The Minerals, Metals & Materials Series

Stephan Broek
Editor

Light Metals 2023

Editor
Stephan Broek
Kensington Technology Inc.
Burlington, ON, Canada

ISSN 2367-1181 ISSN 2367-1696 (electronic)
The Minerals, Metals & Materials Series
ISBN 978-3-031-22531-4 ISBN 978-3-031-22532-1 (eBook)
https://doi.org/10.1007/978-3-031-22532-1

This Springer imprint is published by the registered company Springer Nature Switzerland AG
The registered company address is: Gewerbestrasse 11, 6330 Cham, Switzerland

Preface

I have the great honor and privilege to present the *Light Metals 2023* proceedings. After several challenging years, these proceedings finally take us in a positive direction where conferences like TMS's Annual Meeting can finally be what they should be again: a gathering of accomplished industry professionals sharing science and knowledge in a social environment harnessing the power of human interaction. Our industry is resilient, as illustrated by the number and quality of manuscripts and oral presentations by engineers, scientists, and others represented in this well-referenced bundle of technical articles.

The *Light Metals 2023* proceedings embody the work of the authors, symposium chairs, session chairs, peer reviewers, and TMS staff who made invaluable contributions to this collective publication. A first thank you goes to the authors of technical papers and posters. It is their engagement and active participation that ultimately make the Light Metals program at the Annual Meeting and this publication an ongoing success. Special acknowledgement goes to the symposium chairs of the Light Metals program and of the five special symposia held in 2023. I thank Pierre Marcellin (Aluminum Reduction Technology), Roy Cahill (Electrode Technology for Aluminum Production), Halldór Guðmundsson (Cast Shop Technology), Julie Lévesque (Aluminum Alloys, Characterization and Processing), and Errol Jaeger (Alumina & Bauxite) for organizing the subjects in the Light Metals program. A further thank you to Mark Dorreen (honorary symposium for Prof. Barry J. Welch), Pernelle Nunez (Aluminum Waste Management & Utilization, and for Aluminum Industry Emissions Measurement, Reporting & Reduction), and Timothy Langan (Scandium Extraction and Use in Aluminum Alloys). All these chairs undertook the bulk of the organization, and that is much appreciated.

A constant force behind these efforts is the staff at TMS. In addition to ensuring that deadlines were met, the significant help from Patricia Warren and Trudi Dunlap also safeguarded the society's guidelines. I extend my sincere appreciation to you both for your hard work and dedication.

Finally, I'm grateful to past editors Dmitry Eskin and Linus Perander for their important support throughout the year and to Aluminum Committee colleagues over the years. To be granted the opportunity by my peers to be the editor of the *Light Metals 2023* proceedings is a special honor.

It has become clear over this past year that incredible changes in our industry lie ahead. The push for sustainability is firm and must deliver solutions within the next few decades that we only can imagine today. This will require scientific breakthroughs, new technologies, and different ways of working together. Platforms like the Light Metals program at the TMS Annual Meeting are invaluable opportunities for sharing the insights and knowledge necessary to spur critical innovations. Students, scientists, researchers, engineers, and other professionals must contribute and collaborate to achieve the crucial changes that will propel our industry, save our climate, and ultimately ensure our collective future.

In 2023 we gathered in San Diego, California for the TMS Annual Meeting. At the time of writing, we have largely overcome the COVID-19 pandemic and travel restrictions have been lifted. But other global events had a profound impact on the 2023 conference since many of our peers have been affected. I can only offer sincere wishes that you all are safe and that we will see each other again soon.

Stephan Broek

Contents

Part VII Cast Shop Technology

About the Editor

Stephan Broek is the president and principle consultant of Kensington Technology Inc. in Burlington, Ontario, Canada. He has a B.Sc. in Chemical Engineering from the University of Applied Sciences in Amsterdam, with a postgraduate diploma in Process Technology from Twente University in Enschede, The Netherlands. For twelve years, he worked at Tata Steel Europe in the engineering group of Danieli Corus. His assignments included process engineering, technology management, and product development. In 2003, Stephan joined Hatch, Mississauga, Canada, where he worked for six years on commercialization and development of technologies. For the next twelve years, Stephan went on to dedicate his work to the global aluminum industry where he is recognized for his know-how in environmental engineering and technologies. He was also actively working on smelter feasibility studies and project management, including the rebuild of potline 2 at PT Inalum. In October of 2021, he joined Boston Metal to provide leadership in the industrial development and deployment of inert anode technology for metal production. This is an exciting and game-changing development that can have a real impact on climate change linked to global metal production. As of March 2023 he is the president and principle consultant of Kensington Technology Inc., a specialized consulting firm primarily focused on the global aluminum industry.

Stephan has co-authored over 40 articles and papers and is a regular speaker at international conferences. Over the years, he has been actively involved with TMS, and has served as secretary of the Aluminum Committee for six years. In addition, he has been a subject chair and session chair in several of the TMS annual meetings. Stephan also shares his knowledge within our industry. He created and lectured in several TMS short courses and is a lead instructor of the Potline Scrubber & Fugitive Emissions (PSFE) industrial training course that is part of professional development by TMS.

60 Years of Taking Aluminum Smelting Research and Development from New Zealand to the World: An LMD Symposium in Honor of Barry J. Welch

Mark Dorreen has his Bachelor and Ph.D. degrees in Chemical and Materials Engineering, and a broad range of leadership experience in both R&D and innovation commercialization. Since late 2022 he has been Research Group Leader–Electrochemical Processing at CSIRO's Mineral Resources division in Australia. Prior to that he was the Director of the Light Metals Research Centre (a commercial unit within the University of Auckland, focused on technical innovation within the global aluminum industry), followed by some time as the founding CEO of Enpot Ltd (a spinout company from the LMRC). Earlier in his career, Dr. Dorreen had several technical and commercial roles within the aluminum and steel industries in New Zealand.

Alumina & Bauxite

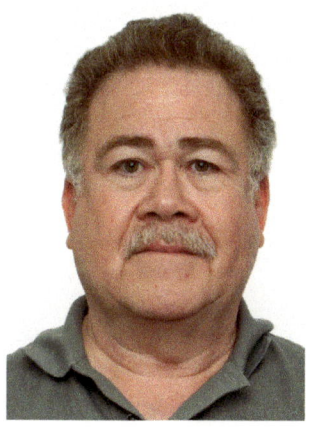

Errol Jaeger has more than 35 years of experience in the aluminum industry. His background has been primarily in the bauxite and alumina sectors and includes basic and conceptual engineering studies of Bayer process plants, greenfield and brownfield commissioning and start-up planning, project management, and due diligence. Errol has worked with Alcoa, BHP-Billiton, Kaiser Aluminum, Ma'aden Aluminium, Vimetco Management, and SNC-Lavalin.

Aluminum Alloys, Characterization and Processing

Julie Lévesque (P.Eng., Ph.D.) is an R&D project manager at Quebec Metallurgy Center (CMQ), where she is the leader of the metals forming and assembly axis. As such, she is responsible for projects related to welding and mechanical forming operations, including rolling, deep drawing, wire drawing, stamping, forging, and other solid state forming processes. She holds a master's degree from Laval University, a Ph.D. from the University of Sherbrooke, and has close to 15 years of experience in the optimization of forming processes in collaboration with industries in the automotive, aerospace, and packaging industry, amongst others. She is an active member of the Aluminium Research Center – REGAL and the director of the Innovation Catalyst in Advanced Manufacturing Processes (college-university-industry cluster). Julie is also the 3rd VP of the Metallurgy and Materials Society of the Canadian Institute of Mining, Metallurgy and Petroleum (MetSoc of CIM).

Aluminum Industry Emissions Measurement, Reporting & Reduction and Aluminum Waste Management and Utilization

Pernelle Nunez joined the International Aluminium Institute in 2015 and is the Deputy Secretary General and Director of Sustainability. The IAI is the leading global association for the aluminum industry with members involved with bauxite mining, alumina and aluminum production across all major producing regions. In addition to being a spokesperson for the IAI, Pernelle is responsible for the Institute's broad sustainability work program including its Greenhouse Gas Pathways Working Group and Environment and Energy Committee. Pernelle has led many collaborative industry projects across topics including environmental footprint analyses, sustainable waste management, responsible mining practices, and GHG emissions accounting and pathways. Pernelle holds a MSci in Geology from The Royal School of Mines, Imperial College London, and a Post Graduate Certificate in Sustainable Value Chains from the University of Cambridge.

Aluminum Reduction Technology

Pierre Marcellin is Principal Advisor in the Rio Tinto Aluminium Technology Solutions (ATS) department in France. This department designs, delivers, and supports aluminum process control supervision and regulation systems, particularly ALPSYS pot control system. Pierre oversees the technical development of ATS solutions and provides assistance to ATS solutions users and delivery projects.

Pierre graduated from the Ecole Normale Supérieure (Paris, France) with a Civil Engineering Degree and a post graduate research study in computer science.

Pierre has been working for Rio Tinto and its predecessors since 1989 and has held various positions in aluminum reduction process control (in R&D in LRF/France, in ALBA smelter/Bahrain, and St Jean de Maurienne smelter/France), in the computer department (Dunkerque smelter/France), and managing major projects in the St Jean de Maurienne smelter and in ALUVAL (Technical Sales projects).

Cast Shop Technology

Halldor Gudmundsson received his B.Sc. in Physics in 1986 from the University of Iceland, M.Sc. in Materials Science in 1989 from the University of Virginia, and Master in Light Metals Reduction Technology 2009 from the University of Auckland. Halldor has taught engineering courses on and off for the past 30 years at the University of Iceland. He was an adjunct lecturer in the Mechanical Engineering department from 1993 to1998. He joined Nordural in 1998, and prior to that spent about 10 years doing various consulting work and testing in the field of metallurgy and materials science. His main area of expertise is microstructural analysis using optical and electron microscopy. At Nordural he has had various responsibilities; quality control of products and raw materials, environmental management, process control in potrooms (2005–2010), technical manager of reduction (2010–2014), and product development manager in the casthouse.

Electrode Technology for Aluminum Production

Roy Cahill has 26 years of experience in the aluminum industry, spanning 3 companies, and has worked across various areas of the industry such as alumina refining, anode production, coke calcination, smelting, and spent potliner treatment systems. He currently is the Carbon Manager for the Smelter Technical Support Team within Rio Tinto Aluminium's Pacific Operations. The Team is responsible for development of data analysis and visualization tools, process stability, technical training of both new engineers and operators, process technical support, selection and utilization of anode raw materials (coke, pitch, and refractories), and future technical strategies. Roy Cahill has a Ph.D. in Inorganic Chemistry from Texas A&M University.

Scandium Extraction and Use in Aluminum Alloys

Timothy Langan has extensive experience in all aspects of the commercialization and development of advanced material technologies, ranging from fundamental research up through product engineering and launch. His primary area of technical expertise involves modifying microstructures and surfaces of advanced materials to optimize corrosion resistance and mechanical properties. Dr. Langan is currently working with Sunrise Energy Metals to develop applications for scandium that will be produced from its Sunrise Project in New South Wales, Australia. In this role, Dr. Langan is working with industrial partners to guide, develop, and focus research efforts on aluminum-scandium alloys at universities including Deakin University, Michigan Technological University, Monash University, and Chongqing University.

Dr. Langan has extensive experience with aluminum alloy development. Working at Surface Treatment Technologies (ST2), he developed a family of advanced corrosion resistant weldable scandium containing aluminum alloys. Dr. Langan joined ST2 after working as a Technical Director of Ashurst Technologies Corporation. In this capacity, he was instrumental in developing US markets for materials technology from the Former Soviet Union, particularly Ukraine. Before joining Ashurst, Dr. Langan was Group Leader, Special Programs Group, in the Advanced Alloys Department at Martin Marietta Laboratories in Baltimore, Maryland. In this position he co-invented a family of aluminum-lithium alloys known as Weldalite™ alloys and worked with engineers at Martin Marietta, NASA, and Reynolds Metals to successfully use the alloy to build the Super-lightweight tank, which first flew as the space shuttle main fuel tank in 1998.

Aluminum Committee 2022–2026

Executive Committee 2022–2023

Chairperson
Dmitry Eskin, Brunel University, Middlesex, UK

Vice Chairperson
Stephan Broek, Kensington Technology Inc., Burlington, ON, Canada

Past Chairperson
Linus Perander, Yara International, Sandefjord, Norway

Secretary
Kristian Etienne Einarsrud, Norwegian University of Science & Technology, Trondheim, Norway

***JOM* Advisor**
Anne Kvithyld, SINTEF, Trondheim, Norway

Light Metals Division Chair
Eddie McRae Williams, Arconic, Pennsylvania, USA

Members-At-Large Through 2023

Houshang Alamdari, Laval University, Quebec, Canada
Corleen Chesonis, Metal Quality Solutions LLC, Pennsylvania, USA
Mark Dorreen, Auckland, New Zealand
Marc Dupuis, Quebec, Canada
Les Edwards, Rain Carbon Inc, Louisiana, USA
John Grandfield, Grandfield Technology Pty Ltd., Victoria, Australia
John Griffin, ACT LLC, Pennsylvania, USA
Ali Jassim, Emirates Global Aluminum, Dubai, United Arab Emirates
Lorentz Petter Lossius, Hydro Aluminium AS, Øvre Årdal, Norway
Eric Nyberg, Kaiser Aluminum Trentwood, Washington, USA
Arne Ratvik, SINTEF, Trondheim, Norway
Barry Sadler, Net Carbon Consulting Pty Ltd. Victoria, Australia
David Sydney Wong, Queensland, Australia

Members-At-Large Through 2024

Stephan Broek, Kensington Technology Inc., Burlington, ON, Canada
Dmitry Eskin, Brunel University, Middlesex, UK

Duygu Kocaefe, University of Quebec, Quebec, Canada
Johannes Morscheiser, Aleris Rolled Products Germany GmbH, Koblenz, Germany
Jayson Tessier, Alcoa Corporation, Quebec, Canada
Alan Tomsett, Rio Tinto Pacific Operations, Queensland, Australia

Members-At-Large Through 2025

Linus Perander, Yara International, Sandefjord, Norway
Derek Santangelo, Hatch, Quebec, Canada
Samuel Wagstaff, Oculatus Consulting, Georgia, USA

Members-At-Large Through 2026

Kristian Etienne Einarsrud, Norwegian University of Science & Technology, Trondheim, Norway
Mertol Gökelma, Izmir Institute of Technology, Urla, Turkey
Stephen Instone, Speira GmbH, Bonn, Germany
Martin Iraizoz, Aluar Aluminum, Puerto Madryn, Argentina
Anne Kvithyld, SINTEF, Trondheim, Norway
Julien Lauzon-Gauthier, Alcoa Corporation, Quebec, Canada
Pascal Lavoie, Alcoa Canada, Quebec, Canada
Olivier Martin, Rio Tinto, Saint Jean, France
Ray Peterson, Real Alloy, Tennessee, USA
Andre Phillion, McMaster University, Ontario, Canada
Daniel Richard, Hatch, Quebec, Canada
Gudrun Saevarsdottir, Reykjavik University, Reykjavik, Iceland
Andre-Felipe Schneider, Hatch, Quebec, Canada
Dimitry Sediako, University of British Columbia, British Columbia, Canada
Camilla Sommerseth, SINTEF, Mo i Rana, Norway
Eddie McRae Williams, Arconic, Pennsylvania, USA
Andrey Yasinskiy, Siberia Federal University, Krasnoyarsk, Russia

60 Years of Taking Aluminum Smelting Research and Development from New Zealand to the World: An LMD Symposium in Honor of Barry J. Welch

What Makes TMS Special? Let Us Consider a Case Study in Volunteer Excellence: Barry J. Welch

James J. Robinson

Abstract

The Minerals, Metals & Materials Society (TMS) classically serves as an intersection point of academia and industry, giving experts opportunities to expand their expertise while simultaneously helping other experts expand their own expertise in kind. It is a complexly intertwined yet beneficial set of relationships that supports technology transfer, that improves or reinvents the state of the art, and that thrives on collegiality. Via his volunteer work through TMS, Professor Barry Welch has consistently and tirelessly exemplified all of these qualities and has routinely and positively influenced the work and workforce of the worldwide aluminum industry. This paper presents personal reflections on some specific instances of Prof. Welch's volunteer activities that have made TMS a better professional society and that have elevated the field and the good of the order in the process.

Keywords

JOM • Light Metals Division • TMS • Volunteerism

Introduction

The Minerals, Metals & Materials Society, a.k.a. TMS, has a lengthy and distinguished history that stretches back to 1871 with the founding of what is today called the American Institute of Mining, Metallurgical, and Petroleum Engineers. Over the 152 years since the founding of AIME and the 65 years since TMS was formed within it, much has changed in the metals community, not the least of which is the development of the entire aluminum industry. While the metals industries have invented and sometimes reinvented themselves many times since 1871, one fact has remained constant: Professional societies have thrived because of volunteer engagement. Within the TMS community, generation after generation of volunteers have lent their energy, ingenuity, experience, and reputations to advance the collective good of the order by leveraging TMS' power to assemble networks of like-minded volunteers and amplify their collective capabilities.

Great associations are great because they have great volunteers, and those great volunteers infectiously inspire other volunteers toward greatness. TMS is a great professional society because of this virtuous cycle.

Within TMS, Barry Welch has been a great volunteer. I feel very confident in this assertion because of my first-hand observation of Barry's volunteer work from the perspective of a professional staff member. I have witnessed him innovating for the Society, giving generously of his time and resources, and inspiring others to engage as well. He has helped me grow as a professional, and he has helped TMS grow as a professional society. What he has not helped is my memory, so apologies in advance for any misremembering on my part. I remember it all like it was yesterday, but my memory of actual yesterday is not always perfect!

The *JOM* Years

I first became acquainted with Barry Welch in the early 1990s as he was taking on the volunteer role of *JOM* Advisor from the Aluminum Committee. In *JOM* parlance, an "advisor" is a volunteer from a technical committee who gathers a suite of four to six papers for publication in *JOM*; these papers should represent the technical interests of the committee to the full TMS membership. The papers might be invited or could be "over the transom" submissions. It is the job of the advisor to arrange a review of the papers and decide which ones to publish. Still used robustly today, the

J. J. Robinson (✉)
The Minerals, Metals & Materials Society (TMS), 5700 Corporate Drive Suite 750, Pittsburgh, PA, USA
e-mail: robinson@tms.org

© The Minerals, Metals & Materials Society 2023
S. Broek (ed.), *Light Metals 2023*, The Minerals, Metals & Materials Series,
https://doi.org/10.1007/978-3-031-22532-1_1

advisor system was less than a decade old when Barry became part of the advisor team. He joined at a very meaningful time.

During the latter half of the 1980s, *JOM* (which was then still named *Journal of Metals*) and the Aluminum Committee had a lively and sometimes contentious relationship. The contentions generally centered on the journal's too-limited attention to primary aluminum production and cast shop technologies—the stuff that has powered the *Light Metals* volume year after year since 1971. Passions on the Aluminum Committee to see more representative and expanded coverage in *JOM* ran deeply. The question came down to this: Why is the technical committee that produces the most technical programming for the Society and producing the most sought-after proceedings volume so proportionately underrepresented in the journal that is supposed to be reflective of the interests of the entire membership? Many philosophical and logistical debates ensued within the volunteer community and among volunteers and staff.

How do I know about this? I participated in many of these discussions as I joined TMS to work on the *Journal of Metals* in 1984; the initial job was to edit manuscripts and transition our manuscript preparation from typewritten papers to digital files. More responsibility soon followed, and it didn't take long to become Associate Editor, then Managing Editor, and then Editor. These transitions occurred in parallel with the how-to-better-represent-aluminum discussions that were underway with the volunteers, and I became more imbued with each title change.

In my experience, solutions to problems come from people working together rather than acting dogmatically. That was certainly the case with *JOM*. The competing perspectives were harmonized by two bridge-building volunteers representing the Aluminum Committee to the journal as advisors: Wayne Hale and Rod Zabreznik. Rod and Wayne served at times individually as advisors and at times as a duo. They worked hard on behalf of the Aluminum Committee, advocated on behalf of TMS being the global home of the aluminum industry, and expanded my understanding of primary aluminum considerably.

What was the harmonious solution? The journal would publish aluminum industry papers quarterly—an aluminum topic in the February, May, August, and November issues. We went from a plan of 1–5 aluminum papers annually to 20 or more. This was a big publishing commitment for the volunteers to fulfil as so many more quality papers would have to be acquired annually. And it was a big commitment from *JOM* as this meant bigger issues and a bigger expense to TMS as the publisher. (At the time, *JOM* operated at a deficit to the Society. So, publishing more papers meant increasing the deficit operation.)

Having helped build a new aluminum publishing model, Rod and Wayne began to step back from their focus on *JOM* and were advisors no more. Rod would go on to chair the Light Metals Division, and Wayne would not only chair the division but serve as TMS President in 2001.

With Rod and Wayne's *JOM* retirements from collecting and reviewing *JOM* papers, the question became who should actualize the new publishing model as the next *JOM* advisor from the Aluminum Committee? Would the model be more vulnerable to failure without the participation of the volunteer architects? Could we actually realize the expanded commitments?

The answers took form in the person of Barry Welch. I do not know if any cajoling or arm-twisting was performed in persuading Barry to accept this assignment, but Barry immediately presented enthusiasm and energy in being the next advisor from the Aluminum Committee. He recognized that the new commitments would not be satisfied effectively without effort. We needed quantity *and* quality. Barry embraced the challenge! And, he set the bar high as he has very high standards for want constitutes a publication-worthy paper. Could the criteria be satisfied?

The answer turned out to be "yes" as Barry knew that we could not just wait for more papers to come flooding in unsolicited. Instead, he devised a three-pronged action plan.

First, he invited people in his network to contribute papers. Barry has a large network, and he is an exceedingly difficult person to whom to say "no." By my admittedly subjective impression, the number of high-profile and up-and-coming contributors of papers to the journal went on the ascend.

Second, Barry agreed to occasionally author or co-author papers for *JOM*. Barry writes an exceptional technical paper and overview. This would be leadership by example.

Third, Barry thought it important that publication in *JOM* provides a prestigious opportunity beyond the standard recognition that comes with being accepted for publication. An inspirational model could be taken from the aluminum community, where subject awards are issued for papers published in *Light Metals*. He set to work with the Aluminum Committee and the Light Metals Division to create a new award: The Light Metals Division *JOM* Best Paper Award.

It was this final development that left me amazed: The journal had accomplished a long and difficult journey from being something that held little to no appeal to members of the aluminum community to being one where the journal was publishing enough quality content that multiple papers could be worthy of award consideration. And speaking of multiple papers, the first award recipient, Manaktala, was recognized for a set of three:

- Nov. 1992: "The Primary Aluminum Industry in the Commonwealth of Independent States—Part I."
- Feb. 1993: "The Primary Aluminum Industry in the Commonwealth of Independent States—Part II."
- May 1993: "The Primary Aluminum Industry in the Commonwealth of Independent States—Part III."

It was a fascinating set of papers and very timely at this moment in history. It was well-deserving of the award, as have the many papers that have followed. The second award recipient was Alton T. Tabereaux for November 1994: "Anode Effects, PFCs, Global Warming, and the Aluminum Industry"—some topics never go out of fashion. Interestingly, Alton's paper is one of only two bylined papers in *JOM* history to use the phrase "global warming" in the title. (The other was "Global Warming and the Primary Metals Industry" by David Forrest and Julian Szekely in December 1991.)

- Nov. 2008: "Applying Fundamental Data to Reduce the Carbon Dioxide Footprint of Aluminum Smelters."

He was also co-author of another of the award-receiving papers: "The Multivariable Model-Based Control of the Non-alumina Electrolyte Variables in Aluminum Smelting Cells" by Fiona J. Stevens McFadden, Barry J. Welch, and Paul C. Austin (February 2006). Fiona was one of Barry's many excellent students, and he clearly mentored her well as she followed him as *JOM* advisor from the Aluminum Committee. She ably kept all of the wheels turning and the progress continued.

With these initiatives, it seemed that *JOM* had cracked the code of robustly engaging primary aluminum. No question that many, many volunteers played important roles in this progress, but I will always think of Barry as the anchor runner for the relay team that led to this success. He was the

The final award of the afternoon was the new *JOM* Best Paper Award, which recognizes a light metals article of excellence from the previous volume year of *JOM*. The award was developed as a result of *JOM*'s increased editorial emphasis on developments in the aluminum industry. The winner of the first award is S.C. Manaktala, who wrote the three-part series "The Primary Aluminum Industry in the Commonwealth of Independent States." Mr. Manaktala, who is a widely acknowledged expert on aluminum production in the former Soviet Union, is currently the director of technical services for the Primary Products Division of the Kaiser Aluminum & Chemical Corporation.

Announcement of Manaktala receiving the first Light Metals Division *JOM* Best Paper Award at TMS1994. (From the *LMD Edition* newsletter insert in the May 1994 *JOM*.)

Alton Tabereaux (right) receives the second Light Metals Division *JOM* Best Paper Award from Light Metals Division Vice Chair Rodney Zabreznik (left) at TMS1995. (From the *LMD Edition* newsletter insert in the July 1995 *JOM*.)

Barry himself would be the primary author on three papers to receive the award (after his service as *JOM* advisor had concluded, of course):

- Feb. 1998: "Cathode Performance: The Influence of Design, Operations, and Operating Conditions."
- May 1999: "Aluminum Production Paths in the New Millennium."

propagator of the good outcomes and the blunting agent of potential problems. He was remarkably generous of himself. Most importantly, he had established a replicable model that future *JOM* advisors could employ and build upon in their own unique ways.

And this all took place before smartphones, low-cost international phone calls, text messages, websites, and email. We used the fax machine as our "miracle" technology!

A Passion for Industry and for TMS

Barry's departure from the advisor role did not mean that Barry ceased to volunteer for TMS. The title of this paper cites "volunteer excellence" for a reason.

While Barry is well-known and much revered as an educator and shaper of multiple generations of scientists and engineers, I often think of Barry as an "industrial guy" within TMS. Like many professors in the Society's ranks, he speaks with passion about the interests and needs of the industrial community and has numerous industrial ties. I see him as walking in both worlds and being highly respected in both. I have observed him manifest this duality via his volunteerism in TMS with *JOM* and beyond.

For example, Barry has gone straight to the cell line and taught countless shop-floor and plant-management professionals about optimizing anodes, cathodes, reduction cells, emissions, and the best practices of all of those elements that constitute modern aluminum reduction. I've never asked him, but I often wondered if there was a primary aluminum production facility in the world that he has neither visited nor lectured at. I hope that he never disabuses me of my impression that he has been everywhere. Whenever I think of Barry, I am put in mind of American singer/songwriter Johnny Cash who sang in the song "I've Been Everywhere,"

> *I've been everywhere, man*
> *Crossed the deserts bare, man*
> *I've breathed the mountain air, man*
> *Of travel I've had my share, man*
> *I've been everywhere*

Very notably as a world traveller while volunteering for TMS, Barry engaged with Halvor Kvande, Alton Tabereaux, and other aluminum industry luminaries in teaching the much-esteemed TMS short course: Industrial Aluminum Electrolysis Course: The Definitive Course on Theory and Practice of Primary Aluminum Production.

This course is an exemplar of how TMS volunteers can come together and advance the industrial community. Its instructors have travelled to almost every continent to teach on site at a smelter operation. Some course participants hail from the host operation's site; others travel in from all points of the compass for the unique experience. All are provided wonderful opportunities to witness plant operations while the instructors present the latest techniques in electrolysis, link theory, and practice and provide tools and information to help the course participants save costs, improve productivity, and manage emissions. Thanks to the volunteer leadership, this course is looked upon as the model continuing education activity within the Society. It both strengthens the workforce and inspires future advancement.

The Industrial Aluminum Electrolysis Course very much advances the TMS mission, which is "to promote the global science and engineering professions concerned with minerals, metals, and materials." It does this in two ways—the first way is obvious: It provides a platform for our volunteers to develop and deliver a curriculum needed by the community, it helps participants be more informed professionals and more effective metal producers, and it serves as a model of success that other volunteers can emulate in developing their own courses through TMS. The second way is less obvious: The course provides revenue to the Society greater than the cost that is required to conduct the work. That surplus is then used to subsidize the activities by the society that are not self-sustaining (which is most of them).

Another way that Barry supported the mission of TMS was by helping to establish a limited-duration award within TMS: "The Vittorio de Nora Prize for Environmental Improvements in Metallurgical Industries." Vittorio de Nora was a pioneer in the materials processing field and a great Italian technologist. Barry facilitated communication between the de Nora Family and TMS leadership to create an endowment within TMS that would allow the issuance of a $20,000 prize. It was presented six times in the 2010s with the purpose of recognizing outstanding materials science research and development contributions to the reduction of environmental impacts, and particularly greenhouse gas emissions, as applied in global metallurgical industries, especially focused on extractive processing. I thought back on this prize while listening to the recipient of the 2012 award, Jim Yurko of Apple, who presented the all-conference plenary talk at TMS2022. In the talk, he referenced Apple's collaboration with ELYSIS on employing aluminum produced without greenhouse gasses. What splendid symmetry with de Nora, I thought!

One more example of volunteerism in action? Barry has served on the Aluminum Committee many times (of course!) and was its Chair leading up to TMS1998. Many know that the chair of the committee serves as Editor of the annual *Light Metals* volume, and he was Editor of *Light Metals 1998*. A big job.

It is more than fitting that Barry is a TMS Fellow (Class of 2015). It is the highest honor that TMS bestows on a member. His spot-on citation reads,

> *For significant contributions to the advancement of aluminum smelting technology through pioneering research in aluminum electrolysis cell reactions and fundamental processes followed by outstanding teaching to students and engineering practitioners.*

Well said!

2014 TMS President Hani Henein (left) presents the TMS Fellow Award to Barry Welch (right).

Closing Thoughts

Beyond the grand efforts that I recount in this brief paper, Barry has benefited the aluminum community and TMS through his volunteerism in countless ways. Still, I don't think that his greatest impact within TMS has been through the formalisms of these high-visibility activities. Similarly, I also don't think that it has been by presenting at a microphone, or in writing papers, or in building a consensus among peers, or in chairing a discussion amongst volunteers. Instead, I suspect that his greatest impact as a volunteer has been in the hallways of conferences and in those informal moments after a session's conclusion—those one-on-one impromptu times when you can seek his advice or opinion, hear his candid thoughts, benefit from some mentorship, kick around an idea, or just chat about how the family is doing.

Barry's impact has been strong in TMS over the decades and his standing as a role model will surely influence current and future generations of volunteers for many decades to come.

Meeting the Requirements of Potline Customers: The Largely Unmet Challenges Set by Barry Welch to Carbon Anode Producers

Barry Sadler and Alan Tomsett

Abstract

Barry Welch is well recognized for his contribution to advancing the science and practice of smelting Alumina to Aluminium Metal. He has also made a significant contribution to advancing anode technology, in particular through the work of his students. This is explored further in an accompanying paper. What is probably less well recognized are the significant challenges and opportunities Barry has laid out to anode producers to improve anode quality to meet the increasingly stringent requirements of the Potlines customer. These challenges and opportunities will be outlined and their potential impact described.

Keywords

Barry Welch • Carbon anodes • Aluminum smelting • Improvement • Problems

Introduction

The contribution made by Barry Welch to improving the understanding of the science and practise of smelting alumina to aluminium is well known and covered by other speakers at this Honorary Symposium. Barry has also made very significant contributions to advancing the understanding of the key issues related to improving anode performance with some significant practical achievements along the way. The advances achieved through the students Barry has supervised are covered in a companion paper also presented at this Honorary Symposium [1]. In the present paper, some of Barry's work over the past 10–15 years on helping improve the understanding of the principles underpinning good anode quality and performance are presented. This includes better defining "what good should look like" when it comes to anodes and some ways to get there—either from Barry or suggested by the Authors. The length of the list of topics (See below) covered in this more recent work may be a surprise to those who only associate Barry with reduction cell operations.

In a number of cases, Barry has presented the issues and opportunities he has identified as challenges to anode technologists and manufacturers to basically "lift their game" in order to meet the ever-increasing demands of the potline customer. Indeed one of the challenges he has presented is for anode manufacturers to genuinely see potlines as a customer, and not just the people that take away the rodded anodes (of a quality that producers can "get away with") and replace them with consumed anode butts (And complain on the odd occasions when quality does not meet the current specifications).

As an observation, many of the challenges presented by Barry (which are based on an understanding of the underpinning science and impact on the customer) have not been addressed by the industry, and indeed a significant number appear to have not been given much attention at all, i.e. they have been put in the "too hard basket" or just ignored. In the following, a number of the challenges identified by Barry will be listed and briefly outlined, and in some cases, potential pathways proposed to capture the opportunity or resolve the issue.

B. Sadler (✉)
Net Carbon Consulting Pty Ltd, Melbourne, Australia
e-mail: barry.sadler@bigpond.com.au

A. Tomsett
Rio Tinto Pacific Operations, Brisbane, Australia
e-mail: Alan.Tomsett@riotinto.com

© The Minerals, Metals & Materials Society 2023
S. Broek (ed.), *Light Metals 2023*, The Minerals, Metals & Materials Series,
https://doi.org/10.1007/978-3-031-22532-1_2

Challenges to Improve Anode Properties and Performance

The following list of challenges presented by Barry to anode carbon technologists and producers has been organised into four broad groupings: Mindset, Quality/performance, Design, and Operations.

a. The necessary mindset of anode manufacturers:
  i. Really recognise that "*Potlines is the customer*" [2] and stop "*blame shifting*". (This is discussed further in Sect. 3)
  ii. Anodes need to be seen as "value add to aluminum production—an essential component that should be designed for performance, not operating convenience" [2].
  iii. Anode producers and potlines should "*have the data that quantifies and proves the most common anode problems*" [3]. (Anode data systems are discussed further in Sect. 4)
  iv. (Continually) Ask "How can anodes help the cells perform better?" [4].
b. The required anode quality and performance:
  i. Barry has consistently and persistently pushed (or in his words, "hammered") the following aspects of anode quality with a strong theme of producing anodes with a low dusting tendency:
  • (Very) Low Sodium in anodes achieved by excellence in butt cleaning [2, 5]. The importance of butt cleanliness is well known, however, it is generally not well monitored. It is now possible to get commercially available devices to do this online, but these are not widely installed. The conventional approach of manual visual observations of butts after fine cleaning (shot-blasting), and maybe daily analysis of crushed butt samples, is insufficient to monitor cleaning effectiveness in a way that reflects the importance of excellence in cleaning butts. Barry's challenge is for a maximum anode sodium level of 200 ppm. In the experience of the Authors, there may only be a handful of smelters that con-sistently achieve this level, despite it generally being possible to meet this target as long as anodes/butts have not been impregnated with sodium in the potlines. Failure to meet Barry's target can be due to operational (e.g. Operator care and attention) or plant equipment limitations, but it is likely that new approaches to butt cleaning will be required to consistently meet the challenge. Innovation is required.
  • (Very) High and consistent baking temperatures to reduce the differential in reactivity between coked binder pitch and filler coke/butts and to reduce anode carboxy reactivity [2, 5]. There is a limit to how high baking temperatures can be which is set by the onset of significant desulfurisation. Using low-sulfur raw materials is necessary to reduce this limitation. (This is discussed further in Sect. 8.)
  • *High density, but without microcracking that will increase resistivity and affect other properties* [2]. This will give less open porosity which will reduce anode consumption rates [5].
  • *No "Free dust" on anodes sent to Potlines* [2], e.g. eliminate Packing Material Accretions (PMA), damaged anode surfaces (mechanical or slumping damage during baking), particle segregation from anode forming, damaged vertical corners, and packing coke in slots (e.g. Fig. 1). All of these anode defects contribute to dusting in cells but can be fixed, i.e. there is no technical constraint on meeting this challenge from Barry. It requires the determination to not accept these defects as unavoidable and to do the work needed to eliminate them.

  ii. Reduce the electrical resistance of anode connections by, for example, not having any distortion or attack of rod/yoke surfaces that make contact in the electrical circuit [2].

Fig. 1 Examples of free dust on anodes. The cause of all of these defects is known, as are the solutions. Unfortunately, it is still common to see anodes set in cells with these problems. From [2]

iii. Barry recommended a specification for anode quality [5] that is intended for all anodes, not just for samples from selected locations in a baking furnace. A target outcome from the specifications is to have less anode dust and help reduce cell energy consumption through excellent anode quality as "*Producing anodes with raw materials, formulations, and baking conditions that minimize carbon dust will also help cell efficiency gains by eliminating dust generated spikes*" [2]:

Baked apparent density	>1.6 g/cm^3 (with commensurate low open porosity)
Air permeability	<0.5 nPm
Carboxy reactivity residue	>93%
Carboxy reactivity dust	<1%
Air reactivity residue	>80%
Air reactivity dust	<5%
Electrical resistivity	<55 μΩm
Sodium	<200 ppm

This is a tight specification compared to many in use at present, especially considering that it is aimed at every anode sent to potlines, not just a selected sample. While some well-run plants may be able to meet this specification for a period of time, it is felt highly unlikely that this compliance is sustainable especially as baking furnace conditions change over time. Innovation is required to consistently achieve this anode quality standard instead of just writing off the specifications as "impossible".

iv. Barry has also outlined elements of current anode quality that if improved could decrease cell energy consumption by more than 0.4 DCkWh/kgAl [5] through reductions in Gross and Net Carbon Consumption:

- *Desulfurise coke/anodes before setting in the cells (Reduces anode preheat and conversion energy demand, lowers gross carbon).* A relatively simple way of doing this is thermal desulfurisation during calcining; however, this degrades coke quality [6], so innovation will be required to find successful alternatives to reduce sulfur in anodes.
- Reduce carboxy reactivity (Reduce anode preheat and conversion energy demand, reduce net carbon consumption, minimize dust formation, and lower CO_2e emissions).
- *Deliver preheated rodded anodes for setting (Reducing cell operating disturbances at the cell, improving Current Efficiency,* will also reduce in-cell

anode cracking). (Anode preheating is discussed further in Sect. 7.)

c. Anode design opportunities, including slot details:

i. Anode slots must be optimised based on considerations including changes to the mass transfer induced by bubble flow [2, 3]. Slot size, number, depth, location, orientation, and cut vs formed, all need to be considered so that Current Efficiency and Power Efficiency gains with slots are balanced with mixing, alumina dissolution, and local Perfluorocarbon (PCF) emissions. (Slots are discussed further in Sect. 5.)

ii. The top shape of the anode should be designed to minimise gross carbon consumption, allow the anode to be positioned during setting, retain anode cover to protect from airburn, and act as a surge reservoir for bath to help to avoid flux wash of stubs [2, 3]. (Anode top shape is discussed further in Sect. 6.)

iii. The anode rod assembly needs to be redesigned so that it does not encumber potline operations, enables effective butt cleaning, and reduces rodded assembly resistance [2]. The yoke design should facilitate anode covering and butt cleaning [3] and allow some "flex" to reduce anode cracking and stub bending ("toe-in") but must be able to fit under the cell fume plate during setting (for example, see Fig. 2 below). Barry's view is that there is a need to start with a blank sheet of paper and relook at the anode assembly design including how electrical contact is made with the anode carbon [2].

d. Opportunities to improve anode production operations.

i. "*Preheat anodes prior to setting to reduce the negative impact anode setting has on cell condition and performance*" [2, 5]. (Anode preheating is discussed further in Sect. 7.)

Fig. 2 Example of a yoke design that gives advantages of ease of butt cleaning and some flex to reduce stress on the anode resulting in less butt cracking and stub "toe-in". From [5]

ii. "Better manage baking furnace maintenance so the focus is on maintaining anode quality" [2, 5] and this includes maintaining higher baking temperatures than are commonly used at present to reduce anode dusting. (Anode baking is discussed further in Sect. 8.)

A selection of these topics will now be covered in more detail.

Potlines Is the Customer

In his 2020 TMS Annual meeting Electrode symposium keynote address [2], Barry covered a wide range of topics. One of the challenges he set was for anode producers to make "*Potrooms is the customer*" more than just a slogan by having carbon plants supply rodded anodes "*of such a design and quality that will enable the smelter to approach benchmark standards for gross and net carbon consumption as well as metal quality*" [2]. Sadly, in many plants, it only takes a walk down potlines past anodes arranged ready for setting to see that this anode quality challenge is not always being met. All too often the appearance of these anodes shows examples of defects such as packing coke accretions, inadequate cast iron pour quality including over-pour, carry-over shot on the anode top, damage from anode baking, surface cracking, yoke and rod damage, and misalignment of the rod and/or anodes, and the anode assembly used will probably be a legacy design with little innovation to address recognised issues. These are some of the defects evident from visual inspection of the anode surface, if internal inspection systems that are (finally) being developed for anodes are used, extensive internal cracking in many anodes would be found. There are also likely to be stubholes and stubs that do not meet the current relatively modest rejection criteria but these defects are hidden by cast iron.

These defects would not be present in anodes that were about to be set if Potlines was really recognised as the customer. There are readily known fixes to all of the defects listed (the design opportunities need a bit more work). To some degree, it seems that the acceptable quality is partially defined by reject criteria (Which may or may not reflect the needs of potlines), but also by "what we can get away with before Potlines will complain". To meet Barry's challenge, this must change, anode defects must be seen as failures, not something "to get away with", and the work done to eliminate them.

Having the Data Needed to Quantify Anode Problems (and Guide Problem Solving)

Barry has presented a challenge to anode producers and potlines to "*have the data that quantifies and proves the most common anode problems*" [3]. While process data from green anode manufacture are generally available and well used, data from anode baking and rodding are less available. Anode quality data are usually even sparser, as are anode performance data. In many plants, the only online data available about each anode, beyond visual observations, are limited to green anode properties at forming, baked/rodded anode mass, and low-frequency anode core test results often collected using anode sampling approaches that limit the usability of the data. Anode performance data are commonly restricted to rod + butt mass, and whether (and in some cases, how) an anode failed prematurely in operation.

This scarcity of online data which are critical for efficient and effective anode improvement is disappointing given that the tracking systems designed to deliver these data have been available for the best part of 20 years (e.g. Fig. 3). While some plants have installed these systems and the associated instrumentation such as butts imaging devices, their adoption across the industry has generally been quite limited. There may be site- or company-specific reasons for this; however, a common issue seems to be that these systems are "enablers" and only generate a return on investment when they are used to deliver improvement. This means they fail to meet the accountant's Rate of Return requirements for project approval. So the projects are rejected, despite the tremendous capability of these systems to deliver improvement and financial benefit when the data they deliver are effectively and widely used. In the cases where tracking systems have been approved and installed, it appears to have been an outcome from forward-thinking management and/or a specific target (e.g. Stub repair costs) to provide the benefits side of the project cost/benefit analysis.

Anode tracking systems that extend into potlines (i.e. individual anode performance is measured by cell and stall) have the potential to provide complete transparency about anode quality and performance across the smelter. Integrating anode and cell operating data addresses another of Barry's challenges listed above, i.e. "Blame shifting" about anode performance problems becomes virtually impossible (e.g. Fig. 4).

Anode tracking systems can also substantially improve the capacity of increasingly scarce anode Process Engineers to deliver necessary improvements (and meet more of

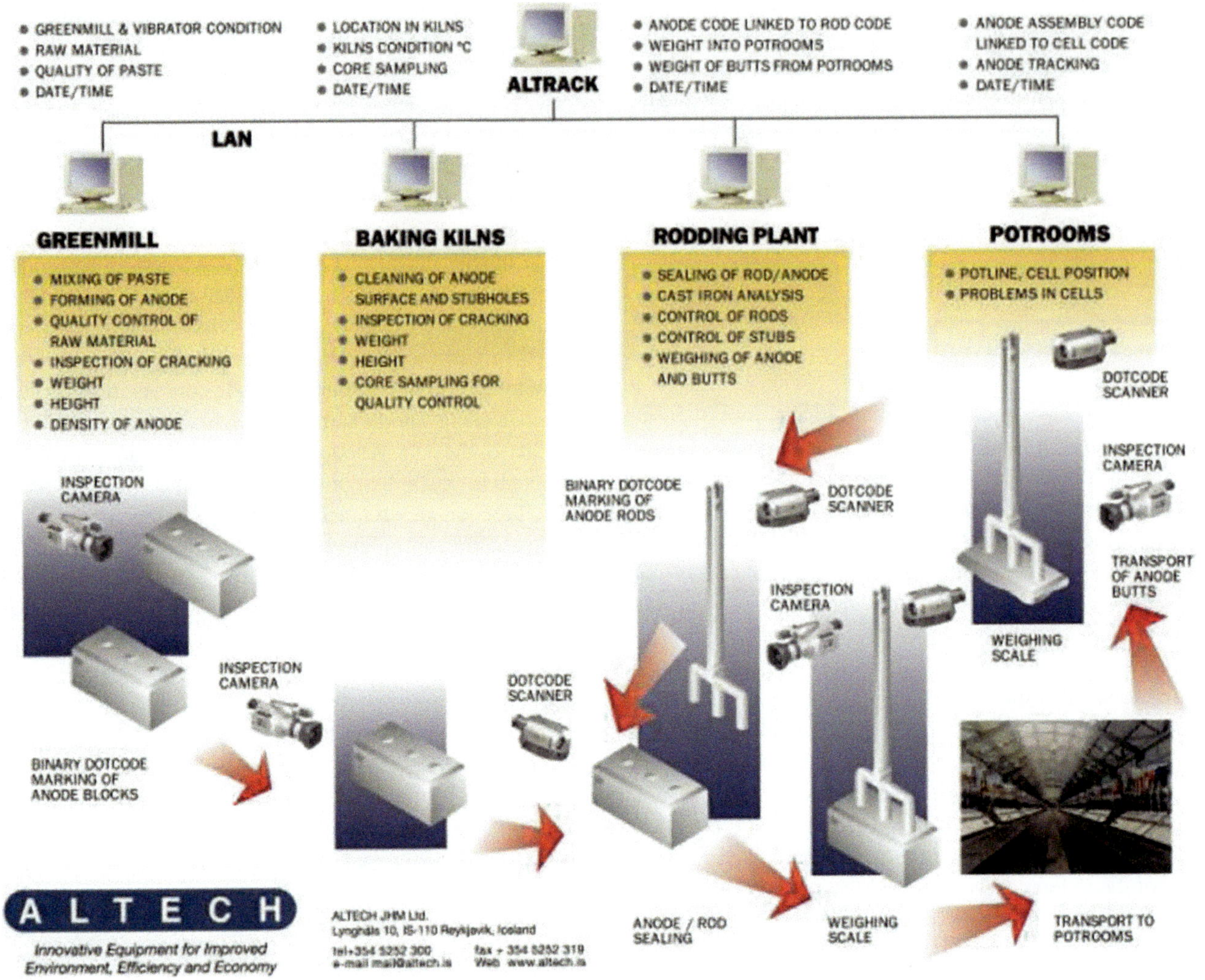

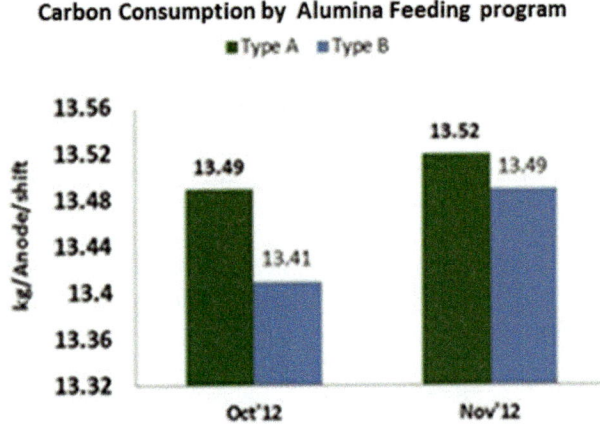

Fig. 3 Schematic example of an early anode tracking system design. From [7]

Carbon Consumption by Alumina Feeding program

■ Type A ■ Type B

Fig. 4 Example of clarity in understanding the drivers of anode consumption gained through the use of an anode tracking system. In this case, it can be seen that alumina feeding program type B gave a lower anode consumption rate than feeding program type A. Reproduced from [8]

Barry's challenges). Hopefully, this, and the focus driven by "Industry 4.0" initiatives, such as big data analytics, will see anode tracking and online anode measurement systems become accepted as fundamental and necessary components of anode plants in the near future.

Optimising Slot Design

Barry has made it clear for some time now that while slots have delivered significant quantifiable gains in Current Efficiency (CE) and Power Efficiency (PE), they have downsides associated with their impact on electrolyte mixing within the cell. His challenge is presented as "*to optimise the size, depth, and location (Not transverse) of slots to balance gains in CE/PE with the greater spatial variation of mixing which is leading to spikes, alumina dissolution issues, and local PFC evolution*" [2, 3]. This challenge is particularly relevant as a caution to ensure that the reasonably common efforts to maximise slot height, with the maximum set when the slots become "full height" (meaning the slots remain in the anodes for the full anode life), take into account all of the effects slots have on cell operations and performance.

Anode slots were first implemented to disrupt the stress patterns at the base of anodes during setting and thereby reduce thermal shock cracking. They have been very successful at doing this and have reduced restrictions on anode filler coke quality (and hence availability) for anode manufacture by allowing the use of more isotropic cokes in blends. As important as this benefit from slots is, their rapid adoption to an almost universal level across the industry was due to the positive impact they had on CE and PE. Slots provide an escape path from the inter-electrode gap (or Anode-Cathode Distance (ACD)) for the gas bubbles generated during electrolysis, thereby reducing the proportion of the ACD filled with bubbles and lowering cell resistance. The motion of these bubbles through the electrolyte, however, provides the driving force for essential mixing within the cell, so the reduction in bubble volume has reduced this mixing giving rise to the issues identified by Barry.

Higher slots reduce bath movement further and give less of the mixing required to maintain consistent electrolyte concentration and avoid localised variation and alumina dissolution issues [2]. Full height slots maximise this effect as during early anode life the slots are exposed above the bath level and hence minimise any bubble mixing. Barry's challenge is to ensure that all factors, especially those based on an understanding of the science and physics involved, are considered when making changes regarding slots [9]. Rigour is required when assessing the merits of design changes to slots and more fundamental work would seem appropriate to understand and then accommodate the impact of slots on electrolyte mixing.

Barry has indicated a preference for cutting slots rather than forming [2] as the latter impacts anode carbon quality in the vicinity of the slots and this contributes to dust [3]. He also reminds us that all slots increase the anode surface area accessible to attack from electrolytically generated CO_2 and this can generate dust. These factors also need to be considered when making changes to slots.

Redesigning the Anode Top Profile

Barry has demonstrated how anode top design/shape can impact anode and cell performance, challenging the industry to adopt a design *that helps to minimize Gross Carbon, makes positioning of the anode in the cell easier and more effective, and helps retain cover on the anode to provide insulation and reduce air ingress and airburn attack* [2, 3].

Fig. 5 Anode top profile recommended by Barry showing how a surge reservoir for bath overflow is formed; this reduces stub flux wash. From [2]

An example of this design is provided in Fig. 5, showing a profiled top with a sloped area forming a step that gives a flat area around the anode top to retain cover and provide a reservoir on the anode top to help cope with surges of bath from anode motion during anode effect termination. This reduces stub attack and iron in aluminium [2]. While some anode designs already have these features, others have flat tops or chamfers that extend to the anode edges that leave room for improvement.

The recommended design gives better protection to the anode from the improvement in cover, better top shape (less exposed vertical surfaces), and if combined with an optimised (sequential) anode setting pattern has been shown to deliver potential savings in net carbon by limiting airburn and associated dust to ≤ 30 kg C/t Al [2]. There would appear to be minimal downsides to adopting an anode profile like that recommended by Barry, and as changes to the top profile are relatively straightforward to make, it appears that this challenge is not difficult to meet. The benefits of easier anode setting, lower net and gross carbon, less dust, and lower iron in metal would seem to make this worthwhile.

Implement Anode Preheating

The setting of "cold" anodes is a major disruption to cell operations and performance, increasing energy consumption and contributing to the formation of spikes associated with the bath freeze that forms on the cold anodes [2, 5]. Barry's challenge is to "preheat" anodes to reduce these impacts.

Ideally, hot anodes (Temperature at about 400 °C to avoid airburn) would be transported straight from the baking furnaces to anode rodding, then immediately delivered in an insulated carrier to the cell for Just In Time (JIT) setting. To make this work, the following would be required:

- Baking furnace operations would need to operate to a very tight drumbeat so that the anode temperature on removal from the furnace is consistent and on target.
- Conveyors and other transfer equipment would need to be able to resist the anode temperatures and be insulated to reduce heat losses.
- Cast Iron thimble details, Cast Iron metal specification including casting temperature, and stub preheating are optimised to ensure a suitably tight stub-carbon connection with the "hot" anodes.
- Although not essential, robotic casting and anode delivery by autonomous vehicles would help to control cycle times and achieve a more consistent anode temperature at the cell.

Meeting Barry's challenge by full implementation of this process means a suitable redesign of the equipment involved and would benefit from locating baking furnace(s) and the rodding room in close proximity to the potlines, but there is potential to implement a degree of preheating using existing equipment (with some modifications to give better heat resistance) now. A high degree of equipment reliability and much greater rigour in scheduling operations /managing cycle times are required but these should be possible in a step toward full implementation. Given the benefits of setting preheated anodes, this challenge would seem to be worth tackling.

Improve Anode Baking

As part of the challenge to reduce dusting, Barry has recommended anode properties that are likely to need higher than generally accepted final baking temperatures. This will require limitations on anode sulfur level to avoid deleterious desulfurisation during baking and require that anode baking furnaces are capable of sustaining these temperatures on an ongoing basis. While ultimately this may require a rethink on how anodes are baked, it should immediately prompt changes to how many existing furnaces are maintained.

Dusting has long been recognised as being important, and yet baking furnace condition is frequently allowed to deteriorate to the point it seriously impacts anode quality and cell performance before the furnace is rebuilt [2]. In the experience of the Authors, inadequate baking furnace condition is the single most common reason for serious dusting excursions in smelters. This indicates that the challenge set by Barry to not allow furnace conditions to increase anode dusting is appropriate.

Baking furnace rebuilds are generally undertaken according to one of three different methods: (1) Continually replacing fluewalls as needed during normal operations, (2) Replacing full section(s) of the furnace according to a tight schedule while maintaining (reduced) furnace operations ("rebuild on the run"), or (3) Shutting the furnace down for a more complete rebuild. Shutting down furnaces tends to be cost-prohibitive if anodes have to be purchased to replace the lost production capacity, but rebuilding on the run is a reasonably common approach. It has the advantages of reducing losses in furnace production, allowing limited changes in refractory details, headwalls can be replaced, and the furnace structure can be re-aligned to correct refractory movement during operation. Rebuilding on the run does require significant forward planning, and once the plans are set, they are not easy to change. If the criteria used to determine the timing of the rebuild prove to be flawed, there is a high likelihood that poor anode quality will result in dusting as a likely outcome. This is not a rare situation in practise.

Rebuilds on the run are commonly planned based on criteria such as visual appearance of the furnace refractories, measurements of deviations in furnace dimensions, data on furnace issues such as blocked/deformed fluewalls, or other criteria such as benchmarks for fluewall life. These criteria are then used to project the furnace refractory life and plan/schedule the rebuild. While anode quality may be considered to some degree, using detailed monitoring of anode quality and performance to set the timing of rebuilds is rarely done, with the outcome mentioned previously— anode quality deteriorates as the furnaces age before the rebuild, and dusting excursions can result.

The challenge set by Barry is to change the way furnaces are maintained so the key criteria for planning are to sustainably supply anodes to potlines that continue to meet a demanding specification including low dusting. Invariably this will mean more furnace maintenance than is done now and will likely drive changes in refractory specifications and baking furnace designs so that acceptable anode quality is consistently achieved for longer periods of time.

While anode sampling, coring, and testing is one approach to getting the data needed to properly project baked anode quality, a better approach is to use a full anode tracking system (see Sect. 4). This will enable anode quality and performance from each baking location within the furnace to be monitored and furnace maintenance to be planned based on statistical projections of the actual anode performance data.

An alternative to this could be to forgo some of the benefits of rebuilds on the run and adopt the continuous fluewall replacement approach, ideally with a tracking system and furnace operating data to help indicate when fluewalls need to be replaced and avoid any deterioration in anode quality.

Meeting Barry's challenge to use delivered anode quality as the key criteria to determine furnace maintenance activities will increase furnace maintenance costs and will likely justify higher quality refractories [5]; however, these can be assessed against the very high cost of anode dusting excursions that are an all too frequent outcome from the current approaches.

Longer term, with tighter criteria for anode quality, revisiting the current concept for anode baking is appropriate.

Conclusions

Barry Welch has outlined numerous science-based challenges and improvement opportunities to anode carbon technologists and producers. A number of these have not been addressed by the industry. Some are simply a matter of adopting a new mindset with existing processes and raw materials, to not accept anode properties or defects that affect anode and cell performance. Others require innovation to meet Barry's challenges. Perhaps one of the most difficult aspects of the challenges from Barry is the question "who is going to do this innovation on carbon anodes"? The answer to this question is compounded by the reportedly relatively short time until inert anodes/cathodes become available, even though it is likely that carbon anodes will continue to prevail for years to come.

References

1. Tomsett AD, Sadler BA (2023) Anode Quality Optimisation – Industry Learnings from the Research Supervised by Barry Welch. Light Metals 2023, TMS.
2. Welch BJ (2020) The development of anode shape, size and assembly designs – past present and future needs. Light Metals 2020, TMS, p 1151–1160.
3. Welch BJ (2012) Good Anodes - Some considerations & Trends. Presented at an internal plant seminar Future leaders Workshop 2012.
4. Welch BJ (2017) Prebake Anodes. Presented at TMS Industrial Aluminium Course. Iceland October 2017.
5. Welch BJ (2014) The design and production of the anodes – "Do they match 21st century requirements for the aluminium smelting Industry" Presented at the 6th Electrodes conference, Iceland, 2014.
6. Edwards L (2017) Thermal Desulfurization of Petroleum Coke for Anode Use, Proceedings of 35th International ICSOBA Conference, Hamburg, Germany, 2017, ICSOBA, p 625.
7. Magnusson J Integrated anode and rod tracking and quality control system. Extracted from Alltech Company brochure, circa 1980's.
8. Sahu TK et. al. (2017) Improving net carbon consumption at EGA with the support of anode tracking and butt analyser, Proceedings of 35th International ICSOBA Conference, Hamburg, Germany, 2017, ICSOBA, p 743.
9. Welch BJ (2023) The need to respect the interlink between science, physics, and cell design in an environmentally responsible manner – The next big challenge for aluminium smelting. Light Metals 2023, TMS.

The Need to Respect the Interlink Between Science, Physics, and Cell Design in an Environmentally Responsible Manner: The Next Big Challenge for Aluminium Smelting

Barry J. Welch

Abstract

By building on the scientific and practical knowledge outstanding smelting cell designs have been developed that are energy efficient and environmentally responsible. With constraints arising from regional availability of energy and the substantial cost of supporting infrastructure, capacity creep by increasing amperage on the installed capacity provides a path for satisfying the growth in demand for the metal. Invariably design and operating retrofits become desirable or necessary because of cell heat balance and other constraints and these can lead to adverse performance and operating characteristics, such as a substantial reduction in cell life or reduced efficiencies. Root-casue contributors to such undesireable consequences are spatial and temporal changes of necessary transfer processes—the physics of getting the consumables or energy to or from the reacting sites at the rate required in order to maintain the desired uniformity in the cell. Hitherto this has been ignored—avoiding or minimising is the next challenge.

Keywords

Electrolyte dynamics • Spatial variations • Capacity creep

The Status of Modern Point Fed Pre-bake Anode Technology

Thanks to a wide range of excellent fundamental laboratory studies over the last 130 years, including the publication of more than 5000 papers directly related to the efficient extraction of aluminium from bauxite in the TMS Light Metal series alone, today all smelters have similar cell design principles, property specifications for the solid electrode materials, and electrolyte composition.

For example, today almost all technologies operate with a modified molten cryolite based solvent that usually has 9–11 wt.% AlF_3, and 4–6 wt.% CaF_2 to both lower the operating temperature and solubility of metal in the electrolyte while maintaining controlled and adequate alumina solubility.

The physicochemical properties of electrolyte and electrode products coupled with the constraint on operating temperature to prevent corrosion of the materials of construction, necessitate a cell design with horizontally oriented electrodes and to be self-heating. However, the knowledge generated has enabled the design and operation of cells that are more than four times bigger than they were a half century ago while simultaneously they incorporate design variations that utilize modern laboursaving automation technology. When the cells are being operated in self-heating mode (without heat recovery) but otherwise targeting minimum energy consumption, all modern cell-designs are capable of exceeding 95% faradaic efficiency while achieving minimum energy consumption at or below 13 DC kWh/kg, provided the temperatures are within the band 955–965 °C. Simultaneously the advanced control systems have dramatically reduced the tendency for the cells to evolve the environmentally harmful perfluorocarbons.

Increasing Cell Productivity

The continuous growth in demand for aluminium requires a steady growth in production, a requirement that cannot be easily or economically satisfied by building a completely new potline. Capacity creep by increasing line current provides an option, but that inevitably introduces changes in the "physics". Increased rates of materials and energy inputs automatically demand changes in the speed required for

B. J. Welch (✉)
Welbank Consulting Ltd., 100B Coates Ave., Orakei, 1071, New Zealand
e-mail: barry@barry.co.nz

© The Minerals, Metals & Materials Society 2023
S. Broek (ed.), *Light Metals 2023*, The Minerals, Metals & Materials Series,
https://doi.org/10.1007/978-3-031-22532-1_3

interactions such as alumina pre-heat and dissolution, thermally equilibrating newly set anodes, and for the electrolyte to mix well throughout the cell. Some of the changed demands are spatially confined, while others are temporal/time-dependent. The changed physics that results encompass secondary chemical reactions and mass and heat transfer processes. Minimizing the variability within the cell while simultaneously maintaining the required self-heating balance is necessary for optimizing the process control and efficiency. Minimising the changed physics following capacity creep presents challenges.

Different "retrofits" (i.e., changed operating materials specifications, design features, and control practices) have been used, to minimise the problems introduced, but deterioration in performance indicators and significant reductions in cell life are common consequences.

Common needs that should be focused on to avoid costly performance changes arising from the changed physics include:

- The provision of adequate thermal energy and electrolyte in the alumina feed-zone to dissolve the increased rate and amount of cold Al_2O_3 powder being added before it can sink through the liquid metal to form sludge or freeze on the cathodic carbon.
- Ensuring the electrolyte flow velocity and mixing pattern enables constraining any spatial concentration or temperature variations to within the limits acceptable for the cell control logic.
- Limiting the secondary reactions that are enhanced by the changed electrode potentials that result from the changed current density to within the environmental economic, and controllable limits.

Strengths and weaknesses of different actions taken to increase productivity and maintain controllability are discussed below.

Increased Potline Amperage

Subject to maintaining current efficiency, productivity is proportional to line current, so increased amperage is both a necessary and a simple step for higher productivity. *Or is it??*

Commonly, we describe the Hall–Heroult process for aluminium extraction according to the following reaction:

$$2Al_2O_3 + 3C \Rightarrow 4Al(l) + 3CO_2(g) \qquad (1)$$

yet no one has ever successfully operated the cell and only formed those two products.

Thermochemical and electrochemical laws highlight that it is impossible to avoid co-evolution of $CO(g)$ even if the cell was operating at 100% current efficiency. Thermodynamics show CO formation is the primary product from the structural rearrangement at the electrode surface. Practically, however, its removal from the electrode surface layer requires the product to transform to the gaseous phase, and the rate of this is constrained by the rate of the heat transfer necessary to satisfy the deficit in energy requirement for the phase transformation that is not provided by the operating electrode potential. In retrospect, anode overpotential measurements [1, 2] gave compelling proof that at low current densities $CO(g)$ is the product with the Tafel curve giving a visible accelerating current inflection at current densities above $0.1\ Acm^2$ but when the current or voltage is being reduced slowly on the same electrode from a high overpotential, the measured current density in the region of the inflection is always significantly higher. The inflection was erroneously interpreted as being due to an impurity even though high-quality materials had been used in the experiment.

Practically at fixed current density surface coverage by the CO intermediate automatically increases the current density and hence electrode potential lifting it to a level that enables CO_2 co-evolution. Much less energy and heat transfer is required for CO_2 release as a gas for the prevalent amount of oxygen evolution once thermodynamically enabled by the very small electrode potential increase necessary. Besides the degree of disorder in the structure of the anode carbon, the physics of heat transfer between the electrolyte and anode surface also influence the proportions of $CO(g)$ and $CO_2(g)$ at any current density. The range of variation can be quite substantial as illustrated by the studies of Hume et al. [3] and Dorreen et al. [4].

Likewise, the cathode product is never pure Al but always a Na-Al alloy. Although pure sodium has a higher deposition potential than aluminium from the electrolyte, its deposition potential is lowered by alloying with aluminium, and some are always co-deposited. The formation can be represented by the reaction:

$$6NaF + Al \Rightarrow Na(in\ Al) + Na_3AlF_6(l) \qquad (2)$$

The minimum Na concentration formed has been determined by Tingle et al. [5] who carefully equilibrated controlled masses of an extensive range of NaF/AlF_3 ratios—Al mixtures over a range of temperatures and analysed the product compositions. From this, we can deduce that the normal smelter grade electrolyte operating at a weight ratio of approximately 1.12 and a temperature of 960 °C would have an approximately 50 ppm sodium in the aluminium in the reversible state.

However, the Na content increases according to exponential laws for increases in both temperature and sodium fluoride concentration of the electrolyte.

In the operating cell, the actual amount of Na co-deposited will always be significantly higher because of such factors as:

- the over potential necessary to sustain the Al deposition rate potential which increases with current density,
- the interfacial electrolyte being enriched in Na^+ through its role transporting current through the electrolyte.

Since the alloy is in contact with the carbonaceous cathode block and "above equilibrium Na concentration" the sodium content has an extra thermodynamic force for chemical reactions with both the carbon of and the electrolyte filling the open pores of the cathode block. In both instances the reaction products grow as solids, thus generating destructive forces through the molar-volume increase. Consequently, cathode life is adversely impacted by increased line current when no other design change occurs.

The increased interfacial sodium in the aluminium also increases the metal solubility in the electrolyte—which is the dominant cause of current efficiency loss. Even though many people refer to the dissolved metal causing the loss in current efficiency through the back reaction as being "dissolved aluminium" it was rigorously proven that the dissolved metal is sodium [6], with the solubility increasing with the NaF content of the electrolyte and being a maximum in molten NaF.

Another adverse consequence of increasing amperage is the magnification of the spatial increases in the rate of heat generation at the extremities of the cathode blocks. Von Kaenel et al. [7, 8] and others have drawn attention to the asymmetric variation in the current density distribution at the cathode block—metal pad interface across the smelting cells. This is generated by the higher collector bar resistances of an equal-length cross-sectional segments when compared to cathode carbon block. Essentially, the current flow takes the path of least resistance. The asymmetry became much worse when the industry replaced the anthracitic cathode blocks with low-resistance graphitized blocks. Figure 1 illustrates the current distribution variation along cathode block surface for a collector bar design and size typical of that in a common technology assuming it has a perfectly balanced magnetic field. The consequential percentage variation in heat generation is also annotated in the figure. At the end of the cathode block, it will be approximately four times the rate of heat generation near the centre channel, the location where

the alumina is fed and, consequently, has the highest thermal energy demand.

This increased thermal asymmetry automatically changes the "physics" of some secondary reactions and processes in the cell. The increase in temperature and superheat of the electrolyte can result in melting the protective side-ledge, while the higher temperature and electrochemical processes drive the sodium deeper through the cathode block and side walls. Since pure sodium has a boiling point of 883 °C with the heat of a vaporization only being 88.6 kJ mol^{-1} the sodium has a strong driving force for migration through open or closed pores within the carbon to continue destructive reactions throughout the cathode block. Reactions with electrolyte are particularly prevalent at the upper interface between the collector bar and cathode block and the growth of NaF crystals becomes a major driver for "wing-cracking" for the common collector bar slot design.

Retrofitting Modified Shape and Size of Anodes

Enlarging the Anode Cross-sections

Retrofitting anode with a larger footprint is an option used for both reducing cell's ohmic resistance and keeping the anode overpotential low. Both changes are beneficial for reducing unit energy consumption or minimising the need to reduce the anode–cathode distance to maintain heat balance. These changes are also beneficial for maintaining current efficiency, the reduced anode overpotential is also beneficial for alumina feed control, giving greater flexibility and minimising the risk of transgressing to initiating co-evolution of fluorocarbons.

Countering the benefit, each newly set anode requires approximately 340 kWh per tonne of carbon to achieve the state where it is able to draw its balanced share of potline current. Initially, the heat is extracted from the electrolyte existing underneath and on the sides of the newly set anode with the electrolyte freezing quantitatively [9]. The physics of electrolyte flow changes, resulting in the cell being in the current imbalanced state for an extended time. Inevitably, a significant portion of the anode footprint increase encroaches on the centre channel, reducing the electrolyte volume.

Adverse physics associated with alumina dissolution and electrolyte mixing have become common.

The thermochemical requirement for alumina dissolution without sludge formation is that for every kilogram of alumina added it needs to be able to quickly mix with 100 kg of electrolyte at 10 °C superheat. This can satisfy the high heat

Fig. 1 The variation in surface current distribution across cathode block for various carbon qualities (7, 8) and flagged annotations for heat generation variation of the common design

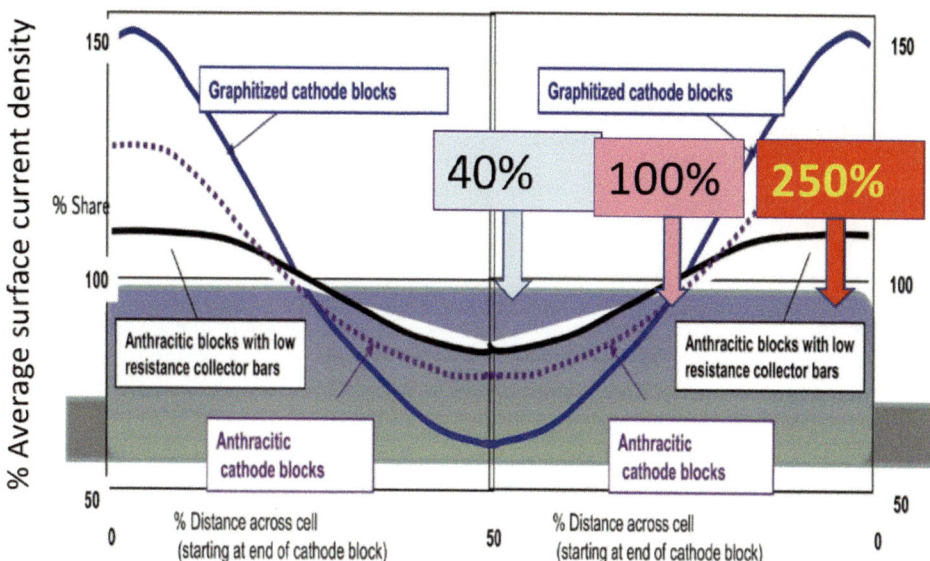

Fig. 2 The changed conditions impacting alumina dissolution with capacity creep [10]

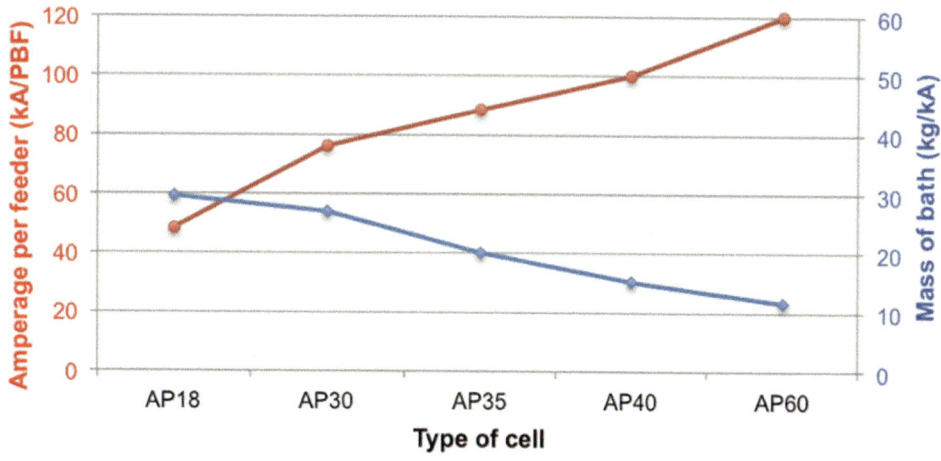

demand for the alumina pre-heating and dissolution thus avoiding electrolyte freezing on the grains surface and sludge formation. Magnifying the alumina dissolution problem in cells that have undergone capacity creep is the need for it to dissolve proportionately faster to match the line current increase! But then there is also the need to flow faster because of both the greater travel distance and the need to compensate for the electrolyte lost through the reduced anode cathode distance (ACD). Martin et al. [10] have analyzed the trends resulting from these changes in the developed technologies they have been involved with and it is seen in Fig. 2 that the dissolution time for each point feeder is substantially reduced, while even though the mass of electrolyte is more than halved.

Slots in the Anode

Slots were introduced to lower the cell resistance by reducing the gas travel path and hence the average bubble volume of the electrochemically formed gas that displaces the conducting electrolyte. However, the large bubbles rising from the electrolyte are the main driving force for the transverse electrolyte mixing—a crucial action needed to maintain uniform conditions in cells where the alumina is only fed to the centre channel [9]. Transverse slots retard the heat flow in newly set anodes which, because of the frozen electrolyte formed on setting, are not conducting their normal current. The remelting of this material is retarded by transverse slots, and they sometimes result in an increase in anode spike formation. Deep longitudinal slots dramatically reduce the ability for the anode gas release to assist transverse mixing.

Collective Consequences of Capacity Creep on the Physics of Cells

The consequential changed physics resulting from retrofits have several have consequences on cell operation:

- *Spatial variations in cell operating conditions have been magnified.* This impacts the traditional approach for cell control which assumes that a single spot measurement is representative of all the cell's condition.
- *The driving forces for electrolyte flow have been hindered significantly more both spatially and temporally.* This also impacts control logic, resulting in the deterioration of operating efficiency
- *The location and proportions of the secondary reactions that occur at the anode, the aluminium cathode, and the aluminium- cathodic carbon interface all change. These accelerate the degradation of the materials, as well as harming the energy and environmental efficiencies.*

The collective effect is the variations lower the quality of all mathematical models used for designing and operating the cells. There is a need to minimise spatial variability—either by design or changed control and operating practice.

Transverse Mixing of the Electrolyte in Cells

With the reduced ACD, and higher amperage, it becomes even more important for getting the dissolved alumina from the feed point to the extremities of the anode.

Mixing measurements performed in low energy, high-efficiency cells with a high bath volume/kA ratio [11] highlighted that while the total circulating rate, which incorporates reaction to the magnetohydrodynamic forces, is slow, and typically takes 1 hour for total homogenization it was adequate for alumina feed control. As a consequence of multiple feeders and with the assistance of gas-driven forces the electrolyte composition is within the band necessary for feed control.

However, in the last few years, several people from smelters around the globe who were operating technologies that have undergone capacity creep, have forwarded to me a blue-coloured material they have found in the side channel of tapped-out failed cells. This crystalline material has been identified as the mineral diaoyudaoite, a structural form of β-alumina. As such it can only crystallise from alumina saturated electrolytes that have cryolite molar ratios greater than 3. It has been proven this condition exists in operating cells that had undergone capacity creep and incorporating enlarged anodes that reduced the volume of electrolyte around the periphery. The limited cooling after switching off the current flow of a cell, before carefully removing the anodes after a short duration resulted in crystallisation of diaoyudaoite on the peripheral facing anodes surface. An example is shown in Fig. 3.

Since the electrolyte composition in these cells is controlled by analysis of the centre channel electrolyte and the makeup AlF_3 is fed feeding into the centre channel it demonstrates the cell has extremely poor transverse mixing—more than a 10 wt.% variation in excess AlF_3 concentration.

The design of slots needs to be changed so that there are adequate driving forces imparted for adequate transverse mixing.

As the freeze formed on newly set anodes is also a major barrier to transverse mixing, problems would be reduced by introducing preheated anodes. While this approach has received limited trials, more innovation is needed. It would contribute to energy conservation as well as reducing the transverse mixing barriers. Changed work practices relating to the anode removal and avoiding the need for cavity cleaning would also be beneficial for shortening the duration that newly set anodes present barriers to transverse mixing. Perhaps shortening the time of delivery of hot anode blocks from the baking furnace via rodding en-route to being set in the operating cells presents an opportunity.

Spatial Variation in Current Density

The intensity of the variation in transverse current density along the cathode block (refer Fig. 1) will be reflected by a similar variation in the current density profile at the metal

Fig. 3 Diaoyudaoite crystallised on a disconnected anode that had undergone limited cooling prior to removal from the cell

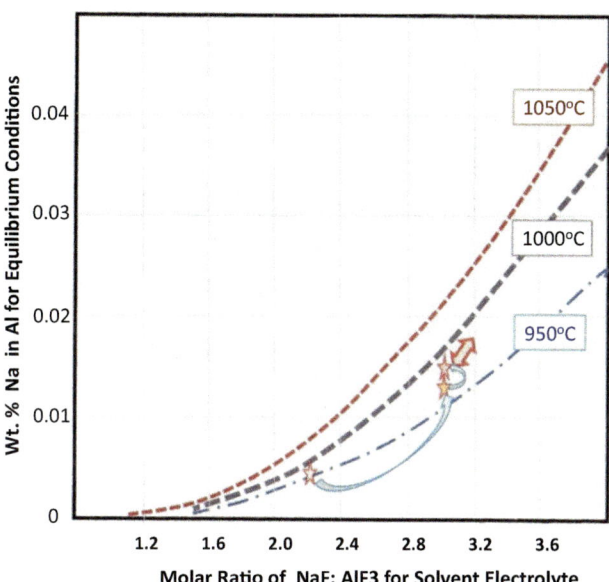

Fig. 4 Tracking the spatial increase in sodium co-deposition for the hotter NaF-enriched electrolyte against the reversible values measured by Tingle et al. [5]

pad-electrolyte interface. Therefore, the proportion of sodium being co-deposited in the high current density zones is increased due to three aspects of the electrochemical science and electrolyte properties:

- the overpotential necessary to sustain the higher current density automatically increasing the interfacial potential. This increases the minimum/reversible Na concentration in the aluminium that will be deposited.
- the increase in metal and electrolyte temperature further lowers the deposition potential of sodium relative to that for aluminium.
- The enrichment of the interfacial electrolyte composition through sodium ions being the dominant transporter of the current between electrodes further enriches the minimum deposition amount.

When all these considerations are combined with the realisation that the high current density coincides with the region where the electrolyte is alkaline and therefore has a molar ratio in excess of 3.0, we see the sodium co-deposition in that zone can have a fourfold increase above the reversible value.

This increase in sodium in the metal has three harmful chain reactions!

(a) with Na concentration being above equilibrium, it will have a higher solubility in the adjacent electrolyte, and this dissolved sodium is the cause of the increased

lowering of current efficiency observed through capacity creep.

(b) The above equilibrium Na of the metal has a greater interaction with the cathode carbon forming more intercalation compounds—and faster. This results in greater expansion stress forces causing cracking and cathode wear in the hotter higher Na concentration zone at the ends of the cathode block. This is evident in Fig. 5.

(c) The migration of the sodium through the cathode carbon is also accelerated by the increased temperature generating a high vapour pressure of sodium—the cell only being 60 °C above its boiling point! This also leads to sodium permeating through the sidewall blocks and also out collector bar windows corroding aluminium flexes.

Meanwhile, the centre channel, which is deficient in heat generation, yet has the highest heat demand for dissolution of alumina, readily develops problems arising from sludge formation and alumina solubility! The adverse consequences of both the asymmetric heat generation and sodium co-deposition can also be seen in Fig. 5.

The spatial effect of increased sodium co-deposition through higher current density caused by current flow taking the shortest path through the cathode carbon is not confined to the ends of the cathode block. Cell have been constructed with a "wave breaker" cathode block design on the erroneous assumption it might reduce the metal pad velocity, flatten it, and also lower the current density. One such design is illustrated in Fig. 6, for a cell that only operated in the potline for approximately 40 days. The higher current density between metal and cathode block at the stepped base of raised section gave increased sodium transport and expansion causing delamination of part of the block. Most, if not all the cells, in the potline failed within 400 days.

Changed Physics and Cell Design

Historically, a weak point in the cell construction has been the need to have an insulating material that is resistant to corrosion or attack by the metal, the anode gases and electrolyte. Silicon nitride-bonded silicon carbide has proven to be the material that satisfies the requirements best, although it does undergo oxidation by the cell gases and that facilitates other degradation processes. Therefore, the practical approach has been to attempt to operate the cell with a controlled sidewall heat loss that ensures a frozen cryolite ledge remains on the inner face in contact with the metal and electrolyte. The high heat generation that is consequential to the high current density at the ends of the cathode blocks limits the effectiveness of this approach and harmful

Fig. 5 A failed cell showing the consequences of excessive heat generation and sodium uptake at the cathode block ends, and insufficient heat generation in the centre channel

Fig. 6 Cathode block design giving extremely early failure as a consequence of ignoring the science of sodium uptake in high current density zones

secondary consequences have arisen through the absence of the protective ledge. Existing models on which cell designs are based need refining!

Changed Physics and Cell Feeding and Control Logic

For good performance, keeping the electrolyte in the cell within a narrow band is crucial for good energy efficiency and productivity. Typically, for the high productivity retrofitted cells, the increased loss in efficiency as a consequence of capacity creep approaches 3%. Based on the science associated with current efficiency the contributing factors are primarily linked with the changed current distribution at the cathode surface.

Monitoring both the cell voltage and the rate of voltage change has sufficient accuracy to prevent transgression to an anode effect—but not necessarily to prevent PFC co-evolution, since the root cause of anode effects is electrode passivation through insufficient heat. The extremely high spatial heat generation in the electrolyte will overcome the heat deficit in the outer zones, so there is a risk of spatially co-evolving low levels of fluorocarbons in selected zones without any sign of it occurring.

The spatial and temporal notion nature of the signal changes means the control logic is not responding to the right magnitude of process change.

Overcoming the Problems Introduced by Capacity Creep—Remaining Challenges!

The dominant causes of the substantial loss in cell life, and reduction in current and energy efficiency are operating with process parameters that exceed acceptable limits because of constraints the cell physics introduce to operating temperature, current distribution, and electrolyte mixing. They highlight design weaknesses.

1. The asymmetry in cathode current distribution is directly linked with the design and materials selection of the cathode block-collector bar assembly. The physico-chemical properties of these materials require the collector bar electrical resistance to be low and have a significant increase in resistance as you progress from the centre channel of the cell to the extremities if the design persists with introducing the current from the sides of the pot shell. A range of options have been proposed [5, 6] but the only openly disclosed designs focused on reducing the resistance in total, without regard to distribution of the current at the carbon surface. Indeed, some have magnified the problems of current asymmetry.

2. Co-deposition of sodium is undesirable at any time, but scientific principles mean it cannot be prevented. Besides having a reduced overall metal pad temperature (and hence cell operating temperature) having a good circulating metal pad flow and mixing combined with an enlarged surface area are the only options for lowering the total sodium being co-deposited.

3. Innovation for increasing the electrolyte transverse mixing is needed. This is primarily influenced by anode length and gas release slot designs. Imaginative thinking is needed.

4. With the harm caused by the anode setting practice introducing spatial and temporal problems, it also raises the question on priorities in anode setting and its pattern (e.g., 2 anode assemblies v's 1 in same zone of the cell).

Finally, has the industry reached the stage where "just in time" technology for the removal of blocks from baking furnaces, rodding and delivery to setting site is appropriate? After all, total energy efficiency is a global responsibility.

But for further advances, it is clear that it is time for the industry to respect to the interlink between science, physics, cell design, and environment responsibility!

Acknowledgements Special thanks to all the outstanding postgraduate students and postdoctoral research workers that I have had the privilege of motivating, guiding, and having an input into their experimentation. Collectively, the valuable information they have generated leaves the industry in a sound position for resolving the problems it creates when attempting to translate it to the operating mode! Also special thanks to companies associated with the industry that not only sponsored the projects but also made their special laboratory facilities available and hosted the students when doing experimental work in their facilities—especially R&D Carbon. All the industry sponsors showed trust in my ill-defined needs for sponsorship, commonly agreeing to fund proposal prior to terminating the phone call. And thanks to my academic colleagues who joined me on the journey and added the extra dimension through their own specialty and expertise.

References

1. Richards N. E. & Welch B.J. "Anode overpotentials in the electrolysis of alumina." In "Extractive metallurgy of aluminum, Vol. 2 Aluminum" ed. Gerard G. Interscience. New York (1963) pp 15–30.
2. Dewing E.W. and van der Kouwe E.T., *J Electrochem Soc.* Vol 122, 358,(1975)
3. Hume S.M, Utley M.R, Perruchaud R.C, & Welch B,J. "The influence of low current densities on anode performance" In Light metals 1993. The Minerals Metals and Material Society, Pittsburgh; Springer, New York pp 525–531.
4. Dorreen Mark M.R., Richards Nolan E., Tabereaux Alton. T., and Welch Barry J. "Role of Heat and Interfacial Phenomena for the Formation of Carbon Oxides In Light metals 2017. The Minerals Metals and Material Society, Pittsburgh; Springer New York pp 659–667.
5. Tingle W. H, Petit J. & Frank W.B. "Sodium content of aluminium in equilibrium with $NaF – AlF_3$ melts", Aluminium 57, (1981) pp 286–288.
6. Gjrotheim K. "Contributions to the theory of the aluminium electrolysis" D. KLG Norsk. Vid. Sel. Skrifter 1956 NR 5, 90 pages.
7. von Kaenel R., Antille J., "Energy savings by using new cathode designs", *Proc. 5ᵗʰ International Electrodes Conference*, Reykjavik, Iceland, 10–12 May 2011.
8. *von Kaenel R, Antille J.,& Bugnion L.. "THE IMPACT OF CATHODE AND COLLECTOR BAR DESIGNS ON CELL PERFORMANCE."* Proc 11ᵗʰ AASTC, Dubai 2018 Kenote#9
9. Jassim, Ali, Akhmetov S, Welch B, Skyllas-Kazacos M, Jie Bao and Yuchen Yao, "Studies on anode preheating using individual anode signals and haul her rolled reduction cells."
10. Martin O., Gariepy R., Girault G. "*APXe AND AP60: "THE NEW REFERENCES FOR LOW ENERGY AND HIGH PRODUCTIVITY CELLS."* Proc 11ᵗʰ AASTC Conf. 2018 Th9.
11. Purdie J.M., Bilek, Taylor M.P.,Zhang W.D., Chen J.J.J., & Welch B.J. ' Impact of anode gas evolution on electrolyte flow and mixing in aluminium electrowinning cells. In Light metals 1993. The Minerals Metals and Material Soc., Pittsburgh; Springer New York pp 355 –361

Anode Quality Optimisation: Industry Learnings from the Research Supervised by Barry Welch

Alan Tomsett and Barry Sadler

Abstract

Barry Welch has a long history of supporting carbon anode research from the perspective of both the anode producer and the potroom customer. The outcomes of this research from Barry and his students continue to be essential reading for anode producers and technologists. The key learnings of the research will be summarised to provide a single entry point to the huge body of work that has been produced, to remind experienced technologists of the source of the work and to inspire new scientists and engineers to explore the world of anode carbon.

Keywords

Anode carbon • Student research • Aluminium smelting

Introduction

Carbon anodes are continuously consumed in the Hall-Héroult process for the production of aluminium largely via the reaction:

$$2Al_2O_3 + 3C \rightarrow 4Al + 3CO_2 \tag{1}$$

The theoretical anode consumption is 0.33 t carbon/t Al which would lead to CO_2 emissions of ~ 1.22 t CO_2/t Al. The actual consumption, however, is typically between 0.4 and 0.45 t carbon/t Al with emissions of 1.45–1.6 t CO_2/t Al due to current efficiency losses and excessive reaction to air and CO_2 in the cell. Anode producers and technologists are continually working on methods to minimise excess consumption, through a reduction in the intrinsic reactivity of the anode carbon, improved anode formulation and baking, and improved covering of the anodes in the reduction cell. Much of the basic understanding of the performance of anode carbon that supports the improvement work comes from the work of students supervised by Barry Welch. This paper will summarise key aspects of this research work conducted at the University of New South Wales, Sydney, Australia (UNSW); The University of Auckland, New Zealand; and R&D Carbon, Sierre, Switzerland.

Anode Reactivity Fundamentals

Carbon anodes are produced by the compaction of a mixture of calcined petroleum coke, recycled spent anodes and coal tar pitch as binder into a solid mass. These "green" anodes are then baked to temperatures up to 1200 °C to achieve the desired properties. Within the reduction cell, the main mechanisms for carbon consumption are

- Electrolytic consumption—the consumption of carbon to produce aluminium metal via Eq. (1),
- Carboxy consumption—Reaction with the CO_2 generated in the cell via the Boudouard reaction

$$C + CO_2 \rightarrow 2CO \tag{2}$$

- Oxidation—Reaction with air in the cell, and
- Dusting—Physical loss of carbon particles from the anode surface.

Barry's students have made significant contributions to our understanding in at least the first three areas.

Sources of Excess Carbon Consumption

Equation (1) is the dominant reaction for aluminium production at the current density typically used in the industry.

A. Tomsett (✉)
Rio Tinto Pacific Operations, Brisbane, Australia
e-mail: alan.tomsett@riotinto.com

B. Sadler
Net Carbon Consulting, Kangaroo Ground, Australia

© The Minerals, Metals & Materials Society 2023
S. Broek (ed.), *Light Metals 2023*, The Minerals, Metals & Materials Series,
https://doi.org/10.1007/978-3-031-22532-1_4

Farr-Wharton [1] confirmed that at low current density, the production of CO becomes favoured via the reaction:

$$Al_2O_3 + 3C \rightarrow 2Al + 3CO \qquad (3)$$

A comparison of Eqs. (1) and (3) shows that the theoretical carbon consumption doubles at a lower current density. While this theoretical change was not measured, a significant increase in consumption was seen (Fig. 1). This was confirmed and further investigated by Hume et al. [2], to show that one source of excess consumption is the generation of CO during electrolytic consumption at the low current density areas on the side faces of the anode and in slots, and is also in turn impacted by the depth of immersion of the anode in the bath.

Farr-Wharton [1] also studied the impact of porosity on consumption. Electrolytic consumption was found to take place in pores >3 μm radius. The Boudouard reaction (2), on the other hand, takes place in pores <1 μm radius. Anodes with high porosity, and a wide pore size distribution range, such as those with a lower than optimum pitch

content (14% in Fig. 2), had the highest penetration of CO_2 into the anode structure.

The work also showed that the pore size distribution became broader, and the pore volume increased during the reaction (Fig. 3).

This work, which remains relevant today, highlighted the importance of defining and maintaining the optimum pitch level in anodes, with data to show why underpitched anodes are more reactive and the spent anode butts are soft and dusty.

The carboxy reactivity of a range of petroleum and pitch cokes, as well as crushed anode cores were tested in a fluidized bed reactor by Janette Wood at the University of Auckland [3]. Two key conclusions were

1. The carboxy reactivity of anode samples was, as expected, dependent on baking level. However, the results did show that the rate of improvement in reactivity decreases at very high baking temperatures (Fig. 4). The results support an industry rule of thumb that the optimum heat treatment of the anode during baking is to match the heat treatment seen by the filler coke during calcination. This optimum is based on a balancing of anode quality and baking costs. As discussed in the companion paper [4], if the focus is solely on anode quality the use of higher baking temperatures up to the point, where desulfurization occurs, may be preferred.
2. The total porosity of the coke sample has the largest impact on carboxy reactivity (Fig. 5).

The small number of samples made it difficult to separate the impact of catalytic metals and calcination temperature. This problem was resolved by the use of a large number of samples in a study by Sheralyn Hume at R&D Carbon [5].

In this work, a statistical analysis and experimental study of the reactivity of a wide range of petroleum coke samples were conducted. Almost 500 different cokes, with a wide range of properties, were analyzed.

Models of petroleum coke reactivity were developed, based on impurity levels and structure. For CO_2 reactivity,

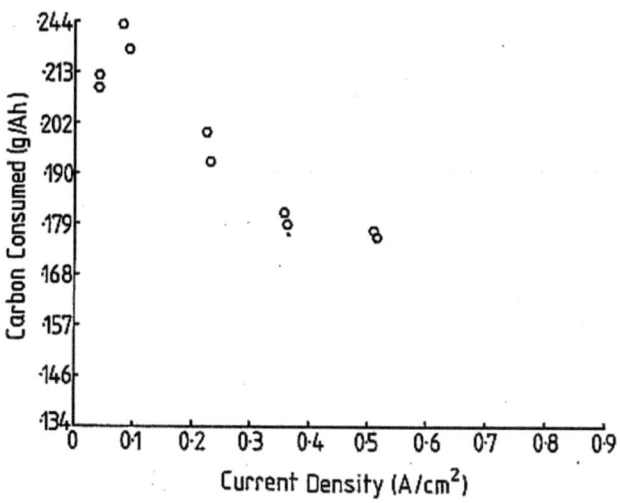

Fig. 1 Impact of current density on anode consumption—anode produced with optimum binder baked at 1150 °C [1]

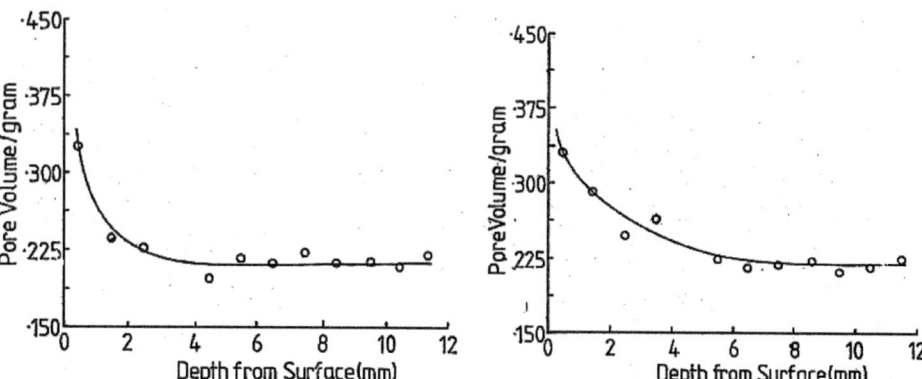

Fig. 2 Extent of subsurface oxidation in carbon anodes—Optimum binder (LHS) and low binder content (RHS) [1]

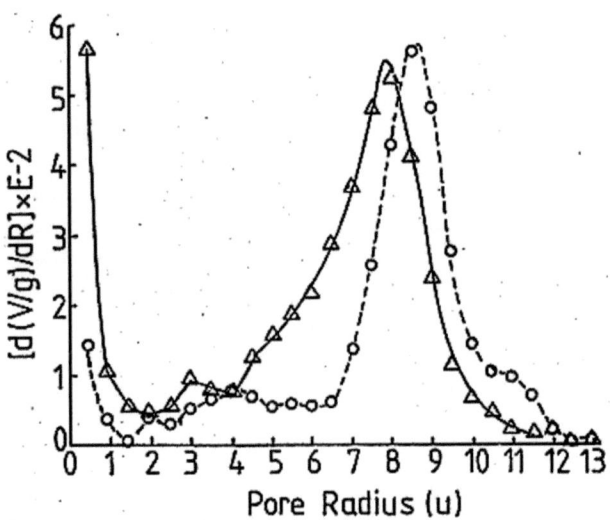

Figure 4.2. Oxidation of a Carbon Composite
              100% CO₂: Δ Surface; O Unoxidized Blank.

Fig. 3 Pore size distribution at the surface of an oxidized sample (Δ)
and the original unoxidized sample (O) [1]

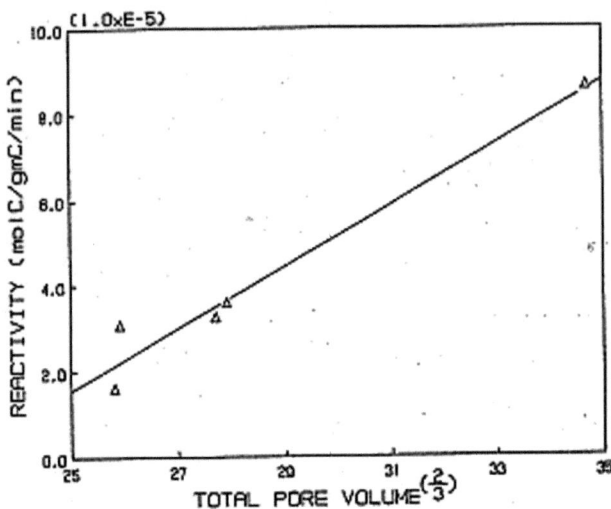

Fig. 5 Cokes with higher total pore volume had the highest reactivity
[3]

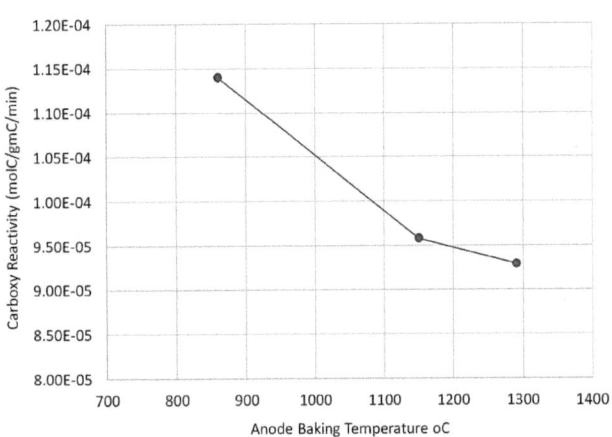

Fig. 4 Anode reactivity reduces with increased baking temperature

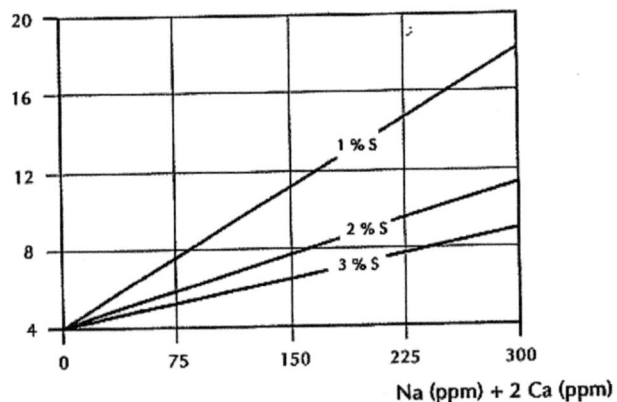

Fig. 6 Influence of coke purity on carboxy reactivity [5]

the bulk density and calcination level have some impact, but
the main drivers of the reactivity were found to be sodium,
calcium, and sulfur:

$$RCO_2(\%) = 4.0 + (0.0411\ Na\ (ppm) + 0.101\ Ca(ppm))/S\ (\%) \quad (4)$$

The models for oxidation were more complex because
purity, calcination level, and porosity all have a significant
impact on air reactivity. Further details on the air reactivity
models can be found in Reference [5].

The CO_2 reactivity model is shown graphically in Fig. 6.
A key finding in this work was that the sulfur content in coke
has an inhibiting effect on the catalytic activity of sodium
and calcium.

This becomes particularly important for the reactivity of
industrial anodes, where additional sodium and calcium are
added due to electrolyte carryover with recycled butts. As an
example, in Fig. 7, is the carboxy reactivity rate of two
cokes, along with the reactivity of laboratory anodes with
extra sodium and calcium contamination. The reaction rates
of the coke and anode are similar in the higher sulfur coke
(Coke S—2.1% S), whereas the anode reactivity rate is
higher than the coke when using a low-sulfur coke (Coke
M2—0.82% S). The environmental drive to reduce SO_2
emissions through the use of low-sulfur coke introduces a
need to continually strive to reduce sodium and calcium in
the anode.

Secondary Ion Mass Spectroscopy was used to show that
there is a strong binding action between sulfur and sodium
which restricts both the migration of sodium throughout the

Fig. 7 Arrhenius plots of petroleum cokes and binder matrix electrodes in CO_2 [5]

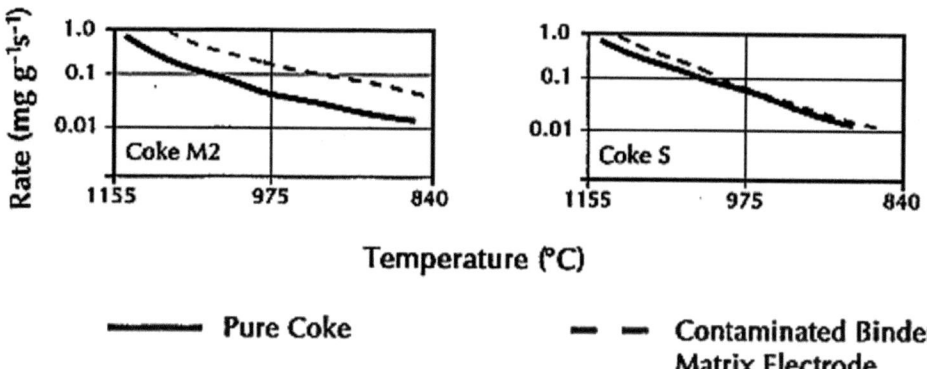

binder matrix, and the catalytic activity of sodium in high-sulfur composites. For the low-sulfur samples, there was insufficient sulfur to bind all the sodium, allowing it to migrate throughout the composite. The broad distribution of free sodium also increases the catalytic activity of the sodium. The relevance for anode manufacturers is that there needs to be a balance between the environmental requirements of using lower sulfur cokes with the need for higher sulfur to minimise carbon reactivity and excess consumption. This study was the first to give clarity about this mechanism of sodium sensitivity of coke and anodes.

Anode Oxidation

In addition to the studies on carbon carboxy reactivity, a study on anode oxidation was completed by Fitchett at the University of Auckland, which included both laboratory studies and plant trials at the New Zealand Aluminium Smelter (NZAS) [6]. The study confirmed the importance of anode protection when in the reduction cell.

In laboratory tests, the anode carbon oxidation rate was found to be minimal at temperatures below 420–460 °C, but then increased rapidly as the temperature increases to about 650 °C. Above 650 °C, the reaction becomes mass transfer controlled, and the rate stabilizes (Fig. 8) but at a high level. The rate of oxidation varies based on the structure, porosity, and trace metal content of the anode (good vs. bad anode in Fig. 8b).

While the quality of the anode has an impact on the oxidation rate, the study showed that unprotected carbons exposed to air will oxidize once the temperature reaches around 450 °C, and that the oxidation rate can be stabilized when the mass transfer of oxygen to the carbon surface is limited. Fitchett also studied the impact of cover depth on oxidation rate. The results in Fig. 9 show that the oxidation rate of poor-quality carbon can converge to the oxidation rate of good-quality carbon when the anode is adequately protected from air in the cell.

The results reinforce the need not only to adequately cover the anodes in the cell with crushed bath and alumina, but also to redress/recover during the anode life, particularly

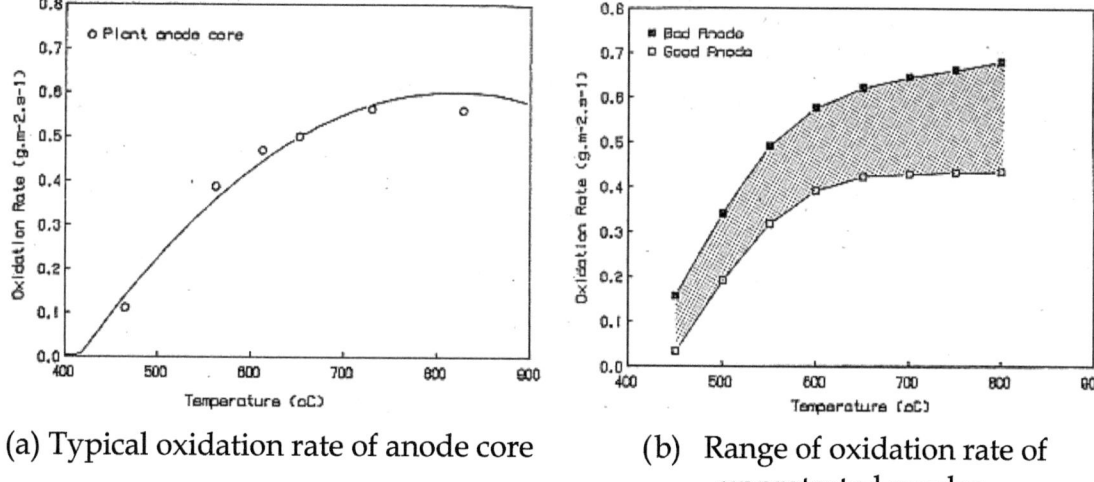

(a) Typical oxidation rate of anode core

(b) Range of oxidation rate of unprotected anodes

Fig. 8 Unprotected anode oxidation rate in the temperature range seen within the cell [6]

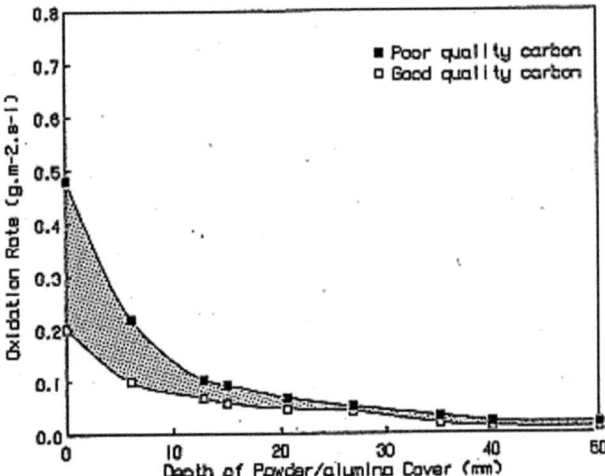

Fig. 9 Oxidation rate is impacted by carbon quality and level of protection [6]

if there has been a process disturbance in the cell that may have damaged the cover. The essential message of this study remains relevant and critical to good anode performance. It provides data to show where to put efforts to reduce airburn.

In addition to bath/alumina cover, NZAS uses aluminium spray to protect the vertical surfaces of the anode. Fitchett also qualified the impact of the spray on the carbon loss from different anode faces (Fig. 10). The results confirmed that the use of aluminium spray or an alternative coating successfully reduces anode oxidation in cells where it is difficult to protect the vertical anode faces, particularly in the centre of the cell.

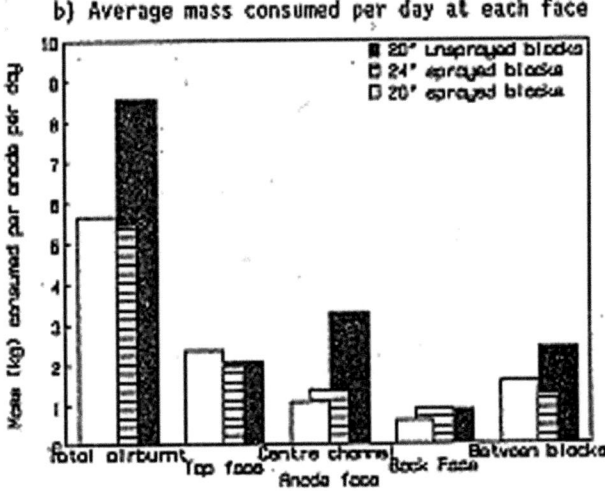

Fig. 10 Aluminium spray can be used to reduce oxidation loss from anodes [6]

Anode Production

The importance of binder properties, binder content, and binder matrix was further investigated by Smith [7] and Hulse [8] at R&D Carbon under the supervision of Barry Welch, Werner Fischer, and Raymond Perruchoud.

Smith produced carbon composites from various pitches and carbon fines to simulate the performance of the binder-fines matrix in the anode and confirmed the impact of sodium and calcium, and baking temperature, on the composite reactivity. Of particular interest was that sodium and calcium in pitch have a strong catalytic activity on the composite reactivity. The level of these impurities needs to be considered in pitch purchasing decisions, and the ability to adequately clean the spent anode butts.

Smith also conducted an industrial survey of the ball mill and filter dusts used by anode manufacturers. There was a wide range of particle sizes reported, as measured by Blaine Index (Table 1). Filter dust in particular was highly variable with the Blaine Index ranging from 2000 to 28,000.

Binder-fines matrix composites were produced using the range of dust granulometry identified in the survey. A dust fineness of 3000–4000 Blaine, with a dust content in the anode of between 30 and 40% was found to provide the optimum reactivity behaviour. This was one of the first works to clarify the concept and importance of the binder-fines matrix as a component of the anode structure.

Hulse [8] conducted a more detailed follow-up survey of industrial anode producers, requesting details of raw materials, paste plant and baking furnace parameters, and anode core properties. The data received was analyzed and linked to the literature to provide a detailed report on the sensitivity of raw materials and production parameters on the quality and performance of green anodes.

It is not possible to do full justice to Hulse's work in a short review paper. The thesis remains a reference book for anode production especially the critical issues in green anode manufacture.

The key points for anode manufacturers and anode users are, however:

- There is significant interdependence between variables that determine anode quality.
- Based on models from other industries, a straight line recipe (granulometry) provides the best anode quality.
- The highest possible packing density of aggregate does not always provide the optimum anode quality.
- The choice of anode recipe needs to be adjusted to meet the requirements of the specific process equipment

Table 1 Average particle size (Blaine index) of industrial carbon dusts [7]

Material	Average Blaine index $\pm 2\sigma$
Mill dust	3033 ± 1384
Filter dust	9573 ± 14,498
Combined dust	3337 ± 1325

installed and the flexibility of that equipment. For example, vibrated anodes were found to be more sensitive to changes in granulometry than pressed anodes.

- A combination of dust fineness and dust content needs to be considered when adjusting granulometry. An example is provided in Fig. 11 which shows the results of a plant trial where improved properties were achieved with anodes produced with a lower content of finer fines (25–29% of 5500 Blaine) compared to the reference of 34% of 3500 Blaine. In addition, adjustments to mixing and forming parameters are required.
- Paste mixing parameters are important to achieve optimum anode mechanical properties. The quality of mixing also depends on the equipment used. Mixing parameters

need to be adjusted when changing granulometry (e.g. finer fines).

- Paste rheology is a critical parameter and provides the theoretical underpinning of the use of paste cooling between mixing and forming in anode production.
- Optimum anode quality is typically achieved at a baked anode density just below the maximum value. This reduces the likelihood of anode over compaction which has a negative impact on electrical resistivity and mechanical properties.

Many of these findings should be revisited to reduce the impact of deteriorating raw material quality.

Conclusions

Barry Welch supervised students working on anode reactivity and anode production for over 20 years. The results of these studies have made a significant contribution to the understanding of anode performance in the reduction cell.

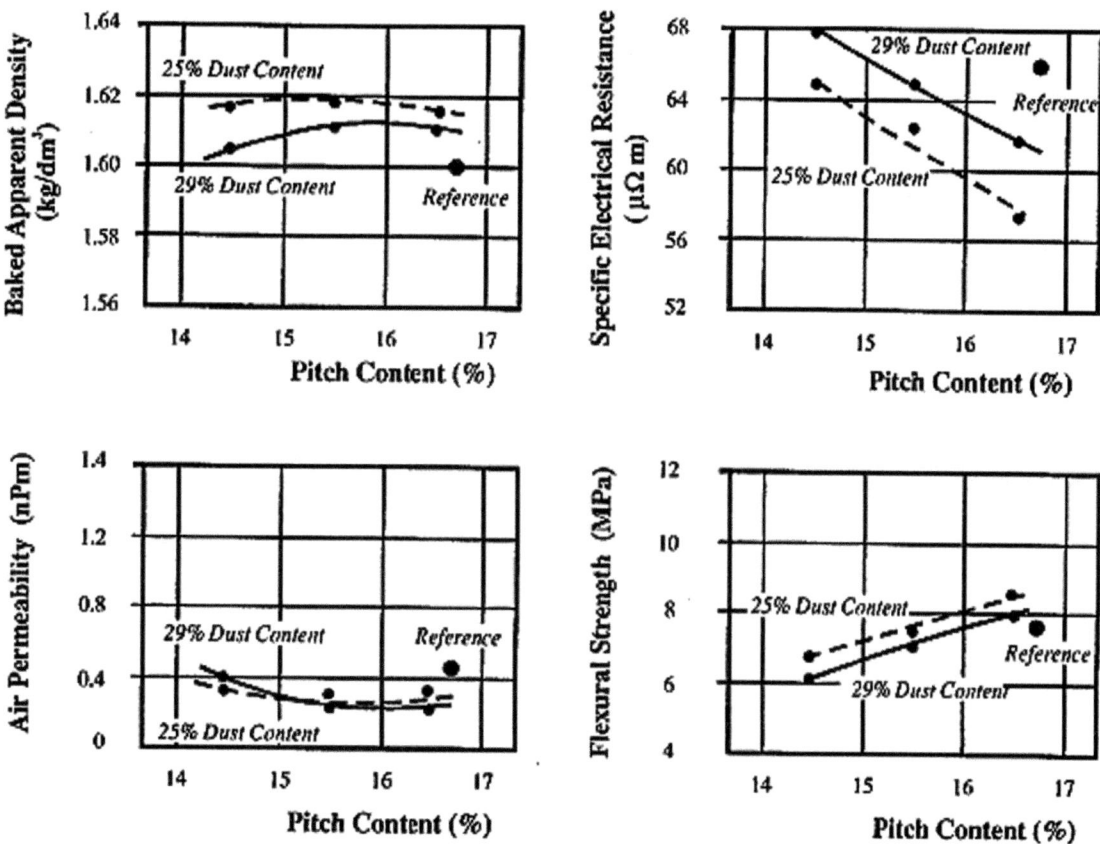

Fig. 11 Influence of formulation on anode quality [8]

Much of the basic understanding of the production of anodes and anode performance is based on this body of work. These studies remain highly relevant today.

References

1. Farr-Wharton R. (1980) An Investigation into the Oxidation of Carbon Anode Composites While Subject to Electrolysis in Fused Cryolite-Alumina, PhD Thesis, University of NSW
2. Hume, S, Utley, M, Welch, BJ and Perruchoud, R (1992), The Influence of Low Current Densities on Anode Performance; In Cutshall, ER (Ed), Light Metals 1992, The Minerals, Metals and Materials Society, Pittsburgh p 687–692
3. Wood, JM (1988), Fluid Bed Carboxy Reactivity of Cokes, Report to Comalco Limited, The University of Auckland
4. Sadler, BA and Tomsett, AD (2023) Meeting the Requirements of Potline Customers – the Largely Unmet Challenges set by Barry Welch to Carbon Anode Producers, Submitted to Broek, S (Ed), Light Metals 2023, The Minerals, Metals & Materials Society, Pittsburgh; Springer, New York
5. Hume, SM (1999), Anode Reactivity – Influence of Raw Material Properties; R&D Carbon Ltd, Sierre.
6. Fitchett, AM (1988) The Oxidation and Protection of Heterogeneous Carbon Anodes Used for Aluminium Smelting, Report to Comalco Aluminium Limited, The University of Auckland.
7. Smith, MA (1991) An Evaluation of the Binder Matrix in Prebaked Carbon Anodes used for Aluminium production, PhD Thesis, The University of Auckland
8. Hulse, KL (2000), Anode Manufacture – Raw Materials, Formulation and Processing Parameters, R&D Carbon Ltd, Sierre

Process Recovery to Unlock Power Efficiency Improvement at BSL

Evan Andrews, Thomas Booby, Murray Ure, and Hao Zhang

Abstract

In February 2019, Boyne Smelters Limited (BSL) established a project team to recover Reduction Line 3 from a process excursion. The Line 3 operation had deteriorated progressively through 2018, with increasing levels of cell instability, anode spiking, and work handed over between shifts had reached critical levels. By January 2019, average current efficiency had dropped to 91.7% and power efficiency had climbed to 13.45 DCkWhr/kg Al. The recovery project ran for 2 months and quickly achieved stable operation through a series of project clusters targeting exception cell recovery, new cell control, and management of intergenerational cell designs. An average of 64 mV was saved in 2019 compared to 2018, representing a power efficiency improvement of 0.24 DCkWhr/kg Al. Further stabilisation into 2020 resulted in full-year performance of 94.2% and 12.96 DCkWhr/kg Al current efficiency and power efficiency respectively. This paper describes the fundamental approach taken to problem-solve and deliver value quickly.

Keywords

Excursion recovery • Power efficiency • Specific energy • Current efficiency

E. Andrews (✉) · T. Booby · M. Ure
Boyne Smelters Limited, Hadley Dr, Boyne Island, 4680, Australia
e-mail: Evan.Andrews@riotinto.com

T. Booby
e-mail: Thomas.Booby@riotinto.com

M. Ure
e-mail: Murray.Ure@riotinto.com

H. Zhang
Transformation and Technical Support, Pacific Operations, 155 Charlotte St, Brisbane, 4000, Australia
e-mail: Hao.Zhang@riotinto.com

Introduction

Boyne Smelters Limited (BSL) is located in Queensland and is the second largest smelter in Australia, with an annual output of approximately 500kT. BSL commenced operations in August 1982 with two reduction lines utilising 480 Sumitomo 170kA end riser cells. There has been an extensive modification of the Sumitomo Technology with the most recent development described in [1]. In 1997, the third line of 264 cells was started using magnetically compensated side riser AP30 technology.

Line 3 was operating at 370kA and in the process of transitioning from the third to fourth generation (G3 and G4) cell designs during 2018, when performance progressively deteriorated. Carbon dust in cells increased and anode spiking became widespread across Line 3 from early September 2018. There was an associated drop in current efficiency as reported by alumina shot mass as shown in Fig. 1.

The number of cells with extreme metal pad noise increased substantially from October 2018, as shown in Fig. 2. A similar trend was also seen for high-frequency anode noise. The noisy resistance signal resulted in alumina feed control issues with aborted tracks (Fig. 3), sludge and anode effects.

The control system and operations team responded to quell noise by the addition of additional power to cells, as shown in Fig. 4, which further inflamed the situation. Thermally, the cells were becoming overpowered leading to increased liquid level variability, loss of anode cover integrity, air burn, and doubling of average Line 3 Iron levels to 1800 ppm.

By January 2019, the situation had become critical with substantial amounts of work being handed over between shifts due to insufficient people and crane resources to manage the situation. Current efficiency had dropped to 91.7% and power efficiency had climbed to 13.45 DCkWhr/kg Al. The full extent of the situation was difficult

© The Minerals, Metals & Materials Society 2023
S. Broek (ed.), *Light Metals 2023*, The Minerals, Metals & Materials Series,
https://doi.org/10.1007/978-3-031-22532-1_5

Fig. 1 Indication of cell performance based on alumina shots mass normalised with amperage

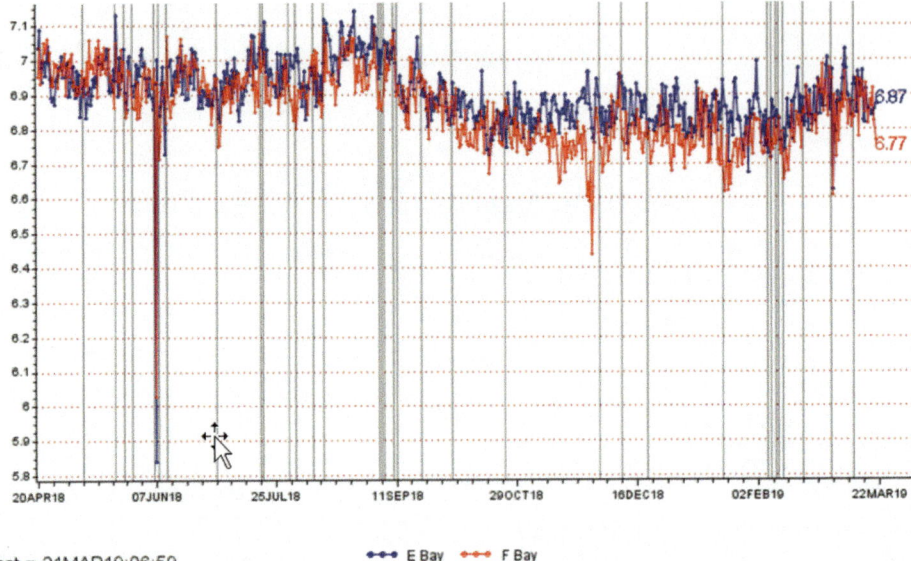

Fig. 2 Number of cells with metal pad noise >0.022 nano-ohms

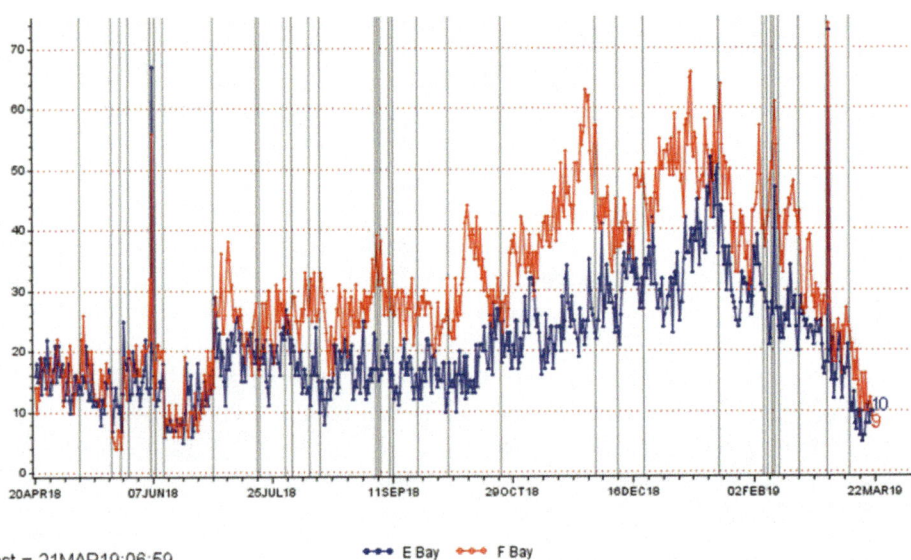

to assess due to poor tracking and management of exception work. There were also underlying critical concerns with people resources, crane reliability, and longer term anode supply with depletion of stocks.

Formation of Recovery Team

At the start of February 2019, a recovery team was formed to quickly regain control of the process over a defined 8-week period. The team was comprised of a technical leader, 3 senior engineers with reduction experience, and an expert cell controller. The recovery team worked alongside the existing Line 3 Operations team and was supported with

weekly onsite help from the Brisbane Smelter Technical Support Team.

Daily interactive briefing and training sessions were held with the process and operations team members to brief on the process recovery and the next steps. These sessions proved valuable in building trust and rapidly identifying critical issues and/or roadblocks that needed immediate support.

A daily update meeting was also held with the General Manager, Reduction Line 3 Manager, operational leaders, and the recovery team to review action progress against the plan, recovery metrics, and approve change management requests. A summary of the daily highlights, key performance indicator (KPI) graphs, and action plan was issued to

Fig. 3 Number of aborted tracks
per shift per cell

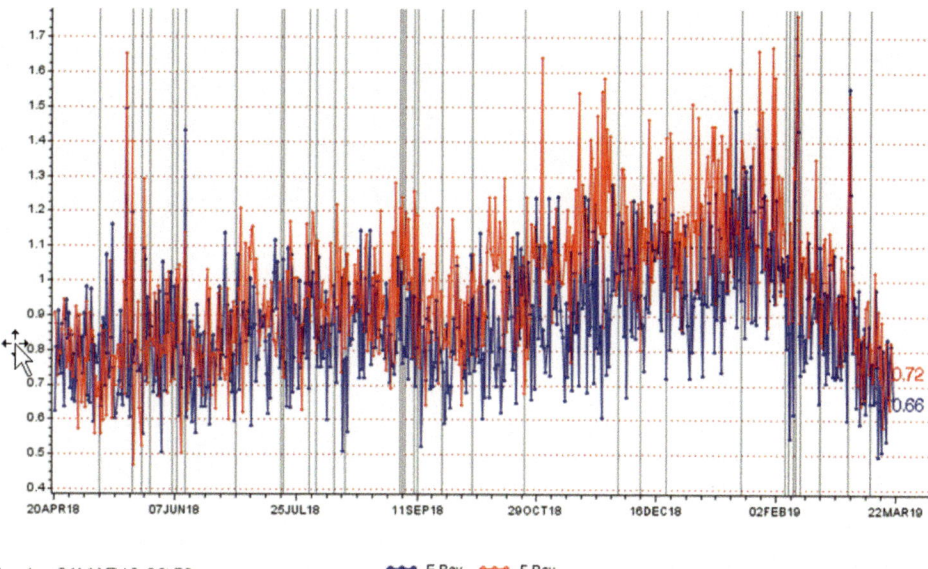

Last = 21MAR19:06:59

Fig. 4 Operator and Noise
resistance (micro-ohms)

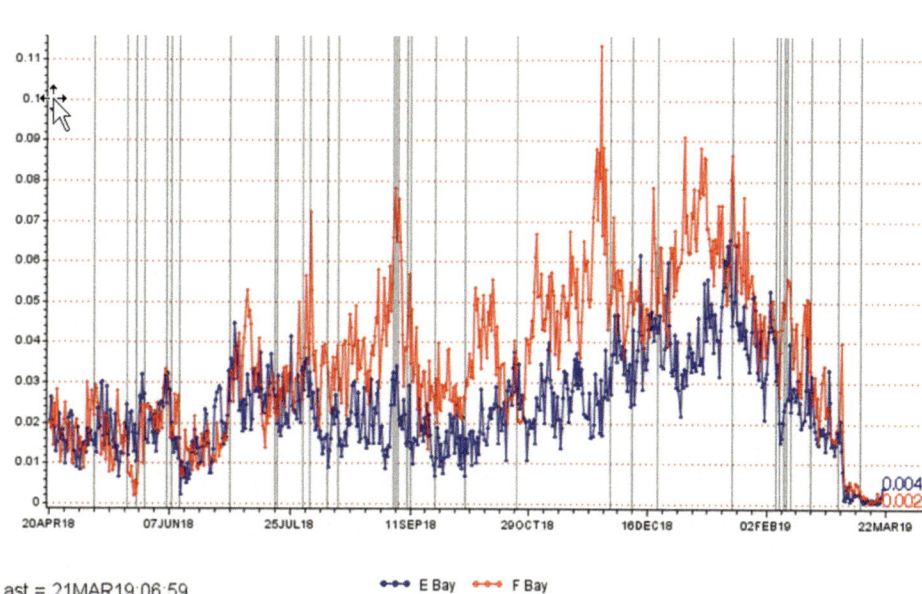

Last = 21MAR19:06:59

all stakeholders at the end of each day. This daily ritual kept everyone informed and maintained tension and urgency in the process recovery.

Immediate Actions

On the formation of the recovery team, the process was rapidly losing stability and control, with an immediate focus necessitating a triage approach to halt the deterioration in stability. Of particular initial concern was the opacity of operational work being handed over. Initial anecdotal reports of the number and criticality of unscheduled anode changes being handed over were unreliable, and significant gaps existed in the record-keeping of these anode issues. Furthermore, it was clear that the volume of corrective actions required far exceeded the resource capacity to complete said actions.

A war room was established to enable tracking and prioritisation of exception work on individual cells. There was limited capacity to action work, and so the focus was given to critical cell exceptions resulting in spikes and anode burn-offs.

The Cell Controller shift log books were updated to ensure that the status of all work was known and captured. Once this was done, it was identified that the numbers of unscheduled anode changes handed over between shifts was still growing, before peaking at 160 anodes per shift.

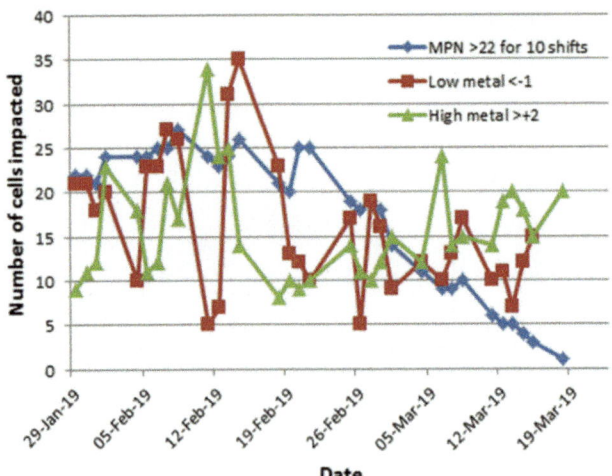

Fig. 5 Number of cells >0.022 Noise for >10 shifts and number of cells with high and low metal

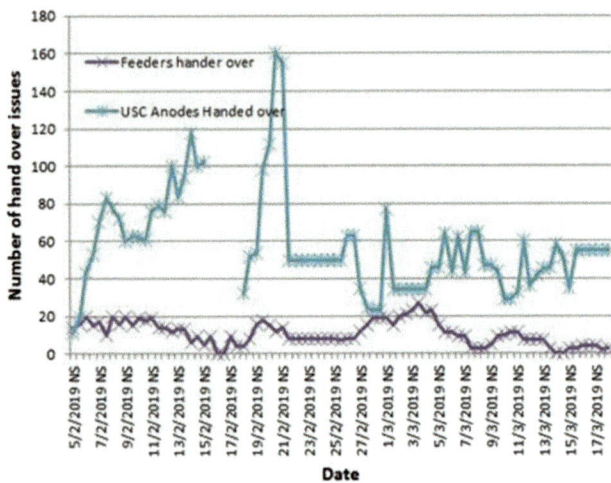

Fig. 7 Number of non-functioning alumina feeders and unscheduled anode changes handed over between shifts

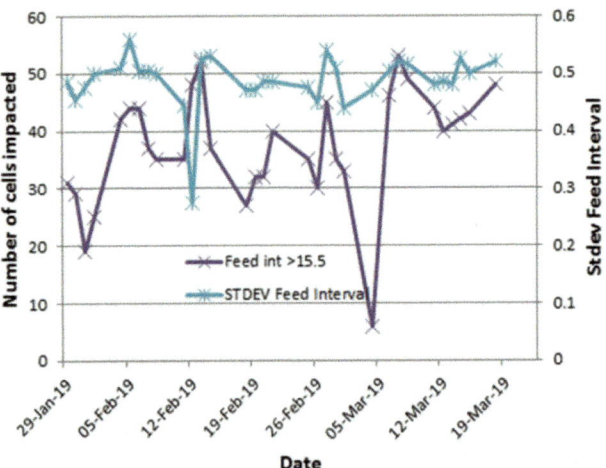

Fig. 6 Alumina feed interval and standard deviation

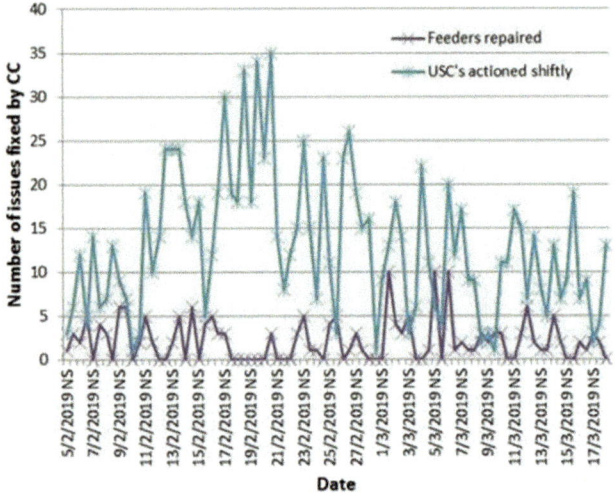

Fig. 8 Number of alumina feeders repaired and unscheduled anode changes actioned per shift

An agreed set of critical KPI's were established to track the recovery process. Figures 5, 6, 7, 8, and 9 show the basic KPI's used which included

- Number of cells >0.022 nano-ohm Noise for >10 shifts.
- Number of cells with high and low metal.
- Alumina feed interval and standard deviation.
- Number of non-functioning alumina feeders and unscheduled anode changes handed over between shifts.
- Number of alumina feeders repaired and unscheduled anode changes actioned per shift.
- Number of suspected spiked anodes per cell based on sample current draw measurements.

Recovery Strategy

With the situation changing quickly, it was important to understand the root cause(s) and key drivers to stabilise and then recover Line 3. Figure 10 shows the strategy that was developed from a fundamental analysis of the data and situation with key stakeholders.

High levels of cell exceptions, including unscheduled anode changes, unstable new cells, and failed alumina feeders, were hypothesised to be caused by an insufficient understanding of how to operate the G4 cell. Furthermore, insufficient exception cell priority and standards were

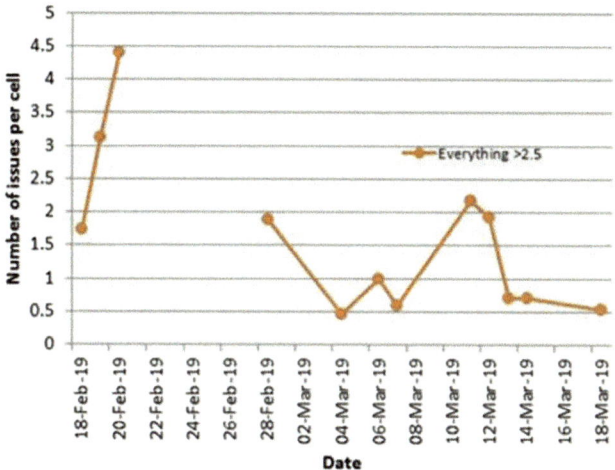

Fig. 9 Number of suspected spiked anodes per cell based on sample current density measurements

hypothesised to be contributing to the high levels of exception. It was important to break both cycles to reduce the overall generation of exceptions and a summary of the approaches taken are given in the sections below.

Exception Cell Priority and Standards

Prior to the implementation of the war room, a large number of exceptions were not being actioned and this helped perpetuate the cycle of exception generation. The war room was critical in providing a system to action exceptions and break this cycle.

The war room had three key objectives:

- Transparency—have a system to track exception work. This had to be highly visual. At the time, exception work that was not completed (i.e. handed over) was not tracked effectively.
- Prioritization—have criteria to organize exceptions into levels of priority. There was an imbalance in the importance placed on scheduled versus unscheduled work, to the point where unscheduled work had decreased to practically nothing on some shifts. Figure 11 shows the method that was used to improve prioritization of exceptions under excursion recovery conditions.
- Resource allocation—have a method to allocate resources to exception work. The amount of exceptions being generated far exceeded the resources available for action.

With the war room providing a means to manage exceptions, a number of other work standards and process control initiatives were implemented to reduce the rate of exception generation. These were generally "coarse" control type initiatives, and this largely meant moving back to the last known good or reverting to fundamentals or best practice. Some examples are listed below that are representative of this approach:

Fig. 10 Line 3 recovery strategy

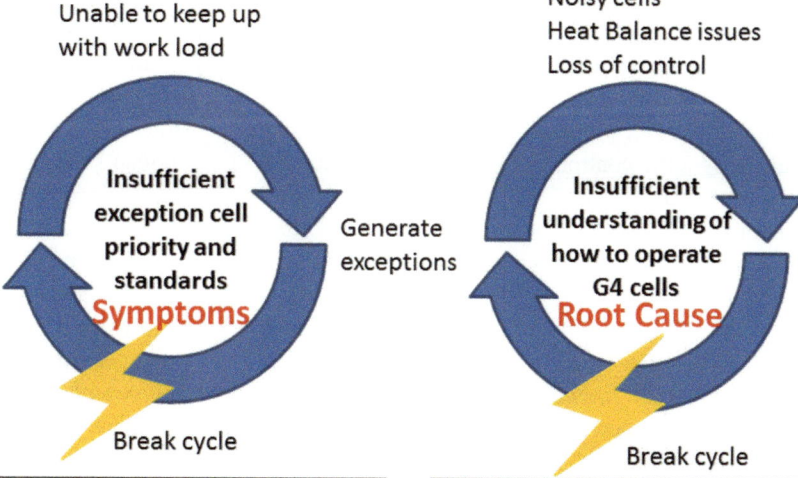

Issue:	Category:	When to Action:
Unstable Cell CD findings More than 990 deg C Feeder Issues and Multiple AEs Spikes Transition joint problem Burn off	Cat. 1	Before scheduled work where possible
Liquid Level Issue Exposed Stubs Multiple Cathode Noise alarms Required Raises - stable cell	Cat. 2	During scheduled set
Deformation Air burn	Cat. 3	After scheduled work is complete

Fig. 11 Method used for prioritization of exceptions

- Amperage—reduced amperage by 3 kA to help stabilize Line 3.
- Turnaround activity—cell replacement activity was temporarily paused in order to increase the ability to action exceptions.
- Bath Control—historically reduction Line 3 had used three levers to control bath levels with great effect:

  - Immediate control by removing or adding liquid bath;
  - Mid-term "fine" control by adding solid bath from feeders in cell; and
  - Long-term "coarse" control by adjusting alumina:bath blend in the anode cover.

This had drifted with too much emphasis on immediate control and long-term control, with the mid-term control through the bath feeders basically turned off. The reliance on long-term control to keep the line in bath surplus also had some unintended consequences with the percentage of alumina in the cover dropping ~10% over a number of years (approx. 55→45%). Observations of "sintering" and "bridges" forming in the crust were made, meaning air was able to get underneath the cover and attack the anodes. Changes were made to increase the percentage of alumina in the cover back to the desired target and increase the use of solid bath additions to get an overall better balance between the 3 levers:

- Carbon Dust Control—high levels of carbon dust (70% of all cells as shown in Fig. 12) was resulting in anode-

cathode distance (ACD) squeezing and generation of spikes. At the time, the number of resources allocated to hand skimming was not consistent, and the effectiveness was unclear as the amount of carbon being removed was not tracked. During the set operation if carbon was identified during the cavity cleaning process ("pac-manning"), then the expectation was that the operators identify the skip where the material was placed and it be segregated. These segregated skips and hand-skimming carbon were not returned to the bath process but removed through waste streams. No specific strategy was in place to remove carbon through extra pac-manning. Changes were implemented to ensure consistent skimmer resources, tracking of carbon being removed, targeted approach for hand skimmers, and pac-man skimming of corner stalls.

- Current Density Measurements—a strategy was implemented to complete current density measurements (voltage drop on anode rod) on all anodes for a rolling 10-cell group using the project resources. This helped the operations team in finding exception work, but more importantly was used as a barometer for process health. The number of high reading anodes gave a true indication of the spike rate when extrapolated, rather than measuring the whole line. Figure 13 shows that the most prolific type of anode failure were carbon and bath spikes (DEF), and addressing this issue would form a large part of the next phase of the recovery work (100-day plan). Routine current density measurements were also reviewed and

Fig. 12 Percentage of G3 and G4 cells with carbon dust by Line 3 Bay

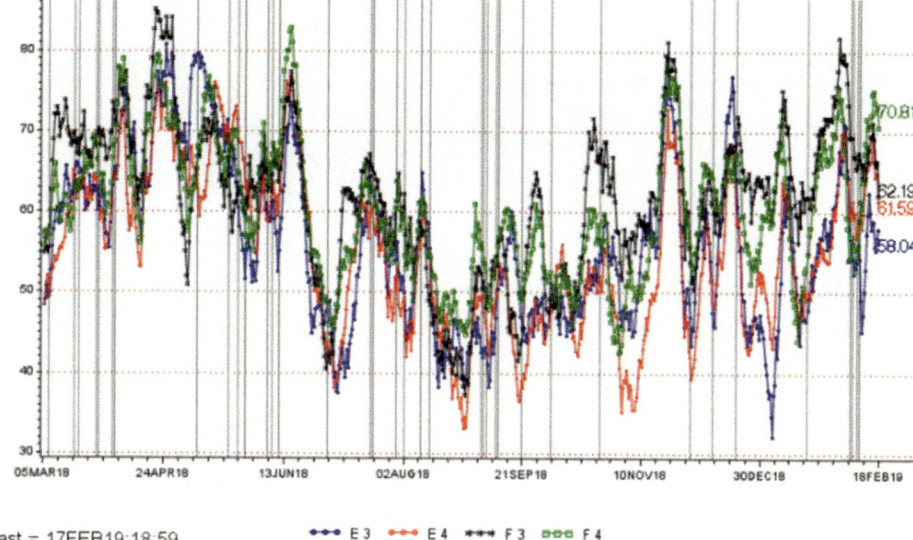

Fig. 13 Pareto chart of why anodes have been replaced

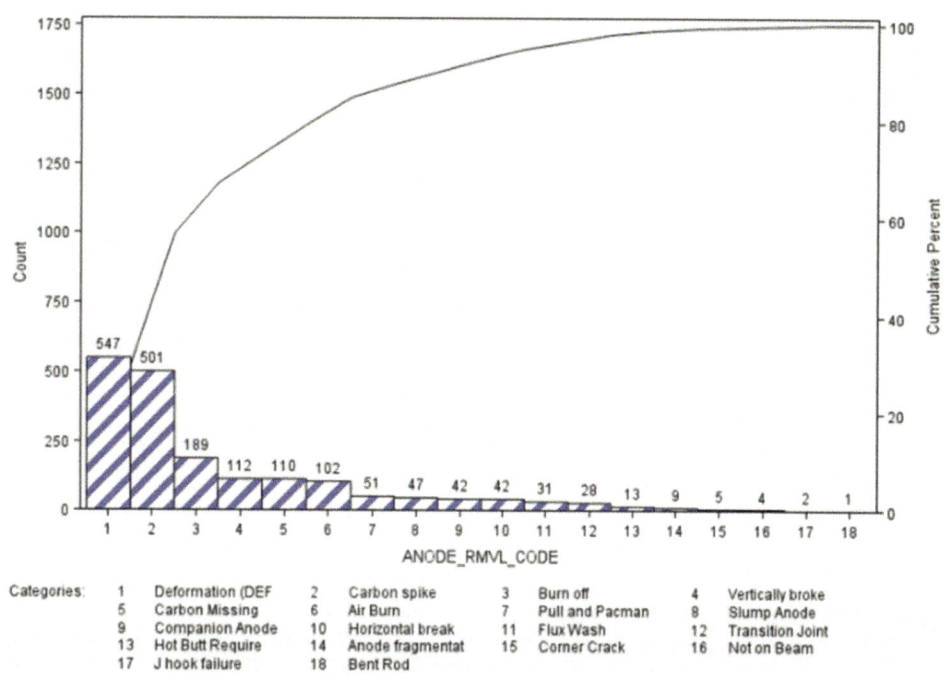

reverted to 24 h post setting operation versus 12 to allow more effective actioning of setting accuracy issues.

- Setting Increment—the data from the current draw measurement campaigns also had the additional benefit of providing data to better understand anode problems. Observations had been made from cell controllers that a large number of anode raises were taking place for anodes soon to be replaced. The current draw data indicated that new anodes were not picking up current draw as quickly as desired, meaning that the adjacent anodes, soon to be set, were drawing more current (resulting in raises). The

setting increment for new anodes had been increased in a previous project to combat anodes shattering (21–27 mm). Consequently, a reduction in the setting increment to 24 mm was made in order to get the desired current draw for new anodes (∼80% at 24 h). For a line 3 cell with 40 anodes, each anode carries approximately 2.5% of the current load. Figure 14 shows that after 1 day only 20% of anodes were pulling >80% of full load or 0.02 (2%).

- Cover Window—reduction Line 3 had recently changed from 70:30 delayed cover application to 100% cover

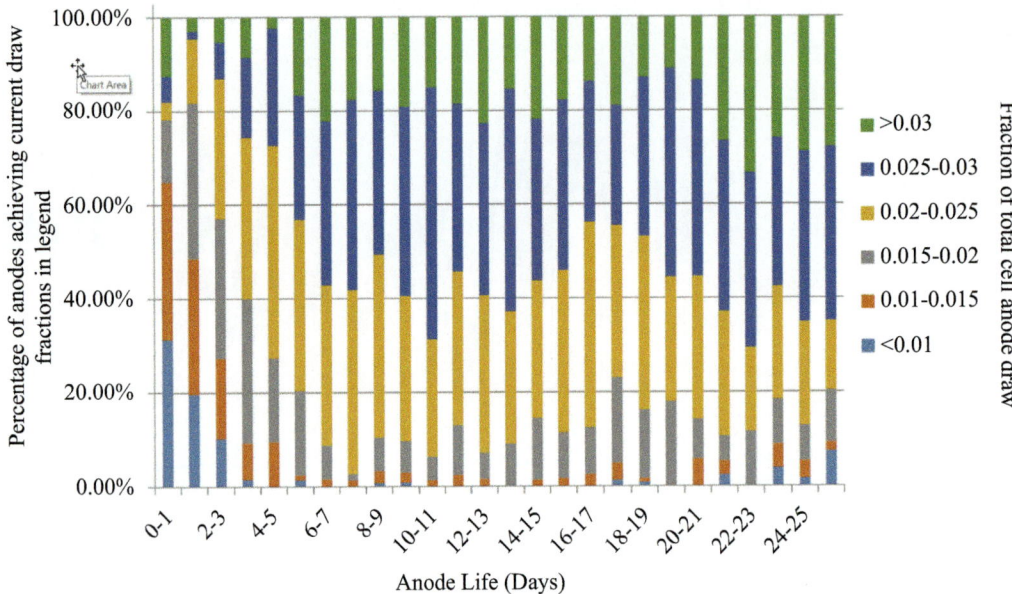

Fig. 14 Plot of Anode current draw versus time during the rota before adjustment of setting offset

being applied 2 to 4 h after set to improve productivity. It was observed that the 2–4 h window did not allow for sufficient freezing over and crust to support the cover, resulting in sludge and noise. Compliance to the 2–4 h cover window standard (70%) driven heavily by the need for conformance and feedback from operators was that even if crust had not formed, they would cover the anode to meet the compliance metric. Field observations were made, in combination with a review of noise and alumina feed data. Consequently, the anode cover window was changed to 3 to 6 h to allow better freeze-over and discretion for operators. Changes from a single anode cover to a partial anode cover approach were not considered given operational constraints.

- Noise Kicker—work was done to take advantage of developments in the cell control system to automate previously manual functions and free up process advisor time. A lot of the tasks process advisors were performing involved manual changes to control parameters based on certain criteria, whereas this could be done automatically by the control system if configured to do so. One such example was the noise kicker which "kicked in" when cells reached certain noise thresholds and then applied different sets of control parameters, e.g. Automatic modifications to alumina feeding logic and power added to cells.

- New cell control—Fig. 15 shows that new cells were typically remaining noisy for the first 400 days of their life. Key changes for new cells included aluminium fluoride and soda ash additions, bath sampling frequency, and parameter updates.

Management of Intergenerational Cell Designs

In order to further improve power efficiency as well as potentially operate at higher amperage, the G4 design was introduced to BSL L3 in 2014 after a period of trial and validation to replace the existing G3 design, which was introduced around 2003. The objective of the G4 design is to deliver a PE benefit of >0.3 kWh/kgAl compared to the G3 design at 370 kA. To achieve this, substantial changes were made to the cathode linings, cathode and collector bar designs. Target operating points were also changed in order to capture the full benefit.

Prior to the 2019 excursion, there were a mixed population of G3 and G4 cells in operation. A review was conducted to examine how the cells were operated and alternative operating strategies were assessed and trailed. It was found that the G3 and G4 cells were effectively being run the same way and therefore the advantages of the low energy cell design were not being realised and in fact the heat balance was detrimental to cell performance and cell life.

Figure 16 shows the metal pad noise in G4 cells was considerably worse than in G3 cells prior to the commencement of the recovery work in February 2019. Anode cover height and metal level were the two key parameters requiring correction on the G4 cells. Lowering the metal level and increasing the cover heat allowed the G4 cells to operate at much reduced total heat input with appropriate heat balance. Monthly cathode voltage drops (CVDs) were

Fig. 15 Metal pad noise by cell age and generation type for G3 and G4 cells

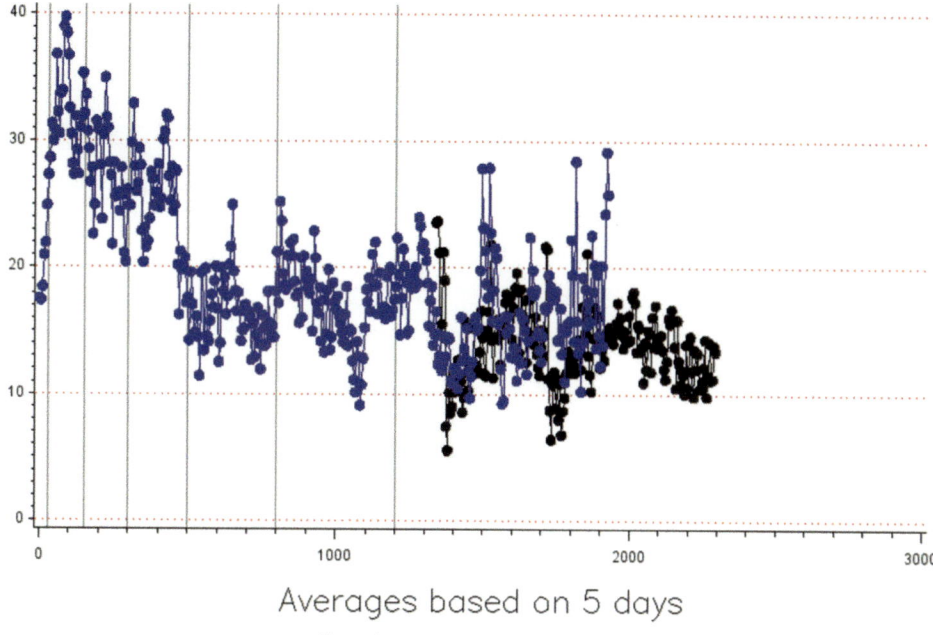

Averages based on 5 days

Generation ●━● G3 ●━● G4

Fig. 16 Average metal pad noise by cell design generation

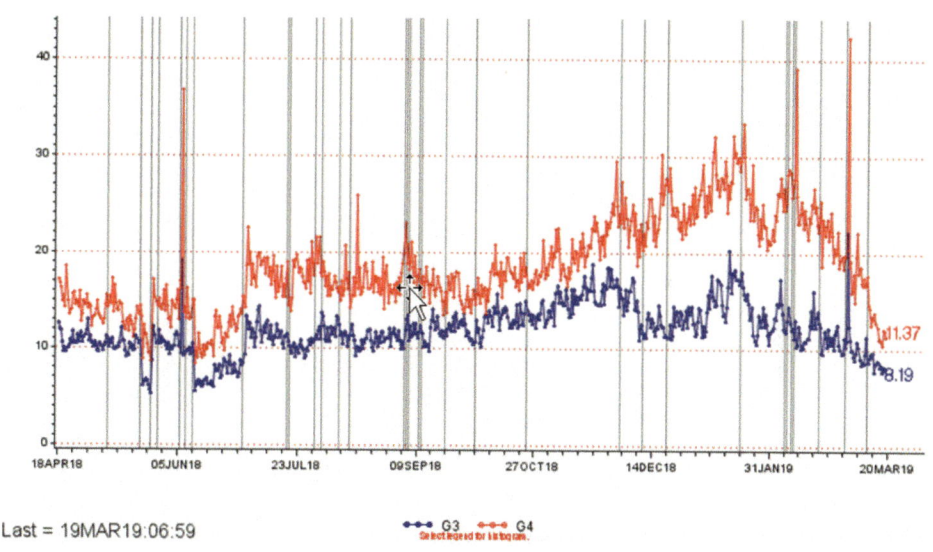

Last = 19MAR19:06:59

●━● G3 ●━● G4
Select legend for histogram.

monitored to ensure the CVDs had not deteriorated and that sufficient anode-cathode distance was maintained.

Figure 17 illustrates the overall recovery strategy with the use of the operating window. The concept of the operating window and its application have been discussed in previous papers [2, 3].

As shown in Fig. 17, amperage was reduced from January 2019 to help with the stabilisation of the process. The adjustment of anode cover and metal heights effectively lowered the thermal limit, which enabled substantial reduction in cell voltage. As soon as process stability was regained and appropriate heat balance established, amperage was increased to the target level with minimum disruption to internal heat. An 80 mV reduction in G4 cells was achieved

in June 2019 compared to January 2019 at the same amperage with correction of cell heat balance.

Outcome

The process recovery achieved over the defined eight-week period was substantial with Line 3 stabilised and a clear pathway forward to all metrics back in control. A key highlight was the reduction in metal pad noise as shown in Fig. 18, all whilst the power applied to cells was reduced as indicated in Fig. 4. Furthermore, Fig. 19 shows the length of time critically noisy cells were remaining noisy was also reduced from typically 5 shifts to under 1 shift. The

Fig. 17 G4 Cell operating
window

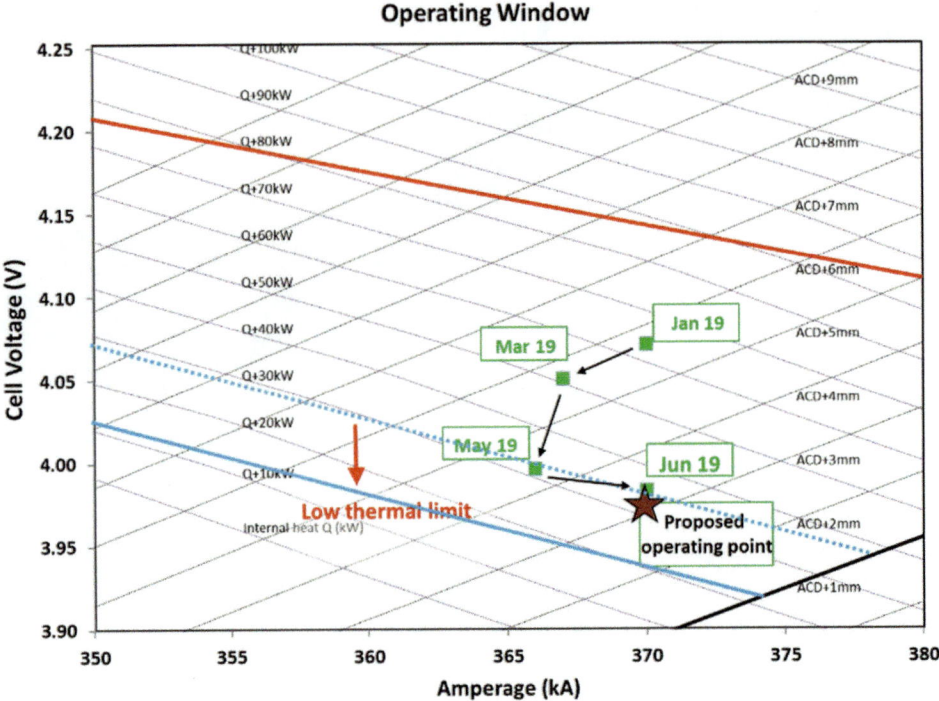

Fig. 18 Average metal pad noise
by Bay for Line 3 nano-ohms

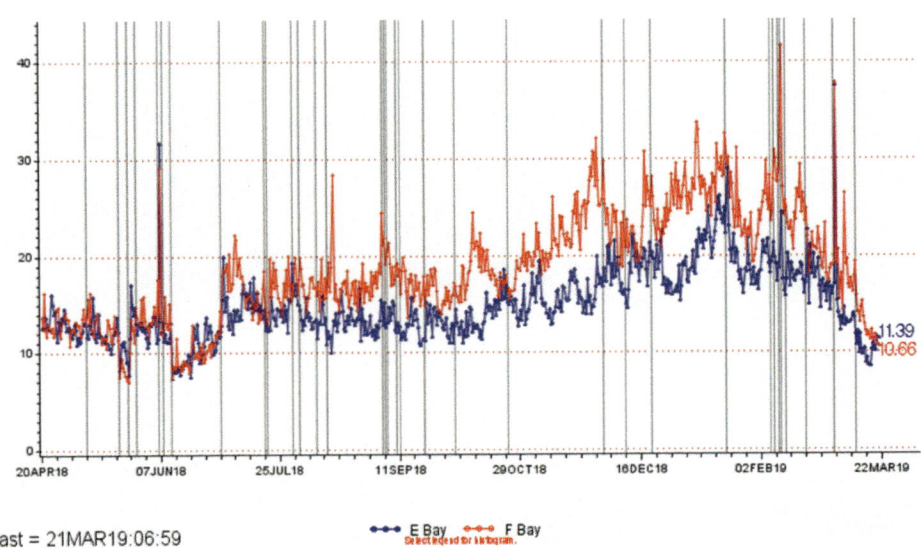

process improvements also had a noticeable impact on
workload, a boost in the morale of the workforce, and work
standards.

Figure 20 shows the 100-day recovery plan that was
developed to support the operations team. The critical few
initiatives were identified with key metrics for tracking.

There were numerous longer term actions that were effectively parked in stage 1 of follow-up work.

Looking back, the recovery work enabled an average cell
voltage reduction of 64 mV in 2019 compared to 2018,
representing a power efficiency improvement of 0.24
DCkWhr/kg Al. Further stabilisation and fine-tuning of Line

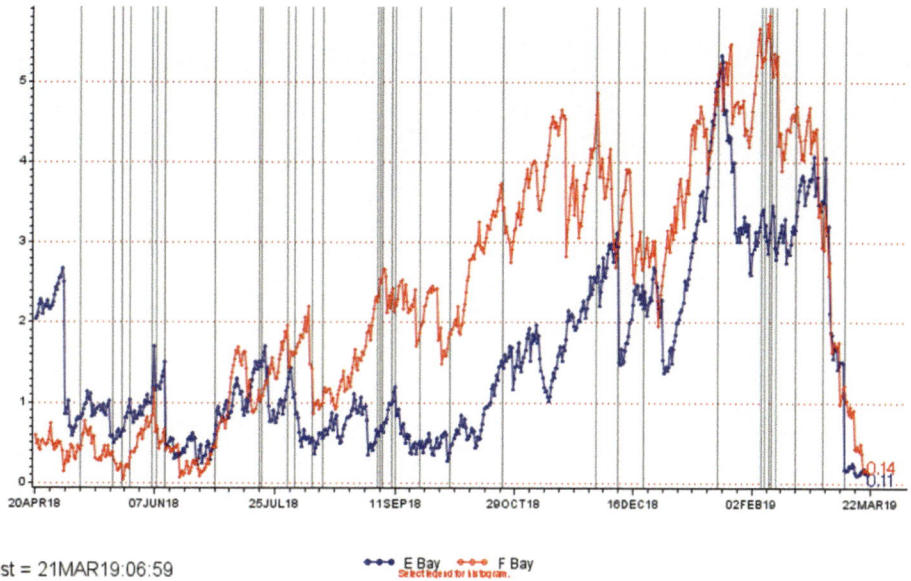

Last = 21MAR19:06:59

Fig. 19 Number of shifts for Metal Pad Noise >0.022 nano-ohms

Line 3 Recovery plan

Critical initiative	4-Mar	11-Mar	18-Mar	25-Mar	1-Apr	8-Apr	15-Apr	22-Apr	29-Apr	6-May	Longer term actions
Cover Strategy		SE1 Roll-out			Timing to be defined for conversion of remaining 4 sections to new cover standard						Setting resistance
		Key metrics: BACI summary for CE/PE/C									Setting Increment
											Alarms/deadbands
War Room		Continue focus on exception cell priority									2 Stage Cover
Exception Priority	key metrics: USC & Feeders actioned per shift, USC & Feeders hander over per shift, number of CD's out of range per shift										Alumina concentration in Cover
											Bath control - Feeders
											G4 Parameters
Carbon Skimming		R14 Skimmers - Targeted skimming 3 months									Noise Kicker v2
		key metrics: Cat 1&2 C dust, amount of C Dust skimmed per day									New Cell Control
											100 to 60 to 45 Days
Noisy Cells	Individual targeting of noisy cells										CaF2 - Long term trial
	Embed and refine Noise Kicker										CCS /Fundamentals Training
	key metrics: number of cells > 22 microhms noise										CVD Strategy
											AlF3 Mass Balance refine
New G4 Cells	Robust Strategy										Chemical Analysis (AlF3)
				Track progress							
	key metrics: MPN, Bath temperature (must be above 970deg C first 100 days whilst stabilising potlin)										
Amperage		Go/No go									
	key metrics: Decision to increase based on above KPI's										
Line 3 team Day											

Fig. 20 Forward 100-day recovery plan for operations

3 into 2020 permitted a full-year performance of 94.2% and 12.96 DCkWhr/kg Al current efficiency and power efficiency respectively.

Key Learnings

The Line 3 recovery in 2019 demonstrates the importance of getting the fundamentals right in terms of understanding and operating the process. Furthermore, when faced with a multi-faceted problem, it is important to get back to basics, identify the root causes, and then support operations with the execution of the plan. It also demonstrates the resource intensiveness and extended time required to recover a Reduction Line that loses stability and control, thereby highlighting the value in focusing first and foremost on stability and reducing variation.

The recovery team consisted of highly experienced and motivated team members who worked collaboratively with operations and the Brisbane Smelter Technical Support team to identify quick wins and implement recovery initiatives quickly. At all times, careful diligence was given to change management systems, risk assessments, and syndication of changes. The daily support provided by the site General Manager and senior leaders was pivotal in aligning the recovery work and demonstrating both the urgency and criticality of the situation.

Acknowledgements Special thanks to Mansel Ismay, Andrew Karbowiak, Mark P. Taylor, the BSL Operations and Technical teams, and the Rio Tinto Smelter Technical Support team in Brisbane, Australia.

References

1. Corby C, Zhang H, Lewis M and Roberts, J (2019) Modernisation of Sumitomo S170 Cells at Boyne Smelters Limited, Proceedings of Light Metals 2019. pp. 543-552.
2. Andrews EW, Tomsett A, Hamilton S (2014) Improving Power Efficiency in Power Constrained Aluminium Smelters, 11th Australasian Aluminium Smelting Technology Conference (AASTC), Dubai, 6–11 Dec.
3. Heithcoate K, Tinnoch S, Moratti S, Andrews EW, Illingworth, M (2018) NZAS Step Change Power Efficiency Improvement, 12th Australasian Aluminium Smelting Technology Conference (AASTC), Queenstown, New Zealand, 2–7 Dec.

A Smart Individual Anode Current Measurement System and Its Applications

Choon-Jie Wong, Jing Shi, Jie Bao, Barry J. Welch, Maria Skyllas-Kazacos, Ali Jassim, Mohamed Mahmoud, and Konstantin Nikandrov

Abstract

This paper discusses a new individual anode current measurement scheme and its applications in real-time monitoring and control of the Hall-Héroult process. While anode current can be directly measured from the anode rod, this approach takes measurements from the anode beam allowing the sensors to remain intact through various cell operations, including anode change. This instrumentation scheme employs smart sensors that are daisy-chained on a common bus for digital data transfer. This approach limits electromagnetic interferences and offers system self-configuration and self-diagnosis, thus allowing for easy maintenance. The system can be configured to work across a broad range of cell technologies. Monitoring anode current distributions helps improve process operation and allows early detection of process faults such as perfluorocarbon co-evolution and blocked feeders. This also offers the ability to monitor process states such as local alumina concentration and bath temperature, along with potential improvements to cell operation and current efficiency.

Keywords

Individual anode current measurement • Online monitoring • Current distribution

Introduction

Measuring individual anode current signals in a Hall-Héroult cell is an attractive option to overcome the lack of observability of local cell variables. Conventionally, only the line current and cell voltage are being continuously measured for monitoring and control purposes [1, 2], but these indicate only the overall cell conditions, such as the average alumina concentration level. The line current and cell voltage cannot detect local cell variations or faults, which are becoming more critical especially as modern cells are designed to operate at higher currents. With anode size growth outpacing the increase in molten electrolyte volume, the assumptions of homogeneity in cell conditions are becoming questionable. Additionally, the anode-cathode distance (ACD) is typically squeezed in modern high-current cells to limit energy consumption, which further exacerbates the uneven cell conditions, due to limited bath mixing arising from the restricted flow.

The distribution of anode currents both affects and is affected by the local condition of each anode-cathode path (e.g., bath composition and temperature). This is because, on the one hand, the regulated line current must distribute among all the anode-cathode paths according to the local reversible potential, cell overpotentials, and ohmic potential drop of each path; these are functions of the local path conditions. On the other hand, anode currents drive the local electrolytic reactions and local ohmic heating. Therefore, the measurements of individual anode currents reveal not only the spatial cell information, but also how it will change.

This paper first presents a review of measurement schemes found in the literature, followed by a discussion of the new developed smart individual anode current measuring system, along with its applications in monitoring and control of smelter cells.

C.-J. Wong · J. Bao (✉) · B. J. Welch · M. Skyllas-Kazacos
School of Chemical Engineering, The University of New South Wales, Sydney, NSW 2052, Australia
e-mail: j.bao@unsw.edu.au

J. Shi · A. Jassim · M. Mahmoud · K. Nikandrov
Emirates Global Aluminium, Jebel Ali Operations, P.O. Box 3627
Dubai, UAE

© The Minerals, Metals & Materials Society 2023
S. Broek (ed.), *Light Metals 2023*, The Minerals, Metals & Materials Series,
https://doi.org/10.1007/978-3-031-22532-1_6

Measurement Principles

There are generally two methods of measuring anode currents as described in the literature. One method, which the development described in this paper employs, is based on the 'isometric voltage drop' principle. In this method, current I flowing through a conductor is determined by measuring the voltage drop V across a known, short distance L on said conductor. For a conductor with a resistivity that changes linearly with temperature T, this is given by

$$V = \frac{(\rho_1 T + \rho_2)L}{A} I, \qquad (1)$$

where A is the surface area through which current flows, and ρ_1 and ρ_2 are the coefficients specific to the conductor material for calculating resistivity.

Table 1 summarises the developments of an anode current measurement system employing the isometric voltage drop principle. In its simplest form, the voltage drop on each anode rod is directly measured. The raw data can be immediately converted into anode currents by a local microprocessor and/or sent to a remote computer for later processing and analysis. However, due to the nature of the Hall-Héroult process where the carbon anodes are continuously consumed, one or two anode assemblies must be replaced every few days. This necessarily means that the measuring probes must be detached and re-attached to the new anode rods, which subjects the wirings and the rods to damages due to excessive handling in addition to inconveniencing the process operators.

Keniry and Shaidulin [4] suggest taking the measurements from the anode beam. However, in this approach, the potentials (cf. potential difference) on the beam were measured, which Yao [15] argues that this can lead to unreliable measurement of all the anode currents even if only one potential measurement is faulty. Nonetheless, this scheme inspired Li et al. [19], Yang et al. [20], and Yao [15] to measure the voltage drops on the anode beam, from which beam currents can be determined. The anode current is then the difference in the two neighbouring beam currents.

Another method of measuring anode currents is based on the 'Hall effect' principle. In this method, the anode current I is determined by measuring the magnetic field strength B around the anode rods induced by the current, with [24]

$$B = \frac{\mu_0}{2\pi r} I, \qquad (2)$$

where μ_0 is the magnetic field constant and r is the distance between the centre of the conductor to the Hall sensor. However, this method is susceptible to magnetic field interference from currents flowing in neighbouring anodes, risers, and crossovers, as well as those in neighbouring cells. Hence, to minimise this interference, researchers have proposed the use of multiple Hall sensors for each anode [25, 26] (e.g., the

Table 1 Review of instrumentation systems based on isometric voltage drop

Author	Smelter	Year	Methodology summary
Holmes et al. [3], Keniry and Shaidulin [4]	Alcoa, United States	1969	P-225 cells were designed for a high degree of automation. The anode currents were monitored by six Mod Comp II computers and controlled by independently adjusting the elevation of the 16 anode pairs
Potocnik and Reverdy [5]	Reynolds Metals, Germany	1973	Connectors were plugged into the anode rods to measure voltage drops, which were sent to a local microcomputer at each cell. The processed results were sent to the main computer system. The system was abandoned after 6–9 months due to excessive damages to anode rods and wires
Langon and Varin [6]	Aluminium Pechiney, France	1975	Four prototype 280 kA cells were constructed, of which two can adjust each of 20 anode pairs automatically (presumably with the use of anode current readings), in addition to conventional main beam adjustments. Individual anode adjustment was later abandoned due to high superstructure cost, and the other two conventional cells performed just as well by improving operation and control
Huni [7]	Alcan, Canada	1987	A-275 prototype cells were designed with individual anode drives. Voltage drops on the flexible current leads which connected the anode rods to the busbar were measured and processed by a computer at each cell
Barnett [8]	Reynolds Metals	1988	Voltage drop and temperature on each anode rod were encoded to ASCII characters for transmission on a RS-485 multidrop network. The data was then converted for a RS-232 multidrop network for radio-transmission to a remote computer
Keniry et al. [9]	Comalco, Australia	1992	Voltage drops on anode rods were filtered, conditioned, and amplified. A local microprocessor then digitalised and transmitted the voltage signals to a computer via the fibre-optic link, which then calculated the current. The system could sample every two minutes for 30 days, or at 50 Hz for 20-s bursts

(continued)

Table 1 (continued)

Author	Smelter	Year	Methodology summary
Rye et al. [10]	Elkem Aluminium, Norway	1998	The anode rods were individually connected to the central beam with a flexible. The voltage drops on these flexibles were measured and converted to anode currents. During normal operation, the anode current distribution was represented as a bar graph with a LED matrix, but not recorded
Panaitescu et al. [11, 12]	ALRO SA, Romania	2000	For visualising metal pad waves, the voltage drops between the bimetal plate and stubs on each anode were measured with the acquisition system DAQ642. On a cell with 16 anodes, 64 stub voltage readings were recorded on a hard disk as well as visualised in real time
Shaidulin et al. [13, 14]	RUSAL, Russia	2005	Each anode rod was fitted with an electric load measuring transducer, and the signals were sent and collected by a control and data acquisition unit. The anode current can then be calculated
Keniry and Shaidulin [4], Yao [15]	RUSAL, Russia	2008	Due to frequent damages and inconvenience arising from the frequent disconnection of the sensor from the anode rods, this design measured the potentials on the anode busbar, relative to a common ground. The anode currents could be determined simultaneously accounting for the cell superstructure design
Cheung et al. [16], Cheung [17]	DUBAL, United Arab Emirates	2013	Voltage drops across spring-loaded contact pins on anode rods were amplified and transmitted to a NI CompactRIO 9024 acquisition system at 10 Hz, while the temperature was also measured and transmitted once every minute. These data were then transferred to a remote computer via an Ethernet connection
Qiu et al. [18]	Yunnan Aluminium, China	2013	Voltage drops on anode rods were measured and transmitted to a converter module which transfers the data to a remote system for data processing and analysis
Li et al. [19], Yang et al. [20]	Jinlian Aluminium, China	2015	Voltage drops and temperatures on the beam were collected at 1 Hz by wires riveted to the beam. A collection unit with a browser then transmitted and displayed the anode currents
Bao et al. [21]	DUBAL, United Arab Emirates	2017	Voltage drops on the anode busbar and their corresponding temperatures were sent to a NI CompactRIO 9082 acquisition system at up to 200 Hz. A new file was created for these raw data every 3 min, which were then uploaded to a cloud storage (Dropbox) for further analysis
Huang et al. [22], Yin et al. [23]	Zunyi Aluminium, China	2021	Voltage drops and temperatures on anode rods were digitalised with AD7705 and transmitted via RS-485 bus to a main controller powered by STM32 for display and control. The data can be transmitted to the factory production process control system

system described by Evans and Urata [27, 28] has five Hall sensors per anode), which drives up the capital costs of the system. It also requires solving a complicated model describing the cell design for good measurement accuracy [25], which can be computationally costly—experiences at TRIMET [29] and Alouette [30] show that the data has to be archived and processed with a cloud computing service (AWS). Hence, the system described in this work focuses on the isometric voltage drop principle.

New Measurement Scheme with Smart Sensors

Based on years of experience and learning from successful deployments of anode current monitoring systems in our previous projects [15, 16], this work further develops an in-house, improved individual anode current measurement system with superior signal quality and system robustness.

A schematic of this measurement system is shown in Fig. 1. Measurements of the voltage drop and temperature are taken from the anode beam, and so the installation will remain intact through cell operations such as anode change and beam raising. The sensors are daisy-chained together in a bus topology network, while a communication unit collects the data and transmits it to a computer server. This system samples data at 2 Hz under normal operation, and a higher sampling frequency of up to 100 Hz can be enabled on-demand for special studies or for diagnostic purposes. The measurement probes can be installed on either the top or the side of the anode beam, as the scheme is flexible.

Exceptional Signal-to-Noise Ratio

Designed to scale for future large cells with more anodes, distributed processing forms part of the core design principle. Each sensor installed at each signal pickup location has

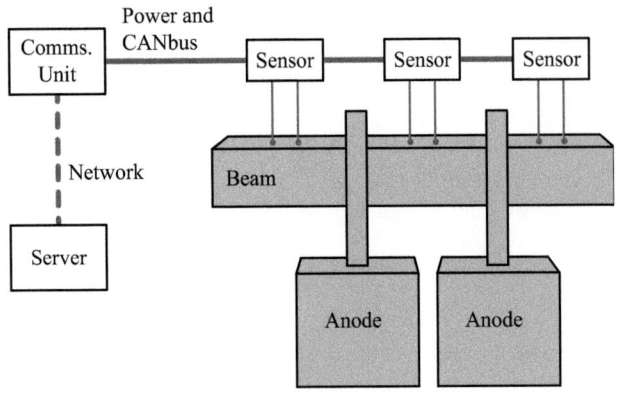

Fig. 1 Schematic of measurement scheme with smart sensors

its own microprocessor and can function autonomously under the instructions of the communication unit. All the sensors can individually and automatically calibrate their readings to reject voltage offset and gain offset errors, as well as independently adjust their instrument amplifier gains. This achieves a high signal-to-noise ratio since the signal magnitude varies greatly depending on the locations on the beam.

The sensors and the communication unit transfer signals digitally on the bus line. The analogue-to-digital converter of each sensor uses 'oversampling and decimation' method [31, 32] to increase the effective resolution—the sensors achieve a better representation of the input signal by measuring at above 250 Hz before subsequent filtering and down-sampling to the required sampling frequency.

System Robustness

The digital signals are transmitted through shielded twisted pair cables to minimise the impacts of electromagnetic interference. The multi-node bus topology also reduces the cabling required, in contrast to other designs with a star topology, where analogue signals from all the measurement points are transmitted to a single cell-level microprocessor with long cables which inadvertently work like antennas. Using a single multiplex bus for all nodes reduces the cross-talks between cables, while also allows easier maintenance of the system as sensors can be added or removed easily from the daisy-chain.

This bus uses the controller area network (CAN) interface protocol, which has seen increased adoption in industrial applications partly due to its robustness in harsh electrical environments [33]. As CANbus protocol transmits data with voltage differential between two bus lines (see Fig. 2), the bus is not susceptible to inductive spikes or other noises due to interference. Additionally, CANbus has built-in error detection capabilities, with each node tracking transmission

errors to prevent faulty devices from interfering with the bus traffic [34]. CANbus also has a major advantage over other protocols, such as the RS-485, in how it handles messages. In a RS-485 system, if multiple nodes are transmitting data simultaneously, the signal performance will suffer degradation possibly leading to hardware damages [33]. In contrast, the CANbus protocol can resolve this issue by queuing messages according to the priority ranking of each message.

To ensure timely data sampling and delivery, especially during the on-demand high-frequency sampling, all the sensors and the communication unit are time-synchronised. Timer-based interrupts are used to precisely control the sampling interval, while timestamping of all data packets accounts for transmission delays.

Ease of Maintenance

The measurement system has self-configuration and self-diagnosis capabilities. Each sensor in the CANbus has a unique addressable identifier, which is used by the communication unit to both identify the source of data packets and target an instruction. As such, the system can automatically detect and report changes in identifiers observed (for example, a missing identifier signifies sensor failure). In some cases, the system can be automatically configured (for example, observing a missing identifier followed shortly by a new identifier signifies that a sensor has been manually replaced).

With the use of identifiers, each sensor can send and receive diagnostic information and instructions. To indicate system health, each sensor also measures and sends its board temperature to the communication unit, in addition to the beam temperature. Each sensor also has an integrated watchdog timer [35] for monitoring microprocessor responsiveness and for ensuring that the CANbus connection is alive.

The firmware of the communication unit and the sensors can also be remotely managed by an IT department, owing to the bootloader-program architecture. The bootloader checks the integrity of the firmware at boot time, and if it fails (such as when a firmware update process was interrupted), an older known-good version of the firmware can be restored. The communication unit also has a physical reset switch for on-site interactions—a quick press reboots the unit and a long press factory-resets the system.

Field Test of Proposed Measurement System

The system was deployed in *Emirates Global Aluminium, Jebel Ali Operations* to verify the reliability and ease of use of the smart individual anode current measurement system

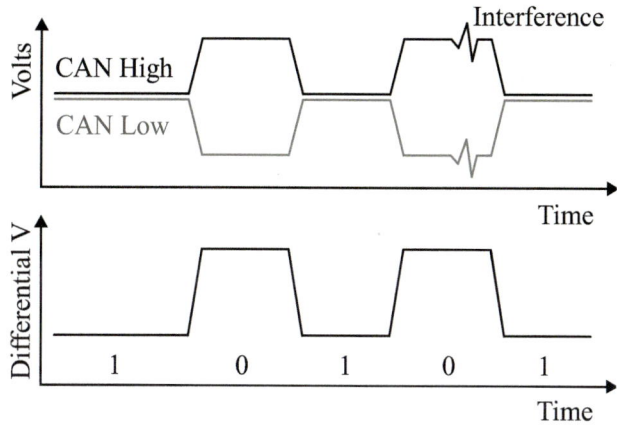

Fig. 2 CANbus data transmission with voltage differential

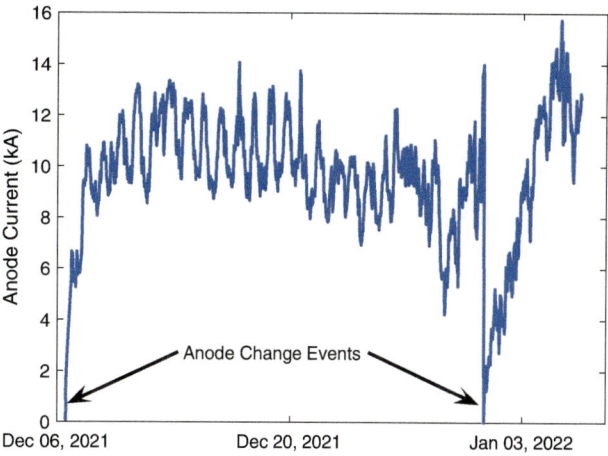

Fig. 4 Anode current measured continuously through cell operations

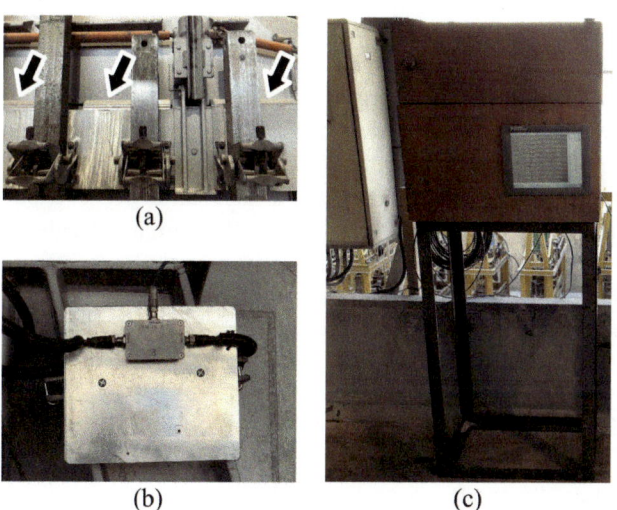

Fig. 3 **a** Sensors installed on the anode beam (enclosed in the box for protection), **b** communication unit mounted at cell duct-end, **c** a prototype human-machine interface beside the cell

developed in this paper under real conditions. For the 400+ kA cell design, each cell has 44 sensors or nodes in its CANbus, in addition to a communication unit that can simultaneously connect and transmit data to both a production server and a development server, via standard TCP/IP on a dedicated virtual local area network (vLAN) set up by the EGA IT department.

The installation (see Fig. 3) was finished in late 2021 and the system was continuously tested for more than 8 months. Data can be collected continuously as the sensors remain intact during the numerous cell operations that were conducted in this period, including for anode change operations. For example, Fig. 4 shows the current of a particular anode which is measured by the system as it undergoes two anode change operations in the span of a month. Throughout the anode service life, the fluctuations caused by cell operations including the changing of other anodes, material feeding patterns, and anode beam adjustments are also visible.

Applications of Anode Currents Measurements

As previously explained, the anode current individually reveals information on local cell conditions. Additionally, since all the anode currents are coupled due to the parallel electrical connection, the distribution of anode currents also provides information on the relative path resistance of all anode-cathode paths, which allows monitoring and control of spatial variations in cell conditions.

Individual Anode Drives

As summarised in Table 1, individual anode currents were utilised to independently adjust the elevation of anodes since at least 1969. It has been experimentally shown that the low setting of an anode causes current overloading, which also results in uneven overheating of anode stubs, leading to uneven thermal expansion and possibly carbon cracking [36]. As the anode currents determine the rate of local material consumption and the rate of local heat generation, this current imbalance also exacerbates the spatial non-homogeneity in the cell.

Overloading or underloading of anode currents can be determined by comparing the measured anode current with the target anode current (total line current divided by the number of anodes). The individual anode drives on the anodes can then be independently adjusted to bring the anode current closer to target values [3, 6, 7]. For automatic controls, the elevation adjustments must consider the nature of anode current re-distribution and the state the cell is operating in, such that the anode adjustments do not lead to instability.

Limiting Perfluorocarbon Emissions and Preventing Anode Effects

Perfluorocarbons (PFCs), particularly CF_4 and C_2F_6, are greenhouse gases that start to co-evolve when the electrical potential across the anode exceeds the reversible potentials for PFC reactions of 1.83–2.75 V [37]. PFCs are generated in large quantities during an undesirable phenomenon associated with the depletion of alumina, called the 'anode effect.' In this event, the anodic gas production reaction changes, leading to dewetting and partial passivation of anodes, as the anode currents fall [37, 38]. As anode current imbalance grows, the PFC co-evolution propagates to all other anodes. For this reason, anode current measurements can be used to characterise and allow early detection/mitigation of anode effects [4, 10, 38–41]. The studies also extend to detecting faults in feeders or crust-breakers [30, 42], since they also contribute to alumina depletion.

In recent decades, it was also established that PFCs are emitted even under typical operating conditions; these are termed 'low voltage PFCs' (LV-PFCs) since they are emitted in the absence of anode effects and the associated characteristic high cell voltage. Studies found that LV-PFCs contribute a significant portion towards total PFC emissions [43–45], with estimates ranging as high as 93% of total smelter emissions [46]. It may be possible to identify LV-PFCs formation with anode current signals.

Identifying Anode Faults

Both the time and frequency domain analyses of anode current signals can be used to identify local cell conditions, including fault identification in the region of the carbon anode. Experiments [10, 42] show that the anode current can indicate the formation of an anode spike, which lowers the cell current efficiency.

The frequency domain can be obtained by performing a windowed Fourier transform of the anode current signals. Low-frequency noises (0.02–0.05 Hz) are typically caused by thermal imbalance and magnetohydrodynamics instability, while high-frequency ones are typically related to the anode-metal shorting arising from anode faults such as slippage, cracking, and spikes [4]. The study of the bubble dynamics (0.5–2 Hz) also yields information on the presence of slots as anode ages, the onset of anode effects, and if an anode slippage occurred [40].

Understanding, Modelling, and Improving Anode Setting

Carbon anodes are regularly replaced in commercial cells as their service lives end. This operation, called the 'anode setting,' introduces significant disturbances to the mass and thermal balance, as the bath freezes around the cold new anodes and impedes the flow of anode currents. The gradual warming up of the new anodes and the dissolution of bath freeze contribute to the local energy deficits introduced by this operation.

Anode current uptake trajectories of young anodes were used to study the freeze dissolution process and anode conditions at various stages while the anodes are heating up [17, 40]. The measurement of anode currents also allows coupled mass and thermal balance dynamic models to be proposed based on the first principles [47, 48] for monitoring/control studies of a cell undergoing anode setting operation. The use of anode current signals in preheating studies [49, 50] has been proposed to minimise the mass and thermal perturbations introduced by the operation.

Spatial Distribution Monitoring and Control

While the anode currents individually reveal the local cell conditions, the distribution of anode currents as a whole indicates the extent of spatial variations in the cell. The monitoring and control of cell homogeneity have become increasingly important as designs of modern cells and retrofitting of existing cells focus on increased productivity; the changes include reduced bath volume per line current amperage, reduced number of alumina feeders per line current amperage, impeded bath mixing due to reduced ACD/cell voltage for energy savings, and bigger anodes taking longer to heat up after replacements [37].

As the anode current distribution also indicates how local electrolytic reactions and local ohmic heating will vary in the future, it can be used in conjunction with model-based state/parameter observers (e.g., Kalman Filters) to estimate local process variables such as alumina concentration, ACD, bath flow velocity, and bath temperature [51–56]. Controllers, either model-based or data-based, have also been developed to improve cell homogeneity by minimising the imbalance of anode currents [57–60].

Conclusion

It is becoming increasingly critical to maintain the homogeneity in cell conditions as modern cell designs and retrofitting of existing cells pursue higher production rates and lower energy consumptions at the cost of cell stability (e.g., squeezing of ACD leading to increased shorting risks). There is wide recognition that an anode current measurement is a promising approach for revealing spatial cell conditions, for the purpose of process monitoring/control and fault detection/mitigation. A successful application requires a stable, reliable instrumentation system for accurate and continuous signal analysis.

In this paper, a new individual anode current measurement scheme with smart sensors is proposed. The signals are sampled continuously from the anode beam, uninterrupted by cell operations, and are amplified by each sensor to achieve a high signal-to-noise ratio. The system was designed to be robust and easy to maintain, with features including self-configuration, self-diagnosis, and remote system management. An industrial field test was conducted to verify the system's performance under environments of elevated temperature, strong magnetic field, and EMI conditions, as well as corrosion and dust. As the system is designed in-house, the sensors can be mass manufactured at a reasonable cost without having to purchase costly off-the-shelf data acquisition systems, such as a CompactRIO. This instrumentation system is designed to work with various cell designs, with the option to retrofit it onto existing cells.

Acknowledgements The authors acknowledge the financial and technical support from Emirates Global Aluminium and Jebel Ali Operations, as well as thank Mr. John Lam for the technical expertise.

References

1. Homsi P, Peyneau J-M, Reverdy M (2000) Overview of Process Control in Reduction Cells and Potlines. Paper presented at the Proceedings of the TMS Light Metals, Nashville, Tennessee, USA.
2. Gao YS, Taylor MP, Chen JJJ, Hautus MJ (2011) Operational and control decision making in aluminium smelters. Advanced Materials Research. 201–203(2011):1632–1641.
3. Holmes GT, Fisher DC, Clark JF, Ludwig WD (1980) Development of Large Prebaked Anode Cells by Alcoa. Paper presented at the Proceedings of the TMS Light Metals, Las Vegas, Nevada, USA.
4. Keniry J, Shaidulin E (2008) Anode signal analysis—the next generation in reduction cell control. Paper presented at the Proceedings of the TMS Light Metals, New Orleans, Louisiana, USA.
5. Potocnik V, Reverdy M (2021) History of Computer Control of Aluminum Reduction Cells. Paper presented at the Proceedings of the TMS Light Metals, Orlando, Florida, USA.
6. Langon B, Varin P (1986) Aluminium Pechiney 280 kA Pots. Paper presented at the Proceedings of the TMS Light Metals, New Orleans, Louisiana, USA.
7. Huni JPR (1987) A-275—Individual anode control. Paper presented at the Proceedings of the TMS Light Metals, Denver, Colorado, USA.
8. Barnett WM (1988) Measuring Current Distribution in an Alumina Reduction Cell. Patent number 4786379.
9. Keniry JT, Barber GC, Taylor MP, Welch BJ (2001) Digital processing of anode current signals: an opportunity for improved cell diagnosis and control. Paper presented at the Proceedings of the TMS Light Metals, New Orleans, Louisiana, USA.
10. Rye KA, Königsson M, Solberg I (1998) Current Redistribution Among Individual Anode Carbons in A Hall-Heroult Prebake Cell at Low Alumina Concentrations. Paper presented at the Proceedings of the TMS Light Metals, San Antonio, Texas, USA.
11. Panaitescu A, Moraru A, Panaitescu I (2000) Visualisation of the Metal Pad Waves in the Aluminum Reduction Cell with Pre-baked Anodes. Paper presented at the Proceedings of the TMS Light Metals, Nashville, Tennessee, USA.
12. Panaitescu A, Moraru A, Panait N, Dobra G, Munteanu N, Cilianu M (2001) Experimental Studies on Anode Effects by the Visualisation of the Molten Aluminium Surface Oscillations. Paper presented at the Proceedings of the TMS Light Metals, New Orleans, Louisiana, USA.
13. Shaidulin E, Gusev A, Vabischevich P (2005) Method of controlling aluminum reduction cell with roasted anodes. Patent number 2303658C1.
14. Shaidulin E, Gusev A, Vabischevich P (2007) Method of controlling aluminum reduction cell with prebaked anodes. Patent number 11/592557.
15. Yao Y (2017), Process Monitoring, Modelling and Fault Diagnosis in Aluminium Reduction Cells. Ph.D. thesis, School of Chemical Engineering, Faculty of Engineering, the University of New South Wales, Australia.
16. Cheung CY, Menictas C, Bao J, Skyllas-Kazacos M, Welch BJ (2013) Frequency response analysis of anode current signals as a diagnostic aid for detecting approaching anode effects in aluminum smelting cells. Paper presented at the Proceedings of the TMS Light Metals, San Antonio, Texas, USA.
17. Cheung C-Y (2013), Anode Current Signals Analysis, Characterization and Modeling of Aluminum Reduction Cells. Ph.D. thesis, School of Chemical Engineering, Faculty of Engineering, the University of New South Wales, Australia.
18. Qiu Z, Ji F, Li Y, Yu Q, Li L, Li C (2013) Online measurement and data analysis device for anode current distribution of aluminum electrolysis cell. Patent number CN203080085U.
19. Li J, Yang S, Zou Z, Zhang H (2015) Experiments on Measurement of Online Anode Currents at Anode Beam in Aluminum Reduction Cells. Paper presented at the Proceedings of the TMS Light Metals, Orlando, Florida, USA.
20. Yang S, Zou Z, Li J, Zhang H (2016) Online anode current signal in aluminum reduction cells: measurements and prospects. JOM. 68(2):623–634.
21. Bao J, Welch BJ, Akhmetov S, Yao Y, Cheung C-Y, Jassim A, Skyllas-Kazacos M (2017) Method of monitoring individual anode currents in an electrolytic cell suitable for the Hall-Heroult electrolysis process. Patent number WO 2017/141135 A1.
22. Huang R, Li Z, Cao B (2021) Design and Implementation of Online Detection System for Anode Current Distribution in Aluminum Reduction Cell. Paper presented at the China Automation Congress (CAC).
23. Yin Y, Wang J, Cui J, Xiao G, Xu Z, Zhang S, Wang F (2016) Aluminium cell positive pole distribution electric current precision

measurement appearance with self calibration function. Patent number CN205501431U.

24. Wieser C, Helmbold A, Glzow E (2000) A new technique for two-dimensional current distribution measurements in electrochemical cells. Journal of Applied Electrochemistry. 30(7):803–807.

25. Urata N, Evans J (2010) The determination of pot current distribution by measuring magnetic fields. Paper presented at the Proceedings of the TMS Light Metals, Seattle, Washington, USA.

26. Hung OK (2000) Anode and Cathode Current Monitoring. Patent number 6136177.

27. Evans JW, Urata N (2011) Technical and operational benefits of individual anode current monitor. Paper presented at the Proceedings of 10th Australasian Aluminium Smelting Technology Conference, Launceston, Tasmania, Australia.

28. Evans JW, Urata N (2012) Wireless and non-contacting measurement of individual anode currents in Hall-Heroult pots; experience and benefits. Paper presented at the Proceedings of the TMS Light Metals, Orlando, Florida, USA.

29. Lützerath A, Evans JW, Victor R (2014) On-line monitoring of anode currents: Experience at TRIMET. Paper presented at the Proceedings of the TMS Light Metals, San Diego, California, USA.

30. Dion L, Lagac C-L, Evans JW, Victor R, Kiss LI (2015) On-line monitoring of individual anode currents to understand and improve the process control at Alouette. Paper presented at the Proceedings of the TMS Light Metals, Orlando, Florida, USA.

31. Baker B (2011) How delta-sigma ADCs work, Part 1, in Data Acquisition, Texas Instruments Incorporated. https://www.ti.com/lit/an/slyt423a/slyt423a.pdf?ts=1660200202417. Accessed 11 August 2022.

32. Microchip Technology (2019) SAM D/L/C—Oversampling and decimation feature. https://microchipsupport.force.com/s/article/SAM-D-L-C—Oversampling-and-decimation-feature. Accessed 11 August 2022.

33. Gee R (2020) CAN vs. RS-485: Why CAN is on the move, in Delivering Robust Communications, Maxim Integrated, Editor. https://www.maximintegrated.com/content/dam/files/design/technical-documents/white-papers/can-wp.pdf. Accessed 11 August 2022.

34. Total Phase (2019) What is CAN Bus Protocol? https://www.totalphase.com/blog/2019/08/5-advantages-of-can-bus-protocol/. Accessed 11 August 2022.

35. ABLIC (2020) What is a watchdog timer (WDT)? https://www.ablic.com/en/semicon/products/automotive/automotive-watchdog-timer/intro/. Accessed 11 August 2022.

36. Peterson RW (1976) Temperature and voltage measurements in Hall cell anodes. Paper presented at the Proceedings of the TMS Light Metals, Las Vegas, Nevada, US.

37. Wong D, Welch B, Nunez P, Dion L, Spirin A (2019) Latest progress in IPCC methodology for estimating the extend of PFC greenhouse gases co-evolved in the aluminium reduction cells and challenges in reducing these emissions. Paper presented at the Proceedings of the International ICSOBA Conference, Krasnoyarsk, Russia.

38. Tarcy G, Tabereaux A (2011) The initiation, propagation and termination of anode effects in Hall-Heroult cells. Paper presented at the Proceedings of the TMS Light Metals, San Diego, California, US.

39. Zarouni A, Reverdy M, Zarouni AA, Venkatasubramaniam KG (2013) A study of low voltage PFC emissions at DUBAL. Paper presented at the Proceedings of the TMS Light Metals, San Antonio, Texas, USA.

40. Cheung C-Y, Menictas C, Bao J, Skyllas-Kazacos M, Welch BJ (2013) Characterization of individual anode current signals in aluminum reduction cells. Industrial & Engineering Chemistry Research. 52(28):9632–9644.

41. Zarouni A, Al Zarouni A (2011) DUBAL's experience of low voltage PFC emissions. Paper presented at the Proceedings of 10th Australasian Aluminium Smelting Technology Conference, Launceston, Tasmania, Australia.

42. Puzanov II, Zavadyak AV, Klykov VA, Makeev AV, Plotnikov VN (2016) Continuous monitoring of information on anode current distribution as means of improving the process of controlling and forecasting process disturbances. Journal of Siberian Federal University. Engineering & Technologies. 9 (6):788.

43. Marks J, Bayliss C (2012) GHG measurement and inventory for aluminum production. Paper presented at the Proceedings of the TMS Light Metals, Orlando, Florida, USA.

44. Marks J, Nunez P (2018) Updated factors for calculating PFC emissions from primary aluminum production. Paper presented at the Proceedings of the TMS Light Metals, Phoenix, Arizona, USA.

45. Wong DS, Fraser P, Lavoie P, Kim J (2015) PFC emissions from detected versus nondetected anode effects in the aluminum industry. JOM. 67(2):342–353.

46. Li W, Chen X, Yang J, Hu C, Liu Y, Li D, Guo H (2012) Latest results from PFC investigation in China. Paper presented at the Proceedings of the TMS Light Metals, Orlando, Florida, USA.

47. Wong C-J, Yao Y, Bao J, Skyllas-Kazacos M, Welch BJ, Jassim A (2021) Modeling Anode Current Pickup After Setting. Paper presented at the Proceedings of the TMS Light Metals, Orlando, Florida, USA.

48. Wong C-J, Yao Y, Bao J, Skyllas-Kazacos M, Welch BJ, Jassim A, Mahmoud M, Arkhipov A (2021) Modelling of Coupled Mass and Thermal Balances in Hall-Heroult Cells During Anode Change. Journal of the Electrochemical Society. 168(21):123506.

49. Jassim A, Akhmetov S, Welch B, Skyllas-Kazacos M, Bao J, Yao Y (2016) Studies on anode pre-heating using individual anode signals in Hall-Heroult reduction cells. Paper presented at the Proceedings of the TMS Light Metals, Nashville, Tennessee, USA.

50. Wong C-J, Yao Y, Bao J, Skyllas-Kazacos M, Welch BJ, Jassim A (2020) Study of heat distribution due to ACD variations for anode setting. Paper presented at the Proceedings of TMS Light Metals, San Diego, California, USA.

51. Jakobsen SR, Hestetun K, Hovd M, Solberg I (2001) Estimating alumina concentration distribution in aluminium electrolysis cells. IFAC Proceedings Volumes. 34(18):303–308.

52. Hestetun K, Hovd M (2005) Detecting abnormal feed rate in aluminium electrolysis using extended Kalman filter. IFAC Proceedings Volumes. 38(1):85–90.

53. Yao Y, Cheung C-Y, Bao J, Skyllas-Kazacos M (2015) Monitoring local alumina dissolution in aluminum reduction cells using state estimation. Paper presented at the Proceedings of the TMS Light Metals, Orlando, Florida, USA.

54. Yao Y, Cheung C-Y, Bao J, Welch BJ, Skyllas-Kazacos M, Akhmetov S, Jassim A (2017) Method for estimating dynamic state variables in an electrolytic cell suitable for the Hall-Heroult electrolysis process. Patent number WO 2017/141134 A1.

55. Yao Y, Bao J (2018) State and parameter estimation in Hall-Heroult cells using iterated extended Kalman filter. IFAC-PapersOnLine. 51(21):36–41.

56. Wong C-J, Yao Y, Bao J, Skyllas-Kazacos M, Welch BJ, Jassim A, Mahmoud M (2021) Discretized Thermal Model of Hall-Heroult Cells for Monitoring and Control. IFAC-PapersOnLine. 54 (11):67–72.

57. Shi J, Yao Y, Bao J, Skyllas-Kazacos M, Welch BJ (2020) Multivariable Feeding Control of Aluminum Reduction Process Using Individual Anode Current Measurement. IFAC-PapersOnLine. 53(2):11907–11912.

58. Shi J, Yao Y, Bao J, Skyllas-Kazacos M, Welch BJ, Jassim A, Mahmoud M (2021) A New Control Strategy for the Aluminum Reduction Process Using Economic Model Predictive Control. IFAC-PapersOnLine. 54(11):49–54.

59. Shi J, Yao Y, Bao J, Skyllas-Kazacos M, Welch BJ, Jassim A, Mahmoud M (2022) Advanced Model-Based Estimation and Control of Alumina Concentration in an Aluminum Reduction Cell. JOM. 74(2):706–717.

60. Wang R, Bao J, Yao Y (2019) A data-centric predictive control approach for nonlinear chemical processes. Chemical Engineering Research and Design. 142:154–164.

Light Metals Research at the University of Auckland

J. B. Metson, R. Etzion, and M. M. Hyland

Abstract

The University of Auckland has had a more than 40-year history of research contributions supporting the light metals industries. This research programme was initiated with the appointment of Professor Barry Welch in Chemical and Materials Engineering but subsequently broadened to embrace academics across a range of disciplines, particularly in chemical sciences. Work was initially focused on the aluminium and alumina industries but has extended at times into lithium, magnesium and titanium. A defining characteristic of this work has always been close engagement with the light metals industries. However, over time, the nature of this interaction has evolved. Of note has been convergence on a range of industry-defining and challenging issues where cross-sector collaborations, often working closely with government agencies, have progressively displaced individual projects addressing technical advantage. Environmental footprint and response to a dynamic energy environment are prominent within this common ground. These trends will be examined along with potential future directions for this research relationship.

Keywords

Universities • Light metals research • Professor Welch

Introduction

The light metals occupy an increasingly important position in the periodic table. Location towards the top of the table makes them abundant, while their utility and recyclability make them increasingly important materials in a range of applications. Aluminium and Magnesium have been widely used in construction, packaging etc. and are critical in light-weighting of transport fleets, while titanium has been a key metal in aerospace, transport and bio-medical application.

The ores from which raw materials for smelting are sourced are abundant and widely distributed across the globe leading to similarly widely distributed industry centres. Although the embedded energy in their manufacture is high, for aluminium and magnesium in particular most of this embedded energy is recovered when the metal is recycled. This leads for example to the remarkable statistic that an estimated 75% of aluminium ever made is still in use [1]. In an energy constrained world, a singular focus on the energy of production, while important, ignores the more significant issue of the lifetime energy cost of these materials. Provided target compositions can still be achieved, and recycling the metals ensures maximum utilization of embedded energy and these materials become a critical contribution to a greener economy and environment.

Despite these advantages, the metals still present major challenges in manufacture and utilization in an energy-constrained and increasingly environmentally conscious world. For aluminium, the relative maturity of the Hall-Heroult and Bayer processes' central technologies, dating from the 1880s, presents a particular problem. At a time of incremental optimization of established technologies operating at very large scale, there are pressing demands for step-change improvements in energy efficiency, environmental footprint and resource utilization. This has presented fertile ground for a range of related research themes both in support of these current technologies and in the exploration of more radical approaches to aluminium and alumina production. Research and innovation are key to the future of the light metals.

J. B. Metson (✉) · R. Etzion · M. M. Hyland
The University of Auckland, 22 Princes St, Auckland, 1010, New Zealand
e-mail: j.metson@auckland.ac.nz

© The Minerals, Metals & Materials Society 2023
S. Broek (ed.), *Light Metals 2023*, The Minerals, Metals & Materials Series,
https://doi.org/10.1007/978-3-031-22532-1_7

The University of Auckland

Research at the University of Auckland in support of the light metals industries was largely initiated by the arrival of Professor Barry Welch's in 1980, the same year the first edition of the seminal "Aluminium Smelter Technology" by Grjotheim and Welch was published [2]. This began a sustained period of growth in fundamental and applied research, largely in support of the international aluminium and alumina industries.

The 1970s and 80s saw a major expansion of the aluminium industry in Australasia. In New Zealand, Comalco's construction of the Tiwai Point Aluminium Smelter, completed in 1971, had made the primary aluminium industry strategically important to New Zealand [3]. This was part of major growth in the aluminium industry in the wider region, in part attributed to the contraction of the industry in Japan following the oil shock. This generated both a significant demand for skills, including graduate engineers and a need for research and technology support to sustain this growth.

The research model established by Professor Welch at the University was distinguished by a combination of excellence in basic research, often with a direct line of sight to application. Projects were frequently coupled with significant time in smelters and, on occasion, alumina refineries, for most of the many students and their supervisors engaged with these research programmes. This combination nurtured strong industry relationships, depth and flexibility in researchers and potentially short pathway to the application of knowledge.

Of note was the breadth of experimental capability established at the University during this time, with major furnace and electrochemical capabilities, able to simulate aspects of the aluminium reduction industrial environment at a sufficient scale to provide meaningful insights. This capability was supported by a range of leading-edge characterization facilities. The Research Centre for Surface and Materials Science was established in 1988 and also hosted by the Department of Chemical and Materials Engineering. The Centre housed both electron microscopy facilities and New Zealand's first X-ray Photoelectron Spectrometer (XPS) instrument, offering a unique surface analytical capability very quickly harnessed in support of such industry-connected research programmes. The supporters of this XPS purchase included Comalco, NZ Steel, and New Zealand's Defence Scientific Establishment.

As the Auckland program grew, it embraced an increasingly diverse group of academic researchers, including Professors Chen, Hyland, Metson, Muller-Steinhagen, Taylor and a steadily expanding range of interests and capabilities. Work covered a remarkable span of research topics, including:

- Alumina dissolution and the resistance curve.
- Bubble dynamics and impacts on current efficiency.
- Anode reactivity and carbon consumption.
- Mechanisms of cathode degradation.
- Degradation of sidewall refractories.
- Heat transfer through cell cover.
- Operational and control decision-making for smelters.
- Alternative technologies for aluminium production.
- Emissions and the mechanisms of HF generation and adsorption on alumina.
- Dry scrubbing and the performance of aluminas.
- Sulfur gases and impacts on dry-scrubbing.
- Impacts of impurities in reduction cells.
- Alumina calcination and resulting performance in the smelter.
- PFC generation.
- Potroom dust and the implications for environmental emissions.
- Power modulation of reduction cells.
- Catalytic destruction of PFCs.

More than 30 Ph.D. students have completed their degrees on these projects. The experimental capabilities have also been harnessed in commercial materials testing, particularly in cathode degradation, alumina dissolution and general alumina characterization.

The close relationship of Professor Welch and others with Comalco, the operator of the Tiwai Point smelter, was highlighted by the very significant role University researchers played in support of the major upgrade of the smelter completed in 1996. A major catalyst for the growth of research at Auckland through the 1980s and 90s was undoubtedly the technological ambition of Comalco in aluminium reduction. Work on drain cathode cells [4, 5], the development of Torbed dry-scrubbers [6, 7], amongst other initiatives, provide some of the more challenging examples. This close relationship also saw the appointment at the University of Auckland of now Professor Margaret Hyland as the Comalco Lecturer in Materials Science and Engineering in 1991.

Of note also has been long-term relationships with a range of other industry players such as Aluminium Pechiney in the formative years, Hydro Aluminium, Trimet Aluminium, EGA, Rio Tinto, Outotec, SAMI, GAMI and others. This also generated particularly strong relationships with other University based research providers including NTNU, UNSW and the REGAL consortium.

The formation of the Light Metals Research Centre in 2002 was led by Professors Welch and Metson and triggered by the retirement of the former and the desire to build on the momentum already established. The subsequent recruitment of the former Auckland graduate Professor Mark Taylor as the Director saw a major build in activity, with a particularly fruitful relationship with the rapidly emerging Chinese industry and addressing the increasing survival challenges of European based smelters. Dr Mark Dorreen took over in this role from 2014, however the years following the Global Financial Collapse saw increasingly tight conditions in the industry, with flat metal prices prompting limited research activity, contraction of the industry in Australasia and refocusing the University of Auckland efforts.

Education

As with other universities engaged in research with the industry, part of the Auckland model has also been to support the development and training of industry personnel. In addition to contributions to numerous TMS short courses, the University of Auckland Post-Graduate Certificate in Aluminium Reduction Technology and the follow-on Masters programmes have been offered in a succession of venues including New Zealand, Australia, several venues in the Middle East and Germany.

These courses are centred on research-based teaching and learning, drawing on research leaders and expertise from across the international industry. The format of the courses reflected a pre-Covid world and has typically been based around a three-week residential block course located close to a smelter to allow observation and practical experience of a range of processes. This practical approach has been used both in formal post graduate qualifications such as the PG Cert. and in training smelter teams worldwide on production process fundamentals, benchmarked operation and process control strategies.

Evolution of Research Needs in the Global Industry

While Light Metals research at the University of Auckland has now passed through several generations of researchers, the industries themselves have also seen a major structural change through this same period. Of particular note has been the consolidation of the historical industry through mergers and, with this, the contracting numbers of major industry centres of research excellence, particularly in North America, Europe and Australia. In parallel, the industry has seen expansion elsewhere driven by power costs, with the emergence of accompanying major research and technology capabilities, particularly in China and the Middle East. This period has seen the rise of a new geographical distribution of both producers and technology suppliers.

This change has resulted in a more limited and concentrated research base with significant dependence on research institutions offering complementary skills and facilities. This breadth has also reflected a wider range of industry challenges and access to the more diverse range of disciplines needed to address them. Industry challenges such as environmental impact, reduced energy consumption, and managing power insecurity also represent "common good" research where industry partnerships, frequently with government involvement, are needed to make real progress.

The collaboration between the University of Auckland and governmental agencies has led to several projects that addressed fundamental issues within the aluminium industry. A US EPA-funded initiative has resulted in developing a fluoride emissions manual [8] produced in Mandarin and English and working on understanding PFC emissions and ways to reduce them. A collaboration of Auckland University with Australian Universities funded by the Australian Government via CSIRO led to a cluster of projects that aimed to find innovative breakthrough solutions to significant problems of the production process by focusing on three themes: cell design and operation, alternate processes/breakthrough technologies, and process control. The New Zealand Government's support has also helped test innovative power modulation technology [9] in the Tiwai Point smelter and ultimately deployment in Europe.

The strategy of sourcing innovation and complementary capabilities outside of the industry has followed a similar pathway to developments in other sectors. Perhaps the lead has been shown by the pharmaceuticals sector, where the cost of in-house research capacity, particularly in drug discovery, has become near unsustainable. Industry instead seeks to harness the power and breadth of smaller, nimbler innovator companies and particularly Universities [10]. Tralau-Stewart et al. [11] highlight the role of universities in this drug discovery process and argue for new partnership models that leverage public funding where common areas of interest allow a win–win outcome. It could be argued that the initiatives above, particularly in emissions control, already demonstrate the application of this model to the aluminium industry. There is little doubt that in parallel with company specific initiatives, the collaborative model will be increasingly prominent in the industry.

Conclusions

The University of Auckland has had a long and distinguished history of engagement with the light metals industries, particularly in the smelting of aluminium and the manufacture and properties of metallurgical alumina. Built on the pioneering work of Professor Barry Welch, this has provided more than 30 years of fundamental insights, research support and innovative technologies to a steadily evolving industry. This evolution has been structural in terms of the major companies, technological, including in terms of models of research collaboration and increasingly geographic, such that in some parts of the world the industry continues to face major challenges to its very existence.

References

1. https://international-aluminium.org/work_areas/recycling/ accessed 31/07/2022.
2. K. Grjotheim and B.J. Welch (1980) Aluminium Smelter Technology: A Pure and Applied Approach, Aluminium-Verlag. Second edition published in 1987.
3. Grant, David. Aluminium smelter, Tiwai Point. teara.govt.nz. Te Ara - The Encyclopedia of New Zealand.
4. B J Welch JOM, 51 (5) (1999), pp. 24–28. Aluminum Production Paths in the New Millennium.
5. G.D. Browne et al., "TiB2 Coated Aluminium Reduction Cells: Status and Future Directions of Coated Cells in Comalco," Proc. 6th Aust. Al Smelting Tech. Workshop, ed. B.J. James, M. Skyllas-Kazacos, and B.J. Welch (Sydney, Australia: U. of N.S.W. and RACI, 1998), pp. 499–508.
6. Torbed dry scrubber-Register, J. (1995). Tiwai's Upgrade - Thank Sweat and the ECA. New Zealand Engineering, 50(10), 6–9.
7. https://www.nzas.co.nz/files/1235_20160718112004-1468797604.pdf.
8. Nursiani Tjahyono, Yashuang Gao, David Wong, Wei Zhang & Mark P. Taylor. Fluoride Emissions Management Guide (FEMG) for Aluminium Smelters. Light Metals 2011 pp 301–306.
9. Lavoie, P., Namboothiri, S., Dorreen, M., Chen, J.J.J., Zeigler, D. P., Taylor, M.P. (2011). Increasing the Power Modulation Window of Aluminium Smelter Pots with Shell Heat Exchanger Technology. In: Lindsay, S.J. (eds) Light Metals 2011.
10. Alexander Schuhmacher, Oliver Gassmann & Markus Hinder (2016) Changing R&D models in research-based pharmaceutical companies. Journal of Translational Medicine volume 14, Article number: 105.
11. Cathy J. Tralau-Stewart, Colin A. Wyatt, Dominique E. Kleyn and Alex Ayad. (2009) Drug Discovery Today Volume 14, Numbers 1/2 January 2009 REVIEWS.

Impact of Aluminium Reduction Cell Parameters on Feeder Hole Condition

Pascal Lavoie and Mark P. Taylor

Abstract

Aluminium reduction requires alumina to be fed and dissolved into electrolyte. The feeder hole condition may impact severely the ability to dissolve the added alumina. In this study, two sets of observations were made on a large sample of aluminium reduction cells to understand the feeder hole condition progress during feed events and determine which reduction cell parameters affect the feeder hole condition and the progression pathway of the feeding event sequence. The results show that feed rate has a major impact on the initial feeder hole condition and is determinant to the pathway of the feeding sequence. Higher superheat, lower excess AlF_3 and longer time since the last anode change increased the probability of finding feeder hole conditions conducive to dissolution. The observations of blocked feeder holes were found to be linked to abnormal process conditions or a mechanical issue. Changes made to the feeding strategy between the two observation sets led to a significant improvement of the feeder hole condition. The proportion of opened feeder holes increased from 12 to 41%, resulting in a more than 50% reduction of the anode effect frequency smelter-wide over an 18 month period.

Keywords

Alumina • Dissolution • Feeder • Superheat • Feeder hole • Sludge

The work reported in this paper is part of independent research from Pascal Lavoie conducted during completion of his Ph.D. at the University of Auckland, prior to joining Alcoa.

P. Lavoie (✉)
Alcoa Corp., Quebec, Canada
e-mail: pascal.lavoie@alcoa.com

M. P. Taylor
The University of Auckland, Auckland, New Zealand
e-mail: mark.taylor@auckland.ac.nz

Introduction

In modern electrolysis cells alumina is fed through point feeders which periodically add a volumetric dose equivalent to between 1 and 2 kg into the cell. Most of these feeders consist of two devices, as shown in Fig. 1. The first device, a "breaker", is an air actuated steel shaft coming down to break the electrolyte crust that may have formed between feeding events. The second device, a "feeder", delivers a fixed volumetric dose of alumina to the feeder hole [1].

The condition of the feeder hole may impact the dissolution rate of the alumina added. Point feeders are normally designed to feed into molten bath in order to maximise the effective contact area between the alumina particles and the electrolyte [3]. However, feeder holes have been observed to operate in a push feed mode [4] where the alumina from the last feeding event lodges in the feeder hole and is pushed into the bath by the breaker at the next feeding event. This has the potential to greatly limit the effective contact area between the alumina and bath, and hinder alumina dissolution leading to sludging and operational difficulties [5]. Some smelters may intentionally operate the feeder hole in such closed state to reduce gaseous emissions or conserve heat but at the detriment to dissolution.

In this article, the authors describe the factors and mechanisms that may affect the feeder hole condition. The results of a study based on experimental observations of feeder hole conditions and selected cell parameters are presented. Finally, the implications of controlling cell parameters to maximise alumina dissolution are discussed.

Factors That May Affect Feeder Hole Condition

In a previous article, a list of factors that may affect the feeder hole condition and the mechanisms that in turn would affect alumina dissolution have been presented [4].

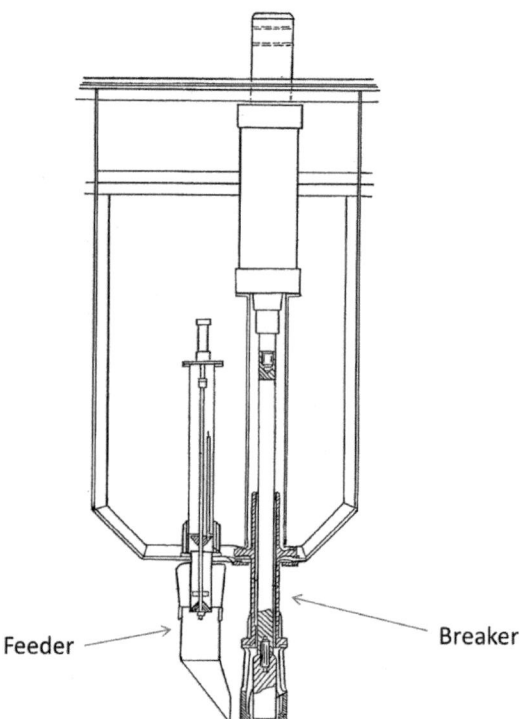

Fig. 1 Point feeder schematic. Adapted from [2]

The hypothesis developed in [4] is that by carefully controlling the feeder hole condition, through the cell parameters management, alumina dissolution can be aided.

This first linkage in Fig. 2 is the focus of the present paper, while the second linkage was also investigated but will be subject of further publications from the Ph.D. thesis of the primary author [6]. The actual relationship between these cell parameters previously hypothesised as factors impacting dissolution coupled with observations of feeder hole conditions is therefore examined in the present paper.

The cell parameters studied are the following:

- Bath level.
- Metal level.
- Total liquid level.
- Bath Temperature. In the present study, a Heraeus FiberLab.[1]
- Liquidus Temperature. In the present study, also the FiberLab synchronously with bath temperature.
- Superheat. As defined in the aluminium industry, the difference between bath temperature and liquidus temperature.
- Excess AlF_3. In the present study, also using the FiberLab.

---

[1] Measurement of bath temperature by optic fiber and analysis of the cooling curve inferring bath composition. www.heraeus.com.

In addition to the cell parameters measured on the day of the observations, the following elements were noted:

- Last anode changed
  An aluminium reduction cell consists of a multitude of anodes. Anodes changed in the vicinity of the feeder hole have the potential to cool the bath locally as they are introduced at room temperature and will absorb sensible and latent heat of freezing from the bath as they heat up to the operating temperature.
- Feed phase
  A cell undergoes a feed cycle imposed by the cell controller to regulate the alumina concentration [7]. As a consequence, the sensible heat in the feeder hole region gradually depletes during overfeed (enriching of the alumina concentration in the bath), as more cool alumina is introduced at a higher rate than stoichiometrically required for aluminium production. Conversely, the sensible heat of the bath in and around the feeder hole increases during the underfeed phase, depending on how far above and below these feed rates are from the theoretical requirement of the cell the feed rate [7]. This variation in temperature due to the feed cycle can be significant and is observed globally throughout the cell as well as locally in the feeder holes, as is represented in Fig. 3. In the present study, the feed cycle is schematically represented in Fig. 4.

Feeder Hole Behaviour Description and Characterisation

Observation Sequence

For each feed event observed, the feeder hole condition and behaviour throughout that event was recorded following semi-quantitative, ordinal characterisation scales describing the state or behaviour of the feeder hole during the event time sequence below:

1. Initial state/Type of break. The state of the feeder hole immediately before initiation of a feeding event (0–10 s).
2. Response to breaking. The behaviour of the feeder hole during the first phase of a feeding event, just after the breaker is actuated (0–5 s).
3. Response to feeding. The behaviour of the feeder hole during the second part of a feeding event, just after the alumina dose is delivered by gravity to the bath surface (0–10 s).
4. Closing state. The state of the feeder hole, approximately 10 s after the alumina dose has been delivered.

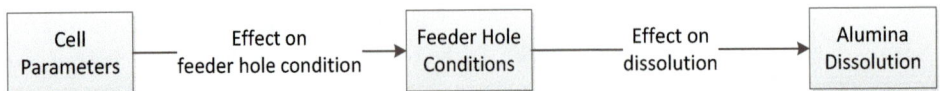

Fig. 2 Link between cell parameters and alumina dissolution

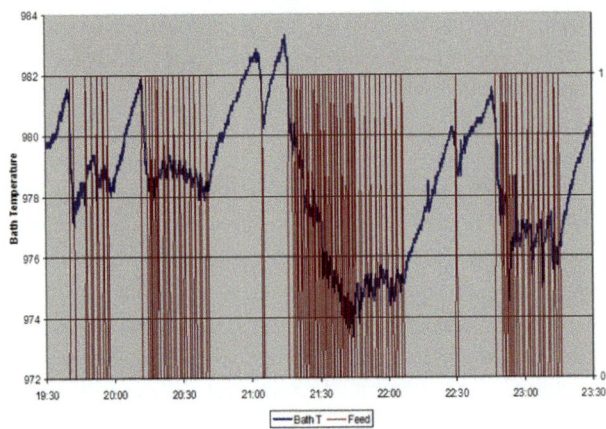

Fig. 3 Example of extensive bath temperature variation measured at the end of a cell throughout the feeding cycles [7]

Feeder Hole Condition State Definition

The following feeder hole condition states have been defined for the purpose of the study. They consist in an ordinal scale of the feeder hole condition. Under the hypothesis concerning the feeder hole condition affecting dissolution, the scale ranges from the least favourable to the most favourable for alumina dissolution. These states can be visualised in Fig. 5.

1. Blocked feeder hole
   A blocked feeder hole consists of a crusted electrolyte with a significant accumulation of alumina in the hole resulting in the inability of the breaker to break the crust and push the material in molten bath.
2. Push feeder hole
   A push feeder hole consists of a crusted electrolyte surface with a piled quantity of alumina. As opposed to the blocked feeder hole, the breaker action successfully pushes the material in the molten bath, with material present in the hole after the breaker retraction.
3. Break feeder hole
   Molten bath surface cannot be seen in the initial state, its surface is covered by crusted electrolyte with powder alumina on top or not. The breaker operation breaks the crust and pushes the material in the molten bath underneath. The molten bath surface can be seen after the breaker retraction.

4. Open feeder hole
   In the open feeder hole state, the molten surface of the electrolyte can be seen and is free of solid material in this initial stage.

Response to Breaking

Four behaviours have been defined to describe the response of the feeder hole following the breaker actuation

1. No Response
   The breaker comes down and may push the material inside the molten bath or not. In either case, there is no dynamic change notable within the feeder hole once the breaker retracts.
2. Particle Aeration
   The breaker action leads to the evacuation of process gases throughout a small opening, leading to alumina particles being aerated upwards.
3. Bubbling
   A bubbling feeder hole consists in a semi-solid electrolyte surface with a small quantity of alumina present on the top. Gaseous emissions can be observed creating a bubbling effect in the alumina present on the top.
4. Open
   Same definition as with the feeder hole state.

Response to Feed

Three behaviours have been defined to describe the reaction in the feeder hole when the alumina dose is delivered. Note that number 2 has been omitted in the sequence to conserve a quantitative scale consistent with other states or behaviours.

1. On Material
   The alumina dose falls on the crust or alumina already present in the feeder hole. No gas evolution or movement of the material can be seen.
3. Bubbling
   Same definition as with the feeder hole condition.

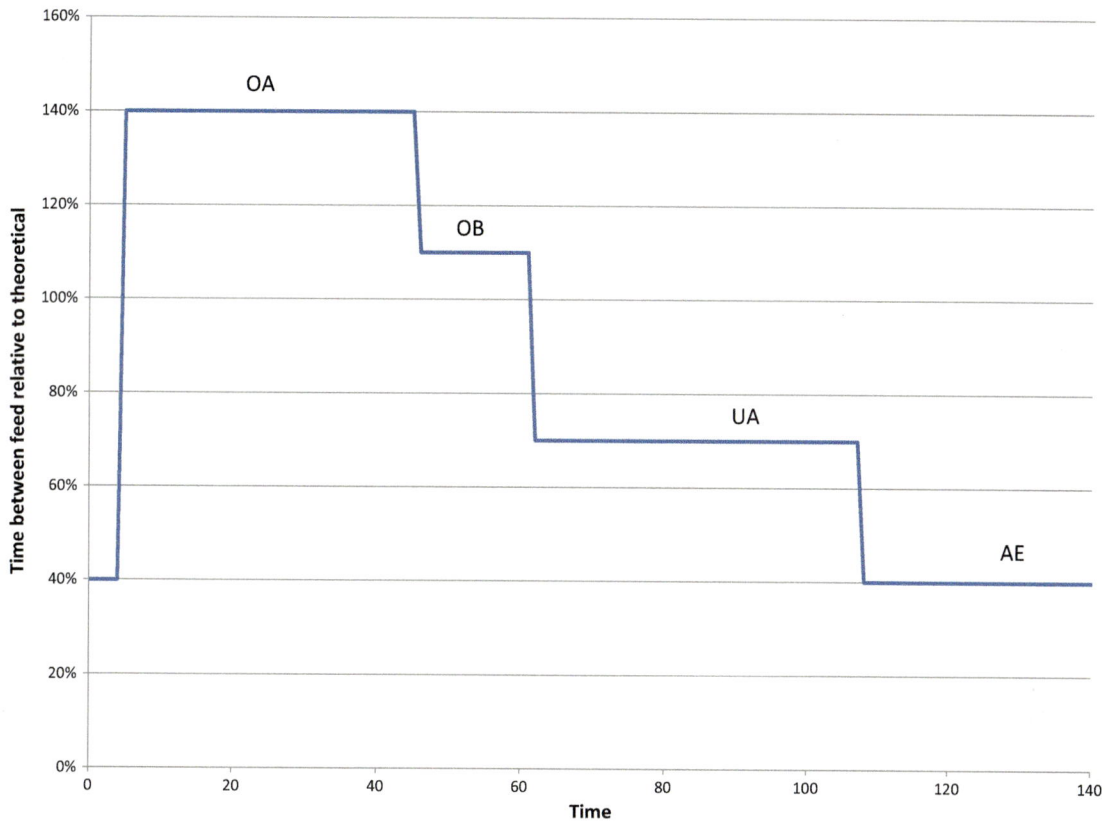

Fig. 4 Feed cycle diagram showing the four feeding rates used by the controller for the cell technology under study (OA, the highest federate; OB, at slight overfeed; UA, at slight underfeed; AE, the lowest feed rate)

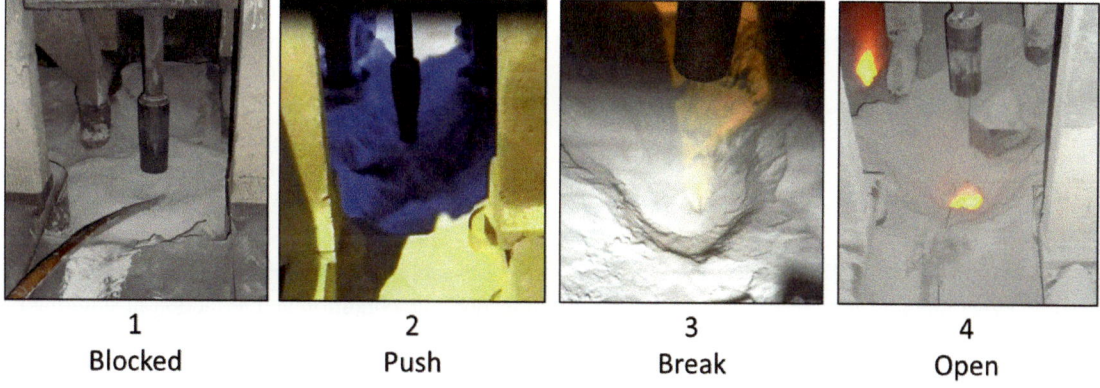

Fig. 5 Ordinal scale of feeder hole states. Going from the least (1) to the most (4) favourable to alumina dissolution

4. Open

The molten surface of the electrolyte can be seen but a floating alumina raft could be present. However, the wave movement is noticeable and the alumina raft can be seen dissolving or floating away.

Experimental Observations

In two separate series of experiments, a relatively large number of feeding events in reduction cells of the same technology were observed, 66 and 70 events respectively. For each set of experiments the breaking/feeding states and responses defined above were characterised. The series of observations in each experiment occurred over a timeframe of approximately two hours.

Between these sets of process characterisations, some of the operating conditions were changed. Adjustments were made to the feeding strategy and a specific abnormality detection algorithm implemented to detect and prevent sludging of the cells similar to that described by Mulder et al. [8]. The potline was however operated with the same amperage, working cycle and other parameters targets remained similar.

The observations were made at random timeframes during the feed cycle. The cells were observed sequentially. They were not selected randomly but were within a group where the cell parameters (Temperature, Liquidus, $XsAlF_3$) had been measured within the 8-h shift, and where no other physical cell operations were occurring.

Results

State and Behaviour Progression Throughout the Feeding Sequence

During a feeding event, the feeder hole condition can evolve along different pathways, progressing from the initial state to the closing state. In order to assess the progression likelihoods of the feeder hole state and behaviour throughout the feeding sequence, a probability tree was built. The

observations were compiled to represent the likelihood of progression to each branch of the next step, schematically shown in Fig. 6.

The evaluation of individual links along the feeding sequence is detailed in the primary author's Ph.D. thesis [6]. Examples of the probability tree evaluation are provided in this paper to illustrate the methodology and the conclusions presented below.

As a first example, differences can be seen on Fig. 7, when comparing the proportion of initial state by feed phase between the two observation sets. During the first set of observations, the likelihood of finding a feeder hole state detrimental to dissolution (1 "blocked" or 2 "push") was the highest (A), except during the lowest feeding rate phase AE (B), where it was more likely to be beneficial to dissolution (3 "bubbling" or 4 "open"). However, it was equally likely to find a feeder hole state beneficial to dissolution in all feed phases during the second set of observations (C), except when feeding at the highest rate OA (D), where the likelihood tended slightly towards a detrimental state.

This improvement in the feeder hole condition from observations set 1 to set 2 can be attributed to the adjustments made to the feed strategy by the smelter management, aimed at improving dissolution. These included restricting feed rates during overfeed and the addition of an online abnormality detection of excessive alumina feed against normal demand through time.

Although the occurrence of detrimental feeder hole states was marginally reduced in set 2 from 56 to 50%, the increase in open feeder holes from 12 to 41% mostly came from improvement in the "Break" state condition (Not explicitly shown on Fig. 7).

As a consequence, a significant reduction in the sludging behaviour is observed in the cells under study. The alumina cusum amplitude, a soft sensor for the extent of undissolved alumina settled at the bottom of the cell [9] reduced by approximately 500 kg (99% confidence). This improvement in feeder hole condition and reduction in sludging behaviour has yielded a large decrease in anode effect frequency of approximately 45% observed smelter-wide and sustained over one year. This last observation is significant and will be discussed later in the paper.

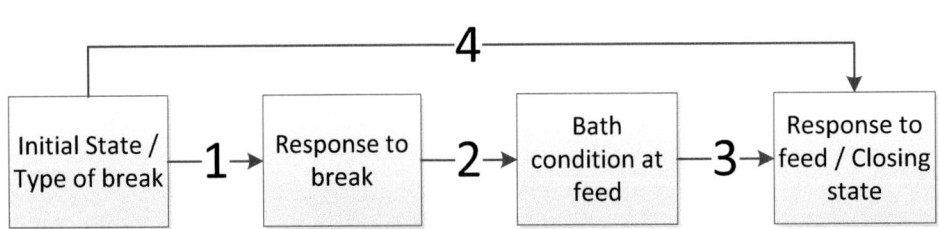

Fig. 6 Analysis of progression of state and behaviour through the feeding sequence

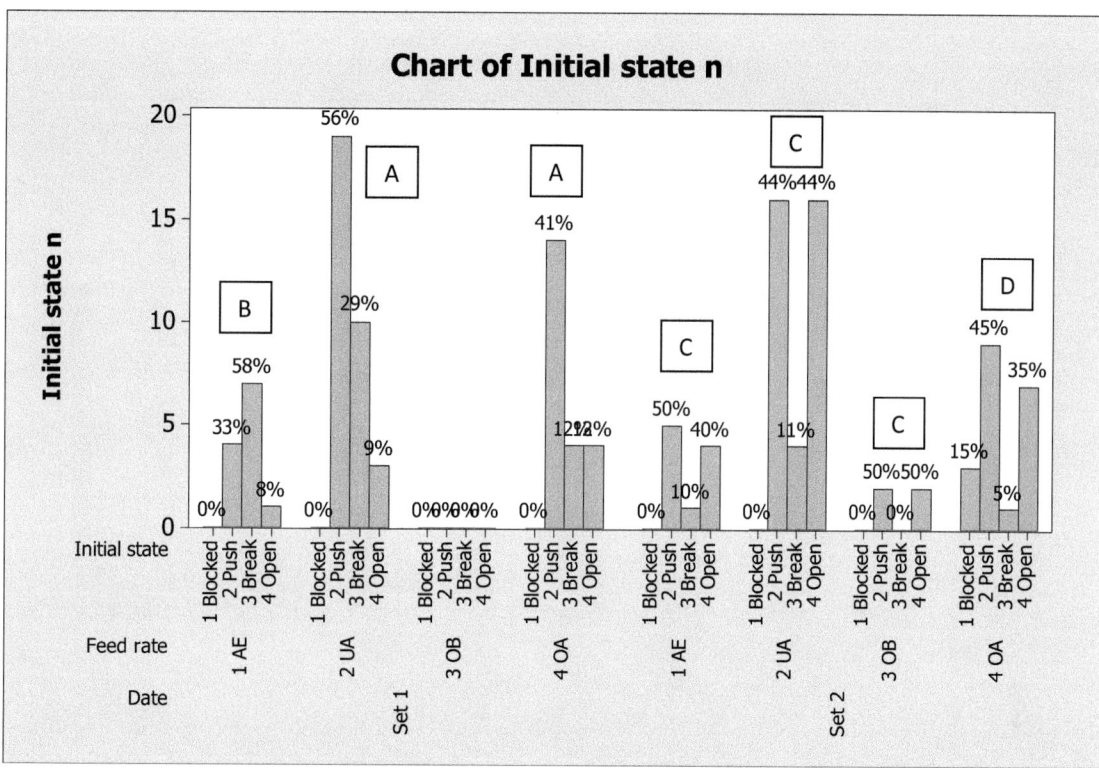

Fig. 7 Initial feeder hole state distribution between feed phase and observation sets

Progression from Initial State to Response to Break (Link 1, Fig. 6)

As a second example, Fig. 8 shows the effect of the feed rate and the initial state on the response to break. Because of low occurrence, the initial condition "Blocked" and the feed rate "OB" were omitted from this graph. As can be expected, having an open feeder hole condition initially was never affected by the breaker actuation.

When starting with a detrimental state ("push" in this case), the likelihood of obtaining a detrimental response to break is high (77%). However, the response to break improves with lower feed rates, as the modal behaviour moves from "1 no response" in OA to "2 particle aeration" in UA and "1 no response" becomes exceptional in AE feed, the lowest feed rate. Feeder hole condition can therefore improve during lower feed rate phases, probably because there is more time to replenish and increase sensible heat in the feeder hole between each feed event. Such behaviour has also been demonstrated in a previous study and can be observed in Fig. 9 [7].

Finally, when the initial condition is "3 Break", there is by definition a response to break. In the majority of observations (85%), the response to break is favourable, having modes at either "3 Bubbling" or "4 Open". Contrary to the general trend found within the study, with the initial

condition "3 Break" the lowest feed did not lead to the most favourable response. This could be due to differential bath temperature and liquidus temperature evolutions or to the time between breaks allowing a stronger crust to form, getting a higher proportion of "2 Particle Aeration" than with higher breaker operation frequency.

Progression from Response at Break to Bath Condition at Feeding (Link 2, Fig. 6)

As a final example, Fig. 10 shows the progression from the response at break to the bath condition at feeding. In all cases where no response occurred at break (A), the alumina dose was delivered onto material without perceivable movement of alumina in the feeder hole, maintaining the detrimental state. Similarly, a feeder hole that was open (B) invariably had the alumina dose delivered into the molten bath.

When the breaker operation was partially successful (C), resulting in aeration of particles, the condition at feed was mostly detrimental. The alumina would most likely be delivered onto a pile of material (62%). The feeder hole condition had however improved between breaking and feeding, moving into bubbling 33% of the time and even opening completely in 5% of the cases. Performing ordinal

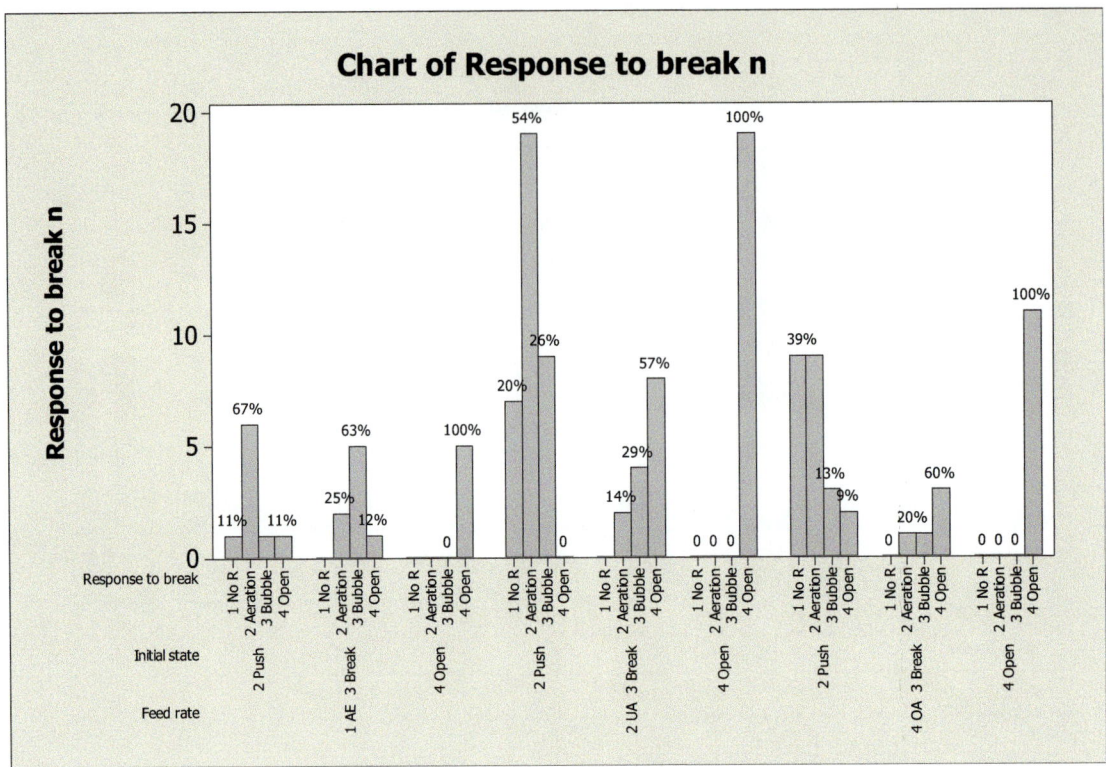

Fig. 8 Effect of feed rate and initial state on response to break. Overfeed "3 OB" and initial condition "1 Blocked" omitted due to low occurrence

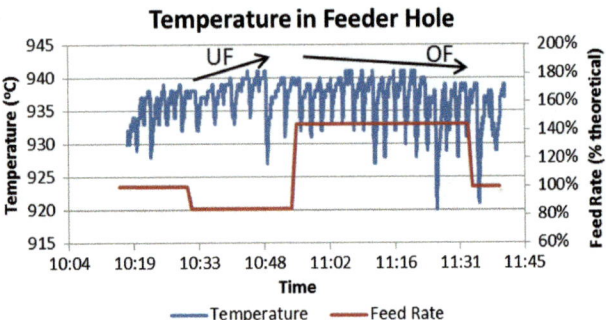

Fig. 9 Sensible heat increasing during underfeed and decreasing during overfeed [7]

logistic regression allows identification of cell parameters that affect the outcome likelihoods [10]. Four factors were determined to be statistically significant (90% confidence) leading to a better feeder hole condition:

- Superheat. For each additional degree of superheat, the odds of feeding on material are reduced by 40%.
- Lower $XsAlF_3$. For each 1% increase in $XsAlF_3$, the odds of feeding on material double.

- Bath level. Surprisingly, as bath level increases, the odds of feeding on material increased as well. Closely examining the parameters however, revealed that the trend had been driven by three observations of low bath levels that also had low $XsAlF_3$ and enough superheat. This suggests that the other favourable parameters helped overcome low bath levels to maintain the feeder hole condition at a better state.
- Feed rate. The statistical significance of this factor is only marginal as the number of data points are small, especially at OB (feed rate 3). Therefore, there may be a trend that as the feed rate increases, and the probability of feeding on the material increase as well. The statistics cannot however quantify the likelihoods accurately.

Finally, when the breaker was successfully operated (D), obtaining a bubbling response to break this state remained until feeding in the vast majority of observations. In exceptional cases (2 occurrences or 9%), the bubbling behaviour had stopped after feeding occurred. In one case, the superheat was only 4 °C which could explain the issue. In the other case, the parameters do not seem to explain the outcome.

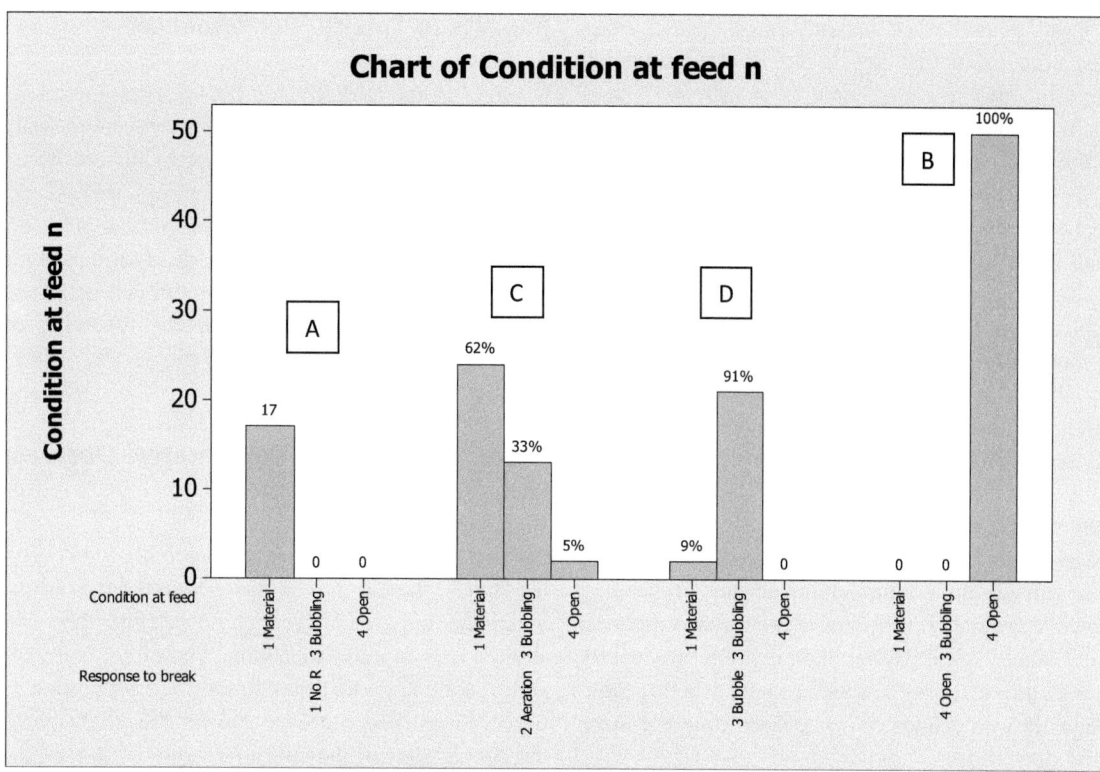

Fig. 10 Effect of response to break on bath condition at feeding

Discussion

Effects

Cell Parameters Affecting the State Progression

The complete analysis [6], too long for this publication, shows that the alumina feeding rate does have a significant influence on the initial feeder hole condition. As the feed rate is increased, the likelihood of feeding alumina into a feeder hole in a detrimental state increases. This has repercussions throughout the feeding sequence as the initial feeder hole condition influences greatly the pathway of the condition through the subsequent feeding steps. The mechanism by which this occurs is probably the gradual depletion of sensible heat during the overfeed period through feeding before local sensible heat has had time to recover [4]. This suggests that restraining the range of feed rates to ensure the local sensible heat is replenished before the next feed event is a key to maintaining the feeder hole condition.

In contrast, the study found marginal statistical influence of very low feed rates to drive a detrimental condition at feeding and response to feed. It is possible that as the time between the crust break increases, the thickness and strength of the crust formed between feed events also increase, creating more difficulty in breaking. In that case, restricting the

range of aggressive underfeed (i.e. AE feed in this study) or actioning the breaker at a higher frequency than the feeder would help maintaining the breaking ability.

The statistical analysis showed that two cell parameters were clearly impacting the feeder hole condition throughout the feeding sequence. The first is superheat, governing the available sensible heat for dissolution. Maintaining enough superheat is therefore critical in maintaining the feeder hole condition. The second is Excess AlF_3, impacting the alumina solubility and therefore the mass transfer gradient away from the dissolving alumina is needed for its dissolution.

Finally, anode change also impacted the ability to maintain feeder hole condition. In the case of this study, the anode position itself did not appear to have statistical significance, but the time since the last anode change did. Optimisation of additional energy input to the cell after anode change could be necessary to compensate for this effect.

Blocked Feeder Holes

Obtaining a blocked feeder hole condition should be considered a process failure as it is abnormal to get the condition and it stems from assignable causes of variation. In all 3 recorded occurrences, the condition was observed during OA, the over feed phase where sensible heat in the feeder hole is likely to be at its lowest as was shown in Fig. 9. Examining the cell parameters of these 3 cases leads to the conclusion that low liquid levels, preventing the breaker from reaching the liquid (low bath in 1 case, low metal in another case) and mechanical issues (leaky feeder in the last case) provoked a blocked feeder hole condition.

A blocked feeder hole remained blocked in the 3 cases recorded, all observed during an overfeed phase. It is possible that the hole condition could improve during an underfeed phase, but from experience and other observations [11], it is more likely to require physical attention from a worker to unblock or possibly a special breaking treatment triggered by the automatic controller if this condition can be detected automatically.

Improving the Hole Condition to an Open Hole

During the two sets of observations, no observations were made where a feeder hole condition terminated its feeding sequence in an open state if the initial condition was different than open. Since open feeder holes were frequently observed, there are certainly situations under which a non-open feeder hole would improve to an open state but these are yet to be determined.

Impacts to Smelter Performance

Changes to the feed strategy between set 1 and set 2, made partially as a result of the data collected here in set 1, yielded a significant performance increase to the smelter, with the Anode Effect Frequency reduced from 0.094 AE/cell/day to a level of 0.042 AE/cell/day. This represents a drastic reduction of more than 50%. There are always multiple factors contributing to the outcome of overall cell performance, but the anode effect frequency is directly related to the effectiveness of the alumina feeding process (Fig. 11).

Limits of Study and Comparative Observations with Other Technologies

As the observations were made randomly with regards to the feed cycle, there are cases which would have been classified under the subsequent feed phase because they were made shortly after a phase transition. Therefore, there could possibly be a categorical bias for a small proportion (<5%) of observations. It is not expected for this slight bias to affect the conclusions of the study.

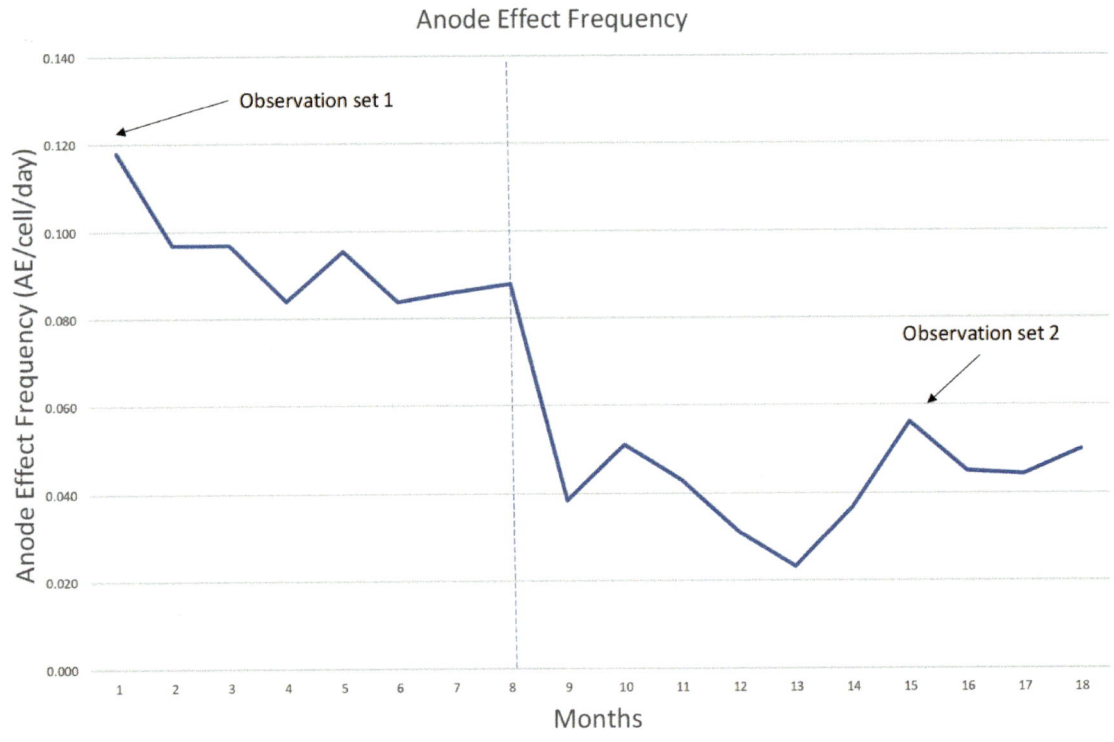

Fig. 11 Smelter-wide anode effect frequency reduction

The observations gathered for this study and the ensuing analysis can be considered representative for this particular smelter and can probably be extended to similar reduction technologies. However, smelters with different technologies, operating practices or operating parameters may still obtain different results.

Although published mainly in internal reports, the authors have made similar observations across a multitude of aluminium smelters with differing reduction technologies and production intensity (amperage) [12–15]. There appears to be consistency with regards to the general trends across technologies, and the specific observations made in the present study. The only exception to this was one smelter with a much larger reduction cell where feeder holes were experiencing blockage at a high frequency within hours of a corner anode being changed [16].

Conclusions and Next Steps

This work developed a rigorous methodology to quantify the feeder hole condition and measure the progression pathway throughout the feeding sequence.

From the analysis of the feeding event observations performed during this study, we can draw the following conclusions:

- The likelihood of finding a feeder hole condition detrimental to dissolution increases with feed rate.
- Blocked feeder holes within this study (3 occurrences) all had specific attributable physical causes (low liquid levels, mechanical problem with breaker, leaking feeder).
- Feeder hole condition evolution through the feeding sequence is highly dependent on the initial condition and is impacted by the feed rate. The condition is more likely to worsen at high feed rates. Although the extreme lowest feed rates may cause difficulty in crust breaking.
- Increasing superheat and decreasing excess AlF_3 increases the odds of finding feeder hole conditions promoting dissolution.
- It is more likely that feeder hole conditions detrimental to dissolution will be found in the hours following an anode change.
- The feeder hole condition appears to have a profound impact on the outcome of the cell performance, as evidenced by the improvement in the feeder hole condition giving rise to a reduction of the anode effect frequency for the whole smelter in observation set 2.

The present study focused on reduction cell parameters affecting the feeder hole conditions. In the future, other factors affecting the feeder hole conditions should also be studied. Rigorous observations using this methodology across technologies could help identify the impact of dissolution zone volume, turbulence (with/without anode slots, current density, bath circulation rate/pattern and alumina feed mass. Also, experiments with different alumina properties (alpha, gibbsite and particle size) and heat content (pre-heating) could be performed.

A question remains concerning what conditions can lead the feeder hole state to improve to an open hole. This was not observed during the present study.

Finally, the central hypothesis brought forward in the introduction of this work (refer to Fig. 2) postulated that the feeder hole condition was central to the dissolution of alumina in reduction cells. This hypothesis is strongly supported by the results of the present study. Further experiments confirming this link but also studying the link between the feeder hole condition and alumina dissolution (alumina concentration) were conducted. These results will be subject of further publications, based on the Ph.D. thesis of the author [6].

Acknowledgements The authors thank TRIMET Aluminium SE for the opportunity to gather observations in their smelter, especially Till Reek, Roman Düssel, and Albert Mulder. The content of this paper was previously reported in [6].

References

1. Andrews, E. W. (1998). A Controllable Continuous Mass-Feed System for Aluminium Smelters. PhD thesis, The University of Auckland
2. Bonny, P., Gerphagnon, J.-L., Laboure, G., Keinborg, M., Homsi, P., & Langon, B. (1984). Process and apparatus for accurately controlling the rate of introduction and the content of alumina in an igneous electrolysis tank in the production of aluminium. U.S. Patent 4,431,491
3. Bagshaw, A. N., Kuschel, G., Taylor, M. P., Tricklebank, S. B., & Welch, B. J. (1985). Effect of operating conditions on the dissolution of primary and secondary (reacted) alumina powders in electrolytes. Light Metals 1985, 649–659
4. Lavoie, P., Taylor, M. P., & Metson, J. B. (2016). A review of alumina feeding and dissolution factors in aluminum reduction cells. Metallurgical and Materials Transactions B, 47(4), 2690–2696
5. Stam, M. A., Taylor, M. P., Chen, J. J. J., Mulder, A., & Rodrigo, R. (2008). Common behaviour and abnormalities in aluminium reduction cells. Light Metals 2008, 309–314
6. Lavoie, P. (2021) On the Factors Affecting Alumina Dissolution in Industrial Reduction Cells. PhD thesis, The University of Auckland
7. Lavoie, P., Metson, J.B. and Taylor, M.P. (2016). Alumina Feed Control, course notes of the Post-graduate certificate in Light Metals Reduction Technology, The University of Auckland
8. Mulder, A., Gao, Y., Zhou, D., Wong, D. S., Ming, L., Lavoie, P., … Yang, X. (2014). New generation control for daily aluminium smelter improvement generation 3 Process control for potlines. Light Metals 2014, 835–840

9. Tandon, G. (2010). Causal Factors in the Variation of the Cell Voltage in Industrial Aluminium Smelting Cells. PhD thesis, The University of Auckland

10. Ordinal Logistic Regression Table – Minitab Online Help, https://support.minitab.com/en-us/minitab/18/help-and-how-to/modeling-statistics/regression/how-to/ordinal-logistic-regression/interpret-the-results/all-statistics/logistic-regression-table/

11. Lavoie, P. (2013). Project communication with Taylor, M.P. HB review 12/12/2013. Confidential Research, Light Metals Research Centre

12. Lavoie, P. (2007). 1118 Feed Test. Confidential research, Light Metals Research Centre, The University of Auckland, Original *redacted* Trial, 06/2007

13. Lavoie, P. (2011). Alumina feeding project summary. Confidential research, Light Metals Research Centre, The University of Auckland, 29/11/2011

14. Lavoie, P. and Depree, N. (2011). Visit Feedback. Confidential research, Light Metals Research Centre, The University of Auckland, *redacted* Feed Control, 06/05/2011

15. Lavoie, P., and Taylor, M. P. (2016). Alumina Concentration Gradients in Aluminium Reduction Cells. Proceedings of the 10th International Conference on Molten Slags, Fluxes and Salts 2016

16. Lavoie, P. et al (2013). Project communication with Mulder, Confidential research, Light Metals Research Centre, The University of Auckland, Feeder Holes, 29/12/2013

A Dynamic Coupled Mass and Thermal Model for the Top Chamber of the Aluminium Smelting Cells

Luning Ma, Choon-Jie Wong, Jie Bao, Maria Skyllas-Kazacos,
Barry J. Welch, Nadia Ahli, Mohamed Mahmoud,
Konstantin Nikandrov, and Amal Aljasmi

Abstract

The cell top chamber is an important region for exchanging material and energy with the bath and the surrounding environment. A considerable heat loss of the aluminium smelting cells is lost through this space. The heat transfer behaviour in the cell top is complex—ambient air and gases generated in the bath enter the cell top, undergo chemical reactions, and are finally collected by the exhaust duct. This paper presents a dynamic coupled mass and thermal model for the top chamber of the aluminium smelting cell. Based on the model, the dynamics of cell top temperature under different operations are studied. This model can be used to investigate the influence of cell operations on the cell top heat balance, especially during power modulation.

Keywords

Dynamic modelling • Mass balance • Thermal balance

Introduction

The Hall-Héroult aluminium smelting process is energy intensive, with Australia's smelting industry consuming 29.5 TWh of electricity in 2007. However, in a typical smelter, about 50% of the energy in the production cost is dissipated as heat from the top, sidewall, and bottom of the cell to the surrounding environment [1, 2]. In order to operate with higher energy efficiency, the smelters now attach great importance to researching the heat loss at the top of the electrolytic cell, which is around 55% of the total heat loss [3–7].

The top heat loss from the bath to the air under the hood is mainly transmitted through two paths: (1) conduction through the shell and anode cover, and convection to the air under the hood. (2) The gas generated in the bath carries heat into the chamber under the hood. However, when calculating the heat flux from the cell body to the cell top, the existing literature mainly put their work on the first path [8, 9]. In order to obtain more accurate heat flux, both paths should be considered in calculating the energy balance of the cell. On the other hand, in the cell top, up to 76% of this heat is carried by gases into the exhaust duct and the rest is lost to the potroom through the hoods and superstructure [7, 10]. To get a better understanding of the heat flow between cell body, cell top, and surrounding environment, the research on the cell top chamber is worthwhile. In addition, in our previous works [11, 12], a detailed cell body model has been established to describe the heat and mass behaviour of the cell body. As an extension of this research, a detailed coupled mass and thermal model for the cell top chamber is required.

Flexible power modulation methods are proposed for improving the energy efficiency of aluminium smelting process [13, 14]. To ensure the safe operation of cells, cell monitoring becomes essential. The temperature dynamics in the cell top can be used as an indicator of cell states under cell operations, e.g., the duct gas temperature is low during the anode setting due to the removal of partial hoods [7]. There have been efforts on the modelling of the cell top chamber: Taylor [15] studied the simplified mechanism of heat transfer from the bath to the chamber and calculated the heat flow between them. Dupuis [16] presented a way to calculate the cell top temperature and showed the relationships between it and cells' other steady states. Abbas [2] and Gusberti [17] both presented a more detailed mechanism description for the cell top heat loss and proposed different

L. Ma · C.-J. Wong · J. Bao (✉) · M. Skyllas-Kazacos · B. J. Welch
School of Chemical Engineering, The University of New South Wales, Sydney, NSW 2052, Australia
e-mail: j.bao@unsw.edu.au

N. Ahli · M. Mahmoud · K. Nikandrov · A. Aljasmi
Emirates Global Aluminium, Jebel Ali Operations, P.O. Box 3627 Dubai, United Arab Emirates

© The Minerals, Metals & Materials Society 2023
S. Broek (ed.), *Light Metals 2023*, The Minerals, Metals & Materials Series,
https://doi.org/10.1007/978-3-031-22532-1_9

CFD models to study the temperature distribution in the cell top. However, the above studies are limited to steady-state modelling and analysis. In this paper, a dynamic coupled mass and thermal model of the cell top is developed, which can be potentially used for real-time cell monitoring (e.g., the bath temperature), especially during power modulation where the operating point of the cell is changed frequently, a dynamic cell top model is needed.

This paper is organised as follows: the mass balance of the cell top model is first presented, followed by the heat balance of the cell top model. Finally, a series of simulation studies are presented to analyse the differences between cell top dynamics and cell body dynamics under different operations.

Modelling

For the modelling of the cell top, mass conservation and energy conservation are applied. The control volume for the coupled mass and thermal model is represented by the area surrounded by dash dot line in Fig. 1. For the mass flow, air from the potroom and gases generated from bath enter the chamber. After fully mixing and undergoing a series of reactions, all gases will leave the chamber through the exhaust duct. For the energy flow, the enthalpies associated with mass flow need to be considered. Apart from this, heat transfers (such as conduction and convection) and the heat generated by under-hood reactions also need to be

considered. The real situation in the cell top chamber is very complex, and it is difficult to describe it fully and exactly. In order to ensure that the model can produce results within a reasonable framework, the following assumptions are made.

Assumptions

1. The temperature $T_{celltop}$ in the cell top is uniform, and the duct gas temperature is equal to the cell top temperature, i.e., $T_{duct} = T_{celltop}$.
2. The pressure in the cell top is lower than atmospheric pressure and always stable.
3. All $NaAlF_4$, CO and COS are separately converted to Na_3AlF_6, CO_2, and SO_2 in the cell top.
4. The cell top is full of gases, so the total gas volume is equal to the volume of the chamber under the hood.

Mass Balance

The cell top obeys the principle of mass conservation; the difference between mass inputs and output is the mass accumulation. The mass inputs are the mass of the air drawn into the cell, bath emissions, and the reaction products under the hood. The mass outputs are the mass of the air leaving the cell and the reaction reactants under the hood. The mass equation is listed as follows:

Fig. 1 Control volume of cell top chamber (mass flow and heat flow are presented)

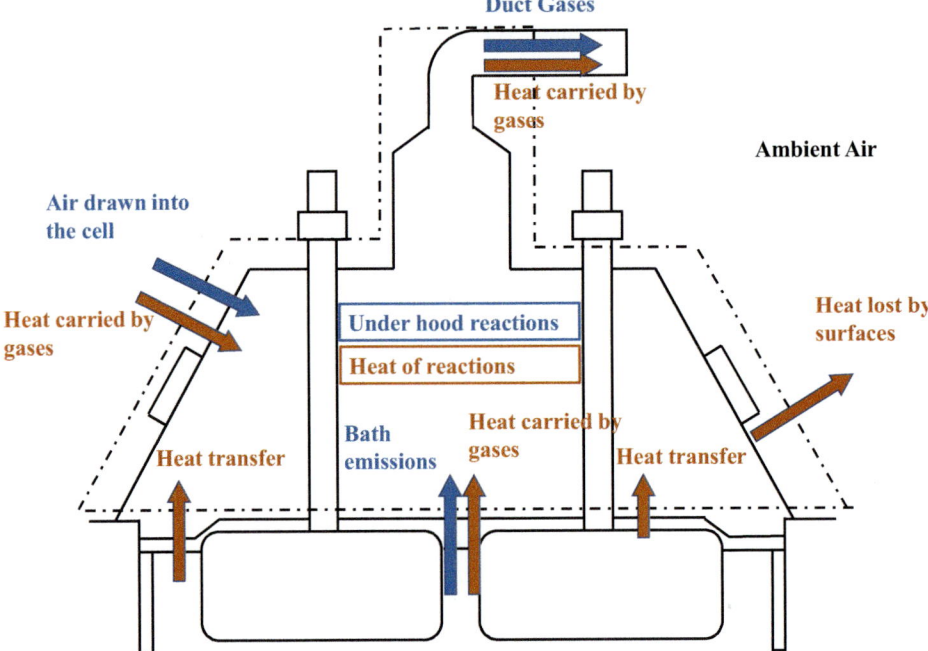

$$\frac{dM_{total,celltop}}{dt} = m_{airin} + m_{bath\,emission} + m_{generation} - m_{duct}$$

$$= m_{airin} + m_{bath\,emission} + m_{generation} - \rho_{duct} v_{duct}$$

$$(1)$$

where $M_{total,celltop}$ is the total mass of all gases in the cell top, m_{airin} is the total air mass flow entering the cell hood, $m_{bath\,emission}$ is the mass flow of gases evolved from the bath, $m_{generation}$ is the net mass flow of gases involved in reactions in the cell top, m_{duct} is the mass flow of gases leaving the cell through the duct, ρ_{duct} is the density of duct gas at standard pressure and temperature, and v_{duct} is the volume flow rate of duct gases.

For each gas, the mass balance is given by

$$\frac{dM_{i,celltop}}{dt} = m_{i,airin} + m_{i,bath\,emission} + m_{i,generation} - m_{i,duct}$$

$$(2)$$

where $M_{i,celltop}$ is the total mass of a specific gas in the cell top, $m_{i,airin}$ is the air mass flow of a specific substance entering the cell hood, $m_{i,bath\,emission}$ is the mass flow of a specific substance evolved from the bath, $m_{i,generation}$ is the net mass flow of a specific substance involved in reactions in the cell top, and $m_{i,duct}$ is the mass flow of a specific substance leaving the cell through the duct. More details are given in the following sections.

Air Drawn into the Cell

The composition of air entering the cell is considered to be atmospheric standard and is composed of the substances: N_2, O_2, CO_2, Ar, and H_2O. Therefore, the total mass flow of air drawn into the cell is given by

$$m_{airin} = m_{H_2O,airin} + m_{N_2,airin} + m_{O_2,airin} + m_{CO_2,airin} + m_{Ar,airin}$$

$$(3)$$

where the water vapour content is evaluated according to potroom relative humidity and temperature as given by

$$m_{H_2O,airin} = Sat_{H_2O}(T_{amb}) \frac{HumF}{100} m_{airin} \quad (4)$$

For other substances, the mass inlet flow is given by

$$m_{i,airin} = proportion(i)\left(m_{Airin} - m_{H_2O,airin}\right) \frac{Mr_i}{Mr_{Air}} \quad (5)$$

where $m_{H_2O,airin}$ is H_2O mass flow entering the cell as air humidity, $Sat_{H_2O}(T_{amb})$ is the humidity saturation concentration, $HumF$ is the fraction of saturation humidity, and Mr_i is the molar mass of a specific gas.

The proportions of N_2, O_2, CO_2, and Ar in the atmospheric air are 78.074, 20.954, 0.038, and 0.934%, respectively. The contribution of other minor inert gases present in

the atmosphere is not considered to be relevant for the process mass and energy balance, because except for argon, other inert gases are trace elements [18].

The duct gas flow rate is normally set as a technical parameter, and according to our assumptions that the pressure in the cell top is constant, so the mass flow rate of inlet air will be adjustable, and it is

$$\frac{dM_{airin}}{dt} = \left(\frac{V_{chamber}}{V_{Mr}(T_{celltop})} - n_{total,celltop}\right) Mr_{air} \quad (6)$$

where $V_{chamber}$ is the volume of the cell top chamber under the hood, $V_{Mr}(T_{celltop})$ is the molar volume of air at $T_{celltop}$, $n_{total,celltop}$ is the total number of moles of all gases in the cell top chamber, and Mr_{air} is the molar mass of air.

Bath Emissions

The mass flow of gases from the bath entering the cell top is composed of the flowing substances:

$$m_{bath\,emission} = m_{CO,bath\,emission} + m_{CO_2,bath\,emission} + m_{COS,bath\,emission}$$

$$+ m_{NaAlF_4,bath\,emission} + m_{HF,bath\,emission}$$

$$(7)$$

For the detailed value of each gas, inspired by [17], they can be calculated by the following equations:

$$m_{CO,bath\,emission} = m_{Al}(1 - SC_{Factor}) \frac{100 - CE}{CE} \frac{3Mr_{CO}}{2Mr_{Al}} \quad (8)$$

$$m_{CO_2,bath\,emission} = m_{Al}\left(1 - (1 - SC_{Factor})\frac{100 - CE}{CE} - \frac{m_{S,in}}{m_{Al}}\frac{2Mr_{Al}}{3Mr_{Al}}\right)\frac{3Mr_{CO_2}}{4Mr_{Al}}$$

$$+ m_S \frac{Mr_{CO_2}}{Mr_{Al}}$$

$$(9)$$

$$m_{COS,bath\,emission} = (m_{S,in} - m_{S,out}) \frac{Mr_{COS}}{Mr_S} \quad (10)$$

$$m_{NaAlF_4,bath\,emission} = (m_{Na_2O,s} - m_{Na_2O,p}) \frac{2Mr_{NaAlF_4}}{Mr_{Na_2O}} \quad (11)$$

$$m_{HF,bath\,emission} = \left(m_{AlF_3,s} - m_{Na_3AlF_6,duct}\frac{2Mr_{AlF_3}}{Mr_{Na_3AlF_6}}\right)\frac{3Mr_{HF}}{Mr_{AlF_3}}$$

$$- 2m_{NaAlF_4,bath\,emission}\frac{Mr_{HF}}{Mr_{NaAlF_4}}$$

$$(12)$$

where SC_{Factor} is the fraction of current efficiency lost by short-circuiting between the anodes and the metal pad, CE is the current efficiency, $m_{S,in}$ is the mass flow of sulphur entering the cell with the new anode, $m_{S,out}$ is the mass flow of sulphur leaving the cell with the anode but, $m_{Na_2O,s}$ is the mass flow of Na_2O entering the cell with secondary alumina feeding, $m_{Na_2O,p}$ is the mass flow of Na_2O entering the cell with primary alumina feeding, $m_{AlF_3,s}$ is the mass flow of AlF_3 entering the cell with secondary alumina, and m_{Al} is the aluminium production rate.

The aluminium production is given by Faraday's law, which is given by

$$m_{Al} = \frac{1}{F} \cdot \frac{I \cdot Mr_{Al} \cdot CE}{1000 \cdot e} \tag{13}$$

where I is the cell line current, e is the number of free electrons per atom of aluminium, and F is the Faraday's constant 96485.3383 ± 0.0083 C/mol.

Reactions

The reactions under the hood are considered, and the mass change is

$$
\begin{aligned}
m_{generation} = {} & m_{CO,reactions} + m_{CO_2,reactions} + m_{O_2,reactions} + m_{COS,reactions} \\
& + m_{H_2O,reactions} + m_{NaAlF_4,reactions} + m_{HF,reactions} \\
& + m_{Na_3AlF_6,reactions}
\end{aligned}
\tag{14}
$$

For the detailed mass changes of different components, inspired by [17], their calculating equations are listed as follows:

$$m_{CO,reactions} = -m_{Al}(1 - SC_{Factor})\frac{100 - CE}{CE}\frac{3Mr_{CO}}{2Mr_{Al}} \tag{15}$$

$$m_{CO_2,reactions} = m_{C,airburn}\frac{Mr_{CO_2}}{Mr_C} - m_{C,Boud}\frac{Mr_{CO_2}}{Mr_C} \tag{16}$$

$$
\begin{aligned}
m_{O_2,reactions} = {} & -m_{S,in}\frac{3M_{O_2}}{2M_S} - m_{C,airburn}\frac{Mr_{O_2}}{Mr_C} \\
& - m_{C,Boud}\frac{Mr_{O_2}}{Mr_C}
\end{aligned}
\tag{17}
$$

$$m_{COS,reactions} = -\left(m_{S,in} - m_{S,out}\right)\frac{Mr_{COS}}{Mr_S} \tag{18}$$

$$m_{H_2O,reactions} = -m_{HF,duct}\frac{Mr_{H_2O}}{2Mr_{HF}} \tag{19}$$

$$m_{NaAlF_4,reactions} = -\left(m_{Na_2O,s} - m_{Na_2O,p}\right)\frac{2Mr_{NaAlF_4}}{Mr_{Na_2O}} \tag{20}$$

$$m_{HF,reactions} = 2m_{NaAlF_4,reactions}\frac{Mr_{HF}}{Mr_{NaAlF_4}} \tag{21}$$

$$m_{Na_3AlF_6,reactions} = \left(m_{Na_2O,s} - m_{Na_2O,p}\right)\frac{2Mr_{Na_3AlF_6}}{Mr_{Na_2O}} \tag{22}$$

where $m_{C,airburn}$ is the carbon consumed by air burn and $m_{C,Boud}$ is the carbon consumed by the Boudouard reaction.

Duct Gases

The cell duct emissions are composed of the air drawn into the cell, including humidity, plus the gases generated inside the control volume minus the gases consumed by the cell inside the volume. The mass of all the duct gas components is given as follows.

$$
\begin{aligned}
m_{duct} = {} & m_{CO_2,duct} + m_{CO,duct} + m_{H_2O,duct} + m_{O_2,duct} + m_{HF,duct} + m_{NaAlF_4,duct} \\
& + m_{SO_2,duct} + m_{COS,duct} + m_{N_2,duct} + m_{Ar,duct}
\end{aligned}
\tag{23}
$$

The mass flow of different gases in the duct is given as

$$m_{i,duct} = n_{duct}\,ratio(i)Mr_i \tag{24}$$

where the ratio of different gases in the duct is equal to the ratio in the cell top,

$$ratio(i) = \frac{n_{i,celltop}}{n_{total,celltop}} \tag{25}$$

and the total number of moles is

$$n_{total,celltop} = \sum n_{i,celltop} \tag{26}$$

For each gas in the chamber, it has its own dynamic as follows:

$$\frac{dn_{i,celltop}}{dt} = \frac{d\left(n_{i,airin} + n_{i,bath\,emission} + n_{i,reaction} - n_{i,duct}\right)}{dt} \tag{27}$$

It should be noted that, due to the above assumptions, $m_{CO,duct} = 0$, $m_{NaAlF_4,duct} = 0$, and $m_{COS,duct} = 0$. In addition, there is no consumption of inert gases, so $m_{N_2,duct} = m_{N_2,airin}$ and $m_{Ar,duct} = m_{Ar,airin}$.

Heat Balance

The cell top obeys the principle of energy conservation. The difference between heat inputs and output is the heat accumulation. The heat inputs are expressed by the enthalpy associated with the gases entering the chamber, heat transfer, and heat generation. The heat outputs are expressed by the enthalpy associated with the duct gases and the heat dissipation. The equation for the temperature in the cell top chamber can be expressed as follows:

$$
\begin{aligned}
\frac{dT_{celltop}}{dt} & = \frac{q_{ac}}{\sum\left(M_{i,celltop}\frac{1}{Mr_i}\right)C_{p,celltop}} \\
& = \frac{q_{ac}}{V_{celltop}\frac{1}{V_{Mr}(T_{celltop})}C_{p,celltop}}
\end{aligned}
\tag{28}
$$

and

$$q_{ac} = q_{airin} + q_{bath\,emission} + q_{bath-celltop} + q_{anode-celltop} + q_{reactions}$$
$$- q_{duct} - q_{celltop-amb}$$
(29)

In (28), the average specific heat capacity of gases in the cell top is

$$C_{p,celltop} = \frac{\sum\left(M_{i,celltop}\frac{1}{Mr_i}C_{p,i}\left(T_{celltop}\right)\right)}{\sum\left(M_{i,celltop}\frac{1}{Mr_i}\right)\times 1000}$$
(30)

where q_{ac} is the power accumulated in the cell top, q_{airin} is the total enthalpy of the air drawn into the cell at T_{amb}, $q_{bath\,emission}$ is the total enthalpy of gases evolved from the bath at T_{bath}, $q_{bath-celltop}$ is the power transferred from the bath to the cell top, $q_{anode-celltop}$ is the power transferred from the anode to the cell top, $q_{reactions}$ is the power generated from reactions under the hood, q_{duct} is the total enthalpy of gases leaving the cell through the duct at $T_{celltop}$, $q_{celltop-amb}$ is the power transferred from the cell top to the ambient air, and $C_{p,i}\left(T_{celltop}\right)$ is the specific heat capacity of one specific gas in the cell top at $T_{celltop}$. More details are given in the following sections.

Heat Carried by the Air Drawn into the Cell at T_{amb}

The total heat of this part is represented by the sum of the enthalpies of all substances entering the chamber from the ambient air, and the heat enthalpy is separately represented

by $H_{CO_2}(T_{amb})$, $H_{O_2}(T_{amb})$, $H_{H_2O}(T_{amb})$, $H_{N_2}(T_{amb})$, $H_{Ar}(T_{amb})$.

Therefore, the total power can be calculated by the following equation:

$$q_{airin} = \sum\left(m_{i,airin}\frac{1}{Mr_i}H(T_{amb})\right)$$
$$= m_{O_2,airin}\frac{1}{Mr_{O_2}}H_{O_2}(T_{amb}) + m_{CO_2,airin}\frac{1}{Mr_{CO_2}}H_{CO_2}(T_{amb})$$
$$+ m_{N_2,airin}\frac{1}{Mr_{N_2}}H_{N_2}(T_{amb}) + m_{Ar,airin}\frac{1}{Mr_{Ar}}H_{Ar}(T_{amb})$$
(31)

The enthalpy associated with a certain mass of a substance at temperature T is

$$H_i(T) = \Delta_f H^0_{298} + \int_{298}^{T} C_p(T)dT$$
(32)

where $\Delta_f H^0_{298}$ is the formation enthalpy of the substance at 298.15 K and the second integral term is the enthalpy difference from 298.15 K to the target temperature.

From the HSC5 [19], $C_p(T)$ is presented in the form

$$C_p(T) = A + 0.001B + 10^5 CT^{-2} + 10^{-6}DT^2$$
(33)

where the parameters A, B, C, D can be found in Table 1.

Heat Carried by Gases Evolved from Bath at T_{bath}

The total heat of this part is represented by the sum of all substances' enthalpy entering the chamber from the bath

Table 1 Detailed parameters for calculating $C_p(T)$

Formula	A J/(mol K)	B J/(mol K)	C J/(mol K)	D J/(mol K)
Al(s)	32.974	−20.677	−4.138	23.753
H$_2$O(l)	186.884	−464.247	−19.565	548.631
H$_2$O(g)	28.408	12.477	1.284	0.36
O$_2$(g)	29.78	−6.177	−0.021	15.997
CO$_2$(g)	22.226	56.2	0.105	−22.518
CO(g)	29.304	−2.905	0	7.925
COS(g)	19.573	91.896	0.148	−65.664
SO$_2$(g)	29.134	37.222	0.058	−2.885
N$_2$(g)	29.298	−1.567	−0.007	3.419
Ar(g)	20.786	0	0	0
NaAlF$_4$(g)	44.15	290.428	−1.1	−287.031
Na$_3$AlF$_6$(s)	192.569	122.838	−11.825	0.1
HF(g)	29.105	0.068	0.002	0.121
C(s)	−0.299	11.491	0	83.49
Al$_2$O$_3$(s)	9.776	294.725	−2.485	−198.174

emission, and the heat enthalpy is separately represented by $H_{CO_2}(T_{bath})$, $H_{CO}(T_{bath})$, $H_{COS}(T_{bath})$, $H_{NaAlF_4}(T_{bath})$, $H_{HF}(T_{bath})$. Therefore, the total power can be calculated by the following equation:

$$q_{bath\,emission} = \sum \left(m_{i,bath\,emission} \frac{1}{Mr_i} H_i(T_{bath}) \right) \quad (34)$$

Heat Transferred from Bath to Cell Top

There are two ways for heat transfer from bath to cell top: heat convection and heat conduction. It can be calculated by using the following equations:

$$q_{bath-celltop} = \frac{T_{bath} - T_{celltop}}{(R_{conduction,bath-celltop} + R_{convection,bath-celltop}) \times 1000} \quad (35)$$

where

$$R_{conduction,bath-celltop} = \frac{\delta_{crust,bath}}{k_{crust,bath} A_{crust,bath}} \quad (36)$$

$$R_{convection,bath-celltop} = \frac{1}{h_{cover-celltop} A_{crust\,bath}} \quad (37)$$

$k_{crust,bath}$ is the thermal conductivity of the crust on the bath surface, $A_{crust,bath}$ is the surface area of the crust on the bath, $\delta_{crust,bath}$ is the thickness of the crust on the bath, and $h_{cover-celltop}$ is the heat convection coefficient from the bath crust cover to the cell top.

Heat Transferred from Anode to Cell Top

There are two ways for heat transfer from anode to cell top: heat convection and heat conduction. The total heat can be calculated by using the following equations:

$$q_{anode-celltop} = \frac{T_{anode} - T_{celltop}}{(R_{conduction\,anode-celltop} + R_{convection\,anode-celltop}) \times 1000} \quad (38)$$

where

$$R_{conduction,anode-celltop} = \frac{\delta_{crust,anode}}{k_{crust,anode} A_{crust,anode}} \quad (39)$$

$$R_{convection,anode-celltop} = \frac{1}{h_{cover-celltop} A_{crust,anode}} \quad (40)$$

$k_{crust,anode}$ is the thermal conductivity of the crust on the anode surface, $A_{crust,anode}$ is the surface area of the crust cover, and $\delta_{cover,anode}$ is the thickness of the crust on the anode surface.

Under Hood Reactions $q_{reactions}$

There are four reactions under the hood:

(1) CO Oxidation ($q_{CO\,oxidation}$)

$$2CO(g) + O_2(g) \rightarrow 2CO_2(g) \quad (41)$$

(2) Carbon air burn ($q_{C\,airburn}$)

Part of anode surface is exposed to air and is hot enough to combust. The carbon quantity loss by air burn is dependent on the conditions of anode exposure.

$$C + O_2(g) \rightarrow CO_2(g) \quad (42)$$

(3) COS Oxidation ($q_{COS\,oxidation}$)

$$2COS(g) + 3O_2(g) \rightarrow 2CO_2(g) + 2SO_2(g) \quad (43)$$

(4) NaAlF$_4$ dissociation ($q_{NaAlF_4\,dissociation}$)

$$3NaAlF_4(g) + 3H_2O(g) \rightarrow Al_2O_3 + Na_3AlF_6 + 6HF(g) \quad (44)$$

The total power is given by their sum.

$$q_{Reactions} = q_{CO\,oxidation} + q_{C\,airburn} + q_{COS\,oxidation} + q_{NaAlF_4\,dissociation} \quad (45)$$

where

$$q_{CO\,oxidation} = -\left(m_{CO,bath\,emission} + m_{CO,Boud} \right) \frac{1}{2Mr_{CO}} \Delta H_{CO\,oxidation} \quad (46)$$

$$q_{C\,airburn} = -m_{C\,airburn} \frac{1}{Mr_C} \Delta H_{C\,airburn} \quad (47)$$

$$q_{COS\,oxidation} = -m_{COS,bath\,emission} \frac{1}{2Mr_{COS}} \Delta H_{COS\,oxidation} \quad (48)$$

$$q_{NaAlF_4\,dissociation} = -m_{NaAlF_4,bath\,emission} \frac{1}{3Mr_{NaAlF_4}} \Delta H_{NaAlF_4\,dissociation} \quad (49)$$

Heat Carried by Cell Emissions at $T_{celltop}$

The total heat of this part is represented by the sum of enthalpy leaving the chamber through the exhaust duct for all gas species. The heat enthalpies considered are $H_{CO_2}(T_{celltop})$, $H_{O_2}(T_{celltop})$, $H_{HF}(T_{celltop})$, $H_{Na_3AlF_6}(T_{celltop})$, $H_{H_2O}(T_{celltop})$, $H_{SO_2}(T_{celltop})$, $H_{N_2}(T_{celltop})$, $H_{Ar}(T_{celltop})$. Therefore, the total power can be calculated by the following equation:

$$q_{duct} = \sum \left(m_{i,duct} \frac{1}{Mr_i} H_i \left(T_{celltop} \right) \right) \qquad (50)$$

Heat Transferred from Cell Top to Ambient Air

There are two ways for heat transfer from bath to cell top: heat conduction and heat convection. They can be calculated by using the following equations:

Heat conduction:

$$R_{conduction,celltop-amb} = \frac{\delta_{hood}}{k_{hood}A_{hood}} \qquad (51)$$

Heat convection:

$$R_{convection,celltop-amb} = \frac{1}{h_{hood-amb}A_{hood}} \qquad (52)$$

Total power of this part

$$q_{celltop-amb} = \frac{T_{celltop} - T_{amb}}{\left(R_{conduction,celltop-amb} + R_{convection,celltop-amb} \right) \times 1000} \qquad (53)$$

where k_{hood} is the thermal conductivity of the hood, A_{hood} is the surface area of the hood, δ_{hood} is the thickness of the hood, and h_{hood} is the surface convective heat transfer coefficient.

Results and Discussion

Using the cell top model presented in section "Modelling", in conjunction with the dynamic model of the smelting cell [12], the relationship between bath temperature and duct gas temperature is studied, under different scenarios, including changes in the line current change, ACD, and duct gas flow rate.

Case 1: Changes in Line Current

For this scenario, the line current was increased and decreased by 10% at the 4th hour, with the operation lasting for 12 h. Then, the line current was returned to the nominal value for a further 12 h. The results are as follows.

It can be seen from Figs. 2 and 3 that when the line current is increased or decreased, the bath temperature and the duct gas temperature have a large response. This is because when the line current increases, more heat is generated in the bath and anode, and thus, more heat is transferred to the cell top

through the bath cover and anode cover. For the fast change of duct gas temperature, this is because when the line current changes, a large amount of gases are suddenly generated from the bath and carry heat into the cell top space. For the subsequent slow change, it is mainly due to the increase in the bath temperature. On the other hand, at the 28th hour, both temperatures did not return to the initial value, because the operation of increasing the line current made the cell store more energy. Therefore, both temperatures have a large increase in response to an increase in line current and vice versa.

Case 2: Changes in ACD

For this scenario, the ACD was separately increased and decreased by 0.2 cm at the 4th hour, with the operation lasting for 12 h. Then, the ACD was returned to the initial value for a further 12 h.

It can be seen from Figs. 4 and 5 that when the ACD is increased or decreased, the bath temperature and the duct gas temperature exhibit a similar response, with different ranges. It can be seen that the effect of ACD change on the cell top is less than that on the cell body. Besides, compared with the operation of line current change, there is no sudden change. This is because the cell top heat balance is mainly affected by the bath temperature change under this operation. When the ACD increases, it mainly affects the cell body heat balance: more heat is generated in the bath and the bath temperature starts to increase slowly.

Case 3: Changes in Duct Gas Flow Rate

For this scenario, the duct gas flow rate was increased and decreased separately by 2000 Nm³/h at the 4th hour, with the operation lasting for 12 h. Next, the duct gas flow rate was returned to the initial duct gas flow rate for a further 12 h.

It can be seen from Figs. 6 and 7 that whether the duct gas flow rate is increased or decreased, the duct gas temperature has a large variation and the bath temperature has a

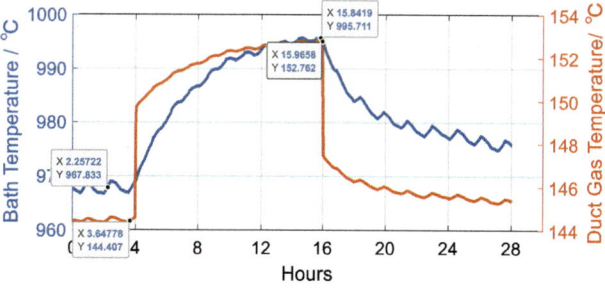

Fig. 2 Response of bath temperature and duct gas temperature to line current (increase line current by 10% at 4 h and it lasts for 12 h)

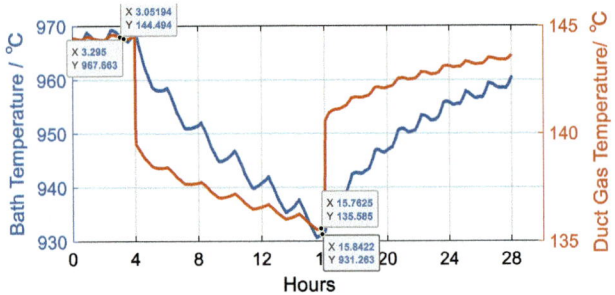

Fig. 3 Response of bath temperature and duct gas temperature to line current (decrease line current by 10% at 4 h and it lasts for 12 h)

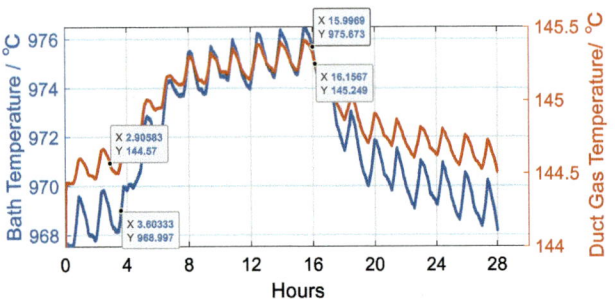

Fig. 4 Response of bath temperature and duct gas temperature to ACD (ACD increases by 0.2 cm at 4 h and it lasts for 12 h)

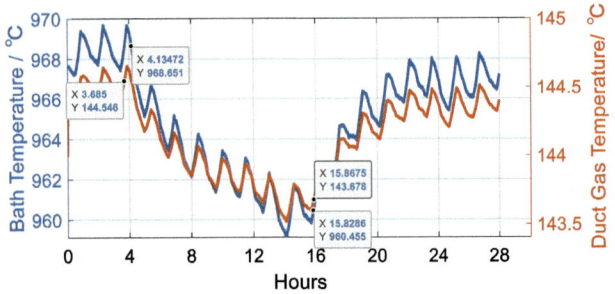

Fig. 5 Response of bath temperature and duct gas temperature to ACD (ACD decreases by 0.2 cm at 4 h and it lasts for 12 h)

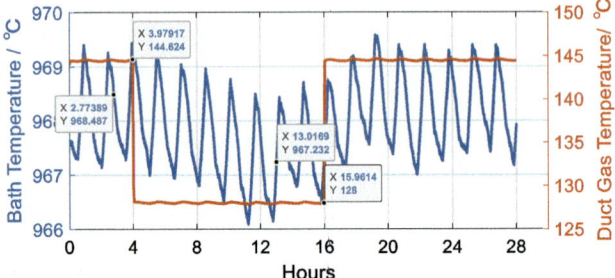

Fig. 6 Response of bath temperature and duct gas temperature to duct flow rate (duct flow rate increases from 8000 to 10,000 Nm^3/s at 4 h and it lasts for 12 h)

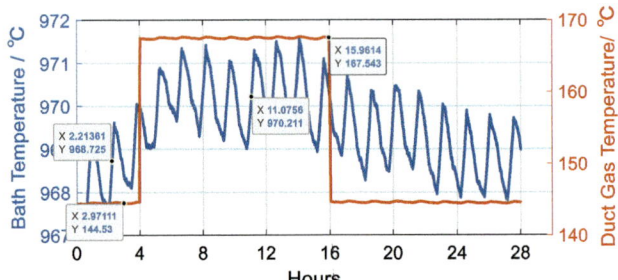

Fig. 7 Response of bath temperature and duct gas temperature to duct gas flow rate (duct flow rate decreases from 8000 to 6000 Nm^3/s at 4 h and it lasts for 12 h)

Conclusions

This study selects the cell top as the research object, using mass conversation and energy conversation to do the modelling. Detailed mass flow and heat flow are included, and the effect of anode temperature is also considered. With the extension of the cell model in the cell top chamber, the dynamic relationship between the cell top and cell body under different operations is studied. Three cases are given to present it. In these cases, although bath temperature and duct gas temperature have different responses to cell operations, the model proves that the duct gas temperature is related to bath temperature. In addition, the model of this paper is based on the general mechanism of mass and heat balance in the cell top chamber, so it can be applied to a variety of prebaked cell technologies. On the other hand, with the dynamic description provided by the model, the states of the cell top chamber can be used for state estimation. Since the temperature of the cell top is far lower than the bath temperature, it enables us to do continuous measurements in practice for estimation work with this model, such as using duct gas temperature to estimate the bath temperature. This will be included in our further work.

small variation. On the other hand, the duct gas temperature can quickly respond to the duct flow rate change. These are because when the duct gas flow rate increases, it mainly affects the heat balance of the cell top; when more heat is carried out by the air flow, the duct gas temperature decreases a lot and vice versa.

In addition, by combining Figs. 2, 3, 4, 5, 6, and 7 and comparing three different cases, this model presents that the bath temperature and the duct gas temperature are not a simple static relationship and they have different dynamics in response to different operations. An extended Kalman filter [20] is required to estimate the bath temperature in real time from the duct gas temperature, duct gas flow rate, line current, and ACD.

Definition of Symbols

Symbol	Meaning	Unit
$A_{crust,anode}$	The surface area of the crust cover	$[m^2]$
$A_{crust,bath}$	The surface area of the crust on the bath	$[m^2]$
A_{hood}	The surface area of the hood	$[m^2]$
CE	The current efficiency	[%]
$C_{p,celltop}$	The average specific heat capacity of gases in the cell top	[J/ (mol K)]
$C_{p,i}$	The specific heat capacity of a gas	[J/ (mol K)]
e	The number of free electrons per atom of aluminium	[–]
F	Faraday's constant	C/mol
$h_{cover-celltop}$	The heat convection coefficient of crust cover	[W/ $(m^2$ K)]
h_{hood}	The surface heat convection coefficient of hood	[W/ $(m^2$ K)]
$H_i(T_i)$	The enthalpy of a specific substance at a specific temperature	[kJ/mol]
$HumF$	The fraction of saturation humidity	[%]
I	The cell line current	[A]
$k_{crust,anode}$	The thermal conductivity of the crust on the anode surface	[W/ (m K)]
$k_{crust,bath}$	The thermal conductivity of the crust on the bath surface	[W/ (m K)]
k_{hood}	The thermal conductivity of the hood	[W/ (m K)]
m_{Al}	The aluminium production rate	[kg/s]
m_{airin}	The total air mass flow entering the cell hood	[kg/s]
$m_{bath\,emission}$	The mass flow of gases evolved from the bath	[kg/s]
m_{duct}	The mass flow of gases leaving the cell through duct	[kg/s]
$m_{generation}$	The net mass flow of gases involved in reactions in the cell top	[kg/s]
$m_{i,airin}$	The air mass flow of a specific substance entering the cell hood	[kg/s]
$m_{i,bath\,emission}$	The mass flow of a specific substance evolved from the bath	[kg/s]
$m_{i,generation}$	The net mass flow of a specific substance involved in reactions in the cell top	[kg/s]
$m_{i,duct}$	The mass flow of a specific substance leaving the cell through duct	[kg/s]
$m_{AlF_3,s}$	The mass flow of AlF_3 entering the cell with secondary alumina	[kg/s]
$m_{C,airburn}$	The carbon consumed by air burn	[kg/s]
$m_{C,Boud}$	The carbon consumed by the Boudouard reaction	[kg/s]

Symbol	Meaning	Unit
$m_{Na_2O,s}$	The mass flow of Na_2O entering the cell with secondary alumina feeding	[kg/s]
$m_{Na_2O,p}$	The mass flow of Na_2O entering the cell with primary alumina feeding	[kg/s]
$m_{S,in}$	The mass flow of sulphur entering the cell with the new anode	[kg/s]
$m_{S,out}$	The mass flow of sulphur leaving the cell with the anode butt	[kg/s]
$M_{i,celltop}$	The total mass of a specific gas in the cell top	[kg]
Mr_{air}	The molar mass of air	[g/mol]
Mr_i	The molar mass of a specific gas	[g/mol]
$M_{total,celltop}$	The total mass of all gases in the cell top	[kg]
n_{duct}	The number of moles of all gases leaving the cell through the duct	[mol]
n_i	The number of moles of a specific gas in the cell top chamber	[mol]
n_{total}	The total number of moles of all gases in the cell top chamber	[mol]
q_{ac}	The power accumulated in the cell top	[kW]
q_{airin}	The total power of the air drawn into the cell at T_{amb}	[kW]
$q_{anode-celltop}$	The power transferred from anode to the cell top	[kW]
$q_{bath\,emission}$	The total enthalpy of gases evolved from bath at T_{bath}	[kW]
$q_{bath-celltop}$	The power transferred from bath to the cell top	[kW]
$q_{celltop-amb}$	The power transferred from cell top to the ambient air	[kW]
q_{duct}	The total power of gases leaving the cell through the duct at $T_{celltop}$	[kW]
$q_{reactions}$	The power generated from reactions under the hood	[kW]
$Sat_{H_2O}(T_{amb})$	The humidity saturation concentration	[–]
SC_{Factor}	The fraction of current efficiency lost by short-circuit	[–]
$V_{chamber}$	The volume of cell top under the hood	$[m^3]$
$V_{Mr}(T_{celltop})$	The molar volume of air at $T_{celltop}$	[L/mol]
$\Delta_f H_{298}^T$	The formation enthalpy of the substance at 298.15 K	[kJ/mol]
$\delta_{crust,bath}$	The thickness of the crust on the bath	[m]
$\delta_{cover,anode}$	The thickness of the crust on the anode surface	[m]
δ_{hood}	The thickness of the hood	[m]
ρ_{duct}	The density of duct gas at standard pressure and temperature	$[kg/m^3]$
v_{duct}	The volume flow rate of duct gases	$[Nm^3/s]$

(continued)

References

1. Tandon, S. C., & Prasad, R. N. (2005). Energy Saving in Hindalco's Aluminium Smelter. Light Metals, 303–309.
2. Abbas, H. (2010). Mechanism of Top Heat Loss from Aluminium Smelting Cells. Doctoral dissertation, The University of Auckland.
3. Robinson, T. P. (2005). Evaluating and Funding New Technologies to Support the U.S. Aluminium Industry, Light Metals.
4. Tsukahara, H., Ono, N., & Fujita, K. (1982). Establishment of Effective Operation of Prebaked Anode Pots. Light Metals, 82, 471.
5. Arai, K., & Yamazaki, K. (1975). Heat Balance and Thermal Losses in Advanced Prebaked Anode Cells. Light Metals, 1, 193.
6. Peacey, J. G., & Medlin, G. W. (1979). Cell Sidewall Studies at Noranda Aluminium. Metallurgical Soc. of AIME.
7. Gadd, M. D., Welch, B. J., & Ackland, A. D. (2000). The Effect of Process Operations on Smelter Cell Top Heat Losses. Light Metals, 231.
8. Dupuis, M., & Haupin, W. (2003). Performing Fast Trend Analysis on Cell Key Design Parameters. In Light Metals Warrendale Proceedings. (pp. 255–262). TMS.
9. Taylor, M. P., Johnson, G. L., Andrews, E. W., & Welch, B. J. (2004). The Impact of Anode Cover Control and Anode Assembly Design on Reduction Cell Performance. In Light Metals Warrendale Proceedings. (pp. 199–206). TMS.
10. Gadd, M. D. (2003). Aluminium Smelter Cell Energy Flow Monitoring. Doctoral dissertation, The University of Auckland, Chemical and Materials Engineering.
11. Wong CJ; Yao Y; Bao J; Skyllas-Kazacos M; Welch BJ; Jassim A; Mahmoud M; Arkhipov A. (2021). Modelling of Coupled Mass and Thermal Balances in Hall-Heroult Cells during Anode Change, Journal of the Electrochemical Society, vol. 168.
12. Wong CJ; Yao Y; Bao J; Skyllas-Kazacos M; Welch BJ; Jassim A; Mahmoud M, (2021). Discretized thermal model of Hall-Héroult cells for monitoring and control, IFAC-PapersOnLine, vol. 54, pp. 67–72.
13. Eisma, D., & Patel, P. (2016). Challenges in Power Modulation. In Essential Readings in Light Metals (pp. 683–688). Springer, Cham.
14. Depree, N., Düssel, R., Patel, P., & Reek, T. (2016). The 'Virtual Battery'—Operating an Aluminium Smelter with Flexible Energy Input. In Light Metals 2016 (pp. 571–576). Springer, Cham.
15. Taylor, M. P. (2007, November). Anode Cover Material–science, Practice and Future Needs. In Proceedings of 9th Australasian Aluminium Smelting Technology Conference (pp. 4–9).
16. Dupuis, M., & Haupin, W. (2003). Calculating Temperatures Under Hood of a Prebake Anode Cell. In Light Metals (Metaux Legers) 2003: International Symposium on Light Metals as held at the 42nd Annual Conference of Metallurgists of CIM(COM 2003) (Vol. 2003).
17. Gusberti, V. (2014). Modelling the Mass and Energy Balance of Aluminium Reduction Cells. Doctoral dissertation, The University of New South Wales.
18. Lutgens, F. K., Tarbuck, E. J., & Tusa, D. (1995). The Atmosphere. Englewood Cliffs, NJ, USA: Prentice-Hall.
19. HSC Chemistry 5, https://www.hsc-chemistry.com/.
20. Simon, D. (2006). Optimal state estimation: Kalman, H infinity, and nonlinear approaches. John Wiley & Sons.

Following Alumina Dissolution Kinetics with Electrochemical and Video Analysis Tools

Daniel Marinha, Astrid J. Meyer, Marián Kuchařík, Sylvie Bouvet, Miroslav Boca, Michal Korenko, Vladimir Danielik, and Frantisek Simko

Abstract

The rate of alumina dissolution in cryolitic melts is critical for the management of aluminium production pots. The kinetics of the dissolution reaction depends on the combined effects of physical and chemical characteristics of alumina. The literature mentions several methods to follow the rate of dissolution, with variable degrees of complexity and success. We report a combination of analytical tools to evaluate the dissolution rates of various industrial aluminas in an industrial cryolite bath collected shortly before the anode effect. Batches of powders were sequentially added to an electrochemical cell specifically designed for this test. The dissolution rate was measured electrochemically, and an automatic script was developed to measure the flotation time of the alumina rafts from digital recordings. We correlate the alumina dissolution rate and flotation times, with the initial characteristics of the alumina powder.

Keywords

Alumina • Dissolution • Alumina rafts

D. Marinha (✉) · S. Bouvet
Rio Tinto Aluminium Pechiney Aluval, 725 Rue Aristide Berges, BP-7 38341 Voreppe, France
e-mail: daniel.marinha@riotinto.com

A. J. Meyer · M. Kuchařík
Hydro Aluminium AS, 3905 Porsgrunn, Norway

M. Boca · M. Korenko · F. Simko
Slovak Academy of Sciences, Institute of Inorganic Chemistry, Dubravska Cesta 9, Bratislava, 845 36, Slovakia

V. Danielik
Institute of Inorganic Chemistry, Technology and Materials, Faculty of Chemical and Food Technology STU in Bratislava, Radlinského 9, 812 37 Bratislava, Slovak Republic

Introduction

The progressive increase of the current in the electrolytic Hall-Héroult process is followed by a need to increase alumina feed rates, and/or by increasing the size of the anodes and decreasing the quantity of bath in the pot. The kinetics of the alumina dissolution is increasingly critical since they can determine the quality of cell operation. A collaboration project between Hydro Aluminium and Rio Tinto Alcan has been established to develop a method to measure the alumina dissolution rate and identify key characteristics of alumina for an optimal dissolution. In this paper, we present the results of a method developed by Slovak Academy of Sciences. These tests were used to evaluate the ability of the method to distinguish and benchmark different aluminas.

Substantial effort is required in designing the experimental apparatus. Conditions in industrial cells are difficult to replicate in a laboratory cell. Lab crucibles typically contain 1 kg of bath or less, however, a larger crucible might improve heat capacity and give more controllable conditions and hence more reproducible results. Furthermore, the choice of stirring method, alumina feeder, cell size, batch size, bath chemistry, bath temperature, and superheat have a significant impact on the dissolution kinetics that may be greater than the physical and chemical characteristics of the alumina. Here, we preferred practical solutions that target reproducibility to guarantee internal consistency and allow comparison of results across different tests.

Several methods to measure alumina dissolution have previously been reported and reviewed [1, 2]. These can be classified into electrochemical [3–5], chemical, and visual [6, 7] methods. Electrochemical methods perform in situ measurements of electrochemical reactions occurring at the interface between a working electrode and the bath. Examples include chronopotentiometry, sweep voltammetry, impedance spectroscopy, or use of oxygen sensors. These techniques are extremely sensitive and relatively fast compared to the alumina dissolution rates; they are thus able to

© The Minerals, Metals & Materials Society 2023
S. Broek (ed.), *Light Metals 2023*, The Minerals, Metals & Materials Series,
https://doi.org/10.1007/978-3-031-22532-1_10

track the chemical changes occurring during dissolution. Because of this high sensitivity, the reproducibility of the experimental conditions is critical both in practical terms of the setup (e.g., design, nature, and position of the electrodes), and in terms of the chemistry of the bath itself where minor impurities may greatly impact the electrochemical signal. The presence of impurities also hinders the deconvolution and interpretation of the signal since they may overlap, or chemically interact, with the species of interest. Additionally, electrochemical methods require a reference electrode, but only quasi-reference electrodes with different degrees of stability are currently available because of the high reactivity of cryolitic melts.

Here, we report on the development of an electrochemical method based on Square Wave Voltammetry (SWV) and experimental validation using 6 different alumina sources. A combination of temperature measurements and video recordings can provide complementary information to evaluate electrochemical results, especially regarding evaluations of the results and understanding why there may be deviations in seemingly identical experiments.

Experimental

Alumina Sources

A total of 7 alumina samples were used for this study, from 6 different sources. The focus was placed on secondary aluminas. The validation was based on the ability to differentiate the dissolution behavior of aluminas reported by smelters to have very different pot room behaviors and hence were expected to have markedly different dissolution rates. For this, we compare the dissolution rates of primary (A_prim) and secondary (A_sec), from the same shipment of source A. Further, we tested and compared the remaining aluminas from different refinery sources marked C to E. Lastly, we compare the dissolution of a 1:1 mixture of aluminas B and C, known to have widely different properties both in terms of shipment certificate values and reported smelter performance. All secondary alumina source samples have been taken "as is" with automatic samplers or at pot feeding at various smelters and are such spot samples representing themselves.

Alumina Characterization

All samples were split using a Retch rotating splitter. This ensures minimal variation in properties between each sample split. All samples for dissolution testing were split into the correct batch size (9 g) before delivery.

All the conventional alumina quality parameters are included in the characterization, as well as some additional parameters. However, only the most relevant parameters are included in this paper.

The analytical methods used were as follows:

– Particle Size Distribution: Coulter laser, laser diffraction using Mie optical model.
– Flow time: 10 mm Pechiney Funnel, in-house equipment.
– BET: Multi-point BET Tristar 11 Plus 3030, ISO 8008.
– Trace elements: XRF, fused beads.
– Alpha and gibbsite: XRD (area, mean of three peaks for alpha, and one peak for gibbsite).
– MOI/LOI: TGA, temperature intervals from room temperature to 120 °C, from 120 °C to 300 °C and from 300 °C to 1100 °C. Based on ISO 806.
– F content: Skalar.
– C content: Leco, ASTM E 1941.

The conditional formatting shows low values in blue and high values in red (Table 1). The MOI RT-120 °C measurement was introduced since the samples are heated and stored at 120 °C before the dissolution testing. The MOI 120–300 °C values have been calculated based on the differences between the conventional MOI RT-300 °C and the new MOI RT-120 °C measurements.

The flow analysis is carried out by taking the time for a defined amount of alumina to run through a specific funnel. In some cases, the flowability is limited and the material plugs the funnel to a complete stop.

Experimental Setup, Electrode Materials, and Test Conditions (Bath Composition, Alumina)

Experiments were carried out in a closed alumina cell under a constant argon flow (Fig. 1). 3.6 kg of the industrial bath is introduced into a graphite crucible with 140 mm × 150 mm (internal diameter x height), installed inside the furnace. The bath was extracted from an industrial cell shortly before an anode effect. The composition was determined by XRF, XRD, and LECO: 11.9 wt% AlF_3—5.13 wt% CaF_2— 0.97 wt% Al_2O_3—82 wt% Na_3AlF_6. The liquidus temperature is 964 °C as determined from thermal analyses. This corresponds well with the calculated value of 968 °C, using the Solheim equation [8] (ca. 964 °C when impact of impurities is accounted for). The working temperature was fixed at 980 °C. The bath is agitated by a BN stirrer generating bath speeds near 4 cm/s. The bath temperature is continuously measured with a thermocouple (Pt/PtRh10, type S) immersed 4 cm into the bath. The temperature gradient from the bottom of the crucible to the surface of the

Table 1 Summary of the most relevant alumina quality parameters of the tested samples

Quality Parameter	Unit	A_sec	B	C	D	E	A prim.	B/C mix.
Na2O	%	0.66	0.5	0.59	0.69	0.64	0.42	0.56
F	%	1.31	2.00	1.59	1.57	1.38	0	1.98
C	%	0.12	0.31	0.09	0.13	0.15	0.07	0.22
Alpha alumina	%	1.4	9.3	3	2.5	1.6	1.5	5.8
Gibbsite	%	<0.5	<0.5	<0.5	2	<0.5	<0.5	<0.5
BET	m²/g	73.4	64.8	72.3	70.2	106.9	76.4	63.9
MOI RT-120°C	%	2.36	2.46	1.53	1.04	3.4	1.76	1.64
MOI RT-300°C	%	3.56	4.13	2.62	2.6	5.15	2.41	2.96
LOI 300-1100°C	%	1.98	2.83	2.01	2.24	2.52	0.83	2.44
MOI 120-300 °C (calc.)	%	1.20	1.67	1.09	1.56	1.75	0.65	1.32
Flow time	sec	108	202	84	122	plugged	87	108
PSD Laser								
+150 µm	%	8.81	0.05	6.56	3.15	0.55	11.2	3.2
+106 µm	%	39.8	2.18	41.8	27.4	10.7	48.7	22.9
+75 µm	%	69.4	29.3	79.7	61.2	40.4	76.7	52.8
+53 µm	%	86.2	59.8	92.6	80.5	73.9	91.2	74.7
-45 µm	%	9.26	29.2	5.36	14.7	17.6	6.1	18.4
-20 µm	%	2.19	5.35	1.63	5.02	3.7	1.04	3.67

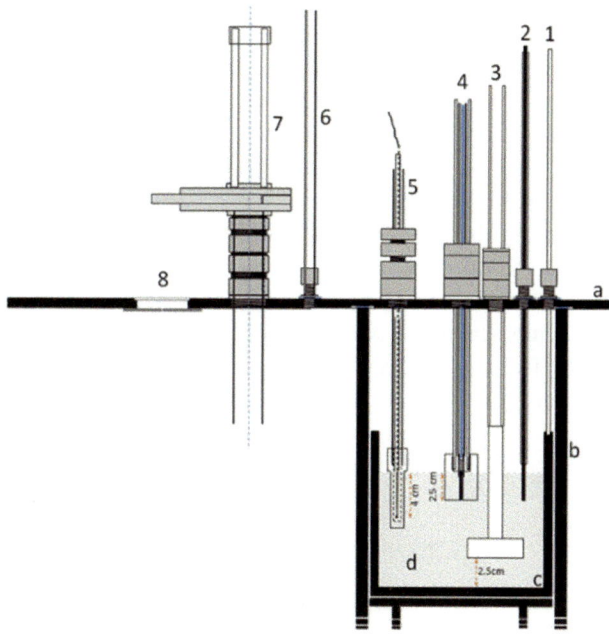

Fig. 1 Cross-section of the experimental setup; counter electrode (1), comparison W electrode (2), Boron Nitride stirrer (3), disc working electrode (4), Pt/Pt10Rh thermocouple with protective Boron Nitride shaft (5), outlet gas tube (6), feeder (7), aperture (8), brass flange with sapphire window (**a**), alsint tube (**b**), carbon crucible (**c**), and bath (**d**)

bath is 4 °C. A digital camera was installed above a 5 cm sapphire window in the upper flange with visibility to the surface of the bath.

The electrochemical measurements are carried out using a 3-electrode setup. The rotating working electrode is a 2.8 mm diameter disc made of vitreous carbon with a rotation frequency of 100 RPM. The graphite crucible is the counter-electrode and quasi-reference. The alumina feeding system is composed of a knife valve, a bottom stainless tube, and a top tube with removable lid. The alumina batches of

9 g (0.25 wt% of the bath) were pre-heated at 120 °C for at least 2 h before being added into the bath every 50 min.

Square Wave Voltammetry

The electrochemical response is measured using the square wave voltammetry technique (SWV). The square wave is characterized by a pulse height (P_H), step height (S_H)a, and pulse width (P_W). The pulse width can be expressed in terms of the square wave frequency $f = 1/(2P_W)$. The scan rate v is $P_H/(2P_W)$. The following parameters are used for the tests: $f = 5$ Hz, $P_H = 0.5$ V, $S_H = 0.005$ V, and Potential range = 0 to 3.8 V/ref. Each alumina batch was measured with 10 scans. Three scans are done before the alumina addition and seven afterwards, totalling 45.2 min/batch, where the duration of each scan was 151 s, and the time interval between two following scans was 120 s.

Video Analysis

A procedure was developed to analyse still images extracted from the video recordings and determine the floating time of alumina rafts formed after the addition of the batch to the melt. The procedure calculates the average brightness from each still image and plots it as a function of time. This strategy relies on the fact that the bath is much brighter than the alumina powders and the fixed parts of the furnace and setup (Fig. 2). The complete disappearance of the alumina raft corresponds to a plateau of the brightness.

The resulting plots are fitted using a simple combination of two linear curves for the initial transition stage during raft formation and disappearance, and the steady state after all alumina disappears. The model is fitted based on an ordinary least squares model.

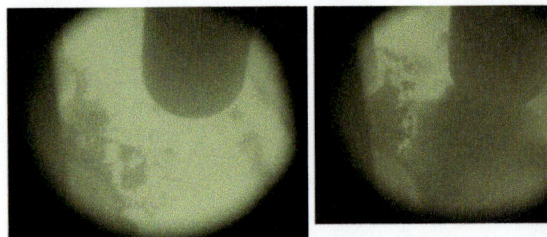

Fig. 2 Images from video recordings showing the bath before (left) and moments after alumina addition (right)

Results and Discussion

Method Evaluation

Three experiments were carried out for each of the aluminas A_sec and B to determine reproducibility of the results.

Interpretation of SWV Records

A complete description of the reaction mechanism at the anode in the Hall-Héroult process has remained elusive [9]. This was not the focus of this work. Instead, we interpreted the electrochemical results based on a reasonable hypothesis and validated the method empirically. The following sections are dedicated to the validation process.

Experimental SWV curves can be fitted by a combination of several Gaussian or Lorentzian curves or peaks, each assigned to a different electrode reaction as a function of the electrochemical potential, including the anodic alumina reaction. Due to the obvious complexity of the system,

different hypotheses were tested, and fitting parameters were adjusted accordingly. We are currently developing an automated procedure to evaluate SWS results, which should bring significant improvement since data processing is currently very time-consuming.

Reproducibility Tests

For the reproducibility evaluation, Aluminas A_sec and B were tested 3 times each, in identical conditions. The area of the peak assigned to alumina was extracted during the fitting procedure (Fig. 3). The relative change of the peak assigned to alumina area as a function of time and alumina batches highlights the different behaviour between the two aluminas. For alumina A_sec, the alumina concentration in the bath increases almost instantly after addition. A plateau corresponding to the end of the dissolution and an equilibrium state is quickly reached for the first 5 batches. Thereafter, the time to reach equilibrium increases progressively with the alumina concentration previously dissolved in the melt. Alumina B takes longer to reach equilibrium compared to alumina A_sec, averaging about 27 min for initial batches. After 6 batches, the complete dissolution is reached after more than 50 min, i.e., minutes before the next addition. For alumina A, the plateau becomes increasingly difficult to distinguish during the last 3 batches. However, equilibrium was not reached during the final 2 batches of alumina B.

An empirical model was developed to fit the experimental results. This model was inspired by the Nernst and Brunner equation (1): the variation of dissolved material versus time is proportional to the area, diffusion coefficient, and the difference of concentration between the surface and the bulk:

Fig. 3 Evolution of the peak area of the SWV with time for alumina A_sec additions, and the modelled behaviour (solid line). Dotted lines mark each batch addition to the melt

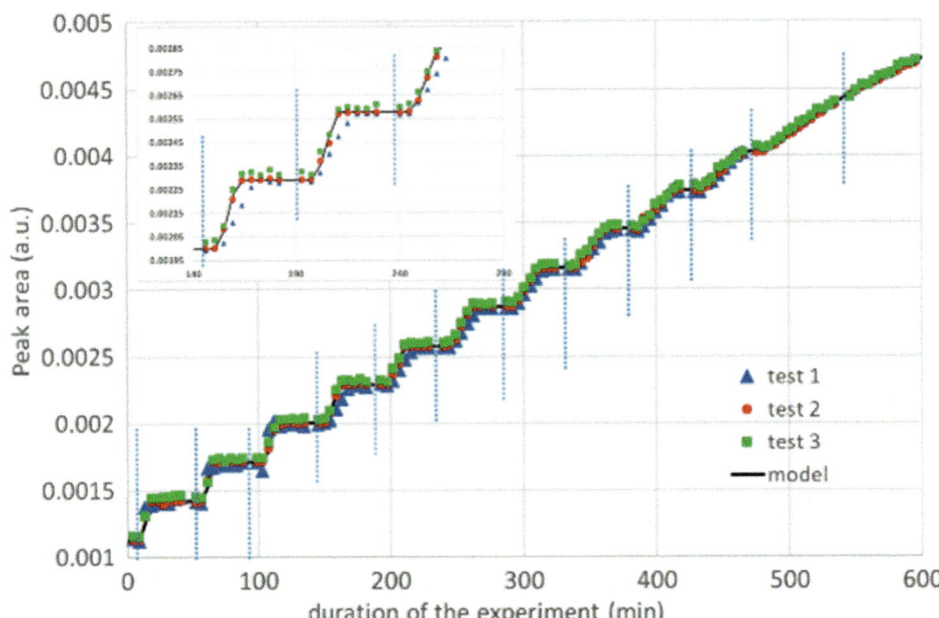

$$\frac{dm}{dt} = A\frac{D}{d}(c_s - c_b) \qquad (1)$$

where

m   Mass of dissolved material.

t   Time.

A   Area of the interface between the dissolving substance and the solvent.

D   Diffusion coefficient.

d   Thickness of the boundary layer of the solvent at the surface of the dissolving substance.

C_s Mass concentration of the substance on the surface.

C_{BS} Mass concentration of the substance in the bulk of the solvent.

In the case of the alumina dissolution, the equation becomes

$$\frac{dn}{dt} = \frac{k}{V^{a-1}} \cdot (n_s - n)^a = K \cdot (n_s - n)^a \qquad (2)$$

where

a   Empirical constant depending on the alumina

K   Empirical constant depending on the alumina

n   Number of moles of alumina with time.

n_s Number of moles of alumina in the alumina saturated melt at the working temperature according to Solheim equation

t_0 The initial condition for the numerical solving of the differential equation: $t = t_0$ is the time when the first alumina molecule from the current addition gets to the electrode surface

The Runge-Kutta method of fourth order is used to determine the empirical constants a and K and t_0 (Table 2). Figure 3 shows that the experimental results are very well fitted with this model.

The time t_0 can be used as a marker of the ability of alumina to dissolve in the cryolite electrolyte. Another way to describe the dissolution rate of alumina is the average dissolution rate until equilibrium is reached. For example, the average rate for alumina A_sec in the first batching is 0.042 wt% of alumina per minute. The average dissolution rate for the B in the first batching is 0.0071 wt% of alumina per minute. It is quite visible that the dissolution rate of A_sec at lower concentration is more than 5 times higher than the dissolution rate of alumina B. At higher concentrations, the dissolution rate of alumina A_sec at the 9th

batch is 0.00962 wt% of alumina per minute. Alumina B has at the same condition as the dissolution rate of 0.0049 wt% per minute. Consequently, the difference between dissolution rates decreases between both aluminas as the concentration of alumina in the bath increases. Regardless, the alumina A_sec dissolution rate is still almost 2 times higher than the dissolution rate of alumina B at higher alumina bath concentration.

A different way to compare the kinetics of the dissolution is to plot the time needed to reach the complete dissolution of the batch for each addition, i.e., the time to reach the plateau or the steady state, which is a function of the concentration of the alumina in the bath (Fig. 4).

This presentation of the results shows a high discrimination between aluminas A_sec and B. The duration of the complete dissolution of each batch increases with the concentration of the alumina previously added. A quadratic equation fits the trend of the dissolution rate versus the number of batches. The increase of dissolution duration with the concentration of alumina in the bath is more important as the dissolution is slow. The increase is non-linear, and the extent of non-linearity appears to be characteristic of the type of alumina and can be quantified by the percentual difference between the dissolution time of the first and last batches, $(t_f - t_0)/t_0$, which expresses the relative impact that alumina concentration in the melt has on the dissolution rate for a given alumina.

Primary Versus Secondary Alumina

Another method validation test consisted in testing the dissolution rate of an alumina from the same source, but before (A_prim) and after (A_sec) being used for scrubbing the smelter gases. The results confirm that the dissolution of the primary alumina is relatively slower when compared to secondary alumina [10]. Nevertheless, the primary alumina is still quicker to dissolve than alumina B (Fig. 5).

Benchmark of Different Alumina Sources

Finally, we compare the results of the measurements of the complete set of alumina sources (Fig. 6). The method discriminates between all aluminas, and they can be grouped in terms of their dissolution rate as fastest (A_sec, C, and D), and slowest (A_prim, B, and E), in accordance with reported industrial smelter experience.

Table 2 Parameters of the dissolution model for alumina A_sec and B

Alumina	K	a	t_0 (min)	Std. deviation
A_sec	1.04E−04	5.18	0	0.0082
B	9.55E−04	1.12	6.6	0.0053

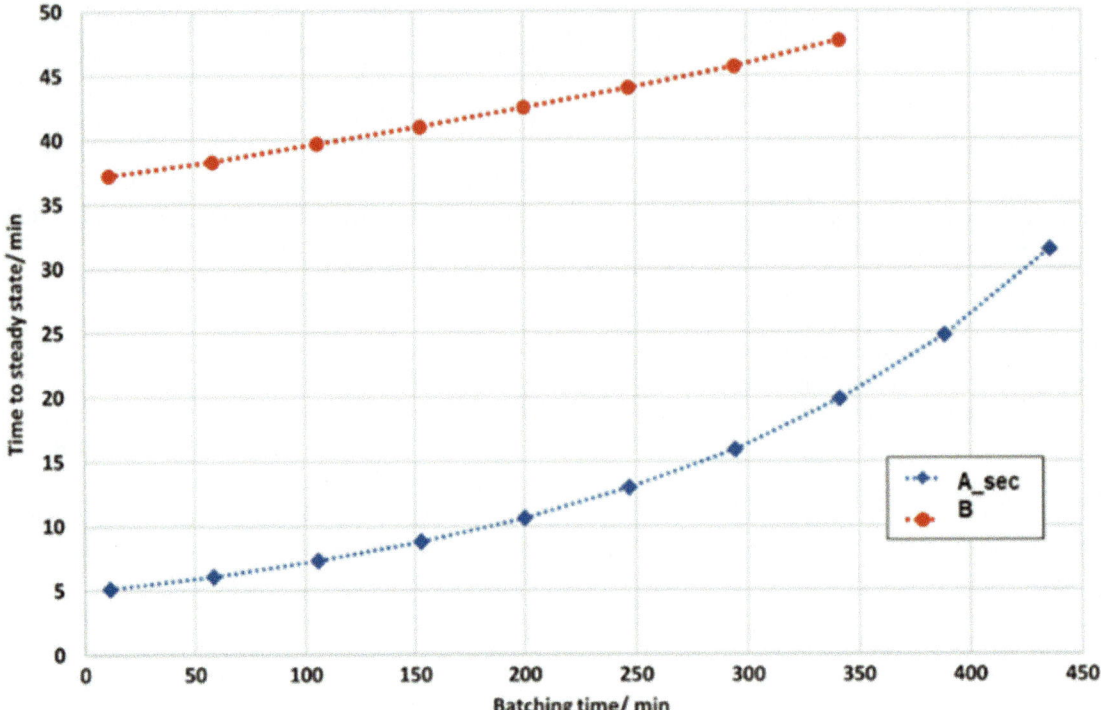

Fig. 4 Comparison of the time needed to reach the complete dissolution of a batch with the number of additions already done for aluminas A_Sec and B

Fig. 5 Dissolution kinetics for primary (A_prim) and secondary (A_sec) aluminas

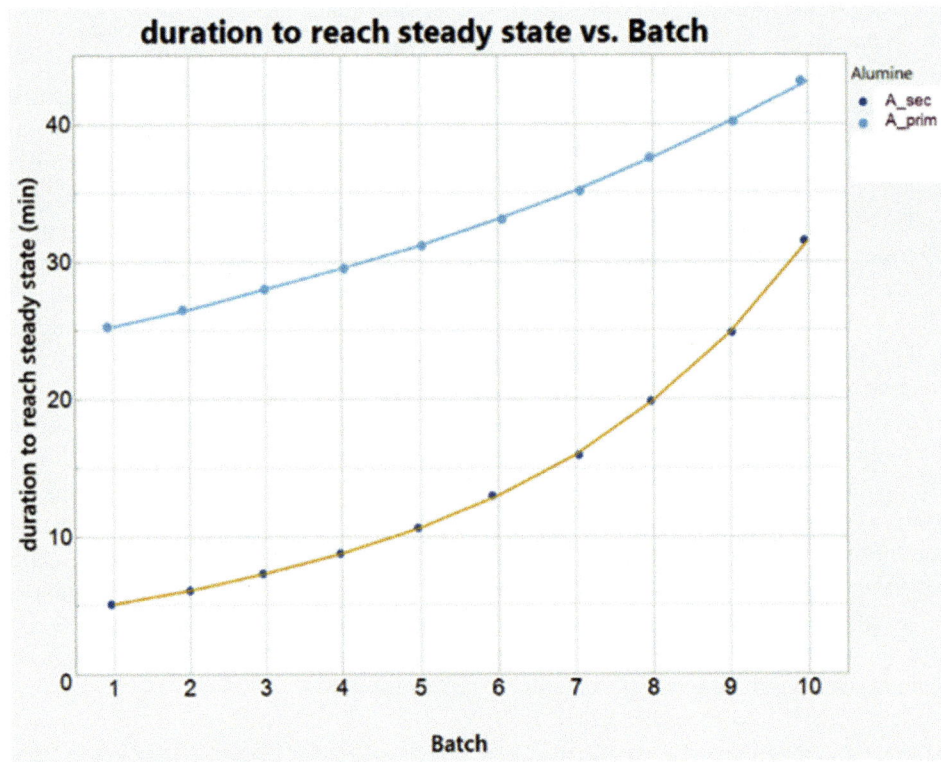

Fig. 6 Time to steady state extracted from the models as a function of batching time for all 6 alumina sources

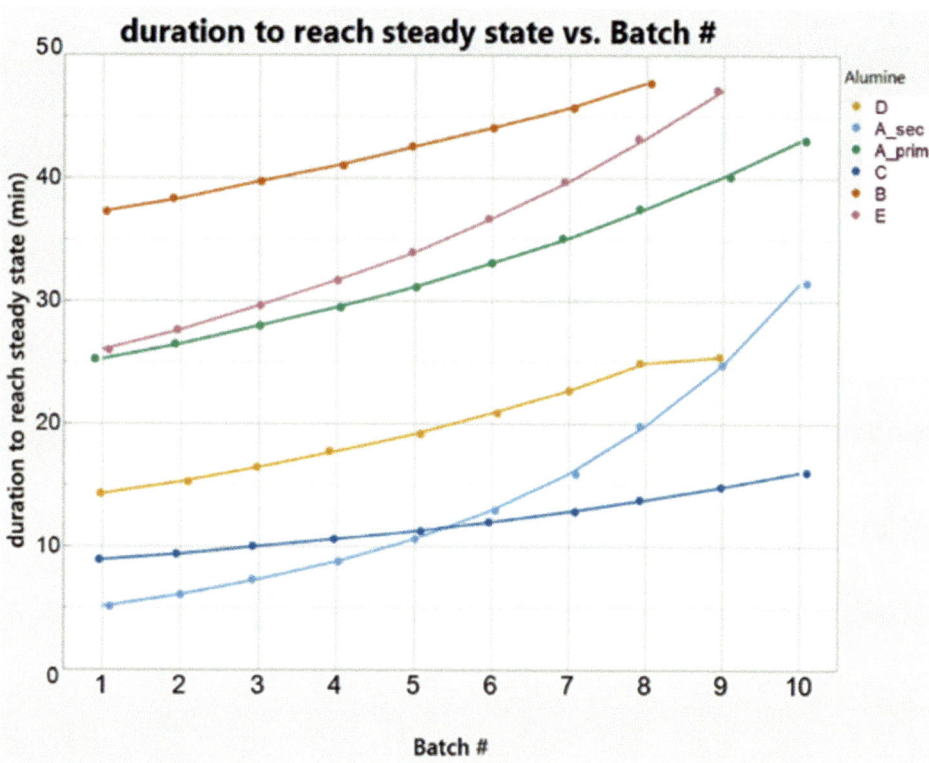

Dissolution of a Mixture of Aluminas

Alumina B was mixed with alumina C with a 1:1 ratio, and the duration to completely dissolve each addition of the mixture (Fig. 7).

For the first 3 additions, the duration to dissolve the whole batch of the B/C mixture is nearly the average of the value of the two aluminas. With subsequent additions, the observed duration to dissolve the mixture increases and tends towards the values of the B alumina. The non-linearity is exacerbated as the slowest dissolving alumina dominates the dissolution behaviour. It is not clear why it is only observed at higher alumina concentrations and additional tests are considered.

Video Analysis and Temperature Drop

It is known that there is a direct correlation between the duration of the alumina rafts atop the melt and the dissolution rate. Therefore, a second method based on the video analysis of the rafts provides a complementary result to the electrochemical method. Videos from aluminas B, C, D, and E were analyzed.

The variability of the floating time is relatively large for any given alumina. From the video observations, this may be a consequence of the random way that alumina batch falls into the melt and how much it spreads or concentrates along the surface. The ability to discriminate aluminas is low, so we calculate the average "floating time" from all the batches for each alumina and plot it versus the alumina dissolution time measured electrochemically (Fig. 8). Despite only using 4 data points, the statistical correlation measured by the R^2 is high (0.91), confirming initial expectations. This method focuses on the ability of the aluminas to form rafts, and how long they take to submerge. For the tested aluminas, it shows that alumina rafts that take longer to submerge also take longer to dissolve. However, this may not be true for every alumina (to be confirmed), in which case it would be interesting to understand which alumina characteristics lead to raft formation, independently of the dissolution rate.

In some cases, alumina dissolution is characterized by the initial temperature drop after addition to the melt [11–13]. With the same experimental condition for each tested alumina and batch addition (bath temperature, superheat and bath stirring/velocity), a trend resulting from the endothermic nature of the dissolution should be expected. Faster dissolving aluminas should be represented by a deeper initial temperature drop opposite to the slower dissolving aluminas. For the present experimental setup, recorded temperature drops varied between 0.5 °C and 2.0 °C in most cases. This is a quite low-temperature response and considering the measurement accuracy of the thermocouple of ±1 °C, one

Fig. 7 Comparison of the duration to reach the complete dissolution of additions of the 1:1 mixture (mix) of aluminas B and C

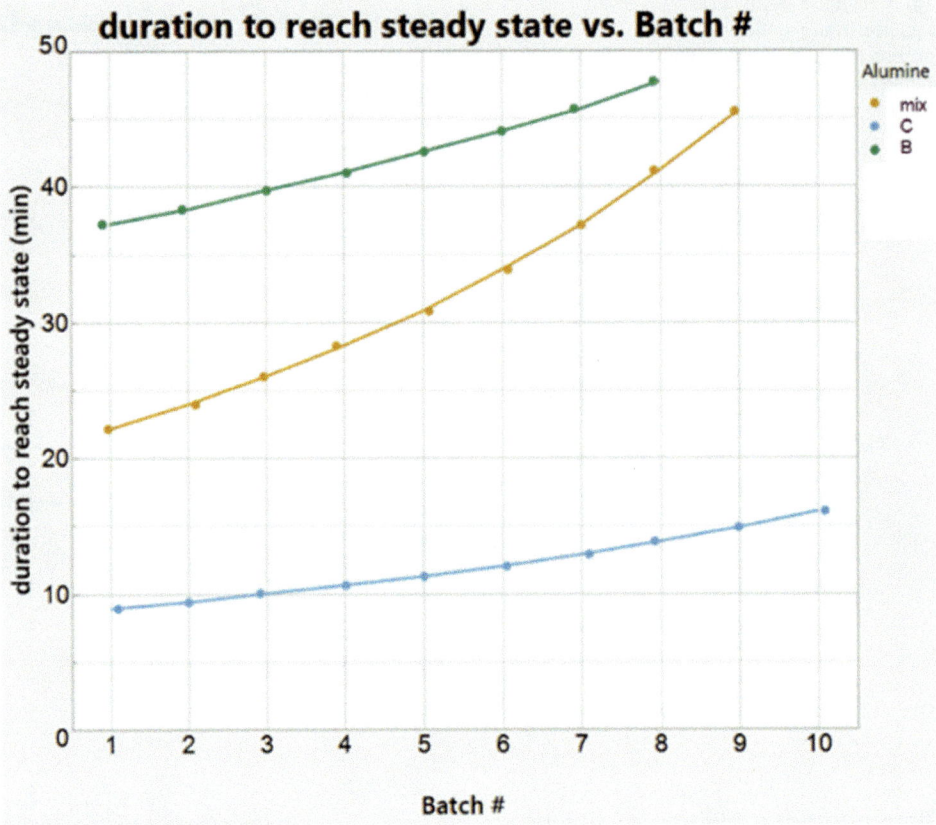

Fig. 8 Alumina dissolution time measured electrochemically versus the average raft floating time for aluminas B, C, D, and E. The error bars are standard deviations

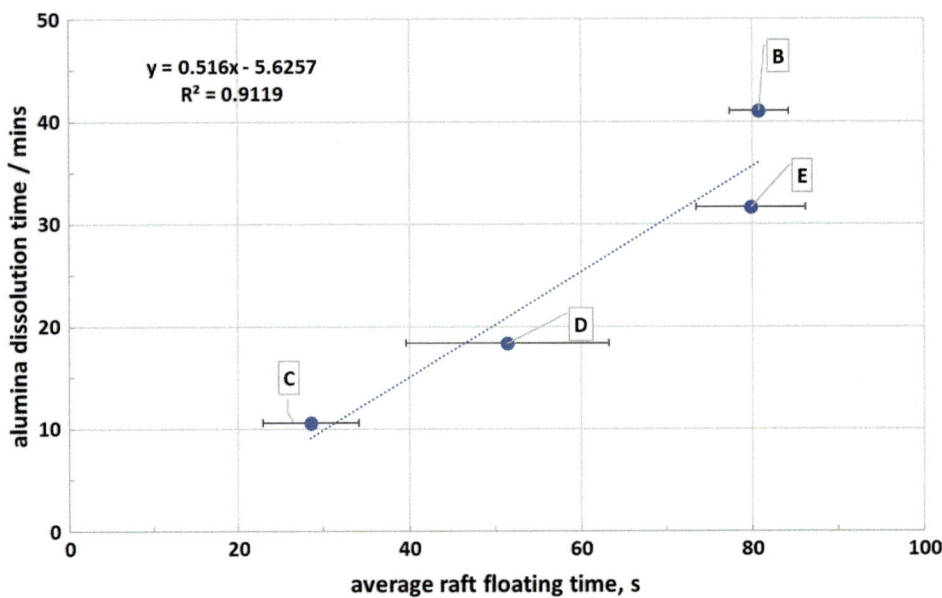

can doubt the value of the evaluation of the temperature drop in this case. Nevertheless, there is an expected trend as can be seen in Fig. 9, where the alumina dissolution time versus the average temperature drop is plotted. The small temperature response is also due to a relatively small alumina batch of 0.25 wt%.

Alumina Properties in Terms of Dissolution Rate

The measured dissolution rate shows an overall agreement with smelter experiences, where alumina A is considered a good quality with good dissolution properties, whereas alumina B is considered a poor alumina quality with poor

Fig. 9 Alumina dissolution time measured electrochemically versus the average temperature drop for aluminas A_sec, B, C, D, and E. The error bars are standard deviations

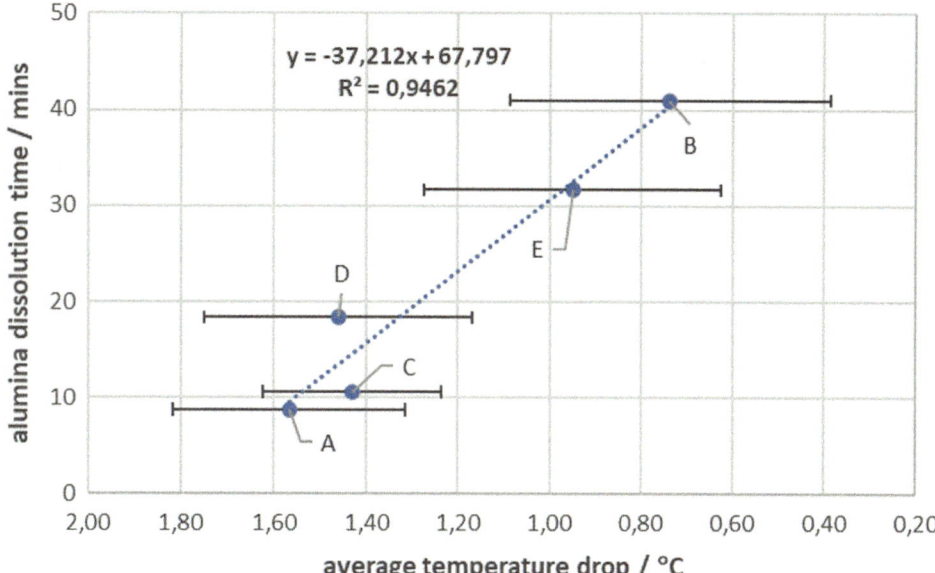

dissolution properties. As can be seen in the characterization results, the samples represent a wide range of different parameter levels.

The MOI and LOI are thought to have a strong influence on the dissolution rate, as well as the alpha and gibbsite content [12]. Other relevant parameters may be the fluoride (F) and carbon (C) content. High levels of MOI/LOI, F, and C content are thought to aid dissolution through breakage and spreading of the particles, or through changed wettability and spreading on the surface. The gibbsite phase is thought to enhance dissolution as well, whereas the thermodynamically stable alpha phase is slower to dissolve.

Of the MOI/LOI parameters, the MOI 120–300 °C and the LOI 300–1000 °C are the most relevant. Since the samples are pre-heated before dissolution testing, any humidity that is released when heating to 120 °C is not relevant.

Aluminas with a high fines content are known to have poor smelter performance and cause several operational disturbances. The fines often have a significantly different chemical composition than the rest of the size fractions. Impurities from pot room dust are brought back to the alumina through the gas scrubber. In addition, the alumina fines could consist of larger portions of either over- or under-calcined material.

Aluminas A_sec and C have a much coarser PSD than alumina B. Alumina A has lower alpha content and flow time, whereas alumina B has higher MOI/LOI and content of the impurities F and C which are thought to aid dissolution. Mixing alumina B and C gives a sample with approximately average levels of the quality parameters of the two pure sources. The dissolution rate also partly falls in between

samples B and C, as expected. This also shows that the method is sensitive enough and that the results make sense.

Comparing the A_prim and A_sec, the secondary alumina dissolves quicker. The PSD is not significantly different; hence, the chemical properties must play a role. However, as alumina A_prim seems to have similar dissolution properties as alumina E and the B/C mix, which both have a much finer PSD than alumina A_sec and A_prim, it complicates the picture. Many attempts have been published to explain the observed dissolution properties based on the alumina quality parameters, however, no clear conclusions have been made regarding which is the dominant factor, and the impact of the different parameters have not been quantified. It is also known that two aluminas with seemingly similar Certificates of Analyses may give very different smelter responses. There are important properties that cannot be explained by the conventional alumina quality parameters, which is why the development of a method to measure the dissolution rate is important. From the number of tests and the complexity of the alumina characterization parameters of the given samples, it is difficult to pinpoint the reasons behind the resulting range in dissolution rate. However, the impact of PSD, alpha, and MOI/LOI is subject to further testing.

Conclusions

An electrochemical method based on square wave voltammetry was successfully developed to measure the dissolved alumina in a cryolite-based bath, able to distinguish between different aluminas.

The main results are summarized as follows:

- The method allows ranking alumina sources according to their dissolution rates with good reproducibility.
- The measured dissolution rate shows an overall agreement with smelter experiences, where aluminas A_sec, C, and D are considered of "good quality", meaning that they were reported to have a fast dissolution, whereas alumina B was reported to dissolve more slowly.
- The method also distinguishes between a primary and a secondary alumina from the same source, the former having a faster dissolution rate.
- The dissolution rate of a mixture of "slow" and "fast" dissolving aluminas results in an intermediary dissolution rate between the two individual aluminas. This means that in principle the dissolution rate of an alumina can be improved by mixing it with a second alumina with faster dissolution kinetics.
- A complementary video analysis determines the floating time of the alumina rafts formed during batch additions and shows a strong correlation with the dissolution rates. The same is valid for the temperature response.

References

1. Richards, N. E., Rolseth, S., Thonstad, J., & Haverkamp, R. G. (1995). Electrochemical analysis of alumina dissolved in cryolite melts. Light Metals (WARRENDALE PA), 391–404.
2. Kuschel, G., & Welch, B. J. (1990). Effect of alumina properties and operation of smelting cells on the dissolution behavior of alumina. Pasmico Research Center, Newcastle, Australia. Department Chemical and Material Engineering, University of Auckland, New Zealand.
3. Haverkamp, R. G., Welch, B. J., Bouvet, S., & Homsi, P. (1997). Alumina quality testing procedure. LIGHT METALS-WARRENDALE, 119–126.
4. Wang, X. (2009). Alumina dissolution in aluminum smelting electrolyte. Light Metals, 2009, 383–388.
5. Molin, A. (2015). Étude expérimentale sur les mécanismes de dissolution de l'alumine (Doctoral dissertation, Université du Québec à Chicoutimi).
6. Yang, Y., Gao, B., Wang, Z., Shi, Z., & Hu, X. (2013). Mechanism of dissolution behavior of the secondary Alumina. Metallurgical and Materials Transactions B, 44(5), 1296–1303.
7. Kan, H. M., Zhang, N., & Wang, X. Y. (2012). Dissolution rate determination of alumina in molten cryolite-based aluminum electrolyte. Journal of Central South University, 19(4), 897–902.
8. Solheim, A., Rolseth, S., Skybakmoen, E., Støen, L., Sterten, Å., & Støre, T. (1996). Liquidus temperatures for primary crystallization of cryolite in molten salt systems of interest for aluminum electrolysis. Metallurgical and Materials Transactions B, 27(5), 739–744.
9. J. Thonstad et al. (2011) "Aluminium Electrolysis: Fundamentals of the Hall-Héroult Process" 3rd edition, Beuth Verlag GmbH.
10. Rye, K., Rolseth, S., Thonstad, J., & Zhanling, K. (1990). Behaviour of alumina on addition to cryolitic baths. In Alumina Quality in a Highly Dynamic Market Environment. Second International Alumina Quality Workshop (pp. 24–37).
11. Thonstad, J., Solheim, A., Rolseth, S., & Skar, O. (2016). The dissolution of alumina in cryolite melts. In Essential readings in light metals (pp. 105–111).
12. Kuschel, G. I., & Welch, B. J. (2016). Further studies of alumina dissolution under conditions similar to cell operation. In Essential Readings in Light Metals (pp. 112–118).
13. Liu, X., Purdie, J. M., Taylor, M. P., & Welch, B. J. (1991). Measurement and modelling of alumina mixing and dissolution for varying electrolyte heat and mass transfer conditions. TMS-AIME Light Metals, 289–298.

Monitoring Cell Conditions and Anode Freeze Dissolution with Model-Based Soft Sensor After Anode Change

Choon-Jie Wong, Jie Bao, Maria Skyllas-Kazacos, Ali Jassim, Mohamed Mahmoud, and Alexander Arkhipov

Abstract

Carbon anodes in Hall-Héroult cells are consumed continuously and changed periodically. The anode change operation introduces a major energy deficit which adversely perturbs the mass and heat balance of the cell. The freeze that formed underneath the newly replaced anodes impedes normal bath circulation and mixing, removes local heat as it dissolves, and disrupts the normal anode current flow which deforms the metal pad via magnetohydrodynamic forces. These may lead to increased process faults and loss of process efficiency. This paper proposes the use of individual anode current measurements to monitor the cell conditions and the dissolution of anode freeze in real time. This helps to return the cell to nominal operation conditions quickly, by identifying issues such as low local superheat, along with providing information to enable further improvements to operational procedures such as ACD adjustments for new anodes. This improves process and energy efficiencies following anode change operations.

Keywords

Anode change • Online monitoring • Current distribution

Introduction

The prebaked anodes used in the Hall-Héroult process for aluminium smelting are consumable carbon blocks that must be replaced at the end of their service life of about a month.

C.-J. Wong · J. Bao (✉) · M. Skyllas-Kazacos
School of Chemical Engineering, The University of New South
Wales, Sydney, NSW 2052, Australia
e-mail: j.bao@unsw.edu.au

A. Jassim · M. Mahmoud · A. Arkhipov
Emirates Global Aluminium, Jebel Ali Operations, P.O. Box 3627
Dubai, UAE

In a cell, typically one or two anodes are replaced every 30–36 h in a cell operation known as "anode change," which is the most frequent manual routine in a smelter cell. As a result, this regularly introduces significant perturbations to the cell mass and thermal balances.

With the typical size of modern anodes, more than 50 kWh of energy stored in spent anode butts are removed from a local zone, and more than 160 kWh of energy are needed to reinstate the optimum operating conditions, including to preheat the new anodes from room temperature [1]. As the newly replaced anodes have low temperatures, it also causes a layer of freeze to form underneath the anodes, extending down to the metal pad. This impedes the flow of current through the new anodes and instead redistributes the current to the remaining anodes, altering the rates of local material consumption/production and rates of local heat generation. The freeze also impedes normal bath circulation and mixing, thereby increasing the spatial non-homogeneity of cell conditions. This freeze can take 10–15 h to redissolve [2]. Therefore, anode change operations have been attributed with numerous cell abnormalities, including improper feed dissolution, sludge formation, anode effect occurrence, and anode spike formation [1, 3, 4]. These have all contributed to losses in process and energy efficiencies.

In order to help provide insights and improve anode change operations, the authors have published elsewhere [1] a detailed spatially discretised model with integrated mass and thermal balance to describe the dynamics in local cell conditions during the operation. In this paper, we propose the use of individual anode current measurements, in conjunction with a simplified anode change cell model, to monitor the cell conditions along with the anode freeze dissolution in real time. This is possible as the line current is regulated, and thus the anode currents must react and redistribute according to the local path conditions, allowing information of the anode freeze to be derived. This work

© The Minerals, Metals & Materials Society 2023
S. Broek (ed.), *Light Metals 2023*, The Minerals, Metals & Materials Series,
https://doi.org/10.1007/978-3-031-22532-1_11

contributes towards increasing the observability of the process variables during anode change operations, to improve cell monitoring, operations, and control.

Anode Change Model

While our previous work [1] involves a model with fine spatial discretisation for describing the local cell dynamics during anode change operation, this work considers a coarser, reduced-order model which serves a different purpose: the structure and complexity of this anode change model must be suitable for use in an online model-based soft sensor, such as an Extended Kalman Filter (EKF) [5]. Therefore, each of the cell conditions (state variables) considered for inclusion in this estimation model must be significant to the process operation, and these are elaborated in this section.

Figure 1 is a schematic diagram of a cell illustrating the model structure and the state variables considered. The cell is divided into two subsystems connected in a parallel electrical network: one containing the newly replaced anodes, and the other containing all the remaining anodes. This is because these two groups of anodes are considered to have vastly different local conditions:

- In the new-anode subsystem, the state variables considered are the average anode–cathode distance (ACD), D_{new}, and the ratio of anode area that can conduct current, A_{new}.
- As for the older-anode subsystem, the state variable considered is the average ACD, D_{old}.
- Other cell conditions such as the alumina concentration, C, the bath temperature, T_{B}, and the ledge thickness, L_{L}, are considered lumped-element variables.

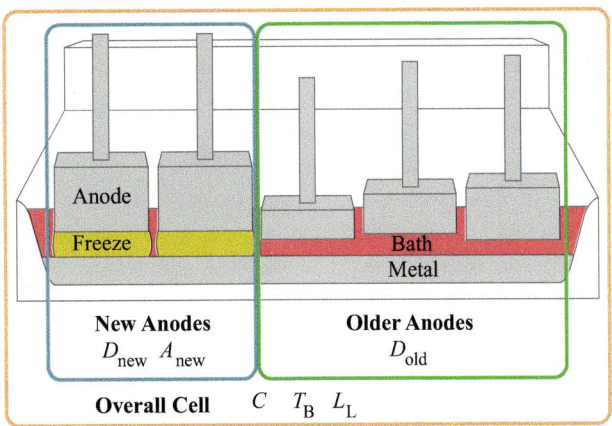

Fig. 1 Model structure and state variables considered in the model

Alumina Concentration

Alumina concentration is an important variable as low values lead to excessive formation of perfluorocarbon gases and occurrence of anode effects. Industrial observations indicate that there is a higher tendency for anode effects to occur after the first underfeeding following anode change operations [4]. Additionally, as alumina concentration level is cycled according to the demand feed strategy [6], the liquidus temperature of the bath also changes, which alters the heat available for freeze dissolution. For this model, the details of the individual feeder actions are not required, and so the change in alumina concentration level C can be modelled with the difference in the rates of alumina feeding and consumption, with

$$\frac{dC}{dt} = \frac{I_{\text{line}}\eta M_{\text{Al}_2\text{O}_3}}{2Fzm_{\text{B}}}(u_{\text{FR}} - 1) + \omega_1, \tag{1}$$

where I_{line} is the line current, η is the current efficiency, $M_{\text{Al}_2\text{O}_3}$ is the molecular mass of alumina, F is Faraday's constant, m_{B} is the total mass of the bath, and u_{FR} is the feeding rate. When u_{FR} has the value of 1, it corresponds to base feeding rates. Finally, ω represents the additive, white, Gaussian, zero-mean process noise.

Anode–Cathode Distance (ACD)

The ACD plays a role in cell thermal balance control, as it alters cell voltage by changing bath resistance [7]. The ACD can be raised before an anode change event to preheat the cell and to supply the energy required to reinstate the cell to optimum operating conditions [2]. The rate of change in ACD is modelled by the difference in rates of carbon consumption (functions of anode currents) and metal accumulation (function of line current), with

$$\frac{dD_{\text{old}}}{dt} = \delta_c(I_{\text{old}}) - \delta_a(I_{\text{line}}) + u_{\text{BM}} + \omega_2, \tag{2}$$

$$\frac{dD_{\text{new}}}{dt} = \delta_c(I_{\text{new}}) - \delta_a(I_{\text{line}}) + u_{\text{BM}} + \omega_3, \tag{3}$$

where D_{new} and D_{old} are the average ACD of new anodes and remaining anodes, respectively, δ_c is the anode consumption rate, δ_a is the metal accumulation rate, u_{BM} is the beam movement distance, I_{new} is the total anode current passing through the newly replaced anodes, and I_{old} is the total anode current passing through the remaining anodes. Additionally, ACD also changes as the metal pad shape shifts due to magnetohydrodynamic forces, and as new anodes expand as they heat up. Although these process uncertainties can be static or slow-evolving, they can be treated as process noise in the state estimator, similar to parameter estimation techniques [5].

Bath Temperature

The bath temperature provides thermal energy for material dissolution [8] and to maintain the molten state of the electrolyte for ionic mobility [9]. The heat generation in a cell occurs almost entirely via the ohmic heating process, with minor contributions from some exothermic reactions in the cell [10]. Thus, the thermal balance around the bath can be given by

$$\frac{dT_B}{dt} = \frac{Q_{B,gen}(I_{line}, V) - Q_{B,feed}(I_{line}, u_{FR}) - Q_{B,topbtm}(T_B, A_{new})}{m_B c_B + m_M c_M}$$
$$+ \frac{-Q_{L,gain}(C, T_B) - Q_{F,gain}(C, T_B, A_{new})}{m_B c_B + m_M c_M} + \omega_4, \quad (4)$$

where T_B is the average bath temperature, $Q_{B,gen}$ is the net heat generation rate in the bath region, $Q_{B,feed}$ is the rate of heat energy used for alumina preheating and dissolution, $Q_{B,topbtm}$ is the rate of heat loss from the top and bottom of the cell (① in Fig. 2), and $Q_{L,gain}$ and $Q_{F,gain}$ are the rates of heat transfer from the bath to the ledge (②) and to the freeze (③), respectively. As the bath and the metal are assumed to share the temperature dynamics in this model, both these layers contribute towards the total heat capacity, with m_B and m_M being the bath mass and metal mass, respectively, and c_B and c_M being the specific heat capacity of the bath and metal, respectively.

In Eq. 4, $Q_{B,gen}$ refers to the net heat generated within the cell, as heat generated along the external superstructures does not contribute towards cell heating. A portion of the input electrical energy (5.474 kWh/kg Al) is used for enabling and completing the electrowinning chemical process, and some energy is used for preheating the carbon anodes and other additives. The net heat generation is then [11]

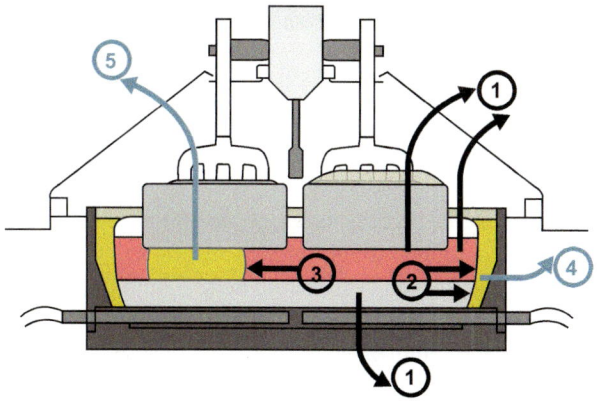

Fig. 2 Some important heat fluxes out of a cell

$$Q_{B,gen} = (V - I_{line} R_{external}) I_{line} - 5.474 \cdot 3600 \cdot \frac{I_{line} \eta M_{Al}}{Fz},$$
$$(5)$$

where $R_{external}$ is the total electrical resistance outside of the cell cavity, dependent on the cell design. There is an additional 1.126 kWh/kg Al energy requirement for preheating and dissolving the alumina feed (with impurities) [11], which is equivalent to 2,146 kJ/kg Al_2O_3. Thus, $Q_{B,feed}$ in Eq. 4 depends on the rate of alumina addition, and thus is given by

$$Q_{B,feed} = 2,146 \cdot u_{FR} \cdot \frac{I_{line} \eta M_{Al_2O_3}}{2Fz}. \quad (6)$$

In Eq. 4, $Q_{B,topbtm}$ (① in Fig. 2) represents the rate of heat loss from the bath region through the top and bottom of the cell. It has three components in parallel:

- heat loss upwards through the anodes and crust layer to the ambient potroom air,
- heat loss upwards through the air gap between the bath and the crust layer, and through the crust layer to the ambient potroom air, and
- heat loss downwards through the layers of refractory materials to the ambient potroom air.

Thus, $Q_{B,topbtm}$ can be calculated as

$$Q_{L,gain} = \frac{T_B - T_{ambient}}{R_{B\rightarrow ambient}}, \quad (7)$$

where $T_{ambient}$ is the temperature of the potroom air and $R_{B\rightarrow ambient}$ is the total effective thermal resistance from the bath region to the potroom air, through the three parallel thermal pathways discussed. Conductive and convective heat transfers are considered, and radiative heat transfers are included in the effective heat transfer coefficients. This effective thermal resistance can be computed for any cell design, for given cell dimensions and materials' thermal properties.

In Eq. 4, $Q_{L,gain}$ (②) and $Q_{F,gain}$ (③) both represent heat transfer from the bath/metal to frozen bath material. These can be modelled with convective heat transfer, with varying heat transfer coefficients depending on local thermal conditions [1]. For example, the heat transfer coefficient is larger in the centre channel than at the back wall due to increased bath volume and bath mixing in the centre channel. Additionally, experiments [12] show that the interface between the ledge/freeze and bath remains at the liquidus temperature of the bath. The liquidus temperature is considered to be a time-varying parameter that changes depending on bath composition, of which the relation has been published elsewhere [13]. These heat transfers can be modelled with, for example,

$$Q_{L,gain} = (T_B - T_{liquidus}) \cdot (h_{B-L}A_{B-L} + h_{M-L}A_{M-L}), \quad (8)$$

where $T_{liquidus}$ is the liquidus temperature, h_{B-L} and h_{M-L} are the bath-ledge and metal-ledge heat transfer coefficients, respectively, and A_{B-L} and A_{M-L} are the corresponding heat transfer areas. It is also noted that $(T_B - T_{liquidus})$ is the superheat, which is widely considered to be the thermal driving force.

Ledge Thickness/Conducting Anode Area

The ledge is a thin frozen bath layer on the cell sidewalls, the thickness of which should be maintained as it protects the sidewall materials from the corrosive bath and metal. The dissolution/formation of the ledge and the freeze are both driven by the difference in heat transferred from the bath/metal to the ledge/freeze surface (Eq. 8), and the heat leaving that surface (④ and ⑤ in Fig. 2), with

$$\frac{dL_L}{dt} = \frac{Q_{L,gain}(C, T_B) - Q_{L,loss}(C)}{l_f \rho (A_{B-L} + A_{M-L})} + \omega_5, \quad (9)$$

$$\frac{dA_{new}}{dt} = \frac{Q_{F,gain}(C, T_B, A_{new}) - Q_{F,loss}(C)}{l_f \rho D_{new} A_{an}} + \omega_6, \quad (10)$$

where L_L is the ledge thickness, A_{new} is the ratio of anode area that can conduct current, $Q_{L,loss}$ is the heat loss rate from the ledge surface through the sidewall materials to the ambient potroom air, $Q_{F,loss}$ is the heat loss rate from the freeze through the anodes and crust layer to the ambient potroom air, l_f is the heat of fusion of cryolite, ρ is the cryolite density, and A_{an} is the total surface area of the new anodes that can conduct electricity when they are at full current capacity.

Discrete-Time State-Space Model

With the use of digital computers in monitoring and control applications, data are sampled and processed at fixed time intervals. For use in model-based soft sensors, the model Eqs. 1–10 are rewritten in discrete-time format with timestep Δt (1 min for this work) using backward difference approximation technique:

$$C(k+1) = C(k) + \Delta t \cdot \frac{I_{line} \eta M_{Al_2O_3}}{2Fzm_B} (u_{FR}(k) - 1) + \omega_1, \quad (11)$$

$$D_{old}(k+1) = D_{old}(k) + \Delta t \cdot [\delta_c(k) - \delta_a(k)] + u_{BM}(k) + \omega_2, \quad (12)$$

$$D_{new}(k+1) = D_{new}(k) + \Delta t \cdot [\delta_c(k) - \delta_a(k)] + u_{BM}(k) + \omega_3, \quad (13)$$

$$T_B(k+1) = T_B(k) + \Delta t \frac{Q_{B,gen}(k) - Q_{B,feed}(k) - Q_{B,topbtm}(k)}{m_B c_B + m_M c_M} + \Delta t \frac{-Q_{L,gain}(k) - Q_{F,gain}(k)}{m_B c_B + m_M c_M} + \omega_4, \quad (14)$$

$$L_L(k+1) = L_L(k) + \Delta t \frac{Q_{L,gain}(k) - Q_{L,loss}(k)}{l_f \rho A_{B-L}} + \omega_5, \quad (15)$$

$$A_{new}(k+1) = A_{new}(k) + \Delta t \frac{Q_{F,gain}(k) - Q_{F,loss}(k)}{l_f \rho D_{new} A_{an}} + \omega_6. \quad (16)$$

Converting to state-space representation, the states and inputs can be defined as $x = [C \quad D_{old} \quad D_{new} \quad T_B L_L A_{new}]^T$ and $u(k) = [I_{old}(k) \quad I_{new}(k) \quad u_{BM}(k) \quad u_{FR}(k) V(k-1)]^T$. It is worth noting that while individual anode currents were measured from the cell, I_{new} in the EKF refers to the sum of currents flowing through the newly replaced anodes, and I_{old} is the sum of the remaining anode currents. In the EKF algorithm, which is further elaborated in the next section, the measurements (cell voltage) should be made for the model outputs. Since both the new-anode and older-anode subsystems are connected in parallel and must have the same cell voltage, this constraint is achieved by the EKF algorithm by defining model outputs as $y = [V_{old} \quad V_{new}]^T$. Thus, the model can be rewritten succinctly as

$$\begin{cases} x(k+1) &= f(x(k), u(k), \omega(k)) \\ y(k) &= h(x(k), u(k), v(k)) \end{cases}, \quad (17)$$

where $f(\cdot)$ is a vector function consisting of the right-hand side of Eqs. 11–16, and $h(\cdot)$ is the cell voltage output equation which has been published elsewhere [14]. ω and v are the process and output noise with covariance matrices Q and R.

Model-Based Soft Sensor

The Hall-Héroult process lacks observability as the harsh operating conditions (e.g., high temperatures, corrosive electrolyte, and strong magnetic fields) limit the use of physical sensors. The Extended Kalman Filter (EKF) algorithm [5] can be implemented as a model-based soft sensor to monitor the important process variables described in the previous section. The setup of the soft sensor based on EKF

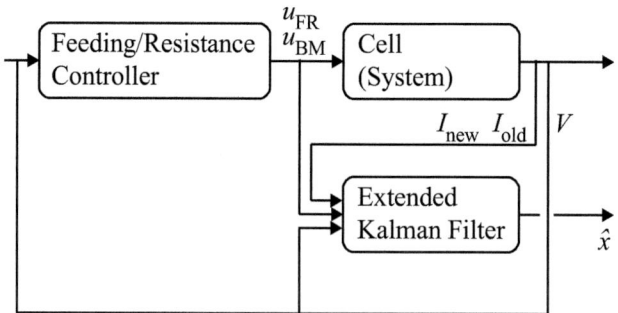

Fig. 3 Block diagram showing Extended Kalman Filter

is shown in Fig. 3, where control actions and measurements of anode currents and cell voltage are used to estimate those process variables.

The EKF is implemented as follows. First, the initial estimated states, $\widehat{x}_0^+$, and error covariance, P_0^+, are given values within an expected or reasonable range. Next, the state variables, $\widehat{x}_k^+$, are estimated for each timestep $k = 1, 2, \ldots$ with [5]

$$F_{k-1} = \left.\frac{\partial f_{k-1}}{\partial x}\right|_{\widehat{x}_{k-1}^+}, \tag{18}$$

$$L_{k-1} = \left.\frac{\partial f_{k-1}}{\partial \omega}\right|_{\widehat{x}_{k-1}^+}, \tag{19}$$

$$P_k^- = F_{k-1} P_{k-1}^+ F_{k-1}^\top + L_{k-1} Q L_{k-1}^\top, \tag{20}$$

$$\widehat{x}_k^- = f_{k-1}\big(\widehat{x}_{k-1}^+, u_{k-1}, 0\big), \tag{21}$$

$$H_k = \left.\frac{\partial h_k}{\partial x}\right|_{\widetilde{x}_k}, \tag{22}$$

$$M_k = \left.\frac{\partial h_k}{\partial v}\right|_{\widetilde{x}_k}, \tag{23}$$

$$K_k = P_k^- H_k^\top \big(H_k P_k^- H_k^\top + M_k R M_k^\top\big)^{-1}, \tag{24}$$

$$P_k^+ = (I - K_k H_k) P_k^-, \tag{25}$$

$$\widehat{x}_k^+ = \widehat{x}_k^- + K_k\big(y_k - h_k(\widehat{x}_k^-, 0)\big), \tag{26}$$

where $F_{k-1}, L_{k-1}, H_k, M_k$ are the Jacobian matrices used to linearise the system in Eq. 17, K_k is the filter gain, and Q and R are the covariance matrices of ω and v in Eq. 17.

Experimental Results

The actual measurements of anode currents and cell voltage from a commercial cell, along with control decisions from the industrial controller, were provided to the model-based soft sensor (Extended Kalman Filter) described in the previous section, as the cell underwent the replacement of a pair of anodes (Anodes 29–30 in experiment 1 and Anodes 17–18 in experiment 2). This allows the cell conditions (process states) to be estimated shortly after the anode change operation. For validation purposes, one of the new anodes was lifted out of the cell at scheduled times to check the freeze underneath it. The alumina concentration was measured from the tap-end on an hourly basis, while the bath temperatures were measured at a frequency ranging from 15 min to 1 h at four locations:

- between the pair of new anodes (• in Figs. 4 and 6),
- between the pair of anodes opposite to the new anodes (×),
- at tap-end (+), and
- at duct-end (▷).

Experiment 1

Figure 4 shows the process inputs (alumina feeding rates, beam movements, and anode currents) and cell voltage provided to the EKF soft sensor. The EKF was initialised shortly after the anode change with initial values that are arbitrary but within the expected range; there is little value in starting the estimation prior to the operation, since the operation introduces huge perturbations that would increase the uncertainties in the process variables. The output (cell voltage) estimation error shows that the EKF took just about an hour to converge successfully, well within the time needed for complete anode freeze redissolution.

In general, the estimations reflect well the trends in the measurements, although a higher bath sampling rate would have been ideal to vigorously validate the trends caused by the demand feed algorithm. For the alumina concentration, the discrepancies between the estimated average concentration level and the tap-end measurements are attributed to the non-homogeneity in cell conditions. Meanwhile, for the bath temperature, the estimated average values match well with the measurements at three cell locations, whereas the new

Fig. 4 State estimation results for experiment 1 (anodes 29–30 replaced at 0 h, anode 30 photographed at 2.6 h)

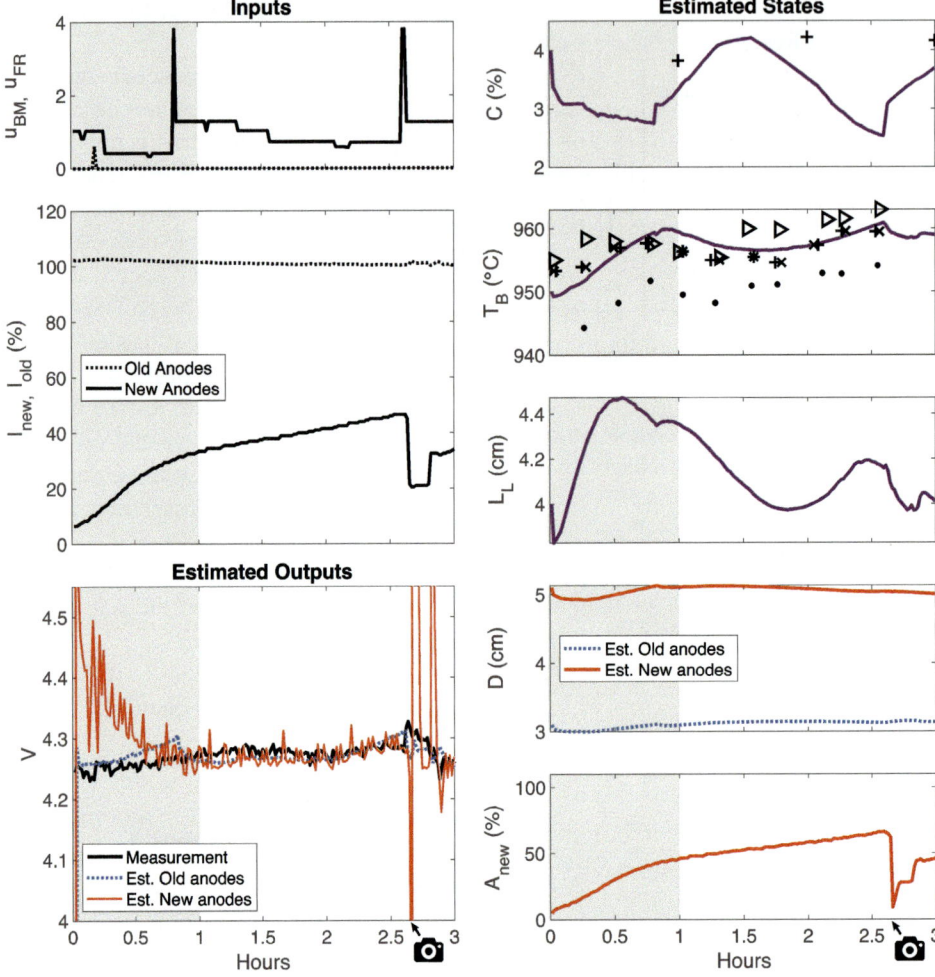

anode region has lower temperature due to the localised energy deficit. The overall fall in estimated ledge thickness (after EKF convergence at 1 h) is reasonable because the cell was gradually restoring its operating temperature. The ACD estimations also highlight the importance of discretising the cell into two subsystems: due to the lower consumption rate of the new anodes resulting from their low anode currents, new anodes are typically replaced at higher elevation as their ACD decreases quickly. Thus, the new anodes have different ACD values and dynamics.

At the time when anode 30 was pulled out for freeze observation, it was estimated that around 35% of its working area would be covered by the freeze. Unfortunately, the freeze appeared to have fallen off due to excessive agitation as the anode was removed from the cell with overhead cranes. However, the dark shadow observed in Fig. 5 could infer the size and location of the freeze, as the grey-white parts are anode areas that were previously exposed to the bath, and this bath froze into a coating as the anode was removed from the cell.

Fig. 5 Photograph of bottom of anode 30 at 2.6 h

Experiment 2

The experiment was repeated for anodes 17–18 for a longer period, and the estimation results are shown in Fig. 6. Similarly, the EKF took only about an hour to converge, while the estimated dissolution time was around 8 h. In this experiment, anode 18 was pulled up twice to check the freeze underneath. For these time points, it was estimated

Fig. 6 State estimation results for experiment 2 (anodes 17–18 replaced at 0 h, anode 18 photographed at 2.65 and 7.8 h)

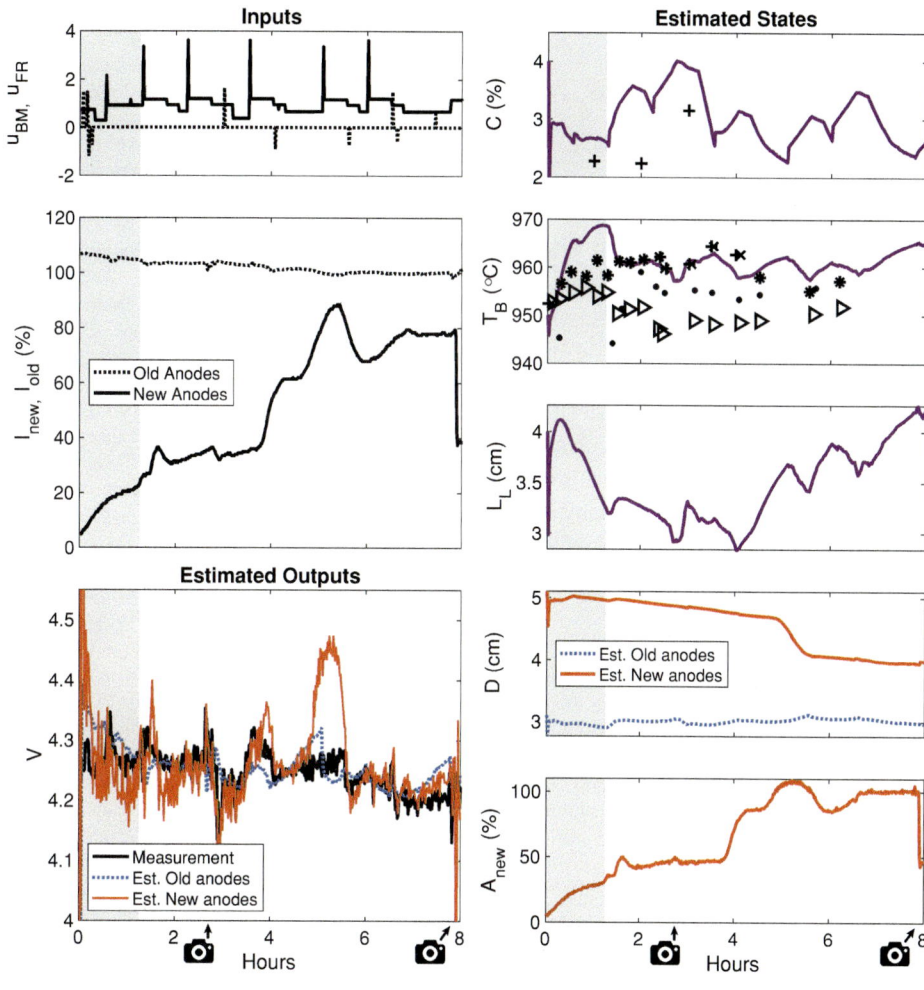

Fig. 7 Photograph of bottom of anode 18 at 2.65 and 7.8 h

that around 50 and 1% of the freeze remains, which is corroborated by the photos in Fig. 7.

This experiment demonstrates the robustness of the soft sensor as the cell was not functioning well. On the day of the experiment, the breaker closest to the tap-end was failing, which limited the delivery of alumina into the bath. The effect on the cell was evident from the low concentrations measured from the tap-end, as well as the premature termination of several underfeeding windows. As this was automatically reported by the actuators, this information was

provided to the EKF as well. There was an unscheduled anode check at 3 h for an anode near the duct-end, causing additional heat deficit in that region and leading to low duct-end bath temperatures as observed. This information was not provided to the EKF as it is unmodelled, but a resulting growth in ledge thickness was still being estimated. The cell issues, unscheduled operations, and a long underfeeding window at 4.3 h finally led to a huge estimation output error, from which the EKF was able to recover automatically. This also shows that a huge output error can be used for detecting faults and deviations from normal operating conditions, but further studies will be needed for a more rigorous solution.

Conclusion

This article presents a simplified model, consisting of two subsystems, for describing the conditions of a cell undergoing an anode change operation, and proposes its use in the online model-based soft sensor by adopting the Extended Kalman Filter algorithm. Two sets of results from experimental studies are presented, validating the estimations of alumina concentration, bath temperature, and freeze dissolution. This work contributes towards increasing the observability of the process variables during anode change operations, which can be used to further improve cell operation and process control, such as ACD adjustments for new anodes.

Acknowledgements The authors acknowledge the financial support from ARC Research Hub for Integrated Energy Storage Solutions, and Emirates Global Aluminium Jebel Ali Operations for their technical support especially from the Technology Development and Transfer team and Operations team. The authors also gratefully thank Prof. Barry Welch for providing numerous helpful insights and guidance for this work.

References

1. Wong C-J, Yao Y, Bao J, Skyllas-Kazacos M, Welch BJ, Jassim A, Mahmoud M, Arkhipov A (2021) Modelling of Coupled Mass and Thermal Balances in Hall-Heroult Cells During Anode Change. Journal of the Electrochemical Society. 168(21):123506.

2. Jassim A, Meintjes G, Blasques J, Sadiq M, Kumar A, Al-Jallaf MM, Al Zarouni AHAM (2012) Improved Energy Management during Anode Setting Activity. Paper presented at the Proceedings of the TMS Light Metals, Orlando, Florida, USA.

3. Wong C-J, Yao Y, Bao J, Skyllas-Kazacos M, Welch BJ, Jassim A (2020) Study of heat distribution due to ACD variations for anode setting. Paper presented at the Proceedings of TMS Light Metals, San Diego, California, USA.

4. Zarouni A, Al Zarouni A (2011) DUBAL's experience of low voltage PFC emissions. Paper presented at the Proceedings of 10th Australasian Aluminium Smelting Technology Conference, Launceston, Tasmania, Australia.

5. Simon D (2006) Optimal state estimation: Kalman, H infinity, and nonlinear approaches. 1 ed.: John Wiley & Sons.

6. Robilliard KR, Rolof B (1989) A Demand Feed Strategy for Aluminum Electrolysis Cells. Paper presented at the Proceedings of the TMS Light Metals, Las Vegas, Nevada, USA.

7. Potocnik V, Reverdy M (2021) History of Computer Control of Aluminum Reduction Cells. Paper presented at the Proceedings of the TMS Light Metals, Orlando, Florida, USA.

8. Haupin W (1992) The liquidus enigma. Paper presented at the Proceedings of the TMS Light Metals, San Diego, California, USA.

9. Taylor MP, Welch BJ, McKibbin R (1986) Effect of convective heat transfer and phase change on the stability of aluminium smelting cells. AIChE journal. 32(9):1459-1465.

10. Grjotheim K, Welch BJ (1980) Aluminium smelter technology. Aluminium-Verlag.

11. Wong C-J (2022), Dynamic Mass and Heat Balance Model of Hall-Heroult Cells: A Discretised Approach. Ph.D. thesis, School of Chemical Engineering, Faculty of Engineering, University of New South Wales, Australia.

12. Fallah-Mehrjardi A, Hayes PC, Jak E (2014) Investigation of freeze-linings in aluminum production cells. Metallurgical and Materials Transactions B. 45(4):1232-1247.

13. Solheim A, Rolseth S, Skybakmoen E, Støen L, Sterten ÅA, Støre T (1995) Liquidus temperature and alumina solubility in the system Na_3AlF_6–AlF_3–LiF–CaF_2–MgF_2. Paper presented at the Proceedings of the TMS Light Metals, Las Vegas, Nevada, USA.

14. Haupin W (1998) Interpreting the components of cell voltage. Paper presented at the Proceedings of the TMS Light Metals, San Antonio, Texas, USA.

EGA's First Holistic Mobile Application for Smelter Operations

Ahmed Al Haddad, Dinesh Kothari, Ghalib Al Falasi, Yousuf Ahli, Sergey Akhmetov, Najeeba Al Jabri, Abdulla Karmustaji, Mustafa Mustafa, and Mahmood Al Awadhi

Abstract

Artificial Intelligence has the potential to generate significant value and unlock numerous business opportunities while supporting Emirates Global Aluminium's (EGA) strategic ambitions and values of innovation and continuous improvement. In order to build a high performing machine learning model we need to fuel it with a large amount of quality data. Based on an initial survey done to determine our paper consumption within the Potlines, we have approximately 1.6 million sheets of paper consumed annually in Reduction. This is where Smart Assistance and Quick Response (SAQR)—an advanced mobile application can unlock new value via digitization, automation and enable future AI implementation. The SAQR application is expected to enhance EGA's data collection, increase overall accuracy and provide transparent tracking throughout our operations, enabling faster decision making. The team behind SAQR is now focused on extending its reach within EGA. SAQR is currently being used in potlines, crucible tracking, reduction process control and busbar monitoring. Today we have full visibility of every crucible journey within EGA and in the future we aspire to have full visibility of every process, action, and activity within the entire production chain.

Keywords

SAQR • Mobile application • Digital transformation • Artificial intelligence

A. Al Haddad (✉) · A. Karmustaji
Product Owner, Industry 4.0, Emirates Global Aluminium (EGA), Abu Dhabi, UAE
e-mail: aalhaddad@ega.ae

D. Kothari
Senior Project Leader - Applications, IT, Emirates Global Aluminium (EGA), Abu Dhabi, UAE

G. Al Falasi
Senior Manager, Reduction, Emirates Global Aluminium (EGA), Abu Dhabi, UAE

Y. Ahli
Vice President, Reduction, Emirates Global Aluminium (EGA), Abu Dhabi, UAE

S. Akhmetov
Executive Vice President, Midstream, Emirates Global Aluminium (EGA), Abu Dhabi, UAE

N. Al Jabri
Vice President, Technical, Emirates Global Aluminium (EGA), Abu Dhabi, UAE

M. Mustafa
Manager, Reduction Technology Development & Transfer, Emirates Global Aluminium (EGA), Abu Dhabi, UAE

M. Al Awadhi
Senior Manager - Lead Translator, Industry 4.0, Emirates Global Aluminium (EGA), Abu Dhabi, UAE

Introduction

During the first industrial revolution data was purely analog based [1], however, over the past few decades and since the introduction of smart devices such as the computer, mobile and cloud-based technologies the volume of data being collected has increased significantly and the storage capacities in parallel have grown. Today we are going through the fourth industrial revolution where we see connectivity and integration of multiple smart technologies which stream data to the cloud in quantities which are too large and complex for humans to process, therefore resulting in missed opportunities and inaccurate and delayed decision-making. However, this large data can be processed by powerful computers and consumed by machine learning models to predict failures, cluster features, recommend operating parameters and detect new patterns. Artificial intelligence provides us with an opportunity to optimize and automate and accelerate our decision-making while unlocking new insights and paving the way for process and technology innovation and

S. Broek (ed.), *Light Metals 2023*, The Minerals, Metals & Materials Series,
https://doi.org/10.1007/978-3-031-22532-1_12

ultimately maximizing value creation within the current supply chain.

To build a high-performing machine learning model with acceptable accuracy our data has to be clean and of sufficient size and scale. The current challenge stems from the inability to acquire clean data and conduct live data analysis since the majority of the data is analog based. The second challenge involves data aggregation and the lack of integration where data is split across multiple legacy systems and applications.

To address this challenge, we have introduced a mobile application which has been integrated with various production applications. Moreover, Internet of Things (IoT) can be introduced and integrated to instantaneously capture and store missing information within the Aluminium manufacturing process which can support EGA's digital transformation journey and enable future technologies.

Current Data Collection and Reporting Process in Reduction at EGA

The current method of collecting data is divided into automated versus manual approaches. In the automated approach, data is automatically captured and streamed via smart sensors and programmable logic controllers to measure key pot parameters such as pot voltage, resistance, noise etc. The data is continuously collected and displayed without the need for human intervention or manual work.

However, there are many other processes and activities which currently follow a manual data collection approach, namely, metal and bath height measurement, bath temperature, anode current distribution etc. These measurements require an operator to manually collect the measurements from the shop floor and return back to the control room to transfer these measurements from paper to computer via *Microsoft Excel* and finally summarized in a report and shared with the management team via *Microsoft Outlook*. The pain points with the current approach are

- Multiple data entry interfaces
- Lack of visibility since data reporting is delayed
- Inefficiency due to repetitive steps within the process
- Difficulty in tracking historical data since it is paper based
- Lower data accuracy due to the chance of human error and lack of validation

Having the right data at the right time is key to making the optimum decision. As mentioned earlier the final destination of this data stream is the machine learning model which will output data in the form of a recommendation, prediction, or cluster. However, the quality of our model is highly dependent on consuming data which is accurate and complete.

For example, when a process control engineer recommends a parameter modification or an action to be taken on a specific pot, they will only come to know what manual actions were taken on a pot at the end of a shift once the report is shared. Therefore, their decisions may result in unwanted results due to not having full visibility on the status of the pot which poses the risk of having certain biases in our model and ultimately leads to inefficiency.

Introduction of SAQR Mobile Application

Mobile technology has improved drastically over the past decade and is part of our daily lives and can be seen as an extension of human consciousness. Today innovative mobile apps are being developed and deployed to close the existing gaps, disrupt business models, and generate value in a multitude of industries around the world.

To address the existing gaps within our current data collection and reporting process, a mobile application was proposed and developed in-house as a solution to

A. Provide shop floor workers with a platform to collect and consolidate data as well automate the reporting activity
B. Present the data via intuitive reports and live dashboards to enhance decision making

SAQR was developed over the past few years with the vision of being a one-stop shop for all EGA operations to increase overall visibility, accountability, and collaboration across all areas within the production chain. Figure 1 shows a screenshot of the homepage of the SAQR mobile application.

SAQR was not created to replace the current systems or workforce, but rather as a technology to augment the current systems by consolidating various datasets as well as the shopfloor workers and decision makers by making fact-based decisions which can lead to higher levels of productivity.

In the scenario shown in Fig. 2, the shop floor operator is manually collecting data and uploading it into the SAQR database which is automatically generating a live shift report that is visible to all relevant stakeholders instantaneously. Additionally, information from several databases is being consolidated in SAQR to provide full visibility on all operating parameters, processes and actions taken to make better decisions and result in faster response times. Data is also streamed via IoT's and presented in SAQR in the form of a table, alerts and notifications eliminating the need for manual measurements and constant monitoring but also providing the shift supervisors with full visibility on the

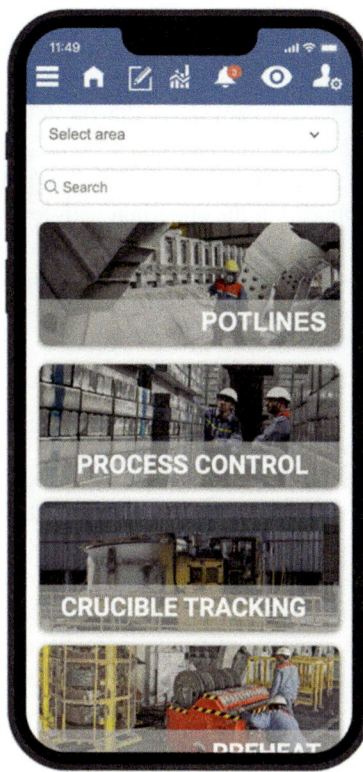

Fig. 1 SAQR mobile application homepage user interface (UI)

operating parameters which can possibly influence or act as a root cause to other operational issues.

Once the data is collated and centralized into a single dataset then we can introduce a data scientist or engineer who will clean, normalize and reduce the data into a form which can be consumed by a machine learning model in the future. Then based on how the problem is framed, we can apply machine learning to several use cases and derive new insights which may perhaps require us to change our current processes, procedures, or parameters to conduct our manufacturing operations at peak productivity and efficiency.

In terms of current functionality, SAQR currently offers its users the following:

- Data entry of measurements, operation status and actions taken
- Conducting job quality and safety audits
- Tracking status and location of assets
- Capturing images of anode problems and vehicle or equipment breakdowns
- Automated generation of shift reports and dashboards
- Export function to download and share reports in multiple formats such as PDF, .csv etc.
- Voice recognition to write and submit data hands-free
- Geo-location to map routes or locate personnel during emergencies and incidents

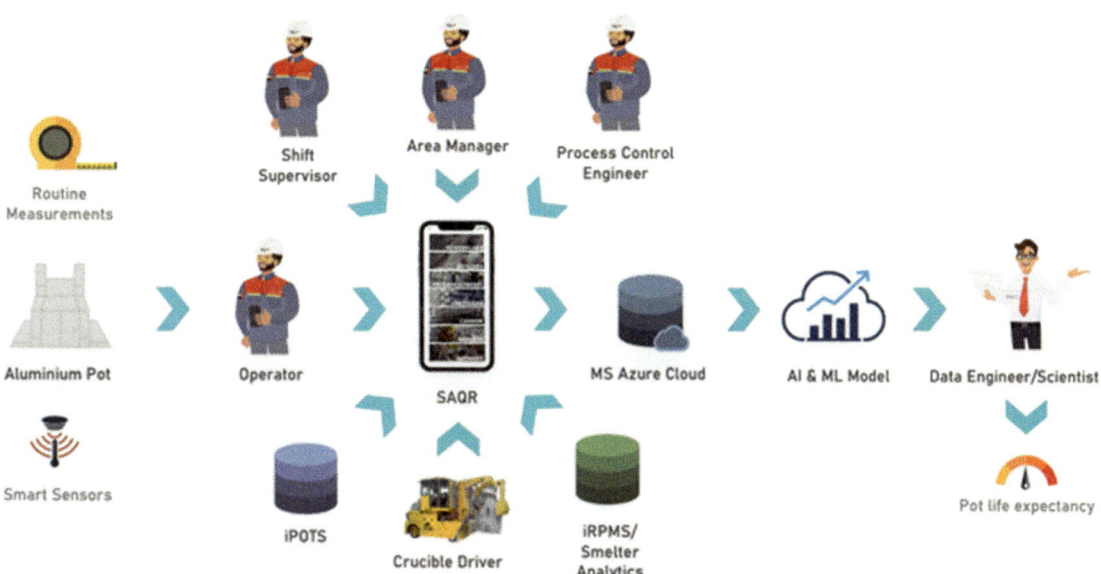

Fig. 2 Example of information flow and data consolidation via SAQR

- Push notification system to send alerts to users
- User access management to onboard users and manage varying access levels
- Live feed to track all actions and data entered within a shift
- Full integration with *Microsoft PowerBI* to create and display dashboards
- Integration with IoT's to view live data
- Integration with multiple Application Programming Interfaces (APIs) to acquire relevant data

These features and plugins can be replicated across the application to create a custom and tailored user experience for any department with minimal development time.

Digital Transformation of Reduction Operational Processes via SAQR

As a starting point, SAQR was focused on the heart of the aluminum smelter—the Potlines. Many processes and activities were heavily reliant on paper use and also relied on multiple interfaces resulting in poor data quality, visibility and historical tracking.

Following the agile methodology, we compiled a backlog of user stories and tasks that stemmed from current operational pain points and prioritized them based on business value and operational impact.

In the potlines, the main bottlenecks were related to data collection of pot measurements, actions taken and shift reports. Other challenges involved heavy consumption of paper due to metal tap ticket and daily work schedule creation. In the past liquid metal and bath height measurements were collected manually by a technician or supervisor and noted on paper and a chalkboard as shown in Fig. 3.

Furthermore, after the measurement activity was complete, they had to manually calculate the total liquid bath available within a section and specify the quantity to be poured or tapped within a pot as per the target bath level.

After communicating the abnormal bath-level pots to the bath-tapping vehicle operator, they would then proceed with the bath level pouring and tapping activity and restore it to equilibrium. The technician or supervisor would in parallel physically make their way to the potline control room to manually register the measurements into Reduction Potlines Management System (iRPMS)—EGA's current production database. This would allow them to calculate the total averages for the section and generate metal tap tickets based on the previous measurements for the incoming shift to execute.

Today the operations team can use SAQR to enter data directly from the shopfloor into the production database and automatically receive a bath pour/tap recommendation, generate a shift report, and create a tap ticket all with the touch of a button. An example of the metal and bath height measurement user interface is shown in Fig. 4. This transformation has halved the current process steps from 6 to 2 saving thousands of man-hours per year as well as reducing human error in collecting data by applying restrictions and validations within the application.

The metal tap ticket creation process which was mentioned earlier is the highest paper-intensive task within EGA operations today. More than 300,000+ sheets of paper are consumed annually to communicate the details of Aluminium tapping at the Potlines and distribution within the furnaces at the casthouse. An example of the current hot metal route in Al Taweelah is shown in Fig. 5.

Fig. 3 Current metal and bath height measurement activity workflow in Potline 2, Al Taweelah, EGA

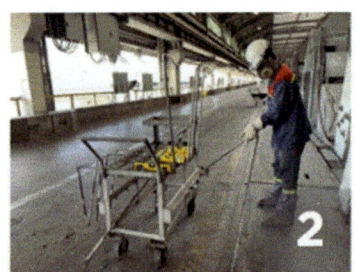

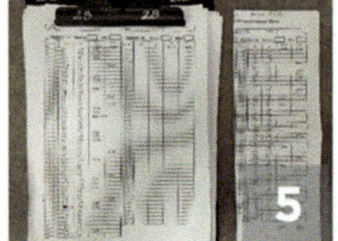

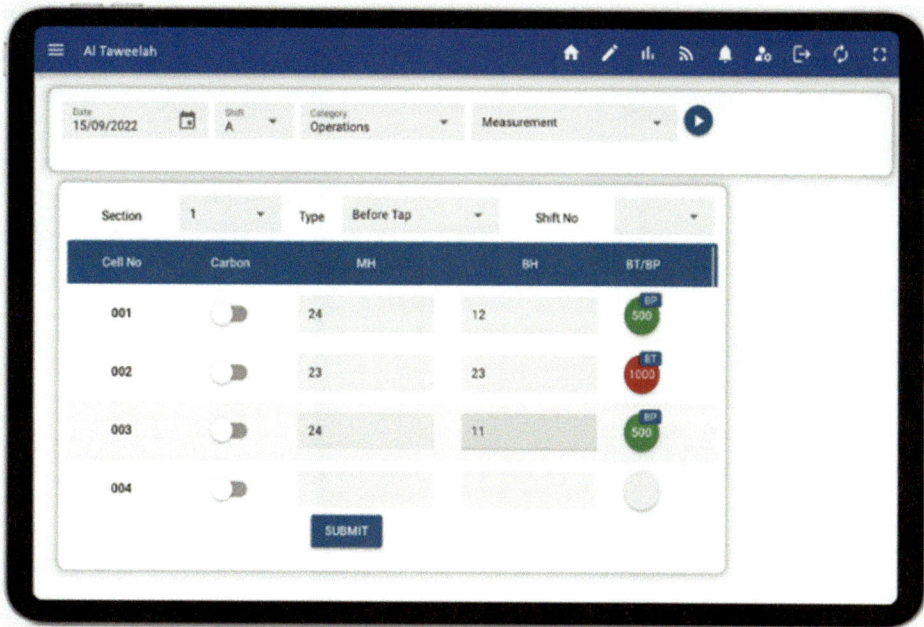

Fig. 4 Metal and bath height measurement user interface on SAQR

Fig. 5 Current hot metal route in Al Taweelah, EGA

Another challenge is that the hot metal coordinator (HMC) is only aware of the actual metal specification on route after the crucible has reached the first station where the analog data on the paper is transferred to a digital environment via a computer.

SAQR tackled this challenge by importing all tap ticket specifications from the production database and presenting them on a mobile user interface for the operator to view and execute. There is also a new feature on the pot programmable logic controller/human machine interface (PLC/HMI) to enter the metal crucible no. which will be mapped to the respective tap ticket and the driver simply has to collect the crucible and transfer it to the next station within the route.

Fig. 6 Crucible tracking application UI in operation

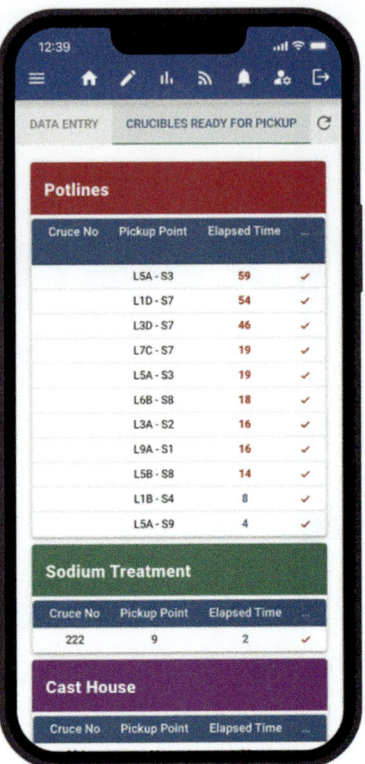

Fig. 7 Crucible tracking dashboard UI on SAQR

The metal-tapping operator has instant access to the work schedule and can easily track the status of each ticket.

Now, the HMC can immediately view which metal tap ticket is on route and begin the furnace allocation planning process ahead of time. Moreover, the previous requirement of manually creating tap tickets via paper and delivering them to the operator is redundant.

SAQR has also tackled the logistics and vehicle tracking pain points related to hot metal crucibles via a fully fledged crucible tracking application. In the past, there was no process or technology in place to track the location of crucibles other than experience on the job. This resulted in lots of bottlenecks via pending jobs, traffic jams and ultimately negatively impacting overall production. Conflict was another major side-effect which stemmed from lack of visibility or fact-based data.

As shown in Fig. 6, today, all crucible transfer vehicles (CTV) have been retrofitted with iPads where the driver can easily enter crucible no. information via voice recognition and the application will track the location at each station within the route and present the crucible no., location, expected time of arrival as well as the driver information.

The metal crucible dashboard can be seen in Fig. 7 and is currently visible to all relevant stakeholders which promotes visibility, accountability and collaboration while consequently eliminating conflict due to the facts being presented and documented.

Many pain points were also present in the process control itself, for example:

- It takes too long to get all required and available data to make informed decisions on efficiently managing the pots (recorded on site using a paper template, then transferred to systems when back in office)
- It requires a lot of effort to enter data digitally (multiple entries of the same data required in different platforms)
- It requires lots of paper and printing ink to record data (on site measurements still done on paper-based forms and templates)
- Many data entries are found to contain errors (e.g. typing errors, invalid measurements, missing data, data entered in wrong location etc.)

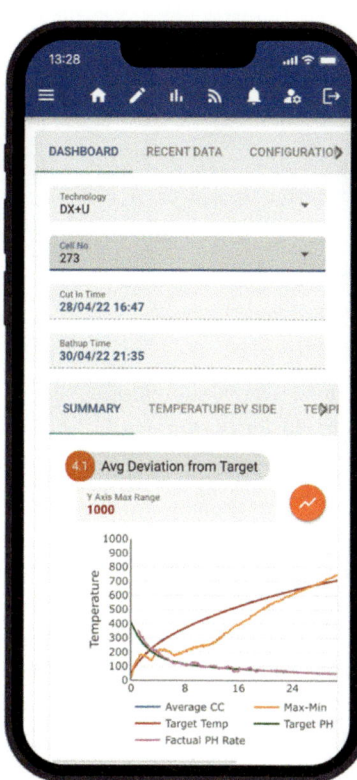

Fig. 8 Pot preheat status dashboard UI on SAQR

To address these challenges the following solutions were proposed:

- To think of new ways that will improve our way of work without compromising on safety, quality and productivity (i.e. minimizing waste by utilizing latest available technologies)
- To reduce data entry and effort (e.g. automated transfer, data entry validation screen)
- To gather the needed resources (form correct team with guided and focused efforts that will address the pain points)

- To develop a system that can offer on spot data capture, with automated data validation, and automated data transfer to all existing databases

A critical activity which affects a pot's life [2] and productivity is pot preheating. The pot's life can either extend or reduce either by optimum preheating or poor preheating.

One of the previous challenges with pot preheating was the lack of visibility and instant access to the latest temperature data. This data is crucial in monitoring the preheat progress and stability as well as determining when to introduce a liquid bath into the pot. The quality of preheated cathode is evaluated by many factors like the final average cathode surface temperature, the final cathode surface temperature distribution etc.

A preheat module was developed on SAQR which can be seen in Fig. 8 to address this gap and now the process control technicians are able to enter measurement data directly from the shop floor and the decision makers are able to have instant access to the latest temperature data resulting in optimum planning.

The data analysis and visualization process which was previously done manually by the process control engineer is now fully automated via SAQR and the final report can be exported into various formats such as PDF to share with relevant stakeholders or .csv to do some further exploratory data analysis and apply machine learning if required.

References

1. Sharma, Ashwani & Singh, Dr. (2020). Evolution of Industrial Revolutions: A Review. International Journal of Innovative Technology and Exploring Engineering. 9. 66–73. https://doi.org/10.35940/ijitee.I7144.0991120.
2. Ali, Mohamed. (2013). Preheating and start-up of prebaked aluminium reduction cells. JES. Journal of Engineering Sciences. 41. 2354–2364. https://doi.org/10.21608/jesaun.2013.114987.

Testing Feeding Alumina in Three Channels in a Wide Cell

Marc Dupuis and Valdis Bojarevics

Abstract

For years, Barry Welch has been promoting the idea of designing wider cells. Preliminary modelling work (Dupuis and Welch in Aluminium 93:45–49, 2017; Dupuis in Proceedings of the 37th international ICSOBA conference, 2019, 849–859) already demonstrated some of the advantages of using wider cells like reducing the external potshell area per electrolysis area ratio that helps the design of low energy consumption cells and reducing the potshell length to width ratio that helps designing even higher amperage cells without further increasing the potroom width and the crane span. A third advantage is to help reduce the formation of dissolved alumina concentration gradient in the bath by allowing the addition of two new alumina feeding channels. The MHD cell stability code MHD-Valdis including its alumina dissolution module has been used to study the impact of the addition of two extra longitudinal feeding channels in order to use three channels to feed alumina into the cell. Results obtained will be presented.

Keywords

Aluminium reduction • Electrolysis • Wide cell • Alumina dissolution • Modelling

Introduction

There is a kind of universal tendency to design bigger and bigger items like buildings, ships and planes to make them more efficient. This tendency equally applies to processes and the aluminium reduction cell is no exception, considering only the last 40 years, the maximum cell amperage about doubled from 300 to 600 kA. Since the anodic current density stayed more or less the same in the range of 0.75–0.95 A/cm^2, this means that the anode electrolysis surface area also about doubled from 35 to 70 m^2.

Before looking at how this increase of electrolysis surface area was achieved, let's look first on how bigger items are typically designed: they are made bigger by increasing uniformly all three spatial directions keeping more or less the same global aspect ratio. Since the blockage of the Suez Canal by the Ever Given mega containership in 2021 [1], the history of the evolution of the containership size became very well documented. Figure 1 from [2] presents the evolution from the very beginning of that type of ship in the 50s. It is clear that the containership capacity was augmented by increasing at the same time the ship length, width and height. Similar exercises could be done with other big items like planes leading to the same constatation.

Yet, when you look at the case of the aluminium electrolysis cell, it is not what happened in the last 40 years. The anode maximum length was kept more or less the same in the range of 1.6–1.8 m. Since the anode panel configuration was also kept the same with 2 rows of anodes and anode electrolysis surface area doubled, this means that the length of electrolysis area had to double from about 10.3 m to about 20.6 m. Assuming an anode width of 0.74 m, this translates to 28 anodes per side for a total of 56 anodes blocks per cell.

Barry Welch was the first to recognize that this strategy of increasing the productivity of aluminium electrolysis cell by increasing only its length had several limitations and he started to promote the idea of designing wider cells. This

M. Dupuis (✉)
Genisim Inc., 3111 Alger St., Jonquière, Québec G7S 2M9, Canada
e-mail: marc.dupuis@genisim.com

V. Bojarevics
School of Computing and Mathematics, University of Greenwich, 30 Park Row, London, 10 9LS, UK

© The Minerals, Metals & Materials Society 2023
S. Broek (ed.), *Light Metals 2023*, The Minerals, Metals & Materials Series, https://doi.org/10.1007/978-3-031-22532-1_13

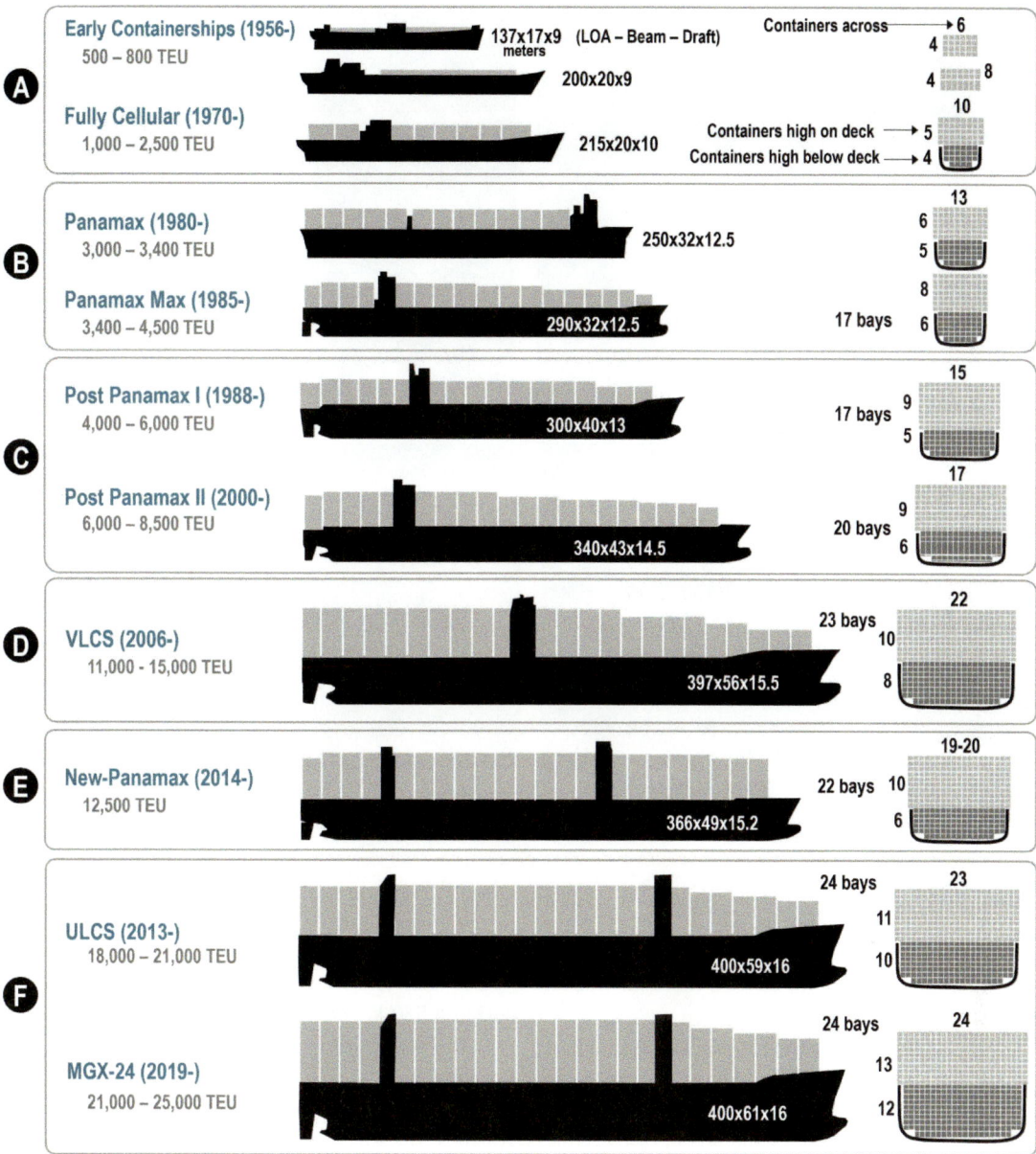

Fig. 1 Evolution of the containership size over the years from [2]

triggers among other things the production of the exploratory modelling work presented in [3]. That initial exploratory modelling work produced two wide cell designs, one operating at 762 kA and a second one operating at 1000 kA. This exploratory modelling work was continued eventually leading to a wide cell design operating at 530 kA and 10 kWh/kg [4]. Those wide cell designs will be briefly represented next.

Cell Layouts

HHCellVolt software [5] can be used to quickly generate cell layouts of different cell design proposals. The first one presented in Fig. 2a is from the dimensions of the generic regular width 600 kA cell presented in the introduction so a cell operating at 0.85 A/cm^2 using 56 anode blocks of

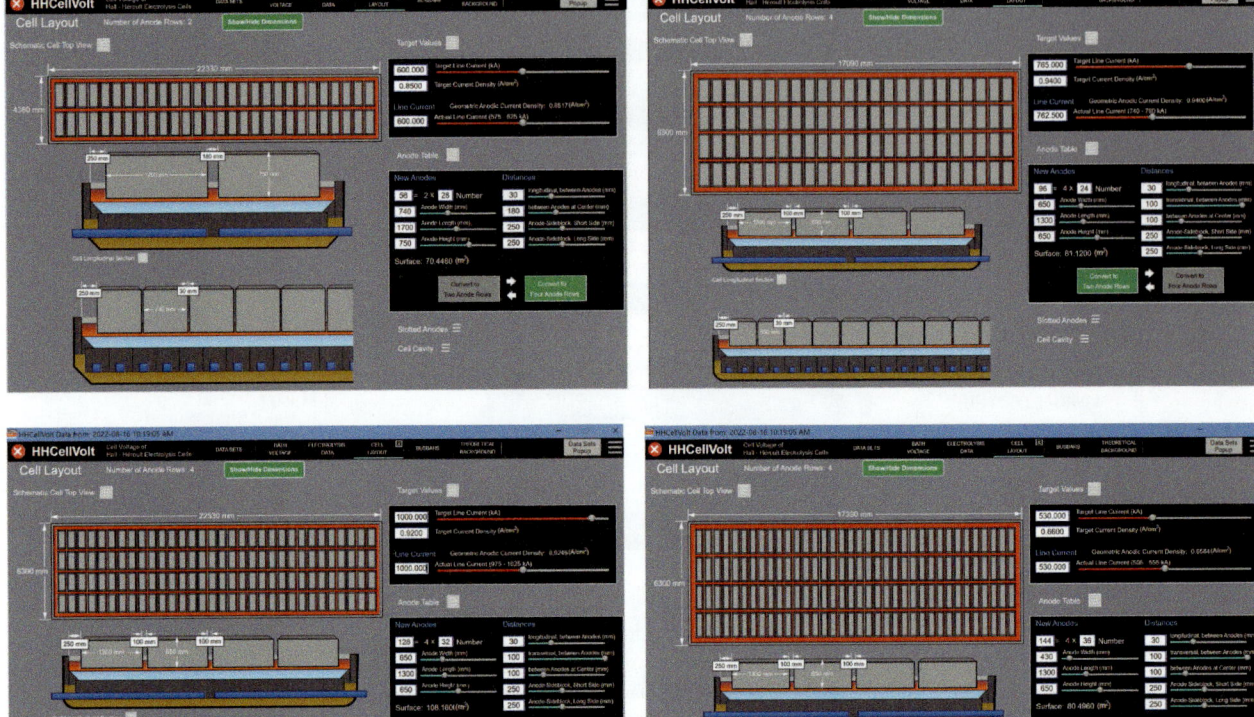

Fig. 2 **a** Generic regular width 600 kA cell layout, **b** Wide 762 kA cell layout as presented in [3], **c** Wide1000 kA cell layout as presented in [3], **d** Wide 530 kA, 10 kWh/kg cell layout as presented in [4]

1.7 m × 0.74 m. Such a cell would be a bit more than 22 m long in total.

Figure 2b presents the layout of the wide 762 kA cell first presented in [3]. The length to width aspect ratio of that wide cell is 2.71 while the aspect ratio of the generic regular width 600 kA cell is 5.1. Figure 2c presents the layout of the wide 1000 kA cell first presented in [3]. The length to width aspect ratio of the wide 1000 kA cell is 3.6.

The cell length to width ratio is important for several reasons, it affects the cell stability characteristics but most importantly it has an effect on cell heat loss characteristics and the bath homogeneity. As the main subject of this paper is the alumina feeding, that last aspect will be treated further down.

A second aspect ratio can be used to characterize the cell heat loss: the cell perimeter to cell electrolysis area ratio. That ratio is 0.76 for the regular width 600 kA cell ($2 \times (22.33 + 4.38)/70.45$) while it is reduced to only 0.58 for the wide 762 kA cell and 0.53 for the wide 1000 kA cell. Clearly, it is easier to reduce the specific cell energy consumption of a wider cell having proportionally less wall surfaces dissipating heat. This point was clearly expressed in [3] already.

For that reason, the wide cell concept was selected to design a 530 kA cell operating at 10 kWh/kg as presented in [4]. The anode design of that wide 10 kWh/kg cell is very different from the above two wide cell designs so the cell layout is different. It is presented in Fig. 2d. There are 144 1.3 m × 0.43 m anode blocks in that anode panel layout. Each anode is made of four anode blocks so there are 36 individual anodes in that cell design as presented in the next section.

Anode Designs

What is special about the anode design of wide cells is that the anode has to be made of at least two anode blocks, one in front and one in the back separated by a new type of channel that will be used to feed alumina as we will see below. This is a key part of Barry Welch's wide cell design concept. The original anode concept presented in [3] is presented in Fig. 3 which is Fig. 3 of [3].

The channel width between the two anode blocks was set to 6 cm in the anode model, while it was set to 10 cm in the

Fig. 3 Wide 762 kA anode design and predicted voltage drop, Fig. 3 in [3]

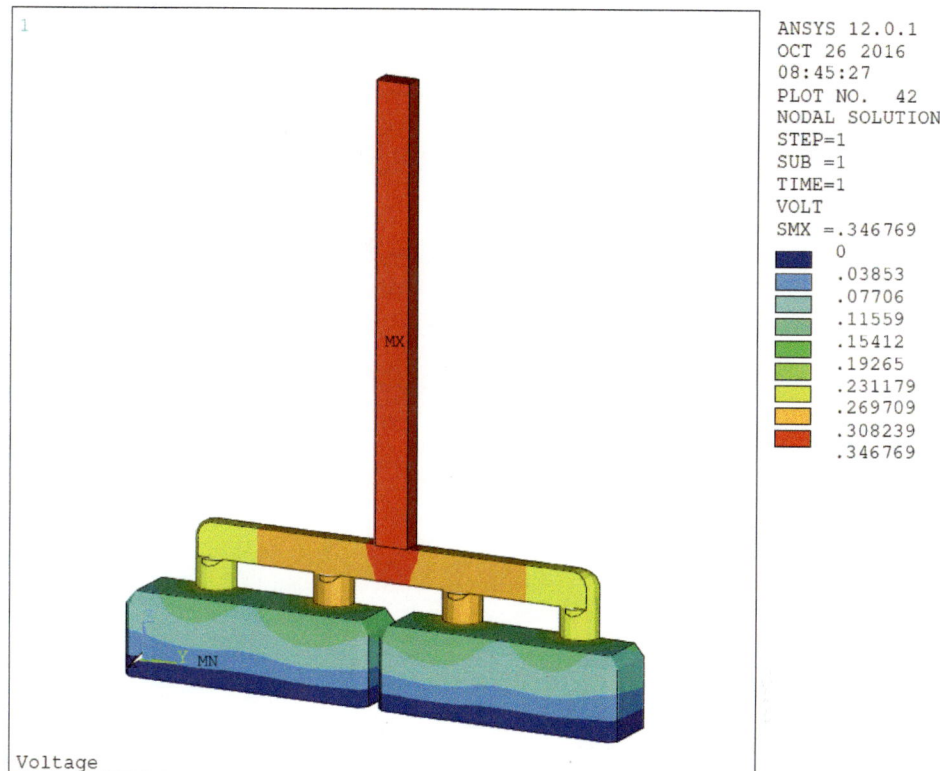

layouts presented above. The width of the 2 new longitudinal channels created to allow alumina feeding in new locations has not been optimized as this cannot be done using existing models. Wider channels help alumina dispersion but further increase the cell width and hence the pot-to-pot distance.

The anode concept used in [3] was good enough to produce cell designs operating at about 12.8 kWh/kg but a completely new anode topology was required to be able to operate a cell at 11.0 and eventually 10.0 kWh/kg. That new anode design is presented in Fig. 4, each anode rod feeds current to four anode blocks through a very long anode yoke having a significant section of copper in it and a total of 12 stubs.

Cathode Designs

One of the main factors limiting the design of very wide cells is the intense horizontal current in the metal pad that it would generate using conventional cathode assembly with regular steel collector bars. A more conductive collector bar must be used in order to address that issue. This is fortunately not a problem anymore since the introduction of copper insert and 100% copper bar cathode assembly designs [6, 7]. The cathode design used in [3, 4] rely on a massive copper insert to very significantly reduce the

horizontal current in the metal pad even in the case of this very long cathode assembly.

The analysis of the massive copper insert cathode assembly was first preformed in [8], where Fig. 5 presents the cathode assembly design showing the massive copper insert. Figure 6 shows the full cell quarter thermo-electric model mesh used to calculate the metal pad current density for the case of a regular width 500 kA cell presented in [8]. Figure 7 unveils the resulting 3 components of the horizontal current showing that very little transversal horizontal currents (CDY) are present.

The same cathode assembly concept was used for the wide cell designs presented in [3, 4], where Fig. 8a presents the obtained cathode current density for the 762 kA wide cell. The metal pad current density is not available for this cell design but the current density on the cathode surface gives a good enough indication of the efficiency of the cathode assembly design to reduce metal pad horizontal current density.

For the 530 kA, 10 kWh/kg wide cell design, the requirements for the cathode design are even more stringent as the cathode voltage drop and the cathode heat loss must be reduced to levels way below what is obtained using conventional cathode designs. Figure 8b presents the obtained cathode current density for the 530 kA, 10 kWh/kg wide cell design.

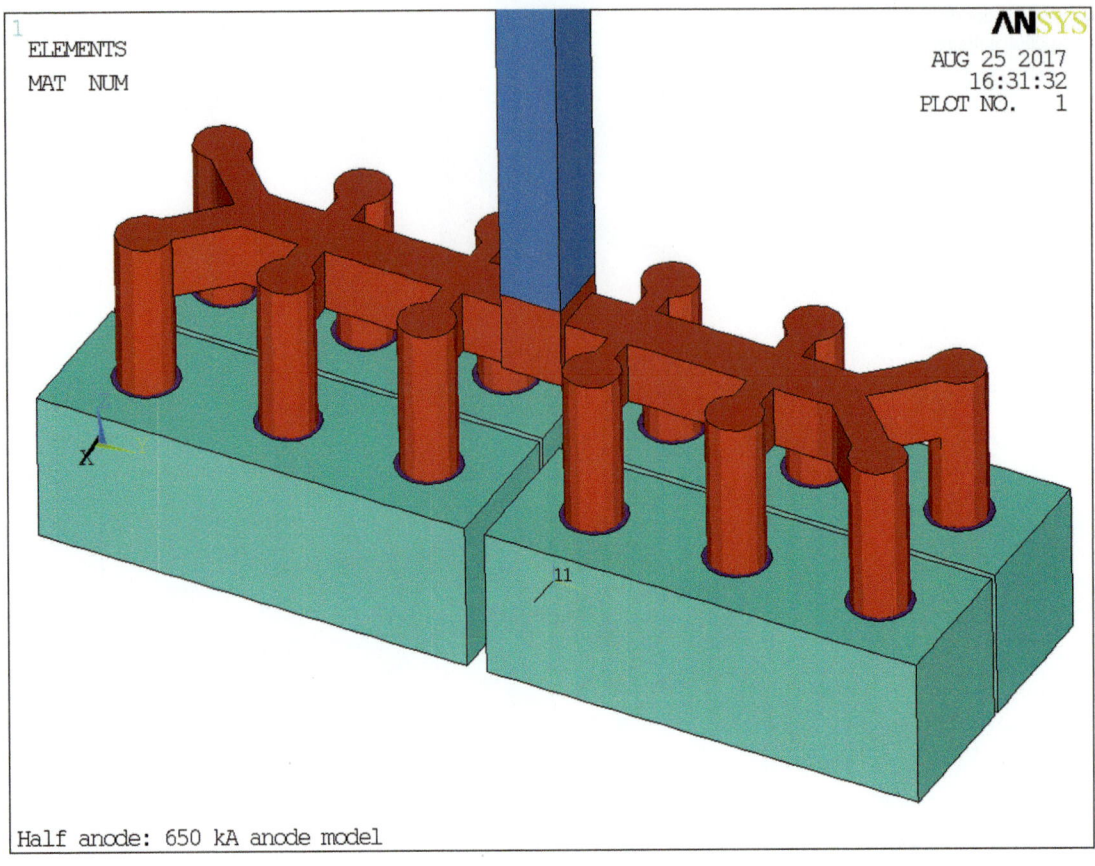

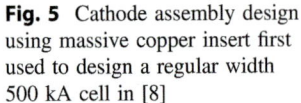

Fig. 4 Four anode blocks and a total of 12 stubs anode design for the 650 kA, 11 kWh/kg cell and later the 530 kA, 10 kWh/kg cell

Fig. 5 Cathode assembly design using massive copper insert first used to design a regular width 500 kA cell in [8]

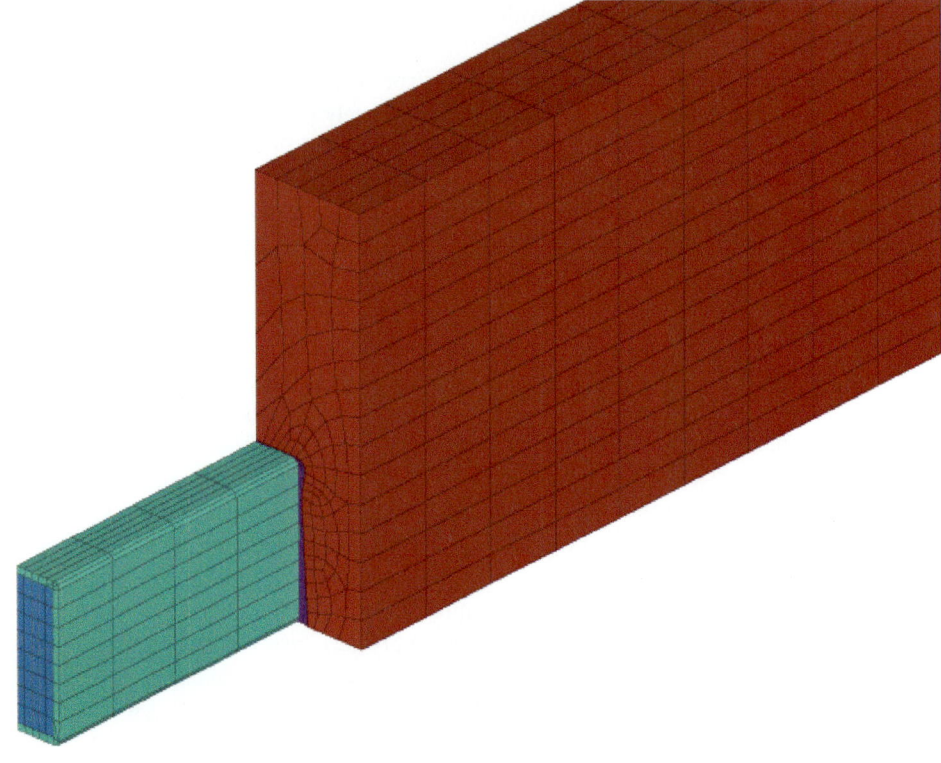

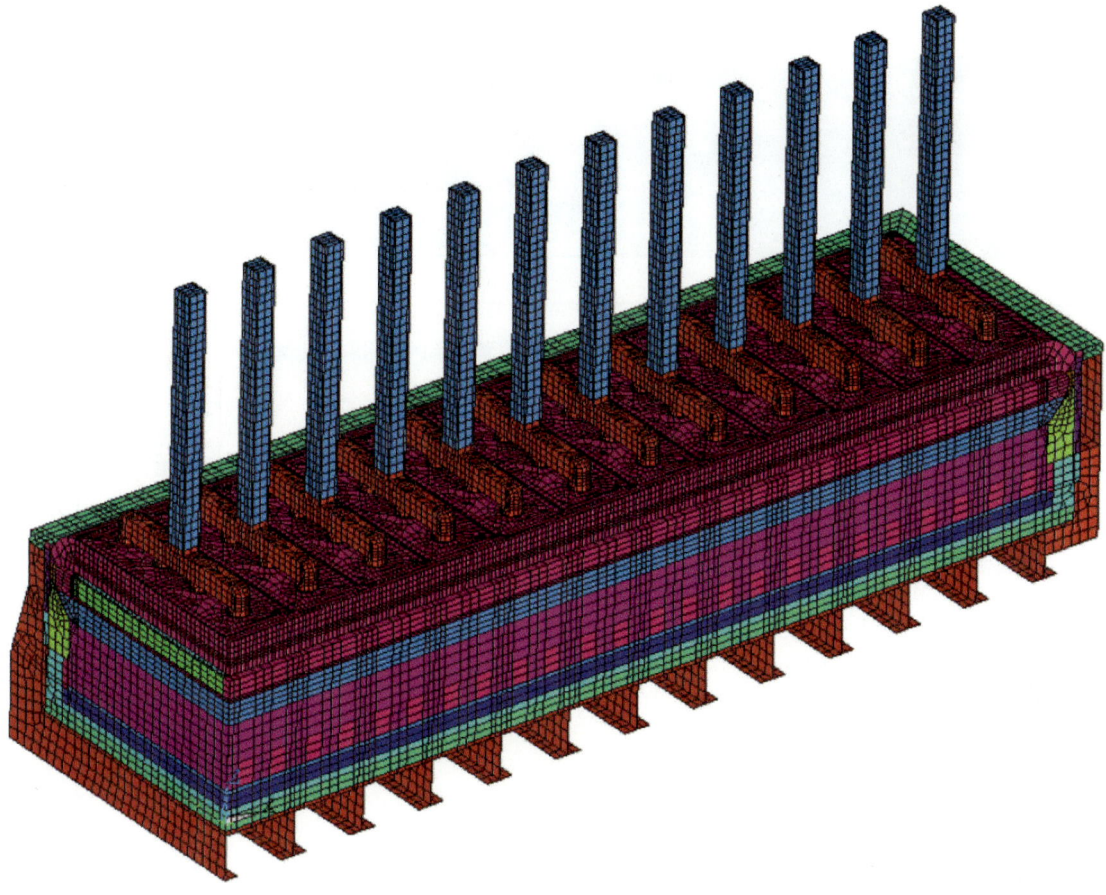

Fig. 6 Full cell quarter thermo-electric model mesh of a regular width 500 kA cell, Fig. 8 in [8]

Busbar Designs

The final part of the wide cell design, the busbar design, will be reviewed here. Several busbar design concepts could have been used for the wide cells, namely ICC, ECC, CCC and RCC concepts that have all been reviewed in [9]. The RCC (Reversed Compensation Current) busbar design concept was selected in [3] because it is easy to implement and it is modular, and the 1000 kA wide cell design using one extra module as compared to the 762 kA wide cell design. The 650 kA, 11 kWh/kg wide cell presented in [10] was used instead of the 530 kA, 10 kWh/kg wide cell to carry on the new alumina feeding study. Both cells are retrofits of the original 762 kA wide cell so they are using the same RCC busbar design concept with three modules. Figure 9 is presenting the 650 and 1000 kA cell busbar layouts. The busbar sections are much bigger in the 650 kA cell in order to minimize the external busbar voltage drop, which in turn helps to reduce the cell voltage and hence ultimately the cell energy consumption.

That RCC busbar design concept is using alternating upstream and downstream risers that produce a globally flat but wavy Bz (vertical component of the magnetic field) as seen in Fig. 10, an antisymmetric Bx (longitudinal horizontal component of the magnetic field) and an upstream/downstream symmetric bath/metal interface deformation, as seen in Fig. 11. Finally, the busbar configuration produces a very unique metal flow pattern characterized by multiple small recirculations as displayed in Fig. 12.

Regular Center Channel Alumina Feeding with and Without Physics of Raft Formation

The two bath flow solutions corresponding to the metal flow solutions presented in Fig. 12 will be used to model the alumina dispersion and dissolution using MHD-Valdis dissolution module previously developed [11–14]. The physics of the raft formation, displacement in the centre channel and eventual dissolution was added in [13] and that new physics was already tested on the 650 and 1000 kA wide cell designs in [14]. The present work is the continuation of that investigation.

Since the August 2022 code version of MHD-Valdis used in this study does not support the addition of two extra

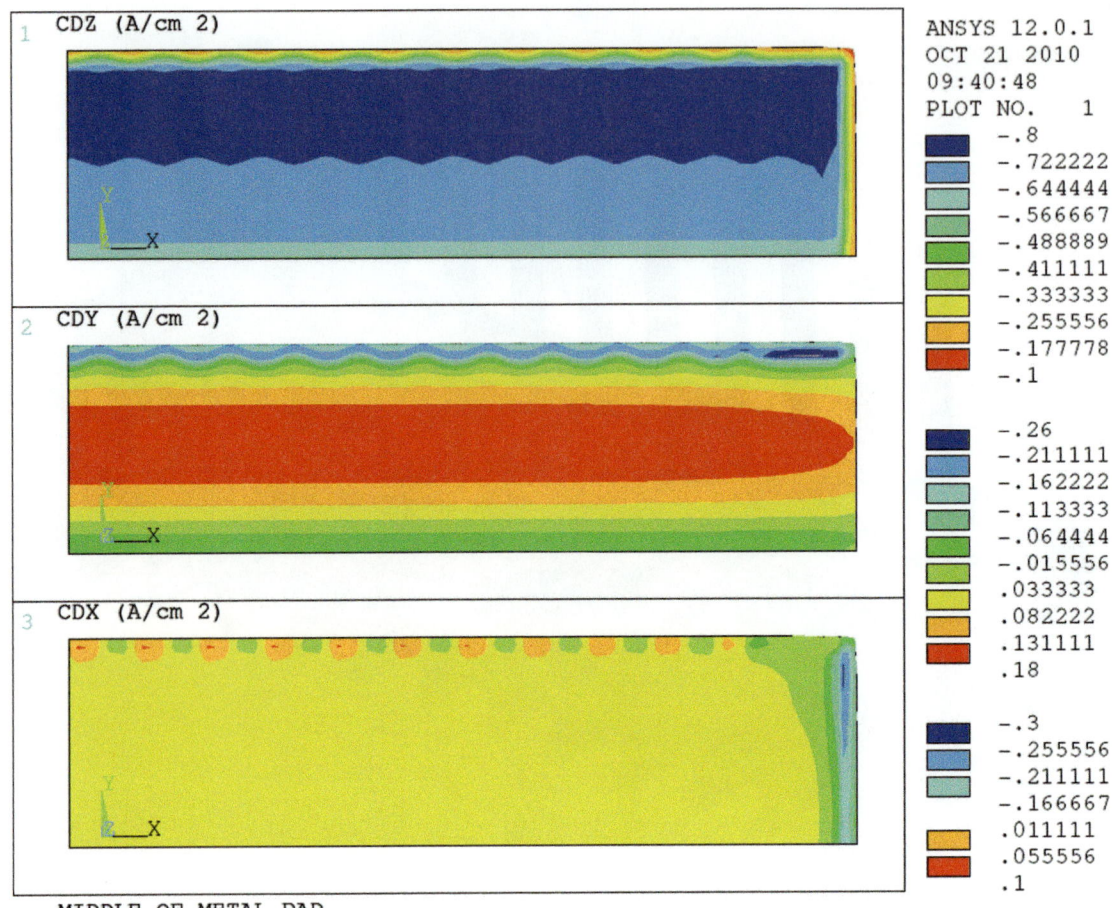

Fig. 7 Resulting horizontal current density in the middle of the metal pad for the 500 kA cell design presented in [8]

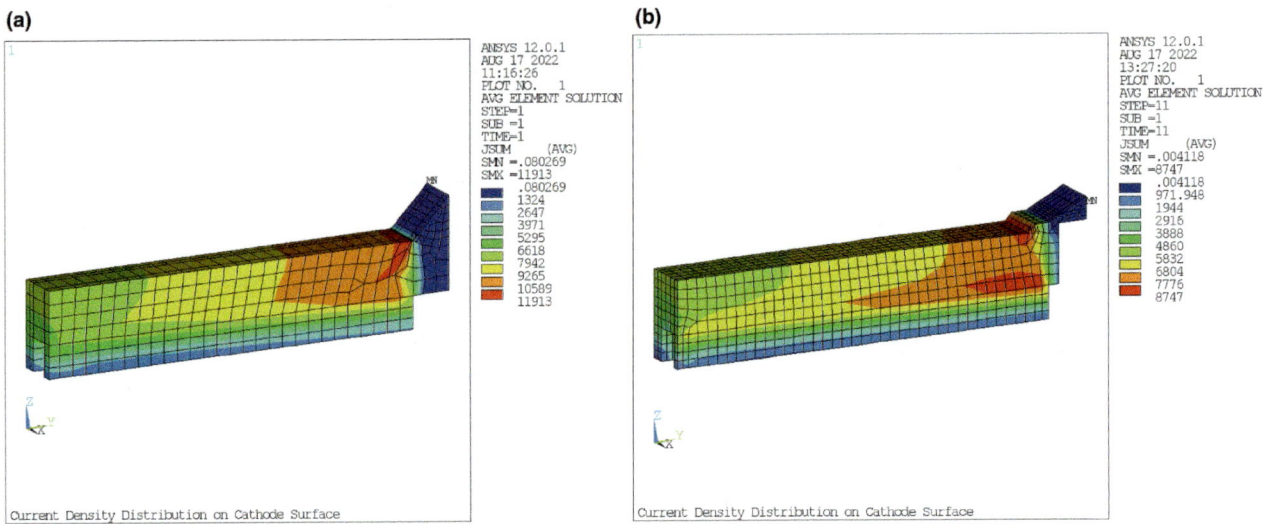

Fig. 8 **a** Current density in the cathode block for the 762 kA wide cell design presented in [3]. **b** Current density in the cathode block for the 530 kA wide cell design presented in [4]

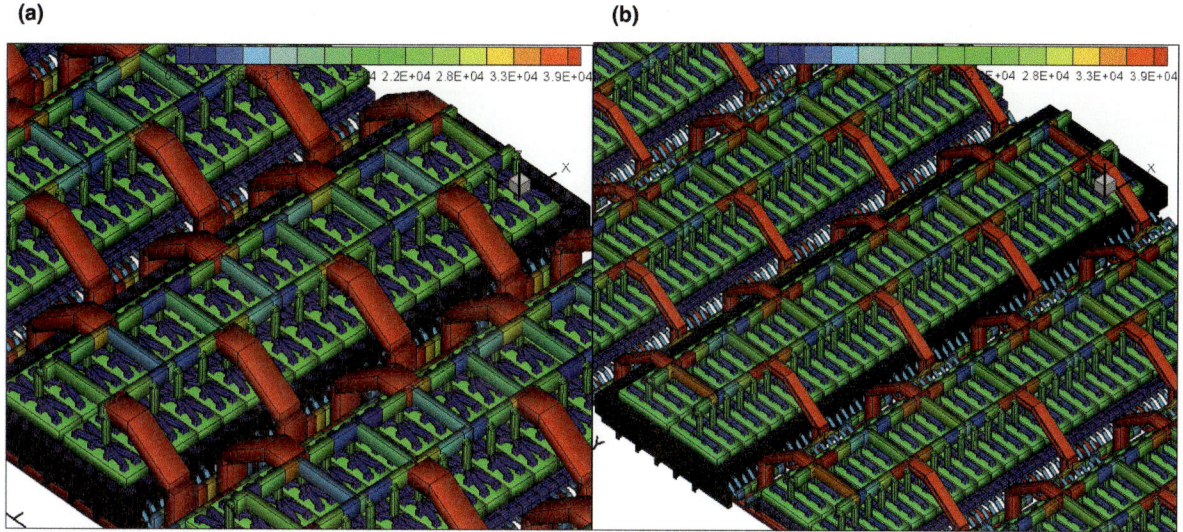

Fig. 9 **a** RCC busbar design layouts for the 650 kA cell, **b** RCC busbar design layouts for the 1000 kA cell

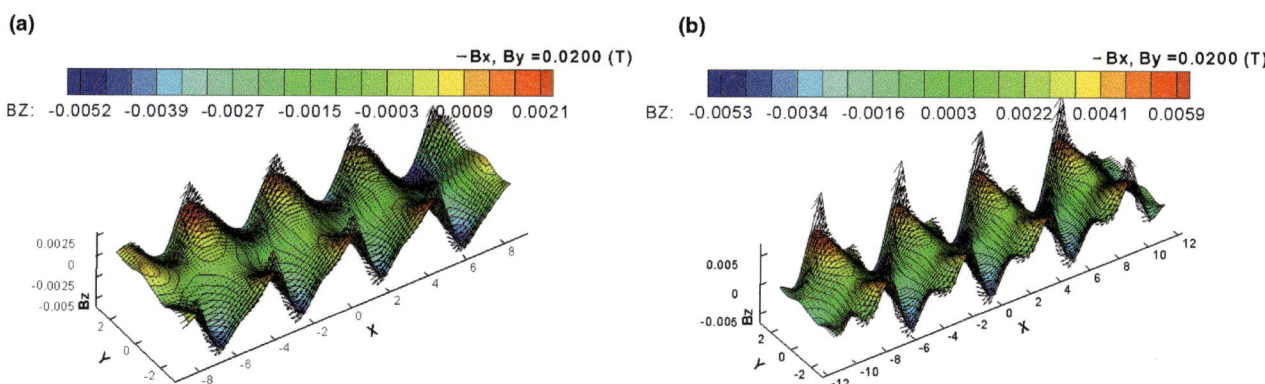

Fig. 10 **a** Vertical component of the magnetic field for the 650 kA, **b** Vertical component of the magnetic field for the 1000 kA cell

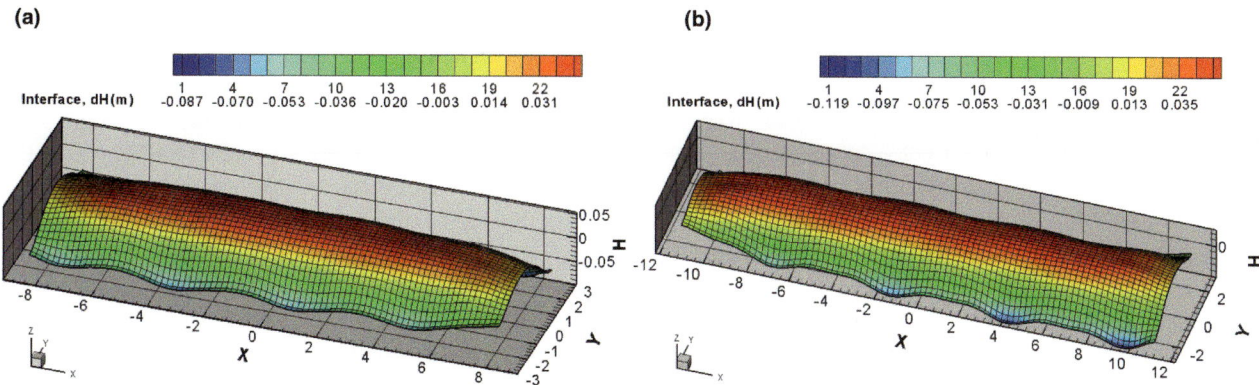

Fig. 11 **a** Bath/metal interface deformation for the 650 kA cell, **b** Bath/metal interface deformation for the 1000 kA cell

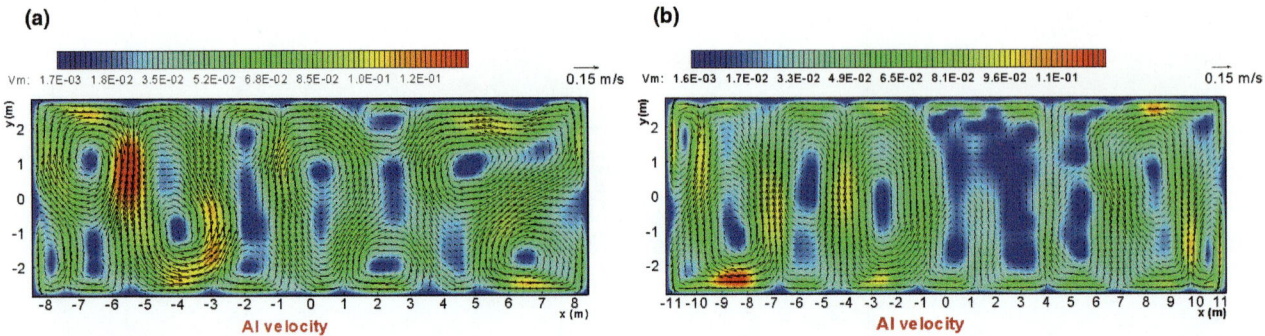

Fig. 12 **a** Metal flow for the 650 kA cell, **b** Metal flow for the 1000 kA cell

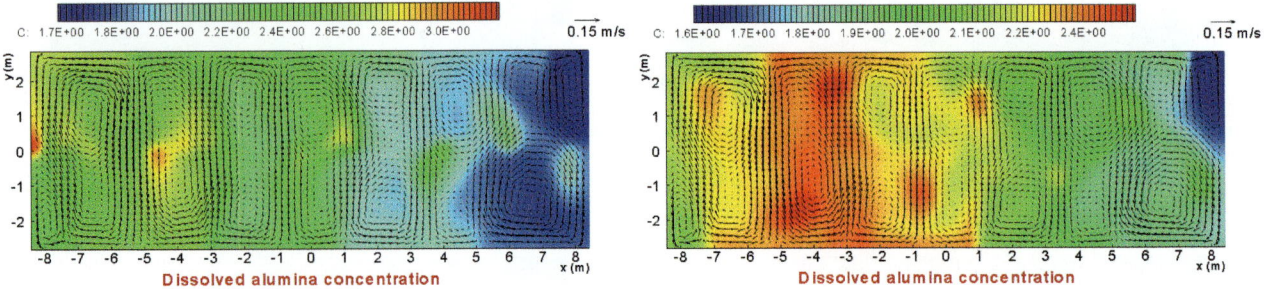

Fig. 13 Dissolved alumina concentration after 1 h of nominal feeding from 8 feeders in the centre channel for the 650 kA wide cell, with raft formation on the left and without raft formation on the right

longitudinal channels where rafts could form and move along before dissolving, the first step of the current study is to compare the dissolved alumina gradients in the bath obtained with and without raft formation. Figure 13 presents the results after 1 h of nominal alumina feeding using 8 feeders located in the centre channel starting from an homogeneous concentration with and without raft formation for the 650 kA, 11 kWh/kg wide cell design.

With physics currently implemented, the rafts are added at the feeder locations, they can move away from those locations in the centre channel following the bath flow and they eventually directly dissolve. Smaller alumina particles fed at the same time are free to follow the bath flow under the anodes, and they also gradually dissolve as they move around. Without the raft formation, 100% of the alumina particles added are free to follow the bath flow under the anodes from the feeding locations.

Results presented in Fig. 13 demonstrate that raft dissolution at fixed locations produces higher alumina concentration gradients. Yet other features are similar: the global feeder distribution produces a global alumina concentration gradient as previously observed in [13] and depending on bath recirculation positioning relative to feeder locations, where some bath recirculation can receive very little alumina supply and hence experiencing very low alumina concentration. It is the case of the downstream right corner zone for the 650 kA wide cell.

The same exercise was repeated for the 1000 kA wide cell. Figure 14 presents the results obtained after 1 h of nominal alumina feeding using 8 feeders located in the centre channel starting from a homogeneous concentration with and without raft formation for the 1000 kA wide cell.

Similar observations can be made for the 1000 wide cell, where greater alumina dissolution gradients are present when raft formation is activated and zones of higher and lower concentration are present in both solutions.

When comparing the 650 and 1000 kA solutions without raft formation, the 1000 kA case is a bit more homogeneous, possibly due to a more appropriate feeders positioning in the centre channel. Notice that very little effort has been made to optimize the feeder's positioning.

Three Channels Alumina Feeding Without Physics of Raft Formation

Since the current version of MHD-Valdis does not support the option to add two extra feeding channels, it is only possible to model the impact of adding feeders away from the centre channel without activating the physics of raft formation. This was done in the second step of the current study and results are reported in this final section for both the 650 ka and the 1000 kA wide cells.

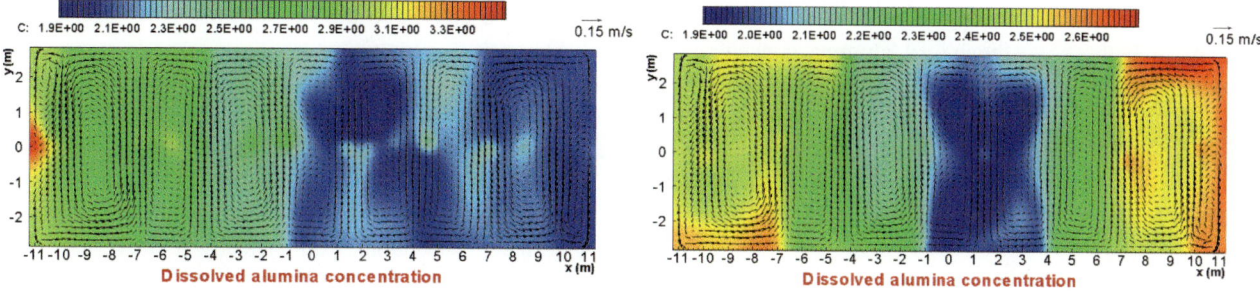

Fig. 14 Dissolved alumina concentration after 1 h of nominal feeding from 8 feeders in the centre channel for the 1000 kA wide cell, with raft formation on the left and without raft formation on the right

In both cases, 14 feeders were used, five were located in the new downstream channel, four were located in the centre channel and five were located in the new upstream channel. Notice that as the current version of MHD-Valdis does not take into consideration the presence of those two extra channels, feeders could have been located anywhere. Some efforts have been done to optimize those 14 feeder locations but not an extensive one. In the selected 14 feeders configuration, there are 14 feeder locations to move around and each try requires a bit less than one hour CPU to simulate one hour of nominal feeding.

Figure 15 presents the last two feeder positioning optimization trials for the 650 kA wide cell case. Very small feeder repositioning can have a significant impact on the obtained alumina concentration solution. Yet, it is important to remember that the solution obtained is not static and that the solution is highly dependent on the bath flow used and on the physics of alumina dispersion and dissolution selected, so spending time on a more rigorous optimization procedure would be premature.

A similar partial 14 feeders localization optimization was done for the 1000 kA wide cell case, Fig. 16 presents the last two feeder positioning optimization trial solutions.

Globally, comparing the alumina concentration results obtained using one row versus three rows of feeders without the physics of raft formation, that comparison clearly indicates that Barry Welch was quite right to speculate that adding two additional rows of feeders would greatly

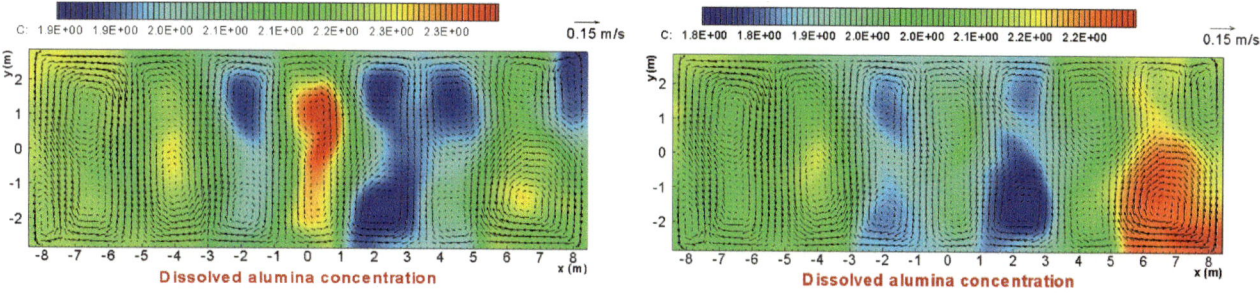

Fig. 15 Dissolved alumina concentration after 1 h of nominal feeding from 14 feeders in total (5–4–5) for the 650 kA cell without raft formation for 2 different sets of feeder positions

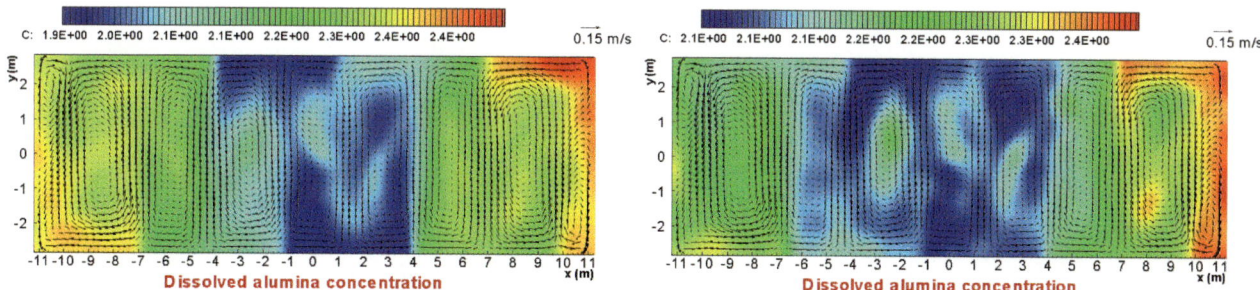

Fig. 16 Dissolved alumina concentration after 1 h of nominal feeding from 14 feeders in total (5–4–5) for the 1000 kA cell without raft formation for 2 different sets of feeder positions

contribute to reduce the alumina concentration gradients in the bath. In both cases, the global alumina concentration variation was reduced about by half from about 0.8% to about 0.4%.

Future Work

The present results are useful to give an idea of the potential of adding two extra channels to be able to feed alumina in three channels instead of one, but they cannot be considered definitive. Pending the development of a new version of MHD-Valdis that supports multiplefeeding channels anode panel geometry, it is not possible to activate the physics of raft formation. Furthermore, as presented in [13], in reality, cell rafts do not directly dissolve, instead they dismantle releasing the small alumina particles initially dumped by the feeders. A new release of the code MHD-Valdis that supports a different physics for the raft dismantling is also pending. Finally, as it is well known that the bath flow is also affected by the bubble release drag force, a new release of the code supporting the bubble release drag force is also pending.

Eventually, when a MHD-Valdis code release supporting all that extra physics becomes available, the current study of the impact of adding two extra channels in order to feed alumina in three channels instead of only one in the centre of the cell should be repeated to have a better evaluation of how it would impact the bath dissolved alumina concentration homogeneity.

Conclusions

For the first time, a modelling investigation of Barry Welch's idea to feed alumina in three channels in a wide cell has been performed.

Two wide cells have been investigated, one operating at 650 kA and 11.0 kWh/kg and another one operating at 1000 kA and 12.8 kWh/kg. In both cases, 8 feeders located in the centre channel are replaced by 14 feeders, 5 feeding in the new upstream side channel, 4 still feeding in the centre channel and 5 feeding in the new downstream side channel (5–4–5).

In both cases, the global concentration variation has been reduced about by half from about 0.8% to about 0.4% when the physics of the raft formation is deactivated.

This preliminary study is hence confirming Barry Welch's intuition that adding two extra feeding channels to disperse feeding locations would greatly improve bath dissolved alumina concentration homogeneity.

References

1. "2021 Suez Canal Obstruction, Wikipedia, https://en.wikipedia.org/wiki/2021_Suez_Canal_obstruction
2. Evolution of Containerships, The Geography of Transport Systems, https://transportgeography.org/contents/chapter5/maritime-transportation/evolution-containerships-classes/
3. Marc Dupuis and Barry Welch, Designing Cells for the Future – Wider and/or even Higher Amperage?, ALUMINIUM, 93 (1/2), 2017, 45-49.
4. Marc Dupuis, Second Attempt to Break 10 kWh/kg Energy Consumption Barrier Using a Wide Cell Design, Proceedings of the 37th International ICSOBA Conference, 2019, 849–859.
5. Peter Entner, About HHCellVolt, https://peter-entner.com/apr18_files_index/hhcellvolt_pamphlet.pdf
6. Mustafa Anwar Mustafa and al., Design and Performance of a Full Copper Collector Bar Pot at EGA, Proceedings of the 39th International ICSOBA Conference, 2021, 803–817.
7. Markus Pfeffer and al., Ready-to-Use Cathodes in High-Amperage Technologies, TMS Light Metals, 2022, 946–963.
8. Marc Dupuis and Valdis Bojarevics, Retrofit of a 500 kA Cell Design into a 600 kA Cell Design, ALUMINIUM 87 (2011) 1/2, 52–55.
9. Marc Dupuis, A New Aluminium Electrolysis Cell Busbar Network Concept, Proceedings of 33rd International ICSOBA Conference, 2015, 699–708.
10. Marc Dupuis, Breaking the 11 kWh/kg Al Barrier, ALUMINIUM, 94 (1/2), 2018, 48-52.
11. Valdis Bojarevics, Dynamic Modeling of Alumina Feeding in an Aluminium Electrolysis Cell, TMS Light Metals, 2019, 675–682.
12. Valdis Bojarevics, In-Line Cell Position and Anode Change Effects on the Alumina Dissolution, TMS Light Metals, 2021, 584–590.
13. Valdis Bojarevics and Marc Dupuis, Advanced Alumina Dissolution Modelling, TMS Light Metals, 2022, 339–348.
14. Marc Dupuis and Valdis Bojarevics, Testing the New MHD-Valdis Alumina Dissolution Module, ALUMINIUM 98 (2022) 1/2, 60–62.

A Pragmatic Model for Bath Temperature Evolution During Alumina Feeding

Kurian J. Vachaparambil, Stein Tore Johansen, Asbjørn Solheim, and Kristian Etienne Einarsrud

Abstract

In this paper, we extend the pragmatic model for alumina feeding, presented in Johansen et al. Light metals 2022, to treat the associated evolution of bath temperature. The underlying idea of the framework is to reduce its computational overhead to allow for very fast simulation of the dissolved alumina, particle alumina, and bath temperature evolution within aluminium reduction cells. Two main advances presented in this work: (1) a generic method to read velocity data from any CFD simulation to the pragmatic model and (2) inclusion of governing equation to simulate the bath temperature evolution. The physics associated with the various heat loss modes from the bath as well as the temperature dependence of the alumina solubility in the bath is represented in the model. The demonstration cases discussed in this work show that the proposed framework runs much faster than real time (up to 75 times). The average bath and metal temperature in the demonstration case is observed to depend on the feeding pattern used to add alumina into the bath. The proposed framework can be used to observe the spatial and temporal changes of dissolved and particle alumina as well as the bath temperature within a cell.

K. J. Vachaparambil (✉) · S. T. Johansen · A. Solheim
SINTEF Industry, 7465 Trondheim, Norway
e-mail: kurian.vachaparambil@sintef.no

S. T. Johansen
e-mail: stein.t.johansen@sintef.no

A. Solheim
e-mail: asbjorn.solheim@sintef.no

K. E. Einarsrud
Department of Materials Science and Engineering, Norwegian University of Science and Technology (NTNU), 7491 Trondheim, Norway
e-mail: kristian.e.einarsrud@ntnu.no

Keywords

Pragmatic model • Hall–Heroult cell • Alumina distribution • Coarse graining • Real time • Digital twin

Introduction

Industrial production of aluminium is based on the Hall–Heroult process. This is an electrochemical process that results in the production of molten aluminium at the cathode and CO_2 gas at the anode. In this process, particle alumina, which is fed via feeders, dissolves in the molten bath (which is essentially molten cryolite). The dissolution process of alumina in the bath is very important—as the very low value can lead to low/high-voltage anode effects [3], whereas high amounts of dissolved alumina lead to sludging [9]. Additionally, maintaining the operational temperature (950–970 °C [9]) of the bath is complex as bath temperature can vary locally as a result of the alumina dissolution process, heat loss to the surroundings, and the ohmic heating of the bath due to current flow in addition to melting/freezing [6, 8]. With the increase in the amperage of the cells used in industrial applications, which is associated with larger anodes to maintain the current densities and reduced distance between anode and cathode, understanding the dissolution and transport of alumina in the bath and its coupling to the energy balance of the cell is critical for improving productivity and cell stability.

Although CFD can be used to study this in detail, cell scale simulations can become prohibitively expensive—due to the coupled multiscale and multiphysics nature of the phenomena. A pragmatic model makes use of physics-based modelling approaches along with domain-specific assumptions to reduce the computational overhead for a cell scale system simulation. In the domain of aluminium reduction cells, several works in literature have explored the pragmatic modelling approaches. [10] developed a pragmatic dynamic model to simulate the energy balance of the aluminum reduction cell.

© The Minerals, Metals & Materials Society 2023
S. Broek (ed.), *Light Metals 2023*, The Minerals, Metals & Materials Series,
https://doi.org/10.1007/978-3-031-22532-1_14

The work by [1, 2] used a shallow water approach to develop a model that can simulate alumina dissolution which is decoupled from the bath temperature evolution. And our previous work [5], which developed a coarse grained model to simulate alumina dissolution decoupled from bath temperature.

This paper will describe and demonstrate an extension of the model presented in [5] to simulate the alumina dissolution and its coupling to the bath temperature with low computational overhead. The underlying idea behind the development of this model is to use it as a hybrid digital twin of the cell for control and fast look-ahead applications which typically requires the framework to be as high as 10–1000 times faster than real time.

Model Description

The model assumes a steady velocity field $\vec{U}$ which can be obtained from CFD simulations (or from measurements if good quality data is available). This velocity field is the time averaged bath flow velocities considering effects of magneto-hydrodynamics, bubble evolution, and turbulence. Since the mesh used by CFD can vary in the type of cells used, we have developed a generic approach to read this data. The interpolation is based on the Inverse Distance Weighted (IDW) approach which computes value at an unknown position based on known values in the 'neighbourhood'. For example, the value of the velocity component along the X direction (U^*) at an arbitrary position $\vec{x}$ is given by

$$U^*(\vec{x}) = \frac{\sum_{p=1}^{p=N} U_p w_p}{\sum_{p=1}^{p=N} w_p}, \qquad (1)$$

where w_p is calculated as $1/d(\vec{x}, \vec{x}_p)^m$ with $d(\vec{x}, \vec{x}_p)$ defined as the distance between the arbitrary position ($\vec{x}$) and the known data point ($\vec{x}_p$), and m is known a the power parameter (larger the value of m, larger is the effect of the points positioned nearby on the point at which data is being interpolated to). At any cell face where the velocity field is stored, the value is assigned as an average between four points that lie on the same plane which for the X component of velocity is computed as

$$U^*(x, y, z) =$$
$$\frac{U^*(x, y + \frac{\Delta y}{2}, z + \frac{\Delta z}{2}) + U^*(x, y - \frac{\Delta y}{2}, z + \frac{\Delta z}{2})}{4}$$
$$+ \frac{U^*(x, y + \frac{\Delta y}{2}, z - \frac{\Delta z}{2}) + U^*(x, y - \frac{\Delta y}{2} y, z - \frac{\Delta z}{2})}{4}.$$

where Δy and Δz are the grid spacing along the Y and Z directions. Based on the averaged interpolated velocity field obtained, the dispersion or deviation velocities can be computed as the root mean square of the deviation of the averaged interpolated velocity field and the four points in the same plane from which the interpolation is performed, which for the deviation velocities in the X direction is

$$U^+(x, y, z) = \sqrt{0.25 \sum_{i=1}^{i=4} (U_i^* - U^*)^2}. \qquad (2)$$

The averaged and the deviation velocities for a given grid are computed just once and saved for use by the framework when required.

As a result of the interpolation of the CFD velocity field to the coarse grid of the pragmatic model, the obtained velocity field is not mass conserving. In order to obtain the divergence-free velocity ($\langle \vec{U} \rangle$), we introduce a correction potential (ϕ) which is defined as

$$\langle \vec{U} \rangle = \vec{U}^* - \nabla \phi. \qquad (3)$$

Since the mass conserving velocity field should satisfy $\nabla \cdot \langle \vec{U} \rangle = 0$, applying divergence operator on both sides of Eq. 3 gives

$$\nabla^2 \phi = \nabla \cdot \vec{U}^* \qquad (4)$$

Once ϕ is computed, the mass conserving velocity ($\langle \vec{U} \rangle$) is computed based on Eq. 3.

The transient evolution of particle alumina is governed by

$$\frac{\partial}{\partial t}(\rho_b X_p) + \nabla \cdot (\rho_b \vec{U} X_p) = \nabla \cdot (\rho_b \Gamma_T \nabla X_p) + S_{feed} - S_{MT}, \qquad (5)$$

where ρ_b is the mass density of the bath, S_{feed} accounts for the particle alumina added via feeders into the bath, Γ_T is a turbulent diffusivity, and X_p is the mass fraction of the particles, see [5]. In Eq. 5, S_{MT} is the dissolution rate, based on single particle of alumina (of size D_p) [4], whose computation is described based on the solubility of alumina in the bath, Sherwood number (Sh) correlation and terminal velocity of the settling particle; see [5] for details. In this work, the solubility of alumina in the bath (X_s) is calculated as a function of temperature and bath composition based on [7]

$$X_s = \frac{1}{100} A \left(\frac{T - 273.15}{1000} \right)^B \qquad (6)$$

where T is the temperature of the bath (in K), A and defined as

$$A = 11.9 - 0.062[AlF_3] - 0.0031[AlF_3]^2 - 0.5[LiF]$$
$$- 0.2[CaF_2] - 0.3[MgF_2] + \frac{42[LiF][AlF_3]}{2000 + [LiF][AlF_3]}$$

$$B = 4.8 - 0.048[AlF_3] + \frac{2.2[LiF]^{1.5}}{10 + [LiF] + 0.001[AlF_3]^3},$$

where terms in the square brackets indicate the weight percent of the compound. The dissolved alumina mass fraction (X_d) is governed by

$$\frac{\partial}{\partial t}(\rho_b X_b) + \nabla \cdot (\rho_b \vec{U} X_b) = \nabla \cdot (\rho_b \Gamma_T \nabla X_b) + S_{MT} - S_{RE} \tag{7}$$

where S_{RE} describes the consumption of alumina due to electrochemical reactions at the anode surface (based on Faraday's law) with current efficiency of 95%, see [5] for more details.

The evolution of the bath temperature (T) in the aluminium reduction cells is governed by

$$\frac{\partial}{\partial t}(\rho_b C_{p,b} T) + \nabla \cdot (\rho_b C_{p,b} \vec{U} T) = \nabla \cdot (\langle \lambda_b \rangle \nabla T) + S_T, \tag{8}$$

where $C_{p,b}$ is the specific heat capacity of the bath, $\langle \lambda_b \rangle$ is the effective thermal conductivity of bath (with contributions from turbulence and bath conductivity, i.e. $\langle \lambda_b \rangle = \rho_b C_{p,b} \Gamma_T + \lambda_b$) and S_T is the source terms which is a sum of contributions from

- Joule heating: The passage of current through the molten cryolite between the anode and cathode results in heating the bath. In the proposed model, the current is assumed to flow vertically between the anode and cathode, a reasonable assumption for the bath since the gap between the anode and cathode is in the order of a few centimeters (when compared to the electrodes) and the electrical conductivity of the bath is relatively low (compared to aluminium). The corresponding source term, which is active only on the cells below the anode, is computed as

$$S_o = \frac{j^2}{\sigma_b} \tag{9}$$

where σ_b is the electrical conductivity of the bath.
- Heat lost through the top: The heat lost from the top part of the bath due to heat transfer with the surroundings are driven by two 'modes': (1) heat lost from the top canals of the bath and (2) heat lost via anodes.
The heat lost from the top canals of the bath considers the heat lost due to radiative and convective heat transfer to the crust which will be in equilibrium with the surrounding air. This is accounted via source term active only at the top channels as

$$S_{t1} = -\frac{\mathcal{H}_{t1}(T - T_{air})}{\Delta h_1}, \tag{10}$$

where $\mathcal{H}_{t1}$ is the user defined overall heat transmission coefficient for this mode of heat transfer, T_{air} is the temperature of surrounding air and Δh_1 is the length scale of the cell along the normal to the face contributing to the top canal surface.

The heat lost via the anodes considers the conductive and convective heat transfer from the bath to the electrodes. Assuming the electrode system is in equilibrium with the surrounding air allows us to express the source term corresponding to this heat transfer mode as

$$S_{t2} = -\frac{\mathcal{H}_{t2}(T - T_{air})}{\Delta h_2}, \tag{11}$$

where $\mathcal{H}_{t2}$ is the user defined overall heat transmission coefficient for this mode of heat transfer and Δh_2 is the length scale of the cell along the normal to the face contributing to the anode face. This source term, which is active on all the cells that have a bath-electrode interface.
- Heat lost through sidewalls: The heat loss from the bath through the sidewalls is in reality a complex process involving side-ledge (which is frozen cryolite) melting/freezing governed by heat and multi-species transport, see [8] for more information. Due to the complexity of the side-ledge dynamics, modelling this phenomena is not within the scope of this work. Assuming that side-ledge dynamics does not substantially influence the bath temperature and composition, as well as the rate of freezing/melting of side-ledge is negligible, an overall heat transfer, which take into account both heat transfer from bath to side-ledge and side-ledge to out of the cell, can be defined. The corresponding source term is estimated as

$$S_s = -\frac{\mathcal{H}_s(T - T_{air})}{\Delta h_s}, \tag{12}$$

where $\mathcal{H}_s$ is the user defined overall heat transmission coefficient through the side of the cell and Δh_s is the length of cell in the direction normal to the cell face which is part of the side wall.
- Heat lost into metal: The source term corresponding to the heat transfer through bath-metal interface (as molten aluminium floats over the cathode) is calculated as

$$S_m = -\frac{\mathcal{H}_m(T - T_{metal})}{\Delta h_m}, \tag{13}$$

where $\mathcal{H}_m$ is the user-defined overall heat transfer coefficient between the bath and metal, Δh_m is the length of the cell in the direction which is direction normal to the cell face which is part of the metal-bath interface, and T_{metal} is the metal temperature which is computed as

$$\rho_m C_{p,m} V_m \frac{dT_{metal}}{dt} = \overline{S}_m - \mathcal{H}_{m,a}\left(T_{metal} - T_{air}\right) A_{mo} \tag{14}$$

where $\overline{S}_m$ is the total heat transferred via the metal-bath interface (computed as $\sum -S_m V$ where S_m and V are computed at each cell which has a face contributing to the interface), $\mathcal{H}_{m,a}$ is the user defined overall heat transmission coefficient between the metal and surrounding, ρ_m is the density of molten aluminium, $C_{p,m}$ is the specific heat capacity of aluminium, A_{mo} is the total surface area of the metal (excluding the surface area of metal-bath interface). Assuming that the metal region is equivalent to a rectangular cuboid $A_{mo} = 2W_c d_m + 2L_c d_m + W_c L_c$ with W_c as the width of the cell, L_c as the length of the cell and d_m as the user-defined value of the depth of metal region in the cell, and corresponding V_m is the volume of metal which is equal to $d_m L_c W_c$.

- Heat lost via heating of fed alumina: The alumina which is fed into the bath is relatively colder, so heat is drawn from the bath (resulting in local cooling of the bath) to heat these particles. This process of heating up of alumina consists of two steps according to [9]: endothermic step of increasing the temperature of fed alumina (which consists of $\gamma - Al_2O_3$) and exothermic step of transformation from $\alpha - Al_2O_3$. This is accounted via source term which is calculated as:

$$S_{heat} = -S_{feed}\left(\Delta H_1 + \Delta H_2\right) \tag{15}$$

where S_{feed} is the feeding rate of alumina used in Eq. 5. Enthalpy changes corresponding to heating of $\gamma - Al_2O_3$ (ΔH_1) and $\gamma - \alpha$ transformation of Al_2O_3 (ΔH_2) are computed based on [9] as

$$\Delta H_1 = 1000\left(1.483(T - T_0) - 199.3\ln\left(\frac{T}{T_0}\right)\right),$$

$$\Delta H_2 = 1000\left(-185 - 0.067(T - 298) + 9.6\ln\left(\frac{T}{298}\right)\right),$$

where T_0 is the temperature of the alumina fed via the feeders.

- Heat lost due to alumina dissolution: The dissolution of the heated alumina occurs as a result of reactions with the bath. The corresponding source term, due to dissolution, is treated as

$$S_{diss} = -S_{MT}\Delta H_{diss} \tag{16}$$

where ΔH_{diss} is enthalpy change due to the alumina dissolution.

Implementation

The governing equation discussed earlier is solved using finite volume method which is implemented in Python, similar to [5]. The velocity is stored at the cell faces whereas remaining flow parameters are stored in the cell center—known commonly as a staggered grid. To illustrate the implementation of

the governing equation, we consider an arbitrary advection-diffusion equation of Ψ (which is similar in form to Eqs. 5, 7, and 8)

$$\frac{\partial}{\partial t}\left(C\Psi\right) + \nabla \cdot (C\vec{U}\Psi) = \nabla \cdot (\zeta\nabla\Psi) + S$$

which can be written as

$$\int_V \frac{\partial}{\partial t}(C\Psi)dV + \int_V \nabla \cdot (C\vec{U}\Psi)dV = \int_V \nabla \cdot (\zeta\nabla\Psi)dV + \int_V SdV$$

$$\frac{(C\Psi)^{n+1} - (C\Psi)^n}{\Delta t}\Delta V + \sum_f (C\vec{U}\Psi)^{n+1} \cdot \vec{A}_f = \sum_f (\zeta\nabla\Psi)^{n+1} \cdot \vec{A}_f + S^n\Delta V$$

$$\frac{(C\Psi)^{n+1}}{\Delta t}\Delta V + \mathcal{F}^{n+1} = \frac{(C\Psi)^n}{\Delta t}\Delta V + S^n\Delta V$$

where Δt is the time step used by the solver (which is calculated based on the user defined maximum Courant number), $\mathcal{F}^{n+1}$ consists of the flux from convection and diffusion terms which is computed as

$$\mathcal{F} = \sum_f (C\vec{U}\Psi)^{n+1} \cdot \vec{A}_f - \sum_f (\zeta\nabla\Psi)^{n+1} \cdot \vec{A}_f$$

$$\mathcal{F} = \sum_f \left(C(\langle\vec{U}\rangle + \vec{U}^+ + \vec{U}^-)\Psi\right)^{n+1} \cdot \vec{A}_f - \sum_f (\zeta\nabla\Psi)^{n+1} \cdot \vec{A}_f$$

where $\langle\vec{U}\rangle$ is the average velocity which is the divergence-free flow field obtained after the velocity correction step in Eq. 4, $\vec{U}^+$ and $\vec{U}^-$ are the positive and negative dispersion/deviation velocities. It should be noted that $\vec{U}^+ \cdot \vec{A}_+ + \vec{U}^- \cdot \vec{A}_- = 0$, due to mass conservation, which gives us: $\vec{U}^- \cdot \vec{A}_- = -\vec{U}^+ \cdot \vec{A}_+$ and the positive deviation velocities are known from Eq. 2.

The discretized form of the governing equation can be written into the form $\mathbf{AX} = \mathbf{B}$ from which $\mathbf{X}$, at $n+1$, is obtained by direct solving $\mathbf{A}^{-1}\mathbf{B}$. It should be noted that the source terms in Eqs. 5, 7, and 8 are computed explicitly. The results form the model are visualized using Paraview—an open source visualization software.

Test Case

In order to showcase the proposed framework, we consider an hypothetical 160 kA aluminium reduction cell consisting of 16 anodes in two rows. Each anode is submerged in the bath by 0.12 m and distance between the anode-metal (commonly referred to as anode-cathode distance) is 4 cm. Each anode has a cross sectional area of 0.72×1.51 m^2 which gives a current density of 9198 A/m^2. Along the shorter edge of the cell, the gaps between the anodes and the cell boundaries is 0.1 m whereas the gaps between the two rows of anodes is 0.22 m. Along the longer size of the cell, the gap between the anodes and the cell walls is set to 0.12 m but the gaps between the anodes (in the same row) are equal to 0.08 m. The coarsest

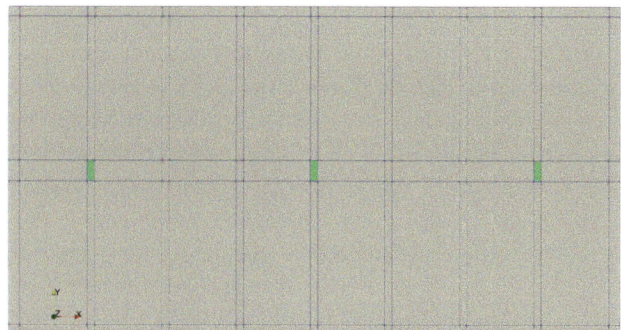

Fig. 1 The mesh used for the simulation along with the position of the three feeders (highlighted in green) used to feed alumina

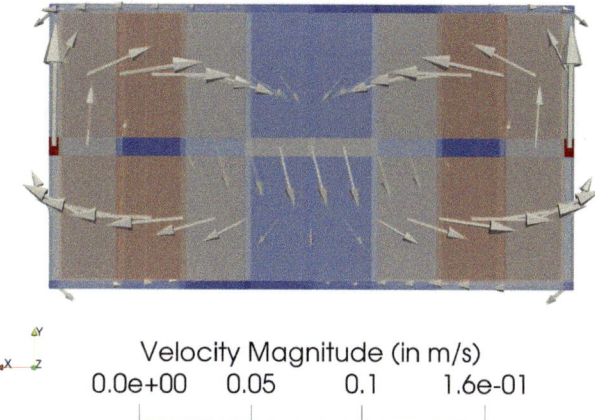

Fig. 2 Velocity field used for the simulations in this work. It should be pointed out that for visualization, the velocities, which are defined at the cell faces, is interpolated to cell center

possible mesh, shown in Fig. 1, is used for the simulation. The depth of the metal (d_m) is assumed to be 0.15 m.

The cell has three feeders, which are located in the gap between the two rows of anodes, as indicated in Fig. 1 with the left, middle and right feeder adding 0.5 kg of alumina at each feeding instance via each feeder. The simulation is setup such that all the feeders become active together which means every time feeding occurs 1.5 kg of alumina is added into the bath. Two scenarios of feeding is used in the simulation: Case 1—uniform feeding pattern which means that feeding occurs every 75 s, and Case 2: non-uniform feeding pattern which means that the feeding time gap between the 5 consecutive feeding instances is 100 s then 50 and 100 s and so on. The velocity field used to generate the flow field used for the pragmatic model is based on

$$U_p = -0.2 \sin \frac{2\pi y}{W_C} \sin \frac{\pi(x+1.64)}{L_C} - 0.2 \sin \frac{2\pi y}{W_C} \sin \frac{\pi(x+4.92)}{L_C}, \quad (17)$$

$$V_p = -0.2 \sin \frac{2\pi y}{W_C} \sin \frac{\pi(x+1.64)}{L_C} - 0.2 \sin \frac{2\pi y}{W_C} \sin \frac{\pi(x+4.92)}{L_C}, \quad (18)$$

$$W_p = 0.0, \quad (19)$$

where W_C is the total cell width in y-direction and L_C is total cell length in x-direction. The corrected velocity obtained from the above velocity field generates two counter-rotating vortices which is visualized in Fig. 2. The velocity used in this work does not aim to represent a realistic flow field for a Hall–Heroult cell—it is constructed purely as an example. Fluid and particle properties used for the simulation are:

- Bath density: $\rho_b = 2070$ kg/m^3
- Particle density: $\rho_p = 2250$ kg/m^3
- Bath viscosity: $\mu_b = 0.0027$ Pa s
- Diffusivity of alumina in bath: $\mathcal{D}_{d,b} = 1 \times 10^{-9}$ m^2/s
- Average alumina particle diameter: $D_p = 1 \times 10^{-4}$ m

Velocity Magnitude (in m/s)
0.0e+00   0.05   0.1   1.6e-01

- Turbulent diffusivity: $\Gamma_T = 0.001$ m^3/s
- Electrical conductivity: $\sigma_b = 215$ S/m
- Density of metal: $\rho_m = 2300$ kg/m^3

The thermal properties used in the simulation are:

- Specific heat capacity of bath: $C_{p.b} = 1900$ J/kg K
- Thermal conductivity of bath: $\lambda_b = 0.8$ W/m K
- Specific heat capacity of metal: $C_{p.m} = 1127$ J/kg K
- Temperature of surrounding air: $T_{air} = 290$ K
- Temperature of the fed alumina: $T_0 = 373.15$ K
- Enthalpy change due to the alumina dissolution: ΔH_{diss} = 1.275×10^6 J/kg of alumina

The composition of the bath used in the simulation to calculate the saturation level of alumina is $[AlF_3] = 23$, $[LiF] = 7$, $[CaF_2] = 3.5$, and $[MgF_2] = 0$. Additionally, the overall heat transfer/transmission coefficients corresponding to various modes of heat loss from the cell are set to:

- Heat transfer coefficient between bath and metal ($\mathcal{H}_m$): 575 W/m^2 K
- Heat transmission coefficient between metal and surroundings ($\mathcal{H}_{m,a}$): 3.95 W/m^2 K
- Heat transmission coefficient between bath and sidewalls ($\mathcal{H}_s$): 23.1 W/m^2 K
- Heat transmission coefficient between bath and anode ($\mathcal{H}_{t2}$): 2.3 W/m^2 K
- Heat transmission coefficient from top canals of the bath ($\mathcal{H}_{t1}$): 2.3 W/m^2 K

It should be noted that these heat transfer/transmission coefficients are set to give a reasonable amount of heat lost through the boundaries. Simulations are run for 5000 s flow time with a time step of 0.72 s, initialized with a mass fraction of alumina of 0.02, bath temperature of 960 °C, and metal temperature of 958 °C.

Results

The bath temperature and the particle alumina distribution in the aluminium reduction cells have been observed to be highly non-uniform irrespective of the feeding pattern, see Fig. 3. The channels between the anode through which the particle alumina is convected show a relatively cooler bath as a result of the endothermic nature of alumina dissolution. In comparison, the regions of the cell where the alumina concentration is low the bath temperature is observed to be comparatively higher. Both the feeding cycles used in the model are observed to under feed the cell resulting in decrease of the dissolved

alumina over time. The framework is also able to conserve mass. This is represented by *Numerically accumulated alumina mass*, described in [5], and here is equal to zero for both cases. The average bath and metal temperature predicted by the proposed framework is observed to almost reach a steady value with the bath being warmer than the metal (by around 5 K) for Case 1 when using a uniform feeding pattern, see Fig. 4. Interestingly, when the non-uniform feeding pattern is used, like in Case 2, the average bath and metal temperature is observed to vary based on the alumina feeding pattern, see Fig. 4. As the total alumina added to the cell over 5000 s is slightly higher for the Case 1 when compared to Case 2, the average bath temperature is relatively lower for Case 1—as more heat is lost due to particle dissolution. Higher temperature of the bath is also observed to result in a larger value of the metal temperature. The metal temperature variation seems to lag behind the bath temperature evolution which is a result of the high heat capacity of the molten aluminium. The average bath temperature, which is around 961–963 °C, predicted by the model, is within the typical operating temper-

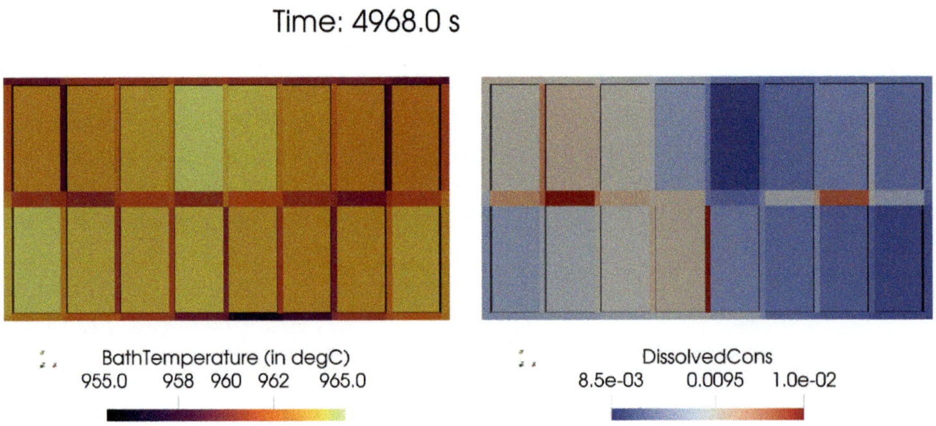

Fig. 3 Visualization of bath temperature (left) and dissolved alumina mass fraction (right) distribution in the cell during the simulation for Case 1 (with uniform feeding)

Time: 4968.0 s

BathTemperature (in degC)
955.0   958 960 962   965.0

DissolvedCons
8.5e-03   0.0095   1.0e-02

Fig. 4 Temporal evolution of average bath and metal temperature for Case 1—uniform feeding, and Case 2—non-uniform feeding

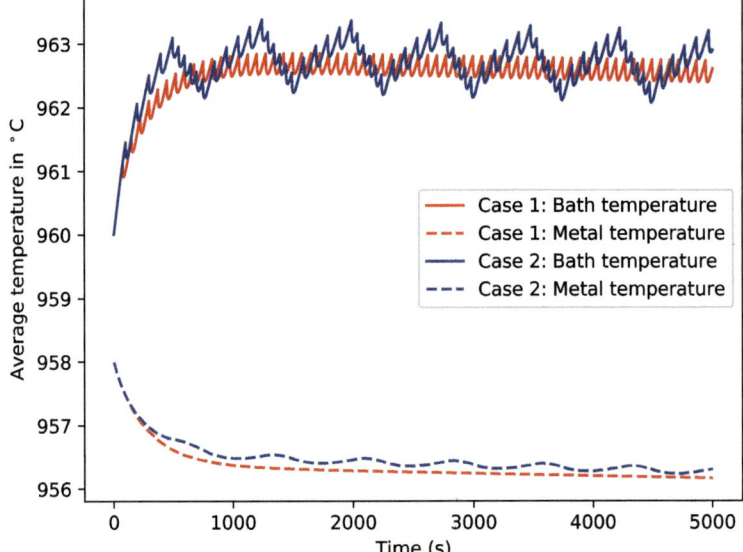

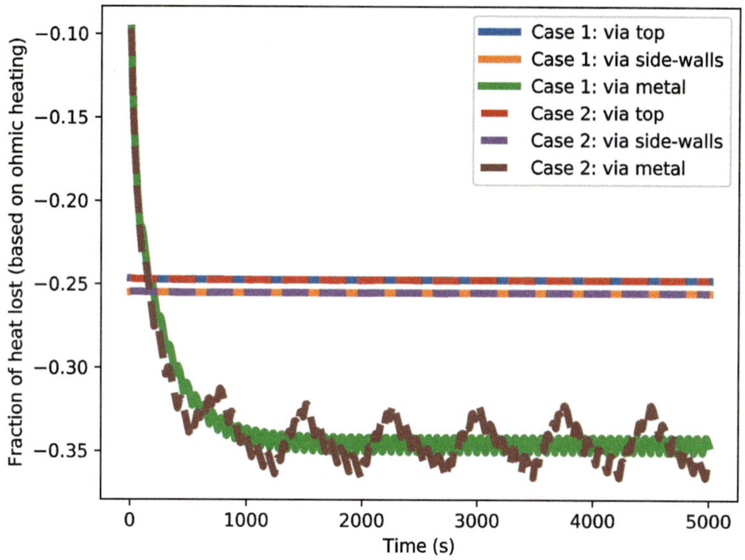

Fig. 5 Fraction of the heat lost from the bath through the top (top canals + anodes), sidewalls and into the molten metal for Case 1—uniform feeding, and Case 2—non-uniform feeding

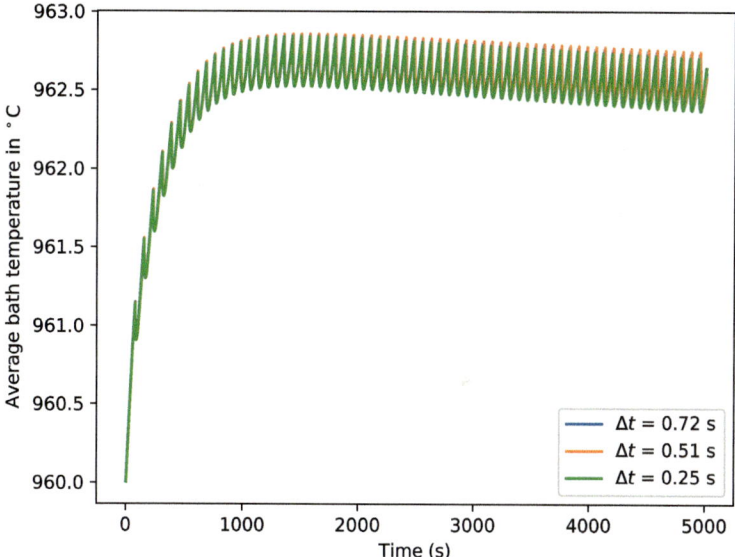

Fig. 6 Transient evolution of the average bath temperature predicted by the proposed framework for various time steps for Case 1 (uniform feeding)

ature of an industrial aluminium reduction cell (in the range of 950–970 °C [9]). The heat lost from the bath through the sidewalls, via the top (via both anode and top canals), and into the molten metal is around 25, 25, and 35% (of the heat added due to ohmic heating), respectively, see Fig. 5. As expected, the heat lost into the metal from the bath is dependent on the feeding pattern. Interestingly the heat lost through the sidewalls and top (top canals and anodes) is relatively steady and independent of the feeding pattern used (although the bath temperature can change with time), see Fig. 5.

Effect of Time Step

As the pragmatic model would typically be run at the largest possible timestep but as the source terms are computed explicitly, there is a constraint on the maximum time step used based on the Courant number. To investigate it, we simulate the case with uniform feeding with three different time steps used by solver, see Table 1. The effect of time step on the bath temperature is not substantial as seen in Fig. 6 which allows the use of a large time step to obtain reliable results.

Table 1 Runtime of the model for the various the time steps used by the solver

Time step used for the simulation (Δt)	Runtime
$\Delta t = 0.72$ s (Courant number = 0.99)	$\approx$66 s (76$\times$ faster)
$\Delta t = 0.51$ s (Courant number = 0.7)	$\approx$96 s (52$\times$ faster)
$\Delta t = 0.25$ s (Courant number = 0.35)	$\approx$191 s (26$\times$ faster)

Conclusions

A pragmatic model framework for alumina dissolution presented in [5] is extended in this work to simulate the associated evolution of bath temperature. The proposed framework is demonstrated for a hypothetical cell with a uniform and non-uniform feeding cycle. Although the framework becomes slower, which is a result of solving additional governing equation for bath temperature when compared to [5], it is still 25–75 times faster than real time. In addition, a generic method, which is independent of the mesh used in the CFD simulation, to generate the velocity (as well as the deviation velocities) for the pragmatic model is also presented in this work.

Acknowledgements The current work has been funded by SFI Metal Production (Centre for Research-based Innovation, 237738). The authors gratefully acknowledge the financial support from the Research Council of Norway and partners of the center.

References

1. Bojarevics, V.: Dynamic modelling of alumina feeding in an aluminium electrolysis cell. In: C. Chesonis (ed.) Light Metals 2019, pp. 675–682. Springer International Publishing, Cham (2019)
2. Bojarevics, V., Dupuis, M.: Advanced alumina dissolution modelling. In: D. Eskin (ed.) Light Metals 2022, pp. 339–348. Springer International Publishing, Cham (2022)
3. Cassayre, L., Palau, P., Chamelot, P., Massot, L.: Properties of low-temperature melting electrolytes for the aluminum electrolysis process: A review. Journal of Chemical & Engineering Data **55**(11), 4549–4560 (2010). https://doi.org/10.1021/je100214x
4. Haverkamp, R., Welch, B.: Modelling the dissolution of alumina powder in cryolite. Chemical Engineering and Processing: Process Intensification **37**(2), 177–187 (1998). https://doi.org/10.1016/S0255-2701(97)00048-2. https://www.sciencedirect.com/science/article/pii/S0255270197000482
5. Johansen, S.T., Einarsrud, K.E., Solheim, A., Vachaparambil, K.J.: A pragmatic model for alumina feeding. In: D. Eskin (ed.) Light Metals 2022, pp. 503–511. Springer International Publishing, Cham (2022)
6. Kovács, A., Breward, C., Einarsrud, K., Halvorsen, S., Nordgård-Hansen, E., Manger, E., Münch, A., Oliver, J.: A heat and mass transfer problem for the dissolution of an alumina particle in a cryolite bath. International Journal of Heat and Mass Transfer **162**, 120232 (2020). https://doi.org/10.1016/j.ijheatmasstransfer.2020.120232. https://www.sciencedirect.com/science/article/pii/S0017931020331689
7. Skybakmoen, E., Solheim, A., Sterten, Å.: Alumina solubility in molten salt systems of interest for aluminum electrolysis and related phase diagram data. Metallurgical and Materials Transactions B **28**(1), 81–86 (1997). https://doi.org/10.1007/s11663-997-0129-9
8. Solheim, A.: Some Aspects of Heat Transfer Between Bath and Sideledge in Aluminium Reduction Cells, pp. 381–386. Springer International Publishing, Cham (2016). https://doi.org/10.1007/978-3-319-48160-9_68
9. Solheim, A., Skybakmoen, E.: Mass- and heat transfer during dissolution of alumina. In: A. Tomsett (ed.) Light Metals 2020, pp. 664–671. Springer International Publishing, Cham (2020)
10. Taylor, M., Zhang, W., Wills, V., Schmid, S.: A dynamic model for the energy balance of an electrolysis cell. Chemical Engineering Research and Design **74**(8), 913–933 (1996). https://doi.org/10.1205/026387696523094. https://www.sciencedirect.com/science/article/pii/S0263876296714818. Heat and Mass Transfer

A New Strategy for Transient Heat Transfer Models with Phase Change for the Aluminum Electrolysis Industry

Bastien Pansiot, Marc Lebreux, Martin Désilets, Francis Lalancette, Jean-Francois Bilodeau, and Alexandre Blais

Abstract

The heat balance of an Aluminum Electrolysis Cell (AEC) represents a very sensitive aspect of its operation. Steady-state heat balance is quite well understood, thanks to industrial expertise and widely used steady-state finite elements models. With the increasing interest in power modulation, these tools become insufficient to predict the time-dependent heat balance of the cell. Above all, they are not suited to predict the melting or formation of the side ledge, a protective solid layer formed on the cell to avoid the corrosive action of the liquid electrolytic bath on the sidewalls. The authors developed a new thermal–electrical model coupling ANSYS® Mechanical software and Python, able to compute the transient heat balance of an AEC, including the phase change dynamic of the side ledge and thermal contact resistance. It has been found to give both accurate results and improved computational time, which becomes a crucial aspect for transient simulations of such complex and heavy models.

Keywords

Heat transfer • Simulation • Aluminum electrolysis • Side ledge • Phase change

Introduction

Industrial Context

Since it has been patented in 1886 by C.M. Hall and P.L. Héroult, the Hall–Héroult process has been improved to reach better efficiency mainly through the design of the Aluminum Electrolysis Cell (AEC) and process operation.

The thermal balance in the aluminum electrolysis process is a very sensitive aspect of process control and optimization. In order to protect the sidewalls of the AEC, it is necessary to form a protective solid bath layer onto it. Usually, a few centimeters thick, this layer is very sensitive to heat balance perturbations. Indeed, the process operates around 965 °C, that is only a few degrees higher than the liquidus of the bath. For instance, to prevent the side ledge to completely disappear, it is known that the lateral heat flux must be high enough as compared to the top ones [1]. The heat balance is very complicated to manage. Unwanted perturbations can occur, such as anode effect or energy shortage for instance. On the other hand, other perturbations are planned, such as anode replacement, aluminum tapping, anode position adjustment or alumina feeding. As energy becomes more and more valuable, it is in the interest of aluminum producers to optimize its consumption. In order to achieve it, two strategies arise: the first consists of increasing the cell electric current intensity along with anodes size in order to achieve better productivity of each cell. The second strategy is to increase and decrease power input according to power availability and cost variations.

Two different parts of the AEC are affected significantly by its heat balance: the top part including the Anode Cover Material (ACM) and the side part with the side ledge. The whole heat balance depends on the ACM "state". Allard et al. [2] studied its thermal and chemical behaviour and showed that this material undergoes chemical and structural transformations under the influence of temperature and acid gas from the bath. It changes from powder to a dense solid called crust at about 700 °C. When this crust is submitted to temperatures higher than 930 °C, it melts and forms a cavity between the top of the liquid bath and the bottom of the crust material. Unlike the side ledge, this transformation is irreversible and will thus change the heat balance until the anode is changed. In this case, the side ledge will act as a thermal buffer and will adapt itself to the new equilibrium [1]. Allard et al. [2, 3]

B. Pansiot (✉) · M. Lebreux · M. Désilets · F. Lalancette
Department of Chemical Engineering and Biotechnological Engineering, Université de Sherbrooke, 2500 Boulevard de l'Université, Sherbrooke, Qc J1K 2R1, Canada
e-mail: bastien.pansiot@usherbrooke.ca

J.-F. Bilodeau · A. Blais
Rio Tinto Aluminium (Arvida Research and Development Centre), 1955 Boulevard Mellon, Jonquière, Qc G7S 4K8, Canada

© The Minerals, Metals & Materials Society 2023
S. Broek (ed.), *Light Metals 2023*, The Minerals, Metals & Materials Series,
https://doi.org/10.1007/978-3-031-22532-1_15

improved a thermal–electrical model of the AEC in order to take into account the behaviour of the ACM.

On the other hand, the behaviour of the side ledge is difficult to assess and is still not well understood. First, the direct ledge thickness monitoring relies on a very difficult and inaccurate measurement [4]. In addition, its chemical composition is not constant, neither in time nor in space. Indeed, due to the difference in heat flux during the bath solidification, the ledge formed in front of the liquid bath is more porous than the one formed in front of the liquid aluminum, as reported by Poncsák et al. [5, 6] and Liu et al. [7]. Experimental observations lead to conclude that the composition as well can vary according to the solidification rate. At a low solidification rate, the ledge is formed at a close to equilibrium composition because species have time to migrate to the bath as cryolite solidifies (macrosegregation). At higher rates, this segregation is more difficult and other species such as AlF_3, CaF_2 or Al_2O_3 are trapped as liquid pockets into the solid cryolite phase. As the temperature decreases, these liquid pockets will solidify eventually. During melting, these species have a lower liquidus temperature and will decrease the whole ledge liquidus which will make the ledge melt much faster than expected. This makes the ledge thermal and chemical characterization very difficult because it depends on the bath composition and thermal conditions during which it has been formed. If several authors have tried to use numerical models to understand the behaviour of the side ledge, a gap in the literature seems to appear regarding the modelling of the side ledge phase change dynamics.

Because the whole heat balance involves an intimate relationship between the top, side and in a lesser manner the bottom heat losses, the thermal dynamic appears to be difficult to evaluate online. In order to understand it more accurately, numerical models combined with industrial measurements are required.

Thermal–Electric Models for AEC

Several kinds of models have been developed to study the aluminum electrolysis process. Magneto Hydro Dynamic models (MHD) are used to study fluids movements generated by the magnetic field, mechanical models to study mechanical stresses inside the AEC's components like the pot shell, and thermal–electrical models to study its heat balance. In this paper, the focus will be put on the latter. Typically, 1D and 2D models are used for quick dynamic operation study and 3D models help to investigate the steady state in AEC design studies.

The 3D, thermal–electrical, steady-state model of an AEC slice is a very popular tool to explore new cell designs. Recently, this model has been made transient to study the cell dynamics. Allard et al. [2] exposed an improved representation of the top part of the AEC by introducing the ACM dynamics detailed previously. One of the missing pieces remains the phase change dynamics of the ledge.

Solheim developed a model to describe the ledge formation in front of the liquid aluminum based on the liquid bath film theory [8]. He also defined a phase change model based on composition and heat transfer rate [1]. Marois [9] was one of the first to work on a time-dependent AEC slice model. He compared two different methods for phase change using a MATLAB® solver. The first does not take into account the latent heat and is called the "Single phase method" and the second uses the "Enthalpy method" (also implemented in ANSYS®). He concluded that the first method is accurate enough to describe small perturbations, but the time to reach a new steady state can be quite under estimated (33 %) for more significant events.

It is thus obvious that a representative numerical model describing the phase change behaviour of the side ledge is needed. In order to be used in the industry, this model needs to be accurate as well as computationally efficient. This is precisely the purpose of the work exposed in this article.

Phase Change Modelling

In numerical methods, different strategies exist to model a phase change problem, also known as "Stefan problem". The main idea is to determine the time dependent position of a solid/liquid interface. In this kind of problem, energy transfer occurs along with mass transfer, which will change the thermal and physical properties of the numerical domain. This introduces some numerical instabilities which will be further discussed later. Lame and Clapeyron and then Stefan have first studied this kind of problem and determined the interface speed also known as Stefan condition (Eq. 1).

$$\lambda_s \frac{\partial T_s}{\partial x} - \lambda_l \frac{\partial T_l}{\partial x} = \rho H_f \frac{dX}{dt} \qquad (1)$$

Here, λ denotes the thermal conductivity, T the temperature with the subscript s and l for "solid" and "liquid" state, H_f is the latent heat of fusion, X the solid/liquid interface position, and finally, t represents the time and x the space variable.

Two analytical methods allow to solve this kind of problem in very simple cases (heat transfer only, 1D and constant properties), the "Neumann's method" [10] and the "Integral heat conservation method" [11]. For more complex cases (geometry, boundary conditions, etc.), numerical methods are required.

Several reviews on this subject were published to present and compare the main numerical methods [12, 13]. The "fixed grid" approach (as opposed to the "moving grid" approach) has been selected here due to its computational performance. Only the "source-based method" as developed by Voller and Prakash [14], as well as the "enthalpy method" presented by Eyres and Hartree [15], were studied here due to their accuracy, stability, computing efficiency, and finally because they are widely used and trusted. The theoretical aspects of each method will not be detailed here as it is already well done in the open literature.

The ANSYS® Model

Solving phase change problems has already been implemented in ANSYS® and, because this software is widely used in this field, it will be tested first. A simplified model of the cell slice thermal–electrical model has been created in order to reduce the computation time and facilitate debugging. From now on, the strategy using only ANSYS® will be referred to as "strategy 1" as opposed to the new method described later ("strategy 2").

Geometry

The simplified model consists of a rectangular parallelepiped taken at a representative horizontal location on the AEC, going from the bath to the side wall (Fig. 1). The size of this parallelepiped is 2.040 m length by 0.01 m height and 0.01 m depth. This geometry is 3D, but is similar to a 1D case due to the boundary conditions detailed later (heat transfer is taken only in the horizontal direction). All the materials are the same as those found inside the AEC slice model.

Boundary Conditions

The following boundary conditions are considered: the top and bottom part are adiabatic as well as the front and back faces, the right boundary is a symmetrical condition, and finally, for the left part, convection is applied on the side wall ($h = 80$ W/m^2 K and $T_{amb} = 40$ °C). Joule effect taking place inside the liquid bath due to the high electrical

current will be simulated through a volumetric heat source added only in the right part of the liquid bath (where no phase change is occurring). It is important to note that thermal contact resistances are set at the shell/SiC slab interface ($h_{cont} = 200$ W/m^2 K) and at the liquid bath/solid bath interface ($h_{cont} = 1500$ W/m^2 K).

Material Properties

The material properties of the phase change material are summarized in Table 1. Properties of other materials must remain confidential.

Modelling Method

The enthalpy method for phase change is already implemented in ANSYS® Mechanical, which is the software typically used by industrial engineers for 3D cell slice models. Thus, it will be used here as well. In this method [13], the enthalpy that includes latent heat is solved as the main degree of freedom of the model using Eq. 2. c_s and c_l denote respectively the solid and liquid specific heat capacity. The liquid fraction field (noted f) can also be computed from the temperature field according to Eq. 4.

$$\frac{\partial H}{\partial t} = \nabla(\lambda \nabla T) \tag{2}$$

where

$$H = (1 - f) \int_{T_{ref}}^{T} \rho c_s dT + f \rho H_f + f \int_{T_{ref}}^{T} \rho c_l dT \tag{3}$$

$$f = \frac{T - T_s}{T_l - T_s} \tag{4}$$

The tricky part here is the thermal contact resistance at the solid/liquid interface. In ANSYS®, in order to set this parameter, a real interface is needed between two distinct materials. Because this interface will move according to the phase change dynamics, the geometry and mesh have to be updated at every time step, thanks to an iterative solving scheme (Fig. 2). The main issue with this method, as it will be exposed later, is that changing the geometry and re-meshing at each iteration is time consuming. A new strategy has been developed to tackle this problem and will be detailed in the next section.

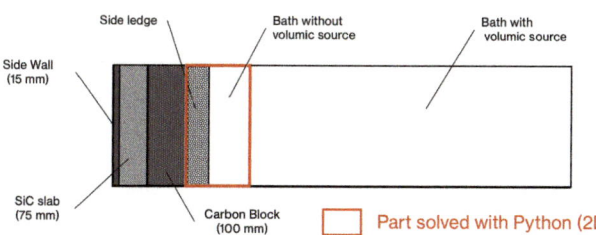

Fig. 1 Geometry of the simplified model

Table 1 Thermophysical properties of the different materials [16, 17]

Property	Value	Unit
ρ	2100	kg/m^3
c	1800	J/kg K
λ_{sol}	1.2	W/m K
λ_{liq}	10000	W/m K
H_f	510.10^3	J/kg
T_s	900	°C
T_l	950	°C

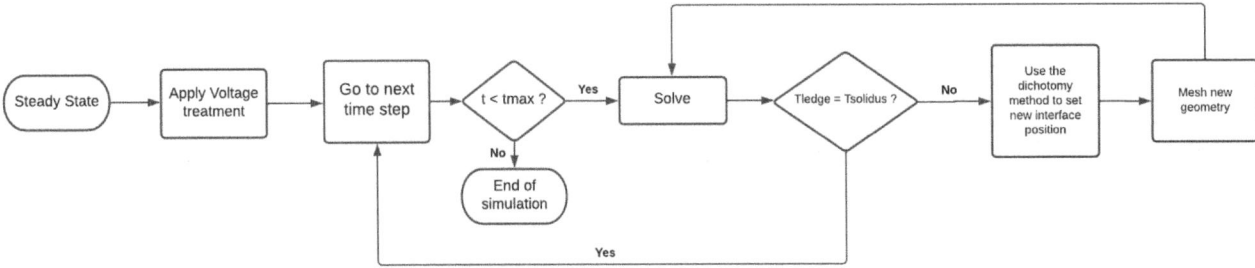

Fig. 2 Phase change strategy implemented in ANSYS® (strategy 1)

The PyAPDL Model

The strategy 1 is easy to implement in ANSYS® and is the basis of the transient model already used in the industry. However, the iterative strategy involving re-meshing at each iteration is time consuming and limits the application of the transient model to short time periods. Here, a new strategy (strategy 2) linking ANSYS® APDL language and Python is detailed and will be compared to strategy 1.

Python Solver

In order to study the phase change modelling methods and compare them, a solver has been developed in Python based on the Finite Volume Method [18]. Python has been chosen because it does not require any licence, and it can easily be linked to other software such as ANSYS®. Geometry and meshes are achieved with SALOME platform and are then imported into Python. The heat balance equation is then solved in 2D with phase change using the source based method [14], accounting for phase change using a source term as detailed in Eq. 5.

$$\rho c \frac{\partial T}{\partial t} = \nabla \cdot (\lambda \nabla T) + S_h \qquad (5)$$

$$S_h = -\partial H \frac{\partial f}{\partial t} \qquad (6)$$

$$\partial H = \rho(c_l - c_s)T + \rho H_f \qquad (7)$$

A different approach has been implemented to take into account the thermal contact resistance and to avoid the excessive iterative problem detailed earlier. The solid/liquid interface is first located in the phase change material (if existing). The thermal diffusivity for the control volumes forming the interface (α_f) is then modified to account for the thermal contact resistance noted Rc (Eq. 8). Here, λ_f represents the thermal conductivity for the control volumes forming the interface and d_f the distance between the two control volumes centres. This strategy allows the solver to compute the temperature and liquid fraction fields in the phase change material as one unique material with thermophysical properties depending on temperature/liquid fraction.

$$\alpha_f = \frac{1}{\rho c(1/\lambda_f + Rc/d_f)} \qquad (8)$$

Python and ANSYS® APDL Coupling

This Python solver only solves the bath part with phase change as illustrated in Fig. 1. To create a solver for the whole cell, slice model with Python is the next step of this project. Figure 3 describes this algorithm written in Python in order to link the ANSYS® and Python solvers.

First, the algorithm checks if a steady state exists or if it needs to be updated for initial conditions. If not, the steady state is calculated with ANSYS® and boundary conditions, as well as temperature field for the bath, are imported in Python. The Python solver runs for some time steps to check if the

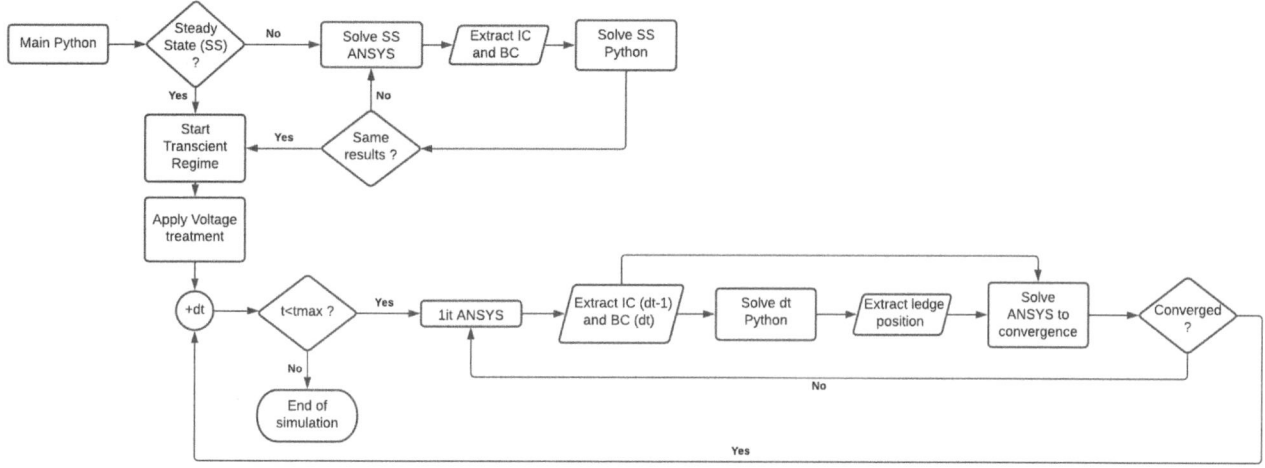

Fig. 3 New strategy for phase change with Python and ANSYS® (strategy 2)

steady state is similar to the one found in ANSYS®. Once this check is done, the true dynamic simulation can begin. The perturbation is applied (here input power variation) and a single iteration of the ANSYS® solver is performed. Boundary conditions at bath boundaries are extracted from ANSYS to Python and the Python solver uses them to perform the time step calculation. The solid/liquid interface position is found and imported in ANSYS®. ANSYS® now solves the same time step until convergence, counting on a better initial guess for the interface position that will save iterations and thus computation time. When the solution is found, the simulation can switch to the next time step.

Results and Discussion

Results Comparison

First, a steady-state solution is calculated and compared for both solvers with an input power of $P_{nom} = 9500$ W/m³. The predictions obtained with Python only differ by 0.6% from the result given with ANSYS®, which is acceptable. Thus, it can be concluded that both give similar results at steady state.

Then two transient scenarios have been simulated. The first is a power boost at $P_{boost} = 22,300$ W/m³ during 7200 s s from the steady state and then going back to P_{nom} until reaching steady state. The second is similar but with a power decrease at $P_{minus} = 6000$ W/m³ during 7200 s.

The two scenarios were run for 90000 s with a time step of 600s. Figures 4 and 5 give the results of the strategy 1 as compared to the strategy 2 for both scenarios. Here, the position of the solid/liquid interface (ledge thickness), the external shell temperature, as well as the external shell heat flux, are presented for each time step. Those variables are chosen among others because they are either important to control or easy to

measure online and thus to compare to model results. One can easily observe that both give approximately the same response and thus are equivalent in accuracy.

Computation Time Comparison

Because both models give similar results, it is interesting to check their computational performance. Strategy #2 saves iterations in ANSYS® preventing excessive re-meshing, thus it is expected to save time. Meshing computation time increases with the number of nodes in the model and preventing it will become more advantageous with larger models. To underline this aspect, different geometry sizes have been compared in order to come closer to a real cell slice model in terms of the node number. Only the depth is increased, thus keeping the domain solved in Python constant. Figure 6 gives the computation time for the cooling scenario with both strategies as a function of the total node number.

To put things in perspective, the cell slice model is around 70,000 nodes. If the results from Fig. 6 are extrapolated, this would take 32,179 s for the strategy 1 to compute and only 20,318 s for the strategy 2. The use of this new solver could enable to save 37% in computation time with similar results.

Conclusion

A new strategy has been developed in order to add phase change in the existing cell slice transient model representing aluminum electrolysis. The classic method, already implemented in ANSYS®, works quite fine, but the presence of the thermal contact resistance at the solid/liquid interface imposes an iterative strategy with re-meshing at each time step. Thus,

Fig. 4 Results for the "Pboost" scenario

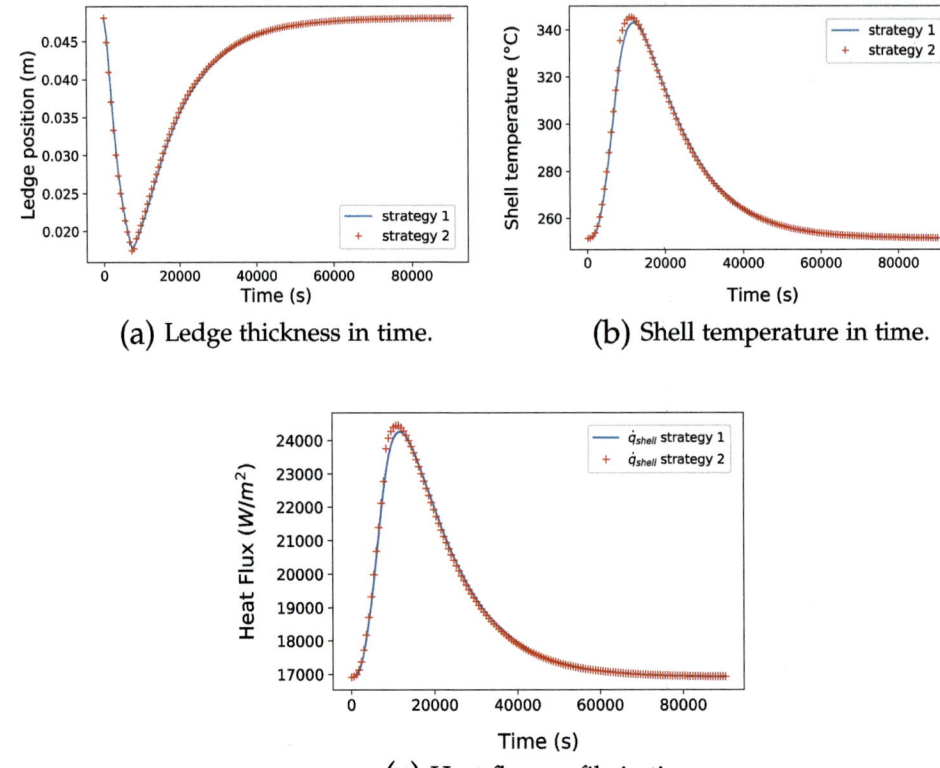

(a) Ledge thickness in time.

(b) Shell temperature in time.

(c) Heat flux profile in time.

Fig. 5 Results for the "Pminus" scenario

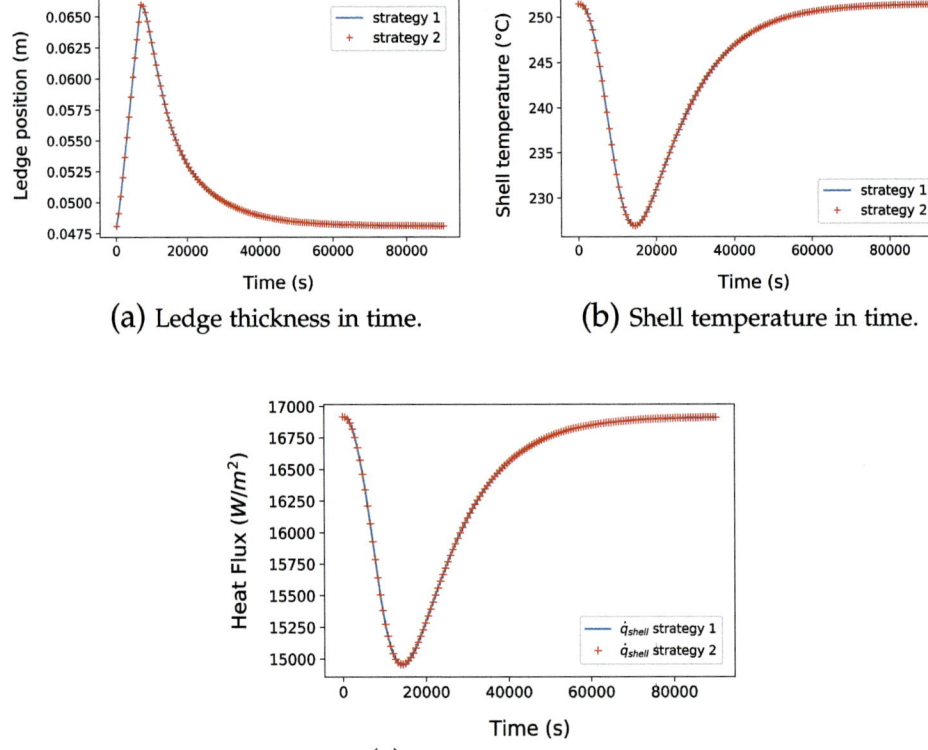

(a) Ledge thickness in time.

(b) Shell temperature in time.

(c) Heat flux profile in time.

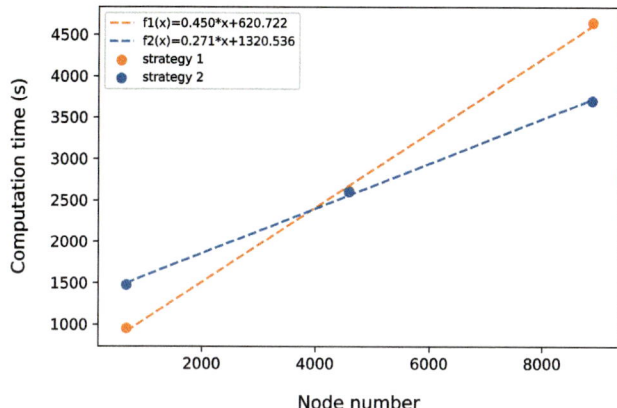

Fig. 6 Comparison of computation time for both strategies in function of node number

computational time becomes prohibitive for long simulations including high number of nodes. The idea behind the new strategy is to use an external solver that allows computation of the phase change part without meshing at each iteration in order to give a better initial guess for the ANSYS® solver and thus save iterations.

A simplified model has been developed in order to study this problem more efficiently. This approach facilitates debugging and analysis as well. First, both strategies give similar results for steady state and transient scenarios. This gives confidence in the accuracy of the new method and its capacity to model transient events. Both of them have stable convergence and the heat balance is conservative as well. Then, computation time analysis showed that the new strategy allows to save computation time for a number of nodes higher than 4000. Extrapolation to the cell slice model size would give a decrease in computation time of 37% with this new method.

These promising results obtained with a simplified model must now be confirmed with the cell slice model. Indeed, the dynamic simulations with this model can take between 6 h, up to days to solve. Thus, a decrease in the computation time is more than welcome.

Acknowledgements The author would like to thank Yves Caratini from Rio Tinto Alcan for his precious help concerning the ANSYS® model. The authors are also very grateful to the Natural Sciences and Engineering Council of Canada (NSERC) and to Rio Tinto Alcan for their financial support.

References

1. Solheim A. "Some aspects of heat transfer between bath and side leedge in aluminium reduction cells." In: *TMS (The Minerals, Metals & Materials Society)* (2011), pp. 381-386.
2. Allard F. et al. "Thermal, chemical and microstructural characterization of anode crust formed in aluminum electrolysis cells". In: *Thermochimica Acta* 671 (2019), pp. 89-102.
3. Francois Allard et al. "Improved heat transfer modeling of the top of aluminum electrolysis cells". In: *International Journal of Heat and Mass Transfer* 132 (2018), pp. 1262-1276. https://doi.org/10.1016/j.ijheatmasstransfer.2018.12.062.
4. P. Boily. "Application des capteurs thermiques implantés pour la détection du profil de gelée dans la cuve d'électrolyse." In: Mémoire de maîtrise: Université du Québec à Chicoutimi (2001).
5. Poncsák S. et al. "Structural characterisation and thermophysical properties of the side ledge in Hall-Héroult cells." In: *TMS (The Minerals, Metals & Materials Society)* (2014), pp. 585-589.
6. Poncsák S. et al. "Impact of the Heat Flux on Solidification of Cryolite Based Bath." In: *TMS - Light Metals* (2016).
7. Liu J. et al. "The Structure of the Smelting Cell Ledge Under Variable Sidewall Heat Flow Conditions." In: *The Journal of The Minerals, Metals & Materials Society (TMS)* 71 (2019), pp. 514-521.
8. Solheim A. et al. "Sideledge facing metal in aluminium eletrolysis cells: freezing and melting in the presence of a bath film." In: *TMS (The Minerals, Metals & Materials Society)* (2016), pp. 333-338.
9. M. Marois et al. "Comparison of two different numerical methods for predicting the formation of the side ledge in an aluminium electrolysis cell." In: *TMS Annual Meeting* (Jan. 2009), pp. 563-568.
10. J. Stefan. "Ueber die Theorie der Eisbildung, insbesondere über die Eisbildung im Polarmeere". In: *Annalen der Physik* 278.2 (1891), pp. 269-286.
11. T.R. Goodman. "The heat-balance integral and its application to problems involving a change of phase". In: *Transactions of the ASME* 80 (1958), pp. 335-342.
12. Hu H. and Argyropoulos S. A. "Mathematical modelling of solidification and melting: a review". In: *Modelling and Simulation in Materials Science and Engineering* 4.4 (1996), pp. 371-396.
13. V. R. Voller, C. R. Swaminathan, and B. G. Thomas. "Fixed grid techniques for phase change problems: A review". In: *International Journal for Numerical Methods in Engineering* 30.4 (1990), pp. 875-898. doi: https://doi.org/10.1002/nme.1620300419
14. V. R. Voller and C. R. Swaminathan. "General source-based method for solidification phase change". In: *Numerical Heat Transfer, Part B: Fundamentals* 19.2 (1991), pp. 175-189. https://doi.org/10.1080/10407799108944962.url: https://doi.org/10.1080/10407799108944962
15. Eyres N. R. and Hartree D. R. "The calculation of variable heat flow in solids". In: *Philosophical Transactions of the Royal Society of London. Series A, Mathematical and Physical Sciences* 240.813 (1946), pp. 1-57.
16. Marc LeBreux et al. "On the Prediction of the Crust Evolution Inside Aluminum Electrolysis Cells". In: *TMS Light Metals* (Jan. 2014), pp. 655-660. https://doi.org/10.1007/978-3-319-48144-9_110.
17. Allard F. et al. "A Modeling Approach for Time-Dependent Geometry Applied to Transient Heat Transfer of Aluminum Electrolysis Cells". In: *Metallurgical and Materials Transactions* B 50 (2019), pp. 958-980.
18. Hrvoje Jasak. "Error Analysis and Estimation for the Finite Volume Method With Applications to Fluid Flows". In: *Thesis - Imperial College* (Jan. 1996).

A Discussion on Thermal Impact of Anode Change in Aluminum Reduction Cell

Zhibin Zhao, Wei Liu, Yafeng Liu, Michael Ren, and Zhaowen Wang

Abstract

Anode change is the most important operation in aluminum electrolysis industry. Influence of new anodes is mainly concentrated in two aspects: thermal impact and magneto-hydrodynamic (MHD) impact. This paper only focuses on the thermal impact part. The paper theoretically calculates the heat required of new anodes and heat generation of the anode set modifier, heat release of metal/bath temperature decrease and electrolyte solidification. There may be a lack of more than 50% of energy or heat during the process of anode change. The anode change in one pot has an individualized characteristic. There are some 'sensitive anodes' with smaller anode current pick-up rate, of which the location is closely related to bath flow. A 'customized technology for anode set modifier' is proposed for anode changes in different regions. Both numerical and industrial works have proven that this technology played a positive role in improving the heat and electric behavior of 'sensitive anodes'.

Keywords

Anode change • Thermal stability • 'Sensitive anodes' • Anode set modifier

Introduction

The size and amperage of modern aluminum reduction cells were continuously increased during the past decades. CHINALCO has successfully developed SY600 ultra large capacity aluminum electrolysis cell technology in 2012 [1]. The first 600 kA commercial line was designed and put into production later in 2016. Even today, ultra large cell technology is still one of the most important development trends for electrolytic technology in the world, such as AP60 [2], NEUI600 [3], RA550 [4], AP50 [5]. There is even clear evidence that some scholars have discussed the possibility of 1000 kA cells [6].

Another development trend of aluminium smelting is to reduce greenhouse-gas emissions. Based on this background, some alternative technologies compared to traditional electrolysis technologies have become hot topics, such as inert anode electrolysis, carbonless electrolysis and ionic liquid electrolysis. But these technologies are still at the stage of experiment or semi-industrial verification. It is expected that the traditional Hall-Heroult process would still be a main-stream industrial production method for primary aluminum in a long period of time.

As the heart of aluminum reduction cells, carbon anodes need to be periodically replaced in 28 to 33 days due to their inherent consumption of anodes. Therefore, anode change becomes one of the most necessary operations in modern aluminum smelting industry. With the enlargement of the pot size and capacity, the anode also shows a trend of continuous enlargement in dimensions and mass.

The influence of new anodes (cold anodes) is mainly concentrated in two aspects: (a) the huge impact on the

Z. Zhao (✉) · W. Liu
Science and Technology Management Department, Shenyang Aluminium and Magnesium Engineering and Research Institute Co., Ltd., Shenyang, 110001, China
e-mail: Zhibin.Zhao@sami.com.cn

W. Liu
e-mail: liuwei@sami.com.cn

Y. Liu
Aluminum Reduction Department, Shenyang Aluminium and Magnesium Engineering and Research Institute Co., Ltd., Shenyang, 110001, China
e-mail: lyf@sami.com.cn

M. Ren
Sunlightmetal Consulting Inc, Ontario, Canada
e-mail: michael.ren@sunlightmetal.ca

Z. Wang
Northeastern University, Shenyang, 110819, China
e-mail: wangzw@smm.neu.edu.cn

© The Minerals, Metals & Materials Society 2023
S. Broek (ed.), *Light Metals 2023*, The Minerals, Metals & Materials Series,
https://doi.org/10.1007/978-3-031-22532-1_16

thermal balance of aluminum reduction pots. A new anode normally requires 16 to 28 h of heating to gradually recover to the electrolysis temperature [7]; (b) newly settled anodes wrapped by a layer of insulating solidified electrolyte generate remarkable horizontal current in aluminum pad [8]. The magneto-hydrodynamic (MHD) stability of the cell is strongly impacted by this operation, and it also needs about 24 h to pick-up the full anode current. The adverse impact of anode change is embodied in the loss of current efficiency. In one such study, a current efficiency loss of 2.2% was reported for each anode change [9].

With the fast development of electrolytic technology, the research on anode change has become a new hot topic and interest point. Hershall [10] calculated the total energy required to bring the new anode to a steady state from the views of pot resistance and heat required. EGA and Alcoa developed anode preheating technology using an external heat equipment [11, 12], and got very positive results. EGA also tried to improve the recovery rate of the new anode current by optimizing the sequence of anode change [13]. CHINALCO Qinghai Branch made similar attempts [14]. Very recently, some scholars began to investigate the anode change process by coupling multi-field and bath solidification and melting together [7, 15].

The first part of this paper theoretically calculates the heat absorption of new anodes and heat generation or release of other parts, such as the anode set modifier, metal/bath temperature decrease, and electrolyte frozen. The second part illustrates that the anode change in one pot has an individualized characteristic, for which different modifiers should be applied. The third part discusses the effect of anode preheating both from the view of numerical simulation and some published industrial tests.

Theoretical Calculations of Heat Absorption and Release

A 500 kA cell with 48 anodes was chosen as our physical geometry. The anode size and other physical parameters were used to analyze the theoretical heat absorption and heat generation or release of several parts after anode change.

Heat Absorption by New Anodes

The weight or mass of one anode in a 500 kA cell is 1,161 kg. The specific heat of anode is 889 J/kg-K. The related parameters and physical properties are listed in Table 1.

Table 1 Physical properties of anodes

One Anode	Value	Unit
Mass	1,161	kg
Specific heat	889	J/kg-K

The temperature of new anodes before anode change equals to the ambient temperature of 30 °C and is picked here.

After the current of the new anodes is completely recovered, the temperature of the bottom part immersed in the electrolyte is close to bath temperature of 952 °C. The temperature of the upper part out of the cover materials is around 190 °C [16]. If we assume that the temperature of the anode carbon is distributed linearly along the height, the weighted average temperature of the total anode is 571 °C.

In this case, the total heat absorption can be calculated as:

$$\text{Specific heat (Cp)} * \text{anode mass (m)} * \text{temperature increase } (\Delta t)$$
$$= 889 \,\text{J/kg-K} * 1,161 \,\text{kg} * 541 \,\text{K} = 558 \,\text{MJ}$$

For convenience of comparison, we convert the heat to equivalent voltage for a 500 kA cell in 24 h:

$$\text{Equivalent Voltage} = 558 \,\text{MJ}/500,000 \,\text{A}/(24 * 60 * 60) \,\text{s} * 1000$$
$$= 12.9 \,\text{mV}$$

Two anodes need to absorb 25.8 mV energy.

Heat Generation by Anode Set Modifier

In the industry practice for anode change, the anode set modifier is always applied to increase heat generation and anode cathode distance (ACD). Each smelter has its unique modifier based on its operating situation. Table 2 lists three kinds of modifiers.

In Smelter A, the initial anode set modifier increases the pot voltage by +45 mV for 40 min in the first stage, then decreases to +30 mV for 40 min in the second stage, and finally decreases to +15 mV for 40 min. Smelter B and Smelter C have similar strategies, but the detailed values are different.

Most Chinese Smelters choose similar modifiers.

It should be noted that most Chinese smelters apply pot resistance in another form (resistance ohm * current I) instead of pot resistance (ohm) in western smelters.

For comparison purpose, the generated energy by the anode set modifier is converted to equivalent voltage for a 500 kA cell in 24 h:

Table 2 Heat generation by the anode set modifier in different Smelters

Smelter A		Smelter B		Smelter C	
Modifier magnitude (mV)	Modifier duration (min)	Modifier magnitude (mV)	Modifier duration (min)	Modifier magnitude (mV)	Modifier duration (min)
45	40	75	45	90	60
30	40	50	45	60	60
15	40	25	45	30	60

$$\text{Current (I)} * \text{anodeset modifier magnitude (V)} * \text{anodeset modifier duration(t)}/\text{Current (I)}/24 \text{ hours}$$
$$= (500,000 \text{ A} * 45 \text{ mV} * 40 \text{ min} + 500 \text{ kA} * 30 \text{ mV} * 40 \text{ min} + 500 \text{ kA} * 15 \text{ mV} * 40 \text{ min})/(500,000 \text{ A} * 24 * 60 \text{ min})$$
$$= 2.50 \text{ mV}$$

Using the same method, the equivalent voltage of the anode set modifier in Smelter B is 4.69 mV, and the equivalent voltage in Smelter C is 7.50 mV.

In the summary section, the data of Smelter B is chosen for comparative analysis.

Heat Release by Temperature Decrease of Bath/Metal

The introduction of cold anodes would destroy the original thermal balance in the electrolytic cell. The first and most obvious observation is the decrease of melt temperature and super-heat.

We measured the temperature (average value of tap-end and duct-end) variations of 10 cells before and after anode change. The data included cases of 4 corner anodes and 6 internal anodes. The bath temperature decreased to the lowest point about 2 h after anode change. The average temperature drop was 4.8 K. Figure 1 selects four representative curves of bath temperature variation.

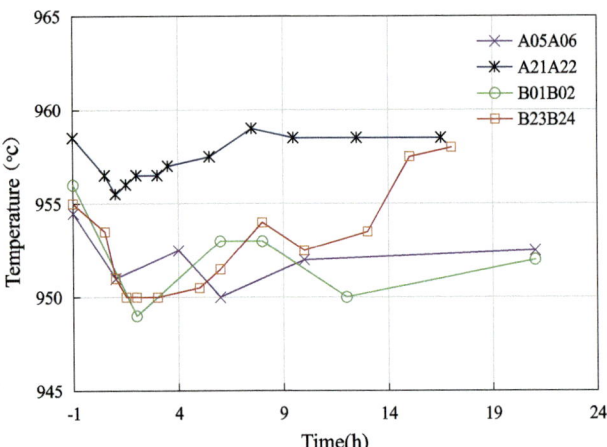

Fig. 1 Bath temperature variation before and after anode change

The total mass of the bath in a 500 kA cell is about 10,000 kg. The specific heat of the bath is 1,760 J/kg-K. The total mass of molten aluminum is about 35,000 kg, and its specific heat is 1,080 J/kg-K. The physical properties of the bath and metal are shown in Table 3.

Based on data in Fig. 1, the bath temperature or super-heat is decreased by 4.8 K. The theoretical released energy can be calculated as:

$$\text{Specific heat (Cp)} * \text{bath mass (m)} * \text{temperature decrease (}\Delta t)$$
$$= 1,760 \text{ J/kg-K} * 10,000 \text{ kg} * 4.8 \text{ K} = 84.5 \text{ MJ}$$

For comparison reason, the released energy by temperature decrease of the bath is converted to equivalent voltage for a 500kA cell in 24 h:

$$\text{released energy (E)}/\text{Current (I)}/24 \text{ hours}$$
$$= 84.5 \text{ MJ}/500,000 \text{A}/(24 * 60 * 60) \text{ s} * 1,000 = 1.96 \text{ mV}$$

Because of the excellent thermal conductivity of molten aluminum, the temperature of the metal decreases as the electrolyte temperature decreases. Similar to bath calculation, the theoretically released energy by temperature decrease of the metal can be calculated as:

$$\text{Specific heat (Cp)} * \text{metal mass (m)} * \text{temperature decrease (}\Delta t)$$
$$= 1,080 \text{ J/kg-K} * 35,000 \text{ kg} * 4.8 \text{ K} = 181.4 \text{ MJ}$$

For comparison reason, it also converts to equivalent voltage:

$$\text{released energy (E)}/\text{Current (I)}/24 \text{ hours}$$
$$= 181.4 \text{ MJ}/500,000 \text{ A}/(24 * 60 * 60) \text{ s} * 1,000 = 4.20 \text{ mV}$$

Heat Release by Electrolyte Solidification

Taylor [17] have published their findings that the maximum freeze mass given corresponds to 7 cm of solid bath around

Table 3 Physical properties of bath and metal

Bath and metal	Value	Unit
Mass of bath	10,000	kg
Specific heat of bath	1,760	J/kg-K
Mass of metal	35,000	kg
Specific heat of metal	1,080	J/kg-K

Table 4 Physical properties of solid electrolyte

Solid electrolyte	Value	Unit
Volume	0.236	m^3
Density	2,130	kg/m^3
Mass	503	kg
Melting heat	520,000	J/kg

anodes. Odegard et al. [18] report that the thickness of the frozen layer reached the maximum value in 1–3 h. In this direction, the freezing electrolyte around two new anodes of 500kA pot can be estimated as 0.236 m^3 (immersion depth 0.15 m, and ACD 0.05 m). During anode change process, these solid electrolytes will re-melting again.

Table 4 shows some physical properties of the solid electrolyte.

Based on calculation, the theoretical heat released by solidification is:

$$\text{Mass of solid electrolyte (m)} * \text{Melting Heat (L)}$$
$$= 503 \text{ kg} * 520,000 \text{ J/kg} = 261.6 \text{ MJ}$$

where it can be converted to equivalent voltage:

$$\text{released energy (E)/Current (I)/24 hours}$$
$$= 261.6 \text{ MJ}/500,000 \text{ A}/(24 * 60 * 60) \text{ s} * 1,000 = 6.06 \text{ mV}$$

Internal Summary

Table 5 summarizes the heat absorption and generation or release of each part in Sects. 2.1–2.4.

After two anodes were replaced, the anode set modifier was applied, the bath and metal temperature were reduced by 4.8 K, the surrounding electrolyte obtained 7 cm of solidification, and there is still 34% lack of heat or energy for one anode change.

It should be noted here that the above calculation is based on a thermal efficiency of 100%. In this direction, the extra energy generated by the anode set modifier is all converted into the temperature increase of new anodes, the heat releases from bath and molten aluminum are all transferred to new anodes, and the heat release by electrolyte solidification is also converted into temperature increase of the new anodes.

In real-world situation of aluminum production, the energy or heat gap may be more than 50%. This should attract the attention of aluminum smelters, especially for some Chinese smelters.

In industrial production, the bath temperature returns to the level before anode change within 24–48 h. But the incremental trend of temperature is not in a linear form. Figure 1 shows four representative curves of temperature recovery, which illustrate an aperiodic fluctuating growth pattern (increase—decrease—increase).

Here we propose a conjecture of the thermal behavior in aluminum reduction cell after anode change:

bath temperature or super-heat decrease—side ledge solidification—insulation effect increase—bath temperature or super-heat increase—side ledge melting—bath temperature or super-heat decrease—temperature slowly increase.

After anode change, the heat effects of the anode set modifier, super-heat decrease and electrolyte solidification are not enough to provide enough heat for new anodes. It is possible that the heat lost to the surrounding atmosphere is reduced by the solidification of the side ledge, and the reduced heat lost compensates the cold effect of new anodes. This speculation needs further test verification, which would be another piece of work.

Table 5 Heat absorption and release of each part after anode change

500 kA pot	Anode absorption	Anode set modifier	Temperature decrease of bath	Temperature decrease of metal	Bath solidification	Heat gap
Voltage (mV)	−25.80	4.69	1.96	4.20	6.06	−8.89

Individualized Characteristic of Anode Change in Same Pot

In our measurement of temperature variation before and after the anode change, we found an unique phenomenon that the maximum temperature drop in the corner anode change was 5.7 K (4 pots) and that of the internal anode change was 4.3 K (6 pots).

It is possible that this happened because the measurement points were close to new anodes. In order to get a better understanding, we made more statistical analysis on the current recovery rate of all anodes. Statistics were done on the anode current recovery of 8 pots within 2 months.

Investigation of Individuation on Anode Change

Figure 2 shows the pick-up rate of anode current within 24 h after anode change in different regions of the electrolytic cell. The data of anode current is linear fitted, of which the slopes of linear fitting lines can approximately represent the recovery rate of the anode current.

In order to get a better comparison, the slopes of 24 fitting lines are plotted as an histogram in Fig. 3. It can be seen that the slopes of current recovery in the middle region of A11 to A13, corner region of B01 to B02, B21 to B24 and A23 to A24 are relatively smaller than those in other regions. These anodes can be named 'sensitive anodes'.

Combined with the flow pattern of the bath in Fig. 4 (cited from Ref. [19]), it is speculated that the distribution of 'sensitive anodes' is closely related to the flow pattern. The faster bath flow would bring more energy or heat to new anodes for better recovery of temperature and current, while this effect is not obvious in these 'sensitive regions'.

A Proposal of 'Customized Technology for Anode Set Modifier'

Aimed to the problem mentioned in Sect. 3.1, this section proposes a customized technology for anode change. The current pick-up rate of 'sensitive anodes' could be improved by increasing the anode set modifier, which means the anode change in different regions could be applied with different voltage modifiers. We named this technology 'customized technology for anode set modifier'.

The following research took corner anodes as representative of 'sensitive anodes'.

Numerical Works

Our previous research developed a numerical model coupled Solidification and Melting model to investigate the coupled thermo-electric-flow behavior after anode change [20]. The numerical results agreed with the phenomenon shown in Fig. 3. The model was applied here to simulate temperature recovery after corner anode change in different anode set modifiers.

Figure 5 lists two curves of bath temperature recovery for corner anodes with different anode set modifiers. Case 3 is the case with a larger modifier magnitude, and Case 4 is the case with a longer modifier time. Case 1 is the base case which chose the real anode set modifier from an operating smelter.

The numerical simulation shows that both modifiers play a positive role on the recovery of liquid temperature. The effect in Case 3 is very obvious in a short time. As the time continued, the effect of Case 4 became more obvious, and the electrolyte below the new anodes recovered to a higher temperature than that in Case 3 after 4 h.

Considering the anode set modifier is applied to the whole pot in a short period of time, it is very possible that some local or global thermal issues would be triggered by the melting of the side ledge. It is necessary to conduct industrial trials.

Industrial Trials

Figures 6 and 7 show the current pick-up lines of corner anodes under different modifiers. The smelter usually makes some adjustments for the anodes with abnormal pick-up current after 16–20 h of anode change, thus the data in 16 h after anode change are chosen in this section.

It can be seen that the current pick-up rates of the test cells with two modifiers are higher than that of reference cells. According to the fitting lines, the pick-up rate of the anode current of the test cells in Fig. 6 is 8% higher than that in reference cells. The current pick-up rate of the test cells in Fig. 7 is 11% higher than that in reference cells. The measured bath temperature shows there is no clear difference between the test cells and reference cells. The bath temperature varies in reasonable values and returns to the level before anode change after 24 h.

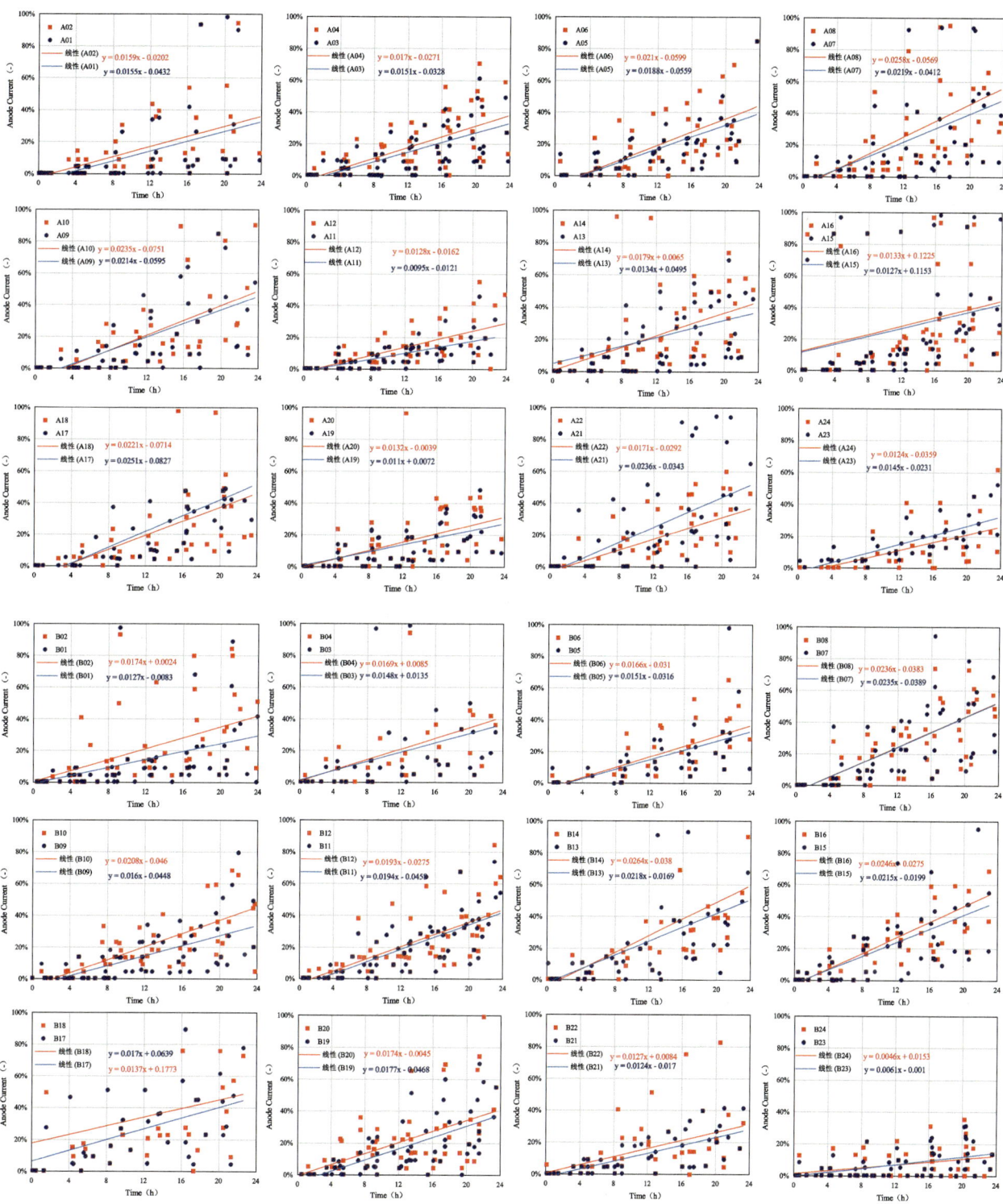

Fig. 2 Anode current recovery of all anodes

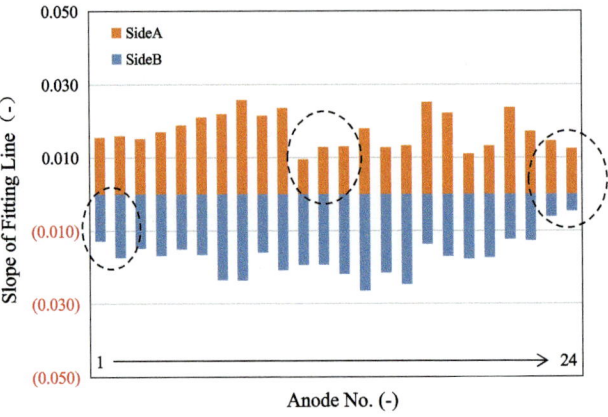

Fig. 3 Slopes of fitting lines of current pick-up rate

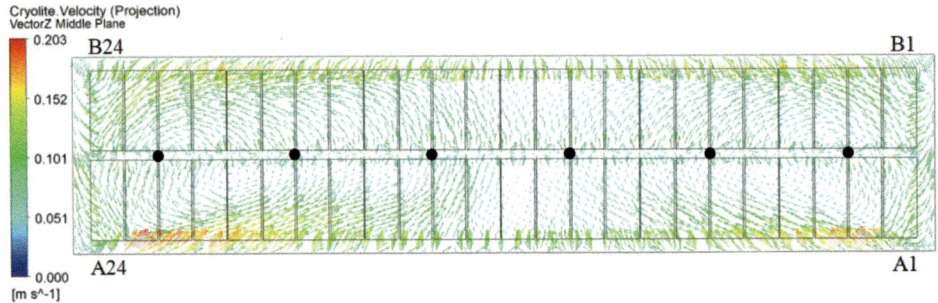

Fig. 4 Bath flow in ACD (Cited form Ref. [19])

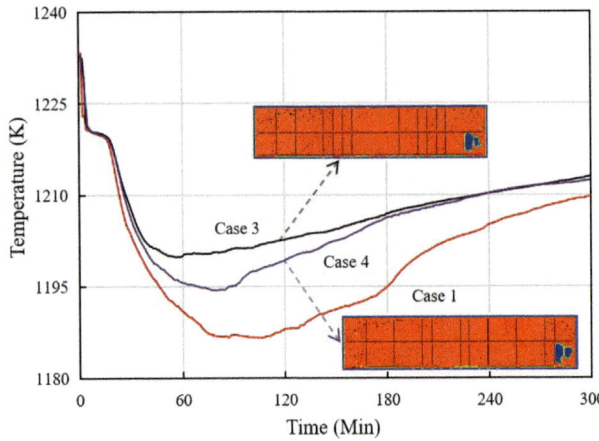

Fig. 5 Temperature recovery for corner anodes with different anode set modifiers

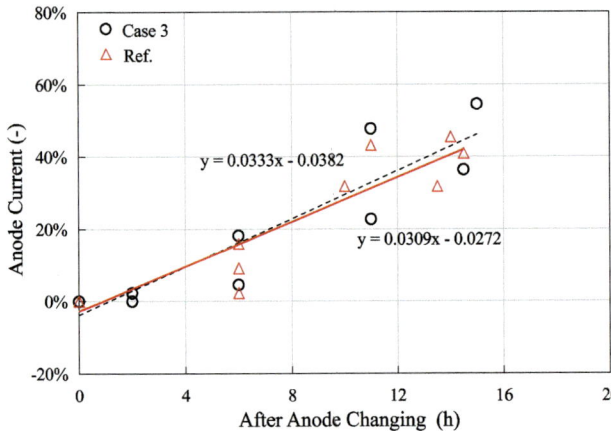

Fig. 6 Current pick-up lines of corner anodes similar with Case 3 (measured data)

The aluminum reduction cell is very sensitive to the instantaneous large heat input. The bath temperature rises sharply after anode effect, and the surrounding side ledge is very easy to melt during this process [21]. It should be more advantageous to use a relatively gentle energy increase scheme such as prolonging the modifier time of the anode change.

After the industrial test, all pots in this smelter were applied to the second anode set modifier for their corner anode change. The current recovery of corner anodes was significantly improved.

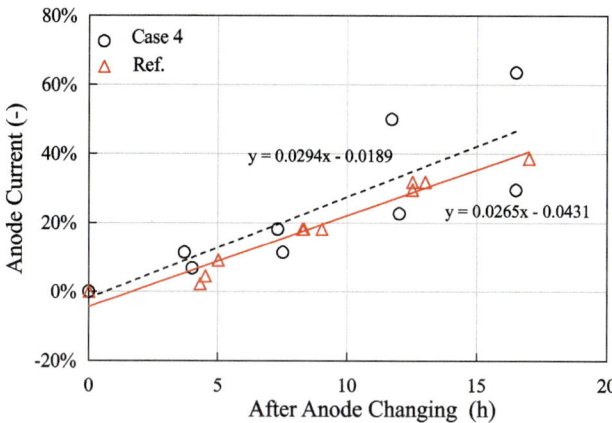

Fig. 7 Current pick-up lines of corner anodes similar with Case 4 (measured data)

Attempts of Anode Preheating

In order to reduce the huge thermal impact caused by anode change, the most popular technology is anode preheating.

Numerical Works

A previous model [20] is also used here to simulate the effect of anode preheating. Figure 8 shows the temperature recovery in two cases with two kinds of preheated anodes (Case 5—initial temperature 120 °C and Case 6—initial temperature 300 °C). It can be seen that the effect of anode preheating is very clear. The maximum temperature drop is significantly smaller than that of cold anodes. The recovery time of the temperature is also shorter than that of the cold anode change.

The adjustment of the anode set modifier is to perform additional energy supplement in the whole pot, which will inevitably cause some heat loss. Setting preheated anodes can heat the new anodes in advance, and its effect is bound to be better than the modifier adjustment in Case 3 or Case 4.

In addition, it is also possible to explore new energy-saving schemes from the perspective of reducing the amplitude or duration of the anode set modifier for all preheated anode changes.

Industrial Trials

Alcoa [11] and EGA [12] made some industrial attempts of replacement of preheated anode change.

Alcoa preheated six anodes per day, and the maximum temperature reached 500 °C. These preheated anodes were used to supply two test pots for anode change. Compared with the reference cell using cold anodes, the current recovery rate was 25% faster in the first hour, more than 40% faster in the second hour, and 11–15% faster after four hours. After the test, the current efficiency was increased by 0.5–1.0%, and the energy consumption was reduced by 40 kWh/t-Al [11].

EGA preheated the anodes by adding gas combustion pre-heat stations. The target temperature was 470–500 °C, and the final temperature for the anode change was 300 °C. Within 6 h after anode change, the recovery rate of the anode current was 42% faster than that of cold anodes [12].

Marc Dupuis and scholars in SINTEF [22] proposed a new concept of heating new anodes with waste heat from butts. In the lab experiments and numerical simulation, a cold anode block was put in contact with the bottom surface of a butt. From the thermal contact, it was expected approximately 0.2 kWh/kg Al would be contributed by the heated anode.

However, there are no further public reports on an industrial scale application. The extra investment of heating furnace may be a possible reason.

Conclusion

This paper theoretically calculated the heat absorption of new anodes and heat generation or release of other parts, such as anode set modifier, metal/bath temperature decrease, and electrolyte solidification. Even based on thermal efficiency of 100%, the energy gap may be more than 50% for each anode change.

The anode change in one pot has an individualized characteristic. This results in some 'sensitive anodes' with lower anode current pick-up rate. The distribution of 'sensitive anodes' is closely related to the flow pattern.

Aimed to 'sensitive anodes', a 'customized technology for anode set modifier' is proposed for different anode changes.

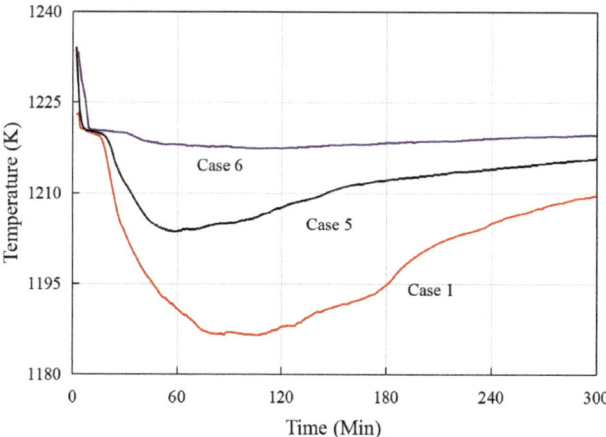

Fig. 8 Temperature recovery for two kinds of preheated anodes

The numerical simulations have proven that this technology played a positive role in improving heat behavior of 'sensitive anodes', and industrial tests have proven that it played a positive role in improving electrical behavior without any thermal issues.

References

1. Zhou Dongfang, Yang Xiaodong, Liu Ming, Liu Wei. Chinalco 600kA High Capacity Low Energy Consumption Reduction Cell Development[J]. Light Metals. 2015, 483–487.

2. Martin Forte, Martin Robitaille, Nicolas Gros, Rene Gariepy, Jean-Pierre Figue. Arvida Aluminum Smelter-AP60 Technological Center, Start-Up Performance and Development of the Technology [J]. Light Metals. 2015, 495–498.

3. Yungang Ban, Jihong Mao, Yu Mao, Jing Liu, and Gaoqiang Chen. Development and Industrial Application of NEUI600 High Effificiency Aluminum Reduction Cell[J]. Light Metals. 2018, 705–713.

4. Viktor Mann, Andrey Zavadyak, Iliya Puzanov, Vitaly Platonov, and Vitaly Pingin. RA-550 cell technology: UC RUSAL's new stage of technology development[J]. Light Metals. 2018, 715–719.

5. B. Benkahla, Y. Caratini, H. Mezin, S. Renaudier, S. Fardeau. Last development in AP50 cell[J]. Light Metals. 2008, 451–455.

6. Marc Dupuis, Barry Welch. Designing cells for the future–Wider and/or even higher amperage[J]. Aluminium 2017(1-2): 45-49.

7. Qiang Wang, Baokuan Li, Mario Fafard. Effect of anode change on heat transfer and magneto-hydrodynamic flow in aluminum reduction cell[J]. JOM 2016(68): 610-622.

8. Valdis Bojarevics, Sharnjit Sira. MHD stability for irregular and disturbed aluminium reduction cells[J]. Light Metals. 2014, 685–690.

9. R.T. Poole, C. Etheridge. Aluminum reduction cell variables and operations in relation to current efficiency[J]. Light Metals. 1977, 163–182.

10. Hershall Wayne Cotten. Understanding the basic requirements of the anode set modifier. Light Metals. 2016, 577–581.

11. Otavio Fortini, Srinivas Garimella, Edwin Kuhn, et al. Experimental studies of the impact of anode Pre-heating[J]. Light Metals. 2012, 595–600.

12. Ali Jassim, Sergey Akhmetov, Barry Welch, Maria Skyllas-Kazacos, Jie Bao and Yuchen Yao. Studies on anode Pre-heating using individual anode signals in Hall-Héroult reduction cells[J]. Light Metals. 2016, 623–628.

13. Ali Jassim, Sergey Akhmetov, Barry Welch. Studies on background PFC emission in Hall-Hearoult reduction cells using online anode current signals[J]. Light Metals. 2015, 545–550.

14. Zhou Hong. Application of 'Four low and one high' anode changing method to production management of aluminum reduction[J]. Light Metal, 2013(7): 33-35. (In Chinese)

15. Hongliang Zhang,Ling Ran,Jinding Liang,Tianshuang Li,Jie Li. Study on 3D full cell ledge shape calculation and optimal design criteria by coupled thermo-flow model[J]. 2018, 587–596.

16. SAMI internal Report 2018.

17. M. P. Taylor, B. J. Welch. Bath/Freeze heat transfer coefficients: experimental determination and industrial application[J]. Light Metals. 1985, 781–789.

18. R. Odegard, A. Solheim and K. Thovsen. Current pickup and temperature distribution in newly set pebaked hall-heroult anodes [J]. Light Metals.1992, 457–463.

19. Zhu Jiaming, Li Jie, Zhang Hongliang. CFD Investigation of Bath Flow and Its Related Alumina Transmission in Aluminum Reduction Cells: Slotted Anodes and Busbar Designs[J]. Metals, 2020, 10(6): 805-820.

20. 21. Zhibin Zhao, Wei Liu, Xiaodong Yang. Numerical Simulation of Thermo-electric-flow Multi-field after Anode Change and Analysis of Additional Voltage Strategies in Aluminum Electrolysis Cells[J]. Light Metal, 2020(05): 26-30. (In Chinese)

21. Ketil A. Rye, Trygve Eidet and Knut Torklep. Dynamic ledge response in Hall-Heroult Cells[J]. Light Metals. 1999, 347–352.

22. Marc Dupuis, Henrik Gudbrandsen and Kristian Etienne Einarsrud. Heating New Anodes Using the Waste Heat of Anode Butts Establishing the Interface Thermal Contact Resistance[J]. Light Metals. 2021, 676–689.

Development and Deployment Measures in PLC-Based Pot Control System at Low Amperage Aluminium Reduction Cell

Rajeev Kumar Yadav, Shanmukh Rajgire, Md. Imroz Ahmad, Goutam Das, Ravi Pandey, Mahesh Sahoo, and Amit Gupta

Abstract

Hirakud 85 kA, end-to-end potline was converted to prebake with GAMI technology during 2006–2009. Later, it was observed that, in the existing pot control system, the application software and control modules are encrypted and hence difficult to modify in line with process requirements. Therefore, control logic and PLC-based system were developed in-house, in correspondence to the Hirakud pot design and process. During the implementation of the new control system, the primary challenges were the acceptability on shop floor and high pot instability due to sludge forming tendency. Also, preventive and corrective measures are delayed due to high manually intensive operations. To mitigate the above challenges, the following actions were taken. (a) Technical training was provided to the shop floor personals. (b) Developed control logic was modified based on pot conditions. (c) SCADA features, additional reporting, and auto mailing were developed. This article would provide detailed insight into the development and deployment measures.

Keywords

Control logic • Control system • Operations • Pot instability • SCADA

R. K. Yadav (✉) · S. Rajgire · A. Gupta
Aditya Birla Science and Technology Company (P) Ltd., Mumbai, India
e-mail: rajeev.yadav@adityabirla.com

Md.I. Ahmad · G. Das · R. Pandey · M. Sahoo
Hindalco Industries Ltd., Mumbai, India

Introduction

Aluminium (Al) is extracted from its oxide, alumina (Al_2O_3), by the Hall-Héroult process. In this process, Al_2O_3 is fed to the system through point feeders, it gets dissolved in molten cryolite (Na_3AlF_6), and is electrolyzed to produce Al (l) and CO_2(g). This process operates in the range of 940–980 °C and produces 99.5–99.8% pure Al. A few additives such as AlF_3, MgF_2, CaF_2, and LiF are added to the molten cryolite (Na_3AlF_6) to lower its liquidus temperature and for more efficient electrolysis of dissolved alumina (Al_2O_3) [1]. This process is highly energy-intensive with electricity comprising 30–40% of the cost of production; the schematic diagram of this process is shown in Fig. 1. Modern smelters run at current efficiencies (CE) close to 94% with a benchmark of 96% [2]. Higher CE can be gained by regulating and controlling alumina in the optimum range provided that other operational practices must be followed according to the standard operating procedure (SOP).

Nowadays, most aluminium smelters have point-feed technology for alumina feeding, which involves a cylinder-type volumetric feeder, operating in coordination with a crust breaker [3]. In general, alumina addition is supplied in a batch-wise manner by the feeder, where a volumetrically measured amount of alumina (1–3 kg) is delivered into the bath, through a feeder hole at certain time intervals. Initially, feed control strategies were based on the voltage, which is a function of the alumina concentration, anode-cathode distance, electrolyte electrical conductivity, and operating anode current density, as reported by Welch [3].

Different feeding systems such as batch feeding, demand feeding, and distributed feeding have evolved to regulate the alumina in the system. In the batch feeding system, Alumina is used to feed once in a shift into the pot. However, in the demand feeding system, the feed strategy is based on resistance response, as the resistance increases by a set value, which initiates the breaking and feeding into the pot. Distributed, feeding system consists of theoretical (Normal)

© The Minerals, Metals & Materials Society 2023
S. Broek (ed.), *Light Metals 2023*, The Minerals, Metals & Materials Series,
https://doi.org/10.1007/978-3-031-22532-1_17

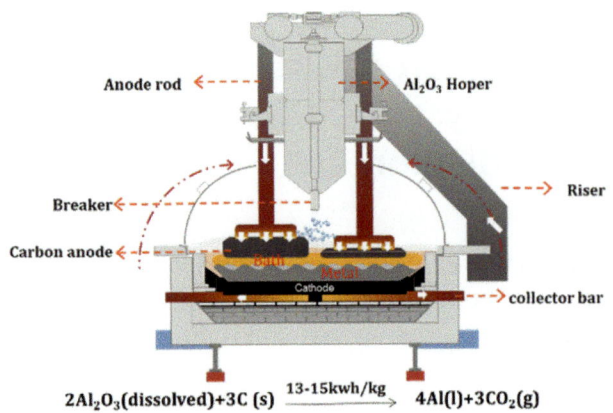

Fig. 1 Schematic cross-section of Hall-Héroult cell

process. These gases form a highly electrically resistive layer underneath the anode, that suddenly increases the pot voltage resulting in an anode effect (AE), which causes thermal disturbance in the pot. At the other extreme, operating at higher alumina concentrations can lead to the formation of "sludge" that is difficult to dissolve and causes instability and high noise. Sludge has a potential detrimental impact on pot life as it can enhance the erosion of the cathode [4]. This paper talks about the scope for improvement in the existing control system and different measures taken to mitigate the challenges faced during the development and deployment of the new PLC-based pot control system.

Scope for Improvement in Pot Control System at Hirakud 85 kA

Since Hirakud 85 kA is the oldest smelter in India, it has been running with very old technology with few outdated components. Therefore, there is a lot of scope for further upgradation with respect to process requirements. A few scopes for improvements related to design, pot control logic, software, hardware, and pot connections are discussed below.

Pot Design and Control Logic

Firstly, the hardware competence in accordance with alumina feeding and distribution was analysed. The number of point feeders was compared to the aspect ratio (ratio of length to width) of the pot, which is the determining factor for alumina distribution to all the anodes. During analysis, it was observed that the number of feeders to the aspect ratio at Hirakud 85 kA is quite similar to the modern smelter as shown in Fig. 2. Further, the quantity of alumina discharge

feeding, underfeeding (UF), and over-feeding (fast feeding) system, and it is performed based on the change in the pseudo-resistance slope (dR/dt), which determines the change in the feed rate. These feed strategies are defined as follows: when the feeding rate of alumina is the same as its theoretical consumption rate as per Faraday's law, it is known as theoretical feeding (TF); for alumina feed rate lower than the theoretical consumption rate, it is known as underfeeding (UF); and for the feeding rate is higher than theoretical consumption, it is described as over-feeding (OF). Distributed feeding systems maintain the pot with leaner Al_2O_3 concentration, consequently increasing the CE and improving the decision criterion for feeding; therefore, modern smelters have adopted this feed system.

Modern smelter approaches are based on the slope of the resistance curve, using the response of pot voltage to perturbations of the feed flow of alumina into the pot. At very low alumina concentrations (<2%), cryolite decomposition occurs and perfluorocarbon (PFC) gases evolve in the

Fig. 2 Number of feeders to the aspect ratio of pot at modern and Hirakud 85 kA pot

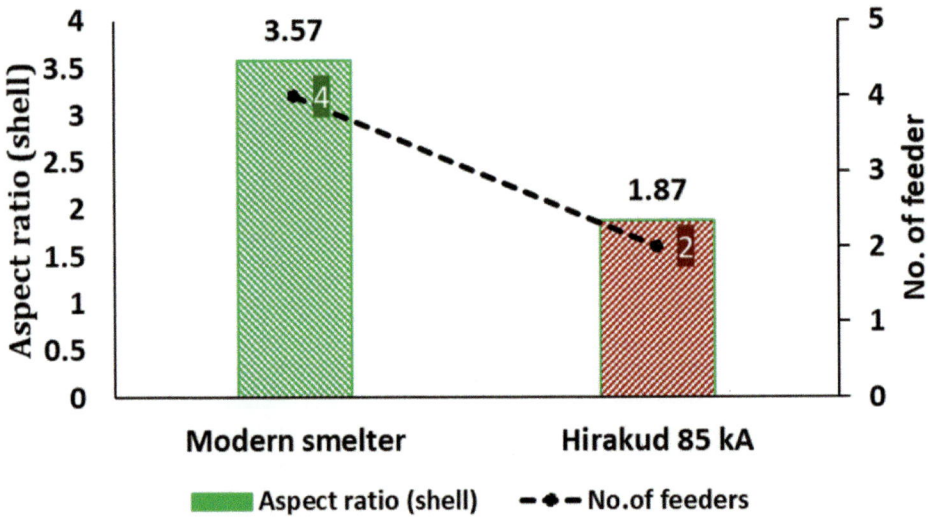

Fig. 3 Ratio of alumina discharge to electrolyte mass at modern and Hirakud 85 kA pot

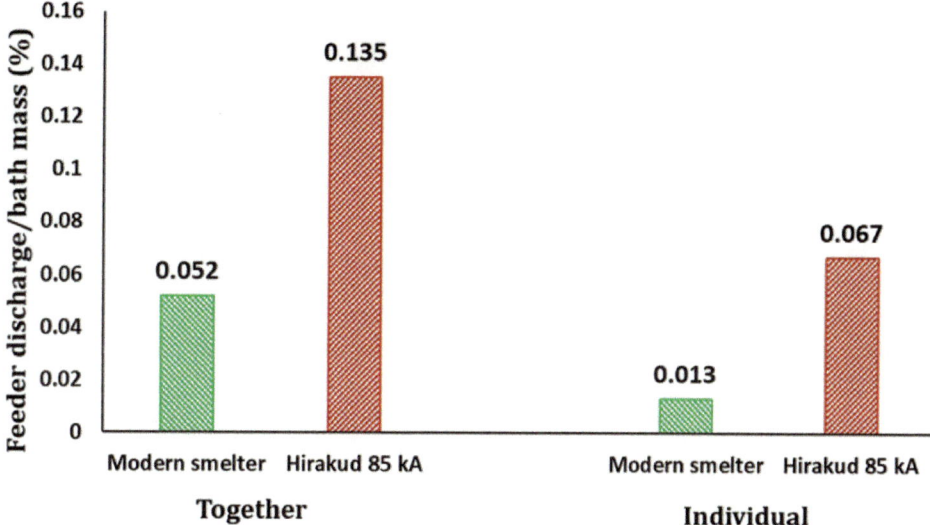

was compared to the electrolyte mass present in the pot. The heat required for the heating and dissolution of alumina is obtained from the adjoining electrolyte. Hence, this factor is an indicator of the dissolution rate of alumina and the tendency for sludge formation. This ratio is compared for the feeders operating together and individually (one at a time). The ratio for Hirakud 85 kA is nearly double of the modern smelter as shown in Fig. 3, which indicates a slow dissolution rate at 85 kA smelter. However, it can be addressed by modifying feed control logic and maintaining optimum superheat in the electrolyte.

Moreover, dynamic feeding systems were also observed in the existing control system, which may cause high variation in the alumina concentration and pot voltage. Therefore, to reduce this variation, the constant over-feeding duration would help to monitor and control the variation of alumina concentration and voltage [5]. The feed rates can be intensified compared to existing ones, and this would increase the number of feed cycles from 5–8 per day to 18–24 per day, helping in maintaining pot at leaner alumina with improved decision-making. It would also help to control the alumina concentration, resulting in reducing the sludge-forming tendency and increasing productivity. In addition to using the pot voltage, the pseudo-resistance would also help in better decision-making for feed strategy by eliminating the impact of fluctuation of line current. The decision criterion would be based on the threshold values of slope (derivative) and curvature (double derivative) of pseudo-resistance.

During the Anode change process, energy loss can be computed, and this would be used to compute the amount and duration of voltage buzz required. Further, these buzzes should be decreased in a stepwise manner by controlling the pot voltage in the operating control band after specific intervals. Stepwise reduction in voltage buzz would help to

reduce the chance of hardware component damage, thermal shock, and pot instability.

Software, Hardware, and Pot Connection System

Due to encrypted application software in the existing control system, it was difficult to modify control logic as per requirement. Moreover, the SCADA software designed in Delphi 7.0 runs over the operating system MS Windows 2000, which has been declared obsolete by Microsoft. Existing hardware components are very old and have turned obsolete, restraining the OEM from providing service and spare support, also the frequent components malfunction. Due to the linear connection of the pots, there was a high possibility of frequent communication interruption and loss of data. The connection break or loss at one pot leads to communication interruption at all other pots.

Development of the New Pot Control System

After identifying the scopes for improvements in the existing system, the number of improvements related to pot control logic, software, and hardware was done, which are discussed in detail.

Pot Control Logic

Pot control logic performs based on the operations being done on the pots. Al_2O_3 concentration plays a key role to control anode effect frequency (AEF), voltage variation, and sludge-forming tendency. Therefore, distributed feed logic has been developed to control alumina concentration in the pot. Moreover, other major control logic such as metal tapping, anode change, anode beam raising, auto anode effect termination, instability control, and automatic voltage control have also been developed and incorporated. The details of the alumina feeding logic are discussed in the following section.

Alumina Feeding

As discussed in the previous section, optimum alumina concentration in the pot has a positive impact on pot performance. Initially, the requirement of alumina per day was identified based on the line current and CE. Further, the theoretical feeding (NB) interval was calculated. Whereas the time interval of other feed strategies (UF and FF) depends on the NB interval. This logic is designed, based on the slope and curvature of the pseudo-resistance. It starts with underfeeding, and continuously checking slope, deviation, and curvature, based on the voltage and alumina concentration control, threshold values for FF trigger have been identified. Figure 4 presents the different feeding rates of alumina feeding in one feed cycle into the pot, which includes underfeeding, ultrafast feeding, and fast feeding, also the impact of feeding on pot voltage along with pot noise, as well as it also shows the number of dumps.

Software, Hardware, and Pot Connectivity

PLC-based, new pot control system has been developed, where the control logic is developed in ladder logic language. Its SCADA software design in FactoryTalk® View, which has unique features such as fast data capturing rate for analog and digital, auto data fetch to SCADA, programming editing, and modification that can be done online even when the pot process is in operation. Besides, it improved the reliability against the obsolescence of old control system hardware and operating systems. A new control system with a fiber-optic network was installed with PLC-based control logic. It consists of two redundant communication buffers (Chassis A and B) systems as shown in Fig. 5. Data from the SCADA is communicated with the communication buffers, which are connected to all the pots in ring architecture as shown in Fig. 6. Due to ring topology, this connection ensures 100% network and data availability, because, in case of a cable disconnection from one side, it communicates from the other side. It also enhanced reports, graphics, and fault diagnostics tools for better operational and maintenance control.

Network Architecture of Pot Control System

Initially, pot control logic was developed and incorporated into the PLC-based control system. The network architecture of the pot control system is shown in Fig. 7. The network system has been arranged in such a way that the input signal is coming from the pot to pot controller, and further is being communicated with the process control system. The process control system further takes the action based on the input signal and sends it to the pot controller for further execution. The process control system is also connected to the SCADA system, where output/pot performance is observed, and based on the performance, certain manual input can also be provided from SCADA to the control system.

Fig. 4 Historical trend with alumina feed strategy, voltage, noise, and dumps

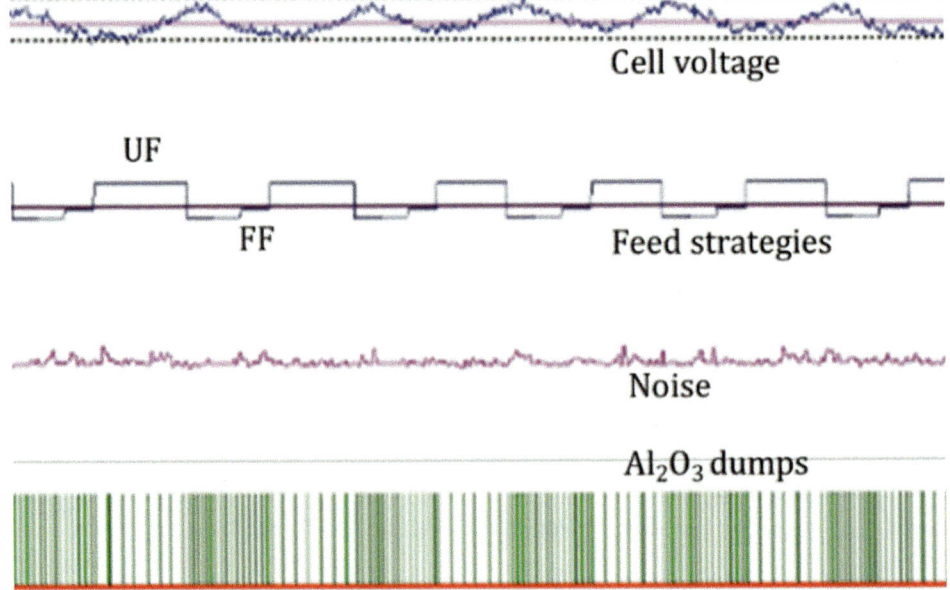

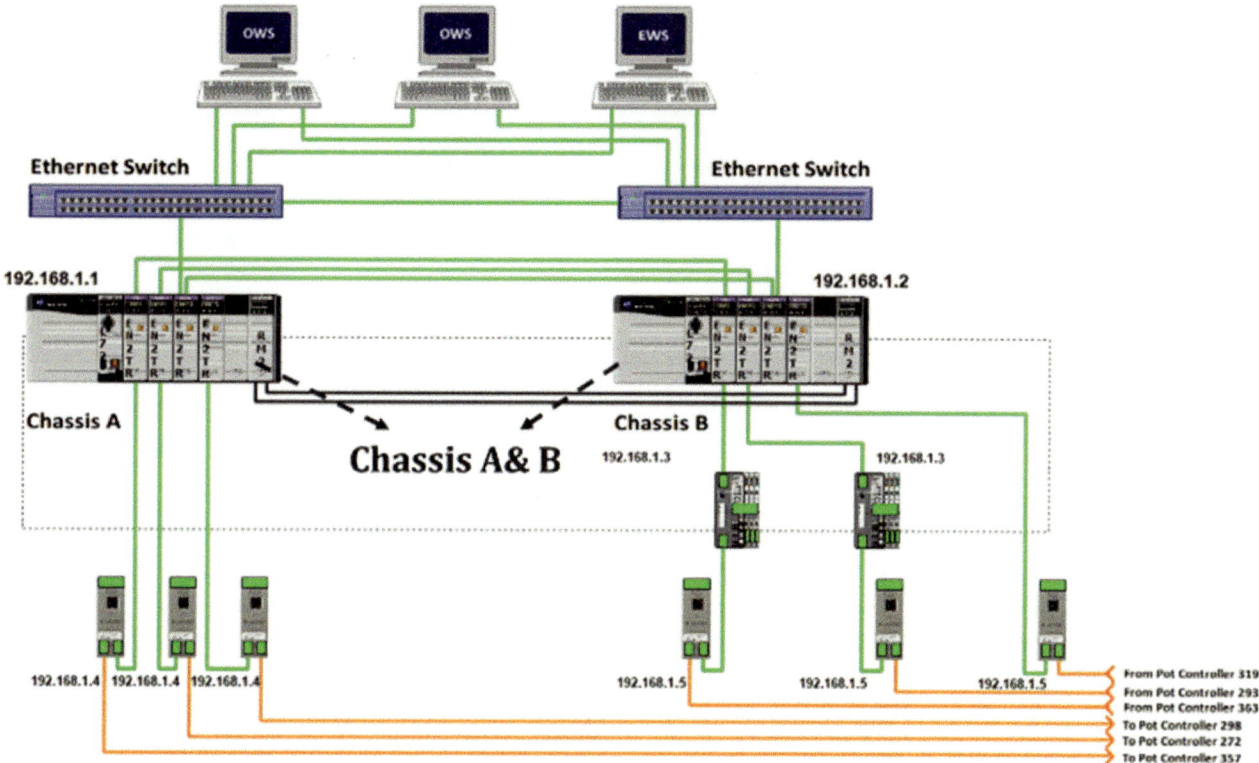

Fig. 5 Buffer PLC system architecture

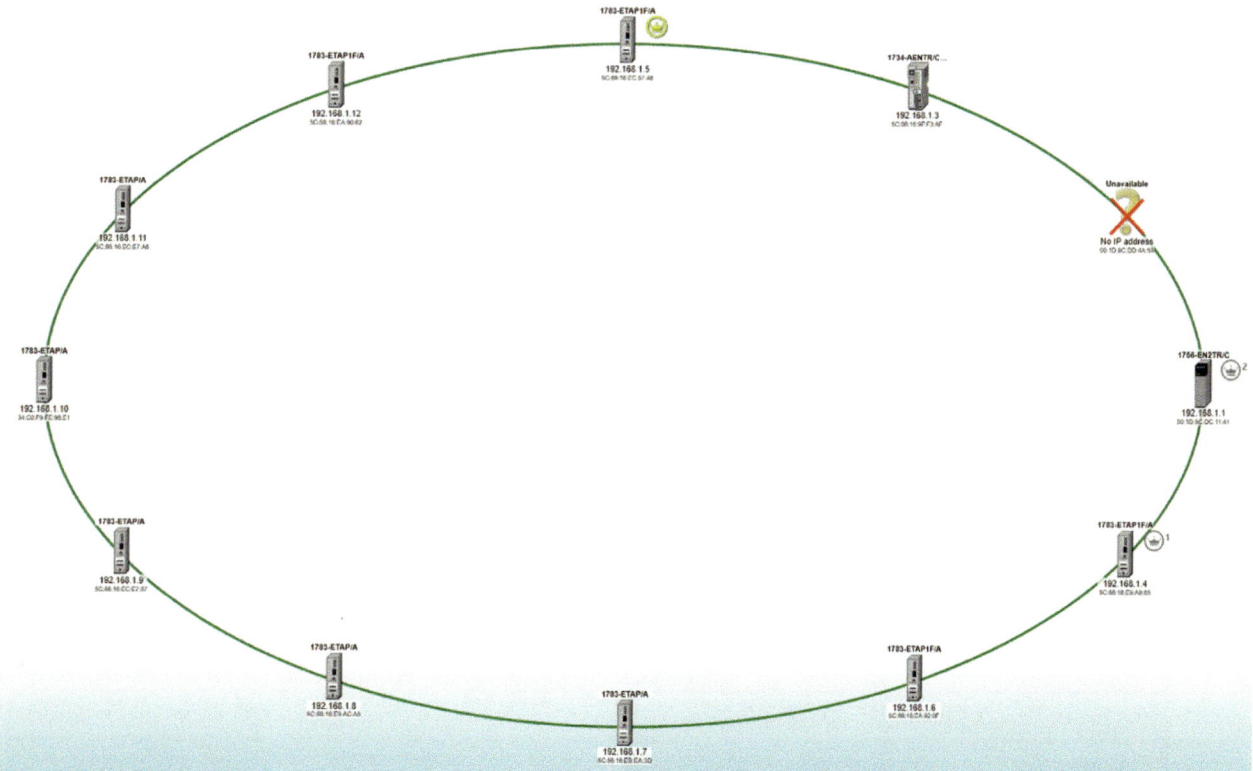

Fig. 6 Ring architecture connected with all pots

Pot Design and Anode Layout

In the modern smelter, the central channel varies about from 150 to 250 mm. The sufficient gap in the central channel helps to reduce the chance of Al_2O_3 feeder hole choke and reduces the chance of breaker damage due to impact with the anode. At Hirakud 85 kA, the central channel is 50 mm, which is very small compared with the modern smelter, and the dump weight is also high as discussed earlier section. Due to high dumps weight and small central channel, there is a localized decrease in superheat, and frequent hole choke tendency was observed, which results in multiple anode effects due to scarcity of Al_2O_3 concentration in the bath. To reduce hole choke conditions, hole choke preventive features (alarm and empty breaking) were incorporated, and also for further improvement, feeder size has also been reduced, tested, and installed in a few pots.

Control System Implementation Challenges and Mitigations

After developing the main control logic as discussed in the earlier section, initially, the trial was started in a pot, and for a period of 4 months ,the avg. AEF was observed close to 0.2, which has a significant reduction compared with plant AEF (~ 0.5 per pot/day). Based on the trials, the new control system was extended to more pots. During implementation, a few challenges were observed such as hardware faults, pot abnormalities, maintaining parameters such as bath height

(BH) and metal height (MH), delays in scheduled operations (mainly metal tapping), frequent feeder hole choke, and occurrence of sludge forming tendency in few pots. These challenges and their mitigation have been discussed in the following sections.

Hardware-Related Challenges and Mitigations

The major challenges associated with hardware are frequent anode slip incidents and alumina feeding quantity deviating from the theoretical requirement. This leads to an increase in the noise and disturbs the feeding cycle. Actions taken to mitigate these challenges are discussed in the following section.

Anodes Slip Alarm

Anode-cathode distance (ACD) is the main operational parameter that can control the pot voltage. Anodes are clamped to the anode beam in such a way that the bottom surface of all the anodes in the electrolyte is at the same level with uniform current density. This controls the heave at the bath-metal interface, thus improving the pot's stability. As an anode slips, higher current flows at the corresponding anode and it disturbs the bath-metal interface leading to an increase in pot instability. Therefore, anode voltage drop (AVD) measurement is planned, when observed in historical trend, to identify the anode slip condition. Due to the delay in identifying the anode slip condition, pot instability increases and reduces current efficiency significantly.

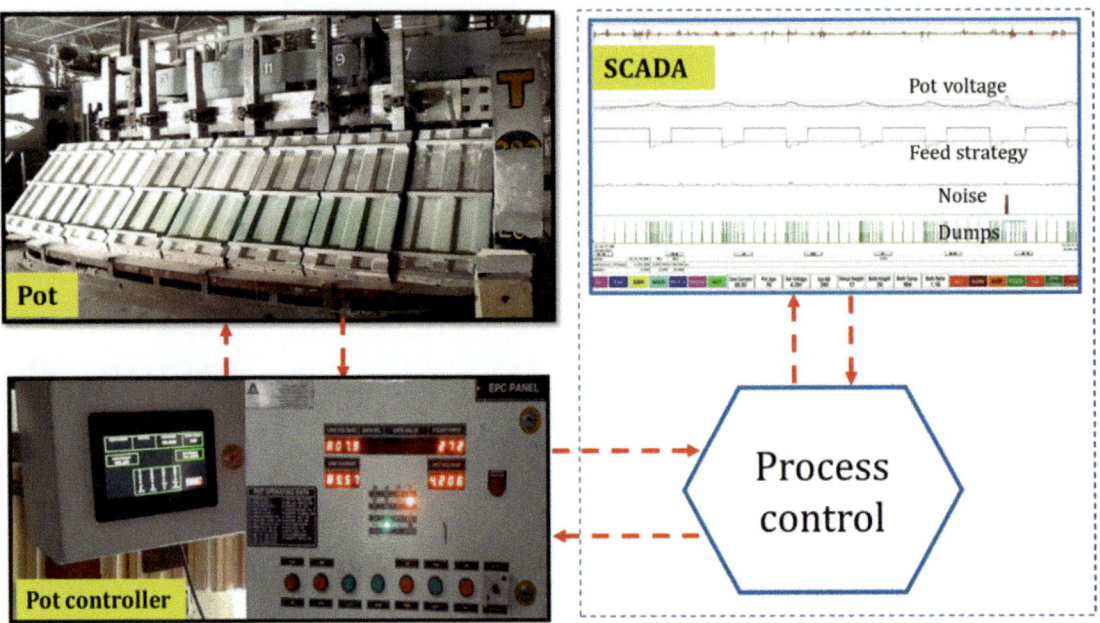

Fig. 7 Network architecture of PLC-based pot control system

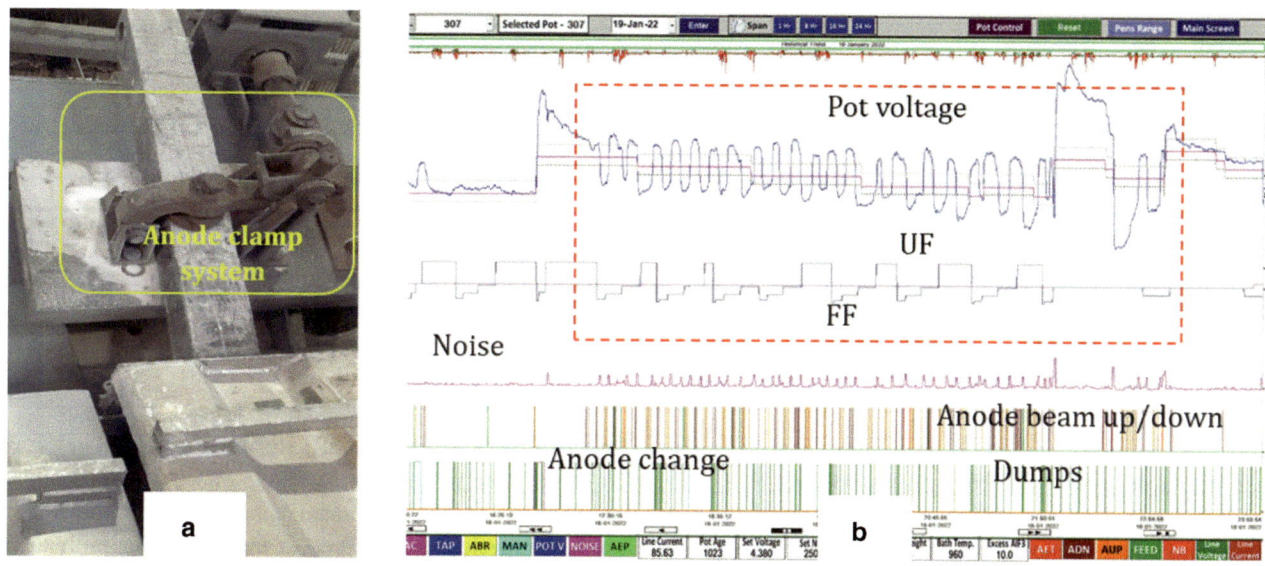

Fig. 8 **a** Anode clamping system; **b** Pot voltage fluctuation after anode clamp gets loose

Therefore, based on the "curvature" response, an anode slip alarm algorithm has been developed to predict and generate an alarm from the anode slip. A prediction alarm helps to immediately address the pot with anode slip and takes corrective action, thus reducing the pot's instability.

Anode Clamp System

Unlike modern smelters, anode clamps are being tightened manually at Hirakud 85 kA. When the anode clamp gets loose, there is high fluctuation in the pot voltage, resulting in poor decision criterion and pot instability as shown in Fig. 8. To overcome this challenge, control logic has been developed based on the voltage response. As this condition occurs, anode beam movement is restricted and an alert to check the anode clamp is generated.

Dump Test Alarm

Sometimes the number of dumps is consistently more or less than the theoretical requirements. It may happen because of mechanical wear of the volumetric feeding system used at Hirakud 85 kA. This leads to variation in the alumina discharge quantity from the feeder. Hence, the dump test is carried out in a scheduled manner every month. But for better timely correction, a "dump test alarm" has been introduced, based on the deviation in the designed feed cycle and the theoretical number of dumps per day. This alarm helps to identify the pot with high or less feeding, and further, it can be corrected after the dump test.

Shop Floor Challenges and Mitigation

During implementation on the shop floor, the major challenges were with acceptability, because it was a new system as well as operational and visualization differences. Also, due to high manual dependency, there was a delay in identifying pot abnormalities; therefore, preventive and corrective actions were getting delayed, which results in hampering the pot's performance.

Pot Performance Monitoring and Actions

The software application-based SCADA (supervisory control and data acquisition) system is being used in the industry to control/visualize the pot performance. It collects real-time data and performs the corrective actions and displays them as an output on the screen (monitor). It also helps to modify the technical parameters for individual pots or lines. Since, in the old system, there was no provision for auto sludge control, time-dependent temporary voltage, control tracking (no feeding), and the latest measure and lab data display, therefore, an in-house developed new control system was installed with customized and advanced features in the SCADA system. To observe the pot performance and take corrective action, a historical trend has been modified with the necessary parameters. Initially, it was developed similarly to an existing system, with the following parameters such as potline current, pot room voltage, pot voltage, feed strategies, and Al_2O_3 feeding which are being displayed.

are similar to the old control system; however, for acceptance at the shop floor few advanced critical alerts and essential reports have been developed, which help to address the abnormalities and performance monitoring. Further, for parameters, corrective actions and advancement in SCADA systems with pot control features have also been developed. These features help to control the impact of parameter deviation and pot abnormalities, resulting in improved pot performance.

Acknowledgements The authors gratefully acknowledge the Hindalco Hirakud smelter and ABSTC team, which have facilitated this new development and deployment successfully.

References

1. George E. Totte et al., Handbook of Aluminum Alloy Production and Materials Manufacturing, 1st Edition, Volume 2, 2003, 736 pages.
2. Geir Martin Haarberg (2015) Effects of electrolyte impurities on the current efficiency during aluminium electrolysis, Proceedings of 33rd International ICSOBA Conference, Dubai, Paper AL27, Travaux 44, 29 November - 1 December page-761–767.
3. JING SHI et al. (2021) Advanced Model-Based Estimation and Control of Alumina Concentration in an Aluminum Reduction Cell, The Minerals, Metals & Materials Society, January 2022
4. H. Heidari, Development of Wettable Cathode for Aluminium Smelting, Ph.D. Dissertation (Universite´ Laval, 2012), p. 14.
5. Shanmukh Rajgire et al. (2019) Advancement in Control Logic of Hindalco Low Amperage Pots. 148th TMS Annual Meeting, San Antonio, Texas, USA, 10–14 March 2019

Process Simulation with Tertiary Cyclone for Kaolinite Removal from Amazonian Bauxite

Allan Suhett Reis, Geraldo Magela Pereira Duarte, Eslyn Neves, Geovan Oliveira, and Thiago Jatobá

Abstract

Bauxite is the main ore for metallic aluminum production, consisting of aluminum and iron oxides and kaolinite, a clay mineral commonly found in Amazonian bauxites, as the main carrier of reactive silica. In the process, due to the small particle size, kaolinite is usually removed by attrition and washing of coarse material followed by desliming using hydrocyclones. In the Bayer process, kaolinite reacts with sodium hydroxide, increasing reagent consumption in the process. The beneficiation process at Mineração Paragominas is based on the separation of coarser fractions with higher gibbsite content from the clay minerals, where kaolinite is more concentrated. The separation takes place in hydrocyclones, equipment that inherently presents a by-pass of fine particles for the underflow, consequently, contaminating the concentrate with kaolinite, and increasing the operating cost in the Bayer process. Additional desliming stages were evaluated for a potential increase in kaolinite removal. Industrial surveying campaigns were carried out and mass balances obtained after laboratory characterization and statistical analysis of operational databases. Nageswararao model for hydrocyclones was calibrated and used in tertiary stage simulation. A black-box model was developed for predicting the benefits of bauxite chemical composition. Five routes for bauxite silica reduction were analyzed considering the inclusion of tertiary stages of desliming in cyclones. Process simulations pointed to a potential solution for the reduction of reactive silica in Paragominas' bauxite, setting tertiary cyclone in fine classification circuit (26″ cyclones) as the most attractive solution, with a potential silica reduction of 4.9%, with 0.4% loss in metallurgical recovery of alumina.

Keywords

Process simulation • Bauxite beneficiation • Clay removal • Silica reduction

Introduction

Bauxite is the main ore to produce metallic aluminum. Production is generally carried out by the Bayer process to form alumina, followed by the Hall-Héroult process to carry out the reduction of alumina into metallic aluminum. Some factors that interfere with this process are available alumina (av. Alumina) and reactive silica (re. silica) contents in bauxite [6].

Bauxite is a rock composed of aluminum oxides. Bauxites can be "lateritic" or "karst" types. Lateritic bauxites are formed in equatorial regions and are mostly composed of gibbsite as the main mineral and gangue composed of kaolinite, iron oxides, titanium oxides, and quartz. The Amazonian bauxite is of the lateritic type and is mostly located in Paragominas, Juruti, Trombetas, and Almerim [1, 9].

Kaolinite is a clay mineral commonly found in Amazonian bauxite and Brazilian iron ores. Due to its particle size, kaolinite is usually removed from the process in desliming steps using hydrocyclones. Kaolinite is the source mineral for re. silica. In the Bayer process, kaolinite reacts with sodium hydroxide, increasing the consumption of the reagent in the process and forming a desilication product [3, 8].

Hydrocyclone is a well-known mechanical classifier used for separation of particles of different sizes. Hydrocyclone performance characteristics such as coarse particle recovery to underflow, cut size, and by-pass reduction are related to

A. S. Reis (✉) · G. M. P. Duarte · E. Neves · G. Oliveira
Mineração Paragominas, Est. da Mineração, Km 30, a partir da BR 010, Platô Miltônia, Paragominas, Pará 68625-970, Brazil
e-mail: allan.reis@hydro.com

T. Jatobá
MinPro Solutions, Rua Dona Ana Neri, 379, Mooca, São Paulo, São Paulo 03106-010, Brazil
e-mail: tjatoba@minpro.com.br

© The Minerals, Metals & Materials Society 2023
S. Broek (ed.), *Light Metals 2023*, The Minerals, Metals & Materials Series,
https://doi.org/10.1007/978-3-031-22532-1_18

operating parameters: pressure, solids percentage, apex diameter, vortex diameter, and cone angle [4, 5].

Mineração Paragominas processes lateritic Amazon bauxite. To remove kaolinite, the beneficiation plant has two cycloning stages in two different circuits, fine particles classification circuit (FC) and superfine particles classification circuit (SFC). Seeking to reduce the content of re. silica in bauxite, the work aims to evaluate the installation of tertiary cycloning in the beneficiation plant of Mineração Paragominas to reduce by-pass of fine particles to underflow and consequently reduce re. silica content in Paragominas' bauxite.

Materials and Methods

Paragominas' Beneficiation Plant

Bauxite processing circuit at Mineração Paragominas includes three main classification steps to separate coarser fraction, with higher gibbsite content, from the finer fraction, where most of the kaolinite is concentrated. The first step, also responsible for separating fine particles from pebbles, which follows re-crushing process, is carried out on vibrating screens. The second step, FC, is carried out in 26″ diameter hydrocyclones and separates clay and finer bauxite particles, from midsize particles that feed the ball mill. The third step, SFC, is carried out in 10″ diameter hydrocyclones and is the final step for separation of clay, beneficiation process tailings, to fine bauxite particles, that forms the product. A simplified flowsheet of the processing plant is presented in Fig. 1.

More than 80% of Paragominas' bauxite is processed in hydrocyclones, equipment that presents a relative low classification efficiency regarding the presence of fine particles on the coarser flow being carried by respective water partition [2]. As a solution to mitigate this factor, the current

plant design is composed of classification circuits in two sub-steps, where the secondary sub-step processes the product (coarser flow) from the primary one, reducing the residual fraction of deleterious fine particles on the classification circuit product.

Mass Balance, Model Fit, and Process Simulation

The applied method consisted of carrying out an industrial surveying campaign, followed by laboratory treatment, technical and historical data gathering through statistical treatment on beneficiation plant instrumentation and control data, model fitting of the Nageswararao hydrocyclone model [7], definition of an empirical model for estimating grades based on granulochemical vectors and performance simulations for different circuit configurations, and operational conditions.

An industrial plant surveying campaign was carried out on secondary steps of classification on both fine's and superfine's circuits. Cyclone feed, underflow, and overflow samples were taken in 7 increments, with an interval of 15 min between each increment.

Laboratory characterization included solids percent, wet screening, cycloclassification of a fraction below 74 μm, and chemical analysis, which made it possible to carry out granulochemical analyses up to 13 μm. A total of 178 chemical analyses were carried out to support project development, including av. alumina and re. silica grades and weight percentage of oxides, from X-Ray Fluorescence.

The raw data generated by sample analysis were mass-balanced and used for performance assessment and model fitting the Nageswararao hydrocyclones model which was originally structured to facilitate scale-up, with parameters being dependent essentially on feed solids characteristics. The model comprises equations that may be used to predict the effect of changing one or more design variables,

Fig. 1 Simplified flowsheet of beneficiation plant at Mineração Paragominast

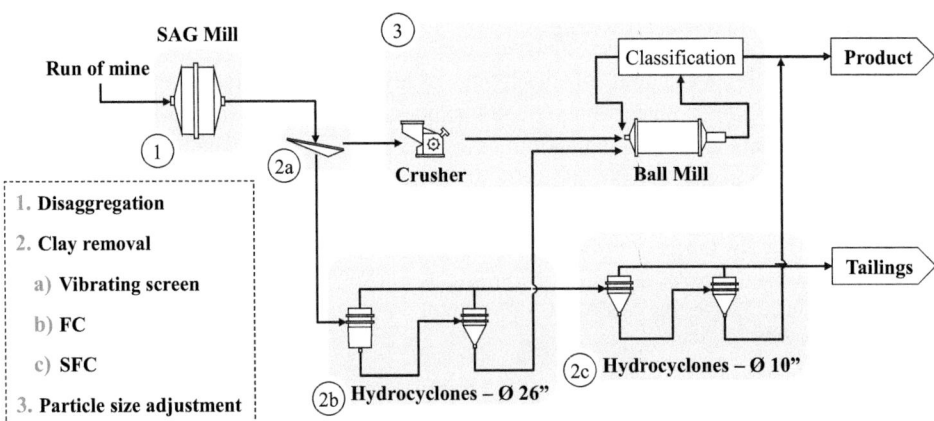

1. Disaggregation
2. Clay removal
   a) Vibrating screen
   b) FC
   c) SFC
3. Particle size adjustment

such as the cyclone diameter and length, or operation variables, such as apex and vortex diameters or, also, the influence of pressure drop in the efficiency curve.

The mass balance and performance analysis for each beneficiation plant equipment was performed with JKSimMet, a software developed by Julius Kruttshnitt Mineral Research Centre (JKMRC), University of Queensland, Australia.

Base Case for mass and metallurgical balances were defined for Mineração Paragominas' beneficiation plant. Besides the main purpose of modeling plant base case, which was allowing integrated analysis of tertiary cyclones on the plant, other key outcomes were obtained. An important one was the diagnostics of contribution from each re. silica source to the product. FC underflow was identified as the main source of contamination to the product, representing more than 50% of the total (Fig. 2).

Process Simulation

Before defined base case, the mass balance and performance analysis for each beneficiation plant equipment were performed with JKSimMet, a Simulator developed by Julius Kruttshnitt Mineral Research Centre (JKMRC), University of Queensland, Australia.

Seeking the most attractive route to enhance fine particle removal/silica reduction on beneficiation process, five alternative scenarios were evaluated and compared to the base case, without tertiary cyclones:

1. Tertiary classification on FC, with tertiary cyclone fed with an existing secondary step of fine particle classification underflow;

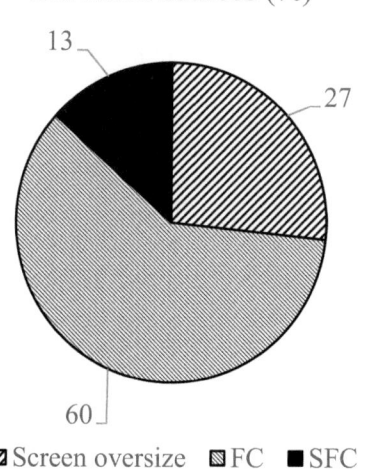

Re. silica sources (%)

13

27

60

☒ Screen oversize   ▨ FC   ■ SFC

Fig. 2 Re. silica sources on base case

2. Tertiary classification on SFC, with tertiary cyclone fed with an existing secondary step of superfine particle classification underflow;
3. Tertiary classification on both FC and SFC, with tertiary cyclone fed with a secondary step of classification underflow, on each circuit;
4. Tertiary classification on FC, similar to scenario 1, with tertiary cyclone overflow recirculation;
5. Tertiary classification on both FC and SFC, similar to scenario 3, with tertiary cyclones overflow recirculation.

Schematic flowsheets representing the scenarios evaluated are presented in Figs. 3 and 4.

The listed alternative scenarios mainly evaluate three conditions: the implementation of tertiary steps of classification individually in one of the classification circuits, FC or SFC; implementation of tertiary classification in both FC and SFC; and finally, scenarios with tertiary cyclone overflow recirculation to the upstream classification step slurry tank, avoiding impacts on downstream equipment caused by the incremental slurry flowrate associated with the new tertiary cyclones.

The main indicators selected for technical evaluation and discussion in the present work were av. alumina increase, re. silica reduction and impact over metallurgical recovery.

Results and Discussion

Comparing with the base case (46.30% of av. alumina), higher av. alumina increases were observed in scenarios with tertiary classification in both FC and SFC, without (46.90%) and with overflow recirculation (46.81%), respectively. Comparing the tertiary step of classification individually on the two circuits, a better performance was observed on FC, reaching 46.72% av. Alumina grade versus 46.44% on SFC (Fig. 5a).

Ranking the alternative scenarios from the better performance to the worse, a similar behavior was observed on re. silica reduction, when compared to av. alumina. Higher silica reductions were observed on scenarios with tertiary step in both classification circuits (Fig. 5b). Base case re. silica grade was 4.80% and project scenarios without and with overflow recirculation presented 4.51 and 4.54% re. silica grades, respectively.

Better results in terms of av. alumina increase and re. silica reduction on the scenarios with tertiary cyclones on both circuits were obtained as those scenarios sum the benefits of tertiary classification individually on each circuit.

Comparing scenarios with and without overflow recirculation, scenarios with recirculation present lower product chemical quality as tertiary cyclone overflow has high silica

Fig. 3 Schematic flowsheet of scenarios 1–3

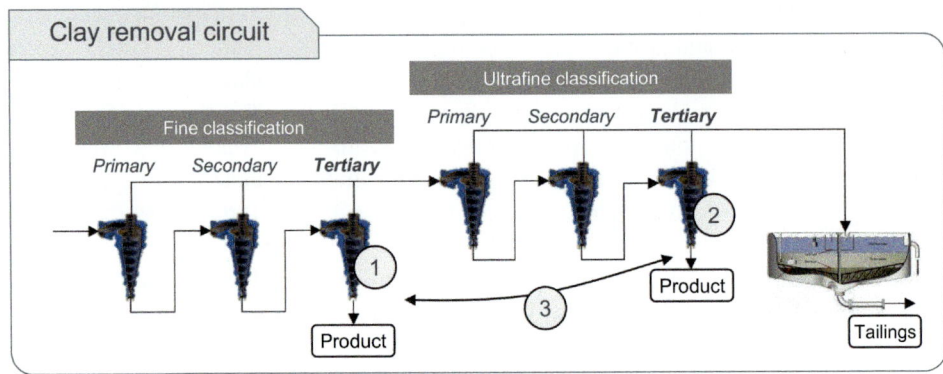

Fig. 4 Schematic flowsheet of scenarios 4 and 5

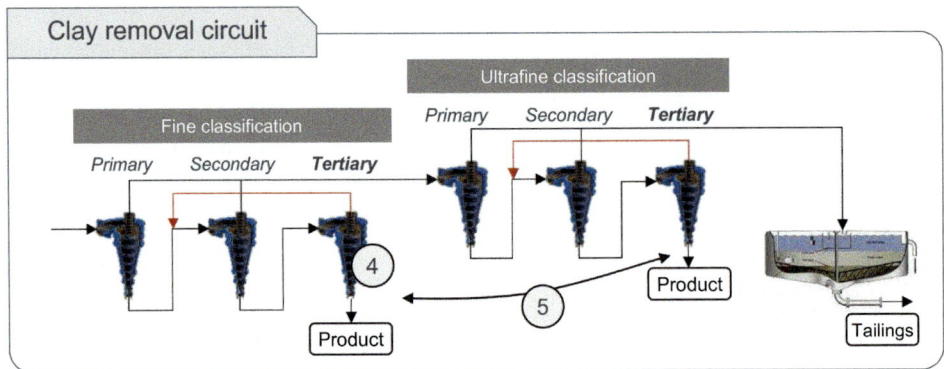

grade and when recirculated, brings lower silica removal efficiency to the circuit.

Better results on FC, when compared to SFC, are explained by chemical characterization of each process flow. Particles below 13 μm on FC have a re. silica grade almost two times higher than on SFC (Fig. 6).

In terms of metallurgical recovery of av. alumina recovery (Fig. 7), higher impacts were observed on SFC, when compared to FC, what can be explained by the lower av. alumina content on the removed particles from FC. The base case, SFC, and FC metallurgical recovery of av. alumina were respectively 84.00, 83.66, and 83.24%.

Lower efficiency in fine particle removal was observed in scenarios with overflow recirculation, besides the negative impact on product quality, resulting in a higher metallurgical recovery when compared to the scenarios without recirculation.

As it is a size classification equipment (hydrocyclone), the inclusion of a tertiary step promotes a reduction in the re. silica content, but it is possible to see a drop in metallurgical recovery. Gibbsite (main mineral) and kaolinite (gangue mineral) particles that are of the same size tend to follow the same flow. To reduce the impacts on the production of Mineração Paragominas with tertiary cycloning, projects to increase production capacity must be evaluated. In addition, it is important to carry out pilot scale tests to evaluate and confirm the results obtained in process simulation.

Conclusion

A simulation model was developed and calibrated for the beneficiation plant at Mineração Paragominas. The simulation of processes with tertiary cyclone steps proved to be a potential solution for the reduction of re. silica in the bauxite of Paragominas. From the 5 alternative scenarios studied, the tertiary step of classification in FC, without overflow recirculation, was considered the most attractive solution, reaching a 4.9% re. silica reduction, from a 4.80% content to 4.56%, with an impact of 0.4% on metallurgical recovery.

Fig. 5 **a** Av. Alumina increase, relative to base case. **b** Re. silica reduction, relative to base case

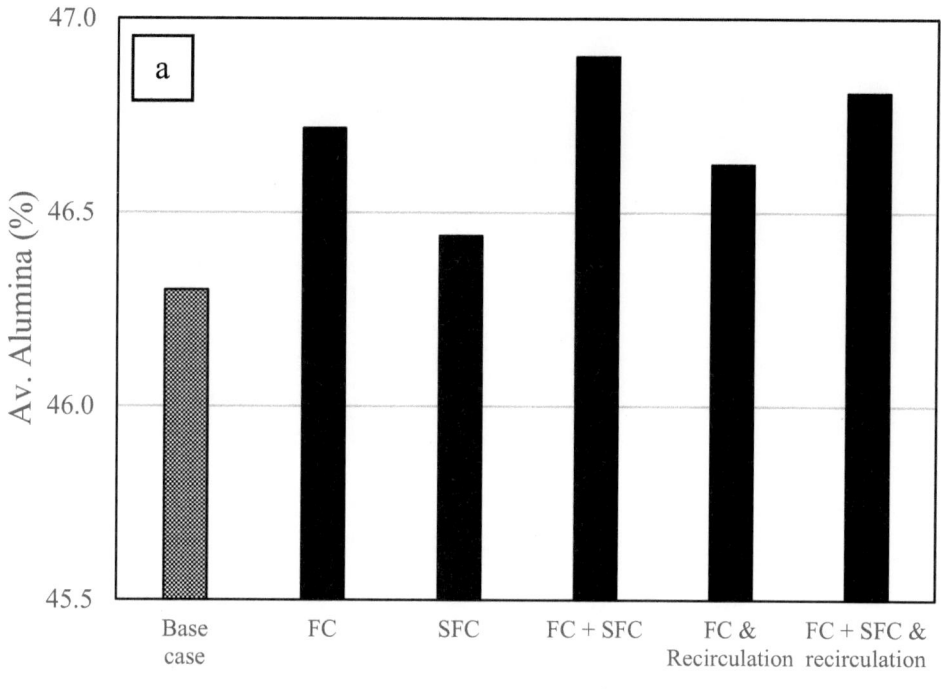

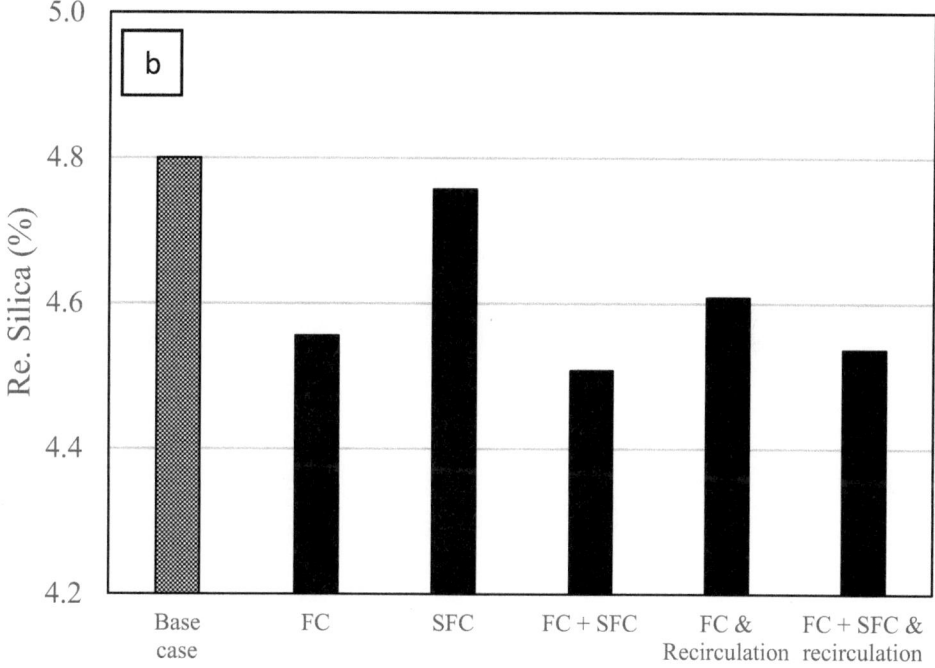

Fig. 6 Chemical characterization of fine particles

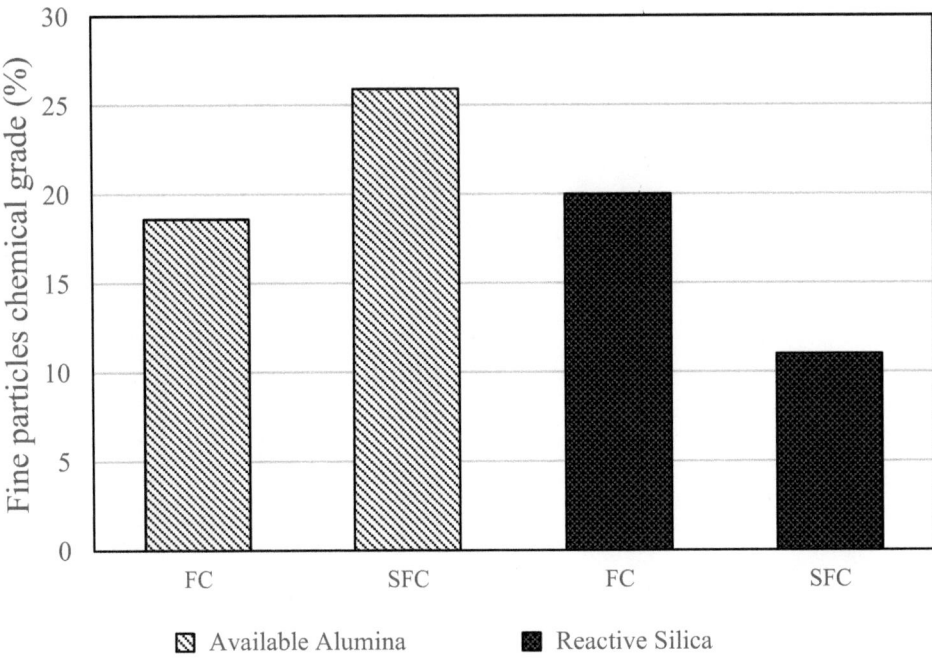

Fig. 7 Av. alumina recovery, relative to base case

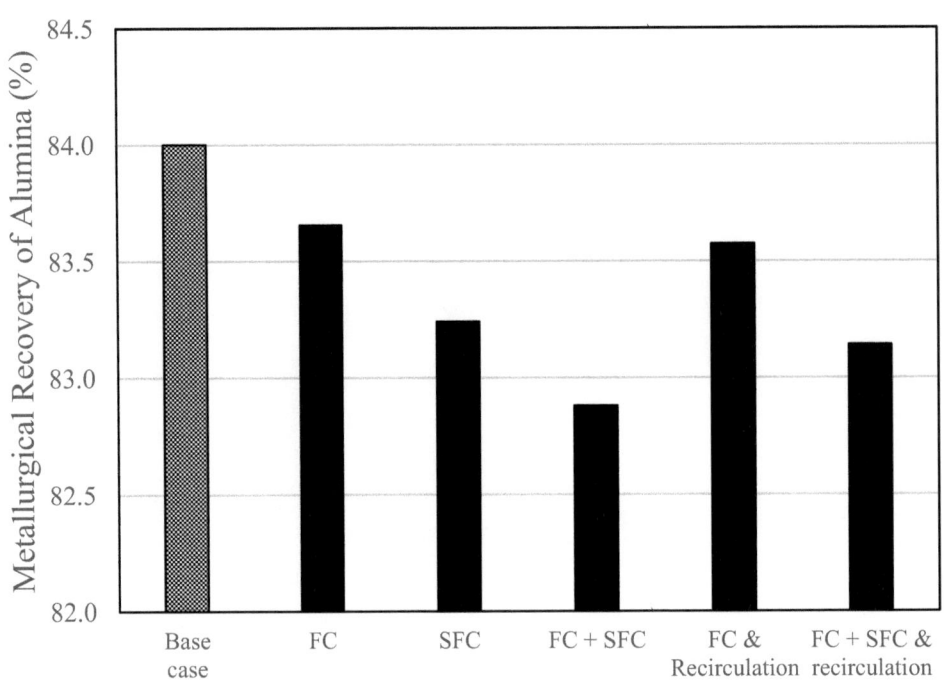

References

1. Costa, ML et. Al (2014) On the geology, mineralogy and geochemistry of the bauxite-bearing regolith in the lower Amazon basin: Evidence of genetic relationships. Journal of Geochemical Exploration, v. 146. ELSEVIER BV. https://doi.org/10.1016/j.gexplo.2014.07.021

2. Frausto, JJ et. al (2021) The effect of screen versus cyclone classification on the mineral liberation properties of a polymetallic ore. Minerals Engineering, v. 169. Elsevier BV. https://doi.org/10.1016/j.mineng.2021.106930

3. Gomes, JF et. al (2021) The formation of desilication products in the presence of kaolinite and halloysite – The role of surface area. Hydrometallurgy, [S.L.], v. 203. Elsevier BV. https://doi.org/10.1016/j.hydromet.2021.105643

4. Luz, AB, Sampaio, JA, Almeida, SLM (2010) Classificação e Peneiramento. In: Tratamento de Minérios (ed), Brasil, Rio de janeiro, p 257–296

5. Mandakini, P, Vakamalla, TR, Mangadoddy, N (2022) Iron ore slimes beneficiation using optimised hydrocyclone operation. Chemosphere, V. 301. ELSEVIER BV. https://doi.org/10.1016/j.chemosphere.2022.134513

6. Meyer F M (2004) Availability of Bauxite Reserves. Natural Resources Research, [S.L.], v. 13, n. 3, p 161–172. Springer Science and Business Media LLC. https://doi.org/10.1023/b:narr.0000046918.50121.2e

7. Nageswararao, K. A generalized model for hydrocyclone classifiers. AusIMM Proceedings, Parkville, December 1995. v. 2, n. 300, 21 p

8. Rodrigues, OMS et. at (2013) Kaolinite and hematite flotation separation using etheramine and ammonium quaternary salts. Minerals Engineering, v. 40. ELSEVIER BV. https://doi.org/10.1016/j.mineng.2012.09.019

9. Smith P (2009) The processing of high silica bauxites – Review of existing and potential processes. Hydrometallurgy, [S.L.], v. 98. Elsevier BV. https://doi.org/10.1016/j.hydromet.2009.04.015

Granulometry Impact on Digestion Efficiency and Cost-Economics in Alumina Refinery for East Coast Bauxite (India)

Suchita Rai, M. J. Chaddha, Prachiprava Pradhan, K. J. Kulkarni, M. Panchal, and A. Agnihotri

Abstract

Due to the fast depletion of bauxite resources, beneficiation is a crucial step, mainly to reduce silica and iron content. Crushing, grinding, and sieving the bauxite in various size fractions are essential for determining suitable feed size during alumina production. East Coast bauxite was grounded and crushed into different granulometry ranging from +2.8 to −0.045 mm. Al_2O_3 enriched in the coarser fractions, while SiO_2, Fe_2O_3, and TiO_2 content increased in finer fractions. There is a decrease in digestion efficiency from coarser to finer size, increasing the specific bauxite and caustic soda consumption. Consequently, up-gradation can be accomplished by discarding a few finer sizes. With an improvement in bauxite quality, for a 1-million-tonnes alumina/annum plant capacity, caustic soda consumption can be reduced by approximately 2160 tonnes of NaOH with a savings of 18,000 tonnes of process bauxite.

Keywords

Bauxite • Granulometry • Upgradation • Digestion efficiency • Specific bauxite • Caustic soda consumption

S. Rai (✉) · M. J. Chaddha · P. Pradhan · K. J. Kulkarni · M. Panchal · A. Agnihotri
Jawaharlal Nehru Aluminium Research Development and Design Centre (JNARDDC), Amravati Road, Wadi, Nagpur, 440023, Maharashtra, India
e-mail: suchitarai@jnarddc.gov.in; ; suchitarai1968@gmail.com

M. J. Chaddha
e-mail: mjchaddha@jnarddc.gov.in

P. Pradhan
e-mail: prachi@jnarddc.gov.in

K. J. Kulkarni
e-mail: kjkulkarni@jnarddc.gov.in

M. Panchal
e-mail: meghapanchal@jnarddc.gov.in

A. Agnihotri
e-mail: director@jnarddc.gov.in

Introduction

Bauxite is the most important mineral resource to produce alumina by using suitable mineral processing techniques which mainly consist of gibbsite, boehmite and diaspore etc. [1]. The chemical, mineralogical and physicomechanical properties of bauxite widely vary depending upon the parent rock composition, mode of origin, geomorphological position and duration and age of bauxite formation. However, the quality of some of the bauxites is typical for the individual deposits and therefore it is necessary to exactly know the characteristics of ore in each deposit.

Beneficiation and upgrading of bauxite are very important steps for the subsequent extraction of alumina in the Bayer process. The important aspect is to consider the suitability of ore for alumina production. Therefore, apart from simple chemical analysis of bauxite, mineralogy becomes an important consideration that influences the technology for processing. Bauxite comprises various major minerals such as trihydrated gibbsite ($Al_2O_3 \cdot 3H_2O$), monohydrate boehmite ($Al_2O_3 \cdot H_2O$) and oxide-hydroxide diaspore (alpha-AlO(OH)) and a few other minerals like iron-bearing minerals as hematite, goethite and silicate minerals mainly kaolinite and quartz etc. Production of alumina through the Bayer process requires finely ground bauxite ore for carrying out digestion using sodium hydroxide solution at a temperature range of 100–250 °C. During the process, the residue generated called "red mud" mainly depends on the quality of bauxite being processed. Apart from the alumina dissolution in the digestion process, some amounts of silica such as kaolinite and quartz also get dissolved. Here the high percentage of reactive silica consumes more caustic soda as compared to quartz and the bauxite with more than 5% reactive silica is not suitable in the Bayer process due to excessive caustic soda consumption. Economical extraction of alumina through the Bayer process can be possible at a higher Al_2O_3 to SiO_2 mass ratio of greater than 10 in the bauxite [2]. Bauxite containing more than 5% reactive silica cannot be

© The Minerals, Metals & Materials Society 2023
S. Broek (ed.), *Light Metals 2023*, The Minerals, Metals & Materials Series,
https://doi.org/10.1007/978-3-031-22532-1_19

directly processed by the Bayer process and high alumina production cost is not preferable. Thus, it is necessary to remove the aluminosilicate minerals from the bauxite to increase the Al_2O_3 to SiO_2 mass ratio [3]. Mineralogical and petrographic characteristics of bauxite are very crucial for the better economic evaluation of the Bayer process because the presence of various phases such as reactive silica which is associated with different Ti and Fe oxides can reduce the efficiency of the process [4].

It is often possible to mine and selectively sort out bauxites of various kinds including metallurgical grades from the same deposit and these can also be produced by adopting simple physical beneficiation processes such as wet scrubbing and sieving. Ore characteristics can be determined from the bauxite composition in the natural state and based on that it can be used for various applications [5]. Therefore, beneficiation is often considered for the up-gradation of ore to make it suitable for the Bayer process in the alumina refinery [2, 5–7]. Before any ore processing and digestion studies at certain parameters, grinding and crushing is the principal operation that attracts attention due to the associated mineral phases such as Al (alumina), and Fe (iron) and their distributions [6].

Comminution plays an important role during the beneficiation of many ores due to the liberation of valuable and gangue minerals in fine and ultrafine particle sizes [8]. Numerous researchers have revealed that the difference between aluminium-containing minerals and silicon-containing minerals in the bauxite can be achieved through various selective grindability studies and the A/S ratio can be increased by using hydro cyclone [9] and flotation method [10]. Different researchers have conducted systematic research on the effect of grinding on the flotation process and it is investigated that the grinding stage is one of the most important steps in mineral processing. Crystalline structure and comminution properties can be related through different types of grinding mediums and their effects. However, it is found that this process shows a great effect on bauxite beneficiation in terms of the quality of concentrate [11]. Selective grinding has an important role in obtaining a high Al_2O_3 to SiO_2 ratio and alumina recovery in the coarser size material than in finer size fraction. This is a very good technique for improving the alumina to silica ratio in the product through better separation of alumina and silica in the ore from the initial step only [12]. However, all these beneficiation techniques reported by various authors are not cost-effective. Even the removal of clay from bauxite by washing is also a simple technique but it requires a large quantity of water during the process [13].

In this view, a study of granulometry's impact on digestion efficiency and cost-economics in alumina refinery for east coast bauxite was conducted. The ROM (Run of Mine) bauxite and ground bauxite samples (13 in no.) have been carried out to improve alumina content by decreasing silica and iron in the ore through beneficiation. In the current examination, a detailed characterization of ROM bauxite and thirteen different bauxite samples after classification has been stated. The interpretation of different size fractions is done for the optimum utilization of ore and to explore the potential beneficiation options that affect the distribution of alumina and iron-containing phases. A digestion study of the ROM bauxite and all the classified bauxite samples have been conducted. Using the better-quality bauxite after beneficiation will promote a higher digestion efficiency at a low specific bauxite consumption and low caustic consumption.

Materials and Methods

Materials

About 1 tonne of bauxite sample in lump form (about 25 mm size) was procured from an alumina refinery located on the East Coast of India (longitude: 18.769496 and latitude: 82.865392). The major chemical composition of representative ROM east coast bauxite is shown in Table 1. Table 1 also shows the trihydrate alumina (THA%) and reactive silica ($RSiO_2$%) of ROM bauxite. Mineralogical phases along with the chemical formula are given in Table 2. Figure 1 shows the XRD diffractogram. Two different scales

Table 1 Major chemical constituents of ROM bauxite

Constituents	Weight (%)
Al_2O_3	44.65
Fe_2O_3	25.15
SiO_2	3.85
TiO_2	2.40
LOI	23.30
THA	41.48
$R(SiO_2)$	3.04

Table 2 Mineralogical phases of ROM bauxite

Mineralogical phases		Chemical formula	Value (%)
Al$_2$O$_3$ as	Gibbsite	Al$_2$O$_3$·3H$_2$O	40.20
	Kaolinite	Al$_2$O$_3$·2SiO$_2$·2H$_2$O	2.76
	Alumogoethite	FeAlOOH	1.00
SiO$_2$ as	Kaolinite	Al$_2$O$_3$·2SiO$_2$·2H$_2$O	3.26
	Quartz	SiO$_2$	0.50
TiO$_2$ as	Rutile	TiO$_2$	ND
	Anatase	TiO$_2$	1.50
	Ileminite	FeTiO$_3$	0.53
Fe$_2$O$_3$ as	Hematite	Fe$_2$O$_3$	15.00
	Alumogoethite	FeAlOOH	9.17
	Ileminite	FeTiO$_3$	0.53

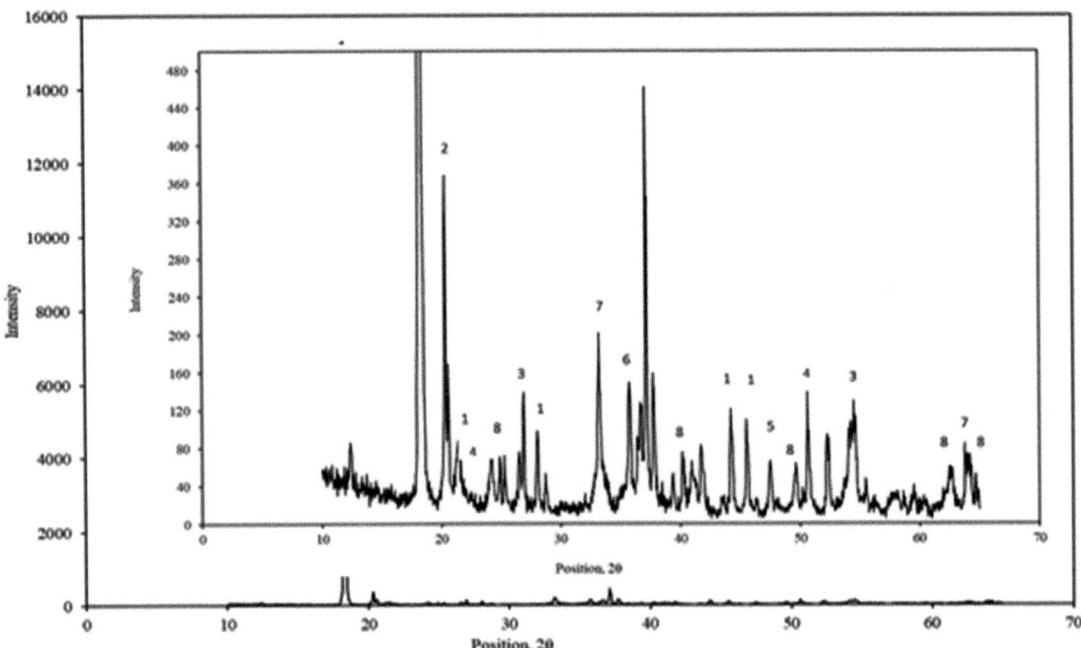

Fig. 1 XRD patterns of ROM bauxite (1: Gibbsite; 2: Kaolinite; 3: Alumogoethite; 4: Quartz; 5: Rutile; 6: Anatase; 7: Ileminite; 8: Hematite)

are being used in the diffractogram of Fig. 1 to show the major peak of gibbsite on one scale (0–16,000 counts per second intensity) and other mineral phases (0–480 counts per second intensity).

Synthetic Bayer liquor of desired caustic (Na$_2$O) concentration of 182.13 g/L, alumina (Al$_2$O$_3$) concentration of 108.63 g/L and silica (SiO$_2$) concentration of 0.73 g/L was prepared by dissolving the caustic soda flakes in distilled water along with the product hydrate at a high temperature of 1 h holding time.

Methods

Sample Preparation of ROM Bauxite and Different Bauxites Fractions

The east coast bauxite after 10–15 days of sun-drying was crushed in a jaw crusher to obtain a size of 10–12 mm and then further crushed inside the roll crusher to get a representative sample of size of 3.36 mm. After proper mixing, coning and quartering through a refill sampler about 50 kg of a representative sample of 3.36 mm size fraction was generated.

From this 50 kg of ROM bauxite sample, 1 kg was taken and ground to 145 μm (100 mesh) size for chemical analysis of major components. Further, the rest ROM bauxite sample (about 48 kg) was classified using sieves in Rotap Sieve shaker in 10 cycles for generating thirteen different size fractions such as +2.8 mm, −2.8 + 2.00 mm, −2.0 + 1.4 mm, −1.4 + 1.0 mm, −1.0 + 0.7 mm. −0.7 + 0.5 mm, −0.5 + 0.25 mm, −0.25 + 0.15 mm, −0.15 + 0.106 mm, −0.106 + 0.075 mm, −0.075 + 0.063 mm, −0.063 + 0.045 mm, and −0.045 mm for executing different digestion experiments. A detailed sample preparation flowsheet for different size fractions of east coast bauxite is given in Fig. 2. Size distribution of particles in −6 mesh (3.36 mm) size bauxite in each cycle has been given in Fig. 3. It presents the cumulative weight percent of the bauxite sample on each sieve size.

The most commonly used matrix for particle size distribution is $(d_{10}, d_{50}$ and $d_{90})$ which are intercepts for 10, 50 and 90% of the cumulative weight percent. The cumulative weight percentage of the bauxite on the successive sieves increases with the sieve sizes as observed in Fig. 3. Successive sieving indicated that d_{10} for this sample lies in between 2.8 and 2 mm size fraction. Similarly, d_{50} lies between 1 and 0.710 mm whereas d_{90} is below 0.074 mm. The remaining 10% of the particles are below 0.074 mm.

Analytical Instruments and Experimental Set-up

The ROM bauxite and its classifications into different size fractions were carried out by sieving using Rotap Sieve shaker (Scientific Corporation, Nagpur) and sieves (Jayant Scientific IND., Mumbai ASTM). These bauxite samples

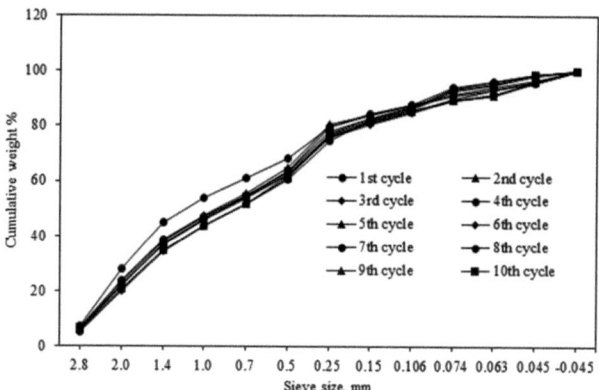

Fig. 3 Cumulative weight % of bauxite retained on different sieve sizes

after classification were characterized by wet chemical analysis and through different analytical instruments. Mineralogical phases for the samples were analysed through an X-Ray diffractometer (X-Pert Pro MPD, Netherland) using Cu Kα radiation ($\lambda = 1.541$ Å). X-Ray Fluorescence (PANalytical, Axios[mAX], Netherland). The wet chemical method was used for the chemical analysis of major constituents such as Al_2O_3, SiO_2, Fe_2O_3, TiO_2, and LOI (Loss on Ignition) of the bauxite. Synthetic liquor preparation and digestion tests were carried out using a 200 ml capacity Bomb digestor (Noble Polymech Corporation, Mumbai, India). The reactor was rotated horizontally to ensure proper mixing of the reactants. Solids and liquor obtained after the digestion tests were separated using Centrifuge (REMI KPR-70). Liquor was analysed for measuring the Al_2O_3, and

Fig. 2 Sample preparation flowsheet for ROM bauxite

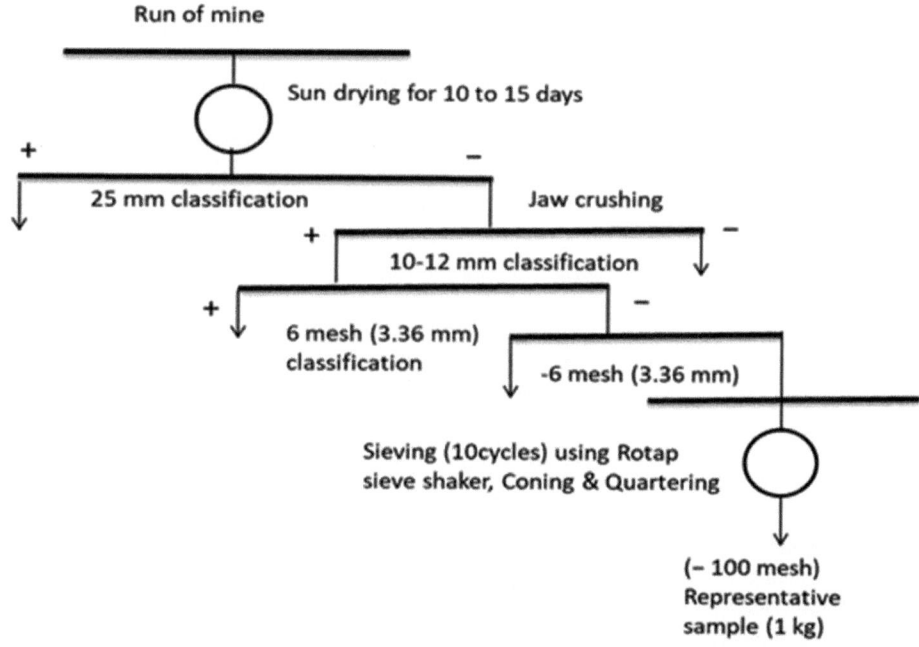

Na$_2$O concentration using the volumetric titration method and SiO$_2$ concentration using a spectrophotometer before and after the digestion test. Similarly, the maximum dissolution of the alumina in the liquor was measured in terms of RP (Ratio Ponderal) i.e. the ratio of Al$_2$O$_3$ concentration to Na$_2$O concentration in the digestion liquor as well as in the liquor after digestion.

Digestion Tests

Digestion experiments for ROM bauxite and different size fractions were carried out using synthetic liquor at a fixed initial caustic concentration (Na$_2$O$_c$) of 182 g/L, alumina (Al$_2$O$_3$) 108.63 g/L, total caustic (Na$_2$O$_t$) 202.27 g/L, RP as 0.596, Silica (SiO$_2$) 0.73 g/L and C/S 0.911. A set of digestion experiments was performed at a temperature of 145 °C for 45 min by keeping the target RP of 1.15 with a target caustic concentration (Na$_2$O$_c$) of 155 g/L. Digestion efficiency is calculated based on total alumina (TA) by the following formula:

$$D_{eff}(TA)\% = \frac{Al_2O_3\%(TA)\,in\,bauxite - M.L \times Al_2O_3\%(TA)\,in\,mud}{Al_2O_3\%(TA)\,in\,bauxite} \times 100$$

(1)

$$Mud\,load(M.L) = \frac{(Fe_2O_3\% + TiO_2\%)\,content\,of\,bauxite}{(Fe_2O_3\% + TiO_2\%)\,content\,of\,digested\,mud}$$

Similarly, specific bauxite consumption (t/t of alumina), and caustic consumption in terms of NaOH (kg/t of alumina) have been calculated for different size fractions.

Results and Discussion

ROM Bauxite Characterisation

From Table 1, it is seen that the Al$_2$O$_3$ content of the ROM bauxite is 44.65% having the maximum extractable alumina in the form of trihydrate alumina (THA) at 41.48%. It contains a high iron phase with 25.15% Fe$_2$O$_3$ and a low titanium oxide phase (TiO$_2$) of about 2.4%. It is also noted that the total silica is in the range of 3–4% i.e. 3.85 weight % with reactive silica content of about 3%. The loss of ignition (LOI) is strongly dependent on the mineralogical composition of bauxite [14] and with the east coast ROM bauxite, it is approximately 23.30%. This higher value for LOI is due to the variation in mineralogical compositions and that aluminium oxide is hydrated with three molecules of water in the respective sample. A high Al$_2$O$_3$/SiO$_2$ ratio of 11.59 and Al$_2$O$_3$/Fe$_2$O$_3$ ratio of 1.775 is observed in the ROM sample.

Mineral phases in the ROM bauxite and their chemical composition is shown in Table 2. From the Table, it is noticed that the mineral proportions for the ROM bauxite are around 40% gibbsite, 3.26% kaolinite, 15% hematite, and 9.17% alumogoethite, 0.5% quartz and 1% anatase respectively. These results present a higher quantity of alumina-containing minerals of around 44%, iron-containing minerals of approximately 25% and very low mineralogical phases of 4% silica and 2% titanium oxide in the ROM bauxite. Alumina is present primarily as gibbsite. Silica is present mainly as kaolinite and quartz. Hematite is the main iron mineral with alumogoethite and ileminite in small quantities. Anatase and ileminite are the main titanium-bearing minerals. It is also seen that the hematite to alumogoethite ratio is around 1.63. Figure 1 shows the distribution pattern of various minerals such as gibbsite, kaolinite, alumogoethite, quartz, rutile, anatase, ilmenite; hematite as shown from their respective X-ray reflection peak.

Variation of Chemical Analysis Based on Granulometry of Bauxite and Average Weight % Retained on the Sieves

Table 3 shows the major chemical compositions of thirteen different bauxite samples after classification (sieving) along with THA%, R(SiO$_2$) % and an average of the weight retained on the sieves in all the ten cycles of sieving. Similarly, Figs. 4 and 5 show the variation of alumina to silica ratio and alumina to iron ratio in different size fractions.

From Table 3, it is seen that interesting results were obtained for all the samples generated through different size-wise classifications. Al$_2$O$_3$% decreased with a decrease in size fractions whereas Fe$_2$O$_3$% and SiO$_2$% increased from coarser size fractions to finer size fractions. The contents of Al$_2$O$_3$ (47.79–38.50%), Fe$_2$O$_3$ (22.18–32.63%), SiO$_2$ (2.94–4.95%), TiO$_2$ (1.92–3.1%) are observed from coarser to finer fraction. Silica gets enriched in the finer fractions and it starts depleting in the coarser size fractions. There is an increase in silica content up to 68.36% in the finer fraction (−0.045 mm) from the coarser fraction (+2.8 mm). Enrichment of silica is reciprocal to that of alumina particles in the ore and is found in finer fractions which are as per the research which states that kaolinite is more abundant in the friable material [15]. Iron increases up to 47.11% in the finer fraction (−0.045 mm) from the coarser fraction (+2.8 mm). There is a decrease of about 20% in alumina content from the coarser (+2.8 mm) to the finer fraction (−0.045 mm size).

Available alumina in terms of (THA) decreases with a decrease in size fractions. Trihydrate alumina (THA %) is 40.75% in the coarser fraction and then it gradually decreases to 31.28% in the finer fraction. R(SiO$_2$) % also increases from 2.33 to 4.07% in the finer fraction. Similarly,

Table 3 Major chemical compositions of different bauxite samples after classification

Bauxite size fraction (mm)	Al$_2$O$_3$%	Fe$_2$O$_3$%	SiO$_2$%	TiO$_2$%	LOI %	THA %	R (SiO$_2$)%	Average weight % on sieve
+2.8	47.79	22.18	2.94	1.92	24.57	40.75	2.33	6.12
−2.8 + 2.0	47.29	22.75	3.10	1.96	24.45	40.46	2.38	16.54
−2.0 + 1.4	46.95	22.82	3.32	2.04	24.30	40.12	2.37	14.91
−1.4 + 1.0	46.47	23.20	3.55	2.10	24.07	39.10	2.85	8.96
−1.0 + 0.7	45.35	24.37	3.83	2.12	23.68	38.76	2.93	7.94
−0.7 + 0.5	44.47	25.33	4.14	2.09	23.30	37.40	3.23	8.75
−0.5 + 0.25	42.89	27.37	4.52	2.29	22.21	36.04	3.52	14.18
−0.25 + 0.15	41.46	28.73	4.79	3.04	21.22	34.50	3.82	5.10
−0.15 + 0.106	39.47	29.69	4.78	4.06	21.20	33.32	3.76	3.80
−0.106 + 0.074	39.18	30.38	4.72	4.29	20.61	32.30	3.31	5.18
−0.074 + 0.063	39.07	31.12	4.81	3.39	20.70	31.62	3.73	2.24
−0.063 + 0.045	39.07	31.23	4.90	3.23	20.80	31.50	3.56	3.42
−0.045	38.50	32.63	4.95	3.10	20.52	31.28	4.07	2.86

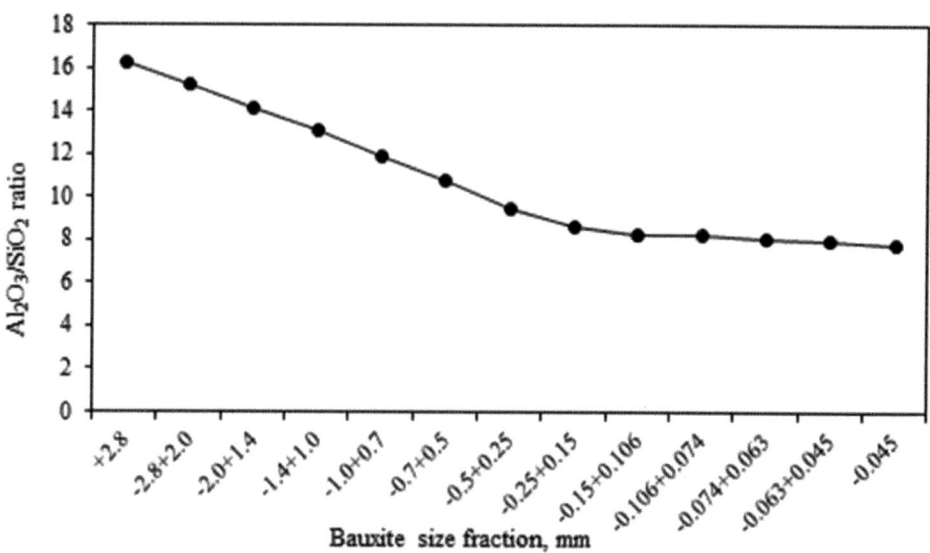

Fig. 4 Variation of alumina to silica ratio in different bauxite size fractions

in Figs. 4 and 5, the increase in alumina content and decrease in silica and iron content in the coarser fraction results in higher alumina to silica ratio of 16.26 and alumina to iron ratio of 2.15 in the coarser size fraction. Alumina to silica ratio of 7.97 and alumina to iron ratio of 1.25 is found in the finer fraction.

Digestion Study of ROM Bauxite

Digestion study with ROM bauxite show digestion efficiency of 86.01% with specific bauxite consumption of 2.741 t/t of alumina while the caustic consumption is found to be 91.59 kg/t of alumina. The mineralogy of bauxite residue is shown in Fig. 6.

Fig. 5 Variation of alumina to the iron ratio in different bauxite size fractions

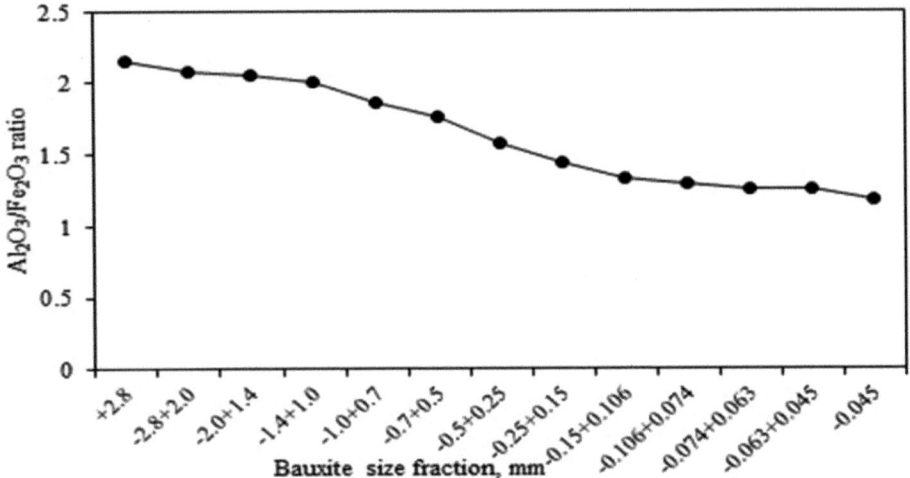

Fig. 6 XRD pattern of bauxite residue after digestion of ROM bauxite (1: Gibbsite; 2: Boehmite; 3: Alumogoethite; 4: Sodalite; 5: Kaolinite; 6: Quartz; 7: Rutile; 8: Anatase; 9: Ilmenite; 10: Hematite)

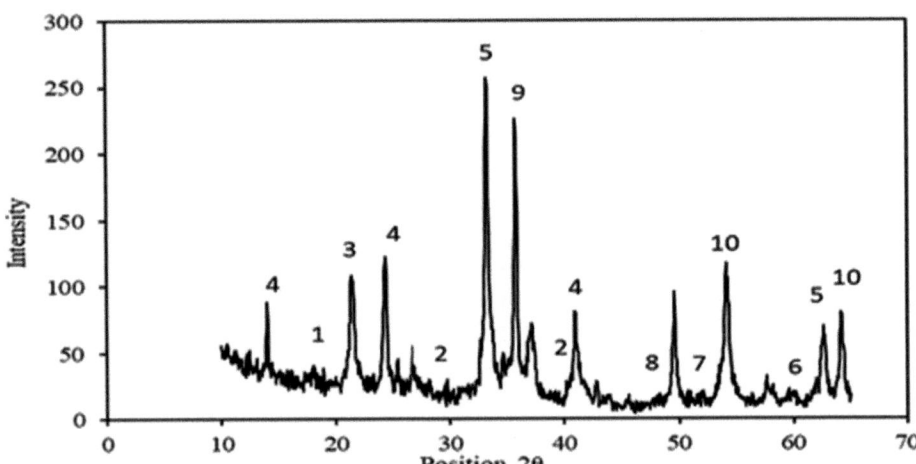

The XRD patterns of bauxite residue of ROM bauxite as listed in Fig. 6 show various minerals. These are sodalite ($Na_2O \cdot Al_2O_3 \cdot SiO_2$), gibbsite ($Al_2O_3 \cdot 3H_2O$), boehmite ($Al_2O_3 \cdot H_2O$), alumogoethite (FeAlOOH), kaolinite ($Al_2O_3 \cdot 2SiO_2 \cdot 2H_2O$), quartz ($SiO_2$), rutile ($TiO_2$), anatase ($TiO_2$), ilmenite ($FeTiO_3$), hematite ($Fe_2O_3$). Undigested gibbsite mineral remains in the bauxite residue.

Digestion Study of Bauxite Fractions with Different Granulometry

The digestion efficiency graph for thirteen different bauxite samples is shown in Fig. 7. Similarly, digestion efficiency, specific bauxite consumption and caustic consumption are represented in Table 4.

Digestion efficiency for the coarser size fraction is found to be 87.34% whereas it reduces to 81.29% in the finer size fraction as seen in Fig. 7. This is about a 7% reduction in

overall digestion efficiency from a coarser to a finer fraction. From Fig. 7 it is seen that the digestion efficiency follows a decreasing trend from coarser to finer size fraction and there is a further decrease in digestion efficiency from 1 mm size fraction onwards. This shows that in an alumina refinery, bauxite quality plays an important role in achieving the desired digestion efficiency while keeping the other parameters constant [16]. Here the increase in digestion efficiency above 1 mm particle size is due to the enrichment of alumina and reduction in silica content as seen in Table 3.

Table 4 shows that specific bauxite consumption (t/t of alumina) and caustic consumption (kg/t of alumina) increased with a decrease in bauxite size. Specific bauxite consumption is found to be 2.522 for the coarser size fraction and then it gradually increased to 3.364 for the finer fraction. Similarly, caustic soda consumption is 54.34 for the coarser size fraction and then it increased to 155.93 for the finer fraction. Nearly a difference of 1 t/t of alumina is found between the coarsest and finest fraction in terms of bauxite

Fig. 7 Digestion efficiency of various bauxite size fractions

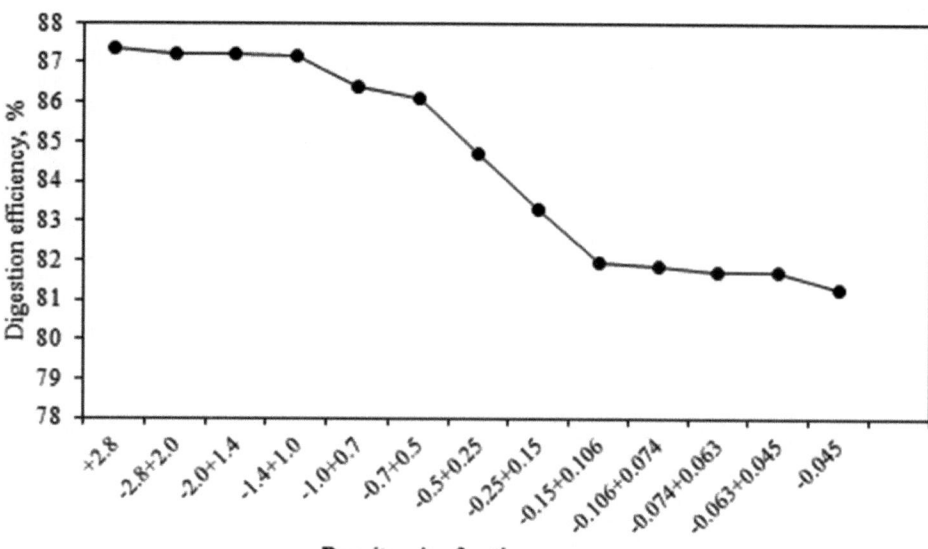

Table 4 Digestion efficiency, Specific bauxite consumption and caustic consumption with different granulometry of bauxite size fractions

Bauxite size fraction (mm)	Digestion efficiency (%)	Specific bauxite consumption (t/t of alumina)	Caustic soda consumption (kg/t of alumina)
+2.8	87.34	2.522	54.34
−2.8 + 2.0	87.23	2.552	60.30
−2.0 + 1.4	87.21	2.571	64.62
−1.4 + 1.0	87.18	2.598	68.46
−1.0 + 0.7	86.38	2.687	86.73
−0.7 + 0.5	86.11	2.749	98.84
−0.5 + 0.25	84.70	2.898	119.36
−0.25 + 0.15	83.32	3.047	136.09
−0.15 + 0.106	81.97	3.253	145.99
−0.106 + 0.074	81.86	3.282	144.65
−0.074 + 0.063	81.73	3.297	148.33
−0.063 + 0.045	81.69	3.298	151.41
−0.045	81.29	3.364	155.93

consumption while the caustic consumption increases threefold in the finest fraction compared to the coarser one due to a decrease in alumina and increase in silica content.

Impact of Bauxite Granulometry on the Digestion Efficiency

The impact of bauxite granulometry is seen from the characterisation and digestion study of ROM and the bauxite fractions. It is noted that digestion efficiency reduces with a decrease in alumina content in the finer fractions. Likewise, silica and iron content increased in the finer fraction from the coarser fraction. In all the thirteen different bauxite size fractions, the average alumina, iron and silica content is 42.92, 27.06 and 4.18% respectively. Hence, if all the last three or five bauxite fractions are discarded, the quality of bauxite will improve. This seems to be advantageous to discard finer fractions before wet ball milling to reduce the silica in the process bauxite by increasing the alumina content. This increase in alumina content and decrease in silica content will lead to a higher digestion efficiency in the ore. Consequently, this will decrease the specific bauxite and caustic soda consumption per tonne of alumina. This has been explained in Table 5 along with the alumina, iron and silica content of ROM bauxite and after discarding 3 or 5 fractions.

Presently in the alumina refinery, the process bauxite of the east coast, after the primary crusher, of size 30 mm is fed

Table 5 Specific bauxite consumption and caustic soda consumption for bauxite samples

Bauxite sample	Al_2O_3 (%)	Fe_2O_3 (%)	SiO_2 (%)	Specific bauxite consumption, (t/t of alumina)	Caustic soda consumption (kg/t of alumina)
ROM	44.65	25.15	3.85	2.741	91.59
After discarding the last three finer fractions	44.95	24.92	3.80	2.723	88.98
After discarding the last five finer fractions	45.57	24.36	3.71	2.686	83.75

to the ball mill and ground to around 80% passing through −63 microns using caustic soda (digestion liquor). If the process bauxite is crushed without using digestion liquor to a size of +2.8 to 0.045 mm size, and then subjected to sieving to remove −0.15 + 0.016 mm (3 or 5 finer fractions) there will be an improvement in the quality of bauxite based on alumina and silica content as seen from Table 5. Later, the above-sieved process bauxite can be fed to the ball mill. Since the feed size of the process bauxite has reduced, the throughput from the process bauxite through the ball mill will be higher. Due to the reduction in silica in the sieved process bauxite and a slight improvement in the alumina content of the sieved process bauxite, the specific bauxite consumption and the specific caustic soda consumption will decrease. If the last three finer fractions such as −0.074 + 0.063 mm, −0.063 + 0.045 mm and −0.045 mm are discarded, then the specific bauxite consumption (t/t of alumina) and caustic consumption (kg/t of alumina) reduces to 2.723, 88.98 from 2.741 and 91.59 respectively. Similarly, for discarding last five finer fractions such as −0.15 + 0.106 mm, −0.106 + 0.074 mm, −0.074 + 0.063 mm, −0.063 + 0.045 mm and −0.045 mm results in further decrease in specific bauxite consumption (t/t of alumina) i.e. 2.686 from 2.741 in ROM bauxite and caustic consumption (kg/t of alumina) i.e. 83.75 from 91.59 in ROM bauxite.

Cost Economics for 1 Million Tonne Capacity Alumina Production Refineries

When the last three finer fractions are discarded,

Specific bauxite consumption reduction = 2.741 − 2.723 = 0.018 tonne.

For a capacity of 1 million tonne alumina production, bauxite consumption would reduce by 0.018 × 10,00,000 = 18,000 tonne.

Specific caustic soda consumption reduction = 91.59 − 88.98 = 2.61 kg.

For a capacity of 1 million tonne alumina production, caustic soda consumption would reduce by 2.61 × 10,00,000 = 2,610,000 kg = 2610 tonne.

If the cost of NaOH flakes is Rs. 30,000–Rs. 40,000/- per tonne.

2610 × 35,000 = Approximately Rs. 915 lakhs will be saved.

However, bauxite loss in this context needs to be explained.

If we discard the last three fractions with granulometry i.e. −0.045, −0.063 + 0.045 and −0.074 + 0.063 mm, the weight loss from 1 tonne of bauxite will be 2.86 + 3.42 + 2.24 = 8.52% (Table 3). i.e. from 1000 kg bauxite, we will be losing 85.2 kg of finer fraction resulting in a bauxite yield of 914.8 kg having Al_2O_3 content of 44.95, SiO_2 −3.80 and reactive SiO_2 of 2.91%.

Conclusion

There is a presence of high Al_2O_3 content of 47.79% with low SiO_2 content of 2.94% in the coarser size fraction (+2.8 mm) and low Al_2O_3 content of 38.5% with high SiO_2 content of 4.95% in the finer size fraction (−0.045 mm) of east coast bauxite (India). As the silicate minerals containing alumina elements are closely associated with the aluminosilicate clay minerals in the finer fractions, there is an increase in SiO_2 content and a decrease in Al_2O_3 content in these bauxite fractions. Granulometry impacts the digestion efficiency and the specific bauxite consumption and caustic consumption increase with a decrease in bauxite size. In general, it can be concluded that the digestion efficiency is reduced with the decrease in alumina content in the finer fractions. Upgradation of bauxite can be done by discarding the last three or five finer fractions which would result in improving the digestion efficiency with a reduction in specific bauxite and caustic consumption in an alumina refinery. However, in translating these results to an industrial scale, some points must be taken into consideration:

(1) Digestion studies carried out with samples on the laboratory scale were of uniform mesh size but when it comes to an industrial scale these vary in the granulometry.

(2) Laboratory and plants have different agitation levels.

(3) Flashing of digested slurry can be done in the plant, however, it is not the case for laboratory-scale experiments.

(4) The amount of process bauxite lost due to the removal of 3 fine size fractions/5 fine size fractions should also be studied while calculating the techno-economics of the process.

Acknowledgements The authors acknowledge the financial support from the Ministry of Mines, Gov. of India for this research work. The authors would also like to thank Sabhajeet Yadav, Technical Assistant, Alumina Department, JNARDDC for his consistent support during the entire experimentation.

Conflict of Interest Statement On behalf of all authors, the corresponding author states that there is no conflict of interest.

References

1. He J, Bai Q, Du T (2020) Beneficiation and upgrading of coarse sized low-grade bauxite using a dry-based fluidized bed separator. Adv Powder Technol 31(1):181–189. https://doi.org/10.1016/j.apt.2019.10.009

2. Rao D.S, Das B (2014) Characterization and beneficiation studies of lowgrade bauxite ore. J Inst Eng India Ser D 95 (2): 81–93. https://doi.org/10.1007/s40033-014-0050-8

3. Ou L, Feng Q, Chen Y, Lu Y, Zhang G (2007) Disintegration mode of bauxite and selective separation of Al and Si. Miner Eng 20: 200–203. https://doi.org/10.1016/j.mineng.2006.09.009

4. Boni M, Rollinson G, Mondillo N, Balassone G, Santoro L (2013). Quantitative mineralogical characterization of Karst bauxite deposits in the Southern Apennines, Italy. Econ Geol 108:813–833. https://doi.org/10.2113/econgeo.108.4.813

5. Fernandes de Aquino T, Gracher Riella H, Michael Bernardin A (2011). Mineralogical and physical chemical characterization of a bauxite ore from Lages, Santa Catarina, Brazil, for refractory production. Min Proc Ext Met Rev 32(3): 137–149. https://doi.org/10.1080/08827508.2010.531069

6. Petrakis E, Bartzas G, Komnitsas K (2020) Grinding behaviour and potential beneficiation options of bauxite ores. Minerals 10 (314):1–21. https://doi.org/10.3390/min10040314

7. Kumar M, Senapati B, Sateesh Kumar C (2013) Beneficiation of high silica bauxite ores of India an innovative approach. Light metals. 187–190

8. Santosh T, Angadi S.I, Dash N, Eswaraiah C, Tripathy S K (2019). Characterization and comminution studies of low-grade Indian Iron Ores. Mining Metallurgy & Exploration. https://doi.org/10.1007/s42461-019-0051-0

9. Shu-ling G, Xia-an L, De-zhou W, Ping F, Chun-yun J, Wen-gang L, Cong H (2007). Beneficiation of low-grade diasporic bauxite with hydrocyclone. Trans Nonferrous Met Soc China 18: 444–448. https://doi.org/10.1016/S1003-6326(08)60078-3

10. Wang P, Wei D (2012). Study of beneficiation technology for low-grade bauxite. Advanced Materials Research. 454: 299–304. https://doi.org/10.4028/www.scientific.net/AMR.454.299

11. Ou L.M, Feng Q.M, Zhang G.F, Chen Y (2008). Comminution property of bauxite and selective separation of Al and Si in bauxite. Min Proc Ext Met 117(3):179–184. https://doi.org/10.1179/174328508X283487

12. Zhu Y, Han Y, Tian Y, Hong W (2011). Medium characteristics on Selective grinding of lowgrade bauxite. Advanced Materials Research 158:159–166. https://doi.org/10.4028/www.scientific.net/AMR.158.159

13. Ahmad I, Hartge E-U, Werther J, Wischnewski R (2014). Bauxite washing for the removal of clay. Int J Min Met Mater 21(11): 1045–1051. https://doi.org/10.1007/s12613-014-1008-4

14. Ostojic G, Lazic D, Skundric B, Skundric Jelena P, Sladojevic S, Keselj D, Blagojevic D (2014). Chemical mineralogical characterization of bauxites from different deposits. Contemp Mater 1: 84–94. https://doi.org/10.7251/COMEN1401084O

15. Anand R.R, Gilkes R.J, Roach G.I.D (1991). Geochemical and mineralogical characteristics of bauxites, Darling Range, Western Australia. Appl Geochem 6: 233–248. https://doi.org/10.1016/0883-2927(91)90001-6

16. Raghavan P.K.N, Kshatriya N K, Dasgupta S (2011). Digestion studies of central Indian bauxite. Light metals. 29–32

Effect of Thermal Activation Temperature on Pre-desilication of Low-Grade Bauxite

Chaojun Fang, Tianrui Cai, Bo Lv, Xiaowei Deng, Jinming Zhang, Zeya Zhao, and Bobing Dong

Abstract

With the increase of requirements for environmental protection, the sintering process with high carbon dioxide emission was gradually reduced in the application. However, for low-grade bauxite, it was difficult to be directly economically utilized by the Bayer process, and a pre-desilication process was often required. This paper researched the influence of different thermal activation temperatures on the pre-desilication process of low-grade bauxite. The results showed that the phase transformation of low-grade bauxite was not much different at different thermal activation temperatures, but the pre-desilication behavior was different. Compared with 650 and 700 °C, the pre-desilication results of low-grade bauxite after thermal activation at 600 °C had a significant improvement. Only after 30 min of leaching, the ratio of alumina to silica of low-grade bauxite can be increased from 2.39 to 6.78. This study may have a certain significance for selecting a suitable thermal activation temperature to promote the high-efficiency pre-desilication of low-grade bauxite.

Keywords

Low-grade bauxite • Pre-desilication • Thermal activation • Phase transformation

Introduction

Aluminum is one of the most important metals, which has been widely applied in aerospace, aircraft, automobiles, trains, ships, electrical appliances, military, and national defense fields [1], while alumina, as an important raw material for aluminum, plays an indispensable role in aluminum industry [2].

Generally, alumina is extracted from bauxite resource. The bauxite resources are relatively abundant on earth, however, the reserves of bauxite resources in China are insufficient, and the majority is diasporic bauxite. The alumina-to-silica (A/S) ratio of diasporic bauxite is low, and it is difficult to be directly treated by the Bayer process [3]. Therefore, desiliconization is required to increase the alumina-to-silica ratio in advance [4, 5].

According to the desilication mechanism, the pre-desilication methods for low-grade bauxite were divided into physical desilication [6, 7], chemical desilication [8–10], and biological desilication [11–13]. Flotation desilication was one of the most important methods of physical desilication, which had been successfully applied in industrial production. Diasporic bauxite with an A/S ratio of 5–6 can be increased to above 11 by flotation desilication [14, 15]. Biological desilication can deal with low-grade bauxite which was difficult to be treated by physical desilication and chemical desilication; it had the advantages of low cost and environmental friendliness [16]. However, flotation

C. Fang · T. Cai · B. Lv (✉) · X. Deng · Z. Zhao · B. Dong
College of Chemistry and Chemical Engineering, Henan Polytechnic University, Jiaozuo, 454000, China
e-mail: bo_lv@21cn.com

C. Fang
e-mail: fangchaojun1009@126.com

T. Cai
e-mail: tianrui_cai@21cn.com

Z. Zhao
e-mail: zeya_zhao@21cn.com

B. Dong
e-mail: bobing_dong@21cn.com

C. Fang · T. Cai · B. Lv · X. Deng · Z. Zhao · B. Dong
Collaborative Innovation Center of Coal Work Safety and Clean High Efficiency Utilization, Jiaozuo, 454000, China

C. Fang · J. Zhang
Zhongzhou Aluminum Co, Ltd. of CHALCO, Jiaozuo, 454171, China
e-mail: jinming_zhang1@21cn.com

C. Fang
State Key Laboratory of Complex Nonferrous Metal Resources Clean Utilization, Kunming, 650093, China

© The Minerals, Metals & Materials Society 2023
S. Broek (ed.), *Light Metals 2023*, The Minerals, Metals & Materials Series,
https://doi.org/10.1007/978-3-031-22532-1_20

desilication relied on bauxite with simple mineral composition and simple ore source. Biological desilication had the disadvantage of a low desiliconization rate. Consequently, chemical desilication was increasingly becoming an important direction to meet the high-efficiency desilication of low-grade bauxite in large quantities [17].

To achieve the goal of pre-desilication, the silicon-containing minerals in low-grade bauxite should be preferentially dissolved in the chemical desilication process. The technique had a significant application potential and was ideal for processing low-grade bauxite with tiny embedded particle size or near cohabitation of alumina and silica-bearing minerals [18]. In practical application, thermal activation and the addition of NaOH or lime were frequently used to carry out chemical desilication, while the thermal activation temperature played an important role [9]. Through thermal activation, the desiliconization efficiency of low-grade bauxite was improved, and the loss of Al_2O_3 during desiliconization was reduced. However, higher temperature was usually required in traditional thermal activation, which consumed a large quantity of energy and emitted more carbon dioxide, which limited the large-scale industrial application of chemical desilication [10].

Therefore, this paper focused on the effective pre-desilication of low-grade diasporic bauxite under lower thermal activation temperature. The phase transformation of minerals in low-grade bauxite was explored, and the pre-desilication effect under different thermal activation temperature was investigated, which may serve as a foundation for the efficient pre-desilication and comprehensive utilization of low-grade diasporic bauxite.

Experiments and Methods

Experiment

The low-grade bauxite sample used in the experiment was the same as that in the previous report [19], and the A/S ratio of the sample was 2.30. Thermal activation made up the first portion of the experiment, and pre-desilication made up the second. A muffle furnace was used to thermally activate a five-gram bauxite sample at a predetermined temperature. Then, the activated bauxite sample was filled in a reactor containing 100 ml of sodium hydroxide solution at a concentration of 4 mol/L. The pulp was filtered after a predetermined amount of reaction time, and the residue was dried and ground for XRF and XRD measurement. Analytical purity grade sodium hydroxide was employed in the experiment, and distilled water was used during the entire experiment.

Sample Measurement

An X-ray diffraction (XRD) device was used to detect the variation of mineral composition of low-grade bauxite under different thermal activation temperatures. The incident X-ray wavelength was 1.5406 nm, and the 2θ spectra varied from $5°$ to $70°$, while an energy-dispersive X-ray fluorescence (XRF) facility was used to evaluate the variation of chemical components.

Results and Discussion

Effect of Thermal Activation Temperature on Chemical Composition of Low-Grade Bauxite

In general, the thermal activation process will have a certain impact on the chemical composition of low-grade bauxite. In which the thermal activation temperature played a significant role due to the difference in volatilize or decompose temperature of particular components. Table 1 showed the variation in the chemical composition of low-grade bauxite under different thermal activation temperatures.

It can be seen from Table 1 that the content of alumina and silica were 55.75 and 24.21%, respectively. The A/S ratio of the low-grade bauxite sample was about 2.30, and it was a typical low-grade bauxite with high alumina, high silica, and low A/S ratio. After thermal activation, the content of alumina and silica had a certain increase, but the A/S ratio of low-grade bauxite only had a little fluctuation. The possible reason was that some components in bauxite volatilized or decomposed under high temperatures, while alumina and silica were relatively stable, resulting in the content increase of alumina and silica.

However, it should be noted that the chemical composition of bauxite was not significantly affected by the thermal activation temperature in the experimental temperature range. When thermal activation temperature was controlled at 600 °C, the content of alumina presented a slight increase, while the content of silica had an insignificant decline. Otherwise, the chemical composition of bauxite was comparable at all three thermal activation temperatures.

Effect of Thermal Activation Temperature on Phase Transformation of Low-Grade Bauxite

It was shown in a previous study that diasporic low-grade bauxite was composed of diaspore, kaolinite, anatase, illite, hematite, and other minerals, in which part of minerals may undergo phase transformation due to thermal activation. The

Table 1 Effect of thermal activation temperature on chemical composition of low-grade bauxite (%)

Temperature	Al_2O_3	SiO_2	Fe_2O_3	TiO_2	CaO	MgO	K_2O	Na_2O
Before thermal activation	55.75	24.21	3.15	2.27	0.36	0.15	0.53	0.04
600 °C	63.13	26.44	1.91	2.73	1.40	0.10	0.48	0.69
650 °C	60.89	27.49	3.18	2.47	0.37	0.15	0.60	0.02
700 °C	60.96	27.68	3.21	2.51	0.38	0.14	0.61	0.05

phase transformation results of diasporic low-grade bauxite under thermal activation at 600 and 700 °C were shown in Figs. 1 and 2, respectively.

It is suggested from Fig. 1 that after thermal activation at 600 °C, diaspore, the main alumina-containing mineral in low-grade bauxite, was transformed into dialuminium trioxide. For silica-containing minerals, the diffraction peaks of illite became more prominent, while the diffraction peaks of kaolinite disappeared. Additionally, minerals like anatase and hematite did not go through phase transformation, although the intensity of the diffraction peaks had a slight increase compared with that before thermal activation.

The results in Fig. 2 were similar to that in Fig. 1. Compared with the thermal activation at 600 °C, the diffraction peak of specific minerals was slightly different from that activated at 700 °C, in which the diffraction peak intensity of illite had an insignificant decrease, and the explanation might be that as the thermal activation temperature rises from 600 to 700 °C, a very small portion of illite started to gradually dissolve while the diffraction peak intensity of other minerals such as dialuminium trioxide, anatase, and hematite had a slight increase on the contrary. The principal reason might be that more compounds decomposed at higher temperatures, which caused minerals

that difficult to disintegrate showing more significant diffraction peaks.

However, it is worth noting that whether thermally activated at 600 °C or 700 °C, there are bulge peaks among 18–28 degree in bauxite XRD pattern, indicating that some minerals transform into amorphous phase after thermal activation. Combined with the previous reports and the results in Figs. 1 and 2, it was generally believed that kaolinite in low-grade bauxite was transformed into amorphous metakaolin, which had a better chemical activity and was easier to be leached. In addition, the thermal activation temperature had little effect on the pattern of diffraction peak, indicating that kaolinite can already be transformed into a more soluble amorphous metakaolin even at a thermal activation temperature of 600 °C.

Therefore, it can be concluded that both diaspore and kaolinite underwent phase transformation after thermal activation, and diaspore was converted into dialuminium trioxide which was not easily soluble in alkali solution, while kaolinite was converted into amorphous metakaolin that was easy to dissolve in alkali solution. The phase transition of diaspore and kaolinite was less affected by the thermal activation temperature, and 600 °C was enough for realizing the phase transformation of diaspore and kaolinite.

Fig. 1 XRD pattern of the low-grade bauxite under thermal activation at 600 °C

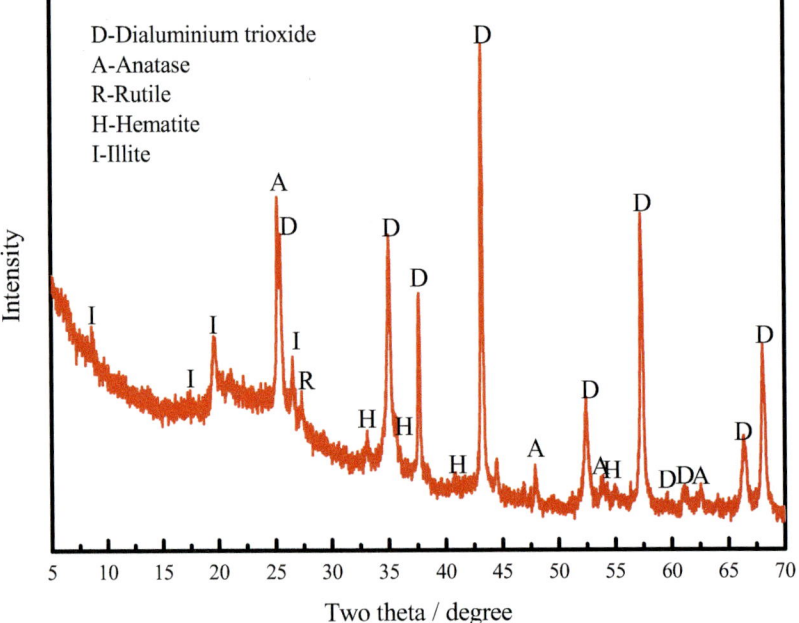

Fig. 2 XRD pattern of the low-grade bauxite under thermal activation at 700 °C

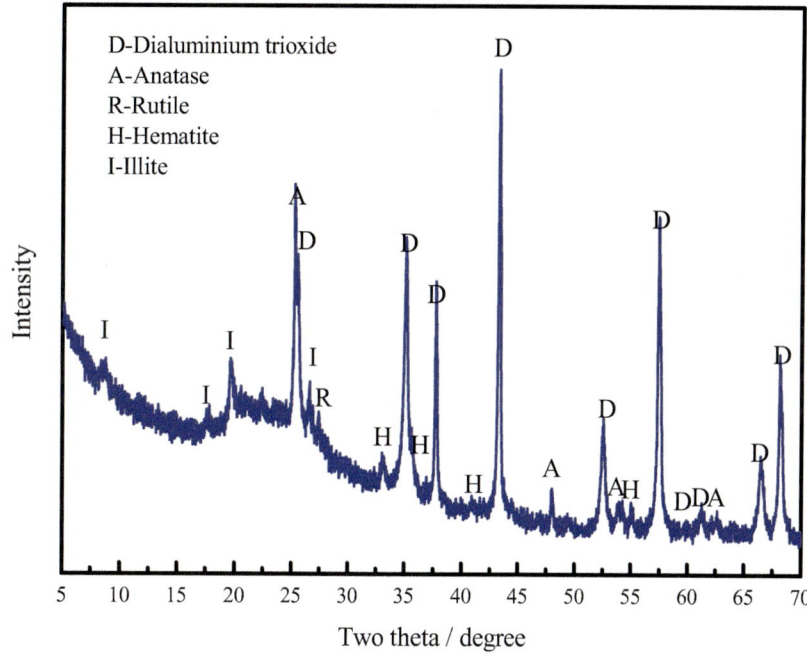

Effect of Thermal Activation Temperature on the Dissolution of Alumina and Silica

According to the previous research, when the thermally activated low-grade bauxite was added to the NaOH solution, the alumina component was still difficult to be dissolved, while the silica component was easy to leach. Figure 3 showed the content of alumina and silica in leaching residue, attaching with the A/S ratio.

It is presented in Fig. 3 that the contents of alumina and silica in low-grade bauxite both had a variation by the leaching of NaOH solution. The content of alumina was increased to around 65%, while the content of silica decreased dramatically from over 25% to about 10%, which resulted in the increase of A/S ratio and realized the pre-desilication of low-grade diasporic bauxite. Especially when thermal activated at 600 °C, the A/S ratio of low-grade bauxite can be improved from 2.30 to 6.78.

However, it was worth noting that the thermal activation temperature had a significant effect on the pre-desilication of low-grade bauxite. Although the content of alumina in the leaching residue was similar, the content of silica had an obvious increase with the increase of the thermal activation temperature, which resulted in the decrease of the A/S ratio

Fig. 3 The contents of alumina and silica, and the A/S ratio of leaching residue

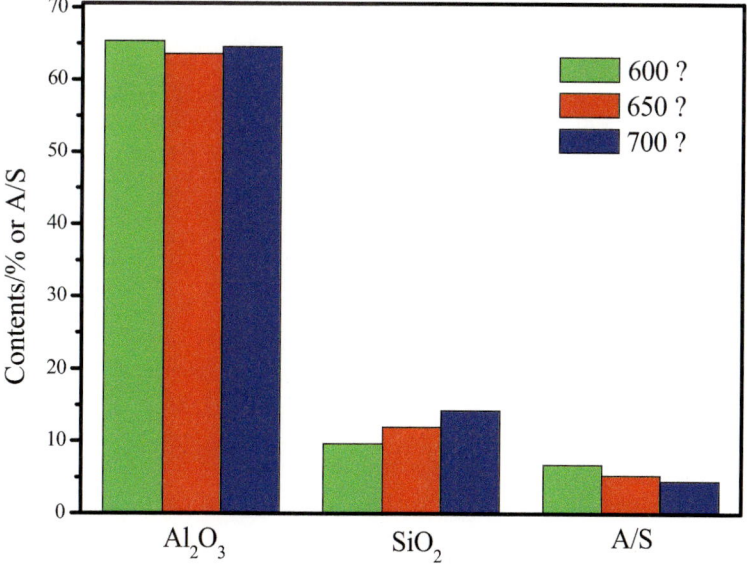

in leaching residue, and weakened the pre-desilication effect of low-grade bauxite. Although the A/S ratio can be improved from 2.30 to 6.78 when low-grade bauxite was thermally activated at 600 °C, it dropped to 4.51 if the thermal activation temperature raised to 700 °C. Therefore, in the pre-desilication process of low-grade bauxite, the thermal activation temperature was not as high as possible.

Conclusions

A single thermal activation process had a rare effect on the chemical composition and A/S ratio of low-grade bauxite, but it induced the phase transformation of low-grade bauxite. The alumina-containing mineral diaspore was converted to dialuminium trioxide, while the primary silica-containing mineral kaolinite was converted to amorphous metakaolin. Then the pre-desilication of low-grade bauxite was achieved by selective and preferential dissolution of amorphous metakaolin by NaOH solution.

The chemical composition and phase transformation of low-grade bauxite were slightly impacted by thermal activation temperature within a given temperature range, but the thermal activation temperature had a significant effect on the pre-desilication of low-grade bauxite. Compared with 650 and 700 °C, the pre-desilication effect of low-grade bauxite was better when thermally activated at 600 °C. The A/S ratio of low-grade bauxite can be increased from 2.30 to 6.78, which meets the requirements of the Bayer process. Therefore, the thermal activation temperature of low-grade bauxite was not as high as possible; controlling the thermal activation temperature in an appropriate range was of great significance for the pre-desilication of low-grade bauxite.

Acknowledgements The authors acknowledge the financial support of the National Natural Science Foundation of China (52004086, U1932129, 51904096, and 51964024), China Postdoctoral Science Foundation (2021M703620), The Key Project of Science and Technology of Henan Province (202102310543), Open Foundation of State Key Laboratory of Complex Nonferrous Metal Resources Clean Utilization (CNMRCUKF2009), Key Research Project of Colleges and Universities in Henan Province (21A440007), and Natural Science Foundation of Henan Polytechnic University (B2020-27).

References

1. Li, H. Research on sustainable development of bauxite resources industry in China. China University of Geosciences (Beijing), China, **2010**.

2. Fang, C.J.; Yu, S.C.; Wei, X.Y.; Peng, H.; Ou, L.M.; Zhang, G.F.; Wang, J. The Cation Effect on Adsorption of Surfactant in the Froth Flotation of Diaspore and Kaolinite. *Miner. Eng.* **2019**, 144, 106051.

3. Liu, C.M.; Feng, A.S.; Guo, Z.X. Investigation and optimization of use of anionic collectors in direct flotation of bauxite ores. *Physicochem. Probl. Miner. Process.* **2016**, 52(2), 932–942.

4. Barbosa, F.M.; Bergerman, M.G.; Horta, D.G. Removal of iron-bearing minerals from gibbsitic bauxite by direct froth flotation. *Technol. Metal. Mater. Miner., São Paulo* **2016**, 13(1), 106–112.

5. Gibson, B.; Wonyen, D.G.; Chelgani, S.C. A review of pretreatment of diasporic bauxite ores by flotation separation. *Miner. Eng.* **2017**, 114, 64–73.

6. Fang, C.J.; Chang, Z.Y.; Feng, Q.M.; Xiao, W.; Yu, S.C.; Qiu, G. Z.; Wang, J. The influence of backwater Al³⁺ on diaspore bauxite flotation. *Minerals* **2017**, 7, 195.

7. Deng, L.; Wang, S.; Zhong, H.; Liu, G. A novel surfactant 2-amino-6-decanamidohexanoic acid: Flotation performance and adsorption mechanism to diaspore. *Miner. Eng.* **2016**, 93, 16–23.

8. Peng, H.; Kim, T.; Vaughan, J. Acid Leaching of Desilication Products: Implications for Acid Neutralization of Bauxite Residue. *Ind. Eng. Chem. Res.* **2020**, 59, 8174–8182.

9. Pan, X.L.; Yu, H.Y.; Dong, K.W.; Tu, G.F.; Bi, S.W. Pre-desilication and digestion of gibbsitic bauxite with lime in sodium aluminate liquor. *International Journal of Minerals Metallurgy and Materials* **2012**, 19(11), 973–977.

10. Xu, Y.P.; Li, J.Q.; Chen, C.Y.; Lan, Y.P.; Wang, L.Z. Desilication and Recycling of Alkali-Silicate Solution for Low-Grade High-Silica Bauxite Utilization. *JOM* **2020**, 72(7), 2705–2712.

11. Dopson, M.; Lövgren, L.; Bostrm, D. Silicate mineral dissolution in the presence of acidophilic microorganisms: Implications for heap bioleaching. *Hydrometallurgy* **2009**, 96(4): 288–293.

12. Mockovčiaková, A.; Iveta, Š.; Jiří, Š.; Ivana, K. Characterization of changes of low and high defect kaolinite after bioleaching. *Appl. Clay Sci.* **2008**, 39(3), 202–207.

13. Vasan, S.S.; Modak, J.M.; Natarajan, K.A. Some recent advances in the bioprocessing of bauxite. *Int. J. Miner. Process.* **2001**, 62(1): 173–186.

14. Lu Y.P. Research on bauxite desilication by selective grinding-aggregation flotation. Doctor, Central South University, Changsha China, **2012**.

15. Fang, C.J.; Yu, S.C.; Peng, H.; Deng, X.W.; Wang, J. Ionic effect of NaCl and KCl on the flotation of diaspore and kaolinite using sodium oleate as collector. *Light Metals* **2020**, 40–46.

16. Wang, Q.; Sheng, X.F.; He, L.Y.; Shan, Y. Improving bio-desilication of a high silica bauxite by two highly effective silica-solubilizing bacteria. *Minerals Engineering* **2018**, 128, 179–186.

17. Xu, B.; Smith, P.; Wingate, G.; Silva, D.L. The effect of calcium and temperature on the transformation of sodalite to cancrinite in Bayer digestion. *Hydrometallurgy* **2010**, 105(1–2), 75–81.

18. Xu, Y.P.; Chen, C.Y.; Lan, Y.P.; Wang, L.Z.; Li, J.Q. Desilication and recycling of alkali-silicate solution seeded with red mud for low-grade bauxite utilization. *J. Mater. Res. Technol.* **2020**, 9(4), 7418–7426.

19. Fang, C.J.; Cai, T.R.; Gao, L.J.; Deng, X.W.; Lv, B.; Peng, H. Effect of Thermal Activation on Phase Transformation and Pre-desilication of Low-Grade Bauxite. *Light Metals* **2022**, 65–69.

Application of Repeatability and Reproducibility Analysis in the Determination of Available Alumina and Reactive Silica in Bauxites

Paula Lima, Walter Santana, Danielle Matos, Jaqueline Melo, and Janyne Ramos

Abstract

Bauxite is a rock constituted of minerals and impurities, which are mainly kaolinite, gibbsite, hematite, and anatase. Gibbsite is one of the useful minerals for aluminum production. In mining industries, the quantification of useful minerals and damaging impurities is necessary, with gibbsite reported in levels of available alumina and kaolinite, one of the main impurities, as reactive silica. The accuracy of these results influences the measurements of KPI's (Key Performance Indicators) necessary to determine productivity, helping in decision-making, so the application of a robust methodology to guarantee the quality of determinations, such as R&R (repeatability and reproducibility), is necessary. R&R analysis of these measurements was performed at Hydro Alunorte (Barcarena/PA/Brazil) using 3 different bauxites and statistical treatment by ANOVA. Precision/Tolerance values of 6.5 and 10% were reached for the analysis of available alumina and reactive silica, respectively. Possibilities for improvement were implemented, with a significant reduction in reproducibility deviations. With the improvements and the data found, it was possible to guarantee the reliability and quality of the results involved.

Keywords

Bauxite · Repeatability · Reproducibility · Gibbsite · Kaolinite

P. Lima (✉) · W. Santana · D. Matos · J. Melo · J. Ramos
Alunorte Alumina do Norte do Brasil, Barcarena, Brazil
e-mail: paula.lima@hydro.com

Introduction

Bauxite is an aluminum ore, formed by a mixture of aluminum hydroxide or oxy-hydroxide minerals and impurities such as kaolinite, hematite, goethite, quartz, and anatase. The three useful forms of aluminum that can constitute bauxite are gibbsite, boehmite and diaspore [1, 2].

In the alumina production process carried out at Hydro Alunorte (Barcarena/PA), the largest metallurgical alumina refinery in the world, gibbsite bauxite from the region of Oriximiná (West of Pará) and Paragominas (Northeast of Pará) is used. For gibbsite bauxites, available alumina is associated with gibbsite and reactive silica with kaolinite [3]. The most conventional process for extracting alumina from gibbsite is the Bayer Process, where bauxite is digested with NaOH (leaching agent) at temperatures between 130 and 160 °C; the alumina is dissolved in the form of sodium aluminate, and under specific conditions it precipitates in the form of aluminum hydroxide (hydrate) which after the calcination process is transformed into alumina (Al_2O_3). In this process, kaolinite acts as a potential contaminant, as it also reacts with NaOH forming a desilication product of the sodalite type, which results in irreversible NaOH [4–7].

For the quality control of bauxites destined for aluminum production, from the prospection of the ore to its entry into the Bayer process, the main quality and process parameters are the available alumina content ($av.Al_2O_3$) and silica reactive ($r.SiO_2$) determined according to a procedure that simulates the Bayer digestion process on a laboratory scale [8], and these parameters are fundamental for calculating the KPI's necessary to determine productivity helping in decision-making.

Thus, due to the importance of these measurements, an evaluation of their repeatability and reproducibility was carried out with the identification of improvements in order to guarantee the quality and reliability of these analyses carried out in the Research and Development laboratory of Norsk Hydro Brasil.

© The Minerals, Metals & Materials Society 2023
S. Broek (ed.), *Light Metals 2023*, The Minerals, Metals & Materials Series,
https://doi.org/10.1007/978-3-031-22532-1_21

Methodology

Analysis of Available Alumina and Reactive Silica

In this methodology, 3.25 g of bauxite is digested with 25 mL of 2.5 N NaOH. After digestion, all this content is placed in a 500 mL flask with NaCl, and 50 mL of this solution is subjected to filtration. The filtrate and the fraction retained on filter paper follow the processes below

– Filtrate: The filtrate goes to the analysis of available alumina. In 10 mL of filtrate is added methyl red indicator, HCl to acidify, and CDTA (Cyclohexylendinitrile-1,2 Tetra acetic Acid), in addition to hexamethylenetetramine, for complexation. The excess added CDTA reacts with the aluminum oxide (alumina) present. The titration is performed with zinc sulfate that reacts with excess CDTA, making it possible to determine the amount of Al_2O_3 that complexed.
– Fraction retained on filter paper: The fraction retained goes to the determination of reactive silica. The filter paper used in the filtration is placed inside each flask corresponding to the digested sample. Hydrochloric acid and organic flocculant are added. After about 5–10 min of decantation, the sample is filtered, 15 mL of the filtrate is transferred to a 500 mL volumetric flask, made up of distilled water, and homogenized. 10 mL of the sample has its pH adjusted with hydrochloric acid or sodium hydroxide to the range of 0.7–0.8 followed by the addition of ammonium molybdate, tartaric acid, reducing solution, and oxalic acid. Finally, measurements are carried out in a spectrophotometer (Figs. 1 and 2).

Statistical Treatment: ANOVA

To obtain the necessary data for the study of repeatability and reproducibility through the application of ANOVA [9],

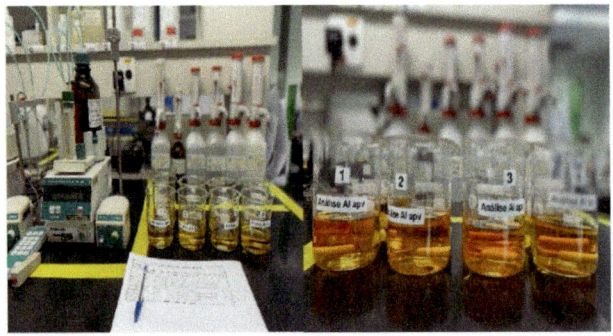

Fig. 1 Titration process in available alumina analyses

Fig. 2 Filtration process in reactive silica analysis

three different bauxite samples were used. The analyses were performed by two different operators. The same methodology was applied, starting from the same materials and using the same equipment. The analysis was performed in triplicate for each bauxite. The same glassware was used.

To apply the ANOVA, the following structural model was considered:

$$Xijk = u + Ai + Bj + Abij + Zk(ij)$$

Ai = Sample (3) random variable.
Bj = Operator (2) random variable.
K = Repetitions (3).
α (significance level) = 5%.

As a measurement quality parameter, the Precision/Tolerance ratio was calculated, which relates the measurement system precision estimate and specification tolerance. For P/T values less than 10%, the measurement system is considered adequate, for values between 10 and 25% acceptable, and greater than 25% is considered inadequate.

Results and Discussion

Available Alumina: Measurement System Parameters

The results obtained for the analysis of available alumina are shown in Table 1, in which samples 1, 2, and 3 are bauxite samples and X the respective repetitions.

Bartlett's test was performed for analysis of equality of variance (Fig. 3). The P-value was equivalent to 0.092, which means that the hypothesis of equality of variance between the populations involved was not rejected. In Fig. 4, it is possible to see that the data fit reasonably well into a straight line, so it can be said that the behavior of the data follows a normal probability distribution.

According to the application of ANOVA, the samples show differences between them which are supported by the

Table 1 Available alumina results for the analyzed samples

Sample	Operator A			Operator B		
	X1	X2	X3	X1	X2	X3
1	48.68	48.84	48.68	48.99	48.89	48.91
2	47.06	47.02	47.03	46.92	46.81	46.83
3	40.14	40.15	40.15	40.33	40.26	40.28

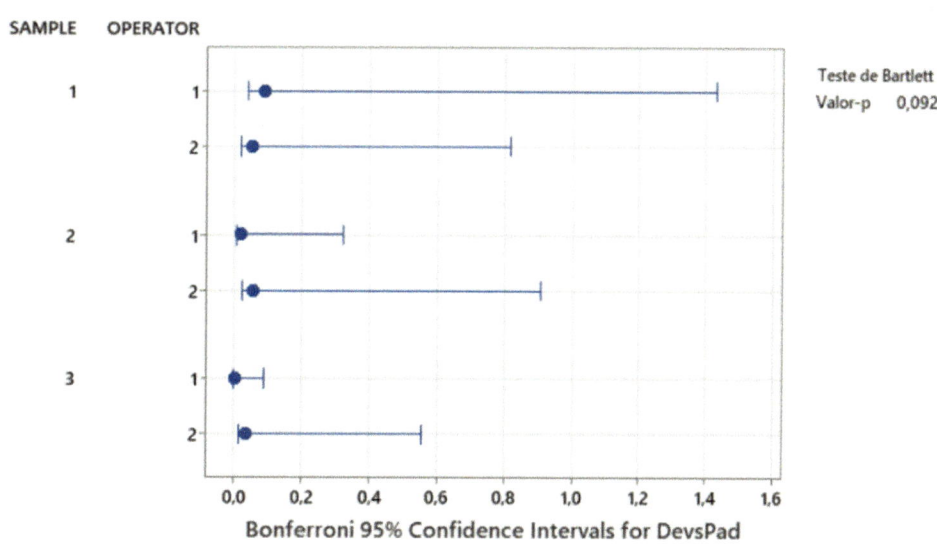

Fig. 3 Bartlett's test—test of equality of variances

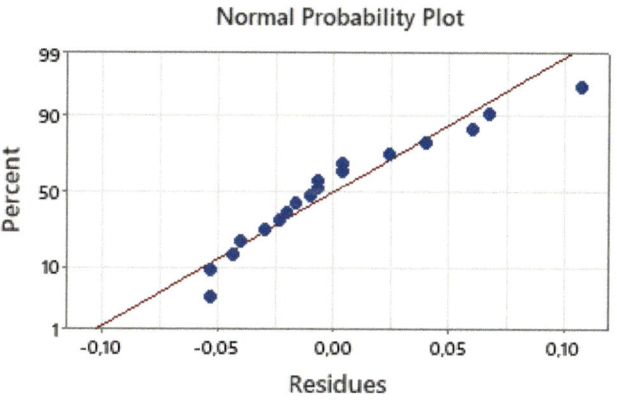

Fig. 4 Normal probability plot

fact that 3 different bauxites were used, with a repeatability deviation equivalent to 0.05 and reproducibility equivalent to 0.14. There is interaction between the sample and operator. The variation between operators was not significant (Fig. 5). The R^2 obtained was equivalent to 99.98%. For the results obtained in the analysis of available alumina, the P/T ratio was 6.5%, so the measurement system is considered adequate (Tables 2 and 3).

With these results, to improve reproducibility, some actions were taken, such as the elaboration of specific glassware kits for each analysis and carrying out blind sampling. With the implementation of these improvements, there was no interaction between the factors, and there was a reduction in reproducibility to 0.03 (78% reduction) with the maintenance of the Precision/Tolerance ratio quality parameter.

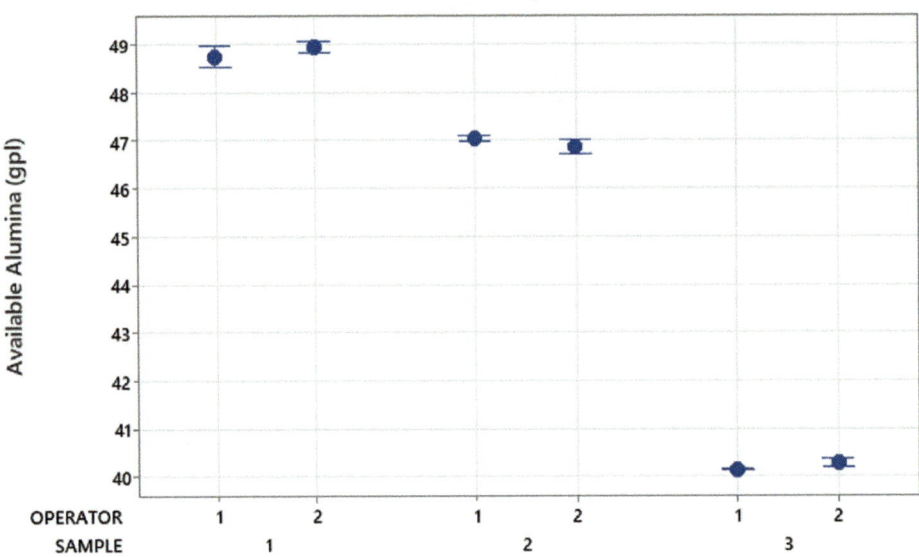

Fig. 5 Confidence interval graph (95% IC for average)

Table 2 Analysis of variance—available alumina analysis

	Degrees of freedom	Squared Sum	Mean squared	Observed factor	P-value
Sample	2	245.99	122.99	1937.30	0.001
Operator	1	0.012	0.012	0.19	0.703
Sample * operator	2	0.127	0.063	23.04	0.000
Repeatability	12	0.033	0.003		
Total	17	246.17			

Table 3 Standard deviations of measurements

	Standard Deviation (SD)
Total measurement R&R	0.1516
Repeatability	0.0525
Reproducibility	0.1423
Piece by piece	4.5264
Total variation	4.5290

Table 4 Reactive silica results for the analyzed samples

Sample	Operator A			Operator B		
	X1	X2	X3	X1	X2	X3
1	4.39	4.40	4.31	4.52	4.42	4.54
2	3.45	3.64	3.57	3.69	3.70	3.84
3	0.93	0.90	0.93	0.95	0.99	0.98

SAMPLE   OPERATOR

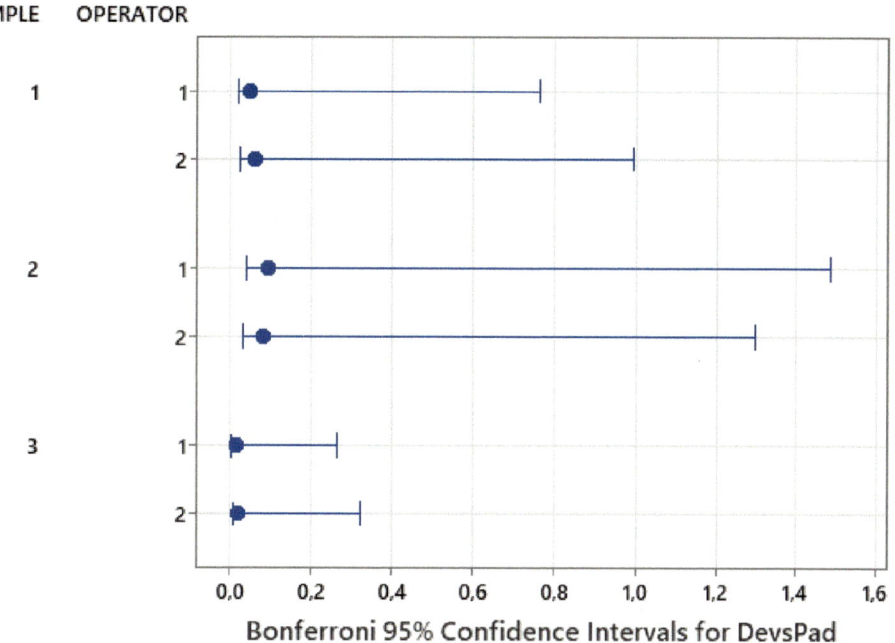

Fig. 6 Bartlett's test—test of equality of variances

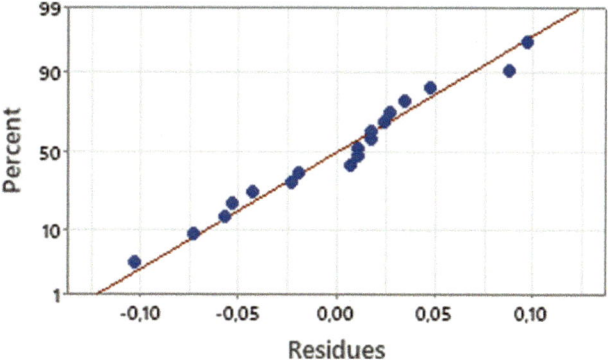

Fig. 7 Normal probability plot

Reactive Silica: Measurement System Parameters

The results obtained are shown in Table 4, in which samples 1, 2, and 3 are bauxite samples and X the respective repetitions.

Bartlett's test was performed for analysis of equality of variance (Fig. 6). The P-value was equivalent to 0.277, which means that the hypothesis of equality of variance was not rejected. The data follows a normal probability distribution (Fig. 7).

According to the application of ANOVA, the samples were different (3 different bauxites were used), with a repeatability deviation equivalent to 0.06 and a deviation of reproducibility equivalent to 0.08 with $R^2 = 99.88\%$. The variation between operators was not significant (Fig. 8). For the results obtained in the analysis of reactive silica, the P/T ratio was 10%, so the measurement system is considered acceptable (Tables 5 and 6).

Improvements were identified for the analysis of reactive silica in order to reduce the deviations obtained for reproducibility. With the implementation of specific glassware kits for each analysis and with the implementation of blind sampling, there was no interaction between the factors and the reproducibility was reduced to 0.003, a 96% reduction.

Fig. 8 Confidence interval graph
(95% IC for average)

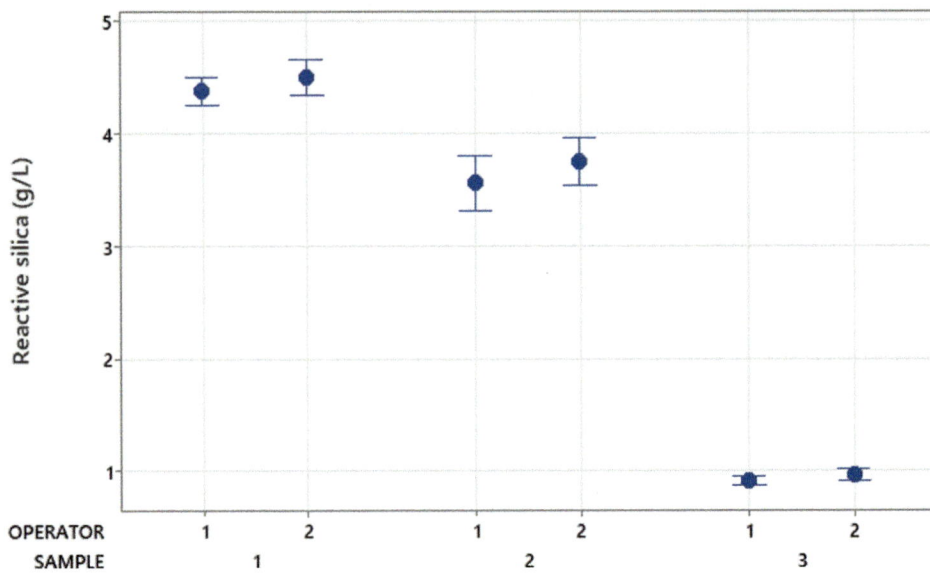

Table 5 Analysis of variance—reactive silica analysis

	Degrees of freedom	Squared sum	Mean squared	Observed factor	P-value
Sample	2	40.087	20.043	2856.57	0.000
Operator	1	0.068	0.0684	9.76	0.089
Sample * operator	2	0.014	0.0070	1.79	0.209
Repeatability	12	0.047	0.0039		
Total	17	40.22			

Table 6 Standard deviations of measurements

	Standard Deviation (SD)
Total measurement R&R	0.1072
Repeatability	0.0661
Reproducibility	0.0843
Piece by piece	1.8275
Total variation	1.8307

Conclusions

The results showed an adequate Precision/Tolerance ratio for the required levels of the respective analyses. In order to improve reproducibility, some actions were taken, such as the creation of glassware kits for analysis and blind sampling, to avoid possible interference in the deviation. These actions have been shown to have a significant impact on reproducibility with a reduction of 78 and 96% of deviations for available alumina and reactive silica analyses, respectively. The knowledge of the R&R (repeatability and reproducibility) and quality deviations of the methodologies allows the guarantee of the accuracy of the measurements of KPI's necessary to determine the industrial productivity, helping in the decision-making in the Hydro group.

References

1. Andrews, WH (1984) Uses and Specifications of Bauxite. In: Jacob Jr., L. Bauxite. Bauxite Symposium, Los Angeles, California: American Institute of Mining, Metallurgical and Petroleum Engineers, New York, p. 49–66.
2. Barrand P, Gadeau R, Dumas A (1967) Enciclopedia del Aluminio: Produccion del Aluminio. Ediciones URMO: España, 300 p.
3. Rayzman VL, Pevzner IZ, Sizyakov VM, Ni LP, Filipovich IK, Aturin AV (2003) Extracting Silica and Alumina from Low-Grade Bauxite. Extractive Metallurgy: 47–50.
4. Barnes MC, Addai-Mensah J, Gerson AR (1999). The kinetics of desilication of synthetic spent Bayer liquor and sodalite crystal

growth, Colloids and Surfaces A: Physicochemical and Engineering Aspects, 147: 283–295.

5. McCormick PG, Picaro T, Smith PA (2002). Mechanochemical treatment of high silica bauxite with lime. Minerals Engineering, 15: 211–214.

6. Croker D, Loan M, Hodnett BK (2008) Desilication Reactions at Digestion Conditions: An in Situ X-ray Diffraction Study. Crystal Growth & Design, 8 (12): 4499–4505.

7. Smith P (2009) The processing of high silica bauxites – Review of existing and potential processes. Hydrometallurgy, 98: 162–176.

8. Ostap S (1986) Control of silica in the Bayer process used for alumina production. Canadian Metallurgical Quarterly, 25: 101–106.

9. Montgomery DC (2009) Design and Analysis of Experiments. John Wiley & Sons, Hoboken, New Jersey.

Zero Waste Alumina Production

Casper van der Eijk and Camilla Sommerseth

Abstract

The Bayer process holds an exclusive status as almost all the world's metallurgical grade alumina is produced by this method. The Pedersen process is an alternative process for the production of alumina where bauxite ores, or other aluminous materials, are smelted with lime to produce pig iron and a slag of mainly calcium aluminate composition. The latter is subsequently leached in Na_2CO_3 solutions to extract alumina hydrate which is calcined to alumina. The Pedersen process can handle a wide range of bauxite qualities and even bauxite residue from the Bayer process as raw material. Moreover, by-products from the Pedersen process can be used in other industries, thereby eliminating the need for land deposits. An EU-financed project, called EnsureAl, has investigated the possibility of the commercialization of an improved version of the Pedersen process. This paper will summarize the main findings of this project concerning the quality of the produced alumina and the by-products. It will also assess a comparison with the Bayer process concerning the estimated costs of the process.

Keywords

Aluminium • Recycling and secondary recovery • Pyrometallurgy

Introduction

Production of alumina produces large amounts of bauxite residue. This represents a potential environmental issue, since the bauxite residue is strongly alkaline, and a potential land use issue, as land availability is often becoming limited in many regions. Moreover, it is an issue of resource inefficiency since the bauxite residue contains many potentially useful materials, among them the oxides of iron and aluminium. The Pedersen process is an alternative process for the production of alumina. Initially, bauxite ores, or other aluminous materials, are smelted with lime to produce pig iron and a slag of mainly calcium aluminate composition. The latter is subsequently leached in Na_2CO_3 solutions to extract alumina hydrate. The process has been well described in the literature [1–8].

Experimental

Bauxite and bauxite residue were used as raw materials to produce alumina hydrates, iron, and grey mud residue via the Pedersen process at the pyro- and hydro-metallurgical pilot units located within the industrial area of the MYTILINEOS Agios Nicolaos plant in Greece. The normalised composition of the raw materials is summarized in Table 1. The Loss of Ignition is not included in the analysis because the first part of the process is the smelting which means that the hydroxides, moisture, and carbonates will disappear. The process parameters for the production are as described in reference [3]. Seeding was used during the production of the precipitates from bauxite. This seeding was not used during the precipitation of the hydrates produced from bauxite residue. The obtained hydrates were characterised by X-Ray Diffraction with a Bruker D8 Focus powder XRD equipped with a Cu k-alpha radiation source and LynxEye. Micrographs were taken with a Hitachi S3400N Scanning Electron Microscopy. Moreover, Differential Thermal Analysis (DTA) and Thermal Gravimetric Analysis (TGA) were performed in a Setaram SetSys 2400 apparatus up to 1200 °C.

Aluminum hydrate calcination tests were performed at 950 °C in the 200 mm fluidizing bed pilot plant at Metso Outotec's R&D Center in Frankfurt am Main, Germany.

C. van der Eijk (✉) · C. Sommerseth
Department of Metal Production and Processing, SINTEF, Trondheim, Norway
e-mail: casper.eijk@sintef.no

© The Minerals, Metals & Materials Society 2023
S. Broek (ed.), *Light Metals 2023*, The Minerals, Metals & Materials Series,
https://doi.org/10.1007/978-3-031-22532-1_22

Table 1 Normalised composition, as measured by XRF, of the bauxite and bauxite residue which were used as raw material during the pilot trials

%wt, normalised	Fe_2O_3	SiO_2	CaO	Al_2O_3	TiO_2	Na_2O
Bauxite	23.0	2.8	0.8	70.1	3.3	0.0
Bauxite Residue	49.1	8.0	10.8	22.7	6.1	3.4

Two different feedstock materials were calcined: (1) standard aluminium hydrate (Bayer process route) and (2) aluminium hydrate produced at the EnsureAl (Pedersen) pilot.

After calcination, the alumina was characterized by XRF, XRD, SEM, PSD, the loss on ignition (LoI) at 1000 °C, and by the surface area that was measured with the Brunauer-Emmett-Teller technique (BET). In addition, alumina dissolution tests were conducted to establish if there were differences between the rate of dissolution of industrial grade Bayer alumina and the Pedersen alumina. The method is based on having a certain amount of industrial cryolite melt in a graphite crucible and, through sampling and using a probe, by tracking the increase in alumina content after adding a certain amount of alumina. The bath used was an industrial bath from Norsk Hydro's plant in Sunndalsøra. The AlF_3 content was 12.1 wt%. The alumina content was measured at SINTEF, using the same Leco instrument mentioned above, to be 2.07 wt%. The temperature was around 980–983 °C at the time of the alumina addition.

The batches of grey mud received from the pilot plant were analysed as material towards the building industry. In this study, the pozzolanic reactivity was assessed for the grey mud as screening tests prior to the cement mortar tests.

Additionally, cement mortar experiments according to EN 196-1 were also conducted.

Results and Discussion

Alumina

XRD of the alumina hydrate shows that the precipitates from the bauxite residue are mainly bayerite. The ones produced from bauxite are gibbsite.

A SEM micrograph of a precipitate produced from bauxite with the Pedersen method is shown in Fig. 1. One can observe that particles seem to grow in a rectangular prism-shaped manner. The precipitates produced from the bauxite residue are all very small. The precipitates produced from bauxite contain a few large particles and many small particles. Probably, the large particles have grown on the seeds that were added, while the small particles were nucleated in the precipitation stage.

TGA and DTA measurements analyse the weight loss and energy consumption during phase transformations. As shown in Fig. 2, the bayerite produced from bauxite residue

Fig. 1 SEM micrograph of a precipitate (gibbsite) produced from bauxite with the Pedersen process

Fig. 2 The weight change and heat flow of the hydrates during calcination in Ar

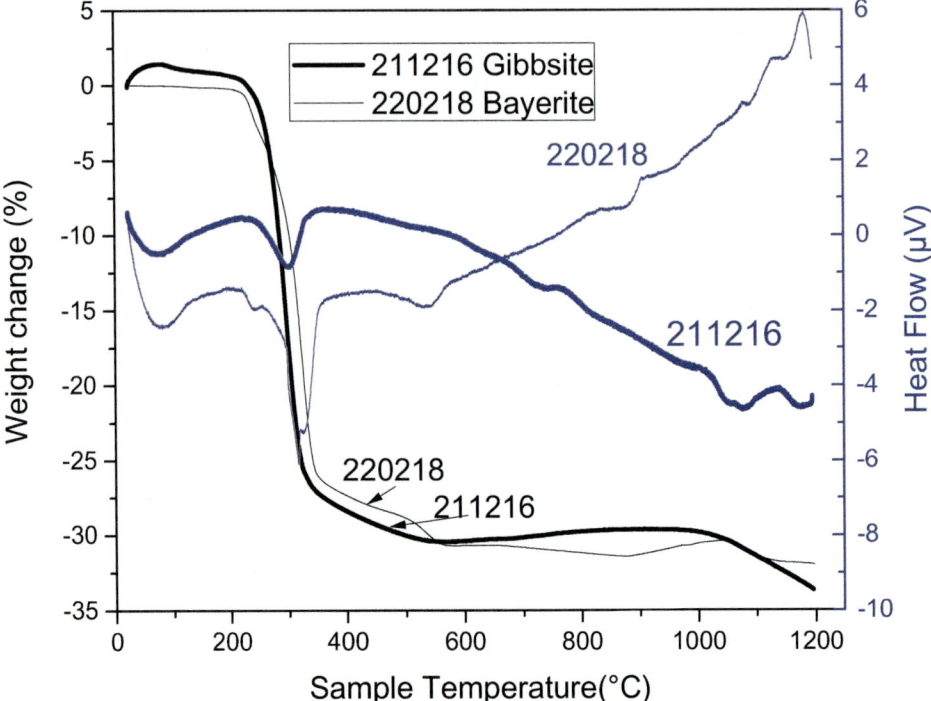

and the gibbsite produced from bauxite with the Pedersen method have a weight loss of ca. 26% until around 340 °C. Further weight loss is 4.0–5.0% until 1030 °C. During the last stage from 1030 to 1200 °C, the weight loss is around 2.0–3.0%. The final weight loss of gibbsite produced from bauxite (33.5%) and bayerite produced from bauxite residue (32%) approaches the theoretical weight loss of calcination, 34.6 wt%. The energy consumption during heating of the gibbsite seems to be higher from 600 to 1100 °C. This means that the calcination of the gibbsite will require more energy than the calcination of the bayerite.

The hydrates from the EnsureAl (Pedersen) pilot have a smaller particle size ($d_{50} \approx 67$ μm) than the typical hydrates from the Bayer process. This is likely due to these hydrates being precipitated in a small pilot-scale tank and having less time to grow as the precipitates from the commercial-sized Bayer process. SEM investigations performed after calcination, shown in Fig. 3, prove that the rectangular prism shape is still intact after calcination.

The particle breakage is measured by the increase of fines (here defined as fraction minus 45 μm) in the calcine (total discharge, bottom + filter) compared to fines in the hydrate feed. The calcination tests in the circulating fluidized bed showed more particle breakage for the Pedersen than the Bayer material. A possible explanation of the elevated particle breakage could be the particle structure which is agglomerates of crystals. If the crystals separate from each other as a result of mechanical forces, then the fines fraction is increased.

Fig. 3 SEM micrograph of alumina produced from bauxite with the Pedersen process

The angle of repose of the Pedersen alumina is higher than the angle of repose of the Bayer alumina. This indicates that the Pedersen material does not flow as easily as the Bayer material. This is most likely caused by the particle shape.

Table 2 lists the typical range of parameters for aluminium oxide used for aluminium production and compares it with the experimental results. The content of α-Al_2O_3 in the Pedersen alumina is below the detection limit of the XRD measurements. The minus 45 μm particle fraction was

Table 2 Characteristic parameters of aluminium oxide used for aluminium production

Parameter	Alumina from the Bayer process	Alumina from the Pedersen process produced from bauxite	Typical commercial range for alumina
α-Alumina, %	0.9	<0.6	1–10
LoI 1000, dried, %	0.4	0.42	0.5–1.0
Fraction <45 μm, %	3.5	22	5–25
BET surface area, m^2/g	62	57.5	50–80

Table 3 XRF analysis of the alumina produced (wt%)

wt%	Alumina from bauxite residue	Alumina produced from bauxite
CaO	0.61	0.20
MgO	0.01	
SiO_2	1.24	1.4
Na_2O	3.81	0.9
SO_3	0.04	
Fe_2O_3	0.04	0.0
Cl	0.02	

measured to be 22% of the total particle distribution. This is a very high number for the calcine from Pedersen process material compared to the standard Bayer alumina. The loss on ignition (LoI) 1000 °C and surface area (BET) of the calcine from Pedersen alumina are comparable to the values obtained from the pilot fluidized bed test with Bayer alumina

Table 3 lists the results of XRF analysis of the alumina. It is obvious that there is a large amount of silicon that is also being dissolved during leaching. The slag from the bauxite residue contained more silicon than the slag produced from bauxite so there does not seem to be a relationship between the silicon content in the raw materials and the precipitates. Recent work [9] has shown that it is possible to remove up to 90% of the silicon in a desilication step. Such a desilication process was not included in the pilot scale runs that produced the current materials.

The alumina dissolution tests performed in a cryolitic melt showed that the alumina produced from bauxite residue is more difficult to dissolve in cryolite than Bayer alumina. The alumina produced from bauxite behaves like commercial alumina during dissolution tests.

Grey Mud

The composition of the residue from the leaching process (grey mud) is varying depending on the raw materials used. The main phase is $CaCO_3$. When bauxite is used, the SiO_2 and TiO_2 contents are about 2 and 1 wt%, respectively.

When bauxite residue is used as raw material, the content of SiO_2 and TiO_2 can reach up to 11 and 6 wt%, respectively. This is due to the higher SiO_2 and TiO_2 contents in the bauxite residue.

In order to test the application of the grey mud as a substitute for cement, the pozzolanic reactivity was determined for both grey mud from bauxite and bauxite residue. The highest reactivity was found for the sample with the highest amount of calcium aluminate phases. The results indicated the performance of grey mud as a supplementary cementitious material. By further optimisation of the fineness and the final chemistry of the binder mix, a replacement level of minimum 10% of the CEM I binder is possible. The strength development for both mortars was either close to or above 90% of the 28 days reference strength.

Scandium and other rare-earth elements which are present in the raw materials typically end up in the grey mud. In the case of scandium, the bauxite residue used in the pilot contained 74 ppm, while the corresponding grey mud contained 125 ppm of scandium.

Iron

The iron produced in the process contains typically up to 5 wt% carbon. It can be used in the foundry industry. Depending on the composition of the raw materials, there is usually Ti in the metal present as Ti-carbides. These can be removed with slag refining. Moreover, Ti-carbides can sometimes also be beneficial like in applications where wear resistance is a requirement.

Techno-Economic Evaluation

A techno-economic evaluation has been performed [10] where various scenarios for alumina production have been evaluated. The main findings of the study were that the Pedersen process cannot compete with the Bayer process in terms of techno-economics. The process is too energy consuming, and the need for the electric arc furnace drives the CAPEX cost of the industrial site to such an extent that it is not possible to replace the Bayer process. This is in line with a recent LCA study [8] that also points out that the high energy consumption of the pyrometallurgical phase is a major constraint for industrialization.

However, the study did find that a scaled-down Pedersen process is feasible in terms of techno-economics, using bauxite residue as the raw material. This way, the CAPEX costs are greatly reduced, and the aluminium deposits in the bauxite residue are utilized to produce smelter grade alumina and iron. The need for depositing bauxite residue will potentially disappear. Placed next to an existing Bayer plant, the Pedersen process can treat the bauxite residue efficiently giving an environmental positive effect but is also economically feasible due to the revenues from iron sales and increased production of alumina. If scaled slightly up, it is also possible to start treating existing bauxite residue deposits and hence slowly clean up these deposits. Further research work in that direction is now ongoing [11].

Conclusions

It is possible to produce alumina with the Pedersen process and find useful applications for its by-products. There are some parts of the process that need to be improved to be able to produce commercially grade alumina, such as desilication and growth of larger precipitates.

The main drawback of the Pedersen process in competition with the Bayer process is on the cost side because the high-temperature smelting phase of the process is costly in regards to energy consumption and CAPEX. There could however be a profitable application of the process for the processing of bauxite residue from the Bayer process.

Acknowledgements The authors would like to thank Sarina Bao, Henrik Gudbrandsen, Julian Tolchard, and Christian John Engelsen from SINTEF and Jochen Grünig from Outotec for their contributions. The project leading to this application has received funding from the European Union's Horizon 2020 Research and Innovation Programme under grant agreement No 767533.

References

1. Miller J, Irgens A (1974) Alumina Production by the Pedersen Process-History and Future. Light Metals: 789–800.
2. Vafeias M et al. (2018) From red to grey: revisiting the Pedersen process to achieve holistic bauxite ore utilisation. Paper presented at 2nd International Bauxite Residue Valorisation and Best Practices Conference, Athens, Greece 7–10 May 2018 https://doi.org/10.5281/zenodo.3235325
3. Konlechner D, Koenig R, Preveniou A, Davris P, Balomenos E. (2021) First Industrial Scale Process Concept for the Reengineered Pedersen Process within ENSUREAL. Materials Proceedings 5 (1):8. https://doi.org/10.3390/materproc2021005008
4. Lazou A, van der Eijk C, Balomenos E, Kolbeinsen L, Safarian J. (2020) On the Direct Reduction Phenomena of Bauxite Ore Using H_2 Gas in a Fixed Bed Reactor. J. Sustain. Metall. 6: 227–238. https://doi.org/10.1007/s40831-020-00268-5
5. Azof FI, Vafeias M, Panias D, Safarian J (2020) The leachability of a ternary $CaO-Al_2O_3-SiO_2$ slag produced from smelting-reduction of low-grade bauxite for alumina recovery. Hydrometallurgy 191 https://doi.org/10.1016/j.hydromet.2019.105184
6. Lazou A, van der Eijk C, Tang K, Balomenos E, Kolbeinsen L, Safarian J (2021), The Utilization of Bauxite Residue with a Calcite-Rich Bauxite Ore in the Pedersen Process for Iron and Alumina Extraction. Metallurgical and Materials Transactions B 52: 1255–1266 https://doi.org/10.1007/s11663-021-02086-w
7. Manataki A, Mwase JM, van der Eijk C (2021) Simulating the Use of a Smelter Off-Gas in the Precipitation Stage of the Pedersen Process. Materials Proceedings 5(1):53. https://doi.org/10.3390/materproc2021005053
8. Ma Y, Preveniou A, Kladis A, Berg Pettersen J (2022) Circular economy and life cycle assessment of alumina production: Simulation-based comparison of Pedersen and Bayer processes. Journal of Cleaner Production 366 https://doi.org/10.1016/j.jclepro.2022.132807
9. Mwase JM, Safarian J (2022) Desilication of Sodium Aluminate Solutions from the Alkaline Leaching of Calcium-Aluminate Slags. Processes 2022, 10 https://doi.org/10.3390/pr10091769
10. Sommerseth C, Eldrup NH, Konlechner D, van der Eijk C (In press) A Techno-Economical Study Comparing the Bayer and Pedersen Process for Alumina Production and Bauxite Residue Treatment. Conference Proceedings ICSOBA 2022.
11. Van der Eijk C, Dalaker H (2022) Recovery of Copper, Iron, and Alumina from Metallurgical Waste by Use of Hydrogen REWAS 2022: Developing Tomorrow's Technical Cycles (TMS), Anaheim, USA, February 27–March 3 2022 https://doi.org/10.1007/978-3-030-92563-5_75

Statistical Modelling of Operating Parameters on Bauxite Slurry Hyperbaric Filtration

Clara Souza, Antonio Silva, Eduardo Moreira, Enio Laubyer, Fabricia Ferreira, and Raimundo Neto

Abstract

The Hydro Alunorte refinery receives bauxite slurry used in the alumina production from Mineração Paragominas, both sites located in Pará—Brazil. This bauxite slurry, which is dewatered in hyperbaric filters located in Alunorte, has been affecting the lifetime of the pipeline through erosion and corrosion processes since the beginning of its transportation in 2008. To increase the lifetime of the pipeline and preserve the dewatering plant production capacity, this work evaluated the influence of filter rotation, basin level, and feed slurry density on filter productivity by applying a factorial design. The experiment, which was carried out on the industrial filters, complemented the outcome of a previous investigation executed on a laboratory scale using slurry density, top size, and pH as control factors to assess the solids throughput through Box Behnken Design statistical technique. Results indicated that a top size reduction and pH elevation by the addition of milk of lime decreases the productivity by 13.3%. However, an increase in density, filter rotation, and basin level proved to offset and surpass this predicted loss, making viable the pipeline lifetime increase project and preventing alumina production losses in Alunorte.

Keywords

Hydro Alunorte • Pressure filtration • Bauxite slurry • Design of experiment

Introduction

About 69% of the bauxite consumed at Hydro Alunorte for the alumina production comes from Mineração Paragominas (MPSA), both sites located in Pará, Brazil. Since 2008, the bauxite transport is carried out by pumping a slurry with 50% solids through a pipeline 244 km long. Once at Alunorte, the slurry is dewatered in a plant with 13 hyperbaric filters (HBF) whose product is bauxite with 86% solids content. The bauxite's filtrate is treated to reduce its solids content and recycled to incorporate the refinery's industrial water supply. Out of specification bauxite with coarser particles and the transport of untreated river water between slurry batches have caused premature wear and metal thickness loss of the pipeline due to erosion and corrosion problems. For this reason, two studies were carried out, one on a laboratory scale and the other on an industrial scale, to study the influence of pH modification with the addition of milk of lime, particle size (top size), and bauxite slurry density on the pipeline's wear rate and Alunorte's HBF process. In addition, other operational parameters, such as filter speed and basin slurry level, also had their impacts quantified to compensate for possible losses in filtration due to changes in slurry characteristics. After complete optimization, it is expected that the pipeline's lifetime, initially designed to operate until 2032, will be extended to 2049.

Milk of lime was the product initially chosen to raise the pH of water and slurry batches due to its low cost and wide industrial applications, including the Bayer process, used in Alunorte for alumina production. The milk of lime has benefits in the alumina extraction, controlling impurities formation from ore, aiding in the pregnant liquor filtering process, and caustic soda loss minimization by bauxite residue [1]. Preliminary studies indicated a consumption of approximately 700–1,000 g CaO per tonne of bauxite in order to raise slurry pH from 7.0 to 10.5. This prior CaO addition in slurry would probably replace or reduce the 800 g/t milk of lime dosage currently used for bauxite's

C. Souza (✉) · A. Silva · E. Moreira · E. Laubyer · F. Ferreira · R. Neto
Hydro Alunorte - Alumina do Norte do Brasil, 481 PA Highway Km 12, 68.447-000, Barcarena-Pará, Brazil
e-mail: clara.de.souza@hydro.com

© The Minerals, Metals & Materials Society 2023
S. Broek (ed.), *Light Metals 2023*, The Minerals, Metals & Materials Series,
https://doi.org/10.1007/978-3-031-22532-1_23

digestion depending on the quantity of CaO that remains in the solid phase after bauxite slurry filtration. This issue will be approached in parallel studies along with the review of possible impacts that industrial water containing milk of lime may offer to its users, such as cooling tower make up and flocculant preparation.

According to the fundamentals of the filtration theory in Eq. 1 [2], the solids throughput (m_s) is proportional to the square root of filter rotation (n), pressure difference (Δp), and cake formation angle (α_1). While the cake formation angle is constant after filter design definition, filter speed and pressure difference are adjustable parameters during filter operation. The basin level control of the filter should always guarantee that all the cake formation zone of the rotating disc is submerged in the slurry, otherwise the filter cake will be thinner, negatively affecting throughput and air consumption. Although the basin level is not present in the filtration process basic formula, the hydrostatic pressure increases linearly with the slurry level, and this minor effect is added to the pressure difference applied against the filter cloths as the driving forces for the filtration process. The hydrostatic pressure is particularly effective in filters with big disc diameters and a high submergence in the slurry [2].

$$\dot{m}_s = \rho_S \times (1 - \varepsilon) \times \sqrt{\frac{2 \times k \times \Delta p \times n \times \alpha_1}{\eta_L \times r_c}} \qquad (1)$$

where:

m_s	Specific solids throughput, kg/(m^2 h)
ρ_S	Solid density, kg/m^3
ε	Cake porosity,
k	Concentration parameter,
Δp	Filtration pressure difference, bar
n	Filter speed rotation, rpm
α_1	Cake formation angle, °
η_L	Filtrate viscosity, kg/(m s)
r_c	Cake resistance, 1/m^2

Another parameter that influences filtration throughput is cake resistance (r_c), which is affected mainly by the particle size and cake porosity (ε). If these parameters decrease, cake resistance increase, once the permeability of the cake reduces. In Al-hydrate coarse seed filtration, it was found that the cake formation time was 15–20% higher for particles with d50 = 90 μm compared to particles with d50 = 100 μm. As the throughput increases with the square root of the cake formation time quotient, the throughput rate rises about 10% if d50 increases from 90 to 100 μm. Increasing slurry density also increases the productivity since thicker cakes are formed. For the same industrial application with coarse seed, a 15% higher filtration performance was found by increasing the slurry density only by 2.5% [2].

Laboratory Test

The first stage of studies involved testing different bauxite slurry specifications relating to density, pH, and particle top size. Currently, up to 12% mass of the particles exceed 300 μm, and typical values for d90, d50, and d10 are 197, 75, and 4 μm, respectively. Higher pH levels above the current value of 7.0 were tested by adding milk of lime with 176 g/L solids content and d90, d50, and d10 equal to 75, 15, and 5 μm, respectively.

The statistical technique of Box Behnken Design (BBD), introduced by Box and Behnken in 1960 [3], was applied for this test with slurry top size, density, and pH as control factors with two levels and a center point and productivity as the response variable, as shown in Table 1.

The design with 3 replicates was executed on benchtop equipment that simulates pressure filtration. The experimental procedure and the productivity calculation methodology are described by Souza et al. [4] and Hahn et al. [5], respectively. The regression model generated from the analysis of variance with R^2 (adj) of 80.56% is described by Eq. 2 [4].

$$P = -0.1322TS - 26,532D + 128.5pH + 9,527D^2 + 1.214pH^2 \\ + 0.02099TS \times pH - 106.9D \times pH + 18,668$$

$$(2)$$

where:

P	Productivity predicted by model, t/h.
TS	Top size, μm
D	Density, g/cm^3
pH	pH

Table 1 Design factors and levels of BBD [4]

Design factor	Levels			Response variable
	−1	Center point	+1	
Top size	100% < 355 μm	100% < 425 μm	100% < 500 μm	Productivity
Density	1.44 g/cm^3	1.46 g/cm^3	1.49 g/cm^3	
pH	7.0	9.25	11.5	

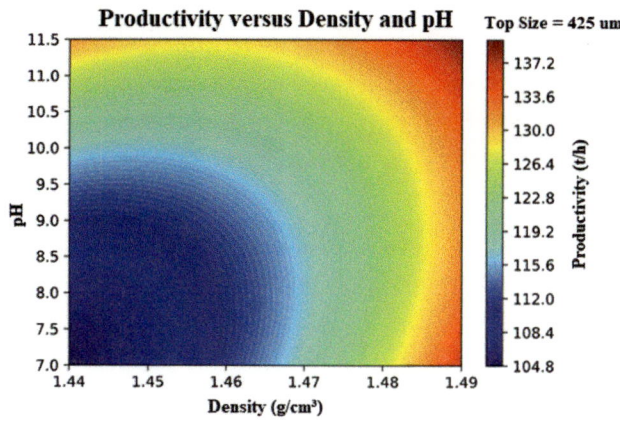

Fig. 1 Contour plot of productivity versus density and pH

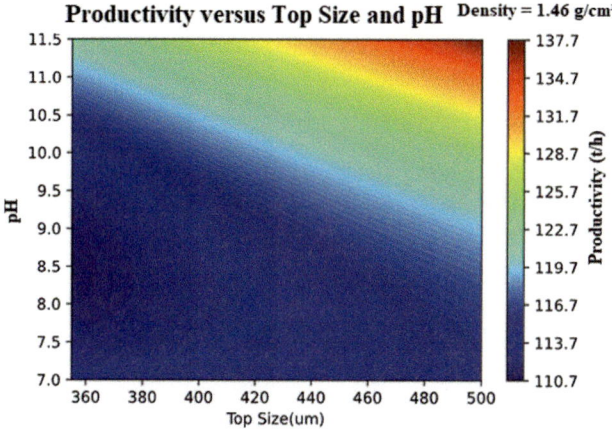

Fig. 2 Contour plot of productivity versus top size and pH

As described by Souza et al. [4], the model in Eq. 2 shows that slurry density is the factor with the greatest influence on productivity. In general, the increase in pH, top size, and density impacts positively on the response variable. However, at higher pH and density levels, the negative effect of reducing the top size is enhanced, as seen in the contour plots of Figs. 1, 2 and 3.

Besides the BBD experiment, two other experiments were performed using the same laboratory procedure described by Souza et al. [4] for two blank cases of the bauxite slurry, named "laboratory base cases". Since the current top size of the slurry is unknown, it will be denoted as "original" in this paper. Thus, both base cases had slurry at pH 7.0 and the original top size. The only difference between these cases was the density, which was 1.45 g/cm³ for the first case and 1.48 g/cm³ for the second one. The throughputs obtained for these base cases were 136.7 and 155.3 t/h, respectively. These experiments will be essential to scale the impacts observed in laboratory concerning top size and pH to the actual productivity of the industrial HBF, which Sect. 3 will cover in further details.

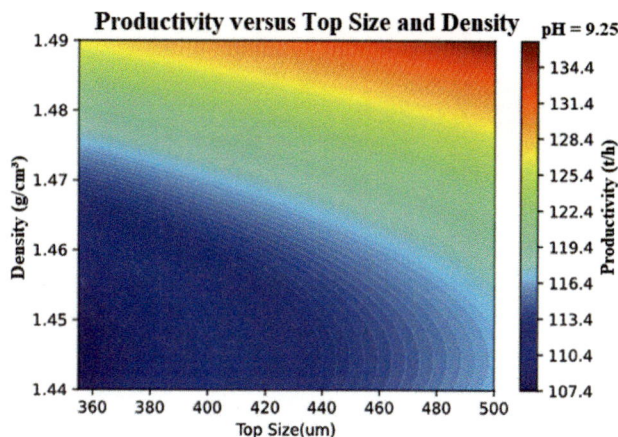

Fig. 3 Contour plot of productivity versus top size and density

Industrial Test

Methodology

In addition to the laboratory tests, field tests were also performed with the industrial HBF with slurry density, rotation speed, and basin level as control variables. These results were useful to complement those obtained in the laboratory for the complete optimization of the bauxite dewatering plant operating parameters, aiming to preserve its full production capacity after the modification of slurry top size and pH to mitigate pipeline metal thickness loss.

The HBF operates continuously while filter cake is discharged onto conveyors through reciprocating knife gate valves that maintain the pressure vessel, which remained constant at 5.9 bar. Cake formation angle was fixed at 122°. Once the desired filter operating parameters were set, the filter was allowed to operate for one hour. Feed slurry was sampled three times for each run at the beginning, then after 30 min and the last one at the end of the run along with the filter cake sampling. Cake moisture was determined by oven drying at 100 °C for 12 h and the feed slurry was filtered before oven drying to determine the solid content. Unfortunately, it is not possible to measure the cake thickness, since a completely crushed bauxite cake reaches the conveyor at the sampling point.

A 2^3 factorial design was performed with productivity as the response variable. Table 2 shows the high and low levels adopted for each factor. The significance level of the test was $\alpha = 0.05$.

The experiments, listed in Table 3, were replicated twice, totalling 16 randomized runs.

The HBF throughput for each run was calculated using the average slurry feed flow rate registered by the flow transmitter instrument, the slurry solids content analysed in

Table 2 Design factors and levels of 2^3 factorial design

Design Factor	Levels		Response variable
	−1	+1	
Density	1.45 g/cm^3	1.48 g/cm^3	Productivity
Speed rotation	1.5 rpm	2.0 rpm	
Basin level	85%	95%	

Table 3 Non-randomized experiment matrix

Experiment	Density	Speed rotation	Basin level
1	−1	−1	−1
2	−1	−1	+1
3	−1	+1	−1
4	−1	+1	+1
5	+1	−1	−1
6	+1	−1	+1
7	+1	+1	−1
8	+1	+1	+1

laboratory, and the bauxite solid density, whose value of 2.65 g/cm^3 was known a priori. For the slurry solids content, the average of the 3 feed slurry samples was collected during each one-hour experiment was used.

Model

The analysis of variance of the productivities obtained from the field test generated the regression model of Eq. 3 with R^2 (adj) of 88.70%. It is worth mentioning that the model is applicable only within the levels on which it was built.

$$P = 409.8D + 89.8R + 1.74L - 0.716R \times L - 669 \quad (3)$$

where:

P   Productivity predicted by model, t/h
D   Density, g/cm^3
R   Speed rotation, rpm
L   Basin level, %

Figures 4 and 5 show the main effects and interaction plots for the response variable and Figs. 6, 7 and 8 present the contour plots of productivity.

The outcomes obtained for productivity related to feed slurry density and filter rotation are in line with the expected according to the filtration theory. Besides, the positive effect of raising the filter basin level confirmed the minor effect that the slurry hydrostatic pressure adds on filter throughput and the importance of keeping the cake formation zone of the disc as immersed as possible in the slurry to guarantee the maximum designed cake formation time.

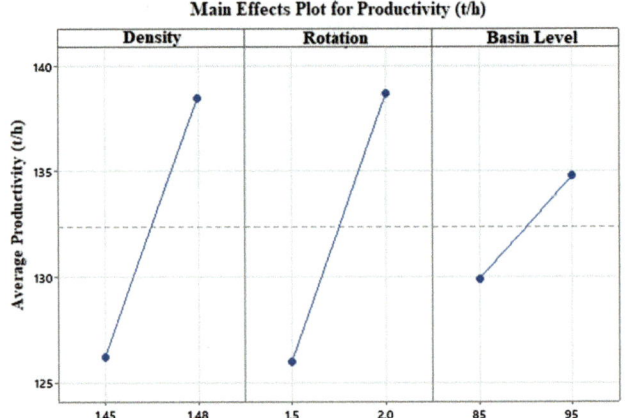

Fig. 4 Main effects plot for productivity

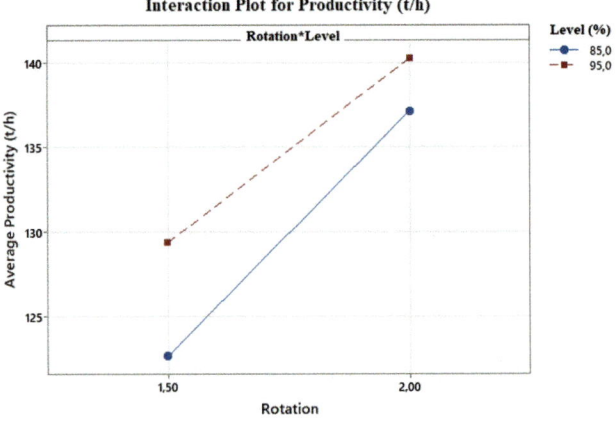

Fig. 5 Interaction plot for productivity

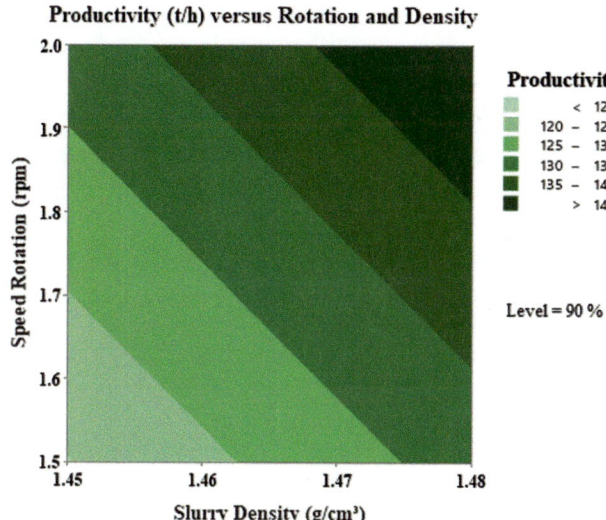

Fig. 6 Contour plot of productivity versus rotation and density

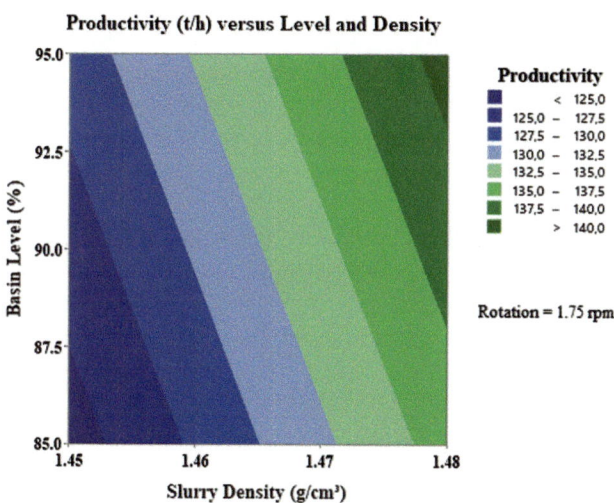

Fig. 7 Contour plot of productivity versus level and density

Model Adequacy Checking

The model from the analysis of variance is valid when residuals are independent and distributed according to a normal distribution, with zero mean and constant variance, according to homoscedasticity criteria. Therefore, this verification is fundamental for a complete analysis, making decisions based on the experiment more technical and valid [6]. Three types of tests should be performed: normality verification, graphical analysis of the residuals, and the homoscedasticity test, as shown in Figs. 9, 10, 11 and 12.

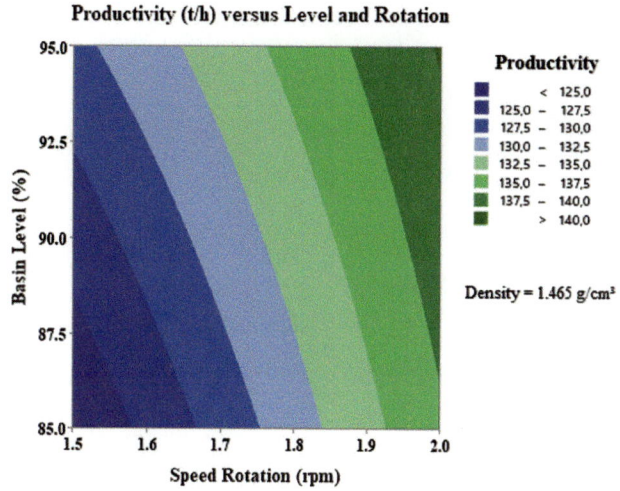

Fig. 8 Contour plot of productivity versus level and rotation

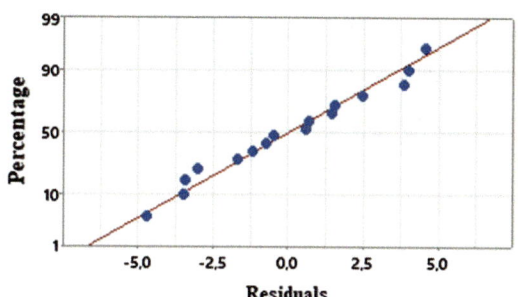

Fig. 9 Normal probability plot of residuals

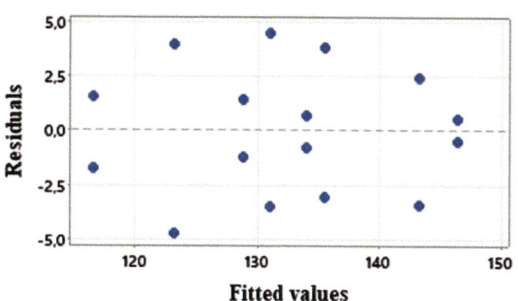

Fig. 10 Plot of residuals versus fitted values

The normality test of the residuals from the model of Eq. 3 resulted in a p-value equal to 0.841, higher than the significance level $\alpha = 0.05$, showing that the distribution of residuals follows a normal distribution. Figures 10 and 11 show no trend or anomalous behavior in the arrangement of residuals versus fitted values and order of observation of the experiments, respectively, ensuring the homoscedasticity criteria. Finally, Fig. 12 shows the histogram of the residuals with zero mean.

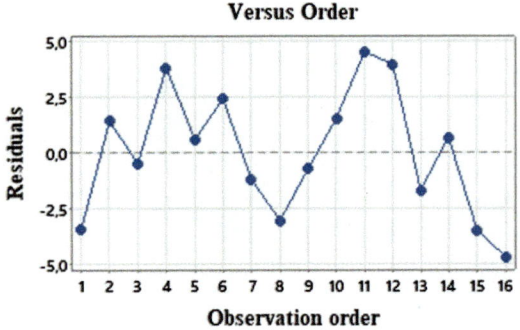

Fig. 11 Plot of residuals versus observation order

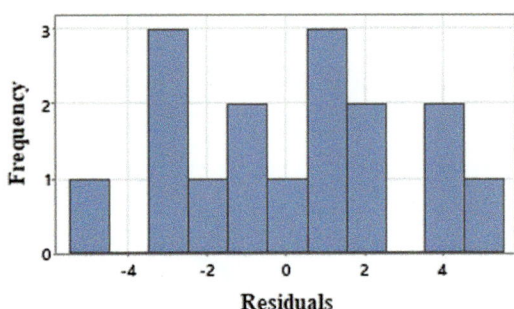

Fig. 12 Histogram of residuals

Results and Discussion

In this section, both outcomes obtained from the laboratory and field tests are integrated. To simplify the understanding of the discussion, the model of each experiment will be identified as follows:

- Model 1 = generated by the laboratory test, with productivity as response variable and slurry density, top size, and pH as control variables, as described by Eq. 2.

- Model 2 = generated by the field test, with productivity as response variable and slurry density, filter rotation, and basin level as control variables, as described by Eq. 3.

As mentioned in Sect. 2, there are two base cases for the laboratory tests, one for each slurry density (1.45 and 1.48 g/cm^3) and both with the original top size and pH at 7.0. The throughputs of both base cases are shown in Table 4.

The productivities for the two density cases with 500 μm top size and pH 9.5 were simulated using Model 1, as presented in Table 5.

Therefore, comparing Tables 4 and 5 for the case of 1.45 g/cm^3 slurry density, there is a 13.3% productivity loss (from 136.7 to 118.5 t/h) if the top size is reduced from original to 500 μm and the pH is raised from 7.0 to 9.5. Similarly, there is a 16.6% productivity decrease (from 155.3 to 129.5 t/h) if the 1.48 g/cm^3 slurry undergoes the same top size and pH changes.

These laboratory-scale productivity losses are important for predicting industrial-scale loss scenarios, since operational and cost constraints prevented the modification of slurry top size and pH for the execution of field tests.

It is worth mentioning that the benchtop equipment used for the laboratory tests evaluates the influence of the factors only from the cake formation aspect. Therefore, the absolute productivity results (t/h) obtained from Model 1 and shown in Tables 4 and 5 do not represent the real HBF productivity, since other phenomena such as cake discharge efficiency and cloth lifetime were not considered in the laboratory. Only the relative (percentage) impacts between the laboratory cases will be useful to predict productivity impacts in field.

Table 6 shows two base-case simulations using Model 2.

The "industrial base case 1" refers to the HBF current operating conditions, i.e., with the slurry presenting the original top size, 1.45 g/cm^3 density, pH 7.0, and filters

Table 4 Base cases for the laboratory test

Parameter	Laboratory base case 1	Laboratory base case 2
Top size	original	original
Density	1.45 g/cm^3	1.48 g/cm^3
pH	7.0	7.0
Productivity	136.7 t/h	155.3 t/h

Table 5 Scenarios predicted from Model 1

Parameter	Scenario 1	Scenario 2
Top size	500 μm	500 μm
Density	1.45 g/cm^3	1.48 g/cm^3
pH	9.5	9.5
Productivity	118.5 t/h	129.5 t/h

Table 6 Base cases for the industrial test

Parameter	Industrial base case 1	Industrial base case 2
Top size	original	original
Density	1.45 g/cm^3	1.48 g/cm^3
pH	7.0	7.0
Rotation	1.5 rpm	2.0 rpm
Level	85%	95%
Productivity	116.5 t/h	146.4 t/h

operating at 1.5 rpm and 85% basin level. For this case, the current reference HBF productivity is 116.5 t/h.

According to the industrial-scale test and the effects of each control factor described by Model 2, the higher the slurry density, filter rotation, and basin level, the higher the productivity observed in the filter, as shown in Figs. 4, 5, 6, 7 and 8. With this, the "industrial base case 2" indicated in Table 6 refers to the HBF maximum potential productivity (146.4 t/h) considering all control variables at their high levels (density, rotation, and basin level).

Note that top size and pH remained constant in the description of these two base cases in Table 6 because these variables were not considered in the field test and, thus, cannot be simulated by Model 2. The impact of changing these two parameters on the HBF real productivity will be calculated through the simulations obtained from Model 1 through the same way as presented in Tables 4 and 5. Recapitulating what has been previously mentioned, Table 5 shows that by reducing top size from original to 500 μm and raising the pH from 7.0 to 9.5, the productivity has a 13.3 and 16.6% drop, respectively, for the 1.45 and 1.48 g/cm^3 slurry densities. Therefore, from the application of these percentages to the industrial base case productivities in Table 6, two other scenarios are generated in Table 7.

Scenario 3 describes how the HBF productivity would be affected if top size and pH were changed at MPSA without any adjustment at Alunorte to compensate for the expected loss. In this case, the average productivity would decrease from 116.5 t/h (industrial base case 1 in Table 6) to 101.0 t/h (scenario 3 in Table 7). This 13.3% filtration capacity loss would represent a loss of more than 63,000 tons in Alunorte's annual alumina production.

Scenario 4 in Table 7, whose productivity is 16.6% lower than the industrial base case 2 in Table 6, represents the compensation of the productivity loss caused by top size and pH alteration, by increasing slurry density, filter rotation, and basin level of the HBF. The 122.0 t/h productivity of scenario 4 in Table 7 should be compared to the 116.5 t/h of the industrial base case 1 shown in Table 6, which represents the area's current (initial) condition before any change in bauxite slurry and filter operational parameters. These results show that the expected productivity loss caused by top size reduction and pH elevation will be compensated (and even overcome) by adjusting the operational parameters of filter rotation and level and increasing the slurry density.

Conclusions

The experiments performed in laboratory and field satisfactorily described the productivity impacts of Alunorte's hyperbaric filters that dewater bauxite slurry pumped by MPSA if it were modified to mitigate the pipeline metal thickness loss caused by the slurry transportation since 2008.

The laboratory experiment proved that there would be a 13.3% productivity loss of the hyperbaric filters (from 116.5 to 101.0 t/h) if the top size were reduced from the original level to 500 μm and the slurry pH raised from 7.0 to 9.5 by adding milk of lime if all other filter and slurry parameters remained constant at current operating levels. However, both the laboratory and industrial tests were able to demonstrate together that it is possible to adjust filter operating parameters and slurry density to compensate for this predicted loss. More specifically, by increasing filter speed rotation from

Table 7 Scenarios predicted from Model 1 for the industrial filters

Parameter	Scenario 3	Scenario 4
Top size	500 μm	500 μm
Density	1.45 g/cm^3	1.48 g/cm^3
pH	9.5	9.5
Rotation	1.5 rpm	2.0 rpm
Level	85%	95%
Productivity	101.0 t/h	122.0 t/h

1.5 to 2.0 rpm, basin level from 85 to 95%, and bauxite slurry density from 1.45 to 1.48 g/cm^3, the resulting 122.0 t/h predicted productivity exceeds the current level of 116.5 t/h even after the unfavorable modification of top size and pH.

These results make it feasible to modify the specification of the bauxite slurry pumped by MPSA to extend the pipeline lifetime from 2032 to 2049, in addition to avoiding the loss of more than 63,000 tons per year in Alunorte's alumina production.

Future laboratory and industrial tests will be performed to evaluate the influence of slurry top size, pH, density, filter speed, and basin level on bauxite cake moisture. Furthermore, experiments will be performed to study the impacts on using the dewatered filtrate with milk of lime as the industrial water supply for the Bayer process.

References

1. Whittington BI (1996) The chemistry of CaO and Ca(OH)$_2$ relating to the Bayer process. Hydrometallurgy 43:13-35
2. Bott R, Langeloh T, Hahn J (1960) Necessary fundamentals for successful alumina seed filtration. Bokela Ingenieurgesellschaft für Mechanische Verfahrenstechnik mbH
3. Box GEP, Behnken DW (1960) Some new three level designs for the study of quantitative variables. Technometrics 2(4):455-475
4. Souza C, Lima P, Miotto P (2022) Study of influences on the productivity of bauxite slurry hyperbaric filtration through Box Behnken design. 40th International Conference and Exhibition ICSOBA, Athens, Greece, 10–14 October 2022
5. Hahn J, Bott R, Langeloh T (2012) Theory and practice of filtration in the alumina industry. Proceedings of the 9th International Alumina Quality Workshop 2012, Perth, Australia, p 103–109
6. Montgomery DC (ed) (2001) Design and analysis of experiments. John Wiley & Sons, New York

Reduction of GHG Emissions and Increase Operational Reliability Using Immersed Electrode Boiler in an Alumina Refinery

Rodrigo Neves, Fernando Melo, Everton Mendonça, Erinaldo Filho, and Jeferson Carneiro

Abstract

The Hydro Alunorte refinery, located in Pará—Brazil, uses heavy oil, mineral coal and electrical energy boilers in its energy matrix to meet the steam demand of the Bayer process. Aligned with Hydro's environmental strategy to reduce greenhouse gas (GHG) emissions, Hydro Alunorte started operating the first electric boiler with immersed electrode technology as one of the initiatives to increase the share of electricity in the steam generation energy matrix. The new immersed electrode boiler has higher operational reliability with an availability of approximately 99.5% and with load control by adjusting the level of the electrode chamber, thus loads variation can be done in minutes to control the consumption demand of the refinery. Therefore, it is expected to reduce fuel oil consumption consequently reducing CO_2 emissions by approximately 150,000 tons per year.

Keywords

Sustainability · Electrode boiler · Emissions

Acronyms and Abbreviations

CO_2	Carbon dioxide
CO_2E	Carbon dioxide equivalent
GHG	Greenhouse gas
HAP	Hazard air pollutants
Mt	Million tonnes
MW	Megawatt
Nox	Nitrogen oxides
PLC	Programmable logic controller
SO_2	Sulphur dioxide
VSD	Variable speed drive

Introduction

One of the most important challenges in terms of the environment today is global warming. Fossil fuels are still the most used as sources of steam and energy generation, which cause a very significant impact due to the high emission of carbon dioxide. According to Moradikhou and Ravanshadnia, "Greenhouse gas emissions are the main contributing factor to global warming, with CO_2 having the greatest share (65%) among other greenhouse gases" [1]. In Fig. 1 it can be seen that energy supply sources are still led by fossil fuels such as coal and oil, however it is encouraging to see that there is an increase also in the use of renewable energy from 2012 onwards.

According to the Enerdata 2022 yearbook [3], Brazil is the third in the ranking of countries with highest share of renewables sources in electricity production with 78.4% of its energy deriving from renewables. Norway and New Zealand are the first and second, respectively. Taking advantage of this, Hydro is investing in projects in Brazil that can replace fossil fuels in all its value chain i.e. increase greener production. Silvia Madeddu et al. 2020 states that "replacing fossil fuels with low-carbon electricity has become the core climate change mitigation strategy as supported by many climate change mitigation scenarios" [4], therefore, with the current need to reduce the emission of greenhouse gases, especially carbon dioxide, to contain the advance of global warming, Hydro has established goals and actions that will sustain the reduction of the carbon footprint to less than a quarter per kilo of aluminum in relation to the global average. For this to be possible, it is necessary to

R. Neves (✉) · F. Melo · E. Mendonça · E. Filho · J. Carneiro
Hydro Alunorte - Alumina do Norte do Brasil, 481 PA Highway Km 12, 68.447-000, Barcarena-Pará, Brazil
e-mail: rodrigo.vieira.neves@hydro.com

© The Minerals, Metals & Materials Society 2023
S. Broek (ed.), *Light Metals 2023*, The Minerals, Metals & Materials Series,
https://doi.org/10.1007/978-3-031-22532-1_24

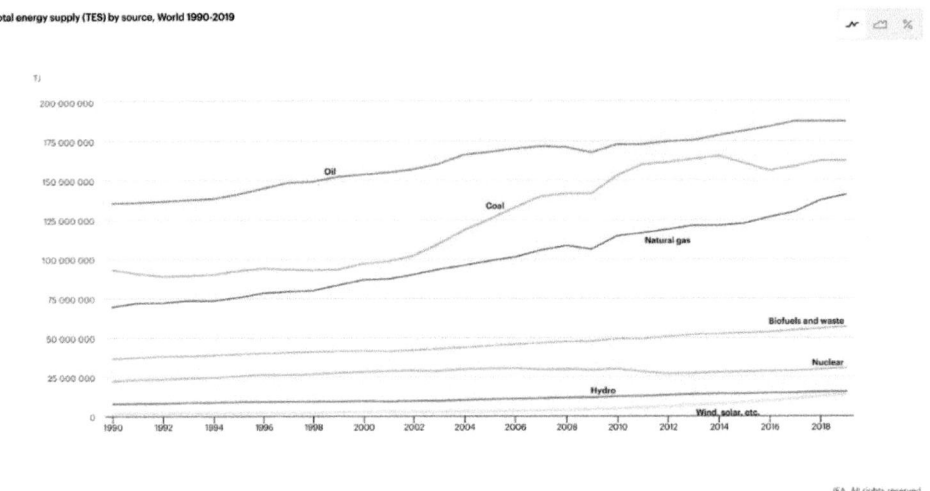

Fig. 1 Total energy supply by source [2]

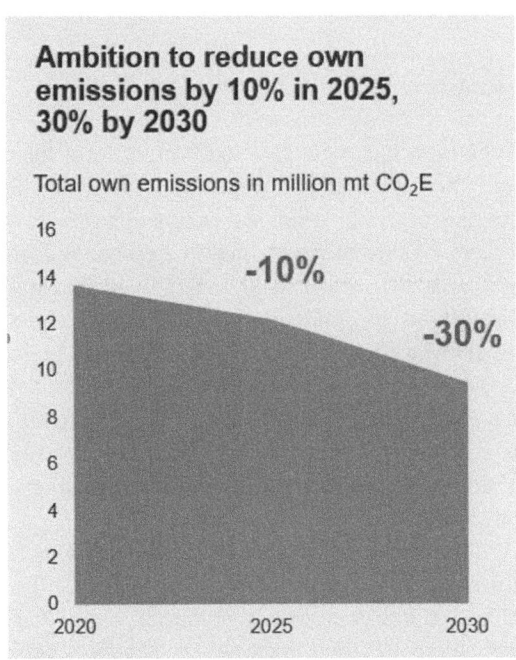

Fig. 2 Hydro's greenhouse gas emissions reduction ambitions [5]

reduce emissions by up to 30% in the entire aluminum chain, including the division of extruded products, as shown in the Fig. 2.

Context

In the alumina refining process from bauxite, the utilities area is responsible for supplying compressed air, electricity and steam to the entire facility. To produce steam, at Hydro Alunorte, a refinery located in Barcarena in the state of Pará,

Brazil, the boilers operate with 3 types of fuel: coal, heavy oil and electricity, thus composing its energy matrix. According to Tomei, Gregory L. "air emissions control with emphasis on particulate, NOx, SO_2 and hazardous air pollutants (HAPs) is perhaps the most significant environmental concern for fossil fuel-fired systems" [6]. One of the actions in line with Hydro's carbon footprint reduction guideline was the installation, at the Alunorte refinery, of a new, higher-capacity electric boiler with immersed electrode technology. The immersed electrode electric boiler allows renewable sources to be used, achieving near zero greenhouse gas emissions, in addition to contributing to an increase in the share of electric energy in Alunorte's steam matrix. This technology also increases the security of steam supply to the refinery due to its high availability and reliability, in addition to allowing operational and commercial flexibility in the energy matrix. This paper discusses how this technology is helping Alunorte to create a platform for refinery electrification supporting the achievement of the target of 30% reduction in CO_2 emissions by 2030.

Opportunity Scenario at Alunorte

Alunorte's boilers area is currently composed of 12 boilers, three of which are coal fired, three electric and six runs with heavy fuel oil, as shown in Fig. 3.

Electric boilers are normally used as an opportunity to reduce Alunorte's operating costs by taking advantage of low electricity prices in Brazil during the rainy season. As showed in Fig. 4, during dry season (hatched area) electricity prices are too expensive, however it is worth to stay with electrical boilers maximized due to CO_2 emissions reduction [7].

Fig. 3 Alunorte steam system

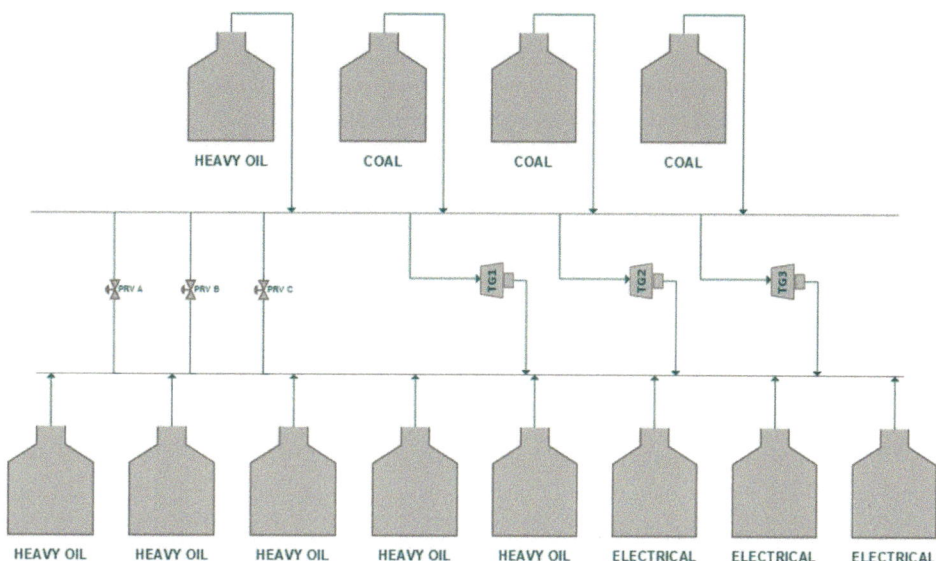

Fig. 4 Brazil energy price

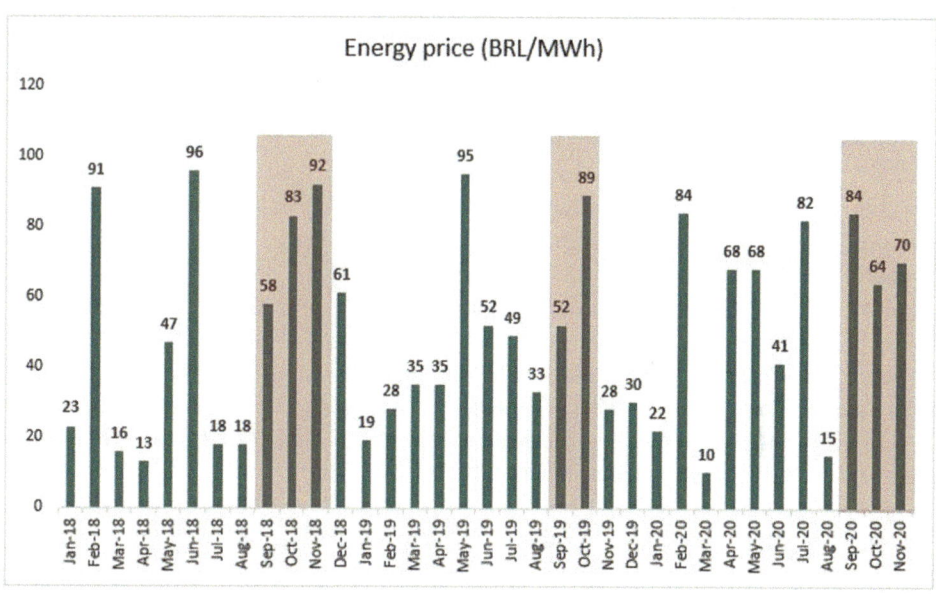

Between 2016–2018 two retrofit projects were executed and delivered at existing jet-type electric boilers at Alunorte. The two boilers did not achieve the desired performance after the retrofit and are currently limited to 70% of the maximum capacity and suffering frequent shutdowns. Due to this, a third retrofit was reevaluated and replaced by a project for installing a new electrical boiler of the immersed electrode type, a different technology that is more reliable, robust and flexible than the jet type [8]. Electrical boilers have been identified as an important decarbonization technology for steam generation at different countries in the world, so Alunorte piloting this technology was important to assess the actual feasibility of immersed electrode boilers at the refinery. The installation of the new boiler potentialized the savings on GHG emissions also allowed refinery to take advantage of the seasonal operation of the electrical boilers without any changes to the current Alunorte operating paradigms and maintenance procedures.

Operation Principle

In the electrode boiler electrical current is added to water through several electrodes. Water is used as a conductor enabling current to flow between the electrodes and the electrical current added to the boiler water is converted to heat for steam production. The electrodes penetrate the top side of the boiler shell and are insulated by ceramic insulators. The ceramic insulators are mounted in a way that secures very low electrical and steam leakage. Steam production on Immersed Electrode Boilers is usually done by circulating the boiler water through the upper chamber where the electrodes are suspended. Steam is produced in the upper chamber and released on the top of the boiler. The water inside is controlled by a throttle valve and recirculation pumps (using VSDs) that regulates the level in the upper boiler chamber. The water conductivity is continuously monitored to ensure that the boiler gives the correct output. When the conductivity exceeds the programmed set point, an automatic blowdown is initiated.

The main components of an electrode boiler are:

– Outer chamber and upper chamber;
– Electrodes;
– Ceramic insulators;
– Circulation pump;
– Control valve for level control in the upper chamber;
– Blowdown system;
– Chemical dosing system;
– Conductivity control with sampling station;
– Pressure safety valve.

The electrodes of the new boiler were connected to Alunorte's 13.8 kV system at the substation. The boiler has a power of 60 MW, producing up to 95 tons per hour of medium pressure steam at 15 bars. The steam is generated by the contact of the energized electrodes with the water contained in the upper chamber. When the water in contact with the electrodes, it evaporates, and the generated steam is released through the pipe located at the top of the boiler and sent to the steam header. Steam production is proportional to the water level in the upper chamber, that is, the more submerged the electrodes are in the water, the greater the contact surface and the greater the production of steam. The circulation pumps are controlled by the local Programmable Logic Controller (PLC), which, through the determined setpoint and the current level of the upper chamber, determines the amount of water to be transferred. The steam produced by the boiler is sent to Alunorte's 15-bar saturated steam system, which receives steam from the other low-pressure boilers as well as the exhaust from the turbo generators and pressure-reducing valves. This system supplies steam to all consumers of the plant, composed of several heating processes, the main one being alumina digestion. Part of the steam generated, after passing through the heat exchangers in the areas, returns as condensate to storage tanks and is mixed with demineralized make-up water. From the condensate tanks, the water passes through the deaerators to reduce the amount of oxygen and is later supplied to the low-pressure boilers. The water supplied has a low conductivity and the electric boiler works with a conductivity controlled through a setpoint. The conductivity control is done through chemical dosing and blowdown

continuously, which makes this a critical parameter for the performance and reliability of the boiler and must be automatically monitored by the control system.

Results and Conclusion

Since it was commissioned, the electric boiler has reached an average availability of 99.3%, which guarantees Alunorte operational reliability and clean production from renewable energies, with the possibility to explore its maximum potential for CO_2 reduction. The boiler operating at maximum capacity allows a heavy oil boiler with the same capacity to be turned off, which brought the benefit of approximately 64,000 t/CO_2 reduced in 5 months of operation. Figure 6 shows the stable operation of the boiler installed at Alunorte shortly after start-up, both in terms of power (MW) and in terms of steam flow (tons per hour), corroborating the initiative to keep electric boilers maximized ahead of the others that operate with coal and heavy oil. With the successful implementation of this technology, it is possible to replace the other 2 electric boilers that still operate with jet technology for the immersed electrode technology, which will support Hydro to achieve the proposed emission reduction targets (Fig. 2).

The emissions reduction targets at Alunorte will enable Hydro to deliver aluminum that can compete with other materials in different market segments and reduce the threat of substitution based on environmental footprint. It will also secure Hydro's position in the forefront of sustainable aluminum production and help maintain an essential competitive advantage against industry peers.

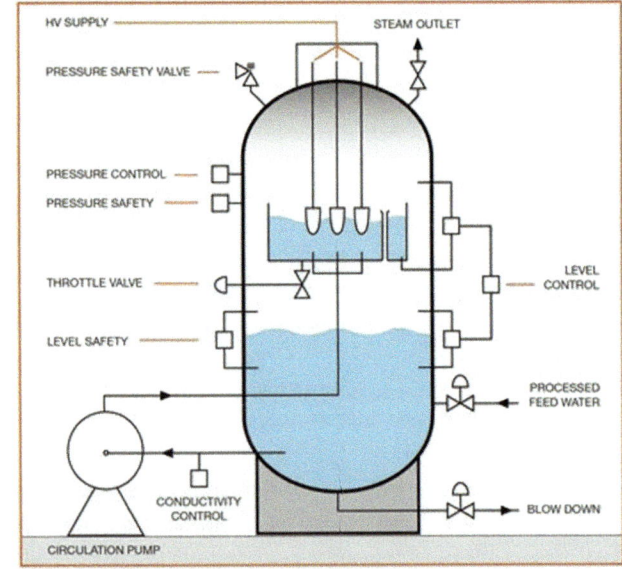

Fig. 5 Immersed electrode boiler diagram [8]

Fig. 6 Alunorte electrical boiler data

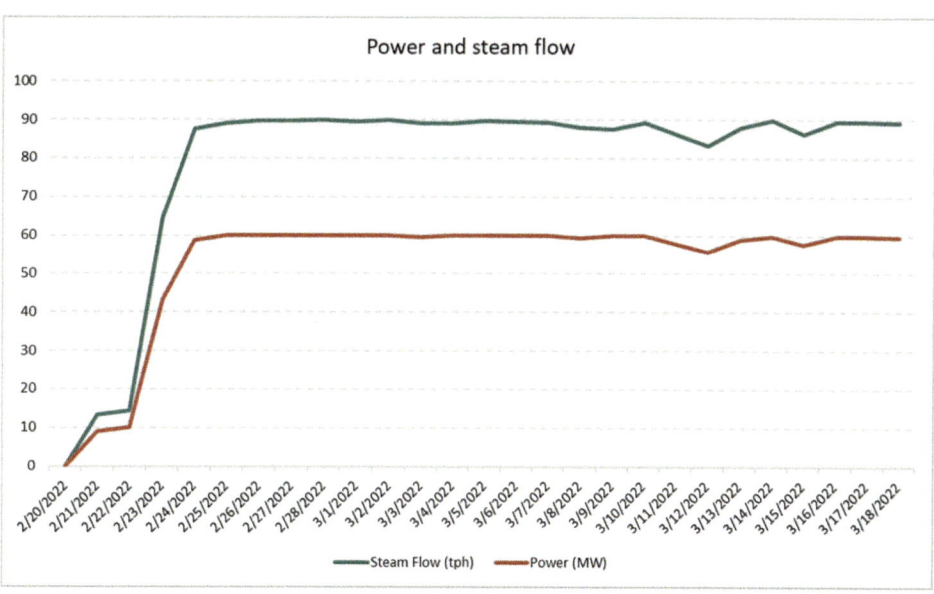

References

1. Amir Moradikhou (2021). Evaluation of CO_2 emissions reduction strategies in the Iranian Cement Industry. Journal of Civil Engineering and Materials Application p 107–114.
2. International Energy Agency (2022). https://www.iea.org/data-and-statistics/data-tools/energy-statistics-data-browser?country=WORLD&fuel=Energy%20supply&indicator=TESbySource. Acessed 25 August 2022
3. Enerdata Yearbook (2022). In: The dictionary of substances and their effects. Royal Society of Chemistry. https://yearbook.enerdata.net/renewables/renewable-in-electricity-production-share.html. Accessed 30 August 2022
4. Silvia Madeddu et al 2020 Environ. Res. Lett. 15 124004. In: The CO_2 reduction potential for the European industry via direct electrification of heat supply (power-to-heat)
5. Hydro's climate ambitions. https://www.hydro.com/en/sustainability/our-approach/environmental/climate. Accessed 28 August 2022
6. The Babcock & Wilcox Company, Steam 42 (2015). Ch31 p31-1
7. Brazil Energy Price Panel. https://www.ccee.org.br/web/guest/precos/painel-precos. Accessed October 03
8. PARAT Halvorsen AS (2022). https://www.parat.no/en/products/industry/parat-ieh-high-voltage-electrode-boiler/ Acessed 25 August 2022

Steam Grid Stability Using Advanced Process Control and Real Time Optimization in an Alumina Refinery

Daniella Costa, Rodrigo Neves, Fernando Melo, João Freitas, Juvenal Sousa, Ediciano Junior, and Danilo Arcodaci

Abstract

The Hydro Alunorte refinery has a complex steam system composed by boilers and power generation turbines that use three types of fuel (heavy oil, coal, and electrical energy) to produce steam for the Bayer process. Due to the complex nature of the refinery and the variable price of fuels, two programs integrated directly with the plant Distributed Control System (DCS), aiming to improve energy efficiency and steam system stability, reduce greenhouse gas (GHG) emission and avoid cascaded plant shutdowns. Advanced Process Control (APC) controls boilers load, steam headers pressure and minimizes relief valves opening. Real Time Optimization (RTO) is above APC and is responsible for giving the best configuration to the steam matrix, based on fuel price. This paper will describe how both programs can help steam grid to be stable and financially optimized for the refinery.

Keywords

Process technology • Process control • Optimization

Acronyms and Abbreviations

DCS	Distributed Control System
GHG	Greenhouse gas
APC	Advanced Process Control
RTO	Real Time Optimization
SGA	Smelter grade alumina
PRV	Pressure reducing valve
MFT	Master Fuel Trip
PID	Proportional-Integral-Derivative control
MPC	Model-predictive control
TG	Turbo generator
MV	Manipulated Variable
CV	Controlled Variable
OPC	OLE for Process control
R&D	Research and Development department

D. Costa (✉) · R. Neves · F. Melo · J. Sousa · E. Junior
Hydro Alunorte - Alumina do Norte do Brasil, 481 PA Highway
Km 12, 68.447-000, Barcarena-Pará, Brazil
e-mail: daniella.costa@hydro.com

R. Neves
e-mail: rodrigo.neves@hydro.com

F. Melo
e-mail: fernando.melo@hydro.com

J. Sousa
e-mail: juvenal.sousa@hydro.com

E. Junior
e-mail: ediciano.junior@hydro.com

J. Freitas · D. Arcodaci
Barcarena-Pará, Brazil
e-mail: joaopaulojprf@yahoo.com.br

D. Arcodaci
e-mail: danilo.arcodaci@gmail.com

Steam System Presentation

Hydro Alunorte is a refinery that processes bauxite, and its final product is smelter grade alumina (SGA), a raw material for aluminum and for other industrial applications. Located in the municipality of Barcarena, state of Pará, and with a total production capacity of 6.3 Mtpy (millions of tons of SGA per year), it is currently the largest alumina refinery in the world in terms of industrial production capacity, outside China. The Norwegian group Norsk Hydro holds most of its public traded shares.

The Bayer process is used to produce alumina in the refinery, and steam is considered one of the main inputs for the operation of this process, together with bauxite and caustic soda.

The company has a steam generation and distribution system with cogeneration, responsible for supplying 1,240 t/h of saturated steam at 14.5 kgf/cm^2 for the process and

© The Minerals, Metals & Materials Society 2023
S. Broek (ed.), *Light Metals 2023*, The Minerals, Metals & Materials Series,
https://doi.org/10.1007/978-3-031-22532-1_25

generating about 70% of the electrical energy needed for operation. The remaining 30% of the electrical power demand of the plant is imported from the energy market.

The system consists of steam headers that store and distribute steam to the main consumers (process areas). The distribution of main steam demand occurs in two major levels from its sources of generation, the boilers, which produce steam at high pressure (superheated steam at 87.5 kgf/cm^2 of pressure and 450 °C of temperature) and at medium pressure (saturated steam at 15.0 kgf/cm^2 at 209 °C temperature).

The high-pressure system consists of a heavy oil boiler and three coal fired boilers in a circulating/bubbling fluidized bed. This steam is sent to a main header and distributed to three back-pressure turbogenerators, which are responsible for generating up to 90.0 MW of active power for the refinery. After the turbogenerators and the de-superheating system, the steam, already in saturated condition, is sent to a medium pressure header and later channeled to the consumer areas that use it in the production process.

The medium pressure system is composed of five heavy oil boilers and two electric boilers, generating saturated steam, which is directed to a main collector. From this header the steam is distributed to different processes of the refinery.

In the medium pressure header, steam relief valves to the atmosphere are installed to protect the system against overpressure effects resulting from cuts in consumption by customers (process areas).

Part of the refinery's steam system are three pressure reducing valves, called PRVs (Pressure Reducing Valve) that are interconnected between the high and medium pressure headers. PRVs operate as by-pass valves for turbogenerators, that is, they are closed most of the time, causing all the steam generated at high pressure to be used to generate electricity. Alternatively, PRVs also act as a system protection mechanism by diverting steam flow to the process in the absence of a turbine. The following figure provides a simplified representation of the described process (Fig. 1).

The existing steam load rejection system has the function of cutting (disposal) off consumers, to maintain the stability of the plant, when there is a loss of a source of steam generation (boilers), and can operate in automatic or manual mode.

In case of an event of loss of any generation source, the system discards consumers (closes the steam supply valve in the customer's area) following a priority order that is defined by the operators, as it depends on the plant's operating scenario.

The load-shedding system can check the type of power source shutdown to define whether the cut of consumers will be total or partial. The reason for this cut-off criterion is because coal-fired boilers admit hot starts when shutting down by MFT (Master Fuel Trip—"fuel trip"), which cuts only coal feeders. This type of occurrence allows an immediate return to operation because the furnace remains at the

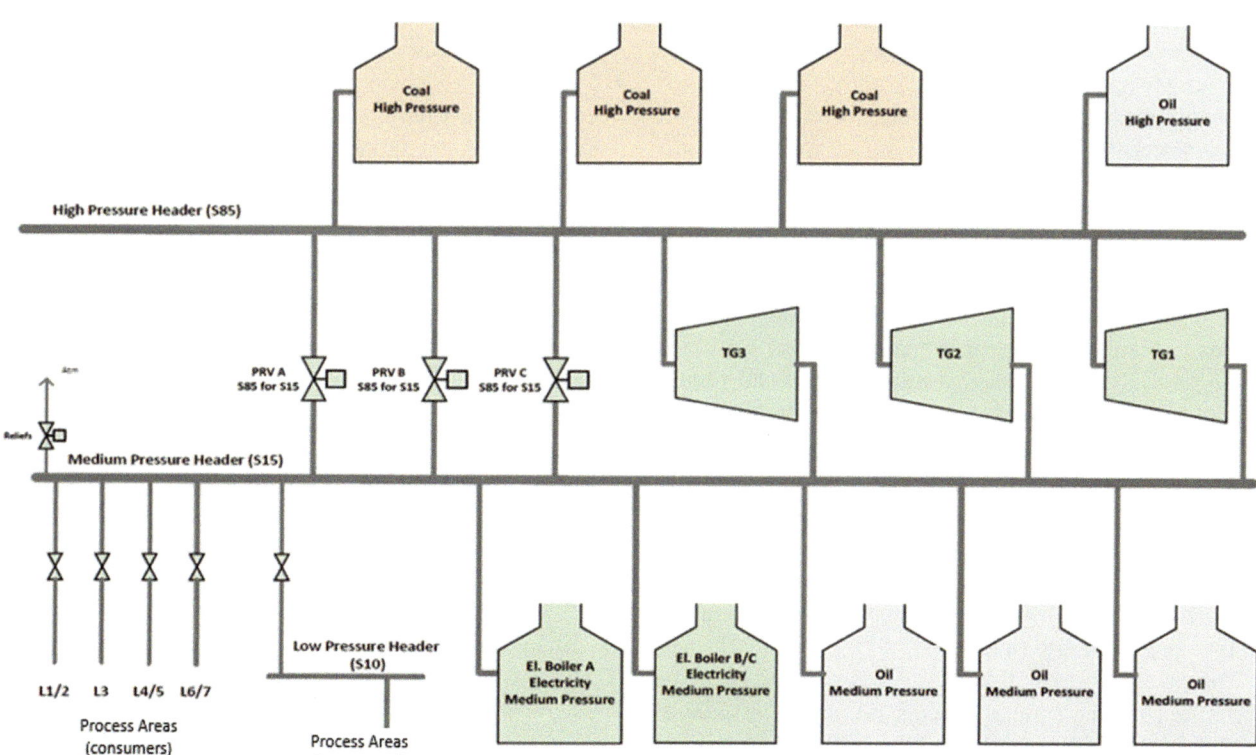

Fig. 1 Hydro Alunorte steam system overview

burning temperature, enabling a quick resumption of the process. Due to this possibility, it is assumed that a consumer can be partially cut off considering that the source can be resumed before the consumer is totally affected.

The steam load rejection system operates satisfactorily and in automatic mode most of the time. Few manual consumer disposal interventions are reported by the operation [1].

Main Difficulties in Regulatory Control

The existing system, despite having instrumentation for checking its main variables, is quite dependent on the operators for its main maneuvers such as: boiler load modulation, steam demand controls, disturbance and transients.

In the steam generation system, each boiler has a main or "master" control loop, responsible for modulating the boiler load through the crossed limits strategy, which modulates the stoichiometric ratios of fuel and oxidizer according to the required load demands, respecting the specific combustion curve.

As the boilers have their steam outlets interconnected to a main collector, sudden load variations in one of them cause disturbances in the others, which if not properly controlled can lead to unforeseen shutdowns and thus cause losses in steam generation and production. of alumina from the refinery.

The pressure control in the high collector was carried out through a simple feedback control loop, which modulates the steam reducing valves (PRV's A/B/C) to regulate the upstream pressure.

For the medium pressure manifold, the existing control loop connects the manifold pressure measurement to 3 vapor-to-atmosphere relief valves installed at the outlet of the turbogenerators in "split-range" mode (split-range control). In this way, the control only works in the event of collector overpressure, protecting the turbogenerators' back pressure from the high protection limit that causes them to shut down. In these occurrences, the steam flow is diverted to the atmosphere, until the operator takes manual control action to reduce the load of some medium boiler. In cases of sudden consumption and occurrence of under pressure, the control action occurs manually, also by the operator, increasing the steam production in some medium pressure boiler, until the system operating pressure level is recovered. If this action is not immediate and the permissible low back pressure limit is reached, the turbogenerators can be turned off by a protection trip (interlocks).

The control of steam distribution among customers in the operational routine currently occurs manually: as steam is needed for the process, a verbal request is made by the consumer and, manually, the operator of the steam area increases production in some boiler to supply the demand.

The turbogenerators, in turn, have a local control system dedicated to each machine that operates independently, with only supervision through the refinery's control and general automation system (a DCS) and the possibility of receiving set-points load remotes. Whenever it is necessary to increase or decrease the load of each machine, load modulations of the high-pressure boilers are first carried out manually until the steam flow necessary to increase or decrease the load of the turbogenerator in question is reached. These operations also cause pressure variations in the high-pressure header.

Even with the use of control loops responsible for regulating the pressure of the collectors, through regulatory control techniques, some aspects intrinsic to the configuration of the Hydro Alunorte steam system do not allow the achievement of the desired level of stability using this type of technology, are they:

- The boilers with different fuels and different (dynamic) response capacities, operating in parallel on the same collector. Thus, gains from a master controller in the high collector, for example, that modulates an oil boiler are not satisfactory for a coal boiler due to the different response capacities to load demands.
- Occurrences at one pressure level affect the other by virtue of the high and medium pressure systems being coupled and dependent on each other.
- Steam consumers (process areas) with different load take-off capacities due to the particularities of their processes, which require different performances for the regulation of a single master PID controller.
- System to be considered as multivariable, where PID controls do not have controllability effectiveness of more than one variable in the occurrence of transient events.
- Different elements of the steam system (boilers, turbogenerators and reducing valves) can compete with each other in an attempt to control an occurrence, causing instability events.
- The lack of modeling of the steam system makes the regulation of the PID parameters of the controllers, to achieve good performance indexes, difficult and ineffective, since they are performed empirically, by trial and error methods.
- The aspects listed above made it difficult to develop solutions to improve performance and desirable stability only using regulatory control strategies (PID control) [1].

RTO and APC Presentation

The powerhouse Optimization System includes a coordinated over-all control strategy designed to minimize the purchased power and fuel costs, while improving the control responsiveness and operational flexibility of the power plant. The overall optimization and control involve all the high pressure and medium pressure boilers, the turbogenerators, pressure reducing valves and steam vent valves. The strategy also provides coordinated control of the steam header pressures.

The solution is based on a multi-layered approach, coupling two separate optimization layers:

- APC (advanced process control)
- RTO (real time optimization)

The above items were complemented by modifications in existing DCS strategies and logics as needed.

The advanced optimization and control strategy is a combination of base level control logic changes and model-predictive control. The model-predictive control (MPC) application is implemented using ABB's Optimize^{IT} Predict & Control software.

Because of the complex nature of the plant and the variable pricing for purchased electrical power, a second application layer sits on top of the MPC. The next figure shows the general concept of the integrated solution (Fig. 2).

The figure above provides a general indication of the main interactions between the layers.

The APC and RTO interact with the existing plant DCS to pursue the optimization objective, in a multi-layered control strategy. This general approach uses real-time optimization (RTO), connected to advanced process control (APC).

At the top level, the optimization program can monitor the current situation in the plant and identify pricing data to make timely decisions about loading individual equipment in the plant.

Loading decisions will cover, for example, balancing steam production over the various boilers so to minimize production costs consider individual fuel type costs and, at the same time, boiler specific efficiency. Boiler efficiency is influenced by boiler type and construction parameters but, also, by boiler load.

The main APC targets are:

- Keep PRV valves in the control range and minimize them
- Keep medium pressure header at a desired Setpoint
- Keep vent valves as closed as possible
- Maximize Power Production with TGs

The main operator interface for APC system functions was configured on standard DCS consoles to make it easier for plant operators to monitor the controller. DCS consoles were used to activate/deactivate APC and to enter APC targets and constraints/limits.

During project execution and commissioning some protection logics were developed in the DCS, to have some permissive for APC activation and some automatic deactivations.

The controller controls the process through the setpoints of the manipulated variables (MVs), which sends directly, periodically and automatically to the basic regulation.

These setpoint values are calculated, every cycle, by the controller itself. The calculation is carried out to achieve the desired values for the controlled variables (CVs), which indicate what the operators want from the process. There are 3 CVs in the APC: total vents flow, total PRVs flow, medium pressure header.

The manipulated variables (MVs) of the APC are basically the set points that the multivariable controller uses to optimize the process. These set points were moved by the controller with the following constraints:

- That there is agreement to operations by the plant process team (in the form of an ON status for the variable).
- That the set point to be sent is within the minimum/maximum interval defined by the operator for each variable (absolute limits).
- That the set point to be sent is within a maximum variation from the current set point (speed limits).

These three aspects are very important for understanding how the controller works. The first aspect substantially requires that there is a request from the operator so that a variable can work under advanced control [2]. Every boiler master control is an MV in the APC system.

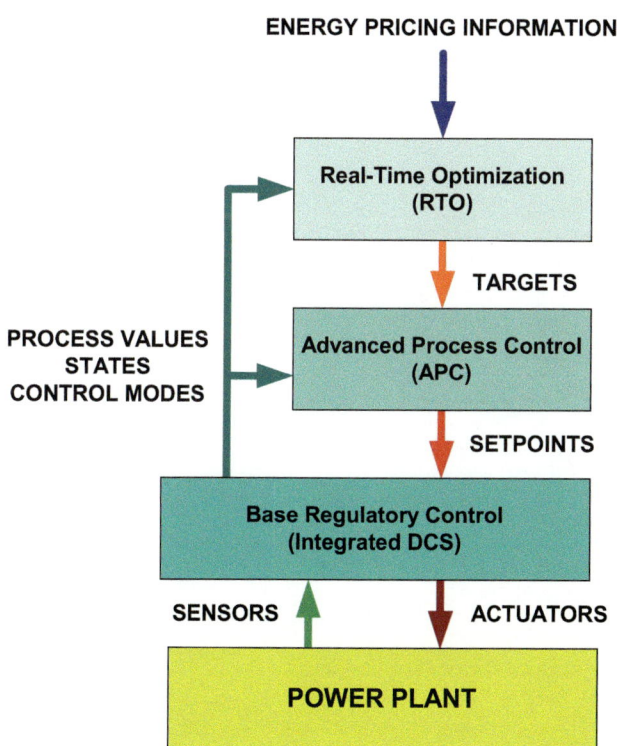

ENERGY PRICING INFORMATION

Real-Time Optimization (RTO)

TARGETS

Advanced Process Control (APC)

SETPOINTS

Base Regulatory Control (Integrated DCS)

PROCESS VALUES STATES CONTROL MODES

SENSORS   ACTUATORS

POWER PLANT

Fig. 2 APC/RTO integrated solution

The APC controller is configured to work with all the variables active. Il will be possible to deactivate some manipulated variables but at the expense of performance.

To help the operators to check if all the conditions to turn each MV on were met, an individual checklist process display were created to each boiler/MV.

The APC system is configured to automatically shut down in case of whatever anomaly, for example communication failure between DCS and APC (Watchdog Alarm), or if the operator does not turrn on at least 2 MVs (1 on the high pressure header and 1 on the medium pressure header). Additionally, a process display to show the APC automatic shutdown reasons was also created (Fig. 3).

The scope of RTO is a global steam and power generation plant optimization, through turbine optimization and better boiler load allocation.

The RTO shall contain the overall energy and steam cost as objective function, which depends on:

- The steam produced by each boiler, which in turn depends on the fuel flowrate sent to each boiler;

- The power generated by the turbogenerators, which in turn depends on the steam flow rate sent to each turbine.

The final cost of the Energy matrix is the sum of the cost to produce steam plus the cost of the purchased energy. The RTO's main goal is to minimize this cost maintaining the overall steam production.

RTO calculates the best target for fuel flow, PRVs total flow, turbo generators and electrical boilers generated power and suggests them to the plant operator for manual implementation (that is called advisory mode, when the system does not act directly to the APC and neither to the plant).

In case of one boiler or turbine unavailability (e.g. because it is in manual or stopped for maintenance purposes) the relevant variable is removed by the optimization problem and put in tracking to the actual process values [3].

The RTO can be also connected to the APC (closed loop mode), but it was not commissioned at Hydro Alunorte yet (Fig. 4).

The APC and RTO servers are hosted in a network out of DCS client server network. Between these 2 networks there

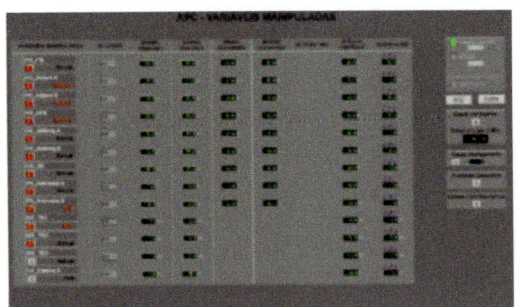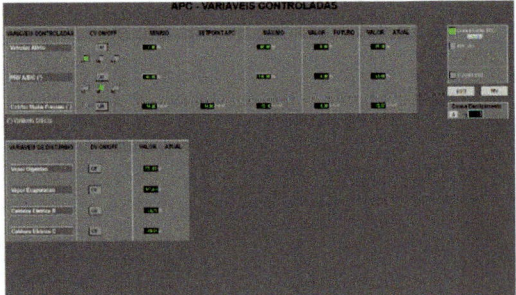

Fig. 3 APC process displays in the DCS

Fig. 4 RTO process display

are firewalls with specific rules to allow the OPC communication between a DCS dedicated interface server (OPC server) and the system servers (OPC clients). An OPC Tunneller software is used in the communication, which has a cycle of 15 s.

Benefits and Savings

Both RTO and APC were commissioned in 2021s semester. The potential benefits that can be achieved thanks to the APC + RTO combined solution were divided in three different categories:

- Global steam and power generation plant optimization, through turbine maximization and better boiler load allocation;
- Minimization of the steam losses to vent; and
- Backpressure minimization with a higher production of electric power by the TGs.

Moreover, the solution will provide pressure stabilization mainly for the medium pressure header [4].

Due to some plant restrictions, the project engineering team has defined a minimum scenario to the APC operation, which keep coal Boilers out of scope, thus benefit coming from coal boilers maximization is not included in this calculation.

The real benefits achieved in the first 8 months of APC operation, based on a baseline of 2021, are described below:

- Vent minimization—Saving of 138 tons/day (reduced waste)

- No enough data to calculate backpressure minimization yet and load minimization

Based on this data, project payback will be around 1 year.

Conclusions

Even with some operational restrictions, APC has shown great potential to improve operation at the Hydro Alunorte steam production plant.

The main challenges of system implementation were to perform step tests (during project phase), keep the boilers in automatic mode (before system commissioning) and change operational culture. All of them have been overcome with care, courage and collaboration of all teams involved in the project (automation, process engineering, operation, maintenance, engineering and R&D).

References

1. Hydro Alunorte/Norsk Hydro. 2021. Especificação Técnica para fornecimento de produtos e serviços. Unpublished internal company document.
2. Hydro Alunorte/Norsk Hydro. 2021. APC Detailed Design Specification – APC and RTO Engineering. Unpublished internal company document.
3. Hydro Alunorte/Norsk Hydro. 2021. RTO Detailed Design Specification – APC and RTO Engineering. Unpublished internal company document.
4. Hydro Alunorte/Norsk Hydro. 2021. Alunorte Benefit assessment report – APC and RTO Engineering. Unpublished internal company document.

Effects of Impurities on the Boiling Point of Bayer Liquor

Erwei Song and Erqiang Wang

Abstract

Bayer liquor (or named sodium aluminate solution), is a kind of mixed solution from Bayer process, whose boiling point will change while heating according to the thermodynamics properties of multiple components. Bayer liquor obtained from the reaction between bauxite and alkaline solution, whose components vary with the bauxite characteristics and Bayer process condition. Based on the analysis of Bayer process, the boiling point of Bayer liquor has a significant effect on the energy consumption and evaporation efficiency during evaporation process one of the most important sections in the Bayer process, in which realizing alkaline solution concentration, water balance and carbonate removal. However, it is hard to calculate the boiling point using the theory formula directly. To overcome above problems, the effects of impurities on the boiling point of Bayer liquor were investigated using simulation method, which can bring us useful guidance to setup criterion for the impurities management during Bayer process.

Keywords

Bayer liquor • Boiling point • Simulation method

Introduction

Bayer process, widely used to produce alumina, could be traced down two patents proposed by Bayer between 1887 and 1892, which was a breakthrough in mineral processing that enabled alumina to be extracted from bauxite ore economically on a large scale, and has been used to produce gibbsite from bauxite for over one hundred years [1]. The basic mechanism of Bayer process can be explained using below reaction equation:

$$Al_2O_3 \cdot xH_2O + 2NaOH + aq \leftrightarrow 2NaAl(OH)_4 + aq$$

During above reaction equation, it can be seen that alumina-containing minerals in bauxite are dissolved in hot sodium hydroxide solution in the digestion process, followed by precipitation, decomposition, evaporation, and calcination process, producing alumina product and cycling liquor. From the Bayer process sketch as shown in Fig. 1, it can be seen that alumina production process is a cycled process, in which the spent liquor for digestion remains in a same composition after every cycle, that means its caustic and alumina concentrations will keep the same. However, the composition of cycling liquor will change according to the bauxite difference, thus the impurities in the liquor will change, which can affect the physical and chemical properties of the liquors.

John Vogrin et al. [3] investigated the effects of chloride on sodium aluminosilicate solubility in Bayer liquor under equilibrium conditions in synthetic Bayer liquors, which demonstrated that the presence of NaCl promoted the formation of spherical nano-sized particles instead of polyhedral micron-sized particles, and increased the thermal stability of the desilication product. Malito [4] proposed equations to calculate the solubility of Na_2CO_3 and Na_2SO_4 in Bayer liquor during evaporative crystallization, predicting impurity equilibrium concentrations from the free Na^+ ion concentration, based on the liquor molality and assume the presence of univalent ions in caustic solutions as proposed by Dewey.

Boiling point elevation, an important property of numerous industrial solutions [5–7], is calculated based on the difference of boiling temperatures between a real solution and its pure solvent under the same pressure. Evaporation means heating liquor contained non-volatile materials to

E. Song · E. Wang (✉)
School of Chemical Engineering, University of Chinese Academy of Sciences, Beijing, China
e-mail: wangerqiang@ucas.ac.cn

E. Song
e-mail: songerwei15@mails.ucas.ac.cn

© The Minerals, Metals & Materials Society 2023
S. Broek (ed.), *Light Metals 2023*, The Minerals, Metals & Materials Series,
https://doi.org/10.1007/978-3-031-22532-1_26

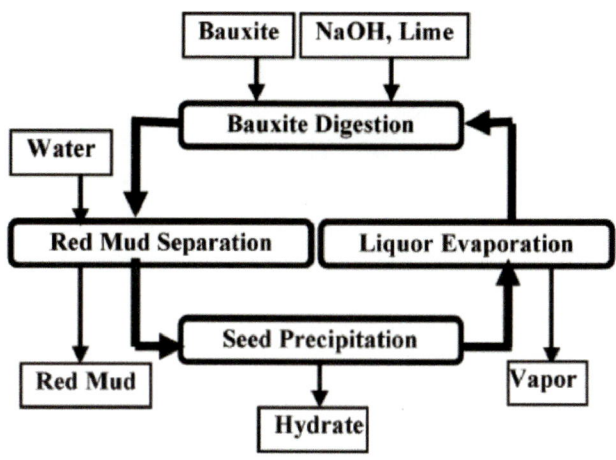

Fig. 1 Typical Bayer Cycle [2]

boiling condition, removing low boil point solvent, to realizing the liquor concentration. Evaporation process, an essential sector in the Bayer process, plays an important role in realizing alkaline solution concentration, water balance and carbonate removal, thus its performance directly affect the quality of product and profit, et al. the value of boiling point rising changes with different properties in different liquor, and the boiling point elevation may make the evaporator heat transfer area increasing [8], which should be taken into consideration during evaporator design and operation process. Bayer liquor comes from the reaction between bauxite and alkaline liquor, thus its components change with the components and type of the bauxite, which makes Bayer liquor complex, leading Raoult's law cannot be effective, and the mechanism of boiling point elevation for Bayer liquor is also complex.

Given the important role of boiling point elevation during evaporation process in the alumina refinery and above reasons, we try to explore the inherent mechanism of boiling point elevation of Bayer liquor with the help of modeling and simulation methods, which can bring us some useful information and guidance during evaporator design and operation process.

Simulation Module and Methods

Boiling points of different liquors were calculated using the modules designed in Aspen Plus, which is a modular process simulation software with an extensive database for many chemical systems including electrolyte solutions. In particular, the electrolyte non-random two liquids (ENRTL) model, based on Chen's work [9], was used to fit experimental data for numerous electrolyte systems. Figure 2 shows the modules designed to transfer mass concentration to mass flow of each component in the liquor, which consists of one heater and flash modules in Aspen Plus. Then the boiling point of liquor was calculated using mixture and stream property analysis tools.

Results and Discussion

Boiling Point of Water

The effects atmosphere pressure on the boiling point of water was simulated, results are shown in Fig. 3.

From Fig. 3, it can be seen that the value of boiling point for water increase with atmosphere pressure increasing, and can be explained that with external pressure increasing, the vapor to approach boiling will need elevated temperature. Besides among the setup range (0.6–2.0 atm), the corresponding math model formulation can be expressed as Eq. 1, where x and y represent atmosphere pressure and boiling point respectively.

$$y = 62.10 + 49.06x - 8.43x^2 \qquad (R^2 > 0.99). \qquad (1)$$

Boiling Point of Caustic Soda Solution

The effects atmosphere pressure on the boiling point of caustic soda solution was simulated, results are shown in Fig. 4.

Fig. 2 The sketch for transferring mass concentration to mass flow of each component in the liquor

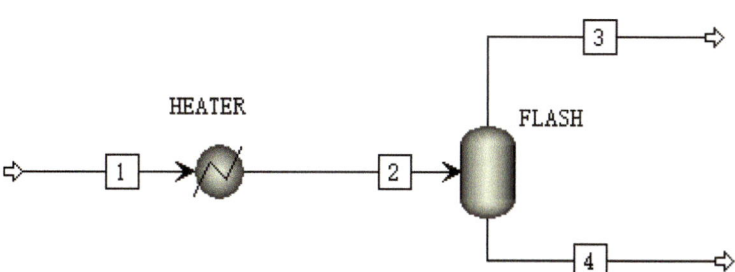

Fig. 3 Boiling point of water at different pressure

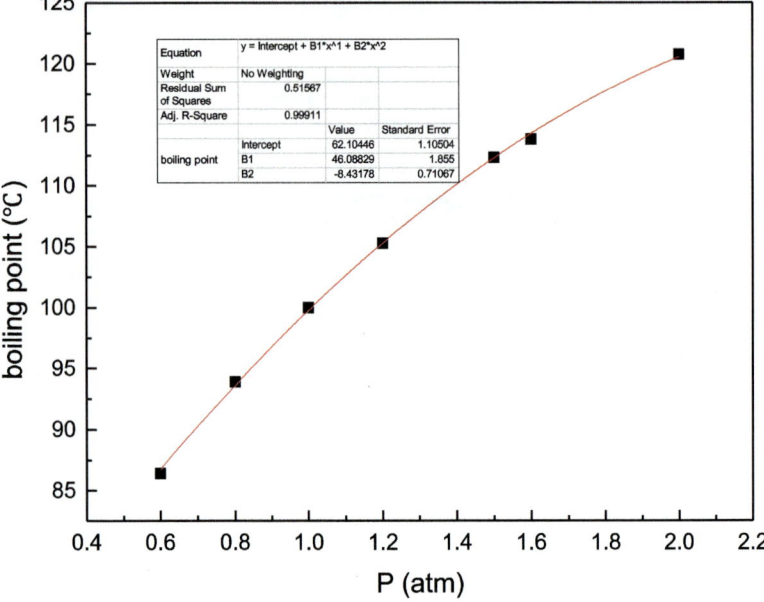

Fig. 4 Boiling point of caustic soda solution at different mass concentration

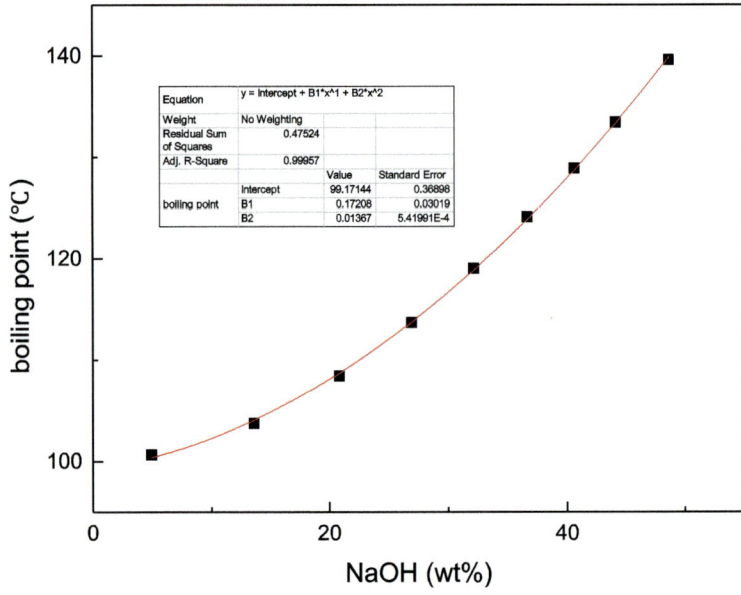

From Fig. 4, it can be seen that the value of boiling point for caustic solution increase with the content of caustic soda in the solution increasing, and can be explained that sodium hydroxide is easily soluble in water, decomposing into Na^+ and OH^- at the same time, and diffusing inside the liquid. If water molecules are to become water vapor, they need to be freed from the bondage of Na^+ and OH^- while being freed from the attraction of other water molecules, so it needs more energy, and higher temperature is needed to produce water vapor. Besides among the setup range (5–50 wt%), the corresponding math model formulation can be expressed as Eq. 2, where x and y represent mass fraction of NaOH and boiling point respectively.

$$y = 99.17 + 0.17x + 0.014x^2 \qquad (R^2 > 0.99). \qquad (2)$$

Boiling Point of Sodium Aluminate Solution

The effects atmosphere pressure on the boiling point of sodium aluminate solution was simulated, results are shown in Fig. 5.

Fig. 5 Boiling point of sodium aluminate solution at different mass concentration

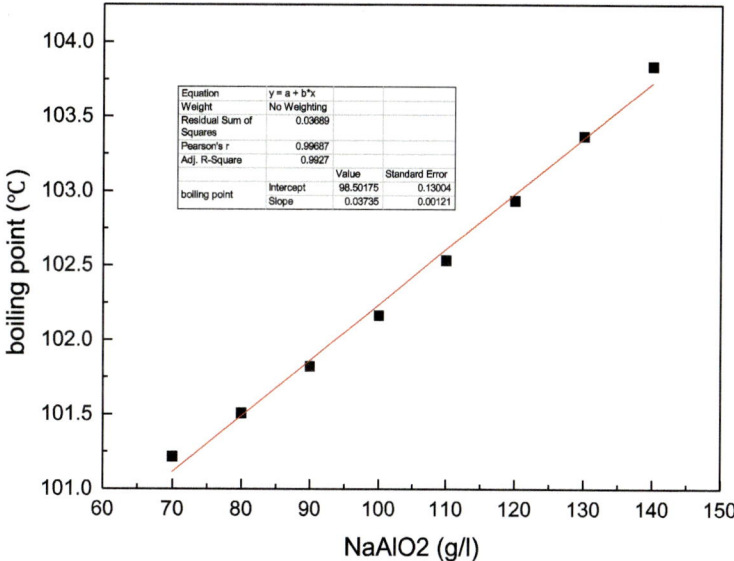

From Fig. 5, it can be seen that the value of boiling point for sodium aluminate solution increase with the content of sodium aluminate in the solution increasing, and can be explained that sodium aluminate is easily soluble in water, decomposing into Na^+ and AlO_2^- at the same time, and diffusing inside the liquid. If water molecules are to become water vapor, they need to be freed from the bondage of Na^+ and AlO_2^- while being freed from the attraction of other water molecules, so it needs more energy, and higher temperature is needed to produce water vapor. Besides among the setup range (70–140 g/l, as Al_2O_3), the corresponding math model formulation can be expressed as Eq. 3, where x

and y represent mass content of $NaAlO_2$ and boiling point respectively.

$$y = 98.50 + 0.037x \qquad (R^2 > 0.99). \qquad (3)$$

Effects of NaOH

IN order to understand the effects caustic content on the boiling point of sodium aluminate solution, different solution (Na_2O 100, 120, 140, 160, 180, 200, 220 g/l) with fixed mole molecular ratio (a = 1.4) was simulated, results are shown in Fig. 6.

Fig. 6 Effects of NaOH on the boiling point of sodium aluminate solution (a = 1.4, Na_2O 100, 120, 140, 160, 180, 200, 220 g/l)

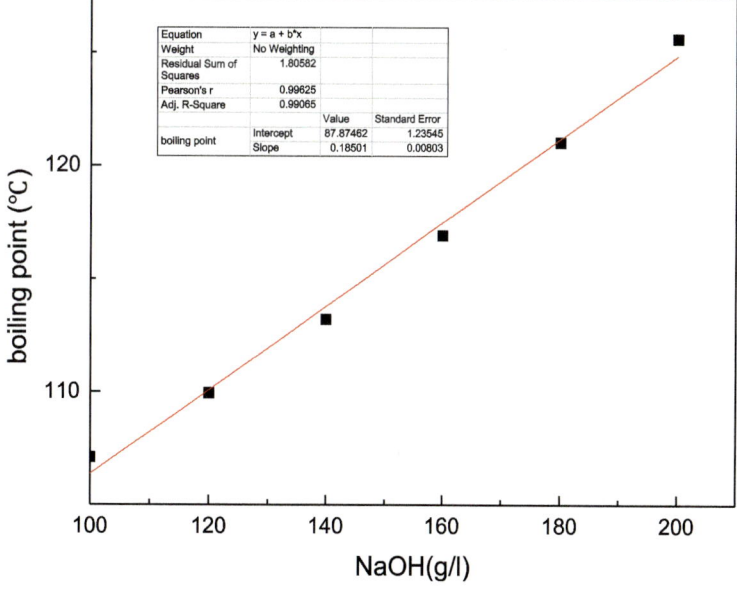

From Fig. 6, it can be seen that the value of boiling point for sodium aluminate solution increase with the content of NaOH in the solution increasing when a = 1.4, and the trend shows approximately linear relation, which can be expressed using one simple model formulation as below, where x and y represent mass content of NaOH and boiling point respectively.

$$y = 87.87 + 0.185x \qquad (R^2 > 0.99). \qquad (4)$$

Effects of Na$_2$CO$_3$

IN order to understand the effects Na$_2$CO$_3$ content on the boiling point of sodium aluminate solution, different solution (Na2O 100, 120, 140, 160, 180, 200 g/l) with fixed mole molecular ratio (a = 1.4) was simulated, results are shown in Fig. 7.

From Fig. 7, it can be seen that the value of boiling point for sodium aluminate solution increase with the content of Na$_2$CO$_3$ in the solution increasing under the condition of a fixed mole molecular ratio (a = 1.4), and the trend shows approximately linear relation, which can be expressed using one simple model formulation as below, where x and y represent mass content of Na$_2$CO$_3$ and boiling point respectively.

$$y = 99.74 + 2.84x \qquad (R^2 > 0.99). \qquad (5)$$

Effects of Na$_2$SO$_4$

To understand the effects Na$_2$SO$_4$ on the boiling point of sodium aluminate solution, different solution (Na$_2$O 100, 120, 140, 160, 180, 200 g/l) with fixed mole molecular ratio (a = 1.4) was simulated, results are shown in Fig. 8.

From Fig. 8, it can be seen that the value of boiling point for sodium aluminate solution increase with the content of Na$_2$SO$_4$ in the solution increasing with other conditions keeping constant.

Conclusion

Based on the foundation of Bayer Process and modeling theory, with the help of Aspen Plus simulation software, the effects of impurities on the boiling point of Bayer liquor were investigated. From the simulation results, it can be seen that impurities indeed have effects on the boiling point of sodium aluminate solution, besides the effects vary with their concentration. For the complexity of Bayer liquor changes with the Bayer process and bauxite, thus there is no one simple mathematical model that can calculate the boiling point of Bayer liquor, but this research proposed one simple but effective method that using modeling and simulation tools to calculate the boiling point of Bayer liquor, which can bring us useful guidance to setup criterion for the impurities management during Bayer process.

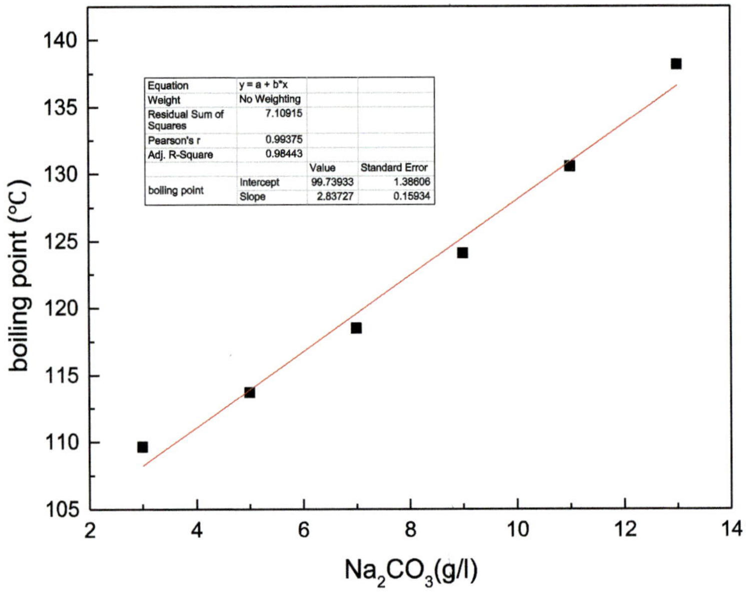

Fig. 7 Effects of Na$_2$CO$_3$ on the boiling point of sodium aluminate solution (a = 1.4)

Fig. 8 Effects of Na$_2$SO$_4$ on the boiling point of sodium aluminate solution (a = 1.4, Na$_2$O 100, 120, 140, 160, 180, 200 g/l)

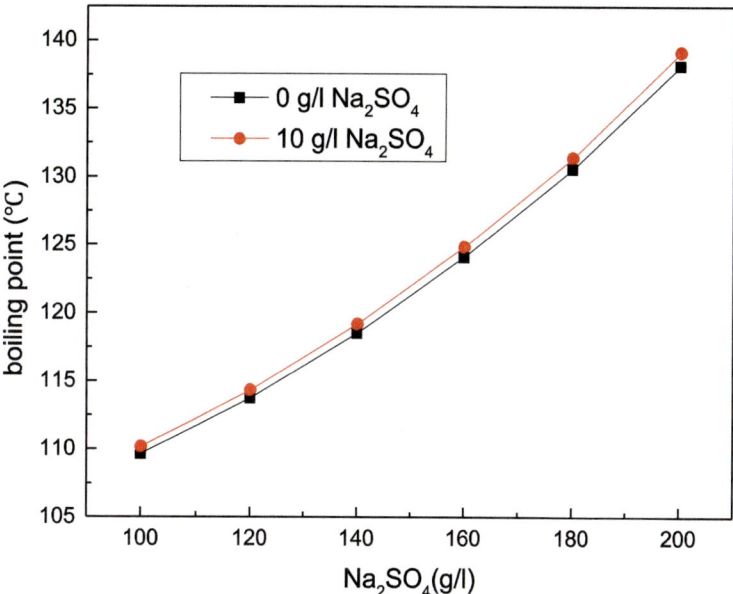

Acknowledgements This work was supported by the Ministry of Education Industry-University Cooperation Collaborative Education Project [Research on energy saving and consumption reduction technology during the Bayer process] and project [GYY-DTFZ-2022-016], funded by Binzhou Weiqiao National Higher Technology Research Institute.

References

1. Yang Zhongyu. Alumina production technology [M]. Metallurgical Industry Press, 1982.
2. Gu, Songqing. Bayer Process Efficiency Improvement [J]. Light Metals, 2013:163–167.
3. Vogrin J, Santini T, Peng H, et al. Influence of chloride on sodium aluminosilicate solubility in Bayer liquor[J]. Microporous and Mesoporous Materials, 2020, 299(110086):1-8.
4. Malito, J. T. Equilibrium solubility of sodium sulfate in Bayer liquor. Light Met. 1984, 147–156
5. Bujanovic B, Cameron J H. Effect of Sodium Metaborate on the Boiling Point Rise of Slash Pine Black Liquor[J]. Industrial & Engineering Chemistry Research, 2001, 40(11):2518-2524.
6. Ilangantileke S G, Jr A B R, Joglekar H A. Boiling point rise of concentrated Thai tangerine juices[J]. Journal of Food Engineering, 1992, 15(3):235-243.
7. Meranda D, Furter W F. Elevation of the boiling point of water by salts at saturation: data and correlation[J]. Journal of Chemical & Engineering Data, 1977, 22(3):315-317.
8. Yizhong Yuan. Influence of Material Liquid Boiling Point Elevation on Evaporator Heat Transfer Area[J]. Liaoning Chemical Industry, 2013,42(9):1140–1142.
9. Song Y, Chen C C. Symmetric Electrolyte Nonrandom Two-Liquid Activity Coefficient Model[J]. Industrial & Engineering Chemistry Research, 2009, 48(16):7788-7797.

Effects of Different Precursors on the Preparation of β''-Al$_2$O$_3$

Hongsheng Che and Yang Zhang

Abstract

β''-Al$_2$O$_3$ was prepared at 1280 °C using α-Al$_2$O$_3$, γ-Al$_2$O$_3$, and boehmite as alumina precursors and NaHCO$_3$ as sodium oxide source. The effect of different alumina precursors and amounts of NaHCO$_3$ on β''-Al$_2$O$_3$ formation behavior were studied. The phase composition and microstructure of the as-prepared β''-Al$_2$O$_3$ were analyzed, respectively, by X-ray diffraction and scanning electron microscopy. β''-Al$_2$O$_3$ co-existing with NaAlO$_2$ and α-Al$_2$O$_3$ was obtained when using α-Al$_2$O$_3$ and γ-Al$_2$O$_3$, whereas phase-pure β''-Al$_2$O$_3$ was prepared by calcining the mixture of boehmite and 22 wt% of NaHCO$_3$. The phase-pure β''-Al$_2$O$_3$ particles did not agglomerate and showed a sheet-like morphology with a thickness of about 1 μm.

Keywords

β''-Al$_2$O$_3$ • Boehmite • Sodium bicarbonate

Introduction

Sodium–sulfur battery is an excellent electrochemical energy storage power source with the advantages of high safety, high specific energy, high charge–discharge efficiency and low cost. β''-Al$_2$O$_3$ electrolyte ceramic is the core material of sodium–sulfur battery, which directly determines the performance, production process, and cost of sodium–sulfur battery [1]. The ideal chemical formula of β''-Al$_2$O$_3$ is Na$_2$O·5Al$_2$O$_3$, which belongs to the hexagonal crystal system. The crystal structure of β''-Al$_2$O$_3$ is similar to that of β-Al$_2$O$_3$, with similar lamellar structure and interlamellar

distance. The unit cell of β-Al$_2$O$_3$ contains two spinel blocks and two sodium ion layers, and there is a loose Na–O layer with equal amounts of Na and O ions perpendicular to the c-axis. Na ions can diffuse in the two-dimensional space of the Na–O layer [2]. With the migration of Na ions, there is a charge movement. Therefore, layered β-Al$_2$O$_3$ has excellent Na ion conductivity. The unit cell of β''-Al$_2$O$_3$ contains three spinel blocks and three sodium ion layers, and the c-axis is about 1.5 times of β-Al$_2$O$_3$. Therefore, β''-Al$_2$O$_3$ has better conductivity, which is about 10 times of β-Al$_2$O$_3$ [3].

The main preparation method of β''-Al$_2$O$_3$ includes the solid phase reaction method, sol–gel method, and co-precipitation method. The solid-state reaction process requires a higher sintering temperature, which results in a partial loss of sodium oxide through evaporation. Although the β''-Al$_2$O$_3$ prepared by sol–gel method and co-precipitation method has smaller particle size and lower sintering temperature, the precursors react randomly with each other in solution, and the phase cannot be controlled accurately, and the production process is more complicated. With the development of research, Barison et al. [4] prepared β''-Al$_2$O$_3$ with different purity using α-Al$_2$O$_3$, γ-Al$_2$O$_3$, and Al(OH)$_3$ as precursors, Na$_2$CO$_3$ as sodium oxide sources, and MgO and Li$_2$CO$_3$ as stabilizers. Zhang et al. [5] synthesized electrolyte materials with high Na-β''-Al$_2$O$_3$ phase by solid phase reaction using one-dimensional γ-AlOOH prepared by the hydrothermal method as precursor. Wang et al. [6] prepared a mixture of β-Al$_2$O$_3$ and β''-Al$_2$O$_3$ at 1170°C using α-Al$_2$O$_3$ as precursor and sodium carbonate as a sodium oxide source.

Boehmite (AlOOH) has a regular spinel structure with oxygen atoms cubic tightly packed and aluminum atoms intercalated [7], which is similar to the spinel-like structure with oxygen atoms cubic tightly packed of β''-Al$_2$O$_3$.

H. Che (✉) · Y. Zhang
Zhengzhou Non-ferrous Metals Research Institute Co. Ltd of CHALCO, Zhengzhou, 450041, China
e-mail: zyy_lcl@rilm.com.cn

© The Minerals, Metals & Materials Society 2023
S. Broek (ed.), *Light Metals 2023*, The Minerals, Metals & Materials Series,
https://doi.org/10.1007/978-3-031-22532-1_27

Moreover, boehmite has a certain degree of continuity to the morphology during calcination. Therefore, the lamellar boehmite is very suitable for the preparation of lamellar β''-Al$_2$O$_3$ precursor.

In this work, boehmite with good crystallinity and regular morphology was synthesized by hydrothermal method using alumina hydroxide, aqueous solution as reaction medium with appropriate template agent. At the same time, α-Al$_2$O$_3$ and γ-Al$_2$O$_3$ were selected as precursors to prepare β''-Al$_2$O$_3$ for comparison, and the influence of different precursors on the preparation of β''-Al$_2$O$_3$ was studied, and the influence of the amount of sodium oxide source on the preparation of β''-Al$_2$O$_3$ was explored.

Experiment

Materials

α-Al$_2$O$_3$ was produced by Zhengzhou Fine Alumina Material Factory, γ-Al$_2$O$_3$ was industrial alumina produced by Henan Branch of Aluminum Corporation of China LTD., and Sodium bicarbonate was produced by Fuchen (Tianjin) Chemical Reagent Co., LTD. The preparation process of boehmite used in the experiment was as follows: aqueous solution was used as the reaction medium, and aluminum hydroxide (chemical composition as shown in Table 1) and appropriate template were added into the reaction vessel. After heating the reaction vessel to 220 °C and holding for 4 h, the hexagonal lamellar boehmite with a particle size of about 0.5–1 μm was prepared, with microstructure morphology of boehmite as shown in Fig. 1, and chemical composition as shown in Table 2.

Method

Sodium bicarbonate was weighed according to the calculated proportion, added to pure water, and stirred. After completely dissolved, it was added to α-Al$_2$O$_3$, industrial alumina and boehmite respectively. After stirring for 1 h at room temperature, it was poured into a stainless steel tray and placed in an oven at 110 °C for drying for 24 h. The samples were then placed in a 500 ml crucible and calcined at 1280 °C for 4 h. The calcined samples were passed through a 60 mesh quasi-sieve.

Characterization

The particle size distribution of the samples was analyzed by laser particle size analyzer (Bettersize 2000, Dandong Baiter, China), the phase composition of the samples was analyzed by X-ray diffraction (Empyream, Malvern Panalytical Ltd., the Netherlands), and the microstructure of the samples was observed by scanning electron microscopy (JSM-6360, JEOL Ltd., Japan).

Results and Discussion

Effect of Precursors on the Preparation of β″-Al₂O₃

β''-Al$_2$O$_3$ was prepared with α-Al$_2$O$_3$, γ-Al$_2$O$_3$ and boehmite as alumina precursor and 20 wt% of NaHCO$_3$ as sodium oxide source. The XRD patterns of the prepared β''-Al$_2$O$_3$ are shown in Fig. 2. As can be seen from the figure, when

Table 1 Chemical composition of aluminum hydroxide

Chemical composition (wt%)			
Al₂O₃	SiO₂	Fe₂O₃	Na₂O
65.036	0.006	0.007	0.210

Fig. 1 SEM of hexagonal flake boehmite prepared by hydrothermal method

Table 2 Chemical composition of boehmite prepared by hydrothermal method

Chemical composition (wt%)		
Al_2O_3	SiO_2	Fe_2O_3
82.590	0.046	0.038

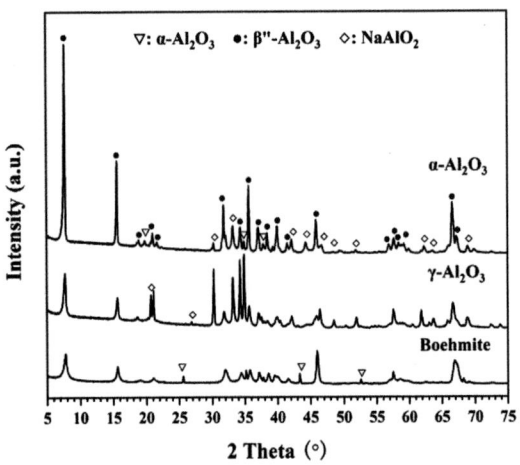

Fig. 2 XRD patterns of $\beta''-Al_2O_3$ prepared by different precursors

$\alpha-Al_2O_3$ was used as precursor, a large number of $\beta''-Al_2O_3$ diffraction peaks appeared, which were high and sharp, and the rest were the diffraction peaks of $NaAlO_2$ and $\alpha-Al_2O_3$. The results showed that when $\alpha-Al_2O_3$ was used as precursor, the main phase after calcination was $\beta''-Al_2O_3$, but a small amount of $NaAlO_2$ and residual $\alpha-Al_2O_3$ appeared at the same time. When $\gamma-Al_2O_3$ was used as precursor, the diffraction peaks of $\beta''-Al_2O_3$ and $NaAlO_2$ were the main ones, and there were a small amount of $\alpha-Al_2O_3$ diffraction peaks too, indicating that there was a large amount of $NaAlO_2$ in the main product after the reaction between $\gamma-Al_2O_3$ and sodium bicarbonate reaction. When boehmite was used as precursor, the diffraction peaks of $\beta''-Al_2O_3$ existed except for a small amount of $\alpha-Al_2O_3$.

Compared with $\gamma-Al_2O_3$ and boehmite, $\alpha-Al_2O_3$ had the highest crystallization degree and the lowest reactivity. Additionally, the complete crystallization of $\beta''-Al_2O_3$ produced by the reaction of sodium bicarbonate with $\alpha-Al_2O_3$, part of sodium oxide decomposed from sodium bicarbonate did not react with $\alpha-Al_2O_3$ and volatilized at high temperature, and the remaining sodium oxide dissolved on the surface of $\alpha-Al_2O_3$, forming a small amount of $NaAlO_2$. $\gamma-Al_2O_3$ was more active than $\alpha-Al_2O_3$, so it was more favorable to the formation of $\beta''-Al_2O_3$. However, the crystallized $\gamma-Al_2O_3$ was not conducive to the further

diffusion of sodium oxide, resulting in the enrichment of sodium oxide and the formation of the $NaAlO_2$ phase. Boehmite decomposed during heating, and the alumina formed had higher activity and lower crystallization degree than $\gamma-Al_2O_3$ and $\alpha-Al_2O_3$. Therefore, it was favorable for the solid phase reaction with sodium oxide, and was favorable for the further diffusion of sodium oxide, so that the sodium oxide was evenly dispersed in the lattice of alumina, and the $\beta''-Al_2O_3$ formed.

The microstructure of $\beta''-Al_2O_3$ prepared with different precursors is shown in Fig. 3. It can be seen from the figure that the morphology of $\beta''-Al_2O_3$ prepared by different precursors was quite different. When $\alpha-Al_2O_3$ was the precursor, the prepared $\beta''-Al_2O_3$ had a layered structure with a thickness of 0.2–0.8 μm and a diameter of about 25 μm. When $\gamma-Al_2O_3$ was used as the precursor, the $\beta''-Al_2O_3$ was layered and thinner than prepared with $\alpha-Al_2O_3$, which was stacked into agglomerates layer by layer, but the grain boundary between grains was not obvious. When boehmite was selected as precursor, $\beta''-Al_2O_3$ had a sheet structure of about 1 μm, and the grains were well dispersed without agglomeration.

It can be seen from the above results that when $\alpha-Al_2O_3$ was selected as precursor to prepare $\beta''-Al_2O_3$, in addition to $\beta''-Al_2O_3$, there was a small amount of $NaAlO_2$ and residual $\alpha-Al_2O_3$. The $\beta''-Al_2O_3$ had a layered structure, formed with agglomerates layer by layer, and the grain boundary interface was obvious. When $\beta''-Al_2O_3$ was prepared by $\gamma-Al_2O_3$ precursor, $\beta''-Al_2O_3$ and $NaAlO_2$ existed at the same time. The $\beta''-Al_2O_3$ also had a lamellar structure, and the particle size was smaller than $\beta''-Al_2O_3$ prepared by $\alpha-Al_2O_3$, but the grain boundary was not obvious. The $\beta''-Al_2O_3$ prepared by boehmite precursor had high crystalline and complete crystal morphology, with a sheet structure of about 1 μm and no agglomeration between grains.

Effect of Sodium Bicarbonate Addition on the Preparation of $\beta''-Al_2O_3$

According to the results in Fig. 2, when $\beta''-Al_2O_3$ was prepared with boehmite as precursor and 20% sodium bicarbonate, there was still a small amount of $\alpha-Al_2O_3$ in the sample. In order to prepare $\beta''-Al_2O_3$ with high purity, the influence of sodium oxide source addition on $\beta''-Al_2O_3$ was further studied. Specifically, the XRD patterns of $\beta''-Al_2O_3$ prepared by 15, 20, 22, and 25% sodium bicarbonate addition are shown in Fig. 4. The diffraction peaks of $\beta''-Al_2O_3$ and $NaAl_5O_8$ were observed in the sample with 15% sodium bicarbonate after secondary calcinations, so did the diffraction peaks of $\alpha-Al_2O_3$.

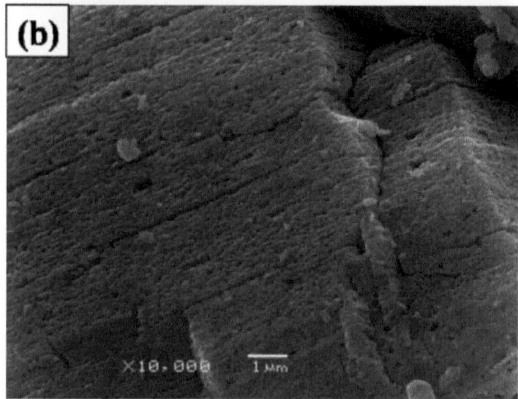

Fig. 3 SEM of β″-Al₂O₃ prepared by different precursors (a, α-Al₂O₃, b, γ-Al₂O₃, and c, boehmite)

The above results showed that β″-Al₂O₃ can be obtained by secondary calcinations of boehmite with 15% sodium bicarbonate addition, but due to the low addition of sodium oxide source, γ-Al₂O₃ decomposed from boehmite converted to α-Al₂O₃ in large quantities. The diffraction peak of α-Al₂O₃ decreased and that of β″-Al₂O₃ increased with the addition of sodium bicarbonate to 20%, indicating that the transformation from γ-Al₂O₃ to α-Al₂O₃ decreased and a

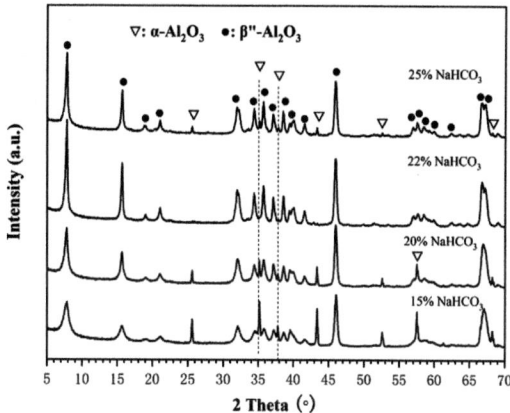

Fig. 4 XRD patterns of β″-Al₂O₃ prepared by different addition of sodium bicarbonate

large amount of β″-Al₂O₃ formed after the secondary calcinations. When the amount of sodium bicarbonate was 22%, the diffraction peak of α-Al₂O₃ disappeared, and the phase in the sample was β″-Al₂O₃ pure phase. However, when the addition of sodium bicarbonate was further increased to 25 wt%, a small amount of α-Al₂O₃ diffraction peak appeared in the sample, which may be because the excess Na₂O and Al₂O₃ formed a liquid phase at high temperature, and α-Al₂O₃ precipitated from the liquid phase during cooling. In conclusion, a pure β″-Al₂O₃ phase sample can be prepared when the addition of sodium bicarbonate was 22%.

Conclusions

When β″-Al₂O₃ was prepared with α-Al₂O₃ as precursor, in addition to β″-Al₂O₃, there was a small amount of NaAlO₂ and residual α-Al₂O₃. β″-Al₂O₃ with layered structure was piled into aggregates with an obvious grain boundary. β″-Al₂O₃ prepared with γ-Al₂O₃ as precursor had more NaAlO₂, and the particle size of β″-Al₂O₃ with lamellar structure was smaller than that prepared with α-Al₂O₃, but there were more agglomerations and grain boundary was not obvious. β″-Al₂O₃ prepared with boehmite as precursor had high crystalline and complete crystal morphology, with a sheet structure of about 1 μm and no agglomeration between grains.

When the addition of sodium bicarbonate was 22%, pure β″-Al₂O₃ phase product can be prepared, without other miscellaneous phases, and the hexagonal sheet of boehmite precursor was maintained.

References

1. Kim I,Park JY, Chang HK, et al (2016) A room temperature Na/S battery using a β″-alumina solid electrolyte separator, tetraethyleneglycol dimethyl ether electrolyte, and a S/C composite cathode. Journal of Power Sources 301: 332–337.
2. Beevers C A, Brohult S (1936) The Formula of "β Alumina", $Na_2O·11Al_2O_3$. ZeitschriftfürKristallographie-Crystalline Materials 95(1–6): 472–474.
3. Fanjat N, Lucazeau G, Bates J, et al (1989) Dynamics of Na^+ in a single crystal of Na-beta"-Al_2O_3. Physica B Condensed Matter 156:342-345.
4. Barison S, Fasolin S, C Mortalò, et al (2015) Effect of precursors on β-alumina electrolyte preparation. Journal of the European Ceramic Society35(7):2099-2107.
5. Zhang X, Yang Y, Wu YG, Liang S, Shen HF, Sun WZ (2019) Preparation of Na-β″-Al2O3 Electrolyte Material with High Phase Content from One-dimensional Boehmite.Journal of the Chinese Ceramic Society 47(6): 758-763.
6. Wang X (2014) Studies of preparation of β″-alumina powders and β″-alumina ceramic wafers. Central South university, Changsha.
7. Halbach TS, Thomann Y, Mülhaupt R (2008) Boehmite nanorod-reinforced-polyethylenes and ethylene/1-octene thermoplastic elastomer nanocomposites prepared by in situ olefin polymerization and melt compounding. Journal of Polymer Science Part A: Polymer Chemistry 46(8): 2755-2765.

Determination of Unit Cell Parameters of α-Alumina Reference Material

Lin Zhao, Hongsheng Che, Bo Li, and Shuchao Zhang

Abstract

This paper briefly introduces a reference material for the determination of α-phase in alumina powders (aluminum oxide) with X-ray diffraction methods, and focuses on the measurement and characterization of the crystal cell parameters. The α-phase alumina powder reference material was made of boehmite as a precursor through high-temperature calcination, screening and homogenization, and it had no preferred orientation in the detection, and its diffraction peak intensity was high and stable, which was a challenge with other precursor materials. According to the requirements of GB/T 15,000 "Directives for the Work of Reference Materials", the statistical method was adopted by 8 testing institutions to complete the determination of the crystal cell parameters. After evaluation and statistical analysis, the crystal cell parameters of the alumina reference material were determined to be: $a = 0.47592 \pm 0.00002$ nm, $c = 1.29921 \pm 0.00008$ nm, which met the Joint Committee on Powder Diffraction Standards (JCPDS) standard card 00–043-1484.

Keywords

X-ray diffraction • α-Al_2O_3 • Cell parameters • Reference material • Detection method

Introduction

Alumina is widely used in chemical industry, medicine, ceramics, electronics and other fields. At present, there are more than 10 kinds of homogeneous crystals of alumina, the common ones are β, δ, η, θ, κ, α, etc. X-ray powder diffraction methods can be used to qualitatively/ quantitatively analyze the composition of the different phases. As early as 1964, Visser and deWolff [1] proposed the use of α-Al_2O_3 (corundum) as an internal standard for quantitative phase analysis by powder diffraction method. The use of α-Al_2O_3 as the reference material has the following advantages: Isotropy, obtaining strong spectral lines in a wide d-space range, good chemical stability, and obtaining powder samples with the desired size and morphology in commercial products easily.

The α phase content of alumina is an important quality parameter for use of alumina in ceramics, catalysts and other industries/applications. The internal standard sample used is an α-Al_2O_3 powder [2–3]. In order to meet the needs of detection and analysis, our company has developed a reference material for the determination of α-Al_2O_3 powder with diffraction method. This paper briefly introduces the reference material and focuses on the measurement and characterization of cell parameters of the alumina reference material.

Characterization

Instrument

The certified value of the alumina reference material was characterized by X-ray powder diffraction method. The certified value was measured using diffractometers by different makers and of different models, including: The Netherlands PANalytical Empyrean X-ray diffractometer, The Netherlands PANalytical X 'pert PRO X-ray diffractometer, the Japan Rigaku MinFlex600 X-ray diffractometer, the Japan Rigaku Ultima IV X-ray diffractometer, Germany Bruker D8 ADVANCE/YQ03 X-ray diffractometer.

L. Zhao · H. Che (✉) · B. Li · S. Zhang
Zhengzhou Non-Ferrous Metals Research Institute Co. Ltd of CHALCO, Zhengzhou, 450041, China
e-mail: zyy_lcl@rilm.com.cn

© The Minerals, Metals & Materials Society 2023
S. Broek (ed.), *Light Metals 2023*, The Minerals, Metals & Materials Series,
https://doi.org/10.1007/978-3-031-22532-1_28

Parameters

For determining the cell parameters the detection operated with a Cu target, a scanning range of 129°-140°, a step size of 0.008°, 50 s/step, the divergence slit at 1° and the scattering slit at 1°, a receiving slit of 0.5 mm, an acceleration voltage of 45 kV, and an electrical current of 30 mA. The scanning range was 5°-75° for the phase identification.

Sample

Alumina Reference Material

α-Al_2O_3 can be prepared from a variety of precursors and with different processes. It is more common to thermally decompose industrial aluminum hydroxide and gibbsite precursors [4]. However, these products are not ideal as reference materials for diffraction detection. When the calcination temperature is low, there will be undecomposed transitional alumina phase. When the calcination temperature is high, although it is pure α phase, the product is usually composed of coarse particles, and the morphology is mostly flaky. Due to the influence of crystal habit, these powders will show a preferred orientation, which is not desired when examined by powder diffraction.

After studying the influence of precursor types, particle size, morphology, calcination process and so on, α-Al_2O_3 reference material for determination by powder diffraction method has been developed under the optimum production equipment and technological parameters, through thermal decomposition, sieving and homogenization at 1450°C with a specific boehmite as precursor.

Characterization

In order to eliminate the influence of instrument zero displacement and deviation between sample and goniometer center, a high purity silicon powder (99.99999%) and alumina reference material were mixed at a ratio of 1:3, and the obtained sample was analyzed by X-ray diffraction.

Because of the crystal structure relaxation [5], there were some new reaction products and absorbed substances (usually water) on the surface, so there was some kind of disordered amorphous surface layer in the crystal material, and the amorphous composition cannot be measured by X-ray diffraction. Although the alumina reference material was calcined at high temperature, it still absorbed water due to its high specific surface area. It was found in the experiment [6] that the reference material was placed in different environments for different periods of time, and it would absorb water in different degrees, resulting in a decrease in the intensity of the diffraction peak, and this was more obvious with the increase of water content. Even if the sample was placed in a dryer, the moisture change was not significant, but its diffraction peak intensity was still lower than the newly prepared sample. Therefore, alumina reference material need to be dried at 110°C ± 5°C for 2 h before X-ray powder diffraction test.

Results and Discussion

Phase Identification

Figure 1 shows the X-ray diffraction pattern of the reference material, the existing crystal phase is shown in Fig. 1a, and the crystal plane index is indicated in Fig. 1b. It can be seen that the alumina reference material is comprised of only α phase, which was consistent with the phase identification by the testing institutions.

Determination and Calculation of Cell Parameters

Before the determination and calculation of cell parameters, the reliability of the pattern index should be evaluated [7]. The evaluation was based on the M_{20} factor, proposed by deWolff [1], and the F_N parameter, proposed by Smith [1], which can be calculated according to formulas (1) and (2).

Fig. 1 X-ray diffraction pattern of alumina reference material

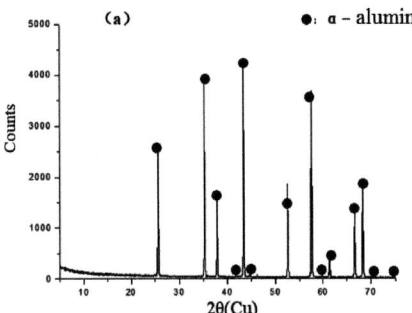
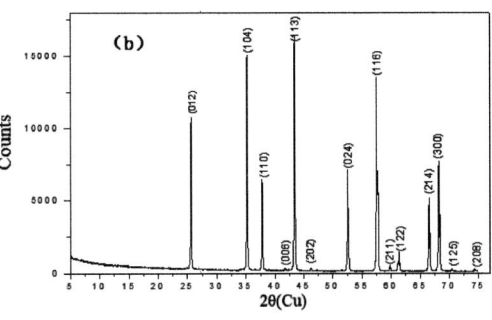

$$M_{20} = \frac{Q_{20}}{2\bar{\varepsilon}N_{20}} \qquad (1)$$

Here,

Q_{20} is the Q value of the 20th diffraction line observed.

N_{20} is number of Q values from Q to Q_{20}.

$\bar{\varepsilon}$ is Q Average deviation between observed value and calculated value

$$F_N = \frac{1}{|\Delta 2\theta|} \times \frac{N}{N_{pass}} \qquad (2)$$

Here,

$|\Delta 2\theta|$ is the absolute deviation between the observed value and the calculated value of 2θ.

N is the diffraction number. It was fixed as 30 here.

N_{pass} is the number of possible diffraction lines before the nth observed diffraction line.

According to the calculation, most $M_{20} > 20$, $F_{30} > 10$, and the index results of scanning patterns were reliable.

In order to obtain a more accurate and stable d value, scanning was conducted at a high angle and the range of 2θ was $129°$ to $140°$. The obtained characteristic peaks of (2012) and (146) were calculated according to Formula (3). Figure 2 showed the X-ray diffraction pattern for determining cell parameters.

$$\frac{1}{d^2} = \frac{4(H^2 + K^2 + HK)}{3a^2} + \frac{L^2}{c^2} \qquad (3)$$

Here, d is the interplanar spacing(Å), a、c are the lattice parameters(Å), and H、K、L are indices of crystal plane.

Uniformity of Reference Material

The boehmite used as a precursor for the alumina reference material was prepared by a hydrothermal method in high pressure reaction kettle with good uniformity. During

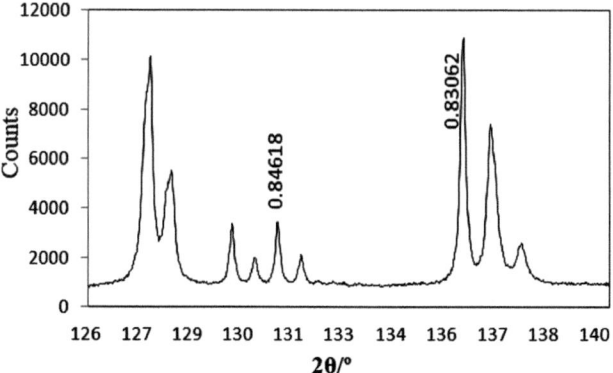

Fig. 2 Cell parameters of alumina reference material by X-ray diffraction

high-temperature calcination, boehmite was placed in several corundum crucible and homogenized, so that the developed alumina reference material had a good uniformity. The inspection process was divided into two steps. First, 5 small samples were taken equidistantly along the longitudinal direction of the reference material of 20 kg in two large bags before packaging, the X-ray pattern of these 10 samples was scanned and calculated, and the t-test method was used to carry out the initial uniformity inspection. Then the samples were packaged according to the specification of 50 g/bottle, and 20 bottles were randomly selected and numbered in sequence, the X-ray pattern was scanned and calculated, and the uniformity of the calculation results was rechecked by the variance F method. After inspection, the uniformity of alumina reference material was qualified.

Characterization

In accordance with GB/T 15,000 "Directives for the Work of Reference Materials" and YS/T 409 "Specification for certified reference materials of non-ferrous metal product for analysis", a statistical method was adopted for the determination of the cell parameters of the alumina reference material. Test samples (silica powder: alumina = 1:3) were uniformly provided by our laboratory and sent to the 8 testing institutions invited to conduct X-ray diffraction pattern scanning according to the defined testing parameters, and 4 patterns were provided from each institution. The cell parameters were calculated in our laboratory according to the crystal plane spacing and the corresponding crystal plane index marked on the pattern.

All the cell parameter results were tested for outliers, normality and equal accuracy. Finally, the qualified data were evaluated to obtain the average value of the cell parameter of the alumina reference material. Meanwhile, the uncertainty was evaluated with the uniformity test. The results and process parameters are shown in Table 1.

The characterization results of the alumina reference material are in accordance with the values in the Joint Committee on Powder Diffraction Standards (JCPDS) standard card 00–043-1484 (where a = b = 0.47592 nm and c = 1.29920 nm).

Comparison with NIST SRM 676a

This Standard Reference Material NIST (SRM) 676a consists of an alumina powder (corundum structure) intended primarily for use as an internal standard for quantitative phase analysis using powder diffraction methods QPA. The first generation of NIST SRM 676 was certified in 1992. Later, with the continuous improvement of alumina

Table 1 Statistics of unit cell parameter certified value and process data

		Unit cell parameter a	Unit cell parameter c
	Certified value /nm	0.47592	1.29921
	Sum of between-unit variation Q1	2.04×10^{-9}	3.157×10^{-8}
	Sum of within-unit variation Q2	3.64×10^{-9}	4.306×10^{-8}
	The variance between-unit homogeneity S_1^2	5.45×10^{-12}	1.95×10^{-10}
	Single measurement standard deviation	7.08×10^{-6}	4.77×10^{-5}
	Total mean standard deviation $S_{\bar{x}}$	2.50×10^{-6}	1.69×10^{-5}
	Stability uncertainties u_s	0.00001	0.00003
	Uncertainty u_c/nm	0.00001	0.00004
	An expanded uncertainty U/nm	0.00002	0.00008
	Number of data groups	8	8

an expanded uncertainty $k = 2$.

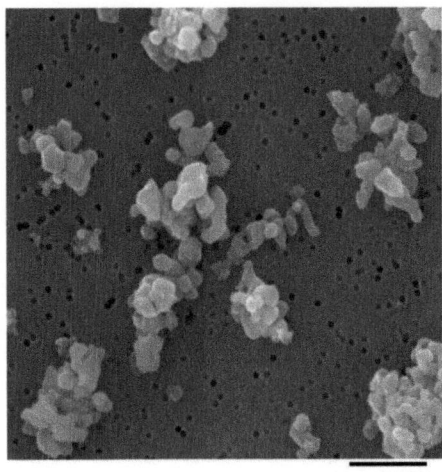

2μm 10000X

Fig. 3 SEM morphology of SRM 676a α-Al$_2$O$_3$ reference material (×10,000) [5]

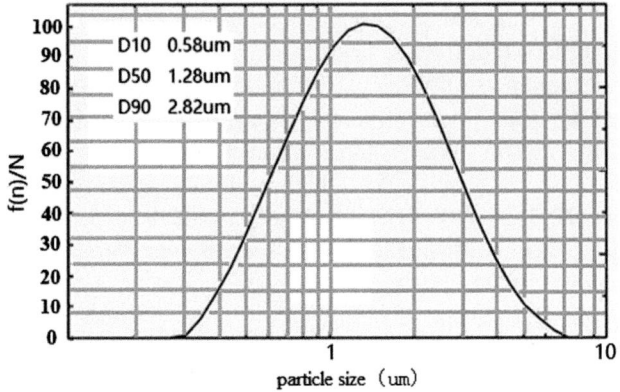

Fig. 4 Laser particle size distributor image of SRM 676a [5]

production and the in-depth development of powder diffraction technology, the second generation SRM 676a was derived [5]. The developed alumina reference material was compared with that of NIST SRM 676a.

Figures 3 and 4 show the morphology and laser particle size of SRM 676a and the reference material developed in this study. Preparation of SRM 676a uses alum as a precursor, and is decomposed at 1400°C and dispersed by jet grinding to get the product. D50 was 1.28um, SPAN = 2.66.

Figures 5 and 6 show the morphology and laser particle size of the developed α-Al$_2$O$_3$ reference material. Without crushing and dispersion treatment, it was directly taken from the product under 60 mesh after high-temperature calcination for analysis. The D50 was 1.27um and the distribution width SPAN = 1.51. Compared with SRM 676a reference material, it had better dispersibility, narrower particle size distribution and more consistent morphology, which can well satisfy the quantitative analysis of α phase content in alumina by X-ray diffraction.

Shaft samples (such as spherical) are ideal powders for diffraction analysis. By grinding products with good sphericity and with small granularity (as shown in Fig. 7) can be obtained. However, during grinding the grain boundary surface of the sample can be destroyed due to the external forces, and the intensity of the diffraction peaks is reduced [6]. Therefore, the grinding process was not applied in the preparation of the reference material. Although the reference material retained its original appearance it did not achieve the desired sphericity, there was no preferred orientation in the detection, and the diffraction peak intensity was high and stable, meeting the needs of X-ray powder diffraction detection.

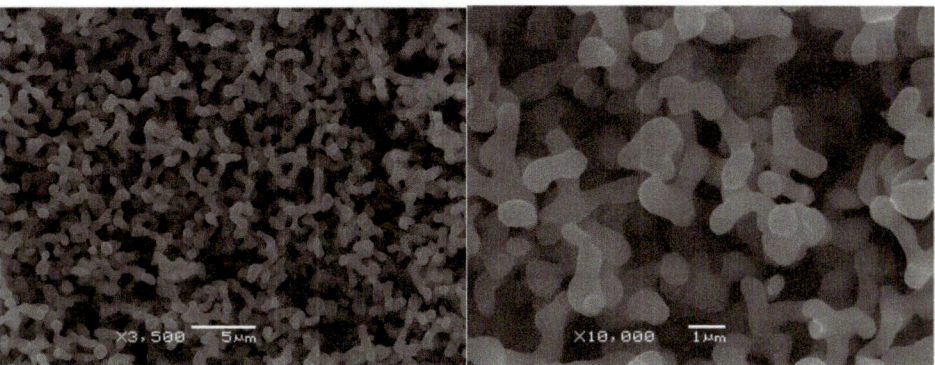

Fig. 5 SEM morphology of the α-Al₂O₃ reference material developed

Fig. 6 Laser particle size distribution of the α-Al₂O₃ reference material developed

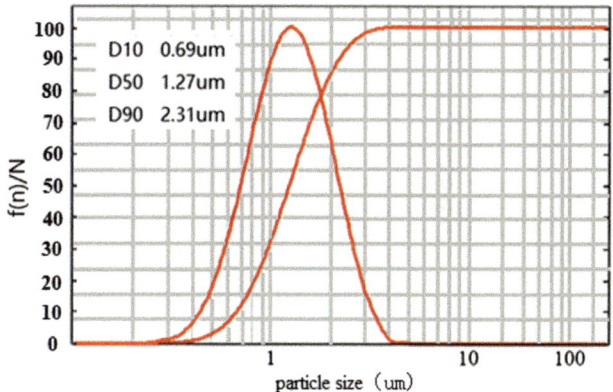

Fig. 7 SEM morphology of alumina reference material milled by mortar in laboratory

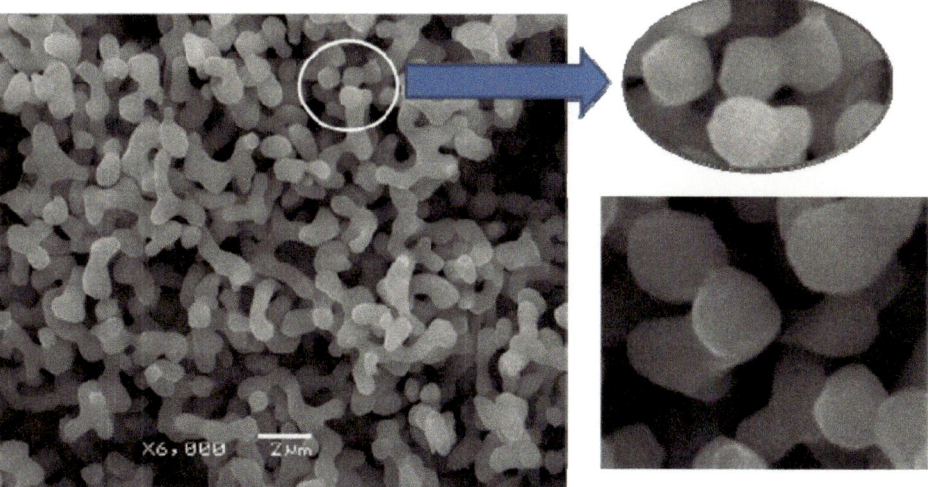

Conclusions

An α phase alumina powder reference material for diffraction method was prepared from a boehmite precursor through high-temperature calcination, screening and homogenization. X-ray diffraction confirmed that only α phase existed in the reference material produced.

According to the guidelines of GB/T 15,000 reference material, the cell parameters of alumina reference material $a = 0.47592 \pm 0.00002$ nm, $c = 1.29921 \pm 0.00008$ nm were calculated by applying statistical methods. These findings are in accordance with the joint Committee on Powder Diffraction Standards (JCPDS) standard card 00–043-1484. The reference material had no preferred orientation and had high diffraction peak intensity. Compared with NIST SRM 676a, the developed reference material had narrower particle size distribution, good dispersion and consistent morphology, which well met the application of X-ray powder diffraction method for quantitative analysis.

Acknowledgements The authors acknowledge the financial support of the Standardization Administration of China (S2021011).

References

1. Ma L D (2004) Modern X-ray polycrystalline diffraction - experimental technology and data analysis. Chemical Industry Press, Beijing.
2. China National Standardization Technical Committee of non-ferrous metal (2009) Administration of China. Chemical analysis methods and determination of physical performance of alumina-Part 32:Determination of α-alumina content by X-ray diffraction: GB/T 6609.32–2009. Standards Press of China, Beijing. 2009:5
3. China National Standardization Technical Committee of non-ferrous metal (2015)
4. Determination of α-alumina content in calcined alumina by X-ray diffraction:YS/T 976–2014. Standards Press of China, Beijing
5. Chen SF, Huang J, Li JW (2006) Development trend and current market situation of non-metallurgical alumina. Refractories 40(3): 225-230.
6. Cline JP, Von Dreele RB, Winburn R et al(2011) Addressing the amorphous content issue in quantitative phase analysis: The certification of NIST SRM 676a. Acta Crystallographica Section A 357–367.
7. Zhao L, Li B, Zhang SC, Che HS(2020) Discussion on several problems in the detection of alpha phase content in alumina. Light Metals (3):48-52.
8. Chi GC, Xiao G, Wu Y(2014) Detecting of Fluorite Crystal Cell Parameters by X-ray Powder Diffraction. Geological Survey and Research 37(1):77-80.

Improvement Seminars: Continuous Improvement and People's Engagement to Support Sustainability

Nathalia Martins, Bruna Dias Cabral, Renan William Costa da Cruz, Raphael Costa, Silene Vendrasco, Jaise Carvalló, Gustavo Silva, Guilherme Brazão, and Karina Trindade

Abstract

A robust management system guarantees the stability and predictability of a company's processes. The focus is on continuous improvement, supported by people, in structured methodologies, through a recognition program to leverage initiatives to contribute to sustainable business. Coordinating different facets, integrating areas, evolving its structure, and using Lean principles, focusing on people makes the challenge even more complex. This project presents the technical evolution of a recognition and improvement program in the bauxite and alumina business in Pará, Brazil. In 2022, it had more than 600 improvement projects that generated an engagement of more than 670 people with proven gains in dimensions such as health, safety, environment, quality of life in the work environment, innovation, and current income in the millions level (US$). It indicates a growing potential for the involvement of teams in the development of improvements. The consequence is sustainability in achieving the global company's goals.

Keywords

Engagement • Continuous improvement • Recognition • Management system • Sustainability

N. Martins (✉) · R. Costa
Norsk Hydro, Rio de Janeiro, Brasil
e-mail: nathalia.cordeiro@hydro.com

R. Costa
e-mail: raphael.costa@hydro.com

N. Martins · S. Vendrasco · J. Carvalló
Norsk Hydro, Belém, Brasil
e-mail: silene.vendrasco@hydro.com

J. Carvalló
e-mail: jaise.carvallo@hydro.com

B. D. Cabral · G. Brazão · K. Trindade
Hydro Alunorte S.A., Barcarena, Brasil
e-mail: bruna.cabral@hydro.com

G. Brazão
e-mail: guilherme.brazao@hydro.com

K. Trindade
e-mail: karina.trindade@hydro.com

R. W. C. da Cruz
Norsk Hydro, Paragominas, Brasil
e-mail: renan.cruz@hydro.com

R. W. C. da Cruz · G. Silva
Mineração Paragominas S.A., Paragominas, Brasil
e-mail: gustavo.lopes@hydro.com

Introduction

Hydro is a leading industrial company committed to a sustainable future, aiming to create more viable societies and develop natural resources in products and solutions in an innovative and efficient way. Founded in 1905, it is a global company with more than 31,000 employees and is present in 40 countries worldwide. The company's business areas include extraction and refining of Bauxite ore for Alumina production, metal aluminum production, extrusion, and energy (Fig. 1).

The extraction of bauxite and refining of alumina occurs in the state of Pará, in the Amazon region of Brazil. Bauxite is extracted from Mineração Paragominas, one of the largest bauxite mines in the world, located in the city of Paragominas. Every processed product is shipped through a pipeline to Alunorte, the world's largest alumina refinery outside China, located in the town of Barcarena. This business area, Hydro Bauxite & Alumina (Hydro B&A), covers the two operating units and administrative offices of Norsk Hydro Brasil, located in the cities of Belém and Rio de Janeiro.

The system used to conduct the management processes in these units is BABS (Bauxite & Alumina Management System), sharing best practices and based on the company's values; BABS is used as part of the strategy of achieving global sustainability and profitability goals, implementing in the day-to-day processes, principles, rules in use, and tools

© The Minerals, Metals & Materials Society 2023
S. Broek (ed.), *Light Metals 2023*, The Minerals, Metals & Materials Series,
https://doi.org/10.1007/978-3-031-22532-1_29

Fig. 1 Hydro's indicators
(*Source* Hydro.com, 2022)

derived from the *lean manufacturing philosophy*, originating from the Toyota Production System (TPS).

TPS is a management philosophy that seeks to optimize the organization to meet the needs of the customer in the shortest possible time, in the highest quality, and at the lowest cost while increasing the safety and morale of its employees, involving and integrating not only manufacturing but all parts of the organization [2].

All the best practices observed in the application of Lean Manufacturing were adapted to hydro's management system and gave rise to the five principles we used in BABS.

The Bauxite & Alumina Management System approach consists of the practical and daily implementation of its principles and fundamental tools aiming at the predictability and stability of processes to have a world-class performance. Its main focus is the continuous improvement of processes focused on people and eliminating waste.

The five principles of BABS that guide the processes are as follows:

- 1st principle: Standardized work processes
- Principle 2: Well-defined relationships between customers and suppliers
- 3rd principle: Optimized flow
- 4th principle: Dedicated teams
- Principle 5: Visible leadership

In practice, each principle is associated with several tools derived from the Toyota Production System, such as the 5S, PDCA, Kaizen, and Value Stream Mapping.

According to rule no. 4 of the Toyota Production System, improvements need to be made in accordance with a scientific method, under the guidance of a technical leader, and at the lowest possible hierarchical level of the organization. Given this, the 4th principle of BABS, which deals with dedicated teams, seeks to foster the creation of multidisciplinary teams focused on developing improvements and competencies at all levels of the organization.

The practical application of this principle provides the stimulus to the feeling of "process owner" in each person on the team so that they are focused on continuous improvement

to achieve results in accordance with the company's overall objectives and goals. The work of these teams is organized, using methods and tools of the Management System to ensure the sustainability of improvements.

Hydro has implemented a systematic monitoring of the work developed by these teams, and a recognition structure was created to reward and value the effort and dedication of employees even more. Hydro Bauxite & Alumina's recognition program has been named CONECTA since 2019. These are seminars, managed by the Bauxite & Alumina Management System (BABS), aimed at presenting the improvement projects developed by dedicated teams throughout the year and providing awards and recognition for the outstanding teams and their leaders.

Accordingly, this article addresses the technical evolution of this recognition program as a way of engaging people in search of continuous improvement of processes and obtaining real gains for the company, contributing to the achievement of local and global goals and objectives, with the whole process based on the practical application of the five principles of BABS.

Methodology

The need to create a recognition program for dedicated teams was born in 2005, with the creation of the Integrated Seminar on Opportunities for Improvement (SIOM), in Alunorte. At the time, the seminar did not cover all areas of the company, and the teams presented their projects to a local jury responsible for choosing the highlights. The history of improvements and records of the event were not maintained satisfactorily, compared to the present day, due to the little development of communication networks and technology.

In 2014 the initiative arrived at Mineração Paragominas with the SOMAR Award (Seminar of Opportunities for Improvement of Areas). Despite having the same objective as SIOM, the SOMAR Award followed a dynamic created for the reality of the place. The records were already more consolidated at the time but still without standardization.

In 2017, the SOMAS (Seminar of Opportunities for Improvement of Support Areas) was created in Belém to recognize the improvement projects developed by the support areas that are part of Norsk Hydro Brasil (NHB), such as Supplies, Legal, HSE, Compliance, among others. At the time, it was possible to keep records of the event with the aid of cloud computing.

As an initial part of a standardization journey in 2019, all Hydro B&A employees could vote for the name that would be used in all units for the Hydro B&A Improvement Seminar, and the name chosen was CONECTA. This initiative allowed to unite other enterprises and also a gradual process of integration, standardization, and monitoring of the evolution of the indicators of the success of events.

For this study, the evolution of the numbers available on record since the year 2005 was analyzed. However, focusing on the results presented as of the creation of CONECTA, having as initial base the works shown since 2020 with the creation of CONECTA's dedicated team.

For obtaining and governing CONECTA information, systems were used for document storage, analysis, and visualization of data, such as SharePoint, Microsoft Forms, Microsoft Power Automate, Microsoft Excel, and Power BI. Through these digital tools, it was possible to evaluate the technical evolution of the event's results.

To assertively follow up on the event data, in 2020, was implemented as a test, the form of the capture of gains, made in Microsoft InfoPath, which allowed improve the accounting and visualization of all the gains obtained with the improvement work presented in CONECTA, started in 2019, but in an exploratory way and with little data.

The continuous improvement and application of lean manufacturing principles and tools led to the evolution in the management and conduct of the seminars. They allowed the analysis of the results presented in this article.

Results

Aiming at the standardization of improvement seminars in Hydro's B&A's business unit, the Management System team has built a dedicated team to bring together and standardize best practices between the three units and ensure a process of continuous improvement with each CONECTA cycle. This dedicated team is formed by a representative of the Paragominas Mining Management System team, one from Alunorte and one from NHB.

The standardization process began in 2019, with the change and unification of the event's name, starting to be called CONECTA. For this decision, a voting form was put together and available to all employees of the B&A business unit. This evolution of entries submitted can be seen in the chart below (Fig. 2):

Fig. 2 Evolution of registrations in improvement seminars

The year 2021 marked a record of improvement projects enrolled in CONECTA, which also generated a greater need for the Management System team to ensure that the projects and results presented are valid. This need allowed an advance in the standardization of the stages of the event until the presentation. In 2021, this recognition process had the following steps improved:

- Review of the procedure unifying the voting rules and weights of the notes of the works according to their category;
- Single-form sign-ups using systems such as Microsoft Forms;
- Validation of the projects registered, using criteria of the CONECTA regulation of each unit;
- Filling out the earnings capture form in a single system (Microsoft InfoPath);
- Validation of the gains presented, involving representatives from various company areas, validated the information, such as Tecnologia, IT, Controllership, Health, Safety, and Environment, among others.
- Finalization of presentations in standard format in Microsoft PowerPoint;
- Presentations at CONECTA improvement seminars in Paragominas Mining, Alunorte and NHB;
- Recognition and award of outstanding improvement projects of the year.

As the standardization process progressed, it was possible to retain and maintain all records regarding team participation in CONECTA Seminars in a single SharePoint database in 2020. Based on this, and despite a pandemic scenario in COVID-19, it is possible to see maintenance in the number of participants between 2020 and 2021, as illustrated in Fig. 3, in absolute numbers.

With the maintenance of the number of participants and the validation system of the papers increasingly robust, it was also possible to notice an increase in the number of

Fig. 3 Absolute numbers of participation in CONECTA

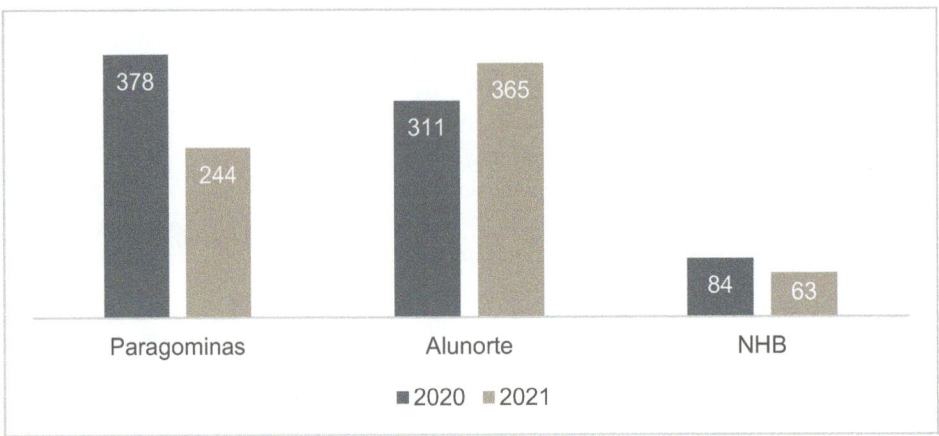

Fig. 4 Absolute numbers of participation in CONECTA

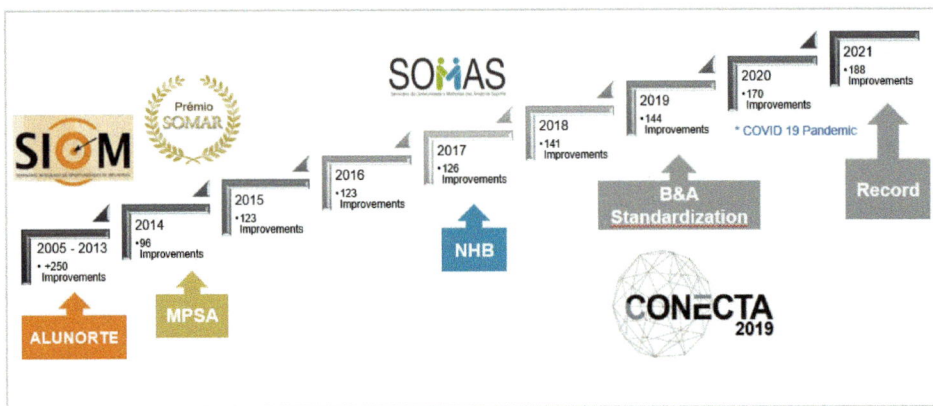

papers presented in 2021, in relation to the year 2020, in absolute numbers. This evolution is also increasing when compared to the years before standardization. Such evolution is explained in the following figure (Fig. 4).

In addition, in 2020, 1224 prizes were delivered, and the following year 1970 awards were awarded, divided into three categories:

- participation awards: provided to all team participants with submitted papers;
- 2nd place awards: delivered to all participants of the teams ranked 2nd place during the presentation of the seminars;
- 1st place awards: provided to all participants of project teams ranked 1st in CONECTA seminars.

In addition, to these groups, leaders, judges, and support teams are also recognized with awards. Certificates and honorary plaques are also delivered to the teams to recognize their participation in the seminars.

With the implementation of a unique form for capturing the gains in 2021, it was possible to accurately measure and visualize all the real gains obtained from the improvement projects presented in CONECTA. The gains were classified in the spheres of Health, Safety and Environment, Financial Gains, Gains in Performance, and Gains in Innovation and are represented in the figures below Figs. 5, 6, 7.

The construction of these boards created another moment of recognition for the teams participating in CONECTA seminars. In 2021, a meeting was held with the presence of the EVP of Hydro Bauxite & Alumina and directors from all areas, in which a team representative had a moment of positive exposure in which he could explain the improvement he has developed and be recognized by it.

Recognition directly impacts the productivity of the employee and the team because he must know that the results of his work have positively impacted the business results and feel fulfilled by it. Data from the year 2015 from the consultancy Towers Watson indicates that employee recognition leads to a more than 60% increase in team engagement. Public recognition—which is visible to the rest of the company—contributes to this and generates a positive effect by reinforcing efficient practices and behaviors [6].

Companies that invest in an effective recognition program can increase engagement and have 31% fewer voluntary layoffs [5].

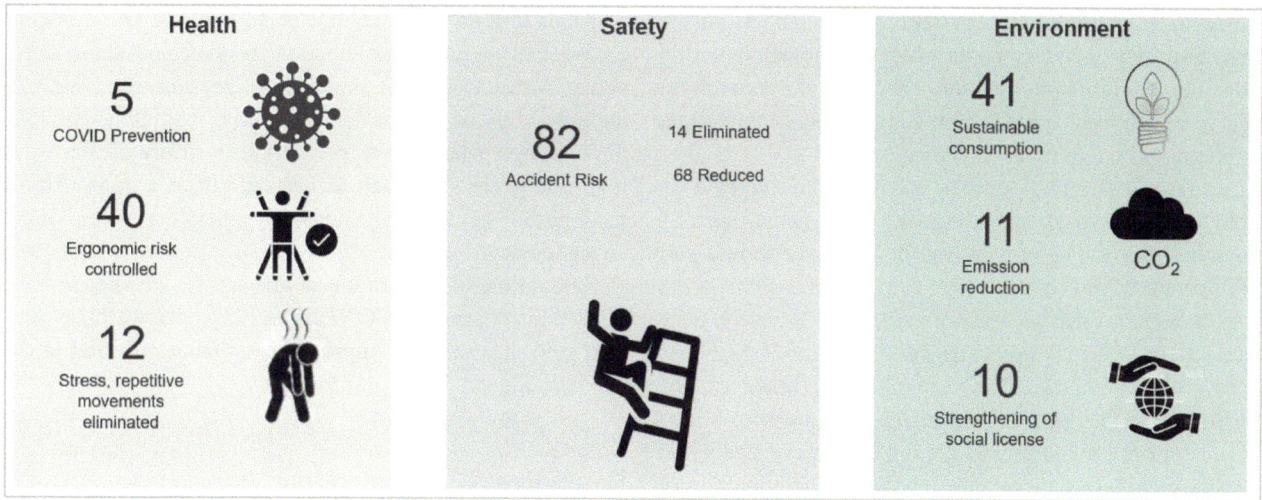

Fig. 5 Numbers of works with gains evidenced in Health, Safety, and Environment

Fig. 6 Gains in Performance and Financial

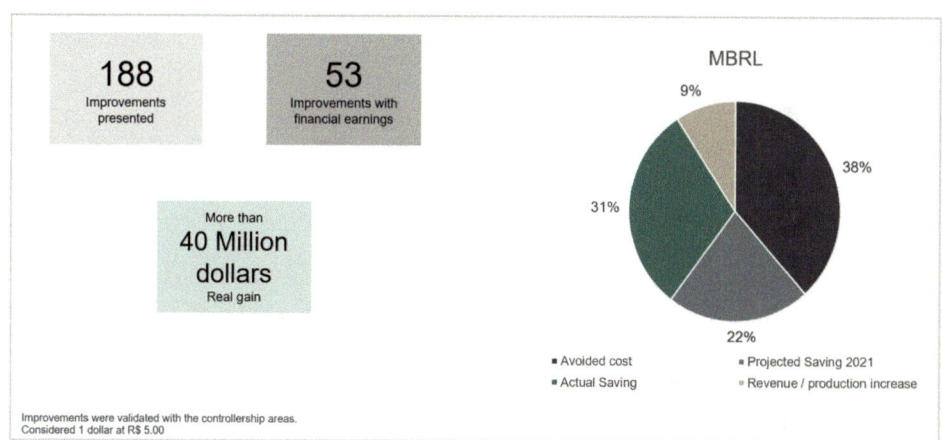

Fig. 7 The number of studies with gains evidenced in different thematic axes

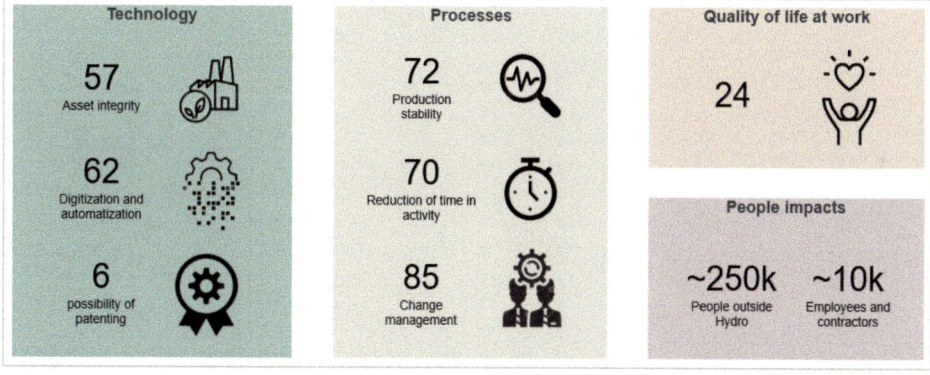

It is undeniable that all this work, which reflects on the gains, now evidenciable, contributes to innovation within Hydro. In addition, recognition programs are proven to improve employees' organizational climate and motivation, stimulating productivity and innovation.

During CONECTA, we can measure the increase in employee engagement, including on social media, such as LinkedIn. It's a source of pride for the employee to be a part of this moment.

All these items can be contemplated when we analyze the success story of the Hydro Bauxite & Alumina (CONECTA) Improvement Seminars. The recognition is based on the valorization of the company's primary asset, people, guided by the company's values and aiming to achieve goals and stimulate continuous improvement based on the principles of the Management System. All this builds an increasingly competitive company with employees increasingly engaged in search of the reach of their targets.

Conclusion

Analyzing the technical evolution of Hydro B&A's Improvement Seminars shows that the need for a structured recognition system has always been part of the company and is critical to engage employees further to achieve the goals.

It was also observed that the evolution of the seminars followed a growing curve in parallel with the growth of available technologies. At the same time, applying tools and concepts derived from the lean manufacturing philosophy became possible, the basis of the company's Management System.

Analyzing the statistics related to CONECTA, it is concluded that people's participation in dedicated teams competing with CONECTA has been increasing, motivated by the possibility of recognition. Similarly, with the creation of the earnings panel, it was possible to measure all the work involved in the dedicated teams and create a new form of recognition at a higher level, generating even more engagement.

At the time of preparation of this article, we already have 1032 registrations for CONECTA 2022, and we have, as a next step, the gains of improvements being reflected in the annual budget.

References

1. ABOUT HYDRO (2022).) Key facts. Available in: https://www.hydro.com/en/about-hydro/this-is-hydro/facts/. Accessed August 25, 2022.
2. GHINATO, P. (2000). Published as 2nd. Cap. In: Book Production & Competitiveness: Applications and Innovations, Ed.: Adiel T. de Almeida & Fernando M. C. Souza, Edit. of UFPE, Recife.
3. NEGREIROS RR et al. (2020). Application of Toyota Production System Tools in a Port Logistics Company. Paper presented at the XL National Meeting of Production Engineering, Foz do Iguaçu, Paraná, 20–23 October 2020.
4. SPEAR S, HK BOWEN (1999). Decoding the DNA of the Toyota Production System. September/October.
5. OSORIO A (2022). 8 actions to create an effective professional recognition program. Available in: https://orienteme.com.br/blog/programa-de-reconhecimento/. Accessed August 25, 2022.
6. ARAUJO R (2022). Professional recognition: understand the importance and how to improve. Available in: https://www.qulture.rocks/blog/reconhecimento-profissional. Accessed August 25, 2022.

Turning Bauxite Residue to Metal Adsorption Materials Through a Low-Cost Approach

Hong Peng, James Vaughan, Shengchun Ma, Sicheng Wang, and Xinyu Tian

Abstract

A low-cost approach to bauxite residue (BR) utilization has not been achieved due to the complicated mineralogy. In this study, BR has been treated by dilute acid to reduce the alkalinity of residue by dissolving sodalite phase. The acid washed BR samples are of high iron grade (>80% Fe_2O_3) with low sodium content (<0.3% Na_2O) which can be potentially used as ironmaking feed. The washed acid solution was treated by caustic NaOH to synthesise the zeolite-like adsorption materials. The materials have been used for adsorption of copper and lithium metal salts. The Cu adsorption performance of the synthesised product shows the exhibits fast loading kinetics as within 15 min over 99% of adsorption efficiency is observed, compared with commercial grade zeolite LTA.

Keywords

Bauxite residue • Acid neutralization • Zeolite LTA • Metal adsorption • Sodalite

Introduction

The production of alumina has a major dependency on the use of bauxite ore through the Bayer process. This process generates a waste slurry known as bauxite residue (or red mud), approximately 1–2 t of bauxite residue is generated for every t of alumina [1]. About 2.7 Bt of residue is contained within disposal areas worldwide, expected to increase by 120 Mt per annum [2]. Bauxite residue is alkaline due to the desilication product in the residue with aqueous slurries having pH values of 10.5 even after 1000 times washing [3].

In addition to the high pH, another environmental concern is the mobilisation of metals residual radioactivity [4]. BR contains multiple elements classified as major in the form of: Fe, Ti, Al, Si and Na. As well as minor in the form of: Sc and rare earth elements (REEs). Much research effort has focused on direct utilization of bauxite such as in cement production or the extraction, through hydrometallurgy or pyrometallurgy of a single to two elements of interest such Fe or Sc due to the high price of its high purity oxide [5–8]. Nevertheless, there is still a need for delving into this research topic due to the following reasons. The effect of acid treatment on BR mineral composition and the treatment mechanism have not been thoroughly understood. Besides, some studies applied complicated and energy-consuming procedures, such as high-temperature calcination and concentrated acid to modify BR. These could act as a barrier to develop cost-effective approaches to BR utilization.

However, to truly achieve zero-waste in alumina refineries, there is a need to develop a process that utilizes all the components within bauxite residue with low cost and energy consumption. From all major and minor constituents, this method may also more economical as the extraction of single elements have high energy requirements (ironmaking) which cannot be supported by the production and sale of the single element, and existing BRs may require chemical and physical processing steps to meet ironmaking feed requirements. The object of this research work is to demonstrate a proof of concept for the complete utilization of BR in a simple and low-cost process as shown in Fig. 1. The aluminium and silica gel were recovered from the filtrates of acid-washed BR then was used for zeolite synthesis by caustic precipitation. Both hematite and anatase were concentrated in acid-washed BR which can be potentially used as steelmaking clinker. The synthesized zeolite samples were tested for use as metal adsorbents. The results show the fast kinetics for copper adsorption from aqueous solution. The main advantages of this process can be summarised as follow: (1) minimal acid consumption as acid is effectively

H. Peng (✉) · J. Vaughan · S. Ma · S. Wang · X. Tian
School of Chemical Engineering, The University of Queensland, Brisbane, Australia
e-mail: h.peng2@uq.edu.au

© The Minerals, Metals & Materials Society 2023
S. Broek (ed.), *Light Metals 2023*, The Minerals, Metals & Materials Series,
https://doi.org/10.1007/978-3-031-22532-1_30

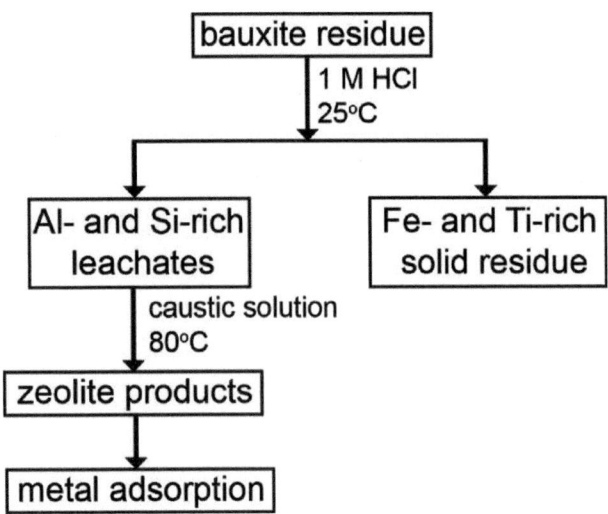

Fig. 1 A conceptual process flow-diagram for low-cost BR utilization

used in the leaching step; (2) selective leaching with limited leaching of Fe and Ti; (3) production of zeolites as adsorption materials; The research could provide an understanding of utilizing BR as an applicable adsorbent for environmental purposes, and how to obtain a cost-effective approach of using BR.

Experimental

Materials and Reagents

Bauxite residue samples were synthesized from bauxite under Bayer process conditions (150 °C for 2 h with 20% NaOH) [9]. Commercial grade Zeolite LTA powder (Sigma_LTA), sodium hydroxide pellets (98%), copper sulfate pentahydrate ($CuSO_4*5H_2O$, 98 wt. %), concentrated hydrochloride acid (HCl, 37%) and lithium nitrate ($LiNO_3$, > 99%) were sourced from Sigma-Aldrich. Concentrated solutions of up to 12 M NaOH were prepared by dissolving analytical grade chemicals in DI water (< 0.5 µS cm-1). Experimental solutions were prepared by combining concentrated caustic solutions with washed liquor to achieve a synthetic liquor composed of 3 M NaOH.

Acid Wash, Zeolite Synthesis and Metal Adsorption Tests

Bauxite residue samples were washed using 1 M HCl with a solid loading of 100 g/L at room temperature for 30 min using a magnetic stirrer based on previous optimised experimental conditions [3]. The washed solid residue was separated from wash liquor with centrifuge (3000 rpm) for 5 min then dried at 60 °C in an oven overnight for further analysis.

For zeolite synthesis, the separated wash liquor was mixed with 12 M caustic solution with volume ratio of 3:1 at room temperature then the temperature was increased to 80 ° C while stirring on a hot plate for 2 h. The kinetic samples were periodically withdrawn at 15, 30, 60, 90 and 120 min. The solid samples collected were washed with DI water to remove any associated solution before drying at 105 °C for further characterisation. The final 2 h solid samples are used for further metal adsorption test.

For metal salt adsorption tests, 0.2 g of zeolite solids (synthesised_LTA and Sigma_LTA) were added into 25 ml of metal salt solution with initial metal concentrations of 100 mg/L (Li^+ or Cu^{2+}), corresponding to approximately 8 g/L solid loading. The tests were conducted at room temperature (23 ± 1 °C) with magnetic stirring over 120 min. The initial pH value for both solutions after adding solids is under 6 to minimize the alkalinity of zeolites. This pH value adjustment avoids precipitation of metal hydroxide precipitation. Solution samples were isolated from the solids by syringe filtering through 0.22 µm nylon membranes. The samples were diluted (20X) into dilute nitric acid prior to assay by ICP. The residual solids at the end of the experiment were isolated vacuum filtration. These solids were also dried overnight at 105 °C before sending for further characterisation.

Characterization

The solid samples were sent for XRD analysis with a Bruker D8 Advance X-ray diffractometer with a LynxEye detector and Cu Kα irradiation (λ = 1.5406 Å) at 40 kV with a scanning speed of 0.02 deg/s over the 2θ angle range of 5 − 40°. The 2017 PDF database from BRUKER was used for reflection identification. The solid powder samples were also disbursed on the carbon tape and then were coated with carbon. The coated samples were imaged by scanning electron microscopy (SEM, HITACHI SU3500).

Results and Discussion

Acid Washing of the Bauxite Residue

The main chemical compositions of the BR sample were determined by X-ray fluorescence (XRF) as shown in Table 1. The main elements are Fe, Na, Al, Si and Ti which is consistent with other BRs reported where Fe_2O_3 typically comprises the highest mass fraction followed by Al_2O_3, SiO_2 and Na_2O. As shown in Fig. 2, its XRD pattern reveals different crystalline phases which include sodalite ($Na_6(AlSiO_4)_6*2NaOH*2H_2O$), quartz ($SiO_2$), hematite ($Fe_2O_3$) and anatase ($TiO_2$). We did not identify the unreacted boehmite

Table 1 XRF data of initial bauxite residue sample and acid wash residues (wt.%)

Sample	Al_2O_3	SiO_2	Fe_2O_3	TiO_2	Na_2O	LOI
Original BR	19.81	21.0	37.4	2.74	12.3	4.66
Acid wash residue	2.14	2.38	84.8	5.06	0.26	6.21

Fig. 2 XRD patterns of bauxite residue and acid washed sample (H): Hematite; (A): Anatase; (S): Sodalite;(Q): Quartz

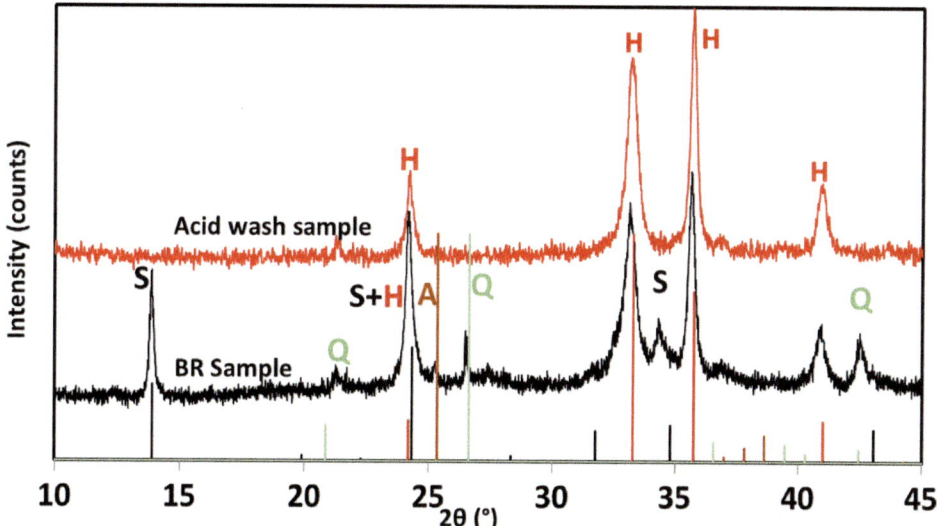

(γ-AlO(OH)) phases as with previous bauxite residue samples [2, 3, 7, 10, 11]. After acid washing at room temperature, sodalite completely disappeared based on the patterns of the residual, while other mineral phases were still present but reacted to some extent (Fig. 2). These observations above suggest that the silica gel does not form under conditions of low temperature with dilute acid concentrations despite of the dissolution of silica-rich sodalite.

The iron grade has been significantly increased from 37 to 85% Fe_2O_3. In the same time, the total sodium oxide content has been significantly reduced from 12% to 0.26%. The washed bauxite residue with much lower alkalinity and high content of iron oxide which can be potentially used as feed for steelmaking industry [8, 12].

Synthesis of Zeolites and Metal Adsorption

When the acid washed liquor contacted with NaOH solution, the Al concentration shows a significantly drop from initial 160 mM to 120 mM resulting from precipitation of aluminosilicate as shown in Fig. 3. The Al concentration keeps decreasing until 90 min, with final concentration of ~ 60 mM. The XRD patterns of precipitates as a function of time were presented in Fig. 4. It can be seen in this figure that amorphous sodium aluminosilicate solid was distinctive from the broad XRD array exhibiting a board hump at 2θ in the range of 15–35°[13] at 30 min sample. For the one-hour sample, zeolite LTA phases appear as the pattern with

decreasing hump intensity for the amorphous phase. For the 2-h sample, the XRD pattern shows a highly crystalline Zeolite LTA phase with minor phase of sodalite. The SEM images of solid samples at time of 15 min, 60 min and 120 min were shown in Fig. 5. The images show that irregular shape of amorphous zeolites formed at 15 min, round shaped Zeolite LTA crystals co-existed with small particles of amorphous zeolites at 60 min, and cubic shaped Zeolite LTA formed at 120 min. The final 120 min samples were collected and dried for metal adsorption test along with commercial grade Sigma LTA. As shown in Fig. 6, the copper adsorption by both zeolites is within 15 min and thus very fast, with the concentration of copper dropping to values under 1 ppm from original 100 ppm. However, for lithium, the adsorption kinetics were slower for both zeolite samples. This is attributed to the lithium ion being strongly hydrated cations, which inhibit the adsorption process [14].

Conclusions

Dilute acid and low temperatures resulted in selective sodalite dissolution against other mineral phases for bauxite residue. Without formation of silica gel, acid-washed BR samples were easily filtered and the resulting solids were greatly enriched in iron (>80% Fe_2O_3) and anatase phases with low content of sodium (<0.3% Na_2O) suitable for the steelmaking industry. In addition, the acid-washing liquid was treated with caustic solution was used for synthesis of

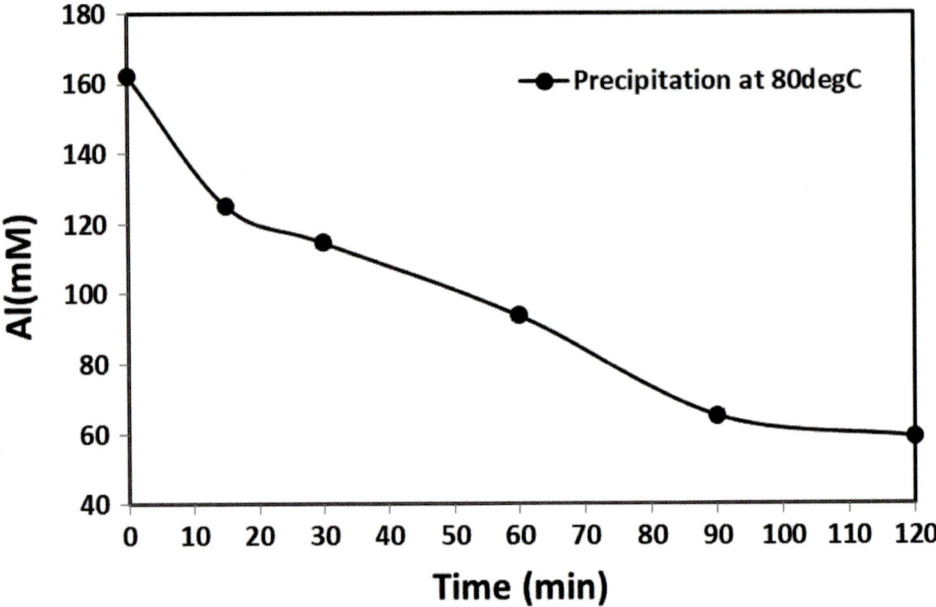

Fig. 3 Al concentration in solution as a function of reaction time during acid leaching liquor treated with 3 M NaOH at 80 °C for 2 h

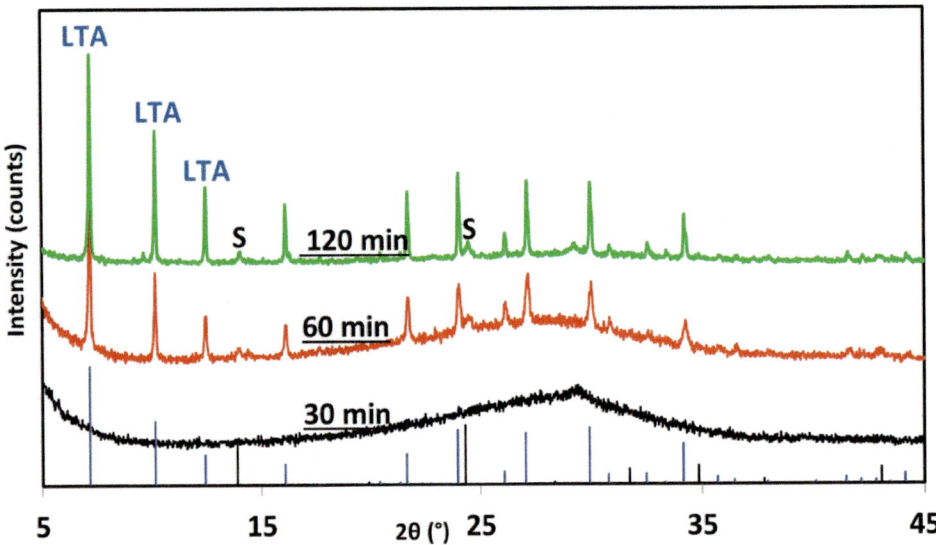

Fig. 4 XRD patterns as a function of reaction for zeolite synthesis. (S): Sodalite;(LTA): Zeolite LTA;

Fig. 5 SEM images of solid products at sampling time of 15 min, 60 min and 120 min

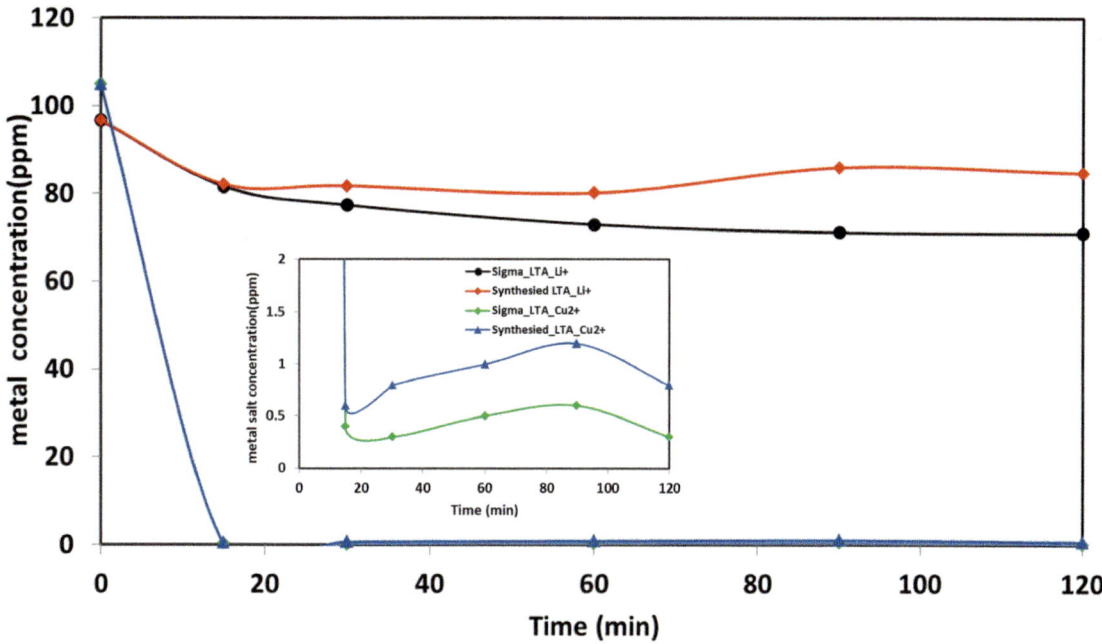

Fig. 6 Li and Cu concentration when treated by Sigma_LTA and Synthesized LTA within testing time of 120 min with initial metal concentration of 100 ppm at room temperature

Zeolite LTA, and tested for use in metal adsorption. In comparison with commercial grade of zeolite LTA, the synthesised zeolite products show the comparable adsorption performance for both copper and lithium. The stepwise design targeted a zero-waste process for the utilization of BR which appears promising for producing high performance adsorption materials and an enriched hematite solid residue from bauxite residue.

Acknowledgements The authors gratefully acknowledge the funding for this research from the Queensland Government through Advance Queensland Industry Research Fellowship (AQIRF014-2019RD2). We acknowledge the facilities, and the scientific and technical assistance, of the Australian Microscopy & Microanalysis Research Facility at the Centre for Microscopy and Microanalysis, The University of Queensland.

References

1. Power, G., M. Gräfe, and C. Klauber, Bauxite residue issues: I. Current management, disposal and storage practices. Hydrometallurgy, 2011. **108**(1): p. 33–45
2. Peng, H., et al., Advanced characterisation of bauxite residue. 2018
3. Peng, H., T. Kim, and J. Vaughan, *Acid Leaching of Desilication Products: Implications for Acid Neutralization of Bauxite Residue.* Industrial & Engineering Chemistry Research, 2020. **59**(17): p. 8174-8182.
4. Kong, X., et al., *Natural evolution of alkaline characteristics in bauxite residue.* Journal of Cleaner Production, 2017. **143**: p. 224-230.
5. Anawati, J. and G. Azimi, *Recovery of scandium from Canadian bauxite residue utilizing acid baking followed by water leaching.* Waste Management, 2019. **95**: p. 549-559.
6. Borra, C.R., et al., *Selective recovery of rare earths from bauxite residue by combination of sulfation, roasting and leaching.* Minerals Engineering, 2016. **92**: p. 151-159.
7. Borra, C.R., et al., *Leaching of rare earths from bauxite residue (red mud).* Minerals Engineering, 2015. **76**: p. 20-27.
8. Klauber, C., M. Gräfe, and G. Power, *Bauxite residue issues: II. options for residue utilization.* Hydrometallurgy, 2011. **108**(1): p. 11–32.
9. Peng, H., S. Peters, and J. Vaughan, *Leaching Kinetics of Thermally-Activated, High Silica Bauxite*, in *Light Metals 2019.* 2019, Springer. p. 11-17.
10. Ahmed, M.H., et al., *Red-mud based porous nanocatalysts for valorisation of municipal solid waste.* Journal of Hazardous Materials, 2020. **396**: p. 122711.
11. Liu, Z. and H. Li, *Metallurgical process for valuable elements recovery from red mud—a review.* Hydrometallurgy, 2015. **155**: p. 29-43.
12. Ning, G., et al., *Large-scale consumption and zero-waste recycling method of red mud in steel making process.* Minerals, 2018. **8**(3): p. 102.
13. Peng, H., D. Seneviratne, and J. Vaughan, *Role of the amorphous phase during sodium aluminosilicate precipitation.* Industrial & Engineering Chemistry Research 2018. **57**(5): p. 1408–1416
14. Ozdemir, O., et al., *Adsorption and surface tension analysis of concentrated alkali halide brine solutions.* Minerals Engineering, 2009. **22**(3): p. 263-271.

Hematite and Anatase Conversion to Magnetic Phases During Reductive Re-digestion of Gibbsitic Bauxite Residue

Paula de Freitas Marques Araújo, Patricia Magalhães Pereira Silva, Andre Luiz Vilaça do Carmo, Fernando Gama Gomes, Adriano Reis Lucheta, Raphael Vieira da Costa, and Marcelo Montini

Abstract

Bauxite residue (BR) is a by-product of the Bayer process for alumina production. The physical and chemical characteristics of BR from gibbsitic bauxite such as the fine particle size distribution and mineralogical phases (hematite, sodalite, goethite, anatase, gibbsite and quartz) prevent its wide use and limit the recovery of elements by conventional physical operations. As an example, the magnetic recovery of hematite from gibbsitic RB is unfeasible. An alternative to enable a magnetic concentration operation would be to obtain higher magnetic susceptibility phases, whose magnetic properties could overcome the processing limitation determined by the fine size of the RB particles. In this work, a Brazilian BR from gibbsitic bauxite processing with an iron and titanium concentration of 47.03% Fe_2O_3 and 5.45% TiO_2 was re-digested in presence of iron(II) sulphate 7-hydrate ($FeSO_4 \cdot 7H_2O$), at high temperature (230°C) and high caustic concentration (370 g/l Na_2CO_3). Magnetite, ulvite (titanium-magnetite) and cancrinite minerals were obtained. The results demonstrated that this route may be promising for magnetic iron recovery, especially if side reactions were minimized. For future studies, one should evaluate the threshold for the conversion when the severity of conditions was diminished, as well as, the assessment of magnetic properties and the evaluation of potential recovery of the magnetic product.

Keywords

Bauxite residue • Iron • Titanium • Magnetite • Recovery

Introduction

Bauxite residue (BR) is the solid mixture remaining after the lixiviation of aluminium minerals from bauxite added to other compounds formed during the Bayer process [1]. Chemical composition, mineralogy, and other characteristics of bauxites led to a classification of bauxite from the standpoint of the Bayer plants, such as gibbsitic, bohemitic, and diasporic bauxite [2]. The leaching conditions (pressure, temperature, and caustic content) of gibbsitic bauxite are milder than the conditions applied to bohemitic and diasporic bauxite. Therefore, the bauxite residue generated in each case has differences [1–3].

Regardless of the bauxite type (i.e., diasporic, gibbsitic, or boehmitic), usually the major constituent of bauxite residue is iron (30–50% Fe_2O_3). Iron is commonly present in two main phases, hematite (α-Fe_2O_3) and goethite (α-FeOOH) [3, 4]. Once that iron is one of the main constitutes of BR, concentration of the iron-bearing minerals is promising for reducing BR inventory [1, 5]. However, the fine particle size (< 100 μm) distribution of the BR evaluated in this study makes it unfeasible for straightforward concentration of iron minerals by conventional mineral concentration methods [6, 7]. For overcoming the particle size limitation, chemical transformation prior to concentration may increase the differentiability among the individual phases enabling concentration processes. Since magnetite has properties that makes it more responsive for magnetic

P. de Freitas Marques Araújo · P. M. P. Silva · A. L. V. do Carmo · F. G. Gomes · A. R. Lucheta (✉)
SENAI Innovation Institute for Mineral Technologies, Tv. Dr Moraes, 78, Belém, PA 6035-080, Brazil
e-mail: adriano.isi@senaipa.org.br

P. M. P. Silva
e-mail: patricia.isi@senaipa.org.br

P. de Freitas Marques Araújo · P. M. P. Silva
Federal Institute of Education, Science and Technology of Pará-IFPA, Belém, PA 66093-020, Brazil

R. V. da Costa · M. Montini
Norsk Hydro Brasil, Av. Gentil BittencourtEd. Torre Infinito, Batista Campos, Belém, PA 54966035-340, Brazil
e-mail: marcelo.montini@hydro.com

© The Minerals, Metals & Materials Society 2023
S. Broek (ed.), *Light Metals 2023*, The Minerals, Metals & Materials Series,
https://doi.org/10.1007/978-3-031-22532-1_31

separation; the synthesis of magnetite from hematite (para-magnetic) present in BR has the potential to enable magnetic iron concentration operations [8].

Magnetite (Fe_3O_4 or $FeO.Fe_2O_3$) was obtained from hematite (α-Fe_2O_3) of diasporic bauxite in reductive digestion conditions, achieved using iron powder as reducing agent [9–12]. Hematite was also transformed in magnetite by reductive re-digestion of bauxite residue from a diasporic bauxite processing [8]. In all studies the reducing caustic medium was resulted from addition of metallic iron or iron (II) sulphate 7-hydrate ($FeSO_4 \cdot 7H_2O$), occasionally calcium oxide (CaO) was used to minimize side reactions and to improve diaspore dissolution [8–12].

It is interesting to note that the bauxite residue studied by Pasechnik and colleagues [8] came from diasporic bauxite, so it has a chemical and mineralogical composition considerably different from the gibbsitic BR, proposed to be studied in this work [6–8]. However, both have in common, hematite phase. The other phases identified in the diasporic bauxite residue studied by the authors were katoite, diaspore, sodium silico—aluminate, bayerite, and chamosite [8]. While in the gibbsitic bauxite residue under study, the phases usually found are Al-goethite, anatase, quartz, gibbsite and sodalite [6, 7]. Regardless of both materials being bauxite residue, they are a complex mixture of different minerals and solids and may result in different products when submitted to severe synthesis condition, as high temperature, high free alkali concentration adding an unconventional reducing agent for a Bayer process standpoint.

In this study, a BR from gibbsitic bauxite was submitted to reductive digestion using $FeSO_4 \cdot 7H_2O$ as reducing agent. Here we show a preliminary investigation of magnetic solids syntheses, as well as, how the phases present in gibbsitic BR behaves in such harsh reaction conditions.

Materials and Methods

Norsk Hydro Alunorte Refinery provided the bauxite residue studied in this work. The refinery is located in the state of Pará, northern Brazil. The study was conducted at the SENAI Innovation Institute for Mineral Technologies (ISI-TM), also located in the state of Pará, Brazil.

Analytical Methods and Characterization

Chemical composition was determined by X-ray fluorescence (XRF). The XRF was carried out in an Epsilon 3^{XLE} / PANalytical spectrometer equipped with Rh ceramic X-ray tube operating at maximum power level of 15 W. The powder samples were pressed. BR sample (0.5 g) were calcinated at 1.000°C for 1 h in an oven for loss on ignition analysis (LOI). The products of the reaction were not submitted to LOI analysis due to the possible presence of magnetite that would be oxidized, gaining mass, which would result in a higher experimental error.

The mineralogy determinations were made by X-ray diffraction (XRD). The XRD was carried out in a divergent beam diffractometer Empyrean/PANalytical (XRD) with θ—θ goniometer and a Co X-ray tube ($K\alpha 1 = 1.78901$ Å) with a fine long focus of 1.800 W, a Fe kβ filter, and a PIXcel3D 1 × 1 area detector, operating in a linear scanning mode (1D), with active length of 3.3473° 2θ (255 active channels). The samples were prepared according to the powder method with backloading preparation.

Morphological micrographs were obtained by Scanning Electron Microscopy (SEM: Vega 3 LMU, Tescan), operated at 20 kV and 10 µA with a focal distance between 8 and 15 mm. The samples were dried at 105 °C and coated by a thin gold layer (Desk V metallizer/Denton Vacuum) before SEM. The punctual chemical analysis was performed with EDX detector, Act-X model, from Inca Instruments.

Particle size distribution (PSD) was determined by laser diffractometry in a Malvern Mastersizer 3.000 using water as dispersant. The refraction coefficient for BR was accepted as 1.568 and for the products, 2.340 due to possible presence of magnetite (Fe_3O_4).

Reductive Caustic Digestion of Gibbsitic BR

The digestion test was carried out in a stirred autoclave reactor (Parr Instrument Company, Series 4520, 1L) at 230°C and 100 rpm. The pressure was consequence of the vapour-liquid equilibrium at the reaction conditions. The digestion time was 90 min after temperature setting.

The reagents were gibbsitic BR, caustic soda solution, and iron (II) sulphate 7-hydrate (Synth, reagent grade > 98%). BR was dried at 100°C for 24 h. Caustic solution at 370 g/l Na_2CO_3 (7 mol/l NaOH) was prepared by diluting a 50% NaOH ACS solution (Sigma-Aldrich). The mass fraction for the reaction was 30% RB and 8% iron (II) sulfate 7-hydrate [9].

After digestion, the autoclave was depressurized and cooled down to room temperature naturally. The contents of the reactor were poured into a polypropylene beaker. After the sedimentation of the solids, the supernatant was siphoned off. A washer procedure in batch was applied where distilled water was fed to the beaker and siphoned off after 24 h of resting time, until pH stabilization at 10–11 (pH indicator strips).

Results and Discussion

The chemical and the mineralogical composition of the gibbsitic BR are given in Table 1 and Fig. 1. The major elements were iron, aluminium, silicon, sodium, and titanium and the mineral associated phases were hematite, sodalite, gibbsite, aluminous goethite, anatase, quartz, and calcite. The results were similar to previous determinations obtained in the past for the BR from Norsk Hydro Alunorte Refinery [6, 7]. Gibbsitic BR is a complex mixture of minerals and solids and when submitted to high temperature, to strong caustic solution (high free alkali concentration) and adding a new reagent that chances the redox potential of the system one would expect many reactions.

Immediately after autoclave reactor quenching, it was observed a fast settlement of the solids and a prompt response to the approach of a magnet bar. After 24 h of the last washer, many suspended brown/reddish solids were still observed, while the dark solids settled in the bottom of the beaker. Figure 2 presents (a) initial response to a magnet and (b) different colours of reaction products. Figure 3 presents the original BR, the underflow solids (US), and the suspended solids (SS) after drying.

The strong caustic environment makes possible the oxidation of the reducing agent, $FeSO_4 \cdot 7H_2O$ to $HFeO_2^-$, as consequence, the conditions to convert hematite from reducing Fe^{3+} into Fe^{2+} are favourable. In terms of crystal chemical structure, magnetite has an inverse spinel structure $(Fe^{3+}(Fe^{2+}Fe^{3+})O_4)$ where half of the Fe^{3+} occupies tetrahedral sites and the other half, together with Fe^{2+} occupy octahedral sites [13]. Therefore, it is not required a reduction of all iron present in the system. The mechanism proposed by Li and colleagues [9–12] is, regardless of the reducing agent, first occurs the generation of $HFeO_2^-$ and then diffusion of this specie to hematite surface, where the reaction takes place and magnetite is formed [8–12]. Table 2 presents the chemical composition of the products. It is remarkable the difference in iron content between SS and US, as well as

aluminium, silicon, and sulphur contents. Side reactions with sulphur have not been mentioned in the literature found. Based on the XFR results, it is possible to suggest that sulphur from $FeSO_4 \cdot 7H_2O$ took part in the reaction once that it was incorporated in the products.

Figure 4 shows XRD diffractograms of RB, SS, and US. Hematite and Al-goethite were observed in all materials, which confirms that a completely conversion of hematite is difficult [8–12]. Sodalite from BR was transformed to cancrinite, and it was present in SS and US. Sodalite and cancrinite have the same general chemical formula, $Na_6[Al_6Si_6O_{24}] \cdot 2NaX \cdot 6H_2O$, where X can be OH^-, Cl^-, NO_3^-, $1/2CO_3^{2-}$, or $1/2SO_4^{2-}$; but their crystal structures are different. Sodalite is cubic and cancrinite is hexagonal [14, 15]. High temperature, calcium (Ca) and carbonate (CO_3^{2-}) promote the transformation of sodalite in cancrinite [11, 14]. As have been reported in several previous work, reductive Bayer digestion, resulting from iron or iron sulphate addition, replaces the role of Ca in improvement of diasporic digestion [9, 11, 16]. Therefore, the present results suggest that the iron from iron sulphate is playing the same role as Ca in cancrinite formation, and the anion sulphate is replacing carbonate in the cancrinite structure. Anatase was not detected in the products in spite of high titanium content in SS and US. The reaction between titanium-containing phases and other elements depends on temperature, reaction time, and the free alkali concentration in the Bayer process, [16–19]. In special, anatase suppress the dissolution of diaspore by formation of a sodium titanate covering on the surface of the diaspore particle and it also reacts to iron and silicon forming Ti-/Si-bearing solids [16, 17]. Besides this reaction, titanium-magnetite or ulvite (Fe_2TiO_4) was one of the products of the reaction between anatase and $FeSO_4 \cdot 7H_2O$ in synthetic Bayer liquor [10]. Ulvite was detected in US but any titanium bearing mineral was identified in SS in spite of high titanium content what suggests a presence or malformation crystals or amorphous [10, 18]. Another finding is that gibbsite was completely gone from BR, but

Table 1 Chemical composition of gibbsitic BR		BR
	$\%Fe_2O_3$	47.03
	$\%Al_2O_3$	14.49
	$\%SiO_2$	10.74
	$\%Na_2O$	10.87
	$\%TiO_2$	5.45
	$\%CaO$	0.98
	$\%ZrO_2$	0.91
	$\%V_2O_5$	0.15
	$\%LOI^*$	9.15

* LOI = Loss on ignition

Fig. 1 XRD pattern of gibbsitic BR

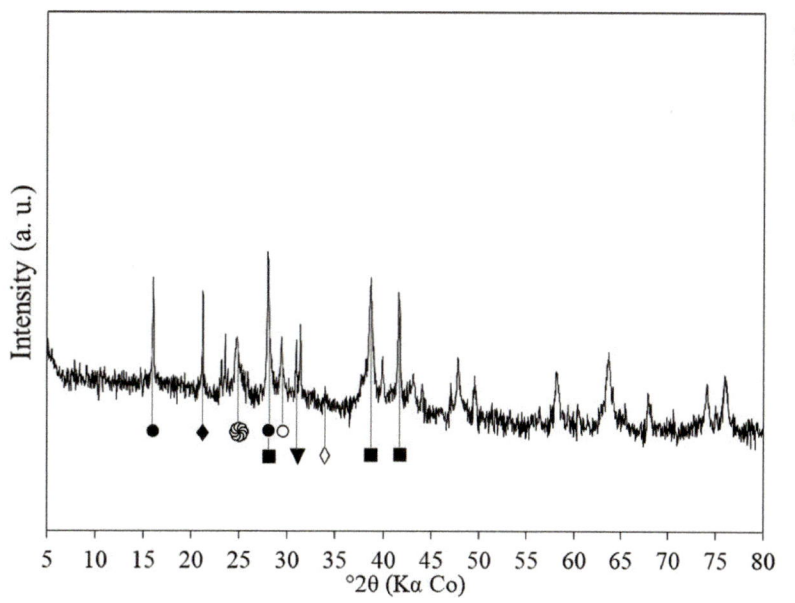

BR:
- ● Sodalite
- ◆ Gibbsite
- ✿ Al-Goethite
- ■ Hematite
- ○ Anatase
- ▼ Quartz
- ◇ Calcite

Fig. 2 Initial observations of the reaction products

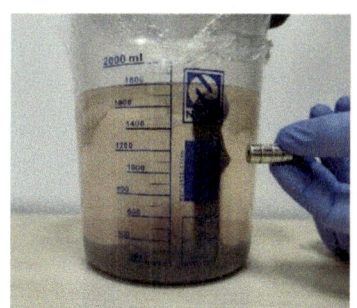

(a) Product answering to a magnet

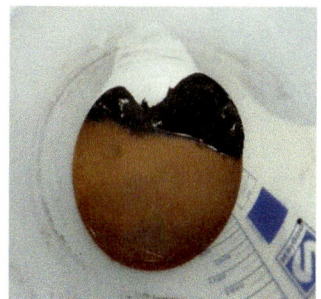

(b) Colours of the products after washing.

(a)   (b)   (c)

Fig. 3 Gibbsitic BR (**a**), underflow solids-US (**b**) and suspended solids-SS (**c**)

the aluminium content was still high in SS and US, only cancrinite formation may be not enough to justify the incorporation of aluminium in the solids, what also suggests side reactions between aluminium and titanium [18]. Finally, magnetite was detected in US. Therefore, the addition of $FeSO_4 \cdot 7H_2O$ and the reaction conditions resulted in the partial conversion of Fe_2O_3 into Fe_3O_4.

Figure 5 presents the BR micrographs (a, b, c, and d), EDX analysis, and particle size distribution (PSD). It is possible to observe uniform particle sizes of rounded aspects

(Fig. 5a, b). PSD indicated that 80% of particles (P80) were below 40 μm. Due to Bayer process characteristics related to minerals lixiviation and new solids formation, it is unlikely to find a particle where just one phase exists, so usually the presence of the major components such as Al, Fe, Na, Si, Ca, and Ti are detected by EDX. Therefore, gibbsite and sodalite were the suggested phases identified in Fig. 5c, d.

Figure 6 presents the US micrographs. There was not enough mass of SS to perform SEM and PSD determinations. The micrographs showed that US has formed large aggregates and their surface has a roughness aspect (Fig. 6 (a) and (b)). PSD indicated that 80% of particles (P80) were below 86 μm, a remarkable incremental when compared to BR; same findings were discussed in previous works [12, 18]. Figure 6c shows an octahedral magnetite etched on a mineral particle which confirms that the conversion process of hematite to magnetite takes place on the hematite surface [9–12]. On the surface of the same particle, EDX indicated the presence of Fe, Na, Al, Si, S, and Ti and the deposited material looks amorphous. As have been discussed before, anatase reacts in high temperature and free caustic

Table 2 Chemical composition of gibbsitic BR and products

	BR	SS	US
%Fe$_2$O$_3$	47.03	14.36	49.77
%Al$_2$O$_3$	14.49	27.82	14.77
%SiO$_2$	10.74	34.03	15.09
%Na$_2$O	10.87	8.34	10.31
%TiO$_2$	5.45	4.28	5.11
%CaO	0.98	0.90	0.92
%ZrO$_2$	0.91	–	0.87
%V$_2$O$_5$	0.15	–	0.11
SO$_3$	–	7.53	2.83
P$_2$O$_5$	–	0.56	–
ZnO	–	1.29	–
K$_2$O	–	0.76	–
%LOI*	9.15	-	–

* LOI = Loss on ignition

Fig. 4 XRD pattern of gibbsitic BR and the products of reductive digestion (SS and US)

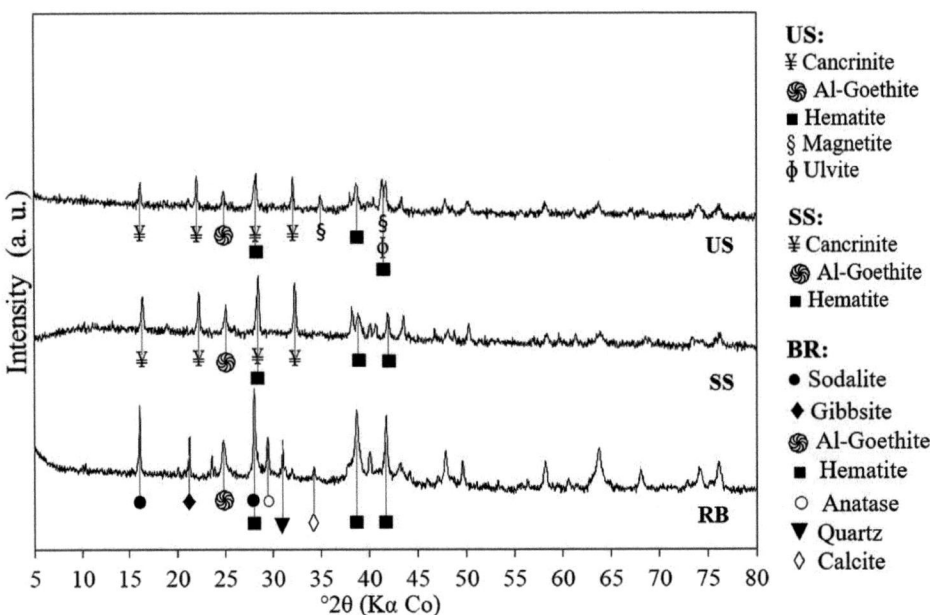

conditions, so is likely that this layer of solids is the result of Ti side reactions with these elements [11, 16, 19]. However, Ti also formed ulvite as have been detected by XRD patterns and a well-shaped crystal is shown in Fig. 6c. Another phase identified by SEM (Fig. 6d) was cancrinite of hexagonal tubular shape similar to the forms of cancrinite studied by Jayalatharachchi [15]. The presence of the cancrinite elements was confirmed by EDX (Fig. 4d).

Conclusions

A preliminary evaluation of magnetite syntheses from hematite present in gibbsitic BR was performed. After washing the solids obtained by reductive digestion, two different materials were obtained: Dark underflow solids (US) and brown-reddish suspended solids (SS). US

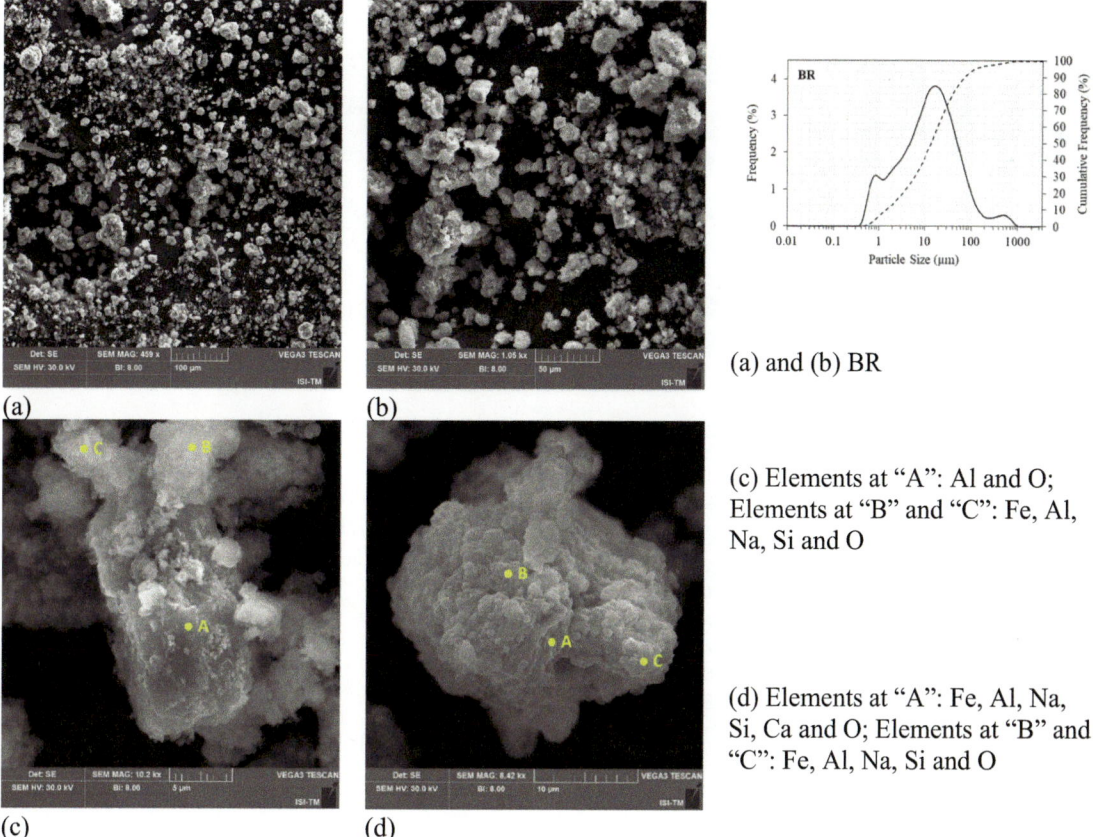

(a) and (b) BR

(c) Elements at "A": Al and O; Elements at "B" and "C": Fe, Al, Na, Si and O

(d) Elements at "A": Fe, Al, Na, Si, Ca and O; Elements at "B" and "C": Fe, Al, Na, Si and O

Fig. 5 BR morphology, EDX analysis, and particle size distribution

presented as main phases hematite, magnetite, ulvite (titanium-magnetite), cancrinite, and Al-goethite; while in SS was identified hematite, cancrinite, and Al-goethite. As observed in previous work [8–12], micrograph of US indicates that magnetite was formed on the surface of the particles. In spite of the successfully magnetite formation, several side reactions were observed specially related to titanium. Elementary characterization for SS and US indicated high content of elements such as Al, Na, Si, and Ti which suggests formation of amorphous complex compounds involving Ti, knowing by its high reactivity in such reaction conditions [16–19].

The findings of this work indicated that it was possible to obtain a higher magnetic phase from hematite from a gibbsitic BR; however, side reactions may carry out incorporation of undesired elements in the magnetic particles. Looking to Fe concentration as a next step, if the magnetic particles have a high mineral combination, i.e., several heterogeneous aggregations, the grade of Fe in the concentrated stream will still be poor. So, as future work, one should seek the balance between formation of magnetic phases and side reaction minimization.

Acknowledgements The present paper is part of the first author's Professional Master's at the Graduate Program in Materials Engineering (PPGEMAT), from Federal Institute of Education, Science and Technology of Pará-IFPA. The authors are grateful for the financial support of this study by Hydro Alunorte S/A and all technical and logistical support from Hydro's R&D team.

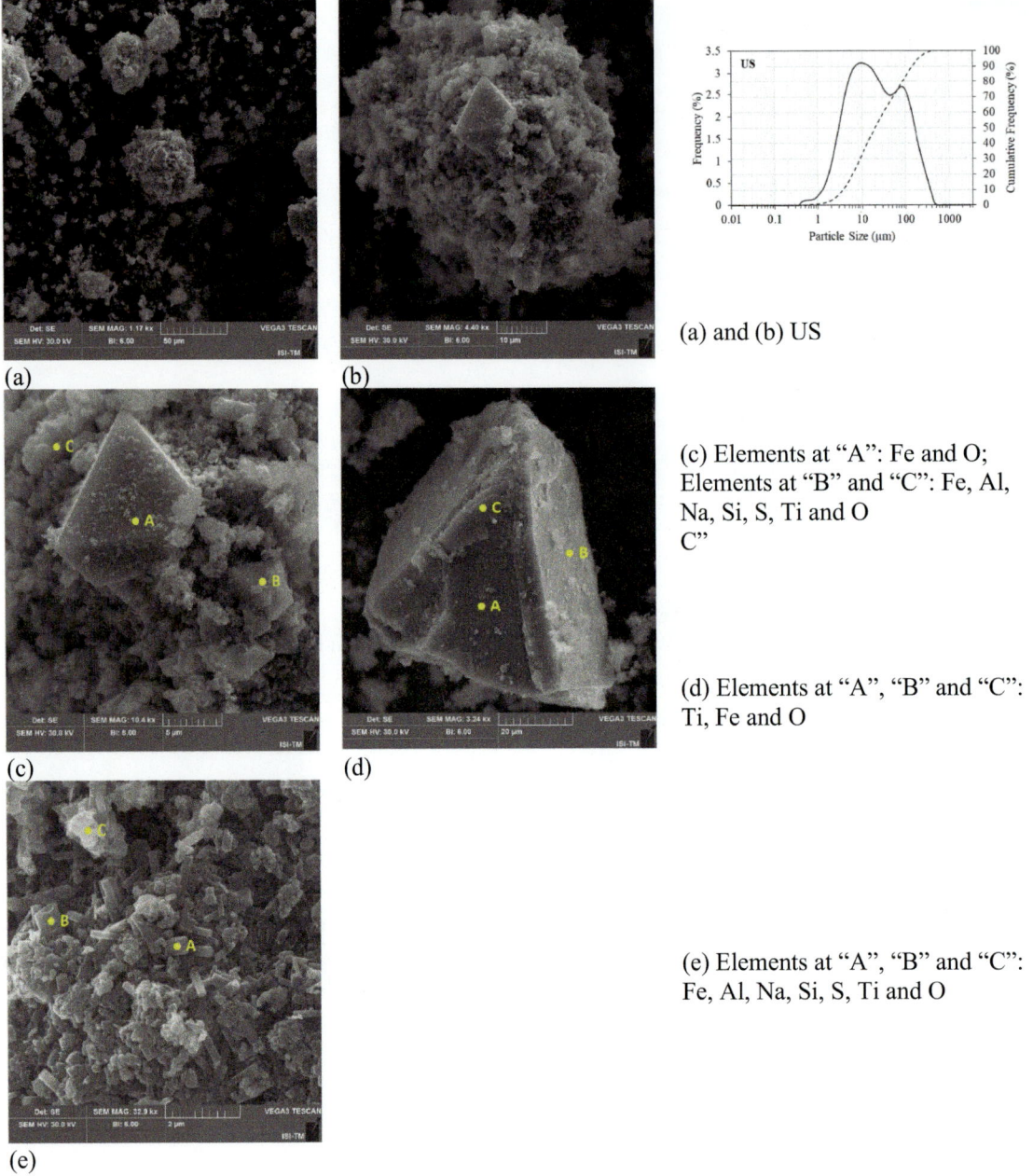

(a) and (b) US

(c) Elements at "A": Fe and O;
Elements at "B" and "C": Fe, Al, Na, Si, S, Ti and O
C"

(d) Elements at "A", "B" and "C": Ti, Fe and O

(e) Elements at "A", "B" and "C": Fe, Al, Na, Si, S, Ti and O

Fig. 6 US morphology and particle size distribution

References

1. Raahauge BE, Williams FS (2022) Smelter grade alumina from bauxite history, best practice, and future challenges, Springer, Switzerland.
2. Hill VG, Robson RJ (1981) The classification of bauxites from Bayer plant standpoint. In: Light Metals 1981. The Minerals, Metals & Materials Society, Pittsburgh; Springer, New York, p 30–66
3. Gräfe M, Power G, Klauber C (2011) Bauxite residue issues: III. Alkalinity and associated chemistry. Hydrometallurgy (108): 60-79.
4. Safarian J, Kolbeinsen L (2016) Sustainability in Alumina Production from Bauxite. Sustainable Industrial Processing Summit and Exhibition (5): 75 – 82.
5. Joyce PJ, Björklund A (2019) Using Life cycle thinking to assess the sustainability benefits of complex valorization pathways for bauxite residue. J. Sustain. Metall. (5): 69–84.
6. Araujo PFM, Silva PMP, Carmo ALV, Gomes FGG, Santos AM, Costa RV, Melo CCA, Lucheta AR, Montini M (2020). Bayer Process towards the circular economy —Metal recovery from bauxite residue. In: Tomsett A (ed) Light metals 2020: The Minerals, Metals & Materials Society, Pittsburgh; Springer, New York, p 98–106
7. Araujo PFM, Silva PMP, Carmo ALV, Gomes FGG, Santos AM, Costa RV, Melo CCA, Lucheta AR, Montini M (2021) Gravity

Methods Applied to Bauxite Residue for Mineral Pre-concentration. In: Perander, L. (eds) Light Metals 2021. The Minerals, Metals & Materials Series. Springer, New York, p. 68–73.

8. Pasechnik LA, Skachkov VM, Bodgdanova EA, Yu Chufarov A, Kellerman DG, Medyankina IS, Yatsenko SP (2020) A promising process for transformation of hematite to magnetite with simultaneous dissolution of alumina from red mud in alkaline medium. Hydrometallurgy (196): 1-13.

9. Li X, Yu S, Dong W, Chen Y, Zhou Q, Qi T, Liu G, Peng Z, Jiang Y (2015) Investigating the effect of ferrous ion on the digestion of diasporic bauxite in the Bayer process. Hydrometallurgy (152):183-189.

10. Li X, Liu N, Qi T, Wang Y, Zhou Q, Peng Z, Liu G (2015) Conversion of ferric oxide to magnetite by hydrothermal reduction in Bayer digestion process. Trans. Nonferrous Met. Soc. China (25): 3467-3474.

11. Li X, Wang Y, Zhou Q, Qi T, Liu G, Peng Z, Wang H (2017) Transformation of hematite in diasporic bauxite during reductive Bayer digestion and recovery of iron. Trans. Nonferrous Met. Soc. China (27): 2715-2726.

12. Li X, Wang Y, Zhou Q, Qi T, Liu G, Peng Z, Wang H (2018) Reaction behaviors of iron and hematite in sodium aluminate solution at elevated temperature. Hydrometallurgy (175): 257-265.

13. Klein C (2012) Manual de ciência dos minerais. Bookman, Porto Alegre.

14. Xu B, Smith P, Wingate C, De Silva L (2010) The effect of calcium and temperature on the transformation of sodalite to cancrinite in Bayer digestion. Hydrometallurgy (105): 75-81.

15. JAYALATHARACHCHI V (2016) Understanding desilication products in bauxite refinery residues. Masters by research thesis, Queensland University of Technology.

16. Wang Y, Li X, Zhou Q, Wang B, Qi T, Liu G, Peng Z, Zhou K (2020) Observation of sodium titanate and sodium aluminate silicate hydrate layers on diaspore particles in high-temperature Bayer digestion. Hydrometallurgy (192): 1-7.

17. Li X, Shunwen Y, Liu N, Chen Y, Qi T, Zhou Q, Liu G, Peng Z (2014) Dissolution behavior of sodium titanate in sodium aluminate solutions at elevated temperatures. Hydrometallurgy (147-148): 73-78.

18. Wang Y, Zhang T, Lv G, Zhang W, Zhu X, Xie L (2017) Reaction behaviors and amorphization effects of titanate species in pure substance systems relating to Bayer digestion. Hydrometallurgy (171): 86-94.

19. Wang Y, Li X, Wang B, Zhou Q, Qi T, Liu G, Peng Z, Zhou K (2019) Interactions of iron and titanium-bearing minerals under high-temperature Bayer digestion conditions. Hydrometallurgy (184): 192-198.

Digestion Efficiency Improvement of Gibbsite-Boehmite Bauxite

Fengqin Liu, Zegang Wu, Songqing Gu, and Michael Ren

Abstract

Bayer digestion performance of gibbsite-boehmite bauxite is studied in this paper. The chemical and mineral compositions of a gibbsite-boehmite bauxite from Australia are analyzed and tested by XRF, XRD, and SEM. Test results show that the bauxite contains $\geq 50\%$ of alumina (gibbsite and boehmite) and 9% of silica (kaolinite and quartz). Its digestion efficiency is studied in detail with various digestion conditions such as temperatures, holding times and additives. A medium temperature digestion technology has been developed based on the study results and industrial tests in Chinese refineries: boehmite in the bauxite will be totally digested in just a few minutes with bits of lime addition for a high alumina recovery efficiency and most of the unreactive silica minerals in the bauxite are at a stable state with caustic consumption reduced by over 10%, alumina recovery efficiency increased by 2–3% compared with traditional digestion technologies. There is a great application prospect with this medium temperature digestion technology for the Australian bauxite.

Keywords

Gibbsite-boehmite bauxite • Bayer digestion • Digestion efficiency • Caustic consumption • Medium temperature • Holding time • Additive

F. Liu (✉) · Z. Wu · S. Gu
State Key Laboratory of Advanced Metallurgy, University of Science and Technology, Beijing, Beijing, 100083, China
e-mail: liufq@ustb.edu.cn

F. Liu · Z. Wu · S. Gu
School of Metallurgical and Ecological Engineering, University of Science and Technology, Beijing, Beijing, 100083, China

M. Ren
Sunlightmetal Consulting Inc, Beijing, China
e-mail: Michael.ren@sunlightmetal.ca

Introduction

Chinese alumina output reached 77.45 million tons in 2021, which exceeded half of the global total output by about 54.4%. Imported bauxite has become the major raw material for Chinese alumina refineries, in which the imported Guinea, Australia, and Indonesia bauxites are the main resource for Chinese alumina industry. One of the largest reserves of bauxite resource in the world is from Australia and its bauxite export has been relatively stable [1–4].

Most of the imported Australia bauxite is from north-east Australia with both gibbsite and boehmite as the major alumina minerals. Australian bauxite usually has a complex mineral composition. A study on digestion behaviors of gibbsite-boehmite bauxite has been carried out by some researchers. Qi [5] studied the digestion performances of boehmite in the bauxite at 180–240 °C for 30 min and the mineral compositions of the bauxite residues formed at 143 °C and 245 °C were also discussed. The study results show that boehmite and quartz cannot react at 143 °C. Both boehmite and quartz were completely reacted at 245 °C for 30 min. Jiang [6] studied the bauxite digestion at 270 °C and 280 °C and showed that the best digestion rate of the bauxite could be obtained at 270 °C for 10 min or 280 °C for 5 min. In addition, the relationship between the reaction rate of quartz and the recovery rate of alumina was also studied. Zhang [7] found that the total available alumina in a bauxite at 150 °C and 250 °C were 36.25% and 43.33% respectively, which showed that the high-temperature digestion process for the bauxite should be more suitable by the comparison between alumina lost by the quartz reaction and alumina recovery from the boehmite at high digestion temperature. It is well known that alumina loss can be caused by the reaction of quartz in the high-temperature digestion process. However, a detailed study should be carried out on what kind of digestion process including digestion temperatures, holding times and additives will be beneficial to obtain the best alumina recovery for the gibbsite-boehmite bauxite?

First, it is necessary to study the reaction behavior of different minerals in the gibbsite-boehmite bauxite during the high-temperature digestion process for its best Bayer digestion performance, which is of significance for the Chinese refineries to obtain better production benefit.

A study on the digestion efficiency of the Australia gibbsite-boehmite bauxite has been carried out and a medium temperature digestion technology has successfully been developed based on the study results.

Gibbsite-Boehmite Bauxite from Australia

Common Description of Gibbsite-Boehmite Bauxite Digestion

Based on the test results the north-east Australian bauxite is a type of gibbsite-boehmite bauxite, in which the major effective aluminum-bearing minerals are gibbsite and boehmite and the main silica-containing minerals are kaolinite and quartz. Gibbsite can react rapidly with caustic liquor at a lower temperature of 145 °C [8–10], while the higher digestion temperatures of more than 220 °C are needed for boehmite digestion [11, 12]. To ensure maximum dissolution of boehmite in the bauxite the higher temperatures above 220 °C are needed in the digestion process. But at the relatively high digestion temperatures and long holding times, more quartz dissolves, which consumes alumina and caustic in the liquor. Therefore, the key to the study on the digestion process of gibbsite-boehmite bauxite is to ensure the complete digestion of boehmite and to minimize the reaction of quartz [13–16].

The Australian bauxites from different origins have quite different chemical and mineral compositions due to the different mineralization characteristics. The chemical and mineral compositions and digestion performances of multiple bauxite samples from different origins were investigated in this paper. The effects of different digestion temperatures, holding times and additives on alumina to silica ratio (A/S) and caustic to silica ratio (N/S) in the bauxite residues were studied, as well as the bauxite and caustic consumption factors in the digestion process.

Chemical Compositions

The chemical composition of multiple samples of gibbsite-boehmite bauxite from the Australian origins A and B were analyzed by X-ray fluorescence spectroscopy (XRF) and chemical titration analysis. The results are shown in Table 1.

It can be seen in Table 1 that for the bauxite samples 1–3 from the origin of A the Al_2O_3 content is above 51.6%, the SiO_2 content is over 8.6%, and the alumina to silica ratio is in the range of 5.5 to 6. But the bauxite from the B origin has an alumina content of about 51%, a lower SiO_2 content of less than 8.3%, and a higher Fe_2O_3 content. It can be seen that there are some differences in the chemical compositions of the bauxite from the different sources.

Mineral Compositions

The mineral compositions of the bauxite A1# and B1# were analyzed by X-ray diffraction analyzer (XRD) and the results are shown in Fig. 1.

It can be seen in Fig. 1 that there is almost no obvious difference in the mineral types in the different bauxite samples and only the difference in the individual mineral content in the bauxites. In the gibbsite-boehmite bauxite samples the major mineral compositions are basically the same, such as gibbsite, boehmite, kaolinite, quartz, hematite, anatase, etc. However, the peak intensity of hematite in the B1# sample is much higher than that in the A1# and the peak intensity of quartz is weaker. It can be expected that the reaction performances of the different samples may have certain differences in the digestion process.

Test Procedures and Analysis Instruments

Bauxite Grinding

The bauxite samples used in the experiments were ground by using vibrating milling. The test process had the following characteristics such as very short grinding time each, multiple sieving, and re-grinding of large particles. This grinding method is mainly to ensure that the particle size of the bauxite samples for digestion tests is less than 20 mesh, and the particles larger than 250 microns are more than 50%. This is just for a size distribution of the samples to be courser and more homogeneous.

Bomb Digestion

The bomb digestion equipment was used for bauxite digestion tests, which consists of three parts: molten salt heating furnace, bomb digesters, and operating control system. 6 separate bomb digesters with 150 ml each are configured in and heated by the molten salt. This equipment can ensure not only the same reaction temperature for the different bomb digesters but also independent operation for

Table 1 The chemical compositions of multiple samples of gibbsite-boehmite bauxite

Name	SiO$_2$	Fe$_2$O$_3$	Al$_2$O$_3$	A/S	CaO	TiO2
A1#	9.33	9.72	51.83	5.56	0.07	3.13
A2#	9.25	10.77	51.68	5.59	0.08	2.97
A3#	8.63	10.39	52.09	6.01	0.07	3.09
B1#	7.92	13.97	51.16	6.46	0.09	2.38
B2#	8.28	15.02	51.11	6.17	0.06	2.61

Fig. 1 The mineral compositions of the bauxites A1# (a) and B1# (b)

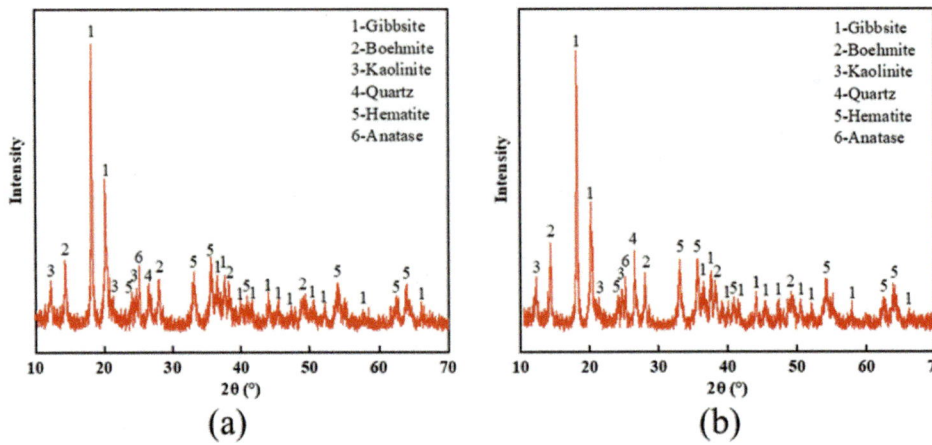

them. The bomb rotating device and precise temperature control system is installed for the uniform stirring of the bomb digesters and accurate digestion temperature (±1 °C).

Materials Characterization

The chemical compositions of the bauxite and residue samples were determined by XRF (mAX, AXIOS, Alamelo, The Netherlands). The X-ray diffractometer (PW1710, Philips, Amsterdam, The Netherlands) was used with Cu Kα-radiation at 40.0 kV and 30.0 mA for the mineralogical study. The XRD tests were conducted within the 2θ range of 10–60° using a scanning speed of 0.1°·min⁻¹. The samples for XRF and XRD tests were ground to pass through a 200-mesh sieve and dried in an oven at 105 °C for 2 h. The micrographs of the samples were observed using a scanning electron microscope (Regulus 8100, Hitachi, Japan) at an accelerating voltage of 15.0 kV. An EDS (ultra-DLD, Shimadzu, Kyoto, Japan) connected with SEM was used to perform elemental analysis on the sample particles. The elements with the content of 3% ~20 wt.% have the accuracy of relative error < 10%. And the elements with content greater than 20 wt.% have the higher accuracy of relative error <5%.

Test Program and Procedure

Test Procedure

The caustic liquor used in the digestion experiments was prepared with industrial caustic liquor adjusted by the analytically pure sodium hydroxide. The chemical compositions of the caustic liquor and bauxite samples were analyzed and the liquid-to-solid ratio is calculated according to their compositions.

The formula for calculating the dosage mass of bauxite (M) required per 100 mL of digestion liquor is as follows:

$$1.4 = \frac{N_k\text{-}M \times S_B \times (N/S)_M \times 10}{M \times A_B \times 10 - M \times S_B \times (A/S)_M \times 10 + AO}$$

where N_K is the caustic concentration in the digestion liquor, (g/L); M is the dosage mass of bauxite required for per 100 mL of digestion liquor, (g); S_B is the SiO$_2$ mass fractions in the bauxite, $(N/S)_M$ is the theoretical caustic to silica ratio in the bauxite residue, A_B is the Al$_2$O$_3$ mass fractions in the bauxite, $(A/S)_M$ is the theoretical alumina to silica ratio in bauxite residue, and AO is the alumina concentration in the digestion liquor (g/L), 1.4 is the expected caustic ratio of the liquor after digestion.

According to the calculation results a quantitative amount of bauxite samples, lime additives and caustic liquor for the tests were added to the bomb digesters respectively. When the molten salt heating furnace reached the preset temperature, the bomb digesters installed on the rotating frame will be put into the molten salts to preheat, turn on the rotary stirring system and start timing after the preheating stage of just a few minutes.

The bomb digester can be taken out from the molten salts when the digestion time is reached and is put immediately into cold water for cooling. Then the bomb digester is opened and the bauxite slurry is filtered for liquor analysis by chemical titration and the digestion residue is further filtered, washed, and dried for chemical composition analysis by XRF.

Test Program

In this paper, the digestion tests were carried out for the bauxite samples A and B. The digestion test program is shown in Table 2.

Test Targets

In order to study the digestion performances of the bauxite samples, the alumina to silica ratio (A/S) and caustic to silica ratio (N/S) in the bauxite residues were tested by XRF and the chemical analysis of the caustic liquor was analyzed by titration analysis. The A/S is the mass ratio of alumina and silica in the bauxite residues, and the N/S is the mass ratio of sodium oxide and silica in the bauxite residues.

According to the chemical compositions of bauxite samples and their digestion residues, from which the alumina digestion rate, bauxite and caustic consumption in the digestion process can be calculated. The calculation formula is as follows.

$$\eta_{AI} = \left[1 - \frac{(A/S)_M}{(A/S)_B}\right] \times 100\% \; (\%) \qquad (1)$$

where η_{AI} is the alumina digestion rate, $(A/S)_M$ is the mass ratio of Al_2O_3 and SiO_2 in the bauxite residue and $(A/S)_B$ is the mass ratio of Al_2O_3 and SiO_2 in the bauxite sample.

$$MC = \frac{1}{A_B - S_B \times (A/S)_M}, (t/t - Al_2O_3)$$

$$CC = \frac{(N/S)_M}{(A/S)_B - (A/S)_M} \times 1.2903 \times 1000, (kg/t - Al_2O_3)$$

where A_B and S_B are the mass fractions of Al_2O_3 and SiO_2 in the bauxite sample respectively, $(N/S)_M$ is the mass ratio of Na_2O and SiO_2 in the bauxite residue, 1.2903 is the mass conversion factor of Na_2O converted to NaOH, MC and CC are the bauxite and caustic consumptions respectively.

In order to study the digestion behavior of the bauxite samples during the digestion process the mineral compositions of bauxite samples and their residues were analyzed by XRD. In addition, SEM observation was carried out on the morphology of the different minerals in the bauxite samples and their residues.

Test Results and Discussion

A/S and N/S of the Bauxite Residues vs Digestion Temperatures at Different Times

The digestion performances of A1# bauxite at different temperatures were studied. The A/S and N/S of the bauxite residues at the different digestion times at 230–250 °C have been shown in Fig. 2.

The A/S (A/S = 1.11) in the bauxite residues digested at 230 °C for 5 min was very high and the digestion rate of alumina is only 78.95%. The main reason is that the boehmite in the bauxite fails to react completely at 230 °C. With the longer digestion time the A/S in the bauxite residues decreased and the highest digestion rate of alumina is 80.57% at 230 °C, which means the boehmite still remained in the residue. The A/S within 10 min at 240 °C and 250 °C is lower (<0.98), but with longer digestion time the digestion rate of alumina gradually decreased. It seems that the

Table 2 The digestion conditions of digestion experiments	No	Bauxite	Temperature	Time	Lime additives
	1	A1#	230–250 °C	5 min–45 min	2%
	2	A2#	240 °C	5 min–60 min	2%
	3	B1#	240 °C	5 min–60 min	2%
	4	B1#	240 °C	10 min, 60 min	0–2.5%
	5	A2#	240 °C	10 min	0%, 1%, 2%

Fig. 2 The A/S (a) and N/S (b) in the residues at the different digestion times at 230–250 °C

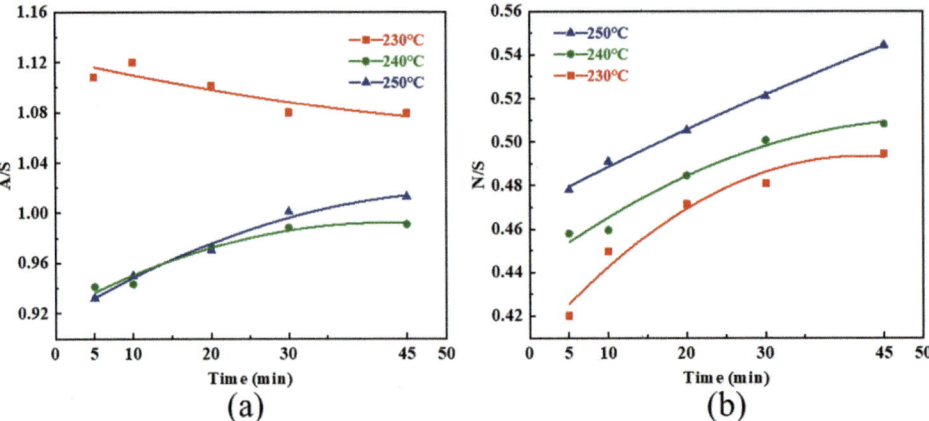

(a)

(b)

boehmite can almost be completely reacted at temperatures above 240 °C for 10 min.

N/S in the bauxite residues is mainly determined by the mineral compositions in the bauxite residues and the attack content of quartz. N/S in the bauxite residues will be obviously increased as the quartz is attacked in large quantities. It can be seen in Fig. 2 that N/S in the bauxite residues is increased with the higher digestion temperature and longer digestion time. N/S in the bauxite residues digested at 240 °C for 10 min is still kept at a lower level (less than 0.46).

Therefore, it can be seen by the test results that when the gibbsite-boehmite bauxite is digested for a short time at 240 °C, a better digestion rate of alumina can be obtained than longer time and N/S in the bauxite residues can be kept lower at the same time.

The tests were also carried out for A2# and B1# bauxite at 240 °C for the different digestion times. It can be seen in Fig. 3 that both A/S and N/S in the bauxite residues show lower values within 10 min. This means that the bauxite digestion is carried out at the medium temperatures for a relatively short time is more suitable for the gibbsite-boehmite bauxite.

A/S and N/S of the Residues vs Lime Additives

The effects of lime addition on the digestion results have been studied for the gibbsite-boehmite bauxite samples. Figure 4 shows the influences of the different lime additions on A/S and N/S in the bauxite residues at the medium digestion temperature for a short time.

It can be seen in Fig. 4 that with the increase of lime addition the A/S and N/S in the residues decrease at the digestion temperature of 240 °C for 10 min. When the lime addition exceeds 2% the digestion rate of alumina will decrease and N/S in the bauxite residues will increase.

Moreover, A/S and N/S in the bauxite residues can be reduced by the definite addition of lime in the different bauxite digestion processes. It can be found from the mentioned above that the lime addition generally has two effects. One is to eliminate the negative influence of titanium dioxide on the bauxite digestion to improve alumina digestion rate; the second is to change the DSP mineral composition and to replace a part of the sodium oxide in DSP for reducing N/S in the bauxite residues.

Fig. 3 A/S and N/S in the bauxite residues at different digestion times at 240 °C (a) A2# bauxite; (b) B1# bauxite

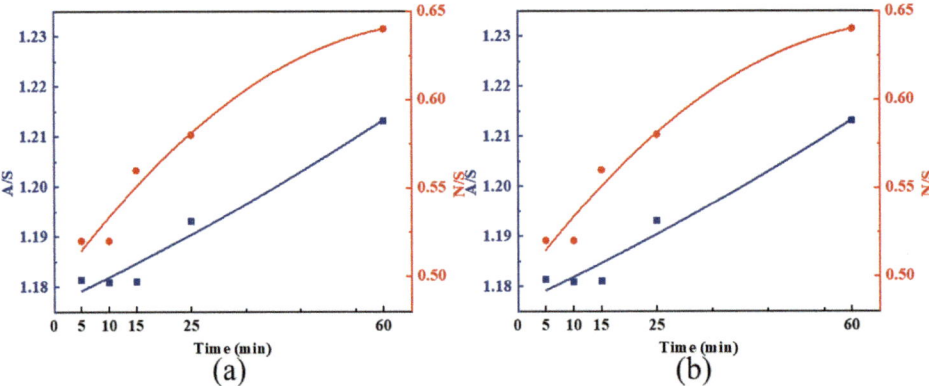

(a)

(b)

Fig. 4 The effect of different lime additions on the A/S and N/S in the bauxite residues at 240 °C for 10 min (a) B1# bauxite; (b) A1# bauxite

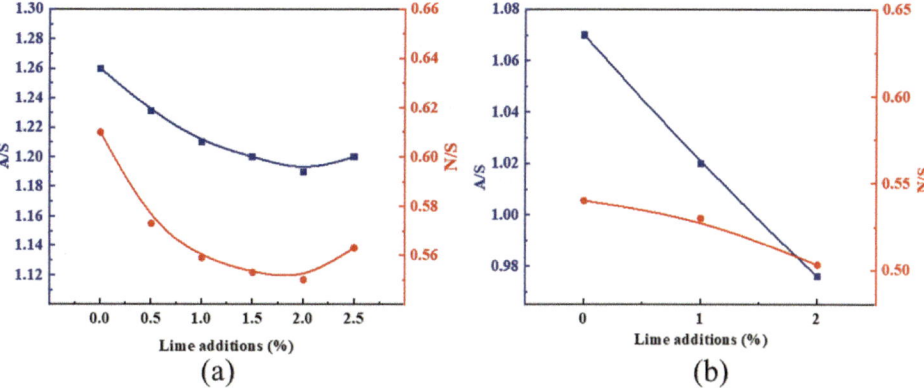

(a)   (b)

It can also be found that the alumina digestion rate can be improved by lime addition during short-time digestion and N/S in the bauxite residues can be reduced as well. The alumina in the bauxite can be completely digested for a long digestion time without adding lime. It is suggested [17–21] that N/S in the bauxite residues decreases with the addition of lime. However, hydrogarnet will be produced instead of common sodium aluminosilicate hydrate when lime is added in excess during the digestion process, resulting in the loss of alumina and caustic. Therefore, the optimum lime addition in the digestion process of gibbsite-boehmite bauxite should be determined by tests.

Mineral Composition Changes During Digestion

The mineral compositions in the bauxite residues at 240 °C for different digestion times are shown in Fig. 5.

It can be seen in Fig. 5 that the bauxite residues are basically the same in terms of mineral types and the differences between the bauxite residues at the different digestion conditions are the content of quartz and boehmite in the residues. It can be found combined with the results in Fig. 2 that boehmite in the bauxite does not completely react for 5 min at 240 °C so its A/S is relatively high. But when digestion time is extended to 10 min the boehmite is digested completely to reduce A/S in the bauxite residues.

It is shown by comparison of the peak intensity of quartz in the bauxite residues in Fig. 5 that the quartz remained in the residues is gradually reduced with the time extension. The reaction equation for quartz can be described as follows:

$$SiO_2 + 2NaOH = Na_2SiO_3 + H_2O$$

$$xNa_2SiO_3 + 2NaAl(OH)_4 + (x + n - 4)H_2O = Na_2O \cdot Al_2O_3 \cdot xSiO_2 nH_2O + 2xNaOH$$

It is seen from the equation that the reaction of quartz in caustic liquor will cause alumina and caustic loss to form sodium aluminosilicate hydrate in bauxite residues. At medium temperatures the major reaction carried out in 10 min is the rapid digestion of the boehmite and the caustic attack to quartz is relatively slow. After 10 min. digestion the boehmite digestion will be completed and only quartz continue to react with caustic liquor consuming Al_2O_3 and Na_2O in the liquor continuously. Therefore, both A/S and N/S in the bauxite residues will be increased with the time extension after 10 min digestion.

It is concluded that the alumina recovery and caustic consumption during the bauxite digestion process can be changed by the different digestion conditions due to the digestion behaviors of boehmite and quartz.

Fig. 5 The mineral compositions in the bauxite residues for the different digestion times (a) A1#, 5 min; (b) A1#, 10 min; (c) A1#, 20 min

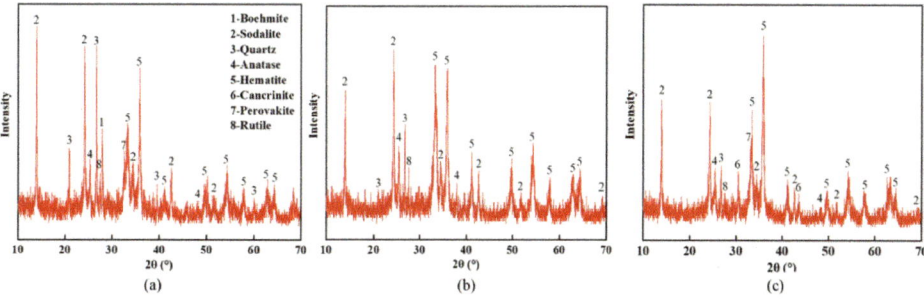

Fig. 6 The SEM images of quartz in the gibbsite-boehmite bauxite

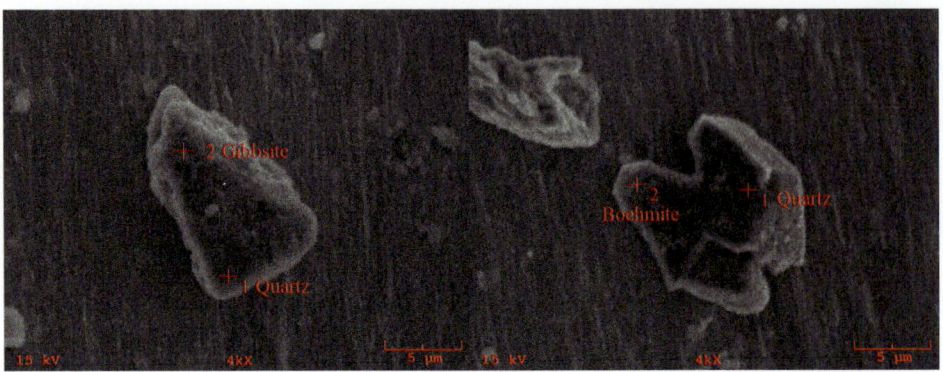

SEM Test Results and Discussion

The quartz morphology and its mosaic situation in the bauxite were observed by SEM to find the reason why there will be the different attack extent for the quartz in the residues at the different digestion conditions.

The SEM images of quartz in gibbsite-boehmite bauxite are shown in Fig. 6.

It can be seen in Fig. 6 that the two kinds of quartz particles present a relatively regular polyhedral shape and the particle size is larger than 10 microns. It seems that there is a thick mineral coating layer on the surface of the quartz particles.

EDS analysis was performed on the position where the coating was thicker, and the elemental composition of the coating material was determined to be Al and O. It can be determined that the encapsulated materials on the quartz surface are gibbsite and boehmite respectively [22]. The existence of this coating layer containing boehmite is beneficial to reducing caustic attack to quartz. In the digestion process, the caustic attack to the quartz can be stopped or delayed to reduce the alumina and caustic loss due to the wrapping of boehmite at the beginning of digestion until the boehmite is completely digested.

Influences of Digestion Parameters on the Consumption of Bauxite and Caustic

The bauxite consumption and caustic consumption refer to the mass of bauxite and the mass of sodium hydroxide required for each 1 ton of alumina produced in the digestion process. The bauxite and caustic consumptions are the most important factors to influence the alumina production cost, which is mainly related to the bauxite grade, A/S and N/S of the bauxite residues. The only way to reduce alumina production costs is to develop the new processes to reduce A/S and N/S in the bauxite residues since bauxite grade is an uncontrollable factor.

It can be seen from the study results mentioned above that the bauxite and caustic consumptions can be changed by the digestion parameters such as temperatures, holding times and additives.

The new digestion processes have been studied in the Chinese alumina refineries using the gibbsite-boehmite bauxite from Australia. Table 3 shows the differences in the digestion results and bauxite and caustic consumptions for the different digestion conditions in the Chinese alumina refinery.

It can be found in Table 3 that both bauxite and caustic consumptions has been reduced by changing the digestion conditions. The bauxite consumption is reduced by 50 kg per ton of alumina and caustic consumption is reduced by 34 kg per ton of alumina. Total alumina production cost is decreased by more than RMB 100.

Conclusions

The gibbsite-boehmite bauxite from Australia contains about 50% of alumina and 9% of silica with alumina to silica ratio of more than 5.

Gibbsite and boehmite are the major alumina minerals and kaolinite and quartz are the main silica minerals existing in the gibbsite-boehmite bauxite.

Table 3 The different consumptions for the different digestion conditions

Different digestion conditions	A/S in residues	N/S in residues	Bauxite consumption	Caustic consumption
250 °C, 60 min, no additive	1.07	0.66	2.42 t/t	185 kg/t
235 °C, 15 min, 2% of lime	0.98	0.55	2.37 t/t	151 kg/t

It can be seen that A/S and N/S of the bauxite residue will be less than 1 and 0.55 respectively after digestion at the medium temperatures of 230–250 °C for a few min. with a little lime addition. And the boehmite in the bauxite can be digested and most of the quartz remains in the bauxite residue at the digestion conditions mentioned above.

A Bayer digestion technology has been developed based on the test results with a medium digestion temperature of 230–250 °C, a short holding time of a few minutes, and bits of lime addition.

References

1. Meyer, F.-M. Availability of Bauxite Reserves[J]. Natural Resources Research, 2004, 13(3),167–172. https://doi.org/10.1023/B:NARR.0000046918.50121.2e.

2. Pan, Z.-S.; Zhang, Z.-Z.; Zhang, Z.-N.; Feng. G.-Q.; Cao, X.-S. Analysis of the import source country of the bauxite in China [J]. China Mining Magazine, 2019, 28(02), 13–17.

3. Gao, X.-T.; Guo, S.; Li, H.-L.; Summary and study of the imported bauxite in China [J]. Light Metals, 2016, (07), 4–11. https://doi.org/10.13662/j.cnki.qjs.2016.07.002.

4. Zhao, B.-Y.; Men, C.-S. Discussion on bauxite supply prospect for domestic alumina refineries [J]. Light Metals, 2016, (08), 8–12. https://doi.org/10.13662/j.cnki.qjs.2016.08.003.

5. Qi, L.-K. Study on the treatment of monohydrate gibbsite by Bayer process [J]. Nonferrous Metals, 1991, 6, 21–23. https://doi.org/10.3969/j.issn.1007-7545.

6. Jiang, Y.-H. Optimization of the digestion technology of gibbsite-boehmite bauxite [J]. Light Metals, 2011, (S1), 57–59.

7. Zhang Z.-Y. Experimental study on digestion of gibbsite and boehmite mixture [J]. Light Metals, 2019, (01), 10–13.

8. Bi, S.-W. et al. The alumina production by Bayer process [M]. Metallurgical Industry Press, 2007.

9. Dillinger, B.; Batchelor, A.; Katrib, J.; Dodds, C.; Suchicital, C.; Kingman, S.; Clark, D. Microwave digestion of gibbsite and bauxite in sodium hydroxide[J]. Hydrometallurgy, 2020, 192. https://doi.org/10.1016/j.hydromet.2020.105257.

10. Pereira, J.-A.-M.; Schwaab, M. Dell'oro, E.;, Pinto, J.-C.; Monteiro, J.-L.-F.; Henriques, C.-A. The kinetics of gibbsite dissolution in NaOH [J]. Hydrometallurgy, 2008, 96(1) 6–13. https://doi.org/10.1016/j.hydromet.2008.07.009.

11. Authier-martin, M.; Forte, G.; Ostap, S.; See, J. The mineralogy of bauxite for producing smelter-grade alumina[J]. JOM, 2001, 53 (12), 36–44. https://doi.org/10.1007/s11837-001-0011-1.

12. Hudson, L.-K. Alumina production[M]. 1987.

13. Gan, B.-K.; Taylor, Z.; Xu, B.-A.; Riessen, A.; Hart, R.-D.; Wang, X.-D.; Smith, P. Quantitative mineral analysis of bauxites and their dissolution products[J]. International Journal of Mineral Processing, 2013, 123, 64–72. https://doi.org/10.1016/j.minpro.2013.05.005.

14. Raghavan, N.-S.; Fulford, G.-D. Mathematical Modeling of the Kinetics of Gibbsite Extraction and Kaolinite Dissolution/Desilication in the Bayer Process[J]. Light Metals, 1998, 29–36. https://doi.org/10.1002/9781118647868.ch34.

15. Roach, G.-I.-D.; White, A.-J. Dissolution Kinetics of Kaolin in Caustic Liquors[J]. Light Metals, 1988, 240–246. https://doi.org/10.1007/978-3-319-48176-0_32.

16. Wu, Y.; Pan, X.-L.; Han, Y.-J.; Yu, H.-Y. Dissolution kinetics and removal mechanism of kaolinite in diasporic bauxite in alkali solution at atmospheric pressure[J]. Transactions of Nonferrous Metals Society of China, 2019, 29(12), 2627–2637. https://doi.org/10.1016/S1003-6326(19)65169-1.

17. Zhao, Q.-J.; Yang, Q.-F.; Chen, Q.-Y.; Yin, Z.-L.; Wu, Z.-P.; Yin, Z.-G. Behavior of silicon minerals during Bayer digestion [J]. Transactions of Nonferrous Metals Society of China, 2008, 20, S1–S9. https://doi.org/10.1016/S1003-6326(10)60002-7.

18. Pan, X.-L.; Wu, F.; Yu, H.-Y.; Bi, S.-W. Precipitation of desilication products in CaO-Na2O-Al2O3-SiO2-H2O system based on the Bayer process [J]. Hydrometallurgy, 2020, 197. https://doi.org/10.1016/j.hydromet.2020.105469.

19. Whittington, B.; Fallows, T. Formation of lime-containing desilication product (DSP) in the Bayer process: factors influencing the laboratory modelling of DSP formation [J]. Hydrometallurgy, 1997, 45(3), 289–303. https://doi.org/10.1016/S0304-386X(96)00085-0.

20. Whittington, B.-I.; Fletcher, B.-L.; Talbot, C. The effect of reaction conditions on the composition of desilication product (DSP) formed under simulated Bayer conditions [J]. Hydrometallurgy, 1998, 49(1), 1–22. https://doi.org/10.1016/S0304-386X(98)00021-8.

21. Xu, B.-A.; Smith, P.; Wingate, C.; Silva, L.-D. The effect of calcium and temperature on the transformation of sodalite to cancrinite in Bayer digestion [J]. Hydrometallurgy. 2010, 105(1), 75–81. https://doi.org/10.1016/j.hydromet.2010.07.010.

22. Chen, Z.-H.; Chen, X.-Q.; Li, S.-S.; Ma, J.-W. Research on Process Mineralogy for a Bauxite Ore in Queensland, Australia [J]. Multipurpose Utilization of Mineral Resources, 2013, 2013(5), 50–54. https://doi.org/10.3969/j.issn.1000-6532.2013.05.014.

Decanter Centrifuge for Dewatering Bauxite Tailings

Camila Botarro Moura, Rafael Alves de Souza Felipe, and Roberto Seno Junior

Abstract

Among the challenges faced by the mineral industry, issues related to reduction on water consumption, tailings disposal, and environmental management are highlighted. It is fundamental to study dewatering methods to improve the sustainability of the mineral activity. Different methods can be seen worldwide, and decanter centrifuges are already an option for dewatering tailings on different operations. Regarding bauxite mines, it is estimated that tons of tailings are disposed every year, mainly in settling ponds or tailings dams. Since typical bauxite concentration consists basically in washing the ore and removing fine particles from the product, the bauxite tailings consist of fine kaolinite, what makes the dewatering operation even more challenging. In this scenario, the purpose of the current work is to analyze the bauxite tailings dewatering by decanter centrifuge, with the tailings from bauxite processing fed by ore from Minas Gerais, Brazil.

Keywords

Decanter centrifuge • Dewatering • Bauxite • Tailings

Introduction

With the challenges faced by the mineral industry, especially in matters related to the reduction of consumption and water recovery, waste disposal, and environmental management, it's importance study dewatering methods that make the sustainability of mineral activity essential.

As a result of bauxite beneficiation, it is estimated that tons of ore are disposed of every year, mainly in tailings dams or disposal ponds. These fractions have large amounts of kaolinite, which makes it difficult to dewater this material. Currently, most tailings are discharged by natural sedimentation in tailings dams. The environmental impact of these large is huge, not to mention the area requirement and risk involved.

With the search for new disposal methods, centrifuges, specifically the decanter type, are versatile equipment and traditionally used in industries such as food, pharmaceutical, petrochemical, sanitation, among others. In mining, they are generally used when gravity sedimentation is very slow or when you want to reduce the amount of water in the densified phase [1].

Discussion

Bauxite

Lateritic bauxite is the most abundant on earth, responsible for about 85% of the deposits, and it is the only type of bauxite that occurs in Brazil. The first deposits discovered in southern France and central Europe were associated with carbonate rocks. In tropical regions, aluminum rich materials have been found on igneous rocks and metamorphic and in the form of sedimentary deposits [2].

That regardless of the geological age of the deposits of bauxite, the formation process will depend on five main factors:

- Climate: expressed by the seasonal variation in temperature and in the distribution of rain;
- Relief: influences the system of infiltration and drainage of water;

C. B. Moura (✉) · R. A. de Souza Felipe · R. S. Junior
CBA, Moraes Do Rego Str 347, Alumínio, 18125-000, Brazil
e-mail: camila.moura.cm1@cba.com.br

R. A. de Souza Felipe
e-mail: rafael.felipe.rf1@cba.com.br

R. S. Junior
e-mail: roberto.seno@cba.com.br

© The Minerals, Metals & Materials Society 2023
S. Broek (ed.), *Light Metals 2023*, The Minerals, Metals & Materials Series,
https://doi.org/10.1007/978-3-031-22532-1_33

- Fauna and flora: supply organic matter for chemical reactions and remobilize materials;
- Mother rock: which, according to its nature, presents differentiated resistance to the processes of weathering alteration; and
- Time: period of exposure of the rock to all the agents described.

The world reserve of bauxite is in the order of 33.4 billion tons. Brazil accounts for 3.5 billion tons of these reserves, of which 95% is metallurgical bauxite [3].

Bauxite Processing

The main objective of bauxite processing is to adjust the granulometry, increase the usable alumina content, and reduce the reactive silica content. As the ore fines concentrate the source of reactive silica, usually clay minerals, the bauxite beneficiation processes consist of washing and classifying the ore, to discard the finer fractions. In Fig. 1, we can see an example of a bauxite ore processing plant.

The fraction that is not economically viable within the bauxite beneficiation process is called tailings and is rich in kaolinite, with a fine granulometric distribution and generally presenting in the form of dilute pulp with solids concentration that can vary according to the beneficiation process.

Solid–Liquid Separation

Most mineral and metallurgical beneficiation processes take place in the presence of water due to the inherent advantages of wet processing (separation, transport, among others). However, this demand for water has become an important concern in the sector, especially in terms of conservation, treatment, and reuse. In this context, dewatering operations gain greater importance, being considered essential in a mineral processing flowchart, because, in addition to the complexity of some separation operations, there are also high capital and maintenance investments associated with the equipment [4].

In a tailings dam, solids are settled by sedimentation, so the force on the sediment is the earth's natural gravity. However, when the presence of fines in the tailings is between 90 and 100% and the percentage of solids in the pulp is approximately 30%, the sedimentation speed is very low, due to the high viscosity that this pulp presents. In these cases, dewatering becomes an interesting option. Bauxite tailings fit in this case, given that in its process, all the fine fraction fed is considered tailings.

The dewatering operation may also, in the future, enable the creation of co-products from bauxite tailings, in civil construction areas and as agricultural inputs. Examples of potential applications: cement production, road construction, brick production, soil quality improvement, among others.

Dewatering by Decanter Centrifuges

The use of mechanical equipment to promote dewatering favors the solid–liquid separation in such level that makes it possible to reach a tailings moisture low enough that enables a dry disposal, further, increasing the reuse of water and, thus, eliminating inconvenient risks in dams. This means that tailings do not need to be stored in ponds. There is a reduction in infiltration rates and in the area used, in addition to the elimination of long-term environmental liability.

The centrifuge is a dewatering equipment whose main characteristic is the use of centrifugal force, which acts on solid particles with an intensity much greater than the gravitational force and that can be multiplied by increasing the rotation speed [1]. Its mechanical separation technology was developed approximately a century ago, and allows the continuous processing of the pulp, which is fed through the central axis, while the products are discharged through outlets in the equipment body, as shown in Fig. 2.

Centrifugal force causes the pulp to form a ring inside the cylinder. As the density of solids is greater than that of

Fig. 1 Bauxite processing

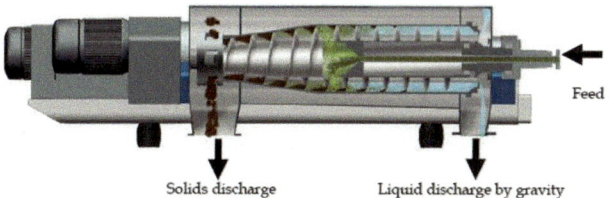

Fig. 2 Cutting of a solid bowl decanter centrifuge

liquid, solids settle on the inner wall of the rotor. To remove solids, a helical screw rotates inside the cylinder with a differential speed in relation to the cylinder and transports the accumulated solids towards the solids discharge.

Results

To assess the best route for dewatering the tailings from bauxite processing, laboratory and field tests were carried out, the results of which are explained below.

Laboratory Tests

The Spin Test is performed to predict the effectiveness of the decanter centrifuge in separating the liquid and solid phases of a given pulp. It is done with the pulp samples being placed in transparent and graduated tubes, which are inserted in a support that will undergo rotation with speed between 1,000 and 4,000 rpm, and times from 1 to 30 min, simulating the effect of a centrifuge.

The tests were carried out in a laboratory centrifuge, with a timer, variable angle, and maximum rotation of 4,500 rpm, with and without the use of flocculant. After carrying out the tests, it was possible to observe in Fig. 3 a high rate of sedimentation in the sample, with good clarification and solid capture.

After the tests, a high sedimentation rate can be seen in the sample, and the clarified phase after the centrifugation test without the addition of flocculants has a clear appearance and a low fraction of suspended solids. For the tests carried out, the objective was to evaluate the performance of the bauxite tailings in the centrifuge without the addition of any type of product.

Pilot Tests

The evaluation and critical analysis of the behavior of the bauxite tailings from the Zona da Mata of Minas Gerais was carried out after the processing of the "run of mine" (ROM) in a pilot plant, and its subsequent dewatering using a decanter centrifuge and evaluating the results considering equipment characteristics and operational parameters.

Pilot plants are operated to generate information on system behavior for use in designing larger installations. The objective when designing, building, and operating a pilot plant is to obtain information about a given process in continuous operation. Based on the data obtained, we can determine whether the process is technically and economically viable, as well as establish the optimal operating parameters for such a process ("scale up"), for the subsequent design and construction of the plant on an industrial scale.

For the proposed work, the ROM was processed in a pilot plant with the unitary operations common to a bauxite beneficiation, as we show in Fig. 4, without processing in a fines circuit. The pilot plant has a processing capacity of 20 t/h of ROM, has an average mass recovery of 40% and is composed of the following main equipment:

- A sizer crusher, with a 50 mm opening;
- A pulping chute that uses high-pressure jet technology to promote breakdown between clay particles and bauxite particles;

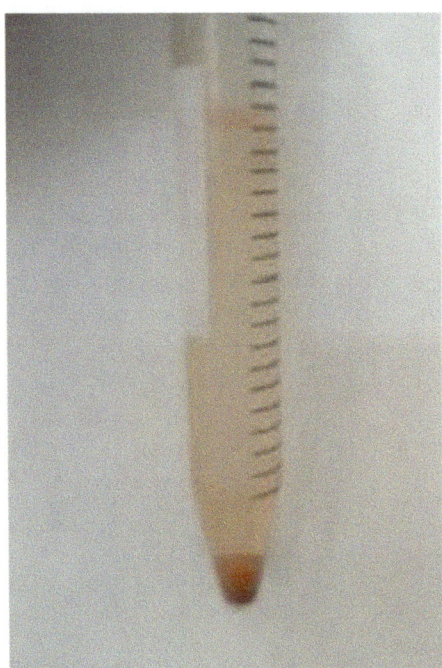

Fig. 3 Sample centrifuged after 30 s

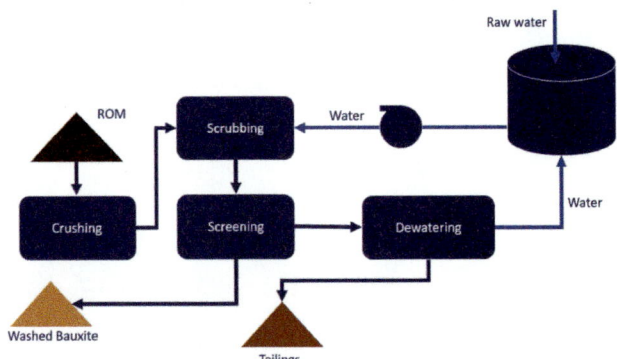

Fig. 4 Flowchart of pilot plant

- A sieve with an opening of 1 mm; and
- A decanter centrifuge.

The main process control variables are the solids flow from the plant, product granulometry, product solids percentage, product washing efficiency, centrifuge feed granulometry, centrifuge feed solids percentage, centrifuge cake solids percentage, and centrifuge clarified solids percentage.

Results

To obtain more reliable data regarding the use of the centrifuge to dewater bauxite tailings, CBA promoted the installation and operation of a pilot dewatering plant at the unit in Itamarati de Minas-MG. The equipment tested was the Decanter 620 MX-V2, supplied by Lindner Techno Systems LTDA, with a bowl diameter of 620 mm and 2.800 mm of length.

The centrifuge parameters during the pilot test presented the best results were 1.500 G and a variation between 35 m³/h and 50 m³/h of volumetric feed rate. For determination of total solids content in the underflow, overflow and in the feed of the centrifuge a moisture analyzer was used. The results of the main tests are shown in Fig. 5.

With the evaluation of the results, we have the average values for feed, 12.5% of solids; overflow 0.5% of solids; and underflow, 81.7% of solids, which represents an excellent result for bauxite tailings and suggest that can be handled in a conveyor belt after the dewatering.

The results above 70% of solids in underflow indicated that the material can be in a dry disposal without challenges, be more sustainable, and with less risks than a traditional wet disposal. In Fig. 5, we can see the appearance of the material after the dewatering (Fig. 6).

Fig. 6 Dewatered solids fraction

Conclusions

The results were used to evaluate a new tailings disposal strategy whereby many technical, economic, and environmental problems associated with current disposal technology can be minimized. The applicability of the equipment for the dewatering of bauxite tailings from the Zona da Mata of Minas Gerais allows for further studies on the disposal of dewatered tailings and the use of the material as co-products.

Based on the results of the tests performed ("spin test" and tests in the pilot plant), it is concluded that the use of a centrifuge decanter for the dewatering of the tailings of the bauxite beneficiation process is a technically acceptable solution.

References

1. CHAVES, A. P. Desaguamento, Espessamento e Filtragem. Teoria e Prática do Tratamento de Minérios. 4. ed. São Paulo: Ed. Signus, 2013. v. 2.
2. CARVALHO, A. 1989. As Bauxitas no Brasil: Síntese de um programa de pesquisa. Universidade de São Paulo - Instituto de Geociências. São Paulo, 1989.
3. SAMPAIO, J.A.; ANDRADE M. C.; DUTRA A. J. B. Bauxita. In Luz, A. B., Lins F. A. F. Rochas e minerais industriais, 2. Ed. Rio de Janeiro: CETEM/MCT, 2008. P.311–377.
4. FRANÇA, S. C. A.; MASSARANI, G. Separação Sólido-Líquido. In: Tratamento de Minérios, 6. Ed. Rio de Janeiro: CETEM, 2018. Cap. 14.

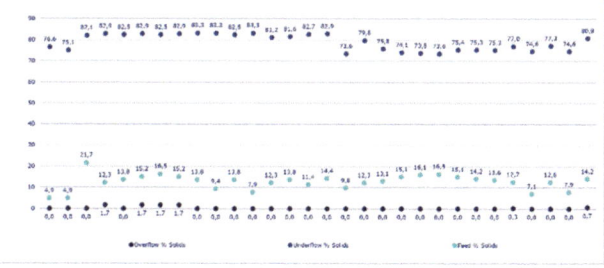

Fig. 5 Results of solids concentration in centrifuge flows

## Part III

Aluminum Alloys, Characterization, and Processing

Comparison of TiB2 and TiC Grain Refiners' Impact on Surface Quality, Edge Cracking, and Rolling Performance of AA5182 DC-Cast Ingot

Josh Lawalin, Pascal Gauthier, and Tao Wang

Abstract

Titanium diboride (TiB2) and titanium carbide (TiC) are both widely used as grain refiners in the production of DC-cast aluminum ingot for rolled products. This study evaluated industrially produced DC-cast aluminum ingot of alloy AA5182 with TiB2 and TiC and the grain refiners' impact on ingot surface condition, as-cast microstructure, rolling performance, grain size distribution, as well as final product mechanical properties. Through optical microscopy and SEM-EDS metallographic examination of the sheet post-hot rolling, the study found that the use of TiC grain refiner resulted in fewer surface stress risers and improved surface quality, rollability, and edge cracking performance of AA5182. It also improved grain size uniformity and created less grain refiner agglomeration versus the same alloy cast with TiB2 grain refiner.

Keywords

Grain refiner • AA 5182 • Microstructure • Rolling performance • Downstream attributes

J. Lawalin
Commonwealth Rolled Products, 1372 St. Hwy 1957, Lewisport, KY 42351, USA

P. Gauthier
Arvida Research and Development Centre, Rio Tinto Aluminium, Jonquiere, QC G7S 4K8, Canada

T. Wang (✉)
Rio Tinto Aluminium Technical Marketing, 200 E Randolph St., Suite 7100, Chicago, IL 60601-7329, USA
e-mail: Tao.Wang@riotinto.com

Introduction

Aluminum alloy flat-rolled products have a wide range of applications in the packaging, aerospace, and automotive markets [1, 2]. The most common manufacturing method for Al-based sheet products is direct chill (DC) casting followed by downstream rolling and heat treatment processes. Many applications mentioned above require the flat-rolled product to have a high surface quality which is directly dictated by the DC-cast ingots' microstructure [3–6].

In the aluminum industry, titanium diboride (TiB2) and titanium carbide (TiC) are both widely used as grain refiners in the production of DC-cast aluminum ingot for rolled products. For the TiB2 grain refiner, there are 3 major types of compositions used which are 5:1 (5wt% Ti: 1wt% B), 3:1, and 5:0.2, while TiC only has 3:0.15 composition as the predominant option. The synthetic TiB2 particle itself has no grain refiner ability without Ti addition which helps to form the Al_3Ti on the surface of refiner and reduce the lattice mismatch with σ-aluminum matrix. In contrary, the TiC refiner doesn't require any doping to be functional. The grain refiner particles, when added to the liquid metal, serve as heterogenous nucleation sites during the solidification process and prevent the non-uniform grain coarsening and thus reduce the tendency of ingot cracking and micro-shrinkage porosity.

Even though the functionality of different grain refiners has been well studied, there are a few research focused on the effect of grain refiner on ingot rollability and microstructure. This study evaluated industrially produced DC-cast aluminum ingot of alloy AA5182 with TiB2 and TiC and assessed the grain refiners' impact on ingot surface condition, as-cast microstructure, rolling performance, grain size distribution, as well as final product mechanical properties. Through optical microscopy and SEM-EDS metallographic examination of the sheet post-hot rolling, the study found that the use of TiC grain refiner resulted in fewer surface stress risers and therefore improved surface quality, rollability, and edge cracking performance of AA5182. It also improved the grain size uniformity and created less grain refiner agglomeration versus the same alloy cast with TiB2 grain refiner.

Experimental

Ingot Casting

In this paper, the 5xxx alloys for packaging and automotive applications were selected for the tests. The trial ingots were cast using Direct Chill (DC) casting technology provided by Wagstaff®. The grain refiner was added to the liquid metal after the tilting furnace but before the degasser system in the form of rod as illustrated in Fig. 1. The TiB2 (5/1) and TiC (3/0.15) grain refiners were added with the same rate and metallic Ti was also added in the furnace to keep the same level Ti in the final chemistry for all ingots. The steady-state portion of the DC-cast ingots were cut into smaller slices for metallographic analysis. Lastly, the final sheet microstructure and mechanical properties were evaluated for all the products produced with different grain refiners.

Pre-Heating, Rolling, and Post-Heat-Treatment Processes

Industrial scale, hot-rolling trials were conducted to better understand the effect of different grain refiners on ingot surface quality, microstructure, and rolling performance, especially the edge cracking performance. Prior to the rolling process, the rolling faces of each ingot were scalped to eliminate the surface imperfections, undesirable microstructure, and any compositional segregations resulted from the initial non-steady solidification.

After scalping, each DC ingot was preheated (6 h of gradual heating to over 500 °C), and then soaked for 10–18 h to achieve a uniform microstructure and suitable rolling surface temperature. Subsequently, the samples were hot rolled and cold rolled. The ingot thickness was reduced from 660 mm to approximately 3 mm and 1 mm after hot rolling

and cold rolling, respectively. The hot rolled sheet edges were closely monitored and rated based on cracking severity. The coil was completed edge trimmed before sending to cold rolling mill to eliminate all the edge defects. The trimmed edge samples were then sent to research lab for further metallurgical analysis using optical microscope, scanning electron microscope with energy dispersive spectrometer.

Results and Discussion

Grain Refiner's Effect on Hot Rolled Sheet Edge Cracking

The ingots cast with TiC and TiB2 grain refiners were prepared and hot rolled with the same procedure. It is found that the TiC refined sheet consistently demonstrated less edge cracking than the TiB2 refined sheet. As shown in Fig. 2, the average edge cracking was about 25 mm into the width for the TiB2 refined sheet while it is only 5 mm deep for the TiC refined product. It is also noticed that the frequency of edge cracking occurrence is significantly reduced on the TiC refined product with its edges remained close to a straight line. This phenomenon is explained by the transverse elongation pattern of the ingot edge material during rolling, the straight-line edge indicates the even material expansion during rolling.

Effect on Ingot Surface Topography

The surface topography was monitored closely on the test ingots based on the meniscus bandwidth between the peaks and depth of the valleys as illustrated below in Fig. 3. The TiC refined ingot demonstrates an overall smoother surface as TiC has less particle agglomeration which affects the ingot surface solidification pattern.

The surface smoothness measurements were listed in Fig. 4, and the results are consistent from the rolling faces and short faces. Compared to the TiB2 refined ingot, the TiC ingot presents very shallow meniscus band, and the severity of ingot surface roughness was reduced by approximately 50%.

Fig. 1 Grain refiner addition system in the aluminum slab cast house

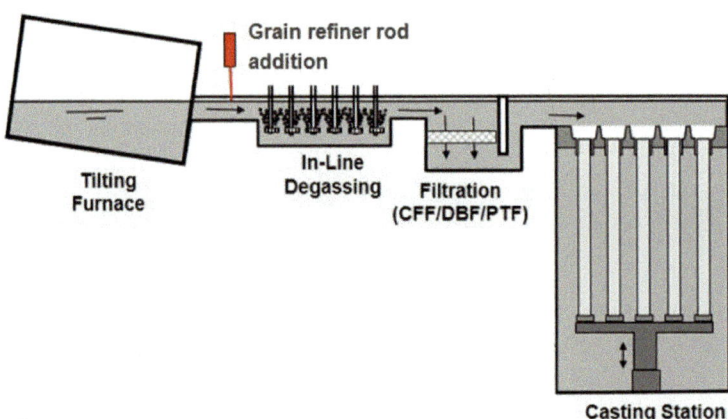

Fig. 2 Edge cracking on sheet samples rolled from **a** TiB2 refined ingot **b** TiC refined ingots

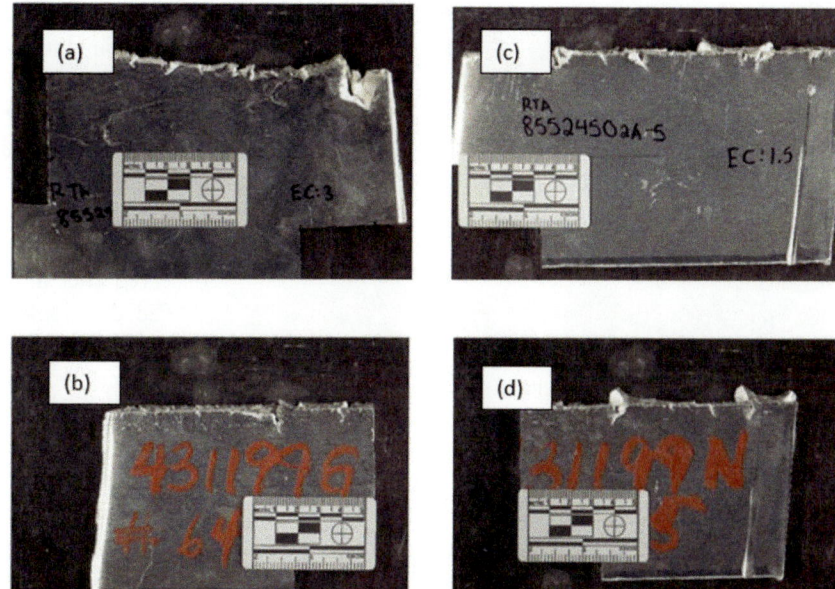

Fig. 3 Surface quality of as-cast ingot with **a** TiC grain refiner and **b** TiB2 grain refiner

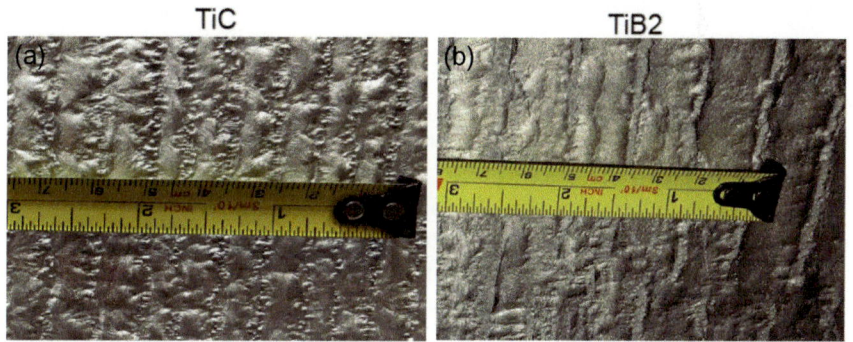

Fig. 4 Meniscus band measurement on ingots refined with **a** TiC and **b** TiB$_2$

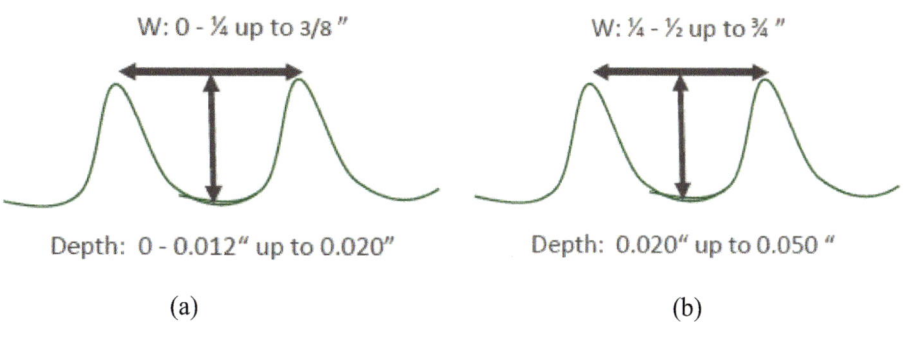

The effect of the surface topography has been studied and it is found that the subsurface below meniscus band dictates the surface rolling and edge cracking performance as these bands drive the inverse segregation deeper into the ingot body and occasionally beyond the scalping depth. Lab heat treatment was conducted on one of the test ingot slices with severe meniscus band as shown in Fig. 5 to determine if the heat treatment would modify the microstructure enough to reduce the band depth. After 30 h soaking at 525 °C, the meniscus bands are still visible, even though the intensity was slightly reduced. Since the dissolution is diffusion-controlled process, by applying heat over time, only a small part of the Mg- and Fe-bearing intermetallics were dissolved. In summary, it is very critical to prevent the formation of surface banding by using the suitable grain refiner such as TiC during cast as the banding is a persistent defect.

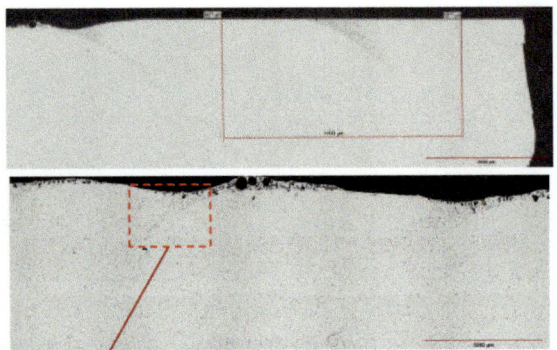

Fig. 5 Meniscus band on the test ingot slice **a** before heat treatment and **b** after heat treatment

Effect on Grain Structure and Intermetallic Phases

One of the main purposes of adding grain refiner in the casting process is to achieve an even microstructure by suppressing the undesirable grain coarsening. The effectiveness of the grain refiner is assessed by both the grain size and the evenness of the size distribution. Even though the sheet samples exhibit similar overall grain structure, it is observed that the grain size distributions around the hot rolled sheet edges are different between the TiC and TiB2 refined products. As illustrated in Fig. 6, the grains of the hot rolled sheet samples are highly deformed and showed sign of recrystallization. The grains of the TiC refined samples are slightly more equiaxed which indicated even grain growth during recrystallization.

As shown in Fig. 7, there was a small island of large grains around the edge cracking areas from the TiB2 refined sample while the TiC sample showed very evenly distributed and overall smaller grain sizes. It is believed the grain size difference around the sample edge led to the uneven force distribution and provoked the edge cracking during hot rolling. This microstructural disadvantage existing in the TiB2 product explains its severe edge cracking occasionally observed after finishing hot rolling process.

The dispersoids were detected in both types of rolled samples and the compositional analysis confirmed they are predominantly Fe-phase and Mg2Si. Despite being the similar phases, the dispersoids exhibit different morphologies from different samples as shown in Fig. 8. It is measured the Mg2Si and Fe-phase to be 3.6 µm and 3.4 µm, respectively, in the TiC samples compared to 4.7 µm and 4.1 µm in TiB2 samples. Also, the dispersoids in the TiC samples are found to be more evenly dispersed in contrast to those in the TiB2 samples which agglomerate in the rolling direction. It is believed that the smaller and well-distributed dispersoids in TiC samples contribute partially to the finer grain size by better pining the grain boundary and preventing directional grain coarsening.

All in all, it is believed that the TiC refined samples present an overall better microstructure and therefore have better edge cracking performance during hot rolling.

Effect on the Mechanical Properties

The mechanical properties of the final rolled products were tested by the uniaxial tensile testing, bending tests, and stamping tests, and the results were compared between sheet

Fig. 6 Grain structure after rolling from **a** TiC refined sample and **b** TiB2 refined sample

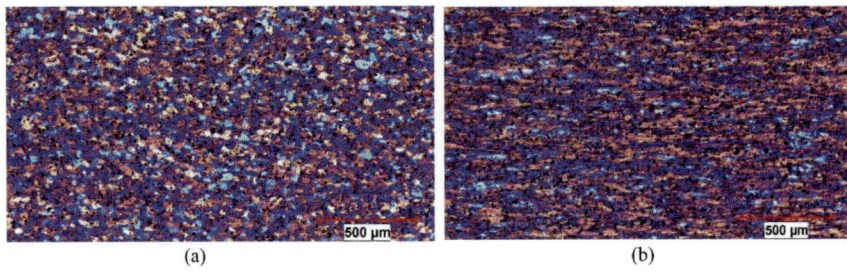

(a)

(b)

Fig. 7 Grain structure analysis around edge cracks from **a** TiC refined sample and **b** TiB2 refined sample

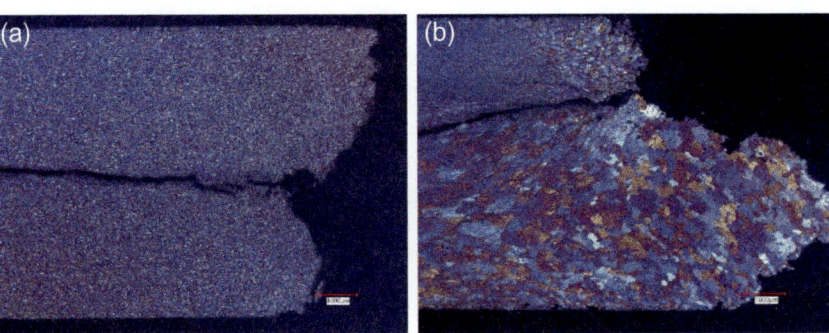

Fig. 8 Intermetallic phases in **a** TiC refined sample and **b** TiB2 refined sample

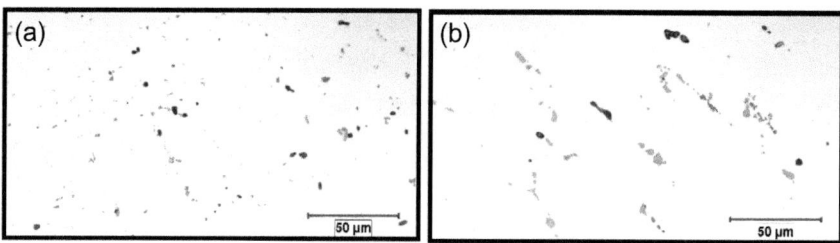

products rolled from both TiC and TiB2 refined ingots. No significant difference was noticed in the final products' downstream attributes as expected.

desirable microstructure and surface condition from the TiC refined product improves its overall rollability without sacrificing mechanical properties.

Conclusion

In this study, the ingots cast with different grain refiners were hot rolled, cold rolled, and heat treated. The effects of grain refiners on surface condition, as-cast microstructure, rolling performance, grain size distribution, and final product mechanical properties were studied using optical microscopy, and SEM-EDS. It is found that TiC grain refiner resulted in fewer surface stress risers and produced an overall smoother surface, which contributes to a better edge cracking performance from the standpoint of both frequency and severity. Also, after hot rolling, the slab cast with TiC exhibits a more uniform grain shape and even size distribution. The dispersoids in the TiC samples are found to be more evenly dispersed and the smaller dispersoids contribute partially to the finer grain size by better pining the grain boundary and preventing directional grain coarsening. The

References

1. Khangholi, S.N., et al., *Optimization of mechanical properties and electrical conductivity in Al–Mg–Si 6201 alloys with different Mg/Si ratios.* Journal of Materials Research, 2020. **35**(20): p. 2765-2776.
2. Grandfield, J.F., D.G. Eskin, and I.F. Bainbridge, *Direct-Chill Casting of Light Alloys.* 9781118022658. 2013, The Minerals, Metals & Materials Society.
3. O'Dette, J.H., *Blister formation in rolled aluminium,* in *The Minerals, Metals & Materials Society (TMS).* 1957.
4. Gali, O., *Micro-mechanisms of Surface Defects Induced on Aluminium Alloys during Plastic Deformation at Elevated Temperatures,* in *Department of Mechanical, Automotive & Materials Engineering.* 2017, University of Windsor.
5. Ito, H., *STUDY ON BLISTER OF ALUMINIUM AND ITS ALLOY.* Materials Science, Journal of Japan Institute of Light Metals, 1954.
6. Mofarrehi, M., M. Javidani, and X.-G. Chen, *On the intermetallic constituents in the sodium-induced edge cracking of hot-rolled AA5182 aluminium alloys.* Philosophical Magazine, 2021: p. 1849–1870

The Influence of Crystallographic Texture Gradients on the Deformation Response of Aluminum Extrusions

W. J. Poole, Y. Wang, A. Zang, M. A. Wells, and N. C. Parson

Abstract

The use of 6xxx series aluminum extrusions in automotive applications is increasing due to the requirement to reduce vehicle weight. Extruded profiles can be subjected to severe and complex plastic strain paths during fabrication or during crash loading. During the extrusion process, deformation gradients develop due to (i) friction between the tooling and the extrudate, (ii) redundant work, and (iii) splitting of the extrusion and welding in porthole dies when fabricating hollow profiles. The current study examined these effects in materials with high dispersoid densities at relatively low extrusion temperatures ($\sim$400 °C) where recrystallization was deliberately suppressed. Local textures were characterized using electron backscatter diffraction (EBSD) and the local stress–strain response was predicted using the visco-plastic self-consistent (VPSC) polycrystal plasticity model. The results were used to rationalize (i) the inhomogeneity of through thickness deformation during tensile testing and (ii) the localization of plastic deformation at the weld seam.

Keywords

Aluminum extrusion alloys • Crystallographic texture gradients • Plastic response • Porthole dies

W. J. Poole (✉) · A. Zang
Department of Materials Engineering, The University of British Columbia, 309-6350 Stores Road, Vancouver, BC V6T 1Z4, Canada
e-mail: warren.poole@ubc.ca

Y. Wang · M. A. Wells
Mechanical and Mechatronics Engineering, University of Waterloo, 200 University Avenue West, Waterloo, ON N2L3G1, Canada

N. C. Parson
Rio Tinto Aluminium, 1955 Mellon Blvd, Jonquière, PQ G7S 4K8, Canada

The application of 6xxx series aluminum extrusions is increasing for both traditional internal combustion engine (ICE) and battery electric vehicles (BEV) [1]. Lightweighting of vehicles is beneficial for ICE vehicles to reduce fuel consumption and to extend the range of BEV vehicles. There are challenges for the application of aluminum alloys in terms of their processing and final performance. In this study, the effect of microstructure gradients on the plastic response of extruded products has been examined with the objective of relating variations in local microstructure to the bulk response of the product.

Extrusion is particularly prone to producing microstructure gradients due to the combination of (i) friction between the die and the material being extruded and (ii) redundant work within the die. The effect of friction has been examined for simple rectangular strips which results in a variation of microstructure through the thickness of the strip, i.e. between the surface of the extrudate and its interior [2]. On the other hand, porthole dies are used to produce hollow extrusions (potentially with multiple cavities). The material fed into the porthole die is split into multiple streams which pass through ports separated by bridges that support a mandrel (or mandrels) which define the internal cavity dimensions. In this scenario, after the material passes through the ports, it rejoins a common stream in the welding chamber of the die, resulting in longitudinal weld seams in the final product. The presence of the ports/bridges results in redundant work and additional friction between the extrudate and the die. In this study, a simplified geometry was used to produce a simple rectangular strip with a weld seam along the centre of the extrudate.

There have been some studies on weld quality (incomplete welding and voids at the weld seam [3, 4]) and various criteria have been proposed for use in numerical simulations, e.g. see reference [5], but relatively little research has been done on the effect of microstructure gradients in the thickness of the extrudate or near weld seams where sound weld lines are present. In these situations, local variations in

Table 1 Alloy Compositions (wt %)

Alloy	Grain structure	Mg	Si	Mn	Fe	Cr	Al
AA3003	Recrystallized	–	0.10	1.27	0.54	–	Bal
AA6082	Unrecrystallized	0.71	1.04	0.520	0.23	0.15	Bal

(i) grain size, (ii) second phase particle distributions, and (iii) crystallographic texture could affect the localization of plasticity leading to premature failure at weld seams. For example, the recent work of Wang et al. identified significant differences in strain localization at weld seams in extrusions with recrystallized and unrecrystallized microstructures [6]. As such, the current study will provide an overview of recent results on the plastic response of materials with (i) through thickness microstructure gradients and (ii) microstructures variations near weld seams.

Alloys of AA3003 and AA6082 were direct chill cast, homogenized, and extruded on an instrumented pilot scale extrusion press by Rio Tinto Aluminium. The alloy compositions are shown in Table 1. The homogenization conditions were 24 h at 600 °C for AA3003 and 2 h at 550 °C for AA6082.

The AA3003 was extruded at a billet temperature of 400 °C and a ram speed of 32 mm/s while for AA6082 a billet temperature of 480 °C and a ram speed of 5 mm/s was employed. The die orifice had dimensions of 3 × 90 mm for AA3003 and 2.5 × 50 mm for AA6082 (with a streamlined bridge in the centre of the die, see [7] for more details). The grain structure was characterized on metallographically prepared samples using electron backscatter diffraction (EBSD) mapping with a Zeiss Σigma field emission gun microscope with an EDAX/TSL OIM configuration. Finally,

in the case of the porthole extrusions, the sample was deformed in tension perpendicular to the weld seam and the local strain was measured by digital image correlation (DIC) using the DaVis software, note: the Fe rich constituent particles in the alloy were used as the fiducial markers for the DIC.

Effect of Through-Thickness Microstructure Gradients

Figure 1a is an inverse pole figure (IPF) EBSD map showing the grain structure variation from the surface (bottom) to the centre of the extrudate (top) for the AA3003 extrusion. The 001 pole figures shown in Fig. 1b, c illustrate the significant difference in the crystallographic texture between the centre and surface, respectively. The difference in average orientation of the grains at the centre (a combination of cube and Goss orientations) and those near the surface (cube and Goss rotated around ND direction) probably arises due to the large surface shear strains parallel to the surface caused by friction between the extrudate and the die [7].

The impact of the through-thickness texture variation is complicated but an example of an effect is illustrated in Fig. 2 where cross-sections perpendicular to the tensile axis close to the fracture surface were prepared for tensile

Fig. 1 **a** EBSD inverse pole figure map of grain structure from the surface to the centre, **b** 001 pole figure illustrating orientation of grains within ∼0.3 mm of the surface and **c** 001 pole figure for grains near the centre

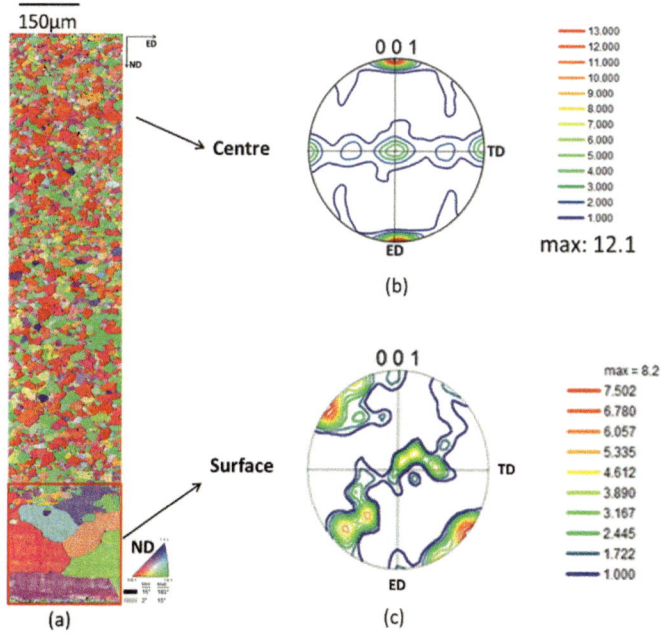

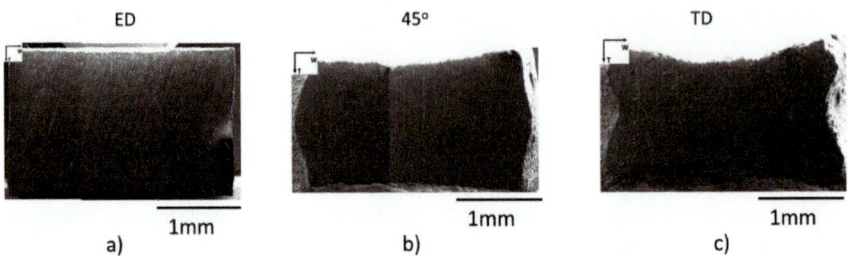

Fig. 2 Secondary electron images showing the polished cross-section from tensile samples tested **a** parallel to ED, **b** at 45° to ED, and **c** TD (i.e. 90° to ED). A true strain of ∼0.7 was applied

samples deformed parallel to ED, at 45° to ED and parallel to TD. Here, it can be observed that the cross-section is rectangular for the case of testing parallel to ED (Fig. 2a), but the sides of the tensile sample are convex for the 45° sample and concave for the TD sample. The shapes of the cross-sections can be rationalized by the difference in the anisotropic plastic response of the surface layer and the interior of the sample (fundamentally related to the differences in crystallographic texture shown in Fig. 1). See reference [2] for a more detailed explanation.

Microstructures Variations Near Weld Seams

Figure 3a shows a plane section view of the porthole die taken at the centre of the extrudate. It shows the splitting of the incoming billet into the two ports, their recombination in the weld chamber and the resulting weld seam in the final extrudate. Figure 3b illustrates an end-on view of the emerging microstructure around the longitudinal weld seam. Finally, Fig. 3c gives examples of 001 pole figures for selected regions showing that the local texture varies

significantly. Figure 4 shows a map of the plastic strain component, ε_{xx}, after the sample was stretched in tension perpendicular to the weld seam. Careful comparison of Figs. 3b and 4 shows that there appears to be correlation between the local texture and the level of strain in a given region. Given that it is well known that the stress–strain response of a polycrystal depends on its crystallographic texture, it is speculated that the local texture determines the local stress–strain response. This produces regions that are relatively harder or softer resulting in different levels of local strain during loading.

In summary, the current work illustrates that the nature of the extrusion process (friction between the extrudate and the die and redundant work) leads to spatial variations in crystallographic texture. These manifest as through thickness microstructure gradients and as local variations in microstructures near weld seams in porthole die extrusions. Plastic deformation of extrudates with these spatial variations of microstructures (in particular crystallographic textures) leads to a complex interplay between microstructure and strain localization. It is proposed that the resulting experimentally observed plastic strain variations can be

Fig. 3 a section through porthole die, **b** end on EBSD IPF map of grain structure near the weld seam and **c** 001 pole figures illustrating local crystallographic texture from selected regions suggest that this would explain the observed local strain distribution measured experimentally [8]

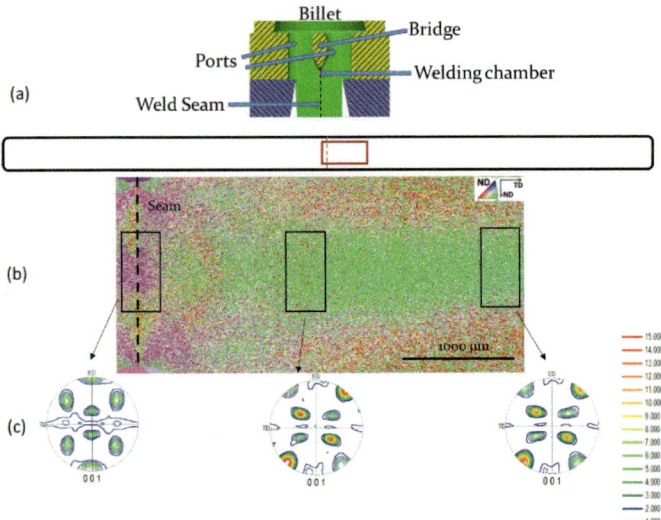

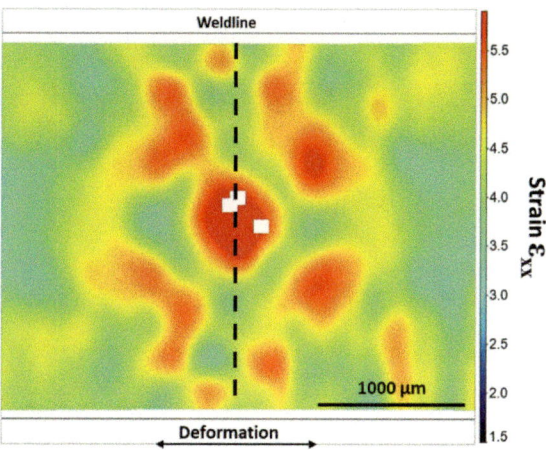

Fig. 4 DIC map of the strain component, ε_{xx}, (%) which corresponds to the extension strain perpendicular to the weld seam

rationalized by relating local texture to the local plastic response.

Acknowledgements This work was undertaken, in part, thanks to funding from the Canada Research Chair program (Poole). Additional support of Rio Tinto Aluminium and the Natural Sciences and Engineering Research Council of Canada (NSERC) is gratefully acknowledged.

References

1. G. Scamans, *Light Metal Age*, Electric vehicle spike demand for high strength aluminium extrusions, 2018. **76**: pp. 6–12.
2. J. Chen, W.J. Poole, and N.C. Parson, *Mater. Sci. Eng. A.*, The effect of through thickness texture variation on the anisotropic mechanical response of an extruded Al-Mn-Fe-Si alloy, 2018. **A730**: pp. 24–35.
3. J.Q. Yu, G.Q. Zhao, and L. Chen, *J. Mater. Proc. Tech.*, Analysis of longitudinal weld seam defects and investigation of solid-state bonding criteria in porthole die extrusion process of aluminum alloy profiles, 2016. **237**: pp. 31–47.
4. J. Yu and G. Zhao, *Mater. Charaterization*, Interfacial structure and bonding mechanism of weld seams during porthole die extrusion of aluminum alloy profiles, 2018. **138**: pp. 56–66.
5. L. Donati and L. Tomesani, *J. Mater. Proc. Tech.*, The prediction of seam welds quality in aluminum extrusion, 2004. **153–154**: pp. 366–373.
6. Y. Wang, A. Zang, M. Wells, W.J. Poole, and N. Parson, *Mat. Sci. Eng.*, Strain localization at longitudinal weld seams during plastic deformation of Al–Mg–Si–Mn–Cr extrusions: The role of microstructure, 2022. **A849**: pp. 143454 (12 pages).
7. Y. Wang, A. Zang, Y. Mahmoodkhani, M.A. Wells, W.J. Poole, and N.C. Parson, *Metall. Mater. Trans. A*, The Effect of Bridge Geometry on Microstructure and Texture Evolution During Porthole Die Extrusion of an Al–Mg–Si–Mn–Cr Alloy, 2021. **52A**: pp. 3503–3516.
8. A. Zang, J.F. Beland, N.C. Parson, and W.J. Poole, *Strain Localization Effects in AA6082 Extrusion Seam Welds*, in *18th Int. Conference on Al Alloys*. 2022: Toyama, Japan.

Mechanical Properties, Microstructures, and Textures of Cold Rolling Sheets Made from a Low-cost Continuous Cast Al-1.5Cu Alloy with Potential Application in Auto Sheets

Xiyu Wen, Yan Jin, and Wei Li

Abstract

Mechanical properties, microstructures and textures of cold rolling sheets from a low-cost continuous casting (CC) Al-1.5Cu alloy were characterized and investigated thoroughly using tensile testing machine, Optical & Scanning Electron microscopes (SEM), and X-ray diffractometer (XRD). Compared with AA5xxx and AA3xxx aluminum alloys, ascribing to the precipitation and solid solution of Al-Cu phase, material characteristics of the Al-Cu alloy after cold rolling at different gauges are significantly different with some interesting findings. The findings from the cold rolling hardening would potentially affect the subsequent processing, such as solid solution heat treatment, quenching, forming, and aging, for application (replace AA5xxx-O sheet) in auto sheets.

Keywords

Aluminum copper alloy sheet • Microstructure • Mechanical properties and annealing • Textures • Continuous cast processing

Introduction

Aluminum-Copper (Al-Cu) alloy is one of the high strength aluminum (Al) alloys. Currently, it has been extensively used in the aircraft and military industries, such as aircraft skin, spars, forging parts, and rivets because of combination of high strength and good toughness. In general, Al-Cu alloys commonly contain 3.8–4.9 wt.% Cu with addition of other alloying elements of Mn and Mg. By going through a regular manufacturing process, including DC casting, hot rolling, solid heat treatment, quenching, and aging, the plates or sheets with different gauges were ultimately applied in fields of aircraft and military. With fast-paced development of automotive weight reduction, high strength aluminum alloys have showed a tendency to replace 5000 and 6000 alloys due to their excellent specific strength. For the Al-Cu alloys containing low Cu, i.e., 1 to 3.8 wt.%, the relevant study and literatures can barely be searched. Given that the Al-Cu alloys from special low-cost processing have potential application in automotive industries, there were also several papers published regarding low Cu (1-2wt%) Al-Cu alloys, which were only focused on the aging precipitation and texture evolution after homogenization and cold rolling [1–6].

Compared to conventional casting method, i.e., direct chill (DC) casting of commercial aluminum sheets, low-cost continuous cast (CC) processing routes started to be used more frequently. The CC processing have remarkable advantages because of its low cost and high efficiency with omission of large ingot casting equipment, scalping machine, homogenization furnace, and breaking-down rolling. Currently, the continuous cast technology can produce 1000, 3000, 5000 series low strength alloys perfectly as well as some 8000 series alloys; however, there are still some difficulties in casting ingot slabs for 2000, 6000, and 7000 series medium or high strength alloys. Auto manufacturers all over the world have focused on automotive parts manufacturing from the high strength aluminum alloys rather than low strength aluminum alloys owing to increasing requirement of auto lightweight. More importantly, compared to the conventional manufacturing process, i.e., DC route, of these high strength aluminum alloys, CC route significantly varies in cooling rate during casting, hot rolling reduction, and production cost. Therefore, the mechanical properties, microstructures, and textures of CC Al sheets show remarkable

X. Wen (✉) · Y. Jin
Center for Aluminum Technology, University of Kentucky, 1505 Bull Lea Road, Lexington, KY 40511, USA
e-mail: Xwen2@uky.edu

W. Li
Department of Mechanical and Aerospace Engineering, University of Kentucky, 351 Ralph G. Anderson Building, Lexington, KY 40506, USA

© The Minerals, Metals & Materials Society 2023
S. Broek (ed.), *Light Metals 2023*, The Minerals, Metals & Materials Series,
https://doi.org/10.1007/978-3-031-22532-1_36

difference in the final products, although their mechanical properties are sometimes comparable [7–15].

In the paper, mechanical properties, microstructures and textures of cold rolling sheet of a low-cost continuous casting (CC) Al-1.5Cu alloy were thoroughly investigated using tensile testing machine, optical & scanning electron microscopes, and X-ray diffractometer (XRD). Compared with AA5xxx and AA3xxx alloys, material characteristics of the Al-Cu alloy after cold rolling at different gauges are significantly different with some interesting findings due to its precipitation and solid solution of Al-Cu phase. The findings from the cold rolling hardening would potentially affect the subsequent processing, such as solid solution heat treatment, quenching, forming, and aging, for application with the purpose of replacing AA5xxx-O sheet in automotive body sheets.

Design of the Al Alloy, Material, and Experiment

Material of Experiment

An Al-1.5Cu alloy was used as shown in Table 1. The composition of the Al-Cu alloy with very low wt.% Si and Fe met with AA2037 alloy certification. Cu is at low level (fall in 1.4–2.2%) due to requirement of twin belt continuous cast processing. The hotband with 2.5 mm gauge of Al-1.5Cu alloy is from a twin belt CC processing.

Annealing and Cold Rolling of the Hotband

Heat treatment of the hotband samples was carried out in the Ney Vulcan 3–1750 furnace at different temperatures (149 °C, 232 °C, 354 °C, 427 °C, 482 °C, and 511 °C) for 12 h (1.0 °C/min heating rate), and then the hotband samples were cold rolled to final gauges with 40%, 60%, and 80% reductions, respectively.

Microstructures

Particle microstructures were characterized using Scanning Electron Microscopes (SEM—JOEL JSM-5900LV and FEI

Helios Nanolab 660) and the grain structure analysis was performed by Optical Microscope (OLYMPUS GX51).

Mechanical Property Test

Tensile specimens were made using a press machine for "dog-bone" tensile specimens of 50.8 mm gauge length and 12.5 mm width in the rolling direction. The tensile tests were performed utilizing the calibrated Material Test System MTS 810 with calibrated axial extensometer (50.8 mm gauge length). A 2267 kg load cell was used in the 0–2267 kg calibration range. The speed of cross head was kept constant at 2.54 mm/minute throughout the test. The yield strength was at 0.2% offset. The elongation recorded is the plastic elongation until fracture. The tests were performed in accordance with ASTM E8/E 8 M.

Texture Measurement

Samples for texture measurement were prepared on the rolling plane, i.e., L–T (rolling direction–transverse direction) plane. The surfaces for the measurement were carefully polished to reduce the surface roughness and meanwhile minimize the effect of surface residual stresses due to aggressive grinding.

The Rigaku D/Max X-ray diffractometer unit with an Eulerian cradle was used to conduct the pole figures measurement and analysis. Four {111}, {200}, {220}, and {311} pole figures for each sample were measured. The alpha rotation angle was from 15° to 90°, and alpha step angle was 5° by the Schulz back-reflection method using CuK_α radiation. The three-Dimensional Orientation Distribution Functions (ODF) and volume fractions of textured components were calculated by use of the Tex Tools software developed by ResMat Corporation. The orientation intensity is expressed in form of a triple of Euler angles (φ_1, Φ, φ_2) according to Bunge's notation. The calculation of volume fraction from the different components was based on the well-accepted definition system, i.e., Cube-{001} < 100 >; Goss-{011} < 100 >; Brass-{011} < 211 >; S-{123} < 634 >, Copper-{112} < 111 > and R-cube-{001} < 110 >. The tolerance angle used in the calculation is 15° that was widely used for each sample.

Table 1 Chemistry of hotband of the Al-Cu alloy (*American Association Teal Sheets 2018)

Wt%	Si	Fe	Cu	Mn	Mg	Cr	Zn	Ti	V	Al
Hotband	0.06	<0.01	1.48	0.24	0.45	<0.01	0.02	0.02	0.01	Rem
AA2037*	0.50	0.50	1.4–2.2	0.10–0.40	0.30–0.80	0.10	0.25	0.15	0.05	Rem

Results and Discussion

Microstructures

Figure 1 shows the particle structures of the hotband samples after annealing at different temperatures and cold rolling to different gauges. When annealing temperature was less than 427 °C, the primary particle structures did not change a lot. When annealing temperature was higher than 427 °C, the number density of the bright Al-Cu primary particles decreased significantly. In Fig. 2, a series of high magnification SEM images showed the very fine precipitates or dispersoids with bright contrast, i.e., Al-Cu or Al-Mn, which were not uniformly distributed in the cold rolling samples via four different heat treatment processes (hotband without annealing, hotband + annealed at 149 °C, 354 °C, and 482 °C for 12 h). Direct cold sheet has about 50 nm in diameter fine precipitates from casting process, while for the samples annealed at 149 °C, 354 °C and 482 °C for 12 h and cold rolled, the average diameter of precipitates is approximately 90 nm, 70 nm, and 100 nm, respectively. The result from above observation is similar to Zhai's findings on dispersoids using TEM [16]. The fine precipitates or dispersoids at different sizes may influence the deformation behavior in the process of cold rolling.

Figure 3 shows the optical micrographs of grain structures of the hotband samples after annealing at different temperatures and cold rolling to different gauges. Severe deformation of grains was observed when annealing temperature is lower than 354 °C, which is a typical cold rolling grain structure. When annealing temperature is higher than 427°F, the recrystallized grain structure can be seen with micro shear-bands in grains.

Evolution of Textures

Figures 4, 5, 6 show the ODFs of the hotband samples annealed at different temperatures and cold rolled to different gauges. It was revealed that there is strong β fiber texture from Brass through S to Copper with weak Goss and Cube textures in the sheets when annealing temperature is less 232 °C. With cold reduction increasing, the intensity of β fiber texture increases. When annealing temperature reaches 354 °C or higher temperature, the combination of β fiber and Cube textures becomes main components with cold reduction increasing. Figure 7 clearly shows the change of volume fractions of six typical texture components in the hotband samples after annealing at different temperatures and cold rolled to different gauges. The result from Fig. 7 confirmed measurement from Figs. 4, 5, 6.

Evaluation of Mechanical Properties

Figure 8 shows the tensile curves of cold rolling sheets from CC hotband samples after annealing at different temperatures and cold rolling. Figure 9 shows the curves of UTS vs Cold reduction from hotband samples after annealing at different temperatures and cold rolling. It was found that after annealing at 323 °C and 354 °C, the cold rolling sheets have lowest strength and work hardening (deformation hardening) rate than other cold rolling sheets. In general, the Al-Mn precipitations would precipitate at annealing temperature range of 316 °C–371 °C. This implied that the fine Al-Mn dispersoids precipitated after annealing at 232 °C and 354 °C might have contributed to the low deformation hardening rate. Direct cold rolling sheet and sheet after annealing at 300°F

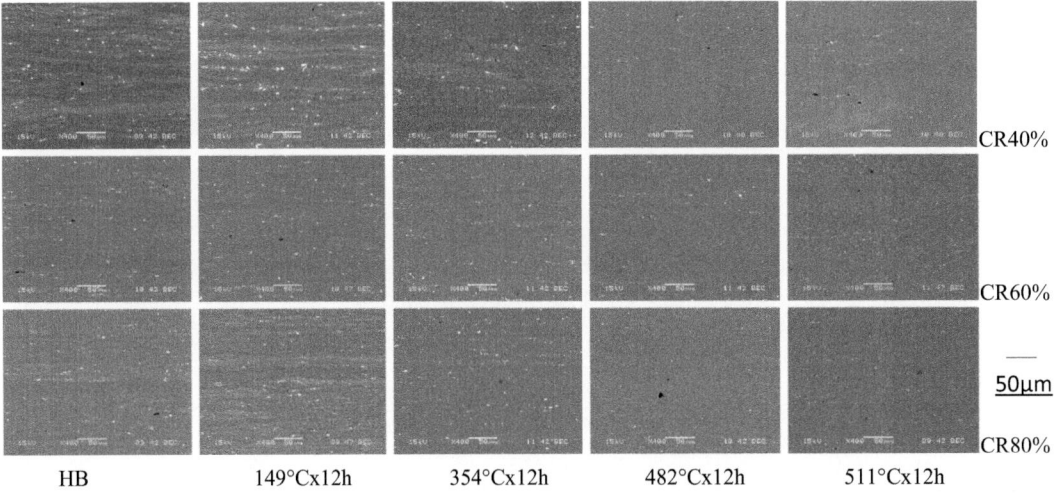

CR40%

CR60%

50μm

CR80%

HB   149°Cx12h   354°Cx12h   482°Cx12h   511°Cx12h

Fig. 1 SEM images of particle structures of the hotband samples after annealing and cold rolling

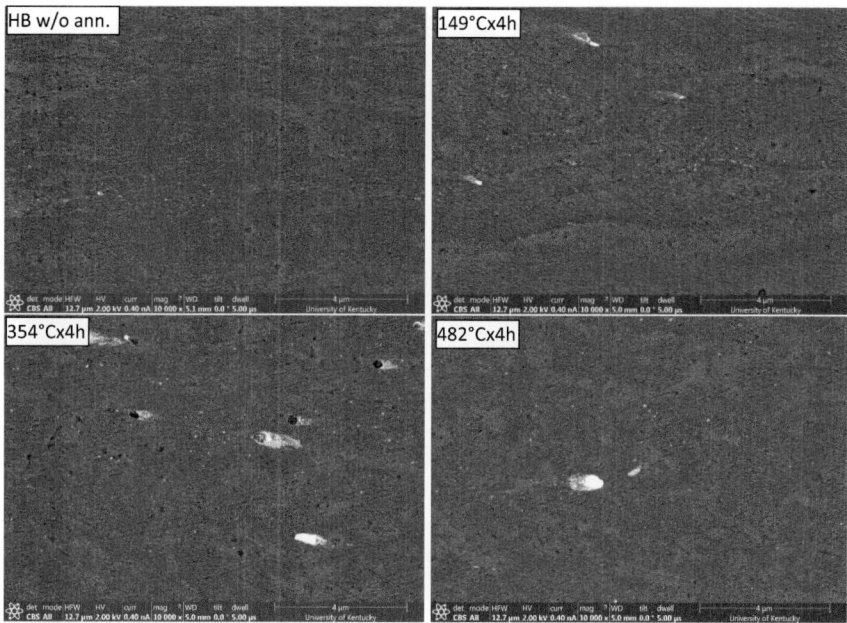

Fig. 2 SEM images of fine precipitations of the hotband samples after annealing and cold rolling

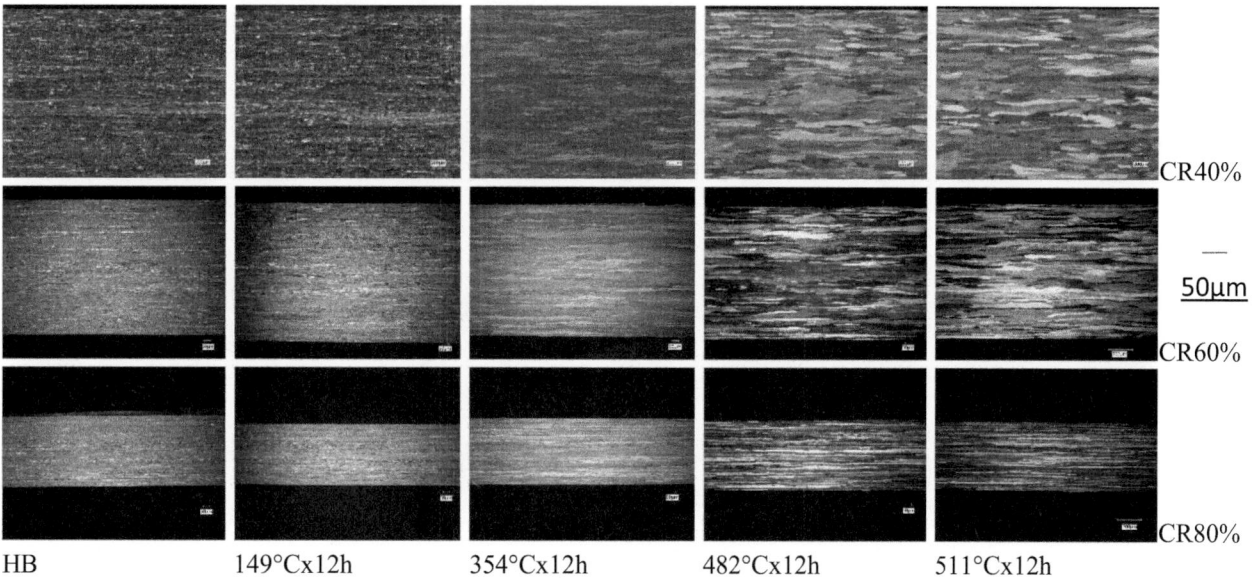

Fig. 3 Optical photograph of grain structures of the hotband samples after annealing

have almost the same strength with ones after annealing at 427 °C–511 °C and cold rolling. This indicated that they have similar solid solution level. Hence, it was indicated that the effect of Al-Mn precipitation was lower than solid solution hardening on the deformation hardening during cold rolling. With temperature increased from 427 °C to 511 °C, solid solution hardening increased. UTS of the alloy after 511 °C and cold rolling 80% can rise to 376 MPa. Compared to 5000 series alloys, it was found that the alloy has higher strength because of solid solution of Cu atoms hardening at 427 °C–511 °C or without annealing. This could apply to auto parts for lightweight because of high strength.

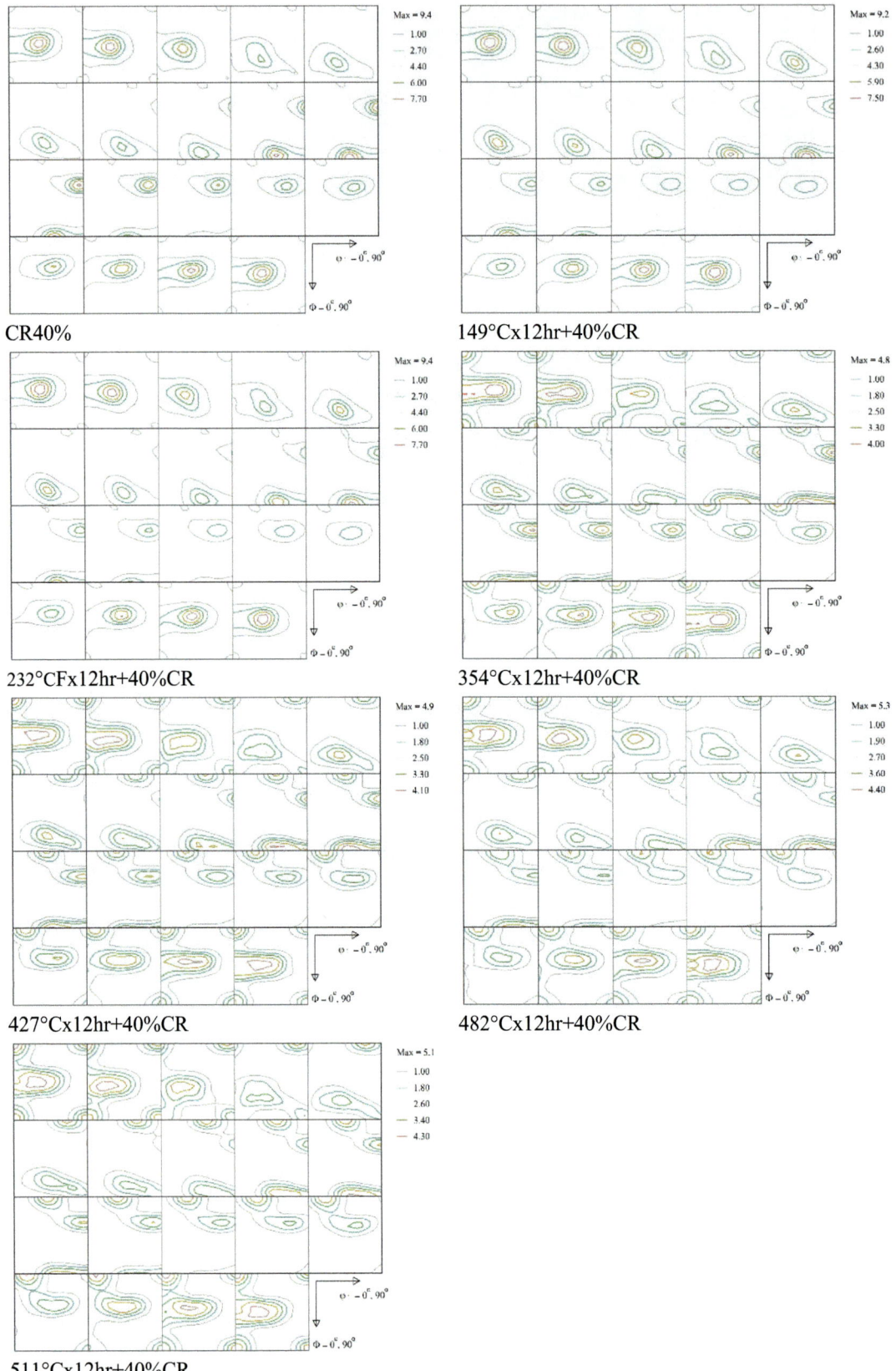

Fig. 4 ODFs of the sheet samples after annealing at different temperatures and 40% cold rolling

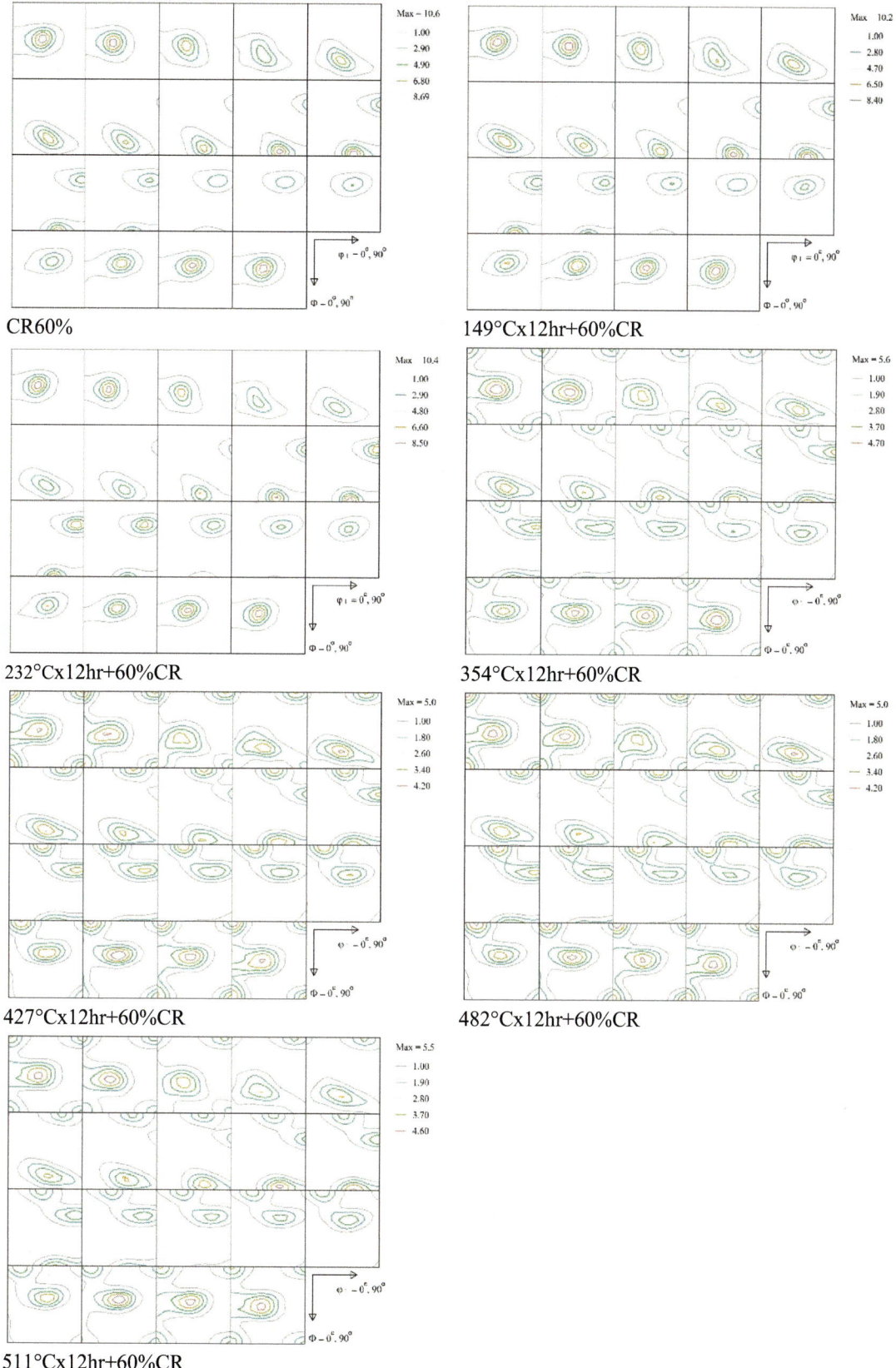

Fig. 5 ODFs of the sheet samples after annealing at different temperatures and 60% cold rolling

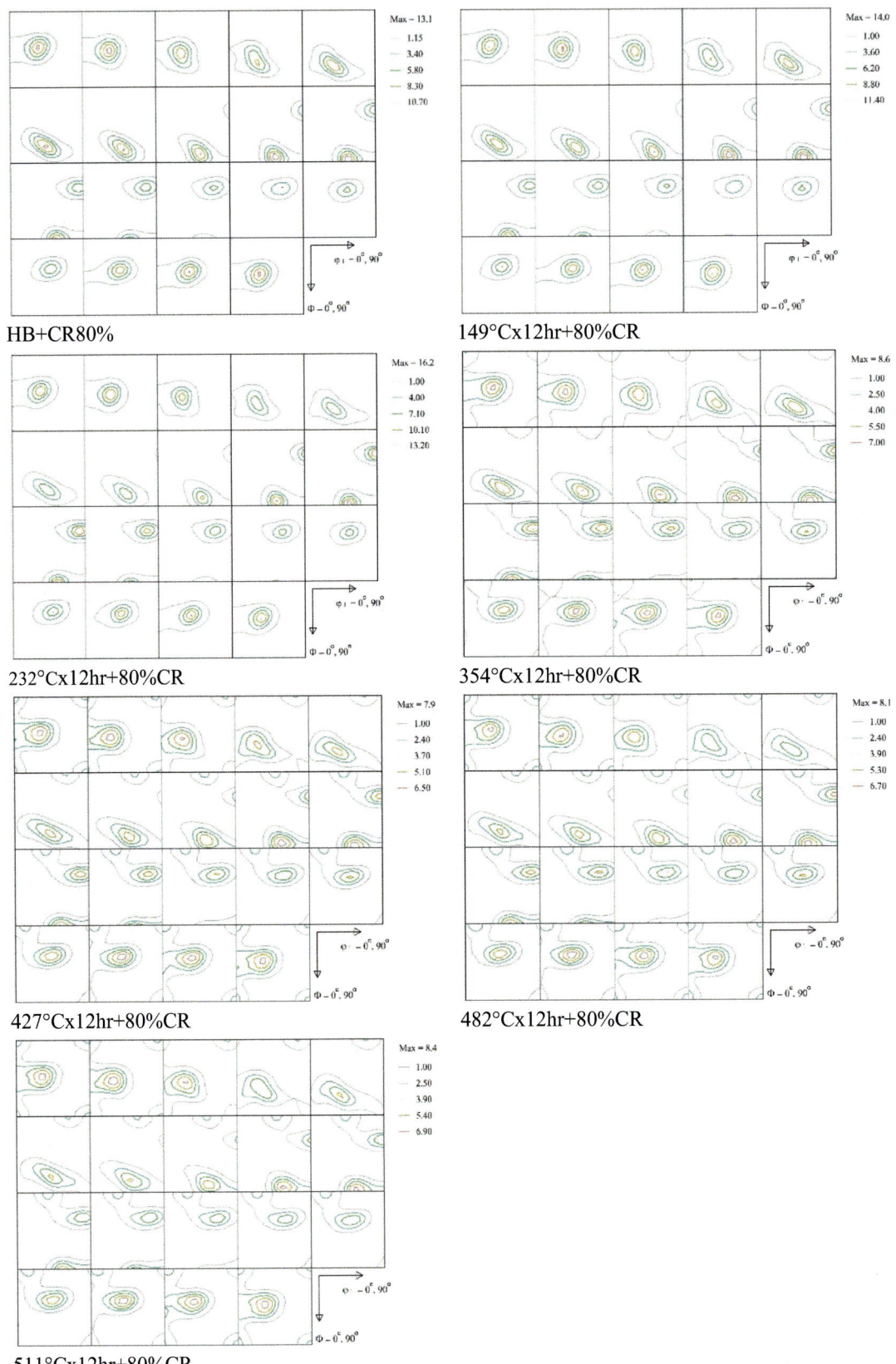

Fig. 6 ODFs of the sheet samples after annealing at different temperatures and 80% cold rolling

Fig. 7 Volume fractions of six typical texture components in the hotband samples after annealing and cold rolling to different gauges (1: HB, 2: 149 °C×12 h, 3: 232 °C× 12 h, 4: 354 °C × 12 h, 5: 427 °C × 12 h, 6: 482 °C × 12 h and 511 °C × 12 h)

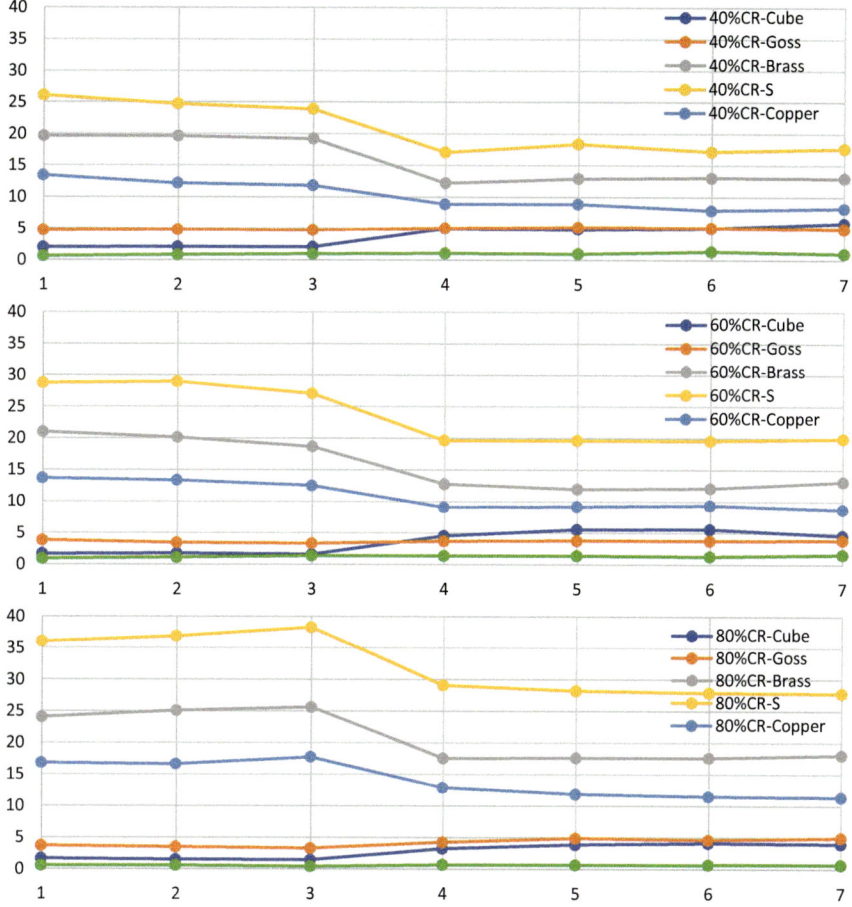

Fig. 8 Tensile curves of cold rolling sheets from CC hotband samples after annealing at different temperatures and cold rolling (1ksi = 6.8947 MPa)

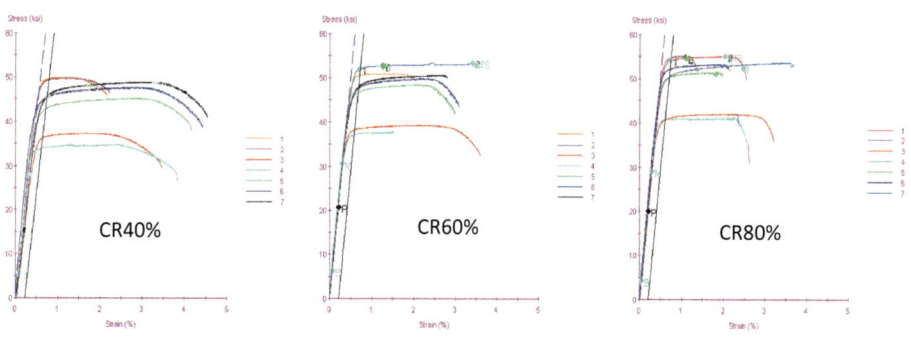

Fig. 9 UTS vs Cold reduction from hotband samples after annealing at different temperatures and cold rolling (1ksi = 6.8947 MPa)

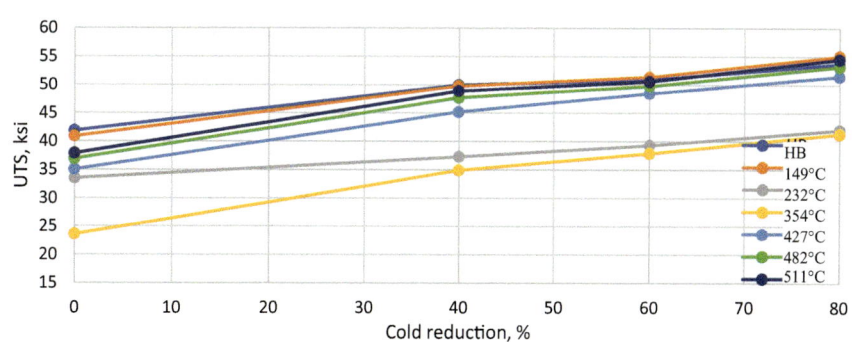

Conclusions

In the present work, mechanical properties, microstructures and textures of cold rolling sheets from a low-cost continuous casting (CC) Al-1.5Cu alloy were studied. The main conclusions are drawn as follows:

(1) When annealing temperature is less 427 °C, the primary particle structures did not change a lot. The bright Al-Cu primary particles decreased dramatically when annealing temperature is higher than 427 °C.

(2) Direct cold sheet has about 50 nm in diameter fine precipitates or dispersoids from casting process, while for the samples annealed at 149 °C, 354 °C and 482 °C for 12 h and cold rolling, the average diameters of precipitates are approximately 90 nm, 70 nm, and 100 nm, respectively.

(3) Severe deformation of grains was observed when annealing temperature is lower than 354 °C, which is a typical cold rolling grain structure. When annealing temperature is higher than 427°F, the recrystallized grain structure can be seen with micro-shear bands in grains.

(4) There is a strong β fiber texture from Brass through S to Copper with weak Goss and Cube textures in the sheets when annealing temperature is less than 232 °C. With cold reduction increasing, the intensity of β fiber texture increase. When annealing arrived at 354 °C or higher temperature, β fiber and Cube textures are stronger with cold reduction increasing.

(5) After annealing at 232 °C and 354 °C, the cold sheets have low strength and work hardening (deformation hardening) rate. Direct cold rolling sheet and sheet after annealing at 149 °C have almost the same strength as ones after annealing at 427 °C–511 °C and cold rolling.

(6) Compared to 5000 series alloys, it was found that the alloy has higher strength because of solid solution of Cu atoms hardening at 427 °C–511 °C or without annealing.

References

1. Zaki Ahmad (ed) (2011) Recent Trends in Processing and Degradation of Aluminum Alloys, InTech Europe, Slavka Krautzeka, ISBN 978-953-307-734-5, www.intechopen.com
2. Zhang JX, Sun HY, Li J, Liu WC, "Effect of precipitation state on recrystallization texture of continuous cast AA 2037 aluminum alloy",
3. Zhang JX, Liu C, Liu WC, Man C.-S (2017), "Effect of precipitation state on texture evolution in cold-rolled continuous cast AA2037 aluminum alloy", Journal of Alloys and Compounds, PII: S0925-8388:33043-8, https://doi.org/10.1016/j.jallcom.2017.09.014
4. [4]Li PH, Wu YC, Jiang J, Xue XD, Liu YF, Wang SJ and Zhai T (2012), "Coincidence Doppler Broadening Study of the Precipitates of a Continuous Cast AA2037 Al Alloy after Annealing and Ageing", Physics Procedia 35: 28 – 33, doi: https://doi.org/10.1016/j.phpro.2012.06.006
5. E.A. Kenik, Q. Zeng and T. Zhai, "Characterization of Continuous Cast AA2037 Al Alloy", Microsc Microanal 15(Suppl 2), 2009
6. Wen XY (2022), "Micro and Mechanical Characteristic of Hotband Annealed of a Continuous Cast Al–1.5Cu Alloy with Potential Application in High Strength and Low Cost Auto Forming Parts or Sheets", Light Metals, Anaheim, CA, USA, Feb. 27-Mar 3, 212–220
7. Liu Y, Cheng XM, Liu J and Morris JG (2000): Materials Science Forum,331-337: 191-196
8. Liu Y, Liu YL, Liao G and Morris JG (1999), Proceeding of 12th International Conference on Textures of Materials, Montreal, Quebec, Canada, 9–13 Aug., 2: 1148–1153
9. Li Z, Ding SX and Morris JG (1995), Light Metals, Las Vegas, NV, USA, 12–16 Feb., 1149–1154
10. Ding SX, Fan XY and Morris JG (1993), Aluminum Alloy for Packaging, Chicago, Illinois, USA, 1–5 Nov. 237–250.
11. Li Z, Li CX and Morris JG (1993), Aluminum Alloy for Packaging, Chicago, Illinois, USA, 1–5 Nov., 200–207.
12. Wen XY, Wang YL, Zheng YZ, Bao E and Li XQ: Materials of Mechanics and Engineering (China), 18: 27–28, 36
13. Wen XY, Wang YL, Zheng YZ, Sun SH (1994): Transactions of Metal Heat Treatment (China), 1994, vol. 15, No.3, pp. 44–49.
14. Yu XF, Wen XY, Zhao YM, Zhai T, Mater. Sci. Eng. A, 2005, 394: 376-384
15. Wen XY, Liu Y, Tong LR, Zeng Q, Zhai T, Li Z (2016), "Effect of iron content on recrystallization texture evolution and microstructures of the cold rolling continuous cast AA5052 alloy sheet", Metallurgical and Materials Transactions A, 47A: 1865-1880
16. Kenik EA, Zeng Q and Zhai T, "Characterization of Continuous Cast AA2037 Al Alloy" Microsc Microanal, 15(Suppl 2), 1072–1073 (2009)

Challenges in the Production of 5754 Automotive Alloy Sheet via Twin Roll Casting Route

Dionysios Spathis, John Tsiros, and Andreas Mavroudis

Abstract

An industrial trial to produce 5754 alloy automotive application sheets through the twin roll casting route was carried out in ELVAL's rolling plant. The paper describes several challenges faced to produce final product within the automotive customer specifications using the twin roll casting process. Microstructural control, as-cast surface quality (level lines), and final mechanical properties are the major challenges in 5754 sheet production via twin roll casting. Microstructural control refers mainly to the final soft temper grain size and the amount of the central line segregation phase. Casting two different 5754 alloy versions with respect to Mn content (high and low Mn alloy) and rolling the coils to final gauge using the same thermomechanical process shows clearly the significant influence of alloy Mn content in the final grain size and the final mechanical properties.

Keywords

TRC • 5754 • Automotive

Introduction

The automotive market has an increased demand for 5xxx series aluminum alloys sheets due to the new lightweight projects coming from the manufacture of fully electrical vehicles (battery closures, battery crash management systems).

In the past, several aluminium rolled sheet producers carried out industrial trials for 5xxx series alloy production via twin roll casting route. Past industrial trials reported data focus on final gauge material formability properties and very few papers focus on the production and quality challenges of twin roll cast 5754 alloy [1–3]. The novelty in the present paper is the influence of 5754 alloy Mn content in final gauge recrystallized grain size and final gauge formability properties (uniform and total elongation) of roll cast 5754 alloy.

Twin roll casting results in a short solidification time in the roll gap causing a high solidification rate. Twin roll casting is suitable for short solidification range alloys like 1xxx, 8xxx and 3xxx series with a maximum solidification range up to 20 °C. According to Yucel Birol [4], alloy 5754 has a very wide non-equilibrium solidification range of 98 °C for a cooling rate of 40 °C/min.

Twin roll casting products exhibit centerline segregation, formed by the deformation of the mushy zone and it is mainly control by casting conditions (speed, cast gauge, and separating forces) and cast alloy [5]. Usually alloys with a wide solidification range like 5754 exhibit a significant amount of centerline segregation phases and an increased risk of surface segregate formation due to extended mushy zone length [5]. Twin roll casting skips initial as-cast surface scalping therefore a tight control of cast surface quality is very critical. The cast surface must be free of level lines and surface segregation because roll cast surface is a part of final gauge product surface. Microstructural control of the as-cast plate is also very critical due to limited structure refinement induced by low cold rolling deformation degree from as-cast gauge down to final product gauge.

ELVAL's twin roll caster produced as-cast coils of two versions of alloy 5754 (high and low Mn content alloy). 5754 alloy version coils were rolled under the same thermomechanical process at the final gauge of 1 mm which is a typical automotive product gauge. The coils after final cold rolling were annealed for O temper product.

Figure 1 shows the thermomechanical process of the two coils (intermediate full anneal at gauge of 3,5 mm).

D. Spathis (✉) · J. Tsiros · A. Mavroudis
Technology, Quality and Innovation Department, Aluminium Rolling Division of ElvalHalcor, Elval Hellenic Aluminium Industry, 61St Km Athens-Lamia National Road, 32011 Oinofyta, Greece
e-mail: dspath@elval.com

© The Minerals, Metals & Materials Society 2023
S. Broek (ed.), *Light Metals 2023*, The Minerals, Metals & Materials Series, https://doi.org/10.1007/978-3-031-22532-1_37

Fig. 1 Thermomechanical process of the 5754 alloy coils

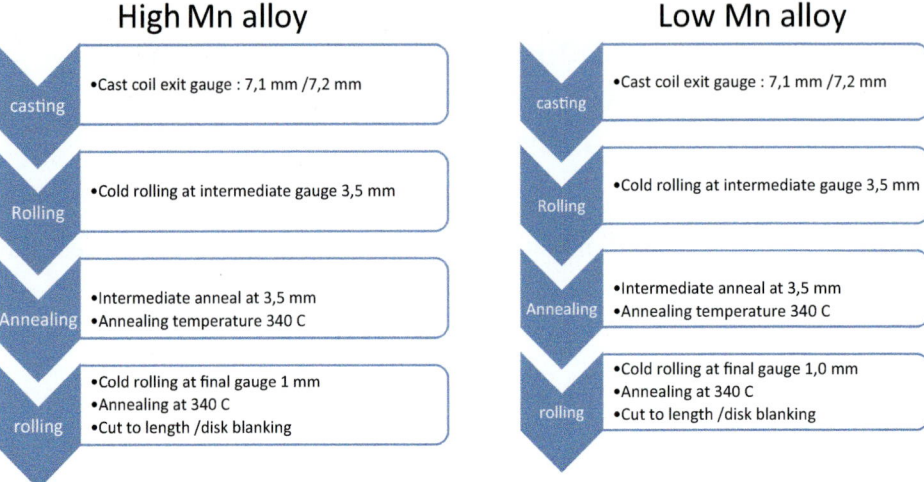

Experimental Procedures and Equipment

Microstructural analysis (grains and intermetallic particles) of as-cast strip and final gauge strip was performed in metallographically prepared cross sections parallel to the working direction (casting or rolling). Metallographical cross sections were etched with 10% HF solution to reveal intermetallic particles and anodized with Barker's reagents to reveal grain structure. Surface grains in final gauge sheet of 1 mm were revealed by macroetching using mixture of acids (Tucker's reagent).

All microstructural observation was done using Nikon Epiphot 200 inverted Metallurgical optical microscope and Nikon Stereo microscope SMZ 1100. The average grain size was determined using line intercept method described in the ASTM standard E112. Mechanical properties of final gauge sheet material (YS, UTS, A%, n, and r) were determined using dog bone shape tensile test specimen in testing direction transverse to the rolling direction. The tensile testing was performed according to ASTM B557-15 (standard test method for tension testing of wrought and cast aluminum and magnesium products).

Full Anneal Grain Structure at Final Product Gauge

In Fig. 2, stereoscope images show the surface grains of high and low Mn content 5754 alloy at final gauge product of 1 mm after the industrial full anneal for O temper. Low

Mn 5754 alloys show a significant refinement of the final fully recrystallized grain size relative to high Mn alloy. High Mn content alloy exhibits an average surface grain size of 250 μm and the low Mn content alloy exhibits a surface grain size below 50 μm.

Figure 3 shows optical microscope micrographs of fully recrystallized grains in cross sections parallel to the rolling direction at a final gauge of 1 mm. The measured average grain size of the high Mn content alloy is around 80 μm and the average grain size of the low Mn content alloy is 17 μm.

Final Gauge O Temper Mechanical Properties Data

Table 1 illustrates the final gauge mechanical properties after industrial full anneal of low and high Mn content 5754. The tensile testing direction was transverse to the rolling direction. Alloy 5754 high Mn version exhibits a total average elongation A80 of 17.7% and uniform average elongation A_g of 15.8% well below the minimum values set by the automotive industry. Alloy 5754 low Mn version exhibits elongation values well above automotive specifications (A80% of 26.6% and A_g of 17.7%) but with a significant reduction of ultimate tensile strength (almost 20 MPa reduction of UTS). Uniaxial tensile test fractured specimens exhibit significant surface roughening (orange peel) in the case of high Mn alloy 5754. Severe surface roughening appears also on the surface of a drawn cup from the high Mn content alloy final gauge O temper sheet (Fig. 4).

Fig. 2 Stereoscope images of surface grains after Tucker's reagent etching/High Mn (**a**) versus Low Mn (**b**) 5754 alloy

Center Line Segregation and Final Gauge Bending Properties

Alloy 5754 due to the wide solidification range exhibits significant amount of centerline segregation under certain casting conditions. Figure 5a optical microscope micrograph show the typical centerline segregation morphology observed in the as-cast structure of both 5754 alloy versions. The center line segregation particle types are Al–Fe-Mn, Al-Si-Fe–Mn (also called α phase), and Mg$_2$Si based on Scanning Electron Microscope EDS analysis resulting spectra (Fig. 6a).

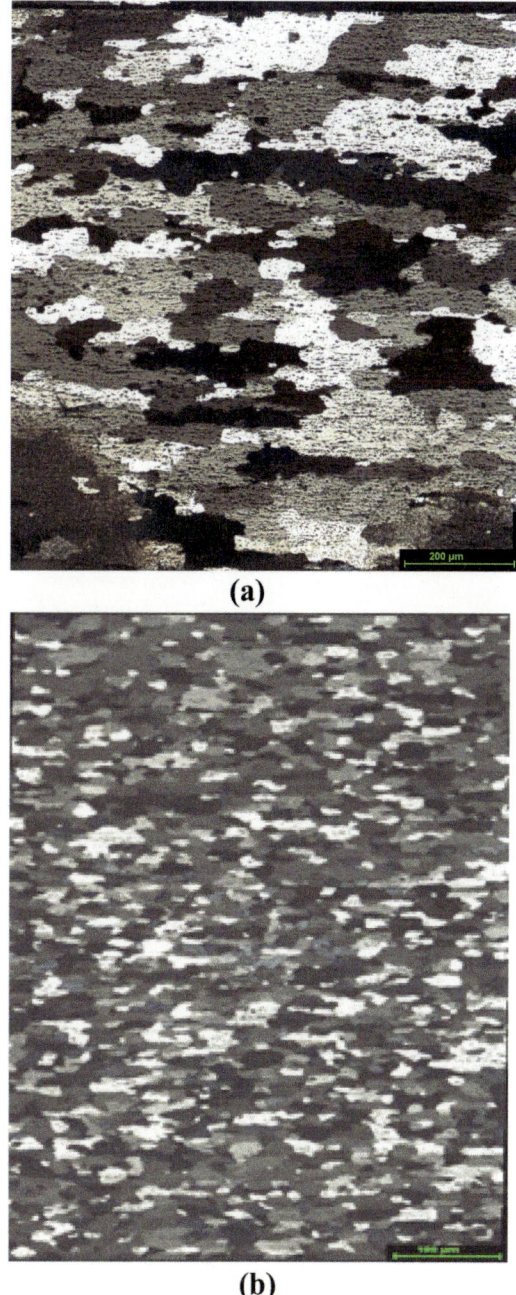

Fig. 3 Optical microscope images of cross section grain structure of High Mn (**a**) versus Low Mn (**b**) content 5754 alloy

The large size segregation intermetallic agglomerates cause strain localization during the cold rolling process in the form of shear bands around the segregation agglomerates

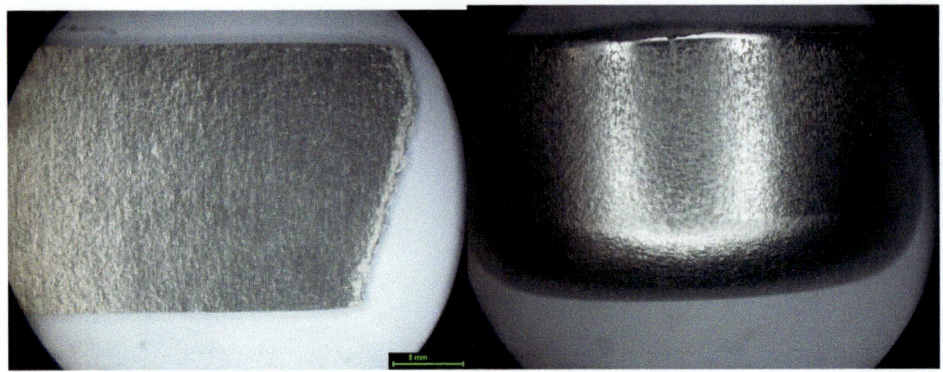

Fig. 4 Surface roughening of tensile fractured specimen and drawn cup test surface

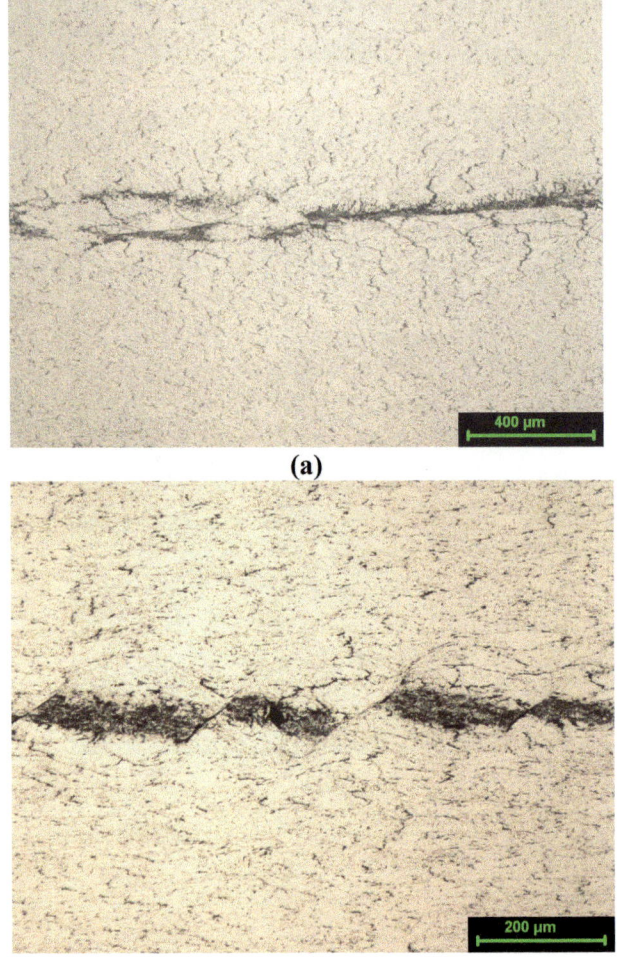

(a)

(b)

Fig. 5 As-cast center line segregation (**a**) and (**b**) shear bands/cracks developed in center line segregation after cold rolling

and finally lead to shear micro-cracks in the centerline region (Fig. 5b). The centerline segregation bands seem do to have any negative impact in the 180 0XT bending performance. Figure 7 optical microscope micrographs of 180° 0XT bend test cross section did not show any cracking evidence in the centerline segregation region.

Surfaces Issues (Ripples and Severe Segregation)

The as-cast surface at the bottom side exhibits ripple marks. Figure 8 shows as-cast surface ripples on the bottom side of a full width strip, which remains on the final gauge material surface of 1 mm. Figure 9 stereoscope image shows the ripple marks in the cold rolled surface at intermediate gauge of 3,8 mm which appears like alternative dark and light bands aligned transverse to rolling direction. The cold rolling process did not eliminate the ripple marks even at the final gauge of 1 mm. Figure 10 micrographs show cross section of ripples area versus normal surface appearance area in the as-cast gauge material. The ripple mark region seems to have a coarse shell size and higher number of intermetallic particles relative to the normal surface appearance area. Ripples could be a periodic disruption of the meniscus oxide skin leading to periodic variation of surface cell size and number of intermetallic particles (light surface segregation). In some cases, severe surface segregation of bottom side cast strip was observed under the light optical microscope (Fig. 11 micrographs). The large size surface segregates are brittle under cold rolling deformation stress forming metal crack bands aligned transverse to the rolling direction in the first cold rolling pass (Fig. 12 micrographs). Figure 13

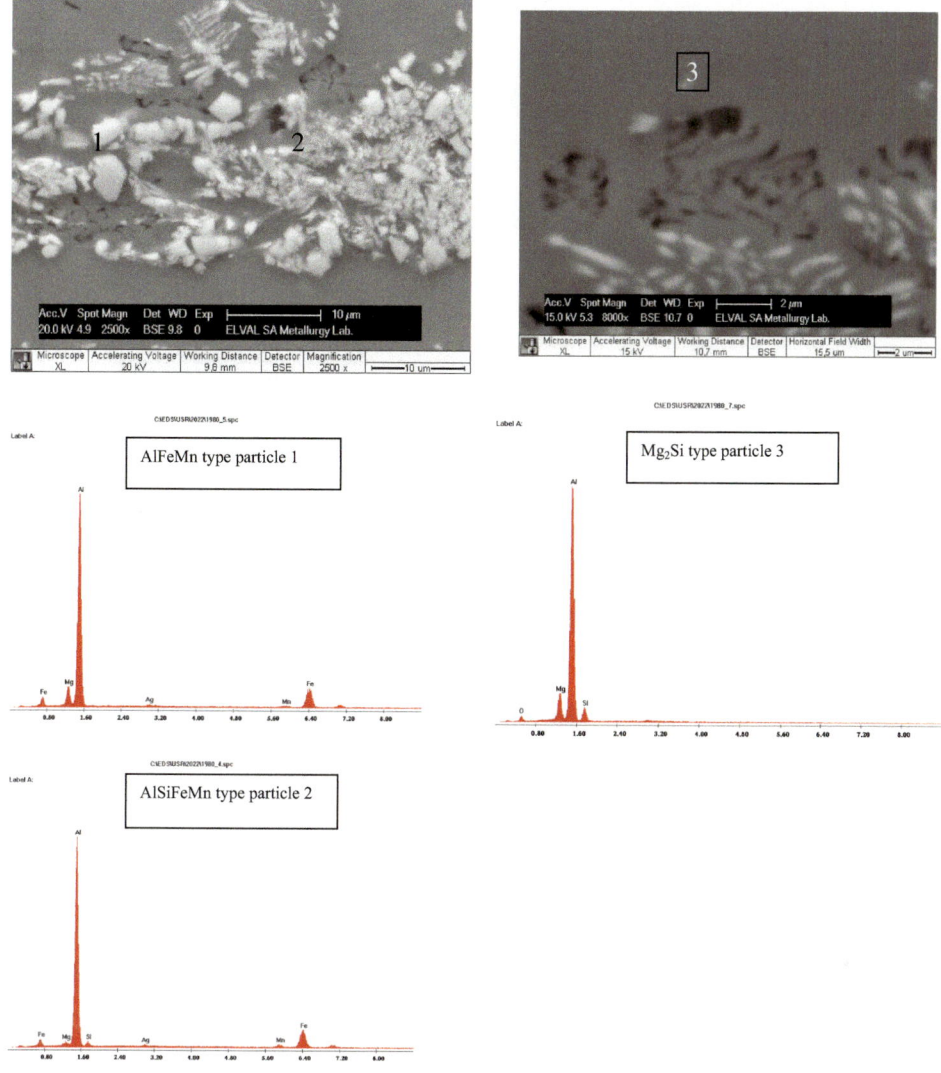

Fig. 6 EDS analysis spectra of center line segregation particles

optical microscope micrograph shows surface cracking cross section with severe surface segregation near the fracture area.

Summary and Conclusions

High Mn content twin roll cast 5754 alloy has final product gauge total (A80%) and uniform elongation (A_g) outside automotive specifications due to coarse recrystallized grains after final anneal process size (average grain size 80 μm). The average final gauge grain size of 80 μm cause severe surface roughening (orange peel) under a different forming process (drawing test and uniaxial tensile testing).

On the contrary low Mn content twin roll cast 5754 exhibits very fine grain size (average grain size 17 μm) after final gauge full anneal with total elongation and uniform elongation values within automotive specifications. The significant reduction of O temper UTS for low Mn content alloy version compared to the high Mn content version could be mainly attributed to different alloy Mn content.

Mn element is a dispersoid particle forming element and therefore it is possible that high Mn content alloy has higher density of dispersoid particles relative to low Mn alloy.

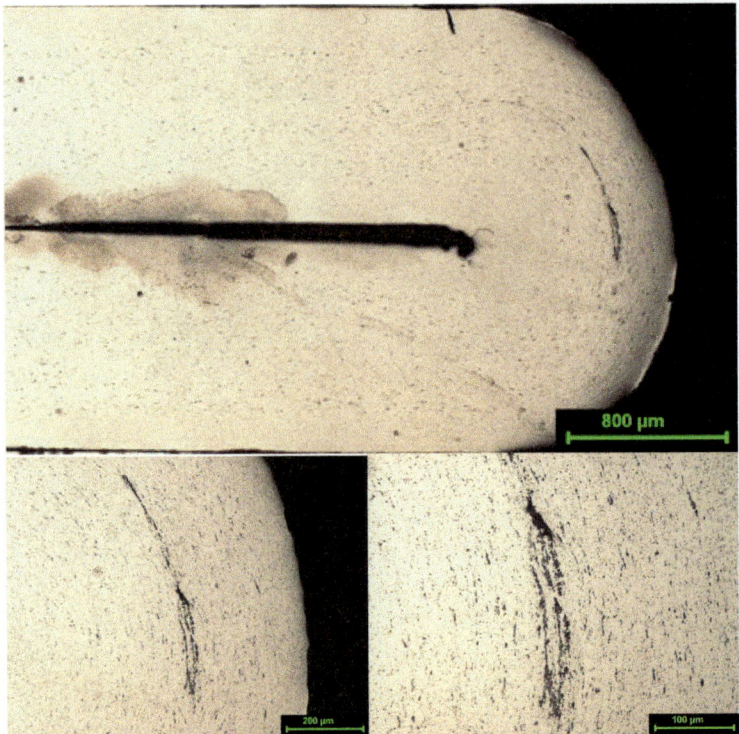

Fig. 7 Cross sections of 180 0XT bending test coupons

Fig. 8 Level lines in the bottom side cast strip and final gauge product level lines at the same side

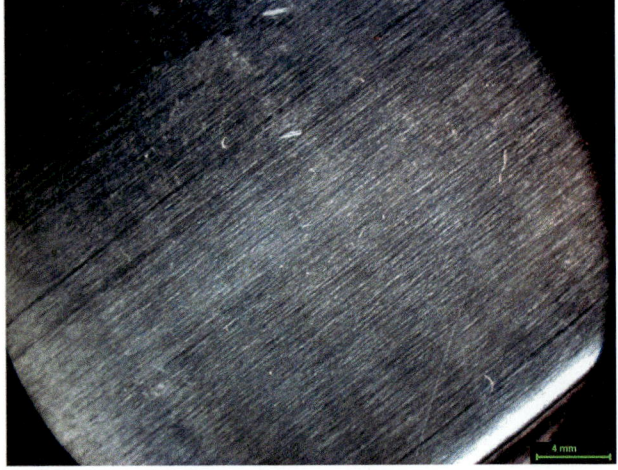

Fig. 9 Stereoscope image of ripples (alternative dark and light zone). Gauge: 3.8 mm

Fig. 10 Cross section of bottom side as-cast gauge ripples (**a, c**) area versus normal surface area (**b, d**)

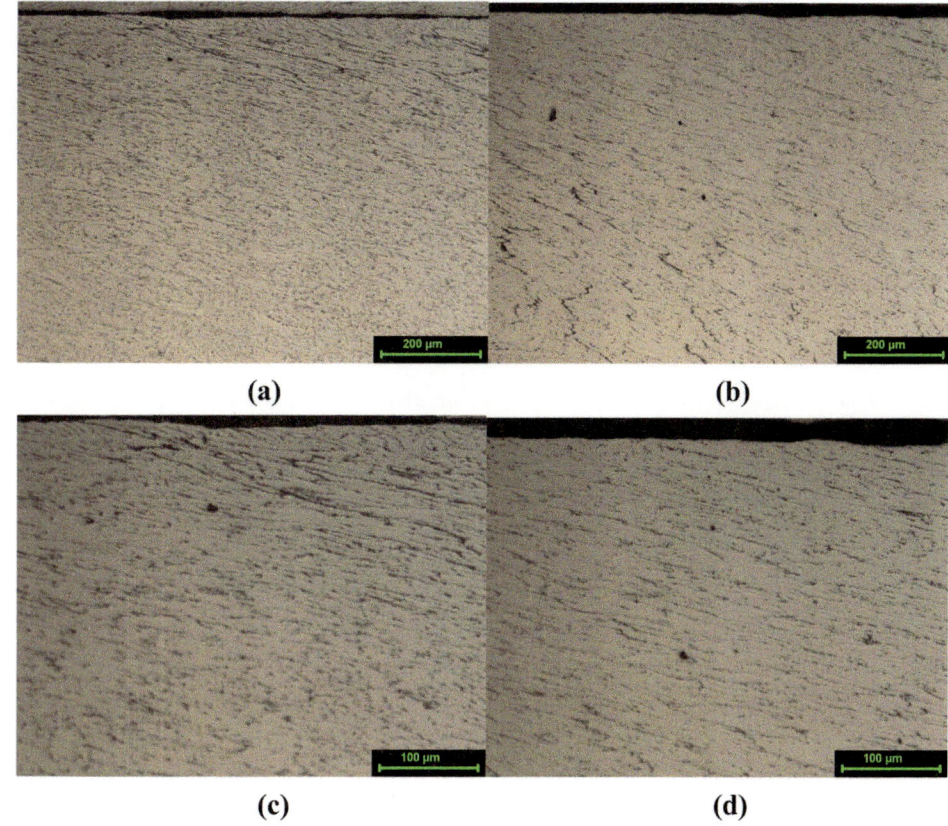

(a)   (b)

(c)   (d)

Fig. 11 Severe bottom side surface segregates which cause surface microcracks in the cold rolling

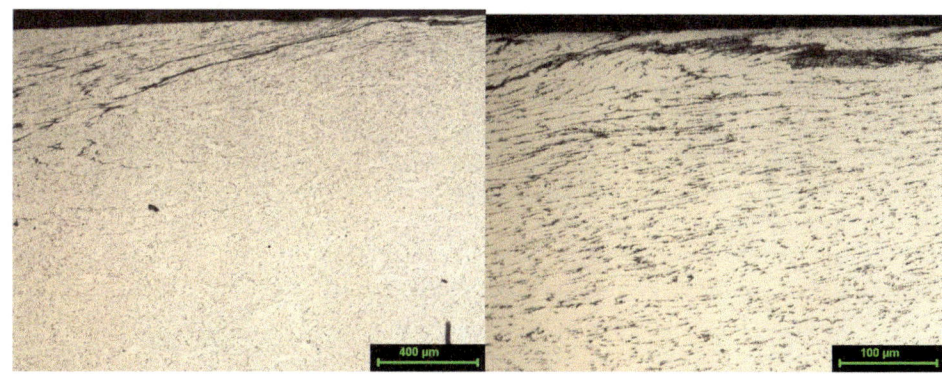

Fig. 12 Stereoscope and optical microscope image of transverse cracks band and optical microscope of cracks band

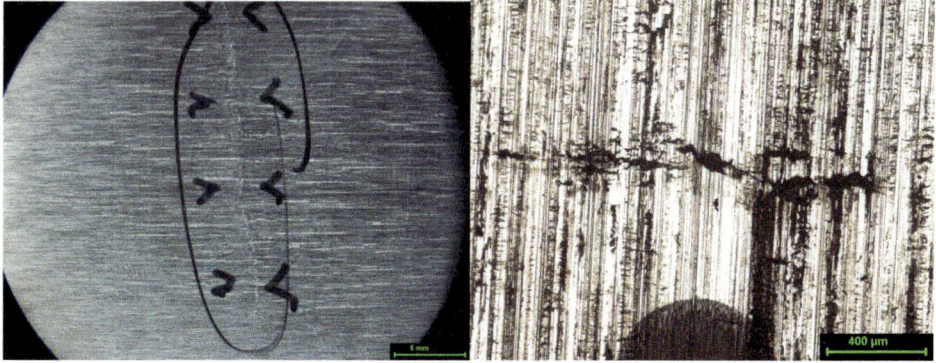

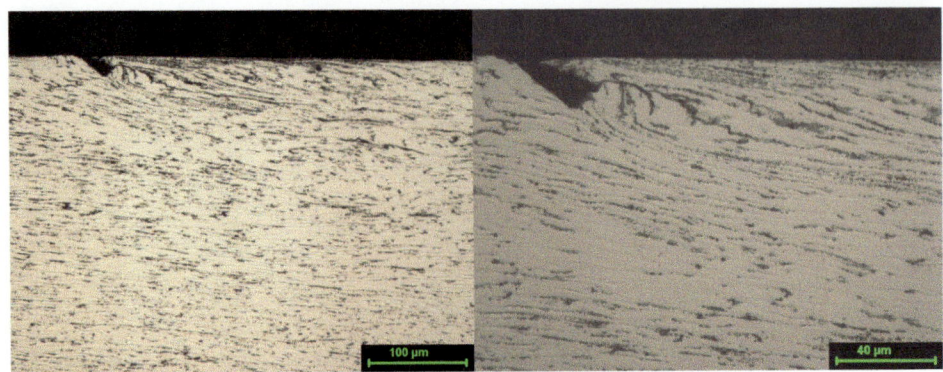

Fig. 13 Optical microscope image of cracks cross sections showing heavy surface segregation residues

Higher density of Mn containing dispersoids prior final full anneal pins the recrystallized grain nucleation leading to coarse-grained structure [7]. Both alloy versions of 5754 exhibit the typical centerline segregation pattern. Segregation intermetallic clusters did not influence negative the 180 0XT bending forming process. Bending has the maximum tension stress in the outer bend surface and the centerline segregation zone area exhibits almost zero bend stress. On the contrary, shear cracks around centerline segregates could reduce significantly metal formability in a biaxial stretching due to strain localization.

Surface issues like ripple marks on the bottom side due to a light surface segregation remains on the final cold rolled product at final gauge of 1 mm. In some cases, severe surface segregation appears on the cast surface bottom side of 5754 alloy causing metal surface transverse cracks in the cold rolling due to the brittle nature of the surface intermetallic particles. The heavy surface segregates are not typically isolated segregation pockets on the surface with clear boundaries between the segregate and the matrix [6]. The segregates have boundaries with the matrix and appear to be an entrapment-frozen meniscus.

Figure 14 shows a mechanism of ripples formation in twin roll casting process [8]. Unstable meniscus buckles under metallostatic pressure and entraps rich in alloying elements frozen meniscus causing ripples and surface segregation. The meniscus stability is very important in order to

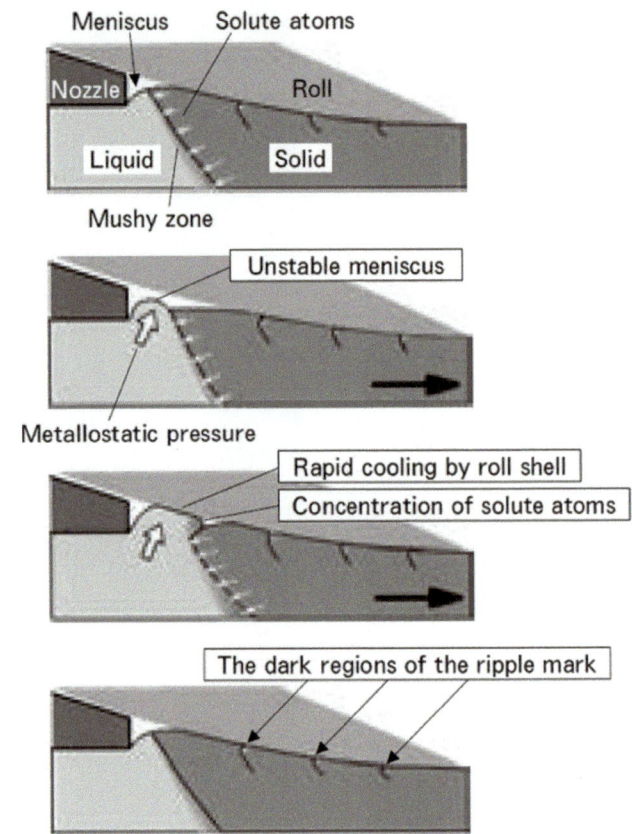

Fig. 14 Ripples formation mechanism due to unstable meniscus [8]

Table 1 Final gauge O temper mechanical properties (testing direction: transverse to rolling)

Low Mn alloy	UTS (MPa)	YS (MPa)	A80 (%)	Ag	n5%	r10%
1	206.04	97.94	27.9	17.8	0.32	0.86
2	208.11	97.68	25.6	17.5	0.32	0.89
3	206.83	98.76	26.9	17.9	0.33	0.84
Average	207.0	98.1	26.8	17.7	0.33	0.86
St div	0.9	0.5	0.9	0.2	0.01	0.02
High Mn alloy	UTS (MPa)	YS (MPa)	A80 (%)	Ag	n5%	r10%
1	228.49	94.67	18.5	15.5	0.33	0.64
2	226.74	94.2	17.8	15.1	0.34	0.68
3	228.34	93.71	19.3	18.5	0.34	0.77
4	228.96	94.93	19	16.5	0.34	0.74
5	230.26	94.16	15.9	14.6	0.33	0.74
6	227.8	95.04	18.4	16.3	0.35	0.67
Average	229.0	94.7	17.8	15.8	0.34	0.72
St div	1.0	0.4	1.3	0.9	0.01	0.03
Auto specs	Min 190	80–120	Min 23	Min 17	Min 30	Min 0.65

achieve as-cast surface free of level line marks and severe surface segregation. The casting condition must be fine tune in order to maintain a stable meniscus during twin roll casting.

Acknowledgements The authors thank Dr. Filippos Patsiogiannis, Bridgnorth Aluminum Ltd Head of Technology & Manufacturing Excellence, for reviewing the manuscript.

References

1. Characterization of Surface Defects encountered in Twin Roll Cast Aluminun Strip. Murat Dundar, Ozqul Kelles. TMS Light metal 2007.
2. Formability performance of 5XXX series aluminum alloys produced with twin roll castig technology. Murat Dundar, Kemal Sariogloy, Yucel Birol, Sonner Akurt, C. Romanowski. TMS light metal. 2002
3. Press formability of twin roll cast (TRC) alloys. Sooho Kim and Anil K. Sachdev. TMS 2008
4. Analysis of Macrosegragation in Twin Roll Casting Aluminium Strip via solidification curves. Yucel Birol. Journal of alloys and compounds. November 2009
5. Twin rolling casting of Alumiun alloys – an overview. N. S. Barekar, B. K. Dhindaw, Material and Manufacturing process 29. Page 651–661. 2014
6. The formation of surface segregates during twin roll casting of aluminium alloy. B.Forbod and others. Material science and engineering. 2006
7. The effect of dispersoids on the recrystallization behavior in a cold rolled AA 3103 Aluminium alloy. Stian Tangen, Hans Bjerkaas, Trond Furu and Erik Nees. Proceding of 9th international conference on alumiunm alloy 2004.
8. Effects of strip casting condition on the ripple marks of twin roll cast Al-Mg-Si. Hiroki Esaki, Yoshi Watanabe, Kaoru Ueda, Hideyuki Uto, Kazuhisa Shibue . Sumitomo Light Metal Technical report volume 47. 2006

Fabrication of Bright-Rolled Aluminum Suitable for Design Elements in the Automotive Industry

Anita Gründlinger, Peter Johann Uggowitzer, and Josef Berneder

Abstract

The present work deals with the fabrication of rolled aluminum sheets based on 5xxx alloys, striving for a superior appearance and formability. Due to the material's exceptional surface properties, processability and corrosion resistance, the developed material may be employed to manufacture sophisticated design elements dedicated to the automotive industry. Beyond optical investigations conducted on brightened and anodized material, phase studies are carried out, suggesting thermal processing close to the solidus temperature prior to cold rolling. This is backed up by thermodynamic simulations, dealing with the impact of precipitates on the material's appearance. Moreover, the grain structure as well as mechanical properties in temper H2X are tackled, addressing the demand for decorative elements with a sufficient formability and enhanced corrosion resistance.

Keywords

Aluminum • Bright • Gloss • Design • Decoration • Automotive • 5505

Introduction

Beyond safety, functionality and comfort the main requirement for today's automobiles is an elegant design, aiming to attract the end customer's attention. This can be achieved by employing design elements, so-called automotive trims.

A. Gründlinger (✉) · J. Berneder
AMAG Rolling GmbH, Post Office Box 32,
Lamprechtshausenerstraße 62, 5285 Ranshofen, Austria
e-mail: Anita.gruendlinger@amag.at

P. J. Uggowitzer
Department of Materials, ETH Zürich, Wolfgang-Pauli-Str. 10,
8093 Zürich, Switzerland

The perfect materials to fabricate the same are rolled aluminum alloys, which offer a wide range of decorative applications such as window surrounds, entry sills, or emblems. While muted finishes are common for sports cars, high-end and middle-class vehicles may be equipped with bright automotive trims. Through polishing, brightening, and anodizing high-purity alloys, the surface of the same can be tuned in such a way that a superior aspect can be accomplished. Combined with aluminum's exceptional forming properties, sophisticated designs and complex geometries can be achieved. Since beauty lies in the eye of the beholder, there are limited means to compare the aspect of decorative elements. One method, that does enable a quantitative comparison of different surfaces, is the gloss measurement. When executing a gloss measurement two major influential factors shall be considered. One of them is the surface roughness, the other, that shall be taken into account, is the presence of intermetallic phases within the anodizing layer of the material. In addition to the gloss level, the impact of intermetallic phases on the corrosion resistance, when exposed to a corrosive environment (e.g. acid rain or detergents), shall be addressed.

In this work, it is evidenced that the minimization of the surface roughness and intermetallic phases within the material are crucial to achieve the highest possible gloss level. While mechanical, chemical, and electro-chemical brightening are common methods to reduce the surface roughness, there is only limited expertise related to the processing of 5xxx alloys aiming to reduce gloss minimizing phases and achieve an appealing appearance without compromising formability. Therefore, this work addresses this issue and shows the impact of thermal processing post and prior to cold rolling on the gloss level and correlates it to simulated as well as actual phase fractions. Furthermore, the microstructure and mechanical properties are evaluated, addressing the demand for design elements featuring small bending radii free of orange peel and fatigue cracks.

S. Broek (ed.), *Light Metals 2023*, The Minerals, Metals & Materials Series,
https://doi.org/10.1007/978-3-031-22532-1_38

Materials and Methods

Given the requirements for a bright appearance, sufficient corrosion resistance, and formability, the material of choice for fabricating automotive trims is the high-purity alloy EN-AW 5505 in temper H2X. The EN standards for the chemical composition are given in EN 573-3. The actual chemical compositions of the investigated materials are listed in Table 1.

Table 1 EN standards and the chemical composition of the analyzed samples (Reference and samples 1–5) are given in weight-%

	Si [%]	Fe [%]	Mg [%]	Zn [%]	Ti [%]	Each [%]
Min			0.8			
Max	0.06	0.04	1.1	0.04	0.01	0.01
Reference	0.03	0.03	0.83	0.01	0.01	
Sample 1	0.03	0.03	0.83	0.01		
Sample 2	0.03	0.03	0.83	0.01		
Sample 3	0.03	0.03	0.83	0.01	0.01	
Sample 4	0.02	0.04	0.86		0.01	

In the first step, the impact of surface roughness and foreign phases on the material's aspect shall be discussed. For comparison, the material's appearance may be quantified by means of gloss level evaluations [1]. In general, the gloss level is at a maximum if the angle of incident and reflected light are equal. As stated by Fresnel [2] this specular reflection applies to perfectly flat surfaces and homogenous material compositions, which are free of foreign phases. Considering that aluminum either comes with a natural or an artificial oxide layer, it is assumed that the reflection of light happens at the interface between the aluminum and the aluminum-oxide layer. Given aluminum's absorption properties [3], the transmission of light within the aluminum alloy shall be neglected. Therefore, the highest gloss level is achieved, if the surface roughness is at a minimum, as shown in Fig. 1a [2]. Nevertheless, in real metal systems the surface roughness is finite. As a result, diffuse reflection of light takes place, leading to a reduction in gloss, as shown in Fig. 1b. Furthermore, foreign phases functioning as scattering agents within the oxide layer may cause diffuse reflection of light within the aluminum-oxide layer, which holds for phase diameters exceeding or corresponding to the wavelength of visible light, known as Mie-Scattering [4]. This is depicted in Fig. 1c, resulting in a decrease in gloss. As a matter of fact, the lowest gloss level is observed for rough surfaces in presence of a large quantity of phases, illustrated in Fig. 1d. In terms of corrosion resistance, not only prevalent phases [5], but also the grain structure need to be taken into account.

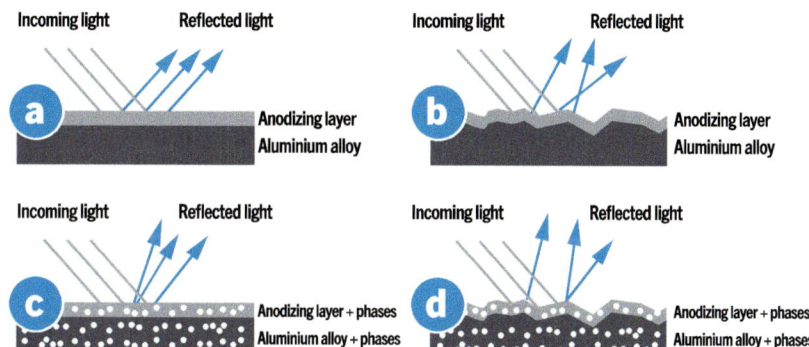

Fig. 1 Schematic sketch of **a** specular reflection for a perfectly flat surface, as well as **b** diffuse reflection for a rough surface, **c** a smooth surface in presence of phases, and **d** a rough surface with phases affecting the gloss

Considering globular grain, it is evident, that the path length for intergranular corrosion ($l_{ic,a}$), indicated in Fig. 2a, is at a minimum. An effective way to increase the path length is to alter the surface-to-volume ratio. Through mechanical processing, e.g. rolling, the grain becomes elongated, which allows reducing the susceptibility to intergranular corrosion along the S-L direction, as shown in Fig. 2b. A comparison between the path lengths of elongated ($l_{ic,b}$) and globular grain ($l_{ic,a}$, shown in Fig. 2a) is given in Eq. (1). The ratio of the mean grain width and length are referred to as elongation, which shall be defined in Eq. (2) according to ASTM E112. The variable AI_l represents the grain elongation, $\bar{l}_{l(0°)}$ the mean grain length and $\bar{l}_{l(90°)}$ the mean grain thickness.

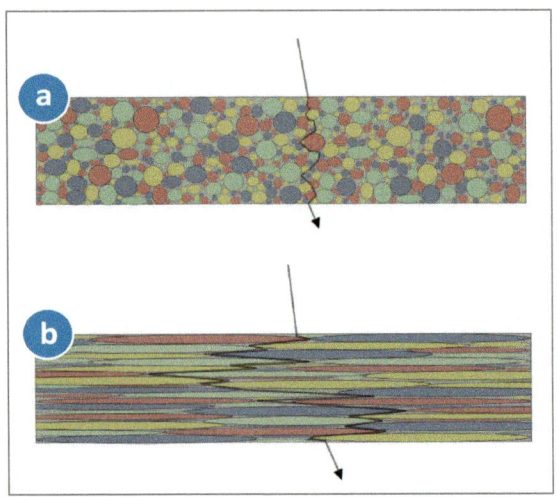

Fig. 2 Pathways of intergranular corrosion for **a** globular and **b** elongated grain shown in S-L direction. Different colors indicate different grain orientations

$$l_{ic,a} < l_{ic,b} \tag{1}$$

$$AI_l = \frac{\bar{l}_{l(0°)}}{\bar{l}_{l(90°)}} \tag{2}$$

Understanding that the minimization of the surface roughness as well as phase fraction is essential to obtain the maximum gloss level and exhibit the best corrosion resistance, a process route shall be defined combining those two requirements. While adjusting the surface topology is a task, that can be overcome by mechanical processing (e.g.

utilization of rollers with gradually decreasing roughness) and subsequent etching, the phase control turns out more challenging. Starting the fabrication process with mechanical and thermal processing of slabs, i.e. hot rolling, cold rolling and annealing, necessary to adjust the material thickness and mechanical properties, it is evident that precipitation is promoted during processing. Focusing on the phases Mg_2Si, $Al_{13}Fe_4$, and $AlFeSi$, the solvus temperature of these phases, evaluated by means of the software Pandat by CompuTherm, is illustrated in Fig. 3.

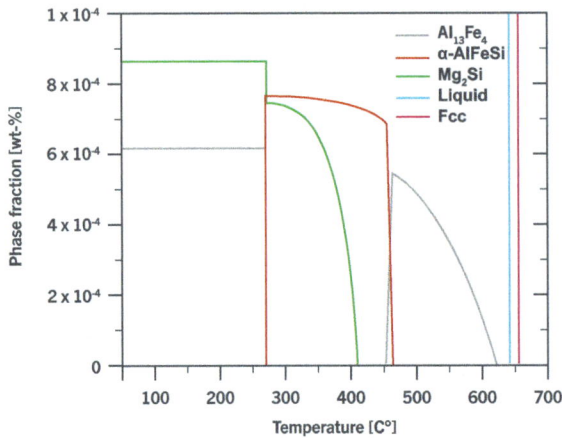

Fig. 3 Phase fraction over annealing temperature, indicating the solvus temperature of Mg_2Si, $AlFeSi$, $Al_{13}Fe_4$ as well as liquidus and solidus temperature. The chemical composition corresponds to sample "Reference" according to Table 1

Given the fact that the Fe phases are present even at temperatures close to the solidus temperature of the aluminum alloy and high-temperature processing results in a limited grain size control, the annealing, which will be referred to as intermediate annealing hereafter, shall be executed closer to the solvus temperature of Mg_2Si than to the solidus temperature of the alloy. To reduce the impact of the Fe phases on the gloss, the phases' diameters shall be reduced to a range below the wavelength of visible light spectrum [6]. This may be accomplished by mechanical processing, i.e. cold rolling, leading to the fragmentation of these phases. Therefore, the process route for the as-cast slabs is defined as follows: hot rolling, intermediate annealing with subsequent quenching, cold rolling and final annealing, as indicated in Fig. 4.

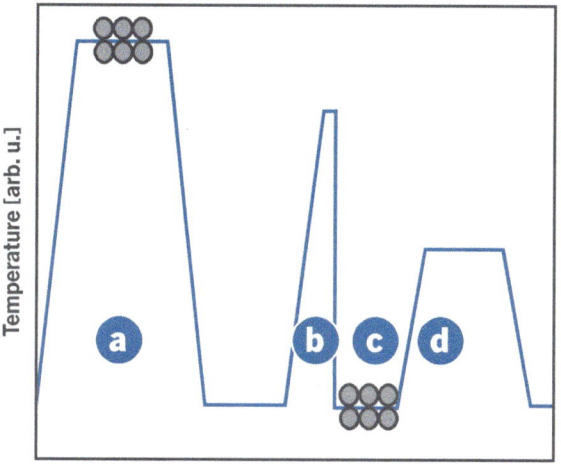

Fig. 4 Schematic sketch of the production process covering **a** hot rolling, **b** intermediate annealing and subsequent quenching, **c** cold rolling and **d** final annealing

It shall be highlighted that hot rolling takes place at temperatures greater than 400 °C and below the solidus temperature of the alloy to prevent incipient melting. The intermediate annealing may be performed above the solvus temperature of Mg_2Si followed by quenching. By cold rolling to the final material thickness of 1.20 mm with a gradually decreasing roughness of the rollers smooths the aluminum's surface. In the final production step, the desired temper is adjusted through balanced recovery at 250 °C for 12 h, ensuring a sufficient material formability combined with enough strength required for the according application. Hence, the goal may be to perform the final annealing below the recrystallization temperature to maintain the elongated grain structure and prevent precipitation of phases. In the last step, the individual lots are chemically brightened to equalize the surface roughness followed by the application of an anodizing layer. The oxide layer growth shall be deliberately limited to a maximum of 10 μm in thickness, delivering a sufficient resistance of the material against corrosive media, allowing a direct comparison of different processing variants.

Results and Discussion

In the following section, the findings linked to the microstructure and the mechanical properties as well as the surface roughness and gloss level are discussed. Considering that the chemical composition is similar for all samples, the impact of the chemical composition is neglected. In the first step the evolution of the yield strength $R_{p0.2}$, ultimate tensile strength R_m, and elongation at break A_{50} are investigated for material intermediate annealed at a thickness of 8 mm. The tensile tests were performed according to EN ISO 6892-1 B. The results are shown in Fig. 5. Comparing the as-rolled

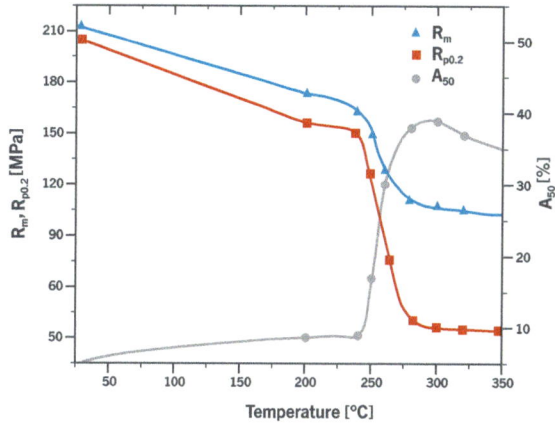

Fig. 5 Mechanical properties after annealing for 12 h and varying temperatures. Measured data and fits are represented by data points and lines, respectively

state (corresponding to Fig. 4a–c), the tensile and yield strength are reduced by ca. 10 and 8%, respectively, when annealed below 250 °C. Furthermore, the elongation at break is increased by ca. 100% through recovery. For final annealing temperatures beyond 250 °C, the material significantly loses strength through the initiation of grain recrystallization, which is confirmed by the investigated microstructures, shown in Fig. 6.

Fig. 6 Microstructure along S-L for sample 1 in the **a** as-rolled condition (no final annealing) as well as final annealed for 12 h at **b** 250 °C, **c** 260 °C, **d** 280 °C and **e** 350 °C

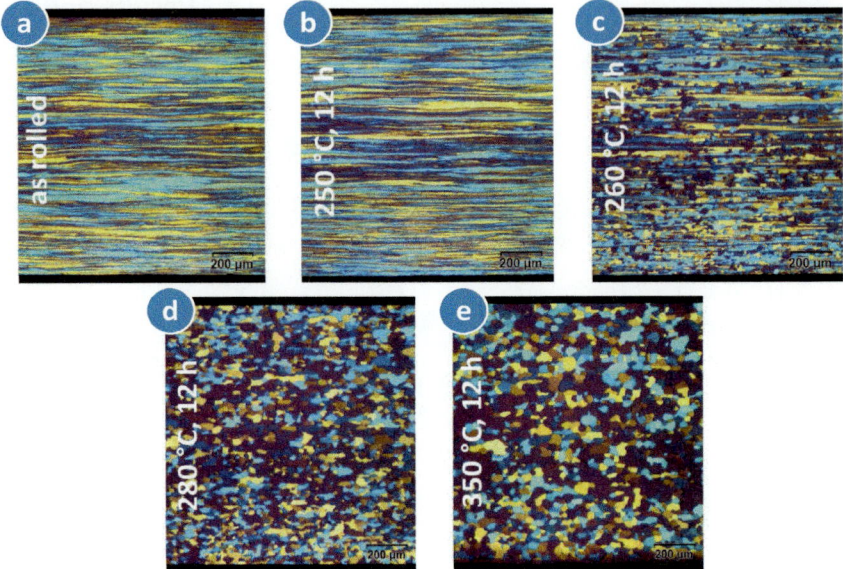

At the peak of the elongation at break between 280 and 300 °C the grain is fully recrystallized. Thereafter, higher final annealing temperatures result in a larger grain size, as shown in Fig. 6e. The sample "Reference" lacks an intermediate annealing, while samples "1" and "2" were intermediate annealed at 8 mm and 5 mm, respectively (given in Table 4). In Fig. 7 the microsections of the polished and etched samples, which were treated with Barker etchant are shown in the *S-L* direction.

The etching was performed for 50 s. According to the shown microsections, the majority of the grains are elongated, while single globular grains indicate beginning recrystallization for sample 2 and the reference sample. The overall microstructures for the sample "Reference" and sample 1 are similar, despite a higher fraction of recrystallized grain in Fig. 7a compared to b, which may be justified by particle stimulated nucleation of Mg_2Si particles.

Notably thicker grains can be observed for sample 2, resulting from a lower degree of cold deformation compared to sample 1. The lower reduction rate by cold rolling after the intermediate annealing for sample 2 is also manifested in the grain elongation measured i. a. by the linear intercept method according to ASTM E112 given in Eq. 2. The elongation is significantly smaller than for sample 1, as evidenced in Table 2. The deviation in grain elongation AI_l

for sample "Reference" and 2 might be explained by partially recrystallized grains.

Table 2 Elongation of grain measured of the reference sample, samples 1 and 2

Sample	AI_l [arb. u.]
Reference	18
1	21
2	12

Regarding the mechanical properties, the tensile strength R_m, yield strength $R_{p0.2}$, and elongation at break A_{50} for the investigated samples are listed in Table 3. It shall be emphasized that increased grain elongations result in a decrease in elongation at break, while the strength stays balanced.

Table 3 Mechanical properties measured according to EN ISO 6892-1 B

Sample	R_m [MPa]	$R_{p0.2}$ [MPa]	A_{50} [%]
Reference	147	117	22
1	148	123	18
2	145	116	15

Fig. 7 Microstructure along S-L for **a** reference sample, **b** sample 1, and **c** sample 2

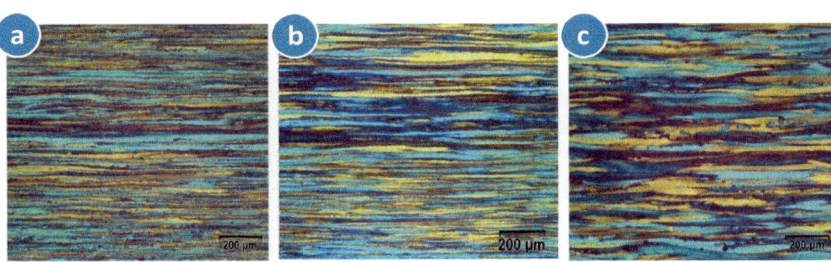

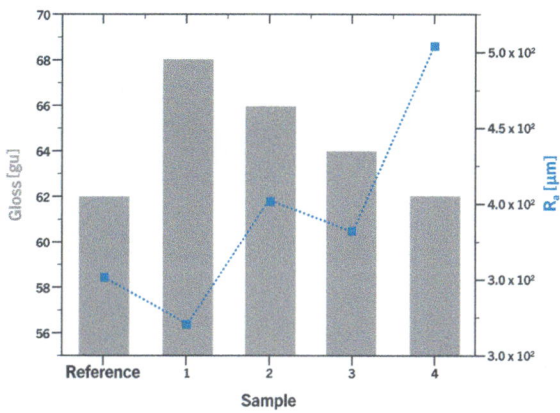

Fig. 8 Correlation of gloss level and surface roughness per processing variant

Table 4 List of surface roughness, hot rolling and intermediate annealing thickness per sample. The samples "Reference" lacks an intermediate annealing, samples 1–5 were intermediate annealed at a thickness ranging from 5 to 8 mm

Sample	Hot rolling thickness [mm]	Intermediate annealing thickness [mm]	Roughness [μm]
Reference	9.5	–	3.5×10^{-2}
1	8	8	3.2×10^{-2}
2	5	5	4.0×10^{-2}
3	8	8	3.8×10^{-2}
4	8	8	5.0×10^{-2}

In the next step the appearance is evaluated through gloss measurements, which are correlated to the surface roughness. The gloss levels and surface roughness after brightening and anodizing of the samples are depicted in Fig. 8. The intermediate annealing thicknesses are listed in Table 4. As shown in Fig. 8, the surface roughness and gloss level are indirectly proportional. This is confirmed when comparing samples 1, 3, and 4, which were intermediate annealed at the same thickness and solely differ in terms of surface roughness from one another. Therefore, it is evidenced that the highest gloss level is achieved for the lowest surface roughness (measured according to DIN EN 149). Furthermore, when comparing the reference sample, which lacks an intermediate annealing step, but shows a low roughness, the positive effect of the intermediate annealing and subsequent quenching is further supported. When comparing samples 2 and 3, a minor discrepancy in terms of gloss level of 2 GU (gloss unit) and surface roughness R_a of 0.002 μm is shown. On the one hand, this discrepancy in gloss may solely result from a finite accuracy of the roughness measuring unit. On the other hand, the intermediate annealing thickness of sample 2 was 5 mm, for which an enhanced heat transfer took place. Therefore, in comparison to sample 3, which was annealed at 8 mm, an improved phase control was achieved for sample 2, resulting in an overall lower phase fraction. To further support the positive effect of the intermediate annealing and adjacent quenching, the phase fraction and microstructure of the reference sample, as well as 1 and 2 are inspected. Therefore, for each sample a polished microsection in *S-L* direction was evaluated by optical microscopy. As suggested by the results in Fig. 9, the reference sample shows the highest phase fraction, while the phase fractions for samples 1 and 2 are reduced. Even though sample 1 exhibits more phases than sample 2, the discrepancy in gloss of the samples may be justified by the increased surface roughness of sample 2. As a result, it shall be emphasized that, for a comparable surface roughness, intermediate annealing of the hot band is a sustainable solution to minimize intermetallic phases, which allows enhancing the gloss level.

Fig. 9 Microsections of **a** the
reference sample, **b** sample 1, and
c sample 2 along S-L plane,
allowing a qualitative comparison
of the individual phase fraction

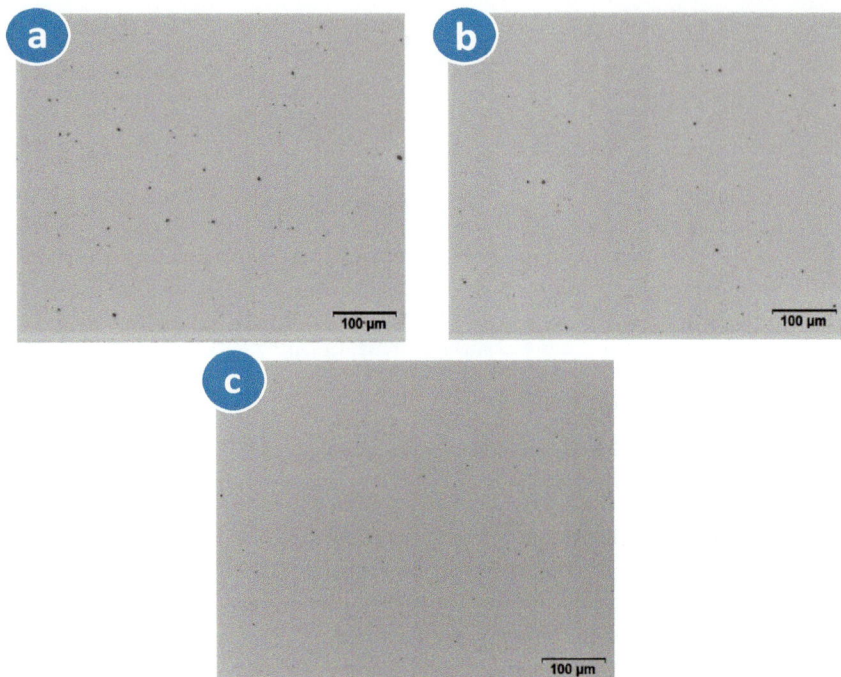

Conclusion

In this paper, the successful fabrication of rolled aluminum
suitable for the application of automotive design elements is
addressed. Through intermediate annealing and subsequent
quenching prior to cold rolling, the gloss level was
enhanced. Additionally, the impact of the surface roughness
on the gloss level was demonstrated. As suggested by the
results, the highest gloss level could be achieved for a
minimal surface roughness and the lowest tested intermediate annealing thickness. Nevertheless, it shall be emphasized
that in terms of intergranular corrosion resistance and
formability the reduction of the intermediate annealing
thickness is limited. This is due to stimulated grain growth
for reduced intermediate annealing thicknesses. Aiming for a
final thickness of 1.2 mm and considering all requirements

(superior gloss, formability and corrosion resistance), the
preferred intermediate annealing thickness lies between 8
and 5 mm.

References

1. Silvennoinen R, Peiponen KE, Myller K (2007) Specular Gloss.
   Elsevier
2. Rahlves M, Seewig J (2009) Optisches Messen technischer
   Oberflächen: Messprinzipien und Begriffe. Beuth, Germany
3. Ostermann F, (2007) Anwendungstechnologie Aluminium.
   Springer, Meckenheim
4. Lockwood D J (2016) Rayleigh and Mie Scattering. In: Luo, M.R.
   (eds) Encyclopedia of Color Science and Technology. Springer,
   New York. https://doi.org/10.1007/978-1-4419-8071-7_218
5. Richardson T J A (2009) Shreir's Corrosion. Elsevier
6. Tipler P A, Mosca G (2ßß7) Physics for Scientists and Engineers.
   Springer

Effects of Aging Conditions on Fracture Characteristics of Al–Mg–Si Alloys

Zeynep Tutku Özen, İlyas Artunç Sarı, Anıl Umut Özdemir,
Abdullah Kağan Kınacı, Emre Çankaya, Alptuğ Tanses,
and Görkem Özçelik

Abstract

The automotive industry has started using lightweight materials like aluminum with the aim of increasing fuel efficiency by supporting carbon neutral production. Structural parts of an automobile like bumper and chassis requires not only tensile strength but also crashworthiness. One of the main requirements of an aluminum alloy is to increase crashworthiness by over aging of the alloys. In this study, 6005 aluminum alloy production has been followed with automotive parts production route of billet production by direct chill casting, homogenization, extrusion and heat treatment. Different natural and artificial aging conditions have been applied to 6005 alloy. Strengths, crashworthiness, elongations and fracture characteristics have been evaluated according to the results of characterization studies including tensile tests and SEM Analysis.

Keywords

Aluminum • Billet casting • Extrusion • Crashworthiness
• 6005 alloy

Introduction

Aluminum alloys are preferred for several usage areas due to their low density, high specific strength and corrosion resistance. One of the main usage areas of aluminum alloys is automotive industry with respect to performance, safety, fuel efficiency, durability and environmental benefits. When safety issues are considered, aluminum has high crashworthiness as it is able to absorb twice as much energy than the same weight of steel. This feature of aluminum enables it to be used for the manufacturing of front and back crumple zones of a vehicle [1, 2].

Aluminum extrusion is a process which produces aluminum profiles. In order to produce profiles, an extrusion equipment applies force to the billet material which has previously heated, causing plastic deformation of the material. The mentioned deformation of the pre-heated billets occurs in an extrusion die which has crucial parts such as bolster, bracket and dummy block to sustain the required force. As a result of billet motion and the applied force of the extrusion press, production of certain cross-sectioned profiles is possible. Aluminum profiles are preferred in structural applications to produce parts for automotive, railway, aerospace and architecture sectors. The technology that enables the production of closed sections in different thickness values is an integral part of the mass production of automotive parts [2, 3].

Figure 1 shows the aluminum alloys mainly used in different parts of the cars as a component [10]. Main aluminum alloys are 5xxx and 6xxx series Body-in-White parts production. Table 1 shows the aluminum alloys and their main usage areas for the automotive industry, with respect to their properties have been summarized. Generally processing of different wrought alloys is different as well. 1xxx, 3xxx, 5xxx alloys are generally cold rolled, 2xxx, 6xxx and 7xxx alloys are generally extruded, 4xxx alloys are generally pressure casted and some 5xxx and 6xxx alloys may be hot rolled [3–7].

In order to produce a crash management system for an automobile, 6xxx series aluminum alloys are used mainly since they have properties like high compression strength, moderate tensile strength and good impact resistance. Also, design optimization freedom of the extrusion process makes extrusion a good manufacturing method for lightweight constructions. Also, 6xxx alloys have good formability which is desired with respect to extrusion process. Especially, 6005 and 6063 alloy are used for crash management

Z. T. Özen (✉) · İ.A. Sarı · A. U. Özdemir · A. K. Kınacı ·
E. Çankaya · A. Tanses · G. Özçelik
ASAŞ Aluminium San. Ve Tic. A. Ş, Sakarya, Turkey
e-mail: tutku.ozen@asastr.com

© The Minerals, Metals & Materials Society 2023
S. Broek (ed.), *Light Metals 2023*, The Minerals, Metals & Materials Series,
https://doi.org/10.1007/978-3-031-22532-1_39

Fig. 1 Different materials used
in vehicles [10]

Table 1 Usage areas of wrought aluminum alloys for automotive industry [5–7]

Aluminum alloy	Components of car	Features
1xxx	Heat insulators	Excellent corrosion resistance
2xxx	Pistons, break components, rotors, cylinders, wheels, gears	High strength, high fatigue resistance
3xxx	Piping, paneling, radiators, body sheet, fenders, doors, floor paneling	Good formability, good workability, good drawing characteristics, excellent corrosion resistance,
4xxx	Pistons, compressor scrolls, engine components	Excellent weldability, abrasion resistance, good castability
5xxx	Body paneling, fuel tanks, steering plates, piping, disk and drum breaks	Highest strength of non-heat treatable alloys. readily weldable
6xxx	Cross members, brakes, wheels propeller shafts, truck and bus bodies, crashbox	Dent resistance, excellent surface finishing characteristics, corrosion resistance, high strength, impact resistance
7xxx	Impact beams, seat sliders, bumper reinforcement, motorbike frames, rims	Highest strength, good welding character

systems as a crash box material due to their compression behaviors under stress. The decision on aluminum alloys for crash box production is done with respect to the final usage area and process efficiency approaches. One of the main variables that affects properties of these alloys is heat treatment process and its efficiency. In this study, the effects of natural aging caused by industrial delays of the profiles to be heat treated and different artificial aging durations have been investigated to increase process efficiency with respect to the results [3, 4].

Study Flowchart

In this study, the process has begun with the direct chill casting of 14 inches (355 mm) billets of 6005 aluminum alloys in the prototype casting facility of ASAS Aluminum. After the casting operation, chemical spectrometer analysis of the billets has been done via mass spectrometer. Compositions of the casted alloy have been shown at the Table 2 with its related standard.

Table 2 Chemical composition of 6005 alloys

	Si	Fe	Cu	Mn	Mg	Zn	Cr	Ti
EN 573-3	0.5–0.9	0–0.35	0–0.3	0–0.5	0.4–0.7	0–0.2	0–0.3	0–0.1
ASAS	0.65	0.3	0.1	0.1	0.5	0.1	0.1	0.1

Fig. 2 Flowchart of this study

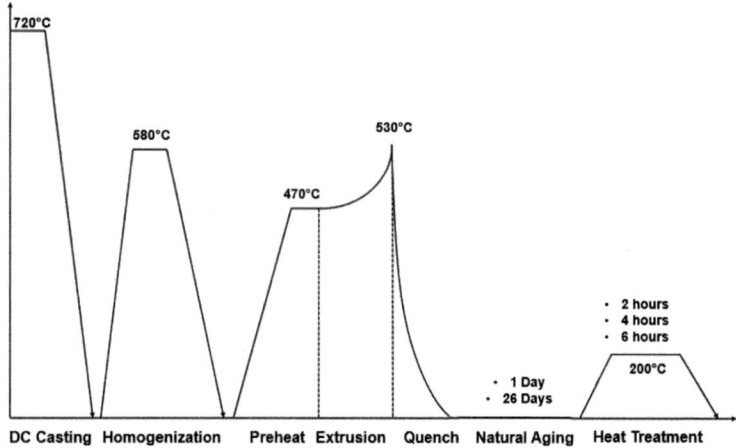

Table 3 Experimental design of the study

	AA6005—200 °C—2 h	AA6005—200 °C—4 h	AA6005—200 °C—6 h
Profiles in 1 day	Tensile test, compression test, fractography	Tensile test, compression test, fractography	Tensile test, compression test, fractography
Profiles in 26 day	Tensile test, compression test, fractography	Tensile test, compression test, fractography	Tensile test, compression test, fractography

Billets were homogenized in a batch type homogenization furnace at 580 °C for 3 h. The extrusion process has been carried out in an extrusion press with 29 MN force capability. The inlet temperature of the extrusion is 470 °C and the outlet temperature is 530 °C. The procedure was followed by quenching through water spraying at 80% capacity. Following the quench, 1.4% stretching was applied and profiles were cut into requested the length of 2250 mm. Natural aging is applied to the profiles for 1 day and 26 days separately. Artificial aging studies are carried out to both naturally aged and non-aged profiles at the temperature of 200 °C with the duration of 2, 4 and 6 h. The flowchart of this study can be seen in Fig. 2.

Characterization and test are carried out after the production route. Compression tests are carried out with a 160 tons hydraulic press. Tensile tests are applied with a Zwick universal testing device. Analysis of fracture types has been carried out with fractured tensile test specimens via SEM, Zeiss EVO MA15 brand. Table 3 demonstrates the design of the experiments respectively.

Results and Discussion

Results of this study have been investigated with respect to mechanical properties, crash properties and fracture characteristics of the alloys with different aging conditions.

Figures 3 and 4 illustrate the effects of applying prolonged natural pre-aging (NA) and subsequent three different artificial aging (AA) durations on mechanical properties. In general, there is a decreasing trend in yield and tensile strength after conducting NA and AA. The yield strength of the NA 1 day and AA 2 h at 200 °C sample is 275 MPa and the tensile strength of it is 298 MPa, respectively. On the other hand, when the NA time is increased to 26 days and the same AA condition is applied, the yield tensile strength decreased to 267 MPa. The tensile strength difference of naturally aged sample at 2 h of artificial aging conditions is very slight with 295 MPa. The yield strength of NA 1 Day and AA 4 h at 200 °C sample is 272 MPa its tensile strength is 296 MPa. Same AA conditions results in a higher

Fig. 3 Yield strength curve of naturally and artificially aged profiles

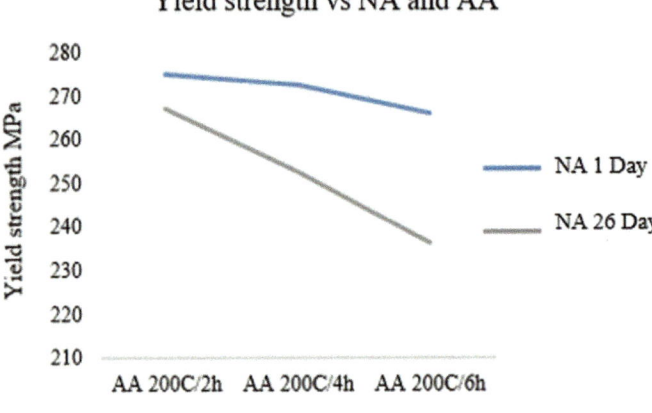

Fig. 4 Tensile strength curve of naturally and artificially aged profiles

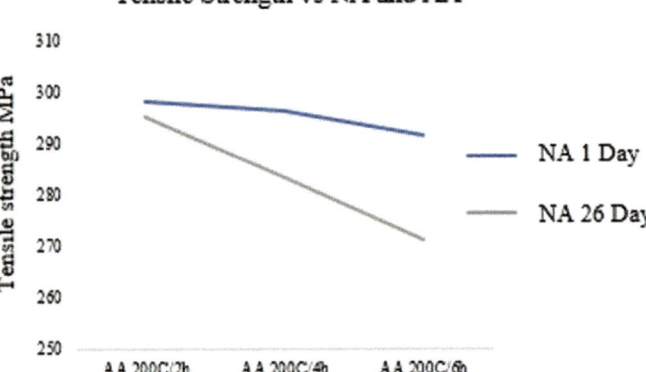

decrease for naturally over aged samples. Yield strength of it is 252 MPa and the tensile strength of it is 283 MPa. Finally, AA 6 h condition results in 266 MPa yield strength and 291 MPa tensile strength for NA 1-day samples and 236 MPa yield strength and 271 MPA tensile strength for NA 26 days samples. As a result, both yield and tensile strength values have been decreased with increasing aging time and with the presence of natural aging. Also, yield strength of minimum duration artificially aged and naturally aged samples shows similar mechanical properties with 6 h of artificial aged and non-naturally aged samples. Thus, natural aging conditions affect negatively the mechanical properties of 6005 alloys. One can claim that, 26 days of natural aging has similar mechanical properties to 6 h artificial aging at 200 °C.

When the AA time is increased from 2 to 4 h at the same AA temperature, yield strength has decreased by approximately 1% and the tensile strength around 0.5% for NA 1 day. These ratios increased to 5.5% yield difference and 4% tensile strength difference at NA 26 days. While evaluating decrease of the strength with respect to artificial aging duration both for naturally aged and non-aged samples from 4 to 6 h, non-aged samples showed 2% decrease for yield strength and 2% decrease for tensile strength. Naturally aged

samples values are 6% decrease for yield strength and 4% decrease for tensile strength respectively.

As long as NA is prolonged the effect of AA increases significantly on mechanical properties. The effects of longer NA might be due to the accumulation of β″ precipitates as the natural aging time increases. Besides, dissolution kinetics of these clusters and the growth of β″ needles slows down at subsequent AA [8].

Figures 5, 6, 7, 8, 9 and 10 illustrate the SEM images of tensile fracture morphology of artificial aging specimens with different natural aging durations. Secondary electron (SE) images have been taken to observe shapes of dimples. Also, frequency of the dimples has been analyzed respectively. Backscattered electron images (BSE) have been taken to interpret fracture characteristics with respect to depth of dimples. Finally, the numbers, distributions and morphologies of dispersoids have been investigated with × 3000 magnified BSE Images of fractured surfaces. As a general observation for all specimens, the width of dimples is reduced while the depth and density of the dimples are increased as exposed to longer natural aging.

According to Fig. 5a, frequency of the dimples is lower than others. Also, dimension of the dimples relatively higher. This is caused because of the relatively lower

Fig. 5 SEM images of 1 day NA and 2 h AA at 200 °C specimens; **a** SE image with 1000 ×magnification, **b** BSE image with 1000 ×magnification, **c** BSE image with 3000 × magnification

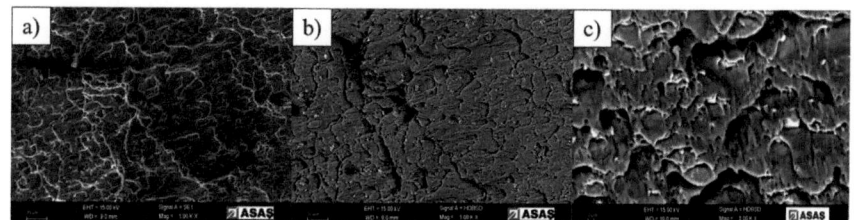

Fig. 6 SEM images of 1 day NA and 4 h AA at 200 °C specimens; **a** SE image with 1000 × magnification, **b** BSE image with 1000 × magnification, **c** BSE image with 3000 × magnification

Fig. 7 SEM images of 1 day NA and 6 h AA at 200 °C specimens; **a** SE image with 1000 × magnification, **b** BSE image with 1000 × magnification, **c** BSE image with 3000 × magnification

Fig. 8 SEM images of 26 Day NA and 2 h AA at 200 °C specimens; **a** SE image with 1000 × magnification, **b** BSE image with 1000 × magnification, **c** BSE image with 3000 × magnification

Fig. 9 SEM images of 26 day NA and 4 h AA at 200 °C specimens; **a** SE image with 1000 × magnification, **b** BSE image with 1000 × magnification, **c** BSE image with 3000 × magnification

Fig. 10 SEM images of 26 day NA and 6 h AA at 200 °C specimens; **a** SE image with 1000 × magnification, **b** BSE image with 1000 × magnification, **c** BSE image with 3000 × magnification

ductility of the material. Shapes of the dimples are generally rounded. Figure 5b shows the BSE images with × 1000 magnification. Depths of the dimples are low which supports the brittle fracture characteristics and mechanical properties. Figure 5c, shows the dispersoids shape and distribution. The dispersoids located at the edges of the dimples and they have round and spherical shapes. It can be stated that, fracture caused mainly around dispersoids.

Figure 6a demonstrates the SE Images of 1 Day NA and 4 h AA specimens. The frequency of the dimples is lower than others. Also, the size of the dimples is relatively higher. This is because of the relatively lower ductility of the material. However, it should be in between of 2 and 6 h results. There should be surface abrasion caused from laboratory conditions. Shapes of the dimples are generally rounded for this investigation. BSE images with × 1000 magnification has been showed at the Fig. 6b. Depths of the dimples are higher than 2 h artificially aged samples and lower than 6 h aged samples which supports the mechanical properties. Figure 6c, shows the dispersoids shape and distribution. The dispersoids located at the edges of the dimples and they have more sharp edges than 2 h aged specimens.

The most ductile fracture characteristic is observed for 1 day naturally aged samples as shown in Fig. 7a. The frequency of the dimples is higher than others. Also, dimension of the dimples is relatively lower. Shapes of the dimples are generally rounded for this investigation. Depths of the dimples are higher than other non-naturally aged samples with respect to Fig. 7b. Shapes of the dispersoids have needle like structure which has been illustrated at Fig. 7c.

Figure 8a illustrates the dimples shape and distribution for 26 days naturally aged and 2 h artificially aged samples. The sizes of the dimples are lower than non-naturally aged ones but higher than naturally aged and 6 h artificially aged samples. The fracture characteristic has similar topology with 1 day naturally aged and 6 h artificially aged samples. The depth of the dimples are higher than other 2 h AA sample but lower than 26 days NA and 6 h AA samples which is shown in Fig. 8b. Dispersoid shapes are rounded as it is demonstrated at Fig. 8c. This can be caused by the effect of natural aging on intermetallic precipitates growth.

Figure 9a shows the SE image of 26 days NA and 4 h at 200 °C AA samples with × 1000 magnification. As it can be observed, frequency of dimples is higher than others and the size of them lower than the others. This results also supports the ductile fracture and low strength values. Depths of the dimples are higher than the other specimens except the one with 26 days NA and 6 h AA. This can be seen from the Fig. 9b. Dispersoids have more sharp edges than the samples with 26 days NA and 2 h AA. The shapes of these dispersoids are illustrated at the Fig. 9c.

Figure 10a demonstrated the most ductile fracture behavior of all with highest frequency of dimples and lowest sizes of them. Results are supporting the increased aging conditions for aluminum alloys yields in increasing the ductility and decreased strength phenomena. The deepest fracture dimples are observed at the Fig. 10b which is belong to 26 days NA and 6 h AA at 200 °C samples. Surprisingly, dispersoids are spherical and located at the middle of dimples. This situation can be seen in Fig. 10c. The reason why dispersoids have greater size and rounded shape is the over aging conditions with respect to study of Alexopoulos et al. [9].

Conclusion

In this work, effects of natural aging and different artificial aging conditions on mechanical properties of 6005 aluminum alloys are studied. As a general result, natural aging has negative effects on strength values of the 6005 alloys, the strength also decreases when artificial aging time increases to 4 h at 200 °C and more. It can be claimed that 4 h and 6 h samples are over-aged. Furthermore, it can be stated that samples with a 26 days natural aging duration and 2 h artificial aging at 200 °C has similar mechanical properties with 1-day natural aging duration (non-naturally aged) and 6 h at 200 °C artificially aged samples. According to SEM results, existence of dimples demonstrates that the fracture type is ductile. Increased dimple frequency and small size of dimples refer to enhanced ductile behavior on the other hand the lesser frequency and bigger dimples refer to lower ductile behavior compared to the higher frequency and smaller size dimple containing material. The opposite situation where dimples are bigger and their frequencies are lower, implies that the strength of the material is higher. In this study, non-naturally aged and 2 h AA samples showed highest strength with brittle fracture characteristics. The samples with 26 days NA and 6 h AA showed a great example for over-aging where dispersoids are agglomerated and showed spherical behavior in the middle of the dimples. Investigation of crash properties and natural aging mechanism would be fruitful for further research.

References

1. Mallick, P. K., & Graf, A. (2010). Aluminum alloys for lightweight automotive structures. In Materials, design and manufacturing for Lightweight Vehicles (pp. 97–123). Essay, CRC Press.
2. Miller, W. S., Zhuang, L., Bottema, J., Wittebrood, A. J., De Smet, P., Haszler, A., & Vieregge, A. (200). Recent development in aluminium alloys for the automotive industry. Materials Science and Engineering: A, 280(1), 37–49. https://doi.org/10.1016/s0921-5093(99)00653-x
3. Lakshmi, A., Buddi, T., Subbiah, R., & Bandhavi, C. (2020). Formability studies of Automotive Aluminium Alloy Sheet Series: A Review. E3S Web of Conferences, 184, 01036. https://doi.org/10.1051/e3sconf/202018401036
4. Konar, M., Aslanlar, S., İlhan, E., Kekik, M., Özçelik, G., Güner, M. B., Yiğit, A. F., & Demirkıran, T. (2021). Investigation of weld quality for friction stir welding of extruded 6xxx series aluminium alloys. The Minerals, Metals & Materials Series, 220–226. https://doi.org/10.1007/978-3-030-65396-5_32
5. Sudhanwa M. Kulkarni, Priyal Vemu, Kiran D. Mali, Dhananjay M. Kulkarni, Effect of geometric irregularities induced during manufacturing of a crash-box on its crashworthiness performance, Materials Today: Proceedings, Volume 57, Part 2, 2022, Pages 715–721, ISSN 2214–7853, https://doi.org/10.1016/j.matpr.2022.02.179.
6. Montijo, S. (2022, September). Aluminum in cars: What aluminum alloys are common in aluminum car bodies? Kloeckner Metals Corporation. Retrieved September, 2022, from https://www.kloecknermetals.com/blog/aluminum-in-cars/
7. Understanding the alloys of aluminum. (n.d.). Retrieved September 2022, from http://www.alcotec.com/us/en/education/knowledge/techknowledge/understanding-the-alloys-of-aluminum.cfm#:~:text=5xxx%20Series%20Alloys%20E2%80%93%20(non%2D,the%20non%2Dheat%20treatable%20alloys.
8. DING, X.-fei, SUN, J., YING, J., ZHANG, W.-dong, MA, J.-jun, & WANG, L.-chen. (2012). Influences of aging temperature and time on microstructure and mechanical properties of 6005a aluminum alloy extrusions. Transactions of Nonferrous Metals Society of China, 22. https://doi.org/10.1016/s1003-6326(12)61677-x
9. [9] Zhang, L., He, H., Li, S., Wu, X., & Li, L. (2018). Dynamic compression behavior of 6005 aluminum alloy aged at elevated temperatures. Vacuum, 155, 604–611. https://doi.org/https://doi.org/10.1016/j.vacuum.2018.06.066
10. Audi A3 - materials in the body structure. Car Body Design. (n.d.). Retrieved October 2022, from https://www.carbodydesign.com/gallery/2012/05/audi-a3-design-story/37/

Evaluation of EN AW 3003 Aluminium Alloy Homogenization with Specific Electrical Resistivity Measurement

Maja Vončina, Mitja Petrič, Sebastjan Kastelic, Tilen Balaško, Stanislav Kores, and Jožef Medved

Abstract

The starting point of the research project was the control of the homogenization process of EN AW 3003 aluminium alloy. For this purpose, thermodynamic simulations of the studied alloy were carried out using the Thermo-Calc program to determine the temperature stability of the different phases and the equilibrium concentrations of these phases vs. temperature. Homogenization annealing was simulated using differential scanning calorimetry (DSC) and electrical resistance measurement to determine the suitability of such measurement for optimizing the homogenization process. The efficiency of the homogenization time was investigated using a standard homogenization process of the as-cast samples from the slab in the furnace, followed by a detailed microstructure analysis. Very good agreement was obtained between the simulation of homogenization with DSC analysis and the electrical resistivity curve, confirming the suitability of the electrical resistivity measurement method for monitoring the homogenization process of aluminium alloys.

M. Vončina (✉) · M. Petrič · S. Kastelic · T. Balaško · J. Medved
Faculty of Natural Sciences and Engineering, Department of Materials and Metallurgy, University of Ljubljana, Aškerčeva 12, 1000 Ljubljana, Slovenia
e-mail: maja.voncina@ntf.uni-lj.si

M. Petrič
e-mail: mitja.petric@ntf.uni-lj.si

S. Kastelic
e-mail: sebastjan.kastelic@ntf.uni-lj.si

T. Balaško
e-mail: tilen.balasko@ntf.uni-lj.si

J. Medved
e-mail: jozef.medved@ntf.uni-lj.si

S. Kores
TALUM d.d, Tovarniška C. 10, 2325 Kidričevo, Slovenia
e-mail: stanislav.kores@talum.si

Keywords

EN AW 3003 aluminium alloy • Thermodynamic equilibrium • Homogenization • Electrical resistivity • Differential scanning calorimetry

Introduction

Detailed knowledge of the correlation of microstructure and processing parameters is required to adequately control the final material properties. For relatively pure Al foil alloys, the microchemistry of the material, i.e., the solution/precipitation state of the alloying elements Fe, Mn, and Si and other impurities, primarily affects both the processing and the foil properties at the final gauge. Preheating or homogenising annealing prior to hot rolling also affects processing and final foil properties.

Due to their relatively high strength combined with sufficient ductility, the non-hardenable alloys Al–Fe–Mn–Si are widely used in the automotive industry, in cooling systems, in construction, … [1]. Commercial Al alloys will always contain large Fe-containing phases due to the very low solubility of Fe in Al. The nature, volume, size and especially the morphology of these constituents affect ductility and formability. Preheating or homogenising annealing prior to hot rolling, affect both processing and final foil properties by leading to the formation of fine secondary intermetallic phases or so-called dispersoids [2, 3]. The solute content during hot working must be kept low, while solidification by solid solution hardening is limited, so the predominant solidification mechanism in dilute Al foil alloys is dispersion hardening with some contribution from solutes. Therefore, the volume fraction and size distribution of dispersoids is an important factor in dispersion hardening. Higher hardness leads to lower formability and fatigue resistance because of the acicular particles [4].

© The Minerals, Metals & Materials Society 2023
S. Broek (ed.), *Light Metals 2023*, The Minerals, Metals & Materials Series,
https://doi.org/10.1007/978-3-031-22532-1_40

Electrical resistivity is a physical property of a metal that is used profitably in the field of metallurgy. It is used to determine the amount of inclusions, to monitor the solidification process of metals and the formation of phase diagrams, solid precipitates and dissolution, to follow the repair process and monitor the process and to determine the conditions of homogenization [5].

By measuring the electrical resistivity, the process of homogenization in the material can be followed, since it depends strongly on the scattering of electrons. Scattering occurs when electrons do not travel through the crystal lattice without interference, scattering causes interference [5]. Electron scattering occurs due to thermal oscillations of the lattice disturbing its periodicity, voids and the presence of impurity atoms. Thus, electrical resistivity (ρ) can be written as the sum of the resistance (Eq. 1) due to thermal oscillation of the network (ρ_f) (described by phonons) and the residual or residual resistance (ρ_r), which is the sum of the resistance due to impurities (ρ_i), the resistance due to defects in the crystal lattice and anisotropy (ρ_d), and the resistance due to magnetism (ρ_M) [5].

$$\rho = \rho_f + \rho_i + \rho_d + \rho_M \qquad (1)$$

Electrical resistivity increases with temperature due to thermal oscillation of the network, and atoms in the network move away from each other at elevated temperature, resulting in less overlap of valence orbitals and thus an increase in electrical resistivity [5]. The electrical resistivity of alloys is higher than the resistivity of pure metal. As the proportion of impurity atoms in the alloy increases, e.g. B atoms in metal A, the scattering of electrons increases and hence the resistivity increases. However, as the solubility of element B in A in the alloy decreases and B is separated from A, thus forming phases, the electrical resistivity decreases in this case. This is due to less electron scattering when the B atoms are in the form of phases in the metal than when they are dissolved in metal A [5–8]. For wrought aluminium alloys of group 3xxx, the amount of dissolved manganese in the solid solution increases the electrical resistivity by $2.94 \cdot 10^{-8}$ Ωm per wt. % dissolved Mn. However, when the amount of manganese exceeds the solubility in aluminium of 1.84 wt. %, the electrical resistivity increases by only $0.34 \cdot 10^{-8}$ Ωm per wt. % Mn [9, 10]. The amount of impurities and defects of the crystal lattice also causes an increase in electrical resistivity. As the temperature increases, the electrical resistivity of the metal increases [11].

Aluminium is generally considered to be a very good conductor of electric current, which means that the electrical resistance of pure aluminium is very low. Nevertheless, it is affected by even smaller impurities in aluminium. The electrical resistance depends mainly on the purity of the aluminium. The electrical resistivity is lowest for high-purity aluminium (99.9960–99.9990 wt. %) and is $2.63 \cdot 10^{-8}$ Ωm at 20 °C. As the impurity content increases, the electrical resistivity of aluminium also increases, so that it is $2.68 \cdot 10^{-8}$ Ωm for high-purity aluminium (99.9500–99.9959 wt. %), $2.74 \cdot 10^{-8}$ Ωm for high-purity aluminium (99.8000–99.9490 wt. %), and $2.8 \cdot 10^{-8}$ Ωm for commercially pure aluminium (99.5000—99.7900 wt. %). When measuring the electrical resistance of high-purity aluminium, the temperature at which the resistance is measured also has a great influence [12]. The effect of the presence of silicon in aluminium up to 0.006 wt. % and Fe/Si ratio between 0.8 and 3.8 on the electrical resistivity of aluminium is very small. This effect is drastically enhanced by increasing the silicon content in aluminium to 0.15–0.16 wt. % [12]. The elements chromium, tungsten, manganese and titanium have the greatest effect on increasing the electrical resistivity of aluminium. For aluminium intended for the electrical industry, the total sum of all four elements must not exceed 0.015 wt. % precisely because of this effect. For silicon content between 0.12 and 0.16 wt. %, the sum of chromium, tungsten, manganese and titanium must not exceed 0.01 wt. % [12]. Binary intermetallic compounds are formed in the aluminium corner of the Al-Mn-Fe ternary system, of which the Al_3Fe and $Al_6(FeMn)$ phases are stable. In certain cases, Fe can replace the Mn atoms in the Al_6Mn phase, so that the composition of the $Al_{12}FeMn$ intermetallic phase containing about 12.85 wt. % Fe and 12.64 wt. % Mn can be formed. The addition of Mn to the Al_6Fe phase, known as the metastable Al–Fe phase formed by non-equilibrium solidification, makes the Al_6Fe phase more stable. Al-Mn-Fe-Si do not form a quaternary phase with each other, but a stable phase of $Al_{15}Si_2Mn_4$ and other phases of the Al–Fe-Si system. Metastable Al_4Mn and $Al_{10}Mn_3Si$ phases can form during non-equilibrium solidification [11]. Most aluminium ingots need to be homogenised before further processing or hot working. Homogenization makes a crucial contribution by allowing the precipitation of excess alloying elements in the solid solution prior to subsequent thermomechanical processing. However, the high solidification rate not only promotes supersaturation and metastable state of alloying elements, but also hinders their microsegregation [13, 14]. The process of homogenization achieves a uniform distribution of alloying elements throughout the microstructure. The inhomogeneity of the solid solution can adversely affect corrosion or oxidation resistance, strength, operating temperature (due to artificial melting of the interdendritic spaces), and hot workability, and can lead to the formation of undesirable phases due to segregation. Homogenization annealing is intended to eliminate inhomogeneities, but numerical determination of time and temperature to adequately eliminate undesirable segregation is difficult [5].

In this case, a new device for electrical resistance measurements was built to plan and monitor the homogenization process in wrought aluminium alloys.

Materials and Methods

The aim of the study was to optimize the homogenization annealing of the alloy slab EN AW 3003 by measuring the electrical resistance. The standard composition of the alloy studied is given in Table 1.

Thermodynamic analysis was performed using Thermo-Calc software (TCAL6 database) to obtain the stability of the different phases, as well as the equilibrium concentrations of these phases.

For the measurements of electrical resistance (ER), special samples of 50 mm length and cross-section 5 mm × 5 mm were prepared. Because of the special shape of the specimens, a steel mould was built. When all conditions for casting and solidification were fulfilled according to the simulation programme ProCAST, a mould was made for simultaneous casting of six bars for the electrical resistance measurements. Before casting, the aluminium must be heated to 750 °C and the mould to 400 °C.

Measuring electrical resistance at high temperatures can present some problems. Thermal stresses can occur at the electrode contacts with the sample, which is best avoided by using an AC voltage with a square wave signal and a low frequency as the voltage source, measuring the voltage drops across the sample at negative and positive voltages. Thermal stress is also measured, which is eliminated by summing the positive and negative voltages. Oxidation of the electrodes can also cause a problem in the measurements which can interfere with the measurement. This is best avoided with a protective (inert) atmosphere or with oxidation-resistant electrodes [5, 15–22].

In this case, four contacts were welded to each specimen, to which current and voltage wires were later connected to measure electrical resistivity. The weld surface of the specimen was thoroughly cleaned with sandpaper. The contact was made of technically pure aluminium wire, then it was placed on the specimen where we wanted to weld it using the welding machine's pliers, and a direct current was passed through it. The specimen for which electrical resistivity was measured has contacts welded to the top of the specimen. The contacts were 20 cm long.

In order to carry out the measurements in the horizontal furnace, it was first necessary to make a sample support to ensure that the sample was in the centre of the furnace and did not move. A base measurement carrier measuring 40 × 23.8 × 15.18 mm³ was made from the insulating material Promasil 1000 (calcium silicate), and a hole was drilled in the middle of the base through which we placed the type K thermocouple. A new measuring system was set up, the sketch of which can be seen in Fig. 1. The measurement system includes a furnace and furnace controller, a sample stand with a thermocouple in the centre, insulating wool, a voltage and current (power) generator, a known resistor (1 Ω), and a computer with LabView 8.5 program.

The measurement prepared in the horizontal furnace was carried out through a connection with the sample electrodes and the current and voltage cables, designed to connect directly from the 20 cm long electrodes to the current and voltage cables. The sample was first placed in the center of the support and carefully placed in the center of the horizontal furnace. Then the thermocouple terminals of the furnace control and measurement unit were connected to the thermocouple of the sample. The sides of the horizontal furnace were carefully sealed with insulating wool to prevent contact between the electrodes and heat leakage from the furnace. The sample's electrodes were then connected to the generator's power and voltage cables. The experimental

Table 1 Chemical composition of investigated EN AW 3003 aluminium alloy

Element	Fe	Si	Mn	Cu	Mg	Ti	Zn	V	Ga	Al
wt. %	0.538	0.102	1.056	0.07	0.019	0.028	0.008	0.015	0.013	rest

Fig. 1 Sketch of the measuring system for electric resistance measurements

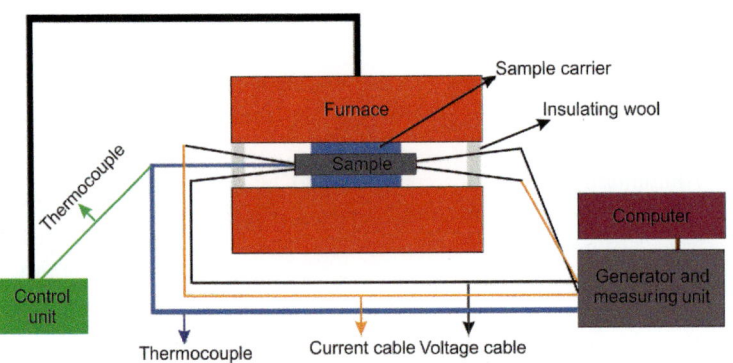

temperature to which the sample was heated (580 °C) was set in the furnace control and computer, and the LabView 8.5 program was started. The results of this experiment are presented as electrical resistance versus time at a given temperature.

Samples for homogenisation simulation on the DSC (differential scanning calorimetry) apparatus were taken from the middle of the slab, using STA Jupiter 449c from Netzsch apparatus. On the DSC device, homogenization of the as-cast sample was carried out, which lasted 12 h at 580 °C. By using the tangent method, the change in the slope of the DSC curves was determined, which are also attributed to changes in the course of homogenization.

To verify the results obtained, samples for the heat treatment tests were cut from the centre of the front surface of the slab in the shape of a cube. The dimensions of the samples were approximately 15 mm (L) × 15 mm (W) 15 mm (H). The one-step solution heat treatment (homogenization treatment) was carried out using an electric chamber furnace that had been previously calibrated. For the study of the solution heat treatments, the samples were kept at a temperature of 580 °C for different times 4, 6, 8, 10 and 12 h to study the effect of homogenization on the dissolution of the particles. The samples were removed from the furnace after a certain time and cooled in air to simulate industrial conditions. All metallographic examinations were carried out on half of the samples after homogenization annealing, while the other samples were ground and polished according to the standard procedure for metallography of aluminium alloys. Using the optic microscope (Zeiss Axio Observer 7) and the scanning electron microscope (SEM—JEOL JSM 6610LV) with energy dispersion spectrometer (EDS with detector EDXS Ultim® Max. at accelerating voltage 10 kV) the microstructural components were analysed.

Results and Discussion

Figure 2 shows the results of the thermodynamic calculations, which indicate that the temperature range for dissolution is 500–640 °C. As can be seen from Fig. 2, the equilibrium phases, and the temperature ranges in which they occur are as follows: the Al_6Mn phase up to 650 °C, the $Al_{15}Si_2Mn_4$ phase up to 590 °C, and the $Al_9Fe_2Si_2$ phase up to 400 °C. This means that when heated to 580 °C for a long time, Al_6Mn partially dissolves and $Al_{15}Si_2Mn_4$ and $Al_9Fe_2Si_2$ completely dissolve. It should also be noted that the homogenization treatment for these alloys should not be carried out at temperatures above 640 °C, as the Al_6Mn eutectic phase starts to melt.

According to the Thermo-Calc simulations, the homogenization temperature was planned, and Fig. 3 shows the electrical resistance measurement combined with the DSC homogenization simulation. The graph in Fig. 3 shows that at a constant temperature of 580 °C electrical resistivity decreases with time, which means that homogenization processes have taken place in the material, reducing the electrical resistivity of the material. In the first few hours of homogenization, the transformation has a greater effect on the resistance of the material than the homogenization of the chemical composition. With the continuation of homogenization, the influence of phase transformation due to a homogenization process decreases and the resistance of alloys decreases due to homogenization of chemical composition. These results coincide with described in Ref. [5–8]. In 12 h, the electrical resistivity decreases from 107 nΩm to 103.8 nΩm. The DSC curve drops sharply in the first 6 h at 580 °C. However, after approximately 6 h, the slope of the curve decreases and finally disappears. From these results, it can be concluded that the homogenization processes at 580 °C are completed after approximately 6 h. Similar was reported in a doctoral dissertation [23] in which research was conducted on an alloy from group 8xxx.

A comparison of the stationary part of the specific ER curve and the stationary part of the DSC curve shows a great similarity in the shape of the curves. With such similarity between the falling DSC curve and the ER curve, the success of the electrical resistivity measurement method for monitoring the homogenization of aluminium alloys can be confirmed.

Typical microstructural elements are marked with coloured circles in the photomicrographs in Fig. 4. Red circles indicate areas in the microstructure of the samples that are characteristic of the non-homogenized structure. Areas of major intermetallic phases are marked in blue. Areas of transitional state of homogenization are marked in yellow, and areas where the microstructure is fully homogenized are marked in green. Figures 5, 6, 7 and 8 show the EDS results of the investigated samples, as-cast and homogenized at 580 °C for 6, 8, and 12 h, respectively. Checking the results of EDS analysis shows that all intermetallic phases in all samples contain Al, Fe and Mn, which is in agreement with the theory [11], since Fe and Mn are substitution elements. It also shows that a lower proportion of Mn is found within the α-Al matrix, which is consistent as Mn has a higher solubility than Fe at homogenisation temperatures. The primary phase present in the studied alloy is stoichiometrically most similar to the $Al_6(FeMn)$ phase [24]. In all homogenized samples, the occurrence of accumulations of Si-containing phases is observed, which are stoichiometrically closest to the

Fig. 2 Thermo-Calc predicted
phase stability in aluminium alloy
EN AW 3003 regarding the
temperature

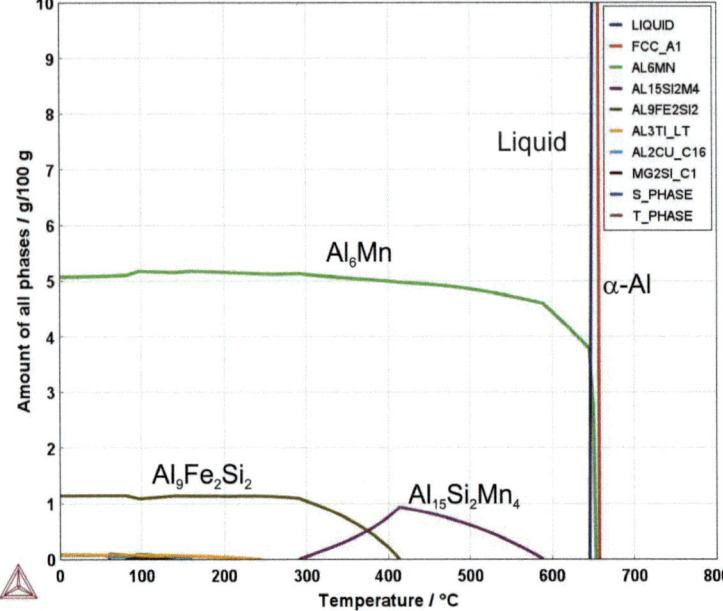

Fig. 3 DSC and ER
measurement of homogenisation
annealing of aluminium alloy
EN AW 3003 at 580 °C

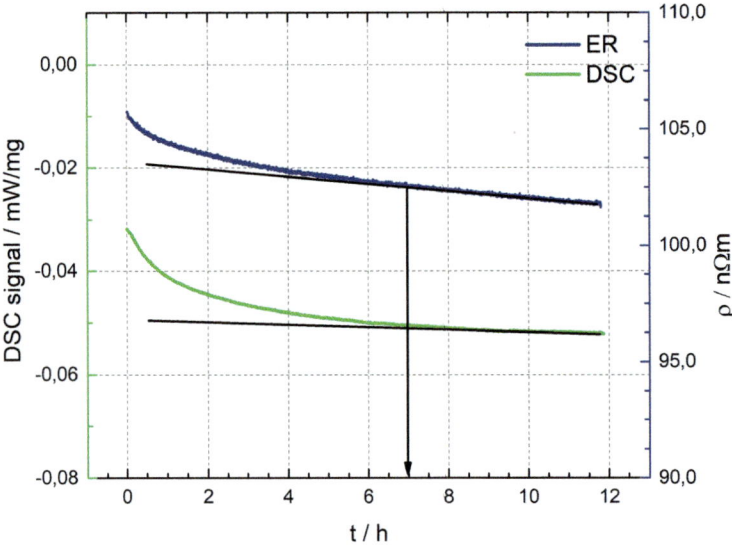

$Al_{15}(Mn,Fe)_4Si_2$ phase. In the non-homogenized sample, there was difficulty in finding domains or Si-containing phases, but these phases contained a higher proportion of Si than the Si phases in the homogenized samples. It is assumed that the Si dissolved and conglomerated on the existing AlMnFe phases due to homogenization, as also reported in [25]. Small precipitates within the α-Al matrix are also observed in homogenized samples as also indicated in [26], but they are too small for analysis from EDS. Some researchers indicate to be α-Al(Mn,Fe)Si [27]. Based on these results of micrography (shape and elemental composition of analysed phases) and DSC and ER measurement of homogenisation annealing, the main part of homogenization annealing time is concluded in 7 h.

Conclusions

Very good agreement was obtained between the simulation of homogenization with DSC analysis and the ER curve, confirming the suitability of the electrical resistivity measurement method for monitoring the homogenization process of aluminium alloys. After 6–8 h of homogenization, there is a change in the slope of the DSC and ER curves, which may indicate a change in the course of homogenization or an end to certain homogenization processes.

According to TC and EDS as-cast microstructure analysis, the alloy EN AW 3003 consists of primary dendrites of α-Al, eutectic $Al_6(Mn,Fe)$ phase and $Al_{15}(Mn,Fe)_4Si_2$ phase

Fig. 4 Microstructure of as-cast and homogenized specimens at 580 °C at 500 × magnification: as-cast (**a**) and after 4 h (**b**), 6 h (**c**), 8 h (**d**), 10 h (**e**), and 12 h (**f**) homogenization. Red circles indicate areas of the non-homogenized structure (presumably $Al_6(FeMn)$ phase), blue indicate areas of major intermetallic phases, yellow indicate areas of transitional state, green indicate areas where the microstructure is fully homogenized (presumably $Al_{15}(Mn,Fe)_4Si_2$ phase)

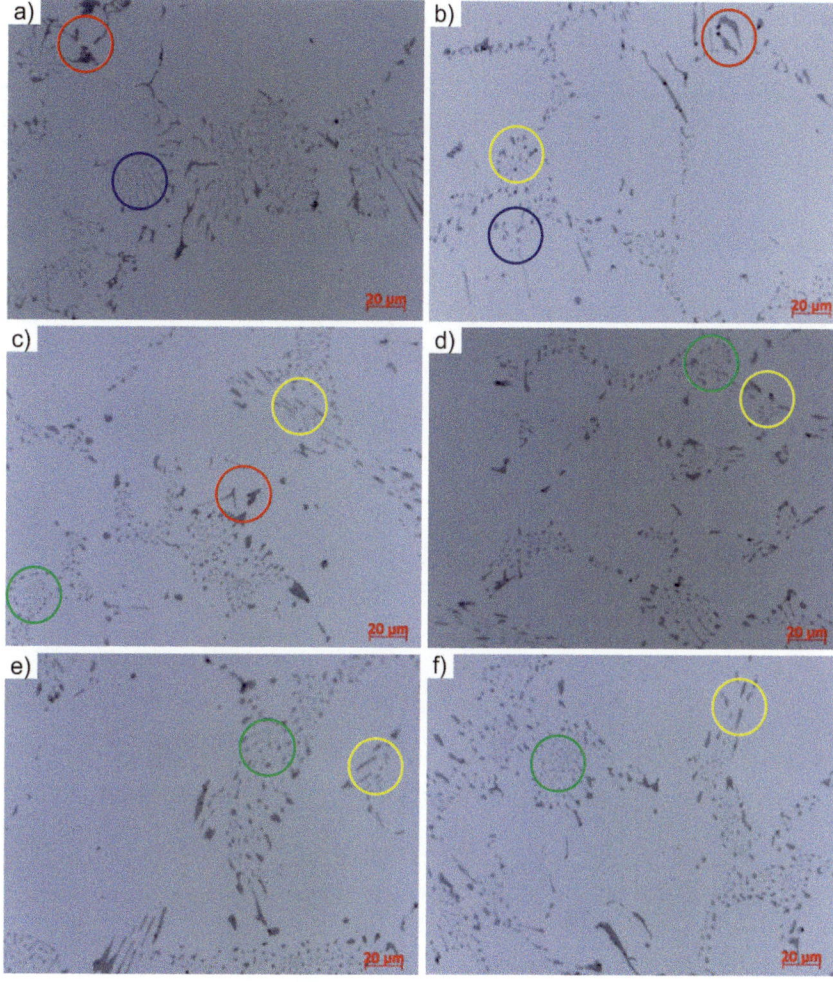

Fig. 5 SEM micrographs of as-cast sample, whereas corresponding EDS analysis are presented in at. %

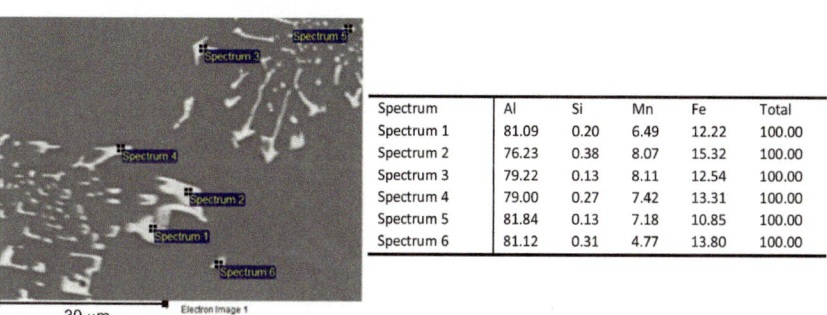

Spectrum	Al	Si	Mn	Fe	Total
Spectrum 1	81.09	0.20	6.49	12.22	100.00
Spectrum 2	76.23	0.38	8.07	15.32	100.00
Spectrum 3	79.22	0.13	8.11	12.54	100.00
Spectrum 4	79.00	0.27	7.42	13.31	100.00
Spectrum 5	81.84	0.13	7.18	10.85	100.00
Spectrum 6	81.12	0.31	4.77	13.80	100.00

Fig. 6 SEM micrographs of sample homogenized at 580 °C for 6 h, whereas corresponding EDS analysis are presented in at. %

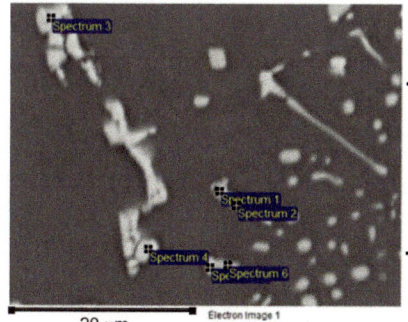

Spectrum	Al	Si	Mn	Fe	Total
Spectrum 1	72.07	3.45	10.40	14.08	100.00
Spectrum 2	81.26	0.08	8.95	9.70	100.00
Spectrum 3	67.88	3.95	10.20	17.96	100.00
Spectrum 4	73.72	3.53	7.92	14.83	100.00
Spectrum 5	73.71	3.40	9.54	13.35	100.00
Spectrum 6	82.17	0.16	8.02	9.65	100.00

Fig. 7 SEM micrographs of sample homogenized at 580 °C for 8 h, whereas corresponding EDS analysis are presented in at. %

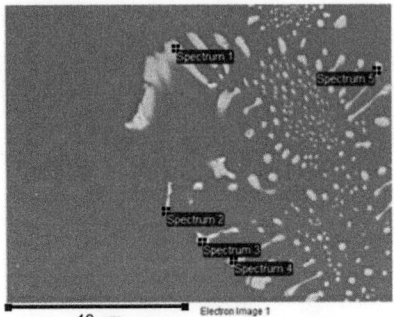

Spectrum	Al	Si	Mn	Fe	Total
Spectrum 1	81.16	0.22	6.30	12.32	100.00
Spectrum 2	74.99	4.20	8.03	12.77	100.00
Spectrum 3	74.33	4.65	8.34	12.69	100.00
Spectrum 4	80.81	0.09	7.23	11.86	100.00
Spectrum 5	81.22	0.06	8.09	10.63	100.00

Fig. 8 SEM micrographs of sample homogenized at 580 °C for 12 h, whereas corresponding EDS analysis are presented in at. %

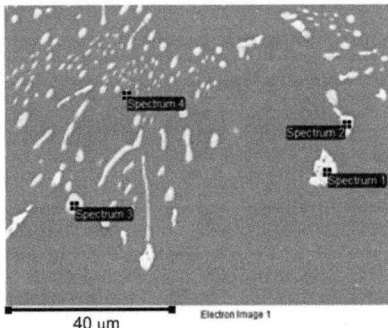

Spectrum	Al	Si	Mn	Fe	Total
Spectrum 1	75.61	5.21	6.35	12.84	100.00
Spectrum 2	72.79	5.47	9.75	12.00	100.00
Spectrum 3	81.57	0.21	6.84	11.38	100.00
Spectrum 4	83.94	0.16	8.56	7.34	100.00

in very small proportions. The homogenization annealing mainly changes the geometrical shape of the intermetallic phases. With a longer homogenization time, the microstructure still consists of primary dendrites of α-Al, eutectic $Al_6(Mn,Fe)$ phase and $Al_{15}(Mn,Fe)_4Si_2$ phase conglomerated on the intermetallic phases and more uniformly distributed throughout the microstructure. EDS analyzes also show that some Mn is dissolved in the primary α-Al crystals. Mn and Fe are stable in the microstructure of all samples in approximately equal proportions in all intermetallic phases. Si-phases are more abundant in homogenized samples than in a non-homogenized sample. The Si-phases in the non-homogenized sample contain a higher proportion of Si than in the homogenized samples.

Currently, the homogenization annealing time is carried out on alloy EN AW 3003 at 580 °C for 12 h. According to the results of this study, the homogenization annealing time could be reduced to 7 h, and the homogenization process would be sufficient for further processing.

Acknowledgements We gratefully acknowledge the financial support of the Republic of Slovenia, the Ministry of Education, Science and Sport, and the European Regional Development Fund. The work was carried out in the framework of the project "Modelling of Thermomechanical Processing of the Aluminium Alloys for High Quality Products: (MARTIN, Grant No.: OP20.03531). The authors would like to thank also to the Slovenian Research Agency (ARRS) for funding under program grant P2-0344.

References

1. Lentz M, Laptyeva G, Engler O (2016) Characterization of second-phase particles in two aluminium foil alloys. Journal of Alloys and Compounds 660:276–288. https://doi.org/10.1016/j.jallcom.2015.11.111
2. Engler O, Laptyeva G, Wang N (2013) Impact of homogenization on microchemistry and recrystallization of the Al-Fe-Mn Alloy AA 8006. Materials Characterization 79:60–75. https://doi.org/10.1016/j.matchar.2013.02.012
3. Shakiba M, Parson N, Chen XG (2014) Effect of homogenization treatment and silicon content on the microstructure and hot work ability of dilute Al–Fe–Si alloys. Material Science and Engineering A 619:180–189. https://doi.org/10.1016/j.msea.2014.09.072
4. Vončina M, Kresnik K, Volšak D, Medved J (2020) Effects of Homogenization Conditions on the Microstructure Evolution of Aluminium Alloy EN AW 8006. Metals 10(3): 419. https://doi.org/10.3390/met10030419
5. Petrič M (2013) Changes of dimensions and electrical resistivity during solidification of alloys from Al–Si system. Ph.D. Thesis, University of Ljubljana
6. Closset B, Drew R, Gruzleski J, Pirie K (1984) Etude par resistivite electrique de la microstructure des alliages de fonderie Al–Si–Mg traites thermiquement. Memoires et etudes scientifiques revue de metallurgie. 81(9): 439
7. Mulazimoglu MH, Drew RAL, Gruzleski JE (1991) The effect of strontium on the electrical resistivity and conductivity of aluminum–silicon alloys. Metallurgical transactions A. 18:941–947. https://doi.org/10.1007/BF03325703
8. Mulazimoglu MH, Drew RAL, Gruzleski JE (1989) The electrical conductivity of cast Al–Si alloys in the range 2 to 12,6 wt pct silicon. Metallurgical transactions A 20A(3):383–389.

9. Hatch JE (1984) Aluminum: Properties and Physical Metallurgy. Aluminum Association Inc. and ASM International

10. McGeer JP (1984) Light Metals: Proceedings of the Technical Sessions. Metallurgical Society of AIME

11. Belov NA, Aksenov AA, Eskin DG (2002) Iron in Aluminum Alloys: Impurity and Alloying Elemen, Taylor & Francis, New York

12. Totten GE, MacKenzie DS (2003) Handbook of Aluminum: Vol. 1: Physical Metallurgy and Processes, no. let. 1. CRC Press

13. Chen Z, Li S, Zhao J (2012) Homogenization of twin-roll cast A8006 alloy. Trans. Nonferrous Metal Soc. China 22:1280–1285. https://doi.org/10.1016/S1003-6326(11)61316-2

14. Sadeghi I, Wells MA, Esmaeili S (2017) Modeling homogenization behavior of Al-Si-Cu-Mg aluminum alloy. Mater. Des. 128:241–249. https://doi.org/10.1016/j.matdes.2017.05.006

15. Brandt R, Neuer G (2007) Electrical Resistivity and Thermal Conductivity of Pure Aluminum and Aluminum Alloys up to and above the Melting Temperature. Int. J. Thermophys. 28:1429–1446. https://doi.org/10.1007/s10765-006-0144-0

16. Van Zytveld JB (1980) Electrical Resistivities of Liquid Transition Metals. Le J. Phys. Colloq. 41:C8–503–C8–506. https://doi.org/10.1051/jphyscol:19808126

17. Monaghan BJ (1999) A Four-Probe dc Method for Measuring the Electrical Resistivities of Molten Metals. Int. J. Thermophys. 20:677–690. https://doi.org/10.1023/A:1022625609033

18. Gui MC, Jia J, Li QC, Guo WQ (1997) Influences of trace additions of strontium and phosphorus on electrical resistivity and viscosity of liquid Al-Si alloys. Trans. nonferrous Met. Soc. China. 7:67–71. http://ir.imr.ac.cn/handle/321006/37963

19. Kokabi HR, Provost J, Desgardin G (1993) A new device for electrical resistivity measurements as a function of temperature (86–700 K) under controlled atmosphere by the four-probe method. Rev. Sci. Instrum. 64(6):1549. https://doi.org/10.1063/1.1144025

20. Riontino A, Zanada G (1994) A precipitation study of two rapidly solidified Al–Fe alloys. Materials science and engineering A 179–180(1):323–326. https://doi.org/10.1016/0921-5093(94)90219-4

21. Brunčko M, Kneissl AC, Kosec B, Lojen G, Anžel I (2007) Identification of liquid/solid transformations in eutectic Pb–Sn alloy. Int. J. Mater. Res. 98(2):112–116. https://doi.org/10.3139/146.101448

22. Çadırlı E, Kaya H, Gümüş A, Yılmazer I (2006) Temperature-Dependence of Electrical Resistivity of Cd-Sn, Bi-Sn, and Al-Si Eutectic and Al-3wt.%Si Hypoeutectic Alloys. J. Mater. Eng. Perform. 15(4):490–493. https://doi.org/10.1361/105994906X124578

23. Arbeiter J (2021) Modelling of phase equilibrium and homogenization kinetics of Al-Fe and Al-Fe-Si alloys. Ph.D. thesis, University of Ljubljana

24. Zhipeng Y, Yiyou T, Ting Y, Yaohua H, Yunhe Z (2021) Effect of the interaction between Fe and Si on the precipitation behavior of dispersoids in Al–Mn alloys during homogenization. Vacuum 184:109915. https://doi.org/10.1016/j.vacuum.2020.109915

25. Dehmas M, Weisbecker P, Geandier G, Archambault P, Aeby-Gautier E (2005) Experimental study of phase transformations in 3003 aluminium alloys during heating by in situ high energy X-ray synchrotron radiation. Journal of Alloys and Compounds 400:116–124. https://doi.org/10.1016/j.jallcom.2005.03.062

26. Huang HW, Ou BL (2009) Evolution of precipitation during different homogenization treatments in a 3003 aluminum alloy. Materials & Design 30(7):2685–2692. https://doi.org/10.1016/j.matdes.2008.10.012

27. Lia YJ, Muggerud AMF, Olsen A, Furu T (2012) Precipitation of partially coherent α-Al(Mn,Fe)Si dispersoids and their strengthening effect in AA 3003 alloy. Acta Materialia 60(3):1004–1014. https://doi.org/10.1016/j.actamat.2011.11.003

The Effect of Octagonal Ingot Shape on AA6xxx Hot Rolling Performance

Joshua Lawalin, Pascal Gauthier, and Tao Wang

Abstract

Edge cracking during hot rolling of aluminum 6000-series ingots presents a challenge to rolling mills, resulting in costly rework, scrap loss, and downtime. The sodium element is known to be a critical contributing factor to edge cracking, while silicon, magnesium, and titanium content, within typical limits, does have a noticeable effect. The inverse segregation zone and meniscus bands are found to serve as potential sites for edge crack initiation and propagation and must be minimized with appropriate casting practices. This study was performed to identify key processes and metallurgical factors of using DC-cast octagonal AA6016 ingots to reduce edge cracking during hot deformation. A microstructural investigation was conducted with an optical microscope and SEM-EDS to understand the impact of using eight-sided ingots. An industrial trial was realized to evaluate the rolling performance and metallurgical characteristics of octagonal ingots compared to the standard rectangular form.

Keywords

6xxx series • Aluminium alloy • Al–Mg–Si • AA6016 • DC-cast ingots • Octagonal shape • Hot rolling • Laminated products • Edge cracking • Failure analysis

J. Lawalin
Commonwealth Rolled Products, 1372 State Road 1957, Lewisport, KY 42351, USA

P. Gauthier
Arvida Research and Development Centre, Rio Tinto Aluminium, Jonquiere, QC G7S 4K8, Canada

T. Wang (✉)
Rio Tinto Aluminium Technical Marketing, 200 E Randolph St., Suite 7100, Chicago, IL 60601-7329, USA
e-mail: Tao.Wang@riotinto.com

© The Minerals, Metals & Materials Society 2023
S. Broek (ed.), *Light Metals 2023*, The Minerals, Metals & Materials Series,
https://doi.org/10.1007/978-3-031-22532-1_41

Introduction

For automotive applications, the heat treatable 6xxx series aluminum alloys are commonly used to produce medium–high strength parts with high formability and good corrosion resistance. However, aluminum alloys are prone to edge cracking defects when they are rolled [1–3]. It compromises the quality of the part because edge cracking cannot be predicted from the design stage and cracks appear during production. This defect affects the product quality by reducing yield as a result of edge trimming, surface damages caused by residual fragments, and the efficiency of production. Most of the sheet shows little lateral flow. Figure 1 demonstrates a bad edge cracking situation during the hot rolling process.

In the centre of the sheet, all of the thickness reduction stretches the material, but there is less constraint against the lateral flow at the edge of the sheet. Some of the thickness reduction causes lateral spread, and the length increase is less than the thickness decrease. The edge tends to flatten during rolling, which causes tension stress in this zone, as opposed to the middle (Fig. 2).

Most of the middle of the slab width (about eighty percent) is in plane strain [4, 5]. This suggests that the horizontal, lateral, and compressive stresses are one half of the through-thickness stresses. The lateral compressive stresses have two sources: friction, which opposes spread, and unrolled metal outside the roll gap which has not been reduced in thickness, and thus has no tendency to spread. This gives a large, negative hydrostatic stress in the centre of the slab. The edge of the slab has no imposed lateral stresses to oppose spread, except for friction at the slab surface. Deformation occurs at constant volume and, in the centre of the sheet, all of the thickness reduction elongates the material. Because of lateral spread at the edge of the sheet, not all of the thickness reduction causes an increase in length so that the edge flattens more during rolling as compared to the middle, which causes tension in the free side surface (Fig. 3).

Fig. 1 Severe edge cracking defect on hot rolled aluminum alloys

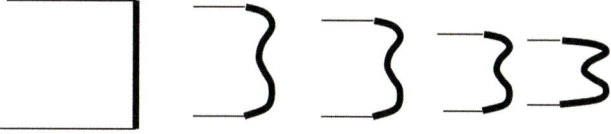

Fig. 2 Lateral cross-section view showing the evolution of the strip edge during hot roughing mill rolling

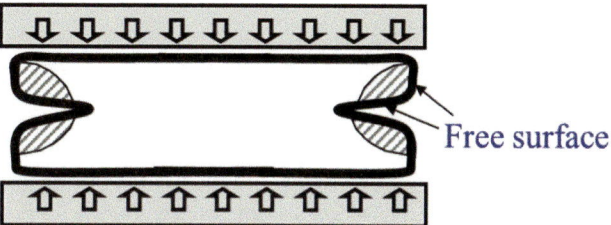

Fig. 3 Stress states profile in rolling

Two reasons explain long edge cracks in rolled aluminum material: (1) edge cracks occur upstream (in the hot roughing mill) and grow longer in subsequent roll passes, or (2) edge cracks originate in the hot finishing mill (HFM) and are long because of high tension applied near the strip edge. Different variables influence the edge cracking sensitivity: existing cracks in transfer slabs, bad ingot quality with sodium contamination, deep inverse segregation/meniscus bands, and ingot transition zones near butt features [7, 8].

Three types of cracks, designated "segregation", "major edge", and "massive" edge cracks, were recognized [6]. Initiation of cracks was usually associated with grain-boundary precipitate particles. Major- and massive-cracking were identified with sodium and hydrogen in association with lack of homogenisation of the ingot. The combination of these bad factors can happen on the short side surface or near edges by creating oxidized particles and crack paths [9]. They will propagate perpendicularly to rolling forces. Figure 4 shows the edge cracking propagation through the edge microstructure or from the short side of the sheet [10].

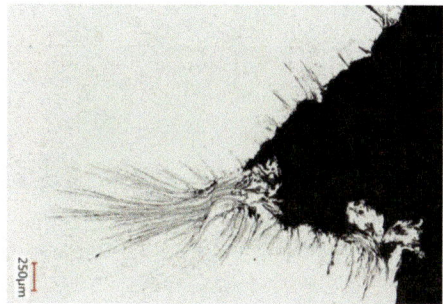

Fig. 4 Edge cracking initiation due to a severe oxidation on the short face and propagation through the rolled aluminum material

To avoid excessive tension in the free surface area during rolling and edge cracking, ingots were designed with eight sides where the short faces have a slight taper. This may lead to slabs with less net deviation than a vertical-sided slab through the hot rolling mill (HRM).

Methodology

The aim of this study was to validate the surface quality and benefits of rolled sheets made from initial DC-cast octagonal ingots. This trial was used to potentially reduce the edge cracking and improve the rolling performance for future 6xxx opportunities.

DC-Casting Octagonal Ingots

AA6016 aluminum alloy was used to produce 610 mm 2032 mm DC-cast octagonal ingots (Fig. 5). The AA chemical composition analyzed by optical emission spectroscopy is shown in Table 1. Metallographic slice analysis

Fig. 5 As-cast octagonal ingot shape and dimensions of 610 mm 2032 mm ingot (or 24 inches × 80 inches) for industrial rolling trial

Table 1 Typical AA6016 alloy designation and its chemical composition limits

AA6016	Si	Fe	Cu	Mn	Mg	Cr	Zn	Ti	Other	Al
Range	1–1.5	0.50	0.20	0.2	0.25–0.6	0.1	0.20	0.15	0.15	Bal

was performed by optical microscopy and acid etching to ensure good surface quality and metallurgical features of the as-cast material. Constituents were characterized by the SEM-EDS technique.

Hot Rolling

An industrial trial was realized with these octagonal ingots to validate the impact of chamfered edges. Five ingots were hot rolled during this trial with eight-sided ingots at a 7.5 mm final sheet thickness. Trimmed sample surfaces were used to evaluate the appearance of edge cracking defect and its severity degree. Finally, microstructural analysis allowed to reveal internal defects, heterogeneities, intermetallic particle distribution, and metallurgical grain size (Fig. 6).

Results

Different metallurgical features were investigated for as-cast octagonal ingots such as cast surface appearance, defect detection, grain structure, inverse segregation thickness, shell zone thickness, dendrite arm spacing, and constituent features.

Octagonal Ingot Characteristics

The surface topography of the material was smooth, without any significant defect or colour variation on the ingot sides. Macrography analysis revealed a clear shell zone at the surface, followed by mixed grain structure types (feathery type on the first 100 mm and equiaxed at the centre) with a

shell zone thickness of 13 mm on the short side and 11–14 mm at the rolling face. Figure 7 shows the solidification pattern near the ingot corner.

Microscopic analysis revealed an inverse segregation zone (ISZ) with a mean value of 990 μm and no more than 1300 μm, as shown in Fig. 8a. Four types of constituents were discriminated in octagonal slices (Fig. 8b). A higher concentration of Al–Mg and Mg–Si phase particles was found at the surface compared to other intermetallics. Iron-related alpha particles (mainly AlMnFeSi) were found in higher concentrations in the middle and the centre of the ingot.

Finally, dendritic arm spacing measurements were done after a 60 s etch with 1% HF. The results show average DAS measurements of 28.5, 56, and 88.6 μm, respectively, for the surface, the midway, and the centre of the ingot.

Rolling Feasibility with Octagonal Ingots

The surface appearance of the five hot rolled sheets was evaluated to observe the topography and defects. Edges showed opened and severe transversal cracks (Fig. 9a). Edge profiles obtained on the second and the fourth rolling ingot demonstrated minor edge cracks but many surface scratches parallel to the rolling direction (Fig. 9b).

The tapered edge was rolled into the short face instead of the rolling faces. Edge alligatoring was reduced significantly from ¾ to 1 inch (approx. 25.4 mm) to ¼ inch (approx. 6.35 mm), as shown in Fig. 10a. However, the head alligatoring defect was higher during this trial (Fig. 10b).

Fig. 6 Rolled sheet output after hot rolling with octagonal-shaped ingots

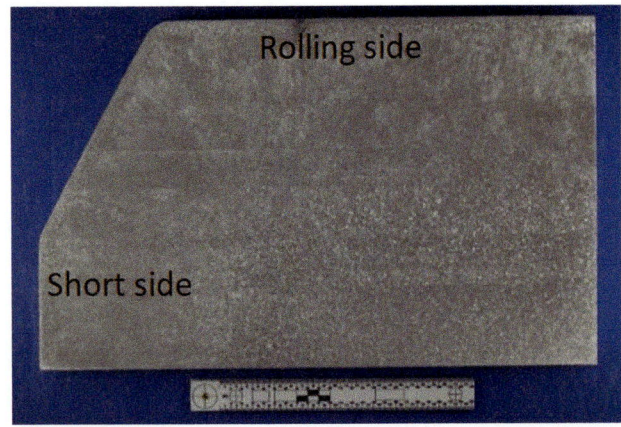

Fig. 7 Macroscopic observations and grain structure of octagonal ingot slices etched with acid Tucker's etchant

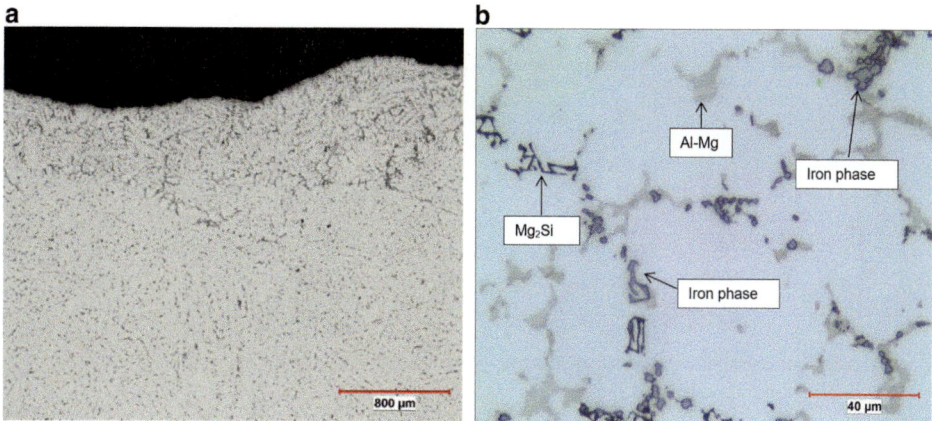

Fig. 8 Microstructural observations of octagonal ingot slice surface

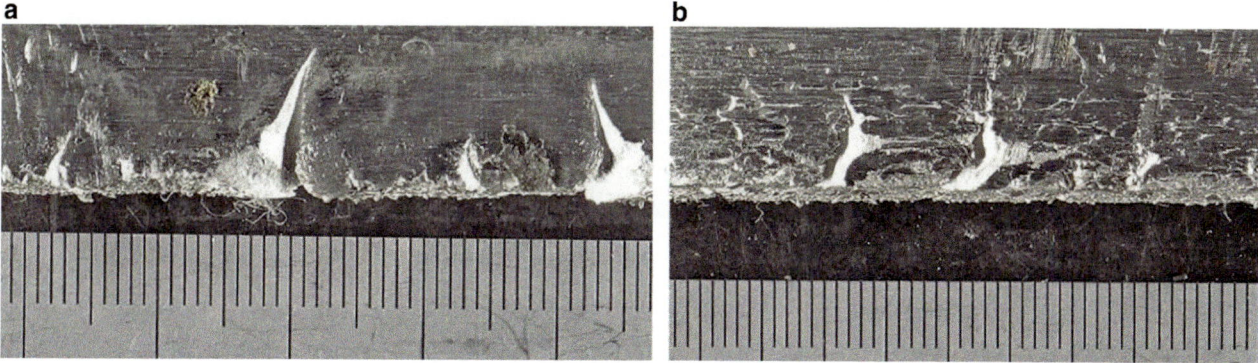

Fig. 9 Edge cracking appearances after the AA6016 octagonal ingots rolling trial

Fig. 10 Surface quality of AA6016 rolled sheets made from octagonal ingot trial

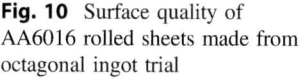

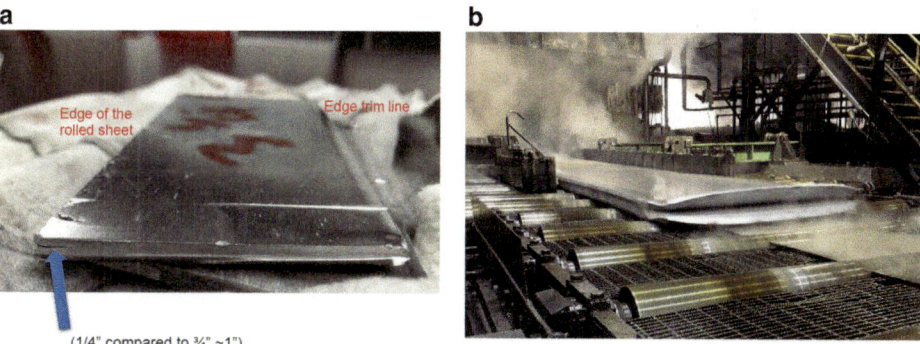

Surface scratches and peel-off defects were found on the surface of some edge-trimmed samples (Fig. 10a). Metallographic observations of octagonal rolled material revealed a mean edge crack depth of 9 mm from the sheet side (Fig. 11a). Trimmed samples revealed a similar edge crack pattern and morphology.

At higher magnification, edge cracks follow intermetallic particles, as seen in Fig. 11b. These second phase particles can fracture, creating voids, and, ultimately, can contribute to sheet failure. Also, a small amount of oxidized particles has been found near internal cracks. They also affect the propagation of cracks into the material.

Finally, the grain structure of the material is mainly fibrous in all samples. High deformation on the sides explains the recrystallization phenomenon near the edge cracks. These fractured lines follow an intergranular path following grain boundaries (Fig. 12).

Fig. 11 Metallographic cross-section of edge trimmed AA6016 samples with typical crack morphology for hot rolled octagonal ingot material

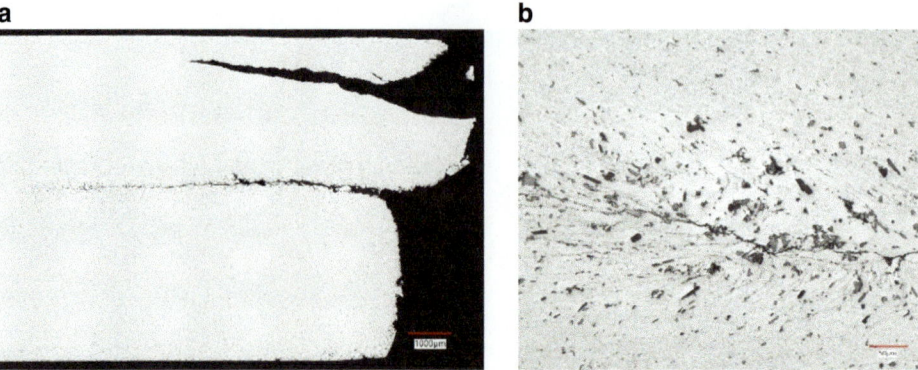

Fig. 12 Metallurgical grains revealed by the 3% HBF$_4$ electrolytic Barker's etchant for the AA6016 hot rolled aluminum sheet specimen made from the octagonal ingot trial

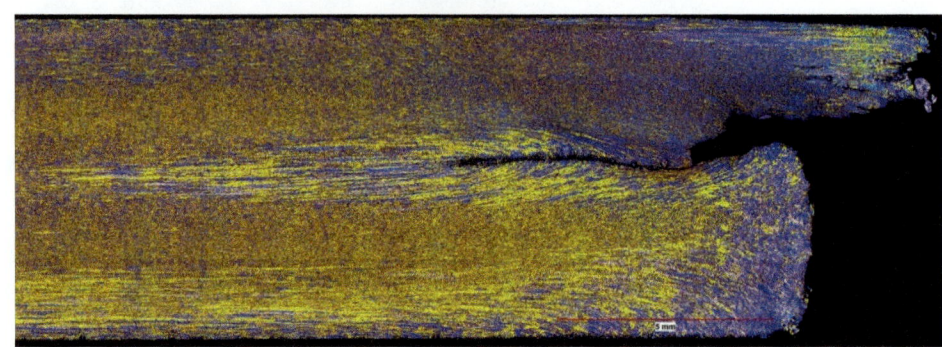

Discussions on the Rectangular and Octagonal Shape Ingots

The microstructural features of octagonal ingots were compared with those of standard rectangular ingots. Surface properties such as shell zone (10–13 mm) and inverse segregation thickness (693–1300 μm) were found similar to those of standard sizes (Fig. 13).

Grain size measurements were found different between cast ingots. It depends on how much grain refiner was added into the AA6016 product, where the grain pattern and size can range from larger grains (a few mm) to small equiaxial grains (around 100–450 μm). See Fig. 14.

The edge cracking pattern of the reference samples has normally no oxide particles/oxide lines within cracks. Edge cracks were found close to the surface, 4–9 mm in depth for rectangular and octagonal rolled ingots (Fig. 15a). The grain morphology and size near the grain boundaries can range from recrystallized grains to highly deformed grains (Fig. 15b). Octagonal rolled ingots showed a typical fibrous matrix microstructure, with some recrystallized grains. In the case of edge cracks in the reference samples, we have a uniform fibrous texture.

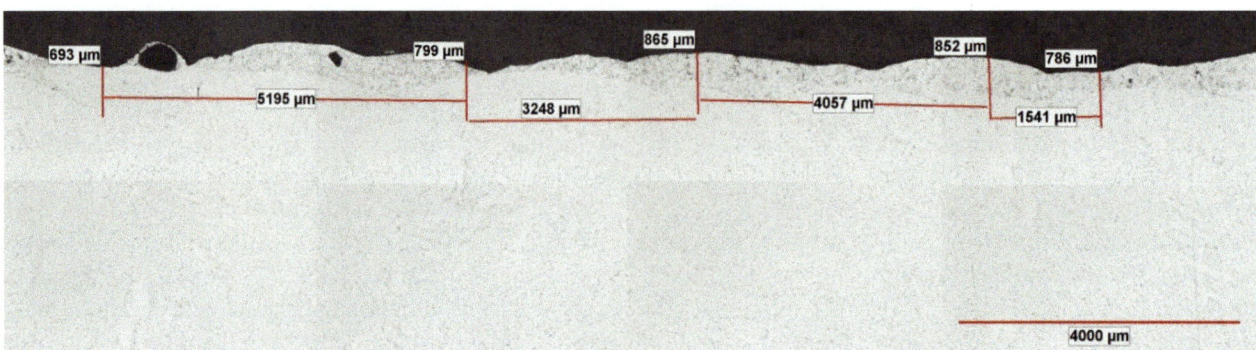

Fig. 13 ISZ thickness measurements of rectangular DC-cast ingot material

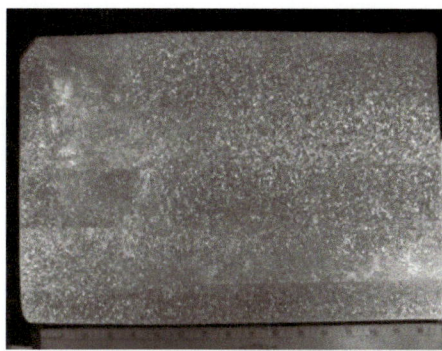

Fig. 14 Grain size evaluation on slightly refined rectangular DC-cast ingot material

Conclusions

The main findings of this work are summarized as follows:

- The cast surface appearance, the shell zone thickness, and inverse segregation layer characteristics were found to be similar between octagonal and rectangular ingots.
- A hybrid grain microstructure was revealed for the octagonal ingots, with feathery grains near the surface and equiaxial in the centre. It can vary from other rectangular ingots due to possible higher grain refiner addition rate, which gives a finer grain structure.

- Metallurgical constituent types and related levels of Al–Mg, Al–Fe–Mn–Si, and Mg–Si were similar compared to the standard rectangular ingot form.
- The hot rolling feasibility of octagonal ingots demonstrated edge cracks on trimmed rolled samples for specific rolling conditions. However, metallographic sections on hot rolled sheets revealed 9-mm cracks that are less deep than when using rectangular ingots (~ 19–25 mm). So, edge alligatoring and cracking were reduced.
- Edge crack failure analysis revealed an intergranular morphology following intermetallic particles at grain boundaries.
- Scratches and peel-off defects were found on some octagonal rolled trimmed samples.
- Head alligatoring was found to be higher during the rolling of octagonal ingots, which could be explained by the special corner configuration and the rolling in the short side of the sheet.

The general microstructure of rolled octagonal ingots is fibrous in all conditions. Metallurgical grains near the edge cracks are recrystallized and small in the same direction of the edge cracks. The final microstructure can differ from rectangular rolled material, where rolling parameters and grain refiner additions play an important role in final sheet properties.

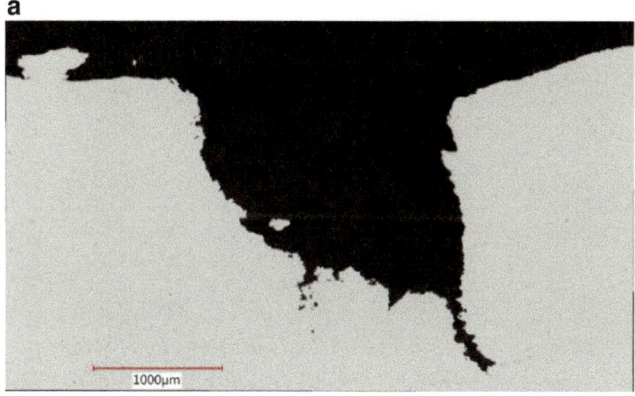
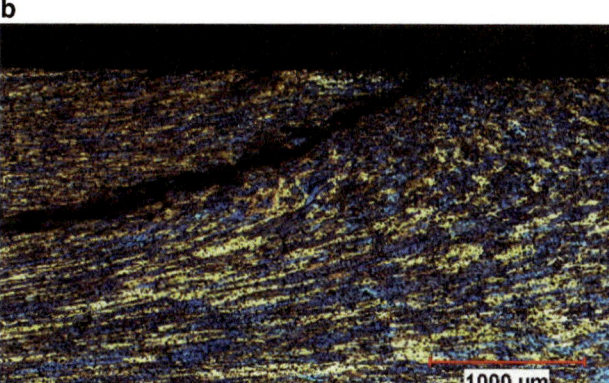

Fig. 15 Typical edge cracking pattern in rectangular rolled ingots

References

1. Dodd Bradley and Boddington Philip, "The causes of edge cracking in cold rolling", Journal of Mechanical Working Technology, 3, 1980, pp. 239–252

2. H. Riedel, F. Andrieux, T. Walde and K. Karhausen, "The formation of edge cracks during rolling of metal sheet", Steel Res. Int 78 (2007), pp. 818–824.

3. Wen Zhe Chen and Xi Peng Xu. "Analysis of Edge Cracks in Hot Rolled Aluminium Alloy Sheets Contained of Magnesium for Can Usage". Advanced Materials Research, Feb. 2012, Vol 472-475, pp. 683–687. https://doi.org/10.4028/www.scientific.net/AMR.472-475.683

4. Alexander Kainz, Sergiu Ilie, Erik Parteder and Klaus Zeman. "From Slab Corner Cracks to Edge-Defects in Hot Rolled Strip – Experimental and Numerical Investigations". Steel research int., 79 2008, No. 11, pp. 861–867.

5. Bo Wang, Jiong-ming Zhang & Yan-zhao Luo. "Research on the Evolution Behavior of Surface Transverse Cracks During the Hot Rolling Process of Medium Plates". Metallography Microstructure and Analysis, January 2015, 4 (1). DOI: https://doi.org/10.1007/s13632-014-0186-9

6. N.M.Burman and P.F.Thomson. "Edge-cracking and profile of hot-rolled aluminium slabs", Journal of Mechanical Working Technology. Volume 13, Issue 2, June 1986, pp. 205–217. https://doi.org/https://doi.org/10.1016/0378-3804(86)90066-5

7. Mohammadreza Mofarrehi, Mousa Javidani & X.-Grant Chen. "On the intermetallic constituents in the sodium induced edge cracking of hot-rolled AA5182". PHILOSOPHICAL MAGAZINE, 2021, VOL. 101, NO. 16, pp. 1849–1870. https://doi.org/10.1080/14786435.2021.1937368

8. S. Zhang, Q. Han & Z.-K. Liu, "Fundamental understanding of Na-induced high temperature embrittlement in Al–Mg alloys", Philosophical Magazine, 2007, 87:1, pp. 147–157. DOI: https://doi.org/10.1080/14786430600941587

9. Hoang, Ruiyin, Guo Y.Y, Zhu Yuanzhi & Liu Wang. "Analysis of edge cracks in hot rolled aluminum alloy sheets contained of magnesium for can usage". Trans Tech Publications ltd. 2022. Advanced Material Research. pp. 683–687.

10. Gali, Olufisayo, "Micro-mechanisms of Surface Defects Induced on Aluminum Alloys during Plastic Deformation at Elevated Temperatures" Electronic Theses and Dissertations, 2017, 5936. 435 pages. https://scholar.uwindsor.ca/etd/5936

The Low-Carbon Production of Wrought Aluminum Alloys Based on Post-consumer Scrap

Varuzan M. Kevorkijan and Sandi Žist

Abstract

The recycling of wrought aluminum alloys from end-of-life products is only carried out to a limited extent. The obstacles are the lack of: (i) an appropriate chemical composition of the recycled melt and (ii) sufficient reliable sources. Both can be eliminated through the consistent traceability of alloys in end-of-life products at all stages of circular aluminum management. This paper describes the process of ensuring the universal traceability of aluminum alloys by type of alloy, by manufacturer, and by location in the end-of-life product. Tracking begins with the disassembly of end-of-life products into their components or components made from wrought aluminum alloys. We have built an innovative concept of ensuring the traceability of Al alloys using barcodes, which can be improved with more advanced versions. The solution enables the consistent separation and extraction of the highest-quality scrap from end-of-life products, comparable to the quality of the return material. This is, as an option, usually returned to the alloy manufacturer, and can be melted directly into wrought alloys. The solution enables significant decarbonization of production and an increase in the added value of the products.

Keywords

Recycling and secondary recovery • Wrought aluminum alloys • Sustainability • Post-consumed scrap

Introduction

The recycling of wrought aluminum alloys (WAAs) from end-of-life products (ELPs) is practiced in the aluminum industry only to a very limited extent. The obstacles are the difficulty, or even the impossibility, of obtaining: (i) the appropriate chemical composition of the recycled melt and (ii) sources with sufficient reliability [1–3].

To eliminate both these obstacles, we designed and prototyped a closed loop that makes it possible to obtain scrap suitable for the production of WAAs. We found that the first obstacle can only be eliminated through the high-quality separation of scrap [4] or with complete traceability of the alloys in the ELPs—during every phase of the circular management of aluminum [5, 6]. Additionally, we found that such traceability and high-quality separation of alloys can only be achieved by consistent disassembly of the ELPs into components. Therefore, in our prototype loop, described in this paper, we are exclusively disassembling rather than shredding the ELPs.

In order to eliminate the second obstacle, it was essential to motivate the owners of the ELPs to keep the material, as far as possible, within a closed loop. As will be described in more detail later, we took care of this with a properly designed closed-loop business model that properly rewards the owners of the ELPs for consistently keeping their products within the loop. The prototype closed loop, which enables the separation of PCS from ELPs in terms of alloys and producers, and ensures more than two-thirds of the return of the WAAs from the ELPs to their producers, we called the Advanced Circular Loop for Wrought Aluminum Alloys (i.e., the ACL for WWAs).

The universal traceability of aluminum alloys by type of alloy, by manufacturer, and by location in the ELP was constructed by introducing our own Digital Product Identification (DPI) concept. The tracking begins already at the stage of disassembling the ELP into its component parts or components made from WAAs. The traceability concept

V. M. Kevorkijan (✉) · S. Žist
Impol Aluminum Industry, Partizanska 38, 2310 Slovenska Bistrica, Slovenia
e-mail: varuzan.kevorkijan@impol.si

© The Minerals, Metals & Materials Society 2023
S. Broek (ed.), *Light Metals 2023*, The Minerals, Metals & Materials Series,
https://doi.org/10.1007/978-3-031-22532-1_42

ensures that the WAAs contained in the ELP within the loop are known by the alloy designation, the manufacturer, and the location of the alloy in the product, allowing them to be consistently separated. It can also contain rules for disassembling the ELPs into components and more detailed or specific instructions for separating them. The existing DPI in the prototype ACL for WWAs was built using QR codes. However, it can be supplemented with more advanced codes if necessary.

This innovation makes it possible to obtain scrap of the highest quality from ELPs, comparable to the quality of the return material. This material, with the right of first refusal given to the original producers, is generally returned to the original alloy producers, and it can be directly cast into wrought alloys with, as an added bonus, a low carbon footprint. A feature that was not possible until now. The solution in the form of an advanced loop for the circular management of aluminum makes it possible to decarbonize aluminum-alloy production and create considerable added value.

With an advanced loop, we effectively solve the problem of increasing the proportion of aluminum scrap based on WAAs from products that have come to the end of their useful life (known as post-consumer scrap or PCS). The target increase in the share of PCS is a minimum of 50%, with an expectation of achieving up to 70%. Since the aluminum in an ELP has practically no carbon footprint, this will result in considerable decarbonization of the aluminum industry (by at least 40%) and large savings in terms of resources and logistics, as well as reducing the dependence of aluminum-alloy producers on primary aluminum (PA) and alloying elements (AEs).

Materials and Methods

We designed the advanced loop for the circular management of aluminum as a prototype with selected buyers of our semi-finished and finished products who were interested and willing to participate in this project. The focus was on semi-finished products (e.g., forging bars) and other forged components for the automotive industry. Our goal was to establish a loop that would allow the alloys we incorporated into our semi-finished and other products to be returned to us at the end of their useful lives. In addition to the direct buyers of our semi-finished products and other products, we also included in the loop all others who process our products to the extent that enables their installation in passenger vehicles. We completed this first part of the loop with passenger-vehicle manufacturers who responded to a call for cooperation. In the second part of the loop, we included: (i) companies engaged in the sale of passenger vehicles and (ii) buyers of passenger vehicles. Since vehicles can change

owners several times during their useful lifetimes, we made sure that all the owners of the vehicles were included in the loop, including the final owner of the vehicle as an ELP. While the cooperation of all the intermediate owners is the key to ensuring traceability, the cooperation of the end-of-life owner is critical to the successful return of the alloys to collection centers, sorting facilities, and end-of-life dismantling facilities. All of the above form the third part of the closed loop, in which end-of-life products are disassembled into components and separated according to the types of alloys and their manufacturers. The components, separated by manufacturer and alloy, are then returned to their manufacturers, who are given the right of first refusal within the advanced loop. The rest of the end-of-life products are offered to other scrap processors.

We can say that the first part of the loop is made up of the producers of alloys, components, and products; the second part is the owners of these alloys, components, and products; and the third part is the scrap processors. If necessary, primary aluminum and alloying elements are fed into the loop, and shredded remains from ELPs are discharged from it. The loop is schematically presented in Fig. 1.

Our innovation solves the problem of separating the WAAs in ELPs and returning them consistently to the alloy producers with the help of full product traceability. Technically, the tracking of WAAs allows the coding of components made from these alloys, Fig. 2, which is nothing new. The key innovation is the rules that we have created, between participating business entities in the advanced loops, for the complete transfer of traceability from one phase of the circular flow to the next. In fact, if the goal is to return to us as much as possible of the wrought aluminum alloys that we have produced and incorporated into products

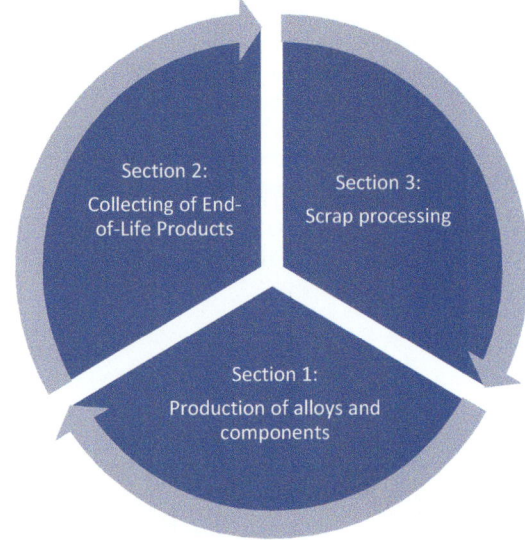

Fig. 1 Schematic of a closed loop

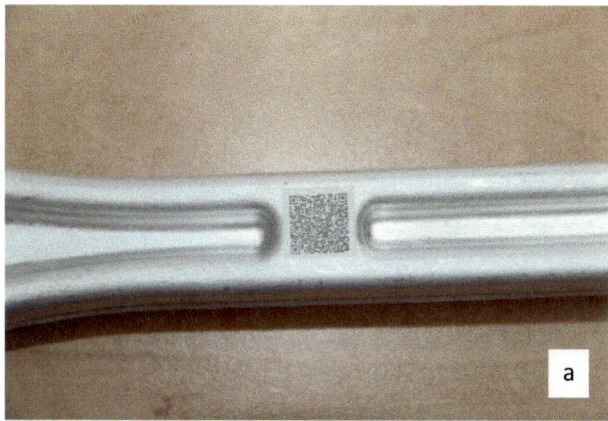

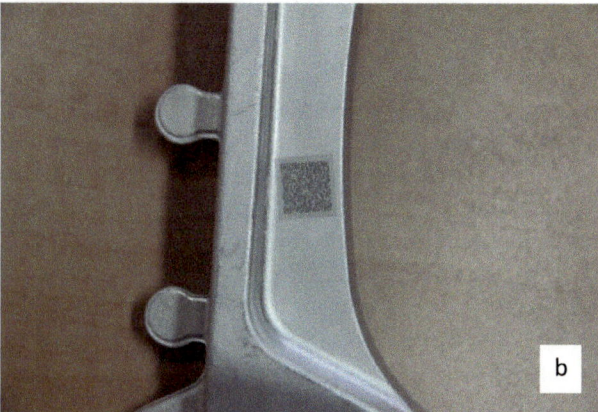

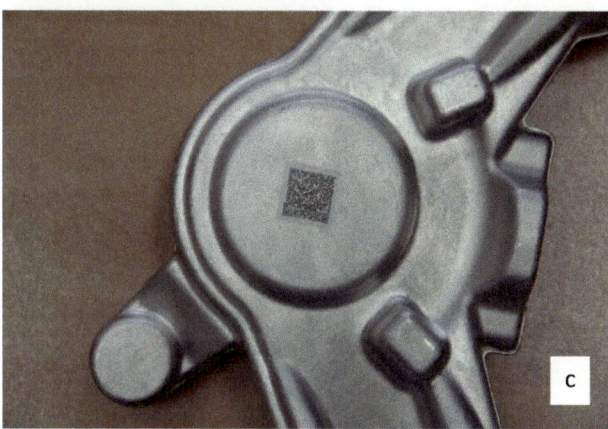

Fig. 2 a, b, c Various aluminum forgings equipped with a barcode. In a specific case, the coded record makes it possible to read the name of the manufacturer of the alloy as well as the standard and internal designation of the alloy, which enables the highest possible degree of subsequent separation

when their useful life expires, we must always keep them within the closed loop. As part of this, the traceability and separation of the aluminum alloys by type and manufacturer are ensured with the advanced DPI tools. Barcodes provide traceability, but we also need some marketing tools to achieve our goal. It is necessary to motivate the owners of the products to sell them to the collection centers that

dismantle the products, which is a prerequisite for their return to a new life cycle (according to the principle of "recycling the same into the same").

Our rules motivate all the participants and so ensure the continuous flow of the WAAs within advanced loops. This ensures full traceability for all phases of the circular flow within the advanced loop, which conventional loops do not have. In the case of semi-finished products (e.g., forged bars), the code contains information about: (i) the manufacturer of the alloy (e.g., Impol PCP, Impol Aluminum Industry), (ii) the standard designation of the alloy (e.g., EN AW-6082), and (iii) designations of the alloy according to the internal nomenclature (e.g., AC 30). If necessary, the data set can be expanded (e.g., by adding the batch code, etc.). In the case of finished products (e.g., automotive components), the code also contains information about the location of the component in the finished product (e.g., in a passenger vehicle). If this information is available before shipping the vehicle component, we will arrange for it to be recorded, but if it is not available, we will agree with the buyer, based on a contract, to enter this information later (e.g., using the buyer's code). In a similar way, we contractually agree with the buyers of our semi-finished products that they transfer data about the alloy, its manufacturer, batch, lot, etc., to all the products made from these semi-finished products.

Our innovation ensures the separation of WAAs from ELPs by individual alloys and alloy manufacturers, which is a major breakthrough. This highest degree of separation makes it possible to: (i) obtain the highest-quality PCS with a completely defined chemical composition, and, consequently, (ii) recycle WAAs to WAAs. The latter means replacing a significant proportion of the PA and AE in WAAs—up to 50% and, in some cases, even up to 70%—without affecting the quality of the final product. As it is obtained from end-of-life products, this scrap is approximately 95% decarbonized and can be used as an efficient and inexpensive source of "green" aluminum, green alloy elements, and green energy, or as a means of decarbonizing our own production of WAAs and the products that result from them. The high quality of the separation, which does not depend on the type of ELP, is ensured by the comprehensive traceability of the WAAs at every phase of their life cycle, i.e., circulation within loops, which is guaranteed by the use of codes. It is this kind of complete traceability, guaranteed by new rules that motivate all the participants to be involved in and maintain the loop, that is the key innovation with which we have succeeded in creating advanced loops for the circular management of aluminum. These ensure a high percentage (up to 90%) of the return of WAAs to their producers (specifically the Impol group), which conventional loops do not provide. The rules we built into the loops also give alloy producers the right of first refusal when it comes to PCS based on their alloys, allowing the "same to be recycled to the same and from the same manufacturer". The latter is also a great innovation and makes it possible to achieve top-quality recycled WAAs, even when it comes to the most demanding applications (as dictated, for example, by the automotive industry).

Results and Discussion

The advanced loop contains a relatively long and complex process chain, which can have a major impact on its effectiveness. The key parameter we pursued in the closed loop is the degree of retention of the components containing wrought aluminum alloys within the process chain, which ensures the return of the alloys to their manufacturers. We tracked this parameter, P, along the entire process chain and in this way identified the critical points in the chain that contribute the most to reducing the efficiency of the process.

The parameter P is calculated as the ratio of the quantity or the number of components, N2, that entered the second part of the loop to the first amount of those, N1, that entered the first part of the loop from production:

$$P_1 = {N_2}/{N_1} \qquad (1)$$

The value of the parameter P1 at the output of the first part of the closed loop was in the range of 0.90, which is well above the target value of 0.70.

The value of the parameter P decreased the most (by approx. 20%) in the second part of the closed loop, especially in the part where the intermediate owners of the vehicle appear, as well as the final owner at the end of the vehicle's useful life. The observed value of P2 was only 0.70, which means that even in this part of the closed loop we managed to reach the target value, although the possibilities for significantly improving the value of the parameter P are obvious.

Such improvements will require the introduction of even more advanced marketing tools, with which it will be necessary to further motivate the owners of products whose useful life has expired, to keep them within the closed loop.

Losses in the third part of the closed loop, where end-of-life products are disassembled into components and separated according to the alloys and their manufacturers, are actually the lowest, amounting to just 3%. The value of the P3 parameter at the exit from the third part of the closed loop varied around 0.67. This means that by organizing a prototype closed loop we managed to achieve an approximately 67%, or two-thirds, return of our aluminum alloys from end-of-life products.

Since the lifetime of components installed in passenger vehicles is from 5 to 12 years (and in some cases longer), in this study, we limited ourselves to components that were installed in vehicles and have the shortest lifetime (approx. 5 years). The achieved result might have been slightly worse if the components installed in vehicles with a much longer service life were also included in the study. With these vehicles (or end-of-life products in general), there is a much higher probability of there being multiple intermediate owners, where tracking and maintaining the product within a closed loop is much more difficult, and therefore, less effective.

However, in this prototype phase of the advanced closed-loop operation, we were not so much interested in achieving the highest possible values for the parameters P1, P2, and P3, rather we wanted to check the correctness of the concept or the closed-loop designs and the performance of the marketing tools, i.e., those with which we motivate the product buyers to keep the products within the closed loop.

We are satisfied with the results we achieved, but there remains a lot of work to be done in the future.

On the prototype advanced loop that was used to create the post-consumer scrap for the production of wrought aluminum alloys, we also monitored the operating costs and the economic benefits, since the price, in addition to the quality of the scrap, is the second-most important factor that determines its widespread use in the production of alloys. The absolute values that we determined must remain confidential and so cannot be disclosed here, but we are permitted to comment on them. In order to ensure the circulation of aluminum alloys, separated in terms of alloys and their producers within the advanced loop, it is absolutely necessary to achieve a higher level of motivation in the second part of the loop (with the owners of semi-finished products and products). It is imperative that they, through their decisions and actions, keep the products within their own loop. We can increase the motivation of the product owners to keep the products in the loop by rewarding their behavior accordingly (and, if necessary, additionally). At the same time, it is our goal to ensure that this higher motivation of the product owners does not lead to an increase in the operating costs of the advanced loop. In other words, as the operating costs of the loop increase in the second part, they must decrease proportionally in the first and third. The loop is organized in the manner of proportional distribution of the newly created value among all the participants in the loop, which gives us the possibility to keep the operating costs of the advanced loop unchanged. In practice, this means that when market conditions dictate, we give participants in part 2 of the forward loop (intermediate and end-of-life product owners) more rewards for handling the products properly and keeping them within the forward loop, and give slightly less to participants in part 1. After all the participants in the loop cover their costs, the profit generated is divided according to the key determined by each participant's contribution to ensuring the circulation of the aluminum.

Without citing specific numbers, we can say that the advanced loop designed in this way creates considerable added value, as it produces post-consumer scrap of the highest quality. Obtaining such scrap makes it possible to produce wrought aluminum alloys containing more than 80% of post-consumer scrap, which significantly reduces the carbon footprint of the alloys and makes them green.

In practice, the most important thing is to ensure that the manufacturer of finished products made from properly coded, semi-finished products correctly transfers the initial coded record to each of the finished products. This is the only way to ensure the complete traceability of alloys. This, in other words, means that anyone who is interested can read the recorded data and, on this basis, take care of the proper separation of the alloy from the other materials in the ELP and return it to the manufacturer. These are the rules, on the basis of which we build advanced loops for the circular management of aluminum, which the participants are obliged to strictly respect. Key among these rules is that the alloy manufacturer has the right of first refusal to purchase its own alloy. The coding of the products, in itself, is nothing new. The innovation of the solution lies in the separation concept, which allows not only separation by individual alloys (which is the highest level of separation), but also by alloy manufacturer, which is a completely new step. Why is this important? Those who deal with the composition of WAAs know that when achieving the desired composition for these alloys, problems often arise with trace elements, the type, and concentration which differ from manufacturer to manufacturer. By separating them in terms of manufacturer, these problems are eliminated or at least reduced to a minimum. Everyone who recycles knows that it is best when "the same is recycled into the same", and with this innovation, we ensure that "the same is recycled into the same and with the same manufacturer". The separation of alloys based on the reading of coded records (specifically, laser-engraved QR codes) also eliminates the need for an on-the-spot determination of their chemical composition, which is expensive, often not entirely reliable, as well as being time consuming. When separating alloys in the way proposed in our innovation, their chemical composition is determined in the laboratory and, if needed, is already available during the phase of their separation. In most cases, we do not even require it, because we are only interested in the standard or, even more so, the internal marking of the alloy, which is completely sufficient for the highest-quality separation. This reduces the costs and improves the separation productivity many times over. According to the proposed methodology, WAAs are separated at the level of components of the ELP, which means that these must be pre-dismantled accordingly.

The alternative to dismantling ELPs is to shred them into chips, and then apply mechanical separation of the chips according to their composition, with processes that are also already developed and commercially available. Unfortunately, when an ELP is crushed, traceability to the alloy producer is completely lost, which can affect the quality of the final products when recycling WAAs for the most demanding purposes. Our technology for "mining" WAAs in end-of-life products rests on ensuring a high degree of traceability for the individual components or the components of products, which takes place from the beginning to the end of their lifecycle, or of their new lifecycle created by recycling. Traceability is achieved by coding or DPI-ing these components, which is impossible to achieve with chips that are created by shredding.

Separation or the sorting of components leads to the highest quality of PCS, which is a prerequisite for the production of WAAs for the most demanding purposes (e.g., for the automotive industry). Another completely new aspect is the rules for the operation of advanced loops for the circular management of aluminum, with which we achieve high transparency for the operating costs of loops and ensure their competitiveness compared to existing ones. One such key rule is the right of first refusal that alloy producers have to purchase PCS based on their own WAAs. This rule allows the PCS from an ELP to be preferentially installed in recycled WAAs by those manufacturers who also produced this alloy at the beginning of its life, which additionally ensures, not only for optimizing the operation of loops, but also, first and foremost, that we can achieve the highest quality of recycled WAAs, i.e., recycling WAAs into WAAs".

Conclusion

The advanced loop for circular aluminum management, the prototype design of which is described here, has many advantages. Its key advantage, with which it stands out from the others, is that it enables a very accurate separation of wrought aluminum alloys from products that have reached their end of life. It is possible to separate alloys by individual types and manufacturers, which is not feasible with other loops and separation procedures. This is ensured by: (i) disassembling end-of-life products into individual components and (ii) full traceability of the individual components made of wrought aluminum alloys, which maintains a high proportion of these components within the closed loop. One of the advanced tools that makes this possible is digital product identification, but a barcode can also be used to take care of this.

However, traceability alone is proving to be insufficient as a tool to keep components made of wrought aluminum alloys within the advanced loop. It is also necessary to introduce very sophisticated marketing tools. The owners of end-of-life products, who are expected to sell a high percentage of them to the collection centers involved in the advanced loop, need to be properly motivated to do so. We've built a number of marketing tools into the advanced loop that makes it possible to achieve this goal.

Dismantling is a prerequisite for the high-quality separation of aluminum alloys from end-of-life products. Dismantling is much more expensive than crushing an ELP, but at the same time, it also creates much greater benefits, through the efficient return to a new lifecycle (according to the principle of "recycling the same into the same").

Advanced loops for the circular management of aluminum are open to all those who wish to participate, as they are built on the principles of respecting the origin and market value of alloys. The respect-of-origin rule provides aluminum-alloy producers with a pre-emptive right to buy back their own alloys, but only at market value, which is taken care of by another rule.

The basic premise from which we proceed is that wrought aluminum alloys in products that have expired retain their economic and functional value permanently. Their market value is slightly lower after each recycling, due to the need to cover the operating costs of the advanced loop (mining costs, including disassembly and separation costs and margins), but it is important that all the above deductions are transparently defined in the operating rules of the advanced loop and written into the relevant contracts.

It is a concept that focuses on the owner of the end-of-life product, but before that, also on the owners of all the semi-finished products, in the intermediate stages of the creation of the final product. The owner is the one who decides to whom and at what price it will offer this product or source of aluminum alloys. With the right motivation, it will offer the product whose useful life has expired to those who contributed the alloys and who will take care of the alloys re-birth or return to a new lifecycle.

Since an advanced loop designed in this way is a source of high-quality post-consumer scrap based on wrought aluminum alloys, its introduction contributes to the decarbonization of production and the creation of considerable added value.

References

1. Fick G., (2021) Feasability of aluminium component dismantling from ELV, IRT-M2P EAStudy-Report-20210225, https://www.european-aluminium.eu/media/3172/irt-m2p-executive-sum-20210412-final.pdf.
2. Norgate T. (2013) Metal recycling: The need for a life cycle approach. EP135565, CSIRO, Australia; https://publications.csiro.au/rpr/download?pid=csiro:EP135565&dsid=DS2.
3. Paraskevas D., Keliens K., Dewulf W. and Duflou J. (2013) Sustainable Metal Management and Recycling Loops: Life Cycle Assessment for Aluminium Recycling Strategies, https://www.researchgate.net/publication/271523601
4. Environmental Affairs Division, Toyota Motor Corporation, (2017) Vehicle Recycling https://global.toyota/pages/global_toyota/sustainability/report/kururisa_en.pdf
5. Schindler R., Schmalbein N., Steltenkanp V., Cave J., Wens B., and Anhalt A. (2012) SMART TRASH, Study on RFID tags and the recycling industry, Rand Europe, I.A.R and P3; https://www.rand.org/content/dam/rand/pubs/technical_reports/2012/RAND_TR1283.pdf
6. Green Deal: New proposal to make sustainable products the norm and boost Europe's resources independence https://ec.europa.eu/commission/presscorner/detail/en/ip_22_2013

Reduce Inclusion Level Study in Aluminum Slab Products 3XX and 5XX

Abdullah Al-Qarni and Bader Dhawi AlMuhana

Abstract

Reducing the inclusion levels in Aluminum is quite challenging considering the metal cleanliness methods starting from TAC (Treatment Aluminum in Crucibles) until the Aluminum been charge into the furnace which then contains a degasser unit to remove the hydrogen from going into the metals a long with A 94 (deep bed filter). Our study was conducted to reduce inclusion level in Aluminum which supply slab. The inclusion depends upon the percentage of (CRU (Can Recycling Unit), reduction, remelt). Our aim was to study the inclusion level with different product mix and minimize the inclusion levels with specified metal input mix (CRU, remelt and reduction metal). This study will be done on MRC slab product. (Ma'aden Rolling Mill Company) with slab ingot of 3xxx, and 5xxx, and it has to go through metal cleanliness test such as Podfa, Limca, and Alscan to verify the metal cleanliness. The study will provide the inclusion levels with different product mix and will suggest the best metal input mix to minimize the inclusion levels in Slab product.

Keywords

Aluminium • Inclusions • Reduce

Introduction

The inclusion level in Aluminum industry is very critical because it will affect the product quality for the end user. The Aluminum industries around the world established metallurgical techniques to evaluate the molten metal cleanliness. Hence, the Aluminum metal source has an impact on the percentage of the inclusion levels since most of the Aluminum smelters takes the metals from reduction, CRU, remelt. Each Aluminum products has different limit when it comes to the inclusion as for example in slab product in 3xxx the inclusion limit is around 0.1 mm^2 on the other hand for 5xxx is 0.2 mm^2 as per Ma'aden standard. However, in some Aluminum smelters the inclusion level varies. In some cases, the inclusion levels depend on the efficiency of the CFF (Ceramic foam Filter) as a good industry practice is to take a Podfa measurements before, and after CFF to look for the CFF efficiency as well as to see the inclusion deviations. Factors may lead to more inclusion on the metals such as amount of grain refiner added, excessive patching in the launders, metal flows, and skimming practice in the furnaces.

Study Approach

The study has been done on slab products for Ma'aden Aluminum company with respect to the metals from four sources: Remelt, UBC "used beverage cans", scrap, and reduction. The total number of 10 samples were taken for each metals input. The metals details are as follows:

Reduction Metal: The metal received from pot line is called reduction metal and it is pure Aluminum 99%. It is treated in TAC (treatment in Crucible) to remove the Alkaline metals before transfer to Slab furnaces.

A. Al-Qarni · B. D. AlMuhana (✉)
Ma'aden Aluminum company, 11342, Ras Al-Khair, 31961, Saudi Arabia
e-mail: garniaa@Maaden.com.sa

A. Al-Qarni
e-mail: muhanab@maaden.com.sa

© The Minerals, Metals & Materials Society 2023
S. Broek (ed.), *Light Metals 2023*, The Minerals, Metals & Materials Series,
https://doi.org/10.1007/978-3-031-22532-1_43

Re-Melt Metal: Scrap received from Ma'aden Rolling mill (scalper and rolling mill process scrap area) that are melted in furnaces and then transferred to slab furnaces by crucibles.
UBC Metal: Scrap from UBC (Used beverage Cans) are melting in melting furnaces and metal is transferred to Slab furnaces.
Scrap: Processed scrap from Slab production (Scrap slabs, drain metal and launder skulls) are directly added to the furnace charge.

The sources of the metals above were checked, and analyzed at the A94 exist using Podfa "Porous disk filtration apparatus" techniques, with correlation to each metal sources which been charge to the slab furnaces for 3xxx as well as 5xxx. Figures 1, 2 pictures of Podfa, pre-heater, and a podfa sample.

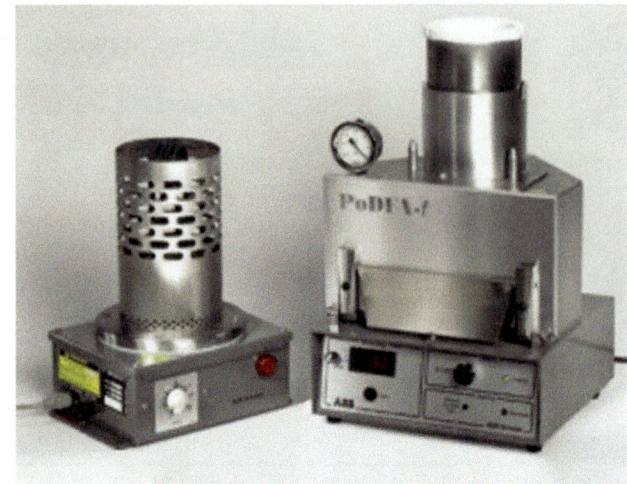

Fig. 2 Podfa device

Analysis Stages

We compared the four input metals sources with the inclusion levls according to Ma'aden standard as for 3xxx the inclusion limit is 0.1 mm^2 on the contrary 5xxx is 0.2 mm^2. The approach of the study was done in each metal input source vs multiple input sources. Starting with reduction metals in Fig. 3 there is a relation between the inclusion level, and the reduction metals hence the plotline metal is having high inclusions which results to high differences is noticed in regression model is around 1.18%

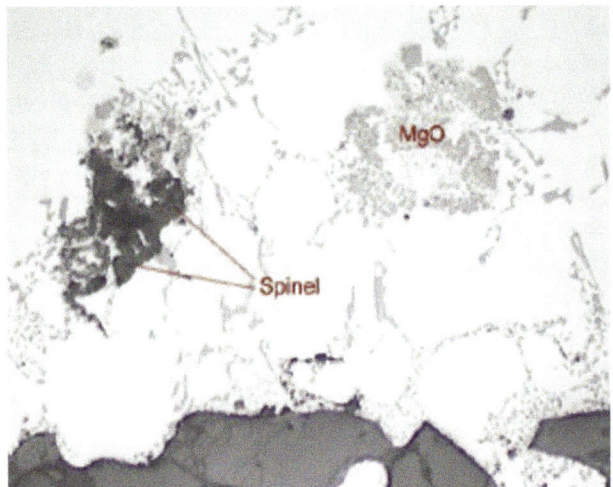

Fig. 1 Podfa sample under microscope

The re-melt metal showed that the inclusion is not significant, and almost 0% which indicate is the best option in order to have less inclusion in our slab products (Fig. 4). Remelt metal is using the scrap generated from rolling mill and slab scrap. There scrapes are passed through all filtration media during casting hence the inclusion levels are low.

The study also indicating that the inclusion levels are very low with remelt metal. The inclusion level in the UBC metal showed that there is a relation with 2.33% which impact the inclusion level (Fig. 5).

The scrap metals were showing an inclusion level with 6.95% variation which indicate significant relation (Fig. 6).

Comparing the total metals sources all together with the inclusion levels were indicating strong relation, and high impact. The variation is 8.61% (Fig. 7).

In the table below is the comparison with each metal sources as in the red color showing the more impact in the inclusion levels, and the green one is showing low impact on the inclusion levels. (Fig. 8).

Conclusion

From the study we came to conclusion the best metals option to charge our slab furnaces are the re-melt metal, and reduction metals then UBC metal or combination of Re-melt and UBC—(CRU metal (Can recycling Unit). Reduction metal is coming from plotlines which is having high amount of aluminum carbides, oxides and some amount of bath. Reduction metal is pure form of metal and doesn't contain

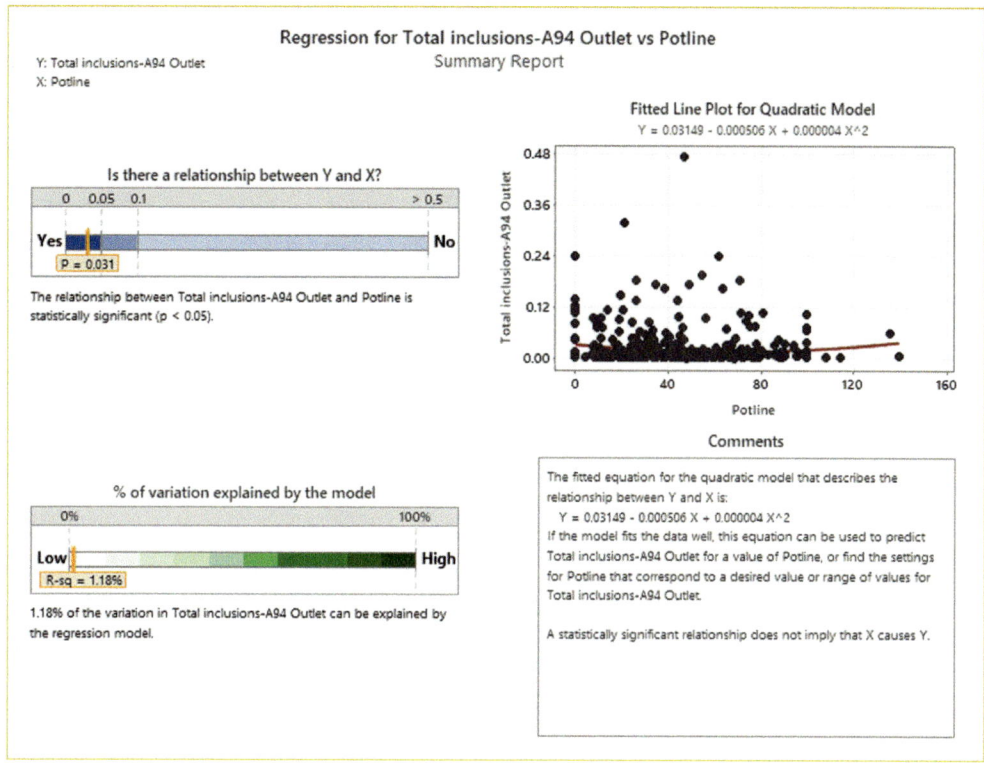

Fig. 3 Relation between inclusion level and reduction

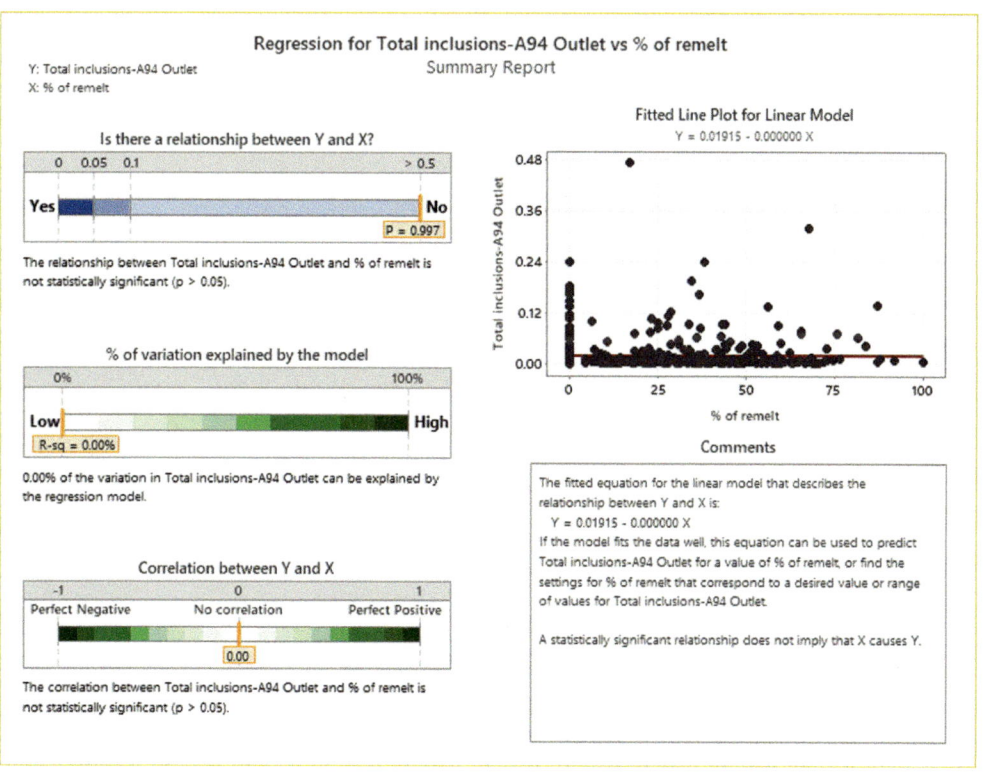

Fig. 4 Remelt metal vs scrap

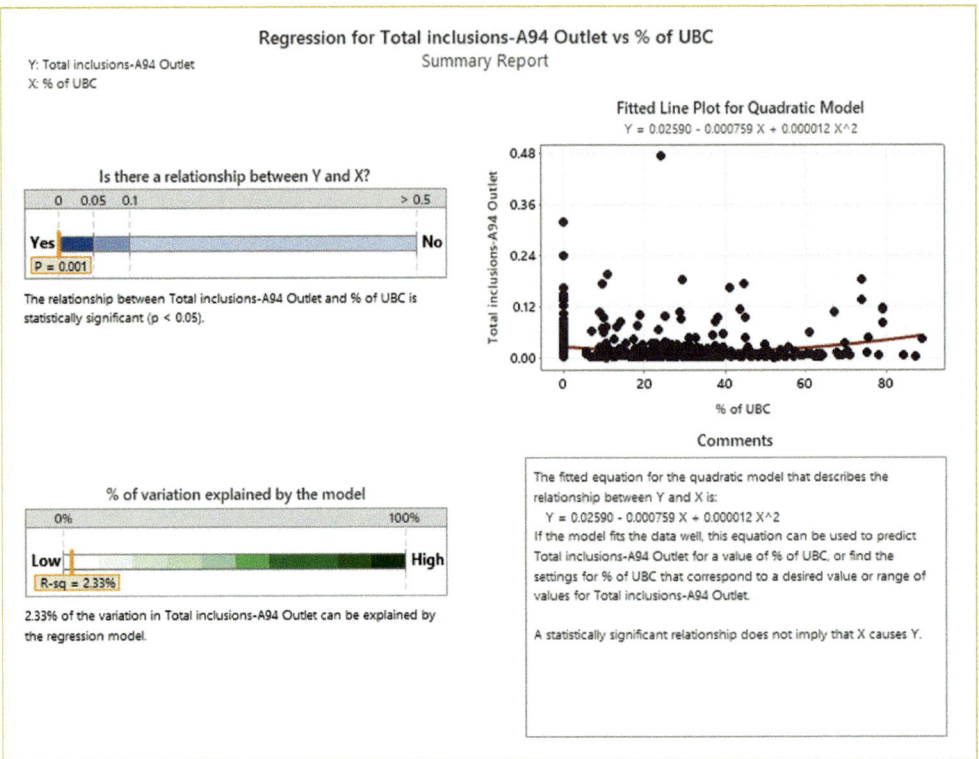

Fig. 5 UBC metal vs remelt metal

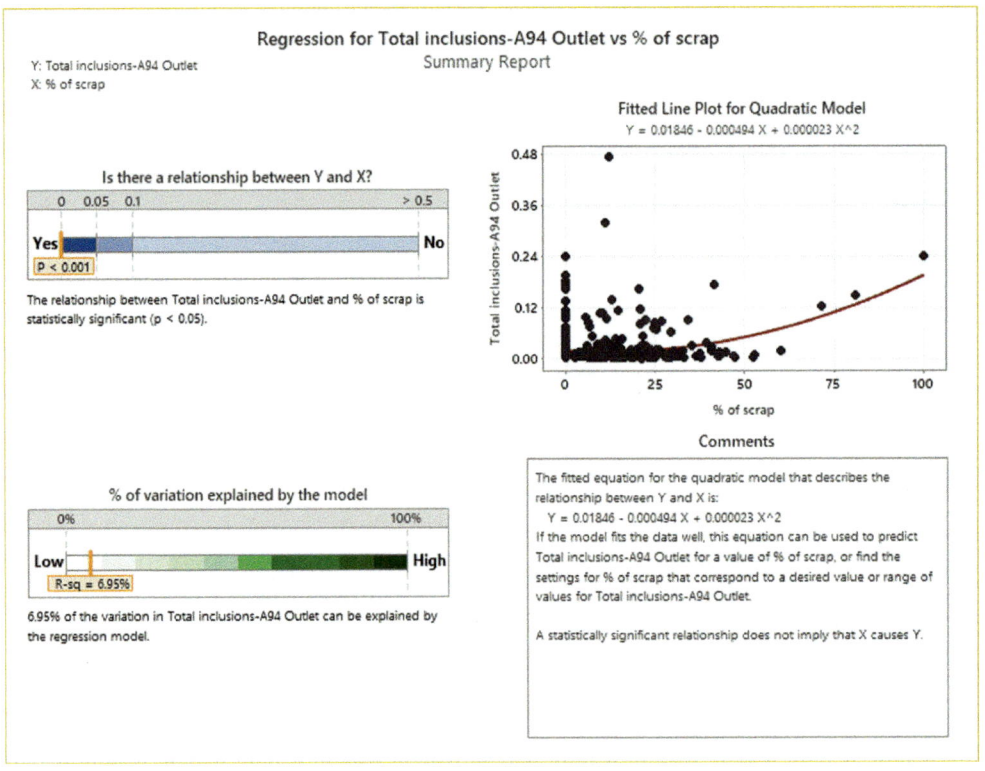

Fig. 6 Scrap metals

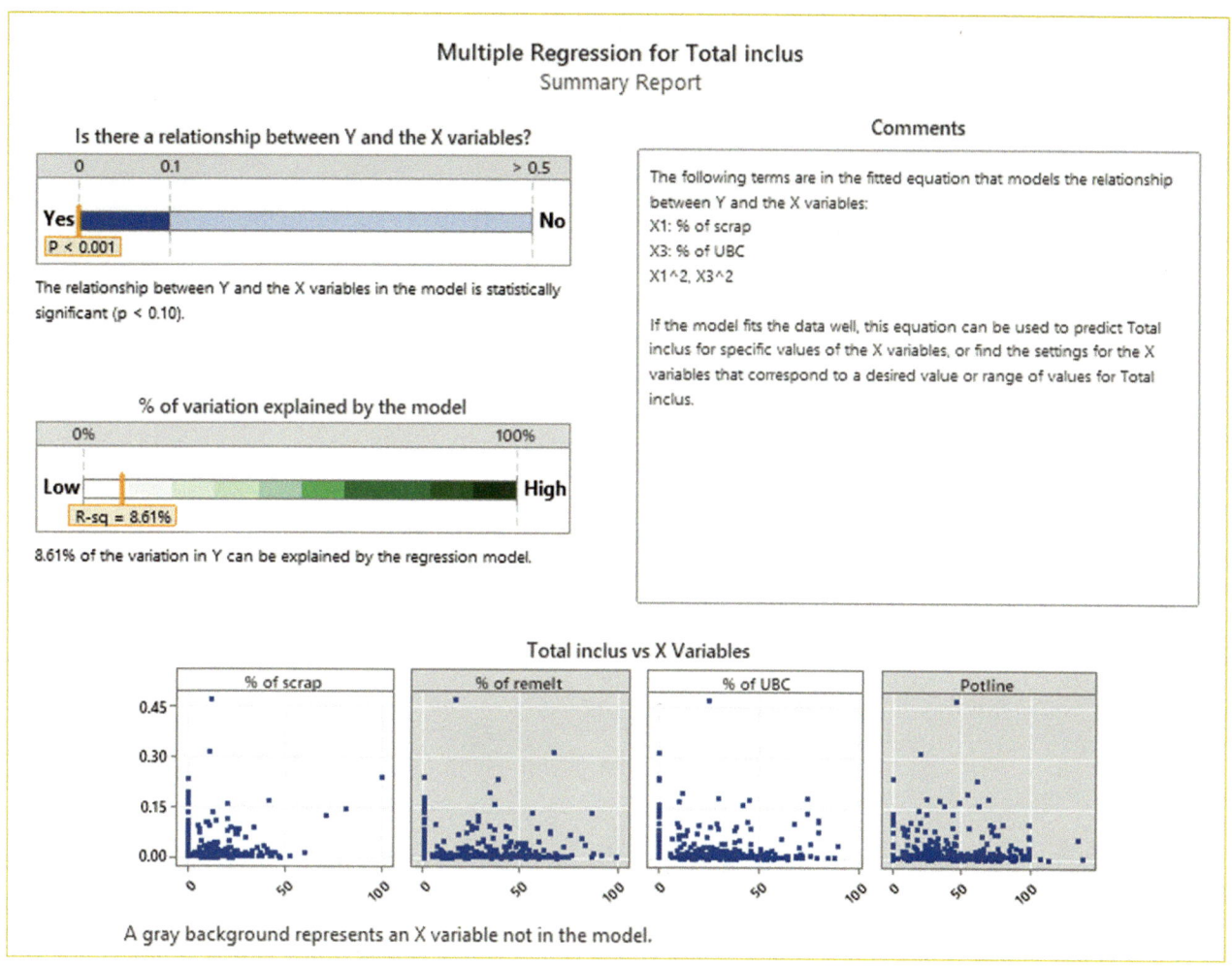

Fig. 7 Total metal sources with variation 8.61%

Input Source	Variation
Scrap	6.95
Re-melt	0
UBC	2.33
Potline	1.18
Scrap+potline	6.95
Remelt+UBC	2.33
Scrap+remelt	6.95
Scrap+UBC	8.61
Potline+remelt+ubc	4.77
Remelt+ubc+scrap	8.61
Ubc+scrap+remelt+potline	8.61

Fig. 8 Comparison with each metal sources

any alloys. To make 3XXX or 5XXX alloys require high amount alloy additions. While adding alloying elements lots of inclusions will be formed during metal preparation which is increasing the inclusion levels at casting. Whereas remelt and CRU metal is having very clean metal as it was passed through filtration media during casting hence inclusions are very less. Remelt and CRU metal is already having all alloying elements so there is no need of addition of alloying elements are very less amount of alloy additions requires hence the inclusion formation is less. CRU and remelt metal is cleaner than reduction metal, hence the best metal to use to have less inclusions at slab is remelt and CRU.

Acknowledgements The study was carried out in the Ma'aden Ras Alkhair Aluminum plant more specifically in the Metallurgy lab. Thanks to our Director Mr. Abdullah Al-Qarni for giving us the opportunity to do the study, and present it to TMS (The Minerals, Metals & Materials Society).

References

1. Alexandre Vianna da Silva, A.M.e.a., *Filtration Efficiency and Melt Cleanness Evaluation using LAIS Sampling at Valesul Aluminio S. A.* Light Metals, 2005: p. 957–960.

2. Voigt, C., et al., *Filtration Efficiency of Functionalized Ceramic Foam Filters for Aluminum Melt Filtration.* Metallurgical and Materials Transactions B, 2017. **48**(1): p. 497-505

3. Syvertsen, M., Frisvold, Frede et al, *Development of a compact deep bed filter for aluminum.* Light Metals: Proceedings of Sessions, TMS Annual Meeting (Warrendale, Pennsylvania), 1999: p. 1049–1055.

4. C. J. Simensen, and G. Berg, "A Survey of Inclusions in Aluminum", Aluminium, vol 56, (1980) pp335-340

5. D. Apelian, "Tutorial: Clean Metal Processing of Aluminum", ASM International, Conference Proceedings from Materials Solutions, (1998) pp 153–162

Effect of Iron and Manganese Content on Microstructure and Mechanical Properties of AlSi11 Alloy in Wheels Produced by LPDC-Process

Sergey Matveev, Dmitry Moiseev, Tatyana Bogdanova, Roman Vakhromov, and Aleksandr Krokhin

Abstract

The increased use of aluminium scrap in the production of cast wheels is considered as a condition to satisfy the global trend for carbon footprint reduction. AlSi11 alloy is widely used in the production of wheels by LPDC-process. Typically, the iron content in this alloy does not exceed 0.19wt.%. The increase of the percentage of aluminum scrap leads to higher iron content in the melt. In the Al-Si alloys, iron tends to form the needle-like β-phase Al5FeSi, presence of which in the structure can significantly reduce the ductility of the alloy. The addition of manganese as an alloying element is used in practice in order to transform the needle-like β-phase Al5FeSi into a less harmful "Chinese script"-like α-phase Al15(Fe, Mn)3Si2. The authors investigated the effect of various Fe/Mn content on the structure and mechanical properties of the AlSi11 LPDC-wheels to ensure the amount of Fe tolerated in the alloy.

Keywords

Aluminium wheels • LPDC • Iron • Manganese

S. Matveev (✉) · D. Moiseev · R. Vakhromov
Light Materials and Technologies Institute UC RUSAL, 6 Leninsky Avenue, Building 21, 119049 Moscow, Russia
e-mail: Sergey.Matveev4@rusal.com

T. Bogdanova
LMZ SKAD LLC, 42 Pogranichnikov Street, Building 12, 660111 Krasnoyarsk, Russia
e-mail: Tatiyana.Bogdanova2@rusal.com

A. Krokhin
Park Pobedy—Victory Park Business Center, JSC RUSAL Management, 1 Vasilisy Kozhinoi St., 121096 Moscow, Russia
e-mail: Aleksandr.Krokhin@rusal.com

Introduction

Cast aluminum alloys occupy a significant share in the structure of world consumption of aluminum alloys. The main consumer of cast aluminum alloys is the automotive industry. At the same time, aluminum cast wheels make up a significant proportion of aluminum in the vehicle structure. It should be noted that unlike the engine and structural components, wheels could be replaced several times during the life cycle of a vehicle. The annual global production of aluminum cast wheels is about 300 million pieces (about 3 million tons). Halving the carbon footprint of cast wheels would reduce the carbon footprint by 12 million tons per year.

Sustainable development requires extensive use of recyclable resources. This is especially true for aluminum alloys, as the global average carbon footprint of 1 ton of primary aluminum is now over 8 tons. European consumers of aluminum alloys are currently aiming to reduce the carbon footprint of their aluminum alloys to less than 4 tons per ton of aluminum alloy. It is possible to reduce the carbon footprint of aluminum alloys by increasing the share of aluminum alloys obtained from waste processing.

However, it should be noted that secondary aluminum alloys contain significantly more impurities than primary aluminum alloys. This requires an assessment of the influence of various impurities on the structure and properties of manufactured products. The main impurities that enter aluminum alloys from scrap are iron, copper and zinc.

The main method for the production of automotive cast wheels is the low-pressure casting method. As alloys for the production of wheels, AlSi7Mg and AlSi11 alloys are mainly used. For these alloys, the most detrimental impurity is iron, which leads to the formation of the needle-like β-phase Al5FeSi in the cast structure and significantly reduces elongation and toughness. Various approaches can be taken to reduce the harmful effect of the needle-like β-phase Al5FeSi.

© The Minerals, Metals & Materials Society 2023
S. Broek (ed.), *Light Metals 2023*, The Minerals, Metals & Materials Series,
https://doi.org/10.1007/978-3-031-22532-1_44

A number of researchers [1, 2] note the positive effect of manganese and strontium in Al-Si alloys with high iron content to transform the needle-like β-phase Al5FeSi into a less harmful "Chinese script"-like α-phase Al15(Fe, Mn)3Si2. Other researchers [3] note the positive effect of elevated (above 820 °C) melt temperature during casting on the formation of "Chinese script"-like α-phase Al15(Fe, Mn)3Si2. However, the high casting temperature in LPDC leads to rapid degradation of the mold material, which makes it impossible to use this approach to control the structure in the wheel.

In [4], authors have shown influence of cooling rate on the changes needle-like β-phase into the fishbone-like δ-Al3FeSi2 phase with the improved tensile properties. It should be noted that different zones of the wheel are cooled during casting at different rates [5]. The highest cooling rate is in the rim area and the lowest in the hub area. The cooling rate in the spoke area is average between the rim and the spoke. Thus, it is difficult to control the structure with the cooling rate.

Based on the above, the only reasonable way to control the structure in the wheel casting is to optimize the chemical composition of the alloy in order to transform the needle-like β-phase Al5FeSi into a less harmful "Chinese script"-like α-phase Al15(Fe, Mn)3Si2.

In this work, a study was made of the influence of various contents of iron and manganese in a ratio of 1 to 1 on the structure and properties of the AlSi11 alloy in the casting of a wheel disk.

Methods

For the preparation of experimental alloys, an induction melting furnace with a capacity of 2.5 tons was used. The charge for test alloys was: Recycled aluminum alloy, magnesium and silicon. The prepared test alloys were subjected to nitrogen degassing and flux modification according to the process, the chemical composition of the alloys was determined on a Q4 TASMAN emission spectrometer.

Next, the melts were poured into the holding furnaces of four casting machines for subsequent casting. After overflow, samples were taken for thermal analysis and samples for determining the density index in the IDECO Thermo—Analysis system.

Experimental casting was carried out for at least 12 h, while a new melt was added to the holding furnaces every two hours so as not to stop casting.

Before the start of casting, the molds were warmed up to 350 °C. The production of castings from experimental alloys was carried out in accordance with the standard technological charts of casting parameters on GIMA casting machines. In the process of casting from experimental alloys, casted parts were subjected to 100% X-ray inspection using X-ray machines YXLON MU231, where castings with internal defects exceeding the requirements of ASTM 155 6 class were rejected. Good wheels were transferred further down the process for machining and painting.

Determination of the level of mechanical properties was carried out on samples from the most loaded areas of the wheel: Outer side flange, spoke, hub (according to UNECE rules [6]). The tests were carried out on three samples from each zone according to ISO 6892–1:2019 and ISO/CD 6506–1, respectively, on a WDW-10 electromechanical machine and on a HB3000B Brinell hardness tester. Metallographic studies were carried out using an OLYMPUS GX51 optical microscope and an OLYMPUS SZX7 stereoscopic microscope using the SIAMS PHOTOLAB program.

Torsional bending tests, impact tests and alternating torque tests were carried out on accredited stands in accordance with UNECE rules [6].

Results and Discussion

The chemical composition of the prepared alloys is shown in Table 1, the result of the study of thermal analysis and density index—in Table 2.

Table 2 shows that a change in the iron content within 0.2 −0.5% does not have a noticeable effect on temperature transformations in the alloy. The density index of all investigated alloys was at the same level.

The results of the level of mechanical properties are presented in Table 3 and Fig. 1.

Figure 1 shows the following trends:

1. DAS increases from the outer flange to the hub, due to an increase in the geometric modulus and does not depend on the iron content in the alloy;
2. Elongation drops with increasing DAS;
3. Plastic limit slightly increases with increasing DAS;
4. Elongation drops catastrophically on an alloy containing 0.5% iron, due to the release of primary iron compounds (Table 4).

Microstructure in different parts of wheels shown are in Tables 4 and 5. Despite the fact that a significant amount of needle-like β-phase was found in the structure, this did not affect the performance properties of the wheels in any way and all wheels passed bench tests (see Fig. 1 Table 6).

Table 1 Chemical composition of experimental alloys

# of alloy	Mass fraction in [%]										
	Si	Mg	Fe	Mn	Ti	Sr	Cu	Cr	Ni	Zn	Al
# 1 (0.2)	11.23	0.049	0.2	0.22	0.114	0.021	0.009	0.004	0.003	0.016	Est
# 2 (0.25)	11.31	0.049	0.25	0.26	0.116	0.02	0.007	0.004	0.003	0.02	Est
# 3 (0.3)	10.75	0.057	0.3	0.352	0.12	0.019	0.002	0.004	0.003	0.012	Est
# 4 (0.35)	10.76	0.046	0.35	0.393	0.143	0.016	0.002	0.005	0.003	0.013	Est
# 5 (0.5)	10.9	0.061	0.5	0.59	0.143	0.016	0.002	0.005	0.003	0.013	Est

Table 2 Results of thermal analysis of experimental alloys and density index of alloy

# of alloy	Theore-tical liquidus, °C	Primary under-cooling, °C	Recales-cence, K	Theoretical eutectic temperature, °C	Eutectic temperature, °C	Recales-cence, K	Liquidus-Solidus, K	DI, %
# 1 (0.2)	590.2	587.0	0.6	576.5	575.3	2.5	14.9	10.0
# 2 (0.25)	589.5	586.7	0.3	576.5	575.4	2.7	14.1	10.2
# 3 (0.3)	590.4	587.2	0.6	576.5	574.6	1.7	15.8	10.1
# 4 (0.35)	589.4	585.9	0.9	576.6	573.2	1.1	16.2	10.0
# 5 (0.5)	590.4	584.5	3.3	576.6	574.0	2.1	16.4	10.4

Table 3 Mechanical properties in different parts of experimental wheels

# of alloy	Mechanical properties				
	Part of wheel	YS, MPa	UTS, MPa	Elongation, %	Hardness, HB
# 1 (0.2)	Outer Rim	95.5	187.0	11.8	57.5
	Spoke	99.8	179.5	8.5	57.0
	Hub	101.5	182.5	7.9	58.5
# 2 (0.25)	Outer Rim	107.2	194.4	9.7	60.2
	Spoke	108.1	194.4	9.3	60.8
	Hub	110.9	194.5	8.4	61.7
# 3 (0.3)	Outer Rim	111.3	198.5	9.5	62.3
	Spoke	111.5	192.8	7.9	62.5
	Hub	111.8	199.0	7.5	64.0
# 4 (0.35)	Outer Rim	107.0	194.0	10.2	60.0
	Spoke	109.7	191.0	8.8	59.3
	Hub	110.7	200.7	8.2	60.7
# 5 (0.5)	Outer Rim	101.0	178.0	5.2	57.0
	Spoke	101.5	169.0	3.5	56.0
	Hub	101.8	168.0	3.3	57.0

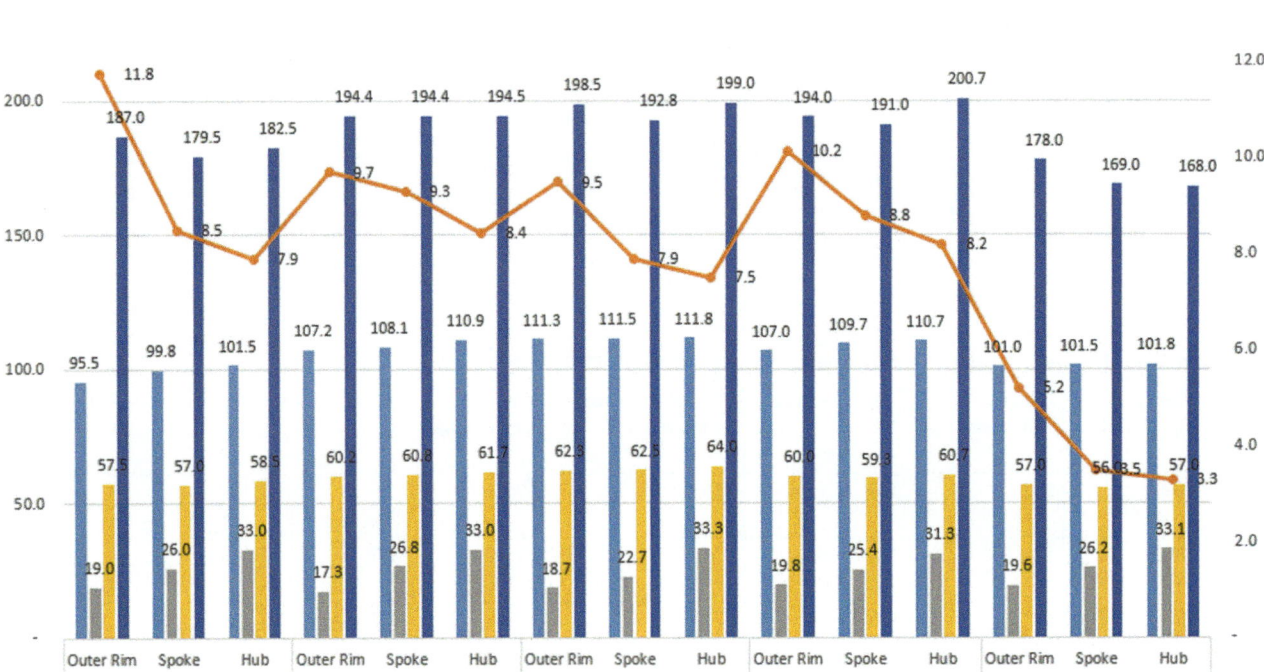

Fig. 1 Mechanical properties comparison

Conclusions

An increase in the strength properties of the alloy in wheels by 10% was revealed due to an increase in the content of iron and manganese from 0.2 to 0.25% and higher.

The possibility of using recycled aluminum alloys with a high content of iron and manganese up to 0.5% in AlSi11 wheel alloy without a noticeable deterioration in ductility has been confirmed. Nevertheless, it is necessary to study in more detail the corrosion behavior of such alloys in the composition of painted wheels assembled with other suspension elements. It should be understood that the corrosion behavior of a wheel as a finished product depends not only on the composition of the alloy, but also on the quality of the painting.

It has shown that an increase of the iron and manganese content up to 0.5% leads to a drop in ductility due to the release of primary intermetallic compounds.

Table 4 Microstructure in different parts of wheels

# of alloy	Outer Rim	Spoke	Hub
# 1 (0.2)			
# 2 (0.25)			
# 3 (0.3)			
# 4 (0.35)			
# 5 (0.5)			

Table 5 Dendrite arms spacing. size of β-phase

# of alloy	Outer Rim		Spoke		Hub	
	DAS, mkm	β-phase, mkm	DAS, mkm	β-phase, mkm	DAS, mkm	β-phase, mkm
# 1 (0.2)	17	25	29	26	35	31
# 2 (0.25)	18	27	26	28	32	37
# 3 (0.2)	21	31	24	37	33	45
# 4 (0.35)	19	48	26	54	33	60
# 5 (0.5)	22	54	26	55	31	72

Table 6 Bench tests

# of alloy	Bench tests			
	Rotational bending fatigue resistance	Impact test		Fatigue resistance under dynamic radial load
		Spoke	Vent hole	
# 1 (0.2)	No cracks	No cracks	No cracks	No cracks
# 2 (0.25)	No cracks	No cracks	No cracks	No cracks
# 3 (0.2)	No cracks	No cracks	No cracks	No cracks
# 4 (0.35)	No cracks	No cracks	No cracks	No cracks
# 5 (0.5)	No cracks	No cracks	No cracks	No cracks

References

1. Mathew J. (2022) Srirangam. P. Microstructure and mechanical properties of Mn–Sr modified Iron containing Al–Si Alloys. Light Metals. 135–141
2. L.B. Otani et al. (2017) Predicting the Formation of Intermetallic Phases in the Al-Si-Fe System with Mn Additions. National Engineering Research Center of Near-net Shape Forming for Metallic Materials. Journal of Phase Equilibria and Diffusion volume 38 298–304
3. R. Haghayeghi et al. (2022) Effects of casting temperature and iron content on the microstructure of hypoeutectic A380 Aluminium Alloy. Light Metals. 790–797
4. X. Shen et al. (2022) Effect of cooling rate on the microstructure evolution and mechanical properties of Iron-Rich Al–Si Alloy. Materials. 15. 411
5. J. Ou et al. (2020) Advanced process simulation of low pressure die cast A356 aluminum automotive wheels. Metals. 10. 563 and 1418
6. Regulation No 124 of the economic commission for Europe of the United Nations (UN/ECE) —Uniform provisions concerning the approval of wheels for passenger cars and their trailers. Official Journal of the European Union L 375 of 27 December 2006

Shear Assisted Processing and Extrusion of Unhomogenized Aluminum Alloy 6063 Castings with High Iron Content

Scott Whalen, Nicole Overman, Brandon Scott Taysom,
Md. Reza-E-Rabby, Timothy Skszek, and Massimo DiCiano

Abstract

Shear Assisted Processing and Extrusion (ShAPE) was used to fabricate aluminum alloy 6063 tubing directly from secondary scrap. Trimmings from an automotive manufacturing facility were cast into billets and spiked with iron (0.3 wt.% Fe) to explore the tolerance of ShAPE to Fe contamination. Billets were extruded in the unhomogenized condition to form tubing with an outer diameter of 12 mm and wall thickness of 2 mm. Tensile properties reached 0.2% YS = 206 MPa, UTS = 238 MPa, and U.E. = 16.3% in the T6 temper. Properties exceed the ASTM minimum standard and are on par with ASM typical values for conventional extrusion of primary aluminum billets in the fully homogenized condition. Property equivalence between conventional extrusion of homogenized primary aluminum billets with nominal Fe, and ShAPE extrusion of unhomogenized secondary aluminum billets with high Fe, are attributed to the highly refined microstructure achieved with ShAPE. Microstructural characterization shows an extensive refinement of grain size along with dispersion, refinement and homogenization of FeAlSi and MgSiO intermetallic phases. This study suggests that ShAPE extrusion may offer a lower carbon manufacturing pathway through direct recycling of secondary aluminum scrap with high Fe content and elimination of the billet homogenization step.

Keywords

Aluminum • Extrusion • 6063 • ShAPE • Recycling • Secondary scrap

Introduction

Although aluminum alloy 6063 (Al 6063) is classified as a ternary Al–Mg–Si alloy, it can be considered a quaternary alloy of Al–Fe–Mg–Si since Fe is a controlled impurity [1]. During casting solidification, Fe, Si, and Mg migrate to inter-dendritic regions and grain boundaries of the Al matrix [2]. This leads to the formation of α_c and β_c-AlFeSi inter-metallic compounds (IMC) [3]. α_c-AlFeSi is dendritic in structure while β_c-AlFeSi is plate-like [4] and more detrimental to ductility [5] which impacts extrudability and product quality [6]. As a result, castings are always homogenized prior to extrusion [7] to transform β_c-AlFeSi into α_c-AlFeSi thereby improving extrudability [8]. Fe also depletes Si away from the Mg_2Si strengthening phase which reduces the extent of precipitation hardening [9]. As a result, unhomogenized Al 6063 billets with high Fe content are not extruded by the industry.

An extrusion process that tolerates high Fe content would be a significant advancement for recycling since a wider variety aluminum scrap streams could be utilized. Additionally, the ability to extrude unhomogenized billets would eliminate the cost and energy associated with the billet homogenization step. This would require a scalable extrusion process that extensively refines and disperses the AlFeSi-enriched IMC phases that form during re-casting. In this research, Shear Assisted Processing and Extrusion (ShAPE) [10] is used to extrude unhomogenized billets made from Al 6063 secondary scrap with Fe content intentionally spiked near the upper composition limit. During ShAPE, the homogenization step is accomplished in situ during extrusion in contrast to the hours long

S. Whalen (✉) · N. Overman · B. S. Taysom ·
Md. Reza-E-Rabby · T. Skszek
Pacific Northwest National Laboratory, 902 Battelle Blvd,
Richland, WA 99354, USA
e-mail: scott.whalen@pnnl.gov

M. DiCiano
Magna International, Inc., Aurora, ON, Canada

© The Minerals, Metals & Materials Society 2023
S. Broek (ed.), *Light Metals 2023*, The Minerals, Metals & Materials Series,
https://doi.org/10.1007/978-3-031-22532-1_45

homogenization step required prior to conventional extrusion. A review of the peer-reviewed literature did not reveal any studies where unhomogenized billets, in any 6xxx aluminum alloy, were extruded with high Fe content. As stated in [7], "It is well established that homogenized [Al 6063] billets extrude easier and faster and give better surface finish and higher tensile properties than as-cast billets. Hence, the production of Al–Mg–Si [Al 6063] extrusions from DC-cast billets almost always starts with a homogenization cycle." This is because homogenization performs the necessary β_c-α_c transformation, and subsequent spheroidization and fragmentation of α_c [11]. With conventional extrusion unable to effectively extrude unhomogenized Al 6063 billets, it is thus noteworthy that ShAPE can not only extrude unhomogenized billets; but can do so with higher than nominal Fe content while still meeting industry standards for mechanical properties.

Experimental Methods

ShAPE extrusion is described elsewhere for Al 2024 [12], Al 6063 [13] and Al 7075 [14], and the reader is referred to these citations for details on the process. In short, the ShAPE process adds a rotational component to conventional extrusion by spinning the die as shown in Fig. 1. This extensively deforms the billet and results in highly refined microstructures akin to those achieved with severe plastic deformation (SPD) techniques. For some aluminum alloys, the extreme deformation at elevated temperature during ShAPE enables in situ homogenization in just seconds. This has been demonstrated for ShAPE of Al 7075 [15] where unhomogenized castings were extruded into tubing having 0.2% yield strength (YS) = 522 MPa, ultimate tensile strength (UTS) = 610 MPa, and elongation = 17%. All of which are above the ATSM minimum [16] and ASM typical values [17].

Billets were cast from Al 6063 scrap collected at one of Magna International's manufacturing facilities. During casting, Fe was added to achieve the upper end of the allowable limit resulting in the composition listed in Table 1.

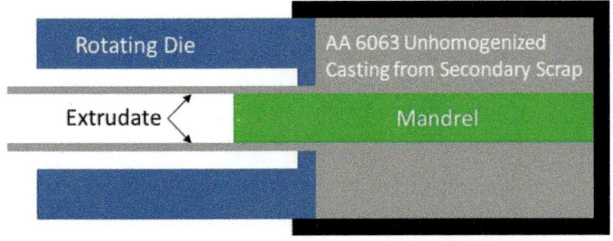

Fig. 1 Schematic of ShAPE for extruding thin-walled tubing

Table 1 Chemical composition by wt.% of unhomogenized Al 6063 castings. The standard Al 6063 composition specification is shown for reference [18]

Al 6063	Al	Mg	Si	Fe	O
High Fe Billet	98.1	0.6	0.4	0.3	0.6
Specification [18]	Bal	0.45–0.9	0.2–0.6	<0.35	–

Fig. 2 Example of Al 6063 tubing extruded by ShAPE with 12 mm OD and 2 mm wall thickness

Castings were supplied in the unhomogenized condition and subsequently machined into billets having a 31.8 mm outer diameter, 8.1 mm inner diameter, and a length of 100 mm. An 8 mm diameter mandrel and 12 mm die orifice were used to extrude tubing with 2 mm wall thickness at an extrusion ratio (ER) of 11.4. An example tube is shown in Fig. 2. Die temperature was measured using a type-K thermocouple spot-welded on the die face at a radius of 10 mm. Extrusions were performed on a ShAPE machine manufactured by BOND Technologies, Inc.

Extrusions were performed with a steady-state die temperature of 530 °C and ram speed of 360 mm/min (4.1 m/min extrudate velocity). This required a die rotation speed of 250 rpm and ram force of 185 kN (259 MPa pressure). Extrusions were spray quenched with 25 °C water immediately after exiting the die. Prior to tensile testing, extrusions were sectioned into 150 mm lengths and heat treated to T6 [19] by solution heat treating at 530 °C for 1 h, water quenching at 25 °C, and artificial aging at 178 °C for 10 h. Tensile testing was performed per ASTM E8 Sect. 6.9.1 [20] using an Instron 8802 load frame pulling at 0.02 mm/s. Displacement of the 25 mm gauge length was measured using a mechanical extensometer per ASTM B557-15 [21]. Tubes were sectioned transverse and longitudinal to the extrusion direction and prepared for microstructural characterization by successively grinding and polishing to a 0.05 μm finish with a colloidal silica suspension. Scanning Electron Microscopy (SEM) was performed in Backscatter Electron (BSE) mode with a JEOL IT500 equipped with an Oxford Instruments Nanoanalysis Package and dual Oxford Ultim Max 100 mm^2 Energy-Dispersive X-ray Spectroscopy (EDS) cameras. EDS mapping occurred at an accelerating voltage of 20 kV, probe current setting of 75, working distance of 10 mm, and 60× magnification. Forescatter Diode Images were acquired using an JEOL IT800 SEM equipped with an Oxford Symmetry S2 EBSD camera.

Results

Mechanical Properties

Tensile properties for ShAPE extruded Al 6063-T6 tubing made from unhomogenized scrap Al billets spiked with 0.3 wt.% Fe are shown in Table 2. A total of 16 specimens were tested. The ASTM minimum [18] and ASM typical values [22] for Al 6063-T6 extruded from homogenized primary Al billets with nominal Fe are shown for comparison. It is observed that YS, UTS, and elongation exceed the ASTM minimum values and are on par with ASM typical values. Note (*) in Table 2, the ASTM and ASM elongation values are for Total Elongation (T.E.), whereas the ShAPE data is for Uniform Elongation (U.E). This suggests that extensive refinement and dispersion of second phase IMCs has occurred since U.E for ShAPE exceeds T.E. for conventional extrusion. This observation is consistent with high elongations observed for ShAPE extrusion of unhomogenized Al 7075 castings [15] and gas atomized Al-12.4TM powders [23]. These results are remarkable given that ShAPE extrusion was performed on unhomogenized billets where plate-like β_c-AlFeSi IMCs are known to limit extrudability and tensile properties [5]. This speaks to the ability of ShAPE to homogenize the composition in-situ during extrusion and suggests that ShAPE also has a higher tolerance to elevated Fe compared to conventional extrusion.

Microstructure Analysis

Figure 3 shows low magnification (a)–(c) and high magnification (d)–(f) SEM-BSE images of microstructures observed in the unhomogenized cast billets (a and d), transverse ShAPE extrusion (b and e), and longitudinal ShAPE extrusion (c and f). This analysis shows the extent of grain refinement following ShAPE processing, quantified in Table 3. The extruded grain structure is highly equiaxed with no observable difference in the longitudinal direction compared to the transverse. The inter- and intra-granular second phases observed in Fig. 3 are analyzed in Fig. 4.

Figure 4 depicts Energy Dispersive Spectroscopy (EDS) maps showing the presence of FeAlSi-enriched IMCs along grain boundaries in the unhomogenized cast billets (Fig. 4a). Figure 4b, c show that these Fe-enriched IMCs have been significantly refined and evenly dispersed throughout the Al matrix. Following ShAPE processing, IMCs exhibit a reduced propensity to decorate grain boundaries, and exhibit a more random distribution, and appear both inter- and intragranularly. This is corroborated by comparing the bright second phases in Fig. 3d with Fig. 3e, f. Figure 4b shows MgSiO-enriched IMCs which result from the billet casting process where the melt was not skimmed or degassed. Note, the magnification of 4b was increased by a factor of 10 to clearly visualize this phase for comparison to the cast material. The size of the MgSiO cast defects was significantly reduced during ShAPE which is advantageous since large defects can have a deleterious effect on mechanical properties.

Conclusions

In this research, we demonstrate that ShAPE is capable of extruding unhomogenized Al 6063 billets, cast from secondary scrap, with Fe content of 0.3 wt.%. Mechanical testing gave YS, UTS, and elongation exceeding the ASTM

Table 2 Tensile properties for ShAPE extruded Al-6063-T6 from unhomogenized secondary billets with 0.3 wt.% Fe. Mean values are given ± one standard deviation

Al 6063-T6	Billet type	Fe Content	0.2% YS (MPa)	UTS (MPa)	U.E. (%)
ShAPE	Unhomogenized secondary Al	High 0.3 wt.%	206.0 ± 4.5	237.8 ± 5.4	16.3 ± 1.1
ASTM minimum [18]	Homogenized primary Al	Nominal	170	205	8*
ASM typical [22]	Homogenized primary Al	Nominal	214	241	12*

High Fe Feedstock Casting   **ShAPE Processed (Transverse)**   **ShAPE Processed (Longitudinal)**

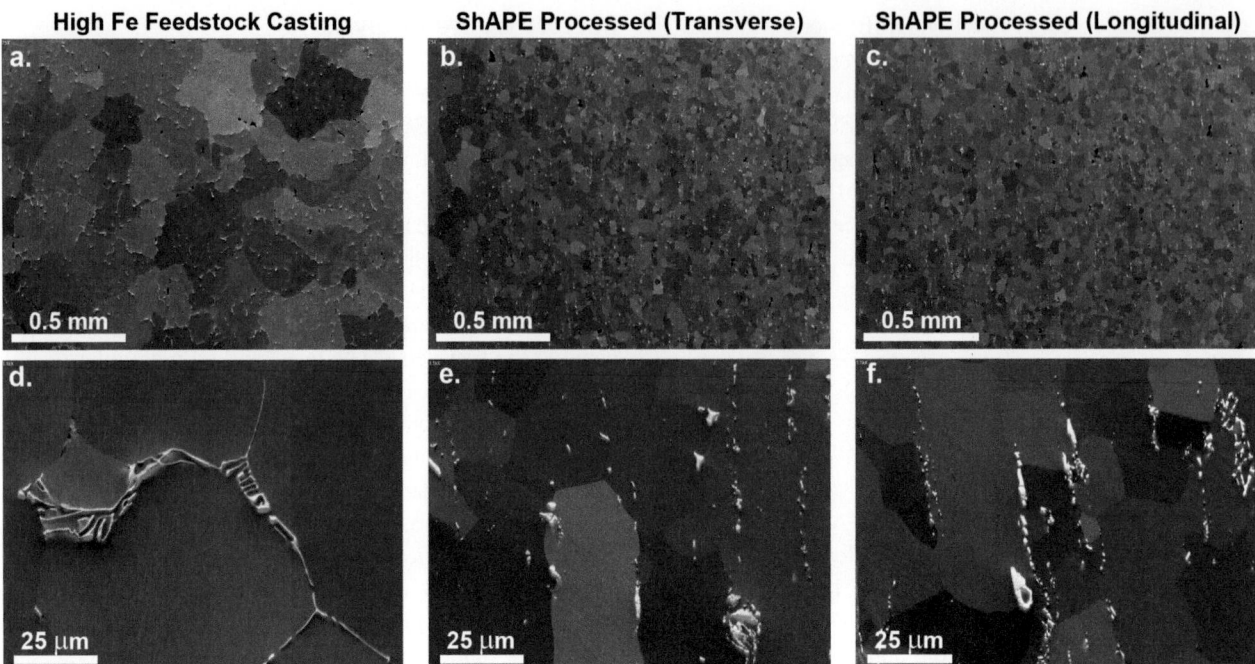

Fig. 3 SEM-BSE images of the unhomogenized cast billet (**a** and **d**), transverse ShAPE extrusion (**b** and **e**), and longitudinal ShAPE extrusion (**c** and **f**)

Table 3 Grain size analysis with Feret diameter reported as an area-weighted mean

Sample	Maximum feret diameter (μm)	Standard deviation (μm)
Unhomogenized cast billet	342.7	137.4
ShAPE transverse	73.3	24.7
ShAPE longitudinal	61.5	21.1

minimum standard for Al 6063-T6 and on par with the ASM typical values for extrusion of homogenized billets, cast from primary Al, with nominal Fe content. Property equivalence between conventional extrusion of homogenized primary Al billets with nominal Fe, and ShAPE extrusion of unhomogenized secondary Al billets with high Fe, are attributed to the novel microstructure achieved with ShAPE. Microstructural characterization shows an extensive refinement of grain size along with dispersion and refinement of FeAlSi and MgSiO intermetallic phases. These microstructural features mitigate property depression despite high Fe content and are responsible for property equivalence with conventional extrusion of homogenized billets having nominal Fe. These results suggest that low-value Al scrap streams with high Fe content may one day be recycled directly into extruded profiles.

Acknowledgements The authors thank the U.S. Department of Energy Vehicle Technologies Office (DOE/VTO) LightMAT Program for supporting this work. The authors are grateful for the dedication of Anthony Guzman for the excellent preparation of specimens for microstructural characterization and numerous Magna Technical Staff for tooling and mechanical testing. The Pacific Northwest National Laboratory is operated by the Battelle Memorial Institute for the United States Department of Energy under contract DE-AC06-76LO1830.

Declaration of Competing Interests The authors declare that they have no known competing financial interests or personal relationships that could have appeared to influence the work reported in this paper.

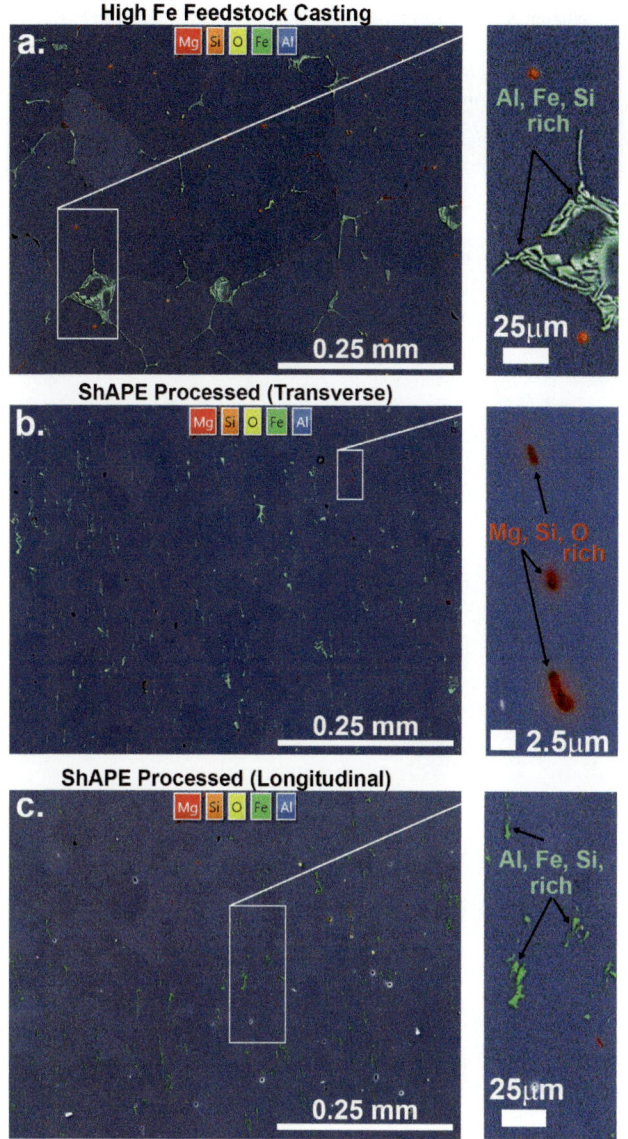

Fig. 4 EDS analysis of unhomogenized cast billets (**a**), transverse ShAPE extrusion (**b**), and longitudinal ShAPE extrusion (**c**). *Note the callout presented in (**b**) is a 10X increase in magnification to better show the refined MgSiO-enriched phase

References

1. Sheppard T, Extrusion of Aluminum Alloys, Springer Science & Business Media Dordrecht, 1999, pg. 78.
2. Zhu H, Couper M, Dahle A, "Effect of Process Variables on Mg-Si Particles and Extrudability of 6XXX Series Aluminum Extrusions," JOM, 2011, pg. 66–71.
3. Sha G, O'Reilly K, Cantor B, Worth J, Hamerton R, "Growth Related Metastable Phase Selection in a 6XXXX Series Wrought Alloy," Materials Science and Engineering A, vol. 304–306, 2001, pg. 612–616.
4. Kumar S, Grant P, O'Reilly K, "Fe Bearing Intermetallic Phase Formation in a Wrought Al-Mg-Si Alloy," Transactions of the Indian Institute of Metallurgy and Materials Engineering, vol. 65, no. 6, 2012, pg. 553–557.
5. Zajac S, Hutchinson B, Johansson A, Gullman L, "Microstructure Control and Extrudability of Al-Mg-Si Alloys Microalloyed with Manganese," Materials Science and Technology, vol. 10, 1994, pg. 323–333.
6. Kumar S, Grant P, O'Reilly K, "Effect of Fe Bearing Intermetallics During DC Casting and Homogenization of an Al-Mg-Si Al Alloy," Metallurgical and Materials Transactions A, vol. 47A, 2016, pg. 3000–3014.
7. Birol Y., "The Effect of Homogenization Practice on the Microstructure of AA6063 Billets," Journal of Materials Processing Technology, vol. 148, 2004, pg. 205–258.
8. Tanihata H, Sugawara T, Matsuda K, Ikeno S, "Effect of Casting and Homogenization Treatment Conditions on the Formation of Al-Fe-Si Intermetallic Compounds in 6063 Al-Mg-Si Alloys," Journal of Materials Science vol. 34, 1999, pg. 1205–1210.
9. Kumar S, O'Reilly K, "Influence of Al Grain Structure on Fe Bearing intermetallics During DC Castings of an Al–Mg–Si Alloy," Materials Characterization, vol. 120, 2016, pg. 311–322.
10. Lavender C, Joshi V, Grant G, Jana S, Whalen S, Darsell J, Overman N, "System and Process for Formation of Extrusion Products," US 10189063, 2019.
11. Rivas A, Munoz P, Camero S, Quintero O, "Effect of the Microstructure on the Mechanical Properties and Surface Finish of an Extruded AA-6063 Aluminum Alloy," Advanced Materials Science and Technology, vol. 2, no. 1, 1999, pg. 15–23.
12. Reza-E-Rabby Md, Wang T, Canfield N, Roosendaal T, Taysom BS, Graff D, Herling D, Whalen S, "Effect of Post-Extrusion Heat Treatment on Performance of AA2024 Tubes Fabricated by Shear Assisted Processing and Extrusion," CIRP Journal of Manufacturing Science and Technology, vol 37, 2022, pg. 454–463.
13. Taysom BS, Overman N, Olszta M, Reza-E-Rabby Md, Skszek T, DiCiano M, Whalen S, "Shear Assisted Processing and Extrusion of High-Strength Aluminum Alloy 6063 Tubing," International Journal of Machine Tools and Manufacture, 169, 103798, 2021.
14. Whalen S, Olszta M, Reza-E-Rabby Md, Roosendaal T, Wang T, Herling D, Taysom BS, Suffield S, Overman N, "High Speed Manufacturing of Aluminum Alloy 7075 Tubing by Shear Assisted Processing and Extrusion (ShAPE), Journal of Manufacturing Processes, vol. 71, 2012, pg. 699–710.
15. Wang T, Atehotua J, Song M, Reza-E-Rabby Md, Taysom BS, Silverstein J, Roosendaal T, Herling D, Whalen S, "Extrusion of Unhomogenized Castings of 7075 Aluminum via ShAPE," Materials and Design, vol. 213, 2022, 110374.
16. ASTM B241-16, Standard Specification for Aluminum and Aluminum-Alloy Seamless Pipe and Seamless Extruded Tube, ASTM International.
17. ASM Handbook Volume 2b - Properties and Selection of Aluminum Alloys, "7075 and Alclad 7075," edited by Anderson K, Weritz J, Kaufman G, ASM International, 2019.
18. ASTM B221M-21, Standard Specification for Aluminum and Aluminum Alloy Extruded Bars, Rods, Wire, Profiles, and Tubes, ASTM International.
19. ASM Handbook Volume 2a – Aluminum Science and Technology, "Heat Treatment Practice of Wrought Age-Hardenable Aluminum Alloys," edited by Anderson K, Weritz J, Kaufman G, ASM International, 2018.

20. ASTM E8-15a, Standard Test Methods for Tension Testing of Metallic Materials.
21. ASTM B557-15, Standard Test Methods for Tension Testing Wrought and Cast Aluminum and Magnesium Alloy Products.
22. ASM Handbook Volume 2b - Properties and Selection of Aluminum Alloys, "6063 Extrusion Alloy," edited by Anderson K, Weritz J, Kaufman G, ASM International, 2019.
23. Whalen S, Olszta M, Roach C, Darsell J, Graff D, Reza-E-Rabby Md, Roosendaal T, Daye W, Pelletiers T, Mathaudhu S, Overman N, "High Ductility Aluminum Alloy made from Powder by Friction Extrusion," Materialia, vol. 6 2019, 100260.

Solutionization via Severe Plastic Deformation: Effect on Natural Aging in an Al–Mg–Si–(Mn) Alloy

Brian Milligan, B. Scott Taysom, Xiaolong Ma, and Scott Whalen

Abstract

Shear Assisted Processing and Extrusion (ShAPE), a severe plastic deformation technique that is fast and scalable, was used to produce thin-wall tubing from alloy 6082 (Al-0.8 Mg-0.9Si-0.7Mn) with in-situ solutionization during processing quickly followed by quenching. Quench medium and input material heat treatment were varied and natural aging behavior (T1 heat treatment) was evaluated using tensile testing. Post-ShAPE, the as-cast material was found to have high initial strength but weak natural aging, while the homogenized material was found to have lower initial strength and greater natural aging. Compared to air quenching, water quenching gave greater strength both early in and after natural aging. These observations along with microscopy suggest that: 1. air quenching was too slow; 2. as-cast material was not fully solutionized but plastic deformation broke up the coarse intermetallic particles, which provided strengthening; and 3. homogenized material was solutionized well, which led to good natural aging behavior.

Keywords

Aluminum • Aging • Precipitation • Processing

B. Milligan (✉) · B. S. Taysom · X. Ma · S. Whalen
Pacific Northwest National Laboratory, 902 Battelle Boulevard, Richland, WA 99354, USA
e-mail: brian.milligan@pnnl.gov

B. S. Taysom
e-mail: brandon.taysom@pnnl.gov

X. Ma
e-mail: xiaolong.ma@pnnl.gov

S. Whalen
e-mail: scott.whalen@pnnl.gov

Introduction

Natural aging is nearly unavoidable yet sometimes useful phenomenon that occurs in certain precipitationstrengthened aluminum alloys, including the 6XXX-series (Al–Mg–Si) alloys [1]. Three things are required for natural aging: fast-diffusing alloying elements which can form atomic clusters or Guinier–Preston (GP) zones, a relatively high-concentration supersaturated solid solution that provides the driving force for clustering and precipitation, and a high concentration of quenched-in vacancies which accelerate the diffusion kinetics [1–3]. 6XXX-series alloys often show this behavior after a solutionizing and quench heat treatment, as both Mg and Si are relatively fast-diffusing elements, and the solutionization and quench form a supersaturated solid solution and trap quenched-in vacancies. In these alloys, Mg and Si form clusters at room temperature over the course of days to weeks [2]. The natural aging eventually saturates because the diffusivity of the alloying elements is reduced after the quenched-in vacancies anneal out over time [4]. These clusters not only increase the strength of the alloy, but can also have a detrimental effect on the precipitation of other strengthening precipitates, often increasing the amount of time required to reach the peak strength during artificial aging (i.e., T5 or T6 heat treatment) and/or reducing the peak strength that can be reached [5]. However, aluminum alloys are sometimes used in the natural aged state (i.e., T1 or T4 heat treatment), mostly in order to reduce cost [1, 6].

Another topic of recent industrial interest is the severe plastic deformation (SPD) of metals. SPD processes can have a range of benefits including refined microstructures [7–9], increased solid solubility [10, 11], and reduced aging times [12]. However, most SPD processes are very difficult to scale up to larger than the laboratory scale. One SPD process that does not have this scalability problem is Shear Assisted Processing and Extrusion (ShAPE). A schematic of the process is shown in Fig. 1. This process essentially

© The Minerals, Metals & Materials Society 2023
S. Broek (ed.), *Light Metals 2023*, The Minerals, Metals & Materials Series,
https://doi.org/10.1007/978-3-031-22532-1_46

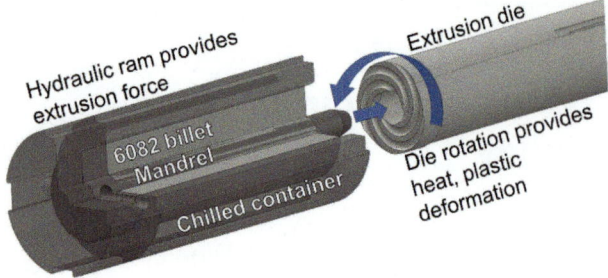

Fig. 1 Schematic of the process of manufacturing tubing via ShAPE

combines conventional indirect extrusion with a rotating die that allows pre-heating of the input material to be avoided and the heat instead comes from friction and adiabatic heating [13]. In addition, it allows for a quench system to be added to the die exit, which can be used for press quenching, allowing a separate solutionization heat treatment to also be avoided [14]. The severe plastic deformation can also act to homogenize the alloy and break up large intermetallic particles, potentially allowing for a separate homogenization heat treatment step to be avoided [15]. Together, avoiding these three thermal treatment steps can reduce the total energy cost for production by a large margin [16].

In this study, ShAPE will be used to produce thin-walled tubing out of aluminum alloy 6082, which is a medium-strength age-hardened alloy. The input material will be either as-cast or homogenized to study the effect of pre-treatment and will be quenched in-situ with either water or air to study the effect of cooling rate. The material will then be naturally aged for time periods ranging from 7 to 90 days, after which point natural aging will likely saturate. The material's mechanical properties will be determined by tensile testing.

Methods

The input material was AA6082, the nominal composition is listed in Table 1. The material was provided by Rio Tinto in both the as-cast and homogenized state.

This material was fed into a Bond Technologies ShAPE machine with a feed rate of 360 mm/min, a rotation rate of 275 RPM for the homogenized and 225 RPM for the as-cast material to control the temperature to $515 \pm 5\ ^\circ$C. The as-cast material had a higher flow stress, which lead to more adiabatic heating, requiring the RPM to be set lower for an

equivalent extrusion temperature. A three-ring quench system was attached behind the tool, which quenched the material approximately 3 s after processing. The flow rates of the water and air quench mediums were 4.7 L per minute and 520 standard liters per minute, respectively.

Four processing conditions were produced with two quench methods and two input material heat treatments. The four conditions are as-cast air quench (AC-AQ), as-cast water quench (AC-WQ), homogenized air quench (H-AQ), and homogenized water quench (H-WQ). Specimens from each of these extrusions were cut out and naturally aged for 7, 14, 30, 60, and 90 days (T1 heat treatment) as well as artificially aged for 10 h at 180 °C after 30 days of natural aging (T5 heat treatment). Each of these specimens was then evaluated using tube tensile testing according to ASTM E8 Sect. 6.9.1 [17]. There were a few outliers that fractured prematurely, likely due to brittle defects or porosity, which were not evaluated.

Results and Discussion

Results from the tensile tests of the AC-AQ specimen are shown in Fig. 2. Mechanical property data from the tensile tests are shown in Table 2. Note that there was some increase in the yield strength with natural aging, but no naturally aged condition reached close to the T5 strength. The yield strength as well as the UTS and the elongation were within the minimum specifications for mechanical properties of 6082-T4 from ASM even after only 7 days of natural aging [18]. There are two potential explanations for the low degree of hardening during natural aging: poor solutionization and low quenched-in vacancy concentration. Both these potential causes can be related to the quench. A slow quench will cause precipitation upon quenching, which limits the amount of alloying elements in a supersaturated solid solution. A slow quench will also allow for non-equilibrium quenched-in vacancies to anneal out during cooling. However, only solutionization is likely to depend on the input material heat treatment, as the as-cast material may not have all the potential alloying elements dissolve during processing because they are trapped in large intermetallics formed during solidification. Therefore, the low T5 strength signifies that the solutionization is poor, and the low hardening during natural aging signifies that the quenched-in vacancy concentration may also be poor.

	Si	Mg	Mn	Fe	Cr	Zn	Cu	Ti	Al
Table 1 Nominal composition of aluminum alloy 6082, given in weight percent	0.7–1.3	0.6–1.2	0.4–1.0	<0.5	<0.25	<0.2	<0.1	<0.1	Bal

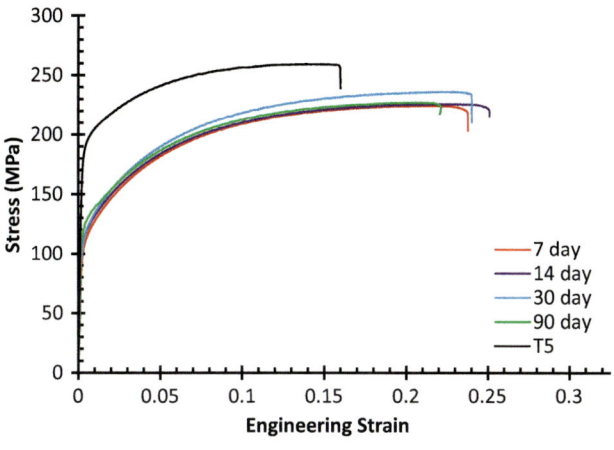

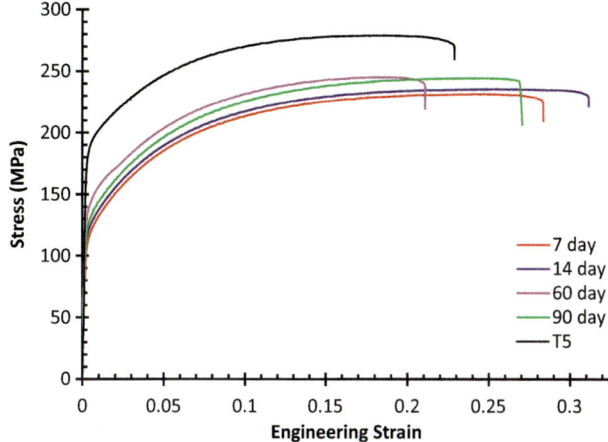

Fig. 2 As-cast, air quench (AC-AQ) tensile curves for natural aging periods ranging from 7 to 90 days, as well as one artificially aged specimen. Note that there was only a small increase in the yield strength during natural aging but the artificially aged specimen was still much stronger than any naturally aged condition

Fig. 3 As-cast, water quench (AC-WQ) tensile curves for natural aging periods ranging from 7 to 90 days, as well as one artificially aged specimen

Table 2 Mechanical properties of each condition from the tensile tests. Blue shading indicates naturally aged specimens and green shading indicates artificially aged specimens. Two specimens fractured prematurely due to microstructural defects, which is why the 60 day AC-AQ and 30 day AC-WQ data are not reported

Run	Feedstock	Quench	Aging temp. (°C)	Aging time	YS (MPa)	UTS (MPa)	Total elongation
34	As-cast 6082	Air	25	7 day	110	225	24
				14 day	113	226	25
				30 day	113	236	25
				90 day	126	228	22
			180	10 h	185	260	16
39	As-cast 6082	Water	25	7 day	116	232	28
				14 day	121	236	31
				60 day	124	245	29
				90 day	130	245	27
			180	10 h	198	277	21
37	Homogenized 6082	Air	25	7 day	115	206	19
				14 day	113	205	19
				30 day	124	217	20
				60 day	122	217	22
				90 day	124	212	18
			180	10 h	197	258	14
38	Homogenized 6082	Water	25	7 day	137	240	24
				14 day	139	240	21
				30 day	142	243	24
				60 day	142	243	20
				90 day	143	242	23
			180	10 h	247	299	17

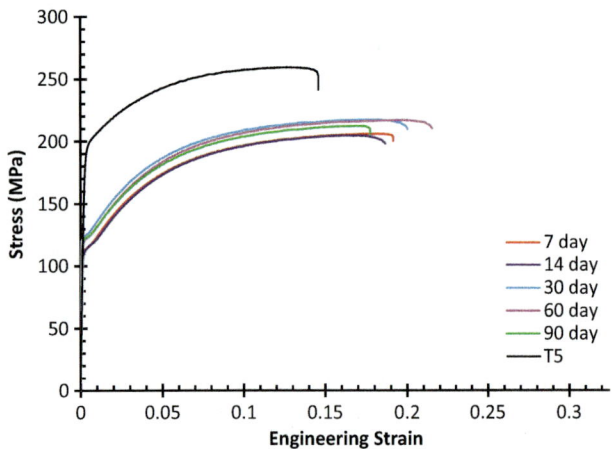

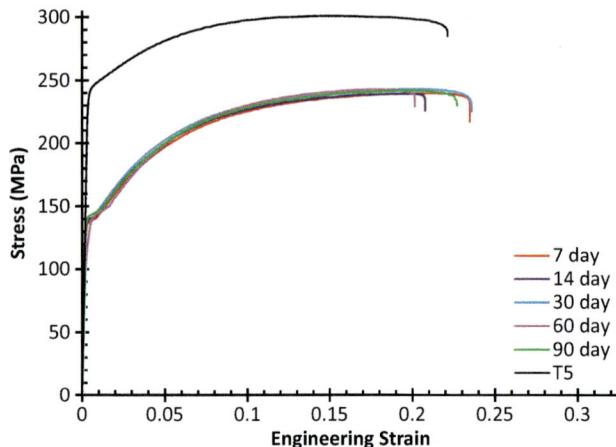

Fig. 4 Homogenized, air quench (H-AQ) tensile curves for natural aging periods ranging from 7 to 90 days, as well as one artificially aged specimen. The high T5 strength signifies that the solutionization was good

Fig. 5 Homogenized, water-quenched (H-WQ) tensile curves for natural aging periods ranging from 7 to 90 days, as well as one artificially aged specimen

Results from tensile tests on the AC-WQ specimen are shown in Fig. 3 and Table 2. Note that the degree of hardening is much larger in this specimen than in the AC-AQ specimen. The moderate T5 strength signifies that the solutionization is moderately good, and the high degree of strengthening during natural aging signifies that the quenched-in vacancy concentration is high.

Results from tensile tests on the H-AQ specimen are shown in Fig. 4 and Table 2. Note that the degree of hardening during natural aging is higher than the AC-AQ specimens but lower than the AC-WQ. The moderate T5 strength signifies that the solutionization is moderately good. If the quenched-in vacancy concentration was good, this moderately good solutionization should allow for a high degree of strengthening during natural aging. However, the natural aging behavior is poor, so it is concluded that the quenched-in vacancy concentration is low in this condition.

Results from tensile tests on the H-WQ specimen are shown in Fig. 5 and Table 2. This condition shows a quite high natural aged strength in the H-WQ specimens, although the degree of strengthening during natural aging is low. High T5 strength signifies that the solutionization was very good, and the water quench has been shown to be sufficiently fast for a high quenched-in vacancy concentration. It can, therefore, be concluded that due to the high- concentration

supersaturated solid solution that gives a high driving force for clustering and the high quenched-in vacancy concentration accelerating the diffusion kinetics, a large amount of the natural aging occurred before the 7 day time window. This observation is also supported by the relatively high natural aged yield strength even at 7 days. A comparison of the natural aging curves for each condition is also shown in Fig. 6.

Two additional relevant observations can be made on Figs. 2, 3, 4 and 5. The first is that there is a considerably higher degree of strain hardening in the AC-AQ and AC-WQ specimens (average of 92 and 100 MPa for the H-AQ and H-WQ, respectively). The second is that the H-AQ and H-WQ specimens both have a plateau in the stress right after the material yields. Both these observations point towards there being different primary strengthening mechanisms based on the input material heat treatment. There likely exist large intermetallic particles in the AC-AQ and AC-WQ specimens, which would have to be bypassed via Orowan looping. Orowan looping is a precipitate bypass mechanism that is often associated with high strain hardening rates. In contrast, the H-AQ and H-WQ specimens will not have many large intermetallic particles, and the main bypass mechanism for dislocations around their strengthening precipitates is shearing, which is associated with lower strain hardening rates. A summary of these observations is shown in Table 3.

Fig. 6 Natural aging curves from 7 to 90 days for each of the four conditions

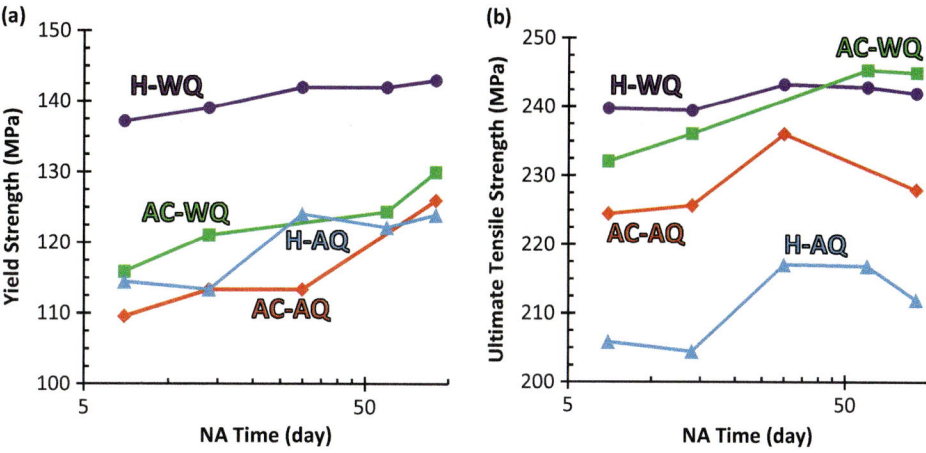

Table 3 Summary of the conclusions made on each of the specimens

	Air quench	Water quench
As-cast	• Poor solutionization • Low quenched-in vacancy concentration • Primarily strengthened by large intermetallics from solidification	• Moderate solutionization • High quenched-in vacancy concentration • Primarily strengthened by large intermetallics from solidification
Homogenized	• Moderate solutionization • Low quenched-in vacancy concentration • Primarily strengthened by solute clusters	• Good solutionization • High quenched-in vacancy concentration • Primarily strengthened by solute clusters

Conclusions

Four different processing condition combinations were tested to evaluate the effectiveness of solutionization via ShAPE on the natural aging behavior in the aluminum alloy 6082. The following conclusions were drawn from the analysis:

1. ShAPE processing followed by quenching was insufficient for solutionization of as-cast material, and intermetallic particles from solidification remained in the microstructure even after processing, limiting the strength in both the naturally aged and artificially aged conditions.

2. Air quenching from the processing temperature of 515 °C was insufficient for proper solutionization of the alloying elements, and also resulted in a low quenched-in vacancy concentration, both of which prevented a high degree of natural aging and limited the artificially aged strength.

3. The primary precipitation strengthening mechanism in the as-cast versus homogenized input material differed after ShAPE processing and natural aging. Precipitate bypass in the as-cast input material specimens was dominated by Orowan looping, while precipitate shearing dominated the homogenized input material specimens.

Further work on this topic is recommended to solidify these conclusions, primarily in microscopy or X-ray diffraction to identify the phases present and their volume fractions, particularly in the as-cast input material specimens.

Acknowledgements The authors thank the U.S. Department of Energy Vehicle Technologies Office (DOE/VTO) Lightweight Metals Core Program for supporting this work and Rio Tinto for providing feedstock materials. Pacific Northwest National Laboratory is operated by the Battelle Memorial Institute for the DOE under contract DE-AC05-76RL01830.

References

1. M. Werinos, H. Antrekowitsch, T. Ebner, R. Prillhofer, W.A. Curtin, P.J. Uggowitzer, S. Pogatscher, Design strategy for controlled natural aging in Al–Mg–Si alloys, Acta Materialia. 118 (2016) 296–305.https://doi.org/10.1016/j.actamat.2016.07.048

2. J. Banhart, C.S.T. Chang, Z. Liang, N. Wanderka, M.D.H. Lay, A. J. Hill, Natural Aging in Al–Mg–Si Alloys – A Process of Unexpected Complexity, Advanced Engineering Materials. 12 (2010) 559–571.https://doi.org/10.1002/adem.201000041

3. A. Cuniberti, A. Tolley, M.V.C. Riglos, R. Giovachini, Influence of natural aging on the precipitation hardening of an AlMgSi alloy, Materials Science and Engineering: A. 527 (2010) 5307–5311. https://doi.org/10.1016/j.msea.2010.05.003

4. Z. Yang, X. Jiang, X. Zhang, M. Liu, Z. Liang, D. Leyvraz, J. Banhart, Natural ageing clustering under different quenching

conditions in an Al-Mg-Si alloy, Scripta Materialia. 190 (2021) 179–182.https://doi.org/10.1016/j.scriptamat.2020.08.046

5. T. Abid, A. Boubertakh, S. Hamamda, Effect of pre-aging and maturing on the precipitation hardening of an Al–Mg–Si alloy, Journal of Alloys and Compounds. 490 (2010) 166–169.https://doi.org/10.1016/j.jallcom.2009.10.096

6. Y. Chen, B.Q. Lu, H.A. Zhang, Hardening and precipitation of a commercial 6061 Al alloy during natural and artificial ageing, IOP Conf. Ser.: Mater. Sci. Eng. 770 (2020) 012065. https://doi.org/10.1088/1757-899X/770/1/012065.

7. J. Kawałko, M. Wroński, M. Bieda, K. Sztwiertnia, K. Wierzbanowski, D. Wojtas, M. Łagoda, P. Ostachowski, W. Pachla, M. Kulczyk, Microstructure of titanium on complex deformation paths: Comparison of ECAP, KOBO and HE techniques, Materials Characterization. 141 (2018) 19–31.https://doi.org/10.1016/j.matchar.2018.04.037

8. B. Beausir, J. Scharnweber, J. Jaschinski, H.-G. Brokmeier, C.-G. Oertel, W. Skrotzki, Plastic anisotropy of ultrafine grained aluminium alloys produced by accumulative roll bonding, Materials Science and Engineering: A. 527 (2010) 3271–3278.https://doi.org/10.1016/j.msea.2010.02.006

9. J.R. Croteau, J.-G. Jung, S.A. Whalen, J. Darsell, A. Mello, D. Holstine, K. Lay, M. Hansen, D.C. Dunand, N.Q. Vo, Ultrafine-grained Al–Mg–Zr alloy processed by shear-assisted extrusion with high thermal stability, Scripta Materialia. 186 (2020) 326–330. https://doi.org/10.1016/j.scriptamat.2020.05.051

10. E. Botcharova, M. Heilmaier, J. Freudenberger, G. Drew, D. Kudashow, U. Martin, L. Schultz, Supersaturated solid solution of niobium in copper by mechanical alloying, Journal of Alloys and Compounds. 351 (2003) 119–125.https://doi.org/10.1016/S0925-8388(02)01025-3

11. M. Komarasamy, X. Li, S.A. Whalen, X. Ma, N. Canfield, M. J. Olszta, T. Varga, A.L. Schemer-Kohrn, A. Yu, N.R. Overman, S.N. Mathaudhu, G.J. Grant, Microstructural evolution in Cu–Nb processed via friction consolidation, J Mater Sci. 56 (2021) 12864–12880.https://doi.org/10.1007/s10853-021-06093-9

12. S. Farè, N. Lecis, M. Vedani, Aging Behaviour of Al-Mg-Si Alloys Subjected to Severe Plastic Deformation by ECAP and Cold Asymmetric Rolling, Journal of Metallurgy. 2011 (2011) 1–8.https://doi.org/10.1155/2011/959643

13. S.A. Whalen, D.R. Herling, X. Li, M. Reza-E-Rabby, B.S. Taysom, G.J. Grant, Devices and Methods for Performing Shear-Assisted Extrusion and Extrusion Processes, U.S. Patent 20210053100A1. (2021). https://patents.google.com/patent/US20210053100A1/en (accessed August 16, 2021).

14. T. Sheppard, Press quenching of aluminium alloys, Materials Science and Technology. 4 (1988) 635–643.https://doi.org/10.1179/mst.1988.4.7.635

15. T. Wang, J.E. Atehortua, M. Song, M. Reza-E-Rabby, B.S. Taysom, J. Silverstein, T. Roosendaal, D. Herling, S. Whalen, Extrusion of Unhomogenized Castings of 7075 Aluminum via ShAPE, Materials & Design. 213 (2022) 110374. https://doi.org/10.1016/j.matdes.2021.110374

16. Brandon Scott Taysom, Nicole Overman, Matt Olszta, Md Reza-E-Rabby, Tim Skszek, Massimo DiCiano, Scott Whalen, Shear assisted processing and extrusion of enhanced strength aluminum alloy tubing, Internal Journal of Machine Tools and Manufacture. 169 (2021). https://doi.org/10.1016/j.ijmachtools.2021.103798.

17. ASTM International, ASTM Standatd E8/E8M: Standard Test Methods for Tension Testing of Metallic Materials, (2016).

18. Kevin Anderson, John Weritz, J. Gilbert Kaufman, ASM Handbook, Volume 2B: Properties and Selection of Aluminum Alloys, ASM International, 2019. https://www.sciencedirect.com/science/article/pii/0378380481900255 (accessed April 19, 2022).

Manufacture of Nano-to-Submicron-Scale TiC Particulate Reinforced Aluminium Composites by Ultrasound-Assisted Stir Casting

Guangyu Liu, Abdallah Abu Amara, Dmitry Eskin, and Brian McKay

Abstract

Aluminium alloys reinforced with ceramic particles have been extensively investigated for automotive applications. Presently, growing attention has been drawn to their application within electric vehicles due to their high strength and lightweight. Among various reinforcements, titanium carbide (TiC) is of particular interest due to its superior hardness, high wear resistance, excellent elastic modulus, and relatively low coefficients of thermal expansion. This study aims to explore the effect of nano-/submicron-sized (<2 μm) TiC particles on the mechanical properties of a commercial aluminium alloy by incorporating Al-45 wt.% TiC master-alloy powders and Al-6 wt.% TiC master ingots. Stir casting with the aid of ultrasound processing was applied to facilitate the mixing and dispersion of the TiC particles. Composites with additions of 0.5, 1, 2, and 5 wt.% TiC were prepared by gravity casting using a permanent steel mould. The effect of the TiC particles on the microstructure, with respect to casting defects, particle distributions, and particle/matrix bonding has been investigated.

Keywords

Aluminium MMCs • TiC particles • Stir mixing • Ultrasound • Mechanical properties

Introduction

Aluminium metal matrix composites (AMMCs), which now constitute over 45% of the global metal matrix composites (MMCs) market, have been widely developed in automotive and aerospace industries [1, 2], due to a combination of the ductility and toughness of the aluminium matrix and the superior strength and stiffness of the ceramic reinforcements. The incorporated reinforcements are diverse in structure and chemistry and can be divided into continuous or discontinuous fibres, whiskers and particulates. Particulate reinforced AMMCs are of special interest because they deliver isotropic mechanical properties, are easier to manufacture, and are often cheaper in comparison with other types of reinforcing fillers [3]. These reinforcements are typically ceramic particles including oxides, carbides, and nitrides, such as Al_2O_3 [4], TiC [5], SiC [6], and AlN [7]. Among them, TiC is known to exhibit superior hardness, high elastic modulus, high wear resistance, excellent thermal conductivity and thermal stability, making Al/TiC composites popular for the structures and applications where the abovementioned characteristics are needed, such as pistons, brake rotors, and the propeller shaft [8].

The high mechanical performance of AMMCs is correlated with a uniform dispersion of the reinforcements with small-scale dimensions (nano-, or sub-micron) and strong reinforcement/matrix bonding [9]. The strengthening mechanism is various, including load transfer mechanism [10], Orowan strengthening [11] and coefficient of thermal expansion (CTE) mismatch strengthening [12]. The load transfer mechanism, through carrying the loading along with the aluminium matrix, is principally significant in the presence of high-volume reinforcements with coarse sizes. Orowan strengthening and coefficient of thermal expansion (CTE) mismatch strengthening are particularly dominant when the reinforcing particles are fine (<1 μm) and distributed inside the grains. Specifically, the Orowan strengthening is based on the resistance of small hard particles against the motion of dislocations, the dislocation movement proceeds with their passing of these obstacles by bowing, reconnecting, and forming a dislocation loop around the particles, leading to high work-hardening rates with improved strength [13]. CTE mismatch strengthening

G. Liu (✉) · A. A. Amara · D. Eskin · B. McKay
Brunel Centre for Advanced Solidification Technology, Brunel University London, Uxbridge, UB8 3PH, Middlesex, UK
e-mail: guangyu.liu2@brunel.ac.uk

© The Minerals, Metals & Materials Society 2023
S. Broek (ed.), *Light Metals 2023*, The Minerals, Metals & Materials Series,
https://doi.org/10.1007/978-3-031-22532-1_47

originates from volumetric strain mismatch and the thermal mismatch between the monolithic matrix and reinforcing particles during solidification and cooling of the AMMCs, thereby producing geometrically necessary dislocations (GND) around reinforcing particles to accommodate the CTE difference [14]. In addition, the grain refinement effect, caused by the addition of particles that might potentially introduce more nucleation sites and/or provide agglomerates as obstacles to grain growth during solidification, improves the performance and overall properties of the as-cast material [15].

To facilitate the introduction and enhance dispersion of the reinforcements, master alloys that contain reinforcing particles are often externally synthesised as additives and are subsequently incorporated into the molten aluminium. These master alloys are typically produced using two methods: (1) solid-state in-situ powder processing technique, using industrial high energy ball milling (HEBM); and (2) liquid-based in-situ reaction. The addition of the same reinforcing particles produced by the two different methods could cause different results due to differences in interfacial features between the reinforcing particles and the matrix, and the difference in the concentration of the particles in the master alloys produced by the two manufacturing methods. For example, the particles may exhibit reaction layers on the surface formed during the liquid-based in-situ reaction, increasing the wettability and adhesion of the particles to the matrix, thus enhancing the bonding. Whereas solid-based in-situ reaction promotes the synthesis of more primary TiC particles in the nano-range and a much higher concentration of reinforcements can be achieved in the master alloys that are produced using in-situ HEBM, targeting a cost-effective and flexible solution for the incorporation of reinforcements and enabling easier integration in Al manufacturing processes than using free nanoparticles. In this study, the master-alloy powders and/or ingots containing TiC particles produced by solid-based in-situ HEBM and a liquid-based in-situ reaction are employed to investigate the effects of the TiC particles on the microstructure and mechanical properties of the aluminium alloys.

Stir casting combined with ultrasonic processing is considered as an effective method for good dispersion of particles in the molten matrix [16]. During ultrasonic treatment, acoustic waves generate tensile stress in the molten metal, leading to the formation of tiny cavities. The cavities (bubbles) grow and collapse alternately during the expansion-compression cycles, producing transient micro "hot spots" within an extremely short duration (in the order of microseconds), where extremely high temperatures and pressures are generated [17]. This process of so-called "acoustic cavitation" can generate high-intensity shock waves to break the agglomerates and disperse them through acoustic streaming [18]. Besides, the local transient high temperature and pressure could significantly improve the wettability of particles by removing or desorbing the gases from the surface of the particles. At a very high temperature reached upon cavitation bibble collapse (5000 °C), the surface tension of liquid with vapour is significantly decreased, further enhancing the wettability of the nanoparticles [11].

In this study, TiC reinforced aluminium composites are prepared using ultrasound-assisted stir casting processing, to produce the AMMCs with homogenously dispersed reinforcements and improved properties. Al-45% TiC master-alloy powders and Al-6% TiC master ingots made by solid-based in-situ high-energy ball milling and liquid-based in-situ reactions, respectively, are also employed in an attempt to improve the incorporation and dispersion of the TiC particles. The distribution and effects of TiC particles on the resultant microstructure and mechanical properties are thoroughly investigated and the findings are discussed herein.

Experimental

Sample Preparations

Commercial aluminium alloys (supplied by Constellium, UK) were used as the raw materials for remelting to produce the base alloy. The Al-45% TiC master-alloy powders and Al-6% TiC master ingots, as shown in Fig. 1a and b, were used as the source of reinforcements to prepare the composites. The Al-45% TiC master-alloy powders produced via solid-based in-situ high-energy ball milling (HEBM) were provided by MBN Nanomaterialia (Italy) as part of their current development of master-alloys powder for the production of aluminium metal matrix nanocomposites and Al-6% TiC master ingots by liquid-based in-situ reactions were supplied from Qinhuangdao Fengyue Science & Technology Co., Ltd (China). The alloying elements of the Base alloy include Si, Mg, Cu, Fe and other tracing elements. The composite materials were defined as Base-0.5TiC (L), Base-0.5TiC, Base-1.0TiC, Base-2.0TiC, and Base-5.0TiC, and the specifications are shown in Table 1.

Figure 1c–e show the schematic experimental setup for making the TiC reinforced AMMCs using the ultrasound-assisted stir casting technique. It includes an electric resistance furnace, impeller, and ultrasonic unit. A total amount of ∼3 kg of the base alloy was placed in the crucible and melted in an electrical resistance furnace (Carbolite) at 750 °C, with a later subsequent addition of Mg to prepare the molten aluminium matrix. After the melt was homogenised, the preheated (at 200 °C) Al-45TiC master-alloy powders wrapped in Al foil, or the Al-6TiC master ingots were fed into the melt at the side of the vortex created by a

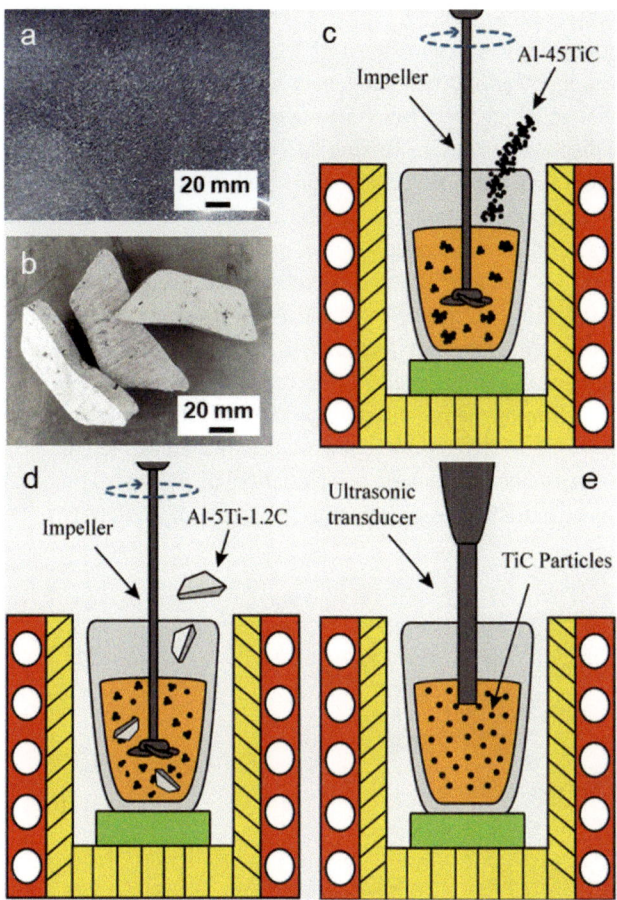

Fig. 1 Photos and schematics showing **a** the Al-45TiC master-alloy powders, **b** Al-6TiC master ingots, **c, d** the stir mixing process and **e** ultrasound processing

four-bladed titanium impeller (coated with boron nitride), to achieve a pre-mixing and preliminary introduction of particles in the melt, as shown in Fig. 1c and d. In the stir mixing process, nominal 0.5, 1.0, 2.0, and 5.0 wt.% TiC particles were introduced into the melt, producing the composite melt, with the melt temperature decreasing by 30–60 °C in the time range of 5–10 min, depending on the amount of the particle additions. Then, the pre-mixed composite melt was reheated up to 750 °C before it was ultrasonically processed. A 20 kHz ultrasonic wave transmitted by a 19 mm diameter niobium probe generated with a 3.5 kW capacity ultrasonic

generator unit was applied to the melt for 15 min to achieve a further dispersion of particles (Fig. 1e). After ultrasonic processing, the composite melt was poured at 700 °C into a permanent steel mould to produce cylinder-shaped casting bars of 20 mm in diameter and 165 mm in length. The Base alloy was separately produced using the same process except that no TiC reinforcements were added.

Microstructural and Mechanical Characterisations

The metallurgical microstructural examination was conducted on the cross-section of Al-45TiC and Al-6TiC master alloys and the casting bars. The surface to be examined was ground using SiC abrasive papers and then polished using silica suspension (OPS, 0.05 μm water-based SiO_2 suspension). Quantitative analysis of the microstructure was performed using AxioVision Rel. 4.8 software. Detailed information on intermetallic phases, reinforcing particles and grain structures was obtained using the Zeiss Supra 35 field-emission scanning electron microscope (FESEM), equipped with energy-dispersive X-ray spectroscopy (EDS). The mechanical test was performed on the as-cast specimens with a circular cross-section, extracted from the central part of the cylinder-shaped casting bars. The tensile test was conducted using Instron 5500 universal Electromechanical Testing System equipped with Bluehill software and a 50 kN load cell. Room temperature tensile tests were performed according to ASTM E8/E8M [19]. A gauge length of 50.0 mm and a gauge diameter of 10.0 mm were applied. Each data reported with standard deviation was based on the mechanical properties attained from 6 samples.

Results

Microstructure of the Al-45TiC and Al-6TiC Master Alloys

Figure 2a–c present back-scattered SEM micrographs showing the size, morphology, and distribution of TiC particles in the microstructure of the Al-45TiC master-alloy powders that were produced by in-situ HEBM. The XRD

Table 1 Specifications of the aluminium composites, wt.%

Materials	TiC additions	Resources
Base-0.5TiC (L)	0.5	Al-6TiC
Base-0.5TiC	0.5	Al-45TiC
Base-1.0TiC	1.0	Al-45TiC
Base-2.0TiC	2.0	Al-45TiC
Base-5.0TiC	5.0	Al-45TiC

pattern (Fig. 2d) revealed that the microstructure mainly consists of α-Al and TiC phases. No Al₃Ti phase was identified. It is noted that the unlabelled XRD peaks correspond to the resin that was used to mount the powders. The TiC particles displayed round-shaped morphologies and were either evenly dispersed or severely agglomerated, thus exhibiting two distinguishing features: individual TiC particles and agglomerates. The TiC particles and the agglomerations showed a Feret diameter in the range of 0–2.0 μm and 0–20.0 μm, with a mean Feret diameter of 0.4 and 3.2 μm, respectively, as shown in Fig. 2e and f.

Figure 3a–c present SEM micrographs showing the size, morphology, and distribution of TiC particles in the microstructure of Al-6TiC master ingots synthesised by liquid-phase in-situ reactions. The corresponding XRD pattern (Fig. 3d) revealed that the microstructure mainly

comprised α-Al and TiC phases. The particles showed similar morphology to that in the Al-45TiC. Whilst the distribution of particles was more homogeneous, despite the formation of clusters which were much less agglomerated and were easier to disperse during the mixing and ultrasound processing. It is noted that the size distribution of the particles was more homogeneous compared to the particles in the Al-45TiC (Fig. 2e), as evidenced by the narrower distribution curve (Fig. 3e), showing the mean Feret diameter of 1.0 μm, instead of 0.4 μm. TiC particles in the Al-6TiC were more evenly distributed, which was expected to improve the distribution of the particles in the melt during the stir mixing process. The more homogeneous distribution of the TiC particles in the Al-6TiC master ingots, compared to that in Al-45TiC, is a result of the much less concentration of TiC particles, the bigger particle size and the pre-dispersion in liquid Al.

Fig. 2 **a**, **b**, **c** BSD-SEM images and **d** XRD pattern for the Al-45TiC master-alloy powders, indicating the presence of α-Al and TiC phases; **d**, **e** the size distribution of the TiC individual particles and agglomerations

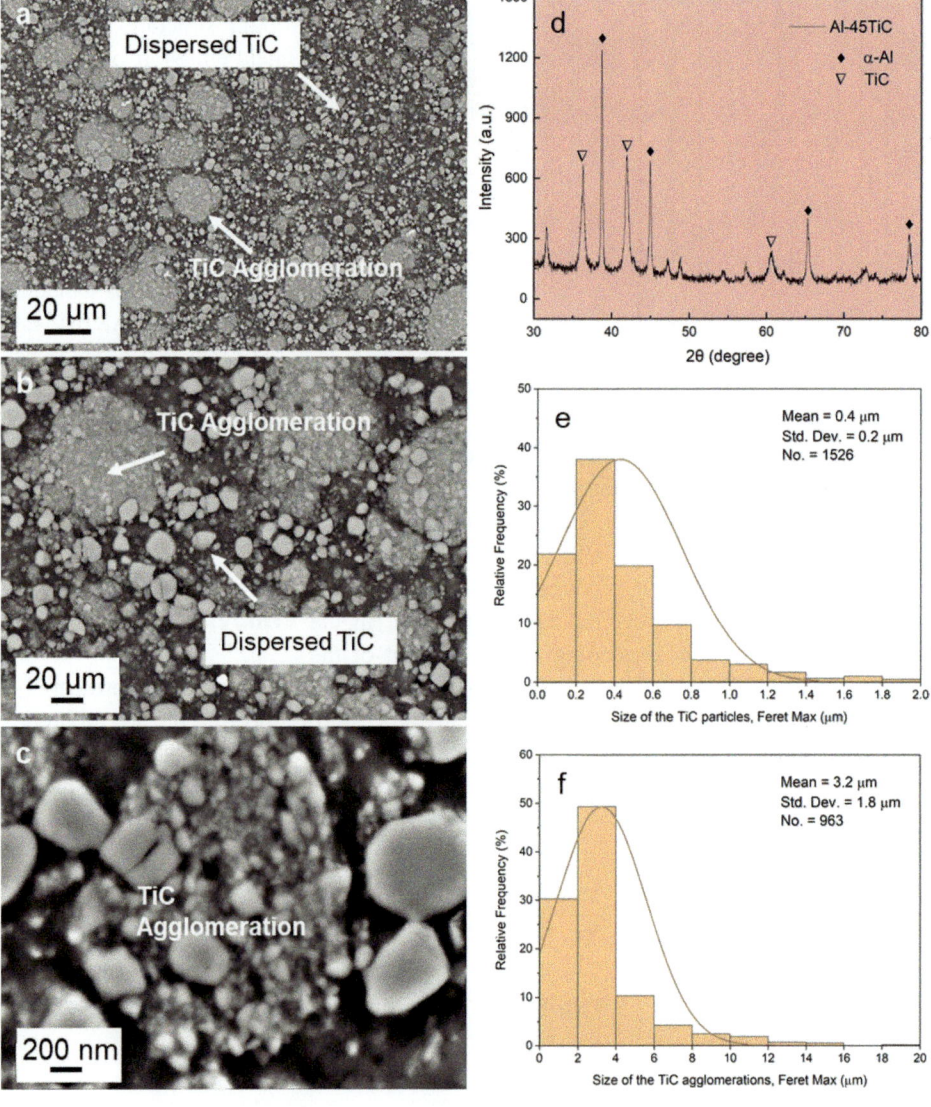

Fig. 3 **a**, **b**, **c** SE-SEM images and **d** XRD pattern for the Al-6TiC master ingot, indicating the presence of a-Al and TiC phases; **e** the size distribution of the TiC particles

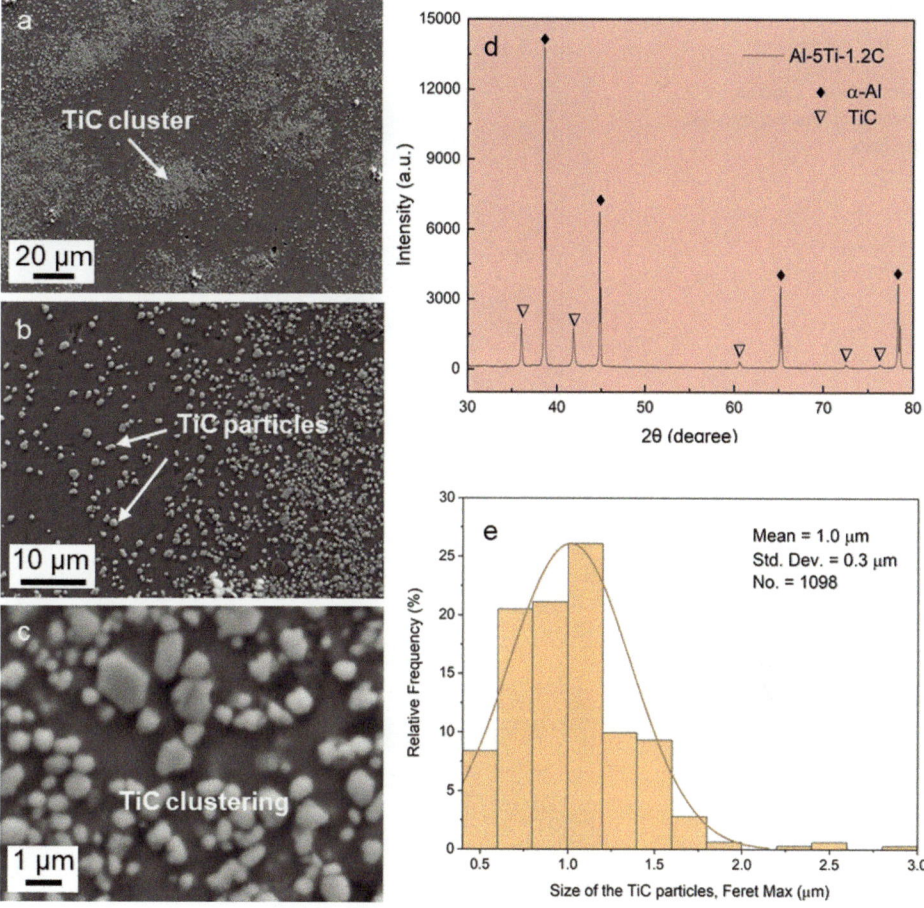

Microstructure of the Base Alloy and Composite

Base Alloy and Al-0.5 TiC (L) Composite

Figure 4 shows the optical and SEM images of the Base and Base-0.5 TiC (L) samples, indicating that the microstructure of the Base alloy comprised primary α-Al phase, Cu-rich and Fe-rich intermetallics, and pores (Fig. 4a and b). TiC particles and/or clusters (particles that tend to be located in the vicinity of each other but are not severely agglomerated) were seen in the Base-0.5 TiC (L) and were found homogeneously distributed in the matrix, as indicated in the region highlighted by the circles/ellipses as shown in Fig. 4c and d. The dimension of the TiC clusters was in the range of 5–20 μm. These TiC clusters tended to segregate to the boundary regions (Fig. 4e). Also, oxides were occasionally observed in the vicinity of the TiC clusters, as shown in Fig. 4d, those could be formed during the mixing process. It is seen from Fig. 4f that the TiC particles showed a similar size and morphology to the TiC in the Al-6TiC master ingots. It is worth noting that some interfacial features between TiC and Al matrix were observed, as shown in Fig. 4f.

Aluminium Composites Reinforced by Al-45 TiC

Figure 5 shows optical micrographs presenting the distribution of TiC particles and the formation of Ti–rich intermetallic phases (IMCs) in the microstructure of the composites reinforced by 0.5, 1.0, 2.0, and 5.0 wt.% additions of TiC that were from the Al-45TiC master-alloy powder. The proportion of TiC particles/clusters was consistently increased when the addition of TiC was increased from 0.5 to 5.0 wt.%. With increasing the TiC additions, more TiC clusters were observed, as indicated by the regions highlighted by the circles/ellipses. When the TiC addition was at 1.0, 2.0, and 5.0 wt.%, a blocky-shaped IMC phase was observed, particularly, in the case of Base-5TiC, the blocky-shaped IMCs were coarse, reaching 50 μm (see Fig. 5d). The large amounts of TiC agglomerates and the formation of coarse IMCs should be detrimental to the mechanical properties.

Figure 6 presents the SEM micrographs and the XRD patterns showing the distribution of TiC particles and intermetallic phases in the microstructure of the Base-5TiC. The coarse block-like intermetallic phase observed in Fig. 6a was identified as the $Al_3Ti(Si)$, as indicated by the EDS

Fig. 4 Optical and SEM images of **a**, **b** Base and **c**, **d**, **e**, **f** Base-0.5 TiC (L), showing the microstructure including α-Al, intermetallic phases, and TiC particles

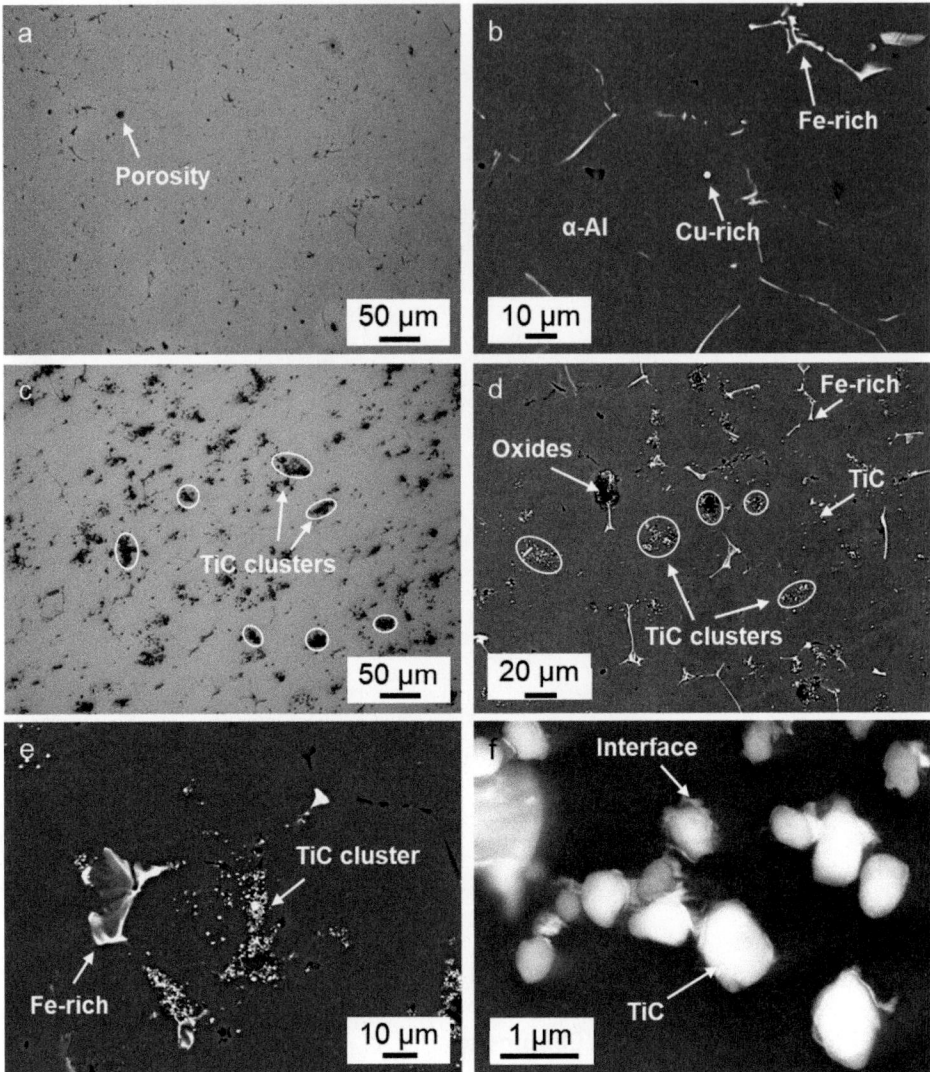

results (Fig. 6a inset). The TiC particles displaying the round morphology showed a similar size and morphology to the TiC particles observed in the Al-45TiC master-alloy powders (Fig. 2). XRD further confirmed the presence of Al₃Ti and the TiC phase, as shown in Fig. 6c.

Mechanical Properties

Figure 7 shows the tensile properties including the YS, UTS, and elongation for the Base and composite materials. The YS of the Base was 87 ± 1 MPa, which increased by 24% to 108 ± 2 MPa for the Base-0.5TiC (L). The YS of the Base-0.5TiC and Base-1TiC was 96 ± 1 MPa and 94 ± 1 MPa, seeing increases of 10% and 8% compared to

the Base, respectively. When the addition of TiC increased to 2 wt.%, the YS displayed only a marginal increase (2%), to 89 ± 2 MPa. The YS of Base-5TiC was 82 ± 2 MPa, experiencing a drop compared to the Base alloy. This was ascribed to the increased TiC agglomerations and the porosity especially when the addition of TiC particles was over 2 wt.%. At the same time, the UTS displayed a similar variation in trend to the YS, with the respective value being 196 ± 2, 216 ± 3, 194 ± 2, 195 ± 3, 193 ± 2, 128 ± 3 MPa. The respective elongation was 16, 13, 11, 11, 10, and 3%, indicating decreases after the addition of TiC particles due to the formation of coarse Al₃Ti phase and the increased porosity originating from the extra air that was introduced and entrained during the mixing processing.

Fig. 5 Optical micrographs showing the distribution of TiC particles or clusters and the coarse Ti–rich IMCs in the microstructure of **a** Base-0.5TiC, **b** Base-1TiC, **c** Base-2TiC, and **d** Base-5TiC

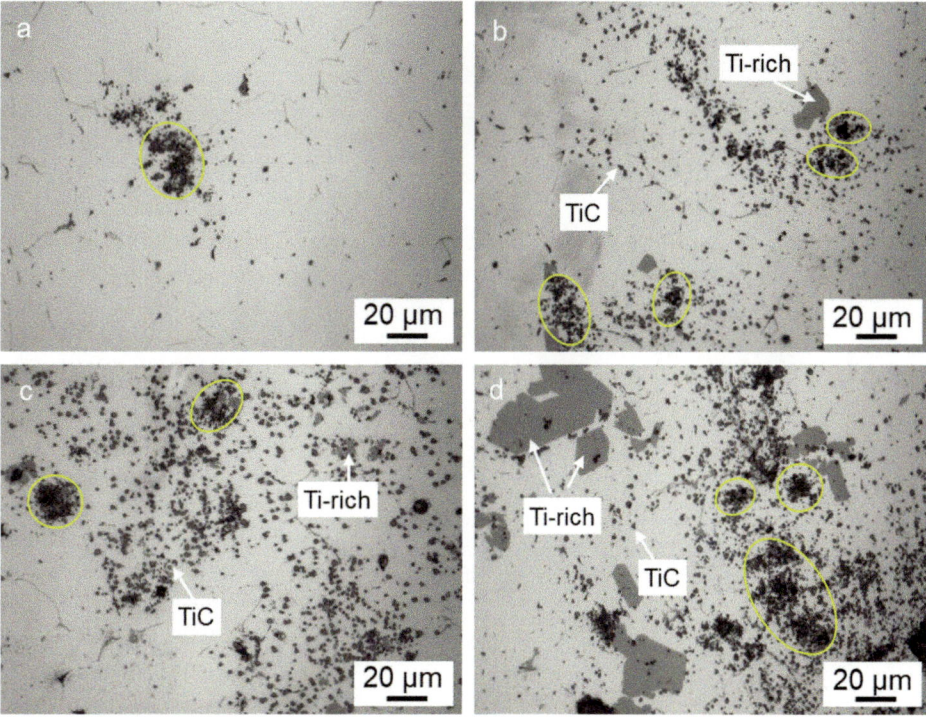

Fig. 6 **a**, **b** BSE-SEM images and **c** XRD patterns for the Base-5TiC sample, indicating the formation of Al₃Ti phases

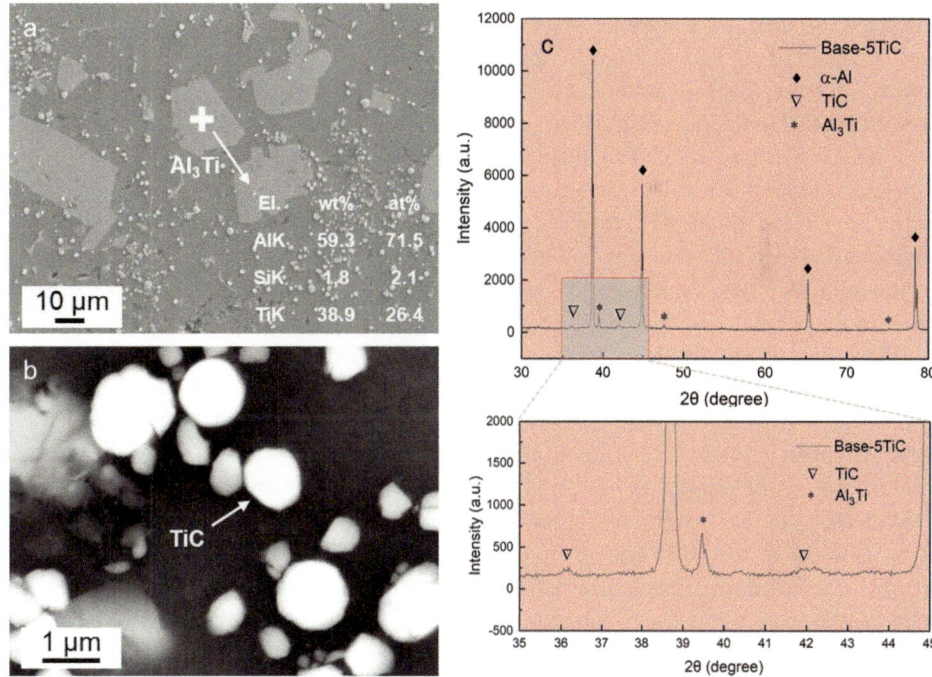

Discussion

Distribution of the TiC Particles

A relatively good distribution of TiC particles or clusters was achieved in the composites reinforced by introducing

Al-45TiC master-alloy powders (by in-situ HEBM) and Al-6TiCmaster ingots (by liquid-based in-situ reaction), in the samples reinforced with lower TiC additions (0.5 wt.% and 1.0 wt.%). This, to a large extent, suggested that a good dispersion of the particles in the molten metal was obtained under the combined effects of the physical pre-mixing and ultrasonic treatment and the use of master alloys including

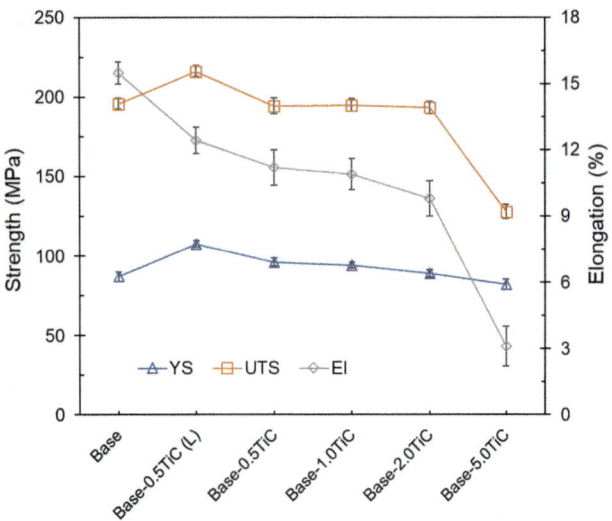

Fig. 7 The tensile properties showing the yield strength (YS), tensile strength (UTS) and elongation of the Base alloy and the TiC reinforced AMMCs

Al-45TiC and Al-6TiC. On the other hand, it has been confirmed that the composite reinforced with Al-6TiC delivered a more homogenous distribution of TiC particles, compared to that introduced by Al-45TiC, as shown in Figs. 4 and 5. This was ascribed to the fact that the TiC particles within Al-6TiC were more homogeneously dispersed than that in the Al-45TiC, (Figs. 2 and 3), thanks also to a less concentration of TiC particles within Al-6TiC compared to that in Al-45TiC. This aided and facilitated the dispersion of individual TiC particles in the molten aluminium using mechanical mixing and ultrasound processing, resulting in the more homogeneous redistribution of TiC particles during the solidification process.

When the aluminium melt containing suspended TiC particles solidified, interactions between the solidification front and the TiC particles occurred. The TiC particles were either engulfed by the solidification front, and thus distributed within the interior of the grain, or pushed by the solid–liquid interface toward the grain boundaries. For a given velocity of the solidification front, it was suggested [20, 21] that the critical particle size can be calculated, which is the particle size above which particles are engulfed by the interface and below which they are pushed. Whether the particle pushing or engulfment occurs is an extremely complicated phenomenon. To put it simply, the repulsive force from interfacial energy change and the dragging force resulting from the flow of liquid are considered the two main parameters [22] determining the critical particle size in a given melt for a certain velocity of the solidification front. The repulsive force and the dragging force compete with each other. If the repulsive force outstrips the dragging force, particles will be pushed, inversely, the particles will be

engulfed. The critical particle size will be calculated supposing the two opposite forces are equal. It has been largely reported that the critical particle size is increased with the decrease of velocity [23]. In this sense, larger-sized particles are more likely to be engulfed and thus cause a more homogeneous distribution of particles at the same velocity of the solidification front. It has been measured that the size of TiC particles in Al-6TiC was over two times larger (1.0 μm) than that in Al-45TiC (0.4 μm), which could be the result of the improved distribution of the TiC particles in the Base-0.5TiC (L) composite in line with increased occurrence of engulfment of TiC particles during solidification.

In addition, the TiC particles produced by liquid-based in-situ reaction could show a better wettability compared to those synthesised by in-situ HEBM and this could deliver stronger bonding between the TiC and Al matrix and thus deliver more improvements in the mechanical properties. It has been reported that the orientation relationship between Al and TiC can be inherited from the Al-TiC master alloy that was synthesised by the liquid-based in-situ reaction to the inoculated aluminium [24]. This could also result in ease of engulfment of the TiC particles when Al-6TiC was externally added to the aluminium melt, due to the existing interfacial adhesion/affinity between TiC and Al. The detailed in-depth microscopic examination of TiC/Al matrix interfacial features and behaviours associated with mechanical performance will be conducted and investigated in future work.

Strengthening Mechanisms

For the composites reinforced with the TiC particles originating from Al-45TiC, a 0.5–2 wt.% addition of TiC increased the strength, whilst a 5 wt.% addition of TiC decreased the strength. The corresponding ductility was found to decrease consistently with increasing the addition of TiC. This was associated with the strengthening mechanisms in line with the addition of TiC reinforcements, including the CTE. The Al alloy matrix and TiC particles have different coefficients of thermal expansion (CTE), during cooling of the composite melt in the solidification process, a great number of dislocations will be generated due to thermal mismatch between the matrix and the TiC particles to accommodate the inconsistency of geometry variations. This will cause a high work hardening rate and a great strengthening effect. However, the strengthening effects introduced by the TiC particles were offset or even diminished with the formation of large-sized TiC agglomerates, Al_3Ti intermetallics, and porosity such as observed in the Base-5TiC. These structure features acted as stress-concentrators and crack origins decreasing the elongation and accelerating the fracture.

For the composites reinforced with the less-reinforcement-concentrated Al-6TiC master alloys, a more significant improvement in the strength (by 24% in YS) was achieved when comparing the YS of Base-0.5TiC (L) with that of Base-0.5TiC. This was due to more diluted TiC particles within Al-6TiC master alloys thus the more homogeneous distribution of TiC particles in Base-0.5TiC (L) (Fig. 4), which enhanced the particle strengthening effects. Also, the increased YS could be attributed to the improved bonding between the TiC and Al matrix. This will be further investigated in future work. On the other hand, it is worth noting that the Al-45TiC delivered a great advantage of the extremely-higher concentration of TiC reinforcements in the nano-range compared to Al-6TiC master ingots, which provided significant ease of manufacturing of the composites by using stir casting. Whilst there was a trade-off that the highly-concentrated TiC particles in Al-45TiC increased the potential for the formation of TiC agglomerations (see Fig. 2a) which increased the difficulty in dispersing the TiC particles in the melt. In this sense, the Al-TiC master alloys with the concentration of TiC being lower than 45 wt.% and higher than 6 wt.% could be potentially developed to achieve fewer agglomerations within the master alloy and ease of incorporation of the reinforcements during the stir mixing process.

Conclusions

The present work produced the aluminium composite by incorporating external nano-/submicron-scale TiC particles (<2 μm) that were in the form of Al-45TiC master-alloy powders and Al-6TiC master ingots, via the ultrasound-assisted stir mixing technology, in attempts to explore the effect of varying addition of TiC particles on the microstructure and mechanical properties of the aluminium alloy. The following conclusions can be drawn:

- For the composites reinforced with the TiC that originated from Al-45TiC, a relatively uniform distribution of TiC particles or clusters was obtained with 0.5 and 1.0 wt.% addition of TiC. However, with the addition of TiC increasing to 2.0 wt.% and 5.0 wt.%, more TiC agglomerates, pores, and the coarse block-like Al_3Ti intermetallics formed.
- For the composites reinforced with the 0.5 wt.% addition of TiC that was in the form of Al-6TiC master ingots, a more homogeneous distribution of TiC clusters was obtained. Fewer TiC agglomerates were observed. This was due to the initial uniform dispersion of the TiC in the Al-6TiCmaster ingots which was more diluted compared to the TiC particle within Al-45TiC.

- The YS was improved by 10% after 0.5 wt.% addition of TiC particles from Al-45TiC, due to a strengthening effect of different coefficients of thermal expansion (CTE), whilst the strength decreased when the additions of TiC were increased to 2 and 5 wt.%. This was attributed to the increase in porosity, agglomerates, and formation of coarse block-like Al_3Ti phase.
- The YS was increased by 24% after a 0.5 wt.% addition of TiC particles from Al-6TiC, with a much higher improvement in the YS compared to the composite with the same addition of TiC from the Al-45Ti master-alloy powder. This was due to a more homogeneous dispersion of TiC particles/clusters that was a result of the less-concentrated and more homogeneous distribution of TiC particles within Al-6TiC.

Acknowledgements Financial support from the European Union (FLAMINGo Grant Agreement No. 101007011) is gratefully acknowledged. The authors are also grateful to Constellium, MBN Nanomaterialia (Italy), and Qinhuangdao Fengyue Science & Technology Co., Ltd (China) for providing commercial aluminium alloys, Al-45TiC, and Al-6TiC master alloys, respectively. The Experimental Technical Centre (ETC) is gratefully thanked for providing access to the equipment for microstructural characterization.

References

1. Ramnath BV, Elanchezhian C, Annamalai RM, Aravind S, Atreya TS, Vignesh V, Subramanian C (2014) Aluminium metal matrix composites–a review. Rev. Adv. Mater. Sci. 38(5): 55–60.
2. Chawla NC, Chawla KK (2006) Metal-matrix composites in ground transportation. JOM 58(11): 67–70.
3. Miracle DB (2005) Metal matrix composites–from science to technological significance. Compos. Sci. Technol. 65(15–16): 2526–40.
4. Alaneme KK, Sanusi KO (2015) Microstructural characteristics, mechanical and wear behaviour of aluminium matrix hybrid composites reinforced with alumina, rice husk ash and graphite. Eng. Sci. Technol. Int. J. 18(3): 416–22.
5. Krasnowski M, Gierlotka S, Kulik T (2015) TiC–Al composites with nanocrystalline matrix produced by consolidation of milled powders. Adv Powder Technol 26(5): 1269–72.
6. Liu, G., Karim, M., Wang, S., Eskin, D., & McKay, B. (2022) Processing of SiC nano-reinforced AlSi9Cu3 composites by stir mixing, ultrasonication and high pressure die casting. J Mater Res Technol 18: 2384–2398.
7. Jia L, Kondoh K, Imai H, Onishi M, Chen B, Li SF (2015) Nano-scale AlN powders and AlN/Al composites by full and partial direct nitridation of aluminum in solid-state. J Alloys Compd 629: 184–7.
8. Mavhungu ST, Akinlabi ET, Onitiri MA, Varachia FM (2017) Aluminum matrix composites for industrial use: advances and trends. Procedia Manuf. 7: 178–82.
9. Dieter GE (1986) Mechanical Metallurgy, third edition, McGraw-Hill (NY), New York.
10. Tjong SC, Mai YW (2008) Processing-structure-property aspects of particulate-and whisker-reinforced titanium matrix composites. Compos. Sci. Technol. 68(3–4): 583–601.

11. Srivastava N, Chaudhari GP (2016) Strengthening in Al alloy nano composites fabricated by ultrasound assisted solidification technique. Mater Sci Eng A 651: 241–7.

12. Habibnejad-Korayem M, Mahmudi R, Poole WJ (2009) Enhanced properties of Mg-based nano-composites reinforced with Al2O3 nano-particles. Mater Sci Eng A 519(1–2): 198–203.

13. Nguyen QB, Gupta M (2008) Enhancing compressive response of AZ31B magnesium alloy using alumina nanoparticulates. Compos. Sci. Technol. 68(10–11): 2185–92.

14. Dai, L. H., Ling, Z., & Bai, Y. L. (2001) Size-dependent inelastic behavior of particle-reinforced metal–matrix composites. Composites Science and Technology, 61(8): 1057–1063.

15. Ferguson JB, Sheykh-Jaberi F, Kim CS, Rohatgi PK, Cho K (2012) On the strength and strain to failure in particle-reinforced magnesium metal-matrix nanocomposites (Mg MMNCs). Mater Sci Eng A 558: 193–204.

16. Cao G, Konishi H, Li X (2008) Mechanical properties and microstructure of SiC-reinforced Mg-(2, 4) Al-1Si nanocomposites fabricated by ultrasonic cavitation based solidification processing. Mater Sci Eng A 486(1–2): 357–62.

17. Suslick KS, Didenko Y, Fang MM, Hyeon T, Kolbeck KJ, McNamara III WB, Mdleleni MM, Wong M (1999) Acoustic cavitation and its chemical consequences. Philosophical Transactions of the Royal Society of London. Series A: Mathematical, Physical and Engineering Sciences 357(1751): 335–53.

18. Eskin, DG (2017) Ultrasonic processing of molten and solidifying aluminium alloys: overview and outlook, Mater. Sci. Technol. 33 (6): 636–645.

19. A. CommitteeStandard Test Methods for Tension Testing of Metallic Materials, 2003.

20. Sen S, Dhindaw BK, Stefanescu DM, Catalina A, Curreri PA (1997) Melt convection effects on the critical velocity of particle engulfment. J Cryst Growth 173(3–4): 574–84.

21. Youssef YM, Dashwood RJ, Lee PD (2005) Effect of clustering on particle pushing and solidification behaviour in TiB2 reinforced aluminium PMMCs. Compos. Part A Appl. Sci. Manuf. 36(6): 747–63.

22. Chen XH, Yan H (2016) Solid–liquid interface dynamics during solidification of Al 7075–Al2O3np based metal matrix composites. Mater Des 94: 148–58.

23. Schultz BF, Ferguson JB, Rohatgi PK (2011) Microstructure and hardness of Al2O3 nanoparticle reinforced Al–Mg composites fabricated by reactive wetting and stir mixing. Mater Sci Eng A 530: 87–97.

24. Yang, H., Qian, Z., Chen, H., Zhao, X., Han, G., Du, W., Liu, X. (2022) A new insight into heterogeneous nucleation mechanism of Al by non-stoichiometric TiCx. Acta Materialia, 233: 117977.

Effect of Mn Content on Quench Sensitivity of 6082 Alloys

Emrah F. Ozdogru, Aleyna Gumussoy, Hilal Colak, and Isık Kaya

Abstract

EN AW 6082 alloys are used in many different areas from construction to automotive industry. EN AW 6082 alloys have one of the highest mechanical properties in the 6XXX group. Studies on the quench rate of this alloy have significant importance to observe its behavior in different conditions industrially. The present work was conducted to investigate the manganese modification effect on microstructure and mechanical properties, depending on the quench rate of this alloy. Low and high Mn contained samples homogenized industrial practical and end quench test method was applied. Then, samples were aged for different durations between industrially chosen temperature levels. Depending on quenching sensitivity, mechanical properties and microstructure differences were determined on the minimum and maximum Mn contained aluminum alloys.

Keywords

EN AW 6082 • Quench sensitivity • Alloy modification • End quench test method

Introduction

EN AW 6082 is an Aluminum–Magnesium–Silicon family (6XXX series) alloy. It has been widely used in many sectors such as automobile, aircraft, marine, and construction, owing to its low cost, high strength to weight ratio, good formability, weldability, excellent corrosion resistance, and higher thermal conductivity [1]. EN AW 6082 alloys have the highest strength of the 6XXX group alloys. The principal alloying elements in EN AW 6082 alloys are Silicon (Si),

Magnesium (Mg), and Manganese (Mn) which play a significant role in the structure and properties of the alloy [2]. Artificial aging is used as the hardening mechanism in these alloys. Depending on precipitation hardening, an increase in hardness is observed in the materials. First, solution heat treatment was applied, then cooled in the Jominy end quench test rig. For artificial aging, which is the second stage, aging is applied to the material at the optimum hour between 150 and 190°. The transformation that occurs during these processes is considered as follows. The precipitation sequence for 6xxx alloys, which is generally accepted in the literature, is: SSSS $\rightarrow$ atomic clusters $\rightarrow$ GP zones $\rightarrow$ β'' $\rightarrow$ β' $\rightarrow$ β (stable), where SSSS is the super saturated solid solution [3].

One potential way to improve aluminum alloys is through the addition of alloying elements. For Al–Mg–Si 6XXX alloys, Mn is an important alloying element to increase the strength and control the grain structure [4]. The addition of Mn in EN AW 6082 alloys could significantly enhance the precipitation during the homogenization and increase the quench sensitivity [5]. Revealing the change in mechanical properties depending on the cooling rate after homogenization helps to better understand the material hardening mechanism.

The Jominy end quench test, well known as a method of measuring hardenability in steels, offers a method for studying samples. Jominy end quench test for aluminum alloys is a simple test that can provide information regarding quench sensitivity, microstructural characterization, and alloy development. End quenching test, not very common for aluminum alloys but provided a large range of cooling rates. Some studies show that this method has been used to determine the quench sensitivity of Al alloys and to improve the parameters [6].

The aim of the present work concerns the relationship between the quench sensitivity and hardness test results of commercial EN AW 6082 alloy by using the Jominy end quench test. This study will be helpful to develop this type of Al alloy. The Jominy end quench test offers a method for

E. F. Ozdogru (✉) · A. Gumussoy · H. Colak · I. Kaya
TRI Metalurji A.S., Istanbul, Turkey
e-mail: Emrah.ozdogru@trimetalurji.com

© The Minerals, Metals & Materials Society 2023
S. Broek (ed.), *Light Metals 2023*, The Minerals, Metals & Materials Series,
https://doi.org/10.1007/978-3-031-22532-1_48

studying many quenching conditions with EN AW 6082 samples. The potential for developing a new understanding of the complex response of nonferrous alloys to processing conditions, especially cooling rate, and artificial aging, have presented. The present work identified the effect of Mn content on the quench sensitivity of the EN AW 6082 alloys. [7].

Materials and Methods

In this study, two EN AW 6082 alloys with different Mn content, melting, and alloying were performed in the electric melting furnace at 750 °C, after that, the melt was subsequently poured into a steel mould to form Jominy bar. The chemical compositions of these EN AW 6082 alloys were measured by arc spark optical emission spectroscopy and given in Tables 1 and 2, respectively.

The end quench specimen was prepared in accordance with ASTM 255. Figure 1 shows the end quench test specimen. 5 mm diameter holes were drilled at 10 and 90 mm for K type thermocouples for the temperature record during the quench. The specimen was placed in the furnace and the solution was heat treated for 3 h at 560 °C. The specimen was then removed from the furnace and placed into the Jominy quench rig. The typical transfer time between the furnace door opening and the start of the quench was approximately 5 s. The specimen remained in the Jominy quench rig for 8 min to allow sufficient time to cool to room temperature. After the specimen has been quenched, parallel flats were cut equally on the specimen and hardness measurements were done at the center of the bars. Brinell hardness test was applied with DURASCAN 50G5 EMCOTEST to 6 points as described in Fig. 1 as well. After the Jominy end quench test, the samples taken from the 1st and 6th regions were grinded, polished, and etched (5%HF) for metallographic investigation. Microstructure investigation was performed by NIKON LV150N optical microscope Then, other samples were heated at 180 °C and held between 2 and 14 h. After aging heat treatment, a hardness test was applied to all samples, and microstructure investigations were made for underaged and aged from the 1st and 6th regions.

Results and Discussion

Cooling Curves

Quenching is an important part of aluminum manufacturing. The cooling rate can play a major role in the overall mechanical properties of the alloy. End quench tests were applied to homogenized bar samples at 560 °C for 3 h. Cooling curves were recorded and taken from the datalogger. Due to the nature of these alloys, important transformations occur at the cooling period from 560 °C to 200 °C during quenching. Based on this information, the average cooling rate was calculated between this temperature range. It can be found that the cooling rates decrease during the quenching process. The temperature of the quenched end is at a lower range in a short time, while it takes a longer time for the alloys away from the quenched end. In Fig. 2a, the cooling rate calculated for low Mn EN AW 6082 alloy for the 1st region (fast cooling area) is 6.53 °C/s and the 6th region (slow cooling area) is 2.59 °C/s. On the other hand, in Fig. 2b, the cooling rate calculated for high Mn EN AW 6082 alloy for the 1st region (fast cooling area) is 6.54 °C/s and the 6th region (slow cooling area) is 2.67 °C/s.

The cooling rate difference in the 1st region (fast cooling area) was calculated as approximately 1%. However, this difference is around 8% when compared to the 6th region (slow cooling area). In the 1st region, the mechanical properties (hardness) are expected to be close to each other, whereas the 6th region shows the opposite.

Hardness Measurement

To further understand the mechanical behavior of these alloys with respect to quench sensitivity, a hardness test was applied. Hardness curves of the end quench test for specimens with low content Mn and high content Mn are given in Fig. 3. The higher the Mn content, the greater the decrease of hardness along the direction of the distance far from the end quenched surface. It is clearly seen from the diagrams that the alloy with high Mn exhibits a more quench sensitive alloy than the low Mn alloy.

Table 1 Low content Mn EN AW 6082

Si (%)	Fe (%)	Mn (%)	Mg (%)
0,80–0,85	0,2	0,45	0,70–0,75

Table 2 High content Mn EN AW 6082	Si (%)	Fe (%)	Mn (%)	Mg (%)
	0,80–0,85	0,2	0,95	0,70–0,75

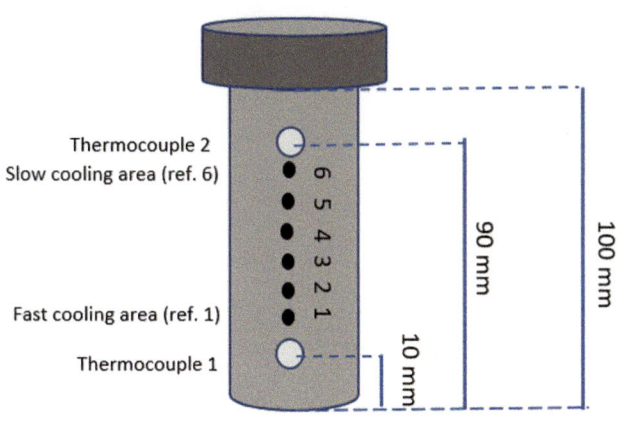

Fig. 1 Jominy end quench test bar schematic representation

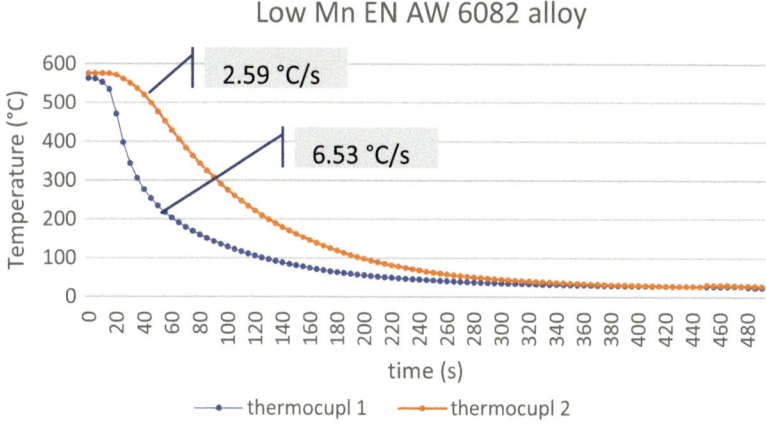

Fig. 3 Hardness test results after Jominy end quench test (before aging test)

Fig. 2 Cooling curves of high and low Mn EN AW 6082 alloys

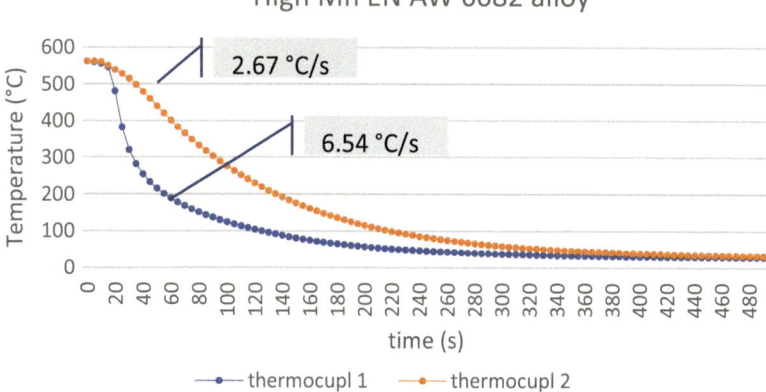

(a) Low Mn EN AW 6082 alloy cooling curve

(b) High Mn EN AW 6082 alloy cooling curve

There is a direct correlation between cooling rate and mechanical properties after T6 treatment. End quenching aging specimens aged between 2 and 14 h at 180 °C. The cooling rate decreases with distance from the quenching face. Hardness measurements were applied in the aging heat treatment to the 6th region whose schematic representation was given in Fig. 1. Hardness values after aging heat treatment of EN AW 6082 alloys with low and high Mn content are given in Fig. 4.

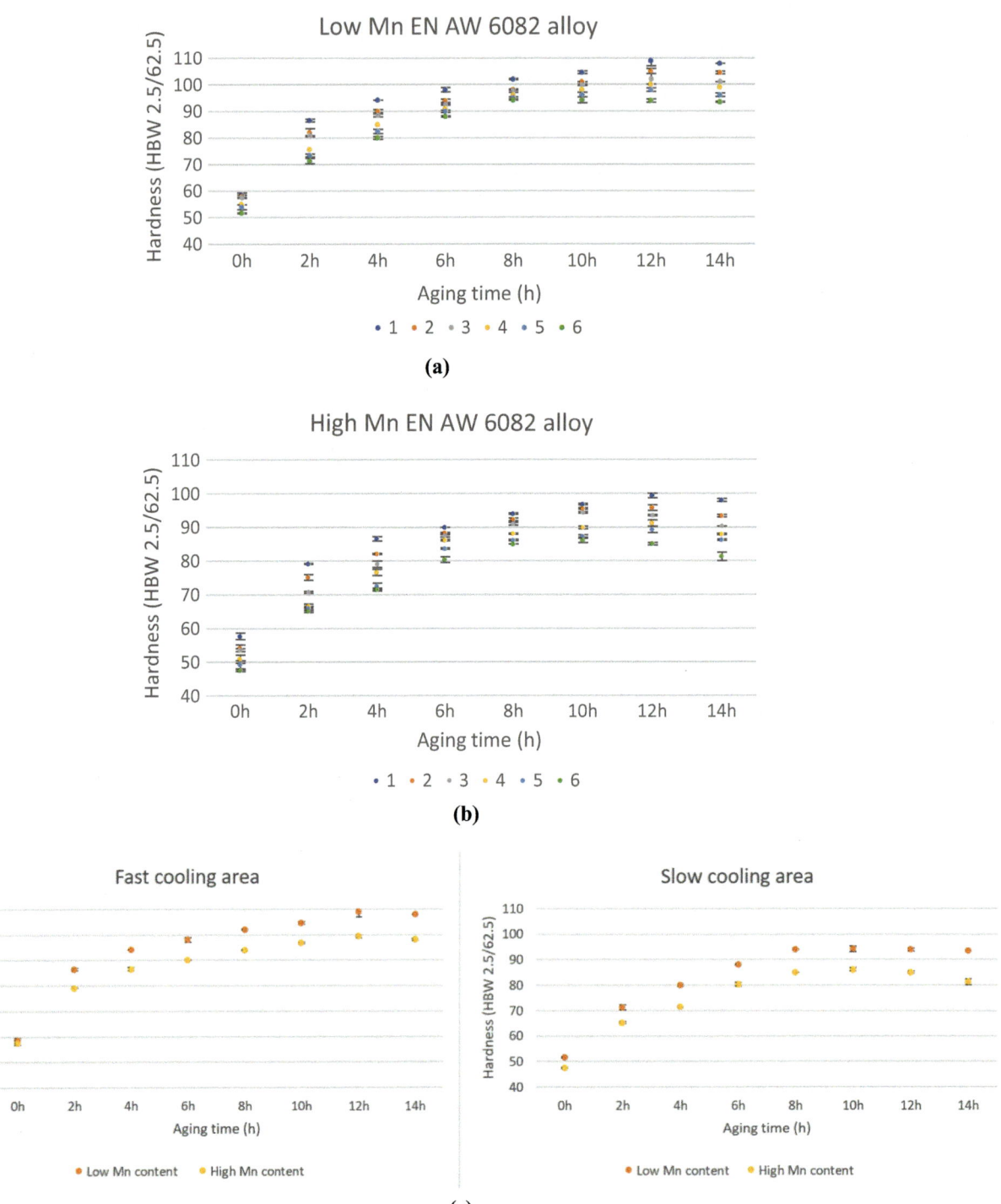

Fig. 4 180 °C 2–14 h artificial aging hardness test results, **a** Low Mn EN AW 6082 alloy artificial aging hardness test, **b** High Mn EN AW 6082 alloy artificial aging hardness test, **c** Hardness comparison of the fast cooling area and slow cooling area

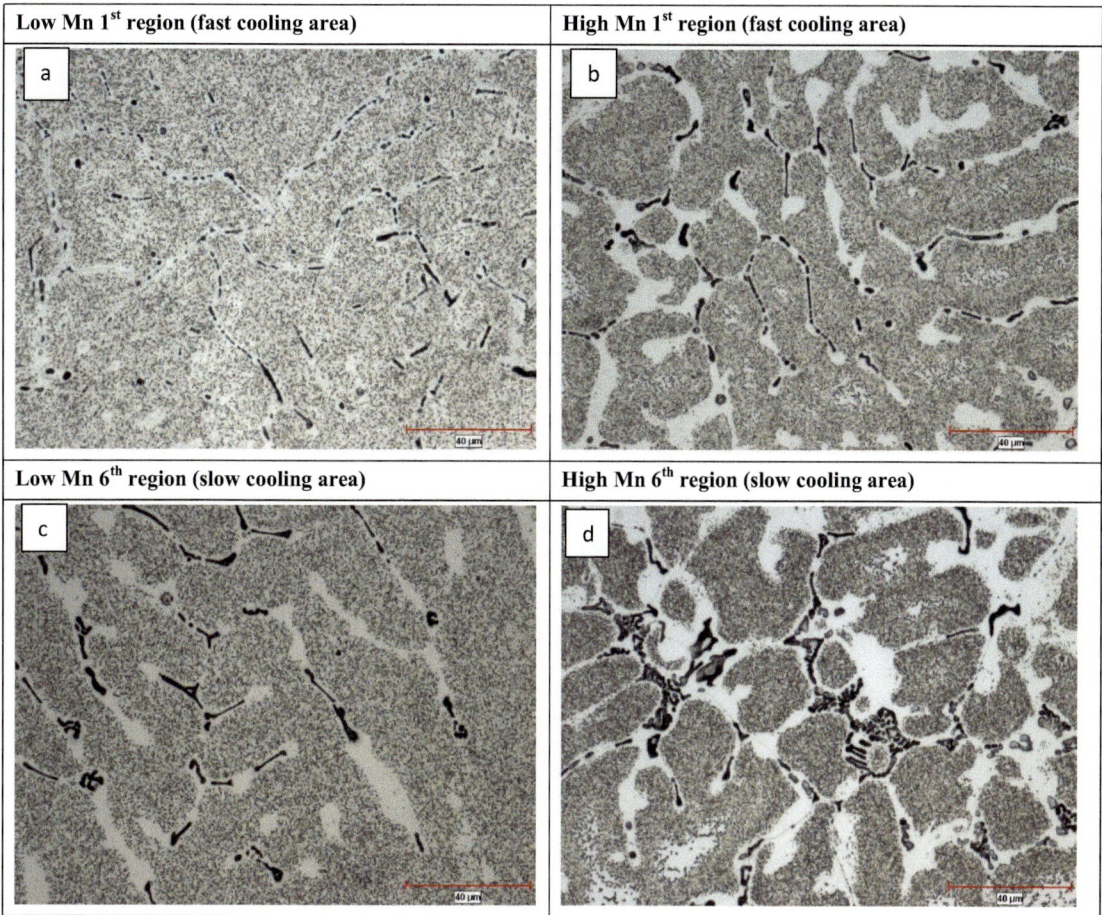

Fig. 5 Microstructures of low and high Mn EN AW 6082 alloys after aging, **a** Low Mn 1st region (fast cooling area), **b** High Mn 1st region (fast cooling area), **c** Low Mn 6th region (slow cooling area), and **d** High Mn 6th region (slow cooling area)

Figure 4a shows that, after aging, hardness measurements were made in low manganese alloy. The hardness was measured at the highest in the 1st region (fast cooling area). At the 12th hour, the alloy reached the highest hardness value.

Figure 4b shows that, after aging, hardness measurements were made in high manganese alloy. According to the test results, the highest hardness was measured at the 12th hour for this alloy after aging. On the other hand, comparing the hardness, increases according to high and low manganese content, the hardness values of the 6082 alloy with low manganese content were higher.

The hardness of the specimens decreases as it moves away from the quenching surface (Fig. 4c). The highest hardness was reached in the aging heat treatment performed in the region closest to the quenching surface. Figure 4c shows that high Mn EN AW 6082 alloy hardness measurement is lower than the low Mn alloy. This is because, at slower quench rates, Mn-based dispersoid phases are able to act as nucleation sites for coarse, nonhardening Mg-Si precipitates. The reduction in the amount of Mg and Si

remaining for precipitation reduces the alloy's response to age hardening treatments; resulting in lower achievable strength.

Transition metals such as Mn play a large role in quench sensitivity as higher levels of these alloy additions can lead to larger dispersoid densities for a given homogenization treatment and are known to be one of the key factors that affect the quench sensitivity of an alloy.

Microstructural Analysis

Additions of small or medium amounts of manganese (Mn) to AlMgSi-alloys are normally used to modify the microstructure. This results in a decrease in the strength that can be obtained with the aging process if the cooling rate after quenching is reduced. This happens because the addition of Mn increases the number of dispersoids. Dispersoids will form during the homogenization process. The Mn will be tied up in the constituent iron (Fe)-rich phases (α-Al(Mn, Fe)Si) during solidification, and generate dispersoids (containing Fe, silicon (Si) and aluminum (Al)) during high-temperature treatments such as homogenization, and some of the Mn may also remain in solid solution. The level of Mn in these different primary and secondary phases depends on the casting and homogenization practices, as well as the thermomechanical processing parameters. Figure 5 shows the microstructure of both alloys at the fast and slow cooling areas after aging practices.

It is shown that the intermetallic transformations observed at the grain boundaries of the low manganese alloy in the 1st region (fast cooling area) after homogenization are better compared to the high manganese alloy. In high manganese containing alloy, the density of the precipitate free zone can cause low mechanical properties in the material.

On the other hand, a precipitate free zone was observed in both low and high manganese containing alloys in the slow cooling region. Slow cooling adversely affects the microstructural properties of the material before aging and, accordingly, its mechanical properties. In both low and high Mn content alloys, slow cooling areas have the minimum hardness values.

Conclusions

- The cooling rate decreases with distance from the quenching face. Fast cooling area has a bigger cooling rate.
- The higher the Mn content, the greater decrease of hardness along the direction of the distance far from the end quenched surface. Increasing the Mn content refers to higher quench sensitivity.

- The addition of Mn increases the number of dispersoids. Increasing the number of dispersoids decrease the mechanical properties (hardness) after T6.
- Precipitation free zones were detected more intensely in the high Mn alloy. On high Mn content alloy, the nucleation sites on the grain boundary are higher than the nucleation sites of low Mn alloy, so more precipitate formation occurs on the grain boundary on High Mn alloy and these precipitates drain the adjacent matrix from the solute, and therefore, more clear PFZ arises.
- Mn is one of the most effective elements for the quench sensitivity. Therefore, the addition of Mn appears to increase the ability of Mg_2Si to precipitate during the quenching operation, which is probably due to the increased number of dispersoids and nucleation sites for the Mg_2Si to precipitate on.

References

1. Kumar N, Goel S, JayaganthanR and Brokmeier H-G 2015 Effect of solution treatment on mechanical and corrosion behaviors of 6082-T6 Al Alloy Metall. Microstruct. Anal. 4 411–22.
2. Kolachiov, B.A., Elagin, V.I., Livanov, V.A. Metallurgy and heat treatment of nonferrous metals and alloys. -Moscow: MISIS, 2001, 413 p.
3. M. Murayama, K. Hono, M. Saga, M. Kikuchi, Mater. Sci. Eng. A 250 (1998) 127–132.
4. Y. Birol, J. Mater. Process. Technol. 173 (2006) 84–91.
5. C. Liu, Q. Du, N. Parson, W. Poole, Scr. Mater. 152 (2018) 59–63.
6. X.M. Zhang, Y.L. Deng, Y. Zhang. Chin Pat, 200710034410.8, 2007.
7. Newkirk J W, Mackenzie D S. The Jominy end quench for light-weight alloy development [J]. Journal of Materials Engineering and Performance, 2000, 9: 408–415.

Characterization of Aluminum Conductors Steel Reinforced in Overhead Transmission Lines

M. Hassanipour, M. Diago, D. Valiquette, F. Guay, and A. Leblond

Abstract

Extensive laboratory tests have been performed on 20 Aluminum Conductor Steel Reinforced (ACSR) conductors that were sampled from Hydro-Québec overhead transmission lines. A series of tensile tests along with metallographic analyses were carried out to reveal the mechanical properties and the remaining thickness of the coated zinc on steel strands. The age, geometry, current, and the type of environment corresponding to each ACSR conductor, were considered to quantify the ageing and degradation behavior of conductors. It is shown that as the conductor ages, the zinc loss on steel strands triggers the decrease in mechanical properties. An aggressive environment induces a higher decrease in the remaining zinc thickness. However, a higher number of aluminum wire layers and diameters can retard this degradation. It was revealed that a low current conducted through the wires might lead to higher degradation.

Keywords

ACSR conductor • Degradation • Overhead transmission lines • Tensile strength • Galvanized steel • Zinc coating

Introduction

The aluminum conductor steel reinforced (ACSR) conductors are the most commonly used conductors on Hydro-Québec TransÉnergie's overhead transmission lines (OHTL) [1]. These conductors are aging over time, thus there is an essential need to prioritize their maintenance and periodic inspections by understanding their degradation mechanisms. These conductors are designed with different layers of aluminum and steel strands. The external aluminum strands conduct the electric current and the steel strands in the core are designed to endure the mechanical loads induced such as wind and ice. However, the interaction of steel and aluminum strands with the environment will lead to the corrosion of the strands over time. This corrosion may be in the internal strands and not visible on the external surface. Thus, it may lead to a decrease in the mechanical properties of conductors leading to unexpected failure [2].

In order to have a slower corrosion rate on steel and aluminum strands, the steel strands in the core are galvanized and protected by zinc [3]. However, zinc gradually corrodes as a function of time, so a total zinc loss will be the initiation period for the higher corrosion rates of unprotected steel and aluminum strands. Corrosion in strands can be in the form of loss of section, local pitting, or corrosive deposits that leads to the failure of conductors [4].

It has been reported that the zinc corrosion in ACSR depends on a variety of factors such as their interaction with the environment and their geometry [5, 6]. The environment in which the conductors are operating is mainly categorized as rural, urban, saline, and industrial [5]. Many studies have shown that conductors can have a lower residual life in saline and industrial environment as compared to those installed in the rural areas [6, 7].

Moreover, the geometry of the conductor has an effect on its degradation with respect to the surrounding environment. It was shown that conductors with a lower number of steel strand layers are more prone to corrosion and they have lower residual life [6]. Some studies have shown that a higher number of aluminum layers leads to a lower level of corrosion in conductors [7, 8]. It was revealed that the internal layer of aluminum that is in contact with steel strands corrodes faster than the outside layers [8]. However,

M. Hassanipour (✉) · M. Diago · D. Valiquette · F. Guay
Institut de Recherche d'Hydro-Québec, Varennes, Québec J3X 1S1, Canada
e-mail: Hassanipour.meysam2@hydroquebec.com

A. Leblond
TransÉnergie, Hydro-Québec, Montreal, Québec H2L 4P5, Canada

© The Minerals, Metals & Materials Society 2023
S. Broek (ed.), *Light Metals 2023*, The Minerals, Metals & Materials Series,
https://doi.org/10.1007/978-3-031-22532-1_49

most of the aforementioned studies were conducted in the laboratory under controlled test conditions that are different from those in the field under different environments.

On the other hand, the effect of high current on conductors has been mainly addressed on the mechanical properties, such as tensile strength [9–11]. It was shown that high current passing through the conductor leads to an increase in the temperature of the conductor leading to recrystallization of aluminum layers. It has been put in evidence that large holding time at high temperature can induce a decrease in the mechanical properties of aluminum strands in conductors [9]. However, the effect of current on the corrosion of conductors has not been studied yet.

It should be noted that the abovementioned studies were performed in the laboratory under controlled and accelerated conditions. Thus, the induced mechanisms of corrosion in the field can be different from those in the laboratory. In this study 20 conductors were selected from the field with different ages at different environments. Afterwards, a wide range of tensile tests on the steel and aluminum strands of those conductors were performed. On the other hand, cross-sectional metallographic analysis was conducted in order to measure the zinc thickness on steel strands. Moreover, the effect of current on the zinc loss of these conductors was investigated in their corresponding operational environment.

Experimental Procedure

A detailed visual examination was carried out and recorded for each conductor followed by a series of tensile tests and metallographic analyses. A series of 20 conductors were tested, 8 samples are located in the aggressive environment with lower age and the rest of the samples are older and in rural environment.

Mechanical Tests

A series of tensile tests were conducted on the aluminum and steel strands according to the ASTM B230-07 and B498-08, respectively [12, 13]. An Instron hydraulic with a 20 KN load cell was used in order to measure the stress–strain corresponding to the strands in conductors. The measured values are normalized as compared to the values for new conductors assigned in the aforementioned ASTM.

Metallographic Analysis

The conductors were cut in the transversal direction in order to polish and reveal the aluminum, steel, and zinc thickness in the ACSR conductor. A systematic procedure was employed in order to polish the conductors and measure the

remained zinc thickness. The maximum and minimum values of zinc thickness on each steel strand were recorded.

Current Analysis

The corresponding current to each overhead line was extracted and quantified with respect to the conductors. The utilization ratio was quantified in order to investigate the effect of current in different conductors. This ratio is the current conducted by the conductor as compared to the maximum current that can be conducted throughout a year of utilization. For the details of calculation, the readers can be referred to [14, 15].

Results and Discussion

The cross-sectional image of the Curlew conductor is shown in Fig. 1a. As it can be seen, the 7 galvanized steel strands are encircled by 3 layers of aluminum strands. The zinc thickness on the steel strands is measured and shown in Fig. 1b.

The mechanical properties and zinc thickness were quantified for each of the conductors. These properties were normalized and compared to new ACSR; thus, values close to 1 show that the ACSR properties are similar to new ones and the values close to 0 reveal a high degradation of the aging conductors.

The rated tensile strength of conductors is a measure of the mechanical strength of steel and aluminum strands in conductors. It can be written as

$$RTS = (n_{St} \times A_{St} \times \sigma_{St}) + (n_{Al} \times A_{Al} \times \sigma_{Al}) \qquad (1)$$

where n and A, are the number of strands and surface area, respectively; the subscripts St and Al refer to steel and aluminum, respectively. The σ_{st} is the tensile stress at 1% of strain in steel and σ_{Al} is the ultimate tensile strength of aluminum.

It can be seen in Fig. 2 that the local zinc loss in samples leads to a decrease in the average mechanical properties of the conductors. It can be seen from Fig. 2a and b that as the zinc thickness decreases gradually towards zero, there is a decrease in the rated tensile strength and tensile strength, respectively. This clearly shows that the loss in mechanical properties initiates with the loss of zinc in the conductors. This shows that a local zinc loss can be related to a gradual decrease in steel and aluminum strength.

The zinc loss can lead to a localized galvanic cell between aluminum and steel strands that can induce local loss of aluminum. Thus, this loss can induce a decrease in the mechanical properties of those strands and an overall decrease in the mechanical properties of the conductor [7].

The cumulative distribution function of maximum and minimum values of zinc measured on steel strands are

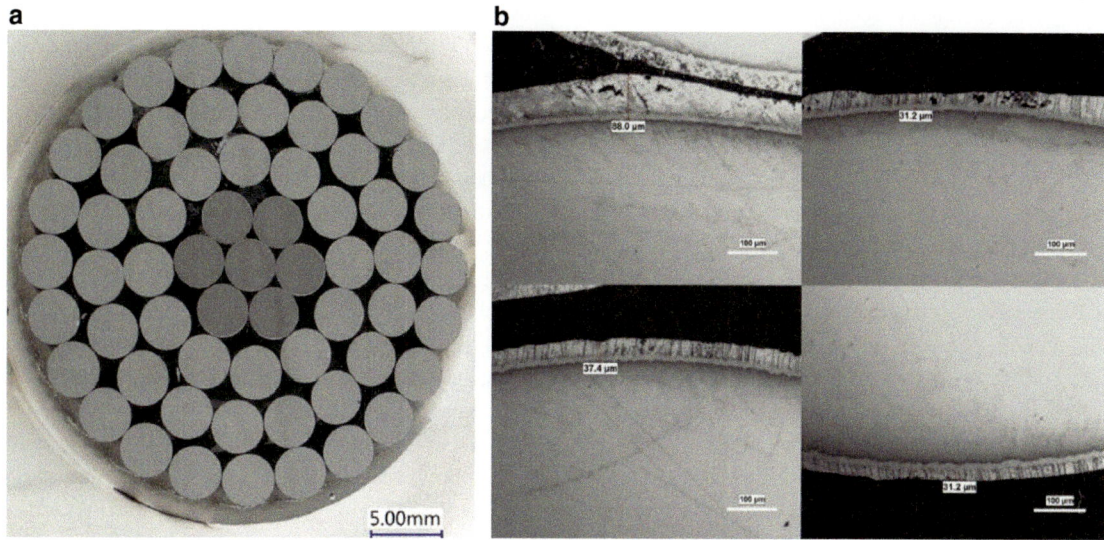

Fig. 1 Cross-sectional images of Curlew conductor. **a** Global image of aluminum and steel strands, **b** local images of galvanized steel strands with the measured zinc thickness

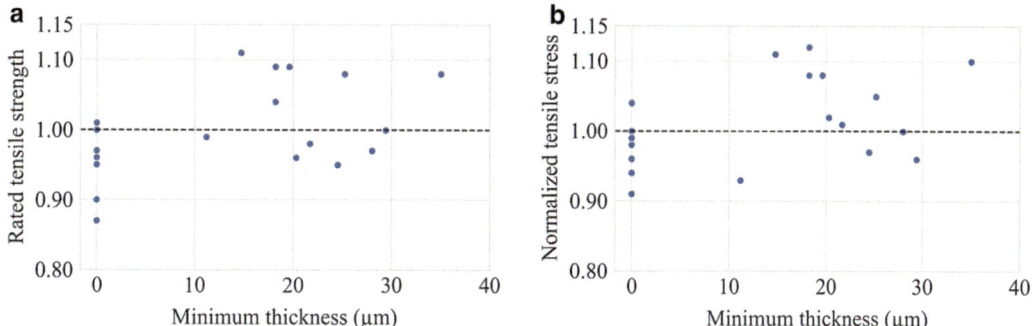

Fig. 2 Mechanical properties as a function of zinc thickness. **a** Average rated tensile strength, and **b** average steel strength

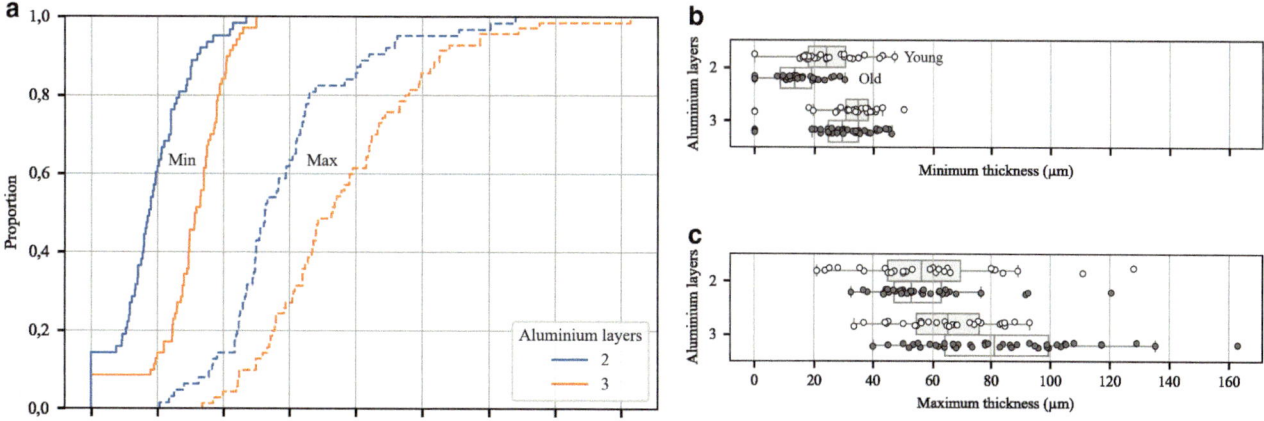

Fig. 3 Minimum and maximum zinc thickness in 2 and 3 layers of aluminum conductors. **a** cumulative distributive function of minimum and maximum zinc thickness, young and older conductor for the **b** minimum and **c** maximum zinc thickness

shown in Fig. 3a. It can be seen that the minimum and maximum values of thickness for a 2 layers aluminum conductor are lower than the 3 layers aluminum conductors. For instance, 20% of zinc values have minimum values below 10 and 24 microns followed by maximum values of 44 and 58 microns for the 2 and 3 layers of conductors.

The conductors with lower age are compared to the older ones in Fig. 3b and c by showing the minimum and maximum zinc thickness values in box plots. The minimum and maximum values are shown by the two extreme horizontal lines and the 25 and 75th percentile is shown by the box ends.

The older conductors have lower values of minimum zinc thickness for 2 and 3 layers aluminum conductors. This difference becomes less significant for the 3 layers aluminum alloys which shows that they are less prone to degradation. It can be said that a higher number of aluminum layers protects the conductor from pollutants that can penetrate into aluminum and steel strands. This will retard the degradation of

zinc loss and the potential to have a galvanic corrosion between aluminum and steel strands as reported [7, 16].

For the maximum values of zinc thickness, it can be seen that older conductors have lower values for 2 layers aluminum. However, this is not the case for 3 layers aluminum conductors that show a large dispersion in the zinc thickness. This may imply that zinc thickness in 3 layers aluminum conductors has higher variation as compared to 2 layers aluminum conductors.

The utilization ratio of each conductor is quantified and then classified into two groups of conductors with high and low utilization ratios. This can aid to investigate the effect of utilization ratio on the zinc loss conductors in their operational environment.

As it can be seen from Fig. 4a, the utilization ratio is quite similar, however, it can be said that at a lower utilization ratio, there is a higher variation in the zinc loss which can be seen from the box plot. As the zinc thickness decreases,

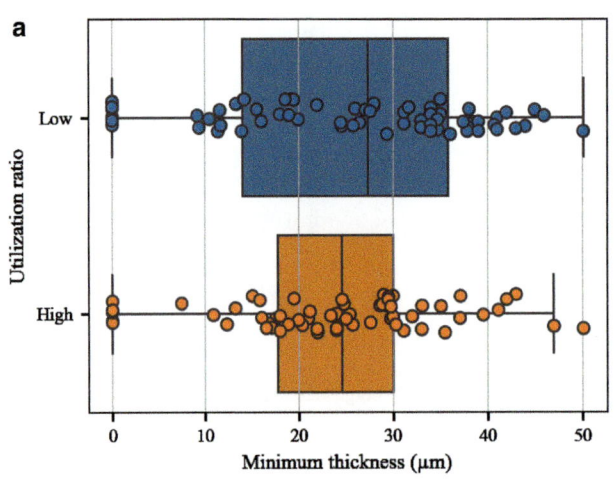

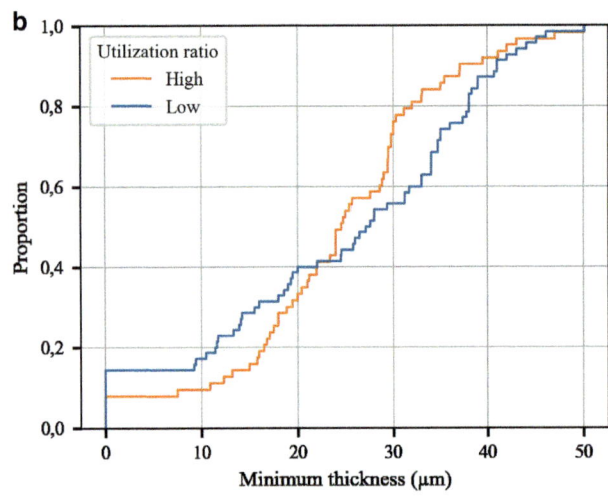

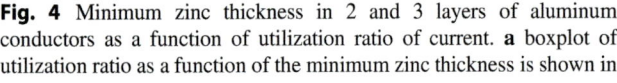

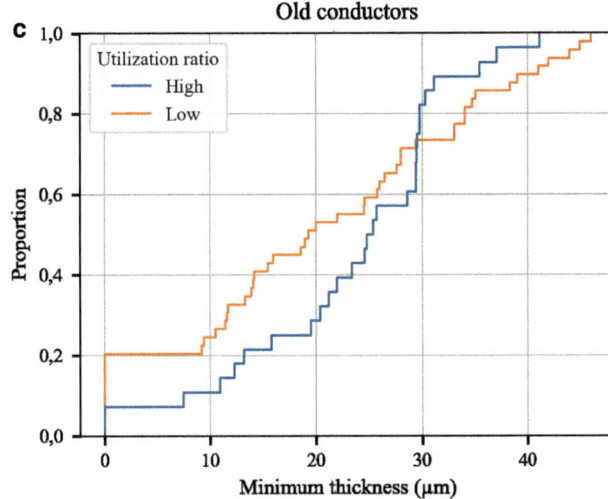

Fig. 4 Minimum zinc thickness in 2 and 3 layers of aluminum conductors as a function of utilization ratio of current. **a** boxplot of utilization ratio as a function of the minimum zinc thickness is shown in **a** as a box plot and **b** as a cumulative distributive function. **c** Zinc thickness as a function of utilization ratio in old conductors

there is a higher proportion of zinc values in the overhead lines with lower utilization ratios (see Fig. 4b).

The conductors with higher and lower utilization ratios can be divided into subgroups of old conductors. These conductors consist of a combination of 2 and 3 aluminum layers within a similar range of age, so a specified comparison can be achieved. Figure 4c shows indeed that there is a higher difference in terms of zinc thickness in the case of old conductors, and it can be seen that lower utilization ratio results in lower zinc thickness on the conductors.

This implies that the utilization ratio may be one of the parameters that can affect the loss of zinc and the degradation of the conductors. Some studies have shown that at lower temperature, there is an increase in the adsorbed water as monolayers can induce corrosion [17, 18]. It may be said that the lower current induces a lower temperature on the surface and inside the conductor. Thus, it can be said that these absorbed monolayers may lead to a decrease in zinc thickness on the steel strands of conductors.

Conclusions

In this study, 20 conductors that are selected from the field at different ages in different environments are tested. A wide range of tensile tests on the steel and aluminum strands are conducted with cross- sectional metallographic analysis in order to reveal the zinc thickness. Moreover, the effect of current in this different conductor is coupled with the zinc loss and the environmental effect. The age, geometry, current, and the type of environment corresponding to each ACSR conductor, were considered to quantify the ageing and degradation behavior of conductors.

It is shown that as the conductor ages, the zinc loss on steel strands triggers the decrease in mechanical properties. The aggressive environment induces a higher decrease in the remaining zinc thickness. However, a higher number of aluminum wire layers and diameters can retard this zinc loss and degradation.

It was revealed that a lower utilization ratio of conductors throughout a year may lead to zinc loss and higher corrosion in conductors. This is associated with a decrease in the conductor's temperature that leads to an increase in the adsorbed water as monolayers inside the conductors causing the degradation of zinc and conductors.

References

1. Pouliot N, Rousseau G, Gagnon D, Valiquette D, Leblond A, Montambault S, et al. An Integrated Asset Management Strategy for Transmission Line Conductors, Based on Robotics, Sensors, and Aging Modelling. In: Liyanage JP, Amadi-Echendu J, Mathew J, editors. Engineering Assets and Public Infrastructures in the Age of Digitalization, Cham: Springer International Publishing; 2020, p. 30–8. https://doi.org/10.1007/978-3-030-48021-9_4.
2. Murray N. Assessment of Phase Conductors. EPRI Report 2009:162.
3. Jackson R, Ferguson J, Lewis G, Gibbon R. Supergrid condition assessment. Electron Power UK 1987;33:641.https://doi.org/10.1049/ep.1987.0384.
4. Forrest JS. Service Experience of the Effect Of Corrosion On Steel-Cored-Aluminium Overhead-Line Conductors. Proceedings of the IEE-Part II: Power Engineering 1954:13.
5. Harvard DG, Bellamy G, Buchan PG, Ewing HA, Horrocks DJ, Krishnasamy SG, et al. Aged ACSR conductors. I. Testing procedures for conductors and line items. IEEE Trans Power Delivery 1992;7:581–7. https://doi.org/10.1109/61.127052.
6. Havard DG, Bissada MK, Fajardo CG, Horrocks DJ, Meale JR, Motlis JY, et al. Aged ACSR conductors. II. Prediction of remaining life. IEEE Trans Power Delivery 1992;7:588–95. https://doi.org/10.1109/61.127053.
7. Fujii K, Miyazaki K. Corrosion Characteristics Based on an Investigation of Sampled OHTL Conductors and a Probabilistic Lifetime Estimation Method. CIGRE 2012:8.
8. Calitz J. Overhead conductor corrosion study n.d.:120.
9. Morgan V. The Loss of Tensile Strength of Hard-Drawn Conductors by Annealing in Servicec. IEEE Trans on Power Apparatus and Syst 1979;PAS-98:700–9. https://doi.org/10.1109/TPAS.1979.319273.
10. Morgan VT. Effect of elevated temperature operation on the tensile strength of overhead conductors. IEEE Trans Power Delivery 1996;11:345–52https://doi.org/10.1109/61.484034.
11. Morgan VT, Zhang B, Findlay D. Effects of temperature and tensile stress on the magnetic properties of a steel core from an ACSR conductor. IEEE Trans Power Delivery 1996;11:1907–13. https://doi.org/10.1109/61.544275.
12. ASTM B498. Specification for Zinc-Coated (Galvanized) Steel Core Wire for Use in Overhead Electrical Conductors. ASTM International; n.d. https://doi.org/10.1520/B0498_B0498M-08.
13. ASTM B230. ASTM B498-Standard Specification for Zinc-Coated (Galvanized) Steel Core Wire for Use in Overhead Electrical Conductors. ASTM International; n.d.
14. Morgan VT. The thermal rating of overhead-line conductors Part I. The steady-state thermal model. Electric Power Systems Research 1982;5:119–39. https://doi.org/10.1016/0378-7796(82)90033-5.
15. Morgan VT. The thermal rating of overhead-line conductors part II. A sensitivity analysis of the parameters in the steady-state thermal model. Electric Power Systems Research 1983;6:287–300. https://doi.org/10.1016/0378-7796(83)90040-8.
16. Fujimoto K, Nishida K, Wakabayashi N, Chiba H, Watabe M, Ozaki T. Development of Estimation and Diagnostic Method of ACSR Inner Corrosion Based on the Corrosion Mechanism Study n.d.:8.
17. Thierry D, Persson D, Le Bozec N. Atmospheric Corrosion of Zinc and Zinc Alloyed Coated Steel. Encyclopedia of Interfacial Chemistry, Elsevier; 2018, p. 55–78. https://doi.org/10.1016/B978-0-12-409547-2.13431-6.
18. Esmaily M, Mortazavi N, Svensson JE, Johansson LG. Evidence for an Unusual Temperature Dependence of the Atmospheric Corrosion of Zinc. J Electrochem Soc 2016;163:C864–72.https://doi.org/10.1149/2.0751614jes.

Mechanical and Electrical Properties of Permanent Steel Mold Cast Eutectic Al-1.8Fe Alloy

S. Liu, A. Hu, A. Dhaif, W. Shen, and H. Hu

Abstract

A eutectic Al alloy containing 1.8 wt.% Fe (Al-1.8%Fe) for electric vehicles was prepared by permanent steel mold casting (PSMC) along with commercial-purity (CP) Al (99.7%). The mechanical properties including ultimate tensile strength (UTS), yield strength (YS), and elongation (e_f) were evaluated by tensile testing. The electrical conductivity was measured by the phase-sensitive eddy current method. The addition of 1.8 wt.% Fe increased both the UTS and YS of the cast CP Al significantly to 86.5 and 28.3 MPa from 34.5 to 12.3 MPa. However, the e_f and electrical conductivity of the cast alloy decreased to 19.8% and 48.4%IACS from 33.8% and 58.5%IACS. The large area fraction of the eutectic Al–Fe phases in the PSMC Al–1.8%Fe alloy should be responsible for the difference in mechanical and electrical properties between the PSMC Al-1.8%Fe alloy and the PSMC CP Al.

Keywords

Al–Fe alloy • Mechanical properties • Electrical conductivity • Electric vehicle • Permanent casting

S. Liu · A. Hu · A. Dhaif · W. Shen · H. Hu (✉)
Department of Mechanical, Automotive and Materials Engineering, University of Windsor, Windsor, ON N9B3P4, Canada
e-mail: huh@uwindsor.ca

S. Liu
e-mail: liu1114z@uwindsor.ca

A. Hu
e-mail: hu14l@uwindsor.ca

A. Dhaif
e-mail: dhaifa@uwindsor.ca

W. Shen
e-mail: shen12a@uwindsor.ca

Introduction

Due to environmental concerns and government regulations, battery-powered electric vehicles (BEVs) are picking up steam on the market. To keep growing their market, it is essential for the automotive industry to develop inexpensive, lightweight, highly efficient induction motors, since the average weight of the BEVs is significantly greater than that of the gasoline-powered vehicles (GPVs). As a key component in the induction motor, squirrel cage rotor as one of the popularity types in the industry plays an important role in the field of electromechanical energy conversion [1–3]. Currently, pure aluminum has been used to replace copper as the rotor bar material in induction motors by casting due to its relatively high electrical conductivity, lightweight, and low price. However, the mechanical properties, in particular, the ultimate tensile strength (UTS) and yield strength (YS), of the pure aluminum are very low. To satisfy the engineering performance of Al rotor bars, a large cooling system has to be installed in the induction motor, which increases the size and weight of the electric motor considerably, and consequently the weight of the BEV [1]. To lower battery energy consumption for the same range, the BEV weight must be reduced [3]. The implementation of high-strength Al alloys for the rotor bar can eliminate the large cooling system in the motor. But, the introduction of traditional alloying elements such as Si, Mg, and Cu adversely affects the electrical conductivity of these alloy alloys owing to their solubility in aluminum [4–11]. Fe appeared as an attractive alloying element, since its solubility in Al was as low as 0.0052 wt.% at room temperature [12]. Wang et al. [13] found that the Fe content varying from 0.75 to 2 wt.% affected the tensile properties of rheo-extruded Al–Fe alloys due to the presence of nanosized Al_3Fe phase. Cubero et al. [14] studied the electrical conductivities of Al–2.0 wt.% Fe alloys processed by high-pressure torsion (HPT) in wrought wire form and treated by subsequent aging. The resulting yield stress was up to 600 MPa. The relatively poor

© The Minerals, Metals & Materials Society 2023
S. Broek (ed.), *Light Metals 2023*, The Minerals, Metals & Materials Series,
https://doi.org/10.1007/978-3-031-22532-1_50

electrical conductivity (40% of IACS) after high-pressure torsion (HPT) processing also improved above 50% of IACS after aging treatment due to the recovery of lattice defects and the precipitation of solute Fe dissolved in the matrix. However, simultaneous studies on both of mechanical properties and electrical conductivities of castable Al–Fe alloys are very limited.

In the present study, a eutectic Al-1.8Fe alloy and CP Al (99.7%) were cast in a permanent steel mold (PSMC) to produce rectangular casting plates. The microstructure of the Al-1.8Fe alloy and CP Al was analyzed. The mechanical properties and the electrical conductivities of the prepared castings were evaluated.

Experimental Procedure

Materials and Casting

The eutectic Al-1.8Fe alloy with the chemical composition in Table 1 was selected for investigation. Commercial-purity (CP) Al (99.7%) were also used for the purpose of comparison, of which composition is given in Table 1. In each casting run, about 1 kg of the alloy melt was prepared in an electric resistance furnace using a graphite crucible. The melt was held at 750 °C ± 10 °C for about 20 min, stirred for 10 min to homogenize its chemical composition, and then poured into a permanent steel mold (PSM) to produce a rectangular casting plate with the dimensions of 150 mm × 125 mm × 25 mm.

Porosity Measurement

Porosity was quantitatively determined by density measurements. Based on the weight measurement of both the permanent steel mold cast (PSMC) Al-1.8Fe and PSMC CP Al specimens in air and water, the actual density (D_a) of each specimen was determined using Archimedes principle based on ASTM Standard D3800 [15]

$$D_a = \frac{D_w w_a}{W_a - W_w} \tag{1}$$

where W_a and W_w are the weight of the specimen in the air and in the water, respectively, and D_w the density of water.

The porosity of each specimen was calculated by the following equation. The porosity of each specimen was calculated by the density values through the following equation (ASTM C948) [16, 17]

$$\%Porosity = \left[\frac{D_t - D_a}{D_t}\right] \times 100\% \tag{2}$$

where Dt is the theoretical densities of the pure Al, with the density of 2.7 g/cm^3 [16], and the Al-1.8Fe alloy calculated based on the weight percentages of Al (2.7 g/cm^3) and Fe (7.8 g/cm^3), with the density of 2.8 g/cm^3.

Tensile Testing

The mechanical properties of both the PSMC Al-1.8Fe and PSMC CP Al specimens were evaluated by the tensile testing, which was carried out at ambient temperature on a MTSTM Criterion (Model 43) Tensile Test Machine (Eden Prairie, MN, USA) equipped with a data acquisition system. According to ASTM B557 [18], subsize flat tensile specimens (0.025 m in gage length, 0.006 m in width, and 0.010 m in as-cast thickness) were machined from the SC and C-HPDC coupons. The strain rate during tensile testing was 0.5 mm/min with a sampling rate of 10 Hz. The tensile properties, including 0.2% yield strength (YS), ultimate tensile strength (UTS), elongation to failure (e_f), and elastic modulus (E) were obtained based on the average of three tests.

Electrical Conductivity

The electrical conductivity is an important electrical property of both the PSMC Al-1.8Fe and PSMC CP Al specimens. The handhold device SIGMASCOPE with FS40 probe was employed the electrical conductivity measurements of the squeeze cast alloys based on the phase sensitive eddy current method. This type of signal evaluation enables non-contact measurement. It also minimizes the influence of surface roughness. The measuring range of the device is 0.5–108% IACS (% International Annealed Copper Standard), and the accuracy at room temperature is ±0.5% of the measured reading. The minimum measurable radius of the specimen

Table 1 Chemical compositions of Al-1.8Fe alloy and commercial-purity (CP) Al

Materials	Chemical composition (wt.%)							
	Fe	Si	Mg	Cu	Ni	C	Al	Others
Al-1.8%	1.82	0.029	<0.01	<0.01	<0.01	<0.01	Remain	None
CP Al	0.09	0.05	<0.01	<0.01	<0.01	<0.01	99.70%	0.13

was 7 mm. The electrical conductivity data were obtained based on the average of three tests.

Results and Discussion

Tensile Properties

The representative engineering stress–strain curves from tensile testing of the PSMC Al-1.8%Fe alloy and PSMC CP Al (99.7%) with a cross-sectional thickness of 20 mm are shown in Fig. 1. Table 2 lists the tensile properties of the PSMC Al-1.8%Fe alloy and PSMC CP Al (99.7%). As shown in Fig. 1, the slope of the linear portion of the engineering curve for the PSMC Al-1.8%Fe alloy had a large increase tendency than that of the PSMC CP Al (99.7%). The UTS of the PSMC Al-1.8%Fe alloy and PSMC CP Al (99.7%) were 86.48 and 34.52 MPa, respectively, which signifies an improvement of 151% over that of the PSMC CP Al (99.7%) for the as-cast conditions. The yield strength of the PSMC Al-1.8%Fe alloy was 28.33 MPa on average, which was 130% higher than that of the PSMC CP Al (99.7%) (12.32 MPa). However, the elongation of the PSMC Al-1.8%Fe alloy was 19.8%, whereas the elongation of the PSMC CP Al (99.7%) was 33.76%, which had a decreasing of 41% over that of the PSMC CP Al (99.7%).

Resilience

The ability of a material to absorb energy is referred to as resilience when it is deformed elastically, and releases that energy upon unloading. The resilience is usually measured by the modulus of resilience which is defined as the maximum strain energy absorbed per unit volume without creating a permanent distortion. It can be calculated by integrating the stress–strain curve from zero to the elastic limit. In uniaxial tension, the strain energy per unit volume can be determined by the following equation [19]:

$$U_r = \frac{(YS)^2}{2E} \tag{3}$$

where U_r is the modulus of resilience, YS is the yield strength, and E is Young's modulus. The calculated modulus of resilience for PSMC Al-1.8%Fe alloy and PSMC CP Al (99.7%) are given in Table 3. In comparison between the PSMC Al-1.8%Fe alloy and PSMC CP Al (99.7%), the modulus of resilience in the PSMC Al-1.8%Fe alloy was 6.37 kJ/m^3 higher than that of the PSMC CP Al (99.7%) (1.18 kJ/m^3). This implied that the PSMC Al-1.8%Fe alloy was much more capable of resisting energy loads in engineering application during service, in which no permanent deformation and distortion are allowed.

Toughness

The tensile toughness of a ductile alloy is its ability to absorb energy during static loading condition, i.e., static deformation with a low strain rate. The ability to bear applied stresses higher than the yield strength without fracturing is usually required for various engineering applications. The toughness for ductile alloys can be considered as the total area under the stress–strain curve for the amount of the total energy per unit volume. To evaluate the deformation behavior, the energy expended in deforming a ductile alloy per unit volume given by the area under the stress–strain curve can be approximated by [20]

$$U_t = U_{el} + U_{pl} = \frac{(YS + UTS) \times e_f}{2} \tag{4}$$

where U_t is the total energy per unit volume required to reach the point of fracture, U_{pl} is the energy per unit volume for elastic deformation, U_{el} is the energy per unit volume for plastic deformation, and e_f is the elongation at fracture. Table 3 lists the calculated U_t for the PSMC Al-1.8%Fe alloy and PSMC CP Al (99.7%). The PSMC Al-1.8%Fe alloy had a U_t value of 11.37 MJ/m^3 higher than that (7.91 MJ/m^3) of the PSMC CP Al (99.7%). This indicated that the PSMC Al-1.8%Fe alloy was tougher than the PSMC CP Al (99.7%) because the PSMC Al-1.8%Fe alloy had the much higher ultimate tensile strength and yield strength, despite the PSMC CP Al (99.7%) having a high elongation. As a result, the total area under the engineering stress and strain curve was greater for the PSMC Al-1.8%Fe alloy.

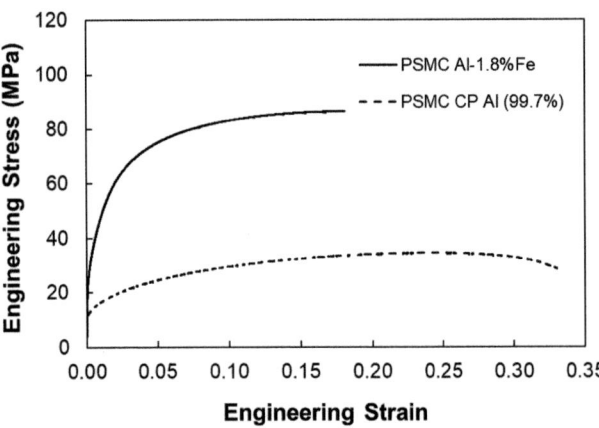

Fig. 1 Representative engineering stress versus strain curves of the PSMC Al-1.8%Fe alloy and PSMC CP Al (99.7%)

Table 2 Tensile properties of the PSMC Al-1.8%Fe alloy and PSMC CP Al (99.7%) at room temperature

Material	UTS (MPa)	YS (MPa)	e_f (%)	Modulus (GPa)
PSMC Al-1.8%Fe alloy	86.48	28.3	19.8%	63.0
PSMC CP Al	34.52	12.3	33.8%	64.4

Table 3 Tensile toughnesses and resiliences of PSMC Al-1.8%Fe alloy and PSMC CP Al (99.7%) at room temperature

Material	Toughness (MJ/m^3)	Resilience (kJ/m^3)
PSMC Al-1.8% Fe alloy	11.37	6.37
PSMC CP Al	7.91	1.18

Electrical Conductivities

The electrical conductivity of both the PSMC Al-1.8% Fe alloy and PSMC CP Al are displayed in Fig. 2. The electrical conductivity of the PSMC Al-1.8% Fe alloy was 48.4% IACS, while it was 58.5%IACS for the PSMC CP Al (99.7%). There was a decrease of 17% in the electrical conductivity of the PSMC Al-1.8% Fe alloy over the PSMC CP Al (99.7%). Therefore, the addition of 1.8 wt.% Fe affected the electrical conductivity of the PSMC CP Al, but insignificant. The microstructure analyses indicated that the addition of 1.8 wt.% Fe was primarily consumed in the formation of the micron and nano-eutectic Al–Fe phases. As a result, a limited amount of Fe was dissolved in the primary Al, which made the Al electron movement almost unaffected.

The preliminary microstructure analyses by optical and scanning microscopies revealed that, compared to that of the pure Al, the massive presence of Al–Fe eutectic phases in the Al-1.8%Fe alloy should be responsible for the resultant mechanical and electrical properties, as also indicated by Wang et al. [13]. The detailed results of microstructure analyses will be reported in future publication.

Summary

The eutectic Al-1.8%Fe alloy and commercial-purity (CP) Al were cast in a permanent steel mold. The results of tensile testing showed that the UTS and YS, toughness and resilience of the PSMC Al-1.8%Fe alloy were 86.5 and 28.3 MPa, 12.98 MJ/m^3, and 6.37 kJ/m^3, respectively, which were higher than those (35.4 and 12.3 MPa, 9.41 MJ/m^3, and 1.18 kJ/m^3) of the PSMC CP Al. But, the elongation of the PSMC Al-1.8%Fe alloy was only 19.8%, which was lower than that (33.8%) of the PSMC CP Al. The difference in mechanical properties between the PSMC Al-1.8%Fe alloy and the PSMC CP Al should be attributed to the fact that the large area fraction (9.5%) of the micron and nano-eutectic Al–Fe was present in the PSMC Al-1.8% Fe alloy, compared to only 0.5% of the secondary phases in the PSMC CP Al. The addition of 1.8 wt.% Fe affected the electrical conductivity of the PSMC CP Al, but insignificant.

References

1. Joachim Doerr, Nikolai Ardey, Günther Mendl, Gerhard Fröhlich, Roman Straßer, Thomas Laudenbach, The new full electric drivetrain of the Audi e-tron, In: Liebl, J. (eds) Der Antrieb von morgen 2019. Proceedings. Springer Vieweg, Wiesbaden. https://doi.org/10.1007/978-3-658-26056-9_2.
2. Ishikawa, H.; Takashima, Y.; Okada, Y. Squirrel-Cage motor Rotor and Squirrel-Cage Motor. U.S. Patent No. 9,935,533, 4 March 2018.
3. Yuxian Li, Anita Hu, Yintian Fu, Sufeng Liu, Wutian Shen, Henry Hu and Xueyuan Nie, Al Alloys and Casting Processes for Induction Motor Applications in Battery-Powered Electric Vehicles: A Review, Metals. 2022, 12, 216.https://doi.org/10.3390/met12020216.
4. Yuxian Li, Yintian Fu, Anita Hu, Xueyuan Nie and Henry Hu, Effect of Sr and Ni Addition on Microstructure, Tensile Behavior and Electrical Conductivity of Squeeze Cast Al-6Si-3Cu Al Alloy, Key Engineering Materials. 2022, 921, 3–14.
5. Sivanesh, P.; Charlie, K.; Robert, S.J.; Ethan, F.; Paul, E. Aluminum Alloys for Die Casting. U.S. Patent No. WO 2020/028730 A1, 6
6. February 2020.

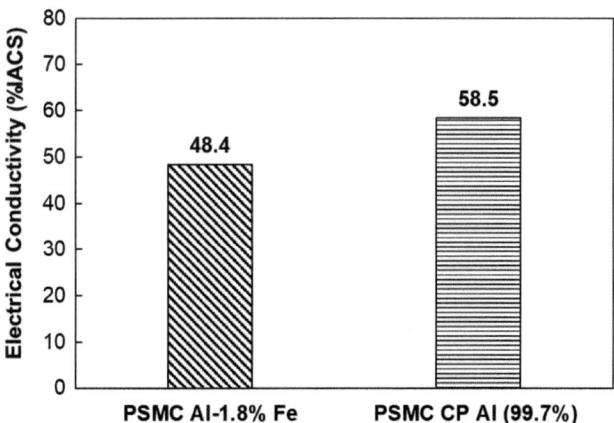

Fig. 2 Electrical conductivities of PSMC Al-1.8%Fe alloy and PSMC CP Al (99.7%)

7. Chen, X.H.; Lu, L.; Lu, K. Electrical resistivity of ultra-fine grained copper with nanoscale growth twins. J. Appl. Phys. **2007**, 102, 083708.

8. Aksöz, S.; Ocak, Y.; Maraşlı, N.; Çadirli, E.; Kaya, H.A.S.A.N.; Böyük, U. Dependency of the thermal and electrical conductivity on the temperature and composition of Cu in the Al based Al–Cu alloys. Exp. Therm. Fluid Sci. 2010, 34, 1507–1516.

9. Plevachuk, Y.; Sklyarchuk, V.; Yakymovych, A.; Eckert, S.; Willers, B.; Eigenfeld, K. Density, viscosity, and electrical conductivity of hypoeutectic Al-Cu liquid alloys. Metall. Mater. Trans. A **2008**, 39, 3040–3045.

10. Kaya, H.A.S.A.N. Dependence of electrical resistivity on temperature and composition of Al–Cu alloys. Mater. Res. Innov. **2012**, 16, 224–229.

11. M. Y. Murashkin, I., Sabirov, X., Sauvage, & R. Z. Valiev, Nanostructured Al and Cu alloys with superior strength and electrical conductivity, J Mater Sci. 2016, 51(1), 33–49.

12. S. Liu, A. Hu, H. Hu, X. Nie, G. Byczynski, and N. C. Kar, Potential Al-Fe Cast Alloys for Motor Applications in Electric Vehicles: An Overview, Key Engineering Materials. 2022, in press.

13. X. Wang, R.G. Guana, R.D.K. Misra, Y. Wang, H.C. Li, Y.Q. Shang, The mechanistic contribution of nanosized Al$_3$Fe phase on the mechanical properties. Mater. Sci. Eng A. 2018, 724, 452–460.

14. J.M. Cubero-Sesin, H. In, M. Arita, High-pressure torsion for fabrication of high-strength and high-electrical conductivity Al micro-wires. J Mater Sci. 2014, 49, 6550–6557.

15. Standard Test Method for Density of High-Modulus Fibers, D3800-99, *ASTM Standards,* ASTM, Vol 15.03, 2002, 186–187.

16. Standard Test Method for Dry and Wet Bulk Density, Water Absorption, and Apparent Porosity of Thin Sections of Glass-Fiber Reinforced Concrete," C948-81, *ASTM Standards,* ASTM, Vol 04.05, 2002, 588–589.

17. J.R. Davis, *Aluminum and Aluminum Alloys-ASM Specialty Handbook,* ASM International, 2002, Materials Park, OH, USA, 645.

18. Standard Test Methods for Tension Testing Wrought and Cast Aluminum-and Magnesium-alloy Products," B557M, *ASTM Standards,* ASTM, Vol 02 02, 2002, 424–439.

19. F.C. Campbell, Elements of metallurgy and engineering alloys. ASM International, Ohio, 2008.

20. B. Larson, Toughness, NDT Education Resource Center, Iowa, 2001.

Effects of the Friction Stir Welding Sliding and Sticking Mechanisms on the Microhardness, Texture, and Element Concentration

Nicholas Sabry, Joshua Stroh, and Dimitry Sediako

Abstract

Friction stir welding has gained importance as an effective way to join dissimilar materials due to its solid-state nature. However, the effects of plastic deformation with increased tool rotation, traverse speed, and plunging force on the mechanical properties of the weld material are poorly understood. In the current study, electron backscatter diffraction, microhardness, and inductively coupled plasma emission analyses were used to characterize two distinct groups of welds conducted in a lap weld configuration joining 6061 and A365 aluminum alloys. The grain structure reveals clear distinctions between the expected dynamically recrystallized zone, the thermomechanical affected zone, and the heat-affected zone with varying microhardness between weld groups. The first weld group was conducted at high rotational speed, traverse speed, and plunge force compared to the second weld group. Furthermore, this study finds that the distribution and generation of precipitates is not the main factor leading to increased surface hardness between the weld groups. Instead, the additional texture resulting from increased plastic deformation as the operation's rotational speed, traverse speed, and plunging force increase is found to cause the mechanical property variation between the weld operations.

Keywords

Friction stir welding • Solid state welding • Aluminum alloy • Microhardness • Texture

Introduction

Owing to their low density, the high specific properties of Al alloys make them appropriate for numerous aerospace and automotive industry applications. For example, the wrought Al–Mg–Si alloy is strong and lightweight, has high corrosion resistance, high thermal and electrical conductivity, and hot/warm formability. In contrast, the fluidity of cast Al-Si alloys is essential to producing intricate components utilizing the high-pressure die casting technique (HPDC) [1]. However, joining dissimilar material components (i.e., wrought Al–Mg–Si and cast Al–Si) is costly and time-consuming for traditional fusion welding, therefore, impractical for mass production [2]. This problem in fusion welding dissimilar material components is principally due to the differences in the joined materials' mechanical, physical, chemical, and metallurgical properties [3]. Nevertheless, to meet the ever-growing demand for high strength-to-weight ratios, joining cast aluminum components with wrought materials is becoming an efficient strategy in the automotive industry, allowing for lighter and more robust designs. Therefore, there is a growing necessity for proper manufacturing operations that join dissimilar material components. Friction stir welding (FSW) is efficient at joining materials with significant differences in chemical composition that lead to substantial differences in thermal conductivity, heat capacity, and coefficient of thermal expansion/contraction [4–6]. This welding is possible as FSW is a solid-state joining method that utilizes a pin-shoulder tool to join two materials below their melting temperatures. FSW joins materials via heat generation by the rotation and plunging action of a tool into the workpiece

N. Sabry (✉) · J. Stroh · D. Sediako
High-Performance Powertrain Materials Laboratory, University of British Columbia, 1137 Alumni Ave, Kelowna, BC V1V 1V7, Canada
e-mail: nsabry@mail.ubc.ca

J. Stroh
e-mail: joshua.stroh@ubc.ca

D. Sediako
e-mail: Dimitry.Sediako@ubc.ca

© The Minerals, Metals & Materials Society 2023
S. Broek (ed.), *Light Metals 2023*, The Minerals, Metals & Materials Series,
https://doi.org/10.1007/978-3-031-22532-1_51

material. After which the tool is advanced down the material creating a weld joint, thereby mechanically mixing and forging plasticized material together [7]. The solid-state nature of FSW avoids bulk melting of the workpiece material, allowing dissimilar materials with different melting temperatures to join, though with the added result of creating several unique welding zones. The near-pin welding results in a dynamically recrystallized zone (DXZ) comprising reduced grains sizes that are more refined compared to the thermo-mechanically affected zone (TMAZ), heat-affected zone (HAZ), and base-metal (BM). Slightly afar the pin and shoulder is the TMAZ region. Owing to the fact that aluminum can withstand significant plastic strain without recrystallization, the welding action creates a distinct boundary of elongated grains between the DXZ and the HAZ region called the TMAZ [8]. Not only does the friction stir welding process significantly change the microstructure through heat and plastic deformation, but FSW also changes the type and density of the precipitates in the weld regions. For example, when joining A365 and 6061, the heat and plastic deformation soften the material causing the disappearance of β'' precipitates and the evolution of β' and β-Mg_2Si. Though aluminum 6061 contains many strengthening precipitates (i.e., GP-I zones, GP-II zones or β'', and β'), the high density of fine needle-shaped β'' precipitate is the primary strengthening mechanism for AA6061-T6. Therefore, the type, size, and distribution of precipitates are particularly important to the mechanical properties of the friction stir weld joint. Because high-strength Al alloys consist of high solute levels and several phases are dependent on thermo-mechanical history, this strengthening mechanism is susceptible to welding conditions, as these conditions dictate the local temperature and the heating/cooling rates reached during welding. Despite that, many factors that control the resulting microstructure and properties in friction-welded joints are inadequately understood. A study by Dong and Sun [9] discussed the post-weld natural aging of FSW 6005A-T6 Al alloy. They found the loss in hardness for the DXZ and TMAZ an hour following welding occurred due to the loss of the original β'' precipitates from high temperatures and non-optimal cooling rates. However, further post-welding hardness testing revealed a hardness recovery mechanism; attributed to the formation of independent Mg and Si clusters (at day one) and GP zone precipitation (at week four of natural aging). Another study by Li et al. [10] described the butt configured joint of AA6061-T6 sheets utilizing stationary shoulder FSW and found that small, dotted GP-I zones mainly represent the DXZ region. However, seldom discussed in the literature are the effects or contribution of plastic deformation with different welding parameters on the resulting mechanical property distribution. Furthermore, since the torque required to rotate the tool inside the material depends on a combination of the plunging

force, coefficient of friction (sliding friction), and the flow strength of the material neighbouring the tool (sticking friction), these forces and thermal input are responsible for deformation in the workpiece material [11]. These inputs will subsequently cause the generation of the residual stresses in the weldment and determine the resulting mechanical properties of the weld. Therefore, this study applies industry-focused FSW to create complex weld profiles in proximity to each other and evaluates their mechanical properties and microstructural characteristics. This analysis is accomplished utilizing Vickers indent testing to identify hardness properties across each weld. Electron backscatter diffraction (EBSD) was utilized to determine the plastic deformation and texture differences across each weld. Finally, inductively coupled plasma atomic emission spectroscopy (ICP-AES) was utilized to account for accurate compositional differences across each weld.

Experimental Details

Materials and Weld Process

This study analyzes three weld joints from an electric vehicle battery tray A365-casting where several 6061-T6 plates are friction stir welded in a lap weld configuration to the casting (see Fig. 1a).

The main structure of the battery tray, along with the coolant channels, is manufactured utilizing high-pressure-die-cast (HPDC) A365 aluminum. The HPDC process is limited in creating internal geometries, therefore, requiring FSW to seal all exposed cooling channels with 6061-plates (Fig. 1a). Welding is divided into two groups, the 56 mm welds and then the 0- and 112-mm welds (see Fig. 1b), with different traverse speeds, rotational speeds, and plunging forces. The transverse and rotational speeds, as well as the plunging force, were almost doubled for the 56-mm pass, whereas the dwell time was much extended (6X) for 0-mm and 112-mm joints.

EBSD and ICP-AES Spectrometer Sample Preparation

EBSD was conducted on a metallurgical sample extracted from the top surface of the 0-, 56-, and 112-mm weld strictly from the 6061 material. First, the sample was mounted and polished following the principles outlined in the ASTM E3-11 standard. Then, to ensure reasonable grain detection rates for EBSD, the sample surface was prepared via a vibrating polisher, with colloidal silica blue (10 pHs) as the suspension, for 4–5 h. The sample preparation for inductively coupled plasma atomic emission spectroscopy

Fig. 1 **a** Battery tray coolant channels/cast structure and plate position. **b** FSW locations on battery tray shown by the magenta weld path

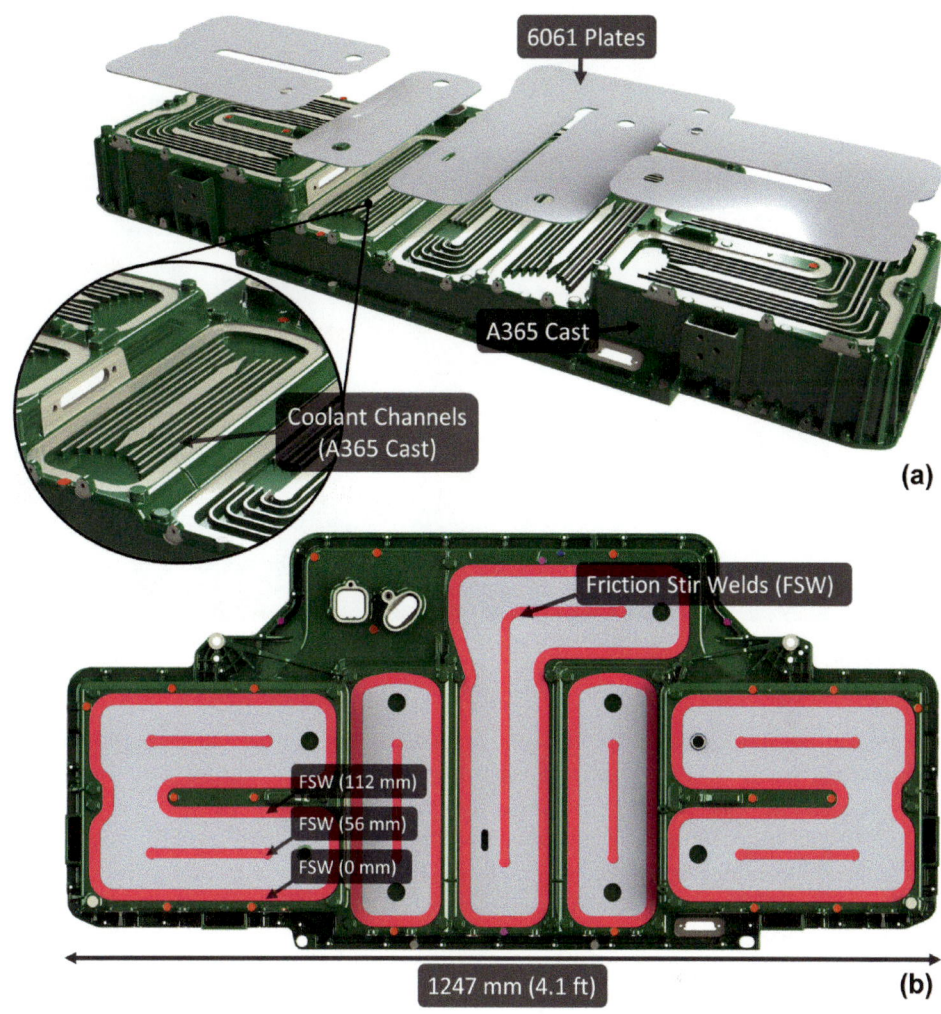

(ICP-AES) requires the dissolving of a 0.1 g sample of aluminum through heating and NaOH to dissolve the Si in the alloy, followed by the addition of peroxide (H_2O_2) to clear dissolved gas. Then, a mixture of HNO_3 and HCl is added to complete the digestion and retain the dissolved elements in the solution. Finally, dilution of the solution is required to finalize sample preparation for analysis [12].

Microhardness Sample Preparation

The Vickers hardness testing is conducted on a metallurgical sample extracted from the battery tray extending across each of the welds, as shown in Fig. 2.

First, the sample was polished to increase the contrast between the indents and material, then mounted in the sampling stage and tested with a load of 500 g for a 10-s duration, following the principles outlined in the ASTM E384-17 and ASTM E92-17 standards for Vickers micro indentation testing.

Results and Discussion

Microhardness

Several micro-hardness profiles were created utilizing the indent measurements across the three welds, FSW (0 mm), FSW (56 mm), and FSW (112 mm), along the surface and into the depth of the material (see Fig. 3). The top surface data in Fig. 4 contains relatively low hardness values for the 112- and 0-mm weld, ranging between 90 and 80 HV. In contrast, the middle 56 mm weld reaches 105 HV on the top surface. This variation in hardness between the 56 mm weld and the 112- and 0-mm weld may be attributed to the significant increase in rotational speed utilized for the 56 mm weld. Similar to the present paper, the study by Li et al. [10] on AA6061-T6 using stationary shoulder FSW, observed a significant decrease in hardness at the DXZ, with the most severe softening to 50–75 HV for the lowest tested rotational speed (750 RPM). However, as the rotational speed

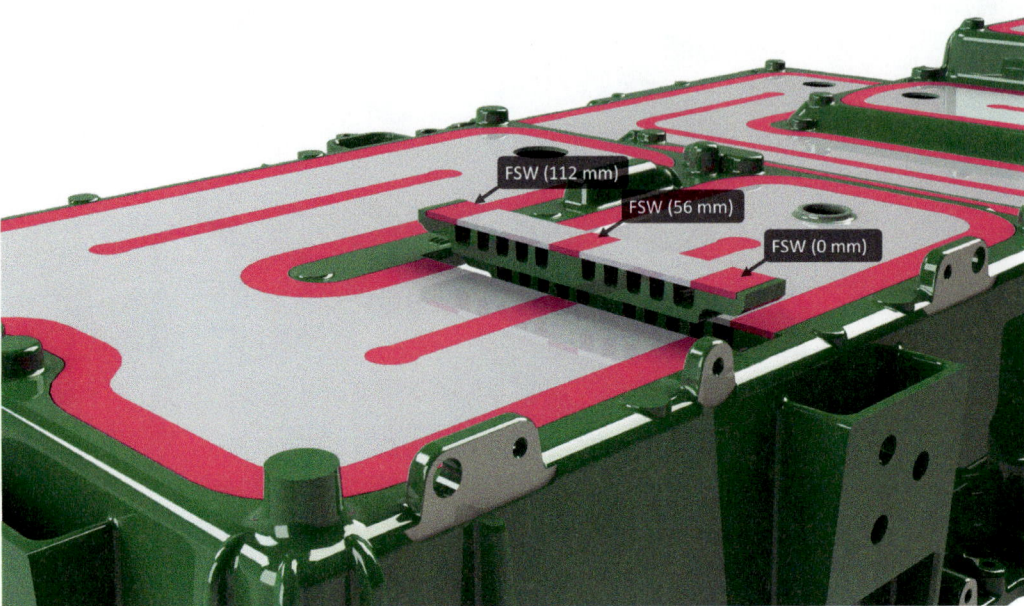

Fig. 2 Hardness sample location

increased at constant welding speeds, the hardness in the DXZ reached nearly 80 HV for a rotational speed of 1500 RPM. This hardness magnitude is consistent with the 0- and 112-mm welds tested at similar rotational speeds for the present study. Additionally, the study by Li et al. [10] states that more strengthening precipitates are generated at high tool rotation speeds, supporting the result of increasing hardness as the rotational speeds increase. Yet, in this study, the maximum measured hardness for the greatest rotational speed on the 56 mm weld is only near-surface, as seen by the maximum peak on the 0.1 and 0.15 mm curves for the 56 mm weld in Fig. 4. However, further down into the material, at 0.6 and 1 mm, the hardness values remain relatively consistent along the entire measurement line between the 112-, 56-, and 0-mm weld (see Fig. 4). The near-surface hardness variations may suggest an entirely different mechanism contributing to hardness increasing with increasing rotational speeds. For example, it is known that aluminum has excellent heat dissipation properties, creating a consistent temperature environment throughout the 1 mm depth directly below the tool [13]. Therefore, the generation of precipitates should be consistent near the surface and at the 1 mm depth. However, what is inconsistent is the amount of applied plastic deformation by the tool. Plastic deformation occurs near the material's surface during FSW. Therefore, if higher rotational speeds and plunging forces are utilized, the material may experience more plastic deformation and grain randomness, increasing hardness at the surface. This may explain why the 0.6 mm profile has a consistent hardness profile below the surface throughout the three welds conducted at vastly different rotational speeds. This concept is

further investigated and supported by EBSD work performed for this study on the identical welds tested for hardness.

EBSD

Analysis with EBSD is based on the band contrast map created by the beam's interactions with the crystal structure. These interactions create Kikuchi patterns or bands that are detectable through EBSD [14]. The position between the Kikuchi bands determines the interplanar angles, while the bandwidths determine the interplanar spacings. To index the grain, the angles and widths of the Kikuchi bands are compared with theoretical values known in advance for the actual crystal structure, in this case, aluminum face-centered cubic (FCC). Only the first three families of {hkl} planes are required to solve a backscatter Kikuchi pattern for FCC as it is cubic and of high crystal symmetry. After each grain is indexed, the crystallographic orientation or misorientation is described by (hkl) [uvw] notation or by three Euler angles (ϕ_1, Φ, ϕ_2). Consequently, EBSD deformation analysis employs grain pattern rotation maps referencing local misorientation, average misorientation, or quantifications through calculated dislocation densities. These methods depict plastic deformation around defects, cracks, or distorted grains in polycrystalline materials. Bulk dislocations that gather in the lattice during plastic deformation are indirectly detectable through EBSD. The indirect detection method is due to the fact that EBSD only observes the changes in the polycrystals resulting from plastic

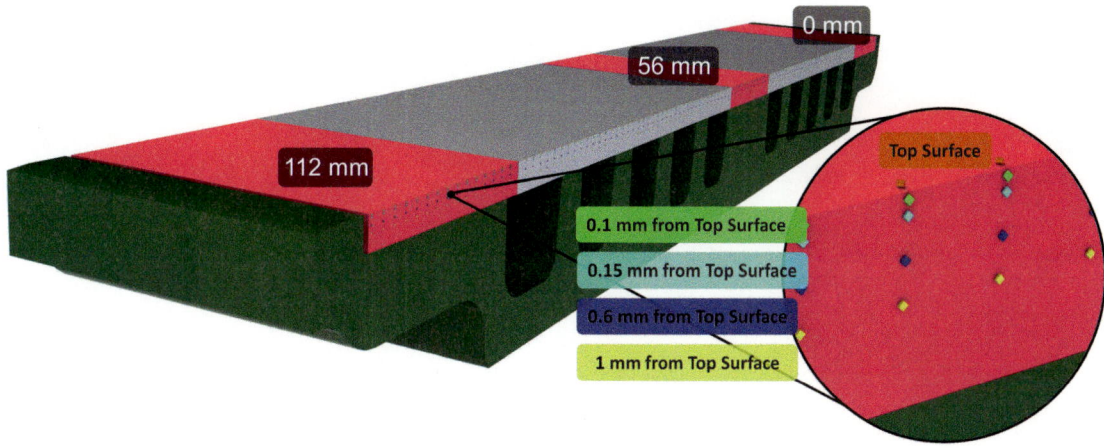

Fig. 3 Location of each hardness profile

deformation. The observed changes in polycrystals due to deformation are achieved by measuring grain misorientation as a spatial gradient in orientation. This measurement is then related to lattice curvature (deformation) through the Nye tensor [15]. The Nye tensor approximates the geometrically necessary dislocation (GND) density of a crystalline sample based on measurements of individual crystal grain orientation variations obtained from EBSD; therefore, with more estimated GND, more plastic deformation must have taken place. Essential to note that misorientation must be about some other orientation; accordingly, the reference orientation will drastically change the visualization of plastic deformation. For example, if only a highly deformed section of the material is measured, it would not appear deformed to the misorientation map, as the reference is another equally deformed grain. Therefore, a large area indicative of the non-deformed and plastically deformed material is measured for an accurate assessment. One effective way to produce a misorientation map is by kernel average misorientation (KAM) [16]. KAM calculates the arithmetic mean of the scalar misorientation between groups of pixels and maps them; this map is produced for the three welds in Fig. 5. In Fig. 5, a comparison of angular grain misorientation utilizing the KAM map is made between each weld. The HAZ is in low misorientation for each weld, signified by the blue-coated grains. The green-coated grains illustrate the TMAZ is at a misorientation of 12° for each weld. Finally, the region directly in contact with the tool shoulder is the DXZ with the most misorientation. The map shown in Fig. 5 contains 66 individual EBSD scans stitched together utilizing a scanning step size of 4 μm. Sometimes the grains will not index as the DXZ may have refined grains below 4 μm in size. Therefore, the stitching software revealed difficulties connecting the images over the map, more so in the DXZ, resulting in black strips of non-indexed grains. Additionally, with high dislocation density (more deformation), detection

quality is reduced during EBSD [15]. Nevertheless, the KAM map measures significant grain misorientation in the DXZ, indicating plastic deformation throughout each weld. However, the 56 mm weld has significantly higher surface hardness (see Fig. 4), indicating more plastic deformation; therefore, grains should be at higher misorientation in the KAM map. However, through the KAM map, the visual distinction between the 56 mm weld and the 0- and 112-mm weld is difficult. Therefore, the center of each weld, the area of the yellow box in Fig. 5, is quantified to analyze and compare the welds numerically (see Fig. 6).

In Fig. 6, the grains are grouped in bins where each bin contains the grains of the associated misorientation in degrees. For example, the fraction of grains on the low angle side, between 4.5° and 28.5°, is dominated by the 0- and 112-mm weld. In comparison, the fraction of grains on the high angle side, between 34.5 and 58.5°, is for the majority dominated by the 56 mm weld conducted at nearly twice the plunge force, rotational speed, and traverse speed. Therefore, the 56 mm weld should have more GND density, resulting in more plastic deformation and higher surface hardness. Furthermore, severe plastic deformation could be considered the primary source of texture variation during FSW. Therefore, if the crystallographic texture is significantly different between the 56 mm weld and the 0- and 112-mm weld, this may further illustrate the increasing plastic deformation with increasing rotational speed. Moreover, the mechanical properties between the welds may also be severely affected depending on the final crystallographic texture.

Fortunately, EBSD texture analysis is readily available by use of the inverse pole figure (IPF) map. The IPF map colour codes specific crystal planes perpendicular to the rolling, transverse, or normal axes (see Fig. 7b). The colour code key is based on the cubic crystal's stereographic projection (see Fig. 7a). This projection retains only the [001], [101],

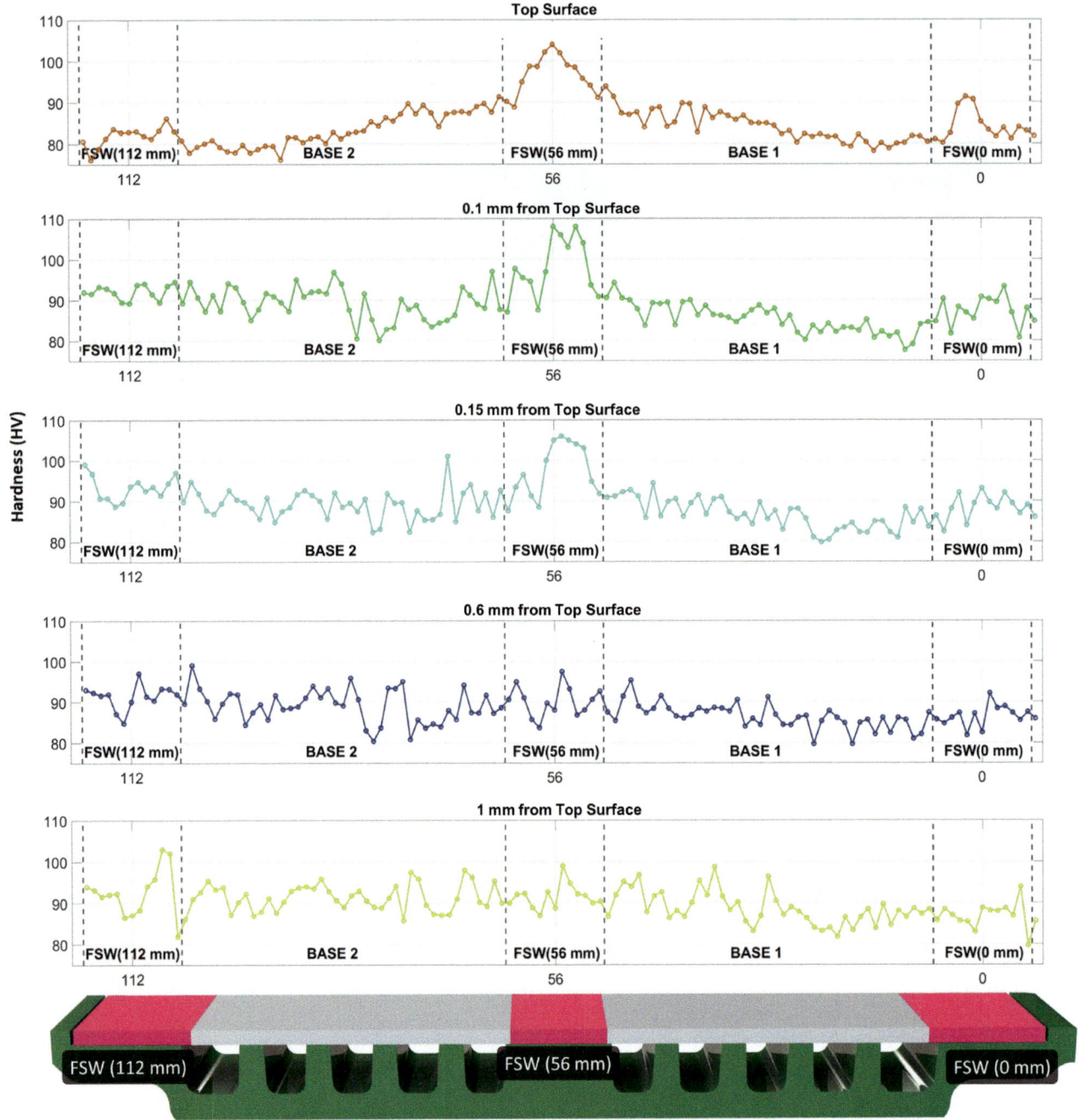

Fig. 4 Measured hardness across the sample

and [111] directions from the aluminum crystal. A specific axis is chosen, and all crystals on the IPF map are then coated depending on how perpendicular they are to the selected global axis. For example, in Fig. 7c, a small portion of the overall map is viewed, where the selected global axis is the normal direction, and the crystals are colour assigned based on which plane is perpendicular to the normal direction. However, Fig. 7c is an ideal case; not all the crystals align with one of the chosen directions. Depending on a

crystal's perpendicularity to the selected axis, it can be coloured between one of the colour key planes (yellow, magenta, or cyan—see Fig. 7a). In the IPF map for the 0- and 112-mm weld in Fig. 7, a strong ordered texture is seen across the weld corresponding to the location of the pin where the [101] planes are perpendicular to the normal direction; and in the shoulder contact region more of the [001] planes are perpendicular to the normal direction. As the plasticized material sticks to the tool surface and is

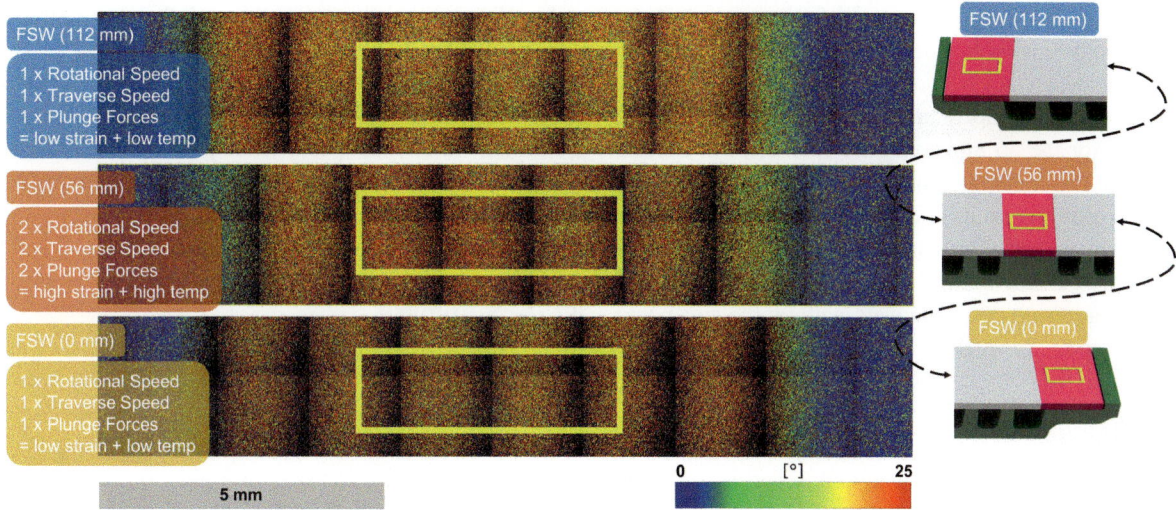

Fig. 5 KAM map of each weld along with the quantified region (yellow box)

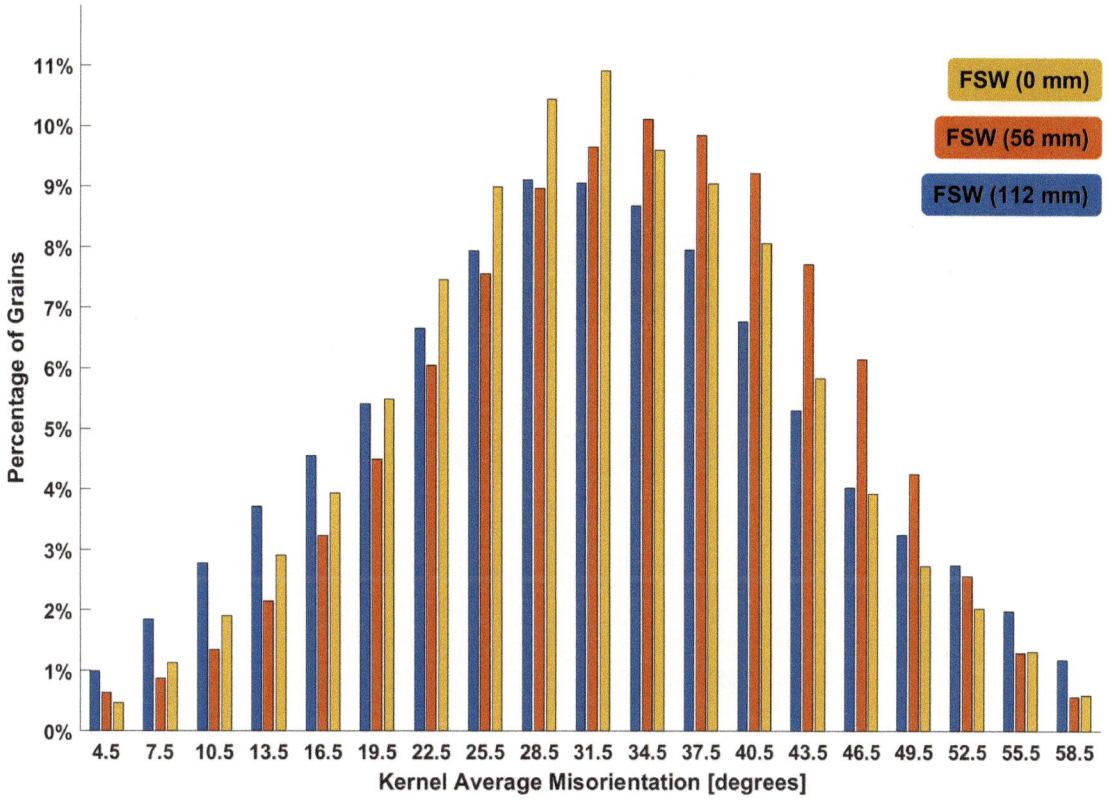

Fig. 6 Quantified region of the KAMA map (yellow box from Fig. 5)

subsequently left behind, the reason should follow when the tool's surface direction changes from pin to shoulder; the texture direction should change in the respective areas. Therefore, producing a different texture direction for the pin and shoulder is expected; studies on flow velocity support this reasoning [17]. On the other hand, the IPF map for the 56 mm weld has a much more unique texture with a laminar structure of many different planes layered over each other. This layer textured structure may be indicative of a change in which point the material is deposited by the tool. Consider the coefficient of friction with respect to temperature; lower temperatures result in a larger coefficient of friction, creating more material sticking. Hence, the material is held on the tool's surface for extended periods resulting in arranged material deposits ordering the material from one end of the weld to the other, as seen in the 0- and 112-mm welds in Fig. 7.

However, when temperature increases, with increasing tool rotation for the 56 mm weld, the coefficient of friction reduces, and the material more readily slips off the tool surface depositing randomly, creating a layered texture of different grain planes. This random and layered texture is visible for the 56 mm weld in Fig. 7; the [111], [001], and [001] planes alternate down the weld path on the advancing side of the tool. Supporting the changing coefficient of friction, a study by Nie et al. [18] found that with increasing temperature, the coefficient of friction reduces, and the slip rate increases during FSW. This increase in temperature is a direct result of increasing rotational and traverse speed [19, 20]. Therefore, given that the 56 mm weld is conducted at nearly double the rotational and traverse speed of the 0- and 112-mm weld, more slipping is occurring, causing the layered texture. Though there is less sticking and an increase in the slip rate of material resulting in a layered texture in the 56 mm weld, the 56 mm weld is still conducted at nearly double the rotational speed compared to the 0- and 112-mm

weld. With increasing rotational speed, the plastic strain also increases [21, 22]. Therefore, even though there is less sticking in the 56 mm weld, the material that does stick experiences much more plastic deformation due to the increase in rotational speed, leading to more plastic deformation and a harder surface for the 56 mm weld (see Fig. 4 for hardness distribution).

ICP-AES Analysis

To further support the increase in material slip rate with increasing rotational speed, the segregation of the elements can be analyzed. For example, it is known that the dragging effect of the rotating threaded tool pin can concentrate elements within the weld [23, 24]. Therefore, if there truly is a reduction in the sticking mechanism and an increase in the sliding mechanism during FSW with increasing rotational speeds, there may also be an associated reduction in the concentration of elements between welds that have more material sticking (more concentration) and welds that stick less and slid more (less concentration in comparison). In this

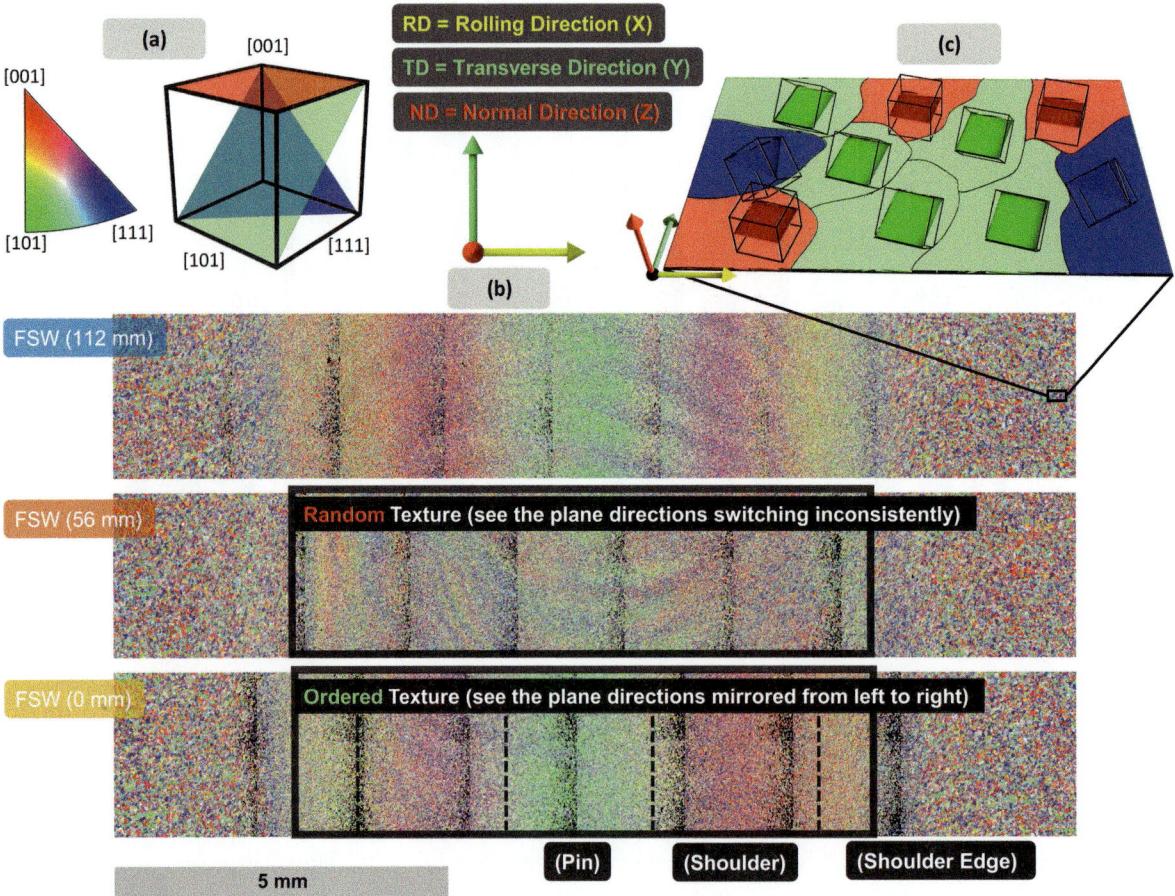

Fig. 7 Inverse Pole Figure (IPF) map

study, ICP-AES analysis tests this concept as it can measure the concentration of elements within each weld to the accuracy of a few parts per million (PPM). ICP-AES achieves element detection utilizing an emission spectrophotometric technique. This technique exploits the fact that excited electrons emit energy at a particular wavelength once the energy level it typically occupies is restored to its lowest energy or ground state. This exploit is beneficial as each element emits several unique wavelengths based on its chemical nature. The ICP-AES technique most frequently selects a single wavelength for a given element during detection. Following this, the energy intensity emitted at the selected wavelength is proportional to that element's concentration. Thus, by determining which wavelengths are emitted by a sample and their intensities, the sample's elemental composition can be precisely obtained in reference to a standard. This detection requires the metals to be in solution through the digestion of their solid form. The digestion is achieved through a series of acids to dissolve the sample completely [25]. Before digestion in this study, the A356 cast material was removed entirely from the welded sections of the 6061 material. The 6061 material was then sectioned into nine samples separating the welded area from the surrounding material (see Fig. 8).

Immediately evident in Fig. 8 is the noticeable change in silicon (Si) concentration between the welds, whereas other elements, like copper (Cu), iron (Fe), and manganese (Mn), still have visible but relatively minimal changes between the welds and base material. The minimal changes in the non-Si elements are most likely due to differences in diffusion rates of Cu, Fe, and Mn in Al compared to Si in Al at welding temperatures. For example, at 500 °C, the diffusion coefficient of Si (1.5×10^{-9} cm^2/s) in Al is much faster than Cu (5.9×10^{-10} cm^2/s), Fe (3.9×10^{-11} cm^2/s), and Mn (7.5×10^{-13} cm^2/s) in Al [26]. Henceforth, utilizing Si as a reference point, Si is, in fact, concentrated in the weld (see Fig. 8). As suggested by other researchers, the dragging and rotating tool action may be the mechanism for the concentration of elements in the weldment during FSW [23, 24]. Furthermore, the degree of the Si concentration is distinct when comparing the 56 mm weld to the 0- and 112-mm weld (see Fig. 8). This distinction suggests that more sliding is occurring due to the higher rotational speeds of the 56 mm weld concentrating less silicon as less material is adhering to the tool surface. Compared to the 0- and 112-mm welds at lower rotational speeds, more material sticks to the tool, causing ordered texture and more Si accumulation in the weld (see Fig. 7 for texture variations and Fig. 8 for Si accumulation results). The discussion on the sliding and sticking mechanism is important when considering material flow; as noted by other researchers, the grip or stickiness of the tool shoulder on the plasticized materials primarily establishes the material flow [27]. Nevertheless, the sliding

and sticking mechanisms generate heat to soften the material; however, material flow is strictly caused by sticking. Therefore, to create a sound weld with good adhesion to the material below, there should be an adequate adherence of plasticized material to the tool surface utilizing the sticking mechanism. Otherwise, the material does not stick and deposits in an unordered-layered manner, as shown by this study, possibly altering the effectiveness of the joint. For example, a study by Arora et al. [28] suggested that the optimum shoulder diameter should correspond to the maximum sticking torque for a given set of welding parameters and workpiece material. Any further increase in the tool rotation and traverse speed with constant diameter could result in the decreased grip of the tool on the material (more sliding and less sticking).

Conclusions

In this study, a friction stir welding process is used to seal coolant channels of an aluminum battery tray utilizing several 6061-plates welded to an A365 HPDC structure in a lap weld configuration. Different processing parameters are used for the 56 mm weld (higher plunge forces, rotational, and traverse speeds) as compared to the 0- and 112-mm weld (half the plunge forces, rotational, and traverse speeds in comparison) throughout the battery tray resulting in varied hardness properties for the two weld groups. The effect of varied processing parameters on microstructural evolution and hardness properties was investigated in detail. Based on the observations and analysis, the following conclusions were drawn:

1. ICP results indicate an accumulation of silicon during the FSW operation for both weld groups. However, as the coefficient of friction reduces with increasing temperature proportional to the increasing rotational speed, silicon accumulation reduces in the weld. This may be because the sticking mechanism is less effective at higher temperatures as the coefficient of friction is reduced. Therefore, with less sticking, a reduction in the silicon accumulation effect is seen in the 56 mm (high-temperature weld) compared to the 0- and 112-mm weld (low-temperature weld).

2. Distribution and generation of precipitates are not found to be the main factor resulting in increased surface hardness between the 56 mm weld to the 0- and 112-mm weld. Instead, the additional texture resulting from increased plastic deformation as the operation's rotational speed, traverse speed, and plunging force increase is found to cause the mechanical property variation between the weld groups.

Fig. 8 ICP data for the 6061 material

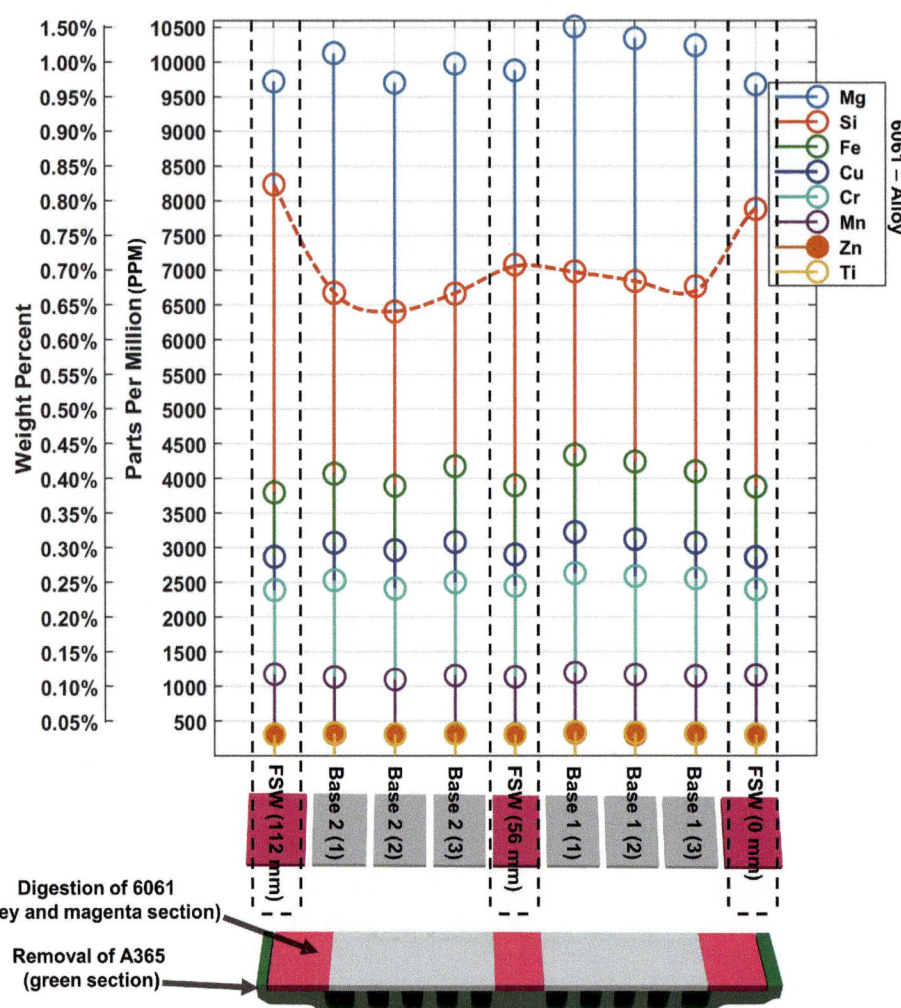

3. It was found that due to the increase in rotational speeds, traverse speeds, and plunging forces, the KAM results show an increase in the density of dislocations. Additional dislocation accumulation indicates more severe plastic deformation as a result. Although there is less sticking with increased rotational speed, the material that does stick experiences much higher plastic strain. The competing factors of increasingly less ordered texture and higher plastic strain with increased rotational speeds may be why the welds conducted at higher rotational speeds have increased hardness even though there is less ordered texture.

References

1. L. Wang, M. Makhlouf and D. Apelian, "Aluminium die casting alloys: alloy composition, microstructure, and properties-performance relationships," *International Materials Reviews*, vol. 40, pp. 221–238, 1995.

2. Noor Zaman Khan, Arshad Noor Siddiquee and Zahid A. Khan, Friction Stir Welding Dissimilar Aluminum Alloys, Boca Raton London New York: CRC Press, 2017.

3. Nilesh Kumar, Wei Yuan and Rajiv S. Mishra, "Chapter 2 - A Framework for Friction Stir Welding of Dissimilar Alloys and Materials," in *Friction Stir Welding of Dissimilar Alloys and Materials*, Elsevier, 2015, pp. 15–33.

4. J. Mohammadi , Y. Behnamian, A. Mostafaei, H. Izadi, T. Saeid, A. H. Kokabi and A.P. Gerlich, "Friction stir welding joint of dissimilar materials between AZ31B magnesium and 6061 aluminum alloys: Microstructure studies and mechanical characterizations," *Materials Characterization,* no. 101, pp. 189–207, 2015.

5. E. G. Col, A. Fehrenbacher, N. A. Duffie, M. R. Zinn, F. E. Pfefferkorn and N. J. Ferrier, "Weld temperature effects during friction stir welding of dissimilar aluminum alloys 6061-t6 and 7075-t6," *Int J Adv Manuf Technol,* no. 71, pp. 643–652, 2014.

6. P L Threadgill, A J Leonard, H R Shercliff and P J Withers, "Friction stir welding of aluminium alloys," *International Materials Reviews,* 2013.

7. Mostafa M.El-Sayed, A.Y.Shash, M.Abd-Rabou and Mahmoud G. ElSherbiny, "Welding and processing of metallic materials by using friction stir technique: A review," *Journal of Advanced Joining Processes*, vol. 3, 2021.

8. Rajiv Sharan Mishra, Partha Sarathi De and Nilesh Kumar, Friction Stir Welding and Processing, Switzerland: Springer, 2014.

9. P. Dong, D. Sun and H. Li, "Natural aging behaviour of friction stir welded 6005A-T6 aluminium alloy," *Materials Science & Engineering A,* vol. 576, pp. 29–35, 2013.

10. D. Li, X. Yang , L. Cui, F. He and H. Shen, "Effect of welding parameters on microstructure and mechanical properties of AA6061-T6 butt welded joints by stationary shoulder friction stir welding," *Materials and Design,* vol. 64, pp. 251–160, 2014.

11. B. R. Singh, A Hand Book on Friction Stir Welding, UK: LAP Lambert Academic, 2012.

12. K. Neubauer and C. Jiao, "Analysis of Aluminum Alloys with the Avio 220 Max ICP-OES Following London Metal Exchange Guidelines," PerkinElmer, Inc, Waltham, Massachusetts USA, 2021.

13. D. G. Andrade, C. Leitão, N. Dialami, M. Chiumenti and D. M. Rodrigues, "Modelling torque and temperature in friction stir welding of aluminium alloys," *International Journal of Mechanical Sciences,* vol. 182, pp. 1–11, 2020.

14. S. Nishikawa and S. Kikuchi, "The Diffraction of Cathode Rays by Calcite," *Proceedings of the Japan Academy,* vol. 8, pp. 475–477, 1928.

15. S. I. Wright, M. Nowell and D. P. Field, "A Review of Strain Analysis Using Electron Backscatter Diffraction," *Microscopy and Microanalysis,* vol. 17, pp. 316–329, 2011.

16. A. J. Schwartz, M. Kumar, B. L. Adams and D. P. Field, Electron Backscatter Diffraction in Materials Science, New York: Springer, 2000.

17. P. Zhang, N. Guo, G. Chen, Q. Meng, C. Dong, L. Zhou and J. Feng, "Plastic deformation behavior of the friction stir welded AA2024 aluminum alloy," *The International Journal of Advanced Manufacturing Technology,* vol. 74, pp. 673–679, 2014.

18. N. L, W. Y. X and G. H, "Prediction of temperature and residual stress distributions in friction stir welding of aluminum alloy," *The International Journal of Advanced Manufacturing Technology,* vol. 106, pp. 3301–3310, 2020.

19. E. Sharghi and A. Farzadi, "Simulation of temperature distribution and heat generation during dissimilar friction stir welding of AA6061 aluminum alloy and Al-Mg2Si composite," *The International Journal of Advanced Manufacturing Technology,* vol. 1, pp. 1–13, 2021.

20. Arbegast WJ and Hartley PJ, "Proceedings of the fifth international conference on trends in welding research," Pine Mountain, GA, USA, 1998.

21. T. T. Feng, X. H. Zhang, G. J. Fan and L. F. Xu, "Effect of the rotational speed of on the surface quality of 6061 Al-alloy welded joint using friction stir welding," in *Global Conference on Polymer and Composite Materials,* 2017.

22. M. M. Hadi, M. M. H, A. Khafaji and A. D. Subhi, "A Numerical Study of Friction Stir Welding for AA5754 Sheets to Evaluate Temperature Profile and Plastic Strain," *Engineering and Technology Journal,* vol. 40, pp. 1–11, 2022.

23. M. M. Attallah and H. G. Salem, "Friction stir welding parameters: a tool for controlling abnormal grain growth during subsequent heat treatment," *Materials Science and Engineering A,* vol. 391, pp. 51–59, 2005.

24. H. Salem, A. Reynolds and J. Lyons, "Structural Evolution and Superplastic Formability of Friction Stir Welded AA 2095 Sheets," *Journal of Materials Engineering and Performance,* vol. 13, pp. 24–31, 2003.

25. J. L. Todolí and J. M. Mermet, "Sample introduction systems for the analysis of liquid microsamples by ICP-AES and ICP-MS," *Spectrochimica Acta Part B,* vol. 61, pp. 239–283, 2006.

26. C. Zhang, J. D. Robson, S. J. Haigh and P. B. Prangnell , "Interfacial Segregation of Alloying Elements During Dissimilar Ultrasonic Welding of AA6111 Aluminum and Ti6Al4V Titanium," *Metallurgical and Materials Transactions A,* vol. 50, pp. 5143–5152, 2019.

27. G. G. Yeshitla and B. Singh, "Friction Stir Welding and its Applications: A Review," *High Technology Letters,* vol. 26, pp. 682–705, 2020.

28. A. Arora, M. Mehta, A. De and T. DebRoy, "Load bearing capacity of tool pin during friction stir welding," *The International Journal of Advanced Manufacturing Technology,* vol. 1, pp. 1–10, 2011.

Experimental Investigation of the Effect of High-Temperature Shot Peening on the Surface Integrity of 7010-T7452 Aluminum Alloy

Abouthaina Sadallah, Benoit Changeux, Hong-Yan Miao, Anindya Das, Sylvain Turenne, and Etienne Martin

Abstract

The aluminum alloy 7010 is widely used in the aerospace industry due to its comparatively lightweight and high strength. Shot peening is an important surface modification procedure for structural materials meant for improving the fatigue life. The surface properties can be further enhanced by changing the temperature of the surface during the process of peening. In this study, shot peening at room and high temperatures ranging from 100 to 300 °C was performed at an Almen intensity of 0.15 mmA, on cylindrical specimens of AA7010-T7452. The influence of the process on surface roughness, residual stresses, hardness, and microstructure was investigated. The results showed that the surface roughness increased with the temperature of shot peening and an optimized peening temperature of 250 °C was identified from the variation of hardness profiles.

Keywords

High-temperature shot peening • Almen intensity • 7010-T7452 aluminum alloy • Surface roughness • Surface hardness

Introduction

The aluminum alloy 7010 is widely used in the aerospace industry due to its comparatively lightweight and high strength [1]. Components made of AA7010 are shot peened to increase their fatigue lives [2]. Shot peening (SP) is a surface treatment that consists in impacting a metallic surface with high-velocity shot to introduce plastic strains. These plastic strains induce compressive residual stresses (CRS), cold work, and grain refinement that can delay the crack initiation at the surface and crack propagation beneath the surface, and therefore increase the fatigue life of the component [3]. However, the induced CRS in shot peened alloys relax under cyclic loading [4], which reduces the effectiveness of the SP process at increasing fatigue life.

When compared to SP at room temperature (RT), SP performed at higher temperatures leads to better fatigue strength for steels [3–6], titanium and aluminum composites [7, 8], as well as for magnesium alloys [9]. The fact that the CRS are more compressive over a larger depth and relax less than those induced by room temperature shot peening (RTSP) under cyclic loadings [3–6] could explain this increased fatigue performance. The publicly available literature related to the mechanical performance after high-temperature surface treatments of aluminum alloys focusses on other high-temperature surface treatments such as high=temperature laser shock peening (HTLSP) and high-temperature deep rolling (HTDR), which increase the surface hardness of aluminum alloys by 17% [10]–73% [11], respectively. This surface hardening phenomenon is due to the dynamic strain aging (DSA) that occurs during the high-temperature deformation of aluminum alloy. The DSA combines both strain hardening effect through the surface plastic deformation and the precipitation hardening effect through the generation of nanoprecipitates in the surface of the treated materials. No study was reported in the publicly available literature on the effect of HTSP on the surface integrity of aluminum alloys, especially the AA7010.

In this research, the effects of the HTSP on AA7010-T7452 cylindrical specimens were investigated. The HTSP at an Almen intensity of 0.15 mmA was carried out at temperatures between RT and 300 °C. The temperature profiles extracted during the HTSP process are analyzed and the peened specimens are characterized in terms of

A. Sadallah · H.-Y. Miao · A. Das · S. Turenne · E. Martin (✉)
Department of Mechanical Engineering, École Polytechnique, 2500 Chemin de Polytechnique, Montreal, QC H3T 1J4, Canada
e-mail: etienne.martin@polymtl.ca

B. Changeux
Materials and Processes Department, Safran Tech, Rue des Jeunes Bois, 78772 Magny-Les-Hameaux, France

© The Minerals, Metals & Materials Society 2023
S. Broek (ed.), *Light Metals 2023*, The Minerals, Metals & Materials Series,
https://doi.org/10.1007/978-3-031-22532-1_52

surface roughness and hardness. The relationships between the temperatures and the resulting surface properties were obtained.

Experimental Procedure

The material used for this study was AA7010-T7452, supplied by Safran in form of cylinders of 20 mm in diameter and 40 mm in length. The T7452 thermo-mechanical processing of the material involves the following steps: solution heat treatment, quenching, cold compression, and artificial aging at 120 °C for 15 h, followed by an over-aging at 175 °C for 12 h. The chemical composition of AA7010-T7452 is listed in Table 1.

AA7010-T7452 specimens were shot peened at an Almen intensity of 015 mmA with full coverage, at RT and at high temperatures ranging from 100 to 300 °C with an interval of 50 °C with two repetitions, which leads to 12 tests. The AA7010-T7452 cylindrical specimens were fixed to an aluminum holder. A high-temperature setup for performing HTSP tests was installed inside a conventional SP machine, including an induction coil. The induction coil was used to heat the specimens and a pyrometer was used to measure and record the surface temperature during the tests.

The determination of SP Almen intensity consisted in performing several SP tests based on previous experience with the SP machine. In our test, ceramic media Z425 (0.425 mm diameter), an air pressure of 69 kPa, a mass flow of 2 kg/min, a stand-off distance of 152.4 mm, and a peening angle of 90° between the nozzle and the specimen holder were selected to achieve Almen intensity of 0.15 mmA.

The specimens were heated and shot peened simultaneously during the HTSP process using the induction coil and the temperature profiles were recorded using the pyrometer. Figure 1 shows a typical surface temperature profile for a target temperature of 150 °C. To obtain a stable temperature during the SP process, the induction heating was programmed at a temperature (induction temperature) of 10 °C higher than the target temperature to avoid under heating during the blowing of air associated with the media.

The HTSP process includes three main steps as shown in Fig. 1. Step 1 is the pre-heating phase where the induction is activated to heat the specimen at the target peening temperature. At the end of this step, the nozzle moves above the specimen to activate the media at the pre-defined mass flow and the table starts rotating, which explains the increase in the

temperature in that second. Step 2 is the HTSP process. In this step, the induction heating is always activated, and the media are impacted onto the surface of the rotating specimen. At the end of this step, the nozzle is moved above the specimen to deactivate the media. Step 3 is the cooling phase. The induction heating and the mass flow are deactivated. However, the air still blows to cool down the specimen.

For the extracted temperature profiles at various temperatures, a simple calculation was made to obtain the average surface temperature during the period of HTSP (step 2) to make sure that the average value is in the range of $\pm 10\,°C$ the target temperature. A Mitutoyo profilometer was used to measure the surface roughness profiles along the longitudinal direction of the samples. A Clemex automated Vickers microhardness machine was applied to test the surface micro-hardness of specimens using a load of 100 gf and a 10 s dwell time. Five measurements were performed for each specimen. The measurements were averaged in order to count for the material heterogeneity and measurements errors.

Results and Discussion

Figure 2 presents the average surface temperatures measured during the HTSP of specimens for the target high temperatures of 100, 150, 200, 250, and 300 °C. The results show that when using an induction temperature of 10 °C higher than the target temperature, the specimens' surface temperature was $\approx \pm 5\,°C$ of the target temperature.

The change of surface roughness after shot peening is critical for the fatigue properties. Arithmetic average (Ra) was extracted from the roughness profiles to characterize the mean depth of the peening indentations. Figure 3 shows the Ra of the AA7010-T7452 specimens peened at different temperatures. The average surface roughness increases with the increase of temperatures from RT to 200 °C after that it reaches a saturation plateau. When the temperature increases, the yield point and the hardening mechanisms of the material decrease. Therefore, the media indentation increases, which relates to larger surface roughness.

Figure 4 shows the relationship between the average surface hardness and the SP temperature. When compared to the untreated specimen, SP at RT increases the average hardness of the specimens. The increase of the process temperature from RT to 150 °C decreases the average hardness of specimens. A significant increase in the hardness

Table 1 Chemical composition of AA7010-T7452 (weight %)

Element	Al	Zn	Mg	Cu	Fe	Zr	Si	Mn	Ti	Cr	Ni
Min	Bal	5.7	2.1	1.5	–	0.1	–	–	–	–	–
Max	Bal	6.7	2.6	2	0.15	0.16	0.12	0.1	0.04	0.05	0.05

Fig. 1 Temperature profile of a target temperature of 150 °C including the three steps of the HTSP program, step 1: heating the specimen by induction, step 2: HTSP process, and step 3: cooling using the machine air pressure

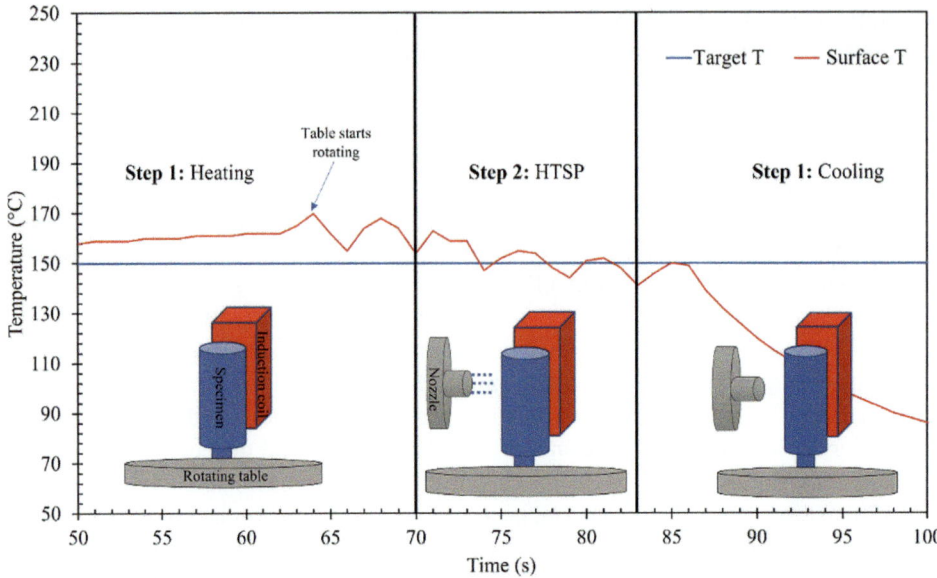

Fig. 2 Average surface temperature measured during the HTSP process period at an Almen intensity of 0.15 mmA for the target temperatures: 100, 150, 200, 250, and 300 °C

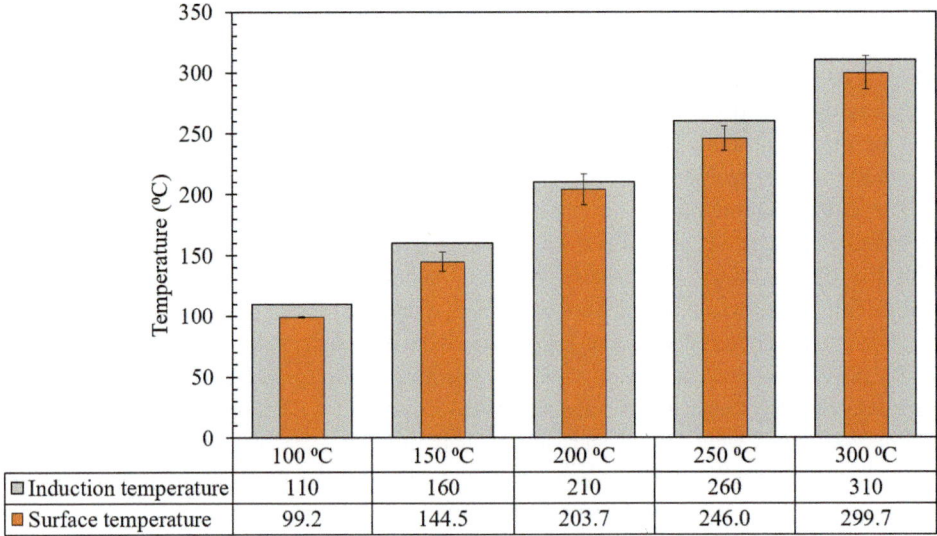

	100 °C	150 °C	200 °C	250 °C	300 °C
Induction temperature	110	160	210	260	310
Surface temperature	99.2	144.5	203.7	246.0	299.7

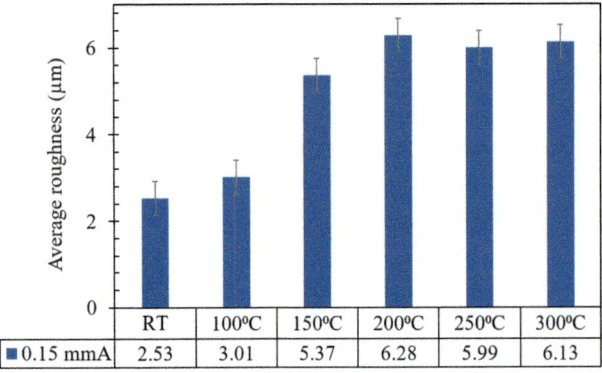

	RT	100°C	150°C	200°C	250°C	300°C
0.15 mmA	2.53	3.01	5.37	6.28	5.99	6.13

Fig. 3 The average surface roughness of AA7010-T7452 specimens shot peened at different temperatures: RT, 100, 150, 200, 250, and 300 °C and at an Almen intensity 0.15 mmA

is observed from 150 to 250 °C. The peak of hardness was observed at 250 °C. It is assumed that the hardness evolution in the case of HTSP of AA7010-T7452 comes from the two competing mechanisms: hardening and softening [12, 13] as shown in the case of HTLSP of AA6160 [10], HTLSP of AA7075 [14], and HTDR of AA6110 [11]. The decrease in the surface hardness is due to the softening mechanism activated in the temperature range of RT to 150 °C. Furthermore, the increase in hardness from 150 to 250 °C may be due to the hardening mechanism activated at this temperature as seen in the case of AA7075 after HTLSP at 250 °C [14], where the authors explained that this hardening mechanism is due to the DSA activated at this temperature range.

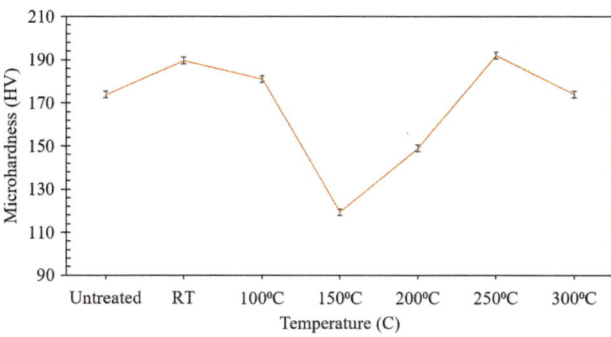

Fig. 4 The average surface microhardness of the AA7010-T7452 specimens peened at 0.15 mmA and at temperatures: RT, 100, 150, 200, 250, and 300 °C

Conclusions

SP and HTSP treatments on AA7010-T7452 specimens were carried out to reveal the effects of temperatures ranging from RT to 300 °C on the surface roughness and hardness of the aluminum alloy. The new HTSP setup yielded a controlled surface temperature during SP. The average surface roughness increased with the increase of the peening temperature that can be related to the large surface plastic deformation induced during the HTSP process, which led to larger dents. The surface roughness reached a plateau after 200 °C. When compared to the untreated specimen and the SP at RT, HTSP at 250 °C led to higher material hardness. CRS measurements and further microstructural observations will be conducted on the AA7010-T7452 specimens peened at RT and at high temperatures to understand the material behavior after SP at different temperatures.

References

1. R.R. Cervay, Mechanical property evaluation of aluminum e DTIC, (1980).
2. G.H. Fair, the Effect of Shot Peening on Fatigue Fretting-Fatigue of Aluminium Alloys, (1988) 255.
3. C. Bianchetti, Analytical fatigue life prediction of shot peened 7050-t7451 aluminum alloy, (2018).
4. M. Benedetti, V. Fontanari, B.D. Monelli, Numerical simulation of residual stress relaxation in shot peened high-strength aluminum alloys under reverse bending fatigue, J. Eng. Mater. Technol. Trans. ASME. 132 (2010) 110121–110129. https://doi.org/10.1115/1.3184083.
5. A. Wick, V. Schulze, O. Vohringer, Influence of the shot peening temperature on the relaxation behaviour of residual stresses during cyclic bending, 7th Int. Conf. SHOT Peen. 4140 (1999) 102–109.
6. R. Menig, V. Schulze, O. Vöhringer, Optimized warm peening of the quenched and tempered steel AISI 4140, Mater. Sci. Eng. A. 335 (2002) 198–206. https://doi.org/10.1016/S0921-5093(01)01915-3.
7. M. Schilling-Praetzel, F. Hegemann, P. Gomez, G. Gottstein, Influence of temperature of shot peening on fatigue life, in The 5th International Conference of Shot Peening, (1993), pp. 227–238.
8. A. Tange, H. Koyama, H. Tsuji, Study on warm shot peening for suspension coil spring., SAE Trans. (1999) 463–467. https://www.jstor.org/stable/44650641.
9. Y. Huang, W.C. Liu, J. Dong, Surface characteristics and fatigue performance of warm shot peened wrought magnesium alloy Mg-9Gd-2Y, Mater. Sci. Technol. (United Kingdom). 30 (2014) 1481–1487. https://doi.org/10.1179/1743284713Y.0000000450.
10. C. Ye, Y. Liao, G.J. Cheng, Warm laser shock peening driven nanostructures and their effects on fatigue performance in Aluminum Alloy 6160, Adv. Eng. Mater. 12 (2010) 291–297. https://doi.org/10.1002/adem.200900290.
11. P. Juijerm, I. Altenberger, Effect of high-temperature deep rolling on cyclic deformation behavior of solution-heat-treated Al–Mg–Si–Cu alloy, Scr. Mater. 56 (2007) 285–288. https://doi.org/10.1016/j.scriptamat.2006.10.017.
12. X. Quelennec, E. Martin, L. Jiang, J.J. Jonas, Work hardening and kinetics of dynamic recrystallization in hot deformed austenite, J. Phys. Conf. Ser. 240 (2010). https://doi.org/10.1088/1742-6596/240/1/012082.
13. É. Martin, W. Muhammad, A.J. Detor, I. Spinelli, A. Wessman, D. Wei, "Strain-annealed" grain boundary engineering process investigated in Hastelloy-X, Materialia. 9 (2020). https://doi.org/10.1016/j.mtla.2019.100544.
14. C. Ye, Y. Liao, S. Suslov, D. Lin, G.J. Cheng, Ultrahigh dense and gradient nano-precipitates generated by warm laser shock peening for combination of high strength and ductility, Mater. Sci. Eng. A. 609 (2014) 195–203. https://doi.org/10.1016/j.msea.2014.05.003.
15. W. Luan, C. Jiang, V. Ji, Y. Chen, H. Wang, Investigation for warm peening of tib2/Al composite using X-ray diffraction, Mater. Sci. Eng. A. 497 (2008) 374–377. https://doi.org/10.1016/j.msea.2008.07.016.
16. W. Luan, C. Jiang, V. Ji, The texture effect of warm peening on tib2/Al composite, Mater. Sci. Eng. A. 504 (2009) 124–128. https://doi.org/10.1016/j.msea.2008.10.035.

Quality Assessment and Features of Microdrilled Holes in Aluminum Alloy Using Ultrafast Laser

Suman Chatterjee, Abhijit Suhas Cholkar, David Kinahan, and Darmot Brabazon

Abstract

In this work, we present the use of an ultrafast laser system for the high aspect ratio micro-drilling of aluminum alloy thin foils. Hole sizes in between 20 and 40 μm were fabricated in arrays with sub-micron level precision in terms of diameter and hole location. The Design of Experiment approach was employed to analyze the influences of the laser process parameters like laser power, frequency, and exposure time on the resulting quality of the produced micro-holes. The outputs measured were hole size, location and the variability in these measures. The metallurgical and geometrical features were examined using a scanning electron microscope and optical microscope. Processing throughput is also important in industrial laser processes. The parametric effect on circularity and taper has been observed to understand the features of the hole. The features of holes help in fabrication in a plethora of industries to produce applications such as fins, filters, microgrid circuits, and biomedical devices.

Keywords

Aluminium alloy • Femtosecond laser • Circularity • Taper • Micro drilling

Introduction

In laser drilling, a high-intensity infrared laser beam is focused on a spot of the workpiece (generally between 0.1 and 2.0 mm in spot diameter) to remove materials to produce a hole. This laser drilling operation takes place in three phases (I) melting of material (II) vaporization and (III) chemical degradation throughout the depth of the material [1]. Laser drilling is gradually becoming a popular machining technique for micro-drilling on various engineering materials and components of intricate shapes. Researchers and scientists are working diligently to develop a new technology to attain the recent demand in the scientific field. There is the widespread application of micro-machined products in heat exchangers for micro-electronic products, micro-nozzle systems, micro-electromechanical systems (MEMS), micro-molds and micro-fluidic systems etc. [2–4]. Literature suggests that the present trend of research is directed toward the micro-machining of a wide variety of work materials ranging from easy-to-cut to hard-to-cut materials for aerospace applications [5–7].

Micro-machining on engineering materials such as aluminum alloys, stainless steels, titanium alloys, nickel alloys and ceramics is quite popular among practitioners because of their widespread applications in aviation, automobile, aerospace, medical and electronic industry as they possess favorable properties like low thermal conductivity, resistance to high temperature, high corrosion resistance and high strength to weight ratio [8–12].

Laser drilling of aluminum alloy is not as effective as it could be because aluminum alloy has a poor absorption rate of laser beam energy and is easily oxidized in the air,

S. Chatterjee (✉) · A. S. Cholkar · D. Kinahan · D. Brabazon
I-Form Advanced Manufacturing Centre, Dublin City University, Dublin 9, Ireland
e-mail: suman.chatterjee@dcu.ie

A. S. Cholkar
e-mail: abhijit.cholkar2@mail.dcu.ie

D. Kinahan
e-mail: david.kinahan@dcu.ie

D. Brabazon
e-mail: dermot.brabazon@dcu.ie

Advanced Processing Technology Research Centre, School of Mechanical and Manufacturing Engineering, Dublin City University, Glasnevin, Dublin 9, Ireland

National Centre for Plasma Science & Technology, Dublin City University, Glasnevin, Dublin 9, Ireland

© The Minerals, Metals & Materials Society 2023
S. Broek (ed.), *Light Metals 2023*, The Minerals, Metals & Materials Series,
https://doi.org/10.1007/978-3-031-22532-1_53

necessitating the use of protective gas during drilling. As a result, a few investigations have been done on the laser drilling technology for aluminum alloy [13–15]. In order to increase the rate of material removal and lessen the taper of the hole, Mishra et al. [4, 16] primarily constructed a numerical model for aluminum alloy laser drilling and predicted and set the best drilling process parameters (pulse width, pulse repetition frequency, pulse peak power, etc.) through the numerical model. The processing of materials using different pulsed lasers was studied by Fujita et al. [17] to determine the dependence of wavelength and pulse width. The research revealed that processing efficiency and thermal damage might be improved using shorter wavelengths and pulse widths. Higher cutting efficiency and a smaller heat-affected zone (HAZ) phenomenon resulted from shorter wavelength and pulse width.

However, micro-machining of aluminum alloy using conventional machining techniques is difficult due to continuous chip production and built-up edges near the drilled holes. As far as micro-drilling is concerned, laser drilling is a preferred drilling technique because a high-intensity laser beam can be focused on a precise location for a small-time interval to achieve the desired hole avoiding thermal degradation of the material. In the present study, laser drilling of aluminum alloy has been performed using a femtosecond millisecond laser in ambient air. The influence of machining parameters such as laser power, repetition rate, and exposure time on the quality characteristics of laser drilled samples is examined. Quality characteristics considered for the study are circularity at entry, circularity at the exit, and taper. The study has adopted the design of experiment (DOE) approach to design the experimental layout so that maximum process-related information can be gathered from fewer experimental trails.

Materials and Methods

Materials

The material considered in the present study is made up of commercially available Aluminum alloy Al 1145 (UNS A91145) in the form of thin foils having 0.03 mm thickness. The chemical composition and mechanical properties of Al 1145 are as mentioned in Tables 1 and 2.

Experimental Procedure

Laser drilling on Al 1145 foil (a thin foil of 0.03 mm) has been carried out to study the effect of process parameters on performance measures. In this study, we used an ultrafast femtosecond laser (NKT One Five Origami 10XP) that

Table 1 Chemical constituents of Al 1145 [18]

Constituents	% Content
Titanium, Ti	0.03
Zinc, Zn	0.05
Magnesium, Mg	0.05
Manganese, Mn	0.05
Copper, Cu	0.05
Aluminum, Al	99.45
Silicon, Si + Iron, Fe	Remaining

Table 2 Physical properties of Al 1145 [18]

Properties	
Density	2.6–2.8 g/cm^3
Elastic modulus	70–80 GPa
Poisson's ratio	0.33
Thermal conductivity	227 W/mK

generates 400 fs pulses with 1030 nm central wavelength at a maximum pulse repetition rate of 1 MHz. The beam diameter of the laser at the focused position was 45 μm. Figure 1 shows the schematic layout for the laser drilling operation. The workpiece is mounted suitably on the vice and the laser beam is focused on the workpiece. Here, a stand-off distance of 221 mm has been considered. The laser processing parameters considered for the drilling operation are laser power, repetition rate, and exposure time. Table 3 shows the parametric setting and their levels. The study has adopted the design of experiment (DOE) approach to design the experimental layout to gather maximum process-related information from fewer experimental trails.

Characterization of Quality Features

Laser drilling of micro holes on aluminum alloy has been performed using the experimental plan as per Box-Behnken designs. Seventeen experimental runs have been performed and each experiment has been repeated more than 20 times. The laser-drilled holes are shown in Fig. 2. The average value of each quality characteristic is calculated and noted for the analysis of aluminum alloy. The quality of the hole can be assessed in terms of circularity (both at entry and exit) and taper. Image of the laser-drilled holes is acquired using a scanning electron microscope to measure the performance measures viz. circularity at entry, circularity at the exit, and taper using image processing software ImageJ. Figure 2 indicates the quality characteristics of the workpiece after the laser drilling process. The circularity of the hole is expressed as the ratio of minimum diameter to maximum diameter

Fig. 1 Schematic of laser drilling setup

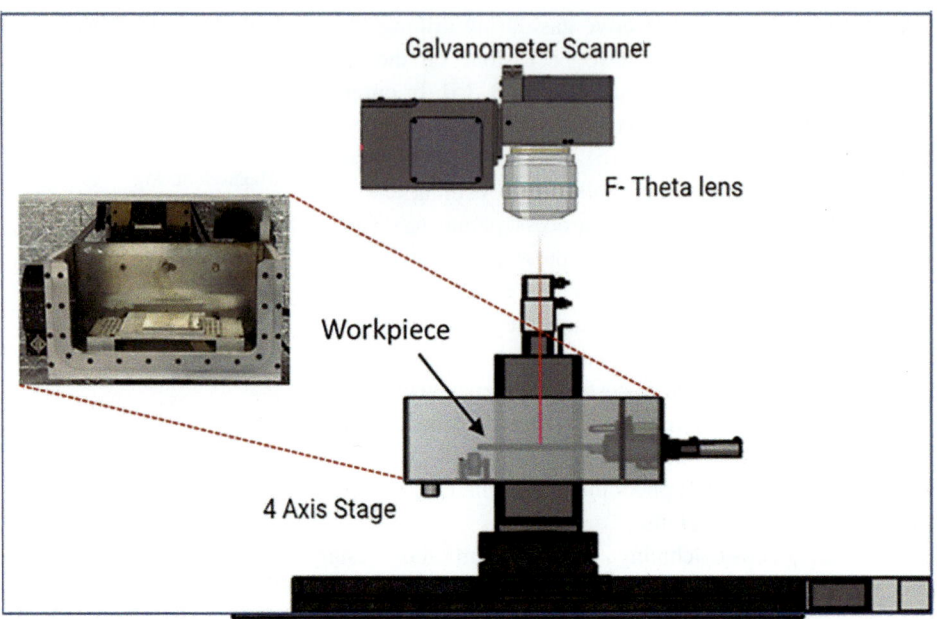

Table 3 Laser drilling setup and the parametric levels

Parameters	Level		
	1	0	−1
Laser power (W)	2	2.5	3
Repetition rate (kHz)	100	200	300
Exposure time (μs)	30	40	50

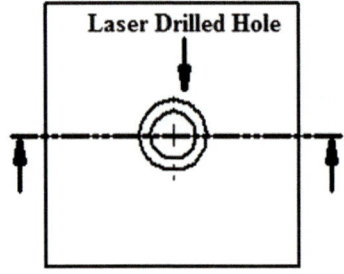

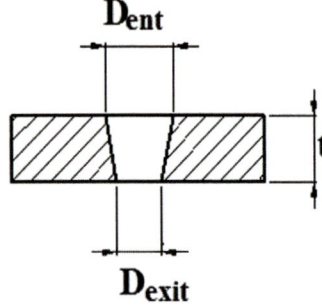

Fig. 2 Schematic diagram of taper [20]

Ferret's diameter of the hole [19, 20]. The taper of the laser-drilled hole is calculated by Eq. (1) as shown below [19, 20].

$$\text{Taper} = \frac{D_{ent} - D_{exit}}{2 \times (\text{thickness of workpiece})} \quad (1)$$

where D_{ent} and D_{exit} represent the diameter of the hole at entry and exit, and t is the thickness of the hole.

Results and Discussion

The measured quality characteristics for laser drilled holes on aluminum alloy Al 1145 has been discussed in the following section. To know the significance of process parameters, analysis of variance (ANOVA) has been performed for each performance measure for laser drilled holes on Al 1145. Co-efficient of determination (R^2) for circularity at entry, circularity at exit, and taper are 0.845, 0.8087 and 0.772, respectively, indicating statistical validity of the analysis. The surface plot for the performance measures shown in Figs. 3, 4 and 5 helps to understand the parametric effect on quality features of laser drilled holes. The main effect surface plot in Fig. 3 shows that circularity (at entry and exit) increases with increase in laser power, repletion rate and exposure time. The surface plot shows that circularity varies linearly with repetition rate and pulse width. Circularity increases with increases in repetition rate and laser power. As the pulse repetition rate increases, pulse off-time gets reduced, materials get melted and solidified with lower agitation and disorder and maximum circularity is attained. Similarly, an increase in exposure time leads to increased heat input, resulting in higher circularity. Similar type of observation has been observed for circularity at exit (Fig. 5).

The surface plot (Fig. 5) discusses the influence of control parameters on taper formation. Figure 5 indicates the combined effect of laser power and repetition rate on the taper of drilled holes. The graph shows that taper of holes

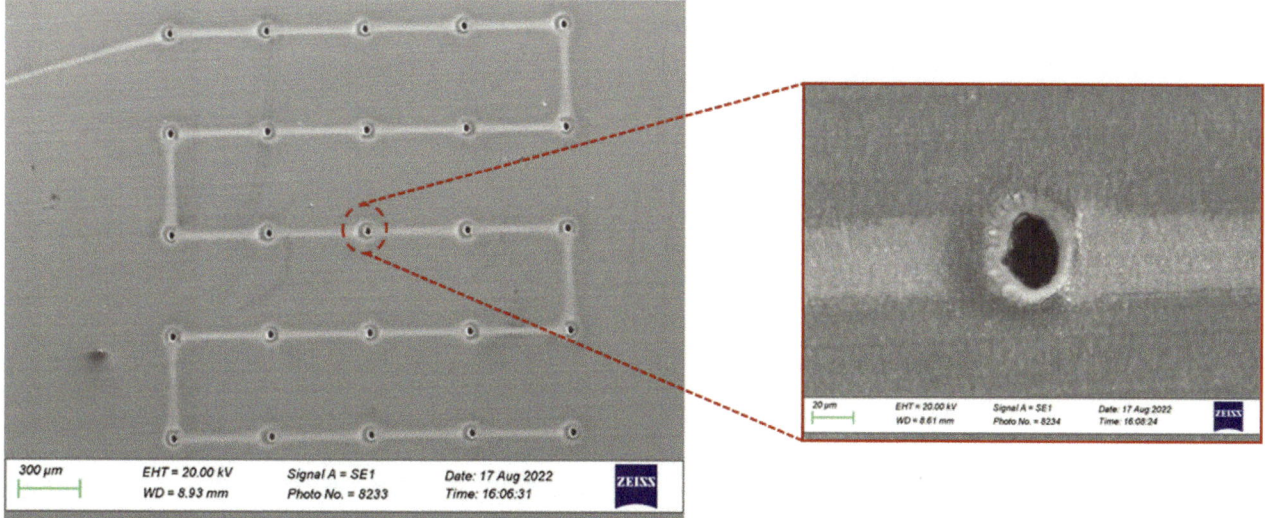

Fig. 3 SEM image showing the array of laser drilled holes at identical experimental setting for experiment number 6 (at top side)

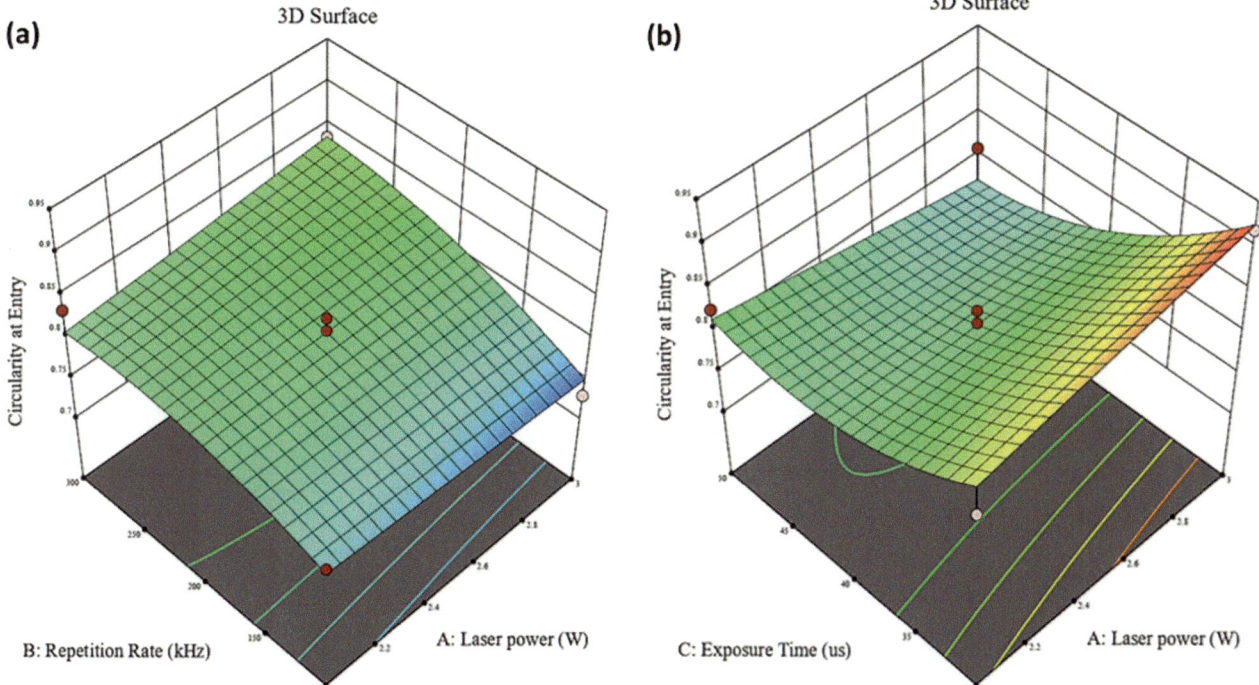

Fig. 4 Response surface plot for circularity at entry with respect to **a** repetition rate and laser power, **b** exposure time and laser power

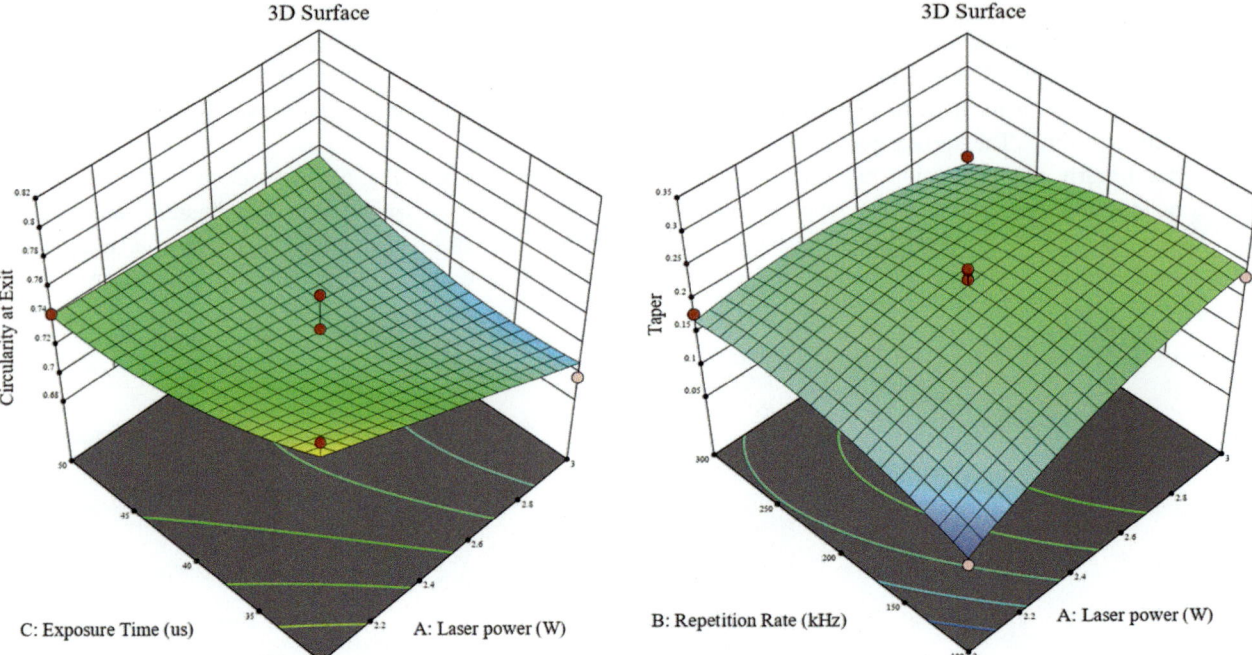

Fig. 5 Response surface plot for circularity at exit with respect to exposure time and laser power

Fig. 6 Response surface plot for taper with respect to repetition rate and laser power

increases simultaneously with the increase in repetition rate and laser power. As the control parameter, viz. laser power current is directly proportional to laser energy [9, 21]; as a result, at a higher value of laser input current thermal energy will also be high. This indicates that at high laser energy, the material's surface gets melted and vaporized instantly; as a result, a huge volume of material is removed from the top surface as compared to the remaining thickness of the material. This may lead to the formation of taper of drilled holes. At the same time the graph also suggests that taper increases with an increase in repetition rate. A similar observation has been reported by Biswas et al. [22].

To analyze the microstructural features of the drilled holes, the drilled holes were observed under a scanning electron microscope. Figure 6 shows the presence of spatter deposition near laser drilled holes for experiment number 6. The microstructural analysis for all the holes were performed at 5000X magnification in SEM. The study suggests that spatter deposition increases on increasing the laser power and exposure time. It may be due to the vapour pressure developed inside the laser-drilled hole which leads to the generation of molten material ejection [23, 24] (Fig. 7).

Conclusion and Future Studies

This paper provides an experimental study of ultrafast pulse laser micro-drilling of commercially available grade aluminum alloy (Al1145). Al1145 surfaces were drilled with a femtosecond laser source using a galvo scanner at different laser power, repetition rates, and exposure times. The study was carried out using the design of experiment approach to determine the influence of laser control parameters on performance characteristics such as the circularity of the hole, and taper. The study shows the significant effect of laser power and exposure on all the performance measures. The microstructural study reveals the impact of process parameters on spatter deposition. It was observed that spatter deposition increases with increasing the laser power and exposure time. Further study will be extended to optimize and select the best laser parametric settings for the desired conditions. The work will be further extended to the use of scanning optics and spatial light modulation (SLM) to contribute toward developing a rapid and scalable ultrafast laser process. The timeframe for conventional galvanometer

Fig. 7 SEM image shows the spatter deposition near laser drilled holes for experiment number 6 at 5000X

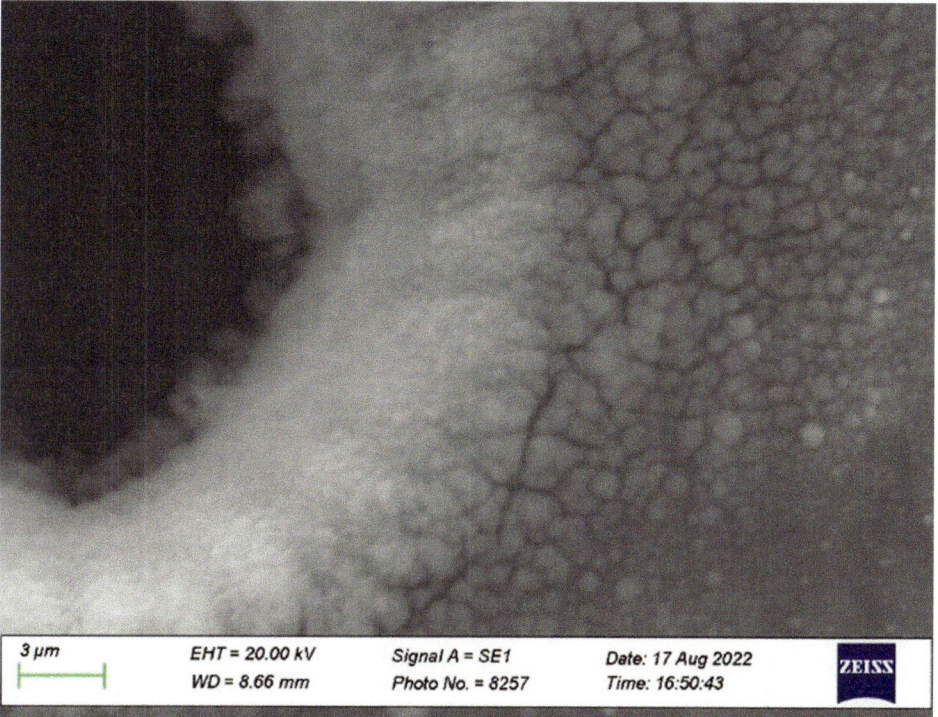

processing versus SLM processing will be compared for different geometric array patterns (number and geometric arrangement of holes) production.

Acknowledgements This publication has emanated from research supported by the European Union's Horizon 2020 Research and Innovation Program under grant agreement No. 862100 (NewSkin), supported by Science Foundation Ireland (SFI) under Grant Numbers 16/RC/3872 and is co-funded under the European Regional Development Fund and by I-Form industry partners.

References

1. Shams, O.A., Pramanik, A. and Chandratilleke, T.T., 2017. Thermal-assisted machining of titanium alloys. In Advanced manufacturing technologies (pp. 49–76). Springer, Cham.
2. Chatterjee, S., Mahapatra, S.S., Sahu, A.K., Bhardwaj, V.K., Choubey, A., Upadhyay, B.N. and Bindra, K.S., 2017, December. Experimental and parametric evaluation of quality characteristics in Nd:YAG laser micro-drilling of Ti6Al4V and AISI 316. In ASME 2017 gas turbine India conference (pp. V002T10A006-V002T10A006). American Society of Mechanical Engineers.
3. Chauhan, A. and Vaish, R., 2012. A comparative study on material selection for micro-electromechanical systems. Materials & Design, 41, pp. 177–181.
4. Mishra, S. and Yadava, V., 2013. Modeling and optimization of laser beam percussion drilling of thin aluminum sheet. Optics & Laser Technology, 48, pp. 461–474.
5. Meijer, J., 2004. Laser beam machining (LBM), state of the art and new opportunities. Journal of Materials Processing Technology, 149(1-3), pp. 2–17.
6. Dahotre, N.B. and Harimkar, S., 2008. Laser fabrication and machining of materials. Springer Science & Business Media.
7. Dubey, A.K. and Yadava, V., 2008. Laser beam machining-a review. International Journal of Machine Tools and Manufacture, 48(6), pp. 609–628.
8. Chen, M.F., Hsiao, W.T., Wang, M.C., Yang, K.Y. and Chen, Y. F., 2015. A theoretical analysis and experimental verification of a laser drilling process for a ceramic substrate. The International Journal of Advanced Manufacturing Technology, 81(9–12), pp. 1723–1732.
9. Kuar, A.S., Doloi, B. and Bhattacharyya, B., 2006. Modelling and analysis of pulsed Nd: YAG laser machining characteristics during micro-drilling of zirconia (ZrO2). International Journal of Machine Tools and Manufacture, 46(12), pp. 1301–1310.
10. Low, D.K.Y., Li, L. and Byrd, P.J., 2003. Spatter prevention during the laser drilling of selected aerospace materials. Journal of Materials Processing Technology, 139(1), pp. 71–76
11. Ng, G.K.L. and Li, L., 2003. Repeatability characteristics of laser percussion drilling of stainless-steel sheets. Optics and lasers in engineering, 39(1), pp. 25–33.
12. Yilbas, B.S., Akhtar, S.S. and Karatas, C., 2011. Laser trepanning of a small diameter hole in titanium alloy: temperature and stress fields. Journal of Materials Processing Technology, 211(7), pp. 1296–1304.
13. Kikin, P.Y., Pchelintsev, A.I. and Rusin, E.E., 2007. Features of the hole shape formation during laser drilling of an ultrafine-grained aluminum alloy. Technical Physics Letters, 33 (11), pp. 917–918.
14. Kikin, P.Y., Perevezentsev, V.N., Pchelintsev, A.I. and Rusin, E. E., 2006. Effect of ultrafine-grained structure on the regime of laser drilling in aluminum alloy 1420. Technical physics letters, 32(10), pp. 845–846.
15. Zhang, Y., Qiao, H., Zhao, J. and Cao, Z., 2022. Research on water jet-guided laser micro-hole machining of 6061 aluminum alloy. The International Journal of Advanced Manufacturing Technology, 118(1), pp. 1–13.

16. Mishra, S. and Yadava, V., 2013. Prediction of hole characteristics and hole productivity during pulsed Nd: YAG laser beam percussion drilling. Proceedings of the Institution of Mechanical Engineers, Part B: Journal of Engineering Manufacture, 227(4), pp. 494–507.

17. Fujita, M., Ohkawa, H., Somekawa, T., Otsuka, M., Maeda, Y., Matsutani, T. and Miyanaga, N., 2016. Wavelength and pulsewidth dependences of laser processing of CFRP. Physics Procedia, 83, pp. 1031–1036.

18. https://www.azom.com/article.aspx?ArticleID=6621.

19. Chatterjee, S., Mahapatra, S.S. and Abhishek, K., 2016. Simulation and optimization of machining parameters in drilling of titanium alloys. Simulation Modelling Practice and Theory, 62, pp. 31–48.

20. Chatterjee, S., Mahapatra, S.S., Bharadwaj, V., Choubey, A., Upadhyay, B.N. and Bindra, K.S., 2019. Drilling of micro-holes on titanium alloy using pulsed Nd: YAG laser: parametric appraisal and prediction of performance characteristics. Proceedings of the Institution of Mechanical Engineers, Part B: Journal of Engineering Manufacture, 233(8), pp. 1872–1889.

21. Padhee S, Pani S, Mahapatra SS. A parametric study on laser drilling of Al/SiCp metal-matrix composite. Proceedings of the Institution of Mechanical Engineers, Part B: Journal of Engineering Manufacture 2011;226:76–91.

22. Biswas R, Kuar AS, Mitra S. Multi-objective optimization of hole characteristics during pulsed Nd: YAG laser microdrilling of gamma-titanium aluminide alloy sheet. Optics and Lasers in Engineering 2014;60:1–11.

23. Chryssolouris G, Bredt J, Kordas S, Wilson E. Theoretical aspects of a laser machine tool. Journal of Manufacturing Science and Engineering 1988;110(1):65–70.

24. Low DKY, Li L, Corfe AG. Characteristics of spatter formation under the effects of different laser parameters during laser drilling. Journal of materials processing Technology 2001;118(1):179–186.

Surface Characterization Methods to Evaluate Adhesive Bonding Performance of 6xxx Automotive Alloys

T. Greunz, M. Hafner, R. Gruber, T. Wojcik, J. Duchoslav, and D. Stifter

Abstract

6xxx series aluminum alloys are increasingly used within the automotive industry due to their lightweight potential. For assembling several joining techniques are common. This paper focuses on adhesive bonding, which is widely employed but its mechanism is by far not fully understood. The sheet metal process, including hot and cold rolling, solution heat treatment, pickling and Ti/Zr conversion treatment, contributes heavily to the final bonding performance. Even minor changes in the rolling process may alter the surface near deformation layer or a variation during pickling can lead to unfavorable oxidic conditions. To evaluate these changes X-ray photoelectron spectroscopy (XPS), which is extremely powerful in gathering elemental, chemical and oxidic information from the topmost surface (<10 nm), was performed. Furthermore, transmission transmission Kikuchi diffraction (TKD) in combination with a focused ion beam (FIB) preparation was employed to investigate the surface near grain structure near grain structure. Additionally, scanning electron microscopy (SEM) provided valuable knowledge concerning the surface topography.

Keywords

Aluminum • 6xxx • Adhesive bonding • XPS • FIB-TKD • SEM

T. Greunz (✉) · M. Hafner
AMAG Rolling GmbH, Lamprechtshausenerstr. 61, 5282 Ranshofen, Austria
e-mail: theresia.greunz@amag.at

R. Gruber · J. Duchoslav
CEST Competence Centre for Electrochemical Surface Technology GmbH, Stahlstraße 2-4, 4020 Linz, Austria

T. Wojcik
Institute of Materials Science and Technology, Vienna University of Technology, Getreidemarkt 9/E308, 1060 Wien, Austria

D. Stifter
ZONA Center for Surface and Nanoanalytics, Johannes Kepler University Linz, Altenberger Straße 69, 4040 Linz, Austria

Introduction

Adhesive bonding of structural elements is widely applied in the automotive industry. A major advantage of this joining technique is that there is no weakening of the substrate material in terms of a thermal or mechanical weakening zone and hence the initial material strength is preserved [1, 2]. However, a crucial point is seen in the long-life durability of the bonded joints which is significantly reduced when exposed to a humid or a corrosive environment [3–5]. The process route for sheet production comprises hot and cold rolling including intermediate annealing and a solution heat treatment. Afterwards a surface pre-treatment in terms of two-stage pickling including a rinsing cascade is common. The standard industrial production process is completed by a conversion treatment. Each step is considered to affect the final product surface in terms of the near-surface deformation as well as the elemental, chemical and oxidic conditions and hence, all these factors enter the lifetime of the bonded joints [6–10].

The current work considers 6xxx aluminum sheet material with a special focus on the influence of the pre-treatment and its contribution to the bonding performance. In particular, the influence of pickling was probed within the first study. In a second one, the effect of subsequent rinsing (after pickling) was separately scrutinized, and the findings were related to the results from tensile shear strength testing of the bonded joints. To simulate the long-time influence of harsh corrosive environment, an accelerated corrosion test was performed within each survey.

The grain structure from the near-surface deformed layers, which was investigated with respect to the pickling test,

© The Minerals, Metals & Materials Society 2023
S. Broek (ed.), *Light Metals 2023*, The Minerals, Metals & Materials Series,
https://doi.org/10.1007/978-3-031-22532-1_54

was visualized by a combination of focused ion beam (FIB) lamella preparation and transmission Kikuchi diffraction (TKD). The benefit of this experimental combination is seen in the significantly higher spatial resolution compared to conventional electron backscatter diffraction (EBSD) [11, 12].

Elemental and chemical information, as well as the determination of the oxide layer, was was obtained by XPS standard and angle-resolved analysis. This method stands out especially due to its high surface sensitivity [13–15] and has proven successful within the study on the changes in the rinsing conditions.

Experimental

All samples presented in this publication were taken from the AMAG rolling GmbH production line. The The Al-Mg-Si alloy composition was adjusted after EN 573-3 with elemental limits as follows: 1.0–1.5 wt.% Si, max. 0.50 wt.% Fe, 0.25–0.6 wt.% Mg and max. 0.20 wt.% Cu, Zn and Mn each.

The material production process is addressed in the subsection below as well as the adhesive bonding procedure, the artificial aging and the corresponding analytical methods.

Material Production and Methods

Ingot casting process via direct chill casting was followed by a homogenization step. The hot rolling procedure comprises two material variations V1 and V2 with material thicknesses below 8.5 mm. The difference refers to a modified pass reduction regime during hot rolling and was intended to alter the surface near microstructure—the so-called deformation layer. Besides, no changes in cold rolling and skin-pass texturing, in this case, texturing, were taken. The final thickness of the aluminum sheets was 1.15 mm for the first sample set (V1 and V2). These variations were used for investigations of the surface near microstructure within the study on the effect of pickling. The material for the second test series was produced analogously to material variation V1 with a final material thickness of 1.00 mm. All further processing steps after cold rolling were carried out in the laboratory, including solution heat treatment, an alkaline and an acidic pickling. Commonly, surface rinsing with deionized water follows each pickling step to remove dissolved surface impurities of organic and inorganic nature. In contrast, the rinsing procedure was completely omitted for the second sample set simulating an extreme case of insufficient rinsing.

Specimen Preparation for Adhesive Bonding

Specimen preparation was done after DIN EN 1465. For that purpose, the samples were cut into dimensions of 100 mm × 25 mm, whereas the short side is parallel to the rolling direction. Surface cleaning with petroleum ether and acetone is progressing the bonding step. Immediately afterwards the specimens were bonded with a 1 K epoxy-based resin adhesive (Betamate BM 1630, Dow Chemical). The resulting adhesive joint covers an area of 10 mm × 25 mm. The specimen preparation is completed by a curing procedure (180 °C/30 min).

Aging and Mechanical Testing

To accomplish artificial aging, the neutral salt spray test (NSST) after DIN EN ISO 9227 was implemented. In respect thereof the bonded specimens were mounted in the salt spray cabinet with an inclination of 70° with respect to the horizontal line and exposed to a neutral mist of 5 wt.% NaCl at 35 °C for 360 h. Afterwards the samples were taken from the chamber and dried under ambient conditions. A minimum time of 72 h was given prior to continuing with the determination of the tensile shear strengths of the bonded joints, including reference and artificially aged specimens. These measurements were executed on a ZwickRoell assembly.

Focused Ion Beam (FIB) Lamella Preparation and Transmission Kikuchi Diffraction (TKD)

The FIB lamella preparation was carried out on a ThermoFisher Scios II system. Prior to preparation, amorphous carbon and a tungsten protection layer were applied by vapor deposition. The obtained lamella was cut parallel to the rolling direction measuring approximately 20 μm in length and around 10 μm into the depth with approximately 50–100 nm thickness.

Transmission Kikuchi diffraction (TKD) was performed on a ZEISS Sigma 500 VP scanning electron microscope in transmission mode that requires a sample thickness of 50–100 nm. The investigations were carried out with an acceleration voltage of 30 kV and a step size of 20 nm. In contrast to common electron backscatter diffraction (EBSD) the benefit of the FIB preparation in combination with TKD analysis lies in the high local resolution of ∼20 nm that is achievable by a small interaction volume of the lamella with 50–100 nm in thickness. Data analysis was achieved with the EDAX OIM 8.6 software package.

Scanning Electron Microscopy (SEM)

Surface topography investigations were done on a ZEISS EVO 40 scanning electron microscope (SEM equipped with a tungsten filament. All data were recorded in the secondary electron image (SEI) mode with an acceleration voltage of 15 kV.

Angle Resolved X-ray Photoelectron Spectroscopy (AR-XPS)

Data collection was was realized on a Theta Probe system from ThermoFisher (UK). The system is equipped with a monochromated AlK_α X-ray radiation (1486.7 eV) and a hemispherical section analyser (HSA) with a multichannel-plate detector that facilitates simultaneous data acquisition of up to 112 energy channels and 96 angle channels at a single position. All measurements were recorded with a pass energy of 50 eV, which was considered as a reasonable compromise between count rate and data resolution for standard and angle-resoled data collection. A dual food gun system providing Ar^+ ions and low kinetic electrons to the sample surface was used for charge neutralization. Data acquisition and assessment were achieved with the supplier's Avantage software package. Angle resolved data enables the determination of a so-called relative depth plot [16] that facilitates an average depth ordering of elemental, chemical or oxide groups within the information depth (<10 nm) of XPS. The information gain is of qualitative nature since no depth information is considered in the calculation. The relative depth of a single element, or in general, of a functional group is determined by the negative logarithm of the ratio of the respective bulk- and surface near concentrations. The bulk near concentrations were recorded under an angle of 24° and the surface near ones under an angle of 61° with respect to the surface normal. Elements or functional groups that are distributed over an extended depth range appear at the average position of the plot.

Results and Discussion

In this section the results from the investigations firstly, reflecting the influence of the pickling and secondly, the impact of subsequent rinsing, on the bonding performance are discussed. The analytical methods are selected in such a way, as to either address the surface near microstructure or the surface chemistry and the oxidic overlayer.

Investigation of the Deformation Layer Region Prior to Pickling and its Impact on the Adhesive Bonding Performance

The following investigation aims the examination of the deformation layer and its potential negative influence on the bonding performance. The visualization of this very fine characteristic grain structure on the surface near zones from samples V1 and V2 (Fig. 1) was facilitated by the combination of TKD analysis on FIB prepared lamellas. TKD probes the crystal structure with a significantly improved resolution compared to EBSD, and as a result, a mapping depicting the crystallographic orientation is obtained, which enables even the evaluation of the ultra-fine grains within the outermost surface region. Isolated white areas within the mapping in Fig. 1 refer to positions where a unique orientation was not identifiable and therefore, the number and the area of the affected grains were not assessable. Accordingly, these regions were excluded from further evaluation. Nevertheless, values of ∼0.1 to ∼0.15 μm in terms of an *equivalent sphere diameter* were found to characterize the

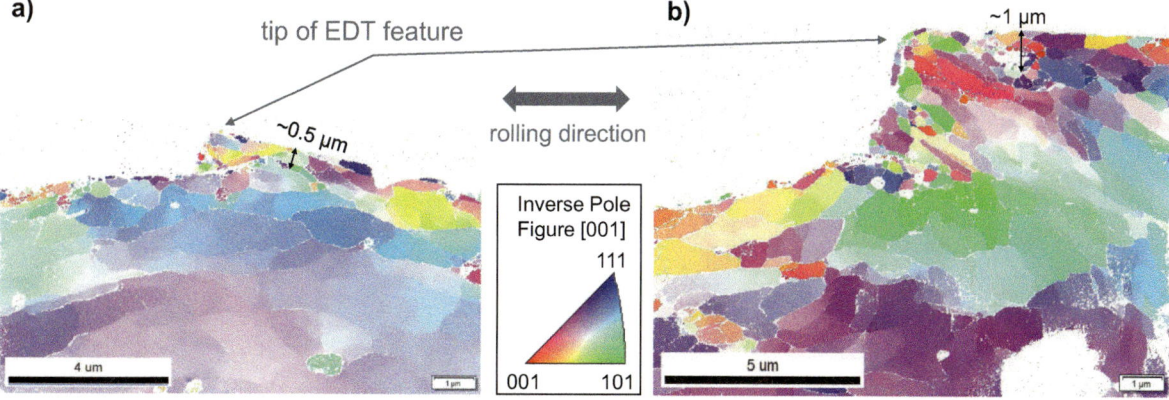

Fig. 1 The characteristic fine grain within the surface/near-surface related regions visualized by means of TKD data analysis on FIB prepared lamellas of V1 in (**a**) and V2 in (**b**). Both mappings include the tip of a single EDT feature representing a zone that experienced a significantly high degree of deformation

grains within the first ~ 0.5 to ~ 1 μm on the topmost surface. The occurrence of this deformed layer that is assigned to the high shear deformation process expands even to depths in the micrometer range. Especially at sample variation V2, this, in many aspects unfavorable zone, is significantly more developed compared to variation V1. In general, we attribute this difference between V1 and V2 to the modified pass reduction regime and secondly to local sample position effects, e.g., the degree of deformation.

Turning to the subject of adhesive bonding, cold roll strips of V1 and V2 were taken as basic material that received a laboratory heat treatment with conditions described, e.g., in Aluminium-Taschenbuch [17]. The following customary two-stage pickling procedure was omitted prior to the bonding step. Finally, five parallel samples each were further artificially aged for 360 h (NSST), whereas the other five reference specimens remained unchanged. The results from tensile shear strength testing are presented in Fig. 2.

The columns represent the mean values of five parallel samples each. The error bars correspond to the standard deviation and are only fully displayed for the V1 data, since for the second sample set, the standard deviation, calculated for the 360 h (NSST) aged specimens, exceeds even the mean value, and was therefore excluded. Furthermore, the remaining strengths are seen on the second y-scale, exhibiting a remaining strength of only 38% for V1 and an even significantly lower value of 11% for V2. To further

classify these findings, a fracture image evaluation after [18] was done. The respective images are given in Fig. 3, revealing a nearly complete adhesive fracture for the V2 joint after debonding. According to the exposed surfaces, this failure is associated with a corrosive behavior of the metallic substrate, indicated by an almost entirely coverage with corrosion products. Unlike this, the surface for V1 appears differently since aside from an adhesive fracture also positions revealing typical cohesive failure characteristics can be observed. Their contribution is accounted for $\sim 25\%$ in total of the overall area proportion and hence, is responsible for the slightly improved remaining strength of 38%.

It can be concluded that the presence of the deformation layer highly impacts the bonding performance in a negative way. Unfavorable conditions of this outermost surface zone are seen, e.g., after the delamination model from Wang et al. [19], which was proposed by the example of AlFeSi cold roll sheets. They suggested a reduced ductility and the tendency for the nucleation sites of sub-surface cracks during forming. These cracks include voids and grain boundaries that are decorated with oxidic species. Accordingly, these prerequisites lead to a progressing crack propagation within the near-surface/bulk interface and finally to delamination. Additional artificial aging as described above is considered to enhance the delamination process yielding a poor bonding performance finally due to corrosion.

Fig. 2 Results from tensile shear strength of samples V1 and V2. A drastic decrease in the remaining strength is observed for artificial *aging* after 360 h (NSST). Especially sample variation V2 performs poorly with only 11% of the original strength remaining

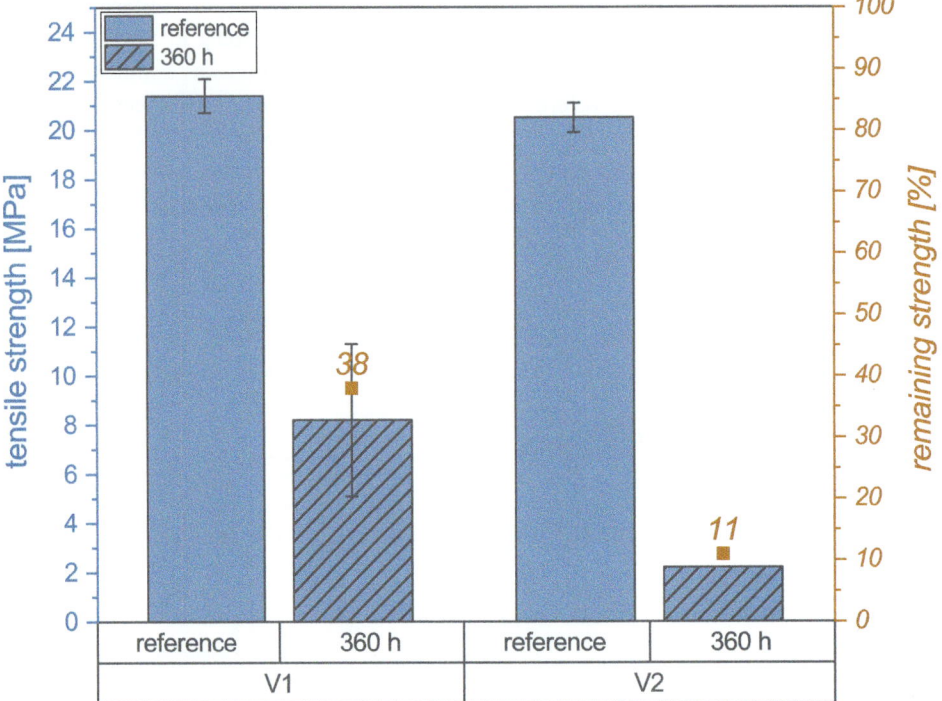

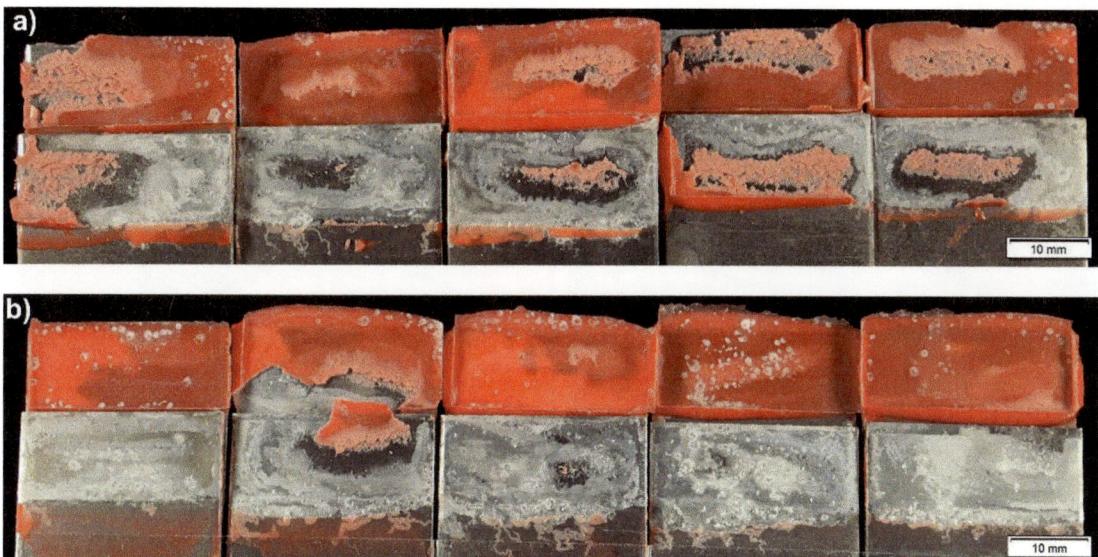

Fig. 3 The exposed surfaces after debonding for V1 are shown in (**a**) and for V2 in (**b**). The corrosive attack of the metallic substrate is responsible for the significantly low performance

Characterization of the Surface Topography, Chemistry, and Oxide Overlayer Thickness with Respect to the Effect of Rinsing after Pickling

The previous investigations aimed at the investigation of the deformation layer, which is commonly removed by two-stage pickling within the industrial process and hence, its negative influences on, e.g., the bonding performance is inhibited. Nevertheless, each pickling step is followed by a proper rinsing to ensure the removal of unwanted residues of organic or inorganic nature. The following analysis is considered as an extreme trial referring, on the one hand, to an insufficient and, on the other, hand to an optimized rinsing condition. By means of XPS analysis in standard and angle-resolved mode as well as SEM imaging, the initial surfaces prior to the bonding are characterized. The results are compared to each other and related to the respective findings from adhesive bonding.

The substrate material was processed in the same way as the previously introduced variant V1 but with a cold roll thickness of 1.0 mm. Both specimen variations, in the following noted as "no rinsing" and "improved rinsing", experienced an immersion into an alkaline and acidic pickling media for 10 s with the difference that the rinsing with deionized water (flow rate: 12 l/min; time: 8 s) after each pickling stage was omitted for the "no rinsing" sample variation. Finally, all surfaces were dried under ambient conditions.

At the beginning, the initial surface topographies were investigated by SE imaging (Fig. 4), from which clear differences were determinable. The surface of the sample without rinsing is characterized by a considerable crack pattern of an overlayer (Fig. 4a) that fully covers the surface and impedes the identification of the typical EDT. In contrast, the improved rinsed surface (Fig. 4b) clearly reveals no peculiarities in terms of undefined buildups related to the previous pickling routine.

Afterwards the adhesive bonded joints were prepared as described in section "Experimental Section"(c) and tested. The results from tensile shear strength testing are presented in Fig. 5 for reference and 360 h (NSST) aged samples. From there, it can undoubtedly be taken that by omitting the rinsing, reduction in the tensile strength must be expected. This effect is already observed for the unaged specimens, where ~18 MPa are determined for the "no rinsing" and ~22 MPa for "improved" sample variations. This trend drastically continues for the artificially aged samples. Finally, a remaining strength of 28% was found for the sample without rinsing and a value of 93% for the sample with improved conditions.

This substantial difference originates to a certain extent from the diverging surface morphologies and topographies as seen in Fig. 4. However, the role of the oxide layer and the respective chemistry shall be considered as well. Therefore, XPS measurements in standard and angle-resolved mode were carried out. Charge referencing was done by adjusting the C–C/C–H functional group to the binding energy of 285.0 eV (Fig. 6a and d). The involved high-resolution binding energy levels from standard measurement mode are presented in Fig. 6, including peak deconvolution. From these spectra, the concentrations were derived and listed in Table 1 together with the corresponding binding energies.

Firstly, the oxide layer thickness is calculated. For that purpose, we follow the approach of Strohmeier [20], which

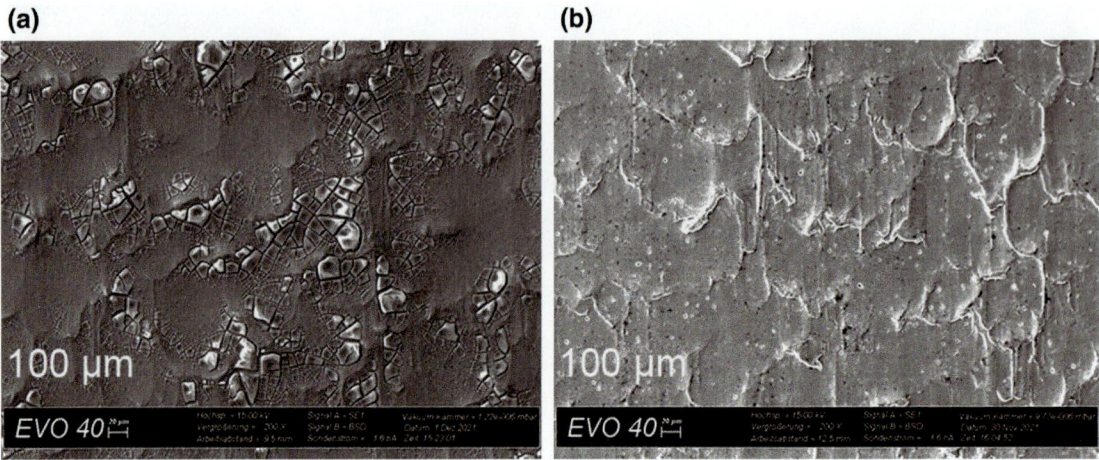

Fig. 4 SE imaging on the surface without rinsing (**a**) versus the improved rinsed sample in (**b**). These surfaces represent the reference state prior to the adhesive bonding

Fig. 5 A significant decrease in tensile shear strengths for shortcomings in the rinsing procedure is noticeable

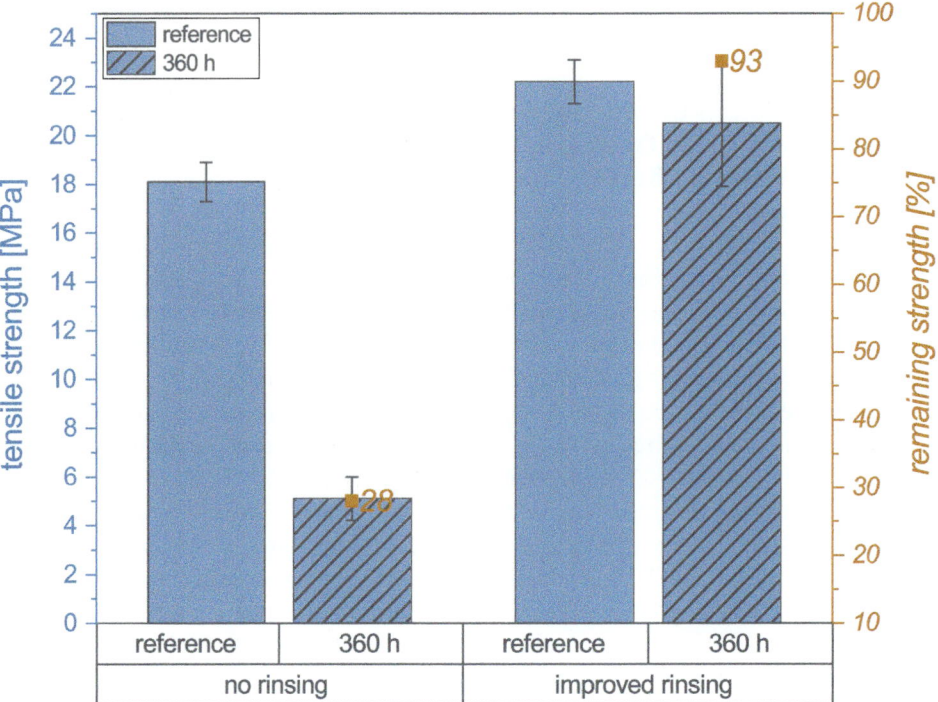

is based on a uniform overlayer model. It represents a feasible way that solely requires the intensity ratios determinable from oxidic and metallic Al2p level taken from standard high-resolution data. The corresponding concentrations are given in Table 1. The overlayer thickness d is now calculated by:

$$d = \lambda_o \; \sin \; \Theta \; \ln\left[\frac{N_m \lambda_m}{N_o \lambda_o} \frac{I_o}{I_m} + 1\right] \qquad (1)$$

where λ_m, and λ_o are the inelastic mean free paths (IMFPs) of the photoelectrons in the metal and the oxide,

respectively. Furthermore, N_m/N_o is the related ratio of the volume densities of the metal atoms in the metal or in the oxide. The electron take-off angle Θ measured with respect to the sample surface refers to the specific experimental setup conditions. Finally, the metal and oxide intensities, I_m and I_o, respectively, from the Al2p photoelectron peak are involved. However, from the latter quantities, it becomes apparent that this method is not applicable for the conditions represented in Fig. 6b since no metallic contribution is observable within the Al2p spectrum. This finding reflects the fact that the oxide overlayer thickness exceeds the information depth of this approach, which is estimated to be

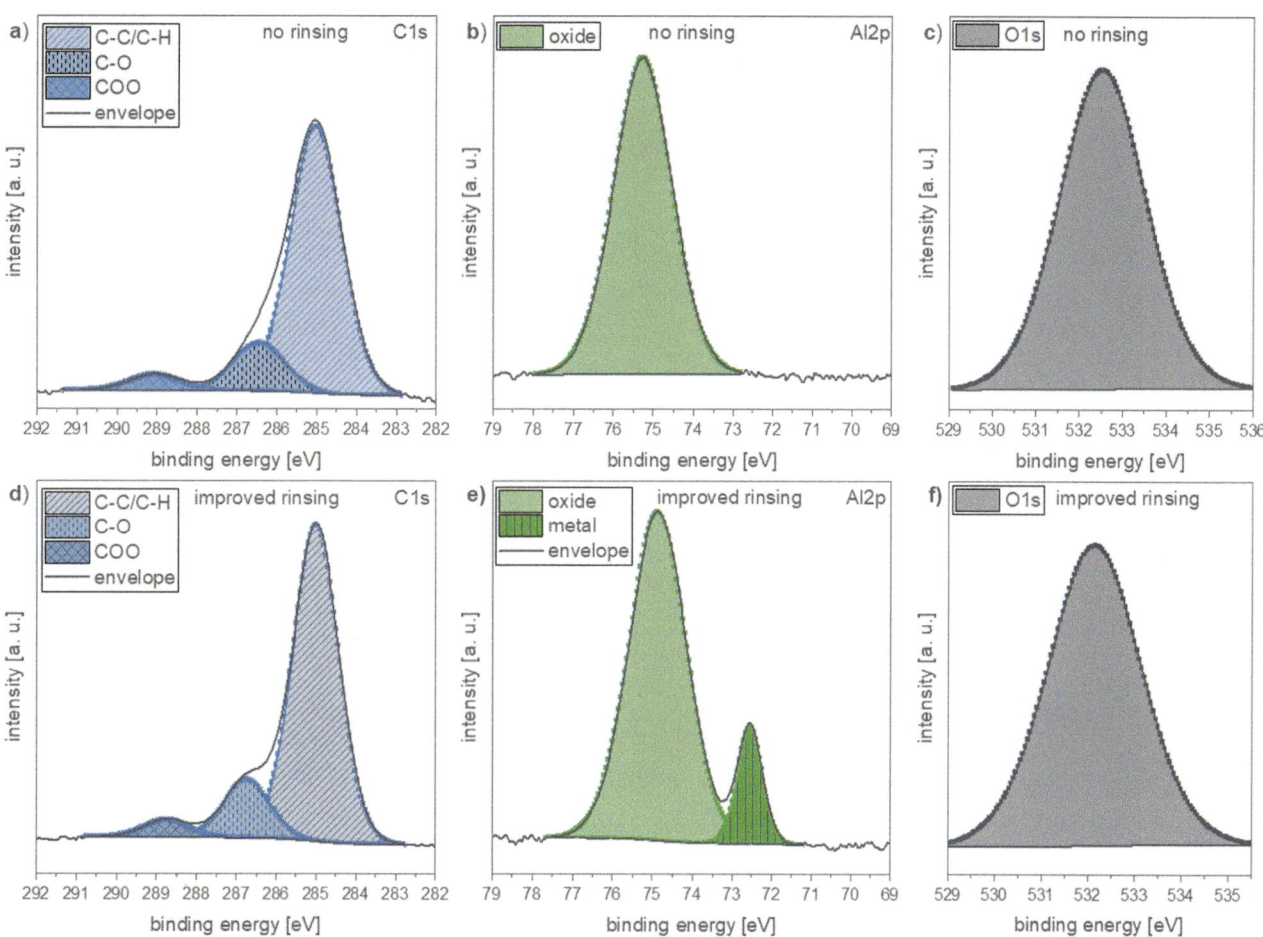

Fig. 6 Deconvoluted XPS high-resolution data from C1s, Al2p and O1s levels. Figures **a–c** refer to the sample without rinsing and the data in **d–f** to the surface with an optimized rinsing

Table 1 Elemental concentrations in at.% and peak binding energies E_b [eV] derived from standard XPS high-resolution data in Fig. 6

Rinsing	Al2p conc. (at.%) E_b [eV]		C1s conc. (at.%) E_b [eV]			O1s conc. (at.%) E_b [eV]	F1s conc. (at.%) E_b [eV]	Residues conc. (at.%) E_b [eV]
	Oxide	Metal	C–C/C–H	C–O	COO			
Without	**17.0** *75.3*		**27.1** *285.0*	**5.1** *286.7*	**1.7** *289.1*	**45.0** *532.5*	**4.1** *686.0*	**0.0**
Improved	**17.3** *74.9*	**2.8** *72.5*	**28.2** *285.0*	**5.3** *286.7*	**1.6** *288.8*	**42.0** *532.1*	**1.6** *686.2*	**1.2**

between ~ 7.5 and 8.5 nm at maximum [20]. Accordingly, the oxide overlayer thickness can only be determined within this approach for the improved rinsed sample surface. Taking the parameters from Table 2 into account, a final value of ~ 5.5 nm for the oxide thickness is calculated from Eq. (1).

Secondly, the evaluation of the standard high-resolution O1s levels Fig. 6c and f to reflect the oxidic nature of the respective overlayer is discussed. A comparison of the two spectra in Fig. 6c (without rinsing) and Fig. 6f (improved

rinsing) exhibit no considerable differences, and due to the highly symmetric peak shapes, a reliable peak deconvolution into organic, hydroxidic and oxidic functional groups was not considered as a reliable practice. However, instead of taking the absolute binding energies for peak identification into account, the binding energy difference ΔE_b from the respective O1s and Al2p$_{oxide}$ levels (Table 1) provides a much more feasible way to accomplish this issue since charge shifting can be neglected and therefore, a comparison with data from literature is facilitated. Now for both investigated

Table 2 The input parameters after [20] and the photoelectron peak intensities from Table 1 for the determination of the oxide overlayer thickness after Strohmeier

I_o (at.%)	I_m (at.%)	$IMFP_o$ (nm)	$IMFP_m$ (nm)	N_m/N_o	sin Θ (rad)	D (nm)
17.3	2.8	2.8	2.6	1.5	0.9	5.5

surfaces a ΔE_b (O1s-Al2p$_{oxide}$) of 457.2 eV was assessed. According to Strohmeier [20] the binding energy differences for the oxide and hydroxide components are found to be 456.8 eV (ΔE_{oxide}) and 458.3 eV ($\Delta E_{hydroxide}$), respectively. Hence, taking the information depth of this method into account, it is highly presumable that both investigated overlayers consist of a mixture of oxide, hydroxide, oxy/hydroxy(-fluoride) and to a certain content of organic species [21] though to a not further specifiable content.

Finally, the measured carbon concentrations that stem from, e.g., contaminates from ambient as well as from residues of the pickling media, are discussed. A high concentration of over 30 at.% independently from the rinsing conditions was measured. Moreover, both samples exhibit nearly equal amounts of C–C/C–H (including adventitious carbon), C–O (hydroxy/ether) and COO (carboxyl/ester) groups (Fig. 6a and d) and the assessed concentrations are provided in Table 1. Taking a carbon concentration of $\sim$30 at.% into account, the question on the depth localization arises, since carbon contaminants, e.g., at the oxidic overlayer/substrate interface may not be beneficial for an improved bonding performance. Whereas carbon species from dry lubricants on the top surface of the overlayer can be absorbed to a certain extent from the adhesive resin and therefore no negative effects are expected regarding adhesive bonding results.

To obtain an estimation on the depth location of the two main carbon functional groups (C–C/C–H and C-O) within the XPS information depth, a so-called *relative depth* plot is given in Fig. 7 for the improved rinsed sample. The necessary surface and bulk concentrations required for the calculation of the relative depths are provided in Table 3.

Within the relative depth plot (Fig. 7) a relative ordering of elemental and chemical groups against each other is enabled. From this plot, it can be taken that the carbon containing groups are situated at the topmost surface sites, followed by an oxidic/hydroxidic zone (O1s and Al2p$_{oxide}$) comprising a lower amount of fluorine. The bulk region is represented by the Al2p$_{metal}$ contribution. To be clearer,

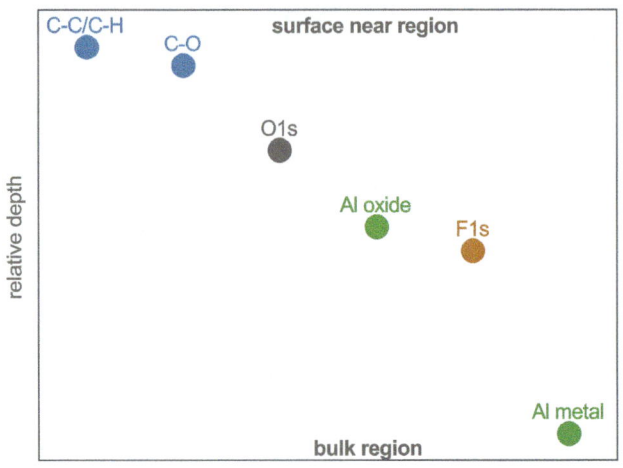

Fig. 7 Relative depth plot derived from angle-resolved high-resolution XPS data revealing the relative ordering of the fitted elemental or chemical components over information depth for the sample with improved rinsing conditions

there is no apparent hint that the interfacial region between the overlayer(oxidic/hydroxidic) and the metallic bulk is affected by a significantly high amount of carbon contamination. Hence, as confirmed by the data of the remaining strength, no negative influence with respect to the adhesive bonding performance could be identified in this certain case.

Finally, from SEM imaging and XPS analysis it is concluded that the morphology, here in particular, the oxide/hydroxide/oxy-hydroxy(fluoride) crack pattern is judged as the largest negative contributor to the poor bonding performance. The crucial point is ascribed to the overlayer that reached in the one case due to omission of the rinsing, an unfavorable high value. Therefore, this overlayer resides as a brittle and undefined structure on the metallic substrate. By means of the XPS results it was shown that both analyzed surfaces (without and with improved rinsing) reveal a quite similar nature with respect to elemental, chemical and oxidic concentrations and therefore, are not considered to be responsible for the extreme difference in the bonding performance.

Table 3 Concentrations in at.% of elemental and chemical species involved in the relative depth plot of the improved rinsed sample

		C1s		O1s	Al_{oxide}	F1s	Al_{metal}
		C–C/C–H (at.%)	C–O (at.%)	(at.%)	(at.%)	(at.%)	(at.%)
	Bulk	24.5	3.8	40.1	25.1	1.5	5.1
	Surface near	35.1	5.1	38.9	18.2	1.0	1.7
	Relative depth	*0.36*	*0.29*	*−0.03*	*−0.32*	*−0.41*	*−1.10*

Conclusion

The focus of this work was set on the examination of surface characteristics that are attributed to highly affect the adhesive bonding performance. In particular, the heavily deformed near-surface layer consisting of an ultra-fine grain structure was investigated by a combination of FIB preparation and TKD. Within this approach the spatial resolution is significantly increased compared to conventional EBSD and hence enables the grain size determination in ranges of 0.15 μm and smaller. Furthermore, these data enable the depiction of the grain distribution and therefore, allow determining the thickness of this deformed region. Moreover, these results could be related to the findings from tensile shear strength tests. From this it is concluded that in case of poor adhesive bonding performance, the pickling procedure with respect to the removal of the deformation layer shall be reconsidered.

By means of standard XPS analysis, the oxide thickness was calculated for the sample with improved rinsing conditions in a manner that solely urges the evaluation of the high-resolution Al2p level. The determined thickness was ~5.5 nm. For the sample without rinsing, representing the extreme case of insufficient rinsing, the oxide thickness was not accessible via this method and SEM image revealed a cracked surface pattern, most likely of oxide/hydroxide mixture on the initial sample surface. This unfavorable surface condition is reflected in the data obtained from tensile strength evaluation. Therefore, proper rinsing after pickling is assumed as a prerequisite for excellent bonding performance. Furthermore, it was shown that the relative depth plot determined from angle resolved XPS data helps to achieve a more differentiated picture of elemental or chemical gradients within the information depth of XPS.

Acknowledgements Financial support from the Austrian Government and the provinces of Upper and Lower Austria within the COMET Program managed by the Austrian Research Promotion Agency (FFG) is gratefully acknowledged.

References

1. Cavezza F, Boehm M, Terryn H, Hauffman T (2020) A Review on Adhesively Bonded Aluminium Joints in the Automotive Industry. *Metals* 10(6):730. https://doi.org/10.3390/met10060730
2. Barnes T, Pashby I (2002) Joining techniques for aluminium spaceframes used in automobiles. J. Mater. Process. Technol. 99:72–79
3. Watts J (2010) Role of Corrosion in the Failure of Adhesive Joints. Shreir's Corros. 3:2463-2481
4. Prakash R, Srivastava VK, Gupta GSR (1987) Behavior of adhesive joints in corrosive environment. Exp. Mech. 27:346–351. https://doi.org/10.1007/BF02330305
5. Silva, LFM, Sato, C (2013) Design of Adhesive Joints Under Humid Conditions. Springer Berlin Heidelberg
6. Zhou X, Liu Y, Thompson GE, Scamans GM, Skeldon P, Hunter JA (2011) Near-Surface Deformed Layers on Rolled Aluminium Alloys. Metall. Mater. Trans. A 42:1373–1385. https://doi.org/10.1007/s11661-010-0538-2
7. Liu Y, Frolish MF, Rainforth WM, Zhou X, Thompson GE, Scamans GM, Hunter JA (2010) Evolution of near-surface deformed layers during hot rolling of AA3104 aluminium alloy. Surf. Interface Anal., 42:180–184. https://doi.org/10.1002/sia.3135
8. Prolongo SG, Ureña A (2009) Effect of surface pre-treatment on the adhesive strength of epoxy–aluminium joints. Int. J. Adhes. Adhes. 29:23–31
9. Özkanat Ö, Wit F, de Wit FM, de Wit JHW, Terryn H, Mol JMC (2013) Influence of pretreatments and aging on the adhesion performance of epoxy-coated aluminum. Surf. Coat. Technol. 215:260–265. https://doi.org/10.1016/j.surfcoat.2012.07.096
10. Saleema N, Sarkar DK, Paynter RW, Gallant D, Eskandarian M (2012) A simple surface treatment and characterization of AA 6061 aluminum alloy surface for adhesive bonding applications. Appl. Surf. Sci. 261:742–748. https://doi.org/10.1016/j.apsusc.2012.08.091
11. Keller, RR, Geiss, RH (2012) Transmission EBSD from 10 nm domains in a scanning electron microscope. J. Microsc. 245:245–251. https://tsapps.nist.gov/publication/get_pdf.cfm?pub_id=912886 (accessed 13th September 2022)
12. Trimby PW, Cao Y, Chen Z, Han S, Hemker KJ, Lian J, Liao X, Rottmann P, Samudrala S, Sun J, Wang JT, Wheeler J, Cairney JM (2014) Characterizing deformed ultrafine-grained and nanocrystalline materials using transmission Kikuchi diffraction in a scanning electron microscope. Acta Mater. 62, 69–80. https://doi.org/10.1016/j.actamat.2013.09.026
13. Cumpson PJ (2012) Angle-resolved XPS depth-profiling strategies. Appl. Surf. Sci. 144–145:16–20. https://doi.org/10.1016/S0169-4332(98)00752-1
14. Oswald S, Reiche R, Zier M, Baunack S, Wetzig K (2005) Depth profile and interface analysis in the nm-range. Appl. Surf. Sci. 252(1):3–10. https://doi.org/10.1016/j.apsusc.2005.01.102
15. Zähr J, Oswald S, Türpe M, Ullrich HJ, Füssel U (2012) Characterisation of oxide and hydroxide layers on technical aluminium materials using XPS. Vacuum 86:1216–1219. https://doi.org/10.1016/j.vacuum.2011.04.004
16. Help Manual for Thermo Avantage software provided from Thermo Fisher Scientific
17. Aluminium-Taschenbuch (1983), 14th edition, Aluminium-Verlag GmbH Düsseldorf
18. Kleben im Karosseriebau: Bewertung von Bruchbildern (DVS Richtline 3302)
19. Wang J, Zhou X, Thompson GE, Hunter JA, Yuan Y (2015) Delamination of near-surface layer on cold rolled AlFeSi alloy during sheet forming. Mater. Charact. 99:109–117. https://doi.org/10.1016/j.matchar.2014.11.011
20. Strohmeier, BR (1990) An ESCA method for determining the oxide thickness on aluminum alloys. Surf. Interface Anal. 15:51–56. https://doi.org/10.1002/sia.740150109
21. Briggs D (1998) Surface analysis of polymers by XPS and static SIMS. Cambridge University Press, England

Investigation of Resistance of Intergranular Attack for Various Heat Treated 2011 Alloys After Hard Anodizing

Ilyas Artunc Sari, Gorkem Ozcelik, Zeynep Tutku Ozen, and Onuralp Yucel

Abstract

The main alloying element of 2011 aluminum alloy is copper and it is disadvantageous from the point of corrosion compared to other aluminum alloys. 2011 alloy also contains bismuth and lead in their chemical composition. These elements have low melting point and enhance machinability of the alloy. In this way, 2011 alloys are frequently used in areas where machinability is important. Also, 2011 alloys are preferred in valve manufacture and therefore understanding its corrosion behavior is critical. In this study, 2011 aluminum billets were produced by direct chill (DC) casting and then homogenization was carried out. As the next step, billets were extruded. Different samples were taken from extruded profiles and heat treated to obtain T4, T6, T79, T73 conditions. Samples were then hard anodized. After anodizing, samples were taken to corrosive environment. Optical microscope, Scanning Electron Microscope (SEM) and Energy Dispersive Spectrometry (EDS) techniques were used for characterization and corrosion behavior of 2011 alloy under different heat treatment conditions was investigated. It was found that T6 condition, which is the most preferred heat treatment condition for 2011 alloys, has the maximum corrosion depth and other experimental results obtained agree well with expected properties.

Keywords

2011 aluminum alloy • Casting • Extrusion • Heat treatment • Hard anodizing • Corrosion

I. A. Sari (✉) · G. Ozcelik · Z. T. Ozen
ASAS Aluminium San. ve Tic. A.S., Sakarya, Turkey
e-mail: artunc.sari@asastr.com

I. A. Sari · Z. T. Ozen · O. Yucel
Istanbul Technical University, Istanbul, Turkey

Introduction

Aluminum and its alloys are one of the most abundant elements in the world and its popularity is increasing day by day in industries such as construction, aerospace and automotive. 2xxx, 6xxx and 7xxx series wrought alloys can be strengthened by precipitation hardening. Copper element in 2011 alloys forms the Al_2Cu phase and this phase precipitates in microstructure, allowing the alloy to reach a high strength. The formation sequence of the Al_2Cu phase is as follows [1].

$$SSS \rightarrow GP \text{ zones} \rightarrow \theta'' \rightarrow \theta' \rightarrow \theta\,(Al_2Cu)$$

2011 alloy, unlike other 2xxx alloys, contains bismuth and lead elements in its chemical composition. These elements have a low melting point and increase the machinability of the alloy. Thanks to these properties, 2011 alloy is preferred in sectors where high machinability is important [2].

As the purity of an aluminum alloy decreases, its corrosion resistance becomes lower. Alloying elements reduce corrosion resistance [3]. There are two types of corrosion progression in metals caused by impurities: transgranular and intergranular. The intergranular corrosion has an electrochemical mechanism due to the potential difference between grains and grain boundaries. Presence of copper increases the intergranular corrosion sensitivity of the alloys [4]. Therefore, 2xxx series alloys are more likely to corrode by intergranular corrosion [5].

In heat treatable alloys, sensitivity of intergranular corrosion varies according to their microstructure and metallurgical conditions. The microstructure of aluminum alloys significantly changes according to the heat treatment, which directly affects the corrosion behavior [6]. The intergranular corrosion sensitivity of T3 and T4 temper is lower than T6. The corrosion resistance of over-aging temper T73 is high [5]. In terms of strength, there is an opposite relationship

© The Minerals, Metals & Materials Society 2023
S. Broek (ed.), *Light Metals 2023*, The Minerals, Metals & Materials Series,
https://doi.org/10.1007/978-3-031-22532-1_55

Fig. 1 Representation of relationship between corrosion and strength [7]

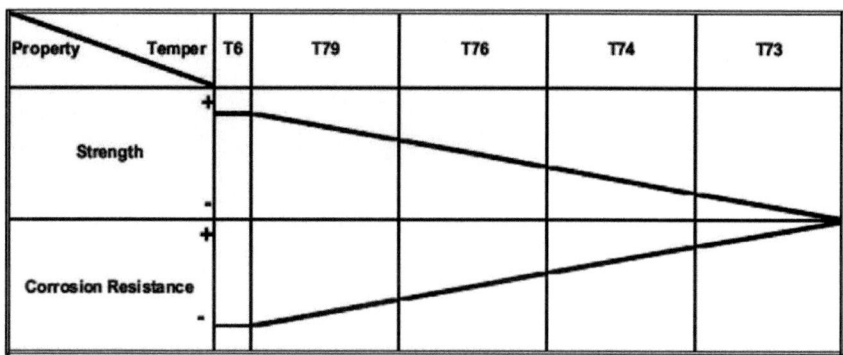

between them [7]. The strength-corrosion relationship according to heat treatments is shown in Fig. 1.

Zheng et al. [6], in their study on corrosion behavior of aluminum alloys, observed that maximum corrosion depths of T4 and T7 tempers were close to each other. It was also noticed that the maximum corrosion depth of the alloy in T6 temper was higher (119 μm) compared to underaged T4 (85 μm) and overaged T7 (83 μm) tempers [6].

In the valve industry, materials with high corrosion resistance are desired. So, the 2011 alloy, which can be used as a valve, should be able to withstand harsh corrosion conditions. There are also rare studies on this subject in the literature. For this purpose, in this study, corrosion behavior of hard anodized 2011 alloy under different heat treatment conditions (T4, T6, T79 and T73) was investigated.

Materials and Methods

14″ billets were casted with DC casting method and casted billets were homogenized at 525 °C, which is the recommended homogenization temperature for 2011 alloy. Chemical composition ranges that 2011 alloy should have ISO EN 573-3 [8] and chemical composition of casted billets in ASAS Aluminum are shown in Table 1. In Fig. 2, optical images before corrosion are given.

Mechanical strength values that 2011 alloy should have under different heat treatment conditions are shown in Table 2 with reference to EN 755-2.

T4, T6, T79 and T73 tempers were obtained by applying different heat treatments to the profiles produced in 62 MN press at ASAŞ Aluminum, which has Turkey's largest extrusion press force. Applied aging processes are shown in Table 3.

Heat treated alloys were cut in 30 × 20 × 2.5 mm dimensions with reference to ISO EN 11,846 standard and subjected to hard anodizing process. With this method, all samples were anodized at ∼25 μm thickness. Anodized samples are shown in Fig. 2.

Hard anodized samples were immersed for 24 h in corrosion solution containing 30 g of NaCl and 10 mL of HCl in 1 L of water.

In order to examine the microstructure, samples were first taken into bakelite. Mounting process was carried out by using Metkon Epocold-R and Metkon Epocold-H in a 5 to 1 ratio. After that, manual sanding was carried out. After sanding process, polishing process was fulfilled. Then, samples were analyzed under optical microscope (Zeiss Scope A1) without etching. After corroded zones were examined under optical microscope for each temper, SEM (Scanning Electron Microscopy) imaging and EDS (Energy Dispersive Spectroscopy) analyzes were performed.

Table 1 Chemical composition of casted 2011 billets (wt%)

	Cu	Fe	Bi	Pb	Al
2011 (EN 573-3)	5.0–6.0	0.7	0.2–0.6	0.2–0.6	Bal.
2011 (ASAS)	5.32	0.47	0.31	0.27	Bal.

Fig. 2 Microstructure of anodized 2011 alloy before corrosion (10x)

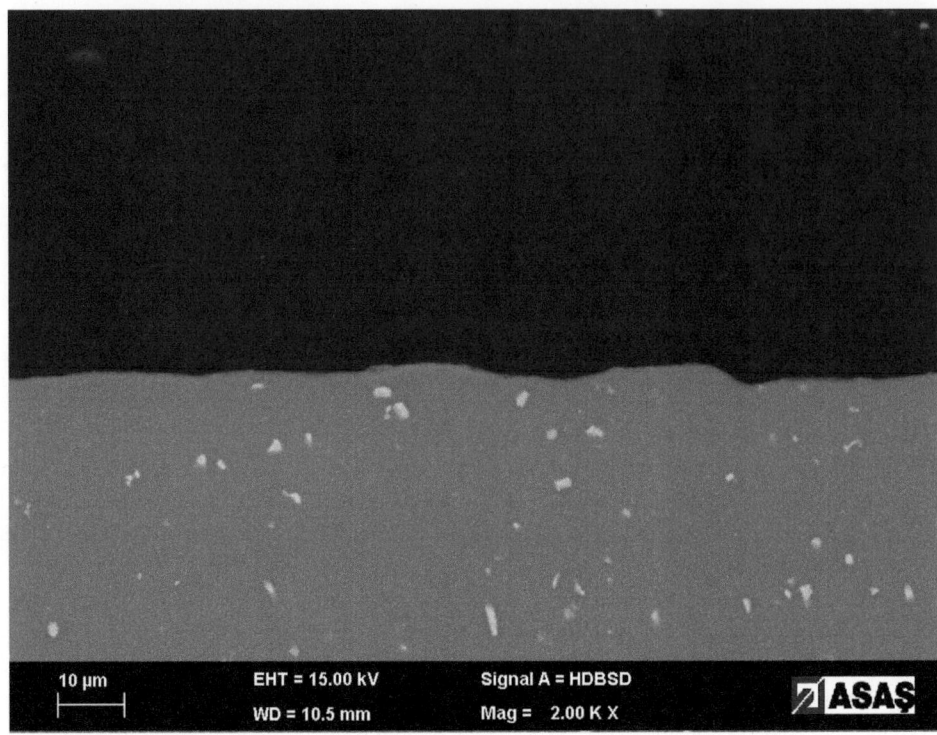

Table 2 Mechanical strengths of 2011 alloy [9]

	Yield strength	Tensile strength	Elongation	Hardness
T4	min. 125 MPa	min. 275 MPa	min. 14%	min. 95 HBW
T6	min. 230 MPa	min. 310 MPa	min. 8%	min. 110 HBW

Table 3 Applied aging treatments to achieve various tempers

Temper	Aging treatment
T4	2 months
T6	180 °C 7 h
T79	210 °C 2 h
T73	215 °C 6 h

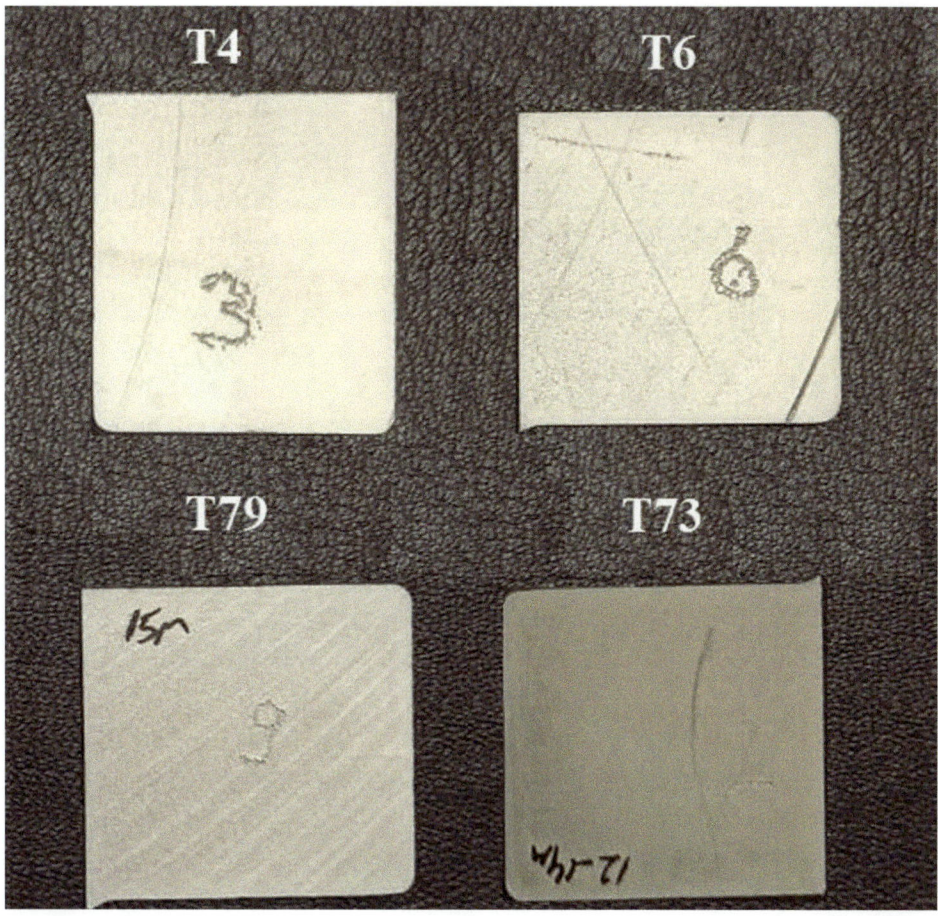

Fig. 3 Anodized 2011 samples

Results and Discussion

Microstructures of metallographically prepared samples were examined under optical microscope. Figure 3 provides the maximum corrosion depths of the 2011 alloy in tempers T4 (a), T6 (b), T79 (c) and T73 (d).

In images taken at 50 × magnification, maximum corrosion depth was measured as 54.16 μm for T4 temper, 57.61 μm for T73 temper, and 172.62 μm for T79 temper. Since the corrosion depth is much higher in T6 temper, optical imaging was performed under 20× magnification and maximum corrosion depth was obtained as 434 μm. It

has been observed that the corrosion resistance of T4 and T73 tempers is much better than others. For all heat treatment conditions, corrosion progression is intergranular.

After optical analysis, samples were examined under SEM (Fig. 4) and elemental analyzes were performed with EDS. Analyze results are in line with the examinations made under the optical microscope.

EDS analyzes were performed on white (EDS Spot 1), black (EDS Spot 2) and gray spot (EDS Spot 3) of the alloy in T4 temper (Fig. 5).

As shown in Fig. 6a, high amount (86.13%) of bismuth element was encountered in white colored intermetallic spot. Aluminum, copper, oxygen and carbon elements were also

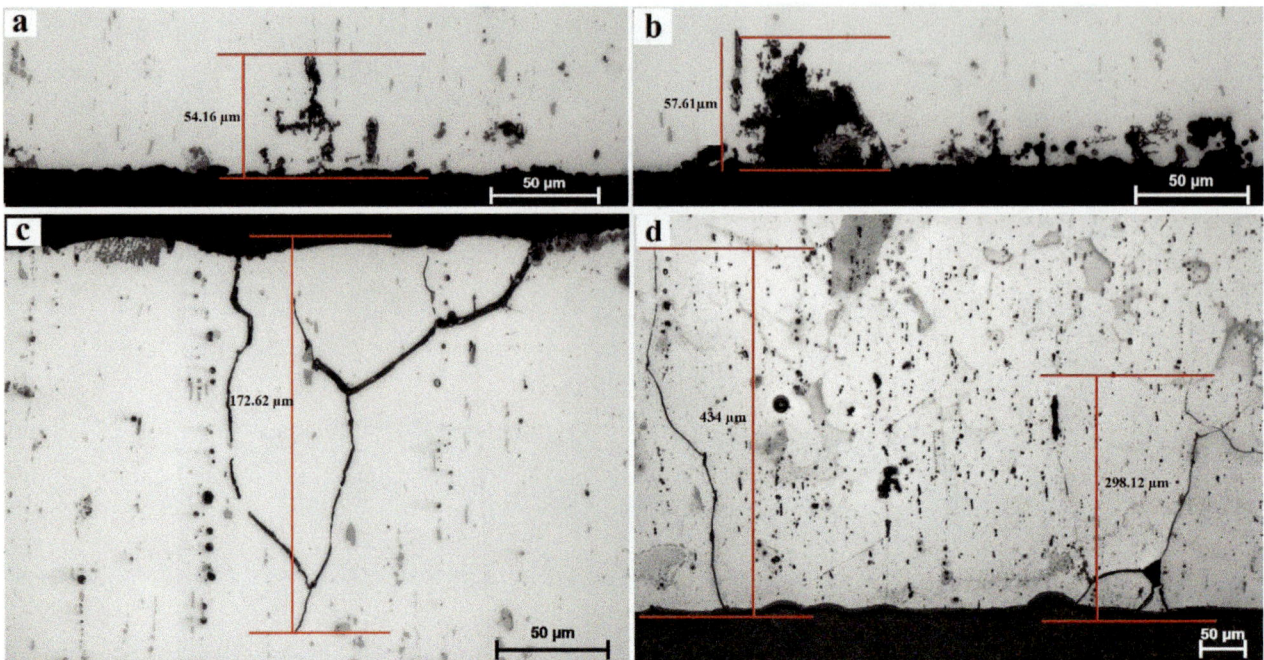

Fig. 4 Optical images of maximum corrosion depth of **a** T4 (50x) **b** T73 (50x) **c** T79 (50x) **d** T6 (20x)

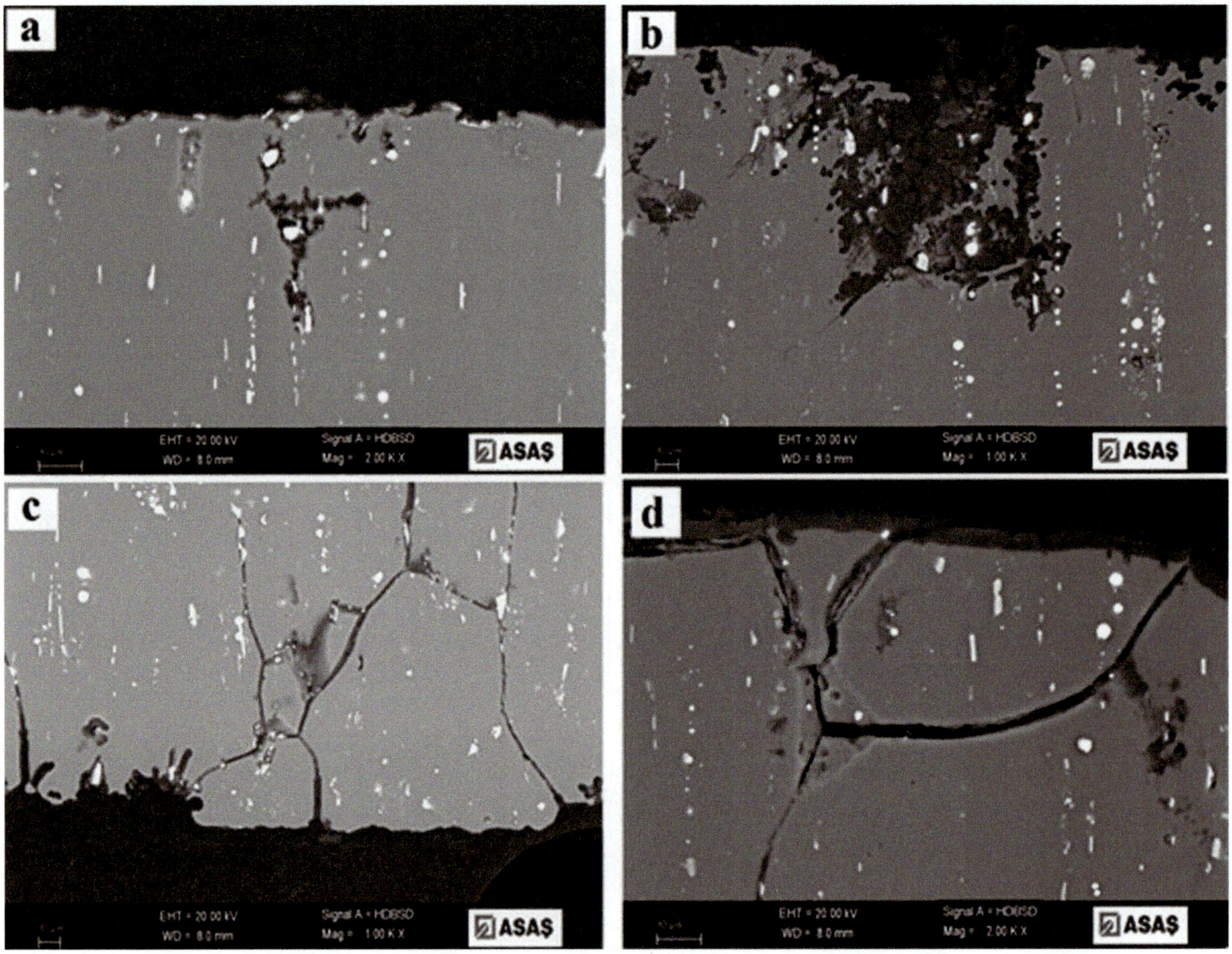

Fig. 5 SEM images of corroded 2011 alloy in **a** T4 **b** T73 **c** T79 **d** T6 temper

Fig. 6 EDS spots of T4 temper

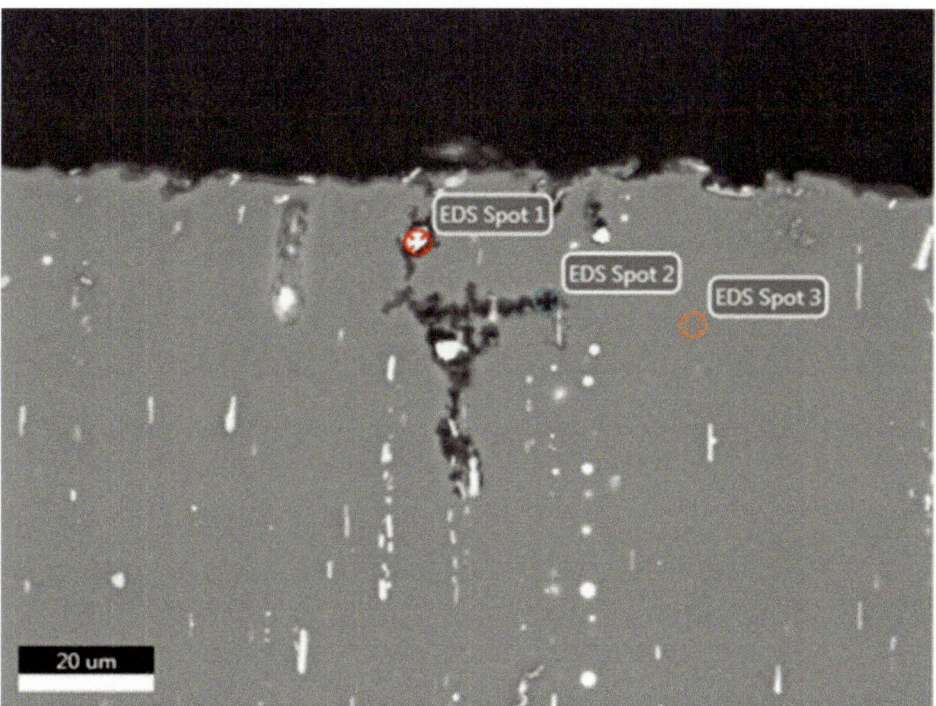

detected in the same region. As indicated in Fig. 6b, aluminum, carbon and copper elements were found in black colored corrosion spot. As an outcome of the analysis of grey spot in Fig. 6c, it was observed that 94.59% of this spot is aluminum and 5.41% is copper. It can be said that thre grey spot represents the matrix.

EDS analysis spots of 2011 alloy in T6 temper are shown in Fig. 7. As one can notice in Fig. 7, analyzes were made from spots with different contrasts.

EDS analysis revealed that weight percent of lead and bismuth elements in 2011 alloy are high in the white spot in Fig. 8a. Figure 8b displays matrix which has 95% aluminum and 5% copper. Reason for the observation of these two elements is that copper is the main alloying element of 2011

alloy. Al_2Cu phase formed in this region increases the strength. Since precipitate sizes are in the nanoscale [10], it is very difficult to detect them by SEM/EDS. Figure 8c presents the results of EDS analysis from fractured region in microstructure. As the failure had happened because of corrosion, different detrimental elements such as oxygen, chlorine, calcium are spotted out in the analysis. In 2011 alloy, aluminum, iron and copper may form Al_7Cu_2Fe intermetallic phase [11]. These elements were detected in spot 4 and can be seen in Fig. 8d.

Figure 9 indicates overall EDS spots in T79 temper. Analyzes were performed on corrosion cracking, matrix and intermetallics at different contrasts.

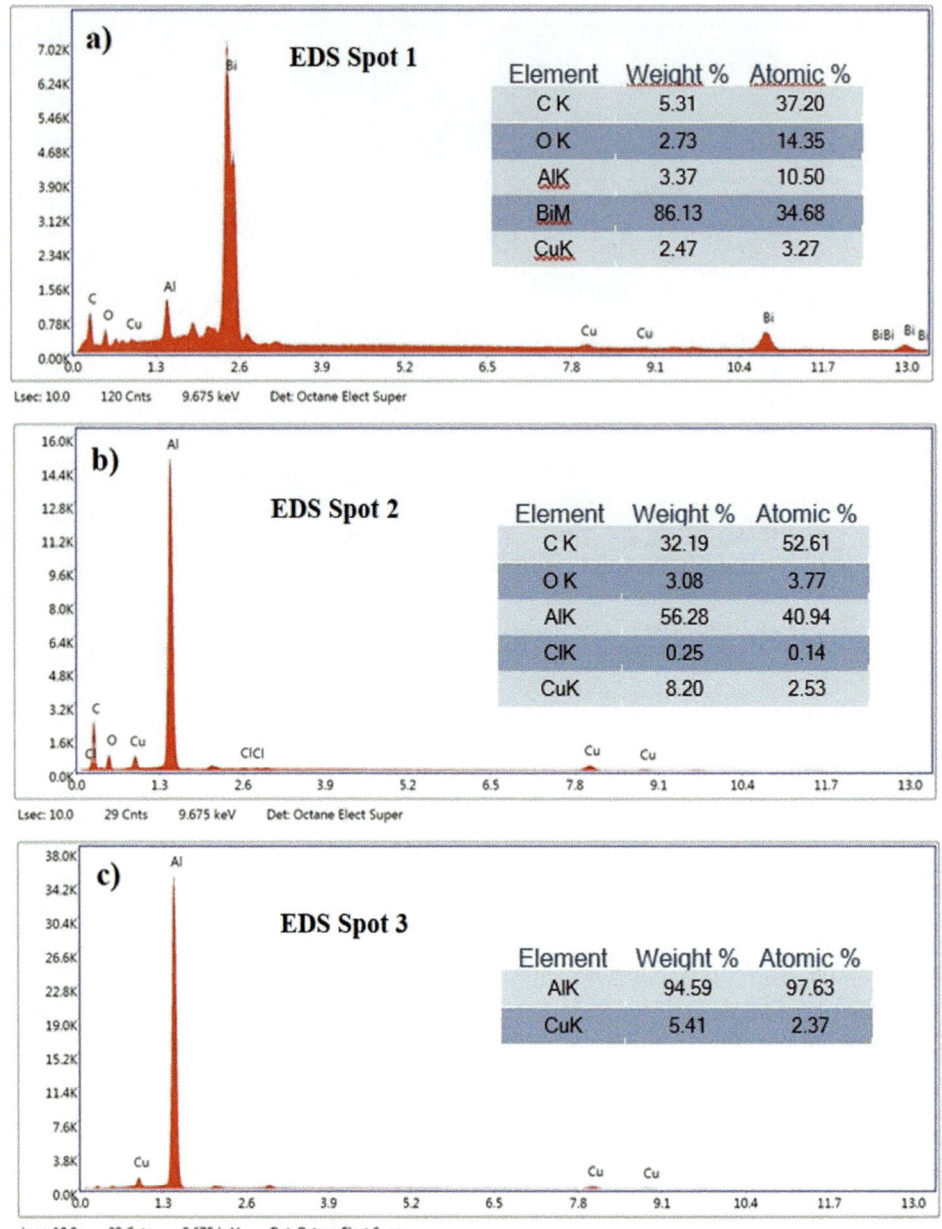

Fig. 7 EDS analysis results of corroded 2011-T4 alloy, **a** EDS spot 1, **b** EDS spot 2, and **c** EDS spot 3

Fig. 8 EDS spots of T6 temper

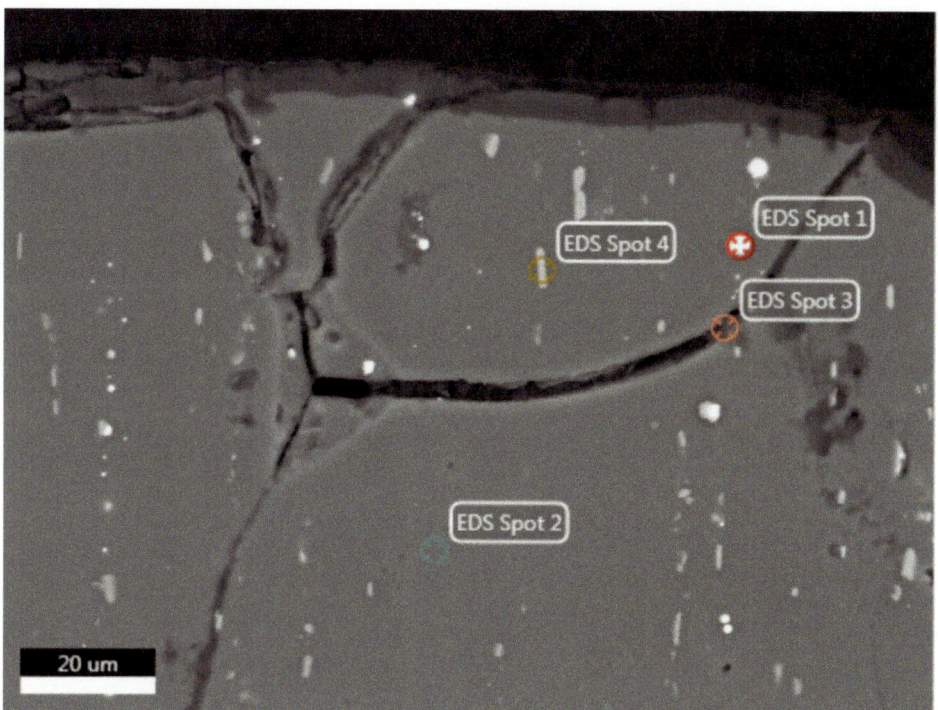

Based on EDS analyzes, Fig. 10a displays the sBiPb$_2$ intermetallic phase due to high lead and bismuth content. Figure 10b presents another intermetallic phase which contain higher bismuth concentration. Different from others, this phase is inside the corrosion crack. Figure 10c provides elemental analysis of the corrosion crack spot. In this area there are aluminum, oxygen, copper, chlorine and carbon elements. Carbon element may exist due to contamination. Figure 10d depicts matrix, similar to previous results.

Presence of 95.11% aluminum, 4.89% copper and the absence of other elements confirm this result.

Figure 11 shows EDS spots in 2011-T3 alloy. Analyzes were performed on spot 1 to find out corrosion product, spot 2 to determine composition of intermetallic phase and spot 3 to specify the matrix.

Analyzes result in Fig. 12a shows that the corroded area mainly contains aluminum and oxygen. This is because Al$_2$O$_3$ compound is formed in a related spot. Figure 12b

Fig. 9 EDS analysis results of corroded 2011-T6 alloy, **a** EDS spot 1, **b** EDS spot 2, **c** EDS spot 3, and **d** EDS spot 4

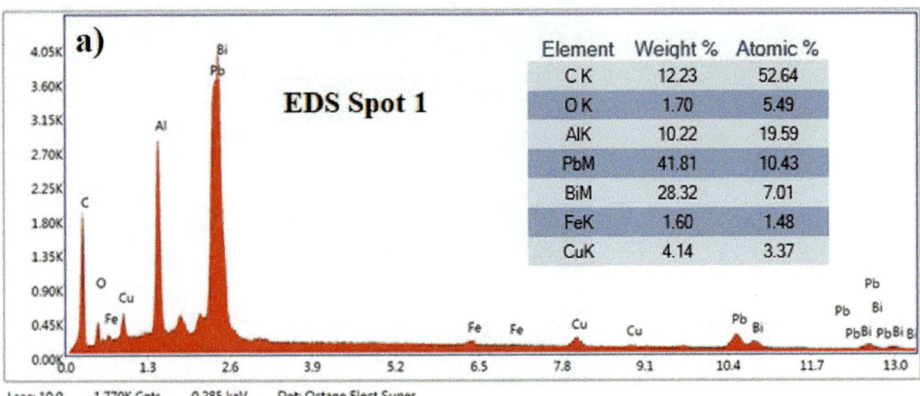

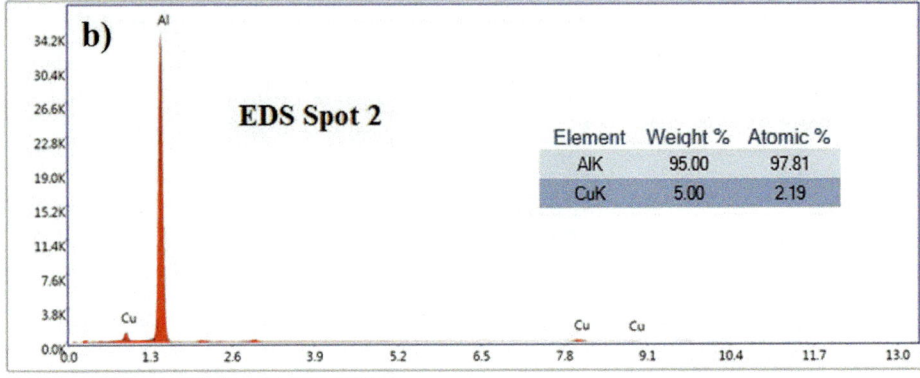

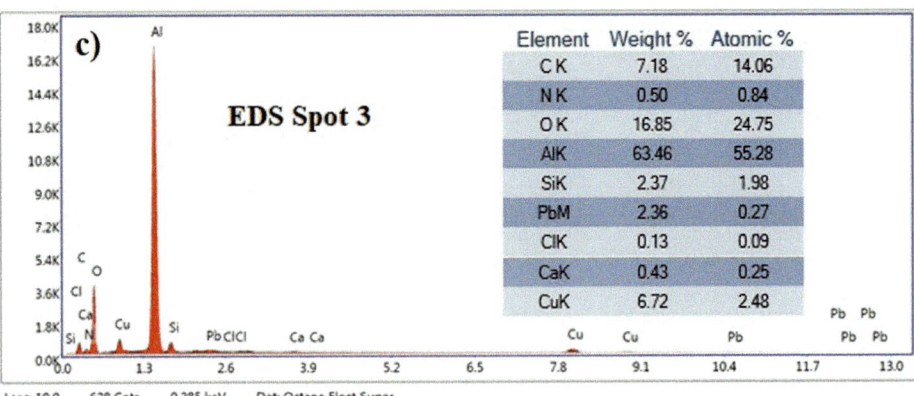

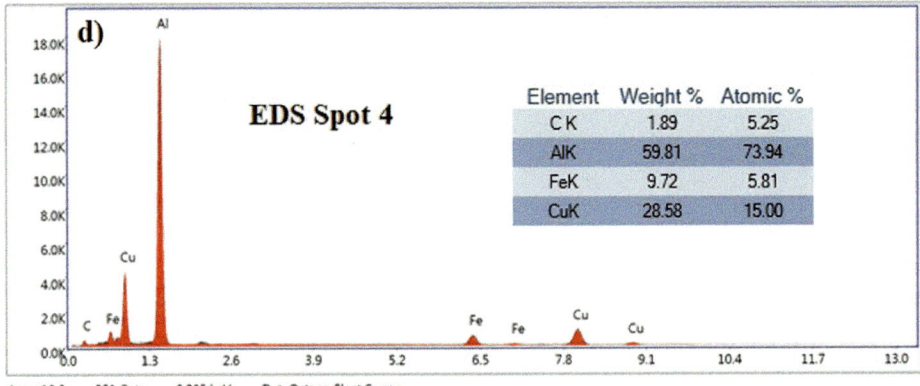

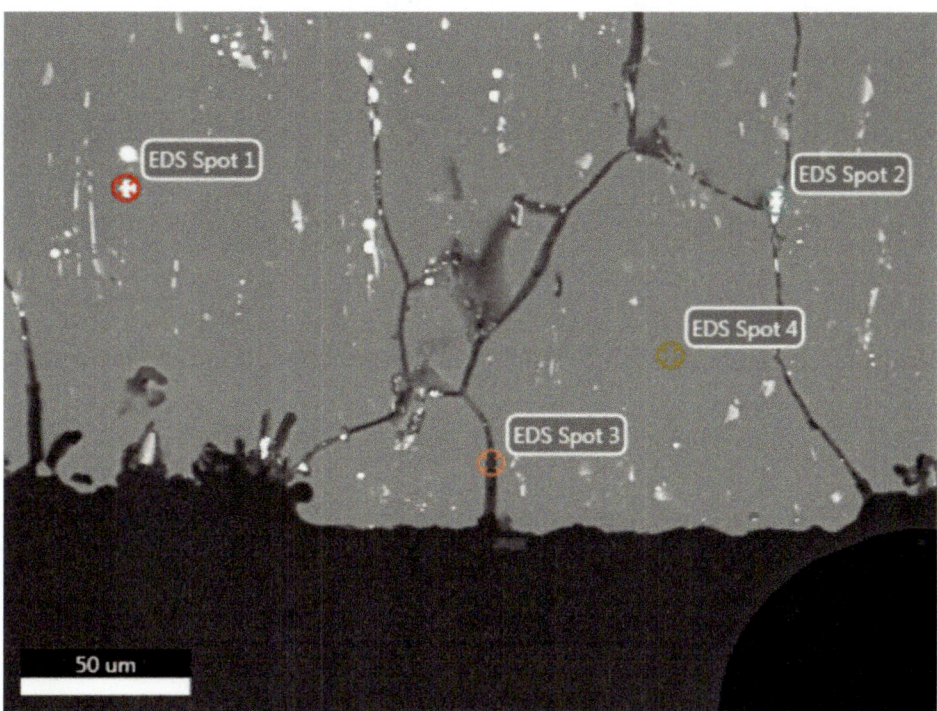

Fig. 10 EDS spots of T79 temper

demonstrates that spot 2 includes principally lead and bismuth. Together with other elements presents, this spot can be said to be of an intermetallic phase. Figure 12c displays the matrix with high aluminum content it contains and the main alloying element copper (Fig. 13).

Microstructure examinations defend that the corrosion propagation depends on heat treatment conditions. Maximum corrosion depth of 2011 alloy under different heat treatment conditions is summarized in Table 4. Obtained results are in line with the studies of Zheng et al. [6] with Al–Mg–Si alloys [6].

Fig. 11 EDS analysis results of corroded 2011-T79 alloy, **a** EDS spot 1, **b** EDS spot 2, **c** EDS spot 3, and **d** EDS spot 4

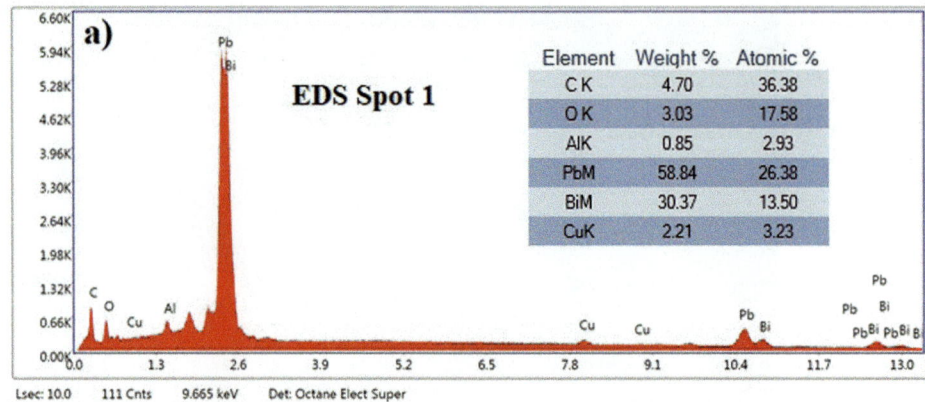

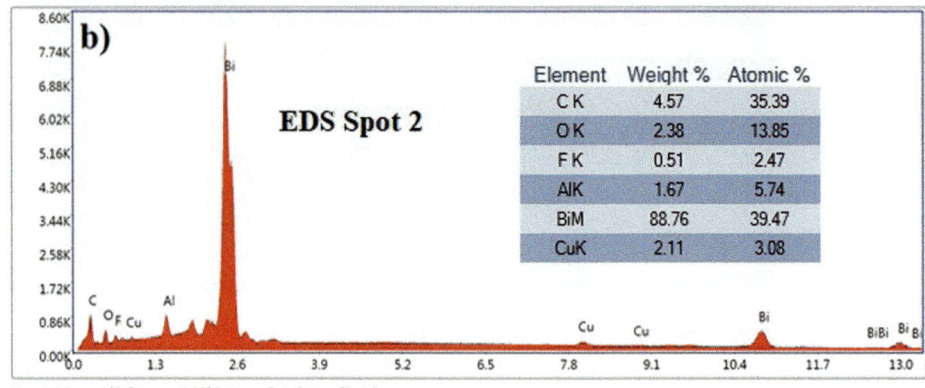

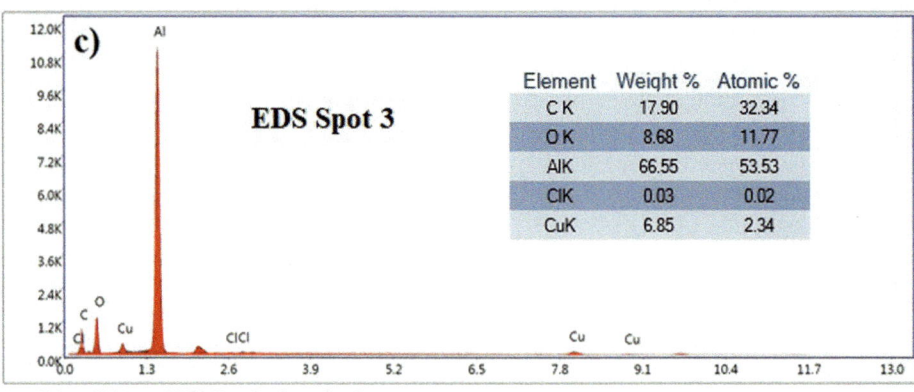

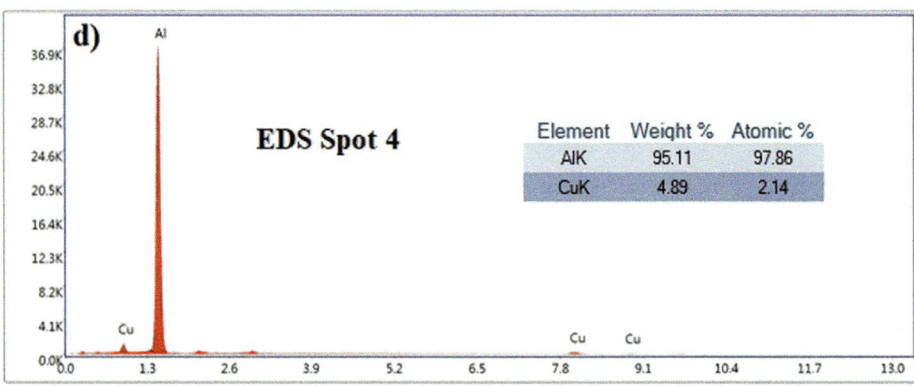

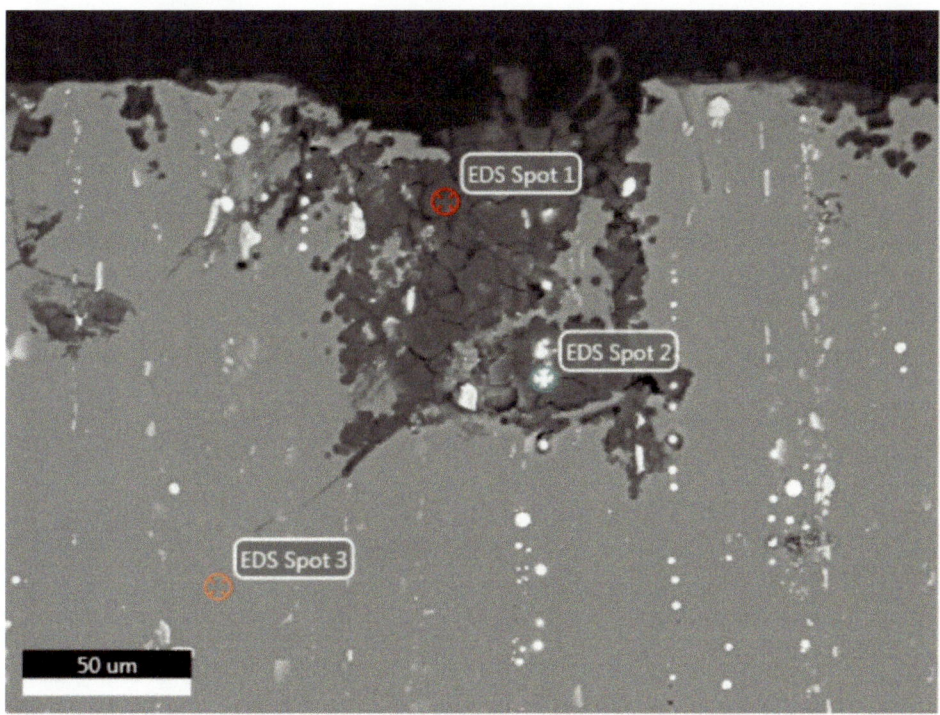

Fig. 12 EDS spots of T73 temper

Fig. 13 EDS analysis results of corroded 2011-T73 alloy, **a** EDS spot 1, **b** EDS spot 2, and **c** EDS spot 3

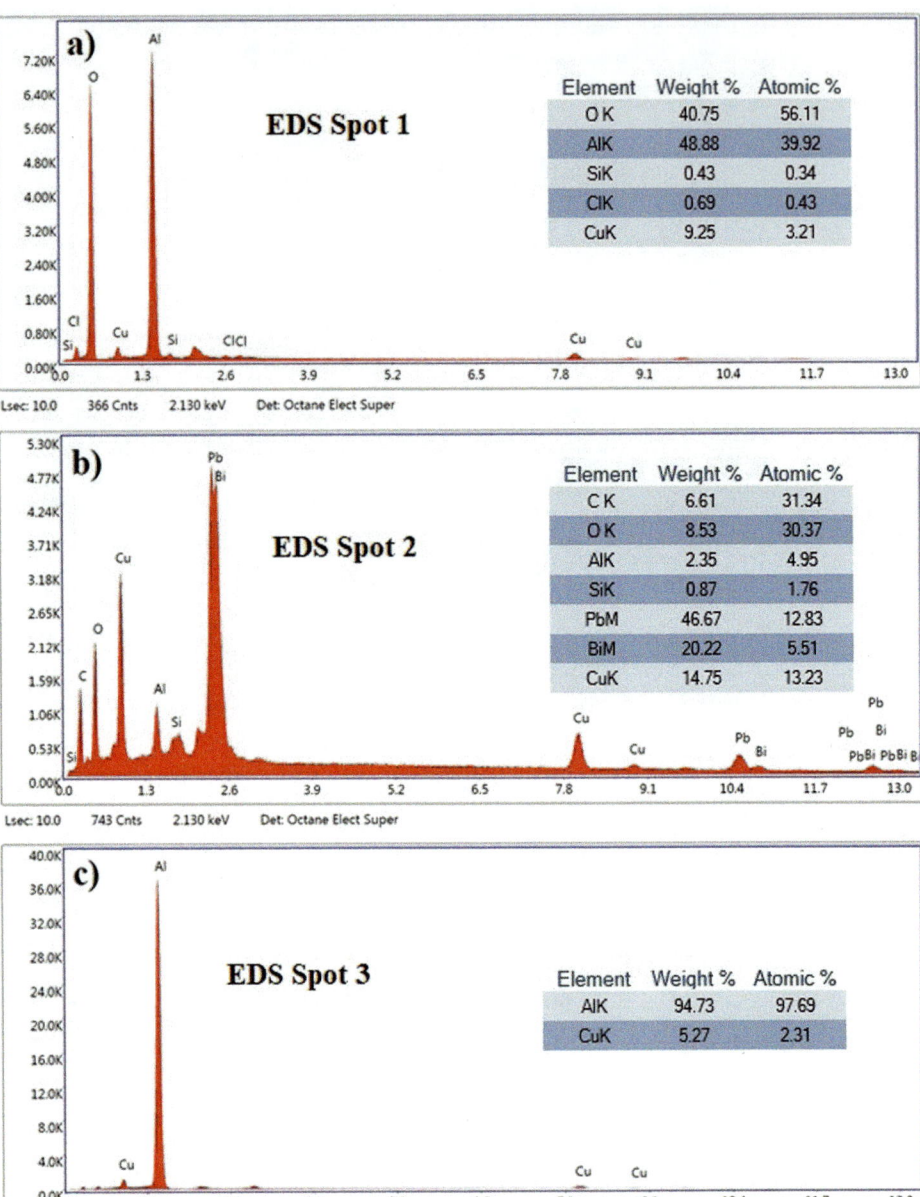

Table 4 Maximum corrosion depth under different tempers

Temper	Max. corr. depth (μm)
T4	54.16
T6	434
T79	172.62
T73	57.61

Conclusion

Since 2011 alloys are used in valves that operate in harsh corrosion conditions, one of the best ways to protect against corrosion for these alloys is hard anodizing. Our previous works show that conventional (technical) anodizing is not adequate to protect 2011 alloy in the harsh corrosive environment [12]. Therefore, in this study, corrosion behavior of hard anodized 2011 alloy under underage (T4), peak age (T6) and overage (T79, T73) conditions was investigated. As a result of optical microscope and SEM imaging examinations;

- Alloys in overaged (T79 and T73) temper have better intergranular corrosion resistance than peak aged (T6) temper.
- Artificial ageing parameters enhance corrosion behavior. Better corrosion resistance was encountered when the heat treatment temperature and time were increased. Corrosion resistance of T73 temper is better than T79.
- Corrosion susceptibility of 2 months natural aged (T4 temper) and overaged (T73) is similar.

References

1. American Society for Metals. (1991). Heat Treating of Aluminum Alloys. In Heat treating (Vol. 4, pp. 841–879) essay.
2. Daviess, J.R., Aluminium & Aluminium Alloys, ASM, Materials Park, Ohio, 1993.
3. Chong, P. H., Liu, Z., Skeldon P. ve Thompson, G.E. (2003) Corrosion behavior of laser surface melted 2014 aluminium alloy in T6 and T451 tempers, The Journal of Corrosion Science and Engineering, 6, 12.
4. K. Nisancioglu, Ø. Strandmyr, Corrosion of AlMgSi alloys with Cu additions: the effect of Cu content up to 0.9 weight percent, Report no. STF34 A78052, SINTEF, Trondheim, Norway, 1978.
5. Vargel, C. (2020). Intergranular corrosion. Corrosion of Aluminium, 185–197. https://doi.org/10.1016/b978-0-08-099925-8.00015-6.
6. Zheng, Y., Luo, B., He, C., & Su, Y. (2019). Effect of aging time on microstructure evolution and corrosion behavior of an al–mg–si alloy. Materials Research Express, 6(11), 116582. https://doi.org/10.1088/2053-1591/ab4bec.
7. Benedyk, J. C. (2010). International Temper Designation Systems for Wrought Aluminum Alloys. Light Metal Age, 16–22.
8. ISO EN 573-3.
9. ISO EN 755-2.
10. Bahl, S., Rakhmonov, J. U., Kenel, C., Dunand, D. C., & Shyam, A. (2022). Effect of grain-boundary θ-Al2Cu precipitates on tensile and compressive creep properties of cast Al–Cu–Mn–Zr alloys. Materials Science and Engineering: A, 840, 142946. https://doi.org/10.1016/j.msea.2022.142946.
11. Hardouin Duparc, O. (2004). Alfred Wilm and the beginnings of Duralumin. Revue De Métallurgie, 101(5), 353–360. https://doi.org/10.1051/metal:2004157.
12. Sarı, İ. A., Özçelik, G., & Yücel, O. (2022). Effect of Anodizing Thickness on Corrosion Behaviour of 2011 Aluminium Alloy. ms, Istanbul.

Fundamental Study on Modified Solidification of 1370 and ALSI7 with and without Commercial Grain Refiners

Robert Fritzsch, Amund Ugelstad, Henrik Gobakken, Silje Li, Shahid Akhtar, Lars Arnberg, and Ragnhild Aune

Abstract

The microstructure of aluminium (Al) and its alloys is a well-investigated key parameter utilized to adjust their mechanical, chemical, and physical properties. The microstructure of Al can be altered by several processes, such as controlled solidification rate, grain refiner additions, and external force fields. The present study has chosen to focus on conventional procedures for grain refinement and compare the results to new low-energy concepts. In view of this, three different commercial grain refiners (2 different AlTi3B1, AlTi5B1) were added to lean 1370 and AlSi7 aluminium alloys at (i) different solidification rates, and (ii) influenced by alternating electromagnetic fields. The results were evaluated based on the effects on grain size and electrical conductivity. The initial results revealed an inverse relationship between the conductivity and the final grain size, which also proved to be independent of the origin of the refined grain structure.

Keywords

Aluminium • Al–Si alloys • Grain refiner • Electromagnetic refining • Microstructure

R. Fritzsch (✉) · A. Ugelstad · H. Gobakken · S. Li · L. Arnberg · R. Aune
Department of Materials Science and Engineering, Norwegian University of Science and Technology, N-7491 Trondheim, Norway
e-mail: Robert.fritzsch@gmail.com

R. Fritzsch
Pyrotek MCR, Faraday House Eastern Avenue, Stretton, Burton upon Trent, DE13 0BB, UK

S. Akhtar
Hydro Aluminium, Karmøy Primary Production, Håvik, Norway

Introduction

Aluminium is a superior material for electrical applications due to its physical properties. Copper can compete with the electrical conductivity of aluminium, while the latter offers lower density with higher mechanical strength, assuming purity and treatment are applied correctly. The unchallenged key advantage of aluminium over copper is the price per unit weight, giving aluminium a unique position in applications for high voltage transmission lines [1].

Low alloyed aluminium alloys, e.g., of the 1xxx series, offer the highest conductivity due to the low amount of impurities [1]. The mechanical properties of aluminium can be improved by altering the grain size of the microstructure to a finer structure. There are different methods adjusting the grain size to achieve the desired mechanical properties: (i) solidification rate, (ii) heat treatments and iii) modified solidification, such as grain refiner (GR), supercooled melt, artificial forces (pressure, vibrations, electromagnetic vibrations/velocity), and agitation.

In the current project, the impact of solidification rate, grain refiner and the effect of alternating electromagnetic fields on commercial 1370 alloy and AlSi7 alloys has been investigated. 1370 alloy, e.g., 99.7% pure aluminium will from now on be shortened to Al.

This work acts as an initiation phase of a fundamental research project, aiming at altering the microstructure, investigating the resulting effects on the electrical conductivity of Al and its alloys.

Theory

Al is suitable for electrical purposes, because it is a good conductor (63.62 %IACS) and has a low density (2.7 g/mm^3). Pure aluminium is soft and requires alloying and microstructural modifications to gain sufficient mechanical strength. By adding other elements to Al, alloys

© The Minerals, Metals & Materials Society 2023
S. Broek (ed.), *Light Metals 2023*, The Minerals, Metals & Materials Series,
https://doi.org/10.1007/978-3-031-22532-1_56

can be generated to meet specific demands. The physical and chemical properties change according to the alloying elements, the amount of these, and how they are present in the final metal microstructure.

The relevant properties and phases of the investigated alloys are given in Table 1. The alloying elements, such as Si and Mg, alter most material characteristics, including electrical [2], thermal [3], and mechanical properties [1].

The electrical conductivity is a key parameter, which is essentially altered by impurities, defects, or other disturbances in the microstructure, and can be correlated by the International Annealed Copper Standard of 100 %IACS, as defined in 1913 [5].

Aluminium normally solidifies in a polycrystalline state, where the metal structure consists of several crystals often referred to as grains. These grains are oriented in different directions and the size can vary depending on the history of the produced metal. Properties of aluminium are greatly affected by grain size. This makes grain size refinement an important aspect of casting and/or manufacturing parts. A key parameter is the yield strength, the stress at which a material begins to permanently deform. This parameter increases as the average grain diameter decreases, known as the Hall–Petch relationship [6].

The electrical conductivity has a similar behaviour. The least conductive area in the metal, acting as a bottleneck or resistance, defines the conductivity of the bulk. The lowest conductivity is usually at the grain boundaries, where intermetallic phases and impurities are concentrated. Fewer grain boundaries lead to a higher concentration and physical dimension of intermetallics and impurities. Samples with large grains and few grain boundaries are therefore expected to have a lower conductivity compared to samples with smaller grains [7, 8].

The first aspect of the solidification process is nucleation, the spontaneous forming of small solid nuclei in the liquid. Particles, such as TiB_2, can act as grain refiners. The grain refiners remain solid in the molten Al, thereby acting as nucleation points. This increases the number of nucleation points and leads to a reduced total grain size. $AlTi_5B1$ and $AlTi_3B1$ are aluminium samples with 5 wt.% and 3 wt.% titanium respectively, and 1 wt.% boron. This means that the grain refiners contain high purity aluminium metal, as well as TiB_2 particles.

In induction coils of any shape with an alternating current flowing in a common direction, a magnetic field B [T] is induced. This magnetic field then induces a current [A/m^2], opposing the main current in accordance with Lenz's law of the conservation of energy. It is known that real induction coils of almost any geometry must be counted as "short" induction coils, where the flux density along the centre axis of the coil has a gradient and can be estimated by the Nagaoka coefficient, or short coil correction factor k_N [9, 10]. The gradient of the magnetic field results in a similar variation of the induced current in the liquid metal, and this is even further altered by the change of the effective electrical conductivity of the metal during solidification.

Lorentz forces F_L [N/m^3] are the resulting "electromotive" forces and can be calculated as the cross product of the induced current density and the magnetic flux. These forces act on all components that possess electrical conductivity inducing a current density. This includes the carbon crucible, the liquid, the solidifying metal, and the mushy zone. The correlation of the Lorentz forces is shown in Eq. (1):

$$F_L = J_\Phi \times B_z \qquad (1)$$

The different coil arrangements were used to obtain the maximal interaction of liquid metal with the alternating magnetic field. The sketch of the apparatus is shown in Fig. 1, giving not only an overview of the used coil geometries, but also the applied position and geometries.

During solidification, aluminium changes its density from approximately 2368 kg/m^3 at 700 °C to 2700 kg/m^3 at 25 °C. This reduction in volume is typically seen as a dimple on top of the cast, forming while the still molten metal moves down to replace the reduced volume of solid aluminium. This shrinkage can also lead to shrinkage pores.

Table 1 Chemical composition, possible intermetallics, solidification temperature in °C, and electrical conductivity in % IACS for Al 99.7% and AlSi7 [4]

Alloy	Chemical composition (wt%)	Possible intermetallics	Solidification temperature (°C)	Electrical conductivity (%IACS)
Al 99.7%	Al 99.7	β–Al_5FeSi	630–660	59.5–63.9
	Fe 0.2	δ–Al_4FeSi		
	Si 0.1	α–Al_8Fe_2Si		
AlSi7	Al balance	β–Al_5FeSi	575–625	38.3–40.1
	Si 6.5–7.5	δ–Al_4FeSi		
	Mg 0.4–0.7	$Al_9FeMg_3Si_5$, Mg_2Si		
	Fe 0.15	π–$Al_8Mg_3FeSi_6$		

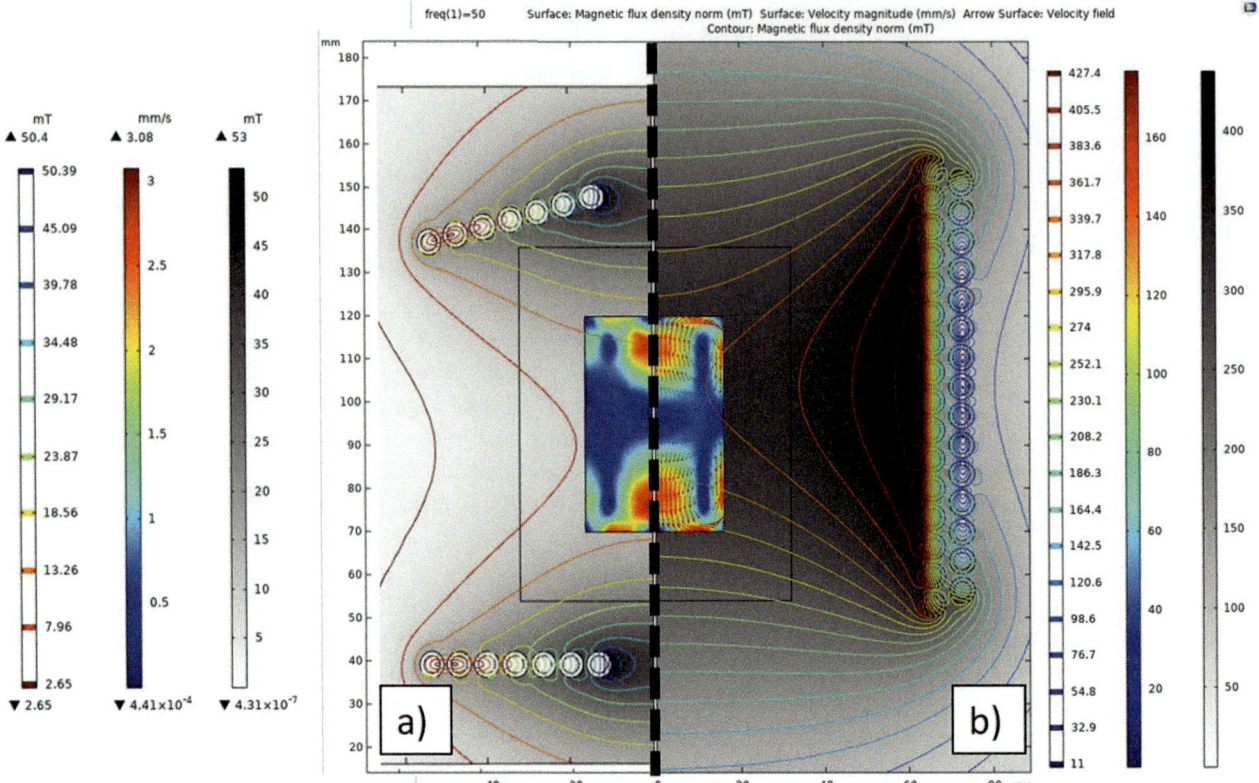

Fig. 1 Sketch of the two coil setups, **a** using the pancake inductors, with the legend to the left and **b** the round induction system, with the legend to the right. The modelling results are showing the generated magnetic field and the resulting velocity field in the melt. The velocity field is in scale on both models

The hydrogen solubility in aluminium increases with temperature and decreases when the metal changes from solid to liquid. During a cast and solidification, most of the dissolved hydrogen recombines to form H_2 gas. This creates small gas bubbles, nucleating at impurities or grain boundaries in the solidifying melt. Dissolved hydrogen mostly adds to the gas volume, hereby generating larger bubbles. These bubbles may break free and rise to the surface or get entrapped in the solidifying structure, creating gas pores [11].

In metallurgy, a dendrite is a tree-like structure of crystals growing as the molten metal solidifies. To attain dendrites, the given metal must be undercooled, which is common for aluminium in regular castings. A slow cooling process may result in less nucleation compared to a rapid cooling process, leading to larger dendrites. Dendritic growth is also more prominent in samples with a larger solidification range. Al has a smaller solidification range than alloys such as AlSi7.

Experimental

The experimental setup consisted of electrical melting furnaces, four different moulds for solidification, as shown in Fig. 2, five different grain refiners and different induction coils for inducing the electromagnetic field during solidification.

After melting the metal and reaching the correct temperature, the dross was removed. For samples with grain refiner, 0.1 wt.% of grain refiner was added at a temperature of 730 °C ± 10 °C. The melt was then stirred with a Boron Nitride (BN) coated graphite rod for 30 s prior to pouring into the four moulds shown in Fig. 2. The procedure was repeated in the same manner for Al 1370 (Al) and AlSi7 alloy, both with and without the addition of the different grain refiners (KBM AlTi3B1, GR AlTi3B1, and AlTi5B1). Al without grain refiner was cast in a similar manner. The

Fig. 2 The four different crucibles, **a** showing the massive cast iron mould, **b** the industrial RSD sampler, **c** the carbon crucible, and **d** the insulated carbon crucible

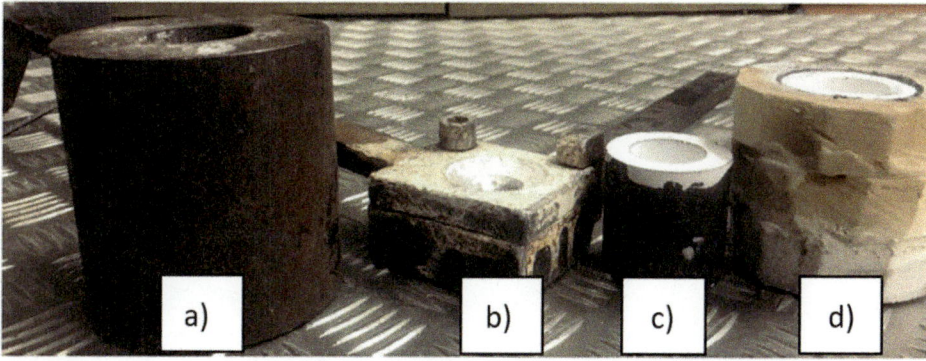

samples have been taken after short dissolution time for the GR because long settling time has a negative impact of GR on the microstructure as discussed by earlier work of M. Hassanabadi at the same labs [12].

For samples using the electromagnetic field, no grain refiner addition was used. The metal, at 730 °C ± 10 °C, was poured into a carbon crucible, which was heat-soaked at 750 °C, and insulated, closing the lid after adding the metal into the crucible. This enabled slower solidification while being exposed to the magnetic fields. Two different types of magnetic fields were applied: A low field intensity (<10 mT) double pancake coil setup and a high field intensity (<200 mT) round coil setup. Both coil setups are shown in Fig. 3a–c.

The power source utilized the mains frequency of 50 Hz; the applied power and labelling of EMC samples are listed in Table 2. The EM field was applied immediately after pouring the metal into the mould and positioning it in the vicinity of the coil. The field was kept energized with the specified power for 10 min in the pancake coil and 20 min in the round coil.

The different setups of reference samples, grain refinement, and electromagnetic modified solidification resulted in 50 samples, which all were analysed by optical inspection and electrical conductivity.

The samples were then prepared by metallographic means, generating a mirror surface finish and finalizing by the anodized step for grain size analysis. All samples were cut to expose a surface where the solidification orientation and front were similar despite the different crucible geometries. This is shown in Fig. 4a–d. The polishing method consisted of aluminium oxide P500 grit sandpaper until plane, 15 μm diamond particle solution on Largan 9 plate for five minutes, 3 μm diamond particle solution on Allegard 3 plate for five minutes, and then OP-S on a cloth disc for five minutes. The samples were cleaned by using an ultrasonic ethanol bath between each step.

The anodizing for visual differentiation of the grains was done in Barker's reagent bath. After the metallurgical preparation, the samples were digitalized using a moving stage on a downlight microscope, scanning the full surface of all samples. The individual images were stitched together generating a full surface scan. The software used for all imaging operations was Image-Pro® and Stage-Pro®. The full methodology of acquiring the images and the whole sample preparation is documented in the thesis report [13].

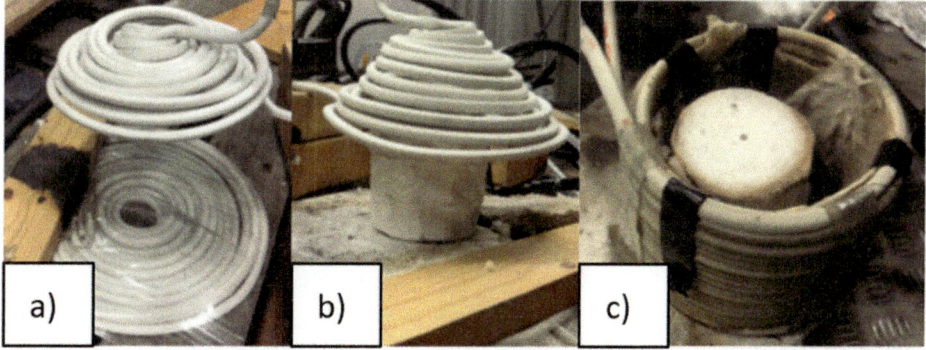

Fig. 3 The different coil geometries are shown here. In **a** the double pancake coil design is shown. In **b** the same coil as in **a** is shown with a protective refractory plate above the bottom inductor and the crucible in the centre of the assembly. In **c** the crucible in the insulation is shown inside the round inductor

Table 2 Labelling, voltage, current, real power, and average magnetic field strength for the applied EM fields during casting

Coil	Label	Voltage (V)	Current (A)	Real power (W)	Average magnetic field (mT)
Pancake	α	1.33	76	90	2.2
Pancake	β	2.76	154	380	4.2
Pancake	γ	4.05	226	816	6.04
Round	R	28	738	11,500	158.56

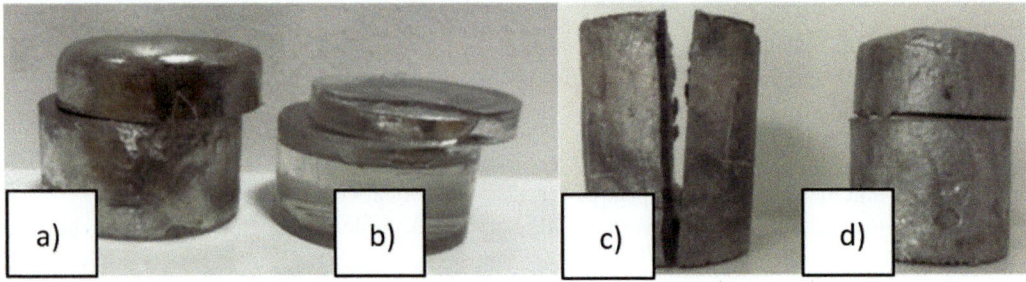

Fig. 4 Examples of castings after being cast with the different moulds as shown in Fig. 2a–d. Here showing the position of the cut and hereby the investigated area for the results

Conductivity data was collected from all samples after digitalizing with the microscope using GE AutoSigma 3000 Electrical Conductivity Meter, with a probe setting of 500 kHz, taking an average from three measured points from each sample.

The grain size analysis was done by using an in-house developed code generating quantitative results of grain size distribution.

Results and Discussion

Before cutting and polishing the samples, each sample was examined visually, noting variations in surface texture, gleam, and shape. The regular samples showed no difference, while the EM-samples had a rougher texture, reduced gleam, and no dimpling at the top, but a narrowing at the sides. This is shown in Fig. 5, where the darker sample is an EM cast and the lighter one a regular sample.

The rougher texture and the reduced gleam are likely a result of the metal solidifying mid movement, caused by the magnetic field. The EM-samples showed no dimpling due to the magnetic field instigating movement towards the middle of the sample.

The microstructure of the samples are shown in Figs. 6 and 7, showing the Al and AlSi7 samples with and without grain refiner. In Figs. 8, 9, and 10, the EM-modified samples of pure and alloyed Al are shown. The key numbers for the sample codes are presented in Table 3.

The rate of solidification produced substantial variations in grain size and structure. The samples indicated that the slower solidification rate generated larger grains with a clearer convergence towards the centre of the sample as shown in the top row of Fig. 6.

The C and D samples in Fig. 6 first row indicate that solidification instigates at the interface between the mould and the liquid Al. The solidification initiates at the edges and the grains grow towards the centre, creating long grains.

Fig. 5 Top and side profiles of one EM-cast (**a**) and one regular cast (**b**)

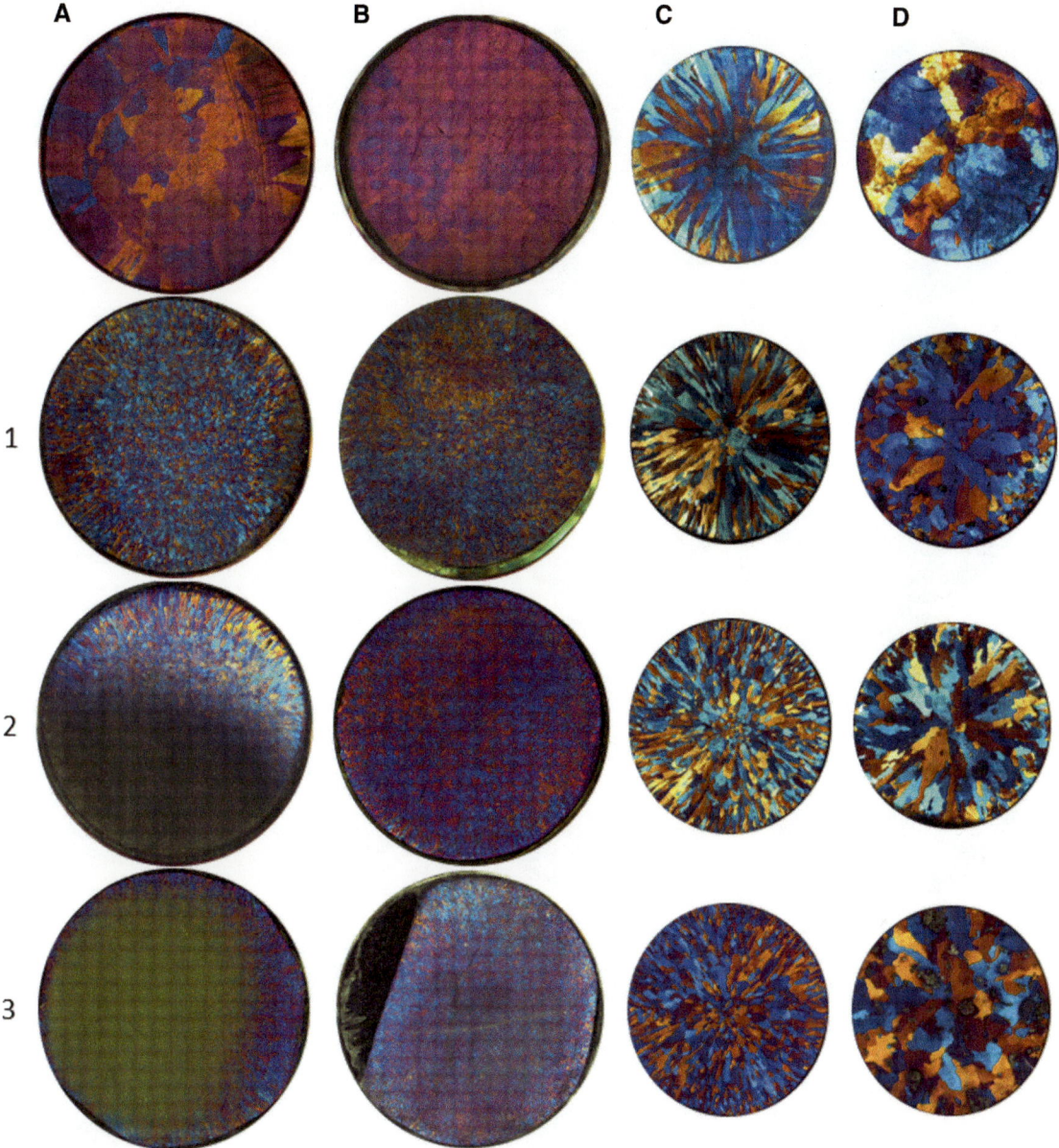

Fig. 6 All non-EMC Al casts. The blank row is Al 1370 without grain refiner, row "1" is with KBM AlTi3B1, row "2" is with GR AlTi3B1, and row "3" is with Aleastur AlTi5B1

The D sample had the lowest solidification rate, and the grains therefore had more time to grow, resulting in large grains. The A sample has indications of grain growth starting from the edges. This could be caused by the mould shape, because the bottom is thicker than the walls, encouraging faster solidification from the bottom. The B sample has a lower solidification rate than A and faster grain growth converging towards the centre was therefore expected. However, no such convergence is seen in the Al-B sample. This is likely that mould B, by creating a disk-shaped sample, has a different solidification pattern, as opposed to the cylindrical-shaped mould A. This entails a larger interface between the sample and mould at the top and the bottom.

Adding the grain refiners to the 1370 Al yielded smaller grains for all solidification rates, as shown in Figs. 1, 2, 3, and 6, having coarser grain shapes with a better-insulated mould. The samples Al-2-A and Al-3-A are unclear, but the visible grain refinement trends are uniform throughout the samples. The C and D samples with grain refiners solidify from the edges and converge towards the centre of the sample, while B samples do not. The grain refiner does not seem to affect the convergence of grains towards the centre of the samples.

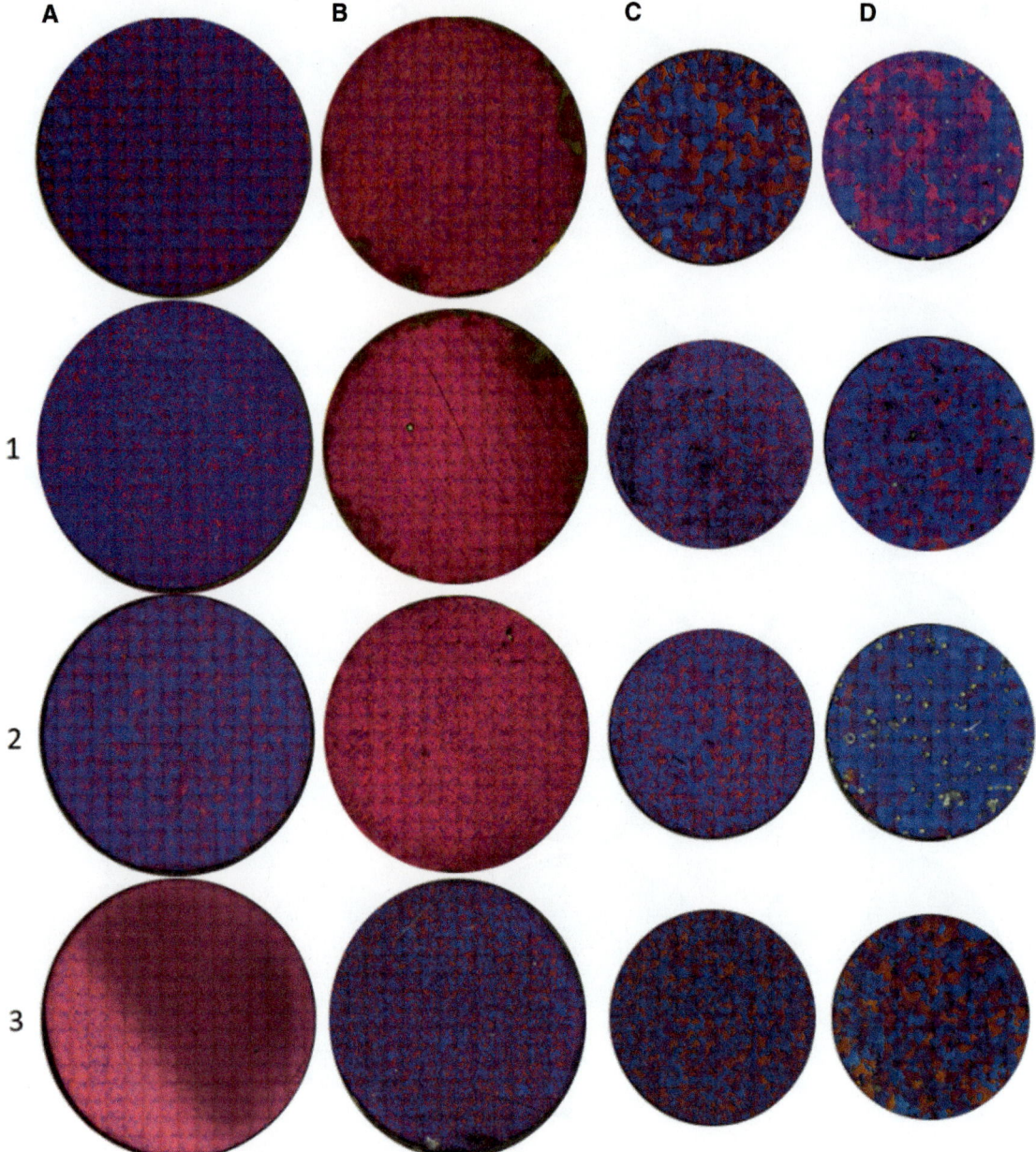

Fig. 7 All non-EMC AlSi7 casts. The blank row is AlSi7 without grain refiner, row "1" is with KBM AlTi3B1, row "2" is with GR AlTi3B1, and row "3" is with Aleastur AlTi5B1

In a similar study by Hassanabadi et al. [12], the effect of Al-3Ti-1B and Al-5Ti-1B grain refiner on commercial pure Al 99.7% cast in an iron mould was studied. Hassanabadi did not observe the same level of grain size reduction as observed in this study, shown in Figs. 6 and 7. The cause of this is likely the TiB_2 particles agglomerating in the melt, as the time between the addition of the grain refiner and the pouring of the metal was substantially longer in the earlier study. They used intervals of 30, 60, and 90 min, while this report used an interval of 30 s.

The top row of Fig. 7 shows the AlSi7 alloy with no added grain refinements solidified at four different rates. The observed growth characteristics in the C and D samples differ significantly from the Al. The grain sizes are smaller at the edges, but there is no convergence towards the centre of the sample. The eutectic solidification process and the mushy zone can be accounted for that. Figure 7 illustrates that AlSi7 will create a partly solidified zone during solidification. These partly solidified areas were spread throughout the sample and act as seeds for solidification, thereby

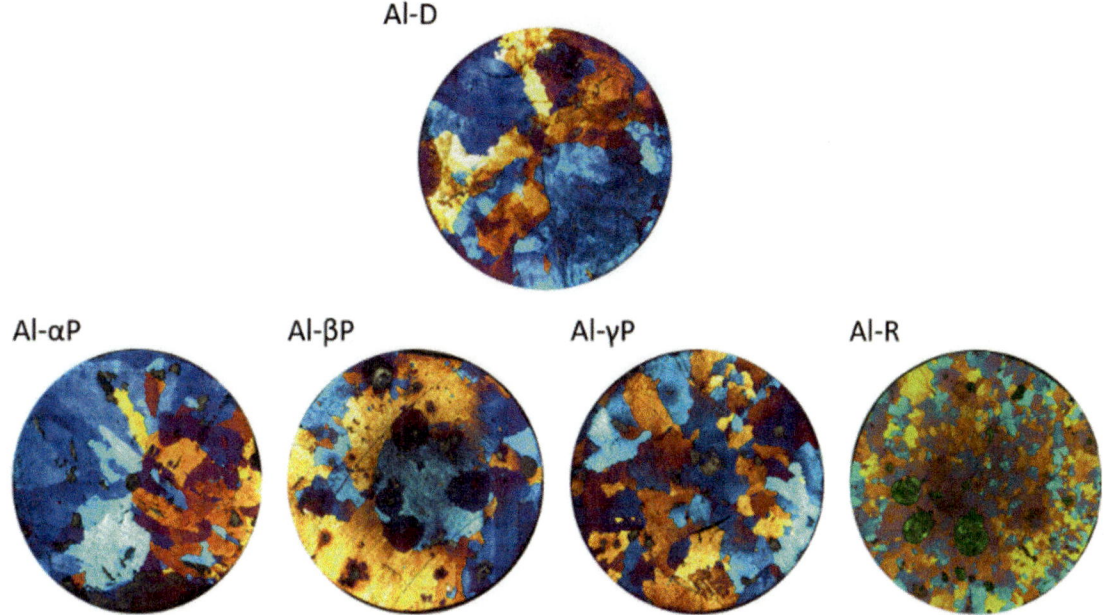

Fig. 8 Circular cross-section of Al-D and Al-EMC samples. The applied field strength is increasing from left to right

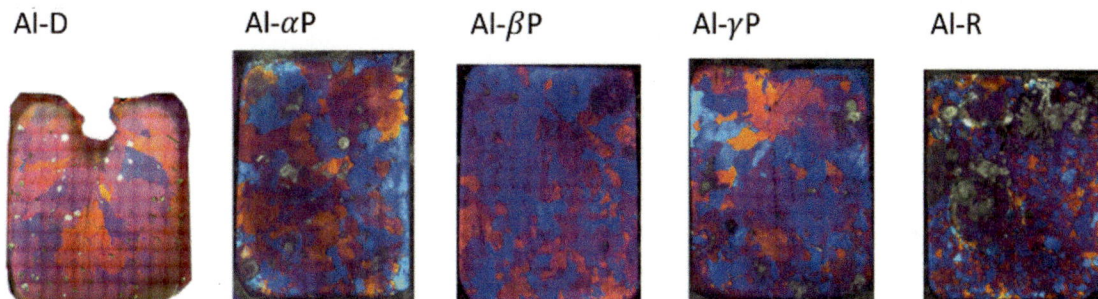

Fig. 9 Rectangular cross-section of Al-D and Al-EMC samples

causing a more uniform grain growth throughout the sample and not just from the edges and inwards.

With decreasing solidification rate, an increasing grain size is seen throughout all grain refinement additions to AlSi7, Fig. 7 rows 1–3. The difference in grain size from adding grain refiner to AlSi7 is not as substantial compared to Al, being in accordance with the recommended application of the specific grain refiners.

The circular cross-sections of Al-D and the Al-EMC samples are shown in Fig. 8. The Al-D sample is used as an unmodified reference sample, containing the same chemical composition as the AL-EMC samples and the longest solidification time of the non-EMC samples. The main difference in grain size for these samples is therefore a result of the EM field applied during casting. In Table 2 the used coils and the power is noted.

The samples in Fig. 8 show that with an increasing EM-field strength, the average grain size reduces

considerably. The most significant reduction is seen in Al-R and Al-γ P, which are the samples with the strongest EM-field. The weaker field, as applied to Al-α P and Al-β P, seems to have a small effect on grain size.

Figure 9 was generated to study the impact of the weak EM-field effects on the grain size. Al-D has a clear dimple formed during directional solidification. The direction of grain growth was shown as they are all oriented from the edges towards the middle of the mould. None of the samples exposed to an EM-field has a dimple or a similar direction of grain growth. The electromagnetic stirring was sufficient for agitating the molten metal and homogenizing the melt temperature. During solidification, nucleated grains form at the edge and are moved by the agitated melt, from the edge of the mould into the centre, acting as nucleation seeds.

A comparison between AlSi7-D and the EMC samples is shown in Fig. 10. AlSi7-D is the reference because they all have the same composition and similar solidification rates.

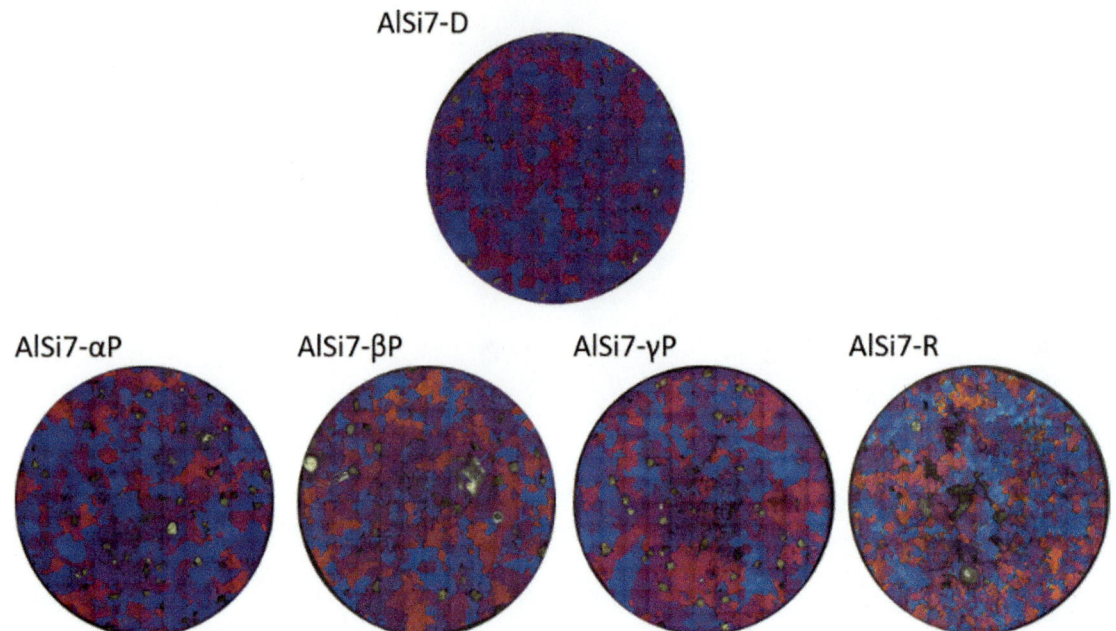

Fig. 10 Circular cross-section of AlSi7-D and AlSi7-EMC samples. The applied field strength is increasing from left to right

Table 3 Summary of all conducted trials as a sampling matrix with given key information

Sample label	Casting type	Mould used	Temperature of mould	Number of samples
A	Regular	Cylindrical iron mould	–	8
B	Regular	RSD iron mould	–	8
C	Regular	Graphite mould	250 °C	9
D	Regular	Insulated graphite mould	650 °C	9
P	EMC pancake coil	Insulated graphite mould	750 °C	12
R	EMC round coil	Insulated graphite mould	750 °C	4

The AlSi7-P samples indicate no significant grain refinement effect, compared to the reference. The higher viscous mushy zone of the alloying system of the AlSi7 alloy is expected to lower the efficiency of the EM stirring by quickly dissipating the stirring movement. However, AlSi7-R shows considerably smaller grains caused by a significant increase in EM-field strength for this sample, as seen in Table 2. A similar effect was observed by Metan [14, 15].

AlSi7-R shows zones with needle-like structures, as shown in Fig. 11. During solidification in AlSi7, α−Al starts solidifying, causing the Si concentration to increase in the remaining melt until it reaches a eutectic concentration and solidifies. This eutectic concentration usually solidifies at grain boundaries, but because of the strong EM stirring, the molten eutectic Al–Si was separated from the solidified α−Al. The molten eutectic solidified as needle-like structures shown in Fig. 11.

The grain size analysis matched the visual inspection of the samples for the C, D, and EMC samples. These samples are used for the grain size–conductivity comparison.

Figure 12 shows the average grain size, plotted against the conductivity measurements for Al C, D, and EMC samples.

The coefficient of determination (R^2) is 0.788. This indicates a strong correlation between conductivity and grain size for the 1370 Al. It is expected, by increasing the grain boundary volume, the existing impurities within the melt will distribute at these grain boundaries in a lower local concentration, resulting in an increase of the conductivity of the grain boundaries and hence an increase of the conductivity of the sample. It is assumed that conductivity increases as grain boundary volume increases, assuming an even distribution [13]. A power regression model was used as the grain size volume [mm^3] is a function of the measured grain

Fig. 11 Eutectic zones created by strong EM stirring in AlSi7-R

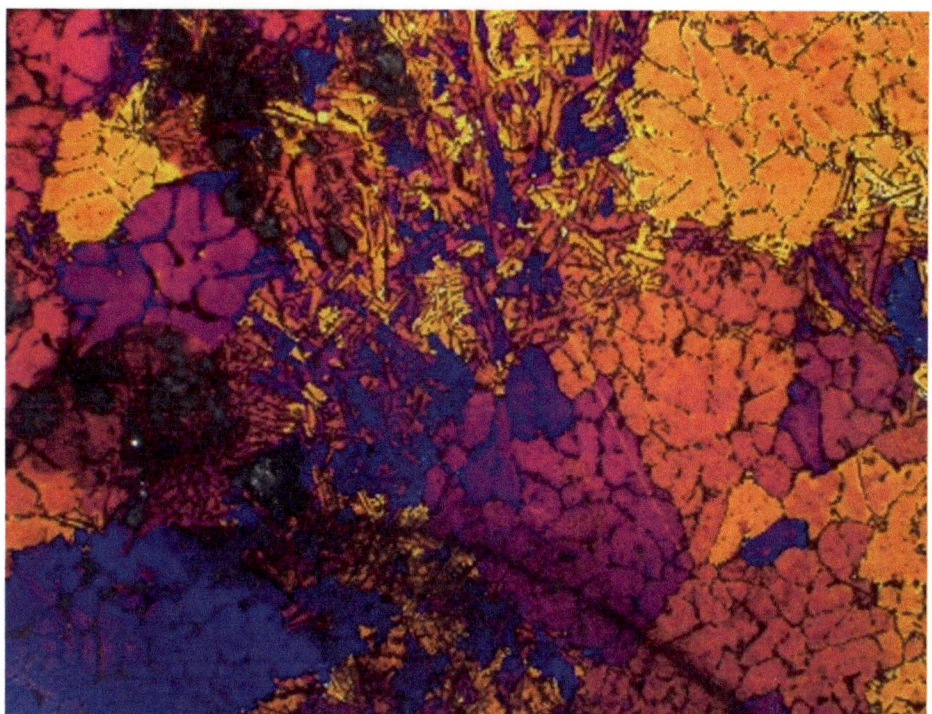

Fig. 12 Conductivity and grain size for Al C, D, and EMC samples

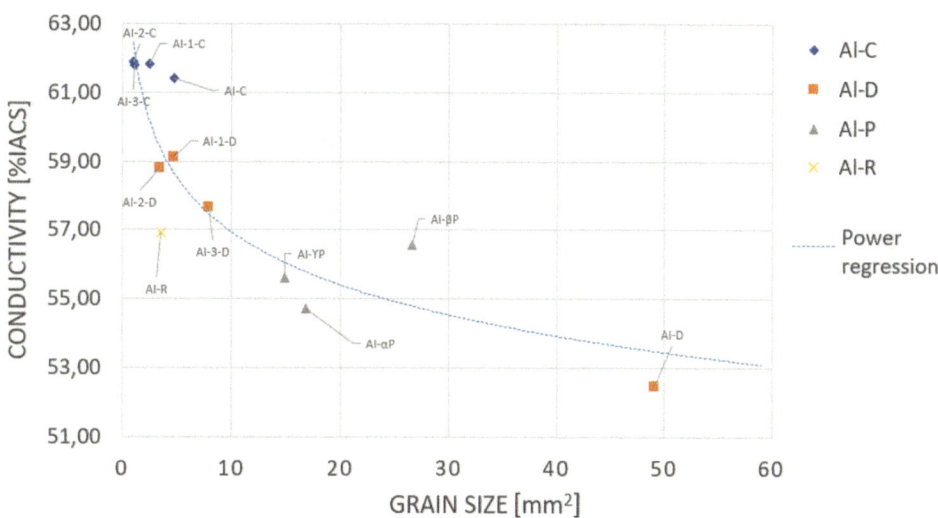

size [mm²], and it is assumed that these are related to a power factor.

Figure 13 shows the average grain size plotted against the conductivity measurements for the AlSi7 C, D, and EMC samples.

R^2 for the plot is 0.398 in Fig. 13, which indicates a weak correlation. Two main factors contributing to the low correlation of average grain size and conductivity for AlSi7 were identified. First, the microscope did not achieve the same level of contrast as for the Al samples. Colour shifting and grid line artifacts from the microscope and image stitching caused further inaccuracies. Second, the high silicon content leads to a high concentration of impurities at the grain boundaries compared to Al. The grain size range for the AlSI7 samples is smaller and narrower compared to Al samples. The dilution of impurities when increasing the grain boundary volume is smaller and less impactful on conductivity for AlSi7 [13].

Fig. 13 Conductivity and grain size for AlSi7 C, D, and EMC samples

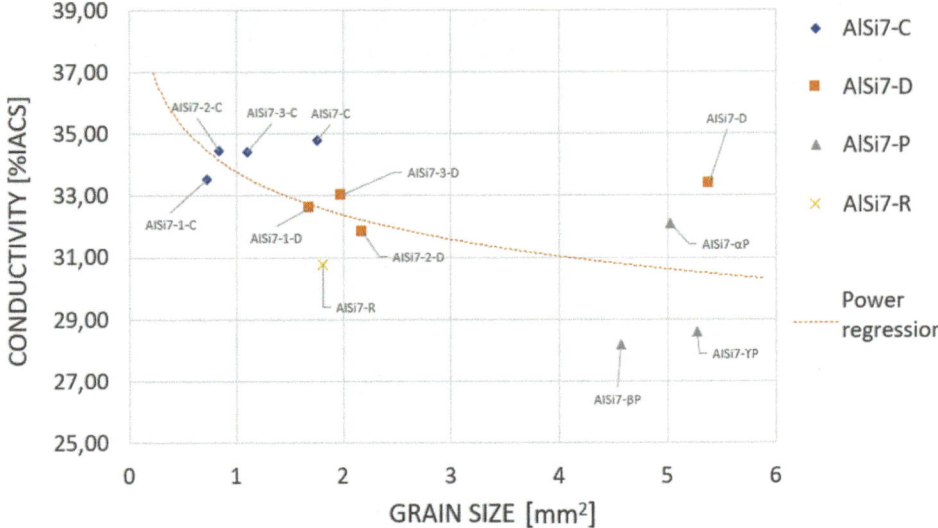

Conclusions

The present work demonstrated that EM fields have the potential to act as a substitute or contributor for grain refining during casting. Several points pertaining to solidification speed, grain refiners, EM casting and conductivity were discovered:

- A slower solidification speed results in larger grains and a more directional growth in Al, while AlSi7 did not exhibit directional growth.
- The addition of grain refiner resulted in a reduction in grain size in Al and to a lesser degree in AlSi7, though it did not appear to influence the directionality of the grains.
- The application of the EM field yielded a greater porosity in the samples while also reducing the size of the grains and causing non-directional growth.
- The results indicated an inverse correlation between grain size and conductivity, with and without the addition of grain refiners.

Future Work

The investigation on the grain refinement using different grain refiners was conclusive and could be related to the alloys, moulds, and the difference efficiencies of the GR types.

Nevertheless, the different EM fields in this study have given an insufficient differentiation about the impact of the magnetic field, especially the implication of weak fields on the true effect of the refinement. Here, further studies with

different coils generating weak EM fields, with alteration of the used frequency, will be conducted.

The effect of the EM fields on microstructure developments at different solidification speed offers an interesting field of research, including comparison with traditional grain refiners.

And, as a key aspect, the degassing of the melt before the application of microstructural adjustment must be taken to account.

Acknowledgements The authors wish to express their gratitude to the good and conclusive discussions with the aluminium alloy and casting experts Prof. Lars Arnberg from NTNU and PhD. S. Akhtar from Hydro Aluminium, and the data analyst Hannes Zedel who has quantified the grain size distribution of all the samples discussed in this report. A special thanks goes also to the Department of Materials Science and Engineering (DMSE) at the Norwegian University of Science and Technology (NTNU) for the intensive usage of the casting laboratory equipment.

This publication is a result of an excellent collaborative thesis work for reaching the level of a BSc of the students Amund Ugelstad, Henrik Gobakken, and Silje Li. They have generated a high-quality report [13], which will be a baseline for the newly founded work group investigating microstructural modifications on Al and its alloys.

References

1. Davis JR (1993), Aluminum and aluminum alloys, ASM International
2. Brandt R, Neuer G (2007), Electrical resistivity and thermal conductivity of pure aluminum and aluminum alloys up to and above the melting temperature, International Journal of Thermophysics 28(5): 1429–1446
3. Leitner M, Leitner T, Schmon A, Aziz K, Pottlacher G (2017), Thermophysical properties of liquid aluminum, Metallurgical and Materials Transactions A 48(6): 3036–3045
4. Ansys GRANTA EduPack 2021 R2 software, ANSYS, Inc., Cambridge, UK, Version 21.2.0 (www.ansys.com/materials)

5. JDavis JR (2001), Copper and copper alloys, ASM International

6. Hall E (1951), The Hall-Petch relationship, Proc. Phys. Soc. Ser. B, 64: 747–753

7. Hansen N (2004), Hall–Petch relation and boundary strengthening, Scripta Materialia 51(8): 801–806

8. Olafsson P, Sandstrom R, Karlsson Å (1997), Comparison of experimental, calculated and observed values for electrical and thermal conductivity of aluminium alloys, Journal of Materials Science 32(16): 4383–4390

9. Nagaoka H (1909), The Inductance Coefficients of Solenoids, Journal of the College of Science 27: 18-33

10. Kennedy MW, Magnetic Fields and Induced Power in the Induction Heating of Aluminium Billets, Licentiate Thesis, KTH University Stockholm, Sweden

11. Campbell J (2003), Castings, Elsevier

12. Hassanabadi M, Akhtar S, Arnberg L, Aune RE (2019), Grain Refinement of Commercial EC Grade 1370 Aluminium Alloy for Electrical Applications, Light Metals 1015–1023

13. Gobakken HL, Li S, Ugelstad A (2022), Comparison of Electromagnetic Fields and Grain Refiners on Microstructure Evaluation of Al and Al-Si Alloys, Bachelor Thesis, NTNU University Trondheim

14. Metan V, Eigenfeld K (2013), Controlling mechanical and physical properties of Al-Si alloys by controlling grain size through grain refinement and electromagnetic stirring, The European Physical Journal Special Topics 220(1): 139–150

15. Metan V, Eigenfeld K, Räbiger D, Leonhardt M, Eckert S (2009), Grain size control in Al–Si alloys by grain refinement and electromagnetic stirring, Journal of Alloys and Compounds 487(1–2): 163–172

Improving the Mechanical Properties of Cast Aluminum via Ultrasonication-Induced Microstructural Refinement

Katherine Rader, Jens Darsell, Jon Helgeland, Nathan Canfield, Timothy Roosendaal, Ethan Nickerson, Adam Denny, and Aashish Rohatgi

Abstract

This study investigates the use of ultrasound to refine the microstructure of cast aluminum alloys during solidification and thus improve their mechanical properties. An A356 aluminum alloy (Al–Si–Mg) with added Fe (to mimic a recycle-grade alloy) was cast in a graphite mold with the simultaneous application of ultrasound via an ultrasound probe inserted in the mold. Tensile specimens were extracted from the castings and heat treated to a T6 temper. Ultrasonication during casting transformed the morphology of primary aluminum grains from dendritic (~ 140 microns in size) to globular (~ 36 microns in size). The ultrasonically refined microstructure had 88% greater ductility, on average, and up to 10% greater tensile strength than the dendritic microstructure. This improvement in strength and ductility demonstrates the potential for ultrasonic processing to improve the performance of cast aluminum alloys without altering their chemistry or additional post-processing.

Keywords

Ultrasound • Grain-refinement • Solidification • Aluminum • Recycle

K. Rader (✉) · J. Darsell · J. Helgeland · N. Canfield · T. Roosendaal · E. Nickerson · A. Denny · A. Rohatgi (✉)
Pacific Northwest National Laboratory, 902 Battelle Blvd, Richland, WA 99354, USA
e-mail: Katherine.Rader@pnnl.gov

A. Rohatgi
e-mail: Aashish.Rohatgi@pnnl.gov

J. Darsell
e-mail: Jens.Darsell@pnnl.gov

J. Helgeland
e-mail: Jon.Helgeland@pnnl.gov

N. Canfield
e-mail: Nathan.Canfield@pnnl.gov

T. Roosendaal
e-mail: Timothy.Roosendaal@pnnl.gov

E. Nickerson
e-mail: Ethan.Nickerson@pnnl.gov

A. Denny
e-mail: Adam.Denny@pnnl.gov

Introduction

Currently, cast aluminum alloys account for 60–70% of the aluminum used in vehicles [1]. As such, cast aluminum alloys offer cost-effective lightweighting opportunities through part consolidation and integration with other product forms. The microstructures of cast aluminum alloys are typically dendritic, inherently less homogeneous, and contain porosity. Consequently, they typically have poor mechanical properties, especially compared to wrought materials. Some enhancement in as-cast properties can be achieved by refining the microstructure. One method of refining the microstructure is the use of chills during casting, which increases the local cooling rate [2]. However, this is not always practical for certain mold designs. Another method is the addition of grain refiners, which enhance the nucleation rate in the melt [3–5]. However, grain refiners are limited in how small of a grain size they can efficiently produce and change the overall composition of the alloy, making it more difficult to recycle. Another method is friction stir processing, which is capable of producing remarkably refined microstructures in cast aluminum alloys [6, 7]. However, this technology requires additional processing steps following casting, which can add to the total part cost.

Ultrasonic melt processing is a casting used for purposes such as degassing, fine filtration, and the production of non-dendritic, globular microstructures [8]. Through mechanisms such as enhanced nucleation and the fracture of dendrites, dendritic primary aluminum grains are transformed into finer globular grains [9–12]. This ultrasonically refined morphology has improved strength, ductility, and fatigue life compared to dendritic microstructures [8, 9]. This

S. Broek (ed.), *Light Metals 2023*, The Minerals, Metals & Materials Series,
https://doi.org/10.1007/978-3-031-22532-1_57

study is part of a larger investigation to incorporate local ultrasonic intensification into permanent mold casting techniques to refine the local microstructure. Rather than applying ultrasound to molten aluminum before/during pouring, ultrasound is applied to the aluminum as it solidifies in the mold. This allows for ultrasound to be applied to targeted locations within a casting casting during the casting process itself rather than as a post-casting step.

In this study, local ultrasonic intensification was applied to an A356 aluminum alloy with added Fe content as it was cast in a graphite mold. Ultrasound was applied via a probe inserted into the mold. The resultant microstructures were characterized using optical microscopy and electron backscatter diffraction (EBSD. The tensile properties of the microstructures were measured using specimens extracted from the castings. X-Ray computed tomography (CT) was used to visualize and measure the pores within the tensile specimens.

Methods and Materials

The material investigated in this study was an A356 aluminum alloy produced by Eck Industries with added Fe content (A356 + Fe). This produced a composition with a relatively high Fe level that mimics a recycled aaluminum alloy The composition of the alloy is listed in Table 1.

During casting, 200 g of A356 + Fe were melted in an aalumina crucible inside a box furnace, heated to ~ 720 °C, and cast in a graphite mold at room temperature. A Type-K thermocouple placed in the mold recorded the temperature of the aluminum as it cooled and solidified. The average cooling rate during solidification was ~ 2 °C/s. This cooling rate is similar to those of permanent mold casting techniques, which are on the order of 0.1–1 °C/s [13]. A 13-mm-dia. ultrasound probe made of Ti6Al4V was inserted into the mold for all casting experiments. For experiments where local ultrasonic intensification was applied during casting, ultrasound was started just before the aluminum was poured into the mold and was stopped once the aluminum was fully solidified. The frequency of the ultrasound was 20 kHz and power varied up to 750 W to maintain a constant amplitude of 33 µm.

Metallography specimens were sectioned from the castings, polished, and etched using Keller's reagent to reveal phases for optical microscopy. The same specimens were repolished for EBSD. Misorientation angles of 15° or more were defined as high-angle grain boundaries separating grains. ImageJ software was used to measure the equivalent grain size and sphericity of the primary aluminum grains [14]. Equivalent grain size is defined at $\sqrt{4A/\pi}$, where A is the area of the grain. Sphericity, sometimes referred to as roundness, is defined as $4\pi A/P^2$, where P is the perimeter of the grain. Sphericity values range from 0 to 1, with values closer to 1 indicating a more circular morphology and values closer to 0 indicating a more needle-like morphology.

Tensile specimens were extracted from the castings using electro-discharge machining. The length and the width of the reduced section were 12 and 2.5 mm, respectively. Each specimen was polished to remove surface damage from machining and the final thickness of the individual specimens was ~ 1.9 mm. The specimens were solution heat treated at 540 °C for 6 h, quenched in water at 80 °C, and aged at 155 °C for 4 h to a T6 condition [7, 15–17]. The specimens were loaded in uniaxial tension until rupture at a constant engineering strain rate of 10^{-3} s^{-1}. Strain was measured using microscopic digital image correlation (micro-DIC). From each casting, 4 tensile specimens were successfully tested in the T6 condition. After testing, X-Ray CT was used to visualize the porosity of the tensile specimens. For consistency across all samples, a $240 \times 400 \times 1200$ voxel region of interest (approximately 15 mm^3) was sub-selected adjacent to the rupture point of each tensile specimen. The smallest volume of pore that could be detected was 1.21×10^{-7} mm^3. From these data, the number density of pores, mean pore volume, largest pore, and porosity of the specimens were calculated. The number density of pores is defined as the number of pores divided by the total volume of the region of interest.

Results

Microstructural Refinement

Figure 1 shows the dendritic microstructure of the specimen that was cast without the application of ultrasound (*i.e.*, the control casting). The mean secondary dendrite arm spacing is 23 µm. Figure 2a shows three distinct microstructural morphologies that were observed in the specimen cast with the simultaneous application of local ultrasonic intensification (i.e., the ultrasonicated casting). Figure 2b shows a subregion of the dendritic microstructure, which was observed to the side of the probe at distances of approximately 2–6 mm. Figure 2c shows a subregion of the microstructure microstructure, which was observed throughout

Table 1 The composition of the A356 + Fe alloy studied

Si	Mg	Fe	Cu	Mn	Ti	Sr	V	Al
6.78	0.35	0.91	0.01	0.01	0.11	0.01	0.02	91.8

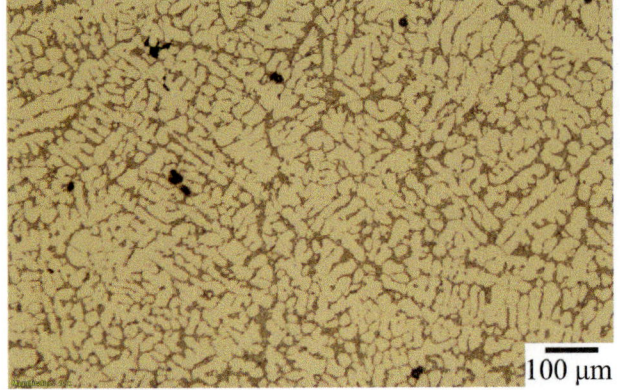

Fig. 1 This optical micrograph depicts the dendritic microstructure of the control casting. The specimen is etched so that the Si eutectic phase appears brown while the primary Al phase appears a lighter yellow. The black spots are porosity

the majority of the specimen. Figure 2d shows a subregion of the very-fine-grained microstructure, which was observed at distances up to 2 mm away from the location of the ultrasound probe.

Electron backscatter diffraction (EBSD was conducted to identify and characterize the primary aluminum grains. Figure 3 is an inverse pole figure (IPF-Z, where Z is the direction of the incident electron beam) map of the dendritic microstructure of the control casting. Figures 4a and b are IPF-Z maps of the globular and very-fine-grained microstructures, respectively, of the ultrasonicated casting. Table 2 lists the mean equivalent grain size and mean sphericity of the primary aluminum grains in each microstructure. Relative to the dendritic microstructure of the control casting, the mean equivalent grain size and sphericity of the globular microstructure are 74% smaller and 23%

Fig. 2 These optical micrographs depict the three microstructural morphologies present in the ultrasonicated casting. The specimen was etched so that Al appears as a light yellow color, Fe-rich phases (such as β-Al$_5$FeSi) appear as an intermediate beige color, and Si appears as a dark brown color. The image in (**a**) is stitched from multiple optical micrographs. Colored lines were added to help distinguish the transition between the different microstructural morphologies. The optical micrographs depict (**b**) the dendritic microstructure, (**c**) the globular microstructure, and (**d**) the very-fine-grained microstructure

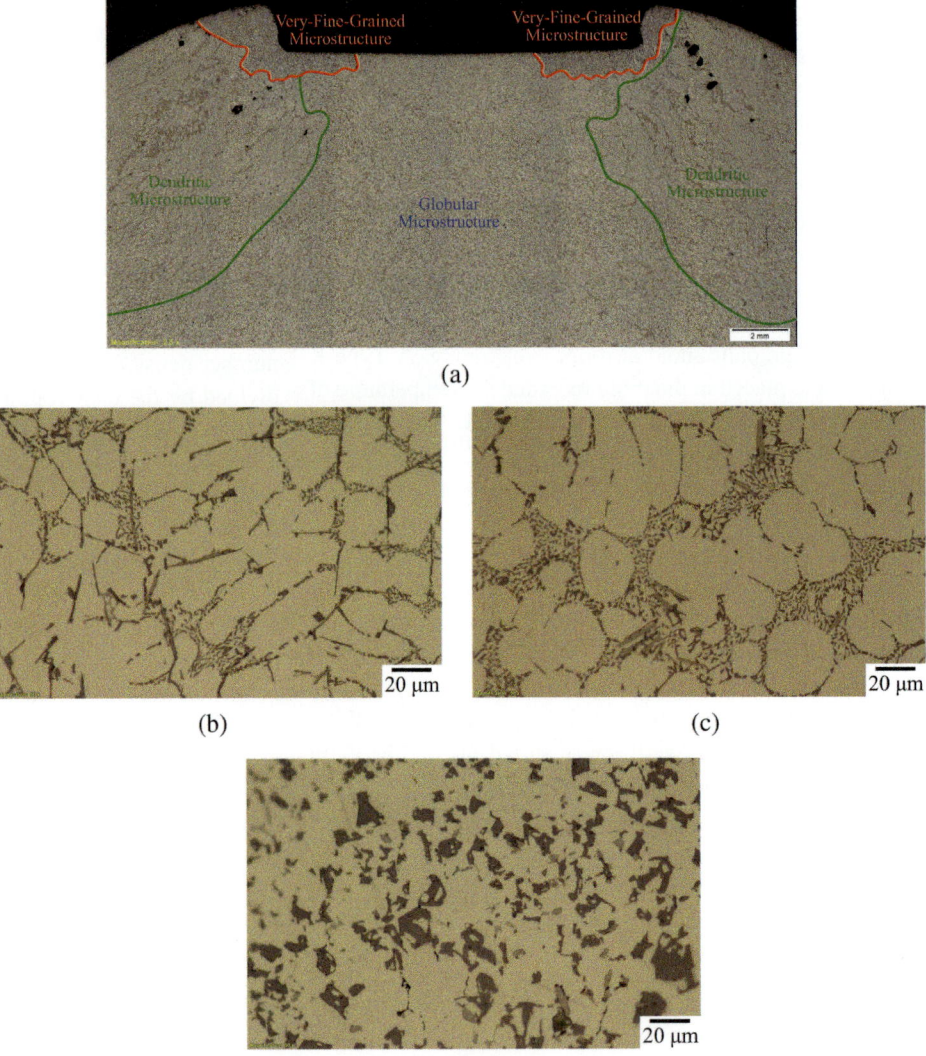

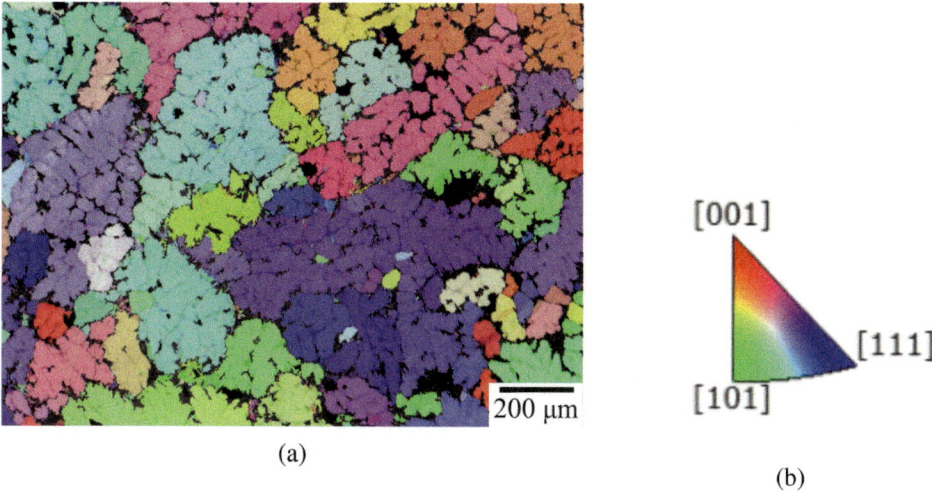

(a)

(b)

Fig. 3 **a** This inverse pole figure (IPF-Z) map identifies Al grains within the dendritic microstructure of the control casting. The crystallographic orientation of each grain relative to the incident electron beam can be identified using the color key in (**b**)

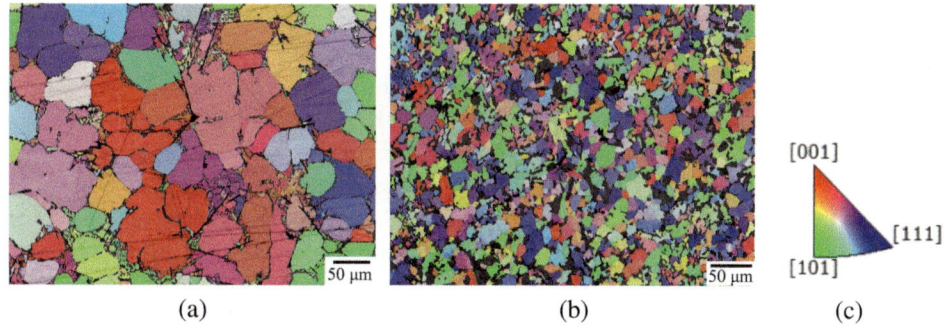

(a)   (b)   (c)

Fig. 4 These inverse pole figure (IPF-Z) maps identify Al grains within the **a** globular and **b** very-fine-grained microstructure regions of the ultrasonicated casting. The crystallographic orientation of each grain relative to the incident electron beam can be identified using the color key in (**c**). Dark colored regions in the IPF-Z maps correspond to non-aluminum phases (e.g., Si, β-Al$_5$FeSi, etc.)

Table 2 Mean equivalent grain size and mean sphericity of the primary Al grains by microstructure and processing method

Microstructure	Equivalent grain size (μm)	Sphericity
Dendritic (no ultrasound)	140 ± 210	0.39 ± 0.17
Globular	36 ± 27	0.48 ± 0.14
Very-fine-grained	9.3 ± 5.3	0.56 ± 0.16

greater, respectively. The mean equivalent grain size and sphericity of the very-fine-grained microstructure are 93% smaller and 44% greater, respectively, relative to the dendritic microstructure of the control casting.

Tensile Tests

Tensile tests were conducted using specimens extracted from the castings. The microstructure of the tensile specimens extracted from the control casting was dendritic while the microstructure of the tensile specimens extracted from the ultrasonicated casting was globular. A summary of the tensile properties are listed in Table 3. The globular microstructure of the ultrasonicated casting has 7% lower yield strength than the dendritic microstructure of the control casting. However, the average UTS between the two microstructures is the same and the globular microstructure has 88% greater elongation at rupture.

X-Ray Computed Tomography

X-Ray CT was conducted on tensile specimens to quantify porosity. The difference in the number and size of the pores before and after testing is considered negligible since the

Table 3 A summary of tensile properties by microstructure and processing method

Microstructure	Yield strength (MPa)	Ultimate tensile Strength (MPa)	Elongation at rupture (%)
Dendritic (no ultrasound)	191 ± 6	204 ± 8	0.75 ± 0.36
Globular	177 ± 13	203 ± 18	1.41 ± 0.23

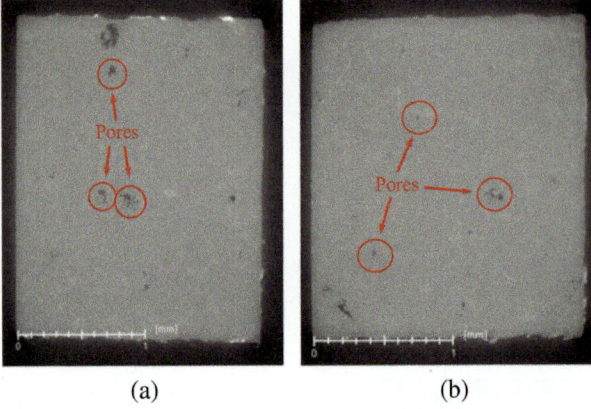

(a)   (b)

Fig. 5 Representative X-ray CT slices taken through the cross sections of tensile specimens show pores within (**a**) the dendritic microstructure of the control casting and (**b**) the globular microstructure of the ultrasonicated casting. Pores appear as black while the aluminum appears as light gray. Red circles and annotations have been added to identify examples of pores within each image

Discussion

Local ultrasonic intensification was successfully applied to an A356 + Fe aaluminum alloyalloy to refine the as-cast microstructure of the alloy as it solidified. Local ultrasonic intensification differs from most other applications of ultrasonic melt processing in that ultrasound is applied to the melt as it solidifies in the mold, rather than applying ultrasound to the melt before/during pouring. The mean equivalent grain sizes of the globular and very-fine-grained microstructures are similar to those of the non-dendritic microstructures produced using friction stir processing [6, 7], and are smaller than the grain sizes produced using chills and grain refiners [2–5]. However, local ultrasonic intensification has the additional benefit of not requiring chemical modifiers or additional post-processing steps. This will make components easier to recycle, since the composition is not changed, and will help reduce the costs of implementing the technology, since the microstructure is refined during the casting process.

Three microstructural morphologies were observed in the specimen cast with local ultrasonic intensification: a globular microstructure, a very-fine-grained microstructure, and a dendritic microstructure.. The globular microstructure was observed throughout the majority of the specimen, and was observed at distances as far as 45 mm away from the location of the ultrasound probe in the specimen. The globular microstructure is similar to other microstructures that have been produced in studies investigating ultrasonic melt processing [8–12]. The very-fine-grained microstructure was only observed within 2 mm of the location of the ultrasound probe, and has not been reported in previous studies investigating ultrasonic melt processing. This microstructure has very fine primary aluminum grains and relatively large,

elongation at rupture of the tensile specimens is very small. Figure 5 shows representative slices from two specimens: one from the control casting (dendritic microstructure), and one from the ultrasonicated casting (globular microstructure). Table 4 lists the number density of pores, mean pore volume, the volume of the largest pore, and porosity (in volume %) for each microstructure. The difference in the mean pore volume and the volume of the largest pore between the dendritic microstructure of the control casting and the globular microstructure of the ultrasonicated casting is negligible. However, the average porosity and number of pores in the globular microstructure are 33% greater than those of the dendritic microstructure. Porosity was not consistent throughout either casting, as the porosity of individual tensile specimens ranged from 0.33 to 2.94%.

Table 4 Number density of pores, mean pore volume, volume of the largest pore, and mean porosity (vol. %) by microstructure and processing method

Microstructure	Number density of pores (mm^{-3})	Mean pore volume (mm^3)	Largest pore (mm^3)	Porosity (%)
Dendritic (no ultrasound)	266	3.79×10^{-5}	1.68×10^{-2}	1.01
Globular	353	3.80×10^{-5}	1.56×10^{-2}	1.34

blocky Si and β-Al$_5$FeSi phase particles whose sizes are on the same order of magnitude as the primary aluminum grains. Computational fluid dynamics (CFD) simulations of ultrasound applied to molten aluminum by Riedel et al. measured a higher volume fraction of cavities near the face of the cylindrical ultrasound probe and tracked a high density of collapsed bubbles near the outer edge of the face of the probe [18]. The pressure exerted by these collapsing bubbles, coupled with additional mixing from ultrasonic side lobes [19], could possibly account for the unique very-fine-grained microstructure that is more refined than the microstructures observed at distances further away from the ultrasound probe. The CFD simulations by Riedel et al. also showed that the shape of the acoustic streaming pattern in molten A356 is tear-drop shaped [18]. However, in a relatively small container, such as the cylindrical mold used in the current work, the acoustic flow is expected to follow the sides of the container until the entire volume of the liquid is mixed [18]. In the specimen cast with local ultrasonic intensification, the shape of the ultrasonically refined zone (i.e., the regions in the specimen that have a very-fine-grained or globular microstructural morphology) is tear-drop shaped. This suggests that the energy applied by the ultrasound probe was not sufficient for the acoustic stream to be redirected by the boundaries of the mold.

Compared to the dendritic microstructure in the control casting, the ultrasonically refined globular microstructure is able to accommodate greater plastic deformation, as evidenced by the 88% greater ductility of the tensile specimens with a globular microstructure. However, the tensile strength and ductility of the globular microstructure produced in this study are less than those measured from similar ultrasonically refined microstructures produced in previous studies [7, 9] and are also less than those of typical 356-T6 permanent mold castings [13]. One possible explanation for this difference is that the alloy in this study had significantly more Fe content (0.91 wt.% Fe, compared to 0.13 wt.% Fe in [7] and 0.66 wt.% Fe in [9]). High Fe content in aluminum alloys alloys can make the alloy weaker and less ductile, particularly when the brittle and needle-shaped β-Al$_5$FeSi phase forms [16]. Another possible reason why the microstructures in this study are weaker and less ductile than similar microstructures in literature is that this study did not use an optimal casting procedure (e.g., the mold was only partially filled, the flow rate was inconsistent, the head height was quite high, etc.). This is evidenced by the large pores and high porosity levels in some tensile specimens, which can decrease strength and ductility. Instead, priority was given to performing the local ultrasonic intensification treatment. Optimization of the casting procedure is expected to increase both the tensile strength and ductility of the

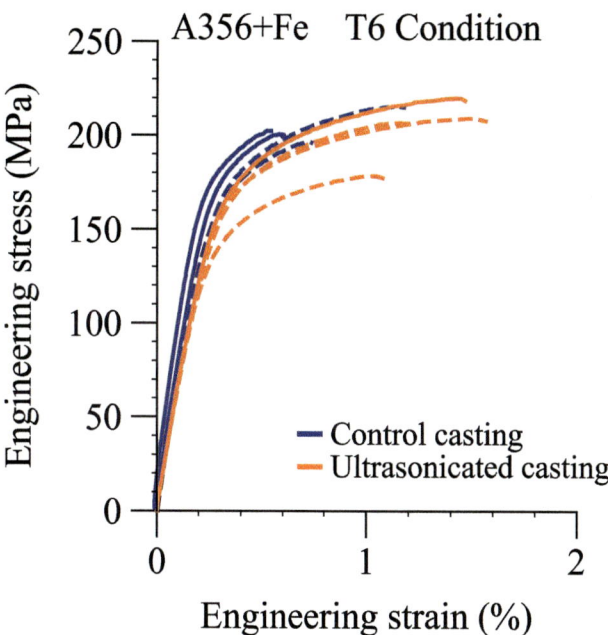

Fig. 6 Engineering stress is plotted as a function of engineering strain for tensile specimens extracted from the control casting (dendritic microstructure) and the ultrasonicated casting (globular microstructure). Solid lines indicate tensile data from specimens with porosity levels <1%

resultant castings. In tensile specimens where porosity was <1%, the globular microstructure had 10% greater UTS and 200% greater elongation at rupture, as shown in Fig. 6. This demonstrates that when porosity levels are sufficiently low, the iimprovement in mechanical properties as a result of ultrasonically induced microstructural refinement is more pronounced. Further development of this casting process could allow local ultrasonic intensification to be used to improve the mechanical properties of aluminum alloys with high Fe content. Greater strength and ductility in Fe-containing aluminum alloys alloys would increase the amount of Fe that can be tolerated in cast aluminum alloys alloys used in automotive applications and would allow cheaper, recycled alloys to be used.

Conclusions

Local ultrasonic intensification was applied to an A356 + Fe aluminum alloy as it solidified, during casting, to refine the as-cast microstructure. The morphologies and mechanical properties of the resultant microstructures were quantified using a combination of metallographic techniques and tensile tests, respectively. Based on the experimental conditions employed in this work, the following conclusions were reached.

1. The application of local ultrasonic intensification during casting produced two morphologies in the as-cast microstructure: globular primary Al grains ~ 36 μm in size and very-fine-grained primary Al grains ~ 9 μm in size. These sizes correlate to a grain size reduction of 74 and 94%, respectively, compared to the dendritic microstructure (140 μm) of the control casting. This reduction in grain size is comparable to the use of friction stir processing and is more effective than the use of chills or grain refiners [2–7].

2. The globular microstructure of the ultrasonicated casting had equivalent UTS and 88% greater ductility, on average, compared to the dendritic microstructure of the control casting. For tensile specimens with <1% porosity, however, the globular microstructure of the ultrasonicated casting had 10% greater UTS and 200% greater elongation at rupture than the dendritic microstructure of the control casting.

3. The globular microstructure of the ultrasonicated casting had 33% more porosity than the dendritic microstructure of the ultrasonicated casting. Additional work is needed to reduce porosity in both castings.

Acknowledgements This work was funded by the Department of Energy Vehicle Technologies Office as part of the Lightweight Metals Core Program. The authors thank Mr. David Weiss at Eck Industries for providing the Al alloy for this work and technical discussions. The authors also thank Anthony Guzman and Michael Blazon of Pacific Northwest National Laboratory for their technical assistance.

References

1. Abraham A, Schultz R, Rakoto B, Murphy J, Ling L, Merta M, Dudley J (2020) 2020 North America Light Vehicle Aluminum Content and Outlook. Ducker Frontier. https://drivealuminum.org/resources/outlooks-and-ducker/
2. Zhang LY, Jiang, YH, Ma Z, Shan SF, Jia YZ, Fan CZ, Wang WK (2008) Effect of cooling rate on solidified microstructure and mechanical properties of aluminum- A356 alloy. J. Mater. Process. Tech. 207:107–111
3. Sigworth GK, Kuhn TA (2007) Grain refinement of aluminum casting alloys. Int. J. Metalcasting Fall: 31–40
4. Yu L, Liu X, Wang Z, Bian X (2005) Grain refinement of A356 alloy by AlTiC/AlTiB master alloys. J. Mater. Sci. 40:3865–3867
5. Peeratatsuwan C, Chowwanonthapunya T (2020) Investigation on the grain refining performance of Al–5Ti–1B master alloy on the recycling process of A356 alloy. Materialwiss. Werkstofftech. 51:1346–1352
6. Nelaturu P, Jana S, Mishra RS, Grant G, Carlson BE (2018) Influence of friction stir processing on the room temperature fatigue cracking mechanisms of A356 aluminum alloy. Mater. Sci. Engr. A 716:165–178
7. Ma ZY, Sharma SR, Mishra RS (2006) Microstructural modification of as-cast Al–Si–Mg alloy by friction stir processing. Met. Mater. Trans. A 37:3323–3336
8. Eskin GI (2001) Broad prospects for commercial application of the ultrasonic (cavitation) melt treatment of light alloys. Ultrasonics Sonochem. 8:319–325
9. Puga H, Barbosa J, Costa S, Ribeiro S, Pinto AMP, Prokic M (2013) Influence of undirect ultrasonic vibration on the microstructure and mechanical behavior of Al–Si–Cu alloy. Mater. Sci. Engr. A 560:589–595
10. Wang F, Eskin D, Mi J, Wang C, Koe B, King A, Reinhard C, Connolley T (2017) A synchrotron X-radiography study of the fragmentation and refinement of primary intermetallic particles in an Al-35Cu alloy induced by ultrasonic melt processing. Acta Mater. 141:142–153
11. Zhang Z, Wang C, Koe B, Schleputz CM, Irvine S, Mi J (2021) Synchrotron X-ray imaging and ultrafast tomography in-situ study of the fragmentation and growth dynamics of dendritic microstructure in solidification under ultrasound. Acta Mater. 209 (116796):1–12
12. Khalifa W, Tsunekawa Y, Okumiya M (2008) Effect of ultrasonic melt treatment on microstructure of A356 aluminum cast alloys. Int. J. Cast Metals Res. 21(1–4):129–134
13. Lampman S (2018) Permanent mold casting of aluminum alloys. In: Anderson K, Weritz J, Kaufman JG (ed) ASM Handbook 2A:209–231. https://doi.org/10.31399/asm.hb.v02a.a0006513
14. Abramoffv MD, Magalhaes PJ, Ram SJ (2004) Image Processing with ImageJ. Biophotonics Int. 11(7), p. 36–42.
15. Standard practice for heat treatment of aluminum-alloy castings from all processes (2020). ASTM B917-12(2020). ASTM International, West Conshohocken, Pennsylvania
16. 356.0 and A356.0: Al-Si-Mg high-strength casting alloys (2019). In: Anderson K, Weritz J, Kaufman JG (ed) ASM Handbook 2B:548–552. https://doi.org/10.31399/asm.hb.v02b.a0006568
17. Menargues S, Martin E, Baile MT, Picas JA (2015) New short T6 heat treatments for aluminum silicon alloys obtained by semisolid forming. Mater. Sci. Engr. A 621:236–242
18. Riedel E, Liepe M, Scharf S (2020) Simulation of ultrasonic induced cavitation and acoustic streaming in liquid and solidifying aluminum. Metals 10(47):1–22
19. Quien MM, Saric M (2018) Ultrasound imaging artifacts: How to recognize them and how to avoid them. Echocardiography. 35:1388–1401

Microstructural Changes on a Ternary Al–Cu–Si Eutectic Alloy with Different Pre-heated Mold Temperatures

Seung-Hwan Oh, Sung-Soo Jung, and Young-Cheol Lee

Abstract

In order to understand the solidification behavior and microstructural evolution of the Al–Cu–Si ternary eutectic alloy system, the evolution on the microstructure of the Al–Cu–Si ternary eutectic alloy with different pre-heating mold temperatures was investigated. When the mold was preheated at 500 °C, the primary Si and Al_2Cu dendrites were mainly observed and (α-Al + Al_2Cu) binary eutectic and needle-shaped Si phase were also developed. When the mold preheating temperature was 300 °C, the primary Si and Al_2Cu dendrites were observed with colonial regions where the (α-Al + Al_2Cu + Si) ternary eutectic phase was also present. When the mold preheating temperature was 150 °C, the bimodal structure which composed of (α-Al + Al_2Cu) binary eutectic phase and (α-Al + Al_2Cu + Si) ternary eutectic phases was observed When the mold preheating temperature was changed from 500, 300, and 150 °C, Si phase undergoes the most distinctive microstructural changes. When it passed the critical cooling rate, the (α-Al + Al_2Cu + Si) ternary eutectic was formed. It was concluded that the growth of Si phase was suppressed by the high cooling rate and the growth of Al and Cu containing eutectic phases, which grow cooperatively, were also suppressed. This resulted in the ternary eutectic alloy with a finer microstructure.

Keywords

Ternary eutectic alloy • Eutectic reaction • Modification of eutectic • Rapid solidification

Introduction

Al–Cu–Si ternary eutectic alloy does not form an intermetallic compound consisting of three alloy elements. However, when 26–31wt.%Cu and 5–6.5wt.%Si are added to pure Al, it has a ternary eutectic point (L $\rightarrow$ α + β) and exhibits a solidification behavior similar to pure metal [1–3]. The solidification of the Al–Cu–Si ternary eutectic alloy has the advantage of growing two or more solid phases in one liquid phase in cooperation with each other, showing various shapes (Lamellar, Fibrous, Rod, etc.) according to the composition and cooling rates mainly. Mechanical properties of the alloys are varied depending on the growth pattern and size of the eutectic structure.

The Al–Cu–Si ternary eutectic alloy has superior corrosion resistance and hot cracking resistance and has superior specific strength among the many Al–Si-based alloys. However, the Al–Cu–Si ternary eutectic alloy is generally brittle and has a limited application as a structural material. However, through alloy design and optimizing the manufacturing process, it is possible to implement a bimodal structure with different scales microstructure, and improve plasticity and toughness significantly. In addition, the microstructure of the Al–Cu–Si ternary eutectic alloy is formed from a bimodal structure composed of micro-sized (α-Al + Al_2Cu) and nano-sized (α-Al_2Cu + Si) microstructure at a high cooling rate (100 K/s or more). This results in the high elongation of over 11% or more and high compressive strength of 1GPa at room temperature [4–7].

Contributing authors S.-H. Oh and Y.-C. Lee contributed equally to this work.

S.-H. Oh · S.-S. Jung · Y.-C. Lee (✉)
Korea Institute of Industrial Technology, Busan, 46938, Republic of Korea
e-mail: yclee87@kitech.re.kr

S.-S. Jung
e-mail: jungss@kitech.re.kr

© The Minerals, Metals & Materials Society 2023
S. Broek (ed.), *Light Metals 2023*, The Minerals, Metals & Materials Series,
https://doi.org/10.1007/978-3-031-22532-1_58

However, previous studies were conducted with rapid solidification and it was difficult to implement rapid cooling in the actual casting industry. For this reason, a practical method is needed to implement the bimodal structure of the Al–Cu–Si ternary eutectic alloy at room temperature.

Experimental Procedure

Pure aluminum (99.7%), pure copper (99.9%), and metal silicon (99.9%) were charged into a graphite crucible to manufacture the Al-27wt.%Cu-5wt.%Si ternary eutectic alloy used in this study. Melting was conducted using a high-frequency induction furnace at 750 ± 5 °C.

In order to remove moisture and secure fluidity, the bottom gating mold (FC25) was preheated to 350 ± 5 °C and the cavity mold (Cu–Be) was preheated to 150, 300, and 500 ± 5 °C in order to obtain different cooling rates, respectively, and the casting conditions are listed in Table 1.

The composition of the eutectic alloys was analyzed using optical emission spectroscopy with an Al-28wt.%Cu-6wt.% Si standard specimen, and the results are shown in Table 2.

Based on the Flite database required for thermodynamic numerical analysis of the aluminum and magnesium alloy system of Factsage v8.1 S/W, prediction of the composition and phase percentage of Al-27wt.%Cu-5wt.%Si three-component eutectic alloys that appear at 1 °C interval under equilibrium conditions were derived. For observation of microstructure, the surface was etched using a Weck etchant (4 g $KMnO_4$ + 1 g NaOH + 100 ml distilled water), and then an optical microscope and SEM-EDS analysis were performed.

Results and Discussion

Growth Mechanism of a Ternary Eutectic Alloy

The binary eutectic alloy solidifies at a specific temperature which is similar to the solidification of pure metal. In a eutectic reaction, in general, two solid phases are formed in a liquid phase and can be expressed by the following:

$$L \text{ (Liquid)} \rightarrow \alpha + \beta \text{(eutectic, solid)}$$

The eutectic reaction has a short solid–liquid coexistence section, and thus nuclei are generated and grown by diffusion of alloy elements in a short time. This growth mechanism is called cooperative growth as shown in Fig. 1 and the same applies to the ternary alloy system.

Different alloy elements A and B constituting the eutectic alloy are released at each interface of α-phase and β-phase as the temperature decreases. Therefore, as shown in Fig. 1a, solute A accumulates at the interface of β-phase, solute B accumulates at the interface of α-phase, and α-phase and β-phase simultaneously grow along the interface.

As each phase grows, the interface of the β-phase between the α phase and the α phase becomes unstable due to the undercooling generated by the non-uniform distribution of the solute of the solid–liquid interface. In addition, as shown in the pictures in Fig. 1b and c, the surface tension formed by three phases of α-phase, β-phase, and liquid phase forms a mechanical equilibrium. However, in the case of a lamellar structure, the larger the distance between α-phase and α-phase, the more interface instability of the center of β-phase appears. This interface instability occurs at both α-phase and β-phase interfaces, and eventually forms an unstable solid–liquid interface [8–10].

The formation of the unstable eutectic interface can be expressed by a primary function of the ratio of the composition (C_0), the growth rate (V), and the temperature gradient (G) at the solid–liquid interface, as shown in Fig. 2a.

In Fig. 2b, if the amount of an added alloying element is small and has a low G/V ratio value, dendrite is preferentially formed in the early stage of solidification, and a eutectic phase is formed in the latter part of the solidification.

Figure 2c shows that an irregular eutectic is formed when the amount of an added alloying element is large and the value of the high G/V ratio is high. When nuclei are formed and grown in a liquid phase, the value of surface energy varies depending on the crystallographic orientation and crystals grow preferentially in the direction of low surface energy. When forming an irregular eutectic, the difference in surface energy between alloy elements affects the growth rate and growth direction of the phase formed during solidification. In addition, the addition of a third alloy element affects the solid–liquid interface and contributes to the formation of an unstable eutectic interface [11–14].

Table 1 Casting conditions

Melting condition		Parameter	Value
	Casting	Alloy	Al-27wt%Cu–5wt%Si
		Melting temp	750 °C
		Mold preheated temp	150, 300, and 500 °C

Table 2 Chemical composition of Al-27Cu-5Si ternary eutectic alloy used (wt.%)

	Si	Cu	Fe	Al
Al-27Cu-5Si alloy	4.93	26.7	0.21	Bal.

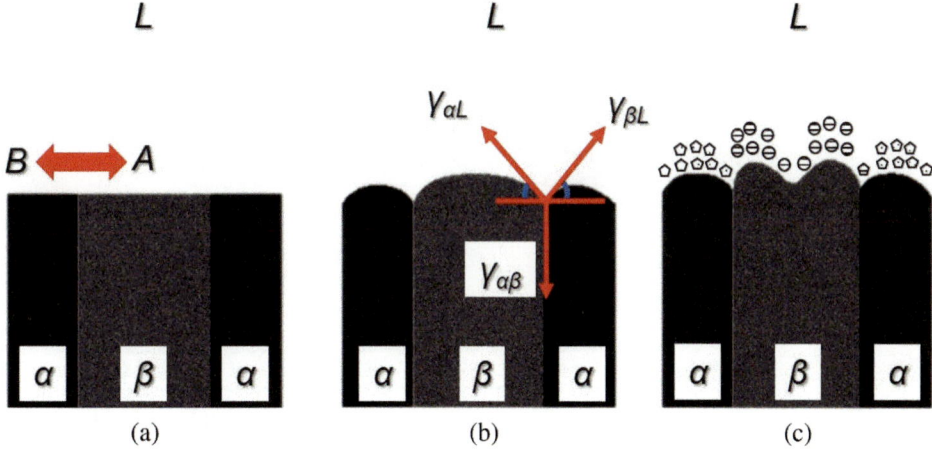

Fig. 1 Interface morphology during solidification; where **a** boundary energy and accumulation of solutes at the interface are overlooked; **b** the excess energy at the grain boundary has been considered; **c** both accumulation of solute and boundary energy has been considered [14]

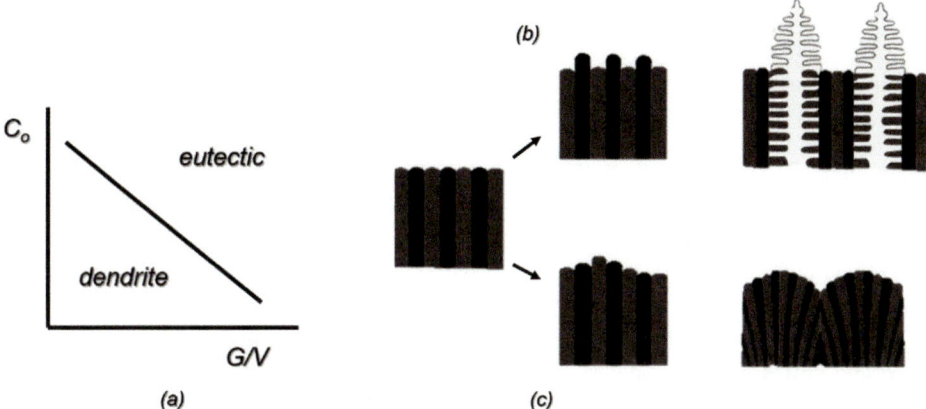

Fig. 2 Instability of the eutectic interface (after Kurz and Fisher 1998); **a** Instability of one phase resulting in dendrites + interdendritic eutectic, **b** Instability of both phases resulting in two phase eutectic cells or dendrites [14]

Observation of Microstructure with Different Cooling Rates

Figure 3 shows the constituent phases and SEM-EDS analysis results of the Al-27wt.%Cu-5wt.%Si ternary eutectic alloy with a mold preheating temperature of 500 °C. The primary Si phase, Al$_2$Cu dendrite, and a (α-Al + Al$_2$Cu) binary eutectic were observed. However, a (α-Al + Al$_2$Cu + Si) ternary eutectic was not observed.

Based on the Al–Cu–Si ternary phase evaluation of the alloy used in this study through computational simulation using Factsage v8.1 S/W (Fig. 6), it is found that the primary Si phase is precipitated first, and then the Al$_2$Cu dendrite is simultaneously formed. The (α-Al + Al$_2$Cu + Si) ternary eutectic is formed in the last stage of solidification.

Since it is an Al–Cu–Si ternary eutectic alloy, the formation of a ternary eutectic phase composed of (α-Al + Al$_2$Cu + Si) is estimated. However, as shown in Fig. 3a, the

primary Si phase, needle-shaped Si, Al$_2$Cu dendrite, and (α-Al + Al$_2$Cu) binary eutectic phases were observed mainly and the (α-Al$_2$Cu + Si) ternary eutectic was not observed.

The Al–Al$_2$Cu eutectic alloy is a typical regular eutectic alloy, and a lamellar structure of α-Al and Al$_2$Cu eutectic phase is observed. From the microstructural analysis, it can be seen that a lamellar eutectic structure is irregularly grown by the addition of a Si element.

In the case of Al$_2$Cu dendrites, they nucleate and grow immediately after the primary Si phase is precipitated. As the mold preheating temperature decreases, Al$_2$Cu dendrites are not observed and (α-Al + Al$_2$Cu) binary eutectic and (α-Al + Al$_2$Cu + Si) ternary eutectic phases are shown.

Si observed in Al–Cu–Si ternary eutectic alloys is recognized as the same primary Si in hypereutectic aluminum alloys and needle-shaped Si in hypoeutectic aluminum alloys. Du et al. reported that the diffusion of Si atoms in the

Fig. 3 Analysis of constituent phases on Al-27wt%–5wt%Si ternary eutectic alloy

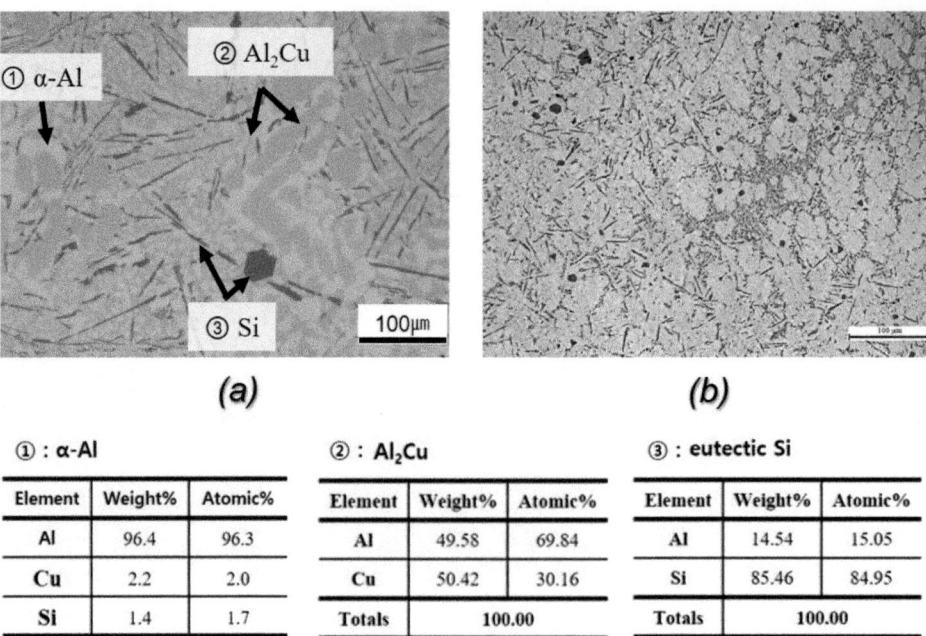

①: α-Al

Element	Weight%	Atomic%
Al	96.4	96.3
Cu	2.2	2.0
Si	1.4	1.7
Totals	100.00	

②: Al₂Cu

Element	Weight%	Atomic%
Al	49.58	69.84
Cu	50.42	30.16
Totals	100.00	

③: eutectic Si

Element	Weight%	Atomic%
Al	14.54	15.05
Si	85.46	84.95
Totals	100.00	

(c)

aluminum molten metal proceeds at a rate 8 times faster than that of Cu atoms [15]. Si as a non-metal has a higher diffusion rate than metal atoms in the aluminum melt, so it is preferentially nucleated over other phases and progressed TPRE (Twining Plane Re-entrant Edge) growth [14].

As previously stated, it can be explained that the Si phase forms and grows in a specific direction that lowers the surface energy, and it is assumed as the Al, Cu, and Si elements do not form a ternary compound. In addition, Ian et al. reported that the surface energy of the molten metal in the atmosphere decreases when Si is added to pure aluminum melt by 4–5wt.%. It is believed that the decrease in surface energy facilitates the movement of atoms in the aluminum melt, which affects nucleation and growth [16].

Figure 4 shows the microstructure of the ternary eutectic alloy made with different mold preheating temperatures. When the mold was preheated at 500 °C, primary Si, needle-shaped Si, and (α-Al + Al₂Cu) binary eutectic were observed. However, ternary eutectic phases composed of (α-Al + Al₂Cu + Si) were not observed.

When the mold preheating temperature was 300 °C, a ternary eutectic structure composed of (α-Al + Al₂Cu + Si) was partially observed and colonial regions of containing the ternary eutectic structure and without was distinctively observed. At this time, the Si phase was observed as needles near the colonies of the (α-Al + Al₂Cu) binary eutectic phase, and the (α-Al + Al₂Cu + Si) ternary eutectic phase was formed

when the critical cooling rate was reached. Figure 5a shows SEM images of the bimodal structure of the (α-Al + Al₂Cu) binary eutectic and the (α-Al + Al₂Cu + Si) ternary eutectic.

When the mold preheating temperature was at 150 °C, a bimodal structure composed of the (α-Al + Al₂Cu) binary eutectic and the (α-Al + Al₂Cu + Si) ternary eutectic, was observed. The bimodal structure refers to a microstructure with phases of different length scales, and Park et al. has reported that the bimodal structure of the Al-Cu-Si ternary eutectic alloy is implemented with a high cooling rate (100 K/s or more) [6]. However, it is estimated that the bimodal structure was formed with a cooling rate of 100 K/s or less in this study.

As the mold preheating temperature decreases, It is assumed that the time for solute diffusion decreases and the evolution of that microstructure take place due to the difference in diffusion rates of Al, Cu, and Si elements.

One interesting fact is that, in the case of the Al–Cu–Si ternary eutectic alloy, the Si phase seems to grow independently. However, morphologies of the Si phase were changed with different cooling rates of the residual liquid in the final solidification stage. In addition, because of the cooperative growth of the Al–Cu–Si ternary eutectic, the growth of the surrounding Al and Cu elements is also inhibited as the growth of the Si phase is suppressed.

Figure 6 shows the phase diagram of the Al-27wt.%Cu–5wt.%Si ternary eutectic alloy derived using Factsage v8.1

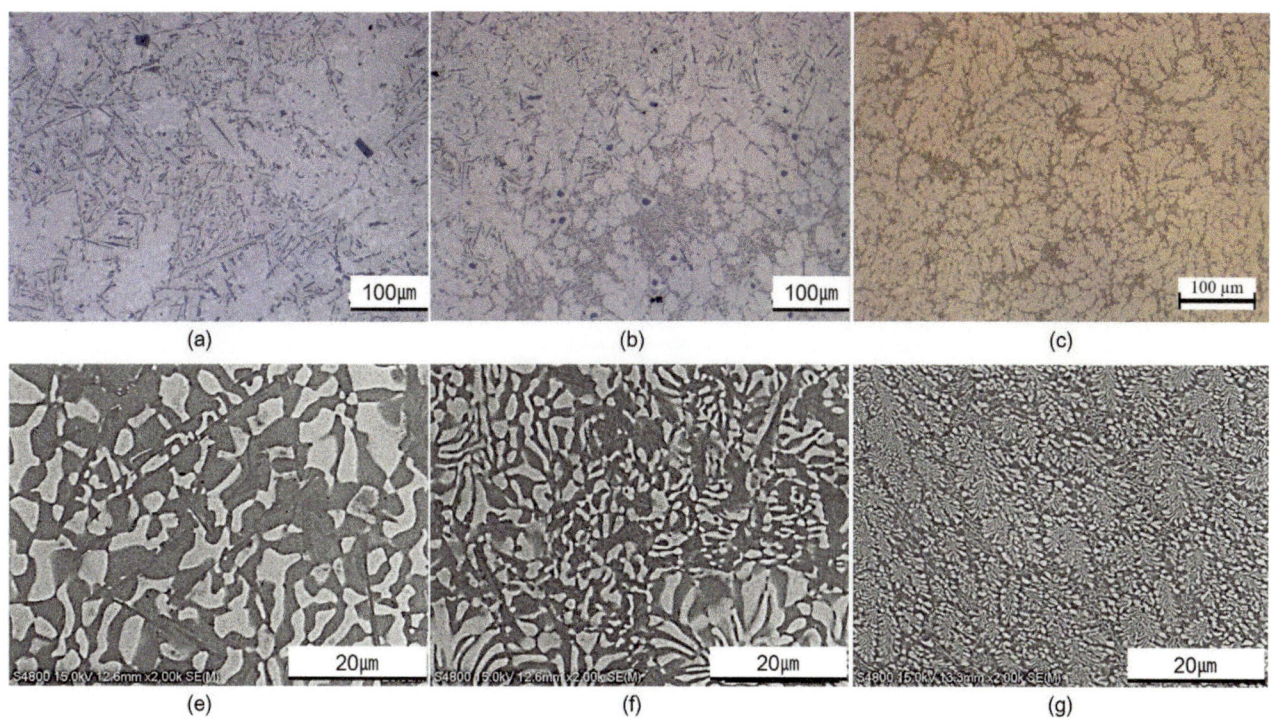

Fig. 4 Microstructures of Al-27wt%–5wt%Si ternary eutectic alloy with different cooling rates; **a** optical image of samples at mold temperature 500 °C, **b** 300 °C, and **c** 150 °C; **e** SEM image of samples at mold temperature 500 °C, **f** 300 °C, and **g** 150 °C (bimodal structure)

Fig. 5 Microstructure of Al–Cu–Si ternary eutectic alloy; bimodal structure of (α-Al + Al₂Cu) binary eutectic and (α-Al + Al₂Cu + Si) ternary eutectic

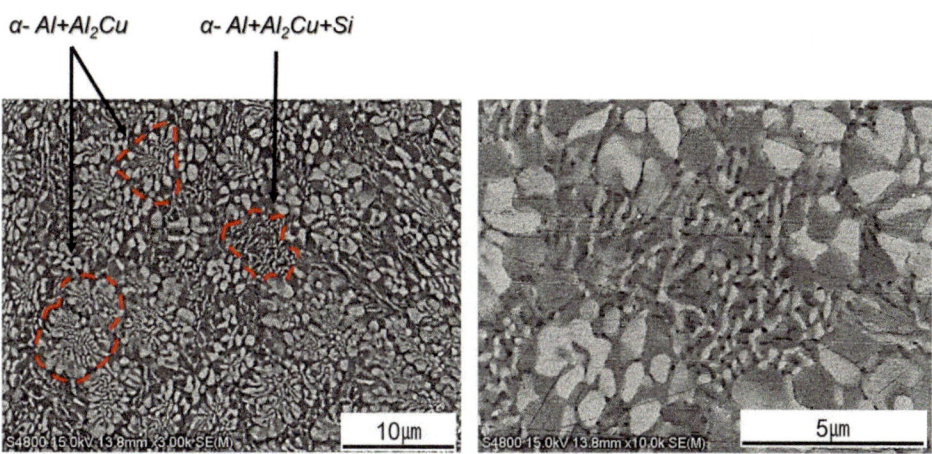

S/W, and the solidification stage of the Al-27wt.%Cu–5wt.% Si ternary eutectic alloy and the mass fraction of each phase are shown in Table 3.

The eutectic temperature of the Al-27wt%Cu–5wt%Si ternary eutectic alloy derived through thermodynamic calculation is 528.1 °C and the first precipitated phase during solidification is expected to be the primary Si phase at 530 °C. When the eutectic temperature is reached, the Al₂Cu phase is formed and the ternary eutectic is formed immediately. Solidification is to be completed at the eutectic temperature, and the volume fraction is expected to be 4.2% Si phase, 48.3% Al₂Cu phase, and 47.5% α-Al phase. It is also confirmed that the simulation results of the Al–Cu–Si ternary eutectic alloy predicted through Factsage v8.1 S/W match the phases observed in the actual casting microstructures.

Fig. 6 Phase diagrams on Al-27wt.%Cu–5wt.%Si ternary eutectic alloy with the Factsage8.1 S/W

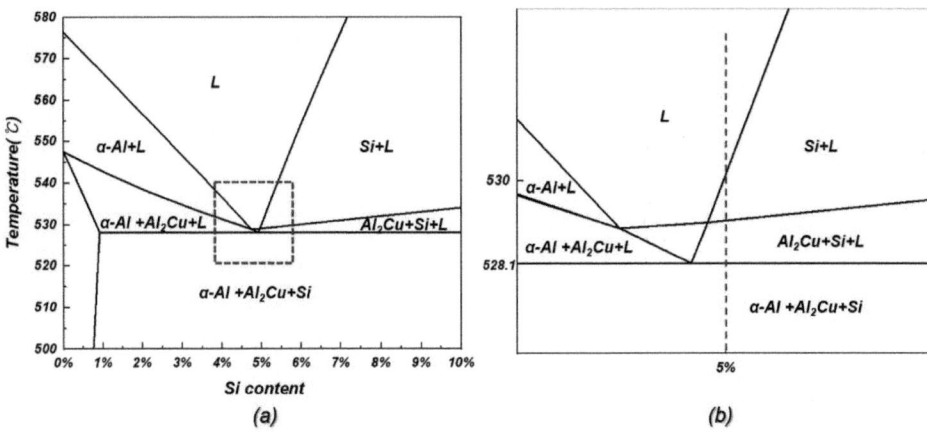

Table 3 Computed reaction and parameter of Al-27wt.%Cu–5wt. %Si ternary eutectic alloy in equilibrium solidification

Solidification path	Reaction	Temperature	Mass fraction at 528.1	Note
1	L → Si	530.0	4.2	
2	L → Si + Al₂Cu	528.1	48.3	
3	L → Si + Al₂Cu + α-Al	528.1	47.5	Ternary eutectic reaction

Conclusions

The following conclusions can be drawn after observing the microstructural evolution on Al-27wt%Cu–5wt%Si ternary eutectic alloy according to different mold preheating temperatures:

(1) When the mold was preheated at 500 °C, the primary Si and Al₂Cu dendrites were mainly observed and (α-Al + Al₂Cu) binary eutectic and needle-shaped Si phase were also developed. When the mold preheating temperature was 300 °C, the primary Si and Al₂Cu dendrites were observed with colonial regions where the (α-Al + Al₂Cu + Si) ternary eutectic phase was also present. When the mold preheating temperature was 150 °C, the bimodal structure which composed of (α-Al + Al₂Cu) binary eutectic phase and (α-Al + Al₂Cu + Si) ternary eutectic phases was observed.

(2) When the mold preheating temperature was changed from 500, 300, and 150 °C, the Si phase undergoes the most distinctive microstructural changes. When it passed the critical cooling rate, the (α-Al + Al₂Cu + Si) ternary eutectic was formed. It was concluded that the growth of the Si phase was suppressed by the high cooling rate and the growth of Al and Cu containing eutectic phases, which grow cooperatively, were also suppressed. This resulted in the ternary eutectic alloy with a finer microstructure.

References

1. S.L. JT Kim, SH Hong, HJ Park, JY Park, NS Lee, YH Seo, WM Wang, JM Park, KB Kim (2016) Understanding the relationship between microstructure and mechanical properties of Al–Cu–Si ultrafine eutectic composite. Materials and Design 92:1038–1045.
2. C.S. Tiwary, S. Kashyap, K. Chattopadhyay (2014). Development of alloys with high strength at elevated temperatures by tuning the bimodal microstructure in the Al–Cu–Ni eutectic system. Scripta Materialia, 93, 20–23. https://doi.org/10.1016/j.scriptamat.2014.08.020
3. L.F. Mondolfo, Aluminum alloys: Structure and properties, Butterworth-Heinemann 1976.
4. S.W. Lee, J.T. Kim, S.H. Hong, H.J. Park, J.-Y. Park, N.S. Lee, Y. Seo, J.Y. Suh, J. Eckert, D.H. Kim (2014) Micro-to-nano-scale deformation mechanisms of a bimodal ultrafine eutectic composite. Scientific reports 4(1):1–5. https://doi.org/10.1038/srep06500.
5. X. Ma, L. Liu (2015). Solidification microstructures of the undercooled Co–24at%Sn eutectic alloy containing 0.5at%Mn. Materials & Design, 83, 138–143. https://doi.org/10.1016/j.matdes.2015.06.010.
6. J.M. Park, N. Mattern, U. Kühn, J. Eckert, K.B. Kim, W.T. Kim, K. Chattopadhyay, D.H. Kim (2009) High-strength bulk Al-based bimodal ultrafine eutectic composite with enhanced plasticity. Journal of Materials Research 24(8):2605–2609. https://doi.org/10.1557/jmr.2009.0297.
7. G.C. Pettan, C.R.M. Afonso, J.E. Spinelli (2015) Microstructure development and mechanical properties of rapidly solidified Ti–Fe and Ti–Fe–Bi alloys. Materials & Design 86:221–229.
8. K.A. Jackson, J.D. Hunt, Lamellar and Rod Eutectic Growth, in: P. Pelcé (Ed.), Dynamics of Curved Fronts, Academic Press, San Diego, 1988, pp. 363–376.
9. T. Sato, Y. Sayama (1974). Completely and partially co-operative growth of eutectics, Journal of Crystal Growth, 22(4), 259–271. https://doi.org/10.1016/0022-0248(74)90170-5.

10. Wattis, A.D. Jonathan (1999) A Becker-Döring model of competitive nucleation. Journal of Physics A: Mathematical and General 32(49):8755-8784. https://doi.org/10.1088/0305-4470/32/49/315.

11. BJ Kim, D. K. Arne, YH Park, Y. Lee (2022) The effect of Sr additions on Al–Cu–Si ternary eutectic alloys for a high-ductility bimodal microstructure. Materials Science & Engineering A 833. https://doi.org/10.1016/j.msea.2021.142547.

12. W. Kurz (2001) Dendrite growth in eutectic alloys: coupled zone. Advanced Engineering Materials 3(7) 443–452. https://doi.org/10.1179/imtr.1979.24.1.177.

13. D. Stefanescu, R. Ruxanda, Fundamentals of Solidification, 2004, pp. 71–92.

14. D.M. Stefanescu, Science and engineering of casting solidification, Springer 2015

15. Y. Du, Y.A. Chang, B. Huang, W. Gong, Z. Jin, H. Xu, Z. Yuan, Y. Liu, Y. He, F.-Y. Xie (2003), Diffusion coefficients of some solutes in fcc and liquid Al: critical evaluation and correlation. Materials Science and Engineering A 363:140–151. https://doi.org/10.1016/S0921-5093(03)00624-5.

16. F.B. Ian, A.T. John (2013) The surface Tension of Pure Aluminum and Aluminum Alloys. Metallurgical and Materials Transaction A 44A:3901–3909. https://doi.org/10.1007/s11661-013-1696-9.

Nanoparticle-Enhanced Arc Welding of Aluminum Alloys

Narayanan Murali and Xiaochun Li

Abstract

Arc welding high-strength aluminum alloys is a great challenge due to characteristic defects upon melting and re-solidification. However, it is an economical method by which these important engineering alloys can be processed. An emerging metallurgical technique known as nano-treating has seen increasing use in tandem with classical metallurgy, with the latest advances being presented here. Incorporating a small volume fraction of nanoparticles into the welding process significantly alters the associated structure/processing/property relationships and enables the joining of difficult-to-weld aluminum alloy systems. From a structural perspective, nanoparticles suppress the formation of dendrites, refine grains, and attenuate large secondary phases that exist in conventional arc welding, thus eliminating hot cracking. The microstructural evolution that nanoparticles permit along with traditional strengthening mechanisms elevates the performance of the weld over its traditional barriers. Thus, the nano-treating approach paves the way for new possibilities in arc welding high-strength aluminum alloys and other "unweldable" systems.

Keywords

Welding • Aluminum alloys • Nano-treating

N. Murali · X. Li (✉)
Department of Materials Science and Engineering, University of California at Los Angeles, Los Angeles, CA 90095, USA
e-mail: xcli@seas.ucla.edu

X. Li
Department of Mechanical and Aerospace Engineering, University of California at Los Angeles, Los Angeles, CA 90095, USA

Introduction

Aluminum alloys (AAs) have become the prime material candidate for engineering applications in the automotive, aerospace, and other industries. These alloys are also an important component in structural lightweighting, where weight reductions from using low-density alloys result in improved energy efficiency and lower emissions [1]. Aluminum alloys possess a wide range of material properties to suit different applications, so industry trends have been pointing toward the use of these alloys wherever possible. For example, AA2024 (Al–Cu–Mg), known for its fatigue resistance and high strength, is normally used in aerospace, AA5083 (Al–Mg), with excellent corrosion resistance in seawater, sees use in marine vessels, and AA6061 (Al–Mg-Si), which can be readily welded or formed, is generally used in structures across industries.

The heat-treatable 2XXX- and 7XXX-series alloys are considered "high-performance" grades because of their high strengths, which allow them to be used in primary structures or in heavily stressed areas. The 2XXX series is alloyed with high Cu contents and may contain other elements like Mg, Mn, or Si. The most widespread alloy in this series is the aforementioned AA2024, which relies on S (Al_2CuMg) and θ' (Al_2Cu) precipitates for strengthening [2]. Previously referred to as Duralumin, these alloys, because of their susceptibility to corrosion and stress corrosion cracking, have been slowly replaced with the higher strength 7XXX series. The principal alloying element for the series is Zn, though most contain significant amounts of Mg and variable amounts of Cu. The key strengthening precipitate in this system is η' ($Mg(Zn, Al, Cu)_2$) [3].

In the manufacturing industry, traditional assembly methods, such as bolting, riveting, etc., are necessary for fabricating large structures consisting of high-strength alloys, such as aircraft and spacecraft, and aluminum alloys are no exception. However, for these purposes, fusion welding is a necessity for creating permanent joints where

© The Minerals, Metals & Materials Society 2023
S. Broek (ed.), *Light Metals 2023*, The Minerals, Metals & Materials Series,
https://doi.org/10.1007/978-3-031-22532-1_59

disassembly is not required. In this vein, the two primary methods for joining aluminum alloys are gas-metal arc welding (GMAW)/metal inert gas (MIG) welding and gas-tungsten arc welding (GTAW)/tungsten inert gas (TIG) welding. In GMAW, the wire itself is the electrode, and the arc is maintained by the continuously fed wire while it deposits. This process can be readily automated because the wire is automatically fed from a spool, but because of the nature of the arc and the thin wire, there is potential for metal spattering that can reduce the quality of the weld bead. In the GTAW process, the electrode is a separate tungsten rod, while the filler metal is fed manually or automatically. Each method has its own advantages: for example, GMAW is preferred for a higher throughput, while GTAW is slower but generally produces higher quality joints because of the greater control over the arc and filler feeding. As high-strength aluminum alloys see wider use in primary structures, multiple issues concerning the creation of reliable joints through arc welding come to the forefront. These topics will be briefly reviewed later in the paper, but among them are hot cracking, shrinkage porosities, inability to create dissimilar joints, and a severe loss in strength from using common filler materials.

A new metallurgical approach known as nano-treating is seeing increased investigation as a way to address several issues surrounding the arc welding process as well as the general casting process. At its core is the dispersion of a low amount of ceramic nanoparticles in metals and alloys to improve their processability and final properties. The objective of this paper is to present a review of the effectiveness of nano-treating on the arc welding process. Though it is still a nascent approach, it shows great promise in improving the overall quality of arc welded joints and paves the way for a more facile processing route for materials that are considered difficult or impossible to process through traditional arc welding processes.

Challenges of Arc Welding High-Performance Aluminum Alloys

Arc welding aluminum alloys involves the melting and re-solidification of the base metal, but because of their varied alloying content, their properties generally differ from grade to grade. Among them, the most relevant to arc welding are the liquidus and solidus temperatures and coefficients of thermal expansion (CTEs). The aforementioned high-performance alloys contain a higher overall alloying concentration that grants them their strength and other properties. However, the disadvantage of melting and resolidifying them during the welding process is that they are prone to hot cracking.

Kou explains that hot cracks occur during the last stages of solidification in intergranular liquid films, when the semi-solid weld metal (WM) is weak, due to tensile stresses exerted upon it by freezing primary grains within the equilibrium solidification range [5], defined as the difference between the alloy's liquidus and solidus temperatures. Furthermore, the dilution of the total alloying content in the WM can place the composition of the region at great risk of hot cracking; for example, high-strength AA7075 can crack more easily than AA6061 because of its Cu, Mg, and Zn content [4]. Another aspect of hot cracking involves the remnant liquid that is present in the WM towards the terminal stage of solidification. For high-performance alloys, significant solute rejection from growing Al grains allows the liquid to flow between grains for much of the weld solidification process. But primary grain formation in the WM is usually heavily dendritic in nature, so the liquid can become trapped between dendrite arms and prevent it from backfilling more solidified zones. Because of a lack of liquid feeding in the semi-solid zone, pores can potentially form and help nucleate and grow hot cracks [6]. On the other hand, too much solute-rich liquid results in continuous films that create easy paths for hot crack propagation [7]. Finally, Kim and Nam observed that a higher equilibrium solidification range correlates to a higher susceptibility to hot cracking [8]; the longer the solidification process takes to complete, which is the case for highly alloyed high-performance grades of aluminum, the more time is available for hot cracks to form and propagate through the WM.

Another problem that can contribute to low-quality arc-welded joints is porosity, specifically the formation of shrinkage pores. The volumetric difference between liquid and solid aluminum can manifest as macropores, while "clusters" of smaller pores can form at the roots of primary dendrites [9].

Shrinkage porosities differ from gas porosities based on their sphericity; the latter tend to be more spherical in three-dimensions and appear more circular [10]. As previously mentioned, hot cracks can be initiated from a pore, but even if the WM can remain intact throughout solidification, these porosities can severely hamper the performance of the joint during service. For example, Li, et al. showed that a repair weldment using GTAW contained shrinkage pores that acted as stress concentration sites and preformed fatigued cracks that contributed to premature fatigue failure [11].

A significant limitation of the arc welding process is the inability to join different grades of aluminum together. The economical nature and design freedom of arc welding in tandem with the ability to combine the properties of two or more alloys at a single joint is of great industrial importance.

Unfortunately, this process is extremely difficult, and the overarching source of this problem is the differing material properties on either side of the joint. For example, during solidification, varying CTEs can more easily cause harmful welding defects because one alloy could completely freeze while the other alloy is in the middle of its equilibrium solidification range, inevitably causing hot cracking. In addition, combining different alloyed aluminum systems can potentially create a WM composition that is more susceptible to hot cracking compared to each base metal. Therefore, due to the multitude of problems and considerations that necessarily arise as a result of the liquid-to-solid phase transition, most processes that require the joining of different high-strength alloy grades have been relegated to solid-state processes such as friction-stir welding [12, 13].

The final problem that arc welding processes pose concerns the final mechanical properties of the weldment. High-strength alloys are considered impossible to arc weld due to the major problem of hot cracking, but it is also important to consider the filler composition with which they could potentially be welded when using their nano-treated counterparts. The most common series that are used to weld a plethora of aluminum alloys are 4XXX (Al–Si) and 5XXX (Al–Mg). These compositions aim to drive the Si or Mg content of the WM outside the range within which the region would be susceptible to hot cracking [14].

Unfortunately, using them dilutes the alloying content of 2XXX- or 7XXX-series base metals that give such high strength. In addition to the fact that 4XXX- and 5XXX-series alloys cannot be heat treated, this means the joint loses much of its as-welded strength and cannot recover much strength through solution heat treatment and artificial aging.

These limitations render arc welding extremely difficult to implement for joining high-performance aluminum alloy grades. However, nano-treating shows distinct advantages that can address these problems to better process these kinds of joints for various applications. The remainder of the paper will discuss the established literature surrounding nano-treatment for the solidification processing of aluminum alloys before surveying the progress made in arc welding.

Nano-treating for Solidification Processing of Aluminum Alloys

Nano-treating is the infusion of a small percentage of ceramic nanoparticles into metals and alloys to improve their processability and properties. Using a low volume of nanoparticles allows for better dispersion in the final product, so nano-treating is typically limited to approximately 1 vol.%. Choosing a suitable nanoparticle for dispersion involves energetic considerations, which were proposed by Xu via energy potentials [15]. Nanoparticle dispersion must occur in the molten metal, where the amount of available energy is far greater than that available in the solid state, so the nanoparticles' interfacial energy with liquid Al, their Brownian potential, and their van der Waals potential all impact their final distribution. Good dispersion occurs when the system's thermal energy, kT, gives nanoparticles enough energy to escape their mutual van der Waals attraction, as seen in Fig. 1.

Alternatively, pseudo-dispersion can occur if their mutual van der Waals attraction is greater than the thermal energy provided by the system. In both cases, it is crucial that the nanoparticles have good wetting with the liquid matrix to discourage clustering and possible sintering. Analyzing the relationship between these three factors within a liquid metal and being able to pick an appropriate nanoparticle for the system ensures that the desired nanoparticle distribution can be achieved.

Fig. 1 Interaction potential for nanoparticle dispersion [15]

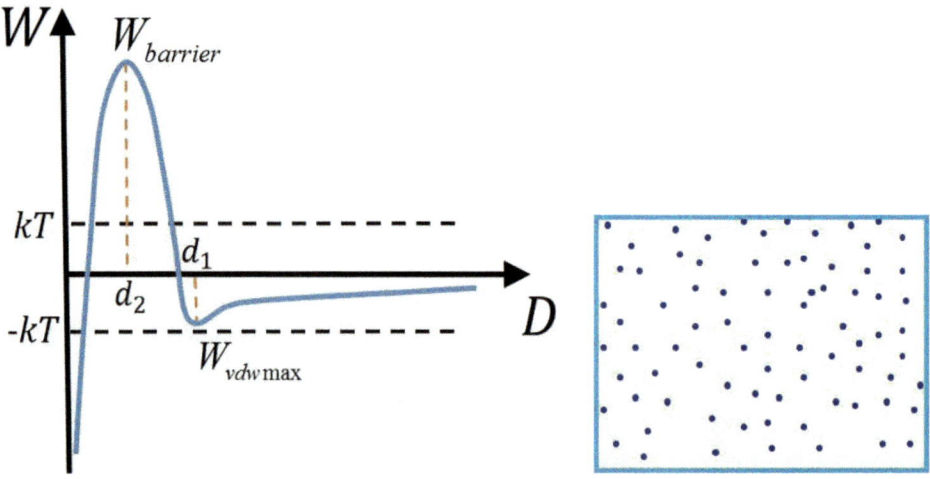

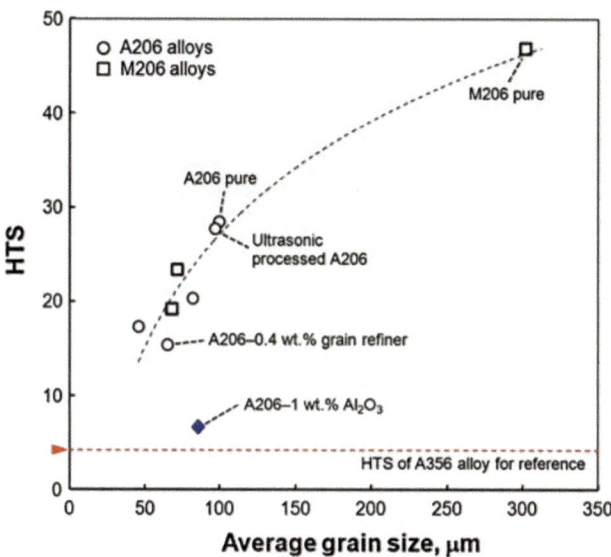

Fig. 2 Hot tear susceptibility of A206 alloys as a function of grain size; M206 denotes the Ti-free version of A206 [16]

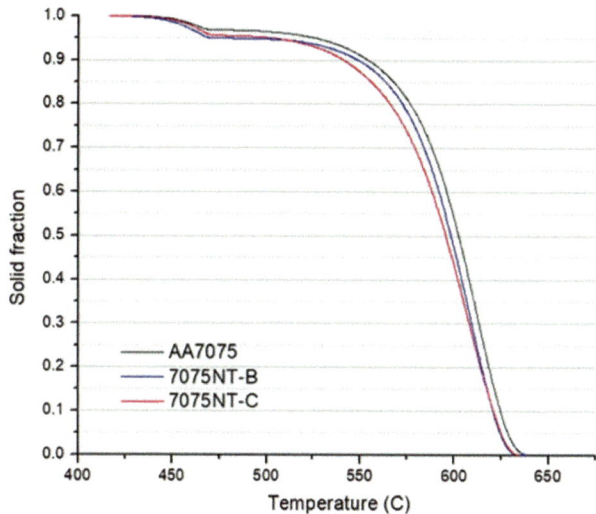

Fig. 3 Solid fraction evolution of AA7075 during solidification when nano-treated with TiB_2 (NT-B) and TiC (NT-C) [19]

Harnessing these principles led to the use of nano-treatment in aluminum solidification processing. Choi, et al. cast ultrasonically processed A206 with 1 wt.% γ-Al_2O_3 nanoparticles and discovered that the nanocomposite's hot tearing susceptibility (HTS) was lowered to a level comparable to that of A356. They observed a heavily refined microstructure in terms of both primary and secondary phases and attributed the reduced HTS to the prevention of dendrite bridging and accommodation of contraction strains by the finer semi-solid structure, observations that were not shown when simply adding the conventional grain refiner [16], as shown in Fig. 2.

De Cicco, et al. further studied the effects of various nanoparticles (Al_2O_3, TiC, TiB_2, and SiC) on the nucleation behavior of A356 and found that TiB_2 and TiC nanoparticles can also act as effective nucleation catalysts for primary Al grains by virtue of their low lattice mismatch [17]. Zuo, et al. investigated the casting of AA7075 with 1 vol.% TiC nanoparticles at different cooling rates and noted that, even with slower cooling, the primary Al grains were heavily refined, to less than 20 µm, relative to an identical casting without TiC [18]. It was also noted that a fine-grained microstructure was retained in the nano-treated castings after a heat treatment, conditions that would necessarily coarsen grains in conventional alloys. Sokoluk, et al. performed thermal analysis on nano-treated AA7075 to compare TiC with TiB_2 and found that, while TiB_2 was less effective than TiC due to clustering in the liquid state, both nanoparticles were able to delay the onset of primary grain coherency, slow down the overall solidification of AA7075, and increase the amount of terminal liquid at the end of solidification [19], as shown in Fig. 3.

The increase in terminal liquid fraction and the delay in grain coherency helped explain the alloy's significantly reduced susceptibility to hot cracking with the aid of nanoparticles. The methods were also applied to high-performance AA2024, and a similar conclusion was reached, demonstrating the wide applicability of nano-treatment and addressing a longstanding obstacle in the way of fabricating high-performance aluminum alloys [20]. Yuan, et al. nano-treated a high-performance 7XXX-series alloy with 1 vol.% TiC but increased the percentage of Zn; they noticed that the nano-treated alloy showed more complete secondary phase dissolution during solution treatment and an improved response to artificial aging with regard to hardness evolution and aging speed [21]. Yuan, et al. further explored the solution treatment aspect in nano-treated AA7034, where nanoparticles again promoted a faster and more complete dissolution of secondary phases while affecting the volume fractions and distributions of eutectic phases [22]. Similarly, the influence of nano-treating on solution treatment was shown in a high-performance 2XXX-series alloy with 1 vol.% TiC, where nanoparticles were shown to increase the solubility limit of Mg and enable a high Mg supersaturation in the alloy after solution heat treatment because of a reduction in interfacial energy between TiC and secondary S phases [23]. Nano-treating was also shown to improve the natural aging response of a high-strength 7XXX-series composition by helping form GPI and GPII zones faster and in higher volume compared to an untreated sample [24]. As shown by these studies, nano-treatment can be used as an effective processing aid in multiple cast and wrought aluminum compositions, and its use can bring about beneficial property increases for various applications.

Progress in Nano-treating for Aluminum Alloy Arc Welding

Since arc welding can be viewed as a high-energy casting process, the use of nano-treatment can be expanded into this area of manufacturing. A key difference between casting and welding is the system's energy being increased significantly in the latter. Aghakhani, et al. explained the behavior of ceramic TiO_2 nanoparticles under an arc in submerged arc welding, stating that the addition of nanoparticles past a certain threshold improved the arc's penetration due to their decreased thermal conductivity [25]. During the welding process, heat is carried by hot electrons from the liquid WM, and eventually they are conducted into the base metal. But because the nanoparticles have reduced thermal and electrical conductivities, they are able to effectively trap the heat in the WM region and increase penetration. Furthermore, ceramic nanoparticles can increase melting efficiency by preventing hot electrons from conducting into the base metal. This behavior was also noted by Tseng and Lin when conducting GTAW on stainless steel [26]. When investigating the use of Al_2O_3 and SiC microparticles and nanoparticles, they observed that adding Al_2O_3 did not show any meaningful change to penetration, whereas additions of SiO_2 nanoparticles increased the weld depth by 524%. Also, the use of nanosized SiO_2 improved the depth-to-width (D/W) ratio of the WM to 1.08, whereas the other particulate inclusions had ratios below 1. This D/W ratio was used to describe the power density of the welding operation, and since the nanosized SiO_2 inclusion had the highest ratio, this led to other benefits, such as 78% lower angular distortion in the assembly and a narrower HAZ. These benefits can be attributed to the lower thermal conductivity of ceramic nanoparticles.

Regarding the use of nanoparticles in arc welding, there is literature showing their incorporation into the arc welding process via the filler material. For instance, for TIG welding, Fattahi, et al. fabricated ER4043 with TiC nanoparticles to join AA6061 [27], while Ramkumar, et al. made 1XXX-series filler with TiO_2 nanoparticles to join AA3003 [28]. Unfortunately, the available literature generally does not report the fusion welding of high-performance alloy series like 2XXX and 7XXX due to the problems mentioned, most often citing their notorious hot cracking tendencies. With the aid of nano-treatment, these grades can be joined, and the current progress in this regard will be reviewed hereafter.

Sokoluk, et al. successfully TIG welded AA7075 by using nano-treated AA7075 filler in what is perhaps the most important showcase of nano-treatment for arc welding [29]. Microstructural and thermal analyses that followed echoed the observations found during casting, showing refined primary grains and secondary phases along with more gradual latent heat release in solidification. AA7075 welds made with different fillers are shown in Fig. 4, along with a high-magnification cross section of the nano-treated filler rod.

T6 heat treatment of the welds showed a significant strength recovery of approximately 60 HV in the WM, outweighing that of that unaffected base metal. Simultaneously, the study also showed the low mechanical potential of the weldment made with commercial ER5356, which gained only between 20 and 25 HV. By using the same high-strength composition as the carrier for nanoparticles, important alloying elements were not diluted in the WM and could plentifully precipitate to effectively strengthen the region. Oropeza, et al. followed with a combined study on GTAW of AA7075 and additive manufacturing using nano-treated AA7075 wire and noted that successful joining

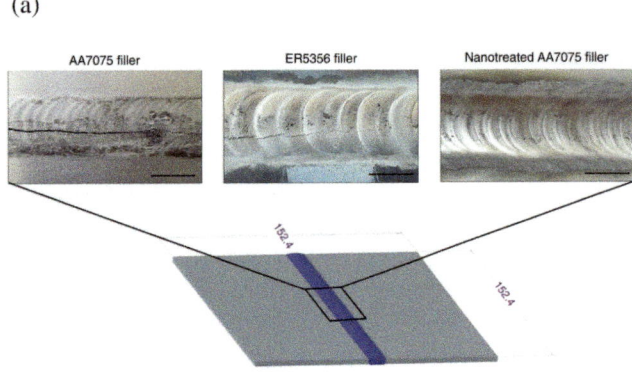

Fig. 4 Welds of AA7075 made with AA7075, commercial ER5356, and nano-treated AA7075 (**a**); TiC nanoparticle band within the nano-treated AA7075 filler rod (**b**). Scale bars are 10 mm in (**a**) and 100 μin in (**b**) [29]

Fig. 5 Grain structure (**a**) and secondary phase distribution (**b**) in the dissimilar AA2024/AA7075 weld [31]

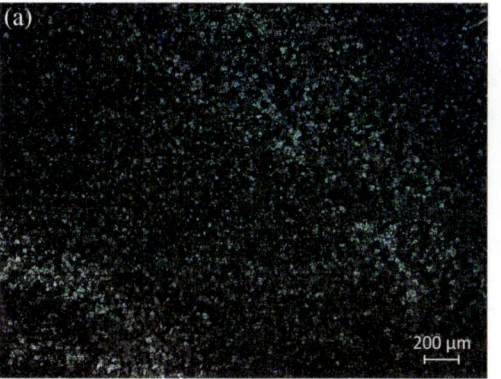
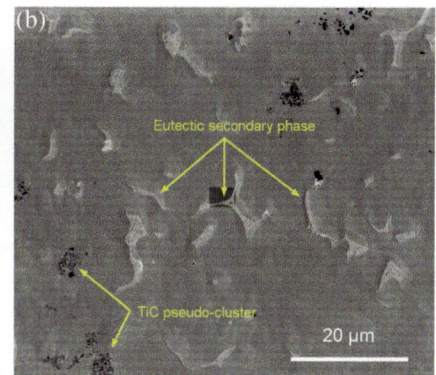

Fig. 6 Hardness changes during natural aging of AA2024/AA7075 joints welded with nano-treated AA2024-TiC (**a**) and nano-treated AA7075-TiC (**b**) fillers [34]

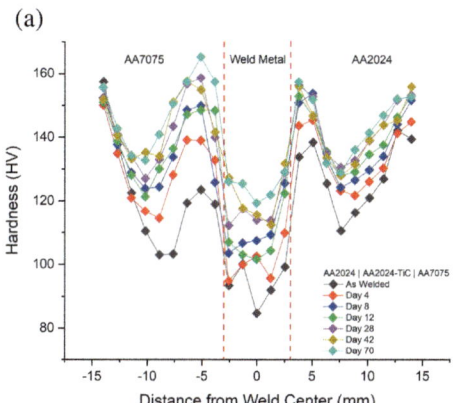
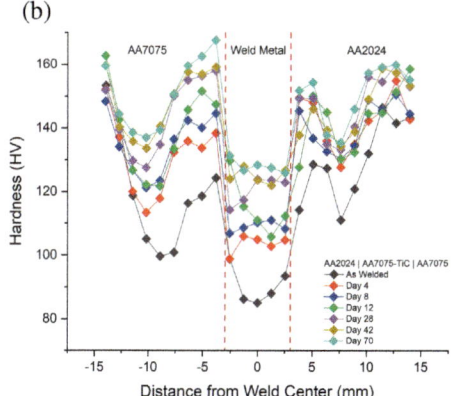

was achieved without cracking or measurable porosity, while the additively manufactured sample was fully densified and free of cracks [30]. Heat treatment of the welded sample returned strength and ductility to nearly match that of the parent material. A few studies were conducted on dissimilar welding of high-performance alloys, whereas the ones reviewed above used a single grade as both the base and filler metals. Murali, et al. nano-treated ER5183 (Al–Mg-Mn) with 1 vol.% of TiC nanoparticles and successfully joined AA2024 and AA7075 [31], as shown in Fig. 5. TiC nanoparticles were observed around intergranular secondary phases, while a multitude of intermetallic compounds were detected in the WM.

Despite the differences in properties between the two high-performance alloys, the nano-treated filler managed to control the solidification of the weld to ensure a crack-free joint. Homogeneous weldments of AA2024 and AA7075 were made in a separate investigation to show that nano-treatment can join each high-strength alloy independently [32]. The feasibility of joining AA7075 to the widely used AA6061 was also examined with nano-treated AA6061 and AA7075 fillers and compared to commercial ER4043 and ER5356 [33]. Study on the AA2024/AA7075 system

was extended to natural aging due to the difficulty in formulating a heat treatment schedule for a dissimilar weldment [34]. The authors fabricated dissimilar joints using nano-treated filler material matched to one of the base metals (i.e., nano-treated AA2024 and AA7075 fillers). After ten weeks of natural aging, the dissimilar joints gained between 30 and 40 HV of strength (as shown in Fig. 6), showing that nano-treatment can promote appreciable natural aging responses in both casting and arc welding conditions. Additionally, the joint made with nano-treated AA7075 as filler registered higher average hardness values towards the end of the aging period, likely due to the WM containing a higher fraction of an AA7075 composition.

The survey of literature regarding nano-treatment in arc welding is mostly concerned with GTAW, though the same principles can be applied to GMAW in the form of conventional joints or additive manufacturing with WAAM, the latter having established research. Since the nanoparticles will behave similarly under the harsh conditions of an arc, weldments and structures made with either GTAW and GMAW will show similar trends in processing and properties when nano-treatment is applied to the filler alloy or wire feedstock.

Conclusion

In the field of modern metallurgy, nano-treatment has emerged as a potent method by which high-performance aluminum alloys can be processed. Regardless of their chemical composition, nanoparticles can be incorporated and dispersed to yield improved properties while avoiding the harmful drawbacks that usually accompany these alloys, such as hot cracking. Moreover, meaningful results stemming from the use of nanoparticles can be observed in arc welding. High-performance aluminum alloys can be successfully joined without hot cracking and other defects upon solidification while displaying significantly elevated microstructures and performance. Furthermore, different high-performance alloy grades can be brought together in a single joint to potentially gain the properties of both. Nano-treatment thus far has shown to be quite promising for welding aluminum alloys that are traditionally difficult or impossible to weld, and with further research and experimentation, it can provide important guidance towards the rational design of diverse yet robust structures.

References

1. Raabe D, Tasan CC, Olivetti EA (2019) Strategies for improving the sustainability of structural metals. Nature 575:64–74. https://doi.org/10.1038/s41586-019-1702-5

2. Huda Z, Taib NI, Zaharinie T (2009) Characterization of 2024-T3: An aerospace aluminum alloy. Mater Chem Phys 113:515–517. https://doi.org/10.1016/j.matchemphys.2008.09.050

3. Lee SH, Jung JG, Baik S Il, et al (2021) Effects of Ti addition on the microstructure and mechanical properties of Al–Zn–Mg–Cu–Zr alloy. Mater Sci Eng A 801:140437. https://doi.org/10.1016/j.msea.2020.140437

4. Cheng CM, Chou CP, Lee IK, Lin HY (2005) Hot cracking of welds on heat treatable aluminium alloys. Sci Technol Weld Join 10:344–352. https://doi.org/10.1179/174329305X40688

5. Kou S (2003) Solidification and liquation cracking issues in welding. Jom 55:37–42. https://doi.org/10.1007/s11837-003-0137-4

6. Coniglio N, Cross CE (2009) Mechanisms for solidification crack initiation and growth in aluminum welding. Metall Mater Trans A Phys Metall Mater Sci 40:2718–2728. https://doi.org/10.1007/s11661-009-9964-4

7. Kah P, Rajan R, Martikainen J, Suoranta R (2015) Investigation of weld defects in friction-stir welding and fusion welding of aluminium alloys. Int J Mech Mater Eng 10:26. https://doi.org/10.1186/s40712-015-0053-8

8. Kim HT, Nam SW (1996) Solidification cracking susceptibility of high strength aluminum alloy weldment. Scr Mater 34:1139–1145. https://doi.org/10.1016/1359-6462(95)00644-3

9. Legait P-A (2005) Formation and Distribution of Porosity in Al-Si Welds. Worcester Polytechnic Institute

10. Gu C, Lu Y, Luo AA (2021) Three-dimensional visualization and quantification of microporosity in aluminum castings by X-ray micro-computed tomography. J Mater Sci Technol 65:99–107. https://doi.org/10.1016/j.jmst.2020.03.088

11. Li L, Liu Z, Snow M (2006) Effect of defects on fatigue strength of GTAW repaired cast aluminum alloy. Weld J 85:264s-270s

12. Cavaliere P, Nobile R, Panella FW, Squillace A (2006) Mechanical and microstructural behaviour of 2024-7075 aluminium alloy sheets joined by friction stir welding. Int J Mach Tools Manuf 46:588–594. https://doi.org/10.1016/j.ijmachtools.2005.07.010

13. Khodir SA, Shibayanagi T (2008) Friction stir welding of dissimilar AA2024 and AA7075 aluminum alloys. Mater Sci Eng B Solid-State Mater Adv Technol 148:82–87. https://doi.org/10.1016/j.mseb.2007.09.024

14. Lippold JC (2014) Welding Metallurgy and Weldability. John Wiley & Sons, Inc, Hoboken, NJ

15. Xu J (2015) Achieving Uniform Nanoparticle Dispersion in Metal Matrix Nanocomposites. University of California at Los Angeles

16. Choi H, Cho WH, Konishi H, et al (2013) Nanoparticle-induced superior hot tearing resistance of A206 alloy. Metall Mater Trans A Phys Metall Mater Sci 44:1897–1907. https://doi.org/10.1007/s11661-012-1531-8

17. De Cicco MP, Turng LS, Li X, Perepezko JH (2011) Nucleation catalysis in aluminum alloy A356 using nanoscale inoculants. Metall Mater Trans A Phys Metall Mater Sci 42:2323–2330. https://doi.org/10.1007/s11661-011-0607-1

18. Zuo M, Sokoluk M, Cao C, et al (2019) Microstructure Control and Performance Evolution of Aluminum Alloy 7075 by Nano-Treating. Sci Rep 9:10671. https://doi.org/10.1038/s41598-019-47182-9

19. Sokoluk M, De Rosa I, Pan S, Li X (2021) Thermal Analysis of the Solidification Behavior of AA7075 Containing Nanoparticles. In: Minerals, Metals and Materials Series. pp 250–256

20. Sokoluk M, Yuan J, Pan S, Li X (2021) Nanoparticles Enabled Mechanism for Hot Cracking Elimination in Aluminum Alloys. Metall Mater Trans A Phys Metall Mater Sci 52:3083–3096. https://doi.org/10.1007/s11661-021-06302-9

21. Yuan J, Zuo M, Sokoluk M, et al (2020) Nanotreating High-Zinc Al–Zn–Mg–Cu Alloy by TiC Nanoparticles. In: Minerals, Metals and Materials Series. pp 318–323

22. Yuan J, Pan S, Zheng T, Li X (2021) Nanoparticle promoted solution treatment by reducing segregation in AA7034. Mater Sci Eng A 822:141691. https://doi.org/10.1016/j.msea.2021.141691

23. Yuan J, Liu Q, Pan S, et al (2022) Nano-Treating Promoted Solute Dissolution for Novel High Strength Al-Cu-Mg Alloys. Materialia 101466. https://doi.org/10.1016/j.mtla.2022.101466

24. Yuan J, Liu Q, Pan S, et al (2022) Nano-Treating Promoted Natural Aging Al-Zn-Mg-Cu Alloys. J Compos Sci 6:114. https://doi.org/10.3390/jcs6040114

25. Aghakhani M, Ghaderi MR, Karami A, Derakhshan AA (2014) Combined effect of TiO2 nanoparticles and input welding parameters on the weld bead penetration in submerged arc welding process using fuzzy logic. Int J Adv Manuf Technol 70:63–72. https://doi.org/10.1007/s00170-013-5180-x

26. Tseng KH, Lin PY (2014) UNS S31603 stainless steel tungsten inert gas welds made with microparticle and nanoparticle oxides. Materials (Basel) 7:4755–4772. https://doi.org/10.3390/ma7064755

27. Fattahi M, Mohammady M, Sajjadi N, et al (2015) Effect of TiC nanoparticles on the microstructure and mechanical properties of gas tungsten arc welded aluminum joints. J Mater Process Technol 217:21–29. https://doi.org/10.1016/j.jmatprotec.2014.10.023

28. Ramkumar KR, Natarajan S (2018) Investigations on microstructure and mechanical properties of TiO2 Nanoparticles addition in Al 3003 alloy joints by gas tungsten arc welding. Mater Sci Eng A 727:51–60. https://doi.org/10.1016/j.msea.2018.04.111

29. Sokoluk M, Cao C, Pan S, Li X (2019) Nanoparticle-enabled phase control for arc welding of unweldable aluminum alloy 7075. Nat Commun 10:98. https://doi.org/10.1038/s41467-018-07989-y

30. Oropeza D, Hofmann DC, Williams K, et al (2020) Welding and additive manufacturing with nanoparticle-enhanced aluminum

7075 wire. J Alloys Compd 834:154987. https://doi.org/10.1016/j.jallcom.2020.154987

31. Murali N, Sokoluk M, Yao G, et al (2021) Gas-Tungsten Arc Welding of Dissimilar Aluminum Alloys with Nano-Treated Filler. J Manuf Sci Eng Trans ASME 143:1–24. https://doi.org/10.1115/1.4049849

32. Murali N, Sokoluk M, Li X (2021) Study on aluminum alloy joints welded with nano-treated Al-Mg-Mn filler wire. Mater Lett 283:128739. https://doi.org/10.1016/j.matlet.2020.128739

33. Murali N, Li X (2021) TIG Welding of Dissimilar High-Strength Aluminum Alloys 6061 and 7075 with Nano-Treated Filler Wires. In: Perander L (ed) Minerals, Metals and Materials Series. The Minerals, Metals & Materials Society, pp 316–322

34. Murali N, Chi Y, Li X (2022) Natural aging of dissimilar high-strength AA2024/AA7075 joints arc welded with nano-treated filler. Mater Lett 322:132479. https://doi.org/10.1016/j.matlet.2022.132479

Phase Equilibria in Al–Fe Alloys

Jožef Medved, Maja Vončina, and Jože Arbeiter

Abstract

Solidification of aluminium alloys is a complex process in which inhomogeneities form. Aluminium and iron form the equilibrium phase $Al_{13}Fe_4$ and various metastable intermetallic phases: Al_6Fe, Al_mFe, and Al_xFe. Three alloys were prepared and analysed in the laboratory: AlFe1, AlFe1Si0.1, and AlFe1Si0.5. Thermodynamic calculations, differential scanning calorimetry, electrical resistivity measurements, and optical and scanning electron microscopy were used to analyse the alloys. The results show that the addition of silicon to the AlFe1 alloy has a great influence on the distribution, amount, and morphology of the phases formed during the solidification process. The addition of 0.1 wt.% Si reduces the amount of metastable Al_6Fe phase. The transformation sequence was defined, starting with the dissolution of the metastable phase and the nucleation of the stable phase $Al_{13}Fe_4$. Due to the increased diffusion lengths of the iron atoms, the transformation takes place in the eutectic range after 4 h of homogenization at 600 °C.

Keywords

Aluminium alloys • Phase transformation • Thermodynamic calculation • Metastable phase • Homogenization

J. Medved (✉) · M. Vončina · J. Arbeiter
Faculty of Natural Sciences and Engineering, Department of Materials and Metallurgy, University of Ljubljana, Aškerčeva 12, 1000 Ljubljana, Slovenia
e-mail: jozef.medved@ntf.uni-lj.si

M. Vončina
e-mail: maja.voncina@ntf.uni-lj.si

J. Arbeiter
e-mail: joze.arbeiter@ntf.uni-lj.si

Introduction

Iron is present in almost all aluminium alloys because it can be introduced into the alloys during electrolysis from ores, by interaction with furnace linings and fluxes, or simply by the dissolution of foundry equipment during casting, but most contamination by iron usually occurs during recycling [1, 2]. Iron-rich intermetallic phases are often associated with lower mechanical properties; on the other hand, iron is used as an alloying element in some wrought aluminium alloys because it provides a good combination of ductility and strength at thinner thicknesses in the presence of small amounts of silicon [3, 4]. Since aluminium alloys do not solidify in equilibrium under industrial conditions, the formation of metastable or non-equilibrium intermetallic phases may occur. Therefore, high-temperature homogenization treatment is required to reduce the effects of metallurgical segregation.

Aluminium and iron form the equilibrium phase $Al_{13}Fe_4$, but non-equilibrium solidification of alloys containing iron as an alloying element or impurity can lead to the formation of various metastable intermetallic phases, such as Al_6Fe, Al_mFe, or Al_xFe [5, 6]. During homogenization, the transformation from metastable to stable phases takes place. The transformation of Al_mFe starts at a lower temperature (at about 390 °C) than that of Al_6Fe (at about 490 °C), but the transformation rate is slower in the case of Al_mFe [7]. Due to the low solubility of silicon in the Al_6Fe phase, the addition of silicon in the alloy decreases the amount of Al_6Fe phase, favouring the formation of the Al_mFe phase [8]. The transformation may also be influenced by the prior presence of stable intermetallic $Al_{13}Fe_4$ particles, which act as nuclei [9].

A combination of thermodynamic calculations and differential scanning calorimetry (DSC) has been successfully used to determine the optimal homogenization parameters and to understand the phase transformations during this process [10, 11]. In particular, the effects of homogenization on the changes in the morphology of Fe constituents, which are known to negatively affect formability, have been studied.

© The Minerals, Metals & Materials Society 2023
S. Broek (ed.), *Light Metals 2023*, The Minerals, Metals & Materials Series,
https://doi.org/10.1007/978-3-031-22532-1_60

The subject of this work was the analysis of metastable phase transformation during the homogenization of an Al–Fe-(Si) alloy. Following the research on binary Al–Fe alloys [12], a study on the influence of silicon addition on the transformation of the metastable intermetallic phase Al_6Fe to the stable intermetallic phase $Al_{13}Fe_4$ was carried out by differential scanning calorimetry (DSC) and metallographic analysis.

Materials and Methods

Two Al–Fe-Si alloys were produced in an induction furnace by combining 99.99 wt.% pure aluminium and 99.99 wt.% pure iron with the master alloy AlSi10. To ensure that the alloy did not contain impurities, all instruments, moulds, and crucibles were coated with boron nitride (BN). Casting was performed in steel moulds with a length of 160 mm and a diameter of 15 mm. A cooling rate between 30 and 40 K/s was selected for the solidification of the metastable Al_6Fe phase. To achieve the desired cooling rate, the mould was preheated to 450 °C. The chemical compositions of the prepared alloys are listed in Table 1, and measurements were performed by inductively coupled plasma spectroscopy (ICP).

To perform the homogenization annealing, the rods were cut into 10 mm sections using a water-cooled circular saw. Homogenization annealing was performed in an electric chamber furnace. The start of the homogenization annealing was recorded when the temperature reached the desired temperature of 600 °C. At this time, the first sample was removed from the furnace, and thereafter each additional sample was removed at the designated time of homogenization. Each sample was immediately quenched in water. The samples were homogenized from t = 0 to 24 h. The results of this work show the samples in as-cast state, and 2, 4, 8, 12, and 24 h after homogenization. All samples are labelled with the alloy composition and homogenization time in hours.

After homogenization, the samples were cut in half along the cross section. One half was used to prepare the specimens for metallographic analysis, and the other was used to prepare the specimens for DSC measurements. A 2.0 mm thick slice was cut from the centre of the sample using a water-cooled precision circular saw, and two identical samples were cut from the centre of the slice for DSC analysis [12].

The samples were characterised using DSC (STA Jupiter 449C from Netzsch). All measurements were performed in an inert atmosphere with argon gas, with a heating rate of 10 K/min from room temperature to 750 °C. An optical microscope ZEISS Axio Imager A1m was used for microstructure analysis, equipped with an AxioCam ICc 3 digital imaging camera and AxioVision image processing and analysis software. For a more detailed analysis of the microstructure elements, the scanning electron microscope FEG-SEM ThermoFisher Scientific Quattro S was used. In addition, thermodynamic calculations were performed using the Thermo-Calc software tool. Theoretical cooling curves and solubility products for the alloys considered were obtained since knowledge of the equilibrium state is essential when considering homogenization.

Results and Discussion

Figures 1 and 2 show thermodynamic calculations for the alloys AlFe1Si0.1 in AlFe1Si0.5. From the diagrams of the phase fractions as a function of temperature (Fig. 1a), the large differences in the phase composition of alloys A [12] and B below 500 °C are apparent when the solubility of silicon in the primary solid solutions of α-Al decreases to the point where it becomes the thermodynamically stable Al_8Fe_2Si phase. From the beginning of the formation of this phase, the fraction starts to increase, while the fraction of the $Al_{13}Fe_4$ phase decreases proportionally. At a temperature of 415 °C, the fraction of Al_8Fe_2Si even exceeds the fraction of $Al_{13}Fe_4$, but at 380 °C the transformation of Al_8Fe_2Si into $Al_9Fe_2Si_2$ occurs. Consequently, due to the higher silicon content in the $Al_9Fe_2Si_2$ phase than in the Al_8Fe_2Si phase, less $Al_9Fe_2Si_2$ phase is present in the microstructure of the alloy, but the proportion of $Al_{13}Fe_4$ phase increases again.

The Scheil diagram of alloy B (Fig. 1b) was calculated using two different functions. In both cases, the same diffusion of the iron atoms was considered, since iron diffuses slowly through the α-Al matrix, while for the diffusion of the silicon atoms the function of the fast-diffusing atoms in addition to the classical one was used. Due to the similarity of the atomic radii of aluminium and silicon, the silicon moves in the aluminium matrix as a substitute, but despite this fact, the results with the mentioned function were more similar to our predictions, since the solidification is completed in an interval of 10.0 °C, while with the classical

Table 1 Chemical composition of the manufactured aluminium alloys [wt.%]

Alloy	Si	Fe	Ti	Cr	Pb	Cu	Mn	Zn	Al
B-AlFe1Si0.1	0.08	1.10	<0.01	<0.01	<0.01	<0.01	<0.01	<0.01	Rest
C-AlFe1Si0.5	0.44	1.17	<0.01	<0.01	<0.01	<0.01	<0.01	<0.01	Rest

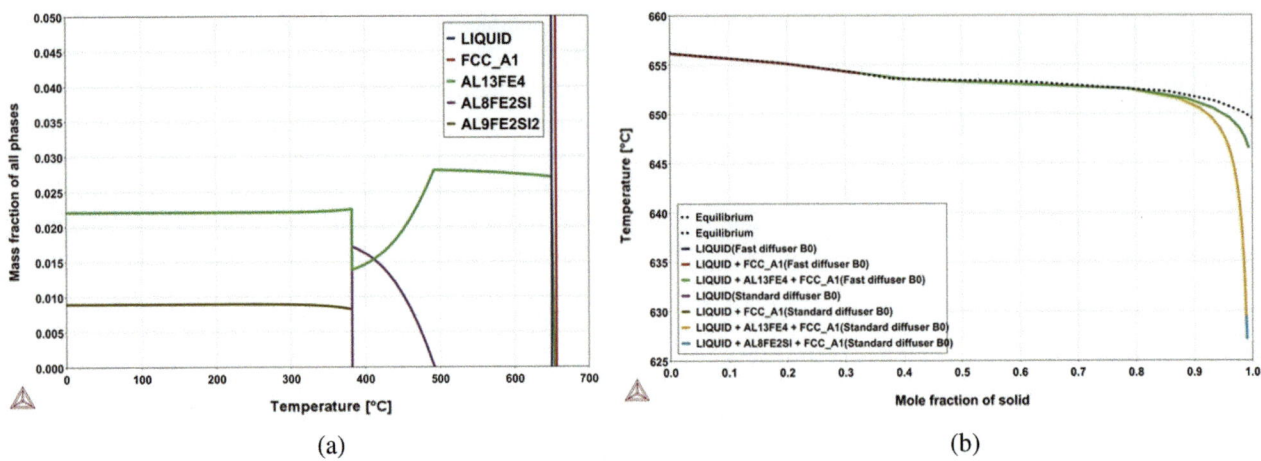

Fig. 1 Thermodynamic calculations of the alloy AlFe1Si0.1: **a** fraction of phases as a function of temperature and **b** Scheil diagram of the theoretical non-equilibrium cooling curve with and without consideration of the diffusion coefficient for Si

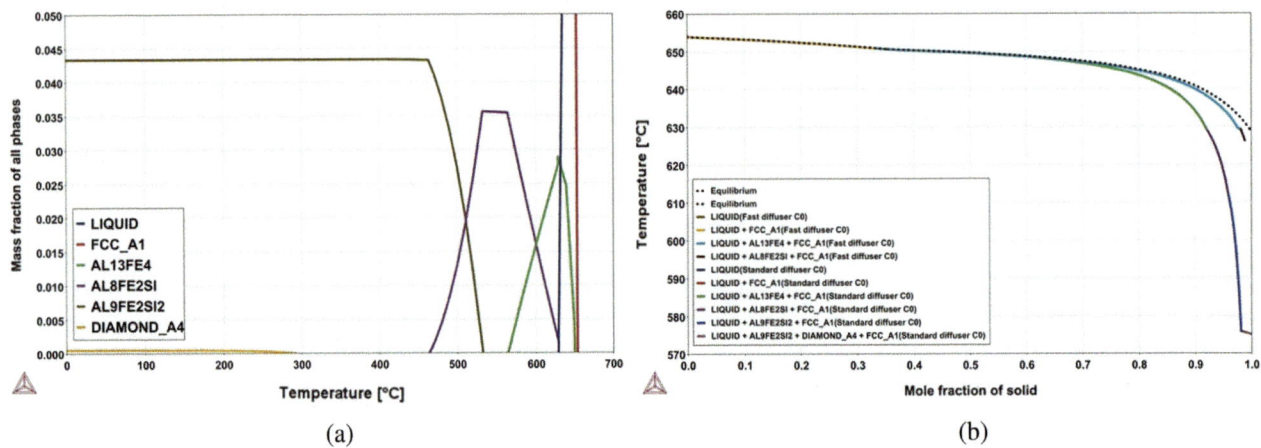

Fig. 2 Thermodynamic calculations of the alloy AlFe1Si0.5: **a** fraction of phases as a function of temperature and **b** Scheil diagram of the theoretical non-equilibrium cooling curve with and without consideration of the diffusion coefficient for Si

diffusion function the solidification took place in an interval almost 30 °C wide. The classical diffusion function also predicts solidification of the α-AlFeSi phase or eutectic (α–Al + Al_8Fe_2Si) in the last 3.0 °C of solidification and is completed at 627.2 °C. Using the fast-diffusing Si atom function, solidification is completed by solidification of the eutectic (α-Al + $Al_{13}Fe_4$) at 646.8 °C.

In alloy C (Fig. 2), the major difference at a temperature of 600 °C is the presence of the Al_8Fe_2Si phase, which contains 60.11 wt.% Al, 32.59 wt.% Fe, and 7.30 wt.% Si. Some silicon is still dissolved in the $Al_{13}Fe_4$ phase, as it contains 0.73 wt.% Si in addition to 60.20 wt.% Al and 39.07 wt.% Fe. The rest of the silicon is dissolved in the primary α-Al solid solutions, which contain 0.326 wt.% Si and 0.025 wt.% Fe, and the rest is aluminium. In total, alloy C contains 96.80 wt.% α-Al, 1.67 wt.% $Al_{13}Fe_4$, and 1.53 wt.% Al_8Fe_2Si.

Magnifications of the DSC curve of the AlFe1Si0.1 alloy in the region of the onset of melting are shown in Fig. 3. Compared with the binary Al–Fe alloy [12], the phenomenon of reduction of the amount of the metastable phase and, at the end, the complete transformation of the metastable phase into the stable phase is obvious. In the AlFe1Si0.1 alloy, the transformation occurs much faster than in the binary Al–Fe alloy, since a very small variation is observed on the DSC curve, which represents the beginning of the melting of the metastable phase.

Figure 4 shows a comparison of the curves of the AlFe1Si0.5 alloy. There are no significant differences between the individual DSC curves in the characteristic temperatures, the enthalpy of fusion, and the shape of the curve itself. The first possible reason for this phenomenon is that the as-cast AlFe1Si0.5 alloy does not contain a metastable phase Al_6Fe. Due to the higher Si content, a

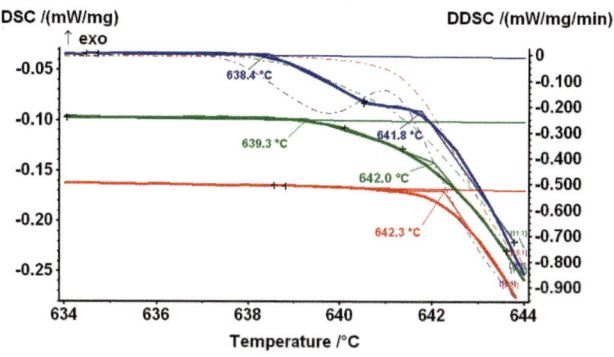

Fig. 3 Comparison of DSC heating curves of the alloy AlFe1Si0.1 (B0—blue, B600-2—green, and B600-24—red). The range of the onset of melting in the temperature range between 643 °C and 650 °C is shown

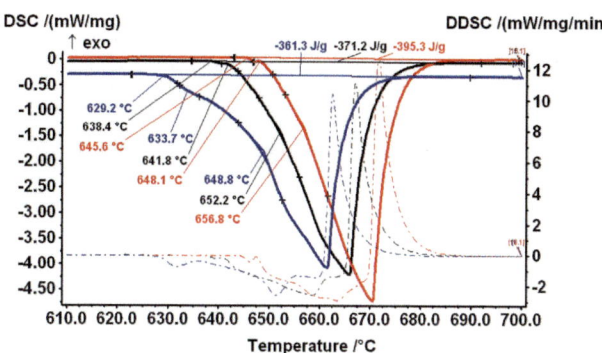

Fig. 5 Comparison of DSC curves of alloys A0—red, B0—black, and C0—blue. The entire melting temperature range of all investigated alloys is shown, from 610–700 °C

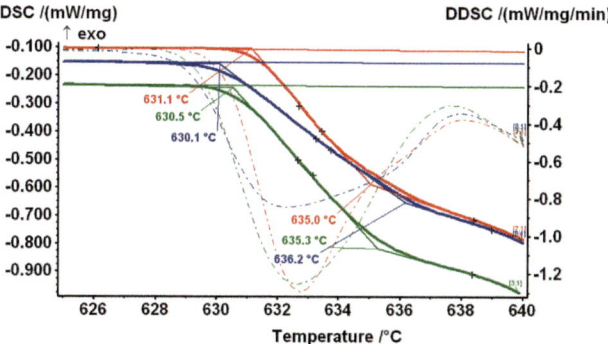

Fig. 4 Comparison of DSC curves of the alloy AlFe1Si0.5 (C0—blue, C600-2—green, and C600-24—red). The range of the onset of melting in the temperature range between 625 °C and 640 °C is shown

metastable phase Al_mFe was formed, which disappeared by the rapid transformation into $Al_{13}Fe_4$ already during the heating process to the homogenization temperature.

A comparison of the DSC curves of samples A0 [12], B0, and C0 (Fig. 5) shows a strong influence of the Si addition on the alloy melting process. In addition to the reduction of the enthalpy of fusion, the temperature of the melting point also decreases, which confirms the results of the thermodynamic simulation with the Thermo-Calc program.

A comparison of the microstructure of the alloy AlFe1Si0.1 is shown in Fig. 6. The sharp, rod-shaped particles that already appear in sample B0 are particles of the stable $Al_{13}Fe_4$ phase. In homogenized samples, the influence of the transformation of the metastable phase into the stable phase is noticeable. In sample B600-1, slightly more rounded particles were observed. In samples B600-4 and B600-12, a phase-less region, and the growth of larger rod-shaped particles at the boundary between α-Al and the eutectic region can be seen. In sample B600-12, the growth of a rod-shaped particle within the eutectic region was also observed. DSC analyses show that in alloy B the transformation of the metastable phase into stable $Al_{13}Fe_4$ has

almost completely occurred after two hours of homogenization and that the samples contain only the stable $Al_{13}Fe_4$ phase and primary solid solutions of α-Al. Therefore, we conclude that the small spherical particles are already transformed particles of the stable phase $Al_{13}Fe_4$ and that after 4 h only the growth of the particles takes place.

A comparison of the microstructure of alloy AlFe1Si0.5 is shown in Fig. 7. In sample C0, there is a thin, branched, rod-like phase that transforms into round particles upon homogenization. Even in sample C600-0, which was only heated to a homogenization temperature of 600 °C, an immediate change in the shape of the intermetallic particles can be seen. Although some particles were still rod-shaped, their edges had already begun to round off. In samples C600-1 and C600-4, the rod-shaped particles gradually disappear or are transformed into a more thermodynamically favourable spherical state. In sample C600-8, rod-shaped particles are rarely observed or are larger than in the less homogenized samples. All other particles have a spherical shape, which is also observed in samples C600-12 and C600-24, although the growth of round particles is also noticeable in the latter. Considering the speed of the initial transformation in sample C600-0, the state made from the DSC results can be confirmed; the entire metastable phase transforms up to the melting temperature of the alloy and the subsequent change in the shape of the particles is not the result of the transformation, but only of the growth of the particles.

Further, Figs. 8–11 show the EDS analyses of some experimental samples, where larger deviations from the stoichiometric values of the analysed phases were expected since the measurement also detects the values of the matrix. The size of the analysed particle has a great influence on the results. In most results, the content of alloying elements is, therefore, lower than the stoichiometric, which for the stable phase $Al_{13}Fe_4$ amounts to 37.3 to 40.7 wt.% Fe, the rest is aluminium, and for the metastable phase Al_6Fe 20.0 wt.% Fe.

Fig. 6 Comparison of the
microstructure images of the alloy
AlFe1Si0.1 at
2500× magnification made with
a scanning electron microscope.
The microstructures of the
non-homogenized sample and the
samples homogenized for 4, 12,
and 24 h are shown

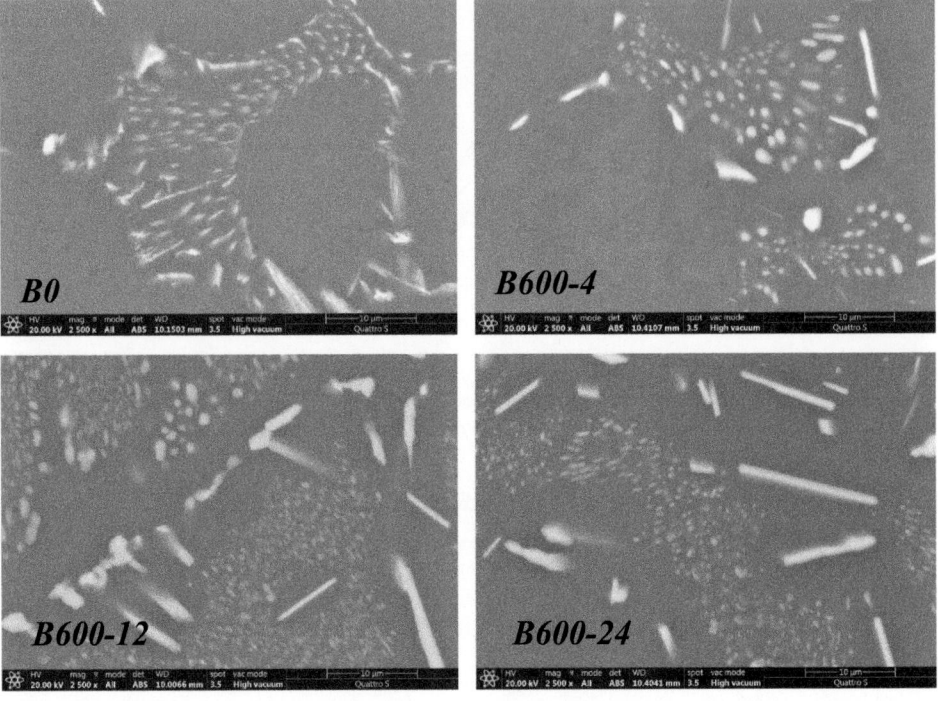

Fig. 7 Comparison of
microstructure images of the alloy
AlFe1Si0.5 at
2500× magnification made with
a scanning electron microscope.
The microstructures of the
non-homogenized sample and the
samples homogenized for 4, 12,
and 24 h are shown

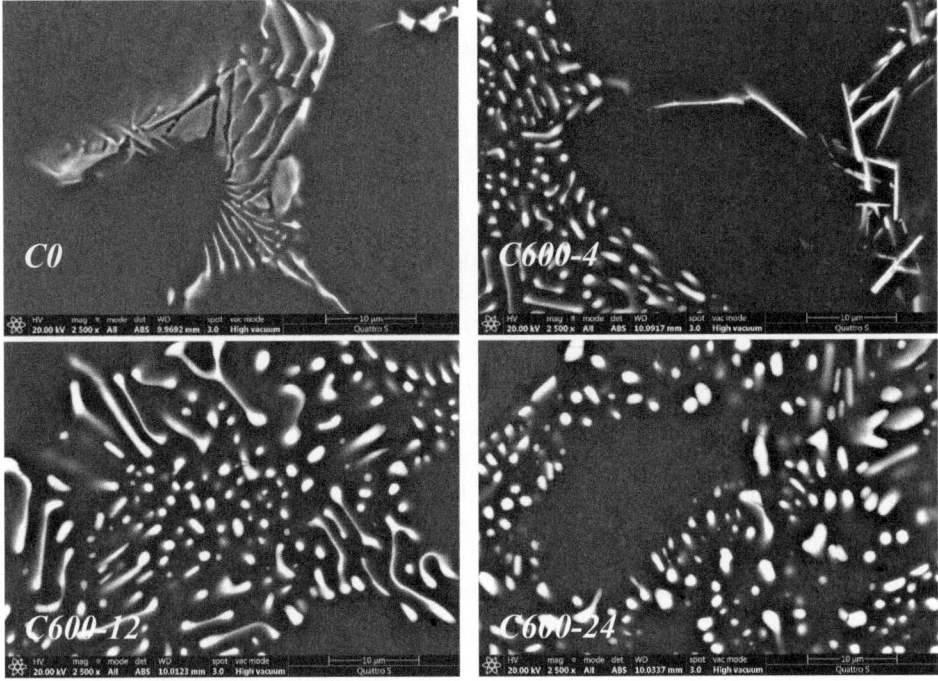

In the analyses of alloys B and C, phases containing silicon are also found in the microstructure, whose stoichiometric values should be 31.6 wt.% Fe and 78 wt.% Si for the Al_8Fe_2Si phase and 25.6 wt.% Fe and 12 0.8 wt.% Si for the $Al_9Fe_2Si_2$ phase. The silicon content never exceeds 1.0% by weight, which means that the samples do not contain phases from the Al–Fe–Si system but are Al–Fe phases in which silicon is dissolved (Figs. 8 and 9). Two distinct phases were observed in sample C0 (Figs.10 and 11). Figure 10 shows the $Al_{13}Fe_4$ phase, which has a distinct rod shape. In all homogenized samples of alloy C, the metastable Al_mFe phase could not be detected.

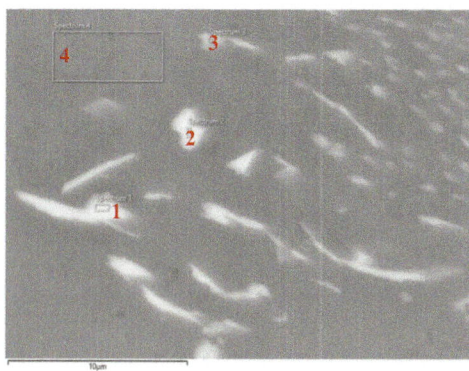

Element/ Wt%	Sp. 1	Sp. 2	Sp. 3	Sp. 4
Al	78.15	77.49	90.14	99.71
Si	0.64	0.55	0.42	0.12
Fe	21.21	21.96	9.43	0.17

Fig. 8 EDS analysis with corresponding results for sample B0. The micrograph was taken using the backscattered electron method at 5000 × magnification. The high Fe contents in the particles from measurements Sp. 1 and 2 represent particles of the stable phase, Sp. 3 represents a particle of the metastable phase, Sp. 4 represents the measurement of the α-Al matrix

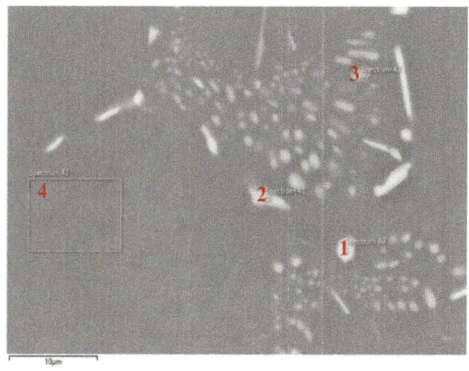

Element/ Wt%	Sp. 1	Sp. 2	Sp. 3	Sp. 4
Al	74.28	79.55	88.35	99.78
Si	0.53	0.49	0.33	0.12
Fe	25.19	19.96	11.33	0.10

Fig. 9 EDS analysis with the corresponding results for sample B600-4. The microphotograph was taken using the backscattered electron method at 2500 × magnification. The measurements Sp. 1, 2, and 3 represent particles of the stable phase, Sp. 4 represents the measurement of the α-Al matrix

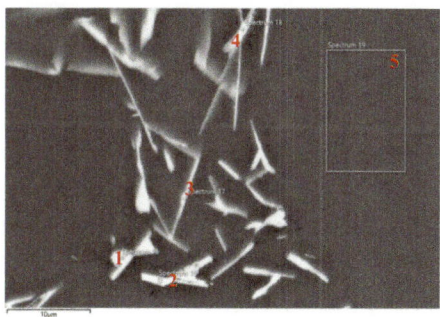

Element/ Wt%	Sp. 1	Sp. 2	Sp. 3	Sp. 4	Sp. 5
Al	72.83	76.03	84.08	79.89	100
Si	1.73	1.51	1.36	2.42	0
Fe	25.44	22.46	14.56	17.69	0

Fig. 10 EDS analysis with corresponding results for sample C0. The micrograph was taken using the backscattered electron method at 2500 × magnification. Sp. 1–4 represent particles of the Al–Fe-Si phase, respectively. Sp. 5 represents the measurement of the α-Al matrix

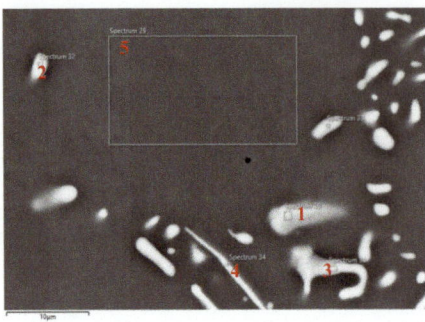

Element/Wt%	Sp. 1	Sp. 2	Sp. 3	Sp. 4	Sp. 5
Al	72.22	78.06	81.54	80.55	100.00
Si	4.90	4.76	3.71	2.84	0
Fe	22.87	17.17	14.75	16.61	0

Fig. 11 EDS analysis with corresponding results for sample C600-4. The microphotograph was taken using the backscattered electron method at 2500 × magnification. Sp. 1, 2, and 3 represent Al–Fe–Si particles. Sp. 4 represents an Al–Fe–Si particle despite its rod shape, and Sp. 5 represents an α-Al matrix

Conclusions

With non-equilibrium solidification at high cooling rates (dT/dt ≈ 35 K/s), metastable phases are solidified in the Al–Fe-Si system. The addition of silicon has a strong influence on the type and amount of metastable phase formed. With DSC, we can detect the presence of a metastable phase. This phenomenon is observed in the as-cast state for the AlFe1Si0.1 alloy. DSC measurements showed that 550 °C is not a high enough homogenization temperature for the transformation of the Al_6Fe metastable phase. When examining homogenized samples, the effect of silicon on the content of the metastable phase is immediately visible in the rate of transformation of the metastable phase into the stable one. The transformation of the metastable phase in the AlFe1Si0.1 alloy is completed after 4 h of homogenization. In the case of the AlFe1Si0.5 alloy, no metastable phase is observed on the DSC heating curves. The reason for this phenomenon is the high rate of transformation of the Al_mFe metastable phase present in this alloy.

During the solidification of the AlFe1Si0.1 alloy with the mentioned cooling rate, the Al_6Fe metastable phase solidifies, but the addition of silicon has an impact on the proportion of the metastable phase since, in the case of the AlFe1Si0.1 alloy, the $Al_{13}Fe_4$ stable phase also solidifies. In the AlFe1Si0.5 alloy, the Al_6Fe phase does not solidify at such cooling rates, but the Al_mFe phase becomes more stable and more likely due to the high addition of silicon. The appearance of the metastable phase is also noticeable at lower cooling rates, which occur in the industrial process of slab casting.

Acknowledgements The authors would like to thank the Slovenian Research Agency (ARRS) for funding under program grant P2-0344.

Funding
This research was funded by the Slovenian Research Agency (ARRS) under program grant P2-0344.

References

1. Belov N.A.; Aksenov A.A.; Eskin D.G. Iron in Aluminum Alloys: Impurity and Alloying Element. Taylor & Francis, London, UK, 2002, p. 360.
2. Chen, J., Dahlborg, U., Bao, C.M. et al. Metall and Materi Trans B (2011) 42: 557. https://doi.org/10.1007/s11663-011-9485-6
3. Pan L, Liu K, Breton F, Grant Chen X. Effect of Fe on Microstructure and Properties of 8xxx Aluminum Conductor Alloys. J Mater Eng Perform. Springer US; 2016;25:5201–8.
4. Liu K, Cao X, Chen XG. Solidification of iron-rich intermetallic phases in Al-4.5Cu-0.3Fe cast alloy. Metall Mater Trans A Phys Metall Mater Sci. 2011;42:2004–16.
5. Belov, N.A.; Eskin, D.G.; Aksenov, A.A. Multicomponent Phase diagrams, Applications for Commercial Aluminum Alloys; Elsevier: Amsterdam, The Netherlands, 2005; p. 424.
6. Vončina, M.; Nagode, A.; Medved, J.; Paulin, I.; Žužek, B.; Balaško, T. Homogenisation Efficiency Assessed with Microstructure Analysis and Hardness Measurements in the EN AW 2011 Aluminium Alloy. *Metals* **2021**, 11, 1211. https://doi.org/10.3390/met11081211
7. Allen CM, O'Reilly KAQ, Cantor B, Evans PV. Intermetallic phase selection in 1XXX Al Alloys. Prog. In Mat. Sci. 1998;43:89-170.
8. Shakiba M, Parson N, Chen XG. Effect of homogenization treatment and silicon content on the microstructure and hot workability of dilute Al-Fe-Si alloys. Mater Sci Eng A [Internet]. Elsevier; 2014;619:180–9. Available from: https://doi.org/10.1016/j.msea.2014.09.072
9. Allen CM, O'Reilly KAQ, Cantor B, Evans PV. A Calorimetric Evaluation of the Role of Impurities in the Nucleation of Secondary Phases in 1xxx Al Alloys. Mat. Res. Soc. Symp. Proc. 1998;481:3.
10. Vončina, M.; Kresnik, K.; Volšak, D.; Medved, J. Effects of Homogenization Conditions on the Microstructure Evolution of Aluminium Alloy EN AW 8006. Metals 2020, 10, 419.
11. Condruz, M.R.; Matache, G.; Paraschiv, A. et al. Homogenization heat treatment and segregation analysis of equiaxed CMSX-4 superalloy for gas turbine components. J Therm Anal Calorim 134, 443–453 (2018). https://doi.org/10.1007/s10973-018-7085-2.
12. Arbeiter, J.;Vončina, M.; Šetina Batič, B; Medved, J. Transformation of the Metastable Al6Fe Intermetallic Phaseduring Homogenization of a Binary Al-Fe Alloy. Materials 2021, 14, 7208https://doi.org/10.3390/ma14237208

Secondary Phase Refinement in Molten Aluminum via Low Power Electric Current Processing

Jonathan Goettsch, Aaron Gladstein, David Weiss, Ashwin Shahani, and Alan Taub

Abstract

Metal matrix composites offer improved mechanical properties at reduced weight compared to monolithic alloys. The reinforcing capabilities of the particulate are dependent on their size, distribution, and bonding with the matrix. The intermetallic compound, Al_3Ti, has excellent bonding with the surrounding matrix but can grow detrimentally large with extended hold times in the molten state. In addition, passing electrical current at low power levels through the melt has been found to reduce the size of the Al–Ti intermetallic phase. This research focuses on the novel processing method of passing an electric current through molten aluminum alloy to refine the reinforcing secondary phase. This technique could have broad impacts on particulate refinement in multiple metal matrix composite systems.

Keywords

Aluminum • Microstructure • Electric current processing • Metal matrix composites

Introduction

Aluminum alloys have become increasingly used in the automotive and aerospace industries for structural components due to their high strength-to-weight ratio. However, monolithic alloys can suffer poor mechanical properties as a result of large grains or inclusions in the microstructure. Metal matrix composites are able to refine the microstructure and improve the mechanical properties of the alloy but can be expensive. The utilization of effective grain refiners allows manufacturers to cast parts that are able to withstand the applied wear and stress. Some research has shown that passing an electric current through the metal as it solidifies can result in grain refinement as well as other changes to the microstructure. Stone showed that increases in DC amperage applied to gallium directly increased its undercooling time [1]. Ivanov and Tsurkin have some of the most extensive work related to electric current processing of molten metal [2–8]. They have extensively discussed the changes in the electric field due to a particle of different electrical conductivity being suspended in the melt. They also discussed the differences between DC, AC, and Pulsed AC treatment methods. Davidson et al. showed the impact that electric arcs have on dielectric fracturing with some similar results to those found in the intermetallic fracture electrified molten metal shown later [9]. Vashchenko et al. demonstrated that passing an electric current through cast iron as it solidifies refined the graphite and primary crystals of the metallic base, stating that the applied current also reduced solidification time [10]. Dvoskin et al. utilized arcing electricity through molten metal in sand or permanent mold castings as the metal solidifies to improve mechanical properties [11]. Li et al., as well as Choi et al., credited forced dendrite fragmentation (increased melt flow resulting in the dendrite tips breaking away) and dissociation of crystal nuclei from the electrode interface to the fine equiaxed grains that present when an electropulse or DC waveform is passed through pure molten aluminum during solidification [12, 13]. Ge et al. discussed the mechanisms by which carbon can be incorporated into the metal lattice when a high electric current is passed through the metal along with precursor carbon powder to create a material called "Covetics" which boasts improvements in electrical conductivity [14]. Zhao et al. and Dai et al. quantified the mechanisms by which particle cluster deagglomeration can occur or state observed inclusion refinement in the microstructure [15, 16]. Table 1

J. Goettsch (✉) · A. Gladstein · A. Shahani · A. Taub
College of Materials Science and Engineering, University of Michigan, Ann Arbor, MI, USA
e-mail: jgoe@umich.edu

D. Weiss
Eck Industries, 1602 N 8Th St, Manitowoc, WI 54220, USA

© The Minerals, Metals & Materials Society 2023
S. Broek (ed.), *Light Metals 2023*, The Minerals, Metals & Materials Series,
https://doi.org/10.1007/978-3-031-22532-1_61

Table 1 Processing conditions for electric current methods in selected literature

References	Voltage	Current	Treatment time	Furnace temperature	Material
[1]	~DC	0, 5, 10, 15, 20 A	300 s	Cycled between 107 °C and −13 °C	Gallium
[2]	~DC	1×10^4–5×10^4 A/cm^2	~	~	~
[4]	10 kV–50 kV Pulse	3.5 kA–80 kA $*I_{max}$	2 Hz micropulse for 1 min	700 °C to air	~
[5]	~AC	10 A	n/a	~	~
[6]	~AC/Pulse	28 A/200 A	n/a	700 °C	~
[7]	~Pulse	Hundreds of amps to dozens of kiloamps	100 μs	~	~
[9]	5 kV Arc	~	~	200 °C, 700 °C, 1600 °C	Various dielectric materials
[10]	120 V DC	4–5 mA/cm^2	"…from the beginning of pouring until the end of solidification"	~	Cast Iron
[12]	~Pulse	480 A, 990 A	50 Hz pulse	900 °C to air	Aluminum
[13]	~DC	300 A, 700 A	14–169 s	665° to air	Aluminum
[14]	10 V DC	150 A	2 min ON 1 min OF	900 °C (1100 °C Peak during processing)	Aluminum – Carbon
[15]	20 V Pulse	6.67×10^6 A/m^2	~	~	~
[16]	~Pulse	1e4 A/m^2	30 min	1550 °C	Steel

includes the processing parameters found in the literature mentioned above. It is noteworthy that, in nearly all of the literature, the metal was processed with electric current as the metal solidified and/or at high currents/voltages. None of the literature discusses the fracturing of the secondary phase as a result of the electric current processing. We have found that passing a lower amount of electric power, less than 300 Watts, through the metal while it is still molten will result in fracturing of the solid intermetallic. Refinement of this phase could lead to significant grain refinement of the alloy without the need for additional grain refiners.

Experimental

Material

An Al–Ti binary alloy was fabricated by two different methods for testing the electric current processing. The first was a remelt of an Al–6wt%Ti master alloy, diluted in 99.99% pure aluminum, to target Al–2Ti final composition. The second method involved fabricating the Al–Ti alloy using a Ti containing flux (K_2TiF_6) that was poured on top of 99.99% pure aluminum melt once it reached the experimental temperature, targeting a final composition of Al–3wt %Ti. The flux-aluminum reaction results in titanium dissolving into the melt and forming the intermetallic

compound, Al_3Ti [17]. This intermetallic can form many different morphologies based on local composition and temperature, the morphologies primarily shown below are the "plate" and "blocky" structures [18].

Electric Current Processing Setup and Conditions

The electrical treatment experiments were split into short- and long-time treatments. The short-time treatments were conducted on both alloys using a variable AC transformer to pass a 60 Hz, 15 V, and < 20 A, AC waveform through the melt at 850 °C for 45–90 s. An inserted graphite rod and the graphite crucible itself acted as electrodes. The melt was manually stirred with the graphite rod, with care taken to avoid touching the crucible walls, during the treatment time in the air. A schematic of the processing setup can be seen in Fig. 1. When fabricating the material using the Ti flux, it was allowed to dwell on the melt for 2–3 min to fully liquify prior to applying the current. After processing the melt was poured into a rectangular graphite crucible and allowed to air cool prior to segmenting and SEM preparation.

The long treatment time experiment was only conducted with the Al-flux alloy. The experiment was done with a waveform generator and high voltage amplifier to pass a 60 Hz, measured 36 V, AC waveform through the melt at 700 °C with the same electrode setup as before, but for these

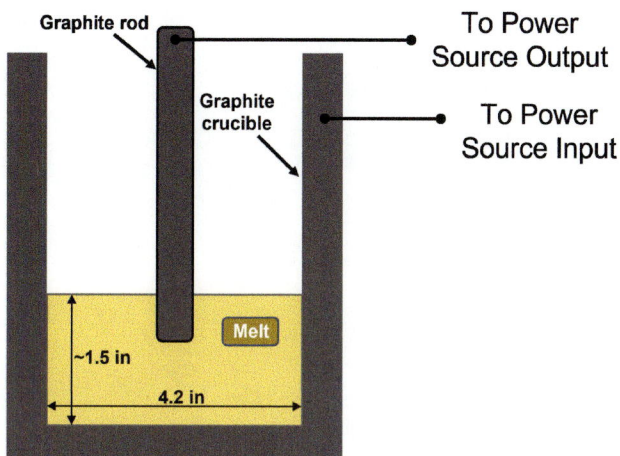

Fig. 1 Schematic of the electric current processing setup: The graphite rod is handled with thermally and electrically insulative material during the manual stirring process. Electrical contact is made with the graphite crucible by inserting a steel rod into the crucible wall with the wire contact fastened to it, this acts as a quick connect so that the wire is held outside of the furnace during treatment. Only the graphite rod and crucible walls contact the melt

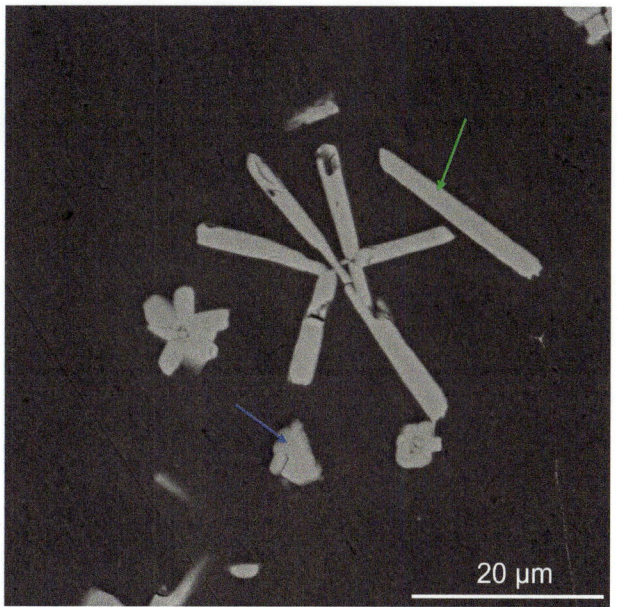

Fig. 2 Unprocessed Sample Microstructure: Under similar cooling conditions to the electrically processed material the untreated material results in a sharp matrix-intermetallic interface (the bright features are the Al_3Ti intermetallic phase). The blue arrow identifies the typical "blocky" intermetallic morphology. The green arrow identifies the typical "plate" intermetallic morphology. Average particle size: 74.3 μm^2

experiments, the graphite rod was suspended in place instead of being used as a stirring rod. The processing began immediately after the Ti-containing flux was added to the surface of the melt, not allowing it time to fully liquefy and react with the molten aluminum, and the current was applied continually for 20 min in the air. Samples were solidified and prepared in the same manner as the short-time experiments.

Results and Discussion

Microstructure of Intermetallic Particles in Unprocessed Material

Figure 2 shows the typical morphology of the Al_3Ti intermetallic phase fabricated without electric current processing. It should be noted that the particle-to-matrix interface is sharp and does not demonstrate much roughness. The unprocessed material has a measured average intermetallic particle size of 74.3 μm^2.

Microstructure of Intermetallic Particles in Electric Current Processed Material

Figures 3, 4, and5 show the impact of the electric current processing on the morphology of the Al_3Ti intermetallic particles. Figures 3 and 4 show the appearance of intermetallic breakage at the interface between the intermetallic and surrounding matrix. The fragments at the particle surface appear to be in an intermediate step of dispersion and have

an average fragment size of 0.12 μm^2. The angular nature of the fractured particle morphology indicates that fracture occurs along the crystallographic planes of the treated intermetallic. Fracture along the crystallographic planes was determined by Davidson et al. to be the result of acute thermal expansion associated with the local Joule heating from an arcing current [9]. However, this method of fracture is unlikely as the processing in this work was performed at a significantly lower voltage.

Figure 5 shows the microstructure of the long-time processing experiments. The limited flux dwell time and extended treatment period led to a reduction in average intermetallic particle size to an average of 3.7 μm^2, or 20 × smaller than the unprocessed intermetallic particles. These fine intermetallic particles could be the result of increased nucleation of the intermetallic phase from the Ti-containing flux since it was not given enough time to fully react with the aluminum before electric treatment started. Li et al. showed that the use of an electric current being passed through molten aluminum as it solidifies can create fine equiaxed grains, as the system free energy is modified by a new term for the electric current [12]. The decrease in intermetallic size might also be the result of the direct fragmentation of the original intermetallic particles along with the dispersion and growth of the particle fragments. Further studies are planned to elucidate the underlying mechanism.

Fig. 3 Electric Current
Processed Master Alloy Sample
Microstructure: **a** Large plate
intermetallic structures are the
sole morphology in the master
alloy **b** enlarged view of the
intermetallic/matrix interface

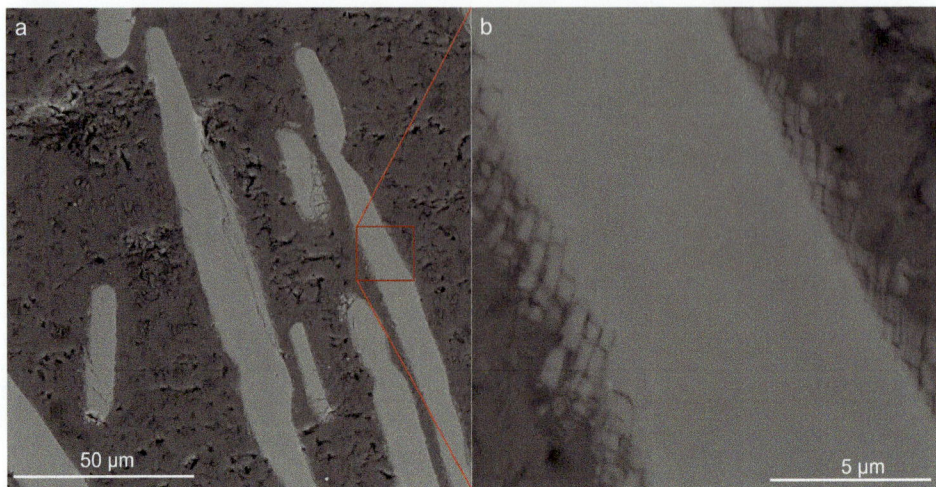

Fig. 4 Electric Current
Processed Flux Formed Sample
Microstructure: (a_1) full-length
view of a processed plate particle
morphology (a_2) enlarged view of
the intermetallic/matrix interface
(b_1) processed blocky particle
morphology (b_2) enlarged of the
intermetallic/matrix interface

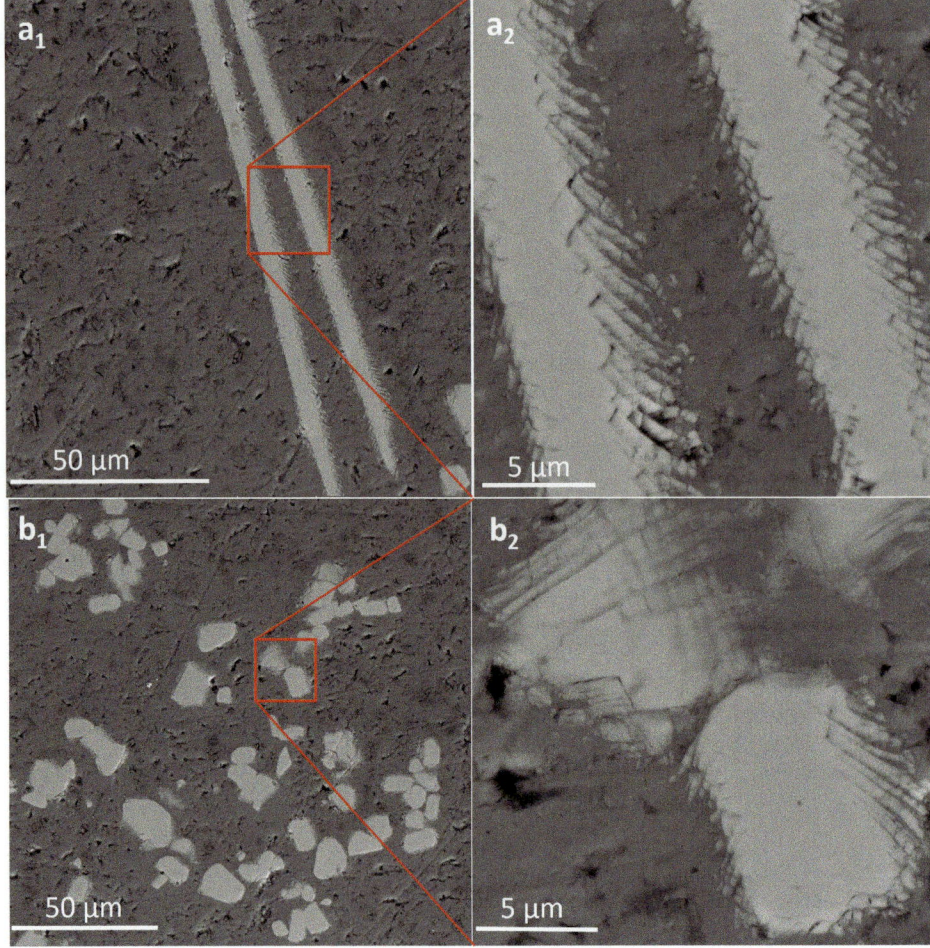

Fig. 5 Electric Current Extended Processed Flux Formed Sample Microstructure: The 20 min electric current processing experiment resulted in Al$_3$Ti particles that are significantly more refined than the untreated or short treatment time samples. Average particle size: 3.7 μm^2

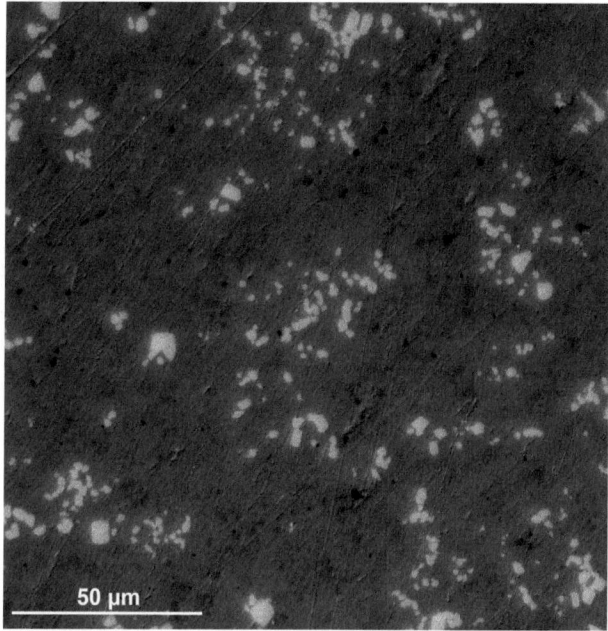

50 μm

Summary

In this work, we applied a novel process to Al–Ti melts using low-power electrical current. The results from this processing method found a significant decrease in particle size, the apparent breakage of intermetallic at the Al-particle interface at short treatment times, and fully refined particles after extended treatment times. The order of magnitude difference in cross-sectional area would result in a significantly finer reinforcing particle being distributed throughout the microstructure. As these intermetallic particles already act as nucleation sites for grains during solidification, it is reasonable to presume that a greater number density of particles would result in a finer grain structure and accompanying improvement in mechanical properties. The low-power nature of this treatment also means that it would have a low economic footprint if implemented into the metal treatment workflow. Future research is planned to determine the mechanism by which the intermetallic breakage occurs the mechanism by which significant particle refinement occurred in the extended treatment experiment, and what the optimal processing parameters are for refining the solid phase particles present in the molten metal for improvement of the material's mechanical properties.

References

1. Stone, Katherine M. Effects of an electric current applied to metals during solidification. Thesis: S.M., Massachusetts Institute of Technology, Department of Materials Science and Engineering, 2019. https://dspace.mit.edu/handle/1721.1/122446.
2. Ivanov, A.V., Sinchuk, A.V. & Tsurkin, V.N. Electric current treatment of liquid and crystallizing alloys in casting technologies. *Surf. Engin. Appl.Electrochem.* **47**, 456–464 (2011). https://doi.org/10.3103/S1068375511050115.
3. Tsurkin, V.N., Kreptiuk, Y.V., Grabovyi, V.M. *et al.* Simulation of the physical processes in liquid metal at electrohydropulse treatment and its complexing with magnetodynamic treatment of melt. *Surf. Engin. Appl.Electrochem.* **44**, 115–122 (2008). https://doi.org/10.3103/S1068375508020075.
4. Ivanov, A.V., Sinchuk, A.V. & Ruban, A.S. Effect of the technological parameters of the melt treatment by a electric pulse current on the mixing process. *Surf. Engin. Appl.Electrochem.* **48**, 180–186 (2012). https://doi.org/10.3103/S106837551202007X.
5. Tsurkin, V.N., Ivanov, A.V. Peculiarities of Redistribution of Electric and Thermal Fields at the Interface When Passing the Electric Current through the Melt. *Surf. Engin. Appl.Electrochem.* **54**, 577–584 (2018). https://doi.org/10.3103/S1068375518060133.
6. Ivanov, A.V., Tsurkin, V.N. Peculiarities of Distribution of Electromagnetic and Hydrodynamic Fields for Conductive Electric Current Treatment of Melts in Different Modes. *Surf. Engin. Appl.Electrochem.* **55**, 53–64 (2019). https://doi.org/10.3103/S1068375519010101.
7. Tsurkin, V.N., Ivanov, A.V. Functional Capabilities of Electromagnetic-Acoustic Transformations in Current Mode in the Metal Melt. *Surf. Engin. Appl.Electrochem.* **58**, 239–247 (2022). https://doi.org/10.3103/S1068375522030139.

8. Ivanov, A.V., Sinchuk, A.V. & Ruban, A.S. Study of the dependences of the characteristics of a pulse discharge current vs the phase state of a liquid-metal conductor. *Surf. Engin. Appl. Electrochem.* **48**, 348–354 (2012). https://doi.org/10.3103/S1068375512040096.

9. Davisson, J. W. and W. H. Vaughan. "ELECTRICALLY INDUCED FRACTURE OF MATERIALS." (1969). https://www.semanticscholar.org/paper/ELECTRICALLY-INDUCED-FRACTURE-OF-MATERIALS-Davisson-Vaughan/c92b8333f6944b5bc4657cfffc4877d50f60c2ad.

10. Vashchenko, K.I., Chernega, D.F., Vorobev, S.L. *et al.* Effect of electric current on the solidification of cast iron. *Met Sci Heat Treat* **16**, 261–265 (1974). https://doi.org/10.1007/BF00663070.

11. Dvoskin P. et al., *Treating molten metals by moving electric arcs* (Pub. No.: US 2005/0098298A1) U.S. Patent and Trademark Office https://patents.justia.com/patent/20050098298.

12. Li, Ning et al. "The Role of Electric Current-Associated Free Energy and Forced Convection on Grain Refinement in Pure Aluminum under Electropulsing." *Materials (Basel, Switzerland)* vol. 12,23 3846. 22 Nov. 2019, doi:https://doi.org/10.3390/ma12233846.

13. Jee Seok Choi, Moonwoo La, DongEung Kim, Kyeong-Hwan Choe, Soong-Keun Hyun, Moon-Jo Kim, Effect of electric current on the microstructural refinement of pure aluminum, Journal of Materials Research and Technology, Volume 12, 2021, Pages 818–830, ISSN 2238-7854, https://doi.org/10.1016/j.jmrt.2021.03.040.

14. X. Ge, C. Klingshirn, M. Wuttig, K. Gaskell, P.Y. Zavalij, Y. Liang, C.M. Shumeyko, D.P. Cole, L.G. Salamanca-Riba, Mechanism studies and fabrication for the incorporation of carbon into Al alloys by the electro-charging assisted process, Carbon, Volume 149, 2019, Pages 203–212, ISSN 0008-6223, https://doi.org/10.1016/j.carbon.2019.04.049.

15. Zhao, Z. C. and Qin, R. S. (2017). Inclusion agglomeration in electrified molten metal: thermodynamic consideration. Materials Science and Technology, 33(12) pp. 1404–1410.

16. W. B. Dai, J. K. Yu, C. M. Du, L. Zhang & X. L. Wang (2015) Refinement of inclusions in molten steel by electric current pulse, Materials Science and Technology, 31:13, 1555-1559, DOI: https://doi.org/10.1179/1743284715Y.0000000015.

17. Z. Liu, M. Rakita, X. Wang, W. Xu, Q. Han, In situ formed al3ti particles in al alloy matrix and their effects on the microstructure and mechanical properties of 7075 alloy, Journal of Materials Research 29(12) (2014) 1354–1361.

18. Gladstein, A., Visualization and Analysis of Nanoscale Microstructure Evolution of *In Situ* Metal Matrix Composites, Materials Science and Engineering, University of Michigan, 2022.

Fluidity and Microstructural Analysis of Al–Ni Alloys with Varied Ni Concentrations

Vigneshwar Hari, Dong Xu, Stuart D. McDonald, Zherui Tong, Dongdong Qu, and Kazuhiro Nogita

Abstract

The manufacturing associated with emerging technologies, including electric vehicles, is impacting demand for cast aluminium alloys, including those based on the Al–Ni system. These alloys typically have a eutectic with a fine lamellar spacing in the as-cast condition, display high thermal stability, high electric conductivity, resistance to hot tearing, good fluidity, and strength. These properties combined make this alloy system suitable for a variety of applications. This research investigates the effect of varying the concentration of Ni on the solidification mode, microstructure, and fluidity of Al–Ni alloys. Hypoeutectic, eutectic, and hypereutectic compositions of Al–Ni from 0 to 10 wt.% were investigated, and it was found that hypereutectic Al–7.7wt.%Ni alloy had the best fluidity. As cast microstructures were compared to investigate microstructure evolution in the Al–Ni system.

Keywords

Aluminium • Nickel • Fluidity • Microstructure • Al–Ni • Eutectic • Casting

Introduction

Aluminium alloys are some of the most versatile and widely used cast alloys. Interest in cast Al–Ni alloys has recently increased for their potential use in high temperature [1] or electro-mechanical components [2], and for their ability to be directionally solidified via thermal gradient/magnetic fields [3–5]. One of the promising properties of Al–Ni alloys is their good fluidity, which is an important consideration for a cast alloy as good fluidity reduces casting defects, and improves mould filling [6]. These properties were noted by Palanivel et al. with their invention of an Al–Ni alloy with Fe and Ti additions [7]. Fluidity, in simpler binary Al-xx systems, is normally the highest for pure Al, drops to a minimum for a hypoeutectic composition, then climbs to another peak at/near the eutectic composition [6]. The exact location of this peak in fluidity varies slightly between each binary system. For the Al–Cu system, the peak fluidity occurs at the eutectic composition, while for Al–Si and Al–Mg, the peak fluidity occurs in the hypereutectic region [6]. Eutectic Al–6.1Ni (wt.%) has been noted as having good fluidity and low hot tearing, compared to Al–Si and Al–Fe [8]. Another work investigated the fluidity of Al–Si–Ni alloys using spiral moulds. Al–(2,4,6,8)Ni alloys were all tested, with near eutectic Al–6Ni found to have the highest fluidity, however, Al–4Ni was also found to have similar fluidity [9]. An alternative work tested Al–xNi fluidity using vacuum pulling methods and found that the peak in fluidity was at 7.1Ni [10]. Other research in binary Al–Ni alloys has mainly focused on its properties and it has been shown that increasing the Ni content increases tensile strength and hardness [11] and hypereutectic Al–8Ni has improved electric conductivity when compared to hypereutectic Al–Si and Al–Fe [12]. This work will investigate the fluidity of binary Al–Ni alloys to clarify the data from the literature. The results are interpreted with reference to cooling curves, along with microstructure, and macrostructure development in Al–Ni samples. There is some noted disagreement in the literature with regards to the exact composition of the eutectic. Earlier works proposed the eutectic equilibrium reaction occurs around 2.58 at.% ($\sim$5.45 wt.%) Ni [13]. For this work, however, 6.1-wt.% Ni addition was used as the eutectic composition, based on modern CALPHAD assessments of the Al–Ni system taken from Thermo-Calc's TCBIN database (Fig. 1) in combination with published literature [14–16]. This composition has been accepted by

V. Hari (✉) · D. Xu · S. D. McDonald · Z. Tong · D. Qu · K. Nogita
Nihon Superior Centre for the Manufacture of Electronic Materials, School of Mechanical and Mining Engineering, The University of Queensland, St Lucia, QLD 4072, Australia
e-mail: uqgharik@uq.edu.au

© The Minerals, Metals & Materials Society 2023
S. Broek (ed.), *Light Metals 2023*, The Minerals, Metals & Materials Series,
https://doi.org/10.1007/978-3-031-22532-1_62

several researchers [1, 5, 17, 18]. Carrara et. al also notes that the Al–Ni system has an asymmetric coupled zone with undercooled hyper eutectic compositions capable of forming eutectic microstructure [5].

Methodology

Roughly 2 kg of each of the compositions shown in Table 1 was prepared by melting 99.95% pure Ni (trace Fe and other elements) into a corresponding amount of 99.96% pure Al (trace Fe, Si, and other elements) using an induction furnace. A sample of liquid was taken from each melt and had its cooling curve characterised to establish the experimental liquidus temperature. The pouring temperatures in Table 1 were determined by adding 60 °C to this measured liquidus temperature. The melt was poured at this temperature to characterise fluidity by pouring it into a room temperature sand-mould prepared using Foseco Coolset (water and 60% sodium metasilicate). The geometry of the spiral shaped pattern used to make the mould is shown in Fig. 2. A ventilation channel was created at the end of the spiral to minimise back pressure during filling. The spiral (Fig. 2b) had a raised 'bump' every 25 mm, to assist in measuring the length of the spiral for each condition. The cross section of the spiral is given in Fig. 2c. Three sand moulds were poured sequentially for each composition and the remaining melt was solidified as ingots and later remelted for detailed cooling curve analysis and quenching experiments.

To perform cooling curve analysis and quenching, ingots were placed into crucibles and heated to 750 °C in an electric resistance furnace. Once ready to perform the experiment, dross was removed and two Nickel cups (see

Fig. 2a), coated with boron nitride and preheated to 150 °C, were used to simultaneously take a sample of the liquid melt. Both cups were quickly placed between two pieces of insulating board. One cup was monitored with a K-type thermocouple and National Instruments datalogger with a capture rate of 4.5 Hz. The other cup was left to solidify until quenched in room temperature water when the desired quench time was achieved.

The quench times varied with the alloy composition, with quench locations chosen to occur during the primary phase solidification, during the start of the eutectic reaction, and the middle of the eutectic reaction (Q1, Q2, and Q3, respectively). This was done to compare how the primary and eutectic phases develop during solidification.

Samples were sectioned vertically and mounted in conductive resin and prepared with conventional metallographic techniques. The quenched samples were then etched with a modified Keller's reagent containing: 1% HF, 1.5% HCl. 2.5% HNO_3, and 95% water for 10–20 s each. Microstructures were investigated using a Hitachi TM1000 Scanning Electron Microscope (SEM). The macrostructures of quenched samples were imaged with a Cannon EOS 5D MkII with a macroscopic lens attached.

Results and Discussion

Microstructure

Figure 3 displays the microstructure for all compositions tested. There were 3 categories of microstructure observed, as shown by the backscattering electron (BSE) SEM images in Fig. 3. Hypoeutectic Al–Ni alloys had large Al dendrites surrounded by eutectic phases (Al–Al₃Ni). Eutectic Al–6.1Ni only had the eutectic phases form, with occasional Al dendrites. The morphology of the eutectic depended on their location within the cellular colonies, as near the cellular boundaries, the eutectic spacing is coarser. Hypereutectic compositions had Al₃Ni needles surrounded by eutectic phases. The hypoeutectic alloys Al–3.1Ni and Al–5.3Ni (Fig. 3a, b) both have Al dendrites that are surrounded by eutectic phases. The Al dendrites have several secondary and tertiary arms that were observed in SEM, and the dendritic spacing would be determined mainly by cooling rate [19]. The higher addition of Ni in Al–5.3Ni resulted in Al dendrites occupying a smaller volume fraction of the microstructure [20]. The morphology of the eutectic was finely spaced and aligned rods of Al₃Ni surrounded by an Al matrix [21]. Some regions of the eutectic were noted to have rods that were larger and spaced further apart. Previous experimentation using directional solidification and varying thermal gradients/growth rates showed that eutectic spacing is mainly dependent on cooling rate, with higher cooling

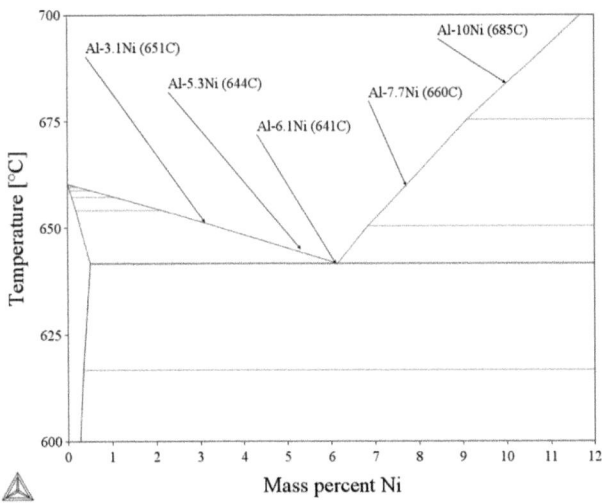

Fig. 1 Al–Ni phase diagram calculated from Thermo-Calc 2022a using the TCBIN database with locations of alloys tested in this work with liquidus temperatures indicated

Table 1 Sample composition, pouring temperature calculated from experimentation, liquidus from Thermo-Calc phase diagram, as well as the average time for complete solidification from cooling curves

Composition (wt.%)	Pouring temperature (°C)	Liquidus from phase diagram (°C) (taken from Fig. 1)	Average Solidification time (sec)
Al–3.1Ni	710	651	177.9 s
Al–5.3Ni	702	644	203.3 s
Al–6.1Ni	702	641	162.0 s
Al–7.7Ni	702	660	148.2 s
Al–10Ni	705	685	164.3 s

a) 38 mm 35 mm 22 mm

b)

c) 5 mm 10 mm

Fig. 2 **a** Diagram of the cup used for quenching experiments (Sigma Aldrich product catalogue). **b** Spiral mould pattern with **c** cross section of the spiral. Down sprue has a 54 mm top diameter with a 20° taper over a 50 mm depth. The centreline of the spiral is 4.5 revolutions with a 17 mm pitch

rates leading to more closely spaced eutectic Al₃Ni rods [2, 5, 22, 23]. It was found that if the composition is varied but the cooling rate was maintained, there was not a significant change in rod spacing for the eutectic colonies [20]. Thus, the eutectic morphology was seen in each sample varied by the location of the micrograph.

Figures 4a, and b show a comparison between the microstructure at the edge vs centre of the sample in Al–5.3Ni. Figure 4a shows Al dendrites and eutectic phases in between the dendrites. There are sections with a fine fibrous eutectic structure, but also other areas with larger Al₃Ni particles. Figure 4b shows SEM images from a region in the centre of the Al–5.3Ni sample. Here the eutectic is highly refined. The eutectic colonies are also easily seen as their boundaries have small regions of primary Al (see Figs. 3c and 4b) with some larger Al₃Ni particles, that have previously been associated with Fe impurities being rejected from the eutectic colonies [5, 24]. Al–6.1Ni (Fig. 3c) was noted to have some very small Al dendrites, but most of the volume is composed of eutectic colonies. The hypereutectic alloys had a microstructure composed of long Al₃Ni needles/particles surrounded by eutectic colonies (Fig. 3d, e). These needles were in the range of 300–3000 μm in length and as wide as 200 μm. Al₃Ni crystals have been shown to have a strong preference to grow in the <010> direction [21], thus

resulting in the long needles noted in the micrographs. The increased addition of Ni in Al–10Ni (Fig. 3e) caused more Al₃Ni needles to nucleate. The eutectic phases in hypereutectic alloys had their morphology similarly affected by cooling rates, but the proximity of Al₃Ni phase also seemed to have an effect. The volume immediately adjacent to Al₃Ni particles appeared to be devoid of eutectic phases and only contain enriched Al phase due to Al segregation during the formation of primary Al₃Ni (See locations labelled as α in Fig. 3d). The eutectic rods were also coarser and spaced further apart in regions with several globular Al₃Ni particles nucleating (Fig. 4c). These regions are likely nucleation regions of Al₃Ni that formed perpendicular to the vertical section imaged. Regions closer to the parallel long needles had the typical eutectic morphology noted in other samples, with variation in rod spacing and size likely due to impinging solute fields during the later stages of solidification.

Fluidity

The fluidity length as measured for each of the compositions is given in Fig. 5. The length of the 3 spirals was measured and averaged to provide the average spiral length for each

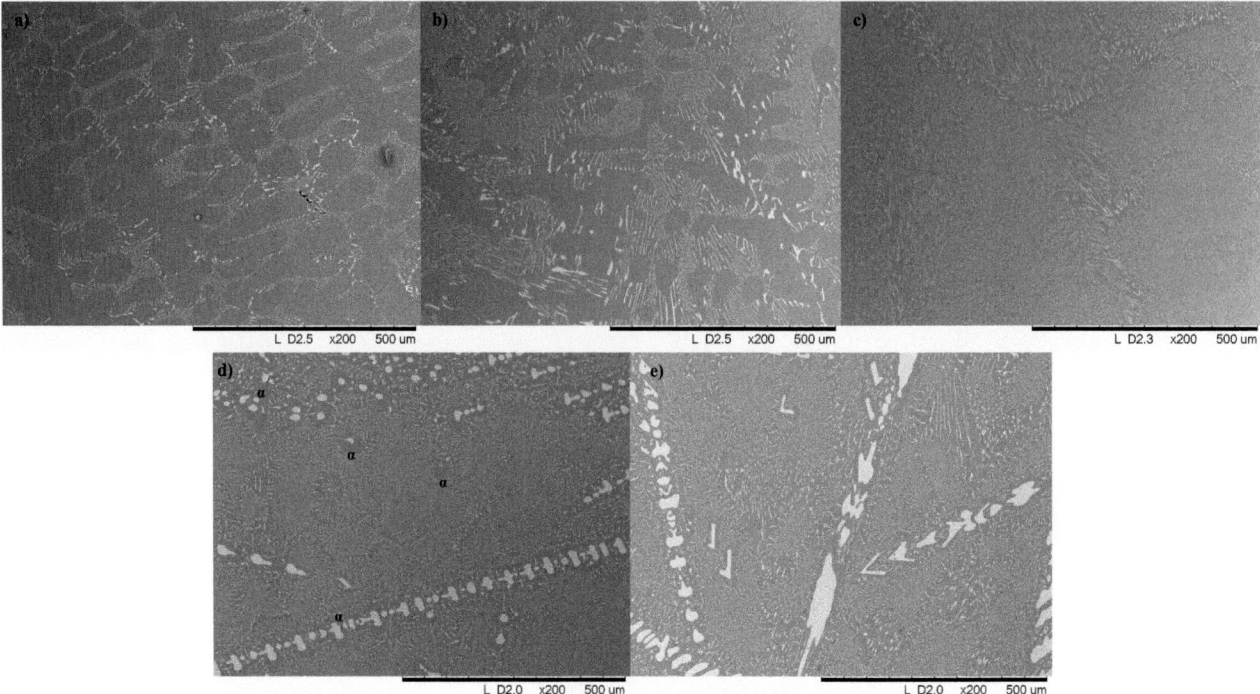

Fig. 3 SEM-BSE images for the microstructures from **a** Al–3.1Ni, **b** Al–5.3Ni, **c** Al–6.1Ni, **d** Al–7.7Ni, **e** Al–10Ni

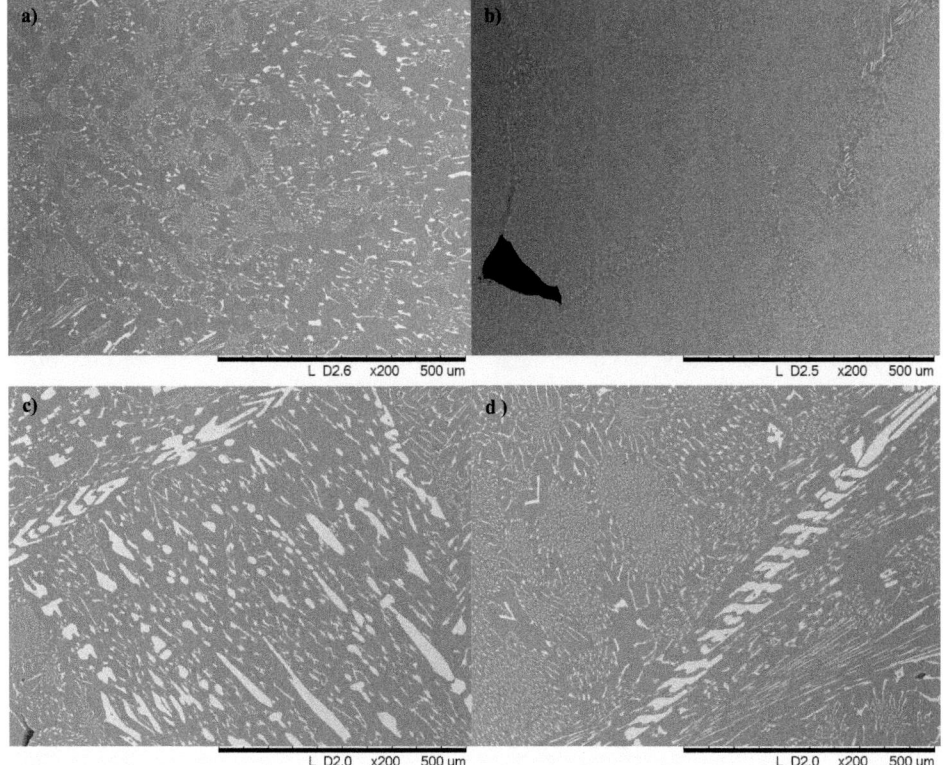

Fig. 4 SEM-BSE images of different locations of samples. **a** Al–5.3Ni scan of the edge of the sample. **b** Al–5.3Ni scan of the sample at the centre. **c** Al–10Ni scan of the coarse region with several rounded Al$_3$Ni particles. **d** Al–10Ni scan of the typical eutectic region

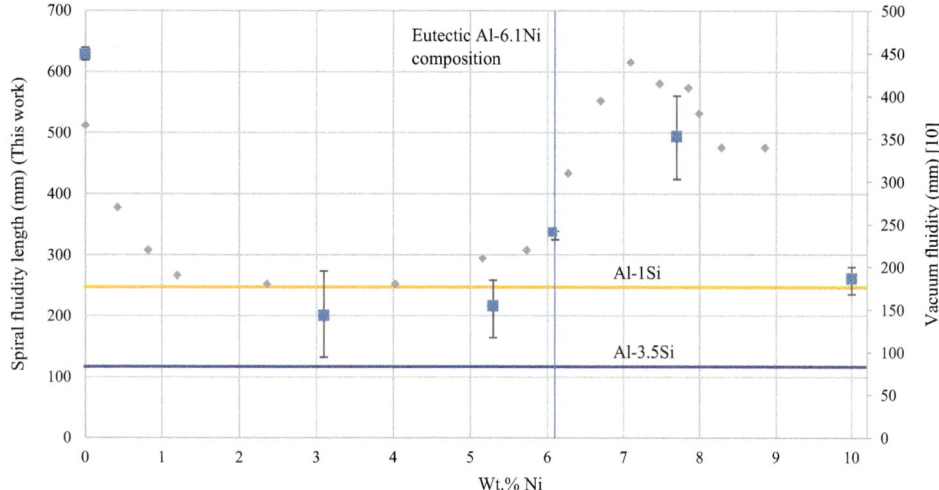

Fig. 5 Fluidity results from this work (■) compared with Al–1Si and Al–3Si also cast with the exact same method as the Al–Ni alloys and used a point of comparison. Data from previous work (◆) was also plotted onto a secondary axis. This work used vacuum pulling in a straight pipe with a pressure head of 40 cm with a constant melt temperature of 1300°F (704 °C) [10]

composition. The error bars show the range for all the measured fluidity. Similar to Al–Si alloys [6], the point of maximum fluidity for the Al–Ni alloys tested in this research was found to be at a hypereutectic composition of Al–7.7Ni, which had an average spiral length of 495 mm. This is an improvement in the fluidity of almost 50% compared to eutectic Al–6.1Ni, which had an average fluidity of 333 mm. However, none of the compositions tested had better fluidity than pure Al. The results of the fluidity tests were compared to previous research in Al–Ni alloys using a fixed vacuum pressure, 40 cm head pressure, and pouring temperature of 1300°F (704 °C) for various Al–Ni alloys [10]. The fluidity behaviour from the current work was found to have a similar pattern to this earlier work, which had a fluidity valley at a hypoeutectic composition between 1–5%, before fluidity rising and peaking at a composition of Al–7.1Ni, before dropping, with Al–8.8Ni being the highest Ni concentration tested. This work extended the composition range to Al–10Ni and confirmed the drop in fluidity past Al–8.8Ni.

The fluidity pattern found in the Al–Ni system is similar to the fluidity of Al–Si alloys, which have a peak in fluidity at around 18-wt.% Si, which is 6% above the eutectic composition et al.-12Si [6]. A possible explanation for this behaviour in Al–Si alloys is that the formation of the Si phase in hypereutectic compositions released large amounts of heat as the enthalpy of formation for primary Si is 4.5–3.7 times higher than Al in the melt [6]. The primary Si particles were also noted to be non-dendritic in shape and were more mobile in the melt, improving fluidity [25]. This is similar to what is noted in this research with Al–Ni. Microstructure investigation has shown that the Al_3Ni phase forms into long needles with a high aspect ratio, instead of dendrites, and the standard enthalpy of fusion for Al_3Ni is −40.7 kJ/mol, [15,

26], which is much higher than Al (−10.7 kJ/mol [27]) and similar to the enthalpy of fusion for Si (-48.31 kJ/mol [28]). Therefore, it is possible that the proposed reasoning for Al–Si alloys having a peak in fluidity at a hypereutectic composition could also apply to the Al–Ni system.

Cooling Curve Analysis

Representative cooling curves from each composition are plotted in Fig. 6. The temperature derivative curves (dT/dt) were also calculated, with a 5-point moving average smoothing applied. Al–3.1Ni was seen to have two clear temperature peaks and with a sharp recalescence of roughly 4.5 °C for the primary Al reaction, and about 0.5 °C for the eutectic reaction. Since the liquidus temperature was 651 °C, there was about 7 °C undercooling before the start of the reaction. This was followed by about a 3 °C difference between the equilibrium liquidus temperature and the temperature measured by the thermocouple. This difference was noted in at least one of the three cooling curves collected for each composition, for example, Al–5.3Ni had a roughly 2–3 °C difference between the liquidus temperature predicted by the thermodynamic database (Fig. 1) and the measured value (Fig. 6). Al–6.1Ni undercooled to 636.3 °C before the solidification reaction started, and recalesced to 641 °C. The behaviour during cooling for Al–7.7Ni and Al–10Ni differed, with a more gradual recalescence of a lower magnitude and a steady climb in temperature throughout the formation of the primary phase. The eutectic phases form later and there is a similar response in the cooling curve, with a slow increase in temperature. Al–10Ni had a recorded drop in temperature around 60 s after the start of

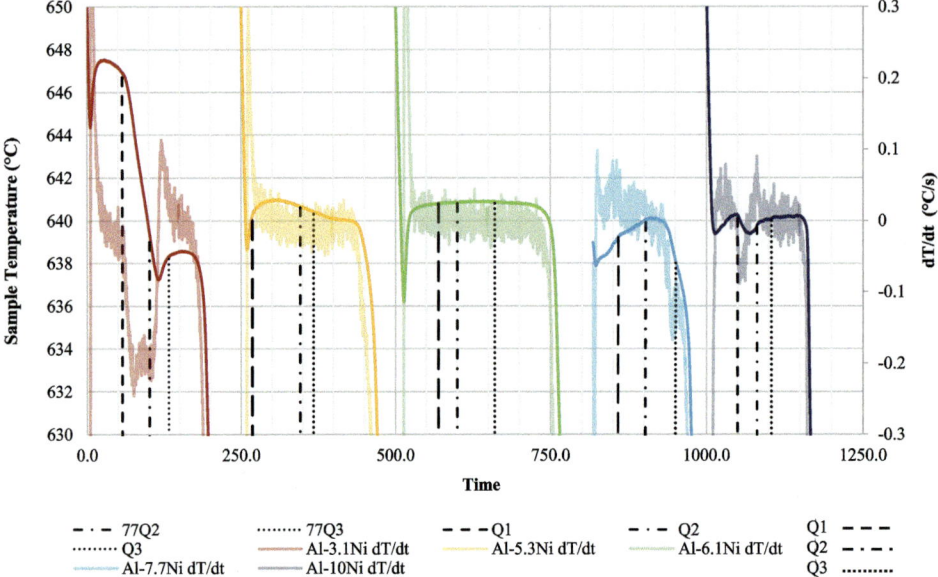

Fig. 6 Representative cooling curves for each alloy (250 s offset for each composition for clarity) with the quench locations indicated by dashed vertical lines

solidification, which marks the clear end of the primary reaction. The cooling curves for Al–7.7Ni and Al–5.3Ni both did not display this distinct drop in temperature to separate the primary and eutectic reactions. Quenched samples reveal that for these two compositions, while the primary phase did nucleate first, there were small amounts of eutectic that also solidified shortly after, thus a clear demarcation cannot be seen in the cooling curves. For hypereutectic compositions, there was a very large difference when comparing the equilibrium liquidus temperature for Al–7.7Ni and Al–10Ni with the thermodynamic database (660 and 685 °C, respectively) with the recorded liquidus temperature (637.9 and 639.4 °C, respectively).

Quenched Samples

The aim of the quenches was to capture the macroscopic progression of the primary and eutectic phases. Firstly, none of the quenched samples showed eutectic phases nucleating in the centre of the sample. This is unlike Al–Si alloys, where the eutectic often nucleates throughout the melt [29] where it may negatively impact fluidity by obstructing flow. Figure 7 observes the formation of microstructure in the Al–7.7Ni sample, the composition which displayed the highest fluidity in this work. Figure 7a–c are each a progressive quench in the Al–7.7Ni series, from Q1-Q3, respectively. Figure 7a from Q1 shows that the Al₃Ni needles nucleate throughout the entire sample. There seem to be more needles near the edge of the sample compared to the centre, which is likely due to the larger undercooling in this region. There are

minimal amounts of eutectic phases on the very edge of the sample, and these could have possibly formed during quenching. Figure 7b from Q2 shows the Al₃Ni needles fully formed in the centre at their appropriate size. In-between the centre and wall of the sample, it is seen the eutectic phases have largely coalesced to form a continuous band that is growing towards the thermal centre of the sample. There are small amounts of eutectic that advance from the Al₃Ni needles near the leading edge of the growth band. When these regions are observed under the SEM, it is seen that this is due to the formation of pockets of Al and eutectic phases surrounding these needles. It is likely that the large Al₃Ni needles have an Al-rich region surrounding them during growth explaining this microstructure. Finally, closer to the wall there are eutectic colonies that have nucleated in the spaces between Al₃Ni needles. In Fig. 7c from Q3, the result looks very similar to the microstructure from Q2, but the eutectic growth front has progressed further into the sample. One thing to note is that while there are still the regions of eutectic surrounding the Al₃Ni needles, these eutectic zones stay confined to this area and the growth of eutectic is from the advancing colonies from the outer edge of the sample. Figure 7d shows Q3 from Al–5.3Ni as a comparison of how the solidification pathway is different for hypoeutectic alloys. The eutectic phases grow from the wall just like in hypereutectic alloys, however, the growth is not as uniform, with areas that have grown much closer to the centre, and even some small areas at the wall that are free of eutectic. Furthermore, the shape and size of the Al dendrites clearly show that the dendrites would occupy a large volume fraction and tend to compact together, limiting fluid flow in a mould.

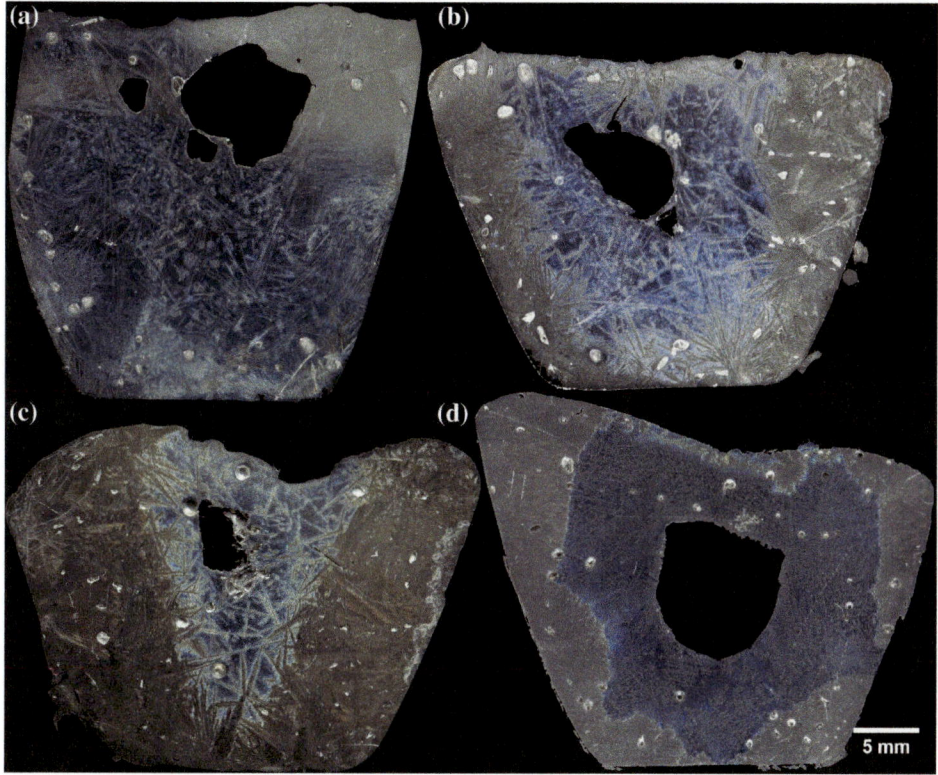

Fig. 7 Optical macroscopic photos of quenched and etched samples (original photos cropped to show metallic surface only). **a** Al–7.7Ni Q1; **b** Al–7.7Ni Q2; **c** Al–7.7Ni Q3; **d** Al–5.3Ni Q2

Conclusions

The Al–Ni alloy system was thoroughly investigated to see what effect varying Ni concentration has on the microstructure and phase evolution in the alloys of Al–xNi (where x = 3.1, 5.3, 6.1, 7.7, 10 wt.%). Results include microstructure, fluidity, cooling curves, and interrupted solidification experiments. The major findings of this work are:

1. The microstructure of hypoeutectic alloys consisted of primary aluminium with ultra-fine eutectic Al–Al$_3$Ni forming around the Al dendrites. The Al–Ni eutectic showed some variation in the eutectic spacing, even within a given sample, which likely reflects changes in the local cooling rate during solidification. Hypereutectic Al–Ni alloys contained many long needles of Al$_3$Ni.
2. The fluidity peak of the compositions tested occurred at a hypereutectic composition of 7.7wt% Ni, largely in agreement with previously published data [10]. The fluidity decreased at higher Ni concentrations.
3. The cooling curves revealed differences in the recalescence behaviour between hypo and hypereutectic alloys. This may be related to differences in available nuclei along with the high enthalpy of fusion of Al$_3$Ni.

4. Interrupting solidification by quenching, revealed that the eutectic phases nucleate in the vicinity of the edge of the sample and progressively grow inwards for all alloys. However, for hypereutectic alloys, the eutectic phases form a more macroscopically planar growth front, whereas hypoeutectic alloys have a more scalloped uneven growth front.

References

1. Suwanpreecha, C., et al., *New generation of eutectic Al–Ni casting alloys for elevated temperature services.* Materials Science and Engineering: A, 2018. **709**: p. 46–54.
2. Kaya, H., et al., *Measurements of the microhardness, electrical and thermal properties of the Al–Ni eutectic alloy.* Materials & Design, 2012. **34**: p. 707–712.
3. Kim, J.-M., et al., *Mold filling ability and hot cracking susceptibility of Al–Fe–Ni alloys for high conductivity applications.* Jurnal Teknologi, 2015. **75**.
4. Wang, C., et al., *Phase alignment and crystal orientation of Al3Ni in Al–Ni alloy by imposition of a uniform high magnetic field.* Journal of Crystal Growth, 2008. **310**(6): p. 1256–1263.
5. Carrara, A.P., et al., *Effect of cooling rate on microstructure and microhardness of hypereutectic Al–Ni alloy.* Archives of Civil and Mechanical Engineering, 2021. **21**(1).

6. Ravi, K.R., et al., *Fluidity of aluminum alloys and composites: A review.* Journal of Alloys and Compounds, 2008. **456**(1–2): p. 201–210.

7. Palanivel, S., et al. (2020) Aluminum alloys for die casting. U.S. P. Office WO/2020/028730. 2020–02–06.

8. Koutsoukis, T. and M.M. Makhlouf, *An Alternative Eutectic System for Casting Aluminum Alloys I. Casting Ability and Tensile Properties,* in *Light Metals 2015,* M. Hyland, Editor. 2016, Springer International Publishing: Cham. p. 277–281.

9. Yang, L., et al., *Effect of Si and Ni contents on the fluidity of Al–Ni–Si alloys evaluated by using thermal analysis.* Thermochimica Acta, 2016. **645**: p. 7–15.

10. Ford, J.B. (1970) The Casting Fluidity Of Some Aluminum–Nickel Alloys. Masters Thesis, The Ohio State University.

11. Hernández-Méndez, F., et al., *Effect of Nickel Addition on Microstructure and Mechanical Properties of Aluminum-Based Alloys.* Materials Science Forum, 2011. **691**: p. 10–14.

12. Chankitmunkong, S., D.G. Eskin, and C. Limmaneevichitr, *Effects of Ultrasonic Melt Processing on Microstructure, Mechanical Properties, and Electrical Conductivity of Hypereutectic Al–Si, Al–Fe, and Al–Ni Alloys with Zr Additions,* in *Light Metals 2021.* 2021. p. 192–197.

13. Du, Y. and N. Clavaguera, *Thermodynamic assessment of the Al–Ni system.* Journal of Alloys and Compounds, 1996. **237**(1): p. 20–32.

14. Davey, T., et al., *First-principles-only CALPHAD phase diagram of the solid aluminium-nickel (Al–Ni) system.* Calphad, 2020. **71**.

15. Chen, H.L., et al., *Thermodynamics of the Al3Ni phase and revision of the Al–Ni system.* Thermochimica Acta, 2011. **512**(1–2): p. 189–195.

16. Okamoto, H., *Supplemental Literature Review of Binary Phase Diagrams: Al–Ni, B–Hf, Ca–Sc, Cr–Sc, Fe–Rh, Hf–Mn, La–Sb, Ni–Re, Ni–Sm, Ni–Zr, Sb–Tb, and Ti–Zr.* Journal of Phase Equilibria and Diffusion, 2019. **40**(6): p. 830–841.

17. Martínez-Villalobos, M., et al., *Microstructural Evolution of Rapid Solidified Al–Ni Alloys.* Journal of the Mexican Chemical Society, 2016. **60**: p. 67–72.

18. Pandey, P., et al., *On the origin of a remarkable increase in the strength and stability of an Al rich Al–Ni eutectic alloy by Zr addition.* Acta Materialia, 2019. **170**: p. 205–217.

19. Júnior, P.F., et al., *Heat-flow parameters affecting microstructure and mechanical properties of Al–Cu and Al–Ni alloys in directional solidification: an experimental comparative study.* International Journal of Materials Research, 2022. **113**(3): p. 181–193.

20. Sankanit, P., V. Uthaisangsuk, and P. Pandee, *Tensile properties of hypoeutectic Al–Ni alloys: Experiments and FE simulations.* Journal of Alloys and Compounds, 2021. **889**.

21. Fan, Y. and M.M. Makhlouf, *The Al–Al3Ni Eutectic Reaction: Crystallography and Mechanism of Formation.* Metallurgical and Materials Transactions A, 2015. **46**(9): p. 3808–3812.

22. Kakitani, R., et al., *Relationship between spacing of eutectic colonies and tensile properties of transient directionally solidified Al–Ni eutectic alloy.* Journal of Alloys and Compounds, 2018. **733**: p. 59–68.

23. Zhuang, Y.X., et al., *Eutectic spacing and faults of directionally solidified Al–Al3Ni eutectic.* Science and Technology of Advanced Materials, 2001. **2**(1): p. 37–39.

24. Bian, Z., et al., *Thermal stability of Al–Fe–Ni alloy at high temperatures.* Journal of Materials Research and Technology, 2019. **8**(3): p. 2538–2548.

25. ASM Handbook Committee, *Pure Metals,* in *Properties and Selection: Nonferrous Alloys and Special-Purpose Materials.* 1990, ASM International. p. 0.

26. Chrifi-Alaoui, F.Z., et al., *Enthalpies of formation of the Al–Ni intermetallic compounds.* Journal of Alloys and Compounds, 2004. **364**(1–2): p. 121–126.

27. Stølen, S. and F. Grønvold, *Critical assessment of the enthalpy of fusion of metals used as enthalpy standards at moderate to high temperatures.* Thermochimica Acta, 1999. **327**(1): p. 1–32.

28. Yamaguchi, K. and K. Itagaki, *Measurement of high temperature heat content of silicon by drop calorimetry.* Journal of Thermal Analysis and Calorimetry, 2002. **69**(3): p. 1059–1066.

29. Nogita, K., et al., *Eutectic Solidification Mode in Sodium Modified Al-7 mass%Si-3.5 mass%Cu-0.2 mass%Mg Casting Alloys.* Materials Transactions, 2001. **42**(9): p. 1981–1986.

Effect of Ti Addition on the Microstructure and Mechanical Properties of Hypo-Eutectic and Eutectic Al–Si Alloys

Chandan Choudhary, K. L. Sahoo, Ashok J. Keche, and D. Mandal

Abstract

To study the effect of Ti–B content on the microstructure and mechanical properties of Al-7Si and Al-12.6Si alloys with Al-5Ti-1B content of 0% and 2%, respectively, were prepared through conventional melting and casting route. The microstructure and mechanical properties of as-cast alloys were investigated, and a correlation has been established between microstructural features (shape, size, and area of eutectic Si, SDAS of primary α-Al, inter-spacing between two nearby eutectic Si particles, and hardness) with strength and ductility of as-cast alloys. Results show that Ti content changes the morphology of primary α-Al grain from dendrite to rosette type in Al-7Si alloy and coarse dendrite to fine columnar dendrite in Al-12.6Si alloy. Eutectic Si particle's shape and size also change with the addition of Ti to the melt. The thermally stable phase Al₃Ti is associated with Si and forms the $Ti_7Al_5Si_{14}$ phase. Lower SDAS values, finer dendrite α-Al grains, reduce thickness of eutectic Si particles, homogeneous structure, and minimization of casting defects in 2 wt.% Al-5Ti-1B added Al-12.6Si alloy plays a vital role in enhancing the hardness (71 HV), yield strength (YS, 126 MPa), ultimate tensile strength (UTS, 198 MPa).

Keywords

Al–Si alloys • Grain refiner • Microstructure • Mechanical properties

C. Choudhary (✉) · A. J. Keche
Department of Mechanical Engineering, Maharashtra Institute of Technology, Satara Parisar, Beed Bypass Road, Aurangabad, Maharashtra 431005, India
e-mail: cc.15mme1101@phd.nitdgp.ac.in

K. L. Sahoo
Materials Engineering Division, CSIR-National Metallurgical Laboratory, Jamshedpur, 831007, India

D. Mandal
Department of Metallurgical and Materials Engineering, National Institute of Technology, Durgapur, 713209, India

Introduction

Over the past few decades, a lot of effort has been put into developing lightweight engineering materials [1]. In the fields of transportation, space exploration, and building, Al–Si alloys are widely employed [2, 3]. The as-cast Al–Si alloy features a coarse needle or acicular type eutectic silicon incorporated in the Al matrix. The qualities of the alloy are significantly influenced by the morphology, shape, size, distribution, and other casting flaws, including shrinkage and gas porosity.

Al–Si alloys have been the subject of numerous studies to improve their mechanical properties by modifying the microstructural morphology. Grain morphology has been modified by altering the chemical composition and cooling rate and applying thermomechanical treatments. The addition of transition elements like Fe, Mn, Ni, Cu, Cr, and Zr could significantly improve the mechanical properties because these elements form thermally stable phases [4]. Due to their ease of use, low cost, and effectiveness in modifying the microstructure of Al–Si alloys, grain refinement using chemical refiners is extensively recognized. The Al–Si alloy undergoes constitutional supercooling as a result of the addition of grain refiners like Ti, B, V, Zr, Sc, and Nb, and this offers adequate nucleation sites for solidification [5, 6]. Grain refiner is added to the molten alloy to restrict dendritic network formation and increase the melt feeding and fluidity to lessen the shrinkage flaws. Additionally, casting reliability and mechanical qualities are enhanced [7]. The Al-5Ti-1B master alloy is the grain refiner that is most frequently used in industry to create finer grains for aluminium alloys [8]. The two factors that play a role in the grain refinement in Al–Si alloys are the heterogeneous nucleation of a-Al grains and the growth restriction factor dependent on undercooling. The grain refinement mechanism through a suitable grain refiner (Al-5Ti-1B) is quite straightforward. Titanium or boron-enriched several aluminide particles serve as a heterogeneous nucleation site for

© The Minerals, Metals & Materials Society 2023
S. Broek (ed.), *Light Metals 2023*, The Minerals, Metals & Materials Series,
https://doi.org/10.1007/978-3-031-22532-1_63

the nucleation of primary grains. The α-Al grains nucleate at the surface of aluminides particles and grow as solidification proceeds [9].

This research aims to study the effect of grain refiner on microstructural features like size, shape, the morphology of phases, casting defects, and eutectic Si particles, which influence the hardness and tensile properties of hypoeutectic and eutectic Al–Si alloys.

Materials and Methods

Commercially pure Al (99.7%), commercially pure crystalline Si (99.5%), and Al-5Ti-1B master alloy were used as raw material to develop Al-7Si, Al 12.6Si, Al-7Si-0.1Ti-0.02 B, and Al-12.6Si-0.1Ti-0.02B alloys. The alloys were developed through the traditional melting and casting route. The melt was held for 10 min after adding a grain refiner to confirm the complete dissolution of Ti and B elements into the melt [10]. After degassing with solid hexachloroethane, the liquid metal was poured into a preheated permanent cast iron mould. The chemical composition of developed Al–Si alloys is presented in Table 1.

Leica DM 2500 M optical microscopy and a scanning electron microscope (SEM, Model: SIGMA Carl Zeiss Germany) were used to examine the microstructural characteristics of developed alloys. X-Ray Diffraction methods (Philips: PANalytical X'Pert PRO diffractometer) were used to identify the various phases in developed alloys. The tensile test of 3 ASTM E8M standard tensile specimens was performed using a Universal Testing Machine (Instron-8800, 100 kN capacity) with a strain rate of 10^{-3} s^{-1} at room temperature. The bulk hardness of developed alloys was obtained using a Vickers hardness testing machine (BV 250 (S), BIE, Miraj, India).

Results and Discussion

Microstructural Characterization

Figure 1a, d shows the typical optical microstructure of developed Al–Si alloys. Figure 1a depicts the dendrite morphology of primary α-Al grain and eutectic Al–Si at the interdendritic region in a typical as-cast Al-7Si alloy. The

addition of 2 wt.% Al-5Ti-1B to Al-7Si alloy could significantly change the morphology of primary α-Al grain from dendrite to rosette, as shown in Fig. 1b. α-Al grains are effectively refined by TiAl3 particles, which are associated with α-Al grains and act as a nucleation site for α-Al grains. TiB2 particles are observed at the eutectic Al–Si and have low potency of nucleation for α-Al grain [11]. Figure 1c depicts the typical eutectic Al–Si in Al-12.6Si alloy. With the addition of Si, more than 10 wt.% and Ti concentration is 0.1 wt.%, there is the formation of τ_2 ($Al_5Si_{14}Ti_7$). The formation of $\tau2$ is due to the peritectic reaction of $liquid + TiAl_3 \rightarrow \alpha - Al + \tau_2$ at 593 °C. It is reported that the formation of τ_2 silicide could easily restrict the epitaxial nucleation of α-Al grains and also reduce the growth restriction factor by consuming nucleating potent ($TiAl_3$, TiB_2) [12, 13]. Figure 1d depicts the microstructure of as-cast Al-12.6Si-0.1Ti-0.02B alloy. It contains mainly α-Al grain and eutectic Al-Si. The formation of α-Al grain in Al-12.6Si-0.1Ti-0.02B alloy is due to heterogeneous nucleating potent sites and undercooling.

The eutectic Si structure of developed Al–Si alloys in SEM microstructure is in Fig. 2a, d. Al-5Ti-1B added alloys depict the regular shape of eutectic Si, while an irregular eutectic Si has been observed in Al-7Si and Al-12.6Si alloys. Mostly TiB_2 particles are segregated near eutectic Al–Si as observed in Fig. 2d and hence have low nucleation potencies towards heterogeneous nucleation of α-Al. It is observed from Fig. 2d that the silicide phase can be developed on the surface of TiB_2 particles (Fig. 3).

The presence of different phases in developed Al–Si alloys has been identified by the analysis of X-Ray diffraction patterns as depicted in Fig. 4a, b. The silicide phase (τ_2) can thermodynamically coexist with $TiAl_3$. However, these silicide phases could not be detected in X-Ray diffraction patterns or SEM/TEM. A slight shifting of the X-Ray diffraction patterns of Al-12.6Si-0.1Ti-0.02B alloy is due to the formation of τ_2 ($Al_5Si_{14}Ti_7$) which increases the lattice misfit of α-Al grain.

Mechanical Properties of Developed Al–Si Alloys

The hardness values and tensile properties of developed Al–Si alloys are tabulated in Table 2. The investigation shows that the hardness value of Al-12.6Si-0.1Ti-0.02B alloy is higher than other developed alloys. This is due to the

Table 1 Chemical composition of developed Al–Si alloys

Sl. no.	Alloys	Si	Fe	Ti	B	Al
1	Al-7Si	7.0	0.06	–	–	Bal.
2	Al-7Si-0.1Ti-0.02B	6.96	0.03	0.1	0.02	Bal.
3	Al-12.6Si	12.43	0.02	–	–	Bal.
4	Al-12.6Si-0.1Ti-0.02B	12.16	0.018	0.1	0.02	Bal.

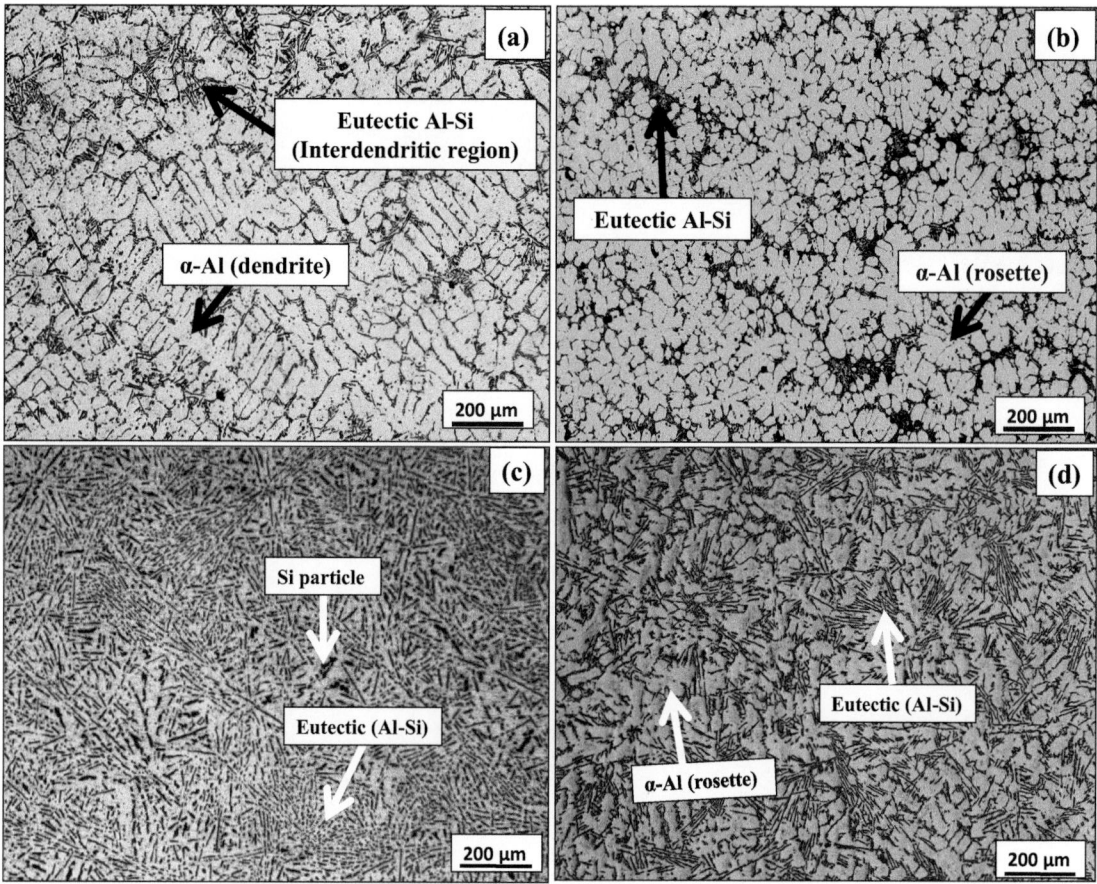

Fig. 1 Optical microstructure of developed Al–Si alloys **a** Al-7Si, **b** Al-7Si-0.1Ti-0.02B, **c** Al-12.6Si, and **d** Al-12.6Si-0.1Ti-0.02B

presence of the silicide phase and higher concentration of Si. The hardness value has increased significantly with the addition of 2 wt.% of Al-5Ti-1B to Al-7Si alloy. This is due to harder aluminide particles at the eutectic (TiB2) and near the α-Al grain (TiAl3). Si concentration in the alloys also plays a vital role in enhancing the hardness of alloys.

The mechanical properties such as YS, UTS, and %El (total elongation, up to fracture) obtained from the tensile test of developed Al–Si alloys and the engineering stress to strain curve are shown in Fig. 4. The 2 wt.% Al-5Ti-1B added alloys show higher strength and ductility than its corresponding alloys (without added Al-5Ti-1B). The YS, UTS, and elongation to fracture (%El) increases linearly with the addition of Al-5Ti-1B to Al-7Si and Al-12.6Si alloy. The elongation to fracture (%El) value of developed Al–Si alloys decreases with increasing Si concentration. This is due to harder Si particles in the matrix restricting the material flow during the tensile test.

The mechanical properties of Al-5Ti-1B added Al–Si alloys are improved due to changes in the size and morphology of α-Al grain and Si particles, formation of

aluminides (TiAl$_3$, TiB$_2$) particles and silicide (τ_2) phase, and reduction in casting defects such as porosity and shrinkage [14]. Adding Al-5Ti-1B to the alloys increases the grain quantity and the grain boundary per unit volume. These grain boundaries absorbed the energy during tensile testing; thus, more energy is required for the failure of the alloys. The finer equiaxed microstructure increases the dislocation interaction at the grain boundary, thus achieving higher strength. Figure 5a, b shows the relationship between UTS and %El to the length of Si particles and porosity. From Fig. 5a, b, it is clear that the porosity (%) of the alloy is inversely proportional to the mechanical properties (UTS, % El) as the porosity decreases with the addition of Al-5Ti-1B to the Al–Si alloy, UTS, and % El of the alloy increases.

A good correlation between the microstructural features of Si particle (size, area, and aspect ratio), porosity, and hardness value was developed using a linear regression equation with a coefficient of determination (R^2) value, and the same is tabulated in Table 3. It is observed that the Si particle's size, area, and porosity are inversely proportional to the strength of the alloy, while the aspect ratio (minor to

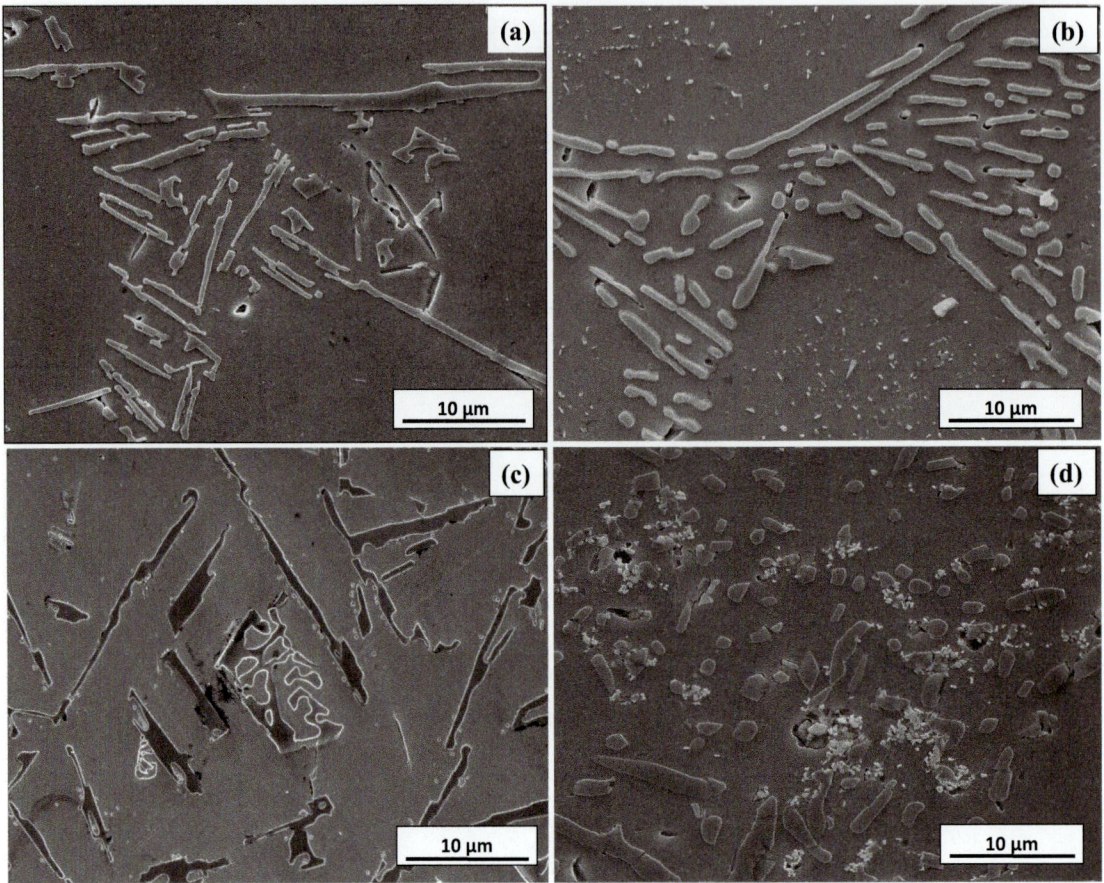

Fig. 2 SEM microphotograph of developed Al–Si alloys **a** Al-7Si, **b** Al-7Si-0.1Ti-0.02B, **c** Al-12.6Si, and **d** Al-12.6Si-0.1Ti-0.02B

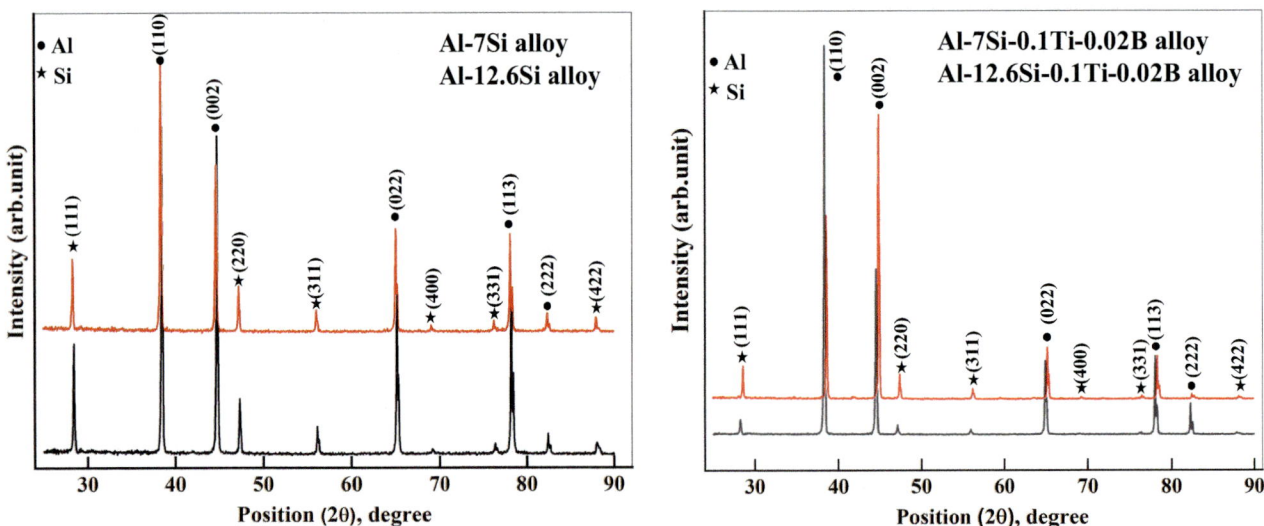

Fig. 3 X-Ray diffraction patterns of developed Al–Si alloys **a** without Al-5Ti-1B, and **b** 2 wt.% Al-5Ti-1B

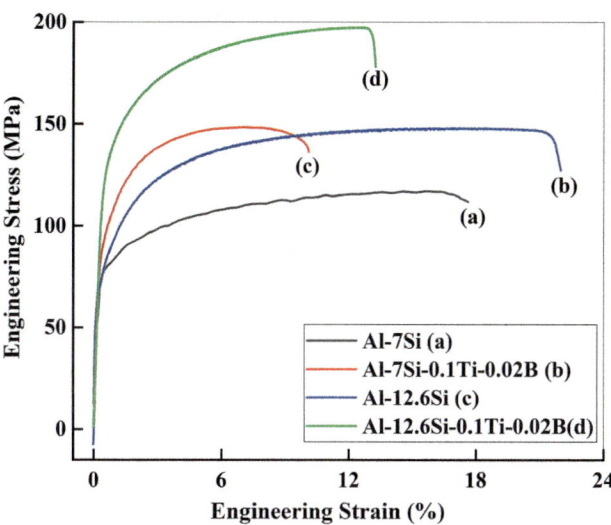

Fig. 4 Engineering stress to strain of developed Al–Si alloys **a** Al-7Si, **b** Al-7Si-0.1Ti-0.02B, **c** Al-12.6Si, and **d** Al-12.6Si-0.1Ti-0.02B

major length) and hardness are directly proportional to the strength of the alloy. This linear regression equation gives a better correlation with the strength of the alloy, which is confirmed in the reported literature [15, 16].

Conclusion

The conclusion drawn from the study of the addition of Al-5Ti-1B to Al-7Si (hypoeutectic alloy) and Al-12.6Si (eutectic alloy) are following:

- Addition of Al–5Ti–1B (2 wt.%) to Al–Si melt significantly changes the morphology of α-Al grain eutectic Si particles. The microstructure of the developed alloy is mainly composed of α-Al, eutectic Si, $TiAl_3$, TiB_2 and τ_2 ($Al_5Si_{14}Ti_7$) particles.
- Addition of Al–5Ti–1B to Al–Si melt, improves the fluidity and feeding characteristics of liquid, thus a reduced value of porosity is observed.
- TiB_2 particles are segregated at the eutectic region, thus having low heterogeneous nucleating potencies than $TiAl_3$ particles.
- Silicides (τ_2) phases developed at the surface of TiB_2 at the eutectic region in Al-Si/Al-5Ti-1B alloy system.
- The hardness (HV) value, YS (MPa), UTS (MPa), and El (%) have been improved significantly with the addition of Al-5Ti-1B to Al–Si alloys.
- A significantly higher value of hardness (71 HV), strength (UTS = 198 MPa), (YS = 126 MPa), and

Table 2 Mechanical properties of developed Al–Si alloys

Properties	Mean ± SD			
	Al-7Si		Al-12.6Si	
	0 Ti	0.1 Ti-0.02B	0 Ti	0.1 Ti-0.02B
Hardness (HV)	38 ± 2	51 ± 5	52 ± 3	71 ± 7
YS (MPa)	84 ± 6	112 ± 5	116 ± 6	126 ± 3
UTS (MPa)	117 ± 7	148 ± 6	150 ± 5	198 ± 5
El (%)	16 ± 2	22 ± 1	10 ± 1	13 ± 2

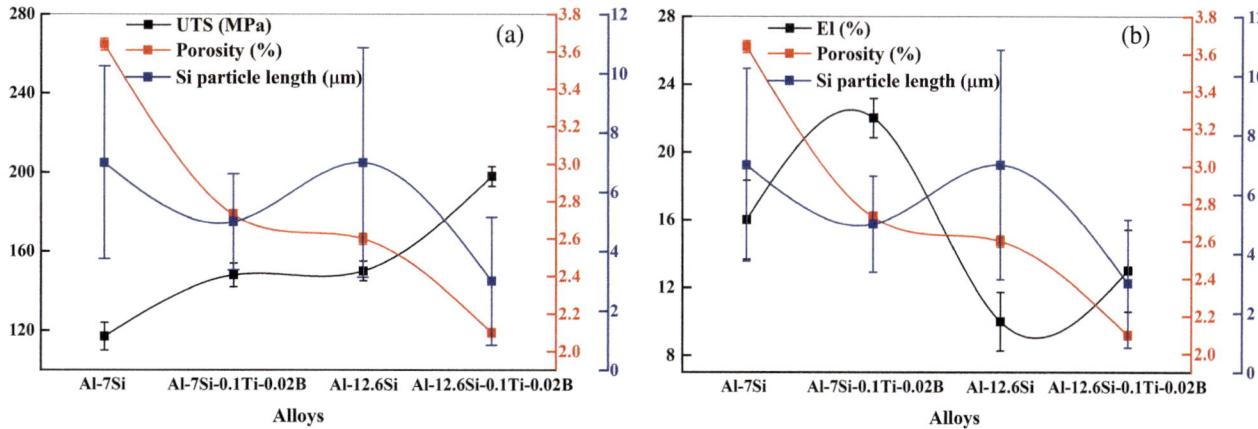

Fig. 5 The relation of Si particle length (μm) and porosity (%) with **a** UTS (MPa), and **b** El (%)

Table 3 Relation of microstructural features and casting defects with UTS of developed Al–Si alloys

Relationship	Linear equation	Coefficient of determination R^2
$UTS = f(D_{Si})$	$-15.318 \times D_{Si} + 237.5$	0.77
$UTS = f(Area_{Si})$	$-7.225 \times Area_{Si} + 245.54$	0.96
$UTS = f(AR_{Si})$	$136.867 \times Aspect\ Ratio_{Si} + 87.89$	0.74
$UTS = f(Porosity)$	$-54.833 \times Porosity + 303.22$	0.92
$UTS = f(H)$	$2.46 \times Hardness + 22.85$	1.0

UTS = ultimate tensile strength (MPa), D_{Si} = Si particle size (μm), $Area_{Si}$ = eutectic Si particles area (μm^2), AR_{Si} = Si Aspect Ratio, Porosity (%) and H = hardness (HV)

ductility (El = 13%) is observed in Al-12.6Si-0.1Ti-0.02B alloy. Hardness value and mechanical properties of developed Al–Si alloy is dependent on microstructural features of cast alloy, porosity (%), and the presence of harder phases present.

Acknowledgements The authors would like to thank the SERB DST and DST-FIST program, Govt. of India, for providing financial and instrumental support and are obliged to the Director of the National Institute of Technology (NIT), Durgapur, and CSIR-National Metallurgical Laboratory (CSIR-NML), Jamshedpur, for providing the facility to carry out this research work.

Compliance with Ethical Statement The authors declare that they have no conflict of interest.

References

1. Yang B, Wang F, Zhang J (2003) Microstructural characterization of in situ TiC/Al and TiC/Al-20Si-5Fe-3Cu-1Mg composites prepared by spray deposition. Acta Mater. 51(17): 4977-4989. https://doi.org/10.1016/S1359-6454(03)00292-1
2. Jorstad, JJ, Rasmussen, WM (1993) Aluminum Casting Technology. American Foundry Society. 2nd ed. Illinois; Chicago, p 76–77.
3. Ammar HR, Samuel AM, Samuel FH (2008) Porosity and the Fatigue Behavior of Hypoeutectic and Hypereutectic Aluminum Silicon Casting Alloys. Int. J. Fatigue. 30(6): 1024-1035. https://doi.org/10.1016/j.ijfatigue.2007.08.012
4. Belov N, Eskin D, Avxentieva N (2005) Constituent Phase Diagrams of the Al-Cu-Fe-Mg-Ni-Si System and Their Application to the Analysis of Aluminium Piston Alloys. Acta Mater. 53 (17): 4709-4722. https://doi.org/10.1016/j.actamat.2005.07.003
5. Rathod NR, Manghani JV (2012) Effect of Modifier and Grain Refiner on Cast Al-7Si Alloy: A Review. Int, J. Emerg. Trends Eng. Develop. 5(2): 574–581.
6. Choudhary C., Sahoo KL, Roy H, Mandal D (2022) Effect of Grain Refiner on Microstructural Feature Influence Hardness and Tensile Properties of Al-7Si Alloy. J. Mater. Engg. Perform. 31: 3262-3273. https://doi.org/10.1007/s11665-021-06413-9
7. Kashyap KT, Chandrashekar T (2001) Effects and Mechanisms of Grain Refinement in Aluminium Alloys Bull. Mater. Sci. 24: 345-353. https://doi.org/10.1007/BF02708630
8. Zeren M, Karakulak E (2008) Influence of Ti Addition on the Microstructure and Hardness Properties of Near-Eutectic Al-Si Alloys. J. Alloys. Compds. 450 (1-2): 255-259. https://doi.org/10.1016/j.jallcom.2006.10.131
9. Perrot, P (1990) Aluminium-Silicon-Titanium Ternary Alloys-A Comprehensive Compendium of Evaluated Constitution Data and Phase Diagram. V Publishers, New York.
10. Ghassemalia E, Riestraa M, Bogdanoffa T, Kumar BS, Seifeddinea S (2017) Hall-Petch Equation in a Hypoeutectic Al-Si Cast Alloy: Grain Size vs. Secondary Dendrite Arm Spacing. Proc. Engg. 207: 19-24. https://doi.org/10.1016/j.proeng.2017.10.731
11. Li Y, Hu B, Liu B, Nie A, Gu Q, Wang J, Li Q (2020) Insight into Si poisoning on grain refinement of Al-Si/Al-5Ti-1B system. Acta Mater. 187: 51–65. https://doi.org/10.1016/j.actamat.2020.01.039
12. Schumacher P, McKay BJ (2003) TEM investigation of heterogeneous nucleation mechanisms in Al-Si alloys. J. Non-Cryst. Solids 317(1–2): 123–128. https://doi.org/10.1016/S0022-3093(02)01992-0
13. Qiu D, Taylor JA, Zhang MX, Kelly PM (2007) A mechanism for the poisoning effect of silicon on the grain refinement of Al-Si alloys. Acta Mater. 55(4): 1447-1456. https://doi.org/10.1016/j.actamat.2006.09.046
14. Iwahashi Y, Wang J, Horita Z, Nemota M, Langdon TG (1996) Principle of equi-channel angular pressing for the processing of ultra-fine-grained materials. Scrip. Mater. 35(2): 143-146. https://doi.org/10.1016/1359-6462(96)00107-8
15. Yajjala RK, Inampudi NM, Jinugu BR (2020) Correlation between SDAS and mechanical properties of Al–Si alloy made in Sand and Slag moulds. J. Mater. Res. Tech. 9 (3): 6257-6267. https://doi.org/10.1016/j.jmrt.2020.02.066
16. Choudhary C, Bar HN, Pramanick AK, Sahoo KL, Mandal D (2022) Effect of Strain Induced Melt Activation Process on the Microstructure and Mechanical Properties of Al-5Ti-1B Treated Al-7Si Alloy. Met. Mater. Int. https://doi.org/10.1007/s12540-021-01154-9

Compatibility Study of Polymeric Binders for Aluminum Binder Jet Parts

Solgang Im, Rasim Batmaz, Arunkumar Natarajan, and Étienne Martin

Abstract

Binder jetting of aluminum powder has progressively gained industry interest especially from automotive and aerospace for the fabrication of lightweight products. However, the interactions between the polymeric binders and oxidative behavior of aluminum powder have not yet been extensively studied by researchers. Hence, this research focuses to understand the impact of different polymeric binders on the final quality of the fabricated sample through performing composition and porosity analyses. The resulting impurity levels and final density were evaluated to identify the most compatible polymer binder for aluminum powder. Five liquid binders were prepared and utilized to fabricate green state samples with the AlSi10Mg powder. The lowest level of impurities was obtained with the alcohol solvent-based binders whereas organic solvent-based binders, such as highly non-reactive thermoplastic fluoropolymer and cyanide-based synthetic polymer, and water-based binders were reported to deposit moderate to high binder residues on the samples after the debinding heat treatment. The alcohol solvent-based binder provided the highest sintered density of 95.1%, while the PVA binder delivered the lowest sintered density. Hence, this research recommends the PVP binder as the most suited binder among the considered binders as it provided the highest densification with minimal chemical interactions with the aluminum powder.

Keywords

Additive manufacturing • Binder jetting • Aluminum powder • Polymeric binder • Densification

Introduction

Recently, binder jetting additive manufacturing (BJAM) technology has gained interest from the automotive and aerospace industries due to the cost-effective and fast printing process that allows to fabricate complex geometries without a need for support structure or special atmospheres [1–5]. In such industries, there has been a demand to investigate the possibility of fabricating aluminum parts using the BJAM process to reduce the overall weight of automobiles and aircraft. However, the aluminum powders are highly susceptible to forming oxide layers on the surface of powder particles, providing challenges for sintering attributed to the high thermal stability of oxides [6]. The impact of polymer binders on the oxidation of aluminum powder has not yet been understood there is currently limited research available in BJAM of aluminum powder.

Thermoplastic polymers, such as polyvinylpyrrolidone (PVP), polyvinyl alcohol (PVA), and polyacrylic acid (PAA), have been widely adopted in the BJAM, metal/ceramic injection molding, and other powder processing industries as a binder. Hence, in the current work, the above-mentioned three thermoplastic polymers alongside the polymers that do not contain oxygen in their polymer chains, such as polyacrylonitrile (PAN), and polyvinylidene fluoride (PVDF), are used to study their compatibility with the aluminum powders. For this study, 2-propanol, dimethylformamide (DMF), and de-ionized (DI) water are selected solvents, and the liquid binders are categorized according to the solvent type, namely alcohol solvent, organic solvent, and water-based binders. The fabricated green state samples with different binders are debound and

S. Im · É. Martin (✉)
Polytechnique Montréal, 2500 Chemin de Polytechnique, Montréal, QC H3T 1J4, Canada
e-mail: etienne.martin@polymtl.ca

R. Batmaz · É. Martin
University of Waterloo, 200 University Avenue West, Waterloo, ON N2L 3G1, Canada

S. Im · A. Natarajan
GE Additive, 8556 Trade Center Drive, West Chester, OH 45011, USA

© The Minerals, Metals & Materials Society 2023
S. Broek (ed.), *Light Metals 2023*, The Minerals, Metals & Materials Series, https://doi.org/10.1007/978-3-031-22532-1_64

sintered to evaluate the effect of different binders on part impurities and densification to determine the most compatible binder for aluminum powder.

Experimental Procedures

Materials

AlSi10Mg was procured from the SLM Solutions to manufacture simulated BJAM green state samples. The obtained powder consists mostly of spherical particles, with a particle size ranging from 20–63 μm. The nominal chemical composition of the acquired AlSi10Mg powder is shown in Table 1.

The PVP, PVA, PAA, PAN, and PVDF with average mol. wt. of 40,000, 31,000, 50,000, 450,000, 150,000, and 534,000 were purchased from the Sigma Aldrich. These polymers were formulated together with DMF, 2-propanol, and DI water to prepare liquid binders for this study. The solvents, DMF (Mw: 73 g/mol, 99.8% purity) and 2-propanol (Mw: 60.10 g/mol, 99.5% purity), were also purchased from the sigma Aldrich.

Sample Fabrication

The liquid binders were prepared by mixing the polymer with solvent in pre-defined ratios. For this study, five polymer binders each consisting of one polymer and one solvent were formulated. The liquid binders were categorized into alcohol solvent, organic solvent, and water-based binders and the corresponding solvent for each category is 2-propanol, DMF, and DI water, respectively. Two alcohol solvent-based binders were prepared with 5 wt.-% of PVP or 0.5 wt.-% of PAA dissolved in 2-propanol when two organic solvent-based binders were composed with 5 wt.-% of PAN or 3.5 wt.-% of PVDF in DMF. Lastly, 2 wt.-% of PVA was dissolved in DI water to prepare the water-based binder.

The simulated BJAM cubic green state samples (~ 1 cm^3) were fabricated by utilizing a silicone mold, where a thin layer of the AlSi10Mg powder and liquid binder were progressively added. The liquid binder was applied with a pipette to adequately saturate the powder bed. Then, the silicone mold was placed on a benchtop vibrator to facilitate the wetting, remove the trapped air bubbles, and smooth the surface. This process was repeated until the silicone mold

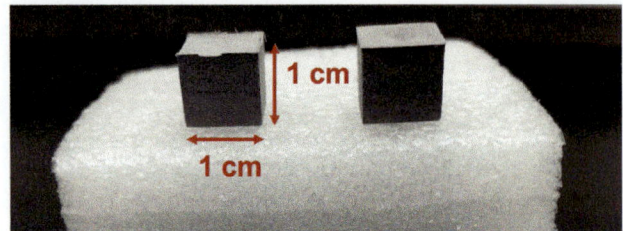

Fig. 1 Simulated BJAM apparent defect-free green state samples studied in this research

was compacted with the AlSi10Mg powder. The filled silicone mold was placed on a hot plate for 12 h at 60 °C to cure the liquid binder and remove the excess solvent. After curing, the green state samples were carefully removed from the silicone mold and visually inspected for surface integrity. Only samples with no significant surface anomalies were chosen for further analysis, and an example of the simulated BJAM green state samples is shown in Fig. 1. The calculated green densities fluctuated between 61 to 64% across all specimens, mainly attributed to the use of different binder systems.

Determination of Debinding Temperatures

Prior to sintering, debinding heat treatments were performed to remove the polymer binders from the green state samples. To determine the debinding temperatures for each binder solution, the thermal gravimetric analysis (TGA) on the polymers was conducted using the TA Instrument Q500. A sample of each polymer powder, except for the PAN, was heated at 10 °C/min to 600 °C to measure the weight loss over an increasing temperature and time. Additionally, the PAN was heated to 800 °C with the same heating rate to record the complete degradation of the polymer. All experiments were performed under flowing air at a flow rate of 60 mL/min.

Debinding and Sintering Conditions

The tube furnace, KSL-1200X-H-UL, from the MTI Corporation was utilized for debinding and sintering heat treatments for all samples. The debinding cycles were conducted under a low vacuum to facilitate the complete pyrolysis of polymer binders. The vacuum was dropped

Table 1 Nominal chemical composition of the AlSi10Mg powder in wt.-% [7]

Si	Mg	Fe	Mn	Ti	Zn	Cu	Ni	Pb	Sn	Other	Al
9–11	0.2–0.45	0.55	0.45	0.15	0.1	0.05	0.05	0.05	0.05	0.15	Balance

below 0.1 Pa absolute pressure on the gauge before initializing the heat treatment. After the debinding heat treatment, a flowing nitrogen gas with 99.9% purity was introduced to the system to promote the sintering of aluminum samples.

Elemental Analysis

Chemical analysis of the debound samples was performed and repeated four times to quantify the impurities such as carbon (CS 844) and oxygen (ONH 836) using the elemental analyzers from the LECO Corporation. The virgin AlSi10Mg powder was heat-treated at 500 °C for 60 min to measure the impurities accumulated from the virgin powder during the heat treatment. This value was later subtracted from the chemical compositions of the debound samples to normalize the oxygen and carbon residues deposited throughout the debinding cycles.

Image Analysis and Porosity Quantification

The green and sintered samples were mounted, ground, polished, and visually investigated under the Keyence VHX7000 digital microscope for porosity quantification. The micrographs of as-polished samples were analyzed using an image analysis software, ImageJ. The image analysis procedure outlined in [8] was modified to determine the total area of porosity [μm^2] to identify the relative sintered density [%]. For each sample, at least eight optical images were collected and analyzed to obtain a better representation of the sample status.

Microstructure Characterization

The sintered samples were etched using the Keller's reagent (5 mL HNO_3, 3 mL HCl, and 2 mL HF) for 10 s prior to the characterization to reveal the morphologies and microstructure. Then, the prepared samples were observed under the JEOL JSM-7600 field emission gun scanning electron microscopy (FEG-SEM) equipped with an Oxford Instruments X-Max 80 mm EDX detector. The secondary electron (SEI) SEM micrographs in varying magnifications were obtained to analyze the morphologies and microstructure of the sample on the grain boundaries. Furthermore, the elemental analysis was performed by using the EDX at an accelerating voltage, beam current, and aperture of 7 kV, 1.9 nA, and 110 μm, respectively to identify the elemental compositions on the grain boundaries.

Results and Discussion

Determination of Debinding Temperatures

The weight loss was differentiated with respect to the time and plotted against to obtain the TGA profiles shown in Fig. 2. The TGA profiles indicate two stages of weight loss on all test specimens. The weight loss of polymers is mainly due to thermolytic polymer degradation, volatiles, and consequent gas release [7]. The pyrolysis of the PVA and PAA began at a relatively lower temperature compared to other polymers, and a significant drop in weight occurred between 240 °C and 400 °C. In contrast, the PVDF and PAN degraded at a higher temperature, having a major weight loss between 400 °C and 500 °C, and 350 °C and 650 °C, respectively.

Fig. 2 TGA profiles of polymers used to identify the debinding conditions of PVP, PVA, PAA, PAN, and PVDF

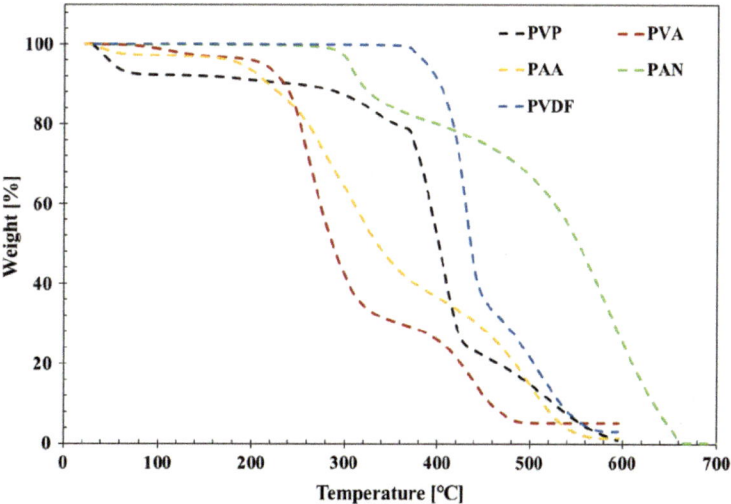

Table 2 Debinding conditions of PVP, PVA, PAA, PAN, and PVDF

Polymers	Debinding temperature 1 [°C]	Debinding temperature 2 [°C]	Dwell time 1, 2 [mins]
PVP	375	485	60
PVA	245	405	60
PAA	265	460	60
PAN	295	500	60
PVDF	410	480	60

Table 3 Normalized binder residues (carbon and oxygen contents) with respect to the heat-treated virgin AlSi10Mg powder

Polymers	Debinding temperature 1 [°C]	Debinding temperature 2 [°C]	Carbon content [%]	Oxygen content [%]
PVP	375	485	0.008	0.001
PVA	245	405	0.096	1.886
PAA	265	460	0.000	0.000
PAN	295	500	0.146	0.066
PVDF	410	480	0.061	0.006

The optimum debinding temperatures were estimated according to EN ISO 11358 [9] by finding the point of intersection of the starting mass and the tangent to the TGA curve at the point of maximum gradients. A step-like pattern shown on the TGA curves in Fig. 2 indicates a multistage decomposition. Hence, to ensure complete pyrolysis, all sections that presented slope change were considered and extrapolated to obtain debinding temperatures for each material. The similar degradation temperatures of each polymer were previously reported by [10–14]. However, there may be a small variation due to the different molecular weights of the polymer and experiment conditions. The determined debinding temperatures for each polymer are summarized in Table 2. For the debinding heat treatments, an isotherm of 60 min was added at each debinding temperature to ensure complete pyrolysis.

Effect of Binder on Oxidation and Carburization

A clean-burning property is a critical characteristic of a successful binder system to minimize the overall impurity level of the final samples. The test specimens in the current study were also evaluated to identify the carbon and oxygen contents to determine the most compatible binder solution for aluminum powders. Prior to conducting the LECO analyses, the green state samples were debound according to the debinding schedules presented in Table 2. The debinding heat treatments were performed under a vacuum with a constant heating rate of 5 °C/min. The impurities of the heat-treated virgin AlSi10Mg powder were subtracted from the debound samples to normalize the oxidation and carburization values as previously explained in section "Determination of Debinding Temperatures".

Table 3 presents the carbon and oxygen contents and the corresponding debinding temperatures of each polymer. The alcohol solvent-based binders (PVP and PAA) provided the lowest impurities, where the PAA binder demonstrated a clean-burning property and left no residues on the fabricated sample. On the other hand, an increase in the oxygen levels was observed with PVDF and PAN binders despite the absence of oxygen in their polymer chains. A substantial level of impurities was noted from the PAN and PVA binders due to carburization and oxidation that occurred during the debinding heat treatment.

Generally, a lower contamination level is often expected with decreasing debinding temperature [15]. However, an interesting result was obtained, where the PVA binder with the lowest debinding temperatures provided the highest impurities. This outcome may suggest that the base materials, such as polymer and solvent, and the composition of the binder are more important in determining the impurity levels of the debound samples than the debinding conditions. Hence, this recommends that the alcohol solvent-based binders are more compatible with the aluminum powders relative to the organic solvent and water-based binders due to the lowest chemical interaction between the alcohol solvent-based binders and aluminum powder.

The photographs of debound samples are shown in Fig. 3. These samples were visually investigated to examine any changes in the physical properties. All samples, except for the PAN, did not present considerable changes compared to the appearances of the green state samples. However, the sample made with the PAN binder displayed a significant discoloration, where the surface of the sample changed from light grey to black upon the debinding heat treatment. This sudden color change was previously reported by [16], who indicated that heat treatment between 200 °C and 300 °C in

Fig. 3 Debound AlSi10Mg samples fabricated with PVP, PVA, PAA, PAN, and PVDF binders

oxidative conditions promotes the formation of ladder structure. This suggests the use of the PAN as a binder in any constituents of the carburizing atmosphere, such as carbon monoxide (CO), carbon dioxide (CO_2), air, or any source of oxygen and carbon [17], needs to be avoided. However, a trace amount of air may be present during the debinding cycles even if the heat treatments were performed under vacuum or reducing atmospheres. This highly limits the application of the PAN as a binder due to the carburizing property near the degradation temperature of the polymer. Hence, in the current work, the PAN as a binder was identified to be less compatible with the AlSi10Mg powder.

Effect of Binder on Part Densification

Sintering heat treatment was also conducted to evaluate the influence of different polymer binders on the densification of fabricated aluminum samples. The debinding cycles were completed according to the conditions provided in Table 2, and the heat treatment was progressed to the sintering at 550 °C for 240 min. The sintering temperature was identified to be approximately 10 °C below the solidus of the AlSi10Mg, which was determined through performing the differential scanning calorimeters (DSC) in pure nitrogen [18].

Figure 4 shows the optical micrographs of the sintered samples and Table 4 presents the corresponding green and

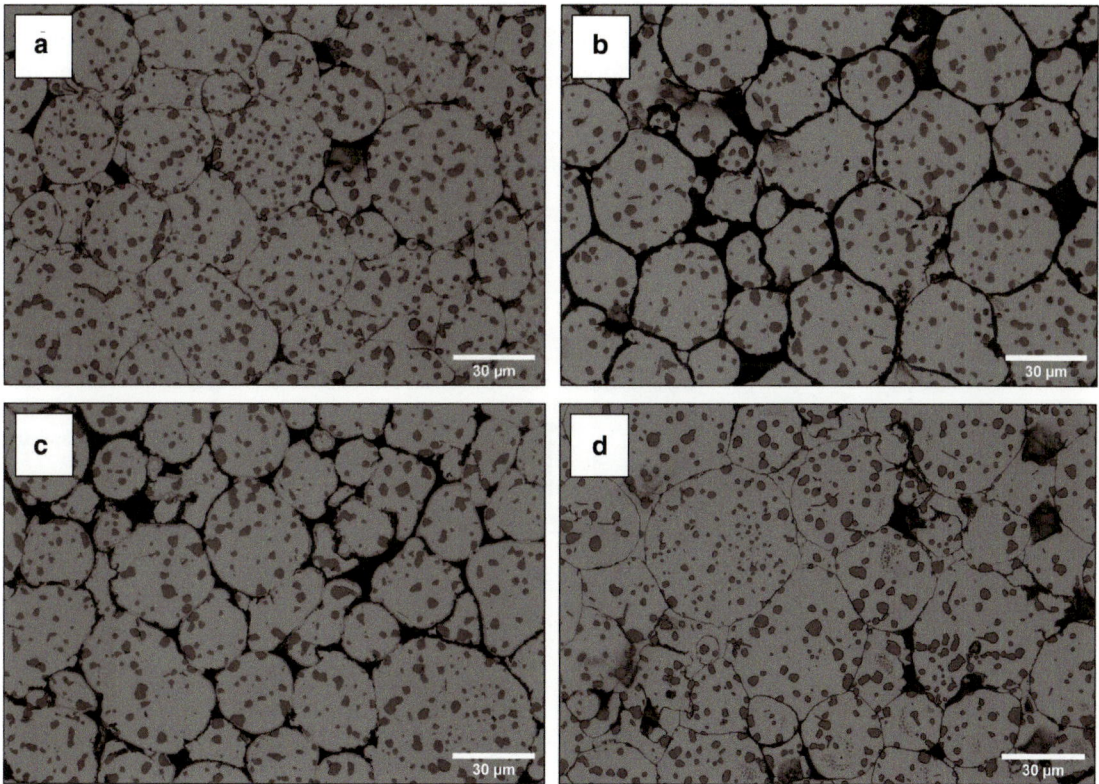

Fig. 4 Optical micrographs of debound and sintered AlSi10Mg samples fabricated using: **a** PVP, **b** PVA, **c** PAA, **d** PAN

Table 4 Green and sintered relative densities of the AlSi10Mg samples fabricated using PVP, PVA, PAA, and PAN binders

Polymers	Green density [%]	Sintered density [%]
PVP	63.5	95.1
PVA	63.1	87.5
PAA	61.3	87.7
PAN	63.9	90.6

sintered densities. Due to the poor structural integrity of the sample prepared with the PVDF binder, this sample set was excluded from the porosity analysis. A high level of impurities was previously reported to hinder the consolidation by inhibiting the sintering between the neighboring powder particles [19]. Similar behavior was found in this study with samples fabricated using the PAN and PVA binders, where low sintered densities of 90.6% and 87.5% were obtained as a result of high contamination levels. This led to believe that polymers with lower impurities, such as PVP and PAA binders, to provide higher sintered densities. However, a relatively high density of 95.1% was only attained with the PVP binder when the PAA binder was not successful in promoting the sintering and delivered the sintered density of 87.7%. Yanez-Sanchez et al. [20] stated that this phenomenon may be due to the different wetting behavior of the PVP and PAA binders. The authors observed the development of effective binding with the PVP binder whereas individually coated powder particle was found using the PAA binder attributed to the high affinity of the PAA to form a covalent bond with the metal powder surface [20]. This may create a larger gap between the powder particles, prohibiting the formation of a sintering bond. Moreover, a lower green density may also have affected the final density as presented by [2]. Therefore, the PVP can be suggested as the most compatible polymer binder for building aluminum parts that are structurally stable relative to other polymeric binders.

Effect of Polymeric Binder on Part Microstructure

The effect of polymeric binder on the final microstructure of the AlSi10Mg part was determined by analyzing the passive layers formed along the surfaces of the aluminum particles. The specimen fabricated by using the PAA binder was selected for this study due to the minimal interactions that it presented with the AlSi10Mg powder (see Table 3). The SEM micrograph in Fig. 5a clearly shows the surface layers, preventing further development of the sintering bond. Moreover, the elemental spectrum analysis discovered that the main constituent of the layers is composed of aluminum oxide as shown in Fig. 5b. This is typical of aluminum powder as it reacts with an oxidizing agent during the powder atomization process and forms the aluminum oxide [21]. Other contaminants, such as carbide, were not observed on the microstructure of the specimen. Hence, this indicates that the carbon from the PAA binder did not deposit any residues during the debinding and sintering cycles. The result is consistent with the LECO analyses discussed in Table 3.

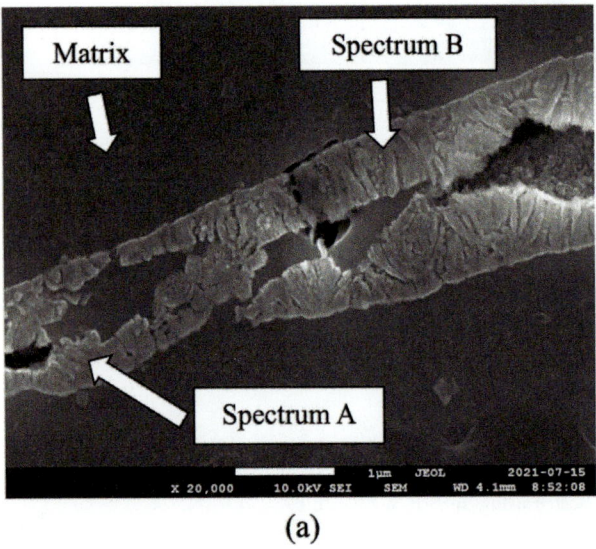

(a)

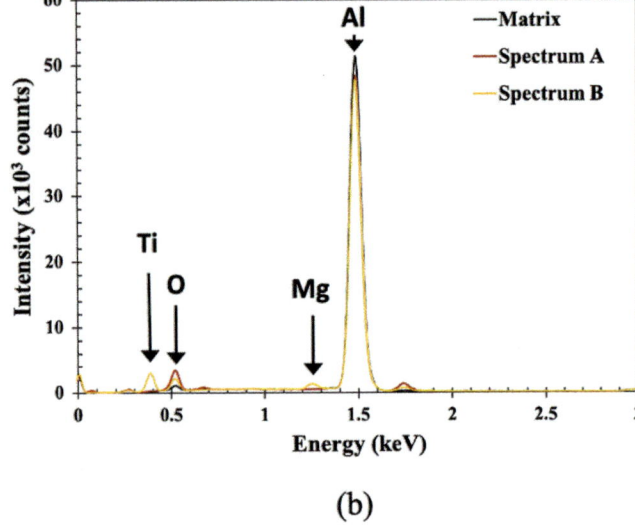

(b)

Fig. 5 High-resolution SEM micrographs of sintered specimen fabricated using PAA binder showing **a** passive layer and **b** spectrums of elemental composition at passive layer

Conclusions

In the study, the AlSi10Mg powder was used to manufacture simulated BJAM green state samples with various polymer binders. The fabricated samples were tested to determine impurity levels and final density to identify the most suitable binder for aluminum powder. The alcohol solvent-based binders (PAA and PVP binders) presented the lowest impurities levels and demonstrated clean-burning properties. On the other hand, moderate to high carbon and oxygen contents were obtained from organic solvent and water-based binders that were composed of PVDF, PAN, and PVA. The highest final density of 95.1% was achieved with the PVP binder whereas the PVA binder, the water-based binder, did not actively facilitate the sintering. In the LECO and porosity analyses, the alcohol solvent-based binders generally presented the most favorable result. However, the low sintered density provided by the PAA binder indicates that this may not be a compatible binder. Therefore, this research suggests the PVP as the most compatible polymer for processing aluminum powder. However, a further study to cross-link polymers, such as PAA and PVP, may be considered as future research to improve the final density.

Acknowledgements The authors are thankful to Natural Sciences and Engineering Research Council of Canada (NSERC) under grant no. CRDPJ 533406-18, (CM)2 Labs at Polytechnique Montreal and GE Additive, for supporting this work.

Compliance with Ethical Statement On behalf of all authors, the corresponding author states that there is no conflict of interest.

References

1. E. Martin, A. Natarajan, S. Kottilingam, and R. Batmaz, "Binder jetting of "Hard-to-Weld" high gamma prime nickel-based super-alloy RENÉ 108," *Additive Manufacturing*, vol. 39, no. 101894, Feb. 2021

2. R. Batmaz et al., "An investigation into sinterability improvements of 316L binder jet printed parts," Metallurgical and Materials Transactions A, Accepted.

3. M. Li, W. Du, A. Elwany, Z. Pei, and C. Ma, "Metal binder jetting additive manufacturing: A literature review," *Journal of Manufacturing Science and Engineering*, vol. 142, no. 090801, Jun. 2020.

4. P. K. Gokuldoss, S. Kolla, and J. Eckert, "Additive manufacturing processes: Selective laser melting, electron beam melting and binder jetting – Selection guidelines," *Materials*, vol. 10, no. 6, pp. 672-684, Jun. 2017.

5. D. Snelling *et al.*, "Lightweight metal cellular structures fabricated via 3D printing of sand cast molds," *Advanced Engineering Materials*, vol. 17, no. 7, pp. 923–932, July. 2015.

6. T. B. Sercombe, "On the sintering of uncompacted, pre-alloyed Al powder alloys," *Materials Science and Engineering: A*, vol. 341, pp. 163-168, Jan. 2003

7. *Material data sheet: Al-Alloy AiSi10Mg*, SLM Solutions, Accessed on: Dec. 07, 2021. [Online]. Available: https://www.slm-solutions.com/fileadmin/Content/Powder/MDS/MDS_Al-Alloy_AlSi10Mg_0221.pdf

8. A. Chakraborty *et al.*, "In-process failure analysis of thin-wall structures made by laser powder bed fusion additive manufacturing," *Journal of Materials Science & Technology*, vol. 98, pp. 233-243, Jan. 2022.

9. *Plastics-Thermogravimetry (TG) of polymers*, ISO 11358, July. 2014. [Online]. Available: https://www.iso.org/standard/59710.html

10. T. H. C. Salles, C. B. Lombello, and M. A. d'Ávila, "Electro-spinning of gelatin/poly (vinyl pyrrolidone) blends from water/acetic acid solutions," *Materials Research*, vol. 18, no. 3, pp. 509-518, Jun. 2015.

11. K. S. Hajeeassa, M. A. Hussein, Y. Anwar, N. Y. Tashkandi, and Z. M. Al-amshany, "Nanocomposites containing polyvinyl alcohol and reinforced carbon-based nanofiller: A super effective biologically active material," *Nanobiomedicine*, vol. 5, no. 5, pp. 1-12, Aug. 2018.

12. V. Datsyuk *et al.*, "In situ nitroxide-mediated polymerized poly (acrylic acid) as a stabilizer/compatibilizer carbon nanotube/polymer composites," *Journal of Nanomaterials*, vol. 2007, pp. 1–12, July. 2007.

13. T. Hsieh, Y. Wang, P. Tseng, K. Ho, and Y. Tsai, "Electrospun highly transparent, conducting Ag@polyacrylonitrile nanofibers prepared by post thermal treatment in the presence of tetraanilines," *Polymer Degradation and Stability*, vol. 144, pp. 146-154, Oct. 2017.

14. A. Kuila, N. Maity, D. P. Chatterjee, and A. K. Nandi, "Temperature triggered antifouling properties of poly(vinylidene fluoride) graft copolymers with tunable hydrophilicity, *Journal of Materials Chemistry A*, vol. 3, no. 25, pp. 13546–13555, May. 2015.

15. T. Ebel, "Metal injection molding (MIM) of titanium and titanium alloys," in *Handbook of Metal Injection Molding*, 2nd ed. D. F. Heaney, Ed. Duxford, United Kingdom; Elsevier, 2019. pp. 435–436.

16. M. S. A. Rahaman, A. F. Ismail, and A. Mustafa, "A review of heat treatment on polyacrylonitrile fiber," *Polymer Degradation and Stability*, vol. 92, pp. 1421-1432, Aug. 2007.

17. R. Singh, "Heat treatment of steels," in *Applied Welding Engineering*, 2nd ed. United Kingdom: Butterworth-Heinemann, 2016. Pp. 111-124.

18. M. Voncina, P. Mrvar, and J. Medved, "Thermodynamic analysis of AlSi10Mg alloy," *Materials and Geoenvironment*, vol. 52, no. 3, pp. 621-633, Oct. 2005

19. D. Li *et al.*, "Effect of oxygen contents on predominant sintering mechanism during initial stage of pure titanium powder," *Powder Technology*, vol. 361, pp. 617-623, Nov. 2019.

20. S. I. Yanez-Sanchez, M. D. Lennox, D. Therriault, B. D. Favis, and J. R. Tavares, "Model approach for binder selection in binder jetting," *Industrial & Engineering Chemistry Research*, vol. 60, no. 42, pp. 15162-15173, Oct. 2021.

21. J. F. Flumerfelt, *"Aluminum powder metallurgy processing,"* Ph. D. thesis, Metallurgy, Iowa State University, Iowa, 1998.

Material Evaluation Framework of Additive Manufactured Aluminum Alloys for Space Optical Instruments

Zachary J. Post, Walter R. Zimbeck, Steven M. Storck,
Floris van Kempen, Gerard C. J. Otter, John D. Boldt,
Ludger van der Laan, Steven R. Szczesniak, Ryan H. Carter,
Robert K. Mueller, Salahudin M. Nimer, Douglas B. Trigg,
Michael A. Berkson, M. Frank Morgan, and William H. Swartz

Abstract

A framework for additive manufacturing aluminum alloy selection was developed to determine the preferred composition and process parameters from which to fabricate topology-optimized optical instrument housings and light-weighted freeform mirrors for the Compact Hyperspectral Air Pollution Sensor (CHAPS). In recent years, a number of high-strength laser powder bed fusion aluminum alloys have become commercially available, which are attractive for aerospace applications due to their high specific strength. Three aluminum alloys were selected for a three-Round experimental comparison. Each Round used a down-selected subset of alloys and parameter sets (candidates) from the previous Round. Round 1 screened a wide range of laser parameter sets for those that produced the highest density and tensile yield strength. Round 2 evaluated build quality using test geometries representative of CHAPS and assessed compatibility with post-processing, including optically black coating for the optical housings and nickelphosphorus plating for the mirrors. Round 3 characterized anisotropy in tensile and thermal properties. A rating system was developed which involved assigning priority weighting for CHAPS-specific criteria and binning test results into scoring categories to give a comparison score for each candidate which was used in the down-selection between Rounds. The framework selection process enabled a comparison of the relative strengths and weaknesses of each candidate and resulted in the selection of Scalmalloy as the preferred alloy for CHAPS. The selected candidate was used to develop design allowables for the topology optimization of CHAPS prototype housings, which were then fabricated.

Keywords

Additive manufacturing • Aluminum • Mechanical properties

Introduction and Background

The Compact Hyperspectral Air Pollution Sensor (CHAPS) is a small imaging spectrometer intended for targeted, science-quality measurements of air pollution at unprecedented spatial resolution (better than 1×1 km^2) from low earth orbit [1]. The compact size and cost-effectiveness of CHAPS make a small satellite instrument or hosted payload feasible. One possible future scenario is a constellation of CHAPS instruments each tuned for particular species of interest.

To achieve size reduction of the instrument, CHAPS leverages freeform optics and additive manufacturing (AM) of topology optimized (TO) designs. Freeform optics is an emerging technology with potentially huge advantages over traditional optics, including fewer optical surfaces, lower mass and volume, and improved image quality [2, 3]. AM, and specifically metal laser powder bed fusion, enables the production of complex parts and enables significant design freedom including improved optimization outcomes by eliminating toolpath clearance (e.g., milling) constraints. AM also enables new lightweight mirror designs via complex internal lattice structures which cannot be fabricated by conventional methods.

To achieve the thermal stability required of an optical instrument with minimal mass, a TO design was created for the optical components housing, including integrated light baffles and optical mounting fixtures. For CHAPS and other

Z. J. Post · W. R. Zimbeck · S. M. Storck · J. D. Boldt ·
S. R. Szczesniak · R. H. Carter · R. K. Mueller · S. M. Nimer ·
D. B. Trigg · M. A. Berkson · M. F. Morgan · W. H. Swartz (✉)
Johns Hopkins University Applied Physics Laboratory, 11100
Johns Hopkins Road, Laurel, MD 20723, USA
e-mail: Bill.Swartz@jhuapl.edu

F. van Kempen · G. C. J. Otter · L. van der Laan
The Netherlands Organization for Applied Scientific Research
(TNO), Stieltjesweg 1, Delft, 2628, CK, The Netherlands

space optical instruments, optimization priorities are dimensional stability through launch (vibrational stresses) and on-orbit (thermomechanical deformation). CHAPS topology optimization sought to limit displacement and rotation of optics during operational orbits to acceptable tolerances, limit stress during launch, and minimize mass. In addition, the mirrors require compatibility with nickelphosphorus (NiP) plating to produce optical-quality mirror surfaces.

Other optimization criteria were informed by TNO's (The Netherlands Organization for Applied Scientific Research) experience developing a TO AM telescope (AMDT—Additive Manufacturing Demonstrator Telescope) for a version of the Ozone Measurement Instrument (OMI). This project highlighted the importance of minimizing the difference in coefficient of thermal expansion (CTE) between the AM component and cubesat structure (6061-T6 Al alloy) and of AM materials being compatible with down-stream processes, such as optically black surface treatment to minimize stray light effects, both of which are applicable to CHAPS [4]. AMDT was designed and fabricated using AlSi10Mg, a widely used AM aluminum alloy. However, the design optimization was constrained by the relatively low yield strength (<250 MPa) of this alloy [5]. Given that the CHAPS design is larger and more complex than AMDT, and will benefit from a high yield strength additive aluminum alloy to minimize mass while maintaining stress margins, a new alloy composition was desirable. Because AM processing and variations therein can yield material properties that differ from those of conventionally manufactured materials, it is not straightforward to select a composition or processing approach. Therefore, there was a need to develop an evaluation framework to enable the identification of an optimal AM aluminum alloy for this space optical application. In this work, an AM material selection framework was developed according to the above considerations. The framework was implemented to select an optimal AM alloy and associated laser processing parameters for the fabrication of a CHAPS prototype and to use that selection to generate design allowables for TO of the CHAPS housing.

Materials and Methods

Candidate Alloy Selection

Candidate AM aluminum alloys were identified by a review of commercially available alloys and a comparison of their properties from technical data sheets, emphasizing high tensile yield strength. Three alloys were selected for experimental evaluation: 7A77 purchased from HRL Laboratories (Malibu, CA); A20X purchased from Aluminum Materials

Technologies (Worcestershire, UK); and Scalmalloy purchased from LPW Technology (Runcorn, UK). A summary of properties for each alloy with reference values for AlSi10Mg is given in Table 1.

AM Fabrication

Test samples were fabricated using a Renishaw AM400, a pulsed, single laser system with an ytterbium fiber laser with 1070 nm wavelength and a maximum power of 400 W. For orientation, "horizontal" means parallel to the plane of the build plate and "vertical" means perpendicular to the plane of the build plate.

Post-Processing

For each build, post-processing consisted of powder removal, heat treatment of samples on the build plate, and electrical discharge machining (EDM) of samples from the plate. Heat treatments for each alloy were:

- 7A77: 2 h at 470 °C, water quench. 14 h at 121 °C, furnace cool [9].
- A20X: AMT Proprietary Heat Treat, solution treatment, water quench, and precipitation hardening to T7 condition [6].
- Scalmalloy: 4 h at 325 °C, water quench (water quench deviated from the datasheet) [10].

Evaluation Criteria and Framework Design

In developing the framework, requirements for the CHAPS application were defined, including high tensile yield strength, CTE matched to 6061-T6 aluminum, high thermal conductivity, low build orientation anisotropy, good build quality of CHAPS geometries (e.g., high aspect ratio), and compatibility with post-processing. Specific properties were identified based on these requirements. Evaluating the identified properties for hundreds of combinations of alloy and laser processing parameter sets (candidates) would have been impractical, so the process was split into three Rounds that progressively narrowed down the processing space. The framework consisted of the following steps:

- Round 1—Parameter Screening—Implement a laser parameter design of experiments (DOE) to identify candidates that produce low porosity and high tensile yield strength.

Table 1 Summary of the properties of the three candidate aluminum alloys compared to the widely used AlSi10Mg alloy [5–8]

Alloy	Vendor	Yield strength (MPa)	Elongation (%)	Chemistry (wt%)	Al series
AlSi10Mg	Renishaw	230	11.5	Si: 9.00–11.00 Mg: 0.20–0.45	–
A20X	Aluminum materials technologies	440	13	Cu: 4.20–5.00 Ti: 3.00–3.80 B: 1.20–1.50 Ag: 0.60–0.90	2000
Scalmalloy	LPW technologies	490	14	Mg: 4.00–4.90 Sc: 0.60–0.80 Mn: 0.30–0.80 Zr: 0.20–0.50	5000
7A77	HRL laboratories	537	10	Zn: 4.50–6.10 Mg: 1.80–2.90 Cu: 1.10–2.10 Zr: 0.50–2.80	7000

- Round 2—Build Quality and Post-Processing Compatibility Assessment—Use candidates chosen in Round 1 to fabricate CHAPS-representative geometries to evaluate the ability to form fine features (holes), unsupported overhangs, high aspect ratio walls, etc., accurately.
- Round 3—Anisotropy Characterization—Use candidates chosen in Round 2 to assess the anisotropy of tensile and thermal properties relative to build orientation (horizontal and vertical).
- Material Property Characterization for TO—Use the final selected candidate to fabricate tensile specimens to calculate design allowables for TO of the housing design.

Requirements Definition

To enable down-selection following each round of the process, material and manufacturing properties of interest to CHAPS were identified. Each property was assigned a weight from one to ten, with one indicating low importance and ten indicating high importance. To convert measured property values into normalized scores, the range of each property was divided into a set of bins, one to ten (best to worst). For each candidate, the bin score for each property was multiplied by the weight to give a weighted score. The weighted scores across all the properties were summed to produce a total score, with the lowest being the best. Both the weight and choice of possible bins acted as measures of importance for each property, with higher values for either leading to a potentially higher (worse) score. Table 2 shows the properties of interest and associated weights and bins.

For example, CTE is a critical material property and was assigned a weight of ten. The target CTE value was within ± 5% of the CTE value for 6061-T6 aluminum (23.6 µm/m/K) [11]. For a given candidate, if the measured

CTE was within the range, the bin score was one and the weighted score was ten. If the CTE was outside the range, the bin score was three and the weighted score was 30.

This scoring system was not utilized in Round 1 and was implemented in Round 2.

Round 1—Parameter Screening

Round 1 AM Sample Fabrication

A DOE was developed to identify candidates with low porosity and high tensile yield strength. The laser parameter variables were power, hatch spacing, pulse exposure time, and pulse point distance. The layer thickness was fixed at 30 µm. A volumetric energy density (VED) metric was calculated using Eq. 1, where point distance divided by exposure time for a pulsed laser is used as an analogue to the speed of a continuous laser [12].

$$\frac{\text{Power (W)} * \text{Exposure Time (s)}}{\text{Hatch Spacing (mm)} * \text{Point Distance (mm)} * \text{Layer Thickness (mm)}} = VED\left(\frac{\text{J}}{\text{mm}^3}\right)$$

(1)

Because the candidate alloys did not have laser parameter sets for the AM400 provided by the vendors, the DOE was built around the standard AlSi10Mg laser parameter set for the AM400. Parameters were chosen to create 75 parameter sets across the parameter space shown in Table 3, which maximized the number of samples in a single build. Ranges for parameter variables considered machine limitations (maximum power 400 W), build rate (low hatch spacing and point distance with high exposure time yield impractically slow build rates), and the likelihood of melt pool overlap (on the order of 100 µm for aluminum alloys) to avoid lack of fusion porosity [13]. All other parameters were standard

Table 2 Properties of interest for selecting an aluminum alloy for CHAPS and respective weights and bins for each property, indicating relative levels of importance for each property, with higher weights and the existence of higher-value bins indicating a more important property

Round	Property	Weight	Bins
1	Porosity (%)	2	**1:** <= 0.1| **3:** > 0.1–0.2| **4:** > 0.2–0.3| **5:** > 0.3–0.4| **6:** > 0.4–0.5| **7:** > 0.5–0.6| **8:** > 0.6–0.7 | **9:** > 0.7–0.8| **10:** > 0.8
	Surface Roughness (μm)	2	**1:** < 20| **2:** 20–25| **3:** > 25–30| **4:** > 30
1,3*	Tensile Yield Strength (MPa)	7	**1:** > 450| **2:** > 400–450| **3:** > 350–400| **4:** > 300–350| **5:** <= 300
	Ultimate Tensile Strength (MPa)	3	**1:** > 550| **2:** > 500–550| **3:** > 450–500| **4:** > 400–450| **5:** > 350–400| **6:** <= 350
	Elongation (%)	3	**1:** > 10| **2:** > 5–10| **3:** <= 5
	Elastic Modulus (GPa)	2	**1:** > 70| **2:** > 60–70| **3:** <= 60
2	30° Facet Overhang Thickness (mm)	3	**1:** 1.0–1.1| **2:** > 1.1–1.5| **3:** > 1.5–2.0| **4:** > 2.0
	Minimum Diameter Horizontal Hole (mm)	3	**1:** 0.5| **2:** 1
	Pre-Heat Treat Stress Comb Curlup (mm)	3	**1:** < 1.78| **2:** 1.78- < 1.91| **3:** 1.91- < 2.03| **4:** 2.03- < 2.16| **5:** 2.16- < 2.29| **6:** >= 2.29
	NiP Plating	7	**1:** Successfully plated| **3:** Cannot be plated
	Black Aeroglaze Z306 Painting	4	**1:** Paint survives scratch test| **2:** Paint removed by scratch test| **3:** Cannot be painted
	Powder Cost ($/kg)	2	**1:** < cutoff| **2:** cutoff
	Powder Availability	1	**1:** no issues in acquiring powder| **3:** Significant delay in acquiring powder
	Build Rate (μm²/μs)	3	**1:** >= 1.60| **2:** > 1.28–1.60| **3:** > 0.80–1.28| **5:** <= 0.80
3	ASTM E8 Tension-Modulus Inomogeneity (GPa)	1	**1:** < 10| **3:** >= 10
	ASTM E8 Tension-Modulus Anisotropy (GPa)	1	**1:** < 10| **3:** >= 10
	Thermal Conductivity (W/m/K)	4	**1:** 125–175| **3:** >= 100–125| **5:** < 100
	CTE (μm/m/K)	10	**1:** 22.42–24.78 um/m/K| **3:** < 22.42 or > 24.78
	CTE-Variability (μm/m/K)	5	**1:** <= 1| **3:** > 1
	CTE-Anisotropy (μm/m/K)	5	**1:** <= 1| **3:** > 1
	Post Heat Treat Baffle Warping	3	**1:** < 0.1 mm| **2:** 0.1–1 mm| **3:** 1–10 mm

Table 3 DOE variable ranges used to explore the processing parameter space in Round 1 to identify candidates with low porosity and high tensile yield strength. Ranges are compared to the AM400 standard AlSi10Mg parameter set as reference, as AlSi10Mg is a commonly used AM aluminum alloy

	Round 1 DOE	AlSi10Mg
Power (W)	87–396	275
Hatch Spacing (μm)	40–110	80
Exposure Time (μs)	30–172	40
Point Distance (μm)	40–100	80
VED (J/mm³)	40–351	57

AM400 AlSi10Mg settings. For each candidate, two sample types were fabricated to enable the characterization of both porosity and tensile properties.

Round 1 Sample Characterization

X-ray computed tomography (XCT) analysis was conducted on Ø6 mm × 6 mm cylinders. A North Star Imaging, Inc. X-50 was used for scanning, and efX-CT and VGSTUDIO MAX v3.4 were used for post-processing to calculate the porosity and surface roughness (R_a) on the sidewall of the sample. Core porosity (>250 μm from the mean surface) and contour porosity (<250 μm from the mean surface) were measured, with pore sizes as small as 48 μm identifiable by the XCT analysis. The maximum value between the two was used as the porosity value for comparison between candidates.

Tensile testing was performed on specialized tensile specimens with a 2 mm × 2 mm gauge cross-section and 13.6 mm gauge length (2 × 2 gauge sample) developed to facilitate high-throughput characterization for measurement of 0.2% offset tensile yield strength (YS), ultimate tensile strength (UTS), elongation (elong%), and elastic modulus (E). Specimens were speckled with a random dot pattern before testing. During testing on an MTS 30/G universal load frame at a constant strain rate of 10^{-3} strain per second, specimens were imaged with a telecentric lens. Strain was measured using two-dimensional digital image correlation.

The best seven candidates for each alloy were chosen for Round 2 by prioritizing high-yield strength but discounting high porosity and low elongation candidates without utilizing the scoring system.

Round 2—Build Quality and Post-Processing Compatibility Assessment

Round 2 AM Fabrication

The top seven candidates for each alloy were used to fabricate samples to evaluate build quality and assess compatibility with post-processing crucial to CHAPS. Samples consisted of stress combs to evaluate the relative level of residual build stress, holed cubes with holes representing small vent holes in CHAPS, geometric samples consisting of features such as triangles, rounded surfaces, and lattice structures, and high aspect ratio thin walls and faceted overhang samples representing geometries and orientations

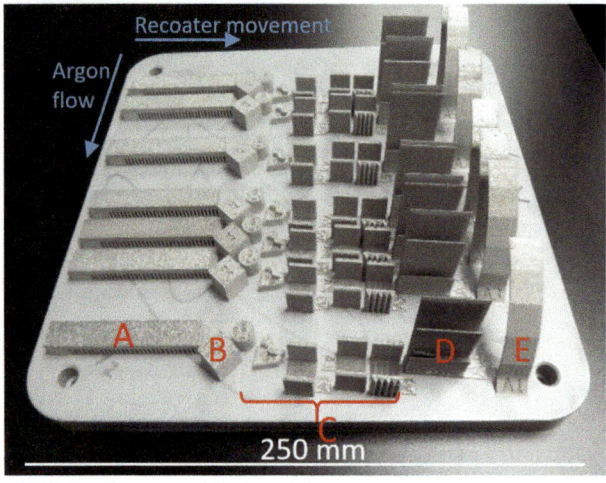

Fig. 1 Build layout of Round 2 samples used to evaluate build quality and compatibility of candidates with post-processing to select candidates for Round 3. Each row of samples was produced with a unique parameter set. Samples from left to right are stress combs (**A**), holed cubes (**B**), geometric samples (**C**, five kinds), high aspect ratio thin walls (**D**), and faceted overhang samples (**E**)

of light-blocking baffles and housing walls for CHAPS. An example build layout is in Fig. 1.

Round 2 Sample Characterization

Stress combs were measured by cutting through the comb teeth with EDM before heat treatment and measuring the height of the highest point of the sample above the build plate relative to the build plate surface with a height gauge. The height is indicative of the level of residual stress in the material, with a larger height indicating higher residual stress which leads to a greater risk of material fracture and geometric distortion during fabrication [14, 15]. The minimum modeled horizontal hole diameter in the holed cube was checked by whether a metal wire with a diameter of 0.2 mm was able to pass through the nominally 0.5 mm and 1.0 mm horizontal holes. If the wire could not pass through a hole, it indicated there was enough material overhang to block the wire, meaning a hole fabricated with that nominal size could not be verified to be a through-hole. Geometric samples were visually inspected to confirm basic geometries could be produced without significant over-melting. High aspect ratio walls (25 mm width, 51 mm height) were used to test the adhesion of black Aeroglaze Z306 paint and the compatibility of the alloy with NiP plating. Paint adhesion was tested by ASTM D3359 Method A [16]. Compatibility with NiP plating was based on the report from the plating vendor (CZL Tilburg B.V., Tilburg, The Netherlands). Faceted overhang samples consisted of five faceted walls at 90°, 75°, 60°, 45°, and 30° relative to the horizontal, with a thickness of 1 mm. Using calipers, the maximum thickness of the 30° facet was measured as an indication of down-facing surface quality of an extreme overhang, where a thinner (closer to 1 mm) indicated better surface quality.

Round 2 Logistics and Manufacturability Assessment

Manufacturability metrics were considered, including powder cost, powder availability, and build rate. Powder cost was recorded from vendor quotes but the values are not public information and are not shared here. Powder availability was evaluated on whether there were delays in acquiring powders. The areal build rate was calculated using Eq. 2 (higher value indicates a faster build rate).

$$\frac{\text{Hatch Spacing } (\mu m) * \text{Point Distance } (\mu m)}{\text{Exposure Time } (\mu s)}$$
$$= \text{Areal Build Rate} \left(\frac{\mu m^2}{\mu s} \right) \qquad (2)$$

Round 2 measurements were evaluated according to the metrics in Table 2 to identify the best candidate for each alloy and to decide which candidates would advance to Round 3.

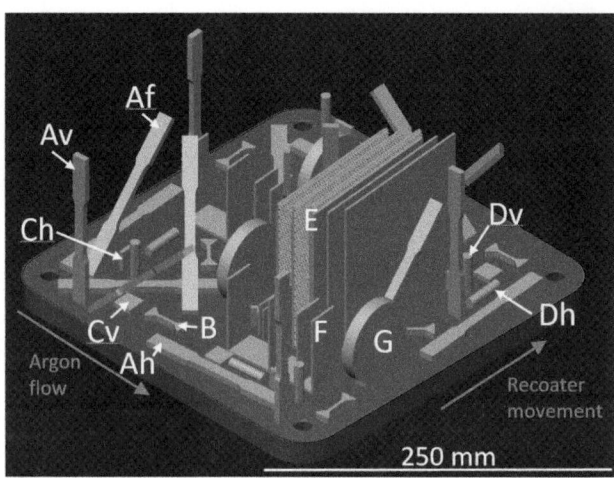

Fig. 2 Isometric view of the Round 3 build layout including ASTM E8 specimens: horizontal (Ah), vertical (Av) and 45° (Af), 2 × 2 gauge specimens (B), Kth samples: horizontal (Ch) and vertical (Cv), CTE samples: horizontal (Dh) and vertical (Dv), high aspect ratio walls (E), U-shaped breadboards (F), and mirror blanks (G). These samples were used to evaluate the anisotropy of tensile and thermal properties in order to select a candidate for Material Property Characterization

Round 3—Anisotropy Characterization

Round 3 AM Fabrication

Round 3 was used to evaluate orientation anisotropy in the key properties of E, thermal conductivity (K_{th}), and CTE. The build layout used for Round 3 is in Fig. 2.

Six subsize plate-type tensile specimens were fabricated according to ASTM E8 in each of three orientations: horizontal, vertical, and 45° from the build plate [17]. Tensile specimens did not undergo any machining after AM fabrication and EDM. Five K_{th} samples (12.5 mm × 12.5 mm area, 2.5 mm thickness) and five CTE cylinders (Ø6.0 mm × 25.4 mm) were fabricated in each of two orientations: horizontal and vertical. High aspect ratio walls (100 mm × 20 mm) were fabricated vertically. 2 × 2 gauge samples, U-shaped breadboards, and mirror blanks were produced for later experimentation.

Round 3 Characterization

Tensile specimens were tested using an MTS 30/G universal load frame with an MTS Model 634 clip gauge extensometer with a 0.5″ gauge length that was removed at 5% strain. Specimens were tested at 0.01 in/min (1.7×10^{-4} strain per second) up to 5% strain, then the rate was increased to 0.06 in/min (10^{-3} strain per second) for the remainder of the test. K_{th} samples were tested by the ASTM E1461 laser flash method [18]. CTE samples were characterized using a Mettler-Toledo DSC-1. The high aspect ratio walls were

evaluated for warping after heat treatment and removal from the build plate by pinning one edge on a flat surface and measuring the height of the opposite edge relative to the height of the pinned edge with a height gauge.

The criteria and weights reported in Table 2 were applied to the Round 3 data to identify the best candidate. The values reported for YS, UTS, elong%, E, K_{th}, and CTE were the values of the mean plus or minus two times the standard deviation, whichever produced a lower bin score. For variability values, which quantified the spread of values of a material property across all samples for a given orientation, the reported value was two times the standard deviation of the orientation with the largest standard deviation. For anisotropy values, which quantified how a material property varied across multiple orientations, the mean of the orientation with the smallest value was subtracted from the mean of the orientation with the highest value.

Material Property Characterization for TO

Material Property Characterization AM Fabrication

The candidate with the lowest score identified in Round 3 was used to fabricate 40 additional ASTM E8 Subsize plate-type tensile specimens in each of the vertical and horizontal directions to generate a larger data set for robust design allowables for TO of the CHAPS housing.

Material Property Characterization Sample Characterization

Samples were tested using the same method and equipment as in Round 2. Design allowables were calculated for YS and E. For YS, a B-Basis value, a design allowable for which "at least 90% of the population of values are expected to equal or exceed the B-Basis mechanical property allowable with a confidence of 95%" and typically requires at least 100 samples for calculation over a minimum of ten lots, was estimated [19]. Typically B-basis values are not calculated for E, so the mean value was calculated.

Results

Round 1—Parameter Screening

The results of the characterization of samples produced with the laser parameter DOE were used to identify a subset of seven candidates for Round 2.

Plots of core and contour porosity versus VED are shown in Fig. 3. Scalmalloy and A20X both had a low-porosity

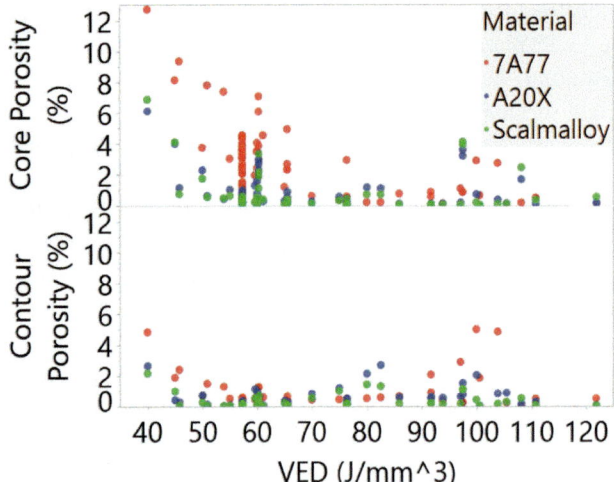

Fig. 3 Core porosity (>250 μm from the sample surface) and contour porosity (<250 μm from the sample surface) versus VED for all candidates, limited to a VED of 25 to 125 J/mm³ for clear visualization. A20X (blue) and Scalmalloy (green) have more similar levels of porosity, and exhibit a low-porosity region near 65 J/mm³ that is likely a region between lack of fusion and keyhole defect formation. 7A77 (red) has a much higher porosity than the other alloys but exhibits a low-porosity region near 80 J/mm³ [12]

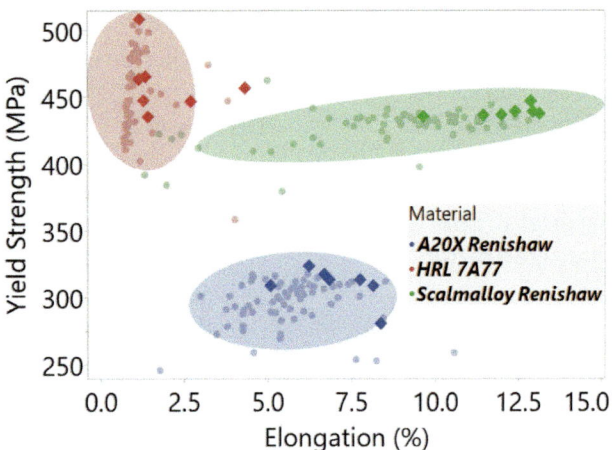

Fig. 4 YS versus elong% for all alloys. Bivariate normal density ellipses indicate the relative region each alloy occupies. Darkly colored parameters were chosen for Round 2 by prioritizing high YS, but rejecting samples with high porosity (near or greater than 1%) and where possible choosing higher elongation values. 7A77 (red) exhibits the highest yield strength, Scalmalloy (green) exhibits the highest elongation and still has relatively high yield strength, and A20X (blue) has relatively low yield strength but superior elongation to 7A77

region near a VED of 65 J/mm³ that is likely a region between lack of fusion and keyhole defect formation, whereas 7A77 exhibited this region closer to 80 J/mm³ [12].

Using a YS versus elong% plot shown in Fig. 4, candidates were selected by prioritizing high YS, but rejecting samples with high porosity (near or greater than 1%) and where possible choosing higher elongation values. The chosen parameters are colored more darkly.

Round 2—Build Quality and Post-Processing Compatibility Assessment

The property values measured in Rounds 1 and 2 were used to identify the lowest score candidates from two alloys (Scalmalloy and A20X) for Round 3 based on the lowest total score. Property values and total scores for the best candidate for each alloy are in Table 4.

Round 3—Anisotropy Characterization

The best (lowest score) candidate from Scalmalloy was selected for Material Property Characterization. Table 5 contains the final scores comparing the best candidate from each of A20X and Scalmalloy after Round 3 data was collected.

Material Property Characterization for TO

The summary of design allowables for the best Scalmalloy candidate is shown in Table 6, separated by build orientation. The estimate of the B-Basis allowable for YS and the mean of E from the vertical orientation were used as more conservative values for the TO of the CHAPS housing.

Discussion

The best candidate for 7A77 had the highest yield strength of the three candidate materials in Round 1 testing (466 MPa), which is expected for a 7000 series aluminum [19]. However, 7A77 suffered from high porosity (all 7A77 parameter sets had porosity > 0.1%), which may be partially attributed to the higher K_{th} of 7000 series aluminum requiring higher energy input to concentrate thermal energy [19]. This is reflected in the low-porosity region for 7A77 being at a higher VED than for the other alloys. This region was sparsely explored in the DOE, so there may exist lower porosity parameter sets near that region. 7A77 had the highest R_a, highest residual stress as measured by the stress comb, and slowest build rate (<50% the speed of the other alloy candidates). The incompatibility of 7A77 with the

Table 4 Measured property values, bins, weights, and total scores for the best candidates for each alloy out of the seven candidates per alloy that were tested through Round 2. Scalmalloy had the lowest total score, as it tends to have the lowest bin scores for the properties with the highest weights. The A20X and Scalmalloy candidates in this table were selected for Round 3

Round	Property	A20X		Scalmalloy		7A77		Weight
		Value	Bin	Value	Bin	Value	Bin	
1	Porosity (%)	0.14	3	0.05	1	0.18	3	2
	R_a (µm)	15.8	1	21.0	2	33.5	4	2
	YS (MPa)	318	4	436	2	466	1	7
	UTS (MPa)	401	4	471	3	478	3	3
	Elongation (%)	6.6	2	11.5	1	1.3	3	3
	E (GPa)	98	1	69	2	74	1	2
2	30° Facet Overhang Thickness (mm)	1.57	3	1.51	3	1.64	3	3
	Minimum Diameter Horizontal Hole (mm)	1.0	2	1.0	2	1.0	2	3
	Pre-Heat Treat Stress Comb Curlup (mm)	1.85	2	1.98	3	2.46	6	3
	NiP Plating	Pass	1	Pass	1	Fail	3	7
	Black Aeroglaze Z306 Painting	Pass	1	Pass	1	Pass	1	4
	Powder Cost ($/kg)	–	2	–	2	–	1	2
	Powder Availability	Delay	3	Ok	1	Ok	1	1
	Build Rate (µm²/µs)	1.60	1	1.35	2	0.40	5	3
Total Score After Round 2		**98**		**82**		**117**		

Table 5 Measured property values, bins, weights, and total scores for the best candidates for each alloy evaluated in Round 3. Scalmalloy had the lowest total score, as it tends to have the lowest bin scores for the properties with the highest weights. This score comparison was used to select Scalmalloy as the alloy for Material Property Characterization and fabrication of CHAPS prototype optical instrument housings and mirrors

Round	Property	Value	Bin	Value	Bin	Weight
1	Porosity (%)	0.14	3	0.05	1	2
	R_a (µm)	15.8	1	21.0	2	2
2	30° Facet Overhang Thickness (mm)	1.57	3	1.51	3	3
	Minimum Diameter Horizontal Hole (mm)	1.0	2	1.0	2	3
	Pre-Heat Treat Stress Comb Curlup (mm)	0.073	2	0.078	3	3
	NiP Plating	Pass	1	Pass	1	7
	Black Aeroglaze Z306 Painting	Pass	1	Pass	1	4
	Powder Cost ($/kg)	–	2	–	2	2
	Powder Availability	Delay	3	Ok	1	1
	Build Rate (µm²/µs)	1.60	1	1.35	2	3
3	ASTM E8 Tension-YS (MPa)	363	3	408	2	7
	ASTM E8 Tension-UTS (MPa)	407	4	433	4	3
	ASTM E8 Tension-Elongation (%)	1.4	3	1.7	3	3
	ASTM E8 Tension-E (GPa)	45	3	57	3	2
	ASTM E8 Tension-Modulus Variability (GPa)	11.0	3	6.0	1	1
	ASTM E8 Tension-Modulus Anisotropy (GPa)	15.00	3	5.00	1	1
	K_{th} (W/m/K)	100.8	3	81.5	5	4
	CTE (µm/m/K)	22.30	3	25.00	3	10
	CTE-Variability (µm/m/K)	0.7	1	0.5	1	5
	CTE-Anisotropy (µm/m/K)	0.45	1	0.38	1	5
	Post Heat Treat Baffle Warping (mm)	7.2	3	0.9	2	3
Total Score After Round 3		**165**		**161**		

Table 6 Design allowables for YS and UTS in each orientation. TO inputs for the CHAPS housing were the vertical YS B-Basis and the vertical E mean as more conservative values

Property	Orientation	Estimated B-Basis
YS (MPa)	Horizontal	440.6
	Vertical	415.1
		Mean ± Standard Deviation
E (GPa)	Horizontal	66.4 ± 3.6
	Vertical	62.9 ± 4.5

standard NiP plating process, required for mirror fabrication, led to its removal from further consideration after Round 2. The high zinc content in 7A77 is a potential cause, as the inhomogeneous distribution of zinc on an aluminum surface can interfere with the plating process [20]. Modification of the plating process to better suit the 7A77 alloy may be possible, but was out of scope for this work.

The best candidates for Scalmalloy and A20X performed similarly. The region of low porosity for both alloys was near a VED of 65 J/mm^3, which is higher than the 57 J/mm^3 VED for AlSi10Mg. Si has high laser absorptivity compared to aluminum at the 1070 nm wavelength and that may account for Scalmalloy and A20X, which are not alloyed with Si, requiring higher VED to obtain sufficient melting [21]. A20X scored better in the categories R_a (25% lower than Scalmalloy), build rate (25% faster), and K_{th} (24% higher), however, these were among the less important parameters. In the majority of other categories, including many higher importance ones, Scalmalloy scored better than A20X. Scalmalloy has significantly lower porosity (0.05% vs. 0.14%), and its tensile properties are typically 10% higher than A20X. Scalmalloy has consistently lower variability and anisotropy in tensile and thermal properties compared to A20X, which is critical for reducing uneven distortion of the housing to prevent optical distortion during instrument operation. Scalmalloy also showed much lower distortion after heat treatment in thin, high aspect ratio walls, which are designed to mimic integrated baffles in the optical housing. This is an important criterion because the integrated baffles would be difficult to access by machining, and distorted baffles could block the optical pathway.

Ultimately, Scalmalloy was selected due to its candidate having the lowest score, which clearly reflected that Scalmalloy was the most well-matched material for this application. For the properties with the highest weights (CTE, NiP plating, YS, CTE variability, and CTE anisotropy), Scalmalloy was equal to or better than A20X, indicating that it met the most critical requirements of the application. Based on the score for A20X, it would be a reasonable backup selection for the CHAPS application and this is supported by looking at individual critical properties such as compatibility with NiP plating and black painting, as well as similar overhang thickness and minimum diameter of horizontal holes.

Conclusion

A framework for additive manufacturing aluminum alloy selection was developed based on CHAPS-specific criteria including high tensile yield strength and compatibility with post-processing. These requirements were translated into measurable properties to facilitate the comparison of different alloys and laser processing parameter sets. Alloy candidates (7A77, A20X, and Scalmalloy) were selected based on available tensile yield strength data. Sequential Rounds narrowed down viable alloy and processing parameter sets. Round 1 identified candidates with high tensile yield strength and low porosity, Round 2 evaluated build quality and post-processing compatibility, and Round 3 evaluated build orientation anisotropy. The framework facilitated a comparison of all candidates that led to the selection of Scalmalloy with a specific processing parameter set that met CHAPS-specific criteria. The best Scalmalloy candidate was used to generate design allowables for topology optimization of the CHAPS housing, for which a prototype was successfully fabricated.

Beyond CHAPS, this framework is broadly useful and can be extended to other applications and material classes. For a novel application, the requirements can be redefined and used to inform the critical properties to measure. Weights and bins that indicate the importance of each property can be adjusted such that final scores will reflect the applicability of each candidate to the chosen application. An initial material selection can then begin and the framework can be implemented for the novel application in order to select a combination of alloy and processing parameter set that matches the intended application.

References

1. Swartz WH et al. (2021) CHAPS: a sustainable approach to targeted air pollution observation from small satellites. Paper presented at SPIE 2021, San Diego, California, 1–5 August 2021
2. Reimers J, Bauer A, Thompson KP, and Rolland JP (2017) Freeform spectrometer enabling increased compactness. Light. Sci. Appl. https://doi.org/10.1038/lsa.2017.26

3. Fang FZ et al. (2013) Manufacturing and measurement of freeform optics. CIRP Annals – Man. Tech. 62(2):823–846. https://doi.org/10.1016/j.cirp.2013.05.003

4. Thiel M et al. (2018) Development and test of a two-mirror telescope using additive manufacturing technology. Paper presented at ECSSMET 2018, Noordwijk, The Netherlands, 5 May-1 June 2021

5. Renishaw, Inc (2015). Data sheet: AlSi10Mg-0403 powder for additive manufacturing

6. Aluminum Materials Technologies, Ltd (2020), A20X Powder Materials Datasheet.

7. APWorks (2019). Data sheet: Scalmalloy

8. HRL Laboratories (2019). Data sheet: Aluminum 7A77.60L

9. HRL Laboratories (2020). Material Process Guide: 7A77.50 and 7A77.60L

10. Carpenter Additive (2021) Scalmalloy Datasheet.

11. Aluminum Association (2000) Aluminum Standards and Data, 2000

12. Montalbano T et al. (2021) Uncovering the coupled impact of defect morphology and microstructure on the tensile behavior of Ti-6Al-4V fabricated via laser powder bed fusion. Journ. of Mat. Proc. Tech. https://doi.org/10.1016/j.jmatprotec.2021.117113

13. Liu B et al. (2019) Numerical investigation on heat transfer of multi-laser processing during selective laser melting of AlSi10Mg. Res. in Phys. 12:454-459. https://doi.org/10.1016/j.rinp.2018.11.075

14. Lyu DD et al. (2020) Numerical prediction of residual deformation and failure for powder bed fusion additive manufacturing of metal parts. Journ. of Mech. 36(5):623-636. https://doi.org/10.1017/jmech.2020.30

15. James, MN (2011) Residual stress influences on structural reliability. Eng. Fail. Anal. 18(8):1909-1920. https://doi.org/10.1016/j.engfailanal.2011.06.005

16. ASTM Standard D3359, 2017, "Standard Test Methods for Rating Adhesion by Tape Test", ASTM International, West Conshohocken, PA

17. ASTM Standard E8, 2016a, "Standard Test Methods for Tension Testing of Metallic Materials", ASTM International, West Conshohocken, PA

18. ASTM Standard E1461, 2013, "Standard Test Method for Thermal Diffusivity by the Flash Method", ASTM International, West Conshohocken, PA

19. Battelle Memorial Institute (2017). MMPDS-12: Metallic Materials Properties Development and Standardization

20. Othman I et al. (2018) Impact of single and double zincating treatment on adhesion of electrodeposited nickel coating on aluminum alloy 7075. Jour. Of Adv. Man. Tech. 12(1):179-192

21. Aversa A et al. (2019) New aluminum alloys specifically designed for laser powder bed fusion: a review. Mat. https://doi.org/10.3390/ma12071007

Comparison of Additively Manufactured and Cast Aluminum A205 Alloy

Heidar Karimialavijeh, Morteza Ghasri-Khouzani,
Apratim Chakraborty, Jean-Philippe Harvey, and Étienne Martin

Abstract

The relationships between the microstructure and thermal behaviour of additively manufactured and cast aluminum A205, a recently developed Al–Cu–Mg–Ag–TiB$_2$ alloy, were investigated. Microstructural characterization performed using scanning electron microscopy (SEM) indicated a significant difference in grain size between laser powder bed fusion (LPBF) and cast specimens. Ultrafine round intercellular precipitates were observed in the LPBF structure, while intergranular precipitates were found in the cast structure. Trans-cellular TiB$_2$ particles were observed in the LPBF structure, while accumulated intergranular TiB$_2$ particles were found in the cast structure. Moreover, X-ray diffraction analysis revealed a higher fraction of precipitates in the LPBF specimen compared to the cast specimen. This was attributed to extremely high cooling rates which extend the solubility of alloying elements in the matrix. The thermo-analytical study demonstrated a strong correlation between the microstructural scale and precipitate dissolution kinetics of A205. The higher diffusion rate inherent in refined microstructures facilitates the dissolution of precipitates in the LPBF A205.

Keywords

Additive manufacturing • Casting • Aluminium alloys • Precipitate dissolution • Microstructural characterization

H. Karimialavijeh · J.-P. Harvey · É. Martin (✉)
Polytechnique Montréal, 2500 Chemin de Polytechnique, Montréal, QC H3T 1J4, Canada
e-mail: etienne.martin@polymtl.ca

M. Ghasri-Khouzani · A. Chakraborty · É. Martin
University of Waterloo, 200 University Avenue West, Waterloo, ON N2L 3G1, Canada

J.-P. Harvey
CRCT, Department of Chemical Engineering, Polytechnique Montréal, Montréal, QC H3T 1J4, Canada

Introduction

The excellent combination of light weight, strength, and corrosion resistance makes precipitation-hardenable aluminum alloys great candidates for the aerospace and automotive industries [1]. The A205 alloy, known as a metal matrix composite, takes advantage of TiB$_2$ inoculants and has been adapted for the LPBF technique due to its resistance against hot cracking, a common defect in precipitation-hardenable aluminum alloys fabricated by LPBF [2, 3]. This alloy exhibits outstanding mechanical properties at elevated temperatures owing to the inclusion of elements such as copper, magnesium, and silver in A205 [4]. Moreover, according to Avateffazeli et al. [5], the microstructure of A205 is noticeably finer compared to the cast structure due to the extremely high cooling rates inherent to the LPBF process. However, the authors did not investigate the relationship between the refined as-printed A205 structure with precipitate dissolution kinetics. In another study, Mair et al. [6] observed that Al$_2$Cu precipitates at the cell boundaries of as-printed A205. The authors also proposed that ultrafine intragranular titanium-boride particles act as grain nucleants, while larger particles agglomerated at cell boundaries control grain growth. Shakil et al. [7] compared the microstructure of cast and LPBF AlSi10Mg showing significant differences between the two conditions. However, the authors did not compare the volume fraction of precipitates between the cast and LPBF conditions, which significantly affects the mechanical properties of the alloy [7]. Zamani et al. [8] reported the relationship between solidification rate and dissolution of precipitates in the A205 cast alloy, showing an increase in cast A205 diffusion rate with the increase in the cooling rate during solidification. According to the authors, this difference is attributed to higher diffusion rates of refined structures created by rapid solidification. A comparison between the microstructure of aluminum alloys fabricated using cast and LPBF has been reported in the literature. However, there exists a knowledge gap on the relationship between the effect

© The Minerals, Metals & Materials Society 2023
S. Broek (ed.), *Light Metals 2023*, The Minerals, Metals & Materials Series,
https://doi.org/10.1007/978-3-031-22532-1_66

of structure refinement on the dissolution kinetics of A205 precipitates. The response of A205 to the precipitation hardening treatment is critical for precipitate dissolution and kinetics. Dissolution kinetics in A205 LPBF alloy has never been investigated experimentally.

Experimental Procedure

Materials

The feedstock material used for both casting and LPBF in this study was the A205 alloy (Aluminum Materials Technologies Ltd). The chemical compositions of the as-printed and as-cast A205 coupons are shown in Table 1. The powder particles for LPBF processing were mainly spherical with a particle size distribution (D10-D90) of 15–63 μm.

Sample Fabrication

The LPBF sample was manufactured using a Concept Laser M2 Series 5 LPBF machine with process parameters shown in Table 2. The specimen was fabricated on a 245 mm × 245 mm build plate. The oxygen content in the printing chamber was kept below 0.1% through nitrogen injection. The cast specimen was fabricated using the investment casting method. Molten A205 (750 °C) was poured into ceramic moulds pre-heated at 250 °C under vacuum (10^{-5} Pa). Both samples were cylindrical with a common diameter and length of 10 mm and 20 mm respectively.

Microstructure Characterization

Following build plate removal, cylindrical specimens with a diameter and length of 8–10 mm were extracted from the printed parts and prepared for metallurgical studies using standard metallographic procedures. Each specimen was mounted in a copper-filled epoxy, ground with silicon carbide papers, polished with 9 to 1 μm diamond suspensions, and a final polishing step was performed using 0.04 μm colloidal silica. Microstructure characterization was carried out using a JEOL JSM-7600F field emission gun SEM (FEG-SEM). The phases were characterized using X-ray diffraction analysis using a Bruker D8 Advance machine equipped with Cu Kα radiation (λ = 1.5407 Å). The analysed spectra were obtained for 2θ within the range of 18° to 100° and step size of 0.02°.

Thermoanalytical Study

Differential scanning calorimetry (DSC) was used to measure the characteristic temperatures and evaluate precipitate dissolution kinetics accordingly. DSC was performed using SETARAM LABSYS EVO equipped with a type-E rod. DSC analysis was performed with a scanning rate of 5 °C/min in a two-step cycle, including heating from room temperature to 750 °C and cooling from 750 °C to room temperature. The analysis was performed under the argon (99.999%) flow of ml/min.

Results and Discussion

Microstructure Characterization

Figure 1 illustrates a comparison of the microstructure between the cast and LPBF-printed A205 alloys. The LPBF-built microstructure (Fig. 1a, b) is composed of a fine cellular network (cell size of ∼1 μm) with cell boundaries containing solute-rich phases and trans-cellular particles. However, the cast alloy (Fig. 1c, d) exhibited a dramatically coarser microstructure, with most intermetallic particles agglomerating at cell boundaries (see Fig. 1d). The coarser grain structure of the cast specimen is explained by the extremely low solidification rate compared to the LPBF specimen.

According to Mair et al. [6] and Avateffazeli et al. [5], intermetallic dispersoids observed at cell boundaries of both as-cast and as-printed A205 include θ (Al_2Cu) precipitates and titanium-boride (TiB_2) particles. According to Mair et al. [6], larger titanium-boride particles are located at grain

Table 1 Chemical composition of A205 alloy in wt.%

	Al	Cu	Mg	Ag	Ti	B	Fe	Si
As-printed	Bal.	4.7	0.29	0.76	3.42	1.48	0.02	0.04
As-cast	Bal.	4.68	0.33	0.77	3.47	1.46	0.03	0.06

Table 2 LPBF process parameters used in the study

Power	Scanning speed	Layer thickness	Hatch spacing
370 W	1250 mm/s	40 μm	140 μm

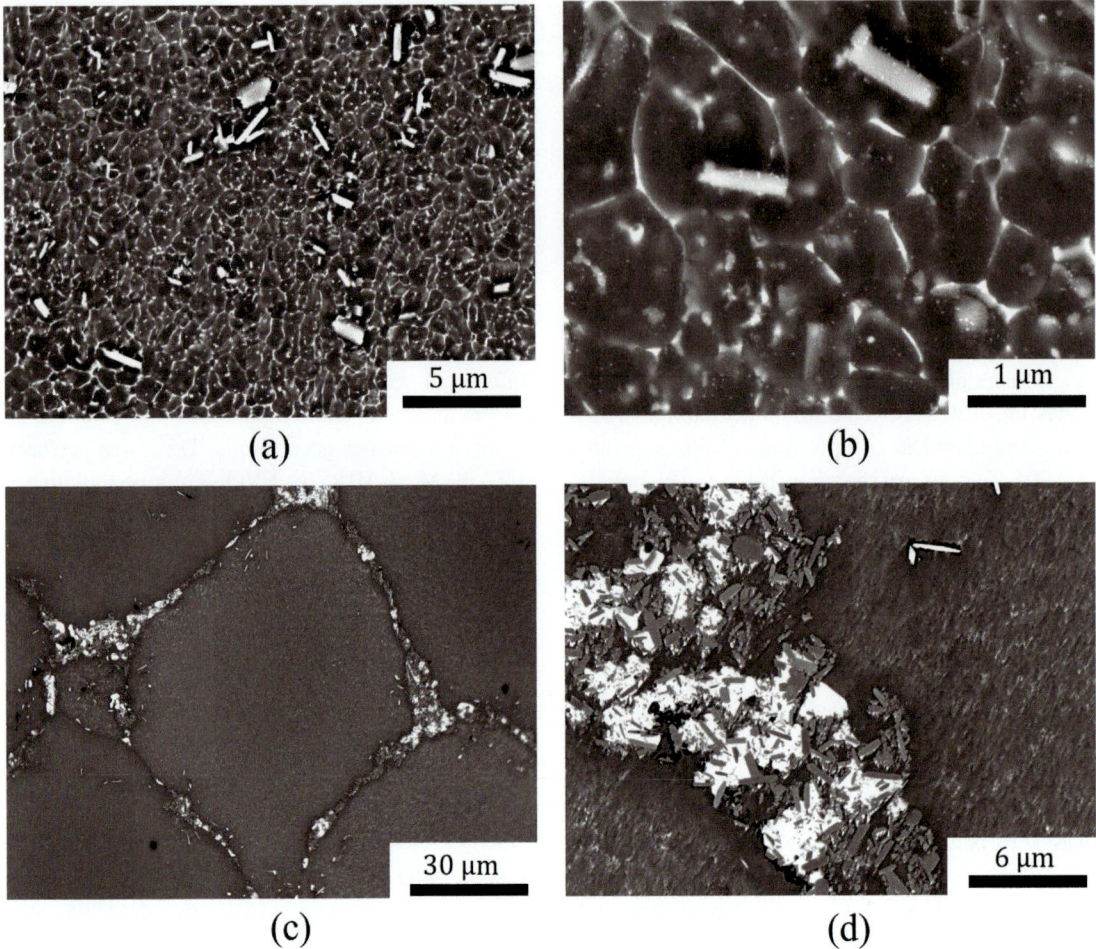

Fig. 1 SEM micrographs showing the microstructure of A205 alloy in as-printed (**a, b**) and as-cast (**c, d**) conditions

boundaries controlling grain growth through the Zener pinning mechanism, while smaller particles act as inoculants during the solidification process [9, 10].

Figure 2 shows the XRD patterns acquired from the LPBF-printed and cast A205 alloys. The FCC aluminum

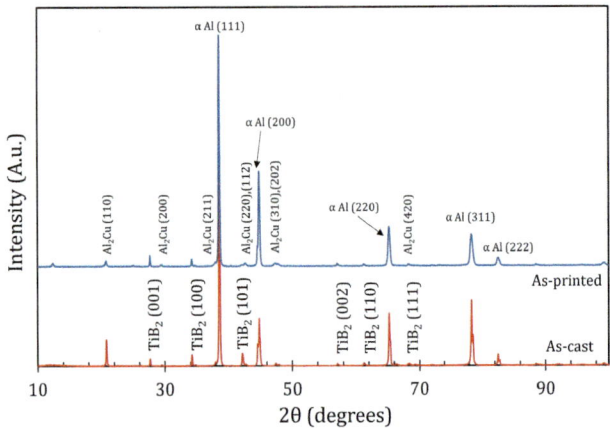

Fig. 2 XRD pattern of the cast and LPBF-built A205 alloy

matrix (α), BCT Al_2Cu (θ), and HCP TiB_2 phases are detected in both as-cast and as-printed specimens. The fraction of Al_2Cu in the as-cast structure is higher than the as-printed structure. This difference is attributed to the difference in solidification rate between the two conditions. The extremely high solidification rate inherent to the LPBF process (10^3–10^7 K/s) [11] enhances the Cu solubility in the Al matrix, which lowers the amount of Cu contribution towards Al_2Cu precipitate formation [12].

DSC Analysis

Figure 3 shows the DSC curves for both the as-cast and as-printed A205 samples. Considering as-printed A205, the only peak detected during the heating cycle is the major endothermic peak corresponding to melting. During the cooling cycle, after the major exothermic peak corresponding to solidification, an exothermic thermal event is observed at around 520 °C that can be attributed to dissolution reaction or incipient melting. The absence of the solvus peak

Fig. 3 DSC curve of aluminum A205 fabricated by casting and LPBF processing obtained at the heating rate of 5 °C/min and argon flow of 15 ml/min

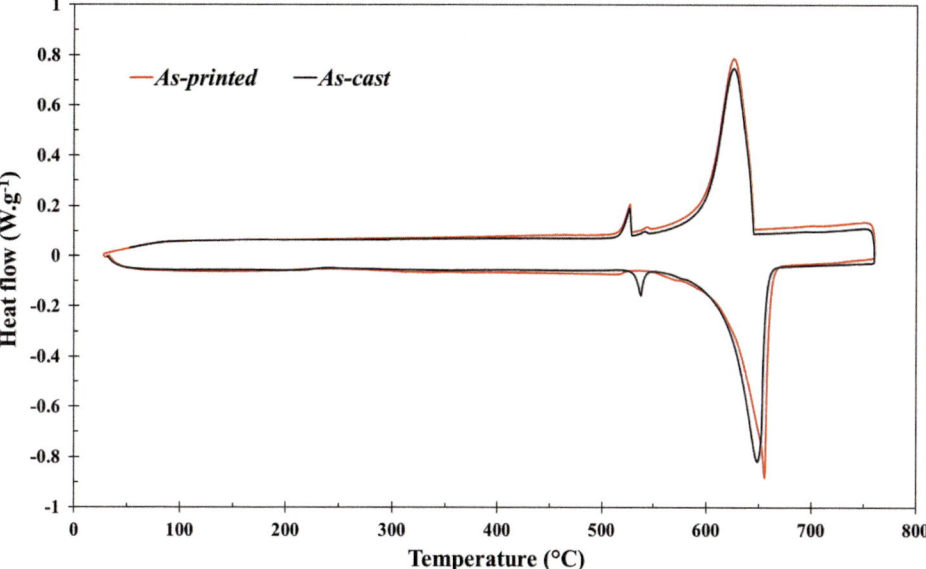

during the heating of as-printed A205 is due to the high diffusion rates present in the ultrafine structures. The highly refined microstructure with a high-volume fraction of grain boundaries in the LPBF-built alloy accelerates the mobility of solute atoms. This results in a smooth dissolution of precipitates, not detectable by DSC [8]. On the contrary, the as-cast A205 exhibited an endothermic event around 535 °C corresponding to the dissolution of precipitates or incipient melting of the eutectic phase. However, no eutectic phase was observed in the as-cast structure during the microstructural characterization. As-cast A205 possesses coarse structures with lower diffusion rates, causing slower dissolution of precipitates easy to capture in DSC. Both as-printed and as-cast A205 exhibit similar precipitate dissolution behaviours after re-melting during the cooling cycle. This occurs because the differences in microstructure are eliminated on complete melting at the end of the heating cycle.

Conclusion

By comparing the microstructure and thermal behaviour between the cast and LPBF A205 aluminum, the following conclusions can be drawn:

1. The extremely high cooling rates during the LPBF process generate finer structures compared to the casting process. Both as-cast and as-printed A205 include solute-rich phases at the cell boundaries. Inter-cellular solute-rich precipitates were observed in the as-printed structure, while most precipitates in the as-cast structure were located near the grain boundaries.

2. Due to the rapid solidification of the as-printed specimen, TiB_2 particles were trapped within the cells during solidification. However, in the cast structure, most TiB_2 particles were pushed to the grain boundaries during the solidification process.

3. A higher fraction of Al_2Cu precipitates was found in the as-cast structure compared to the LPBF structure. This is attributed to the lower solubility of alloying elements during the casting process, which provides a greater amount of the alloying element for precipitate formation.

4. DSC analysis revealed that the dissolution of precipitates occurs faster in the as-printed structure, explained by the faster diffusion rates inherent in fine structures.

Acknowledgements The authors are thankful to Natural Sciences and Engineering Research Council of Canada (NSERC) under grant no. CRDPJ 533406-18, (CM)2 Labs at Polytechnique Montreal and Amber Andreaco, working in the material supply division at GE Additive, for supporting this work.

References

1. W.F. Miao, D.E. Laughlin, Precipitation hardening in aluminum alloy 6022, Scr Mater. 40 (1999) 873–878. https://doi.org/10.1016/S1359-6462(99)00046-9.

2. A. Mehta, L. Zhou, T. Huynh, S. Park, H. Hyer, S. Song, Y. Bai, D.D. Imholte, N.E. Woolstenhulme, D.M. Wachs, Y. Sohn, Additive manufacturing and mechanical properties of the dense and crack free Zr-modified aluminum alloy 6061 fabricated by the laser-powder bed fusion, Addit Manuf. 41 (2021). https://doi.org/10.1016/J.ADDMA.2021.101966.

3. R. Du, Q. Gao, S. Wu, S. Lü, X. Zhou, Influence of TiB2 particles on aging behavior of in-situ TiB2/Al-4.5Cu composites, Mater Sci Eng A Struct Mater. 721 (2018) 244–250. https://doi.org/10.1016/J.MSEA.2018.02.099.

4. I.J. Polmear, The Influence of Small Additions of Silver on the Structure and Properties of Aged Aluminum Alloys, JOM. 20 (1968) 44–51. https://doi.org/10.1007/BF03378722.

5. M. Avateffazeli, P.E. Carrion, B. Shachi-Amirkhiz, H. Pirgazi, M. Mohammadi, N. Shamsaei, M. Haghshenas, Correlation between tensile properties, microstructure, and processing routes of an Al–Cu–Mg–Ag–TiB2 (A205) alloy: Additive manufacturing and casting, Materials Science and Engineering A. 841 (2022). https://doi.org/10.1016/J.MSEA.2022.142989.

6. P. Mair, L. Kaserer, J. Braun, N. Weinberger, I. Letofsky-Papst, G. Leichtfried, Microstructure and mechanical properties of a TiB2-modified Al–Cu alloy processed by laser powder-bed fusion, Materials Science and Engineering A. 799 (2021). https://doi.org/10.1016/J.MSEA.2020.140209.

7. S.I. Shakil, A. Hadadzadeh, B. Shalchi Amirkhiz, H. Pirgazi, M. Mohammadi, M. Haghshenas, Additive manufactured versus cast AlSi10Mg alloy: Microstructure and micromechanics, Results in Materials. 10 (2021). https://doi.org/10.1016/J.RINMA.2021.100178.

8. M. Zamani, I. Belov, E. Sjölander, A. Bjurenstedt, E. Ghassemali, S. Seifeddine, Study on dissolution of Al2Cu in al-4.3cu and a205

9. X.P. Li, G. Ji, Z. Chen, A. Addad, Y. Wu, H.W. Wang, J. Vleugels, J. van Humbeeck, J.P. Kruth, Selective laser melting of nano-TiB2 decorated AlSi10Mg alloy with high fracture strength and ductility, Acta Mater. 129 (2017) 183–193. https://doi.org/10.1016/J.ACTAMAT.2017.02.062.

10. Y.K. Xiao, Z.Y. Bian, Y. Wu, G. Ji, Y.Q. Li, M.J. Li, Q. Lian, Z. Chen, A. Addad, H.W. Wang, Effect of nano-TiB2 particles on the anisotropy in an AlSi10Mg alloy processed by selective laser melting, J Alloys Compd. 798 (2019) 644–655. https://doi.org/10.1016/J.JALLCOM.2019.05.279.

11. S.L. Sing, S. Huang, G.D. Goh, G.L. Goh, C.F. Tey, J.H.K. Tan, W.Y. Yeong, Emerging metallic systems for additive manufacturing: In-situ alloying and multi-metal processing in laser powder bed fusion, Prog Mater Sci. 119 (2021) 100795. https://doi.org/10.1016/J.PMATSCI.2021.100795.

12. G.-F. Rafael, The effect of solidification rate and solutionizing on the mechanical properties and hardening response of aluminum alloys, Michigan Technological University, 2016.

cast alloys, Metals (Basel). 10 (2020) 1–17. https://doi.org/10.3390/MET10070900.

The Role of Ti and B Additions in Grain Refinement of Al–Mn Alloy During Laser Additive Manufacturing

Qingyu Pan, Monica Kapoor, Sean Mileski, John Carsley, and Xiaoyuan Lou

Abstract

In the present work, we investigated the governing mechanism of grain refinement in AA3104 Al–Mn alloy made by laser direct energy deposition (DED) additive manufacturing (AM). The microstructure development and phase transformation of DED AM AA3104 were studied by adding different concentrations of Ti, B, Ti–B mixture, and TiB_2 particles as grain refiners. The transition from large columnar grains to fine equiaxed grains was achieved by Ti or Ti–B addition, but not by B or TiB_2 addition. The study demonstrated that reactive additive manufacturing may be a more effective way to refine grains than directly adding ceramic particles. The formation of Al_3Ti plays the most critical role in grain nucleation during DED AM.

Keywords

Al–Mn alloys • Additive manufacturing • Grain refinement • Second phase particles

Q. Pan (✉) · M. Kapoor · X. Lou
Purdue University, West Lafayette, IN, USA
e-mail: pan398@purdue.edu

M. Kapoor
e-mail: Monica.Kapoor@novelis.adityabirla.com

X. Lou
e-mail: lou49@purdue.edu

S. Mileski · J. Carsley
Novelis Global Research and Technology Center, Kennesaw, GA, USA
e-mail: Sean.Mileski@novelis.adityabirla.com

J. Carsley
e-mail: John.Carsley@novelis.adityabirla.com

Introduction

Grain refinement during Al alloy casting is required to avoid casting defects and promote the uniform microstructure and properties [1–3]. The typical grain refiner used in casting is the master alloy Al–5Ti–1B. During the melting, the formation of TiB_2 and Al_3Ti are considered as grain nucleus [3]. It is believed that TiB_2, which is thermally stable due to its high melting point, serves as the heterogeneous nucleation site. The free Ti atoms form the Al–Ti transition layer on the TiB_2 surface to increase their nucleation potency [3–6]. This theory has been widely accepted, while the roles of Ti and B have not been thoroughly studied.

Laser additive manufacturing (AM) has emerged to offer high manufacturing resolution and excellent part quality for making Al alloy components [7, 8]. Different from conventional methods, large columnar grains and strong crystallographic texture are produced in AM components due to the rapid cooling along the build direction [9–11]. These characteristics produce anisotropic mechanical properties and thermal cracking [12–14]. Therefore, a variety of methods, such as introducing external nucleation sites [15, 16], ultrasound-assisted fabrication [17], have been explored to promote grain refinement during solidification to reduce residual stress and eliminate cracking. Directly adding TiB_2 into the feedstock powder was proven to induce equiaxed grain nucleation and improve the strength [18–20]. It is also reported that adding Zr and Sc into aluminum alloys during laser AM can assist grain refinement by the in situ formation of Al_3Zr or Al_3Sc through peritectic reaction [13, 15].

This study intends to understand the grain refinement mechanism of AA3104 Al–Mn alloy during laser AM. AM AA3104 was fabricated by laser direct energy deposition (DED). The study focused on Ti- and B-based grain refiners, and investigated their roles in promoting heterogeneous nucleation during laser AM. By varying the different types

© The Minerals, Metals & Materials Society 2023
S. Broek (ed.), *Light Metals 2023*, The Minerals, Metals & Materials Series,
https://doi.org/10.1007/978-3-031-22532-1_67

of refiners and their amounts, the study will investigate the role of TiB_2 and Al_3Ti and clarify the grain nucleation path during laser AM.

Experiments

Feedstock materials in this study included gas-atomized spherical pure Al powder (H60, Valimet™) with particle size under 74 μm (−200 mesh), spherical gas-atomized Al–Mg (1:1) alloy powder (Reade™ Advanced Materials) with particle size under 53 μm (−270 mesh), irregular-shaped pure elemental powders including Mn, Fe, Cu purchased from Alfa Aesar™, and Si, Ti, B, TiB_2 from US Research Nanomaterials, Inc. All minor elemental powders were less than 10 μm in size. The composition of the base material, AA3104, is listed in Table .1. Different combinations of grain refiners were added to AA3104 base powder as listed in Table 1. For AA3104-Ti--B, the weight ratio between Ti and B was 5:1 (1 wt.%Ti and 0.2 wt.%B) in this study. For AA3104-TiB_2, 0.67 wt.% TiB_2 was added to AA3104 base powder to study the role of TiB_2. AA3104-1Ti and AA3104-0.6B were also prepared to investigate the effect of single-element addition. Planetary ball milling was used to mix powders uniformly to achieve the targeted composition. Before the mixing, the container was evacuated and back-filled with argon gas multiple times to prevent oxidation during mixing. Ball-milling was carried out for 1 h at 200 RPM with the ball-to-powder ratio (BPR) of 3:1. After the mixing, the ball-milled powder was further sieved to 100 mesh to remove powder agglomerates larger than 150 μm. Before laser AM, the mixed powder was baked at 150 °C for 1.5 h to remove moisture.

Sample Fabrication

Samples were fabricated using an Optomec LENS 500 DED AM system (Optomec Inc, Albuquerque, NM), equipped with a 1 kW continuous wave Nd: YAG laser (IPG Photonics, Inc., Oxford, MA). The deposition was carried out in an argon atmosphere with an oxygen level maintained below 30 ppm. By mixing the powders from different powder feeders, Optomec DED AM system provided the opportunity to achieve numerous compositions by adjusting feeding rates. In this study, AA3104 powder and AA3104 powder with grain refiners were loaded into different powder feeders. Figure 1 shows the assignments to the powder feeders and the alloy compositions generated from each set. The targeted alloy composition was achieved by adjusting the feeding rates of the two feeders.

Samples were fabricated on the sandblasted AA3003 substrates. The laser parameters used in this study were set at 500 W, with 28in/min scan speed. A rotating angle of 90° was set between two adjacent layers. The hatch spacing was 0.015in, and the layer thickness was 0.01in.

The compositional accuracy was verified from the produced AM samples, as shown in Table 2. Three samples (AA3104, AA3104-1Ti, and AA3104-0.5Ti) were measured by inductively coupled plasma atomic emission spectroscopy (ICP-AES). All samples were confirmed to be within the targeted composition range. It is also confirmed that the powder mixing approach produces good repeatability from batch to batch. Mg loss during laser AM was observed in Table 2, due to its low boiling point [11, 21].

Microstructure Characterization

The specimens were all sectioned parallel to the build direction, mounted, and polished up to 0.05 μm colloidal silica suspension. Electrolytic etching was done at 30 V in Barker's reagent (5 mL HBF4, 200 mL DI water) for 2 min to reveal the grain structure. Grain structure was recorded under optical microscopy (OM, ZEISS-AXIOVERT 10) with polarized light. The mean linear intercept method was used to quantify grain size, with at least 30 grains selected. The particle distribution and identification were performed under a JSM-7000 scanning electron microscope (SEM). The composition of constituent particles, measured by electron dispersive spectroscopy (EDS), was averaged from ten particles.

Table 1 Target compositions (wt%) for different types of powders

	Mn	Mg	Fe	Si	Cu	TiB_2	Ti	B	Al
AA3104	0.8–1.4	0.8–1.3	0.4–0.7	0.2–0.4	0.05–0.25	/	/	/	Bal
AA3104-Ti-B	0.8–1.4	0.8–1.3	0.4–0.7	0.2–0.4	0.05–0.25	/	1	0.2	Bal
AA3104-TiB_2	0.8–1.4	0.8–1.3	0.4–0.7	0.2–0.4	0.05–0.25	0.67	/	/	Bal
AA3104-1Ti	0.8–1.4	0.8–1.3	0.4–0.7	0.2–0.4	0.05–0.25	/	1	/	Bal
AA3104-0.6B	0.8–1.4	0.8–1.3	0.4–0.7	0.2–0.4	0.05–0.25	/	/	0.6	Bal

Fig. 1 Schematic showing the mixing method and deposited samples in this study

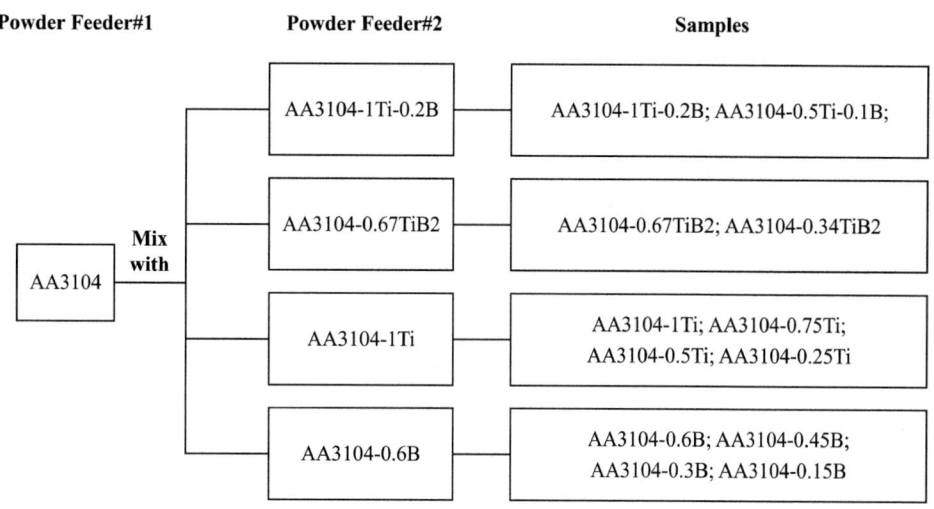

Table 2 Accuracy of chemical compositions of AA3104, AA3104-1Ti, and AA3104-0.5Ti samples measured by inductively coupled plasma atomic emission spectroscopy (ICP-AES)

	Mn (%)	Mg (%)	Fe (%)	Si (%)	Cu (%)	Ti	Al
AA3104	94	61	91	94	89	/	Bal
AA3104-1Ti	94	67	97	92	89	90%	Bal
AA3104-0.5Ti	97	83	91	98	95	99%	Bal

Results and Discussion

Effect of Grain Refinement from TiB$_2$, Ti, and B

Conventional grain refinement in DC casting uses a master alloy (Al–5Ti–1B) as the grain refiner [3]. In laser AM, studies showed that TiB$_2$ could also reach significant grain refinement [22]. We compared the effectiveness of TiB$_2$, Ti, and B. Figure 2 shows the OM images of DED-AA3104 with different grain refiners. Without grain refiner addition, DED AA3104 exhibited large columnar grains which grew epitaxially along the build direction. The grain size reached up to 1–3 mm, following a "zig-zag" pattern. The change in grain growth direction was induced by the alternating scan directions between layers [23, 24]. With the additions of 0.5%Ti and 0.1%B, the columnar grains still dominated the microstructure. However, the average grain size was significantly refined to 165 ± 31 μm, with narrower grain width. Isolated regions with fine equiaxed grains can be observed. A higher loading of grain refiners (1 wt%Ti and 0.2 wt%B) resulted in the fully equiaxed grains, with the average grain size at 18.7 ± 2.6 μm. Based on stoichiometry, 0.5 wt%Ti with 0.1 wt%B would form 0.34 wt%TiB$_2$ and 0.27 wt% free Ti during solidification. As shown in Fig. 2, grain refinement with 0.34 wt% TiB$_2$ was also

evaluated. TiB$_2$ particles were uniformly distributed in the sample during laser AM, pointed by red arrows in Fig. 2. While the average grain size was decreased to 582 ± 82 μm by TiB$_2$ addition, 0.34 wt% TiB$_2$ was less effective than 0.5 wt% Ti + 0.1 wt% B. Doubling the TiB$_2$ content to 0.67 wt %, equivalent to the alloy with 1 wt% Ti and 0.2 wt% B, still did not show significant improvement. Compared to Ti and B addition, TiB$_2$ is not an effective grain refiner in DED AA3104. In literature, TiB$_2$ has been shown effective to promote grain refinement in AM Al-based alloys [22, 25]. The possible reason of our contradictory observation could be due to the less amount of TiB$_2$ used, and its relatively larger particle size. But still, the in situ reaction of Ti (excessive) and B during melting is confirmed more effective than the direct addition of TiB$_2$.

The comparison above suggests that the addition of excessive Ti and B performed better than adding TiB$_2$ for grain refinement in laser AM. Further, the study confirms that Ti is more important than TiB$_2$ and B in promoting grain nucleation. As shown in Fig. 3, 0.25 wt% Ti produced a similar effect as 0.34 wt% TiB$_2$, with the average columnar grain size reduced to 451 ± 68 μm. As the concentration of Ti to 0.5 wt%, ∼50 vol% of the material presented equiaxed grains with the average grain size further dropped to 42 ± 30 μm. We emphasize that the alloy with 0.5 wt% Ti exhibited better grain refinement than the one with 0.5%

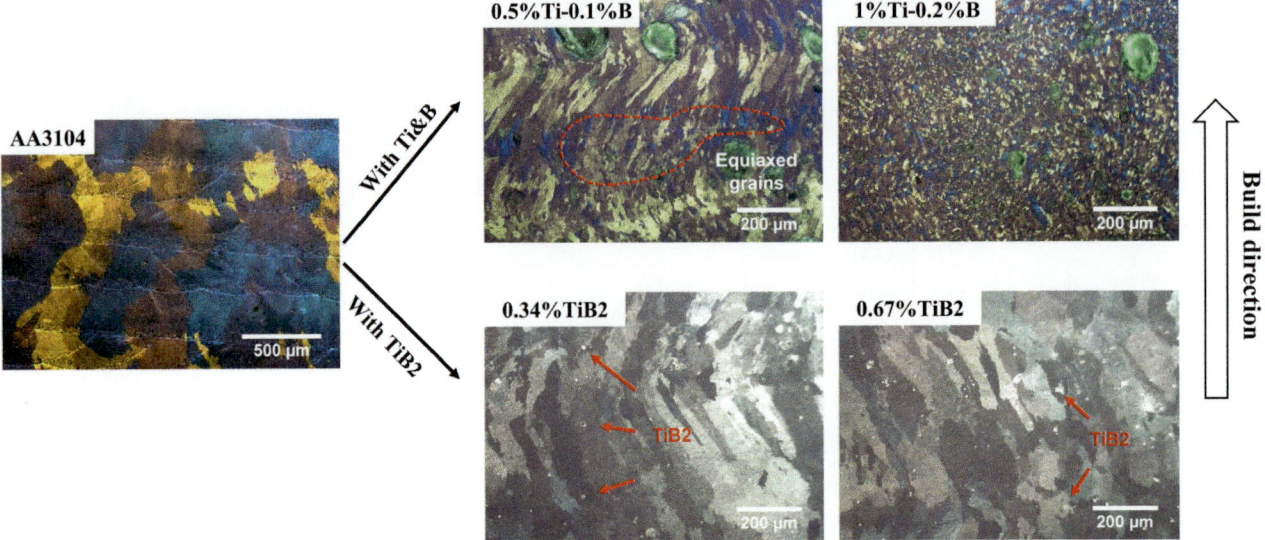

Fig. 2 OM images showing the grain structure of AA3104 and AA3104 with different types and content of grain refiners

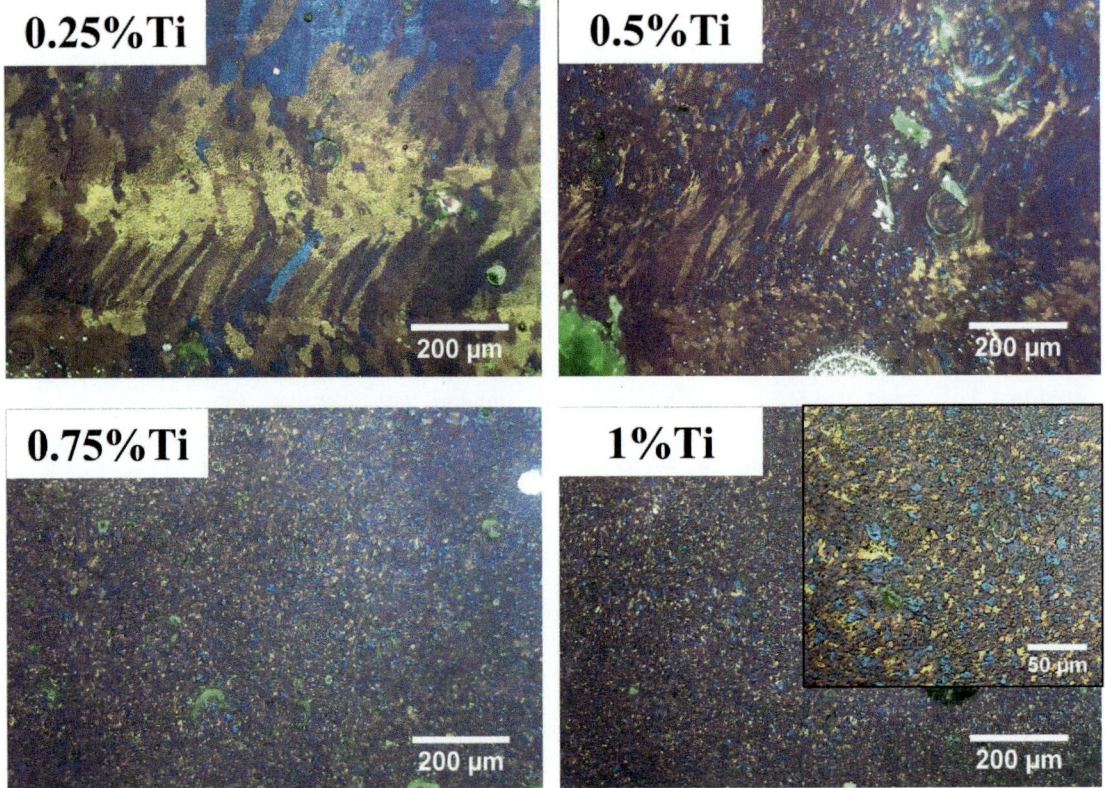

Fig. 3 OM images showing the grain structure of AA3104 with different content of Ti as the grain refiner

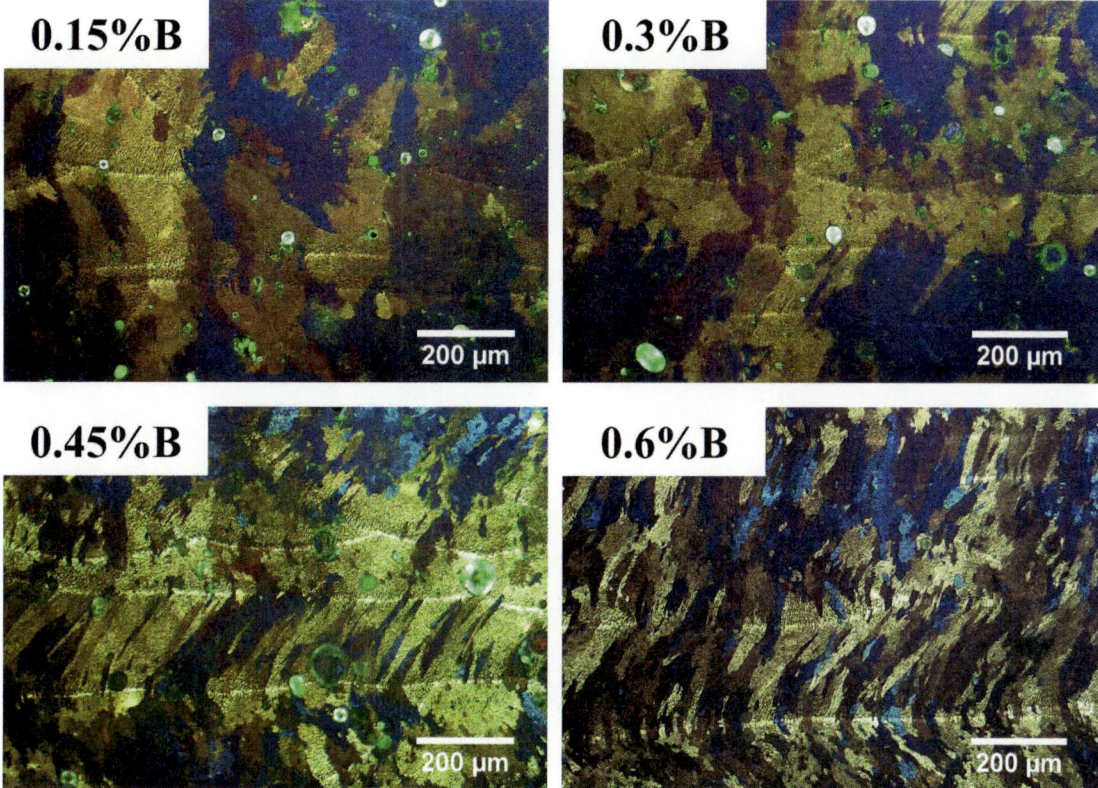

Fig. 4 OM images showing the grain structure of AA3104 with different content of B as the grain refiner

Ti–0.1%B in Fig. 2. Fully equiaxed grains were achieved when 0.75 wt% Ti was added, with the grain size being 9.2 ± 1 μm. 1 wt% Ti further refined the grains to 7 ± 0.8 μm. Compared to adding Ti–B and TiB_2 as shown in Fig. 2, the sole addition of Ti to the alloy reached a smaller grain size, confirming that Ti is the main contributor to grain refinement in DED AA3104. In DED AA3104-0.5% Ti–0.1%B (Fig. 2), the addition of B diluted Ti to form TiB_2, which in fact weakened the refinement effect.

The effect of B on the grain refinement was also studied, as shown in Fig. 4. The sole addition of B, 0.15 wt% and 0.3 wt%, did not change the grain structure compared to the original DED AA3104. In comparison (Fig. 2), the combination of 0.1 wt% B and 0.5 wt% Ti was effective to further decrease in the grain size. With the sole addition of B in DED AA3104, 0.45 wt% B and 0.6 wt% B were able to reduce the grain size to 418 ± 95 μm and 170 ± 20 μm, respectively, but the grains remained columnar. It is known that the addition of B in Al alloys can form AlB_2, which can also act as the grain refiner [26, 27]. However, the stable AlB_2 particles are favorable to form in the Al alloys with high content of Si [28, 29], which could be the reason for the limited contribution of B here in AA3104.

Microstructure of Alloys

Figure 5 summarizes the microstructure of DED AA3104 with TiB_2, Ti–B, and Ti as the grain refiners. For comparison, the microstructure of DED-AA3104 without grain refinement can be found in our previous work [30]. One of the unique microstructural features in AM Al alloys is the formation of cellular structures with second phase particles decorated along the cellular boundaries [31, 32]. The previous studies [30] had identified these bright contrast particles on the cellular boundaries were $Al_6(Fe,Mn)$ and α-Al (Fe,Mn)-Si, both of which are commonly found in conventional DC-Cast AA3104 [33]. Mg_2Si particles were also present as black spots in DED AA3104.

With TiB_2 directly added into DED AA3104, there was no change observed in regard to the distribution and types of constituent particles. As shown in Fig. 5, due to the high melting point of TiB_2 (3225 °C) [34], the TiB_2 particles in the feedstock stock remained in the alloy after laser AM, with the size comparable to its original form. No dissolution or reaction of TiB_2 was observed. In comparison, the addition of Ti–B and Ti led to in situ reaction to form Al_3Ti

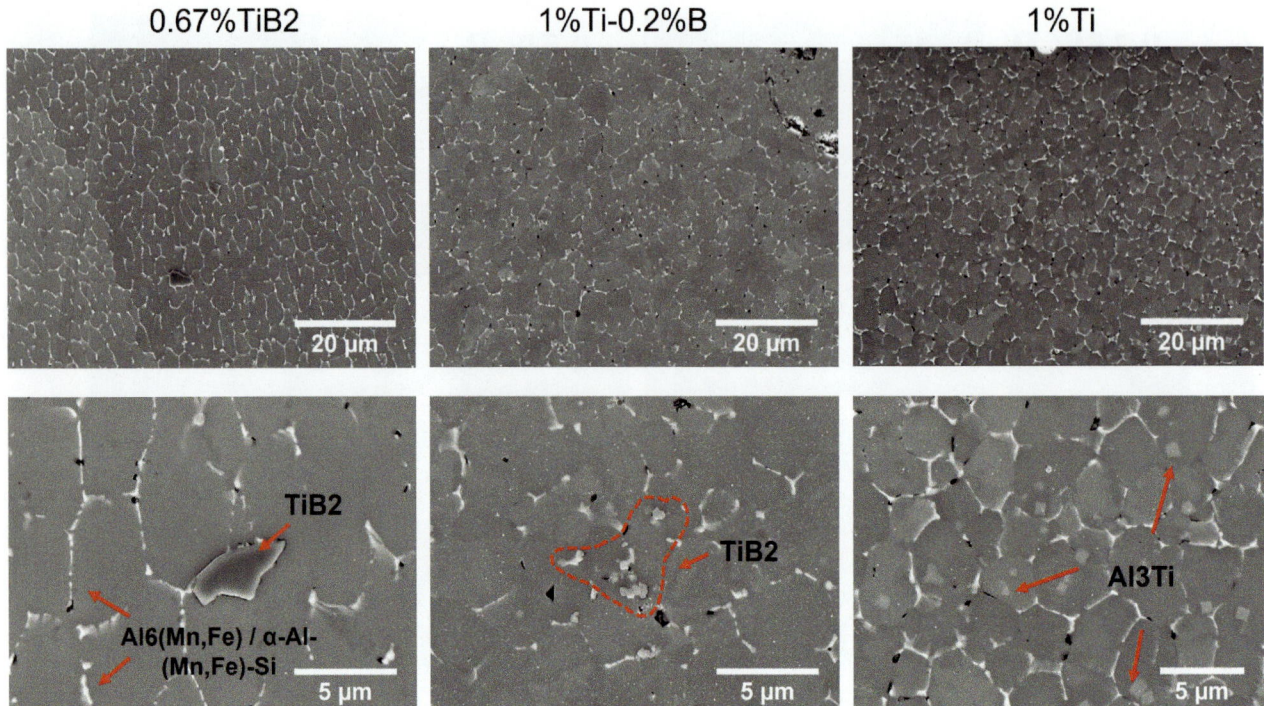

Fig. 5 SEM (SE) images of alloys with different types of grain refiner

during laser AM. The in situ formed TiB_2 clusters presented a much smaller size at ~ 500 nm. It was reported that such clusters can promote grain refinement [5, 6, 35]. Al_3Ti particles in the alloy exhibited the polygon shape with a size close to 1 μm. With 1 wt% Ti addition, the grain size was around 5 μm. Since Al_3Ti particles were mostly located in the center of the grain, we believe they acted here as the main nucleus. It needs to be mentioned that no Al_3Ti was observed in the alloy with 1 wt% Ti and 0.2 wt% B, even though the excessive amount of Ti (0.53 wt% Ti in theory) was available to react with Al (Ti solubility in Al is 0.15 wt.% [3]). We suspect that the in-situ reaction led to the formation of Al–Ti transition layer on the TiB_2 particles. Such process has been experimentally confirmed to reduce lattice mismatch between TiB_2 and α-Al [3], which benefits grain refinement in DC casting.

The beneficial effect of Al_3Ti formation on the grain refinement was revealed in Fig. 6. With the increasing amount of Ti added to DED AA3104, the material exhibited a higher density of Al_3Ti with the reduced grain size. Fan et al. [3] showed the in situ formation of Al_3Ti film on the surface of TiB_2 particle is critical to grain nucleation during casting. However, Al_3Ti is a transient phase which eventually dissolves during DC casting. Our study not only verified the critical role of Al_3Ti in grain refinement but also showed this transient phase can solely act as grain nucleus in AM Al alloys and remains in the alloy after the fast laser solidification.

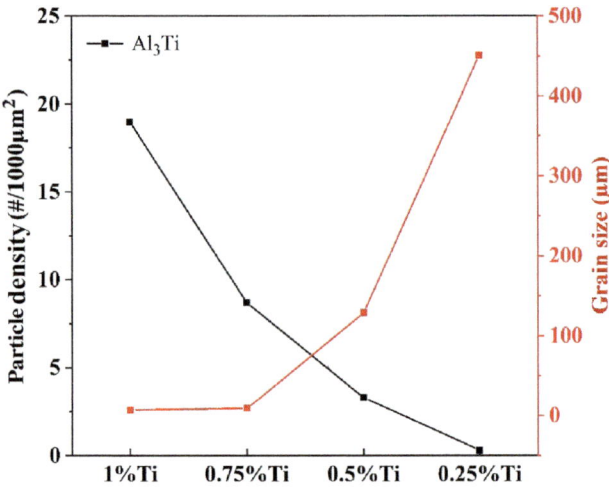

Fig. 6 The evolution of the particle density and the grain size with different content of Ti

Conclusion

This study evaluated the grain refinement of AA3104 by Ti, B, and TiB_2 during laser DED AM. Ti addition was identified as the most effective grain refiner for DED AA3104. The complete transition from columnar grain to equiaxed grain was achieved with 0.75 wt% Ti in the alloy. In comparison, B and TiB_2 showed limited contribution to grain refinement. With solely Ti addition, the in situ formation of

Al$_3$Ti particles during laser solidification served as grain nucleation sites for DED AA3104, and stays as the constituent particles in the final product. With the combined additions of Ti and B, the characterization suggests that Al$_3$Ti may form as the thin film on TiB2 particles to assist α-Al nucleation. This work demonstrated that reactive AM can produce a more effective grain nucleus than the ex situ addition of ceramic particles.

Acknowledgements The research was sponsored by Novelis Global Research & Development Center.

References

1. M. Easton and D. Stjohn, Metall. Mater. Trans. A Phys. Metall. Mater. Sci. **30**, 1613 (1999).
2. R. G. Guan and D. Tie, Acta Metall. Sin. (English Lett. **30**, 409 (2017).
3. Z. Fan, Y. Wang, Y. Zhang, T. Qin, X. R. Zhou, G. E. Thompson, T. Pennycook, and T. Hashimoto, Acta Mater. **84**, 292 (2015).
4. F. Wang, D. Qiu, Z. L. Liu, J. A. Taylor, M. A. Easton, and M. X. Zhang, Acta Mater. **61**, 5636 (2013).
5. P. S. Mohanty and J. E. Gruzleski, Acta Metall. Mater. **43**, 2001 (1995).
6. B. S. Murty, S. A. Kori, and M. Chakraborty, Int. Mater. Rev. **47**, 3 (2002).
7. G. P. Dinda, A. K. Dasgupta, S. Bhattacharya, H. Natu, B. Dutta, and J. Mazumder, Metall. Mater. Trans. A Phys. Metall. Mater. Sci. **44**, 2233 (2013).
8. T. C. Lin, C. Cao, M. Sokoluk, L. Jiang, X. Wang, J. M. Schoenung, E. J. Lavernia, and X. Li, Nat. Commun. **10**, 1 (2019).
9. G. P. Dinda, A. K. Dasgupta, and J. Mazumder, Surf. Coatings Technol. **206**, 2152 (2012).
10. Z. Dong, M. Xu, H. Guo, X. Fei, Y. Liu, B. Gong, and G. Ju, J. Mater. Res. Technol. (2022).
11. D. Herzog, V. Seyda, E. Wycisk, and C. Emmelmann, Acta Mater. **117**, 371 (2016).
12. L. Zhou, H. Hyer, S. Park, H. Pan, Y. Bai, K. P. Rice, and Y. Sohn, Addit. Manuf. **28**, 485 (2019).
13. D. Carluccio, M. J. Bermingham, Y. Zhang, D. H. StJohn, K. Yang, P. A. Rometsch, X. Wu, and M. S. Dargusch, J. Manuf. Process. **35**, 715 (2018).
14. M. J. Bermingham, D. H. StJohn, J. Krynen, S. Tedman-Jones, and M. S. Dargusch, Acta Mater. **168**, 261 (2019).
15. A. Mehta, L. Zhou, T. Huynh, S. Park, H. Hyer, S. Song, Y. Bai, D. D. Imholte, N. E. Woolstenhulme, D. M. Wachs, and Y. Sohn, Addit. Manuf. **41**, 101966 (2021).
16. W. Zhai, W. Zhou, and S. M. L. Nai, Mater. Sci. Eng. A **840**, 142912 (2022).
17. C. J. Todaro, M. A. Easton, D. Qiu, M. Brandt, D. H. StJohn, and M. Qian, Addit. Manuf. **37**, 101632 (2021).
18. B. Jiang, L. Zhenglong, C. Xi, L. Peng, L. Nannan, and C. Yanbin, Ceram. Int. **45**, 5680 (2019).
19. Q. Z. Wang, X. Lin, N. Kang, X. L. Wen, Y. Cao, J. L. Lu, D. J. Peng, J. Bai, Y. X. Zhou, M. E. I. Mansori, and W. D. Huang, Mater. Sci. Eng. A **840**, 142950 (2022).
20. H. Ding, Y. Xiao, Z. Bian, Y. Wu, H. Yang, Y. Li, Z. Chen, H. Wang, and H. Wang, Mater. Sci. Eng. A **845**, (2022).
21. D. Svetlizky, B. Zheng, T. Buta, Y. Zhou, O. Golan, U. Breiman, R. Haj-Ali, J. M. Schoenung, E. J. Lavernia, and N. Eliaz, Mater. Des. **192**, 108763 (2020).
22. S. Y. Zhou, Y. Su, H. Wang, J. Enz, T. Ebel, and M. Yan, Addit. Manuf. **36**, (2020).
23. T. DebRoy, H. L. Wei, J. S. Zuback, T. Mukherjee, J. W. Elmer, J. O. Milewski, A. M. Beese, A. Wilson-Heid, A. De, and W. Zhang, Prog. Mater. Sci. **92**, 112 (2018).
24. G. M. Karthik, E. S. Kim, P. Sathiyamoorthi, A. Zargaran, S. G. Jeong, R. Xiong, S. H. Kang, J. W. Cho, and H. S. Kim, Addit. Manuf. **47**, 102314 (2021).
25. X. Wen, Q. Wang, Q. Mu, N. Kang, S. Sui, H. Yang, X. Lin, and W. Huang, Mater. Sci. Eng. A **745**, 319 (2019).
26. M. M. Guzowski, G. K. Sigworth, and D. A. Sentner, Metall. Trans. A **18**, 603 (1987).
27. G. K. Sigworth and T. A. Kuhn, Int. J. Met. **1**, 31 (2007).
28. Y. Birol, Mater. Sci. Technol. **28**, 385 (2012).
29. Y. Li, B. Hu, B. Liu, A. Nie, Q. Gu, J. Wang, and Q. Li, Acta Mater. **187**, 51 (2020).
30. Q. Pan, M. Kapoor, S. Mileski, J. Carsley, and X. Lou, in *Light Met. 2022* (the Minerals, Metals & Materials Series. Springer, Cham., 2022), pp. 179–185.
31. K. G. Prashanth and J. Eckert, J. Alloys Compd. **707**, 27 (2017).
32. B. Chen, S. K. Moon, X. Yao, G. Bi, J. Shen, J. Umeda, and K. Kondoh, Scr. Mater. **141**, 45 (2017).
33. D. T. L. A. & A. L. Greer, Acta Mater. **50**, 2571 (2002).
34. P. Mair, L. Kaserer, J. Braun, N. Weinberger, I. Letofsky-Papst, and G. Leichtfried, Mater. Sci. Eng. A **799**, 140209 (2021).
35. C. Te Lee and S. W. Chen, Mater. Sci. Eng. A **325**, 242 (2002).

AMAG CrossAlloy®—A Unique Aluminum Alloy Concept for Lightweighting the Future

Florian Schmid, Lukas Stemper, and Ramona Tosone

Abstract

Increasing requirements to reduce CO_2 emissions by using high scrap rates while maintaining the same high material properties represent a major hurdle that modern aluminum alloys must overcome. Since classic aluminum alloy systems are limited to one main alloying element, material improvements to meet the requirements described above are limited. In this context, crossover alloys—a new class of aluminum alloys that is capable of combining beneficial properties of already existing alloys and alloy classes by advanced alloy design—have recently gained increased attention in the scientific community. Industrial trials on 5xxx/7xxx crossover alloys carried out by AMAG (AMAG CrossAlloy.57) additionally reveal an outstanding performance that is not limited to use in any specific industrial sector. This study provides an insight into the potential of CrossAlloys® for a wide range of applications.

Keywords

Aluminum alloy • Crossover alloy • Alloy design

Introduction

The field of material development in the area of aluminum alloys is driven by the improvement of material properties in order to be competitive against other innovative material concepts, for example, modern steel grades. Mechanical properties such as strength or ductility often depict the main research topic but are usually subject to mutual interaction—increased strength at the expense of ductility and vice versa.

In addition, technological properties such as joining properties, corrosion resistance, formability or surface quality play an equally important role in implementing materials to successful industrial use.

For this reason, 5xxx alloys with very good formability are used in automotive construction only for internal components, as they do not sufficiently meet the high requirements in terms of surface quality. Here, 6xxx alloys are used, which, however, are of limited use for complex forming operations. For structural parts with the highest strength requirements, 7xxx alloys can also be used, among others, which offer yield strengths beyond 400 MPa. However, this alloy concept tends to exhibit poor corrosion resistance as well as unfavorable joining properties. In addition, 7xxx alloys can only be processed to a very limited extent at room temperature, which requires very complex and therefore more expensive processes for shaping. To conclude, various alloy concepts must be used to meet the different usage requirements [1].

This multi-material mix, even within one alloy class, leads to a mixture of different alloys with varying element contents considering the collection of process scrap generated by OEMs as well as end-of-life recycling. As a consequence, the production of high-quality rolled products with such mixed scrap is significantly impeded. Still, modern aluminum alloys must be able to meet certain quotas with regard to scrap addition – closed-loop process – in order to reduce the specific CO_2 footprint to a minimum. In most cases, however, this can only be achieved by widening the specific element limits. This results in a field of tension in which narrow alloying limits to achieve specific material parameters are opposed by wide alloying limits for the use of high scrap ratios from partially different alloys [2].

The CrossAlloy® alloy concept developed by AMAG provides a promising answer to this increasingly pressing issue. This concept aims to combine maximum strength with good formability and other advantageous technological properties in order to produce a material with a broad range

F. Schmid (✉) · L. Stemper · R. Tosone
AMAG rolling GmbH, Lamprechtshausener Straße 61, 5282 Ranshofen, Austria
e-mail: florian.schmid@amag.at

© The Minerals, Metals & Materials Society 2023
S. Broek (ed.), *Light Metals 2023*, The Minerals, Metals & Materials Series,
https://doi.org/10.1007/978-3-031-22532-1_68

of applications. In this way, a mix-up of different alloy grades can be effectively countered in the course of recycling. The following provides a detailed description of this innovative alloying concept.

Alloy Concept CrossAlloy®

The decisive factor in the development of this new CrossAlloy® alloy class was to break up the established alloy classifications in order to develop an innovative material concept detached from classic, normative restrictions.

In principle, alloy classes are divided according to their respective main alloying elements, which also determine the main properties. In 5xxx alloys, Mg contributes very strongly to excellent formability, while, in 6xxx and 7xxx alloys, Mg + Si and Zn + Mg, respectively, lead to high and highest strengths. Vice versa, 5xxx alloys show comparatively low strengths, while high-strength 7xxx material shows rather low ductility. These trends are illustrated in Fig. 1 by means of the plotted yield strengths as well as elongations at fracture, although elongation at fracture is not the sole indicator of good or poor formability.

In order to combine these apparently opposite material properties, a first CrossAlloy® concept has been designed by AMAG. AMAG CrossAlloy.57 contains both sufficient Mg for good formability and matched amounts of Zn and Cu to achieve high strength. Simplified, it represents a mixture of a 5xxx and 7xxx alloy and should therefore result in an alloy with a balanced and greatly improved combination of properties—represented by the blue ellipse in Fig. 1 [3].

Material Characterization

The chemical composition of the herein investigated AMAG CrossAlloy.57 is summarized in Table 1. The alloy concept is based on a classic 5xxx alloy with about 5 wt.% Mg and small amounts of Fe, Mn, and Si. In this newly developed alloy, zinc (3.5% by weight) and copper (0.5% by weight) are also included, resulting in age-hardenability after suitable heat treatment similar to 6xxx or 7xxx alloys. Contrary to 7xxx alloys containing Zn and Mg as well, Eta phase does not form and is not responsible for the strength increase during ageing. In this, CrossAlloy® showing an Mg/Zn ratio >1 T-phase (AlnZnCuMg phase) forms. This phase is characterized by a comparatively high thermal stability as well as its very homogeneous distribution throughout the microstructure. In combination with a suitable thermomechanical treatment, a very fine microstructure can be achieved, allowing AMAG CrossAlloy.57 sheets to be processed by means of superplastic forming (SPF) [4, 5].

Mechanical Properties of a CrossAlloy® Sheet

For an initial classification of the mechanical properties, Fig. 2 shows a comparison between the newly developed AMAG CrossAlloy.57 and common alloys from different classes. For the production of outer skin and inner structural parts, 6016-T4 and 5182-soft are mainly used. Due to their high uniform elongation (Ag) ($\sim$22%) and good yield strength ratios, these alloys are very well suited for this purpose. Although this CrossAlloy® has a yield strength of

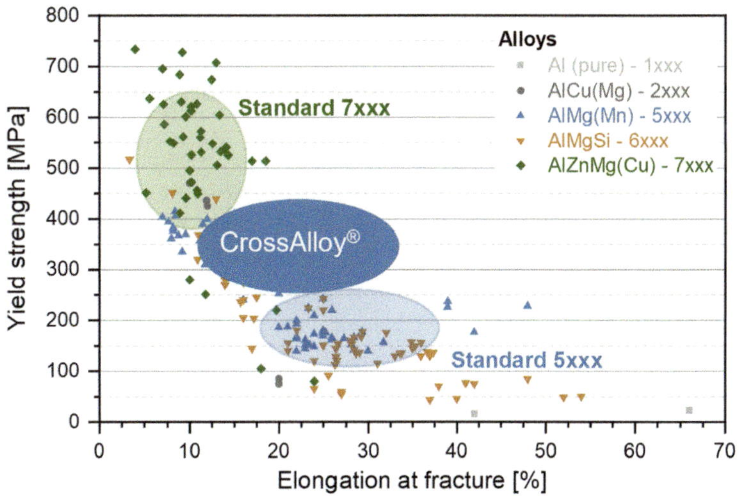

Fig. 1 Plot of yield strength and elongation at fracture of different alloy classes (1xxx, 2xxx, 5xxx, 6xxx and 7xxx) in unspecified conditions (literature data and own data)

Table 1 Chemical composition AMAG CrossAlloy.57 based on a mixture of 5xxx and 7xxx alloy

Chemical composition	Mg	Zn	Cu	Mn	Fe	Si	Al
wt.-%	5	3,5	0,5	< 0,5	< 0,4	< 0,2	Residue

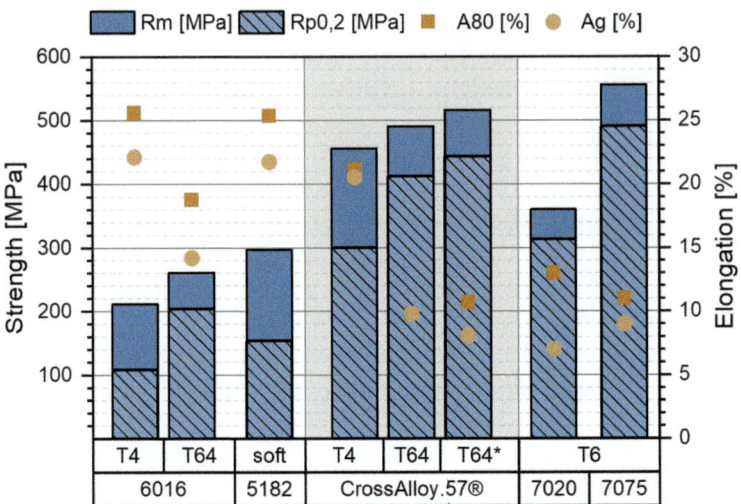

Fig. 2 Mechanical properties (yield strength—Rp0.2, tensile strength—Rm, uniform elongation—Ag, elongation at fracture—A80) of a cold-rolled AMAG CrossAlloy.57 1 mm sheet and other selected aluminum alloys. T64 refers to 2% pre-strain and ageing at 185 ° C/20 min; T64* refers to 2% pre-strain and ageing at 110 ° C/50 min + 165 ° C/35 min + 180 °C/40 min

300 MPa, uniform elongation is also over 20%, which can be regarded as an initial indicator of equally good formability.

In the course of automotive production, the body-in-white undergoes various temperature cycles, among other things for curing the applied adhesive or for drying the various paint layers. During this thermal treatment, age-hardenable aluminum alloys acquire their final product properties, in particular their in-service strength. While 5xxx alloys cannot form any hardening phases during these processes, 6016 experiences a strength increase of about 100 MPa, resulting in a final strength of about 200 MPa Rp0.2. To the same extent, this CrossAlloy® benefits from these temperature cycles and an associated formation of strength-increasing T-phase precipitates. A final strength of over 400 MPa is thus achieved, allowing comparison with high-strength 7xxx alloys.

Mechanical Properties of a CrossAlloy® Plate

A comparison with common alloys from 5xxx, 6xxx and 7xxx series can also be made for hot-rolled plates. On the one hand, a comparison of non-heat-treated 5xxx alloys AlMg3 (e.g. 5754) and AlMg5 (e.g. 5083) with AMAG CrossAlloy.57 in the solution-annealed and cold-aged T4 condition is useful. These three materials show similar values in terms of their measured elongation at break (~25–29%), but the strength values differ significantly. While the naturally hard AlMg3 and AlMg5 show Rp0.2 values of about 110 and 160 MPa, respectively, the CrossAlloy® plate shows a strength of 300 MPa. The measured tensile strengths show similar trends.

After suitable artificial aging to achieve T6 strength, CrossAlloy® outperforms common 6xxx alloys such as 6061 and 6082 or a Cu-free 7020 by more than 150 MPa. In this regard, a comparison with a high-strength 7075 seems fair.

Technological Aspects AMAG CrossAlloy.57

As already mentioned earlier, technological properties are also decisive for industrial use. Table 2 gives a rough schematic overview of the most important properties of this CrossAlloy® in comparison with other alloy classes.

The formability of AMAG CrossAlloy.57 sheets can be compared with products from 5xxx and 6xxx series alloys in terms of maximum degree of forming or feasible component geometries—but at a much higher strength level (see also Figs. 2 and 3). In addition, semifinished products made from AMAG CrossAlloy.57 exhibit very good values with regard to their joining properties such as welding (resistance spot welding) and bonding, which puts them in competition with products typically used in automotive construction. It is also fair to compare this CrossAlloy® type with 7xxx alloys due to their very good mechanical properties. In contrast to these, however, CrossAlloys® exhibit good cold formability even at room temperature, weldability, and corrosion resistance. Similar positive properties in terms of corrosion behavior and weldability (TIG welding) have already been demonstrated for thicker products such as hot-rolled plates too. Good anodizability also enlarges the possible fields of application in the direction of electronics applications and other visible applications.

Summary and Potential Fields of Application

Considering the available results so far, this first AMAG CrossAlloy.57 concept, which is based on a mixture of 5xxx and 7xxx alloys, impresses with its excellent mix of properties. Many of the material properties and processing characteristics investigated are more than competitive with conventional alloy classes and are additionally combined in an exceptional manner in one material. In combination with

Table 2 Schematic comparison of the mechanical–technological properties of 5xxx, 6xxx and 7xxx alloys with AMAG CrossAlloy.57. The arrows shown symbolize very positive (↑↑↑), neutral (~), and very negative properties (↓↓↓)

	5xxx	6xxx	7xxx	AMAG CrossAlloy.57
Strength	~	↑(↑)	↑↑↑	↑↑(↑)
Elongation	↑↑	↑↑↑	↓↓	↑↑
Bendability	↑↑	↑↑↑	↓↓↓	↓
Formability	↑↑↑	↑↑	↓↓↓	↑↑
Corrosion resistance	↑↑↑	↑↑	↓↓	↑↑
Weldability	↑↑↑	↑↑	↑/↓↓	↑(↑)
Adhesive bonding	↑↑↑	↑↑↑	Not tested	↑↑↑
Anodizability	↑↑	↑↑	↑↑	↑↑

Fig. 3 Mechanical properties (yield strength—Rp0.2, tensile strength—Rm, elongation at fracture—A50) of a hot-rolled AMAG CrossAlloy.57 10 mm plate and other selected aluminum alloys

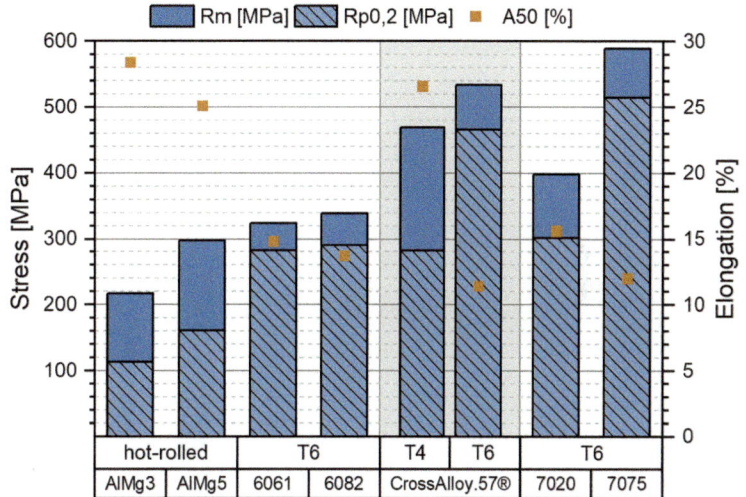

the achievable high strength, this results in a large number of possible fields of application:

- High-strength structural components in automotive engineering.
- Age-hardenable SPF components—complex, high-strength, and joint-free components.
- High-strength panels in the transportation sector.
- Use in specialty applications due to the unique property portfolio.

Finally, a comparison of the specific strengths of various aluminum and steel alloys shows the great potential of AMAG CrossAlloy.57 (see Fig. 4). While the specific strength is close to the level of high-strength steel grades or 7075 alloys, this first CrossAlloy® concept allows component production with conventional processing methods and thus broad use in various application areas.

Fig. 4 Comparison of specific strengths of selected aluminum alloys including AMAG CrossAlloy.57 in different heat treatment states with several steel grades

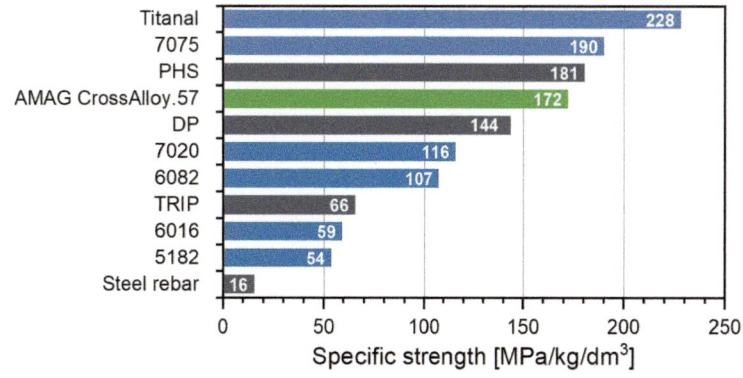

References

1. Ostermann F (2014) Anwendungstechnologie Aluminium. 3. Auflage.

2. Raabe D, Tasan CC, Olivetti EA (2019) Strategies for improving the sustainability of structural metals. Nature 575 (7781): 64–74. https://doi.org/10.1038/s41586-019-1702-5.

3. Stemper L, Tunes MA, Tosone R et al. (2022) On the potential of aluminum crossover alloys. Progress in Materials Science 124 100873.https://doi.org/10.1016/j.pmatsci.2021.100873.

4. Stemper L, Tunes MA, Oberhauser P et al. (2020) Age-hardening response of AlMgZn alloys with Cu and Ag additions. Acta Materialia 195 541–554.https://doi.org/10.1016/j.actamat.2020.05.066.

5. Stemper L, Tunes MA, Dumitraschkewitz P et al. (2021) Giant hardening response in AlMgZn(Cu) alloys. Acta Mater. doi: https://doi.org/10.1016/j.actamat.2020.116617

Effect of Alloying Elements on Corrosion Resistance of Quench-Free Al–Ca Alloys for HPDC

Dmitry Fokin, Sergey Matveev, Roman Vakhromov, Dmitry Ryabov, and Aleksandr Alabin

Abstract

For various products made of cast aluminum alloys, such as automotive components and parts of household appliances contacting with a corrosive environment, increased corrosion resistance is required. One of the ways to increase the corrosion resistance of aluminum alloys is anodizing. However, anodizing increases the cost of products and cannot be applied to bimetallic castings. Al–Ca alloys demonstrate good corrosion resistance in an as-cast state together with excellent castability allowing to produce parts of complex geometry. In the Al–Ca system alloys, various alloying elements are used to achieve a good level of mechanical properties. In this paper, Al–Ca compositions containing Zn, Si, and Mn were analyzed in comparison with reference cast aluminium alloys in salt spray and alkaline media. It is shown that Al–Ca alloys have a good potential for applications where high corrosion resistance is a key characteristic.

Keywords

Cast alloy • Corrosion resistance • Heat treatment free • HPDC

Introduction

Interest in alloys that do not require heat treatment is constantly growing from manufacturers of various products using cast aluminum components. The exclusion of the heat treatment operation makes it possible to reduce the cost by up to 20%. In this regard, new alloys are emerging based on alloying systems alternative to Al–Si alloys, which show a high level of performance in the as-cast condition.

Following this trend, eutectic alloys of the Al–Ca system with Ca content of up to 5% have been developed for high-pressure die casting (HPDC) technology. Al–Ca alloys are used in an as-cast state, have a melting point of 625–640 °C, and a narrow crystallization range of 10–20 °C, which ensures their high casting properties comparable to commercial Al–Si alloys. Thus, such alloys can easily be integrated into an existing serial production process. As shown in [1], alloying the Al–Ca system with manganese, zinc, and silicon leads to the formation of favorable structures in castings and improves the mechanical properties. The combination of properties of these alloys allows them to be considered for a wide range of applications, including the manufacture of automotive components, household appliances, and housings, where special attention is paid to corrosion properties.

In this work, corrosion resistance of Al–Ca alloys in the as-cast state, containing from 1.5 to 4.7% calcium, in comparison with serial HPDC alloys was analyzed. Alloys of the AlSi10MnMg type (in the T6 state) and AlMg5 (in the as-cast state) were taken as references. The first is one of the most common for use in structural automotive components. The second one is also used for casting parts by the HPDC method and has high corrosion resistance typical for the Al–Mg alloys, but relatively lower casting properties. Four Al–Ca alloys were investigated with the calcium content of 1.5, 2.3, 3.4, and 4.7% and with mechanical properties given in Table 1.

The corrosion resistance of the material in many applications related to various aggressive environments often determines the lifetime of the product. According to [2], the pitting corrosion penetration depth h for aluminum alloys is proportional to the cube root of time and is described by Eq. (1):

D. Fokin (✉) · S. Matveev · R. Vakhromov · D. Ryabov
Light Materials and Technologies Institute (LLC LMTI), Leninsky Ave. 6b21, 119049 Moscow, Russia
e-mail: Dmitriy.Fokin@rusal.com

A. Alabin
Russian Aluminum Management (UC Rusal), Vasilisi Kozhinoi, 1, 121096 Moscow, Russia

© The Minerals, Metals & Materials Society 2023
S. Broek (ed.), *Light Metals 2023*, The Minerals, Metals & Materials Series,
https://doi.org/10.1007/978-3-031-22532-1_69

Table 1 Mechanical properties of Al–Ca compositions

Alloy designation	UTS, MPa	YS, MPa	Elongation, %
Al–Ca1.5	192 ± 4	98 ± 3	11 ± 3.0
Al–Ca2.3	213 ± 5	111 ± 4	9 ± 2.5
Al–Ca3.4	244 ± 4	124 ± 5	12 ± 3.5
Al–Ca4.7	267 ± 7	138 ± 4	4 ± 0.5

$$h = kt^{1/3} \qquad (1)$$

where t—time; k—a constant depending on alloy type.

The corrosion rate **r** in this case, obtained from (1), is

$$r = \frac{dh}{dt} = \frac{k}{3\sqrt[3]{t^2}} \qquad (2)$$

From Eq. (2), it follows that the rate of propagation of pitting corrosion decreases with time. Thus, the impact of corrosion on the material is most intense in the initial stage of its operation life. At the same time, as follows from the equations, a decrease in the corrosion rate by 2 times, other things being equal, potentially makes it possible to increase the service life of the product by 8 times.

There are a large number of different tests to determine the characteristics of corrosion resistance in various environments. It should also be noted that in many industries, including the automotive industry and the production of household appliances, there are no uniform standards for corrosion testing, and each manufacturer uses its own methods and sets requirements for the material performance. However, when benchmarking with commercially used alloys, it is possible to use the most common methods or specific tests in environments associated with specific operating conditions to obtain an overall assessment. In this work, studies were carried out in a neutral salt spray environment with a long exposure for 720 h and in a strongly alkaline solution using a shortened method.

Experimental Procedure

The compositions of the alloys studied in this work are presented in Table 2. The samples were made using HPDC and gravity die casting (GDC) techniques. High-pressure die casting (all alloys from Table 2) was carried out on a machine with a locking force of 400 tons, and flat castings were made with a thickness of 3 mm. The pouring temperature in the case of Al–Ca alloys was 690–720 °C, and the temperature of the mold was 150–180 °C. The second group of samples (alloys Al–Ca2.3, Al–Ca4.7, and Al–Si10) was made by gravity casting into a graphite mold without preheating, and the casting temperature did not exceed 750 °C. In this way, samples with a thickness of 12 mm were obtained, and then they were machined to a thickness of 10 mm. The Al–Ca and Al–Mg5 alloys were tested in the as-cast state, and the Al–Si10 alloy was tested in the T6 state. The structure was analyzed using a Zeiss Axio Observer optical microscope. The mass of samples before and after corrosion tests was determined on a Vibra analytical balance with a maximum mass and a measurement resolution of 220 g and 0.0001 g, respectively.

Salt spray tests were carried out in a CC1000ip cyclic corrosion chamber at 35 ± 2 °C by continuous spraying of 5% NaCl solution. The duration of the test was 720 h. The size of the test HPDC pieces (all alloys, Table 2) was 3 × 60 × 60 mm. The size of test specimens made by casting into a graphite mold (Al–Ca2.3, Al–Ca4.7, and Al–Si10 alloys, Table 2) was 10 × 80 × 100 mm. After removal from the chamber, the samples were washed in tap water, dried, and corrosion products were removed from the surface of the samples by immersion in a 30% HNO3 solution for 2 min.

Alkali tests were carried out by immersing samples in a 10% NaOH solution (with a pH value of 13.7). The specimens were 3 × 10 × 20 mm in size. Holding time was 5, 30, and 160 min. After that, the samples were washed in tap water, dried, and corrosion products were removed in a 30% HNO3 solution for 30 s. Corrosion rate **r** [mm/year] was calculated from formula (3):

Table 2 Chemical composition of alloys, wt. %

Alloy designation	Ca	Mn	Zn	Si	Fe	Mg	ρ, g/cm^3
Al–Ca1.5	1.5	1.9	–	0.15	0.2	–	2.71
Al–Ca2.3	2.3	0.9	0.4	0.15	0.2	–	2.67
Al–Ca3.4	3.4	0.85	0.85	0.40	0.2	–	2.66
Al–Ca4.7	4.7	0.85	1.5	0.75	0.2	–	2.65
Al–Si10	–	0.6	–	10.6	0.1	0.3	2.64
Al–Mg5	–	0.6	–	2.1	0.1	5.2	2.65

$$r = \frac{\Delta m \cdot 365 \cdot 1000}{S \cdot T \cdot \rho} \qquad (3)$$

where Δm is specimen mass loss after testing, g; S is specimen surface, m^2; T is duration of the test, days; ρ is alloy density, g/cm^3.

Results and Discussion

From the point of view of evaluating the effect on corrosion properties, in addition to differences in chemical composition, it is advisable to consider how the phase composition changes in the presence of various alloying elements. The considered calcium alloys have the phase composition shown in Table 3. As can be seen, in the presence of zinc, the Al$_4$Ca phase is transformed into (Al, Zn)$_4$Ca, and with the introduction of additional silicon, a new Al$_2$CaSi$_2$ phase is formed. Manganese is completely soluble in aluminum solid solution. The mass fraction of the Al$_4$Ca or (Al, Zn)$_4$Ca eutectic phase is almost proportional to the calcium content in the alloy and is about 3.4 wt.% per 1 wt.% Ca.

Salt Spray Test

Figure 1 shows the corrosion rate values calculated by formula (3) from the results of mass loss measurements in the salt spray chamber for HPDC and GDC samples. Alloys of the Al–Ca system for both types of samples show significantly lower corrosion rates in comparison with the Al–Si10 alloy and are approximately on the same level with the Al–Mg5 alloy. In addition, the Al–Ca4.7 alloy demonstrates the lowest corrosion rate in both cases. The corrosion rate calculated for machined specimens having lower roughness is

predictably lower than for HPDC specimens with the as-cast surface.

Figures 2 and 3 show the appearance of HPDC and GDC samples before and after testing (after removal of corrosion products), as well as a typical structure in the cross-section. As can be seen, there is no visible corrosion damage on all samples. For Al–Ca samples with 3.4 and 4.7% Ca, a slight change in the surface color towards a yellowish tint can be noted. The microstructure pictures also show no pitting corrosion for the Al–Ca and Al–Mg5 alloys. For the Al–Si10 alloy sample, traces of pitting corrosion with a depth of 90 µm were found.

Since the alloys of the Al–Ca system are intended for producing finished geometry castings by HPDC technique without subsequent machining, the most representative are the results presented in Fig. 1a. However, it should be noted that HPDC is characterized by the presence of segregation in the near-surface zone of castings [3], i.e. the difference in the composition of the surface region compared to the metal in the middle of the cross-section. For Al–Ca alloys, as shown in Fig. 4 on the example of Al–Ca3.4 alloy, the surface layer is a eutectic-rich region. In this regard, the data obtained on machined samples make it possible to evaluate the corrosion resistance for the structure in the casting body. In particular, the results from Fig. 1b allow concluding that, in comparison with the Al–Si10 alloy, the mass loss trend for calcium alloys does not change. The corrosion rate is lower if compared to HPDC specimens, but this is obviously due to the low roughness of the machined samples and this is in agreement with the fact that with the decrease in the roughness, the corrosion rate tends to decrease, as was shown for example in [4]. Thus, it can be expected that the rate of corrosion propagation will not depend on the depth of its penetration.

Table 3 Phase composition of Al–Ca alloys

Alloy code	Phases (wt.%)		
Al–Ca 1.5	(Al)–94,5%	Al$_4$Ca–5,1%	Al$_2$CaSi$_2$ < 0.5%
Al–Ca 2.3	(Al)–91,5%	(Al, Zn)$_4$Ca–8,1%	Al$_2$CaSi$_2$ < 0.5%
Al–Ca 3.4	(Al)–87,3%	(Al, Zn)$_4$Ca–11,7%	Al$_2$CaSi$_2$–1,1%
Al–Ca 4.7	(Al)–82,6%	(Al, Zn)$_4$Ca–15,4%	Al$_2$CaSi$_2$–2,0%

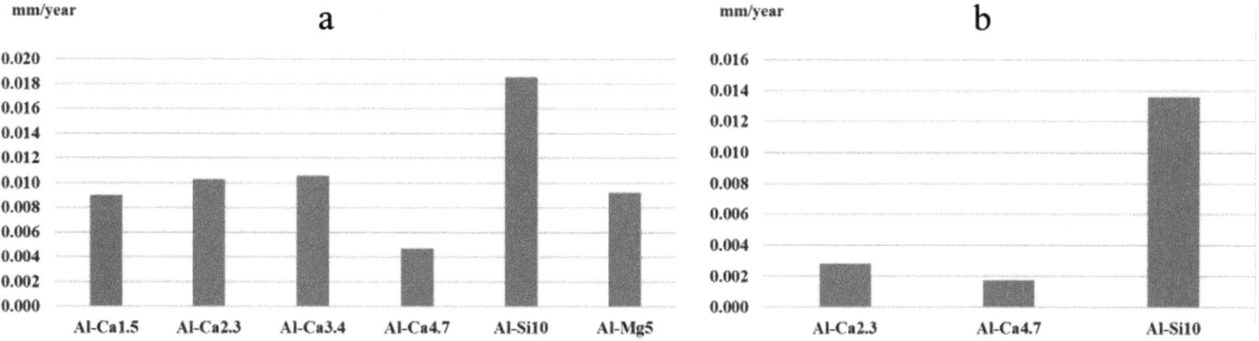

Fig. 1 Corrosion rate in salt spray medium for tested alloys—**a** HPDC specimens (as-cast surface), **b** GDC specimens (machined surface)

Fig. 2 Appearance of the HPDC specimens before and after the salt spray test: **a** Al–Ca1.5 alloy; **b** Al–Ca2.3 alloy; **c** Al–Ca3.4 alloy; **d** Al–Ca4.7 alloy; **e** Al–Si10 alloy; **f** Al–Mg5 alloy

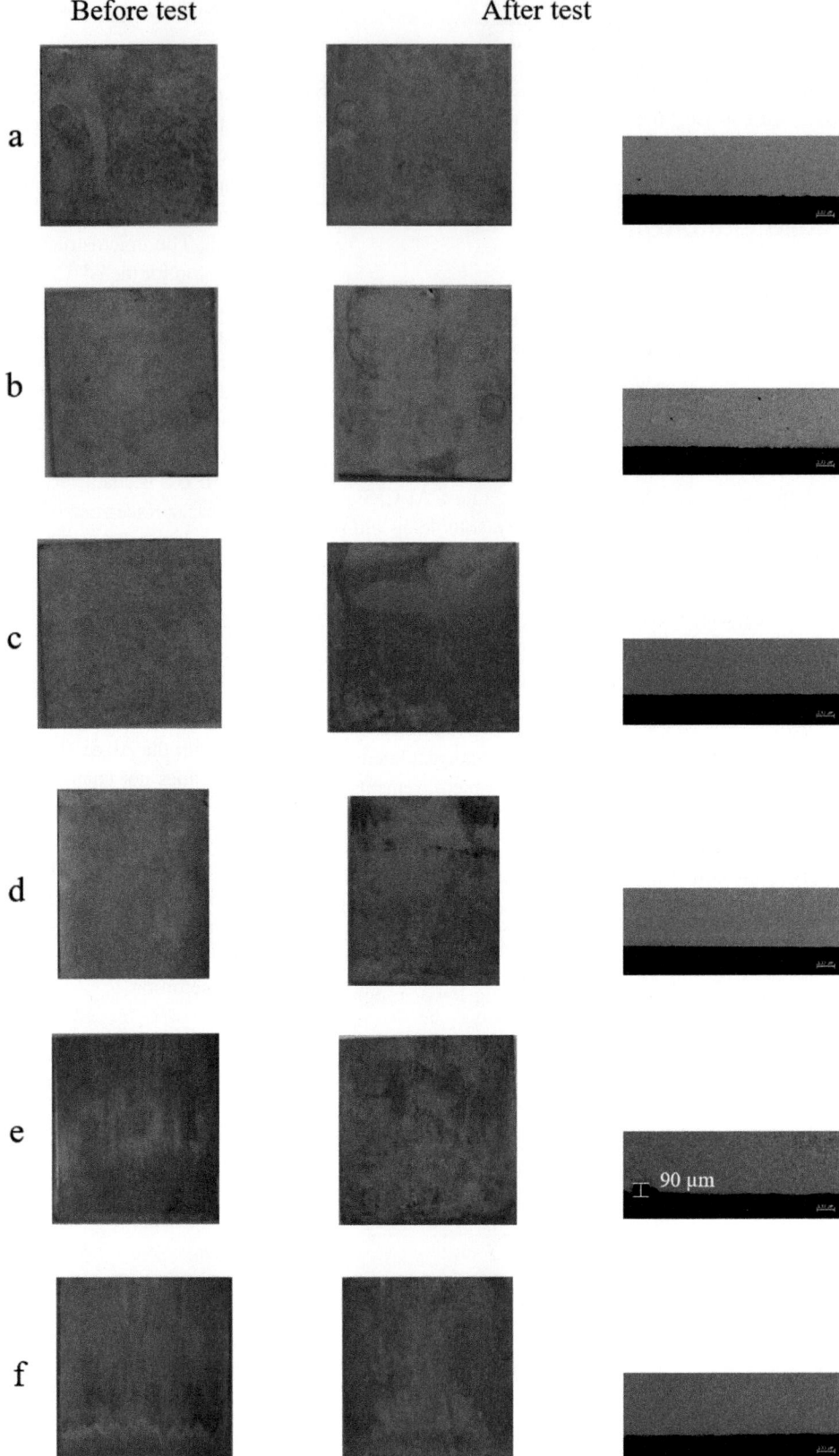

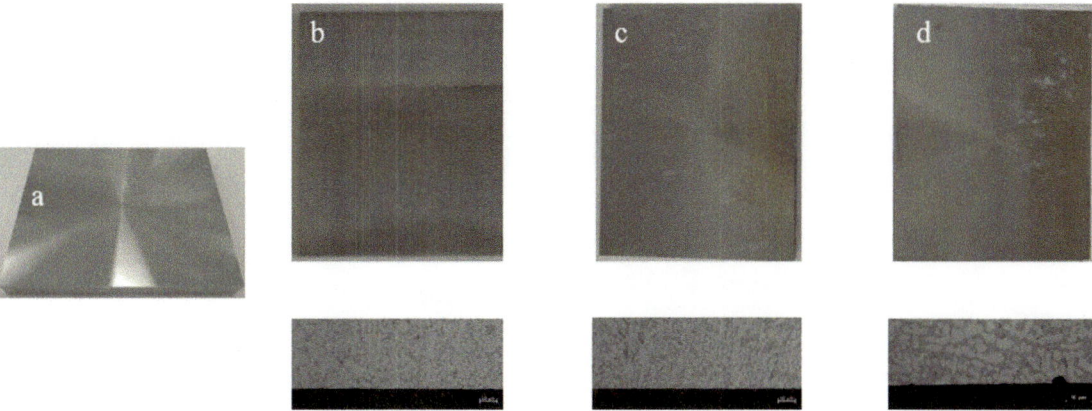

Fig. 3 Appearance of the GDC machined specimens. **a** Specimen before test; specimens after test: **b** Al–Ca2.3 alloy; **c** Al–Ca4.7 alloy; **d** Al–Si10 alloy

Fig. 4 Surface segregation in the 3 mm thick HPDC casting made of Al–Ca3.4 alloy

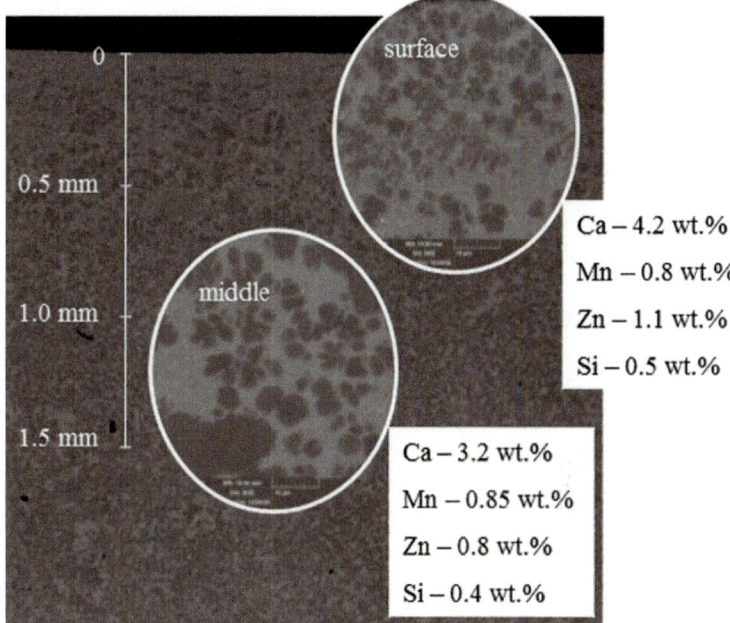

The high corrosion resistance of calcium alloys in a salt medium can be explained in such a way that the potential of the solid solution and eutectic calcium-containing phases is quite close. However, there is no exact data on the values of the potential of calcium phases.

High Alkaline Test

The 10% NaOH solution used in the work is a strong alkali (pH = 13.7) and a very aggressive medium for aluminum. The high dissolution rate of aluminum in NaOH allows testing without long holding times. The results of this test were used to obtain a comparative assessment of alloy resistance in an alkaline environment. For testing, identical samples with dimensions of $3 \times 10 \times 20$ mm (Fig. 5) were

Fig. 5 Specimens $3 \times 10 \times 20$ mm for strong alkali tests

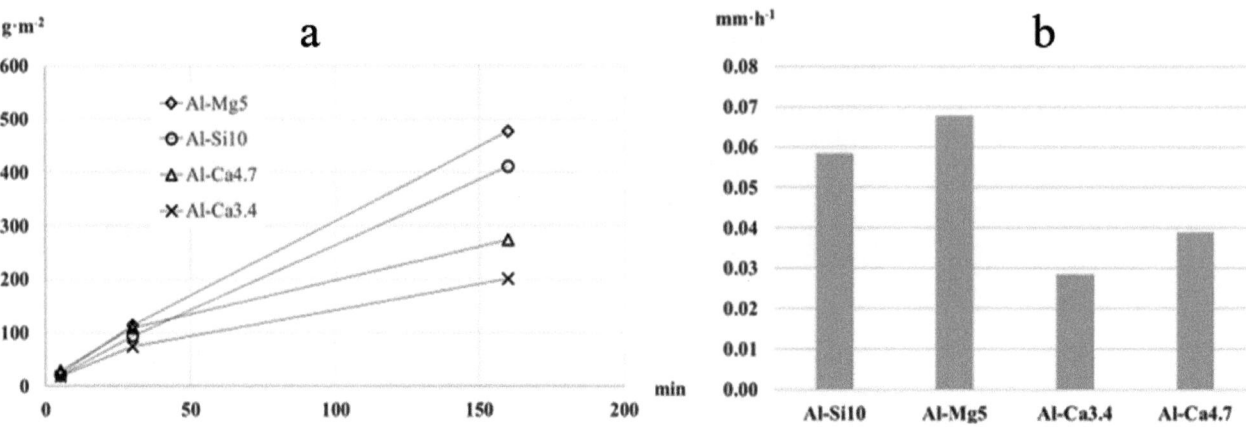

Fig. 6 Mass loss in alkali test, **a** g/m² after 5, 30, 160 min; **b** mm/h

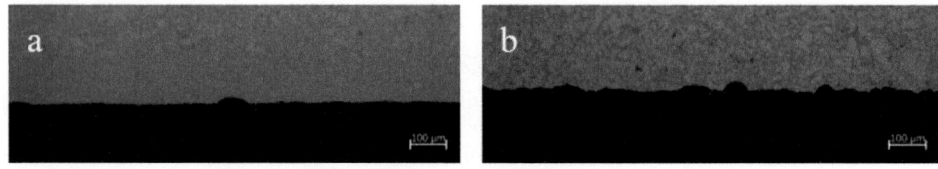

Fig. 7 Micro-sections of the specimens after alkaline test, **a** Al–Ca3.4 alloy, **b** Al–Mg5 alloy

prepared by cutting from 3 mm plates of Al–Ca3.4, Al–Ca4.7, Al–Si10, and Al–Mg5 alloys obtained by HPDC. The weight loss values of the specimens after 5, 30, and 160 min exposure to the solution are shown in Fig. 6a. The corrosion rate was calculated using formula (3) for a test time of 160 min and is shown in Fig. 6b. As in the case of the salt spray test, the corrosion rate samples in NaOH solution is 1.5–2 times lower in comparison with Al–Si10 alloy. At the same time, the Al–Mg5 alloy, which has a higher resistance in a salt medium, is prone to more intense corrosion in an alkali solution compared to the other alloys considered in this work. In addition to corrosion rate values, a comparison of micro-sections of Al–Ca3.4 and Al–Mg5 specimens (Fig. 7) also showed that higher corrosion damage corresponds to Al–Mg5 alloy.

Conclusions

Comparative tests have shown high corrosion resistance of Al–Ca alloys with calcium content from 1.5% to 4.7% in both neutral salt and alkaline media. In a salt spray environment, their corrosion resistance is slightly better than for Al–Mg5 (first of all in case of high Ca) and 1.5–2 times higher than that of Al–Si10. In an alkaline environment, Al–Ca alloys have a 2 times lower corrosion rate in comparison with Al–Mg5 and Al–Si10. The received results indicate the high potential of Al–Ca alloys for applications requiring high corrosion resistance.

References

1. Dmitry Fokin, Sergey Matveev, Roman Vakhromov, Aleksandr Alabin Effect of Alloying Elements on Strength Properties and Casting Properties of Corrosion Resistant Quench-Free Al–Ca Alloys. Light Metals 2022, Springer. P.113–118.
2. Christian Vargel, Corrosion of aluminium, Elsevier 2004
3. S.Otarawanna, C.M.Gourlay, H.I.Laukli, A.K.Dahle. Formation of the surface layer in hypoeutectic Al-alloy high-pressure die castings. Materials Chemistry and Physics. 2011. Volume 130, Issues 1–2. P. 251–258.
4. Ahmad Almansour, Mazen Azizi, Abdul Munem Jesri, Sami Entakly Effect of Surface Roughness on Corrosion behavior of Aluminum Alloy 6061 in Salt Solution (3.5%NaCl), International Journal of Academic Scientific Research. 2015. Volume 3, Issue 4. P.37–45.

Influence of Increased Cu and Fe Concentrations on the Mechanical Properties of the EN AB-42100 (AlSi7Mg0.3) Aluminum Alloy

T. Beyer, D. Ebereonwu, A. Koch, P. Decker, A. Kauws, M. Rosefort, and F. Walther

Abstract

Production of Al alloys from End of Life (EoL) scrap is accompanied by increased Cu and Fe contents, as these cannot be economically removed from the melt. Within the scope of this work, Cu- and Fe-containing alloy variants of AlSi7Mg0.3, based on CALPHAD simulations, were produced at a laboratory scale using gravity die casting. Microstructural analyses and tensile tests of the as-cast state were carried out to characterize the influences of Cu and Fe on microstructural and mechanical properties. It is shown that Fe and Cu contents above the known limits of the standards can result in good mechanical properties. Hardness, yield strength, and tensile strength correspond well to the properties of the base alloy and cause only minor losses in elongation.

Keywords

Recycling • Mechanical properties • Microstructure • EN AB-42100 • AlSi7Mg0.3 • Aluminum • Iron • Copper • Interactions • Cast alloys

Introduction

Global demand for aluminum is expected to increase up to 78 million tons by 2029 [1]. The production of primary aluminum is an energy-intensive process. To reduce the carbon footprint of their products, companies are willing to use aluminum alloys with increased content of recycled material. Increased Fe and Cu contents resulting from the scrap material are a major challenge that need to be solved since no suitable process for removing these alloying elements from the aluminum alloys is available on an industrial scale. Along with the demand for sustainable mobility, the share of conventional combustion engines is expected to decrease significantly [2]. Therefore, increased amounts of Fe- and Cu-containing aluminum scrap will flow back into the recycling loop. Reduced ductility of AlSi(Cu) alloys is observed especially at higher Cu contents [3]. A high Cu content confers high-temperature strength, which is a reason why this type of alloy is often used for engines and pistons [4]. During the production of primary aluminum, Fe occurs as a natural impurity. Molten primary aluminum typically contains up to 0.15 wt% Fe. Further (re-)melting activities lead to increased Fe content [5]. Since such types of scrap differ from the widely used universal casting alloy EN AB-42100 (AlSi7Mg0.3), the influence of increased Fe and Cu contents on the properties of this alloy is investigated in this work. The main goal is to determine if Fe- and Cu-rich scrap materials are a suitable base material to produce Al-casting alloys such as EN AB-42100 that fulfill the mechanical property demand. Therefore, a combined approach of thermodynamic calculation, microstructural investigation, and mechanical testing was chosen to prove the mechanical suitability while simultaneously demonstrating the microstructural origins of the observed effects. Since a broad variance in the Fe and Cu concentrations in scrap material is expected, 18 gravity die cast alloy materials with varying Cu and Fe concentrations were examined.

T. Beyer (✉) · D. Ebereonwu · P. Decker · A. Kauws · M. Rosefort
TRIMET Aluminium SE, Aluminiumallee 1, 45356 Essen, Germany
e-mail: tobias.beyer@trimet.de

A. Koch · F. Walther
Chair of Materials Test Engineering (WPT), TU Dortmund University, Baroper Straße 303, 44227 Dortmund, Germany

© The Minerals, Metals & Materials Society 2023
S. Broek (ed.), *Light Metals 2023*, The Minerals, Metals & Materials Series,
https://doi.org/10.1007/978-3-031-22532-1_70

Experimental

Experimental Design

The experimental matrix, shown in Fig. 1, is based on the evaluation and analysis of literature, patents, and production data and shows the chemical composition of all investigated samples. The numbers correspond to the compositions given in Table 1. Since Fe occurs as a natural impurity during the production of primary aluminum, a base material with a Fe content of 0.12 wt% was used. The maximum Fe content to be considered was set at 1.25 wt% to simulate alloys that tolerate high levels of scrap. The maximum Cu content was chosen to be 3 wt% and is based on existing Cu-containing casting alloys from combustion engine components, as these represent the potential recycling material.

Although AlSi7Mg0.3 is often used in a heat-treated condition, all investigations were carried out on test specimens produced by gravity die casting in the as-cast condition, since a suitable heat treatment would need to be individually redesigned for each set of Fe and Cu contents and the microstructure after casting could be investigated specifically.

Thermodynamic Calculation

For the prediction of the expected microstructural constituents, thermodynamic calculations based on the CALPHAD method were carried out using Thermo-Calc software version 2021b and TCAL8 database. For an ease of computation, only elements with concentrations of > 0.1 wt% were selected in the equilibrium calculation (Al, Si, Fe, Cu, and Mg).

Alloy Production

A resistance-heated furnace with a capacity of 45 kg was used to produce the different alloy variants. The crucible was coated with boron nitride prior to each melting operation. The melt of each alloy variant was cleaned by impeller treatment, resulting in a density index DI < 0.5%. AlTi5B1 master alloy was used for grain refinement. Cylindrical specimens (height 30 mm, diameter 40 mm) were cast for microscopic analysis and hardness tests. The samples for the tensile tests were produced from the cast bars. All steel molds were first heated up to 300 °C, followed by the casting process with a casting temperature of 720 °C. The samples were quenched using water at room temperature. An overview of the composition of all investigated alloy variants is given in Table 1.

Microscopy

The cylindrical specimens were milled down to a height of 15 mm. The cross-sectional area was divided into "center", "middle", and "edge", as can be seen in Fig. 2, to consider varying solidification times within the specimen's cross section. Microstructural analyses included measurement of the length of Fe phases, amount of Cu phases, and qualitative microstructural analysis.

To measure the Fe phase lengths at 500× magnification, three images were taken for every specimen in each of the three areas, with the longest 20 Fe phases being measured in each case. Mean value and standard deviation were calculated from the total of nine images per alloy. The Fe phases were determined based on the typical morphology. A distinction between the Fe phases is not clear for every case,

Fig. 1 Experimental Cu-Fe matrix for all investigated samples

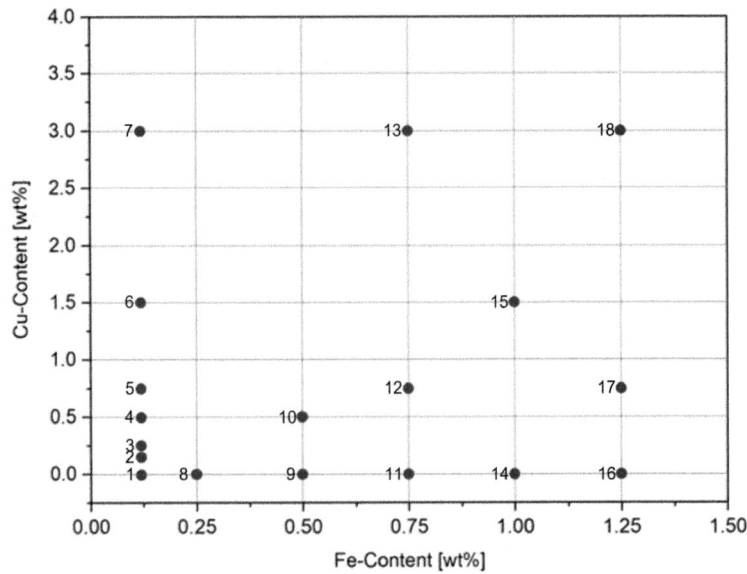

Table 1 Composition of all investigated alloy variants

No	Alloy	Si [wt%]	Fe [wt%]	Cu [wt%]	Mg [wt%]	Others [wt%]
1	Fe0.12	7.08	0.12	0.01	0.31	92.48
2	Fe0.12Cu0.15	7.06	0.12	0.15	0.30	92.37
3	Fe0.12Cu0.25	7.17	0.12	0.25	0.29	92.17
4	Fe0.12Cu0.5	7.01	0.12	0.52	0.31	92.04
5	Fe0.12Cu0.75	7.01	0.12	0.74	0.30	91.83
6	Fe0.12Cu1.5	6.99	0.12	1.56	0.29	91.04
7	Fe0.12Cu3	7.05	0.12	3.05	0.29	89.49
8	Fe0.25	7.18	0.25	0.01	0.31	92.25
9	Fe0.5	7.22	0.49	0.01	0.30	91.98
10	Fe0.5Cu0.5	7.02	0.50	0.50	0.31	91.67
11	Fe0.75	7.10	0.76	0.01	0.31	91.82
12	Fe0.75Cu0.75	7.11	0.77	0.77	0.33	91.02
13	Fe0.75Cu3.0	6.98	0.76	2.98	0.29	88.99
14	Fe1	7.06	0.97	0.01	0.31	91.65
15	Fe1Cu1.5	6.98	0.99	1.48	0.31	90.24
16	Fe1.25	7.32	1.28	0.01	0.30	91.09
17	Fe1.25Cu0.75	7.32	1.28	0.76	0.30	90.34
18	Fe1.25Cu3	6.95	1.23	2.88	0.30	88.64

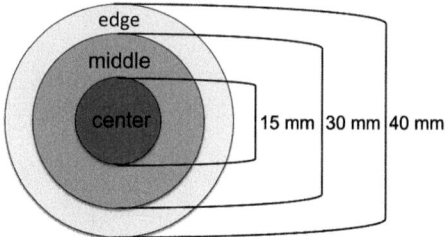

Fig. 2 Cross section of the cylindrical sample, divided into "center", "middle", and "edge"

because it is hardly possible to differentiate between the occurring Fe phases by means of light microscopy, which impedes the categorization of the individual phases in a quantitative analysis. Therefore, only the length of the total Fe phases was measured and analyzed.

Since most of the microscopically detectable Cu phases are visually very distinct from the other parts of the microstructure, an automatic analysis method was chosen to detect the number of Cu phases by calculating the area fraction. Representative images of the microstructure were taken at 500× magnification in the "edge", "middle", and "center" areas. Images were then automatically evaluated in ImageJ software using the "Color Threshold" function. The measure chosen for analysis was "area covered", which is a percentage of the covered area.

Mechanical Properties

Yield strength (Rp0.2), tensile strength (Rm), and elongation at fracture (A) were determined by tensile tests according to DIN EN ISO 6892 using a Zwick/Roell Z250 machine. Bars with dimensions of 250×20 mm^2 were used as initial workpieces to produce tensile specimens according to DIN 50125-A6x30.

Brinell hardness tests HBW5/250 were used to analyze the hardness of the alloy variants. Three specimens per variant were tested. For each specimen, three measurements were taken in the center of the specimen with a spacing of 7.5 mm. Therefore, the mean value and standard deviation were determined from nine measurements.

Results

Thermodynamic Calculation

Equilibrium calculations in Thermocalc for the model AlSi7Mg0.3 with varying Cu and Fe concentrations were performed to predict the phases that occur upon microscopic investigations. Two Fe-rich phases, $Al_{18}Fe_2Mg_7Si_{10}$, also known as π-phase [6], and $Al_9Fe_2Si_2$, also known as β-phase [4, 6], occur over the whole composition range. As expected, $Al_9Fe_2Si_2$ is increasing with Fe concentration and is not affected by Cu. $Al_{18}Fe_2Mg_7Si_{10}$ is only stable at temperatures above 340 °C and might also be found in the microstructure due to solidification and the absence of heat treatment. Both phases are usually described as needle-like when examined in micrographs and are expected to be found close to the grain boundaries [7].

Cu influences the presence of three different phases: Al_2Cu, AlCuMgSi (Q phase) [4], and Mg_2Si. Above 0.2 wt% Cu Mg_2Si cannot be found while the Q phase becomes the dominant Cu-rich phase. The Q phase amount is approx. four times the amount of Al_2Cu independent of the composition. In total, at least four of the five possible phases apart from Si and Al solid solution can be expected in all analyzed samples. Only their phase amount is increasing with the Fe or Cu concentration.

Microstructure

To illustrate the measurement methodology for determining the nominal Fe needle lengths, Fig. 3 shows micrographs of the alloys AlSi7Mg0.3 (a) and AlSi7Mg0.3Fe1.25 (b).

The average Fe needle length as a function of the Fe and Cu content of the individual alloy variants is illustrated in Fig. 4. The Fe needle length increases with Fe content. Cu appears to have almost no effect on the average needle length. Especially variants without an increased Fe content do not seem to follow a clear trend with increasing Cu content.

To illustrate the measurement methodology, Fig. 5 shows typical images used for measuring the Cu phases in ImageJ.

To evaluate the extent to which Cu phases form in each alloy variant, the surface area covered with Cu phases in the respective image section was used. The measured data are shown in Fig. 6. As the Cu content increases, the number of Cu phases also increases. This seems to follow a disproportionate trend. Whether the actual number of Cu-containing phases is affected by Fe, or it rather only affects the measurement, or the allocation of the Cu phases remains questionable.

Mechanical Properties

The influence of Fe and Cu content on tensile strength Rm is given in Fig. 7. Fe itself has almost no influence on the tensile strength of the alloys. Cu shows a major influence on the tensile strength. A negative interaction of Fe and Cu is evident. While Fe has no significant effect on the tensile strength in the absence of Cu, the presence of Cu results in a reduction. Therefore, all Cu-containing alloys with low Fe content show a higher tensile strength than those with increased Fe. The influence of Cu does not appear to be linear but follows a saturation curve. Accordingly, the increase at a low Cu content provides a higher potential for an increase in strength than at an already high content.

Figure 8 shows the influence of Fe and Cu content on yield strength Rp0.2. Fe has almost no effect, whereas Cu has a positive influence on the yield strength. The yield strength increases almost linear for Cu. The slope is significantly higher for Cu compared to Fe, which confirms that Cu has a significantly greater influence on the yield strength than Fe. The maximum appears in the variant with both maximized Fe and Cu content and is therefore positive for the yield strength.

The influence of Fe and Cu contents on elongation at fracture A is illustrated in Fig. 9. A negative interaction between Fe and Cu can generally be observed for both the low-alloyed and the high-alloyed variants. In the case of the variants containing only Fe, above a content of 0.75 wt% a rapid decline with increasing Fe content can be observed. The sole increase in Cu leads to a similar result. However,

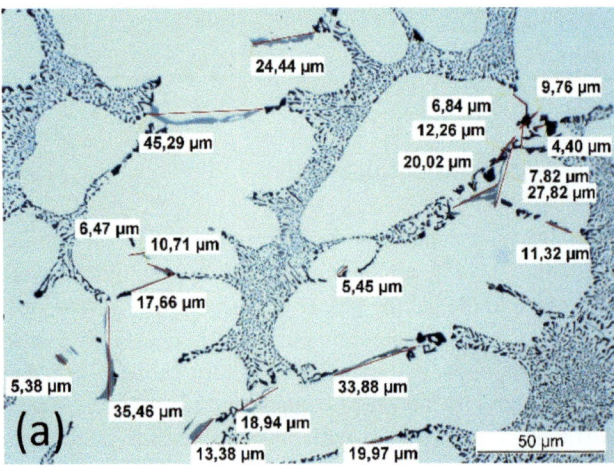
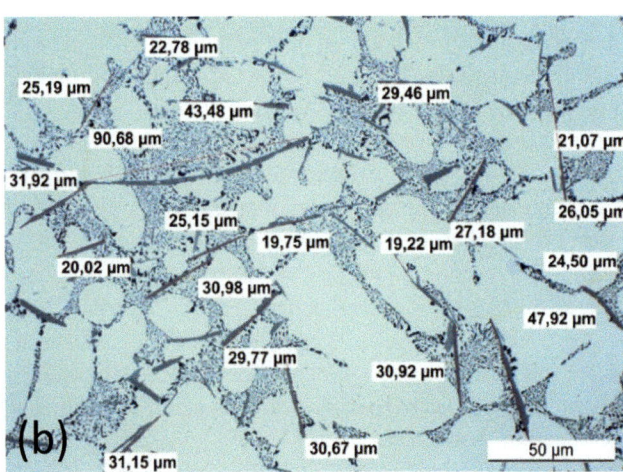

Fig. 3 Nominal Fe needle lengths for **a** AlSi7Mg0.3 and **b** AlSi7Mg0.3Fe1.25

Fig. 4 Influence of Fe and Cu content on average length of Fe phases

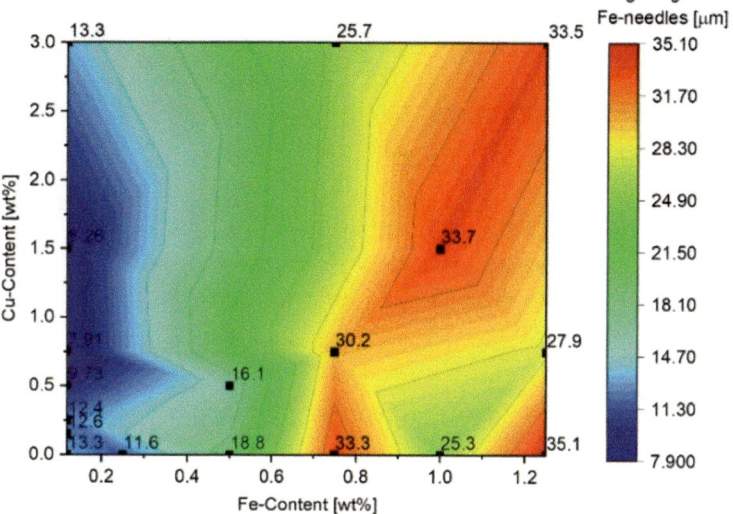

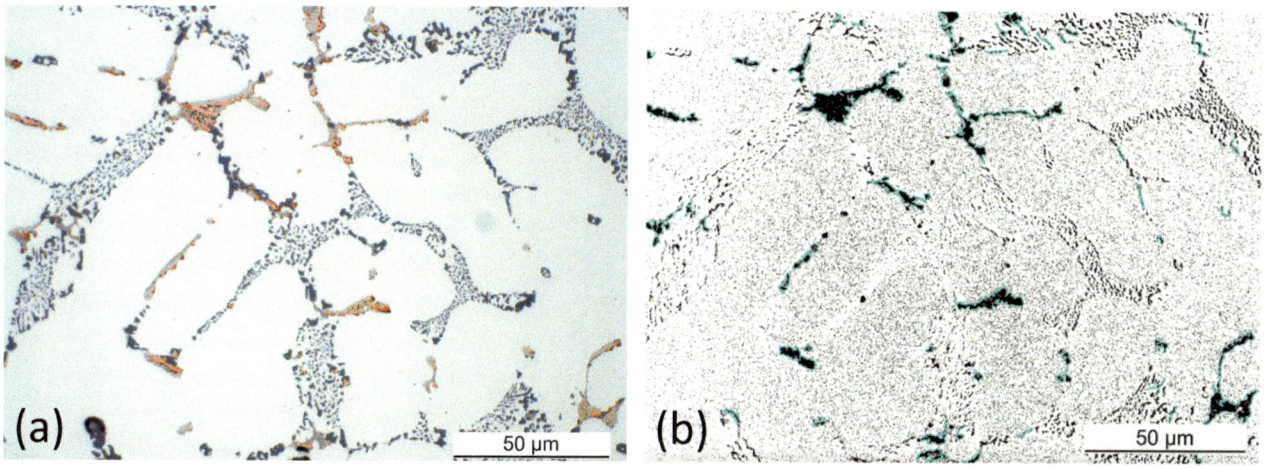

Fig. 5 Typical images of Cu phase measurement, **a** from optical microscope and **b** processed with ImageJ

Fig. 6 Influence of Fe and Cu content on average area covered by Cu phases

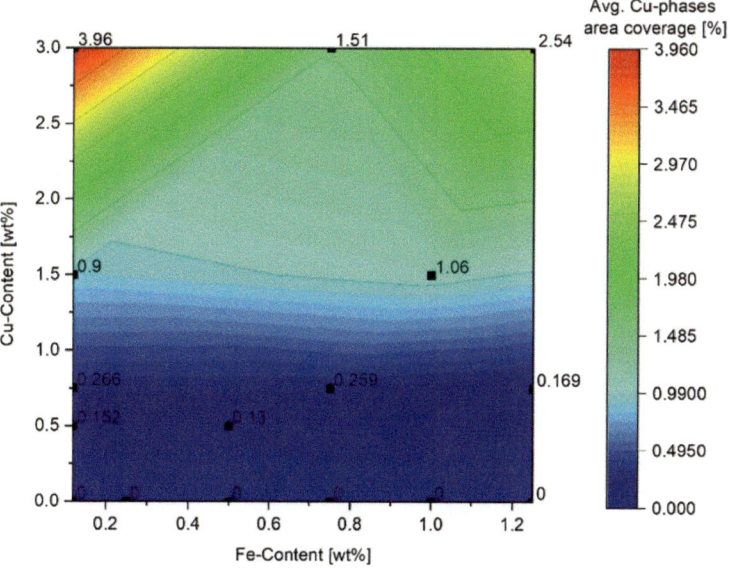

Fig. 7 Influence of Fe and Cu content on tensile strength Rm

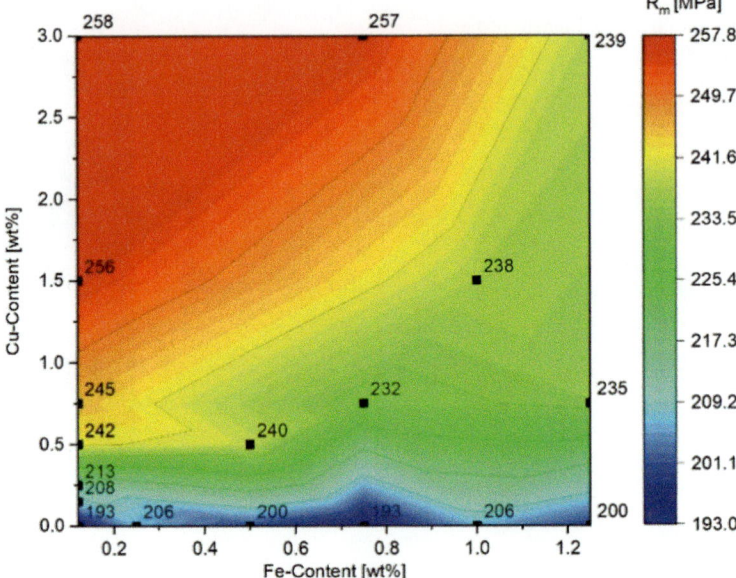

Fig. 8 Influence of Fe and Cu content on yield strength Rp0.2

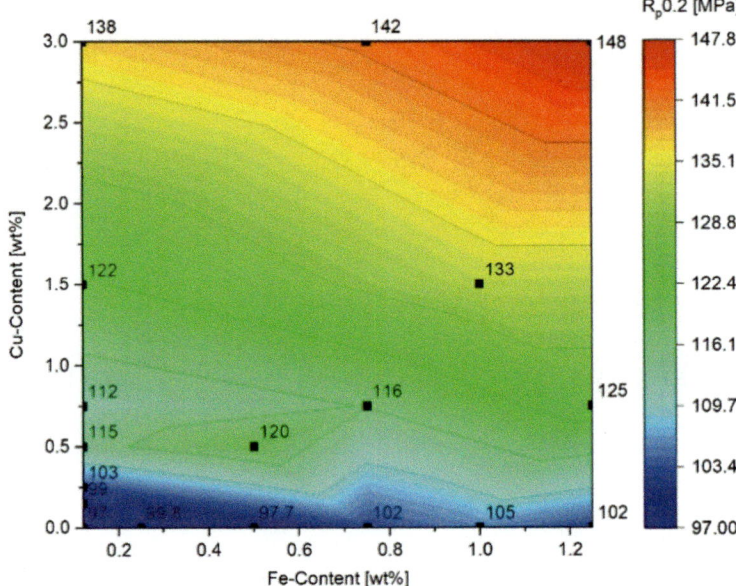

the initial increase in elongation at fracture is quite strong and persists up to a Cu content of 0.75 wt% which exhibits an elongation at fracture of 13.1%, a value well above the base alloy. An increasing Cu content significantly reduces the elongation at fracture starting from a content of 1.5 wt%.

The results of the Brinell hardness tests are shown in Fig. 10. They are similar to those of the yield strength, except the fact that the effect of Fe seems to be more positive. In contrast, as with the yield strength, the positive influence of Cu and the positive interaction between Fe and Cu are evident but seem to be weaker. A linear, albeit weak, relationship can be determined between hardness and Fe content. For Cu, the relationship is in analogy to the tensile strength. There is initially a sharp increase, which then levels off. It can therefore be described as saturation-curve-like.

Fig. 9 Influence of Fe and Cu content on elongation at fracture

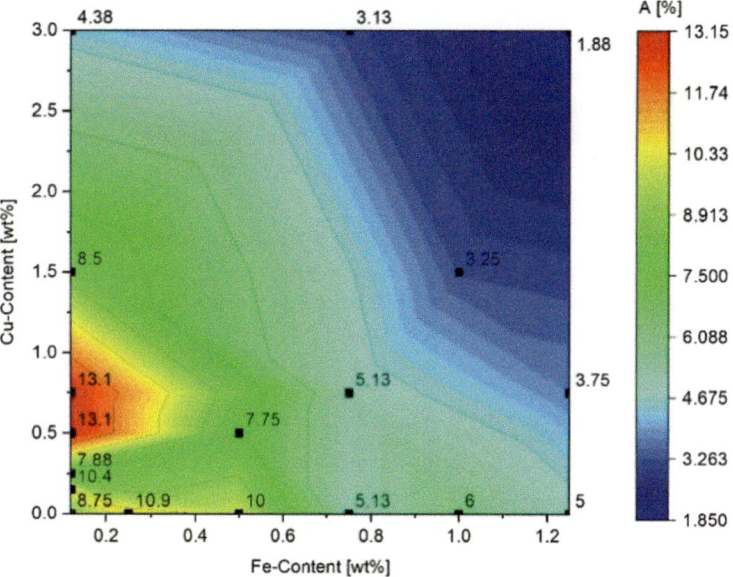

Fig. 10 Influence of Fe and Cu content on hardness HBW 5/250

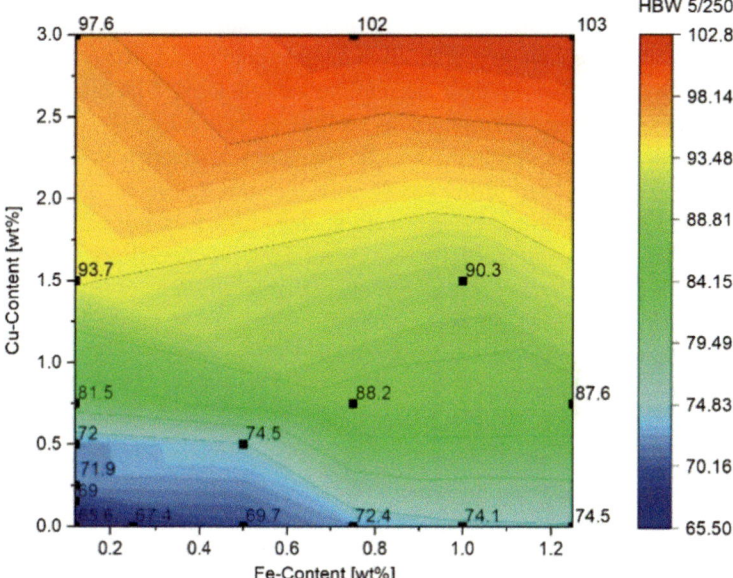

Discussion

Thermodynamic Calculation and Microstructure

The quantification of Fe phases served to evaluate the formation of those phases as a function of Fe content and under the influence of Cu content. Needle-like Fe phases are described as an influencing factor regarding mechanical properties, especially ductility [5, 7]. The objective was to determine to which extent the Fe phases increase and to evaluate their influence on mechanical properties. Initially, the trends described in literature [7, 8] can be confirmed. The length of the Fe phases increases with increasing Fe content.

The trend of the length of the Fe phases can be described as linear. An exception is the Fe0.75 alloy variant. Here, there is a significant increase compared to the Fe0.5 alloy variant. This confirms the critical Fe factor described by Taylor [5]. With a Si content of 7 wt%, the Taylor-factor results in a critical Fe content of 0.475 wt%, which confirms that the critical Fe content is directly related to the Si concentration of the alloy. The reason can be found in the higher amount of Fe that can be tolerated with increasing Si content before the β-phase begins to form in front of the AlSi eutectic. This assumption is supported by means of the CALPHAD simulations. In equilibrium simulation with Fe, it is shown that beginning from a Fe content of about 0.4 wt%, the β-phase starts to precipitate earlier in the solidification process.

Accordingly, the time frame of phase forming during the solidification process is longer which leads to longer grains. The Al$_2$Cu phase forms in AlSi7Mg0.3 at an equilibrium solidification starting at a Cu content of about 0.2 wt%. The maximum solubility of Cu in pure Al is up to 2.5 wt%. During solidification process, phases initially precipitate in the form of clusters and GP-I zones, which are coherent with the matrix, restrict transverse sliding, and promote planar sliding. As a result, a further increase in strength can be observed. In analogy with Fe, this is also supported by the CALPHAD simulation results.

Mechanical Properties

Although the yield strength may increase slightly with increasing Fe content, independent of the Cu content, a negative influence of the Fe content with increasing Cu content was found for the tensile strength. Fe-containing intermetallic phases affect the mechanical properties of the base alloy. They are much more critical against tensile loading than the Al matrix or small Si particles and introduce a notch effect due to their needle-like morphology. The intermetallic phases themselves, as well as at their fracture sites, increase the possibility of the formation of microcracks. When comparing the tensile strength with the yield strength of the alloys examined, Fe slightly improves the yield strength regardless of the Cu content. However, tensile strength in high Fe variants is not significantly affected by increased amounts of Cu. The notch effect of the Fe phases leads to earlier fracture, which is reflected in the measured elongation at fracture. A sustained increase in hardness due to increasing Fe content and a reduction in elongation at fracture were found. In the case of hardness measurement, the notch effect is less. The critical Fe content in AlSi alloys described by Taylor [5] is supposed to describe a limit where losses of ductility are supposed to occur. This value fits the strain curve, since indeed up to an Fe content of 0.5 wt% no significant influence on the strain can be detected. At higher contents, there is a rapid reduction of the elongation at fracture. The described notch effect of the phases explains this effect.

The correlation of Cu content and elongation at fracture appears to be complex. A negative correlation between Cu content, which is found at higher Cu contents, is also confirmed in the literature [9, 10] and is also evident in the elongation at fracture of the high Cu alloys according to the standard alloys described in DIN EN 1706:2020-6 [11]. The behavior of the elongation at fracture as a function of the Cu content can be explained with sliding mechanisms and phase formation in the AlCu system described by Ostermann [4]. Alloys with small amounts of Cu have presumably numerous clusters and GP zones, which would favor planar slip. Alloy variants with higher Cu content have significant amounts of the Al$_2$Cu phase, which is not a particle that cleaves and thus favors transverse sliding. In addition, Al$_2$Cu phases form preferentially at grain boundaries, resulting in precipitationfree zones at the grain boundaries. The grain boundary precipitates and precipitationfree zones in combination cause intergranular fracture [4], which is why the fracture strain decreases at higher Cu contents. It is worth noting that the initial increase in elongation at fracture increases quite sharply when the Cu content is increased separately and continues up to a Cu content of 0.75 wt%, resulting in an elongation at fracture of 13.1%, a value well above that of the base alloy. Regarding the interaction of Fe and Cu, it was found that the fracture strain is decreased by combining the two elements. This is probably due to the interaction effects of Fe and Cu contents. Accordingly, there are more phases which, due to their incoherence to the solid solution, exert a notch effect during tensile loading and thus lead to early fracture initiation. However, interactions between Fe and Cu result in an improvement in yield strength at high amounts, while there is a deterioration in tensile strength.

Conclusions

Based on the work carried out,

- Fe as an impurity was found to have little or no effect on the mechanical properties, except for ductility. Ductility, measured by elongation at fracture, is reduced at high Fe contents (>0.5 wt%). Below that, the elongation at fracture is comparable with EN AB-42100. This limit was found to be important because scrap material with Fe concentrations up to 0.5 wt% can be used without further measures such as the introduction of further alloying elements.
- Cu significantly improves the mechanical properties. The elongation at fracture increases at low Cu contents (up to 0.75 wt%) up to values above that of the base alloy but decreases significantly at high Cu contents. As was seen for Fe, a Cu limit of 0.75 wt% was determined which allows Cu-rich scrap material to be used for EN AB-42100 base material.
- Furthermore, interactions between Fe and Cu were found to result in an improvement in yield strength at high amounts above the determined limits, while tensile strength and elongation are reduced. This effect is linked to incoherent phases rich in either or both Fe and Cu that hinder a ductile response under mechanical load.

Summing up, this work yields two important results:

1. Scrap material can be used to produce alloys comparable to EN AB-42100 for
   a. individual contents of up to 0.75 wt% Fe or 0.75 wt% Cu.
   b. combined contents of up to 0.5 wt% Fe and 0.5 wt% Cu.
2. Above these limits, a further increase in the yield strength and a decrease in tensile strength and elongation are seen. Alloys in this composition range can therefore be used for structural applications where high strength is needed while high ductility is negligible.

Acknowledgements The authors would like to thank Dr. Axel Marquardt, Thien Dang, Alice Siemund, Ines Zerbin, and Chris Schäfer for revealing discussions and their help with sample preparation and evaluation. Many thanks to Nicholas Grant Towsey for proofreading this paper.

References

1. Statista (2021), Worldwide aluminum consumption forecast 2029, https://www.statista.com/statistics/863681/global-aluminum-consumption/, Accessed 18 August 2021.
2. D. Harrison (2019), Powertrain forecast 2020 - 2030, Automotive from Ultima Media, Automotive Logistics.
3. S. Klan (2004), Beitrag zur Evolution von Aluminium-Gusslegierungen für warmfeste Anwendungen, Ph.D thesis, Freiberg University of Mining and Technology.
4. [4]F. Ostermann (2014), Anwendungstechnologie Aluminium, 3rd edition, Springer Vieweg, Berlin.
5. J. A. Taylor (2004), The effect of iron in Al-Si casting alloys, In 35th Australian foundry institute national conference, vol 148, p 157.
6. L. Zhenbang (2015), Effects of Cooling Rate on the Microstructure and Tensile Strength of A356 Alloy Wheels, ICME, doi: https://doi.org/10.2991/ic3me-15.2015.404.
7. D. Závodská (2019) The Effect of Iron Content on Microstructure and Porosity of Secondary AlSi7Mg0.3 Cast Alloy, Periodica Polytechnica Transportation Engineering, vol 47, no 4, art-no 4, doi: https://doi.org/10.3311/PPtr.12101.
8. [8]L. Kucharikova (2019), Analysis of microstructure in AlSi7Mg0.3 cast alloy with different content of Fe, Transportation Research Procedia, vol 40, p 59–67, doi: https://doi.org/10.1016/j.trpro.2019.07.011.
9. R. A. Gonçalves (2015) Influence of Copper Content on 6351 Aluminum Alloy Machinability, Procedia Manufacturing, vol 1, p 683–695, doi: https://doi.org/10.1016/j.promfg.2015.09.014.
10. D. Stanić, (2020) Influence of Copper Addition in AlSi7MgCu Alloy on Microstructure Development and Tensile Strength Improvement, Metals, vol 10, no 12, art-no 12, doi: https://doi.org/10.3390/met10121623.
11. DIN EN 1706:2020–6 (2020).

Temperature Dependence of Lattice Misfit in Determining Microstructural Evolution of High Temperature High Strength Aluminium Alloys—A 3D Phase-Field Study

Dhanish Sidhik and B. S. Sundar Daniel

Abstract

One of the recent advancements in the field of high temperature high strength aluminum alloys is the development of the Al-Sc-Zr alloy. The addition of Sc and Zr in the Aluminium matrix results in trialuminide precipitation that imparts excellent coarsening resistance, making the alloy suitable for high-temperature application. In the early stages of solid–solid phase transformations, the coherent microstructure is developed by maintaining the continuity between lattice planes and directions along the interface. These misfit strains significantly influence the shape and growth of trialuminides. The temperature dependency of misfit strain, which is often neglected in the theoretical simulations of precipitate phases, is considered in the present work to investigate microstructural evolution and resulting strain and concentration field distribution during trialuminide precipitation using the phase-field method. Also, an assessment of the precipitation kinetics during Al_3X (X = Sc, Zr) growth in the Al matrix with the help of particle size and energy variation-time graph is also done to get an insight into the precipitation mechanism.

Keywords

Age-hardening • Multi phase-field modeling • Precipitation • Al alloy • Microstructure evolution • Phase transformation

D. Sidhik (✉) · B. S. Sundar Daniel
Department of Metallurgical and Materials Engineering, Indian Institute of Technology Roorkee, Roorkee, Uttarakhand 247667, India
e-mail: dsidhik@mt.iitr.ac.in

B. S. Sundar Daniel
e-mail: sundar.daniel@mt.iitr.ac.in

Introduction

With the implementation of lightweight aluminium alloys in aerospace and automotive applications, the primary concern in focus during aluminium alloy design has become its ability to retain considerable strength at elevated temperatures. The significant characteristic of aluminium alloy, that makes it a suitable candidate for creep-resistant application, is its FCC closed-packed structure. Also, being a low-density alloy, it has a broad scope in weight-sensitive applications, and also the ability to form a passive layer imparts this alloy high oxidation resistance [1]. The existing high temperature-aerospace alloys such as Ti and Ni are considerably less economical than Aluminium alloy, which raises its' relevance from a research standpoint [2, 3]. Knipling et al. [4], in their pioneering work, have proposed conditions to develop a castable, precipitate strengthened, thermally stable Al-based alloy. He has identified the significance of trialuminide $L1_2$ structured precipitates in imparting superior mechanical properties to Aluminium alloys. The $L1_2$ precipitates are analogous in terms of chemistry to the well-known Ni_3Al ordered phase, which provides high-temperature strength in the Ni-based superalloys. The $L1_2$ precipitates have a similarity in crystal structure with the matrix, and as a result, a coherent interface exists between the phases. This characteristic feature of $L1_2$ precipitates results in its high efficacy to withstand high temperatures without coarsening.

Al-Sc and Al-Zr are the two primary $L1_2$-forming aluminium alloys that have been widely studied in recent years [5, 6]. Sc and Zr belong to the transition metals and come in the periodic table's third and fourth groups. According to the condition put forward by Knipling et al. [7] for a castable creep-resistant alloy, these two alloys can give excellent performance in the high-temperature application providing considerable cost benefits. So far, several people have widely studied the precipitation kinetics of Al_3Sc and Al_3Zr in the aluminium matrix [8, 9]. In most studies concerning

S. Broek (ed.), *Light Metals 2023*, The Minerals, Metals & Materials Series,
https://doi.org/10.1007/978-3-031-22532-1_71

these alloys, the significance of lattice misfit in determining microstructure characteristics and mechanical properties has been well acknowledged [10–13]. However, the significant difference between the lattice thermal expansion of the matrix and precipitate and its possible effect on the microstructural characteristics is an area that is less explored. Currently, there are only a few experimental studies that investigate the microstructural changes by considering the variation of lattice misfit with respect to temperature [14].

The present work investigates the temperature dependency of lattice misfit in alloys Al-Sc and Al-Zr, and their effect on microstructural changes during precipitate growth using the multi phase-field method. The composition is assumed to be 0.5 % for both the alloys and the time and temperature of heat treatment are considered as per the standard industrial processing conditions. The growth of one precipitate and a set of randomly distributed precipitates in the Aluminium matrix will be simulated using the Multiphase field model developed by Ingo Steinbach [15]. Since this study compares and analyses various microstructural features such as composition distribution, strain field, and coarsening behaviour, we believe the work will have a notable impact on the scope of research and analysis of high temperature high strength aluminium alloys.

Model Description

Multiphase Field Model

Multiphase field models are recently developed phase-field models designed to analyze microstructure evolution in multiphase and multi-component systems. Since the present study involves a systematic investigation of the stress, concentration, and growth behaviour involving single and many precipitates, this model is chosen for our study. In this method, diffusion of solute atoms, interface curvature, and elasticity are considered to solve for the interface kinetics. In MPFM, an indicator variable φ (x, y, z, t) [0, 1] determines the presence or absence of a phase at a point, called the phase-field variable. For an individual non-conserved field variable (φ_α), the evolution equation is given below

$$\dot{\phi}_\alpha = \frac{\mu}{N} \sum_{\beta=1}^{N} \left(\frac{\delta F}{\delta \phi_\alpha} - \frac{\delta F}{\delta \phi_\beta} \right) \quad (1)$$

Here α and β are the respective phases of interest, μ is the interface mobility, and N is the total number of phases. The principle applied here is the minimization of free energy F, by which the temporal evolution of the microstructure is obtained. Here the equation has an antisymmetric nature, and

as a result, it must obey a constraint concerning space and time; $\sum_{\alpha=1}^{N} \dot{\varphi}_\alpha = 0$. The phase field φ_α is like the phase fraction, defined in the range $\varphi_\alpha \in [0, 1]$, which leads to another constraint, $\sum_{\alpha=1}^{N} \varphi_\alpha = 1$. The total free energy can have a contribution from several domains based on the type of problem, and in the present study, grain boundary free energy f^{int}, chemical free energy, f^{chem}, and mechanical free energy f^{mech} are added and integrated to obtain total free energy

$$F = \int_\Omega \left\{ f^{intf} + f^{chem} + f^{mech} \right\} \quad (2)$$

Interfacial Free Energy

The interface free energy density over all phases is calculated using the expression below.

$$f^{intf} = \sum_{\alpha,\beta=1,\alpha\neq\beta}^{N} \frac{8\sigma_{\alpha\beta}}{\eta} \left[-\frac{\eta_{\alpha,\beta}^2}{\pi^2} \nabla\phi_\alpha . \nabla\phi_\beta + \phi_\alpha\phi_\beta \right] \quad (3)$$

Here the interfacial width between phases α and β is $\eta_{\alpha,\beta}$, the interfacial energy is $\sigma_{\alpha\beta}$, and N is the total number of phases. The first term in the equation is a cross-gradient term, and the second term is a double obstacle potential. The purpose of a double obstacle function is to maintain an exemplary interface.

Chemical Free Energy Density

The chemical free energy is defined as

$$f^{chem} = \sum_{\alpha=1,\alpha\neq\beta}^{N} \phi_\alpha f_\alpha(c_\alpha^i) + \mu^i \left[c^i - \sum_{\alpha=1}^{N} \phi_\alpha(c_\alpha^i) \right] \quad (4)$$

The expressions for chemical and elastic free energy are obtained by weighing with phase fraction. Here the bulk free energy of phase α is $f_\alpha(c_\alpha^i)$, and it depends on the individual phase concentration of component a. Also, the total concentration of component 'i' is represented as c^i. The parameter, μ^i, used as a Lagrange multiplier, is also called the generalized chemical potential or diffusion potential vector of component i. It serves the purpose of conserving mass balance between the phases [16]. The law of conservation of mass introduces the following constraint, which is utilized to obtain the total concentration vector.

$$c^i = \sum_{\alpha=1}^{N} \phi_\alpha c_\alpha \qquad (5)$$

At any instant, the concentration value of each component is obtained by solving the continuity equation $\dot{c} + \nabla.\left(M\nabla\frac{\partial F}{\partial C}\right) = 0$,

Elastic Free Energy Density

The linearly weighted contribution of the elastic energy density of each phase is expressed as

$$f^{mech} = \frac{1}{2}\sum_{\alpha=1}^{N} \phi_\alpha \left[\epsilon_{\alpha,kl}^{tot} - \epsilon_{\alpha,kl}^*\right] C_{\alpha,klmn}\left[\epsilon_{\alpha,mn}^{tot} - \epsilon_{\alpha,mn}^*\right] \qquad (6)$$

The expression is similar to Hook's Law and $C_{\alpha,klmn}$ is Hook's matrix or elastic constant. The total strain in phase α, $\epsilon_{\alpha,kl}^{tot}$, comprises $\epsilon_{\alpha,kl}^*$, $\epsilon_{\alpha,kl}^{el}$, which are misfit strain and elastic strain, respectively; $\epsilon_{\alpha,kl}^{tot} = \epsilon_{\alpha,kl}^{el} + \epsilon_{\alpha,kl}^*$. The misfit strain or Eigenstrain is described as $\epsilon_{\alpha,kl}^* = \delta_{kl}\varepsilon_0$, where δ_{kl} is the Kronecker delta function and ε_0 is the lattice misfit. An instantaneous mechanical equilibrium is considered, and the local strain at an instant is obtained by solving the force balance equation, which is $\nabla\frac{\partial F}{\partial \epsilon^{ij}} = \vec{0}$.

Phase-Field Governing Equation

The governing time evolution equation for the $L1_2$ precipitate growth is obtained by inserting Eqs. 2,3,4 and 6 in Eq. 1, which is

$$\dot{\phi}_\alpha = \frac{\mu_{\alpha\beta}}{N}\left[\sum_{\gamma=1\neq\beta}^{N}\left[\sigma_{\beta\gamma} - \sigma_{\alpha\gamma}\right]\left[\nabla^2\phi_\gamma + \frac{\pi^2}{\eta^2}\phi_\gamma\right] + \frac{\pi^2}{8\eta}\Delta g_{\alpha\beta}\right] \qquad (8)$$

were

$$\Delta g_{\alpha\beta} = \Delta G_{\alpha\beta}^{chem} + \Delta G_{\alpha\beta}^{mech} \qquad (9)$$

is the total free energy with mechanical and chemical contributions. The chemical driving force can be approximated as

$$\Delta G_{\alpha\beta}^{chem} = \left(c - c_{eq}\right)m_{AB}\Delta S_0 \qquad (10)$$

Here ΔS_0 is the entropy of the formation of the $L1_2$ phase, and m_{AB} is the slope between the matrix and the precipitate region, obtained from the T-X phase diagram. The equilibrium matrix concentration at the heat treatment temperature

is defined as c_{eq}. The mechanical driving force is calculated in the following way:

$$\Delta G_{\alpha\beta}^{mech} = -\left(\frac{\delta}{\delta\phi_\alpha} - \frac{\delta}{\delta\phi_\beta}\right)f^{mech} \qquad (11)$$

To obtain the system's mechanical equilibrium, the quasi-static approximation approach is adopted.

$$0 = \nabla^j\sigma^{ij} = \nabla^j C^{ijkl}\left[\epsilon_\alpha^{kl} - \epsilon_\alpha^{*kl}\right] \qquad (12)$$

The Khachaturyan [15] approach of homogenizing mechanical properties along the interface is used for the present study. Three parameters are homogenized, which are effective stiffness tensor Eigenstrain ϵ^* and total strain, ϵ^*

$$\epsilon^* = \phi\epsilon_\alpha^* - (1-\phi)\epsilon_\beta^* \qquad (13)$$

$$C_{eff} = \phi C_\alpha - (1-\phi)C_\beta \qquad (14)$$

$$\epsilon^e = \epsilon - \epsilon^* = \epsilon - \phi\epsilon_\alpha^* - (1-\phi)\epsilon_\beta^* \qquad (15)$$

A linear homogenization of stiffness tensor C and Eigenstrains, ϵ^*, is adopted by this model. The standard diffusion equation is solved to map the alloying elements Zr and Sc concentration field at each time step, t.

$$\dot{c} = \nabla\left(\sum_{\alpha=1}^{N}\phi_\alpha D_\alpha \nabla c_\alpha\right) \qquad (16)$$

Here and ∇c_α is the concentration gradient of the alloying element in phase α, and D_α is the diffusivity of the alloying element (Sc and Zr) in the matrix.

Simulation Procedure

Parameters for the Phase-Field Simulation

To study the microstructure evolution using the multiphase field method, open source codes Openphase is used in the present study [15]. The boundary condition applied is periodic, and a simulation box of 128^3 is used for many particles simulation, and 48^3 is used for single particle simulation. The time step chosen for the study was 1 s, and the grid size was fixed as 1 nm. The ageing time and temperature chosen for the alloy systems were based on the optimal ageing conditions that gave the best mechanical properties in the previous experimental studies. For the Al-Zr system, 375 and 425 °C were chosen as per the findings of Knipling et al. [13], and for Al-Sc alloy, the ageing temperature chosen was 300 and 350 °C [17]. For many particles system, quasi-random nucleation and growth are employed, and to get mechanical field variables, Fast Fourier Transform (FFT) is used. The total simulation time was fixed as 72 h for many particles and 100 h for single particle analysis. Also, the interfacial mobility of $8.2 \times 10^{-21}\ m^4 J^{-1}\ s^{-1}$ is chosen, which guarantees diffusion-controlled growth according to Karma's thin interface asymptotics [16]. The solution composition was assumed to be 0.5% for both alloys. To obtain chemical driving force, a linearized phase diagram is used with $\Delta G^{ch} = m\Delta S(c - c_{eq})$ (Eq. 10), where the entropy of formation ΔS for Al_3Zr and Al_3Sc was taken as $-13 \times 10^{-5} JK^{-1} m^{-3}$ and $-1.04 \times 10^{-5} JK^{-1} m^{-3}$, respectively [18, 19]. By setting the growth of precipitate from nuclei with a negligible size, any presumption of size and shape is avoided. The elastic constants are expressed with three independent components of the stiffness matrix; C_{11}, C_{12}, and C_{44}, since FCC Al and $L1_2$ phases (Al_3Sc, Al_3Zr) exhibit cubic anisotropy. For the Al matrix, the values taken are 107.1 GPa, 62.9 GPa, and 28.9 GPa, respectively [20]. For Al_3Sc and Al_3Zr, these values were taken as 189, 43, 66 GPa (for Al_3Sc [14]) and 182, 63, 72 GPa [21] (for Al_3Zr). The pre-exponential Diffusion coefficient D_0 is taken as $5.31 \times 10^{-4}\ m^2 s^{-1}$ and $7.28 \times 10^{-2}\ m^2 s^{-1}$ for Sc and Zr, respectively. Similarly, the diffusion activation energy is taken as 173 kJ mol^{-1} and 242 kJ mol^{-1} for Sc and Zr, respectively [4]. The mismatch of Al_3Sc and Al_3Zr with Al is taken as 1.32% and 0.75%, respectively [4]. For interfacial energy, the precise magnitude is unknown and different values are reported in the literature. In the present study, 0.2 J m^{-2} [8] is taken for the Al_3Sc/Al interface, and 0.1 J m^{-2} [6] is taken for Al_3Zr/Al.

Temperature Dependency of Eigenstrain

To allocate the temperature dependency of Eigenstrains of aluminium matrix, in the simulations, the model proposed by Royset et al. [14] is chosen for the present study. Since the present study comes in the temperature range of 300–900 K, the recommended approximation is given by

$$\frac{\Delta L^{Al}}{L_0^{Al}} = 1.8 \times 10^{-4} + 2.364 \times 10^{-5}(T - 300) + 4.164 \times 10^{-9}(T - 300)^2$$
$$+ 8.270 \times 10^{-12}(T - 300)^2 + 8.270 \times 10^{-12}(T - 300)^3$$

$$(17)$$

From the thermal expansion, the temperature-dependent misfit for Al-Sc system $\delta(T)$ is calculated from the following expression:

$$\delta''(T) = \frac{\alpha_{Al_3Sc}\left[1 + 16 \times 10^{-6}\right)(T - 300)\right]}{\alpha_{Al}\left(1 + \frac{\Delta L^{Al}}{L_0^{Al}}\right)} - 1 \quad (18)$$

Also, for the Al-Zr system, the linear thermal expansion coefficient value has been obtained from the studies of Li et al. [22].

$$\delta''(T) = \frac{\alpha_{Al_3Zr}\left[1 + 41 \times 10^{-6}\right)(T - 300)\right]}{\alpha_{Al}\left(1 + \frac{\Delta L^{Al}}{L_0^{Al}}\right)} - 1 \quad (19)$$

These models (Eqs. 17–19) have been integrated into the open-source phase-field code Openphase to get the temperature-dependent misfit of Al-Sc and Al-Zr alloys at the studied temperatures.

Results and Discussion

Composition Distribution

Figure 1 illustrates the composition distribution in the Al matrix after 72 h of heat treatment in Al–Sc and Al–Zr alloy. In the Al-Sc alloy system with an ageing temperature of 300 °C (Fig. 1a), the Sc composition is distributed in a relatively wider range, from 0.14–0.18%. The highest Sc concentration in the matrix from the figure is 0.18%, and we can see that most of the matrix has this concentration. The presence of a concentration gradient in the matrix indicates temporal Sc diffusion and precipitate growth. An interesting observation that can be made from the high- temperature Al–Sc 350 °C (Fig. 1d) composition distribution is the uniform composition distribution. Here the composition is 0.22–0.23% throughout the matrix with a magnitude higher than the low-temperature composition value. From the composition analysis made by Knipling et al. [1], it is evident that there will be a strong partition of Sc towards the precipitates during precipitate growth. From Fig. 1a, it is obvious that out of 0.5 Sc %, 0.32% is partitioned towards the precipitate, and only 0.18% is distributed in the matrix, which supports this observation. But at 350 °C, the precipitates attained a quasi-static equilibrium state indicating termination of further growth. However, in Al–Zr, unlike Al–Sc at low temperature (300 °C), the distribution is narrow but non-uniform. This difference is because of the slower diffusion of Zr atoms in the Al matrix. Thus, to attain a uniform matrix composition like the Al–Sc system at 350 °C, the Al–Zr system at this temperature requires a higher ageing time. Also, at high temperature (450 °C), even though there is a broader distribution range, peak Zr concentration (0.29%) is high compared to other systems.

The activation energy for diffusion is taken as 173 kJ.mol^{-1} for Sc in the Al–Sc system and 242 kJ.mol^{-1} in the present study, and this difference plays a significant role in determining the composition distribution in the matrix. The

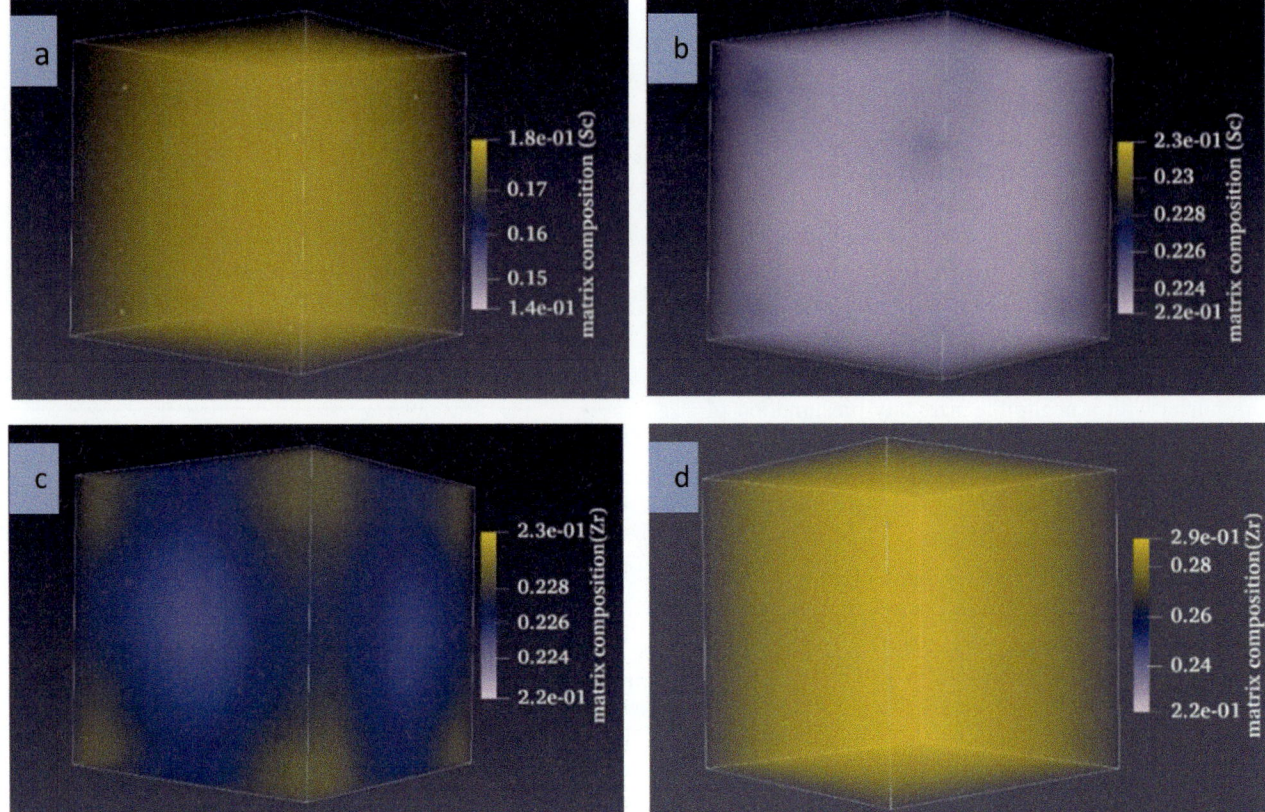

Fig. 1. Composition distribution in **a** Al–Sc −300 °C, **b** Al–Sc −350 °C, **c** Al–Zr −400 °C, and **d** Al–Zr −450°

coarsening of both the precipitates is diffusion controlled [6, 8]. That means, with the advent of reaching equilibrium, there will be a gradual decline of diffusion in the matrix leading to a uniform distribution (Fig. 1b–d).

For instance, consider Fig. 1a which shows Sc distribution at 300 °C heat treatment; there is a non-uniformity in the composition distribution, with a few white spots having 0.14 Sc % and most of the region having 0.18%. But at 350 °C, the system shows uniform Sc distribution indicating a high temperature has increased the diffusion kinetics leading to a uniform distribution. Similarly, an apparent non-uniformity in composition distribution is observed in the Al–Zr system at 400 °C. The corners show yellow, indicating a higher concentration, and towards the centre, there is a minor and gradual decrease. But such a gradient is not observed at 450 °C, pointing to the achievement of equilibrium in this system. Another essential aspect to be considered is the difference in the coefficient of thermal expansion taken for the two systems. The coefficient of thermal expansion of Al_3Sc precipitates is only half the value of Al_3Zr. This difference results in relatively high mechanical free energy for the Al-Zr system affecting the precipitation kinetics and the resulting composition distribution in the system.

Strain Field

The normalized strain field values and their distribution around the precipitates are represented in Fig. 2. To study the quantities that exhibit anisotropy, a normalization method called Frobenius Norm can be used [23]. In our present study, we used this method to normalize strain values at a point since it is more convenient and reduces complexity. The Frobenius norm gives the square root of the sum of squared components of a strain tensor at a point, i.e.

$$||\epsilon_F|| = \sqrt{\sum_{i,j=1}^{3} \epsilon_{ij}^2} \qquad (20)$$

When the precipitate nucleates and grow, the precipitate lattice is subjected to contraction, and the matrix undergoes stretching due to the difference in their lattice parameters. Also, the magnitude of the stress field around the precipitates directly correlates with the particle aspect ratio (thickness/diameter) [24]. Thus, those precipitates with a more significant aspect ratio should have a high-magnitude stress field around them. According to the linear elastic theory, this stress field consequently determines the resulting strain field. In the present study, the highest strain field magnitude is

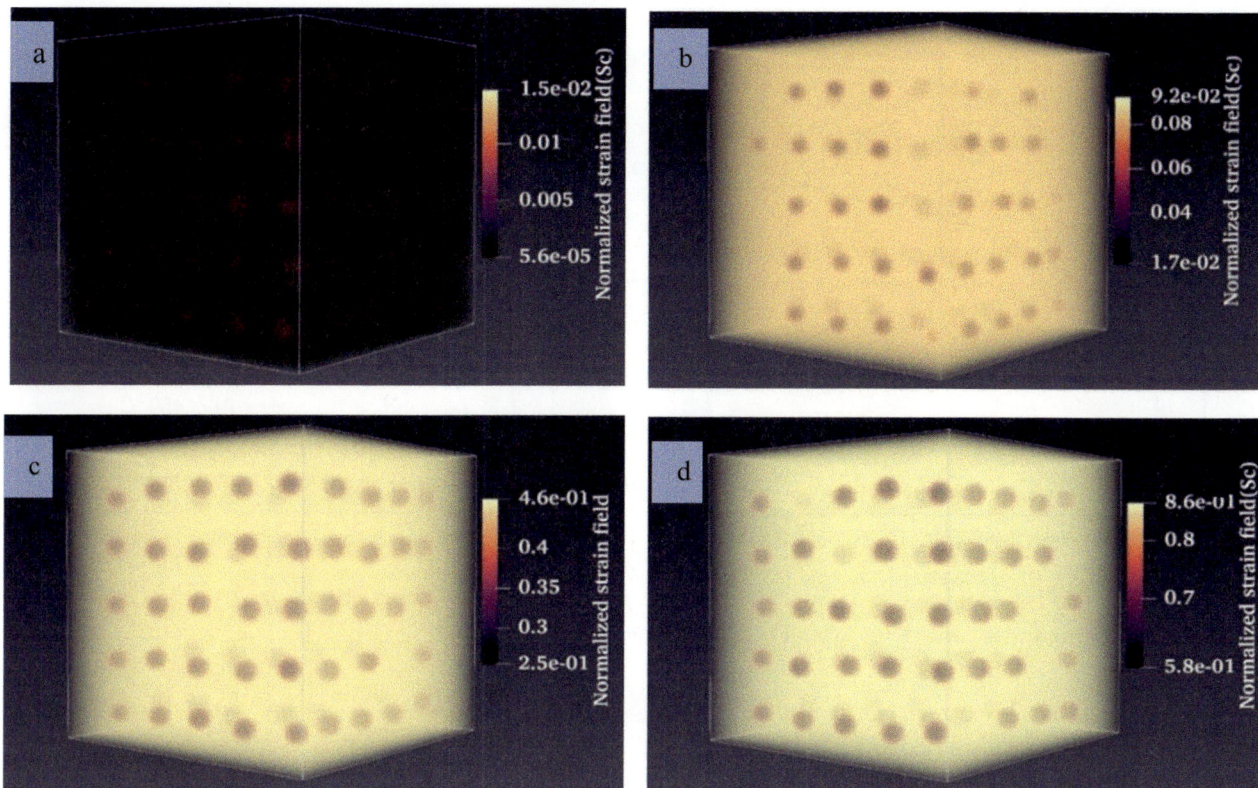

Fig. 2. Normalized strain field distribution in **a** Al–Sc −300 °C, **b** Al–Sc −350 °C, **c** Al–Zr −400 °C, and **d** Al–Zr −450 °C

observed in Al–Zr alloy heat treated at 450 °C Fig. 2d. Similarly, there is a wide distribution of strain in the matrix of Al–Sc heat treated at 300 °C. The strain field varies with an order of 10^{-5} as the lowest value and 10^{-2} as the highest value. This is like the wide concentration field observed in the composition distribution in the previous section (Fig. 1a). Since the precipitates in this system are still in their growth stage, the system is yet to reach an equilibrium state, resulting in the broader strain field distribution. Similarly, in the Al-Sc system heat treated at 350 °C, the normalized strain field magnitude is one order lower than the Al-Zr systems. This difference is an interesting observation since the precipitation kinetics in the Al-Zr system are expected to be lower than in the Al-Sc system. As a result, the Al-Sc system's strain field magnitude is expected to be larger than the Al-Zr system. But in the present study, the reason for such a difference in strain field should be because of the different thermal expansion coefficients. In both systems, the matrix thermal expansion is higher than the precipitate expansion; the overall misfit decreases with temperature [14]. As a result, the overall Eigenstrain value also decreases with temperature. In the Al–Sc system, the decrease of overall Eigenstrain because of temperature is expected to be very high due to its high thermal expansion coefficient, and

as a result, the resulting strain field is also less comparatively. But in Al–Zr, with half the thermal expansion coefficient of Al_3Sc, there is a relatively minor decrease in the overall Eigenstrain. This decrease results in a high-magnitude strain field formation in the Al–Zr system.

Particle Size and Energy Versus Time

To understand the precipitation kinetics of a single particle in the studied systems, the particle diameter and total energy variation with respect to time steps are constructed in Fig. 3. From the image, a considerable size increase can be observed for the Al–Sc system with respect to time, compared to Al–Zr system. This observation supports both the composition and strain field results illustrated above, strengthening the argument that there is higher precipitation kinetics for the Al–Sc system. Also, at 350 °C, there is an initial shoot in the diameter of the precipitate, following which the size remains constant. This indicates the equilibrium in this system in a short time.

As per the LSW theory

$$r^3 - r_0^3 = k_r(t - t_0) \tag{21}$$

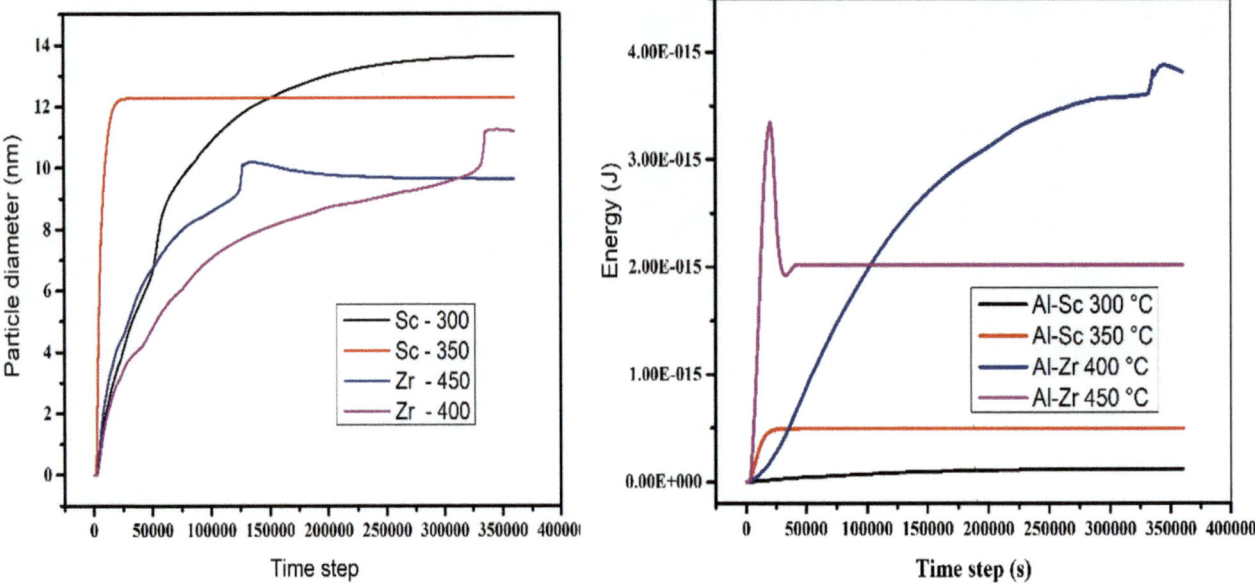

Fig. 3. Variation of diameter and total energy with respect to ageing time in Al–Sc and Al–Zr single particle system

the size and time have an exponential relationship with (1/3) as the power [17]. The slope of the graph will indicate the rate of precipitation, and in the present study, the highest slope is observed for Al-Sc system heat treated at 350 °C, indicating it has the highest precipitation kinetics. Similarly, the least precipitation rate is observed for Al–Zr system heat treated at 400 °C. This observation is expected since Sc's diffusion constant is higher than Zr's, and the activation energy for diffusion for Sc is two orders lower. This implies a high precipitation rate and larger size precipitate formation in the Al-Sc alloy system when compared to the Al–Zr system. It is more obvious from the figure that the size of Al_3Sc precipitates in both conditions is around 12–13 nm, whereas Al_3Zr size is slightly smaller, 8–9 nm. The total energy graph also strengthens the previous results; in Al–Zr at 400 °C and Al–Sc at 300 °C, the graph is monotonously increasing, indicating precipitate is still in the growth stage. Interestingly, the total energy in Al-Sc at 300 °C heat treatment is considerably lower compared to other systems (of the order of 10^{-28}). Also, the largest particle diameter is observed in this system. This observation indicates this system has significant precipitation kinetics compared to other systems. Similarly, in the Al–Zr system, at 400 °C total energy monotonously increases and reaches the highest, whereas this system has the lowest precipitate size indicating the lowest precipitation kinetics. But for Al-Sc at 350 °C and Al–Zr at 450 °C, total energy, and the precipitate size are constant after a point, indicating the attainment of a quasi-static equilibrium in these systems.

Conclusion

The composition and strain field distribution on Al–Sc and Al–Zr alloys, heat treated at their optimum temperature, which gives the best mechanical properties, has been obtained using the multiphase field simulation method. Unlike previous simulations, here, the temperature dependence of the misfit has been incorporated with the model, and the results obtained have the effect of this misfit variation. The following inferences have been made from the study:

(a) The precipitation kinetics is high in Al–Sc system heat treated at 350 °C due to the high temperature, high diffusion constant, and low activation energy of Sc. For the Al–Sc system, in the studied temperature range, slightly larger precipitates are formed due to its high precipitation kinetics. However, rather than the size effect, the thermal strain due to volume expansion contributes more to the strain field distribution resulting in comparatively lower strain field distribution.

(b) A high magnitude of strain field is developed in the Al–Zr system regardless of its ageing temperature and time. The thermal strain mainly contributes to the large misfit strain field due to a high thermal expansion coefficient of Al_3Zr compared to Al_3Sc. Also, the Al–Zr system observes a uniform composition distribution due to the slow diffusion kinetics of Zr and its high mechanical free energy.

An insight into microstructural changes during heat treatment of these high temperature high strength Al alloys can significantly contribute to the research and development wing of aerospace and automobile industries. Thus, we believe the present study has remarkable scientific and commercial significance.

References

1. Knipling KE, Karnesky RA, Lee CP, et al (2010) Precipitation evolution in Al-0.1Sc, Al-0.1Zr and Al-0.1Sc-0.1Zr (at.%) alloys during isochronal aging. Acta Mater 58:5184–5195.
2. Costa S, Puga H, Barbosa J, Pinto AMP (2012) The effect of Sc additions on the microstructure and age hardening behaviour of as cast Al-Sc alloys. Mater Des 42:347–352.
3. Marquis EA, Seidman DN, Dunand DC (2003) Effect of Mg addition on the creep and yield behavior of an Al-Sc alloy. Acta Mater 51:4751–4760.
4. Knipling KE, Dunand DC, Seidman DN (2006) Criteria for developing castable, creep-resistant aluminum-based alloys - A review. Int J Mater Res 97:246–265.
5. Zhang JY, Gao YH, Yang C, et al (2020) Microalloying Al alloys with Sc: a review. Rare Met 39:636–650.
6. Souza PHL, Do Vale Quaresma JM, Silva De Oliveira CA (2017) Precipitation evolution and modeling of growth kinetics of L1$_2$-structured Al3Zr particles in Al-0.22Zr and Al-0.32Zr (wt. %) alloys isothermally aged. Mater Res 20:1600–1613.
7. Knipling KE, Dunand DC, Seidman DN (2006) Criteria for developing castable, creep-resistant aluminum-based alloys – A review. 97:
8. Novotny GM, Ardell AJ (2014) Precipitation of Al3Sc in Binary Al-Sc alloys Precipitation of Al 3 Sc in binary Al – Sc alloys. Mater Sci Eng A 5093:144–154
9. Knipling KE, Dunand DC, Seidman DN (2007) Nucleation and precipitation strengthening in dilute Al-Ti and Al-Zr alloys. Metall Mater Trans A Phys Metall Mater Sci 38:2552–2563.
10. Iwamura S, Miura Y (2004) Loss in coherency and coarsening behavior of Al3Sc precipitates. Acta Mater 52:591–600.
11. Liu S, Wang X, Zu Q, et al (2021) Significantly improved particle strengthening of Al–Sc alloy by high Sc composition design and rapid solidification. Mater Sci Eng A 800:140304.
12. Knipling KE, Dunand DC, Seidman DN (2008) Precipitation evolution in Al-Zr and Al-Zr-Ti alloys during aging at 450-600 °C. Acta Mater 56:1182–1195.
13. Knipling KE, Dunand DC, Seidman DN (2008) Precipitation evolution in Al-Zr and Al-Zr-Ti alloys during isothermal aging at 375-425 °C. Acta Mater 56:114–127.
14. Røyset J, Ryum N (2005) Some comments on the misfit and coherency loss of Al3Sc particles in Al-Sc alloys. Scr Mater 52:1275–1279.
15. Kamachali RD, Borukhovich E, Hatcher N, Steinbach I (2014) DFT-supported phase-field study on the effect of mechanically driven fluxes in Ni4Ti3 precipitation. Model Simul Mater Sci Eng 22:.
16. Steinbach I, Apel M (2006) Multi phase field model for solid state transformation with elastic strain. Phys D Nonlinear Phenom 217:153–160.
17. Seidman DN, Marquis EA, Dunand DC (2002) Precipitation strengthening at ambient and elevated temperatures of heat-treatable Al(Sc) alloys. Acta Mater 50:4021–4035.
18. Li DL, Chen P, Yi JX, et al (2009) Ab initio study on the thermal properties of the fcc Al3Mg and Al3Sc alloys. J Phys D Appl Phys 42:. 407
19. Tamim R, Mahdouk K (2018) Thermodynamic reassessment of the Al–Zr binary system. J Therm Anal Calorim 131:1187–1200.
20. Schwarze C, Gupta A, Hickel T, Darvishi Kamachali R (2017) Phase-field study of ripening and rearrangement of precipitates under chemomechanical coupling. Phys Rev B 95:1–14.
21. Clouet E, Sanchez JM, Sigli C (2002) First-principles study of the solubility of Zr in Al. Phys Rev B - Condens Matter Mater Phys 65:1–13.
22. Li DL, Chen P, Yi JX, et al (2010) Thermal properties of the FCC Al3Zr: First-principles study. Mater Sci Forum 650:313–319.
23. Kochetov M, Slawinski MA (2009) On obtaining effective transversely isotropic elasticity tensors. J Elast 94:1–13.
24. Guo W, Steinbach I, Somsen C, Eggeler G (2011) On the effect of superimposed external stresses on the nucleation and growth of Ni4Ti3 particles: A parametric phase field study. Acta Mater 59:3287–3296.

Microstructure and Mechanical Properties of an Al-Mn-Si Alloy Microalloyed with Sn

Amir R. Farkoosh, David C. Dunand, and David N. Seidman

Abstract

We demonstrate that a small addition of a low-melting point element such as Sn (0.02 at.%), within the impurity tolerances of commercial aluminum alloys, to an *Al-0.5Mn-0.3Si (at.%)* model alloy, converts this non-heat-treatable (with negligible precipitation strengthening) alloy into a heat-treatable (precipitation strengthened) alloy with high strength, creep, and coarsening resistance. The small Sn additions refine significantly the α-Al(Mn,Fe)Si-precipitate distribution, which is related primarily to the formation of Sn-rich nanoprecipitates at intermediate temperatures (~ 200 °C). At higher temperatures, these nanoprecipitates act as heterogeneous nucleation sites for Mn-Si-rich nanoprecipitates—the quasi-crystalline precursors of the α-precipitate. We demonstrate that precipitation hardening by these Sn-modified α-precipitates is a highly efficient approach for designing creep-resistant aluminum alloys.

Keywords

High-temperature aluminum alloys • Tin micro-alloying • Heterogeneous nucleation • Precipitation strengthening • Creep resistance

Introduction

Manganese is the only slow-diffusing element ($D = 6 \times 10^{-19}$ m^2s^{-1} at 400 °C [1]) that has substantial ~ 0.5–0.7 at.% (~ 1–1.4 wt.%) solid solubility in Al, which can form a

A. R. Farkoosh (✉) · D. C. Dunand · D. N. Seidman
Department of Materials Science and Engineering, Northwestern University, Evanston, IL, USA
e-mail: farkoosh@northwestern.edu

D. N. Seidman
Center for Atom-Probe Tomography (NUCAPT), Northwestern University, Evanston, IL, USA

considerable volume fraction of thermally stable precipitates upon aging ($V_f \sim$ 2–3%, when compared to ~ 0.2–0.5% for the coarsening resistant $L1_2$ nanoprecipitates [1–5]). The most common Mn-containing precipitate formed in commercial alloys is the α-Al(Mn,Fe)Si phase, which has a body-centered cubic (b.c.c.) or simple cubic (s.c.) structure depending on its chemical composition with a lattice parameter, a_o, in the range 12–13 Å, corresponding to a cubic approximant phase [6–13]. The Al-Mn-based alloys exhibit, however, poor age-hardening responses due mainly to a high (1.3–1.8 eV) activation energy for nucleation of the Mn-containing precipitates [6, 12, 14–17]. Even in highly supersaturated, rapidly solidified alloys, the hardening increments are much smaller than those of the common age-hardenable aluminum alloys, such as Al-Cu, or $L1_2$ strengthened Al-Sc-Zr [1–3, 6, 18–20]. It is therefore desirable to enhance the precipitation of the Mn-rich precipitates in this alloy system to achieve marked age-hardening and creep resistance.

It is well known that micro-alloying or a trace addition of certain elements can significantly influence the age-hardening behavior of alloys and thereby their mechanical properties [21, 22]. Particularly, in Al-Cu alloys, micro-additions of low-melting-point elements, such as Sn [21, 23–29], In [23, 24], and Cd [6, 24, 26, 27, 30–33] have been employed to promote a finer and more uniform distribution of precipitates. Employing a similar micro-alloying strategy, a 25% increase in the yield strength of an AA3003 alloy was achieved using ~ 0.05 at.% Cd additions, which is related to the formation of Mn- and Si-rich clusters around Al-Cd nanoprecipitates, assisting heterogeneous nucleation of α-Al(Mn,Fe)Si precipitates [12].

Based on the archival literature we hypothesize that Sn micro-alloying additions can be utilized to enhance dramatically the age-hardening response of the Al-Mn system, thereby turning all the non-heat-treatable Mn-containing aluminum alloys (3000 and 4000 series) into heat-treatable alloys with high strength, thermal- and creep resistance. In the present study, we explore the possibility of improving the aging response of an *Al-0.5Mn-0.3Si* (at.%) model alloy by micro-alloying it with Sn, which is known to form nanoscale Sn-rich precipitates in the Al(f.c.c.) lattice [28]. We hypothesize that these precipitates can act as nucleation sites for α-Al(Mn,Fe)Si precipitates, thus increasing their number density and thereby their strengthening efficiency. The effects of Sn micro-alloying on the high-temperature strength of this alloy are also investigated utilizing compressive creep experiments.

Experimental Procedure

The ternary model alloy *Al-0.5Mn-0.3Si* (at.%), with and without Sn micro-additions, Table 1, was prepared in a graphite crucible in an electric resistance furnace using 99.99% pure Al and Sn as well as Al-10Mn and Al-12Si master alloys, all in at.%. The melt was maintained at 900 °C for 1 h, with periodic stirring, and then cast into a graphite mold preheated at 200 °C, which was placed on an ice-cooled copper platen just prior to casting to enhance directional solidification.

The as-cast ingots (cylinders, 12 mm diam.) were cut into smaller samples, which were aged isochronally (with 25 °C steps lasting 1 h or 3 h from 150 to 575 °C) in air, terminated by water quenching. Vickers microhardness measurements (average of ∼10 measurements in five different grains for each sample) were performed on polished samples using a Duramin-5 microhardness tester (Struers), with a load of 200 g and a dwell time of 5 s. For scanning electron microscope (SEM) analyses, specimens were ground with a series of SiC papers and then polished with diamond suspensions followed by a final vibratory polishing step. An FEI Quanta 650 field-emission-gun SEM was used for microstructural observations. For transmission electron microscopy (TEM) analyses, thin foils of the peak-aged samples were prepared by mechanical grinding and electropolished to electron transparency using a Struers Tenupol-5 twin-jet polisher and a solution of 10% nitric acid in ethanol at −10 °C. High-angle annular dark-field scanning transmission electron microscopy (HAADF-STEM) imaging was performed utilizing a cold-field emission S/TEM instrument, JEOL ARM300F, operating at 300 kV. Nanotips for three-dimensional (3-D) atom-probe tomography (APT) investigations were prepared by cutting ∼0.3

0.3 × 10 mm³ blanks from the aged samples, followed by a two-step electropolishing technique [34]. The 3-D atom-probe tomography (APT) experiments were performed utilizing a laser-pulsed LEAP 5000XS tomograph (Cameca Instruments Inc., Madison, WI) at 30 K in ultra-high vacuum (<10⁻⁸ Pa) [35–39]. Picosecond ultraviolet (UV) laser pulses (wavelength = 355 nm) were applied with an energy of 30 pJ per pulse and a pulse repetition rate of 500 kHz, while maintaining an average detection rate (number of ions per pulse) of 4%. Data analyses were performed using the program IVAS 3.8.2 (Cameca, Madison, WI). The proximity histogram methodology [40] was employed to study the compositional variations within the precipitates and Al matrix, after performing background corrections to improve the accuracy of the compositional measurements. Compressive creep experiments on the isochronally peak-aged (1 h –25 °C steps to 475 °C) alloys were performed at 300 °C under step loadings in air. Cylindrical specimens, 11 mm in diameter and 22 mm in height, were utilized, which were placed between boron-nitride-lubricated tungsten carbide platens. Sample deformation was measured with a linear variable differential transducer (LVDT) with a resolution of 10 μm. After the establishment of a steady-state minimum strain rate for a given load, the applied load was increased, and the process was repeated until the total strain reached ∼10% for each specimen.

Results and Discussion

Figure 1 displays isochronal aging curves of the Sn-modified *Al-0.5Mn-0.3Si-0.02Sn* alloy with aging steps of 25 °C for 1 h and 25 °C for 3 h, and the Sn-free *Al-0.5Mn-0.3Si* alloy with an aging step of 25 °C for 1 h. The Sn-free alloy exhibits no significant age-hardening. Two small peaks in the microhardness values, each with ΔHV ∼ 25 MPa with respect to the as-cast microhardness, are observed; the first peak occurs at ∼150 °C, which is attributed to the precipitation of Si (diamond cubic) precipitates as Mn remains in the supersaturated solid-solution due to its extremely small diffusivity at 150 °C: the root-mean-square (RMS) diffusion distance of Mn at 200 °C is less than 0.1 nm. The second broader peak at ∼425–475 °C is attributed to the formation of α-Al(Mn,Fe)Si

Table 1 Chemical composition of the alloys determined by inductively coupled plasma optical emission spectroscopy (ICP-OES) at Genitest, QC

Alloy	Concentration (at.%)[a]				
	Mn	Si	Sn	Fe	Cu
Al-0.5Mn-0.3Si	0.50	0.35	<0.001	0.004	0.004
Al-0.5Mn-0.3Si-0.02Sn	0.51	0.32	0.025	0.004	0.003

[a] Concentrations of other impurities are below 0.001 at.%

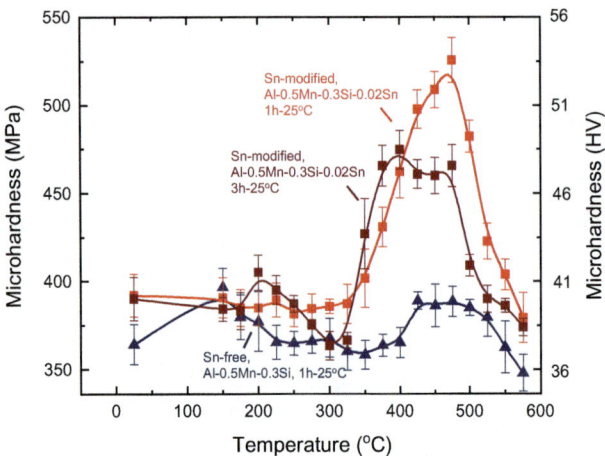

Fig. 1 Vickers microhardness evolution during isochronal aging of the Sn-modified *Al-0.5Mn-0.3Si-0.02Sn* (25 °C for 1 h and 25 °C for 3 h steps) and Sn-free *Al-0.5Mn-0.3Si* (25 °C for 1 h step) alloys

precipitates. In contrast, the Sn-modified alloy exhibits a pronounced age-hardening effect. The 25 °C 1 h aging step achieves a higher peak microhardness value, 525 ± 5 MPa at ~475 °C, than the 25 °C 3 h aging step, 475 ± 6 MPa at ~400 °C.

Upon isochronal aging (25 °C steps of 1 h) of the as-cast alloys to their respective peak microhardnesses at 475 °C, the α-Al(Mn,Fe)Si precipitates form in both alloys, Fig. 2. In the Sn-free *Al-0.5Mn-0.3Si* alloy, Fig. 2a, the precipitates are coarse (D ~ 200 to 1000 nm) and distributed non-uniformly

within the microstructure. In the Sn-modified *Al-0.5Mn-0.3Si-0.02Sn* alloy, Fig. 2b–c, the precipitates are dramatically smaller (D ~ 50 nm) and distributed uniformly.

To study the effects of Sn micro-alloying on the nucleation mechanism of the α-precipitates in the *Al-0.5Mn-0.3Si-0.02Sn* alloy, APT analyses were performed on the specimens aged isochronally (25 °C steps of 1 h) to 300 °C and to 475 °C (peak microhardness). Figure 3 displays the APT reconstruction of a Sn-modified alloy aged isochronally to 300 °C. A segment of a Mn-Si-rich nanoprecipitate associated with a Sn-rich nanoprecipitate is displayed. The association of the Mn-Si-rich nanoprecipitate with a Sn-rich nanoprecipitate is consistent with a heterogeneous nucleation mechanism, through which Mn-Si-rich nanoprecipitates nucleate on Sn-rich nanoprecipitates, which are formed at a lower temperature. The average composition of the Sn-rich nanoprecipitates at 300 °C is ~50 at.% Sn, ~0.2 at.% Si and ~0.1 at.% Mn. The Sn concentrations in the Al (f.c.c.) matrix and the Mn-Si-rich nanoprecipitate are negligible (<10 at. ppm).

The nanostructure of the α-Al(Mn,Fe)Si precipitates formed in an isochronally peak-aged (475 °C) Sn-modified alloy is displayed in Fig. 4a, along with a corresponding proximity histogram (proxigram) in Fig. 4b, which displays the average concentration profiles across the matrix/precipitate heterophase interface. The average composition of the α-Al(Mn,Fe)Si precipitate is $Al_{69.4}(Mn_{18.5}Fe_{0.7})Si_{11.8}$ (at.%), which agrees with the

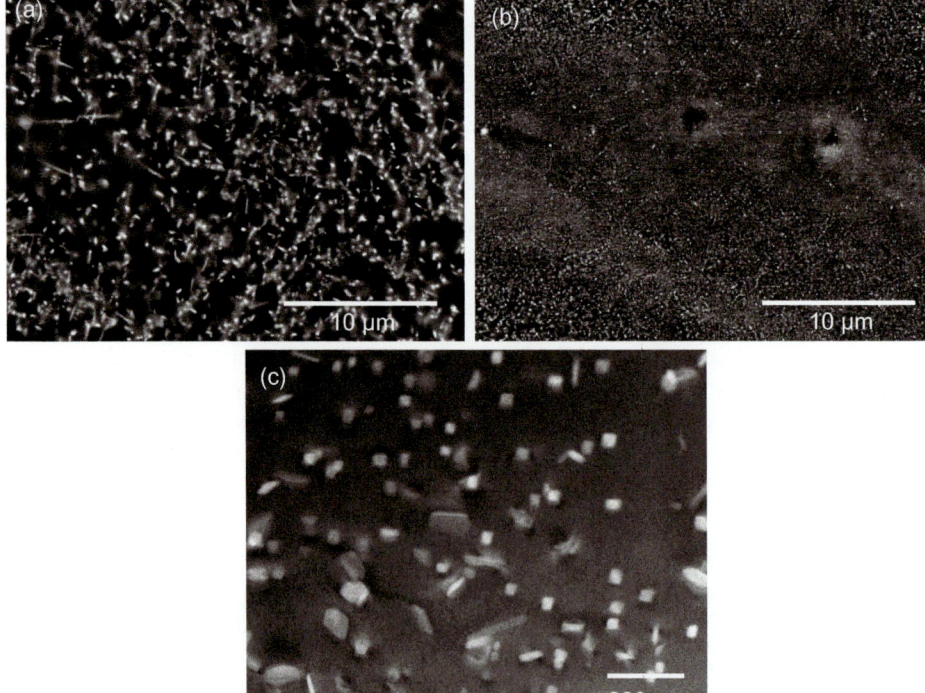

Fig. 2 Backscattered electron-SEM micrographs of: **a** the Sn-free *Al-0.5Mn-0.3Si;* and **b** the Sn-modified *Al-0.5Mn-0.3Si-0.02Sn* alloys peak-aged isochronally to 475 °C in 25 °C steps for 1 h from the as-cast state. **c** High angle annular dark field (HAADF) micrograph of the same alloy as in (**b**) displaying α-Al(Mn,Fe)Si precipitates at a higher magnification

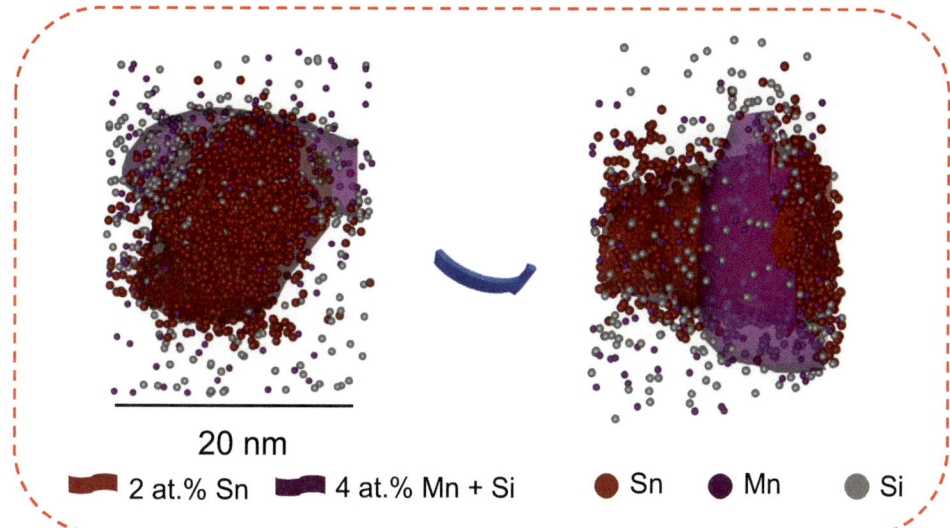

Fig. 3 3-D APT reconstruction of a Mn-Si-rich nanoprecipitate associated with a Sn-rich nanoprecipitate in the Sn-modified *Al-0.5Mn-0.3Si-0.02Sn* alloy aged isochronally (25 °C steps for 1 h) to 300 °C, viewed along two different orientations. The Al atoms are removed completely for the sake of clarity (total number of atoms in this region of interest: 700). The Mn-Si-rich and the Sn-rich nanoprecipitates are delineated with 4 at.% (Mn plus Si) and 2 at.% Sn isoconcentration surfaces, respectively

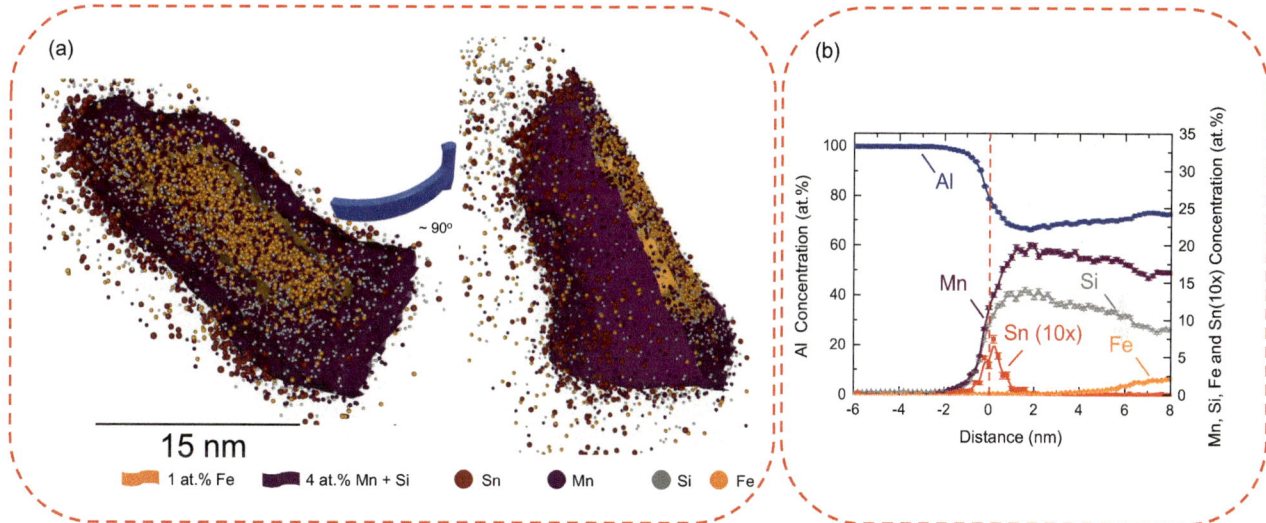

Fig. 4 **a** 3-D-APT reconstructions of an α-Al(Mn,Fe)Si precipitate in the Sn-modified *Al-0.5Mn-0.3Si-0.02Sn* alloy peak-aged isochronally (25 °C steps for 1 h) to 475 °C, viewed along two different orientations. Only a portion of a precipitate is displayed in this 3-D reconstruction. **b** Proximity histogram (proxigram) computed from the dataset displayed in (**a**), showing the average concentration profiles across the matrix/precipitate heterophase interface. The interface (vertical dashed line) is defined as the inflection point of the Al concentration profile. The error bars represent a one-sigma statistical error. A significant Sn segregation at the interface is visible in the proxigram, which is averaged over all the matrix/precipitate interfaces visible in (**a**). The bin size utilized is 0.2 nm

approximate formula of the α-phase without Fe, initially proposed by Cooper and Robinson, $Al_9Mn_2Si_{1.8}$ [13] and later refined by Sugiyama et al., $Al_{69.7}Mn_{17.4}Si_{12.9}$ [41]. The non-uniform distribution of the alloying elements within the α-precipitates indicates different stages of precipitate growth during isochronal aging. The core of the precipitate is enriched in Fe, which is consistent with Fe ($D_{Fe} \sim 5 \times 10^{-18}$ m^2s^{-1} at 400 °C [42]) diffusing faster than Mn ($D_{Mn} \sim 6 \times$

10^{-19} m^2s^{-1} at 400 °C [42]) in the Al(f.c.c.) matrix. Tin segregation at semi-coherent interfaces (the periphery of the precipitate) is observed. Tin partitioning to the α-Al(Mn,Fe) Si precipitates is small (0.03 ± 0.01 at.%).

Compressive creep tests were performed on the peak-aged Sn-free *Al-0.5Mn-0.3Si* and Sn-modified *Al-0.5Mn-0.3Si-0.02Sn* alloys, displaying two significantly different dimensions and distributions of α-precipitates,

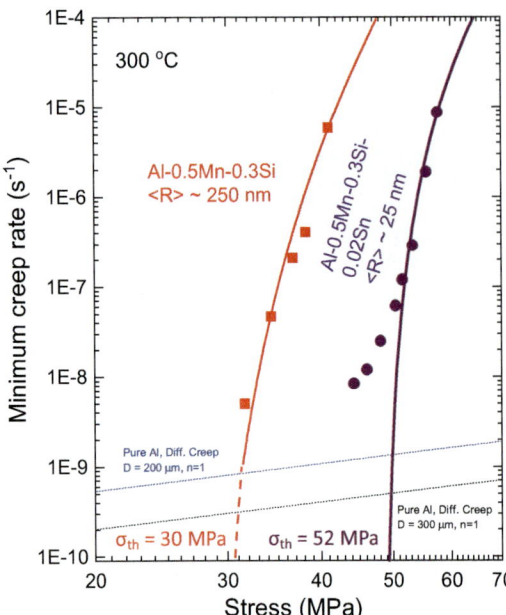

Fig. 5 Double-logarithmic plot of minimum compressive creep strain rate versus applied stress at 300 °C for the Sn-free *Al-0.5Mn-0.3Si* (mean precipitate radius, <R> ~ 250 nm) and Sn-modified *Al-0.5Mn-0.3Si-0.02Sn* (<R> ~ 25 nm) alloys peak-aged isochronally (25 °C steps for 1 h to 475 °C) from the as-cast state. The dotted blue and black lines represent calculated diffusional creep rates (Coble plus Nabarro-Herring creep, D = 200 and 300 μm, D is the grain diameter) for pure aluminum, utilizing data in Ref. [43]

Fig. 2a and b. Figure 5 displays plots of the minimum creep rate, $\dot{\varepsilon}_m$, as a function of applied stress, σ, on a double-logarithmic scale for these alloys. The apparent stress exponents, $n_a(= \partial ln\dot{\varepsilon}_m/\partial ln\sigma)$, in the dislocation creep regimes are much higher than that for pure Al ($n = 4.4$) [43] and vary with stress ($n_a \sim 25 - 30$), which is indicative of a threshold stress, σ_{th}, below which dislocation creep is inhibited [44]. The threshold stresses are attributed to interactions of dislocations with precipitates during the climb bypass process [45, 46]. In the presence of a threshold stress, σ_{th}, the minimum creep rate, $\dot{\varepsilon}_m$, is expressed through a modified power-law equation [44]:

$$\dot{\varepsilon}_m = A(\sigma - \sigma_{th})^n \exp(-\frac{Q}{kT}) \qquad (1)$$

where n is the stress exponent for the aluminum matrix, A is a constant, Q is the activation energy for creep, and kT has its standard meaning. The threshold stresses, determined employing a best-fit procedure given in Ref. [47], are given in Fig. 5. It is apparent that the Sn-modified alloy, with very small α-precipitates, exhibits a threshold stress of ~52 MPa, which is much higher than that of the Sn-free alloy, ~30 MPa, with much larger α-precipitates. This significant improvement in the creep resistance is consistent with the pronounced age-hardening response of the

Sn-modified alloy, Fig. 1, making this alloy one of the most creep-resistant high-temperature aluminum alloys developed to date that is castable and workable.

The Sn inoculation proposed herein for the refinement of the distribution of the α-Al(Mn,Fe)Si precipitates can be employed in most of the commercial Al alloys containing Mn (3xxx, 4xxx, 5xxx, 6xxx and 7xxx) to improve their strengths at both ambient and elevated temperatures. The modified alloys can be utilized at ambient and high temperatures under high stresses for a variety of lightweight applications. This chemical approach is particularly attractive as Sn micro-additions can be integrated into the bulk composition of the alloys, within the impurity tolerances of most of the aluminum alloys. Thus, neither recertification of the alloy systems nor additional processing (such as large deformation steps or complex multistage aging) is necessary. Furthermore, as Sn inoculation appears to be relatively insensitive to the presence of other impurities in Al-alloys, such as Fe and Si, Sn inoculation can be employed for alloys with high recycling contents (with higher Fe and Si concentrations than pristine alloys, such as recycled material from aluminum beverage cans), which leads to significant financial and environmental benefits.

Conclusions

We examined the microstructures and mechanical properties of Sn-free *Al-0.5Mn-0.3Si* and Sn-modified *Al-0.5Mn-0.3Si-0.02Sn* alloys. The following specific conclusions are drawn:

1. Microstructural analyses indicate that α-Al(Mn,Fe)Si precipitates form in both alloys upon isochronally aging to their respective peak microhardness. The α-precipitates in the Sn-free alloy are coarse (D ~ 200 to 1000 nm) and distributed non-uniformly throughout the microstructures, which leads to a negligible hardening increment. In the Sn-modified alloy, however, the α-precipitates are much smaller (D ~ 50 nm) and distributed uniformly, imparting a significant strength to the alloy (ΔHV ~ 125 MPa with respect to the as-cast state, upon isochronal peak aging with 25 °C steps for 1 h).
2. APT results indicate that α-precipitates nucleate heterogeneously on the Sn-rich nanoprecipitates, leading to an enhanced distribution of α-precipitates and a higher strength at ambient temperatures.
3. Compressive creep experiments conducted at 300 °C demonstrate that the Sn-modified *Al-0.5Mn-0.3Si-0.02Sn* alloy exhibits a creep threshold stress of ~52 MPa, which is over 70% greater than that of the Sn-free alloy, ~30 MPa, which translates into over four orders of magnitude slower minimum creep rates of the Sn-modified alloy at a given applied stress. This

significant improvement in the creep resistance of the Sn-modified *Al-0.5Mn-0.3Si-0.02Sn* alloy is attributed to the enhanced dispersion of the α-Al(Mn,Fe)Si precipitates, which agrees with the pronounced age-hardening response of this alloy.

Acknowledgements This research was supported by the Office of Naval Research (N00014-18-1-2550 and N00014-21-1-2782) with Dr. W.M. Mullins and Dr. J. Wolk serving as the grant officers. We are grateful to research associate Prof. Dieter Isheim, Northwestern University (NU), for numerous valuable discussions of the APT experiments. Atom-probe tomography was performed at the Northwestern University Center for Atom-Probe Tomography (NUCAPT). The LEAP tomograph at NUCAPT was purchased and upgraded with grants from the NSF-MRI (DMR-0420532) and ONR-DURIP (N00014-0400798, N00014-0610539, N00014-0910781, N00014-1712870) programs. NUCAPT received support from the MRSEC program (NSF DMR-1720139) at Northwestern's Materials Research Center, the SHyNE Resource (NSF ECCS-1542205), and the Initiative for Sustainability and Energy (ISEN). This research made use of the Materials Characterization and Imaging Facility (MatCI), which receives support from the MRSEC Program (NSF DMR-1720139) of the Materials Research Center at NU. This work made use of the EPIC facility of the NUANCE Center, which receives support from the MRSEC program NSF DMR-1720139) at the Materials Research Center; the International Institute for Nanotechnology (IIN NIH-S10OD026871); and the State of Illinois, through the IIN. DNS and DCD disclose financial interests in Unity Aluminum (formerly Braidy Industries), which is active in aluminum R&D.

References

1. K.E. Knipling, D.C. Dunand, D.N. Seidman, Criteria for developing castable, creep-resistant aluminum-based alloys - A review, Z. Metallk. 97(3) (2006) 246-265.
2. R.A. Michi, J.P. Toinin, A.R. Farkoosh, D.N. Seidman, D.C. Dunand, Effects of Zn and Cr additions on precipitation and creep behavior of a dilute Al-Zr-Er-Si alloy, Acta Mater. 181 (2019) 249-261.
3. A.R. Farkoosh, D.C. Dunand, D.N. Seidman, Effects of W and Si microadditions on microstructure and the strength of dilute precipitation-strengthened Al–Zr–Er alloys, Materials Science and Engineering: A 798 (2020) 140159.
4. A.R. Farkoosh, D.C. Dunand, D.N. Seidman, Solute-induced strengthening during creep of an aged-hardened Al-Mn-Zr alloy, Acta Mater. 219 (2021) 117268.
5. A.R. Farkoosh, D.N. Seidman, D.C. Dunand, Microstructure and mechanical properties of a precipitation-hardened Al–Mn–Zr–Er alloy, Springer International Publishing, Cham (2021) 239-244.
6. L.F. Mondolfo, Aluminum alloys : structure and properties, Butterworths, London, 1976.
7. Y.J. Li, L. Arnberg, Quantitative study on the precipitation behavior of dispersoids in DC-cast AA3003 alloy during heating and homogenization, Acta Mater. 51(12) (2003) 3415-3428.
8. Y.J. Li, A.M.F. Muggerud, A. Olsen, T. Furu, Precipitation of partially coherent α-Al(Mn,Fe)Si dispersoids and their strengthening effect in AA 3003 alloy, Acta Mater. 60(3) (2012) 1004-1014.
9. A.R. Farkoosh, Development of creep-resistant Al-Si cast alloys strengthened with nanoscale dispersoids, McGill Unversity, 2014.
10. A.R. Farkoosh, X.G. Chen, M. Pekguleryuz, Dispersoid strengthening of a high temperature Al-Si-Cu-Mg alloy via Mo addition, Mater. Sci. Eng., A 620 (2015) 181–189.
11. A.R. Farkoosh, X.G. Chen, M. Pekguleryuz, Interaction between molybdenum and manganese to form effective dispersoids in an Al-Si-Cu-Mg alloy and their influence on creep resistance, Mater. Sci. Eng., A 627 (2015) 127–138.
12. F. Qian, S. Jin, G. Sha, Y. Li, Enhanced dispersoid precipitation and dispersion strengthening in an Al alloy by microalloying with Cd, Acta Mater. 157 (2018) 114-125.
13. M. Cooper, K. Robinson, The crystal structure of the ternary alloy α (AlMnSi), Acta Crystallogr. 20(5) (1966) 614-617.
14. Y. Murakami, K. Mori, Highly supersaturated Al-Mn solid solution alloys and their decomposition by heat-treatment, Journal of Japan Institute of Light Metals 18(6) (1968) 339-346.
15. K. Nagahama, I. Miki, Precipitation during Recrystallization in Al-Mn and Al-Cr Alloys, Transactions of the Japan Institute of Metals 15(3) (1974) 185-192.
16. N.J. Luiggi, Isothermal precipitation of commercial 3003 Al alloys studied by thermoelectric power, Metallurgical and Materials Transactions B 28(1) (1997) 125-133.
17. L. Lodgaard, N. Ryum, Precipitation of dispersoids containing Mn and/or Cr in Al–Mg–Si alloys, Materials Science and Engineering: A 283(1) (2000) 144-152.
18. A.R. Farkoosh, D.C. Dunand, D.N. Seidman, Tungsten solubility in L12-ordered Al3Er and Al3Zr nanoprecipitates formed by aging in an aluminum matrix, J. Alloys Compd. 820 (2020) 153383.
19. A. De Luca, D.C. Dunand, D.N. Seidman, Microstructure and mechanical properties of a precipitation-strengthened Al-Zr-Sc-Er-Si alloy with a very small Sc content, Acta Mater. 144 (2018) 80-91.
20. A.R. Farkoosh, D. Dunand, D.N. Seidman, Microstructure and Mechanical Properties of an Al-Zr-Er High Temperature Alloy Microalloyed with Tungsten, Springer International Publishing, Cham, 2019, pp. 379-383.
21. I.J. Polmear, Role of Trace Elements in Aged Aluminium-Alloys, Mater. Sci. Forum 13-14 (1987) 195-214.
22. I.J. Polmear, Light alloys : from traditional alloys to nanocrystals, Elsevier/Butterworth-Heinemann, Amsterdam, 2006.
23. J.M. Silcock, T.J. Heal, H.K. Hardy, The structural ageing characteristics of ternary aluminium-copper alloys wih cadmium, indium, or tin, Journal of the Institute of Metals 84(1) (1955) 23.
24. H.K. Hardy, The ageing characteristics of ternary aluminium copper alloys with cadmium, indium, or tin, Journal of the Institute of Metals 80(9) (1952) 483-492.
25. S.P. Ringer, K. Hono, T. Sakurai, The effect of trace additions of sn on precipitation in Al-Cu alloys: An atom probe field ion microscopy study, Metallurgical and Materials Transactions A 26 (9) (1995) 2207-2217.
26. R. Sankaran, C. Laird, Effect of trace additions Cd, In and Sn on the interfacial structure and kinetics of growth of θ' plates in Al-Cu alloy, Materials Science and Engineering 14(3) (1974) 271-279.
27. J.M. Silcock, H.M. Flower, Comments on a comparison of early and recent work on the effect of trace additions of Cd, In, or Sn on nucleation and growth of θ' in Al–Cu alloys, Scripta Mater. 46(5) (2002) 389-394.
28. T. Homma, M.P. Moody, D.W. Saxey, S.P. Ringer, Effect of Sn Addition in Preprecipitation Stage in Al-Cu Alloys: A Correlative Transmission Electron Microscopy and Atom Probe Tomography Study, Metallurgical and Materials Transactions A 43(7) (2012) 2192-2202.
29. L. Bourgeois *, J.F. Nie, B.C. Muddle, Assisted nucleation of θ' phase in Al–Cu–Sn: the modified crystallography of tin precipitates, Philos. Mag. 85(29) (2005) 3487–3509.

30. B.T. Sofyan, K. Raviprasad, S.P. Ringer, Effects of microalloying with Cd and Ag on the precipitation process of Al–4Cu–0.3Mg (wt %) alloy at 200°C, Micron 32(8) (2001) 851–856.

31. B. Noble, Theta-prime precipitation in aluminium-copper-cadmium alloys, Acta Metall. 16(3) (1968) 393-401.

32. E. Holmes, B. Noble, Resistivity examination of artificial ageing in an aluminium-copper-cadmium alloy, Journal of the Institute of Metals 95 (1967) 106.

33. H.K. Hardy, Aluminum copper cadmium sheet alloys, Journal of the Institute of Metals 83(7) (1955) 337.

34. B.W. Krakauer, D.N. Seidman, Systematic procedures for atom-probe field-ion microscopy studies of grain boundary segregation, Rev. Sci. Instrum. 63(9) (1992) 4071-4079.

35. D.N. Seidman, Three-Dimensional Atom-Probe Tomography: Advances and Applications, Annual Review of Materials Research 37(1) (2007) 127-158.

36. D.N. Seidman, B.W. Krakauer, D. Udler, Atomic scale studies of solute-atom segregation at grain boundaries: Experiments and simulations, J. Phys. Chem. Solids 55(10) (1994) 1035-1057.

37. D.N. Seidman, Subnanoscale Studies of Segregation at Grain Boundaries: Simulations and Experiments, Annual Review of Materials Research 32(1) (2002) 235-269.

38. D.N. Seidman, K. Stiller, An Atom-Probe Tomography Primer, MRS Bull. 34(10) (2009) 717-724.

39. D. Seidman, K. Stiller, A Renaissance in Atom-Probe Tomography, Materials Research Society Bulletin 34(10) (2009) 717.

40. O.C. Hellman, J.A. Vandenbroucke, J. Rüsing, D. Isheim, D.N. Seidman, Analysis of three-dimensional atom-probe data by the proximity histogram, Microsc. Microanal. 6(5) (2000) 437-444.

41. K. Sugiyama, N. Kaji, K. Hiraga, Re-Refinement of [alpha]-(AlMnSi), Acta Crystallographica Section C 54(4) (1998) 445-447.

42. G. Rummel, T. Zumkley, M. Eggersmann, K. Freitag, H. Mehrer, Diffusion of implanted 3d-transition elements in aluminium part I: Temperature dependence/diffusion implantierter 3d-Übergangselemente in aluminium Teil I: Temperaturabhängigkeit, International Journal of Materials Research 86(2) (1995) 122-130.

43. H.J. Frost, M.F. Ashby, Deformation mechanism maps: the plasticity and creep of metals and ceramics, Pergamon press1982.

44. M.E. Kassner, Fundamentals of Creep in Metals and Alloys (Third Edition), Butterworth-Heinemann, Boston, 2015.

45. F.R.N. Nabarro, F. De Villiers, Physics of creep and creep-resistant alloys, CRC press1995.

46. M.E. Krug, D.N. Seidman, D.C. Dunand, Creep properties and precipitate evolution in Al-Li alloys microalloyed with Sc and Yb, Mater. Sci. Eng. A 550 (2012) 300-311.

47. R.A. Karnesky, Jr., Mechanical properties and microstructure of aluminum-scandium with rare-earth element or aluminum oxide additions, 2007.

Innovative Approaches in Development of Aluminium Alloys for Packaging Industry

Stanislav Kores, Simon Strmšek, Maja Vončina, and Jožef Medved

Abstract

Sustainable trends in the packaging industry market are leading us to the development of new aluminium alloys, that enable to achieve higher mechanical properties. Use of circular materials and post-consumer recycled materials are the approaches to reduce the carbon footprint of products. Several aluminium alloys were developed for aluminium narrow strips, cast with a rotary strip caster, to produce slugs for aerosol cans. The newly developed alloys provide constant mechanical properties during the manufacturing process of aerosol cans, a good transformation and high deformable and burst pressures of the aerosol cans. With increasing mechanical properties of aerosol can materials, it is possible to achieve a significant impact on lightweight of the final aerosol cans. Innovative approaches were used with use of a post-consumer recycled material in different proportions in the casting and heat treatment process of aluminium alloys to reduce carbon footprint of products in the packaging industry.

Keywords

Aluminium alloys • Rotary strip casting • Slug • Aerosol cans • Post-consumer recycled material

S. Kores (✉) · S. Strmšek
TALUM d.d, Tovarniška c. 10, 2325 Kidričevo, Slovenia
e-mail: stanislav.kores@talum.si

S. Strmšek
e-mail: simon.strmsek@talum.si

M. Vončina · J. Medved
Faculty of Natural Sciences and Engineering, Department of Materials and Metallurgy, University of Ljubljana, Askerceva 12, 1000 Ljubljana, Slovenia
e-mail: maja.voncina@ntf.uni-lj.si

J. Medved
e-mail: jozef.medved@ntf.uni-lj.si

Introduction

Aluminium aerosol cans are made either by extrusion of aluminium slugs or by deep drawing of discs stamped from aluminium sheet. Aluminium sheets are mainly produced by the direct chill casting process (DC). However, the continuous casting process (CC) offers energy and economy savings while reducing environmental emissions. Compared to the DC casting technology, the CC technology also offers the advantage of high productivity [1].

Approximately 6 billion aluminium aerosol cans are used annually worldwide. Almost 80% of the production is attributable to cosmetics industry. Aluminium aerosol cans are not only user-friendly, but also help to conserve resources. A typical cylindrical aerosol can (38 mm/138 mm) weighs 17 g today, around 30% lighter than in the early 1970s, and there is still further potential for savings. Aluminium is also readily recyclable and can be made into new high-quality products again and again without any loss of quality, and this also applies to aerosol cans [2].

As a durable, lightweight, and infinitely recyclable material, aluminium is important to the transition to carbon neutrality. All the sustainable components–3R (reduce-recycle-reuse) of material are ensured, when aluminium at the end of its life cycle (PCR material) is used in products for the packaging, transportation, and construction industries. Circular economy, infinite recyclability without losing the positive properties of the material, and other advantages make this metal a green and environmentally friendly material.

The manufacturing process of aerosol cans consists of the following steps (Fig. 1):

(1) Slug → (2) impact extrusion → (3) washing and drying → (4) internal lacquering and polymerization → (5) external lacquering → (6) printing → (7) coating with protective lacquer layer → (8) draw in the can shoulder and body in a multi-die necking machine.

© The Minerals, Metals & Materials Society 2023
S. Broek (ed.), *Light Metals 2023*, The Minerals, Metals & Materials Series,
https://doi.org/10.1007/978-3-031-22532-1_73

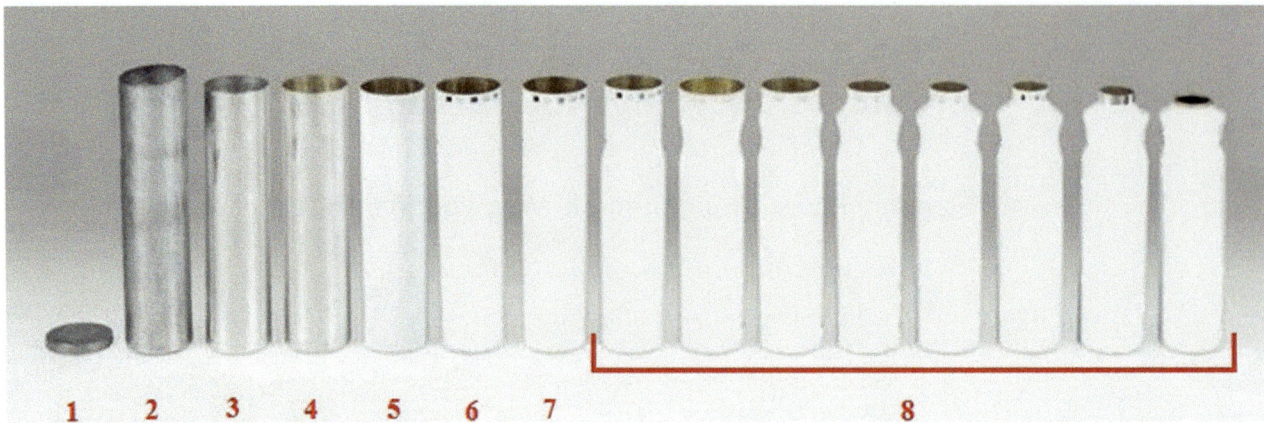

Fig. 1 Sequence of aerosol can manufacture

Drying of washed, externally painted, and printed cans takes place at temperatures of 140–180 °C. The polymerization of the interior coating takes place at temperatures around 250–280 °C. During the manufacturing process of aerosol cans, the mechanical properties of the material decrease by up to 15% after polymerization. This is reflected in the achievement of lower deformation and bursting pressure of aerosol cans.

Aluminium aerosol cans for cosmetics and food industry are produced from aluminium slugs by extrusion. In Talum d.d., the aluminium slugs are semi-finished products stamped from an aluminium narrow strip produced on a rotary strip casting machine (Fig. 2).

An aluminium strip is usually cast by the horizontal casting method, which allows casting of a wide range of aluminium alloys. A narrow aluminium strip cast using the "rotary strip casting system" is limited to casting AA1XXX and AA3XXX series aluminium alloys. The 1XXX and 3XXX series aluminium alloys are widely used in the transportation, food, beverage, and packaging industries. In these applications, control of the sheet's plastic anisotropy is essential to ensure end-product formability and reduce material waste due to earring behaviour [3]. The 1XXX and 3XXX series are characterised by good formability and corrosion resistance, and the AA3XXX series aluminium alloys also have good weldability and relatively good mechanical properties [4]. Good mechanical properties mean that high deformations and burst pressures are achieved [5].

The rotary strip casting machine consists of a casting wheel and a steel belt. The molten metal flows from the casting channel into the area between the endless steel belt and the water-cooled wheel (Fig. 3). The casting wheel is made of a copper or steel alloy.

From the casting machine, the aluminium strip is fed via a roller conveyor to the hot rolling mill and then to the cold rolling mill. In the hot rolling mill, the strip thickness is reduced by 40–70%, while in the cold rolling mill it reaches 30–50%.

The rolled strip, which is made of a narrow aluminium alloy, is then transported to the stamping line, where the slugs are stamped using a stamping machine. From the stamping machine, the slugs are led into annealing furnaces

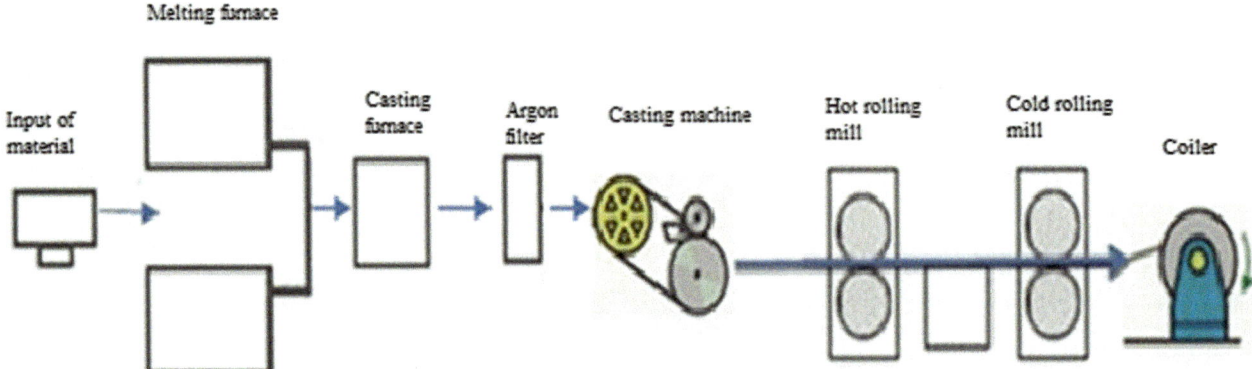

Fig. 2 Production line of aluminium narrow strip

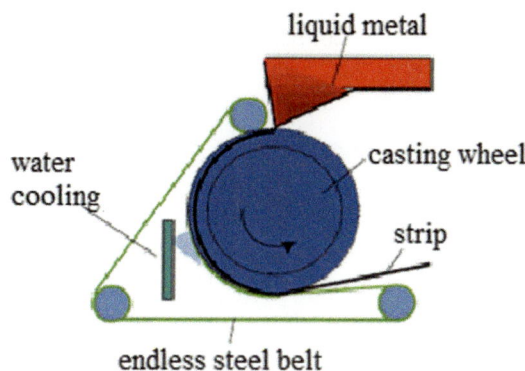

Fig. 3 Rotary strip casting system [6]

where they are softened and the oil remaining from stamping is burned off. After annealing, the slugs are surface treated by sandblasting, vibrating, or tumbling.

To obtain suitable mechanical properties, it is necessary to take into account all influencing factors that affect the mechanical properties of the material. Figure 4 shows the most important influencing factors, such as chemical composition, casting and deformation parameters, and heat treatment parameters. All these factors have a significant influence on the mechanical properties and ultimately on the mechanical properties of the final product.

In this work, we investigated the influence of chemical composition on the casting of aluminium narrow strip in the rotary strip casting process, especially when PCR material is used. With increasing the alloying elements in the chemical composition, the solidification path on the casting wheel changes and some intermetallic phases appear, causing cracks on the narrow strip. By adding some alloying elements to the standard alloys, it is possible to improve the mechanical properties and lower the recrystallization threshold. In the following, the influence of cold

deformation in the cold rolling mill on the mechanical properties, microstructure, and annealing parameters was investigated. When PCR material is used, alloying elements such as Mg, Mn, Cu are present, which accelerate the oxidation of the surface. For this reason, the conditions and the effects on the oxidation rate of alloys with Mg addition were studied.

During high-temperature annealing of Al–Mg alloys in air, magnesium is dissolved out from the interior of the material. This causes an uneven and relatively thick layer of dark oxide to form on the surface. These oxide layers can lead to undesirable effects during product processing. Using Auger electron microscopy (AES), analysis of a sample annealed at 550 °C/90 h with 0.45 wt% Mg found that the magnesium concentration decreases linearly from 45% at the surface to about 5% at a depth of 860 nm. The aluminium concentration at the surface is less than 5%, and at a depth of 860 nm it is about 90%. It was also found that the concentration of MgO decreases from 98 to 86% at a depth of 250 nm, while the amount of spinel ($MgAl_2O_4$) increases from 2 to 14%. The predominant presence of MgO can be explained by faster growth compared to spinel and by magnesium diffusion through local defects that allow magnesium atoms to move toward the surface. Oxide growth is therefore controlled by diffusion of Mg to the surface in MgO and spinel, since oxygen diffusion in aluminium and hence oxide formation is negligible. MgO generally does not form a homogeneous layer, making the base metal depleted of Mg at individual locations and more susceptible to spinel formation. The formation of spinel can also occur through the reduction of Al_2O_3 on the surface of the material [7].

In another study, it was found that the surface of the cold-rolled sheet is covered with a thin protective layer of Al_2O_3 oxide, which has an amorphous structure. They found that when Al–Mg alloys are heated in air to above 350 °C, there is a gradual transformation of the amorphous structure of Al_2O_3 to the crystalline structure of γ-Al_2O_3. The whole transformation process is accompanied by the diffusion of magnesium and aluminium to the surface. Due to the crystalline structure of the converted oxide, it no longer forms a homogeneous layer. As a result, pathways are opened between the individual crystals for the diffusion of magnesium to the surface. Magnesium has a much higher diffusivity than aluminium. Along these pathways, MgO is formed more intensively, forming an inhomogeneous oxide layer on the surface, also called "MgO islands". Over time, a complete transformation of amorphous Al_2O_3 into crystalline γ-Al_2O_3 occurs, and the entire surface is covered with MgO. When the entire surface is covered with MgO, the situation under the oxide layer becomes thermodynamically and kinetically more favourable for the reduction of Al_2O_3 with magnesium. $MgAl_2O_4$ spinel is formed in the lower layers [8].

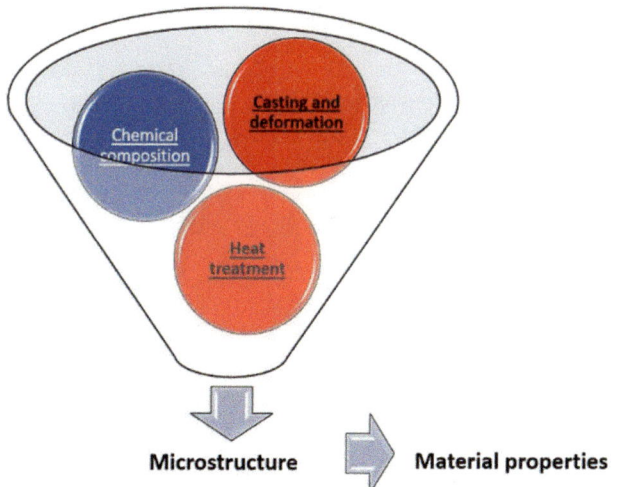

Fig. 4 Influencing factors on material mechanical properties

Table 1 Standard Talum aluminium alloys for aerosol cans

Alloy	Mechanical properties				
	Hardness (HB$_{2.5/15.625}$)	Rm (MPa)	Rp$_{0.2}$ (MPa)	Elongation (%)	Grain number/mm^2
99.7	18.5	70	34	42	60–100
99.5	19.5	75	37	41	60–100
AlMn0.3	22	80	41	40	20–30
AlMn0.6	27	92	55	38	30–60

Materials and Methods

Some basic investigations with the Thermo-Calc software have been carried out at the Faculty of Natural Sciences and Engineering, Department of Materials and Metallurgy: simple thermal analysis, DSC, and observation under light and scanning electron microscope to observe the influence of various chemical elements.

For the development/modification of aluminium alloys, standard aerosol can alloys, that are produced in company Talum (Table 1) were used as the base material. These alloys were modified by combining different alloying elements.

A very important factor in impact extrusion is also the grain number/mm^2. Slugs made from 99.5% and 99.7% pure aluminium have smaller grains than slugs made from AlMn0.3 and AlMn0.6 alloys (Fig. 5).

The reason for modifying aluminium alloys for aerosol cans was the deterioration of mechanical properties during the manufacturing process and the use of PCR material in slug production. The question was how to eliminate the deterioration of mechanical properties and at the same time be able to cast aluminium narrow strip on a rotary strip casting machine.

During the aerosol can manufacturing process, the mechanical properties of the material decrease by up to 15% after polymerization. This is reflected in the achievement of lower deformation and bursting pressures of aerosol cans. Mechanical tests were performed on samples of aerosol cans. Figure 6 shows the mechanical properties of the material of aerosol cans during the manufacturing process—after extrusion and after polymerization [9].

Results and Discussion

Samples of the standard aerosol can material were taken from the cans, and the standard tensile test was used to measure the mechanical properties. After polymerization, the mechanical properties of Al99.7% aerosol cans decreased by 15.9%, those of AlMn0.3 alloy cans decreased by 8.5%, and the mechanical properties of AlMn0.6 alloy aerosol cans decreased by 6.5%. Figure 7 shows the tensile strength of the measured aluminium alloys at the essential stage of the aerosol can manufacturing process.

The aim of the modification of aluminium alloys for aerosol cans was to develop such an alloy from which it is possible to produce an aluminium narrow strip with the rotary strip caster on an existing casting-rolling line.

By combining different alloying elements, higher slug hardness, tensile strength, and yield strength were achieved. The elongation of the modified alloy decreased. In addition, the grain number/mm^2 decreased compared to the standard alloys. The process of aluminium solidification on the

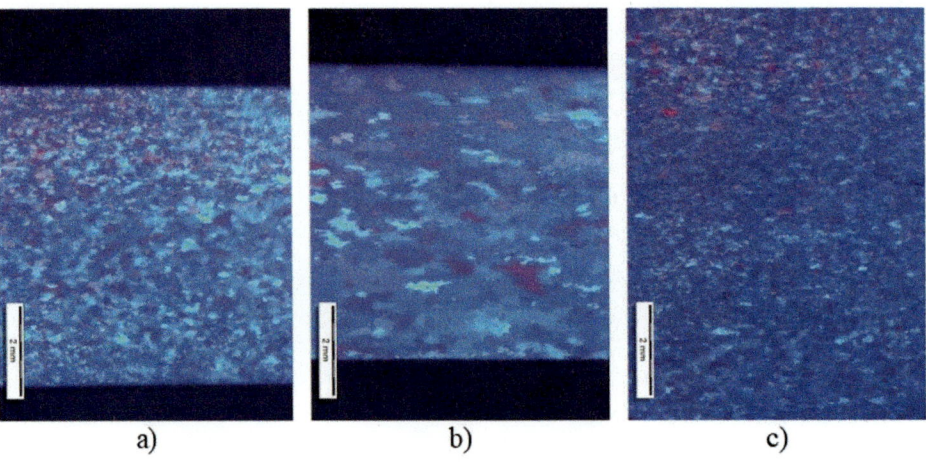

Fig. 5 Grain structure of standard Talum aluminium alloys for aerosol cans: **a** Al99.5%, **b** AlMn0.3, and **c** AlMn0.6 in polarised light

a)   b)   c)

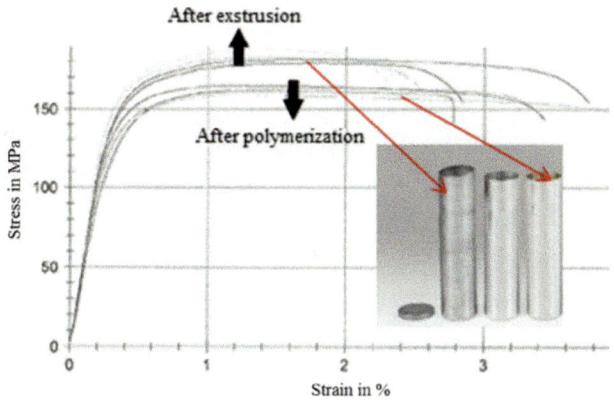

Fig. 6 Mechanical properties of Al99.7% aerosol cans after extrusion and after polymerization

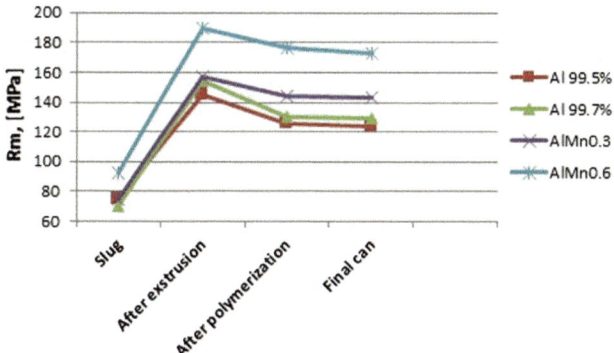

Fig. 7 Mechanical properties of aerosol can material during manufacturing

casting wheel is a very complex process, as it is influenced by many factors. Very important factors are the temperature of the melt, the temperature of the casting wheel, the casting speed, the temperature of a steel strip, the gradient of temperature change of the cooling medium, and the flow rates of water. The casting groove must also be modified.

Figure 8 shows the tensile strength of standard and modified aluminium alloys for aerosol cans. It can be seen from the figure that the drop in mechanical properties after polymerization has been eliminated to 2–3%. The mechanical properties of alloy 1 are in the range of AlMn0.3 alloy, but due to the lower grain number/mm^2, alloy 1 is easier to produce. Alloys 2 and 3 can achieve high deformation and bursting pressure, so this alloy has potential in the market where steel cans are used for packaging.

The comparison of standard alloy Al99.7% with alloy 1 shows that alloy 1 has excellent processing properties, excellent surface of aerosol cans and higher deformation and bursting pressure. The comparison of AlMn0.3 alloy with alloy 2 shows the same properties, while alloy 3 shows a significant increase in deformation and bursting pressure but poorer processing properties compared with AlMn0.6 alloy.

Figure 9 shows the grain structure of the samples of the modified alloys taken from the slugs under polarized light. Alloy 1 has a more or less uniform and homogeneous grain structure, while alloy 2 has a coarser grain structure and alloy 3 has an inhomogeneous grain structure. The mechanical properties of alloy 1 are in the range of AlMn0.3 alloy, but due to the lower grain number/mm^2 alloy 1 can be produced more easily.

Fig. 8 Mechanical properties of aerosol can material during manufacturing

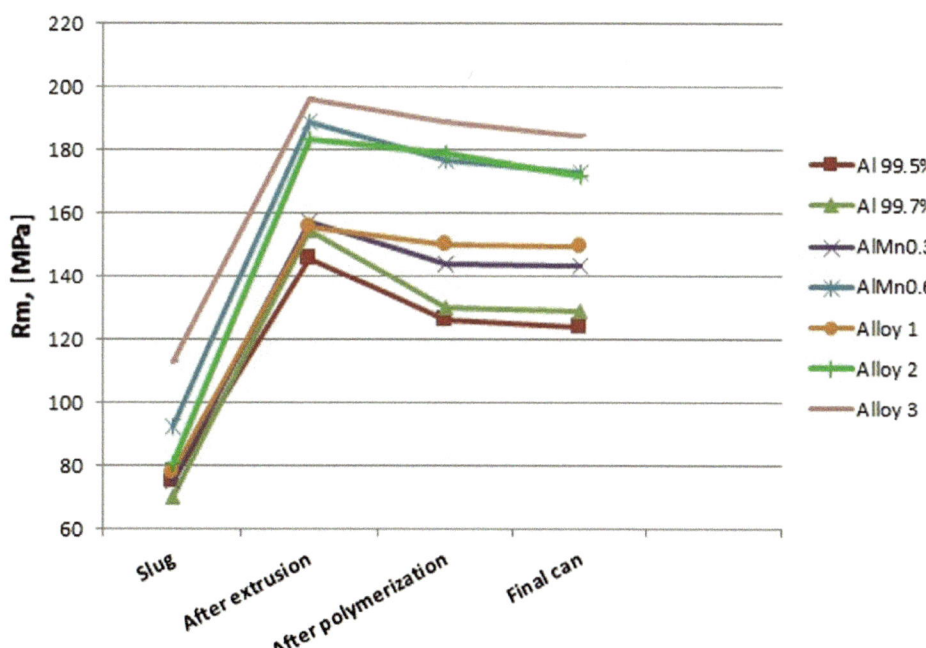

Fig. 9 Grain structure of modified aluminium alloys for aerosol cans under polarised light

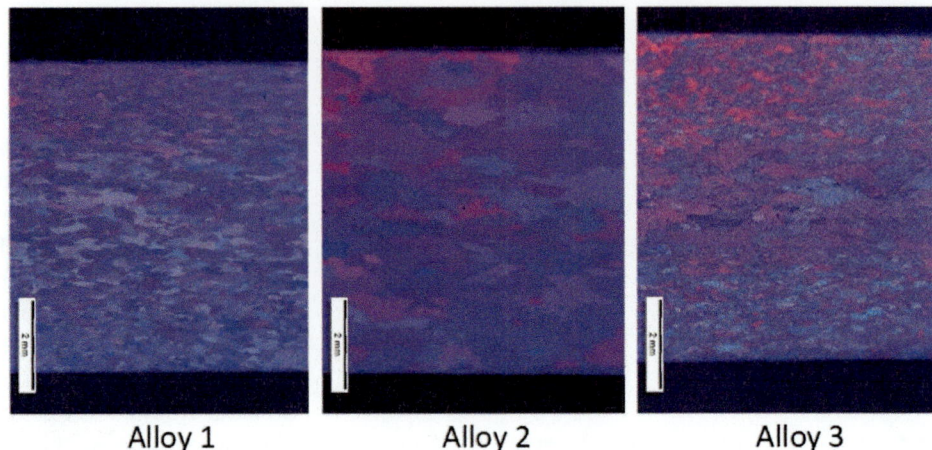

Alloy 1   Alloy 2   Alloy 3

In cooperation with the manufacturers of aerosol cans, the test of alloy 1 has shown a 10–15% higher deformation and bursting pressure of aerosol cans. This means that it is possible to produce aerosol cans with thinner walls, thus reducing their weight and saving material.

The influence of cold forming in the cold rolling mill on the mechanical properties was investigated. Figure 10 shows the influence of the annealing temperature and the reduction rate of the cold rolling mill on the mechanical properties of the narrow strip. It can be observed that the tensile strength values after annealing are very similar, indicating that the soft state of the material has been reached. At the same annealing temperature, the changes in the material properties are different depending on the reduction degree. The higher the reduction degree was, the greater the change was. This indicates that more energy is accumulated in the material during cold working, which together with the increased temperature is the driving force for recrystallization during annealing.

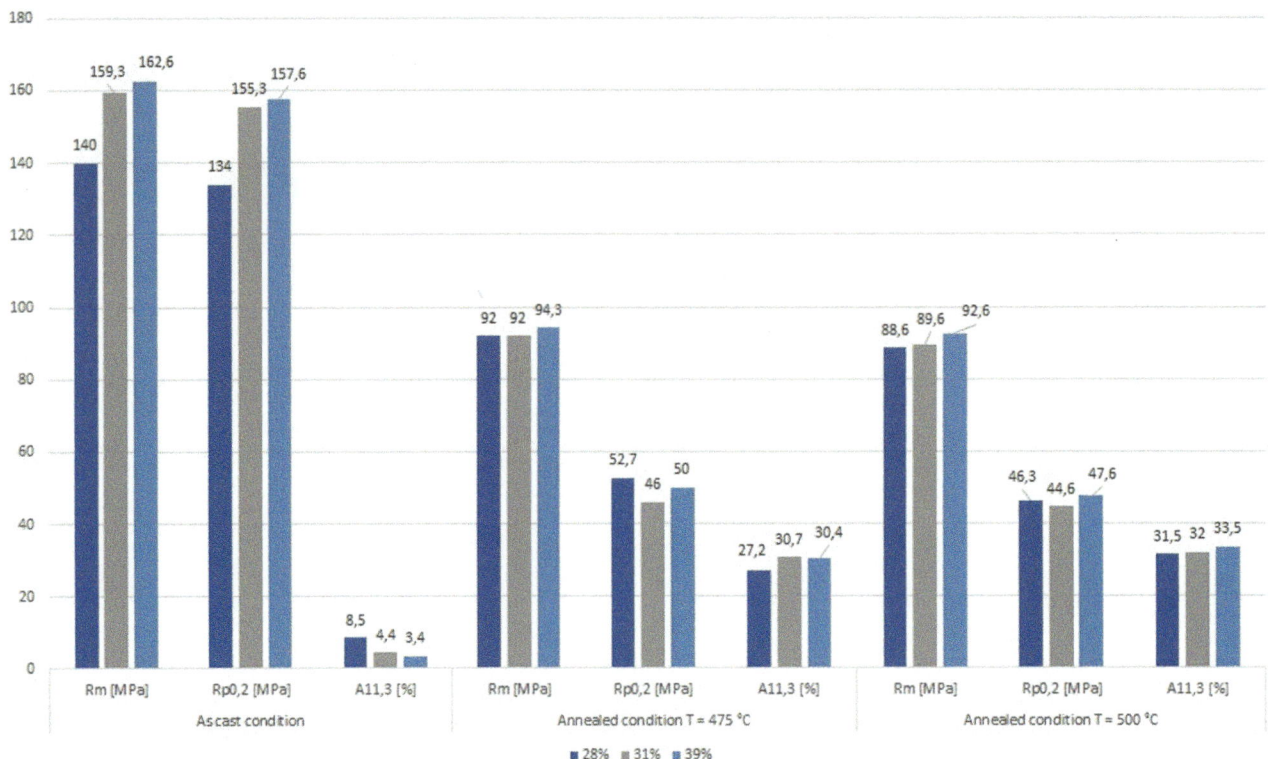

Fig. 10 Influence of annealing temperature and reduction rate of cold rolling mill on mechanical properties of narrow strip

Conclusions

On the existing casting-rolling line, aluminium alloys were modified to produce the aluminium narrow strip for the production of slugs, which allows:

- Casting a narrow aluminium strip at high casting speeds using the "rotary strip casting system" with excellent surface finish and a minimum number of defects,
- constant mechanical properties of the material after polymerization and throughout the manufacturing process of aerosol cans, which is reflected in more than 10% higher burst and deformation pressure of aerosol cans,
- good manufacturability, transformation, and surface of aerosol cans from the developed aluminium alloy slugs,
- the improved mechanical properties of the aerosol can material make it possible to produce aerosol cans with thinner walls and reduce their weight,
- the higher the degree of cold deformation, the higher is the percentage of recrystallized microstructure. With a 39% reduction in the cold rolling mill, compared to a 31% reduction, we achieve a 53% increase in grain size,
- the temperature and annealing atmosphere have a greater effect on the progression and proportion of surface oxidation than the proportion of magnesium in the alloy. The best results were obtained when annealing at 480 °C /50 min.

References

1. Liu J, Morris JG (2003) Macro-, micro- and mesotexture evolutions of continuous cast and direct chill cast AA 3015 aluminium alloy during cold rolling. Materials Science and Engineering A 357 (1–2): 277–296. https://doi.org/10.1016/S0921-5093(03)00210-7
2. http://www.aluinfo.de/index.php/fact-sheets.html (accessed: 7 August 2015)
3. Engler O (2012), Control of texture and earring in aluminium alloy AA 3105 sheet for packaging applications. Materials Science and Engineering A 538: 69–80. https://doi.org/10.1016/j.msea.2012.01.015
4. Mondolfo, LF (1962) Aluminium alloys: Structure and Properties. Butterworth Co, London
5. Medved J, Godicelj T, Kores S, Mrvar P, Vončina M (2012) Contribution of Mn content on the pressure dose properties. RMZ-Materials and Geoenvironment, 59 (1): 41-54
6. Kammer C (1999) Continuous casting of aluminium, Talat Lecture 3210, Goslar
7. ZAYAN, MH, JAMJOOM OM, RAZIK NA (1990) High-Temperature Oxidation of Al-Mg Alloys. Oxidation of Metals, 34: 323-333
8. LEA C, BALL J (1984) The oxidation of rolled and heat-treated Al-Mg alloys. Applications of Surface Science 17: 344-362
9. KORES S, TURK J, MEDVED J, VONČINA M (2016) Development of aluminium alloys for aerosol cans. Materials and technology 50 (4): 601-605

The Role of Microstructure on Strength and Fracture Anisotropy Effects in Al–Mg–Si Extrusion Alloys

S. Kordmir, N. C. Parson, and W. J. Poole

Abstract

Aluminum 6xxx extrusion alloys are attractive candidates for use in automotive applications to decrease vehicle weight. In this study, Al–Mg–Si alloys were extruded into 3 × 90 mm strips on a pilot scale extrusion press with processing conditions designed to produce either recrystallized or unrecrystallized microstructures. The grain shape and crystallographic texture were characterized by electron backscatter diffraction (EBSD) and mechanical anisotropy was measured by tensile and VDA bend testing conducted at 0, 45, and 90 degrees to the extrusion direction. It was found that strength and fracture anisotropy were significantly affected by crystallographic texture and the distribution of second phase particles. The results were rationalized using considerations for the sequence of damage initiation, void growth, and void coalescence.

Keywords

Al–Mg–Si alloys (6xxx) • Strength and fracture anisotropy

Introduction

The increasing use of aluminum extrusions in automotive applications to reduce vehicle weight has drawn attention to consideration of how these materials behave in situations where multi-axial plastic deformation occurs, e.g. in the forming of complex components and during crash scenarios.

This requires characterization of the anisotropic plastic and fracture response of extrusion alloys with different processing histories. There are a number of sources for anisotropy including (i) plastic anisotropy (dependent on crystallographic texture), (ii) morphological anisotropy (related to the shape and size of particles from which the voids nucleate) and topological anisotropy (associated with the spatial distribution of second phase particles [1–7]). This study aims to investigate the sources of anisotropy for two extruded Al–Mg–Si alloys with recrystallized and unrecrystallized microstructures under uniaxial tensile and bending loading paths.

Methodology

Two Al–Mg–Si extrusion alloys were provided by the Rio Tinto Aluminum (RTA) Research and Development Center. Table 1 summarizes the chemical composition measured by optical emission spectroscopy (OES).

The alloys were direct chill cast (DC) as billets 101 mm in diameter, homogenized at 550 °C for 2 h followed by water quenching, and then extruded at a billet temperature of 500 °C and an extrusion ratio of 33:1 to form strips with a cross section of 90 × 3 mm, i.e. almost plane strain deformation. The strips were quenched in a standing wave water tank after exiting the extrusion press. Two different heat treatments were examined after quenching, (i) natural ageing of approximately 1 year (T4) and (ii) artificial aging for 6 h at 180 °C (T6).

Samples from the ND-ED surfaces were prepared for SEM observation and electron backscatter diffraction (EBSD) measurements by grinding with SiC papers up to 4000 grit followed by polishing with 1 μm diamond suspension and finally, a 0.5 μm colloidal silica suspension. EBSD maps were taken using a Zeiss sigma scanning electron microscope equipped with a Nikon high-speed camera and EDAX/TSL OIM Data collection (6th edition)

S. Kordmir · W. J. Poole (✉)
Department of Materials Engineering, The University of British Columbia, 309-6350 Stores Road, Vancouver, BC V6T 1Z4V6T 1Z4, Canada
e-mail: warren.poole@ubc.ca

N. C. Parson
Rio Tinto Aluminium, 1955 Mellon Blvd, Jonquière, QC G7S 4K8, Canada

© The Minerals, Metals & Materials Society 2023
S. Broek (ed.), *Light Metals 2023*, The Minerals, Metals & Materials Series,
https://doi.org/10.1007/978-3-031-22532-1_74

Table 1 Alloy compositions (wt%)

Alloy	Grain structure	Mg	Si	Cu	Mn	Fe	Al
0 Mn	Recrystallized	0.66	0.94	0.002	0.002	0.21	Bal.
0.5Mn	Fibrous	0.74	1.04	0.001	0.520	0.19	Bal.

software, using an accelerating voltage of 20 kV and a 60 μm aperture. Data was captured at a speed of 40 fps with an 8 × 8 bin size. The sample was positioned at a working distance of 13 mm and only the aluminum FCC phase was indexed. A set of overlapping scans were taken at the center area along the extrusion direction and stitched together. Backscatter detector imaging was used to investigate the second phase particle distribution.

To measure mechanical anisotropy, tensile specimens with a 30 mm gauge length were machined with the tensile axis oriented at 0 (ED), 45, and 90° (TD) with respect to the extrusion direction. Tensile testing was conducted for the T4 and T6 conditions at a nominal strain rate of 0.001 s^{-1} using an Instron screw-driven hydraulic test machine. Axial and width clip-on extensometers were used so that the plastic anisotropy could be characterized by the R-value (i.e. the ratio of width to thickness plastic strain). The R-value was characterized at 10% axial strain when possible or as close as possible to 10% when necking occurred before this strain, e.g. in the T6 samples. At least 3 tests were conducted for each condition. The area of the fracture surface was measured from SEM images taking the projection of the fracture cross section using the ImageJ image analysis program. Using this area, the true fracture stress (σ_f) and true fracture strain (ε_f) were calculated with the following formulae:

$$\sigma_f = \frac{F_f}{A_f} \tag{1}$$

$$\varepsilon_f = \ln\left(\frac{A_f}{A_o}\right) \tag{2}$$

where F_f, A_0, and A_f are load at the fracture point, initial area, and fracture area, respectively. A correction for the stress triaxiality in the neck of the tensile sample was employed based on the finite element method calculations of Puydt shown in reference [8]. Specimens for three-point bending were water jet machined with dimensions of 40 mm × 40 mm × 3 mm, where the bend axis was perpendicular to the ED, 45, and TD directions. Bend tests were conducted using a VDA test rig, employing a punch speed of 2 mm/min and a roll gap of 2 times the specimen thickness,

i.e. 6 mm. Force was logged as a function of punch displacement, and the test was stopped when the force dropped by 60 N from the maximum, i.e. the point of fracture. The bend angle at this point was measured manually afterwards using a protractor as well as calculated from the displacement of the punch according to the formula from the VDA 238–100 standard [9].

Results and Discussion

Figure 1 shows EBSD IPF maps illustrating the microstructure of the 0Mn and 0.5Mn extrusions. It can be observed that the 0Mn alloy was fully recrystallized (equiaxed grains) while the 0.5Mn alloy appeared unrecrystallized (highly elongated grains). These results are consistent with the presence of Mn dispersoids which form during homogenization in the 0.5Mn alloy, which are known to suppress recrystallization by the Zener drag mechanism [10–12]. Figure 2 shows 111 and 001 pole figures measured by EBSD with more than 1000 grains in each case. In Fig. 2a, the 0Mn alloy exhibited a typical recrystallization texture after plane strain deformation with cube and Goss texture components [13]. On the other hand, the 0.5Mn material, Fig. 2b showed a typical plane strain deformation texture with Cu, Brass, and S components [14, 15], i.e. the textures were consistent with the EBSD observations on the grain shape.

Figure 3 shows backscatter electron SEM images taken from the 3 perpendicular surfaces aligned with the extrusion, transverse and normal directions of the extrudate which illustrate the spatial distribution of second phase particles. In the case of the 0Mn alloy, the particles were the β-Al$_5$FeSi phase which form as plates during the final stages of solidification [16]. Subsequently, these plates align with the extrusion direction and also fragment due to their fracture during large strain deformation [17]. This leads to an anisotropic spatial distribution of particles as shown in Fig. 3a. In contrast, the 0.5Mn alloy had a more complicated second phase particle distribution. First, the nature of the particles was different with the α-Al(Fe,Mn)Si phase being dominant [18]. This phase appeared in particles of two different length

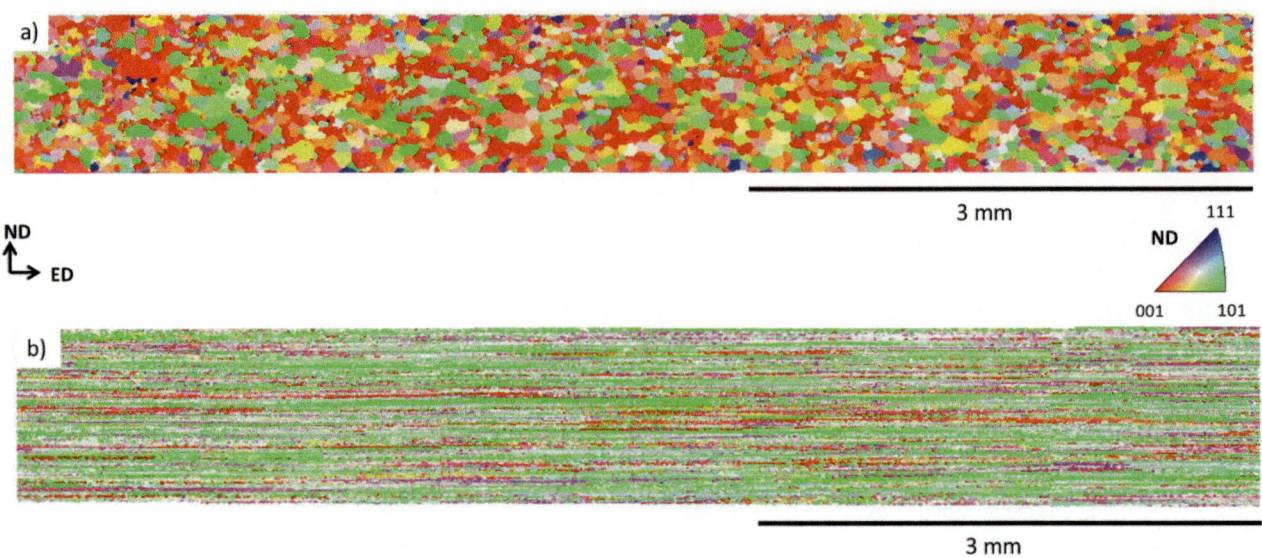

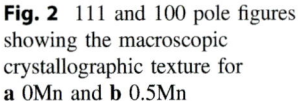

Fig. 1 EBSD inverse pole figure (IPF) grain structure maps in **a** 0 Mn and **b** 0.5Mn extrusions taken near the centre

Fig. 2 111 and 100 pole figures showing the macroscopic crystallographic texture for **a** 0Mn and **b** 0.5Mn

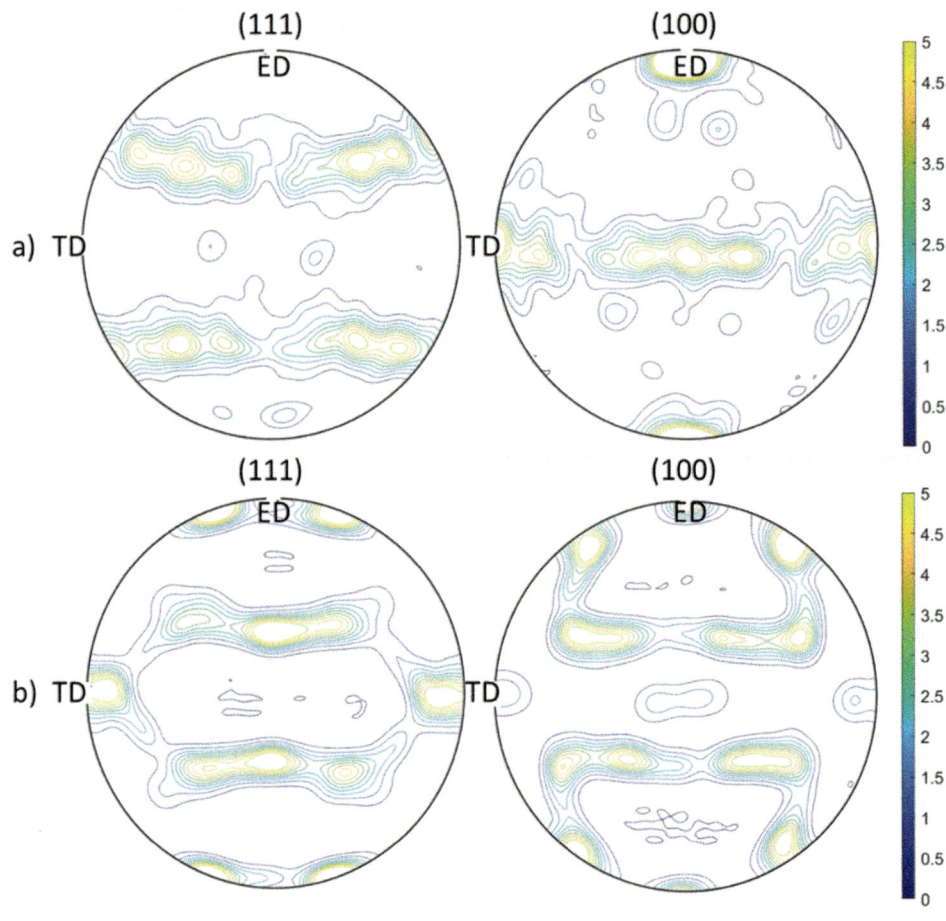

Fig. 3 Backscatter electron images showing the Fe-based particles in the **a** 0 Mn and **b** 0.5Mn

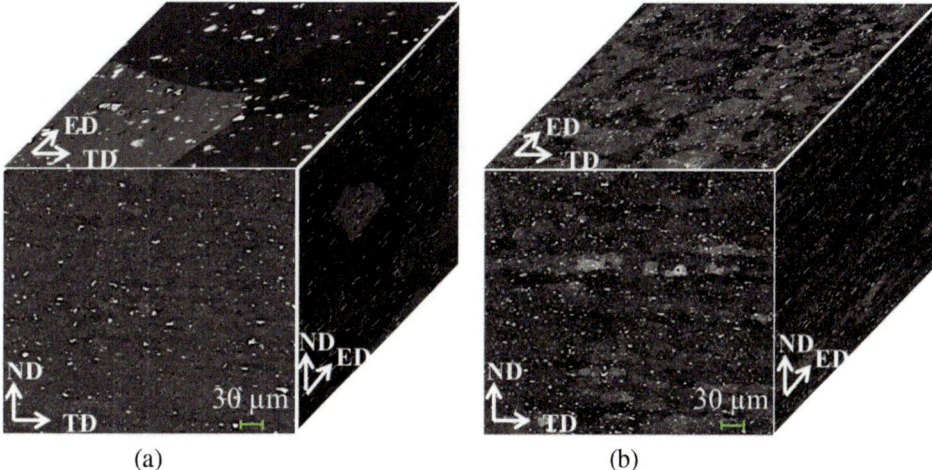

(a)   (b)

scales after homogenization: (i) spheroidized constituent particles 500–1000 nm in size and (ii) dispersoids 20–150 nm in size. By careful examination of Fig. 3b, one can see the larger constituent particles embedded in a fine distribution of dispersoids. In summary, the constituent particles in the 0.5Mn alloy were smaller and more rounded compared to the 0Mn alloy.

Figure 4 shows the true stress-true strain response of the 0Mn and 0.5Mn samples tested in the ED, 45°, and TD directions for both the T4 and T6 tempers. The dashed lines for each test represent the behavior between the necking and final fracture points (assumed to be linear). The results for the yield stress, ultimate tensile strength (UTS), true stress and strain at fracture and the R-value are summarized in Table 2. As expected, both alloys exhibited higher strength and lower uniform elongation in the T6 temper compared to T4. In detail, for the recrystallized extrusion the yield stress

was almost independent of test direction with values of ~ 145 and ~ 305 MPa for the T4 and T6 tempers, respectively. Further, the true strain to fracture was ~ 0.5 for the T4 and ~ 0.1 for the T6 tempers. In the case of the T6 temper, it was noted that there was almost no post-necking strain. The loading direction had very little effect on the entire stress–strain curve (see Fig. 4a). However, the R-value did show a dependence on loading direction with values of ~ 0.43, ~ 0.37, and ~ 1.04 for the ED, 45° and TD loading axes indicating that the recrystallization texture did affect plastic anisotropy (recall, an R-value of 1 would be expected for an isotropic material).

In the case of the unrecrystallised extrusions, the situation was more complicated (see Fig. 4b). The yield stress for the T4 and T6 conditions was similar for the ED and TD directions, ~ 190 MPa and ~ 350 MPa, respectively, (note these values are ~ 50 MPa higher than for the recrystallized

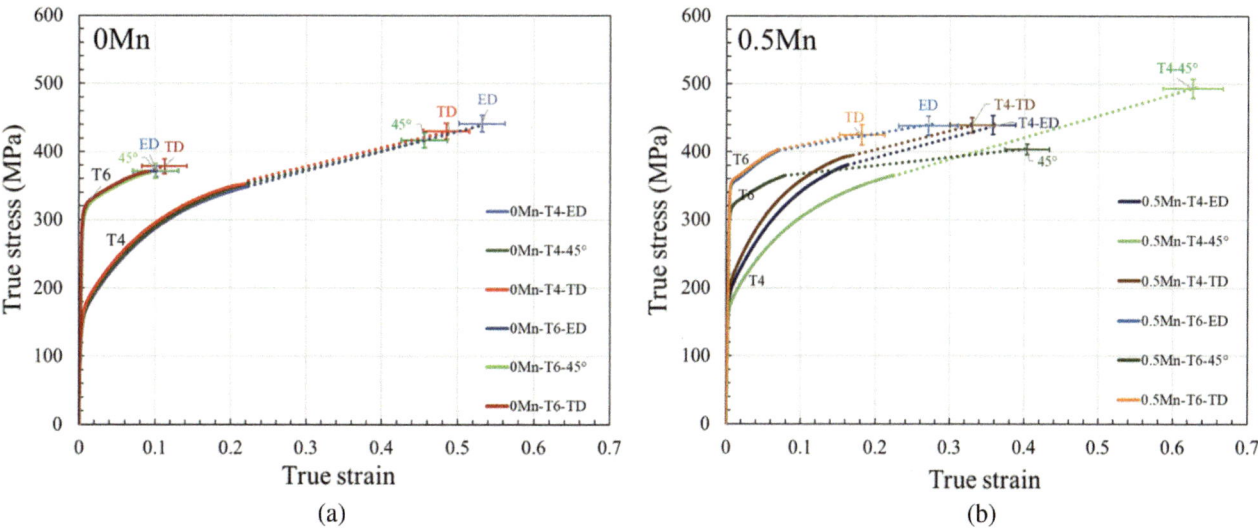

(a)   (b)

Fig. 4 True stress versus true strain plots for the **a** 0Mn and **b** 0.5Mn extrusions in T4 and T6 tempers

Table 2 Summary of mechanical properties from tensile tests

Alloy	Heat treatment	Deformation direction	0.2% offset yield stress (MPa)	UTS (MPa)	True strain at fracture	True stress at fracture (MPa)	R-value*
OMn	T4	ED	144 ± 4	284 ± 6	0.53 ± 0.03	478 ± 12	0.43 ± 0.04
		45°	143 ± 4	283 ± 6	0.46 ± 0.03	446 ± 11	0.35 ± 0.03
		TD	149 ± 4	288 ± 5	0.49 ± 0.03	463 ± 12	0.93 ± 0.06
	T6	ED	305 ± 6	336 ± 5	0.10 ± 0,03	371 ± 8	0.44 ± 0.05
		45°	302 ± 5	338 ± 6	0.10 ± 0.03	372 ± 11	0.39 ± 0.03
		TD	308 ± 4	341 ± 4	0.11 ± 0.03	379 ± 11	1.15 ± 0.07
0.5Mn	T4	ED	184 ± 4	325 ± 6	0.36 ± 0.03	464 ± 14	0.29 ± 0.04
		45°	168 ± 4	294 ± 7	0.63 ± 0.04	541 ± 14	1.71 ± 0.06
		TD	198 ± 4	335 ± 4	0.33 ± 0.03	464 ± 11	1.09 ± 0.11
	T6	ED	346 ± 4	376 ± 4	0.27 ± 0.04	457 ± 14	0.38 ± 0.04
		45°	312 ± 4	339 ± 5	0.40 ± 0.03	429 ± 8	1.97 ± 0.18
		TD	352 ± 4	377 ± 3	0.18 ± 0.03	437 ± 15	1.07 ± 0.10

* R-value was measured at an axial strain of 10% strain for T4 and 6% for T6

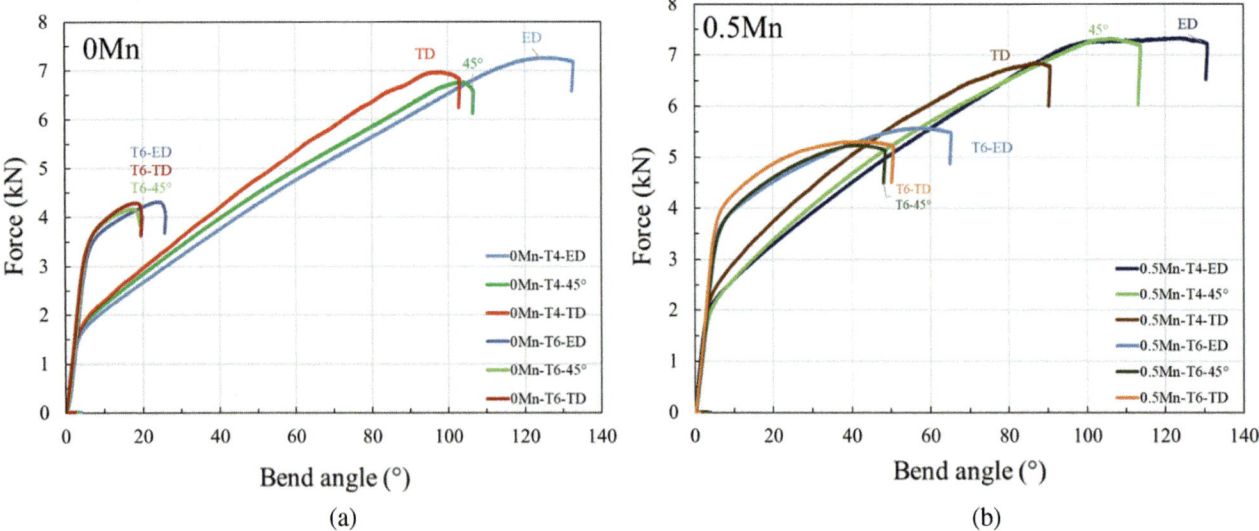

Fig. 5 Force–bend angle curves from VDA bend tests for the **a** 0Mn and **b** 0.5Mn

materials) but were lower for the 45°, ~170 and ~310 MPa for the T4 and T6 cases. For the T4 condition, the true strain at fracture was measured to be ~0.35 for ED and TD directions and ~0.6 for 45°, i.e. ~70% larger. In the case of the T6 material, the true strain at fracture was different for each loading direction with values of 0.27, 0.40, and 0.18 for the ED, 45°, and TD tests. The R-values exhibited the same dependence for both tempers with values of ~0.35, 1.8, and 1.1 for the ED, 45°, and TD directions.

Figure 5 shows the results for the force vs. bend angle determined from the VDA 3-point bend tests. In general, the force was higher for T6 versus T4 materials and the maximum bend angle was correspondingly lower in T6 for both the recrystallized (0Mn) and unrecrystallized (0.5Mn)

materials. Figure 6 summarizes the results in more detail. It can be observed that for the case of the recrystallized alloy (Fig. 6a), the maximum bend angle and the true strain to fracture only showed a weak dependence on test direction, 100–120° and 0.5 for the T4 and 15–20° and 0.1 for the T6 tempers, respectively. In contrast, in the case of the unrecrystallized material (Fig. 6b) the maximum bend angle decreased from ED to 45° to TD for both the T4 and T6 tempers, while the true strain to fracture was larger at 45° compared to ED and TD.

It has been previously reported that the true strain to fracture and the maximum bend angle can be correlated [19, 20] over a range of material yield stresses. Figure 7 examines this for the current data, and it can be seen that this

Fig. 6 Comparison of maximum VDA bend angle and true strain to fracture from tensile tests for **a** 0Mn and **b** 0.5Mn

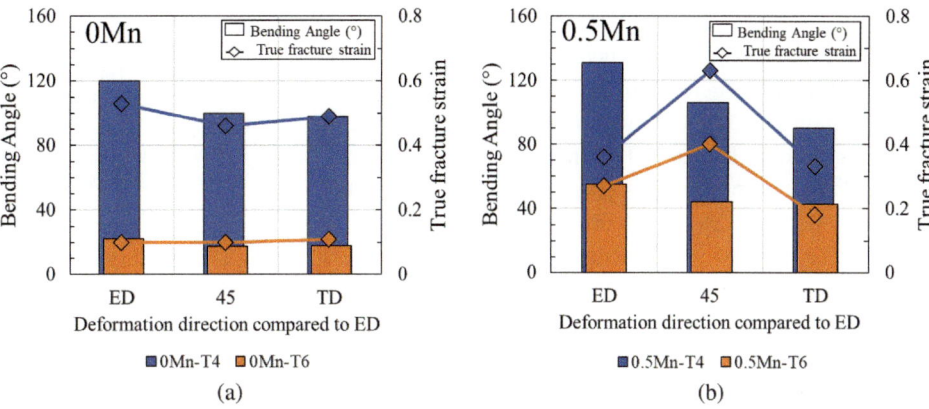

Fig. 7 VDA maximum bend angle versus true strain to fracture from tensile tests for **a** 0Mn and **b** 0.5Mn

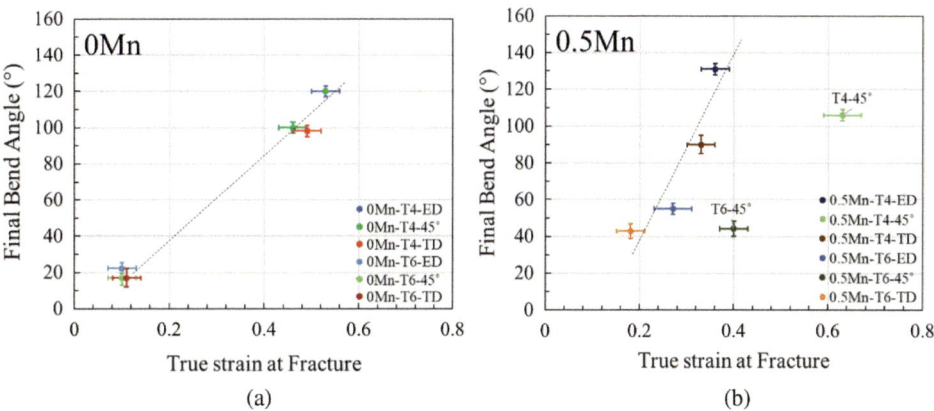

correlation holds reasonably well for the recrystallized material (Fig. 7a). However, in the case of the unrecrystallized material (Fig. 7b), to first approximation, the correlation may hold for the ED and TD directions but follows a different relationship for the 45° tests. Similar results have been reported by Snilsberg et al. [21].

As noted in the introduction, the anisotropy of the fracture process depends in a complex manner on the microstructure features (grain shape, crystallographic texture and size/distribution of second phase particles). Turning first to the case of the recrystallized material (0Mn), the tensile tests showed similar responses in the ED, 45°, and TD directions, although the R-values deviated significantly from isotropic behavior (see Table 2). On the other hand, Fig. 3a showed that the distribution of the β-Al$_5$FeSi second phase particles was highly dependent on the orientation consisting of small elongated plates aligned with the extrusion direction, i.e. morphological and topological anisotropy. However, it appears that these effects combine in a manner to result in relatively little macroscopic anisotropy for both plastic and fracture behavior and bending performance (as measured by maximum bend angle) can be correlated with the true strain to fracture in a tensile test.

The situation is more complicated for the unrecrystallized materials (Fig. 7b). In this case, the grains are highly elongated parallel to the extrusion direction (Fig. 1b), the tensile test response (yield stress, work hardening, and R-value) is strongly orientation dependent (Fig. 4b and Table 2). On the other hand, the distribution of the α-Al(Fe,Mn)Si phase second phase particles appears at least qualitatively to be less orientation dependent, noting that the presence of Mn dispersoids may also play a factor here [22–24]. Taken together, it is not straightforward to rationalize the current results, in particular the anomalous relationship between the maximum bending angle and the true strain to fracture shown in Fig. 7b. Further work is under way to quantify the morphology and topology of the second phase particles and to examine the relationship between the nucleation, growth and coalescence of damage in the microstructure.

Summary

The current study examined the relationship between orientation dependence of mechanical properties measured from uniaxial tensile and 3-point VDA bend tests with the microstructure of Al–Mg–Si aluminum extrusions. Two different microstructures were examined: (i) a 0Mn alloy with a recrystallized microstructure with the β-Al$_5$FeSi second phase constituent particles aligned in the extrusion direction and (ii) 0.5 Mn alloy with an unrecrystallized microstructure which had α-Al(Fe,Mn)Si second phase particles of two different length scales, i.e. dispersoids and constituent particles. Both materials showed aspects of anisotropic plastic response demonstrated by the orientation dependence of the R-values and to a lesser degree, the uniaxial stress–strain response. The results for the recrystallized material exhibited a correlation between the true strain to fracture in a tensile test and the maximum bend angle in a 3-point VDA bend test. However, the unrecrystallized material showed a different dependence, particularly for the samples loaded at 45° to the extrusion direction. Further work is currently underway to attempt to explain this anomaly.

Acknowledgements This work was undertaken, in part, thanks to funding from the Canada Research Chair program (Poole). Additional support of Rio Tinto Aluminium and the Natural Sciences and Engineering Research Council of Canada (NSERC) is gratefully acknowledged.

References

1. F. Hannard, A. Simar, E. Maire, and T. Pardoen, *Acta mater.*, Quantitative assessment of the impact of second phase particle arrangement on damage and fracture anisotropy, 2018. **148**: pp. 456–466.

2. T. Pardoen, F. Scheyvaerts, A. Simar, C. Tekoˇglu, and P.R. Onck, *C.R. Physique*, Multiscale modeling of ductile failure in metallic alloys, 2010. **11**: pp. 326–345.

3. S.-H. Choi, J.C. Brem, F. Barlat, and K.H. Oh, *Acta mater.*, Macroscopic anisotropy in AA5019A sheets, 2010. **48**: pp. 1853–1893.

4. B.H. Frodal, D. Morin, T. Borvik, and O.S. Hopperstad, *Int J. of Solids Struct.*, On the effect of plastic anisotropy, strength and work hardening on the tensile ductility of aluminium alloys, 2020: pp. 118–132.

5. B.H. Frodal, S. Thomesen, T. Borvik, and O.S. Hopperstad, *Int J. of Solids Struct.*, On fracture anisotropy in textured aluminium alloys, 2022. **244–245**: pp. 111563 (12 pages).

6. O. Engler, *Mater. Sci. Eng. A.*, Effect of precipitation state on plastic anisotropy in sheets of the age-hardenable aluminium alloys AA 6016 and AA 7021, 2022. **A830**: pp. 142324 (12 pages).

7. M. Khadyko, O.R. Myhr, and O.S. Hopperstad, *Mech. Materials*, Work hardening and plastic anisotropy of naturally and artificially aged aluminium alloy AA6063, 2019. **136**(136): pp. 103069(12 pages).

8. M. Gaudet, Doctoral thesis, The Tensile Properties and Toughness of Microstructures Relevant to the HAZ of X80 Linepipe Steel Girth Welds, The University of British Columbia, 2015.

9. *VDA 238–100, Verfand der Automobilindustrie.* 2017: Berlin, Germany.

10. R. Hu, T. Ogura, H. Tezuka, T. Sato, and Q. Liu, *Journal of Materials Science & Technology*, Dispersoid Formation and Recrystallization Behavior in an Al-Mg-Si-Mn Alloy, 2010. **26** (3): pp. 237-243.

11. J. Chen, W.J. Poole, and N.C. Parson, *Mater. Sci. Eng. A.*, The effect of through thickness texture variation on the anisotropic mechanical response of an extruded Al-Mn-Fe-Si alloy, 2018. **A730**: pp. 24–35.

12. A.J. den Bakker, X. Wu, L. Katgerman, and S. van der Zwaag, *Mat. Sci. Tech.*, Microstructural and X-ray tomographic analysis of damage in extruded aluminium weld seams, 2015. **31**: pp. 94–104.

13. J. Humphreys, G.S. Rohrer, and A. Rollett, *Recrystallization and Related Phenomena.* 2017: Elsevier.

14. U.F. Kocks, C.N. Tome, and H.-R. Wenk, *Texture and Anisotropy: Preferred Orientation in Polycrystals and their Effect on Material Properteis.* 1998: Cambridge University Press.

15. O. Engler and J. Hirsch, *Mat. Sci. Eng.*, Texture control by thermomechanical processing of AA6xxx Al–Mg–Si sheet alloys for automotive applications—a review, 2002. **A336**: pp. 249–262.

16. C. Liu, *Ph.D. Thesis, The University of British Columbia*, Microstructure evolution during homogenization and its effect on the high temperature deformation behaviour in AA6082 based alloys, 2017.

17. G. Guiglionda and W.J. Poole, *Mat. Sci. Eng. A.*, The role of damage on the deformation and fracture of Al–Si eutectic alloys, 2002. **A336**: pp. 159–169.

18. W.J. Poole, C. Liu, Z. Zhang, Q. Du, L.R. Pan, and N.C. Parson. *Modeling the Precipitation of Dispersoids during Homogenization of 3xxx-Series and 6xxx-Series Extrusion Billets.* in *Extrusion Technology 2022.* 2022. Orlando, Fl.

19. J. Sarkar, T.R.G. Kutty, K.T. Conlon, D.S. Wilkinson, J.D. Embury, and D.J. Lloyd, *Mat. Sci. Eng. A*, Tensile and bending properties of AA5754 aluminum alloys, 2001. **A316**: pp. 52–59.

20. N.C. Parson, J.F. Beland, and J. Fourmann, *Effect of Press Quench Rate on Automotive Extrusion Performance*, in *Extrusion Technology 2022.* 2022: Orlando, Fl, USA.

21. K.E. Snilsberg, I. Westermann, B. Holmedal, O.S. Hopperstad, Y. Langsrud, and K. Marthinsen, *Mat. Sci. Forum*, Anisotropy of bending properties in industrial heat-treatable extruded aluminium alloys, 2010. **638–642**: pp. 487–492.

22. J.M. Dowling and J.W. Martin. *The influence of Mn addition on the deformation and fracture behaviour of an Al-Mg-Si alloy.* in *Third Int. Conf. on Strength of Metals and Alloys.* 1973. Cambridge, UK: Institute of Metals.

23. W.J. Poole, X. Wang, D.J. Lloyd, and J.D. Embury, *Phil. Mag.*, The shearable/non-shearable transition in Al-Mg-Si-Cu precipitation hardening alloys: Implications on the distribution of slip, work hardening and fracture, 2005. **85**(26–27): pp. 3113–3135.

24. W.J. Poole, X. Wang, J.D. Embury, and D.J. Lloyd, *Mat. Sci. Eng. A*, The effect of manganese on the microstructure and tensile response of an Al-Mg-Si alloy 2019. **755**: pp. 307–317.

Comparison of Experimental Test and Finite Element Simulations of Car Crash Boxes Manufactured with Different Aluminum Alloys

Görkem Özçelik and Melih Çaylak

Abstract

In consideration of developments in electric vehicle production, lightweight materials such as aluminum alloys have gained increased interest by the Automotive Industry to diminish total weight of vehicles. Structural parts of automobiles such as car crash boxes require high impact damping energy and high tensile strength properties. To this end, different alloy compositions for aluminum alloys have been used to improve mechanical properties of metals. In this study, different aluminum alloys such as 6082 and 6005 has been considered as test materials and finite element (FE) simulations were carried out with FE software. Crash boxes were manufactured with aluminum extrusion production method and subjected to compression tests. In order to observe mechanical properties, tensile tests were performed according to the extrusion direction. Results of FE simulations have been compared with experimental data. As a result of this study, it has been observed that similar FE results with experimental tests can be obtained.

Keywords

Aluminum • Car crash box • Finite element method

Introduction

Aluminum and its alloys have a wide range of uses especially in automotive industry owing to its specific strength. In order to decrease total weight of vehicles, aluminum alloys are used by automotive designers, particularly crash box parts that requires high impact energy. These mechanical properties are obtained from several mechanical tests. In pursuit of developments in engineering, these tests started to be simulated with finite element (FE) methods and compared with experimental results. In 2018, Mete et. al. carried out some FE studies to investigate folding response of an annular rolled Al 6063-T5 tube [1]. Several experimental tests were conducted in that study and compared with FE studies. It was seen that folding behavior in experimental and simulated results gave similar results when they were compared with each other. Boria and Forasassi investigated the crash behavior of a crash-box with a honeycomb sandwich type geometry in 2008 [2]. Several FE studies were conducted during the study. Piecewise linear plasticity rule and Cooper-Saymonds law were used in order to simulate linear and non-linear behavior of material, respectively. Results of FE studies compared with experimental test data. It is foreseen that FE results and experimental test results matched with each other. Yilmaz and Celik investigated the shock absorbing ability of an aluminum crash-box [3]. Several FE simulations and experimental tests were conducted in the study and it is declared that numerical and experimental results are consistent. There are also several studies investigating car crash box tests in literature [4, 5].

Material and Methods

In the present study, 6005A and 6082 aluminum materials were taken as test materials. The chemical composition of both materials is indicated in Table 1. Crash-boxes were manufactured with aluminum extrusion method and heat-treated after extrusion process. Several tensile tests were conducted in order to determine the mechanical properties of materials. Engineering Stress–Strain curves of both materials are illustrated in Fig. 1.

According to tensile tests that were conducted in the view of this study, mechanical properties of test materials are listed in Table 1.

G. Özçelik (✉) · M. Çaylak
ASAŞ Aluminium San. ve Tic. A. Ş., Sakarya, Turkey
e-mail: gorkem.ozcelik@asastr.com

© The Minerals, Metals & Materials Society 2023
S. Broek (ed.), *Light Metals 2023*, The Minerals, Metals & Materials Series,
https://doi.org/10.1007/978-3-031-22532-1_75

Table 1 Mechanical properties of test materials

		6082	6005A
	Unit	Value	
Density	g/cm³	2.7	2.7
Elastic modulus	GPA	70	70
Engineering yield stress	MPA	307	263
Engineering UTS	MPA	330	286
Strength coefficient	MPA	365	384
Hardening exponent		0.11	0.125

Fig. 1 Stress–Strain curves of test materials

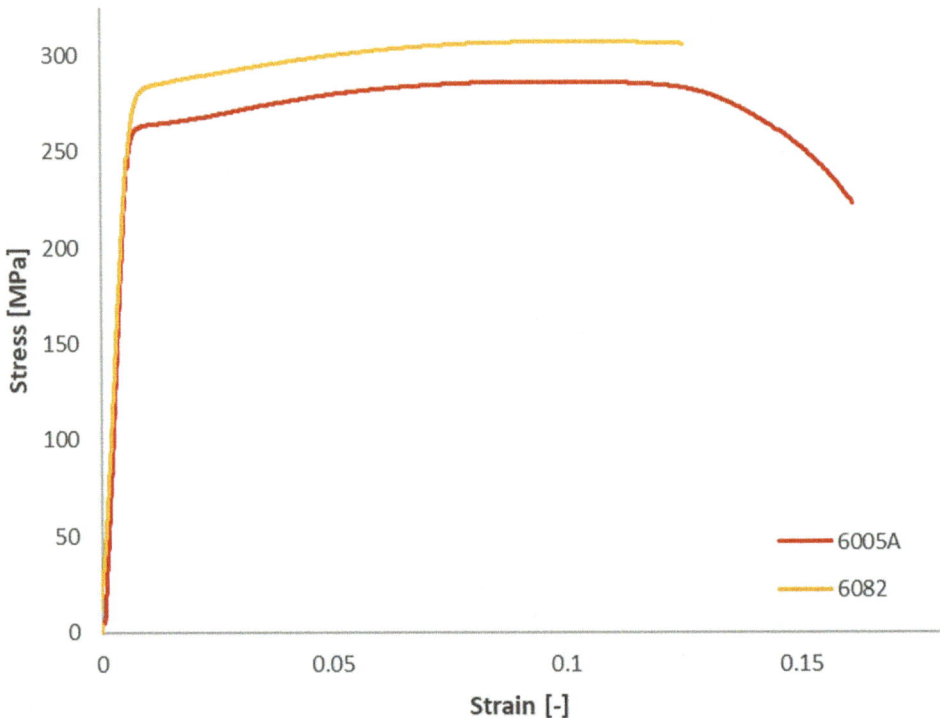

Definition of Material Geometry and Compression Tests

Standard test geometry which has dimensions of 80 × 80 × 2 mm cross-section and 300 mm length were tested and simulated during the study. Cross-section properties of test specimens were figured in Fig. 2.

Various compression tests were carried out to get experimental data for this study. Compression tests were run with Zwick/Roel Z 250 compression test unit. The maximum capacity of this unit is 25 tons and tests were conducted with 10 mm/min compression rate. Figure 3 illustrates the compression test unit. A displacement of 200 mm was applied to test specimen under compression load.

Results of experimental tests are given in Fig. 4 as Reaction Force—Displacement graphics.

FE Model

Within the scope of this study, FE simulations of crash box were carried out in student version of ABAQUS FE Software. In the present study, a power-law hardening rule [4] was employed as hardening rule. In Eq. 1 basic formula of rule was indicated as:

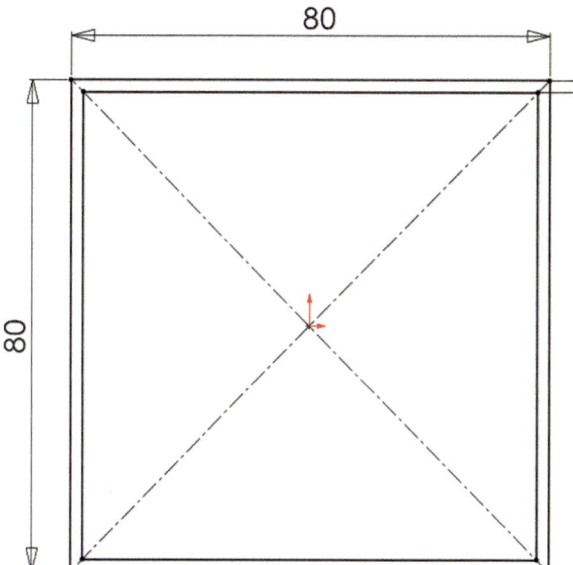

Fig. 2 Cross-section dimensions of test specimen

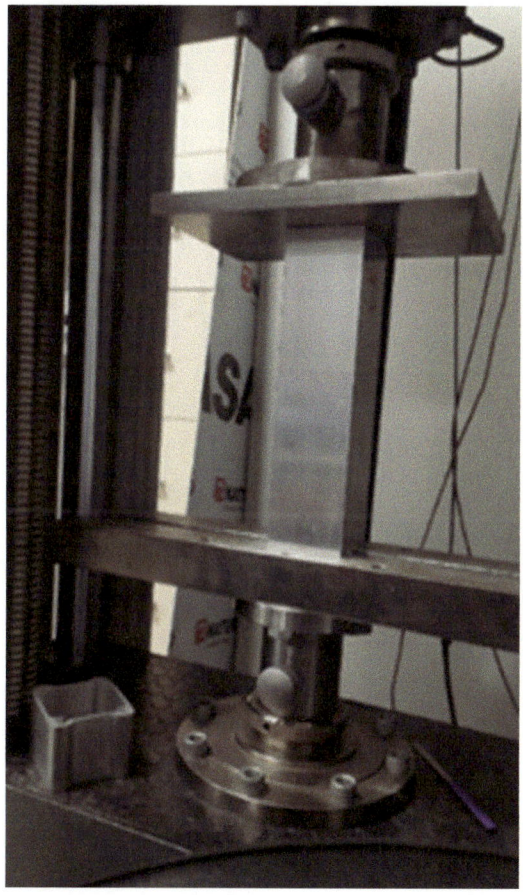

Fig. 3 Compression test unit

$$\sigma = K.\varepsilon_p^n \qquad (1)$$

Here, n is the strain hardening coefficient and K is the hardening constant.

FE model of test specimen was created in ABAQUS software. Shell elements were used as element type and mechanical properties that were obtained from tensile tests of aluminum materials have been used to define the materials properties. Figure 5 shows the FE model of test specimens.

Results

Explicit FE analysis were conducted in this study. Figure 6 indicate the results of FE analysis that were conducted with two different aluminum alloys.

Figure 7 indicates the total strain energy for both analyses and Fig. 8 show the FE results.

Here, it is seen that for 6082 material has been subjected to buckling. The same FE model has been used for both 6005 and 6082 materials. It was considered, buckling is related to material properties.

Figure 9 illustrated the compression tested specimens.

Experimental test and FE results of 6005 were compared. Figure 10 indicates the comparison of results. It is clearly seen that there are some differences between results.

It is clearly seen that there are some differences between experimental data and FE results. Table 2 shows the result values and differences between experimental data and FE results.

There is a 11.98% difference between maximum reaction force values for 6005 material and 49.87% for 6082 material. When it comes to the difference between results of the two different materials, maximum reaction force is different because of the natural behavior of materials. Reaction force values for the different two material and difference between the forces were listed in Table 3.

Conclusion

In the scope of this study, 6005 and 6082 aluminum alloy square crash-boxes were investigated with FE method and results are compared with experimental data. Power-law hardening rule was employed as hardening rule.

During the compression test, the force applied by the press increases continuously until plastic deformation begins on the profile. When a sufficient force value is reached for the deformation, there is a sudden decrease in the force. Then,

Fig. 4 Force—displacement graphics **a** 6005 material **b** 6082 material

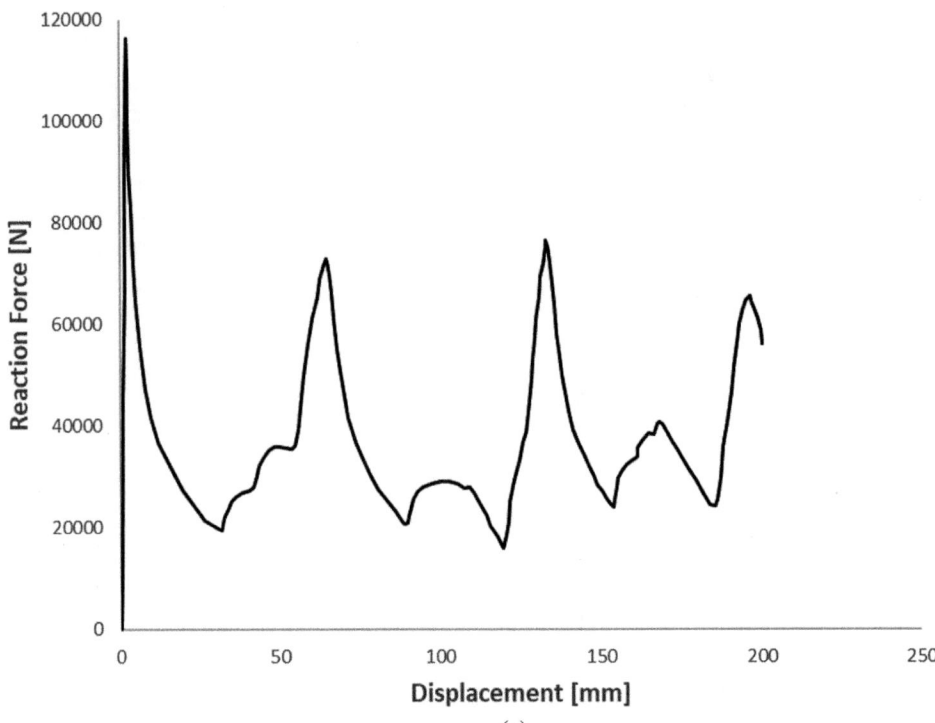

(a)

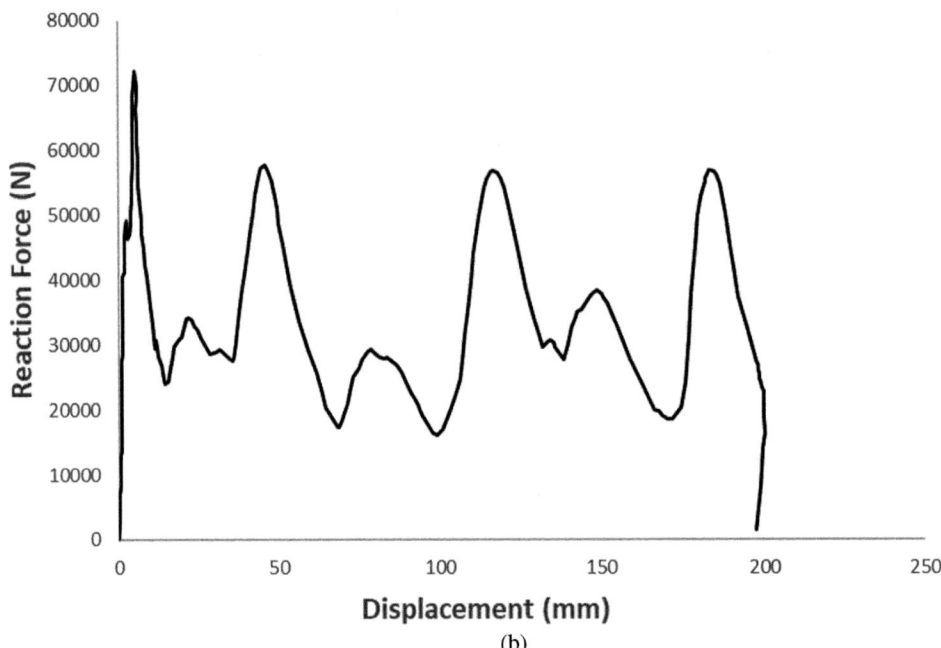

(b)

Comparison of Experimental Test and Finite Element Simulations of Car Crash Boxes ...

553

Fig. 5 FE model of crash boxes

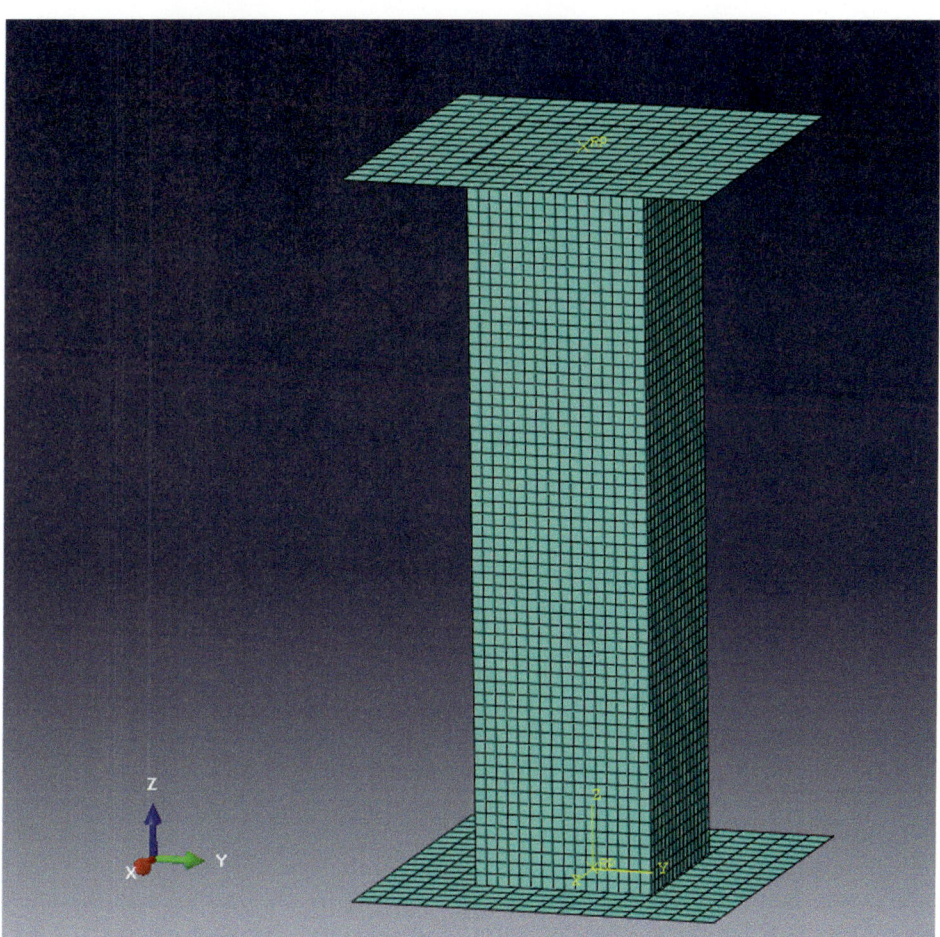

Fig. 6 FE results—reaction forces

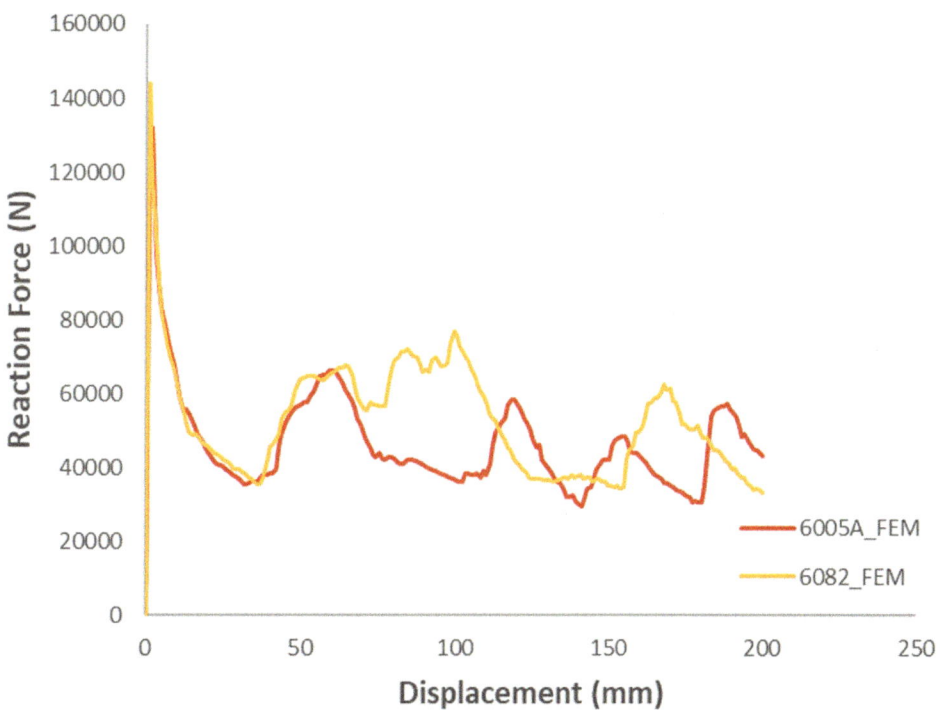

Fig. 7 Total energy—displacement results

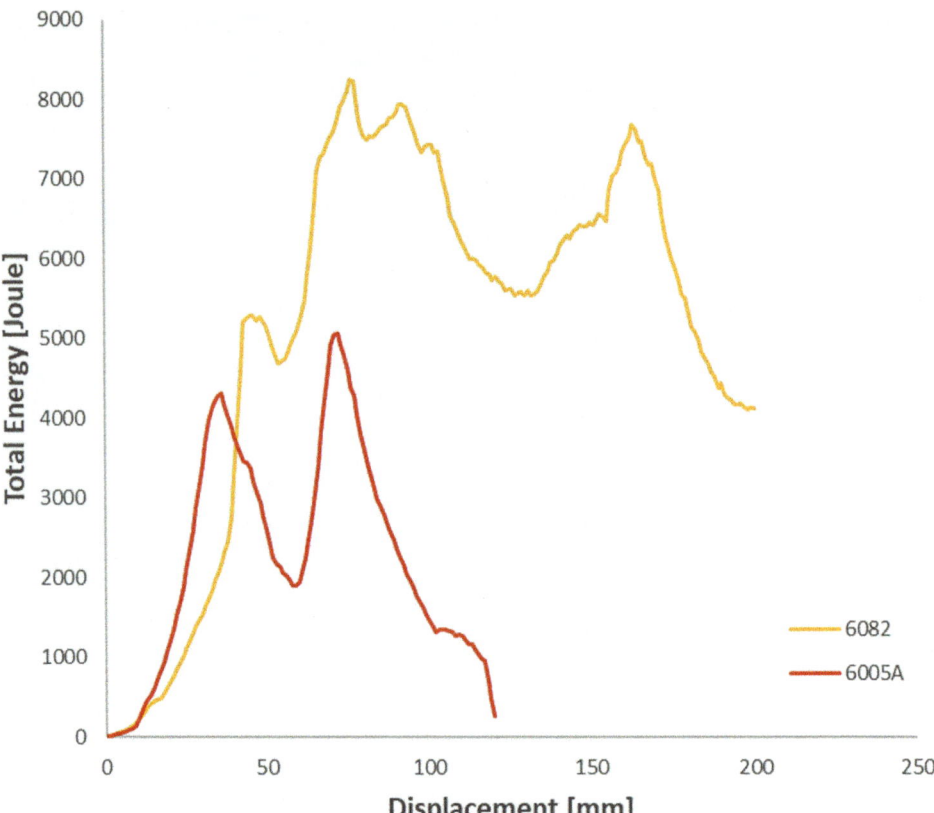

parallel to the folds in the profile, the force oscillates between the two values. When the experimental results and the results obtained in the simulation environment were compared, it was seen that this characteristic was obtained in both cases.

There are some differences between FE results and experimental results. Reasons for thedifferences between results might be the material model that has been used. A different material model should be used to get consistent results. The other reason could be buckling. Alloy 6082

material has buckled during the compression tests. For further studies, forming limit diagrams should be investigated to gather more consistent results.

In FE simulations, both material and geometric modelling were done with the assumption that geometrical and material properties of model is flawless. For the further studies, real defects such as geometrical discontinuities etc. should be examined.

Fig. 8 Results of FE studies
a 6005 material **b** 6082 material

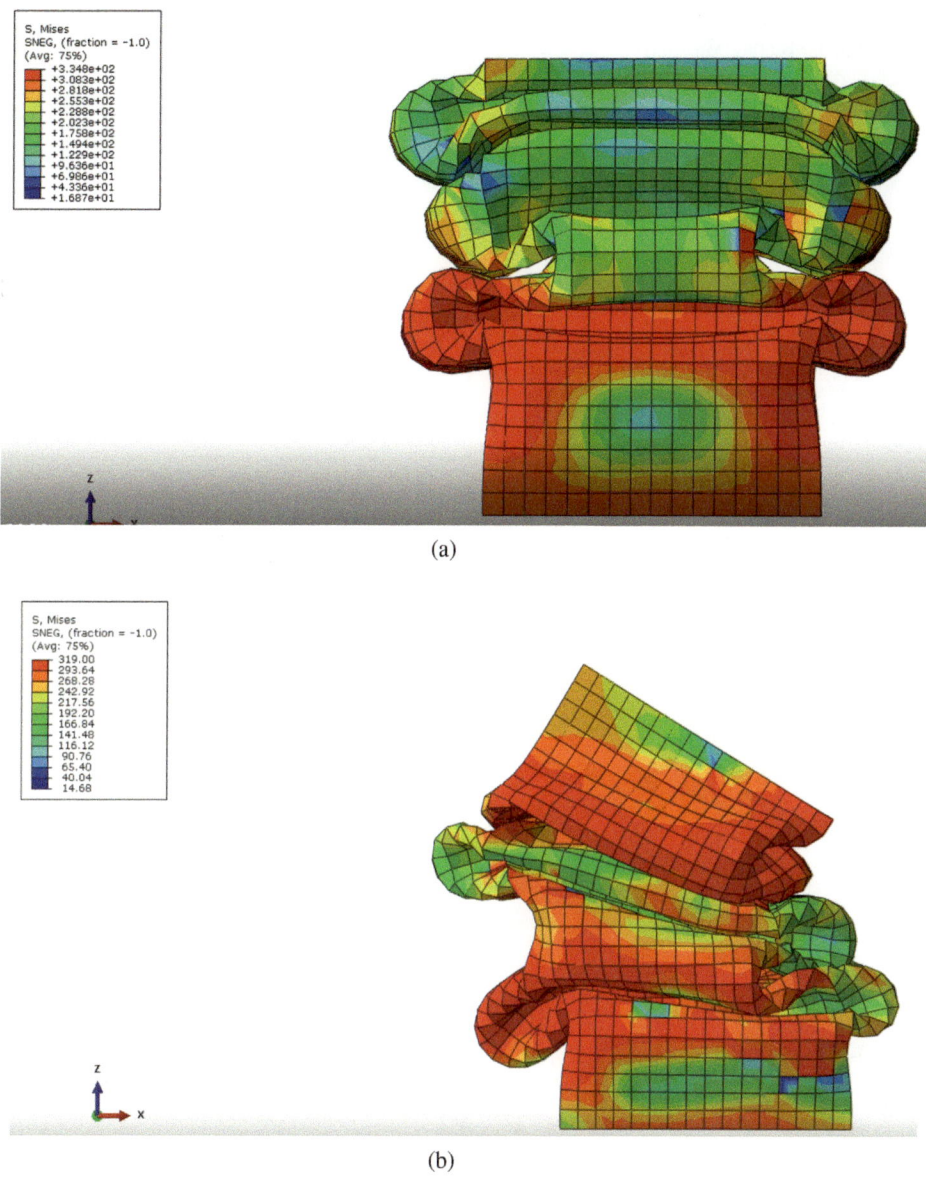

(a)

(b)

Fig. 9 Experimental test results
a 6005 material **b** 6082 material

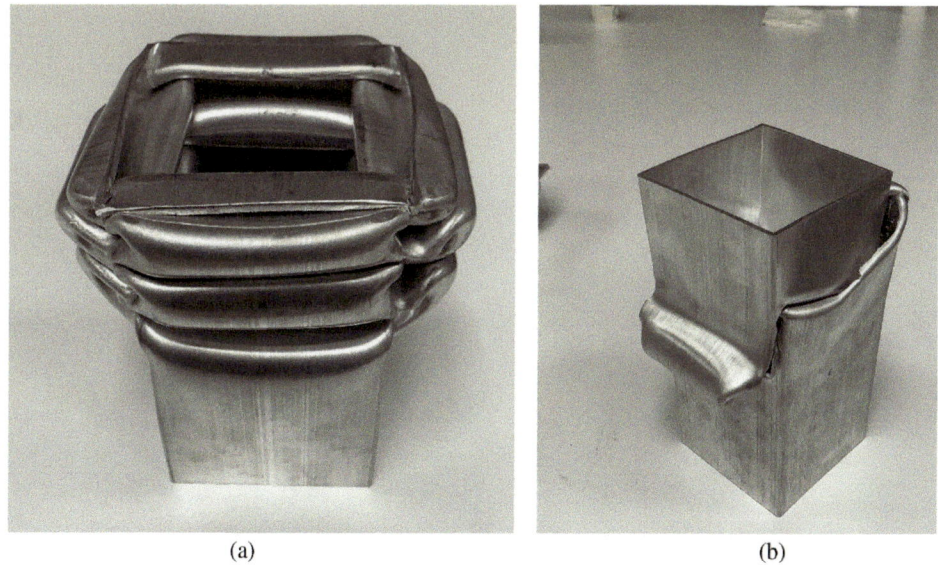

(a)   (b)

Fig. 10 Result comparison
a 6005 Material **b** 6082 Material

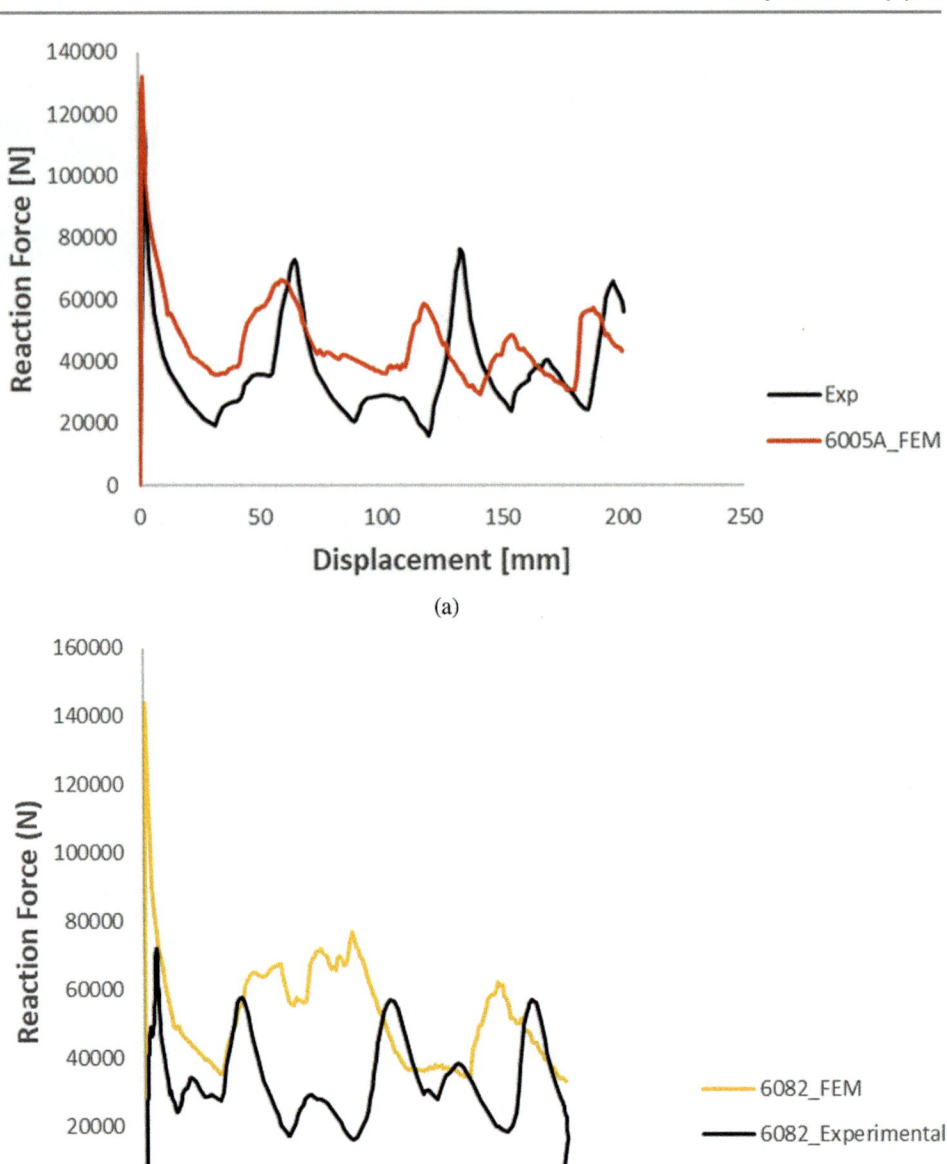

(a)

(b)

Table 2 Experimental and FE
result comparison (a) 6005
Material (b) 6082 Material

Results	Max. force (N)	Min. force (N)
a		
FE	132,069	29,481
Experimental	116,253	12,133
Difference (%)	11.98	58.84
b		
FE	143,818	33,031
Experimental	72,093	24,418
Difference (%)	49.87	35.27

Table 3 FE results comparison of both material

Results	Max. force (N)	Min. force (N)
6082	143,818	33,031
6005A	132,069	29,481
Difference (%)	8.17	10.75

References

1. Mete, H. O., Yalcin, M., Genel, K. (2018). Experimental and numerical studies on the folding response of annular-rolled Al tube. Thin-Walled Structures, 127, 798-808.
2. Boria, S., Forasassi, G. (2008). Honeycomb sandwich material modelling for dynamic simulations of a crash-box for a racing car. WIT Transactions on the Built Environment. Vol. 98.
3. Yilmaz, G et. al (2022) Numerical and Experimental Investigation of Shock Absorbing Ability of Crash Boxes Used in The Automotive Industry. Paper presented at the 10th International Aluminium Symposium, Istanbul, 3–4 March 2022
4. Abdullah, N., A., Z., Sani M., S., M., Salwani, M., S., Husain N., A. (2020). A review on crashworthiness studies of crash box structure. Thin-Walled Structures, 153.
5. Li, Q., Liu, Y., Wang,H.,Yan, S. (2009). Finite Element Analysis and Shape Optimization of Automotive Crash-Box Subjected to Low Velocity Impact. International Conference on Measuring Technology and Mechatronics Automation. 791–794.
6. Abaqus User Help Files

Exploring Semi-solid Deformation of Al–Cu Alloys by a Quantitative Comparison Between Drained Die Compression Experiments and 3D Discrete Element Method Simulations

Te Cheng Su, Meng Chun Chen, Huai Ren Hu, Ying Hsuan Ko, and Ling En Yao

Abstract

Developing computational modeling for semi-solid aluminum alloys with a solid network can help optimize advanced pressurized casting processes such as die-casting, squeeze casting, twin-roll casting, and semi-solid forging. However, a comprehensive numerical approach that can capture the coupled behavior between grain rearrangement and deformation of each individual grain remains a significant challenge. Inspired by recent synchrotron imaging work on deforming equiaxed-globular Al-Cu alloys showing granular deformation mechanisms, this research uses the particulate discrete element method (DEM) in 3D to generate two numerical assemblies of primary aluminum grains. Two-sphere particles and polyhedral grains were adapted in DEM to effectively represent aluminum crystals in 60% solid Al–15Cu alloy sample and 82% solid Al–8Cu alloy sample at 583 °C, respectively. Burgers contact model is introduced to consider viscous interactions between two grains at high temperatures. Contact model parameters are found by an iterative approach to reproduce the rheological response of compression experiments. Developments of 3D DEM simulations verified by compression experiments under controlled drained closed die boundary conditions will be useful for exploring the relationship among deformation process parameters and strain localization phenomena of a bulk semi-solid with enhanced microstructural sensitivity.

Keywords

Aluminum alloy • Semi-solid • Rheological modeling • Solidification microstructures • Discrete element method

Introduction

Cast aluminum alloys are often made by high-pressure die casting (HPDC), which pressurizes a partially solidified alloy in the die cavity during the intensification stage to feed entrapped gas porosities and shrinkage [1]. If equiaxed dendritic grains in the die contact and impinge with each other, this semi-solid mush can have measurable compressive or shear strength, known as dendrite coherency [1–5].

After pressurizing the semi-solid mush containing a solid network, a dilatant shear band can form along the feeding direction [1]. The formation of the dilatant shear band is the result of intensification overcoming the shear resistance of the semi-solid alloy, grains levering one another apart at their contacts during grain rearrangement, and subsequently creating weak shear zones of increased intergranular spaces (dilatancy) and liquid fraction [2, 3]. The localization susceptibility of semi-solid can be the main reason for deformation-induced casting defects such as concentrated porosity [2, 6–8] and macrosegregation [1, 9–12].

In the last decade, in situ synchrotron X-ray imaging studies [5, 13–15] have shown that deforming equiaxed Al–Cu alloys with a solid network has some granular deformation characteristics such as (i) grains mainly rearrange as cohesionless discrete bodies, (ii) local solid fraction can increase due to grains being pushed together by compaction until reaching a critical solid fraction, and (iii) local solid fraction can decrease due to grains pushing one another apart by dilation until reaching a critical solid fraction. Those suggest a granular interpretation of measured semi-solid alloy behavior into a critical state soil mechanics (CSSM) framework [16, 17].

Meanwhile, inspired by recent synchrotron imaging work, some researchers [4, 13, 14, 18] have used the particulate discrete element method (DEM) to produce a virtual semi-solid alloy. Particulate DEM is a microstructure-explicit approach [19, 20], which means that each discrete element with translational and rotational degrees of freedom

T. C. Su (✉) · M. C. Chen · H. R. Hu · Y. H. Ko · L. E. Yao
Department of Materials Science and Engineering, National Taiwan University, Taipei, Taiwan
e-mail: tcterrysu@ntu.edu.tw

© The Minerals, Metals & Materials Society 2023
S. Broek (ed.), *Light Metals 2023*, The Minerals, Metals & Materials Series,
https://doi.org/10.1007/978-3-031-22532-1_76

can directly adapt to the shape and size distributions of real samples. Yuan et al. [4] digitized an equiaxed dendritic crystal assembly to 2D DEM and found the coherency solid fraction of their virtually created grain assembly where shear strength starts to be measurable. Sistaninia et al. [18] developed a 3D hydromechanical granular model for high solid fraction deformation via finite/discrete element coupling to predict the hot tearing of equiaxed polyhedral Al–Cu alloys. The author's past work [13, 14] focused on exploring the rheological transition of equiaxed globular semi-solid Al–Cu alloy undergoing thin-sample shear deformation using coupled 2D lattice Boltzmann method–discrete element method (LBM-DEM) model. There are still quite a few undiscovered combinations of semi-solid microstructure, solid fraction when deformation occurs, and stress modes that DEM and its extensions can simulate.

This research aims to explore bulk semi-solid deformation of Al–Cu alloys using a combined approach between redesigned drained die compression experiments from refs. [17, 21–23] and 3D DEM. The experimental part focused on comparing the rheological behavior of Al–8Cu and Al–15Cu alloys during semi-solid compression at a given temperature and target compressive stress. For the DEM simulation, this research digitized αAl grain morphology in semi-solid at 60% (Al–15Cu at 583 °C) and 82% (Al–8Cu at 583 °C) volumetric solid fractions using two-sphere clumped particles and rounded polyhedral particles. After virtual specimen preparations and contact model installations, the differences in overall stress resistance and grain-level rheological behavior between two-sphere and polyhedral grain assemblies in response to compression are compared. The granular interpretation of the mechanical behavior of semi-solid alloys can help optimize the parameters of advanced casting processes, including applications of combined loading conditions on semi-solid alloys spanning across rheological transitions.

Methods

Specimen Preparation and Globularization

Two castings, Al–8Cu (wt.%) and Al–15Cu (wt.%) with grain refiners were produced using procedures similar to Ref. [5], and their alloy compositions as determined by ICP-MS (Thermo Fisher Scientific iCAP TQ) are shown in Table 1. Ingots of Al–8Cu and Al–15Cu were placed into Al_2O_3 rectangular crucibles and heated to 553 °C ($\sim$5 °C above the αAl + Al_2Cu eutectic temperature) for 72 h. This globularization heat treatment in the semi-solid state is to obtain globular αAl crystals closer to the morphology observed in the interior region of HPDC components. Ingots

were then cooled in air and machined into several $\varnothing25 \times 13$ mm cylinders. The oxide layers were ground with SiC papers, so the dimension of the cylinders became $\varnothing24 \times 12$ mm; the 2:1 diameter-to-height ratio based on one-dimensional drained compression test principles (ASTM-D2435).

Drained Compression Experiment and Microstructure Characterization

The drained compression experiments were conducted using an electric servo press machine (Delta AM-ESP-3000). The experimental apparatus is shown in Fig. 1a. First, the load cell (LM-20 kN) locked by a Delta stamping head detected the compression response from the $\varnothing24 \times 12$ mm semi-solid sample. The mesh size of the detachable filter was 300 μm which is not only for proper drainage but also to prevent globularized and coarsened αAl crystals from passing through the filter. Graphite demolding agent was sprayed to the piston, "filter + filter support," and stainless steel cup to prevent etching of the steel by liquid aluminum. The cup was designed to be separable, and one side of the cup had a small hole for inserting a K-type thermocouple. To inhibit the excessive oxidation of the Al–Cu sample in high temperatures, we applied flux on the surface of the Al–Cu sample before putting it into the cup. In addition, Ar gas was introduced just above the sample through the sleeve to ensure extra protection.

The sample was heated to 583 °C following the heating curve in Fig. 1b using the coil around the cup sleeve. This temperature corresponds to $\sim$60% volumetric solid fraction for Al–15Cu and $\sim$82% volumetric solid fraction for Al–8Cu, referring to thermodynamic calculations and direct measurements from tomographic reconstruction data from Kareh et al. [5]. After holding at 583 °C for 6 min (Fig. 1b), the piston was driven down with a constant rate of 120 μm/s, and the effective compressive strain rate on the sample was 10^{-2} s^{-1}. At the end of the test, when the load cell read 12kN ($\sim$26.5 MPa), airflow was directed towards the cup, and the cooling rate was around 20 °C/min.

As-cast, globularized, and semi-solid compressed specimens for scanning electron microscopy (SEM) observation under backscattered electron (BSE) mode were ground using SiC papers, polishing using diamond suspensions, and 0.05 μm Al_2O_3 powders. Residual Al_2O_3 powders were removed by ultrasonic clear. Note that additional electropolishing by twin-jet electropolisher (Fischione Instruments) with a solution containing 30% nitric acid and 70% methanol at $-$30 °C with a current of 10 mA was conducted for specimens that required electron backscatter diffraction (EBSD) characterization.

Table 1 Compositions of Al–Cu samples tested by ICP-MS

	Cu in wt.%	Ti in ppm	B in ppm	Fe in ppm	Mg in ppm
Al–8Cu	7.94	469.1	163.1	1156	172.3
Al–15Cu	16.66	341.4	99.1	1154	101.3

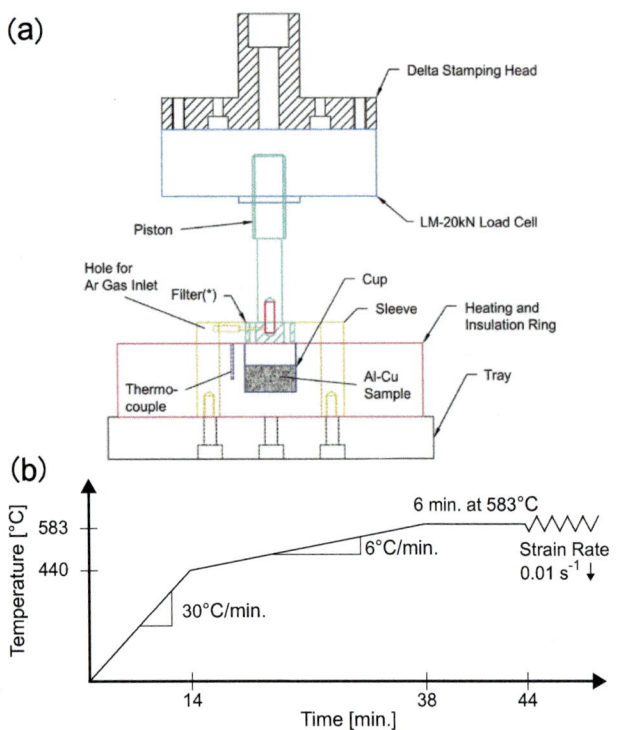

Fig. 1 **a** Schematic of experimental apparatus. **b** The heating curve for drained die compression experiments

Theoretical DEM Framework and Computational Procedure

We generated a virtual assembly "15Cu simulation" comprised of two-sphere grains for reproducing the microstructure of Al–15Cu at 583 °C using DEM software PFC Suite 6.00 developed by Itasca Consulting Group, Inc. [24]. On the other hand, we chose polyhedral rigid blocks in "8Cu simulation" for reproducing the microstructure of Al–8Cu at 583 °C. The discrete elements used in this study were about five times the αAl grain size; this scaling approach, similar to Ref. [25] for 3D DEM, was to reduce computational cost.

The DEM simulation procedure in this research is shown in Fig. 2. First, two virtual walls (2Ws) and two cubic rigid blocks (2RBs) with a side length of 1,410 μm ($\approx 5\bar{d}$) and set mass density of 2,670 kg/m^3, referred to as measurement from Ref. [26], were created for microscale simulation. This "2Ws + 2RBs" system included three contacts: two "RB-W contact" and an "RB-RB contact." The Hertz contact model

[19, 24] was applied to all RB-W contacts to consider computational stability. On the other hand, Burgers contact model [19, 24] was used for the RB-RB contact to capture viscous interactions between two grains at high temperatures. Those three contacts were set to have no overlap initially.

In the normal direction of the RB-RB contact, the values of viscous damping component and spring component in the Maxwell part, $\eta_{1,n}$, $k_{1,n}$ and those in the Kelvin part $\eta_{2,n}$, $k_{2,n}$, were found by an iterative approach to reproduce the past stress–strain measurement from Braccini et al. [27]. The "2Ws + 2RBs" small-scale DEM simulation output a stress–strain curve by letting the top wall move down with an equivalent compressive strain rate of 0.01 s^{-1} for completing an iteration. The required Hertz contact parameters referred to the solid Al close to its melting point [4]. The value of each contact parameter is summarized in Table 2, and this parameter set was also used for another small-scale DEM simulation, "2Ws + 2Ss".

To generate a DEM assembly containing >1,000 RB grains in "8Cu simulation", a ∅24 × 12 mm cylindrical box was created for constraining generated 1,273 spherical discrete elements with a diameter of 1700 ± 10 μm. Those elements were evenly distributed after reaching a dynamic equilibrium state using a simple linear contact model [24]. The centroids of spherical elements then served as points for Voronoi tessellation to generate an equiaxed ∼100% packing fraction assembly comprised of RBs similar to the Refs. [18, 28]. Desired packing fraction (82% for Al–Cu at 583 °C) was reached by applying erosion operation on each RB element. In the "15Cu simulation", 1,289 two-circle grains (represented as a clumped sphere, CS, in Fig. 2) with an equivalent spherical diameter of 1345 ± 368 μm were also constrained inside a virtual ∅24 × 12 mm cylindrical box with a packing fraction of 60%, and the same linear contact model was used to equilibrate this assembly.

The "switching model" step started by replacing the linear contact model with Burgers model for all RB-RB contacts and the Hertz model for RB-W contacts in the "8Cu simulation". Adjustment of $\eta_{1,n}$ shown in Table 2 for compression simulation in 583 °C was due to the consideration of temperature-dependent flow stress of the solid αAl [27]. In addition, the difference in mass densities between the liquid phase and solid phase for semi-solid Al–Cu was accounted for by introducing a buoyancy field $(1 - \rho_L/\rho_S)\mathbf{g}$. Low friction coefficients (0.01 for RB-RB and 0.0 for

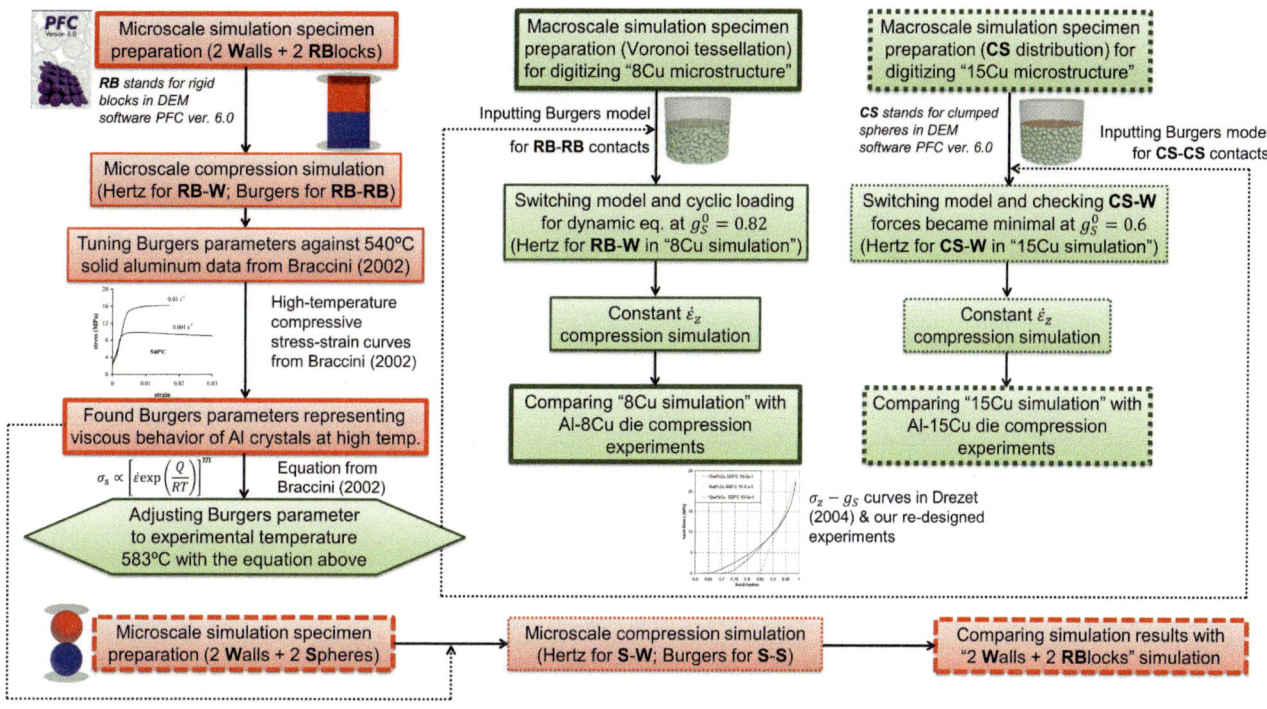

Fig. 2 DEM modeling and simulation flow chart

Table 2 A summary of the contact model parameters used in Hertz and Burgers contact model, including Maxwell and Kelvin parts. Parameters of shear components in Burgers contact model were set the same as those of normal components

Temperature	Parameters	Value	Unit
540 °C and 583 °C	Shear modulus—Hertz contact model (G)	1.71×10^{10}	Pa
	Poisson's ratio—Hertz contact model (v)	0.384	
	Normal contact stiffness—Kelvin part ($k_{2,n}$)	1.55×10^7	N/m
	Normal damping coefficient—Kelvin part ($\eta_{2,n}$)	2.4×10^6	N·s/m
	Normal contact stiffness—Maxwell part ($k_{1,n}$)	2.5×10^7	N/m
540 °C	Normal damping coefficient—Maxwell part ($\eta_{1,n}$)	1.15×10^6	N·s/m
583 °C	Normal damping coefficient—Maxwell part ($\eta_{1,n}$)	8.48×10^5	N·s/m

RB-W) were specified since low μ helps create a stable high packing fraction assembly [4, 19]. After cyclic compression at a slight strain three times similar to Ref. [29], the dynamic equilibrium was again reached. In the "15Cu simulation", Burgers model and Hertz model with the same set of parameters were applied to CS-CS contacts and CS-W contacts, respectively. The total contact forces between the upper layer of the "15Cu simulation" assembly and the top wall drop down very quickly. Virtual "8Cu simulation" and "15Cu simulation" samples were then subject to axial compression by the $\emptyset 24$ mm mobile top wall at a constant displacement rate $du_z/dt = -120\,\mu\text{m/s}$. This corresponds to the engineering compressive strain rate $\dot{\varepsilon} = 0.01\,\text{s}^{-1}$.

Results and Discussion

Microstructural Evolutions of Al–Cu Alloys

Figure 3 shows Al–Cu microstructure under SEM-BSE mode in different processing states. In the as-cast condition, the microstructure of the Al–15Cu specimen is more dendritic due to thermosolutal instability [30] and contains more $\alpha\text{Al} + \text{Al}_2\text{Cu}$ eutectic compared with the Al–8Cu specimen. After globularization heat treatment, grains become coarse and globular due to Ostwald ripening [31]. The grain size of equiaxed αAl also rises significantly from 50–100 μm to

Table 3 A summary of microscale and macroscale simulation settings in this study

	Microscale		Macroscale	
	"2Ws + 2RBs"	"2Ws + 2Ss"	"8Cu simulation" 1,273 grains in a cylindrical box	"15Cu simulation" 1,289 grains in a cylindrical box
Size [μm]	Two cubes in contact $a = 1410$	Two spheres in contact $d = 1750$	$d = 1700 \pm 10$ Voronoi tessellation using sphere centroids	$d = 1615 \pm 368$ two-sphere grain
Eff. Gravity [m/s^2]	None		1.075 (upward)	
Friction Coefficient [–]	$\mu_{RB-RB} = \mu_{RB-W} = 0.1$		$\mu_{RB-RB} = 0.01, \mu_{RB-W} = 0$	

Fig. 3 The SEM-BSE microstructure of Al–8Cu and Al–15Cu alloys in different processing states

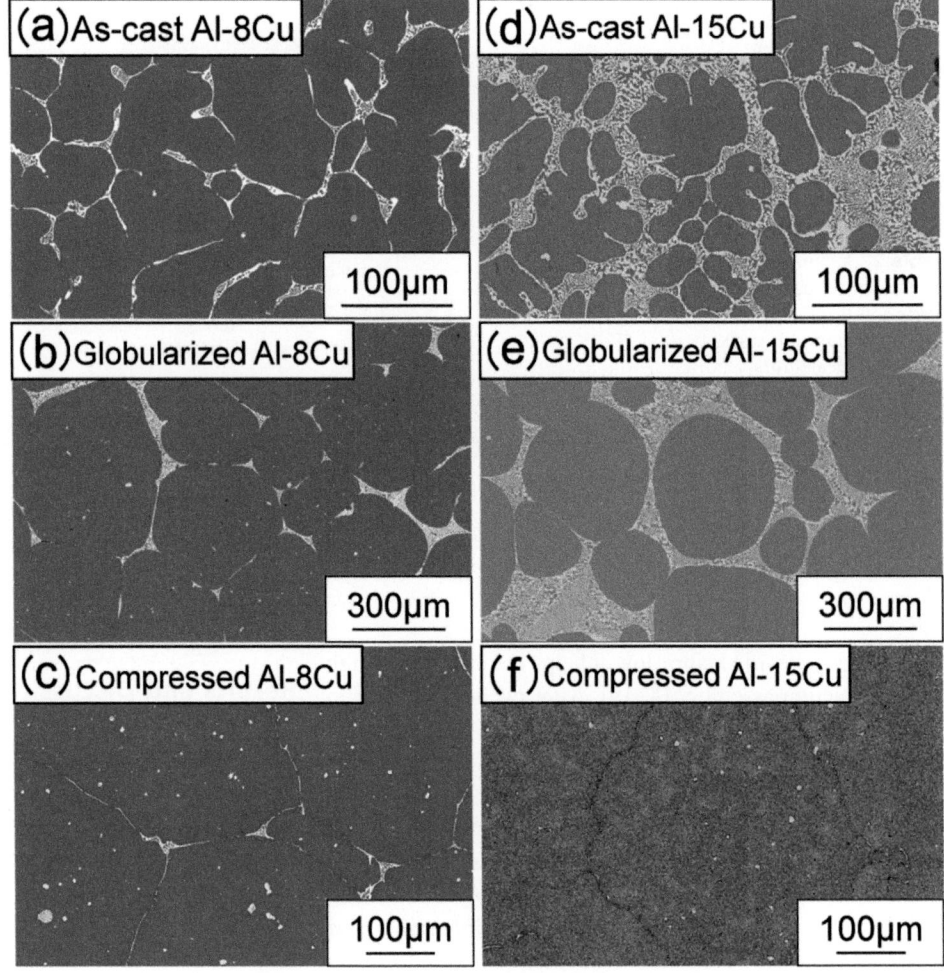

300–450 μm. The globularized equiaxed αAl grains naturally become rounded polyhedral in Al–8Cu. After drained die compression on globularized Al–8Cu and Al–15Cu specimens at 583 °C, liquid expulsion results in the near disappearance of the αAl + Al$_2$Cu eutectic structure that was occupying αAl interstitial space.

Figure 4 shows the EBSD IPF-Z maps of Al–8Cu and Al–15Cu specimens after globularization and drained die compression. Some small αAl grains are distributed along grain boundaries between large αAl grains in compressed samples, which may be explained as the occurrence of multiple metallurgical phenomena during late-stage compression: (i) partial cohesion, (ii) continuous dynamic recrystallization, and (iii) grain growth during cooling. First, some thin liquid films separating contacting grains were replaced with solid–solid interfaces by grain boundary coalescence [32]. The grain boundaries can serve as regions for forming sub-grain enclosed by dislocation cells and increased grain boundary misorientation due to sub-grain rotation known as continuous dynamic recrystallization (CDRX) [33]. The post-compression cooling with a limited cooling rate of 20 °C/min may be responsible for grain growth under the temperature range 480–548 °C [34]. It will be valuable to conduct more EBSD investigations on drain-compressed Al–Cu samples to elucidate the deformed αAl EBSD IPF maps as a function of process parameters and Cu concentrations.

Burgers Contact Behavior in Microscale DEM Simulations

Figure 5a compares stress–strain curves between "2Ws + 2 RBs" microscale DEM compression and previous measurements of compressing high-temperature ($T/T_m \approx 1$) Al–4.5Cu alloy with compressive strain rate 0.01 s^{-1} and 0.001 s^{-1} by Braccini et al. [27]. The stress in this DEM simulation was calculated by the resistance force acting on the top wall divided by a square of the size length, a^2, whereas the strain was defined as the downward displacement of the top wall divided by the initial height of the sample, $2a$. The iterative process and sensitivity analysis have shown that $\eta_{1,n}$ in Burgers model is linearly proportional to the flow stress under the constant $\dot{\varepsilon}$ condition. On the contrary, $k_{1,n}$, $\eta_{2,n}$, and $k_{2,n}$ parameters do not influence the flow stress and slightly affect the yield behavior. Besides, $\eta_{1,n}$ in "2Ws + 2RBs" microscale DEM 0.001 s^{-1} compression was set 7.1×10^6 N·s/m to reproduce the experimental result, which was ~6.3 times higher than "2Ws + 2RBs" 0.01 s^{-1} compression.

Fig. 4 EBSD inverse pole figure of αAl phase in Al–8Cu and Al–15Cu alloys

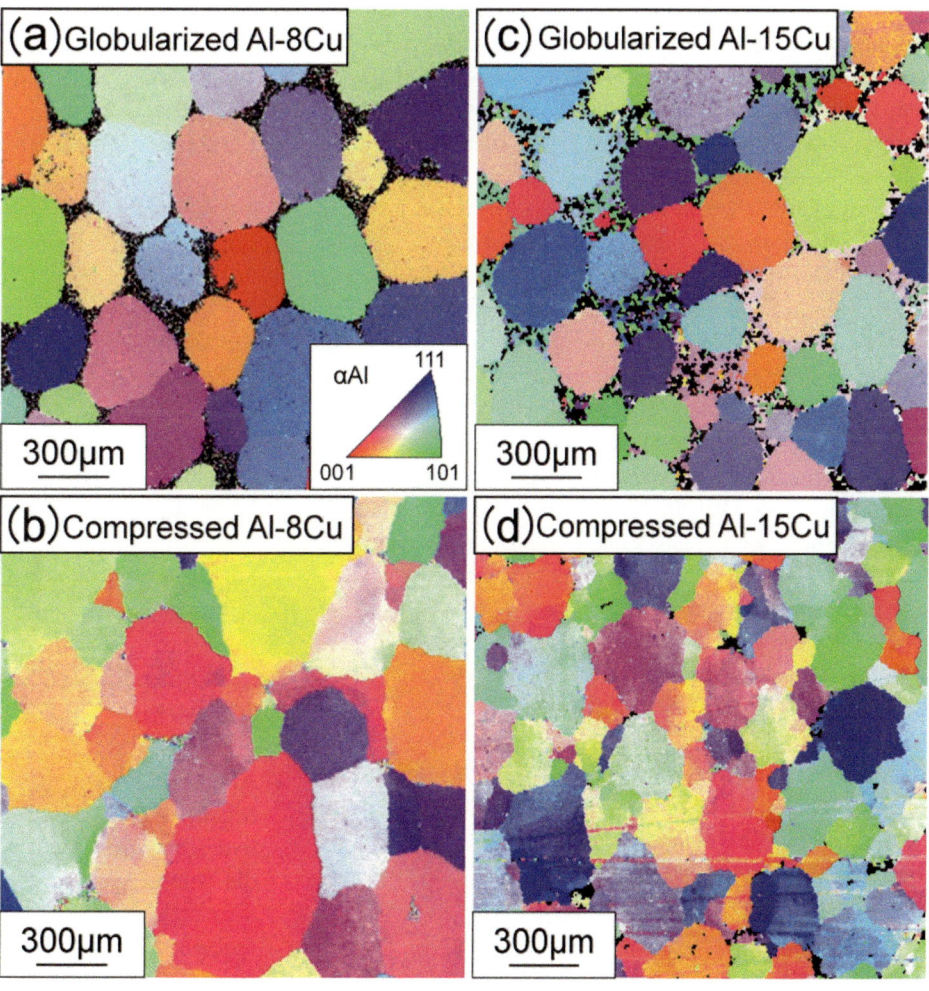

Besides, the contact force–contact overlap relationship during compression is primarily a function of discrete element shape. As shown in Fig. 5b, although the "2Ws + 2 RBs" and "2Ws + 2Ss" were set to have the same simulation loading conditions, the repulsive contact force rises much slower but reaches a higher plateau in "2Ws + 2Ss" than that in "2Ws + 2RBs". The shape sensitivity of discrete elements suggests that DEM can effectively simulate grain rearrangement and grain-level strain localization with various particle shapes.

Visualization of Deformation Microstructure for DEM Simulations

Figure 6 includes snapshots of discrete grain assemblies undergoing constant $\dot{\varepsilon}_z$ compression. Grain displacement along the z-axis is more negative for grains near the top wall. This is because the descending top wall contacted the upper layer of particles, and those particles pushed the lower layer of particles downward. Some grain-level heterogeneous phenomena are also visualized in Fig. 6. For example, grains in the "15Cu simulation" with an initial loose packing fraction of 60% became buoyant and floated. After grains packing into a new solid network, the repacked assembly was then subject to compression from the top wall. However, some particles in "8Cu simulation show high instantaneous velocity magnitude at the later stage of simulation ($u_z = -746\ \mu m$), suggesting the occurrence of calculation instability. Further work should address this issue, possibly by understanding the effects of other Burgers contact parameters ($k_{1,n}$, $\eta_{2,n}$, and $k_{2,n}$).

The bottom row of Fig. 6 shows the contact force chain network. With the increase of the compressive strain, the contact network gradually becomes denser, and the contact force magnitude increases. But at the same time, the contact force network of "8Cu simulation" has several thick main chains and thin branches, which is a typical phenomenon of granular materials [19]. In contrast, the "15Cu simulation" does not show well-defined main chains, which is worth checking its evolution by further compression.

Comparison of Stress–Strain Response in Experiments and Simulations

Figure 7a summarizes stress–solid fraction curves of semi-solid Al–8Cu and Al–15Cu among experimental measurements with a compressive strain rate of 0.001 s⁻¹ by Drezet et al. [21] (Al–Cu alloys were totally remelted and re-solidified to 555 °C, corresponding to Al–8Cu at 84% solid fraction and Al–15Cu at 62% solid fraction), drained

die compression experiments in this study, and macroscale DEM simulations with $g_S^0 = 82\%$ and $g_S^0 = 60\%$. It is found that the Al–8Cu semi-solid compression experiment from Drezet et al. shows higher stress in the solid fraction range 84–95% than measurement in this study. This can be realized because the globularized grains used in this research can rearrange easily, so the grain interlocking and stress resistance developed slower. On the contrary, the stress resistance of Al–8Cu at the late-stage deformation in this study is higher than the measurement from Drezet et al. [21], which can be attributed to the higher strain rate (0.01 s⁻¹) and the development of a solid network in the globularized assembly. On the other hand, the Al–15Cu specimen with an initial solid fraction of 60% in this study has lower stress resistance than Al–8Cu until late-stage deformation. Still, it has higher stress in the solid fraction range of 65–96% than the measurement by Drezet et al. This suggests that the presence of liquid in Al–15Cu may act some pressures on the piston under 0.01 s⁻¹ strain rate compression. Further work includes checking filter mesh size, compressive strain rate, and initial microstructure effects on the stress–volumetric solid fraction curves using the same apparatus.

Figure 7b shows the relationship between interstitial liquid (void) ratio $e = (1 - g_S)/g_S$ and compressive stress is shown in the logarithmic scale known as $e - \log p'$ projection plane in CSSM [17]. We can see that all Al–8Cu (initial solid fraction at 82% or 84%) and Al–15Cu experiments (initial solid fraction at 60% or 62%) tend to converge into a straight one-dimensional normal compression line (1D NCL). It is also noted for the "8Cu simulation" that the facets for each RB are perfectly smooth in DEM. Still, the surface of αAl grains in Al–8Cu can be curved initially, giving rise to even softer contact behavior in the experiment. Nevertheless, the "8Cu simulation" shows yield behavior near the possible 1D-NCL in Fig. 7b. In comparison, the "15Cu simulation" using clumped two-sphere grain approach did not show a significant stress rise in Fig. 7a until 71% solid fraction, and its compression curve in $e - \log p'$ diagram again yields near to the possible 1D-NCL. It will be valuable to couple with the computational fluid dynamics (CFD) approach in 3D to consider the liquid pressure effect, especially for "15Cu simulation".

Conclusions

- After heat treatment, the Al–8Cu and Al–15Cu alloys became equiaxed polyhedral-like and globular, respectively, and the amount of Al₂Cu became few after drained die compression. The change of grain size distribution after drained die compression suggests the occurrence of dynamic recrystallization.

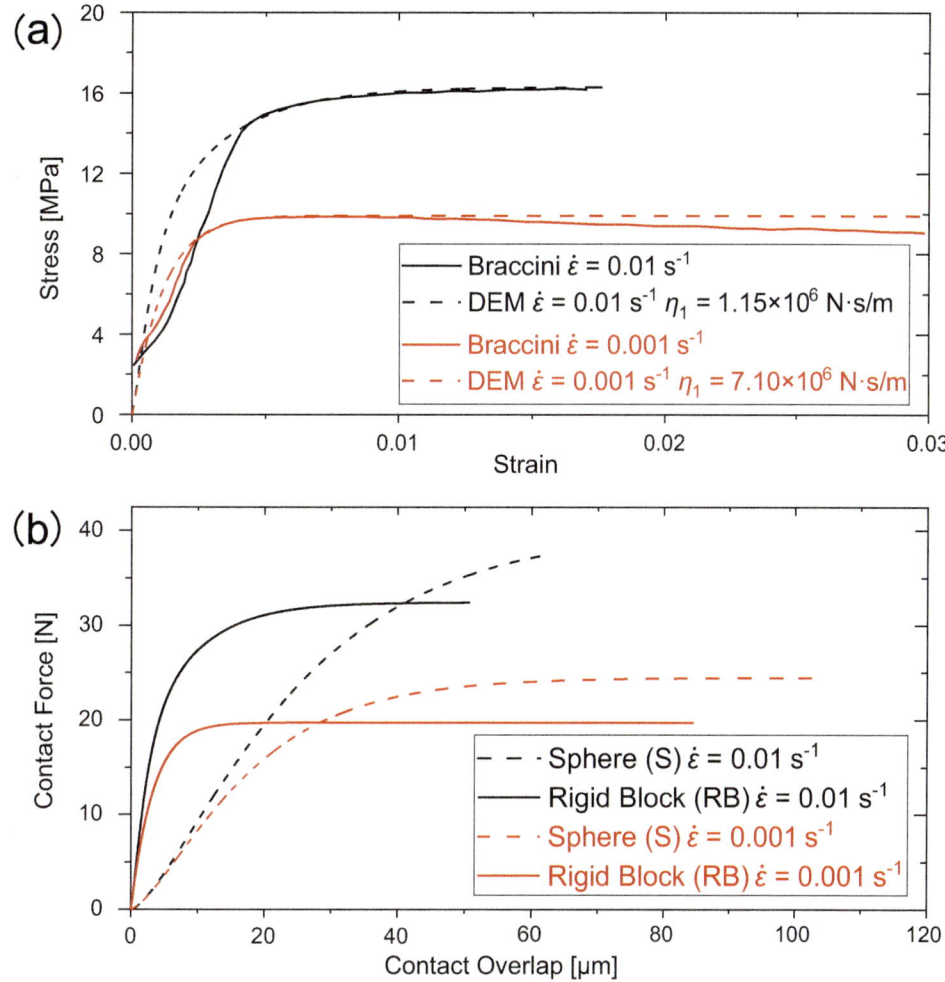

Fig. 5 **a** Stress–strain curves of "2Ws + 2RBs" microscale DEM compression simulation and comparison with experimental measurements by Braccini et al. [27]. **b** Comparison of "2Ws + 2RBs" and "2Ws + 2Ss" microscale DEM compression behavior

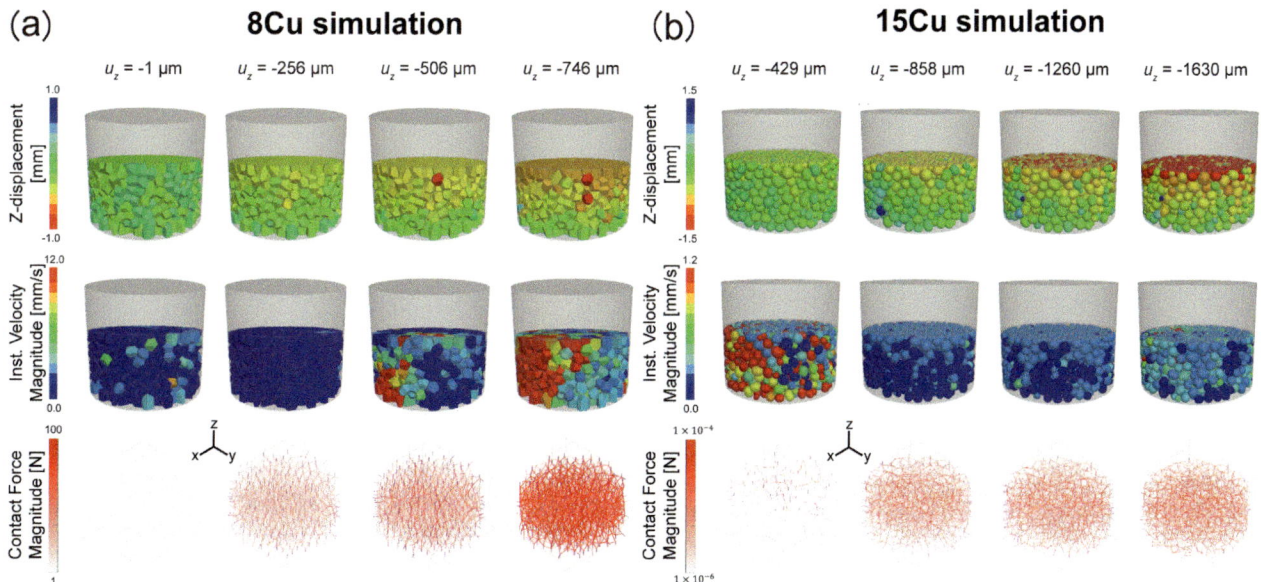

Fig. 6 Snapshots of **a** macroscale DEM compression "8Cu simulation" and **b** macroscale DEM compression "15Cu simulation". The bottom row shows the contact force network: the red color bar levels showing RB-RB or CS-CS contact force magnitude are on a logarithmic scale, and a thicker chain represents the greater force. The RB-W or CS-W contacts are represented by magenta

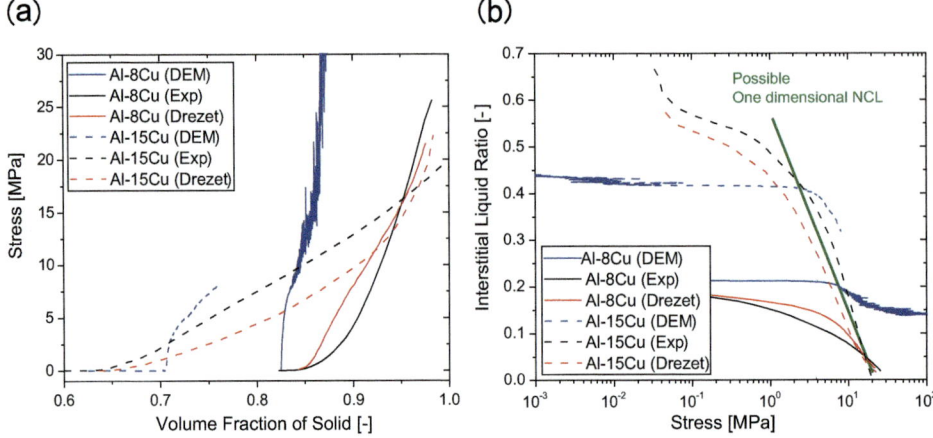

Fig. 7 Comparisons of **a** compressive stress–volumetric solid fraction relationship and **b** interstitial liquid (void) ratio—compressive stress in logarithmic scale of semi-solid Al–8Cu and Al–15Cu alloys among drained die compression tests in this study (Exp), experimental measurements by Drezet et al. [22], and analogs DEM simulations

- The stress–strain curve of constant strain rate compression simulation on two cubic rigid blocks is mainly affected by the viscous damping component in the Maxwell part of Burgers contact model, and the contact force rises slower while compressing on two virtual spheres than on two cubic rigid blocks even when the same contact model parameters are used.

- While visualizing the simulated deformation microstructure, the gradient of grain displacement along the z-axis from downward to upward along the top-to-bottom direction can be found. Meanwhile, the macroscale DEM simulations show discrete behavior, such as uneven instantaneous velocity magnitude and contact force chain.

- The stress resistance of globularized and partially remelted Al–8Cu at the early stage of drained die compression is lower than in the past study using partially solidified Al–8Cu, suggesting the effect of primary αAl morphology on force transmissions. On the other hand, the higher stress resistance of globularized and partially remelted Al–15Cu in this study than in literature at the early stage of deformation may be attributed to the strain rate effect, resulting in some contribution of resistance from liquid pressure. The "interstitial liquid (void) ratio vs. compressive stress shown in the logarithmic scale" diagram shows that all drained die compression experiments tend to converge into a straight one-dimensional normal compression line at the later stage of compression.

Acknowledgements The authors acknowledge the financial support of the National Science and Technology Council (MOST 109-2222-E-002-005-MY3), technical support from the Instrumentation Center at NTU for EPMA experiments (MOST 110-2731-M-002-001, EPMA000300) and the Instrumentation Center at NTHU for ICP experiments (MOST 110-2731-M-007-001, ICP000200).

References

1. Otarawanna S, Laukli HI, Gourlay CM, Dahle AK (2010) Feeding mechanisms in high-pressure die castings. Metall. Mater. Trans. A 41A(7):1836-1846

2. Gourlay CM, Dahle AK (2007) Dilatant shear bands in solidifying metals. Nature 445:70-73

3. Meylan B, Terzi S, Gourlay CM, Dahle AK (2011) Dilatancy and rheology at 0-60% solid during equiaxed solidification. Acta. Mater. 59(8):3091-3101

4. Yuan L, O'Sullivan C, Gourlay CM (2012) Exploring dendrite coherency with the discrete element method. Acta. Mater. 60 (3):1334-1345

5. Kareh KM, Lee PD, Atwood RC, Connolley T, Gourlay CM (2014) Revealing the micromechanisms behind semi-solid metal deformation with time-resolved X-ray tomography. Nat. Commun. 5(4464):1-7

6. Gourlay CM, Meylan B, Dahle AK (2008) Shear mechanisms at 0-50% solid during equiaxed dendritic solidification of an AZ91 magnesium alloy. Acta. Mater. 56(14):3403-3413

7. Gourlay CM, Laukli HI, Dahle AK (2007) Defect band characteristics in Mg-Al and Al-Si high-pressure die castings. Metall. Mater. Trans. A 38A(8):1833-1844

8. Otarawanna S, Gourlay CM, Laukli HI, Dahle AK (2009) The thickness of defect bands in high-pressure die castings. Mater. Charact. 60(12):1432-1441

9. Gras C, Meredith M, Hunt JD (2005) Microdefects formation during the twin-roll casting of Al-Mg-Mn aluminium alloys. J. Mater. Process. Technol. 167(1):62-72

10. Flemings, MC (1974) Solidification Processing. McGraw-Hill, New York

11. Chen CP, Tsao CYA (1997) Semi-solid deformation of non-dendritic structures. 1. Phenomenological behavior. Acta. Mater. 45(5):1955-1968

12. Kim MS, Kim SH, Kim HW (2018) Deformation-induced center segregation in twin-roll cast high-Mg Al–Mg strips. Scripta. Mater. 152:69-73

13. Su TC, O'Sullivan C, Nagira T, Yasuda H, Gourlay CM (2019) Semi-solid deformation of Al-Cu alloys: a quantitative comparison between real-time imaging and coupled LBM-DEM simulations. Acta. Mater. 163:208-225

14. Su TC, O'Sullivan C, Yasuda H, Gourlay CM (2020) Rheological transitions in semi-solid alloys: in-situ imaging and LBM-DEM simulations. Acta. Mater. 191:24-42

15. Cai B, Lee PD, Karagadde S, Marrow TJ, Connolley T (2016) Time-resolved synchrotron tomographic quantification of deformation during indentation of an equiaxed semi-solid granular alloy. Acta. Mater. 105:338-346

16. Schofield, A, Wroth, P (1968) Critical state soil mechanics. McGraw-Hill, New York

17. Altuhafi FN, O'Sullivan C, Sammonds P, Su TC, Gourlay CM (2021) Triaxial compression on semi-solid alloys. Metall. Mater. Trans. A 52:2010–2023

18. Sistaninia M, Terzi S, Phillion AB, Drezet JM, Rappaz M (2013) 3-D granular modeling and in situ x-ray tomographic imaging: a comparative study of hot tearing formation and semi-solid deformation in Al-Cu alloys. Acta. Mater. 61(10):3831-3841

19. O'Sullivan, C (2011) Particulate discrete element modelling: a geomechanics perspective. Spon, London

20. Cundall PA, Strack ODL (1979) A discrete numerical model for granular assemblies. Géotechnique 29(1):47-65

21. Drezet J, Ludwig O, M Hamdi M, Fjaer H, Martin CL (2004) FEM modeling of the compressibility of partially solidified Al-Cu alloys: comparison with a drained compression test. In: Tabereaux, AT (ed) Light Metals 2004. The Minerals, Metals & Materials Society, Charlotte, North Carolina, p 655–660

22. Ludwig O, J.M. Drezet, Martin CL, Suery M (2005) Rheological behavior of Al-Cu alloys during solidification: constitutive modeling, experimental identification, and numerical study. Metall. Mater. Trans. A 36A(6):1525-1535

23. Nguyen TG, Favier D, Suery M (1994) Theoretical and experimental study of the isothermal mechanical-behavior of alloys in the semisolid state. Int. J. Plasticity 10(6):663-693

24. PFC-Particle Flow Code 6.0 (2018) http://docs.itascacg.com/pfc600/pfc/docproject/index.html. Itasca Consulting Group, Inc., Minneapolis, USA

25. Lommen S, Mohajeri M, Lodewijks G, Schott D (2019) DEM particle upscaling for large-scale bulk handling equipment and material interaction. Powder Technol. 352:273-282

26. Ganesan S, Poirier DR (1987) Densities of aluminum-rich aluminum-copper alloys during solidification. Metall. Trans. A 18(4):721-723

27. Braccini M, Martin CL, Tourabi A, Brechet Y, Suery M (2022) Low shear rate behavior at high solid fractions of partially solidified Al-8 wt.% Cu alloys. Mater. Sci. Eng. A 337(1–2):1–11

28. Eliáš J (2014) Simulation of railway ballast using crushable polyhedral particles. Powder Technol. 264:458-465

29. Jiang MJ, Konrad JM, Leroueil S (2003) An efficient technique for generating homogeneous specimens for DEM studies. Comput. Geotech. 30(7):579-597

30. Dantzig JA, Rappaz, M (2009) Solidification. EPFL Press, London

31. Lifshitz IM, Slyozov VV (1961) The kinetics of precipitation from supersaturated solid solutions. J. Phys. Chem. 191:35-50

32. Rappaz M, Jacot A, Boettinger WJ (2003) Last-stage solidification of alloys: theoretical model of dendrite-arm and grain coalescence. Metall. Mater. Trans. A 34(3):467-479

33. Li J, Wu X, Cao L, Liao B, Wang Y, Liu Q (2021) Hot deformation and dynamic recrystallization in Al-Mg-Si alloy. Mater. Charact. 173:110976

34. Dennis J, Bate PS, Humphreys FJ (2009) Abnormal grain growth in Al–3.5Cu. Acta. Mater. 57(15):4539–4547

The Role of Through-Thickness Variation of Texture and Grain Size on Bending Ductility of Al–Mg–Si Profiles

P. Goik, A. Schiffl, H. W. Höppel, and M. Göken

Abstract

Requiring a high strength and concurrently a high ductility in materials is generally a demand for opposing properties in dislocation slip deforming materials, such as Al–Mg–Si wrought alloys. However, these are essential mechanical properties for safety parts in the mobility sector. While the strength of Al–Mg–Si wrought alloys is mainly governed by the state and density of the secondary precipitates, the deformation behavior and ductility are affected by both precipitates and crystallographic texture. The deformation during extrusion leads to the formation of characteristic textures in the bulk, which are distinct to a plane-strain deformation, and a peripheral coarse grain (PCG) layer beneath the surface. This PCG layer can have a detrimental effect on the bending ductility, which assesses the crashworthiness. However, an appropriate texture in the bulk can counteract the detrimental effect of PCG and increases the bending ductility at high strengths. Subsequently, based on EBSD investigations of bending deformed microstructures, a way to enhance bending deformation capability in Al–Mg–Si profiles is proposed.

Keywords

AA-6xxx • Extrusion • EBSD • Crystallographic texture • Peripheral coarse grain • Plate bending • Ductility

P. Goik (✉) · H. W. Höppel · M. Göken
Department of Materials Science and Engineering,
Friedrich-Alexander-Universität Erlangen-Nürnberg, Institute I:
General Materials Properties, Erlangen, Germany
e-mail: philip.goik@fau.de

A. Schiffl
Hammerer Aluminium Industries Extrusion GmbH, Ranshofen,
Austria

Introduction

Profiles from Aluminum alloys are used in mobility applications such as cars, busses, and trains as structural parts for safety components. For these applications, alloys must exhibit a high strength to withstand high forces in case of crash, as well as to utilize maximum lightweight potential. Additionally, to ensure maximum energy absorption in case of crash, safety components must exhibit a ductile deformation behavior [1]. However, strength and ductility are competing mechanical properties, so that for technological applications a best compromise needs to be found [2].

The class of AA-6xxx or Al–Mg–Si describes heat-treatable alloys in which artificial aging leads to precipitation of a sequence of different metastable Mg-Si phases that lead to an increase in strength and hardness, depending on the precipitation state [3, 4]. The characteristic development of strength over aging duration at the beginning consists of an increase of strength during precipitation formation. In this under-aged state, dislocations cut through coherent β''-MgSi. At peak-aging, maximum strength is achieved which decreases with further aging time as precipitates undergo Ostwald ripening, marking the overaged state, where dislocations bow around precipitates by the Orowan mechanism. However, ductility behaves contrary to strength during aging. Consequently, when a structure requires high strength and ductility, a tradeoff between these material parameters is needed [3].

Ductile deformation is further affected by large primary phases (constituent particles) >1 μm at which cracks initiate by braking and decohesion of the matrix under load [5, 6]. In technological Aluminum alloys such constituent particles form intermetallic phases of Al-X-Si and Al-X, where X denotes one or more of the alloying elements Fe, Mn, Cr, Zr, V, and other metallic elements [7–10]. Among these elements Fe is exceptional with a very low solubility beneath solidus temperature of < 0.05 wt.-% [9], thus bound to precipitate in large AlFeSi-phases during casting. Initially large constituent particles (<50 μm) break down during homogenization heat treatment [7] and forming process to 1–10 μm in size. Although detrimental to the mechanical properties, Fe as an impurity can only be avoided by the usage of primary Aluminum. However, energy efficiency and emission reduction require to recycle scrap Aluminum, which increases the Fe contamination. Therefore, the deformation behavior of these secondary Aluminum alloys must be tolerant in regard to the effect of AlFeSi-phases.

The forming of wrought Aluminum alloys leads to the formation of distinct grain structures and crystallographic textures, which is already extensively investigated for rolling of sheet Aluminum [11, 12]. The resulting grain structures and textures have a strong effect on the deformation behavior, which can become crucial for further processing of the sheet material. During rolling, the plane-strain deformation leads to β-fiber orientation texture in the bulk of the sheet [11] and shear deformation beneath the surface of the sheet leads to shear deformation texture components [13]. Subsequent annealing above recrystallization temperature results in the formation of a {100}⟨001⟩ cube orientation texture [12, 14] through a mechanism of compromise growth of small cube-oriented grains into the surrounding grains of β-fiber orientation [15]. Different to sheet rolling, in extrusion deformation is conducted simultaneously at high temperatures and strain rates. Material flows through the die and thus requires low flow strength. Due to high temperatures, varying strain states and friction in the die, disparate microstructures and textures form throughout the profile cross-section and wall-thickness compared to sheet rolling of the same alloy class.

Determining ductile behavior further depends on the load case in testing that is relevant for application. An readily obtained value to represent ductility is the fracture strain from tensile testing [16]. However, this reflects the ductile behavior under a combination of uniaxial load until uniform elongation. Further elongation leads to necking, which describes a localized deformation under complex strain and stress states. Aluminum profiles for safety components must be ductile in the hardened state and withstand complex load cases that lead to buckling of the profile [1, 17]. This buckling is replicated in the three-point plate bending test according to the standard VDA 238–100 [18]. In this bending test, the specimen is a small plate sectioned from the profile, and ductility is determined as the bending angle of the specimen after the bending force drops by a defined value beneath the maximum force.

The aim of this study is to provide insight into the role of texture and grain structure of the Al–Mg–Si alloys during extrusion on the plate bending behavior. To achieve only different textures and grain structures, extrusion speed was varied while alloy composition, extrusion tool, and aging treatments were not altered and kept constant.

Material Production and Characterization Methods

Material Production

The investigation of bending behavior of Al–Mg–Si profiles was performed on an AA-6005A alloy, with the composition of Si 0.5–0.9 wt.-%, Fe < 0.35 wt.-%, Cu < 0.30 wt.-%, Mg 0.4–0.7 wt.-%, Mn < 0.5 wt.-%, and Cr < 0.3 wt.-%, with the additional condition of Mn + Cr 0.12–0.5 wt.-%. The process steps for profile manufacturing consisted of casting, billet homogenization, profile extrusion and artificial aging.

The alloy was cast by Direct Chill casting into 10″ extrusion billets of 7 m length at Hammerer Aluminium Industries (HAI) Casting GmbH, Ranshofen. Consecutive homogenization heat treatment in a gas furnace was conducted at 560 ± 10 °C for 4 h. The extrusion process was performed with a 41 MN servo-hydraulic extrusion press at HAI Extrusion GmbH, Ranshofen. Before loading into the extrusion press, the billet was preheated in a gas furnace, then cut into blocks of 750 mm and brought to a final temperature of 490 °C in an induction furnace. The block was extruded into a thin-walled, two-chamber hollow profile with a wall thickness of 2.5 ± 0.2 mm and an overall cross-sectional area of 1080 mm², shown in Fig. 1a. Profile extrusion speed was 5 and 15 m/min, respectively. The resulting profile temperature upon exit of the die orifice was 528 ± 2 °C and 570 ± 9 °C, respectively. Immediately after exiting the extrusion press, the profiles were quenched in water with a minimum quench rate of 20 K/s. Within 2 h the profile was stretched, cut into smaller parts, and artificially aged in a final heat treatment of 160 °C for 20 h to obtain the T6 state.

Microstructural Characterization

Characterization of grain structure and texture was performed by means of electron back-scatter diffraction (EBSD) at Friedrich-Alexander University Erlangen-Nürnberg (FAU), using a Zeiss 1540 EsB Crossbeam SEM equipped with an Oxford NordlysNano detector. The region of the profile for analysis is shown in Fig. 1a, where EX denotes the extrusion, TD the transversal, and ND the normal direction. Sections in the TD-ND and EX-ND plane were prepared by grinding with SiC-paper, polishing with 6 μm diamond suspension, and electro-polishing using an alcohol-based solution of 3% HClO₄ at −5 °C, with 55 V for 5−10 s. The surface of the profile representing the EX-TD plane was prepared only by electro-polishing.

The distribution of deformation during bending was characterized by EBSD-measurements in the plane normal to

Fig. 1 **a** Thin-walled hollow chamber profile. Red box marks the investigated area for EBSD, tensile and bending analysis. Microstructure cubes of profiles extruded at **b** 5 m/min and **c** 15 m/min in TD-ND- and EX-ND-sections and the EX-TD-surface. IPF-coloring is with respect to the acquisition plane normal

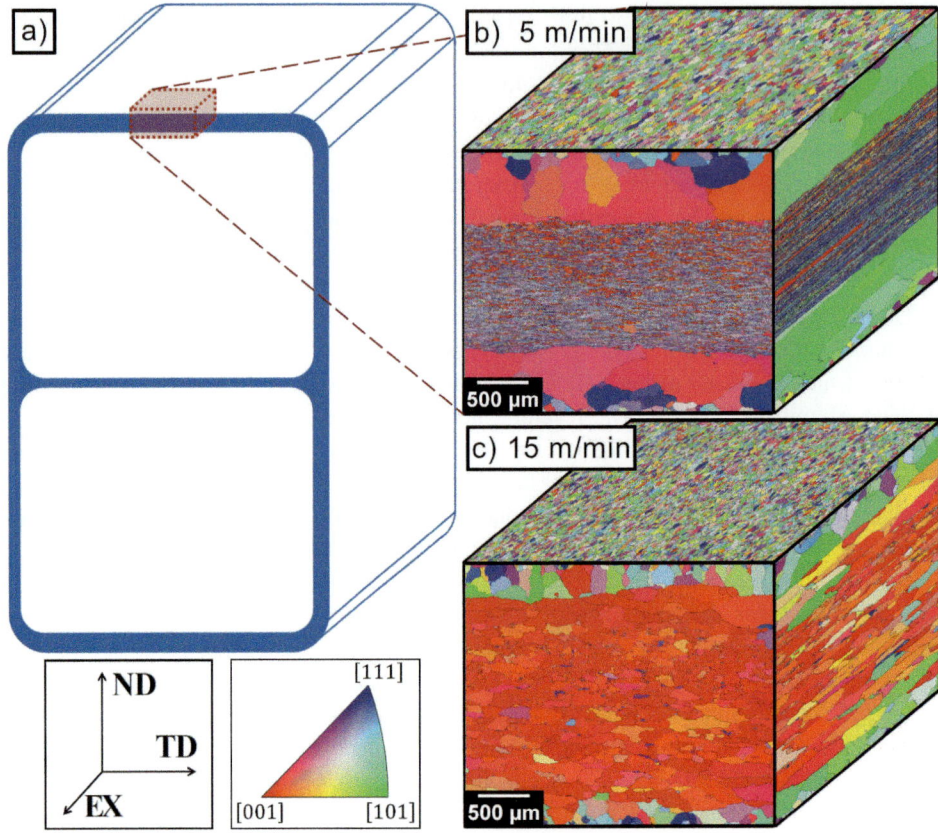

Fig. 1 **a** Thin-walled hollow chamber profile. Red box marks the investigated area for EBSD, tensile and bending analysis. Microstructure cubes of profiles extruded at **b** 5 m/min and **c** 15 m/min in TD-ND- and EX-ND-sections and the EX-TD-surface. IPF-coloring is with respect to the acquisition plane normal

the bending axis. High resolution EBSD mapping was performed at FAU using a FEI Helios SEM equipped with an Oxford Symmetry detector at low acceleration voltage of 5 kV to reduce the electron scatter interaction volume. Orientation determination was conducted in a post-processing routine with the dictionary indexing (DI) method introduced by de Graef et al. [19] in the EMsoft software. Evaluation of all EBSD measurements was performed with customized scripts using the MTEX toolbox for Matlab [20].

Mechanical Testing

Tensile tests were conducted at FAU using an Instron 4505 equipped with a control system from Hegewald-Peschke. Three specimens were taken from the investigated region of each profile to determine the local tensile properties. The specimen's cross-section had a width of 4 mm and the depth of the wall-thickness, with a gauge length of 20 mm. Tensile loading was parallel to the extrusion direction with a strain rate of 10^{-3} s^{-1}.

Using a Hegewald-Peschke Inspekt 50kN at HAI Extrusion GmbH, 3-point plate bending tests according to standard VDA 238–100 [18] were performed. The investigated region of the profile was the same as in the EBSD investigations. The bending axis of the punch was oriented parallel

to TD and perpendicular to EX, resulting in a tensile load parallel to EX. The punch was introduced onto the outer surface of the profile plate at 20 mm/min. The dimension of the specimen had a length of 50 mm parallel to EX and a width of 30 mm parallel to TD.

Four samples for each state were bent until failure, which is defined in VDA 238–100 as a drop of 60 N from maximum bending force. For analysis of the surface grain structure on bending behavior the inner surface of the profile plate was removed in steps of 30−70 μm by milling prior to bending, until a wall-thickness of 2.0 ± 0.1 mm was reached. For analysis of the deformation distribution within the bent area a sample for each extrusion speed was bent to an angle of 80° to obtain macroscopically comparable deformation states.

Results and Discussion

Microstructure and Mechanical Properties

The grain structure of profiles extruded at 5 and at 15 m/min are shown in Fig. 1b and c as microstructure cubes. Depending on the extrusion speed and profile temperature distinct grain structures and corresponding textures form. At 5 m/min the bulk of cross-section consists of grains elongated in EX and TD, while showing small dimension in ND. Those grains consist of equiaxed sub-grains of about 5 µm. Contrary to the fine-grained structure in the bulk a large layer of peripheral coarse grains (PCG) forms underneath the surface, taking up to 25% of the cross-section on each side. The microstructure of the profile extruded at 15 m/min (Fig. 1c) however shows larger grains in the bulk of cube orientation with a spread along the ⟨001⟩‖EX-fiber. A PCG layer can also be identified, but grain size is similar to that of the grains in the bulk. However, it can be distinguished from the bulk by various orientations of poles compared to the ⟨001⟩‖EX-fiber in the bulk.

Exemplary stress–strain curves of a tensile test for each extrusion speed condition are shown in Fig. 2. Yield and ultimate tensile strength as well as uniform elongation fall closely together for both extrusion speeds. However, fracture strain is larger for the sample extruded at 15 m/min.

Table 1 depicts the mean tensile properties and standard deviation of three samples for each extrusion speed condition. Yield strength (YS) is comparable in both microstructures with a difference of only 4 MPa. The sample extruded at 5 m/min shows a slight increase in ultimate tensile strength (UTS), which accounts to 9 MPa difference. This can be explained by a slightly higher uniform elongation strain in this sample. Regardless of the scattering of the data, the fracture strain behaves contrary, with larger values for the coarse-grained microstructure in the bulk of the profile extruded at 15 m/min, indicating that this microstructure sustains larger non-uniform strains than the profile with a fine-grained grain structure in the bulk extruded at 5 m/min.

The force-bending curves of 4 samples for each extrusion speed in Fig. 3a show much larger bending angles for the microstructure of the 15 m/min profile with remaining total bending angles of 123°–124° after unloading while the profile extruded at 5 m/min shows bending angles of only 78 −82°. There is only little scatter between the four samples of

Table 1 Average tensile mechanical properties and standard deviation of three samples of profiles extruded at 5 m/min and 15 m/min

	YS/MPa	UTS/MPa	Uni. Elongation/%	Frac. Strain/ %
5 m/min	278 ± 0.5	308 ± 0.2	8.69 ± 0.89	12.8 ± 1.15
15 m/min	274 ± 0.4	299 ± 0.6	7.71 ± 0.0	15.4 ± 1.16

each microstructure. Figure 3b and c shows a bent sample for each extrusion speed condition. Visible cracks form in both cases; however, the sample extruded at 15 m/min exhibits extensive, cleaving cracks along the bent surface. This indicates that grains in the PCG layer are cracking before reaching the ultimate bending angle. Nevertheless, no premature failure of the bending sample occurs compared to the sample extruded at 5 m/min.

Influence of PCG Layer on Bending Ductility

During bending strains and stresses are not evenly distributed throughout the cross-section but consist of a transient increase from compression to tensile stress perpendicular to the axis of the bending punch. The largest tensile loads occur at the surface of the sample, that is facing away from the bending punch. This promotes crack initiation that can lead to final failure of the sample. To investigate the influence of the PCG layer on the bending behavior, thin layers of 30–70 µm were removed from the plate's surface facing away from the bending punch prior to bending, so that the plates' thickness was between 2.0 and 2.5 mm. Because a thinner plate thickness d leads to smaller strains and stresses at the same bending angles, the resulting bending angles α were corrected to α_{2mm} according to VDA 238–100, projected to a thickness of 2 mm with Eq. (1).

$$\alpha_{2mm} = \alpha \cdot \sqrt{d/2mm} \qquad (1)$$

The corrected bending angles of the profiles extruded at 5 and 15 m/min are shown in Fig. 4. It is to notice that here the bending angles from Fig. 3 of the not thinned samples are corrected by their thickness according to Eq. (1). The bending behavior of the profile extruded at 5 m/min shows an increase in corrected bending angle for plate thicknesses down to 2.14 mm, which correspond to the removal of the first 350 µm of PCG layer. A very strong increase is achieved with PCG reduction between 320 and 350 µm, that leads to a near maximum bending angle of 119°. Further removal of the PCG layer yields only a small increase in bending angle to 120°, until the bending angle drops again to 116°. Although the maximum extent of PCG layer in this microstructure is ∼700 µm, the largest PCG reduction was only of 480 µm because no substantial increase in bending angle was observed beyond 350 µm of PCG reduction.

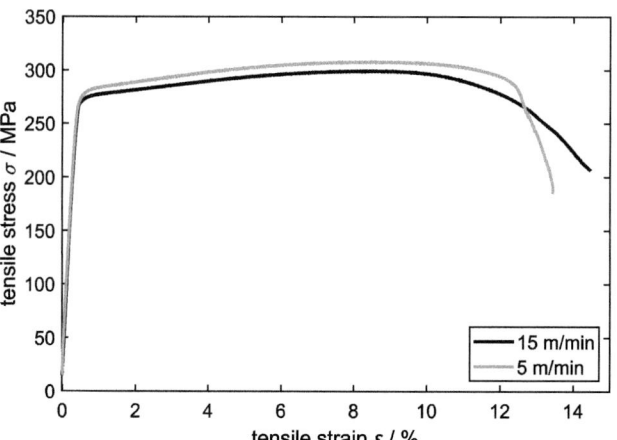

Fig. 2 Exemplary stress–strain curves from tensile tests in extrusion direction of profiles extruded at 5 m/min and 15 m/min

Fig. 3 **a** Force-bending curves of four samples for each extrusion speed. Specimen extruded at **b** 5 m/min and **c** 15 m/min after bending, respectively

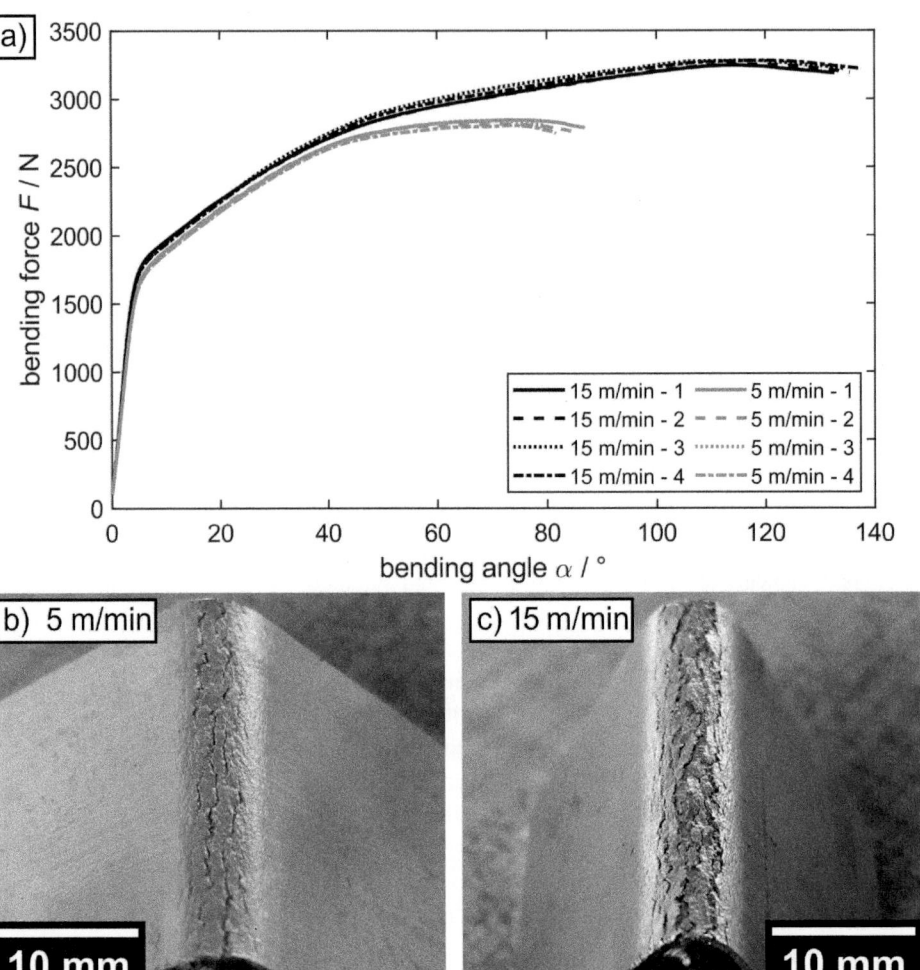

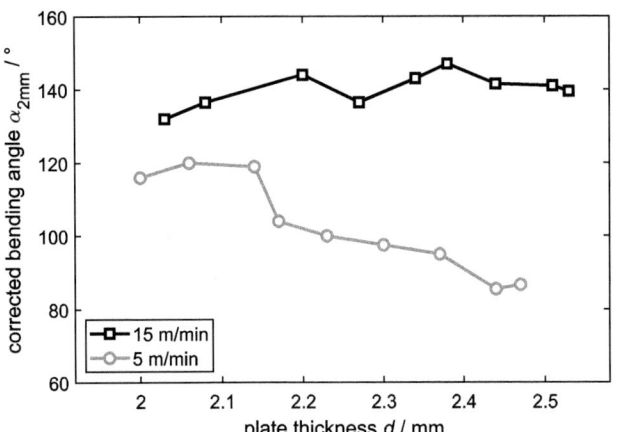

Fig. 4 Corrected bending angle α_{2mm} over reduced plate thickness by reduction of PCG layer of profiles extruded at 5 m/min and 15 m/min

The bending behavior of the profile extruded at 15 m/min in Fig. 4 shows a slight overall decrease from 140° to 132° with reduced plate thickness through increased PCG reduction. A maximum of 146° can be achieved at 2.38 mm, but

this is within the expected scatter. Compared to the profile extruded at 5 m/min the PCG layer does not affect the bending angle negatively. The bending angle in the profile extruded at 15 m/min does not drop below 130° and remains at larger bending angles throughout the investigated plate thicknesses compared to the profile extruded at 5 m/min. This indicates that bending behavior of the sample extruded at 15 m/min is governed by the bulk microstructure of cube orientation texture and a spread along the $\langle 001 \rangle \| EX$-fiber.

Deformation Distribution and Localization

From Fig. 4 it becomes clear, that the maximum bending angle for the profile extruded at 15 m/min is superior to the profile extruded at 5 m/min, irrespective of the amount of removed PCG layer. In this context it should be noted that the chemistry, profile geometry, artificial aging and material strength are identical for both profiles. The main difference between these two extrusion speeds lies in the resulting grain structure, texture, and their respective through-thickness

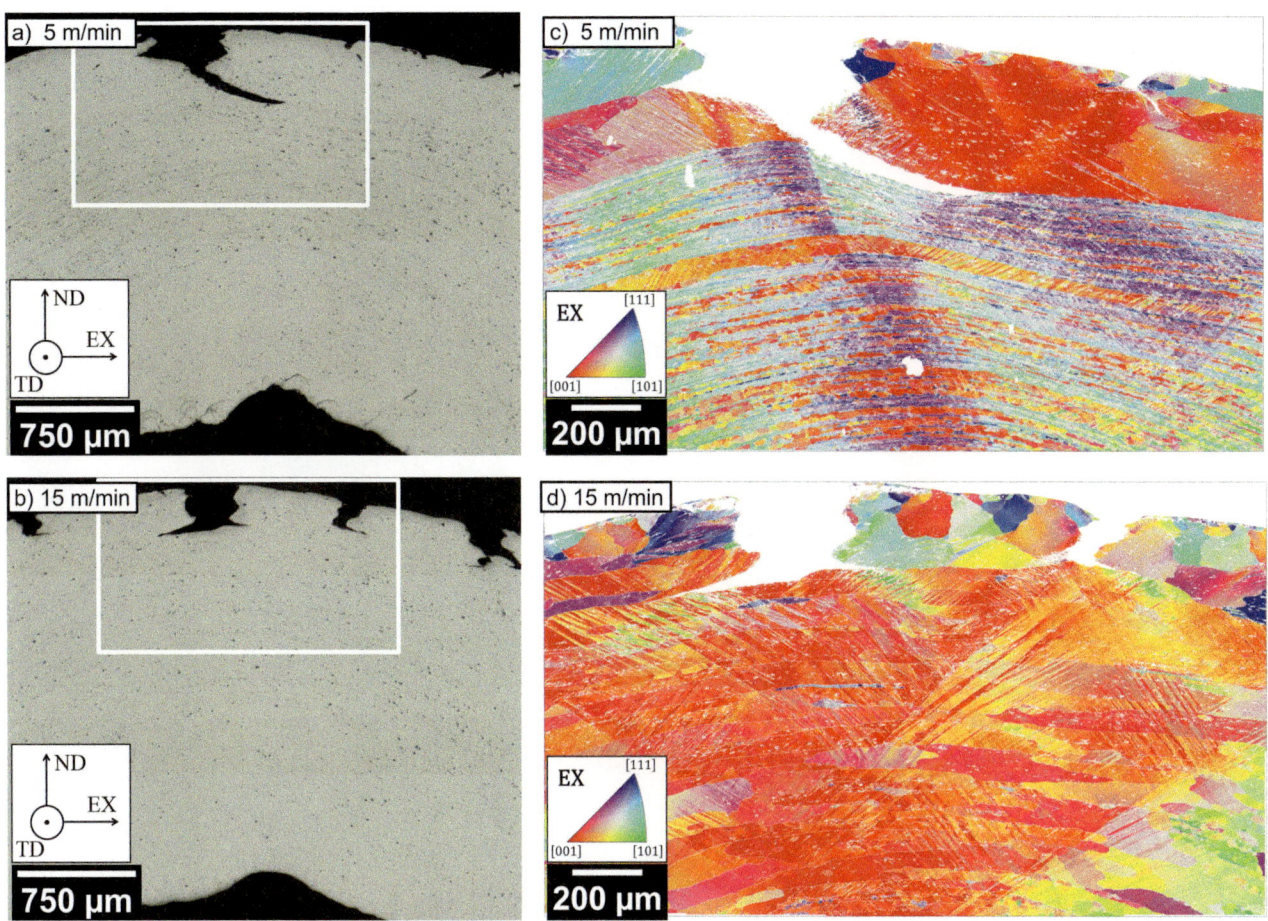

Fig. 5 Crack formation in the cross-section of the samples bent to an angle of 80° of the profiles extruded at **a** 5 m/min and **b** 15 m/min without removing PCG. EBSD measurement of the cross-section of **c** sample extruded at 5 m/min and **d** 15 m/min within the white boxes in **a** and **b**

variation. To investigate the distribution and localization of deformation during bending, samples for each extrusion speed were bent to a bending angle α of 80°, which marks the average failure angle of the sample extruded at 5 m/min. Figures 5a and b show the bent cross-sections of the samples extruded with 5 and 15 m/min, respectively. The outer surface during bending is at the top of Fig. 5a and b, while at the bottom the imprint of the bending punch is visible. Similar to the bending sample shown in Fig. 3b a single large crack forms in the 5 m/min sample at the outside of the sample underneath the bending punch. That crack appears to grow under 45° into the material. The EBSD measurement in Fig. 5c of the area around the crack marked in the box in Fig. 5a, reveals that the crack forms in the PCG layer, grows perpendicular at first until it reaches the boundary between the PCG layer and the fine-grained bulk. There the crack gets deflected and expands further along this boundary. This leads to an effective wall-thickness reduction by the extent of the PCG layer, resulting in a weakened cross-section that ultimately leads to the drop of bending force.

In contrast to the single crack of the profile sample at 5 m/min, several cracks form in the bent cross-section of the sample extruded at 15 m/min in Fig. 5b. These cracks grow perpendicular to the surface to a maximum extent of 300 −400 μm until getting deflected at the boundary between the PCG layer and the ⟨001⟩‖EX-fiber oriented bulk (Fig. 5d). Additionally, bands of periodically changing orientations are visible in the bulk, that expand over several 100 μm, thus over grain boundaries, and distribute homogeneously throughout the bulk of the cross-section. These bands form due to the bending deformation, therefore are referred to as deformation bands. This suggests that in this sample deformation is much more homogeneously distributed.

For further analysis, distribution of deformation is investigated by means of kernel average misorientation (KAM) below 2° misorientation, which describes the local rotation due to slip deformation. The KAM distribution is visualized as a 'heat-map' in Fig. 6a and b for both extrusion speeds. For this, the EBSD-images were sectioned into 40 chessboard-like segments of $4.5 \cdot 10^4$ μm^2 area, within

Fig. 6 Intensity of kernel average misorientation (KAM) as heat-map of the samples bent to an angle of 80° of profiles extruded at **a** 5 m/min and **b** 15 m/min, respectively

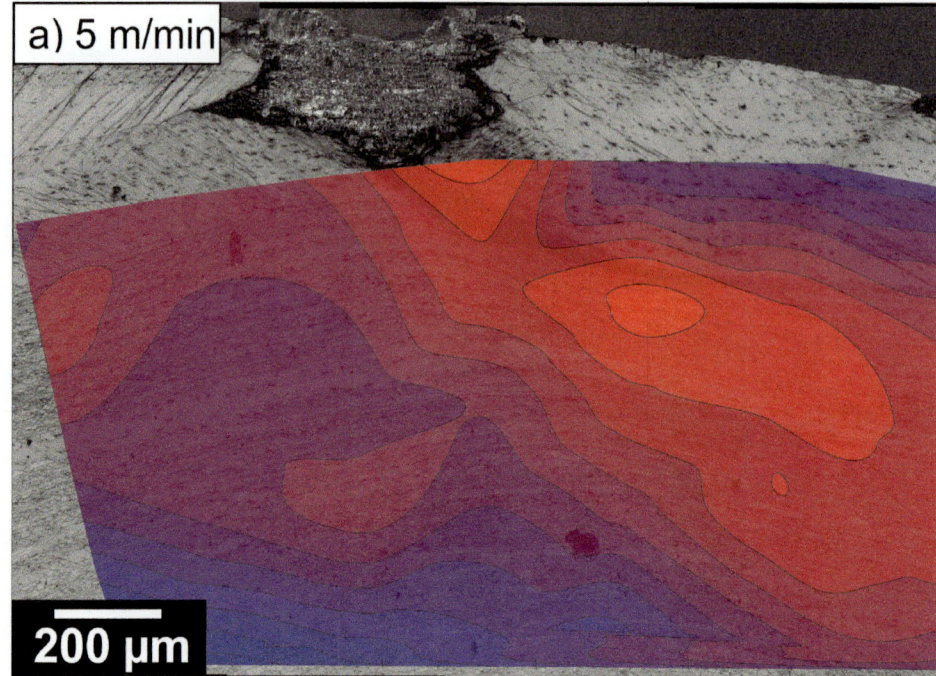

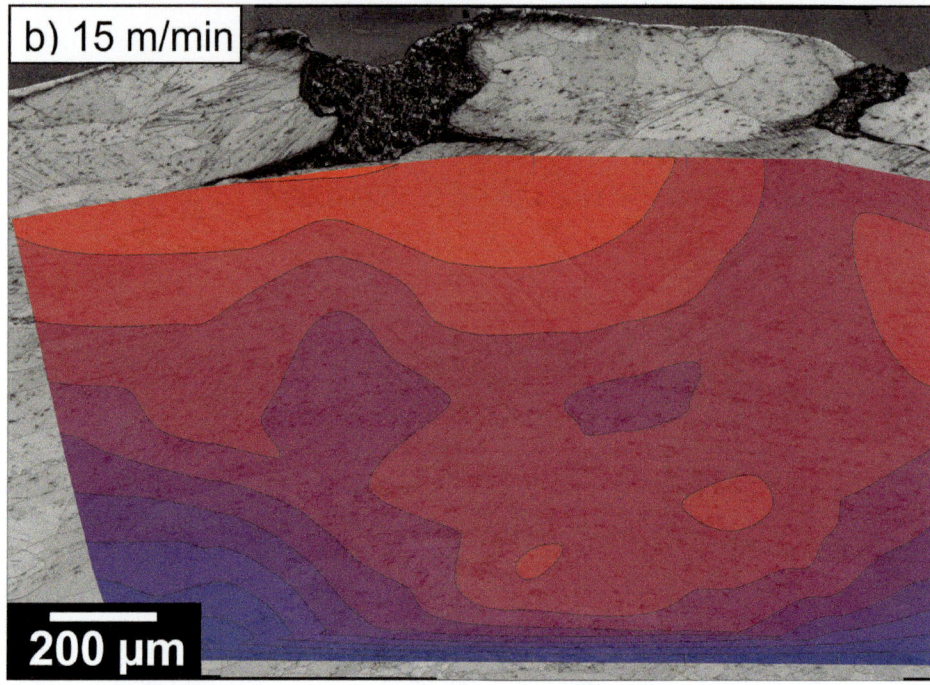

which the KAM values were averaged. The KAM distribution of the sample extruded at 5 m/min in Fig. 6a shows deformation during bending localizes underneath the crack, whilst the areas left and right to the crack carry less deformation. In contrast, the heat-map of the KAM distribution for the sample extruded at 15 m/min (Fig. 6d) shows homogeneous deformation. The deformation increases radially towards the outer surface, as expected for the strain during bending deformation, while tangential distribution shows only little deviation. This emphasizes the homogeneous distribution of deformation bands from Fig. 5d.

The deformed area in the plastic zone of a deflected crack underneath the PCG-bulk interface was investigated by high-resolution EBSD using the EMsoft software package for dictionary indexing by de Graef et al. [19]. This area (Fig. 7a) shows the pronounced formation of deformation bands within a grain, that extend to the underneath lying neighboring grain. Figure 7b shows the misorientation

Fig. 7 a IPF-map over band contrast colored in EX direction from high resolution EBSD close to a crack tip (upper left) in the PCG-bulk boundary from Fig. 5d. Blue arrows indicate the slope of a grain boundary affected by the deformation band. **b** Development of misorientation along the dashed line in **a** relative to the line origin on the left. Arrows denote misorientations close (<5° deviation) to the respective CSL boundary. Different degrees of misorientation for same CSL boundaries are due to the definition referenced to the line origin

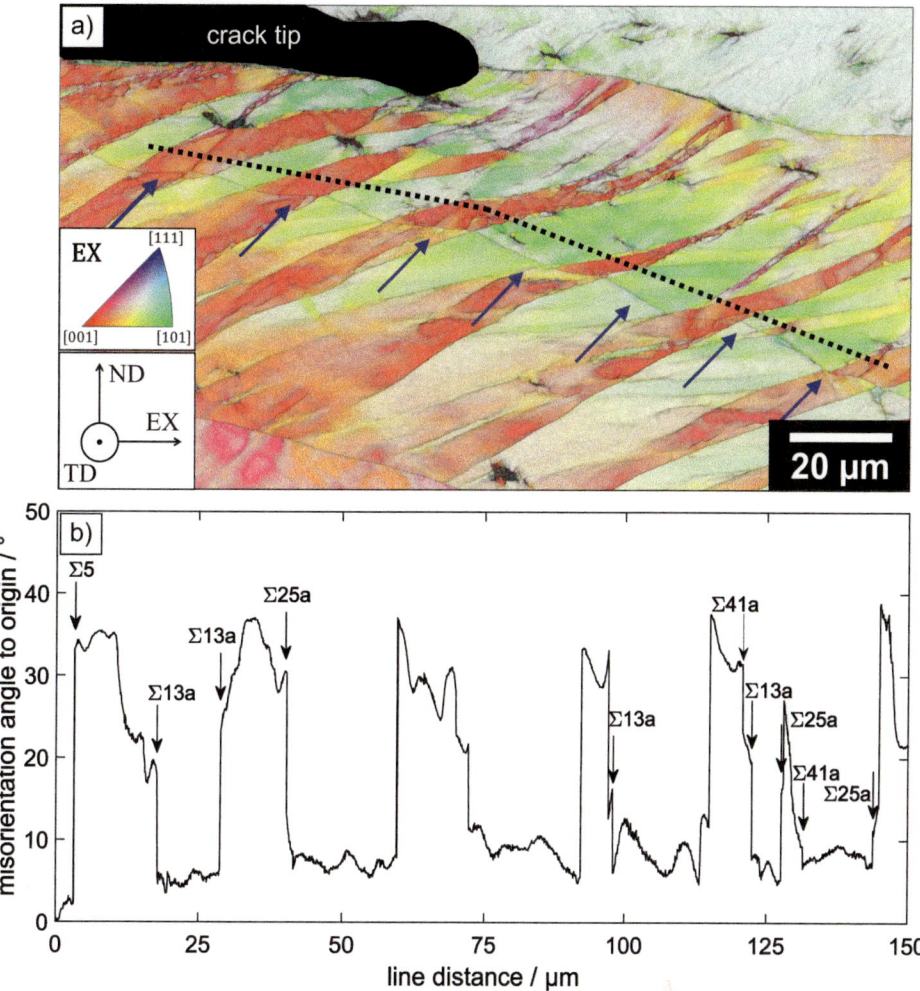

profile along the dashed line from Fig. 7a in respect to the line origin on the left end. The misorientation profile shows alternating change of misorientation in the grain due to the deformation bands, that can be assigned to orientations with ⟨101⟩ poles (green) and ⟨100⟩ poles (red) parallel to EX direction. The misorientation in the green regions along the line from Fig. 7a, that are closely oriented to the first orientation of the line, are showing little variation of misorientation within a region, implying that in these regions only little deformation occurred. The regions that are colored in red, however, show pronounced misorientation change within each zone that is due to dislocation accumulation under bending deformation. Thus, these regions can be identified as the actual deformation bands.

The localized plastic deformation in these bands leads to an increase in geometric necessary dislocation density to facilitate the required macroscopic strain. These dislocations further accommodate the boundaries between the deformation bands and the regions of little deformation. In the misorientation profile in Fig. 7b, these deformation band boundaries can be seen as sharp increase in misorientation.

Misorientation analysis revealed that half of these boundaries show misorientations close (<5° deviation) to the special coincidence site lattice (CSL) boundaries $\Sigma 5$, $\Sigma 13a$, $\Sigma 17a$, $\Sigma 25a$, $\Sigma 37a$, and $\Sigma\ 41a$. Different degrees of misorientations in Fig. 7b for same CSL boundaries are due to the misorientation definition referenced to the line origin in this figure. Common to all these misorientations is the rotational crystallographic axis of ⟨100⟩-type. It can be argued that in Aluminum with its high stacking fault energy special CSL boundaries are energetically not much more favorable than random misorientation, especially at large CSL values. However, twinning has been reported for Al–Mg alloys under high straining [21] as well as cracking for pure Aluminum [22, 23]. While it is not clear from the present investigation, if these special misorientations are generally favorable boundaries in Aluminum, the common rotational axis of ⟨100⟩ parallel to the bending axis appears to be favorable.

This becomes even more important when neighboring grains are considered. In a global texture of cube orientation with a small spread along ⟨001⟩‖EX-fiber most grains are only slightly misoriented, with the common rotational axis of ⟨001⟩ parallel to EX. In other words, in such oriented grains the poles parallel to the bending axis and TD are close to ⟨100⟩, enabling the deformation bands to expand over grain boundaries and maintain the ⟨100⟩ misorientation axis at the deformation band boundaries. This can be seen at the grain boundary in the band contrast of Fig. 7a indicated by blue arrows. The grain boundary shows a stepwise course following the deformation in the deformation band.

In this manner, a higher extrusion speed increases the bending angle (Fig. 3a) through the formation of cube orientation texture with a small spread along ⟨001⟩‖EX-fiber in the bulk. This texture supports the bending ductility independently of the detrimental PCG layer (Fig. 4) by the formation of deformation bands that extent over several grains (Fig. 5), thus can distribute homogeneously across the bent cross-section (Fig. 6). The deformation bands show boundaries to their host grains of a common ⟨100⟩ misorientation axis, that can be sustained in neighboring grains in this texture (blue arrows in Fig. 7a).

Summary and Conclusion

The present investigation on bending ductility shows that different results in bending angle can be achieved only by a variation of grain structure and texture along the cross-section. This was achieved by influencing the recrystallization behavior through variation of extrusion speed and resulting profile temperature, while using constant deformation conditions with the same extrusion tool, the same alloy with constant chemical composition and primary particle content, as well as achieving comparable tensile strength through the same artificial aging heat treatment. At low extrusion speeds a large peripheral coarse grain (PCG) layer forms with a fine-grained microstructure of β-fiber texture in the bulk. At high extrusion speeds, the PCG layer is smaller with large grains in the bulk, which exhibit a cube texture and a small spread along the ⟨001⟩‖ EX-fiber. The bending ductility in terms of bending angle for the profile extruded at low extrusion speed increases with reduction of the PCG layer. However, it never reaches the bending angles of the profile extruded at high extrusion speed. Here, the bending angle is on a high level irrespective of the PCG layer. EBSD investigations of the bent cross-section revealed that bending leads to the development of deformation bands, where deformation localizes, which distribute evenly throughout the cross-section. These deformation bands then form boundaries to the undeformed regions with a common crystallographic misorientation axis of ⟨100⟩, that lies parallel to the bending axis. In the bulk texture of cube orientation with a spread along the ⟨001⟩‖ EX-fiber, the ⟨100⟩-poles of the grains in the bulk are only

slightly misoriented. Due to only small deviation of the misorientation axis, this texture thus enables the development of deformation bands across grain boundaries and macroscopically homogeneous plastic deformation becomes stabilized resulting in increased bending ductility.

References

1. Parson N, Fourmann J, Beland J-F (2017) Aluminum Extrusions for Automotive Crash Applications. SAE Tech Pap 01–16
2. Ritchie RO (2011) The conflicts between strength and toughness. Nat Mater 10:817–822
3. Ryen Ø, Holmedal B, Marthinsen K, Furu T (2015) Precipitation, strength and work hardening of age hardened aluminium alloys. IOP Conf Ser Mater Sci Eng 89:012013
4. Remøe MS, Marthinsen K, Westermann I, Pedersen K, Røyset J, Marioara C (2017) The effect of alloying elements on the ductility of Al-Mg-Si alloys. Mater Sci Eng A 693:60–72
5. Österreicher JA, Schiffl A, Falkinger G, Bourret GR (2016) Microstructure and mechanical properties of high strength Al—Mg—Si—Cu profiles for safety parts. IOP Conf Ser Mater Sci Eng 119:012028
6. Lassance D, Fabregue D, Delannay F, Pardoen T (2007) Micromechanics of room and high temperature fracture in 6xxx Al alloys. Prog Mater Sci 52:62–129
7. Kuijpers NCW, Vermolen FJ, Vuik C, Koenis PTG, Nilsen KE, van der Zwaag S (2005) The dependence of the β-AlFeSi to α-Al (FeMn)Si transformation kinetics in Al–Mg–Si alloys on the alloying elements. Mater Sci Eng A 394:9–19
8. Lodgaard L, Ryum N (2000) Precipitation of dispersoids containing Mn and/or Cr in Al–Mg–Si alloys. Mater Sci Eng A 283:144–152
9. Sundman B, Ohnuma I, Dupin N, Kattner UR, Fries SG (2009) An assessment of the entire Al–Fe system including D03 ordering. Acta Mater 57:2896–2908
10. Shi C, Chen X-G (2015) Effects of Zr and V Micro-Alloying on Activation Energy during Hot Deformation of 7150 Aluminum Alloys. In: Hyland M (ed) Light Metals 2015. Springer International Publishing, Cham, pp 163–167
11. Hirsch J, Lücke K (1988) Overview no. 76 - I. Acta Metall 36:2863–2882
12. Daaland O, Nes E (1996) Origin of cube texture during hot rolling of commercial Al-Mn-Mg alloys. Acta Mater 44:1389–1411
13. Montheillet F, Cohen M, Jonas JJ (1984) Axial stresses and texture development during the torsion testing of Al, Cu and α-Fe. Acta Metall 32:2077–2089
14. Grasserbauer J, Weißensteiner I, Falkinger G, Mitsche S, Uggowitzer PJ, Pogatscher S (2020) Evolution of Microstructure and Texture in Laboratory- and Industrial-Scaled Production of Automotive Al-Sheets. Materials 13:469
15. Molodov DA, Shvindlerman LS, Gottstein G (2003) Impact of grain boundary character on grain boundary kinetics. Int J Mater Res 94:1117–1126
16. Frodal BH, Morin D, Børvik T, Hopperstad OS (2020) On the effect of plastic anisotropy, strength and work hardening on the tensile ductility of aluminium alloys. Int J Solids Struct 188–189:118–132
17. Henn P, Liewald M, Sindel M (2018) Investigation on crashworthiness characterisation of 6xxx-series aluminium sheet alloys based on local ductility criteria and edge compression tests. IOP Conf Ser Mater Sci Eng 418:012125
18. VDA 238–100 (2020) Plate bending test for metallic materials

19. De Graef M (2020) A dictionary indexing approach for EBSD. IOP Conf Ser Mater Sci Eng 891:012009

20. Bachmann F, Hielscher R, Schaeben H (2010) Texture Analysis with MTEX – Free and Open Source Software Toolbox. Solid State Phenom 160:63–68

21. Jin SB, Zhang K, Bjørge R, Tao NR, Marthinsen K, Lu K, Li YJ (2015) Formation of incoherent deformation twin boundaries in a coarse-grained Al-7Mg alloy. Appl Phys Lett 107:091901

22. Li BQ, Sui ML, Li B, Ma E, Mao SX (2009) Reversible Twinning in Pure Aluminum. Phys Rev Lett 102:205504

23. Hai S, Tadmor EB (2003) Deformation twinning at aluminum crack tips. Acta Mater 51:117–131

Anisotropy of Tearing Behavior in AA7075-T6 Sheet at 200 °C

Daniel E. Nikolai and Eric M. Taleff

Abstract

The tearing resistance of AA7075-T6 sheet material was measured at room temperature and at 200 °C along different directions. Tearing resistance is characterized by the energy required to completely tear a specimen, with higher energies indicating greater tearing resistance. Specimens were tested at 200 °C for times no longer than would provide a retrogression heat treatment, from which the full strength of the T6 condition may be recovered by a reaging heat treatment. Tearing resistance is significantly greater at 200 °C than at room temperature, which promises improved deformation processing of AA7075-T6. The tearing resistance at 200 °C varies with direction relative to the rolling direction. The tearing resistance is highest in the L–T orientation and lowest in the T–L orientation, by ASTM E871 specimen orientation designations. The significance of these results to the deformation processing of AA7075-T6 sheet at elevated temperatures will be discussed.

Keywords

Aluminum • AA7075 • Tear resistance • Fracture • Warm deformation

Introduction

There is a desire to use high-strength, heat-treatable wrought aluminum alloys for structural applications in automobiles to reduce vehicle mass and improve performance while maintaining high safety standards. One material of interest is the AA7075 sheet, which provides a yield strength of 505 MPa

and a tensile strength of 570 MPa in the T6 temper [1]. These give AA7075 a strength-to-weight ratio that is an improvement over some ultra-high-strength steels currently used in the automotive industry. These strengths are also significantly greater than those of the 6000-series alloys more commonly used in mass-production vehicles, such as AA6022 and AA6111. The 6000-series alloys are usually formed in the T4 temper or in an alternative proprietary temper condition that provides the ductility and somewhat low strength required for successful room-temperature forming. The parts formed from these 6000-series alloys are then paint baked to increase strength. Both the 6000-series alloy compositions and the difficulties of integrating heat treatments with forming and joining operations limit the part strengths possible. For example, AA6022 can reach yield and tensile strengths of only 272 and 325 MPa after paint baking from the T4 temper [2]. Several barriers to using higher-strength aluminum alloys such as AA7075-T6 include low formability at room temperature, difficulty with joining operations, increased cost, and potential susceptibility to corrosion problems, such as stress corrosion cracking. Recent work on warm forming of AA7075-T6 sheet demonstrates that the concept of *retrogression forming* can be used to form complex components with negligible loss of strength, and a single low-temperature heat treatment can restore strength to equal or surpass the original T6 temper [3, 4]. Figure 1 presents an example of AA7075-T6 sheet formed by retrogression forming [5]. The combination of forming during retrogression at an elevated temperature followed by a reaging heat treatment creates a two-step process that has been termed *retrogression forming and reaging* (RFRA) [3, 4]. The success of the retrogression forming approach is attributed to both a lowering of flow stress at the warm forming temperature, typically 200 °C for AA7075-T6, and an increase in tearing resistance. This investigation explores the tearing resistance of AA7075-T6 sheet at 200 °C, which is anisotropic within the sheet plane.

D. E. Nikolai · E. M. Taleff (✉)
Department of Mechanical Engineering, The University of Texas at Austin, 204 E. Dean Keeton St., Austin, TX 78712, USA
e-mail: taleff@utexas.edu

© The Minerals, Metals & Materials Society 2023
S. Broek (ed.), *Light Metals 2023*, The Minerals, Metals & Materials Series,
https://doi.org/10.1007/978-3-031-22532-1_78

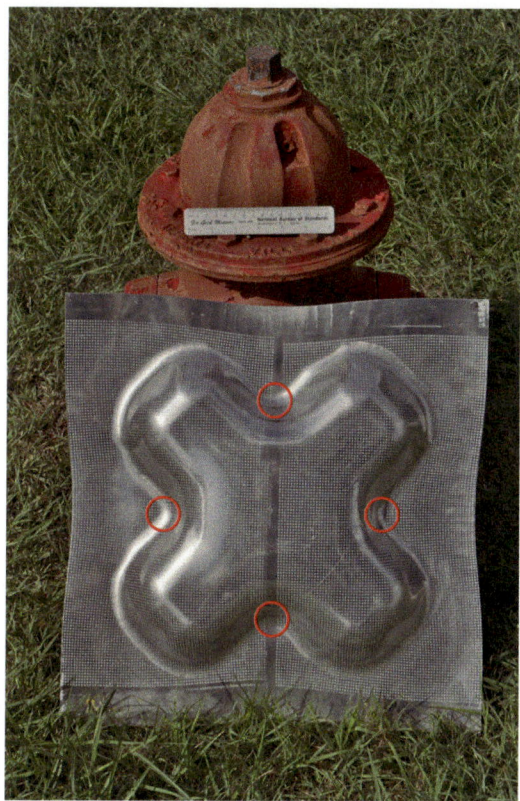

Fig. 1. A part successfully produced by RFRA in an AA7075-T6 Alclad sheet is shown [5]. Parts formed to a greater depth experienced tearing at the locations circled, where a plane strain condition and large strains produced the greatest sheet thinning

Possible sources of this anisotropy are explored, and the most likely are related to material microstructure.

Retrogression forming is based on the concept of simultaneously warm forming an aluminum material in the T6 peak-aged temper while it undergoes a retrogression heat treatment. The primary alloying elements in AA7075 are zinc and magnesium, as shown in Table 1. During a retrogression heat treatment of AA7075-T6, fine η' precipitates either dissolve into the bulk material or coarsen into η phase (Mg_2Zn) precipitates, especially at grain boundaries, while new η phase precipitates may also form [6]. This process slightly decreases strength but increases ductility. Forming during retrogression takes advantage of the increased ductility from retrogression *and* both the increased ductility and lower flow stress at the elevated temperature used for retrogression. A subsequent reaging heat treatment precipitates new η' and coarsens undissolved η' precipitates. This reaging heat treatment can recover strength to a level equal to the T6 temper prior to retrogression.

Prior work measuring tearing energy in AA7075-T6 sheet at an elevated temperature suggested that an increase in tearing resistance is partially responsible for the success of RFRA experiments [5, 7]. Figure 2 depicts two curves of force *versus* displacement for tear test experiments conducted on AA7075-T6 sheet material. Figure 2a presents data from a specimen tested at room temperature (25 °C), and Fig. 2b presents data from a specimen tested at a temperature suitable for a retrogression heat treatment and retrogression forming (200 °C). Both specimens were tested at the same crosshead pulling rate of 2 mm/min, which is a rate recommended in ASTM B871-01 [10]. The total area under each curve is equivalent to the tearing energy of the specimen. The area to the left of the vertical dashed line shown in Fig. 2 is the energy required to initiate a crack in the specimen, and the area to the right of the vertical dashed line is the energy required to propagate a crack completely across the specimen. By comparing the two tests shown in Fig. 2, it is clear that there is a substantial increase in crack propagation energy at the elevated temperature, an approximately 20-fold increase [7]. The slight uptick in force toward the end of the curve denoted by an arrow in Fig. 2b is indicative of a plastic hinge. This is caused by a small ligament of material plastically deforming at the end of the crack propagation path and absorbing energy, resulting in a slight jump in force required to completely tear the specimen at 200 °C. This phenomenon is typical of ductile tearing behavior and is not observed at room temperature for AA7075-T6. Macrophotographs, not shown here, of fracture surfaces provided further visual evidence of increased ductility during tear testing of AA7075-T6 sheet at elevated temperature [7]. The increased tearing resistance in AA7075-T6 sheet at an elevated temperature helps to explain the success of RFRA in forming experiments [5]. It also hints at a potential for additional applications of the RFRA concept to forming and mechanical joining operations that require greater ductility than is available in high-strength aluminum alloys at room temperature.

Experimental Procedures

Kahn-type tear tests [8] were conducted on specimens from 2-mm-thick AA7075-T6 sheet material produced by the Alcoa Corporation, ALCOA-DPW lot 593–042. The nominal chemical composition of AA7075 is shown in Table 1 [9]. The material was supplied and tested in the T6 temper. Tear tests were based on procedures described in ASTM B871-01 [10]. The specimen geometry used for the present

Table 1 The nominal chemical composition of AA7075, wt. pct. [9]

Zn	Mg	Cu	Cr	Fe	Si	Mn	Ti	Al
5.1–6.1	2.1–2.9	1.2–2.0	0.18–0.28	< 0.5	< 0.4	< 0.3	< 0.2	Bal

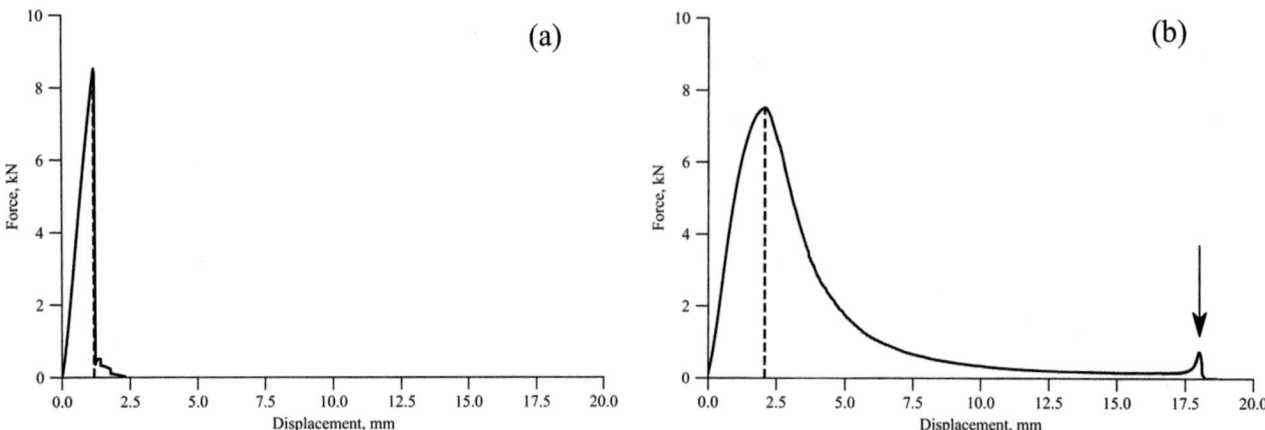

Fig. 2 Force is plotted against displacement for tear tests of AA7075-T6 sheet at **a** 25 °C and 2 mm/min and **b** 200 °C and 2 mm/min. The plastic hinge phenomenon is denoted by the arrow in **b**. Both specimens were tested in the L–T orientation. Data are from Ref. [7]

study was modified slightly from ASTM B871-01 and is similar to that described in the reference [11]. The same specimen geometry was previously used by the present investigators [7].

Figure 3 shows a typical specimen used for tear tests in the present study. Specimens were machined from the AA7075-T6 sheet in two steps using a combination of water-jet cutting and wire electrical discharge machining (EDM). The first step was water-jet cutting of the outer rectangular specimen shape and circular loading-pin holes for specimen mounting. The crack notch was subsequently machined using wire EDM, which provided the precise geometry required for the crack notch radius [10]. Precise machining of the crack notch radius was necessary to create a consistent preferential crack initiation site for all specimens. Specimens were machined in three different orientations: longitudinal-transverse (L–T) with the crack direction transverse to the sheet rolling direction, transverse-longitudinal (T–L) with the crack direction along the sheet rolling direction, and with the crack direction oriented at a 45° angle to the sheet rolling direction (∡ 45° Each

Fig. 3 A typical specimen used for Kahn-type tear tests is shown. This specimen was machined in the T–L orientation. The sheet rolling direction is vertical

specimen was loaded in an MTS FlexTest 40 test frame and held in place by pins with shims to keep the specimen centered along the loading axis. The pin-grips were attached to pull rods with a universal joint on the top grip to allow for free rotation and automatic specimen alignment.

Tear tests were conducted at room temperature (25 °C) and at 200 °C in a computer controlled servohydraulic test frame with computerized data acquisition. A convective oven surrounding the specimen and grips was used to heat the specimen and test fixtures for tests conducted at elevated temperatures. To account for heat lost during specimen insertion into the oven and grips, the oven and fixtures were preheated to approximately 30 °C higher than the desired test temperature prior to specimen loading. After insertion into the grips, the temperature on either end of the specimen was measured with two K-type thermocouples attached to the specimen. A separate K-type thermocouple measured ambient temperature within the convective oven for closed-loop temperature control. Only data from specimens that reached the target test temperature within four minutes are reported. Testing at a constant pulling rate commenced when the average temperature between the two thermocouples attached to the specimen reached within 2 °C of the desired test temperature (200 °C); specimen temperature was monitored throughout each test. Pull rates from 2 to 512 mm/min were applied to determine the effect of pulling rate on material response. Upon rupture, specimens were quickly removed from the test fixtures and quenched in water. Tests conducted at room temperature used the same test fixtures without using the convective oven. Data from completed tear tests were considered valid per ASTM B871-01 only if a crack propagated entirely through the specimen and if the crack propagation path deviated by no more than 10° from parallel with the short edge of the specimen; see Fig. 3 [9]. Tested specimens of particular interest were photographed using a digital single-lens reflex camera with an attached macro lens.

Partial tear tests were also conducted on specimens in the L–T and T–L orientations at 25 °C and 200 °C. The testing protocol was effectively the same as that followed for the regular tear tests, but cracks were stopped after propagating approximately 50 to 80% through the specimen. After testing and quenching, sections of these specimens were excised from regions near the crack tip using a diamond saw. The resulting specimen sections measured approximately 2 mm × 2 mm × 6 mm. These specimen sections were studied using X-ray computed tomography (XRCT) with the assistance of Dr. Philip Noell at Sandia National Laboratories. For XRCT, specimens are subjected to an incident X-ray beam. A detector measures the intensity of the transmitted beam, which depends upon the absorption of X-rays within the specimen. The resulting data are processed to create a three-dimensional rendering of the sample absorptance, which can be loosely interpreted as a density. These three-dimensional data can reveal cavities, cracks, and second-phase particles, all of which have X-ray absorptances different from the base aluminum alloy.

Photomicrographs were taken of as-received material specimens prepared by standard grinding and polishing techniques. Photomicrographs of polished and unetched specimens were taken using an optical microscope to reveal intermetallic particles and cavities. After polishing, some specimens were electrolytically etched to reveal grain structure. A voltage of 25 V was applied for 90s with the specimen immersed in Barker's reagent (95 parts distilled water; 5 parts tetrafluoroboric acid). Photomicrographs were taken using an optical microscope with polarizing filters to reveal grain structure in the as-received condition.

Results, Analysis, and Discussion

A prior study determined that tearing resistance increases from room temperature to 200 °C, an elevated temperature suitable for retrogression heat treatments in AA7075-T6 [7]. This phenomenon is demonstrated by the increased tearing energies shown in Fig. 2. Recall that the area under the curve is the total energy required to tear a specimen, the area to the left of the vertical dashed line is initiation energy, and the area to the right of the vertical dashed line is propagation energy. Force *versus* displacement data for AA7075-T6 specimens tested at 200 °C in the ∡ 45° and T–L orientations are shown in Fig. 4. The data of Fig. 4 demonstrate a significant increase in tearing resistance compared to the room temperature test shown in Fig. 2 for the L–T orientation. Figure 4a shows force–displacement data for a specimen tested at 200 °C in the ∡ 45° orientation and Fig. 4b shows force–displacement data for a specimen tested at 200 °C in the T–L orientation. Both specimens were tested at 2 mm/min, the same rate used for the data shown in Fig. 2. Although tearing resistance is higher at elevated temperature than at room temperature for all orientations tested, scrutiny of the data in Fig. 4 provides evidence for anisotropy in the tearing behavior of AA7075-T6 sheet. The propagation energy shown in Fig. 4a is slightly smaller than in Fig. 2b, and the propagation energy is smaller still in Fig. 4b.

Table 2 provides measurements of tearing resistance as initiation and propagation energies for AA7075-T6 sheet under the different experimental conditions studied. The total tearing energy *E*, initiation energy *IE*, propagation energy *PE*, and unit energies (*UIE* and *UPE*) are presented. *UIE* and *UPE* are the unit initiation and propagation energies, which are normalized for the specimen cross-sectional area as

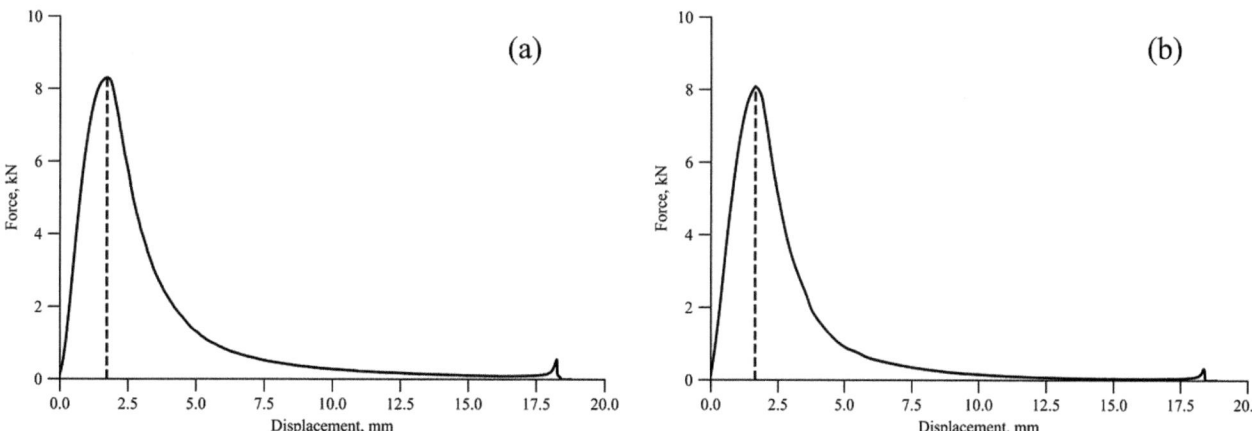

Fig. 4 Force is plotted against displacement for tear tests of AA7075-T6 sheet at 200 °C and 2 mm/min in **a** the ∡ 45° orientation and **b** the T–L orientation

Table 2 Tearing energies at room and elevated temperatures for different specimen orientations

Orientation	T, °C	v, mm/min	Tests	E, J	IE, J	UIE, J/mm²	PE, J	UPE, J/mm²
L–T	25	2	3	6.25	5.35	0.1	0.91	0.02
L–T	200	2	6	28.7	10.7	0.21	18	0.34
L–T	200	64	7	23.1	9.85	0.19	13.3	0.25
∡ 45°	200	2	5	25.5	8.74	0.16	16.8	0.31
∡ 45°	200	64	5	19.3	8.01	0.15	11.3	0.21
T–L	200	2	5	20.9	8.06	0.15	12.9	0.24
T–L	200	64	5	12.8	6.52	0.13	6.25	0.12

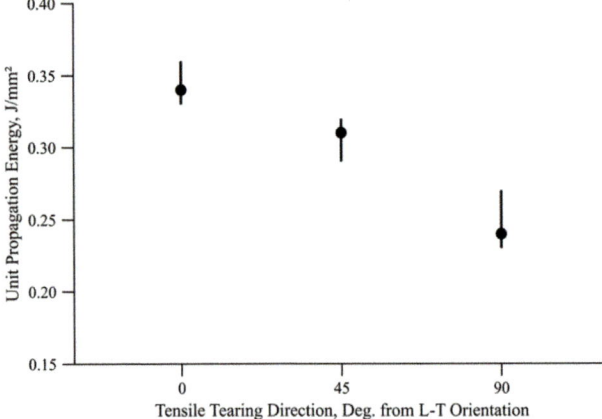

Fig. 5 Average unit propagation energy (*UPE*) is plotted against specimen orientation for specimens tested at 200 °C and 2 mm/min. The error bars represent the maximum and minimum values of *UPE* measured for each orientation

defined by the product of specimen thickness and tear length. The cross-sectional area of the specimens for the present study is b × t = 51.4 mm². The data documented in Table 2 indicate that the L–T orientation has the highest resistance to tearing, the ∡ 45° orientation has an intermediate resistance to tearing, and the T–L orientation has the lowest resistance to tearing. Figure 5 shows the average *UPE* plotted against specimen orientation for specimens tested at 200 °C and 2 mm/min with error bars indicating the minimum and maximum *UPE* values measured for each test condition. It is clear that the *UPE* decreases as the crack propagation direction rotates closer to the sheet rolling direction. This is evident for specimens tested at both slow and fast pulling rates. *UPE* also decreases with an increasing pulling rate for each specimen orientation tested.

The microstructure of the as-received material was investigated for possible mechanisms that cause the observed anisotropy in tearing resistance. Figure 6 shows an optical photomicrograph of the untested material. The horizontal direction is the sheet rolling direction and the vertical direction is the short-transverse direction (through-thickness direction) of the sheet. Optical microscopy of polished

specimens revealed microstructural defects that include cavities and large intermetallic particles. The lightest color in Fig. 6 is the aluminum alloy material. The intermediately colored features in Fig. 6 are intermetallic particles typically formed during commercial material processing and recycling. Slight differences in the color of these particles indicate the different intermetallic compositions present in the as-received material. The intermetallic particles shown in Fig. 6 were examined with energy-dispersive spectroscopy (EDS) in a scanning electron microscope (SEM). The EDS data indicate the presence of three different types of intermetallic particles:

1. copper rich (~ 20 wt% Cu) particles,
2. iron rich (~ 25 wt% Fe) particles, and
3. particles containing both copper and iron (~ 15 wt% Fe; ~ 5 wt% Cu).

Each of these intermetallic particle types contains traces of zinc and magnesium. The darkest regions in Fig. 6 are

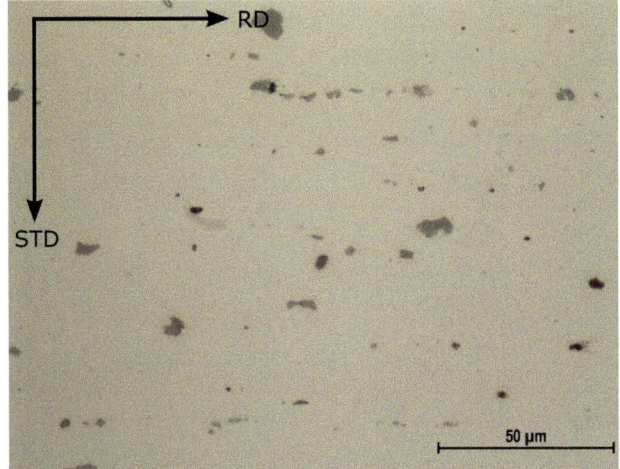

Fig. 6 Microstructure of the as-received AA7075-T6 sheet is shown in an optical photomicrograph. The horizontal direction is the sheet rolling direction; the vertical direction is the short-transverse (through-thickness) direction. Note the presence of intermetallic particles and cavities

cavities. It is unclear whether these cavities form during processing or are the result of intermetallic particles falling out during metallographic preparation. Both intermetallic particles and cavities tend to be aligned along the sheet rolling direction. Figure 6 shows intermetallic particles distributed as stringers along the sheet rolling direction. This is an expected result of large intermetallic particles being broken up and dispersed during sheet rolling. The cavities observed are similarly distributed and are often associated with intermetallic particle stringers. The arrangement of these features provides evidence for microstructural anisotropy in the as-received AA7075-T6 sheet.

Figure 7 depicts an optical photomicrograph from a polished and etched specimen of as-received material viewed using polarizing filters. It is worth noting the tendency for electrolytic etching to preferentially attack cavities, which is why the area fraction of cavities appears much greater in Fig. 7 than in Fig. 6. The horizontal direction is the sheet rolling direction, and the vertical direction is the short-transverse (through-thickness) direction of the sheet. Revelation of the grain structure demonstrates grains elongated along the rolling direction, as is expected in commercial unidirectionally-rolled aluminum alloy sheet material. This elongated grain structure is another anisotropic aspect of the material microstructure.

For a valid test under ASTM B871-01, the tear is required to propagate straight across the specimen, to within a maximum deviation of ±10° [10]. It was difficult for specimens in the ∡ 45° orientation to meet this requirement because the crack propagation path tended to skew toward the sheet rolling direction. This crack deviation effect increased with pulling rates. Figure 8 shows three specimens tested in the ∡ 45° orientation at 200 °C. The specimen on the left (denoted as KTT 078) was tested at 2 mm/min; the specimen in the middle (denoted as KTT 085) was tested at 64 mm/min; the specimen on the right (denoted as KTT 093) was tested at 512 mm/min. The sheet rolling direction in Fig. 8 is denoted by an arrow. Note that as the pulling rate increases, the crack propagation path increasingly deviates to align more closely with the sheet rolling direction. While experimental data were not valid for tests with such extreme crack deviation angles, the specimens shown in Fig. 8 provide strong evidence that tearing resistance is lowest along the sheet rolling direction.

X-ray computed tomography was conducted on small samples of material extracted from near the crack tip in partially torn specimens. Figure 9 shows an image of an XRCT data slice in a specimen partially torn in the L–T orientation at 200 °C and 64 mm/min. In Fig. 9, The horizontal direction is the sheet rolling direction, and the short-transverse (through-thickness) direction is perpendicular to the page. Different microstructural features can be identified in the XRCT slices because of their different X-ray absorption coefficients, such as intermetallic particles and cavities. These features are identified in Fig. 9. Note the meandering nature of the crack path. The crack tip appears to propagate straight through a specimen on the macroscopic scale, but it is apparent that the crack preferentially interacts with microstructural features during propagation, especially cavities. The data in Fig. 9 suggest that anisotropic microstructural features, such as intermetallic particle stringers or cavities or elongated grains, preferentially deflect the crack path. This is useful evidence suggesting that anisotropy of microstructure produces the observed anisotropy in tearing resistance.

Conclusions

Kahn-type type tear tests were conducted on specimens of AA7075-T6 sheet in three different orientations relative to the rolling direction at two different temperatures and several different pulling rates. Microstructures in the material were observed using optical microscopy, SEM with EDS, and XRCT. The following conclusions are drawn from the data produced:

1. The tearing resistance of AA7075-T6 sheet is anisotropic at 200 °C.
2. Cracks propagate most easily along the sheet rolling direction and tend to deviate toward the rolling direction as the pulling rate increases.
3. Tearing resistance is higher at 200 °C than at room temperature for all crack propagation directions.

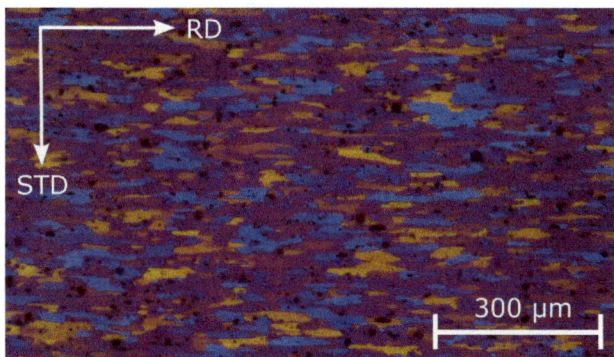

Fig. 7 The microstructure of the as-received AA7075-T6 sheet is shown under polarized light after etching. The horizontal direction is the sheet rolling direction; the vertical direction is the short-transverse (through-thickness) direction. Note that grains are elongated along the rolling (longitudinal) direction

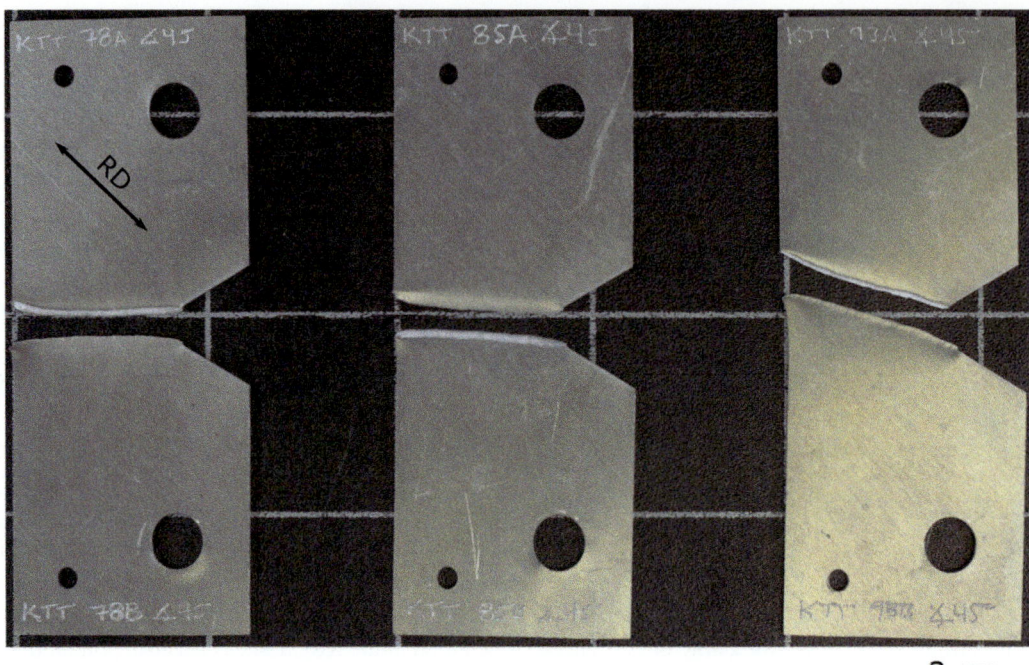

Fig. 8 Specimens tested in the ∡ 45° orientation at 200 °C are shown for pulling rates of (left to right) 2, 64, and 512 mm/min. Note the increasing deviation of the crack path toward the rolling direction as the pulling rate increases

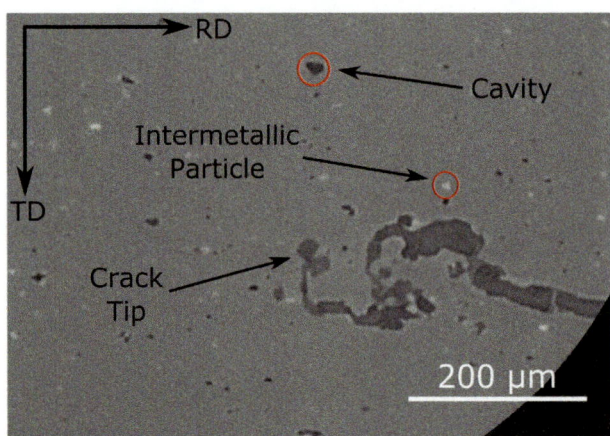

Fig. 9 The crack tip of a specimen tested in the L–T orientation at 200 °C and 64 mm/min is shown in data from X-ray computed tomography. Note the meandering of the crack path

4. Tearing resistance decreases with an increasing pulling rate for all specimen orientations.
5. The crack path interacts with microstructural features.
6. The microstructure of unidirectionally-rolled AA7075-T6 sheet metal is anisotropic.
7. The anisotropy of tearing behavior in AA7075-T6 is likely a result of the anisotropic microstructure.

Acknowledgements This work was supported by the National Science Foundation under GOALI grant number CMMI-1634495. The authors express their gratitude to Dr. Philip J. Noell of the Sandia National Laboratory for acquiring and providing X-ray computed tomography data for this investigation.

References

1. Davis JR (ed) (1993) *ASM Specialty Handbook: Aluminum and Aluminum Alloys*. ASM International, Materials Park, OH
2. Anderson K, Weritz J, Kaufman, JG (eds) (2019) *ASM Handbook, Volume 2B, Properties and Selection of Aluminum Alloys*. ASM International, Materials Park, OH. https://doi.org/10.31399/asm.hb.v02b.a0006712
3. Rader KE, Carter JT, Hector LF, Taleff, EM (2021) Review of Retrogression Forming and Reaging for AA7075-T6 Sheet. Minerals, Metals and Materials Series: Light Metals 2021, Perander, L (ed) The Minerals, Metals & Materials Society, 206–211
4. Rader, KE, Carter, JT, Hector, LG, Taleff, EM (2021) Plastic Deformation and Ductility of AA7075 and AA6013 at Warm Temperatures Suitable to Retrogression Forming. Metallurgical and Materials Transactions A, 52:4003-40017
5. Rader, KE, Carter, JT, Hector, LG, Taleff, EM (2022) Retrogression Forming and Reaging of an AA7075-T6 Alclad Sheet Material. Journal of Materials Engineering and Performance, 31 (7): 5311-5323. https://doi.org/10.1007/s11665-022-06663-1
6. J.K Park and A.J. Ardell (1984) Effect of Retrogression and Reaging Treatments on the Microstructure of Al-7075-T65. Metallurgical and Materials Transcations A, vol. 15A, pp. 1531-43

7. Nikolai, DE, Taleff, EM (2022) Tear Resistance of AA7075-T6 Sheet at Room Temperature and 200 °C. Minerals, Metals and Material Series: Light Metals 2022. The Minerals, Metals, & Materials Society.

8. J.G. Kaufman and A.H. Knoll (1964) Kahn-type Tear Tests and Crack Toughness of Aluminum Alloy Sheet. American Society for Testing and Materials, Materials Research and Standards 4(4): 151-156.

9. Registration Record Series Teal Sheets: International Alloy Designations and Chemical Composition Limits for Wrought Aluminum and Wrought Aluminum Alloys: The Aluminum Association, January 2015, p. 12.

10. ASTM International. Standard Test Method for Tear Testing of Aluminum Alloy Products. Standard Designation B871–01, ASTM International, West Conshohocken, PA, 2001.

11. Waleed M. Al-Mahshi. A Hybrid Experimental-Numerical Approach to Predict Fracture Initiation in AA-6111 Sheets. Master's thesis, University of Michigan-Dearborn, 2014.

Evaluating the Earing Amount of Materials Under Various Chemical Composition and Heat Treatment Processes with Finite Element Simulations of Cup Drawing Tests

Melih Çaylak, Görkem Özçelik, Abdullah Kağan Kınacı, and Koray Dündar

Abstract

Cup drawing is an important sheet metal forming process and is affected by the anisotropic behavior of materials. This anisotropic behavior is influenced by the chemical composition as well as the heat treatment process of materials. Therefore, the selection of proper chemical composition of the material is important for formability studies. In this study, different aluminum alloys such as 6082 and 6005 which were subjected to different heat treatment processes have been considered as test materials and FE simulations were carried out with Marc software. In order to observe the anisotropic properties of test materials, tensile tests were performed with 0, 45, and 90° according to the extrusion direction, respectively. Hill anisotropic yield criterion was used to define the anisotropic behavior of materials during the finite element studies.

Keywords

Aluminum • Deep drawing • Finite element method • Anisotropy • Earing • Formability

Introduction

Aluminum materials have gained great importance with the development of electric vehicles. Vehicle manufacturers started to prefer aluminum instead of steel owing to aluminums specific strength. In order to introduce the anisotropic behavior of aluminum, finite element (FE) studies have been conducted and compared with experimental results. Describing the anisotropic behavior of aluminum comes with some challenges. There are several material models to define the anisotropic behavior of materials. Hill proposed a criterion in 1948 to investigate the anisotropic behavior of materials [1]. Obtaining material parameters and other several anisotropic behaviors were investigated in that study. In this study, Hill r-based yield criteria was used in finite element studies. The experimental and numeric results were compared.

One of the common manufacturing methods is deep drawing. The deep drawing method relies on the substantial plastic deformation of flat metal sheets to produce engineered pieces with precise forms. A metal sheet is subjected to an external force by this plastic deformation. In order to prevent the metal portion from flexing back or changing shape again after being displaced by the external force, the force must be strong enough to move the material into the plastic zone. The ultimate wall thickness, wrinkle-freeness, and unbreaking ability of the pieces created by this procedure determine their quality [2]. Several studies have been carried out in the literature. Aminzadeh et al. investigated forming defects laser-welded blanks during deep drawing process [3]. FE analysis and experimental tests were accomplished with different steel sheets during the study. It was reported that the model that was used in the work to define material behavior is promising and further work can lead to improvements. Sener et al. used FE analysis to study the impact of blank thickness and ear development throughout the din cup drawing process [4]. The anisotropic behavior of a material is defined using the Yld91 yield criterion. Aluminum alloy 2090-T3 was employed as the test substance. The anisotropic behavior of the AA2090-T3 alloy could be sufficiently defined by the Yld91 yield criterion, and the estimated earing profiles and thickness strain distributions from the simulations were consistent with the experiment. Several FE models were run by Aksen et al. to forecast the expansion of the hole [5]. The test material was an AA6016 aluminum alloy, and the yield criterion employed was Hill48. It was determined after comparing the results of experimental tests and FE simulations that there is consistency between the two types of results.

M. Çaylak (✉) · G. Özçelik · A. K. Kınacı · K. Dündar
ASAŞ Aluminium San. ve Tic. A. Ş, Sakarya, Turkey
e-mail: melih.caylak@asastr.com

S. Broek (ed.), *Light Metals 2023*, The Minerals, Metals & Materials Series,
https://doi.org/10.1007/978-3-031-22532-1_79

FE and Material Model Description

In this study, Marc software was used to perform the FE simulations. Tools were modelled based on standard test tools. Dimensions of tools are illustrated in Fig. 1. The blank was modelled using fully integrated shell elements.

All geometries are modelled in Marc software. Die, blank holder, and punch parts were modelled as rigid bodies. In order to decrease analysis time, the FE model was created as one fourth model, using symmetry boundary conditions. Contact regions were defined with the surface-to-surface algorithm.The friction coefficient was determined as 0.1. Extrusion direction was defined on the blank. The punch moves into the die with a punch force of 4000 N and a displacement of 25 mm. Figure 2 illustrates the FE model.

In the scope of this work, the anisotropic behavior of the sheets was defined with Hill48 yield criterion. The hill criterion is given in Eq. 1, where F, G, H, L, M, and N are constants calculated from Eq. 2.

$$2f = F\left(\sigma_{yy} - \sigma_{zz}\right)^2 + G(\sigma_{zz} - \sigma_{xx})^2 + H\left(\sigma_{xx} - \sigma_{yy}\right)^2 \\ + 2L\tau_{yz}^2 + 2M\tau_{zx}^2 + 2N\tau_{xy}^2 = 1 \tag{1}$$

Anisotropy parameters were calculated according to yield stress ratios. Equation 2 indicates the yield stress ratios.

$$F = \frac{1}{2}\left(\frac{1}{R_{22}^2} + \frac{1}{R_{33}^2} - \frac{1}{R_{11}^2}\right), G = \frac{1}{2}\left(\frac{1}{R_{11}^2} + \frac{1}{R_{33}^2} - \frac{1}{R_{22}^2}\right), H \\ = \frac{1}{2}\left(\frac{1}{R_{11}^2} + \frac{1}{R_{22}^2} - \frac{1}{R_{33}^2}\right), L = \frac{3}{2R_{23}^2}, M = \frac{3}{2R_{31}^2}, N \\ = \frac{3}{2R_{12}^2} \tag{2}$$

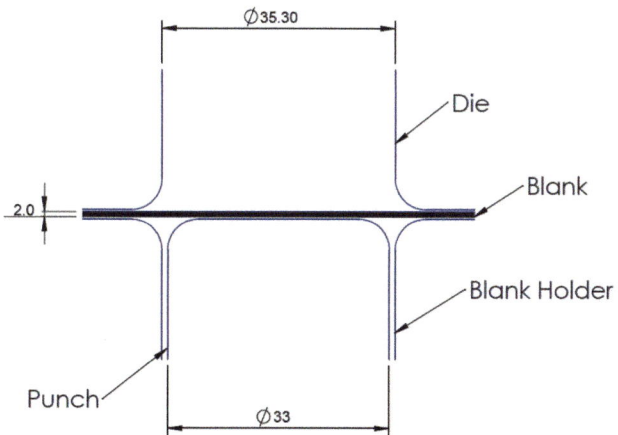

Fig. 1 Tool and blank dimensions

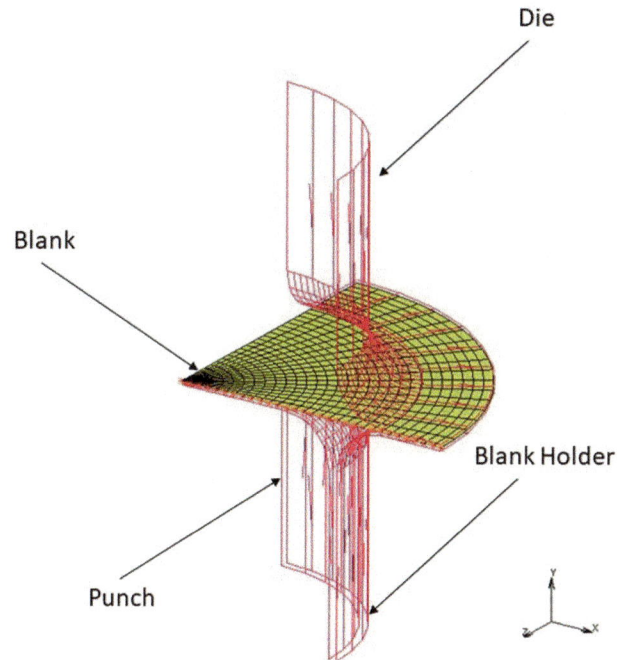

Fig. 2 FE Model, blank with punch, blank holder, and die

Material Properties

Aluminum 6005 and 6082 alloys were chosen as test materials. They were manufactured with aluminum extrusion method. Different heat treatments were applied to the test materials. For 6005 alloys, heat treatment operation was applied at 200 °C for 2 h. On the other hand, for 6082 material, the heat treatment processes were applied with two different methods at 200 °C 2 h and 200 °C 4 h. Several tensile test specimens were prepared at 0°, 45°, and 90° according to the extrusion direction. Material mechanical properties were obtained with several tensile tests. The mechanical properties of test materials were listed in Table 1.

Yield stress ratios of test materials in different directions were shown in Table 2.

Lankford parameters of material is given Table 3.

Results

The earing heights were measured from the FE simulations results at 0°, 45°, and 90° of the extrusion direction of the blank. Table 4 shows those results.

Table 1 Mechanical properties of test materials according to the extrusion direction

Material	Extrusion direction [deg]	σ_y [MPa]	σ_{UTS} [MPa]	Elongation [%]
6005–200° 2H	0°	266	292	9.2
	45°	227	243	7.8
	90°	215	242	5.3
6082–200° 2H	0°	276	302	8
	45°	250	285	12.6
	90°	281	305	7.6
6082–200° 4H	0°	292	317	11.9
	45°	253	278	13.4
	90°	293	309	3.9

Table 2 Mechanical properties of materials

Material	σ_0/σ_0	σ_{45}/σ_0	σ_{90}/σ_0	r0	r45	r90
6005–200°–2H	1	0.853383	0.808271	0.565886	0.467764	0.808255
6082–200°–2H	1	0.905797	1.018116	0.737756	0.486669	0.354521
6082–200°–4H	1	0.866438	1.003425	0.480936	0.749554	0.434471

Table 3 Lankford parameters

Material	F	G	H	L	M	N
6005–200°–2H	0.4471162	0.638616	0.361384	1.5	1.5	1.050733
6082–200°–2H	1.1975167	0.575455	0.424545	1.5	1.5	1.749337
6082–200°–4H	0.7474642	0.675248	0.324752	1.5	1.5	1.777756

Table 4 Earing amount results of FE analysis

Material	Extrusion direction [deg]	Earing amount [mm]
6005–200°–2H	0°	20.45
	45°	16.82
	90°	20.31
6082–200°–2H	0°	21.68
	45°	22.85
	90°	21.82
6082–200°–4H	0°	19
	45°	20.40
	90°	19.15

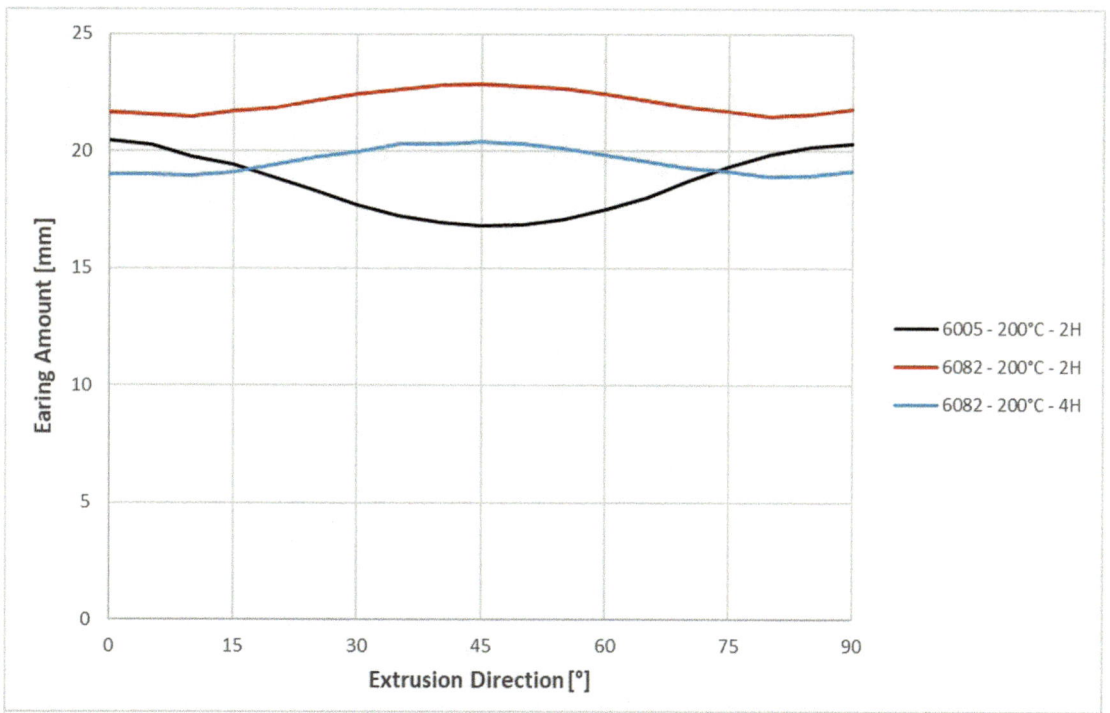

Fig. 3 Earing profile of test materials—FE results

Earing profile according to FE results is given in Fig. 3. It is clearly seen that different alloy compositions result in different anisotropic behavior. For the 6005 aluminum alloy, ear shape shows different orientations than for the 6082 aluminum alloy. When it comes to the behavior of the same alloy with different heat treatment processes, the earing profile was seen in the same direction but the amount was different.

Fig. 4 Illustrates the results of FE cup drawing tests.

Conclusion

In the present study, the anisotropic behavior of different aluminum alloys with various heat treatments was investigated. FE simulations were performed with material data obtained from tensile tests. Hill48 yield criteria was used as yield criterion in the FE studies in order to define material behavior.

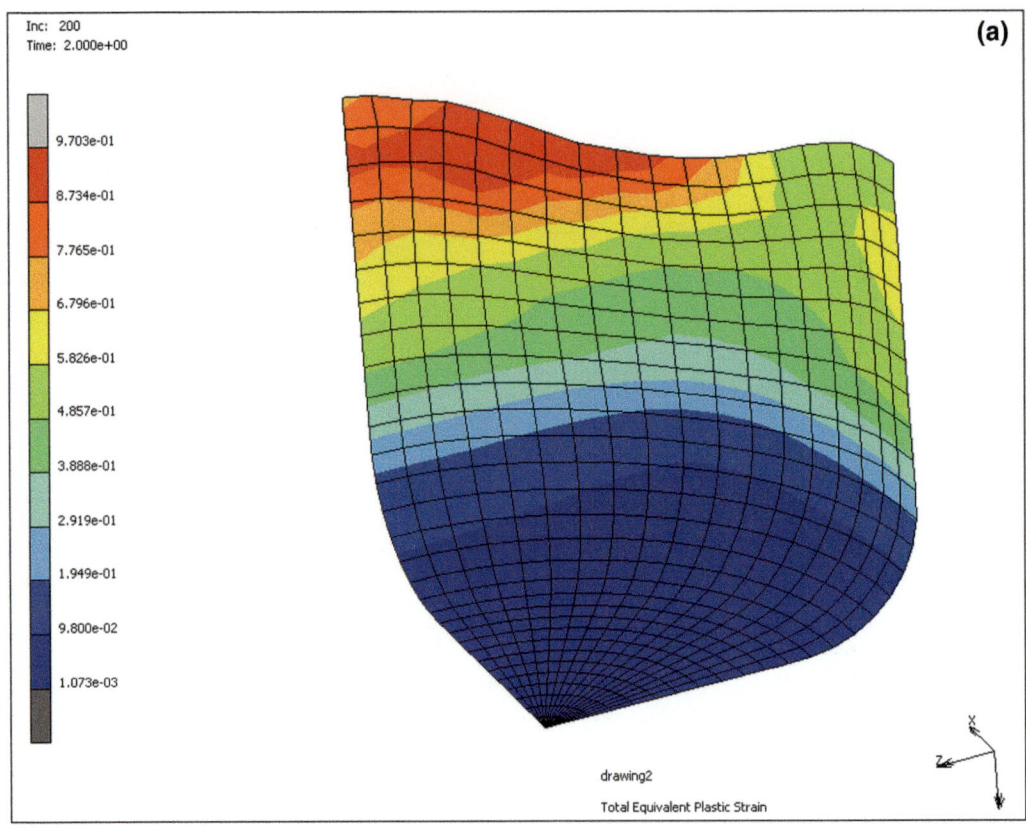

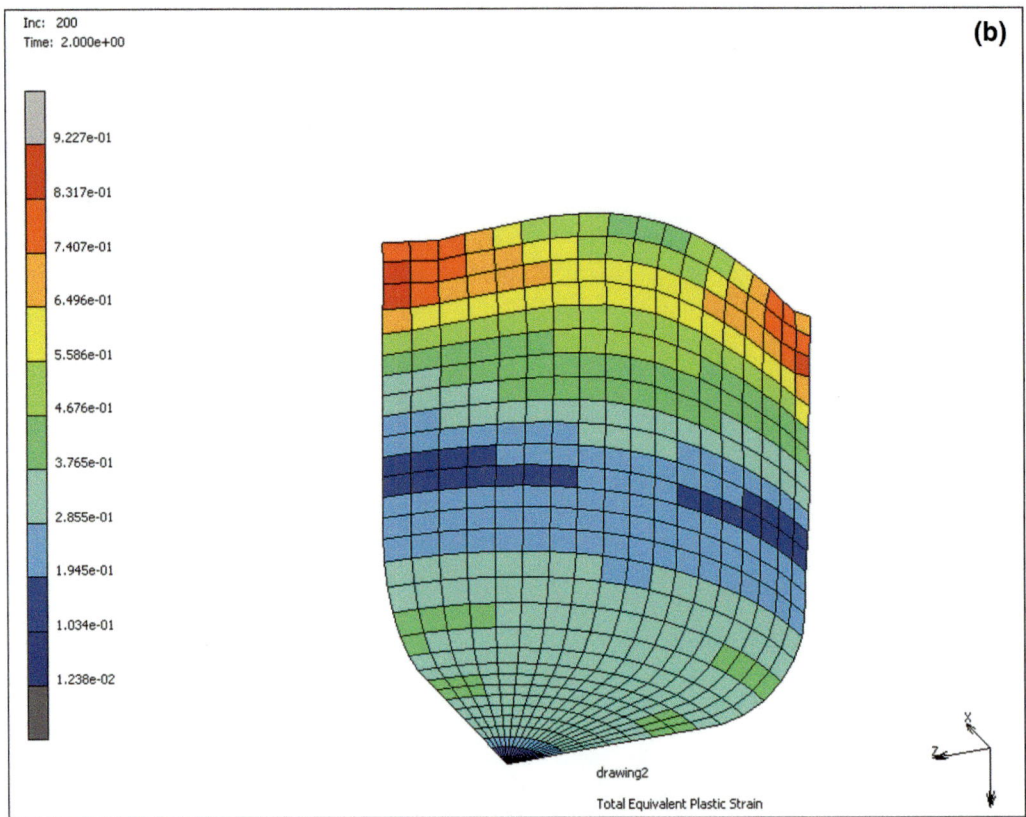

Fig. 4 FE results **a** 6005–200 °C – 2H **b** 6082–200 °C-2H **c** 6082–200 °C-4H

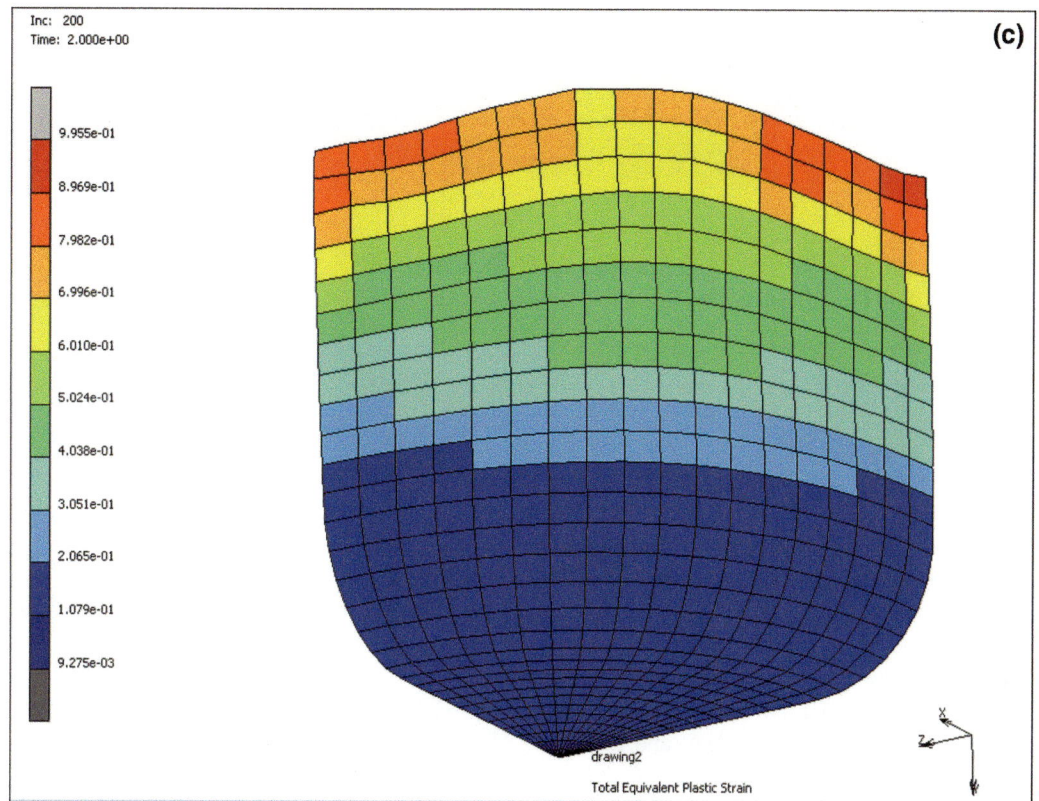

Inc: 200
Time: 2.000e+00

(c)

9.955e-01
8.969e-01
7.982e-01
6.996e-01
6.010e-01
5.024e-01
4.038e-01
3.051e-01
2.065e-01
1.079e-01
9.275e-03

drawing2

Total Equivalent Plastic Strain

Fig. 4 (continued)

As a result of the study, the following results were concluded:

- Earing profile directions are different with respect to alloy composition. Minimum earing amounts were observed at 45° for 6005 material and 90° for 6082 materials. This is an unexpected result. For different alloy compositions, earing profiles are expected in the same direction. The reason for this difference might be the discontinuities in tensile test materials.

- Heat treatment process changes the material's anisotropic behavior. For the 6082 test material, two different heat treatment processes were applied at 2 h and 4 h at 200 °C after the extrusion process. Earing amount of 2 h specimen is 1.17 mm, while it is 1.40 mm for 4 h specimen.

- For further studies, material models will be investigated with different methods to make certain of results. For further studies, different material models such as Barlat-, Yld2004, etc. will be used to compare and verify the results.

References

1. Hill, R. (1948). A theory of the yielding and plastic flow of anisotropic metals. Proceedings of the Royal Society of London Series A Mathematical and Physical Sciences, 193: 281-297.
2. Özçetin Y (2022) Investigation of the deep drawing capability of 3005 aluminum alloy plates. MSc thesis, Yıldız Technical University.
3. Aminzadeh, A., Karganroudi, S., S., Barka, N. (2021). Experimental and numerical investigation of forming defects and stress analysis in laser-welded blanks during deep drawing process. 17, 1193–1207.
4. Şener, B., Akşen, T. A., Fırat, M. (2021). On the Effect of Through-Thickness Integration for the Blank Thickness and Ear Formation in Cup Drawing FE Analysis. European Mechanical Science. 5 (2) , 51-55.
5. Akşen, T. A., Firat, M. (2021). Blank thinning predictions of an aluminum alloy in hole expansion process using finite element method. SN Applied Sciences, 3(3).
6. Marc Mentat User Help Files, Vol A.

Effect of Al-3Ti-1B-1.5Ce Refiner on Microstructure and Mechanical Properties of A356 Aluminum Alloy

Da Teng, Guangzong Zhang, Shuo Zhang, Junwen Li, Yaodong Zhang, and Renguo Guan

Abstract

A356 alloy is widely used in rail transit and aerospace fields due to its excellent comprehensive properties. The refiner is called "flavor element" in aluminum industry, which plays an important role in improving the alloy microstructure and mechanical properties. In this paper, as-cast Al-3Ti-1B-1.5Ce master alloy was prepared by fluoride salt method. The optimum amount and holding time were obtained by comparing the effects of Al-3Ti-1B-1.5Ce and Al-3Ti-1B on the microstructure and properties of A356 alloy. Results showed that the average grain size in A356 alloy modified by 0.6% Al-3Ti-1B was 153.73 μm, which was much smaller than that of A356 alloy with an average grain size of 1750 μm. The introduction of Ce element caused a better refining effect of the Al-3Ti-1B-1.5Ce master alloy. When the addition of Al-3Ti-1B-1.5Ce master alloy was 0.4%, the average grain size, tensile strength and elongation reached 145.96 μm, 213.9 MPa and 11.86%, respectively.

Keywords

Al-3Ti-1B-1.5Ce refiner • A356 aluminum alloy • Grain refinement • Recession resistance

Introduction

A356 alloy is widely used in machinery, automobile industry and other fields upon the advantages of light weight, low cost and good comprehensive performance [1–3]. However, it is easy to form casting defects such as shrinkage porosity and pores due to the coarse Al grains, which reduces the casting compactness and mechanical properties and results in a low safety usage factor. Grain refinement is one of the most effective methods to improve the alloy strengthening and toughness [4–7]. In order to realize casting alloys with a grain-refined structure, chemical inoculation by adding a certain amount of refiners has become a common method in industrial production. The most widely used grain refiner is Al-Ti-B master alloy, which generates Al_3Ti intermetallic compounds and TiB_2 particles in the aluminum matrix. However, the precipitation and aggregation of TiB_2 particles greatly reduces the refining efficiency [8–10]. Thus, seeking an effective method is urgent to improve the refining effect in order to obtain higher A356 alloy properties.

In this study, rare earth Ce element was added into the Al-Ti-B master alloy, the refining effect of Al-3Ti-1B-1.5Ce on A356 aluminum alloy was studied, the existence mode of rare earth Ce in A356 aluminum alloy was discussed, and the effect mechanism of Ce element was determined [11–13]. And the influence and mechanism of the second phase, based on the tensile strength, elongation and fracture morphology of the refined A356 alloy, were studied to determine the optimal Ce addition. The current work is of great significance for the preparation and application of high-performance rare earth A356 aluminum alloy.

D. Teng · G. Zhang · S. Zhang · J. Li (✉) · Y. Zhang · R. Guan
Key Laboratory of Near-Net Forming of Light Metals of Liaoning Province, Dalian Jiaotong University, Dalian, 116028, China
e-mail: joelee0527@163.com

Engineering Research Center of Continuous Extrusion, Ministry of Education, Dalian Jiaotong University, Dalian, 116028, China

R. Guan
School of Materials Science and Engineering, Northeastern University, Shenyang, 110819, China

© The Minerals, Metals & Materials Society 2023
S. Broek (ed.), *Light Metals 2023*, The Minerals, Metals & Materials Series,
https://doi.org/10.1007/978-3-031-22532-1_80

Experimental Materials and Methods

The experimental procedure included the preparation of Al-3Ti-1B and Al-3Ti-1B-1.5Ce master alloys and the refinement of A356. Raw materials used in the preparation of the master alloys include industrial aluminum with a purity of 99.7% (the compositions are shown in Table 1), fluoride salts and Al-10Ce master alloy. And the fluoride salts contain potassium fluorotitanate (K_2TiF_6) and potassium fluoroborate (KBF_4) with purity higher than 99.5%. For refinement of A356 alloy, C_2Cl_6 was used as the degassing and slag removing agent.

Figure 1 shows the experimental process of A356 refinement. A certain amount of A356 alloy was placed in a dried crucible and heated to 750 °C in a resistance furnace. After adding Al-3Ti-1B or Al-3Ti-1B-1.5Ce master alloy with content of 0.2, 0.4, 0.6wt.%, it was stirred for 5 min to make the melt homogeneous. When the temperature was stabilized at 750 °C, the alloy melt was degassed and poured into the ring mold. Before the mechanical property test, the ingot was subjected to heat treatment, i.e., 540 °C solid solution for 2 h and 175 °C aging for 5 h.

After sampling and grinding, the ingots were etched by mixed acid (70wt.% HCl + 25wt.% HNO₃ + 5wt.% HF) to corrode the bottom surface. The macrostructure of samples was observed by an optical microscope, and the average grain size of the alloy was measured by the truncation method. The phases and microstructure of Al-Ti-B-Ce alloy were analyzed by X-ray diffractometer, optical microscope and scanning electron microscope(SEM, Zeiss SUPRA 55,

GER). And based on the GB/T228-2010 standard, the sample was processed into tensile-specimen dimensions as shown in Fig. 2. The samples of Φ8 × 15 mm were measured on the CMT4204 microelectronics universal test tensile machine, and fracture was observed by SEM.

Result and Discussion

Phase Analysis and Microstructure Comparison of Al-3Ti-1B and Al-3Ti-1B-1.5Ce

Figure 3 shows the XRD patterns of Al-Ti-B and Al-Ti-B-Ce alloys. Compared with Al-Ti-B, the diffraction peaks of $Ti_2Al_{20}Ce$ phase appear in Al-Ti-B-Ce except for the typical peaks of Al, $TiAl_3$ and TiB_2. The formation of $Ti_2Al_{20}Ce$ can be explained as that, after Ti and Al form the $TiAl_3$ phase, the element Ce, which has low solid solubility in the aluminum melt, will react with the $TiAl_3$ to form the $Ti_2Al_{20}Ce$ phase.

Figure 4a and b show the distribution of individual particles in Al-Ti-B and Al-Ti-B-Ce alloys, respectively. It can be seen from Fig. 4a that a large number of blocks and cluster particles are distributed around α-Al matrix in Al-Ti-B. In Fig. 4b, dark block particles exist on the off-white particles. To determine the phase composition, EDS analysis was performed on different particles. Table 2 shows the EDS results of points A to E. Combined with the XRD detection, $TiAl_3$ in Al-Ti-B alloy mainly presents as block, while the $Ti_2Al_{20}Ce$ phase in the Al-Ti-B-Ce alloy is in the form of flakes, about 45 μm. The formation of

Table 1 Chemical compositions of A356 aluminum alloy

Compositions	Al	Si	Mg	Ti + Zr	Fe	Mn	Cu	Zn	Others
Content/ wt.%	92.13	7.30	0.41	0.152	0.15	0.01	0.001	0.01	Balance

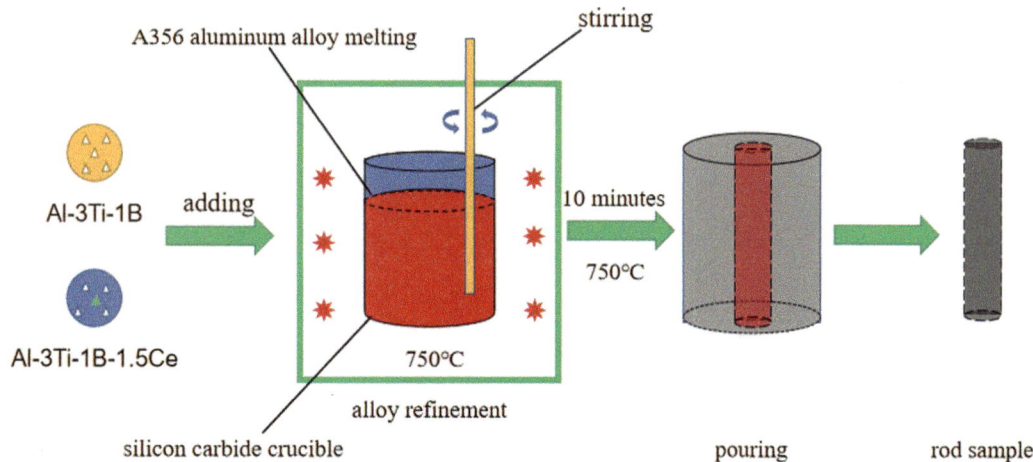

Fig. 1 Experimental process of A356 refinement

Fig. 2 Schematic diagram of
specimens for tensile testing

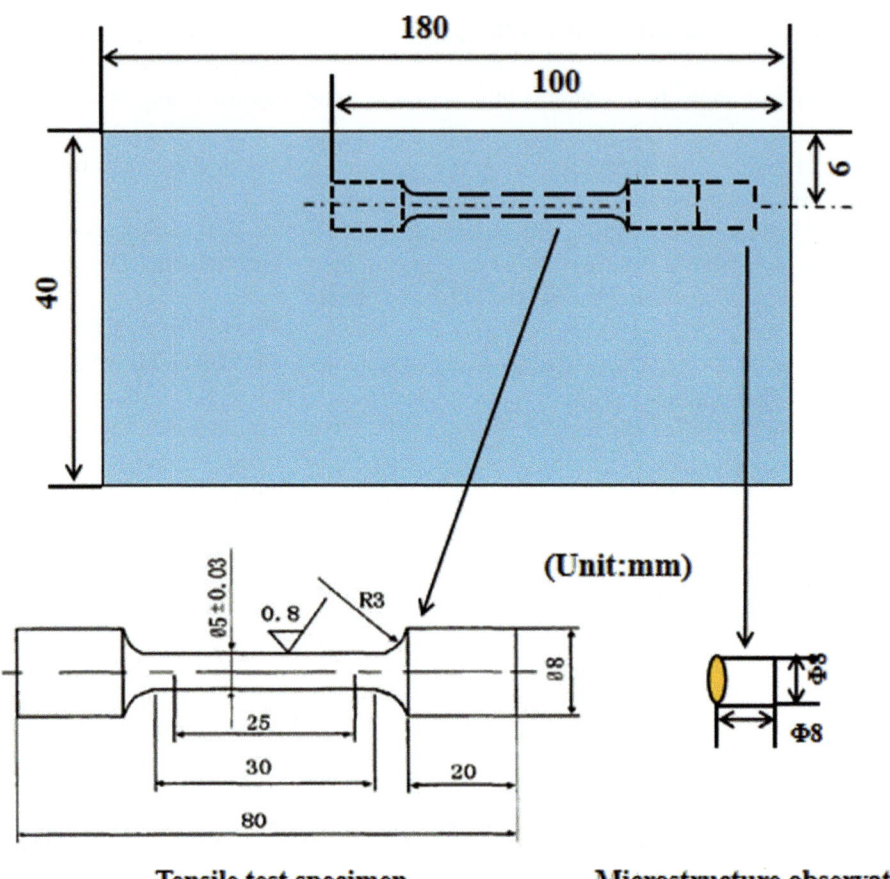

Fig. 3 XRD results of Al-3Ti-1B
and Al-3Ti-1B-1.5Ce alloys

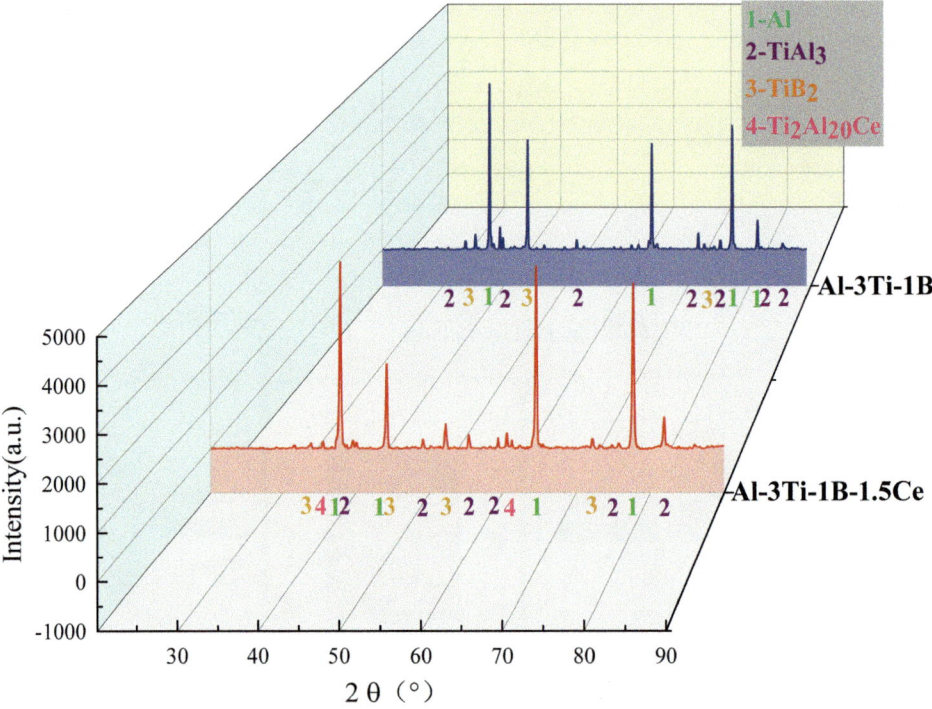

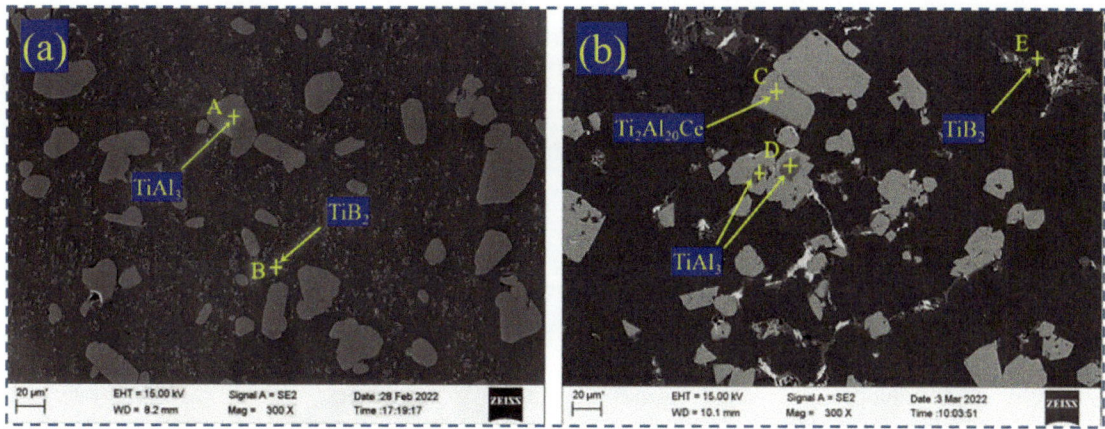

Fig. 4 SEM images of Al-3Ti-1B refiner (**a**) and Al-3Ti-1B-1.5Ce (**b**)

Table 2 EDS results of points A to E

Point	Elements(at.%)				Phases
	Al	Ti	B	Ce	
A	74.41	25.59	0.00	0.00	$TiAl_3$
B	80.67	7.10	12.23	0.00	$TiB_2\backslash TiAl_3$
C	86.15	9.46	0.00	4.39	$Ti_2Al_{20}Ce$
D	74.49	25.51	0.00	0.00	$TiAl_3$
E	34.73	8.18	57.09	0.00	$TiB_2\backslash TiAl_3$

$Ti_2Al_{20}Ce$ reduces the number and size of the remaining $TiAl_3$, the average size of which is less than 20 μm. This is because the presence of rare earth elements in Al-Ti-B-Ce prevents the formation of needle-like $TiAl_3$, hinders the growth and improves the distribution of $TiAl_3$.

Effect of Al-3Ti-1B-1.5Ce on the Microstructure of A356 Alloy

Figure 5 shows the alloy ingots obtained by adding different contents of Al-3Ti-1B and Al-3Ti-1B-1.5Ce into A356 alloys. Figure 6 shows the average grain size of α-Al in A356 alloy measured by the cross-section method.

It can be seen from Figs. 5 and 6, the α-Al grains in initial A356 have an average grain size of 1750 μm, which is obtained by truncated grain counting. After adding Al-3Ti-1B, as shown in Fig. 5a–c, the grain refinement has a certain effect, and the average grain size decreased to 174.5, 156.9, and 153.7 μm under contents of 0.2 wt.%, 0.4 wt.% and 0.6 wt.%, respectively. As shown in Fig. 5d–f, when adding different contents of Al-3Ti-1B-1.5Ce, the grain becomes an equiaxed crystal and the average size decreases to 138.6, 133.2 and 136.4 μm, respectively. Furthermore, Al-3Ti-1B-1.5Ce alloy with addition of 0.4 wt.% has the best grain refining effect, even better than Al-3Ti-1B with addition of 0.6 wt.%.

Effect of Al-3Ti-1B-1.5Ce on the Mechanical Properties of A356 Alloy

A356 is a heat-treatable Al–Si–Mg cast alloy. Therefore, its mechanical properties can be effectively improved after heat treatment. Figure 7a shows the mechanical properties of A356 alloy as a function of refiner addition. The tensile strength of A356 alloy is 170 MPa and the elongation is 2.4%. After adding 0.2 wt.% Al-3Ti-1B and Al-3Ti-1B-1.5Ce master alloy, the tensile strength and elongation reaches 193.94 MPa and 8.5%, and 205.61 MPa and 10.65%, respectively. When the addition is 0.4 wt.%, the tensile strength and elongation reaches 195.68 MPa and 9.72%, and 218.42 MPa and 13.42% respectively, much higher than the former. When the addition of Al-3Ti-1B and Al-3Ti-1B-1.5Ce alloys is 0.6 wt.%, the tensile strength and elongation reach 197.12 MPa, 9.92%, and 213.65 MPa and 13.33%, respectively. Therefore, Al-3Ti-1B-1.5Ce has a better refining effect than Al-3Ti-1B, and the tensile strength and elongation obtain the maximum under Al-3Ti-1B-1.5Ce addition of 0.4 wt.%.

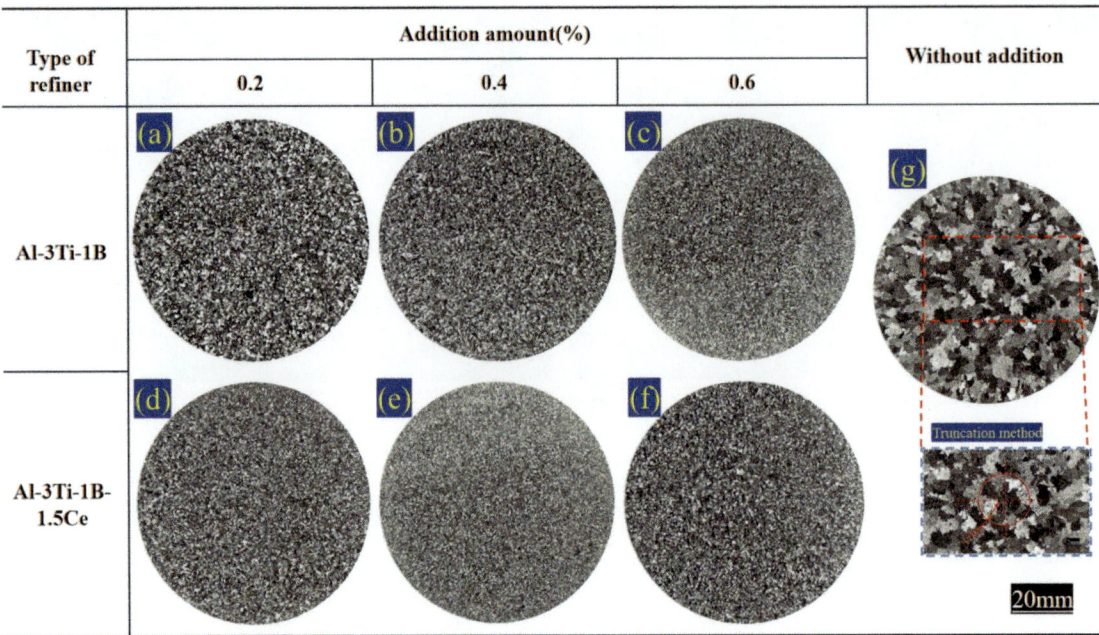

Fig. 5 Macrostructures of A356 alloy under different Al-3Ti-1B and Al-3Ti-1B-1.5Ce additions **a, d** 0.2 wt.%; **b, e** 0.4 wt.%; **c, f** 0.6wt.%; **g** 0 wt. %

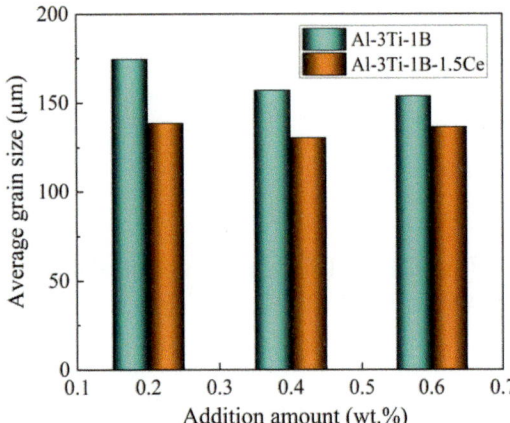

Fig. 6 Average grain size of A356 alloy under different refiner additions

Al-3Ti-1B-1.5Ce Refinement Mechanism Analysis of A356 Alloy

Influence Mechanism of Ce on Grain Refinement

Influence of rare earth element Ce on grain refinement is divided into two aspects. On one hand, elemental Ce can improve the phase compositions and the microstructure of Al-Ti-B-Ce refiner. On the other hand, elemental Ce can improve the grain structure of the aluminum alloy. As a surface-active material, elemental Ce is easy to be adsorbed and separated at the grain boundary and the phase interface

during grain growth, also, it can fill the defects on grain surface. Grain growth is greatly hindered and grain size is thus refined. During the crystallization process, elemental Ce tends to accumulate on the surface and edge of the grain boundary, which hinders the contact of the crystal nucleus with the liquid and residual impurities, and reduces the contact area. As shown in formula (1), due to the presence of elemental Ce, the contact areas S_1 and S_2 decrease. During the grain growth, the surface energy $\sigma_1 S_1 + \sigma_2 S_2$ decreases, and the nucleation work ΔF_k also decreases. The reduction of contact area is thereby beneficial to reduce the grain nucleation work, i.e., it is easier to promote grain nucleation, increase the nucleation rate, and promote grain refinement.

$$F_k = \frac{(\sigma_1 s_1 + \sigma_2 s_2)}{3}$$

where, F_k is nucleation work; σ_1 is specific surface energy between crystal nucleus and impurities, σ_2: specific surface energy between crystal nucleus and liquid, S_1 is the contact area between crystal nucleus and impurities and S_2 is the contact area between crystal nucleus and liquid.

Effect of Ce Addition on Fracture Morphology

In order to better reveal the relationship between the mechanical properties and microstructure of A356 alloy, the fracture of tensile specimens with Al-3Ti-1B and Al-3Ti-1B-1.5Ce under 0.4 wt.% addition was analyzed,

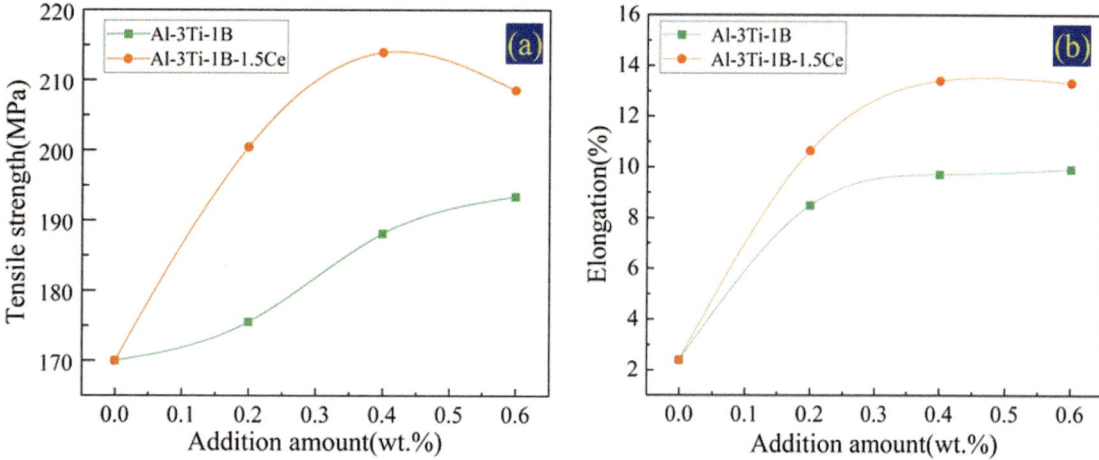

Fig. 7 Mechanical properties of A356 alloy under different refiner additions. **a** Tensile strength; **b** Elongation

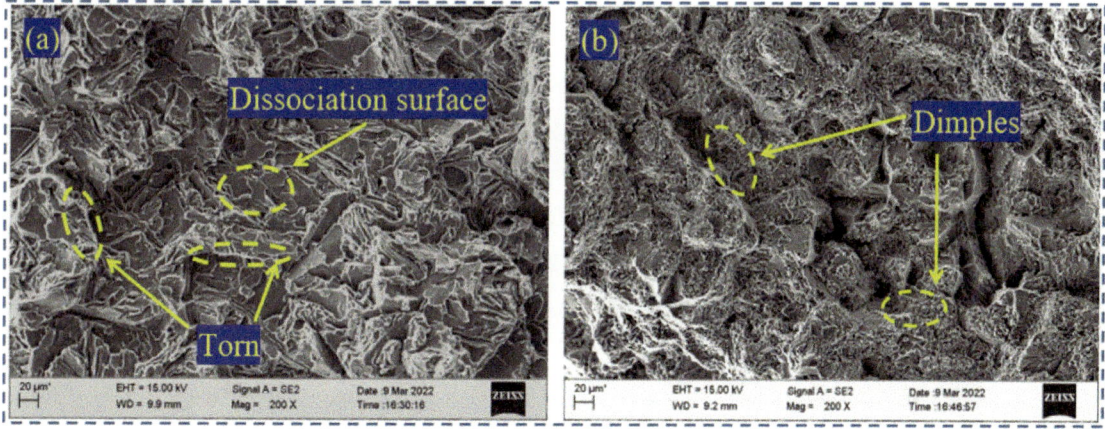

Fig. 8 Fractures of the tensile A356 alloys under different 0.4 wt.% refiner additions. **a** Al-3Ti-1B; **b** Al-3Ti-1B-1.5Ce

respectively. Figure 8a shows the fracture of A356 alloy with 0.4 wt.% Al-3Ti-1B refiner. It can be seen that the fracture has a certain quasi-dissociation plane, which is a typical brittle fracture characteristic. This is because A356 alloy consists of coarse α-Al dendrites and eutectic silicon flakes, leading to premature cracking caused upon tensile stress concentrations.

It can be seen from Fig. 8b that the introduction of elemental Ce causes a significant change in cross-sectional morphology, i.e., the number of uniform and round dimples increases and the number of tear ridges decreases. A356 alloy exhibits better strength and toughness, which signifies the refining effect of 0.4 wt.% Al-3Ti-1B-1.5Ce alloy addition. Fracture investigation is consistent with the results of the microstructure and mechanical property tests.

Conclusion

(1) Al-3Ti-1B-1.5Ce master alloy can effectively refine A356 alloy and improve its microstructure, and its effect is better than that of Al-3Ti-1B. When the addition of Al-3Ti-1B-1.5Ce alloy is 0.4 wt.%, the average grain size of α-Al in A356 alloy is 145.96 μm, and the refining effect reaches the maximum.

(2) Al-3Ti-1B-1.5Ce can significantly improve the mechanical properties of A356 aluminum alloy. When the addition is 0.4 wt.%, the tensile strength and elongation of A356 alloy is 218.42 MPa and 13.42%, respectively, which is 11.62% and 3.7% higher than that of Al-3Ti-1B.

Acknowledgements This work was supported by the National Key Research and Development Program of China [grant No. 2018YFB2001800], National Natural Science Foundation of China [grant No. 51871184] and Dalian High-level Talents Innovation Support Program [grant No. 2021RD06].

References

1. Fan S B, Peng J, Zhou M (2014) Effect of Ce Addition on Microstructure and Mechanical Properties of ZM21 Magnesium Alloys[J]. Materials Science Forum 788: 58-63.
2. Mondal D P, Jha N, Badkul A (2011) Effect of calcium addition on the microstructure and compressive deformation behaviour of 7178 aluminum alloy[J]. Materials & Design, 32(5): 2803-281.
3. Hao S W (2009). Application of density measurement used in A356 aluminum alloy[J]. Light Alloy Fabrication Technology.
4. Masuda T, Sauvage X, Hirosawa S (2020) Achieving highly strengthened Al-Cu-Mg alloy by grain refinement and grain boundary segregation[J]. Materials Science and Engineering: A 793
5. Li H, Lin B, Xu R(2019) Enhanced mechanical properties of Al-Si-Cu-Mn-Fe alloys at elevated temperatures through grain refinement and dispersoid strengthening[J]. Materials Science and Technology, (4):1–13.2020, 793:139668.
6. Xxa B, Yfa B, Hui F C(2019) The grain refinement of 1070 alloy by different Al-Ti-B mater alloys and its influence on the electrical conductivity. Results in Physics 14.
7. Ding W, Zhao X, Chen T (2020) Effect of rare earth Y and Al–Ti–B master alloy on the microstructure and mechanical properties of 6063 aluminum alloy[J]. Journal of Alloys and Compounds,830: 154685.
8. Khaliq A, Mehmood S, Ranjha S A (2018) Microscopic Analysis of TiB2 Formation Mechanism in Al-Ti-B Alloy[J]. Microscopy and Microanalysis, 24(S1): 2262-2263.
9. Chuaypradit S (2018). Quantifying the effects of grain refiner addition on the solidification of Fe-Rich intermetallics in Al–Si–Cu Alloys using in situ synchrotron X-Ray Tomography. Light Metals 2018: 1067-1073.
10. Birol, Y (2008) Production of Al–Ti–B grain refining master alloys from Na_2B4O_7 and K_2TiF_6 Journal of Alloys and Compounds. 458 (1–2): 271–276.
11. Lee M S, Grieveson P (2003) The production of Al-Ti-B grain refining master alloys[J]. Scandinavian Journal of Metallurgy, 32 (6): 769-772.
12. Li PT, Li YG, Nie JF (2012) Influence of forming process on three-dimensional morphology of TiB_2 particles in Al-Ti-B alloys [J]. Transactions of Nonferrous Metals Society of China 22(3): 564-570.
13. Birol Y (2007) Production of Al-Ti-B grain refining master alloys from B_2O_3 and K_2TiF_6 [J]. Journal of Alloys and Compounds 443 (1-2): 94-98.

Effect of Al-Ti-B Refiner on Microstructure and Properties of A356 Alloy by Continuous Rheo-Extrusion

Shuo Zhang, Guangzong Zhang, Da Teng, Junwen Li, and Renguo Guan

Abstract

In this paper, an Al-5Ti-1B grain refiner for continuous rheo-extrusion forming and its effect on the microstructure and properties of A356 alloy is investigated. Phase and microstructure of Al-5Ti-1B were examined by XRD, SEM, and EDS. Furthermore, the effect of Al-5Ti-1B on the microstructure and properties of the A356 alloy was studied. Results showed that the continuous rheo-extrusion can break the $TiAl_3$ phase in Al-5Ti-1B alloy and promote the TiB_2 particle to distribute uniformly. When the content of Al-5Ti-1B is 0.8 wt%, the refining effect was best. The grain size can reach 170 μm and A356 alloy obtained the optimum mechanical properties. The tensile strength, yield strength, and elongation were 290 MPa, 237 MPa, and 8.5%, respectively.

Keywords

Continuous rheo-extrusion • Al-5Ti-1B • Grain refinement • A356 alloy

Introduction

A356 alloy as a hypoeutectic Al-Si alloy is widely used in aerospace, transportation and engineering equipment due to its good casting, machinability and corrosion resistance [1–3]. However, the primary α-Al phase in A356 alloy tends to grow coarsely during the casting, which seriously affects the alloy's mechanical properties [4]. To avoid this undesirable phenomenon, A356 alloy needs to be refined by rapid solidification [5], magnetic field solidification [6], ultrasonic solidification [7] and grain refiner addition [8]. Among these methods, adding grain refiner is the most economical and effective [9].

Grain refiner can effectively improve the grain morphology and size, and improve the mechanical properties of the alloy [10]. At present, the widely used commercial refiner is the Al-Ti-B master alloy, which induces the generation of $TiAl_3$ and TiB_2 particles [11, 12]. In particle theory, $TiAl_3$ and TiB_2 particles act as heterogeneous nucleation cores of α-Al, promoting nucleation of aluminum melt and refining the alloy grains. However, the phenomenon that TiB_2 particles alone have no obvious refining effect on pure aluminum cannot be explained [13]. In peritectic theory, a diffusion zone containing rich Ti elements around $TiAl_3$ exists. When the melt temperature is 665 °C and the Ti element concentration reaches 0.15wt.%, the peritectic phenomenon (L + $TiAl_3$ → α-Al) occurs [14]. This theory can explain the reason why $TiAl_3$ particles refine aluminum alloy, but it cannot explain the contribution of TiB_2. According to the complex phase nucleation theory by Schumacher et al., when Ti element content in aluminum melt exceeds 0.15 wt.%, a layer of $TiAl_3$ phase will be formed on the surface of TiB_2 [15, 16]. Fan et al. obtained TiB_2 particles embedded in the amorphous aluminum matrix by quenching molten Al-5Ti-1B refiner under liquid nitrogen [13]. It was found that there was a $TiAl_3$ layer in nanometers on the surface of TiB_2 particles. Compared with the former two nucleation theories, the complex nucleation theory not only affirms the role of particles but also the role of redundant Ti elements. It is the most comprehensive and convincing nucleation theory at present.

Some studies have shown that [17, 18], the size and distribution of the secondary phase ($TiAl_3$ and TiB_2

S. Zhang · G. Zhang (✉) · D. Teng · J. Li · R. Guan
Key Laboratory of Near-Net Forming of Light Metals of Liaoning Province, Dalian Jiaotong University, Dalian, 116028, China
e-mail: gzzhang@djtu.edu.cn

Engineering Research Center of Continuous Extrusion, Ministry of Education, Dalian Jiaotong University, Dalian, 116028, China

R. Guan
School of Materials Science and Engineering, Northeastern University, Shenyang, 110819, China

© The Minerals, Metals & Materials Society 2023
S. Broek (ed.), *Light Metals 2023*, The Minerals, Metals & Materials Series,
https://doi.org/10.1007/978-3-031-22532-1_81

particles) in master alloy are closely related to its refining effect. When the secondary phase size is small and has a dispersion distribution, the refining effect is the best. Based on hot rolling and heat treatment of Al-5Ti-1B master alloy, Venkateswarlu et al. reduced the phase size of TiAl$_3$ and increased the phase number to improve the refining effect [19]. However, the process is complex and the cost is high, which is not suitable for the continuous production of the refiner. Continuous rheo-extrusion is an advanced short process technology that can realize one-step forming from liquid to solid. According to the previous research, Wang et al. used continuous rheo-extrusion technology to obtain nanosized Al$_{13}$Fe$_4$ phase in Al–Fe alloy [20]. Guan et al. significantly refined the Al–Mg alloy grain by continuous rheo-extrusion [21]. However, reports on applying continuous rheo-extrusion technology to produce Al-Ti-B grain refiner with fine refinement performance can rarely be found.

In this paper, Al-5Ti-1B master alloy was prepared by the fluoride salt reaction method and continuous rheo-extrusion technology. The morphology and size of the alloy phase were analyzed by SEM and EDS. The refining effect of Al-5Ti-1B master alloy with different additions (0, 0.2, 0.4, 0.6, 0.8, 1.0, and 1.2 wt.%) on A356 alloy was investigated, and the optimal dosage was obtained.

Experiment

Material Preparation

In this experiment, Al with purity higher than 99.99% and K$_2$TiF$_6$ and KBF$_4$ with purity of 99% were used as main raw materials to prepare Al-5Ti-1B alloy melt, which was prepared by continuous rheo-extrusion processing. Firstly, the pure aluminum was cut into pieces, subjected to surface treatment, and dried in a dryer at 150~200 °C for 100 min. After aluminum ingot was heated to 750~900 °C, K$_2$TiF$_6$ and KBF$_4$ mixed salt were added. The temperature was kept at 750~1100 °C for 60~150 min. After the complete reaction and degassing treatment, the slag was removed, and the alloy melt was obtained for continuous rheo-extrusion processing. Secondly, the melt was poured into the guide groove of the continuous rheo-extrusion equipment shown as in Fig. 1a. The pouring temperature was 860 °C and the roller speed was 6 rad/min. The alloy wire in Φ9.5 mm was acquired by this way, waiting for the refining experiment. Thirdly, A356 alloy was melted in a resistance furnace at 740 °C. Graphite crucible was used, and 0, 0.2, 0.4, 0.6, 0.8, 1.0 and 1.2 wt.% Al-5Ti-1B alloy wire was added into A356 melt, respectively. After stirring for 1 min, heat preservation for 10 min and degassing treatment by C$_2$Cl$_6$, the melt was poured into the steel casting mold for natural cooling, as shown in Fig. 1b and c.

Microstructure and Mechanical Property Characterizations

Phases in continuous rheo-extruded Al-5Ti-1B were analyzed by X-ray diffractometer (XRD, Panalytical Empyrean, UK). Patterns were obtained at 40 kV and 40 mA with Cu-K α radiation, and the data were collected in the 2-theta range from 10° to 90°. The scanning step was 0.02°, and the scanning speed was 10°/min. Microstructure of Al-5Ti-1B

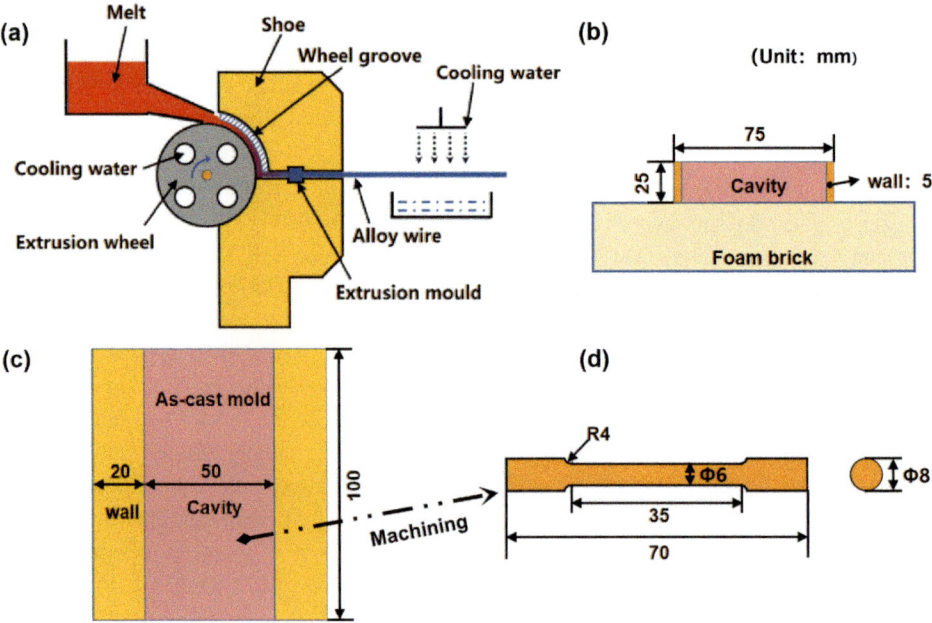

Fig. 1 Schematic diagram of continuous rheo-extrusion equipment **b** annular mold, **c** casting mold, **d** tensile specimen size

alloy was analyzed by scanning electron microscope (SEM, Zeiss supra-55, GER) equipped with energy spectrum analyzer (EDS). After the ingot from annular mold in Fig. 1b was grinded, the mixed acid solution (5 wt.%HF, 25 wt.% HNO_3 and 70 wt.%HCl) was used to uniformly corrode the sample. For the mechanical property, ingot from casting mold in Fig. 1c was solution treated at 535 °C for 3 h and aging treated at 170 °C for 4 h, then the tensile specimen shown in Fig. 1d was tested by a universal testing machine (Shimadzu AG-IC 100 kN, JPN), the tensile strength, yield strength and elongation were included. Each sample was tested at least three times to guarantee the accuracy and repeatability. The tensile rate was 1 mm/min, and the average value was taken as the test value.

Results and Discussion

Microstructure of Al-5Ti-1B Master Alloy in Continuous Rheo-Extrusion

Figure 2 shows the XRD patterns of the continuous rheo-extruded Al-5Ti-1B master alloy. It can be seen that the phases in Al-5Ti-1B mainly includes Al matrix, $TiAl_3$, and TiB_2. Figure 3 shows the SEM and EDS results of the continuous rheo-extruded Al-5Ti-1B alloy. Based on Fig. 3a, the second phases of Al-5Ti-1B master alloy are presented as black and white blocks or cluster particles. From the EDS result in Fig. 3d, the main elements of the black-and-white blocks are Al and Ti, and the atomic ratio of the two elements is about 3:1. In combination with the XRD patterns, the block is identified as $TiAl_3$. Similarly, the particles in Fig. 3a are determined as TiB_2. Moreover, it can be seen that $TiAl_3$ in the alloy has an obvious brittle crack and fracture separation, and the overall distribution of TiB_2

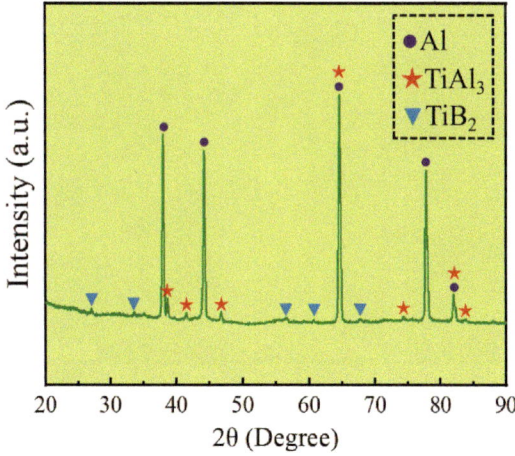

Fig. 2 XRD pattern of continuous rheo-extruded Al-5Ti-1B master alloy

particles is dispersed except for a small amount of aggregation. This should be attributed to the extrusion effect of continuous rheo-extrusion. When the Al-5Ti-1B alloy melt enters the extrusion cavity from the guide groove, the shear stress of the extrusion groove, the angle shear stress of the metal flow path and the extrusion stress of the extrusion die active $TiAl_3$ and TiB_2. The size of refiners reduces, the number increases and the distribution becomes more even, facilitating to improve the refining effect of grain refiners. In this work, the size of $TiAl_3$ and TiB_2 in continuous rheo-extruded Al-5Ti-1B master alloy varies in the range of $20 \sim 50$ μm and $0.5 \sim 1.2$ μm, respectively.

Refining Effect of Master Alloy

Figure 4 shows the macrostructure of A356 alloy under different rheo-extruded Al-5Ti-1B alloy additions, where Fig. 4a–g represents the condition of 0 wt.%, 0.2 wt.%, 0.4 wt.%, 0.6 wt.%, 0.8 wt.%, 1.0 wt.%, and 1.2 wt.%, respectively. It can be seen from Fig. 4a, the grain size of α-Al in initial A356 alloy is coarse, with an average grain size of about 1140 μm, and the macroscopic grain morphology is "irregular quadrilateral". After refining, the grain size of α-Al is reduced, and the macro grain morphology is fine "round", as shown in Fig. 4b–g. This indicates the grain of α-Al has changed from the original strong columnar crystal to fine equiaxed crystal. Figure 5 shows the grain size evolution of α-Al in A356 alloys. With the increase of Al-5Ti-1B addition, the grain size of α-Al varies non-monotonically between 170 and 311 μm. When the content of Al-5Ti-1B is 0.8 wt.%, the average grain size of α-Al obtains the smallest value of 170 μm. As the content further increases, the grain size of α-Al tends to increase. The reason should be explained as that the increasing Al-5Ti-1B content results in excessive $TiAl_3$ and TiB_2 particles in A356 alloy melt, thus the particle aggregation and precipitation cause the number of particles participating in heterogeneous nucleation to decrease. Consequently, the refining effect is affected, and the grain size of α-Al increases.

Mechanical Properties of A356 Alloy

Since the grain size of alloy directly affects its mechanical properties, the mechanical properties of the A356 alloy were studied under different rheo-extruded Al-5Ti-1B alloy additions. The ultimate tensile strength (UTS), yield strength (YS), and elongation were measured, respectively.

Figure 6 shows the mechanical properties of A356 alloy under different rheo-extruded Al-5Ti-1B additions, where Fig. 6a shows the change of UTS and YS and Fig. 6b shows

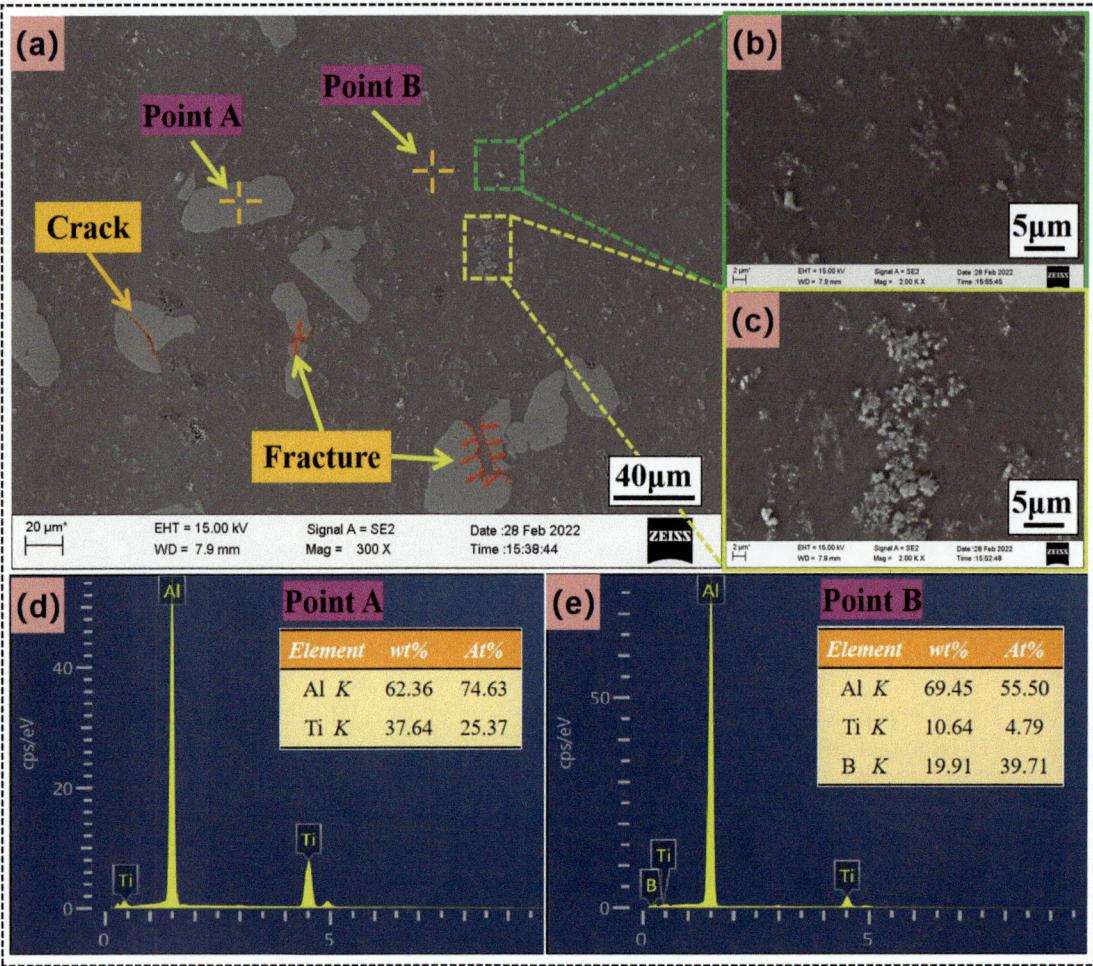

Fig. 3 SEM image and EDS results of continuous rheo-extruded Al-5Ti-1B master alloy

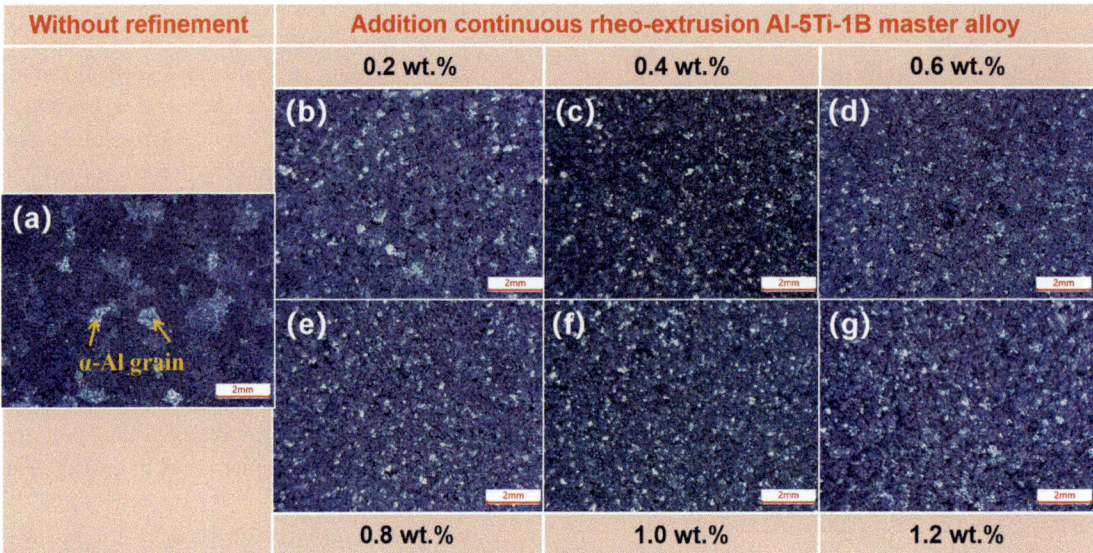

Fig. 4 Macrostructure of A356 alloy under different rheo-extruded Al-5Ti-1B alloy additions **a** 0 wt.%; **b** 0.2 wt.%; **c** 0.4 wt.%; **d** 0.6 wt.%; **e** 0.8 wt.%; **f** 1.0 wt.%; **g** 1.2 wt.%

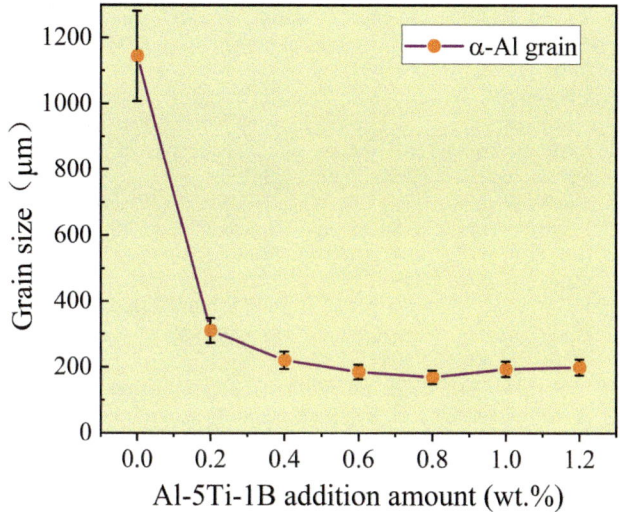

Fig. 5 Grain size evolution of α-Al in A356 alloys

elongation, respectively. It can be seen that the UTS and YS of the initial A356 alloy is 271 MPa and 210 MPa, respectively. After adding 0.2 wt.% continuous rheo-extruded Al-5Ti-1B, the UTS reaches 280 MPa, and the YS increases to 231 MPa. Further increasing Al-5Ti-1B addition, the UTS and YS of the A356 alloy are improved. When the content of Al-5Ti-1B is 0.8 wt.%, A356 alloy obtains the optimum UTS and YS of 290 MPa and 237 MPa, respectively, 7.0 and 12.8% higher than that of the initial A356 alloy. However, as the content of Al-5Ti-1B exceeds 0.8 wt.%, the UTS and YS of A356 alloy decrease. Figure 6b shows the change of elongation of A356 alloy. It can be seen that the elongation of initial A356 alloy is only 4.7%. After adding 0.2, 0.4, 0.6, 0.8, 1.0 and 1.2 wt.% continuous rheo-extruded Al-5Ti-1B, the elongation increases to different degrees.

Also, when the content of Al-5Ti-1B alloy is 0.8 wt.%, the ductility of refined A356 alloy is the best, and the elongation reaches 8.5%, 80.8% higher than that of the initial one.

Conclusion

In this paper, Al-5Ti-1B master alloy wire was prepared by fluoride salt reaction method and continuous rheo-extrusion. The microstructure and the refining effect of rheo-extruded Al-5Ti-1B on A356 alloy were investigated. Following conclusions are drawn:

(1) Microstructure investigation indicated that the rheo-extruded Al-5Ti-1B master alloy contains $TiAl_3$ and TiB_2. After the continuous rheo-extrusion process, the size of refiners reduced, the number increased and the distribution become more even, facilitating to improve the refining effect of grain refiners. The size of $TiAl_3$ and TiB_2 in continuous rheo-extruded Al-5Ti-1B master alloy varied in the range of $20\sim50$ μm and $0.5\sim1.2$ μm, respectively.

(2) Refining result showed that the average grain size of α-Al grains in initial A356 alloy was about 1140 μm. The rheo-extruded Al-5Ti-1B master alloy can significantly refine the grains of A356 alloy. When the content of Al-5Ti-1B was 0.8 wt.%, the average grain size of α-Al obtained the smallest value of 170 μm.

(3) Mechanical properties of A356 alloy were significantly improved after continuous rheo-extruded Al-5Ti-1B addition. When the content of Al-5Ti-1B was 0.8 wt.%, the optimum UTS, YS and elongation were 290 MPa, 237 MPa and 8.5%, respectively, 7.0, 12.8 and 80.8% higher than that of initial A356 alloy, respectively.

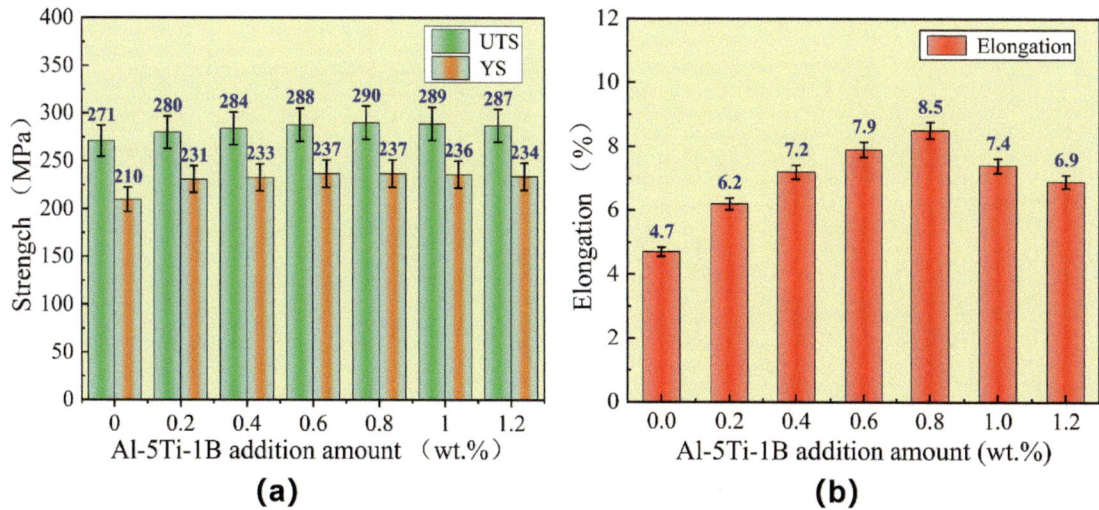

(a)   **(b)**

Fig. 6 Mechanical properties of A356 alloy under different rheo-extruded Al-5Ti-1B additions **a** UTS and YS; **b** Elongation

Acknowledgements This work was supported by the National Key Research and Development Program of China [grant No. 2018YFB2001800], National Natural Science Foundation of China [grant No. 51871184] and Dalian High-level Talents Innovation Support Program [grant No. 2021RD06].

References

1. Zhu M, Jian ZG, Yang GC, Zhou YH (2012) Effects of T6 heat treatment on the microstructure, tensile properties, and fracture behavior of the modified A356 alloys.Materials & Design (1980–2015),36: 243–249.

2. Öztürk İ, Ağaoğlu GH, Erzi E, Dispinar D, Orhan G (2018) Effects of strontium addition on the microstructure and corrosion behavior of A356 aluminum alloy.Journal of Alloys and Compounds,763: 384–391.

3. Li PT, Liu SD, Zhang LL, Liu XF (2013) Grain refinement of A356 alloy by Al–Ti–B–C master alloy and its effect on mechanical properties.Materials & Design,47: 522–528.

4. Zhao CX, Li Y, Xu J, Luo Q, Jiang Y, Xiao QL, Li Q (2021) Enhanced grain refinement of Al-Si alloys by novel Al-V-B refiners.Journal of Materials Science & Technology,94: 104–112.

5. Zhang LY, Jiang YH, Ma Z, Shan S F, Jia YZ, Fan CZ, Wang WK (2008) Effect of cooling rate on solidified microstructure and mechanical properties of aluminium-A356 alloy.Journal of materials processing technology,207(1–3): 107–111.

6. Kang CG, Bae JW, Kim BM (2007) The grain size control of A356 aluminum alloy by horizontal electromagnetic stirring for rheology forging.Journal of Materials Processing Technology,187: 344–348.

7. Zhang SL, Zhao YT, Cheng XN, Chen G, Dai QX (2009) High-energy ultrasonic field effects on the microstructure and mechanical behaviors of A356 alloy.Journal of alloys and compounds,470(1–2): 168–172.

8. Casari D, Merlin M, Garagnani GL (2013) A comparative study on the effects of three commercial Ti–B-based grain refiners on the impact properties of A356 cast aluminium alloy.Journal of Materials Science,48(12): 4365–4377.

9. Zhang DQ, Zhao K, Li DX, Ren L, Liu GL, Liu SD, Liu XF (2022) Microstructure evolution and enhanced mechanical properties in Al-Mn alloy reinforced by B-doped TiC particles. Materials & Design,221: 110906.

10. Tunçay T (2017) The effect of modification and grain refining on the microstructure and mechanical properties of A356 alloy.Acta Phys. Pol. A,131(1): 89-91.

11. Ghadimi H, Nedjhad SH, Eghbali B (2013) Enhanced grain refinement of cast aluminum alloy by thermal and mechanical treatment of Al-5Ti-B master alloy.Transactions of Nonferrous Metals Society of China,23(6): 1563–1569..

12. Qiu K, Wang RC, Peng CQ, Wang NG, Cai ZY, Zhang C (2015) Effects of Mn and Sn on microstructure of Al–7Si–Mg alloy modified by Sr and Al–5Ti–B.Trans. Nonferrous Met. Soc. China,25: 3546-3552.

13. Fan Z, Wang Y, Zhang Y, Qin T, Zhou XR, Thompson GE (2015) Grain refining mechanism in the Al/Al-Ti-B system. Acta Materialia, 84: 292-304.

14. Yuan LU, Chao D, Li YX (2011) Grain refining mechanism of Al-3B master alloy on hypoeutectic Al-Si alloys.Transactions of Nonferrous Metals Society of China,21(7): 1435–1440..

15. Schumacher P, Greer AL (1994) Enhanced heterogeneous nucleation of α-Al in amorphous aluminium alloys.Materials Science and Engineering: A,181: 1335–1339.

16. Schumacher P, Greer AL, Worth J, Evans PV, Kearns MA, Fisher P, Green AH (1998) New studies of nucleation mechanisms in aluminium alloys: implications for grain refinement practice. Materials science and technology,14(5): 394–404.

17. Yu H, Wang N, Guan RG, Tie D, Li Z, An Y, Zhang Y (2018) Evolution of secondary phase particles during deformation of Al-5Ti-1B master alloy and their effect on α-Al grain refinement. Journal of Materials Science & Technology,34(12): 2297–2306.

18. Rokhlin LL, Dobatkina TV, Bochvar NR, Lysova EV (2004) Investigation of phase equilibria in alloys of the Al–Zn–Mg–Cu–Zr–Sc system.Journal of Alloys and Compounds,367(1–2): 10–16.

19. Venkateswarlu K, Murty BS, Chakraborty M (2001) Effect of hot rolling and heat treatment of Al–5Ti–1B master alloy on the grain refining efficiency of aluminium.Materials Science and Engineering: A,301(2): 180–186..

20. Wang X, Guan RG, Tie D, Shang Y Q, Jin HM, Li HC (2018) Microstructural Evolution of Al-1Fe (Weight Percent) Alloy During Accumulative Continuous Extrusion Forming.Metallurgical and Materials Transactions B,49(2): 490-498.

21. Guan RG, Tie D, Li Z, An, YA, Wang X, Li Q, Chen, XB (2018) Microstructure evolution and mechanical property improvement of aluminum alloys with high magnesium content during continuous rheo-extrusion.Materials Science and Engineering: A,738: 31–37.

Effect of Annealing Process on Recrystallization Microstructure and Properties of 1235D Aluminum Alloy Sheet

Wei Tang, Junpeng Pan, Chao Wu, Hongpo Wang, and Zizong Zhu

Abstract

The annealing process during cold rolling significantly affects the mechanical properties of cold-rolled 1235D aluminum alloy sheets and their foil products. The effect of the annealing process on the recrystallized structure and properties of 1235D aluminum alloy cold-rolled sheets was studied. The results show that their tensile strength decreased with the increasing annealing time, from 96 to 89 MPa at 500 °C for 5 to 240 min, the elongation remained unchanged, though. The tensile strength decreased from 93 to 90 MPa with the increasing annealing temperature from 440°C to 560°C, while the elongation also decreased, from 37.4 to 36.8%. The heating and cooling methods of the annealing process have little effect on the tensile strength of cold-rolled sheets after being annealed. In contrast, slow heating and slow cooling are both beneficial for improving their elongation but correspondingly reduced their tensile strength.

Keywords

Aluminum alloy · Annealing · Recrystallization · Tensile strength · Elongation

Introduction

1235 aluminum alloy has a series of advantages such as lightweight, corrosion resistance, low-temperature resistance, airtightness, and suitable cladding. It is widely used in packaging, electromechanics, construction, and electronics field, etc. [1–4]. With the rapid development of aluminum foil production technology, the requirements for the quality and thickness of aluminum foil are gradually increasing. However, the 1235 aluminum foil currently used in China still has problems such as poor tensile strength and elongation [5–7]. Currently, there are two main methods for producing aluminum foil sheets: hot rolling and casting. Compared with hot rolling, continuous casting and rolling is a green and low-cost process with low energy consumption, high efficiency, and low greenhouse gas emissions. Therefore, most aluminum foils in China are produced by continuous casting and rolling [8–11].

The quality of continuous casting slabs and the cold rolling process dramatically affects the mechanical properties of aluminum foils [12, 13]. A good aluminum foil blank is the key to achieving a high yield. Poor quality of aluminum foil blank leads to a lot of impurities in the aluminum foil. The more the content of Si and Fe in the solid solution, the higher the work-hardening rate of the material, resulting in rolling performance deterioration. Therefore, the aluminum foil blank must have an excellent microstructure. Annealing significantly modifies the microstructure, deformation hardening, and composition segregation of cold-rolled aluminum sheets, to improve their deformation and mechanical properties [14–16].

To improve the performance of 1235 aluminum alloy, some enterprises have adjusted its Si and Fe contents to develop a new grade aluminum alloy with good performance namely 1235D. In this work, the effects of the annealing process on the recrystallization structure and mechanical properties of 1235D aluminum alloy sheets, manufactured by continuous casting and rolling, were studied.

W. Tang · H. Wang (✉) · Z. Zhu
College of Materials Science and Engineering, Chongqing University, Chongqing, 400044, China
e-mail: wanghp@cqu.edu.cn

J. Pan · C. Wu
Shangqiu Sunshine Aluminum Co., Ltd, Shangqiu, 476003, China

© The Minerals, Metals & Materials Society 2023
S. Broek (ed.), *Light Metals 2023*, The Minerals, Metals & Materials Series,
https://doi.org/10.1007/978-3-031-22532-1_82

Experimental Materials and Methods

The test material is 1235D aluminum alloy sheets, and their chemical compositions are shown in Table 1. After being smelted and refined, the sheets with the thickness of 6.8 mm were obtained by continuous casting and rolling, followed by cold rolling (6.8 mm → 3.6 mm → 1.8 mm (annealing) → 0.8 mm → 0.36 mm → 0.22 mm). The annealing temperature was designed in the range from 440 to 560 °C, and the holding time was 5–240 min. After the furnace temperature reaches the target temperature, put the sample into it directly or heat the sample with the furnace for annealing; then followed by air cooling, water-quenching or furnace cooling. The size of the sample is 180 mm × 30 mm, as shown in Fig. 1(a). After annealing, it is processed following GB/T 16,865–2013 "Samples and methods for tensile testing of deformed aluminum, magnesium and their alloy processed products" to prepare standard tensile samples, as shown in Fig. 1(b). The experimental protocol for annealing is shown in Table 2. Using the DNS50 electronic universal testing machine to measure its tensile properties along the longitudinal direction, the tensile strength and elongation of each test point is the average of three tests.

When the mechanical property test samples were annealed, 1235D aluminum alloy sheets with a size of 90 × 60 mm (as shown in Fig. 1(c)) were annealed under the same conditions. Then a sample of 8 × 10 mm (as shown in Fig. 1(d)) was extracted. After grinding, mechanical polishing, electrolytic polishing, and film coating, metallographic observation was carried out. The recrystallized structure was observed by scanning electron microscope (JEOL JSM-7800F SEM), and the recrystallized texture was detected by EBSD.

Results and Discussion

Effect of Annealing Time

Figure 2(a) shows the mechanical properties of 1235D cold-rolled sheets after being annealed at 500 °C at different times. It can be seen that the tensile strength of the sheets decreased with the increase of annealing time and gradually stabilized at about 89 MPa after annealing for 120 min. Zhang et al. considered that the residual stress of the samples was reduced after annealing, and α and β phases precipitated at the same time, and the content of solid solution Si was reduced, thereby reducing their tensile strength [17]. After

annealing for 120 min, the content of solid solution Si was stable, so its tensile strength was stable. The annealing process decreased the dislocation density gradually, and the grain size gradually grew, continuously decreasing the tensile strength of the sample.

Theoretically speaking, with the increase of holding time, the motion of atoms in the deformed aluminum plate intensifies, the intragranular dislocations begin to move, subgrain boundaries and grain boundaries begin to appear, the distortion of the deformed aluminum plate decreases, and the elongation increases [18]. Therefore, the elongation of the aluminum sheet should be inversely related to the changing trend of the tensile strength. As the tensile strength decreases, the elongation should increase accordingly. However, as shown in Fig. 2(b), the elongation of the 1235D sheets after annealing decreases first and then increases with the increasing annealing time, which indicates some other factors such as the microstructure uniformity and the distribution of the second phase in the sheets also significantly affect the elongation. The inconsistency of the sample's initial state due to the cold-rolled sheet's uneven structure makes the changing trend of elongation not strictly negatively correlated with the changing trend of tensile strength.

Figure 3 shows the recrystallization structure characteristics of 1235D cold-rolled sheets under different annealing time conditions. It can be seen that, after being annealed at 500 °C, with the increase of the holding time, the recovered grains and the residual deformation structure become less, the equiaxed grains become more, and the recrystallized grain size becomes uniform. When being annealed for 5 min, the samples were dominated by {101} oriented grains; with the prolongation of annealing time, the number of {001} oriented grains increased sharply, and the number of {111} oriented grains increased slowly. When the annealing time was 120 min, the recrystallized grains of the annealed samples were mainly {001} oriented.

Effect of Annealing Temperature

Figure 4 shows the mechanical properties of 1235D cold-rolled sheets under different annealing temperatures. It shows that the tensile strength of 1235D sheets decreases with the increasing annealing temperature. When the annealing temperature reaches 500 °C, the tensile strength is gradually stabilized at about 89 MPa, and the elongation is stabilized at about 36.5%. The recrystallized grains also

Table 1 Chemical compositions of 1235D aluminum alloy (wt%)

Element	Si	Fe	Cu	Mn	Mg	Zn	Ti	Al
Content	0.04	0.39	0.15	0.04	0.002	0.01	0.02	balance

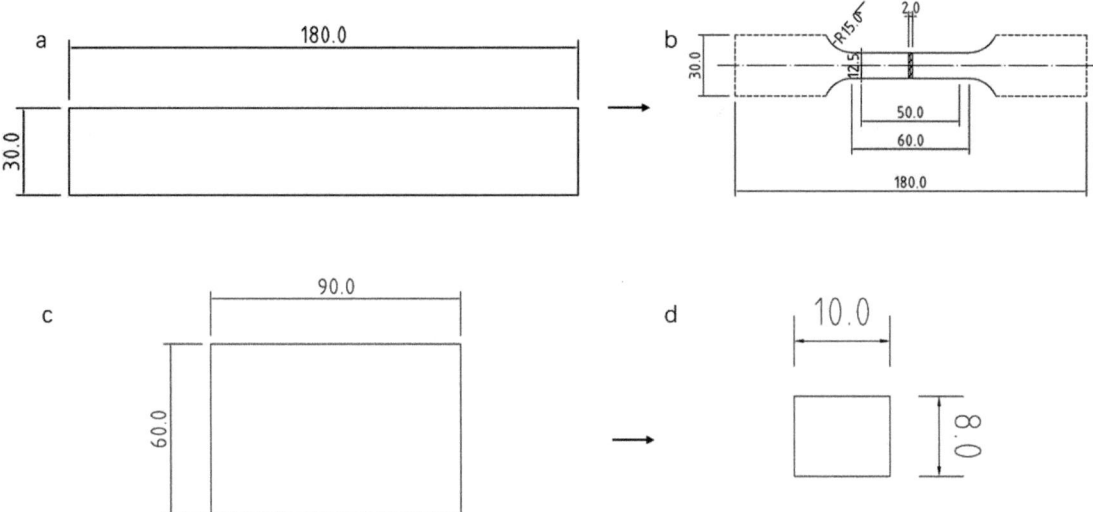

Fig. 1 Schematic diagram of processing sample: **a** Tensile sample wool; **b** Annealed and processed into tensile specimens; **c** Specimen for microstructure observation; **d** Processed into metallographic samples after annealing

Table 2 The designed annealing process

Number	Heating process	Annealing temperature (°C)	Holding time (min)	Cooling process
11–18	Hot fitting	500	5–240	Air cooling
21–27	Hot fitting	440–560	120	Air cooling
31–33	Hot fitting	500	120	Water quenching/air cooling/furnace cooling
41–42	Furnace heating/hot fitting	500	120	Air cooling

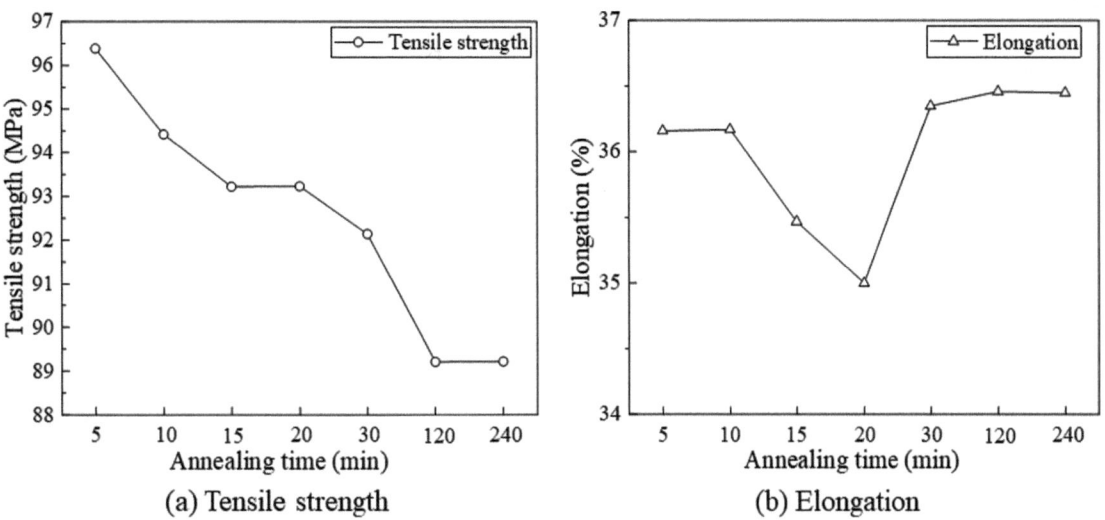

(a) Tensile strength

(b) Elongation

Fig. 2 Mechanical properties of 1235D cold-rolled sheet under different annealing times

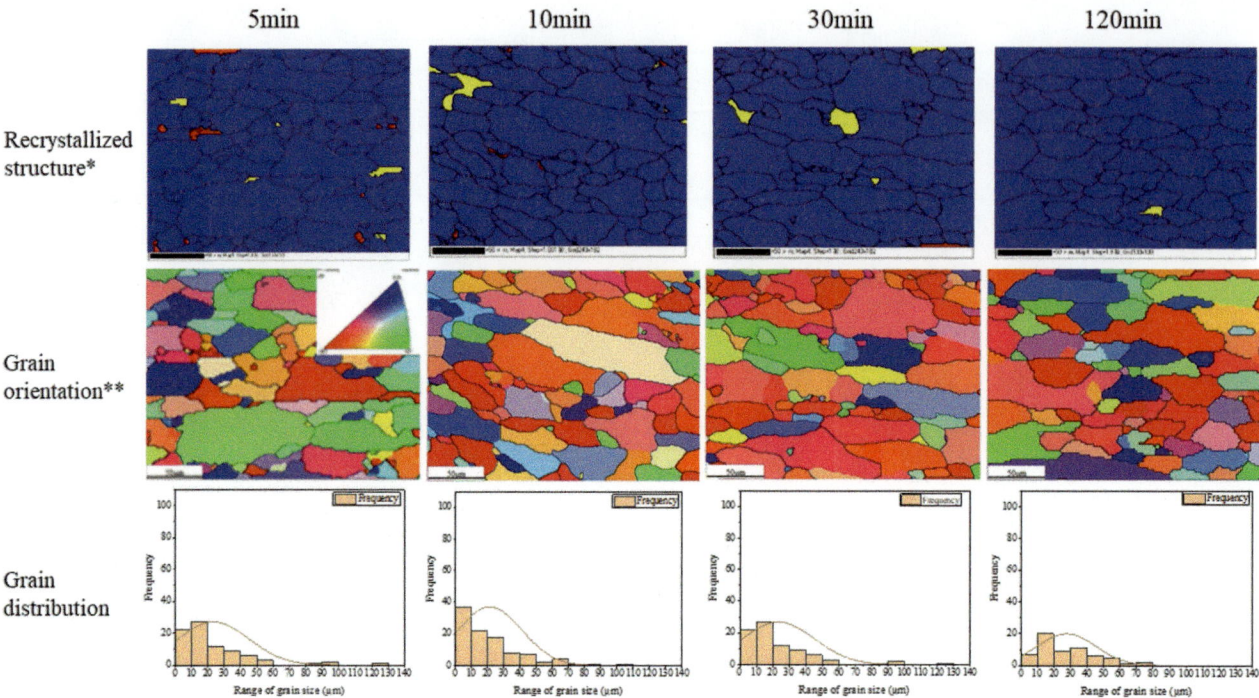

Fig. 3 Recrystallization structure characteristics of 1235D cold-rolled sheets under different annealing time conditions (*blue: recrystallized structure; red: deformed structure; yellow: recovery structure; **{001} orientation is red, {101} orientation is green, and {111} orientation is blue)

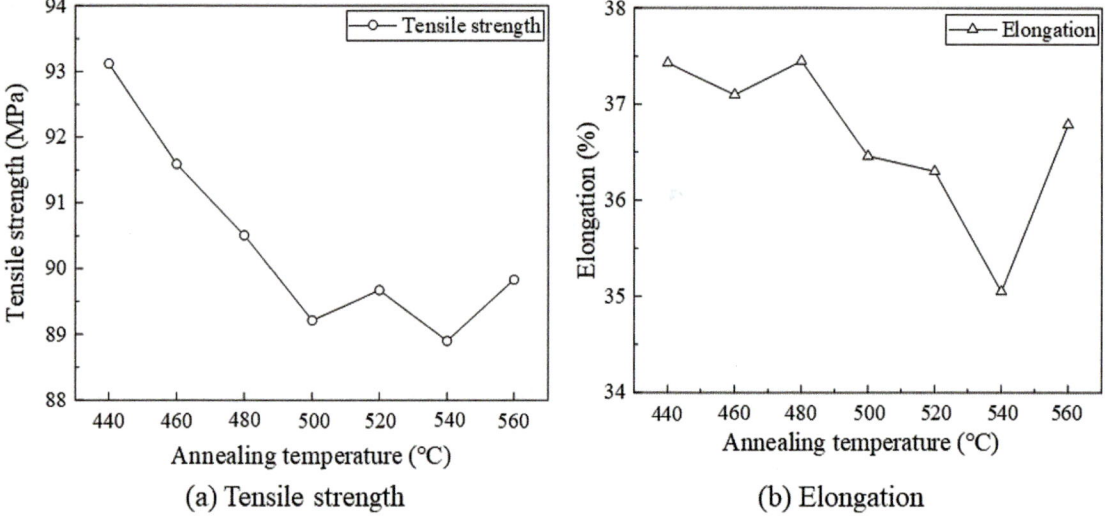

(a) Tensile strength   (b) Elongation

Fig. 4 Mechanical properties of 1235D cold-rolled sheet at different annealing temperatures

continued to grow with the increasing temperature, which reduced the tensile strength and elongation of the samples. For elongation, similar to the results in Fig. 2, the variation trend of elongation is not strictly negatively correlated with the variation trend of tensile strength due to the inconsistency of the initial state of the samples caused by the uneven microstructure of the cold-rolled sheet.

Figure 5 shows the recrystallization structure characteristics of the 1235D cold-rolled sheet under different annealing temperature conditions. With the increase of the annealing temperature, the content of deformed and recovered structures gradually decreases, transforming into equiaxed grains without distortion, and the recrystallized grains grow up. And the recrystallization was almost completed at 500 °C. When the annealing temperature was 460 °C, the recrystallized grains of the samples were mainly {001} oriented; when the annealing temperature went up, the number of {101} oriented grains increased sharply, and

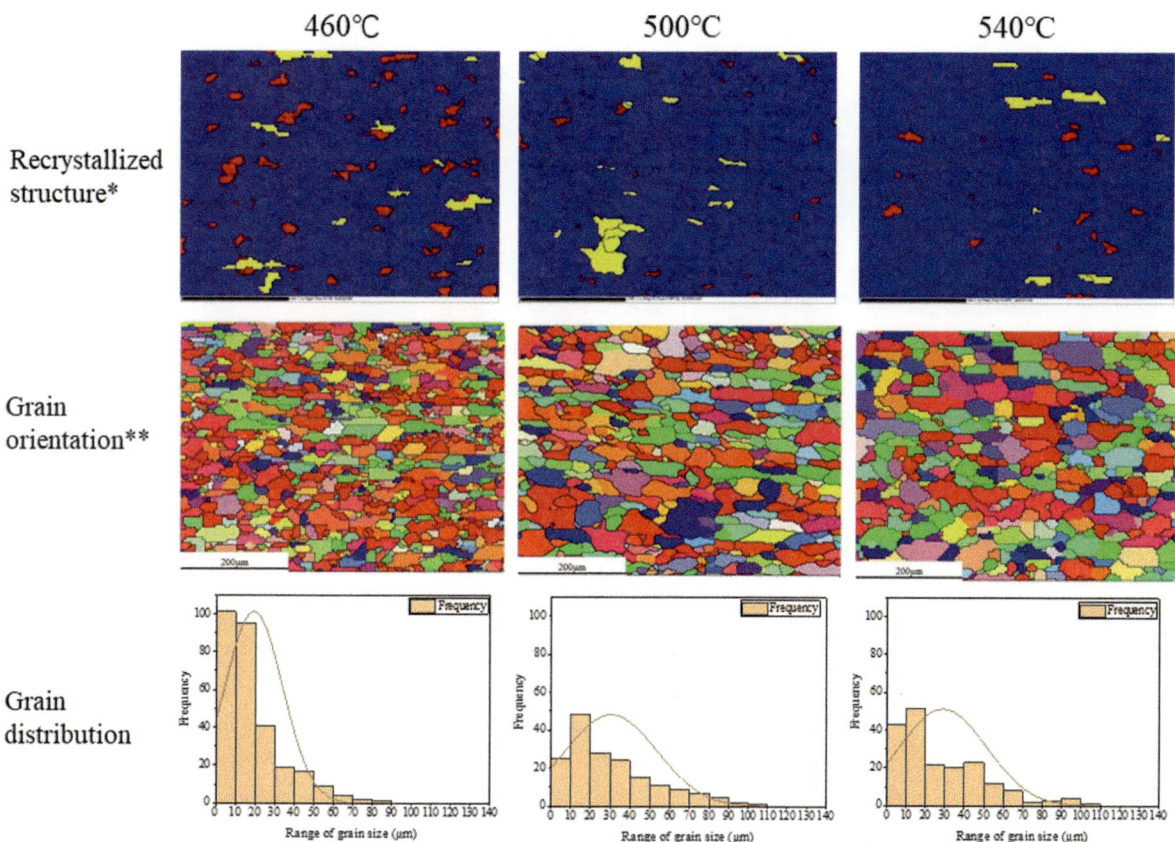

Fig. 5 Recrystallization structure characteristics of 1235D cold-rolled sheet under different annealing temperature conditions (*blue: recrystallized structure; red: deformed structure; yellow: recovery structure; **{001} orientation is red, {101} orientation is green, and the {111} orientation is blue)

the {111} oriented grains slightly increased. When the annealing temperature reached 540 °C, the annealed samples were mainly {101} oriented grains.

The Influence of Heating Mode

Figure 6 shows the mechanical properties of 1235D cold-rolled sheets under different heating methods. It shows that when the sheets were annealed by heating with the furnace and hot charging, respectively, the tensile strength does not change much, but the elongation under the former condition is significantly higher than the latter. This is because the heating rate is fast, the recovery process is insufficient, and the transformation of the deformed structure is not sufficient, so the elongation decreases with the increase of the heating rate. After holding at 500 °C for 120 min, the internal recrystallization of the structure was fully completed, so the heating rate had little effect on the tensile strength. This may be related to the effect of heating speed on the preferred orientation of recrystallized grains, which needs further study.

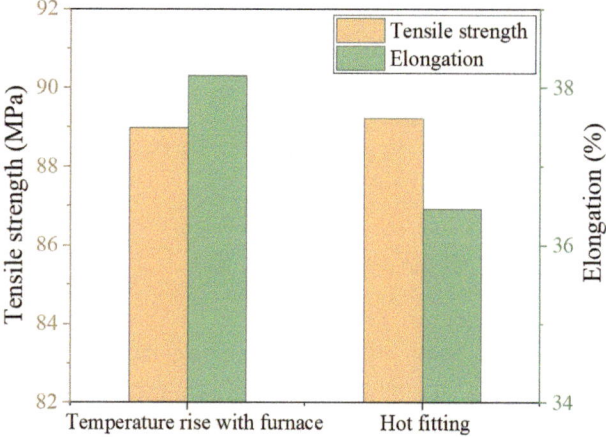

Fig. 6 Mechanical properties of 1235D cold-rolled sheet under different heating methods

Figure 7 shows the recrystallized structure of 1235D cold-rolled sheets after annealing under different heating methods. It can be seen that when the heating rate is slow (the sample is heated with the furnace), and the number of recovered and deformed structures in the sample is relatively

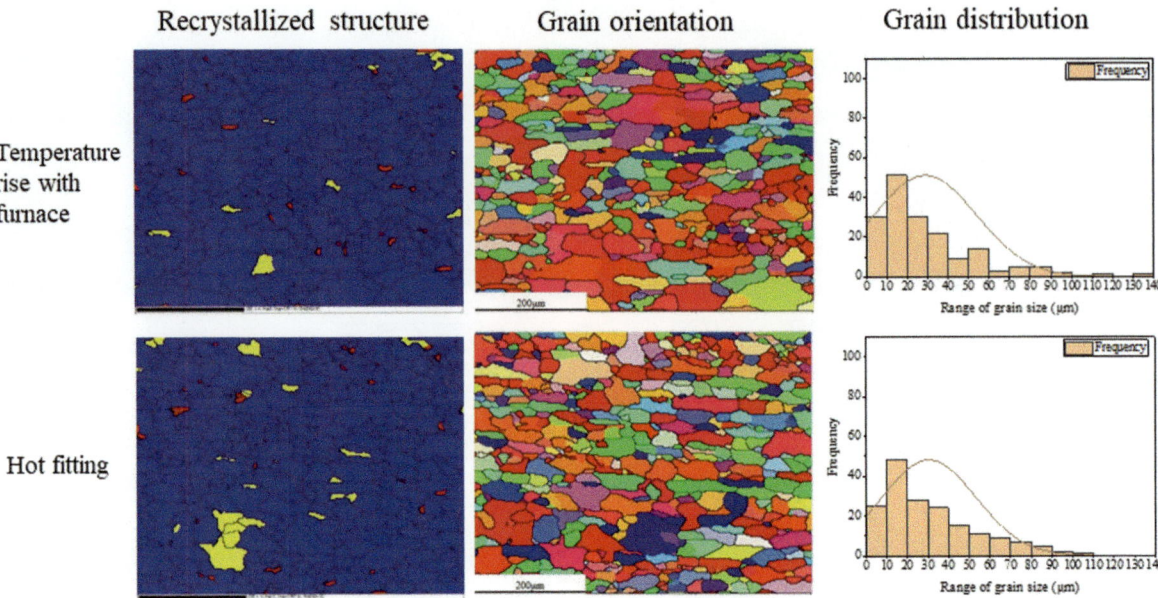

Fig. 7 Recrystallized structure of 1235D cold-rolled sheet after annealing under different heating methods

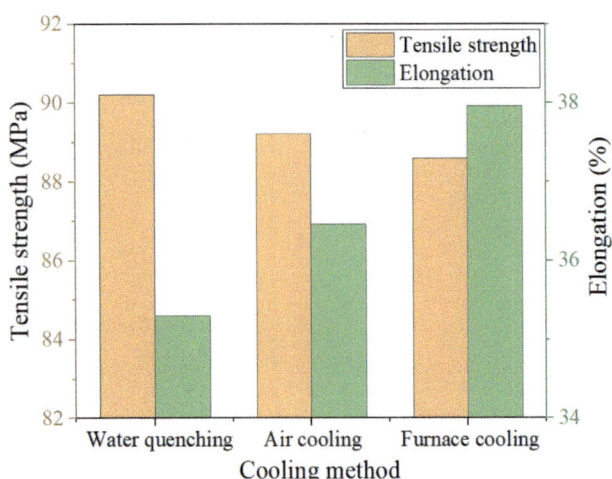

Fig. 8 Mechanical properties of 1235D cold-rolled sheet after annealing under different cooling methods

small. This is because the heating rate significantly affects the release of the deformation storage energy in the sheets.

Influence of Cooling Method

Figure 8 shows the mechanical properties of the 1235D cold-rolled sheet after annealing under different cooling methods. As the cooling rate decreases, the tensile strength of the aluminum plate gradually decreases, the elongation gradually increases, and the sample with furnace cooling has the highest elongation. Combined with the results in the previous section, it can be seen that the slow heating and slow cooling in the annealing process are beneficial to improving the elongation of the aluminum sheet, but correspondingly reduces the tensile strength.

Conclusion

(1) The tensile strength of the 1235D aluminum alloy cold-rolled sheet decreases with the increase of annealing time. When being annealed at 500 °C for 5–240 min, the tensile strength decreased from 96 to 89 MPa; the elongation remained unchanged, however.

(2) The tensile strength of the 1235D aluminum alloy cold-rolled sheet decreases with the increase of annealing temperature. When the annealing temperature increased from 440 to 560 °C, the tensile strength decreased from 93 to 90 MPa; and the elongation also slightly decreased, from 37.4 to 36.8%. The changing trend of elongation is not strictly negatively correlated with the changing trend of tensile strength.

(3) Slow heating and slow cooling during the annealing process are conducive to improving the elongation of the aluminum sheets, but the tensile strength reduces accordingly.

References

1. Guoyin Tan, Gang yang, Youcheng Yue, Liangwei Chen. Influence of Diverter Block on Liquid Flow in Cavity During Casting and Rolling of 1235 Aluminum Alloy. Casting Technology, 2015, 36(2): 394–395.

2. Hong Yang, Weimin Mao. Research Status and Technology Development of Aluminum Foil for Aluminum Electrolytic Capacitors. Material Guide, 2005, 19(9):1-4

3. Huaixi Liu, Dongyang Fang, Jianguo Zhao, Qingxiao Zhang. Research on Common Quality Problems of Double Zero Aluminum Foil. Thermal Processing, 2014, 43(1): 170-172.

4. Youcheng Yue, Gang Yang, Huigang He, Rong Yuan, Ning Ma, Zhugui Rao. Effect of homogenization annealing on the microstructure and properties of ultra-thin double-zero aluminum foil blanks. Metal Heat Treatment, 2013, 38(9): 49-52.

5. Gang Yang, Ning Ma, Huigang He, Yiping Kong, Bo Cao. Research on the Technology of Casting and Rolling Billet for Rolling 0.005 mm Aluminum Foil. Light Alloy Processing Technology, 2007, 35(3): 16–1855.

6. Jun Li, Delin Zhang, Junjie Xia, Jianjun Yan, Wenzhong Liu, Yan Zhao, Guohui Mao. 1235 alloy double-sided smooth aluminum foil for lithium battery and preparation method thereof. China. patent CN201910553753.8. 13 September, 2019

7. Qunli Hong, Deqiang Lu. Research on the production process of 6.0μm negative tolerance aluminum foil from cast and rolled billets. National Aluminum Foil Technology and Marketing Seminar, 2002: 109–112.

8. Bing Ji, Zhutang Wang, Ke Deng. Review of China's Aluminum Foil Industry in 90 Years (2). Light Alloy Processing Technology, 2021, 49(10): 1-12.

9. Guanqi Hu, Huixiao Liu, Ruiqin Zhang. Direct casting and rolling of electrolytic aluminum liquid to produce aluminum foil blanks. Light Alloy Processing Technology, 2007, 35(8): 27-2758.

10. Jing Zhang, Fusheng Pan. Organization Control of Aluminum Foil Wool in Aluminum Foil Production. Material Guide, 2006, 20(5): 108-110116.

11. Meng Li, Haoliang Pan, Qing Zhu. Analysis of black wire defects on the surface of 1235 aluminum alloy cold-rolled sheet. Nonferrous Metal Engineering, 2018, 8(2): 22-25.

12. Guofa Mi, Jianzeng Wang, Peng Geng, Changyun Li. Mechanism and improvement measures of black wire on the surface of cast-rolled coil. Thermal Processing, 2014, 43(23): 83-8589.

13. Meng Li, Guobin Si, Qing Zhu. Influence of melt temperature treatment on the microstructure of billet for casting and rolling ultra-thin double zero aluminum foil. Thermal Processing, 2016, 45(19): 58-61.

14. Aitao Tang, Fusheng Pan, Jing Zhang. Study on Annealing Process and Microstructure of AA1235 Aluminum Foil Billet. Light Alloy Processing Technology, 1999, 27(7): 20-2232.

15. Dongming Song, Hong Yan, Yunkun Wang, Xiaojin Gao, Youcai Zhao. Effect of Annealing Temperature on Microstructure and Properties of Cold Rolled 1235 Aluminum Alloy Sheet. Thermal Processing, 2005, (12): 83-84.

16. Yuqing Fan, Jie Cheng. Mechanical properties of aluminum foil and its improvement. 1990, (9): 11-15.

17. Jing Zhang, Aitao Tang, Fusheng Pan. Comparative study of Korean aluminum foil wool and domestic aluminum foil wool. Light Alloy Processing Technology, 1999, (11): 17-21.

18. Zhengrong Ai, Kai Yu, Hongyan Wu, Janzhong Li. Effect of Rapid Annealing on Microstructure and Properties of Cryogenically Rolled 6063 Aluminum Alloy. Light Metal, 2021, (7): 37-41.

Effect of Thermal Treatment (T5) on Microstructure and Tensile Properties of Vacuum High Pressure Die Cast Al–Si–Mg Alloy

Henry Hu, Ali Dhaif, and Kazi Ahmed

Abstract

In this work, a modified Al–Si–Mg (A356) alloy was prepared by vacuum-assisted high pressure die casting processes (V-HPDC). To release residual stresses, various thermal treatment schemes over a wide range of temperatures between 120 and 350 °C were experimented to the as-cast V-HPDC alloy, in an effort of understanding the effect of thermal treatment on tensile properties of V-HPDC modified Al–Si–Mg (A356) alloy. The morphology of eutectic silicon has a sound effect on the tensile properties of the tested alloy. The content of magnesium-based intermetallic phase, their morphology, and distribution throughout the matrix affect the tensile properties as well. The reduction in the strengths of the alloy treated at 350 °C for two hours should be at least attributed partly to the absence of the magnesium-based intermetallic phase. However, the presence of sufficient amount of magnesium intermetallic phase plays an important role in strengthening the alloy thermally treated at 200 °C.

Keywords

Aluminum alloy • Heat treatment • Vacuum • High pressure die casting

H. Hu (✉) · A. Dhaif · K. Ahmed
Department of Mechanical, Automotive & Materials Engineering, University of Windsor, Windsor, ON N9B 3P4, Canada
e-mail: huh@uwindsor.ca

A. Dhaif
e-mail: dhaifa@uwindsor.ca

K. Ahmed
e-mail: kahmed@uwindsor.ca

Introduction

High pressure die casting is a near-net shape manufacturing process in which molten metal is injected into a metal mould at high speeds and allowed to solidify under high pressures. This technique becomes very popular and cost-effective for mass production of metal components where physical dimensions must be accurately replicated and surface finish is important [1–3]. However, this process creates inherent defects, typified by gas porosity in the produced castings which is mainly due to the entrapment of air in the molten metal as a consequence of the high-speed injection of the molten metal into the die cavity. The presence of gas porosity adversely affects mechanical properties of the cast components. In addition, these formed pores which are located specifically at the casting surface may expand in size during heat treatment, and thus causes blisters [4]. Entrapped gas is always a major source of porosity in conventional die castings and minimization of this problem can improve the mechanical properties to some extent. A vacuum system in high pressure die casting processes is thus a logical necessity to extract the gases from die cavities, runner system, and shot sleeve during processing [1, 3]. Vacuum die casting is such a process which meets this logical requirement, and thus advances the die casting process and makes it more competitive. By creating a pressure lower than atmospheric pressure in the injection chamber and die cavity, the relative absence of air results in high quality of die castings. Back pressures encountered by metal trying to fill the die cavity are also reduced, which ensures a smooth liquid flow through die cavity.

Niu et al. [4] studied the effect of vacuum assistance on aluminum alloys and found that the use of pressure level of 18×10^{-3}–28×10^{-3} MPa during casting, significantly prevents cast components from porosity in comparison with conventional die casting. It was even possible to improve the mechanical properties of die castings by applying solution treatment. Schneider and Feikus et al. [5] studied the effect

© The Minerals, Metals & Materials Society 2023
S. Broek (ed.), *Light Metals 2023*, The Minerals, Metals & Materials Series,
https://doi.org/10.1007/978-3-031-22532-1_83

of solution treatment parameters on tensile properties of vacuum die cast test bars of GD-AlSi9Cu3 (German standard) alloy which is equivalent to A380 in North America. They found that both solution treatment temperatures and times affect the tensile properties of vacuum die cast alloy GD-AlSi9Cu3. According to Hu et al. [6], high pressure vacuum die casting alloy A380 responds to solution treatments. It was shown that solution temperature has a strong effect on tensile properties of solution treated A380. The treatment also changes the morphology of silicon phases which enhance the ductility and UTS. The work by Lumley et al. [7] indicated that blistering is substantially reduced and eventually completely eliminated as the temperature and time of solution treatment are decreased. They have studied aluminum alloy 360 (Al-9Si-0.7Fe-0.6 Mg-0.3Cu-0.2Zn-0.1Mn) and found at temperature below 525 °C for 15 min have no blistering or dimensional instability. Timelli et al. [8] studied the effect of solution treatments on microstructure and mechanical properties of a die cast AlSi7MgMn alloy. According to his study a solution heat treatment of 15 min at 475 °C, or a heat treatment involving even more time at 525 °C, causes spheroidization, coarsening, and an increase in the interparticle distance of the eutectic silicon particles, leading to substantial changes in the microstructure and mechanical properties. The distribution of the silicon particles (eutectic Si) is significantly affected by a short solution heat treatment time. Most of these studies were carried out by solution treatment where morphology distribution of different strengthening phases in the matrix was optimized to improve the mechanical properties. A proper die sealing and vacuum level is always required for approaching high temperature treatment like solution treatment, in die casting components [4]. As recommended in reference 9 that it is not always essential to apply solution heat treatment to die castings. This is because ultra-fine grain structure present in die castings due to severe chill rate often approaches the almost similar microstructural condition of solution heat treatment. Thus, low temperature aging treatment may be good enough for stress relief or dimensional stability. High pressure die cast components always suffer residual stress problem due to the involvement of high pressure and different solidification rates in die cavity. Therefore, a proper aging treatment becomes essential to get relief from this residual stress and to provide a dimensionally stable component without significant sacrifice of mechanical properties.

The present study was carried out to investigate the effect of aging treatment on tensile properties of a vacuum die cast modified aluminum alloy A356, which can meet both strength and ductility requirements for automotive applications. The influence of different thermal treatment schemes on microstructure as well as tensile properties was investigated. An optimum condition was confirmed by analysing the mechanical properties obtained at different thermal treatment temperatures and times.

Experimental Procedures

Thermal Treatment

Low silicon and strontium (200 ppm) modified aluminum alloy A356 (Al-7.75–8.25%Si-0.05–0.15%0.325–0.375% Fe-0.225–0.275%Mg-0.325–0.375%Mn- < 0.04% Zn- < 0.02Ni– < 0.02Sn- < 0.02Cr, wt%) component used in the present study was cast by a 2500-ton vacuum high pressure die casting machine. The alloy was selected due to its relatively high as-cast strength and ductility for structural applications. The cast component was about 0.75 m in length and 0.35 in width with a nominal wall thickness of 0.005 m. For isothermal heat treatments, the components were initially cut into rectangular shapes and then were thermal treated in the temperature range of 120–350 °C. The cast component was sectioned right after the casting operation to avoid natural aging. The duration for thermal treatment was initially selected for 2 h and varied as 0.5, 1, and 1.5 h at optimized temperature. All the treatments were carried out in a muffle furnace of which temperature could be controlled within ±5 °C. All the samples were cooled in air after the thermal treatments.

Tensile Testing

Subsized flat tensile specimens were machined out of as-cast samples and various thermally treated ones according to ASTM standard B557. All samples were loaded to fracture at room temperature at a strain rate of 1×10^{-3}/s on an Instron universal testing machine (Model 8562) with a computerized data acquisition system. During tests, an extensometer with a gauge length of 25.4 mm was attached to test specimens to measure their elongation percentage. Tensile properties, including ultimate tensile strength (UTS), yield strength (YS), and elongation at fracture (E_f%), were computed from the data recorded by a data acquisition system based on the average of three tests for each condition.

Microstructure Analysis

Specimens for microstructural analyses were cut from the same region of the components in each case and prepared following the standard metallographic procedures. After proper polishing and etching (0.5% HF acid solution), microstructural changes were examined on the surface of metallographic specimens obtained from as-cast as well as from various thermally treated samples using optical microscopy and scanning electron microscopy (SEM) which is equipped with an energy-dispersive x-ray (EDX) system for elemental analysis. The densities of specimens measured based on Archimedes principle were used to calculate the porosity levels of tested specimens.

Results and Discussion

Microstructure

Figures 1 and 2 show the microstructures of the alloys at different conditions, namely in as-cast, at 200 °C and at 350 °C thermally treated for two hours. SEM analyses reveal the morphological change of eutectic Si particles and trace of intermetallic phases as the thermal treatment temperatures increased. It is also observed in Fig. 1b that eutectic particles

after aging at 200 °C become more spherical and finer, which could provide high mechanical strengths as indicated in references 10 and 11. Moreover, their inter-spacing (0.26 microns) between eutectic particles tends to be also optimized at the associated temperature. With the excessive increase of temperatures, eutectic particles became coarse at higher temperatures. The inter-spacing between the eutectic silicon increased to about 0.41 microns at 350 °C. Figure 2 shows the presence of different intermetallic phases at 200 and 350 °C. EDS analyses on the specimens treated at different thermal treatment schemes indicated that, after 200 °C thermal treatment, Mg-based intermetallic phase started to dissolve at further higher temperatures. At 350 °C, most of them dissolved in the matrix and were hardly found in the microstructure. Thermal treatment at 200 °C resulted in sufficient amount of all associated intermetallic phases present in the alloy (Fig. 2a).

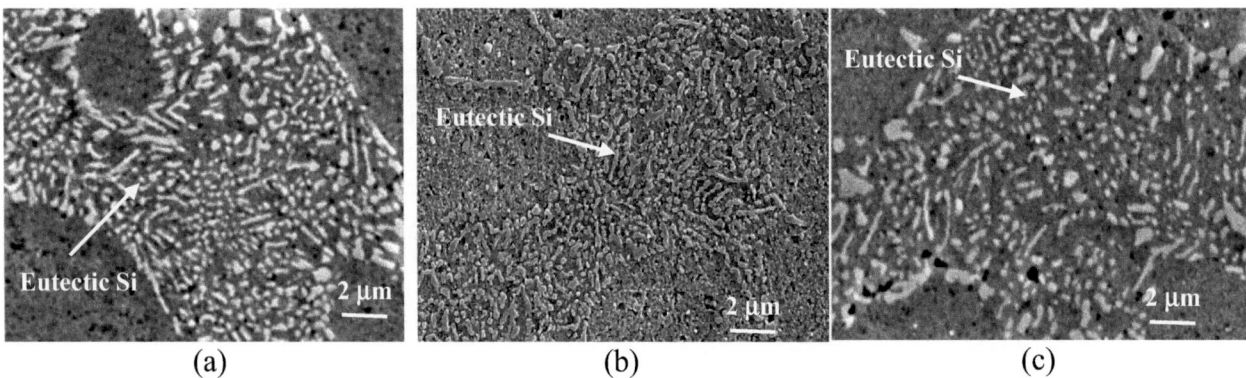

Fig. 1 Eutectic morphology in **a** as-cast, **b** 200 °C, and **c** 350 °C for 2 h thermal treatment

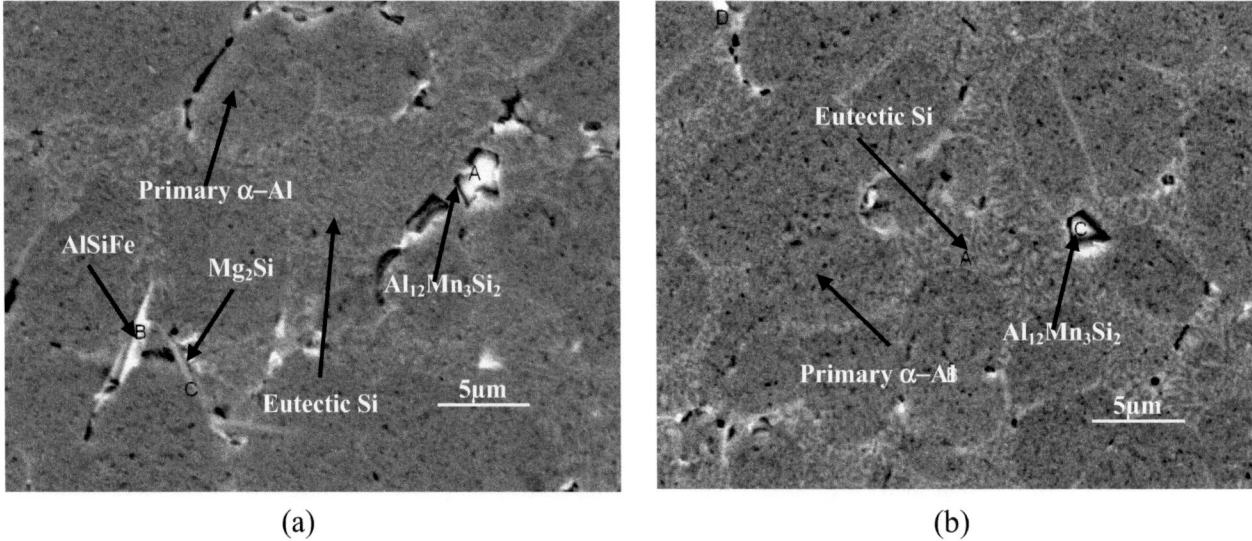

Fig. 2 Intermetallic phases present in the alloy aged at **a** 200 °C and **b** 350 °C for two hours

Table 1 Tensile properties versus aging temperatures for two-hour treatment

Aging temperature (°C)	Yield strength (MPa)	Ultimate tensile strength (MPa)	Elongation (%)
25	133.66 ± 2.5	251.45 ± 16.3	10.7 ± 1.13
120	134.8 ± 2.0	252.6 ± 8.0	10.3 ± 0.21
150	165.9 ± 16.1	275.1 ± 28.3	9.8 ± 1.06
180	190.5 ± 7.5	286.3 ± 22.1	7.4 ± 0.47
200	192.6 ± 7.3	286.3 ± 6.9	7.2 ± 1.27
250	151.3 ± 10.6	224.3 ± 16.1	7.9 ± 1.18
300	150.2 ± 5.8	230.4 ± 22.2	8.7 ± 1.37
350	114.3 ± 3.9	214.6 ± 8.3	9.67 ± 2.46

Table 2 Tensile properties versus aging time at 200 °C

Aging time (minutes)	Yield strength (MPa)	Ultimate tensile strength (MPa)	Elongation(%)
30	160.25 ± 2.5	261.92 ± 15.7	9.43 ± 0.2
60	169.0 ± 8.0	267.87 ± 5.3	7.67 ± 0.3
90	193.8 ± 5.8	286.02 ± 9.3	7.83 ± 0.4
120	192.6 ± 7.3	286.3 ± 6.9	7.2 ± 0.36

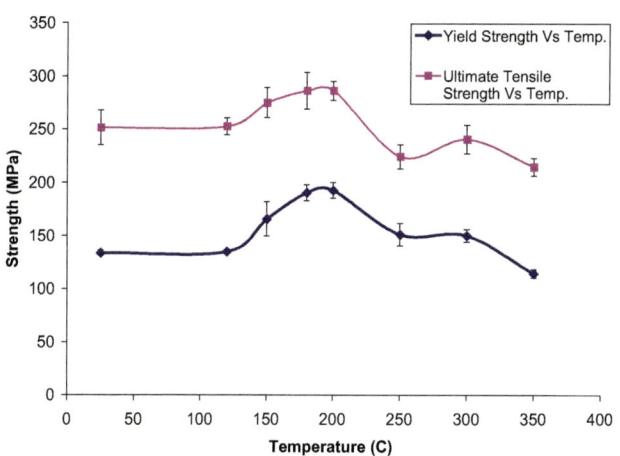

Fig. 3 Strengths (YS and UTS) of the alloy aged at different temperatures for two hours

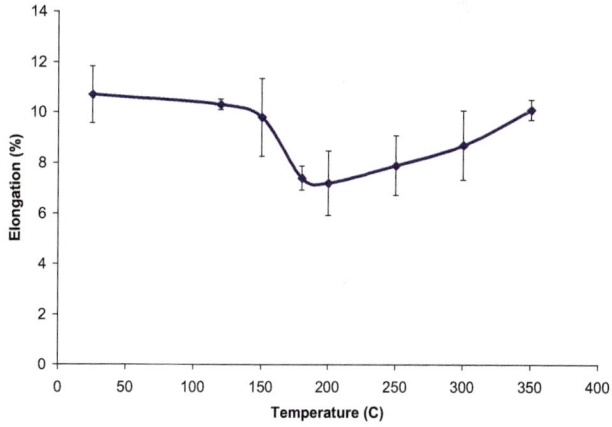

Fig. 4 Elongation (%) of the alloy aged at different temperatures for two hours

Tensile Properties

The tensile properties of the alloy treated at different temperatures and at different times are summarised in Tables 1 and 2. Figures 3 and 4 graphically depict the variation of the strengths and elongation of the alloy with different aging temperatures.

It can be seen from Table 1 as well as from the illustration of Fig. 3, both YS and UTS have increased until 200 °C. However, with increasing the temperature from 200 °C, the strength decreased, but elongation increased. Table 3 represents the time variation with properties at 200 °C. The tensile properties of the alloy appeared to be optimized at

200 °C for 90 min rather than those treated for other different times.

Eutectic silicon morphology has certain effect on increasing tensile strength. Hanna et al. [10] have reported that fine eutectic silicon along with fine primary aluminum grains improves mechanical properties. It is shown in Fig. 2 that eutectic silicon particles distribution changed as well as with their morphology. At 200 °C, eutectic silicon particles appeared to be much more rounded compared with those in the as-cast samples. The inter-spacing between the eutectic particles was also reduced in comparison to that in the as-cast condition. Furthermore, the availability of sufficient amount of Mg_2Si intermetallic phase also provides better

Table 3 Porosity and density variation at different aging times at 200 °C

Treatment	Density (kg/m³)	Porosity (%)	Avg.porosity size (μm)
As cast	2.627 ± 0.07	0.8	13.38 ± 2.52
120	2.623 ± 0.12	1.2	13.59 ± 3.1
180	2.621 ± 0.09	1.4	13.87 ± 2.67
200	**2.618 ± 0.03**	**1.6**	**14.43 ± 1.04**
250	2.614 ± 0.03	2.1	16.78 ± 1.46
300	2.610 ± 0.02	2.4	17.46 ± 0.96
350	2.609 ± 0.02	2.7	19.23 ± 1.43

strength at 200 °C which is shown in Fig. 3. But, all the strength-enhancing factors negatively affect the elongation, of which the samples treated at 200 °C was comparatively lower than that of the as-cast ones. To establish the relation between the eutectic phase and strength, quantitative measurements of size, roundness, and inter-spacing of eutectic silicon based on detailed image analyses are to be performed. The relation is to be reported in future publication.

Successive increase of thermal treatment deteriorates the strength which also can be seen from Table 2. The primary reason should be the coarsening of the eutectic silicon particles and the large inter-spacing between the particles which are shown in Fig. 1. Also, the EDS examination manifests that, at higher temperatures, specifically above 300 °C, the strengthening intermetallic phase Mg_2Si started to dissolve into the matrix. The work of Crepeau et al. [12] on aging treatment of 339 alloy shows that the co-operative precipitation of Al_2Cu and Mg_2Si phases was responsible for alloy hardening. During aging, these elements precipitate in the form of a large proportion of fine particles which in turn render the alloy its strength. Also, the study of Koach et al. [11] has also indicated that an increase in magnesium content of the similar alloy enhanced the yield strength even with low temperature heat treatment. However, it was found that the elongation or ductility of the alloy increased again at 350 °C which might be due to the absence of Mg_2Si intermetallic phase as well as the relief of residual stress.

Effect of Porosity on Tensile Properties

The porosity level at different thermal treatment temperatures is given in Table 3. The measurements of porosity indicate that, due to vacuum assistance during die casting, the porosity level (0.8%) of the as-cast alloy was very low in the cast components compared to that of conventionally die cast aluminum alloys. However, it is almost impossible to completely remove porosity from the casting part in die casting process due to high pressure involvement, leakage at parting line, as well as turbulence effect resulting from high velocity involvement in order to fill relatively thin sections of cast components.

From the experimental values of the porosity and density, it can be seen that, as the thermal treatment temperatures increased, the porosity level raised from 0.8 to 2.7%. This observation indicates that the releasing rate of gas dissolved in the alloy increased with increasing temperature. The increasing trend of the porosity level implies that the high temperature exerted large driving forces to porosity growth. However, the purpose of the aging treatment was to relieve the residual stress from the die cast component, which induced by different solidification rates during die casting. In general, the aging treatment temperature was not very high (maximum temperature was 350 °C), which is far less than solution treatment temperatures. Therefore, it is expected that the applied aging treatment should not provide severe external driving forces to expand gas pore and increase porosity levels. Moreover, at 200 °C, the temperature and time effect on porosity (1.6%) is comparatively much less than samples treated at 350 °C treatment (2.7%), and the overall mechanical properties specifically strength was improved with a moderate value of the elongation.

Conclusions

1. High pressure vacuum die cast modified A356 alloy needs to be thermally treated for the purpose of residual stress relief for dimensional stability of casting. To understand the effect of thermal treatment on tensile properties, different thermal treatment schemes were investigated. It is found that, under a thermal treatment scheme of 200 °C for 90 min, the relatively high YS and UTS of 193.8 and 286.0 MPa respectively were obtained compared to 133.6 MPa for the YS and 251.45 MPa for the UTS of the as-cast alloy. However, this thermal treatment schemes results in a decrease in elongation to 7.83% from 10.7% in as-cast condition.

2. The morphology of eutectic silicon had a sound effect on the tensile properties of the tested alloy. Fine eutectic

particles and their proper distribution in the matrix enhance the mechanical strengths of the alloy.

3. The content of magnesium-based intermetallic phase, their morphology, and distribution throughout the matrix affected the mechanical properties as well. The reduction in the strengths of the alloy treated at 350 °C for two hours should be at least attributed partly to the absence of the magnesium-based intermetallic phase. However, the presence of sufficient amount of magnesium intermetallic phase had played an important role in strengthening the alloy thermally treated at 200 °C for two hours.

4. The tensile properties of the alloy are sensitive to the presence of porosity, their size, and contents. The porosity sensitivity at 200 °C thermal treatment for 90 min was not very significant. However, the high temperature thermal treatment led to an increase in the porosity level of the alloy. The increase in porosity level should arise from the release of some dissolved gas, which may not respond at a relatively low thermal treatment temperature.

5. The application of vacuum assistance systems in die casting process minimizes and reduces gas entrapment in castings. A porosity level of only 0.8% was found in the as-cast condition which was relatively very low in comparison with conventional die casting process (usually about 3%). The thermal treatment at low temperature below 200 °C had no significant effect on the porosity level, rather enhanced the mechanical strengths with a moderate level of ductility.

References

1. H. Kaufmann and P.J. Uggowitzer, Metallurgy and Processing of High-Integrity Light Metal Pressure Castings, Schiele & Schon, Berlin, Germany, 2007.
2. R.N. Lumley, I.J. Polmear, H. Groot and J. Ferrier, Scripta Materialia, Vol 58 (2008), p. 1006.
3. L.Sulley, Die Casting, *Metals Handbook-Casting*, 9th ed., edited by D. Stefanescu, Vol 15, ASM International, Materials Park, OH, (1988), p. 286.
4. X. Niu, B. Hu, I. Pinwill, H. Li, J. Materials Processing Technology, Vol 105 (2000), p. 119.
5. W. Schneider, F.J. Feikus, "Heat treatment of aluminum alloys casting for vacuum die casting", Light Metal Age (1998), p. 22.
6. H. Hu, A.S. Spadafora, R.F. Turchi, A.T. Alpas, "Solution treatment and artificial aging of vacuum die cast alloy A380", Light Metals, edited by T. Lewis, Montreal, Canada, CIM, (2002), p. 475.
7. R.N. Lumley, I.J. Polymer, and P.R. Curtis, Metallurgical and Materials Transactions. A, Vol.40A (2009), p. 1716.
8. G. Timelli, O. Lohne, L. Arnberg, H. I. Laukli, "Effect of solution heat treatments on the microstructure and mechanical properties of a die-cast AlSi7Mg Alloy", Metallurgical and Materials Transactions. A, Vol.39A (2008), p. 1747.
9. NADCA Products Specifications for Die Castings, NADCA, USA, 2009.
10. M.D. Hanna, S.Z. Lu, A. Hellawell, "The effect of SiC particles on the size and morphology of eutectic silicon in cast A356/SiC$_p$ composites" Metallurgical and Materials Transactions A, 15A (1984), p. 459.
11. H. Koch, U. Hielscher, H. Sternau, A.J. Franke, "Silafont 36, the new low-iron high-pressure die-casting alloy", Light Metals, edited by James W. Evans, The Minerals, Metals & Materials Society, (1995), p. 1011.
12. P.N. Crepeau, S.D. Antolovich and J.A. Worden, "Structure-Property Relationships in Aluminium Alloy 339-T5: Tensile Behavior at Room and Elevated Temperature", AFS Transactions, Vol. 98, pp 813-822 (1990).

Numerical Simulation of Flow of Liquid in Molten Pool of Twin-roll Casting Rolling 5182 Aluminum Alloy Strip

Bingxin Wang, Xiaoping Liang, Wenxiong Duan, and Peng Yang

Abstract

The shape and the flow state of the liquid surface in molten aluminum have an important impact on the casting and rolling process of aluminum alloys. Mathematical models of flow of liquid in twin-roll casting and rolling of 5182 aluminum alloy strip was established in this paper. The influence of different nozzle structures of distributors on the shape and the flow state of liquid in the molten pool was studied by numerical simulation. The results show that changes in the inclination of the nozzle of the distributor, the height of the nozzle, and the depth of the distributor into the molten pool will affect the flow of liquid in the molten pool. With the increase of the depth of the distributor into the molten pool, the shape of liquid surface in the molten pool changes obviously, and the turbulent kinetic energy of the liquid surface in the molten pool decreases. The reasonable structural parameters of the distributor nozzle were obtained through research.

Keywords

Twin-roll casting • Aluminum alloy • The state of the flow of the liquid surface • Numerical simulation

Introduction

The twin-roll strip casting technology is a typical near-net shape formation process. The aluminum alloy strip is directly prepared by the twin-roll continuous casting process, and the aluminum alloy strip is obtained by subsequent rolling, which can greatly shorten the production process of plate processing, improve productivity, and reduce production costs [1, 2]. Moreover, it can use a cooling rate higher than traditional semi-continuous casting to improve the strength and plasticity of aluminum alloy sheet and strip, improve corrosion resistance, and obtain high-quality products [3].

The shape and flow state of the molten pool in twin-roll casting directly affects the oxidation of molten aluminum in the molten pool and the quality of the cast strip. Among them, the structure of the distributor of the twin-roll strip casting has an important influence on the liquid surface flow in the molten pool. Wang et al. [4] established a 1:1 water model of a twin-roll strip caster based on the similarity of Froude number and Reynolds number. The effects of molten pool depth and pulling speed on the fluctuation of molten pool surface were studied. Long Chen [5] conducted an optimization study on the structure of the distributor, and mainly studied the influence of parameters such as the distance between the side hole and the bottom of the distributor and the width of the distributor itself on the flow field and the turbulent kinetic energy of the liquid surface.

In this paper, a mathematical model of flow of liquid in twin-roll casting and rolling of 5182 aluminum alloy strip was established. The shape of the liquid surface in the molten pool of the twin-roll casting and rolling process, the flow state law of the liquid surface, and the turbulent kinetic energy of the liquid surface are calculated. The influence of the inclination angle of the nozzle, the height of the nozzle, and the depth of the nozzle on the liquid in molten pool surface was studied.

B. Wang · X. Liang (✉) · W. Duan · P. Yang
College of Materials Science and Engineering, Chongqing University, Chongqing, 400030, China
e-mail: xpliang@cqu.edu.cn

W. Duan
e-mail: duan116984@163.com

© The Minerals, Metals & Materials Society 2023
S. Broek (ed.), *Light Metals 2023*, The Minerals, Metals & Materials Series,
https://doi.org/10.1007/978-3-031-22532-1_84

Mathematical Model of the Flow of Liquid Surface in Two-roll Thin Strip Melt Pool

Geometric Model and Physical Parameters

In the twin-roll thin-strip continuous casting process, the installation structure of the distributor is shown in Fig. 1. The high-temperature molten aluminum flows evenly into the casting and rolling molten pool through the distribution nozzle in the distribution system.

The simulation model takes the casting and rolling molten pool composed of the distributor and the casting roll as the research object. The flow state of molten aluminum in the stable stage of the molten pool is explored. The pouring process does not need to be considered. The model contains two-phase fluids of aluminum liquid and normal temperature air. Since the casting and rolling molten pool is symmetrical along the geometric centerline, the casting and rolling molten pool is appropriately simplified, and 1/2 area is used for modeling. The calculation area of the melt pool is shown in Fig. 2, and the model parameters of casting and rolling molten pool are shown in Table 1.

The parameters of the nozzle of the distributor mainly include the inclination angle of the nozzle (α): the angle between the axis of the side distribution nozzle and the center line of the bottom of the distributor; the height of the nozzle (H): the vertical distance between the center of the side; the immersion depth of the distributor (h): the vertical distance between the inlet end face of the distributor and the center line of the bottom of the distributor.

The simulated material in this paper is 5182 aluminum alloy, which has the characteristics of medium strength, good corrosion resistance and weldability, and is used in automobile inner panels, trunks, doors, etc. Its physical parameters are shown in Table 2.

The physical properties of air are shown in Table 3.

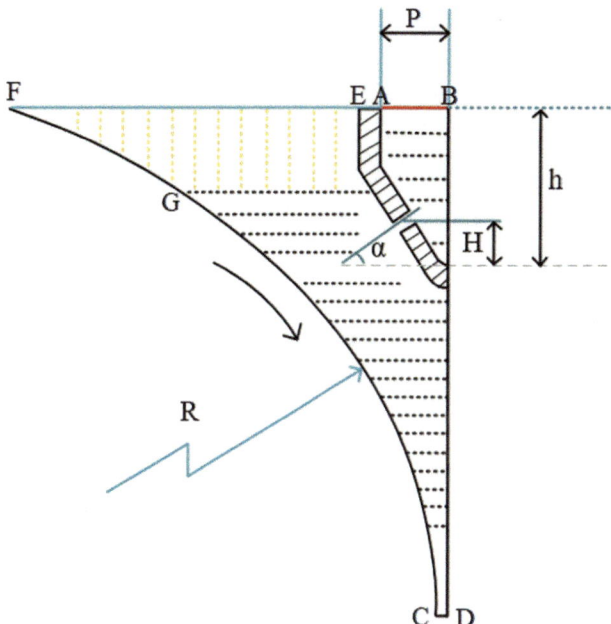

Fig. 2 Pool calculation area

Table 1 Parameters of the casting and rolling model

Parameter	Symbol	Value
Radius of casting rolls	R	400 mm
Half the size of the gate	P	80 mm
The wall thickness of the distributor	t	10 mm

Table 2 Thermal physical parameters of 5182 aluminum alloy

Casting material	Density kg/m^3	Viscosity kg/(m·s)	Surface tension coefficient N/m
5182 aluminum alloy	2700	0.0012	0.873

Table 3 Thermal physical parameters of air

Material	Density kg/m^3	Viscosity kg/(m·s)
air	1.225	1.7894×10^{-5}

Basic Assumptions

The simulation model contains two-phase fluids of molten aluminum and normal temperature air, and the flow situation is very complicated. In order to simplify the calculation, the following assumptions are made:

(1) The molten aluminum in the molten pool is considered to be an incompressible Newtonian fluid, that is, the density of the molten aluminum is constant;

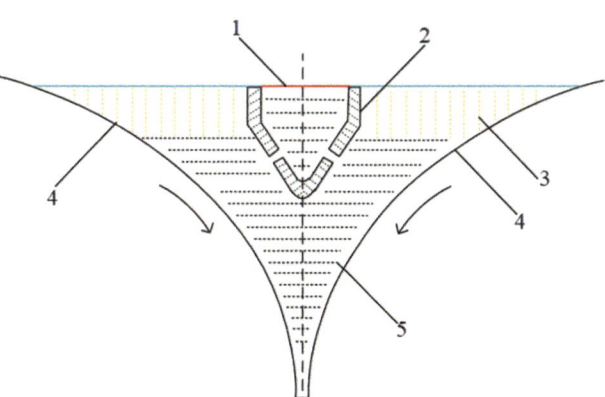

Fig. 1 The installation structure diagram of delivery system 1: Nozzle of distributor 2: Distributor 3: Air 4: Casting roll 5: Molten pool (molten aluminum)

(2) There is no slip between the surface of the casting roll and the molten aluminum in the molten pool;

(3) The influence of the deformation of the crystallizing roller on the flow field is ignored, and the crystallizing roller rotates at a constant speed;

(4) Parameters such as flow field and velocity field are symmetrical along the geometric centerline of the molten pool.

Basic Governing Equations

(1) Treatment of free surface

Describe the transient behavior characteristics of liquid aluminum at the free surface of the twin-roll thin strip molten pool by the VOF (Volume of Fluid) model [6]. In the twin-roll thin strip continuous casting process, only the two-phase flow of molten aluminum and air is considered; Therefore, the VOF (Volume of Fluid) model can be used to reflect the shape of the free liquid surface and the law of the liquid surface flow state.

In the grid unit, the overall integral of molten aluminum and air is 1, and the value of the volume fraction of molten aluminum (α_b) is different, which has different physical meanings:

(1) $\alpha_b = 0$, all is air;
(2) $\alpha_b = 1$, all is aluminum;
(3) $0 < \alpha_b < 1$, contains two phases of molten aluminum and air.

The numerical simulation method selected in this paper is based on the theory of homogeneous equilibrium flow, assuming that the fluid is composed of a single medium mixed by multiphase media. The basic governing equations of mixtures mainly include continuity equation and momentum equation.

(2) Continuity Equation:

$$\nabla \cdot (\rho_m \vec{V}) = 0 \tag{1}$$

In the formula, ρ_m is mixture density (kg/m^3).

When the VOF model deals with free two-phase flow, the values of the density ρ and the molecular viscosity coefficient μ are calculated as follows:

$$\rho_m = \alpha_b \rho_b + (1 - \alpha_b)\rho_a \tag{2}$$

$$\mu_m = \alpha_b \mu_b + (1 - \alpha_b)\mu_a \tag{3}$$

In the formula, ρ_b, ρ_a are the densities of liquid aluminum and air, respectively, (kg/m^3);

μ_b, μ_a are the dynamic viscosities of liquid aluminum and air, respectively, (Pa s).

(3) Momentum equation:

$$\frac{\partial}{\partial x_j}(\rho_m u_i u_j) = -\frac{\partial p}{\partial x_i} + \frac{\partial}{\partial x_i}\left(\mu_{eff}\left[\frac{\partial u_i}{\partial x_i} + \frac{\partial u_j}{\partial x_j}\right]\right) + \rho_m g_i + F_i \tag{4}$$

In the formula, g_i is the component of the gravitational acceleration in the Cartesian coordinate system;

F_i is the component of surface tension in the Cartesian coordinate system;

μ_{eff} is the effective viscosity.

$$\mu_{eff} = \mu_0 + \mu_t = \mu_0 + \rho C_\mu k^2/\varepsilon \tag{5}$$

(4) Turbulence model selection:

In this paper, the standard k–ε double equation model is used for analysis.

k equation:

$$\frac{\partial}{\partial t}(\rho_m k) + \frac{\partial}{\partial x_i}(\rho_m u_i k) = \frac{\partial}{\partial x_j}\left[\left(\mu_m + \frac{\mu_t}{\sigma_k}\right)\frac{\partial k}{\partial x_j}\right] + G_k + G_b - \rho_m \varepsilon + S_k \tag{6}$$

ε Equation:

$$\frac{\partial}{\partial t}(\rho_m \varepsilon) + \frac{\partial}{\partial x_i}(\rho_m u_i \varepsilon) = \frac{\partial}{\partial x_j}\left[\left(\mu_m + \frac{\mu_t}{\sigma_\varepsilon}\right)\frac{\partial k}{\partial x_j}\right] + C_{1\varepsilon}\frac{\varepsilon}{K}(G_k + C_{3\varepsilon}G_b) - C_{2\varepsilon}\rho_m\frac{\varepsilon^2}{k} + S_\varepsilon \tag{7}$$

In the formula, G_k is the turbulent kinetic energy generated by the gradient;

G_b is the turbulent kinetic energy due to buoyancy;

$C_{1\varepsilon}$, $C_{2\varepsilon}$, $C_{3\varepsilon}$ are constants.

$$\mu_t = \rho C_\mu \frac{k^2}{\varepsilon} \tag{8}$$

In the formula, C_μ is constant.

The empirical constants under the standard k-ε turbulence model are shown in Table 4.

Boundary Conditions

(1) Entrance setting (AB)

The inlet adopts the velocity-inlet condition, and the inlet velocity vin is calculated by mass conservation, which is a

Table 4 Empirical constants in the standard k-k-ε equation

$C_{1\varepsilon}$	$C_{2\varepsilon}$	C_μ	σ_k	σ_ε
1.44	1.92	0.09	1.0	1.3

constant; when the casting velocity is given, the inlet velocity and the casting velocity have the following relationship:

$$v_{in} = \frac{V_{pull} \cdot S}{a} \qquad (9)$$

Calculate the turbulence parameters:

$$k_{in} = 0.01 v_{in}^2 \qquad (10)$$

$$\varepsilon_{in} = k_{in}^{1.5}/D \qquad (11)$$

(2) Boundary conditions of free surface (EF, FG)

$$V_y = 0. \frac{\partial V_x}{\partial y} = 0. \frac{\partial T}{\partial y} = 0. \frac{\partial k}{\partial y} = 0. \frac{\partial \varepsilon}{\partial y} = 0 \qquad (12)$$

Free Surface Set to Symmetric Boundary Condition.

(3) Conditions of moving wall (left roller CG)

The speed assignment of each point on the roller surface is realized by UDF, and the speed is

$$v_x = v \cdot \sin\theta \qquad (13)$$

$$v_y = v \cdot \cos\theta \qquad (14)$$

In the formula, v—the rotational speed of the casting roll.

θ—The boundary of the roll surface is the angle between the connection line between the node and the axis of the casting roll and the horizontal line at the minimum roll gap.

(4) Symmetry plane (BD)

The velocity component perpendicular to the plane of symmetry is zero, and the gradients of other variables along the plane of symmetry are set to zero.

(5) Boundary condition of exit (CD)

Outflow boundary outlet condition.

(6) The liquid surface of the molten pool

The liquid surface of the molten pool is processed into a shear plane, which is in contact with the protective gas, and the pressure value of the contact surface is set to 0 Pa, and all other vertical gradients are set to zero.

Results and Discussion

Based on the above mathematical model, the shape of the liquid surface and the flow state in the molten pool under different nozzle structures of the distributor are obtained through numerical simulation. The influence of different nozzle structures of distributors on the liquid surface in the molten pool of twin-roll casting was researched, and reasonable structural parameters of distributor nozzles were obtained.

Influence of the Inclination Angle of the Nozzle of the Distributor on the Liquid Surface

Under the conditions of the parameters set in Table 5, the inclination angles of nozzles of the distributor are 20°, 30°, and 40°, respectively, the shape and the flow state of the liquid surface near the roll surface are calculated, and the results indicated that the change of the inclination angles of nozzles of the distributor had less influence on the shape of the liquid surface of the molten pool after the molten pool reached the stable state. The flow state of the liquid surface of the molten pool near the roll surface is shown in Fig. 3.

As can be seen from Fig. 3, when the inclination angle of the nozzle of the distributor is 20° and 40°, numerous small refluxes are formed on the liquid surface near the distributor in the upper part of the molten pool, and the aluminum oxide film formed by oxidation at the liquid surface is moved above the molten pool by the backflow action without being brought into the lower part of the molten pool. When the inclination angle of the nozzle is 30°, the backflow at the upper part of the molten pool near the distributor decreases, and the oxide film will be brought into the molten pool by the backflow, and the tendency of aluminum oxide inclusions to cause thin strip defects increases.

Figure 4 shows the effect of different inclination angles of the nozzle of the distributor on the turbulent kinetic energy of the liquid surface in the upper part of the molten pool. When the inclination angle of the nozzle of the distributor is 20°, the turbulent kinetic energy of the liquid surface in the upper part of the molten pool is the smallest.

Table 5 Geometric parameters and process parameters used in numerical simulation

Parameter	Symbol	Value
Radius of casting rolls	R	400 mm
Half the size of the gate	P	80 mm
The wall thickness of the distributor	t	10 mm
Casting roll speed	v	$0.50 \ \mathrm{m \cdot s^{-1}}$
Half of the thickness of thin strip	S	4 mm
The width of the nozzle	L	10 mm
The height the nozzle	H	40 mm
Depth of immersion of the distributor	h	140 mm

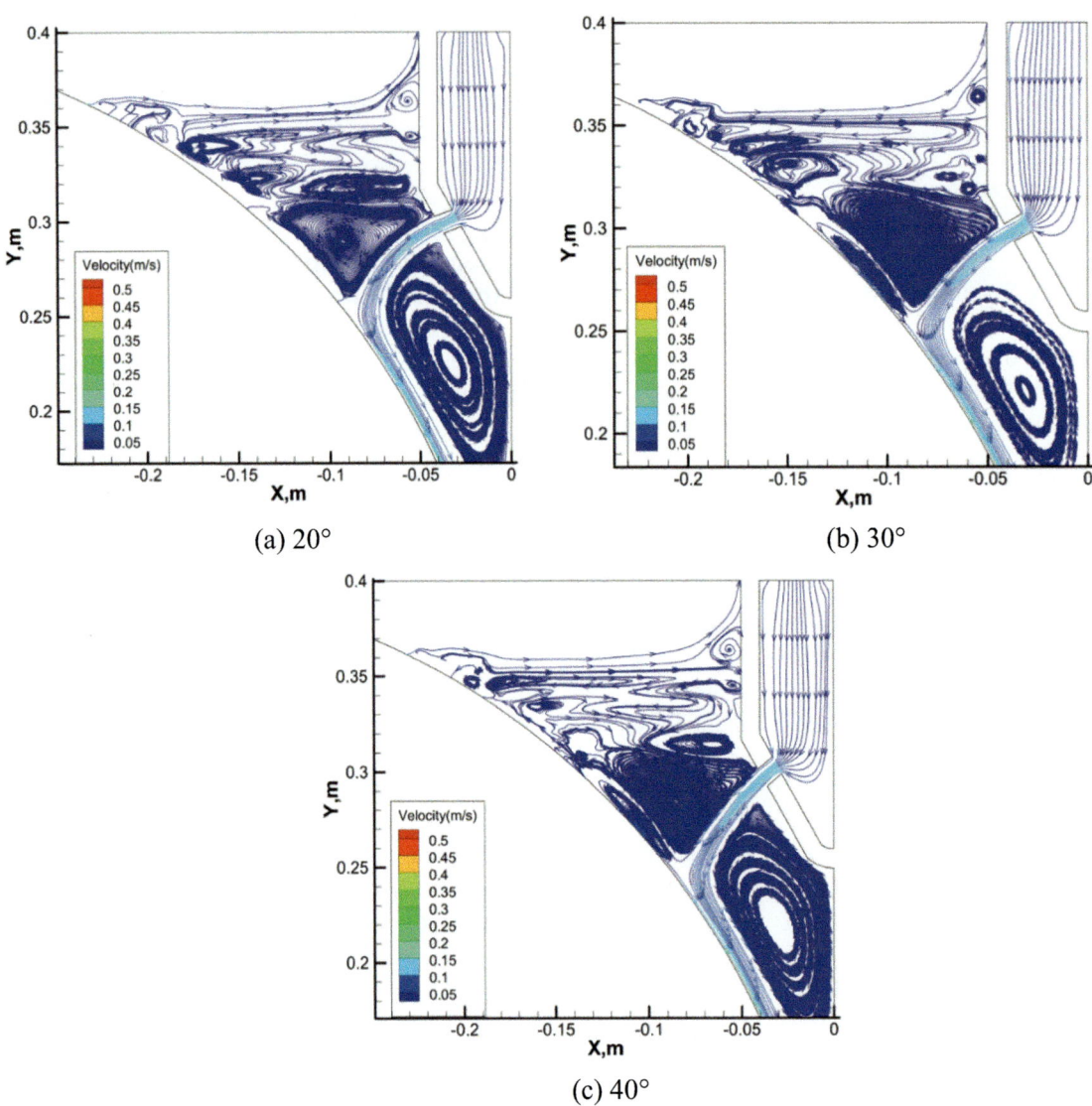

(a) 20°

(b) 30°

(c) 40°

Fig. 3 Flow state of the liquid surface of the molten pool with different inclination angles of nozzle

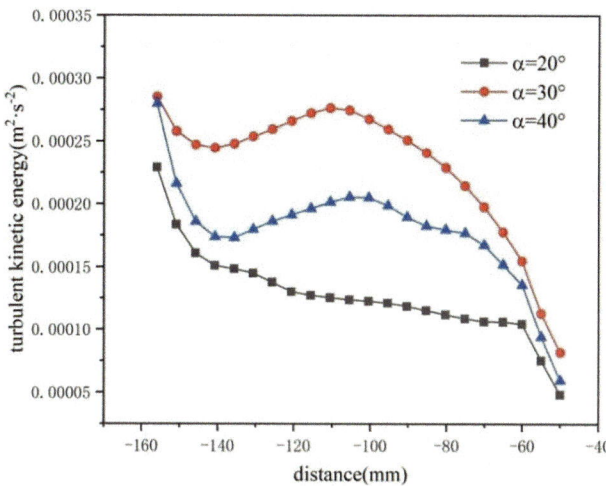

Fig. 4 Effect of different inclination angles of the nozzle of distributor on turbulent kinetic energy of liquid surface in the upper part of molten pool

Influence of the Height of the Nozzle on the Side of the Distributor on the Liquid Surface

Under the conditions of parameters set in Table 6, the height of the nozzle of the distributor is 20, 30, 40, and 50 mm, respectively, the shape and the flow state of liquid surface near the roller surface under different heights of the nozzle of the distributor were calculated. The results showed that the change in the height of the nozzle of the distributor had less influence on the shape of the liquid surface of the molten pool after the molten pool reached the stable state. The flow state of liquid surface near the roll surface is shown in Fig. 5.

It can be seen from Fig. 5 that after the molten pool reaches a stable state, when the height of the nozzle on the side of the distributor increases to 30 and 40 mm, the number of refluxes above the molten pool decreases, and a part of the molten aluminum with a lower volume fraction is brought into the molten pool with the reflux. This increases the tendency of alumina inclusions to cause ribbon defects.

When the height of the nozzle on the side of the distributor continues to increase to 50 mm, the nozzle of the distributor is getting closer and closer to the upper liquid surface of the molten pool. The molten aluminum flowing from the nozzle of the distributor and the molten aluminum in the upper part of the molten pool form a reflux again near the liquid surface of the distributor.

Influence of Immersion Depth of Flow Distributor Nozzle into Molten Pool on Liquid Surface

Under the conditions set by the parameters in Table 7, the depth of immersion of the distributor into the molten pool is 120 mm, 140 mm, and 160 mm, respectively, the shape and flow state of the liquid surface near the roll surface under different immersion depths of the distributor in the molten pool are calculated. The results are shown in Figs. 6 and 7.

It can be seen from Fig. 6 that after the molten pool reaches a steady state, the change in the depth of the flow distributor immersed in the molten pool has a distinct effect on the shape of liquid surface of the molten pool. When the immersion depth of the distributor is 140 and 160 mm in the molten pool, the aluminum liquid fraction near the roll surface is lower than that above the molten pool, and the production trend of alumina inclusions near the casting roll increases. When the flow distributor is immersed in the molten pool at a depth of 120 mm, it can be seen that the fractional distribution of the aluminum liquid near the casting roll is more uniform. And the aluminum liquid fraction at the roll surface is higher than that at the top of the molten pool. The tendency of alumina inclusions to cause ribbon defects is improved.

It can be seen from Fig. 7 that after the molten pool reaches a stable state, when the flow distributor is immersed in the molten pool at a depth of 160 mm, since the nozzle is far from the liquid surface of the molten pool, it can be seen that many small refluxes are formed above the molten pool. These reflows drive the flow of molten aluminum with a

Table 6 Geometric parameters and process parameters used in numerical simulation

Parameter	Symbol	Value
Radius of casting rolls	R	400 mm
Half the size of the gate	P	80 mm
The wall thickness of the distributor	t	10 mm
Casting roll speed	v	0.50 m·s^{-1}
Half of the thickness of thin strip	S	4 mm
The width of the nozzle	L	10 mm
The inclination angle of the nozzle	α	30°
Depth of immersion of the distributor	h	140 mm

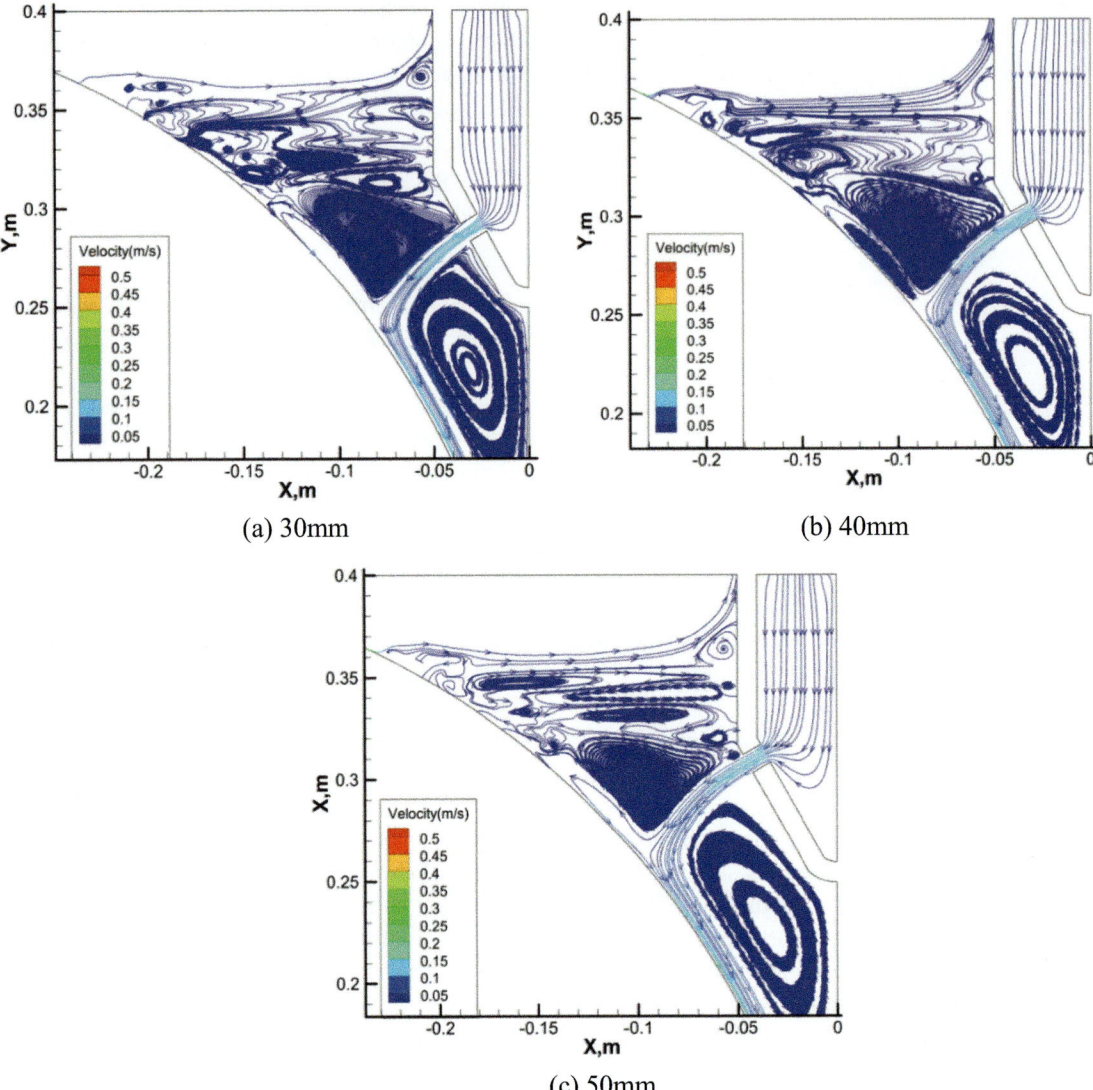

(a) 30mm   (b) 40mm

(c) 50mm

Fig. 5 Flow state of liquid surface of molten pool at different nozzle heights

Table 7 Geometric parameters and process parameters used in numerical simulation

Parameter	Symbol	Value
Radius of casting rolls	R	400 mm
Half the size of the gate	P	80 mm
The wall thickness of the distributor	t	10 mm
Casting roll speed	v	0.50 m·s^{-1}
Half of the thickness of thin strip	S	4 mm
The width of the nozzle	L	10 mm
The inclination angle of the nozzle	α	30°
The height of the nozzle	H	40 mm

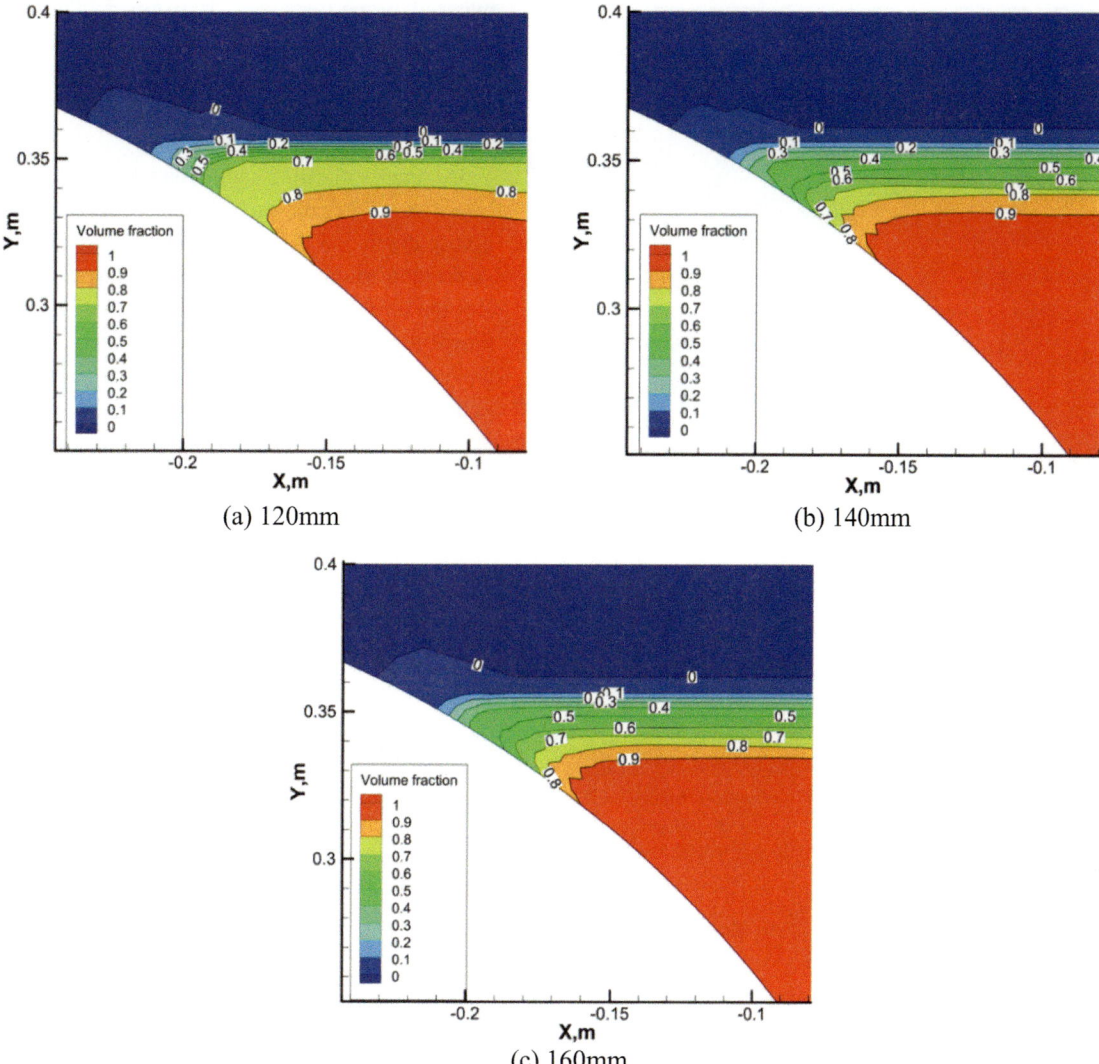

(a) 120mm

(b) 140mm

(c) 160mm

Fig. 6 The shape of liquid surface of molten pool with different diffuser immersion depths

small volume fraction above the molten pool, so that the oxide film generated after the oxidation of molten aluminum can continue to flow above the molten pool to play a protective role.

Figure 8 shows the effect of different distributor immersion depths on the turbulent kinetic energy of the liquid surface in the upper part of the molten pool. It can be seen from the figure that when the flow distributor is immersed in the molten pool at a depth of 160 mm, the turbulent kinetic energy of the liquid surface in the upper part of the molten pool is the smallest.

Conclusion

In this paper, a mathematical model of molten pool flow in vertical high-speed twin-roll casting and rolling of 5182 aluminum alloy strip is established, and the influence of the nozzle structure of the distributor on the flow state of the molten pool is studied. The meniscus shape of the liquid surface at the casting roll and the turbulent kinetic energy of the liquid surface in the upper part of the molten pool were calculated, and the flow state of the molten pool liquid surface was inferred. The conclusions are as follows:

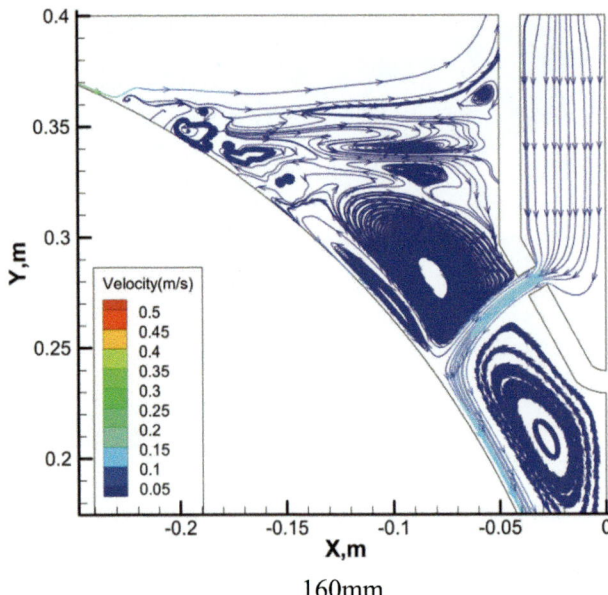

160mm

Fig. 7 Flow state of liquid surface in molten pool at different immersion depths of distributor

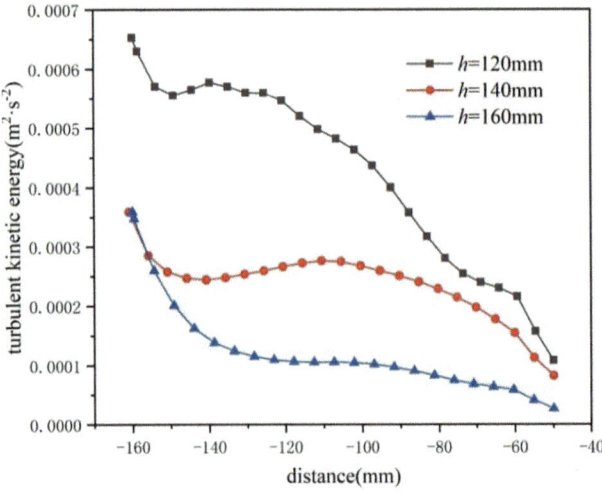

Fig. 8 Effect of different diffuser immersion depths on turbulent kinetic energy of liquid surface in the upper part of molten pool

(1) Mathematical models of flow of liquid in twin-roll casting and rolling of 5182 aluminum alloy strip was established. The calculation area and geometric model are determined, and the boundary conditions and

material properties were determined. The model can describe the shape of the liquid surface in the molten pool and the law of the turbulent kinetic energy of the liquid surface when the two-phase flow of molten aluminum and air was considered.

(2) Using the established mathematical model, the effects of different nozzle inclinations, nozzle heights, and nozzle depths on the molten pool surface were studied. According to the calculation and analysis of the shape and the turbulent kinetic energy of the liquid surface in the molten pool, the reasonable structural parameters of the nozzle of the distributor were obtained: the inclination angle of the nozzle is 20°, the height of the nozzle is 50 mm, and the depth of the distributor immersed in the molten pool is 160 mm.

Conflicts of Interest The authors declare that they have no conflict of interest.

Author Contributions Bingxin Wang contributed to data curation and wrote the paper; Xiaoping Liang contributed to review and editing; Wenxiong Duan contributed to investigation; and Peng Yang contributed to resources.

Funding This research did not receive any specific grant from funding agencies in the public, commercial, or not-for-profit sectors.

References

1. Yuan Fang, Jian Zhang. Development status and future of twin roll strip casting & rolling [J]. Baosteel Technology,2018,(04): 2-6.
2. Xingzhong Zhang, Peng Liao, Minglin Wang. Investigation on Development of Twin Roll Cast Strip [J]. Iron and Steel, 2010, 45 (3): 13-17.
3. P Campbell, W Blejde, R Mahapatra, et al. The castrip process-direct casting of steel sheet at Nucor crawfordsville [J]. Iron and Steel Technology, 2005, 2(7): 56-62.
4. B Wang, JY Zhang, JF Fan, et al. Water modeling of twin-roll strip casting[J]. Journal of Iron and Steel Research International. 2006, 13(1): 14-17.
5. Long Chen. Optimal design and study on delivery system during twin-roll strip casying. Yanshan University,2015.
6. YF Zhang QI Feng-Sheng, F Wang, et al. Modeling of multiphase flow and interfacial behavior between slag and steel in the gas stirring ladle[C]. Asia Steel International Conference B eijing, China,2012: 29–35.
7. Zhiyu Liu, Bo Wang, Qinghua Zhang, Jie Ma, Jieyu Zhang. Numerical Simulation of Filling Process During Twin-Roll Strip Casting. Metallurgical and Materials Transactions B,2014,45(1).

Study of the Solidification Behavior and Homogenization Heat Treatment of the Investment-Cast Al–Cu Foams: Experimental and Modelling Investigations

Waleed Mohammed, Mahan Firoozbakht, and Andreas Bührig–Polaczek

Abstract

Investment casting of aluminum alloy open-pore foams requires high mold temperatures to ensure proper mold filling. The microstructural evolution of these foams is affected by the resulted slow cooling rate and the limited solidification space due to their intrinsic geometry. In our study, solidification and homogenization of investment-cast Al–Cu foams were investigated using cooling curve analysis, optical microscopy, scanning electron microscopy, and energy dispersive spectroscopy. Eutectic Al_2Cu preferably precipitates around Fe-rich needle-like phases and just beneath the surface. The latter suggests that the last melt to solidify in foam casting is near the surface. Solution annealing for 16 h at 535 °C was sufficient to homogenize the Al–3.8 wt.% Cu sample but insufficient for the Al–4.8 wt.% Cu. Moreover, a 1D DICTRA model was proposed to simulate the Cu microsegregation. Simulation results agree with the experiments, and back diffusion was found to contribute slightly to homogenizing the as-cast microstructure during a relatively near-equilibrium cooling.

Keywords

Aluminum Alloys • Investment casting • Metallic foam • Microsegregation • Thermodynamic simulation

W. Mohammed · M. Firoozbakht (✉) · A. Bührig–Polaczek
Foundry-Institute, RWTH Aachen University, Intzestr. 5, 52072 Aachen, Germany
e-mail: m.firoozbakht@gi.rwth-aachen.de

W. Mohammed
e-mail: waleed.mohammed@rwth-aachen.de

A. Bührig–Polaczek
e-mail: office.buehrig-polaczek@gi.rwth-aachen.de

Introduction

Aluminum foams show unique combination of properties, such as low density, high stiffness, impact absorption capability, and superior specific compression strength, which cannot be obtained from their conventional bulk counterparts [1, 2]. The properties of aluminum foams can be modified from different structural hierarchy levels. On the micro-level, the properties can be adjusted by alloying and heat treatment. On the macro-level, they are influenced by the pore size and the cell structure [2, 3]. The latter one can either be open-pore (permeable with the resemblance of a sponge) or closed-pore. The desired cell structure and the physical state of the metal during processing govern the fabrication method of aluminum foams [1, 2].

In the solid-state processing, aluminum foams can be produced from metal powder through sintering [3, 4]. In the gaseous state processing, the metal vapor is allowed to condense on a cold foam precursor through physical vapor deposition [3, 4]. Moreover, investment casting and foaming with blowing agents are techniques to produce open-pore and closed-pore aluminum foams, respectively, as the liquid state processing routes [3, 4]. During the investment casting process, a polymer foam, as a replica of the metal foam, is used as the pattern to create the mold cavity and hence exactly determines the pore size distribution and the strut thickness of the final foam. This polymer pattern evaporates later during the mold firing step. To ensure proper filling of the mold cavity, application of under-pressure on mold, over-pressure on melt and mold heating might be necessary [3, 4]. Therefore, the microstructural characteristics in the as-cast state may be impacted by these casting conditions [5, 6].

Because of their good castability, Al–Cu alloys are potential candidates for the investment casting route of aluminum foams [7]. Another advantage is the heat treatability of Al–Cu alloys, which allows the production of high-strength foams [8]. Heat-treated Al–Cu parts obtain their strengths primarily from precipitation and solid solution hardening mechanisms [5]. These strengthening mechanisms can be impaired by microsegregation during solidification [8]. Moreover, microsegregation impacts the homogenization and age-hardening responses in Al–Cu alloys, and therefore should be studied for different casting conditions.

The microsegregation severity in Al–Cu cast alloys, manifested by the phase fraction of the non-equilibrium Al_2Cu, strongly depends on the solidification cooling rate with an increase at low cooling rates (<1 K/s) followed by a gradual decrease at high cooling rates [9, 10]. High microsegregation level can be attributed to partitioning of solute atoms in the low solidification cooling rate. On the other hand, dendrite arm coarsening during solidification was suggested to explain the decrease at rapid cooling rates [10, 11]. During dendrite arm coarsening, the fine dendrite arms that formed in the early solidification stage and contain a lower solute content get thinner and eventually dissolve, whereas the coarse arms that formed at the later solidification stage get thicker. Solute atoms diffuse from the coarse arms to the fine arms due to the chemical potential difference caused by the various interfacial energies, e.g., curvature difference [10, 12]. Thus, this would lessen the amount of non-equilibrium eutectic Al_2Cu and assist in eliminating microsegregation during solidification [10, 11].

High mold temperatures in investment casting of aluminum foams result in slow solidification and cooling rates. Together with the restricted solidification space in the ultrathin struts, these affect microsegregation profile. However, the extent of this effect is still unknown. To address this knowledge gap, we investigated the microstructural characteristics of open-pore Al–Cu cast foams in the as-cast and homogenized conditions using modeling and experimental techniques.

Experimental Procedures

Molding

We employed block mold casting process, as a modified investment casting technique, in this research to produce aluminum foams [13, 14]. For this purpose, 10 pores per inch polyurethane foams (Foam Partner GmbH, Leverkusen, Germany) with an average strut thickness of 500 μm and size of 30x30x100 mm^3 were used as patterns. The patterns were glued on wax gating systems and then fixed into perforated steel flasks. The flasks were covered with adhesive tapes to prevent slurry leakage. Gypsum-bonded investment powder "Goldstar XXX" (Goodwin Refractory Services Ltd., Staffordshire, UK) was mixed vigorously with 40 wt.% water relative to the weight of the powder. The resulting slurry was poured over the patterns in the flasks under vibration to ensure efficient infiltration of the slurry into the foam patterns. A vacuum chamber was used to reduce the entrapped air pockets inside the slurry. After drying for 2 h at room temperature, molds were dewaxed at 135 °C for 4 h. Afterward, molds were heated to 720 °C at a heating rate of 1 K/min and fired for 4.5 h.

Casting

Two Al–Cu alloys with target compositions of Al–4.8 wt.% Cu and Al–3.8 wt.% Cu were prepared in an induction melting furnace from a commercially pure aluminum ingot (99.8%) and an Al–50 wt.% Cu master alloy. The melt temperature during casting was at 730 °C with around 70 °C superheat. Table 1 presents the chemical compositions of the cast alloys, which were determined by an optical emission spectrometer (Hitachi OE750) through an average of 6 measurements.

The mold temperature was 700 °C during casting to prevent premature solidification. The pouring process was assisted by under-pressure through a vacuum chamber to ensure complete mold filling. The cooling curves of the alloys were recorded by K-type thermocouples that were inserted into the pouring basins, and thermal analyses were performed on them.

Heat Treatment

After casting, the as-cast foams were cut into three cubic samples of approximately 30x30x30 mm^3. The homogenization heat treatment of the alloys was done at 535 °C for 8 and 16 h by an air-circulating chamber furnace. Once the samples were removed from the furnace, they were immediately quenched in water at room temperature.

Table 1 Chemical compositions of the alloys

Target composition	Cu	Fe	Si	Other	Al
Al–4.8 wt.% Cu	4.78	0.073	0.04	<0.01	Bal.
Al–3.8 wt.% Cu	3.71	0.077	0.067	<0.01	Bal.

Microstructure Analysis

The as-cast and the homogenized samples were embedded in the cold and electrically conductive embedding material of KEM 70 (ATM Qness GmbH, Mammelzen, Germany) and standard metallography steps of grinding and polishing were carried out on them. Subsequently, microstructure studies were conducted by an optical microscope of Axio Imager A1 (Carl Zeiss, Oberkochen, Germany) and a scanning electron microscope of Supra 55 VP (Carl Zeiss, Oberkochen, Germany) on the unetched samples. The mean area fraction of the Al_2Cu phase was measured by standard image analysis on 10 micrographs per alloy. The Cu microsegregation between the dendrite arms was measured using energy dispersive spectroscopy (EDS) elemental line scanning to be compared with simulation data. Elemental mapping was used to investigate phase enrichment with specific elements and point analysis was used to determine the composition of specific phases in the microstructure.

Afterwards, samples were anodized with Barker's reagent electrolytic etchant (5 ml HBF_4 diluted in 200 ml of water) at 20 V for 80 s. The grain structures of the anodized samples were studied under the polarized light of Axio Imager optical microscope.

Microsegregation Model

ThermoCalc® 2022a was used to simulate the microsegregation between the dendrite arms using DICTRA module. TCAL5 thermodynamic and MOBAL4 mobility databases were used to perform the simulation. Moreover, the secondary dendrite arm spacing (SDAS) and the cooling rate, which are crucial inputs for the simulation setup, were determined from the microscopic images and the cooling curve, respectively.

Investigation of Cu microsegregation during solidification and further cooling to room temperature was performed using a moving boundary multiphase system model. Figure 1 illustrates the simulation setup. The computational domain is defined as half of the SDAS in the mushy zone due to the symmetry. Simulation started at a temperature of 700 °C with only liquid phase. As the temperature dropped to room temperature, the thermodynamic driving forces for phase formation were calculated by DICTRA. The α-Al and θ-Al_2Cu phases started to form at the left and the right interface boundaries, respectively, when they were thermodynamically stable. A planar geometrical model was used to approximate the shape of the dendrites. Only Cu was considered during the calculations and all other impurity elements and associated phases were excluded from the computational domain for simplification.

Results

Thermal Analysis

Figure 2 demonstrates the cooling curve and its 1st derivatives for Al–4.8 wt.% Cu alloy. The same measurement was done for the other alloy as well. Three inflection points were spotted in each of the first derivative curves. The corresponding temperatures to the inflection points are shown in Table 2.

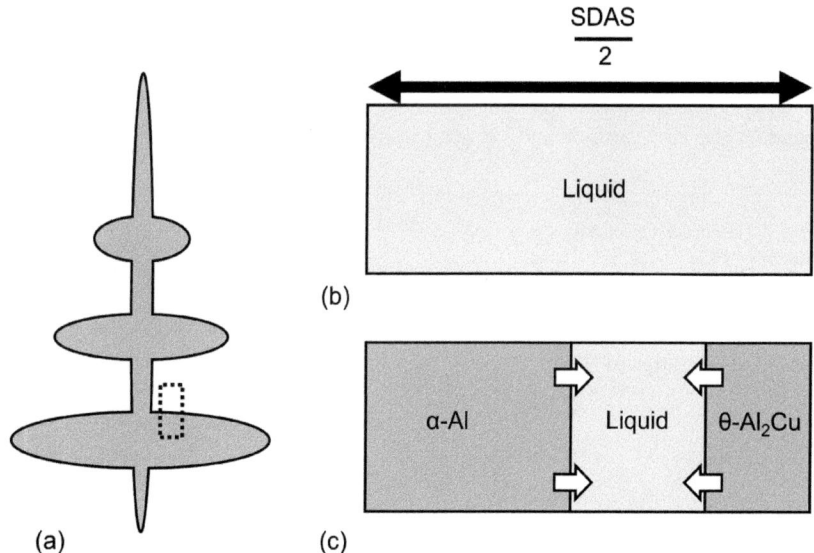

Fig. 1 Schematic illustration of the simulation setup: **a** the microsegregation domain (the dotted rectangle) defined as half of the SDAS, **b** liquid phase stability above the liquidus temperature at the start of simulation, **c** appearance of α-Al (at early solidification stage) and θ-Al_2Cu (at later solidification stage) phases at the domain boundaries

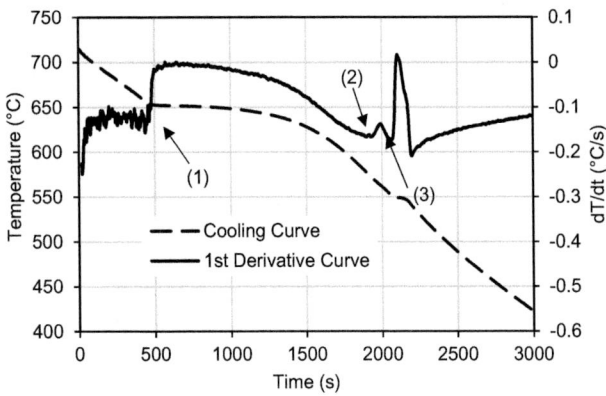

Fig. 2 Thermal analysis from the cooling curve for Al–4.8 wt.% Cu alloy

Table 2 The corresponding temperatures to inflection points for Al–4.8 wt.% Cu and Al–3.8 wt.% Cu

Inflection point	Temperature (°C)	
	Al–4.8 wt.% Cu	Al–3.8 wt.% Cu
1	654.4	658.2
2	576.7	576.7
3	549.4	549.4

Microstructure Study

Figure 3 demonstrates the microstructural characteristics of the as-cast Al–3.8 wt.% Cu struts. The non-equilibrium Al_2Cu phase grew into different morphologies of globular eutectic (GLE), elongated eutectic (ELE), globular blocky (GLB), and elongated blocky (ELB). Figure 4 depicts the measured non-equilibrium Al_2Cu area fractions in the as-cast alloys. Non-equilibrium Al_2Cu content in Al–4.8 wt.% Cu was higher than in Al–3.8 wt.% Cu.

As Fig. 3 presents, the eutectic Al_2Cu phase formed beneath the strut surface and between the dendrite arms. In addition, this phase formed mostly around needle-like phases (Fig. 5). Elemental mapping around the eutectic phase indicated enrichment of the needle-like phases with iron. Moreover, EDS point analysis of the needle-like phases revealed an average composition of 76.7 at.% Al, 17.2 at.% Cu, and 6.1 at.% Fe.

Figure 6 shows the microstructural changes of Al–4.8 wt.% Cu and Al–3.8 wt.% Cu during homogenization at 535 °C for various holding times. Both blocky and eutectic Al_2Cu were still present in the microstructures of the alloys after 8 h of homogenization. After 16 h, there was no visible presence of Al_2Cu in the Al–3.8 wt.% Cu alloy. Blocky Al_2Cu was still present in the microstructure of the Al–4.8 wt.% Cu alloy after 16 h but it was smaller and more spheroidized, however, eutectic Al_2Cu was completely dissolved. The needle-like iron-intermetallic phase showed no signs of dissolution or change in morphology after homogenization even after 16 h.

The grain structure of the as-cast Al–3.8 wt.% Cu is depicted in Fig. 7 from the anodized metallography sample. Differently oriented grains appeared with different colors under polarized light. Each strut was occupied by one single grain and therefore the grain size was comparable to the strut thickness.

Microsegregation of Cu

Figure 8 illustrates the simulated vs. measured Cu microsegregation profile of Al–3.8 wt.% Cu, as well as the region where the elemental line scanning was performed. The brighter areas in the SEM image correspond to the Al_2Cu phase, while the darker areas correspond to the α-Al matrix. Al_2Cu formed in the region between the dendrite arms; therefore, the distance between the bright areas in Fig. 8a was considered as one SDAS, which was equal to 116 μm. The simulated concentration profile agreed well with the experimental results in both qualitative and quantitative manners.

Fig. 3 The microstructural features of the as-cast Al–3.8 wt.% Cu struts under SEM. **a** GLE: globular eutectic, ELE: elongated eutectic and **b** GLB: globular blocky, ELB: elongated blocky

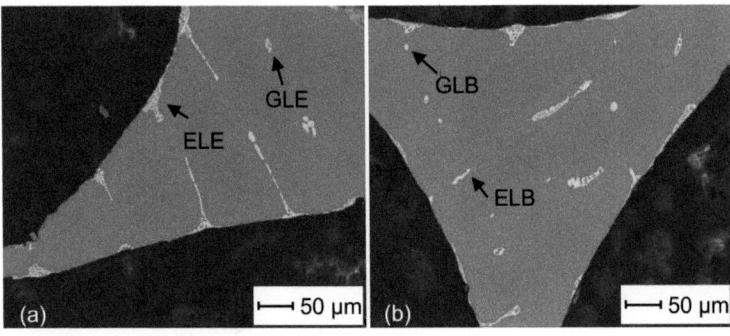

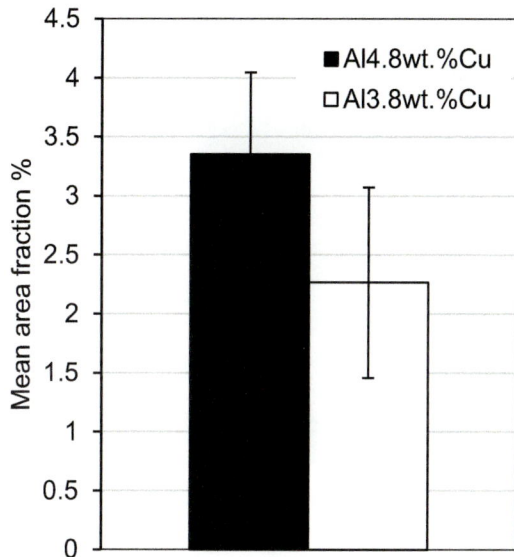

Fig. 4 Mean area fraction of Al$_2$Cu in the as-cast alloys

Discussion

Aluminum foams solidify and cool slowly during the investment casting process as a result of high mold temperatures. This affects the nucleation and growth of different

phases and consequently impacts microstructure refinement, precipitates size, and homogenization behavior. The mechanical properties are influenced by these parameters and it is essential to study them.

Thermal and Microstructural Analysis

The nucleation onsets of different phases can be identified by the inflection points of the first derivative of cooling curves. In our study, the first inflection point corresponds to the liquidus temperature of the alloy and the nucleation of the primary α-Al dendrites. The second point corresponds to the nucleation of the Fe-rich intermetallic phase, and the third point indicates the non-equilibrium eutectic transformation of the Al$_2$Cu phase which occurs at the final solidification stage.

EDS point analysis results and some studies on the bulk Al–Cu alloys [15, 16] suggest that the Fe-rich phase, which appears as needles in metallography, is Al$_7$Cu$_2$Fe. This intermetallic phase is brittle and detrimental to mechanical properties due to the stress concentration at the sharp tips. The harmful effect of the Al$_7$Cu$_2$Fe phase can be mitigated by rapid cooling to reduce the size; however, this is not a viable option during aluminum foam casting due to the process constraints. Slow cooling rate gives the Fe-rich

Fig. 5 EDS elemental mapping around the eutectic phase: **a** SEM image, **b** Fe Kα$_1$, **c** Al Kα$_1$, **d** Si Kα$_1$, **e** Cu Lα$_{1,2}$, **d** Mg Kα$_{1,2}$

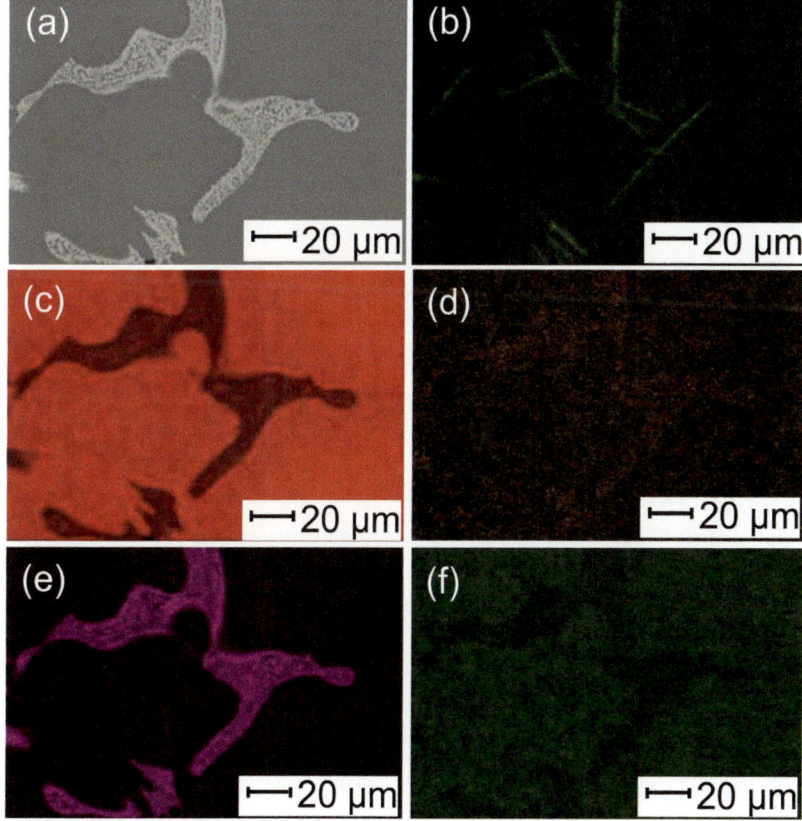

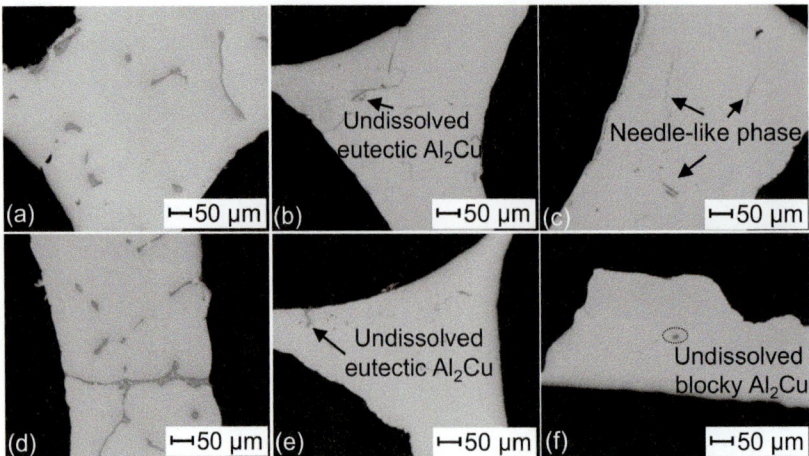

Fig. 6 Optical micrographs for the microstructural changes during homogenization. Al–3.8 wt.% Cu: **a** as-cast, **b** after 8 h: undissolved eutectic Al₂Cu, and **c** after 16 h: complete dissolution of Al₂Cu phase. Al–4.8 wt.% Cu: **d** as-cast, **e** after 8 h: undissolved eutectic Al₂Cu, and **f** after 16 h: undissolved blocky Al₂Cu

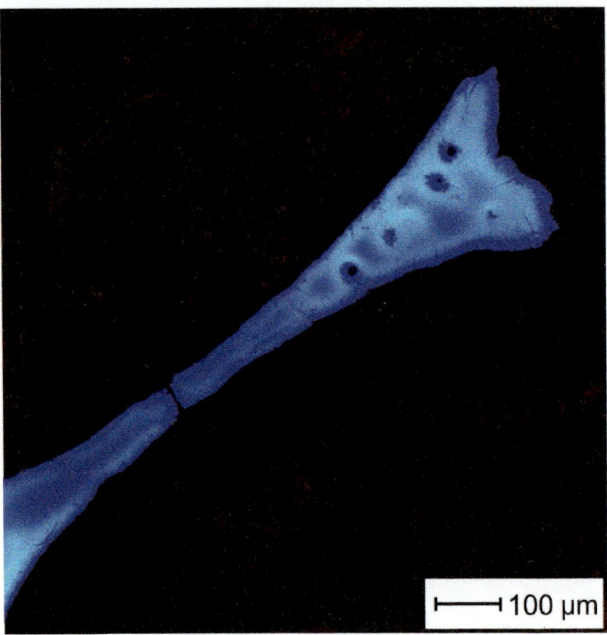

Fig. 7 The etched microstructure of the as-cast Al–3.8 wt.% Cu alloy observed under polarized light revealing the grain size and the dendritic solidification structure

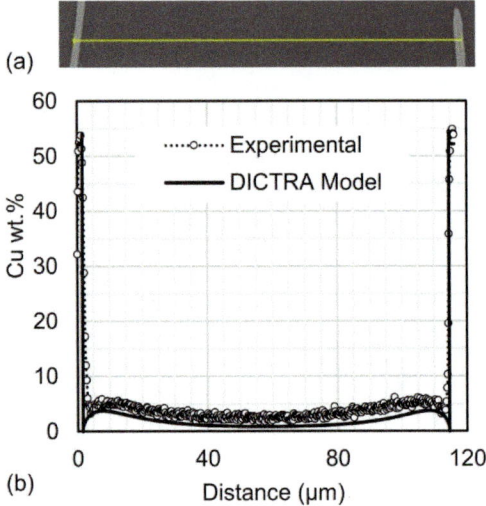

Fig. 8 Cu microsegregation profile inside a secondary dendrite arm of Al–3.8 wt.% Cu: **a** the elemental line scanning region inside one dendrite arm, **b** the measured vs. predicted Cu concentration profiles

phase the opportunity to grow and expand in size in the residual melt. They act afterwards as heterogeneous nucleation sites of the eutectic Al₂Cu phase.

The eutectic Al₂Cu phase forms beneath the strut surface and between the dendrite arms. The superficial formation of this phase can be attributed to high mold temperature during the investment casting process. Both melt and mold are kept at roughly the same temperature during casting. Investment

molding material has high heat content and cools slower than the melt. By start of the solidification, any nucleus that forms at the mold surface re-melts, which causes solidification to proceed endogenously. However, as the mold temperature falls, the final liquid droplets near the mold surface will finally solidify through the eutectic reaction, resulting in the formation of eutectic Al₂Cu.

Slow cooling rate during investment casting induces high growth rate of all the phases leading to the coarse aluminum grains and large precipitates. These microstructural features have a significant impact on the homogenization treatment.

Microstructural Changes During Homogenization

The homogenization rate is strongly influenced by the diffusion distance and the fast diffusion path, e.g., grain boundaries. Finer grains and smaller dendrite arms result in a higher homogenization rate [17, 18]. Therefore, the coarse grains of the cast aluminum foam decrease the rate of the solute atoms diffusion into the matrix. Another factor that controls the duration of homogenization treatment is the dissolution rate of precipitates.

The large size of the precipitates present in the microstructure of our foams has an adverse impact on the dissolution rate, resulting in a longer homogenization treatment. Generally, fine particles with a lower aspect ratio dissolve faster than coarse elongated ones [19, 20]. The dissolution of the eutectic Al_2Cu after 16 h of homogenization, in contrast to the presence of the blocky Al_2Cu, can be attributed to morphology. The blocky Al_2Cu phase is harder to dissolve due to its lower interfacial area with the matrix. Therefore, it dissolves sluggishly via spheroidization. On the other hand, the eutectic Al_2Cu has a higher interfacial area and dissolves faster via fragmentation into smaller segments followed by spheroidization of those tiny segments [21, 22]. Both homogenization and dissolution behaviors are affected by the initial microsegregation profile in the surrounding matrix.

Microsegregation of Cu

The investment cast Al–Cu foams undergo slow cooling to room temperature after solidification which affects the microsegregation behavior. This effect can be realized qualitatively from the microsegregation profile in Fig. 8. The shape of the curve indicates a flux of atoms in the reverse direction, i.e., back diffusion [23]. This contributes to the partial homogenization of the microstructure. However, the dendrite arm spacing similarly increases during the slow cooling leading to a coarser microstructure that can substantially counteract the homogenization effect. Moreover, since Cu is a sluggishly diffusing element in the Al matrix, thus it is expected that the partial homogenization effect from back diffusion would rather be insignificant.

It is crucial to mention that the 1D modeling technique used in this work is not guaranteed to produce accurate results at high cooling rates, where the effect of interface dynamics becomes more pronounced [11]. The model assumes no difference in chemical potentials across phase interfaces. Hence, this simplification neglects many influential factors such as the effects of curved interfaces, solute drag, and elastic stresses [24]. As a result, the model will fail to account for the driving forces and solidification mechanisms that come into play during dendrite arm coarsening such as the difference in interfacial energies (curvature effect) and temperature gradient zone melting (TGZM) which can strongly affect the redistribution of solute atoms and the formation of non-equilibrium phases [12, 25].

Conclusions

In this study, the microsegregation behavior and microstructural features of investment cast open-pore aluminum foams were investigated. The following key points can be drawn from this work:

- Because of the unique casting conditions, investment cast metallic foam generally begins to solidify from the center of the strut to the surface (endogenously).
- The microstructure of the investment cast metallic foam consists of coarse grains with a size comparable to the strut thickness.
- Metallic foams undergo partial homogenization due to the back diffusion during subsequent cooling after solidification; however, the effect of the homogenization appears to be impeded by the coarse microstructure, leaving little or no impact on homogenization heat treatment time.

References

1. Gibson, LJ (1997) Cellular Solids: Structure and Properties. Cambridge University Press, Cambridge
2. Evans AG, Hutchinson JW, Ashby MF (1998) Cellular metals. Current Opinion in Solid State and Materials Science 3 (3):288-303. https://doi.org/10.1016/S1359-0286(98)80105-8
3. Degischer, HP (2002) Handbook of Cellular Metals: Production, Processing, Applications. Wiley-VCH, Weinheim
4. Banhart J (2001) Manufacture, characterisation and application of cellular metals and metal foams. Progress in Materials Science 46 (6):559-632. https://doi.org/10.1016/S0079-6425(00)00002-5
5. Gottstein, G (2004) Physical Foundations of Materials Science. Springer-Verlag, Berlin
6. Ashby, MF (2013) Engineering Materials 2 An Introduction to Microstructures and Processing. Elsevier, New York
7. Lumley, R (2011) Fundamentals of Aluminium Metallurgy: Production, Processing and Applications. Woodhead Publishing, Cambridge
8. Totten GE (ed) (2016) ASM Handbook Vol.4E–Heat Treating of Nonferrous Alloys. ASM International, Materials Park, Ohio
9. Kasperovich G, Volkmann T, Ratke L, Herlach D (2008) Microsegregation during Solidification of an Al-Cu Binary Alloy at Largely Different Cooling Rates (0.01 to 20,000 K/s): Modeling and Experimental Study. Metallurgical and Materials Transactions A 39(5):1183–1191. https://doi.org/10.1007/s11661-008-9505-6
10. Du Q, Eskin D, Jacot A, Katgerman L (2007) Two-dimensional modelling and experimental study on microsegregation during

solidification of an Al-Cu binary alloy. Acta Materialia, 55 (5):1523-1532. https://doi.org/10.1016/j.actamat.2006.10.035

11. Kraft T, Chang Y (1998) Discussion of "Effect of Dendrite Arm Coarsening on Microsegregation". Metallurgical and Materials Transactions A 29: 2447-2449. https://doi.org/10.1007/s11661-998-0120-3

12. A. Turkeli (2006) Simple analytical models for dendrite arm coarsening under high and medium temperature gradients. Materials Science Forum 508:331-336. https://doi.org/10.4028/www.scientific.net/MSF.508.331

13. Fischer SF, Bührig-Polaczek A (2012) Evaluation and Modification of the Block Mould Casting Process Enabling the Flexible Production of Small Batches of Complex Castings. Science and Technology of Casting Processes: 87-114. https://doi.org/10.5772/50621

14. Fischer SF, Schüler P, Fleck C, Bührig-Polaczek A (2013) Influence of the casting and mould temperatures on the (micro) structure and compression behaviour of investment-cast open-pore aluminium foams. Acta Materialia 61(14):5152-5161. https://doi.org/10.1016/j.actamat.2013.04.069

15. Zhou D, Wang J, Lu Y, Bai Z, Li X, Huang Y (2020) Optimization of Homogenization Treatment Parameters and Microstructural Evolution of Large Size DC AA2014 Aluminum Alloy. Materials Transactions 61(7):1210-1219. https://doi.org/10.2320/matertrans.MT-M2019256

16. Haghdadi N, Phillion A, Maijer D (2015) Microstructure Characterization and Thermal Analysis of Aluminum Alloy B206 During Solidification. Metallurgical and Materials Transactions A 46 (5):2073-2081. https://doi.org/10.1007/s11661-015-2780-0

17. Zamani M, Toschi S, Morri A, Ceschini L, Seifeddine S (2019) Optimisation of heat treatment of Al–Cu–(Mg–Ag) cast alloys. Journal of Thermal Analysis and Calorimetry 139:3427-3440. https://doi.org/10.1007/s10973-019-08702-x

18. Liu X, Pan Q, Fan X, He Y, Li W, Liang W (2009) Microstructural evolution of Al-Cu-Mg-Ag alloy during homogenization. Journal of Alloys and Compounds 484(1-2):790-794. https://doi.org/10.1016/j.jallcom.2009.05.046

19. Sjölander E, Seifeddine S (2010) Optimisation of solution treatment of cast Al–Si–Cu alloys. Materials & Design 31(1): S44-S49. https://doi.org/10.1016/j.matdes.2009.10.035

20. Zamani M, Seifeddine S, Jarfors AE (2015) High temperature tensile deformation behavior and failure mechanisms of an Al–Si–Cu–Mg cast alloy — The microstructural scale effect. Materials & Design 86:361-370. https://doi.org/10.1016/j.matdes.2015.07.084

21. Han YM, Samuel AM, Samuel F, Valtierra S (2008) Effect of solution heat treatment type on the dissolution of copper phases in Al-Si-Cu-Mg type alloys. Transactions of the American Foundry Society 116:79-90

22. Li Z, Limodin N, Tandjaoui A, Quaegebeur P, Balloy D (2021) Effects of duration of solution heat treatment on the evolution of 3D microstructure in AlSi7Cu3 alloy: A quantitative X-ray tomography study. Materials Characterization 173:110919. https://doi.org/10.1016/j.matchar.2021.110919

23. Ågren J (2019) The Role of Diffusion in Materials. https://thermocalc.com/products/add-on-modules/diffusion-module-dictra/the-role-of-diffusion-in-materials-a-tutorial/. Accessed 10 March 2022

24. Borgenstam A, Höglund L, Ågren J, Engström A (2000) DICTRA, a tool for simulation of diffusional transformations in alloys. Journal of Phase Equilibria 21(3):269-280. https://doi.org/10.1361/105497100770340057

25. Boussinot G, Apel M (2017) Phase field and analytical study of mushy zone solidification in a static thermal gradient: From dendrites to planar front. Acta Materialia 122:310-321. https://doi.org/10.1016/j.actamat.2016.09.053

Part IV

Aluminum Industry Emissions Measurement, Reporting, and Reduction

The Way Towards Zero Carbon Emissions in Aluminum Electrolysis

Gudrun Saevarsdottir, Sai Krishna Padamata,
Brandon Nicholas Velasquez, and Halvor Kvande

Abstract

The global community has set a goal of carbon neutrality by 2050. Almost one-fourth of the global emissions, attributed to direct emissions from industrial processes, must be addressed by developing zero-carbon alternatives for each process, including the production of aluminum. Several companies and research institutions are working on aluminum electrolysis using oxygen-evolving inert anodes, and recent developments are reported to be quite promising. For existing smelters, carbon capture and sequestration may become a realistic alternative, depending upon the successful adaptation of the flue-gas system enabling increased concentration of CO_2 in the flue gas. A third alternative is electrolysis of aluminum chloride, keeping chlorine and carbon in separate recycling loops. This paper gives a review of the efforts to date, from industry and academia, to decarbonize the electrolysis of aluminum. The development of the largest part of the carbon footprint, arising from the production of the electrical energy used from fossil sources, is also discussed.

Keywords

Aluminum smelting · Greenhouse gas emissions · Specific emissions · Carbon footprint · Sustainability · Decarbonization

Introduction

The demand for aluminum has increased significantly in past decades. In 2021, primary production had increased by a factor of 2.7 since 2000 [1]. In addition comes increased recycling, but recycled material currently supplies around a third of all demand, around 20% being the end-of-life recycling, which captures around 70% of post-consumer scrap, according to the International Aluminium Institute (IAI) alucycle model [2, 3]. Greenhouse gas emissions have increased, not only because of the growing aluminum production but also the limited supply of electrical energy generated from renewable sources. In 2021, 67% of the aluminum production came from electricity from fossil fuels, mainly coal (57%) and natural gas (10%), up from 48% in 2000 [4]. This is down from 71% in 2018 [5], so there are some hopeful signs that the energy mix is starting to shift towards more renewable energy.

The IPCC [6] states in its most recent report that to prevent 1.5 °C of global temperature increase, the total greenhouse gas emissions must be reduced by 45% from 2010 levels by 2030, and reach "net zero" around 2050. The aluminum industry must contribute towards this goal. This paper summarizes the strategies that the industry can pursue to reduce emissions from aluminum electrolysis.

Emissions From Aluminum Electrolysis

The only commercial process for aluminum production is the Hall–Héroult process, where smelter grade alumina, Al_2O_3, produced from bauxite in the Bayer process, is dissolved in a sodium aluminum fluoride molten salt mixture (mainly cryolite, Na_3AlF_6) at about 960 °C in an electrolysis cell. Direct electric current is passed through pre-bake carbon anodes and, predominantly CO_2, forms through electrochemical oxidation. Aluminum containing ionic species are reduced and deposited as metal at the cathode.

G. Saevarsdottir (✉) · S. K. Padamata · B. N. Velasquez
Department of Engineering, Reykjavik University, 101 Reykjavík, Iceland
e-mail: gudrunsa@ru.is

H. Kvande
Previously: The Norwegian University of Science and Technology, Trondheim, Norway

© The Minerals, Metals & Materials Society 2023
S. Broek (ed.), *Light Metals 2023*, The Minerals, Metals & Materials Series,
https://doi.org/10.1007/978-3-031-22532-1_86

Alumina reacts to form molten aluminum and the gases CO_2 and CO according to the overall chemical equation:

$$2\ Al_2O_3(\text{dissolved}) + 3(1+x)\ C\ (s)$$
$$= 4\ Al\ (l) + 3(1-x)\ CO_2(g) + 6x\ CO\ (g) \qquad (1)$$

Although CO_2 is the primary anode product, carbon monoxide starts evolving at a lower potential than CO_2 so it will always be co-evolved, which comes in addition to the Boudouard reaction where CO_2 will react with C to form CO at these temperatures. The CO later reacts to form CO_2. If only CO_2 were formed, the minimum stoichiometric value would be 1.22 t CO_{2eq}/t Al, but the quality of available raw materials for making pre-baked anodes and co-evolution of CO, make it unlikely that these emissions will be brought below 1.33 t CO_{2eq}/t Al [7]. The reversible potential for the CO_2-evolving reaction is 1.22 V, and the total cell voltage to supply the reaction enthalpy is around 2 V, depending on the current efficiency of the process. The total cell voltage is normally above 4 V, so around half of the electric energy supplied to the process is lost as heat, but the sides of the cell are protected by a side ledge, a layer of solid electrolyte which is maintained by the heat loss.

Another mechanism for greenhouse gas emissions occurs when alumina is depleted from the electrolyte. The cell voltage rises until other electrolyte components are oxidized, producing the perfluorocarbon gases CF_4 and C_2F_6. These gases wet the anode and insulate it electrically from the electrolyte, causing a rapid voltage rise known as anode effect. These gases can also be co-evolved in apparently normally operated cells without the onset of an anode effect. They are extremely potent greenhouse gases, 5000–18,000 times more potent than CO_2, depending on the timeframe [8]. According to IAI data, the average PFC emissions in China amount to 1.12 t CO_{2eq}/t Al, while the average outside of China is 0.19 t CO_{2eq}/t Al. The high estimate for China is likely due to non-anode effect PFC emissions being more prevalent in high amperage potlines with many anodes [9, 10]. The global average is 0.75 t CO_{2eq}/t Al [11].

Estimates of the average specific greenhouse gas emissions from aluminum production, from the bauxite mine to cast metal, range between 14 and 17 metric tonnes of CO_2-equivalents per tonne of aluminum produced (t CO_{2eq}/t Al), including both direct emissions from the process and indirect emissions (mainly from energy sources), depending on the assumptions made in the various literature sources [4, 5, 7]. These are significant emissions, and the industry is facing pressure to contribute to global emission reduction goals.

There are four main sources of emissions in aluminum production [5]:

- The Bayer process producing smelter grade alumina from bauxite ore emits an average of 2.7 t CO_{2eq}/t Al with a variability of ±0.5 t CO_{2eq}/t Al. The variability is mostly due to the source of energy used to provide the thermal energy for heating and steam in alumina refining, and higher energy efficiency in newer facilities [12, 13].
- Producing the prebake anodes, including their raw materials, contributes around 0.6 t CO_2e/t Al.
- The Hall-Héroult process, on average, releases 1.5 t CO_{2eq}/t Al from the prebaked anodes during electrodeposition of aluminum, but values down to 1.40 t CO_{2eq}/t Al have been reported by Reny et al. [14]. Additionally, an average of 0.75 t CO_{2eq}/t Al, ranging from 0.13 to 1.3 t CO_{2eq}/t Al, depending on cell technology, arises from (PFC) gas emissions, mainly during anode effects. Total values range between 1.5 and 2.3 t CO_{2eq}/t Al [5, 7, 11, 15].
- The dominating part of the emissions are *indirect* emissions, released in the production of electrical energy used in the electrolysis step, estimated to be on average 9.6 t CO_{2eq}/t Al [4, 16, 17]. These emissions are highly dependent on the energy source, spanning from practically zero when using renewable energy and up to 12 to 18 t CO_{2eq}/t Al when the power is produced using low-grade coal.

There are ongoing efforts to bring down the emissions from the alumina production and anode production, but this paper focuses on the Hall–Héroult process itself and the ways to reduce the emissions from that. Table 1 lists the global average emissions from aluminum electrolysis as well as the lowest emissions achieved using Best Available Technology (BAT), using energy data for 2021.

Efforts to reduce or eliminate the process emissions from aluminum electrolysis can be put in three categories. There are a number of initiatives aiming at replacing the carbon anodes with non-consumable oxygen evolving anodes, or inert anodes. Capturing CO_2 from the Hall–Héroult cell flue gas for sequestration and/or utilization is another route being pursued, and the third is the reaction of the alumina with carbon and chlorine to form $AlCl_3$, which is then electrolyzed in a molten chloride electrolyte.

Decarbonizing aluminum production can, however, not be seriously discussed without addressing the indirect emissions from the electrical energy used. Therefore, the development of the energy mix used for the electrolysis process is presented, and the opportunities are discussed.

Table 1 Data on emissions from aluminum electrolysis, the global average and the best available technology for the CO_{2eq} emission processes in aluminum smelters. Data is from refs. [4, 5, 7, 11, 14, 15, 17–19]

CO_{2eq} emission processes	Chinese average (t CO_{2eq}/t Al)	Rest of the world (t CO_{2eq}/t Al)	Global average (t CO_{2eq}/t Al)	Best available technology (t CO_{2eq}/t Al)
Electricity production	12.4	4.6	9.6	0
Electrolysis	1.5	1.5	1.5	1.40
Perfluorocarbon formation	1.12	0.19	0.75	0.13
Ancillary materials	0.1	0.1	0.1	0.1
Transport	0.3	0.3	0.3	0.3
Total	15.4	6.7	12.2	1.9

Ways to Reduce Emissions from Aluminum Electrolysis

Oxygen Evolving Inert Anodes

One of the influential aspects of obtaining zero-emission during aluminum electrolysis would be to replace the consumable carbon anodes with oxygen-evolving inert anodes in aluminum cells. Adapting inert anodes to the aluminum electrolysis process could eliminate or minimize greenhouse gas emissions, reducing the operating cost related to production and avoid the frequent replacement of carbon anodes, and give a more efficient productivity of aluminum per unit cell [20]. However, finding a suitable anode material that can withstand corrosive electrolytes at high temperatures has been challenging, and the search is ongoing. Primarily, three categories of anode materials have been studied; they are metals, ceramics, and cermets. Metallic anodes are typically an alloy of two or more metals with good electrical conductivity, resistance to thermal shocks, and ease of fabrication, but they are prone to severe corrosion. Ceramics are metallic oxides with high corrosion resistance and are chemically inert, but show poor electrical conductivity. Cermets are a combination of ceramics and metals, possessing properties like good electrical conductivity, low dissolution rate, and high chemical inertness. A good overview of these material types is available in the following papers [20–22].

Inert anodes should show properties such as high electrical conductivity, low dissolution rate, resistance towards fluorination, high mechanical strength, stable potential, low overpotential for oxygen evolution, and low corrosion rate. The primary reaction associated with inert anodes is as follows:

$$Al_2O_3 = 2Al + \frac{3}{2}O_2 \quad E_0 = 2.19V \text{ at } 960\,^\circ C \quad (2)$$

Here, E_0 represents the reversible potential. The reaction enthalpy is around 2.9 V. For comparison, E_0 for the reaction involving the decomposition of alumina and evolution of CO_2 is around 1.19 V at 960 °C, while the reaction enthalpy voltage is around 2 V. The cell voltage would therefore be 1 V higher for an inert anode cell than in the conventional Hall–Héroult process if the process design were similar. The total cell voltage (E_{tot}) can be expressed as below:

$$E_{tot} = E_0 + IR_{electrolyte} + \eta_{bubbles} + IR_{anode} + IR_{cathode} \quad (3)$$

However, studies have shown that the overvoltage of carbon anodes is 0.5 V, which is 0.4 V more than reported or oxygen evolution on Pt anodes [23], while actual candidate anode materials may operate with higher overpotentials, or around 0.4 V [24]. Thus, the cell voltage of the inert anode system would be 0.6–0.9 V higher than for a Hall–Héroult cell. The inert anode cell's total voltage can be further reduced by minimizing the anode–cathode distance. The bubble diffusion overpotential ($\eta_{bubbles}$) caused by the bubbles evolving on the anode can be minimized by selecting an anode material with low bubble wettability behavior. Bubbles generated on the anode surface should detach quickly so that the active anode surface remains constant, and the overpotential caused by the bubbles could be low.

One of the major worries when using inert anodes is the aluminum contamination by anode products. Keller et al. [25] described the mass transfer model to calculate the contamination caused by anode products shown by the equation:

$$\text{Reductionrate} = k_c A_c C_{bulk} \quad (4)$$

Here, reduction rate is the rate at which the anode products are deposited at the anode. k_c, A_c, and C_{bulk} are the anode mass transfer coefficient, the anode's active surface area, and the impurity cation concentration in the bulk phase. Solubility and the mass transfer coefficient of the anode material should be as low as possible for a longer anode lifetime and maximum current efficiency. The most important strategy to counter this is to use an electrolyte that can be used at a lower operating temperature. By replacing sodium in the mixture with potassium, either completely or partly, the cryolite ratio, and thus the liquidus, can be lowered significantly while retaining an acceptable alumina solubility [26]. In this instance, the cryolite ratio becomes CR = (mol% NaF + mol% KF)/mol% AlF$_3$, and, in addition, the potassium ratio, KR = mol% KF/(mol% KF + mol% NaF), is an important process parameter. At high KR, it is therefore possible to operate aluminum electrolysis cells at low CR and temperatures below 800 °C. Operating conditions play a vital role in an inert anode's performance. The operating temperature should be less, or around 800 °C, to protect the anode from corrosion. The electrolyte should be thoroughly saturated with alumina for optimal stability of the anodes [27]. Also, KF is volatile, and vaporization of KF would lower the CR value and make the electrolyte more acidic, risking severe anode corrosion. A solution should be found to maintain the CR values constant, so the cell can operate without voltage fluctuations. Findings show that the anodes have to be homogenized before use to have better corrosion resistance [28].

Solheim argued that inert anode cells would have no significant benefit over a Hall–Héroult cell in terms of economic and environmental aspects [29]. The specific energy consumption in the aluminum production process in 2015 was 13.403 kWh/kg Al, reduced from 14.286 kWh/kg Al in 2006. Due to the higher reversible voltage, the specific energy consumption in inert anode cells would then be 15.35 kWh/kg Al, which is 15% higher energy consumption for inert anode cells. The advantages of inert anodes could therefore only be beneficial if the energy used for the electrolysis process is from renewable sources.

A possible way to avoid this increased energy consumption is to design the inert anode cell as a vertical electrode cell (VEC), rather than using horizontal anodes like in the Hall–Héroult process, thus avoiding complications due to movement and waves in the liquid metal, which is the de facto cathode of the industrial cell. The VEC configuration requires the use of cathodes that are wetted by molten aluminum. TiB$_2$ is the leading candidate material for this application [30]. The thin aluminum film covering wetted cathodes enable a shorter anode–cathode distance, reducing the ohmic resistance in the cell and increasing the electro-active area of the expensive electrodes, while gas bubbles generated on the anode surface could easily escape from the active electrolyte area.

It will, however, always be necessary to provide sufficient voltage to the cell to cover the reaction enthalpy ($E_{\Delta H}$ = 2.9 V) as well as any heat loss. If the cell voltage is maintained at around 4 V, as is normal for Hall–Héroult cells, the heat loss must be 50% lower for the inert anode cell to maintain its bath temperature. This makes it challenging to maintain a frozen lining to protect the sidewalls of the cell, but as VEC enables a more compact design, with higher current per cell volume, the smaller surface area of the cells can help with this issue.

Elysis is developing a VEC technology, which is planned to be commercialized by 2024. They stated that an inert anode's lifespan would be 2.5 years (30 times longer than a carbon anode's lifespan), and operating costs would be 15% less than for conventional cells [31].

Arctus metals along with IceTech are developing a VEC design, based on metallic anodes made of Fe, Ni, and Cu [32]. In fact, out of the three material types mentioned above, metals have been widely studied and are considered promising candidate materials for anodes in inert anode cells [22]. Metals can form a protective oxide layer during the electrolysis process, protecting the anode from oxygen and electrolyte penetration and extending their lifespan. Findings show that the anodes must be homogenized before use to have better corrosion resistance [28]. Pre-oxidation of anodes protects the anode surface from corrosion and fluorination [33]. Another method would be to cover the anode with a protective layer using electrodeposition or vapor deposition processes. One example is the De Nora anodes [34], made from a Ni–Fe-based alloy which was electrodeposited with a Co-based coating to enhance the anode life. The anodes were tested in 100–300 A cells, and results show that the anodes have a lifespan of around 1 year, with an oxide layer dissolution rate of about 3 mm/year.

Rusal, or En$^+$, is also developing an inert anode process and currently produces around 1 t Al per day in a 140 kA inert anode cell [35, 36].

In short, regarding the current efforts to develop an industrial process for aluminum electrolysis with oxygen-evolving inert anodes, and the positive reports thereof, some optimism on successful implementation seems merited. For vertical electrode cell designs, it is likely to be very costly to retrofit older cells. Therefore, alternative solutions, such as CCS, should be explored for existing smelters.

Carbon Capture and Sequestration

If carbon dioxide emissions cannot be prevented in the aluminum production process, a potential method of emission reduction is possible through the capturing of the gas at the production source and subsequent sequestration. Carbon capture and sequestration (CCS) technology has already been shown to be able to work with a broad range of industrial applications, such as in the production of power, fertilizer, hydrogen, iron and steel [37]. This being the case, it is certainly reasonable to think that it can also be applied to the aluminum industry as well. The main obstacle is that the concentration of carbon dioxide in the exhaust gas is very low. The way that modern aluminum cells are designed requires a large amount of air to pass through the system in order to cool process equipment and to prevent fugitive emissions from escaping into the potroom during operation. As a result the concentration of carbon dioxide leaving the system through the ductwork is reduced to about 1 vol%, which is too low to be considered economically and energetically viable for CCS to be implemented. Many current fume treatment plants are usually limited to dry scrubbing and bag filters, which are used to remove HF and recycle fluorides back into the system. Meanwhile, carbon dioxide and SO_2 are generally just released into the atmosphere at low concentrations. Research has been done regarding carbon dioxide, where it has been found that obtaining a concentration of at least 4 vol% or higher would be suitable in order to make the cost of using CCS a worthy investment, as can be seen in Fig. 1 [38].

The most direct way to concentrate the exhaust gas would be to reduce the amount of air pulled into the system. As effective as this would be to increase the concentration, it would also have the negative side effect of disrupting the heat balance in the cell. By reducing the airflow, the amount of heat that is removed from the system would also be reduced, leading to an increase in temperature of all components. In experiments performed by Shen et al. [39] on a 160 kA prebaked cell, it was shown that approximately 76% of the heat leaving the system was through the exhaust gas, with the remaining $\sim 24\%$ being emitted into the potroom via superstructure surfaces. With reduced airflow, more of the heat from the cell will likely flow to other areas of the system, including the sides of the cell, at the risk of melting the solid side ledge. This cannot be allowed to happen, as it

would significantly impact the useful operability of the cell. Research has been done simulating the upper section of a reduction cell with a 50% ventilation reduction while trying to maintain heat transfer levels that were present under normal flow conditions through varied design modifications. These modifications included an increased surface area of the anode assembly (for increased contact with the air), modification of the hood gap geometry (to induce faster flow into the hood), and using different material properties for the anode cover (to increase the emissivity of the surface) [40]. Although none of these modifications were able to fully maintain normal heat loss levels in this case, they do show the necessity for additional modelling and other potential design modifications to be investigated, in conjunction.

Another possibility for increasing the carbon dioxide concentration is to suction the flue gas from the cell separate from the main cooling air. The idea is to construct a suctioning device, dubbed distributed pot suctioning [41], to be placed above the alumina feeding holes, where the majority of the process gases are emitted, see Fig. 2. With this approach, there is far less ambient air being mixed with the process gases, thus allowing the concentration of the collected carbon dioxide gas to be higher than in the traditional system. Tests with this type of device were conducted and seemed to perform well, achieving concentrations approaching 4% while reducing the overall suction rate. The collected gas is also at a higher temperature, as it is suctioned much closer to the heat source in the cell and diluted with less ambient air, which can be troublesome for superstructure integrity and electrical components. Additional testing of devices, such as this, are required before further advancements can be made or have long-term applicability. Several aluminum companies are currently working on cell adaptation to CCS although little is available in the public literature.

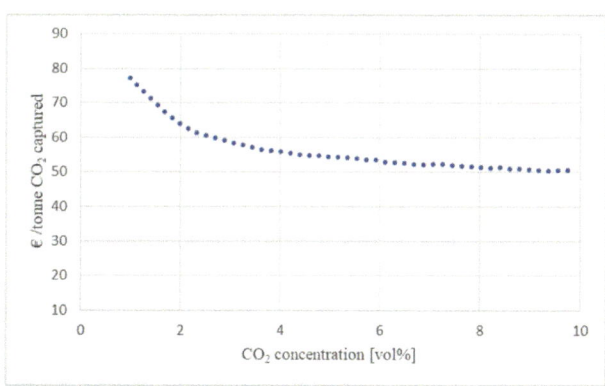

Fig. 1 Cost per tonne of captured CO_2 as a function of CO_2 concentration in exhaust gas. Adopted from [38]

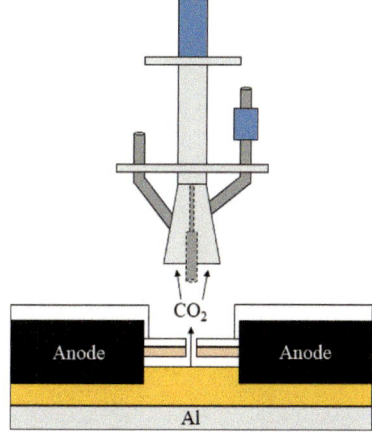

Fig. 2 Distributed cell suctioning device attached to an alumina point feeder. Adopted from [41]

Aluminum Chloride Electrolysis

There is presently an interest in looking at the old aluminum chloride electrolysis process that Alcoa developed in the 1980s. In the temperature range 900–1200 K, alumina can be chlorinated by carbon, by the equation:

$$Al_2O_3 + 3/2 \ z \ C \ + \ 3 \ Cl_2 = \ 2 \ AlCl_3 + 3/2(2-z)$$
$$CO_2(g)_+ 3(z-1) \ CO \ (g) \quad (5)$$

CO can also be used as a reductant at similar temperature [42]. The $AlCl_3$ can then be electrolyzed in a molten chloride electrolyte at around 1000 K to form Al metal and Cl_2 gas. The decomposition voltage for this reaction is around 1.85 V, which is significantly higher than for the Hall–Héroult reaction ($\sim$1.2 V), but lower than for oxygen-evolving inert anodes ($\sim$2.2 V). This approach has been known for a long time, and was the basis of the Alcoa Smelting Process [43]. A molten chloride at 1000 K is less aggressive than cryolite based electrolytes. The chloride electrolysis can be done with inert electrodes and with less heat losses than the Hall–Héroult process, thus claiming to require lower energy consumption. It is also possible to develop the process to use bauxite ore as input, rather than alumina [44, 45].

Aluminium Technologies has developed the so-called CCR process to decarbonize primary aluminum production, replacing both the Bayer process and the Hall—Héroult process. The process uses non-consumable anodes and is claimed to produce high-purity aluminum with no CO_2 emissions while using one-third less energy [46].

Hydro [47] is working on a similar technology, but there are some differences. In this case, CO (g) is used instead of C(s) in the chlorination process of alumina, and the aluminum chloride is then electrolyzed. An interesting solution is that the CO_2 is electrolyzed back to CO (g) again and is recycled to the carbo-chlorination process, releasing O_2 gas while keeping the carbon-containing gases in a closed loop.

Indirect Emissions from Electric Power Used

Aluminum electrolysis is very energy-intensive, with a global average of 14.1 kWh/tonne of aluminum in 2021, including conversion losses [48], and therefore the source of electric energy used for the electrolysis is extremely important for the carbon footprint. A number of papers have been published recently, focusing on the global development of the energy mix for aluminum electrolysis [5, 7, 49]. Many have been curious to explore the development in the carbon footprint during the COVID-19 pandemic and its economic aftermath. According to data from the International Aluminium Institute, which includes data from 2021, there seem to have been some positive developments [48].

Looking at the total energy used for aluminum electrolysis, it can be seen that the fraction of hydropower has significantly increased between 2018 and 2021. This has led to a decrease in the carbon footprint by 0.7 kg CO_{2eq}/kg Al in that period, and is now at 9.6 kg CO_{2eq}/kg Al. Down from a maximum of 10.3 kg CO_{2eq}/kg Al from 2016 to 2018, see data in Fig. 3.

This can be attributed to a significant increase in the use of renewable energy in China, as can be seen in Fig. 4, which shows the energy used for aluminum production in China on the one hand and the rest of the world on the other. The Chinese data for 2019 and 2020 appears to have been updated retroactively, with an increased share of hydropower, which causes a slight discrepancy with previously published data [49], but the important thing is that the Chinese aluminum industry is making a real effort to increase the share of renewable energy used for aluminum

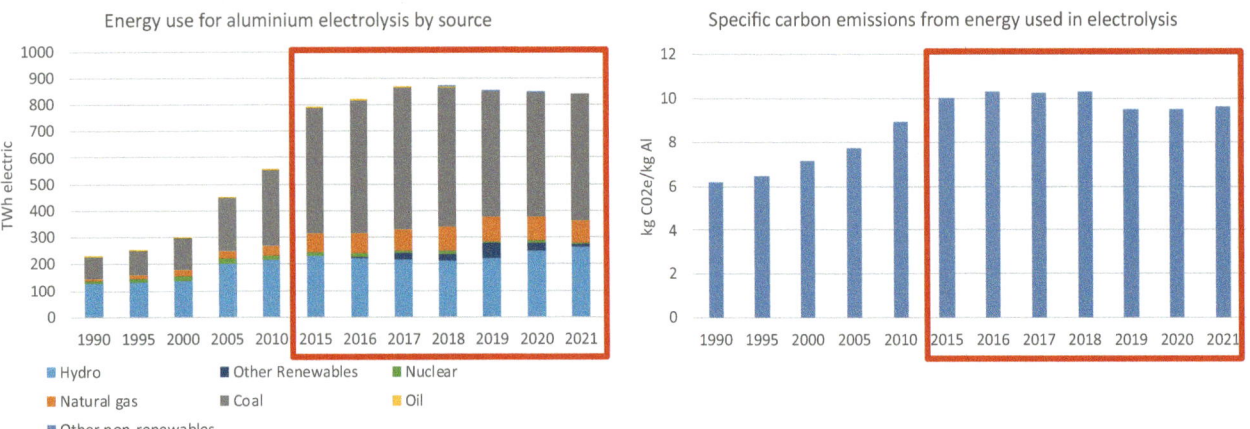

Fig. 3 On the left, the annual global energy used for aluminum production, by source, and on the right, there is the corresponding specific carbon footprint

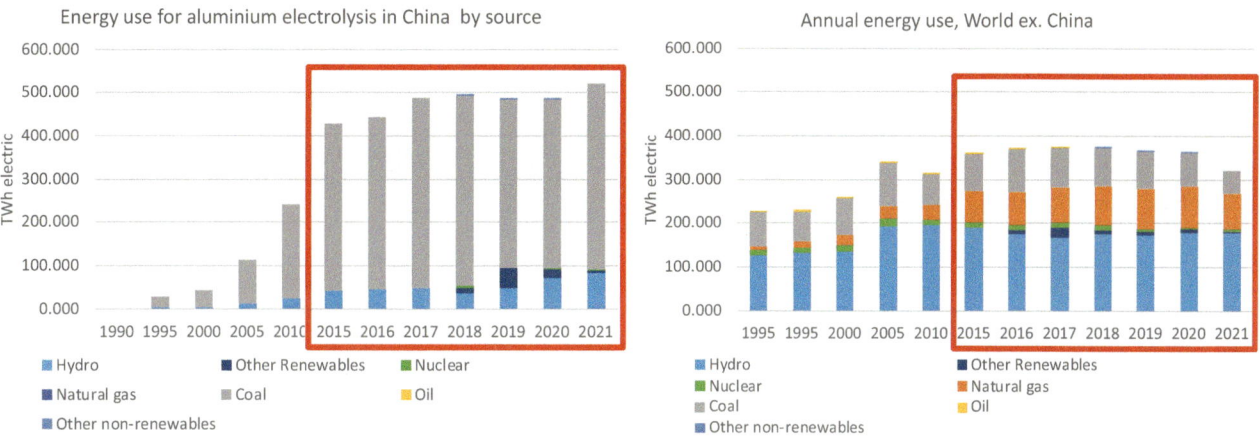

Fig. 4 On the left, the annual energy used in China for aluminum production, by source, and on the right, there is the corresponding energy use in the rest of the world

production. The data shows an increase in the use of hydropower in China, up by around 48 TWh/year from 2018 to 2021, which is equivalent to 3.5 million tonnes of aluminum annually.

This is consistent with declared goals from Chinese authorities, that they wish to shift aluminum production from coal power to renewables, and efforts to transfer aluminum production capacity to southwest China, for example, to Yunnan. The goal for the industry is to be 50% on renewables by 2045, according to an article published on the World Economic Forum website [50]. The Chinese energy carbon footprint was 12.4 kg CO_{2eq}/kg Al in 2021, while it was 13.7 kg CO_{2eq}/kg Al in 2015.

The amount of aluminum produced with electricity from coal was reduced in the rest of the world between 2020 and 2021, so the average energy carbon footprint from aluminum electrolysis for the world, other than China, was 4.6 kg CO_{2eq}/kg Al in 2021, down from 5.4 kg CO_{2eq}/kg Al in 2020. The overall result being a significant improvement in the carbon footprint. This shows that it is possible to get positive development in the carbon footprint, and that each smelter which shifts production from coal-fired power to renewable or nuclear matters for the overall global carbon footprint for aluminum production.

Discussion and Conclusions

As a metal that is important for society moving forward through the green energy transition, it is crucial to reduce the high carbon footprint associated with its production. Emissions from the electrolysis process can be practically eliminated by replacing the CO_2-evolving carbon anodes with oxygen-evolving inert anodes, and there is now increasing

optimism that the initiatives to develop an inert anode process, which is viable on an industrial scale, will indeed be successful. It is important to achieve this without increasing the energy demand for the electrolysis, as the majority of the current emissions come from the production of electric energy used for electrolysis. It is likely that the future of the aluminum industry will be the use of inert anodes, but that requires significant investment, and it is not clear that it is possible retrofit traditional Hall–Héroult cells into vertical electrode configuration without rebuilding the whole potline. Retrofit may be more applicable for horizontal configuration, but that would likely lead to a 15–20% higher energy consumption than for Hall–Héroult. Therefore, there will be demand for applying carbon capture with sequestration on existing smelters during their lifetime, in order to significantly reduce the emissions from the electrolysis process.

Decarbonization of the production of the alumina raw material is also important, as the emissions from that process step are similar to those from the electrolysis itself. There are ongoing efforts to reduce these emissions, mostly by addressing the energy source used, and improving energy efficiency by heat recovery and similar measures.

The most important emission source is, of course, the source of energy used in the electrolysis. It is good to be able to report some positive developments in the overall average global carbon footprint from the production of the energy used for electrolysis. That can be mostly attributed to China's coordinated efforts to shift aluminum production to renewable energy and the reduced production with coal power in the rest of the world. Results can already be reported in averages of the indirect emissions, which has been reduced from 10.3 to 9.6 kg CO_{2eq}/kg Al between 2018 and 2021. This shows that every smelter, which uses low carbon energy rather than coal thermal power, matters for the emissions.

References

1. International Aluminium Institute, "Primary Aluminium Production." Accessed: Aug. 23, 2022. [Online]. Available: https://international-aluminium.org/statistics/primary-aluminium-production/.

2. The International Aluminium Institute (IAI), "IAI MATERIAL FLOW MODEL – 2021 UPDATE." Accessed: Aug. 23, 2022. [Online]. Available: https://international-aluminium.org/resource/iai-material-flow-model-2021-update/.

3. World Aluminium, "Global Aluminium Cycle 2019," Oct. 04, 2021. Accessed: May 08, 2021. [Online]. Available: https://alucycle.international-aluminium.org/public-access/.

4. International Aluminium Institute, "Primary Aluminium Smelting Power Consumption," 2022. [Online]. Available: https://international-aluminium.org/statistics/primary-aluminium-smelting-power-consumption/.

5. G. Saevarsdottir, H. Kvande, and B. J. Welch, "Aluminum Production in the Times of Climate Change: The Global Challenge to Reduce the Carbon Footprint and Prevent Carbon Leakage," JOM, vol. 72, no. 1, pp. 296–308, Jan. 2020, https://doi.org/10.1007/s11837-019-03918-6.

6. Masson-Delmotte, V., P. Zhai, H.-O. Pörtner, D. Roberts, J. Skea, P.R. Shukla, A. Pirani, W. Moufouma-Okia, C. Péan, R. Pidcock, S. Connors, J.B.R. Matthews, Y. Chen, X. Zhou, M.I. Gomis, E. Lonnoy, T. Maycock, and M. Tignor, and T. Waterfield (eds.), "Global Warming of 1.5 °C.An IPCC Special Report on the impacts of global warming of 1.5 °C above pre-industrial levels and related global greenhouse gas emission pathways, in the context of strengthening the global response to the threat of climate change, sustainable development, and efforts to eradicate poverty," © 2019 Intergovernmental Panel on Climate Change., IPCC, 2018.

7. G. Saevarsdottir, H. Kvande, and B. J. Welch, "Reducing the Carbon Footprint: Aluminium Smelting with Changing Energy Systems and the Risk of Carbon Leakage," in Light Metals 2020, Cham, 2020, pp. 726–734.

8. Forster, P., T. Storelvmo, K. Armour, W. Collins, J.-L. Dufresne, D. Frame, D.J. Lunt, T. Mauritsen, M.D. Palmer, M. Watanabe, M. Wild, and H. Zhang, "Chapter 7," in The Earth's Energy Budget, Climate Feedbacks, and Climate Sensitivity. In Climate Change 2021: The Physical Science Basis. Contribution of Working Group I to the Sixth Assessment Report of the Intergovernmental Panel on Climate Change, Cambridge University Press, Cambridge, United Kingdom and New York, NY, USA, 2021, pp. 923–1054.

9. J. Thonstad, S. Rolseth, Й. Тонстэд, and С. Ролсет, "Low voltage Pfc emission from aluminium cells," J. Sib. Fed. Univ., vol. 10, pp. 30–36, 2017.

10. J. Marks and P. Nunez, "Updated Factors for Calculating PFC Emissions from Primary Aluminum Production," in Light Metals 2018, Cham, 2018, pp. 1519–1525.

11. International Aluminium Insitute (IAI), "Perfluorocarbon (PFC) Emissions." [Online]. Available: www.world-aluminium.org/statistics/perfluorocarbon-pfc-emissions/#data.

12. International Aluminium Institute, "Life Cycle Inventory Data and Environmental Metrics for the Primary Aluminium Industry, 2015 data," 2018. [Online]. Available: https://fluoridealert.org/wp-content/uploads/aluminum.life-cycle.2015.pdf.

13. International Aluminium Institute, "Greenhouse Gas Emissions - Primary Aluminium - data for 2021," 2022. [Online]. Available: https://international-aluminium.org/statistics/greenhouse-gas-emissions-intensity-primary-aluminium/.

14. P. Reny et al., "Hydro's New Karmøy Technology Pilot: Start-Up and Early Operation," in Light Metals 2021, Cham, 2021, pp. 608–617.

15. H. Kvande and BJ. Welch, "Kvande H, Welch BJ (2018) How to Minimize the Carbon Footprint from Aluminium Smelters. Light Metal Age 76(1):28–41," Light Met. Age, vol. 76, no. 1, pp. 28–41.

16. T. Faerden, G. Tranell, J. S. Bubetsky, T. Lindstad, and S.E. Olsen, "2006 IPCC Guidelines for National Greenhouse Gas Inventories," [Online]. Available: https://www.ipcc-nggip.iges.or.jp/public/2006gl/pdf/3_Volume3/V3_4_Ch4_Metal_Industry.pdf.

17. World Aluminium, "A life-cycle model of Chinese grid power and its application to the life cycle impact assessment of primary aluminium," World Aluminium, 2017. [Online]. Available: https://www.world-aluminium.org/media/filer_public/2017/06/29/lca_model_of_chinese_grid_power_and_application_to_aluminium_industry.pdf.

18. International Aluminium Institute, "GHG Emission Data for the Aluminium Sector (2005–2019)," 2020. [Online]. Available: https://www.world-aluminium.org/media/filer_public/2020/10/01/ghg_emissions_aluminium_sector_21_july_2020_read_only_25_september_2020. xlsx.

19. P. Nunez and S. Jones, "Cradle to gate: life cycle impact of primary aluminium production," Int. J. Life Cycle Assess., vol. 21, no. 11, pp. 1594–1604, Nov. 2016, https://doi.org/10.1007/s11367-015-1003-7.

20. Galasiu Ioan Rodica Galasiu and Jomar Thonstad., Inert Anodes for Aluminium Electrolysis., 1st ed. Düsseldorf: Aluminium-Verlag., 2007.

21. Y. He, K. Zhou, Y. Zhang, H. Xiong, and L. Zhang, "Recent progress of inert anodes for carbon-free aluminium electrolysis: a review and outlook," J Mater Chem A, vol. 9, no. 45, pp. 25272–25285, 2021, https://doi.org/10.1039/D1TA07198J.

22. A. S. Yasinskiy, S. K. Padamata, P. V. Polyakov, and A. V. Shabanov, "An update on inert anodes for aluminium electrolysis," Non-Ferr. Met. 2020, vol. 1, pp. 15–23.

23. J. Thonstad, "Anodic overvoltage on platinum in cryolite-alumina melts," Electrochimica Acta, vol. 13, no. 3, pp. 449–456, 1968, https://doi.org/10.1016/0013-4686(68)87016-1.

24. J. Thonstad, A. Kisza, and J Hives, "Anode overvoltage on metallic inert anodes in low-melting bath," in Light Metals 2006, 2006, pp. 373–380.

25. R. Keller, S. Rolseth, and J. Thonstad, "Mass transport considerations for the development of oxygen-evolving anodes in aluminum electrolysis," Electrochimica Acta, vol. 42, no. 12, pp. 1809–1817, 1997, https://doi.org/10.1016/S0013-4686(96)00381-7.

26. A. Redkin et al., "Recent Developments in Low-Temperature Electrolysis of Aluminum," 2013.

27. A. Solheim, "On the Feasibility of Using Low-Melting Bath to Accommodate Inert Anodes in Aluminium Electrolysis Cells," in Light Metals 2021, Cham, 2021, pp. 511–518.

28. I. Gallino, M. E. Kassner, and R. Busch, "Oxidation and corrosion of highly alloyed Cu–Fe–Ni as inert anode material for aluminum electrowinning in as-cast and homogenized conditions," Corros. Sci., vol. 63, pp. 293–303, 2012, https://doi.org/10.1016/j.corsci.2012.06.013.

29. A. Solheim, "Inert Anodes—the Blind Alley to Environmental Friendliness?," in Light Metals 2018, Cham, 2018, pp. 1253–1260.

30. S. K. Padamata, K. Singh, G. M. Haarberg, and G. Saevarsdottir, "Wettable TiB2 Cathode for Aluminum Electrolysis: A Review," J. Sustain. Metall., vol. 8, no. 2, pp. 613–624, Jun. 2022, https://doi.org/10.1007/s40831-022-00526-8.

31. A. K. Gupta and B. Basu, "Sustainable Primary Aluminium Production: Technology Status and Future Opportunities," Trans. Indian Inst. Met., pp. 1–16, 2019.

32. G. Gunnarsson, G. Óskarsdóttir, S. Frostason, and J. H. Magnússon, "Aluminum Electrolysis with Multiple Vertical

Non-consumable Electrodes in a Low Temperature Electrolyte," in *Light Metals 2019*, Cham, 2019, pp. 803–810.

33. V. Chapman, B. J. Welch, and M. Skyllas-Kazacos, "Anodic behaviour of oxidised Ni–Fe alloys in cryolite–alumina melts," *Electrochimica Acta*, vol. 56, no. 3, pp. 1227–1238, 2011, https://doi.org/10.1016/j.electacta.2010.10.095.

34. T. Nguyen and V. De Nora, "De Nora oxygen evolving inert metallic anode," in *ight Metals*, 2006, vol. 385.

35. "Rusal Produces Low Carbon Aluminum Using Inert Anode Technology," *Light Metal Age*, Apr. 15, 2021. Accessed: Aug. 05, 2021. [Online]. Available: https://www.lightmetalage.com/news/industry-news/smelting/rusal-produces-low-carbon-aluminum-using-inert-anode-technology/.

36. "Rusal Sets New Low For Carbon Dioxide Output In Primary Aluminium Production," *Aluminium Insider*, Apr. 14, 2021. Accessed: Aug. 05, 2021. [Online]. Available: https://aluminiuminsider.com/rusal-sets-new-low-for-carbon-dioxide-output-in-primary-aluminium-production/.

37. G. Turan, A. Zapantis, and et. Al, "Global Status of CCS 2021, CCS Accelerating to Net Zero," Global CCS Institute, 2021. [Online]. Available: https://www.globalccsinstitute.com/resources/global-status-report/download/.

38. A. Mathisen, H. Sørensen, N. Eldrup, R. Skagestad, M. Melaaen, and G. I. Müller, "Cost Optimised CO2 Capture from Aluminium Production," *Energy Procedia*, vol. 51, pp. 184–190, 2014, https://doi.org/10.1016/j.egypro.2014.07.021.

39. X. C. Shen, M. Hyland, B. Welch, and X. C. Shen, "TOP HEAT LOSS IN HALL-HEROULT CELLS," 2008.

40. R. Zhao, L. Gosselin, M. Fafard, and D. P. Ziegler, "REDUCED VENTILATION OF UPPER PART OF ALUMINUM SMELTING POT: POTENTIAL BENEFITS, DRAWBACKS, AND DESIGN MODIFICATIONS," p. 6.

41. O.-A. Lorentsen, A. Dyrøy, and M. Karlsen, "Handling CO2EQ from an Aluminum Electrolysis Cell," in *Essential Readings in Light Metals: Volume 2 Aluminum Reduction Technology*, G.

Bearne, M. Dupuis, and G. Tarcy, Eds. Cham: Springer International Publishing, 2016, pp. 975–980. https://doi.org/10.1007/978-3-319-48156-2_144.

42. K. Grjotheim and B. Welch, "Impact of Alternative Processes for Aluminum Production on Energy Requirements," *JOM*, vol. 33, no. 9, pp. 26–32, Sep. 1981, https://doi.org/10.1007/BF03339491.

43. K. Grjotheim, C. Krohn, and H. A. Øye, "Aluminiumherstellung aus Aluminumchlorid—eine kritische Betrachtung von Toth- und Alcoa-Verfahren," *Aluminium*, vol. 51, pp. 697–699, 1975.

44. M. A. Rhamdhani, M. A. Dewan, G. A. Brooks, B. J. Monaghan, and L. Prentice, "Alternative Al production methods," *Miner. Process. Extr. Metall.*, vol. 122, no. 2, pp. 87–104, 2013, https://doi.org/10.1179/1743285513Y.0000000036.

45. B. Øye, "Carbochlorination routes in production of Al," SINTEF Materials and Chemistry, 2018:00342, Feb. 2018. [Online]. Available: https://www.sintef.no/globalassets/project/higheff/deliverables-2018/d1.3_2018.01-chloride-process-aluminium.pdf.

46. C. Reilly, "Introducing the Age of Clean Aluminium Technologies." [Online]. Available: https://aluminumtechnologies.us.

47. H.E. Vatne, "Hydros Decarbonization Roadmap," presented at the Greener Aluminium Online Summit, Aluminium International Today, May 24, 2022.

48. International Aluminium Institute, "Primary Aluminium Smelting Energy Intensity," 2021. [Online]. Available: https://international-aluminium.org/statistics/primary-aluminium-smelting-energy-intensity/.

49. H. Kvande, G. Saevarsdottir, and B. J. Welch, "Direct and Indirect CO2 Equivalent Emissions from Primary Aluminium Production," in *Light Metals 2022*, Cham, 2022, pp. 998–1003.

50. Jorgen Sandstrom, Wen Zhang, Shaun Chau, Marc Huang, and Sheila Peng, "How China is decarbonizing the electricity supply for aluminium," *World Economic Forum*, Apr. 21, 2022. [Online]. Available: https://www.weforum.org/agenda/2022/04/how-china-is-decarbonizing-the-electricity-supply-for-aluminium/.

Individual Pot Sampling for Low-Voltage PFC Emissions Characterization and Reduction

Brian Zukas and Julie Young

Abstract

PFC emissions from aluminum smelting are characterized by two mechanisms, high-voltage generation (HV-PFCs) and low-voltage generation (LV-PFCs). HV-PFCs are emissions produced when a cell is undergoing an anode effect, typically >8 V. Modern cell technology has enabled pre-bake smelters to achieve low anode effect rates and durations, thereby lowering their HV-PFC emissions. LV-PFCs are the emissions produced when the cell voltage is below 8 V. Lacking a clear process signal to act upon, LV-PFCs can be difficult to treat. To tackle this issue, Alcoa has conducted sampling on individual electrolysis cells, during which continuous process and emissions data, as well as periodic bath samples, were collected. In the sampled cells, a variety of conditions were observed where LV-PFCs were generated. Understanding what was occurring at the cell level allowed for the identification of opportunities for process improvement, both for the reduction of LV-PFC emissions and cell performance.

Keywords

Low-voltage PFC • Process improvement • Emission measurements

Introduction

Improvements in the aluminum industry's understanding of anode effects and a dedicated effort to prevent their occurrence has led to a sharp decline in high voltage PFC emissions from smelters over recent decades [1]. Focused improvements on process control and potroom operations improvements have contributed to the decline in emissions [2]. When an anode effect does occur, the rapid passivation of multiple anodes by the insulating PFC gas layer serves to drastically increase the overall pot resistance and produces an easy to detect voltage spike. Detection of this voltage spike can then trigger an automatic anode effect termination routine to minimize its duration and resulting emissions. When low voltage PFC generation occurs, the passivated anode surface area is not large enough to significantly influence the overall pot voltage and produce an easily detectable process signal [3, 4]. To add to the challenge, LV-PFC emissions occur through reactions that require less energy than the reactions for HV-PFC emissions that occur during anode effects. The low-voltage reaction for CF_4 has a reversible electrode potential of 1.83 V versus the 2.59 V required for the high-voltage reaction [5].

With the large achievements in reducing anode effect rates and thereby HV-PFCs, LV-PFCs can comprise a large fraction of a pre-bake smelter's PFC emissions. A survey of Alcoa's pre-bake smelters showed that on the high-end, low-voltage emissions could comprise 81% of a smelter's total PFC emissions [6]. Without a clear process signal and multiple generation mechanisms possible in a pot, the LV-PFC fraction of emissions can be hard to treat [7]. The environmental justifications for working to reduce or eliminate the situations that generate these emissions are worthwhile in their own right [1]. There are also process improvements that can be achieved through these efforts. In situations where a pot is operating stably and generating LV-PFCs, the co-evolution of PFCs is partially consuming current that otherwise would be producing aluminum. If PFC generation and pot instability are occurring, identifying the cause of the instability would possibly have the double benefit of emissions reduction and improved pot stability leading to better pot performance. Measurement campaigns on individual pots where both emission and bath sampling are performed can be informative as to which LV-PFC

B. Zukas (✉) · J. Young
Alcoa, Continuous Improvement Center of Excellence, Smelting and Casting Technology, 859 White Cloud Rd., New Kensington, PA, USA
e-mail: brian.zukas@alcoa.com

© The Minerals, Metals & Materials Society 2023
S. Broek (ed.), *Light Metals 2023*, The Minerals, Metals & Materials Series,
https://doi.org/10.1007/978-3-031-22532-1_87

generation mechanisms are occurring at a given smelter. With this knowledge, targeted process improvements can be implemented in an attempt to treat the root cause of LV-PFC generation events.

Experimental Methods

Sampling campaigns were performed at two Alcoa pre-bake smelters, Massena and Bécancour. Massena utilizes P-225 pot technology including anode pairs equipped with continuous current monitoring and jack motors for independent positioning. Bécancour utilizes AP-18 pot technology with a conventional bridge. Both smelters operate in the range of 225–250 kA.

Emission sampling was done from the duct of an induvial pot. During the campaigns, two pots were sampled simultaneously. A stainless-steel tube was inserted into the duct, as seen at Massena in Fig. 1a, which was connected to Teflon tubing which carried the extracted gas stream to the equipment placed nearby in the courtyard, Fig. 1b. Emission measurements were made with ABB MB3000 FTIRs. To protect the FTIRs, the sampled emissions were conditioned by being passed through a particle trap and Drierite canisters. The FTIRs, placed inline, were used to continuously measure the PFC emissions generated by the cell at a frequency of 0.1 Hz. A pump was used to extract the emissions from the duct to an FTIR at a rate of approximately 2 L/min. Sampling was performed on an individual pot for a period of 24–36 h.

When bath sampling was performed, samples were collected every 5 min for 3 h. Samples were collected from the tap hole by using a sampling cup with a retractable cover, which allowed for the collection of a sample only when the cup was immersed in the molten bath. Samples were analyzed for weight% oxygen on a LECO RO600 Oxygen Determinator.

During the period of sampling, pot process data including potential, resistance, noise, feeding, and other related parameters were collected from the process control system as well.

Results and Discussion

Alcoa has previously used emission measurements collected from GTC stacks to improve understanding of the primary factors contributing to LV-PFC emissions and develop strategies for achieving reductions in these emissions. As reported previously, these efforts achieved reductions in anode effect rate, pot noise, and pot voltage [8, 9]. Performing measurements on stack emissions can also help to see large trends in a plant's PFC emissions.

HV-PFC emissions (>8 V) are clearly visible in stack emissions as large spikes in CF_4 levels and can be readily matched to anode effects recorded by the control system. Beyond the HV-PFC emissions, changes in the baseline CF_4 emissions can also be observed [6, 8]. Collecting time stamps for potroom operations such as anode setting can provide a high-level explanation for the observed baseline emissions, but not enough information is available to propose a detailed mechanism for the emissions generation seen. An operation like anode setting can provide multiple opportunities for LV-PFC generation. By moving data collection down to the pot level, a more detailed understanding can be developed of what is occurring in the pot when increases of LV-PFCs are seen. Specific improvements intended to address the root of the LV-PFC can then be proposed and implemented.

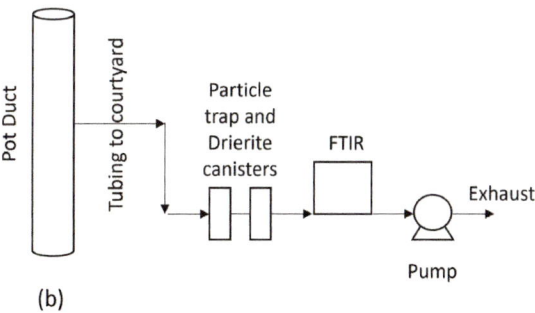

Fig. 1 **a** Tubing for emission sampling connected to the pot's duct at Massena. **b** Schematic of equipment connections for emission measurements

PFC Generation Around Anode Setting

Anode setting results in many temporary changes within a pot that are conducive to LV-PFC generation [4]. Primarily, the average current density of the pot is temporarily increased due to the presence of a cold anode or anode pair that is not drawing current. The increase in current density also tends to be non-uniform, with anodes neighboring a recent set experiencing a larger increase in load. This behavior is also occurring while the alumina and heat distribution and bath flow patterns are disturbed by the new anode(s). Performing measurements at the pot level can provide an understanding of what mechanisms are contributing to anode setting related LV-PFCs.

Figure 2 shows multiple PFC generation events occurring during the time following an anode set. Prior to setting, a period of low emissions was seen. This likely can be attributed to the average alumina concentration being high enough to prevent anode polarization during this time. Leading up to anode setting, the underfeed duration had begun to increase indicating that average alumina concentration had recently trended higher. The first large spike in emissions at point 1 coincides with the end of the first underfeed cycle following the set. From the alumina concentrations seen in Table 1, the expected rise in alumina concentration following an anode set occurs due to the cover material that inadvertently is added to the pot. No CF_4 is seen between set initiation and point 1, this is likely due to the average increase in alumina concentration helping to minimize anode polarization. During this time period, an underfeed is used to consume the additional alumina present in the bath. The spike in CF_4 that occurs indicates that the underfeed produced too low of an alumina concentration and caused localized alumina depletion leading to anode polarization. Previous studies on the impact of anode setting to a pot have shown that the time around point 1 is when the pot is still highly disturbed. Freeze can still be present on the anode, the anode has yet to pick up appreciable current, and the resulting impact to flow patterns changes the alumina distribution in the pot [10–12]. Table 1 shows that for bath samples collected from other pots around the time that point 1 occurs, the average alumina concentration is reduced back to what it would be pre-set. For the still disturbed state that a pot is in that soon after anode setting, maintaining a higher average alumina concentration for longer may help to prevent this type of feeding-related emissions events.

There is also evidence of an additional issue occurring on this pot during the time sampling was performed. Point 2 highlights a period of high pot instability, with direct shorting of an anode possibly occurring. This behavior combined with an overall increase in the variability of the pot potential when the final downward bridge move occurs

at point 3 indicates that an older anode may have been set improperly. An anode set too low relative to its neighbors will draw a higher current density, making LV-PFC generation easier. A decrease in ACD, like that occurring at point 3, would accentuate this issue [13].

Data collected from Massena illustrates how an increase in the current density of an anode or anode pair due to a decrease in the local ACD enables PFC generation. As described earlier, Massena utilizes P-225 pot technology with jack motors on each anode pair. In Fig. 3b, as pair 9 was lowered by the control system, the pair load increased and correspondingly so did CF_4 emissions, Fig. 3a. The increase in current density for the pair increased the anode potential above the overpotential needed to drive low voltage generation of CF_4.

Feeding Generated PFCs

LV-PFCs occurring at the end of underfeeds have been a previously reported phenomenon. The lower overall alumina concentration in the bath and slower feeding rate act to create zones of localized alumina depletion, causing anode polarization and therefore PFC generation. Zarouni et al. reported that 10% of the underfeeds in their study were observed generating PFCs [14]. For the two smelters studied here, Massena had 24% of observed underfeeds generate CF_4, while Bécancour had 53% of underfeeds generate CF_4.

For the 6 pots that were studied at Massena, the underfeeds were not seen to consistently generate CF_4. A series of underfeeds would occur that generated CF_4 and then pot conditions would change and the underfeeds would stop producing CF_4, as seen in Fig. 4a. For Massena, analysis of the pair COV at the end of each underfeed showed that LV-PFC generation was more likely to occur when the pairs were not carrying a balanced load. Pair load data was available as 3-min averaged data and COV was calculated from the 16 pair loads as

$$COV_t = \frac{\sigma_t}{\mu_t} \qquad (1)$$

where μ_t is the average load for the 16 pairs at time t and σ_t is the standard deviation of the pair loads at time t. The COV at the time each underfeed observed terminated was then extracted for the analysis in Fig. 4b. This could possibly indicate that the variation around the average alumina concentration at Massena can temporarily cause anode polarization, particularly when the anode current distribution is unbalanced. In general, though, the average alumina concentration is high enough to prevent anode polarization from occurring during underfeeding.

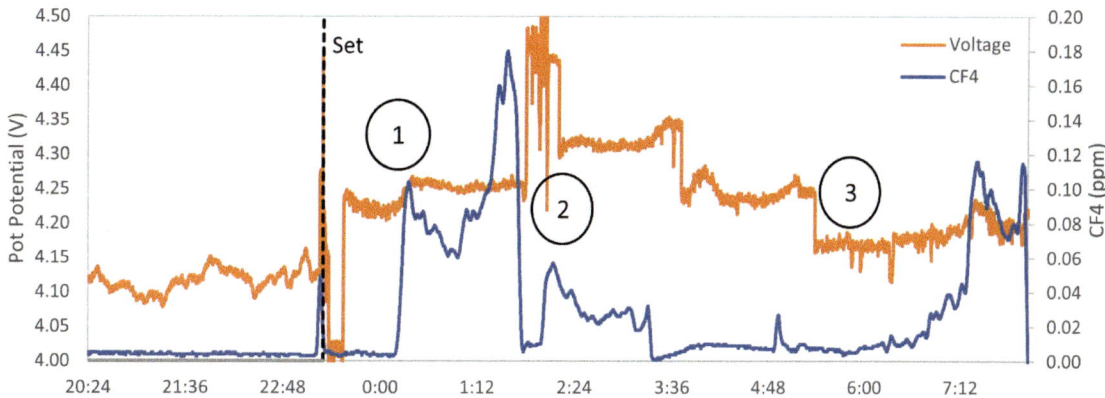

Fig. 2 Pot potential and CF$_4$ concentration versus time during an anode change at Bécancour. Start of anode change is marked by the dashed vertical line

Table 1 Average alumina concentration at different time points near anode setting from 4 pots at Bécancour

		Before set	Between set and 1st underfeed	~3.5 h after set
Avg. %Al$_2$O$_3$		2.5	4.0	2.3
95% CI		[1.7, 3.1]	[3.0, 4.5]	[1.6, 3.0]

Additional analysis of the underfeeds observed at Massena provided a feed control tuning opportunity. A review of Massena's control parameters showed that the delta-resistance target was set too high to ever be triggered by an underfeed. Delta resistance is a measure of the maximum resistance rise that occurs during underfeeds and can be used as a metric for determining underfeed endpoint. For all underfeeds at Massena, only slope was being used to exit. When a comparison was made between the delta-resistance value for underfeeds that generated CF$_4$ and those that did not, underfeeds generating CF$_4$ had a statistically significant higher value, Fig. 5. This analysis provided a value region for a delta-resistance target to trial at the smelter. This parameter adjustment, and other process improvements are currently being trialed on a group of test pots at Massena.

For the 6 pots that were studied at Bécancour, the number of underfeeds generating CF$_4$ was elevated due to two pots in the population that for over 20 h would generate CF$_4$ at the end of each underfeed. Data from one of these pots is shown in Fig. 6. During the 3 h that bath sampling was done on this pot, the alumina concentration near the end of the underfeed went below 2%, insert of Fig. 6. While Fig. 2 shows evidence of Bécancour having anodes on some pots set too low, it appears that increasing the average alumina concentration of the pots would also help to reduce CF$_4$ emissions by preventing those caused by underfeeds. During the sampling campaign at Bécancour, sampling was also done to characterize the resistance versus alumina curve as well. Analysis of that data showed that the critical alumina concentration for anode effect initiation was reached at approximately 1% alumina. As Solheim demonstrated

recently, when the average alumina concentration approaches the critical alumina concentration, the anode area passivated by PFC generation increases [3]. As this passivated anode area increases, the amount of PFCs measured would be expected to increase as well. The combination of bath and emissions sampling aids in developing an appropriate buffer from the critical alumina concentration for a pot to operate. Data from the bath and emissions sampling indicates that a small increase in average alumina concentration would help to reduce low-voltage PFC generation. The alumina concentration would still be low enough to encourage good dissolution and not promote mucking.

An important process difference is also present between the two smelters that should be noted, Massena is using flat (non-slotted) anodes while Bécancour is using slotted anodes. The use of slotted anodes confers many operational advantages to aluminum electrolysis: lower operating voltages, improved pot stability, and increased current efficiency [15]. Due to the change in gas flow patterns with the use of slotted anodes, some recent work has hypothesized that it is easier for localized alumina depletion and therefore LV-PFC generation to occur with slotted anodes [5, 16]. The data collected from Massena and Bécancour indicate that there are other factors to consider as well. One of the main advantages of the slotted anode is the decrease in effective current density that they offer. This is achieved through the more efficient gas release behavior of these anodes which reduces the ability of large gas films to form, exposing more anode surface. This enables slotted anodes to better tolerate the partial passivation that occurs during localized alumina depletion or low average bath concentration without causing

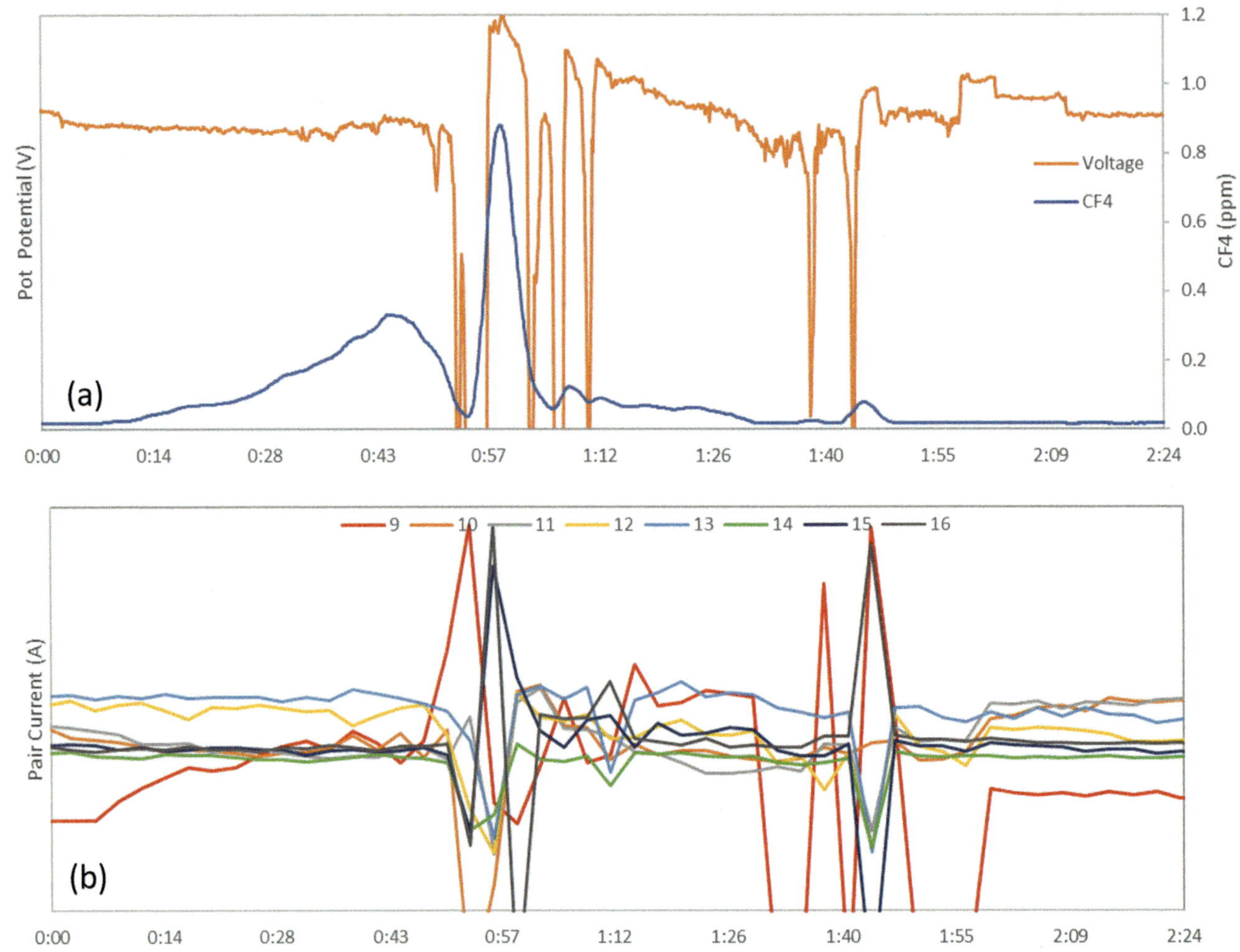

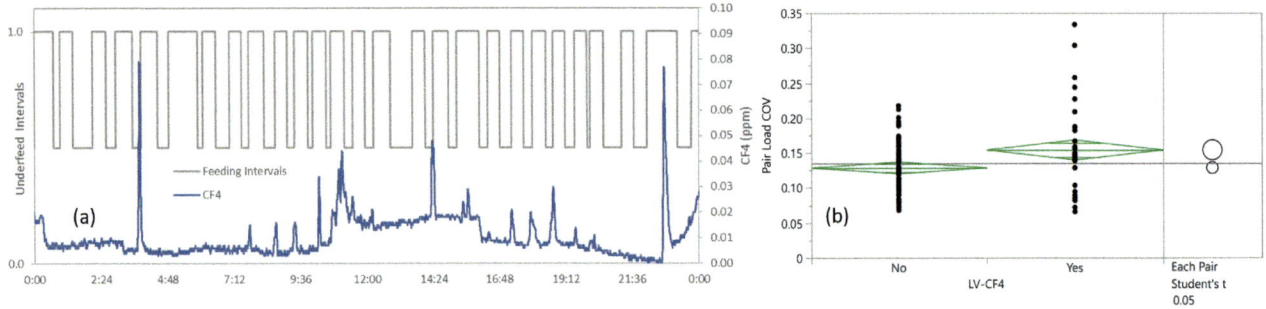

Fig. 3 **a** Pot potential and CF_4 emission versus time during a period of anode pair movement. **b** Pair current versus time during the period of low-voltage CF_4 generation

Fig. 4 **a** Underfeed intervals and corresponding CF_4 emissions versus time at Massena. Values of the feeding interval less than 1 indicate underfeeding. **b** Means testing of the pair load COV at the end underfeeds that do and do not produce LV-CF_4

an anode effect propagation. Kolås et al. observed that older anodes, which would have had their slotted portion consumed, were more likely to passivate than younger anodes [17]. The critical alumina concentration for anode effect initiation at Massena, with flat anodes, was found to be

slightly higher at approximately 1.4%. Figure 6 shows that a pot with slotted anodes can operate stably for an extended time at that concentration range without causing a high-voltage anode effect. With slotted anodes possessing the ability to operate at low alumina concentrations, good

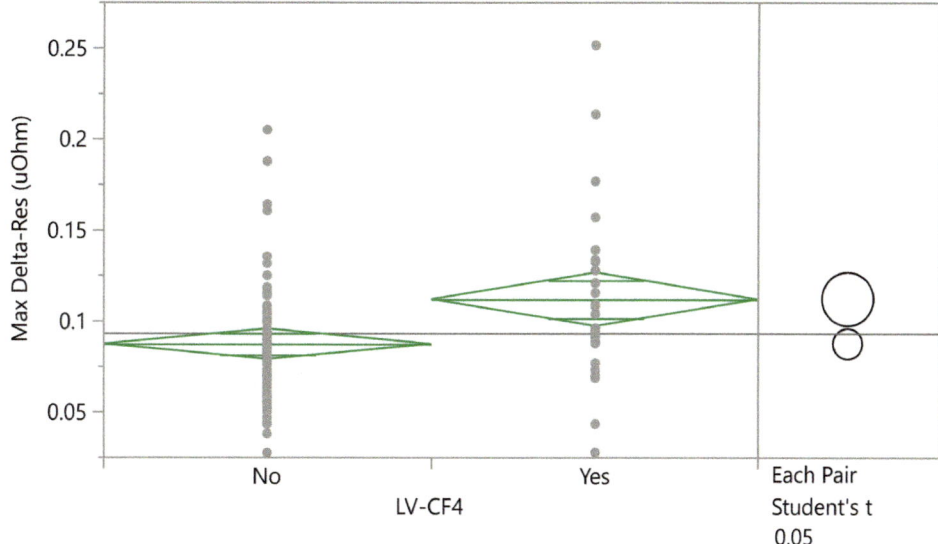

Fig. 5 Means testing of the maximum delta-resistance value measured during underfeeds that did and did not generate LV-CF$_4$

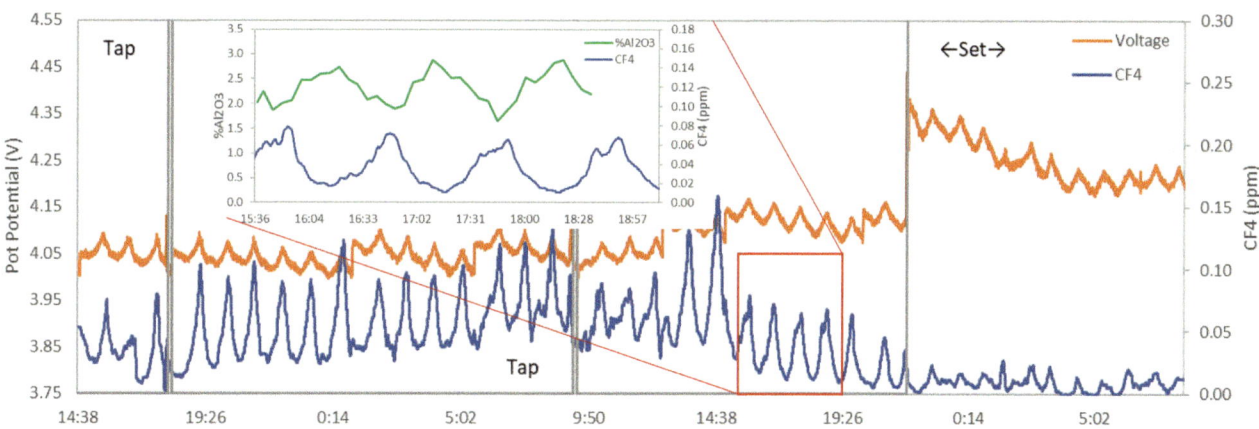

Fig. 6 Pot potential and CF$_4$ concentration versus time, insert shows alumina concentration from bath sampling performed during the highlighted time period

feed control should be a priority to keep alumina concentration high enough to minimize/prevent LV-PFC generation.

Summary

Emission sampling in conjunction with bath sampling on individual pots can provide insight into the mechanisms driving low-voltage PFC emissions, providing opportunities for target process improvements. LV-PFCs generated during anode setting at Bécancour could be attributed to aggressive alumina concentration reduction and possibly improper anode setting. LV-PFCs during underfeeding was seen at both plants. With access to the anode pair load data at Massena, it was possible to demonstrate that unbalanced current distribution was contributing to its generation. At Bécancour, bath sampling indicated that some pots were operating with very low alumina concentrations. Addressing these issues would not only be expected to reduce the LV-PFC emissions at these plants but also improve pot line performance. The findings from the sampling campaigns have been applied to or will be applied to test groups of pots. Follow-up studies will be conducted to determine if these actions were effective at reducing LV-PFC emissions. Sampling campaigns will also be conducted at other Alcoa smelters to determine the mechanisms generating LV-PFC at these plants and identify any areas of improvement.

Acknowledgements The authors would like to thank Xiangwen Wang and Luis Espinoza-Nava for their assistance with this work. In addition, they would also like to thank the technical teams at Massena and Bécancour for their assistance during the setup and execution of these measurement campaigns.

References

1. A. T. Tabereaux and D. S. Wong, "Awakening of the Aluminum Industry to PFC Emissions and Global Warming," *Light Metals*, pp. 554–564, 2021.

2. N. R. Dando, L. Sylvain, J. Fleckenstein, C. Kato, V. Van Son and L. Coleman, "Sustainable Anode Effect Based Perfluorocarbon Emission Reduction," *Light Metals*, pp. 325–328, 2011.

3. A. Solheim, "Reflections on the Low-Voltage Anode Effect in Aluminimum Electrolysis Cells," *Light Metals*, pp. 971–978, 2022.

4. D. S. Wong, A. Tabereaux and P. Lavoie, "Anode Effect Phenomena during Conventional AEs, Low Voltage Propagating AEs & Non-Propagating AEs," *Light Metals*, pp. 529–534, 2014.

5. D. S. Wong and B. Welch, "PFCs and Anode Products-Myths, Minimisation and IPCC Method Updates to Quantify the Environmental Impact," in *Proceedings from the 12th Australasian Aluminium Smelting Technology Conference*, Queenstown, New Zealand, 2018.

6. N. R. Dando, N. Menegazzo, L. Espinoza-Nava, N. Westenford and E. Batista, "Non Anode Effect PFCs: Measurement Considerations and Potential Impacts," *Light Metals*, pp. 551–554, 2015.

7. L. Dion, L. I. Kiss, S. Poncsak and C.-L. Lagace, "Prediction of Low Voltage Tetrafluoromethane Emissions Based on the Operating Conditions of an Aluminum Electrolysis Cell," *JOM*, pp. 2472–2482, 2016.

8. E. Batista, L. Espinova-Nava, C. Tulga, R. Marcotte, Y. Duchemin and P. Manolescu, "Low Voltage PFC Measurements and Potential Alternatives to Reduce Them at Alcoa Smelters," *Light Metals*, pp. 1463–1467, 2018.

9. E. Batista, N. R. Dando, N. Menegazzo and L. Espinoza-Nava, "Sustainable Reduction of Anode Effect and Low Voltage PFC Emissions," *Light Metals*, pp. 537–540, 2016.

10. D. Picard, J. Tessier, D. Gauthier, H. Alamdari and M. Fafard, "In Situ Evolution of the Frozen Layer Under Cold Anode," *Light Metals*, pp. 795–802, 2019.

11. C.-J. Wong, Y. Yao, J. Boa, M. Skyllas-Kazacos, B. J. Welch and A. Jassim, "Modeling Anode Current Pickup After Setting," *Light Metals*, pp. 351–358, 2021.

12. V. Bojarevics, "In-Line Cell Position and Anode Change Effects on the Alumina Dissolution," *Light Metals*, pp. 584–590, 2021.

13. L. Dion, L. I. Kiss, S. Poncsak and C.-L. Lagace, "Simulator of Non-homogenous Alumina and Current Distribution in an Aluminum Electrolysis Cell to Predict Low-Voltage Anode Effects," *Metallurgical and Materials Transcations B*, vol. 49B, pp. 737-755, 2018.

14. A. Zarouni, M. Reverdy, A. A. Zarouni and K. G. Venkatasubramaniam, "A Study of Low Voltage PFC Emissions at Dubal," *Light Metals*, pp. 859–863, 2013.

15. X. Wang, G. Tarcy, S. Whelan, S. Porto, C. Ritter, B. Ouellet, G. Homley, A. Morphett, G. Proulx, S. Lindsay and J. Bruggerman, "Development and Deployment of Slotted Anode Technology at Alcoa," *Light Metals*, pp. 299–304, 2007.

16. A. Jassim, A. A. A. Akhmetov, D. Whitfield and B. Welch, "Understanding of Co-Evolution of PFC Emissions in EGA Smelter with Opportunities and Challenges to Lower the Emissions," *Light Metals*, pp. 829–836, 2019.

17. S. Kolas, P. McIntosh and A. Solheim, "High Frequency Measurements of Current Through Individual Anodes: Some Results From Measurement Campaigns at Hydro," *Light Metals*, pp. 729–734, 2015.

Determination of PFC with Canister Sampling and Medusa GC–MS Analysis in Comparison to General IPCC Estimation Methods

Henrik Åsheim, Morten Isaksen, Ove Hermansen, Norbert Schmidbauer, and Chris Lunder

Abstract

Most aluminum smelters today report on their Perfluorocarbon (PFC) emissions by a method that is derived from the IPCC Guidelines for National Greenhouse Inventories of 2006 or 2019. Often the default industry emission factors have been employed (Tier 1/Tier 2), however, many companies are looking into acquiring plant specific factors (Tier 3). Hydro and Norwegian Institute for Air Research (NILU) have been testing a method of time-integrated air sampling where canisters are used to extract gas from different smelter locations that can later be measured by Medusa GC–MS. Sampling duration is determined by gas flow and canister size and the PFC detection limit is down to ambient air—making the method very applicable to low emitting smelters. Preliminary results from canister sampling deviate from 2006 IPCC Tier 2 slope estimation, whereby the measurements, a total PFC method, show a CO_2e reduction of about 28% compared to the estimation.

Keywords

Environment • Anode effect • PFC • Total PFC emission • GHG • Emission monitoring • CO_2e

Introduction

The primary aluminum industry is one of the main contributors of anthropogenic tetrafluoromethane (CF_4) and hexafluoroethane (C_2F_6) to the atmosphere. These perfluorocarbons (PFCs) come with high global warming potentials (GWP_{100}) of 7380 and 12 400 and estimated atmospheric lifetimes of 50 000 and 10 000 years according to the latest assessment report (AR6) from the Intergovernmental Panel on Climate Change (IPCC) [1]. Any calculation later on will use IPCC AR5 [2] values (GWP_{100} of 6630 and 11 100 for CF_4 and C_2F_6, respectively) as that is what is currently required for formal reporting in Norway.

Estimates from global atmospheric measurements are suggesting periods of both over-accounted emissions (nearly 40% between 1996 and 2002) and under-accounted emissions (about 40% between 2003 and 2010) [3]. Despite significant progress from the aluminum industry in understanding and reducing their emissions over the last three decades, the global model emissions for CF_4 and C_2F_6—using atmospheric measurements as input—continue to rise and are significantly larger than those currently reported by industry and governments. There is still strong evidence for unaccounted PFC-emissions from the aluminum industry as a whole, but also quite large regional discrepancies with special focus on East Asia. Environmental legislation of PFC-emissions may in the near future move towards online-emission monitoring, or time-integrated sampling methods in combination with precise off-line measurement techniques.

Emission of perfluorocarbon gases from aluminum production is a process upset condition that occurs mainly because of insufficient dissolved oxide at the anode/bath interface for the passing current, thus elevating the voltage and introducing other reactions like decomposition of cryolite bath and eventually evolution of PFCs. To which extent this occurs will vary with several factors such as cell technology, feeding and control strategy, metal hydro dynamics,

H. Åsheim (✉)
Hydro Aluminium AS, Verksvegen 1, 6882 Øvre Årdal, Norway
e-mail: Henrik.asheim@hydro.com

M. Isaksen
Hydro Aluminium AS, Hydroveien 67, 3936 Porsgrunn, Norway

O. Hermansen · N. Schmidbauer · C. Lunder
Norwegian Institute for Air Research, NILU, 2027 Kjeller, Norway

S. Broek (ed.), *Light Metals 2023*, The Minerals, Metals & Materials Series,
https://doi.org/10.1007/978-3-031-22532-1_88

quality of operations etc. For a long time, release of PFCs were connected to a specific phenomenon termed the (conventional) anode effect (AE) with elevated voltage (above a set threshold, usually 8 or 10 V), often around 20 V or higher, and irregular current distribution, but newer research clearly shows that PFCs are evolved outside of these events [4–6], and that it may be substantial, in some cases even surpass emissions from conventional AEs [7]. In recent years the two emission routes have been labeled high-voltage (HV) PFC emissions and low-voltage (LV) PFC emission.

The primary aluminum industry in Norway is currently reporting their PFC emissions on rules based on the 2006 IPCC Guidelines for National Greenhouse Gas Inventories [8]. In 2019 IPCC updated the guidelines and produced a refinement to the 2006 guidelines [9], but that is not yet approved for reporting. The main differences are some updated factors for Tier 1 and Tier 2 reporting a few new suggested HV PFC estimation methods and the addition of LV PFC emissions. The changes will require the reporting of total PFC emission, however, not necessary as individual contributions. The most common way to report on PFC emissions is with the 2006 IPCC Tier 2 slope method, whereby process data anode effect minutes (AEM) and metal production is combined with technology specific emission factors. However, some smelters have acquired plant specific factors through measurement campaigns. This is beneficial from a reporting perspective as it should be more accurate. In many cases, if you're charged for emissions there can also be a monetary incentive to acquire smelter specific emission factors.

Measurement campaigns have often been performed with FTIRs (Fourier transform infrared spectroscopy), but canister sampling followed by off-line analysis with Medusa gas chromatography-mass spectrometry (GC–MS) was tested by CSIRO at Hydro Australia's Kurri Kurri smelter in 2008/2009. The study showed that this method was very cost effective, precise, and very suitable for long-time sampling [10]. Little has been reported for some time, but interest in the method has lately been renewed [4, 5].

The aim of this work is to investigate to what extent canister time-integrated sampling together with Medusa GC–MS methodology can be used and further improved as an alternative to the traditional attempts to quantify PFC-emission.

Background

Time Integrated Sampling

"Time-integrated sampling or concentration representative sampling is when a sample is taken over a period of time, in which the concentration of a species is equal to the averaged concentration of an imaginary or real continuous analysis of the same species over the same time period". Since the evolution of PFCs from aluminum production is very variable, longer duration time-integrated sampling should increase accuracy. PFCs are also very volatile, stable, and long-lived compounds, and should as such be ideal candidates to be stored in canisters without analytical impediment.

Analysis with Medusa GC–MS

The Medusa GC–MS is the latest development in analysis equipment from the Advanced Global Atmospheric Gases Experiment (AGAGE) network that NILU is a part of. Medusa is a preconcentration unit that revolves around a cold plate of temperature $-175\ °C$, which together with a resistive heater can control the temperature of two traps independently between $-165\ °C$ and $+200\ °C$. Coupled with trap absorbents this system can effectively separate the desired compounds from more abundant gases that would otherwise interfere with chromatographic separation or mass spectrometric detection, such as N_2, O_2 and H_2O. After separation, distillation and refocusing, the analytes are injected into an Agilent 5973 GC–MS producing sharp and reproducible results. A more thorough description of the analytical system is provided in the literature [11, 12].

A software system has been designed that controls both the "front-end" of the system and the GC–MS in selective ion mode. Blanks and instrument linearities are routinely measured. An important advance in the Medusa is its ability to check its linearity by injecting a wide range of standard gas volumes. Such linearity and composition-independence are critical to accurate calibration, especially when propagating synthetic primary standards or when measuring samples spanning wide concentration ranges. The Medusa system uses a high precision integrating mass flow controller (MFC) for improved measurement of sample volumes. The

Medusa systems are producing exceptional routine precisions. The practice of alternating ambient air and calibration gas analyses obtain the highest precision measurements. By using quantifier (target) and qualifier ions for each measured species, the Medusa also offers improved peak identification and reduced susceptibility to interference by co-eluting species.

The Medusa measurements of the global atmospheric background concentrations of halogenated gases are very precise—usually within less than 1%—and can deal with concentration levels far lower than FTIR or QCL (Quantum cascade laser) instrumentation.

Experimental

Hydro Aluminium Husnes consists of two potlines, each comprising 200 point-fed cells of end-to-end technology. Line B was recently refurbished and restarted in the first half of 2021—after being closed for about a decade. Experiments have been conducted at this plant with canister sampling and analysis by Medusa GC–MS on two occasions over the past couple of years—one before, and one just after the restart.

Spot Samples and First Examination

The plant was first visited in mid-August of 2020 to examine suitable sampling spots and the technical assistance on site. The goal was to gain experience with concentration levels, sample amount, necessary dilution steps and possible interferences with co-emitting compounds like water, SO_2, CO_2, other halocarbons or hydrocarbons, or particles. In mid-September of the same year, plant-personnel took 10 canister samples around the plant, both pre- and post-scrubber as well as on the plant grounds. 3 l canisters were used, and all sampled for 20 min without use of any flow restrictor—only a particle filter on each side of a HF-scrubber filled with smelter grade gamma alumina. The samples were still about 250 mbar under ambient pressure after sampling.

The samples were analyzed by Medusa GC–MS at NILU's laboratory for the whole AGAGE range of halogenated compounds.

Long Term Stack Sampling

CCanister sampling can be performed over several weeks given low flow, large canisters, or both. NILU rented four sets of 30-l stainless steel canisters from CSIRO together with flow restrictors (0.004″ ID tube 1.2 m length) and pressure loggers (Omega OM-CP-PR2000-100-A) for the real sample flow. The same equipment that was successfully used during the Kurri Kurri measurements of 2008/2009 [10] and a good starting point for comparison of sampling procedures.

In June 2021 further measurements were conducted at the Husnes plant. The four 30-l canisters were attached to four of the stack exhausts (RA1a, RA1b, RA2a and RA2b). The canisters were kept in their transport boxes which were later covered with a plastic bag to protect from weather and fastened to a nearby fence. An alumina filter was placed in the pathway of the sampling tube with the tip entering the chimney. There was some uncertainty as to whether the sampling capillary should be facing upwards or downwards so two of each were chosen. When ready the valves were opened, and the canisters started sampling. An image showing the setup is provided in Fig. 1. Six weeks later the valves were closed, and the filled canisters were sent for analysis by Medusa GC–MS.

Medusa Measurements

All in all, 14 samples were collected in relation to this survey. 10 samples with NILU's 3-l canisters and 4 samples with CSIRO's 30-l canisters. The canisters were still at sub-ambient pressure after sampling ended. The pressure was measured, and the canisters were pressurized with Zero-air to an end pressure that allowed several analyses of its content. The end pressure after Zero-air addition determines the dilution factor for the sample. For samples from the roof top or plant grounds low dilution was chosen, whereas a higher dilution was chosen for samples from the chimney.

The sampling volume introduced to the Medusa preconcentration unit varied from one milliliter up to two liters of sample. High-precision loops of 1 and 10 ml volumes were used when high concentrations were expected. The content of the loop was then flushed to the Medusa with a Zero-air stream. The sampling time of the Medusa was set to 20 min —so multiple loop injections could be completed within that timeframe—allowing sample volumes from 1 to 100 ml with 1 or 10 ml steps. In that way, a volume could be chosen to best match the concentrations of the standard runs which were performed before and after each sample run. An image of the instrument running samples at NILU's Kjeller laboratory can be viewed in Fig. 2 [13].

Results and Discussion

Preliminary Spot Samples

Results from the 10-canister preliminary sampling campaign are provided in Table 1 and visualized in Fig. 3. Also included in the table, is a short summary of the two anode

Fig. 1 30-l canister ready to sample the stack. The white and blue cylinder is the alumina filter, and the sampling capillary has already been placed inside the chimney. The insert shows the canister inside its box

effects that were recorded within the timeframe and geographical area of the sampling. Only line A (Hall A) was active during this sampling period. Sample 1 and 2 showed real atmospheric background concentrations—very close to AGAGE background concentration for the northern hemisphere for September 2020 for both CF_4 and C_2F_6. Those samples were taken at ground level close to Hall A and at ground level between Hall A and Hall B, and it was a very small concern that these heavy gases could linger around the plant providing slightly elevated background concentrations, however, that did not materialize. Additionally, a 131 s anode effect occurred during Sample 1, although there is little evidence of it in the measured concentrations.

Samples 3, 4, 5 and 6 were taken at the roof of Hall A with an extension tube down into the roof lamellae. The results show slightly increased levels of both CF_4 and C_2F_6. Those are short-time measurements and thus not representative—but showing small variations of the concentrations. A short anode effect of 15 s was recorded during Sample 3. It is difficult to know if any PFC from that event escaped the cell and were picked up by the canister. Values are only slightly elevated and lower than the other roof samples without the associated anode effect. Those emissions would be undetected with FTIR or QCL instrumentation.

Samples 7 and 8 were samples taken in the stack of Hall A north and south and showed mixing ratios of about 950 ppt for CF_4 and about 70 ppt for C_2F_6. Assuming similar concentration for the whole line and year, and iterating over all the gas treatment centers this would equate to less than 350 t CO_2e/y for about 90 kt Al/y for Hall A.

Samples 9 and 10 were raw-gas taken in the pipe before entering the seawater-scrubber. The mixing ratios are somewhat higher, but those samples were taken one hour later that day. It could also be that some false air is entering the system, thus decreasing the concentrations seen out of the chimney. The exhaust before the wet-scrubber is much richer in sulfur compounds, particles, and HF than towards the end-stack. Since the analysis of the PFCs was not affected by this—sampling at this spot instead of the stack would be a good alternative. The exhaust in the stack is saturated with water which can cause problems to both on-line instruments as well as off-line sampling.

Another slightly puzzling observation is the different ratio between C_2F_6 and CF_4 seen for some of the measurements. It can visually be seen from Fig. 3 that Samples 7 and 8 have a lower C_2F_6/CF_4 ratio than Samples 9 and 10. On a mass basis these ratios span from 0.086 to 0.131.

Long Term Stack Sampling

During this sampling about 200 cells from line B had just been started. It had been a somewhat challenging time in relation to anode effects for the restarted line, with issues related to feeders/breakers, a large new inexperienced workforce and some production parameters that had not yet been fully optimized. Hence, a much higher concentration than during the spot samples was expected.

When the canisters were collected it was immediately noticed that the filter on two of the samples had filled with

Fig. 2 30-l canisters connected to the Medusa at NILU, Kjeller. In the background several 3-l canisters can be spotted

water. It turns out that the two samples with capillary facing upwards had clogged, while the two other samples were as expected. Even though having the capillary pointing downwards worked this time it may be worthwhile looking into other ways of dealing with water and/or particles, and perhaps also other less moist sampling locations.

The canister pressure buildup during sampling at stack RA1a and RA1b is presented in Fig. 4. Small fluctuations in the pressure profile correlate with temperature change that both affect the sample itself as well as the volume of the container it resides in. Overall, the sampling rate is close to linear, especially the first 4 weeks, but it gradually declines over the last two weeks. For accurate results sampling should cease well before ambient pressure is reached, which will markedly reduce the effective volume of the flask. It is, however, possible to make some mathematical fit corrections, like the quadratic fit corrections CSIRO has employed for most of their flow restrictors. Another option is to look at

other types of flow restrictors like critical orifices. They are designed to have a constant volume flow if the pressure difference is above a certain level (0.55 bar). Beyond this limit a completely constant flow is not guaranteed. Thus, for completely linear sampling the effective capacity of a flask will here too be drastically reduced, still performance beyond this pressure might up to a certain point be reasonably good. And even though the volumetric flow is constant, the actual flow does vary with temperature and pressure.

The results from the two working samples are presented in Table 2. The samples were analyzed twice as a 10 ml loop injection from the 30-l canisters with an initial dilution of about 10. Variation between runs on one sample is for CF_4 within fractions of a percent and is a testament to the precision of the method. The variation is somewhat higher for C_2F_6 at about a couple of percent. The dilution is adapted to expected CF_4 levels and thus C_2F_6 might end up outside the linearity of the standard. C_2F_6 could also be the basis for

Table 1 Overview of spot sample location, duration, and their measured PFC concentrations. Any relevant anode effects are also specified

Sample	Location	Period	CF$_4$ (pptv)	C$_2$F$_6$ (pptv)	AE time	AE dur. (s)
1	Outside hall A east	09:15–09:35	86.68	4.73	09:28	131
2	Outside between A and B	08:50–09:10	86.07	4.88		
3	Roof pos. 1	10:05–10:25	89.04	5.29	10:11	15
4	Roof pos. 2	10:25–10:45	106.85	8.49		
5	Roof pos. 3	10:47–11:07	98.09	6.63		
6	Roof pos. 4	11:10–11:30	91.22	5.55		
7	RA1a stack north	12:37–12:57	957.19	79.73		
8	RA1b stack south	12:59–13:20	924.70	73.66		
9	RA1a pre scrubber	13:35–13:55	1333.19	73.85		
10	RA1b pre scrubber	13:55–14:15	1552.48	91.91		

Fig. 3 CF$_4$ and C$_2$F$_6$ concentrations for the spot samples collected at Husnes in September 2020

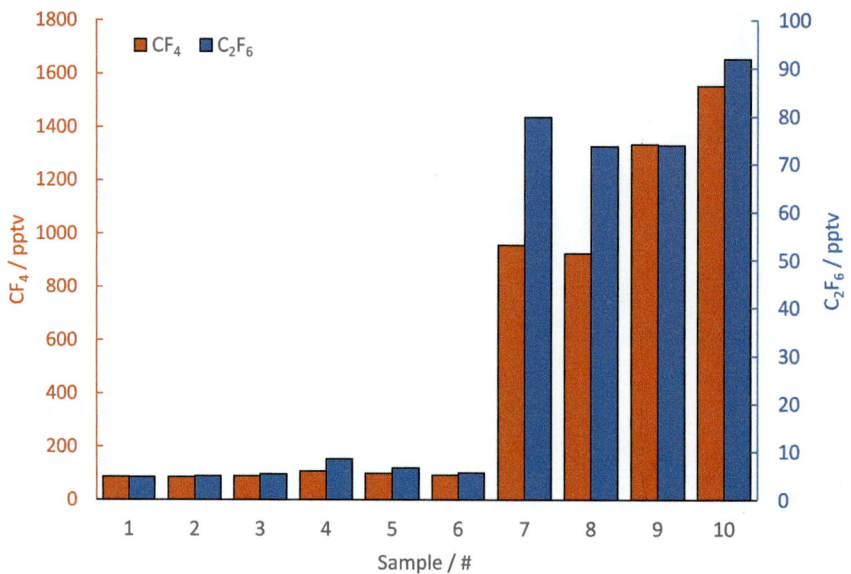

dilution to improve repeatability within runs. On a mass basis the ratio C$_2$F$_6$/CF$_4$ varies from 0.098 to 0.105 between runs, closely resembling previous observations in the literature. The concentrations reported for the long-term sampling was as expected much higher than during the spot samples, more than 100 times higher and is attributed to the mentioned pot control and anode effect related challenges.

A summary of HV anode effect data for cells connected to gas treatment center RA1 is provided in Table 3. These are high numbers for a plant that used to have anode effect frequency (AEF) / anode effect duration (AED) / anode effect minutes (AEM) around 0.15 /1.1/ 0.17. Over the course of the year the anode effect challenges with line B

had much improved and by the end of the year AEM for line B were below 0.1. This development indicates, together with the non-linearity from the sampling (Fig. 4), that a period with likely higher concentration of PFC will contribute more strongly than a period of lower expected PFC emission.

Table 4 shows the summed emissions for the whole 6-week sampling period, both the results from the canister measurements as well the results of the 2006 IPCC Tier 2 slope estimation which is currently used for emissions reporting. There is a considerable discrepancy between the numbers, putting the measured CO$_2$e values about 29% lower than the estimation, especially considering canister sampling is a total PFC method while only HV PFC

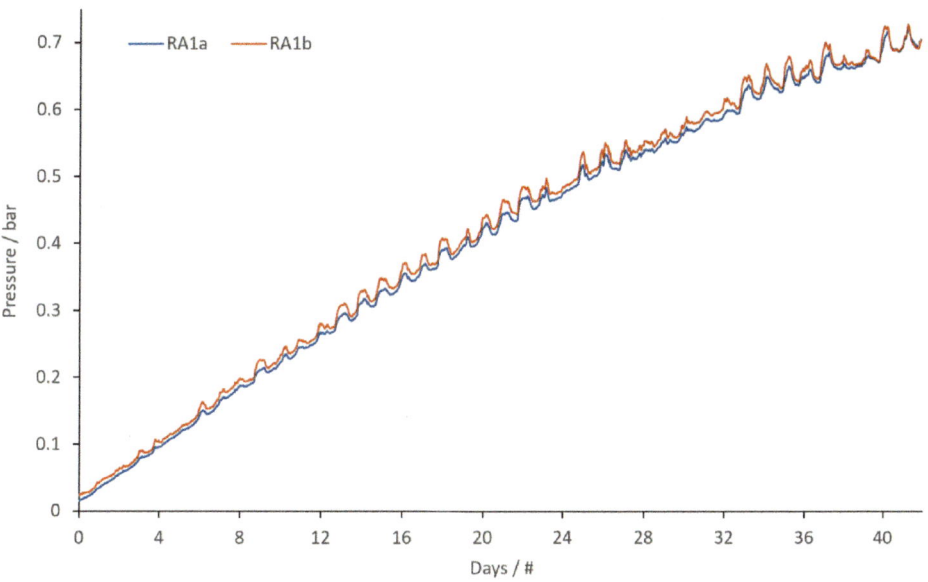

Fig. 4 Canister pressure-buildup during sampling at stacks RA1a and RA1b. The pressure curves are completely analogous, and the smaller fluctuations are mainly stemming from diurnal temperature changes

Table 2 Results from Medusa GC–MS analysis and the flow of the respective stacks during the long-term measurement campaign in 2021

Stack	Run	CF_4 (pptv)	C_2F_6 (pptv)	Flow (kNm^3/h)
RA1a	1	189 140.29	12 642.01	342
RA1a	2	189 086.29	12 482.97	342
RA1b	1	161 829.52	10 158.66	377
RA1b	2	161 825.06	10 428.91	377

Table 3 Anode effect and related parameters for emissions calculation for RA1 cell-section

Parameter	Value
Duration (d)	42
cell-days	4944
# AE	1518
AEF	0.307
AED (min)	2.318
AEM (min/cell-day)	0.712

emissions were treated when the 2006 IPCC guidelines were described. The measurements do, however, currently only cover the stacks and emissions through the roof were not measured during the long-time sampling campaign. Additionally, it can be difficult to get representative samples—maybe especially at Husnes where the roof is a ∼1 km long exhaust-line, per line. Some spot samples were taken at the roof on the preliminary campaign (Table 1) and using the highest as basis (Sample 4) would put roof emissions at 0.56 kg CF_4 and 0.15 kg C_2F_6. Nonetheless, the sample would likely not be representative of the higher anode effect rate during the 2021 campaign, even if cells are mostly closed during anode effects (apart from those needing manual intervention to be extinguished).

Another way to assess the roof emissions is to evaluate gas collection efficiency, which for Husnes and PFCs is estimated to be about 99% , making predicted roof emissions 5 kg CF_4 and 0.5 kg C_2F_6. With this last addition CO_2e emissions from the measurements would still be more than 28% lower than IPCC 2006 Tier 2 estimation.

The "2019R T2a (HVAE) + T1 (LVAE)" method would be the equivalent to the 2006 Tier 2 method in the new guidelines. In the table it estimates lower emissions than Tier 2 of 2006 due to the very high AEM number. For AEM below about 0.3 (depending on PFC GWP values) it will produce a higher estimate. The 2019 refinement also includes a couple of non-linear "T2b" methods that treat each anode effect individually [14, 15]. Part of the goal of these methods is to try to account for the decreasing emissions intensity observed over the duration of an anode effect. With a high average AED (Table 3) both emissions estimates

Table 4 Emissions of CF_4, C_2F_6 and CO_2e from section RA1 during the canister sampling period. The canister data are based on measured concentration and flow. Tier 2 and Tier 1 data based on slope factor and HV AE data. CO_2e conversion performed with IPCC AR5 GWP_{100} values. "Legacy" (amperage < 350 kA) factors are used for all 2019R methods where applicable

Method	Metal prod. (t)	CF_4 (kg)	C_2F_6 (kg)	CF_4 (kg/t Al)	C_2F_6 (kg/t Al)	CO_2e (t)	CO_2e (t / t Al)
Canister (RA1a + RA1b)	6646	490.3	50.1	0.07378	0.00755	3 807	0.57
2006 IPCC Tier 2 slope	6646	676.4	81.8	0.10178	0.01231	5 393	0.81
2019R T2a (HVAE) + T1 (LVAE)	6646	636.8	56.0	0.09583	0.00842	4 844	0.73
2019R T2b (HVAE, Marks) + T1 (LVAE)	6646	381.0	31.2	0.05734	0.00469	2 872	0.43
2019R T2b (HVAE, Dion) + T1 (LVAE)	6646	564.2	42.2	0.08490	0.00635	4 209	0.63

end up well below their linear counterpart, and the estimate of "2019R T2b (HVAE, Marks) + T1 (LVAE)" is even far below the canister measurements. The difference between the two non-linear methods is much influenced by the line current and would at higher currents be switched around.

General Remarks

Observations during the experiments showed that compounds like CFC-11, CFC-12 and SF_6 (and many other halocarbons) can be used as conservative tracers—their concentrations levels were identical to the real background levels and can therefore be used as a true measure for the accuracy of the dilution steps.

Conclusions

Two measurement campaigns with canister sampling and subsequent analysis with Medusa GC–MS were conducted at Husnes. The first comprised spot samples from stacks, the roof and around the plant to assess PFC concentration levels and scout sampling locations. The second campaign aimed at getting good long-term average concentration data that could be used for formal emission reporting. Measuring over a long period ensures both HV and LV PFC emission data are captured—meaning everything from regular anode effect, cell start-ups, cell operations and everything in-between.

The analysis with Medusa GC–MS worked without issues and it tackled the high variation in sample concentration well. There is some improvement potential on the sampling side, both regarding improved sampling linearity as well as issues related to moisture and clogging.

The measured PFC emissions are about 27% lower than suggested from 2006 IPCC Tier 2 slope estimation when weighted as CO_2e. This includes roof emissions estimated from a gas collection efficiency of 99% as that was not measured specifically. The Tier 2 method does not include LV PFC emissions, whereas canister sampling is a total PFC method.

Outlook

Overall, the results of the project are very encouraging. The methods for time-integrated sampling and off-line Medusa GC–MS analysis are promising. There are still some issues to work on—like the prevention of clogging the air flow during stack sampling due to high water content and the final decision on size and design of the HF-scrubber and the pipes within the stack. Sampling at other sites than the stack, did not reveal any problems.

NILU is confident in finding good solutions for those issues and that "true" emission values can be obtained from long time-integrated stack-sampling combined with offline analysis. Time periods of one month could be sampled with a canister size of about 15 l and a flow rate of 0.15 ml/min.

Measurements at individual cell outflow can be performed with short time-integrated sampling ranging from minutes to several hours in order to monitor PFC emissions from conventional HV anode effects or LV anode effects.

Time-integrated sampling of airflows leaving the top of the halls will give a good indication of the diffusive emissions within the halls. Those emissions will vary in time but will all in all be much closer to atmospheric background levels over long time periods. For those mixing ratios, FTIR or QCL are not suitable. Offline sampling and GC–MS would be far superior and cheaper compared to quite insensitive online measurements.

The total cost of sampling and off-line analysis will be very competitive compared to long-time online measurements using FTIR or QCL. The sampling can be performed by the staff—not involving site visits by NILU personnel. Such measurements could be done at all stacks at the same time and over the same time periods. Compared to that, a scenario with multiple online equipment seems very unrealistic.

Acknowledgements The authors would like to thank Husnes and its personnel for their assistance.

References

1. IPCC (2021) Climate Change 2021: The Physical Science Basis. Contribution of Working Group I to the Sixth Assessment Report of the Intergovernmental Panel on Climate Change. Cambridge University Press, Cambridge, United Kingdom and New York, NY, USA https://doi.org/10.1017/9781009157896

2. IPCC (2014) Climate Change 2013–The Physical Science Basis. Cambridge University Press, Cambridge, United Kingdom and New York, NY, USA https://doi.org/10.1017/CBO9781107415324

3. Wong DS, Fraser P, Lavoie P, Kim J (2015) PFC Emissions from Detected Versus Nondetected Anode Effects in the Aluminum Industry. JOM 67:342–353 https://doi.org/https://doi.org/10.1007/s11837-014-1265-8

4. Espinoza-Nava L, Dubois C, Batista E (2020) Method Development to Estimate Total Low Voltage and High Voltage PFC Emissions. Minerals, Metals and Materials Series 758–765 https://doi.org/10.1007/978-3-030-36408-3_102

5. Espinoza-Nava L, Young J (2022) Sampling and Analysis Methodology Review to Report Total PFC Emissions. Minerals, Metals and Materials Series 964–970 https://doi.org/10.1007/978-3-030-92529-1_125

6. Åsheim H, Aarhaug TA, Ferber A, et al (2014) Monitoring of Continuous PFC Formation in Small to Moderate Size Aluminium Electrolysis Cells. In: Light Metals 2014. John Wiley & Sons, Inc., pp 535–539 https://doi.org/10.1002/9781118888438.ch91

7. Marks J, Bayliss C (2012) GHG Measurement and Inventory for Aluminum Production. In: Light Metals 2012. John Wiley & Sons, Inc., Hoboken, NJ, USA, pp 803–808 https://doi.org/10.1002/9781118359259.ch139

8. IPCC (2006) Metal Industry Emissions. In: Eggleston HS, Buendia L, Miwa K, et al (eds) 2006 IPCC Guidelines for National Greenhouse Gas Inventories, Prepared by the National Greenhouse Gas Inventories Programme. IGES, Japan https://www.ipcc-nggip.iges.or.jp/public/2006gl/pdf/3_Volume3/V3_4_Ch4_Metal_Industry.pdf

9. IPCC (2019) Metal Industry Emissions. In: Buendia E, Tanabe K, Kranjc A, et al (eds) 2019 Refinement to the 2006 IPCC Guidelines for National Greenhouse Gas Inventories. IPCC, Switzerland https://www.ipcc-nggip.iges.or.jp/public/2019rf/pdf/3_Volume3/19R_V3_Ch04_Metal_Industry.pdf

10. Fraser P, Steele P, Cooksey M (2013) PFC and Carbon Dioxide Emissions from an Australian Aluminium Smelter Using Time-Integrated Stack Sampling and GC-MS, GC-FID Analysis. In: Minerals, Metals and Materials Series. Springer International Publishing, pp 871–876 https://doi.org/10.1007/978-3-319-65136-1_148

11. Miller BR, Weiss RF, Salameh PK, et al (2008) Medusa: A sample preconcentration and GC/MS detector system for in situ measurements of atmospheric trace halocarbons, hydrocarbons, and sulfur compounds. Anal Chem 80:1536–1545 https://doi.org/10.1021/AC702084K

12. Arnold T, Mühle J, Salameh PK, et al (2012) Automated measurement of nitrogen trifluoride in ambient air. Anal Chem 84:4798–4804 https://doi.org/10.1021/AC300373E

13. Schmidbauer N, Hermansen O, Lunder CR (2022) Hydro Aluminium AS. Measurements of CF4 and C2F6 emissions from Norsk Hydro's aluminium smelter at Husnes, Norway https://hdl.handle.net/11250/2981742

14. Marks J, Nunez P (2018) Updated factors for calculating PFC emissions from primary aluminum production. Minerals, Metals and Materials Series Part F4:1519–1525 https://doi.org/10.1007/978-3-319-72284-9_198

15. Dion L, Gaboury S, Picard F, et al (2018) Universal Approach to Estimate Perfluorocarbons Emissions During Individual High-Voltage Anode Effect for Prebaked Cell Technologies. JOM 2018 70:9 70:1887–1892 https://doi.org/10.1007/S11837-018-2848-6

Heavy Metal Emissions through Particulate Matter from Aluminium Electrolysis

Fride Müller, Thor Anders Aarhaug, and Gabriella Tranell

Abstract

Heavy metal emissions from the aluminium industry are mainly carried from the plant through fugitive particulate matter (PM) originating from the aluminium electrolysis pot room. To evaluate the behaviour of metal-carrying PM, both airborne and settled PM from two different primary aluminium smelters have been characterized and analyzed for composition and particle size distribution, with special emphasis on heavy metals and carbon. In addition, optical particle sensors have been placed at different elevations in one of the plants to determine the concentrations of different particle sizes in fugitive PM. Metals such as Fe and Ni were primarily found as particles together with S and P on partly combusted carbon PM. Settled PM from both plants were generally coarser (mean = 32–39 μm) and had a higher Al:Na ratio compared with airborne PM, with a mean PM of 21–22 μm. The optical sensors measured PM100 concentrations at roof level in the plant 5–6 times higher than the PM10 concentration during fuming events such as anode shift operations.

Keywords

Aluminum • Heavy metals • Fugitive emissions • Particulate matter

F. Müller · G. Tranell (✉)
Department of Materials Science and Engineering, Norwegian University of Science and Technology (NTNU), Trondheim, Norway
e-mail: gabriella.tranell@ntnu.no

T. A. Aarhaug
SINTEF Industry, Trondheim, Norway

Introduction

Through international agreements, the primary aluminium industry has, together with other industries, committed to reducing greenhouse gases, as well as managing and monitoring hazardous waste and emissions of heavy metals in relation to their processes. In Norway, a significant fraction of certain reported heavy metal emissions from the land-based industries originates from aluminium production, as summarised in Table 1. These heavy metal emissions are mainly carried from the plant through fugitive particulate matter (PM) from the pot room with electrolysis cells and calculated based on regular measurements using PM collection filters over a period of weeks, with a known air throughput [1]. The PM typically originates from daily operational processes such as anode change, anode covering processes and metal tapping.

The heavy metals are primarily introduced to the electrolysis process through the carbon anodes and will distribute between metal (main path), bath and air/dust [3]. Carbon anodes are typically made of a mixture of 60–70% calcined petroleum coke (CPC), 15–20% recycled anode butts, and 12–17% coal tar pitch binder. The CPCs may contain different amounts of impurities, such as heavy metals, due to mixing different quality cokes [4]. In Table 2, some of the reported trace elements found in anode coke and their associated concentrations are given. In comparison, the metal content of the alumina raw materials is significantly lower than that of the coke [5].

As seen from Table 2, sulfur is a major impurity in the anode coke. The sulfur can be found as a part of the carbon lattice, attached to chains on the surface of clustered molecules or on surfaces and pores bound by capillary condensation, adsorption, or chemisorption [6]. A recent study [7] showed that V, Ni, and Fe are most likely present in high-sulfur coke mainly as hexagonal sulfides: V was found mainly as V_3S_4, Ni as hexagonal NiS and Fe as hexagonal FeS. These authors found that the metal was well distributed

© The Minerals, Metals & Materials Society 2023
S. Broek (ed.), *Light Metals 2023*, The Minerals, Metals & Materials Series,
https://doi.org/10.1007/978-3-031-22532-1_89

Table 1 Reported Norwegian heavy metal emissions from land-based industries and emissions from primary aluminium production in 2020 (kg) [2]

Element	Emission all land-based industries (kg/y)	Emission aluminium industry (kg/y)	Percentage emission from Al industry
Ni	10,143	2880	28
Pb	2410	260	11
V	733	96	13
Cd	141	26	19
Zn	22,158	898	4

Table 2 Representative concentration of trace elements in anode coke (ppm) [5]

Element	Low	High
Fe	50	350
Ni	50	500
V	30	500
Cu	20	50
Cr	1	50
P	5	30
Pb	3	10
S (wt%)	0.5	5
Mo	10	20
Na	20	140
Al	20	250
K	10	20
Zn	2	150
Mg	50	200

in the carbon matrix and not present as large crystalline inclusions. Previous studies of collected PM in potrooms [8, 9] have found it to consist of bath related compounds (cryolite, chiolite ($Na_5Al_3F_{14}$) and calcium chiolite ($Na_2Ca_3Al_2F_{14}$)), aluminas (γ-, γ'-, θ-, and α-Al_2O_3) and graphitic carbon with metallic and light metal impurities. Reported concentrations of heavy metal (oxides) in these dusts were in the order of 0.2–3.2%. Pot fumes are reported to be mainly fine particles with at least 50% of the particles <20 μm [10] in size. This corresponds with prior extensive studies of PM from electrolysis raw gas [11].

While previous studies have provided information on typical compositions and particle sizes of potroom PM, a more complete picture of the presence of heavy metals (chemical form, in which particle size bracket, coupling to operational events, etc.) in the fugitive PM emissions is needed to predict, for example, the metal-carrying particle

dispersion in air from the plant. Hence, in the current work, the composition and particle size distribution of settled and airborne PM collected in electrolysis halls of two aluminium plants (denoted Plant A and Plant B), have been analysed. Special emphasis has been placed on the relationship between heavy metals and carbon in these materials. In addition, small, inexpensive optical sensors monitoring the in situ concentration of PM emissions were placed at different elevations in the potroom of Plant B to measure the relative concentrations of particle sizes and relate these to operational events. These sensors have previously shown promise in collecting useful data in the primary aluminium industry [12]. Put together, this data will provide a more "dynamic" understanding of metal emissions and thereby a better starting point for metal-containing PM dispersion estimates and modelling.

Experimental Procedure

Sampling of PM for Characterization

For both plants, PM deposited on the mezzanine floor above the electrolysis cell rows were collected by simply filling sample bottles, and denoted "settled". In Plant A, an "airborne" PM sample was collected over a 3-week period on a 50 mm filter in a probe connected to a vacuum pump placed at the regular sampling point at one of the gas ventilation points of the pot room. In Plant B, two airborne PM samples were collected through gas vacuum pumps fitted with 150 mm Whatman filters and neoprene hose and probe. Each sample was collected over a 3-week period: one above the mezzanine floor close to the roof opening (see sketch in Fig. 1) and the second, placed on a frame near floor level.

Collected PM Characterization

Imaging/Individual Particle Analysis
The collected PM was imaged using a Zeiss ULTRA 55 scanning electron microscope (SEM) with a field emission gun (FEG) and chemical composition analyses was done by means of energy dispersive spectroscopy (EDS). The PM was placed on carbon tape and additionally coated with carbon to give the best possible images and element analyses. The primary electron energy was set to 15 keV with a working distance (WD) of 10–15 mm. Individual sizes of carbon particles in the PM were additionally determined in selected samples via image analysis software in the SEM. For these measurements, PM was placed on copper tape rather than carbon tape.

Fig. 1 Schematic of sensor
location placements in Plant B

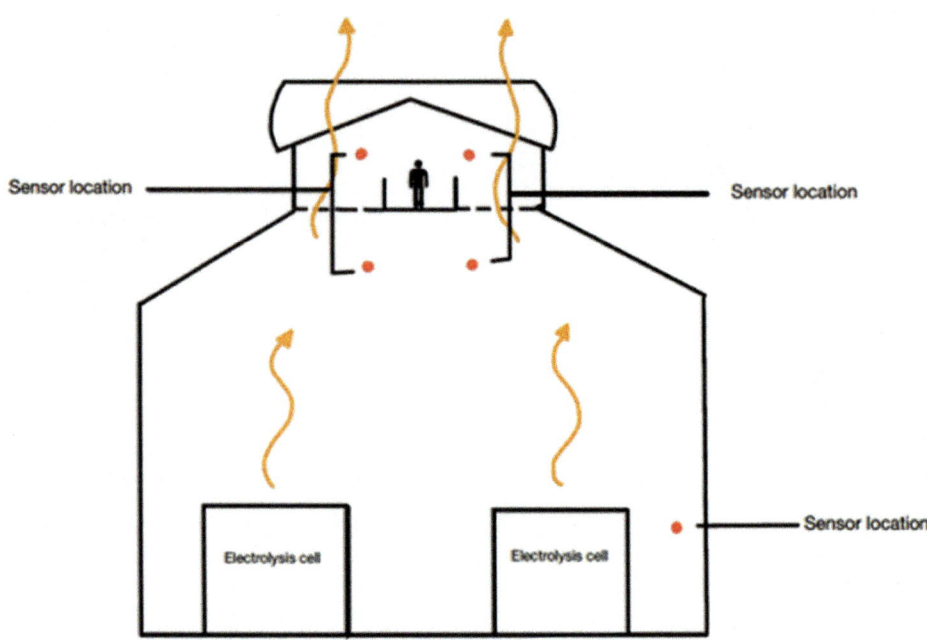

Particle Size

For all collected samples, at least three parallel particle size distribution analyses were carried out using a Horiba LA-960 laser scattering particle size distribution analyser, operating over a particle size range of 0.1–5000 μm. The analyser was operated in dry mode to be able to measure the size of the full sample, including water-soluble particles.

PM Composition

Full dissolution and subsequent Inductively Coupled Plasma–Mass Spectroscopy (ICP-MS) analysis of an airborne and a settled sample from Plant A was carried out by ALS Scandinavia in Sweden to determine the Al:Na ratio in the samples. The compositions of all collected samples, with emphasis on metals, P and S were analysed by Elkem Technology where two parallel samples were dissolved in an HNO_3-HCl-HF solution and analysed by Inductively Coupled Plasma-Optical Emission Spectroscopy (ICP-OES). It should be mentioned that the method did leave a fraction of the sample undissolved which leaves some uncertainty in the measurements. However, the reported concentrations were based on the original sample weight and hence, reported values would potentially be under-estimating rather than over-estimating concentrations. Additionally, the carbon content of the samples was determined by LECO with SINTEF Industry.

In Situ Measurements of Fugitive Dust Emissions

A set of Sensirion SPS30 optical PM sensors were used to gather in-situ data on mass and number concentrations of $PM_{1.0}$, $PM_{2.5}$, PM_4, and PM_{10} at three different locations/elevations in Plant B. Additionally, Nova SDS198 PM100 sensors were placed in the same locations to cover a wider particle size range. One of the sensor systems was placed at floor level, while the other two were placed at roof level near the rooftop outlet and hanging down over the cell rows, respectively. The sensor locations are illustrated in Fig. 1. Temperature and relative humidity were monitored simultaneously. For each location, three parallel sensors were connected to a small "Raspberry pi" computer, collecting data which can be accessed by an external computer, giving continuous measurements.

Results and Discussion

PM Imaging

In Fig. 2, typical images of PM from airborne and settled PM from Plant A are shown. As described by earlier studies, pot room PM consists of a mixture of condensed bath, alumina particles and carbon particles. As depicted in Fig. 3, a closer examination of the PM reveals that heavy metals largely appear as small inclusion clusters on partly combusted carbon particle surfaces.

Another example particle is displayed in Figs. 4a and b where Fe and Ni are observed on the carbon surface by use of Backscattered Electron (BSE) imaging, revealing their typical co-existence with S. The presence of Ni and Fe together with S is aligned with the observations by Jahrsengene [7] of the presence of Ni- and Fe-sulfides in anode coke, confirming that the carbon particles in fugitive PM relate to dusting from the carbon anodes.

Fig. 2 Depiction of airborne (left) and settled (right) PM from Plant A as imaged through secondary electrons in SEM using identical magnification (X1000). The pictures illustrate clearly the difference in particle size between the two types of dust. At the chosen magnification, heavy metal particles cannot be clearly distinguished

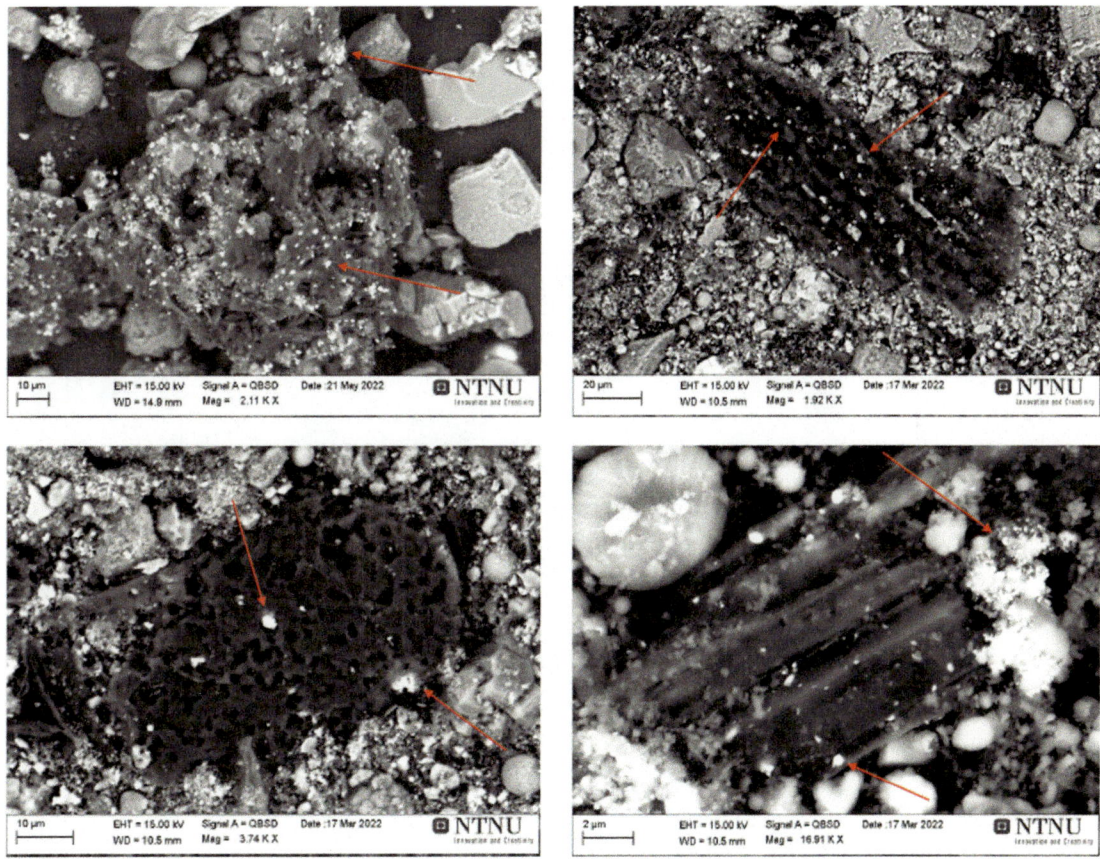

Fig. 3 Typical images of carbon particles in PM. Red arrows mark heavy metal particles

Similar elemental mapping of PM from electrolysis raw gas has also identified the co-existence of Ni and P [13], which corresponds well with a study performed by Haugland et al. [14] where the behaviour of phosphorus impurities in aluminium electrolysis cells were studied. These authors found that dissolved P species can also be reduced by impurities in the bath, where small carbon particles may act as nucleation sites. They also found that loss of phosphorous from the cell was due to evaporation of gaseous elemental phosphorous and that the phosphorus is attached to carbon dust.

Fig. 4 **a** Secondary electron SEM image of a carbon particle in a matrix of alumina and bath fume. **b** Backscattered electron imaging of the carbon particle in **a** illustrating the composition of the bright spots on the carbon surface

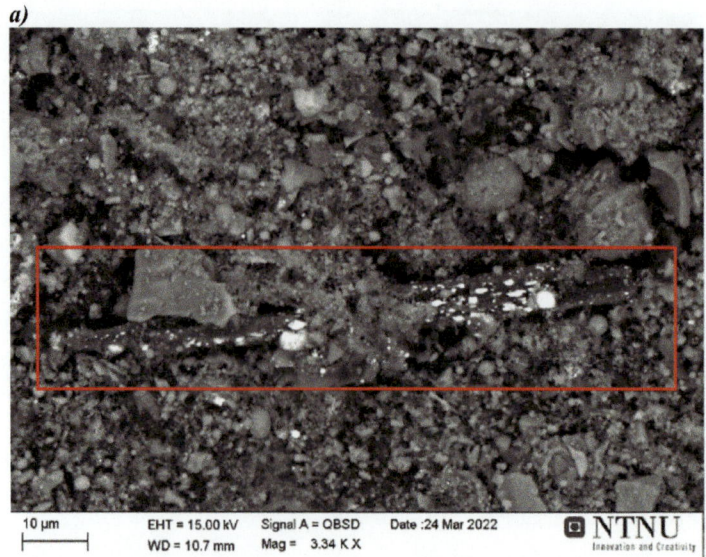

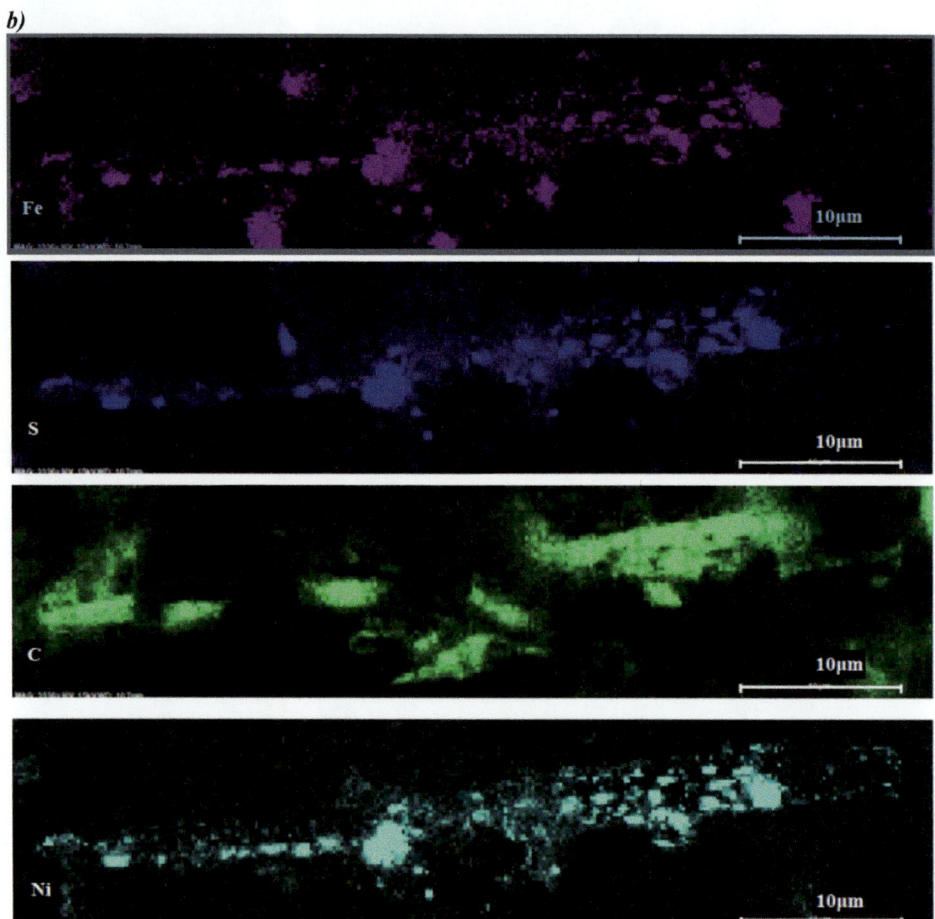

Particle Size

Figure 2 is a visual illustration of the difference in particle size appearance between airborne and settled PM from Plant A. Laser scattering measurements confirm the visual observations as detailed by the particle size distribution curves for different airborne and settled samples from both Plant A and B in Fig. 5. The collected particle size data is summarized in Table 3, showing that the mean particle size of settled PM lies between approximately 32–39 μm, while the mean particle size of airborne PM lies between 21 and 22 μm. Interestingly, when manually measuring the size in the SEM of 100 randomly selected carbon particles in airborne and settled PM, respectively (Fig. 6), the mean size of these particles came out almost double the size of the measured mean of all particles using laser diffraction [15]. While this difference may reflect a selection bias of particles measured manually, it is possible that carbon particles in fugitive emissions are generally larger and more buoyant than other, higher density particles, such as alumina, and consequently preferentially report to the gas escaping the plant.

Particle Composition

The Al concentration of settled PM from Plant A was measured to 72.4% while that of airborne PM was on average 59.6%. However, the Na content was higher in the airborne PM (30.7%) than the settled PM (21.9%). This indicates a higher concentration of bath fumes in the airborne PM and a higher concentration of alumina in the settled PM, which is rather expected. In Table 4, selected metallic, P, S, and C contents of the settled and airborne samples from the two plants are summarized.

Two main trends can be observed in Table 4. The first is that both the metal and light element contents are on average slightly higher in the airborne than in settled PM samples. Since the metal content appears to be mostly associated with carbon particles, this is a logical outcome. However, the compositional uncertainty in the carbon content makes this difference statistically insignificant. The composition of airborne PM from roof and floor level, respectively, in Plant B are not significantly different. The second trend is that both metal, carbon and phosphorous contents are higher

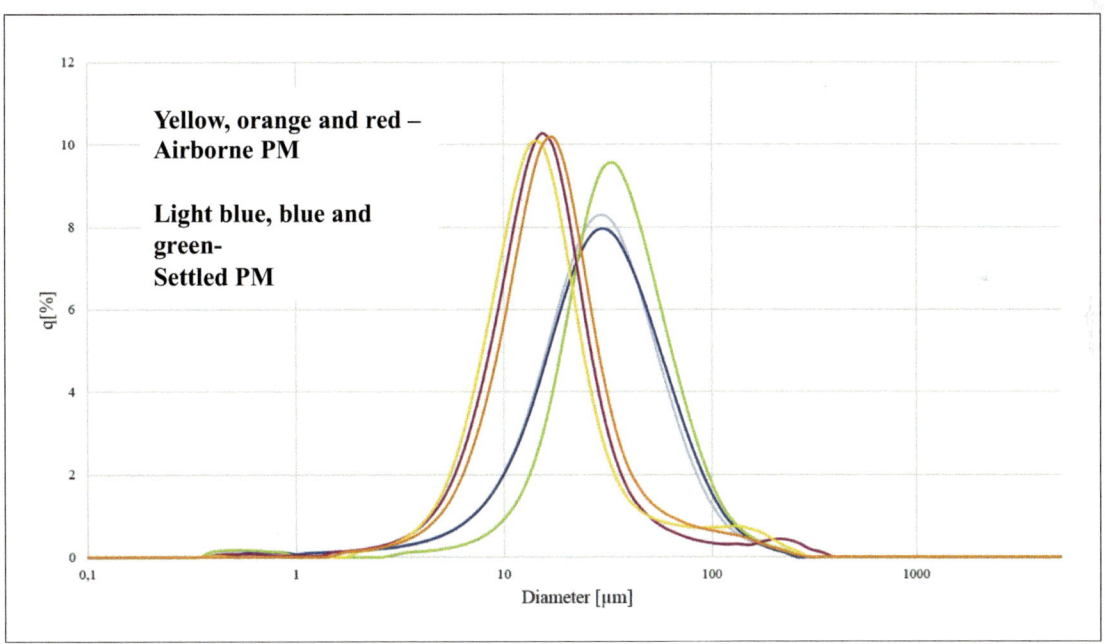

Fig. 5 Particle size distribution of airborne and settled PM samples

Table 3 Particle size distribution for airborne and settled PM (μm)

Particle size (μm)	Airborne			Settled		
	A-1	A-2	A-3	A-1	A-2	B-1
Median	14.3	13.6	15.7	26.6	27.4	31.9
Mean	21.4	21.7	21.9	32.8	34.1	38.6
D_{10}	6.63	6.57	7.37	10.0	10.1	11.0
D_{90}	47.3	46.0	38.4	61.9	65.6	49.4

Fig. 6 Illustration of carbon particle measurements in a settled PM sample from Plant A using SEM imaging

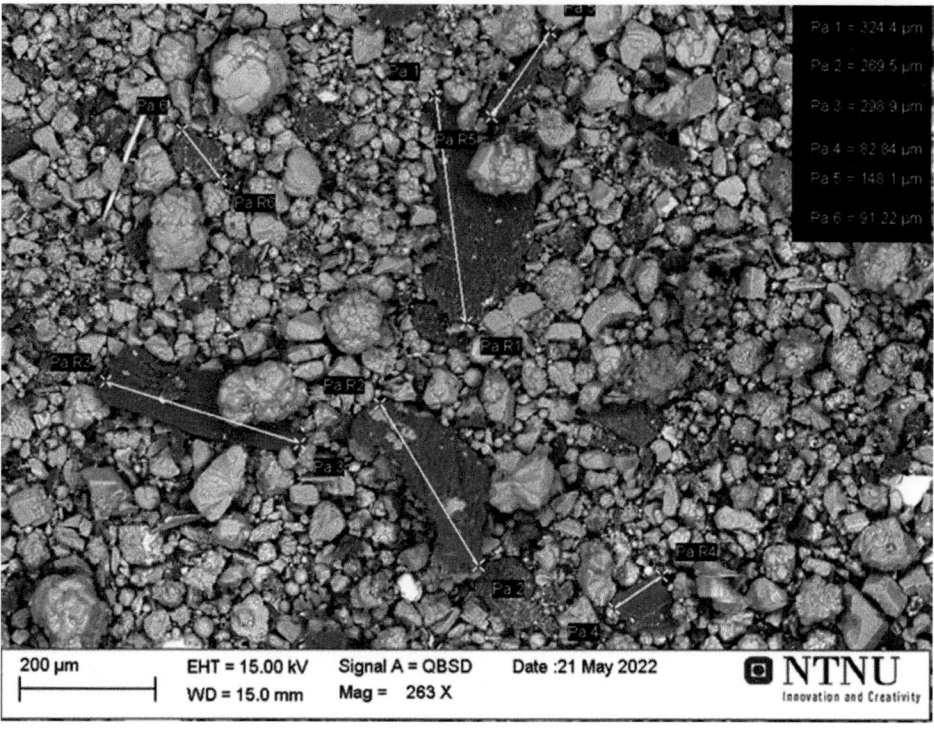

Table 4 Selected metal, P and S contents of settled and airborne PM samples [µg/kg] as measured by ICP-OES. Wt% (C) measured by LECO. All ICP samples except airborne floor PM from Plant B (very little sample) were analyzed in two parallels. The reported variation is that between the two reported parallels. The LECO analysis for airborne PM in Plant B had no parallel

Element	Plant A		Plant B		
	Settled	Airborne	Settled	Airborne (roof)	Airborne (floor)
Fe	3500±0	6100±100	1700±0	2450±150	3400
Ni	3600±100	3550±50	1700±0	3000±200	2900
V	188±2	203±3	84±1	154±66	171
Pb	48±4	116±4	38±2	457±10	216
Mo	39±3	29±4	9±0	10±8	<10
S	3750±50	5100±600	3150±50	9650±350	8000
P	477±16	857±160	225±0	434±122	452
C (wt%)	9.5±1.4	9.8±1.2	7.07±0.7	7.3	8.2

in PM from Plant A than Plant B. The differing amount of carbon in the plant-specific PMs would likely be explained by operational reasons or the use of different carbon materials with different behavior in terms PM generation. The higher S content in airborne PM from Plant B than that in other PM samples is internally consistent with respect to settled PM but not consistent with respect to the lower C content than Plant A and its origin should hence be further investigated.

Particle Sensor Measurements

Figure 7 illustrates the online measurements of the PM10 particle sensor at roof level over a 10-day-long measurement period. It can be seen from the figure that regular "events" account for the bulk of PM emission with spikes reaching concentrations up to 2500 µg/m³. These events coincide with anode shifts, allowing bath evaporation and carbon dust to leave the cell house as has previously been reported in the work of Wong [16] and Myklebust et al. [12].

While the PM 10 concentrations were similar at roof and cell levels, the floor-level PM10 concentrations are generally significantly lower than those above the cells and at roof level, as shown in Fig. 8.

In Table 5, the average mass concentrations of the different particle size bins are summarised. It is found that PM1.0 amounts to 86% of PM10 at roof level, while at floor level, the same fraction amounts to 77% of PM10. PM2.5 amounts to 95% of PM10 at roof level and 89% of PM10 at

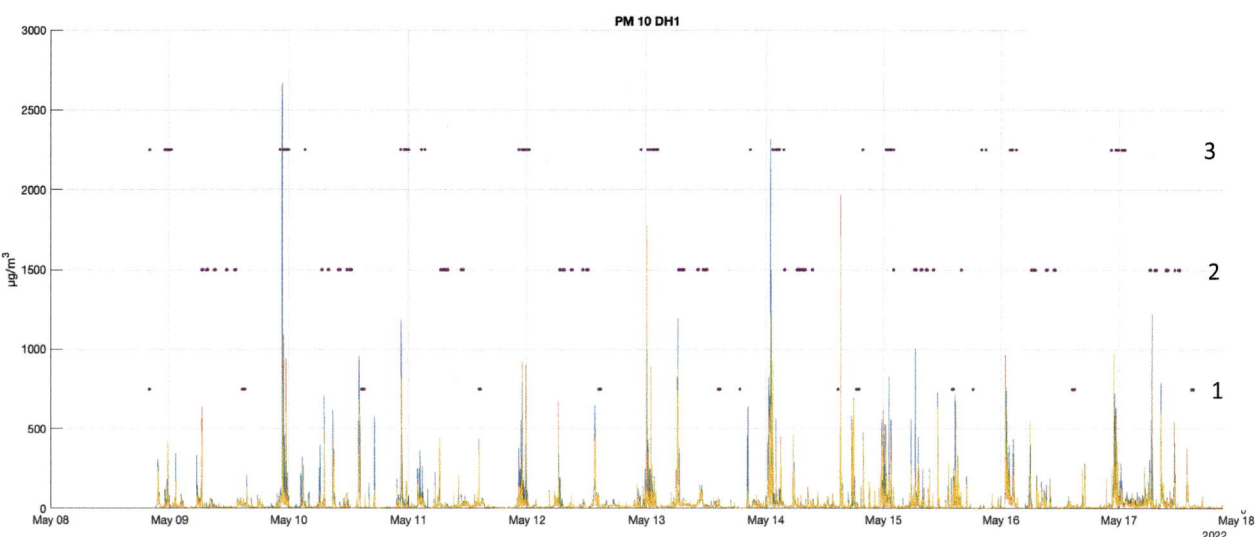

Fig. 7 PM concentration of the 3 parallel PM 10 sensors at roof level (different colored signals). Production activities marked on lines 1 = temperature measurements, 2 = tapping and 3 = anode change

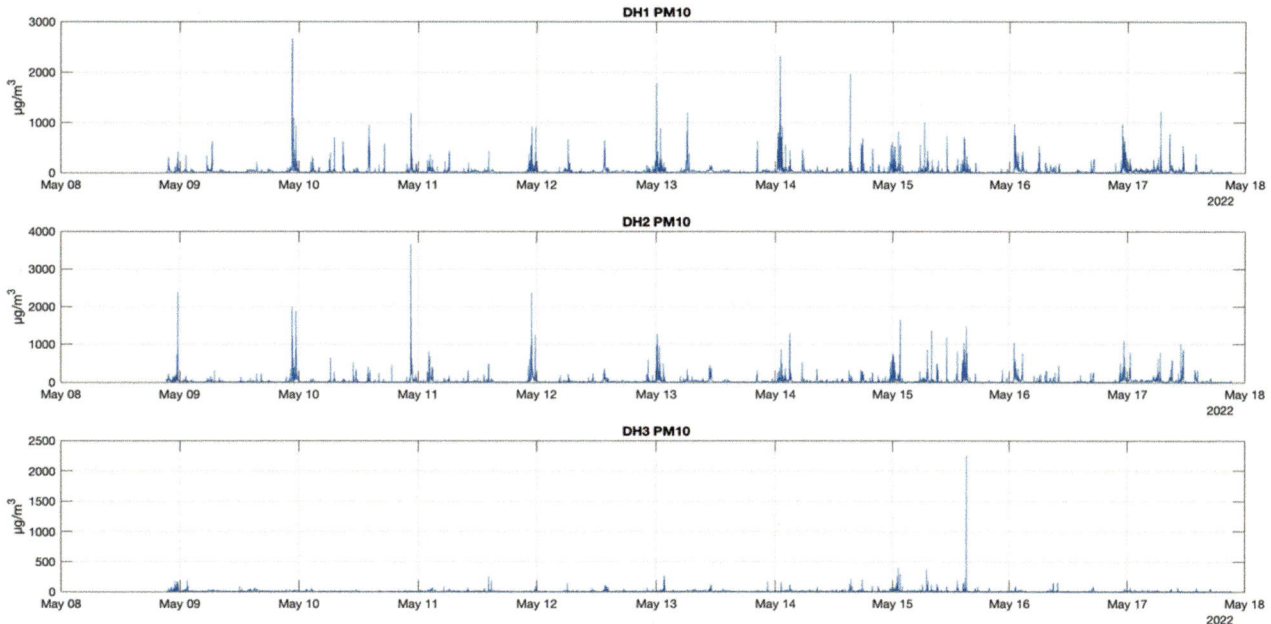

Fig. 8 From the top: PM10 measured at (DH1) roof level, (DH2) cell level and (DH3) floor level

Table 5 Average PM mass concentrations for the low particle size (10PM) sensors in roof, over cell, and floor positions at Plant B

Av. mass concentration ($\mu g/m^3$)	Roof	Over cell	Floor
PM_1	26.4	29.6	6.7
$PM_{2.5}$	29.1	32.6	7.9
PM_4	30.1	33.7	8.5
PM_{10}	30.6	34.2	8.8

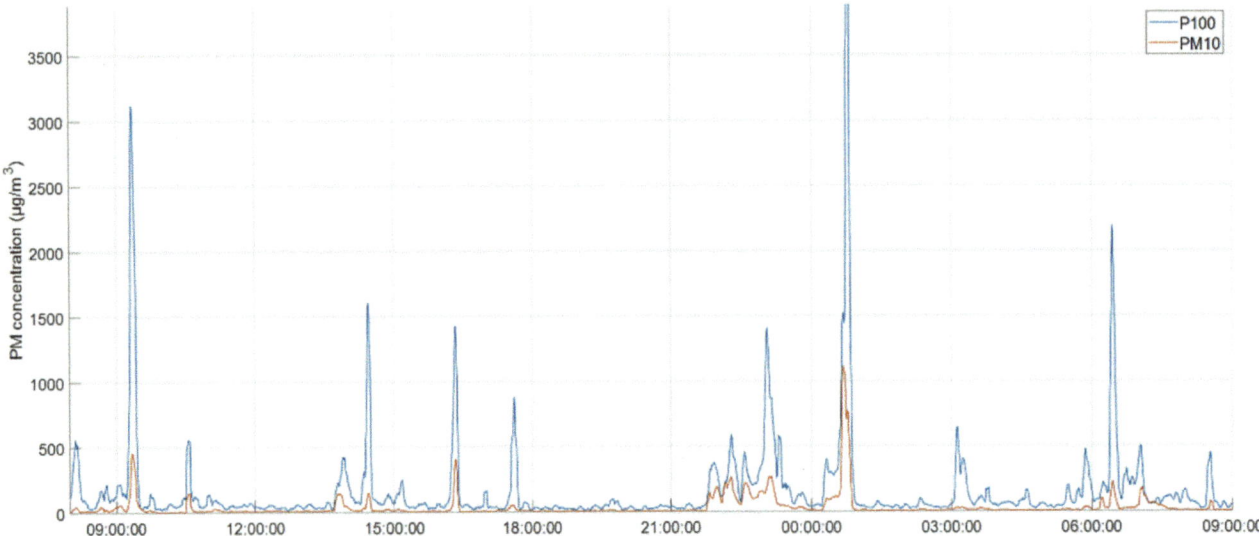

Fig. 9 Comparison of PM10 and PM100 concentration at roof level in Plant B over a 24 h period

floor level. Lastly, PM4.0 amounts to 98% and 96% of PM10 at roof and floor levels, respectively.

The PM10 sensors are rated to reliably measure a maximum of 1000 µg/m^3 while typically reporting up to 2500 µg/m^3 and are often "saturated" during process events. Given the particle size distribution measurements of collected PM in Fig. 5 and Table 3, PM100 sensors measuring at roof and floor levels were used to monitor the concentration of larger particles. These sensors are rated to a concentration up to 20.000 µg/m^3. A comparison between the PM10 and PM 100 measured at roof level is illustrated in Fig. 9. As seen in the figure, the measured concentration ratio between PM100 and PM10 during major events is between 5 and 6. This is in line with the measured particle size distribution of the collected PM samples. The relative background PM concentrations measured by the two different sensor types may, however, be somewhat different. Therefore, one should be careful when comparing measured absolute PM100 and PM10 concentrations.

Conclusions

Airborne and settled PM from two different primary aluminium smelters (Plants A and B) have been characterized and analyzed for composition and particle size distribution with special emphasis on heavy metals and carbon. In addition, optical particle sensors have been placed at different elevations in Plant B to determine the concentrations of different particle sizes in fugitive PM. The following conclusions are drawn:

- Both settled and airborne PM generally consists of bath fume, alumina particles, and carbon particles.
- Heavy metals such as Fe and Ni are largely coexisting with S on the surface of partly combusted carbon particles.
- The S, P and heavy metal contents are on average slightly higher in airborne than in settled PM for both plants. The Na content is higher while the Al content lower in the airborne PM, indicating a higher bath content and a lower alumina content in the airborne PM.
- The heavy metal content, C and P content is higher in PM from Plant A than B but the S content lower.
- Settled PM in both plants were generally coarser that airborne PM with a mean PM measured by laser scattering, lying between approximately 32–39 µm for settled PM while the mean particle size of airborne PM was measured between 21 and 22 µm. Carbon particles were typically larger than other dust particles.
- High particle concentrations (>1000 µg/m^3) were measured by the optical sensors during anode shift at the roof level of Plant B. The mean PM10 particle concentration over the 10-day period for the roof level and over the electrolysis cells were between 30 and 40 µg/m^3, while that at floor level was close to 6–9 µg/m^3. PM1 accounted for 77–86% of the measured PM10. The measured PM100 concentration was between 5 and 6 times higher than the PM10 concentration during fuming events at roof level.

Future Work

As a continuation of the current work, a three-stage impactor will be used in both Plant A and B over an extended period of time in order to collect and measure PM online. This will allow for more specific correlation between composition analysis of different size fractions and their counts/masses. Calibration of different particle sensor "background" concentration measurements should be performed in order for a more reliable comparison between intensities.

Acknowledgements This publication has been funded by the SFI Metal Production (Centre for Research-based Innovation, 237738). The authors gratefully acknowledge the financial support from the Research Council of Norway and the partners of the SFI Metal Production. The authors also wish to thank Ole Kjos at SINTEF industry for aiding the placement of PM100 sensors in Plant B.

References

1. Norwegian standards NS 4861 and NS 4863.
2. Miljødirektoratet og Statistisk Sentralbyråd. Landbasert industri, 2022. Accessed: 28.04.2022. [Online]. Available: https://www.norskeutslipp.no/no/Landbasert-industri/?SectorID=600.
3. G. Jahrsengene, H. C. Wells, A. P. Rørvik, S.and Ratvik, R. G. Haverkamp, and A. M. Svensson. "A XANES study of sulfur speciation and reactivity in cokes for anodes used in aluminium production", Metallurgical and Materials Transactions B, volume 49, pages 1434–1443. 2018.
4. L. Edwards, N. Backhouse, H. Darmstadt, and M. Dion. "Evolution of anode grade coke quality", Light Metals, pages 1207–1212. 2012, Springer International Publishing.
5. T. A. Aarhaug and A. P. Ratvik, "Aluminium primary production off-gas composition and emissions: An overview", JOM, volume 71, pages 2966–2977. 2019.
6. G. Jahrsengene. "Coke impurity characterisation and electrochemical performance of carbon anodes for aluminium production", PhD Thesis, NTNU, ISBN: 9788232643004, 2019.
7. G. Jahrsengene, H. C. Wells, C. Sommerseth, A. P. Ratvik, L. P. Lossius, K. H. Sizeland, P. Kappen, A. M. Svensson, and R. G. Haverkamp "An EXAFS and XANES study of N, Ni, and Fe speciation in cokes for anodes used in aluminum production", Metallurgical and Materials Transactions B, volume 50, pages 2969–2981, 2019.
8. B. L. W. Höflich, S. Weinbruch, R. Theissmann, H. Gorzawski, M. Ebert, H. M. Ortner, A. Skogstad, D. G. Ellingsen, P. A. Drabløs, and Y. Thomassen. "Characterization of individual aerosol particles in workroom air of aluminium smelter potrooms", Journal of Environmental Monitoring, volume 7, pages 419–424, 2005, Royal Society of Chemistry.
9. D. S. Wong, N. I. Tjahyono, and M. M. Hyland "The nature of particles and fines in potroom dust", Light Metals, pages 553–558. 2014 Springer International Publishing.
10. M.M. Hyland and M.P. Taylor, "Origins and effects of potroom dust", Light Metals, pages 141–145. 2005, Springer International Publishing
11. H. Gaertner, "Characteristics of particulate emissions from aluminium electrolysis cells", PhD Thesis, 2013. NTNU, ISBN:978-82-471-4764-1.
12. H. A. H. O. Myklebust, T. A. Aarhaug, and G. Tranell, "Measurement system for fugitive emissions in primary aluminium electrolysis" Light Metals, pages 735–743, 2020, Springer International Publishing.
13. F. Müller, "Heavy metal emissions from primary aluminium production" Technical report, 2021. NTNU.
14. E. Haugland, G. M. Haarberg, E. Thisted, and J. Thonstad, "The behaviour of phosphorus impurities in aluminium electrolysis cells", Essential Readings in Light Metals: Volume 2 Aluminum Reduction Technology, pages 229–233, 2016, Springer International Publishing.
15. F. Müller, "Heavy metal emissions from primary aluminium production" MSc thesis 2022. NTNU.
16. D. S. Wong, M. M. Hyland, N. I. Tjahyono, and D. Cotton, "Potroom operations contributing to fugitive roof dust emissions from aluminium smelters" Light Metals, pages 905–912, 2019 Springer International Publishing.

Verification of Open-Path Dust Laser for Continuous Monitoring of Diffuse Emissions

Lars Moen Strømsnes, Heiko Gaertner, Steinar Olsen, Peter Geiser, and Bernd Wittgens

Abstract

Today, quantification of diffuse dust emissions from large production halls are often estimations based on relatively few spatially and temporally constrained manual measurements in ventilators, wall- and roof openings. Results are used to extrapolate an average operation-related discharge throughout the year where considerable variation in both the quantity, duration, and location will rarely give a representative picture of the emission situation over the year. A new generation of open path instruments for continuous dust monitoring are being developed by NEO Monitors AS. Its improved measuring technology based on a divergent laser aims to significantly increase path length capability while keeping the detection limit low. Stability, robustness, and general ease of use are also significantly enhanced. However, as with current laser-based solutions, calibration and verification relies on reliable gravimetric reference measurements, either in lab or at site installation. Data from existing laser technology for shorter paths is used to identify limitations and challenges related to emissions monitoring that occur through the building's ventilation openings in the ceiling. Gravimetric reference measurements have been carried out to verify the dust levels reported by the laser installation.

Keywords

Dust • Fugitive • Diffuse • Emission • Laser • Transmission • Measurement • Verification • Calibration • Open • Long • Path

Introduction

The potential detrimental health effect of particulate emissions is of considerable concern for mineral and process industry. Comprehensive efforts are undertaken to reduce emissions through defined stacks. However, diffuse or fugitive emissions which do not pass-through controlled stacks, ducts, filters, or other monitored emission channels, are currently challenging to capture or measure with available instrumentation. The urgent need to comply with increasingly stringent emission regulations has motivated industry partners in Norway to join a research project to develop monitoring technology capable of open path measurements of diffuse dust emissions for path lengths that exceed the capabilities of current systems.

In the project "DustDetect" funded by the Norwegian Research Council, Elkem ASA as project owner is collaborating alongside a broad industrial consortium, including NEO Monitors AS for laser technology, on developing and testing next-generation laser technology designed for measuring diffuse dust emission in an industrial environment. As part of the project, methods for calibration and verification, both at lab and at site, are studied and applied. The work presented in this paper focus on experiences from field test of current laser technology along with flow monitoring, and how manual reference measurements may be used for verification at site.

S. Olsen · P. Geiser
NEO Monitors AS, Skedsmokorset, Norway

H. Gaertner (✉) · B. Wittgens
SINTEF Industry, Metal Production & Processing, Trondheim, Norway
e-mail: heiko.gaertner@sintef.no

L. M. Strømsnes
SINTEF Helgeland, Mo i Rana, Norway

© The Minerals, Metals & Materials Society 2023
S. Broek (ed.), *Light Metals 2023*, The Minerals, Metals & Materials Series,
https://doi.org/10.1007/978-3-031-22532-1_90

Transmission Measurements

Common optical methods for dust measurements are based on transmitting light through a volume while measuring the reduction of light intensity or scatter of light. When transmitting light through aerosols, the light intensity that can be measured at a receiver/detector will be reduced due to the absorption and scattering effects that occur when the light hits dust particles. By simply transmitting a beam of light through the dust containing aerosol, the intensity reduction by light deflecting and absorbing particles can be measured. This is referred to as transmission measurements or direct light measurements.

Less light received means more dust particles present. The light intensity will be reduced according to Beer–Lambert law:

$$I = I_0 \cdot e^{\frac{-C \cdot l}{k}} \qquad (1)$$

where I is the light intensity, I_0 is the light intensity without dust present, C is the dust concentration, and l is the optical path length. Factor k is the dust dependent calibration constant, which might vary depending on the type of dust measured. The light transmission, T, is defined as

$$T = \frac{I}{I_0} \qquad (2)$$

Combining and rearranging, this gives the dust concentration, C:

$$C = -\frac{k}{l} \cdot ln(T) \qquad (3)$$

Processing transmission signals will measure dust over the optical path l, while the sensitivity will be mainly dependent on the path length. This implies that high sensitivity (low detection limits) for low dust levels will require longer path lengths.

Commonly used instruments are based on both forward scattered and direct light transmission measurements. Typically, a collimated laser beam is emitted from the transmitter over a defined path length. The receiver unit consists of a direct light detector in the front and a scattered light detector behind it. The instruments are operated in either direct or scatter light mode, depending on duct dimension, homogeneity, and the expected dust concentration range.

Instruments are primarily made and optimized for in-stack measurements with limited path lengths and there are only a limited number of viable options available for monitoring diffuse emissions. To address this, a new generation of Open Path (OP) instruments is under development thereby aiming at a detection limit of values below 0.1 mg/m^3 with instrument stability under 1%, and at path lengths exceeding 100 m. Coupled with that, the new LaserDust OP instrument combines both the laser transmitter and the light receiver in a single unit. A reflector is used to redirect the light beam back to the receiver.

In a transmitter/reflector configuration, lower detection limits are achieved by doubling the effective path length. However, when measuring over long paths, alignment of the system is critical. Therefore, instead of a collimated laser beam, a divergent light beam is used giving an improved alignment offset tolerance without any loss of the returned signal. That said, the diverging beam does not come without its challenges with signal intensity as the returned signal is weakened when the proportion of reflected light decreases with longer path lengths. These challenges are overcome with sophisticated signal/noise processing software. The configuration of the new instrument is illustrated in Fig. 1.

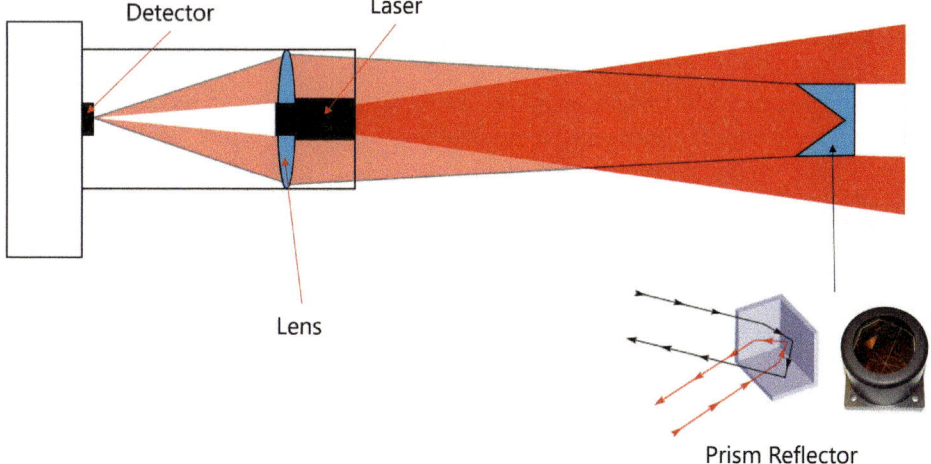

Fig. 1 Alignment of divergent beam reflector setup

Reference Measurements

Monitoring diffuse emissions of dust is a challenging task due to the variety of emission source and their diverse nature. The Industrial Emissions Directive 2010/75/EU [1] describes diffuse emissions as non-channeled emissions to the environment, which usually concern volatile or dust-like substances. Diffuse emissions can be point, linear, surface, or volume sources. Examples include storage facilities during loading and unloading, open-air storage of solid matter, separation pools in oil refineries, doors in coke plants, and electrolysis cells in chlor-alkali plants. Multiple emissions inside a building are normally considered diffuse emissions.

The German VDI 4285 Richtlinien [2, 3] is a series of guidelines dedicated to the determination of diffuse emissions by measurement. According to these guidelines, diffuse emissions are waste streams that are not conducted in channels, and therefore cannot be measured on the basis of the guidelines VDI 2066 Part 1 [4], VDI 4200 [5], and NS-EN 15259:2007 [6].

While the principles for reference measurements in channelled emissions are well defined in standards (i.e. EN 13284, EN 14181), this is not the case for diffuse sources and emissions. For reliable monitoring solutions, laser measurements must be calibrated in a standardized reference regime. To do this, standardized sampling and measurements procedures must be adapted to give representative laser measurement with respect to spatial and temporal variations in the outlet cross section, as well as variations in dust concentration and properties along the laser line of sight. The design of reference measurements with respect to sampling points, sampling duration, and the number of samples, is of significance.

Experimental

Test Site and Installation

To gain experience with diffuse emission monitoring and gravimetric reference measurements, a test pilot was installed in a harsh environment at the Elkem smelter in Mo i Rana. The monitoring path is located in a section of the *Lyre*-outlet of the production hall, with dimensions of approximately 82×1.9 m, and an estimated annual dust emission of 100 tons. The system consists of one platform, a commercial laser, and two ultrasonic weather stations for measuring velocity, direction, and temperature of the air in the outlet. An additional weather station was installed in the production hall close to the pilot installation for supplementary information on local weather conditions. The positioning and alignment of the laser and the two ultrasonic weather stations are outlined in Fig. 2. The laser setup covers a measurement section of about 10 m, located directly above several sources of diffuse dust emissions that included ladle pouring, refining, and bed casting activities. The unit operations located below were expected to provide large variations in dust concentrations that provide both opportunities and challenges in planning and execution of representative reference measurements.

Instrumentation

The largest version of NEO Monitors LaserDust XLP is installed diagonally across a 10 m section. The effective path length was 8.5 m as a result of shielding tubes being installed both at the transmitter and receiver end.

Two sonic weather stations of type Windmaster™ (Gill Instruments) monitor the flow rate at the outlet (hereafter called *Sonic B* and *Sonic C*). The anemometers measure three-axis ultrasonic velocities (wind speed) in the range of 0–50 m/s with an RMS accuracy of 1.5% at 12 m/s. Sonic temperatures are measured in the range of −40 °C to +70 °C, with a ±2 °C (between −20 °C and +30 °C) accuracy.

Manual Reference Sampling Setup

For reference filter sampling, portable work environment sampling pumps of type SKC AIRCHEK ESSENTIAL with adjustable flow rate 5–5000 ml/min were used. Dust is collected in 25/37 mm filter holders on MCE (mixed cellulose ester) filters of equivalent size. The gravimetric detection limits reported by the laboratory (SINTEF Norlab, Mo i Rana) are 0.05 and 0.1 mg for 25 and 37 mm filters, respectively. Sampling was conducted at a fixed flow rate of 2.0 l/min, corresponding to gas velocity of approximately 2.5 m/s in the filter holder opening.

The laser test section has eight rails mounted across the outlet, evenly distributed along the roof section. Each rail has adjustable brackets for the placement of filter holders, directing these at a reasonable orientation in the direction of the flow. The brackets are aligned along the line of sight of the laser, ensuring accurate spatial representativeness with the laser, as shown in Fig. 2.

Sampling Program

Preliminary measurements with the laser showed periods with relatively low dust concentration in the range of 0–10 mg/m^3, with clearly defined peaks reaching 200–300 mg/m^3, and each peak lasting for 5–20 min. Each peak is typically

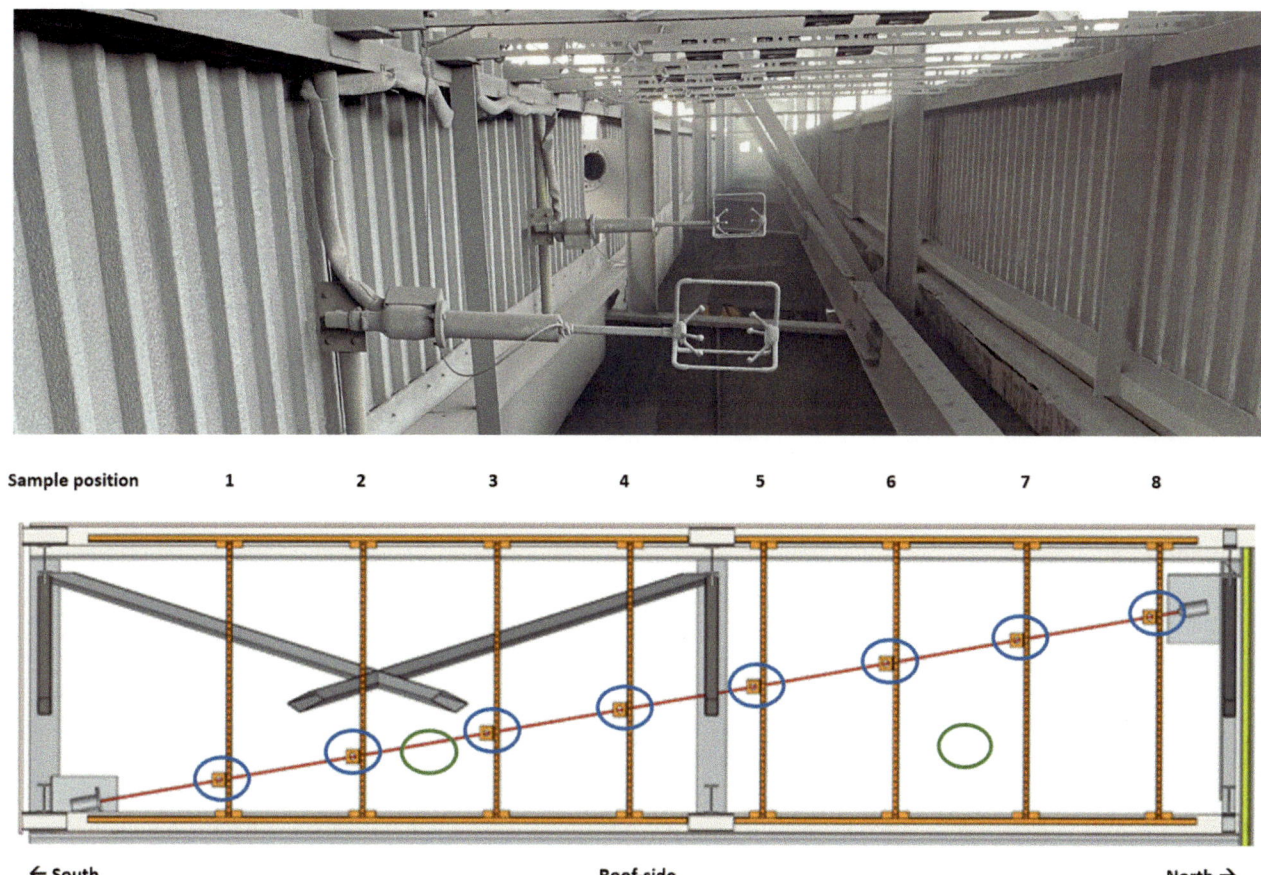

Fig. 2 Placement and alignment of laser in the outlet section. Green circles indicate the approximate placement of wind instruments Sonic B and C, while blue circles indicate the gravimetric filter sampling positions

associated with one specific dust generating activity, either ladle pouring, refining, or casting of refined/unrefined as the most important sources of dust generation.

While the main objective of the sampling program is to verify the laser measurements, the campaign also had a goal of using the results for calibrating the laser in the case that satisfactory agreement of the measurements is not observed. To do this, one ideally has access to some sort of calibration curve with sufficient variation in measured concentration levels. To achieve this, the following sampling strategy was proposed: (i) five series of short-term samples evenly distributed along the lasers monitoring path, and in parallel; and (ii) a long-term sample to give an average for comparison. The idea of the shorter series is to fit these sampling periods with specific activities in the production process, such as refining and casting. This would create sample series with dust concentration expected to be high, with at least some variation between series, as each activity/peak doesn't have the same intensity and emission associated with it.

During each run, a series of four filters is sampled. Overall, eight filters are evenly distributed over the lasers line of sight, of which four filter positions are used for short-term measurements. The remaining four filter positions

are used for the long-term measurements. The choice is a trade-off where more samples give better spatial representativeness of both dust concentration and variations in dust properties, at the expense of costs, handling, and general ease of operation of sampling equipment. Although four samples may seem low, it is important to have in mind that this only involves a smaller section of the actual outlet, and that the sample density is likely to be smaller in a real setup, where the laser is measuring the whole outlet.

Results and Discussion

Long-Term Flow Measurements

Long-term measurements of flow have been evaluated for the period 27 October–23 November 2021 and 14–25 January 2022, 37 days in total. The prevailing wind directions in the dataset are mainly southern and western, with good correlation to official registrations at weather stations *Skamdal* and *E6 Yttervika*.

Figure 3 shows hourly data of the vertical velocity of the two outlet wind units along with the roof units' total wind

Fig. 3 Hourly velocity for Sonic B (orange), Sonic C (blue), and Sonic A—roof monitor (green)

speed. The measured windspeeds are clearly related, as can be seen when comparing the outlet flow direction that tends to shift downwards when wind speed across the production hall is high.

Figure 4 shows scatter plots of the variation in vertical velocity versus roof wind speed and direction, respectively. Negative velocities clearly correlate with wind blowing from south-western directions and exceeding 4–5 m/s. The dataset is somewhat limited, in that wind direction and speed measured on the roof are strongly correlated, meaning that wind speeds exceeding 5 m/s almost only occur in south-western directions. The cause for negative vertical flow might therefore attribute to the speed alone, or to a combination of both speed and direction. For the complete dataset, 18 and 16% of the vertical velocities of Sonic B and Sonic C, respectively, are negative readings. A noteworthy observation that stresses the importance of a representative flow profile determined with enough flow measurement

devices and good time resolution. The flow rate varies or even can change direction over the long measurement section. Negative flow will give a faulty dust emission rate as the dust is not emitted but contributes to a concentration overestimate as it enters and leaves the measurement path at different locations almost at the same time.

Gravimetric Reference Measurements

The reference sampling program consists of a total of six series as outlined earlier. The five short-term series were sampled over a time span of 4–5 h. Series 6 being the long-term sample was sampled continuously in parallel to these.

Each of the series consists of four samples. The filters for the short-term sampling (Series #1–5) were placed along the light path in the positions 2, 4, 6, and 8, (see Fig. 5), while the long-term filter (Series #6) was placed in positions 1, 3, 5, and 7.

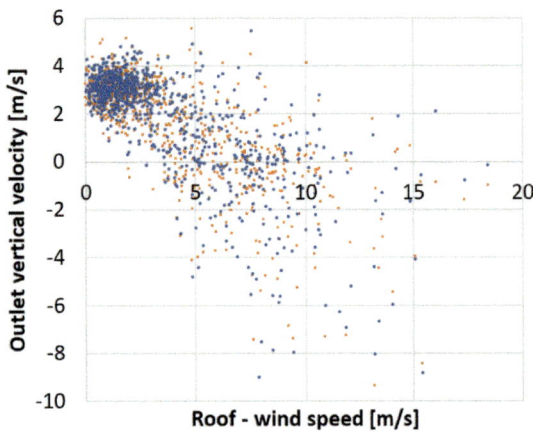

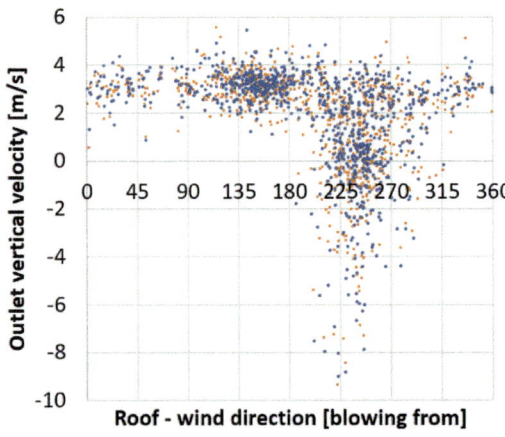

Fig. 4 Scatter plot of hourly outlet vertical velocity (Sonic B and Sonic C) as a function of roof wind speed (left) and roof wind direction (right)

Fig. 5 Reference measurement concentration profiles (distribution along laser stretch)

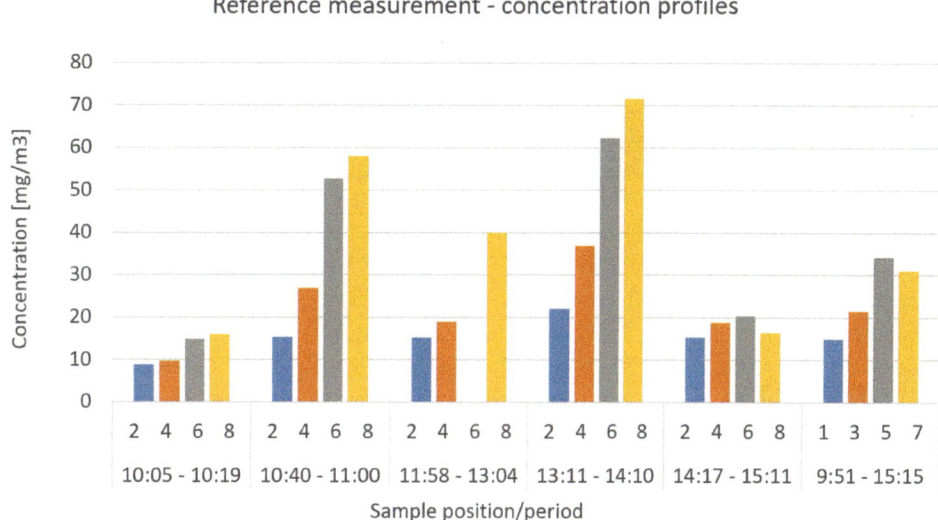

Series 1 and 2 were successfully sampled while casting took place at a sampling time of 15–20 min. Communication to obtain precise information about timing of casting activities was challenging and therefore the sampling strategy for series 3–5 were adjusted to 1-h averages with no specific timing with process activities.

Figure 5 summarizes the distribution of dust concentrations obtained along the laser path, clearly indicating the potential for variations in dust concentration over relatively short distances. As the samples are taken along the diagonal of the outlet section, concentration gradients both along and across the outlet may affect the obtained profiles. This illustrates the importance of a proper sample procedure with respect to the number of samples and that they are placed properly in relation to the laser.

Laser Calibration

Table 1 lists the concentration determined with help of the gravimetric reference measurements and compares with the concentration estimates given by the optical measurement device.

The reference sample series indicate that the optical measurement device underestimates the dust concentration for both average of short-term samplings series #1–5 and long-term series #6, although the elevated background level in the range of 5 mg/m^3 due to optical contamination is not subtracted. Furthermore, the larger deviation between gravimetric and optical measurements during short-term series implies underestimates especially during periods with height dust contents that might be attributed to an incorrect dust-dependent calibration constant k.

The standard deviations for short-term filter samples (Table 1) are significantly higher compared to the long-term samples. This indicates that the dust was unevenly distributed during short emission events. It is worth mentioning that minor deviations in recorded filters mass obtained during short-term series will have a larger effect on measurement errors and concentration estimates compared to minor deviations recorded during long time sampling as the effect of a potential measurement error is relativized by the measurement duration (filter weights are divided by the volume flow).

The measurements overall are found to have a good correlation, with one clear outlier observed at series 4. The

Table 1 Reference concentration and laser transmission and corresponding dust concentrations

Series	Average reference concentration (mg/m^3)	Standard dev (mg/m^3)	Laser transmission [%]	Laser concentration [mg/m^3]
1	12.4	3.1	91.2	12.2
2	38.2	17.7	81.0	28.9
3	27.1*	10.9	87.4	17.9
4	48.2	19.7	87.8	17.2
5	17.7	2.0	86.7	19.1
6	25.5	7.7	83.3	18.1

* Corrected for missing value at sample point 6, assuming equal concentration profile as the other series

laser can therefore be calibrated using a simple regression, either including or omitting series 4. As the series is so clearly deviating from the remaining results, the regression is performed without it. As long-term drift effects are observed during the test period at the pilot installation. adjustment of the baseline (zero level) is also considered while doing the regression analysis.

The slope of the regression curve corresponds to the combined factors, $k/l\cdot N$, according to Eq. (4). The factor N represents temperature and pressure normalization performed by the laser instrumentation and must be corrected for when estimating the calibration constant. Using the reference concentrations and laser transmission from Table 1, and current instrument settings of $l = 8$ m and $N = 1.113$, k is estimated at 1525 using the following equation:

$$C = -N \cdot \frac{k}{l} \cdot ln(T) \qquad (4)$$

In addition to the calibration constant, the intercept of the regression curve is used to estimate the baseline laser intensity (I_0), and the relative influence on T according to Eq. (5). The adjustment of the baseline corresponds to a decrease in concentration of 4.5 mg/m^3 at zero.

$$T = T_0 \cdot e^{-6.9394/212.2} = T_0 \cdot 0.968 \qquad (5)$$

Challenges Related to Optical Measurement Equipment

The reference samples and calibrated laser data correlate well with a good fit for the long-term sample series, as indicated in Fig. 6 However, standard deviations between reference samples (within each individual treatment) are found to be high. Possible reasons can be related to deviations in local flow rate and dust-specific feature with influence on transmission and scatter of light.

Reference Sampling—Flow Rate

Reference concentrations for series 4 are consistently elevated for all sampling positions (Fig. 5). Random sampling errors are unlikely to be the cause of the deviation observed, as the result of each series is the average of 4 individual samples. However, it is more likely that sampling at a flow rate, which differs from the isokinetic rate at the sampling location at the given time caused the systematic sampling errors. The sample flow rate was adjusted according to the standard based on a common reference that was measured with a handheld anemometer at several locations in the roof opening. The sampling was performed at a rate of 2 l/min, corresponding to a gas velocity of 2.5 m/s in the filter holder inlet. The sonic weather stations recorded an average outlet gas velocity of 3.3 m/s, which gives an isokinetic sampling deviation of −24%. A negative deviation in general means that measured concentrations will be overestimated. The standard reference method for the measurement of low dust concentrations in ducted gaseous streams (EN 13,284–1:2017) provides estimates for the effect of isokinetic deviations on the sampling efficiency (representativeness), defined as the ratio of over-/underestimation. An isokinetic deviation of −24% could potentially overestimate the dust concentration of 10–20%, depending on the gas stream conditions and dust properties, especially particle size. This might explain some of the deviation observed between reference and laser data. For series 4, however, the deviation is most likely too high for this to be the only explanation.

Laser Measurement—Calibration Constant and Morphology

The dust-dependent calibration constant for direct mode measurements is highly dependent on particle size, shape, and mass density. Variations in dust properties could explain the deviation observed for sample series 4. The dust

Fig. 6 Comparison of reference measurements with estimated error intervals and calibrated laser data

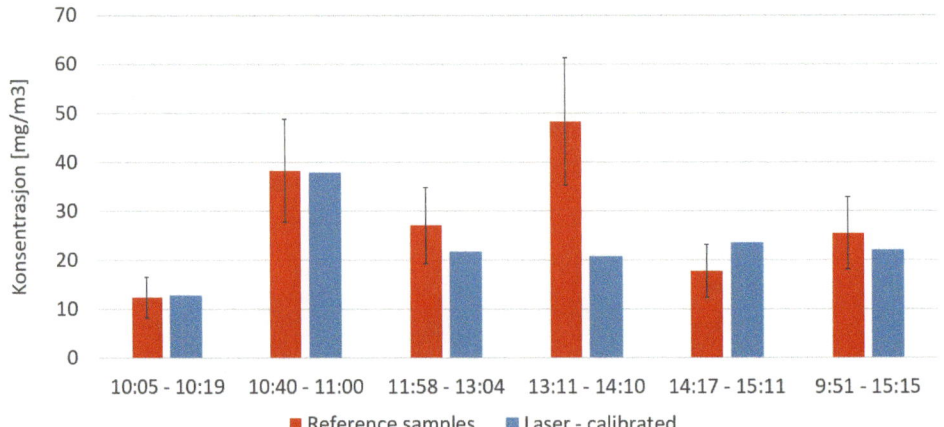

emission has several sources (tapping, refining, casting, and crushing of casted metal), which all emit dust of varying sizes and shapes. This was quantitatively investigated with SEM to compare the dust morphology of selected filters from series 3 and 4. The filters are all heavily loaded with dust and the interpretation of the obtained images is not straightforward. However, the images made it possible to group the dust into smaller fume particles with size <1 μm and spheric shapes, and larger, more irregularly shaped particles with sizes in the range of 10–100 μm. The filter from series 4 contained significantly more of the larger irregularly shaped particles when compared to the filter dust from series 3. It was therefore concluded that the observed deviation in concentration estimates between these series of filters is most likely attributed to variations in dust characteristics, which indicates that the calibration constant might only be valid for a specific type or mixture of dust present at the time of sampling.

Conclusions

The current installation is set up to gain experience in operating an open-path laser measurement system in harsh environments. The weather stations proved to be robust and suitable in an environment with high dust loads. *Sonic B* and *Sonic C* show a high degree of correlation, with 24-h average vertical velocities within 3% of each other. This is in the same order as the specified measuring uncertainty of the weather stations. Long-term evaluation of the data currently shows 18 and 16% negative hourly vertical velocities, respectively, which correlates highly with the roof (*Sonic A*) wind speed and direction. This must be considered when planning and performing reference sampling, as well as when estimating total emission.

Already in its early stage, the pilot installation provided useful data that can be used to tailor mitigation of emissions and target-oriented improvement activities where step changes in cumulated dust emission rates could be correlated to specific operations in production. It became clear that the refining and casting activities are among the largest contributors of diffuse emissions that leave the production hall through the roof openings.

Reference measurements show considerable variation in dust concentrations along the laser line of sight although the monitoring path was limited to a fraction of approx. 8 m of the overall outlet. The calibration factor, *k*, is determined to be 1525 based on regression analysis with series 1, 2, 3, and 5. The long-term series lies within the range of this calibration constant, while series 4 does not. Simplified qualitative SEM analysis shows that this is likely caused by deviating dust characteristics such as size and shape for these samples. These findings illustrate the challenge and inherent uncertainty of measuring diffuse emission from sources of varying dust characteristics with laser technology.

The research works continue because there is an urgent need for a reliable and cost-efficient way for calibration of open-path lasers. Further activities will address specific dust characteristic features like size distribution and morphology, and their influence on laser scatter and absorption. The next-generation Laser Dust Open-Path instrument is currently being tested in a calibration rig designed by SINTEF and installed at an NTNU laboratory in Trondheim, Norway. The instrument is installed in a wind tunnel that allows for dry dispersion and reference measurements under controlled conditions. This means that the rig constitutes a unique opportunity for controlled testing of scatter and transmission measurements over a wide range of dust types, size distributions, and concentration ranges.

Acknowledgements This research was supported by the Norwegian Research Council as part of the BIA/IPN *DustDetect project no.: 309466*. The authors thank Elkem Rana as the project owner and for facilitating the pilot installation at site. We would also like to thank NEO Monitors AS as the technology provider in the project, and for continuous support during installation, testing, and evaluation of results.

References

1. T. Brinkmann og et al, «JRC Reference Report on Monitoring of Emissions to Air and Water from IED Installations,» Integrated Pollution Prevention and Control (European IPPC Bureau), 2018.
2. VDI 4285 Part 1: 2005 "Determination of diffusive emissions by measurements - Basic concepts", Berlin: Beuth Verlag, 2005.
3. VDI 4285 Part 2: 2011 "Determination of diffusive emissions by measurements - Industrial halls and livestock farming", Berlin: Beuth Verlag, 2011.
4. VDI 2066 Blatt/Part 1: 2021-5 "Particulate matter measurement - Measuring particulate matter in flowing gases - Gravimetric determination of dust load – Fundamentals", Berlin: Beuth Verlag, 2021.
5. VDI 4200: 2000-12 "Realization of stationary source emission measurements", Berlin: Beuth Verlag, 2000.
6. NS-EN 15259: 2007 "Air Quality Measurement of stationary source emissions, Requirements for measurement sections and sites and for the measurement objective, plan and report", European Committee for Standardization, 2007.

Characterization of Industrial Hydrocarbon Samples from Anode Baking Furnace Off-Gas Treatment Facility

Kamilla Arnesen, Alexandre Albinet, Claudine Chatellier, Nina Huynh, Thor Anders Aarhaug, Kristian Etienne Einarsrud, and Gabriella Tranell

Abstract

Polycyclic aromatic hydrocarbons (PAHs) are naturally present in raw materials used as a binder in prebaked anodes for electrolysis of aluminum. Green anodes are baked to about 1200 °C through a cycle of 14–17 days where organic hydrocarbon volatiles contribute to the carbonization process. Off-gases contain volatile and semi-volatile organic components, which are further treated to reduce environmental emissions by techniques such as regenerative thermal oxidizers and dry or wet scrubbers. Still, prebaked anode production contributes to a noticeable part of the reported PAH emissions in Norway. Samples of condensed hydrocarbon-based residues have been collected from an off-gas treatment facility and a set of analytical methods applied to determine the presence of different categories of aromatic hydrocarbons in these samples. Based on the results of the characterization, it was determined that the residue contains sulfur-substituted and polar aromatic hydrocarbons in addition to the PAH-16 routinely reported. It was concluded that future research should be dedicated to extending the range of PAHs that can be reliably determined, with particular emphasis on substituted species.

Keywords

PAH • Sulfonates • Sulfates • Pre-baked anodes • Off-gas • Aluminum

K. Arnesen (✉) · K. E. Einarsrud · G. Tranell
Department of Materials Science and Engineering, Norwegian University of Science and Technology (NTNU), Trondheim, Norway
e-mail: Kamilla.arnesen@ntnu.no

A. Albinet · C. Chatellier · N. Huynh
National Institute for Industrial Environment and Risks (INERIS), Verneuil-en-Halatte, France

T. A. Aarhaug
SINTEF Industry, Trondheim, Norway

Introduction

Primary aluminum is produced by electrolysis of alumina with carbon anodes in the Hall-Héroult process. Most modern pot lines use prebaked anodes which are a mixture of petroleum coke, anode butts, and coal tar pitch [1]. Anodes are baked in an open or closed-top baking furnace, where green anodes are stacked in pits, covered with packing coke and, heated to 1200 °C through a cycle of 14–17 days [2]. Petroleum coke originates from a crude oil refinery feedstock which will define the coke quality and structural properties. A typical anode grade calcined petroleum coke has a "sponge" structure, with up to 60% aromatic structure, containing 1.5–3.5 wt% sulfur, and traces of vanadium, nickel and iron [3, 4]. The sulfur is usually present as organic sulfur, five-membered heterocyclic rings named thiophenes, on their own, or as a part of thiophene-containing polycyclic aromatic hydrocarbons (PAHs) structures, as described by Xiao et al. [5]. Coal tar pitch has a composition of about 97% aromatic and hetero-aromatic compounds, with a great variety of structures from 128 to >2500 Daltons. The presence of derivatives such as phenols, ketones, amides, and amines are also reported, contributing to the complexity of the mixture. The off-gas from a baking furnace has been analyzed and is reported to contain CO_2, CO, CH_4, H_2, SO_X, and other volatile organic compounds, which originate from the coke and the coal tar pitch [6, 7]. Currently, in Norway, emissions from plants to air and water, are reported to the government and the priority EPA-16 PAH emissions in 2019 were 67 metric tons, where pre-baked anodes for aluminum production is one of the main contributors [8]. Off-gas treatment plants clean the gas to reduce emissions. Typical equipment to reduce airborne industry emissions are regenerative thermal oxidizers (RTO), which reduce organic matter using combustion chambers operated at about 900 °C. Other technologies used for volatile organic compounds, particulate matter and SO_2 are bag house filters, wet scrubbers, and electrostatic

S. Broek (ed.), *Light Metals 2023*, The Minerals, Metals & Materials Series,
https://doi.org/10.1007/978-3-031-22532-1_91

precipitators, often in combination [9]. While the reported PAH-16 emissions are effectively reduced in the gas, some condensed hydrocarbon residues from anode baking furnaces build up in the off-gas treatment plants and are observed in the pipes and filters.

As there is reason to believe that these condensed residues contain aromatics soluble in aqueous solution, more specifically polar and/or sulfur-substituted aromatic compounds, a more in-depth analysis of the condensed residues was initiated. In the current work, the aim has been to identify reliable methods to firstly establish the presence of various polyaromatic compounds (PACs), and, secondly, determine the type of specie(s) associated with the solubility in various solvents mixtures and their properties.

Experimental Procedure

Industrial Facility and Sample Collection

In Fig. 1, a picture of the condensed hydrocarbon residue from an anode baking furnace off-gas cleaning facility is presented. The residue is semi-hard and tar-like, and was collected from a filter that is located after an RTO and a wet scrubber. An aqueous sample was also collected from the wet scrubber. The same residue sample was used for all characterization performed using the different techniques described below.

^{13}C-NMR Analysis

To investigate the carbon structure and functional groups present in the hydrocarbon residue nuclear magnetic resonance (^{13}C-NMR) analysis was performed. Samples were dissolved in acetone ($\geq$ 99.0% Ph.Eur., VWR Chemicals) and methanol ($\geq$ 99.5% GPR Rectapur, VWR Chemicals), and the extracts were filtered using white ribbon filter paper (Schleicher & Schuell Micro Science, 589/2 grade) before the solvent was removed using a rotavapor. Samples and 1-naphtalenesulfonic acid ($\geq$ 70–<90%, Sigma Aldrich, Germany), used as a standard, were dissolved in deuterated

Fig. 1 Picture of the condensed hydrocarbon residue from an anode baking furnace off-gas cleaning facility

chloroform, and analyzed by a ^{13}C-NMR spectrometer (600 MHz Burker Avance).

ICP-MS

To quantify the sulfur content in the hydrocarbon residue, ICP-MS was used. Triplicates of the residue were prepared together with two samples of coal tar pitch, as a reference material with known sulfur content (0.53 wt%). The precise mass was measured on Sartorius balance (Krugersdorp, South Africa). Ultrapure 50% v/v nitric acid was added using a 5 mL bottle-top dispenser (Seastar Chemicals, Sidney, BC, Canada). The ultrapure nitric acid was produced at NTNU from p.a. grade nitric acid (Merck, Darmstadt, Germany) using a quartz sub-boiling distillation system (Sub-Pur, Milestone, Shelton, CT, USA). The samples were digested using a high-performance microwave reactor (UltraClave, Milestone, Italy). The digested samples were decanted into pre-cleaned 50 mL polypropylene vials (VWR, PA, USA) and diluted to approximately 26.5 mL with ultrapure water to achieve a final acid concentration of 0.6 M. The final weight of the diluted samples was measured with an analytical balance and converted to volume (density 0.6 M HNO_3: 1.0167 g/mL). The samples were analyzed on an Agilent 8800 Triple Quadropole ICP-MS (ICP-QQQ) with SPS 4 Autosampler. They are quantified against standards from Inorganic Ventures.

Fluorescence Spectroscopy

Fluorescence spectroscopy is a widely used technique for the detection and quantification of PACs due to the rigid molecular structure and delocalized electrons of aromatic hydrocarbons.

Calibration

1-Naphtalenesulfonic acid standard ($\geq$ 70–<90%, Sigma Aldrich, Germany) was used as a reference material for the fluorescence measurements. The standard was dissolved in water to produce a stock solution with a concentration of 4120 ppm. Further, standards were prepared by dilution (500, 1000, and 2000 ppm). Each standard, including water and the stock solution, was analyzed by laser induced fluorescence (405 nm) using fluorescence spectroscopy (HD-1000, Advanced Sensors, Carrickfergus, UK) at the emission wavelength range between 400 and 1100 nm. The reported fluorescence signal was an average of five measurements.

Sample Preparation and Analysis

To investigate sample solubility and phase affinity of the PACs in different solvents, residue samples of 10–13 mg were mixed with equal amounts (30 mL) of either of three different organic solvents and aqueous solutions, i.e. combinations of n-hexane, cyclohexane, and dichloromethane (HPLC, Sigma-Aldrich, Germany) and distilled water, with or without sodium hydroxide (Reag.Ph.Eur. VWR Chemicals) (NaOH) at various concentrations (1.0 M, 2.0 M, 3.0 M, and 5.0 M). In Fig. 2, an example of samples dissolved in n-hexane and increasing concentrations of NaOH are presented. Solutions were mixed for 10 or 60 min before being transferred to a separation funnel. The phases were separated, and fluorescence was measured in both phases using the HD-1000 instrument. The sample dissolution time of 10 min was performed in triplicates and 60 min dissolution was performed once.

UHPLC-Fluorescence/UV and UHPLC-HRMS Analysis

Sample Preparation and Analysis

The analysis of 22 PAHs in the residue and wet scrubber samples has been performed by ultra-high-performance liquid chromatography (UHPLC)-fluorescence, while the presence of potential sulfate- or sulfonate-PAHs (SO$_3$-PAHs) has been evaluated using a non-targeted screening (NTS) approach and based on UHPLC-high-resolution mass spectrometry (HRMS) analyses.

Prior to extraction internal standard (6-methlychrysene) was added to the samples to evaluate the recovery rate of PAHs. Solid samples were extracted by sonication in toluene (HPLC, Aldrich), for PAHs, or methanol (HPLC, Aldrich) for NTS evaluation of SO$_3$-PAHs. Aqueous samples were filtrated before liquid/liquid extraction by toluene for PAH analyses. An additional pre-concentration step using an HLB (hydrophilic–lipophilic balanced) solid-phase extraction (SPE) cartridge (Waters, Oasis, HLB SPE), eluted in two

steps with methanol and acidified methanol, was applied of SO$_3$-PAHs. The resulting extracts (Fig. 3) were diluted in acetonitrile (HPLC, Aldrich) before injection on UHPLC-fluorescence (Ultimate 3000, ThermoFisher) for PAH analyses using a C18 UPLC column (Zorbax). Sample extracts were diluted in Milli-Q water (Millipore) before NTS analysis by UHPLC-HRMS of SO$_3$-PAHs (UHPLC 1290 Infinity coupled to a Quadrupole-Time of Flight, IFunnel 6550 (Agilent)) using electrospray ionization (ESI) in negative mode and acquisition in auto MS/MS. Separation was achieved using an HSS T3 column (C18, Waters). The presence of potential SO$_3$-PAHs was based on specific fragmentation patterns and characteristic fragments, namely SO$_3^-$ (m/z 79.9559) and HSO$_4^-$ (m/z 96.9607) fragments [10]. Thirteen commercially available sulfate or sulfonate organic analytical standards were analyzed to study their fragmentation patterns. Additional suspect screening analyses were also performed targeting these compounds as well as based on theoretical m/z of sulfates and sulfonates corresponding to the quantified PAHs in the samples.

Results and Discussion

^{13}C-NMR Analysis

Figure 4 shows the ^{13}C-NMR chemical shift for the sample dissolved in acetone. Three areas in the spectra are noticeable, where one coincides with deuterated chloroform, at 77 ppm, used as a solvent for the extract during analysis. The signals between 0 and 40 ppm indicate the presence of aliphatic carbon and signals at 110–150 ppm originate from aromatic carbon species [11]. It is not possible to determine if the aliphatic carbons are separate components or attached as side chains on an aromatic structure.

Fig. 2 A series with sample dissolved in n-hexane and increasing concentration of NaOH (left to right: Water, 1.0 M, 2.0 M, 3.0 M, and 5.0 M). Organic phase is on the top and aqueous is on the bottom of the flask

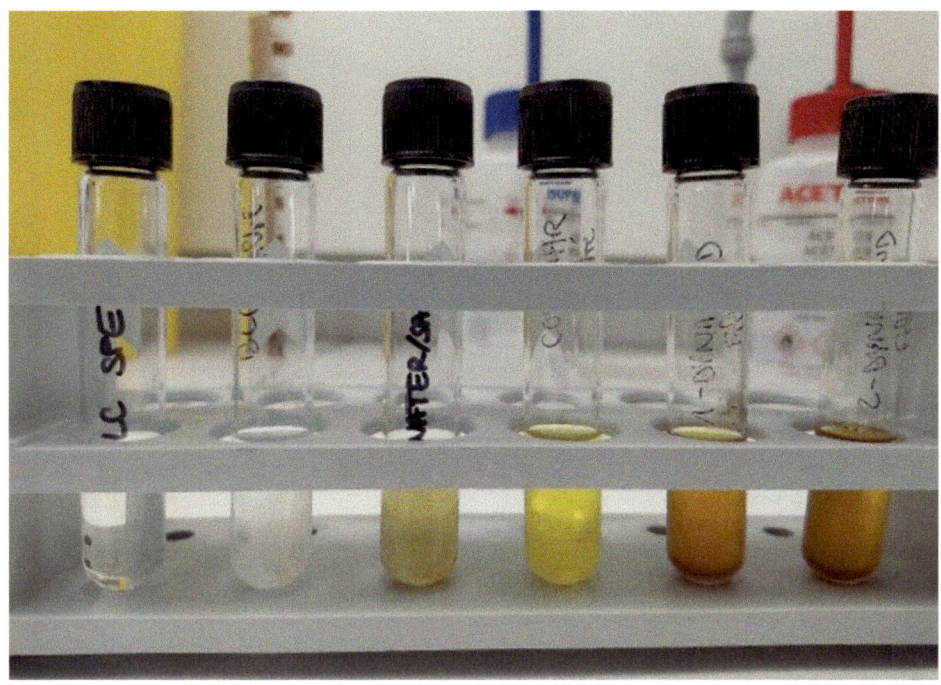

Fig. 3 Extracts for LC-HRMS analyses of the different samples (left to right: solid-phase extraction blank, sonication blank, aqueous, coal tar pitch, Residue-1, Residue-2)

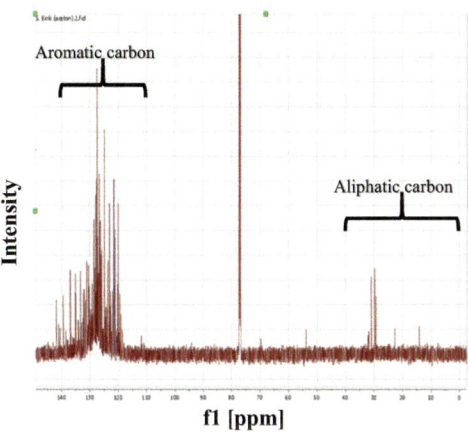

Fig. 4 ^{13}C-NMR chemical shift (ppm) for the condensed sample dissolved in acetone. Deuterated chloroform is used as a solvent during analysis, displayed as the signal at 77 ppm

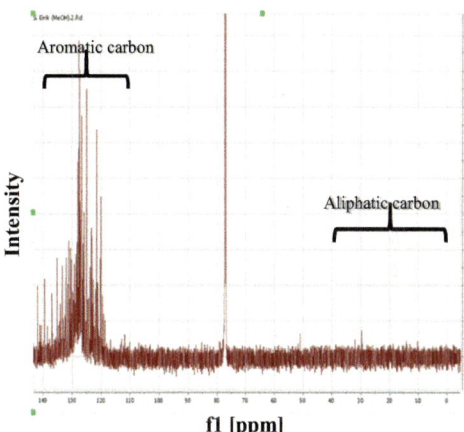

Fig. 5 ^{13}C-NMR chemical shift (ppm) for the condensed sample dissolved in methanol. Deuterated chloroform is used as a solvent during analysis, displayed as the signal at 77 ppm

Figure 5 shows the chemical shift for the sample dissolved in methanol. In addition to the deuterated chloroform at 77 ppm, only the signals from aromatic carbons are present at 110–150 ppm. This may indicate the different properties of the organic solvent influence on which hydrocarbon groups are dissolved. Methanol is a more polar solvent than acetone, which would indicate why the aliphatic carbon is absent from the methanol extract. Aromatic carbon species such as PAH are nonpolar compounds and are generally lipophilic.

Compounds such as benzo[a]pyrene are reported to be sparingly soluble in methanol [12]. This could indicate the presence of other polyaromatic compounds with polar properties, such as heterocycles, alkylated, and substituted PACs [13]. In Fig. 6, the chemical shift from 1-naphtalenesulfonic acid is presented. A signal is visible in the area for aromatic carbon, but as the compound has low solubility in chloroform, the signal is not particularly strong.

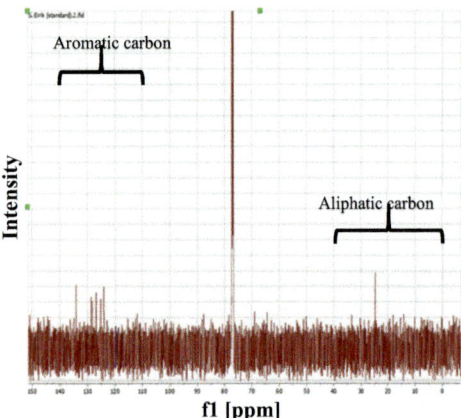

Fig. 6 ^{13}C-NMR chemical shift (ppm) for 1-naphtalenesulfonic acid standard (≥ 70–$< 90\%$). Deuterated chloroform is used as asolvent during analysis, displayed as the signal at 77 ppm

Table 1 Sulfur concentration in residue sample and coal tar pitch from ICP-MS analysis

Sample	S (mg/g)	RDS (%)
CTP 1	4.94	0.9
CTP 2	4.86	1.9
Residue 1	9.42	2.1
Residue 2	9.27	2.0
Residue 3	9.77	1.1
Residue average	9.49	

ICP-MS

Table 1 shows the total concentration of sulfur in the residue sample analyzed by ICP-MS. The average concentration of sulfur is 9.49 mg/g, which is more than in the coal tar pitch raw material used in the anode and could be an indication of an accumulation of sulfur species in the off-gas treatment plant.

Fluorescence Spectroscopy

Calibration

The calibration curve was made using 1-naphtalenesulfonic acid (Fig. 7), as a reference for a possible aromatic species present in the residue sample. The curve is linear in the area between 0 and 4000 ppm, and the fluorescence intensity is between 0.25 and 0.85% based on the pre-programmed calibration of the instrument. In comparison, the intensity of the fluorescence signal from the residue sample could be up to 90 times higher than the concentrations measured for the calibration curve. Not being able to guarantee the continued linearity of the calibration curve, the fluorescence signal was

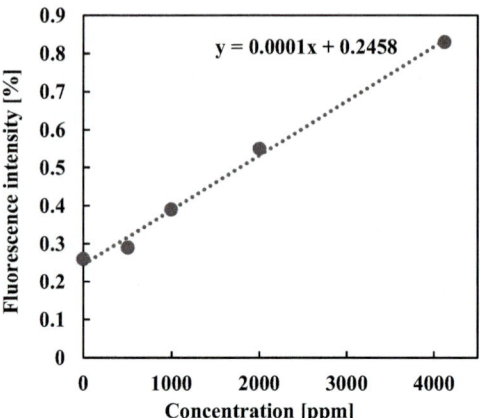

Fig. 7 1-Naphtalenesulfonic acid standard (≥ 70–$< 90\%$) fluorescence emission calibration curve. Excitation wavelength at 405 nm and emission wavelengths between 400 and 1100 nm

normalized based on high and low values within the dataset and not reported as an absolute concentration.

Measurements Using the Industrial Residue Sample

The fluorescence spectra were obtained continuously after the aqueous and organic phases were separated for each sample. The fluorescence signal is an average of five measurements in each solution and has been normalized to compare solubility of the residue sample and the interactions of the dissolved PACs with the organic and aqueous solvents. Figure 8 shows the fluorescence signal in the aqueous phase with different concentrations of NaOH after the

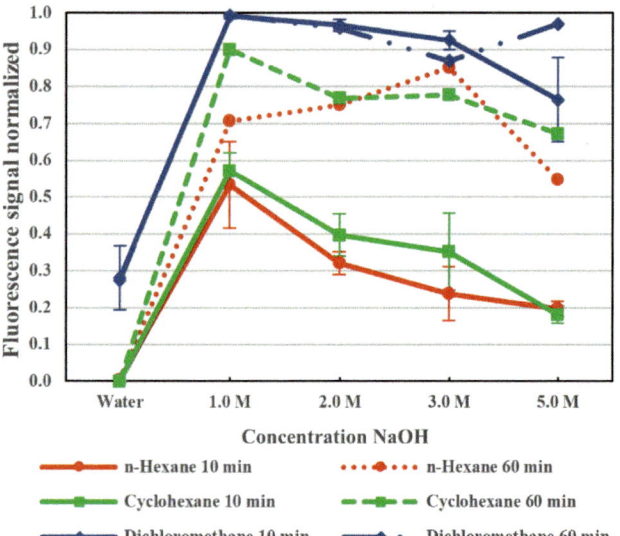

Fig. 8 Normalized fluorescence signal from PACs in the aqueous phase after 10 and 60 min sample dissolution time. Excitation wavelength at 405 nm and emission wavelengths between 400 and 1100 nm. Measurements after 10 min are an average of triplicate tests

sample solution has been mixed for 10 and 60 min.

Both the type of organic solvent and the concentration of sodium hydroxide influence the fluorescence signal from PACs distributed between the two phases. The intensity varies with the type of organic solvent, indicating that the polar dichloromethane is a more effective solvent for the condensed hydrocarbon sample. N-Hexane and cyclohexane have comparable solvent properties, which may result in similar interaction with the dissolved hydrocarbons and explain the overlapping signals for the two solvents. Another overall trend is an increase in fluorescence when the sample is dissolved in an aqueous solution containing NaOH. The addition of hydroxyl ions can react with already dissolved aromatic hydrocarbons, creating aromatic hydrocarbon salts by oxidation, or an acid–base reaction [14]. This reaction would increase the water solubility of the aromatic compounds explaining the increased fluorescence compared to pure water. In addition to being an oxidizing agent, the high-energy vibrations of the hydroxyl ion have been reported to have a quenching effect on the fluorescence emissions, which could explain the decreasing fluorescence signal with increasing NaOH concentration above 1.0 M [15]. Griffiths and Mama also reported on the shift in the fluorescent absorption band when tri-anionic species was produced with increasing pH when looking at the pH-dependency on hydroxyaryl-squarylium dyes [16]. The tests with 60 min dissolution time were performed once, so some uncertainty within the result series may be expected. Overall, the trends are similar to the shorter solubility time, except the signals are somewhat increased. This indicates, especially for n-hexane and cyclohexane, that solubility of the residue could increase with time.

Figure 9 shows the corresponding measurements to Fig. 8, where fluorescence is measured in the organic phase of the sample dissolved in organic solvents and aqueous solutions with varying concentrations of NaOH for 10 and 60 min. Also here, dissolution in dichloromethane results in a stronger fluorescence signal from the aromatic compounds, compared to both cyclohexane and n-hexane. The concentration of NaOH does not seem to affect the distribution of compounds between the two phases.

UHPLC-Fluorescence/UV and UHPLC-HRMS Analysis

Total PAH concentrations (Σ_{22} PAHs) observed were similar in both residue samples (about 160 µg/g) and about 1.6 times higher than in the coal-tar-pitch raw material (about 95 µg/g). The aqueous sample showed concentrations of about 210 ng/ml. Chemical profiles obtained were similar for both residue samples, while higher contributions of heavier PAHs (≥ 5 cycles) were observed for the coal tar

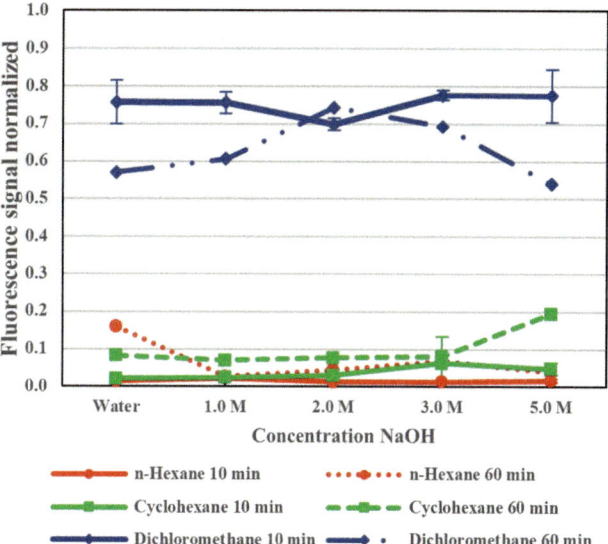

Fig. 9 Normalized fluorescence signal from PACs in the organic phase after 10 and 60 min sample dissolution time. Excitation wavelength at 405 nm and emission wavelengths between 400 and 1100 nm. Measurements after 10 min are an average of triplicate tests

pitch sample. On the opposite, the aqueous sample is dominated by lighter PAH compounds (≤ 4 cycles) which are known to be more water-soluble (Fig. 10).

NTS analyses showed characteristic SO_3^- and/or HSO_4^- fragments highlighting the presence of sulfate or sulfonates organic species in all the samples. The same potential organic sulfate was detected in all the solid samples. For the aqueous sample, two organic sulfonates and one organic sulfate were detected (Fig. 11). Molecular formula calculations (including C, H, O, and S atoms) and confrontation with an in-silico fragmentation tool highlighted potential aromatic sulfonates structures (30/43 and 9/12 of the structures proposed were aromatic sulfonates) [17]. Based on suspect screening analyses, none of the standard sulfate or sulfonate organic compound available was detected in any of the samples. Suspect screening based on theoretical m/z of SO_3-PAHs highlighted the presence of methylnaphthalene sulfate in the aqueous sample.

In summary, the methods used to characterize the industrial residue sample each give different insights into the properties of the condensed off-gas treatment residue. ^{13}C-NMR analysis showed the presence of both aromatic and aliphatic hydrocarbons, in addition to a dependency on solvent properties, as the aliphatic hydrocarbons were only present in the extract using a non-polar solvent. Further characterization using fluorescence spectroscopy confirmed a phase affinity of the aromatic species. The most effective environment for dissolving the sample, both for the aqueous and organic solution with 1.0 M NaOH and dichloromethane, respectively. Dichloromethane is a polar solvent, confirming that the sample contains polar aromatic species,

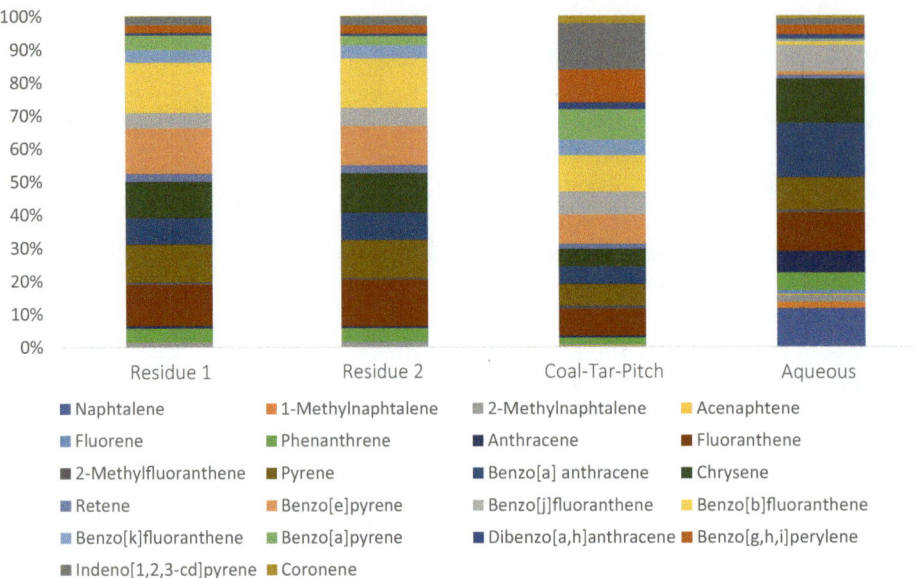

Fig. 10 Distribution of the quantified parent PAH in the different samples

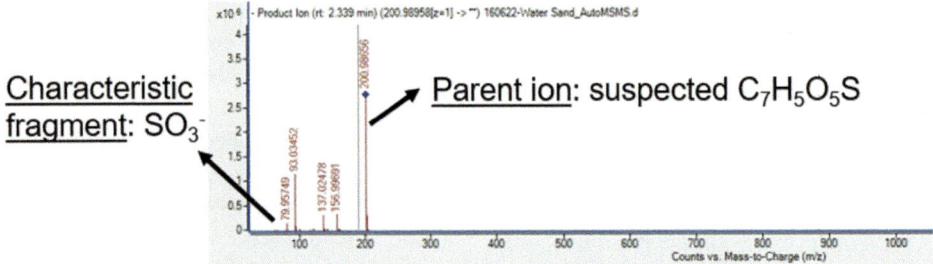

Fig. 11 Example of aromatic sulfonate (probable SO$_3$-PAH) detected in the aqueous sample

which are not present in coal tar pitch and coke, reflecting the complexity of the condensed residue, compared to the raw materials used in the production of pre-baked anodes. The presence of sulfate and sulfonate organic species was confirmed in the hydrocarbon residue and aqueous sample. Polar aromatic hydrocarbons are outside the scope of conventional PAH-16 analysis, but to understand the nature of complex PAH emissions, further research should be dedicated to extending the range of standard methods and techniques to analyze for such species.

Conclusions and Further Work

A multi-method study of condensed hydrocarbon residues from an aluminum anode baking off-gas cleaning facility was carried out in order to determine their content with particular emphasis on polar and/or sulfur-substituted PACs outside the conventional PAH-16 range. The following conclusions were drawn:

- ^{13}C-NMR and fluorescence spectroscopy analysis of the industrial residue confirmed the presence of both polar aromatic and aliphatic compounds in the sample.
- ICP-MS analysis found sulfur species in the residue related to the anode raw materials.
- Substituted PAH (at least aromatic species) with sulfate and/or sulfonate moieties (SO$_3$-PAHs) were highlighted by NTS approach based on ULPC-HRMS analyses.

Given the analytical evidence seen together, it was suggested that the presence of polar aromatic hydrocarbons in the condensed residues, with properties unlike traditional PAH, are due to alkylation and/or substitution reactions occurring during the anode production or in the off-gas cleaning facility. Further experimental work to shed light on their formation, in parallel with analytical method development is suggested.

Acknowledgements This work has been funded by the Norwegian Research Council and the Center for Research-based Innovation, SFI Metal Production (NFR Project number 237738), and supported by the French Ministry of Environment. Furthermore, the authors would like to acknowledge Morten Isaksen in Norsk Hydro ASA and Eirik Sundby at NTNU for their valuable input and discussion.

References

1. E. H. M. Moors, "Technology strategies for sustainable metals production systems: a case study of primary aluminium production in The Netherlands and Norway," *J. Clean. Prod.*, Jan. 2006, doi: https://doi.org/10.1016/j.jclepro.2004.08.005.

2. T. Brandvik, Z. Wang, A. P. Ratvik, and T. Grande, "Autopsy of refractory lining in anode kilns with open and closed design," *Int. J. Appl. Ceram. Technol.*, 2019, doi: https://doi.org/10.1111/ijac.13108.

3. L. Edwards, "The History and Future Challenges of Calcined Petroleum Coke Production and Use in Aluminum Smelting," *JOM*, Feb. 2015, doi: https://doi.org/10.1007/s11837-014-1248-9.

4. M. Legin-Kolar and D. Ugarković, "Petroleum coke structure: Influence of feedstock composition," *Carbon*, Jan. 1993, doi: https://doi.org/10.1016/0008-6223(93)90043-A.

5. J. Xiao, Q. Zhong, F. Li, J. Huang, Y. Zhang, and B. Wang, "Modeling the Change of Green Coke to Calcined Coke Using Qingdao High-Sulfur Petroleum Coke," *Energy Fuels*, May 2015, doi: https://doi.org/10.1021/acs.energyfuels.5b00021.

6. T. Brandvik, H. Gaertner, A. P. Ratvik, T. Grande, and T. A. Aarhaug, "In Situ Monitoring of Pit Gas Composition During Baking of Anodes for Aluminum Electrolysis," *Metall. Mater. Trans. B*, Jan. 2019, doi: https://doi.org/10.1007/s11663-018-1500-8.

7. Q. Zhong, J. Xiao, H. Du, and Z. Yao, "Thiophenic Sulfur Transformation in a Carbon Anode during the Aluminum Electrolysis Process," *Energy Fuels*, Apr. 2017, doi: https://doi.org/10.1021/acs.energyfuels.6b03018.

8. Norwegian Environmental Agency, "Polysykliske aromatiske hydrokarboner (PAH)," *Miljøstatus.* https://miljostatus.miljodirektoratet.no/tema/miljogifter/prioriterte-miljogifter/polysykliske-aromatiske-hydrokarboner-pah/ (accessed Oct. 01, 2021).

9. C. Behrens, O. Espeland, and B. Nenseter, "EMISSIONS OF DIOXINS AND VOC'S FROM THE ÅRDAL CARBON PLANT," p. 7, 2007.

10. M. Riva *et al.*, "Evidence for an Unrecognized Secondary Anthropogenic Source of Organosulfates and Sulfonates: Gas-Phase Oxidation of Polycyclic Aromatic Hydrocarbons in the Presence of Sulfate Aerosol," *Environ. Sci. Technol.*, Jun. 2015, doi: https://doi.org/10.1021/acs.est.5b00836.

11. S. Farmer, T. Soderberg, C. P. Schaller, and L. Morsch, "6.8: Principles of ^{13}C NMR Spectroscopy," *Chemistry LibreTexts*, Jul. 09, 2020. https://chem.libretexts.org/ (accessed Sep. 13, 2022).

12. M. Windholz, S. Budavari, R. F. Blumetti, and E. S. Otterbein, *The Merck index: an encyclopedia of chemicals, drugs, and biologicals*, 10th ed. Rahway, N.J: Merck & Co, 1983.

13. J. T. Andersson and C. Achten, "Time to Say Goodbye to the 16 EPA PAHs? Toward an Up-to-Date Use of PACs for Environmental Purposes," *Polycycl. Aromat. Compd.*, Mar. 2015, doi: https://doi.org/10.1080/10406638.2014.991042.

14. I. J. Keyte, R. M. Harrison, and G. Lammel, "Chemical reactivity and long-range transport potential of polycyclic aromatic hydrocarbons – a review," *Chem. Soc. Rev.*, vol. 42, no. 24, pp. 9333–9391, Nov. 2013, doi: https://doi.org/10.1039/C3CS60147A.

15. J. Maillard, K. Klehs, C. Rumble, E. Vauthey, M. Heilemann, and A. Fürstenberg, "Universal quenching of common fluorescent probes by water and alcohols," *Chem. Sci.*, 2021, doi: https://doi.org/10.1039/D0SC05431C.

16. J. Griffiths and J. Mama, "pH-dependent absorption and fluorescence spectra of hydroxyaryl-squarylium dyes," *Dyes Pigments*, Dec. 1999, doi: https://doi.org/10.1016/S0143-7208(99)00073-X.

17. C. Ruttkies, E. L. Schymanski, S. Wolf, J. Hollender, and S. Neumann, "MetFrag relaunched: incorporating strategies beyond in silico fragmentation," *J. Cheminformatics*, Jan. 2016, doi: https://doi.org/10.1186/s13321-016-0115-9.

CFD Modelling of Solidification and Melting of Bath During Raft Formation

Sindre Engzelius Gylver and Kristian Etienne Einarsrud

Abstract

The dissolution of alumina in cryolite is a complex process, and better understanding is needed to ensure stable cell conditions and high energy efficiency. Additions of cold powder result in freezing of bath that hinders dissolution, and creation of rafts. The current work aims to develop and demonstrate a CFD framework in OpenFOAM for freezing of bath on a fed dose of alumina, based on the volume of fluid (VOF) method, where appropriate source- and sink terms are applied. Essential features have been verified by comparison with a Stefan problem, while simulating the dose as a floating rigid object demonstrate that a larger layer of freeze increase the damping of its movement. When simulating the dose as an immiscible fluid, spreading will hinder enough freeze to be formed around the dose. Hence, the added source terms behave as intended, but improvements on the alumina-bath interactions are needed.

Keywords

CFD modelling • Alumina feeding • Rafts

Introduction

Alumina is, together with carbon, the raw materials used for producing aluminum in the Hall-Héroult process and is added regularly to the electrolyte through point feeders, located at one or several positions in the cell. Modern cells tend to increase in size without a proportional increase of feeder locations, and one feeder must therefore distribute alumina over a higher surface area than earlier. In addition, the anodic-cathodic-distance is decreasing, resulting in lesser available volume for alumina to dissolve in [1], which means that efficient dissolution becomes even more critical.

When a batch of alumina is added, some of the grains might not be dissolved immediately, but form so-called rafts, being a rigid porous body consisting of frozen bath and alumina [2]. Dissolution of single grains in stagnant fluids are however found to be mainly diffusion controlled [3]. However, agglomerates might consist of several thousand grains and this assumption will not hold. Solheim and Skybakmoen [4] propose that heat transfer will restrict the dissolution when a coat of frozen layer is present around the raft, and their calculation showed that the freezing and melting only accounts for about 10% of the total dissolution time. Even though the solidification and melting itself account for only a small fraction of the dissolution process, it cannot be neglected, as demonstrated in the work by Bardet et al. [5], who simulated the alumina concentration in an industrial cell and concluded that the temperature equation must be considered as high superheat is an important driving force for the alumina dissolution. In addition, the freezing of bath can affect the initial geometry of the formed agglomerates, hence determine the available contact area between surface and bath. A model that can predict the formed raft is therefore needed in order to understand the complete picture of dissolution.

The time for bath to melt and freeze around alumina will vary depending on size and geometry. For a single grain with $d = 50\,\mu$m in a stagnant liquid, bath will have frozen and melted within 0.1 s [3], while larger studies by immersing premade agglomerates into the melt showed a total freezing-melting time of approximately 50 s for 5 g [6] and 150 s for 280 g [6].

Dassylva-Raymond et al. [7] base their work on experimental results [6] when developing a model that describes the whole agglomeration process, which also include infiltration of bath into the raft, sintering and dissolution, but do not consider the effect of convection in the bath. Kovács [3] also

S. E. Gylver (✉) · K. E. Einarsrud
Department of Material Science and Engineering, NTNU, Alfred Getz' veg 2, Trondheim 7034, Norway
e-mail: sindre.e.gylver@ntnu.no

K. E. Einarsrud
e-mail: kristian.e.einarsrud@ntnu.no

© The Minerals, Metals & Materials Society 2023
S. Broek (ed.), *Light Metals 2023*, The Minerals, Metals & Materials Series,
https://doi.org/10.1007/978-3-031-22532-1_92

investigates the freezing, melting infiltration and dissolution, assuming the alumina dose to be a porous lump.

Roger et al. [8] developed a coupled model in a Lagrangian framework, by using discrete element method (DEM) to determine the movement of the alumina particles, while smoothed particle hydrodynamics (SPH) was used to compute the flow of bath. A heat transfer model based on the heat conduction equation was also included, where phase change of the bath was modeled using an enthalpy method. Raft formation was modeled by a cohesion force between alumina particles where bath solidification could occur. They have also implemented the effect of natural convection due to temperature dependent density changes successfully.

In Computational Fluid Dynamics (CFD), a regular approach is to use an enthalpy-porosity technique [9–11], achieved by applying additional source terms to the different balance equations and is implemented in several CFD-software [12, 13].

The current work investigates how solidification can be modeled for a multiphase system, where one of the fluids can solidify and melt. This is followed by a demonstration of the models for a system where a raft is subjected to freezing. The said models have been realized in OpenFOAM [14], an open-source framework for numerical simulations.

Governing Equations

Solidification by the enthalpy-porosity method is modeled by adding appropriate source terms to the momentum and energy equation. Assuming that density differences only affect the buoyancy by Boussinesq approximation [15], balances of mass, momentum and energy are written as

$$\nabla \cdot \boldsymbol{u} = 0, \tag{1}$$

$$\frac{\partial}{\partial t}(\rho \boldsymbol{u}) + \nabla \cdot (\rho \boldsymbol{u}\boldsymbol{u}) = -\nabla p + \nabla \cdot \boldsymbol{\tau} + \rho \boldsymbol{g}\beta(T - T_{ref}) + \boldsymbol{S}_d, \tag{2}$$

$$\frac{\partial}{\partial t}(\rho c_p T) + \nabla \cdot (\rho c_p \boldsymbol{u} T) = \nabla \cdot (\kappa \nabla T) + S_h. \tag{3}$$

$\boldsymbol{u}$ and p are respectively the velocity and pressure, ρ is the density, $\nabla \cdot \boldsymbol{\tau}$ is the viscous stress tensor and $\boldsymbol{g}$ is the gravity. $\beta, T, T_{ref} c_p$, and κ are respectively the coefficient of thermal expansion, temperature, a reference temperature, specific heat capacity and thermal conductivity.

$\boldsymbol{S}_d$ dampens the velocity towards zero when the fluid solidifies and can be modeled based on Darcy's law [9]:

$$\boldsymbol{S}_d = -C\frac{(g_s)^2}{(1 - g_s)^3 + q}\boldsymbol{u}, \tag{4}$$

where C is a constant that express the "strength" of the source term, normally in the order of 10^5. q is a small constant needed to avoid singularity and g_s expresses the fraction of solidified liquid in a cell volume, and will have a material dependent relationship with temperature [10]. Assuming an isothermal process, mostly applied for pure metals, g_s can be expressed as

$$g_s = \begin{cases} 1 & \text{for } T < T_M, \\ 0 & \text{for } T > T_M, \end{cases} \tag{5}$$

where a discontinuity in g_s occurs at the melting point, T_M. $\boldsymbol{S}_d$ will then become the dominating term in Eq. (2) as g_s approaches one.

For the energy equation, the term S_h accounts for the enthalpy change due to change of state [11]:

$$S_h = -\rho L \left[\frac{\partial g_s}{\partial t} + \nabla \cdot (\boldsymbol{u}g_s) \right] \tag{6}$$

where L is the heat of fusion. The convective term can in case of isothermal process be neglected [11].

Establishment of a Multiphase Model

The system investigated will contain multiple phases: bath, air, and alumina. The Volume of Fluid (VOF) method [16] is used to distinguish the different phases from each other. Considering a system with N fluids, the amount of a fluid k is expressed by volumetric fraction, α_k, in each computational cell and will have the following balance:

$$\frac{\partial \alpha_k}{\partial t} + \boldsymbol{u} \cdot \nabla \alpha_k = 0. \tag{7}$$

This equation will be solved for $N - 1$ of the fluids, while the last one is solved by the constraint

$$\sum_{i}^{N} \alpha_i = 1. \tag{8}$$

In areas where only one component is present, the single phase (Eqs. (1)–(3)) will in principle be solved. When multiple phases are present in the same area, the equations must be solved using volume-averaged values, which in the case for density will be:

$$\rho_m = \sum_{k}^{N} \alpha_k \rho_k, \tag{9}$$

where the subscript m denotes that it is a mixture. For a system without reactions where the bath phase α_b will solidify and melt, the balance equations can be written as

$$\nabla \cdot \boldsymbol{u} = 0, \tag{10}$$

$$\frac{\partial}{\partial t}(\rho_m \boldsymbol{u}) + \nabla \cdot (\rho_m \boldsymbol{u}\boldsymbol{u}) = \nabla \cdot \boldsymbol{\tau} - \nabla p + \rho_m \boldsymbol{g} + \boldsymbol{F}_{ST} + \boldsymbol{S}_d. \tag{11}$$

$$\frac{\partial}{\partial t}(\rho c_{p,m} T) + \nabla \cdot (\rho_m c_{p,m} \boldsymbol{u} T) = \nabla \cdot (k_m \nabla T) + S_h, \tag{12}$$

$\boldsymbol{F}_{ST}$ expresses the interactions between phases as surface tension. S_h will still have the same form as earlier, while the definition of g_s is slightly changed:

$$g_s = \begin{cases} \alpha_b & \text{for } T < T_M, \\ 0 & \text{for } T > T_M, \end{cases} \tag{13}$$

which ensures that solidification only will occur in the bath phase. The compressible buoyancy effects are also neglected, since density changes in liquid cryolite is small as a function of temperature [17].

The source term presented in Eq. (4) will in principle reduce the velocity to zero. However, in a multiphase system, solidified bath should be affected by movement of the surrounding phases, having a non-zero relative velocity. A possible approach is to apply a temperature dependent viscosity that gain a relatively large value when the fluid becomes solidified [11]. In the current work, the "solid viscosity" is not implemented by varying the actual viscosity of the fluid, but rather by modifying the source term from (4):

$$\boldsymbol{S}_d = \nu_{sol} \cdot \nabla^2 \boldsymbol{u}, \tag{14}$$

where ν_{sol} is dependent on temperature,

$$\nu_{sol} = \begin{cases} \alpha_b C_\nu & \text{for } T \leq T_M, \\ \alpha_b C_\nu \exp[A \cdot (T_M - T))] & \text{for } T_M \leq T \leq T_L, \\ 0 & \text{for } T > T_L. \end{cases} \tag{15}$$

C_ν is and A are user defined constants. In practice, the source term will add an imposed viscosity to the bath phase that increase exponentially from T_L and down to T_M, from zero to C_ν, where the exponential formulation is used in order to avoid numerical issues that can arise when a viscosity is rapidly increased.

Implementation

The fvOptions utility in OpenFOAM provides an opportunity to include source terms to equations, and apply them on multiple solvers without needing to modify the code. The model described is already implemented for single phase solvers, and a detailed description is already published [12].

For a time step $n + 1$, the values of g_s is calculated by the algorithm

$$g_s^* = g_s^n - \frac{\gamma c_p}{L}(T - T_M)$$

$$g_s^{n+1} = \max[0, \min(\alpha_b, g_s^*)], \tag{16}$$

where γ is an under-relaxation factor. The initial value of g_s is by default zero.

The solvers applied in the current cases are either inter-Foam or multiphaseInterFoam, depending of the number of phases. It should be noted that these solvers by default are isothermal, and that the energy equation—written for temperature—as such has been implemented in order to simulate sought effects. The solver is based on the PIMPLE algorithm [18] for pressure-velocity coupling to solve the momentum- and mass equation, followed by the energy equation. Source terms are calculated prior to solving the momentum equation.

Cases

Three cases were investigated with the following purposes:

1. A Stefan case, to verify that the temperature equation and fraction of solidified bath were implemented correctly.
2. A case with a rigid body floating on the surface, to investigate the effect on the developed momentum source term.
3. A three-phase case with cold alumina dose entering a bath on cryolite, to investigate the consequences of alumina being a phase.

The setups are shown in Fig. 1.

Stefan Case

A slab of liquid with infinite length and temperature T_L and a freezing point at $T_m < T_L$ is considered. The left wall of the domain will have a temperature $T_s < T_m$ at $t = 0$, thus creating moving phase front driven by diffusion, where the position of the solidifying front, $X(t)$, can be expressed as [19]

$$X(t) = 2\lambda\sqrt{g_s t}. \tag{17}$$

λ is found solving the Stefan equation:

$$\lambda\sqrt{\pi} = \frac{St_s}{\exp(\lambda^2)\text{erf}(\lambda)} - \frac{St_l}{\vartheta \exp(\vartheta^2 \lambda^2)\text{erfc}(\vartheta\lambda)}, \tag{18}$$

where ϑ expresses the ratio of thermal diffusivity between the liquid and solid part, while Stefan number expresses the ratio of sensible heat to latent heat for respectively the solid and the liquid part, written as:

Fig. 1 Overview of the cases investigated: **Case 1** Stefan case. **Case 2** Rigid body case, with boundaries and phases present (left) and Deformation of mesh with inner and outer distance (right). **Case 3** Cold alumina dose falling into cryolitic bath

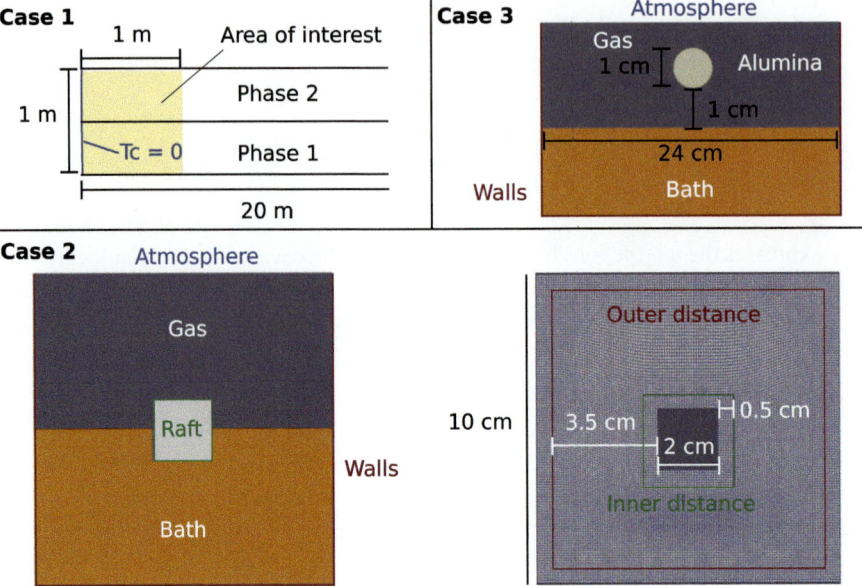

$$St_s = \frac{c_p(T_M - T_s)}{L}, \tag{19}$$

$$St_l = \frac{c_p(T_L - T_M)}{L}, \tag{20}$$

where it is assumed that the solid and the liquid part have the same heat capacity.

The case was set up as a 2D case consisting of four walls, where freezing occurs within a distance of 1 m from the left wall in the investigated time interval, marked as the area of interest. The right wall is 20 m away and while the distance between upper and lower wall is 1 m. The long geometry was used in order to obtain conditions similar to that of the analytical solution, i.e. an infinite domain. The mesh in the area of interest is uniform with rectangular cells that is 5 mm wide and 33.33 mm high, and the width increase to the right after 1 m. Properties of the phases are shown in Table 1, and is set such as the Stefan number and thermal diffusivity of the phases are equal, thus resulting in an equal movement of freezing front. The system has an initial temperature of 100 °C, while left wall has a temperature of 0 °C. Since the case is without any convection and isothermal, only the first term in Eq. (6) was considered.

Time was discretized with a Crank Nicolson scheme, while gradients and Laplacian terms were discretized with a linear scheme and an upwind scheme was used for divergence.

Floating Body

To investigate the pure effect of the damping source term, a raft is considered to be a cubic rigid object floating on a bath of cryolite with air above, as seen in Fig. 1a. The cases are limited to 2D and the body is only allowed to move up and down. When movement of the rigid object occurs, the mesh is being deformed in the area that is set by an inner and outer distance away from the object, illustrated in Fig. 1b with dimensions.

The raft is partly immersed in the melt, with center of mass located 0.4 cm above the surface and has a constant temperature below the melting point of the bath, denoted as T_{raft}. Boundary conditions are given in Table 2, while Table 3 states the properties for the system, where an asterisk marks that the value set is not based on literature value. It is mainly due to that the values cannot be found, except from heat capacity of air, which was set to a high value to avoid that bath freezes due to faster cooling of air. A factorial design was performed, where the effects of bath- and raft-temperature, bath heat capacity and the raft's initial velocity, to illustrate the effect of different drop height, were varied, with values given in Table 4. An additional case with reduced damping strength was also run where all factors besides the velocity were low.

The initial mesh consisted of 153 600 quadratic cells with 0.25 mm length. Time was discretized by a Crank Nicolson scheme, while a linear scheme was applied on the convection and Laplacian terms. For divergence, Van Leer was used for velocity and the phase fraction, while the rest was solved with a upwind scheme. The movement of the rigid object was solved by the Newmark-beta method.

Feeding of a Dose

The third case simulates a dose of alumina falling into a liquid bath, where the effect of solidification and its interactions with several phases are investigated. For simplicity, alumina

Table 1 Physical properties for case 1

Property	Symbol	Phase 1	Phase 2	Unit
Density	ρ	1000	1	kg m^{-3}
Heat capacity	c_p	1	1	$\text{J kg}^{-1}\,{}^{\circ}\text{C}^{-1}$
Viscosity	ν	0.01	0.01	m^2s^{-1}
Thermal conductivity	κ	10	0.01	$\text{Wm}^{-1}\,{}^{\circ}\text{C}^{-1}$
Latent heat of fusions	L	10	10	J kg^{-1}
Melting point	T_M	10	10	${}^{\circ}\text{C}$

Table 2 Boundary conditions for case 2 and 3

Field	Atmosphere	Walls	Raft (case 2 only)
Velocity	Inlet-outlet	No slip	Moving wall velocity
Pressure	Fixed 0	Zero gradient	Zero gradient
Phase fraction	Fixed air	Zero gradient	Zero gradient
Temperature	Zero gradient	Zero Gradient	Fixed, case dependent

is assumed to be a Newtonian fluid. The Boundary conditions and properties are given in Tables 2 and 3, respectively. The time was in this case discretized with a forward Euler scheme, while gradients and divergence were respectively discretized by a cubic scheme and Fromm's scheme. The walls at the sides are placed a long distance away to avoid any unwanted boundary effects. The mesh is uniform with a cell size of 0.125 mm in 4 cm to the left and right, while being graded to become more coarser out. A total of five cases were run, four where C_ν was set to be respectively 0. 0.01, 0.1 and 1, while one case was run without the applying the extra source terms.

Results

Stefan Case

A plot of the phase front as a function of time is shown in Fig. 2, where both the analytical and simulated solution are shown. The current numerical solution was on average 0.014 m below the analytical one, which were not found to increase nor decrease throughout during time.

Floating Body

A selection of cases is presented in Fig. 3. A factorial regression analysis, where the response was set as the lowest value the center of mass had was conducted in Minitab. In that case, individual parameters and their two-way interactions were considered. Each of the individual parameters had a significant positive impact on the depth which the body penetrated, i.e. a higher T_i increased the depth. U_0 and T_i had the highest relative impacts, with respectively 32 and 31%.

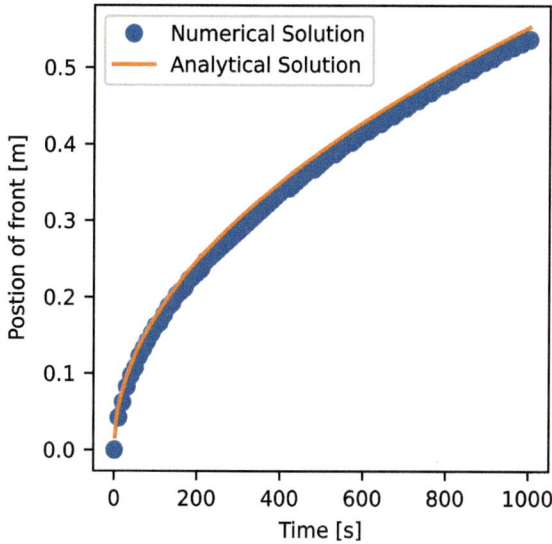

Fig. 2 The analytical solution of the Stefan problem compared with the numerical one

Feeding of a Dose

Figure 4 shows the distribution of different phases, including frozen bath, and the temperature in parts of the domain, when $C_\nu = 0.01$. Images of the phases and freeze at $t = 0.1$ for the different values of C_ν are shown in Fig. 5.

Discussion

The results from the Stefan case, Fig. 2, demonstrate that temperature equation in interfoam (Eq. (12)) and the source term for energy are implemented correctly.

Table 3 Properties of system for case 2 and 3. An asterisk denotes that the value is set, and not a literature value

Property	Symbol	Value case 2	Value case 3	Unit
Alumina density	ρ_a	–	1200	kg m^{-3}
Bath density	ρ_b	2000	2000	kg m^{-3}
Gas density	ρ_g	1	1	kg m^{-3}
Raft density	ρ_r	1200	–	kg m^{-3}
Alumina viscosity	ν_a	–	10^{-6}*	m^{2}s^{-1}
Bath viscosity	ν_b	10^{-6}	10^{-6}	m^{2}s^{-1}
Gas viscosity	ν_g	$1.48 \cdot 10^{-5}$	$1.48 \cdot 10^{-5}$	m^{2}s^{-1}
Alumina heat capacity	$c_{p,a}$	–	1200*	J kg^{-1} °C^{-1}
Bath heat capacity	$c_{p,b}$	Varies	2200	J kg^{-1} °C^{-1}
Gas heat capacity	$c_{p,g}$	10000*	700	J kg^{-1} °C^{-1}
Alumina Thermal Conductivity	κ_a	–	8	W °C^{-1}m^{-1}
Bath Thermal Conductivity	κ_b	0.8	0.8	W °C^{-1}m^{-1}
Gas Thermal Conductivity	κ_g	$2 \cdot 10^{-2}$	$2 \cdot 10^{-2}$	W °C^{-1}m^{-1}
Latent Heat of Fusion	L	530 000	530 000	J kg^{-1}
Temperature constant	T_L	959*	959*	°C
Melting point	T_M	950*	950*	°C
Damping strength	C_ν	10*	Varies	m^{2}s^{-1}
Constant	A	1*	1*	–
Gravity	g	9.81	9.81	m s^{-2}
Initial temperature	T_i	Varies	960	°C
Initial temperature alumina	T_a	–	100	°C
Initial velocity of raft	U_0	Varies	–	m s^{-1}

Table 4 The low and high value of the parameters investigated for case 2

Property	Low	High	Unit
T_i	965	980	°C
T_{raft}	100	500	°C
$c_{p,b}$	1600	2200	J kg^{-1} °C
U_0	0	0.03	m s^{-1}

The rigid object study showed that the initial velocity had the highest effect, followed by temperature and heat capacity. Decreasing the temperatures and the heat capacity will increase the layer of solidified bath and promote damping of the object. The initial velocity increases the penetration depth, although it does not have any effect on the layer formed around the object. When reducing C_ν for the rigid object (right plot in Fig. 3), a larger layer of freeze must form before the effect becomes visible.

When a cold dose hits the bath, a layer of freeze will form, as the bath becomes colder. By assuming the dose to be spherical, it corresponds to add about 0.5 g into the melt. Based on earlier findings [6], it is therefore not expected that melting of the frozen layer has started within the investigated time interval.

As seen in Fig. 5, the added momentum source term does not affect the formation of raft in any large degree. The layer of freeze will become thin as the dose spreads out, resulting in limiting damping effect. C_ν cannot be higher than 1, as this results in numerical issues, for example the strange waves seen in Fig. 5. With $C_\nu = 1$, an even thicker layer of freeze is needed to observe any damping effect (see Fig. 3). In addition, the dose's velocity while entering the bath were around 0.4 m s^{-1}, which is 10 times larger than for the rigid object, and the movement will in these cases therefore in a small degree be affected by heat transfer.

The current cases demonstrate that modelling solidification of bath in the context of the volume of fluid is a possible approach. The issues with the current cases might be improved by allowing for bath mix into the alumina, thus

When considering the added source terms, solidification and melting of bath is not an isothermal process and a more advanced model, such as a linear dependence between solidus-liquidus or Scheil equation should be considered [10].

Fig. 3 Left: The center of mass as a function of time for the floating object at four different configurations. Solid line: no freezing, dashed line: all factors had high values, dash-dot line: all factors had low values, dotted line: all factors except initial velocity had high values. Value of the factors are given in Table 4. Right: comparison of having $C_\nu = 10$ (full) versus 1 (dashed) for the case where the initial velocity is at its high value, while the other factors are low

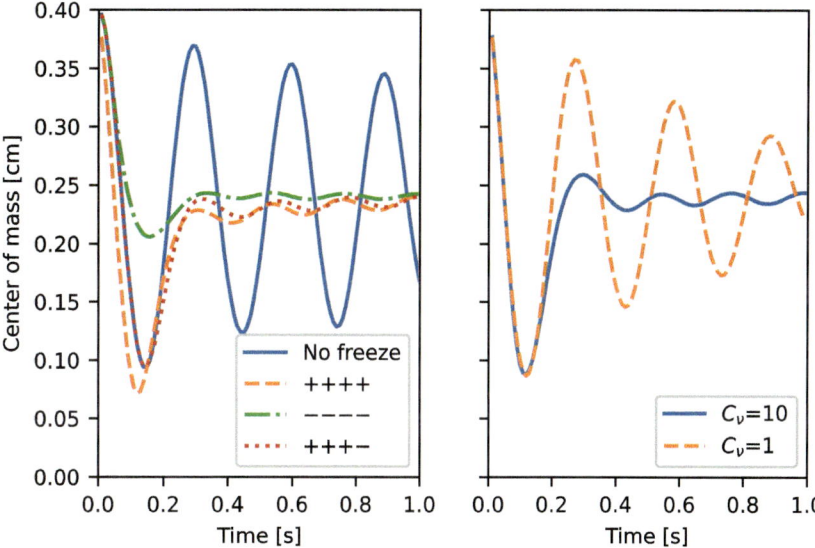

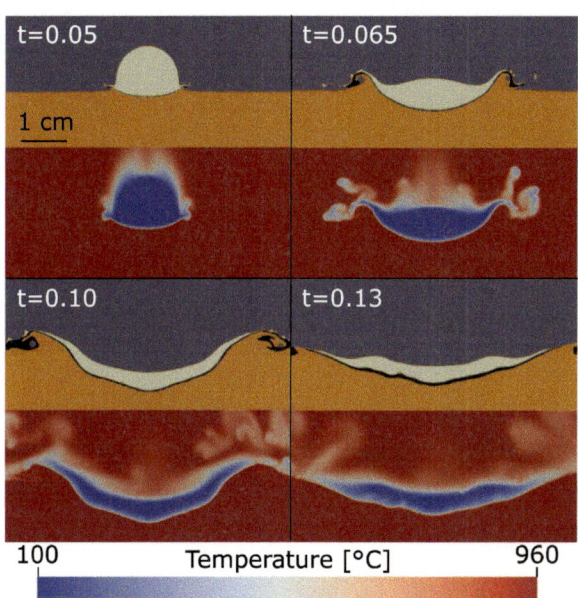

Fig. 4 The phase compositions of molten bath (orange), alumina (beige), air (dark gray) and frozen bath (black), and the temperature distribution in the same area (from cold blue to hot red) for four selected times, when $C_\nu = 0.1$

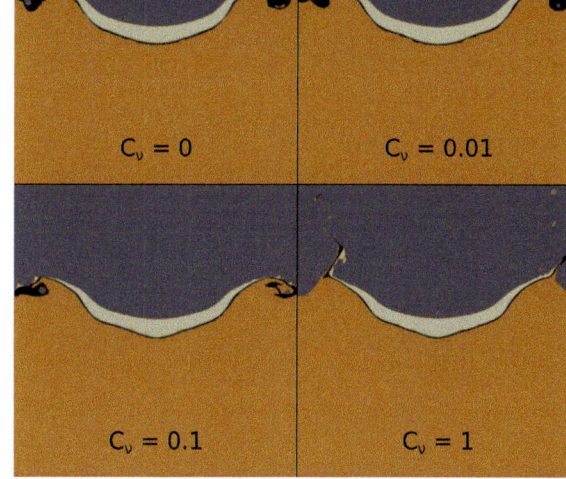

Fig. 5 The phase composition of molten bath (orange), alumina (beige), air (dark gray) and frozen bath (black) amount for the cases with varying damping strength at $t = 0.1$ s

The variation of density of liquid is relatively small [17] and will be important for convection in a stagnant bath [8], which is not the case for a real cell. Frozen bath samples are found to have densities about $500 \, \text{kg} \, \text{m}^{-3}$ above liquid bath [20], which might affect the momentum balance, and should be considered when developing the model.

Conclusion and Further Work

A framework for solidification and melting has been implemented in OpenFOAM in the context of the Volume of Fluid method. A parametric study on a rigid object highlights the importance of thermal masses in raft formation and saw an increase in damping as the freeze layer thickness increased. If alumina is assumed to be a fluid, the effect of freezing was not visible, due to a too thin layer of freeze.

The current model provides a foundation for further work. An improved model should therefore allow bath to infiltrate into the powder, which eventually will allow more bath to freeze. In addition, a more suitable rheology for alumina should be found. The model should also consider a more complex temperature dependency and density changes due to solidifying, and validation by studying a freezing object in water should also be considered.

Acknowledgements The current work has been funded by SFI Metal Production (Centre for Research-based Innovation, 237738). The authors gratefully acknowledge the financial support from the Research Council of Norway and partners of the center.

References

1. Lavoie, P., Taylor, M.P., Metson, J.B.: A Review of Alumina Feeding and Dissolution Factors in Aluminum Reduction Cells. Metall Mater Trans B 47(4), 2690–2696 (2016). https://doi.org/10.1007/s11663-016-0680-3.

2. Gylver, S.E., Omdahl, N.H., Rørvik, S., Hansen, I., Nautnes, A., Neverdal, S.N., Einarsrud, K.E.: The Micro- and Macrostructure of Alumina Rafts. In: Chesonis, C. (ed.) Light Metals 2019. The Minerals, Metals & Materials Series, pp. 689–696. Springer, Cham (2019).

3. Kovacs, A.: "Modelling the feeding process for aluminium production". PhD Thesis, University of Oxford, Oxford (2021).

4. Solheim, A., Skybakmoen, E.: Mass- and Heat Transfer During Dissolution of Alumina. In: Tomsett, A. (ed.) Light Metals 2020. The Minerals, Metals & Materials Series, pp. 664–671. Springer, Cham (2020).

5. Bardet, B., Foetisch, T., Renaudier, S., Rappaz, J., Flueck, M., Picasso, M.: Alumina Dissolution Modeling in Aluminium Electrolysis Cell Considering MHD Driven Convection and Thermal Impact. In: Williams, E. (ed.) Light Metals 2016, pp. 315–319 (2016).

6. Walker, D.: Alumina in aluminum smelting and its behaviour after addition to cryolite-based electrolytes, Phd thesis, University of Toronto, Toronto, 1993.

7. Dassylva-Raymond, V., Kiss, L.I., Poncsak, S., Chartrand, P., Bilodeau, J.-F., Guérard, S.: Modeling the behavior of alumina agglomerate in the Hall-Héroult process. In: Grandfield, J. (ed.) Light Metals 2014, pp. 603–608. Wiley, Hoboken, (2014).

8. Roger, T., Kiss, L., Dion, L., Guérard, S., Bilodeau, J.F., Bonneau, G., Santerre, R., Fraser, K.: Modeling of the Heat Exchange, the Phase Change, and Dissolution of Alumina Injected in Electrolysis Cells. In: Eskin, D. (ed.) Light Metals 2022. The Minerals, Metals & Materials Series, pp. 363–370. Springer, Cham (2022).

9. Voller, V.R., Prakash, C.: A fixed grid numerical modelling methodology for convection-diffusion mushy region phase-change problems. International Journal of Heat and Mass Transfer 30(8), 1709–1719 (1987). https://doi.org/10.1016/0017-9310(87)90317-6.

10. Swaminathan, C.R., Voller, V.R.: A general enthalpy method for modeling solidification processes. MTB 23(5), 651–664 (1992). https://doi.org/10.1007/BF02649725.

11. Voller, V.R., Cross, M., Markatos, N.C.: An enthalpy method for convection/diffusion phase change. International Journal for Numerical Methods in Engineering 24(1), 271–284 (1987). https://doi.org/10.1002/nme.1620240119.

12. Torabi Rad, M.: solidificationMeltingSource: A Built-in fvOption in OpenFOAM® for Simulating Isothermal Solidification. In: Nóbrega, J.M., Jasak, H. (eds.) OpenFOAM®: Selected Papers of the 11th Workshop, pp. 455–464. Springer, Cham (2019). https://doi.org/10.1007/978-3-319-60846-4_32.

13. Muhammad, M.D., Badr, O., Yeung, H.: Validation of a CFD Melting and Solidification Model for Phase Change in Vertical Cylinders. Numerical Heat Transfer, Part A: Applications 68(5), 501–511. Taylor & Francis, (2015). https://doi.org/10.1080/10407782.2014.994432.

14. OpenCFD: OpenFOAM®- Official home of The Open Source Computational Fluid Dynamics (CFD) Toolbox. http://www.openfoam.com Accessed 2019-05-06.

15. Bird, R.B.: Transport Phenomena, 2nd ed. edn. Wiley, New York (2002).

16. Hirt, C.W., Nichols, B.D.: Volume of fluid (VOF) method for the dynamics of free boundaries. Journal of Computational Physics 39(1), 201–225 (1981).

17. Grjotheim, K., Kvande, H. (eds.): Introduction to Aluminium Electrolysis-Understanding the Hall-Heroult Process. Aluminium-Verlag, Düsseldorf (1993).

18. Greenshields, C., Weller, H.: Notes on Computational Fluid Dynamics: General Principles. CFD Direct Ltd, Reading, UK (2022). https://doc.cfd.direct/notes/cfd-general-principles/.

19. Alexiades, V., Solomon, A.D.: Mathematical Modeling of Melting and Freezing Processes, 1st edn. CRC Press, Washington, DC (1993).

20. Poncsák, S., Rakotondramanana, L., Kiss, L.I., Roger, T., Guérard, S., Bilodeau, J.-F.: Evolution of Mechanical Resistance of Alumina Raft Exposed to the Bath in Hall-Héroult Cells. In: Chesonis, C. (ed.) Light Metals 2019. The Minerals, Metals & Materials Series, pp. 667–673. Springer, Cham (2019).

Experimental Investigation of the Alumina Cloud During Alumina Injections in Low- and High-Temperature Conditions

T. Roger, L. Kiss, L. Dion, S. Guérard, J. F. Bilodeau, and G. Bonneau

Abstract

Alumina injections are the most frequent discrete events occurring in aluminum reduction cells. During each feeding, a significant fraction of the mass injected will float and create a raft composed of alumina and frozen bath which hinders the dissolution rate of alumina. However, a small fraction of the alumina sinks in the form of a cloud in the electrolyte which establish idealized dissolution conditions. Specific investigations were performed to understand the fraction of particles in each specific state and the cloud patterns. An analog experimental setup is presented to observe the cloud at low temperature. Organic particles, cooled with liquid nitrogen, were injected in water. Each experimental injection performed was analyzed during the formation of the ice-particle raft to determine the surface and the density of the cloud. The results are compared with experimental injections performed in molten cryolite using a see-through cell to pinpoint the similitude and disparities.

Keywords

Aluminum electrolysis • Alumina injection • Cloud formation

T. Roger (✉) · L. Kiss · L. Dion · G. Bonneau
GRIPS-UQAC, Régal, Université du Québec à Chicoutimi, 555 Boulevard de l'Université, Chicoutimi, QC G7H 2B1, Canada
e-mail: thomas.roger1@uqac.ca

S. Guérard · J. F. Bilodeau
Arvida Research and Development Center, Rio Tinto Aluminium, 1955 Boulevard Mellon, Jonquière, QC G7S 4K8, Canada

Introduction

In the ultimate quest for a more efficient aluminum production, the complexity of phenomenon's surrounding the alumina injection has inspired numerous researchers in their investigation towards process improvements. In this initial step necessary to provide alumina to the electrolysis cell, the particles of alumina, colder than the bath, create a raft with the frozen bath. Nonetheless, as the phase change of the liquid bath occurs rapidly but not instantaneously, a fraction of the alumina injected is allowed to disperse further than the raft. This investigation will focus primarily on these particles as they sink and create a cloud with optimal dissolution conditions. Due to the complexity of observations of alumina injections in electrolysis bath, only a few publications are available in the literature. Out of the few authors that developed experimental rigs [1–3] to observe the injection of alumina, the main topic of their research did not consider this very specific phenomenon in the analysis that was published.

First, this paper will analyze two videos of alumina injection into electrolysis bath, the first video was presented by Bracamonte et al. [2] and the second video was captured by A. Molin [3]. Theses analyses will allow for a better understanding of the injection and will serve as a base for comparison with the next part of this paper.

Secondly, this paper will present an analogue method to understand the injection of a granular material into a liquid to study the cloud formation. The advantage to use analogue method is to understand the physics with some simplification but under more friendly conditions than in the electrolysis cells. Organic particles are close to particles, mustard grain, used for the papers at the TMS conference in 2020 and 2022 [4, 5], but a little bit smaller. The name of the organic granular material is not divulged for confidentiality reasons. Then, the particles are injected into water at 8 °C. The mass of particles injected for a referenced injection is 1 kg. During the experiments, a camera captures the side view of the injection and the dispersion of the particles. The background is illuminated by a white screen and a projector light for easier post-treatment processing. The particles are injected with the same injector as used by the industry.

Alumina Injection Analysis

The alumina injection presented by Bracamonte et al. [2] is made with the addition of 2.5 g of alumina in 250 g of bath. LiF was added to allow a low liquidus point and have a longer lifetime of the transparent quartz crucible. The pictures captured are shown in Fig. 1.

The authors present the injection of the alumina in frame (b) with a formation of a finger. This finger is the penetration in the liquid of the agglomerate formed with alumina and frozen bath. Frames (c) and (d) show the main cloud of alumina sinking in the bath. It is made with a high fraction of particles limiting the observation. Frames (e) and (f) show the presence of small agglomerates, most probably a limited number of alumina particles in cohesion with frozen bath. Frame (g) shows bigger agglomerate sinking on the bottom of the crucible. After this frame, the authors mentioned that the alumina is "almost completely dissolved".

As part of previous research realized by Molin [3], our research group developed a methodology to visually capture the injection of alumina with a larger crucible and mass of alumina. The size of the anthracite crucible was a cube of 39 cm per side and was able to hold nearly 60 kg of electrolyte. The injection of the alumina was 1 kg. The images were captured using a high-speed camera and are presented in Fig. 2.

Figure 2 shows the surface of the liquid (a), the formation (b), and the downward movement (c) of the cloud during the injection to cover all le volume of liquid (d). At 2 s, the cloud is disappeared (e) and gives way to the disintegration of the raft by small agglomerate (f). The line on image (e) shows the bottom of the raft.

The injection presented by Bracamonte et al. is similar, but the cloud disappears more quickly on our injection due to the bigger volume of liquid and the different chemical and thermal conditions. The crucible impacts the intensity of the free convection and influences the disintegration of the raft.

The particles on the cloud are too small on both videos to compute the fraction of particles and the video is too low quality to extract quantitative information. But the different step present during the injection can be compared to the injections with the analogue method and show the potential to understand the cloud formation with an analogue method. The two videos clearly demonstrate that the dissolution of the cloud is very fast compared to the raft life.

Analogue Injection Method

The analogue injection allows to easily create controllable and repeatable conditions to observe the injection from its side without a specific need for an important organization and setup. This part of the paper shows the similarity and difference between low- and high-temperature injections.

To study cloud formation, an organic granular material is used with a 1 mm of diameter. It was cooled previously at −190 °C with liquid nitrogen. The injector used by the aluminum industry is used to inject the particles. It is cooled with a freezer at −20 °C to limit the particles' heating during the injection. The tank (50 cm × 180 cm) is filled with 28 cm of water. The filmed zone is limited by the camera to nearly 80 cm to optimize the pixels per millimeter resolution and improve the quality of the analysis. The movie is taken with a Nikon 5300 at 60 frames per second.

In our investigation on the cloud formation, nine injections were performed with different parameters as summarized in Table 1.

The chosen set of parameters allows to understand the effect of the injection height (position of the injector), the mass injected, and the flow of the liquid. The flow of the liquid is created with two pipes, one creates a vertical flow and a second one for a horizontal flow as shown in Fig. 3. To create the flow, water is injected into the pipe and ejected by the holes made on the pipe. The horizontal flow is to simulate the magnetohydrodynamic flow present in the electrolysis cells and the vertical flow to simulate the flow of bubbles coming from electrolysis. The reference condition (injections 1, 2, and 3) is an injection at 40 cm, with 1 kg of particles and without flow.

To analyze the video of the injection, an algorithm was developed to detect the position of the raft and of each particle present in the cloud. Figure 4 represents a comparison of a frame taken by the camera and the results as computed by the algorithm.

To compare the results, the algorithm gives the surface of the cloud, the fraction of particles present in the cloud, and the surface of the horizontal projection of the raft behind the liquid surface.

The limit of the method developed is the 2D analysis of a 3D phenomenon. A 3D analysis was more complicated and not necessary at this step of the study. This analysis is a first step to understand the behavior of alumina cloud and aims at pinpointing the principal parameters influencing the cloud formation.

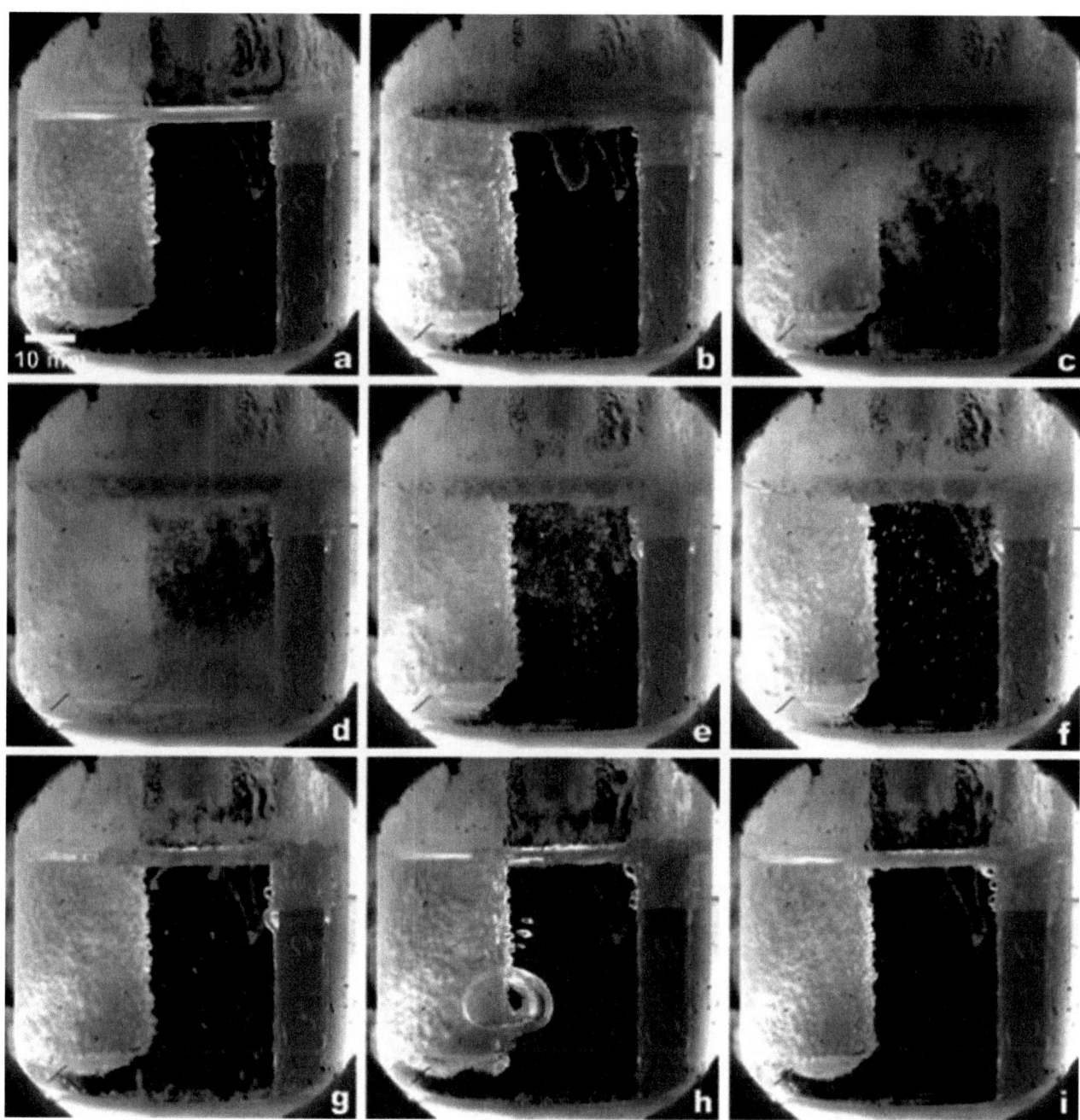

Fig. 1 Images from the dissolution process of the first addition of alumina (**a**—before addition; **b**—upon addition; **c**—1 s; **d**—20 s; **e**—30 s; **f**—40 s; **g**—1 min; **h**—2.5 min; **i**—3 min) [2]

Image produced by NTNU

Comparison Between the Injection of Alumina and the Analogue Method

The first observation is the formation of the raft and a cloud. The geometry of the raft is similar to the one observed by Gylver [6] during the injection of the alumina on the electrolysis cells. The granular material forms a long raft at the surface of the liquid.

During the injection of the organic particles with the analogue method, the cluster of particles create a rotation to form the raft. The density of the granular material is under the density of the liquid. So, the particles try to float at the liquid surface. But the neighbor particles push the particle,

Fig. 2 Images taken by the side of an alumina injection with the limit of the cloud of the aggregate (**a**—before the injection; **b**—during the injection; **c**—at 0.5 s; **d**—at 1 s; **e**—at 2 s; **f**—at 4.5 s). Solid lines were added manually for easier observations

Table 1 Analogue injection parameters

Test	Injection height (cm)	Mass injected (kg)	Flow	Injection temperature (°C)
1	40	1	No	−190.1
2	40	1	No	−188
3	40	1	No	−190.1
4	40	1	**Vertical**	−193.5
5	40	1	**Horizontal**	−192
6	**20**	1	No	−188.2
7	**60**	1	No	−191.1
8	40	**0.5**	No	−190
9	40	**1.5**	No	−182.4

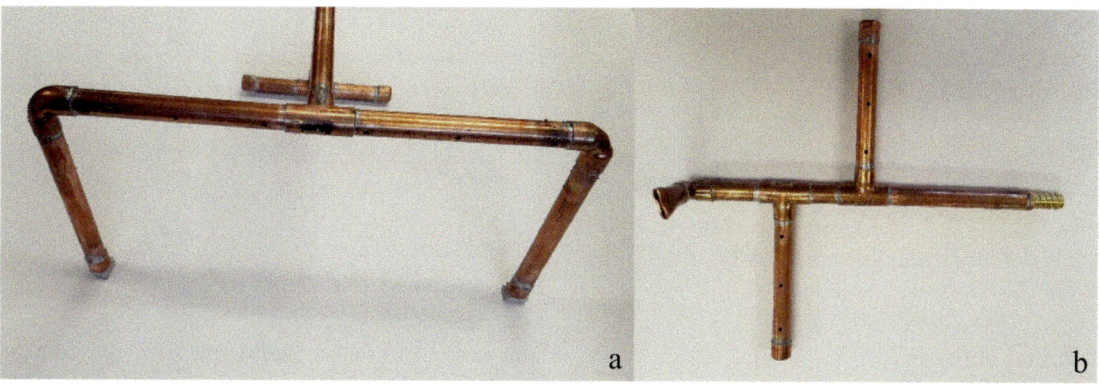

Fig. 3 Water pipe to create flow close to the injection (**a**—horizontal flow; **b**—vertical flow)

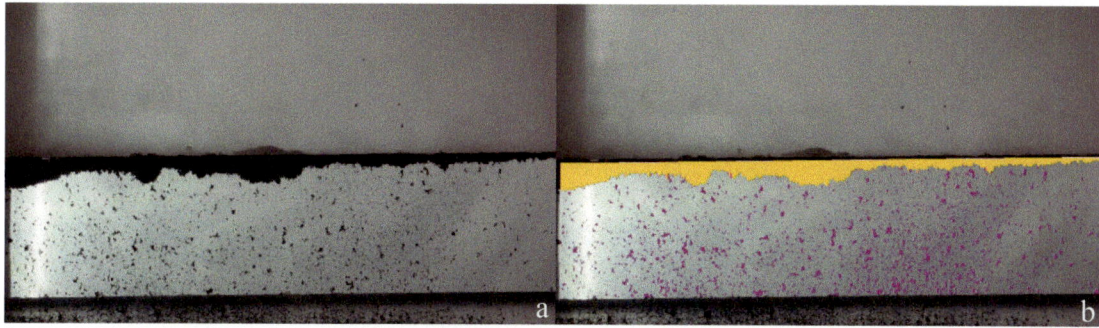

Fig. 4 Frames took at 30 s after the injection before the analysis (**a**) and after (**b**) (yellow: raft; pink: particles in the cloud)

Fig. 5 Comparison between the aggregate sinking created with the analogue model with the high-temperature model

and the rotating movement is created. This phenomenon can be observed in the videos published online by Luis Braca-monte [7] showing the alumina injection.

During the warming of the raft, some aggregate sink (Fig. 5). This agglomerate is the source of sludge formation on the cathode and influence the efficiency of the cells [8]. The phenomenon is present also with the analogue method but creates smaller aggregates. The cohesion between the particles is smaller due to the lower difference in temperature between the injection tem-perature and the phase change temperature of the liquid.

Analogue Injection Analysis

Repeatability of the Measure

As indicated previously, a specific analogue experimental setup at low temperature was developed and an algorithm was designed for data processing. Because of the variability of such sequential processes, the repeatability of the results provided from this work was examined. In this case, three experiments were performed using the reference conditions and the results are compared.

The same metrics will be used for the comparison between the results from all the different injections. The first graph shows the evolution of a two-dimensional cloud surface. It is calculated with the vertical position of the last particles. The second graph shows the number of pink pixels between the bottom of the raft to the bottom of the cloud compared to the original number of pixels, which gives the surface fraction of cloud particles in image. A cloud filled with particles would give a fraction leaning towards a value of "1". The last graph shows the horizontal projection surface of the raft following the algorithm's automatic differentiation between the raft and the cloud. Hence, it represents the surface of the shadow created by the raft as measured for every frame (Fig. 6).

The results demonstrate a good behavior agreement between the repetitions regarding the measured cloud surface and the fraction of particles in the cloud. However, there is more uncertainty regarding the repeatability of the projected surface measurements. No specific cause for the disparity between injection one and the two others was identified.

The noise (error) as a function of time can be calculated with the Eq. 1 to make sure that the error is not important compared to the variation observed in the rest of the analysis (Fig. 7).

$$\text{Error}[\%] = \frac{\max - \min}{\text{average}} \cdot 100 \quad (1)$$

As we can see further in the analysis, the effect of the parameters is dominant and overcomes the noise variability once the injection is complete (after approximately 10 s). To compare the injection method, all the next results will be compared to the average value curve.

Influence of the Injection Height

The second analysis shows the influence of the height on the cloud formation. The height took three values: 20 cm, 40 cm (the reference), and 60 cm (Fig. 8).

This analysis of the injection height shows that the height impacts the fraction of particles present in the cloud. The height increases the movement of the particles during the injection and impacts the link between the particle created by the ice. At the end, the proportion of the particles injected sank will be increased for a higher injection.

Influences of the Mass Injection

The analysis of these conditions shows the influence of the mass injected on the cloud formation. The mass took three values: 0.5 kg, 1 kg (the reference), and 1.5 kg (Fig. 9).

The first graph shows that the mass injected is in close correlation with the cloud formations due to the higher number of particles. But it is not possible to compute the fraction of particles present on the raft and on the cloud. So, the increase of the efficiency could not be computed.

The second graph shows that the mass increases the fraction of particles in the cloud and the duration of a dense cloud is beneficial for the alumina dissolution. More particles are in contact with the liquid. The duration allows to give time to dissolve the particles and avoid the saturation of the bath.

The third graph shows an increase in the raft's surface. The phenomenon that influences the raft formation is the phase change. It creates a link between the solid particles. The thermal energy is increasing with the mass of particles injected. It created more ice and created a bigger agglomerate.

Influence of the Flow

The analysis of these conditions shows the influence of the flow in the cloud formation. Two types of flow are studied and compared to the reference: the horizontal flow and the vertical flow. On the electrolysis cells, the two flows influence the raft and cloud formation at the same time. But to understand the influence of each one, they are studied separately. The horizontal flow is created by the magnetohydrodynamics (MHD). The vertical flow simulates the movement in the liquid generated by the bubbles created at the bottom surface of the anode. Like presented before, the flow is created with an input of water and not with bubbles. The bubbles can create light interference and decrease the performance of the algorithm developed to analyze the data (Fig. 10).

Unfortunately, due to increased movement in the tank, the flow disperses the particle in the tank and the camera is not adapted to this situation and adds another layer of uncertainty to the results. Nonetheless, it is possible to observe

Fig. 6 Result of the repeatability on the cloud surface (**a**), the fraction of particles in the cloud (**b**), and the surface of the projected area of the raft (**c**)

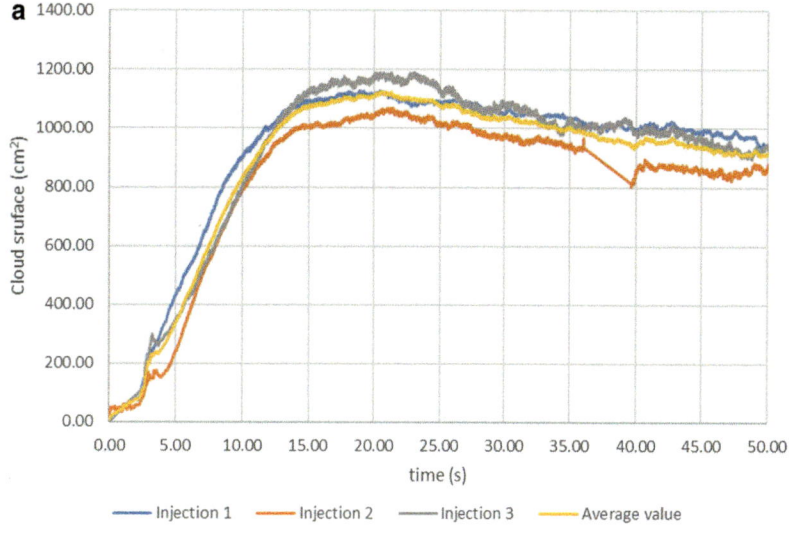

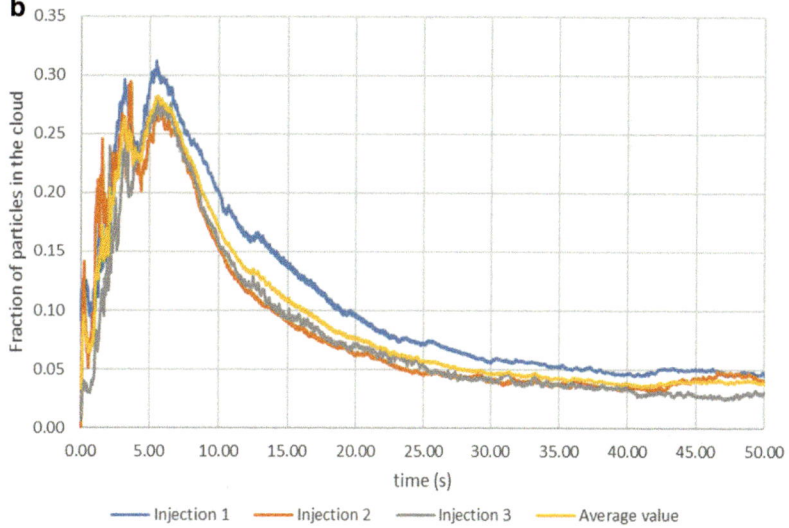

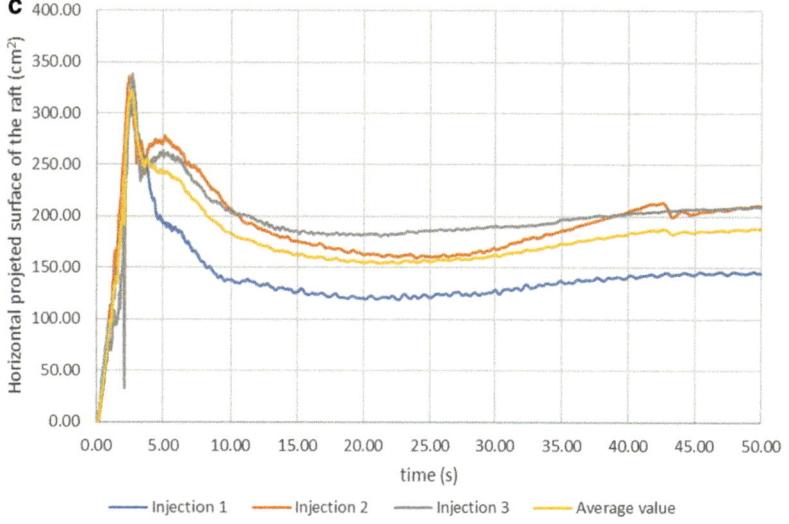

Fig. 7 Noise as a function of the time

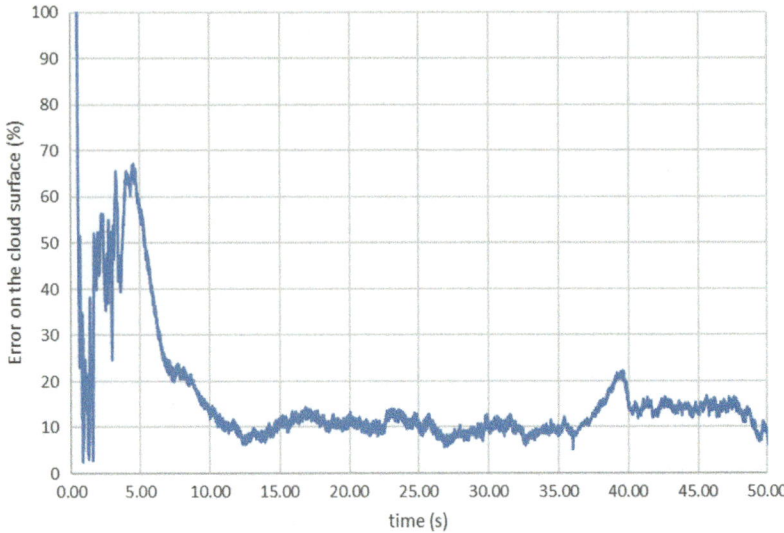

some detailed behaviour. The horizontal flow will provide a more important dispersion of the particles and will improve the cloud surface. However, in the initial phase of the injection (5–10 s), we can observe that the horizontal flow disperses the particles and reduces the fraction of particles in the cloud. This effect is rapidly overcome by the overall mixing in the tank. Finally, the horizontal and vertical movement of the fluid appears to help the propagation of the raft particles during the solidification of the fluid, which increases the surface of the raft.

Conclusion

This paper presented a study of the cloud formation during and after the injection of the alumina in the electrolysis cells. The cloud represents the alumina particles sinking in the bath under the raft. The first part shows two analyses of alumina injection captured by the side at high-temperature conditions. The second part shows the potential to use an analog methodology at low temperature to further improve our understanding of the cloud formation. The two were compared and show some resemblance like the rotation of the raft, the cloud formation, and the aggregate separation. After showing the similarities between the two parts, a study is realized to understand the influence of the mass injected, the injection height, and the flow on the cloud formation using an algorithm and an automated video analysis. The

study is limited to two-dimensional measurements even if the problem is three-dimensional.

The results consider three metrics for the comparisons between the different injections: the surface of the cloud, the fraction of particles in the cloud, and the surface of the horizontal projection of the raft. The first part of the results shows the repetitiveness of the analysis method. The error narrows down close to 10% between the three tests and the average once the injection is complete. Nonetheless, the overall variation needs to be important to see a significant impact on the cloud formation. Most of the result shows a too small variation between the parameter to be considered as a potential influencer for the cloud formation. The mass injected is the only parameter that impacts the most the cloud formation. A bigger mass increases the surface of the raft and the fraction of particles inside it.

Finally, the methodology developed in this study demonstrates good potential. The first improvement is more fundamental with the learning about the cloud formation during the granular material injection. The second benefit is to find the parameters to increase the cloud formation and the duration to improve the alumina dispersion and the dissolution. Fine-tuning is necessary to improve the reproducibility of the metrics considered. Once complete, further low-cost and low-temperature experimentations may be executed for an enhanced understanding of the detailed step-by-step variations and mechanisms occurring during the alumina injection process.

Fig. 8 Results of the influence of the injection height on the cloud surface (**a**), the fraction of particles in the cloud (**b**), and the surface of the projected area of the raft (**c**)

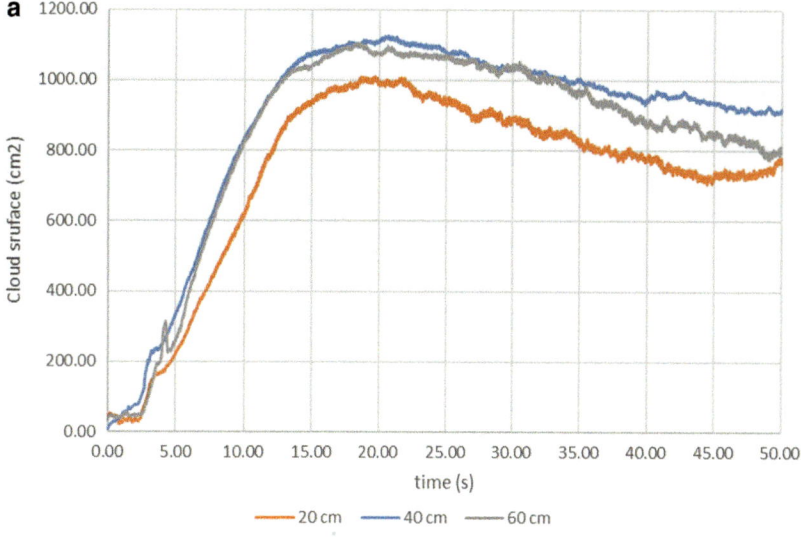

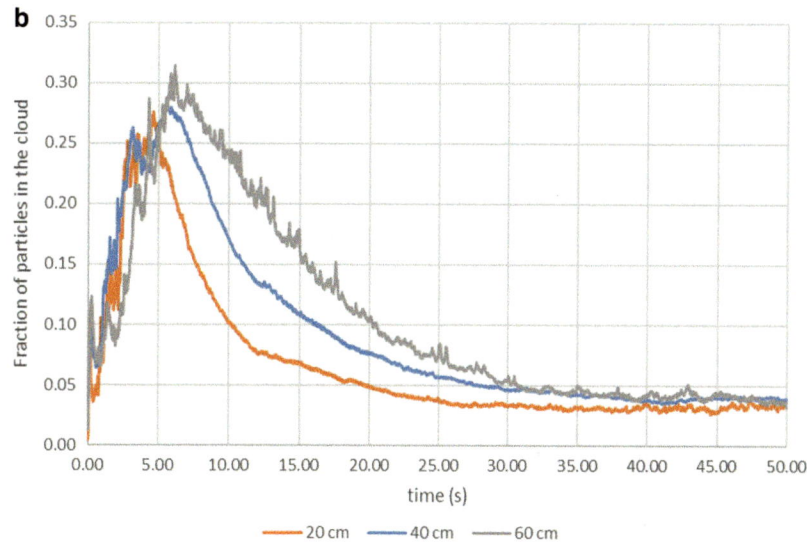

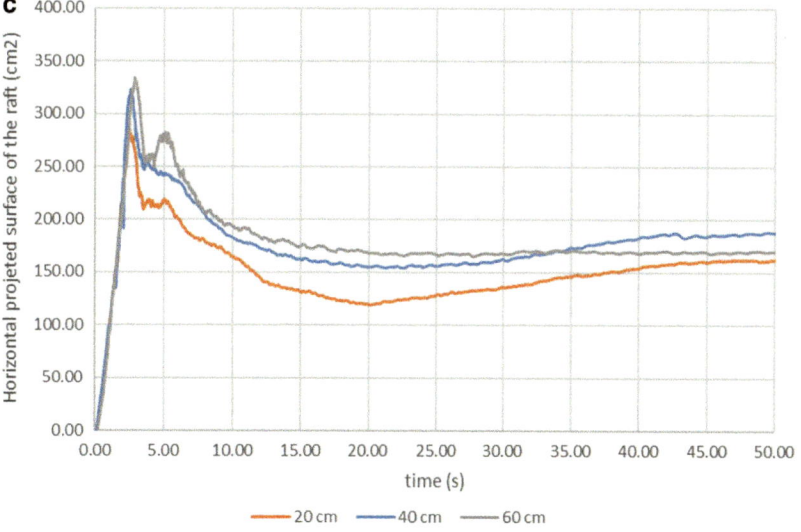

Fig. 9 Result of the influence of the mass of particles injected into the cloud surface (**a**), the fraction of particles in the cloud (**b**), and the surface of the projected area of the raft (**c**)

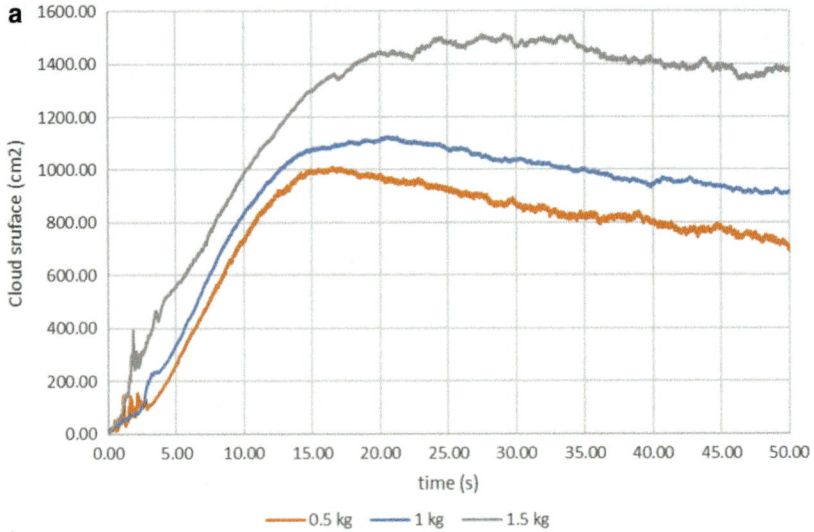

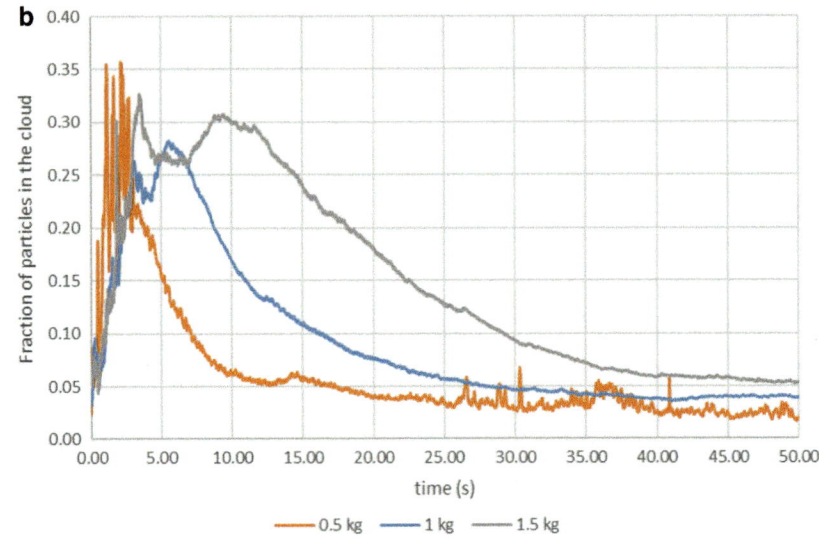

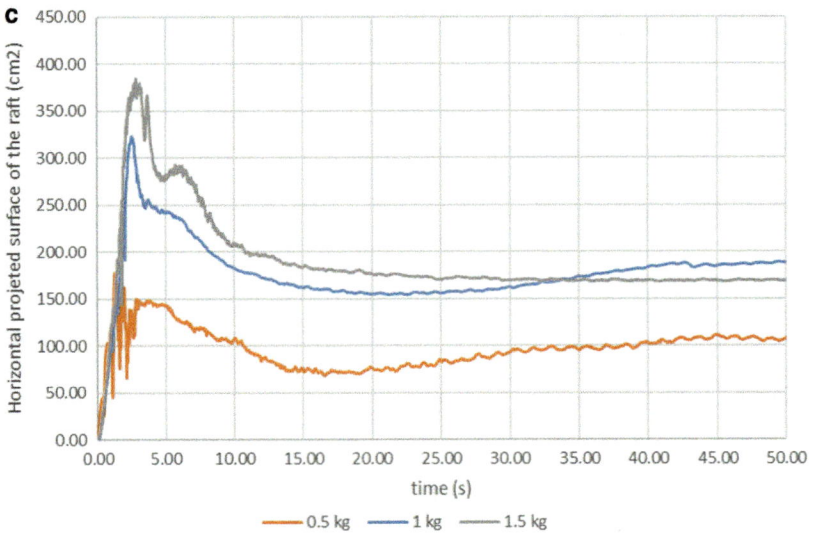

Fig. 10 Result of the influence
of the liquid flow on the cloud
surface (**a**), the fraction of
particles in the cloud (**b**), and the
surface of the projected area of
the raft (**c**)

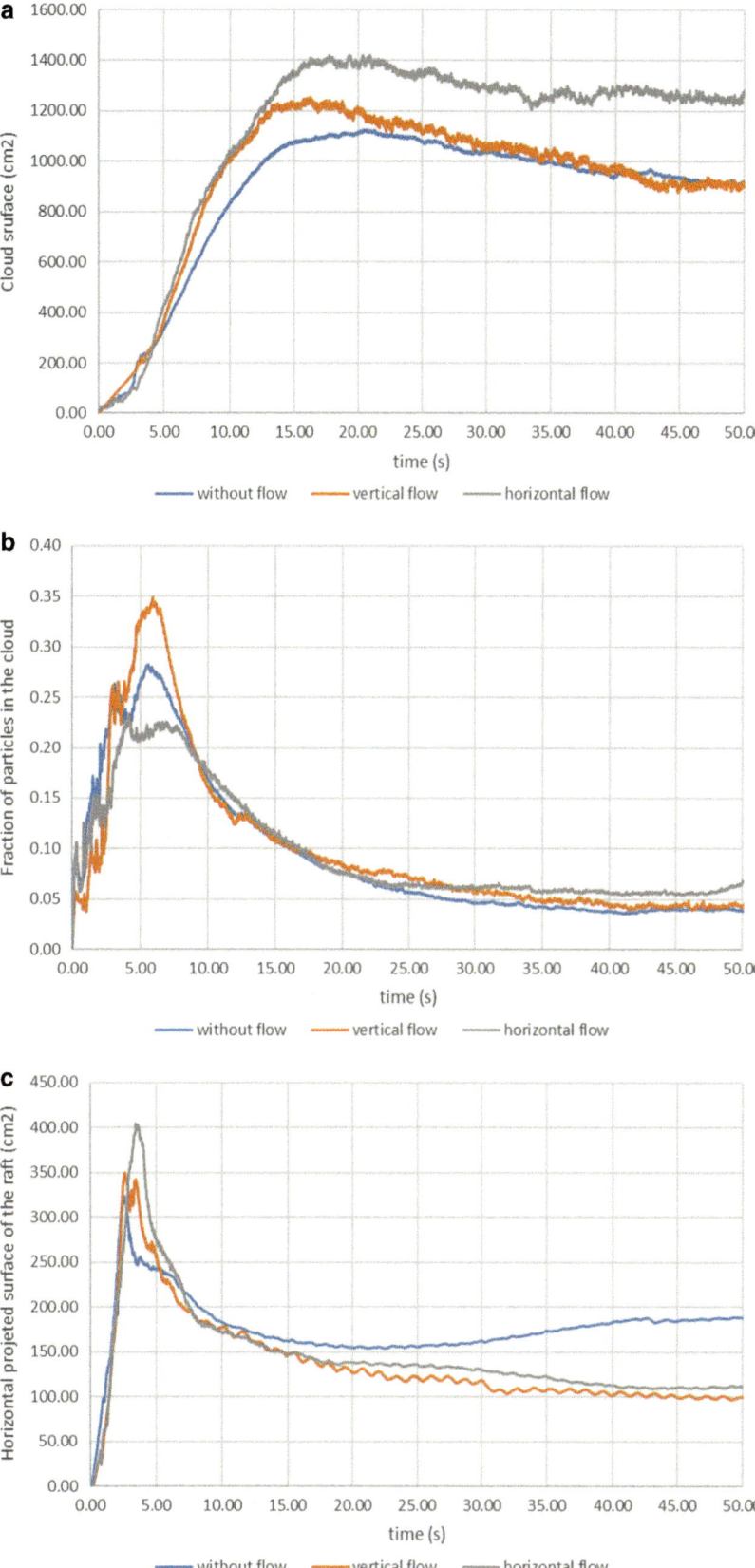

Acknowledgements The research described above was made possible by the financial support of Rio Tinto Aluminum and the Natural Sciences and Engineering Research Council of Canada.

References

1. L. Bracamonte, V. Aulie, C. Rosenkilde, K. E. Einarsrud, et E. Sandnes, "Dissolution Characteristics and Concentration Measurements of Alumina in Cryolite Melts," Dans *Minerals, Metals and Materials Series*, 2021, vol. 6, pp. 495–503.
2. L. Bracamonte, K. E. Einarsrud, C. Rosenkilde, et E. Sandnes, "Oxide Sensor Measurements and Simultaneous Optical Observations During Dissolution of Alumina in Cryolite Melt," Dans *Minerals, Metals and Materials Series*, 2022, pp. 381–391.
3. A. Molin, "Étude expérimentale sur les mécanismes de dissolution de l'alumine," 2015. [En ligne]. Disponible: https://constellation.uqac.ca/3725/.
4. T. Roger *et al.*, "Development of a mathematical model to simulate raft formation," *TMS Light Metals 2020*, 2020.
5. T. Roger *et al.*, "Modeling of the Heat Exchange, the Phase Change, and Dissolution of Alumina Injected in Electrolysis Cells," Dans *Minerals, Metals and Materials Series*, 2022, pp. 363–370.
6. S. E. Gylver, N. H. Omdahl, A. K. Prytz, A. J. Meyer, L. P. Lossius, et K. E. Einarsrud, "Alumina Feeding and Raft Formation: Raft Collection and Process Parameters," Dans *Light Metal 2019*, 2019, pp. 659–666.
7. L. Bracamonte. *Alumina fines first addition.* (Consulté le 2021-12-29.) Disponible: https://www.youtube.com/watch?v=SiTZ5hoWEjM.
8. R. Keller, "Alumina dissolution and sludge formation revisited," Dans *TMS Light Metals*, 2005, pp. 147–150.

Fundamental Mass Transfer Correlations Based on Experimental and Literature Data

Jonathan Alarie, László I. Kiss, Lukas Dion, Sébastien Guérard, and Jean-François Bilodeau

Abstract

Using a specific description of the heat transfer and diffusion coefficients, general mass transfer theory is applied to data available in the literature to identify their respective dissolution rate. The following calculations using data from the literature are then compared to experimental work performed under laboratory conditions using a gravimetric method to evaluate the dissolution rate of alumina disks. The contrast between the data from our experimental work and the validation provided by the literature is assumed inherent to the morphology of the sample and the adequate description of the flow around it. The following discussions highlight the dominant factor affecting the mass transfer coefficients and pinpoint the theoretical challenges to overcome to achieve more precise relations for future works.

Keywords

Aluminum electrolysis • Alumina dissolution • Cryolite properties

J. Alarie (✉) · L. I. Kiss · L. Dion
GRIPS, University of Quebec at Chicoutimi, 555 Boul Universite, Saguenay, QC G7H 2B1, Canada
e-mail: jonathan.alarie1@uqac.ca

L. I. Kiss
e-mail: Laszlo_Kiss@uqac.ca

L. Dion
e-mail: Lukas1_Dion@uqac.ca

J. Alarie
REGAL, Aluminium Research Centre, 2325 Rue de l'Université, Québec, QC G1V 0A6, Canada

S. Guérard · J.-F. Bilodeau
Arvida Research and Development Centre, Rio Tinto, 1955 Mellon Boulevard, Jonquière, QC G7S 4K8, Canada
e-mail: Sebastien.Guerard@riotinto.com

J.-F. Bilodeau
e-mail: Jean-francois.bilodeau@riotinto.com

Introduction

Alumina dissolution in the Hall-Héroult process is a complex phenomenon that limits the uniformity of alumina distribution in industrial electrolysis cells. An uneven distribution may cause instability in the process that leads to undesired anode effects or to an increase of muck under the metal pad. In all cases, environmental and security issues, not to mention productivity, ask for an increased comprehension of the alumina dissolution to prevent such problems. For this purpose, a study on the effect of the bath composition and temperature on the dissolution rate of solid discs has been conducted. The results of this experimental work, reproduced here, can be found in Alarie et al. [1] and represent the foundation of the present work. Basically, a mass transfer model for a flat plate in a perpendicular flow has been applied to describe the dissolution of the alumina discs. However, the sample structure used in this work is different from what is formed by an injection of alumina in electrolysis cells. It is well known that the alumina upon injection in the melt form solid aggregates that are a hard to dissolve. Then, the present work try to apply the same analysis used for the solid discs to data from the literature, with different alumina raft structure.

Since nearly 50 years, researchers tried to explain the parameters that influence the alumina dissolution rate. Despite the many names that could be mentioned here, only a few works are reported in this paper due to the lack of details available in the literature for the purpose of this work. The first one is the work of Jain et al. [2] who gave a good description of their experimental setup and dissolution curves over time. They tested the effect of a few chemical compositions, superheat and alumina properties on the alumina dissolution behavior. With a similar experimental setup, Bagshaw and Welch [3] tried to identify the effect of the alumina properties on the dissolution rate. Isaeva et al. [4] gave few details on their experimental setup to describe the effect of the alumina content in the bath and few alumina properties. Finally, the work of Gerlach et al. [5] are of first interest because they

© The Minerals, Metals & Materials Society 2023
S. Broek (ed.), *Light Metals 2023*, The Minerals, Metals & Materials Series,
https://doi.org/10.1007/978-3-031-22532-1_94

studied pressed tablets (geometry similar to the solid discs) with different chemical compositions and alumina properties.

Method

Before applying the mass transfer model previously developed to the data from the literature, there are some points to mention:

1. To compare the dissolution rate obtained by Jain et al. [2] in their work to other authors, the average dissolution rate was used.
2. Where the authors studied the alumina properties, such as Gerlach et al. [5], Jain et al. [2], and Bagshaw and Welch [3], the average dissolution rate that they observed for the same Reynolds number was used for comparison.
3. The dissolution rates measured by Isaeva et al. [4] are assumed to come from stirred melt.

The next analysis relies on Colburn's analogy to apply dimensional analysis to the alumina dissolution (Eq. 1). Most of the concepts presented here can be easily taken from a general reference book on heat and mass transfer, such as Çengel and Ghajar [6] and more details are available in Alarie et al. [1]. This kind of analysis needs a good description of the flow around the system studied and the fluid properties to find adequate Reynolds and Schmidt numbers (Eqs. 2 and 3, respectively), that leads to the Sherwood number (Eq. 4) and the h_m mass transfer coefficient (Eq. 5). These relations require the use of the fluid velocity (U_l), the characteristic length of the sample (L_c), the melt kinematic viscosity(v), the diffusivity of the alumina in the melt (D) and finally, the concentration gradient ($w_s - w_\infty$), expressed here in mass fraction instead of the common weight percent. Since this information is mostly unavailable from all considered authors, the following assumptions were made:

1. Tests are assumed to has been conducted in round crucibles with perfect insulation on the bottom face, radiative heat input from the sides and pure radiative heat loss from the top.
2. The stirring of the melt, unless otherwise specified, should create a flow with a velocity of about 10 cm/s.
3. In stirred melt, the diameter of the stirrer is one fifth of the crucible diameter.
4. Powder injection creates a raft that has a conical shape, with an equivalent circle diameter as the characteristic length. This length is assumed to stay the same all over the dissolution time studied.

$$\phi \left[\frac{kg}{m^2 s} \right] = \frac{\dot{m}}{A} = h_m \rho (w_s - w_\infty) \tag{1}$$

$$Re = \frac{U_l L c}{v} \tag{2}$$

$$Sc = \frac{v}{D} \tag{3}$$

$$Sh = a Re^m \cdot Sc^n \tag{4}$$

$$Sh = \frac{h_m L_c}{D} \tag{5}$$

These assumptions were used to calibrate the parameters of the physical relationship between the experimental setup and the dissolution rate, such as to choose between the use of the diameter or the radius.

One of the important information in such analysis is the contact area between the raft and the melt. Without a deep knowledge of the geometries involved in the tests performed by other authors; the following simplification was made. The raft is considered to have a conical shape with an angle of repose (θ) of 4° as presented in Fig. 1. Experimental observation of the shape of the heap formed by the injected particles on the surface of a liquid can be as small as 2°. Also, Roger et al. [7] observed angles around 4° in an analog model of organic grains injected on water. In perfect conditions, the diameter of such a raft (d) can then be approximated with Eq. 6, the value used in this work. This relation relies on the geometric proportionality of injected volume of alumina, obtained by the injected mass (m) and its assumed bulk density (ρ), and the diameter of the cone obtained. However, it is also important to mention that the inner diameter of the crucible limits the maximal diameter of the raft. For the subsequent analysis, the diameter found with Eq. 6 will be used as the characteristic length for most of the calculations involved. Another issue with this assumption is that the diameter of the raft changes over time with the dissolution. Ideally, such behavior must be tracked, reducing the characteristic length and the contact area between the melt and the sample. Unfortunately, the information needed to do this update throughout the whole dissolution are mostly unavailable. Also, it can be cumbersome to track the contact area of a dissolving raft. Then, the initial diameter of the raft has been chosen as a reference in the present analysis. Thus, the dissolution rate found with the model presented in this work should be slightly below the real dissolution rate in the first moment of the dissolution and slightly above at the end. However, the average dissolution rate will be well represented by the model.

$$d = \sqrt[3]{\frac{24m}{\rho \pi \tan \theta}} \tag{6}$$

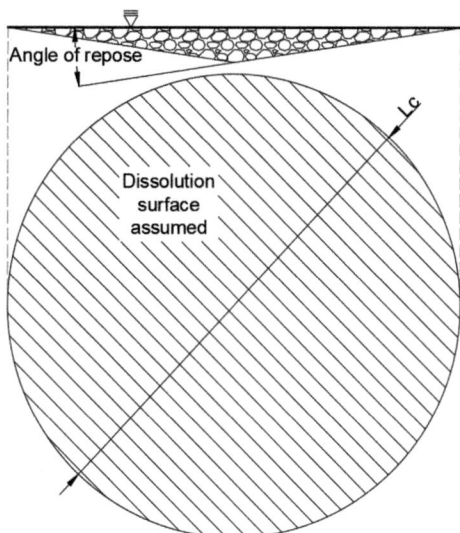

Fig. 1 Sketch of the conical raft assumption to obtain the surface of contact and the characteristic length of the dissolution process

To calculate the Reynolds number, it is important to know the internal properties of the liquid. Most of the publications presented above try to relates the dissolution rate to the mass fraction of the additives in the melt. However, the dissolution process is too complex to describe it by linear relations with the additives content. So, the use of the fluid properties integrates the interactions between the additives and their non-linear behavior. Fortunately, relations between the composition and the melt dynamic viscosity (μ, Hertzberg et al. [8]), density (ρ, Kvande and Rørvik [9]), cinematic viscosity ($\nu = \mu/\rho$), and alumina solubility (w_s, Skybakmoen et al. [10]) are easily available in the literature. Once the material properties of the bath are known, the convection in the melt needs to be described. There are four sources of convection identified in the different experimental setups:

1. Natural convection caused by concentration gradients.
2. Natural convection caused by thermal gradients.
3. Bubbling.
4. Forced convection.

It is important to mention here that the real flow in contact with the sample need to be precisely known to accurately describe the convective dissolution. However, the different works that studied the dissolution of the alumina hardly mention the flow pattern in the experimental setup and barely less their dimensions. So, without a detailed knowledge of the different flow patterns, the following analysis will be considered for comparison purpose only. The first two sources of convection are mostly negligible in the works with stirred melt. The third source applies only in the work with solid discs. These convection sources are detailed in Alarie et al. [1] and are neglected in this analysis.

The last source of convection is the stirring involved in some experimental setups. Most authors considered adding stirring to an intensity similar to that seen in industrial cells, which is estimated to few tenth of meter per second (See Perron [11] and Kobbeltvedt and Moxnes [12]). To obtain the Reynolds number associated with such stirring, the velocity of the fluid is required. So, with the help of the diagram in Jain et al. [2], the diameter of the stirrer was estimated of the fifth of that of the crucible. This value seems coherent from one paper to another, since there is not much physical space in a crucible for a larger stirrer, and a smaller one will require higher rotation speed to achieve the desired stirring. In a related matter, the information provided by Isaeva et al. [4] on her experimental setup does not include any input on the design of stirring. In that case, it was assumed that the stirrer has a rotation speed of 200 rpm, just like the one used by Jain et al. [2] and Bagshaw and Welch [3]. Note that the design of the stirrer can be very different from one author to the other and they can provide different stirring patterns. So, their results are here used to confirm that the assumption made with the works of Jain et al. [2] and Bagshaw and Welch [3] are reasonable. To obtain the velocity of the melt, the rotation speed was used in concert with the radius of the stirrer. The velocity obtained from this method gives congruent results with every author with values between 9.4 and 20.9 cm/s. Now, using Eq. 2 with the diameter of the raft as the characteristic length, the Reynolds number for the convection source can be obtained.

The Reynolds number is used to describe the ratio of the inertial forces on the viscous forces in the fluid [6] that interact with the sample. This interaction is reflected by the amplitude of the Reynolds number and its exponent in the Sherwood number equation (4). In the case of stirred melt, as presented by most of the authors, the Reynolds number is under 5×10^5, suggesting a laminar flow around the samples. As shown in Hess and Miller [13], the flow in an heated round crucible rises along the wall of the crucible then it turns downward in the center. Even with stirring, the melt is assumed to follow a similar flow pattern. In these conditions, the model of a flat plate perpendicular to the flow, with an exponent of 0.748 as described is Alarie et al. [1], will be used for all authors.

Once the flow pattern has been considered, the Schimdt number (Eq. 3) will dictate the fluid properties influence on the mass transfer in the medium. Its value characterizes the relative thickness of the velocity boundary layer over the diffusive boundary layer. So, with the viscosity and the diffusivity found in Alarie et al. [1], it is possible to find the Schmidt number for every data point. The Schmidt number usually comes with an exponent of $\frac{1}{3}$ from the exact analysis of the boundary layer, as presented by Blasius and Pohlausen, in Welty [14]. This leads the analysis to the complete description of the Sherwood number, except for the coefficient (a) in Eq. 4. This coefficient is found from experimental data in the next section.

Results

From the theoretical analysis presented above, the experimental dissolution rate was predicted by the Sherwood number using Eqs. 1–5. Note that the coefficient "a" in the Sherwood formulation (Eq. 4) is found with the help of the regression lines presented in Fig. 2.

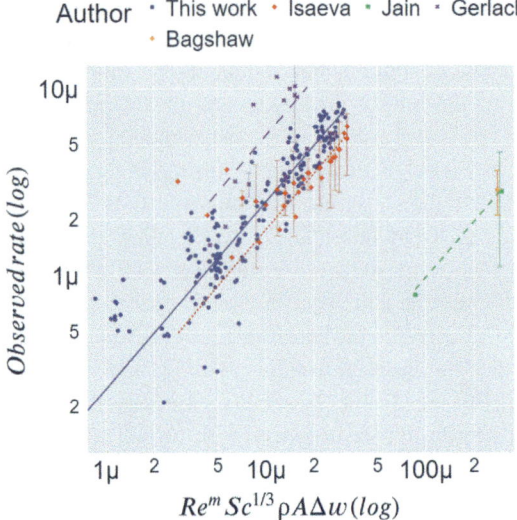

Fig. 2 Correlation between the observed dissolution rate values (vertical axis), and the one predicted by the mass transfer theory (horizontal axis). The error bar represents the amplitude of variation of the dissolution rate caused by the alumina properties, not taken into account in the model of this work. The lines are the predictions of the model for the data from this work (solid), Isaeva et al. [4] (dot), Jain et al. [2] (dash) and Gerlach et al. [5] (long dash)

Fig. 2 show a good accordance between the model and the observed dissolution rates for each author. However, differences can be seen from one author to another. As explained above, the conditions in which the tests were performed are of first importance in these differences. For example, an increase in the stirring or in the diameter of a sample (the characteristic length) will drag the results to the right, thus lowering the coefficient needed to predict the dissolution rate. These causes explain the differences between the data from Isaeva et al. [4] and the other injections performed by Jain et al. [2] and Bagshaw et al. [3], the last two of them using the same experimental method. As stated in the method description, it is difficult to compare the results obtained by different methods, without a deep knowledge of the flow pattern of the melt and the shape of the sample. Here, the assumption made for the stirring in the crucible seems to differ from the reality of the tests of Jain et al. [2] and Bagshaw et al. [3] Though, assuming that these unknowns are fulfilled, the model presented here proved to be able to adequately predict the dissolution rate of powder injection.

It is much relevant to mention that one of the questions that many authors studied, such as Isaeva et al. [4], Gerlach et al. [5], Jain et al. [2], and Bagshaw et al. [3], is the influence of the alumina properties on the dissolution rate. As it can be seen on Fig. 2, the model presented here cannot take these parameters in the account, represented by the error bar in the figure. Still, the influence of the bath internal properties, inherent to its composition and superheat, are larger than most of the alumina properties variations observed by the cited authors, as depicted by the data from this work.

From Fig. 2 it can also be seen that the results obtained by Isaeva et al. [4], Gerlach et al. [5] and this work gives a fairly similar correlation. This means that the geometries and flow patterns involved in these works are similar to each other, or at least that the differences compensate for the difference in behaviors. These differences include the roughness of the surface of the injected raft compared to the smooth surface of a disc or a pressed tablet of alumina.

Sensitivity Study

For added confidence in the previous model, a sensitivity analysis was performed on the results from the literature. This analysis shows the influence of the assumed parameters on the correlation coefficient between the model and the different authors and different forms (powder or pressed tablets). Table 1 shows the parameter's amplitude of variation. Parameters that influence only the natural convection are neglected, since all powder injections were assumed to be made with a similar stirring. The influence of the variation of the parameters are shown in Fig. 3.

Correlation Accuracy

From the data from Isaeva et al. [4], Gerlach et al. [5], it can be seen that the assumptions have a minimal effect on the correlation coefficient between and our model. For Isaeva et al. [4], this is easily explained by the small size of the raft created compared to the crucible used. In the case of Gerlach et al. [5], since they used compressed tablets of alumina, only the Reynolds exponent and the crucible-to-stirrer ratio are considered and they do not have a significant influence. On the other hand, the data from Jain et al. [2], presented as a example in the Fig. 3, show a slight dependence on the parameters value. Despite the very small change in the values of the correlation coefficient, these dependences can help to find better value for the parameters. That is the case for the Reynolds number exponent and the cone angle. The Reynolds number exponent seems to reach an optimum of accuracy near 0.748, which is expected if the flow around the sample corresponds

Table 1 Range of the sensibility study of the model

Parameter	Minimum	Maximum	Reference
Reynolds exponent (m)	0.05	1	0.731
Crucible-to-stirrer (diameter ratio)	1	20	5
Angle of cone (°)	1	10	4
Bulk powder density ($\frac{kg}{m^3}$)	950	1150	1000

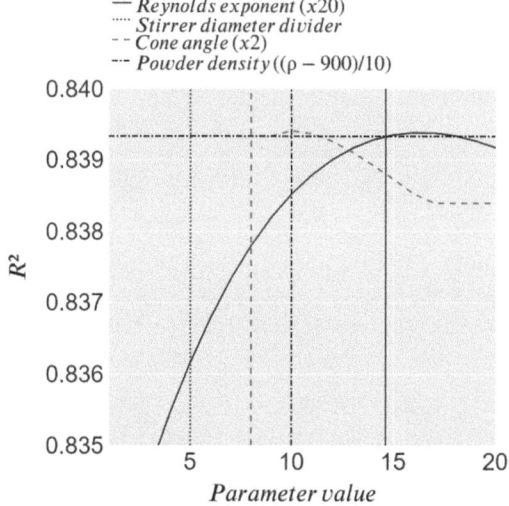

Fig. 3 Sensibility study of the assumption on the value of the correlation coefficient compared to the prediction using the Sherwood number formulation for the data from Jain et al. [2]. 20 different value for each parameters were tested. Vertical lines represent values from Table 1

Coefficient Value

The coefficient "a" in the Sherwood formulation (Eq. 4) is also influenced by the assumption made in the method section, as presented as an example in Fig. 4 from the data of Jain et al. [2]. As mentioned, two models in the mass transfer theory can represent an approximative target for the value of this coefficient; the model for a flat plate, with a coefficient of 0.664 ($0.664 Re^{0.5} Sc^{1/3}$), and the one with a perpendicular flow to an infinite plate, with a coefficient of 0.257 ($0.257 Re^{0.731} Sc^{1/3}$) (see Çengel and Ghajar [6]). These values represent the maximum that could be achieved, if the sample behaves like a perfect plate and if the flow is adequately described by the exponent of the Reynolds number. For this reason and to keep a good readability, Fig. 4 is limited to the range from 0 to 0.5 for the coefficient values. Also, a turbulent flow parallel to a flat plate, that is to say at higher Reynolds number than that observed in this work, will see its coefficient to be around 0.037 ($Sh = 0.037 Re^{0.8} Sc^{\frac{1}{3}}$, see Çengel and Ghajar [6]). This case is not considered in this work, and thus the coefficients observed should be above this other limiting value.

to the assumption previously made. However the rafts, as created by Jain et al. [2], are in theory still touching the walls of the crucible with an angle below 4°, explaining the constant value of the correlation coefficient up to this value. Increasing the angle of the cone to prevent the raft of touching the wall bring the angle near 8°, while the optimum on Fig. 3 is near 6°. This modification also changes the value of the best fitting exponent which is now about 0.5 for a 6° angle. This results suggests a dissolution behavior more like a flat plate with a parallel flow instead of a perpendicular one. Changing the exponent of the Reynolds number in the Sherwood equation (Eq. 4) from 0.748 to 0.5 and running another sensitivity study, shows that the most likely angle should be of 8°, the angle at which the raft stop touching the walls of the crucible. This behavior should indeed be an ideal scenario for a raft in an electrolysis cell. Also, as mentioned previously, the shape of the raft is strongly influenced by the injection method. This means that the scenario of a flat plate describing the data from Jain et al. [2] is more plausible. The last two parameters (the stirrer diameter and powder density) seems to be of little effect on the variation of the correlation coefficient.

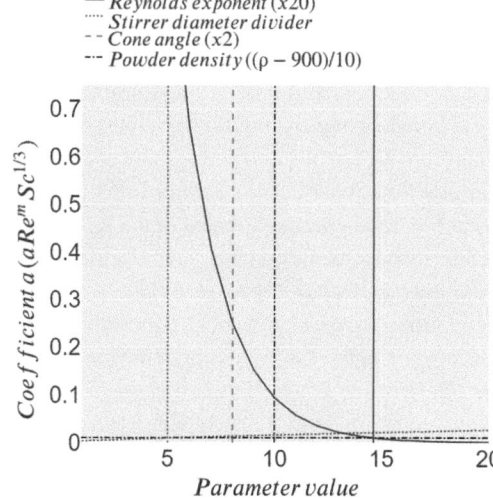

Fig. 4 Sensibility study of the assumption on the value of the coefficient "a" in the Sherwood number for the data from Jain et al. [2]. 20 different value for each parameters were tested. Vertical lines represent values from Table 1. 0.75 represent the highest plausible value (see text)

The most sensitive parameter for the coefficient value is the exponent associated with the Reynolds number, as predicted in the mass transfer theory. A change of the flow pattern and in the intensity of the turbulence is indeed a strong factor that modifies the dissolution behavior of a sample, and thus the coefficient of the Sherwood number formulation. Therefore, the value of the Reynolds number exponent should be selected from a deep knowledge of the flow pattern around the sample. From Fig. 4, it can be seen that the value fixed for the different parameters allows the coefficient to be within the limits identified above. The sole exception comes from the data from Gerlach et al. [5], which show a higher value than the expected 0.257. However, the use of fragile samples, such as compressed tablets of alumina, can favor the disintegration of the sample and thus easily explain this result. The second more sensitive parameter is the crucible-to-stirrer diameter ratio, that directly changes the velocity of the fluid. As this ratio increase, the diameter of the stirrer decrease, and thus does the velocity of the fluid. This will lower the Reynolds number and a greater coefficient in the Sherwood formulation will be needed to fit the data. However, as stated in the method section, a stirrer of one fifth of the crucible diameter at 200 rpm should give melt velocity in the order of 10 cm/s and represents a plausible value.

The last two parameters, the powder density and the angle of the conical raft, have weak effects on the coefficient, since their variation range does not affect much the diameter of the raft, with less than 10% of variation between the limits studied. Also, while the raft is touching the wall, a change in the density of the powder or the angle has no influence on the diameter of the raft, unless they lower the volume of the raft to the point that the side of the raft could not touch the wall of the crucible.

This analysis confirms the adequacy of most of the assumptions and the model used to predict the alumina dissolution rate. Only the conical angle should be modified from 4° to 8° and the data from Jain et al. [2] and Bagshaw et al. [3] benefit from a parallel flow model. The results of this modification are shown in Fig. 5, where the exponent of the Reynolds number in the Sherwood formulation (Eq. 4) for the data from Jain et al. [2] and Bagshaw et al. [3] is set to 0.5.

Once the model is well defined, it is possible to see the effect of some parameters on the output of the model. Figure 6 shows the effect of the convection and characteristic length on the Sherwood number value. As the Reynolds number increases, usually due to its correlation with fluid velocity,

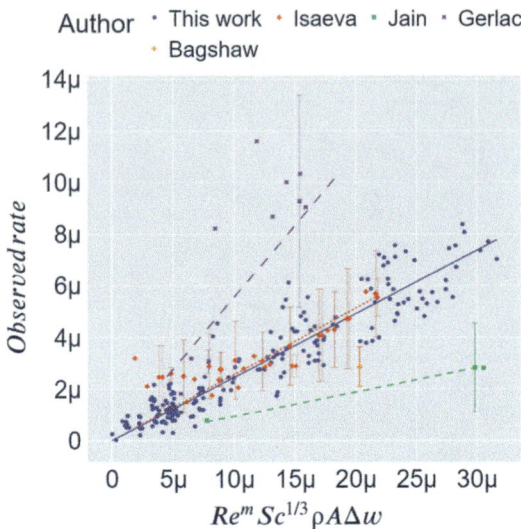

Fig. 5 Correlation between the observed dissolution rate values and the revised model. The lines are the predictions of the model for the data of this work (solid), Isaeva et al. [4] (dot), Jain et al. [2] (dash) and Gerlach et al. [5] (long dash)

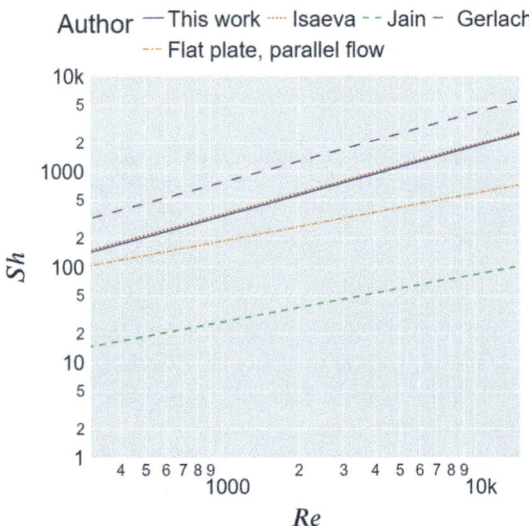

Fig. 6 Variation of the Sherwood number with the Reynolds number as predicted by the model obtained from the data of each author. Dash dot line from Çengel and Ghajar [6], dash line from Jain et al. [2], dot line from Isaeva et al. [4], long dash from Gerlach et al. [5] and solid lines from this work

the Sherwood number will also slowly increase. However, the Reynolds number can also be increased following an augmentation of the characteristic length. These two parameters are foremost inherent to the electrolysis cells design from which the melt and bubble velocity depend. Next, the injector design can largely influence the shape of the raft and thus its characteristic length. On the other hand, the dynamic viscosity of the bath is more a consequence of the bath composition and temperature than set at a desired value. In every case, Fig. 6 represents a valuable tool to compare the results from the different sets of data used in this work, which have different stirring values. This figure also presents that the chosen idealized case represented by the parallel flow above a flat plate is sufficiently accurate to describe the alumina dissolution rate under a different set of conditions. However, to satisfy this ideal/perfect model, it is necessary to accurately describe the evolution of the surface area over the lifetime of the raft. Along with the measure of the dissolution rate at each instant, it is expected that the pairing of the instantaneous surface and dissolution rate will indeed follow a flat plate model with parallel flow. Since the information about the instantaneous raft surface is actually unavailable, the analysis here represents the average dissolution rate over the initial surface of contact. Then, it is natural that the dissolution rates predicted are different than that of the ideal/perfect model. Also, as mentioned in Alarie et al. [1], the flow pattern perpendicular to the raft, with an exponent of 0.748, is similar to that of a piece of raft sinking in the melt, and thus the dissolution rate in this case is higher than with a flat plate in parallel flow.

Conclusion

In this work, a fundamental analysis of mass transfer based on data from experimental tests has been applied to other data from the literature. The results obtained show that the model previously developed for a perpendicular flow to the sample surface can describe most of the dissolution behavior observed. However, at higher Reynolds numbers, a model using a parallel flow over a flat plate is more suitable to predict the dissolution rate.

Also, the model demonstrated the strong sensitivity of the results to the description of the flow around the sample and the sample itself. Namely, the pattern of the flow and the characteristic length are of first importance to accurately find the dissolution rates of the sample. A sensitivity study of the flow pattern to the bath internal properties should be part of every dissolution study. Also, special consideration should be taken to avoid the raft from reaching the wall of the crucible and minimize its proximity in future experimental work. Finally, the models presented here describe the mean dissolution rate of the sample involved. The analysis of the Sherwood number and its relation to the raft surface of contact at each instant

should give results closer to the mass transfer theory over a flat plate with a parallel flow.

Acknowledgements The authors want to thank Rio Tinto, the Natural Sciences and Engineering Research Council of Canada, and the Fonds de recherche Nature et technologies of Quebec for their technical and financial support for this project.

References

1. Alarie, J. *et al.* Determination of the mass transfer coefficient for the dissolution of alumina samples immersed in a cryolitic bath. *Acta Materialia* (Submitted) (2023).
2. Jain, R., Tricklebank, S., Welch, B. & Williams, D. Interaction of aluminas with aluminum smelting electrolytes. *Light Metals 1983* 609–622 (1983).
3. Bagshaw, A. & Welch, B. *The influence of alumina properties on its dissolution in smelting electrolyte*, 783–787 (Springer, 2016).
4. Isaeva, L. A., Braslavskii, A. B. & Polyakov, P. V. Effect of the content of the alpha-phase and granulometric composition on the dissolution rate of alumina in cryolite-alumina melts. *Russian Journal of Non-Ferrous Metals* **50** (6), 600–605 (2009). https://dx.doi.org/10.3103/S1067821209060078. 10.3103/s1067821209060078.
5. Gerlach, J., Hennig, U. & Kern, K. The dissolution of aluminum oxide in cryolite melts. *Metallurgical and Materials Transactions B* **6** (1), 83–86 (1975). https://dx.doi.org/10.1007/BF02825681.
6. Çengel, Y. A. & Ghajar, A. J. *Heat and mass transfer : fundamentals & applications* Fifth edn (McGraw-Hill Education, New York, N.Y., 2015).
7. Roger, T. *et al.* Experimental investigation of the alumina cloud during alumina injections in low and high temperature conditions. *Light Metals 2023* NA (2023).
8. Hertzberg, T., Tørklep, K. & Øye, H. *Viscosity of Molten NaF-AlF3-Al2O3-CaF2 Mixtures: Selecting and Fitting Models in a Complex System*, 19–24 (Springer, 1980). https://onlinelibrary.wiley.com/doi/abs/10.1002/9781118647851.ch3.
9. Kvande, H. & Rørvik, H. *Influence of bath density in aluminum electrolysis*, Vol. 36, 62–62 (Minerals metals materials soc 420 commonwealth dr, warrendale, pa 15086, 1984).
10. Skybakmoen, E., Solheim, A. & Sterten, Å. Alumina solubility in molten salt systems of interest for aluminum electrolysis and related phase diagram data. *Metallurgical and materials transactions B* **28** (1), 81–86 (1997). https://link-springer-com.sbiproxy.uqac.ca/content/pdf/10.1007/s11663-997-0129-9.pdf.
11. Perron, A. *Transfert de quantité de mouvement et augmentation de la résistance électrique causés par la présence des bulles dans une cuve Hall-Héroult*. Thèse de doctorat, Chicoutimi (2006). https://constellation.uqac.ca/464/.
12. Kobbeltvedt, O. & Moxnes, B. *On the Bath Flow, Alumina Distribution and Anode Gas Release in Aluminium Cells*, 257–264 (1997). https://onlinelibrary.wiley.com/doi/abs/10.1002/9781118647851.ch37.
13. Hess, C. F. & Miller, C. W. Natural convection in a vertical cylinder subject to constant heat flux. *International Journal of Heat and Mass Transfer* **22** (3), 421–430 (1979). https://dx.doi.org/10.1016/0017-9310(79)90008-5.
14. Welty, J. R. *Fundamentals of momentum, heat, and mass transfer* 4th [updated]. edn (J. Wiley, New York, 2001). http://catdir.loc.gov/catdir/toc/onix07/00039278.html http://catdir.loc.gov/catdir/description/wiley035/00039278.html.

Potential of Production Al–Si Green Alloys in AP18 Aluminium Reduction Cell

Haris Salihagić Hrenko, Anton Verdenik, Branko Juršek, Dragan Mikša, Maja Vončina, and Jožef Medved

Abstract

The market situation and environmental requirements suggest that the direct synthesis of Al–Si alloys in aluminium electrolysis cell can help to increase added value and reduce a total CO_2 footprint of alloys. The aim of this paper was to determine the influence of SiO_2 additives on the process control during the production of Al–Si alloy. The regulation of the process strongly depends on the electrical resistance of the electrolysis cells, so we determined the electrical conductivity of the electrolyte using the DC four-point method. The measured electrical conductivity was a guide for determining the dosing rate of SiO_2. Optimization of the reduction process has been done with the process computer data and measuring the properties of the electrolyte with STARprobe™. The promising results of direct Al–Si alloys synthesis in AP18 industrial cells could open a path to a large variety of greener Al alloys produced by electrolysis process.

Keywords

Aluminium • Sustainability • Aluminium electrolysis • Al–Si alloys • Feed rate

Introduction

Corporate energy efficiency is becoming more and more self-evident, which is why we are paying more and more attention to so-called green aluminium. The type of energy used, which is not always green, must also be considered in the overall carbon footprint of the aluminium alloys produced. Companies do not have much influence on the choice of energy resources, so they are looking for other ways to produce aluminium and aluminium alloys with the least possible impact on the environment.

Primary aluminium has been produced for over 100 years by a well-know Hall–Héroult electrolysis process in aluminium reduction cells. During the whole period, more than 50 different types of prebake cell technologies have been developed worldwide [1]. In the aluminium reduction cell, aluminium oxide (Al_2O_3) is dissolved in a molten electrolyte where it decomposed to aluminium and oxygen during electrolysis process. The reduced liquid aluminium collects at the bottom of the reduction cell, while oxygen reacts with carbon anode to form CO_2.

Choosing the appropriate electrolyte is very important for the optimal reduction cell performance. Manufacturers around the world use a classic electrolyte consisting of cryolite (Na_3AlF_6), aluminium fluoride (AlF_3), and calcium fluoride (CaF_2). It is very important that the additives to Na_3AlF_6 have a higher decomposition potential than Al_2O_3, prevent loss of electrolyte by evaporation, increase the electrical conductivity of the electrolyte, do not cause secondary reactions in the electrolyte and do not form insoluble substances [2].

H. S. Hrenko (✉) · A. Verdenik · B. Juršek · D. Mikša
TALUM d.d, Tovarniška c. 10, 2325 Kidričevo, Slovenia
e-mail: haris.salihagichrenko@talum.si

A. Verdenik
e-mail: anton.verdenik@talum.si

B. Juršek
e-mail: branko.jursek@talum.si

D. Mikša
e-mail: dragan.miksa@talum.si

M. Vončina · J. Medved
Faculty of Natural Sciences and Engineering, Department of Materials and Metallurgy, University of Ljubljana, Aškerčeva 12, 1000 Ljubljana, Slovenia
e-mail: maja.voncina@ntf.uni-lj.si

J. Medved
e-mail: jozef.medved@ntf.uni-lj.si

© The Minerals, Metals & Materials Society 2023
S. Broek (ed.), *Light Metals 2023*, The Minerals, Metals & Materials Series,
https://doi.org/10.1007/978-3-031-22532-1_95

The control of chemical composition of the electrolyte and temperature is very important for the stability and efficiency of electrolysis process [3]. The resistance of the electrolyte decreases with increasing temperature of the electrolyte and decreases with a lower concentration of AlF$_3$ in electrolyte, while, at low and high concentrations of Al$_2$O$_3$, the resistance of electrolyte increases [4]. Due to rapid changes in concentration, Al$_2$O$_3$ is the most important component in electrolyte. Modern aluminium reduction cells tend to operate with concentrations of Al$_2$O$_3$ in electrolyte of up to 4 wt% [5]. The alumina feeding strategy determines the Al$_2$O$_3$ concentration in the electrolyte and directly affects pseudo-resistance of the aluminium reduction cell [6, 7]. Setting the correct target resistance is important for the optimal aluminium reduction cell regulation [7].

In the past, feeding materials such as white tan clay (47.09 wt% SiO$_2$) and sand (98.66 wt% SiO$_2$) were used to produce Al–Si alloys in industrial aluminium reduction cells [8]. The addition of SiO$_2$ is limited by its solubility in the electrolyte. At the temperature of 1010 °C, the solubility of SiO$_2$ in synthetic cryolite is less than 5 wt% but can increase up to 14 wt% if Al$_2$O$_3$ is present in the system [9]. The addition of SiO$_2$ to the system of the classical electrolyte affects its properties. The electrical conductivity of the electrolyte is one of the most important parameters for the aluminium reduction cell regulation. Equations for calculating the electrical conductivity exist for the most common commercial electrolytes, but they do not consider SiO$_2$ in the electrolyte [10–12]. The electrical conductivity of two-component system containing Na$_3$AlF$_6$ and SiO$_2$ is the most commonly discussed. It was found that the addition of SiO$_2$ to the pure cryolite reduce the electrical conductivity of the melt, since Si in cryolite tends to form large ions [13]. It was found that some of the SiO$_2$ in the electrolyte may react with the liquid aluminium in aluminium reduction cell, leading to reoxidation of aluminium to Al$_2$O$_3$ [14]. The reoxidation of aluminium and the formation of complex ions in electrolyte can negatively affect the current efficiency of the electrolysis process [15].

In this work, the electrical conductivities of classical electrolytes with different additions of SiO$_2$ were measured with aim of determining an acceptable dosage rate of SiO$_2$ and Al$_2$O$_3$ into aluminium reduction cell. The work examines the effects of SiO$_2$ addition on aluminium reduction cell regulation.

Experimental

Electrical conductivity of electrolytes was performed using DC four-point method, in which the electrical resistance of the electrolyte was measured during solidification process. Croning measuring cells were used, which were developed

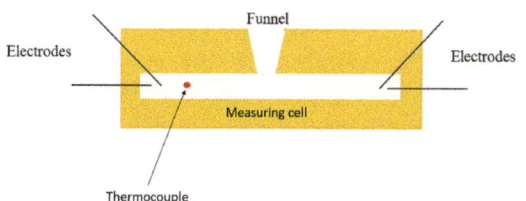

Fig. 1 Croning measuring cell

for "in-situ" electrical resistivity measurements of molten metals [16]. A setup of the Croning measuring cell is shown in Fig. 1. The length of the cell cavity is 216 mm, which ensures the stability of the measurement. 1 mm molybdenum (99.95) wire was used for the electrodes. The temperature during the solidification process was measured with K-type thermocouple. Because of the aggressiveness of the molten electrolytes at high temperatures, inner part of the measuring cell and K-type thermocouple were coated with boron nitride.

A measuring system can be seen in Fig. 2. The current source DC was 150 W power supply, and the known 2.5-Ω resistor (R$_R$) was connected in series with sensing element representing the unknown resistor (R$_c$). The voltage was measured across the known resistor (U$_R$) and the sensing element (U$_C$). The voltages and resistivity of the known resistor were used to calculate the resistivity of the sensing element.

To obtain better results, the cell constant (Z) was determined to be 7.88 cm^{-1}. From Eq. 1, the electrical conductivity (K) can be derived from the measured resistance of the measuring cell (R$_c$) and the cell constant (Z):

$$K = \frac{Z}{R_c} \tag{1}$$

The measurement of electrical conductivity was performed on five different electrolytes. Chemical compositions of electrolytes are presented in Table 1. For each electrolyte, three series of measurements were done. The samples were melted in induction furnace at temperature 1250 °C, and they were poured into an insulated measuring cell. Results of electrical conductivity of electrolyte were presented as average values of series in the temperature range between 900 °C and 1000 °C.

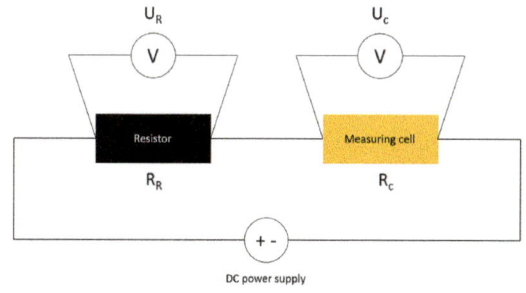

Fig. 2 Measuring system

Table 1 Chemical composition of the electrolytes

Electrolyte	Component (wt%)				
	Na_3AlF_6	AlF_3	CaF_2	Al_2O_3	SiO_2
ELEC	78.5	13.5	5	3	/
6SiO2_E	75.5	13.5	5	/	6
E_1SiO2	77.5	13.5	5	3	1
E_3SiO2	75.5	13.5	5	3	3
E_6SiO2	72.5	13.5	5	3	6

Since the determination of the excess AlF_3 in electrolyte is important for the thermal regulation of the aluminium reduction cell, the chiolite peak at 30.8°2 Θ was investigated by XRD analysis. When the electrolyte samples cool down, excess AlF_3 crystallises as chiolite, which affects the intensity of chiolite peak. The excess AlF_3 in electrolyte can be calculated by the equation proposed by Abramović and Homšak [17].

Talum's potline has 160 AP18 reduction cells. An industrial test to produce Al-Si alloy was carried out in an AP18 aluminium reduction cell. In normal operation, there are 10,000 kg of molten aluminium and 5000 kg of molten electrolyte in the aluminium reduction cell. The average DC current of potline was 187.08 kA. One out of four Al_2O_3 feeders was used to feed of SiO_2. During the test, different SiO_2 feed rates were used. The 0.8 kg SiO_2/10 min, 1.6 kg SiO_2/10 min, 1.6 kg SiO_2/12 min, or 1.6 kg SiO_2/15 min was feed to aluminium reduction cell. The purity of the SiO_2 used was 98.5 wt% SiO_2. The effects of SiO_2 addition on electrolysis process were observed using process control data. The number of alumina feed, number of feeding cycles and duration of tracking were analysed. Measurements of excess AlF_3 and temperature were made every 32 h using STARprobe™ measurements. Al–Si production was performed for 84 days in an aluminium reduction cell. The produced alloy was analysed with XRF. The performance of the cell in which Al–Si alloy was produced was compared to the performance of the potline.

Results and Discussion

Figure 3 shows the results of electrical conductivity measurement of the sample ELEC. The conductivity of the electrolyte was measured during the solidification process, reference values were calculated based on the chemical composition of the electrolyte ELEC using the program AlWeb [18] at temperature 916 °C and from 920 °C to 1000 °C in 10 °C steps. The correlation coefficient between reference values and the electrolyte ELEC electrolyte was higher than 0,99. At the temperature of 1000 °C, the measured electrical conductivity of ELEC was 2.17 S/cm, while

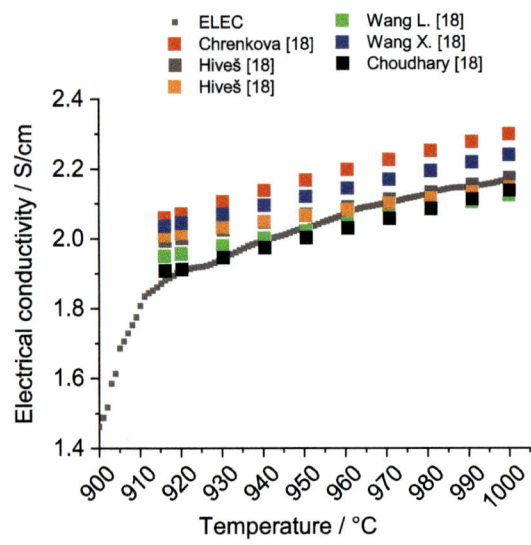

Fig. 3 Electrical conductivity of electrolyte ELEC compared to calculated reference values

the highest reference value was 2.30 S/cm and the lowest 2.12 S/cm. Temperature around 950 °C is optimal for aluminium reduction cell performance, at this temperature, the measured electrical conductivity was 2.03 S/cm. The measured electrical conductivity showed a good correlation with the references [18] for the temperatures from 1000 to 940 °C. Bellow this temperature, similar results were observed by Choudhary [18]. Due to the formation of solid phases during solidification process, a difference between measured electrical conductivity and calculated values was observed at lower temperatures.

The electrical conductivities of electrolytes with addition of SiO_2 are shown in Fig. 4. The electrical conductivity of ELEC was used as a reference value. At the temperature 1000 °C, the highest electrical conductivity (2.21 S/cm) was measured for the sample E_6SiO2, while the lowest electrical conductivity was measured for 6SiO2_E (2.07 S/cm). The decrease in electrical conductivity of E_6SiO2 and 6SiO2_E was very rapid. At high concentrations of SiO_2 and Al_2O_3 in the electrolyte, Al_2O_3 and SiO_2 tend to form complex phases in electrolyte, which decrease the electrical

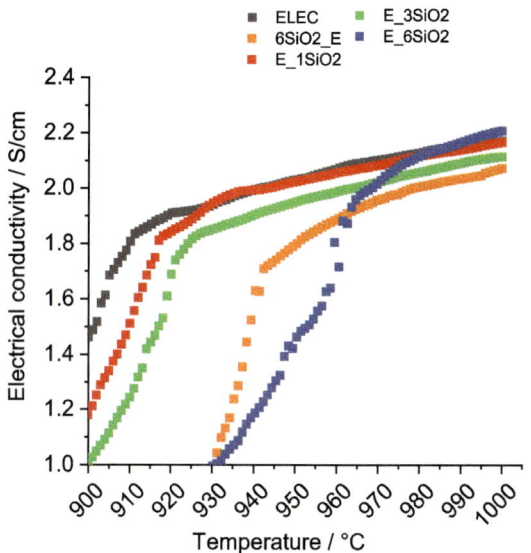

Fig. 4 Electrical conductivity of samples ELEC, 6SiO2_E, E_1SiO2, E_3SiO2, and E_6SiO2

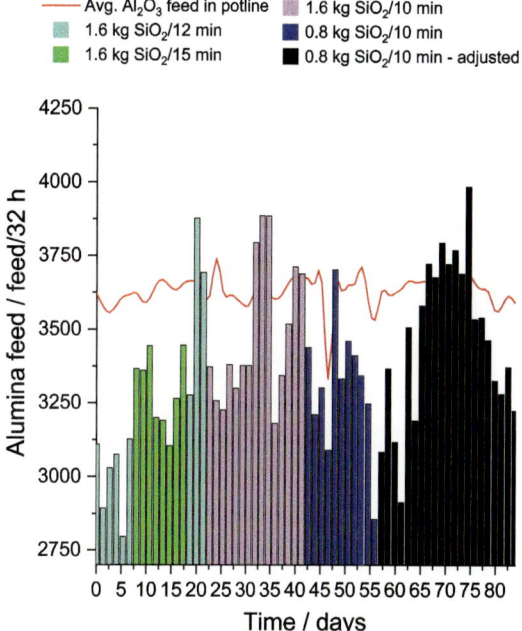

Fig. 5 Average alumina feed in potline compared to Al–Si aluminium reduction cell

conductivity of the electrolytes. As shown in Fig. 4, the addition of Al_2O_3 has a positive effect on electrical conductivities in the Na_3AlF_6–AlF_3–CaF_2–Al_2O_3–SiO_2 system when the SiO_2 concentration was below 6 wt%. At a temperature of 950 °C, the electrical conductivity of E_1SiO2 was 2.02 S/cm, which was 0.5% lower than the electrical conductivity of the sample ELEC. The electrical conductivity of the sample E_1SiO2 was higher compared to E_3SiO2 in the observed temperature range and showed a good correlation with the sample ELEC up to the temperature 926 °C, where the difference between measured electrical conductivities starts to exceed 1%.

From solid electrolytes, samples for XRD analysis were prepared. Average values of peak intensity of the electrolytes at 30.8°2 Θ were observed. It was found that the addition of SiO_2 decrease the intensity of the cyolite peak. It is known that, in classic electrolyte, AlF_3 tends to form $NaCaAlF_6$ and $Na_2Ca_3Al_2F_{14}$, when calcium is "consumed", chiolite ($Na_5Al_3F_{14}$) begins to form [17]. New phases with SiO_2 are not visible in the diffractogram, due to amorphous structure of the new phases.

At the beginning of the Al–Si alloy production in the aluminium reduction cell, the Si concentration in liquid aluminium was at 0.0627 wt% Si. As shown in Fig. 5, the average amount of alumina feed for the potline was stable with 3630 doses of Al_2O_3 per 32 h with a standard deviation of 53.1 doses. When SiO_2 was added to the electrolysis process, the average amount of alumina feed was reduced to 3387 doses per 32 h.

The addition of SiO_2 to the electrolysis influenced the control of alumina feeding control in the aluminium reduction cell where Al-Si alloy was produced. Therefore, the alumina feeding parameters were adjusted. The aluminium reduction cells in potline had an average of 13.49 feeding cycles per 32 h. The aluminium reduction cell where Al–Si alloys were produced, had an average of 10.03 feeding cycles per 32 h. The rate and mass of SiO_2 feed to the aluminium reduction cell affect the occurrence of the feeding cycles. As shown in Fig. 6, cell was fed up to 307.2 kg of SiO_2 per 32 h, resulting in a significant decrease in alumina feeding cycles. The feeding rate of 1.6 kg/10 min resulted in a very inconsistent occurrence of alumina feeding cycles. When the cell resistance exceeds an upper deadband and lowering the ACD cannot reduce the cell resistance, the feeding cycle starts. We hypothesised that the reduction in the number of alumina feeding cycles would be due to the formation of alumina.

One kilogramme of SiO_2 can react with liquid aluminium to form 1.13 kg of Al_2O_3 [14]. At the fastest feeding rate of SiO_2 during a 32-h cycle, an additional 347.14 kg of Al_2O_3 can be formed. The average alumina feed presents 95.63% of the theoretical alumina consumption (standard deviation of 1.4%) during the 32-h cycle, while the average alumina feed rate when SiO_2 was fed to process average alumina is 89.23% of the theoretical alumina consumption (standard deviation of 7.12%). When the reoxidation of aluminium

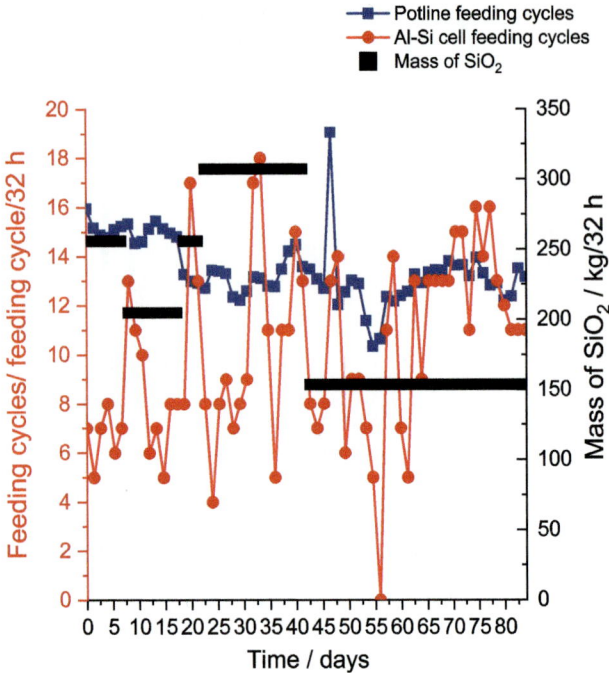

Fig. 6 Feeding cycles of potline, Al–Si cell and mass of added SiO$_2$

was considered, the average alumina feed increased to 95.50% of theoretical alumina consumption. The sum of added alumina and alumina from reoxidation reached up to 111.57% of the theoretical alumina consumption in 32-h periods. Overfeeding of the aluminium reduction cell can lead to accumulation of the sludge at the bottom of the cell. The tracking procedure, in which no alumina was added to electrolysis cell, prevented the sludge accumulation. Therefore, the duration of tracking was analysed. The average duration of tracking in the potline was 37.7 min. The average duration of tracking during SiO$_2$ addition was 62.42 min. During last 27 days, alumina feeding strategy parameters were changed to reduce the effects of alumina formation due to reoxidation of aluminium. When feeding rate was 0.8 kg SiO$_2$/10 min, the underfeed rate time (when all feeding units were activated) was increased by 14% and duration of overfeed rate time was reduced by 10%. A similar approach to adjust underfeeding rate time was presented in the previous study [19]. With the adjusted feeding parameters, the average tracking time was reduced to 50.23 min.

When SiO$_2$ was added, the maximum concentration of 7.2 wt% Si in Al metal was reached, and the average concentration was 4.4 wt% Si. During the production of Al–Si alloy, the average temperature of the electrolyte was 942.2 °C, with a standard deviation of 13.5 °C. The average AlF$_3$ excess in electrolyte was 14.2 wt%, with a standard deviation of 1.5 wt%. The average superheat during the electrolysis production of Al–Si alloy was 14.8 °C.

Conclusions

We have observed the electrical conductivity of electrolytes using DC four-point for measurement of electrical conductivity of liquid metals. The method is suitable for the measurement of electrical conductivity of industrial electrolytes. An electrolyte with 1 wt% has the greatest potential for producing Al-Si alloys in aluminium reduction cells. The electrical conductivity of the electrolyte with 1 wt% SiO$_2$ is comparable to the 78.5 wt% Na$_3$AlF$_6$_13.5 wt% AlF$_3$_5 wt % CaF$_2$–3 wt%Al$_2$O$_3$ system commonly used in industrial aluminium reduction cells. Higher concentration of SiO$_2$ in the electrolyte limits the use of electrolytes at lower process temperatures.

The production of the Al–Si alloy with up to 7.2 wt% of Si is possible in aluminium reduction cell. In the industrial production of Al-Si, the effects of lower electrical conductivity were not noticeable. The reoxidation of aluminium to Al$_2$O$_3$ has a greater impact on the electrolysis process. Feeding of alumina to aluminium reduction cell in the production of Al-Si alloys must be reduced. The parameters of overfeed and underfeed strategies must be adjusted. The temperature of electrolyte and excess AlF$_3$ in electrolyte were not affected by the addition of SiO$_2$, showing that thermal regulation of the cell is able to regulate the system; a minor adjustment is required to reduce superheat temperature bellow 10 °C. The results show that production of Al–Si alloy in AP18 aluminium reduction cell is possible with changes of process parameters.

References

1. Taberaux A (2000) Prebake cell technology: A global review. JOM 52(2):23–29. https://doi.org/10.1007/s11837-000-0043-y
2. Paulin A, Banko Z, Gulin V (1983) Optimizacija elektroliznega postopka s poudarkom na sestavi elektrolita. FNT – VTO Montanistika, oddelek za metalurgijo, Ljubljana
3. Salihagić Hrenko H, Medved J (2018) Managing the Electrolysis Process by Integrating In Situ Measurements of the Bath's Properties. JOM 70(9):1883–1886 https://doi.org/10.1007/s11837-018-2999-5
4. Kolås S, Støre T (2009) Bath temperature and AlF$_3$ control of an aluminium electrolysis cell. Control Engineering Practice 17 (9):1035–1043 https://doi.org/10.1016/j.conengprac.2009.03.008
5. Taylor M P (1997) Challenges in Optimising and Controlling the Electrolyte in Aluminium Smelters. Paper presented at the Molten Slags, Fluxes and Salts 1997 Conference, p 659–674
6. Potocnik V, Reverdy M (2021) History of Computer Control of Aluminum Reduction Cells. In: Light Metals 2021. The Minerals, Metals & Materials Society, Pittsburgh; Springer, Cham, p 591–599
7. Verdenik A (2004) Sinteza vodenja procesa elektrolize primarnega aluminija. MSc thesis, University of Maribor
8. Tebereaux A T, McMinn C J (2016) Production of Aluminium-Silicon Alloys from Sand and Clay in Hall Cells. In

Bearne G, Tarcy G (eds) Essential Readings in Light Metals; Springer, Cham, p 1082–1088

9. Weill D F, Fyfe W S (1964) The 1010 °C and 800 °C Isothermal Sections in System $Na_3AlF_6-Al_2O_3-SiO_2$. Journal of the Electrochemical Society 111(5):582–585

10. Choudhary G (1973) Electrical Conductivity for Aluminium Cell Electrolyte between 950 °C-1025°C by Regression Equation. Journal of the Electrochemical Society 120(3):381–383

11. Wang X, Peterson R D, Tebereaux A T (1993) A multiple regression equation for the electrical conductivity of cryolitic melts. In: Light Metals 1993. The Minerals, Metals & Materials Society; TMS, Warrendale, p 247–255

12. Hiveš J, Thonstad J, Sterten Å, Fellner P (1996) Electrical conductivity of molten cryolite-based mixtures obtained with a tube-type cell made of pyrolytic boron nitride. Metall Mater Trans B 27(4):255–261. https://doi.org/10.1007/BF02915051

13. Sokhanvaran S, Thomas S, Barati M (2012) Charge transport properties of cryolite-silica melts. Electrochimica Acta 66(1): 239–244

14. Grjotheim K, Matiašovský (1971) Chemical Reactions in the System $Na_3AlF_6-SiO_2-Al$. Chemical Papers 25(4):249–252

15. Keller R, Welch B J, Tebereaux A T (1990) Reduction of silicon in an aluminum electrolysis cell. In: Light Metals 1990. The Minerals, Metals & Materials Society, Pittsburgh; AIME, Warrendale, p 333–340

16. Petrič M, Kasteli S, Mrvar P (2013) Selection of electrodes for the "in situ" electrical resistivity measurements of molten aluminium. Journal of Mining and Metallurgy, Section B: Metallurgy 49 (3):279–283. https://doi.org/10.2298/JMMB130118029P

17. Abramović G, Homšak M (2014) Analitika fluorovih soli v proizvodnji aluminija. Paper presented at Slovenski kemijski dnevi, Maribor, Slovenia, 11–12 September 2014

18. Etner P., AlWeb application: https://peter-entner.com/E/ElProp/ElProp-Frame.aspx

19. Salihagić Hrenko H (2020) Mehanizem sinteze aluminijevih zlitin v elektrolizni celici. Ph.D. thesis, University of Ljubljana

Effect of Sulfur Content of Carbon Anode on Measuring Current Efficiency of Aluminum Electrolytic Cell by Gas Analysis Method

Kaibin Chen, Shengzhong Bao, Fangfang Zhang, Guanghui Hou, Huaijiang Wang, Lifen Luo, and Xu Shi

Abstract

In this paper, the gas analysis method was used to measure the current efficiency of multi cells in different aluminum electrolysis potlines, the measurement operation was standardized, and the influence of sulfur content of carbon anodes on the measured values was studied. The results show that the sulfur content in carbon anodes should also be introduced into the gas analysis method as a correction factor in addition to the scaleless volume of Orsat gas analyzer. A systematic approach of gas analysis method and a comprehensive correction coefficient formula were proposed. Using the standardized measurement approach and the comprehensive correction coefficient formula proposed in this paper, the difference between the measured current efficiency by gas analysis method and the statistical current efficiency from metal tapping was found to be less than 0.3%.

Keywords

Aluminum electrolysis • Current efficiency • Gas analysis method • Sulfur content • Correction coefficient

Introduction

Current efficiency is one of the important indicators in the aluminum electrolysis process [1–3]. Measuring and analyzing the current efficiency of aluminum reduction cells is an important means to evaluate the operation quality of workers and economic indicators of aluminum reduction cells. Current efficiency is usually measured by anode gas analysis method, statistical method of metal tapping, and tracer element analysis method [1–5]. Among them, the anode gas analysis method is the most rapid and convenient method.

The theory of anode gas analysis method was based on the hypothesis that the primary anode gas was 100% CO_2, and the CO was all generated by the reoxidation reaction of aluminum and the primary CO_2. That means, the reduction of current efficiency was entirely due to the reoxidation reaction of aluminum and the primary CO_2 gas. According to the reoxidation reaction equation of aluminum and CO_2, and considering other factors influencing current efficiency, the calculation formula of current efficiency is as follows [1–6]:

$$\eta = \frac{1}{2}\varphi(CO_2) + 50\% + K \qquad (1)$$

where η is current efficiency, %; $\varphi(CO_2)$ is volume percentage of CO_2 in anode gas (mixture of CO_2 and CO), %; K is correction coefficient.

The correction coefficient has a significant effect on the calculation result of current efficiency, and many researchers had different values on it. Feng [2], according to a long-term anode gas analysis and the statistical current efficiency of aluminum melt tapping of electrolytic cells at that time, ascertained that the value of K was 3.5%. Shi et al. [4], by measuring the current efficiency of modern large-scale aluminum reduction cells, believed that the value of K was 0, which was more in line with the actual situation. Zhang et al. [5] measured the instantaneous current efficiency to evaluate the rationality of technical conditions of aluminum reduction cells by using a current efficiency analyzer, in which the correction coefficient K was taken as 3.5%. Ma et al. [6] measured and analyzed the relationship between instantaneous current efficiency and process parameters of aluminum electrolytic cells by using a current efficiency analyzer, in which the correction coefficient K was taken as

K. Chen (✉) · S. Bao · F. Zhang · G. Hou · H. Wang · L. Luo · X. Shi
Zhengzhou Non-Ferrous Metals Research Institute Co., Ltd of CHALCO, Zhengzhou, 450041, China
e-mail: Zyy_ckb@rilm.com.cn; zyy_bsz@rilm.com.cn

F. Zhang · G. Hou · H. Wang · L. Luo · X. Shi
National Aluminum Smelting Engineering Technology Research Center, Zhengzhou, 450041, China

© The Minerals, Metals & Materials Society 2023
S. Broek (ed.), *Light Metals 2023*, The Minerals, Metals & Materials Series,
https://doi.org/10.1007/978-3-031-22532-1_96

3%. Cao et al. [7] studied the effect of absorption liquids, gas sampling position, and carbon dust content in electrolyte melt on the accuracy of instantaneous current efficiency test results.

The above research results showed that the value of the correction coefficient had a closer relationship with the measurement operation itself and cell conditions. However, the influence of the measurement operation itself and abnormal conditions of cells could be avoided by using standardized measurement. So it is particularly important to standardize the measurement works and take a reasonable value of the correction coefficient.

In this paper, the gas analysis method was used to measure the current efficiency of multi cells of different aluminum electrolysis potlines. The influence of sulfur content in carbon anode on the measured results was studied. The accuracy of the gas analysis method in measuring current efficiency could be improved through a more reasonable correction coefficient and a standardized measurement scheme.

Current Efficiency Measurement Scheme

Measurement Specifications

Measuring device. Orsat gas analyzer was used to analyze the CO_2 content in anode gas. The scale capacity of the measuring tube was 100 ml, and the absorbent was potassium hydroxide solution with a mass concentration of 30%. The gas sampling pipe was made of steel or stainless pipe with a nominal diameter of 20–25 mm, the total length was 2–2.5 m, the length of the elbow at the intake end was 25–30 cm, and the angle of the elbow was 90–120°. A latex tube with a length of 15–20 m was used to connect the Orsat gas analyzer and the gas sampling pipe, as shown in Fig. 1.

Setting and opening of gas sampling holes. At a position where was on the gap between two carbon anodes, and close to the middle seam of the cell, a circular hole was opened on the

thermal insulating crust with a shell-breaker of an overhead crane. Before opening the hole, remove the loose alumina on the crust surface to expose the hard shell. The gas sampling hole was required to have a diameter of 60–100 mm, have a regular shape with flames emerging and exhaust unobstructed, as shown in Fig. 2. The anodes at the gas sampling hole were 3–10 days away from replacement and had no abnormal current distribution. Air leakage and PFCs (CF_4, C_2F_6) gas generation caused by local anode effects could be effectively avoided by the above standard operations.

Gas sampling and analysis. Put the elbow of the gas sampling pipe into the gas sampling hole, and ensure that there was no alumina or electrolyte blocking the pipe inlet. Keep the pipe inlet 2–5 cm away from the bath level. Under the "Chimney Effect", anode gas would automatically enter the gas sampling pipe and be discharged from the other end (visible to human eyes). Then, connect the gas sampling pipe and Orsat gas analyzer with a latex tube. Conduct gas sampling and CO_2 absorption analysis according to the operational requirements of the Orsat gas analyzer. For each sampling, the volume of gas was 100 ml. The sample gas was passed through the absorption solution repeatedly, CO_2 was absorbed, and the rest was CO. As CO_2 was absorbed, the volume of sample gas gradually decreased. Until the difference of residual volume between the previous and the current absorption was no more than 0.2 ml, indicating that the absorption was completed, the absorption operation could be stopped. According to experience, the times of repeat absorption could be fixed at 8. In addition, the absorption solution should be replaced after taking 20 samples to ensure a good absorption effect.

Measurement requirements. Each cell was measured twice, and different gas sampling holes were used for each measurement. For each measurement, the number of sampling should not be less than 3 times, and the average value was taken as a result. Anode effect, voltage swing, and other abnormal conditions were unaccepted within 4 h before each measurement. The number of measuring cells should be reached more than 10% of the total aluminum electrolysis potlines.

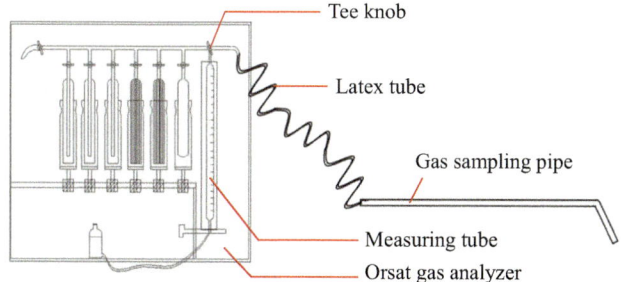

Fig. 1 Schematic diagram of the device for measuring current efficiency by gas analysis method

Fig. 2 Photo of a gas sampling hole

Calculation of current efficiency. After the volume ratio of CO_2 was measured, the current efficiency was calculated with formula (1).

Initial Value of Correction Coefficient

There was an un-scaled connecting pipe between the tee knob of the Orsat gas analyzer and the zero scale line of the measuring tube, so there was an unmeasured volume V_0 when anode gas was sampled and the remaining gas volume was read, as shown in Fig. 3. The existence of the unmeasured volume V_0 caused the actual volume of the sampling gas to be 100 mL + V_0, and the V_0 to be ignored when reading the remaining CO gas volume, which leads to a higher calculated result of current efficiency. Therefore, there was an initial correction coefficient K_0 when measuring the current efficiency with an Orsat gas analyzer.

Assuming that the actual percentage of CO_2 in anode gas was $\varphi(CO_2)$, the actual percentage of CO was $\varphi(CO), \varphi(CO_2) + \varphi(CO) = 1$. When Orsat gas analyzer was used to measure the percentage of CO_2 in the anode gas, the volume of sample gas inhaled in the measuring tube was actually $100 + V_0$ mL. When CO_2 in the sample gas was fully absorbed by the absorption liquid, the remaining CO volume read out from the scale of the measuring tube was V_0 smaller than the actual volume. So the calculated current efficiency was $\frac{1}{200} V_0 \varphi(CO_2)$ higher than the actual current efficiency value. Therefore, the initial correction coefficient K_0 was shown in formula (2).

$$K_0 = \frac{1 - \varphi(CO)}{2} - \frac{100 - [(100 + V_0) \times \varphi(CO) - V_0]}{2 \times 100}$$
$$= -\frac{1}{200} V_0 \varphi(CO_2)$$

(2)

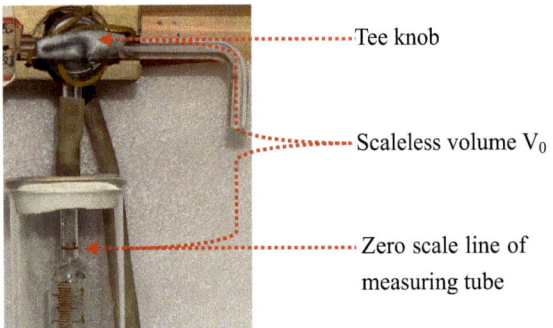

Fig. 3 The unmeasured volume V_0 between the three-way cock and the zero scale line

Tee knob

Scaleless volume V_0

Zero scale line of measuring tube

According to the actual device for measuring current efficiency by gas analysis method, the unmeasured volume V_0 between the tee knob and the zero scale line of the measuring tube was measured and calculated to be 0.75 ml. The actual current efficiency of the aluminum electrolysis potlines to be measured was between 90 and 92%, and the percentage content of CO_2 in anode gas was usually about 80%. Therefore, the $\varphi(CO_2)$ in formula (2) could be taken as 80% and the initial correction coefficient value is $K_0 = -\frac{1}{200} \times 0.75 \times 0.8 = -0.3\%$.

Measurement Objects

The above measurement scheme was used to measure the current efficiency of 131 electrolytic cells of five aluminum electrolysis potlines in three aluminum smelters. Among them, in smelter A, 24 cells were randomly selected from 212 cells of the 200 kA potline, 17 cells were randomly selected from 144 cells of the old 350 KA potline, and 17 cells were randomly selected from 144 cells of the new 350 potline. In smelter B, 34 cells were randomly selected from 300 cells of the 500 kA potline. In smelter C, 39 cells were randomly selected from 300 cells of the 500 kA potline.

The statistical current efficiency of aluminum melt tapping of the five potlines in the past one month (30 days) was retrieved from production reports of the three smelters, and used for comparative analysis with the measured current efficiency by gas analysis method. For the potlines with average aluminum melt level changed during the past month, the statistical current efficiency of metal tapping should be modified according to the change of aluminum metal level. When the aluminum metal level increased, the statistical value of current efficiency needed to be added by $\Delta\eta$, while the aluminum metal level decreased, the statistical value of current efficiency needed to be subtracted by $\Delta\eta$. The calculation formula of $\Delta\eta$ is as follows:

$$\Delta\eta = \frac{\Delta h \times \Delta m_0}{0.3356 \times I_0 \times 24 * 30}$$

(3)

where $\Delta\eta$—the modified value of current efficiency, %; Δh—the difference value of aluminum metal level before and after 30 days of electrolytic cells, cm; Δm_0—the weight of aluminum liquid corresponding to each centimeter of aluminum melt level in an electrolytic cell (usually supplied by smelter), kg/cm; I_0—current intensity of aluminum electrolysis potline, kA.

The sulfur content values of carbon anodes used in the 131 cells selected above were provided by quality inspection departments of corresponding smelters, and no separate sampling and analysis was required.

Results and Discussion

Measurement Results of Current Efficiency

From June to July 2021, the current efficiency of the 131 cells was measured by the measurement scheme mentioned above. The scatter diagram of measurement results is shown in Fig. 4.

It can be seen from Fig. 4 that the measured current efficiency of the randomly selected cells was different, but most of them were within ±1.5% of the average value. The situation of each potline was basically the same.

Figure 5 shows the difference values between the two measurements of the 131 cells. These difference values reflected the reproducibility of the gas analysis method. For a stable cell, the variation of actual current efficiency was small in one day. The two measurements of the same cell were completed in one day, and the deviations of the two measurement results were basically within ±0.6%, and most of them were within ±0.5%.

The current efficiency measured by the gas analysis method was an instantaneous value, but the average value of multi cells could basically represent the actual current efficiency of the potline. In order to check the accuracy of the gas analysis method, the average measured values of multi cells in each potline were compared with the statistical current efficiency of aluminum metal tapping of these pot-lines in one month, as shown in Table 1.

From the above results, the following could be drawn:

(1) The current efficiency measured by the gas analysis method had good reproducibility. With standardized measurement operations, such as setting and opening of gas sampling holes, avoidance of abnormal cell conditions, and gas analysis operations, the impact of air leakage and PFCs (C2F6, CF4) gas generation could effectively be avoided, and the influence of the measurement works itself and the effects of cell conditions could be greatly reduced.

(2) Although the influence of unmeasured volume V_0 of the Orsat gas analyzer was considered, the value of current efficiency measured by the gas analysis method was still slightly higher than the statistical current efficiency of aluminum melt tapping, and the difference values were in the range of 0.2% to 0.6%.

(3) In addition to the unmeasured volume V_0, there must be a common factor needs to be considered in the correction coefficient.

Morphological Changes of Sulfur in Carbon Anode

Sulfur in carbon anodes mainly comes from petroleum coke and pitch coke and exists in the form of organic sulfur. At present, the sulfur content in carbon anodes is in the range of 0.7–3.7% in China [8], and most of them are above 1.5%. In the process of aluminum electrolysis, in addition to the normal decomposition reaction of alumina with carbon, the sulfur in carbon anodes will also generate Sulfur oxide carbon COS, also known as Carbonyl Sulphide, which has the smell of rotten eggs. The reaction equations are as follows [8, 9]:

$$Al_2O_3 + 3/2C = 3/2CO_2 + 2Al \qquad (4)$$

$$Al_2O_3 + 3C + 3S = 3COS + 2Al \qquad (5)$$

$$S + CO_2 + C = COS + CO \qquad (6)$$

$$S + CO = COS \qquad (7)$$

Among them, the reaction of formulas (4) and (5) are the main reactions that occur on the electrolysis reaction interface. The reaction of formulas (6) and (7) are side reactions that occur on the anode surface.

It can be seen from the above reaction equations that the sulfur in carbon anodes will first generate COS during the electrolysis process. The COS is stable at room temperature,

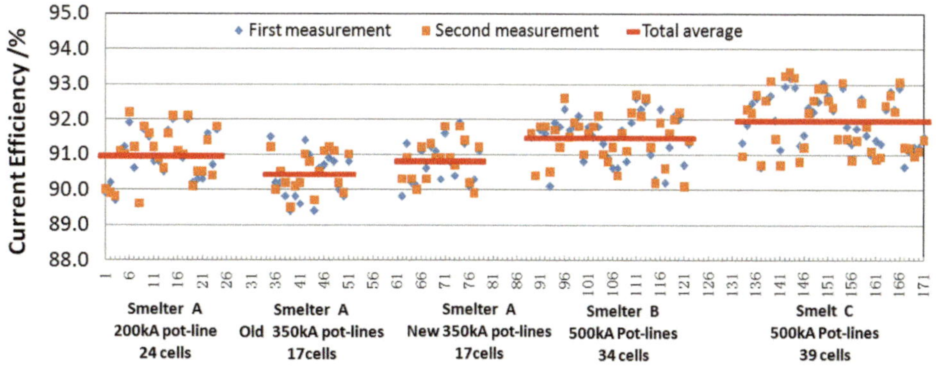

Fig. 4 Scatter diagram of the current efficiency of the 131 cells measured by gas analysis method

Fig. 5 Difference values between the two measurement results

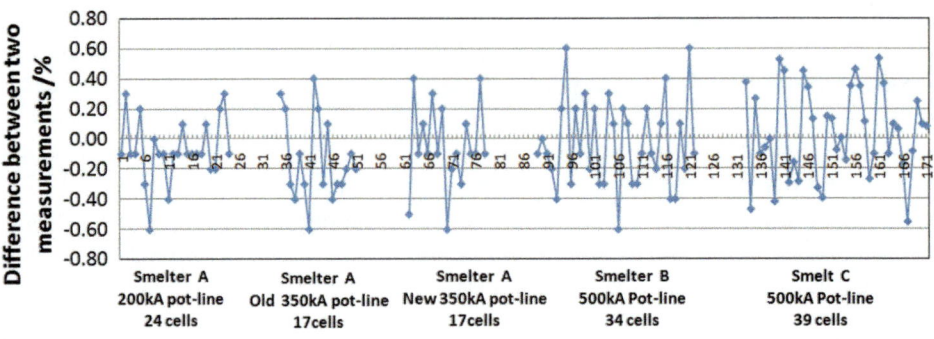

Table 1 Comparison of current efficiency between the measurement results and the statistical values

Aluminum potlines	Smelter A 200kA potline	Smelter A old 350kA potline	Smelter A new 350kA potline	Smelter B 500kA potline	Smelter C 500kA potline
Number of cells	24	17	17	34	39
Average value of current efficiency measured by gas analysis method ($K_0 = -0.3$)/%	90.94	90.41	90.80	91.46	91.96
Average statistical current efficiency of metal tapping of potlines in recent one month/%	90.4	90.01	90.45	91.22	91.67
Difference value/p. p	0.54	0.40	0.35	0.24	0.29

but it will be converted to CO_2 and SO_2 when encountering strong oxidants such as oxygen in the air at high temperature. Therefore, after the COS gas is discharged from the electrolytic cell, about 95% of the gas is finally converted into SO_2 after passing through the dry scrubbing system. However, in the process of sampling anode gas by using an Orsat gas analyzer, there was no air leakage, so it could be considered that the COS gas would not be oxidized basically.

It can be seen from the reaction of (4) and (5) that the oxygen generated by the decomposition of the same mass alumina combines with sulfur and carbon together to generate COS gas, which is twice the mole number of CO_2 generated by combining with carbon alone. That means, to produce the same mass of aluminum, twice as much COS gas is produced as CO_2 gas. As known, COS is easily soluble in water. While CO_2 was absorbed by the absorption liquid in the Orsat gas analyzer, COS was absorbed too. Finally, the measured volume percentage of CO_2 was higher than the actual one.

Assuming that the sulfur content in the anode is γ, and all the sulfur generates COS during electrolytic reactions, then the volume percentage of COS in anode gas would be $\frac{\gamma/32}{\frac{1-\gamma}{12}+\gamma/32}$. When the Orsat gas analyzer was used to measure the volume percentage of CO_2 in anode gas, the measured value would be $\frac{1}{2} \times \frac{\gamma/32}{\frac{1-\gamma}{12}+\gamma/32}$ higher than the actual one. Then

the calculated current efficiency was about $\frac{1}{4} \times \frac{\gamma/32}{\frac{1-\gamma}{12}+\gamma/32}$ higher, which was $\frac{3\gamma}{32-20\gamma}$. Therefore, considering the influence of sulfur content in carbon anodes, the correction coefficient should be $K_s = -\frac{3\gamma}{32-20\gamma}$. If the sulfur content in carbon anodes is 2%, then $K_s = -0.001899$, or -0.1899%.

Comprehensive Correction Coefficient

It can be seen from the above analysis results that the unmeasured volume V_0 of the Orsat gas analyzer and the sulfur content in the carbon anode were the main factors of the correction coefficient in the current efficiency calculation formula. Therefore, the comprehensive correction coefficient of current efficiency measured by the Orsat gas analyzer is

$$K = K_0 + K_s = -[\frac{1}{200}V_0\varphi(CO_2) + \frac{3\gamma}{32-20\gamma}] \quad (8)$$

The current efficiencies of the 131 electrolytic cells were recalculated using the comprehensive correction factor of formula (8). Among them, the value of K_0 was -0.3%, and the actual analysis result of sulfur content in carbon anodes given by the smelters was adopted. The calculation results are shown in Table 2 and Fig. 6.

Table 2 Comparison of current efficiency between the measured results calculated by the comprehensive correction coefficient and the statistical values of metal tapping

Aluminum potlines	Smelter A 200kA Potline	Smelter A old 350kA potline	Smelter A new 350kA potline	Smelter B 500kA potline	Smelter C 500kA potline
Number of cells	24	17	17	34	39
Sulfur content in carbon anode/%	3.2	2.8	2.4	1.84	2.05
comprehensive correction coefficient $K = K_0 + Ks$/%	−0.6061	−0.5672	−0.5284	−0.4745	−0.4947
Average value of current efficiency measured by gas analysis method ($K = K_0 + Ks$)/%	90.64	90.15	90.57	91.29	91.76
Average current efficiency of metal tapping of potlines in recent one month/%	90.4	90.01	90.45	91.22	91.67
Difference value /p. p	0.24	0.14	0.12	0.07	0.09

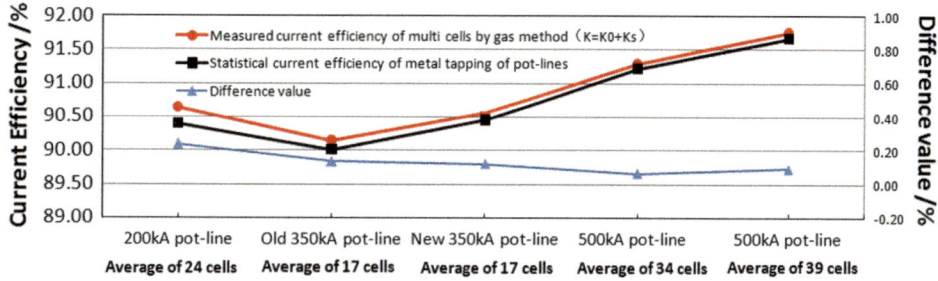

Fig. 6 Comparison of current efficiency between the measured values of multi cells by gas analysis method and the statistical values of metal tapping of pot-lines

It can be seen from Table 2 and Fig. 6 that after the comprehensive correction coefficient was adopted, the difference between the measurement values of the gas analysis method and the statistical values of metal tapping was significantly reduced, which is less than 0.3%.

Conclusions

The gas analysis method is relatively simple and quick to determine the CE (current efficiency) of a cell. However, in order to obtain more accurate measurement results, the standardization of the measurement itself and the value of the correction coefficient is very important.

In this paper, a systematic approach was proposed to minimize possible errors during gas sampling. The sulfur content in carbon anodes should also be introduced into the gas analysis method as a correction factor in addition to the scaleless volume of the Orsat gas analyzer. The correction factor of sulfur was estimated based on the molar ratio of COS and CO_2 formed during the alumina reduction reaction. The correction factor of scaleless volume was derived based on the actual internal volume of the connecting pipe between the zero scale line and the tee knob of the Orsat gas analyzer. Once the two correction factors were included in the calculations, there was a good match between the CE values derived from the formula and CE values from conventional CE measurements based on the metal produced over a 30 day period.

References

1. Qiu ZX (2008). Aluminum production with prebaked electrolysis-cell (3rd edition). Metallurgical industry press, Beijing.
2. Feng NX (2006). Aluminumelectrolysis [M]. Chemical Industry Press, Beijing.
3. Liu YX, Li J, et al. (2008). Modern aluminum electrolysis [M]. Metallurgical industry press, Beijing.
4. Shi ZR, Yang WJ, Zhang G, Bao SZ, et al. (2009). Study on measurement technology of instantaneous current efficiency of large aluminum reduction cell [J]. *Light Metals, 2009* (7):23–26. (Chinese Publication)
5. Zhang SJ, Yue SB, Tan HT, et al. (2010). Application of current efficiency analyzer in electrolytic cell [J]. *Light Metals, 2010*(5):28–33. (Chinese Publication)

6. Ma BY, Sun ZB, Dai YF, et al. (2011). Testing practice of instantaneous current efficiency of aluminum reduction cell [J], *Light Metals, 2011*(6):30–32. (Chinese Publication)

7. Cao YF, Zhou YF, Hou GH, et al. (2017). Study on measurement method of instantaneous current efficiency of aluminum reduction cell [J], *Light Metals, 2017*(5):22–24. (Chinese Publication)

8. Sun XX, Sun Y, Liu CD. Study on the migration behavior of sulfur in anode raw materials for aluminum during production [J], *Light Metals, 2015*(4):45–48. (Chinese Publication)

9. F. M. Kimmerle, L. Noel, J. T. Pisano (1997). COS, CS2 and SO2 emissions from prebaked Hall Heroult Cells [C]. *Light Metals. TMS 1997*: 153–158.

KF Content on Physical and Chemical Properties of Aluminum Electrolysis Electrolyte

Changlin Li, Shengzhong Bao, Fangfang Zhang, Gang Li, Shilin Qiu, Fei He, Guanghui Hou, and Huaijiang Wang

Abstract

Based on the step cooling curve method, continuously varying cell constant (CVCC) method, observation method, and carbon block sodium expansion rate determination method, the effects of KF content on liquidus temperature, conductivity, alumina solubility, and electrolysis expansion rate of cathode were studied. The results showed that the liquidus temperature decreased with the increase in KF content. For every 1% increase of KF, the liquidus temperature decreased by about 1.83 °C. The conductivity increased with the increase of superheat, and the average conductivity increased by 0.009 S/cm for every 1 °C increase in superheat. Under the same superheat, the conductivity increased by 0.033 S/cm for every 1% decrease in KF. Under the same superheat, KF content had little effect on the dissolution rate of alumina. The electrolysis expansion rate of the cathode carbon block increased with the increase in the KF content. The purpose of this work was to study the effect of KF content on the physical and chemical properties of aluminum electrolyte, and to provide a theoretical basis for the selection of appropriate electrolyte composition in aluminum electrolysis industry.

Keywords

Molten salt • Electrolysis • KF • Liquidus temperature • Conductivity • Solubility

Introduction

The physical and chemical properties of electrolytes are an important research issue in the field of molten salt electrolysis, which can directly affect and determine the technical parameters of electrolytic cells and play a crucial role in the electrolysis production process [1–3].

Alumina with high potassium content was used in one smelter for a long time, and potassium was enriched and KF was generated in the electrolyte, and KF content reach above 4%, forming unique high-potassium electrolyte. The concentration of CaF_2, MgF_2 and LiF in the electrolyte system was about 4.8, 0.5, and 1.8%, respectively. The concentration of Al_2O_3 was about 2.2%, and the molecular ratio was controlled at 2.4–2.5. In this paper, according to the characteristics of electrolyte components in the smelter, the influence of KF content on liquidus temperature, conductivity, alumina solubility, and cathode electrolytic expansion rate was investigated by single factor test, in order to establish the relationship between KF content and physical & chemical properties of aluminum electrolytes. It provided a theoretical basis for selecting suitable electrolyte composition.

Experimental

Materials

Before the experiment, the required reagents were dried in a vacuum drying oven at a constant temperature of 200 °C for more than 48 h and proportioned according to the ingredients described in Table 1. Then the electrolyte was fused in a Muffle furnace at a certain temperature with superheat of 15 °C. Finally, the samples were ground and mixed evenly after cooling and put into a dryer for later use.

C. Li (✉) · S. Bao · F. Zhang · S. Qiu · G. Hou · H. Wang
Zhengzhou Non-Ferrous Metals Research Institute Co. Ltd of Chalco, Zhengzhou, 450041, China
e-mail: zyy_lcl@rilm.com.cn

G. Li · F. He
Zunyi Aluminum Co. Ltd, Zunyi, 563000, China

© The Minerals, Metals & Materials Society 2023
S. Broek (ed.), *Light Metals 2023*, The Minerals, Metals & Materials Series,
https://doi.org/10.1007/978-3-031-22532-1_97

Table 1 Electrolyte composition/wt %

No	Molecular ratio	KF (%)	CaF$_2$ (%)	MgF$_2$ (%)	LiF (%)	Al$_2$O$_3$ (%)
1#	2.5	1	4.8	0.5	1.8	2.2
2#	2.5	2	4.8	0.5	1.8	2.2
3#	2.5	3	4.8	0.5	1.8	2.2
4#	2.5	4	4.8	0.5	1.8	2.2
5#	2.5	5	4.8	0.5	1.8	2.2
6#	2.5	6	4.8	0.5	1.8	2.2
7#	2.5	7	4.8	0.5	1.8	2.2

Methods

Liquidus Temperature

The liquidus temperature was measured by step cooling curve method according to national standard GB/T 33908-2017. Firstly, NaCl and NaF, two standard reagents with known liquidus temperature, were selected to determine the temperature deviation of the thermocouple and the accuracy of the measurement system. The measured liquidus temperature was compared with the known literature to determine the system error, as shown in Table 2. The analysis showed that the error of the liquidus temperature measured by this method was in the range of 1–1.6 °C, which indicated that this method was accurate and feasible. Then the liquidus temperature of the electrolytes was measured in the same way.

Conductivity

The conductivity of molten salt was tested according to the CVCC method [4–6]. Boron nitride tube was used as a conductance cell [7, 8], platinum wire was used as the working electrode, graphite crucible connected with a conductive rod was used as the reference electrode and counter electrode, and computer-controlled electrochemical workstation applied frequency to the circuit and recorded data. The device for conductivity determination by CVCC method was shown in Fig. 1.

The Solubility of Alumina in Electrolyte

The solubility of alumina in electrolyte was studied by observation. In order to guarantee the accuracy of the experiment, the electrolyte temperature was corrected before each experiment, and then 1% alumina was added at a certain superheat after the electrolyte was melted for 15 min. A flashlight was used to observe the dissolution of alumina in the electrolyte and a stopwatch was used to record the floating time and dissolution time of alumina in the electrolyte, respectively.

Electrolysis Expansion Rate of the Cathode

The electrolysis expansion rate of the cathode carbon block in the electrolyte melt was measured with the electrolysis expansion rate tester produced by R&D Carbon of Switzerland, and the influence of KF content on the electrolysis expansion rate of the cathode carbon block was studied. The experiment was carried out according to standard YS/T 63.5-2006 aluminum carbon materials test method—Part 5: Under pressure bottom carbon block sodium expansion rate determination.

Results and Discussion

Effect of KF Content on the Liquidus Temperature

The liquidus temperature of electrolyte with different content of KF was tested. The results were shown in Fig. 2. As can be seen from the figure, when CR was 2.5, the liquidus temperature of electrolyte generally decreased with the increase of KF content. When the KF content increased from 1 to 7%, the liquidus temperature decreased by 11.1 °C in total, that was, the average liquidus temperature decreased by about 1.83 °C with the increase of 1% KF.

Effect of KF Content on Electrolysis Conductivity

Considering the influence of KF content on the liquidus temperature of electrolyte, the conductivity of superheat in the range of 3–12 °C was tested in the experiment. The

Table 2 Error analysis

Standard reagent	Measured value T/°C	Literature value T/°C	Error/%
NaCl	800	801.0	0.125
NaF	993.4	995.0	0.160

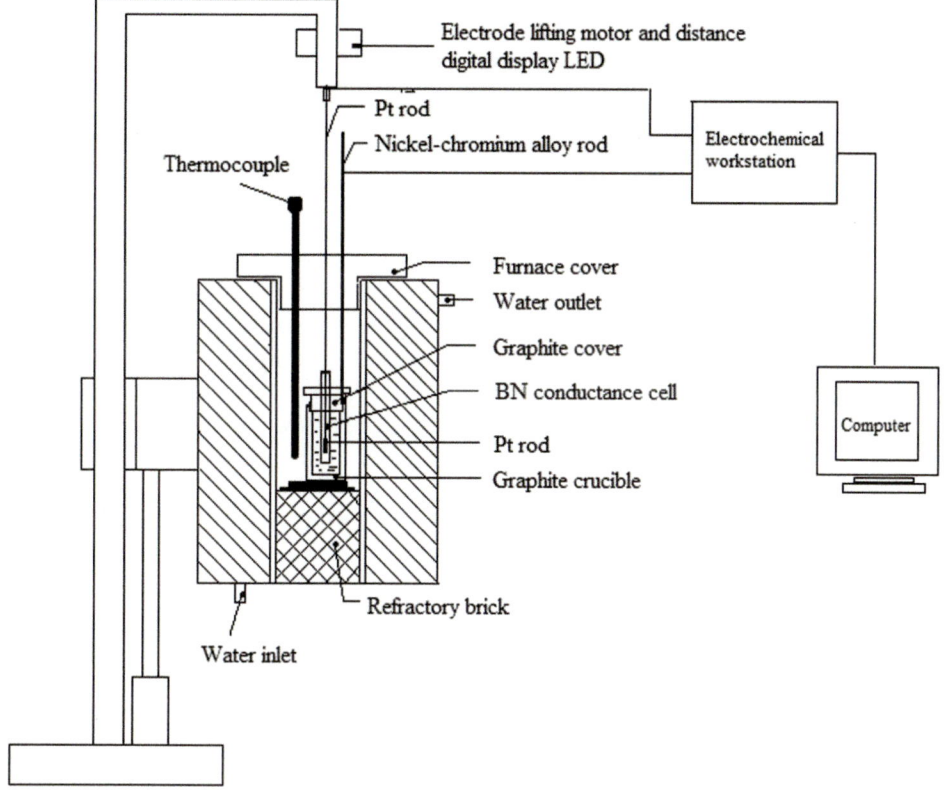

Fig. 1 Test device diagram of electrical conductivity

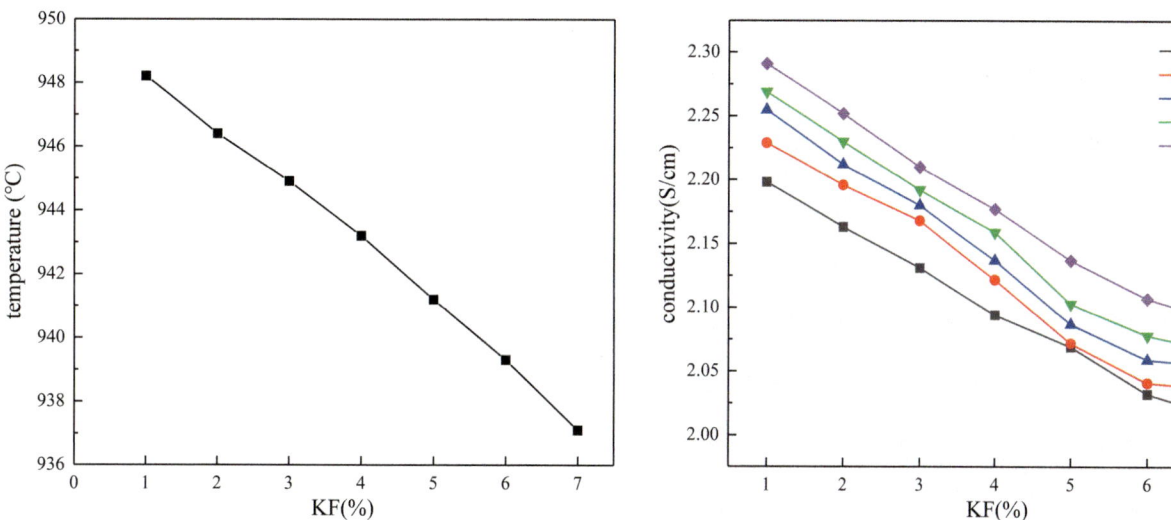

Fig. 2 Liquidus temperature at different KF content (CR was 2.5)

Fig. 3 Electrolysis conductivity at different KF content (CR was 2.5)

electrolysis conductivity results at different superheat were shown in Fig. 3.

It can be seen from Fig. 3 that at the same superheat, when KF decreased by 1%, the electrolysis conductivity increased by 0.033S/cm on average. At the same KF content, the conductivity increased with the increase of superheat.

This was mainly because with the increase in temperature, the kinetic energy of ions increased, making it easier for ions to overcome the attraction between ions and move under the action of the electric field. Macroscopically, with the increase in temperature, the number of conductive ions per unit volume increased, the viscosity of electrolyte decreased,

Table 3 Electrolyte composition and liquidus temperature at different KF content

No	KF (%)	CaF$_2$ (%)	MgF$_2$ (%)	LiF (%)	Al$_2$O$_3$ (%)	Molecular ratio	Liquidus temperature (°C)
1#	0	4.8	0.5	1.8	2.2	2.4	946.2
2#	1	4.8	0.5	1.8	2.2	2.4	945.3
3#	3	4.8	0.5	1.8	2.2	2.4	942.6
4#	5	4.8	0.5	1.8	2.2	2.4	938.7
5#	7	4.8	0.5	1.8	2.2	2.4	935.1

and the resistance of ion movement decreased, which was manifested as the increase in electrical conductivity. In the experiment, when KF was in the range of 1–7%, the conductivity increased by 0.009 S/cm on average for every 1 °C increase in superheat.

Effect of KF Content on Alumina Solubility

Effect of KF content on the solubility of alumina was studied. The electrolyte composition and liquidus temperature were shown in Table 3.

In order to quantitatively study the effect of KF content on the dissolution of alumina, the dissolution time of adding 1% alumina at different KF content was tested at the electrolyte temperature of 950 °C, as shown in Table 4.

With the increase of KF content, the dissolution time of 1% alumina in electrolyte gradually shortened. When the KF content was 7%, the dissolution time of alumina was 22 min 56 s, which was 35% shorter than that without KF (35 min 27 s).

The above results reflected the dissolution rate of alumina at a fixed temperature, in fact, the superheat of the electrolyte gradually increased with the increase of KF content. Therefore, the dissolution rate of alumina in electrolytes with different KF content at the same superheat was further explored in the experimental, as shown in Table 5.

At the same superheat, the liquidus temperature decreased gradually with the increase of KF content, and the dissolution rate at this time was the result of the joint action of temperature and KF content. The liquidus temperature of the electrolyte without KF and the electrolyte with 7% KF were 946.2 and 935.1 °C, respectively, and the difference in the liquidus temperature was 11 °C. However, compared with the electrolyte without KF, the dissolution rate of alumina in the electrolyte with 7% KF was only 12.6% lower, indicating that the dissolution rate of alumina in the electrolyte was only slightly slower.

Effect of KF Content on the Electrolysis Expansion Rate of Cathode Carbon Block

The electrolyte sample used were self-prepared (1#–5#), and the KF content was 0, 3, 6, 9, and 12% in the electrolyte, respectively. The compositions were shown in Table 6. The influence of KF content on the electrolysis expansion rate of cathode carbon block was investigated.

As can be seen from Fig. 4, the electrolysis expansion rate gradually increased with time going on. At the beginning of electrolysis, the expansion rate of increased faster. After 60 min of electrolysis, the growth of expansion rate gradually slowed down and reached the maximum value at about 120 min. At the same time, it can be seen that the electrolysis expansion rate gradually increased with the increase of KF content, indicating that with the increase of KF content, the corrosion of cathode carbon block by electrolyte gradually intensified.

In Fig. 5, with KF content in the electrolyte increasing from 0 to 12%, the maximum electrolysis expansion rate of the cathode carbon block increased from 0.63% to 1.16%, nearly doubling. The maximum electrolysis expansion rate of 1# and 2# were both relatively small and the maximum electrolysis expansion rate increased by 0.05%, indicating that the influence of KF content below 3% on the electrolysis expansion rate was small. With the continuous increase of KF content, the maximum electrolysis expansion rate gradually increased. KF content in the range of 3–12%, the

Table 4 Single dissolution time of alumina in electrolytes at different KF content (950 °C)

Feeding times	Dissolution time (min:s)				
	1# (KF = 0%)	2# (KF = 1%)	3# (KF = 3%)	4# (KF = 5%)	5# (KF = 7%)
1	31:32	31:36	25:43	24:50	16:51
2	33:30	23:07	29:35	27:12	19:23
3	35:27	33:29	32:26	30:23	22:56

Table 5 Single dissolution time of alumina in electrolytes at different KF content (superheat 10 °C)

Feeding times	Dissolution time (min:s)				
	1# (KF = 0%)	2# (KF = 1%)	3# (KF = 3%)	4# (KF = 5%)	5# (KF = 7%)
1	25:59	27:14	26:53	27:56	28:11
2	27:36	28:27	28:01	30:12	31:04
3	29:58	31:49	33:20	33:34	33:45

Table 6 Electrolyte composition of samples for the electrolysis expansion rate of cathode carbon block investigation

No	Molecular ratio	KF (%)	CaF_2 (%)	MgF_2 (%)	LiF (%)	Al_2O_3 (%)
1#	4.0	0	4.8	0.5	1.8	2.2
2#	4.0	3	4.8	0.5	1.8	2.2
3#	4.0	6	4.8	0.5	1.8	2.2
4#	4.0	9	4.8	0.5	1.8	2.2
5#	4.0	12	4.8	0.5	1.8	2.2

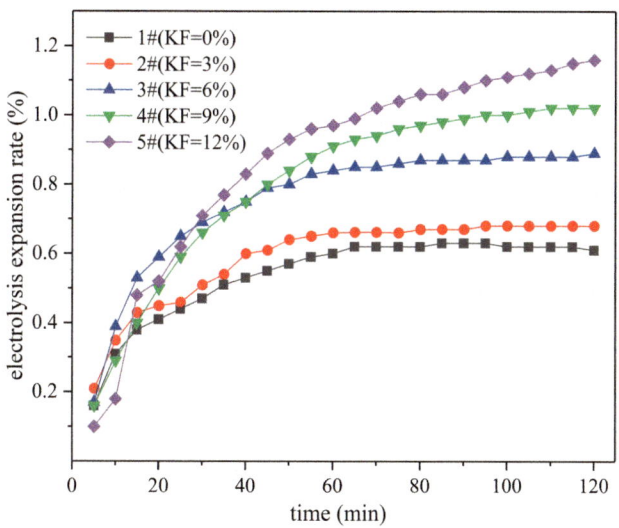

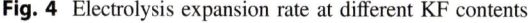

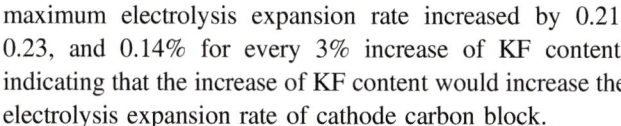

Fig. 4 Electrolysis expansion rate at different KF contents

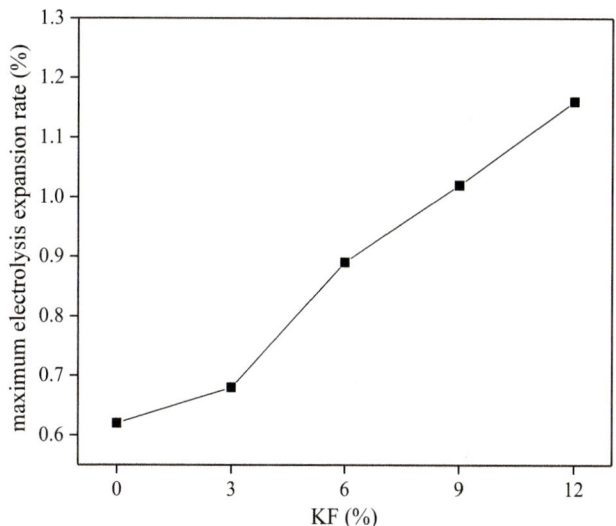

Fig. 5 Maximum electrolysis expansion rate at different KF contents

maximum electrolysis expansion rate increased by 0.21, 0.23, and 0.14% for every 3% increase of KF content, indicating that the increase of KF content would increase the electrolysis expansion rate of cathode carbon block.

Conclusions

(1) The liquidus temperature decreased with the increase of KF content. When the KF content increased from 1 to 7%, the liquidus temperature decreased by 11.1 °C, the average liquidus temperature decreased by about 1.83 °C with the increase of 1% KF.

(2) At the same superheat, when KF decreased by 1%, the electrical conductivity increased by 0.033 S/cm on average.

(3) At the same KF content, the conductivity increased with the increase of superheat. When KF content is in the range of 1–7%, the conductivity increased by 0.009 S/cm for every 1 °C increase in superheat.

(4) At the same temperature, with the increase of KF content, the dissolution time of alumina in the electrolyte gradually decreased. When the KF content was 7%, the dissolution time of alumina was 35% shorter than that without KF. At the same superheat, compared with the electrolyte without KF, the dissolution rate of alumina in electrolyte with 7% KF only reduced by

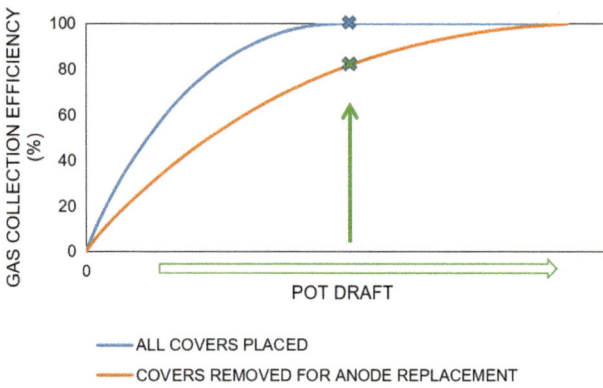

Fig. 1 Gas collection efficiency example

more important for thermal modeling and operations to have consistency in heat loss rates to the fume control systems.

Pot Ventilation Duct Network

The gases drafted from the pots are transported to the GTCs through a network of ducts. The path between the GTC and the most distant pot(s) is called the critical path and the pressure drop (dP) of the gas through this path will determine the static pressure (Ps) demand at the GTC inlet. The duct network is typically designed positioning the GTC in a central location, and the ducting generally includes branches so that the length of the critical path is minimized. The diameter of the ducts is sized so that the gas velocity is sufficient to avoid the deposition of particulate matter and not high enough to cause excessive pressure losses in the system (these would increase demand for energy consumption). Note that the formation of greyscale in the main ducts when gas flow is laminar is generally not of concern.

Figure 2 represents a schema of a sample duct network with 80 pots. The dP of each pot is defined as the difference between the air pressure surrounding each cell at the

potroom (point 0), and the gas pressure upstream of the dampers at the pot discharge (point A). The pressure at the pot discharge for each pot is defined as P_{Ai} where the subscript "i" represents the pot number.

With the same flow evacuated at each pot and assuming the same conditions in terms of flow restrictions and heat transferred to the gas, P_A is ideally the same for all the pots. Point C in Fig. 2 refers to the GTC inlet, and therefore, section A_1-C illustrates the critical path.

Since the static pressure (and total pressure) at points Ai are the same ($P_{A1} = P_{A2} = \cdots = P_{Ai}$), the pressure difference between each of the points A_i and point C shall be the same ($dP_{A1-C} = dP_{A2-C} = \cdots = dP_{Ai-C}$). For this to happen, the dampers at each pot discharge shall create different resistances in order to compensate for the losses among the various paths to point C. The closer that a pot is to the GTC inlet, the more resistance should be created by its damper.

The restrictions at the pots might be actually obtained with dampers (guillotine or butterfly), or by using orifice plates. This manuscript will only refer to dampers for the sake of simplicity.

Figure 3 shows the effect that balancing has on the duct system illustrated in Fig. 2. Note that this represents 40 pots in one potroom with another 40 pots in a second potroom being a mirror image to this duct. As explained, the restrictions at the pots closest to the GTC are the highest, whilst no restriction should be induced at the pots of the critical path(s). The flows represented in this example were calculated with a model built with the software Arrow, considering some typical parameters for the duct system and pot conditions.

Systems that are not balanced are inefficient for fume capture. For example, it would not be unusual for Pot #1 and Pot #40 of Fig. 3 to have fugitive emission losses to the potroom roof that are multiples of the fugitive loss rates of Pot #9 and Pot #31. Pot #20 and Pot #21 in an unbalanced configuration would have lower than average fugitive

Fig. 2 Pressure drops in a sample duct system

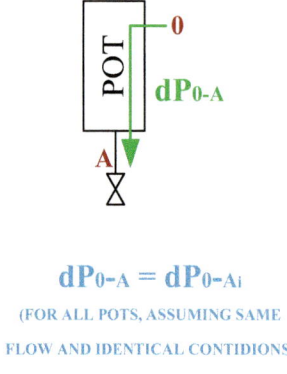

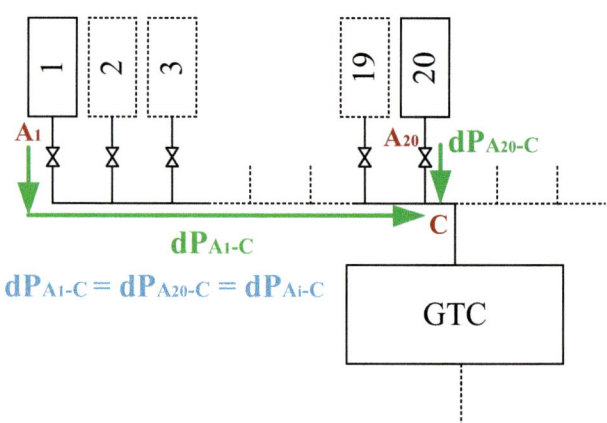

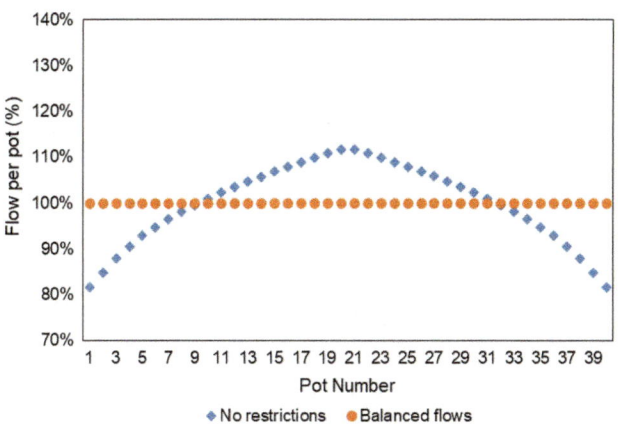

Fig. 3 Flows with no restrictions at the pots vs balanced flows

emission losses, but nowhere near enough to offset the high fugitive loss rates from pots with low exhaust flows.

When balancing a system, the static pressure at the GTC inlet increases as flow increases through the critical path pots at the terminal points of ducting (increasing the system's overall losses). The difference between the maximum and minimum flows per pot in the system with no restrictions will depend on the length and resistance of the critical path.

A patented design of ducts on the ends of duct networks is available. They are known as *Impulse Ducts* and their aerodynamic design minimizes the resistance on the critical path, thus optimizing the overall losses [2].

Many smelters operate with a fixed flow at the pots at a rate that allows for 98% or higher collection efficiency. The majority of fugitive emission losses occur when covers are removed for pot tending activities. Some smelters incorporate High Draft Systems which temporarily allow an increased draft on pots that have had some covers removed during anode changing and other pot tending activities [3].

Figure 4 shows the boost effect created by opening the damper of one of the closest pots to the GTC in a balanced system. The increase in flow rate will depend on the position

of the pot versus the GTC location, and on the ducting system design (critical path). The extra flow drafted from this pot will be deducted from the sum of flows of all the remaining pots connected to the GTC. This should not present a problem as most of these pots will have all their covers in place, maintaining a good collection efficiency even with a slightly lower flow. This is the principle of the Single Line High Draft Systems where pot dampers are temporarily opened to achieve a higher flow. This is also a partial contributor to the flow increase obtained with the Jet Induced Boosted Suction (JIBS) High Draft System [4].

Note that some Single Line High Draft systems are also balanced to limit maximum flow rates for any one pot while other systems fully open the adjustable damper during key, pot tending activities. It has been demonstrated that the use of high draft that is more than double the amount of normal draft is beyond the point of diminishing returns for the capture of fugitive emissions [5].

The boosting capacity of a Single Line High Draft could be raised by consistently increasing the restrictions in all the system dampers. This would even allow some boosting capacity at the pots located in the critical path. It is to be noted that the overall static pressure would be raised in the whole system, thus an evaluation of the GTC fans capacity, fan motors power, and ducting design would be required.

Issues in Ducting Systems

The most important issues of the ducting systems include:

- Lack of maintenance and inspection: ignoring cracks, defective gaskets or expansion joints, and loose flanges that allow ambient air to be drawn into the system stealing exhaust capacity from the pots. Ingress of moisture in the air can also lead to heavy-scale formation. The structural integrity of the ducts may be even compromised to the point of failure.
- Improper flow balancing:
  - Pots with flow rates above the target "steal" capacity from other pots.
  - Pots with flow rates below the target, on the other hand, will have their collection efficiency reduced, which will increase the potroom roof emissions.
  - Excessively low flow rates on individual pots or groups of pots may also lead to accumulations of particulate matter in ducting within pot superstructures or in the ducting leading away from the pots. This may then unbalance the ducting system and may also compromise the structural integrity of the ducts.

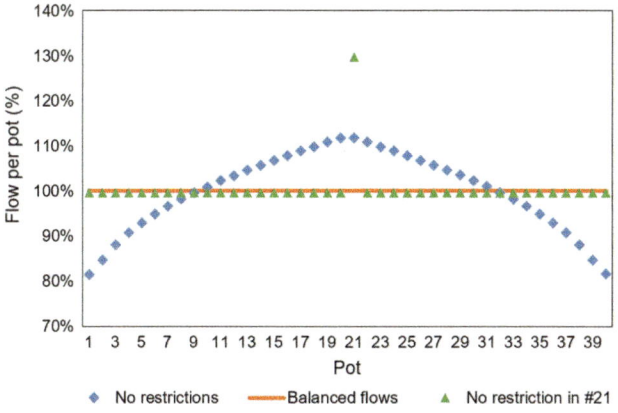

Fig. 4 Effect in pot flows when opening the damper at pot #21

Each time that the superstructure comes out of service for pot repair, it is important to inspect and clean the exhaust ducting of the pots, as well as the fume collection points and superstructure ducting, as these could have been partially or completely blocked. A particulate matter called chiolite tends to stick to surfaces and accumulates in a number of superstructure exhaust designs. Cleaning and inspection are key to achieving similar resistances of the gas flow through the pot. The exhaust pick-up points can often be cleaned out in situ with a shot of compressed air delivered by a hand-held aluminum air pipe that is shaped to perform the task, or by vacuuming if possible.

Measurement of Pot Flows

Prior to the discussion of a preferred method for pot flow measurement, some key points regarding safety are considered as task-related hazards with potentially serious consequences. When working on or around the exhaust duct from the pot it is always important to be alert to the hazard of ground potential. Technicians that balance pot draft systems and their equipment must be made aware to never bridge the gap between potline potential and ground potential of building steel. In addition, if gaining access to the sampling points for pot exhaust requires work that is above floor level, the hazards of different-level falls must also be taken into consideration. The presence of hot surfaces and potential emissions of gas shall be also considered.

Given the high number of pots connected to a single GTC, a simple method of flow measurement is key for obtaining the exhaust flow at all the pots. The most accurate technique is measuring gas velocity with a Pitot tube in the pot exhaust duct, but gas turbulence is often excessive at these locations, thus erroneous measurements may be obtained unless multi-point traversing is precisely done. Also, the procedure to obtain flow values with a Pitot tube can take a considerable amount of time and often requires the use of a specialized lift vehicle.

An easier and much quicker approach can be done by measuring the static pressure at the pot discharges, in a location prior to the balancing damper. This method assumes that fluid resistance and heat transferred to the gas are similar in all pots. As such, the static pressure upstream of the balancing dampers shall be the same for all the pots for a given volumetric flow. Refer to dP_{0-A} in Fig. 2.

The relation between static pressure at the pot discharge and the volumetric flow can be obtained by inducing different flows in a few pots. See an example in Fig. 5 for the actual linear relation obtained in one smelter.

Assuming that a constant amount of heat is transferred to the gas at the pots, another relation can also be obtained

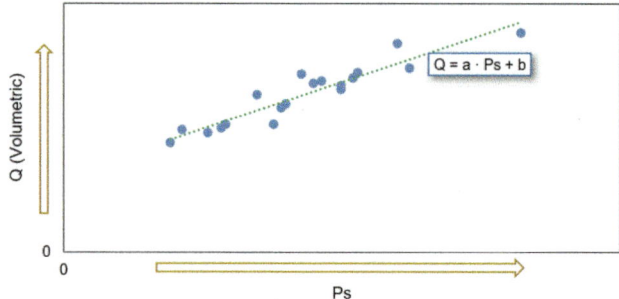

Fig. 5 Relation between static pressure (Ps) and volumetric flow (Q) at pot discharge obtained in one smelter

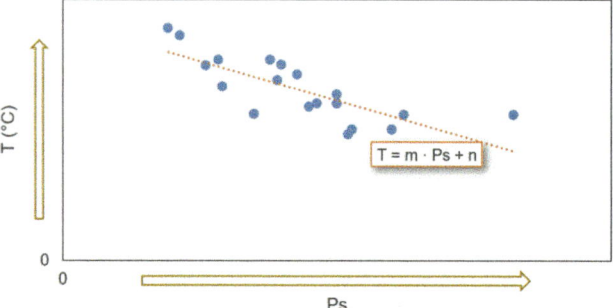

Fig. 6 Relation between static pressure (Ps) and gas temperature (T) at pot discharge obtained in one smelter

between the static pressure and the gas temperature at the pot exhaust. Refer to Fig. 6. Note that gas temperature alone is insufficient for effective balancing of pot drafts. The amount of heat lost through the top crust of individual pots can vary widely in practice.

With the temperature and volumetric flow values as a function of the static pressure measured at the pot discharge, mass flows (or normalized flows) can be calculated. These relations shall be recalculated if any significant parameter is changed such as pot amperage or ambient temperature.

The instrumentation that is used for the measurement of static pressure is also important. The easiest is using a digital manometer but these can be damaged or provide erroneous measurements due to the strong magnetic fields in the potrooms, so they must be tested in advance. Water-type or U-tube manometer with magnetic mounts on the back are well suited to the task, but these might be forbidden from being used in the potrooms at some locations. An inclined manometer is also acceptable but requires attention to proper leveling at each pot, which can be time consuming. Other commonly used instruments such as analog manometers are not suited to the task since the magnetic fields will negatively impact their function.

Pot Flow Balancing Procedure

As the parameters affecting the pot exhaust flow rate are subject to change over time, pot flow balancing shall be done regularly, at least once a year. To this end, the following pot flow balancing procedure is proposed:

1. Obtain the current flows with the method described in Section "Measurement of Pot Flows";
2. Set the value of normalized or mass flowrate (static pressure) for each pot;
3. Modify the positioning of the pot dampers starting with the pots closest to the GTC and sequentially continuing towards the furthest ones. This sequence is important as modifications on the closest pots to the GTC have the most impact on the system balancing. An example of this methodology can be found in Fig. 7.
4. Repeat measuring the pot flows and iterate the balancing procedure as needed. Experience has shown that the preferred approach to reduce iterative passes is to restrict flow on pots with high static pressure measurements on the first balancing passes. On subsequent passes, individual pot dampers can then be increased or decreased until at least 90% of all individual pots are within an acceptable range of target flow rates, often being ±0.1 Nm^3/s/pot.

Note that this procedure should be ideally done while having all the covers installed on all of the pots that are connected to the GTC. Also, the setpoint of GTC fans shall remain unchanged while balancing. Fan setpoints might have to be adjusted once the balancing is finished.

Graphs in Fig. 8 show a simulation of the evolution of the pot flows when applying the balancing procedure in a system with initially no restrictions at the pots. As it can be observed, pot flows are sequentially "pushed" towards the pots at the extremities.

For smelters where orifice plates are installed at the pot discharges, iterative steps can become a daunting and tedious task. One possibility is to add temporary blades that partially obstruct the ducts acting as guillotine dampers. Once the balancing process is completed, orifice plates with the same restriction as the temporary blades can be installed.

A more common practice for smelters with orifice plates is known as patterning. In such cases, iterative steps are taken on just one branch duct of the fume collection system, on perhaps twenty pots. When an acceptable pattern of orifice placements has been achieved on this branch, it is then copied and/or mirror-imaged to other branches of the GTC. Note that this method may not be able to be transferred from one GTC to another unless the two systems are essentially identical in design and in total flow rate. When significant differences exist between GTCs each one will require its own patterning exercise.

Conclusion

Pot flow balancing is a basic practice for an optimum performance of the pot ventilation system. Measurement of the static pressure at the pot discharge can be used as a simple method for obtaining a quick and reliable value of the drafted flow at each pot. These measurements can then be used to get a picture of the balancing condition of the pot

Fig. 7 Proposed sequence for pot flow balancing in a sample smelter

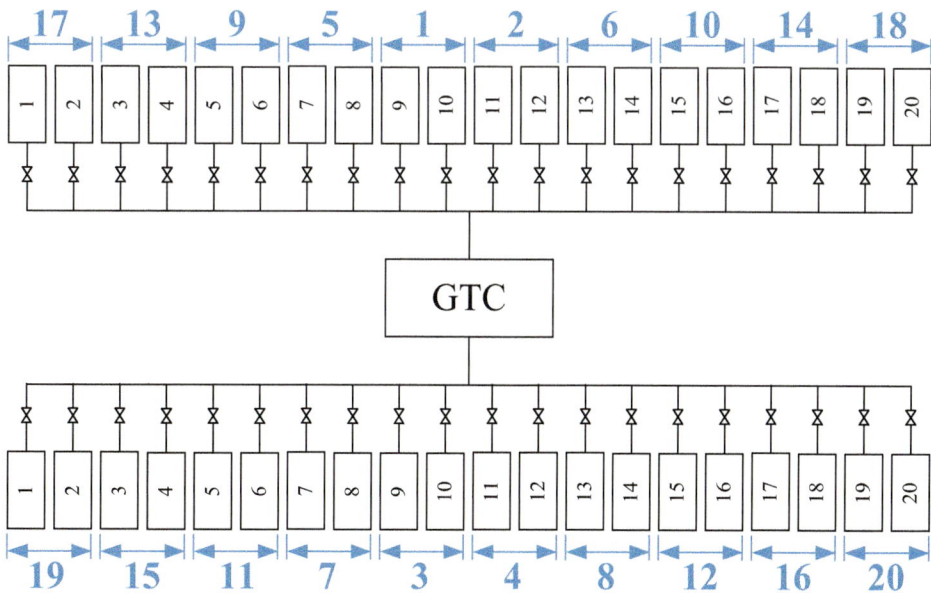

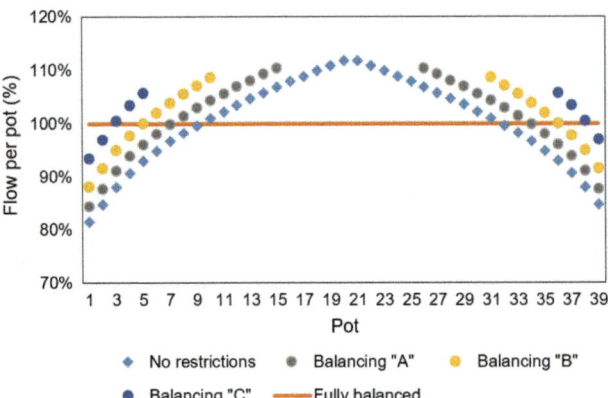

Fig. 8 Representation of pot flows during balancing procedure

flows in a GTC duct system, as well as for balancing the pot flows.

Flow balancing is also an essential practice for being able to obtain the best performance of overall fume capture for any GTC. Improper balancing of individual pot drafts often drives GTC operators to attempt various means to increase system flow rates in order to reduce average roof emission rates. At the very least, this requires additional energy consumption. It can also lead to system modifications and capital expenditures that may not be the wisest way to allocate precious resources.

It must also be noted that changes to GTC system conditions will shift the balance of individual pot exhaust flow rates. In multiple case studies, system operators have pursued higher system flow rates in order to reduce potline roof emissions or to offset some other change to operations such as increased amperage load and thermal input to reduction

cells. Changes in total flow rate and gas temperature affect individual pot exhaust rates and must be considered when exercises such as production de-bottlenecking are undertaken.

Pot draft balancing often comes down to minimizing the impact of the weakest links in the system in order to achieve maximum performance in terms of emissions. It has been observed that individual pots with exhaust flow rates that are below target may emit four times as much emission to the roof compared to the pots with normal draft rates. In extreme cases, individual pots or groups of pots affected by duct accumulations and cracks or gaps in ducting or expansion joints can drive up the total emission rate that is reported for an entire room or potline. Regular checks for acceptable overall balancing remain a necessary aspect of potline operations.

References

1. Edgar Dernedde, "Gas Collection Efficiency on Prebake Reduction Cells", Am. Ind. Hyg. Assoc. Journal 51(1): 44-49 (1990)
2. Michel Sahling, Elmar Sturm, "Improvement of Pot Gas Collecting Efficiency by Implementation of Impulse Duct System", Light Metals 2004
3. Stephan Broek; Stephen J. Lindsay; Dr. Neal R. Dando; Alain Moras, "Considerations Regarding High Draft Ventilation as an Air Emission Reduction Tool", Light Metals 2001
4. Jean-Nicolas Maltais, Michel Meyer, Mathieu Leduc, Guillaume Girault, Hyacinthe Rollant; "Jet Induced Boosted Suction System for Roof Vent Emission Control: New Developments and Outlooks", Light Metals 2012
5. W. Brooks, G. Girault & M. Meyer, "Investigating Potential Solutions to Reduce Fluoride Emissions", 9th Australasian Smelting Technology Conference, 2007

Evaluation of Methodologies for Assessment of SO₃ Concentration in Industrial Off-Gas

Thor Anders Aarhaug, Ole Kjos, Morten Isaksen, and Jan Olav Polden

Abstract

Assessment of acid dew point in industrial off-gas has become increasingly important with strategies for heat recovery and the use of more concentrated gas compositions. Several approaches to acid dew point assessment make use of glassware not compatible with the high fluoride concentrations found in unfiltered gas. In this work, a new and simplistic approach to acid dew point assessment is evaluated side by side with conventional methodology. The new approach captures SO_3 in the gas as sulfate onto a solid NaCl trap. Interference from SO_2 is avoided since it passes through the NaCl trap unreacted. The trap is then dissolved in water for direct analysis by ion chromatography. The importance of isokinetic sampling was evaluated, and no significant difference in results was observed. This suggests that the acid droplets are small. The method is shown to be promising for acid dew point assessment in industrial off-gas.

Keywords

Acid dew point • Off-gas monitoring • Ion chromatography

T. A. Aarhaug (✉)
SINTEF Industry, 7034 Trondheim, Norway
e-mail: taarhaug@sintef.no

O. Kjos
SINTEF Industry, 0314 Oslo, Norway
e-mail: ole.kjos@sintef.no

M. Isaksen
Hydro Aluminium, 3936 Porsgrunn, Norway
e-mail: morten.isaksen@hydro.com

J. O. Polden
Hydro Aluminium, 6600 Sunndalsøra, Norway
e-mail: jan.olav.polden@hydro.com

Introduction

The primary off-gas composition in aluminium production has been extensively studied. Whereas the absolute concentration levels may change with air dilution, the relative concentration between gas constituents remains fairly constant [3]. Less studied is the presence of SO_3 in the off-gas, this is because conventional methods for assessment involve the use of glassware that are not compatible with a high concentration of HF in the off-gas.

Sulfur trioxide in off-gas is believed to be the result of oxidation of SO_2. Conversion is strongly temperature dependent, slow due to kinetics [15], and is often assumed to be in the range of 1 to 5% [14]. Conversion is further known to be catalyzed by impurities like V and Fe known to be present in the carbon anodes [9]. SO_3 is hygroscopic and at temperatures below 400 °C will react with water to form sulfuric acid.

The correlation between SO3 concentration and acid dew point depends on the SO_3 concentration and humidity [13, 18]. More recently, machine learning was applied to improve the prediction [19]

$$\theta_{dew} = 150 + 8.1328 \cdot \ln\left(p_{H_2O}\right) + 11.664 \cdot \ln\left(p_{SO_3}\right) - 0.38226 \cdot \ln\left(p_{H_2O}\right) \cdot \ln\left(p_{SO_3}\right)$$

where partial pressures are in mm Hg and temperature in °C. Traditionally, SO_3 in off-gas has not been a problem since concentrations in the gas translate to acid dew point values lower than the actual temperature of the gas and the duct surfaces: corrosion problems are not common in off-gas ducts. With the introduction of heat exchangers to the off-gas, knowledge of the dewpoint becomes critical if corrosion protection is to be avoided.

Future scenarios for primary aluminium production with carbon as a reducing agent must include efforts to capture CO_2 emitted. With today's concentration in the off-gas in the range of 0.5–1% capture is difficult to apply. Efforts to

© The Minerals, Metals & Materials Society 2023
S. Broek (ed.), *Light Metals 2023*, The Minerals, Metals & Materials Series,
https://doi.org/10.1007/978-3-031-22532-1_99

increase the CO_2 concentration in the off-gas have been demonstrated [10]. More recently, efforts have been made to reduce the air draught by recycling the air and in this way increasing the concentration of CO_2 in the gas. The implications have been discussed [16], and gas recycling will also increase the concentration of HF and SO_2 in the gas, thus exacerbating the problem with acid dew point if heat exchangers are applied to recover the heat. For downstream CO_2 capture with, e.g., amine technology, the presence of sulfuric acid aerosols could lead to losses of absorbent due to the formation of amine mist [11].

The most widely accepted method for sulfur trioxide analysis is the controlled condensation method [4]. The method is described in more detail in [8]. In short, sampled gas is cooled in a condensation column to a temperature higher than the water dew point but lower than the acid dew point. In this way, sulfuric acid is collected as condensate on the column walls and the sintered glass filter is used at the end of the column. By avoiding condensation of water, interference from SO_2 is minimized. After sampling, the glassware is washed with a 5% vol isopropanol solution that is either analyzed by ion chromatography or by titration. Due to the high hydrogen fluoride content in the off-gas, the use of glassware is not suitable for the application discussed in this paper.

Another commonly used method for SO_3 determination is the isopropanol bottle method [12]. Gas is passed through gas bubblers filled with 80% vol isopropanol to capture SO_3. In order to avoid evaporation of absorbent, the gas bubblers must be cooled to 0 °C with ice or even colder temperatures with a cryostatic bath. Most SO_2 will pass through the bubblers, and for concentration estimation, additional gas bubblers filled with 3% vol hydrogen peroxide can be used. Analysis of the SO_3 can be performed with ion chromatography, inductively coupled plasma mass spectrometry or titration. Since there is some solubility of SO_2 in 80% isopropanol, slow oxidation of SO_2 dissolved sulfite to sulfate could interfere with the sulfuric acid estimate. Bubbling of absorbent with inert gas and immediate analysis after sampling are arguments that could minimize the interference from SO_2. The advantage of the ion chromatographic method is that it allows for the evaluation of both sulfite and sulfate concentrations in the absorption solution.

Commercially available condensation probes have previously been available from Breen and Land. The working principle for the probes is to actively cool the probe with, e.g., compressed air so that when it reaches the dew point of the acid, the condensed acid will result in conductance between electrodes embedded on the surface. Thermocouples at the surface are used to estimate the accurate acid dew point. Problematic for application in primary aluminium production is the fact that the probes are made from quartz

glass that will degrade in the hydrogen fluoride atmosphere. SINTEF made an attempt to modify a Land Instruments ADM 200 probe using polymeric glass in the probe [1]. While the replacement of pyrex glass was successful it turned out that the heat conductivity of the polymeric glass was too low for efficient cooling with compressed air: the probe could only be cooled to about 80 °C which was not sufficient to estimate the acid dew point.

More recently, a method using solid absorbent has been revisited [5]. The principle of the method is to use a salt trap where gaseous sulfuric acid reacts with the sodium chloride to form sodium sulfate and bisulfate according to the reactions

$$2NaCl(s) + H_2SO_4(g) \rightarrow Na_2SO_4(s) + 2HCl(g)$$

$$2NaCl(s) + H_2SO_4(g) \rightarrow NaHSO_4(s) + HCl(g)$$

SO_2 in the gas does not react with the salt and will pass through the trap. Since condensation of water would dissolve the salt and potentially solvate SO_2, the temperature of the trap is kept well above the dew point of water. The salt is trapped in a Teflon tube by glass- or Teflon wool if hydrogen fluoride is present. After sampling, salt is dissolved in de-ionized water and analysed by ion chromatography. Evaluation of the use of different salts has been documented [17]. The method was found to be in good agreement with the controlled condensation method [6].

In this work, the salt method has been directly compared with the isopropanol absorption bottle method for the characterization of the acid dew point in the off-gas at the Hydro Aluminium Smelter at Sunndalsøra. The ion chromatograph at the smelter laboratory was used to analyze the samples immediately after sampling.

Experimental

The sampling of gas was performed at a regular sampling point, and a standard isokinetic probe from the smelter was used. Two gas treatment centers were sampled. A sampling pump with a capacity of about 1 m^3 was used, and the total volume sampled was controlled with a calibrated gas meter and a thermometer for temperature compensation. A 142 mm filter (PALL Versapor 1.2 µm) was installed in the probe, but the probe was installed looking away from the gas direction. The justification for this was that particulate buildup on the filter could potentially trap SO_3 onto alumina. On the other hand, omitting filter installation could result in particulate material containing sulfate being included in the analysis and thus constituting a positive error. The "look away" installation of the probe was therefore considered to be the best strategy.

For isopropanol sample bottle method, two (three in the case of using ice) gas washing bottles in plastic material were used in sequence. Isopropyl alcohol (p.a.) was diluted to 80% vol by de-ionized water. A volume of 400 mL was used in the first bottle. The second (third) bottles were left empty. The bottles were either cooled with a cryostat to −20 °C or with water, ice, and salt.

For the salt method, about 2 g of NaCl (suprapur) were trapped in a ¾″ PFA tube by means of Teflon wool. In order to avoid segregation, the trap was installed in a vertical position. The trap was heated with a jacket to a temperature of about 200 °C in order to avoid condensation of water that could dissolve the salt trap as well as to provide temperature for the conversion of the SO_2 to sulfate reactions taking place [17]. After the samples were returned to the laboratory, they were dissolved in 1.00 L of de-ionized water in a volumetric flask. They were also bubbled with inert gas in order to remove SO_2 that could potentially oxidize to sulfate and interfere with the analytical result (Fig. 1).

All samples were analyzed by ion chromatography. A Metrohm 883 Basic IC plus with an 858 Professional sample processor was used for analysis. A MetroSep A Supp 5 150/4.0 column was used with a 3.2 mmol/L sodium carbonate, 1.0 mmol/L sodium bicarbonate buffer. A conductivity detector was used. Analysis was performed as quickly as possible after sampling in order to avoid oxidation of sulfite in the sample. The IC was calibrated with sulfate standards covering the concentration range found in the samples.

The fluoride and sulfate content of the dust collected onto the filters were analysed. For fluoride, an ion-selective electrode was used for analysis [2]. For sulfate, ion chromatography was used as described earlier in this paper after boiling particulate material in sodium carbonate (10% vol) for 20 min.

The SO_2 concentration in the off-gas at GTC3 was estimated with FTIR (ProtIR 204 M, 6.4 m path) to be 119 ppm.

Results

All samples were analysed by ion chromatography. An example of a chromatogram is shown in Fig. 2. For sampling in isopropyl alcohol, a sulfite peak was observable in all samples. This implies that even though the purging of inert gas was performed before analysis, there are still contributions from SO_2 as sulfite in the solution. The contribution from the sulfite already converted to sulfate in the samples at the time of analysis is unknown, but clearly, the time between sampling analysis is important. For sampling with a salt trap, no sulfite peak was recorded. This suggests that the interference from SO_2 in this method is less of a problem.

The results are summarized in Table 1. From GTC3 five samples were collected: two with isopropanol and cryostat cooling, one with isopropanol and ice/salt cooling, and one sample in a salt trap. From GTC 1, two samples in isopropanol and cryostat cooling were collected.

For sampling in isopropanol, a significant loss of absorbent from the original 400 mL was observed. Some were recovered in the downstream bottles but overall losses still significant. Comparing ice versus cryostat, the losses were similar even when considering an additional gas washing bottle used in the case of ice/salt. It was also observed that the concentration of sulfate in the bottles was quite similar.

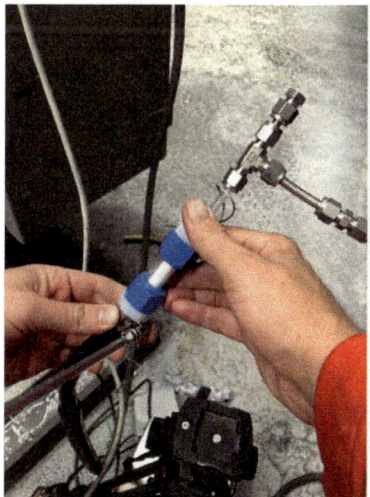

Fig. 1 Sampling of SO₃ with salt and isopropanol. The NaCl trap is shown to the left; connected to the sampling probe and with a heating jacket installed in the middle. To the right is shown a sampling of two gas washing bottles immersed in a cryostat cooled bath with a temperature of −20 °C

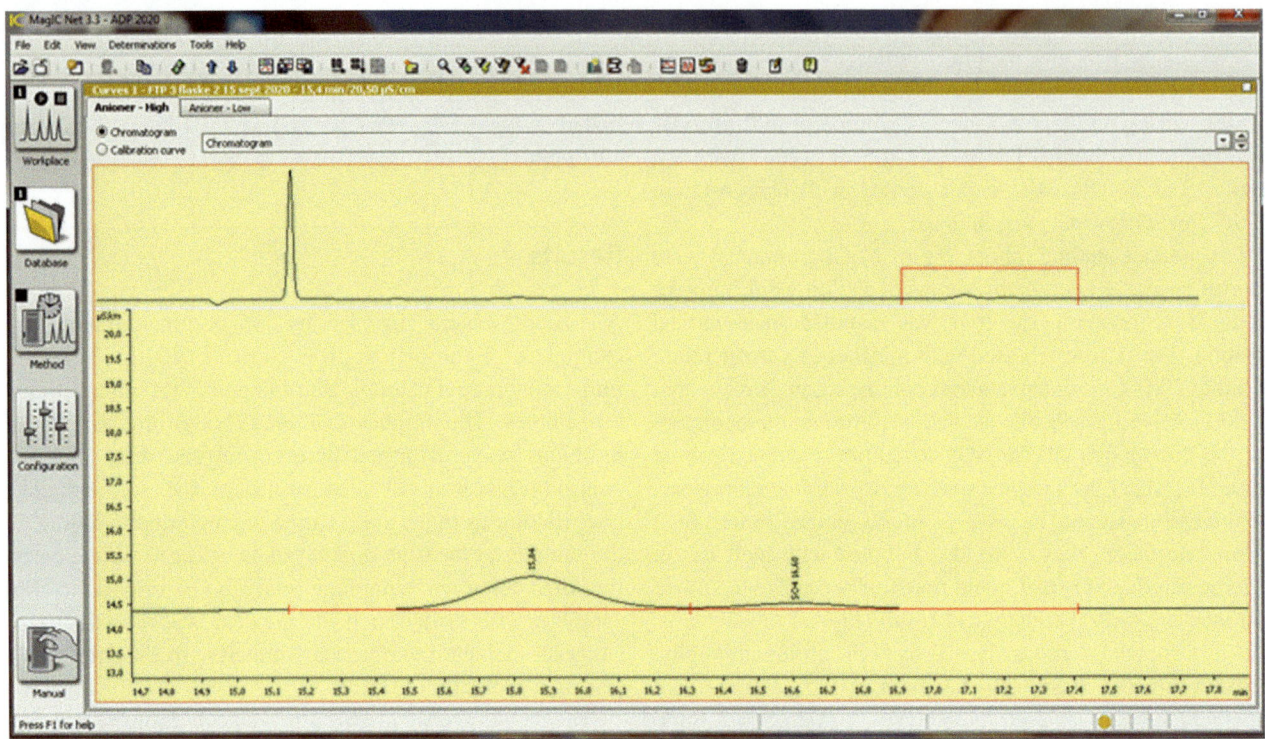

Fig. 2 Example ion chromatogram showing sulfite (ret. 15.84 min) and sulfate (ret. 16.60 min). Baseline conductivity was around 14.4 µS/cm

Table 1 Analytical results of ion chromatographic analysis. Volume is the sampled volume of gas corrected for gas meter calibration and corrected for temperature

Location	Bottle	Cooling	SO4 (mg/L)	Volume (L)	SO4 (mg)	Volume (m3)	SO4 (mg/m3)	SO4 (ppm)	Mean (ppm)
GTC3	2	Cryo	6.58	0.03	0.16				0.37
	1		8.12	0.33	2.68	1.77	1.60	0.41	
GTC3	2	Cryo	4.32	0.12	0.52				
	1		15.39	0.22	3.38	1.75	2.23	0.57	
GTC3	2	Ice	5.01	0.02	0.10				
	1		7.38	0.30	2.18				
	3		5.92	0.01	0.03	1.81	1.27	0.32	
GTC1	2	Cryo	5.26	0.10	0.53				
	1		6.17	0.26	1.57	1.55	1.36	0.35	
GTC1	2	Cryo	2.90	0.06	0.18				
	1		3.84	0.29	1.11	1.55	0.83	0.21	
GTC3			0.47			0.37	1.28	0.33	0.33
			0.49			0.37	1.32	0.34	

The overall concentration of sulfate in the samples with isopropanol sampling was estimated to be 0.37 ppm. The relative standard deviation is 36%, partly owing to the slightly lower average observed for GTC 1 than for GTC 3. No significant difference between using cryostat and ice/salt for cooling was found. For the two parallel analyses of the salt trap sample, a value of 0.33 ppm was estimated. The results indicate a very good correlation between the two methods and are in support of the fact that any error from dissolved SO_2 in the isopropanol is negligible when analyzing samples almost immediately after sampling.

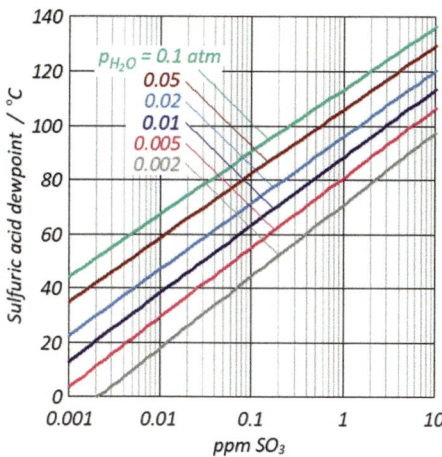

Fig. 3 Acid dew point estimation based on [19] plotted by [15]

estimation is based on [19] plotted by [15]. For the graph, the dew point was estimated to be in the range of 77 to 79 °C.

The dust collected onto the filter was analyzed for fluoride and sulfate according to the standard methodology for the smelter. The fluoride concentration increased with the increasing mass of particulate matter. For sulfate, the amount of sulfur did not appear to be a function of mass. The data is shown in Fig. 4.

This observation reflects the challenge of sampling particulate matter while sampling SO₃: sampling isokinetically would potentially trap SO₃ onto the particulate matter while sampling without a filter could result in contributions from sulfate on alumina in the analysis. The look away probe is a compromise where smaller particulate matter mostly consisting of fluorides [7] are predominantly collected onto the filter. These fluorides do not exhibit absorption capabilities similar to that of smelter-grade alumina for fluoride or sulfur. Therefore, the analysis of the particulate matter shows a mass dependence on fluoride but not on sulfate that is not trapped in the fluoride particles.

Taking an average value of 0.35 ppm as the SO₃ concentration found in the off-gas, and assuming a water content of between 1–1.2% vol, the acid dew point was estimated. Dew point values are plotted in Fig. 3. Acid dew point

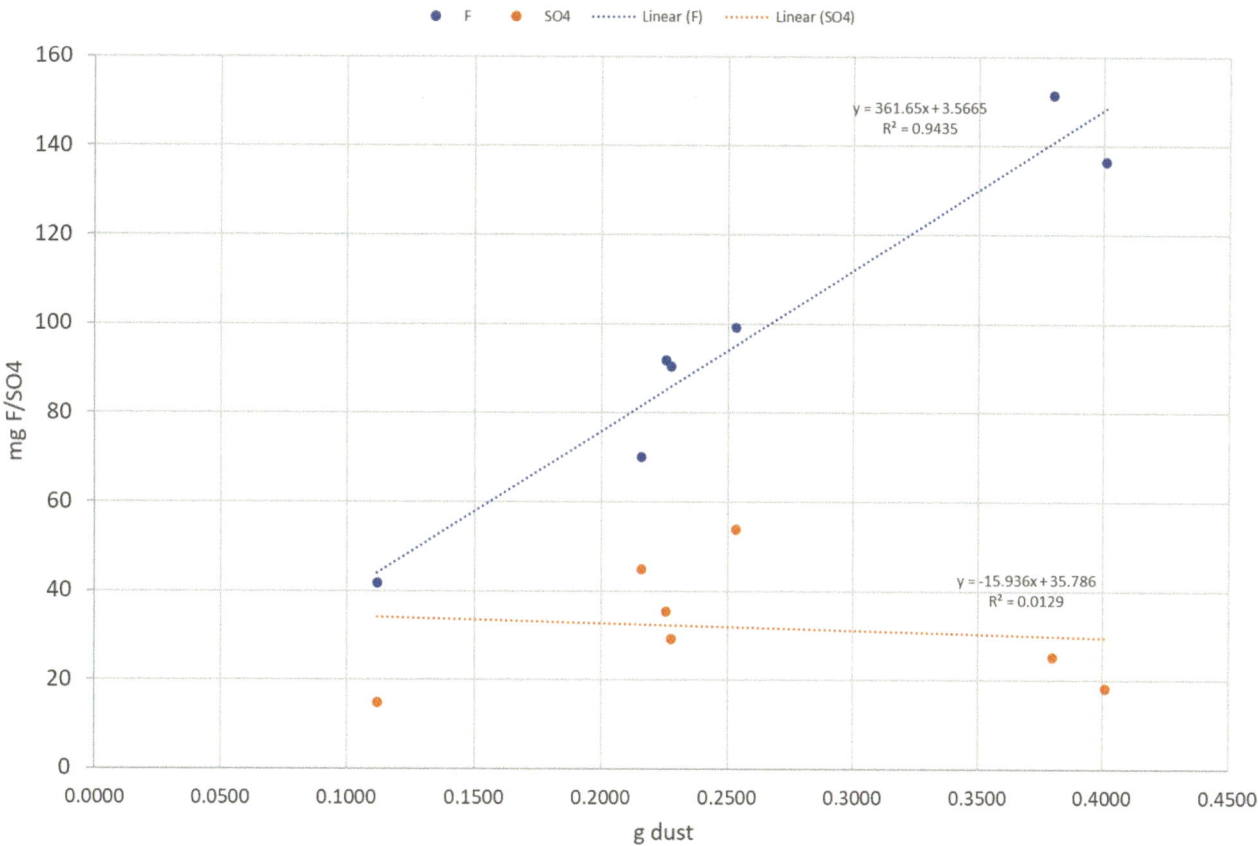

Fig. 4 Collected particulate matter analyzed for fluoride and sulfate

Discussion

Two methods for assessment of acid dew point in off-gas from primary aluminium production have been evaluated. The methods give fairly similar results, with an average of 0.35 ppm. Assuming a water content of 1–1.2% vol in the gas, this translates to an acid dew point temperature between 77 to 79 °C. The salt trap sampling method was found easy to conduct. For sampling in 80% vol isopropanol, some oxidation of SO_2 to sulfate seems to be difficult to avoid unless immediate analysis of the samples is performed. This is despite the purging of the sample with inert gas. Nevertheless, the two methods compared are in very good agreement, suggesting that immediate analysis of isopropanol after sampling minimizes the problem of having contributions to the sulfate concentration from dissolved SO_2 being oxidized in the solution.

When the SO_3 concentration value of 0.35 ppm is compared to the measured value of SO_2 in the off-gas of 119 ppm, this means that the conversion of SO_2 to SO_3 is less than 0.3%. This is much lower than the value of between 1 and 5% often assumed.

Sampling was not performed isokinetically, but using a look-away orientation of the probe. This was done to avoid SO_3 from the particulate matter collected onto the filter. Analysis of the particulate matter indicates that the sulfate concentration is not a function of the particulate mass. This indicates that sulfates are not trapped by the filter, thus affecting the sulfate concentration and acid dew point estimate. Further work should be performed in order to find a good strategy for sampling SO_3 in the presence of particulate matter.

Acknowledgements Hydro Aluminium is acknowledged for providing access to their smelter and laboratory services. This work has been conducted as part of SFI Metal Production, NFR grant 237738.

References

1. Aarhaug, T. A., & Kjos, O. (n.d.). Sulfuric acid determination using condensation probe. *Unpublished Results.*
2. Aarhaug, T. A., Nagy, K., & Smith, K. G. (2012). Potentiometric fluoride analysis with improved analytical performance. *TMS Light Metals.*
3. Aarhaug, T. A., & Ratvik, A. P. (2019). Aluminium Primary Production Off-Gas Composition and Emissions: An Overview. *JOM, 71*(9), 2966–2977. https://doi.org/10.1007/s11837-019-03370-6.
4. *BS 1756–4 Methods for sampling and analysis of flue gases part 4: miscellaneous analyses.* (1977).
5. Cooper, D., & Ferm, M. (1994). *Jämtförelse av mätmetoder for bestämning av SO3- koncentrations ir rökgaser.*
6. Fleig, D. (2012). *Experimental and modeling studies of sulfur-based reactions in oxy-fuel combustion.* Chalmers univeristy of technology.
7. Gaertner, H., Ratvik, A. P., & Aarhaug, T. A. (2013). Trace element concentration in particulates from pot exhaust and depositions in fume treatment facilities. *TMS Light Metals, Light Metals 2013-At the TMS 2013 Annual Meeting and Exhibition.* https://doi.org/10.1002/9781118663189.ch131.
8. Gustavsson, L., & Nyquist, G. (2005). *Värmeforsks mäthandbok* (3rd ed.). Värmeforsk Service AB.
9. Kramer, M., Schubert, M., Lautensack, T., Hill, T., Körner, R., Rosowski, F., & Zühlke, J. (2012). *Catalyst for the oxidation of SO2 to SO3* (Patent No. US 8,323,610 B2).
10. Lorentsen, O. A., Dyrøy, A., & Karlsen, M. (2009). Handling CO2EQ from an aluminum electrolysis cell. *TMS Light Metals,* 263–268. https://doi.org/10.1007/978-3-319-48156-2_144.
11. Mertens, J., Bruns, R., Schallert, B., Faniel, N., Khakharia, P., Albrecht, W., Goetheer, E., Blondeau, J., & Schaber, K. (2015). Effect of a gas–gas-heater on H2SO4 aerosol formation: Implications for mist formation in amine based carbon capture. *International Journal of Greenhouse Gas Control, 39,* 470–477. https://doi.org/https://doi.org/10.1016/J.IJGGC.2015.06.013.
12. Method 8 Determination of sulfuric acid and sulfur dioxide emissions from stationary sources. (2019). *US EPA.*
13. Okkes, A.G.; Badger, B. V. (1987). No Title. *Hydrocarbon Processing, 66,* 53.
14. Schnelle Jr, K., Dunn, R., & Ternes, M. (2015). *Air pollution control technology handbook.* https://books.google.com/books?hl=no&lr=&id=5Hm9CgAAQBAJ&oi=fnd&pg=PP1&ots=27i_wNvaYL&sig=kWBhQcA2J2pvgwCADCN0C5KNBYw.
15. Solheim, A. (2022). An Attempt to Estimate the Sulfuric Acid Dew Point in the Flue Gas from Aluminium Electrolysis Cells. *Minerals, Metals and Materials Series,* 520–527. https://doi.org/10.1007/978-3-030-92529-1_69/COVER.
16. Solheim, A., & Senanu, S. (2020). Recycling of the Flue Gas from Aluminium Electrolysis Cells. *Minerals, Metals and Materials Series,* 803–810. https://doi.org/10.1007/978-3-030-36408-3_107/COVER.
17. Vainio, E., Fleig, D., Brink, A., Andersson, K., Johnsson, F., & Hupa, M. (2013). Experimental evaluation and field application of a salt method for SO 3 measurement in flue gases. *Energy and Fuels, 27*(5), 2767–2775. https://doi.org/https://doi.org/10.1021/EF400271T/ASSET/IMAGES/LARGE/EF-2013-00271T_0013.JPEG.
18. Verkhoff, F.H.; Banchero, J. T. (1974). No Title. *Chemical Engineering Progress, 70*(8), 71.
19. Zarenezhad, B., & Aminian, A. (2011). Accurate prediction of the dew points of acidic combustion gases by using an artificial neural network model. *Energy Conversion and Management, 52*(2), 911–916. https://doi.org/https://doi.org/10.1016/J.ENCONMAN.2010.08.018.

Mathematical Modelling of the Desulfurization of Electrolysis Cell Gases in a Low-Temperature Reactor

Arash Fassadi Chimeh, Duygu Kocaefe, Yasar Kocaefe, Yoann Robert, and Jonathan Bernier

Abstract

SO_2 is one of the main sources of acid rain and air pollution. Semi-dry sorbent injection, using powdered alkaline sorbents, is an effective means of removing SO_2. Since no costly additional equipment is needed, the operating cost is lower, and it is a more economical and efficient process compared to wet and dry desulfurization processes. The reaction between sorbent (hydrated lime, $Ca(OH)_2$) and SO_2 is dominated by the adsorption step. In this study, a mathematical model has been developed to simulate the lab-scale desulfurization reactor employed for the low-temperature gases containing low SO_2 concentration coming from the electrolysis cells used for aluminum production. A parametric study was carried out in order to examine the effects of certain parameters, such as inlet SO_2 concentration, sorbent flow rate, and relative humidity of the gas on the desulfurization efficiency. The model and some of the results are presented in this article.

Keywords

SO_2 removal · Semi-dry desulfurization · Aluminum electrolysis · Computational fluid dynamics (CFD)

A. F. Chimeh · D. Kocaefe (✉) · Y. Kocaefe
Research Chair On Industrial Materials (CHIMI), University Research Centre on Aluminium (CURAL), Aluminium Research Center (REGAL), University of Quebec at Chicoutimi, 555 University Blvd., Chicoutimi, QC G7H 2B1, Canada
e-mail: Duygu_Kocaefe@uqac.ca

Y. Robert · J. Bernier
Arvida Research and Development Centre (ARDC), Rio Tinto, 955 Boulevard Mellon, Jonquière, QC G7S 4K8, Canada

Introduction

The aluminium demand is expected to grow by 4.2% per year till 2050 due to increased demand in the construction, transport, and renewable energy sector [1]. The sulfur-containing gases (SO_2 and SO_3) are emitted from aluminum smelters. These gases may react with water vapor in the air and produce H_2SO_3 and H_2SO_4, which are major contributors to acid rain [2]. In addition to being detrimental to the environment, acid rain is extremely harmful to the health of humans and animals. Also, it affects aquatic life because it contributes to the toxicity of water resources. Another severe consequence of acid rain is the deterioration of historical and ancient buildings [2].

Environmental laws restrict the total emission of contaminants through certain regulations. For instance, 88/609/CEE and 2001/80/CEE aim to control the emissions from fossil fuels coming from industrial plants [3]. Therefore, the industries have to meet the maximum allowable concentration of SO_2 present in the gases exhausted from the stacks [4].

Aluminum is a metal widely used in a large number of applications, including transportation, construction, etc. due to its distinct properties [5]. Canada is one of the major aluminum manufacturing and exporting countries. In 2020, Canada produced approximately 3.2 million tonnes of aluminum and 90% of this was produced in Quebec [6]. The source of sulfur in gases emitted from the electrolysis cells is the raw materials, especially coke, used in anode production. Since the quality of anode raw materials is decreasing, their sulfur content is increasing [7]. Rio Tinto, which has six smelters in Quebec and one in British Columbia, is one of the major aluminum producers in the world. It continuously searches for ways to further reduce its emissions [4].

There are three types of desulfurization processes: dry, wet, and semi-dry [4, 5]. In a wet process, sulfur oxides are scrubbed by passing the gas through a large quantity of solution containing a solute (mostly Na). This process is capital-intensive. Its operation and maintenance costs are high. The solvent has to be neutralized and recovered [8], and the process creates low-quality by-products [7]. In the dry process, the solid particles are injected into the gas stream. The dry process has some disadvantages as well, including low desulfurization efficiency and excessive sorbent utilization [8]. In the semi-dry process, particles suspended in water are injected into the gas. It is reported in the literature that the presence of humidity in gas increases the reaction rate of the sorbent (solid particles) with SO_2. An increase of 14% in SO_2 removal efficiency compared to that of the dry process was observed by some researchers [9].

Contrary to the emissions in thermal plants, gases released from the aluminum smelters contain a much lower SO_2 concentration at somewhat lower temperatures. The elimination of SO_2 at such a low concentration level is complex and requires further attention [4, 10]. Studies focusing on the removal of SO_2 from effluent gases in the smelters are quite rare.

Modelling is an effective tool to determine the design and operational conditions of a process as it reduces the cost. Industrial processes ranging from a simple water air pipe [11] to complex absorbers and reactors [12] mostly contain more than one phase (multi-phase systems), which is the case for the current system.

In this study, a mathematical model of a desulfurization reactor is being developed. SO_2 is removed using hydrated lime $Ca(OH)_2$, therefore, it is a multi-phase system. The main objective is to remove SO_2 from the gas stream as efficiently as possible. The system is isothermal and turbulent. The commercial code Ansys Fluent was used to solve the equations. The kinetic equation was introduced via the UDF (user-defined function). The effect of gas relative humidity on SO_2 removal is demonstrated, which enhanced the removal efficiency by 8% in some cases.

Methodology

System (Laboratory Reactor)

Figure 1 gives the geometry and shows the three domains: filter and the two domains on both sides. Gas inlet, gas outlet, and filter are indicated in the figure as well.

Governing Equations

The model is based on the two-phase Eulerian-Eulerian strategy. Each of the existing phases, i.e. continuous phase (gas mixture of air and SO_2) and dispersed phase (hydrated lime and the reaction product) considered as separate flow domains. The system is isothermal. The temperature is taken as 70 °C. A porous domain is defined to represent the filter.

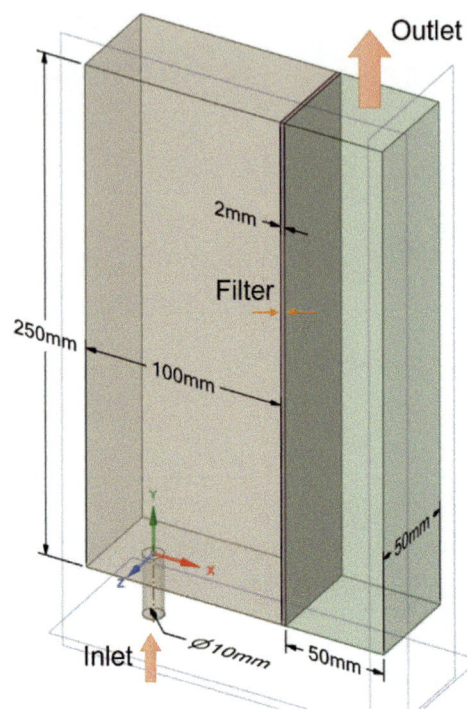

Fig. 1 System: laboratory reactor

The continuity and momentum equations are shown in Eqs. (1) and (2), respectively.

$$\frac{\partial \rho}{\partial t} + \nabla \cdot \left(\rho \, \vec{u} \right) = 0 \tag{1}$$

$$\frac{\partial \left(\rho \, \vec{u} \right)}{\partial t} + \nabla \cdot \left(\rho \, \vec{u}\vec{u} \right) = -\nabla p + \nabla \cdot \left(\overline{\overline{\tau}} \right) + \rho \, \vec{g} + \vec{F} \tag{2}$$

where ρ is the fluid density (kg/m^3), u is the fluid velocity (m/s), and p is the static pressure (Pa). $\rho \, \vec{g}$ and $\vec{F}$ describe the gravitational body forces and external body forces (N), respectively. $\vec{F}$ is the source term representing porous media or any other model dependent source terms. Stress tensor $\left(\overline{\overline{\tau}} \right)$ is defined as shown in Eq. (3).

$$\overline{\overline{\tau}} = \mu \left[\left(\nabla \vec{u} + \nabla \vec{u}^T \right) - \frac{2}{3} \nabla . \vec{u} I \right] \tag{3}$$

where μ is the molecular viscosity and I is the unit tensor [13].

In this process, hydrated lime is injected into the SO_2-containing gas stream. When the hydrated lime mixes with SO_2, the reaction shown in Eq. (4) takes place [4, 12].

$$Ca(OH)_2 + SO_2 \rightarrow CaSO_3 \cdot \frac{1}{2} H_2O + \frac{1}{2} H_2O \tag{4}$$

In reality, part of the calcium sulfite ($CaSO_3$) oxidizes to form calcium sulfate ($CaSO_4$). Thus, the product is a mixture of the two. But, it is taken as sulfite in order to simplify the problem.

The species transport equation shown in Eqs. (5) and (6) is used for simulating the reacting flow in a volumetric reaction model. It was assumed that the reaction takes place in a single phase [13].

$$\frac{\partial(\rho\omega_i)}{\partial t} + \nabla \cdot \left(\rho\omega_i \vec{u}\right) = -\nabla \cdot \left(\vec{J}_i\right) + R_i + S_i \quad (5)$$

$$\vec{J}_i = -\left(\rho D_{i,m} + \frac{\mu_t}{Sc_t}\right)\nabla\omega_i - D_{T,i}\frac{\nabla T}{T} \quad (6)$$

wehere ω_i is the mass fraction of species i participating in the reaction. $\vec{J}_i$ is the diffusion flux of species i, generated due to the concentration gradient. $Sc_t = \frac{\mu_t}{\rho D_t}$ is the turbulent Schmidt number taken as 0.7. μ_t and D_t are the turbulent viscosity and turbulent diffusivity, respectively, used in turbulent Schmidt number formulation. R_i is the net source of chemical species i due to the reaction and defined as given in Eq. (7). There is only one reaction in this process, where $M_{w,i}$ is the molecular weight of the species i, and $\widehat{R}_i$ is the molar rate of the reaction of species i [13].

$$R_i = M_{w,i}\widehat{R}_i \quad (7)$$

The turbulence model used is a two-equation model called standard $k - \epsilon$. The transport equations for both turbulent kinetic energy (k) and dissipation rate ($\in$) are presented in Eqs. (8) and (9), respectively. These equations were proposed by Launder and Spalding [14]. They are applicable to a wide range of turbulent flow systems, especially to those that require simple and accurate enough models to prevent high computational loads. The turbulent (eddy) viscosity term μ_t consists of k and ϵ as shown in Eq. (10) where C_μ is a constant [13].

$$\frac{\partial}{\partial t}(\rho k) + \frac{\partial}{\partial x_i}(\rho k u_i) = \frac{\partial}{\partial x_j}\left[\left(\mu + \frac{\mu_t}{\sigma_k}\right)\frac{\partial k}{\partial x_j}\right] + G_k + G_b - \rho\epsilon \\ - Y_M + S_k$$

$$(8)$$

$$\frac{\partial}{\partial t}(\rho\varepsilon) + \frac{\partial}{\partial x_i}(\rho\varepsilon u_i) = \frac{\partial}{\partial x_j}\left[\left(\mu + \frac{\mu_t}{\sigma_\epsilon}\right)\frac{\partial\epsilon}{\partial x_j}\right] + C_{1\epsilon}\frac{\epsilon}{k}(G_k + C_{3\epsilon}G_b) - C_{2\epsilon}\rho\frac{\epsilon^2}{k} + S_\epsilon$$

$$(9)$$

$$\mu_t = \rho C_\mu \frac{k^2}{\varepsilon} \quad (10)$$

where G_k and G_b represent the generation of turbulence energy due to mean velocity gradient and buoyancy,

respectively. Y_M represents the effect of compressibility on the turbulence model which can be neglected. S_k and S_ϵ are both user-defined source terms that are taken as zero in this study. σ_ϵ and σ_k are as the turbulent Prandtl numbers for k and ϵ, respectively [13]. The model constants are taken as suggested by Launder and Spalding since they apply to a wide range of flows [14].

$$\sigma_\epsilon = 1.3; \ \sigma_k = 1.0; \ C_{1\epsilon} = 1.44; \ C_{2\epsilon} = 1.92; C_\mu = 0.09$$

Filter

There are three different domains in this system. The filter, which, is located between two fluid domains, is defined as a porous medium. It is used to capture the unreacted lime and the reaction product (calcium sulfite) from the gas stream. The porosity of the filter was estimated as 0.6 using an optical microscope. High viscosities are assigned to hydrated lime calcium sulfite (dispersed phase) in order to represent their behavior. The resistivity of the porous filter medium was estimated from $\vec{u} / -\nabla p$, , and the pressure drop was determined using the Ergun equation [13].

SO$_2$-Hydrated Lime Reaction

In this study, a two-phase Eulerian–Eulerian model under turbulent conditions was developed assuming that a reaction with the volumetric species transport model takes place in a single phase as mentioned previously [13]. Hydrated lime (Ca$(OH)_2$) constitutes 5% of the total inlet mass flow. This reacts with SO_2 of the continuous gas phase. The inlet concentration of SO_2 is 300 ppm. The outlet boundary condition is taken as the atmospheric pressure, and the no-slip condition is applied to the walls. The boundary conditions for the base case are presented in Table 1; also the temperature is taken as 70 °C, and there is no humidity in the gas. It should be mentioned here that even if no humidity is injected into the reactor, a certain amount of water forms due to the desulfurization reaction (Eq. (4)). The goal is to remove the SO_2, as calcium sulfite/sulfate, which can be used in the production of valuable by-products such as fertilizer, wallboard, etc.

Rate of Reaction

The kinetic expression for the rate of reaction is introduced to the model based on Eq. (11) using a user-defined function (UDF). This kinetic equation is obtained by Gutierrez and Orello [15] for an in-duct desulfurization process at low temperatures. They have used pilot plant experimental data to derive the equation.

$$\frac{dX}{dt} = 0.0768\left(\frac{BET}{12.9}\right)\exp\left(-\frac{12.9 * 0.0019X}{(RH/100)(BET)Y_{SO_2}}\right) \quad (11)$$

where BET is the specific surface area of the sorbent which is 15 (m^2/g) in this study. Y_{SO_2} is the molar fraction of SO_2,

Table 1 Boundary conditions

Boundary conditions				
Inlet	Parameter		Value	Unit
	SO₂ mass concentration		300	ppm
	Ca(OH)₂ mass fraction		300	ppm
	H₂O mass fraction (Humidity)		0	–
	Air mass fraction		Balance	
	Velocity		1	m/s
Outlet	Gauge pressure		0	Pa

and X is the sorbent molar conversion. RH represents the relative humidity in percent.

Mesh Grid Study

Mesh grid analysis was carried out with a focus on mesh independency and mesh quality. The examination of mesh independency was considered for various cell elements. The number of elements selected based on the verification of the results did not further vary with the number of mesh cells, as illustrated in Fig. 2. The selected mesh has approximately 500 000 elements. The majority of elements are tetrahedral and wedge shaped. The element size is about 4 mm.

The convergence criterion was set as $RMS = 10^{-6}$. Outlet SO₂ concentration was also monitored. Once this parameter no longer changed with subsequent iterations and the condition set for the Root Mean Square (RMS) was satisfied, the results were taken as final. The finer elements were used near the walls to represent the boundary layers appropriately, and certain mesh inflation using ANSYS Meshing was applied near the walls and the filter. The mesh is presented in Figs. 3, 4, and 5.

The mesh quality is another important issue to ensure the reliability of the predictions. It is defined by maximum skewness, minimum orthogonality, and aspect ratio. The maximum skewness, which is suggested to be less than 0.85,

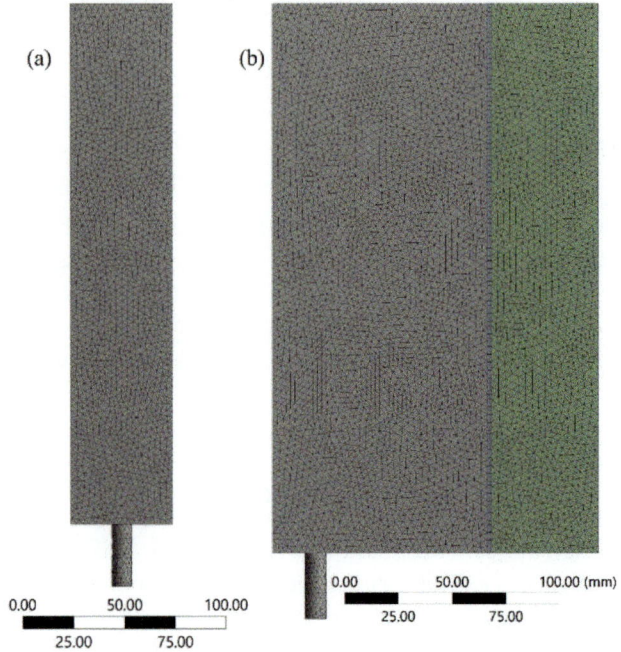

Fig. 3 Mesh used: **a** Side view **b** Front view

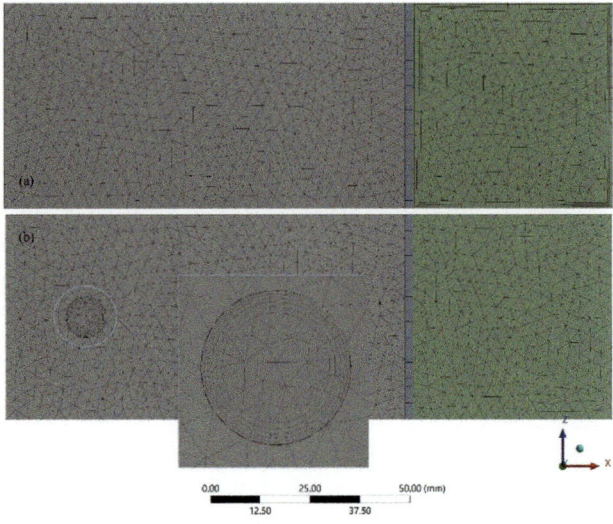

Fig. 4 Mesh used: **a** Top view **b** Bottom view

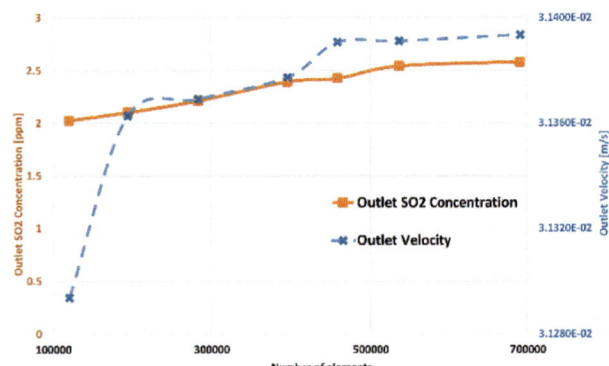

Fig. 2 Mesh independency diagram

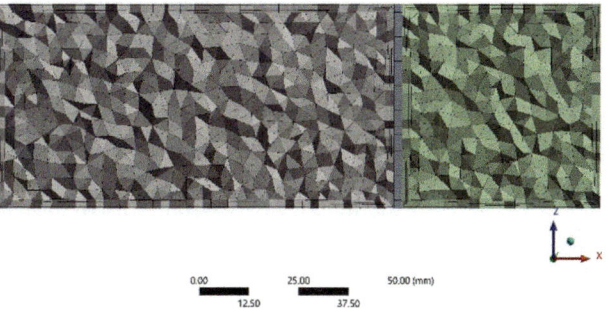

Fig. 5 Mesh used: Cross-section on the central plane

is 0.65. The orthogonality is 0.7, which is greater than the minimum suggested value of 0.25. The maximum aspect ratio is 10, which is recommended to be less than 20 to obtain a smooth convergence [13].

Results and Discussion

Base Case

The results show that SO_2 is efficiently removed under the conditions presented in Table 1. It decreased from 300 to 2 ppm when 5.0 wt.% hydrated lime is used as the sorbent. There is no humidity in the gas for the base case. Also, a parametric study was carried out and the results are presented. The parameters considered are the inlet sorbent (hydrated lime) concentration, inlet SO_2 concentration, and relative humidity of gas on the outlet SO_2 concentration (SO_2 removal).

Fig. 6 Air mass fraction contours

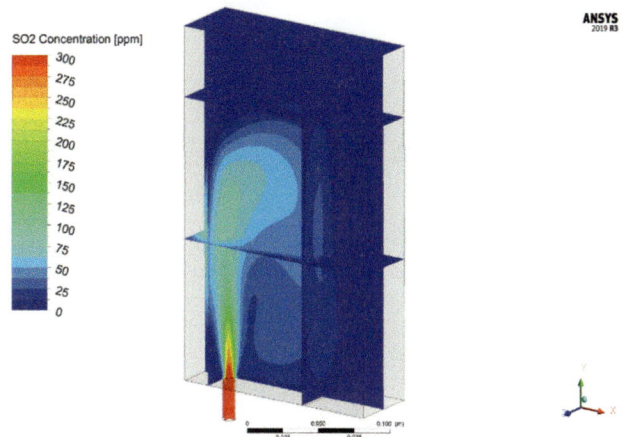

Fig. 7 SO_2 concentration contours [ppm]

Figure 6 shows the air concentration distribution in the reactor, and the variation is negligibly small as expected since it is inert. Its mass fraction is approximately 0.95 in the whole domain when no humidity is injected into the system.

Figure 7 presents the SO_2 mass concentration distribution (in ppm). It decreases from 300 ppm at the inlet to 2 ppm at the outlet under the conditions of the base case as can be seen in the figure. This shows that it is possible to desulfurize the gas efficiently if the reaction rate equation properly represents the actual reaction. The experimental work is underway. The rate expression will be verified based on the experimental results and will be modified if necessary.

The distribution of the lime mass fraction is shown in Fig. 8. The amount of lime is highest at the inlet and reduces as the reaction proceeds in the reactor. The trend is similar to that of SO_2 since they are both reactants. The hydrated lime is mostly retained on the filter as can be seen in the figure. Normally, all particles should be retained, and none should exist after the filter. However, this part is still under development.

The product calcium sulfite forms on/within the lime particles, and the amount formed can be calculated according to the stoichiometry of the reaction (Eq. (4)) and the total amount of SO_2 removed through the reaction. The H_2O is the other product of the reaction (Eq. (4)) that shows the same trend as that of calcium sulfite, but the H_2O concentrations are also very low due to the small SO_2 levels.

Figure 9 presents the velocity contours, showing the flow distribution. A fully developed flow is observed immediately in the inlet section and the centre of the inlet line possesses the maximum velocity. Velocities are higher near the filter which shows that the hydrated lime and the product will be carried by the gas to the zone near the filter surface.

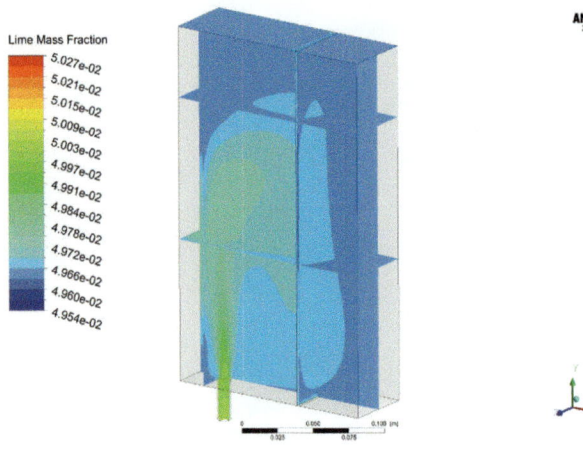

Fig. 8 Lime mass fraction contours

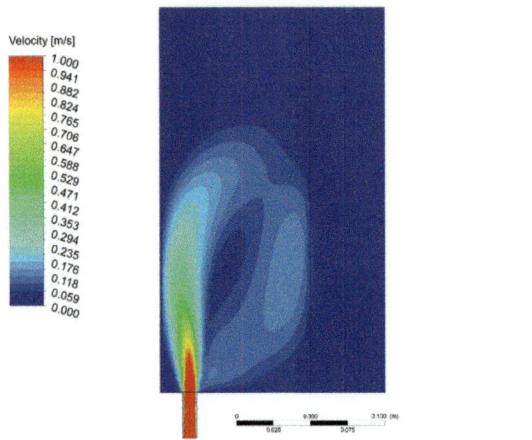

Fig. 9 Velocity contours

Parametric Study

A parametric study was carried out to study the impact of some parameters on the SO_2 removal: inlet sorbent amount, inlet SO_2 concentration, and relative humidity, as shown in Table 2.

Figure 10 shows the effect of the inlet concentrations of sorbent (hydrated lime) and SO_2 inlet concentration on the desulfurization of the gas. Increasing lime content increases the SO_2 removal at a given inlet SO_2 concentration. The higher the inlet SO_2 concentration is, the higher the outlet SO_2 concentration is for the same hydrated lime content. A slight increase in sorbent

(hydrated lime) in the low sorbent ranges has a substantial impact on the removal process as shown in this figure. However, increasing the concentration of the sorbent further affects the removal of SO_2 to a lesser extent, especially when its inlet concentration is low. The results are in agreement with those of Ma et al. [16]. They found that an increase in the amount of calcium-based sorbent leads to a higher SO_2 removal. However, the excessively high Ca/S ratio does not substantially contribute to the removal process [17, 18]. The SO_2 concentration at the inlet affects the desulfurization process as observed by other researches [4, 19, 20].

The presence of humidity also improves the desulfurization process [9, 16] as shown in Fig. 11. The humidity of the gas at the inlet enhances the removal of SO_2 for a given lime mass fraction. The model results have demonstrated that humidity (15% RH, Table 2) can increase the removal efficiency by 3% (92% to 95%) for an inlet lime mass fraction of 2%. The presence of humidity in gas is more influential when the inlet SO_2 concentration is high.

Humidity plays an important role in SO_2 removal when a low amount of sorbent is used as illustrated in Fig. 12. For an inlet SO_2 concentration of 300 ppm, the presence of humidity improves the desulfurization process, resulting in a lower SO_2 concentration at the outlet. For example, it increases the removal efficiency by 8% (77% to 85%) when the inlet lime mass fraction is 0.01.

Conclusions

A multi-phase Eulerian–Eulerian and turbulent CFD model was developed and used to simulate the SO_2 removal from the gas in a lab-scale reactor.

The model solves the continuity, Navier–Stokes, species transport, and $k - \epsilon$ turbulence model equations to determine the flow field and the species distributions in the system. The mesh was assessed in terms of both the quality and independence of the results. In addition, inflation meshing is employed near the walls to well-represent the boundary layers. The process is isothermal (70 °C here, which can be changed) since the impact of the reaction is minimal due to the low SO_2 concentration. A kinetic reaction rate is incorporated into the model using a UDF (user-defined function). The reaction rate was taken from the literature. Based on this

Table 2 Parameters used in the parametric study

	Base case	Parametric study
Inlet $Ca(OH)_2$ concentration, mass fraction	0.05	0.01, 0.02
Inlet SO_2 concentration, ppm	300	100, 200
Relative humidity (RH), %	0	15

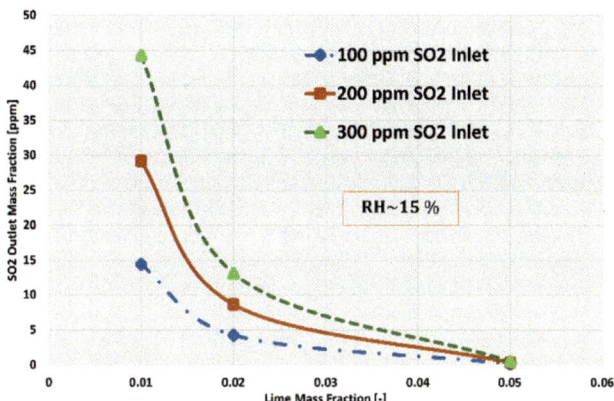

Fig. 10 The effect of inlet lime and SO_2 concentrations

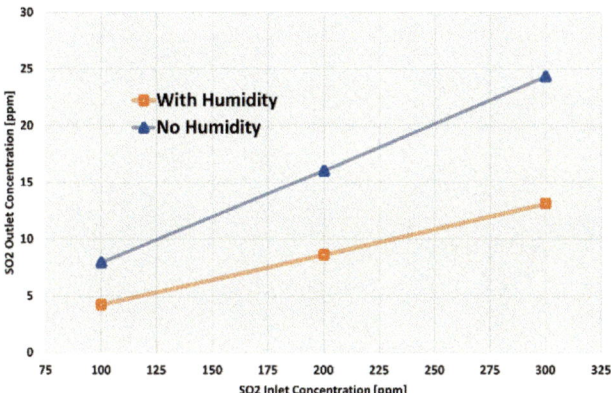

Fig. 11 The effect of gas humidity on desulfurization for an inlet lime fraction of 0.02

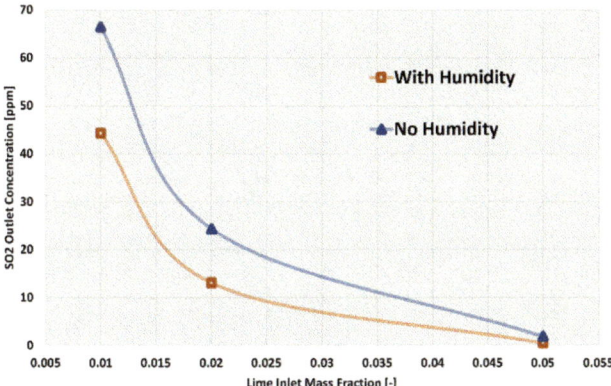

Fig. 12 The effect of gas humidity on desulfurization when the inlet SO_2 concentration is 300 ppm

rate expression, it was possible to remove SO_2 effectively. This will be verified using the experimental system and the necessary modifications will be made depending on the results.

The results showed that the SO_2 mass concentration of dry gas (0% RH) is reduced from 300 to 2 ppm, when 5.0

wt.% hydrated lime is used as the sorbent. Injection of humidity (RH = 15%) into the gas improved the removal efficiency, resulting in lower SO_2 concentration at the outlet (about 0.5 ppm).

The parametric study showed that increasing the inlet SO_2 concentration results in a higher outlet concentration of SO_2 if all the other conditions are kept the same. Furthermore, increasing the sorbent concentration at the reactor inlet without any other change leads to a higher SO_2 removal. The presence of humidity in the inlet gas can increase the removal efficiency, enhancing as much as 8% in some cases depending on the inlet SO_2 and lime concentrations. The results are entirely consistent with those found in the literature. All these findings will be further assessed based on the results of the experimental work currently being undertaken.

Acknowledgements The financial and technical support of the NSERC, Rio Tinto, Graymont, CQRDA, the University of Quebec at Chicoutimi (UQAC), and REGAL is greatly appreciated.

References

1. An assessment of global megatrends and regional and market sector growth outlook for aluminium demand, February 2020, CM group, https://international-aluminium.org/wpcontent/uploads/2021/03/cm_2050_outlook_for_al_demand_20200528_4wycD18.pdf

2. Prasad, D.S.N., Sharma, R., Acharya, S., Saxena, M., and Sharma, A.K., *Removal of sulphur dioxide from flue gases in thermal plants.* Rasayan Journal of Chemistry, 2010. **3**: p. 328–334

3. Gómez, A., Fueyo, N., and Tomás, A., *Detailed modelling of a flue-gas desulfurisation plant.* Computers & Chemical Engineering, 2007. **31**(11): p. 1419-1431.

4. Moran, S.Z.N.F.K., *Fell Feasibility report on technical options to reduce SO2 emissions Post-KMP.* 2013, Rio Tinto Alcan.

5. Charette, A., Kocaefe, Y., and Kocaefe, D., *Le carbone dans l'industrie de l'aluminium.* 2012: PRAL - Press Aluminium.

6. Aluminium Association of Canada, Primary *aluminum production.* Available from: https://aac.metrio.net/indicators/economie/autre/production_aluminium_multi

7. Masterson, L., *Refinery run cuts reduce anode-grade coke supply.* 2020.

8. Maltais, J.-N., Gaudreault, C., Bernier, J., Leclerc, S., and Ross, J., *Development, Proof of Concept and Industrial Pilot of the New CHAC Scrubbing Technology: An Innovative and Efficient Way to Scrub Sulfur Dioxide.* 2016. p. 473–478.

9. Wang, N. and Zhang, X., *Effect of humidification water on semi-dry flue gas desulfurization.* Procedia Environmental Sciences, 2011. **11**: p. 1023-1028.

10. Ghosh, R., Smith, J., and Adams, A., *Horizontal In-Duct Scrubbing of Sulfur-Dioxide from Flue Gas Exhausts.* 2015. p. 595–601.

11. Lotfi, M., FassadiChimeh, A., Dabir, B., and Mohammadi, A.H., *Computational Fluid Dynamics Modeling of the Pressure Drop of an Iso-Thermal and Turbulent Upward Bubbly Flow Through a Vertical Pipeline Using Population Balance Modeling Approach.* Journal of Energy Resources Technology, 2022. **144**(10).

12. Marocco, L. and Mora, A., *CFD modeling of the Dry-Sorbent-Injection process for flue gas desulfurization using*

hydrated lime. Separation and Purification Technology, 2013. **108**: p. 205-214.

13. ANSYS, *ANSYS FLUENT Theory Guide 2022 R2.* 2022.

14. Szablewski, W., *B. E. Launder and D. B. Spalding, Mathematical Models of Turbulence. 169 S. m. Abb. London/New York 1972. Academic Press. Preis geb. $ 7.50.* ZAMM - Journal of Applied Mathematics and Mechanics / Zeitschrift für Angewandte Mathematik und Mechanik, 1973. **53**(6): p. 424–424.

15. Gutiérrez Ortiz, F.J. and Ollero, P., *A realistic approach to modeling an in-duct desulfurization process based on an experimental pilot plant study.* Chemical Engineering Journal, 2008. **141**: p. 141–150.

16. Ma, X., Kaneko, T., Tashimo, T., Yoshida, T., and Kato, K., *Use of limestone for SO2 removal from flue gas in the semidry FGD process with a powder-particle spouted bed.* Chemical Engineering Science, 2000. **55**(20): p. 4643-4652.

17. Xu, G., Guo, Q., Kaneko, T., and Kato, K., *A new semi-dry desulfurization process using a powder-particle spouted bed.* Advances in Environmental Research, 2000. **4**(1): p. 9-18.

18. Bausach, M., Pera-Titus, M., Fité, C., Cunill, F., Izquierdo, J.F., Tejero, J., and Iborra, M., Kinetic modeling of the reaction between hydrated lime and SO2 at low temperature. AIChE Journal, 2005. 51: p. 1455-1466

19. Garea, A., Herrera, J.L., Marques, J.A., and Irabien, Á., *Kinetics of dry flue gas desulfurization at low temperatures using Ca(OH)2: competitive reactions of sulfation and carbonation.* Chemical Engineering Science, 2001. **56**: p. 1387-1393.

20. Harriott, P., *A simple model for SO2 removal in the duct injection process.* Journal of the Air & Waste Management Association, 1990. **40**(7): p. 998-1003.

Improvements to a Mathematical Model Used to Reproduce the Wave and Stream at the Bath-Metal Interface and Assess Their Impact on the Movement of Alumina Rafts

Thomas Richer, Lukas Dion, Laszlo Kiss, Sébastien Guérard,
Jean-François Bilodeau, Guillaume Bonneau, and Martin Truchon

Abstract

A mathematical model has been developed to reproduce the tridimensional interface between bath and metal in an electrolysis cell. In the last year, the mathematical model has been adapted to consider alumina rafts movements in tridimensional coordinates. To reproduce a bath-metal surface, the wave equation was solved around three main phenomena known to occur in operating pot. Specific solution includes strong Lorentz force at the edge of the cell, natural resonance of the geometry, and impulse from perturbation. Among the geometrical challenges inherent to such improvement, it was necessary to properly introduce concepts such as "flow", "interfacial forces", and "buoyancy force". Hence, the model uses interfacial phenomena to reproduce the movement of alumina rafts at the bath-metal interface. The potential of such tracking is shown in different cell conditions. This paper details the scope of the modifications applied to the model, describes in detail the step used to characterize both interface and rafts and shows the raft tracking potential for industrial application.

Keywords

Aluminum electrolysis • Bath-metal interface • Stokes drift • Surface wave • Floating body • Numeric model

Introduction

In an alumina reduction cell, periodic injection of alumina form rafts at the surface of the cryolitic bath. Under specific circumstances, some rafts sink underneath the bath all the way to the metal pad, or alternatively remained caught at the bath-metal interface. Once at the bath-metal interface those rafts begin to move along gravitational wave phenomenon as well as thermal, magnetic, and hydrodynamics flows. There is a high possibility that those rafts eventually sink and further contribute to the formation of sludge at the bottom of the cell. This sludge disrupts the vertical electrical flow of the current, which generates cell instability and may lead to a reduction of the life expectancy of the cell. Consequently, there are two ways for sinking rafts to contribute to the generation of sludge. The first is by direct penetration of the bath-metal interface. The second is by sliding under the metal nap in the outer layer of the cell due to the pronounced angle of the bath-metal interface in these regions. This study will focus on transport phenomena and the conditions leading to this second mechanism.

This study uses the basis developed in a previous study by the same authors [1–3] but greatly expands upon it. To approach the cell condition, the entire surface of the bath-metal interface has been simulated using the standard wave equation [4]. Interface phenomena considered are gravitational mode oscillation [5], disturbance due to bubble formation and pumping effect under the anode [6], and permanent magnetohydrodynamics deformation near the border of the cell [7]. Those phenomena induce multiple superposed waveforms which move the raft by drift and sliding. Moreover, a flow field in the cell has been added to affect the movement of rafts at the interface [8]. Also, the force applies on the raft now includes Stokes drift force [9, 10], buoyancy, and gravitational force. Those modifications mean that it is now possible to use it in different cells state to estimate raft movement contribution to sludge formation.

T. Richer (✉) · L. Dion · L. Kiss · G. Bonneau · M. Truchon
GRIPS-UQAC, Régal, Université du Québec à Chicoutimi, 555 Boulevard de l'Université, Chicoutimi, QC G7H 2B1, Canada
e-mail: Thomas.richer1@uqac.ca

S. Guérard · J.-F. Bilodeau
Arvida Research and Development Center, Rio Tinto Aluminium, 1955 Boulevard Mellon, Jonquière, QC G7S 4K8, Canada

© The Minerals, Metals & Materials Society 2023
S. Broek (ed.), *Light Metals 2023*, The Minerals, Metals & Materials Series,
https://doi.org/10.1007/978-3-031-22532-1_101

Model Description

Overview

The physical model has two main sections, solving the bath-metal interface (BMI) for each temporal increment and applying force derived from it to alumina rafts. In standard operation, the BMI is in an oscillatory mode dependent on the dimension of the cell and the thickness of the bath and metal nap. This mode is bound at the border of the cell by strong magnetic force which induces a permanent deformation of the cell. Local circulating flow made the BMI sink near the corner and border of the cell. In addition, perturbation such as bubble release from the anode can disturb locally the bath-metal interface. Those perturbations have been quantified by Kiss and Klára Vékony in 2008 [6]. Along those liquid phenomena, alumina and liquid bath at the interface are moved by forced magnetic convection and natural thermal convection [4]. All those BMI phenomena combine to form the interface tridimensional profile. From this profile are extracted normal vectors which give direction to apply force on the raft.

As the model was upgraded from previous work, some important hypothesis must be revisited.

1. The raft is always small enough to not disturb the form of the interface.
2. The flow field in a normal operation cell approach a steady state and local disturbance such as bubble or anodic effect is not enough to disturb the overall flow in the cell.
3. The force applied on the raft is of two origins, gravitational and viscous drag. All other forces or phenomena that can affect the raft movement are disregarded.
4. Dissolution is not considered, as such the geometry and density of each single raft is constant if there is no coalescence or fracture of the raft.
5. This study focuses on the movement of the raft and the possible formation of sludge by slipping under the metal pad in "dead zones" located in the perimeter of the BMI. As such the raft can't sink directly through the metal pad.

After all hypothesis considerations, the movement of the rafts can be calculated by a sum of forces along each axis. This movement is dependent on the position and speed of the raft due to Stokes drag from the wave and flow drag from magnetohydrodynamics and thermal flow. A forward Burlisch-Stoer scheme is then used to predict the new position. During each timestep in the simulation, a validation of the rafts' positions is performed with respect to collision and fragmentation before further advancing in time, to maintain the conservation of mass and prevent two rafts from occupying the same physical space. Once confirmed as valid, the mass and speed of each raft are tracked for every time step. The timestep used was 50 ms and 0.1 mm per minute of simulation for convergence.

The model is coded using Python to take advantage of many optimized libraries as the simulation become more complex and more costly in computing time. The interface is simulated using a 1 mm wide mesh in all directions to the respective dimensions of a 300 to 600-kA electrolysis cell. Simulation last 5 min. Further precisions are provided in the next sections for each respective part of the model.

Bath-Metal Interfacial Phenomenon and Modeling

The bath-metal interface form is characterized by the waveform equation:

$$\frac{\partial^2 f}{\partial t^2} - c^2 \Delta f = 0 \tag{1}$$

With c squared proportional to the elasticity module divided by the density of the medium. To solve the wave equation, boundary conditions need to be applied. For simplifications, no slipping between the bath and the metal is considered. In addition, the density difference between the metal and alumina is a small fraction of the density of each liquid. As such, the medium is considered homogenous in flow. Consequently, movement at the bath-metal interfaces are considered as internal gravity wave and continuous across all the surface [4] The specific solution to the wave equation is considered a combination of three distinct and, assumed for simplicity, linearly independent phenomena.

$$f(x,y,t) = deformation(x,y) + mode(x,y,t) + \sum_{1}^{N} perturbation(x,y,t) \tag{2}$$

where *deformation* is a mathematical model for the permanent deformation characterized by previous literature such as Renaudier et al. 2018 [8]. The *mode* function describes the general solution of the wave equation for a surface standing wave in a limited space. At last, *Perturbation* summation describe wave fading with respect to time and distance from an origin impulse within the pot. Each term is detailed below.

The Permanent Deformation

The permanent deformation of the BMI is a known phenomenon in the design of alumina pot. It references a steep inclination of the BMI caused by Lorentz force near the edge of the cell. This incline can plunge as much as half of the metal depth near the corners [7] (Fig. 1).

$$deformation(x, y) = -8 * Pd \left(\frac{x^4}{Lx^4} + \frac{y^4}{Ly^4} \right) \quad (3)$$

where Pd is the maximum depth of deformation in the corners of the cell, Lx is the total length of the cell, and Ly is the total width of the cell. A visualization in a cell simulation with this study model is illustrated below.

Standing Wave and Mode of Oscillation

The wave equation solution for a sustained oscillation of a membrane in a close environment are well known. In an electrolysis cell, the bath-metal interface is not bound to the side of the cell. As such, the boundaries condition for standing wave is free at the edge of the cell. The solutions to the wave equation take the form of Eq. 4 (Fig. 2).

$$mode(x, y, t)_{m,n} = \frac{A}{2} * sin \left(2\pi \frac{(m+1)x}{Lx} \right) \\ * sin \left(2\pi \frac{(n+1)y}{Ly} \right) * sin(\omega t) \quad (4)$$

where A is the amplitude, Lx is the length of the cell, Ly is the width of the cell, and ω is the resonance frequency of the mode. A mode refers to a solution to the wave equation that satisfies boundary condition. As solutions are multiple of each other, only subscripts are used to describe them. Consequently, if m = 0, n = 0 in Eq. 4, the mode is fundamental (0,0) and multiple of this mode are harmonics. In truth, fundamental and harmonics are always present in an oscillating interface. In an electrolysis cell, industrial partners have shown that the cell mode tends to change overtime. In this study, considering short simulation time, only one mode is used for a specific simulation. An example of a mode simulated with the model is illustrated below. Amplitude and frequency data have been taken from Laroche and co. [5]

Process Perturbation

Process perturbations are impulsions on the bath-metal interface originating from normal chemical reaction in the pot during operation. In short, bubbles escape from under the anode, creating a pump effect that perturbs the BMI. Most of this effect happens in front and behind a new anode and transition to the side as the life cycle progress. This has been studied by L.I. Kiss and Klára Vékony [6]. Thus, bubble emission is a natural process perturbation of the BMI. This paper differentiates between process perturbation like bubble and operating perturbation such as alumina injection and anode change. As date of writing, only process perturbations are accounted in the simulation.

Process perturbation is modelized as a collection of impulse across time for each anode in the cell. Each impulse then provokes a fast-decaying wave in time and space, centered around each impulse point. To approach the nature of gas release, a random time delay has been introduced between pulse for each anode. All perturbations are then added to the surface form. This summation takes the form of Eq. 5.

$$\sum_{i=1}^{N} perturbation(x, y, t) = \frac{A}{2} * \ sin(kr * r_i + \omega t + \varphi_i) \\ * e^{-\alpha r_i} e^{-\beta(t-t_0)} \quad (5)$$

where r_i is the vector from the center of the impulse, define as $r_i = \sqrt{(x - x0_i)^2 + (y - y0_i)^2}$, ω is the impulse frequency, φ_i is a random phase between each anode to simulate the nature of gas release and introduce the difference between anode. For decay factors, t_0 is the time of impulse, α is the dispersion constant, and β is a dampening constant. These many waves at the interface form an interference pattern with local variation. This is shown in Fig. 3.

Once all phenomena are combined, we obtain a complex interface such as the ones illustrated in Fig. 4.

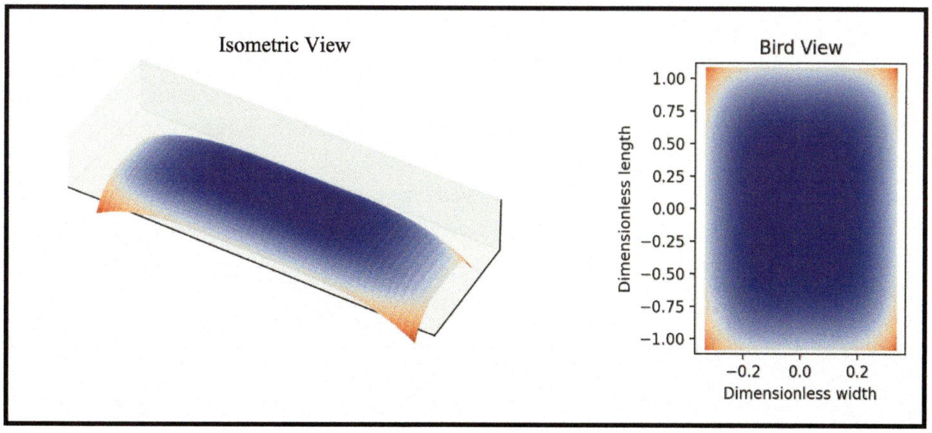

Fig. 1 Permanent deformation of the bath-metal interface. Maximum depth set to 50% of metal depth

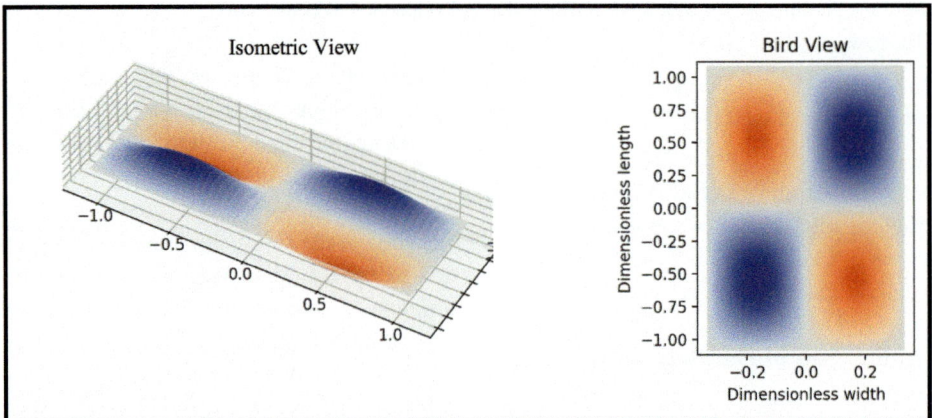

Fig. 2 Mode (2,2) of an electrolysis cell

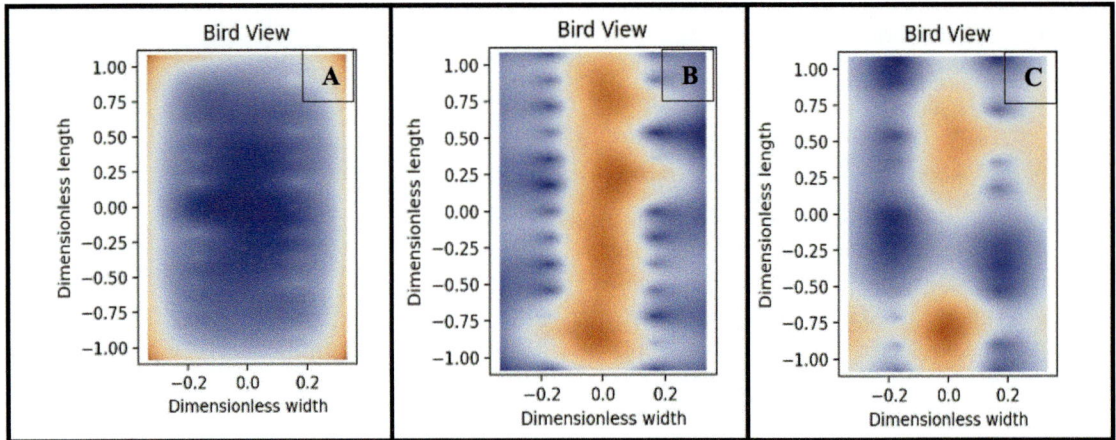

Fig. 3 Interference patron of N dissipating waves center around each anode in a simulated cell. (A) At time equal 0 s. (B) At time equal 30 s. (C) At time equal 60 s

Flow at the Bath-Metal Interface

From past hypothesis and simplification, the flow at the bath-metal interface is considered homogenous across the interface. Consequently, a velocity vector field at each point of the interface mesh is enough to characterized flow in the model. This velocity field has been simulated many times in the industry [8] and used in the development and design of pot.

The flow field is taken as an inertial referential when calculating drag on the raft. As such, the efficient velocity of a raft is taken as the difference between the velocity of the raft, the velocity of fluid particle according to Stokes equation [10], and flow velocity at that point.

Force Applied on the Raft

All rafts are approximated as cylindrical disks in their geometry determined by mass, density, and thickness. Geometry, thickness, and density are parameters that are kept constant throughout the simulation. Hence, without dissolution, the mass and the radius of the rafts are discrete evolving parameters upon coalescence or fracture of the raft. This phenomenon has been explored in a previous study from the same author [12].

Each raft movement is defined by the summation of force on its center mass. Among those forces, apparent weight, hydrostatic pressure and drag effects from Stokes drag and flow field. Figure 5 illustrates how those forces affect the rafts at the BMI.

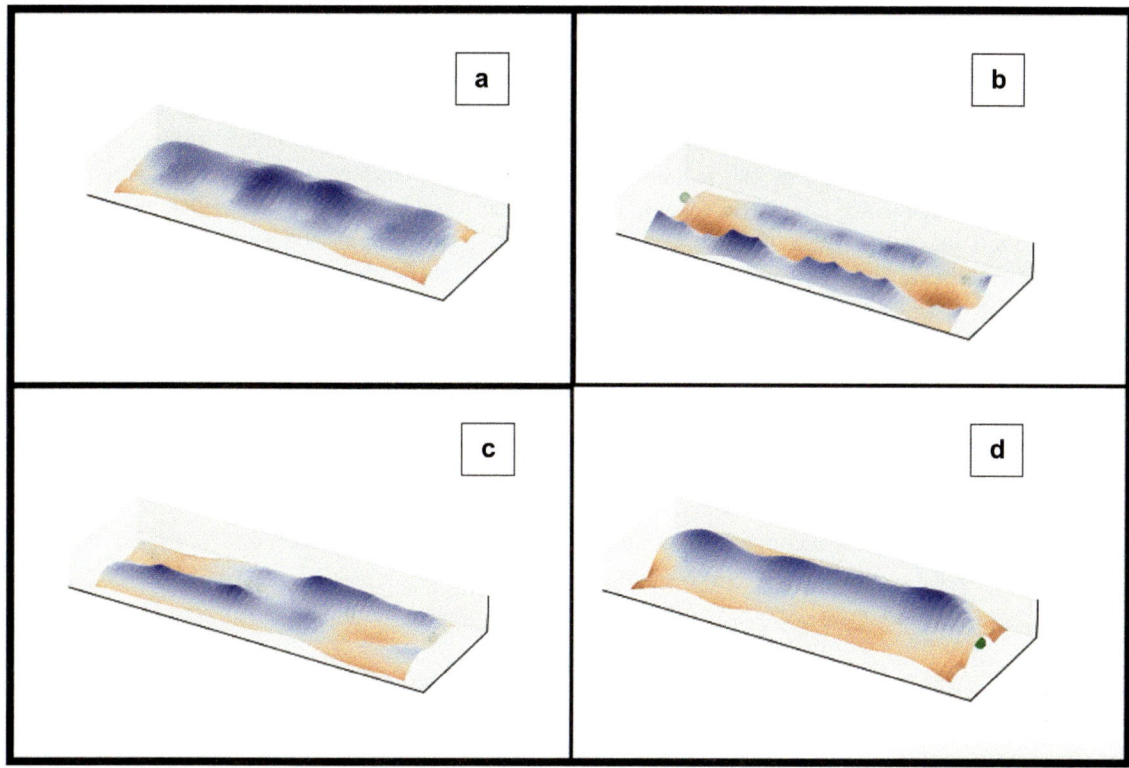

Fig. 4 Example of a full period of the simulated surface. Mode (0,1). **a** Initial form of the interface. **b** A quarter period after initial launch, minimal point in the canal, **c** half period after launch, **d** three-quarter period after launch, maximal height of metal in the cell

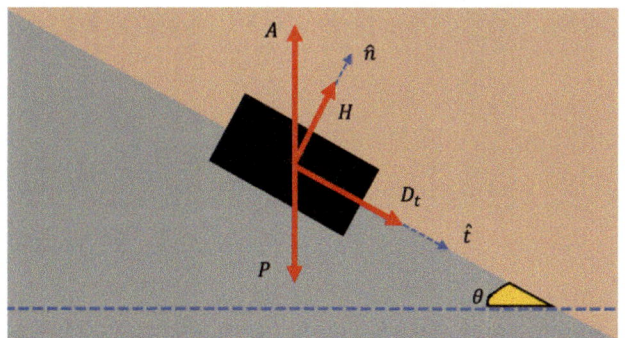

Fig. 5 2D Force diagram on a raft with weight (P), Archimedes' Force (A), Drag along the interface (D), and hydrostatics pressure (H). Normal and tangent axis is represented by n and t

The apparent weight is an important force since it's the only one explicitly dependent on the inclination of the bath-metal interface. Archimedes' force from the two liquid and gravity mostly cancels each other out, and the apparent weight of the rafts leans towards zero. In this state, differential in hydrostatic pressure due to the incline determines how the raft slide on the internal wave. The direction of the

buoyancy force vector is the normal of the surface at the raft position. As such, it is equal to the gradient of the surface at this point.

$$A = 2\pi r g * (E_1 \rho_1 + E_2 \rho_2)\widehat{n} \qquad (6)$$

where r is the radius of the raft, g is the gravitational constant, $E_i \rho_i$ is the thickness and density of the liquid displaced. Also $\widehat{n}$ is the direction of the normal to the interface.

Furthermore, drag forces are defined by the velocity difference between the raft and fluid particle at the interface. The velocity of the fluid is determined by the convection cycle and Stokes drift from perturbation.

$$A = Cd * 2r * (E_1 \rho_1 + E_2 \rho_2) * \left(v_{raft} - v_{fluid} - v_{Stoke}(r_i)\right)^2 \qquad (7)$$

where Cd is the drag coefficient dependent on the geometry (Cd = 0.45 for disk), r is the radius, $E_i \rho_i$ is the thickness and density of the liquid displaced, $v_{Stoke}(r_i)$ is Stokes drift velocity field in the direction of the radius of propagation and v_{fluid} is the velocity field from heat, magnetic, and hydrodynamics effects in the electrolysis cell.

Practical Application

One of the goals of this model is to simulate the displacement of many rafts across the interface in a specific cell condition of operation. At this end, the new model can simulate movement on standard operational regime and across some transitional regime such as anode change and anodic effect. Figure 6 shows the result of a simulation in standard operating condition. This means slow and steady mode with low amplitude and all perturbations under 1 cm.

Results suggest that rafts tend to move along the length of the cell in the central canal. Near the edge they spread out and get trapped by the permanent deformation at this point. Flow and modal oscillation favorize movement "up" the pot (from syphon end). Other scenarios were executed, to simulate a transition state in the low current period that follows an anode change where an anode perturbation has been retired during a simulation.

Figure 7 shows that raft tend to stay closer to the missing anode, a few are even trapped in the « hole» in the pattern created by the changed anode. However, the central canal is still preferential to most of the rafts and they tend to still follow preferred direction of flow and modal oscillation. The last scenario presented in this paper is the unstable cell. To simulate a cell in an unstable condition, the modal oscillation has bigger amplitude and simultaneously, a single anode produces much larger amplitude perturbation from a hypothetical low-voltage anode effect (LVAE). For an easier comparison with Fig. 7, the same anode was used for the anodic effect. Figure 8 shows results of the simulation after 300 s, averaged from 3 simulations.

As shown in Fig. 8, the combination of the LVAE with added instability in the cell produces a lot more rafts trapped on the edges. As such, a cell in those condition is likely to produce sludge along its edge, even for relatively short period of time.

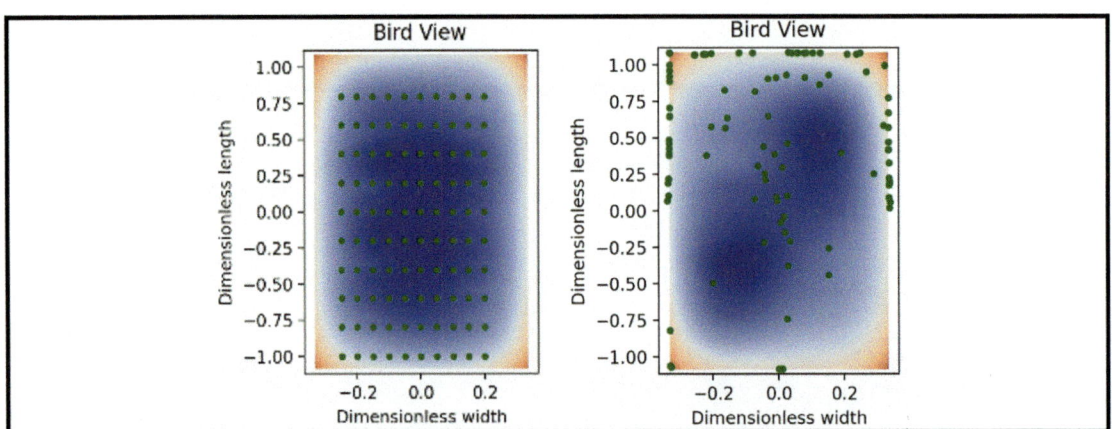

Fig. 6 Position of raft from starting position (left) to their resting position after 300 s of simulation (right) (avg. position from three simulations)

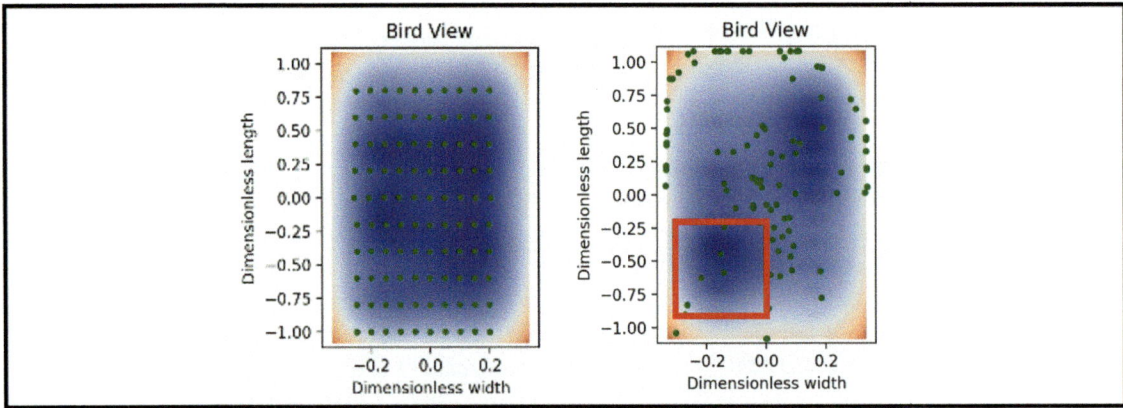

Fig. 7 Position of raft from starting position (left) to their resting position after 300 s of simulation (right) (avg. from three simulations). Red square is the general area of the changed anode

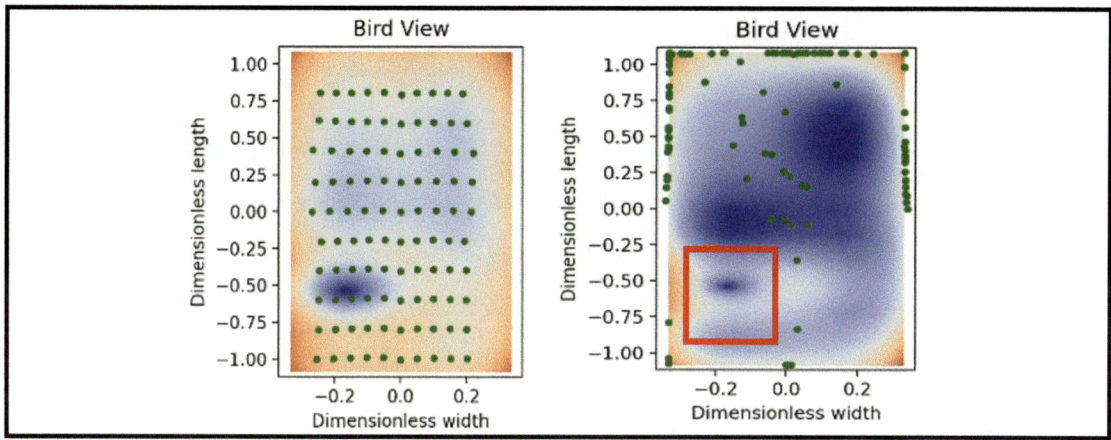

Fig. 8 Position of raft from starting position (left) to their resting position after 300 s of simulation (right) (avg. from three simulations). Red square is the general area of the atypical anode

Simulation with the model has shown that under normal operation, bubble creation from anode mostly cancels out along the length of the cell and interferes with each other to form a preferential area in the central canal. This seems to follow the finding of Mojtaba Fallah Fini and al. [11] in *Sludge Formation in Hall-Héroult Cells: Drawbacks and Significant Parameters*. Furthermore, unequal disturbance in the process, such as anode effect or a recently changed anode causes a change in the displacement pattern of the raft. A changed anode tend to « attract» rafts even if the general behavior is conserved. An unstable cell seems to speed up raft movements and, intensifies its own level of instability via increased sludge creation, which in turn may reduce the lifetime of the electrolysis cell.

Conclusion

In conclusion, an improved model has been refined from a previous version. Updates include tridimensional consideration and overhaul of all force and addition of stream velocity field from literature and buoyancy. In addition, the model can now affect the individual anode effect on the interface. Many 300 s simulations were conducted some with a changed anode or an important low-voltage anode effect. The result has shown the central cell canal is a preferential « highway» for rafts. Also, abnormal cell condition such as unstable cell produces a higher rate of captive raft, likely to become sludge by sliding under the metal nap. At last, a changed anode affects the movement of raft to an extent but do not cause additional sludge formation. This seems to corroborate with literature on the subject. Those results are promising to help predict sludge accumulation on a cell under different sets of conditions. Future refinements

to the process may benefit from coupling these results with additional simulation tools used for cell designs.

Acknowledgements The authors would like to thank our industrial partner Rio Tinto Aluminium (RTA), National Sciences and Engineering Research Council of Canada (NSERC) and REGAL for their financial support. The author also thanks the GRIPS team for their wisdom and support.

References

1. T. Richer, L. Dion, L.I. Kiss, S. Guerard, J-F Bilodeau, L. Rakotondramanana (2022) Mass transport by waves: physical model with coalescence, fragmentation, and displacement on a bath-metal interface. Light Metals.
2. L. Rakotondramanana, L.I. Kiss, S. Poncsak, R.Santerre, S. Guerard, J-F Bilodeau, S.Richer (2021) Mass transport by waves: bath-metal interface deformation, rafts collision, and physical model, Light Metals, 2021.
3. L. Rakotondramanana, L.I. Kiss, S. Poncsak, S.Guérard and J-F Bilodeau (2020) Mass transport by waves on the bath metal interface in electrolysis cell. Ligth Metals 2020, p 510–516.
4. Sutherland, B. (2010). *Internal Gravity Waves*. Cambridge: Cambridge University Press. doi:https://doi.org/10.1017/CBO9780511780318.
5. F. Laroche, "Études des phénomènes d'oscillation régulière de l'interface bain-métal d'une cuve d'électrolyse," Master thesis UQAC, Chicoutimi, 1988.
6. L.I. Kiss, Klára Vékony (2008), Dynamics of the gas emission from aluminum electrolysis cells, Light Metals, 2008.
7. Zhang, H., Yang, S., Zhang, H. *et al.* Numerical Simulation of Alumina-Mixing Process with a Multicomponent Flow Model Coupled with Electromagnetic Forces in Aluminum Reduction Cells. *JOM* **66,** 1210–1217 (2014). https://doi.org/10.1007/s11837-014-1020-1.
8. Steeve Renaudier, Steve Langlois, Benoit Bardet, Marco Picasso and Alexandre Masserey. (2018). Alucell: A Unique Suite of Models to Optimize Pot Design and Performance. Light Metals, 2018.

9. A. Solheim and S. Rolseth, "Some surface and interfacial phenomena encountered in aluminium electrolysis," *Light Metals,* pp. 469–474, 2001.

10. T. S. Van Den Bremer and Ø. Breivik, "Stokes drift," Philosophical Transactions of the Royal Society A: Mathematical, Physical and Engineering Sciences, 2017.

11. Mojtaba Fallah Fini, Jean-René Landry, Gervais Soucy, Martin Désilets, Patrick Pelletier, Loig Rivoaland & Didier Lombard (2020) Sludge Formation in Hall–Héroult Cells: Drawbacks and Significant Parameters, Mineral Processing and Extractive Metallurgy Review, 41:1, 59-74, DOI: https://doi.org/10.1080/08827508.2018.1536658.

12. Richer, T., Dion, L., Kiss, L.I., Guérard, S., Bilodeau, JF., Rakotondramanana, L. (2022). Mass Transport by Waves: Physical Model with Coalescence, Fragmentation, and Displacement on a Bath-Metal Interface. In: Eskin, D. (eds) Light Metals 2022. The Minerals, Metals & Materials Series. Springer, Cham. https://doi.org/10.1007/978-3-030-92529-1_51.

Numerical Investigation of Thermal, Electrical, and Mechanical Behaviour of Aluminium Cell During Preheating Phase

Simon-Olivier Tremblay, Daniel Marceau, Rohini-Nandan Tripathy, Antoine Godefroy, Duygu Kocaefe, Sébastien Charest, and Jules Côté

Abstract

Electrical preheating of aluminium electrolysis cells using coke and/or graphite is a delicate process, which has a significant impact on cell life. Cell preheating is evaluated by many factors such as the heat-up rate, the final cathode surface temperature distribution, anodic current distribution, and longevity of cathode life [1]. Hence, a better understanding of the cell behaviour is required to optimise this critical phase. In this work, electrical, thermal, and displacement data have been measured on a cell during the preheating period. Those measurements were then used to calibrate a quarter-cell model, which includes a transient thermo-electro-mechanical weakly coupled analysis, developed using ANSYS ™. The results obtained are in good agreement with in situ measurements and allow a better understanding of the anodic current distribution, cathode surface temperature, baking level of ramming paste, and stress level distribution in the cathode/lining, which are critical information for further optimisation.

Keywords

Electrolysis cell • Preheating • Cathode • Numerical investigation

S.-O. Tremblay (✉) · D. Marceau · R.-N. Tripathy · D. Kocaefe
University Research Centre on Aluminium (CURAL), Aluminium
Research Centre (REGAL), University of Québec at Chicoutimi,
555, Boul. de l'Université, Chicoutimi, QC G7H 2B1, Canada
e-mail: Simon-Olivier1_Tremblay@uqac.ca

A. Godefroy · S. Charest · J. Côté
Aluminerie Alouette Inc., 400, Chemin de La Pointe-Noire,
Sept-Iles, QC G4R 5M9, Canada

Introduction

It is widely accepted that the preheat, the start-up, and early operation have a strong influence on the performance and life of a Hall-Héroult cell [2]. The main goal of the cell preheating is to create a uniform temperature distribution on the cathode plane with a high enough temperature to prevent thermal shock during start-up, minimize freezing of the bath on the cathode surface, and finally prepare the cell for full production as soon as possible. A good preheating will have a positive impact on the early operation and cell life expectancy [3].

The preheating procedure is controlled by measuring the cathode surface temperature at prefixed positions, the anodic current distribution of all anodes at regular time intervals, and the cathodic current distribution [4]. The goal of the preheating phase is to have homogeneous heat-up rate of the cathode surface as wesll as uniform anodic and cathodic current distributions. These elements will control the baking of the ramming paste and prevent the bath from freezing initially in the cell. These factors are important for avoiding thermal and mechanical shocks to the cathode, which increase cathode life hence decrease cathode production cost and spend pot lining per ton of Al produced [2].

Problem Statement

After the energization of the cell during the preheating phase, the anodic current distribution and the cathode surface temperature are monitored. Often, these data show an important non-uniformity along the cathode surface.

It is observed that the cathode surface temperature in the central zone is much higher than the temperature at the peripheral zone. New cells with a low temperature at the ends are usually unstable and more difficult to operate just after the start-up than more uniformly preheated cells [4]. Similar irregularity is also observed for anodic current

© The Minerals, Metals & Materials Society 2023
S. Broek (ed.), *Light Metals 2023*, The Minerals, Metals & Materials Series,
https://doi.org/10.1007/978-3-031-22532-1_102

distribution. Typically, anodes placed at four corners draw lower currents as compared to the anodes in the middle of the cell.

Freezing of liquid bath on the peripheral cathode surface is also observed during bath pouring or bath generation period, which is not desired for a cell start-up. This affects adversely the cell stability in its early operation and can be a persistent issue in normal operational phase as well.

A cathode failure during the cell start-up and in the early-operational phase is very costly for a smelter. Thermal shock or thermal deformation of lining materials caused by improper preheating and/or start-up may result in open gaps between ramming paste and cathode blocks. Additionally, improper preheating may generate vertical shrinkage cracks in the peripheral ramming paste [4].

Previous Work

Dupuis [5, 6] developed a transient thermo-electric model calibrated by comparing the predictions with the measured temperature evolution curves. According to Dupuis, during preheat phase, the thermal gradients can be minimized by using different coke bed patterns thus the stress on the cathode linings is also minimized.

Recently, Richard et al. [7] presented a thermo-chemo-mechanical slice model which predicts the impact of the preheating on the stress distribution in the cathode block and the baking level of the ramming paste. The results showed that some ramming paste is still not baked properly even after 24 h of preheating.

Marceau et al. [1] modelled a thermo-electro-chemo-mechanical behaviour of the quarter-cell model during its preheating phase. The contact conditions show that sliding and separation occurs mainly for the cathodes at the extremities of the cell and it is most significant for the last cathode block.

In this context, the focus of this paper is on the numerical investigation of the thermo-electro-mechanical behaviour of the cathode/lining during its preheating phase. The identification of the cell behaviour will allow to define the corrective measures to be adopted for the improvement of the preheating phase. This will result in a better cell performance.

Proposed Approach

The proposed approach consists of developing a numerical model of an electrolysis cell using 3D finite element method in order to evaluate the thermal, electrical, and mechanical behavior of a cell during the preheating phase specific to Aluminerie Alouette Inc (AAI) conditions.

To represent the cell preheating behavior properly, the model must include the proper geometry, material properties, and boundary conditions. Cell behavior is particularly sensitive to some of these parameters such as the resistor bed electrical resistance (mix of coke and graphite) and the shell convection coefficient. But they also are difficult to evaluate accurately.

To calibrate these parameters and to validate the model, a measurement campaign was carried out at AAI to obtain thermal, electrical, and mechanical data following a specific instrumentation plan. In parallel, laboratory characterization was done at UQAC to measure the missing and sensitive mechanical properties.

A sequence of numerical analysis is then chosen to reproduce the thermal, electrical, and mechanical behavior of the cell from ambient temperature to the end of the preheating phase of the electrolysis cell.

Some preliminary simulations are first carried out to validate the model by analyzing the independence of the mesh, the calibration parameters, and the *in situ* measurements comparison. Afterward, the model is used to evaluate the preheating performance, more specifically: temperature/ current distribution along the cathode surface and stress distribution in the shell/linings.

Finally, the understanding of the behavior of the cell will allow to define the corrective measures to be adopted to improve the preheating performance.

The Finite Element Model

Considering the double geometrical symmetry of the electrolysis cell, only the quarter of the geometry is considered using the representative boundary conditions.

Figure 1 shows the different components taken into account in the numerical simulation corresponding to the geometry at the end of the preheating cell preparation. Note that the solid bath occupies the totality of the remaining space under the anode yokes. The top part (anode assemblies, cover product, and anode beam) is included in the model to allow a better representation of the heat propagation coming from the joule heating of the coke bed and hence a more realistic temperature distribution of the whole cell. The coke bed pattern/position used by AAI has been respected.

Firstly, a transient thermoelectrical (TE) analysis is used in order to obtain the temperature distribution evolution. Transient effect is needed since, among other reasons, the preheating is a process which lasts around two days before reaching the steady state. Also, considering that the preheating may lead to the melting of a significant mass of the bath, the enthalpy of fusion of this material needs to be considered.

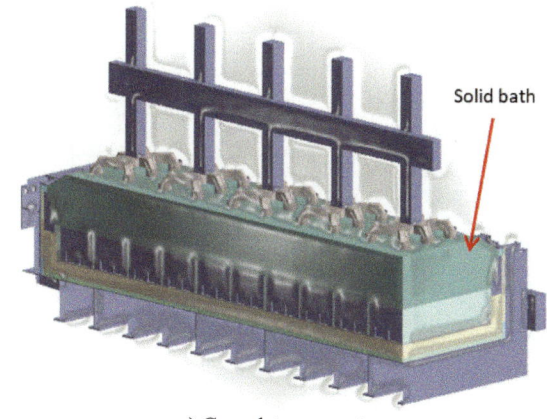

a) Complete geometry

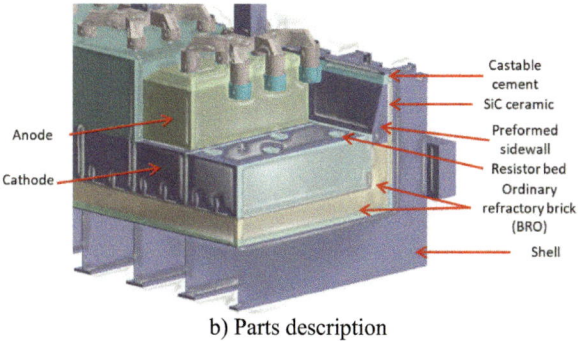

b) Parts description

Fig. 1 Discrete representation of the electrolysis cell during preheating

Once the transient TE analysis is solved, the temperature distribution evolution is exported to a thermomechanical (TM) analysis in order to evaluate the resulting stress distribution from ambient temperature to the end of the preheating phase.

Since it is well known that the shell undergoes permanent deformation, plasticity is considered for the steel shell material. In addition, it was necessary to include the damage of refractory materials, which translate to a lost in stiffness once they reach a stress limit, to represent properly the mechanical behavior of the cell.

Boundary Conditions

Thermo-Electrical Analysis

The boundary conditions used for the TE analysis and some key material properties are as follows:

– Preheat time of 44 h
– Current of 96,000 A at the connection surface between the positive risers and the anode beam

– Zero voltage (0 V) on the lower surface of the bus bar
– Ambient temperature of 37 °C at the exterior of the cell (measured by AAI)
– Initial temperature of 37 °C
– Ambient temperature inside the cell varying linearly from 37 to 140 °C (measured by AAI)
– Convection coefficient on surfaces inside the cell of 13 W/m^{2}K (calibrated previously using AAI measurements)
– Uniform electrical contact resistance (ECR) at the cast iron-carbon interface of 0.5 Ω*mm^2 is used (base on the measured voltage drop between the metal pad and the cathode barre during steady state)
– Electrical resistivity of the bath at preheating: 10 Ω*m below 879 °C (solidus) and 0.004232 Ω*m above 955 °C (liquidus)
– Electrical resistivity of the resistor bed: to be calibrated
– Convection coefficient of the shell: to be calibrated

Thermomechanical Analysis

The thermomechanical analysis only requires the representation of the shell and lining. Therefore, the boundary conditions used are the following:

– Gravitational loading
– Shell supported vertically at its support (second cradle from the end)
– Pressure applied to the cathodic plane representing the weight of the anode assemblies, bath (partially), and resistor bed
– Pressure applied to the surface of the pillar foot representing the weight of the superstructure
– Representative boundary conditions of the symmetries
– All contacts are bounded between components except:
  - the shell with lining (no normal and tangential restriction)
  - the cast iron connectors to cathode (no normal and tangential restriction) which include an initial uniform gap of 0.5 mm

Also, to represent properly the mechanical behavior of the cell, laboratory characterization was done at UQAC for obtaining the missing mechanical properties such as the stress limits of some refractories and the stress as a function of plastic strain of the steel used for the shell.

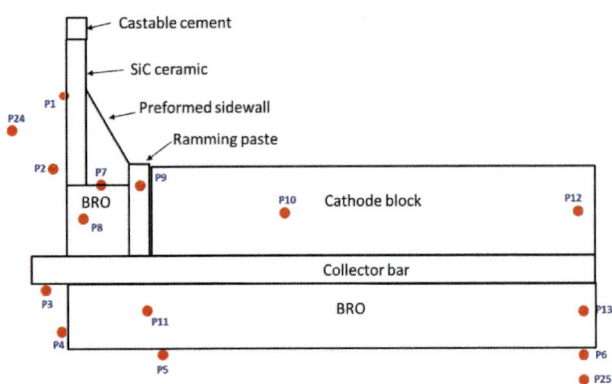

Fig. 2 Location of thermocouples

In Situ Measurements

The main goal of the in situ measurements consists of recording in-plant temperature and displacement of interest to calibrate some of the TE boundary conditions and also to validate the thermal and mechanical behavior predictions of the model.

Figure 2 shows an example of the thermocouple positions chosen strategically to optimise the model validation via the consideration of

– Boundary condition to calibrate
– Higher temperature gradient zone
– Ramming paste baking progression/state
– Current in cathode bar
– Possible thermal contact resistance between the components

Figure 3a shows the ultrasonic sensor installed underneath the shell to evaluate the vertical displacement at three different positions (Fig. 3b). Finally, a total of 92 channels were used each one being recorded at each minute.

Calibration and Validation

The electrical resistance of the resistor bed (mix of granular coke and graphite) needs to be calibrated since it's difficult to evaluate mainly because:

1. Once compressed, the bed height, effective diameter, and density change significantly
2. The electrical resistivity is sensitive to pressure, which varies with time and location, because of the thermal deformation of the cell components

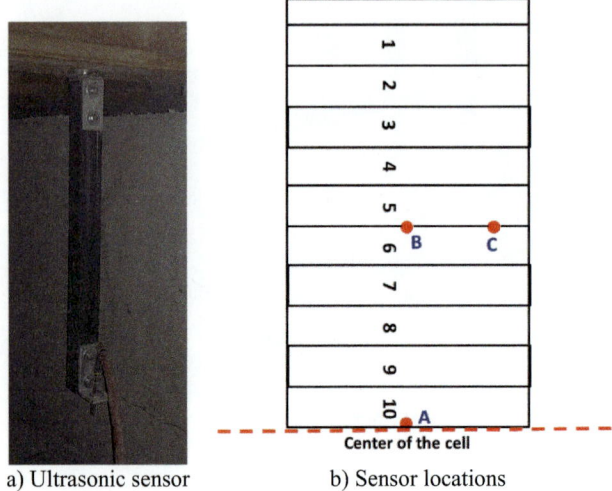

a) Ultrasonic sensor   b) Sensor locations

Fig. 3 Ultrasound used to measure the vertical displacement and the associated positions underneath the shell

Considering a compressed height of 16 mm and an effective diameter of 160 mm, the final calibrated electrical resistivity of the resistor bed is 0.0013 Ω.m. As for the convection coefficient of the exposed surfaces of the shell, the final value is 9.5 W/mm^2.

Figure 4 demonstrates the good agreement between the calibrated TE model results with the measured temperatures at positions P10, P2, and P13 (see Fig. 2).

As for the thermomechanical model validation, Fig. 5 compares the model results with the measured displacements at positions A, B, and C (see Fig. 3b for locations). The model, which takes into account the shell plasticity and the

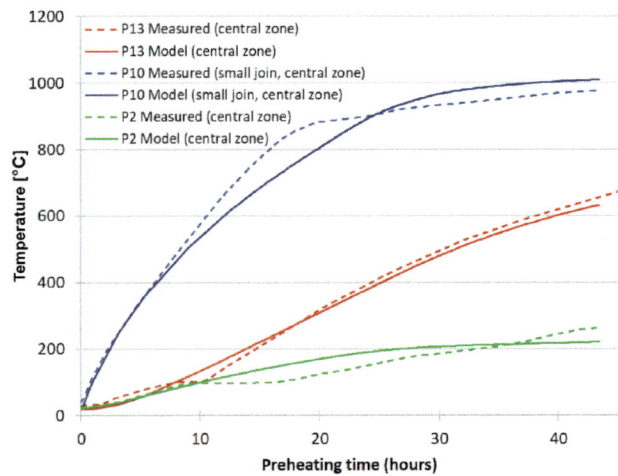

Fig. 4 Comparison of the predicted temperature with the measurements as a function of time

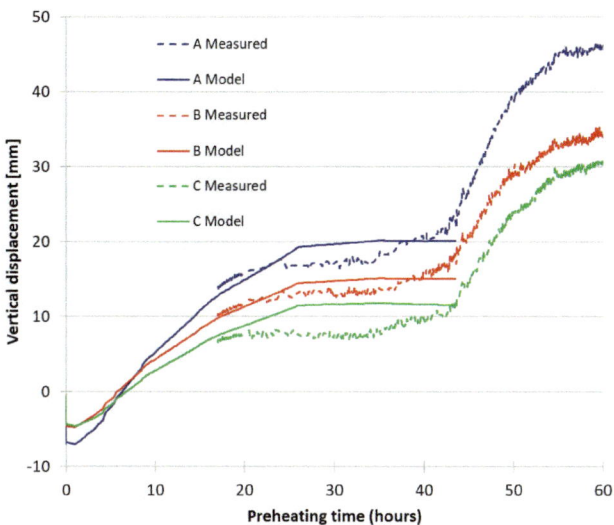

Fig. 5 Comparison of the displacement results predicted by the model with the measurements

refractory damage/lost of stiffness, gives a relatively good quantitative estimation of the measurements at the chosen positions.

Result and Discussion

Transient Thermoelectrical Model

In addition to predicting the temperature distribution needed for thermomechanical analysis, the transient thermoelectrical model allows to evaluate the temperature uniformity along the cathode plane for the actual technology. Also, it allows to have a better understanding of the outputs available in industry, namely, the cell voltage and the anodic current distribution in order to carry out the right intervention.

Figure 6 shows the temperature distribution at the end of the preheating period which is 43 h in this case. There is approximately a difference of 200 °C between the centre and

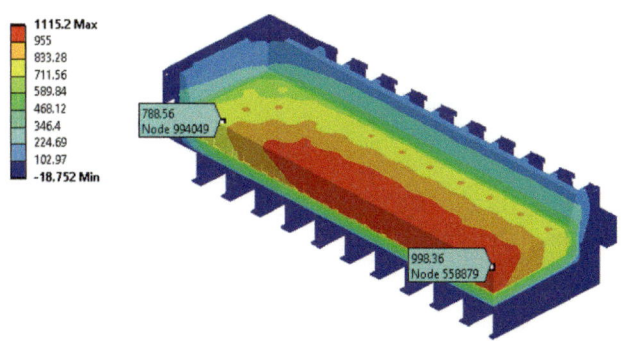

Fig. 6 Temperature distribution at the end of preheating [°C]

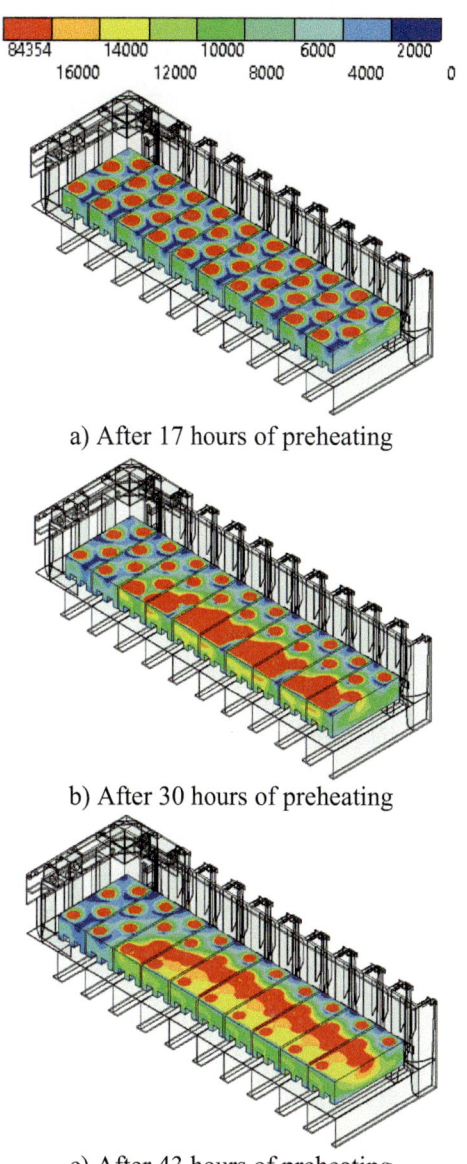

a) After 17 hours of preheating

b) After 30 hours of preheating

c) After 43 hours of preheating

Fig. 7 Current density distribution evolution [A/m^2]

the end of the cell (an offset of 230 °C was measured during the preheat follow-up procedure).

Corresponding to the resulting temperature distribution shown above, Fig. 7 presents the current density distribution evolution considering:

- That the solid bath occupies initially the entire space between the anode base and the cathode surface
- The bath thermal and electrical properties evolution during his phase change are considered (solidus and liquidus temperature of 879 and 955 °C)

Figure 7a shows a relatively uniform current distribution since the resistance of the resistor bed is uniform and the

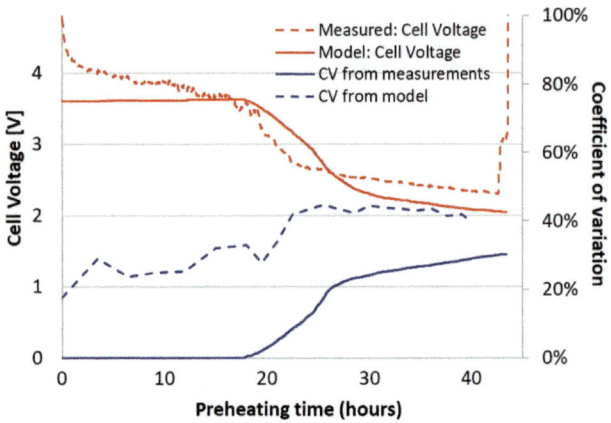

Fig. 8 Comparison of the cell voltage and coefficient of variation predicted by the model with the measurements

bath is not yet melted after 17 h. After 30 h of preheating (Fig. 7b), the bath melts only in the centre of the cell (heat lost coming from the shell side) leading to a less resistive path for the current in the centre even if the bath is 3 times more resistive than the resistor bed. Afterwards, the current distribution remains mostly non-uniform since the current prioritizes the centre/less resistive path (Fig. 7c).

As shown in Fig. 8, the behavior related to the impact of the partial melting of the bath correlates well with the in situ cell voltage evolution (decrease between 17 and 30 h). This behavior is also well represented by the coefficient of variation of the in situ anodic current distribution (used by AAI to evaluate the non-uniformity) which increase after 17 h (model and measurements).

Thermomechanical Model

The main objectives of the cell preheating are to minimize the damage of the lining materials and simultaneously prepare the cell for full production as soon as possible. Hence, parallel to the thermoelectrical model, a thermomechanical model is then used to evaluate the stress and possible damage in the materials.

For a better understanding of the results, the thermomechanical model was firstly used without loss in stiffness of the lining materials reaching their strength limits. In that case, the Mohr–Coulomb criterion is used during post-processing to study the potential damaged regions. This criterion considers the distinct character in tension and compression of the material via the following expression:

$$\sigma^* < 1 \tag{1}$$

where σ^* represents the Mohr–Coulomb ratio (M-C ratio) and is given by

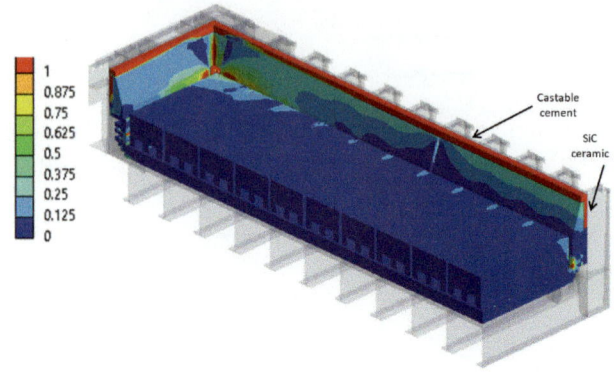

Fig. 9 M-C ratio at the end of preheating without considering the linings damage/stiffness loss

$$\sigma^* = \frac{\sigma_1}{S_{ut}} + \frac{\sigma_3}{S_{uc}} \tag{2}$$

where σ_1 and σ_3 represent, respectively, the maximum and the minimum principal stresses in the material. S_{ut} and S_{uc} represent the tensile and compressive ultimate strengths of the material.

Figure 9 shows the M-C ratio at the end of the preheating period ignoring the loss in stiffness in the lining materials. The results show that the M-C ratio reaches critical values (over 5) in the castable cement and the SiC ceramic (see Fig. 1b for location).

The results presented in Fig. 10 facilitate the understanding of the source of the high M-C ratio at the top of the lining. This figure shows the amplified deformation of the thermal strain (longitudinal direction) when the thermal loading is only applied to a) the shell b) the lining. Results show that the temperature distribution of the shell leads to a

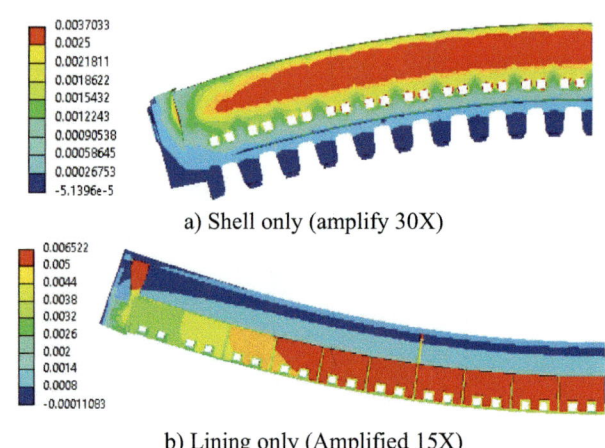

a) Shell only (amplify 30X)

b) Lining only (Amplified 15X)

Fig. 10 Deformation of the thermal strain (longitudinal direction) when only the thermal loading is applied

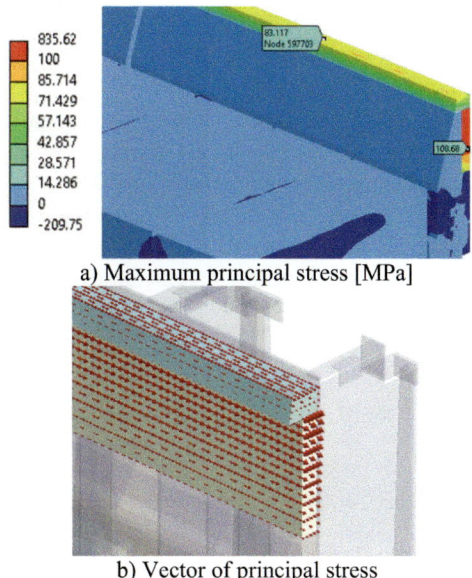

a) Maximum principal stress [MPa]

b) Vector of principal stress

Fig. 11 Vector of principal stress of the high M-C ratio region and the contact pressure between the lining and the shell

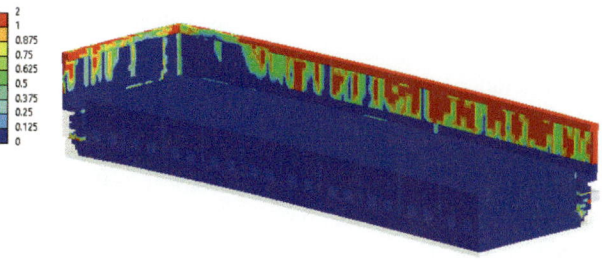

Fig. 12 Damage level of the lining at the end of the preheating phase

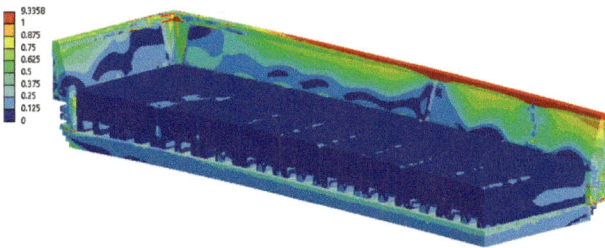

Fig. 13 M-C ratio at the end of preheating considering the lining damage/stiffness loss

positive vertical displacement of 32 mm in the centre of the cell caused by the greater thermal deformation at the top of the shell compared to that at the bottom. Conversely, the temperature distribution of the lining leads to a negative vertical displacement of -61 mm in the centre of the cell caused by the greater thermal deformation at the bottom part of the lining compared to that at the top.

Considering that the contact between the shell and lining has no normal or tangential stiffness (no friction, allow detachment), the shell forces the lining to bend in the opposite direction (or vice-versa). This leads to massive tension on the top part of the lining (castable cement and SiC mostly). Figure 11 shows the maximum principal stress (tension) and the principal tension vector at the top parts of the lining. Results show that the amplitude and orientation (longitudinal direction) correlate with the described behavior and explain the high M-C ratio (tension dominant) of those materials.

Since the high M-C ratio add an unrealistic stiffness to the entire lining (bounded contact), the vertical displacement of the shell at the end of preheating is underestimated by 15.9 mm compared to the measurements shown in Fig. 5 (6.1 mm vs 22 mm). It is then necessary to consider the loss of stiffness of these materials once they get damaged in order to represent properly the mechanical behavior of the cell.

An elastic model with damage is used since the refractory materials behave in a brittle way. The damage appears after reaching a maximum stress in the form of a radical loss of rigidity. These stress limits correspond in the model to the ultimate strength in compression and in tension of the specified material. Also, the shear stress limit is estimated

using the Mohr–Coulomb failure criterion. Once the material reaches one of these limits, the stiffness is instantly reduced by 95% (partially reduced for model stability reasons).

Figure 12 shows the resulting damage status (1 and above is 95% stiffness reduction). Results show that only the castable cement and SiC ceramic reach their damage state leading to a stiffness reduction and stress redistribution in the assembly.

Considering the loss of stiffness of the castable cement and SiC ceramic, Fig. 13 presents the resulting M-C ratio at the end of the preheating phase. Results show that the cathode block remains well under its stress limits. However, the top of the preformed side wall has now a M-C ratio of 0.7 approximately. Also, results show that the carbon paste dilation joint allows a local reduction of the tensile stress of the preformed sidewall. Finally, with the damage consideration, the vertical displacement of the shell at the end of preheating is now underestimated by only 1.4 mm compared to the measurements shown in Fig. 5 (20.6 mm vs. 22 mm).

Conclusion

This paper describes the development of a quarter cell model for the evaluation and prediction of the thermal, electrical, and mechanical behavior of Hall-Héroult cell under electrical preheating. To do so, the transient thermoelectrical analysis, weakly coupled with a thermomechanical analysis, was used. Specifically, the thermomechanical analysis considers the plasticity of the shell and the damage initiation of the refractories leading to a loss of stiffness. Laboratory characterization was done at UQAC to evaluate these key properties.

A calibration/validation of the model was then performed using in situ thermal, electrical, and mechanical measurements at Aluminerie Alouette Inc. Comparison between the thermal (temperature distribution), electrical (cell voltage and anode current distribution), and mechanical (vertical displacement of the shell) results of the model are in good agreements with the in situ measurements.

The evaluation of the transient thermoelectrical model results shows the presence of a non-uniform temperature distribution along the cathode surfaces at the end of preheating. Other than the heat loss from the side shell, the melting of the bath at the centre of the cell (after 20 h) contributes to this non-uniformity since it lowers the resistance path at the centre, thus leading to a non-uniform anodic current distribution accordingly.

The resulting thermomechanical analysis demonstrates the mechanical behavior of the lining and the shell from the thermal loading, which explains the high stress in tension taking place at the top of the lining (castable cement and SiC ceramic mostly). Also, the model shows that the cathode block and the refractories remain under the stress limits during this period.

The actual model has the potential to suggest improved pattern configurations/geometry and procedures to optimise the preheating phase via sensitivity analysis. It can also be used to evaluate the behavior for the start-up, early operation, and steady-state phases.

Note that the model can be significantly improved by considering for instance: a more realistic damage model for brittle materials, the initial gap distribution at the cast-iron/cathode interface caused by the sealing process, the electrical contact conductance evolution of the anode and cathode assembly interfaces (function pressure and temperature), the thermal contact conductance evolution between the shell and lining (function gap), and the pressure variation on the resistor bed leading to a resistance variation along the pattern (due to cell deformation).

Acknowledgements The author acknowledges the financial support of the Fonds de recherche du Québec—Nature et technologies, The Aluminum Research Center—REGAL, and particularly to our industrial partner Aluminerie Alouette Inc.

References

1. Daniel Marceau, Simon Pilote, Martin Désilets, Jean-François Bilodeau, Lyès Hacini & Yves Caratini, "Advanced numerical simulation of the thermo-electro-chemo-mechanical behavior of Hall-Heroult cell under electrical preheating", Light Metals 2011, pp. 1041–1046.
2. Sorlie M., Oye H. (2nd Edition) (1994) Cathodes in Aluminium Electrolysis. Aluminium-Verlag, Germany.
3. Michel Reverdy & Vinko Potocnik, "A historical review of aluminium reduction cell start-up and early operation", Light Metals 2022, pp. 991–997.
4. Carlos Zangiacomi, Victor Pandolfelli, Leonardo Paulino, Stephen Lindsay, Halvor Kvande, "Preheating study of smelting cells", Light Metals 2005, pp. 333–336.
5. Marc Dupuis, "Thermo-electro-mechanical modeling of Hall-Heroult cell coke bed preheating", Light Metals 2001, pp. 757–761.
6. Marc Dupuis, Ghasem Asadi, "Thermal study of the coke preheating for Hall-Heroult cell", Light Metals 1993, pp. 93–100.
7. Daniel Richard, Patrice Goulet, Marc Dupuis & Mario Fafad, "Thermo-chemo-mechanical modeling of a Hall-Heroult cell thermal bake-out", Light Metals 2006.

Simplified 3D MHD Model for Quick Evaluation of Aluminium Electrolysis Cell Design

Ievgen Necheporenko, Alexander Arkhipov, and Abdalla Alzarouni

Abstract

A mathematical model coupling 3D electrical current distribution, magnetic field and magnetohydrodynamic (MHD) behaviour of aluminium electrolysis cell is presented. COMSOL Multiphysics® is used as a software tool to develop the present model. Implementation of electrical current carrying busbars using line current source elements and evaluation of the MHD with the help of shallow-water equations method allow for a quick and robust parameter estimation. Sensitive modelling details parameterization, and automated report generation features of the model provide an easy-to-use solution for preliminary design evaluation. The modelling results were compared with 3D COMSOL model developed before and validated with the measurements. Several simulation scenarios are presented in this paper showing the capability of the model to estimate the influence of the aluminium cell constituents on the MHD behaviour of the cell.

Keywords

MHD • COMSOL Multiphysics® • Aluminium cell design • Magnetic field

Introduction

The list of the models below, along with the challenges described, show the motivation for development of a simplified MHD model, which will increase the efficiency of the cell design process. In general, the models used in EGA can be classified by the physical phenomena. Another distinguishing criterion might be the purpose of the particular model used. Figure 1 shows the classification and mutual dependencies of the existing models by the analysis type, even though there are cases where several different areas are coupled.

Figure 1 shows the coupling required for a smooth and accurate evaluation of the relevant aluminium cell design criteria. This paper will focus on the simulation of the cell magnetic field and MHD behaviour, which rely on the inputs from other models (thermo-electrical busbar model, for instance) according to the dependencies shown in Fig. 1.

To get a better understanding of the motivation for a simplified MHD model development, it is worth looking into the challenges of the existing approach used for the evaluation of the magnetic field and MHD behaviour of the cell. Two approaches have been used in EGA:

- The first one considers the metal and bath as two thin layers by integrating the variables over the thickness of each liquid (shallow-water approach). An example is MHD-Valdis software. Technically, MHD-Valdis package computes electrical current and magnetic field distribution, which are used for MHD calculation. It is a practical tool for cell busbar design [1]. MHD-Valdis gives the solution for the electrical part of the problem, provides the 3D distribution of the magnetic field and evaluates the MHD behaviour of the cell, using the shallow-water approach.

- The second approach uses full 3D geometry of the metal and bath with immersed anodes, and gives 3D MHD behaviour of the metal and bath. Examples of this approach are ESTER-PHOENICS [1] and recently developed MHD model using COMSOL software package [2]. The COMSOL model comprises the thermo-electrical busbar model which produces electrical current distribution used for the magnetic field calculations as an input to the MHD model. Results of this stage are exported into a standalone full 3D MHD model in COMSOL. Depending on the details required by the numerical method used, 3D

I. Necheporenko (✉) · A. Arkhipov · A. Alzarouni
Emirates Global Aluminium, Dubai, United Arab Emirates
e-mail: inecheporenko@ega.ae

© The Minerals, Metals & Materials Society 2023
S. Broek (ed.), *Light Metals 2023*, The Minerals, Metals & Materials Series,
https://doi.org/10.1007/978-3-031-22532-1_103

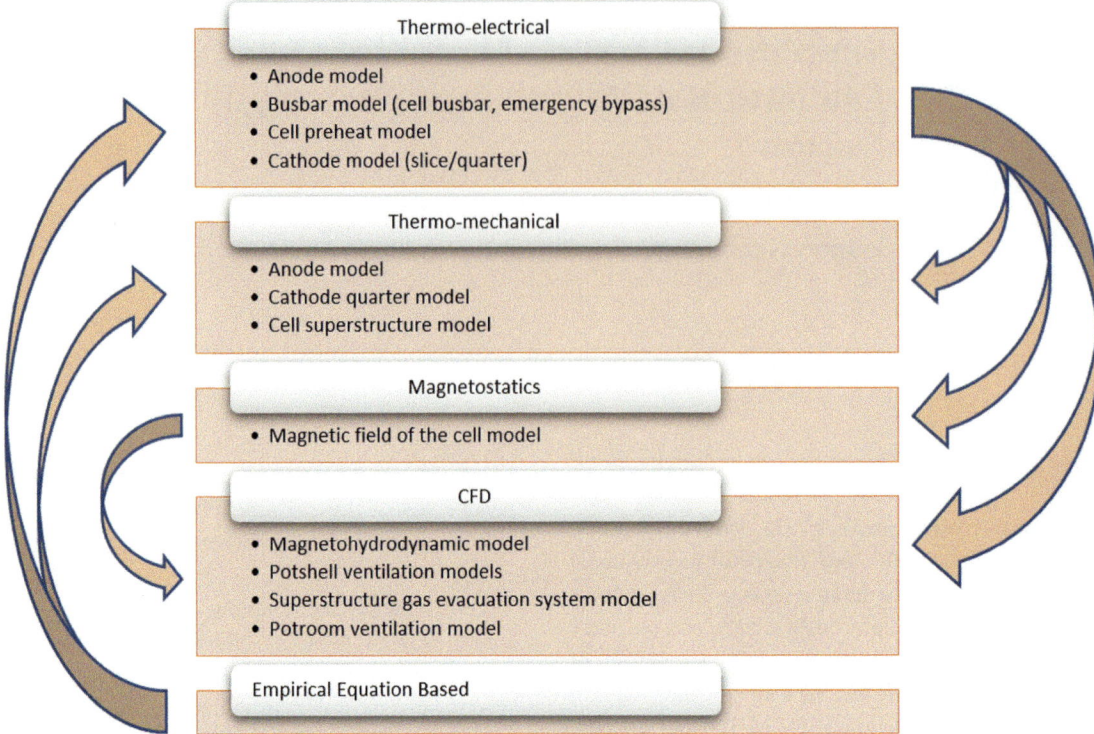

Fig. 1 Classification of the models used for aluminium cell design evaluation

models may be computationally intensive and may require long computer run time. This slows down the design process in which many scenarios have to be evaluated in a short time.

In this paper, the simplified MHD model developed on COMSOL will be compared to the COMSOL 3D MHD model.

Description of the Approach

The aim of the approach presented here was to add the capability of quick MHD evaluation. COMSOL Multiphysics® is used to develop the model. The structure and GUI of the software package allow to combine several models within one framework where variables and results from one model may serve as inputs for another. In general, every COMSOL model is structured with the so-called "Interfaces" [4]. Each interface represents a set of equations and constraints (related to physics) applied to a particular volume, surface, edge, or point. A list of interfaces used within the simplified 3D MHD model is given below

- Electrical currents. Interface for calculating electrical current distribution in 3D model elements such as cathode blocks, collector bars, cast iron, molten liquids, etc.

- Magnetic fields. Interface for calculating magnetic field generated by the electrical currents, evaluated in the previously listed interface along with the electrical currents, defined as line conductors, oriented, and placed as desired.

- Electrical currents on deformed metal-bath interface. Interface for evaluation of the electrical current distribution in the molten electrolyte and aluminium considering metal heave.

- SWE (shallow-water equations [3]) interface is used for embedding electrical currents and magnetic fields for the calculation of metal and bath flow, averaged over the height of each liquid.

Typical computational domain used in the model for the magnetic field evaluation consists of the following geometry entities:

- 3D cathode assembly, including collector bars, cast iron, cathode blocks, and other relevant design details.
- Molten aluminium and electrolyte (volume enclosed by anode-cathode distance, ACD).
- Simplified potshell geometry (3D surfaces prescribed with magnetic shielding boundary condition).
- Simplified busbar geometry (3D edges prescribed with an electrical current passing through the conductors).
- Volume of the air wrapping the modelled cell.

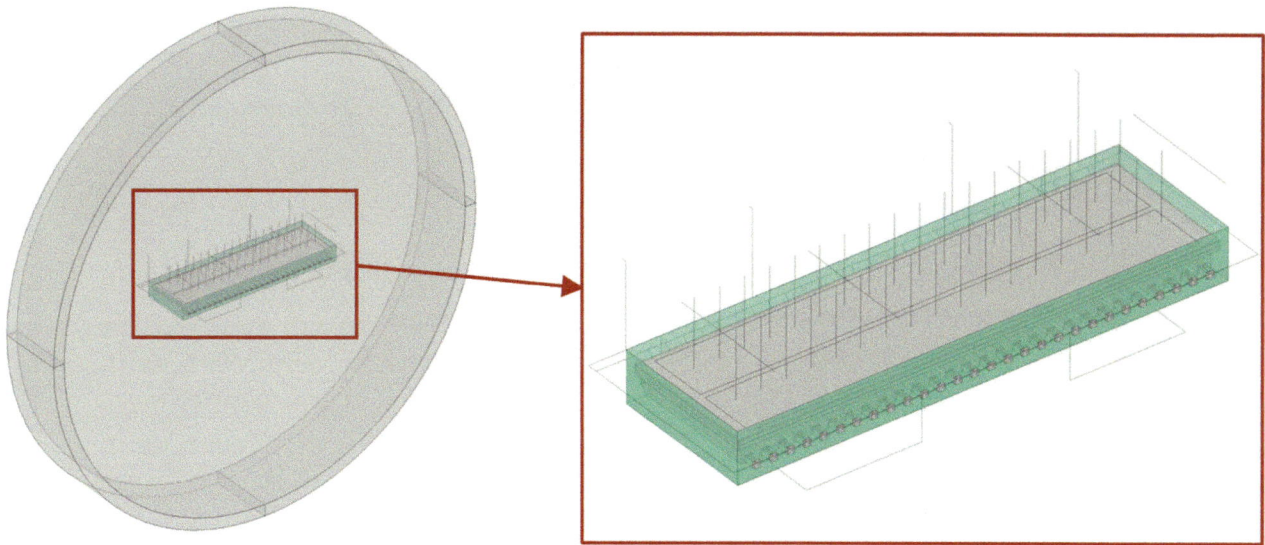

Fig. 2 3D CAD geometry of the computational domain, shown on the left

Figure 2 shows a general view of the geometry, including the computational space around the modelled cell.

Periodic boundary conditions are applied at the computational domain surfaces to account for the influence of upstream and downstream cells of the potline. Typical workflow for the model includes the following steps:

1. Set the desired parameters describing cathode assembly, busbar layout, and anode panel.
2. Assign desired electrical currents in selected busbars/conductors.
3. Run coupled calculations of the electrical currents interface along with magnetic field interface.
4. Run coupled calculations of the deformed metal-bath interface with electrical current interface, along with SWE interface.
5. Post-process magnetic field variables (3D distribution of the B_x, B_y and B_z components of the magnetic field), molten aluminium flow, and metal heave.

One of the important criteria of the model is computational time. For the most common evaluation scenario, step 3 takes approximately 15 min, and step 4 takes another 10–15 min depending on the metal heave. The computational time specified for step # 3 is achieved by running COMSOL in a standard non-parallel mode, solving approximately 2.5 million degrees of freedom (DOFs) using a workstation with Intel Xeon Gold 6146 3.2 GHz CPU and 384 GB of RAM. Step # 4 specified computational time corresponds to solving approximately 50 000 DOFs using the same hardware as in step # 3.

Results

As the initial intention for the simplified MHD model development was to investigate the influence of the electrical current distribution across busbar layout, cathode assembly and molten aluminium on the magnetic field and MHD behaviour of the electrolysis cell, the most relevant results are the magnetic field distribution in the metal pad, metal velocity pattern and the metal-bath interface deformation. Hence, these are the outputs of the model. Even though the method used for evaluating the liquid metal flow does not account for some of the 3D geometry details, the results are reasonable for the early design stage.

The subsequent steps, required to get the MHD results, bring us useful findings in terms of 3D electrical current distribution in the cathode assembly and the current distribution in molten liquids (aluminium and bath). Geometry parameterisation along with the simplified representation of the cathode and anode ring busbar, allow for fast evaluation of the busbar and cathode assembly designs. 3D magnetic field distribution is also calculated across the full model domain.

The magnetic field model was validated by comparing the results with MHD-Valdis, run for the same cell design. Below we compare the distribution of B_x and B_y components of the magnetic field (Figs. 3 and 4).

MHD results were validated by comparing the present SWE model with the previously developed COMSOL 3D MHD model (which had been validated with measurements and compared to MHD-Valdis). Figures 5 and 6 show metal

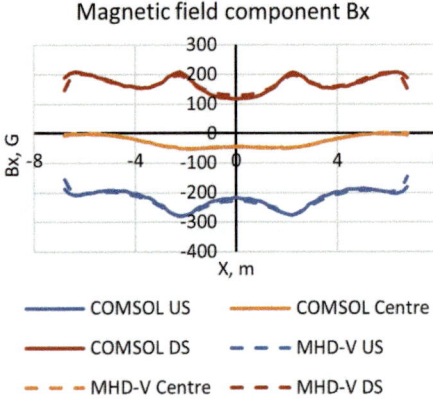

Magnetic field component Bx

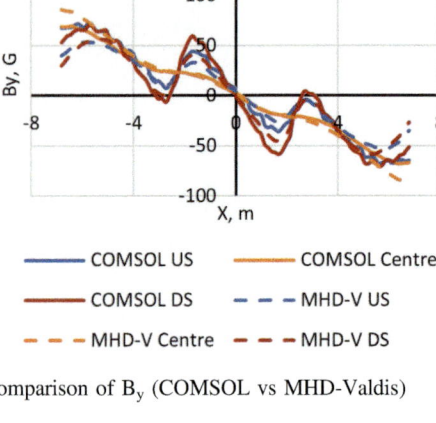

Magnetic field component By

Fig. 3 Comparison of B_x (COMSOL vs MHD-Valdis)

Fig. 4 Comparison of B_y (COMSOL vs MHD-Valdis)

velocities obtained from the simplified MHD model (Fig. 5) and full 3D MHD COMSOL model (Fig. 6).

The new model was used to show the influence of the aluminium electrolysis cell constituents on the magnetic field and MHD of the cell. The following cases were investigated:

- Case 1. The model includes potshell, superstructure, cathode busbar, anode risers, anode beam with cross-beams, 3D cathode assembly, molten aluminium and electrolyte.
- Case 2. Case 1 was modified by removing the superstructure from the model (investigate the influence of the superstructure on the magnetic field and MHD).

- Case 3. Case 1 was modified by removing the potshell (investigate the influence of the potshell on the magnetic field and MHD).
- Case 4. Case 1 was modified by removing the anode beam from the model (investigate the influence of the anode beam on the magnetic field and MHD).
- Case 5. Case 1 was modified by removing the anode rods from the model (investigate the influence of the anode rods on the magnetic field and MHD).
- Case 6. Case 1 was modified by changing the collector bar design (the steel portion of the collector bar was substituted with copper so that the electrical resistance of the collector bar decreased).

Fig. 5 Simplified MHD SWE model, metal velocity patterns, m/s

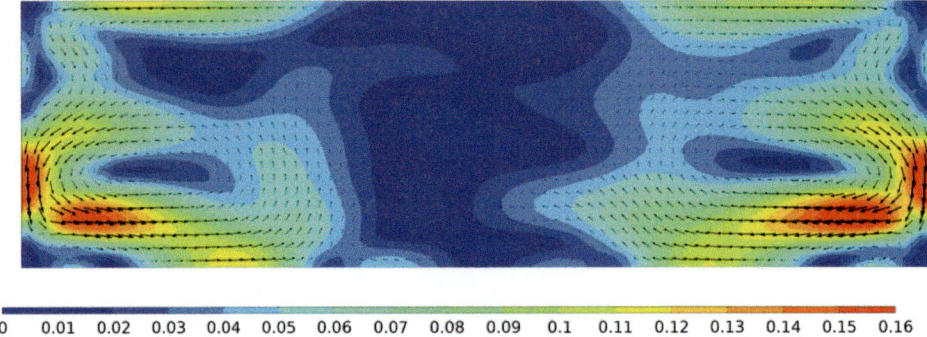

Fig. 6 Full COMSOL 3D MHD model, metal pad velocity pattern, m/s

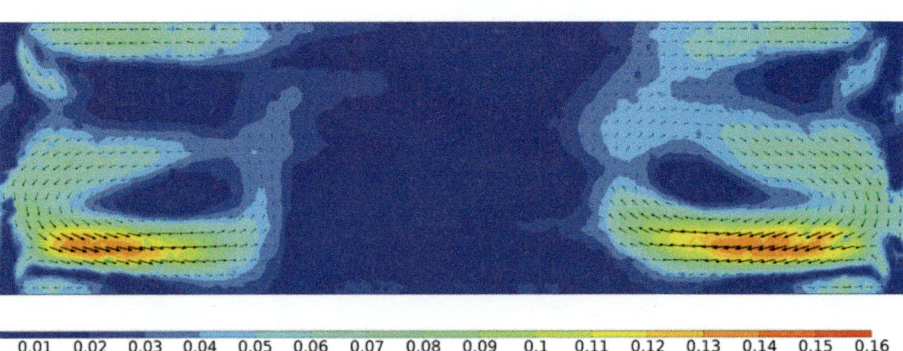

Fig. 7 Case 1 model geometry

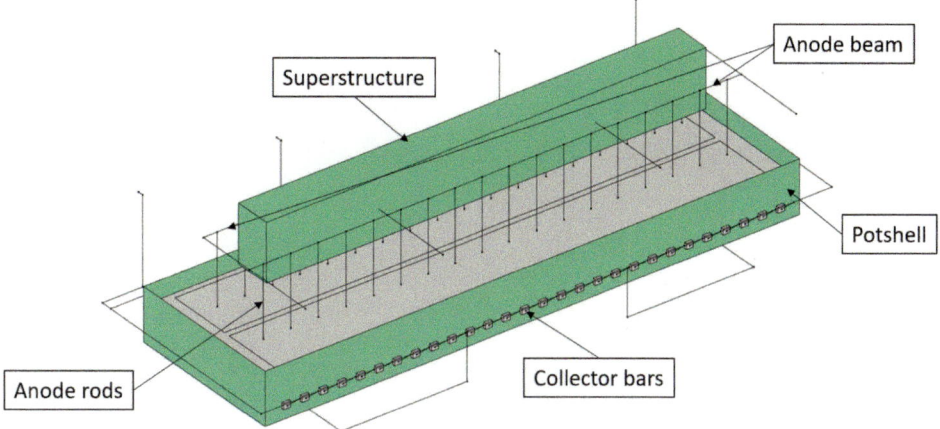

Fig. 8 Locations for data extraction in the middle of the liquid aluminium

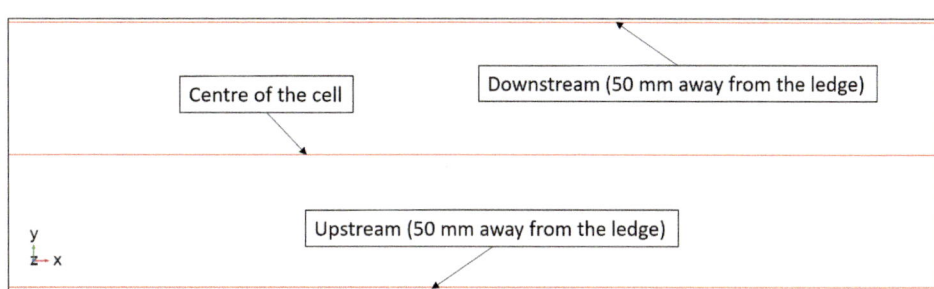

Figure 7 shows the geometry used for Case 1. In other cases, geometry was modified according to the description above.

Due to limited space, it is not possible to present all the comparisons. Besides, some cases do not affect significantly the magnetic field or MHD. Hence, the results presented below correspond to the cases where the influence of the particular design components is clearly seen. Figures 9, 10, 11, 12, 13, and 14 show the magnetic flux densities in the middle of the liquid aluminium at the upstream, centre, and downstream of the cell. Locations of the data extraction are shown in Fig. 8.

It is well-known that the potshell significantly affects the magnetic field distribution and several publications have already reported this result. Nevertheless, case 3 was considered for verification and validation. Figures 9, 10 and 11 show the influence of the potshell on the x, y, and z components of the magnetic field in the middle of liquid aluminium pad.

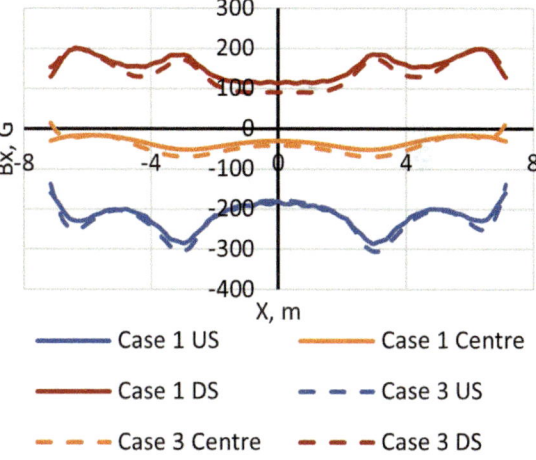

Fig. 9 B_x component of the magnetic field

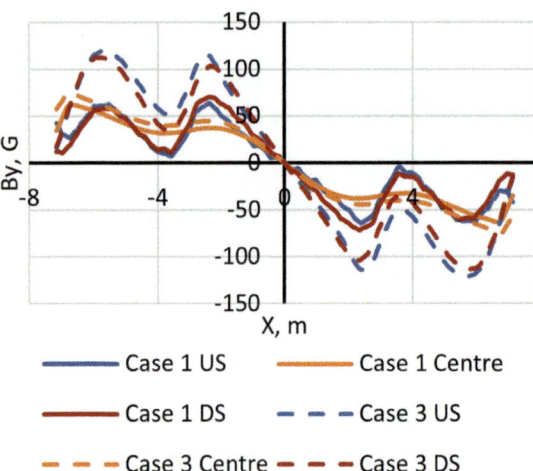

Fig. 10 B_y component of the magnetic field

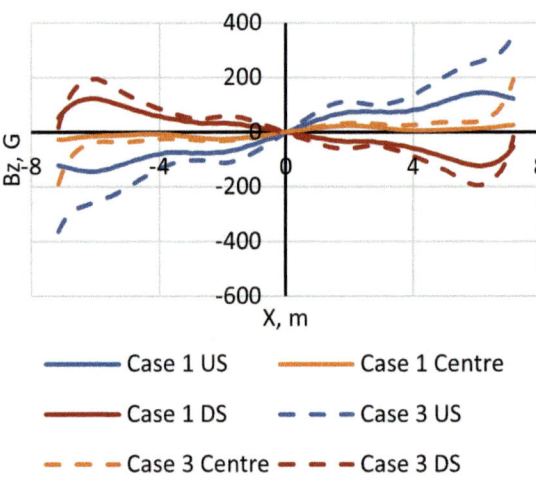

Fig. 11 B_z component of the magnetic field

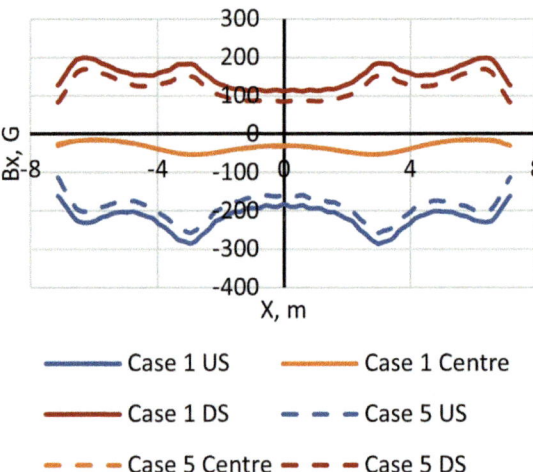

Fig. 12 B_x component of the magnetic field

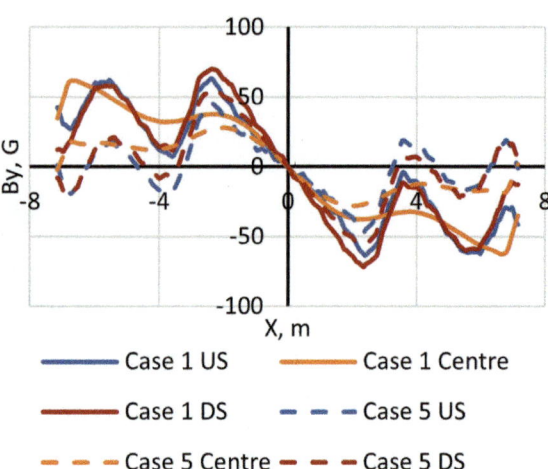

Fig. 13 B_y component of the magnetic field

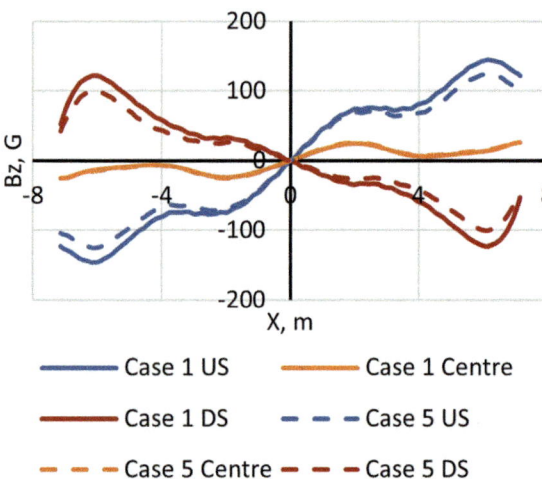

Fig. 14 B_z component of the magnetic field

Anode rods also have a tangible impact on magnetic field distribution. Figures 12, 13, and 14 show that the electrical current, passing through the anode rods, affects all the components of the magnetic field. Case 6 was chosen to show that the model can evaluate electrical current distribution depending on the cathode assembly design changes (collector bars in particular). Figure 15 shows a comparison of the transverse horizontal electrical current density distribution in the middle of the metal pad for case 1 and case 6.

Fig. 15 Horizontal electrical current densities in the central cross-section of the cell in the middle layer of the metal

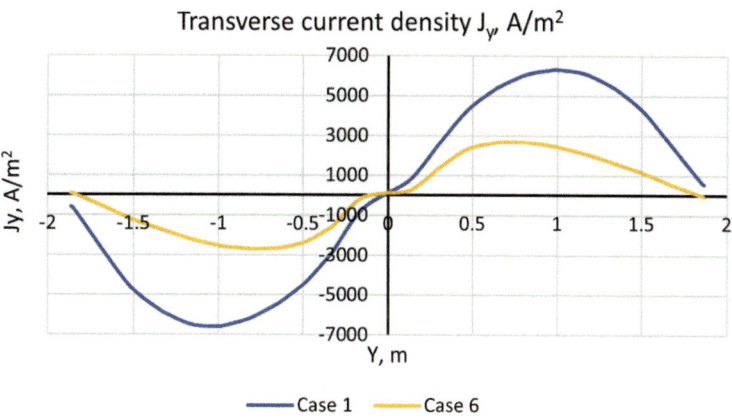

Transverse current density J_y, A/m^2

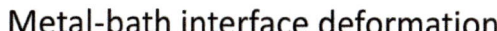

Fig. 16 Metal-bath interface deformation in the centre of the cell

Metal-bath interface deformation

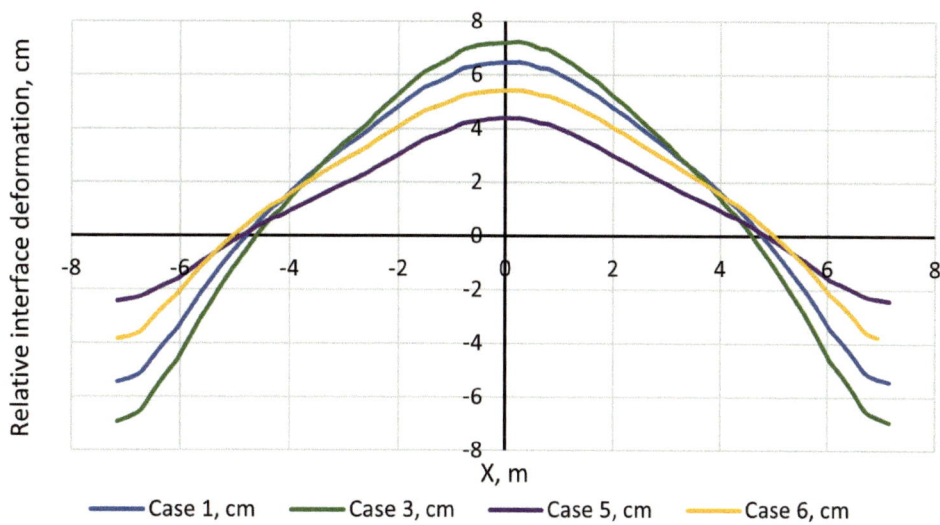

The metal-bath interface relative deformation (with respect to the undeformed flat shape) in the centre of the cell, along the cell, is compared in cases 1, 3, 5, and 6 in Fig. 16. Metal pad velocities are shown in Figs. 17, 18, 19, and 20.

It is difficult to quantitatively compare the influence of different scenarios on the overall magnetic field and electrical current distribution.

Hence, a qualitative analysis of the cases considered was carried out and the results are shown in Table 1.

Fig. 17 Case 1 metal pad
velocity vectors, m/s

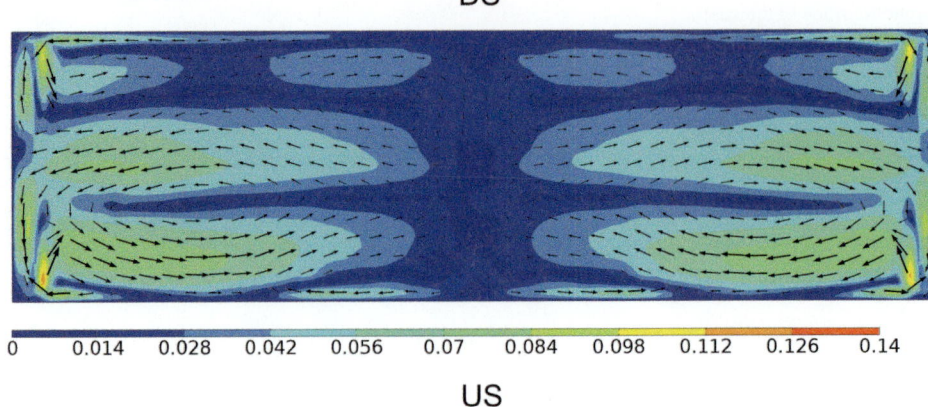

Fig. 18 Case 3 metal pad
velocity vectors, m/s

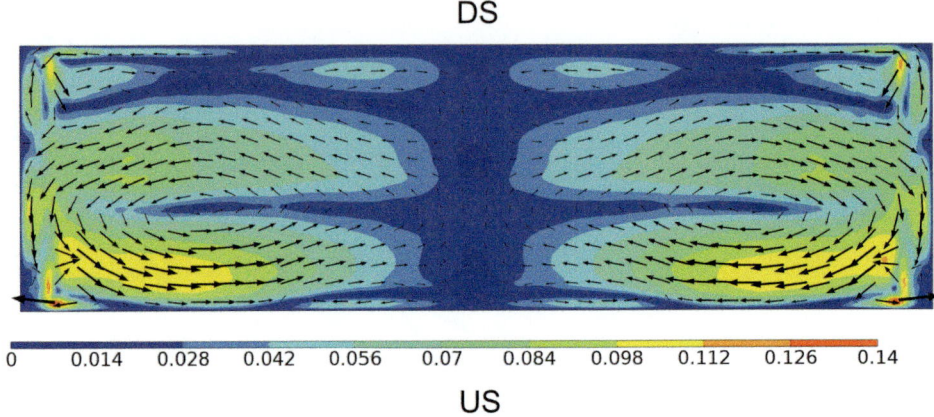

Fig. 19 Case 5 metal pad
velocity vectors, m/s

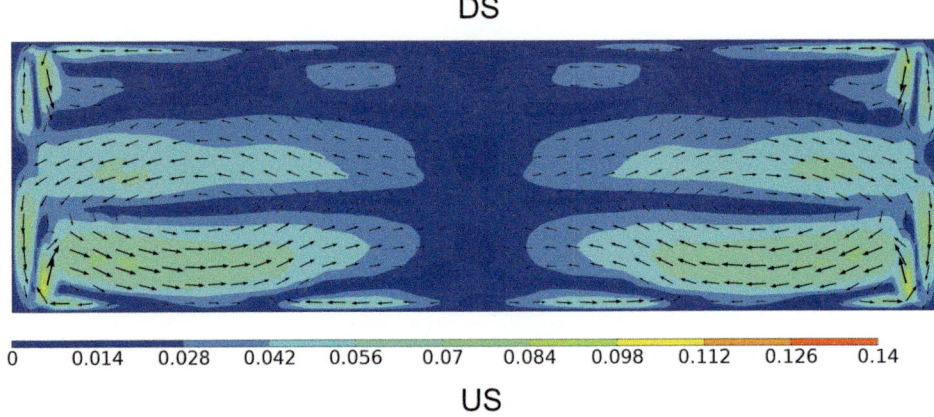

Fig. 20 Case 6 metal pad velocity vectors, m/s

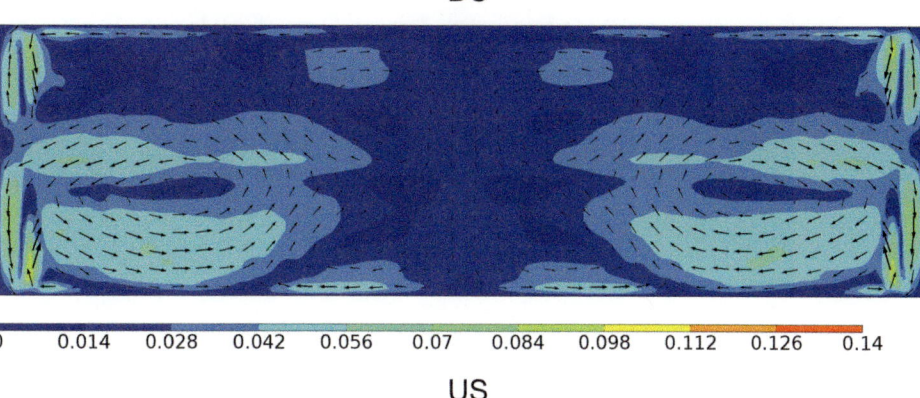

Table 1 Qualitative comparison of the evaluated scenarios. Impact of cell constituents on the magnetic field, metal heave, and metal velocity

Influence of	Magnetic field	Metal heave	Metal velocity
Case 2: Superstructure	Weak	Weak	Weak
Case 3: Potshell	Strong	Medium	Strong
Case 4: Anode beam	Weak	Weak	Weak
Case 5: Anode rods	Medium	Strong	Medium
Case 6: Collector bar design	Weak	Strong	Medium

Way Forward

One of the main drawbacks of the simplified COMSOL model is the absence of cell stability simulation. This is one of the key aluminium electrolysis cell design criteria and its evaluation has to be introduced, which will add significant practical value but also longer computational time. Another feature that will be added as an optional capability is the thermo-electrical calculation, which will give the opportunity to quickly switch the mode where an engineer will be able to balance electrical currents across the busbars after an initial layout has been selected. These improvements will enhance the value of the simplified COMSOL model for quick early design evaluations.

Conclusions

The modelling approach described in this paper can be used for MHD evaluation of the aluminium electrolysis cell designs at early development stages. Parameterization of the busbar elements, position and orientation, 3D cathode assembly dimensions, and desired electrical currents in the busbars makes the design investigation easier and faster. Simplifications used (such as surfaces for the potshell and superstructure instead of 3D geometry, line conductors for the busbar, periodicity boundary conditions, etc.) reduce computation time. It takes approximately 20 min to get magnetic field distribution using this simplified MHD

model, compared to several hours required for the same using the 3D COMSOL magnetic field model.

The simplified MHD model gives the liquid aluminium flow averaged across the thickness of the metal pad (shallow-water equations approach), and the metal-bath interface deformation. These results can be used for rapid preliminary design investigations. Full 3D MHD modelling is still required at later stages of the design development in order to account for complex 3D geometry of the cell liquid zone. This paper shows the use of the simplified COMSOL MHD model to predict the influence of the aluminium electrolysis cell design constituents on the distribution of the electrical current and magnetic field in the liquid metal pad as well as on the metal flow of a cell.

References

1. Zarouni A, Mishra L, Bastaki M, Al Jasmi A, Arkhipov A, Potocnik V. (2013). Mathematical model validation of aluminium electrolysis cells at DUBAL. In: Sadler, B.A. (eds) Light Metals 2013. The Minerals, Metals & Materials Series. Springer, Cham. https://doi.org/10.1007/978-3-319-65136-1_102
2. Necheporenko I, Arkhipov A., Modelling of the magnetic field and magnetohydrodynamic behaviour of a typical aluminium cell using COMSOL. Paper presented at COMSOL conference 2020 Europe section (Online venue).
3. Fe J, Cueto-Felgueroso L et. al., Numerical viscosity reduction in the resolution of the shallow water equations with turbulent term, Int. J. Numer. Meth. Fluids 2000; 00: 1–6.
4. COMSOL Multiphysics® 6.0 User's Guide.

Achieving Low Pot Failure Rate at Aditya Aluminium

Atanu Maity, Venkannababu Thalagani, Deepak Das, Bhanu Shankar, Anish Das, Kamta Gupta, Madhusmita Sahoo, Shanmukh Rajgire, and Amit Gupta

Abstract

At Aditya Aluminium, the average pot life is about 2000 days, pot relining contributes approximately 2.1% of total production cost. To avoid the risk of pot failure, an average of 55–60 pots relining/year are planned out of 360 pots. Further, to reduce failure, an in-house pot life estimation method was developed, which helped in planned pot stoppage. Firstly, pots were prioritized based on the pot age, iron & silica, side shell temperature (SST), Current Efficiency and specific energy consumption. These factors were assigned with individual weightage, which helped in pots shut down, further pot autopsy was performed for validation. A strategy was adopted for measuring regular SST, observing iron/silica pick-up, and Fe/ Mn ratio. Moreover, forced cooling regulation, and selective collector bar cutting were adopted to extend pot life. These methods helped Aditya Aluminium establish a global benchmark among other modern smelters by achieving a pot failure rate ∼0.6% since inception.

Keywords

Pot failure • Pot life • Global benchmark • Side shell temperature • Critical pots

A. Maity (✉) · D. Das · B. Shankar · A. Das · K. Gupta · M. Sahoo
Hindalco Industries Ltd, Lapanga, Odisha 768212, India
e-mail: atanu.maity@adityabirla.com

V. Thalagani · S. Rajgire · A. Gupta
The Aditya Birla Science and Technology Company Pvt. Ltd., Taloja MIDC, Navi Mumbai, Maharashtra 410208, India

Introduction

Aluminium is produced through an electrolytic process in a Hall-Héroult cell or pot, which consists of a steel shell lined with cathodes, collector bars, insulations, and refractories materials as shown in Fig. 1. These cell lining materials offer thermal, mechanical, and chemical stability during cell operation. The cell lining is typically made of layers, comprising a bottom layer of calcium silicate-based insulation, which is protected by alumina silica-based refractory bricks. These refractories house a carbon cathode embedded with steel collector bars, to facilitate the current passage out of the cell.

The cell lining plays an important role in process efficiency and energy optimization. The cell relining cost contributes to around 2–3% of the total production cost as well as it leads to production loss for 5–6 days per pot. The typical life of Aluminium reduction cells is about 5–6 years; however, it may get affected based on certain factors such as cell amperage and technology, lining design, quality of lining materials, preheating & startup procedure, process parameters, operational practices and power disturbances.

Aditya Aluminium smelter is based on the AP 36 technology with 360 pots, these pots are operating above 369 kA, with average life of about 2000 days. All the 360 pots cannot be shut down at once because it involves the design and operational constraints during pot startup, lining resources, and financial constraints. So it should be taken up in a planned manner. Hence it demands precise planning and execution for relining the pots to avoid any sudden risk of failure and safety incidences. Premature pot failure leads to an increase in turnaround time (TAT), production loss, safety risk, and energy loss, in extreme cases it might also lead to the shutdown of the entire potline. Pot failure may occur due to some critical factors like high pot age, power failures, lining and start-up issues, and pot abnormalities. The pot failure may be indicated by a few symptoms like high side and bottom shell temperature, collector bar temperature, high localized cathode current distribution, severe

© The Minerals, Metals & Materials Society 2023
S. Broek (ed.), *Light Metals 2023*, The Minerals, Metals & Materials Series,
https://doi.org/10.1007/978-3-031-22532-1_104

Fig. 1 Aluminum reduction cell lining layout

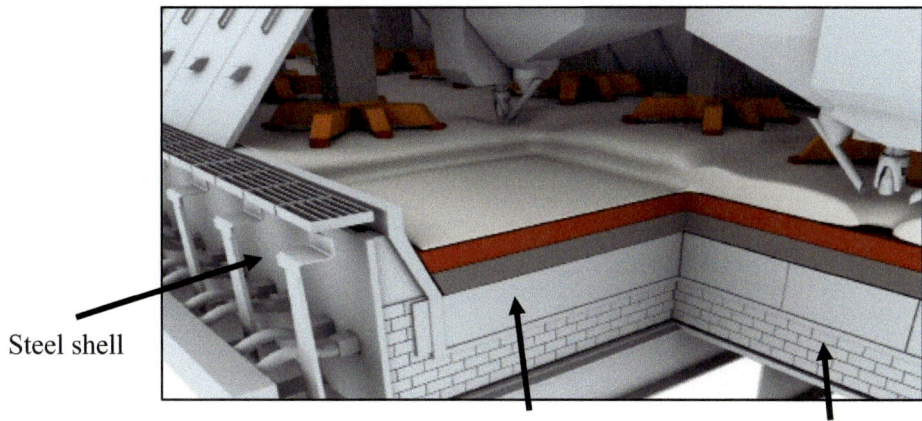

Steel shell

Carbon/Cathode   Refractory Lining

instability, high iron (Fe) and silica (Si) pick up and high Fe/Mn ratio.

During a power outage or blackout for a longer period, the thermal balance of the pot gets disturbed due to reduced heat generation in the pot but continued heat losses from the cell boundaries. This phenomenon decreases the internal heat of the pot, and consequently the bath starts cooling and solidifying from sidewalls toward the center of the pot [1]. After a prolonged power outage, some severe issues can take place during power resumption like high chances of cooling cracks formation due to thermal shocks and liquid metal penetration in the cathode blocks. The cathode deterioration may result in increasing the chances of pot failure [2].

Sometimes pot experiences abnormalities due to deviation in operational or process parameters like low bath temperature, high excess AlF_3 in bath, low superheat, and low anode cover thickness. These abnormalities lead to muck formation and an extended ledge in the shadow of the anode, over the cathode surface. The extended ledge and muck over the cathode make the current flow localized thus current distribution gets disturbed in the cathode. The concentrated cathode current density is one of the detrimental factors for pot life. This localized current increases the formation and dissolution rate of aluminum carbide (Al_4C_3) which leads to cathode wear and cavities or rat holes in the cathodes. A deep enough rat hole is sufficient for penetrating the metal to the collector bar. The collector bar gets corroded eventually leading to pot failure due to high localized current flow [3]. These rat holes are also one of the main sources of the high iron pickup in liquid metal from the collector bar which causes degradation of metal purity.

Pot age is the more considerable critical factor for pot failure. As pot age increases, the degradation of the pot sidelining and cathode wear rate increases due to the occurrence of thermal imbalances inside pot, resulting in depletion of side ledge and exposure of side lining to molten bath. If this chemical erosion continues for a longer time,

sidelining erodes completely, and the steel shell gets exposed to the molten electrolyte thus damaging the side wall and promoting metal leakages. This scenario also leads to high side steel shell temperatures sometimes also indicated by the formation of red shell. This chemical erosion results in the high silica (Si) pickup in the liquid metal and thus degradation in metal quality.

Past Incidences Adversely Affected the Pot Life

During the initial stage, Aditya Aluminium faced some critical challenges due to three major incidents that were water ingression in pot lining, enriched alumina silo collapse at gas treatment center (GTC), and frequent power disturbances which are detrimental to pot life.

Water Ingress in Lining Material During 1st Generation Startup

During the project time, water ingress happened in half of the potline where around 50–60 relined pots were majorly impacted for a long time as shown in Fig. 2. This incident occurred mainly due to a delay in roof sheet work in the pot room during the rainy season. The exposure of cathode lining materials to water might result in deterioration in lining quality in turn resulting in low pot life. After pot stoppage, the autopsy was carried out for those affected pots where metal penetration through small ramming joint at the cathode block (Block no.16, collector bar location 31 & 32) and rat hole in the same cathode was observed as shown in Fig. 3. These incidents might have occurred due to lining quality degradation which leads to low thermal expansion of the cathode during pot startup and might generate a gap between the cathode and ramming resulting in metal penetration and a rat hole.

Fig. 2 Water ingress in pot lining during project stage

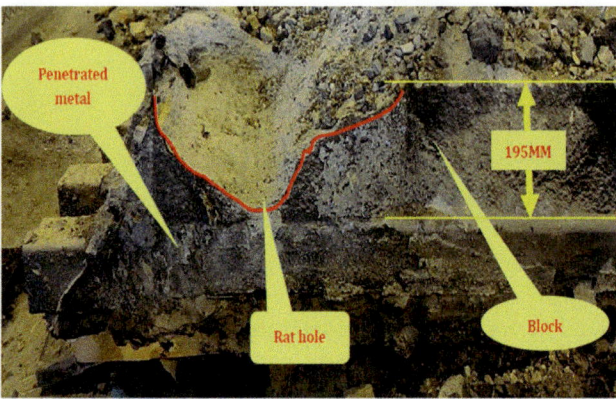

Fig. 3 Pot A092 autopsy observations: **a** Metal penetration through small ramming joint at Cathode block 16 (CB-31 & CB-32), **b** Rat hole in cathode block

Gas Treatment Center (GTC) Alumina Silo Collapse

Another major incident that occurred at Aditya Aluminium in 2015 was the collapse of enriched alumina silo in gas treatment center due to the failure of the supporting structures. This incident resulted in the stoppage of automatic enriched alumina feeding in the pot line, so it required manual feeding initially for 3 days. The bypass arrangements (Fresh alumina filled in the tanker and supplied to pots through hyper dense phase system) were made to supply alumina into the pots for the remaining time till the GTC silo was restored. The fume suction system was also totally disrupted which caused the high clad temperature in several pots, significant anode effect (~ 100 AE per hour in the entire pot line) and pot disturbances that resulted in multiple clad failures.

Multiple anode effect increases the superheat in the pot due to which side ledge melts and pot becomes abnormal. This abnormal situation leads to severe thermal shocks in the cathode thus resulting in multiple cathode cracks and reduced pot life.

Frequent Power Blackouts

Another most common challenge Aditya pot line has faced is a blackout and reduced amperages. The pot line had gone through the zero-power situation about 40 times since inception indicated by star marks in the average amperage graph over the total time period as shown in Fig. 4. These multiple zero-power scenarios reduced the pot life, as described in the earlier section.

To avoid sudden pot failure, Aditya Aluminium adopted a unique methodology, i.e. pot life estimation model with proper corrective actions. The main objective of the pot life prediction model was to calculate the criticality of the pots and to have scheduled planning to stop the pots in advance before it fails.

Pot Life Estimation Model

The concept/methodology adopted is based on the available literature study, and data comparison has been made with the modern smelters. This study provides the information about

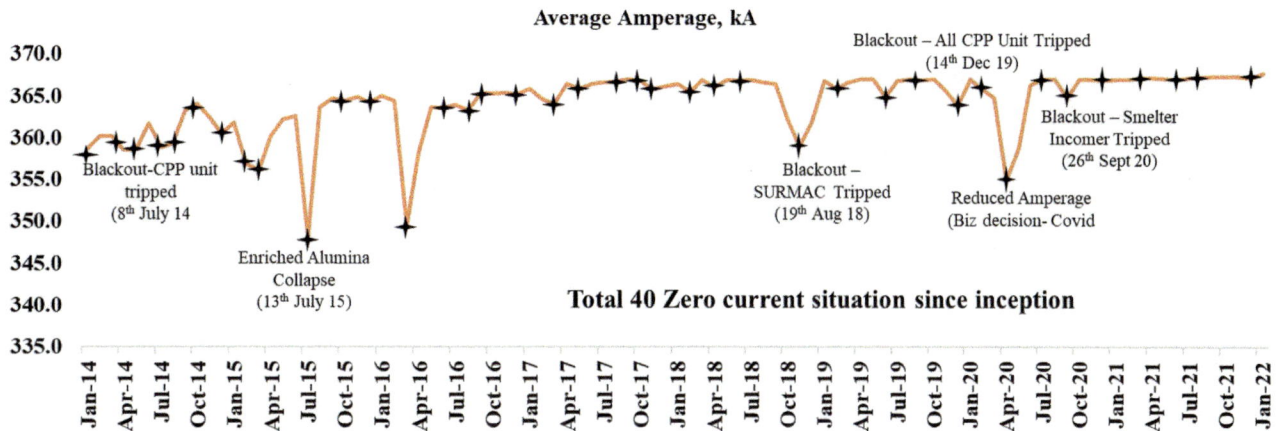

Fig. 4 Pot line amperage fluctuations scenarios since inception

Fig. 5 Methodology approach for pot life prediction

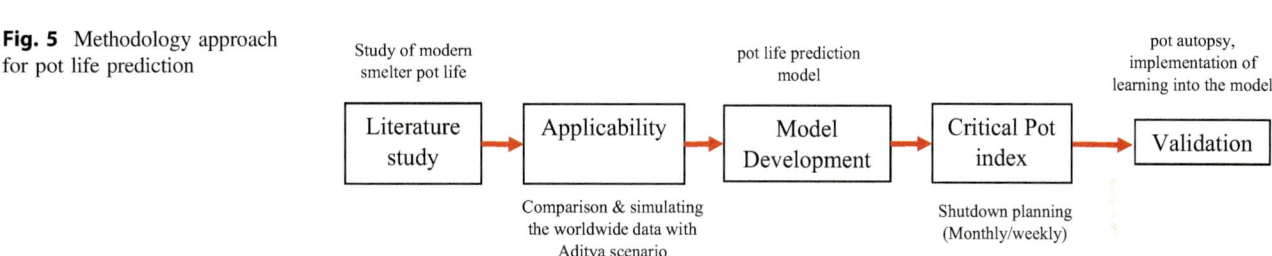

minimum, maximum, and average pot life of modern smelters so it is easy to compare the global benchmarking pot life with the Aditya Aluminium. In Aditya Aluminium, a pot stoppage model was developed which provides the criticality matrix of the individual pots. This criticality matrix helps to plan activities/actions proactively to minimize the chances of sudden failures. This model was also validated through pot autopsy. The approach for developing this methodology at Aditya Aluminium is as shown in Fig. 5.

In this methodology, pots were categorized based on the pot age, high iron & silica, current efficiency, and specific energy consumption and abnormalities like high steel shell temperature (SST), high collector bar temperature (CBT) and cathode flaking, etc. To identify critical pots, an in-house system was adopted which involves a regular SST measurement system above 380 °C, a CBT measurement system above 300 °C, shift-wise analysis of iron and silica pick-up, and Fe/Mn ratio tracking. These factors were assigned with an individual score in the criticality matrix to observe and predict the severity of the pot condition as shown in Table 1. The unique strategy adopted was the consideration of secondary reasons like current efficiency (CE), DC energy, Fe/Mn ratio, and abnormality for better model accuracy. However other plants do predictions based on only primary reasons like pot age, Fe, and Si. In the

Table 1 Pot life prediction model criticality matrix

Rank		Age	CE	DC	Fe	Si	Abn	Total score
		1	2	3	4	5	6	
	Contribution	15	5	5	25	25	25	
1	Pot No 1	1			1		1	65
2	Pot No 2	1				1	1	65
3	Pot No 3		1	1		1	1	60
4	Pot No 4			1	1	1		55
5	Pot No 5			1	1	1		55
6	Pot No 6	1				1		40
7	Pot No 7		1	1	1			35
8	Pot No 8		1	1	1			35

criticality matrix, the primary reasons were given more weightage, whereas secondary reasons were given low weightage. Based on the criticality index score shutdown of pots is planned. The pot obtaining the highest score in the criticality index is given the first priority for stoppage. Pot autopsies are also performed to check the accuracy of this model, and the model accuracy obtained is 70–80%.

To extend the pot life beyond the design value and sometimes below design value due to multiple abnormal conditions, some corrective actions are taken by closely monitoring the critical pots so that sudden failure is minimized. Some of these corrective actions are strengthening the side shell temperature measurement system, in-house fabrication of cooling system for collector bars and side shell, maximizing forced cooling network (FCN) pressure, manual ledge building, collector bar cutting methodology, stringent monitoring of high Fe and Si pots and Fe/Mn ratio tracking.

Process Improvements to Reduce the Pot Failure Rate

Aditya Aluminium adopted the following for preventing sudden pot failure.

1. **Strengthening side shell temperature (SST) measurement system:**

Red shell or high side shell temperature indicates high heat loss from the side wall, and it may be due to side ledge melting or sidelining erosion. Sidelining erosion occurs mainly due to pot age, issues in lining quality, high metal bath movement inside the pot, and ledge melting due to high superheat and process abnormalities for a prolonged time. The sidelining erosion leads to Si pick up in metal and the pot needs to be stopped. The Aditya smelting team is following the below unique procedure to identify the red shell and corrective actions to minimize it.

- Deployed dedicated team for daily SST measurement of critical and high age pots
- Shift-wise monitoring of pot which is having high Si (>600 ppm)
- Training and sensitizing the technicians/operators
- Usage of thermography for accurate temp measurement of side shell
- In-house designed and fabricated external side shell cooling arrangement
- Side breaking along with cryolite addition and manual ledge building

- Maximizing FCN pressure of critical pots by adjusting the damper position
- Removing the adjacent anode for a period of 1–3 h for cooling that area and forming a ledge.
- Weekly review of parameters (superheat, base resistance, bath temp, etc.) of critical pots and necessary actions taken

With the above strengthening of the SST measurement system, the number of high SST pots was reduced as shown in Fig. 6. Also, the pot stoppage plan for most of the critical pots was extended by 30 to 60 days. For a particular pot, the silica has reduced from 1750 to 580 ppm by corrective actions and pot life was extended by ~ 60 days as shown in Fig. 7.

2. **Stringent monitoring of high Fe/Si pots:**

The sources of iron (Fe) pickup in the metal are from collector bar, anode pin, anode cast iron joint, hammer tip, and some external sources. The possible reasons to reach the metal through cathode cracks are the high anode effect (AE), cathode and lining quality, the non-uniform temperature distribution in the cathode during pot startup, and multiple power outage scenarios. The silica pickup (Si) in metal is mainly because of its high pot age, high superheat, absence of side ledge, and poor lining quality. To minimize high Fe/Si pickup and to improve the metal quality, the Aditya smelting team has followed the below procedure in sequence as shown in Fig. 8. The adopted procedure has given a better result and helped to identify the root cause of high Fe/Si pots in advance thus reducing the sudden pot failure. The Aditya team has adopted the below sampling criteria for stringent monitoring of high Fe/Si pots further,

- Changing monitoring criteria for high Fe pots from 1000 to 800 ppm and high Si pots from 600 to 500 ppm
- Measurement and analysis frequency changed from 32 to 8 h for critical pots
- Identification and analysis of all primary and secondary reasons for Fe/Si pickup

The procedure followed to identify the root cause for Fe/Si was the consideration of secondary factors along with the primary factors like bath chisel condition, foreign material addition, bath level, anode block drop in the bath, etc. which helped in corrective actions taken proactively for critical pots to minimize the risk of pot failure. These secondary factors are used in identifying the alternative Fe source apart from the collector bar. The Fe pickup due to these secondary factors would be ignored.

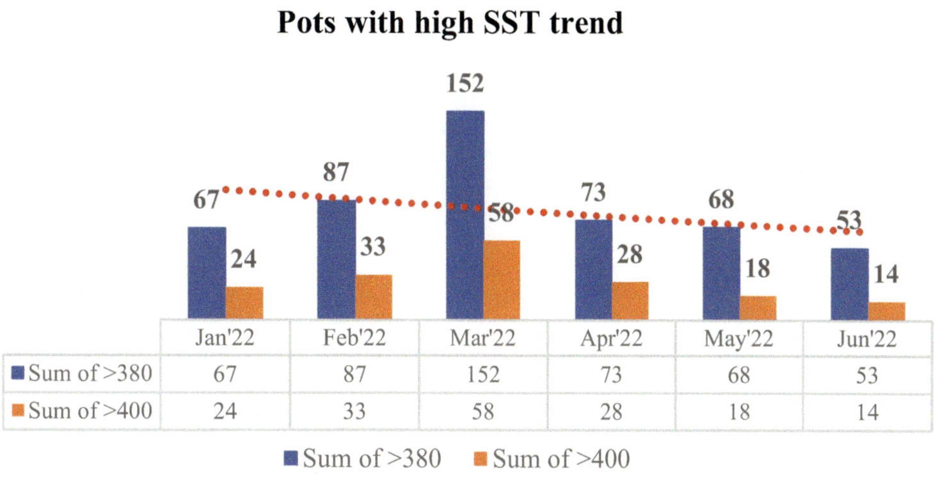

Fig. 6 SST trend before and after strengthening measurement system

Pots with high SST trend

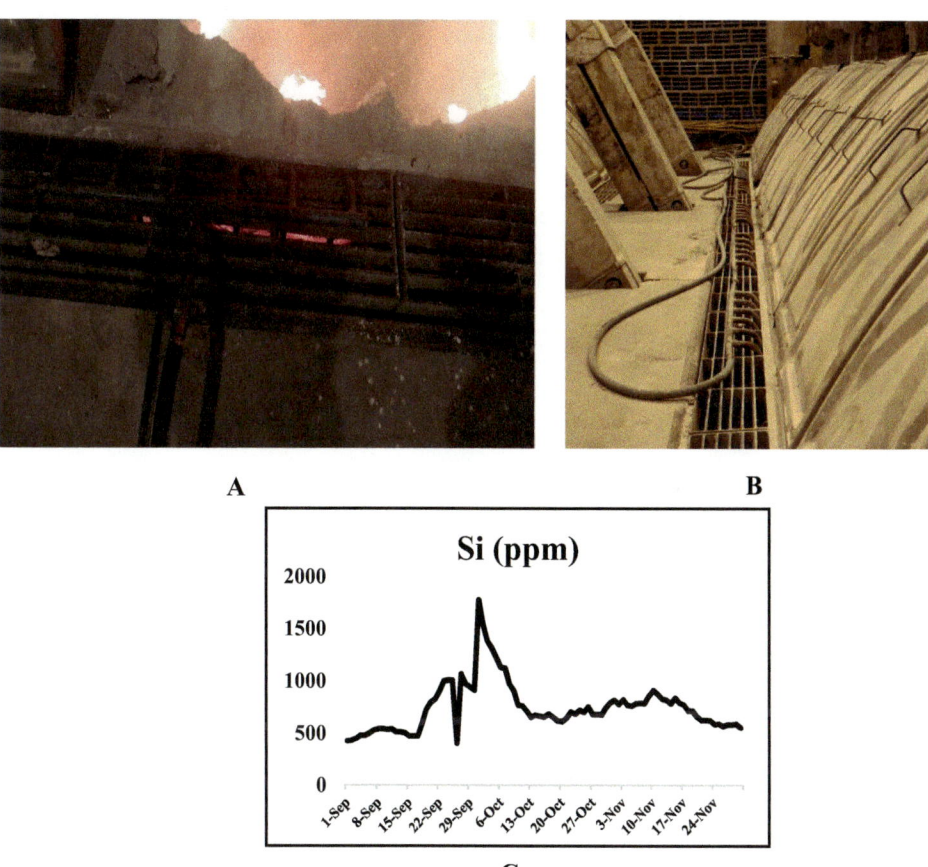

	Jan'22	Feb'22	Mar'22	Apr'22	May'22	Jun'22
■ Sum of >380	67	87	152	73	68	53
■ Sum of >400	24	33	58	28	18	14

■ Sum of >380   ■ Sum of >400

Fig. 7 Side shell strengthening strategy: **a** Removal of anode and manual ledge building, **b** External cooling by compressed air, **c** Impact on the Si content

A

B

C

3. Implementation of monitoring the Fe/Mn ratio:

The Fe/Mn ratio is also known as accumulation ratio. Manganese (Mn) is present in collector bars. It is also present in higher proportion in anode cast iron thimbles so the Fe/Mn ratio can be used to track the source of iron in the pot.

The Fe/Mn ratio above 200 indicates the Fe pickup from the collector bar when molten metal attacks the collector bar, below 200 indicates the Fe pickup is from the anode pin due to anode cast iron thimble exposure as shown in Fig. 9 [4].

After identifying the high Fe/Mn ratio pots, the below actions are taken.

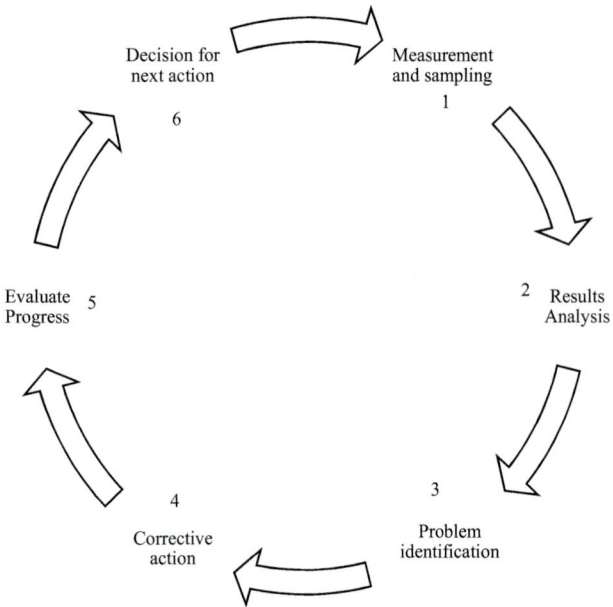

- Increased frequency of collector bar temperature measurement for identification of faulty collector bar
- Use of thermography to identify collector bar temperature in faster way
- Application of in-house designed and fabricated cooling arrangement for the collector bar

If the collector bar temperature consistently persists above 350 °C even after cooling, the defective collector bars are detached from the cathode flex (zero current passage from that collector bar) and allowed to cool. It has helped to avoid the collector bar leakage and sudden stoppages of critical pots thereby extending the pot life by 30 to 60 days. For a particular pot (A142) having the age of 2310 days, initially Fe level had increased up to 3600 ppm, but after collector bars (number: 59 & 60) detachment, Fe level reduced gradually down to 810 ppm while extending life by 26 days as shown in Fig. 10.

Fig. 8 Stringent monitoring process sequence for Fe/Si

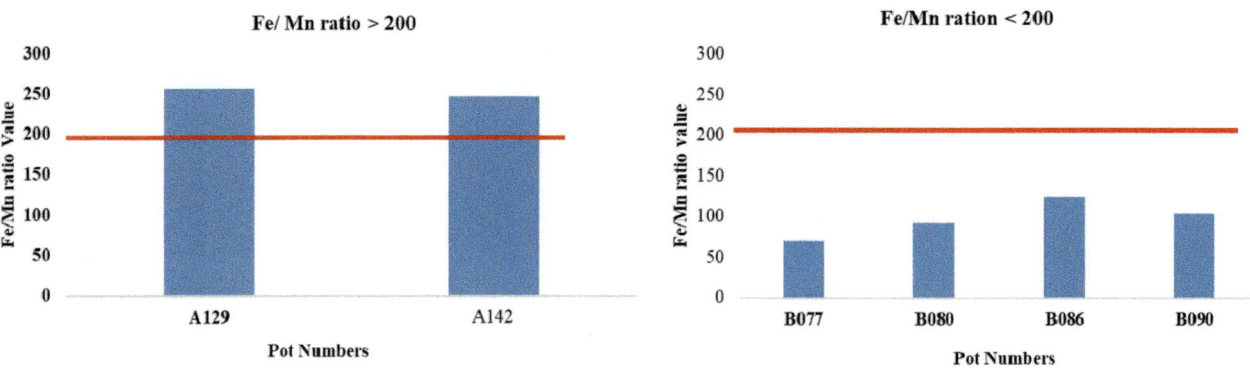

Fig. 9 Fe/Mn ratio analysis

Fig. 10 a In-house collector bar cooling device, **b** Fe trend in liquid metal before and after collector bar detachment

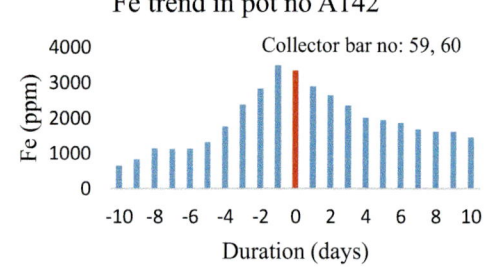

a

b

Fig. 11 Autopsy results of pot no A142: collector bar attack by liquid metal

Validation Through Autopsy

As a regular practice, Aditya Aluminium carries out pot autopsy for every stopped pot. The autopsy process of pot involves various activities such as visual inspection, cavity measurements, cathode wear measurement, imaging from different locations of interest on spent pot lining. For validation, pot autopsy was carried out in the pots stopped in accordance to the prediction model. In pot autopsy, the complete sidelining erosion, collector bar attacked by liquid metal, and metal penetration through cathode cracks were observed. The pot autopsy for pot number A142 showed erosion by molten metal on collector bar number 59 was observed as shown in Fig. 11.

Results/Achievements

With all the adopted systems and the corrective actions taken, the condition and life of the critical pots have improved, and the failure rate has reduced. Despite different challenges, Aditya Aluminium has established a global benchmark in worldwide modern smelters by achieving low pot leakage/ sudden failures as shown in Table 2. The pot

sudden failure rate in FY'2019–2020, FY'2020–2021, and FY'2021–2022 are 0, 1.3, and 0.91%, respectively. The overall pot failure rate is $\sim 0.8\%$ for the last three years and $\sim 0.6\%$ since inception (2 pots failed out of 333 relined pots) with sustainable pot life which can be considered as a global benchmark.

Conclusion

To reduce the risk of sudden pot failure, Aditya Aluminium plant team had developed an in-house pot failure prediction methodology that provides the criticality of the pots. The criticality matrix consists of primary as well as secondary reasons like pot age, side shell temperature, cumulative Fe/Si, Fe/Mn ratio, abnormalities, current efficiency and DC energy. This pot life prediction model helped to identify the critical pots in advance and also in planned pot shutdown. A different strategy was also followed to extend the pot life beyond the design pot age by strengthening of steel side shell temperature system, optimization of FCN pressure, in-house fabricated collector bar cooling system, stringent monitoring of Fe/Si and Fe/Mn analysis and collector bar cutting process. Adopting these measures had helped in extending the pot life by approximately 30 to 60 days and also reduced sudden pot failure. After a comparison of the past data of modern smelters, Aditya Aluminium smelter has achieved the lowest pot failure rate of $\sim 0.6\%$ since inception and established a global benchmark within worldwide modern smelters. Currently Aditya Aluminium is exploring ways that provide the in situ pot metal analysis and digital twin for the remaining useful pot life.

Acknowledgements The authors sincerely acknowledge the contribution of Shashidhar Ghatnatti, Rio Tinto, Brisbane, Australia who guided and supported in the implementation and strengthening of systems. The authors would also like to thank all the pot room operators, technicians, and management of Aditya Aluminium for their support in handling and preventing pot failure/leakage.

Table 2 Pot failure data of Aditya Aluminium since inception

	Fy'2016–2017	Fy'2017–2018	Fy'2018–2019	Fy'2019–2020	Fy'2020–2021	Fy'2021–2022
Total pots relined	9	26	50	61	77	110
Side shell pot leakage	0	0	0	0	0	0
collector bar leakage	0	0	0	0	1	1
leakage/Sudden failure (%)	0	0	0	0	1.3	0.91
Age of leakage pot (months)	0	0	0	0	54	70

References

1. Venkannababu Thalagani et al. Heat Dissipation Study in Hall-Héroult Cells During Power Outage, TRAVAUX 50, Proceedings of the 39th International ICSOBA Conference, 2021, 897–909.
2. Xinliang ZHAO et al. Restart of 300 kA potlines after 5 hours power failure, Light Metals 2011, 405–406.
3. Samuel Senanu et al. Pitting on Carbon Cathodes in Aluminium Electrolysis Cells, TRAVAUX 50, Proceedings of the 36th International ICSOBA Conference, Belem, Brazil, 2018, 643–653.
4. Determination of Iron Source in Molten Aluminium by Using the Accumulation Ratio of Iron to Manganese, TRAVAUX 46, Proceedings of the 35th International ICSOBA Conference, Hamburg, Germany, 2017, 997–1003.

Dissimilar Results in Restarting Two Different Potlines

María Carolina Daviou, María Alejandra Mollecker Rausch, Ricardo Alonso, and María Fernanda Jaitman

Abstract

In March 2020, Aluar reduced its production. Two potlines with different technologies were halted within two weeks. Metal was left in the pots in order to preserve the cathode conditions during stoppage, so the repair procedure for the restart consisted in cleaning the metal surface, sidewall repairing and gas preheating. The restartup of AP22 pots in Potline C was successfully accomplished six months later, with no incidents and a low early failure rate. Two years after, a pre-existed failure mode was identified during the restart of end-to-end al20 pots in Potline B. As a result, a higher amount of pots could not survive. This paper highlights the experience gained from a planned but necessary shutdown of 344 pots together with the challenges involved, not only in restarting two different potlines in a completely diverse context but also in the strategic criterion proposed for a pot restart in each case.

M. C. Daviou (✉)
Researcher, R&D Reduction Process, Aluar Aluminio Argentino, Parque Industrial Pesado, U9120OIA Puerto Madryn, Chubut, Argentina
e-mail: cdaviou@aluar.com.ar

M. A. Mollecker Rausch
Process Engineer, Reduction Process Control, Aluar Aluminio Argentino, Parque Industrial Pesado, U9120OIA Puerto Madryn, Chubut, Argentina
e-mail: mmolleckerrausch@aluar.com.ar

R. Alonso
Head of Reduction Process Control, Reduction Process Control, Aluar Aluminio Argentino, Parque Industrial Pesado, U9120OIA Puerto Madryn, Chubut, Argentina
e-mail: ralonso@aluar.com.ar

M. F. Jaitman
Head of R&D Reduction Process, R&D Reduction Process, Aluar Aluminio Argentino, Parque Industrial Pesado, U9120OIA Puerto Madryn, Chubut, Argentina
e-mail: mjaitman@aluar.com.ar

Keywords

Aluminium electrolysis • Potline restart • Restart with metal • Strategy

Introduction

Aluar is located in Argentina and its aluminium production capacity has tripled since its beginning in 1974. Currently, the installed capacity is 460,000 tpy with 784 pots distributed in 4 potlines: potlines A and B hold 200 end-to-end al20 pots each running at 196 kA; while potlines C and D hold 144 and 240 AP22 pots respectively, running at 222 kA.

When the COVID-19 pandemic was declared in March 2020, the decision was to temporarily cut off 40% of the production. Thus, Potline B (200 pots) and Potline C (144 pots) were stopped within two weeks. Despite the fact that it was a planned curtailment, it was a special situation as personnel at-risk were sent home and others following strict social distance requirements, taking into account the pot conservation and the intention of getting both lines in operation as soon as possible. At that time it was necessary to adequate labour breaks and transport logistics to accomplish current protocols.

The decision of shutting down two lines of different technologies was based in the resources needed to go back to operation.

Potline C began its restart in September 2020 (6 months later), while potline B had been stopped for more than 20 months until the first pot was restarted in December 2021. Even though similar restart procedures were applied, both potlines were involved in unlike circumstances that ended up in a dissimilar performance.

© The Minerals, Metals & Materials Society 2023
S. Broek (ed.), *Light Metals 2023*, The Minerals, Metals & Materials Series,
https://doi.org/10.1007/978-3-031-22532-1_105

Planned Versus Unplanned Shutdown

In 2011, Aluar had to face a similar situation when 50% of the production was cut out due to a blackout [1]. At that time, a sudden flood caused a power interruption that resulted in 208 pots randomly located in potlines A and B, together with the whole potline C out of service. Since auxiliary power was lost too, there was no chance to tap metal out of the pots, so during the conditioning period the metal pad had to be removed in order to have enough room for a later restart, causing (in some cases) a severe damage in cathode blocks. In the end, 132 pots were restarted in potline C: 71 pots without metal, and 61 pots with metal (but presenting an uncontrolled height). In spite of the misfortune, this event helped us identify opportunities for improvements and develop contingency procedures and working instructions that were implemented in the hurried but planned shutdown in March 2020.

In that year, the first decision was to restart pots with metal in order to protect the bottom lining from air infiltration. However, prior to the potline shutdown, the criteria to define which pot would be good enough for the restart was established. This definition bears considerable significance as it will impact straight forward in the final metal height of each pot to be restarted, while pots to be relined should be tapped out firstly. Subsequently, the metal flow rate was coordinated according to the casting house capacity. Based on previous experience, the target in both potlines was 12 cm of metal in every pot to be restarted. This value resulted from restart requirements, such us anode-cathode distance and the liquid bath volume available.

Since power interruption was planned, it was possible to prepare the pots by increasing pot voltage in order to get rid of sludge, tapping enough metal from pots for a potential restart with metal, adding bath to the pots with low bath height, and discontinuing operations like anode change. In addition, other actions were implemented after the shutdown, such as close open holes with cover material to avoid air infiltration up to the bottom lining, raised anodes just above metal height to facilitate restart process, and leave all anodes to ensure a slow cooling down to minimize the presence of cooling cracks. As a result of this, most of the pots to be restarted were stopped in similar process conditions.

Nowadays, the aluminium industry is trying to reduce its specific energy consumption by decreasing pot voltage and/or increasing current efficiency, so many operational strategies have changed in the last decade. In terms of line current and pot voltage, in 2011 the strategy was to restart with high anode-cathode distance and low current; but having less metal pad to dissolve together with the development of a low energy start procedure promoted a new and more controlled operational window in 2020, running close to the set-line current. In order to compare the first 6 months after each potline C restart, the monthly average of these two process variables is shown in Fig. 1.

Bath temperature evolution of restarted pots is presented in Fig. 2. Each box corresponds to an 8-h working shift, so 90 shifts are equivalent to 30 days in operation. Red dashed lines indicate the target, and the grey solid line is the shift average. There is a significant improvement in bath temperature evolution in 2020 when compared to 2011, mainly due to standardized procedures, training and a better control on restart and pots early life.

Restart Criteria

It is well known that a line shutdown may cause irreversible damage to the pots and a loss in life expectancy of 50–150 days for each individual pot [2–5], thus the operational stop strategy was focused on preparing the pots for a subsequent restart with metal.

The criteria to decide whether a pot should be restarted or not was based not only on pot age but also on iron and silicon content in the metal pad accumulated along its life, in cathode current distribution, previous restarts (if any) and the reduction process performance. At that time, the estimated

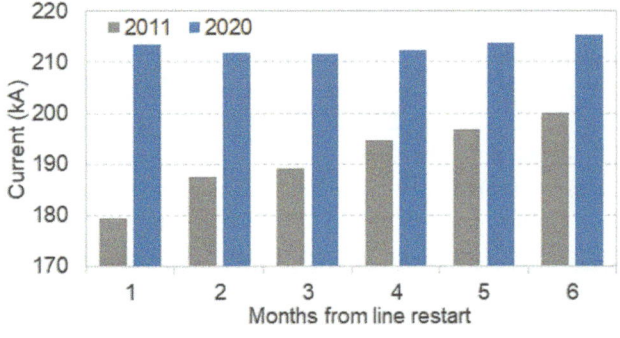
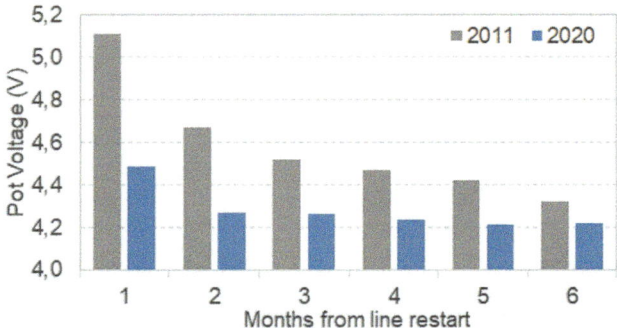

Fig. 1 Comparison between potline C restart in 2011 and 2020

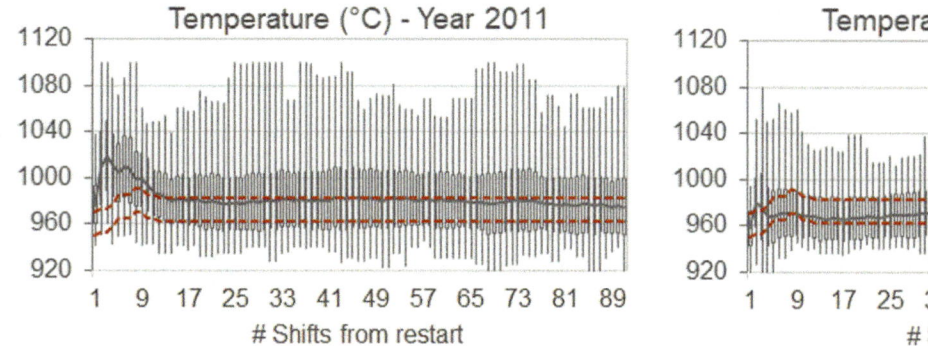
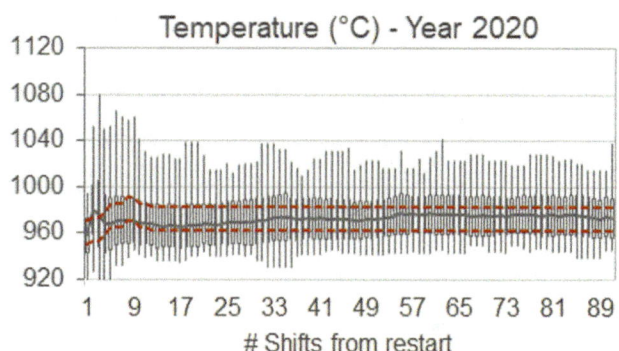

Fig. 2 Temperature evolution of restarted pots in 2011 and 2020 during the first month in operation

lifespan for al20 and AP22 pots was 70 and 82 months respectively, so pots older than 70 months were stopped to be rebuilt, and some pots over 60 months old were conditioned to be restarted.

Repair and Conditioning

The repair and conditioning process is vital for a successful restart. On one hand, pots should be kept as much airtight as possible until restart, but, on the other hand, pots should be prepared for gas preheating and restart. A prolonged exposure and oxidation of weakened sidewalls can lead to early failures. Thus, a well-coordinated work is necessary between electrolysis and relining areas in order to minimize both any delays in starting up the pots and precise task management in the pot rooms. In addition, short circuits were implemented in both potlines in order to reduce specific energy consumption during the restart, and a fuse was developed to restart pots at 50 kA to avoid zero potline current every day.

It was planned to start 1 or 2 pots/day, so repair procedures had to start at least 1 month prior to powering up the potline, taking into account that repairs have an expiration date. Likewise, additional human and material resources were needed to keep up with the restart rate.

Since most of the restarts were carried out with metal, there was no chance to have a look at the cathode surface. Every pot in both potlines was carefully inspected during pot cleaning, and side ledge was left when it was possible. It was preferable to rebuild the upper sides and patch only when the sides were hollow or damaged.

In order to speed up the process, new pots were built in Potlines B and C before having the restart scheduled. In case of a prolonged shutdown, it was decided to delay the ramming stage until the start date was programmed.

Restart Rate

Potline C restart process began in September 2020. As shown in Fig. 3, the average restart rate was 1.3 pots/day, combining new and restarted pots as the strategy was to start neighbouring pots. There was a very good congruity between the original plan and the real restart rate. As usual, new AP22 pots were electrically preheated using graphite plots, but gas preheating was used in restarted pots. Donor

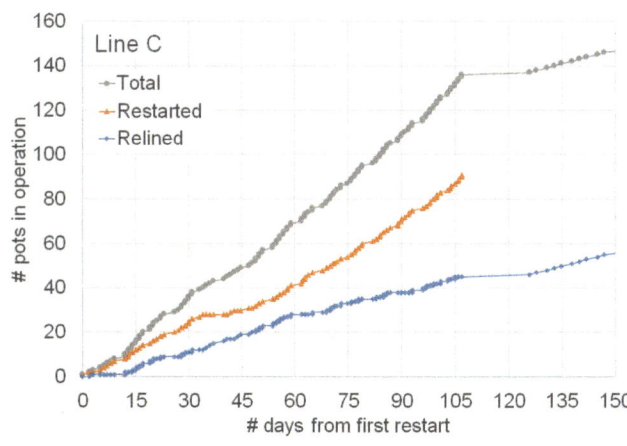
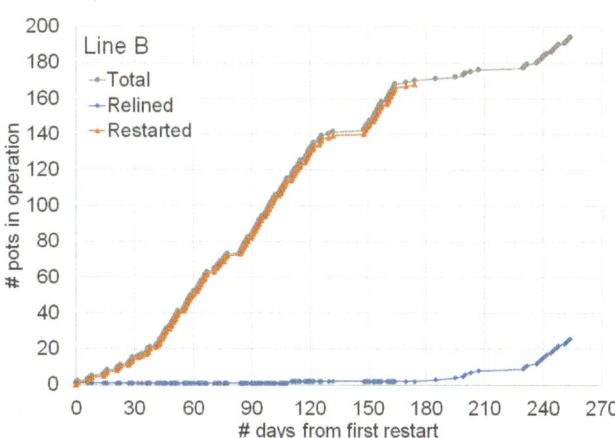

Fig. 3 Restart up rate in Potlines B and C

Table 1 Relined and restarted pot distribution in potlines B and C

Pot Line	Pots		Early failures after restart	
	Condition	Number of pots	After 2 months	After 6 months
B	To relining	35 (18%)		
	To restart	163 (82%)	28 (17%)	53 (33%)
	Already stopped	2		
C	To relining	55 (37%)		
	To restart	89 (63%)	7 (8%)	16 (18%)
	Already stopped	2		

pots were in Potline D, which remained in normal operation during the pandemic. Some butts were used at the beginning, but this strategy failed because they produced significant carbon dust and several time-consuming anode problems. The last restarted pot was put in operation on 18th December, then a short pause was made between Christmas and New Year holidays before completing the potline. It took 150 days to have all 144 pots in operation again.

The restart of Potline B was initiated in December 2021, at least 20 months after the shutdown. In this case, the strategy was to restart all pots before switching on the relined new pots. All pots were gas-preheated and donor pots were in Potline A. It took 180 days to restart 163 pots, as shown in Fig. 3, but after a significant early failure rate, there are still some new pots to be started.

In both cases, a startup team was created and trained before repowering the potline, in search of conforming a teamwork working under a repetition pattern, and achieving a better control of each working instruction in the tasks performed.

Restart Results

Potlines B and C were shut down simultaneously in March 2020, even though they showed a different performance after restart.

First of all and mainly because of the blackout that occurred in 2011, the pot age distribution was older in Potline C, so 37% of the pots were rebuilt. With more new pots than restarts, the process began 6 months later. Only 8% of the restarted pots presented early failures 2 months after the restart, and 18% of the pots were stopped before 6 months in operation.

On the other hand, not only 82% of the pots were restarted in Potline B, but also it was stopped for more than 20 months because its restart began in December 2021. It is not clear yet if this postponed date exacerbated the weakness of cracks already present in the bottom lining. In addition, there were many heated disruptive pots that required extra resources. In order to recover line process control trying not to spend all the resources in fighting troublesome pots, 13 of those pots were stopped and could be restarted again in the near future. As shown below in Table 1, 17% of the restarted al20 pots were stopped before reaching 2 months in operation. At present, not all operating pots have lived longer than 6 months since the restart, so it is not possible to have a successful ratio yet. At least 33% of the restarted pots were stopped.

Figure 4 shows each pot survival against the pot age at shutdown. In Potline C, 89 pots were restarted and there were no early failures in pots younger than 40 months in operation. Today, almost two years after the potline's restart, there are pots older than 90 months still in operation, and the

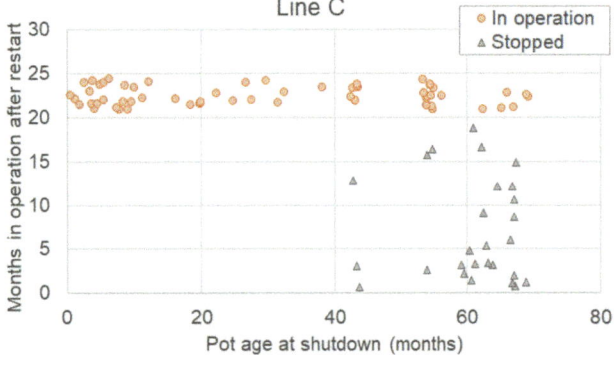
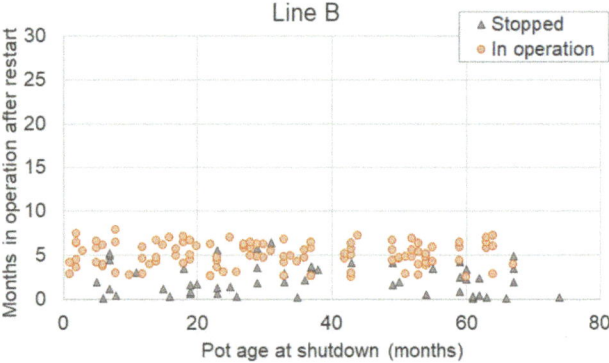

Fig. 4 Pot survival after restart versus pot age at shutdown

Fig. 5 Pot 485, 13 months in operation when potline B was stopped. (Left) First inspection, 11 months after stoppage. (Right) Second inspection, 23 months after stoppage

stopped pots are mostly older than 60 months when the potline was shut down. Thus, we can assume that the restart criteria and adopted operational strategies were successful.

In the case of Potline B, similar criteria and strategies were implemented but results turned out differently. As shown in Fig. 4, several premature failures were registered, even in very young pots. It was found that multiple factors may have influenced in the remaining pot life, such as pre-existent cracks in cathode surface, the duration of the shutdown and the evolution of hot sick pots. It has been a few months since the last pot restart and it may probably be premature to evaluate the potline restart performance; however, we consider it will not be as successful as it was in potline C.

Early Failures in Potline B

Several early failures were registered in potline B shortly after the restart, and it was found that a pre-existed failure mode was already present in pots younger than 40 months in operation. It was essentially metal infiltration in the very early pot life that promoted a longitudinal crack in the cathode surface along the central channel [1]. It is proposed that this phenomena could hasten the final shutdown, mainly because cathode surfaces could not be inspected before restart pots with metal. As a result, young pots failed after potline B restart.

Numerous corrective actions are in progress in potlines A and B to minimize the early metal infiltration and the following central crack on the cathode surface.

In addition, the exposure of cathode surface to air could result in alumina powder formation by reaction between aluminium carbide and moisture in the air, causing a serious damage [6]. The hypothesis is that those cracks already present in the bottom lining had been filled up with aluminium carbide during the two years that the potline remained stopped.

There were at least two attempts to restart the potline in 2020 and 2021, so anodes and bath were removed in some pots, some of them were even repaired, and then the restart

was rescheduled. Despite the fact that pots were covered with alumina, these attempts could promote air infiltration and the subsequent reaction previously described.

In some young pots, a metal pad was pulled out to carry out a surface inspection before restart and determine a possible failure mode, but this procedure worsened the air infiltration when the restart had to be rescheduled. For example, pot 485 was 13 months in operation when potline B was shut down in March 2020, and the metal pad was removed in February 2021 during one of the restart attempts. In this first inspection, the cathode surface presented aluminium carbide on the surface but it was in a good condition. Big and narrow joints were also satisfactory. When restart was rescheduled, this pot was covered with alumina to minimize air infiltration. One year later, in February 2022, a second cathode surface inspection was performed, and after removing all of the free alumina powder produced by the reaction previously described, narrow joints were damaged and several pieces of cathode blocks were loose and easily detached, as shown in Fig. 5.

Summary

Potlines B and C were stopped after the pandemic, and despite the fact that similar operational strategies were implemented, a different behaviour was registered during both restarts. The main differences were the time elapsed until the restart process, the pot age distribution when potlines were shut down, and the pre-existent failure mode found in potline B during the restart stage.

Some advices for a successful restart are:

- Define a criterion to decide whether each pot can be restarted or not, before shutting down the potline.
- Pots with less than 10% of remaining life or registering an uneven cathode current distribution should be relined.
- Restart pots with metal protects the underlying lining in case of a long curtailment. The target metal height has to be planned according to future restart procedures.

- Minimize air access to the pot by leaving anodes and cover material as much time as possible. Remove them only when restart process is imminent.
- If the potline restart date is unknown, ramming stages should be delayed in new pots.

In addition, Potline C was temporarily shut down in 2011 (unplanned) and in 2020 (planned). When both restart periods are compared, the results of improved operational strategies and process control actions could be quantified.

Power loss may occur, sometimes it is planned due to different reasons, and sometimes it is caused by unpredictable events. But restarting a potline is not an easy task, it is time-consuming and involves additional costs, not only in terms of loss in pot life expectancy but also considering the significant human and material resources that need to be applied. The aim of this paper is to share our experience in restart potlines and to humbly contribute to increase the knowledge to minimize future losses in the aluminium industry.

Acknowledgements The challenge of stopping and getting both potlines back in operation would not have been possible without the commitment of each and every employee of the company, and the selfless advices of several aluminium industry experts. Thank you all!

References

1. M.C. Daviou and P. Navarro, "Evolution of the mail failure mechanisms along the pot lifespan", Proc. 11th Australasian Aluminium Smelting Conference, Dubai, UAE, 2014.
2. H. Øye and M. Sørlie, "Power failure, restart and repair", Aluminium International Today Buyers' Directory 2011, pp 4.
3. A. Tabereaux, "Loss in cathode life resulting from the shutdown and restart of potlines at aluminium smelters", Light Metals 2010, pp 1039.
4. A. Tabereaux and M. Dupuis, "Cathode cooling damages due to potline power interruptions", ICSOBA 2020, pp 581.
5. A. Tabereaux, "The survivability of aluminum potlines after lengthy electrical power outages", Light Metals 2022, pp 448.
6. M. Sørlie and H. Øye, Cathodes in Aluminium Electrolysis, Aluminium Verlag, 2010.

Restart of Albras' Potline 2—Improving Performance and Changing Paradigms

Ana Carolina Guedes, Ana Renata Nunes, Bruno Vasconcelos, Flávio Silva, João Vilckas, Johnson Machado, Márcio Souza, and Michel Pena

Abstract

In February 2022, Albras had a potline freeze due to a power outage, shutting down 223 pots (the number of pots operating at the time). After 5 months, Albras was able to return 100% of the pots to operation and production levels are back to normal. The early failure rate was a record low of 5% when considering a restart under such conditions, with only 12 failures. During the restart campaign, there were few thermal excursions above 1000 °C (daily average at 5.5). Average acidity was kept below 4% during the first 5 days of pot operation. Originally expected to end in late November, the work was concluded in July, four months in advance, without any safety incidents. The aim of this work is to describe the steps leading to what has become the most successful restart in Albras' history in terms of schedule completion, success rate, process consistency, operational stability, and human capital development.

Keywords

Line restart • Pot baking • Pot start up • Pot early operation • Pot relining

A. C. Guedes · A. R. Nunes · B. Vasconcelos (✉) · F. Silva · J. Vilckas · J. Machado · M. Souza · M. Pena
Albras, Barcarena, Brazil
e-mail: Ana.Guedes@albras.net

A. R. Nunes
e-mail: ana.nunes@hydro.com

J. Vilckas
e-mail: Joao.Vilckas@albras.net

J. Machado
e-mail: Johnson.Machado@albras.net

M. Souza
e-mail: Marcio.Souza@albras.net

M. Pena
e-mail: michel.pena@albras.net

Introduction

Albras works with four potlines in total, with Line 2 composed of 240 AP13 pots, side by side, with 18 anodes each.

Restarting a production line in any aluminum smelter is a complex large-scale undertaking that requires not only a dedicated team with the proper expertise to conduct the process, but also a highly coordinated interaction between process engineering, operation, maintenance, supply and HSE teams in order to assure high performance results.

The way pots are repaired and restarted will have a major influence in the number of days that they will be able to operate with stability in the future; therefore, the importance of deciding what kind of repair each pot should receive, as well as making sure that the baking, soaking, startup and early operation are carried out without causing damages to either the cathode or the ramming paste cannot be overstated.

During the restart of Albras' Line 2, the required personnel and resources were employed in an effort to bring the line back applying the best process and operational practices. With the support Hydro's Technical and Operational Support (TOS) team, Albras implemented changes to guarantee safety and quality to the restart process. The obtained results reflect how positive the adopted approach was, with the restart of pots finishing four months ahead of the original schedule, without the occurrence of any safety incidents and with a company record low early failure at 5%.

The Line Shutdown

On February 18, 2022 (Friday), at 10:04 pm, an internal power distribution failure occurred in Line 2. It caused a shutdown of 25% of Albras' production capacity. Once production was halted, the organizational effort to restart Line 2 began, aiming at not only bringing the line back to its full production capacity, but also at achieving better operational conditions and stability when compared to the performance before the shutdown, reproducing the best practices to other lines in Albras as well.

The Organizational Efforts

Figure 1 shows the order in which the cleaning process, followed by the preparation, and startup of pots was meant to occur. After the first six pots entered in operation, cleaning, preparation and startup would occur at the same time in the potroom, which led to the need for crossovers installation. Pots would only be started in a section that had finished cleaning, to minimize the interactions between people and industrial vehicles and equipment. After the first half of the potroom was finished, the second half had two sections being started simultaneously, so as to allow for more operational flexibility with the PTMs.

The main kick-starting points, with related activities being carried out on a daily basis were:

- Preheating, Start up and Early life operation SOPs review;
- Mapping of cathodes conditions to guide pots evaluation;
- Support with operation and measurements on shifts;
- Coaching and developing first line leadership on shifts.

The original plan was to complete the restart on November 30th, 2022, assuming there would be a failure rate of 20% of the total pots put in operation. The number of pots initially predicted to undergo total relining was 83.

The Changes and Improvements

Cathode Inspection and Repair

No crash restarts were attempted. In the adopted strategy, one of the first steps was to understand the condition of the cathodes that would be cleaned and restarted in Line 2. It was decided not to restart pots over 1700 days (15 in total) and a group of 10 pots already cut out before the event for total relining. Therefore, excluding these two groups of pots, all the other remaining pots were assessed based on their pre-shutdown operational condition and historical records. The aim was to classify them regarding the risk associated with a restart.

The evaluated criteria were age, Fe and Si content, number of cut collector bars, number of previous restarts and cathode voltage drop. The values for each parameter were compared to a reference value, as follows (Table 1).

After the evaluation, the results for every pot were summed, yielding the final score for it (Fig. 2).

The higher a pot scored, the higher the risk associated with it. The results were used to group the pots into four color groups:

- Green: score under 3. Low risk of failure;
- Yellow: score 3 and 4. Medium risk of failure;
- Orange: score equal or above 5. High risk of failure;
- Red (pots over 1700 days and the classified for total relining before the event): pot should not be restarted.

This approach yielded an initial map, in which it was possible to visualize the condition of the whole potroom (Fig. 3).

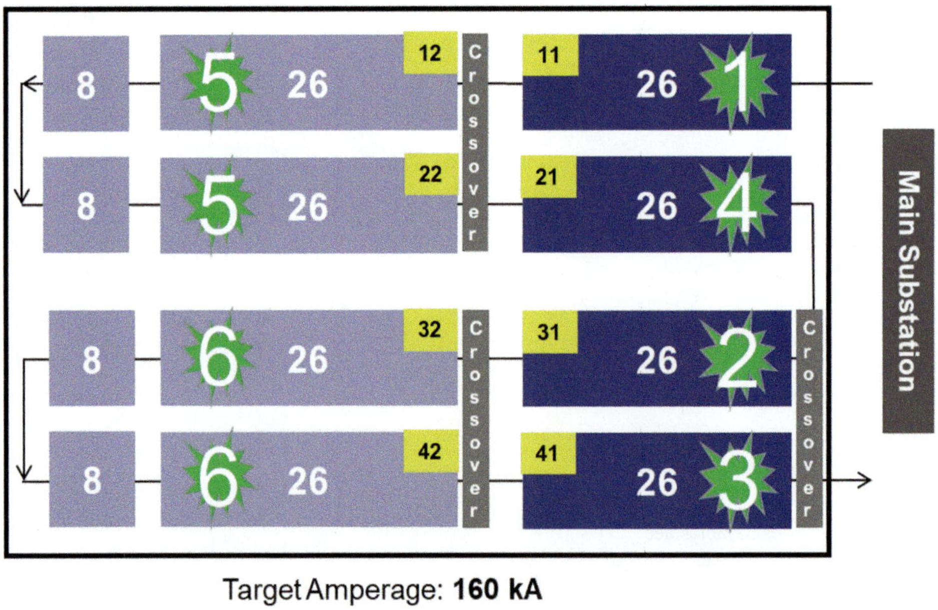

Fig. 1 Planned order of cleaning, preparation and start up (numbers 1 to 6). Numbers in yellow represent the sessions and the mid-sized number in white represent the number of pots in each session

Target Amperage: **160 kA**

Table 1 Treatment of parameters in cathode evaluation

Parameter and reference	Treatment
Age (650 days), Fe (0,35%), Si (0,13%)	The value for the pot is divided by the reference, and the result is rounded
Number of cut bars (#3), number of previous restarts (#3)	The value was kept, up to the reference. Above it, the reference number was used
Cathode Voltage Drop (609 mV)	If the parameter was over the reference, 1 was used, if not, then 0

Device	Age	Fe	Si	Cut bars	Starts	Cathode V-Avg 1M	Age	Fe	Si	Cut bars	Restarts	CVD	Sum
2104	382	0,132	0,052		1	573	1	0	0	0	0	0	1
2105	207	0,125	0,054		1	564	0	0	0	0	0	0	0

Fig. 2 Table of cathodes evaluation

All green, yellow, and orange pots were inspected in the area for cracks, holes, and other damages to the cathodes and/or paste that may hinder operation. The inspection generated a handmade sketch (Fig. 4) and a set of photos for each pot. This data set would later inform decisions such as which bar collector bar to cut in case of Fe contamination and where to apply corundum.

After this evaluation, some pots were sent to total relining, others were restarted as they were, and others had repairs to rebuild damaged sections. This approach was dynamic, meaning that as the cleaning process and the inspections in the area continued, the initial map was revised, turning yellow and orange pots into either green or red ones.

Some pots had to undergo repairs of damaged sections during cleaning and after tap outs occurred after restart. The procedure for repairing pots was improved with the assistance of the ramming paste supplier. Some important points consisted of removing previous ramming paste in the damaged area at least 100 mm under the cathode block level, standardize height of layers (150 mm), pre-compact layers before starting the ramming, ram at least twice or until the paste could not be more compacted (Figs. 5 and 6).

Pot Preparation

Total relining pots were prepared with a 60 mm thick layer of resistive material and restarted pots used 40 mm, which was made using a wooden frame built specifically to form these layers. After anodes were positioned, total relining pots had a layer of crushed bath added, followed by a layer of soda ash, topped by a final layer of crushed bath. This sandwich configuration allowed for excess fluoride in the bath to be lower than 4% in the first hours of operation, since it promoted a quick reaction between the soda ash and the liquid bath added at start up [1, 2]. The low initial acidity

favored the formation of ledge, the cathode protection, and adding a layer of protection to the ramming paste.

For both restarted and total relining pots, the anodes were not completely covered to allow periodic cathode temperature measurements by the process team while monitoring the evolution of the baking phase.

Pot Pre-heating

For total relining pots the desired total pre-heating time was 72 h (between 48 and 62 h for restarted pots, depending on cathode condition). The final pre-heating temperature should be between 850 °C and 950 °C [2, 3]. Temperature of the cathode and current distribution measurements were taken every 2 hours, recorded in a spreadsheet for the technicians and used to make decisions such as if and which anodes to disconnect and whether to use shunt to decrease heating rate and ensure homogenous pre-heating within the desired time. Sometimes, the use of shunt was combined with anode disconnection to reduce the heating rate and favor the heating of specific points on the cathode. Figure 7 demonstrates the pre-heating curve of a pot where anode disconnections were used to homogenize the cathode temperatures and later a pair of shunts was used to extend its heating to 52 h. Figure 8 shows the points on the cathode where temperatures were measured.

Pot Startup

When all points of a pot's cathode reached the temperature of 800 °C, the startup team was called upon by the process technician, beginning the mobilization for a startup (preparation of the pot, crucibles, bath tapping). Some key points were monitored during this activity:

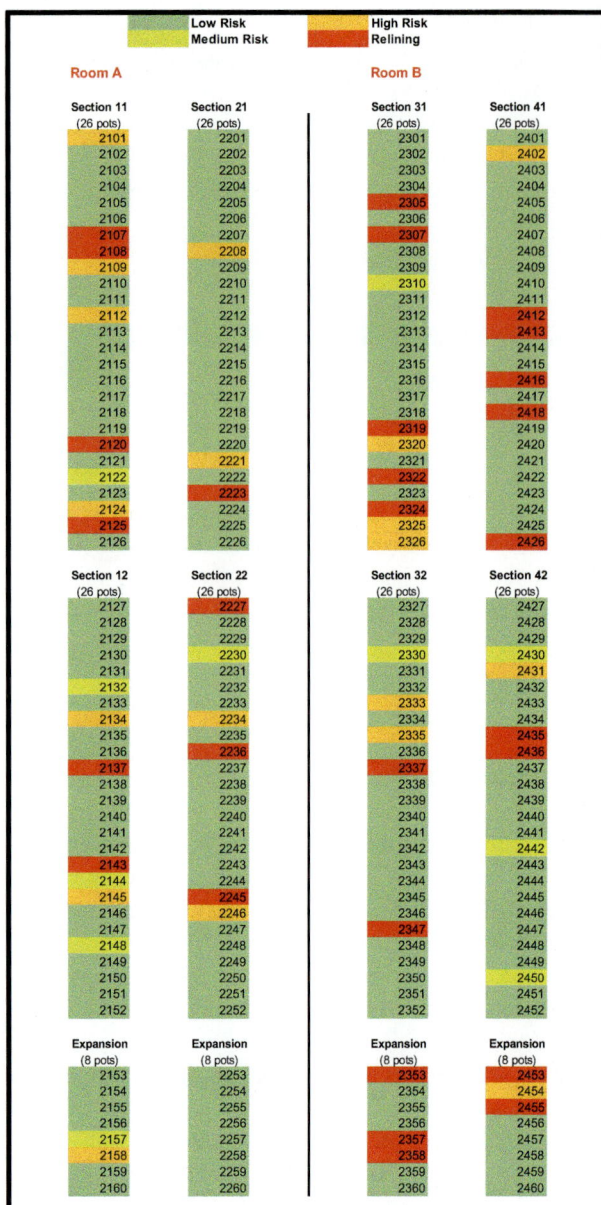

Fig. 3 Map of pot assessment and risk for restart based on cathode conditions

- Absence of metal when pouring bath, to avoid noise and instability.
- All monitored cathode points above 850 °C.
- Time for complete bath addition no longer than 1 h, to avoid bath solidification.
- Maximum voltage of 7,5 V to avoid excessively high temperatures during early operation and down movement of anodes.

These items were part of a checklist for the process technicians.

Soaking and Pot Early Operation

After bath pouring, the pot remained in manual control until the metal addition. In the meantime, for the restarted pots, the temperatures of cathode bars 1 and 14 were taken every two hours and recorded on the technician's control, together with pot voltage and bath temperature. Metal should be added when the temperature of the collector bars reached 300 °C, meaning the cathode would be warm enough for the metal. If the bath temperature reached 1000 °C before the bars, crushed bath would be added to control bath temperature. Using this indirect form of monitoring the cathode temperature, it was possible to avoid dramatic oscillations during this activity. After metal addition, the pot would be put in automatic control. Total relining pots received metal after 24 h, without cathode bar temperature monitoring.

Once the pot received metal (9 tons), the practice of poling was adopted. Green eucalyptus poles were inserted under the anodes in order to retrieve materials from the baking phase, decreasing the likelihood of anode deviations, especially spikes and aggregated material, as well as contributing to temperature control (Fig. 9).

During the first 6 days of operation, pots received no fluoride addition whatsoever to guarantee the sealing and protection of the pot lining with a low acid bath [1]. Fluoride additions took place as needed after the sixth day of operation, respecting the non-addition criteria.

The Results

It is important to point out that the key factor for obtaining the results in Line 2 restart was the training of technicians, supervisors, and operators. With the engineering team making the bridge between the TOS and the rest of the team via training and daily interactions, it was possible to create a chain of knowledge transmission, allowing for quick reactions whenever deviations occurred. That said, some of the highlights in Line 2 were the following.

Pot Start Up Rate

The pace of restart was of 2 pots per day (3 pots when there were total relining pots to be started). With this pace, it was possible to reach the same number of pots in operation of the original plan on May 9th, with 72 pots. From then on, the pace kept the process ahead of the plan until it reached 232 pots in operation on July 30th (the original plan was 146 pots for this date, since the initial idea was to start one pot per day, finishing on November 30, 2022). The good

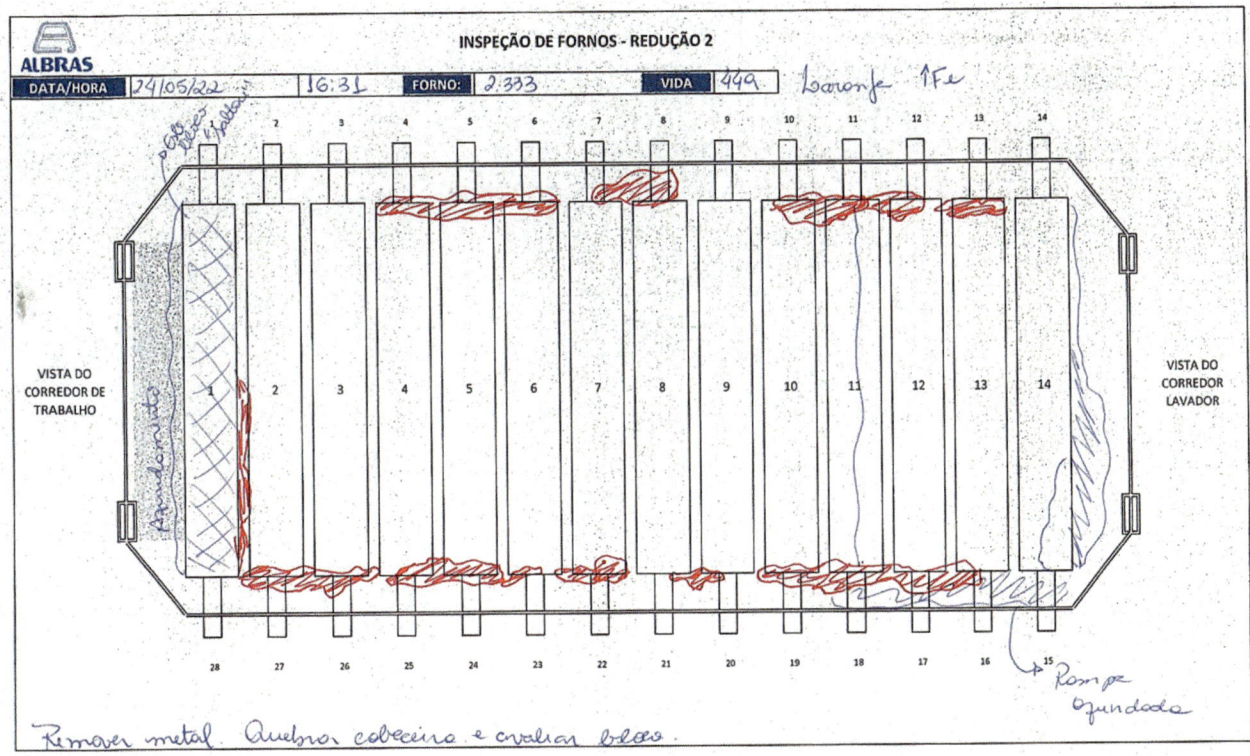

Fig. 4 Handmade sketch of a pot showing metal, cracks, and other damages to a pot after inspection

Fig. 5 Manual ramming of large joint

Fig. 6 Complete manual ramming of large joint in repaired pot

Fig. 7 Example of a pre-heating curve

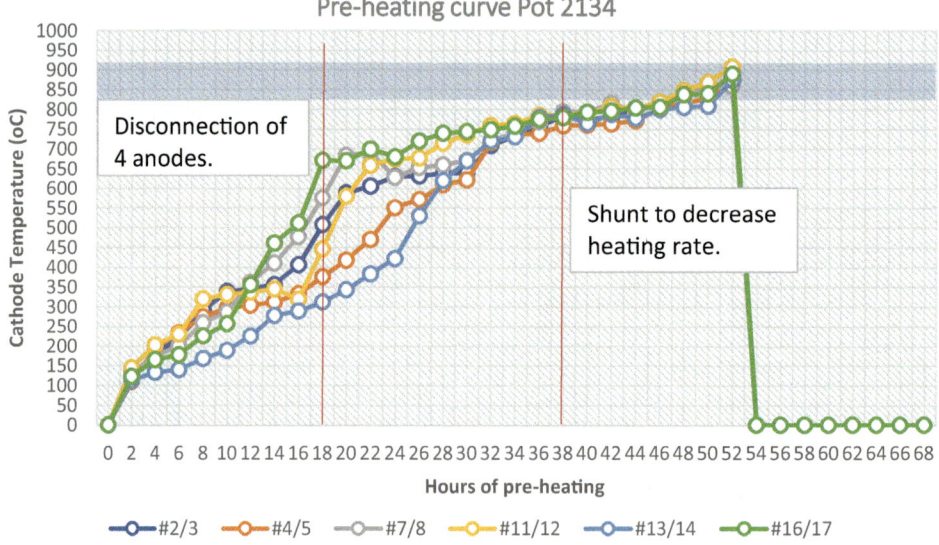

Fig. 8 Points on the cathode where the temperature was monitored

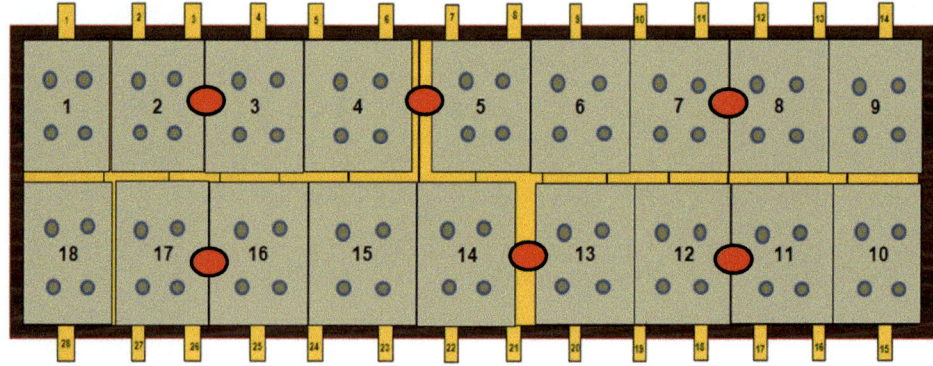

Fig. 9 Poling after metal addition to retrieve carbon and prevent anode deviation

operational stability was the main factor behind keeping the pace. As of September 2022, there are 240 pots in operation (vs 188 in the original plan), its full capacity even before the end of the on-going current ramp up. Figure 10 depicts the evolution of the actual number of pots in operation on a daily basis compared to the original plan.

Anode Effect Rate

The average rate of total anode effect (AE), given by the sum of non-programmed AE (NP AE) and programmed AE (P AE), was kept mostly below the target of 0.22 AE/pot.day through the duration of the pot restart process. The exception was the month of May, when, due to difficulties derived from bath quality, it was necessary to keep a higher number of pots without feeding, resulting in a spike of programmed AE (45% of the total AE rate). Figure 11 showcases that after the best practices to deal with lower quality bath (with higher % of alumina, higher % of silicon, and higher % of fines) were established, the AE rate normalized and as the current ramp up goes on, has produced satisfactory values so far. It is also possible to remark the drastic drop in P AE as a contributor to the total AE rate.

Anode Problems Rate

Mainly as a result of the poling, skimming, and cavity cleaning procedures, it was possible to keep the anode

problems rate well below the expected 40%. Once more the months of April and May concentrated the deviations since they presented factors that made operation more difficult, such as the bath crisis. In June and July, the rate was kept below 10%. Figure 12 showcases the evolution of anode problems rate on a daily basis.

Temperature and Fluoride Within Range

Figure 13 demonstrates that for the most part during the restart process the percentage of pots within their ranges % AlF3 remained above the target of 80%. As for temperature, the number started high, dropped to close to 60% and then recovered, staying between 70 and 90% from week 23 on. The drop in acidity control is a reflection of pots that remained with low % AlF3 even after their early life period had passed. Extra additions are being used to address this issue.

Current Efficiency

Figure 14 illustrates the average monthly values for the current efficiency (CE) of Line 2 (with scrap metal). As the number of pots in operation increased, so too did the CE. Operating at lower amperage contributes to a smoother operation, but even with the amperage ramp ups that happened so far (from initially 160 to 171 kA and increasing), line stability has not suffered severely.

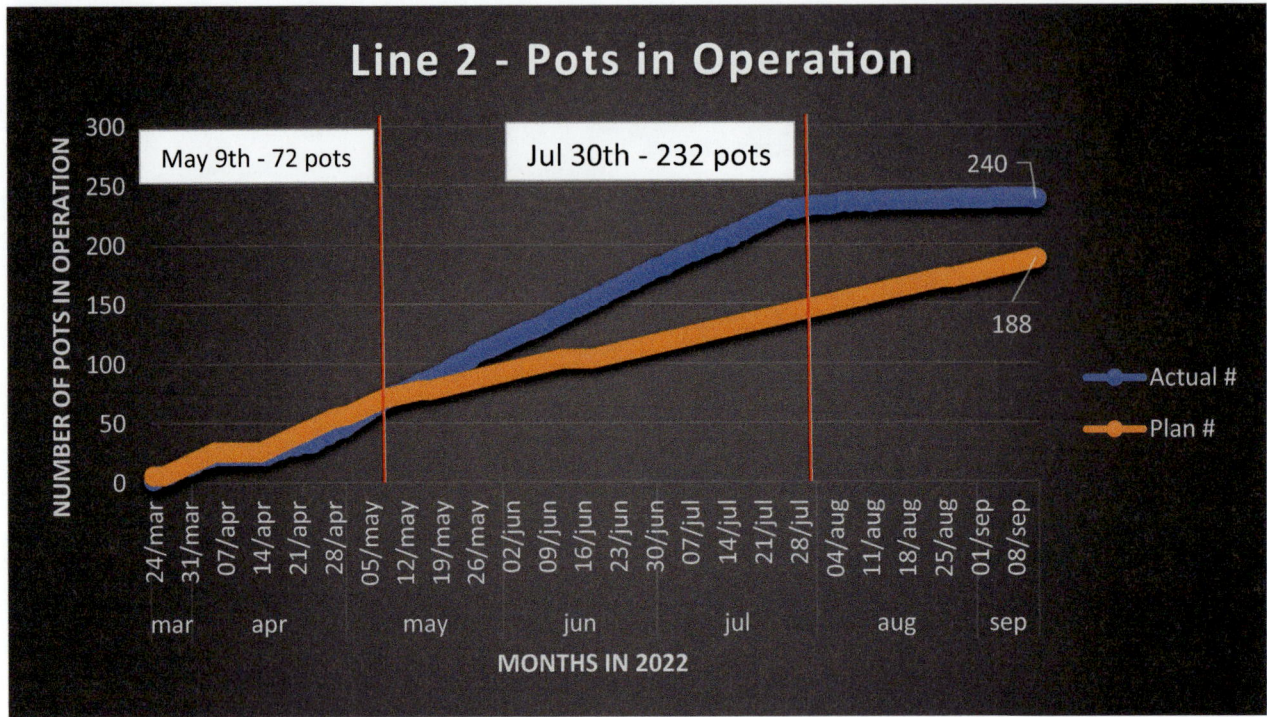

Fig. 10 Pots in operation–Plan versus Actual

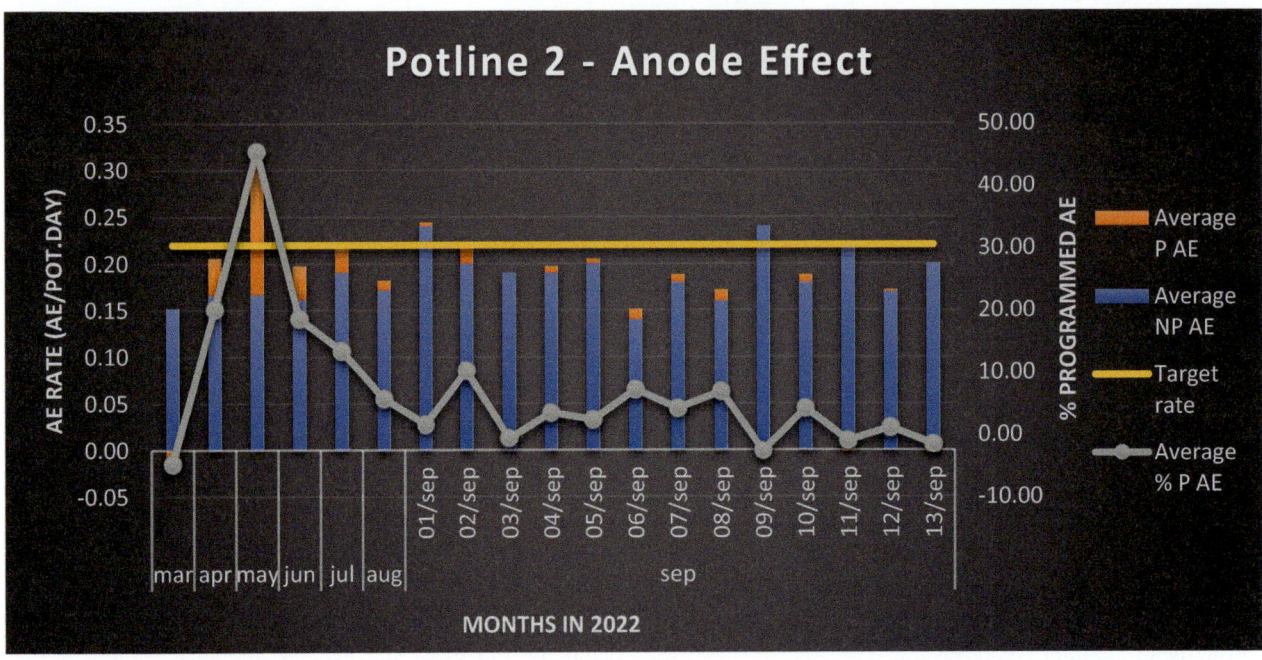

Fig. 11 Non-programmed and programmed AE

Early Failure Rate

The early failure rate is definitely the crowning achievement of the restart process in Line 2. By the end of August, it was of 5%, distributed as follows:

- 5 tap outs during bath phase.
- 3 tap outs after metal addition.
- 3 shutdowns due to high Fe content.
- 1 shutdown due to operational instability.

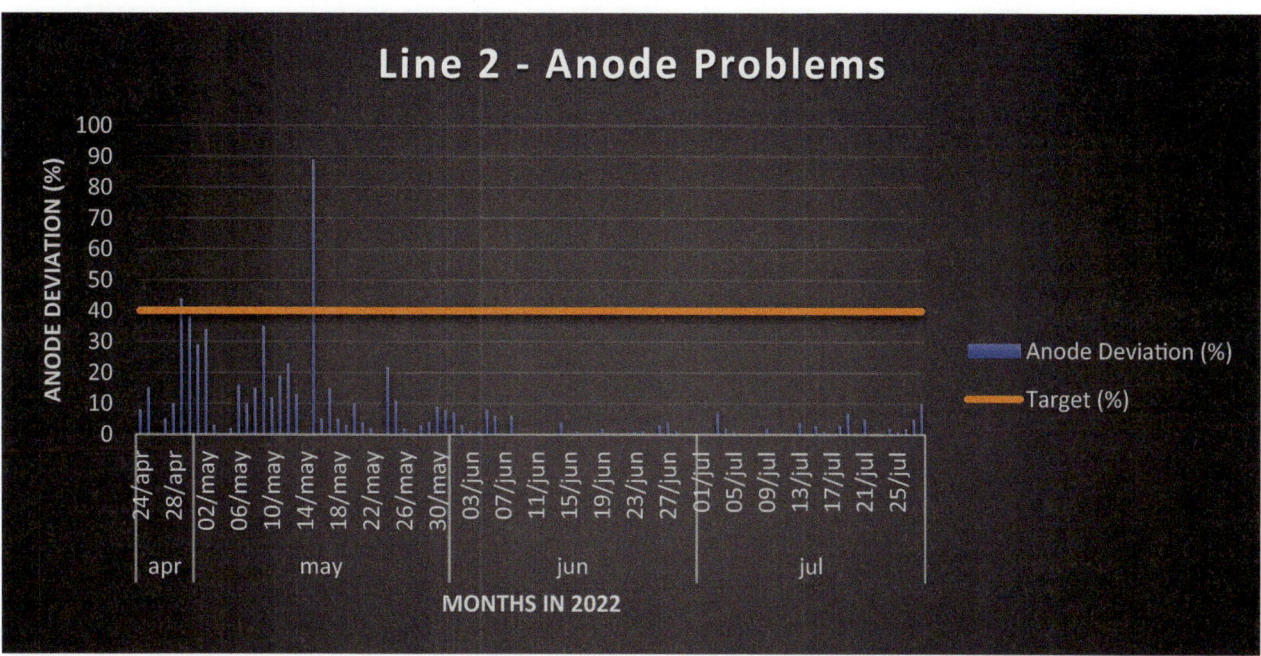

Fig. 12 Anode deviation rate

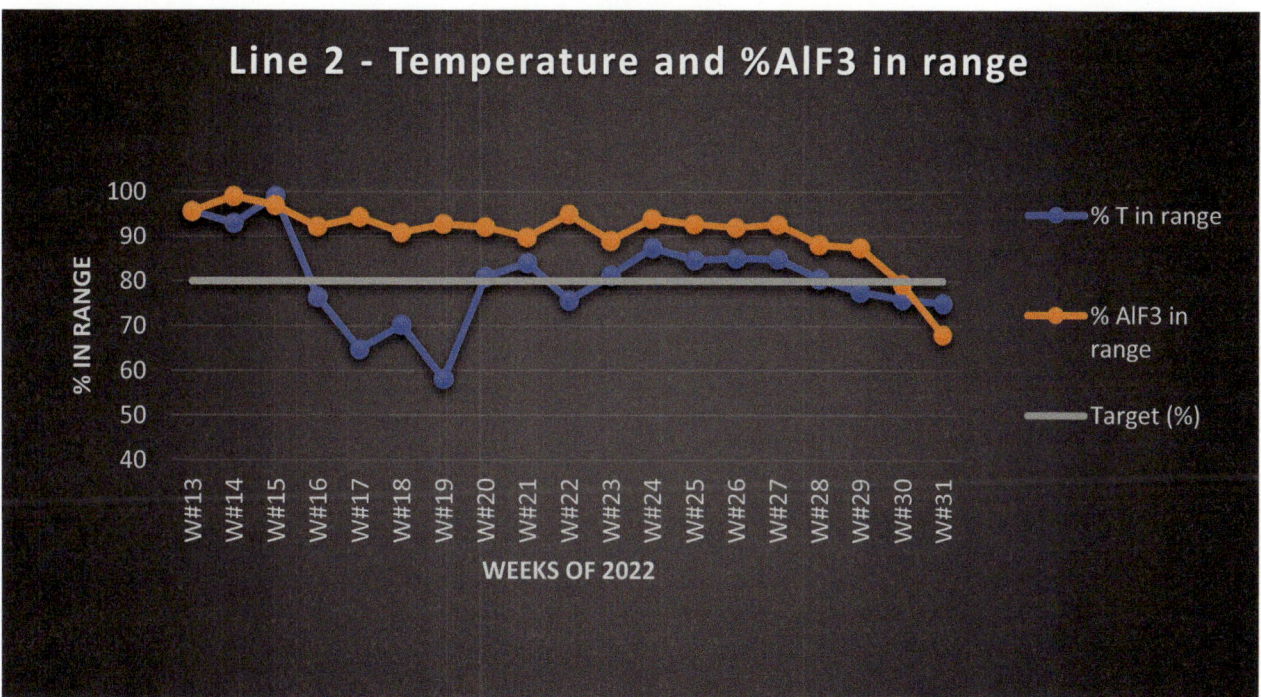

Fig. 13 Temperature and % AlF3 in range

Figure 15 depicts the evolution of the failure rate. It is possible to verify that most shutdowns occurred early in the process, attesting to the team's ability to respond to issues. It is also remarkable that all throughout the process the failure rate remained well below the 20% expected in the initial plan.

Current Ramp Up

Now the focus is to keep operational stability while increasing the current until it reaches 178 kA, the point at which the line was operating before shutdown. As of September 2022, it has been possible to increase 1 kA per

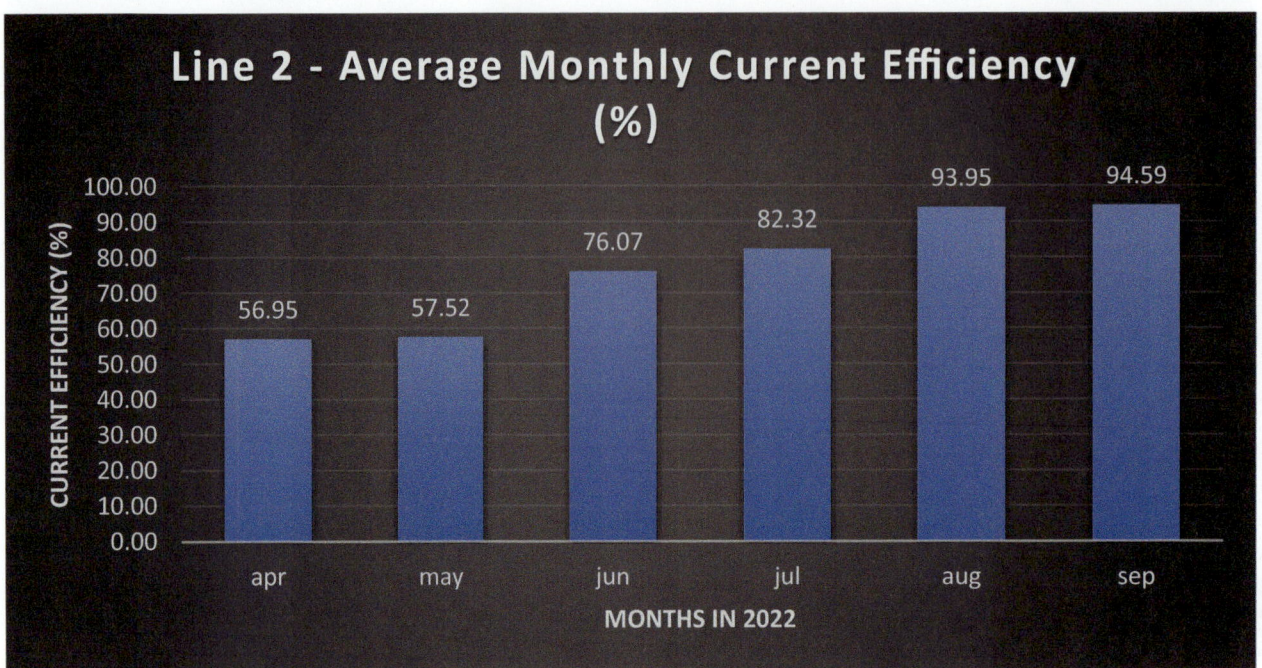

Fig. 14 Current efficiency

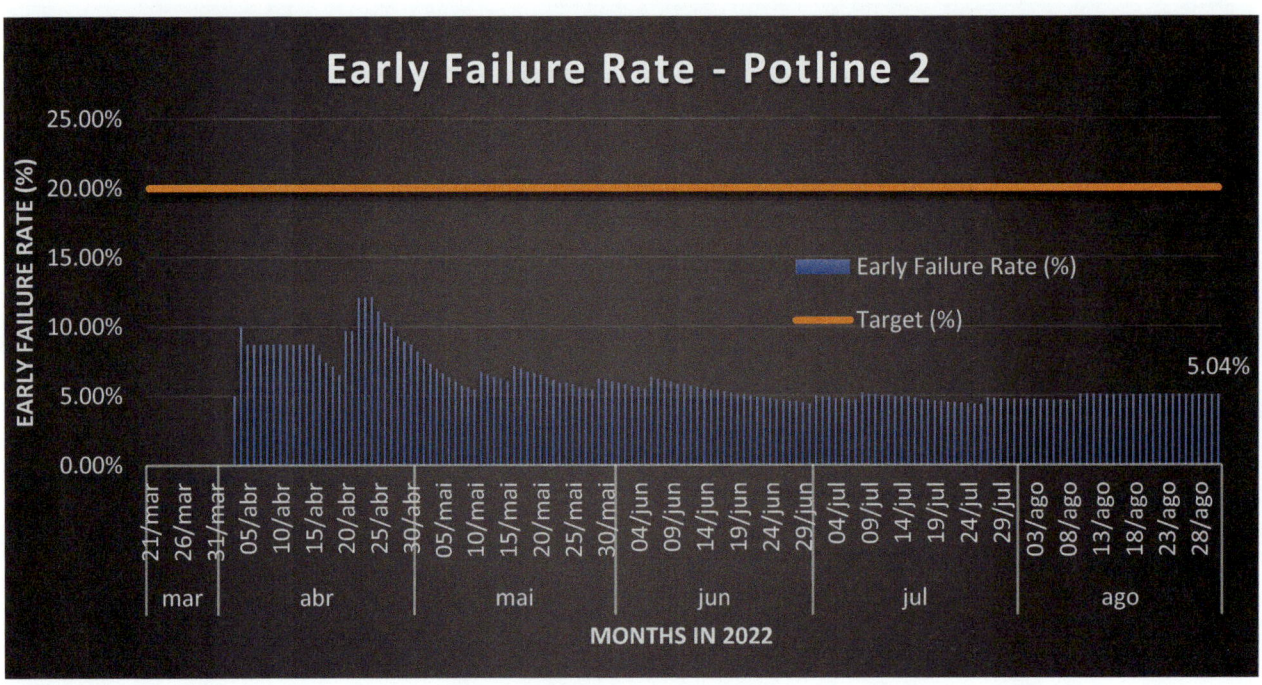

Fig. 15 Early failure rate

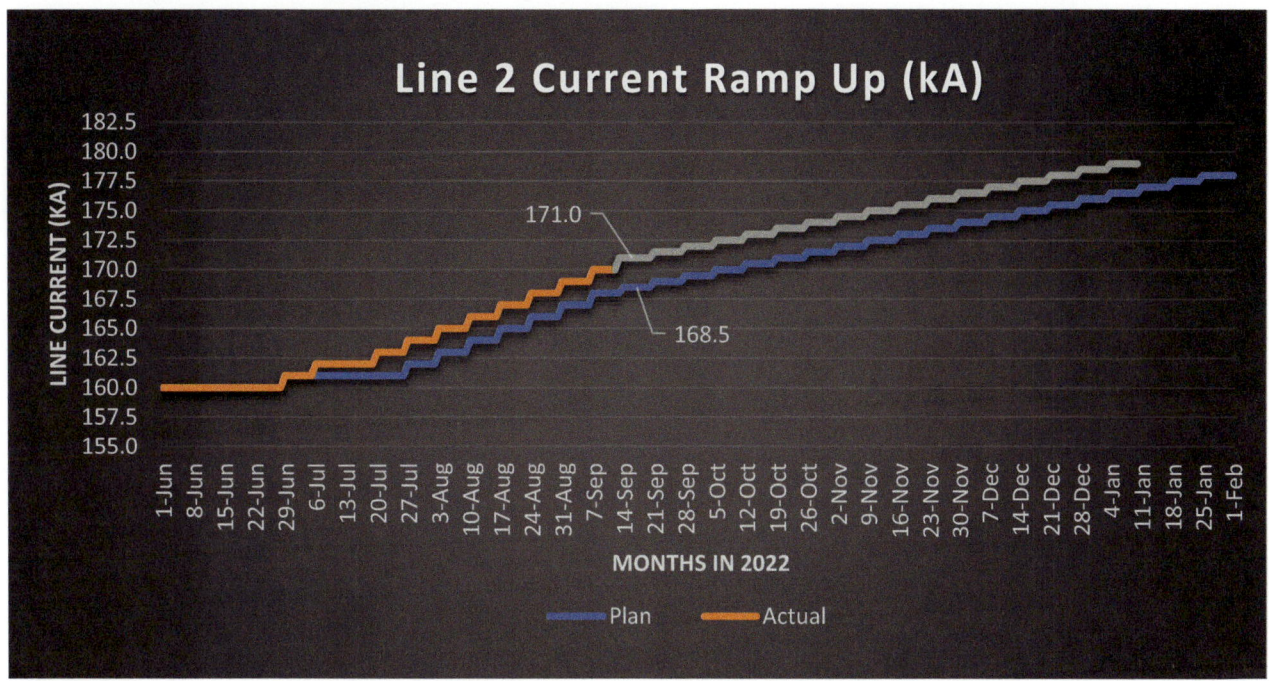

Fig. 16 Line 2 current ramp up

week, without drastic impact on the line. This allows Potline 2 to remain ahead of the original plan, which predicted a 0.5 kA increase every week. If the current pace is kept, the ramp up will be finished in January 2023 (Fig. 16).

Conclusion

The restart of Albras' Line 2 exceeded all expectations, finishing pot cleaning, pot repairs, and pot startup well ahead of schedule and, given the line's operation stability, it is also ahead of schedule in the current ramp up. As of September 2022, there are already more pots in operation than before the shutdown (240 vs. 233) and the current efficiency is already higher the last pre-shutdown value (94,59% vs. 90,47 in January).

This process is a testament that even when issues appear along the way (such as the shortage of bath), they can be overcome if there is transfer of knowledge within a well-integrated team. For Albras, the human capital that was formed as a result of the restart process will be invaluable for improving the operation of the company's other 3 lines.

With revised procedures, dissemination of good practices, a trained and more experienced team and upcoming technical improvements, the company can be optimistic about concluding the current ramp up, sure that it is prepared to act promptly in order to prevent any shutdown inducing failures and that it possesses the know-how to better conduct its normal operations in light of the lessons learned after this successful experience.

References

1. Michel Reverdy and Vinko Potocnik—A Historical Review of Aluminum Reduction Cell Start-up and Early Operation - Light Metals 2022 - 991–997.
2. M. M. Ali, preheating and start-up of prebaked aluminium reduction cells, pp. 2354–2364 (PDF) preheating and start-up of prebaked aluminium reduction cells (researchgate.net).
3. Gomma Abdalla, Khalid Youssif, Elsayed A. Elbadry, Mohamed Ali—Thermal analysis of the baking and start-up stages. Journal of Petroleum and Mining Engineering (PDF) Thermal Analysis of the Baking and Start-up Stages for Hall-Heroult Cells at Egytalum Smelter (researchgate.net).

Application of Cell Retrofit in GP320 Aluminum Reduction Cell Line

Zhuojun Xie, Jian Lu, Weibo Li, Song He, and Xingyu Yang

Abstract

Retrofitting older aluminum reduction cell line is an effective way to extend the productive life of an aluminum smelter. With the modification of busbar system, cathode lining design, and superstructure design, the GP320 aluminum reduction cell can achieve the target of both current creep and energy saving. In July 2020, a retrofitted GP320 cell line was started up (which was initially designed in 2002 and started up in 2003), and the current has been increased to 340kA. Until June 2022, all the cells are operating in stable, high-efficient condition. The retrofit has reached the design target and has shown a great example for the future of older cell lines.

Keywords

Aluminum reduction cell • Cell retrofit • GP320 • Current creep

Introduction

An aluminum smelter consists of 282 GP320 (first generation) aluminum reduction cells operating at 320 kA current. It was originally designed in 2002 and started up in 2003. The smelter is located in Henan province, China, a region with high pollution levels and low environmental carrying capacity. In 2018, the smelter decided to transfer its potline to the hydropower-rich Sichuan province due to local environmental pressure and the smelter's uncompetitive energy consumption index (the average operating DC power consumption in the year before the smelter shutdown was 13,273 kWh/t-Al at 320 kA amperage) [1].

This relocation project was designed by Guiyang Aluminum & Magnesium Design & Research Institute Co., Ltd. (GAMI). The relocation provided a very good opportunity for the cells to be retrofitted. The overall goal of retrofitting the GP 320 cell in this project is to increase the operating current to 340 kA and achieve a more energy-efficient index than the original design. Limited by the investment of the project, some equipment needs to be reused. The retrofitted cell is designed in March 2019, and the project construction started in November 2019. In July 2020, the new line started up at 350 kA. After running for half a year, due to insufficient power supply from the grid, the current has been reduced to a design value of 340 kA since January 2021. The picture of the original and retrofitted GP320 cells is shown in Fig. 1.

General Considerations of Cell Retrofit

As previously described, the project target is to creep the current to 340 kA and achieve a more energy-efficient index than the original design. In order to reach such targets, two main significant aspects must be considered:

• Busbar system and pot stability.

When increasing the current, the busbar system mainly needs to consider two factors: one is the current carrying capacity of the busbar (according to GAMI's criterion, it shall be less than 1.05 A/mm^2 for a retrofitted cell), and the other is whether the layout of the busbar can improve the reduction cell MHD stability.

Z. Xie (✉) · J. Lu · S. He · X. Yang
Guiyang Aluminum & Magnesium Design & Research Institute Co., Ltd, No. 2 Jinzhu W.Rd., Guiyang, 550081, China
e-mail: zhj_xie@chalieco.com.cn

W. Li
Guangyuan Zhongfu High Precision Aluminum Co., Ltd, Yuanjiaba Industrial Park, Guangyuan, 628017, China

© The Minerals, Metals & Materials Society 2023
S. Broek (ed.), *Light Metals 2023*, The Minerals, Metals & Materials Series,
https://doi.org/10.1007/978-3-031-22532-1_107

Fig. 1 Original GP 320 Cell and Retrofitted GP 320 Cell. **a** Original and **b** Retrofitted

- Energy-saving lining design.

To obtain energy-saving targets, the lining design of a cell shall always be highlighted as thermal equilibrium is most relative to the performance. Establishing an energy-saving lining structure is critical in this project. The average cathode voltage drop (CVD) before the cell shutdown was above 350 mV, and some cells even reached more than 400 mV. In the new design, the target CVD is 255 mV.

In addition to the above two main factors, the following challenges were also considered when retrofitting the cell:

- Whether the gas collection of the cell can be more evenly distributed.
- Whether the anode height can be increased to extend the anode changing cycle.
- Due to the long service life, the shells of several cells are seriously deformed, and it is necessary to judge whether the shells can still be used.

Busbar System

As known, the main driving force for the molten metal and bath inside the cell is the electromagnetic force:

$$\mathbf{F} = \mathbf{J} \times \mathbf{B}$$

Therefore, the horizontal current in the metal layer and vertical magnetic field create the wave-driving force, which is mainly generated by the busbar system. Hence, to increase a cell's MHD stability, there must be an optimized busbar system that can guarantee a good magnetic field distribution.

GAMI calculated the magnetic field of the original GP320 design. The magnetic field calculation is based on the same calculation platform established by GAMI when

Table 1 Vertical magnetic field in four quadrants of original design at 320 kA

320 kA	1st quadrant	2nd quadrant	3rd quadrant	4th quadrant	Max
$|Bz|$	19.42	6.43	4.74	16.00	34.85

developing the GP500+ cell. The original design is calculated at the original current of 320 kA. Calculation results of the original design show that the magnetic field is unevenly distributed. Maximum vertical magnetic field $|Bz|$ is 34.85Gs, while the average value of $|Bz|$ is 11.64 Gs. The area with $|Bz| < 10$ Gs only accounts for 52.25%. Values and distribution of the vertical magnetic field are indicated in Table 1 and Fig. 2.

Starting from the research by Sele [2], magnetic field judgement criteria were described. However, those criteria cannot meet the requirements for today's MHD stability. According to the evaluation criteria established by GAMI during the development of large-scale cells, in order to obtain good MHD stability, the vertical magnetic field must meet the following criteria:

- The maximum absolute value is within 20 Gs.
- The average of absolute values is within 5 Gs.
- The area with $|Bz| < 10$ Gs shall be greater than 80%.

Researchers have pointed out that Bz distribution has a much more important role in MHD stability than the absolute maximum Bz value [3]. Even if GAMI does not have a quantitative criterion to judge the distribution of four quadrants, the result of the original design is obviously not good enough. Therefore, it can be concluded that before busbar modification, such a magnetic field cannot guarantee the stable and low-energy operation of the cell.

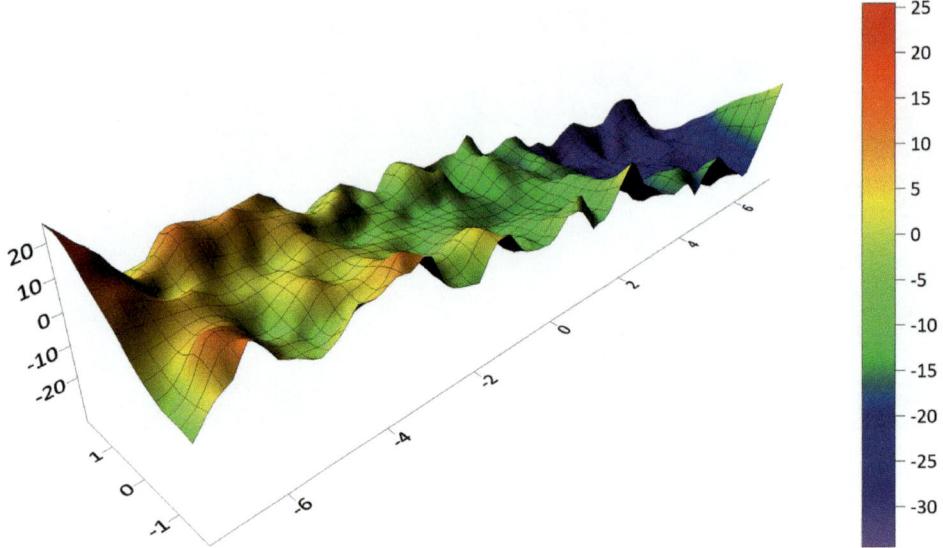

Fig. 2 Vertical magnetic field distribution of original design at 320 kA

Therefore, busbar modification was decided in this project. We have partially modified the underneath busbars and end busbars, and the magnetic field compensation on the duct end was intensified. At the same time, we added the cross section of the busbar where the current density is high. However, due to the limitation of project investment, the risers are completely reused. After the busbar modification, the busbar system has the following capabilities:

- Busbar carrying capacity allows the potline current to creep up to 365 kA.
- MHD stability of the cell is greatly optimized, which allows the pot to operate much more stable.
- The cathode current distribution and riser current distribution of the cell are greatly improved, which is helpful for improving the stability of the cell and reducing the Anode Effect Frequency (AEF).

Compensating busbar is not used in this project. Although adding compensating busbar is a great method to optimize the magnetic field of the existing potline [4, 5], it can neither change the electrical distribution of the busbar system nor increase the busbar carrying capacity. Since this project is a relocation project, it was possible to modify the existing busbar system. So, the final decision is to directly modify the cell busbar. The potroom-to-potroom distance remains unchanged.

A calculation of the modified busbar system was carried out. In terms of boundary conditions, apart from the operating current (new cell target is 340kA), others are the same with the original busbar system calculation. After busbar modification, the magnetic field became evenly distributed. The maximum vertical magnetic field dropped to 19.20 Gs,

Table 2 Vertical Magnetic Field in Four Quadrants with busbar modification at 340 kA

340 kA	1st quadrant	2nd quadrant	3rd quadrant	4th quadrant	Max
|Bz|	3.82	4.11	4.37	3.47	19.20

while the average value of |Bz| dropped to 4.00 Gs. The area with |Bz| < 10 Gs accounts for 96.53%. Values and distribution of the vertical magnetic field are indicated in Table 2 and Fig. 3.

According to GAMI's judgement criteria, after busbar modification, the magnetic field meets the MHD stability requirements. The distribution of four quadrants is also much more improved, which provided the conditions for the cell to operate in stable conditions with low energy consumption.

In addition to magnetic field optimization, the modification of the busbar system also optimizes the current distribution. The deviation of the riser busbar current and cathode current before and after the busbar modification is shown in Figs. 4 and 5, respectively. The results indicated that the current distribution is greatly improved after the busbar system is modified.

Gas Collection System and Superstructure

According to the feedback from the smelter, the ventilation of the original GP320 cell is uneven. Gas collection on the tap end is weak. If the smelter wants to increase the collection efficiency, the negative pressure of the duct shall be increased, which will both increase the power consumption of the Fume Treatment Plant (FTP) and increase the heat loss

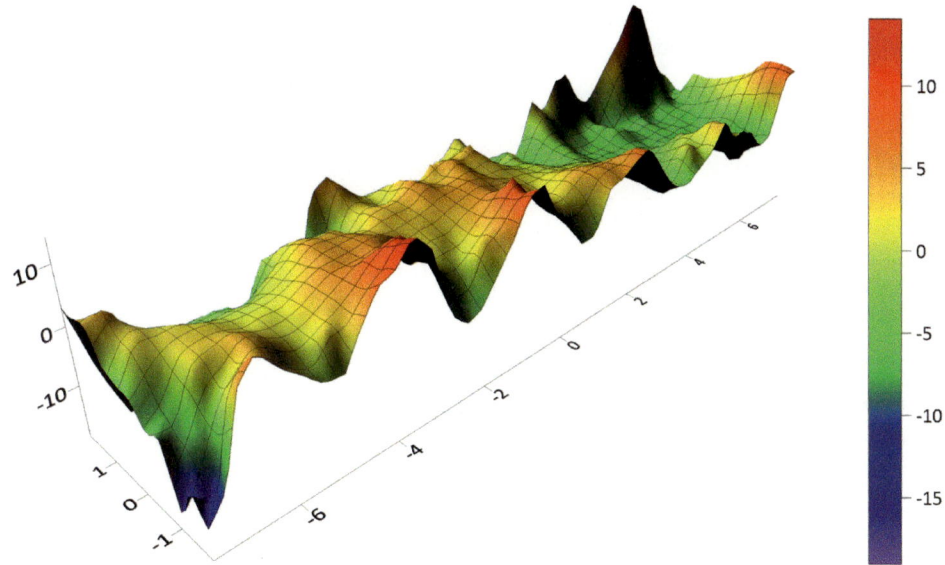

Fig. 3 Vertical Magnetic Field Distribution after busbar modification at 340 kA

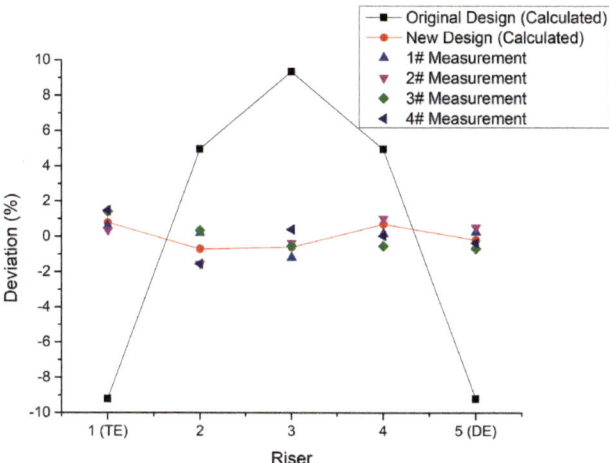

Fig. 4 Deviation of Riser Busbar Current (Calculated value along with 4 retrofitted cells measurement)

of the pot (since sucked gas is hot). In addition, in the original duct design, the distance between the duct and the anode yoke is relatively short. During anode changes, the yoke is easy to collide with the duct, which has a certain impact on the production operation. Therefore, in the project, duct modification will contribute to even ventilation as well as increase the anode change space.

A simulation calculation is developed for the gas collection system. In the calculation, the outlet pressure of the duct is taken as −300 Pa. The calculation result of the original design is shown in Table 3. The calculation results show that the volume flow is larger near the duct end and smaller near the tap end. Meanwhile, the deviation of volume flow is high, ranging from −35.5 to 28.3%. And in actual production operations, the flow in the tap end will be smaller than the calculated value due to the opening of the door (for aluminum tapping, inspection, etc.) [6].

After the redesign of the duct, the calculated value of the volume flow is shown in Table 4. The redesigned duct is changed from the original 6-section structure to 4-section structure. It can be seen from the results that after the redesign, the total volume flow decreased from 13,742.3 to 12,197.2 m^3/h, a decrease of 11.3%, and its uniformity was greatly improved. In addition, the flow of the tap end section is slightly larger than that of other sections, which is beneficial to ensure the effect of collecting fume gas during the door-opening operation.

The new gas collection structure also makes the duct farther away from the anode yoke than before, which extends the space for anode change, and makes it possible to increase the height of the anode. Since the consumption of the anode becomes faster after the current is creeped, increasing the anode height to extend the anode change cycle is beneficial to reduce anode change manpower workload and equipment workload. Besides, the anode gross consumption can be reduced. In the new cell, the cross-sectional size of the anode remains unchanged, but the height is increased from 550 to 620 mm. In doing so, even if the operating current of the cell is increased by 20 kA, the anode change cycle can be extended by 3 days.

Due to increasement in anode height, the superstructure support beam must bear additional loads. Therefore, the support beam with additional load shall be recalculated, and the result is shown in Fig. 6. From the calculation, the maximum deformation in the vertical direction is $Z_{max} = 0.48121$ mm, and the minimum deformation is $Z_{min} = -19.319$ mm.

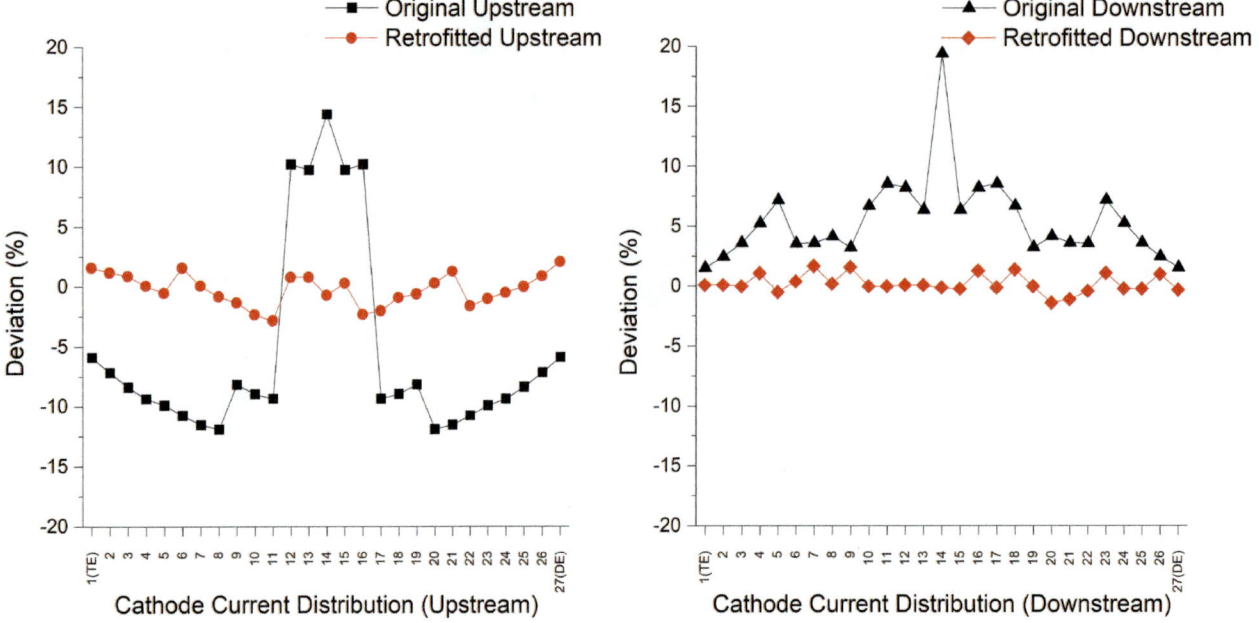

Fig. 5 Deviation of Cathode Current (Values are calculated). *Note* TE–Tap End, DE–Duct End

Table 3 Calculation result of original gas collection system

Section	Volume flow m³/h	Average flow m³/h	Deviation %
1(TE)	1,475.4	2,290.4	−35.5
2	1,728.3		−24.5
3	2,192.0		−4.3
4	2,529.3		10.4
5	2,866.5		25.2
6(DE)	2,950.8		28.3
Total	13,742.3		

Table 4 Calculation result of redesigned gas collection system

Section	Volume flow m³/h	Average flow m³/h	Deviation %
1(TE)	3,141.3	3,049.5	3.2
2	2,964.3		−2.7
3	2,969.9		−2.6
4(DE)	3,123.5		2.4
Total	12,197.2		

According to the criterion of GAMI, the maximum allowable deformation of the support beam of a 300kA level cell is R = 1:400. In this calculation,

$$R = \frac{Z_{\max} - Z_{\min}}{L} = \frac{0.48121 + 19.319}{16,152} = 0.00123 < \frac{1}{400}$$

Therefore, the maximum deformation meets the design requirements.

Energy-Saving Lining Design

In order to achieve the goal of energy saving, the new cell lining design needs to be more insulated than the original design to reduce the sidewall and bottom heat loss. In the selection of cathode material, to balance investment and energy-saving effect, it was finally decided to use 50% graphitic cathode. In fact, according to GAMI's engineering

Fig. 6 Deformation contour of the support beam (deformation exaggerated)

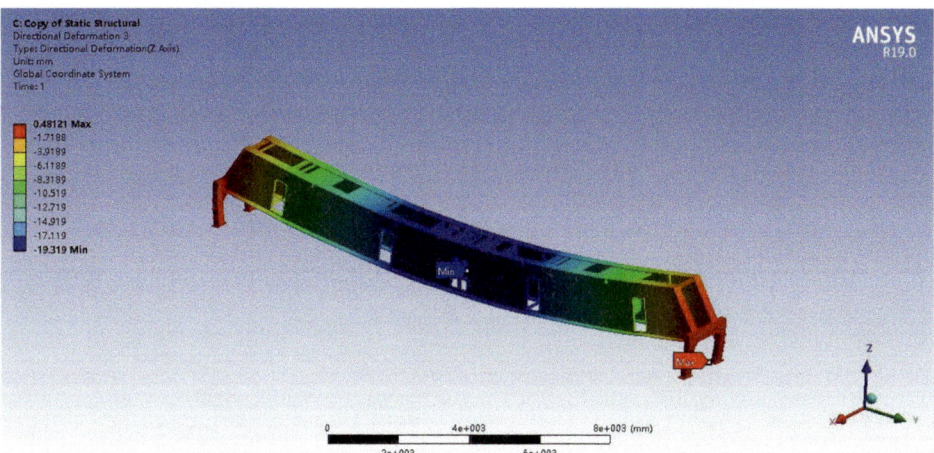

experience, 50% graphitic cathode is the most cost-effective in cells with anode current density lower than 0.8 A/cm^2.

As the depth of the GP320 shell is shallower than that of larger scale cells (GP400, GP500, and GP500+), in order to reduce bottom heat loss, two new refractory materials with better thermal insulation properties have been used. Those two materials were developed during the development of the GP500+ cell. In addition, limited by the insufficient depth of the shell, in order to ensure a sufficient cavity depth, we did not add the cathode height nor the collector bar height. In this case, GAMI's latest horizontal current reduction technology cannot be applied. However, since the magnetic field has been greatly optimized, good MHD stability can still be obtained.

A simulation calculation has been done for the new lining structure, and the key input parameters are shown in Table 5.

The isotherm distribution is indicated in Fig. 7. The result shows that 900 °C isotherm (isotherm of the bath) is located under the cathode carbon block and above the lining insulation layer, which can prevent the insulating materials being reacted with sodium infiltration.

The result of the voltage breakdown calculation is indicated in Table 6. The voltage drop is calculated at 3.984 V, and the voltage composition is reasonable.

In November 2021, a test of the energy balance was conducted. The test was carried out according to the standard YS/T 481–2005 (aluminum cell test for energy balance and its calculating method—five-point feed current and six-point feed current prebaked anode aluminum cell). In brief, the calculation duration is based on 1 h. The energy input is calculated by current and voltage, which can be read from the pot controller. The energy output is calculated based on the following measurement data:

- Volume ratio of CO_2/CO of anode gas (analyzed by austenitic gas analyzer).
- The ambient temperature (measured by thermometer).
- Temperature of the hood cover (side, end, and top), side shell, bottom shell, deck plate of shell, collector bar head, anode rod, yoke and stub, butt, and bath (measured by infrared thermometer and thermocouple).
- Flow, pressure, and temperature of the fume duct (measured by flue gas flow meter).

Table 5 Key input parameters of thermal calculation

Item	Unit	Value
Pot line current	kA	340
Bath temperature	°C	950
Liquidus temperature	°C	942
Ambient temperature	°C	35
Metal level	cm	25
Bath level	cm	18
Anode covering height	cm	20
Cathode type	/	50% graphitic

Fig. 7 Isotherm distribution

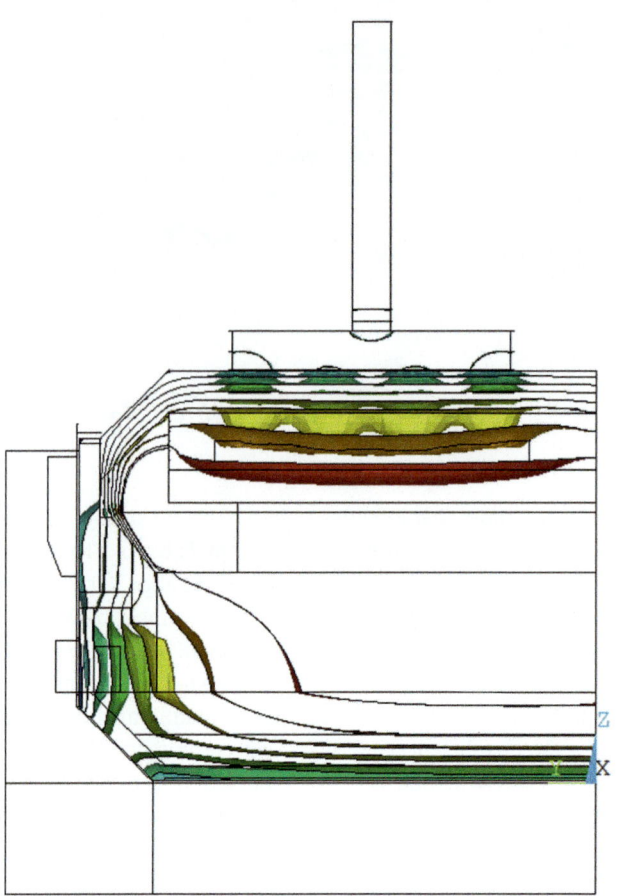

	ANSYS	A	=91.1613
	R14.5	B	=193.58
		C	=295.998
		D	=398.417
		E	=500.835
		F	=603.253
		G	=705.672
		H	=808.09
		I	=910.509

Table 6 Voltage breakdown of retrofitted lining design

Item	Calculated voltage (V)
Anode	0.355
Anode–Cathode	1.420
Cathode	0.255
Back EMF	1.700
Voltage Drop of Busbar Around	0.230
Others (pot line busbar, AE, etc.)	0.023
Avg. Voltage	3.984

Then these measurement data are calculated to energy output, then being converted to equivalent voltage. The average values of the four cells are shown in Table 7. The difference between energy input and output is 0.25%, indicating that the test results are reliable [7]. The energy consumption of aluminum electrolysis reaction is one of the main indicators to judge the energy utilization rate of aluminum reduction cells, and the proportion of 44.24% is among the middle level of Chinese smelters.

Shell

In this project, since the cells have been used for 15 years, some shells have serious deformation. Therefore, in the project, a criterion needs to be established to determine which shells can be reused and which need to be replaced. Below is the detail:

Table 7 Energy balance test result (Four cells' average)

	Item	Equivalent Voltage V	%
Energy Input	1. Electricity	4.0010	100%
	Total	**4.0010**	**100%**
Energy Output	1. Al Electrolysis Reaction	1.7702	44.24%
	2. CO + CO$_2$ gas Heat Loss	0.0099	0.25%
	3. Tapped aluminum Heat Loss	0.1109	2.77%
	4. Butt Heat Loss	0.0038	0.09%
	5. Stub Heat Loss	0.0005	0.01%
	6. Anode Change Heat Loss	0.0257	0.64%
	7. Air Heat Loss	0.9424	23.55%
	8. Cell System Heat Loss	1.1276	28.18%
	a. Side	0.8407	21.01%
	b. Bottom	0.0491	1.23%
	c. Deck Plate	0.0421	1.05%
	d. Hood	0.1106	2.76%
	e. Collector Bar	0.0802	2.00%
	f. Rod	0.0049	0.12%
	Total	**3.9910**	**99.75%**
Difference		0.0100	0.25%

Table 8 Comparison between the structure of the new shell and the old shell

Item	Traditional structure	New structure
Cradle	Independent, bolt connection with the shell	Welded with shell
Inner corner type	Rounded corners	Right-angle corners
Expansion deformation	Larger	Smaller
Weight	46.1 t	45.3 t

- Type A: if shell width expansion deformation is less than 30 mm, then directly reuse it (88 cells).
- Type B: if shell width expansion deformation is between 30–60 mm, then the shell needs to be corrected and meets the criterion of type A without changing the structure (166 cells).
- Type C: if shell width expansion deformation is larger than 60 mm, then replace the shell with a new structure (28 cells).

Note: since the length deformation of the shell is relatively small (maximum deformation among all pots is only 19 mm), the length deformation is not considered in the above criterion.

A new shell structure, which was applied in GP500 + reduction cell, has been used to replace the unqualified shells. The biggest difference in structure between the new shell and the traditional shell is that the new shell welds the cradle and shell together and applied a right-angle design at the corner, which prevents the corner anodes from hitting the ledge. The differences are listed in Table 8.

We carried out a finite element simulation calculation for the new shell. According to the calculation, the vertical deformation of the pot shell at the initial stage of a startup is 5.08 cm, and the longitudinal deformation (expansion) is 7.28 mm during normal production. The expansion deformation of the pot shell is less than 1/200 of the shell's width and the stress is lower than the allowable stress of the material. Therefore, the pot is safe and reliable (Fig. 8).

Cell Performance

Since the cells were started, all the cells are operating at stable, high-efficient conditions. Production key performance parameters have been collected since the cells are normalized (the normalization period is 3 months since startup). The main indicators from the fourth quarter of 2020 till the second quarter of 2022 are listed in Table 9.

It can be seen from the above statistics data that the average DC power consumption of the retrofitted cell since

Fig. 8 Longitudinal deformation (expansion) of new shell structure

Table 9 Key performance parameters of retrofitted GP320 potline

Item	Unit	Original	Retrofitted							
			Q4′2020	Q1′2021	Q2′2021	Q3′2021	Q4′2021	Q1′2022	Q2′2022	Average
Current	kA	320.0	350.8	340.3	340.2	340.8	340.4	339.2	340.1	341.7
CE (tapping)	%	92.5	93.4	93.1	93.2	93.1	93.4	93.1	93.2	93.2
DC Power Consumption	kWh/t-Al	13,273	12,708	12,739	12,741	12,758	12,762	12,768	12,764	12,749

normalization is 12,749 kWh/t-Al. Compared with the original DC power consumption of 13,273 kWh/t-Al, the cell retrofit successfully reduces the power consumption by 524 kWh/t-Al. This proves that the cell retrofit was successful.

Conclusion

This project is a successful case of retrofitting the first-generation GP320 aluminum reduction cell. The current has been increased by 6.25%, and the DC power consumption has been reduced by 524 kWh/t-Al. According to a regulation issued by the Chinese government, the smelter's AC power consumption must be lower than 13,300 kWh/t-Al (desulfurization power consumption will be excluded from total AC power consumption) in 2025 [8]. This means that the DC power consumption of the smelter has to reach the level of 12,750 kWh/t-Al to hit the target. After cell retrofit, the new cell line can reach such a target ahead of schedule. This sets a good example for extending the service life of GP 320 reduction cells (in fact, in the recent decade, due to increasing pressure of energy-saving targets, some 300 kA level potlines were shut down and replaced by higher amperage potlines). The GP 320 reduction cells and the contemporaneous GP 340 reduction cells were widely used in the 2000s. Today, there are still many

smelters using these cell types in China and around the world. The success of this project has explored a path for the future of these aluminum smelters.

References

1. Guiyang Aluminum & Magnesium Design & Research Institute Co., Ltd. Feasibility Study Report of Guangyuan Zhongfu Smelter Relocation Project[R]. 2019.
2. Sele, Thorleif. Instabilities of the metal surface in electrolytic alumina reduction cells[J]. Light Metals, 1977, 8: 613–618.
3. Alexander Arkhipov, Abdalla Alzarooni, Amal Al Jasmi, et al. Improving the understanding of busbar design and cell MHD performance[J]. Light Metals, 2017, 671–677.
4. Wang Xuan, Liu Wei, Li Guangbin, et al. Simulation and Optimization on Magnetic Field of 350 kA Series Aluminum Reduction Cell[J]. Metal Materials and Metallurgy Engineering, 2015, 43 (06): 30–34.
5. Wang Changchang. Influence of Magnetic Field Optimization on Technology Indicators of 350 kA Prebaked Aluminum Reduction Cell[J]. Energy Saving of Nonferrous Metallurgy, 2014, 30 (05): 10–13.
6. Lu Shaoyuan. Application of Numerical Simulation in the Improvement of the Gas Collection Flue of Electrolyzer, 2020 (05): 54–56.
7. Guangyuan Zhongfu Smelter Test Report [R]. 2021.
8. National Development and Reform Commission. Notice of the National Development and Reform Commission on Improving the Ladder Electricity Price Policy for the Aluminum Reduction Industry [EB/OL]. https://www.ndrc.gov.cn/xxgk/zcfb/tz/202108/t20210827_1294888.html.

The Expanded Industrial Pilot of SAMI's NCCT+ Technology

Xi Cao, Yafeng Liu, Hongwu Hu, Xuan Wang, Jinlong Hou, Wei Liu, Kangjian Sun, Michael Ren, and Pengfei Du

Abstract

Shenyang Aluminum and Magnesium Engineering and Research Institute Co. Ltd (SAMI) has carried out a series of technical research and experiments since 2015 to develop a more environment-friendly and energy-efficient aluminum reduction technology. By adopting the technical route of improving MHD stability while reducing cathode voltage drop, SAMI's New Conceptual Cathode Technology namely NCCT Technology has been developed. In 2019, this technology has been applied and verified in the Indonesian INALUM aluminum smelter's upgrade project. Since then, an expanded industrial test was implemented on two SY500 potlines in 2021 and a newer version of NCCT Technology has been developed, called NCCT + Technology. Statistical data over the past year demonstrates that energy consumption of NCCT Technology is <12,300 kWh/t Al, while that of the NCCT+Technology is <2,200 kWh/t Al. Both these technologies provide strong technical support for the aluminum industry to upgrade existing reduction cells for better carbon emission.

Keywords

NCCT+ Technology · MHD stability · Cathode voltage drop · Expanded industrial pilot

X. Cao · Y. Liu (✉) · H. Hu · X. Wang · J. Hou · K. Sun · P. Du
Aluminum Reduction Department, Shenyang Aluminum and Magnesium Engineering and Research Institute Co. Ltd, Shenyang, 110001, China
e-mail: lyf@sami.com.cn

X. Cao
e-mail: caoxi@sami.com.cn

H. Hu
e-mail: hhw@sami.com.cn

X. Wang
e-mail: wangxuan1543@qq.com

J. Hou
e-mail: hou1jin2long3@126.com

K. Sun
e-mail: samiskj@163.com

P. Du
e-mail: 1367872716@qq.com

W. Liu
Director of Science and Technology Management Department, Research Institute Co., Ltd, Shenyang Aluminium & Magnesium Engineering, Shenyang, 110001, China
e-mail: liuwei@sami.com.cn

M. Ren
Sunlightmetal Consulting Inc., Burlington, L7L 6M3, ON, Canada
e-mail: michael.ren@sunlightmetal.ca

Introduction

With the implementation of the "3060" strategic goal of carbon peaking and carbon neutrality in China, China's aluminum reduction industry will continue to develop environment-friendly and highly energy-efficient technologies. Shenyang Aluminum and Magnesium Engineering and Research Institute Co. Ltd (SAMI) has taken the lead in this field. A series of technical research and experiments have been carried out since 2015 [1–3], and a complete set of green and low-carbon aluminum reduction technology system has been formed. The technical route of the system is shown in Fig. 1.

(1) The technical route is based on the essence of the aluminum reduction process, and the core is to improve the MHD stability and reduce the net cell voltage. With the improvement of the raw material quality and upgrade of the control system, the theoretical basis for the stable operation under ultra-low ACD and ultra-low cell voltage is set up.

(2) With the basic conditions for stable operation under the ultra-low ACD, it is also necessary to control the heat balance for the new design and production management technology to realize the stable production of the electrolytic cell under the ultra-low ACD. At the same time,

© The Minerals, Metals & Materials Society 2023
S. Broek (ed.), *Light Metals 2023*, The Minerals, Metals & Materials Series,
https://doi.org/10.1007/978-3-031-22532-1_108

Fig. 1 Roadmap of green low-carbon aluminum reduction technology system

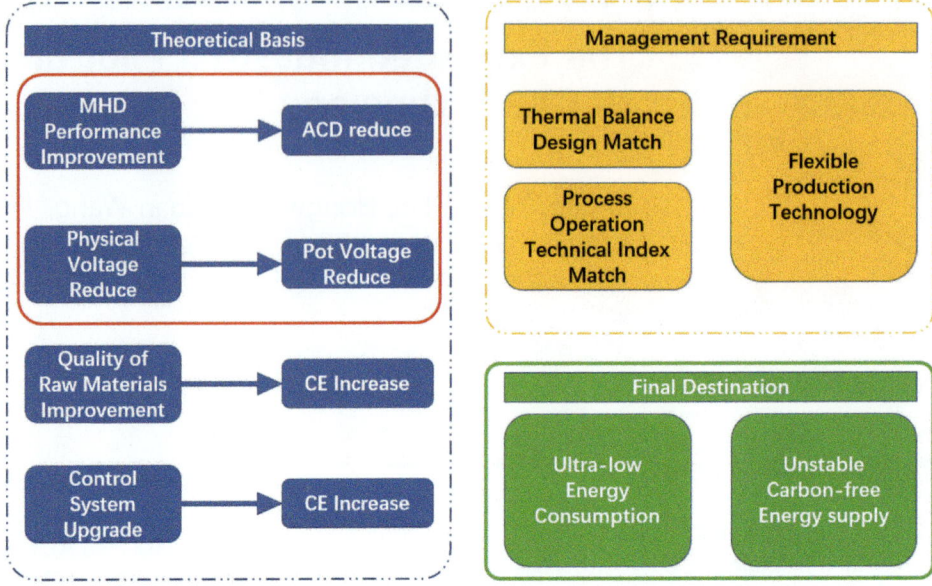

in order to incorporate clean energies such as wind or solar power, flexible production technology is required to adapt to this unstable external power supply form.

(3) Through the above technical measures, it is theoretically possible to realize the production of reduction cells under the ultra-low ACD and ultra-low voltage conditions, and at the same time, it can use the form of clean energies with an unstable power supply.

In this technical system, SAMI has developed a variety of technologies, among which NCCT is a technology that can simultaneously achieve the improvement of magnetic fluid stability and the reduction of net cell voltage drop. The theoretical basis of this technology has two points:

(1) The MHD stability of the pot can be significantly improved by a large reduction of the horizontal current in the metal [4].
(2) The voltage drop of the cathode can be greatly reduced by a special cathode structure design combined with iron casting.

The MHD stability can be summarized as the free surface problem of two-phase flow caused by the difference of electromagnetic force between liquid bath and metal. One of the ways to solve the problem of MHD stability is based on $F = J \times B$. Starting from the magnetic field and the current in the molten aluminum, the source problem is solved: reduce the magnetic field in the vertical direction or reduce the current in the horizontal direction.

The NCCT technology integrates the concept of reducing the horizontal current in the molten aluminum into the design and manufacture of the casting cathode assemblies, which can not only reduce the cathode voltage drop but also suppress the horizontal current in the molten aluminum. Using advanced computer models, the overall design of the cathode structure is optimized.

Firstly, the contact surface and resistance ratio between the cathode carbon block and the steel bar is changed to meet the targeted horizontal current and cathode voltage drop control objectives. Secondly, the design of the cathode groove structure is optimized to further reduce the cathode voltage drop, eliminate local stress concentration, and prevent corner cracks in the groove. A refractory layer is set at the lower part of the cathode carbon block as the first barrier to prevent electrolyte leakage. Another important improvement introduced in the design is changing the cathode assembly sealing methodology from using carbon paste to cast iron. The facility to preheat cathodes and collector bars, prepare cast iron, and seal the cathode with collector bar was developed inhouse.

NCCT technology has been successfully applied and verified in the Indonesian INALUM aluminum smelter's upgrade project [5], and SAMI has further developed this technology on top of that. In order to pursue lower horizontal current and cathode voltage drop, a copper steel bar was adopted, and the impact has been improved to a higher level, thus this technology is called NCCT+ Technology. Compared with NCCT technology, the horizontal current can be further reduced by more than 50%, and the cathode voltage drop can be further reduced by more than 40 mV. The NCCT+ technology can provide even lower ACD and lower cathode voltage drop, which provides better basic conditions for further reduction of DC power consumption.

Reduce Horizontal Current Effect

Sami's advanced thermoelectric coupling calculation model can calculate the horizontal current in molten metal under NCCT and NCCT+ technology. The results comparison in Fig. 2. show that for a typical SY500 cell, the horizontal current density is reduced from 8175A / m^2 to 5539 A / m^2 & 2207 A/m^2, respectively, after adapting NCCT and NCCT + technology, and the horizontal current density of NCCT + technology is reduced by more than 70%, which significantly improves the MHD stability.

At the same time, Fig. 2 illustrates that the cathode voltage drop reductions of the NCCT and NCCT+ technologies are significant. The design value of the cathode voltage drop of a traditional SY500 pot, which is assembled with 50% graphite carbon blocks and pastes, is 328 mV. After adopting graphitized cathode carbon blocks and combining NCCT technology, the cathode voltage drop is reduced to 196mv. When the NCCT+ technology is adopted, the design value of the cathode voltage drop is reduced to 155 mV (Different pot styles may fluctuate slightly). The lower voltage drop is providing the basic conditions for directly reducing the operating voltage of the pot.

MHD Cell Stability Analysis of Reduced Horizontal Current

The MHD performance of the pot is significantly improved due to the reduction of the horizontal current in the metal. MHD model can clearly show the performance improvement [6].

The impact of the reduced horizontal current is studied using the MHD cell stability shallow water model. The metal pad horizontal current is recalculated first in the MHD

model. Figure 3 is presenting the obtained horizontal current density for the SY500 cell before the introduction of NCCT. Figure 4 is presenting the horizontal current density of the new design. As indicated in Figs. 5 and 6, the steady-state metal velocity is reduced by that reduction of the horizontal current intensity.

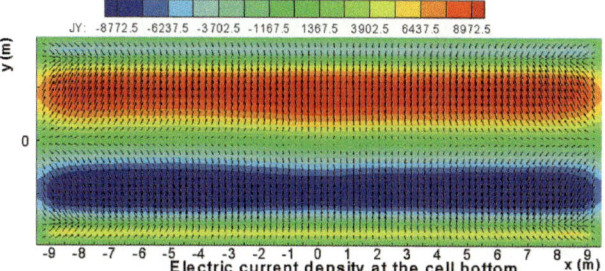

Fig. 3 MHD-Valdis model metal pad horizontal current before NCCT

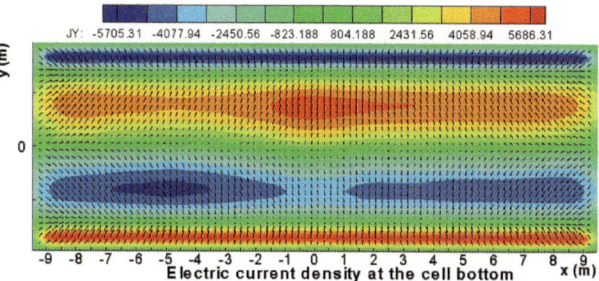

Fig. 4 MHD-Valdis model metal pad horizontal current after NCCT

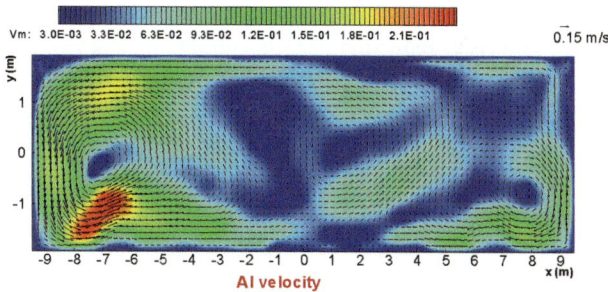

Fig. 5 MHD-Valdis model metal pad steady-state flow before NCCT

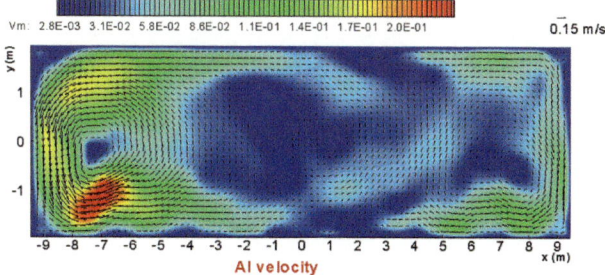

Fig. 6 MHD-Valdis model metal pad steady-state flow after NCCT

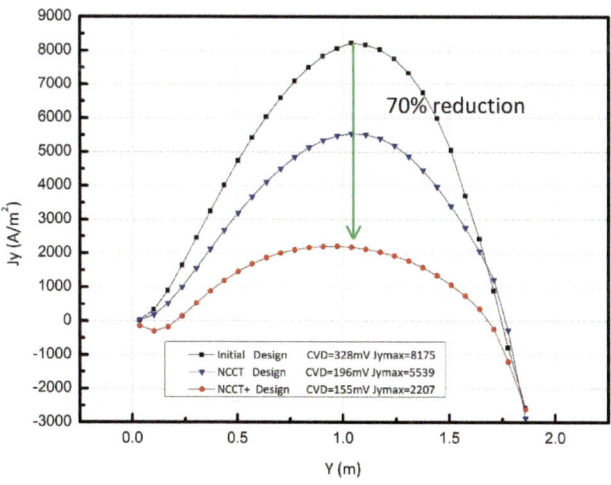

Fig. 2 The metal pad horizontal current of various technology under 500kA current

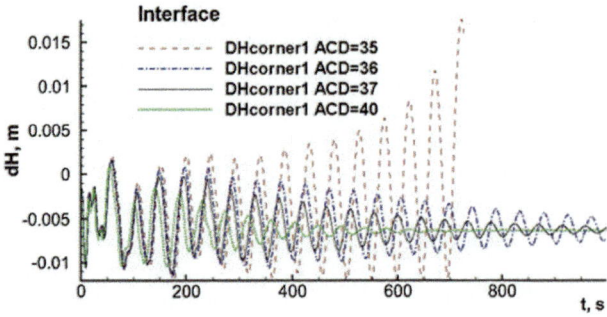

Fig. 7 MHD-Valdis model transient cell stability analysis before NCCT

The model is used to calculate the transient magnetic fluid stability of the cell, and the convergence of the disturbance curve represents that the cell can produce stably at this ACD. The stability analysis was done using a constant metal pad thickness of 18 cm and an initial perturbation amplitude of 5 mm. As shown in Fig. 7, the SY500 cell before NCCT is stable up to 36 mm of ACD but unstable at 35 mm ACD.

But after NCCT, the stability analysis presented in Fig. 8 was performed at a constant 26 mm ACD. To confirm this conclusion, we also investigated the MHD stability at different metal heights at 26 mm ACD. Because the metal height will affect the horizontal current in the metal, the higher the metal, the smaller the horizontal current. The results show that the disturbance also converges when the height of aluminum metal pad exceeds 18 cm. But to be conservative, we consider that 26 cm ACD is already the bottom limit.

It indicates that NCCT technology can provide at least 10 mm safe ACD reduction, but the current efficiency of the cell will not be affected. In the same way, NCCT+ technology can also rely on lower horizontal current to achieve lower safe ACD. Then the purpose of reducing the pot voltage can be achieved by reducing the ACD. Lower cell voltage means lower energy consumption.

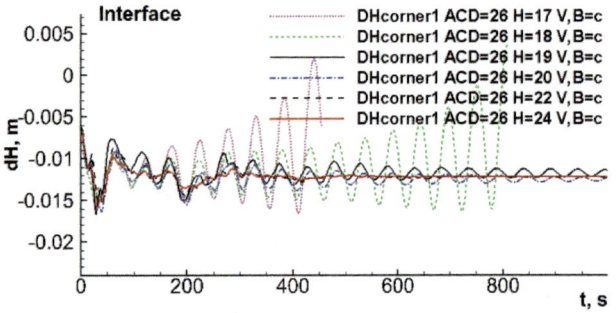

Fig. 8 MHD-Valdis model transient cell stability analysis after NCCT

Thermal Balance Control

NCCT+ technology provides the basis for lower voltage, but whether the pot can operate stably under target voltage is also closely related to the design of thermal balance and operating window. If thermal balance design and thermal balance management deviate, it will be difficult for the pot to operate at the target voltage.

NCCT+ technology greatly reduces the cathode voltage drop, but the paradox is that the heat dissipation of the graphitized cathode is larger, which makes the thermal balance of the NCCT+ pot very different from that of the original design. In order to ensure the thermal balance of the cathode area, the corresponding lining design must reduce the heat dissipation of the cathode area. Good thermal insulation performance and resistance to electrolyte and sodium vapor corrosion materials are chosen to ensure the local and whole thermal balance at the targeted cathode voltage is greatly reduced.

Therefore, SAMI uses an accurate simulation calculation model to optimize the thermal balance of the lining on the basis of ensuring the refractory capability by upgrading the material, structural form, and construction process of the cathode area to prevent the issue of the extended ledge on the cathode surface (ledge toe).

The heat balance calculation results are shown in Fig. 9. Taking Hualei smelter as an example, which uses NCCT + technology, the isotherm distribution of the pot is reasonable for the 800 °C isotherm and the 900 °C isotherm that are both within the refractory layer, while the side isotherms are vertical. The operating voltage of the cell is 3.90 V and the process parameters are within the design window.

The net thinnest thickness of the ledge profile at the metal/bath interface is 7.0 cm shown in Fig. 9, and the shortest distance from the ledge surface to the side block is 16.0 cm. The ledge toe is very short, just about 4 cm below the anode projection. The ledge profile is uniform and there is no longer ledge toe, and the cathode lining is well protected. The calculated ledge profile is also shown in Fig. 9.

In the original lining design, the heat balance calculation results of the test pot at the target voltage are shown in Fig. 10. The thermal isotherm at the bottom of the pot lining moves up significantly. The shortest distance from the ledge surface to the side block increases to 22.2 cm from 16 cm, and the net thinnest thickness of the ledge profile at the metal/bath interface increases to 12.7 cm from 7.0 cm. The ledge toe extends 11.5 cm from 4 cm below the anode projection. The new design form is a great improvement over the original design.

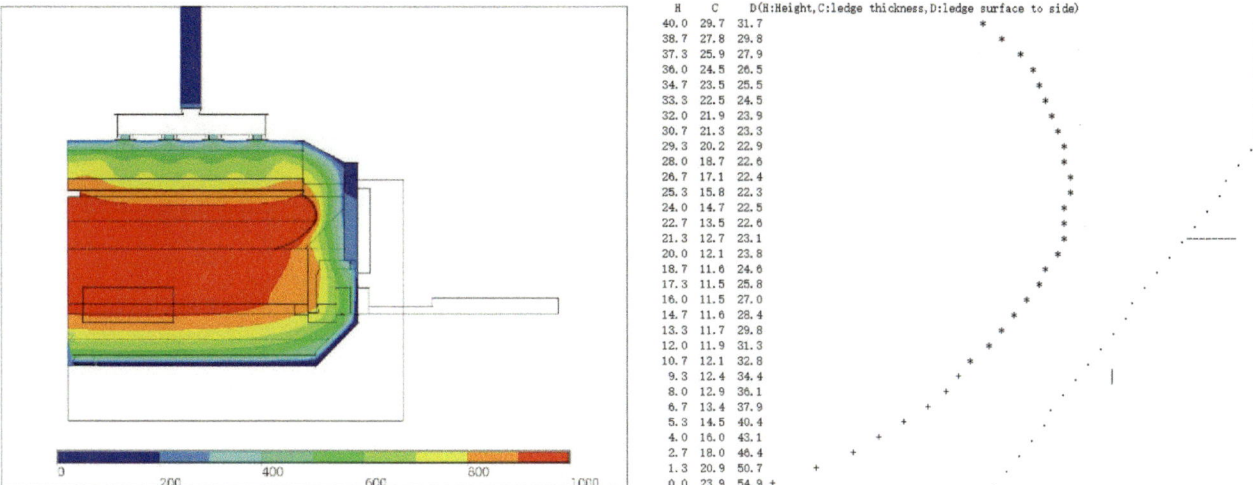

Fig. 9 Thermal isotherm distribution and pot profile shape

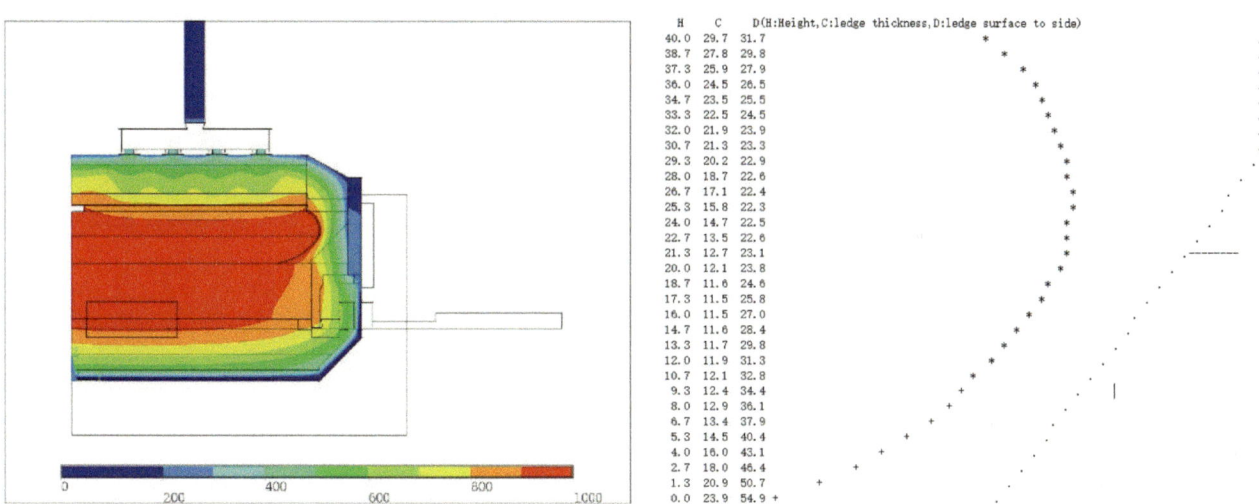

Fig. 10 Thermal isotherm distribution and pot profile shape under the original design

Table 1 lists the voltage breakdown and heat losses of Huayun 500kA NCCT+ technology test cells. The electrolyte chemistry contains 3–4% lithium salt and 2–3% Potassium salt. The calculated anode voltage drop is 0.338 V, bath (including bubbles) voltage drop is 1.244 V, cathode voltage drop is 0.157 V, cell busbar voltage drop is 0.243 V, anode rod to anode beam contact voltage drop is 0.008 V, anode effect average voltage is 0.005 V, and average net cell voltage is 3.850 V.

The Process Technology

This technology puts forward standard requirements for production management. It is necessary to strictly follow the process and operation procedures required by SAMI to carry out the lining and baking start-up of the reduction cells and strictly follow the process parameter adjustment plan in the abnormal period and the process parameter window in the

Table 1 Voltage distribution and energy expenditure of Huayun smelter pilot pot

Voltage breakdown/V		Heat losses/V	
Reversible and overvoltage	1.840	Top heat loss	0.799
Anode voltage drop	0.338	Side heat loss	0.509
Bath (including bubbles) voltage drop	1.293	Heat loss of steel bar	0.110
Cathode voltage drop	0.157	Bottom heat loss	0.150
Anode effect average voltage	0.005	End heat loss	0.066
External busbar voltage	0.189	Busbar and contact surface voltage drop heat loss	0.197
Anode rod to anode beam contact voltage drop	0.008	Material heating	0.220
Noise voltage	0.020	Al Production	1.799
Average net voltage of the cell	3.850	Total heat loss	3.850

Table 2 The process technical conditions requirements of NCCT+

Item	Unit	Huayun	Hualei
Bath temperature	°C	935945	955965
Aluminum level	cm	1820	1921
Electrolyte level	cm	1820	1820
CR	–	2.452.55	2.252.3

normal production period for the Production management of reduction cells. The close combination of production practice and design requirements ensures that the performance of the reduction cells can be achieved.

The electrolyte systems of the two pilot sites are different. The electrolyte of the Huayun smelter contains 3–4% lithium salt and 2–3% potassium salt, and the electrolyte of Hualei smelter contains 1% lithium salt and 1% potassium salt. Therefore, the process technical conditions of NCCT+ pots required by the design are different as follows (Table 2):

Technology Application

In 2021, an expanded industrial test was implemented on two SY500 potlines, in Inner Mongolia Huayun smelter and Guangxi Hualei smelter, respectively, both belonging to Chinalco. In these two aluminum smelters, both NCCT and NCCT+ technologies have been applied.

In each smelter, there are 3 NCCT technology pots with graphitized cathodes and 2 NCCT+ pots, a total of 6 NCCT technology pots with graphitized cathodes, and 4 NCCT + pots. The first set of test pots was started in October 2021 and has been operating for 11 months. The typical key

performance during a normal operation period is shown in Table 3, and the trends of various process parameters since start-up are shown in Figs.11, 12, 13, 14, 15, 16, 17, 18, 19, and 20. The data in Table 3 are statistics from August 1, 2022 to August 31, 2022. Different busbar configuration is applied in the inner Mongolia Huayun smelter and Guangxi Hualei smelter. The process loop busbar configuration is applied in the Huayun smelter designed by SAMI, but not in the Guangxi Hualei smelter, namely New Conceptual Busbar Technology. The process loop busbar also consumes DC power. In order to facilitate comparison with Hualei Smelter, the DC power consumption in Table 3 only includes the net cell and does not include the energy consumption of the process loop busbar.

The cathode voltage drop trend of the pilot pot is shown in Fig. 15. The data clearly shows that the cathode voltage drop of the pot using the NCCT+ technology with copper steel bars is gradually increased and stabilized at about 165 mV, which is around 30 mV and 50 mV, respectively, lower than that of each site NCCT technology pots, which provides more ACD space and heat input.

It should be noted that Inner Mongolia Huayun smelter and Guangxi Hualei smelter not only used different busbar configurations, but also used different cathode group sizes

Table 3 Summary of key performance data[a]

KPI	Unit	Huayun smelter		Hualei smelter	
		NCCT	NCCT+	NCCT	NCCT+
Pot Current	kA	500	500	500	500
Net cell Voltage	V	3.851	3.855	3.929	3.895
Current Efficiency	%	93.34	94.39	95.27	95.21
DC Power Consumption	kWh/t-Al	12,297	12,173	12,292	12,194
Bath Temperature	°C	934	933	956	958
Excessive AlF3	%	7.7	7.7	11.7	11.7
Bath height	cm	18.7	18.3	19.1	19.1
Metal height	cm	20	20	19.1	21
Pot noise	mV	19	20	20	19
CVD	mV	202	168	217	166
Fe content	%	0.053	0.067	0.078	0.066
Si content	%	0.018	0.021	0.029	0.026

[a]The data for each set in Table 3 are statistics from August 1, 2022 to August 31, 2022

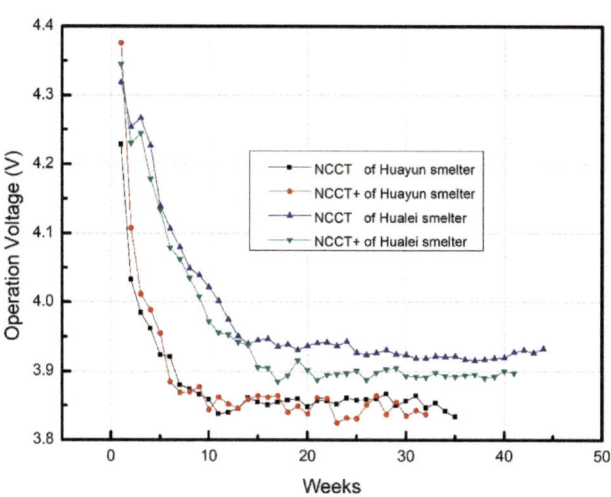

Fig. 11 Net cell voltage

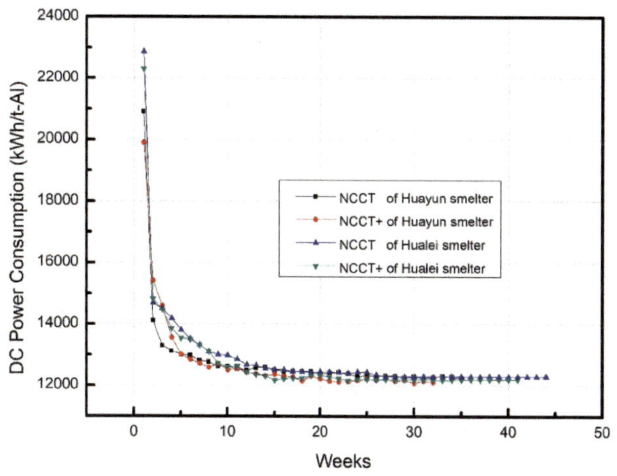

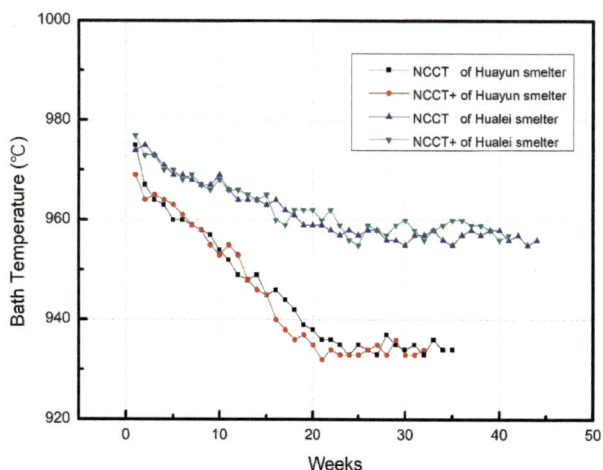

Fig. 13 DC power consumption

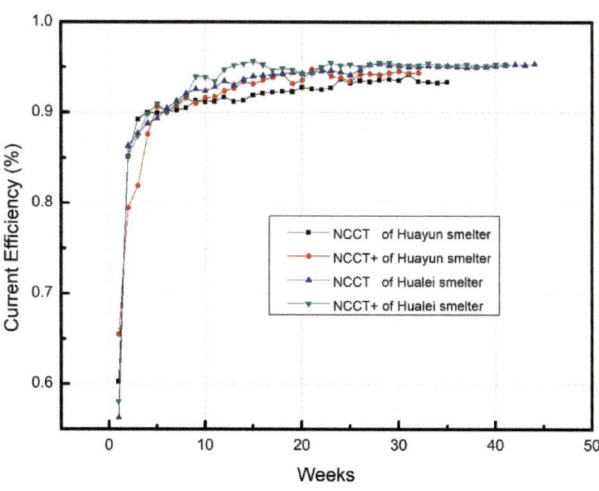

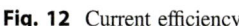

Fig. 12 Current efficiency

Fig. 14 Bath temperature

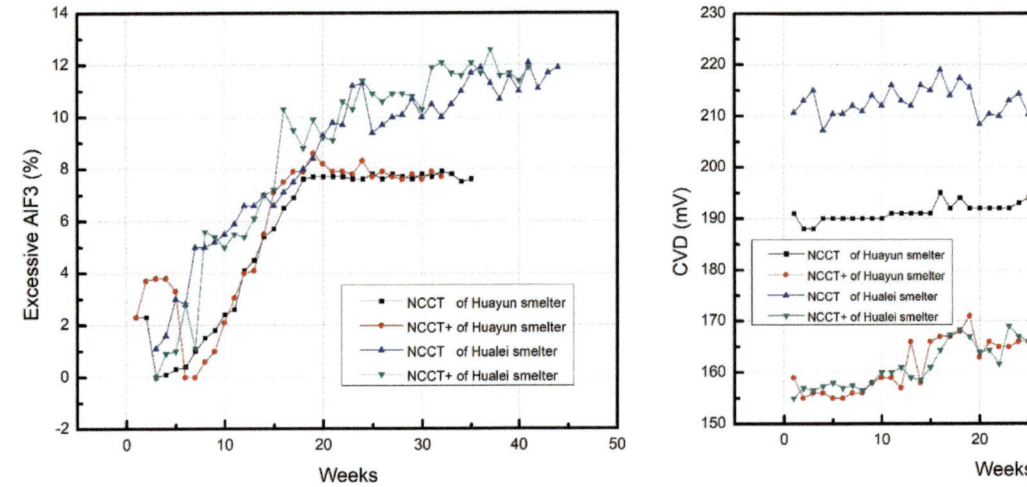

Fig. 15 Excess AlF3

Fig. 18 CVD

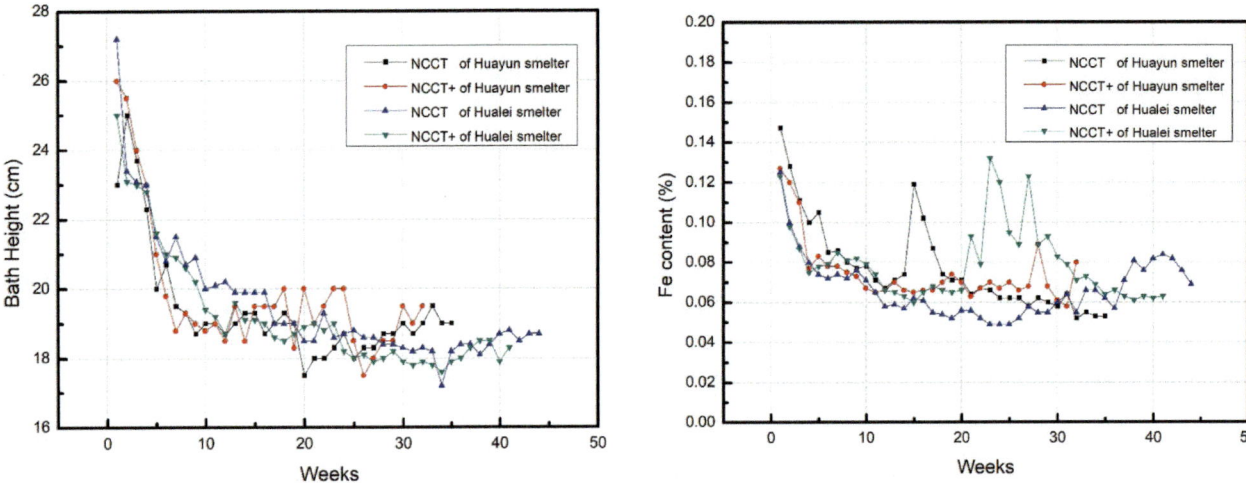

Fig. 16 Bath height

Fig. 19 Fe content

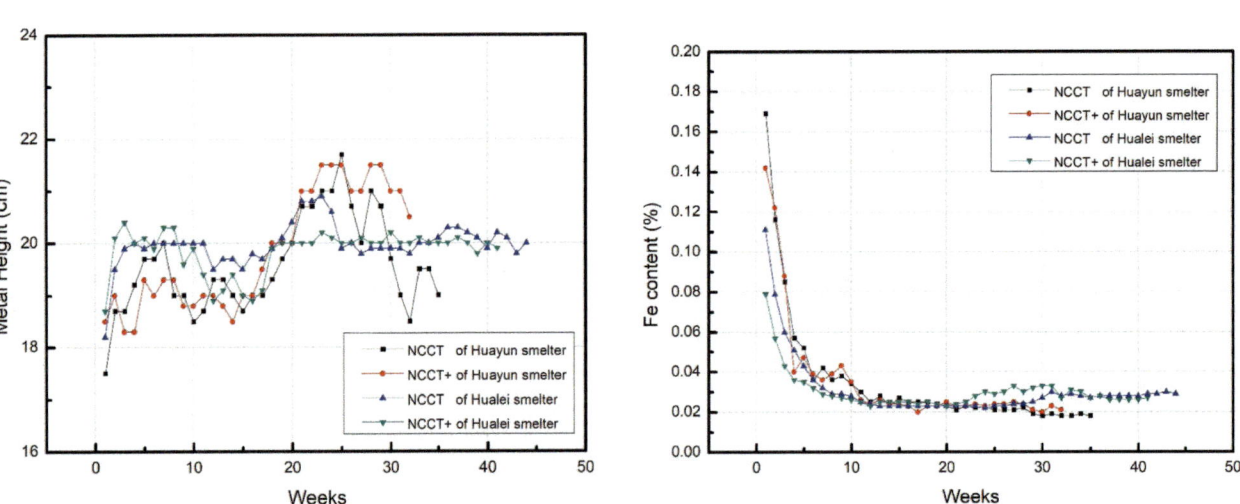

Fig. 17 Metal height

Fig. 20 Si content

and cathode group quantities. Hualei's cathode width is smaller than that of Huayun, and the number of cathode groups is larger. In order to ensure the safety of the lining and the potline of Hualei smelter, the cross-section of Hualei's cathode collect bar is correspondingly smaller than that of Huayun. Therefore, under the NCCT technology, the design value of Hualei's cathode voltage drop is higher than that of Huayun. However, the copper insert collect bar is used by the NCCT+ technology, and we can control the cathode voltage drop by adjusting the amount of copper insert. Therefore, by using different amounts of copper, the cathode voltage drops of the Inner Mongolia Huayun smelter and Guangxi Hualei smelter for NCCT+ technology can be controlled at the same level. The cathode voltage drop of the NCCT technology in Fig. 2 is that of the Huayun smelter.

The content of excess AlF_3 is shown in Fig. 15. It is low initially because the pure cryolite was not used for charging the pots before start-up and instead by the electrolyte combination with sodium carbonate. After the start-up, sodium carbonate is gradually added to the pots to adjust electrolyte composition according to the test results.

It should be noted that the sudden increase of Fe content in Fig. 19 is caused by the bath corrosion of stubs, and the problem is gradually solved after the anode is changed. The Si content has been very stable at a low value, which indicates that there is no problem with the ledge of the pots.

Conclusion

The expanded industrial pilot results show that the NCCT + technical direction of improving the MHD stability by decreasing the horizontal current and reducing the cathode voltage drop by using graphitized cathode and copper steel rod is reliable and demonstrates the targeted reduction of specific energy consumption of the cells. The 500kA pots of two smelters using NCCT and NCCT+ technology can achieve DC power consumption of less than 12,300 and 12,200 kwh/tAl, respectively, and operate stably and safely. The application of the NCCT+ technology not only brings more economic value to smelters, but also greatly reduces carbon emissions, which provides strong technical support for the aluminum industry to upgrade existing reduction cells for better carbon emission.

References

1. Yang Xiaodong, Liu Wei. Discussion on designing high amperage energy saving aluminum reduction pot - busbar, cathode structure and MHD stability[J]. Light Metal, 2016(10): 27–32. (In Chinese)
2. Hu Hongwu, Cao xi, Sun Kangjian. Research and application of new energy saving cathode structure technology in high amperage aluminum reduction pot [J]. Light Metal, 2016(10): 33–37. (In Chinese)
3. Zhang Kun, Cao Xi. Research and Economic Benefit Analysis of Aluminum Reduction Technology with Ultra-low Energy Consumption[J]. Energy Saving of Nonferrous Metallurgy,2020 (4):22-30.
4. Dongfang Zhou, Yafeng Liu, Shaohu Tao. The Research and Trial of the Aluminum Electrolysis Cells with Current Out from the Bottom, *Light Metals 2018*, pp. 1223–1228.
5. Liu Ming, Yang Xiaodong, Liu Yafeng, Lu Yanfeng. Amperage Increase from 195 to 240 kA Through Pot Upgrading, *Light Metals 2019*, pp. 582–591.
6. Wei Liu, Zhibin Zhao, Ming Liu, Xi Cao, Hongwu Hu, Yafeng Liu, Marc Dupuis. Retrofitting of Several Cell Technologies Using a Protruding Collector Bar Cathode Assembly. *ICSOBA 2022, Athens, Greece, 10–14 October 2022.*

The SY500 Pot Technology Development

Kangjian Sun, Yafeng Liu, Hongwu Hu, Xuan Wang, Jinlong Hou,
Wei Liu, Xi Cao, and Michael Ren

Abstract

Since 1996, Shenyang Aluminum and Magnesium Engineering and Research Institute Company Limited (SAMI) has developed a series of aluminum reduction technologies, the SY series, ranging from 160 kA, 190/200 kA,230/240 kA 280 kA, 300 kA, 350 kA to 400 kA. In 2009, SAMI developed the 500 kA reduction technology called SY500, and since then SAMI has been continuously improving this technology and has applied it in the engineering of new aluminum smelters. The paper introduces the development of the first, second and third generation SY500 potline technology, systematically analyzes breakthroughs of new technologies adopted by SY500 pot technology, such as the New Conceptual Cathode Technology (NCCT), New Conceptual Busbar Technology (NCBT), Network Busbar Technology (NBT), etc. The SY series potline technology has been validated for its excellent technical and economic performances and become the most widely used reduction pot technology.

Keywords

SY500 pot technology • 500 kA pot • Reduction

K. Sun (✉) · Y. Liu · H. Hu · X. Wang · J. Hou · X. Cao
Aluminum Reduction Department, Shenyang Aluminum and
Magnesium Engineering and Research Institute Co.Ltd,
Shenyang, 110001, China
e-mail: samiskj@163.com

W. Liu
Science and Technology Management Department, Shenyang
Aluminium & Magnesium Engineering & Research Institute Co.,
Ltd, Shenyang, 110001, China

M. Ren
Sunlightmetal Consulting Inc, Burlington, ON, Canada
e-mail: michael.ren@sunlightmetal.ca

Introduction

Since the beginning of the new century, with the rapid development of China's economy, demand for primary aluminum has been increasing, which has promoted the continuous progress of aluminum reduction technology. The operating current of the reduction pot has gradually increased from 160 kA in the 1990s to 200 kA, 300 kA and 400 kA by early 2000's, and has developed to 500 kA and 600 kA in the recent years.

In recent years, with the continuous improvement and optimization of the 500 kA reduction pot technology, the 500 kA pot developed by SAMI called SY500 pot is now the most widely used reduction technology in the industry. The SY500 upgrades have undergone 3 generation3 in recent years.

Development of First-Generation SY500 Pot

In May 2007, the SY400 electrolytic pot was successfully started, put into operation and reached the advanced level performance. Since then, SAMI developed SY500 pot technology which was more productive, more energy efficient and more environmentally friendly.

The first generation SY500 pot summarizes the development experience of the previous SY400 pots and other SY pot types and greatly improves the stability of the potline by reducing the horizontal current of aluminum liquid. At the same time, the heat balance design of the pot is adjusted according to the optimization results, which maintains low energy consumption. The pot hood efficiency is greatly improved by adopting the superstructure design and the twin duct pipe design. The new cathode structure technology of reducing horizontal current and the optimized MHD design. Table 1 below shows the comparison of typical design parameters of SY400 and SY500 (1st generation) pot.

© The Minerals, Metals & Materials Society 2023
S. Broek (ed.), *Light Metals 2023*, The Minerals, Metals & Materials Series,
https://doi.org/10.1007/978-3-031-22532-1_109

Table 1 Comparison of typical design parameters of SY400 and SY500

Pot		SY500	SY400
Design potline current	kA	500	400
Design pot voltage	V	3.93	4.2
Feeding point	#	6	6
Number of cathode assemblies	#	24	24
Number of anode assembly	group	48	24(double anode assembly)
Anode size (Length × width)	m	1.75X0.74	1.55X0.66
Number of bus risers	#	6	6

In the context of energy and emission reduction, industrial adjustment and revitalization planning, SAMI has studied various mechanism characteristics of aluminum pots through theoretical analysis and a large number of physical field simulation and actual measurements. For 500 kA pots, it has developed a number of energy-saving technologies and applied a number of new technologies such as Low Horizontal Current Cathode Technology, optimized six point riser bus configuration technology to the newly developed SY500 pot.

The Low Horizontal Current Cathode Technology adjusts the height of cathode steel bar and carbon block appropriately. By dividing the steel bar into upper and lower parts according to a certain height ratio, the conductive structure of the cathode steel bar is changed. The assembly structure of cathode steel bar and carbon block is also optimized. The technology greatly reduces the horizontal current in the molten aluminum of the pot. Because of the well-designed magnetic fluid stability, this effectively suppresses the fluctuation of molten aluminum and greatly improves the stability of the pot. The pot anode cathode distance (ACD) is effectively reduced and the voltage can be greatly reduced at

the design level, which effectively reduces the energy consumption of the pot (Fig. 1).

The first generation SY500 pot technology was first applied to CHALCO Liancheng branch, Gansu Dongxing Jiujia project and Xinjiang East hope Co., Ltd. the three potline were designed in 2009 and 2010 and put into production in 2011 and 2012 (Fig. 2).

Several series of the first-generation SY500 electrolytic cells have been successfully put into operation in China. The low horizontal current cathode technology used in the SY500 technology has played an important role in reducing the energy consumption of the potline; at the same time, this technology has also been widely used in the overhaul and modification of the existing potline lining, such as Qingtongxia Aluminum Plant, CHALCO Liancheng 200 kA potline, etc.

The Second-Generation SY500 Pot

At the same time, 500 kA potlines in operation have been tested and evaluated. How to improve the technical production performance of the SY500 pots becomes the next research and development goal.

The most prominent feature of the second generation SY500 pot is the adoption of a new bus design concept and bus configuration mode and the establishment of a new

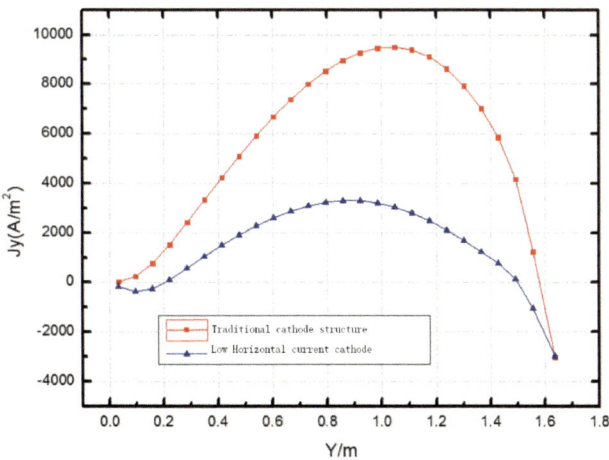

Fig. 1. Comparison of horizontal current of pot between traditional cathode assembly and Low horizontal current cathode technology.

Fig. 2 The first-generation SY500 pot

design standard for magnetic fluid stability which is called New Conceptual Busbar Technology (NCBT). Patent cn105220179a [1].

Compared with the traditional busbar configuration, NCBT technology can reduce the average and maximum flow velocity of liquid aluminum pad and electrolyte pad, reduce the deformation of aluminum liquid electrolyte interface by 30%. The NCBT technology greatly improves the magnetic fluid stability of the pot and achieve energy reduction. At the same time, this technology greatly simplifies the configuration of the bus bar, shortens the pot-to-pot distance by 0.3 m, and reduces the voltage drop of the pot bus bar by 20%.

The comparison of SY500 NBCT busbar design pot and traditional bus configuration is shown in Table 2.

The breakthrough new bus design concept has provided a good design basis for the improvement of the technical indicators of the 2nd generation SY500 pot.

The comparison between the flow field calculation results of SY500 NCBT bus configuration and traditional bus configuration is shown in Table 3 below. It can be seen that compared with the traditional bus bar configuration, the NCBT technology can reduce the maximum flow rate of aluminum liquid by 29%, the average flow rate of aluminum liquid by 26%, the maximum flow rate of electrolyte by 15%, the average flow rate of electrolyte by 26%, and the deformation of the aluminum liquid electrolyte interface by 33% (Fig. 3).

The 2nd generation SY500 pot technology was first applied to Xinjiang Xinfa phase III project, Inner Mongolia Huayun phase I project, Liaoning Zhongwang and other projects. It has achieved good economic and technical performance. Compared with the 1st generation SY500 pot, the stability of the pot has been greatly improved. For example, Table 3 shows the comparison of the average values of

potline high-frequency noise and low-frequency noise recorded by the same pot control system for two adjacent SY500 potline of a plant under the same management operation and raw materials. At the same time, the current efficiency of the potline configured with NCBT busbar potline is 2.54% higher than that of the traditional bus potline, and the power consumption can be reduced by 237 kWh/t-al [2].

The Third-Generation SY500 Pot

The 1st generation and 2nd generation SY500 potlines have been put into production throughout the country. In addition, tests and production evaluations were carried out for the 500 kA potlines already in production.

Further optimization and improvement are also ongoing. How to further comprehensively improve the technical production performance of the pot becomes the next research and development target. In view of the summary of on-site production and operation practice, the third generation SY500 technology has applied a number of more optimized technologies in the previous SY500 pot design and production and operation experience. Including Network Busbar Technology (NBT), New Conceptual Cathode Technology (NCCT), optimized pot shell, optimized superstructure, and optimized new plant ventilation structure.

The NBT technology can greatly reduce the conduction of the uneven distribution of cathode and anode current in the potline under the unstable state of the pot (including the period of pot shutdown, the period of anode effect, and during anode change, etc.), which can reduce the current distribution and magnetic field fluctuation of the unstable pot and its upstream and downstream pots, and further improving the pot stability and anti-interference capability.

Table 2 SY500 NCBT pot and traditional bus configuration data comparison

SY 500 Pot		NBCT busbar design	Traditional bus configuration
Design pot to pot distance	m	6.8	6.5
Design magnetic field Bz	Gs	3.92	2.85
Average liquid aluminum pad flow velocity	cm/s	6.4	8.7
Average liquid electrolyte flow velocity	cm/s	3.2	4.3

Table 3 SY500 NCBT pot and traditional bus configuration noise comparison

Potline	Pot generation	Low frequency noise/mV	High frequency noise/mV
First Potline	First generation SY500	4.5	13.5
Second Potline	Second generation SY500 with NCBT	1.5	9.5

Fig. 3 The second-generation SY500 pot

The current distribution deviation during the anode effect is reduced by 70%, the speed of recovering from an unstable to a stable state is faster. The duration of pot anode effect is shortened which effectively reduces the effect voltage drop and reduces greenhouse effect gas emissions.

The New Conceptual Cathode Technology (NCCT) inheriting the characteristics of the previous low horizontal current cathode technology. The NCCT technology optimizes the structure of the cathode assembly, adjusts the size of the steel bar and structure to improve the stability of the pot. The NCCT technology can improve the magnetic fluid stability and the reduction of the actual net cell voltage drop which effectively reduces the energy consumption of the pot.

The 3rd generation SY500 pot adopts an integrally welded pot shell. The pot shell structure can effectively homogenize the stress distribution, reduce the deformation, ensure the safety of the pot lining, extend the effective life of the high-performance operation of the pot, and facilitate construction, manufacturing and maintenance. As the current of the pot increases, the heat dissipation per unit area will increase correspondingly. In order to improve the operation environment of the potroom and the thermal balance of the 3rd generation SY500 electrolytic pot, the ventilation structure of the original potroom was greatly improved. The ventilation structure can effectively reduce the operating environment temperature, homogenize the air flow around the pots, this not only reduces the temperature of the working area but also improves thermal balance distribution of the pot.

In addition, the ventilation structure of the new plant also fully considers the influence of seasonal changes. By dynamically adjusting the opening of the ventilation window of the potroom, the influence of the environmental difference in winter and summer on the thermal balance of the pot is eliminated.

The 3rd generation SY500 electrolytic pot technology was first applied to Inner Mongolia Mengtai aluminum industry, CHALCO Shanxi Zhongrun, CHALCO Wenshan and other projects, of which CHALCO Shanxi Zhongrun potline was put into operation in May 2018. After the potline was put into normal production, the purity of the primary aluminum which almost all the pot metal purity was above 99.85%, the Fe content in the aluminum was less than 550 ppm, and the silicon content was less than 200 ppm [3]. The current efficiency was higher than 94.18% and the DC power consumption for the whole potline was 12406 kWh/t-al.

The following Fig. 4 shows the third generation SY500 potline, and Fig. 5 shows the potline specific energy consumption after start-up.

Fig. 4 The third-generation SY500 pot

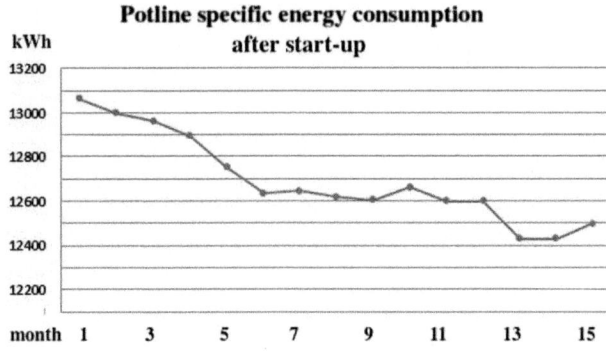

Fig. 5 The third-generation SY500 potline specific energy consumption after start-up

and 25 potline have been put into production, with an operating capacity of more than 12 million tons.

At the same time, related technologies are also constantly moving downward to transform and upgrade the old potlines. This provides a good foundation for the comprehensive progress of domestic aluminum smelting technology.

In addition, in the development process of the new generation of SY pots, new technologies such as digital pot technology, regional control technology and low-carbon emission technology with more advanced concepts are also being developed and will be applied in the next generation of more advanced pots.

Conclusions

Through continuous development and research, design and simulation, on site test and operation, SY500 pot technology have been validated for the good economic performance, low operation cost, low invest cost, low energy consumption, low GHG emission and high potline production.

By the middle of 2020, according to statistics, there were 30 potline designed with SY500 electrolytic pot technology,

References

1. A bus connection method for super large capacity aluminum reduction cell [P]. CN105220179A Shenyang Aluminum and Magnesium Engineering and Research Institute Company Limited
2. Song Shuhong, Fan Jinping. Application Comparison of SY500 NCBT pot and SY 500 traditional bus configuration pot [J]. Light metal, 2021 (1): 23–27.
3. Li Guowei. Practice of quality control in manufacturing and installation of 500 kA pot [J]. Light metal, 2019 (12): 30–35.
4. Marc Dupuis, Kangjian Sun, Review of the SAMI Retrofit Project in QTX Smelter in China; ICSOBA 2022, Athens, Greece, 10–14 October 2022.

Preheat, Start-Up and Early Operation of DX+ Ultra Pots at 500 kA

Mustafa Mustafa, Abdalla Alzarooni, Konstantin Nikandrov, Nadia Ahli, Aslam Khan, Hassan AlHayyas, Marwan AlUstad, and Sajid Hussain

Abstract

In Emirates Global Aluminium (EGA), the first pots were started up in 1979 in its Jebel Ali smelter (then called Dubai Aluminium—DUBAL), using Kaiser P69 cells at an amperage of 150 kA. In 2014 EGA started up its flagship, homegrown DX+ Ultra technology at 450 kA. In 2017 DX+ Ultra was upgraded targeting operations at higher amperages, eventually reaching 480 kA by 2019. Now, in 2022, EGA succeeded in starting up DX+ Ultra technology at 500 kA. This was done on 5 pots in its Jebel Ali smelter research and development demonstration section called Eagle. This paper aims to describe some of the adjustments made from the previous generation of DX+ Ultra technology to enable the additional 20 kA amperage increase and start up the pots at 500 kA. It also details some of the related challenges and results from the preheat, start-up and early operations periods.

Keywords

DX+ Ultra • 500 kA • Start-up • Preheat • Amperage creep • Copper collector bars • Challenges

Introduction

In year 2021 Emirates Global Aluminium (EGA) had the opportunity to start-up a new cell design in its "Eagle" research and development (R&D) section of its Jebel Ali smelter site. Previously Eagle section housed 5 pots of DX+ Ultra reduction cell technology running at 480 kA. The mathematical modelling and engineering work in designing these cells were completed [1]. Thereafter operational tests and demonstrating targeted key performance indicators (KPIs) were completed, giving the chance to upgrade and aspire to new heights. With modifications made to the potlining and potshell EGA intended to creep DX+ Ultra technology up to 500 kA.

In this paper, we will start by describing some of the changes made to enable this creep and detail the experience gained in starting, for the first time, an EGA reduction cell technology at 500 kA. The start-up will focus on setup and performance during pot preheat, bath-up and early operations. We will end the paper by describing some of the challenges faced during these phases and the expected way forward.

Pot Design

The main changes made to DX+ Ultra reduction cell technology, to enable an additional 20 kA creep, were to the potlining and potshell.

- A novel EGA Cathode assembly design—comprising full copper collector bars connected directly to the carbon cathode blocks—was developed and implemented [2]
- The insulation and refractories were adjusted to ensure the right amount of thermal energy was contained (where needed) and extracted (where needed)
- A bolted cathode collector bar flexible connection (to the cathode ring bus) was used to suit the copper collector bar design (the previous DX+ Ultra design employed a welded cathode collector bar flexible connection)
- The potshell was fitted with extra cooling fins in both the side and endwalls to ensure steel temperatures did not exceed critical levels (Figs. 1 and 2)

M. Mustafa (✉) · A. Alzarooni · K. Nikandrov · N. Ahli ·
A. Khan · H. AlHayyas · M. AlUstad · S. Hussain
Emirates Global Aluminium, Jebel Ali, Dubai,
United Arab Emirates
e-mail: mamustafa@ega.ae

© The Minerals, Metals & Materials Society 2023
S. Broek (ed.), *Light Metals 2023*, The Minerals, Metals & Materials Series,
https://doi.org/10.1007/978-3-031-22532-1_110

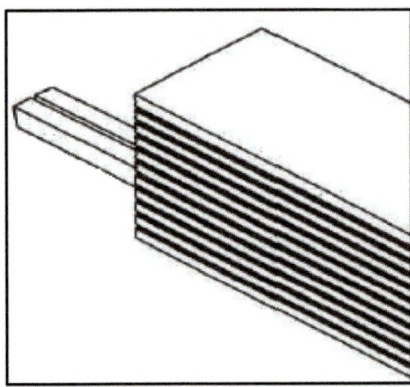

Fig. 1 Sketch of EGA's novel copper collector bar and cathode assembly used in the DX+ Ultra Eagle 500 kA

Fig. 2 Upgrade of potshell with additional cooling fins added to the endwalls of DX+ Ultra Eagle 500 kA

Start-Up Events

The first cell (number 277) was energized on 27 March 2022 whereas the last cell (number 273) completed metal pouring on 1 May 2022 (this is typically the last start-up activity). It was planned to have a minimum of 7 days gap between the start-up of each pot to give the team enough experience and resources to react, learn and improve start-up practices for the next pot. The table below displays the time taken for each activity during the start-up phase for each of the 5 Eagle pots:

Pot Preheat Setup and Practices

The below list describes the general setup and practices adopted in preheating the DX+ Ultra Eagle cells at 500 kA:

- Standard 20 mm thick aluminium fuse plates were used to make cut-out wedge extraction process safer, without electrical arc [3].
- All pots were cut in at 500 kA without the use of shunts.

- All pots used 1880 mm length anodes, with 100 × 100 mm chamfer to the backwall so they could sit on the pot preheat resistor bed and clear the side-lining materials. The anodes were carefully selected, as much as practically possible, to avoid installing anodes with even minor corner chips, to provide best possible contact with resistor bed and to reduce the need of adjusting anodes during pot preheat.
- Graphite was used as the preheat resistor bed material, with a different coverage area between central and corner anodes to better distribute current and achieve least heterogeneity in cathode surface temperatures. The graphite bed dimensions under each anode were quality controlled to ensure most consistent anode current distribution during preheat period. Three different graphite resistor bed dimensions were tested to extend the preheat duration and slow down the preheat rate. The quantity of graphite used for resistor bed increased from 196 to 350 kg accordingly.
- Some pots used pure crushed bath as anode top cover material, whereas others used Anode Cover Recycled Material (i.e., ACRM). This was done to determine if any difference in behaviour was noted during preheat and early operation periods.
- All pots used a sandwich of crushed bath and soda ash between the anodes and the exposed side and endwall lining materials (typically called the backwall location in EGA). The quantity of soda sandwiched in the backwall was optimized based on learnings from first pots bath-up and added manually after bath-up to improve effectiveness and reduce splashing against the pothoods.
- All pots were equipped with several cathode surface temperature probes [4, 5]. The surface thermocouples placed in the center of each cathode were used for preheat control purposes, whereas the peripheral thermocouples were used to study the evenness of temperature across the cathodes.
- A specific online monitoring system was developed as part of EGA's drive towards industry 4.0 for quick and remote referral to the preheat performance of each pot (refer to Fig. 7)
- Two pots had additional voltage drop and temperature probes to more thoroughly study pot preheat performance across the resistor bed: central and corner anodes and collector bar ends.
- Cathode collector bar bolted joints and flexible voltage drops were measured twice throughout preheat period and corrected as needed.
- Anode current distribution and preheat flexible joint drops were measured and corrected as needed

Fig. 3 Tap door view of the inside of an Eagle DX+ Ultra pot setup to start preheat at 500 kA

- Anode preheat flexibles removal started as soon as the target average central cathode surface preheat temperature was achieved (Fig. 4)

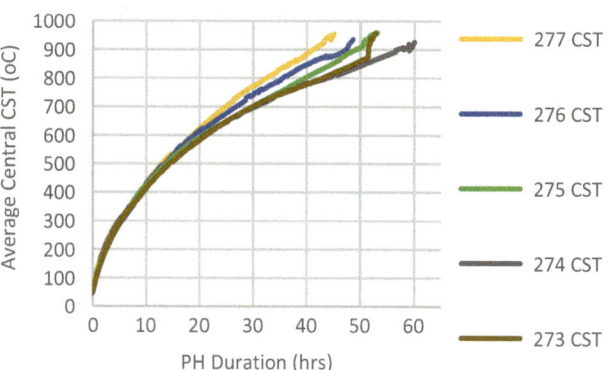

Fig. 4 Average central cathode surface temperatures (CST) per pot of DX+ Ultra 500 kA

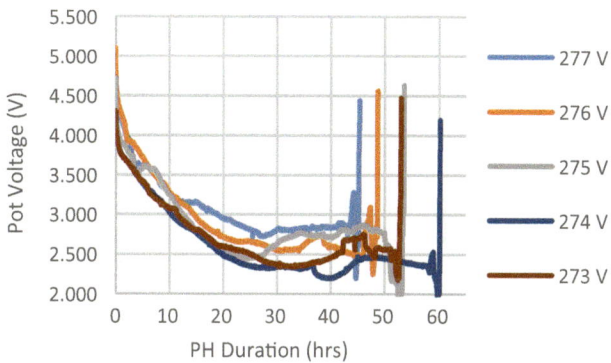

Fig. 5 Pot volts during pot preheat (PH) of DX+ Ultra 500 kA

Pot Preheat Performance

The targeted preheat duration was 65 h to achieve a final average central cathode surface temperature of 850 °C.

The below charts display the general performance for each pot's preheat.

Table 1 and Fig. 4 show that preheat duration was shorter than the target despite attempts to adjust resistor bed dimensions from pot to pot.

Figure 5 shows how the high cut in pot voltage contributed to the fast preheat rate. Even when comparing the maximum cut-in voltage of these pots against other copper collector bar pots (started previously at EGA but in other technologies and lower amperages) the DX+ Ultra pots cut in at 500 kA were higher. The peaks at the end of each trend line, of Fig. 5, signify start of bath-up activity.

One of the plots used to predict the final preheat duration is shown in Fig. 6. After preheating the first 2 pots it became evident that around 70 MWh (dashed blue line above) was required to achieve the final preheat target temperature. The pots showed up to 20% variation in the energy accumulation rate due to the changes made in pot setup after each pot preheat to get closer to the target preheat duration of 65 h. The linear regression equations seen in Fig. 6 show the highest and lowest cumulative energy rate pot preheats).

Bath Up and Early Operations Practices

For Eagle R&D section the requests, to receive liquid bath from planned donor pots of nearby potlines, typically starts with the end of the preheat phase.

An Early Operation Sheet (EOS) was prepared for each pot that listed all the parameters targeted for smooth pot start up and early operations period. This sheet evolved over the

Table 1 Start-up activity durations

Phase/Cell No	277	276	275	274	273
Pot preheat duration (h)	45	48	53	60	53
Preheat flexible removal (h)	1.47	2.55	1.67	1.62	1.72
Bath up duration (h)	1.37	1.12	1.75	0.95	1.03
Metal pouring duration (h)	1.77	2.05	2.22	1.47	2.53

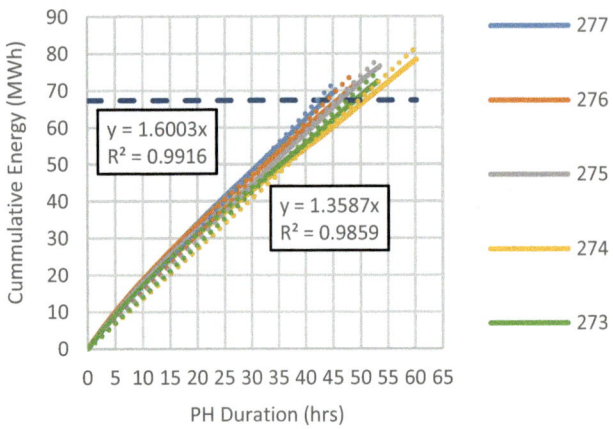

Fig. 6 Cumulative Energy per pot to achieve final preheat temperatures of each DX+ Ultra pot at 500 kA

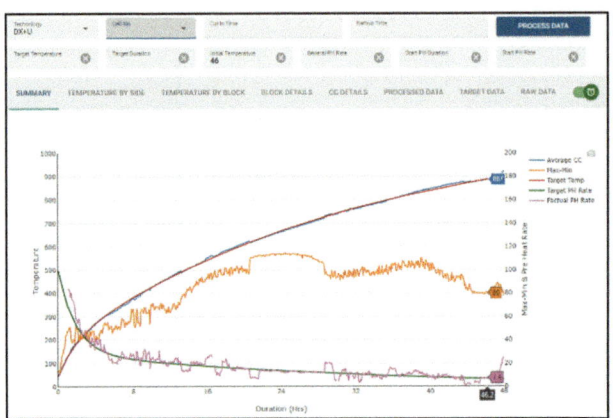

Fig. 7 A snapshot from the preheat online monitoring system developed for DX+ Ultra pot's preheat at 500 kA as viewed in EGA's mobile SAQR application (Smart Assistance and Quick Response)

course of each pot start-up to include lessons learnt. In general, the following practices were performed:

During Bath-up:

- An average of 14 tonnes of liquid bath was poured to achieve a target bath height of 42 cm.
- Central anodes were raised to match the curvature of the cathode surface.
- Carbon skimming was performed from tap end and duct end in every shift.
- Total soda ash addition of 900 kg was added into pots at regular intervals over 3 batches before metal pouring. Bath samples were analyzed regularly to monitor and control the bath chemistry according to the EOS targets.
- Bath temperature was monitored closely because full copper collector bars were used (Fig. 8).

Fig. 8 Close monitoring of bath pouring activity, behind a heat protective shield, to ensure no liquid metal is poured in with the bath

During Metal Pouring:

- Metal pouring was carried out 14 to 16 h after bath up once; bath height reached between 23 to 27 cm and anodes current distribution was within limits.
- Automatic pot voltage control, for metal pouring, was enabled and used for all 5 pots (Fig. 9).
- Finally, anodes were dressed with ACRM to the target level.

Fig. 9 Metal pouring at a moderate rate to ensure the automatic pot voltage control logic works effectively

Thereafter routine operational activities were started in the following sequence:

- Anode setting was carried out as per schedule,
- Metal tapping was carried out once a target metal height of 24 cm was achieved,
- Beam raising was performed as soon as the beam reached the threshold limit for beam raising.

The following conditional activities took place during early life period (the first 28 days of potlife):

- Daily bath samples and carbon skimming activities continued, even after metal pouring, to ensure chemistry control (according to EOS target) and to remove the carbon dust (generated from the graphite bed and air burning of the preheat anodes), respectively.
- Compressed air cooling of the pot shells, which have in built cooling pipes, was continuously applied until the bath temperature reduced below 990 °C.
- Metal sampling was carried out on daily basis to monitor the silicon trend, in the metal. In general, it took around 18 to 20 days to reduce silicon to normal levels (<300 ppm). Much of the silicon comes from the ceramic fibre blanket, used in pot preheat, that cannot be removed easily prior to bath-up.

Early Operations Performance

The graphs below display the overall performance of the 5 DX+ Ultra pots post bath-up. The most striking feature is the bath temperature as shown in Fig. 10. For the first 3 days of pot life the bath temperatures were consistently above 1000 °C, reaching very close to the melting point of pure

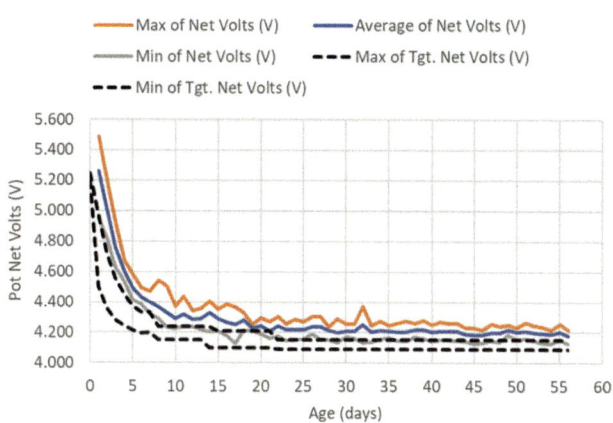

Fig. 11 Pot voltage trend during the start-up period of Eagle DX+ Ultra at 500 kA

copper (~1086 °C). This is obviously concerning for the pot lining since full copper collector bars are used. However, from the 7th day onwards an improvement was seen, where the average bath temperature conforms more closely to the target band. There followed a significant period thereafter (~next 25 days) with the average bath temperature being less than 960 °C.

It is believed that the combination of higher than target pot voltage and excess Aluminium Fluoride (AlF_3) are responsible for this outcome of greater than 1000 °C bath temperature. Figure 11 below shows how the initial pot voltage was a maximum of 300 mV more than the target. Figure 12 shows how excess AlF_3 was a maximum of 3% more than the target. As the pot voltage conformed more closely to the target band the bath temperature dropped to more desirable levels.

One of the reasons it was difficult to reduce pot voltage was due to the disturbance high noise level would have on the auto demand feed logic, which is disabled when noise exceeds 120 mV. As can be seen from Fig. 13 below high

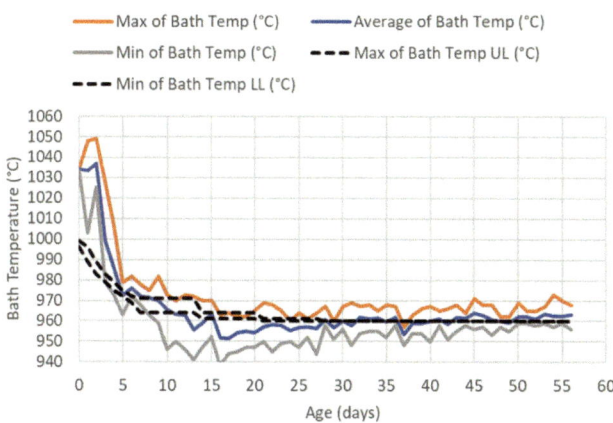

Fig. 10 Bath temperature trend during the start-up period of Eagle DX + Ultra at 500 kA

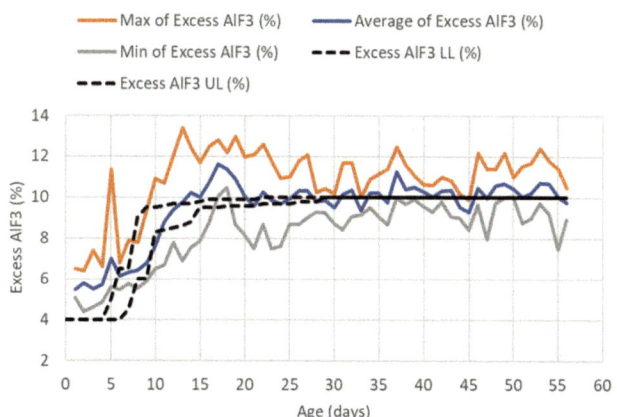

Fig. 12 Excess Aluminium Fluoride trend during the start-up period of Eagle DX+ Ultra at 500 kA

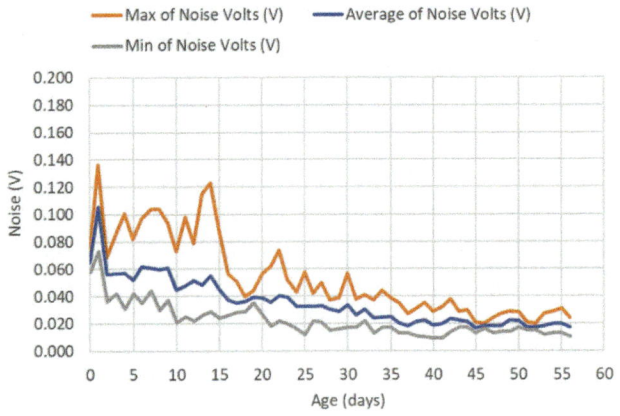

Fig. 13 Noise voltage trend during the start-up period of Eagle DX+ Ultra at 500 kA

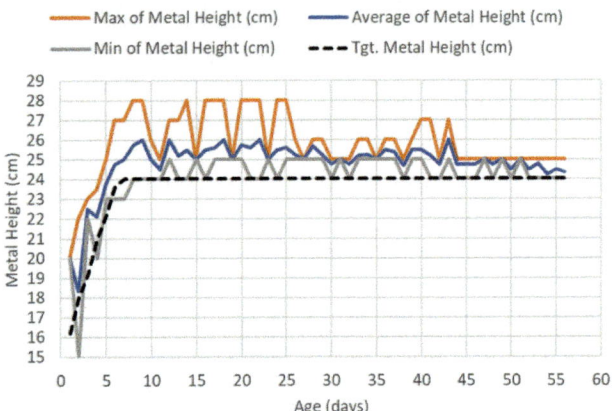

Fig. 14 Metal height trend during the start-up period of Eagle DX+ Ultra at 500 kA

noise was experienced in the first 30 days. Additionally, noise—up to 80 mV—interrupts the underfeed trigger causing an early switch to the overfeed window. The concern with forming muck on the cathode surface by overfeeding the pot, in early operations, combined with the high dependence on auto-feed logic exacerbated the situation.

In general liquid levels (both metal and bath) were found to be higher than the target throughout the early operations period as displayed in Figs. 14 and 15, with bath heights conforming to target near the end of early operations period.

Learnings and Challenges

From the activities, conducted during the cut-in, preheat, bath-up and early operations, EGA demonstrated that 500 kA shuntless graphite resistor bed preheating and start-up is possible with EGA's novel full copper collector bar and cathode block assembly.

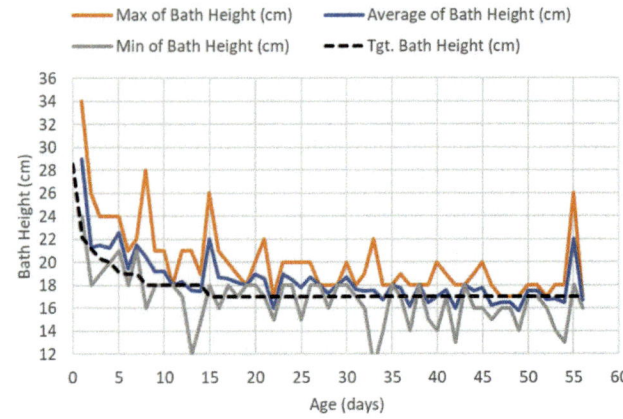

Fig. 15 Bath height trend during the start-up period of Eagle DX+ Ultra at 500 kA

In general, the start-up experience of Eagle 500 kA was deemed satisfactory, despite; high cut-in voltages, faster than desired preheat rate and high bath temperatures in the first few days of start-up.

In facing the above-mentioned challenges some improvement opportunities for the start-up and early operation performance were determined, such as:

- Enhancing the design of cut-in fuses (to provide more time for the fuse to blow after extracting the final cut-out wedge)
- Using shunts during the first few hours of preheat (to lower the cut-in voltage and better control the preheat rate).
- Fine-tuning the auto-demand feed control, in early operations phase (to enable lowering pot voltage and subsequently lowering bath temperature below 1000 °C and overcoming the concern with overfeeding due to typically high start-up noise levels disrupting the effectiveness of the feeding logic).
- Increase frequency of bath sampling, at different locations of the cell (to develop a better understanding of chemistry control during start-up phase and target a slower increase of excess AlF_3 to normal levels).
- Check and adjust anode positions more frequently as pot conditions change (to reduce noise levels during early operations).

Acknowledgements We would like to give special thanks to EGA senior management for their abundant support and continuous encouragement of our research and development activities. Also to all operational, process control and corporate teams involved—whose dedicated efforts and teamwork made it possible to safely demonstrate 500 kA start-up of our homegrown flagship DX+ Ultra technology for the first time in our 40 years history of aluminium reduction. We would also like to acknowledge the support and guidance provided throughout EGA's smelting research and development activities by Professor Barry Welch, Dr. Vinko Potocnik, and Michel Reverdy [6].

References

1. Alzarooni A et al. (2018) DX+ Ultra Industrial Version: Preheat Start Up and Early Operation. Light Metals 2018, 721–729
2. Bernard Jonqua et al., Cathode assembly with metallic collector bar systems for electrolytic cell suitable for the Hall-Héroult process, International Publication WO 2021/240353 A1, Priority Date 26 May 2020.
3. International patent application WO 2015/121796, "Start-up fuse for aluminium reduction electrolysis cell".
4. Arkhipov A et al. (2015) A model-based study of cell electrical preheating practices at DUBAL. Light Metals 2015, 777–782
5. Arkhipov A et al. and Potocnik V (2014) Cell electrical preheating practices at DUBAL. Light Metals 2014, 445–449
6. Reverdy M and Potocnik V (2022) A Historical Review of Aluminum Reduction Cell Start-Up and Early Operation. Light Metals 2022, 991–997

Part VI

Aluminum Waste Management and Utilization

Recovery of Value Added Products from Bauxite Residue

Himanshu Tanvar and Brajendra Mishra

Abstract

The growing stockpiles of bauxite residue and associated environmental hazards require a sophisticated process flowsheet for sustainable residue management and value recovery. Considering the association of multiple elements (Fe, Al, Si, Ca, Ti, V, Sc) within bauxite residue, metal extraction is of prime interest. The complex association of different elements and physical and chemical characteristics makes the extraction and purification process expensive and challenging. The present study focuses on developing a novel hydrometallurgical flowsheet for the subsequent recovery of base metals and critical elements from bauxite residue. The major elements present in bauxite residue are recovered as high-purity magnetite, titanium dioxide, and alumina. At the same time, critical elements (such as V and Sc) are recovered in the liquid stream generated after the recovery of base metals.

Keywords

Bauxite residue • Hydrometallurgy • Magnetite

Introduction

Bauxite residue (red mud) is an industrial waste generated during the extraction of alumina through the Bayer process [1]. More than 160 million tons of bauxite residue is produced annually, with no large-scale recycling or applications [2]. Bauxite residue is characterized by fine particle size (80% passing 10 microns), high alkalinity (pH >11), and moisture (25–30%) content. Disposal, handling, and storage of bauxite residue slurry is expensive and requires a large land area, generating potential health and environmental hazards [3, 4]. Despite the large number of technically viable solutions being developed in recent years, only 2–3% of bauxite residue produced annually is used productively. The billions of tons of bauxite residue being stockpiled are damaging the environment due to its high alkalinity (pH 10.5–12.5) and can result in contamination of ground or surface waters if not managed properly [5, 6]. This is in part due to the large variation in composition, mineralogy, and particle size between bauxite residue from different alumina refineries.

In general, recycling options for bauxite residue consist of utilizing the residue itself and/or recovering various materials. Bauxite residue has been used in road construction, landfill capping, and cement industries in low volume [3]. Material recovery through extractive metallurgy generally focuses on metals such as Fe, Ti, and Al or rare earth elements (REEs) such as Sc, Ce, and La [7]. Bauxite residue consists of mixed phases of Fe (5–60%), Al (5–30%), Ti (1–15%), Ca (2–14%), Si (5–50%), and Na (1–15%) along with Sc, Ga (0.001–0.012%), and V (0.03–0.23%) as critical elements [6, 8]. Pyrometallurgybased processes are focused on the recovery of Fe as pig iron or magnetite concentrate [9–11]. However, economic uncertainty due to high energy consumption is the limiting factor for industrial-scale applications [12]. Hydrometallurgy based processes are mainly concerned with the recovery of Sc from bauxite residue [13–15]. However, a considerably low concentration of the target element (10–100 ppm Sc, Ga) and a high fraction of impurities (Fe, Al, Si, Ti, Na, Ca) results in poor recovery due to multi-element dissolution [16]. These processes are concerned with the dissolution of elemental species; however, separation and recovery of these elements from solution are limited [17, 18]. Only a few literature works are concerned with recycling a high quantity of waste generated; therefore, scaleup efforts and industrial-scale

H. Tanvar (✉) · B. Mishra
Material Science and Engineering, Worcester Polytechnic Institute, Worcester, MA 01609, USA
e-mail: htanvar@wpi.edu

© The Minerals, Metals & Materials Society 2023
S. Broek (ed.), *Light Metals 2023*, The Minerals, Metals & Materials Series,
https://doi.org/10.1007/978-3-031-22532-1_111

processing are questionable for sustainable process development.

The following work is focused on the recovery of high-purity magnetite, titanium dioxide, and alumina from bauxite residue using a hydrometallurgical process. Sc and V oxide are the critical elements present within bauxite residue; however, the low concentration of these elements results in low-volume products at the expense of expensive reagents. Therefore, the recovery of these elements as a byproduct of the base metal recovery process is attempted in the present study. Different stages of treatment are critically evaluated, and a comprehensive flowsheet is presented to recover different metallic oxides from bauxite residue.

Materials and Methods

Sample Information

The bauxite residue sample used in this study was procured from an alumina refinery (Rio Tinto Alcan) in Quebec, Canada. The bulk sample was collected from the dried section of the bauxite residue pond and further dried in an oven to remove residual moisture. The dried bauxite residue was sieved below 250 microns to remove large impurities (wood chips and pebbles) in the bulk sample.

Experimental Procedure

Acid neutralization with HCl was carried out in a 1000 mL Pyrex beaker under continuous stirring using a magnetic stirrer. The desired amount of bauxite residue (10–20%) was added to the acid solution (0.1–1.5 M) to get the required solid to liquid ratio (S/L; pulp density). Acid neutralization was carried out for a short duration of 15 min, followed by filtration of slurry using a vacuum filter. The solid residue was dried in a laboratory oven at 95 °C for 12 h and was further processed to recover iron values.

The leaching experiments with oxalic acid were carried out in a three-necked flat-bottom Pyrex flask using a magnetic stirrer. The reaction flask was placed in a silicone oil bath to regulate the temperature and achieve constant heating. One end of the flask was connected to a condenser, while the other two ends were used to measure the slurry temperature and pH using a thermocouple and a pH probe, respectively. Furthermore, the reacted slurry was filtered using a vacuum filter, and the leach solution was subjected to photochemical reduction to precipitate ferrous oxalate [19].

After leaching with oxalic acid, the resulting residue was subjected to sulfation baking using sulfuric acid to recover Ti and Sc. The solid samples were thoroughly mixed with the desired amount of sulfuric acid (0.5–1.5 mL/g) to form a paste which was further transferred into a ceramic boat and heated in a tubular furnace at 300 °C for 1 h. The sulfated sample was pulverized using a pestle and mortar and leached in 0.5 M sulfuric acid at 65 °C for 90 min. Leaching experiments were carried out 500 mL Pyrex beaker using a magnetic stirrer at 800 rpm stirring speed. The leach residue and leach liquor were separated using a vacuum filter, followed by thermal hydrolysis of leach liquor to recover Ti precipitate. Hydrolysis experiments were carried out by boiling the solution to 100 °C for 4 h in a 1000 mL Pyrex flask connected to a condenser.

Analytical Techniques

The chemical analysis of the samples was carried out using Inductively Coupled Plasma Optical Emission Spectroscopy and Mass Spectroscopy (ICP-OES (PerkinElmer Optima 8000), MS (PerkinElmer NEXION350x)). ICP-OES was used to analyze major elements, while ICP-MS for trace elements. Solid samples were fused with borate flux at 1000 °C for 1 h, dissolved in 25% nitric acid solution, diluted, and further analyzed with ICP-OES and MS. In contrast, the liquid samples were analyzed after dilution with 2% nitric acid. Mineral phases present in the solid samples at different stages were determined using an X-ray diffractometer (XRD; PANalytical Empyrean). The diffraction data were measured using a Cr–K_α radiation in the 2—theta range of 10 to 80°, with a scanning rate of 2°/min and a step size of 0.02°. The morphology of the powder sample was studied using a Scanning Electron Microscope (SEM; JSM 7000F) equipped with Energy-dispersive X-ray spectroscopy (EDS).

Results and Discussion

Feed Characterization

Bauxite residue consists of mixed phases of iron (Fe), aluminum (Al), calcium (Ca), silicon (Si), titanium (Ti), and sodium (Na). The presence of a high amount of Na (7%) and Ca (3%) makes the bauxite residue basic (pH >11) and often makes recycling a colossal challenge. The Canadian bauxite residue used in this work consists of Fe (21.4%), Al (10.2%), Na (6.9%), Ca (2.3%), Si (5.0%), and Ti (3.7%) in the form of hematite (Fe_2O_3), boehmite (Al(OOH)), gibbsite (Al(OH)$_3$), sodium aluminum silicate hydrate ($Na_8Al_6Si_6O_{24}(OH)_2(H_2O)_2$), calcium carbonate ($CaCO_3$), quartz ($SiO_2$), and rutile ($TiO_2$) phases. The chemical analysis and XRD spectrum of the bauxite residue feed sample are shown in Fig. 1a and b. The pH value of the bulk bauxite residue sample was determined as 10.5, reflecting its high basic

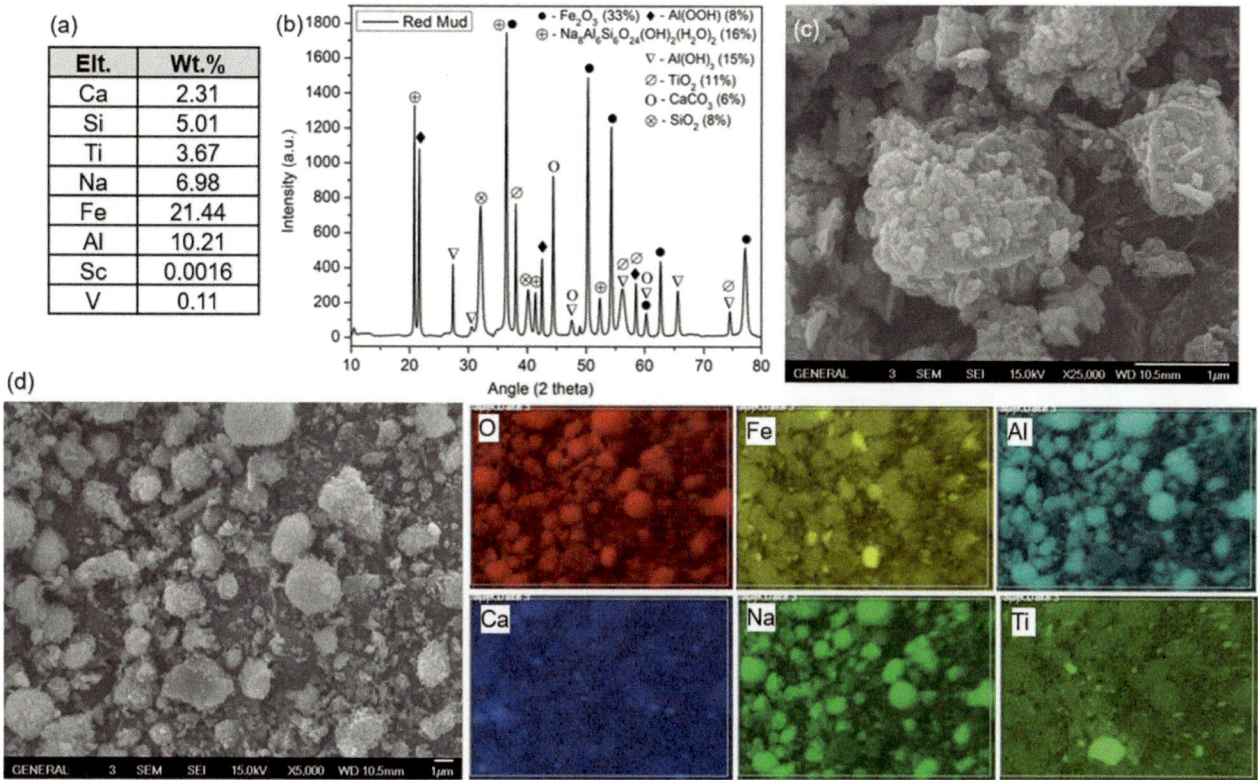

Fig. 1 Feed characterization: **a** Chemical analysis, **b** XRD, **c** SEM (25,000× magnification), **d** SEM-EDS (elemental mapping)

nature. The SEM analysis and EDS elemental mapping of the feed sample shown in Fig. 1c and d depict the presence of sub-micron sized particles with a complex association of different elements in the bulk sample. Mild acid washing was carried out to separate alkali fractions from bauxite residue.

Bauxite Residue Neutralization

Direct leaching of bauxite residue in acid solution requires an additional reagent because a part of it is consumed to neutralize the alkali left behind from the Bayer process. Neutralizing bauxite residue with HCl reduces the acid demand during the second stage of leaching with oxalic acid. Oxalic acid is relatively more expensive than mineral acids; therefore, neutralization with other mineral acids reduces the overall reagent cost. HCl and H_2SO_4 were selected as mineral acids to neutralize bauxite residue. The dissolution value of different elements using HCl and H_2SO_4 during neutralization is shown in Fig. 2a and b. It was found that Na removal depends upon the concentration of H^+ ions in the solution; therefore, 1 M HCl solution resulted in a similar result as with 0.5 M H_2SO_4. The bauxite residue slurry pH was reduced from ~10.5 to ~2.5, along with the dissolution of approx. 38% solid mass with HCl and 27% with

H_2SO_4. Both the acids completely dissociated the sodium aluminum silicate phase at the optimized conditions and separated more than 90% Na, 40–45% Al, 60% Si and less than 10% Fe. The key difference between the two acids is the limited separation of Ca with H_2SO_4 due to the formation of insoluble $CaSO_4$. The leach solution from both HCl and H_2SO_4 wash consists of a high amount of Al (~4500 ppm), along with Si (~4400 ppm), Na (4000 ppm), and Ca (500–3000 ppm). The pH adjustment of the leach solution can be employed to recover silica, alumina, and calcium values, and the remaining solution can be recirculated for acid washing. Considering the high recovery of Na and Ca with HCl, neutralization experiments were performed with HCl. The neutralized bauxite residue was further subjected to leaching with oxalic acid to recover Fe and V values.

Production of Magnetite from Neutralized Residue

The high chelating ability of oxalic acid provides a high affinity for forming an aqueous complex with metal species such as Fe and V [20]. Oxalic acid is a diprotic acid and dissociates into $HC_2O_4^-$ and $C_2O_4^{2-}$ with a pK_a of 5.6×10^{-2} and 6.2×10^{-5}, respectively. The ionization depends on the solution pH, with $C_2O_4^-$ being the dominant

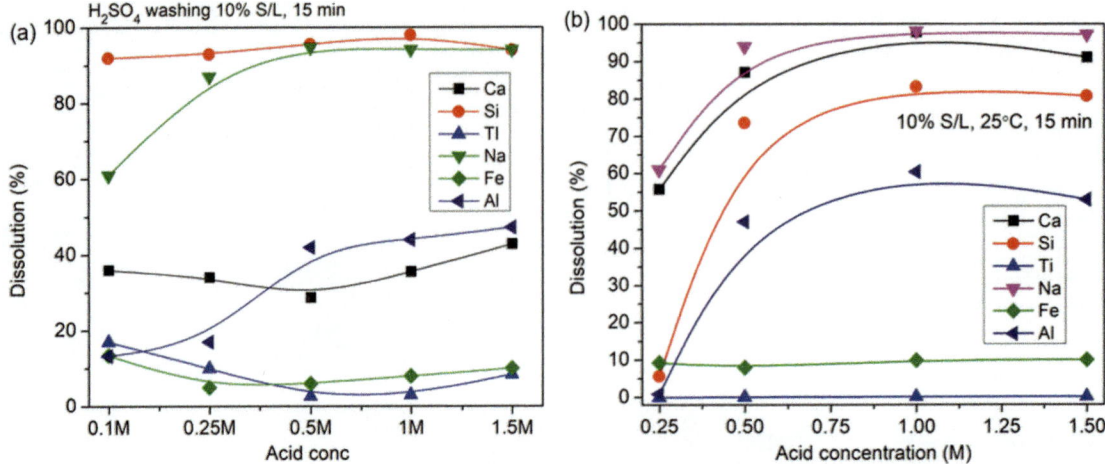

Fig. 2 Dissolution of different elements during neutralization with **a** sulfuric acid, **b** hydrochloric acid

species above pH 4 and $HC_2O_4^-$ in the acidic range [21]. The oxalate ion ($C_2O_4^{2-}$, $HC_2O_4^-$) have (two, one) oxygen atoms with unshared electrons and can form a coordination bond with a metal ion (such as Fe and V) and form a ring, resulting in a complex metal ion species. The chemical reaction for the leaching of Fe and V oxide with oxalic acid is shown in Eqs. (1)–(4). Based on our previous work and literature findings, the activation energy for the dissolution of hematite using oxalic acid was determined as 100–150 kJ/mol in the temperature range of 65 to 95 °C [19, 22, 23]. The leaching kinetics of hematite in oxalic acid is highly influenced by the temperature. The temperature and oxalic acid concentration during leaching was fixed at 95 °C and 2 M, respectively, whereas the leaching time and solid to liquid ratio were varied as shown in Fig. 3. The dissolution value of key elements and concentration of resulting leach liquor is shown in Fig. 3a and b, respectively. High Fe leaching was obtained at a low solid to liquid ratio and

extended leaching duration, where approximately 85.4% Fe, 77.9% V, 16.9% Sc, 4.2% Ti, 12.7% Si, and 48.5% Al was dissolved after 2.5 h using 10% pulp density.

The ferric oxalate leach liquor was further subjected to photochemical reduction using UV light to precipitate and recover Fe as ferrous oxalate. Ferric oxalate ($[Fe(C_2O_4)_3]^{3-}$) is a photochemically active complex that undergoes spontaneous photochemical reduction to form ferrous oxalate [24, 25]. The reduction mechanism proceeds through the formation of carbon dioxide ($CO_2^{\bullet-}$) and oxalate ($C_2O_4^{\bullet-}$) radical anion [24, 25]. Electron transfer from ligand (oxalate) to metal (Fe^{3+}) produces a $C_2O_4^{\bullet-}$ anion in the solution, which further dissociates into CO_2 and $CO_2^{\bullet-}$. The $CO_2^{\bullet-}$ can further reduce another Fe^{3+} ion or combine with $CO_2^{\bullet-}$ anion to form $C_2O_4^{2-}$ in the solution. The Fe(II) complex is further hydrolyzed and precipitated as ferrous oxalate ($FeC_2O_4 \cdot 2H_2O$). The photochemical reduction was performed at an optimized duration of 6 h using a 100 W UV

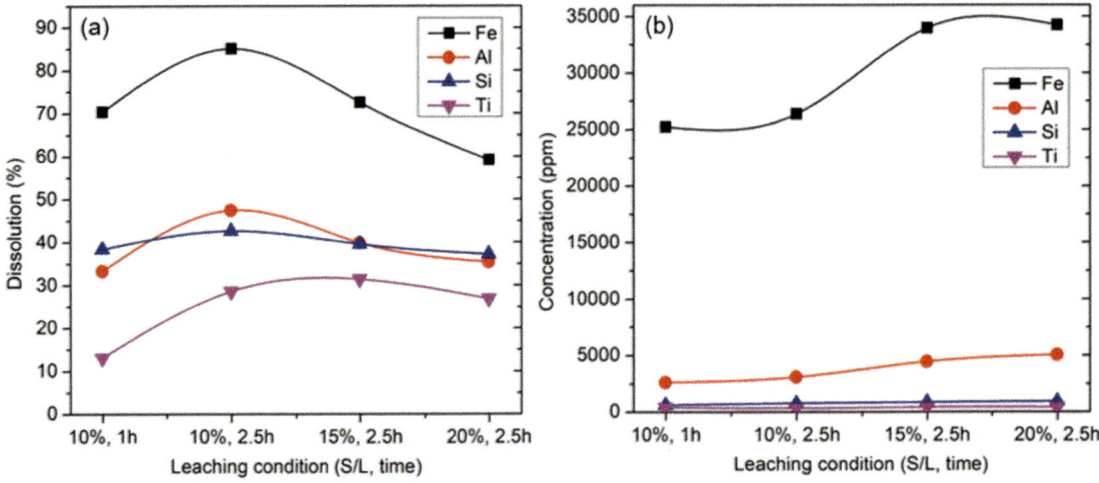

Fig. 3 **a** Dissolution of different elements during leaching with oxalic acid, **b** concentration of solution at different oxalic acid leaching conditions

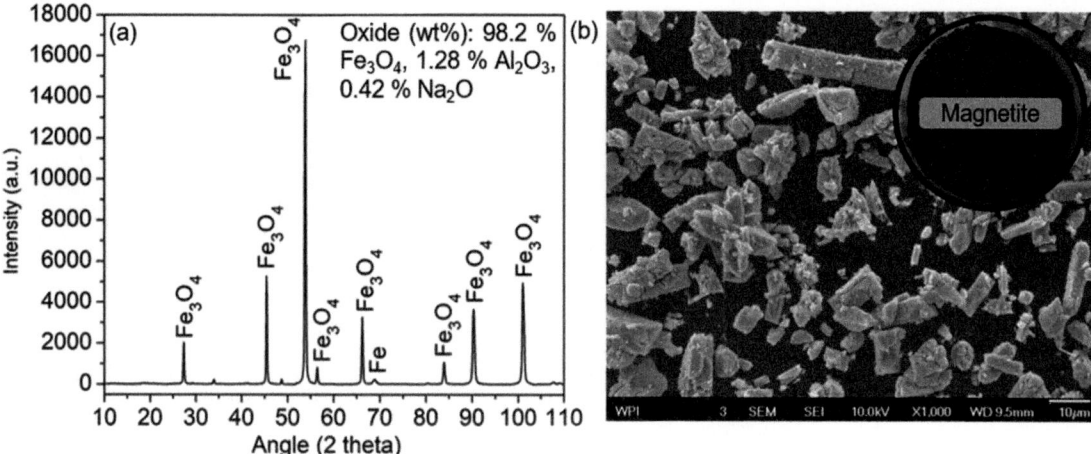

Fig. 4 Characterization of magnetite product: **a** XRD and chemical analysis, **b** SEM micrograph

lamp [19]. After photochemical reduction, more than 99% of the Fe in the solution was recovered as ferrous oxalate precipitate, whereas V was retained in the solution. The filtered solution consists of approx. 58.7 mg/L V which can be recovered through solvent extraction using Cyanex 923, and Aliquat 336 [26].

The ferrous oxalate precipitate was decomposed at a high temperature to produce magnetite [27]. Thermal decomposition was performed in a low oxygen atmosphere at 500 °C for 1 h. A low oxygen atmosphere was maintained using a continuous flow of N_2 in the tube furnace to restrict the oxidation of Fe^{+2} to Fe^{+3}. The XRD spectrum, chemical analysis, and SEM micrograph of the magnetite product are shown in Fig. 4a and b. The magnetite product is characterized by more than 98% purity, with magnetite being the dominant mineral phase. The SEM analysis depicts the presence of porous particles of variable geometry. High purity magnetite possesses application in magnetic resonance imaging, targeted drug delivery systems, photo magnetics, and black pigment material.

$$Fe_2O_{3(s)} + 6C_2O_{4(aq)}^{2-} + 6H_{(aq)}^{+} \rightarrow 2\left[Fe(C_2O_4)_3\right]_{(aq)}^{3-} + 3H_2O_{(l)}\left(\Delta G_{25°C}° = -212.75\ kJ/mol\right) \tag{1}$$

$$Fe_2O_{3(s)} + 2HC_2O_{4(aq)}^{-} + 6H_{(aq)}^{+} \rightarrow 2\left[FeHC_2O_4\right]_{(aq)}^{2+} + 3H_2O_{(l)}\left(\Delta G_{25°C}° = -87.35\ kJ/mol\right) \tag{2}$$

$$V_2O_{5(s)} + 5HC_2O_{4(aq)}^{-} + H_{(aq)}^{+} \rightarrow 2\left[VO(C_2O_4)_2\right]_{(aq)}^{2-} + 3H_2O_{(l)} + 2CO_{2(g)}\left(\Delta G_{25°C}° = -326.05\ kJ/mol\right) \tag{3}$$

$$V_2O_{5(s)} + 5C_2O_{4(aq)}^{2-} + 6H_{(aq)}^{+} \rightarrow 2\left[VO(C_2O_4)_2\right]_{(aq)}^{2-} + 3H_2O_{(l)} + 2CO_{2(g)}\left(\Delta G_{25°C}° = -448.05\ kJ/mol\right) \tag{4}$$

Sulfation Baking for Titanium and Scandium Recovery

The leach residue obtained after leaching with oxalic acid constitutes about 27% of starting bauxite residue weight and consists of approx. 10.3% Si, 15.1% Ti, 8.9% Fe, 13.8% Al, and 0.0048% Sc. It is worth mentioning that Ti and Sc were concentrated in the residue up to fourfold after the recovery of major elements. Separation of Fe before an attempt to recover Sc and Ti presents better selectivity, reduced acid demand, and easy separation in downstream processing.

The oxalic leaching residue was thoroughly mixed with different amounts of sulfuric acid, followed by baking at 300 °C for 1 h. The baking temperature was selected based on the decomposition behavior of sulfuric acid at high temperatures and its reactivity [28]. Sulfuric acid decomposes to gaseous sulfuric acid, SO_3, and H_2O in the temperature range of 127–427 °C. The second stage of decomposition proceeds with endothermic reduction of SO_3 to SO_2 and is observed at a temperature of more than 750 °C [29]. The main advantage of sulfation baking over direct leaching includes reduced acid demand, fast reaction kinetics, and a high reaction rate. The dissolution of different elements at different acid to residue ratio is shown in Fig. 5a. With an increase in acid dosage from 0.5 mL/g to 1 mL/g, the dissolution of all elements increased. The highest dissolution of Ti, Al, and REEs was obtained at acid to residue ratio of 1.3 mL/g, resulting in the dissolution of 42.7% Si, 90.2% Ti, 87.3% Fe, 93.2% Al, and 64.5% RE. The dissolution of RE was limited to 65% even at the high dose of acid, therefore, 1.3 g/mL was considered optimal.

The leach liquor obtained after sulfation baking and water leaching consist of 12.31 g/L Ti, 7.21 g/L Al, and 7.19 g/L Fe as major components. The recovery of Ti was carried out

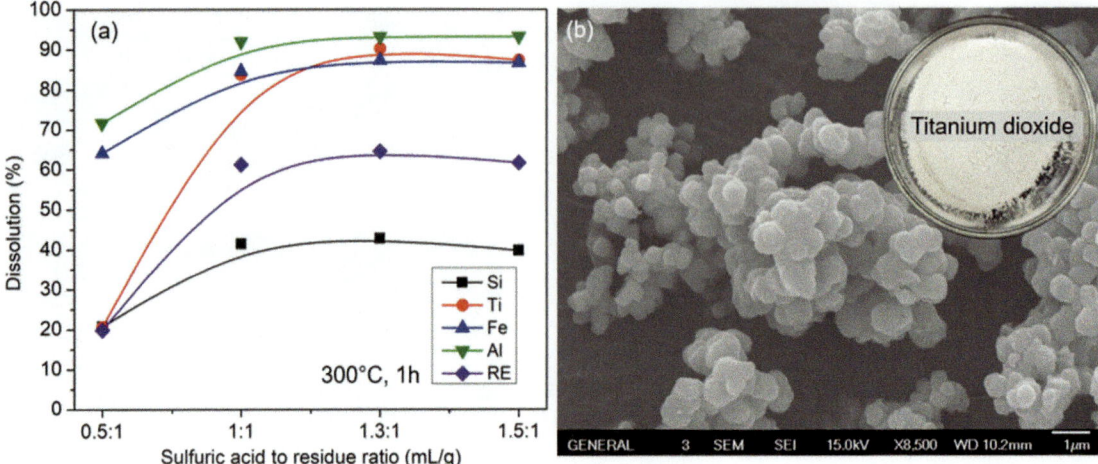

Fig. 5 **a** Dissolution of different elements during sulfation baking and water leaching at different sulfuric acid to residue ratio (300°, 1 h), **b** SEM micrograph and sample photograph of titanium dioxide product

by thermal hydrolysis by boiling the solution at 100 °C for 4 h [30]. Fe(III) was first reduced to Fe(II) with iron powder before thermal hydrolysis to prevent coprecipitation of Fe. The filtered solution after thermal hydrolysis contains 55.10 ppm Ti, 9080.00 ppm Fe, 7190.20 ppm Al, and 4.1 ppm Sc. At the same time, more than 96% of the Ti was recovered as TiO_2 precipitate. The final solution can be further processed through solvent extraction to recover the rare earth concentrate of Sc. The SEM micrograph and sample photographs of the TiO_2 product are shown in Fig. 5b. The purity of TiO_2 precipitate was determined as more than 99%, with minor impurities of Fe and Al.

Comprehensive Flowsheet

The summarized process flow diagram for the recovery of different products from bauxite residue is shown in Fig. 6. The proposed process consists of three-stage processing to selectively recover magnetite, alumina, and titanium dioxide as major products with the high market value from bauxite residue. The critical elements, including Sc and V, are retrieved in the liquid stream after the extraction of major elements and can be further recovered through solvent extraction. Hydrochloric acid, oxalic acid, and sulfuric acid are utilized as chemical reagents for the dissolution of (Al, Si, Ca), (Fe, V), and (Ti, Sc), respectively. Whereas precipitation from solution, photochemical reduction, and hydrolysis are adopted to recover the dissolved species selectively. The following flowsheet provides high recovery and purity of final products. Based on the material balance, the processing of one ton of bauxite residue will result in the production of 254 kg magnetite, 53 kg alumina, and 62 kg titanium dioxide. Critical elements including Sc and V are recovered into the solution containing 58.7 ppm V and 4.1 ppm Sc generating 1200.0 g V_2O_5 and 10.9 g Sc_2O_3, respectively.

Conclusions

Based on the research work carried out in this study, a comprehensive flowsheet for recovering valuable products from bauxite residue is presented. The process includes neutralizing bauxite residue with HCl to separate alkali fractions and obtain neutralized bauxite residue. The leach liquor from the neutralization stage is further processed to recover alumina, silica, and calcium values. The neutralized solid is leached with oxalic acid to selectively dissolve Fe and V. Neutralization of bauxite residue before oxalic acid leaching reduces the acid consumption and provides a high purity product. The oxalic leaching liquor is subsequently reduced with UV light to precipitate ferrous oxalate, further decomposed to magnetite in an inert atmosphere. The residue from leaching with oxalic acid is processed through sulfation baking, water leaching, and hydrolysis process to recover Ti as TiO_2 and generate a liquid stream with dissolved rare earth elements. The major constituents of bauxite residue are recovered in the form of magnetite (98% purity, 95% recovery), alumina (98% purity, 31% recovery), and titania (98% purity, 87% recovery). Based on the experimental work and theoretical analysis, the proposed process results in near-zero waste discharge and presents an excellent opportunity for industrial-scale processing of bauxite residue.

Fig. 6 Summarized process flow diagram for recovery of different products from bauxite residue

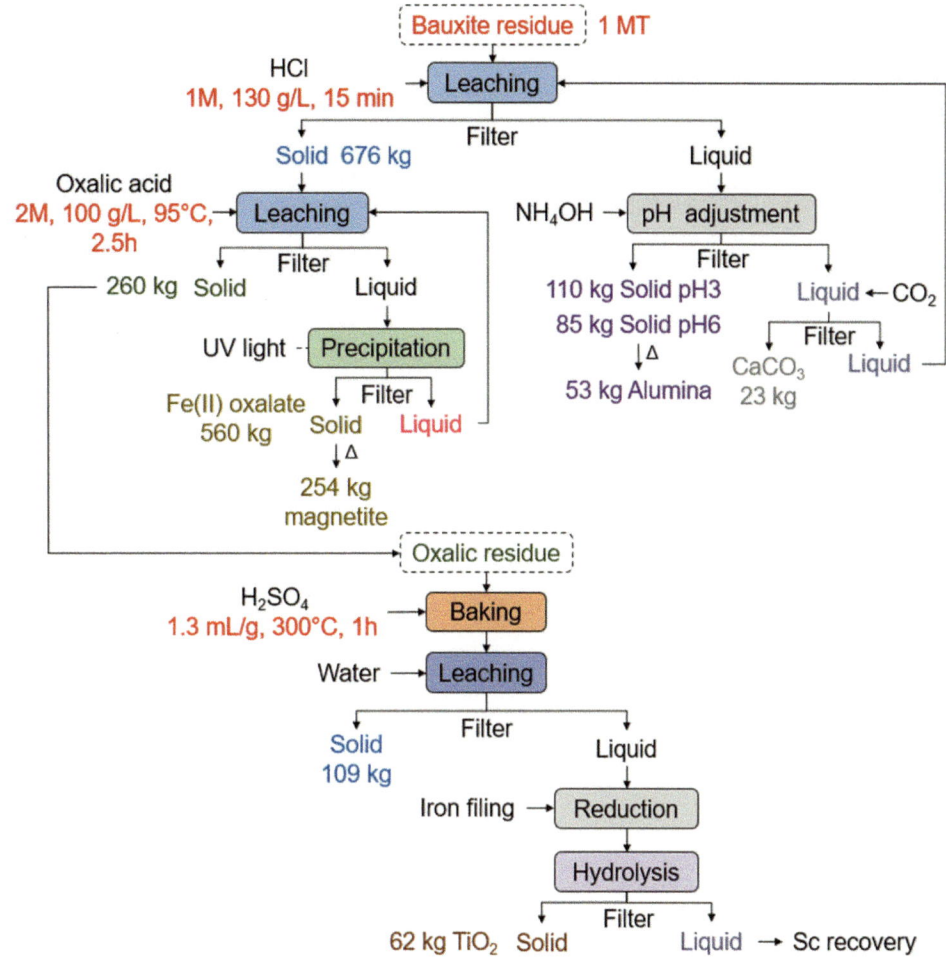

Acknowledgements The authors are thankful to Mr. Glenn Yee for the fellowship he instituted at the Worcester Polytechnic Institute. Thanks are due to the NSF Center for Resource Recovery and Recycling with their technical support through Global Minerals Recovery, LLC.

References

1. Habashi, F. (2016) *A Hundred Years of the Bayer Process for Alumina Production, in Essential Readings in Light Metals.* 85–93.
2. Healy, S. (2022) *Sustainable Bauxite Residue Management Guidance*, S. Healy, Editor., International Aluminum Institute. p. 92.
3. Tsakiridis, P.E., S. Agatzini-Leonardou, and P. Oustadakis (2004) *Red mud addition in the raw meal for the production of Portland cement clinker.* J Hazard Mater 116(1–2) 103–110.
4. Ruyters, S., et al. (2011) *The red mud accident in ajka (hungary): plant toxicity and trace metal bioavailability in red mud contaminated soil.* Environ Sci Technol 45(4) 1616–1622.
5. Hammond, K., et al. (2013) *CR3 Communication: Red Mud—A Resource or a Waste?* Jom 65(3) 340–341.
6. Evans, K. (2016) *The History, Challenges, and New Developments in the Management and Use of Bauxite Residue.* Journal of Sustainable Metallurgy 2(4) 316–331.
7. Borra, C.R., et al. (2016) *Recovery of Rare Earths and Other Valuable Metals From Bauxite Residue (Red Mud): A Review.* Journal of Sustainable Metallurgy 2(4) 365–386.
8. Khairul, M.A., J. Zanganeh, and B. Moghtaderi (2019) *The composition, recycling and utilisation of Bayer red mud.* Resources, Conservation and Recycling 141 483–498.
9. Borra, C.R., et al. (2015) *Smelting of Bauxite Residue (Red Mud) in View of Iron and Selective Rare Earths Recovery.* Journal of Sustainable Metallurgy 2(1) 28–37.
10. Cardenia, C., E. Balomenos, and D. Panias (2018) *Iron Recovery from Bauxite Residue Through Reductive Roasting and Wet Magnetic Separation.* Journal of Sustainable Metallurgy 5(1) 9–19.
11. Mishra, B., A. Staley, and D. Kirkpatrick (2002) *Recovery of value-added products from red mud.* Mining Metallurgy and Exploration 19 87–94.
12. Archambo, M.S. and S.K. Kawatra (2020) *Utilization of Bauxite Residue: Recovering Iron Values Using the Iron Nugget Process.* Mineral Processing and Extractive Metallurgy Review 42(4) 222–230.
13. Reid, S., et al. (2017) *Technospheric Mining of Rare Earth Elements from Bauxite Residue (Red Mud): Process Optimization, Kinetic Investigation, and Microwave Pretreatment.* Sci Rep 7(1) 15252.
14. Ding, W., et al. (2022) *Efficient Selective Extraction of Scandium from Red Mud.* Mineral Processing and Extractive Metallurgy Review 1–9.

15. Ujaczki, É., et al. (2019) *Recovery of Gallium from Bauxite Residue Using Combined Oxalic Acid Leaching with Adsorption onto Zeolite HY*. Journal of Sustainable Metallurgy 5(2) 262–274.

16. Akcil, A., et al. (2017) *Overview On Extraction and Separation of Rare Earth Elements from Red Mud: Focus on Scandium*. Mineral Processing and Extractive Metallurgy Review 39(3) 145–151.

17. Borra, C.R., et al. (2015) *Leaching of rare earths from bauxite residue (red mud)*. Minerals Engineering 76 20-27.

18. Agrawal, S. and N. Dhawan (2021) *Investigation of mechanical and thermal activation on metal extraction from red mud*. Sustainable Materials and Technologies 27.

19. Tanvar, H. and B. Mishra (2021) *Hydrometallurgical Recycling of Red Mud to Produce Materials for Industrial Applications: Alkali Separation, Iron Leaching and Extraction*. Metallurgical and Materials Transactions B 52 3543–3557.

20. Taxiarchou, M., et al. (1997) *Dissolution of hematite in acidic oxalate solutions*. Hydrometallurgy 44(3) 287–299.

21. Panias, D., et al. (1996) *Thermodynamic analysis of the reactions of iron oxides: Dissolution in oxalic acid*. Canadian Metallurgical Quarterly 35(4) 363–373.

22. Lee, S.O., et al. (2006) *Study on the kinetics of iron oxide leaching by oxalic acid*. International Journal of Mineral Processing 80(2–4) 144–152.

23. Salmimies, R., et al. (2012) *Acidic dissolution of hematite: Kinetic and thermodynamic investigations with oxalic acid*. International Journal of Mineral Processing 110–111 121–125.

24. Mangiante, D.M., et al. (2017) *Mechanism of Ferric Oxalate Photolysis*. ACS Earth and Space Chemistry 1(5) 270–276.

25. Ogi, Y., et al. (2015) *Ultraviolet photochemical reaction of [Fe (III)(C$_2$O$_4$)$_3$]$^{3-}$ in aqueous solutions studied by femtosecond time-resolved X-ray absorption spectroscopy using an X-ray free electron laser*. Structural Dynamics 2(3).

26. Liu, Z., et al. (2020) *Separation and recovery of vanadium and aluminum from oxalic acid leachate of shale by solvent extraction with Aliquat 336*. Separation and Purification Technology 249.

27. Angermann, A. and J. Töpfer (2008) *Synthesis of magnetite nanoparticles by thermal decomposition of ferrous oxalate dihydrate*. Journal of Materials Science 43(15) 5123–5130.

28. Narayanan, R.P., N.K. Kazantzis, and M.H. Emmert (2017) *Selective Process Steps for the Recovery of Scandium from Jamaican Bauxite Residue (Red Mud)*. ACS Sustainable Chemistry & Engineering 6(1) 1478–1488.

29. Schwartz, D., et al. (2000) *A Kinetic Study of the Decomposition of Spent Sulfuric Acids at High Temperature*. Industrial & Engineering Chemistry Research 39(7) 2183–2189.

30. Grzmil, B.U., D. Grela, and B. Kic (2008) *Hydrolysis of titanium sulphate compounds*. Chemical Papers 62(1) 18–25.

Current Status and Proposed Economic Incentives for Higher Utilization of Bauxite Residue to Enhance Sustainability of the Aluminum Industry

Subodh K. Das and Muntasir Shahabuddin

Abstract

The three-stage production chain of aluminum from ore to casting is complex and consumes massive amounts of energy and input materials. The process also results in undesirable by-products, such as bauxite tailings and bauxite residues or red mud. From a holistic perspective, the overall sustainability of the aluminum industry is dependent on several factors that go beyond the recent focus on the low carbon drive across the companies and regions. Globally, some 3 billion tons of bauxite residue (red mud) are now stored in massive waste ponds or dried mounds, making it one of the most abundant industrial wastes on the planet. Alumina refining plants generate an additional 150 million tons each year. The industry produces 0.5–2 tons of bauxite residue per ton of aluminum. Although actively globally pursued, the current utilization of bauxite residues is less than 4%. In the author's published analysis, mitigation and utilization of bauxite residue are the number one sustainability-enhancing factor. The objectives of this paper are:

1. Review of bauxite residue utilization status
2. Assess the feasibility and prospects of current efforts underway
3. Suggest economic incentives to encourage higher utilization.

Keywords

Aluminum • Sustainability • Recycling and secondary recovery • Bauxite residue

S. K. Das (✉)
Phinix, LLC, Clayton, MO, USA
e-mail: skdas@phinix.net

M. Shahabuddin
Worcester Polytechnic Institute, Worcester, MA, USA

Introduction

The utility and prevalence of aluminum in modern society belie the century-long impacts of its caustic, crimson byproduct: bauxite residue. Aluminum production begins with the Bayer process: where bauxite, a red ore containing a slew of metal oxides (including iron, aluminum, and titanium), is hydrothermally extracted in sodium hydroxide so the alumina can be selectively removed. From there, the Hall-Héroult process electrolytically reduces the alumina into aluminum metal. The remaining oxides and sodium hydroxide left from the Bayer process leave exorbitant amounts of a high-pH, low-value red slurry dubbed "bauxite residue" or "red mud"—in amounts up to twice the mass of the resulting aluminum produced [1].

The operative "low value" of bauxite residue is caused by its high causticity and iron oxide content—which are expensive to process and result in low-value products. This has led to neither valorization of bauxite residue's component minerals nor its management as a waste product.

Problem Description

Over 3 billion tons of bauxite residues have been amassed in holding ponds which grow by 150 million tons per year [2]. The global alumina industry is expected to have 7 to 8 billion tons of red mud inventory by 2040. These ponds take up significant real estate and require upkeep to prevent spills [3], the damage of which is illustrated by the over 120 casualties caused by drowning and chemical burns during the Timföldgyár alumina plant spill in Ajka, Hungary [4], alongside the resultant, and well-studied, destruction of terrestrial and aquatic ecosystems in its fallout [5, 6]. The most recent disaster of bauxite pond leakage happened in Hydro's Alunorte refinery in Brazil in 2018 [7].

This phenomenon—of large bauxite residue deposits—is not geographically limited. Alumina refineries exist across nearly every continent, with at least 200,000 tons of alumina produced on each one. Their prodigious output of residues necessitates their disposal in accompanying red mud "ponds" and dry stacks, each of which comes with huge maintenance costs and the potential risks of their failure: which can be catastrophic displays, as seen in Ajka, or smaller leakages, as seen in Hydro's Alunorte refinery in Brazil. In either case, the destruction caused by improper waste management has led to quantifiable economic damage and irreparable human impact. Ultimately, mitigating effective disposal and management are key to aluminum's sustainable future.

Current Efforts

Historically, the management of bauxite tailings was completely limited to storage and discarding. The major methods for bauxite residue disposal differed only in their location and solids loading.

1. Aquatic disposal, for example, forgoes the land required to house the tailings, but causes destructive changes in pH and dissolved minerals in aquatic ecosystems. Regulatory pressure has nearly eliminated this option.
2. Land-based disposal has thus become the most popular option, where waste solids content governs cost-benefit analyses. Lower solids content "lagooning" (<30 wt%) does not require capital-intensive moisture removal, but instead involves neutralization and/or long-term storage of the caustic waste. The larger volume is unwieldy and has a larger impact radius in the event of a failure [8].
3. Conversely, higher solids loading "dry stacks" and "dry cakes" lower the land requirement but involve costly filtration processes. In drier climates, the surfaces of these piles can dry and become a dust hazard, while in wet climates, heavy rainfall can prevent high solids loading altogether and cause spillovers [8].

In all situations, the utilized land eventually requires deliberate restoration and needs careful engineering to prevent the failure of enclosures. The high cost and zero to negative value (in the event of failure) of disposal methods clearly illustrate them as stop-gaps on the pathway towards proper valorization and utilization of bauxite tailings.

A number of technology options already exist to supplant storage—some involve use in the chemicals industry as a gas scrubbing agent, catalyst support, or adsorbent, while hydrometallurgical and electrolytic processes seek to neutralize the residue causticity and recover the remaining Fe, Ti, and Al from their oxides [2, 9]. However, to date, there are few efforts to use a notable volume of bauxite residue beyond cementitious aggregates. Even the most common low technology utilization options using it as a bulk aggregate in concrete or structural material require a concerted effort to move the waste material from the alumina plants, integrate it into existing processes, and prove its comparable efficacy and risk against existing options. These factors withheld, cost is still the greatest driver in a market economy. Any given utilization pathway will only be pursued in the presence of economic incentive.

The broad body of literature, intellectual property, and decades of research over the past half-century prove there is no dearth of technological options to manage bauxite residue. And yet, none of them are useful without monetary pressure or incentive established by policymakers. The value of bauxite residue must come from its component minerals, the cost of its storage, and the liability of potential damage borne by it.

Rationale for Suggested Incentives

The author believes that the successful utilization of bauxite residue can be the #1 aluminum sustainability enhancer as discussed recently by the author [10]. Unfortunately, while there has been incremental progress in applying bauxite residue for cement manufacturing, soil modification, and road construction, it is still unresolved, with a grave threat to public health.

In the quest for low-carbon aluminum, the industry is exploring a premium pricing structure for low-carbon aluminum, primarily by using hydropower as the source of electricity. Unfortunately, the availability of hydropower is limited to a few Nordic counties among only a handful of other geographically limited locations, and no new hydro dams are under construction for environmental reasons. Therefore, the hydro-based premium does not apply to most global aluminum production, thus having a minimal and only regional sustainability. Furthermore, this premium structure only applies to companies that already enjoy enormous economic advantages from substantially lower hydropower costs.

Suggested Incentives

We are proposing new pricing premiums for London Metal Exchange (LME) traded primary P1020 aluminum designed to incentivize the effective utilization of newly produced and existing bauxite residue. The suggested premium places a higher incentive for the removal of problematic existing bauxite residue compared to the newly produced residue. The bauxite residue mitigation premiums will apply equitably to all primary aluminum producers using any electrical energy sources (e.g., hydro, other renewables, coal, and natural gas) globally. Increasingly, aluminum consumers are willing to pay higher prices for sustainable aluminum, which is in short supply. Premiums based on "sustainable

Table 1 Suggested premiums—(to initiate discussion on methodology and implementation)

Matrix	Suggested LME premium	
	(% of current LME)	Value at the current LME price of $2,290/MT
Process **NEW** bauxite residue (1 T/per ton of P1020 produced)	0.25	5
Process **NEW** bauxite residue (2 T/per ton of P1020 produced)	0.50	10
Process **NEW** bauxite residue (3 T/per ton of P1020 produced)	0.75	15
Process **NEW** bauxite residue (4 T/per ton of P1020 produced)	1.00	20
Process **EXISTING** bauxite residue (1 T/per ton of P1020 produced)	0.50	10
Process **EXISTING** bauxite residue (2 T/per ton of P1020 produced)	1.0	20
Process **EXISTING** bauxite residue (3 T/per ton of P1020 produced)	1.5	30
Process **EXISTING** bauxite residue (4 T/per ton of P1020 produced)	2.0	40

aluminum", with respect to higher bauxite residue utilization, would provide economic incentives to producers globally to manufacture more primary aluminum (from any electrical energy source) adhering to the LME-defined low-bauxite residue requirements. Table 1 describes suggested aluminum premiums.

References

1. Kurdowski W, Sorrentino F. 6 - Red mud and phosphogypsum and their fields of application. In: Chandra S, editor. Waste Materials Used in Concrete Manufacturing. Westwood, NJ: William Andrew Publishing; 1996. p. 290–351.
2. Klauber C, Gräfe M, Power G. Bauxite residue issues: II. options for residue utilization. Hydrometallurgy. 2011;108:11–32.
3. Liu W, Chen X, Li W, Yu Y, Yan K. Environmental assessment, management and utilization of red mud in China. Journal of Cleaner Production. 2014;84:606–610.
4. TÓTH JI. Key actors of the red sludge disaster in Hungary. Confronting Ecological and Economic Collapse. 1st ed.
5. Winkler D, Bidló A, Bolodár-Varga B, Erdő Á, Horváth A. Long-term ecological effects of the red mud disaster in Hungary: Regeneration of red mud flooded areas in a contaminated industrial region. Science of The Total Environment. 2018;644:1292–1303.
6. Bauxite Residue accident from Norwegian Hydro in Brazil. https://redmud.org/bauxite-residue-accident-from-norwegian-hydro-in-brazil/Mayes.
7. WM, Burke IT, Gomes HI, Anton ÁD, Molnár M, Feigl V, et al. Advances in Understanding Environmental Risks of Red Mud After the Ajka Spill, Hungary. Journal of Sustainable Metallurgy. 2016;2:332–43.
8. Power G, Gräfe M, Klauber C. Bauxite residue issues: I. Current management, disposal and storage practices. Hydrometallurgy. 2011;108:33–45.
9. Lyu F, Hu Y, Wang L, Sun W. Dealkalization processes of bauxite residue: A comprehensive review. Journal of Hazardous Materials. 2021;403:123671.
10. Das S The Quest for Low Carbon Aluminum: Developing a Sustainability Index LIGHT METAL AGE February 2021 edition (pages 44–49)

Aluminium Bahrain (Alba) SPL Sustainable Solution from Landfill to Valuable Feedstock "HiCAL30"

Khalid Ahmed Shareef, Bernie Cooper, Mohsen Qaidi,
Nabeel Ebrahim Mohd Al Jallabi, Fuad A. Hussain Alasfor,
and Vijay Rajendran

Abstract

Spent pot lining (SPL) is one of the largest solid wastes generated from the primary aluminium production process. Around 23 kg of SPL is generated per tonne of aluminum. ALBA is the largest aluminumproducing smelter in the world outside of China with an annual production of 1.56 million tonnes of aluminum (2021), the amount of SPL generated is becoming very significant. SPL is classified as hazardous material contaminated with cyanide and fluorides. The traditional way of handling such waste is either stockpiling around smelters facility or landfilling, which is not the best sustainable solution. ALBA has taken the initiative to construct its own one of a kind SPL treatment plant through its partnership with Regain Technologies to process and detoxify its SPL and convert it into useful feedstock for other industries such as the cement industry. This paper discusses the journey of Alba toward its sustainable solution of converting the SPL into a useful valuable product called HiCAL30. Moreover, this paper will focus on the process flow, detoxification heat treatment process as well as the final product specification and usages.

Keywords

Spent pot lining (SPL) • HiCAL30 • Clinker • Cyanide • Fluoride

K. A. Shareef (✉) · M. Qaidi · N. E. M. Al Jallabi · V. Rajendran
SPL Plant, ALBA, Aluminum Bahrain, P.O. Box 570 Manama,
Kingdom of Bahrain
e-mail: khalid.ahmed@alba.com.bh

B. Cooper
Regain Technologies, Melbourne, Australia

F. A. H. Alasfor
Potlines and Services, ALBA, Aluminum Bahrain, P.O. Box 570
Manama, Kingdom of Bahrain

Introduction

Solid waste generation and management have become a serious issue of concern, especially in aluminum smelters where it represents one of the intractable environmental challenges. Different types of waste are generated; part of which are being hazardous waste. The typical solid waste generated by aluminum smelters are Carbon Dust, Cast Iron Slag, General Waste, Refractory Waste, Spent pot lining (SPL), etc. SPL represents around 50% of the total solid waste generated in the aluminum smelting process.

SPL material has been categorized as hazardous waste by many regulators as it exhibits several characteristics of such waste material. During the life of the reduction cell, the carbon cathode gets saturated with fluorides and cyanides, and these contaminants have the potential to leach into the groundwater if the material disposal is not managed well. Additionally, SPL reacts with water to generate a strongly basic solution that is highly corrosive and emits flammable and explosive gasses during the process such as methane, hydrogen, and toxic gasses such as ammonia. (Fig. 1), SPL removed from the cut out reduction cell.

Disposal of SPL has been carried out conventionally through landfilling. This method is not sustainable and not commercially viable and become unfavorable to most environmental regulators. More sustainable methods are sought that encourage the circular economy and the full recovery and conversion of the material to useful end products with zero-waste output.

Aluminium Bahrain BSC (Alba) is the largest aluminumproducing smelter in the world outside of China with an annual capacity of 1.56 million tonnes of aluminum (2021), the amount of SPL generated is becoming very significant and reach up to 25,000 tonnes of SPL generated per annum.

Alba has always strived to be a leading company in applying the best practices with regard to Environment, Social, and Governance (ESG). To date, Alba has recorded

Fig. 1 SPL removed from the reduction cell

30 million safe working hours without lost time injury (LTI). And protecting the environment with the best Eco-friendly system is one of its top priority.

Alba Journey Towards the Sustainable SPL Treatment Solution

The journey to reach and identify the most suitable SPL treatment process for Alba has taken almost a decade and a common effort with the GAC smelters was started in 2008. In 2009, Alba sent a team to evaluate four different processes by which SPL was being handled. The team concluded that the processes evaluated then were neither mature nor complete. Nevertheless, the effort was not stopped and Alba started again to look into the possibility of collaborating with a local cement plant operating in Bahrain. However, this collaboration wasn't successful due to the low plant capacity and technical difficulties. With the absence of a suitable cement manufacturing plant in Bahrain, it became critical to find an alternative solution to handle Alba SPL. Alba's efforts continued till 2017, when few potential processes were identified again, and the evaluation process started.

Selecting the Right Technology

After shortlisting the potential treatment process and technology suppliers, it was decided to approach three suppliers. Based on the thorough analysis, technology benchmarking, and due diligence, REGAIN technology was selected due to its full recovery and conversion of the material to useful end products with zero-waste output. REGAIN technology has been handling Tomago smelters since 2003.

Building the Plant

During the construction phase, the project has gone through many challenges such as the availability of resources, and delays in supply chain and equipment deliveries as a result of COVID-19 restrictions. To deal with the repercussions of COVID-19, Alba management and project team followed a specific recovery approach to evaluate the overall progress of projects and plan the recovery road map.

Despite the challenges and lockdown due to COVID-19 regulations, the design and construction of the plant commenced in December 2019 and was followed by the commissioning and ramp-up in July 2021. The plan was officially inaugurated in December 2021 by the Chairman of Alba's Board of Directors Shaikh Daij bin Salman bin Daij Al Khalifa (Fig. 2).

The plant was built over a 26,000 square meters site. The plant has an annual capacity of 35,000 tonnes of SPL with provision for future expansion to 70,000 tonnes. The actual capital cost of the plant, including site development and buildings, was US$37.5 million compared with a budget of US$44 million.

Material Characterization

Due to its hazardous classification, the method of disposal poses a major challenge. Also, the export of such hazardous waste for any sort of disposal or usage in other forms is prohibited by almost all regulators worldwide.

Aluminium Bahrain (Alba) considered options for recycling its spent potlining (SPL) and exported its SPL from Bahrain for characterisation of its properties and for recycling by Regain. And for Alba to qualify its SPL and to

Fig. 2 SPL plant during the construction phase

validate the selected technology was the right technology, prior to the plant start-up, Alba has taken the challenge and exported 343 tonnes of SPL to be recycled in Regain's facility in Tomago, Australia. The shipment was sent in sealed shipping containers by sea fulfilling the Basel Convention approvals of Transboundary Movements of Hazardous Wastes.

The shipment consisted of different types of SPL generated from the two different potline technologies namely Hydro/Montecatini and Pechiney AP30 as shown in Fig. 3

The characterization process was divided into two major stages:

1. SPL was divided into four lots and was processed as a separately controlled batch including sorting, metal recovery, crushing, and size reduction
2. Thermal treatment and hydrolysis process

The Alba SPL was processed in January 202. It was processed through the Crushing and Metal Recovery processes in four designated batch lots. A representative sample of the crushed (untreated) SPL for each batch was taken. The analytical results for each lot of testing are set in Table 1.

The second step involved thermal treatment of the different batches with the objectives of:

1. Destroying cyanide in the SPL
2. Eliminating the flammable gas hazard from SPL reacting with water
3. Proving the Alba SPL could be processed using Regain SPL processing technology

Representative samples were taken for each thermally treated batch and the results are shown in Table 2.

The four batches of treated SPL were blended to manufacture a HiCAL 30 product batch. A Representative sample from the batch was taken for the following tests:

1. Flammable gas generation test
2. Elements analytical test (Table 3).

The treated Alba SPL was used in the production of Regain's HiCAL and was shipped from the Tomago facility to be consumed in cement manufacturing outside Australia.

The final product (HiCAL) was consumed by the end customer and the results were satisfactory and accepted by the end user.

Commissioning and Startup

After the thermal treatment plant was pre-commissioned, the kiln was ready to introduce feed. The Thermal Treatment Plant was commissioned as follows:

1. Pre-commissioning verification:
   - Testing of all the automated sequences and key interlocks of the Thermal Treatment Plant.
2. Initial system proving:
   - Charged the feed bin with SPL,
   - Commissioning and prove of the material handling system

Fig. 3 Bulk batch of Alba SPL

Table 1 Analytical results of crushed Alba SPL (untreated)

Description		Unit	Hydro/Montecatini 2nd Cut Batch - BT2774	Hydro/Montecatini 1st Cut Batch -BT2775	Pechiney AP30 1st Cut Batch - BT2776	Pechiney AP30 2nd Cut Batch - BT2777
Silicon	as SiO_2	%	50.6	3.6	6.9	39.4
Aluminium	as Al_2O_3	%	19.8	8.8	9.8	18.3
Iron	as Fe_2O_3	%	2.7	0.6	0.9	3.5
Titanium	as TiO_2	%	0.59	0.06	0.09	0.43
Potassium	as K_2O	%	0.86	0.18	0.22	1
Magnesium	as MgO	%	0.6	0.2	0.4	0.8
Sodium	as Na_2O	%	13.3	14.6	12.2	12.7
Calcium	as CaO	%	1.7	2.2	2.4	2.2
Sulphur	as SO_3	%	<0.1	0.2	0.2	<0.1
Manganese	as MnO	%	0.02	0.01	0.01	0.02
Phosphorus	as P_2O_5	%	0.06	0.01	0.01	0.04
Chloride		%	0.01	0.01	0.01	0.01
Fluoride	Total as F	%	6.6	11.3	9	6.4
Carbon		%	11.5	61.8	60.2	19.5
Calorific Value		MJ/kg	3.9	19.9	18.7	7.4
Cyanide	(CN)	mg/kg	177, 182	198, 194	168, 156	69, 76
Chromium	Total	mg/kg	70	25	25	60

Table 2 Analytical results of Alba thermally treated SPL

Description		Unit	Hydro/Montecatini 2nd Cut Batch - BT2774 A	Hydro/Montecatini 1st Cut Batch -BT2775A	Pechiney AP30 1st Cut Batch - BT2776A	Pechiney AP30 2nd Cut Batch - BT2777A
Silicon	as SiO_2	%	48.2	7	6.6	37.1
Aluminium	as Al_2O_3	%	19	11.1	9.6	23.5
Iron	as Fe_2O_3	%	2.7	1.2	1.1	3.5
Titanium	as TiO_2	%	0.56	0.11	0.09	0.49
Potassium	as K_2O	%	0.83	0.24	0.21	0.65
Magnesium	as MgO	%	0.6	0.4	0.5	0.6
Sodium	as Na_2O	%	13.9	13	11.6	16.6
Calcium	as CaO	%	1.6	2.6	2.4	2.1
Sulphur	as SO_3	%	<0.1	0.3	0.1	<0.1
Manganese	as MnO	%	0.02	0.01	0.01	0.02
Phosphorus	as P_2O_5	%	0.06	0.02	0.02	0.07
Chloride		%	0.01	0.01	0.02	0.01
Fluoride	Total as F	%	7	9.8	8.7	9.2
Carbon		%	13.4	57.5	61.2	15.7
Calorific Value		MJ/kg	4.3	18.5	19.4	5
Cyanide	(CN)	mg/kg	43, 44	42, 47	16, 12	12, 13
Chromium	Total	mg/kg	90	30	20	85

Table 3 Analytical results of Alba blended materials as HICAL 30 product

Description		Unit	Alba Treated Material Blend Batch BT2780	HiCAL 30 Product Specification
Carbon (total)		%	32.5	30 to 35
Calorific Value		Mj/kg	11.0	8 to 11
Aluminium	as Al_2O_3	%	16.7	10 to 25
Calcium	as CaO	%	2.1	1 to 3
Iron	as Fe_2O_3	%	2.6	2 to 7
Sodium	as Na_2O	%	13.7	13 to 18
Potassium	as K_2O	%	0.52	0 to 1
Silicon	as SiO_2	%	28.2	10 to 30
Sulphur	as SO_3	%	<0.1	0 to 2
Fluoride (total)		%	8.1	8 to 12

3. Commissioning Ramp-up:
   - Kiln was dried out, preheated, and then feed was introduce at 3 tonnes per hours. Progressively, commissioning runs were conducted with gradual ramp up to 5 tonnes per hours.

Performance Test

In order to evaluate the performance and complete the final hand over of the plant, Regain the technology provider must demonstrate the SPL processing facility is capable of meeting the performance conditions.

The performance test is separated into two parts: the processing plant and the product qualification:

Cyanide Destruction and Flammable Gas Testing

During commissioning with potentially satisfactory runs at the target processing temperature samples of the treated SPL were taken and sent to Regain for laboratory analysis to check for cyanide levels and the completion of the hydrolysis reaction.

Production Throughput Performance Requirements

The SPL processing plant is expected to perform as per the proposed designed capacity of 35,000 tonnes per annum, based on a thermal treatment plant availability of at least 85%. The performance test was conducted over a period of 7 days of crushing and thermal treatment followed by a curing period to allow the completion of the hydrolysis reaction of the SPL.

After successfully achieving the Performance Test, Regain provided Alba a Certificate of quality with a supporting laboratory testing report for Alba HiCAL 30 that confirmed the quality parameter was met as per the Engineering and Site Support Services Agreement as shown in Table 4.

The Treatment Process

SPL Cyanide and Flammable Gas Hazards

A typical pot life is between 4–5 years, over this period, the lining is penetrated by aluminium metal, sodium metal, and the sodium aluminium fluoride/sodium fluoride electrolyte. Chemical reactions in the lining result in the formation of various carbides, nitrides, and cyanides within the pot lining [1]. Indicative examples of the chemical reactions are:

1. Reaction of cryolite with nitrogen and sodium to form nitrides, e.g.

$$Na_3AlF_6 + 0.5N_2 + 3Na \rightarrow AlN + 6NaF$$

2. Metals such as aluminium react with carbon to form carbides, e.g.

$$4Al + 3C \rightarrow Al_4C_3$$

3. Various carbon–nitrogen compounds are also produced in the forms of cyanides, e.g.

$$1.5N_2 + 3Na + 3C \rightarrow 3NaCN.$$

Table 4 Performance test results of Alba HiCAL30 product

Evidence of Conformance

Description		Unit	Test Results	Specification Range Limits	Test Method
Fluoride	total as F	%	11.0	8 to 12	Ion Selective Electrode (ISE)
Sodium	as Na₂O	%	15.6	13 to 18	
Potassium	as K₂O	%	0.62	0 to 1	
Silicon	as SiO₂	%	21.7	10 to 25	
Aluminium	as Al₂O₃	%	18.7	10 to 25	Inductively Coupled Plasma Spectroscopy (ICP/OES)
Iron	as Fe₂O₃	%	2.4	2 to 7	
Calcium	as CaO	%	2.1	1 to 3	
Sulphur	as SO₃	%	0.2	0 to 2	
Carbon		%	31.5	28 to 35	Liebig technique to Australian Standard AS2434.6
Calorific Value		kcal/kg	2366	Target >2,000	Calorimeter to Australian Standard AS1038.5

SPL is considered to be a hazardous waste due to the presence of leachable cyanide and fluoride as well as its tendency to react with water and produce flammable gases [2]

SPL Treatment Plant Layout and Process Flow

The Alba SPL treatment plant has four main areas; warehouse, process building, thermal treatment plant (TTP), and final product building as illustrated in Fig. 4

The SPL treatment process flow is summarized below and is shown in Fig. 5:

- 50,000 tonne capacity SPL warehouse for receiving raw SPL then sorting and breaking the SPL to a size suitable for feed to the process plant

- Process building housing SPL primary crushing, secondary crushing, and metal recovery process equipment used to separate cathode bars and aluminium metal and prepare the SPL for chemical processing
- Thermal treatment plant that destructs the SPL cyanide and explosive gas hazards in a chemical process
- Products building for blending of thermally treated materials into a chemically homogeneous product, secondary hydrolysis process, storage, and dispatch of the product.

SPL Material Preparation

Alba SPL materials are delivered by trucks from Alba main smelter to the Alba SPL treatment plant. The SPL material is

Fig. 4 SPL plant layout

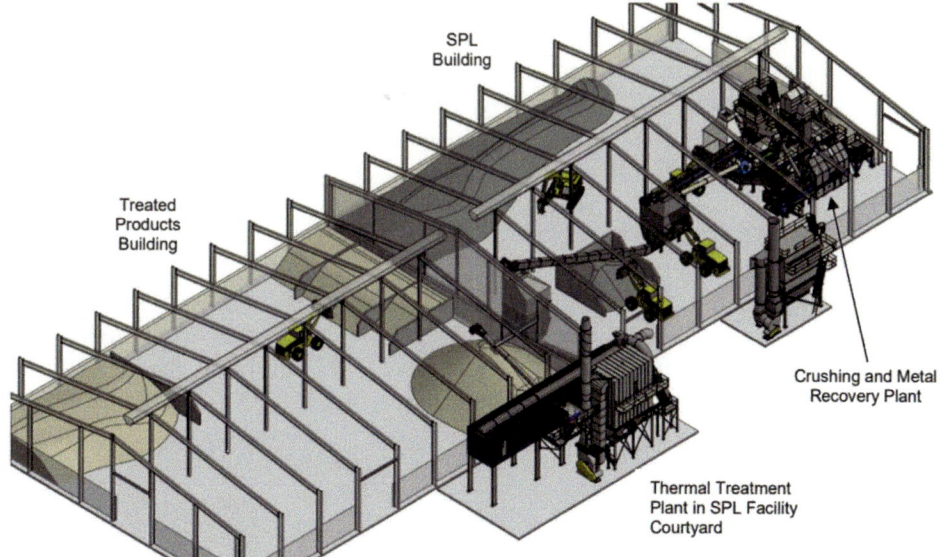

Fig. 5 SPL treatment summary process flow

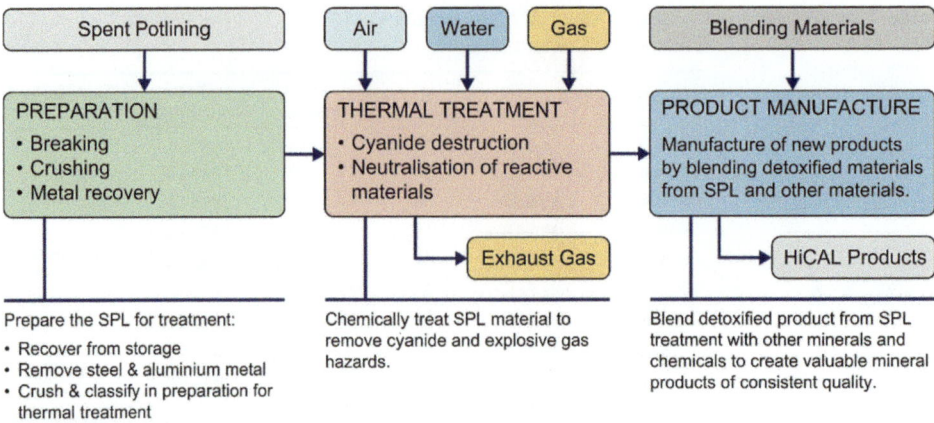

Fig. 6 Alba SPL warehouse

stored in the SPL plant storage shown in Fig. 6. The storage has been designed and built exclusively to handle the preparation process of SPL in Alba.

SPL Preparation involves (a) recovery of material from storage, (b) segregation and removal of metals (aluminum and iron), pulverizing the carbon materials and refractory materials to the required feed size; (c) sorting into like material streams, and (d) crushing and size classification.

Alba Primary Crushing Plant

The SPL material is then conveyed through several transfer conveyors and feeder, which, in turn, are discharged to the primary crusher. The primary crusher crushes the SPL material greater than 200 mm to the nominal 200 mm size (Fig. 7).

Mill Crushing and Metal Recovery

The objective of the crushing and metal recovery plant (C&MRP) is to crush the receiving SPL from the primary crushing plant to minus 12 mm in size and to separate and recover aluminium and ferrous metals prior to feeding the thermal treatment plant.

Prepared SPL is fed to the Alba SPL crushing and metal recovery plant (C&MRP) feed storage bin from the primary crushing plant. The SPL is transferred through several conveyors and feeders which, in turn, discharged into the rod mill, where it is crushed to less than 150 mm in size and delivered to a mill discharge screen which screens a target of 10 mm minus size under-flow goes to feed the thermal treatment plant.

The oversize from the mill discharge screen is passed over the ferrous metal separator which then sees magnetic

Fig. 7 Primary crushing system

material passing to the ferrous metal skip and non-magnetic material passing to the non-magnetic feed conveyor and, in turn, to the non-ferrous metal separator.

Thermal Treatment Process

The cyanide and explosion hazards in SPL are eliminated through the SPL treatment process. Neutralization of the reactive compounds is achieved by bringing on the reactions that generate the explosive gases in a controlled environment such that no more gas can be generated.

The Treatment Process is self-sustaining by re-using the gases generated to destroy Cyanide by thermal oxidation, i.e., heating the material in the presence of oxygen. No residual materials are produced because there of no other chemical processes or additives.

The crushed SPL is fed to the TTP Kiln, reactor, and dust collection system where the SPL is heated to destroy cyanide and then mixed with water in the reactor to generate the hydrolysis reaction (primary hydrolysis) required to neutralize the substances in SPL that react with water and produce flammable gases.

The treated SPL is delivered to a thermally treated material stockpile conveyor and then to the treated SPL building with a typical production rate of 5 tonnes per hour.

Use of Other Smelter Wastes

With the onset of one of its kind of facility in ALBA, the production of HiCAL 30 started to pick up in the SPL treatment plant in CY 2022. However, the produced HiCAL 30 was having slight deviation with respect to targeted specifications in terms of carbon percentage as mentioned in Table 5.

Above deviation to the desired quality demands to add more carbon percentage in the product, which provokes an idea of exploring any other wastes which are rich in Carbon. Shot blast and dust collector dust being generated in Alba Carbon plant (which was found to be having 60% C) are chosen to be blended in an appropriate mix with HiCAL product to increase the carbon percentage to the desired range (Table 6).

Eventually, Alba managed to produce HICAL 30 as per the desired specification by using one of the wastes, i.e. shot bast dust collector dust, which was also disposed of in the landfill till then.

Moving forward, Alba is also making efforts to use another waste from its own carbon plant called contaminated butt to feed along with SPL to increase the carbon percentage in the final product.

Table 5 Carbon percentage in HiCAL30 product

Quality parameter	HiCAL 30 specification	HiCAL from TTP
CN (ppm)	0–60	55 ppm
C (%)	28–35	24.8
F (%)	10–14	12.4
Na (%)	13–18	17.4
Si (%)	10–25	24.0

Table 6 Carbon percentage in HiCAL30 after blending

	HiCAL from TTP	Shot blast dust collector dust	HiCAL 30
Quantity (MT)	750	275	1025
Carbon (%)	23.4	63.9	34.3
F (%)	13.0	9.0	11.9
Na (%)	17.3	6.7	14.5
Si (%)	25.1	0.3	18.4

HiCAL Marketing and Exporting

The HiCAL product has proven valuable to certain cement-producing plants that are geographically located close to the equator. Due to the heavy tropical rain, the quarries used by these plants are usually deprived of the minerals usefully needed in the clinker manufacturing process. Therefore, the addition of HiCAL elevates the concentration of the minerals and improves the clinker throughput and properties, in a cost-effective manner, which also eliminates the toxicity of HiCAL in the chemically irreversible cement clinkering process. In addition, the cement industries realize significant benefits through energy saving, greenhouse gas (GHG) emission reduction, and cost-effective production increases [3].

Cooperation was made with Asian cement producers to ascertain the assessment of this project that there is a high demand for HiCAL and it is fully consumed. In order to ensure the sustainability of the project, a key factor is to gain the regulator's trust on the product and grant their acceptance for the shipping purpose. The importation and use of HiCAL products in cement making have already been accepted by regulatory authorities in eight countries.

Use of HiCAL in Cement Industry

SPL has specific elements that work like mineralizers and fluxes for the clinkering process in cement manufacturing. Flux can be defined as an element that decreases the liquids temperature for a solid substance and/or generates a higher liquid quantity at a certain temperature. Similarly, mineralizer can be defined as an element that helps faster reaction rates where the creation of the required material in the liquid phase for the cement clinker or any similar material is generated. The outcome of such a process is a higher quality product and lower energy usage due to lower kiln resistance time. As far as the cement industry is concerned, sodium as flux and fluorides as minerals are of specific importance coming from SPL [4].

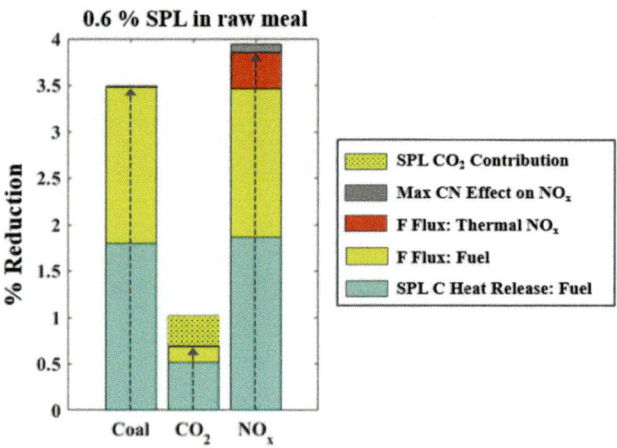

Fig. 8 SPL addition's effect on cement production on fuel consumption, CO_2 and NO_x emissions. The arrow indicates the net reduction (*Source* [7])

Mineralizing clinker using fluorides increases the clinker reactivity and reduces the clinker quantity required for each tonne of cement [5]. The Elite level increases in the clinker due to mineralization that corresponds to less clinker proportion needed to produce a certain strength cement (i.e. improved clinker factor). The clinker factor is improved in the magnitude of 10% or more at the optimum fluoride mineralization. The clinker amount in cement can be reduced by 15 to 20% maintaining the same performance because of fluoride mineralization [6].

According to Al Jawi et al. [7], around 3.8% of coal consumption was reduced in the cement manufacturing plant with the use of 0.6% of SPL in the raw meal. The reduced consumption was equivalent to the energy input provided by the SPL carbon and the flux impact on lowering the needed kiln temperature. At the same time, CO_2 was lower by 1% due to these two factors. Yet, the total CO_2 emission reduction was only 0.65% as part of the CO_2 emission comes from the usage of SPL itself. It was anticipated that NO_x will be reduced by 3.8% due to lower Nitrogen generation associated with lower coal burning and the impact of the flux resulting in reduced fuel and thermal NO_x generation (Fig. 8).

HiCAL as an End of Waste Solution for SPL

Environmental lifecycle assessment of the total system of SPL recovery, treatment process, HiCAL product transport, and consumption in the cement industry shows that each tonne of HiCAL can save:

(a) 4 tonnes of carbon dioxide equivalent (CO_{2e}) greenhouse gas emission
(b) 16 gigajoules of thermal energy
(c) 350 kWh of electrical energy.

Article 6 of the European Commission "*Waste framework directive*" provides guidance as to when certain wastes cease to be waste and become a product or a secondary raw material.

According to Article 6 (1) and (2) of the Waste Framework Directive, certain specified waste ceases to be waste when it has undergone a recovery operation (including recycling) and complies with specific criteria, when:

The substance or object is commonly used for specific purposes:

(a) a market or demand exists for such a substance or object;
(b) The substance or object fulfils the technical requirements for the specific purposes and meets the existing legislation and standards applicable to products; and
(c) The use of the substance or object will not lead to overall adverse environmental or human health impacts.

This criterion is supporting of HiCAL being classified as a product that has ceased to be a waste.

Conclusion

Over ten years of searching was done for a sustainable environmental solution for the recycling of SPL, Alba partnered with Regain Technology to implement an eco-friendly solution as an alternative for landfilling in a sustainable and cost-effective way with zero waste.

The objective of a sustainable zero-waste solution has been realised through Alba's long-term vision and leadership in establishing the SPL Treatment Plant in Bahrain linked with the Regain industrial eco-system of HiCAL product technology, regulatory authorisation, industrial marketing, and a portfolio of end-user cement industry customers.

The investment in the Alba SPL Treatment Plant and the overall approach to dealing with Alba SPL reflects the Company's strategic approach when it comes to ESG. By adopting this strategy, Alba has been able to establish a cost-effective zero-waste SPL solution and has comprehensively addressed the key SPL management aspects identified in the World Aluminium Sustainable Spent Pot Lining Management Guidance document:

- Longterm planning
- Good governance
- Health and safety
- Environmental management
- Treatment
- *Utilisation of the SPL (Eliminating land fill).*

References

1. Shamsili, R., & Oye, H. (1994). Melt Penetration and Chemical Reactionsin Carbon Cathodes During Aluminium Electrolysis. III. *Light Metals 1994*, 731–738.
2. Arpit Agrawal, Chandan Kumar, Arunabh Meshram, Recovery of carbon rich material: Recycling of spent pot lining: A review, Materials Today, Proceedings 46(4), 2021, 1526–1531 DOI: https://doi.org/10.1016/j.matpr.2021.01.143.
3. Kaddatz K.T., Rasul M.G., Rahman A. "Alternative fuels for use in cement kilns: process impact modeling" Central Queensland University, School of Engineering and Built Environment, Rockhampton, Queensland 4702, Australia, Proc Eng. 2013; 56: 413–20
4. B. Cooper, Considerations for dealing with spent potlining. In Proc. 11th Australasian Aluminium Smelting Technology Conference, Dubai, 2–6 December 2014, 6–11.
5. V. Johansen and J. Bhatty, Fluxes and mineralizers in clinkering process, in J. Bhatty, F. Miller, S. Kosmatka and R. Bohan, Eds., Innovations in Portland Cement Manufacturing, (Skokie, Illinois: Portland Cement Association, 2011), 401–438.
6. H. Borgholm, D. Herfort and S. Rasmussen, A new blended cement based on mineralised clinker, World Cement Research and Development vol. 8, 1995, 27–33.
7. M. Al Jawi et al., Environmental benefits of using spent pot lining (SPL) in cement production. In Light Metals 2020, 1251–1260.

Valorization of Treated Spent Potlining in Cement Industry

Laurent Birry, Jean Lavoie, Victor Brial, Claudiane Ouellet-Plamondon, Hang Tran, Luca Sorelli, and David Conciatori

Abstract

Spent potlining (SPL) is a hazardous waste produced by aluminum smelters. It is classified as a hazardous waste due to its contamination with fluorides and cyanides and its reactivity with water, generating explosive gases. After being industrially and hydrometallurgically treated by the Low Caustic Leaching and Liming (LCL&L) process, the refractory part of SPL becomes an inert, non-hazardous material, called LCLL Ash. The cement industry is a major emitter of greenhouse gases. One of the best options for reducing the carbon footprint of concrete is the use of supplementary cementitious materials or fillers to replace part of the cement in the concrete. This article presents the conditions that make LCLL Ash a suitable and value-added material for the cement industry. The results presented were obtained over the past four years as part of an R&D project at the Ecole de Technologie Superieure (Montreal) and Universite Laval (Quebec), with the support of Rio Tinto and Ciment Quebec.

Keywords

Aluminum • Sustainability • Spent pot lining

L. Birry (✉)
Rio Tinto, Arvida Research and Development Centre, 1955 Mellon Boulevard, Bldg. 110, Saguenay, QC G7S 4K8, Canada
e-mail: Laurent.birry@riotinto.com

J. Lavoie
Rio Tinto, Saguenay, Canada

V. Brial · C. Ouellet-Plamondon
Ecole de Technologie Superieure, Montreal, Canada

H. Tran · L. Sorelli · D. Conciatori
Universite Laval, Quebec City, QC, Canada

Introduction

Spent potlining is a hazardous waste produced in the pot rooms of aluminum smelters. This waste is generated from the internal lining of aluminium (Al) cells consisting of carbon and refractory bricks, which are replaced after five to eight years of service life. SPL is classified as a hazardous waste because of its contamination with fluorides and cyanides and its reactivity with water, generating explosive gases. Therefore, transportation, storage and final disposal of SPL are subject to strict environmental regulations.

Each tonne of aluminum produced generates approximately 22 kg of SPL. Several options to treat SPL exist and were reported in the literature [1, 2]. In the early 1990s, Rio Tinto developed the "Low-Caustic Leaching and Liming" hydrometallurgical process (LCL&L) at the Arvida Research and Development Centre (ARDC) in Saguenay, Quebec. This process leaches fluorides and cyanide compounds out of SPL and produces inert by-products having high valorization potential [3, 4]. In 2008, an SPL treatment plant was built in Jonquiere, which processes up to 80 kt of SPL per year [5, 6].

During this hydrometallurgical treatment, the refractory part of SPL becomes an inert, non-hazardous material, called LCLL Ash. Due to its chemical and mineral composition, LCLL Ash is attractive for the clinker chemistry and its use as a raw material for cement production will be presented in this paper. LCLL Ash material has also the potential to be used as a supplementary cementitious material (SCM) in the production of concrete and backfill. Cement production is known to generate 4–8% of global carbon emissions [7]. Efforts are being made worldwide to reduce the use of cement in concrete mixes. Many improvements in the cement production process have been made to reduce the energy consumption of cement plants and make them less polluting. However, the production of one tonne of Portland clinker can produce up to one tonne of CO_2, 60% of which comes from the inevitable decarbonation reaction. The use of a supplementary cementitious material (SCM), to reduce the amount of cement in concrete and its environmental impact, is one of the preferred solutions today. This paper, written in collaboration with the Ecole de Technologie Superieure (ETS, Montreal) and Universite Laval (Quebec), presents the conditions that make LCLL Ash a suitable SCM for applications in concrete. Other valorization options for LCLL Ash, for example in ultrahigh resistance concrete (UHPC), were also studied and summarized in this paper.

The LCL&L Process

The LCL&L process leaches fluorides and cyanide compounds out of SPL and generates inert by-products, which can be valorized more easily than hazardous waste.

As shown in Fig. 1, two separate sectors (one wet and one dry) are needed to treat SPL. The dry sector includes unloading, handling and storage of SPL containers, SPL grinding (less than 300 microns) and ground SPL storage. The wet sector consists of low-caustic serial leaching steps to extract soluble fluorides and cyanide compounds. These inert residues are called carbonaceous by-products (CBPs). Cyanide compounds, contained in the leachate, are subsequently destroyed in pressurized reactors using high temperature hydrolysis, while fluorides are precipitated as inert calcium fluoride after reaction with lime. More details on the recent development of the LCL&L process are available in [8].

Valorization Options for Carbon By-Products (CBPs)

Figure 2 presents the different options for CBP valorization. Since the start-up of the SPL treatment plant, approximately 800 kt of SPL has been treated, half of which comes from SPL stored in warehouses from mid 80 s until the start-up of the plant in 2008. Due to this latter mixed composition (carbon and refractory lining), the produced CBP contains approximately 30–40% of carbon (on a dry basis) and 60–70% of inert materials (mainly SiO_2 and Al_2O_3). All this mixed CBP production has been valorized as a low-density (1250 kg/m^3) civil engineering construction material (called LCLL Geotek) for the construction of dams, at the Rio Tinto bauxite residue disposal site.

Since 2006, Rio Tinto has developed an intensive R&D program to find alternative valorizations for CBP. Due to its mixed nature (carbon and refractory compounds), CBP is essentially limited to be used as alternative fuel and raw material for clinker production in the cement industry. Carbon is attractive for its energy content (approx. 13 GJ/t), while the inert material is attractive for its mineral composition (rich in aluminosilicates), which is very close to that of the clinker chemistry. Depending on the use of CBP in the cement process, it is more interesting to separate the carbon fraction from the refractory fraction. The first cut (SPL fraction above the collector bars, mainly carbon) and the second cut (SPL fraction below the collector bars, mainly refractory brick) can be separated when dismantling electrolysis pots. The carbon concentrate from the first cut

Fig. 1 LCL&L process flow diagram

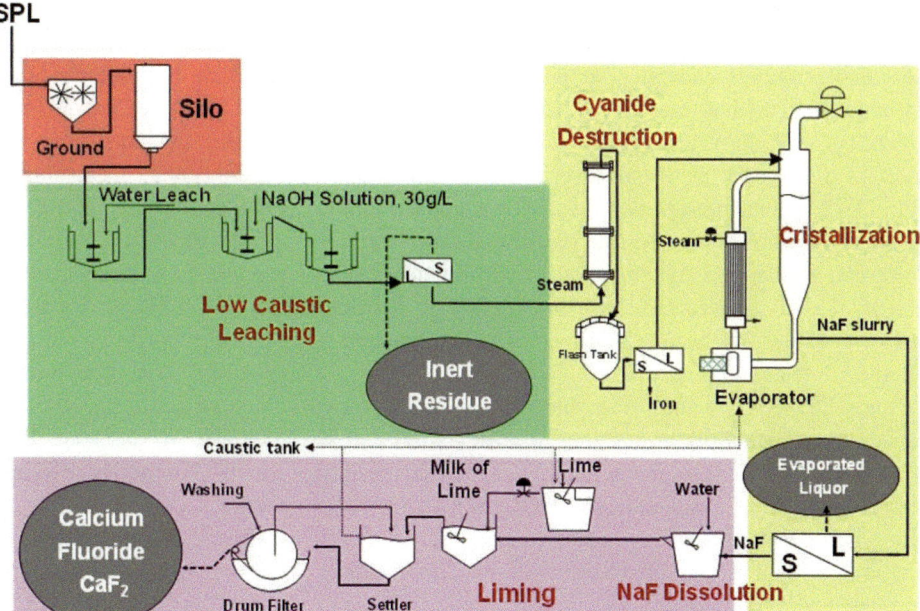

Fig. 2 Valorization options for the carbon by-products (CBPs)

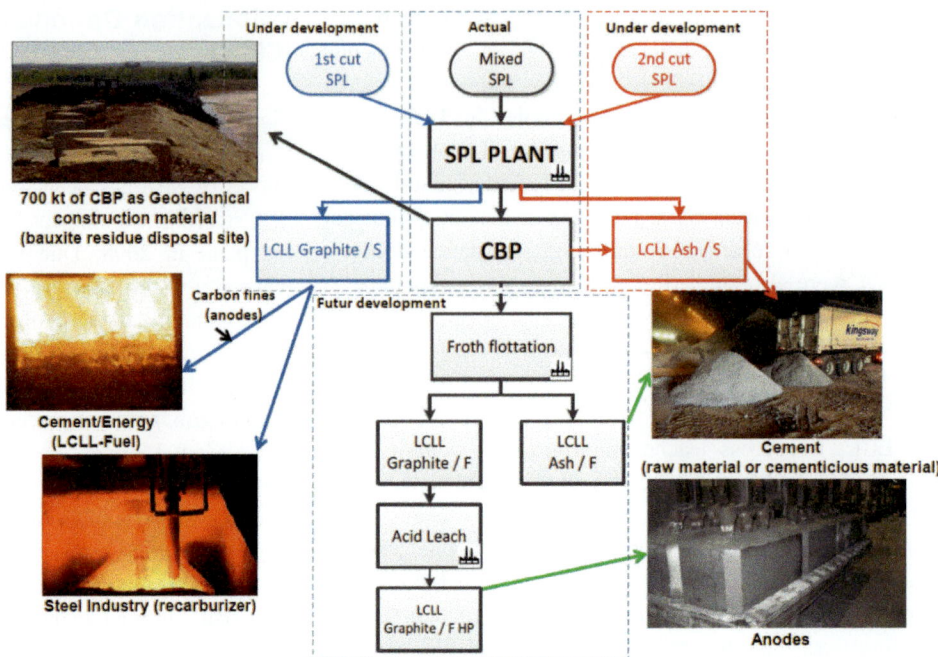

treatment is called LCLL graphite, and the refractory concentrate from the second cut treatment is called LCLL Ash (C < 10%).

Valorization of LCLL Ash in Cement Industry

LCLL Ash as Raw Material for Clinker Production

The cement industry as well as some other industrial pyrometallurgical processes can use SPL as generated, without any pretreatment [2]. However, strict restrictions on the sodium and fluoride content in the final product limit the amount of SPL that can be added to these processes. Nowadays, environmental regulations are also stricter, and it becomes difficult for these industries to accept unprocessed SPL. LCLL ash is attractive for the clinker chemistry due to its mineral composition and can be used as raw material. Compared to the direct valorization of SPL in the cement industry, LCLL Ash contains ten times less total fluorides and lower alkalis (sodium), which enables a higher dosage in the kiln.

Since 2018, Rio Tinto, Geocycle Canada and Lafarge Canada have been working together to reuse LCLL Ash as raw material for clinker production. To meet their specifications, Geocycle Canada and Rio Tinto have developed a new product called Alextra, made from LCLL Ash with additions of alumina or refractory waste materials coming from the aluminum industry. Alextra is the result of years of research and development, aimed at finding new ways to deliver sustainable outcomes and value from used potlining.

Alextra is a good alternative to raw materials such as alumina and silica, which are commonly refined or mined for use in the cement industry. This leads us to also develop new products based on treated mixed SPL and other industrial wastes as sources of silica or iron to meet customer needs.

LCLL Ash as Supplementary Cementitious Material (SCM)

In 2017, an intensive R&D program, for the potential use of LCLL Ash as SCM, was initiated in collaboration with the Ecole de Technologie Superieure (ETS, Montréal), Universite Laval and Ciment Quebec. The aim of this project was to identify the conditions that make LCLL Ash a suitable supplementary cementitious material (SCM) for concrete production.

The procedure for evaluating the reactivity of LCLL Ash for use as a supplementary cementitious material is summarized in Fig. 3. A Frattini test, following the procedures in [9, 10], was carried out to evaluate the pozzolanic reactivity of LCLL Ash. This test evaluates the reactivity of SCM aluminosilicate phases with portlandite to precipitate more hydrates. R^3 tests are based on a mix that recreates the chemical behavior of limestone cement, without cement particles [11]. The heat generated by the precipitation of hydrates was measured by isothermal calorimetry, and the consumption of portlandite was measured by thermogravimetrically analysis. The pastes were measured by X-ray diffraction after the R^3 tests. The reactivity of LCLL Ash (with and without calcination) as well as the other SCMs

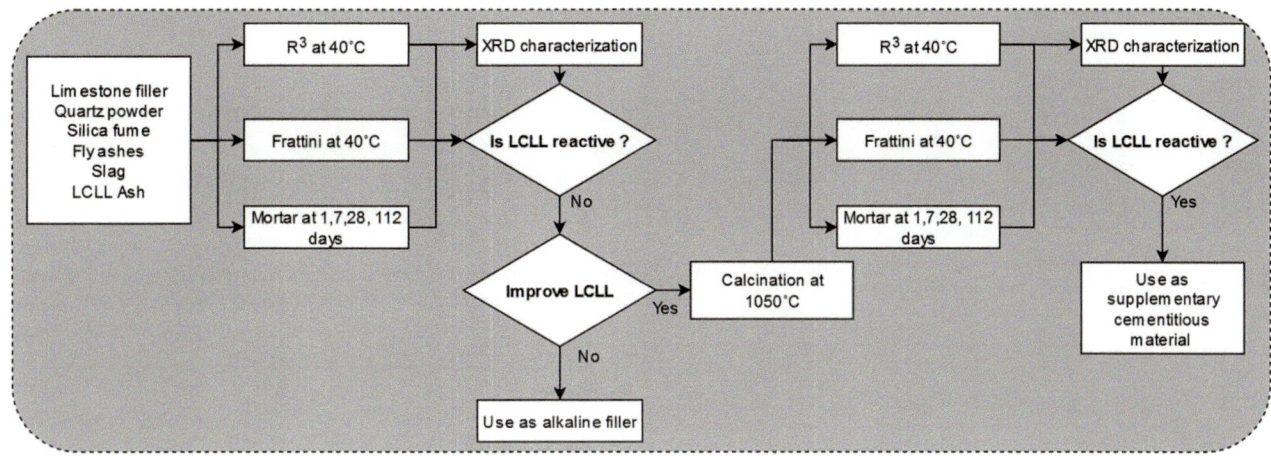

Fig. 3 Schematic of the experimental plan to evaluate the reactivity of LCLL ash [12]

were tested by measuring the compressive strength of mortar containing 20 wt% SCMs.

As shown in Fig. 4, tests on the reactivity of LCLL ash in cement in Frattini and compressive mortar [12] showed similar behaviors to that of fillers such as quartz powder (Q) and limestone filler (FC). For LCLL Ash calcined at 1050 °C, referred to as LCLL-C in Fig. 4, improved reactivity was observed during the Frattini test with higher calcium removal, close to what was observed for fly ash. The same trend was observed on the mortar (Fig. 4) with a relative compressive strength close to that of fly ash (FA) and slag (GGBS) values. This trend was also confirmed by R^3 tests showing higher heat release, higher portlandite

consumption and the presence of new hydrates, identified by XRD, such as monocarboaluminate. The behavior of calcined LCLL Ash is closer to that of supplementary cementitious materials.

LCLL Ash for Ultrahigh Resistance Concrete (UHPC)

This work aimed at developing UHPC mixtures with LCLL Ash powder using their mineral filler effect to improve the competitiveness of UHPC in terms of cost and carbon footprint. More specifically, the effect of partially replacing the cement in UHPC mixtures by LCLL Ash was studied in terms of workability, hydration kinetic, autogenous shrinkage, mechanical properties, and microstructure morphology [13].

Figure 5 shows the compressive strength results and flow table values of the designed UHPC mixtures. The flowability of the mixture is an important parameter enabling the pouring of concrete. LCLL Ash particles reduced the flowability of the mixtures. The spread flow value of samples with 6% LCLL Ash (L6) and 12% LCLL Ash (L12) decreased by 4.3% and 13%, respectively, compared to the reference (L0) without LCLL Ash. This can be explained by the high alumina and silica content of LCLL Ash, which absorbs more water from the total water added, thus reducing the flowability as observed by others [14]. The compressive strength of samples with 6% and 12% of LCLL Ash is comparable to the reference, which reached up to 120 MPa at 28 days of normal curing. This means that 12% of the cement in concrete can be replaced by LCLL Ash with minimal effects on strength and flowability.

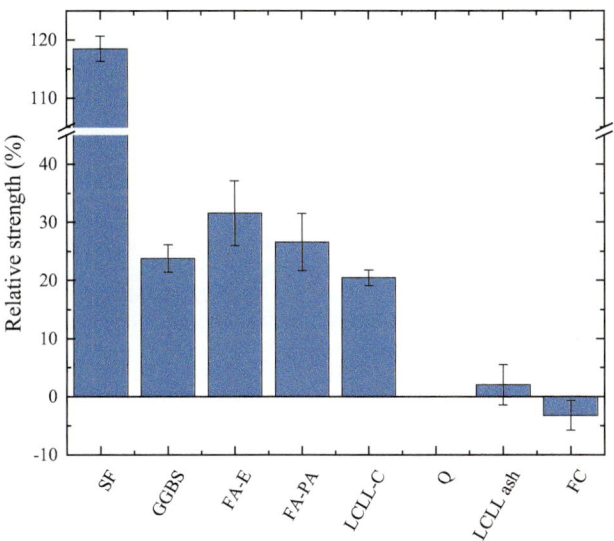

Fig. 4 Relative strength of mortar containing 20 wt% SCMs at 112 days (data normalized to quartz values)

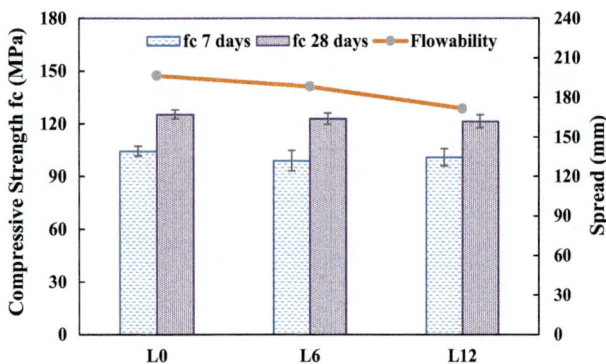

Fig. 5 Compressive strength versus flowability with the standard derivation

Summary

The LCL&L process was developed by Rio Tinto to treat fresh and stored spent potlining generated by the aluminum industry in a sustainable way. It is a proven and robust technology with approximately 800 kt of different types of SPL treated to date. Today, less than 300 kt of SPL remains stored.

Meanwhile, R&D work has been undertaken to develop and implement different ways to valorize LCL&L by-products. As regards the refractory part of SPL (second cut) treated by the LCL&L process, called LCLL Ash, the extensive R&D work carried out over the last five years has enabled great progress to be made in understanding and using this by-product in the cement industry.

For the past two years, LCLL Ash has been used as raw material to produce a manufactured product called Alextra, which is used by Lafarge for clinker production. Regarding the use of LCLL Ash as SCM, R&D studies conducted by ETS have shown that without thermal treatment, the LCLL Ash has low reactivity. However, after further calcination at 1050 °C, LCLL Ash showed a significant improvement of its reactivity in cement, and transforms into a pozzolanic material, similar to fly ash. In addition, due to the high availability of reactive alumina in calcined LCLL Ash, new hydrated phases may precipitate, such as carboaluminate phases. Calcining the material improves its reactivity and makes it potentially a suitable cementitious material to replace Portland cement in concrete. Replacing a portion of the cement with calcined LCLL Ash could reduce the environmental and energetic impacts of blended cement for local use.

For ultrahigh performance concrete, R&D studies conducted by Universite Laval showed that 12% of the cement used in UHPC concrete can be replaced by LCLL Ash without greatly affecting strength and flowability.

Acknowledgements The authors are thankful to NSERC CRD grant program (CRDPJ 515485—17), CRITM consortium, and Ciment Quebec Inc. for their financial support for this project.

References

1. Pawlek R. P. (2012) "Spent potlining: an update", Light Metals, 1313–1317.
2. Holywell G. and Breault R. (2013), "An overview of useful methods to treat, recover, or recycle spent potlining", JOM, 65 (11):1441–1451.
3. Kimmerle F.M. et al. (2001), "SPL treatment by the LCL&L process: Pilot study of two-stage leaching", Light Metals: 199–211.
4. Kasireddy V. et al., Recycling of spent Pot Linings, US Patent no 6,596,252. Jul. 22, 2003.
5. Hamel G. et al. (2009), "From the Low Caustic Leaching and Liming process development to the Jonquiere spent potlining treatment pilot plant start-up, 5 years of process up-scaling, engineering and commissioning", Light Metals:921–925.
6. Hamel G. et al. (2011), "Towards industrial aluminum spent potlining treatment with complete end-product valorization", Light Metals:17–23.
7. Friedlingstein P. et al.(2020), "Global Carbon Budget 2020", Earth Syst. Sci. Data 12(4):3269–3340.
8. Birry L. and Poirier S. (2020), "The LCL&L Process: A Sustainable Solution for the Treatment and Recycling of Spent Pot Lining", Light Metal:1237–1242.
9. Donatello S., Tyrer M., and Cheeseman C.R. (2010), "Comparison of test methods to assess pozzolanic activity", Cem. Concr. Compos. 32(2):121–127.
10. Tironi A. et al. (2013), "Assessment of pozzolanic activity of different calcined clays", Cem. Concr. Compos. 37(1): 319–327.
11. Li X. et al. (2018), "Reactivity tests for supplementary cementitious materials: RILEM TC 267-TRM phase 1", Mater. Struct. Constr. 51(6).
12. Brial V. et al. (2021), "Evaluation of the reactivity of treated spent pot liningfrom primary aluminium production as cementitious materials", Resources, Conservation and Recycling 170:105584.
13. Tran H, et al. (2022) Development of Sustainable Ultra-High Performance Concrete Recycling Aluminum Production Waste. Elsevier Constr Build Mater. (Accepted upon review CONBUILDMAT-D-22–06133)
14. Huang W. et al. (2017), "Effect of replacement of silica fume with calcined clay on the hydration and microstructural development of eco-UHPFRC", Mater. Des. 121: 36–46.

Aluminum Recycling and Recovery of Other Components from Waste Tetra Pak Aseptic Packages

Ilgım Baltacı, Selçuk Kan, Ahmet Turan, and Onuralp Yücel

Abstract

Tetra Pak packages which are multi-layered composite materials (approximately 75% cardboard-cellulose, 20% LDPE and 5% Al by weight) facilitate the distribution of particular food products and aid to preserve some food properties for a long time. In this study, experiments were proceeded to separation of the layers of Tetra Pak aseptic packages efficiently from each other and their recycling. Hydropulping process were carried out for the separation of the celluloses part which is paper for this material; then, hydrometallurgical and pyrometallurgical treatments were followed through for partition of polyethylene and aluminum (PEAl) fractions from each other. In hydrometallurgical pathway, PEAl samples were put in vegetable oil, observed according to increasing temperature, time and solid/liquid ratio parameters, at the end the LDPE and Al phases were purified. The pyrometallurgical studies (pyrolysis) were carried out with various time and temperature combinations, and efficiency of Al recovery from PEAL fraction were investigated.

Keywords

Aluminum • Cellulose • Low density polyethylene • Recycling • Tetra Pak

I. Baltacı · S. Kan · O. Yücel (✉)
Metallurgical and Materials Engineering Department, Faculty of Chemical and Metallurgical Engineering, Istanbul Technical University, 34469 Maslak, Istanbul, Turkey
e-mail: yucel@itu.edu.tr

A. Turan
Materials Science and Nanotechnology Engineering Department, Faculty of Engineering, Yeditepe University, 34755 Atasehir, Istanbul, Turkey

Introduction

Situations such as increase in population, changes in consumption habits, increase in living standards cause an increase in the sales of packaged consumer goods, as well as changes in the composition of solid waste. When the results of this situation are examined in general; packaging wastes including paper-cardboard, plastic, glass, metal, wood, composite packaging, constitute 20% by weight and 50% by volume of this solid waste [1].

Tetra Pak packages which are multi-layered composite materials (approximately 75% cardboard-cellulose, 20% LDPE and 5% Al by weight) facilitate the distribution of particular food products and aid to preserve some food properties for a long time via aseptic production technology. As well, due to its composite structure which is given in Fig. 1, these materials are more complicated to recycle than mono-material packages. When recycling types are examined in detail, primary recycling which includes reprocessing the waste material and using them for their original purpose, clearly is not appropriate for regain of Tetra Pak aseptic packages. Reuse of a material after the recycling process in different domains that have no requirements of the original raw material properties, is secondary recycling. This method is commonly used for separated cardboard material in paper products. Converting the waste material into new various products by breakdown into its chemical components is tertiary recycling. The last sort is quaternary recycling meaning that energy recovery by burning waste product; while this method is the most unwelcomed recycling, regrettably it is a prevalent method in Turkey [2, 3].

In most cases, Tetra Pak waste recycling is only possible with energy recovery or to process wastes into low quality products, if hydropulping is not an option. Here, what is meant by energy recovery could be incineration, gasification or pyrolysis and sometimes it includes municipal solid waste mixture [4].

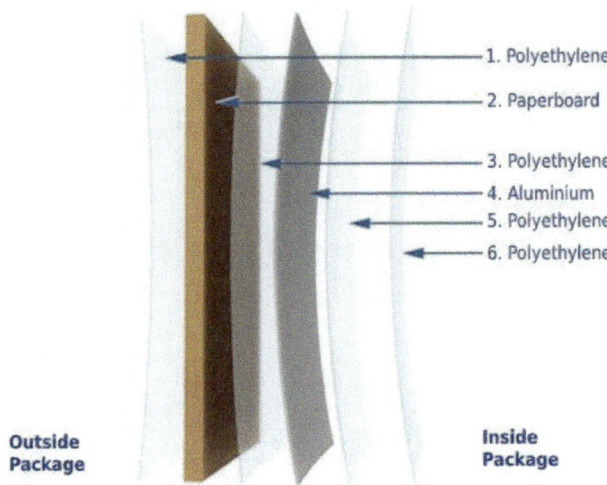

Fig. 1 Tetra Pak packaging layers [3]

The hydropulping process is commonly used to separate the cardboard layer from the low-density polyethylene and aluminum part. When working with smaller amounts to separate the cardboard, blending is also used instead of pulping, as it reduces the time to be spared for cleaning. With the help of the centrifugal power provided by the device after a certain time and temperature, the cardboard part is separated from the PEAl part. Composite PEAL layers, which are generally produced from aseptic cardboard wastes, are currently utilized as roofing material [5].

Following a hydrometallurgical method the solvent creates inflammation by penetrating through the polyethylene molecules and forms a kind of gel as a result. So, polymer chain macromolecules of polyethylene begin to separate and solubilize in the solvent. Plastic waste pyrolysis is the tertiary or raw material recycling route. It is the way to recycle plastic waste materials into basic petrochemicals which can be used as raw materials to make pure plastic or refined fuels [6]. Pyrolysis is advantageous as it is simple, versatile and have low capital investment [7].

In this study, waste Tetra Pak aseptic package materials were collected and, cellulose content was extracted out via hydropulping and wet sieving, then obtained mixed LDPE and aluminum fractions were separated by using hydrometallurgical (plant-based sunflower oil) and pyrometallurgical (pyrolysis) methods. Polyethylene (LDPE) and aluminum fractions were analysed by FTIR and TGA.

Experimental

In this study, experiments were conducted to separate the layers of Tetra Pak aseptic packages efficiently from each other for recycling. Hydropulping process were carried out for the separation of the cellulose part which is paper; then,

hydrometallurgical and pyrometallurgical treatments were followed through for separation of polyethylene and aluminum, which are PEAl fractions, from each other. The hydropulping process was applied in two stages; at first, the separation process was applied on the Tetra Pak aseptic packages. The packages without lids were cleaned of food wastes, cut into small pieces and kept in water, afterwards, pieces were chopped via blender with water, the mass was sieved under water with three-layer sieve system. The sieve system having 3.35, 2, 0.18 mm sieve openings. At the end, segregated fractions on sieves were collected and dried in oven. The second part of the hydropulping process was for PEAl fraction which was collected from the sieves. The fraction was chopped and sieved under water again (with 0.8, 0.18, 0.05 mm sieve openings) in order to decrease the size of PEAl pieces and extinguish all of cellulose; then, the fractions were dried again. Experimental studies were pursued in hydrometallurgical and pyrometallurgical routes to separate LDPE and Al from each other (Fig. 2).

In hydrometallurgical pathway, PEAl samples were put in vegetable oil, appropriate to TS 886 (the edible sunflower oil standard of Turkey) and observed according to increasing temperature, time and solid/liquid ratio parameters. At the end, the LDPE and Al phases were depurated. In leaching experiments, leaching duration of 1 h and 140 °C temperature, 100 rpm stirring rate were determined to be optimum parameters for sun-flower oil having a solid/liquid ratio of 1/4.5, under these circumstances three times leaching were applied, following each other, in conjunction with a cleaning step at the end (Fig. 3).

The pyrometallurgical studies, were carried out under air atmosphere with various combinations of time (2, 3, 4 h) and temperature (400, 500, 600 °C), and efficiency of Al recovery from PEAL fraction were investigated. 1 g of PEAl samples were placed in the furnace at room temperature and then heated. When the desired temperature was reached, the samples were kept for the specified time durations.

At the end of the experiments, the separated fractions which are end products of both methodologies were characterized by means of XRF, XRD and FT-IR methods their thermogravimetric behaviors were investigated by TGA, DTA and, the recovery efficiencies were calculated.

Results and Discussion

In the present study, aluminum recycling routes and recovery conditions were investigated. Furthermore, recovery of the low-density polyethylene and paper (cellulose) fractions, which are present in waste Tetra Pak aseptic package materials were also examined. At the end of the hydropulping, cellulose and PEAl (polyethylene-aluminum) fractions were successfully separated from each other. As a

Fig. 2 Hydropulping process

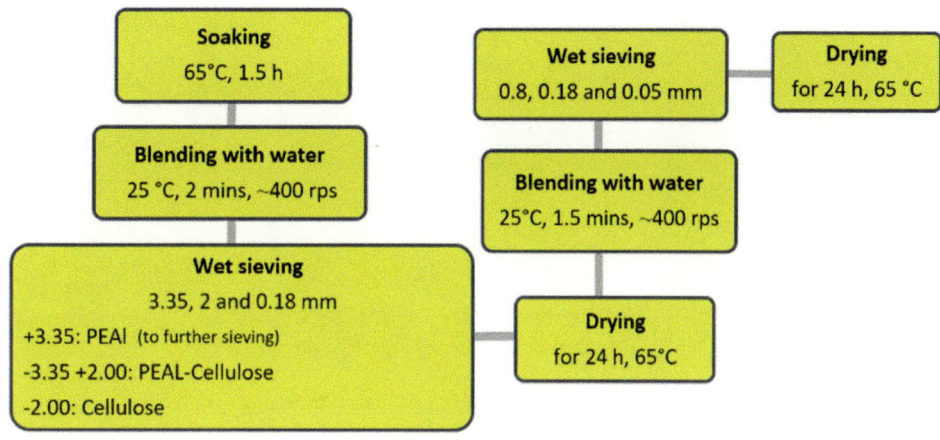

Fig. 3 LDPE Al separation process through hydrometallurgical route

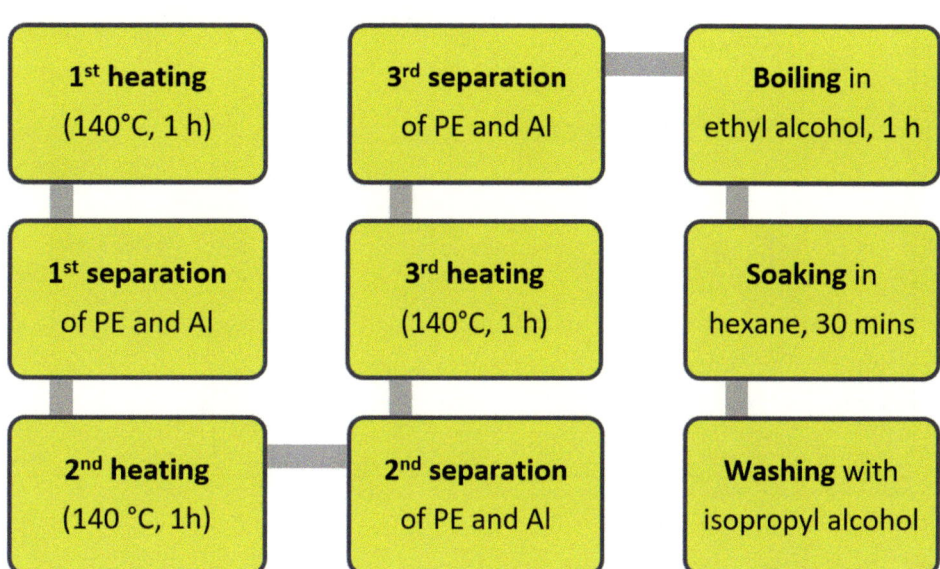

result of hydrometallurgical experiments, it was determined that in the Al phase there is high amount of LDPE remaining while LDPE fraction is almost clean (Figs. 4, 5 and 6).

In thermogravimetric analysis, TGA was applied to understand weight reduction of fractions with removal of polyethylene; DTA shows the energy (heat) amount that sample take to evaporate. At the end, nitrogen atmosphere was used for the characterization of the polymeric fraction. According to the Fig. 7, there were two endothermic peaks on the red line, the first is 100 °C where melting was seen, the second was 400 °C which was the evaporation point.

During the analysis of Al fraction, it was seen in the Fig. 8 that again, there were two endothermic peaks on the red line, the first is 100 °C at which melting was seen and the second one was 400 °C that was the evaporation point.

Also, there were no peaks for plant-based oil in TGA due to the final cleaning of the fractions with hexane and isopropyl alcohol.

The results of pyrometallurgical experiments were given in Table 1. It has been observed that 400 °C is not a sufficient temperature to remove PE. At the end of 2 h at both 500 and 600 °C, it was observed that the weight loss was around 80%. These results agree with the presence of 5% wt. Al in the Tetra Pak package.

The products that were obtained after 2, 3 h at 500 °C and 2 h at 600 °C were analyzed with FT-IR to detect remaining PE but there were no peaks detected. The result is given in Fig. 9. After that, products were characterized with XRD to determine the occurred phase and results were given in Fig. 10. It is seen that all 3 samples are Al and no oxide

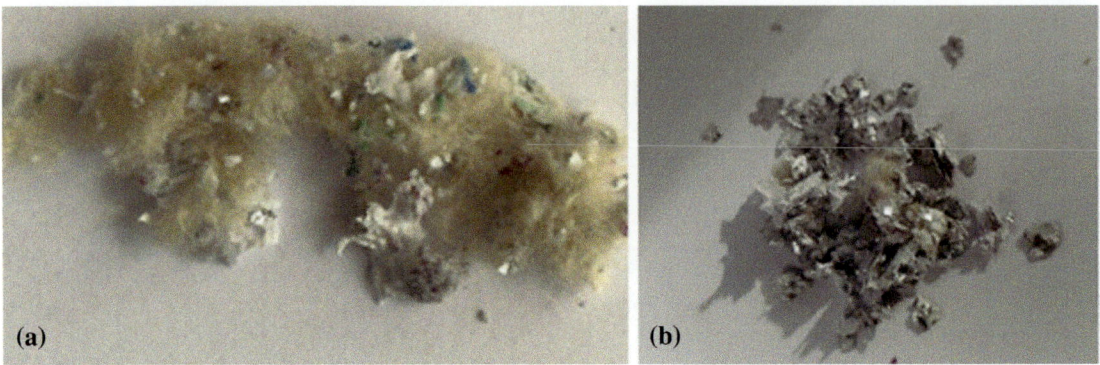

Fig. 4 **a** Polymer fraction and **b** Aluminum fraction of PEAl after oil cleaning with hexane and isopropyl alcohol

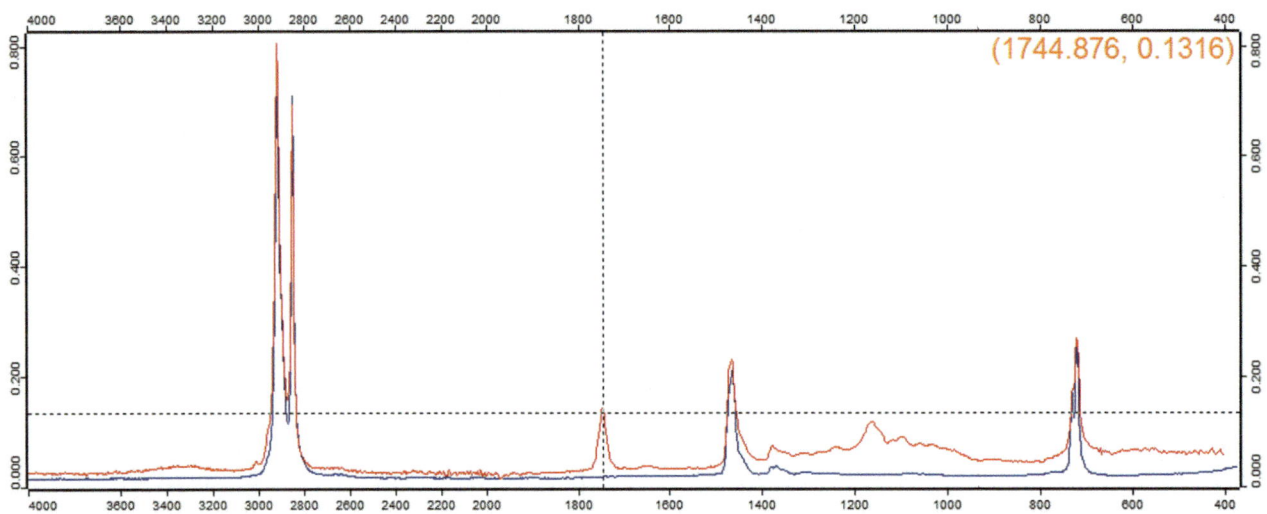

Color	Hit Quality	Compound name	CAS Number	Molecular formula	Molecular weight
	842	Low-Density Polyethylene (PE-LD)			

Fig. 5 PE-FTIR analysis result after oil cleaning with hexane and isopropyl alcohol

phase is observed. These results show that PE was evaporated, in Fig. 11 the aluminum fraction after 3 h under 500 °C pyrolysis is given.

Conclusion

Tetra Pak aseptic wastes were separated from the cellulose part first with the hydropulping process efficiently. As a second step, in hydrometallurgical route PEAl part of the waste was put in a vegetable oil to separate PE and Al fractions at 140 °C, 100 rpm stirring rate and sun-flower oil having a solid/liquid ratio of 1/4.5. At the end of this route, it was seen that in the Al phase there is high amount of LDPE remaining. PEAl samples were pyrolyzed for 2, 3 and 4 h at 400, 500 and 600 °C. PE-free Al was produced successfully by pyrometallurgical method. The optimum conditions were chosen as 2 h at 500 °C. Evaporated PE could be collected by condensation unit as paraffin. Further study is suggested because this paraffin could be a raw material for producing polymers.

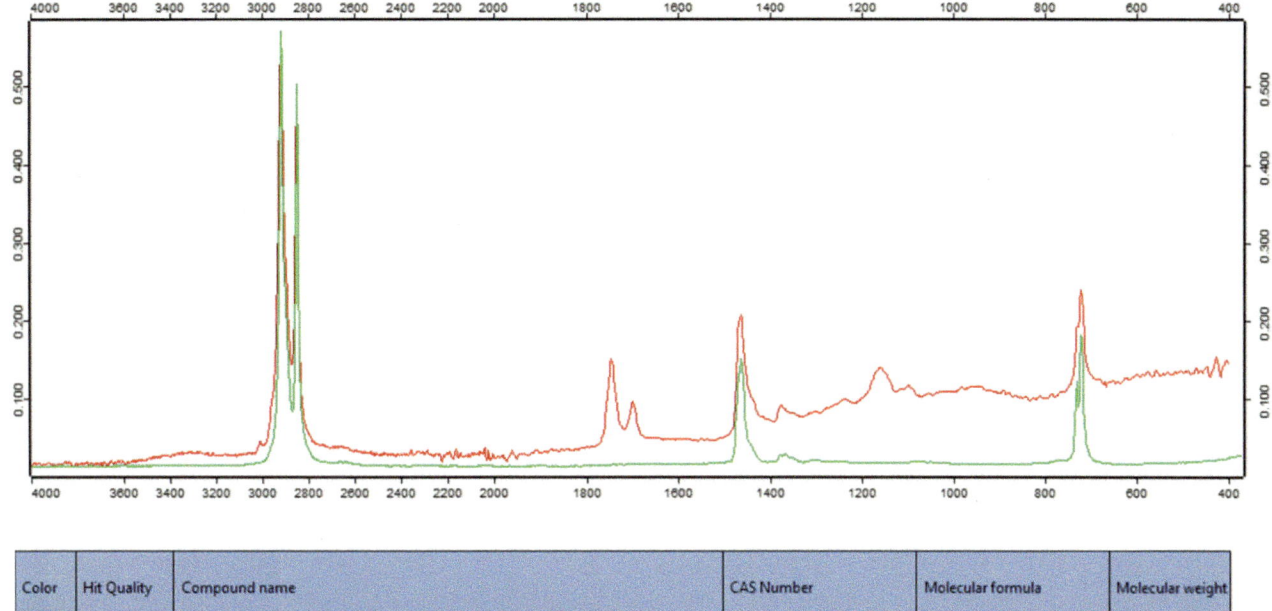

Color	Hit Quality	Compound name	CAS Number	Molecular formula	Molecular weight
(green)	753	Low-Density Polyethylene (PE-LD)			

Fig. 6 Al-FTIR analysis result after oil cleaning with hexane and isopropyl alcohol

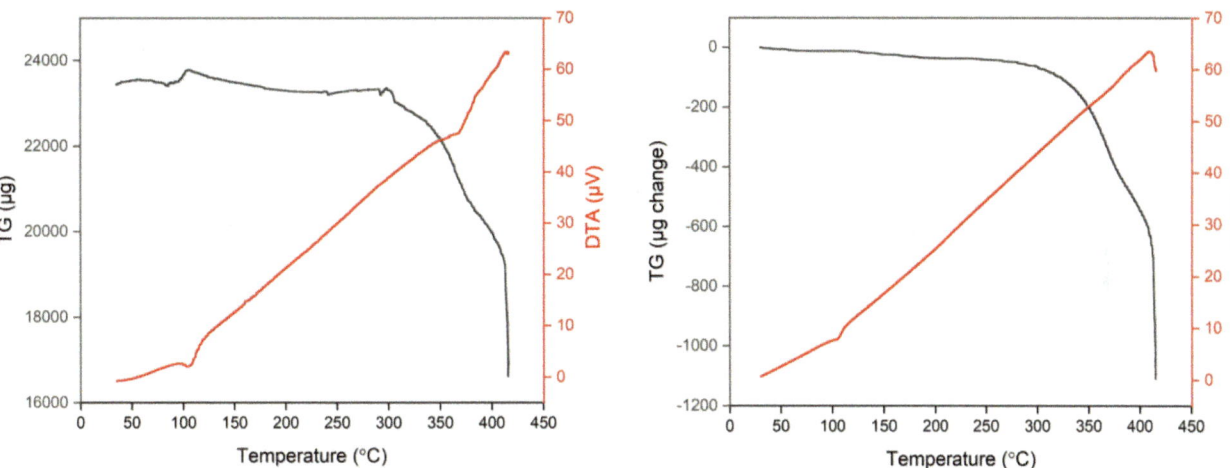

Fig. 7 Thermogravimetric analysis of LDPE

Fig. 8 Thermogravimetric analysis of Al

Table 1 The pyrometallurgical experiments results

Time duration (h)	Weight loss (%)		
	600 °C	500 °C	400 °C
2	82.32	82.31	47.83
3	81.69	81.97	61.58
4	83.60	82.15	71.73

Acknowledgements This study is funded by the Scientific Research Projects Coordination Unit (BAP) at Istanbul Technical University (Project No: 42947).

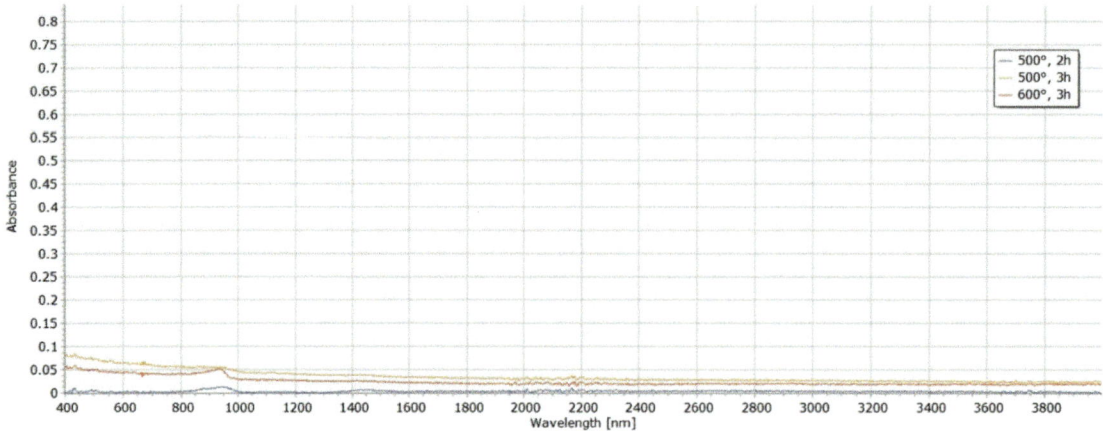

Fig. 9 FT-IR analyze results of the sample obtained after 3 h pyrolysis at 500 °C

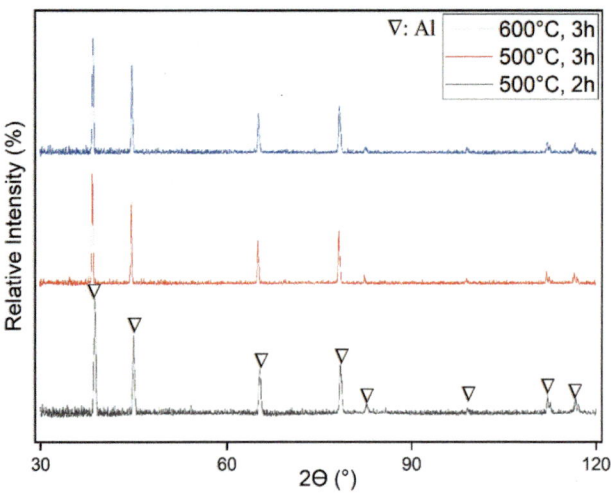

Fig. 10 XRD results of pyrolized samples

Fig. 11 The aluminum fraction after 3 h pyrolysis at 500 °C

References

1. Ş. ve İ. D. B. T.C. Çevre, "Ambalaj Atıkları," 2018. https://cevreselgostergeler.csb.gov.tr/ambalaj-atiklari-i-85757 (accessed Aug. 16, 2022).

2. Robertson, G.L. Recycling of Aseptic Beverage Cartons: A Review. Recycling 2021, 6, 20. https://doi.org/10.3390/recycling6010020.

3. I. Georgiopoulou, G. D. Pappa, S. N. Vouyiouka, and K. Magoulas, "Recycling of post-consumer multilayer Tetra Pak® packaging with the Selective Dissolution-Precipitation process," *Resources, Conservation and Recycling*, vol. 165, Feb. 2021, doi: https://doi.org/10.1016/j.resconrec.2020.105268.

4. Jan Zawadiak, Szymon Wojciechowski, Tomasz Piotrowski, Alicja Krypa. Tetra Pak Recycling – Current Trends and New Developments. *American Journal of Chemical Engineering*. Vol. 5, No. 3, 2017, pp. 37–42. https://doi.org/10.11648/j.ajche.20170503.12.

5. Türkiye Eğitim ve Çevre Vakfı, "Ambalajın Tanımı," May 07, 2022. https://www.tukcev.org.tr/ambalajin-tanimi (accessed Aug. 16, 2022).

6. P. T. Williams and E. A. Williams, "Fluidised bed pyrolysis of low density polyethylene to produce petrochemical feedstock," 1999.

7. M. J. Muñoz-Batista, G. Blázquez, J. F. Franco, M. Calero, and M. A. Martín-Lara, "Recovery, separation and production of fuel, plastic and aluminum from the Tetra PAK waste to hydrothermal and pyrolysis processes," Waste Management, vol. 137, pp. 179–189, Jan. 2022, https://doi.org/10.1016/j.wasman.2021.11.007.

Electromagnetic Priming of Filtration Systems: Pyrotek EM-DF

Robert Fritzsch, Joseph Whitworth, Paul Bosworth, and Jason Midgley

Abstract

The Pyrotek MCR Group has developed a unit, the Electromagnetic Deep Filtration (EM-DF) system, that successfully primes multiple Ceramic Foam Filters (CFF) for filtration in casthouse and foundry operation using patented electromagnetic technology. Regular gravity-primed filter boxes are restricted by metal head for a certain filter grade and the number of filters used; most use a single filter and require a substantial metal head. The EM-DF system enables multiple filters with identical or dissimilar grades to be used at standard launder metal levels. Consistent priming of all filters has also been achieved, expecting to deliver greater and more consistent filtration efficiency, making the filtration process more reliable. This communication presents the results from conducted industrial trials using triple stack 50 grade Pyrotek Sivex CFFs with the size of 24.5″, 23.7″ and 23″. The system's performance, productivity, safety, and reliability are discussed.

Keywords

Aluminium • Filtration • Automation • Ceramic foam filters (CFFs) • State-of-the-art • Engineering solutions

R. Fritzsch (✉) · J. Whitworth (✉) · P. Bosworth · J. Midgley
Pyrotek MCR, EMP Technologies Ltd., Faraday House Eastern Avenue, Stretton, Burton on Trent, DE13 0BB, United Kingdom
e-mail: robfri@pyrotek.com

J. Whitworth
e-mail: joswhi@pyrotek.com

R. Fritzsch
Department of Materials Science and Engineering, Norwegian University of Science and Engineering (NTNU), Alfred Getz vei 2, 7034 Trondheim, Norway

Introduction

Production of high-quality aluminium requires multiple metal treatment stations prior to the casting. After adjusting the alloy within the furnace, the melt can be treated with grain refiner, usually directly before entering a degassing stage, which treats trace elements and dissolved entrainments such as hydrogen. After that, the metal enters the physical filtration stage. The most applied metal filtration system for bulk metal production uses Ceramic Foam Filters (CFFs) [1], offering great flexibility for changing filter grades and alloy compositions with every cast while delivering high filtration efficiencies [2–4]. To enable a filter to "perform", it needs to be "primed", which relates to the exchange of air with liquid aluminium within all the filter volume. Traditional priming methods rely on the use of a static metal head above the filter and it is a known challenge to consistently prime higher grade CFFs using traditional methods.

When relying on gravity as the method of priming, the static metal head can result in inconsistent priming efficiencies since the metal will tend to flow via the paths of least resistance, resulting in a reduction of the filter volume used. The variability of filter volume, caused by the potential inconsistent priming under gravity-priming conditions, leads to localised higher metal velocity, and can alter the filtration efficiency of the used filter grade. This results in a larger filtration efficiency deviation of the CFF grades, often visualised in plots showing higher and lower limits of filtration efficiencies of different filters [5].

The development of the presented technology was in close collaboration with the University of Science and Technology, NTNU, in Norway and has been published [6–11]. After an extensive development phase, the final prototype has now successfully been used repeatedly in an industrial environment, providing key data, benchmarking, and validation information. The result is a market-ready

© The Minerals, Metals & Materials Society 2023
S. Broek (ed.), *Light Metals 2023*, The Minerals, Metals & Materials Series,
https://doi.org/10.1007/978-3-031-22532-1_116

product, called the Electro-Magnetic Deep Filtration system, hereafter known as the Pyrotek EM-DF system.

The industrial unit was designed by Pyrotek MCR engineers to withstand the harsh environment of the cast house or foundry, while reducing human interaction with the filtration process before, during, and after the cast. Health and safety risk reduction and ease-of-use have been prioritised throughout the design process. The experience from our technology experts and design engineers within Pyrotek was used to design this novel system utilizing existing and trusted filtration technology delivered by Pyrotek Sivex with the addition of a novel patented priming method. The patent has been developed and filed by the Norwegian University of Science and Technology (NTNU) and Pyrotek MCR and Sivex have further developed the concept into a reliable industrial technology [12]. The ready-for-market EM-DF unit is shown in Fig. 1.

Ceramic Foam Filters (CFFs) have proven high filtration efficiencies and reliability [4]. Deviation of filtration performance depends mainly on the successful priming during the initial priming stage, the used alloy and temperature, the inclusion load and type in the melt and the metal velocity inside the filter during the cast. The smaller the primed area, the larger the velocity in the remaining filter volume and hence the lower the filtration efficiency. The higher the filter grade, the more challenging it is to successfully prime the CFF. This relates to the tortuous path, defined as the tortuosity, inside the filter, which the metal must follow, ranging from 1.3 for a 30 grade CFF to 3.2 for a 80 grade CFF [13]. This is defining the average path length of the metal inside the filter as the product of the tortuosity and the filter thickness.

The inclusion load of the melt can also alter the filtration performance of the filter. When the melt was highly contaminated and too fine, a filter was used, the filter "blinds", meaning a cake on top of the filter forms, reducing the available area for metal passing through the filter, increasing the velocity of the metal. This results in a head difference on the inlet side to the outlet side of the filter, indicating the pressure head build of the clogging filter. Finally, the filter can fully blind, meaning there is no more metal passing through. If the melt is very clean and a coarse filter is used, the few inclusions are more likely to find a path through the filter. It is therefore highly important to use the right type of filters for the right quality of metal and alloy, to deliver a high melt quality at the casting station.

The key features of the EM-DF system are:

1. The reliable priming and draining of the filters.
2. The improved filtration performance.
3. The repeatability.
4. Less labour for the cast line crew, reducing the human interaction, bias, and error potential.
5. The health and safety improvements.
6. Ability to prime multiple CFFs of different or similar grades and/or a single CFF of a higher grade.
7. Reduced metal head height for priming.
8. Lower running costs and greater flexibility.
9. Ability to provide full automation.
10. Less risk of failed casts due to floating CFFs.
11. Easy to operate and low maintenance.
12. Similar footprint to a standard Sivex CFF filter box.

This document describes the advantages of the EM-DF system and experiences gathered with it on site. The example is the application of a triple stack of 50 grade Non-Phosphorous (NP) Pyrotek Sivex CFF. The stacked filters have a different size, where the top filter is a 24.5″, the

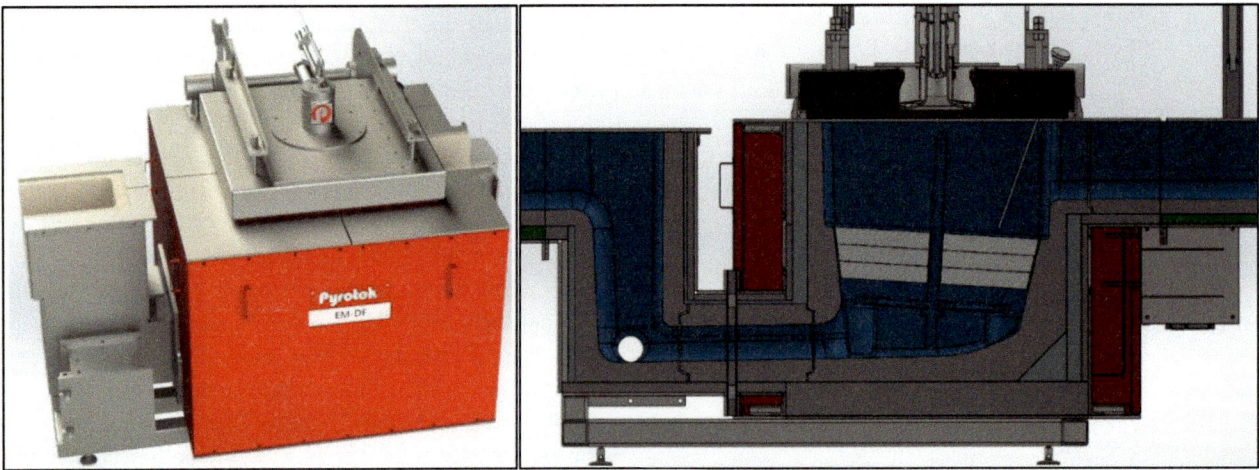

Fig. 1 EM-DF system with closed lid, showing the flat flame burner (interchangeable with an electric heater), the inlet side to the right and the "swan neck" outlet. The draining port is on the backside of the swan neck

middle a 23.7″ and the lowest is a 23″ filter and are positioned with the patented filter handling tool [14]. The EM-DF system can hold any combination of these three filters, as single, double, and triple filter stacks of varying ppi/grade. A typical three-filter stack, as inserted into the EM-DF filter box, is shown in Fig. 2 on the left and a filter mounting and removal station is shown on the right.

A filter calls fully primed when all the volume of the filter has been made available for metal flow during the filtration stage. The high surface tension of aluminium and the aluminium oxide layer are key factors to overcome for achieving fully primed filters. The EM-DF system supports the filter priming by the application of the EM field, by inducing velocity, homogenizing the melt temperature, generating a dynamic pressure, and causing an oscillating, electromotive force, reducing the surface tension and stability of the oxide layer. This together is ensuring full utilization of the filter volume, *e.g.,* full priming of the CFFs.

From empirical experience and earlier research, it is known that there is a metal velocity band the filters should be operated within, depending on the available metal quality, the aimed cleanliness and used production process [5]. This is shown in Table 1, *e.g.,* giving a peak velocity of 10 mm/s for continuous casting and recommending a 50 grade CFF.

Industrial filter segments of regular 50 grade CFF, after being removed from the filter box, are shown as an example in Fig. 3. Figure 3a shows a well-primed and drained 50 grade CF filter. Figure 3b shows a well-primed and partially drained 50 grade CFF, where residual metal close to the filter seat is visible and more of the metal remains inside the filter. And Fig. 3c shows a partially primed filter, where the white areas are the CFF web that has not been in contact with metal.

Experimental

An industrial remelting furnace has been used and all samples have been taken during regular production. The produced metal was with 6XXX and 7XXX series alloy, grain refined prior to the degassing unit. The EM-DF unit was installed directly after the degassing station before the casting pit and has been equipped with a stack of three 50 grade Sivex NP filters for each cast, where the dimensions are given with a square shape and width of 623, 604, and 586 mm (24.5, 23.8 and 23 inch). The casting velocity was similar for each cast and reached from approximately 9 mm/s at the smallest filter (bottom) to 7.9 mm/s at the largest filter (top) with a total casting length of 105 min on average.

The EM-DF filter box has been preheated similarly to a regular filter box via a flat flame burner with 80 kW thermal effect. Electrical heating solutions have been considered, but were not tested in this installation. The box has been preheated following the refractory recommendations and after reaching the thermal soaking temperature, the box was set to 750 °C using the burner control thermocouple. The lid was only opened when the filters were loaded into the box by a crane. The tool, the hook, and the filter seat angle of the filters align when the filters are at the right position. The shaft and base of the tool are further lowered, sliding to the stable resting position at the base of the filter box.

The hook was removed from the shaft after positioning the filters and the tool, which is the only manual labour required during placing the filters. The lid was closed, soaking the filters with residual heat of the box and reducing the energy losses and thermal stress on the refractory/filters. During preheating and soaking of the filters, the flat flame

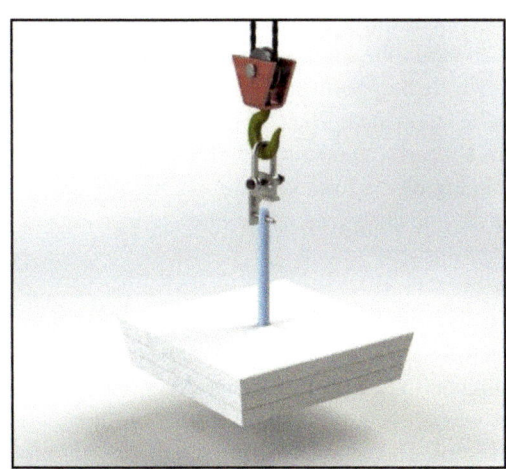
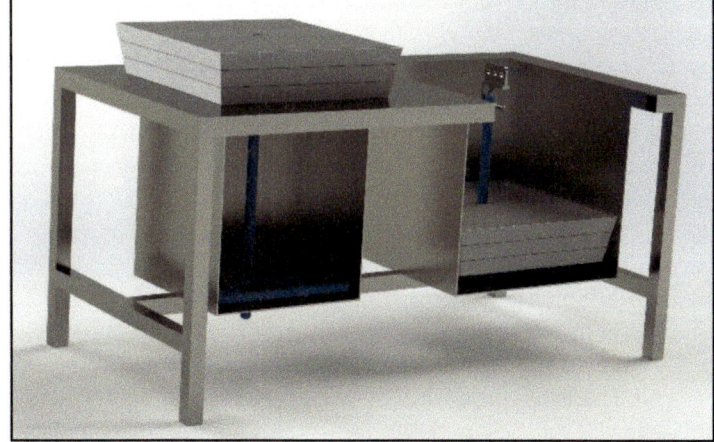

Fig. 2 Patented filter tool [1], as ready for loading a stack of three CF filters into the filter box. The filter tool removal hook is attached to the shaft after the filters are placed on the filter handling tool. A mounting station, shown as an example in (**b**) helps to separate the tool from the used filters (left) while a stack of new filters can be prepared on the right

Table 1 Summary of metal velocity and flow rate for selected casting applications [5]

Type of Casting	Metal Velocity (mm/sec)	Metal Flow (kg/sec/m²)	Metal Flow (lb/min/inch²)	Typical Pore Range
Billet	10 - 19	19 - 36	1.6 – 3.1	Grade 30-40
Slab	6 - 15	17 - 29	1.5 – 2.5	Grade 40-65
Continuous Casting	3 - 10	6 - 19	0.5 – 1.6	Grade 20-50

Fig. 3 Examples of standard industrial single filters primed in a normal filter box by gravity, showing **a** a well primed and drained 50 grade CFF, **b** a well-primed and partially 50 grade CFF and **c** a partially primed and partially drained 50 grade CFF

burner was used on low fire, keeping the temperature above the filter surface at ~750 °C.

The filter handling tool was specifically designed for the insertion and removal of the filter stack into the EM-DF unit. The cast iron for the tool has a bespoke alloy composition and can withstand 800 °C without undesired changes or significant diffusion in the microstructure. The tool has a two-layer protective coating system; the first is a durable hard coat, *ZYP Boron Nitride Hardcoat®*, giving a core protective layer against Fe diffusion into the Al and wear. The second layer is a sacrificial non-wetting material, ZYP *Boron Nitride Lubricoat®*-ZV, which is reapplied after every cast. The filters are sealed by standard alumina wool gaskets, while an expanding gasket is used around the filter tool shaft in the centre.

The *preheating time* of the stack of three 50 grade Sivex NP CFF's has been widely tested and a minimal time of *35* min was sufficient to reduce the danger of damage due to thermal shock to the refractory of the filter box, but also to maintain the integrity of the filters, and getting the filter tool

to temperature, avoiding risk of crystallisation at the tool interface when in contact with liquid aluminium.

The priming was driven by the EM field, generated by the coils, energized when the box is flooded with aluminium and has been subject of multiple publications [6–8, 10–12]. The dam was pulled when the launder reached approx. 150 mm metal depth, which has been done manually during these trials, but can be automated. The coil has been energized for a minimum of 90 and a maximum of 300 s (1.5 to 5 min), giving full priming of the filter volume. During the casting, the coil must stay de-energized to avoid disturbance generated by the coil during filtration.

When a cast was finished, the preparation for the next cast started by draining the EM-DF box. While the metal level sinks, the coil was energized for some trials and for some not, giving different filter draining effects. After draining, the lid of the box was opened, the filter handling tool hook was attached to the tool. The filters were lifted and moved to the disassembly station. The box has been inspected visually after every cast and minimal maintenance has been required during the campaign. The lid was closed, the inlet of the box was sealed with a baffle plate and alumina wool, and the burner was kept on low fire to maintain the 750 °C refractory bulk temperature. Finally, the dam has been placed in the launder and the EM-DF system was ready for a next cast.

Results and Conclusions

The field trials of the EM-DF system were conducted without any major issues and multiple positive experiences have been gained. The main conclusions are as follows:

- The EM-DF demonstrated reliable priming and draining of the triple stacked filters.
- The EM-DF primed reliably all types of CFF with a lower metal head.
- The EM-DF performed repeatedly in a production environment, while requiring less labour and heat exposure for the cast line crew.
- Significantly enhancing the filter charging and removal process and the overall control of the filtration stage.
- Greatly reduced health and safety risks.
- The EM-DF system demonstrated ease of operation, as testified by the workers.

In Fig. 4, the schematic for the different filter layers for the filters as shown in Fig. 5 is displayed. The pictures in Fig. 5 show two triple stacks of 50 grade Pyrotek Sivex NP CFFs, primed in the EM-DF system. In Fig. 5a the system has been primed by using the EM system, but not been drained using the EM. In Fig. 5b the system has been primed and drained with the EM system. For both trials shown in Fig. 5, the EM system has been energized for 90 s during priming and for Fig. 5b the coil has been applied during the draining stage until no metal flow was visible from the draining hole. The filters have been retrieved from the filter box in a single piece by using the filter lifting tool and a crane. The filters drained solely by gravity had some metal

spill during the lifting and transportation to the filter handling station, while the EM-drained filters did not show significant metal dripping.

The inspected filters were all fully primed. This is the key result and conclusion of the EM-DF system, showing its performance as a priming system for filtration solutions. The white section in the centre of Fig. 5a and b, on line 2 and 5 are showing the hole for the filter tool, sealed by the expendable gasket. The white colour indicates that the seal did perform to its expectations.

After draining the filters, a difference is visible, where in Fig. 5a, the filters are drained by gravity alone and (b) the filters have been drained by the support of the EM field from the EM-DF system. The variation of residual metal in the filters was significant. The original weight of the unused triple stack of filters is approx. 27 kg. The gravity only drained filters had an average final weight of 86 kg, whereas the EM-drained filters came to 56 kg. This results in a residual metal of 59 and 29 kg in the filters, depending on the draining method. Theoretically, a stack of 3 filters when filled with Al metal would be 155 kg, resulting in draining figures of 44.5% when gravity drained and 63.9% when drained with the EM field. The data is summarized in Table 2.

The data in Table 2 indicates that the amount of metal that needs to be recovered from the used triple stack of filters can be reduced by ~50%, from 59 to 29 kg. Looking at 3 casts a day, this results already in a difference of approx. 2,500 kg Al per month extra saving, when considering the use of a stack of 3×50 grade CFF's being used for every cast.

The repeatability of the EM-DF system was emphasized as a positive conclusion of the trials, together with the use of

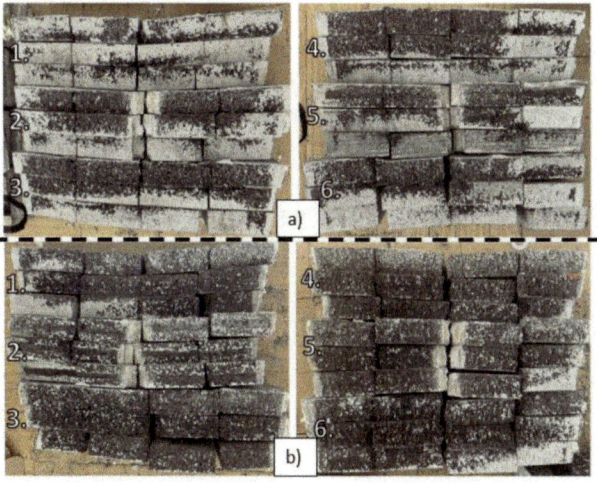

Fig. 5 Example of two filter stacks primed with the EM-DF system after a cast of 45-ton 6xxx Al. The filters in **a** have been drained by gravity only, while the filters in **b** have experienced draining supported by the EM field of the EM-DF system

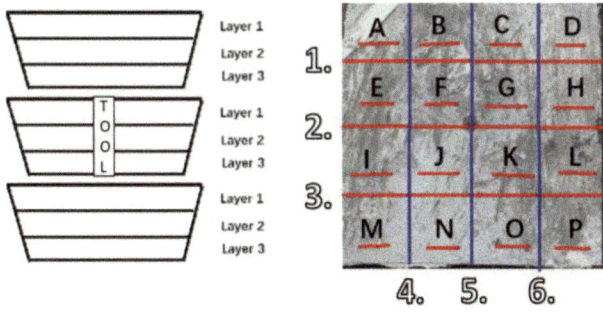

Fig. 4 Stacked filter cut overview and labeling to identify the different layers and the centre of the used filters

Table 2 Summary of the weight information of the triple stack of filters, giving the absolute possible metal weight in the filter, the total weight of all three filters, the drained filter weight with the removal efficiency and the gravity drained weight with the removal efficiency

Abs. total metal weight [kg]	Abs. total filter weight [kg]	Drained filter weight [kg]	Draining efficiency [%]	Gravity only drained weight [kg]	Draining efficiency [%]
127.9	27.1	56	63.9	86	44.5

the filter handling tool for positioning and removing the filters from the box. For a regular CF filter box, the manual placing of the filters is counted as a challenging task, where a small mistake could lead to a bad seal, resulting in metal bypassing the filter during a cast, and is also related to direct exposure of workers to the hot refractory, resulting in worker strain and possible injuries.

Health and safety wise, it is similar when removing filters from a regular CF filter box. With the EM-DF system, the only interaction with hot surfaces the workers will be exposed to, is when attaching the hook to the filter handling tool. There is no need to punch a hole into the used filters, no screw needs to be drilled into the filter face, or any other method of preparing the used filters for removal.

Cooling down the refractory below solidification temperature of aluminium was not required, as the filters are drained by the EM field and lifted when hot by the tool, leaving little to no residual metal, saving a significant amount of energy, and increasing the refractory life.

Another advantage comes from the removal of the filters in one piece, as there are no filter pieces and debris in the filter box, needing cleaning prior the next cast. There was only a small residual of liquid aluminium, which did not drain through the drain hole. Therefore the wear on the filter box refractory is significantly lower, as it sees less thermal cycling than a regular filter box.

Future Work

The EM-DF system has proven itself regarding reliability of priming, its repeatability, the user friendliness, and robustness. This was achieved despite, due to the Covid restrictions over the last years, little to no access to the plant during the installation and the trial/validation phase of the systems.

It had been hoped to validate the Filtration Efficiency of the EM-DF system using Pyrotek's two LiMCA III units as a primary measurement tool during these trials. This was not possible due to unfavorable conditions experienced on the casting line. Only a limited number of PoDFA samples were taken upstream and downstream of the EM-DF system during these trials which did not enable us to generate statistically significant data on the performance of the system.

The next campaign aimed for is the validation of the cleanliness performance of a double and triple 50 grade CFF

by LiMCA units. As a follow-up from that, using different grade CFF's, *e.g.*, 30–50–65, or other combinations, for fine tuning the performance depending on the expected incoming melt quality.

The 30 grade CFF could be used as initial filter face, giving space to distribute the build-up potential of a filter cake and catching larger films and inclusions, without clogging or blinding the finer filters below. Then an intermediate 40–50 grade filter and at the base a fine grade CFF such as a 50–65 or potentially even an 80 grade CFF as the final filter, giving a high probabilistic chance to retain the finest inclusions, without significantly increasing the pressure drop of the filters during the cast length due to build-up and blinding.

Acknowledgements The authors also wish to express their gratitude for the support and conclusive discussions by the Pyrotek MSG group and especially to Dr. Neil Keegan. A special thanks go to all Pyrotek engineers who have contributed to this development through all stages until the final pilot building, the Pyrotek engineers in China, who have given superior support to the pilot trial phase during the difficult times of the Covid pandemic and finally to the team from Pyrotek Sivex and especially Marcin Smorawiński for the extensive experience and knowledge with ceramic foam filters and how to make the EM-DF system a high performing filtration solution.

References

1. K. R. Butcher and D. B. Rogers, "Update on the Filtration of Aluminum Alloys with Fine Pore Ceramic Foam", pp. 797–803, 1990.
2. N. Keegan, W. Schneider, and H. Krug, "Evaluation of the Efficiency of Fine Pore Ceramic Foam Filters", *Light Metals*, pp. 1031–1041, 1999.
3. N. J. Keegan and S. F. Ray, "An evaluation of Industrial Filtration Systems", ed: Alcastek, 2002.
4. H. Duval, C. Rivière, É. Laé, P. Le Brun, and J. Guillot, "Pilot-Scale Investigation of Liquid Aluminum Filtration through Ceramic Foam Filters: Comparison between Coulter Counter Measurements and Metallographic Analysis of Spent Filters", *Metall and Materi Trans B*, vol. 40, no. 2, pp. 233–246, 2009.
5. S. Ray, B. Milligan, and N. Keegan, "Measurement of Filtration Performance, Filtration Theory and Practical Applications of Ceramic Foam Filters", 2005, pp. 1–12.
6. R. Fritzsch, M. W. Kennedy, J. A. Bakken, and R. E. Aune, "Electromagnetic Priming of Ceramic Foam Filters (CFF) for Liquid Aluminum Filtration", *Light Metals 2013*, pp. 973–979, 2013.
7. R. Fritzsch, "Electromagnetically Enhanced Priming of Ceramic Foam Filters", PhD, DMSE, NTNU, 2016.

8. R. Fritzsch, M. W. Kennedy, S. Akbarnejad, and R. E. Aune, "Effect of Electromagnetic Fields on the Priming of High Grade Ceramic Foam Filters (CFF) with Liquid Aluminum", *Light Metals,* pp. 929–935, 2015, doi: https://doi.org/10.1002/9781119093435.ch156. 20 FEB 2015.

9. M. W. Kennedy, S. Akhtar, R. Fritzsch, J. A. Bakken, and R. E. Aune, "Apparatus and method for priming a molten metal filter", ed: Google Patents, 2015.

10. R. Fritzsch, "Filtration of Aluminium Melts using Ceramic Foam Filters (CCF) and Electromagnetic Field", Masters, Department of Materials Science and Engineering, NTNU, Trondheim, 2011.

11. R. Fritzsch, M. W. Kennedy, S. Akhtar, J. A. Bakken, and R. E. Aune, "Electromagnetically Modified Filtration of Liquid Aluminium with a Ceramic Foam Filter", *International Journal of Iron and Steel Research,* no. S1, pp. 72-76, 2012.

12. M. W. Kennedy, R. Fritzsch, S. Akhtar, J. A. Bakken, and R. E. Aune, "Apparatus and Method for Priming a Molten Metal Filter", U.S. Patent Appl. PCT/IB2013/000775, WO 2013160754 A1, 2012.

13. M. W. Kennedy, "Removal of Inclusions from Liquid Aluminium using Electromagnetically Modified Filtration", PhD, Department ofMaterials Science and Engineering, NTNU, Trondheim, ISBN 978-82-471-4537-1, 2013.

14. A. J. Smith, P. Bosworth, R. Fritzsch, and M. Vincent, "Filter handling tool", 2020.

Automated Metal Cleanliness Analyzer (AMCA): Digital Image Analysis Phase Differentiation and Benchmarking Against PoDFA-Derived Cleanliness Data

Hannes Zedel, Robert Fritzsch, Shahid Akhtar, and Ragnhild E. Aune

Abstract

Assessing metal cleanliness of aluminum melts is critical for product quality control, as well as for process optimization. PoDFA is the current standard method for assessing aluminum cleanliness but has limitations in speed and costs due to its manual image processing. The Automated Metal Cleanliness Analyzer (AMCA) method was previously demonstrated to produce cleanliness indicators highly correlating to the main cleanliness indicator of industrial PoDFA analyses on the same samples. In the present work, the features of the AMCA method were expanded, introducing quantitative inclusion characterization and enhanced detection features. The results were systematically benchmarked against industrial PoDFA-derived cleanliness data. The results confirm the equivalence of the total particle area and provide moderate differentiation of inclusion types. Thereby, AMCA shows potential to be used as an alternative to PoDFA, deriving cleanliness data of aluminum samples for generating extensive process data at superior cost-scaling and minimized human biases.

Keywords

Digital image processing • Automated image processing • Particle counting • Metal cleanliness • Quantitative stereology • PoDFA

H. Zedel (✉) · R. Fritzsch
MetIQ AS, Trondheim, Norway
e-mail: hannes.zedel@metiq.no

H. Zedel · R. Fritzsch · R. E. Aune
Department of Materials Science and Engineering, Norwegian University of Science and Technology (NTNU), Trondheim, Norway

S. Akhtar
Hydro Aluminium Karmøy Primary Production, Håvik, Norway

Introduction

The use of both primary and secondary aluminum requires robust cleanliness assessment methods for preventing defects in end products. Ensuring metal quality standards is therefore critical for high quality of aluminum products as well as process optimization and R&D purposes. One of the current standard methods for assessing aluminum cleanliness is the Porous Disc Filtration Apparatus (PoDFA [1]). For PoDFA measurements, a sample of about 1 kg liquid melt is taken and sucked through a ceramic foam filter by a vacuum, creating PoDFA discs of about 5 g. The filter opening is called the "chord length" and typically close to 12.7 mm. Different types of contaminants, mainly inclusions, accumulate on top of the filter and are termed the "filter cake". The discs are cut along the diameter and micrographs are taken from the transverse section of the intersection of the filter and the aluminum bulk material, showing the entire filter cake in approximately 10–15 micrographs per sample at 100× magnification. Each micrograph typically shows a part of the filter grain at the bottom, the filter cake in the middle (inclusion band), and the aluminum bulk at the top. The 2-D total area of the contaminants in the micrographs is used as the main cleanliness indicator for aluminum melt. PoDFA was developed long before the digital age and still relies on manual labor for the quantification of these micrographs. An operator applies a grid of about 100 squares to the microscope for each micrograph, dividing the micrograph into smaller subsections, and estimates the relative amount of up to 21 inclusion types as a percentage for each square. For each sample, the operator therefore estimates about 1000–1500 squares for different contaminant categories. This method is prone to significant errors, partially relating to human biases. Evans et al. reported that the measurement error primarily depends on the overall level of contamination and can range from about 20% in highly contaminated samples to approximately 100% in very clean samples (see Fig. 1).

© The Minerals, Metals & Materials Society 2023
S. Broek (ed.), *Light Metals 2023*, The Minerals, Metals & Materials Series,
https://doi.org/10.1007/978-3-031-22532-1_117

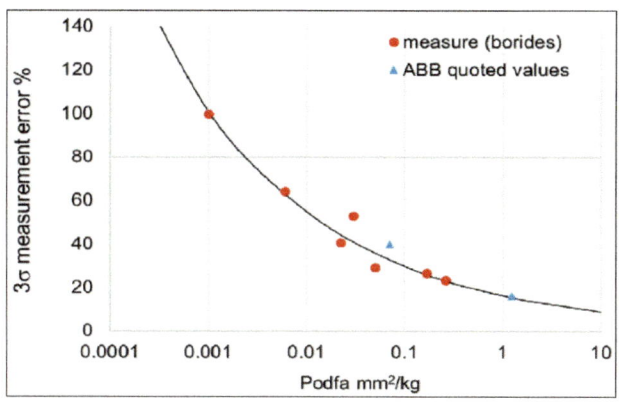

Fig. 1 Measurement error of boride inclusions in PoDFA cleanliness assessments depending on the level of contamination (mm^2/kg) [2]

Because this approach is prone to significant variance, slow, and therefore expensive in operational costs, the industry has a high demand for more robust and automated metal cleanliness assessments. Furthermore, for the assessment of filtration efficiency in aluminum manufacturing, using ceramic foam filters, the quantitative assessment of smaller size fractions of inclusions is of interest [3]. The Automated Cleanliness Analyzer (AMCA) method was previously used for a cost-efficient replication of the main cleanliness indicator of PoDFA, the total particle area, and enabled the detection of small inclusions down to 5 μm Equivalent Spherical Diameter (ESD), limited directly the resolution of the micrographs [4].

Depending on the type of samples, the specific process step, and the specific purpose of the cleanliness data, industrial metallurgical services may use a selection of target inclusion categories in PoDFA analyses or may aggregate categories.

In this study, we use an evolved implementation of the AMCA method to digitally quantify PoDFA micrographs and compare the results to industrial PoDFA-derived cleanliness data of the same samples. The AMCA implementation improves on a previously suggested feature for more differentiated phase detection [5]. The method quantifies these five target classes: Filter grain, aluminum bulk material, particles, titanium boride phases, and intermetallic phases. Previously, phase differentiation was limited to filter grain, aluminum bulk material, particles, and aggregated other inclusions, using an earlier version of the AMCA method.

We here aim to quantitatively demonstrate recent advances of the AMCA method by determining more differentiated inclusion type detections, improved artifact removal strategies, and novel features. This includes improved artifact handling, e.g., by reducing false-positive detection of topological shadows typically seen close to the filter grain, and quantitative automatic isolation of the bulk,

inclusion band, and filter grain areas in each micrograph, enabling more differentiated data sets for all target classes.

Method

Equipment and Materials

The main cleanliness indicator in PoDFA is the Total particle area (also "Total Inclusion Count"—TIC) in mm^2/kg and calculated by the formula

$$TIC = Area_{[mm^2]} \times \frac{Nominal\ Chord\ length_{[12.7\ mm]}}{Measured\ Chord\ length_{[mm]}} \times \frac{Nominal\ Filtered\ Weight_{[1\ kg]}}{Filtered\ Weight_{[kg]}},$$

where Area is the total area of all present contaminants, the chord length is the opening distance of the filter, and the filtered weight is the amount of metal sucked through the ceramic foam filter. The cleanliness is evaluated as shown in Table 1.

In PoDFA cleanliness assessments, the "Area" relates to the sum of the relative content of all grid squares allocated to contaminant classes over all micrographs of a sample.

For this study, Norsk Hydro ASA provided 137 raw micrographs of 11 industrial PoDFA samples from 6xxx alloy melts in secondary aluminum and related PoDFA-derived cleanliness data from the same micrographs by their metallurgical laboratory. The data set covers the filtered weights, the chord lengths, inclusion areas of Mixed oxides, Carbides, TiB$_2$/Ti-rich, MgO, and Spinel phases, as well as the Total particle area for each sample. Carbides, MgO, and Spinel phases were only selectively indicated when present.

Image Processing

The micrographs were imported and processed with Mathworks Matlab R2021a [7]. The algorithm generates labelled masks where each pixel of a micrograph is allocated to a target class. For calculating the TIC, the method provides the

Table 1 PoDFA cleanliness classification [6]

Classification	TIC [mm^2/kg]
(1) Very light	0.0–0.05
(2) Light	0.05–0.10
(3) Moderate	0.10–0.40
(4) Heavy	0.40–1.20
(5) Excessive	≥ 1.20

sum of the number of pixels for contaminant classes over all micrographs of a sample as the "Area", which can be converted to square millimeters. In micrographs that contain a scale bar, a rectangular box covering the scale bar is excluded from the detections.

Color Detection

The segmentation algorithm is based on an AMCA algorithm that was previously used to detect 4 target classes in PoDFA micrographs [4]. Among other cues, that algorithm relied on grayscale thresholding of the green color channel (which was calibrated best in the microscope), complemented by shape metrics and size filters. In the present study, higher quality micrographs are available that enable using color detection. Thereby, information from 3 instead of only 1 color channel can be used to enhance detection precision. In an RGB image. colors represent relative proportions between the three color channel intensities. These proportions are lost when extracting only one of the channels as a gray scale image. Because direct thresholding of the color channels does not maintain the relative proportions of the channels beyond very narrow thresholds that only detect partial features, color thresholding is performed on micrographs that are converted into an HSV format. HSV is a representation of the image by its "Hue" (color tone, describing the relative color channel proportions as a range of the visible spectrum of colors), "Saturation" (color intensity), and "Value" (brightness or luminance). The use of color detection enables better phase differentiation, e.g., between the TiB_2/Ti-rich phases (typically in brown color tones) and the intermetallic phases (typically in gray color tones) which coincide in grayscale representations of the same images.

Target Classes

Target class detection is based on hierarchical elimination, where more robust detections run first and are then removed from a reference image to facilitate further detections and minimize misdetections.

The most robust target class to detect is the filter grain, defined as large dark objects that reach the bottom end of the micrograph. After the filter grain is identified, darker particles and porosity can be detected robustly as well. The remaining structures in our data set relate to TiB_2/Ti-rich phases and intermetallic phases, which then are isolated by color detection layers. Different agglomerate densities and brightness heterogeneity of the micrographs require combining multiple partial color detections, merging them into one target class detection. Brightness homogeneity in our data set shows as a systematic bias between intensities from left to right, where the left was generally slightly darker than the right side of each micrograph. This indicates uneven lighting conditions during image acquisition, e.g., by lamps or sun from an open window. Subsequently, detections may require slightly different HSV ranges for different parts of the micrographs.

Topological Shadows

The dominant source of artifacts in PoDFA micrographs is the occurrence of topological shadows. The filter grain is more resistant to polishing than the bulk material, so the filter grain structures create morphological elevations on samples during polishing that then induce shadows next to the filter grain structures. These shadows gradually cover a wide range of intensities, so they tend to get detected as false-positives in various target phases. In the previous implementation, these shadow detections were partially isolated by slicing gray scale ranges to subranges and applying shape metrics for an acceptable artifact removal. In the current implementation, multiple strategies are combined to minimize artifacts from topological shadows:

- *Color detection* directly reduces the impact of topological shadows for most target phases because they are less likely to coincide in a full color spectrum compared to a grayscale intensity spectrum.
- *Partial target phase detections* facilitate artifact removal functionality, e.g., using size filters. The target phase is then derived by merging multiple color detections of relatively narrow HSV ranges.
- *The relative position* of detected objects is a major cue because topological shadows almost exclusively occur directly next to filter grain structures. Artifact removal functions detect the immediate proximity of an object to the detected filter grains to adjust removal criteria such as size or shape.

Band Isolation

For quantitatively isolating the bulk material at the top of each micrograph, the inclusion band (filter cake) in the middle, and the filter grain at the bottom, the following definitions are applied:

The filter grain is initially defined as a large connected dark structure reaching from the bottom of the micrograph up to a certain y-value of the micrograph. The maximum y-value of that structure is identified before scanning for other large dark objects in the micrograph. Objects that are large, dark, and below the maximum y-value of the initial filter detection, are then allocated to the filter grain as well. Openings in the filter grain and any inclusions within these openings can then be separately counted as being inside the filter grain.

The inclusion band is defined as the area between the filter grain and the top-most y-value of any larger TiB_2/Ti-rich phase detection. The filter grain provides a clear dynamic boundary between the bands, while the TiB_2/Ti-rich phase does not always have a clear upper boundary, so the upper end of the inclusion band is a horizontal line.

The aluminum bulk is then the remaining part of the image above the inclusion band.

Results

For this study, 137 micrographs from 11 industrial PoDFA samples were processing by a deterministic image segmentation algorithm based on the AMCA method. For each sample, there are 10 micrographs showing the filter cake on top of the filter grain. In case of high contaminations, additional micrographs of section further above were taken in addition. An overview of the samples is provided in Table 2.

The AMCA method generates visual overlays of the detections with a color code of the target phases. A typical detection is shown in Fig. 2.

A comparison of the Total particle area between PoDFA reference data and AMCA data is shown in Table 2.

The red (particles/porosity) and blue (TiB_2/Ti-rich) target classes of the AMCA labels are counted into the total particle area. The filter grain, bulk, and intermetallic phases are not indicative of the samples' cleanliness. By exclusion of the excessively contaminated sample 19, PoDFA and AMCA data aligns with a correlation coefficient of 0.98, validating the equivalence of the total particle area (see Fig. 3).

The phase-specific detections of both methods are reported in Table 3. The correlation coefficients relate to the

Table 2 List of PoDFA samples. Sample types A and B indicate sampling from different process steps

Sample	Type	Phase data	Comment	Images
0	Pilot	No	Lower magnification	5
1	A	Yes		10
2	A	Yes		10
3	A	Yes		10
11	A	No	Chord length estimated	10
13	B	No	Chord length estimated	15
16	B	Yes		17
18	A	Yes		10
19	B	Yes		29
20	A	Yes		11
21	A	Yes		10

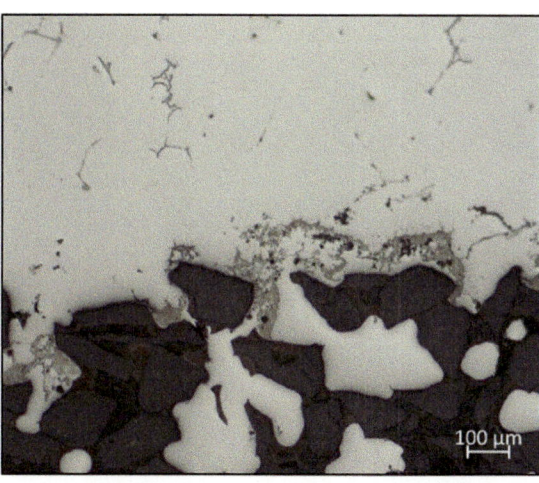 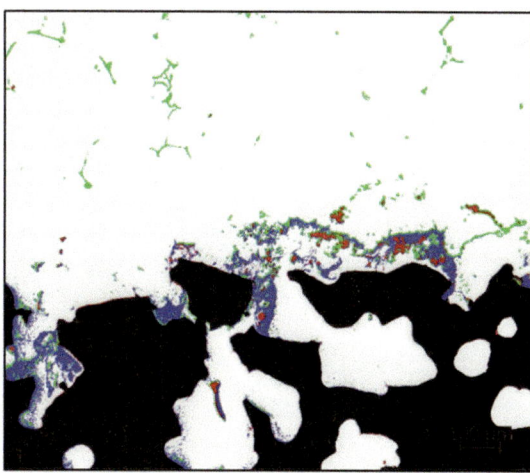

Fig. 2 Example overlay of a typical AMCA detection. Left: PoDFA micrograph. Right: AMCA overlay (filter grain: black, particles and porosity: red, TiB_2/Ti-rich: blue, intermetallic phases: green, bulk: white)

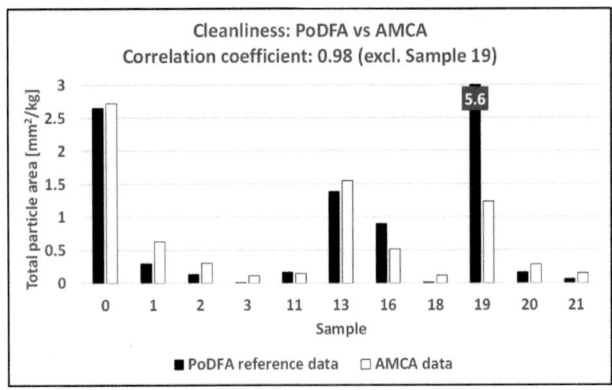

Fig. 3 Main cleanliness data comparison between PoDFA and AMCA

PoDFA reference detections per phase correlated with the different target classes of AMCA. The alignment of TiB_2/Ti-rich detections with adjusted samples 16 and 19 is shown in Fig. 4.

The band isolation is a novel feature, so there is no reference data for comparison. The isolation of the bands is illustrated in Fig. 5. All quantifications can be specifically allocated to the individual bands.

Furthermore, particle size distributions can be derived for all conditions shown in this study, however, these were not in the scope of the present study.

Discussion

Total Particle Area

Compared to previous applications of the AMCA method, the total particle area indicators between PoDFA-derived reference data and AMCA generally have only small deviations. This is largely attributable to the improved artifact handling functions. Previously, there were significantly more false-positive detections from topological shadows and other inhomogeneities. The integrity of the sample 19 reference data could not be verified before finalizing this document, so it remains in question whether the sample 19 data is reliable as a benchmark reference. Excluding it leaves the AMCA results at a 98% correlation with the other PoDFA reference data, validating the AMCA total particle area as an equivalent cleanliness indicator as the industrial PoDFA-derived TIC.

20 μm-rule

In certified PoDFA analyses, clusters of small particles below 20 μm ESD are counted as agglomerates where the small area between agglomerated particles counts into the particle area. In digital analyses, those spaces between small particles are typically not counted. Subsequently, AMCA

Table 3 Quantitative phase differentiation results derived from PoDFA samples and AMCA. Correlations show coefficients for the PoDFA inclusion class at the top correlating with the AMCA class at the bottom right. Intermetallic phases do not count into the total particle area but are reported because they can be indicative of false-positive or false-negative detections in the TiB_2/Ti-rich phase

Sample	Inclusion area [mm^2]							
	PoDFA reference data					AMCA data		
	Mixed oxides	Carbides	MgO	Spinel	TiB_2/Ti-rich	Particles	TiB_2/Ti-rich	Intermetallic phases (excl. from TIC)
1	0.0449	0.0060	0.0060	0.0030	0.2394	0.186	0.406	0.221
2	0.0056	0.0028			0.1305	0.080	0.168	0.235
3	0.0033	0.0016			0.0060	0.052	0.064	0.101
16	0.0180	0.0045		0.0045	0.8750	0.408	0.096	1.451
18	0.0003	0.0002			0.0148	0.049	0.064	0.097
19	0.7248	0.0558		0.0558	4.7391	0.947	0.270	2.896
20	0.0033	0.0017			0.1607	0.094	0.166	0.336
21	0.0013				0.0618	0.049	0.074	0.162
Correlations	0.93	0.94		0.97	0.97	Particles		
	0.40	0.40		0.04	0.36	TiB_2/Ti–rich		
	0.90	0.91		0.90	0.96	Intermetallics		
	0.90	0.91		0.90	0.96	TiB_2/Ti–rich (s16, s19 TiB_2/Ti–rich + intermetallics)		
	0.91	0.92		0.94	0.96	TiB_2/Ti–rich + intermetallics		

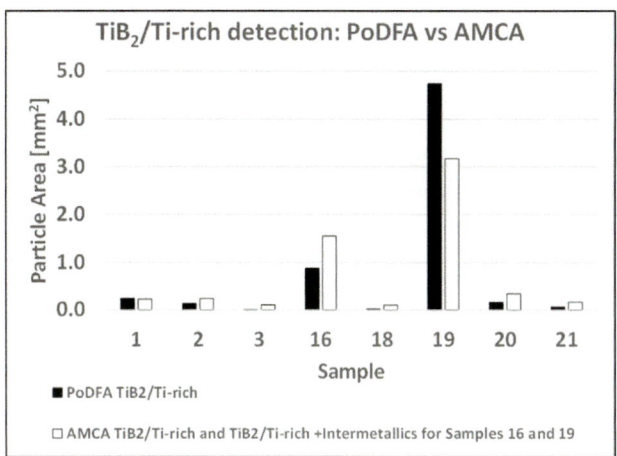

TiB$_2$/Ti-rich detection: PoDFA vs AMCA

- ■ PoDFA TiB2/Ti-rich
- □ AMCA TiB2/Ti-rich and TiB2/Ti-rich +Intermetallics for Samples 16 and 19

Fig. 4 Comparison of TiB$_2$/Ti-rich phase detection between PoDFA-derived cleanliness data and AMCA. For samples 16 and 19 (Sample type B), the sum of TiB$_2$/Ti-rich and Intermetallic phases are used instead

particle areas for small particles can be expected to be systematically lower than related PoDFA counts. This undercounting was considered negligible in this study due to the low overall level of contamination and the low relative amount of particle inclusions (highlighted by red in Figs. 2 and 5). In the present study, deviations were predominantly determined by residual artifacts from topological shadows. The relative impact of the agglomerate detection difference between the methods will be larger for more contaminated samples and needs to be considered for future studies. This also greatly affects particle size distributions, which were not in the scope of the present study but are commonly derived from this type of analysis.

Phase Differentiation

The phase differentiation demonstrated in this study exceeds the level of detail provided in previous studies using the

AMCA method and better aligns with PoDFA standards of assessing the total particle area by improving the isolation of the intermetallic phases from the inclusion categories. Notably, the differentiation between intermetallic phases and TiB$_2$/Ti-rich phases appears to be limited particularly in type B samples (16, 19), which is important because the intermetallic phases do not count into the total particle area (TIC) and thereby are a confounding factor when being misallocated to cleanliness-related target phases. The remaining visual artifacts also affect both phases. That relates to some topological shadows being misidentified as either TiB$_2$/Ti-rich phases or intermetallic phases. Also, inclusion agglomerates and clusters are highly heterogeneous objects that are typically allocated partially to the intermetallic phase and partially to the TiB$_2$/Ti-rich phase, while a human operator may likely prefer making more absolute choices about target class allocations. For the type A samples, the TiB$_2$/Ti-rich phase detection robustly aligns with the reference data.

The correlations of any given PoDFA-derived inclusion class to an AMCA class is very similar in all cases, indicating that the relative distribution between different inclusion types is similar in all samples. The low correlation between any PoDFA-derived class and the AMCA TiB$_2$/Ti-rich phase, unless corrected for by adjusting samples of type B (16 and 19), indicates poor phase differentiation between TiB$_2$/Ti-rich and intermetallic phases by the AMCA method specifically in type B samples.

Band Isolation

The band isolation as demonstrated in Fig. 5 was qualitatively verified by screening the resulting micrographs and was found to be robust for all type A samples that have a distinct TiB$_2$/Ti-rich phase that can reliably be identified. In type B samples, the inclusion band is dominated by oxide films and other inclusions which are too heterogeneous for defining reliable

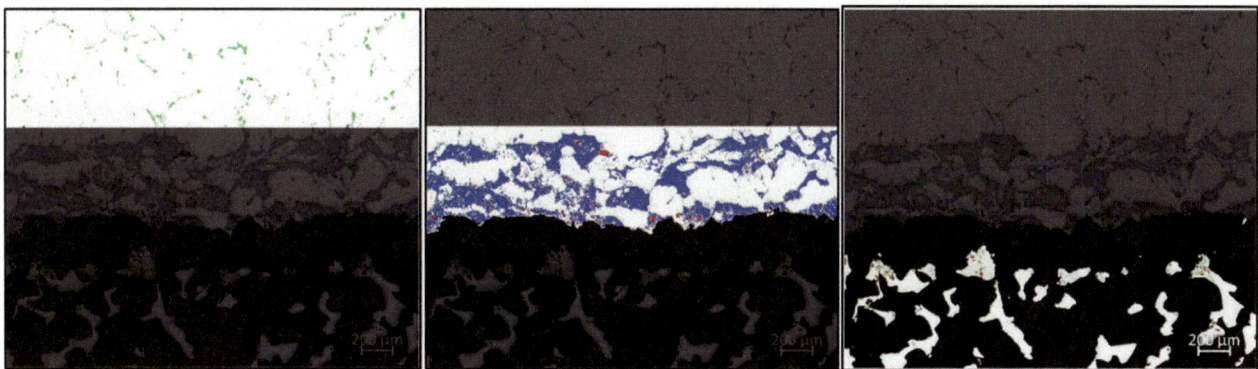

Fig. 5 Example of band isolation of one PoDFA micrograph. Left: Aluminum bulk, Middle: Inclusion Band, Right: Filter Grain. Detection color coding: Black = Filter, White = Bulk, Red = Particles/porosity, Blue = TiB$_2$/Ti-rich, Green = Intermetallic phases

detection criteria for reasonable band boundaries between the inclusion band and the aluminum bulk.

The data can be used for more nuanced analyses of aluminum filtration trials and may contribute to process optimization and control in the future.

Limitations

Deterministic image analysis can be used for batch processing of micrograph sets that were taken from similar samples and with similar image acquisition methods and conditions. The algorithm needs to be tuned to each new batch that originates from conditions not previously encountered. The necessity of adjusting the algorithm currently requires qualitative screening by an operator but may be automatized in the future. The cost-scaling of manual assessments by an operator is a direct function of the surface area to be covered for quantification. In PoDFA assessments that relate to the number of grid squares being estimated, which means the man hours are a function of the number of images. While AMCA scales its man hours as a function of batches of any number of micrographs, it still is not a fully automated method, because it requires initial adjustment to subsets of new data sets. Extensive databases with implementations for different sample types may offset this limitation in the future.

While the application of a tuned algorithm to a batch of micrographs significantly minimizes human biases and makes the resulting data more statistically robust, the tuning of the initial subset of the micrographs is still subject to some degree of human bias. By the Law of Large Numbers, these biases become systematic, making AMCA data more robust for relative comparisons.

Full certified PoDFA assessments provide a high level of inclusion differentiation that is derived not only from the micrographs but also takes contextual information into account, e.g., regarding the alloy type, sample type, sample acquisition method, and process step. The AMCA method currently determines detection criteria solely based on visual distinction and thereby has a limited capacity to provide the same level of inclusion categorization. At the current stage of development, AMCA is validated to be used for a cost-efficient assessment of the total particle area (TIC) of PoDFA micrographs and provides only basic phase differentiations compared to full certified PoDFA assessments.

Conclusion

The present study aimed at evolving the AMCA method to quantitatively analyze PoDFA micrographs and compare the results to industrial PoDFA-derived cleanliness data from the same samples. The goal was to confirm equivalence of the main cleanliness indicator (total particle area, TIC) and demonstrate recent advances of the feature development of AMCA. The implementation of color detection and improved artifact handling functions such as vicinity detection of the filter grain significantly increased detection accuracies. The total particle area between AMCA and PoDFA correlates at 98% at the exclusion of one sample which's data integrity is in question. This suggests that AMCA's total particle area is equivalent to the PoDFA-derived total particle area indicator for assessing metal cleanliness.

In comparison to earlier versions of the AMCA method, more pronounced phase differentiations were demonstrated. These were robust for one sample type but showed significant misallocations between phase detections of the other sample type, which were partially attributed to inhomogeneity of the micrographs, brightness biases in the images due to lighting conditions during image acquisition, generally coinciding features of these two phases, and are otherwise a focus for future improvements of the method.

The band isolation feature is a promising complement to the digital quantification of aluminum manufacturing process data for future studies on filtration efficiency and process optimization.

Overall, our results mark a significant step towards cost-efficient automatization of metal cleanliness assessments by providing an approach that can replace man hour-intensive manual labor with specialized computation for significantly increased cost-efficiency and processing speeds for assessing the overall cleanliness of PoDFA samples and providing basic phase differentiation data.

Future Work

Inclusion Categories/Characterization

Further benchmarking should focus on applying the AMCA method to more types of samples from different process steps during aluminum manufacturing. Also, digital strategies for further improving the phase differentiation are needed. This includes taking contextual information about the samples into account for improving the characterization of inclusion types. In the future, different implementations of the AMCA method for different types of samples may provide higher degrees of phase differentiation to more closely replicate certified PoDFA inclusion categories.

Oxide Film Detection

Full certified PoDFA analyses also provide counts of oxide films as a relevant cleanliness metric. Oxides film detection was not in the scope of the present study. Earlier implementations of the AMCA method suggest that a replication of this is feasible while also providing specific detections of oxide films including their size (area) and distribution. However, this feature needs to be demonstrated and systematically compared to robust reference data in future studies.

Applications Beyond PoDFA

The AMCA method is not specific to PoDFA images and should be applied in other fields to create opportunities for further evolvement. In recent years, the method was adapted to solve similar challenges in adjacent fields such as the quantification of particle size distributions in aluminum production off-gas [8], grain refiner detections, and similar micrograph quantifications documented in other pending publications. These quantitative analyses of metallurgical micrographs may be beneficial for the steel, magnesium, silicon, and zinc industries by streamlining process control and paving the way to big data approaches in the metal industry.

Acknowledgements The authors wish to express their gratitude to the Department of Materials Science and Engineering and the Department of Chemistry at the Norwegian University of Science and Technology (NTNU) for their continuous support, to NTNU Technology Transfer AS (NTNU TTO) for their administrative and strategical support of the project, as well as to Norsk Hydro ASA in Karmøy for the provision of PoDFA micrographs and continuous support of the project.

References

1. 'PoDFA, The complete solution for inclusion measurement, Inclusion identification and quantification analysis'. ABB Inc., 2016. [Online]. Available: https://library.e.abb.com/public/b70691346 2934969befe277d80880795/PB_PoDFA-EN_A.pdf.
2. P. V. Evans, P. G. Enright, and R. A. Ricks, 'Molten Metal Cleanliness: Recent Developments to Improve Measurement Reliability', in *Light Metals 2018*, O. Martin, Ed. Cham: Springer International Publishing, 2018, pp. 839–846. https://doi.org/10. 1007/978-3-319-72284-9_109.
3. H. Duval, C. Rivière, É. Laé, P. Le Brun, and J.-B. Guillot, 'Pilot-Scale Investigation of Liquid Aluminum Filtration through Ceramic Foam Filters: Comparison between Coulter Counter Measurements and Metallographic Analysis of Spent Filters', *Metall. Mater. Trans. B*, vol. 40, no. 2, pp. 233–246, Apr. 2009. https://doi.org/10.1007/s11663-008-9222-y.
4. H. Zedel, R. Fritzsch, S. Akhtar, and R. E. Aune, 'Automated Metal Cleanliness Analyzer (AMCA)—An Alternative Assessment of Metal Cleanliness in Aluminum Melts', in *Light Metals 2021*, L. Perander, Ed. Cham: Springer International Publishing, 2021, pp. 778–784. https://doi.org/10.1007/978-3-030-65396-5_102.
5. H. Zedel, R. Fritzsch, S. Akhtar, and R. E. Aune, 'Estimation of Aluminum Melt Filtration Efficiency Using Automated Image Acquisition and Processing', in *Light Metals 2019*, C. Chesonis, Ed. Cham: Springer International Publishing, 2019, pp. 1113–1120. https://doi.org/10.1007/978-3-030-05864-7_136.
6. C. Stanica and P. Moldovan, 'Aluminium melt cleanliness performance evaluation using PoDFA (Porous Disk Filtration Apparatus) technology', *UPB Sci Bull Ser B*, vol. 71, no. 4, 2009, [Online]. Available: https://www.scientificbulletin.upb.ro/rev_docs_arhiva/full6739.pdf.
7. 'MATLAB'. The MathWorks Inc., Natick, Massachusetts, 2021.
8. D. Perez Clos, H. Zedel, S. G. Johnsen, P. Nekså, and R. E. Aune, 'Particle deposition characteristics in the formation of Hard Grey Scale (HGS) on cold surfaces exposed to aluminium production off-gas', *J. Aerosol Sci.*, vol. 161, p. 105946, Mar. 2022. https://doi.org/10.1016/j.jaerosci.2021.105946.

Automated Image Analysis of Metallurgical Grade Samples Reinforced with Machine Learning

Anish K. Nayak, Hannes Zedel, Shahid Akhtar, Robert Fritzsch, and Ragnhild E. Aune

Abstract

Controlling metal cleanliness in primary and secondary aluminum production is critical for ensuring quality in commercial sales and for effective process optimization. Solidified aluminum melt samples are today typically analyzed using established techniques such as LiMCA and PoDFA, however, these techniques rely on heavy and expensive equipment, extensive running times, and high heterogeneity of the results. The primary bottleneck of PoDFA analyses, the current standard approach, is the manual analysis of melt micrographs by human operators. In the present study, an image analysis platform based on a machine learning algorithm capable of quantifying contaminants in PoDFA micrographs was developed and tested. Machine learning models enable improved performance in heterogeneous datasets compared to common image analysis techniques using minimal computational resources and are envisioned to enable superior cost-scaling in metal cleanliness assessments. Future implementations will expand on the quantitative differentiation of relevant inclusion types.

Keywords

Aluminum • Machine learning • Characterization • PoDFA

A. K. Nayak (✉) · H. Zedel · R. Fritzsch · R. E. Aune
Department of Materials Science and Engineering, Norwegian University of Science and Technology (NTNU), Trondheim, Norway
e-mail: anish.k.nayak@ntnu.no

A. K. Nayak
Institute of Chemical Technology Mumbai (IndianOil Odisha Campus), Bhubaneswar, India

S. Akhtar
Norsk Hydro, Karmøy Primary Production, Hydrovegen, N-4265 Håvik, Norway

Introduction

Quantifying aluminum melt contamination is a priority for the industry for both quality control and process control. Contaminants are categorized into different inclusion categories such as aluminum oxides, magnesium oxides, spinel particles, carbides, titanium, or vanadium-based borides. Other inclusions can originate from bone ash or refractory particles. The type of inclusions to be expected in a melt sample depends on where in the process the sample was taken, as illustrated in Fig. 1.

Contaminations have significant implications for product qualities, even though they are scarce in a melt sample, i.e., in the range from 0 to ~50 ppm. Metal cleanliness assessments for primary and secondary aluminum are therefore carried out on up-concentrated samples. The most established method for this cleanliness assessment is the Porous Disc Filtration Apparatus (PoDFA), where a melt sample of ~1 kg is sucked through a filter using vacuum [1]. After solidification, the filter is cut, and micrographs are acquired from the transverse section as inclusions typically accumulate on top of the filter (the "filter cake"). The micrographs are then systematically assessed by an operator that quantifies the presence of the different types of particles/inclusions in view of the area they take up. As this is a manual step, it would be beneficial for the industry to have automation strategies to reduce operational costs, increase processing speed, and reduce human biases in the assessment.

In the present study, it has been demonstrated, through a preliminary investigation, that combining different digital image segmentation methods is viable for achieving these goals. Machine learning (ML) methods have been used in the past for analyzing metallurgical micrographs, but the labeling of relevant images for training a neural network is still a manual process and too cost-inefficient for widespread use. Therefore, the present study proposes labeling data with semi-automated deterministic image processing methods previously used to quantify PoDFA micrographs [2, 3].

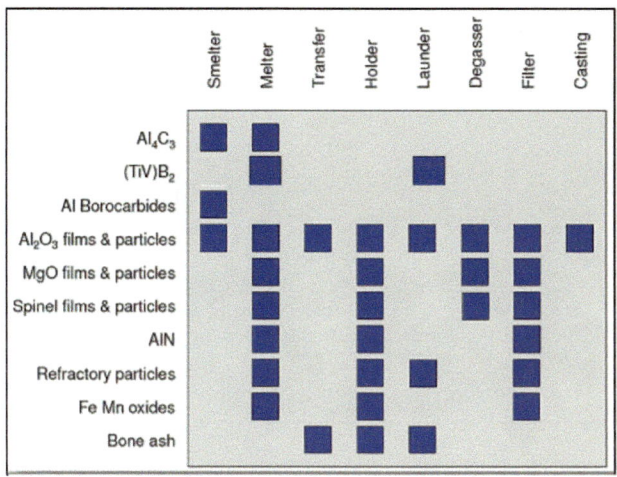

Fig. 1 The type of inclusions in aluminum melt samples depends on the process step it was sampled from. The most common inclusion types for each process step are highlighted [1]

Using such labeled data, a neural network can be trained as a proof of concept. Predictions made by the model in the form of labels can then be validated on micrographs not trained on and, therefore, unknown to the neural network. The goal was to demonstrate the equivalency of the results from both methods for the main melt cleanliness indicator (the total particle area). Depending on the image acquisition conditions and sample type, the deterministic segmentation method must be tuned to micrograph batches. Conceptually, training an ML network with labeled data from different micrograph batches will enable the model to achieve sufficient feature detections in the future without requiring batch-specific tuning and thereby enable a higher degree of automation.

Machine Learning (ML)

It was widely believed until as recently as 1997 that the raw computing power of a machine could never parallel that of the human mind until Deep Blue, the IBM supercomputer, beat the world chess champion at the time, Garry Kasparov. With substantial increases in the processing power and immaculate algorithms of the Graphics Processing Unit (GPU), ML frameworks have seen massive improvements over the past 15 years. The three common types of ML models include supervised, semi-supervised, and unsupervised models. Image analysis algorithms are typically supervised where large amounts of labeled data are fed as training inputs and features, or biases are generated, establishing a correlation between input and output. (E.g., detecting different phases in a metallurgical micrograph).

One specialized type of ML is deep learning. Neural networks used in deep learning consist of individual layers

when pieced together from the framework. A set of layers can be imported from one application and adapted for the required purpose by changing a few layers towards the end. Training algorithms can be modified if insufficient accuracy is reached by varying certain parameters. These accuracy metrics include training loss, accuracy, and base learn rate.

Methods

Equipment and Materials

For the present study, Norsk Hydro ASA (Karmøy, Norway) provided 132 micrographs at $100\times$ magnification from 10 industrial PoDFA samples, as well as the operator's evaluation reports generated by their internal metallurgical laboratory, as reference data. All PoDFA samples utilized are categorized under the 6xxx alloy melts.

Image Processing

Image Processing Reinforced by ML

A deterministic image segmentation method was adopted to generate labeled data suitable for training a neural network. The method that applies a batch-processing approach tunes a deterministic segmentation algorithm to a random subset of a batch of micrographs acquired under the same laboratory conditions, such as sample preparation methods and microscopy settings. The method generates labeled data for each micrograph, allocating each pixel to different target classes (see Fig. 2). A neural network can interpret these labels to train it to similarly detect features in new micrographs. Out of the 132 micrographs, 29 were randomly selected as training data, and 20 different micrographs comprising two entire samples were selected for validation. The validation set was then segmented by the trained ML network and generated pixel labels with the same format as the original labels used for training. The resulting ML prediction could then be compared to the reference data generated from the deterministic image segmentation method, as well as the PoDFA reference data when all micrographs of one or several samples are analyzed.

Predictions made on varied datasets by the ML algorithm can potentially be more accurate than other techniques. Labels generated for the network depend on whether the classification was performed on an object or the segmentation of an image pixel by pixel.

An open-source model of U-Net based on multi-class semantic segmentation [4] was adopted and trained on the 29 images broken down into smaller patches. The labels were generated using the deterministic analysis in the form of

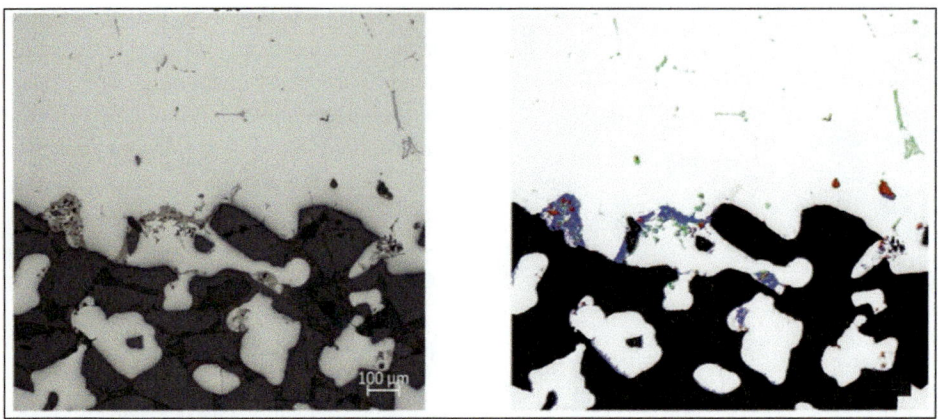

Fig. 2 Masks/labels generated from the raw micrographs using the AMCA (Automated Metal Cleanliness Analyzer) showing filter grains (black), dark particles/inclusions (red), titanium boride phase (blue), intermetallic phase (green), and aluminum bulk material (white). The scale bar is blackened and removed from the detections

Tagged Image Format (TIF) files. Before training, these labels were converted to Numerical Python (NumPy) arrays. The model was optimized using the TensorFlow platform and utilized Keras' neural network library [5]. Basic image operations were carried out using the OpenCV python library [6].

The output of the ML network is generally given as prediction masks for each patch, which can be stitched to the same resolution and transformed to the same format as the original labels from the deterministic segmentation method. The preliminary results showed that creating smaller patches reduced the precision in differentiating filter grains from particles/pores. Because the filter grain detection is very robust in the deterministic approach, the ML prediction masks were post-processed in the MATrix LABoratory (MATLAB) [7] to determine the filter grain and thereby remove artifacts in the filter grain areas commonly seen in the ML predictions. Despite being trained on patches, the U-Net detections were run on larger images with the help of a stitching function. A set of 20 untrained images (two complete samples) were kept separately for quantifying the network performance as validation. The network was trained on five classes, i.e., the bulk aluminum (white in detection overlays), filter grains (black), intermetallic phases (green), TiB_2/Ti-rich phases (blue), and particles/pores (red).

Results

The training of the neural network was initially carried out on a limited number of images, which were taken at a resolution of 2464 by 2056 and rescaled to 308 by 257 prior to generating square patches of 128 by 128 in both conditions. The number of patches generated was about 9000 for the larger images and 100 for the rescaled (smaller) images. Minimal data augmentation barring dimension alteration of images was done.

The U-Net was trained for 300 epochs, with empirical loss and accuracy metrics reduced to a minimum, as seen in Fig. 3.

A set of testing images relating to the different classes and unknown to the algorithm were taken to quantify the network's performance. The training graphs, generated for the neural network when taking approximately 30 images as input, have been displayed for both original and rescaled samples.

The network was trained using an NVIDIA GeForce RTX 3060 Laptop GPU. Training on 9000 patches took approximately 3 h, with 100 patches taking less than 5 min. Average detections taken in the case of larger images ranged between 20 and 30 s and a few seconds for the rescaled images, with some overlays, as can be seen in Fig. 4.

Table 1 shows the phase-specific detections for all target phases between ML on small and large images, as well as the AMCA (Automated Metal Cleanliness Analyzer) reference labels generated along the initial training labels. The red and blue detections (particles/pores and TiB_2/Ti-rich phases, respectively) were counted into the total particle area to determine the cleanliness.

Discussion

Increasing the number of micrographs used to train the network, as well as ensuring a sufficient representation of the TiB_2/Ti-rich phases in the training data set, generally improved ML predictions. For the images at original size, false-positive detections of topological shadows around the filter grains also significantly decreased by increasing the training data, however, topological shadows were still the main source of misdetections and artifacts. In Fig. 5 it is seen that both image analysis methods overcount the total

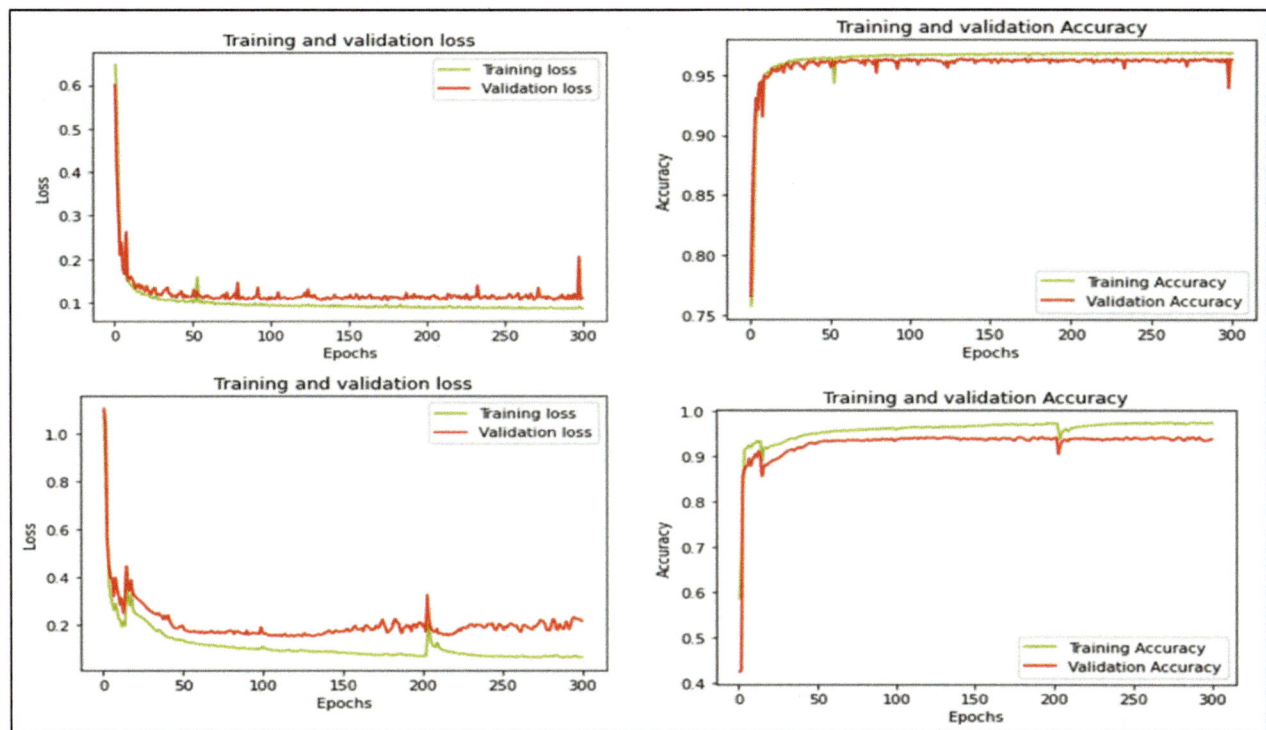

Fig. 3 Training and validation accuracies. The model was trained on an Adam optimization algorithm. Both losses and accuracies were calculated using an empirical formula which were expected to become consistent with the minimum fluctuations with an increase in training time. Top: Model trained on larger images; Bottom: Model trained on rescaled images

particle area in overall clean samples, while the ML trained on large images shows the best performance in predicting the total particle area when compared with the PoDFA reference (manual count). Previous studies [2, 3] suggest that topological shadows relate to a systematic error that is scaled by the amount of filter grain edges present in the micrograph, which cast a shadow during microscopy. Because PoDFA micrographs typically show similar amounts of filter grain areas, this leads to a systematic offset by artifacts induced from the topological shadows that are independent of the cleanliness of the sample, so the relative impact of this absolute offset is larger, the lower the actual particle area is. Evans et al. [8] also concluded that the relative measurement error in PoDFA is higher for cleaner samples. Therefore, the relative deviation from reference data in more contaminated samples is expected to be lower.

On rescaled images, the TiB_2-/Ti-rich phases were better detected than the ML trained on large images, however, the rescaling also led to very significant false-positive detections of particles/inclusions/pores, especially around the filter grain, imposing excessive overcounts, as can be seen in

Fig. 6. The gain of rescaling for detecting the TiB_2-/Ti-rich phase is, therefore, not considered worth the loss in particle/inclusion/pore detection precision. Notably, ML reinforced with labeled data from AMCA deterministic segmentation yielded overall better results for the total particle areas than AMCA detections alone (see Fig. 5). The combined approach is, therefore, a promising path toward automating PoDFA counts.

The main limitations are misdetections, especially of topological shadows, as well as misallocations between the target phases. In particular, the visual features of the TiB_2-/Ti-rich phases tend to coincide with intermetallic phases, which are not relevant to the total particle area. Both challenges can be addressed by improving the deterministic labeling data used as training data for the neural network and by increasing the quality of the micrographs. This includes maximizing the homogeneity of the image acquisition conditions to prevent brightness biases and ensuring consistent contrast and brightness levels.

Furthermore, the present approach aimed at establishing a method that can be flexibly used in different environments

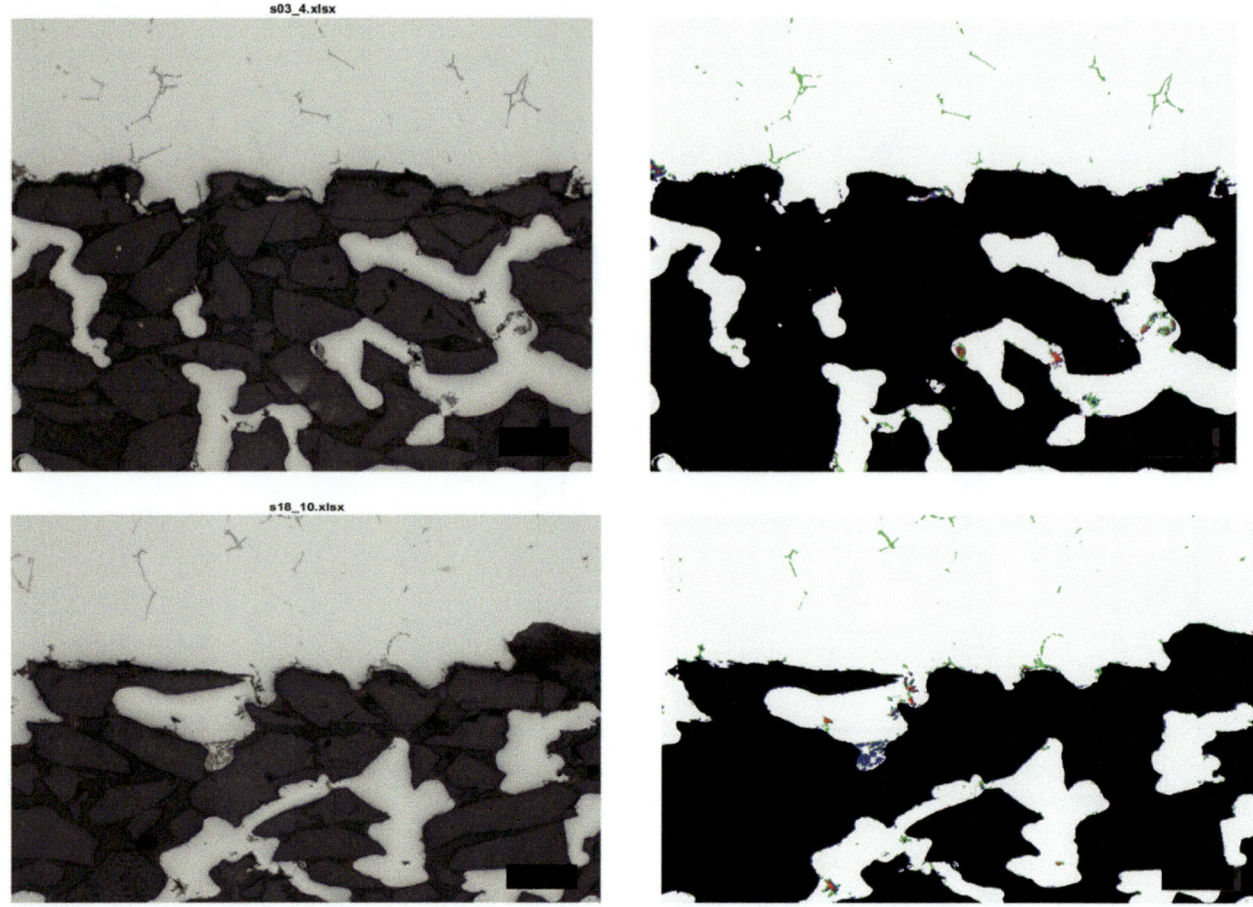

Fig. 4 Overlays generated for s3 (top) and s18 (bottom) when training on images with original dimensions

and should run on a regular laptop. As an increase in training data improves detection precisions, it can be expected that investments in higher computation capacities will increase the neural network's performance, e.g., by utilizing computational clusters for training the neural network on more and larger micrographs.

Conclusion

In the present study, a limited set of data and a cost-effective computational setup were used to segment different types of PoDFA sample micrographs within seconds. In view of comparing the obtained results to similar algorithms, e.g., the AMCA, showed that on certain samples, the ML

approach competed in detecting critical features of the micrographs but had limitations in target phase differentiations. The use of rescaled images led to better results in characterizing TiB_2-/Ti-rich phases but had significant overcounts in the particle layers due to misdetections of topological shadows as particles/inclusions/pores.

Occasional undercounting was still observed when detections were made by the model trained on larger images. When trained on smaller and rescaled images, the network also resulted in inconsistent predictions, most likely due to loss of details. Therefore, the potential for significant improvement exists, and with further optimization, the framework can possibly replace current industrial techniques to quantify contaminants in aluminum melt samples.

Table 1 Phase detections per micrograph. The categories Red and Blue are relevant for determining the particle/inclusion areas. Left columns: ML on rescaled (smaller) images. Center columns: ML detections on original size images. AMCA (Automated Metal Cleanliness Analyzer) column: Cleanliness reference from deterministic analysis

Sample	File	ML detection rescaled images						ML detection original size images						AMCA
		Black (filter grain) (%)	Red (particle, porosity) (%)	Blue (TiB$_2$/Ti-rich) (%)	Green (Intermetallic phase) (%)	White (bulk) (%)	Particle area mm^2 (red and blue detection)	Black (filter grain) (%)	Red (particle, porosity) (%)	Blue (TiB$_2$/Ti-rich) (%)	Green (Intermetallic phase) (%)	White (bulk) (%)	Particle area mm^2 (red and blue detection)	Particle area mm^2 (red and blue detection)
03	'S03_1.jpg'	37.6	2.0	2.0	1.4	57.1	0.055	39.5	0.2	0.5	0.6	59.1	0.011	0.009
03	'S03_10.jpg'	48.4	1.4	1.9	1.3	47.0	0.045	49.4	0.1	0.3	0.7	49.4	0.010	0.006
03	'S03_2.jpg'	40.5	1.6	1.7	1.2	55.0	0.046	42.1	0.1	0.5	0.6	56.6	0.010	0.007
03	'S03_3.jpg'	46.7	1.3	1.5	1.6	48.7	0.040	48.0	0.1	0.4	0.8	50.5	0.012	0.007
03	'S03_4.jpg'	47.6	1.8	1.9	1.3	47.3	0.051	49.2	0.1	0.5	0.6	49.5	0.011	0.008
03	'S03_5.jpg'	44.8	1.9	2.4	1.7	49.2	0.059	46.8	0.1	0.6	0.7	51.8	0.012	0.009
03	'S03_6.jpg'	42.1	2.5	2.4	1.6	51.4	0.068	44.5	0.3	0.5	0.7	53.9	0.010	0.011
03	'S03_7.jpg'	50.1	2.0	2.1	1.6	44.2	0.057	52.1	0.2	0.5	0.8	46.3	0.013	0.009
03	'S03_8.jpg'	47.4	2.4	2.3	1.9	46.0	0.066	49.7	0.4	0.7	0.8	48.3	0.012	0.014
03	'S03_9.jpg'	37.4	1.5	1.2	1.4	58.4	0.038	38.5	0.2	0.5	1.0	59.7	0.016	0.010
18	'S18_1.jpg'	38.6	1.6	1.8	1.6	56.3	0.048	40.4	0.1	0.5	1.0	57.9	0.014	0.008
18	'S18_10.jpg'	44.0	2.0	2.0	1.3	50.7	0.055	45.8	0.1	0.5	0.6	52.9	0.010	0.008
18	'S18_2.jpg'	40.6	1.5	1.1	1.3	55.5	0.036	41.8	0.2	0.4	0.7	56.8	0.010	0.008
18	'S18_3.jpg'	41.3	1.3	1.2	1.3	54.9	0.035	42.4	0.1	0.3	0.8	56.3	0.014	0.005
18	'S18_4.jpg'	41.6	1.4	1.3	1.3	54.3	0.038	43.0	0.1	0.5	0.7	55.6	0.011	0.008
18	'S18_5.jpg'	42.8	2.0	1.9	1.6	51.7	0.054	44.3	0.2	0.4	0.8	54.1	0.012	0.009
18	'S18_6.jpg'	38.3	2.0	1.7	1.5	56.4	0.052	40.3	0.1	0.4	0.8	58.3	0.010	0.006
18	'S18_7.jpg'	36.2	1.9	1.7	1.7	58.5	0.050	38.1	0.2	0.5	0.9	60.2	0.012	0.009
18	'S18_8.jpg'	38.5	1.8	1.8	1.6	56.3	0.049	40.0	0.3	0.7	0.9	58.0	0.013	0.013
18	'S18_9.jpg'	37.8	1.8	1.3	1.3	57.8	0.043	39.5	0.1	0.4	0.5	59.3	0.009	0.008

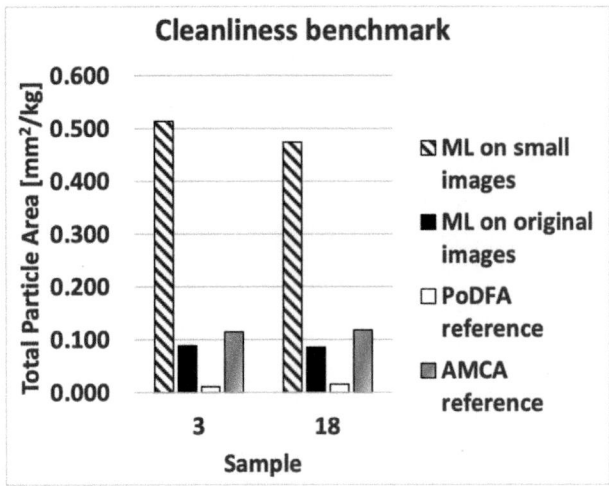

Fig. 5 Cleanliness benchmark comparing the total particle area resulting from ML on downsized and original images, PoDFA reference data, and the deterministic reference data from AMCA (Automated Metal Cleanliness Analyzer) that was generated for training the ML

Future Work

Future studies combining ML and deterministic image analysis for replicating manual PoDFA particle counts will investigate the benefit of increased computation power and, therefore, larger training data sets. Further, the convolutional layers of the neural network should be optimized for taking spatial cues into account, e.g., the relative distance of a potential target object from structural elements like the filter grain, which should decrease confounding artifacts induced from topological shadows and thereby increase the differentiation between filter grains and particles/inclusions/pores. Similar improvements may also be achieved by combining multiple neural networks specializing in certain visual features. The computational setup's performance on more heterogeneous data sets should also be tested. Demonstrating sufficient target phase detections of the same neural network for multiple types of samples would mark a milestone toward automating PoDFA counts.

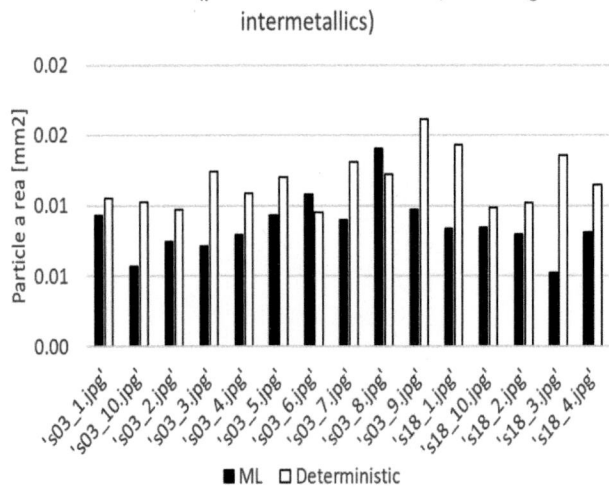

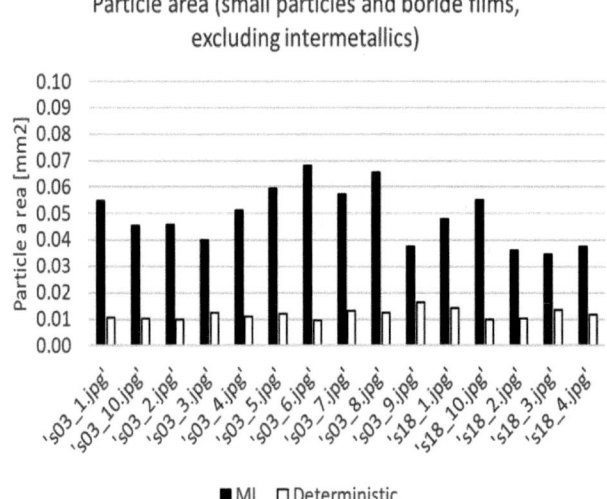

Fig. 6 Particle area distribution per sample image compared to AMCA (Automated Metal Cleanliness Analyzer) results. (Left: Original dimension images, Right: Rescaled images)

Acknowledgements The authors express their gratitude to the Department of Materials Science and Engineering and the Department of Chemistry at the Norwegian University of Science and Technology (NTNU) in Trondheim, Norway, as well as to Norsk Hydro ASA in Karmøy, Norway, for their continuous support of the project. Without the PoDFA micrographs received from Norsk Hydro ASA, the project would not have been possible.

References

1. "PoDFA, The complete solution for inclusion measurement, Inclusion identification and quantification analysis." ABB Inc., 2016. [Online]. Available: https://library.e.abb.com/public/b706913462934969befe277d80880795/PB_PoDFA-EN_A.pdf.

2. H. Zedel, R. Fritzsch, S. Akhtar, and R. E. Aune, "Estimation of Aluminum Melt Filtration Efficiency Using Automated Image Acquisition and Processing," in *Light Metals 2019*, C. Chesonis, Ed. Cham: Springer International Publishing, 2019, pp. 1113–1120. https://doi.org/10.1007/978-3-030-05864-7_136.

3. H. Zedel, R. Fritzsch, S. Akhtar, and R. E. Aune, "Automated Metal Cleanliness Analyzer (AMCA)—An Alternative Assessment of Metal Cleanliness in Aluminum Melts," in *Light Metals 2021*, L. Perander, Ed. Cham: Springer International Publishing, 2021, pp. 778–784. https://doi.org/10.1007/978-3-030-65396-5_102.

4. O. Ronneberger, P. Fischer, and T. Brox, "U-Net: Convolutional Networks for Biomedical Image Segmentation," in *Medical Image Computing and Computer-Assisted Intervention—MICCAI 2015*, vol. 9351, N. Navab, J. Hornegger, W. M. Wells, and A. F. Frangi, Eds. Cham: Springer International Publishing, 2015, pp. 234–241. https://doi.org/10.1007/978-3-319-24574-4_28.

5. "TensorFlow Module: tf.keras." 2022. [Python]. Available: https://www.tensorflow.org/api_docs/python/tf/keras.

6. "Open Source Computer Vision Library." 2015. [Online]. Available: https://github.com/opencv/opencv.

7. "MATLAB." The MathWorks Inc., Natick, Massachusetts, 2021.

8. P. V. Evans, P. G. Enright, and R. A. Ricks, "Molten Metal Cleanliness: Recent Developments to Improve Measurement Reliability," in *Light Metals 2018*, O. Martin, Ed. Cham: Springer International Publishing, 2018, pp. 839–846. https://doi.org/10.1007/978-3-319-72284-9_109.

Characterization of Low- and High Mg-Containing Aluminum White Dross Using Deterministic Image Analysis of EPMA Scans

Cathrine Kyung Won Solem, Hannes Zedel, and Ragnhild E. Aune

Abstract

White dross is a hazardous waste generated during the primary production of aluminum (Al) which consists of a heterogeneous mixture of different oxides and metallic Al in the form of large flakes, lumps, particles, and dust. Due to the heterogeneity of the dross, sampling is challenging. A sampling tool with step-by-step procedures for its use has previously been introduced, with recommendations for how to pulverize and analyze different fractions of the dross using X-Ray Diffraction (XRD). In the present study, Electron Probe MicroAnalysis (EPMA) scans of white dross samples collected from the holding furnace in an Al casthouse during the production of Al alloy 5182 (AlMg4.5Mn0.4) and 6016 (AlSi1.2Mg0.4) have been quantified and referenced to XRD analysis. The obtained results suggest that the image analysis method can quantitatively assess the influence of various process parameters on dross characteristics and thereby contribute to optimizing industrial furnace operations aiming to reduce dross formation.

Keywords

Aluminum • Dross • Characterization • Deterministic image analysis

Introduction

Aluminum (Al) is a diverse metal that has expanded its applications and use over the last decades. Its light weight, formability, and strength have positioned Al as the second most important material in our modern society, and the demand is still increasing [1]. The Al industry is trying to meet the high demand, and even though 40% and 50% of the post-consumer scrap is being recycled in Europe and US, respectively [2], it is still crucial that the Al production is optimized and that the losses through oxidation during production is kept to a minimum.

Oxidation

Oxidation of Al has been thoroughly investigated for decades, and several laboratory studies have reported that the oxidation rate of Al is enhanced by the additions of magnesium (Mg) to the melt and that the oxidation rate increases with increasing Mg concentration [3, 4]. E.g., Surla et al. [5] reported that the oxidation of Mg in AlMg5 alloys initiates the breakaway oxidation of MgO, resulting in the formation of $MgAl_2O_4$ (spinel) when the activity of Mg decreases below 0.023 at 700 °C. Moreover, More et al. [6] concluded that during oxidation of the AlMg alloy 3004, the Mg concentration decreased from 1.0 to 0.86 wt% and in the case of the Al alloy 5182 from 4.5 to 4.12 wt% Mg, indicating that the Mg had oxidized. As early as 1977, Cochran et al. [7] proposed a mechanism for the oxidation of AlMg alloys where amorphous MgO films would be formed and further crystallize and oxidize to MgO and $MgAl_2O_4$ until all of the Mg in the alloy was oxidized. The later studies confirmed in different ways this theory.

Dross Formation

Oxidation is also seen at an industrial scale, known as dross formation, and has been reported to be a challenge, both during primary and secondary Al production, generating white- and black dross, respectively. The dross is generated as an undesirable and hazardous waste [8], and the majority

C. K. W. Solem (✉) · H. Zedel · R. E. Aune
Department of Materials Science and Engineering, Norwegian University of Science and Technology (NTNU), Trondheim, Norway
e-mail: cathrine.k.w.solem@sintef.no;
cathrine.k.w.solem@ntnu.no

© The Minerals, Metals & Materials Society 2023
S. Broek (ed.), *Light Metals 2023*, The Minerals, Metals & Materials Series,
https://doi.org/10.1007/978-3-031-22532-1_119

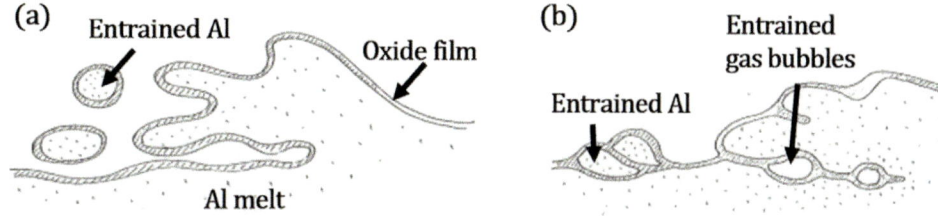

Fig. 1 Molten Al covered by an oxide film that **a** entrains metallic Al lumps and **b** gas bubbles [14, 16]

of the dross is still, on a global level, sent off to be landfilled [9]. From an economic point of view, the dross may contain up to 80% metallic Al, which, in view of this, is discarded rather than being commercially used or recycled [10].

The metallic Al concentrations in dross have been reported to vary [11, 12], and several parameters have, therefore, been addressed to try to influence the dross characteristics, e.g., the chemistry of the melt, injection of primary produced Al, furnace operation, and the skimming process [6, 13]. Even in view of this, <10% of the annual metallic Al produced in a casthouse today is still lost through either oxidization or by pure metallic Al being entrained within oxide films and other non-metallic compounds (NMCs) present in the dross as seen in Fig. 1 [14–16]. It is, in other words, both economically and environmentally beneficial that the dross formation is kept to a minimum.

Before casting the Al melt, removing the dross floating on top of the melt is necessary, which is commonly done by skimming the melt. The state-of-the-art procedure for skimming is based on the manual use of a large rake controlled by a skilled operator from a wheel truck [10]. The dross composition may vary from melt to melt and how it is skimmed from operator to operator, which causes an inhomogeneous morphology of the dross itself. To face some of these issues, the present authors developed and validated the use of a sampling tool and a step-by-step procedure for collecting reproducible dross samples from the holding furnace in the casthouse [17]. The work also tested suitable sampling preparation techniques, and X-Ray Diffraction (XRD) was used to quantify the results. The approach proved that crushing dross particles/lumps >1.25 mm was a challenge due to the high metallic Al content mixed together with oxides/NMCs. A more advanced technique, i.e., cryomilling, was then introduced to pulverize the samples. The cryomilling technique proved to be efficient, however, time consuming due to the requirements of an operating temperature of −196 °C.

Research Aim

In view of the challenges involved in pulverizing dross samples due to their inhomogeneity and thereby obtaining

reproducible analytical results, the present work aims to develop a novel method for quantifying the dross morphology based on deterministic image analysis of Electron Probe MicroAnalysis (EPMA) micrographs. Dross samples from both low- and high Mg-containing AlMg alloys will be collected from the holding furnace in the casthouse and analyzed by the proposed method. The results will later be compared to those obtained using the well-established XRD technique. The impact that the Mg concentration on the morphology of the produced dross will also be addressed.

Experimental Procedure

Sampling of Dross

Dross samples from four different locations in the casthouse holding furnace, i.e., Location 1, Location 2, Location 3, and Location 4 (see Fig. 2), were collected from two different AlMg charges, i.e., Charge A (high wt% Mg) and Charge B (low wt% Mg). The reason for this was that the distance from the injection point of the primary produced Al has been established to influence the dross characteristics [17]. After sampling, the collected dross samples were transported from the sampling tool to stainless steel trays to cool in the ambient air of the casthouse for one hour before being transferred into air-sealed containers. All samples were

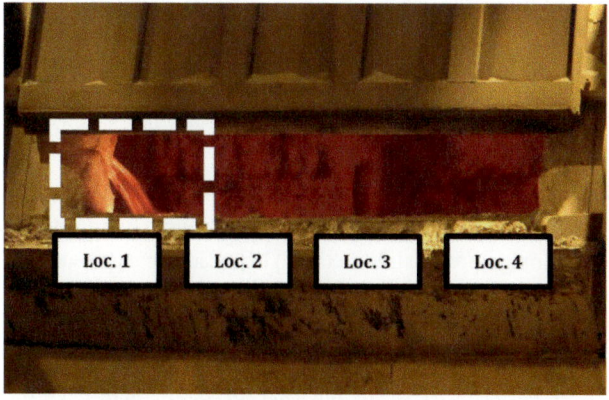

Fig. 2 Injection of primary produced Al from the left side of the casthouse holding furnace and positioning of Location 1 to Location 4

sieved into three fractions, i.e., <1.25, 1.25–4.5, and >4.5 mm, before being prepared for further analysis.

Sample Preparation

X-Ray Diffraction (XRD)

The smallest fraction (<1.25 mm) from each location for Charge A was pulverized using a ring mill, having three cycles with 20 s intervals. The powder was sieved between each cycle, and the powder being <100 μm was put aside, while powder being >100 μm was milled for another cycle, with a maximum of three cycles. The prepared powder <100 μm was then stored in glass containers awaiting further analysis.

Electron Probe MicroAnalysis (EPMA)

The mid-fraction (1.25–4.5 mm) for both Charge A and B was prepared for elemental mapping by Electron Probe MicroAnalysis (EPMA). Eight pieces/lumps of dross from each location were randomly selected, mounted in epoxy (EpoFix Resin), and polished with 1 μm diamonds. The polished samples were placed in a drying cabinet for ∼24 h before being coated with a 10–20 nm thin carbon (C) layer.

Analysis

XRD

The XRD measurements were carried out by a D8 A25 DaVinci Diffractometer (Bruker, Billerica, USA) instrument. Monochromatic CuKα radiation with a wavelength of λ = 1.5406 Å was applied for a scattering angle of 2θ = 6–100°. The results were then quantified by the TOPAS software (version 5.0, Billerica, USA).

EPMA

A JEOL JXA-6500F (JEOL Ltd., Japan) Field Emission EPMA was used for the elemental mapping of the polished dross samples. The same instrumental settings were applied for all mappings, ensuring that the same elements were analyzed from the same detector, as well as giving the same angle between the sample surface and the detector.

Deterministic Image Analysis

The EPMA scans were quantified by a previously introduced digital image processing method [17]. Each scan provided a 2D-grayscale intensity map per element where each pixel represents that element's concentration (Cn) at the pixel's location in percent. The pixel values ranged from an intensity of 0 (0% Cn) to an intensity of 255, relating to the maximal Cn calibrated for that element in a given scan. The calibration was adjusted in each scan for the best visual contrast. In some cases, the maximal Cn calibration may have exceeded 100%, which was considered for all quantitative metrics.

Preprocessing: Each EPMA scan was provided as a collage of a Scanning Electron Microscopy (SEM, here noted as CP) image, as well as up to 15 elemental maps of different elements secured from the same region. All scans were imported to MATLAB version R2021a[x] as 2-dimensional matrices in 8-bit unsigned integer (uint8) format, and each elemental map was individually cut at identical dimensions. The scale bars within each rectangular elemental scan were not part of the quantitative analysis and were therefore subtracted from the sample area. For each elemental map, the maximal Cn calibration was extracted from the EPMA scan color bars, indicating the Cn at the highest pixel intensity of 255. To make the data comparable as absolutes between differently calibrated scans, each elemental map was normalized by its maximal Cn to represent a Cn of 100% at a pixel intensity of 255 for all maps.

Concentrations per scan for each element: For each elemental map, the maximum Cn of 100% was defined as the total number of pixels multiplied by the maximal intensity:

$$Cn_{max} = n \cdot 255$$

where n is the total number of pixels per elemental map. The total concentration of an element was then defined as a percentage derived as the sum of all pixel intensities divided by Cn_{max}:

$$Cn_{element} = \frac{\sum_{i=1}^{n} x_i}{Cn_{max}}$$

where x_i is the intensity value of a given pixel between 0 and 255 (individual pixels may represent more than 100% pixel Cn in the Σ term if the calibration is set to a maximum above 100% for the pixel intensities). In theoretical constructed cases, this could lead to a $Cn_{element} > 100\%$, however, screening of preliminary results showed that $Cn_{element}$ never exceeded 80%. This can be attributed to EPMA scans designed to create visual contrast that includes a certain number of pixels at lower intensities. The resulting concentrations for each element and each EPMA scan can then be pooled to derive statistical metrics from different experimental conditions or series.

Binary mapping of EPMA scans: Due to the penetration depth of the microbeam, EPMA scans represent a volume of the sample displayed as 2-dimensional micrographs. The penetration depth depends on the densities of compounds at different locations of the sample, so the total sample volume represented in one EPMA scan depends on the sample composition and can be neither generalized nor reliably estimated. It should be noted that, at any given pixel location, different materials may coincide at the same location at different depths. For deriving quantitative metrics about the metallics and oxides/NMCs content, the EPMA scan areas were subdivided into target classes by the predominant composition at each pixel location. Locations high in epoxy or pores were detected first. These were considered not representative of the sample composition and therefore excluded from further quantifications. The remaining sample area was divided into metallic-dominant and oxide/NMC-dominant areas. Thereby, each pixel of an EPMA scan was classified as either pore/epoxy, metallic, or oxide/NMC, based on the predominant content at the location. The relative content of each target class per scan can then be described as a percentage of the sample area and be pooled with other scans according to experimental conditions for deriving statistical metrics. The congruence of the elemental maps of the same EPMA scan enabled coincidence mapping, where the detection criteria can be defined based on combined Cns at the same pixel location of different elemental maps. Due to the refraction of the microbeam and sensor angles, the elemental maps for different elements of the same scan had slight offsets considered negligible for the present study.

Pore/epoxy, metallics, and oxides/NMC detections: Based on the Cn at each pixel location for different elements, three criteria were set to define pore- or epoxy-dominant areas:

- Very high C-areas: C Cns above 70 were considered epoxy-dominant areas.

- Moderate C-areas with no oxygen (O): C concentrations >35 in combination with O $Cns < 4$ were also considered epoxy-dominant areas. Preliminary screenings showed that moderate C-areas occasionally coincide with otherwise clear oxide areas, which were higher in O.

- Low C-areas without indications of metallics or oxides/NMCs: Carbon $Cns < 3$ at pixels showing Al $Cns < 24$ and O $Cns < 4$ were considered pore-dominant areas.

Detections from all three categories were excluded from further quantifications so that the derived metrics are based on each sample's non-epoxy/pore areas. The remaining non-pore/epoxy area was then categorized based on each pixel's O Cn, where a $Cn > 4$ was defined as oxide/NMC-dominant and less than 4 as metallic-dominant. All Cn thresholds were iteratively determined through qualitative screenings of the resulting detections, comparing the binary detections of each target phase in the context of all elemental maps and the CP image of each sample.

Quantitative analyses: By applying the same criteria to all normalized EPMA scans, statistically comparable data could be derived. For each experimental condition, there were at least 8 EPMA scans (up to 1 of which were excluded per experimental condition due to excessive cryolite content) that were pooled together for deriving statistical metrics. This includes the mean and median metallic and oxide/NMC content as percentages of the micrographs. These metrics can then be used in relative comparisons between different

Fraction	Phase	Phase distribution/wt%			
		Loc. 1	Loc. 2	Loc. 3	Loc. 4
<1.25 mm	**Aluminum, Al**	**87.9**	**84.3**	**32.9**	**78.0**
	Magnesium oxide, MgO	7.8	8.7	8.4	15.1
	Spinel, $MgAl_2O_4$	–	0.2	1.7	0.1
	Defect spinel, $Mg_{0.388}Al_{2.408}O_4$	1.0	2.4	24.3	0.9
	Aluminum oxide, Al_2O_3	0.9	1.8	27.9	4.7
	Sodium aluminum oxide, $NaAl_{11}O_{17}$	0.5	0.2	0.6	0.1
	Oxide content, total	**10.2**	**13.3**	**62.9**	**20.9**
	Aluminum nitride, AlN	1.2	2.0	4.2	1.1
	Silicon, Si	0.7	0.4	–	–
1.25–4.5 mm		**EPMA Cn, /%**			
	Aluminum, Al	**88.0**	**93.0**	**67.2**	**95.3**
	Oxide content, total	**12.0**	**7.0**	**32.8**	**4.7**

Table 1 Table 1 Results from X-Ray Diffraction (XRD) and mean EPMA quantifications by deterministic image analysis of Charge A, Location 1 to Location 4 for given fractions

conditions and are expected to replicate similar trends as complementary characterization methods such as XRD.

Results and Discussion

Deterministic Image Analysis of EPMA Scans as a Method for Quantitative Analyses of Dross Samples

The image analysis method enabled acquiring quantitative metrics from EPMA scans, which is a novelty in the field. Relevant metrics include the Cn of each element per elemental map, as well as binary maps of target phase detections based on Cn thresholds of one or multiple elements for each pixel location. For Charge A, each location's mean metallic and non-metallic Cns were derived by quantifying the EPMA micrographs. The resulting metrics are shown in Table 1 along with the corresponding XRD results. The metrics for target phase detections used by XRD (wt%) and the EPMA quantification (%) for the 4 locations showed similar trends and strong correlations, i.e., 0.924 and 0.916 for metallic and non-metallic detections, respectively. This suggests equivalence of the indicative value of the novel EPMA metric and thereby marks the first empirical validation of the approach.

Traditionally, EPMA scans are also analyzed qualitatively by expert operators. For further validating the quantification of target phases, a qualitative screening of all images, including all elemental maps and the corresponding CP images, was carried out and verified adequate detection criteria based on traditional EPMA interpretations, i.e.,

interpreting the contrast in the CP image in the context of the relevant elemental maps. The screening confirmed robust detections across all samples in the dataset.

Quantification by XRD and EPMA/Image Analysis

The obtained XRD results confirmed previously reported results [17] and revealed that the metallic Al concentration in the dross decreased from Location 1 to Location 4 simultaneously as the oxide/NMC concentration increased, see Table 1. A deviation from the observed trend was, however, observed for Location 3, where a significant drop in the metallic Al concentration and instead an increase in the oxide/NMC concentration was identified. It is believed that this deviation resulted from inconsistent furnace operations when the floating oxides/NMCs were carefully skimmed to the furnace gate without being removed from the holding furnace. The average concentrations for Charge A for the smallest fraction <1.25 mm were 70.8 wt% metallic Al and 26.8 wt% oxides/NMCs.

The EPMA and deterministic image analysis for the same charge for the mid-fraction 1.25–4.5 mm also showed similar trends as observed from the XRD analysis, having higher metallic Al values for Location 1 than Location 4. Representative results from the EPMA and deterministic image analysis for Charge A, fraction 1.25–4.5 mm, Location 1 are presented in Fig. 3, where it can be seen that the binary maps for metallic detection, oxide detection, and epoxy/pore detection correlate well with the elemental mappings from the EPMA (top line). A combination of the three binary maps is also presented, where the red, yellow, and green

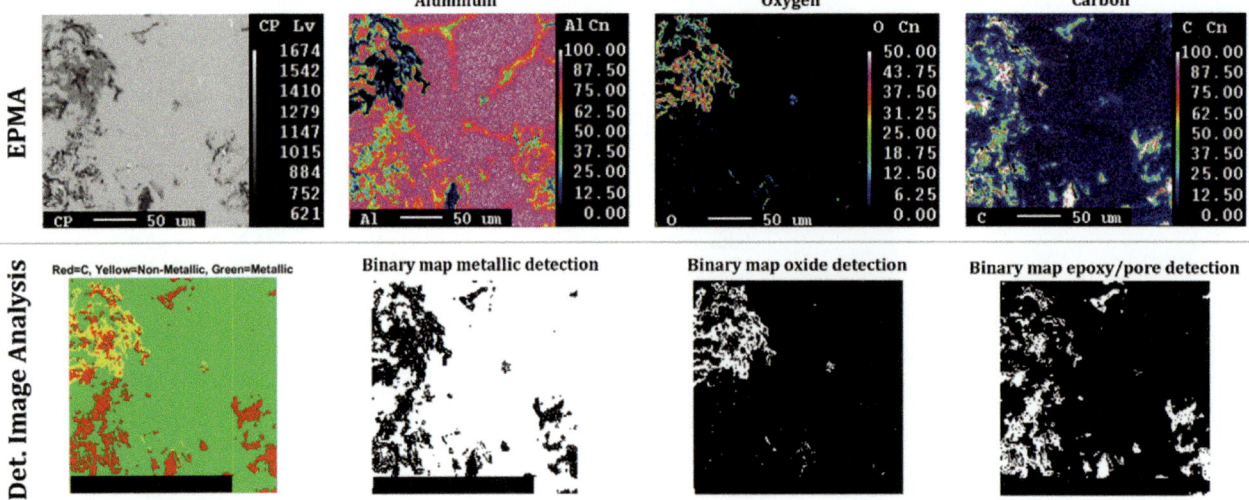

Fig. 3 Representative results of Charge A, fraction 1.25–4.5 mm, Location 1, where the top line show EPMA results of Al, O, and C, and the bottom line shows the deterministic image analysis results, as well as the binary maps of metallic detection, oxide detection, and epoxy/pore detection

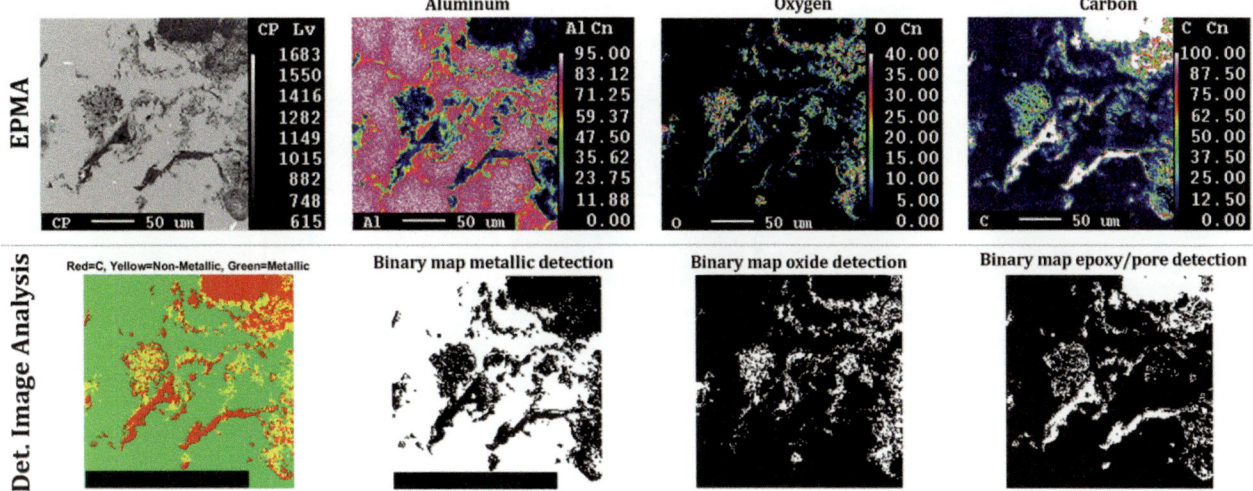

Fig. 4 Representative results of Charge A, fraction 1.25–4.5 mm, Location 4, where the top line show EPMA results of Al, O, and C, and the bottom line shows the deterministic image analysis results, as well as the binary maps of metallic detection, oxide detection, and epoxy/pore detection

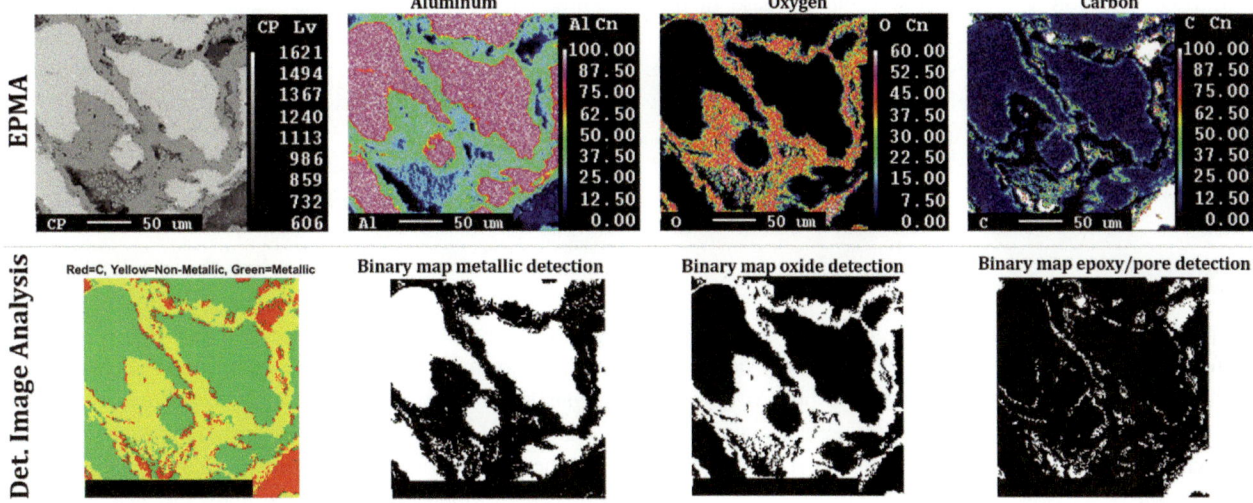

Fig. 5 Representative results of Charge A, fraction 1.25–4.5 mm, Location 3, where the top line show EPMA results of Al, O, and C, and the bottom line shows the deterministic image analysis results, as well as the binary maps of metallic detection, oxide detection, and epoxy/pore detection

represent C/pores, non-metallic, and metallic areas, respectively. A similar analysis is given for Charge A, fraction 1.25–4.5 mm, Location 4 in Fig. 4, where it can be seen that more pores are entrained within the bulk (as seen in Fig. 1b).

On an overall average for Charge A, fraction 1.25–4.5 mm, a metallic Al concentration of 85.6 wt% was obtained, which is higher than for the smallest fraction <1.25 mm obtained by XRD. This correlates well with the literature, where it is proposed that larger dross particles have a higher concentration of metallic Al [18].

EPMA and deterministic image analysis was also carried out for Charge A, fraction 1.25–4.5 mm, Location 3, to investigate the deviation observed from the XRD analysis, see Fig. 5. As seen from the figure, thick oxides are present in the Al matrix confirming the significant drop in the metallic Al concentrations identified by XRD. It can also be seen that the novel deterministic image analysis method provides reproducible quantifying data comparable to the XRD results.

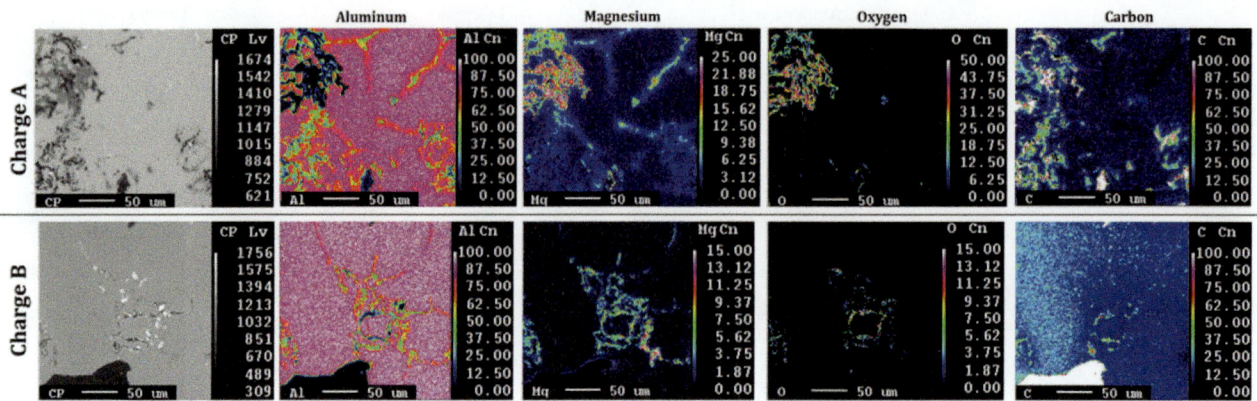

Fig. 6 Representative EPMA results of Al, Mg, O, and C of fraction 1.25–4.5 mm, Location 1 for **a** Charge A, and **b** Charge B. A clear increase in the relative concentration scale bar of Mg and O is seen for Charge A (high Mg-containing Al alloy)

When comparing the EPMA and deterministic image analysis results from Charge A (having high Mg concentrations in the Al alloy) and Charge B (having low Mg concentration in the Al alloy) for the mid-fraction 1.25–4.5 mm for Location 1 to Location 4, an increase in the metallic Al concentration from 85.6 to 94.3 wt% is observed. The increase is believed to be related to the lower Mg concentrations in the melt, see scale bars in Fig. 6, which has been reported to influence the oxidation rate, hence, the dross formation.

Limitations

The quantification of the EPMA scans simplifies the raw data for deriving quantitative metrics. The binary detection criteria for each target phase may lead to both over- and undercounting of the target phase under certain conditions. As previously mentioned, as EPMA elemental maps represent a volume of varying depths, multiple target phases may coincide at the same pixel location. In view of the present definition, the predominant phase is then allocated to being present at that location, whereas other phases are considered not present, which undercounts the neglected target phases. Due to the binary nature of the detection, higher contents of a target phase beyond the detection criteria definition are not reflected in the data, potentially leading to undercounting of the detected target phase. The relevance of these deviations depends on the purposes of the data and the goals of corresponding investigations. It should, however, be noted that the distinction of target phases in EPMA micrographs depends on the detection criteria, i.e., the tuning of elemental thresholding. These thresholds may require retuning to new

datasets depending on the composition of the samples. As EPMA is typically calibrated for each image individually, this leads to a significant labor investment for extracting the relevant data to normalize the elemental maps for quantitative comparability. This step could be prevented by using the same EPMA calibrations for all samples, ignoring the visual contrast of the micrographs so that the raw data is comparable directly without further preprocessing.

Summary and Conclusions

Industrial Al white dross samples have been collected from two charges of AlMg alloys, i.e., Charge A (high Mg-containing Al alloy) and Charge B (low Mg-containing Al alloy), to investigate if a new novel deterministic image analysis of EPMA images is suitable to quantitatively analyze industrial dross samples. The analysis was compared to XRD measurements and evaluated in view of the Mg concentration's influence on the dross characteristics.

Based on the EPMA and deterministic image analysis results, it is concluded that the quantification method provides reliable results in the same range as the XRD results. A higher metallic Al concentration was identified for the dross from Charge B (low Mg-containing Al alloy), indicating that the oxidation rate has been lowered, causing less dross formation.

In view of the present results, quantifying EPMA micrographs using simple deterministic image analysis methods may serve as an alternative to XRD characterizations of dross samples, thereby contributing to expanding the knowledge of dross characteristics. The proposed approach can also be more accessible for a broader range of laboratories compared to the high-resolution instrument and corresponding infrastructure needed for XRD, as well as offer simplified sample processing.

Future Work

Future work will include an extension of the study on the dross formation on an industrial scale. An experimental set-up allowing for cooling under a protective atmosphere will be developed to investigate the impact of the cooling environment on the dross characteristics. Trials with atmospheres containing synthetic air and 5% CO_2 will be performed, as small amounts of CO_2 have proven to inhibit the oxidation of AlMg alloys at a laboratory scale. In view of the proposed novel deterministic image analysis method, the statistical robustness of the EPMA indicator will be confirmed by studies with larger datasets.

Acknowledgements This research has been funded by the SFI Metal Production (Centre for Research-based Innovation, pr. nr. 237738). The authors gratefully acknowledge the financial support from the Research Council of Norway and the partners of the SFI Metal Production.

References

1. Cullen JM, Allwood JM (2013) Mapping the Global Flow of Aluminum: From Liquid Aluminum to End-Use Goods. Environ Sci Technol 47:3057–3064. https://doi.org/10.1021/es304256s
2. Xiao Y, Reuter MA, Boin Udo (2005) Aluminium Recycling and Environmental Issues of Salt Slag Treatment. Journal of Environmental Science and Health, Part A 40:1861–1875. https://doi.org/10.1080/10934520500183824
3. Cochran CN, Sleppy WC (1961) Oxidation of High-Purity Aluminum and 5052 Aluminum-Magnesium Alloy at Elevated Temperatures. J Electrochem Soc 108:322–327. https://doi.org/10.1149/1.2428080
4. Haginoya I, Fukusako T (1983) Oxidation of Molten Al–Mg Alloys. Trans JIM 24:613–619. https://doi.org/10.2320/matertrans1960.24.613
5. Surla K, Valdivieso F, Pijolat M, et al (2001) Kinetic Study of the Oxidation by Oxygen of Liquid Al–Mg 5% Alloys. Solid State Ionics 143:355–365. https://doi.org/10.1016/S0167-2738(01)00861-X
6. More KL, Tortorelli PF, Walker LR, et al (2003) Microstructural Evaluation of Dross Formation on Mg- and Non-Mg-Containing Al Alloys from Industrial Furnaces. Materials at High Temperatures; Leeds 20:453. https://doi.org/10.3184/096034003782748838
7. Cochran CN, Belitskus DL, Kinosz DL (1977) Oxidation of Aluminum-Magnesium Melts in Air, Oxygen, Flue Gas, and Carbon Dioxide. MTB 8:323–332. https://doi.org/10.1007/BF02657663
8. Mahinroosta M, Allahverdi A (2018) Hazardous Aluminum Dross Characterization and Recycling Strategies: A Critical Review. Journal of Environmental Management 223:452–468. https://doi.org/10.1016/j.jenvman.2018.06.068
9. Aluminium Dross Processing: A Global Review. In: alcircle. https://www.alcircle.com/specialreport/306/drossprocessing. Accessed 22 Oct 2020
10. Peterson RD (2011) A Historical Perspective on Dross Processing. Materials Science Forum 693:13–23. https://doi.org/10.4028/www.scientific.net/MSF.693.13
11. Drouet MG, LeRoy RL, Tsantrizos PG (2013) Drosrite Salt-Free Processing of Hot Aluminum Dross. In: Stewart DL, Daley JC, Stephens RL (eds) Recycling of Metals and Engineered Materials. John Wiley & Sons, Inc., Hoboken, NJ, USA, pp 1135–1145
12. Capuzzi S, Timelli G (2018) Preparation and Melting of Scrap in Aluminum Recycling: A Review. Metals 8:249. https://doi.org/10.3390/met8040249
13. Prakash M, Pereira GG, Cleary PW, et al (2010) Validation of SPH Predictions of Oxide Generated during Al Melt Transfer. Progress in Computational Fluid Dynamics, an International Journal
14. Campbell J (2003) Castings. Elsevier
15. Ünlü N, Drouet MG (2002) Comparison of Salt-Free Aluminum Dross Treatment Processes. Resources, Conservation and Recycling 36:61–72. https://doi.org/10.1016/S0921-3449(02)00010-1
16. Solem CKW (2022) Parametric Study of Molten Aluminium Oxidation in Relation to Dross Formation at Laboratory and Industrial Scale
17. Solem CKW, Deledda S, Tranell G, Aune RE (2022) Sampling Procedure, Characterization, and Quantitative Analyses of Industrial Aluminum White Dross. Submitted July 2022 to the Journal of Sustainable Metallurgy
18. Kudyba A, Akhtar S, Johansen I, Safarian J (2021) Aluminum Recovery from White Aluminum Dross by a Mechanically Activated Phase Separation and Remelting Process. JOM 73:2625–2634. https://doi.org/10.1007/s11837-021-04730-x

Assessment of Separation and Agglomeration Tendency of Non-metallic Inclusions in an Electromagnetically Stirred Aluminum Melt

Cong Li, Thien Dang, Mertol Gökelma, Sebastian Zimmermann, Jonas Mitterecker, and Bernd Friedrich

Abstract

Presence of non-metallic inclusions (NMIs) reduces surface quality and mechanical properties of aluminum products. The development of good NMIs removal practices relies on the understanding of inclusion behaviors with respect to separation and agglomeration particularly in the turbulent flow. In the scenario of electromagnetically induced recirculated turbulent flow, the concerned behaviors of inclusions with different sizes have rarely been investigated experimentally. In the presented study funded by AMAP Open Innovation Research Cluster, reference materials were prepared with uniformly distributed NMIs (SiC and $MgAl_2O_4$) via an ultrasound-involved casting route. Reference materials were charged into an aluminum melt where turbulent flow was promoted via electromagnetic force. Microscopical analysis shows non-significant agglomeration tendency of SiC, $MgAl_2O_4$, and TiB_2 inclusion. Time-weight filtration curve, PoDFA, and Spark Spectrometer results suggest a strong dependence of separation rate on particle size. Analytical models were established to estimate the collision rate of particles and to evaluate separation probability of different sized particles.

Keywords

Non-metallic inclusions • Melt cleanliness • Separation • Agglomeration

Introduction and Motivation

During primary and secondary aluminum production, non-metallic Inclusions (NMIs) are formed during smelting, re-melting, and melt handling processes. Without proper NMIs removal process, NMIs pose difficulties in subsequent metal working processes and dramatically deteriorate the surface quality and mechanical property of aluminum products [1].

Conventional NMIs removal processes namely sedimentation, salt fluxing, gas purging, and filtration have been well established in productions [2]. The concept of improving the conventional NMIs removal process based on forced agglomeration of the particle was proposed firstly by Szekely [3]. The author suggested that intensive mixing can promote the coagulation of particles and bigger particles can be easily removed by either floatation or filtration. In simulation works, SiC particles and Al_2O_3 particles were also reported to have a tendency to form agglomerates in the aluminum melt [4, 5]. Marechal et al. [6] reported agglomeration trend of Al_2O_3 and SiC particles with aluminum melt in laminar flow conditions. Nevertheless a robust way to evaluate the size evolution during particle agglomeration was missing. So far the experimental investigation of NMI particles agglomeration behavior in turbulent flow conditions is rare to find. Not to mention the study of agglomeration behavior of practical inclusions namely $MgAl_2O_4$ which is frequently found in 5xxx alloys. The reasons lie mainly in difficulties of introduction of inclusions into the melt and dispersing them homogeneously. Commercial metal matrix composite (MMC) offers a simple way to introduce of NMIs. For instance using Duralcan® MMC, particles like SiC, Al_2O_3 can be readily introduced into the melt. Nevertheless

C. Li (✉) · B. Friedrich
IME Process Metallurgy and Metal Recycling, RWTH Aachen University, 52056 Aachen, Germany
e-mail: cli@ime-aachen.de

T. Dang
TRIMET Aluminium SE, 45356 Essen, Germany

M. Gökelma
Department of Materials Science and Engineering, Izmir Institute of Technology, 35430 Izmir, Turkey

S. Zimmermann · J. Mitterecker
Former student of IME Process Metallurgy and Metal Recycling, RWTH Aachen University, 52056 Aachen, Germany

© The Minerals, Metals & Materials Society 2023
S. Broek (ed.), *Light Metals 2023*, The Minerals, Metals & Materials Series,
https://doi.org/10.1007/978-3-031-22532-1_120

particles in the commercial MMC are often clustered to a different extent and type of particles cannot be customized to meet specific interests.

Electromagnetic (EM) separation, as a relative new separation process, has been proposed since decades yet the application of such process on removal of NMIs from aluminum melt remains in lab-scale [7]. Theoretical [8, 9] works including simulation ones have demonstrated the fundamental principle of EM separation. Meanwhile a simulation work is performed in the heavy metal melt predicted particle size-dependent separation rate [10]. With respect to aluminum melt, several trials demonstrated the effectiveness of EM separation [11]. Nevertheless, to the best knowledge of authors, no trials havre been performed on examining separating rate difference of NMIs with different sizes during EM separation treatment.

In the presented study funded by Advanced Metals and Process (AMAP) Open Innovation Research Cluster, we firstly developed process routes capable of fabricating reference materials with customized inclusion type and meanwhile well-dispersed inclusions. SiC, $MgAl_2O_4$, and TiB_2 particles were charged into the melt via either reference materials or master alloy to represent practical carbide, oxide, and boride inclusions present in 5xxx aluminum alloy. An electromagnetically driven melt was employed to provide intensive stirring conditions for agglomeration study. The treated melt were evaluated via metallography, time-weight filtration curve, PoDFA, and Spark Spectrometer analysis. In the meantime, a separation study was conducted in two different scale furnaces to clarify the size dependent separation rate of particles in the electromagnetically driven flow. Analytical models were established to help interpret experimental findings.

Experimental Methodology

Reference Materials Preparation

Two types of reference materials were fabricated in order to ensure a controlled way for introducing artificial inclusions with a defined concentration as well as particle size distribution.

1st type of reference material A356-SiC was fabricated by diluting Duralcan® Metal Matrix Composite (MMC) which contains ca. 30 vol% SiC particles. The dilution was made under a protective atmosphere in a pre-cleaned parent A356 melt with assistance of a rotor. The melt was mixed for certain time upon charging of MMC and afterwards a final casting was performed to obtain reference materials. 2nd Type of reference material namely 5xxx-$MgAl_2O_4$ was obtained by a combination of stir-casting and ultrasound melt treatment. $MgAl_2O_4$ powders were charged onto the surface of a commercial pure aluminum melt which was agitated via a rotor under a protective atmosphere. Afterwards the melt with suspended particles were casted into a steel mould. The upper part of the obtained ingot enriched with $MgAl_2O_4$ particles (incl. clusters) and pores were sectioned and charged into another prepared 5xxx parent melt. Ultrasound melt treatment was performed for the 5xxx-$MgAl_2O_4$ melt for the sake of degassing and simultaneously break up the big clusters. Treated melt was casted to obtain reference materials (weighing 1–2 kg). Both type of reference materials were sectioned into small pieces for further characterization and usage.

Separation and Agglomeration Trials

The investigation of particle separation and agglomeration was performed using two vacuum induction furnaces (VIF) in difference scales. Table 1 lists dimension, and electromagnetic (EM) parameter of the two used induction furnaces.

For 0.1 L VIF, the set-up snapshot and schematic of the melting facility is shown in Fig. 1. The trials started by first of all preparing base melt (5083 alloy). Afterwards inclusions (SiC, $MgAl_2O_4$) were introduced via the reference materials. Upon charging, the furnace was powered on to generate EM stirring for 60 s. When stirring was completed, the melt along with steel crucible was taken out of the coil and quenched in the water tank.

The trial procedure in 1 L VIF is given in Fig. 2. 5083 alloy was used to prepare the base melt. The main difference comparing with 0.1 L VIF trials is the integration of a melt cleaning step (Fig. 2b). Reference material

Table 1 Dimension, electromagnetic parameter of vacuum induction furnaces (VIF)	Name	0.1 L vacuum induction furnaces	1 L vacuum induction furnaces
	Height of crucible (net value, m)	0.065	0.19
	Diameter of crucible (net value, m)	0.04	0.11
	Frequency (kHz)	10	4
	Number of inductor turns-N_c	6.75	5
	Length of coil (m)-l_c	0.146	0.081
	Inductor current (A)-I_c	540	1423

Fig. 1 0.1 L vacuum induction furnace trial procedures: **a** set-up snapshot; **b** schematic of heating facility; **c** sectioning; **d** assessed columns, blue arrow marked radial direction

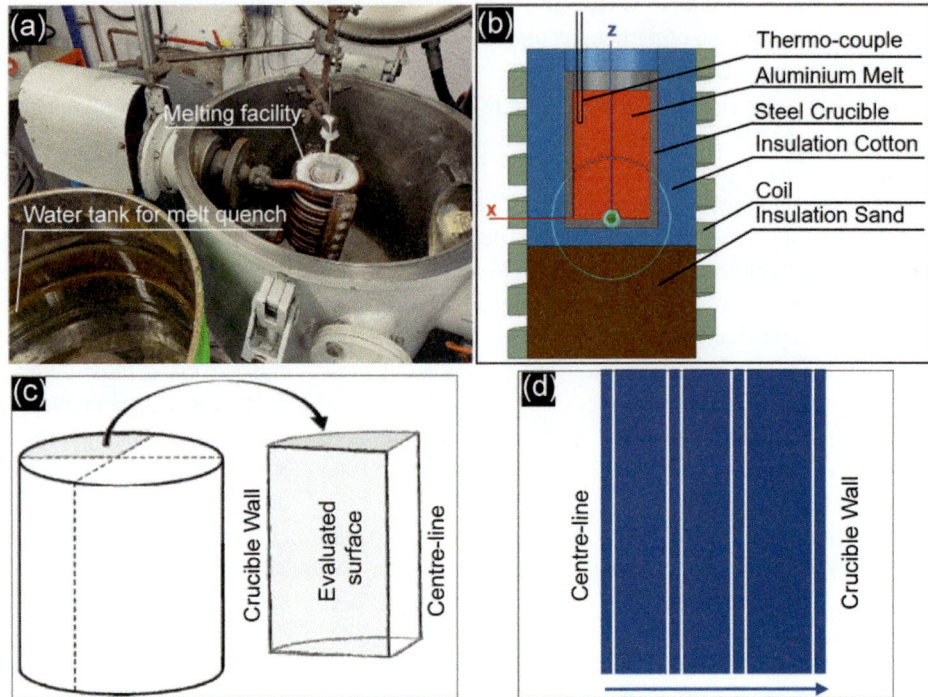

Fig. 2 Procedure of 1 L vacuum induction furnace trial. EM stands for electromagnetic

A356-SiC and master alloy Al-5Ti-1B in the coil form were used in 1 L VIF trials (Fig. 2c, d). Evaluation of inclusions were conducted for melts in non-stirred status and electromagnetically stirred status. Trial parameters for both 0.1 L VIF and 1 L VIF scenarios are listed in Table 2.

Inclusions Evaluation

The quenched ingots from 0.1 L VIF trial were sectioned and subject to metallographic preparation, as is shown schematically in Fig. 1c. Upon metallograpic preparation of

the sample surface, particle spatial distribution and size distribution were assessed and compared with that of the reference materials. Figure 1d shows schematically the evaluation region of the sample in a white column. Optical Micrographs (OMs) were transformed from RGB into binary graphs using Matlab®. The same software was used to quantify particles with respect to area fraction and size distribution within each view field.

The filtration process through a PoDFA set-up was recorded in real-time videos. In each of the 12 trials, the melt was at the same temperature before transferring into the PoDFA crucible and the vacuum level generated beneath the

Table 2 Parameter of trials and evaluation methods

Vacuum induction furnace (L)	Trial Nr.	Inclusion	Conc.	EMS* time (s)	Evaluation method
0.1	1	SiC	0.1 wt %	60	Metallography
	2	$MgAl_2O_4$	0.1 wt %	60	
1	3, 4, 5	SiC	25 ppm	0	Metallography (inc. PoDFA) filtration-curve SS**
	6, 7, 8	SiC	25 ppm	60	
	9, 10, 11	TiB_2	39 ppm	0	
	12, 13, 14	TiB_2	39 ppm	60	

* EMS stands for electromagnetic stirring
** SS stands for spark spectrometer

filter disc during filtration was the same. The data concerning the filtrated mass value (kg) as a function of time (s) was extracted and plotted into time-weight filtration curve. Solidified PoDFA samples containing the filter and possibly the filtration cake were sectioned and prepared metallographically. Afterwards OMs were taken to evaluate the collected inclusions. Spark Spectrometer (SS) analysis were performed for samples taken right before PoDFA filtration.

Results

Characterization of Reference Materials

Figure 3A shows Particle Size Distribution (PSD) of both commercial metal matrix composite (MMC) and A356-SiC reference materials. Note that the normalized particle number fraction within each size sub-range (e.g., 35–40 μm) is represented by a single data point. The graph suggests that with our dilution process, particles were well dispersed. Meanwhile big clusters were broken up with an average particle size reduced from 16.6 to 12.1 μm.

Figure 3b shows PSD of the 5xxx-$MgAl_2O_4$ reference material. The graph demonstrates the effectiveness of the developed process route, i.e., a combination of stir-casting and ultrasound melt treatment. $MgAl_2O_4$ particles were well dispersed with an average size of 21.8 μm.

Agglomeration Behavior of Particles

Agglomeration trials were mainly performed in a 0.1 L vacuum induction furnace (VIF). For SiC particles, normalized particle area fraction for each sub-group of size (e.g., 35–40 μm) was plotted as a function of the particle size (Fig. 4a). The comparison between the sample obtained from 60 s electromagnetically stirred melt and A356-SiC reference material suggest no evidence of agglomeration. On the contrary, a few big clusters seem to be broken up, leading to a shift of particle size to a smaller range and a minor decrease of average of particle diameter. A similar trend of de-agglomeration of particles were observed for the case of $MgAl_2O_4$ (Fig. 4b).

Fig. 3 Particle size distribution in reference materials: **a** A356-SiC; **b** 5xxx-$MgAl_2O_4$. Quantitative results each from at least 40 view fields. Inserts are example optical micrographs

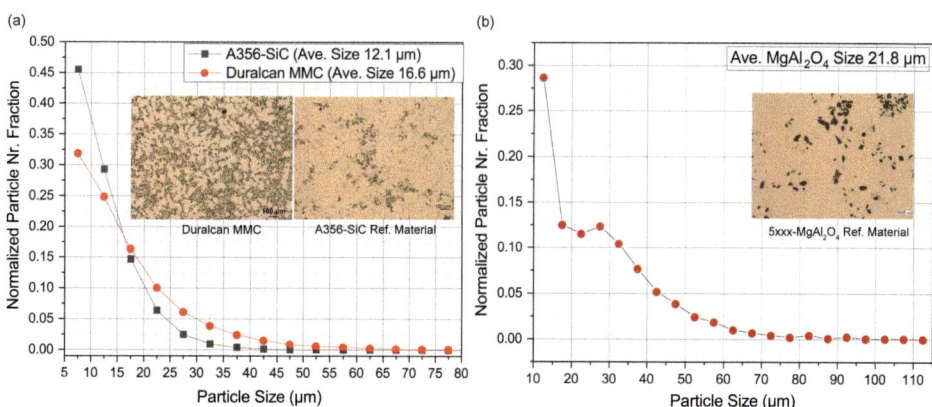

Fig. 4 Normalized particle area fraction within each sub-range of particle size: **a** SiC; **b** MgAl$_2$O$_4$. Quantitative results of A356-SiC and 5xxx-MgAl$_2$O$_4$ each from 160 view fields. Inserts are example optical micrographs

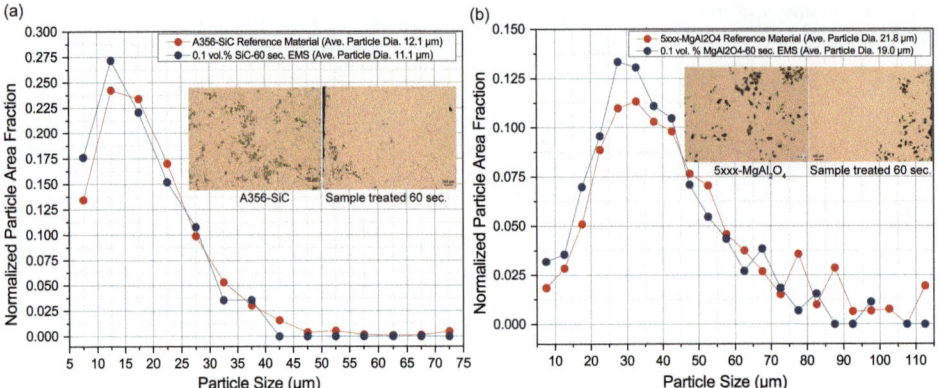

Separation Behavior of Particles

In the 0.1 L VIF scenario, the separation of SiC and MgAl$_2$O$_4$ particles in the near-wall region is shown in Fig. 5. The normalized area fraction of each column (Fig. 1d) for either SiC or MgAl$_2$O$_4$ particle is plotted as a function of radial distance to the center line. The result shown in Fig. 5 suggests that more MgAl$_2$O$_4$ (19 μm) were separated than SiC (11 μm) particles under the same EM parameter.

In the 1 L VIF scenario, the separation of SiC particles was evaluated via time-weight filtration curve and PoDFA metallography. Figure 6a gives the time-weight filtration curve of the melt after 0 s EMS and 60 s EMS for SiC

particle trials. A noticeably higher filtration rate can be observed for the melt after 60 s stirring. PoDFA metallographic result shown in Fig. 7 suggests a nearly 100 vol% reduction of SiC in the melt. This offers a reasonable explanation for the increased filtration rate shown in Fig. 6a.

In 1 L VIF scenario, the separation of TiB$_2$ particles was evaluated via time-weight filtration curve, PoDFA metallography, and Spark Spectrometer (SS) analysis. Figure 6b gives the time-weight filtration curve of the melt after 0 s EMS and 60 s EMS for TiB$_2$ particle trials. Unlike SiC particles, the two curves coincide with each other very well. If one assumes the filtration curve is sensitive primarily to inclusions amount and secondarily to inclusion size distribution, the filtration curve of TiB$_2$ particles does not suggest

Fig. 5 Normalized particle (SiC and MgAl$_2$O$_4$) area fraction in radial direction

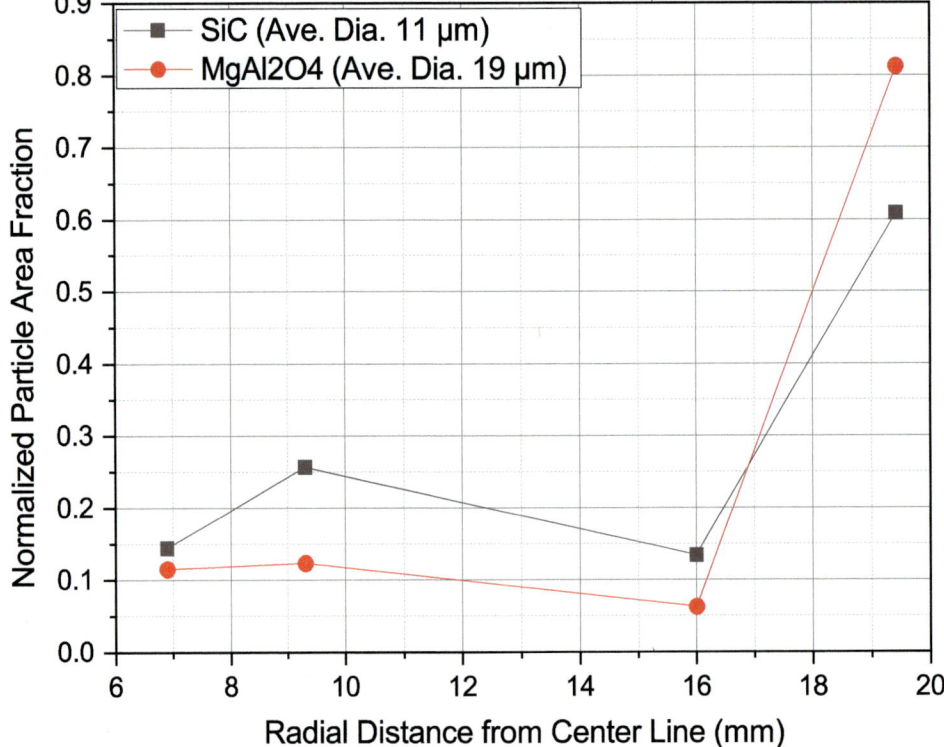

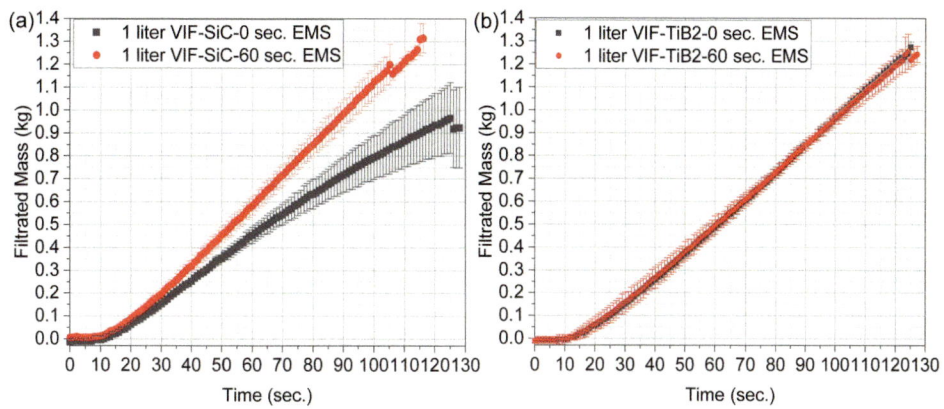

Fig. 6 Time-weight filtration curve from melt after 0 and 60 s electromagnetically stirring (EMS). **a** SiC particle, from trials Nr. 3–8; **b** TiB$_2$ particle from trial Nr. 9–14. All trials conducted in 1 L vacuum induction furnace (VIF)

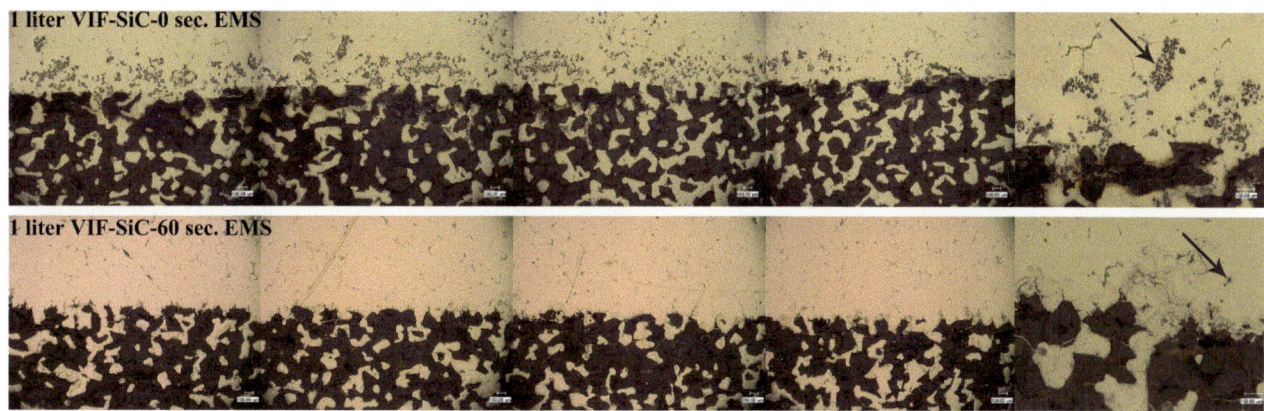

Fig. 7 Micrographs of PoDFA filter (cake) from melt after 0 and 60 s electromagnetically stirring (EMS). Arrows marked SiC particles. Results from trial Nr. 3 and 6 conducted in 1 L vacuum induction furnace (VIF)

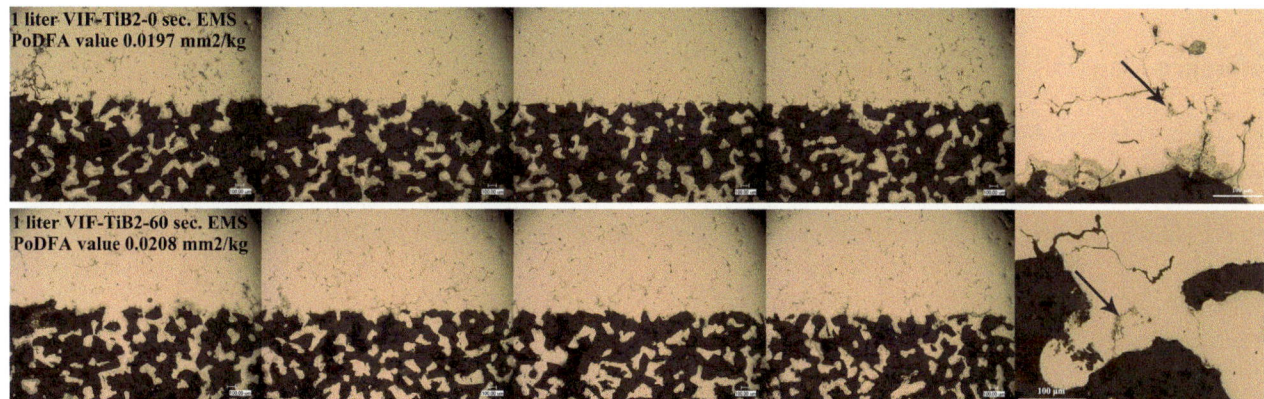

Fig. 8 Micrographs of PoDFA filter (cake) from melt after 0 and 60 s electromagnetically stirring (EMS). Arrows marked TiB$_2$ particles. Quantitative analytical results insert in upper-left. Results from trial Nr. 9 and 12 conducted in 1 L vacuum induction furnace (VIF)

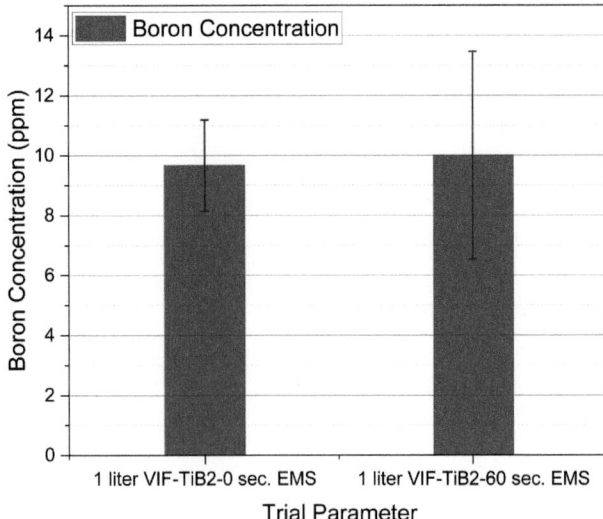

Fig. 9 Spark spectrometer analysis from samples taken from the melt after 0 and 60 s electromagnetic stirring (EMS). Results from trial Nr. 9–14 conducted in 1 L vacuum induction furnace (VIF)

any separation or significant agglomeration of TiB$_2$ particles after 60 s stirring time. PoDFA metallographic result shown in Fig. 8 also suggests insignificant difference in the collected TiB$_2$ particles. Nevertheless, in our study the melt was cleaned before charging TiB$_2$ particles, so most TiB$_2$ particles were in fact not collected by PoDFA filtration. Therefore we resorted to spark spectrometer analysis for determining TiB$_2$ concentration in the melt right before PoDFA filtration. The result is given in Fig. 9, which suggests an insignificant variation of TiB$_2$ concentration in the melt irrespective of melt stirring or not. Such result is in line with time-weight filtration curve and PoDFA results.

Analytical Modelling and Discussions

Reference Materials and Agglomeration

The fabricated A356-SiC and 5xxx-MgAl$_2$O$_4$ reference materials in fact offer already good basis for agglomeration studies. However no experimental evidence of agglomeration for either types of particles were found. For clarifying the reasons, we tried to examine the kinetics factor, i.e., collision constant. The derivation starts from calculation of magnetic flux density in the center line of the virtual melt [12]:

$$B_0 = k_s k_N \frac{\mu_0 \mu_r N_c I_c}{l_c} \qquad (1)$$

where k_s is the shielding factor of the crucible, k_N is the Nagaoka short-coil factor, μ_0 is permeability of free space, μ_r is the relative magnetic permeability of the inductor, N_c is

number of inductor turns, I_c is the current, l_c is the length of the coil. With obtained B_0 value, Alfven's velocity u_A can be calculated. Alfven's velocity can be seen as an indicative value of the characteristic melt velocity in an electromagnetically driven melt [13]:

$$u_A = \frac{B_0}{\left(\mu_0 \mu_r \rho_f\right)^{0.5}} \qquad (2)$$

where ρ_f is the melt density. With the obtained value of u_A, the dissipation rate of turbulent kinetic energy (TKE) ε can be estimated based on following equation [14]:

$$\varepsilon = \frac{u_A^3}{L} \qquad (3)$$

where L is the characteristic length of the melt. According to the classic Saffman–Turner collision theory [15], the collision constant C_t can be calculated by:

$$C_t = \sqrt{\frac{8\pi}{3}} d^3 \sqrt{\frac{\varepsilon}{5v}} \qquad (4)$$

where d is the diameter of the particle, v is the kinematic viscosity of the melt. Assuming that each collision leads to a decrease of a unit number of particle. One may write down [3]:

$$\frac{dn}{dt} = -C_t n^2 \qquad (5)$$

where dn/dt is the particle number density decrease rate, integration of Eq. (5) leads to:

$$\frac{n}{n_0} = \frac{1}{1 + C_t t n_0} \qquad (6)$$

where n/n_0 is the remaining particle number density, t is the EMS time of the melt, n_0 is the initial particle number density.

If we input the initial particle concentration 0.1 vol% and treatment time 60 s, the remains particle number density ratio after melt stirring can be calculated. Results are given in Table 3. It can be seen that the particle number density remains as only 0.65% with our input values. As a consequence the average particle size shall be 5 times larger than the reference materials, which was, however, not observed in any of the 0.1 L VIF trials.

Based on established analytical model, insufficient collision rate of the particles can be excluded as a reason for the particle to not form agglomerates. To account for the non-agglomeration character, other factors such as weak adhesion or swift separation of particles towards the near-wall region of the melt shall be considered.

It is worth mentioning that one should be cautious to extrapolate the non-agglomeration indication to all kinds of

Table 3 Estimated fluid and particle dynamic properties

Crucible shielding factor	Nagaoka shielding factor	Magnetic perm.	Melt density	Chara. length	Melt kinematic viscosity	Alfven's velocity	Diss. rate TKE	Particle density remains
/	/	(H/m)	kg/m^3	m	m^2/s	m/s	m^2/s^3	%
k_s	k_N	$\mu_0 * \mu_r$	ρ_f	L	v	u_A	ϵ	n/n_0
0.34	0.78	1.26E-06	2445	0.02	4.91E-07	0.2	0.51	0.65

MgAl$_2$O$_4$ inclusions. In practical situations, MgAl$_2$O$_4$ inclusions may adopt different morphologies and sizes (e.g., oxide film). Hence their agglomeration behavior (incl. kinetic factor) may deviate from the one we observed in this study.

Size-Dependent Particle Separation

In an electromagnetically driven flow, non-conductive spherical particles are subject to an electromagnetic force f_{EM} whose density is proportional to Lorentz force density F_{EM} [8]:

$$f_{EM} = 0.75 * F_{EM} \tag{7}$$

The force calculated by Eq. (7) will induce a relative velocity of particles in the radial direction which is responsible for the separation of particles. In creep flow (Rep < 1) conditions, the formula of reference radial velocity u_t can be calculated analogously to that of settling velocity [16]:

$$u_t = \frac{1.5d^2 f_{EM}}{36\mu_f} \tag{8}$$

where μ_f is melt viscosity. Eq. (8) gives the first criterion for comparing the separation rate of different sized particles. With the increase of particle diameter, the reference radial velocity also increases.

On the other hand, particles present in the melt tend to follow the streamline because of the drag force whose density F_D is given as [16]:

$$F_D = \frac{18\mu_f}{d^2}\left(u_p - u_f\right) \tag{9}$$

where $u_p - u_f$ is the relative velocity of the spherical particle with respect to melt local velocity. Drag force is responsible for carrying out the mixing of particles and hence counteract the localized enrichment of particles in the melt. Under the same flow conditions, particle relaxation time τ_p justifies the faithfulness of the particles to follow streamlines of the flow and hence the tendency of particle to be well-dispersed. The relaxation time τ_p can be calculated by dividing the relative velocity $u_p - u_f$ with particle acceleration, the latter of

which can be readily calculated using F_D given in Eq. (9) [17]:

$$\tau_p = \frac{\left(u_p - u_f\right)}{\frac{F_D}{\rho_p}} = \frac{\rho_p d^2}{18\mu_f} \tag{10}$$

In the recirculated electromagnetically driven flow, if a particle has a high value of relation time of a particle, it is more prone to deviate from the streamline of the flow pattern. Equation (10) offers the second criterion for comparing the separation rate of different sized particles.

A combination of Eq. (8) and Eq. (10) makes it possible to compare the separation rate of different sized particles under same flow conditions. For instance, in the 1 L vacuum induction furnace scenario, SiC particles will be separated faster towards the wall region than TiB$_2$ particles since they have a higher radial terminal velocity and meanwhile they are inclined to fall out of streamlines.

Based on Eq. (8) and Eq. (10), a comparison of particle separation probability of TiB$_2$, SiC and MgAl$_2$O$_4$ is schematically drawn in Fig. 10. Note that the spherical particle shape is assumed during the derivation of the models. For other shapes of inclusions, e.g., oxide film, model adaption is necessary to give the right prediction of separation probability.

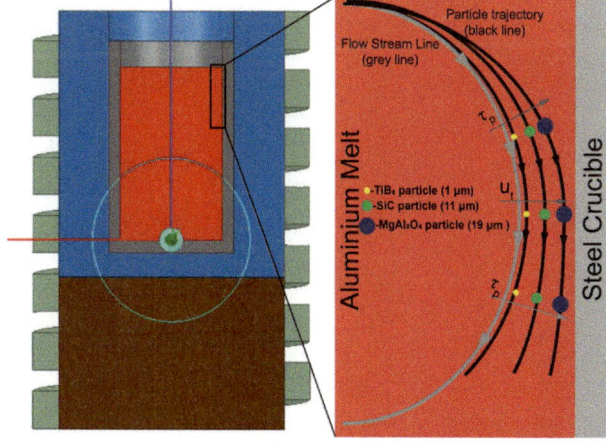

Fig. 10 Schematic of particle separation probability of TiB$_2$, SiC, and MgAl$_2$O$_4$ under same flow condition. Density difference has been accounted for

Conclusions

Agglomeration and separation behaviors of non-metallic inclusions in 5xxx alloy melt were investigated in electromagnetically driven flow. The findings are summarized as follows:

- A combination of stir-casting and ultrasound melt treatment is capable of fabricating 5xxx-$MgAl_2O_4$ materials with well dispersed particles.
- No apparent evidence was found for agglomeration tendency of SiC, $MgAl_2O_4$, and TiB_2 particles with respect to their respective as-charged states. Although established analytical model indicates sufficient collision rate provided by the flow.
- Experimental results demonstrate that particle separation rate is positively related to size. Two separation criteria were proposed to explain size-dependent particle separation rate.

Acknowledgements The research leading to these results was carried out in "Project 4 Continuation (P4C)" within Advanced Metals and Process (AMAP) Research Cluster at RWTH Aachen University, Germany. The authors are thankful for the financial support received from P4C members namely: Constellium, Magma, Nemak, Novelis, Speira, Trimet, and Vesuvius. Special thanks addressed to China Scholarship Council (CSC) for the financial support of Cong Li.

References

1. Li C, Li J, Mao Y, Ji J (2017) Mechanism to Remove Oxide Inclusions from Molten Aluminum by Solid Fluxes Refining Method. China Foundry 14(4):233–243
2. Grandfield J (2017) Developments in Inclusion Removal Technology. Light Metals 2017:1429–1434
3. Szekely AG. The Removal of Solid Particles from Molten Aluminum in the Spinning Nozzle Inert Flotation Process. Metall Mater Trans B 7(2):259–270
4. Tian C, Irons GA, Wilkinson DS (1998) Monte Carlo Simulation of Clustering of Alumina Particles in Turbulent Liquid Aluminum. Metall Mater Trans B 29(4):785–791
5. Johansen ST, Taniguchi S (1998) Prediction of Agglomeration and Break-Up of Inclusions during Metal Refining. Light Metals 1998:855–861
6. Marechal L, EI-Kaddah N, Menet P-Y (1993) Influence of Convection on Agglomeration and Removal of Non-Metallic Inclusions in Molten Aluminum. Light Metals 1993:907–913
7. Zhang L, Wang S, Dong A, Gao J, Damoah L. N. W. (2014) Application of Electromagnetic (EM) Separation Technology to Metal Refining Processes: A Review. Metall Mater Trans B 45(6):2153–2185
8. Leenov D, Kolin A (1954) Theory of Electromagnetophoresis. I. Magnetohydrodynamic Forces Experienced by Spherical and Symmetrically Oriented Cylindrical Particles. J Chem Phys 22 (4):683–688
9. Takahashi K and Taniguchi S (2003) Electromagnetic Separation of Nonmetallic Inclusion from Liquid Metal by Imposition of High Frequency Magnetic Field. ISIJ Int 43(6):820–827
10. Ščepanskis M, Jakovičs A, Baake E, Nacke B (2014) Solid Inclusions in an Electromagnetically Induced Recirculated Turbulent Flow: Simulation and Experiment. Int J Multiph Flow 64:19–27
11. Shu D, Sun B, Li K, Li T, Xu Z, Zhou Y (2002) Continuous Separation of Non-Metallic Inclusions from Aluminum Melt Using Alternating Magnetic Field. Mater Lett 55(5):322–326
12. Kennedy MW, Bakken JA and Aune RE (2015) Impact of Coil Geometry on Magnetohydrodynamic Flow in Liquid Aluminium and Its Relevance to Inclusion Separation by Electromagnetophoresis. J Manuf Sci Prod 15(1):69–78
13. Taberlet E and Fautrelle Y (1985) Turbulent Stirring in an Experimental Induction Furnace. J Fluid Mech 159:409–431
14. Kirpo M (2008) Modeling of Turbulence Properties and Particle Transport in Recirculated Flows. Ph.D. thesis, UNIVERSITY OF LATVIA
15. Saffman PG, Turner JS (1956) On the Collision of Drops in Turbulent Clouds. J Fluid Mech 1(1):16–30
16. Gökelma M, Einarsrud KE, Tranell G, Friedrich B (2020) Shape Factor Effect on Inclusion Sedimentation in Aluminum Melts. Metall Mater Trans B 51(2):850–860
17. Guha A (2008) Transport and Deposition of Particles in Turbulent and Laminar Flow. Annu Rev Fluid Mech 40:311–341

Microalloying of Liquid Al–Mg Alloy Studied In-Situ by Laser-Induced Breakdown Spectroscopy

Kristjan Leosson, Sveinn Hinrik Gudmundsson, Arne Petter Ratvik, and Anne Kvithyld

Abstract

Laser-induced breakdown spectroscopy (LIBS) provides a way to study aluminum melt dynamics. In the present work, liquid-phase LIBS analysis is used to study the behavior of group-II metals in Al–Mg alloys. The effects of Sr and Ca microalloying in high-Mg alloys suggest that these elements have an inhibiting effect on Mg oxidation. It is shown that adding the microelements Sr and Ca in an induction crucible significantly reduces the Mg partial pressure at typical process temperatures. It is believed that Sr and Ca form a barrier that reduces the vaporization of Mg through the surface.

Keywords

Aluminum • Process technology • Aluminum alloys • Chemical analysis • LIBS

Introduction

Understanding the behavior of small amounts of other elements in liquid aluminum is crucial to achieve good quality alloys. Of particular interest are the group-I (Li, Na, K) and group-II (Be, Mg, Ca, Sr) elements. Small amounts of Na and Ca can be introduced into the metal during electrolysis, as may Li in cases where Li is added to the bath. Sr is used to modify the Al–Si eutectic and some amounts can accumulate when remelting scrap. Be is added to the melt in ppm concentrations to inhibit oxidation in melts high in Mg. Mg is a major alloying element in, for instance, the widely used 5XXX alloys.

K. Leosson (✉) · S. H. Gudmundsson
DTE ehf., Arleynir 8, IS-112 Reykjavik, Iceland
e-mail: kristjan.leosson@dte.ai

A. P. Ratvik · A. Kvithyld
SINTEF, Trondheim, Norway

All the above-mentioned elements have been subject to extensive studies. In the 1970's, the Alcan research lab in Kingston, Ontario, showed that Be, Ca, and Sr were found to have a marked beneficial effect in reducing the oxidation of molten 5XXX alloys. Hence, Be has for a long time been used as a microalloying element to inhibit the oxidation of high-Mg alloys [1]. In his Ph.D. thesis, Smith showed (Fig. 1) that a thin $BeO-BeAl_2O_4$ layer was formed between the oxide and the molten metal [2]. Due to the negative health consequences of Be, Sr and Ca are interesting substitutions.

In 1973 Alcan patented 0.01–0.05% additions of Sr and/or Ca to reduce the formation of "pick-up", which are defects in the quality of the surface finish [3]. The upper limit of Ca additions was described in a patent by the Reynolds company in 1994 for electromagnetic casting of 5182 alloys [4], describing maximum additions of 0.05 wt% Ca, preferably between 0.01 and 0.04 wt%. Alcoa patented the use of Ca to improve the quality of the as-cast surface of both 5XXX and 7XXX alloys in 2002 and 2005 [5, 6]. High amounts of Na lead to edge cracking but Ca was not observed to have the same effect when the material contained only Ca. However, Ca had a slight mitigating influence when Na was present [7].

In recent years, significant advances have been made in the direct monitoring of molten metal chemistries by laser-induced breakdown spectroscopy (LIBS) [8]. LIBS provides a unique opportunity to study melt dynamics in real time. For example, the changing concentration of volatile elements and the formation and dissolution of solid phases in molten aluminum have been studied using LIBS measurements on the molten metal [9, 10]. The accuracy of LIBS analysis of aluminum melts was, however, found to be negatively impacted by increasing vapor pressure of the studied elements under standard experimental conditions [10]. In addition, the distribution of electronic energy levels makes group-I and group-II elements particularly susceptible

© The Minerals, Metals & Materials Society 2023
S. Broek (ed.), *Light Metals 2023*, The Minerals, Metals & Materials Series,
https://doi.org/10.1007/978-3-031-22532-1_121

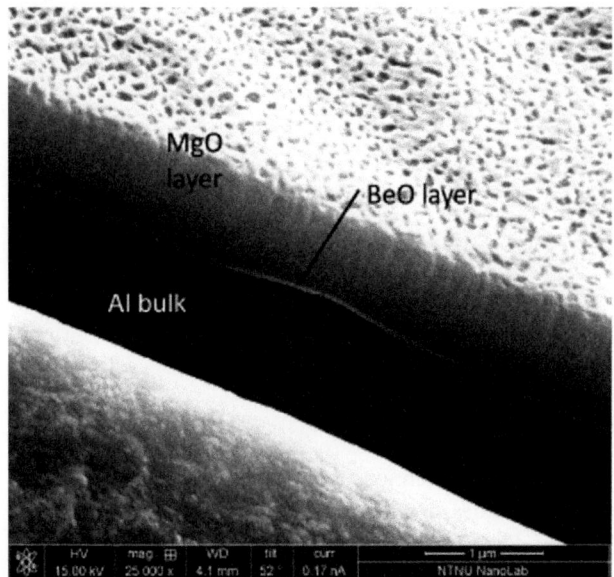

Fig. 1 Thin BeO–BeAl$_2$O$_4$ layer at the oxide-metal interface of a 5182 alloy containing 100 ppm Be after oxidation at 700 °C (from Ref. [2])

to self-absorption, where the atomic emission is reabsorbed by the plasma [11].

In view of their industrial importance, however, the ability to accurately monitor these elements in aluminum melt in real time is of significant value. In this paper, we present introductory LIBS measurements of aluminum-magnesium alloys with Sr and Ca additions. The study demonstrates that although the Mg concentration of the alloy does not change strongly with time, the LIBS analysis suggests that the presence of the microalloying elements causes a significant lowering of the partial pressure of Mg at the melt surface.

Methodology

Liquid aluminum-magnesium with up to 5 wt% Mg was prepared by mixing high-purity aluminum (4N6, Laurand Associates) and Be-free Mg in a boron-nitride-coated zirconium-silicate-reinforced fiberglass crucible (Pyrotek RFM®) using induction heating. Sr and Ca were introduced to the melt by adding weighed pieces of Al master alloy with 0.3% Sr and 10% Ca, respectively, to the molten metal in the crucible. The total amount of alloy in the crucible was approximately 500 g.

Chemical analysis of the melt was carried out at regular intervals using a LIBS molten-metal analyzer (DTE EA-2500), employing 100 mJ pulses of 1064 nm wavelength at 20 Hz repetition rate. In the analyzer, plasma emissions were detected with a cooled CCD camera after dispersion through a high-resolution spectrometer ($\lambda/\Delta\lambda \approx$

9000). Each measurement consisted of an accumulated spectrum from 300 laser pulses. At each time point, the LIBS measurement was repeated three times and the results averaged (total measurement time $\approx$45 s). A flow of argon gas (5 N) was introduced to the sampling point during measurement and the melt temperature was maintained with the induction heater, simultaneously providing continuous stirring of the sample. A mechanical skimming of the melt surface was performed manually, prior to each LIBS measurement.

The behavior of emission peaks in the LIBS spectra and their degree of self-absorption will be affected by the experimental conditions and the selection of emission peaks is therefore specific to those conditions. For the measurements presented here, the Mg I peaks at 552.8405 nm and 333.2146 nm were used for quantification, as they proved to be close to proportional to the Mg concentration of the melt for the range of interest (1–5%). For Ca and Sr, the emission peaks at 315.887 nm (Ca II) and 460.733 nm (Sr I), respectively, were used. All signals were normalized to the Al I peak at 305.0072 nm.

Chemical analysis was also performed using conventional spark optical emission spectroscopy (spark-OES) on solid samples cast from the melt into a preheated ASTM-standard disk mold. The samples were milled to a depth of 1.2 mm from the sprue side and measured using a spark-OES instrument (Thermo Scientific ARL iSpark 8820). In some cases, the concentration of Ca was well outside the calibrated range of the OES spectrometer, but readings were still found to be consistent with the amount of added material.

The tabulated vapor pressures of the pure group-I and group-II elements Na, Mg, Ca, and Sr across a range of temperatures relevant to aluminum processing are shown in Fig. 2. The vapor pressures of Al and Be are several orders of magnitude lower (close to 10^{-7} mbar at 750 °C). For a given alloy, the partial pressure p_M of element M is given by

$$p_M = p_M^0 \gamma_M x_M \tag{1}$$

where p_M^0 is the vapor pressure of the pure element, γ_M is its activity coefficient and x_M is its mole fraction. As a consequence, for a typical magnesium containing alloy the dominating contribution to the total vapor pressure among these elements will be from magnesium.

Time-Dependent Studies of Al–Mg Alloy

A previous thermogravimetric analysis (TGA) using disk samples (mg scale) with 4.5% Mg showed a weight gain close to 13% after oxidation for 1–2 h at 750 °C under air flow, which is consistent with full conversion of Mg from the melt to MgAl$_2$O$_4$ [13]. In this work, a Ca addition of

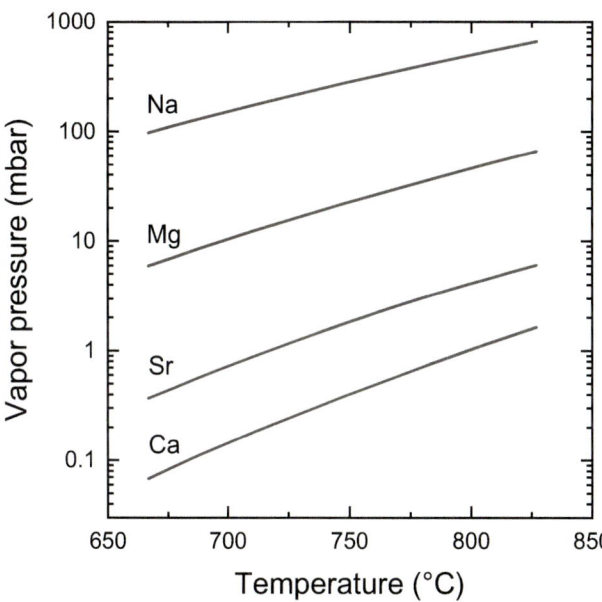

Fig. 2 Tabulated vapor pressures (log scale) of high-vapor-pressure elements (from Ref. [12])

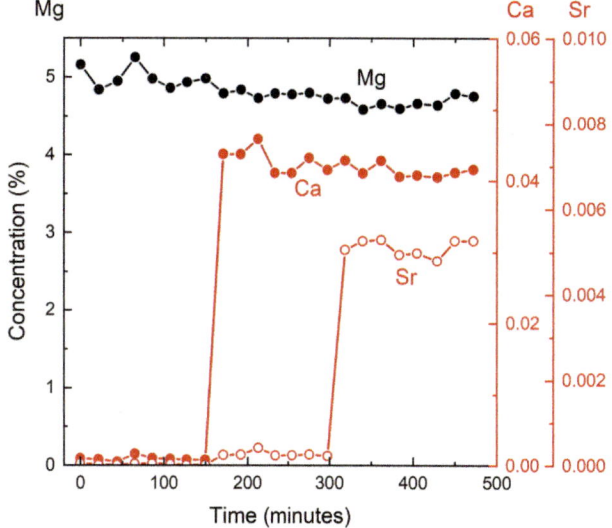

Fig. 3 Concentration of Mg, Ca and Sr in aluminum melt measured with a LIBS analyzer. Ca master alloy was added at t = 150 min and Sr master alloy was added at t = 300 min. The melt was maintained at 750 °C between successive LIBS measurements

1000 ppm and a Sr addition of 500 ppm were found to substantially inhibit breakaway oxidation. TGA experiments on the kg scale, where the volume-to-surface ratio was an order of magnitude higher, were also conducted, both on model Al–Mg alloys and on a commercial alloy. In this case, the relative weight gain was smaller. The impact of Ca (200–300 ppm) and Sr (100–300 ppm) additions were less pronounced, but an inhibiting effect was still noticed.

Motivated by these experiments, the current work was carried out to monitor the content of Mg, Ca, and Sr in a 5% Mg mixture using the EA-2500 LIBS analyzer. LIBS measurements were made at approximately 20 min intervals as shown in Fig. 3. Between measurements, the melt was kept at 750 °C in air for a total time of 8 h. Ca and Sr were added to the melt at t = 150 min and t = 300 min, respectively. As stated above, the oxide layer that formed on the melt surface was skimmed off before each measurement. The results suggest a slight decrease (<5–7% of the original concentration) in Mg and Ca over the measurement period but no measurable decrease in Sr concentration.

In a separate test, molten alloy samples containing approximately 4% magnesium were prepared using the same base materials. Solid samples were cast from the melt as described above and measured using spark-OES. The melt samples were then maintained at 750 °C for a period of several hours. The oxide layer was skimmed off the surface at regular intervals—typically between 10 and 40 min—for further analysis (to be presented elsewhere). Following the oxidation test, a second sample was cast from the melt for spark-OES analysis. Similar tests were done on a melt sample containing 1000 ppm Ca, and a melt sample containing approximately 270 ppm Ca and 160 ppm Sr (see Table 1). The results are consistent with the LIBS results in Fig. 3, showing only a minor (0–15%) reduction in concentration of Ca and Sr during the test period. For Mg, the difference in measured concentration before and after the oxidation tests was of the same order as the variation in repeated spark-OES measurements within each sample.

The total weight of the material removed by skimming places an upper limit on the weight gain of about 1–1.5% during the test (assuming an oxygen weight percentage of 30–45% in the oxide layer). This is lower than in most of the kg-scale TGA test reported in Ref. [13], which may be explained by the fact that the most rapid weight gain in the TGA tests was usually observed after a period of 1–2 h whereas in the present experiments the repeated skimming may have kept the melt in the initial (slower) weight gain regime. If melt oxidation in the present tests was primarily due to formation of MgO and/or $MgAl_2O_4$, a clear difference in Mg concentration should be seen in the LIBS and spark-OES measurements presented above. Conversely, the present study suggests that the oxidation process does not strongly impact the Mg concentration in the melt. It should be kept mind, however, that the volume-to-surface ratio in the present experiments was around 2.3, which is comparable to the kg-scale TGA measurements of Ref. [13], but still about one order of magnitude away from typical industry values (see Table 2 for comparison).

Table 1 Spark-OES results before and after oxidation tests at 750 °C in air

	Before			After			
	Mg (%)	Ca (ppm)	Sr (ppm)	Mg (%)	Ca (ppm)	Sr (ppm)	Duration (hours)
Sample 1	4.2	–	–	4.2	–	–	3.0
Sample 2	4.2	1060*	–	4.1	1066*	–	6.5
Sample 3	3.9	277*	157	3.9	254*	132	6.0

* Value outside the calibrated range of the spark-OES analyzer

Table 2 Comparison of volume-to-surface ratio in different conditions

	Area (cm^2)	Average mass (g)	Volume/area (cm)
This study	80	500	2.3
kg-TGA [13]	200	2200	4
mg-TGA [13]	1.1	0.25	0.1
Industry (typical)	1.2×10^5	2.4×10^7	70

Effects of Microalloying on Mg Vapor Pressure

Considering the above results, the question remains which mechanism is responsible for the reduction in oxidation rate (mass gain) upon addition of the microalloying elements as reported in the kg-scale TGA experiments in Ref. [13]. The temperature-dependent contribution to the Mg signal in the LIBS analysis, arising from the vapor phase (Mg(g)), can be separated from the contribution arising from Mg ablated from the melt (Mg(l)), which does not depend strongly on temperature. Figure 4 shows the temperature dependence of the Mg(g) signal from the molten aluminum, together with the partial pressure of Mg, according to Eq. (1). The temperature-dependent activation coefficient of Mg in Al was obtained from Ref. [14]. It should be pointed out, however, that historically a lack of consensus in the literature regarding the activity of Mg in Al and different experimental methods has yielded different values of the activity coefficient [15].

This analysis was repeated for different Mg concentrations in the pure melt, using a typical process temperature of 730 °C. The result is shown in Fig. 5, displaying the relative Mg(g) signal determined by LIBS analysis (i.e., normalized with the Mg concentration). While Mg(g)/x_{Mg} stays approximately constant for Mg concentrations in the 2–5% range in the pure melt, a significant lowering of the Mg(g) signal was observed upon addition of small amounts of Ca and Sr to the ≈5% Mg alloy (sample of Fig. 3, measured immediately after t = 0 min, t = 150 min and t = 300 min, respectively). We postulate that this reflects changes in the presence of Mg in the vapor phase due to the formation of Ca- and Sr-containing oxides inhibiting the vaporization of

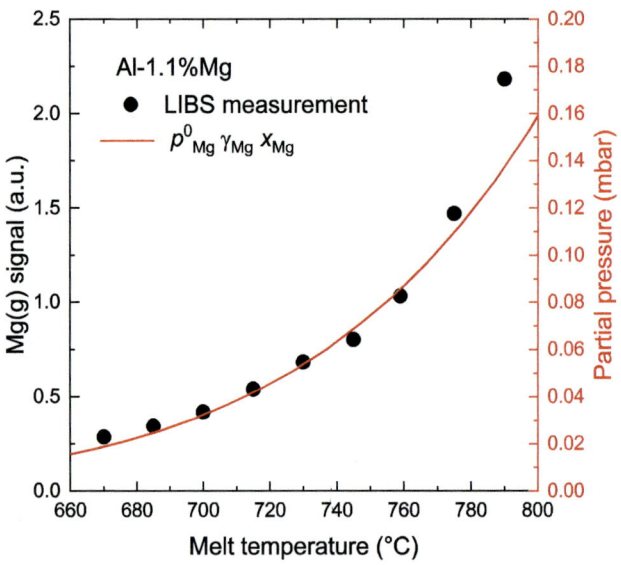

Fig. 4 Temperature-dependent (vapor phase) Mg signal in LIBS measurements of molten aluminum alloy. The partial pressure of Mg is shown for comparison

Mg through the surface. While the present study is still probing only a limited number of concentration values of these microalloying elements, it supports the previous conclusions of Sr and Ca microalloying strongly affecting the surface behavior of Mg in the molten alloy.

Conclusions and Future Work

This work has demonstrated the use of LIBS analysis on molten alloys containing high-vapor-pressure alloying elements. The elements were found to be considerably stable in

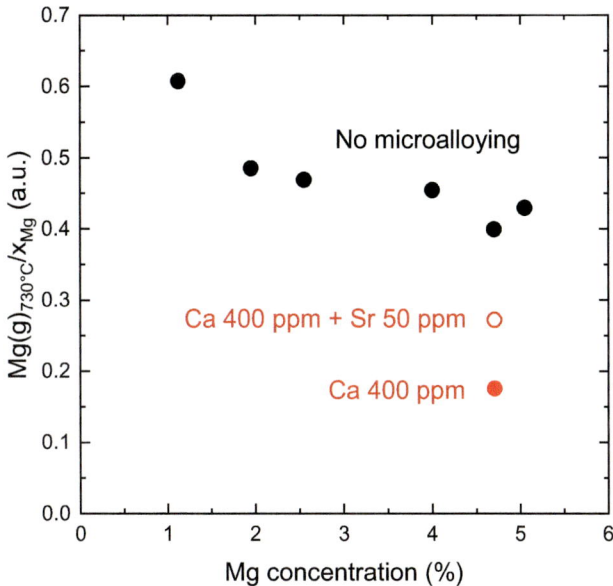

Fig. 5 Normalized Mg(g) signal obtained from LIBS data in pure mixtures of varying Mg concentration (black symbols) and after microalloying with Ca and Sr (red symbols)

the melt over several hours at 750 °C. The vapor pressure of Mg, as measured by the LIBS analysis, was found to be significantly affected by the presence of the microalloying elements Ca and Sr, supporting previous conclusions of their inhibiting effect on the oxidation of the molten alloy. Further analysis will be performed on the skimmed oxides to determine their Ca and Sr concentrations. Also, studies with longer holding times, with longer time between skimming will be conducted.

Acknowledgements This work has been partly supported by the SFI Metal Production (Centre for Research-based Innovation, 237738). AR and AK gratefully acknowledge financial support from the Research Council of Norway and the partners of the SFI Metal Production. The authors also would like to thank Geoffrey Kenneth Sigworth for thermodynamic insights and Don Allan Doutre for being a key contributor to the research question.

References

1. W. Thiele (1962) Die Oxydation von Aluminium- und Aluminiumlegierungs-Schmelzen. Aluminum 38, 707.
2. N. Smith (2019) PhD Dissertation: Methods of oxidation Inhibattion for Al-Mg Alloys. Trondheim Norway: Norwegian University of Science and Technology.
3. Aluminum Alloys (1972) US patent 3926690.
4. Method for improving surface quality of electromagnetically cast aluminum alloys and products therefrom (1994) US patent 5469911.
5. Aluminum alloys having improved cast surface quality (2000) US patent 6412164.
6. Aluminum alloys having improved cast surface quality (2000) US patent 6843863.
7. D. Doutre, P. Wycliffe (2011) The effect of calcium on the rolling behaviour and hot tensile properties of AA5182. Materials Science Forum 693, 256-63.
8. A.K. Myakalwar, C. Sandoval, M. Velásquez, D. Sbarbaro, B. Sepúlveda, J. Yáñez (2021) LIBS as a spectral sensor for monitoring metallic molten phase in metallurgical applications—a review, Minerals 11, 1073.
9. S.H. Gudmundsson, J. Matthiasson, K. Leosson (2020) Accurate Real-Time Elemental (LIBS) Analysis of Molten Aluminum and Aluminum Alloys. In: A. Tomsett (ed.) Light Metals 2020. The Minerals, Metals & Materials Series. Springer, Cham.
10. S.H. Gudmundsson, J. Matthiasson, B.M. Björnsson, H. Gudmundsson, K. Leosson (2019) Quantitative in-situ analysis of impurity elements in primary aluminum processing using laser-induced breakdown spectroscopy, Spectrochim. Acta Part B At. Spectrosc. 158, 105646.
11. F. Rezaei, G. Cristoforetti, E. Tognoni, S. Legnaioli, V. Palleschi, A. Safi (2020) A review of the current analytical approaches for evaluating, compensating and exploiting self-absorption in laser induced breakdown spectroscopy Spectrochimica Acta Part B 169, 105878.
12. Data obtained from https://www.iap.tuwien.ac.at/www/surface/vapor_pressure.
13. N. Smith-Hanssen, M. Syvertsen, A. Kvithyld (2022) Microalloying to Inhibit Oxidation of Al–Mg Alloys. In: D. Eskin (ed.), Light Metals 2022. The Minerals, Metals & Materials Series. Springer, Cham.
14. G.K. Sigworth (2021) Refining of Secondary Aluminum: Important Chemical Factors. JOM 73, 2594.
15. G.K. Sigworth, T.A. Engh (1982) Refining of liquid aluminium—a review of important chemical factors, Scandinavian Journal of Metallurgy 11, 143.

Hydrogen Absorption of Aluminum-Magnesium Melts from Humid Atmospheres

Stefan Tichy, Philip Pucher, Bernd Prillhofer, Stefan Wibner, and Helmut Antrekowitsch

Abstract

In casthouses, water vapor has long been considered a source of high hydrogen contents in aluminum melts. Moisture levels are usually highest in melting and holding furnaces, where H_2O contents of up to 18 vol% occur due to the combustion of natural gas with air. In the case of oxyfuel burners, contents of even more than 60 vol% are theoretically possible. To avoid defects such as annealing bubbles or grain porosity in the subsequent production steps, the absorbed hydrogen must be reduced to a minimum by degassing treatments. To gain a better understanding of the interaction between liquid aluminum and water vapor, the hydrogen absorption of aluminum-magnesium melts under atmospheres with varying water vapor contents of 0–90 vol% at around 700 °C (1300 °F) is investigated in this work. Furthermore, the influence of H_2O on the oxidation behavior is also examined.

Keywords

Aluminum • Hydrogen • Water vapor • Hycal • Oxidation

Introduction

Hydrogen is the only gas with significant solubility in liquid aluminum. Particularly problematic is the sudden decrease in solubility by a factor of 20 during solidification, which leads to the precipitation of hydrogen in the form of H_2 bubbles. The resulting increase in volume leads to various defects in casting products or rolled parts, depending on the time of recombination [1].

The main source of dissolved hydrogen in casthouses is water vapor. This occurs mainly as a reaction product during the combustion of natural gas. In the case of natural gas-air burners, the H_2O content in the off-gas amounts up to 18 vol%. following Eq. (1). For oxyfuel burners, values up to 67 vol% can theoretically be reached, according to Eq. (2) [1–4].

$$CH_4 + 2\,O_2 + 8\,N_2 \leftrightarrow CO_2 + 2\,H_2O + 8\,N_2 \qquad (1)$$

$$CH_4 + 2\,O_2 \leftrightarrow CO_2 + 2\,H_2O \qquad (2)$$

Subsequently, the melt reacts with H_2O according to Eq. (3) to form H_2 and Al_2O_3 [5].

$$\{H_2O\} + \frac{2}{3}\,[Al] \leftrightarrow \frac{1}{3}(Al_2O_3) + \{H_2\} \qquad (3)$$

The hydrogen released in (3) can either be absorbed atomically by the melt (4) or escape back to the atmosphere. Figure 1 illustrates all these steps schematically in a hearth furnace [1, 6].

$$\{H_2\} \leftrightarrow 2\,[H]_{Al} \qquad (4)$$

The hydrogen content (S) of the melt subsequently increases until equilibrium is reached according to Sievert's law (5) [1].

$$S = k \cdot \sqrt{p_{H_2}} \qquad (5)$$

Sievert's solubility constant k depends on temperature and alloy composition. It can be determined using an Arrhenius approach. $p(H_2)$ is the partial pressure of hydrogen above the melt resulting from Eq. (4). In practice, a modified version of Sievert's law (6) is usually applied to calculate the hydrogen solubility in ml/100 g. It contains a separate correction term each for alloy CF(A) (7) and temperature CF(T) (8). S_0 represents the hydrogen solubility of

S. Tichy (✉) · S. Wibner · H. Antrekowitsch
Chair of Nonferrous Metallurgy, Montanuniversitaet Leoben, Franz-Josef-Straße 18, 8700 Leoben, Austria
e-mail: stefan.tichy@unileoben.ac.at

P. Pucher · B. Prillhofer
AMAG Casting GmbH,
Lamprechtshausenerstraße 61, 5282 Ranshofen, Austria

© The Minerals, Metals & Materials Society 2023
S. Broek (ed.), *Light Metals 2023*, The Minerals, Metals & Materials Series,
https://doi.org/10.1007/978-3-031-22532-1_122

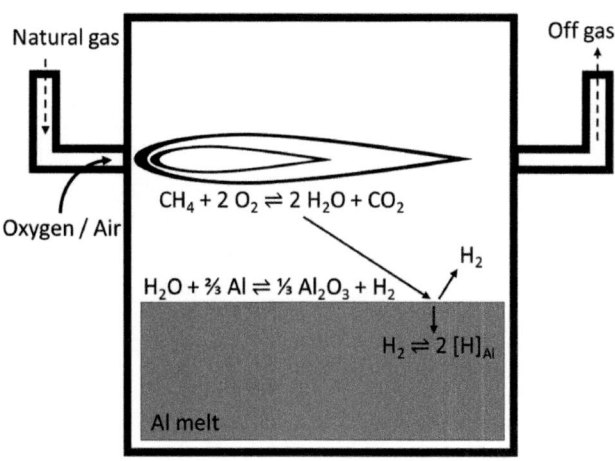

Fig. 1 Schematic for the reaction between water vapor and an aluminum melt in a natural gas-fired furnace [4, 5]

pure aluminum at 700 °C and a hydrogen partial pressure of 1 atm and is 0.92 ml/100 g [6, 7].

$$S = S_0 \cdot \sqrt{p_{H_2}} \cdot CF(A) \cdot CF(T) \qquad (6)$$

$$CF(A) = 10^{0.0170 \cdot \%Mg - 0.0269 \cdot \%Cu - 0.0119 \cdot \%Si} \qquad (7)$$

$$CF(T) = \exp\left(6,531\frac{T - 700}{T + 273}\right) \qquad (8)$$

Measurement of Hydrogen Solubility

The hydrogen dissolved in the melt can be determined by a variety of methods. The indication methods (density index, method of the first bubble, etc.) allow rather a qualitative observation and are therefore unsuitable for scientific measurements. Direct determination methods provide a continuous, quantitative determination of the hydrogen content and are therefore suitable not only for scientific applications but also for process monitoring and quality assurance in the industry. The most important representatives in this area include Chapel, AlScan, and Hycal [1, 8–10].

While Chapel measures the hydrogen partial pressure above the melt directly, AlScan and Hycal determine the hydrogen content of the melt through stored equations. AlScan utilizes Eq. (6), while Hycals software uses a more comprehensive calculation that is not published. In addition to the elements listed in (7), Li, Fe, Ti, and Zn are also included in the calculation, which also influence the hydrogen solubility [7, 9, 11].

AlScan and Hycal also differ in terms of their operating principle. At AlScan, nitrogen is injected into the melt via a probe. Hydrogen accumulates in the rising gas bubbles until they reach equilibrium with the dissolved hydrogen in the melt. The hydrogen content is then determined by measuring the thermal conductivity of the ascending carrier gas. AlScan is only suitable to a limited extent for laboratory-scale investigations since the carrier gas can cause in situ degassing of the melt. This may lead to incorrect measured values, especially at higher hydrogen contents. Hycal uses a thermochemical approach. The hydrogen content is determined by measuring the potential difference between a reference electrode with a known hydrogen concentration and the melt. The probe made of a $CaZrO_3$-In ceramic can conduct hydrogen ions at elevated temperatures and serves as a solid electrolyte. Since purge gas is only required at the beginning of a measurement (approx. 30 s), no significant degassing of the melt takes place even with small amounts of melt.

The third method for measuring dissolved hydrogen in aluminum is hot extraction, which is widely considered the reference standard. The disadvantages here are the low maximum analysis quantity and the large measurement effort, which is why hot extraction is mainly used for scientific purposes. The solidification of the samples and the sample preparation also represent additional sources of error [5, 12].

Since the aim of this work is primarily to simulate industrial conditions and high hydrogen contents (>0.5 ml/100 g) can be expected due to the high concentration of water vapor in the ambient atmosphere, Hycal is particularly suitable for carrying out the hydrogen measurements in the experiments.

Correlation Between the Partial Pressures of Water Vapor and Hydrogen

While hydrogen solubility as a function of hydrogen partial pressure has been extensively studied, limited information is available on the correlation between $p(H_2)$ and $p(H_2O)$. Theoretically, according to Talbot [5], due to the high reactivity of aluminum, all traces of water vapor in contact with the melt are converted into hydrogen, following Eq. (3). This allows the assumption that near the melt surface the partial pressure of hydrogen can be set equal to the partial pressure of water vapor in the furnace atmosphere (9).

$$p_{H_2} = p_{H_2O} \qquad (9)$$

Syvertsen et al. [13] carried out AlScan measurements in parallel with determinations of the H_2O content in the off-gas as part of industrial experiments on a natural gas-fired furnace and determined the correlation according to Eq. (10).

$$p_{H_2} = 0.51 \cdot p_{H_2O} \qquad (10)$$

This is explained by the presence of a diffusion-determining gas boundary layer between the melt/oxide layer and the surrounding atmosphere. This boundary layer is responsible for 49 vol% of the hydrogen formed in reaction (3) being desorbed into the atmosphere and 51 vol% being absorbed by the melt according to reaction (4) [13].

Another theory that explains the deviation of $p(H_2)$ from $p(H_2O)$, but has not yet been investigated in detail, is a potential influence of a formed oxide layer or dross between the melt and the atmosphere on the hydrogen absorption. Talbot and Anyalebechi [14] postulated that a stable oxide layer may delay or even prevent the hydrogen content of the melt to reach equilibrium in this case.

Generally, the hydrogen absorption is maximal when the surface of the melt is blank. In contact with oxygen or water vapor, an amorphous Al_2O_3 layer with a thickness of a few nanometers is formed immediately. After some minutes, a crystalline γ-Al_2O_3 layer forms at the metal-oxide interface and grows into the melt. According to Stephenson [15], once this layer is dense and completely formed, it inhibits further absorption of hydrogen by preventing gas molecules from permeating through it. After several hours, a transformation of γ-Al_2O_3 to α-Al_2O_3 occurs. This modification has a higher density, which leads to a volume contraction of about 24% and thus to the breakup of the oxide layer. This in turn leads to the exposure of blank aluminum melt, which can subsequently absorb hydrogen again. Concerning the influence of Mg, as well as MgO and $MgAl_2O_4$ on the hydrogen absorption of aluminum-magnesium melts, there have been hardly any investigations published so far [14, 15].

Influence of H₂O on the Oxidation Behavior

From reaction (3), it could be assumed that the oxidation of aluminum melts increases at higher H_2O contents. However, various sources [2, 16, 17] report reduced oxidation in environments containing water vapor. Impey et al. [16] attributed this effect to a stabilization of the γ-Al_2O_3 layer and the resulting delayed transformation to α-Al_2O_3. Normally, this transformation occurs due to the continuous reduction of OH^- ions in the oxide layer, as the hydrogen is either absorbed into the melt or escapes into the atmosphere over time. However, the process can be significantly delayed by the continuous resupply from the atmosphere. Thiele [2] made the same observations but explained them with a catalytic effect that H_2O has on corundum formation (α-Al_2O_3). This leads to the rapid homogeneous formation of the oxide layer, which in turn minimizes the diffusion processes required for oxidation. Haginoya et al. [17] observed the oxidation-reducing effect of H_2O also in Al-Mg alloys.

Experimental Setup

All experiments were carried out on a resistance-heated Nabertherm furnace KS20/13/S. The crucible made of clay graphite can be shut with a lid and is thus sealed off against the surrounding air. The furnace atmosphere above the melt is adjusted by a constant gas flow from a gas mixing station outside the furnace. In addition to Bronkhorst mass-flow controllers for argon, nitrogen, and oxygen, a Bronkhorst CEM unit is used for the specific adjustment of the H_2O content (see Fig. 2). To avoid condensation of the water after the evaporator, the transfer pipe is heated to 200 °C in the area between the gas mixing station and the furnace by a heating collar.

Approximately, 25 kg input material of the alloy EN AW-5083 is used for each experiment. The composition of the elements influencing hydrogen solubility is shown in Table 1. Since the same material is used for all tests, deviations resulting from variations in the chemical composition can be neglected. To minimize the input of external moisture, the material was cleaned of any cooling lubricant residues or other contaminants.

During heating and melting, an argon atmosphere is generated in the crucible to keep hydrogen absorption and oxidation as low as possible during this period. The relative humidity and temperature are periodically recorded using a (Testo 608-H2) hygrometer to be able to identify any influence of the surrounding atmosphere, which is primarily weather dependent. When the target temperature is reached, the melt is skimmed to remove the oxide layers and impurities of the input material. Afterward, hydrogen is measured with Hycal to determine the initial hydrogen content. The furnace is then closed and an atmosphere with a defined H_2O, N_2, and O_2 content is applied. While the ratio of H_2O/N_2 varies between experiments, the oxygen content

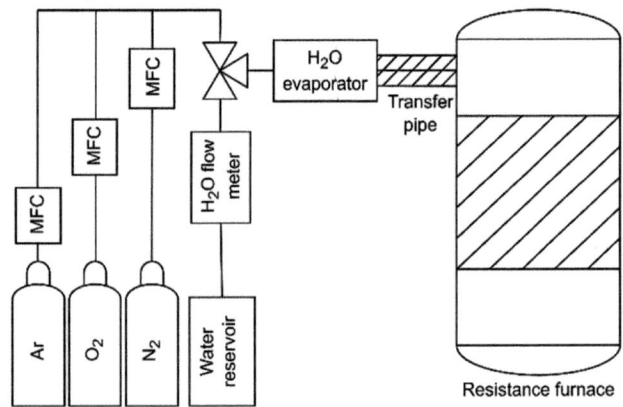

Fig. 2 Schematic of the lab-scale furnace including a gas mixing station

Table 1 Chemical composition of EN AW-5083 [18]

Element	wt%
Mg	4.0–4.9
Si	0.40
Fe	0.40
Cu	0.10
Ti	0.15

always remains constant at 1–3 vol%. This is checked several times by ABB gas analysis using a lance, which can be inserted into the furnace if required. During the measurements, it is shown that the furnace atmosphere also contains a small CO_2 content of 0–1 vol%. This probably originates from the reaction of the clay graphite crucible with oxygen.

The exposure time ranges from 2 to 12 h. During this time, the melt is held isothermally at 720–780 °C with the furnace lid closed.

At the end of each experiment, the melt is skimmed again. The amount of dross is then determined by weighing. In addition, an examination is carried out using a scanning electron microscope. A second Hycal measurement allows the hydrogen absorbed or released during the test to be determined. The statistical experimental design and the evaluation are performed with MODDE 12.0.1 software.

Results Hydrogen Solubility

After melting, there are already significant differences in the hydrogen content of the melt. The values range between 0.08 ml/100 g and 0.36 ml/100 g. The moisture content of the atmosphere in the melting laboratory is between 6.5 and 13.5 g/m^3 H_2O, depending on the weather. This corresponds to an absolute humidity of 0.8–1.8 vol%. Increased hydrogen contents are generally measured at higher melt temperatures. Thus, the hydrogen solubility after melting and heating to 700 °C is 0.08–0.25 ml/100 g, while at 760 °C values in the range of 0.20–0.36 ml/100 g are observed. To eliminate the factor of hydrogen absorption during melting, the results are given in the form ΔS, i.e., final minus initial hydrogen content.

Furthermore, it is important to mention that the melt loses an average of 20 °C in temperature during skimming and Hycal measurementsin total approx. 20 min. The measured hydrogen contents are therefore in the range between 700 and 760 °C.

The results (see Fig. 3) show various trends. For example, the hydrogen absorption of the melt is primarily dependent on the H_2O content of the atmosphere and the exposure time in this environment. Higher water vapor contents always lead to increased hydrogen uptake by the melt. At very low water vapor contents, degassing of the

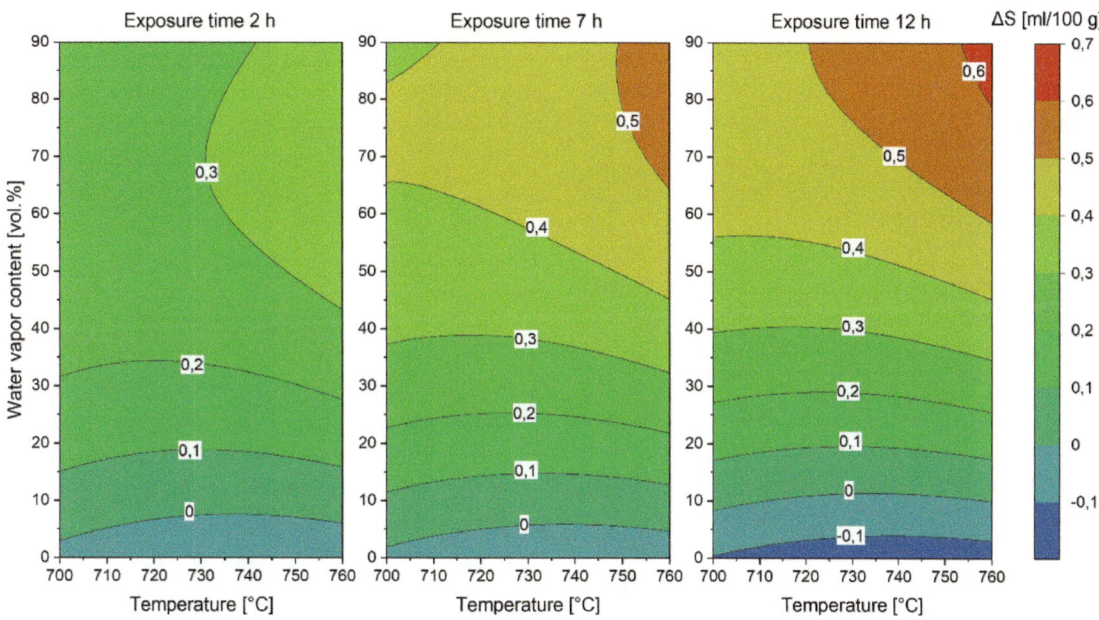

Fig. 3 Statistic analysis of selected experiments regarding the hydrogen absorption of 5083 melts under atmospheres with different humidities

Table 2 Comparison of calculated hydrogen solubilities in equilibrium (S_{eq}) for $p(H_2) = p(H_2O)$ and Hycal measurements in EN AW-5083

p(H₂O) [atm]	700 °C		730 °C		760 °C	
	S_{eq} [ml/100 g]	S (measured) [ml/100 g]	S_{eq} [ml/100 g]	S (measured) [ml/100 g]	S_{eq} [ml/100 g]	S (measured) [ml/100 g]
0.45	0.72	0.47	0.88	0.67	1.06	0.78
0.90	1.04	0.56	1.26	0.77	1.51	0.89

melt occurs. This phenomenon is more pronounced with increasing exposure time. Latter also plays a decisive role in hydrogen absorption. Uptake is highest in the first 2 hours. Nevertheless, up to a holding time of 7 h, a further significant increase takes place. In the range between 7 and 12 h, however, the increase in hydrogen content is small and can be regarded as almost constant. A significant influence of the temperature only becomes apparent at H_2O contents above 40 vol%. In this range, higher temperatures always lead to an increased hydrogen absorption of the melt in the time between the Hycal measurements.

A detailed analysis reveals the highest hydrogen uptake and content of the melt, as expected, at 760 °C and 90 vol% H_2O. However, the value of 0.89 ml/100 g is well below the theoretical equilibrium solubility of 1.51 ml/100 g assuming a hydrogen partial pressure $p(H_2)$ of 0.9 atm. Similarly, at an H_2O content of 45 vol%, the hydrogen content of 0.78 ml/100 g is lower than calculated at $p(H_2) = 0.45$ atm (1.06 ml/100 g). As can be seen in Table 2, this trend is independent of both time and temperature. In other words, when equating hydrogen partial pressure and water vapor partial pressure, the measured hydrogen contents are always below the theoretically determined solubility in thermodynamic equilibrium.

Results Oxidation Behavior

Since the oxygen content is kept constant at 1–3 vol% over the entire series of experiments, the oxidation behavior of the melt can be studied specifically as a function of the parameters, namely, H_2O content of the atmosphere, temperature, and residence time independent of the O_2 content. It is also important to mention that the experiments were carried out without typical industrial bath movement (natural convection due to the burners or electromagnetic pumps), which is why reduced dross occurrence is to be expected.

Regarding temperature and time dependence of the oxidation, the sufficiently known and expected trends can be observed. Both higher temperatures and longer exposure times promote the oxidation of the melt. When investigating the influence of water vapor, reduced oxidation can be observed under humid atmospheres. While the melt surface is

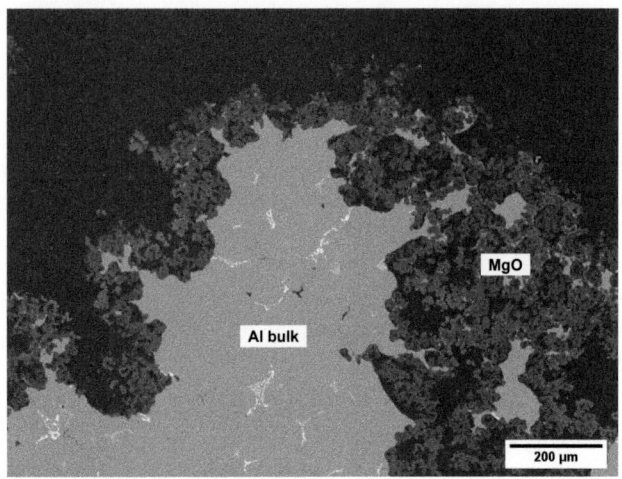

Fig. 4 SEM image of dross found on EN AW-5083 after 2 h exposure at 780 °C in dry atmosphere (98 vol% N_2, 2 vol% O_2)

covered by a thin oxide layer in all tests with $p(H_2O) > 0$ atm, the different phases of break-away oxidation can be observed at water vapor contents of 0 vol%, depending on temperature and exposure time. This is evident both from the amount of dross skimmed and from closer examination of the oxide residues using scanning electron microscopy. At 720 °C, after 2 h there is just a thin oxide layer on the melt (m = 37 g). At 780 °C, after 2 h the melt surface is covered by a layer of MgO several centimeters high (m = 345 g). However, $MgAl_2O_4$ is not formed, according to EDX analysis (see Fig. 4).

After 7 h, all melts kept under dry atmosphere with a water vapor partial pressure of 0 atm show an increase in mass higher than 1000 g. In addition to MgO, MA spinel formation is visible here on the SEM images (Fig. 5).

In a humid atmosphere ($p(H_2O) > 0$), only minor differences are noticeable concerning the oxidation behavior of the melt, as can be seen in Fig. 6. Except for individual small MgO seeds, the dross consists mainly of a thin Al_2O_3 oxide layer, which is hardly detectable by SEM. However, a slight increase in the mass of this oxide residue with increasing water vapor partial pressure can be detected. At 45 vol% H_2O, the amount of oxide is 46–57 g, depending on temperature and exposure time. At 90 vol% H_2O, the dross amount is 46–124 g.

Discussion

As can be expected, hydrogen content increases for the high Mg-containing alloy EN AW-5083 with rising water vapor partial pressure. The kinetics of hydrogen absorption is thereby slower than in the industrial melting processes, presumably due to the lack of bath movement. Since the measured hydrogen contents are in some cases, however, significantly below the values in thermodynamic equilibrium when equating $p(H_2)$ and $p(H_2O)$, it can be assumed that one or more kinetic barriers prevent the setting of the equilibrium state. According to the literature, there are two possible causes:

- The diffusion-determining gas boundary layer above the melt leads to only 51% of the hydrogen formed in reaction (3) being available for absorption (4) into the melt. [13]
- The formed oxide layer shields the melt, as soon as it is dense and stable, from the atmosphere above the melt and thus prevents the setting of equilibrium condition [14].

Figure 7 compares the hydrogen contents measured using Hycal with the two curves for hydrogen solubility at equilibrium for the two states $p(H_2) = p(H_2O)$ and $p(H_2) = 0.51 \cdot p(H_2O)$. The theoretical calculations were performed using formula (6)–(8) for the alloy EN AW-5083. It can be noticed that the measured values are much closer to the lower curve regardless of the experimental temperature, so it can be assumed that in the used experimental setup the hydrogen partial pressure does not equal the present water vapor partial pressure but must be much lower. However, the measurements performed at 90 vol% H_2O are also well below the second calculated curve ($p(H_2) = 0.51 \cdot p(H_2O)$). These values can therefore not be explained with the aid of the diffusion-determining boundary layer.

Fig. 5 SEM image of dross found on EN AW-5083 after 12 h exposure at 780 °C in dry atmosphere (98 vol% N_2, 2 vol% O_2)

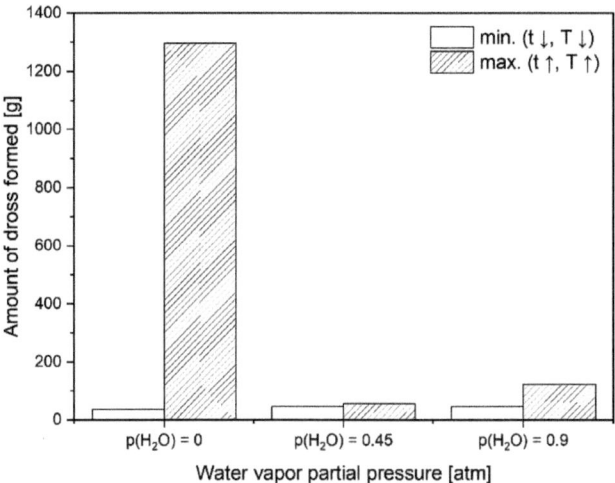

Fig. 6 Comparison of the dross formation depending on the water vapor partial pressure

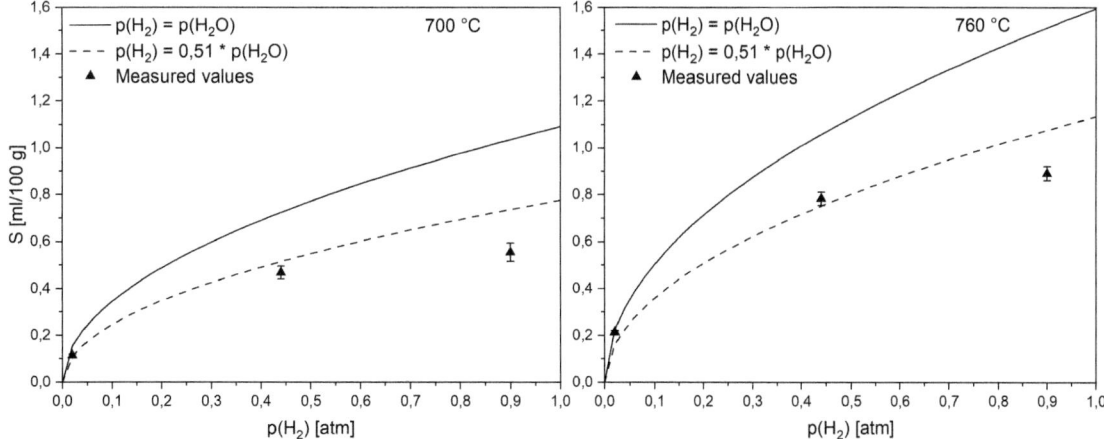

Fig. 7 Plots of measured hydrogen contents of EN AW-5083 as a function of hydrogen partial pressure at 700 °C (left) and 760 °C (right) compared with the calculated solubility at equilibrium for different correlations between $p(H_2)$ and $p(H_2O)$

A possible explanation for the low measured values at 90 vol% H_2O is the stable oxide layer on top of the melt. Due to the lack of bath movement, this represents a barrier to further hydrogen absorption after a certain point of time and prevents reaching a state of equilibrium with the atmosphere. Since the measured values at 45 vol% H_2O are relatively higher, it can also be assumed that the formation of the stable oxide layer requires more time than under an atmosphere with a higher water vapor content. Since the formation of such a stable oxide layer is usually not possible in industrial melting processes due to process-related bath movement, experiments with artificial bath movement are planned in future work, which should ensure the permanent breakup of the oxide layer on the surface of the melt. Within these experiments, it will also be possible to make more detailed predictions regarding the setting of the equilibrium state and/or the presence and effect of a diffusion-determining boundary layer.

Analyzing the influence of H_2O on the oxidation behavior of an EN AW-5083 melt, it becomes apparent that water vapor as part of the atmosphere has a positive effect on the onset of break-away oxidation. While under dry atmosphere (98 vol% N_2, 2 vol% O_2), break-away oxidation starts already after 2 h at 780 °C, it does not occur under humid atmosphere even after 12 h.

Since the oxygen content was kept constant at below 3 vol% over all experiments, this allows the conclusion that H_2O as part of the atmosphere seems to have a similar protective effect as CO_2. This was already presented by Cochran et al. [19] in 1977.

Beyond that, however, no significant differences concerning the influence of various water vapor contents on the oxidation of the melt in the range of $0 < p(H_2O) \leq 0.9$ atm can be observed. The slight mass increase of the oxide residue between 45 and 90 vol% H_2O can be explained by a higher removal of metallic aluminum during the skimming. A detailed evaluation would be possible by remelting the residue under salt addition, as this method allows the aluminum content of the dross to be determined. This can then be subtracted from the originally determined amount of dross. Such investigations are planned for further experiments in the future, analogous to more precise measurements using a thermobalance.

Conclusions

In laboratory-scale melting experiments with EN AW-5083 under atmospheres with constant oxygen and varying water vapor content, Hycal measurements were carried out to determine the hydrogen solubility and the oxidation behavior was investigated by weighing the skimmed dross and by SEM analysis. The planning and, to some extent, the evaluation of the experimental series were carried out software-based via statistical experimental design (MODDE 12.0.1).

The hydrogen solubility of the melt rises with increasing water vapor content of the atmosphere. In addition, the exposure time (up to 7 h) and the temperature also have an impact. Kinetic barriers—such as a diffusion-determining gas layer above the melt or a stable oxide layer—prevent the thermodynamic equilibrium state from being reached. Thus, the hydrogen partial pressure above a quiescent melt cannot be set equal to the water vapor partial pressure of the atmosphere.

Furthermore, water vapor in the atmosphere provides protection against break-away oxidation. The exact hydrogen content required for this inhibition will be determined in future experiments.

References

1. Krone K. (Ed.): Aluminiumrecycling. Vereinigung Deutscher Schmelzhütten, Düsseldorf (2000).
2. Thiele W.: Die Oxydation von Aluminium-und Aluminiumlegierungs-Schmelzen. Aluminium (1962), 707-715, 780–786.
3. Schmitz C. J. (Ed.): Handbook of aluminium recycling. Vulkan-Verlag, Essen (2006).
4. Paul R.: Energieeffizientes Umschmelzen und Recycling von Aluminium mit der Sauerstoffverbrennung. GWI—gaswärme international (2012), 45–49.
5. Talbot D. E. J.: The effects of hydrogen in aluminium and its alloys. In: Book/Institute of Materials, Minerals and Mining, Band: 724. Maney Publ, Leeds (2004).
6. Engh T. A., G. K. Sigworth and A. Kvithyld: Principles of Metal Refining and Recycling. Oxford University Press (2021).
7. ABB: AlSCAN Manual (2018).
8. Kammer C.: Aluminium-Taschenbuch. Aluminium-Verl., Düsseldorf (2002).
9. Marzoli L. et al.: Hydrogen Measurements Comparison in EN-AW 5083 Alloy. In: Chesonis, C. (Hg.): Light Metals 2019. Cham: Springer International Publishing, 973–980.
10. Pelss A. et al.: Benchmark and Practical Application of State of the Art Hydrogen Monitoring. In: Tomsett, A. (Hg.): Light Metals 2020. Cham: Springer International Publishing, 944–950.
11. Anyalebechi P. N.: Analysis of the effects of alloying elements on hydrogen solubility in liquid aluminum alloys. Scripta Metallurgica et Materialia, 33 (1995), 1209–1216.
12. Anyalebechi P. N.: Hydrogen Solubility in Liquid and Solid Pure Aluminum—Critical Review of Measurement Methodologies and Reported Values. Materials Sciences and Applications, 13 (2022), 158–212.
13. Syvertsen M. et al.: Furnace Atmosphere and Dissolved Hydrogen in Aluminium. In: Chesonis, C. (Hg.): Light Metals 2019. Cham: Springer International Publishing, 1051–1056.
14. Talbot D. E. J. and P. N. Anyalebechi: Solubility of hydrogen in liquid aluminium. Materials Science and Technology, 4 (1988), 1–4.
15. Stephenson D. J.: The absorption of hydrogen from humid atmospheres by molten aluminium and an aluminium-magnesium alloy. PhD thesis, Brunel University School of Engineering (1978).

16. Impey S. A., D. J. Stephenson and J. R. Nicholls: Mechanism of scale growth on liquid aluminium. Materials Science and Technology, 4 (1988), 1126–1132.

17. Haginoya I. and Sato K.: Influence of Water Vapor on the Oxidation of Molten Al-Mg Alloys. The Journal of the Japan Foundrymen's Society, 56 (1984), 264–268.

18. Ostermann F.: Anwendungstechnologie Aluminium. In: VDI-Buch. Springer Berlin Heidelberg, Berlin, Heidelberg (2014).

19. Cochran C. N. and W. C. Sleppy: Oxidation of High-Purity Aluminum and 5052 Aluminum-Magnesium Alloy at Elevated Temperatures. Journal of The Electrochemical Society, 108 (1961), 322.

Influence of Cryolite Content on the Thermal Properties and Coalescence Efficiency of NaCl–KCl Salt Flux

Veronica Milani, Alicia Vallejo-Olivares, Gabriella Tranell, and Giulio Timelli

Abstract

Salt fluxes with fluoride additions are necessary for the treatment and recycling of contaminated or oxidized aluminium scrap. This study aims to investigate the effect of cryolite additions on the thermal properties and the coalescence efficiency of a NaCl–KCl salt flux mixture. Thermodynamic calculations were carried out to examine the phase diagram of the salt mixture as the cryolite content increases. The study of the melting properties of the salt was carried out by means of differential scanning calorimetry. The coalescence efficiency of the salts was assessed by re-melting coated aluminium chips. The experimental results show a decrease in the liquidus temperature as the cryolite content increases; this differs from the thermodynamical calculations. For cryolite contents up to 3 wt%, higher cryolite content in the salt leads to higher coalescence of re-melted chips. However, the coalescence differences observed between the recycling products for cryolite contents of 3% and higher were minor.

Keywords

Salt flux • Aluminium recycling • Coalescence • Melting point • Cryolite

V. Milani · G. Timelli (✉)
Department of Management and Engineering, University of Padova, Stradella S. Nicola, 3, 36100 Vicenza, Italy
e-mail: timelli@gest.unipd.it

A. Vallejo-Olivares · G. Tranell
Department of Materials Science and Engineering, Norwegian University of Science and Technology (NTNU), Alfred Getz vei 2, N-7465 Trondheim, Norway

Introduction

Aluminium is the second most used metal in the world. Its high specific strength, good corrosion resistance, high electrical and thermal conductivity, and the versatility of its manufacturing processes have made it a competitive material for several applications: packaging, construction, automotive and aerospace, among others. Despite the benefits arising from fuel and energy savings due to the low weight of aluminium products, aluminium requires a substantial amount of energy for its primary production [1]. Due to its strong affinity with oxygen, a great amount of energy is required for the electrolytic conversion of alumina into pure aluminium metal. The Hall–Héroult process is one of the most energy-intensive industrial processes, and it accounts for approximately 2/3 of the total 3% of global GHGs emissions attributed to aluminium production [2]. Other environmental concerns of aluminium's primary production include the air pollutants and the solid waste arising from the Bayer process, also known as red mud. The demand for aluminium products is predicted to increase to 81% by 2050 [3]. Given the current energy and climate crisis, aluminium recycling is crucial for sustainability since the recycling process requires much less energy and produces significantly fewer waste products and emissions than primary production [1]. The secondary production of aluminium (namely recycling) starts with scrap collection. The scrap is generally categorized according to its origin: *new scrap* originates from the waste of metal-processing industries and is usually characterized by low content of impurities and known composition. *Old scrap* arises from post-consumer utilization, often contains higher concentrations of impurities, and its composition is usually unknown. The scrap collection is followed by comminution and sorting to confer the scrap charge its proper size and eliminate contaminants such as plastic, rubber, glass, and other metals. Some routes of scrap preparation also involve thermal or chemical processes to de-coat the scrap and eliminate paints, paper, moisture, and other contaminants from the scrap surface. The de-coating process may be followed by the compaction of scrap pieces of small size, which tend to oxidate more easily [4]. After the scrap has been prepared, the next step is to re-melt it. Different furnaces are available to melt the scrap according to several parameters such as production volume, energy requirements and cost, scrap characteristics, and final product composition. Rotary furnaces are a suitable choice for processing heavily oxidized and contaminated scrap. Compared to other furnaces used for re-melting aluminium scrap, such as reverberatory furnaces, rotary furnaces have lower emissions, fuel consumption, and higher melting efficiencies. Rotary furnaces consist of a rotating refractory-lined steel cylinder with a flame burner to melt the aluminium scrap. Melting contaminated scrap in rotary furnaces usually requires the addition of salt fluxes to refine the molten metal [5]. Salt fluxes, which generally consist of a NaCl–KCl mixture with small additions of fluorides, perform three main functions: to cover the molten metal to prevent its oxidation, to collect impurities and oxides from the metal bath, and to promote the coalescence of the metal droplets by freeing them from the surrounding oxide layer [6]. Different percentages of NaCl and KCl can be used as a salt flux. NaCl salt is typically cheaper [7] than KCl, but KCl additions decrease the melting point, reducing energy requirements and the cost of production. An equimolar mixture of NaCl–KCl forms a eutectic point with a melting temperature of approximately 657 °C [5]. During the melting process, the salt flux collects oxides and impurities from the molten scrap and forms a slag layer on top of the bath, which sometimes entraps molten metal droplets. It has been discussed in the literature [6, 8, 9] that the small fluoride additions to the salt flux promote the coalescence of the molten aluminium droplets and reduce the metal losses by entrapment in the salt-slag. The fluorides, which often consist of CaF_2 or Na_3AlF_6, strip the oxide layer surrounding the molten metal droplet, which is then free to coagulate with other droplets and coalesce. The mechanism by which fluorides act is not completely understood in the literature. According to Peterson, the rupture of the oxide layer is related to the fluoride's ability to dissolve aluminium oxide [6]. Tenorio and Espinosa suggested that fluoride additions enhance the rupture of the oxide layer according to a mechanism similar to the hot corrosion process, which helps release the molten metal from the surrounding oxide layer [10]. Interfacial tensions between the salt and the metal also play a crucial role in enhancing the coalescence and the coagulation of metal droplets, as suggested by Roy and Sahai [11]. The treatment and disposal of the waste product from secondary aluminium production, also known as *salt cake*, must be addressed properly due to its potential environmental impact. Usually, salt cakes are either treated for recovery of their components or stored in controlled landfills. The latter is forbidden in most of Europe due to the high reactivity of salt cakes to form pollutants and toxic compounds, so its recovery is the most ecological choice [12].

Although the recycling performance and thermodynamical properties of salts with varying NaCl–KCl ratios have been extensively studied [7, 13], there is little available data regarding how varying cryolite additions affect these properties. In this work, a 95 wt% NaCl–5 wt% KCl salt flux with cryolite (Na_3AlF_6) additions ranging from 2 to 15% has been investigated in terms of their melting behaviour and Al coalescence efficiency. The melting behaviour of 70 wt% NaCl–30 wt% KCl and 50 wt% NaCl–50 wt% KCl mixes, both with 2% cryolite content, was also investigated.

Materials and Methods

The salt fluxes studied in this work were composed of sodium chloride, NaCl (99.5% purity, Fisher Scientific), potassium chloride, KCl, (99.5% purity, Sigma-Aldrich), and cryolite, Na_3AlF_6 (97% purity, Sigma-Aldrich). The composition of the analyzed salts mixtures is summarized in Table 1.

The salt mixes with 95/5 ratio NaCl/KCl, and cryolite additions of 0, 2, 3, 5, 7 and 10 wt% were used for the coalescence study. Chips from shredding an aluminium sheet of 8111 alloy and 600 µm thickness, provided by Speira Holmestrand, were re-melted to test the influence of increasing cryolite contents on the degree of coalescence of the recovered aluminium. The aluminium sheet was coated by a polymeric layer of approximately 25 µm on one side and 5 µ on the other. The exact composition of the coating is unknown but based on analysis from a previous study [14], it consists of a polymer resin with oxide particles of TiO_2, SiO_2 and $BaSO_4$.

Sample Preparation

The salt samples were prepared by melting 300 g of each mixture composition in graphite crucibles by means of an induction furnace with Argon flushing. The molten salt mixtures were cast in a rectangular water-cooled copper mould and let solidify. The aluminium sheet was first shredded into chips with a Getecha RS 1600-A1.1.1 with a grate of 8 mm diameter and then sieved using two sieves of square mesh 5 and 2 mm^2 to unify their size. The mean weight of the chips was calculated in a previous study as 48 mg [14].

Table 1 Chemical composition of the analyzed salt mixtures

Mixture	wt% NaCl/KCl	wt% Na_3AlF_6
95-5-0	95/5	0
95-5-2	95/5	2
95-5-3	95/5	3
95-5-5	95/5	5
95-5-7	95/5	7
95-5-10	95/5	10
95-5-10	95/5	15
70-30-2	70/30	2
50-50-2	50/50	2

Thermodynamic Analysis

A preliminary thermodynamic analysis by means of the thermodynamic software *Factsage 8.2*™ [15] was carried out. Since the NaCl/KCl ratio in each mixture is constant, it was possible to obtain binary phase diagrams in which the considered variables were the temperature and the amount of cryolite. NaCl/KCl mixtures with 95/5, 70/30, and 50/50 ratios were used in the equilibrium and phase diagram calculations. For each NaCl/KCl ratio, a binary phase diagram was created, by setting the amount of cryolite (Na_3AlF_6) and the temperature as variables. The FTsalt database (SALTA and solid solutions) was used to describe the salt mixture. The formation of bath and cryolite solutions are ignored in the present calculations to avoid compatibility issues. Thermodynamic description of the pure substances was taken from FactPS database.

Differential Scanning Calorimetry (DSC)

A heat-flux DSC apparatus (Linseis STA PT 1600) was utilized to perform the thermal analysis on samples weighing approximately 60 mg. The DSC tests were carried out with an empty graphite reference crucible as heating and cooling scans (5 °C/min 820; 10 min dwelling at 820 °C; 5 °C/min to 25 °C) in an Argon atmosphere with a gas flow of 0.2 l/min. An empty graphite crucible with a 6.8 mm diameter was utilized for the zero line. The samples were obtained by crushing the previously molten salt mixtures. No calibration was carried out for the experiments. By analyzing the DSC curves, the peak extrapolated onset temperature (T_{onset}) and the peak maximum temperature (T_{peak}) were obtained as the intersection between the ascending peak slope and the baseline and as the intersection point between the fitted lines of the descending and ascending slopes of the peak, respectively. The extrapolated baseline was obtained as a straight line between each peak region's initial and final temperature. The initial and final temperatures of the peak

region were identified as deviations from the baseline of the DSC curve. The extrapolated peak onset temperature was determined as the liquidus temperature of the samples. When two peaks were present, the second peak maximum temperature was determined as the eutectic temperature of the salt mixtures.

Coalescence Study

The re-melting experiments were conducted using ceramic crucibles (Al_2O_3-SiO_2) of 20 Cl volume and a Nabertherm™ resistance furnace. For each trial, 80 g of salt-flux were used to re-melt 20 g of chips (salt/scrap ratio 4). This ratio is higher than the typical industrial ones, but it was chosen so that the molten salt would completely cover the scrap and prevent oxidation reactions. The crucibles containing the mixes of salts were placed into the furnace at 830 °C. After approximately 40 min (once the salts were completely molten) the chips were charged. Instantly, the combustion of the organic components of the coating generated flames and a dark smoke, and the lid of the furnace was kept open until the end of this combustion (approx. 30 s). Then the crucibles were held inside the closed furnace for 7 min, removed and naturally cooled in air. There was no stirring applied. Once the crucibles were at room temperature, the salt was separated from the metal by washing it away with water on a 0.8 mm^2 mesh size sieve. As mentioned in the introduction, the coalescence describes the ability of the individual aluminium pieces to merge, which is critical for a successful re-melting operation without too many metal losses to small metal droplets dispersed in the salt slag. In previous studies, Peterson [6] and Vallejo-Olivares [14, 16] calculated the coalescence efficiency as the percentage of mass recovered which had merged into the largest metal bead per trial. Gökelma [17] and Thoraval [9], as the percentage of mass merged into pieces larger than a specific threshold set depending on the initial size or mass of the scrap. Capuzzi [8] considered in addition the roundness and the diameter of the recovered pieces, and Xiao assessed the coalescence visually in [18] and quantified the size distributions of the metal beads in the salt [19]. In the present study, the average mass of a shredding was just below 50 mg. Thus, it was assumed that the pieces weighting over 0.5 g originated from the coalescence of at least two chips, and the coalescence efficiency calculated according to Eq. (1):

$$\%Coalescence = \frac{m_{sum > 0.5g}}{m_{recovered}} * 100 \qquad (1)$$

where $m_{recovered}$ is the mass sum of all pieces recovered after leaching the salts, and $m_{sum > 0.5g}$ is the mass sum of the pieces heavier than 0.5 g.

Results and Discussion

Thermodynamic Analysis

The phase diagrams obtained with the thermodynamic software FactSage show the presence of a liquid phase above 757 °C for the 95 wt% NaCl–5 wt% KCl, regardless of the cryolite content. In the temperature interval 757–788 °C, a solid rocksalt phase is stable with sodium and aluminum-fluoride solid compounds. For temperatures above 788 °C, the rocksalt phase is not present. At temperatures higher than 788 °C, an increase in the cryolite content leads to the disappearance of AlF_3, which is gradually replaced by $Na_5Al_3F_{14}$ and, for cryolite additions greater than 4%, also

by Na_3AlF_6. The phase diagram for 95 wt% NaCl–5 wt% KCl with cryolite additions up to 15 wt% is reported in Fig. 1.

From the thermodynamic analysis of the 70 wt% NaCl–30 wt% KCl, a liquid phase is present above 800 °C, regardless of the cryolite content. At temperatures higher than 800 °C, an increasing cryolite content leads to the stabilization of sodium-aluminum-fluoride solid compounds (Na_3AlF_6 and $Na_5Al_3F_{14}$), which gradually substitute AlF_3. The phase diagram for the 70 wt% NaCl–30 wt% KCl mixture with cryolite concentrations up to 15 wt% is shown in Fig. 2.

The investigation of the 50 wt% NaCl–50 wt% KCl phase diagram at varying cryolite content reveals the presence of a liquid phase above 650 °C, as can be seen in Fig. 3. The addition of cryolite to the salt mixture for temperature above 662 °C leads to the stabilization of solid Na_3AlF_6 and $Na_5Al_3F_{14}$. However, when cryolite is present below concentrations of 1% at temperatures higher than 775 °C, the presence of a solid potassium-aluminum fluoride compound (KAl_4F_{14}) is also noted, in contrast with the previous phase diagrams.

DSC Analysis

To avoid effects due to superheating, the cooling scans from the DSC were used to determine the characteristic temperatures of the samples. The cooling curves for each mixture

Fig. 1 Phase diagram for the 95% NaCl–5% KCl salt flux with cryolite additions up to 15% weight

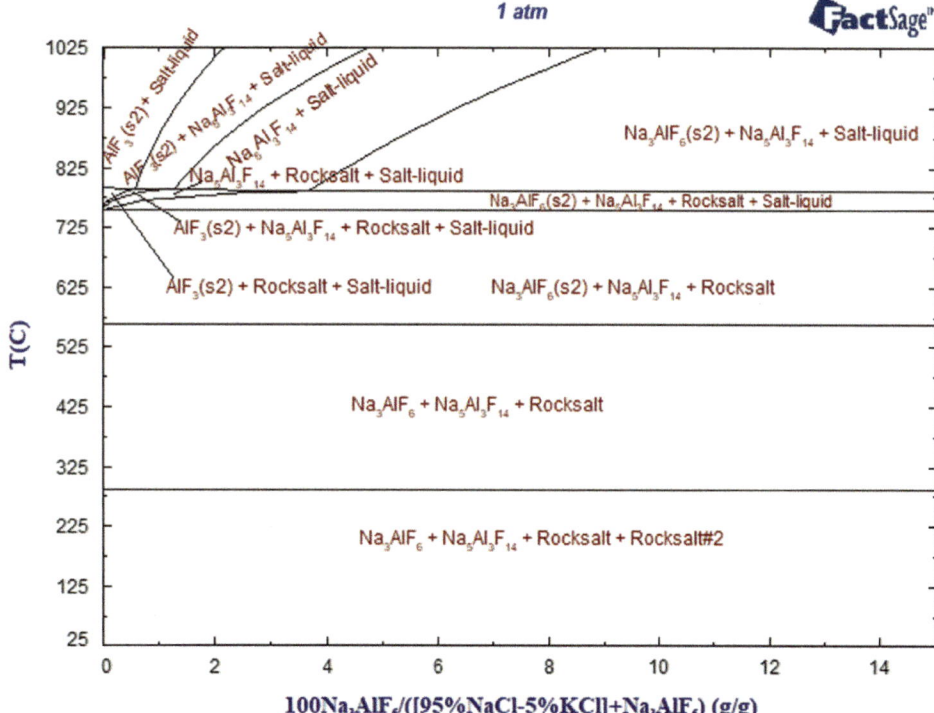

Fig. 2 Phase diagram for the 70% NaCl–30% KCl salt flux with cryolite additions up to 15%

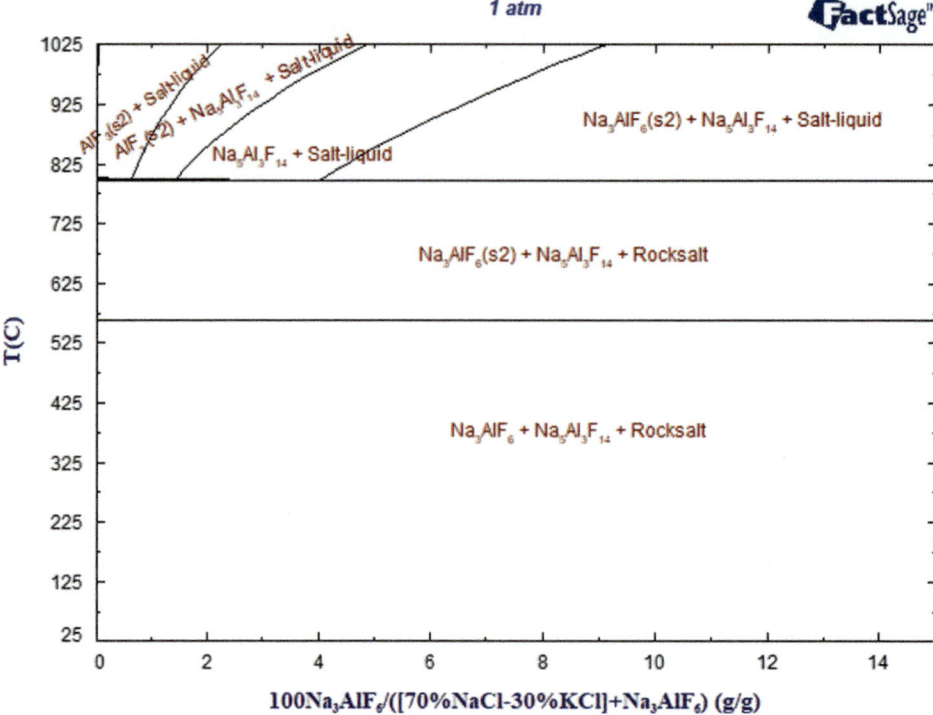

Fig. 3 Phase diagram for the 50% NaCl–50% KCl salt flux with cryolite additions up to 15%

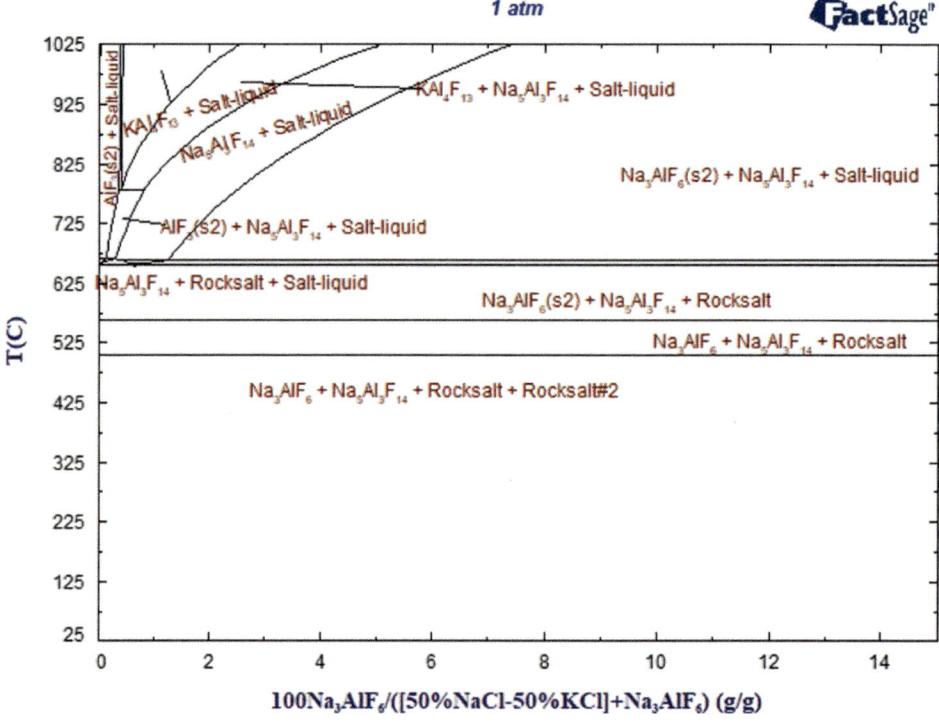

containing cryolite showed two main events: a peak due to the solidification of the eutectic and a peak due to the solidification of the remaining phase, as can be seen in Fig. 4 for the 10% cryolite-containing salt. For cryolite concentrations in the range of 2–10 wt%, minor peaks in the temperature range of 631–619 °C were noted and attributed to crystallization phenomena, since these peaks were not found in the heating curves.

The extrapolated onset temperature of the first peak represents the liquidus temperature, whereas the second peak's

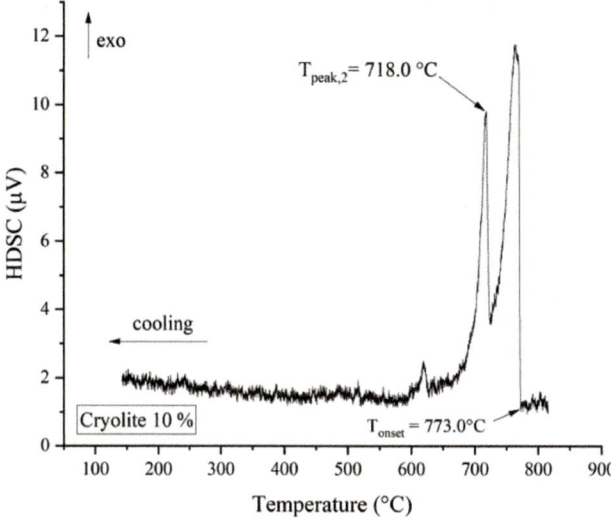

Fig. 4 DSC curve obtained for the 95% NaCl–5% KCl salt flux with 10 wt% cryolite addition

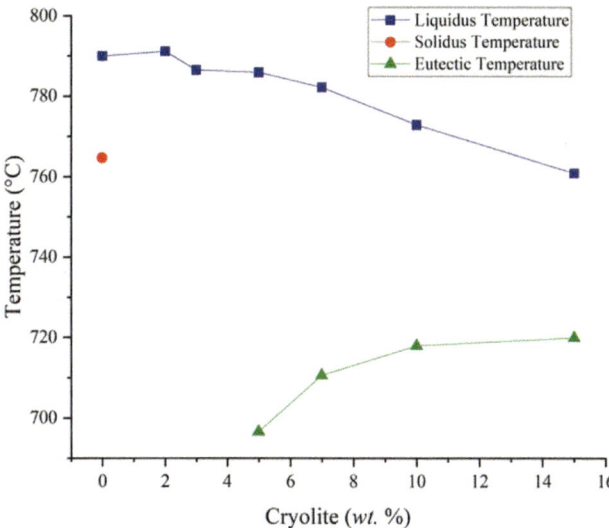

Fig. 5 Plot of the liquidus and eutectic temperatures obtained from the analysis of the DSC curves of the 95% NaCl–5% KCl salt mixtures. The solidus temperature for the mixture without cryolite is also displayed

maximum temperature indicates the eutectic temperature. The mixtures with cryolite contents in the range of 0–3 wt% did not show the presence of the eutectic. The results are shown in Fig. 5.

The 95% NaCl–5% KCl salt without cryolite shows a liquidus temperature of 790 °C, well in agreement with the NaCl–KCl phase diagram from Bolivar and Friedrich [7], whereas the experimental results find the solidus temperature at 765 °C, approximately 10 °C higher than the phase

diagram. The addition of cryolite to the 95% NaCl–5% KCl salt mixture noticeably affects the liquidus temperature of the salt fluxes, especially for cryolite contents above 5 wt%. The liquidus temperature decreases from 790 °C for the salt without cryolite to 761 °C for the 15% cryolite-containing salt. From an industrial point of view, the liquidus temperature is the most relevant: a decrease in the melting point of the salt fluxes is beneficial as it implies lower emissions and reduced energy-related costs. The eutectic temperature for cryolite contents in the range 5–15 wt% increases from 697 to 720 °C.

The liquidus temperatures for the 50% NaCl–50% KCl and 70% NaCl–30% KCl mixtures with 2% cryolite are 661 and 725 °C, respectively. According to the NaCl–KCl phase diagram from Bolivar and Friedrich [7], the liquidus temperatures for 50 and 70% NaCl are 660 and 725 °C, respectively: this implies that 2% cryolite additions to the salt mixtures do not affect significantly their liquidus temperature. This is also true for the 95% NaCl–5% KCl. The liquidus temperatures for the 50% NaCl–50% KCl and 70% NaCl–30% KCl mixtures with 2% cryolite are lower when compared with the 95% NaCl–5% KCl mixtures, regardless of their cryolite content. It is accepted that increasing the KCl/NaCl ratio up to an equimolar mixture decreases the salt's melting temperature [13]. This is also seen in the present study and implies higher energy requirements to melt salt fluxes with higher NaCl content. However, it may be economically viable to decrease the KCl content due to its higher cost compared to NaCl [7]. An additional positive effect by lowering the KCl-content is the decrease of the vapour pressure of the slag and of the salt loss due to evaporation [20]. Furthermore, although the results provide valuable information of the fundamental properties of various salt-flux mixtures, they may not be directly related to their performance during recycling. For instance, neither Pirker's [21] nor Bolivar's [7] recycling studies showed significant differences between the metal yield obtained by salts containing KCl between 10 and 30% when testing in a 160 kg laboratory-scale rotary furnace and 30 kg resistance furnace.

The experimental results from the DSC analysis differ from the preliminary thermodynamic analysis. This may be explained by the fact that the experimental conditions are not those of equilibrium, or to missing information in the used databases. The solubility of $AlCl_3$ and AlF_3 species is not described in the thermodynamic behaviour of the selected liquid salt solution. As such, in the case of solubility of these species in the liquid salt, the overall thermodynamic behaviour of the system might be influenced. As mentioned, the formation of bath and cryolite solutions was ignored due to the compatibility issues with other solutions, which introduces further limitations to describe the system accurately.

Coalescence Ability of Salt Fluxes

According to the coalescence results, presented in Fig. 6a, b, small additions of 2% or 3 wt% cryolite significantly improve the coalescence ability of the salt fluxes, but further additions have a small effect.

Figure 6a, b show the remarkable difference between re-meltings without cryolite, on average 19% coalescence, and the re-melting with salt containing cryolite, which led to coalescences over 83% in all cases. There was a large deviation between the three re-melting trials without cryolite, and the exact values of the three trials were of 4, 5 and 47%. When increasing the cryolite content from 2 to 3% there was an improvement from 85 to 95% average coalescence, but higher cryolite additions increased the coalescence just slightly. Therefore, for the current set up increasing the cryolite additions over 2 or 3% would not pay off considering the extra cost and the environmental risks associated with cryolite. Thoraval [9] also reached the optimal coalescence efficiencies with additions of 2% cryolite for clean and recycled salts contaminated with less than 30 wt% oxides. Similarly, Bolivar concluded that adding 1% CaF_2 increased the metal yield if the NaCl/KCl ratio is 0.85 or higher, but further additions were not necessary. However, comparing the coalescence or recyclability results with the literature is not straightforward, as it was demonstrated that they are influenced by many factors, such as scrap size and composition (e.g. oxides, organic contamination) [17–19], the application or not of thermal or compaction pre-treatments [8, 14, 16], the holding time and temperature [9], the salt/scrap ratio [8] and finally furnace set-up and stirring. For instance, Thoraval [9] used a laboratory scale

rotary furnace to re-melt oxidized chips, Gökelma [17] applied stirring manually to recycle oxidized bottom ash, while Capuzzi [8], Vallejo-Olivares [14], and the present study used a static resistance furnace without stirring to re-melt coated scrap. The present coalescence results for 2% cryolite are higher than those obtained for similar coated chips by Vallejo-Olivares using a salt composition of 68.6/29.4/2.0 $NaCl/KCl/CaF_2$, but lower than those obtained for the de-coated chips. Since CaF_2 is less costly both economically and environmentally, applying a thermal de-coating or mechanical stirring may be more advisable than substituting CaF_2 by cryolite additions to the salt-fluxes to optimize the recycling of aluminium scrap.

Conclusions

This study aimed to investigate the thermal behavior of the 95% NaCl–5% KCl salt flux, as well as its ability in promoting aluminium droplets' coalescence. The following conclusions were drawn:

Thermal Behavior

According to the preliminary thermodynamic analysis performed with FactSage, an increasing cryolite content does not affect the temperature at which the liquid phases appear in the phase diagrams.

The experimental results from the Differential Scanning Calorimetry show a decrease in the liquidus temperature as the cryolite content in the 95% NaCl–5% KCl salt flux

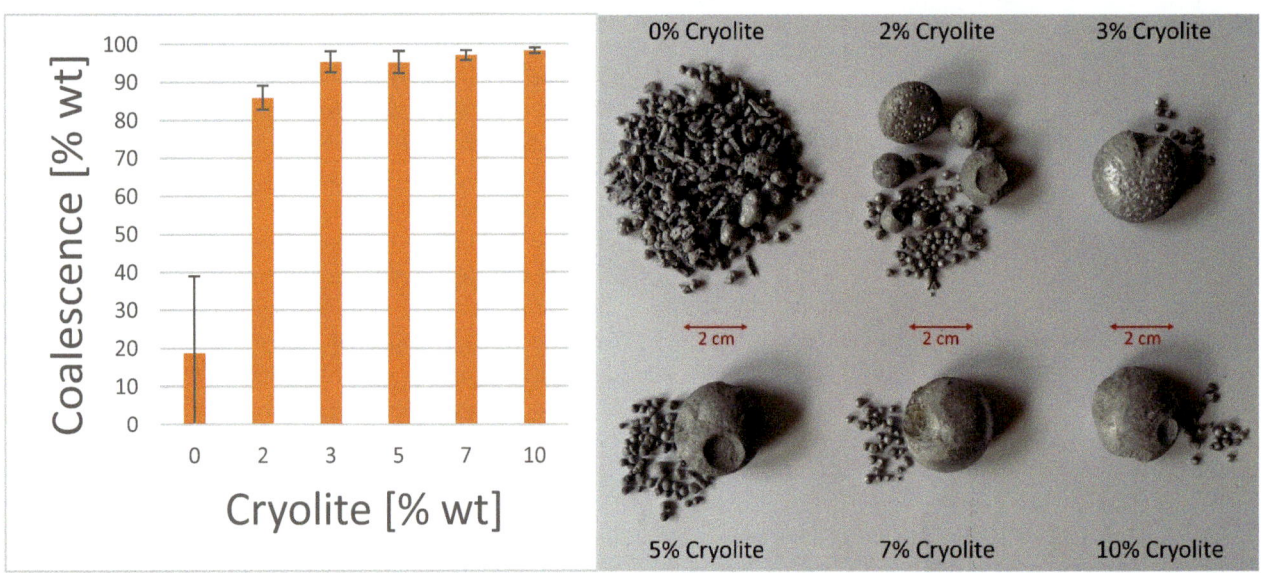

Fig. 6 **a** Left: Plot of the degree of coalescence of the metallic pieces recovered after re-melting and washing away the salts. Average and STD from 3 repetitions. **b** Right: pictures of the recovered metal products

increases, especially for cryolite contents above 5 wt% For cryolite additions in the range 5–15 wt%, the presence of a eutectic was noted in the temperature range of 697–720 °C.

The liquidus temperatures of the 95% NaCl–5% KCl salt mixtures are higher when compared with the 70% NaCl–30% KCl and the 50% NaCl–50% KCl, regardless of their cryolite content.

Based on the discrepancies between experimental and modelled results, it is suggested that the salt-cryolite system is thermodynamically re-assessed.

Coalescence Ability of Salt Fluxes

While small additions of 2% or 3 wt% cryolite greatly improved the coalescence ability of the salt fluxes, further cryolite additions did not have a significant effect on the coalescence of the recovered aluminium.

Acknowledgements The authors gratefully acknowledge the Department of Materials Science and Engineering at NTNU, Trondheim, for the experimental equipment and support, especially to Adamantia Lazou for carrying out and discussing the phase diagrams and to Arman Hoseinpur-Kermani for the help with the induction furnace and DSC equipment.

References

1. J. A. S. Green, Aluminum recycling and processing for energy conservation and sustainability, ASM International, 2007.
2. D. Raabe, D. Ponge, P. J. Uggowitzer, M. Roscher, M. Paolantonio, C. Liu, H. Antrekowitsch, E. Kozeschnik, D. Seidmann, B. Gault, D. De Geuser, A. Deschamps, C. Hutchinson, C. Liu, Z. Li, P. Prangnell, J. Robson, P. Shanthraj, S. Vakili, C. Sinclair, L. Bourgeois and S. Pogatscher, "Making sustainable aluminum by recycling scrap: The science of dirty alloys," *Progress in Materials Science*, vol. 128, 2022.
3. International Aluminium Association—IAI, "Beyond 2 degrees—The outlook for the aluminium sector," 2018. [Online]. Available: https://international-aluminium.org/resource/beyond-2-degrees-the-outlook-for-the-aluminium-sector-factsheet/. [Accessed 20 April 2022].
4. M. E. Schlesinger, Aluminum Recycling, Taylor & Francis Group, 2007.
5. S. Capuzzi and G. Timelli, "Preparation and melting of scrap in aluminum recycling: a review," *Metals*, vol. 8, no. 4, pp. 249–273, 2018.
6. D. R. Peterson, "Effect of salt flux additives on aluminum droplets coalescence," in *Second International Symposium—Recycling of*

7. R. Bolivar and B. Friedrich, "The influence of increased NaCl:KCl ratios on metal yield in salt bath smelting processes for aluminium recycling," *World of Metallurgy—ERZMETAL*, 2009.
8. S. Capuzzi, A. Kvithyld, G. Timelli, A. Nordmark, E. Gumbmann and T. A. Engh, "Coalescence of Clean, Coated and Decoated Aluminium for Various Salts, and Salt-Scrap rations," *Journal of Sustainable Metallurgy*, vol. 4, no. 2018, pp. 343–358, 2018.
9. M. Thoraval and B. Friedrich, "Metal Entrapment in Slag during the Aluminium," in *European Metallurgy Conference*, Aachen, 2015.
10. J. A. S. Tenorio and D. C. Romano Espinosa, "Effect of salt/oxide interaction on the process of aluminum recycling," *Journal of Light Metals*, vol. 2, pp. 89–93, 2002.
11. R. R. Roy and Y. Sahai, "Interfacial tension between aluminum alloy and molten salt flux," *Materials Transactions*, vol. 38, no. 6, pp. 546–552, 1997.
12. P. E. Tsakiridis, "Aluminium salt slag characterization and utilization—A review," *Journal of Hazardous Materials*, Vols. 217–218, pp. 1–10, 2012.
13. D. Coleman and P. Lacy, "The phase equilibrium diagram for the KCl–NaCl system," *Materials Research Bulletin*, vol. 2, no. 10, pp. 935–938, 1967.
14. A. Vallejo-Olivares, S. Høgåsen, A. Kvithyld and G. Tranell, "Effect of Compaction and Thermal De-coating Pre-treatments on the Recyclability of Coated and Uncoated Aluminium," in *Light Metals*, Anaheim, 2022.
15. C. W. Bale, P. Chartrand, S. A. Degterov, G. Eriksson, K. Hack, R. Ben Mahfoud, J. Melançon, A. D. Pelton and S. Petersen, "FactSage thermochemical software and databases," *Calphad*, vol. 26, no. 2, pp. 189–228, 2002.
16. A. Vallejo-Olivares, H. Philipson, M. Gökelma, H. Roven, T. Furu, A. Kvithyld and G. Tranell, "Compaction of Aluminium Foil and Its Effect on Oxidation and Recycling Yield," 2021.
17. M. Gökelma, A. Vallejo-Olivares and G. Tranell, "Characteristic properties and recyclability of the aluminium fraction," *Waste Management*, vol. 130, no. 2021, pp. 65–73, 2021.
18. Y. Xiao and M. Reuter, "Aluminium Recycling and Environmental Issues of Salt Slag," *Journal of Environmental Science and Health*, vol. 40, no. 2005, p. 1861–1875, 2005.
19. Y. Xiao and M. Reuter, "Recycling of distributed aluminium turning scrap," *Materials Engineering*, vol. 15, no. 2002, pp. 963–970, 2002.
20. S. Gül, R. Dittrich and B. Friedrich, "KCl-reduced salt application in aluminium recycling," in *European Metallurgy Conference*, 2013.
21. A. Pirter, H. Antrekowitsh, W. Fragner, H. Suppan and M. Kettner, "Optimization of the Al-recycling Process for Low Grade Scraps," *BHM Berg- und Hüttenmännische Monatshefte*, vol. 160, pp. 320–327, 2015.
22. G. Höhne, W. Hemminger and H. J. Flammersheim, Differential Scanning Calorimetry—Second Edition, Springer, 2003.
23. R. Peterson, "Effect of Salt Flux Additives on Aluminium Droplets Coalescence," in *Light Metals*, 1990.

metals and engineered materials—The Minerals, Metals and Materials Society, 1990.

Oxidation Study of Al–Mg Alloys in Furnace Atmospheres Using Hydrogen and Methane as Fuel

M. Syvertsen, A. Johansson, J. Lodin, A. Bergin, M. Ommedal, Y. Langsrud, and R. D. Peterson

Abstract

By switching from fossil to hydrogen fuel in melting furnaces, the CO_2 footprint of a cast house in the aluminium industry can be reduced considerably. However, using hydrogen instead of fossil fuel will increase the water vapour concentration in the furnace atmosphere and likely change the oxidation behaviour of molten aluminium alloys. A series of tests have been performed where Al–Mg alloys were melted and normally held for 4 h in different atmosphere compositions due to variations in fuel type and burner set-up. The results show that combustion of hydrogen in an air–fuel configuration gives more oxidation on liquid Al–Mg alloys than combustion of hydrogen in an oxy-fuel configuration. The tests also show that as little as 5% CO_2 in the furnace atmosphere significantly suppresses oxidation. The suppressing effect of CO_2 was greater on alloys with 4.7% Mg than on alloys with 3.1% Mg.

Keywords

Aluminium • Environmental effects • Pyrometallurgy • Hydrogen

Introduction

During the last 50–60 years there have been many studies on the oxidation behaviour of aluminium and aluminium alloys. For instance, several studies have shown higher temperatures will increase oxidation rate. Some elements have been shown to decrease oxidation (such as Be), while many others will increase oxidation (Mg, Na, Li etc.) [1]. More recently there has been an increasing focus on reduction of CO_2-emissions from the aluminium industry. In the cast houses the main CO_2 emission comes from the combustion of fossil fuel; NG, LNG, or LPG (Natural Gas, Liquified Natural Gas, or Liquified Petroleum Gas), or even oil for heating in reverberatory furnaces. Therefore, transition to using hydrogen as fuel is interesting to explore.

Little work has been presented on oxidation of molten Al–Mg alloys in high humidity atmospheres. Cochran, Belitskus, and Kinosz [2] published an article in 1977 showing the effects of various atmospheres on oxidation of Al–Mg alloys.[1] They provided rather vague conclusions where both CO_2 and N_2 should postpone the start of the breakaway oxidation where $MgAl_2O_4$ starts forming. They did not make any conclusions with regard to water vapour. Both Haginoya [3] and Smith [4] investigated the effect of CO_2 + air on oxidation of aluminium and aluminium-magnesium alloy samples and saw a clear effect in suppression of oxidation. Haginoya concluded that for Al–10% Mg, they

M. Syvertsen (✉)
SINTEF Industry, Alfred Getz Vei 2, 7034 Trondheim, Norway
e-mail: martin.syvertsen@sintef.no

A. Johansson
Siemens Energy AB, Slottsvägen 2-6, 612 31 Finspång, Sweden

J. Lodin
Linde Gas AB, Varuvägen 2-10, 125 30, Älvsjö, Sweden

A. Bergin
Hydro Aluminium ANS, Romsdalsvegen 1, 6600 Sunndalsøra, Norway

M. Ommedal
Alcoa Norway ANS, Havnegata 40, 8663 Mosjøen, Norway

Y. Langsrud
Benteler Aluminium Systems Norway AS, Grøndalsvegen 2, 2830 Raufoss, Norway

R. D. Peterson
Real Alloy Recycling INC, 388 Williamson Drive, Loudon, TN 37774, USA

---

[1] There is probably an error in the pressure units they used. They write that the total gas-pressure in the experiments were up to 98.6 MPa, it probably was 98.6 kPa (which is nearly 1 bar).

© The Minerals, Metals & Materials Society 2023
S. Broek (ed.), *Light Metals 2023*, The Minerals, Metals & Materials Series,
https://doi.org/10.1007/978-3-031-22532-1_124

needed more than 20% CO_2 at 750 °C to minimize oxidation for oxidation times shorter than 4 h.

Stevens et al. [5] investigated oxidation of Al–4.6% Mg alloy at 800 °C in laboratory scale TGA in dry and humid argon, air and air + CO_2 atmospheres. They found no impact of humidity on the oxidation rate, but clear inhibiting effect from CO_2. One must be aware that the humidity they used gave a maximum vapour concentration of 7% (mol basis) in the atmospheres. The humidity was limited by condensation in the TGA-equipment.

Combustion of hydrogen in oxy-fuel set-up will result in an atmosphere made of only water molecules. Therefore, if we want to perform oxidation studies in realistic furnace atmospheres, it is crucial to use a gas fired furnace with properly controlled atmosphere.

At Linde's lab in Älvsjö close to Stockholm, Sweden, there is a gas-fired furnace with the potential to obtain any concentration of water vapour up to 100%, depending on the fuel and burner set-up. The same furnace was used in earlier studies and is described in publications in previous Light Metals 2020 [6, 7].

Fig. 1 Picture of Linde's furnace used in the experiments

Experimental Set-Up

Linde's lab is equipped with a box furnace designed for testing burners and various combustions of gases. The size of the furnace chamber is about 4 m deep, 2.2 m wide and 2 m high. This furnace is ideal for testing the impact of various combustion gases on various materials. A picture of the furnace with open door is shown in Fig. 1. In the same figure four steel crucibles are shown which each held about 10 kg aluminium metal.

The crucibles are coated with Velvacoat ST 802 from Ask-chemicals which (for most of the experiments) protected the steel from being attacked by the molten aluminium.

The tested aluminium was two alloys from Hydro Aluminium which contained 0.7% Mg and 3.1% Mg and one alloy from Alcoa containing 4.7% Mg. None of the other elements present in the alloys have previously been shown to affect oxidation rate.

Industrial gas fired furnaces are currently running on either Natural Gas or LNG (typically contain >95% methane), or LPG (typically contain >95% propane). The combustion products from these fossil fuels will result in various mixtures of CO_2 and H_2O. A CO_2-free alternative to these hydrocarbon fuels is the use of hydrogen. One other process set-up parameter in the industry is that some furnaces use pure oxygen and other use air as oxygen source. All these furnace burner combinations will give different furnace atmosphere combinations. Figure 2 shows the resulting off-gas atmosphere compositions in the atmosphere

from all six burner set-ups provided there is no leakage[2] and no excess oxygen.[3] It is worth noting that there is very little difference in off-gas composition between the use of CH_4 and C_3H_8 (that is between LNG and LPG).

In this paper the atmosphere compositions are for the most cases positioned within the area defined by the points in Fig. 2. In 2018 we used propane in both oxy- and air–fuel mode, with and without excess air to simulate leakage in furnace. In 2019 we ran two campaigns, this time with methane and various extra additions of extra CO_2, H_2O, O_2, and N_2 to explore more of the ternary diagram, and in 2021 and 2022 we started with combustion of H_2 with addition of extra gas components for further investigations.

Measurements

The Linde test furnace is equipped with mass flow controllers on both the fuel and all other gases which are sent into the furnace chamber through the burner system, or through separate injection lances. The furnace is also equipped with gas measuring devices. They measure dry content of O_2, CO_2, CO and NO, in addition to wet O_2-concentration. From these measurements and input to the

---

[2] An industrial furnace has often unwanted entrainment of excess air, especially when the burner power is low and there is not sufficient over-pressure in the furnace chamber.

[3] When burning hydrocarbons, excess oxygen of 1–2% is always used to avoid formation of CO.

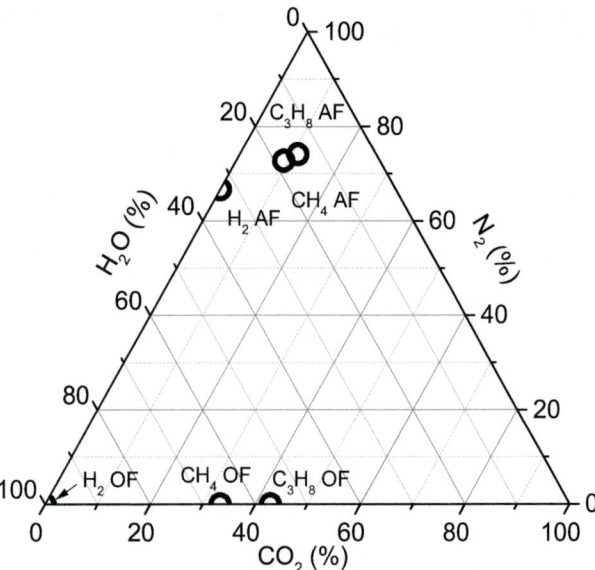

Fig. 2 Off-gas composition from the 6 burner configurations. AF means Air–fuel and OF means Oxy-fuel for combustion of methane (CH_4), propane (C_3H_8) and hydrogen (H_2). No leakage or excess oxygen is assumed

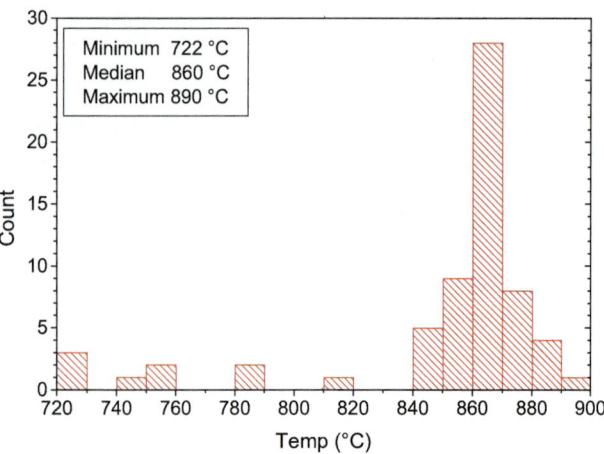

Fig. 3 Distribution of average temperature for all 64 tests

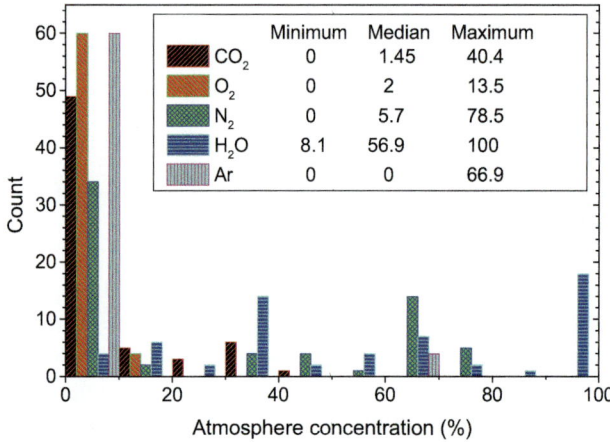

Fig. 4 Distribution of all gas components in 10% bins for all 46 oxidation tests. The variation in the atmosphere concentrations is given in the figure label

furnace it is possible to calculate all wet concentrations in the furnace atmosphere.

In addition to the gases, various temperatures are measured. Since one of the steel crucibles was hanging in a load cell, and we did not want to disturb this, the temperature in this crucible is assumed to be the same as the other crucible with the same alloy (often with 4.7% Mg). The temperature is calculated as an average temperature after melting and stabilising of temperature of each sample. The measured metal temperatures are used to determine when the metal was completely melted, at which point the test time is set to zero. Most of the experiments were stopped after 4 h. For those experiments which lasted for 3 or 8 h, the weight gain is normalized to a 4 h test.

When we experienced high oxidation rate, the coating did not protect the steel crucible completely. Therefore, the weight difference is measured on the full crucible both before and after 4 h of experiment. This weight difference is then divided by the weight of the aluminium metal before melting.

In total 64 (10 kg scale) oxidation tests were performed where 7 had 0.7% Mg, 16 had 3.1% Mg, and 41 had 4.7% Mg. Most tests were performed at about 860 °C as shown in Fig. 3. Variation in atmosphere composition is shown in Fig. 4. Also shown in the label in Fig. 4 is the minimum, maximum, and median value of the atmosphere compositions in all tests.

Figure 5 shows the distribution of the measured weight gain after 4 h. It is seen that about 70% of the tests had a weight gain of less than 0.5%. In other words, for many

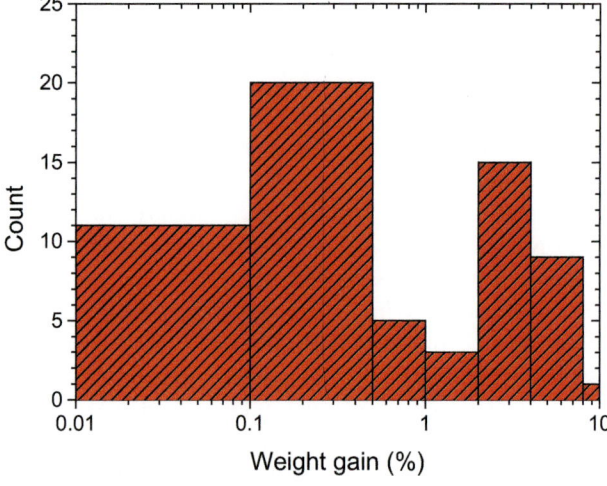

Fig. 5 Distribution of the weight gain measurements for all 46 tests. 32 out of 46 tests (that is 70%) has a weight gain lover than 0.5%

alloy—furnace atmosphere combinations, excessive oxidation was not an issue. In our study we tried to determine which conditions were contributing to high oxidation rates.

Discussion

This study showed there was a clear correlation between weight gain and both measured melt temperature and burner configuration for high-Mg alloy. For the medium high Mg, the difference between the burner configurations was not as clear. This is shown in Fig. 6 where the weight gain is plotted versus melt temperature for the two different burner configurations hydrogen–air and hydrogen-oxy for medium and high magnesium containing alloys. It is clear that the weight gain increases with temperature for all sets of measurements.

Figure 7 shows that for high temperature and low magnesium content in the alloy, the weight gain is close to zero and equal for the two burner configurations. For higher magnesium content, the air-hydrogen set-up showed higher weight gain than oxy-hydrogen. This means that a mixture of H_2O and N_2 for magnesium content greater than 3% gave higher mass gain than pure H_2O-vapour in our tests.

Figure 8 shows that nitrogen is not inert in combination with water vapour. Replacing nitrogen with argon decrease the weight gain by roughly 50% for the high temperature and high magnesium tests. The measurements close to 100% H_2O include both with and without 2% O_2 in the atmosphere. The oxygen content did not seem to affect the weight gain.

At high temperatures, there is a clear effect of CO_2-content in the atmosphere on the measured weight gain. This is

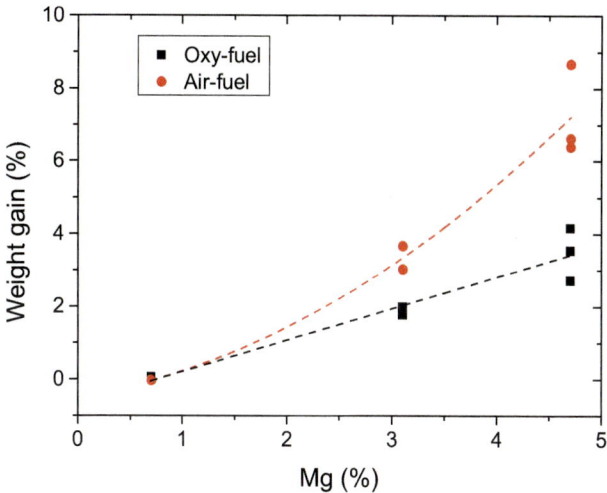

Fig. 7 Magnesium dependency on weight gain for hydrogen combustion in oxy- and air–fuel configuration. Only tests with melt temperatures higher than 856 °C are included

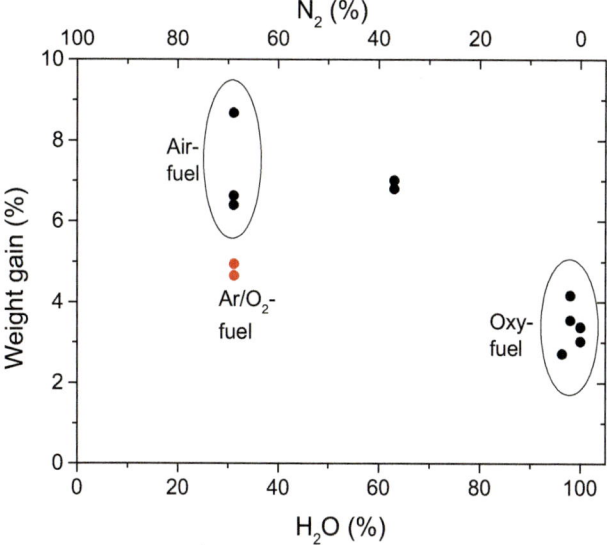

Fig. 8 Weight gain plotted versus water (and nitrogen on top) content in atmosphere for different burner configurations. Only tests with 4.7% Mg, melt temperatures above 855 °C, and hydrogen as fuel is plotted. Two tests were done where nitrogen was replaced with argon

shown in Fig. 9. By adding 5% CO_2 to the atmosphere, the weight gain is reduced to the same level as for shown for the low-Mg alloy.

Figure 10 shows the weight gain measurements as a colour scale in a ternary gas composition diagram. The measurements used are from the high-Mg, high temperature, and 0–2% oxygen tests. If the low oxygen concentration is ignored, the sum of the three gas components can be normalized to 100%. By using OriginPro's routine [8] for plotting ternary contour, the map in Fig. 11 is obtained. This

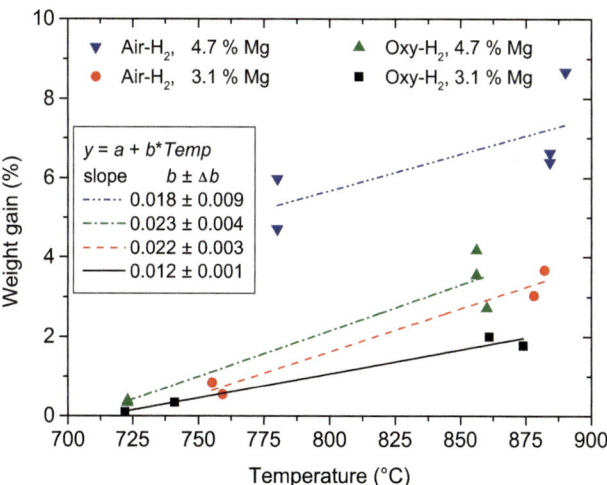

Fig. 6 Weight gain for pure hydrogen combustion in air–fuel and oxy-fuel set-up for the medium and high Mg containing alloys

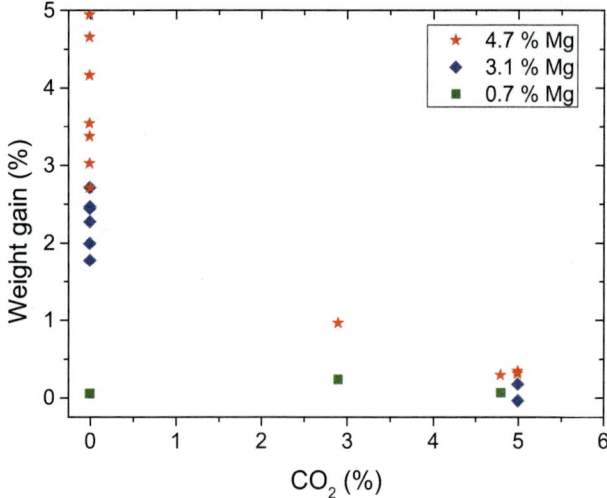

Fig. 9 Weight gain versus added CO_2 in oxy-hydrogen set-up for the three different alloys. Temperature in all measurements is from 846 °C and higher. The rest of the atmosphere is mainly H_2O with up to 2% O_2 and up to 1.7% N_2

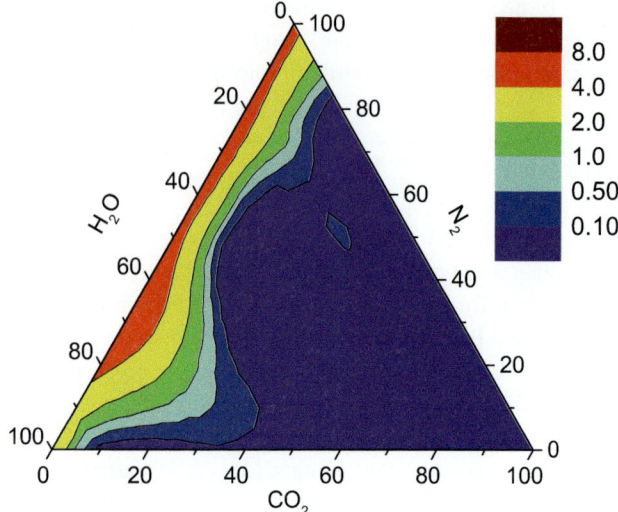

Fig. 11 Calculated weight gain areas in the ternary gas composition diagram. Only data from tests with high Mg and temperatures above 840 °C are used to calculate the regions. Note that the colour scale is logarithmic

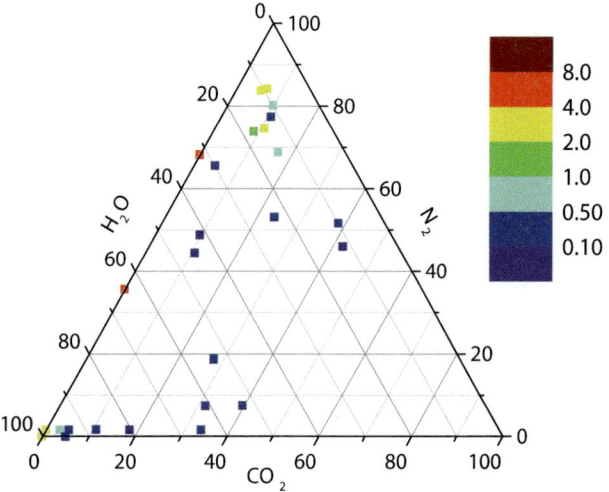

Fig. 10 Weight gain plotted as colour scale in ternary gas composition diagram. Only data from tests with high Mg and temperatures above 840 °C are plotted. Note that the colour scale is logarithmic

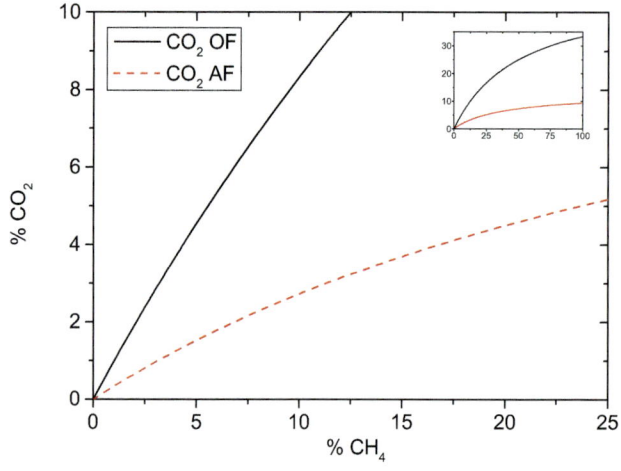

Fig. 12 CO_2 concentration in furnace atmosphere as function of CH_4 in a fuel mixture of $H_2 + CH_4$. The small graph inserted shows the complete range from 0 to 100% CH_4

ternary contour map shows that the weight gain is reduced considerably by introducing 5–10% CO_2 in the atmosphere.

To secure low oxidation rate of a high Mg-alloy, CO_2 can be added to the furnace either directly as CO_2-gas, or by using a mixture of (for instance) methane and hydrogen. It can be calculated that in oxy-fuel setup a $CH_4 + H_2$ mixture with 5.6% CH_4 (+94.4% H_2) will give 5% CO_2 in the furnace atmosphere, while in air–fuel setup 23.4% CH_4 is needed to obtain 5% CO_2. This is illustrated in Fig. 12 where the concentration of CO_2 in the burner off-gas is plotted as function of CH_4 content in the fuel. It can also be

seen from the inserted graph, that for air–fuel configuration it is impossible to reach 10% CO_2 even with 100% CH_4.

While industrial furnaces are designed and built to minimize entrainment of excess air, it is a fact that leaks will occur. Outside air can and will infiltrate into the furnace hearth, especially during low fire when the furnace pressure drops. In addition, furnace burners are normally operated with excess air or oxygen to avoid reducing conditions and the potential to form carbon monoxide, CO. To truly know the atmospheric composition within the furnace it is necessary to measure all components under normal operating

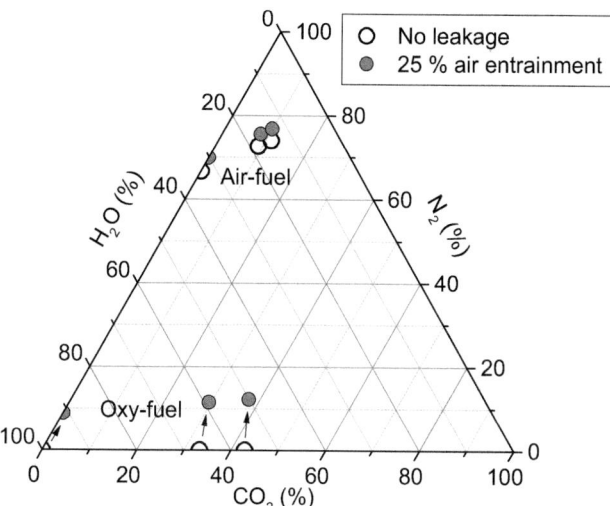

Fig. 13 Gas composition of the atmosphere from same burner configurations as in Fig. 2. Open circles are calculated with no excess air entrainment and filled circles with 25% air entrainment (relative to volume of burner oxidant)

conditions. Figure 13 shows the dilution effect on the gas composition (again excluding any oxygen) when there is 25% excess air entrained into the furnace. This means that when using Fig. 11 for a real furnace, one should either measure the actual gas composition, or at least the relative leakage so that the real point for that specific furnace is located and the necessary CO_2 addition calculated. Figure 13 also shows that any degree of leakage moves all the points towards the upper corner with 100% N_2. It can also be mentioned that leakage will also dilute any CO_2 required for suppressing oxidation, which means that the CH_4/H_2 ratio must be increased to keep the required CO_2 level.

Conclusions

A total of 64 oxidation tests were conducted at Linde's Älvsjö laboratory. The tests show that measured weight gain increase with magnesium content. For low magnesium content (0.7%) there was no dependency on either temperature or atmosphere composition. For medium (3.1%) and high (4.7%) Mg, there was clear correlation with temperature. For high magnesium content alloy, it is also found that:

- Air–fuel hydrogen configuration gave more weight gain than oxy-fuel hydrogen.
- By replacing nitrogen in hydrogen air–fuel with argon, the weight gain was reduced to same level as hydrogen oxy-fuel.

- Addition of 5–10% CO_2 in the atmosphere will reduce oxidation considerably for high magnesium alloys and high melt temperature.
- 5% CO_2 in furnace atmosphere from an oxy-fuel burner can be achieved by using 5.6% CH_4 (+94.4% H_2) while for air–fuel burners 23.4% CH_4 is required. Any false air will dilute the CO_2 further requiring even higher amount of CH_4.

This study showed that hydrogen can be used as a substitute fuel for Natural Gas or LPG in aluminium melting furnaces without increasing the oxidation of most molten aluminium alloys. For the case of high Mg content alloys, it will be necessary to add small amounts of CO_2 to prevent excessive oxidation.

Acknowledgements This research was carried out in cooperation with Linde AB as part of the Research Council of Norway (RCN) funded BIA Projects (No. 269634/O20) BEST and (No. 327564) NoBAl. The projects include the partners: Hydro Aluminium AS, Alcoa Norway ANS, Gränges AB, HYCAST AS, Real Alloy AS, NTNU and SINTEF. The authors also want to acknowledge the staff at Linde's lab and everybody else which has been involved in any discussions during the trials.

References

1. W. Thiele, *Die Oxydation von Aluminium-und Aluminium legierungsschmelzen,* (Aluminium, Vol. 38, 1962), 780–786.
2. Cochran, C.N.; Belitskus, D.L.; Kinosz, D.L. *Oxidation of Aluminum-Magnesium Melts in Air, Oxygen, Flue Gas and Carbon Dioxide.* Metall. Trans. B 1977.
3. Haginoya, I. and T. Fukusako. *Function of CO_2 Gas on the suppression of oxidation of molten Al-Mg alloy.* Journal Japan Insitute of Light Metals 31.11 [1981]: 733–741.
4. Smith, N. PhD Dissertation: *Methods of Oxidation Inhibition for Al-Mg Alloys.* Norwegian University of Science and Technology. 2019.
5. Stevens et.al. *Oxidation of AlMg in Dry and Humid Atmospheres* Light Metals 2011 Edited by: Stephen J. Lindsay. TMS (The Minerals, Metals & Materials Society), 2011 pp. 719-724
6. Johansson A; Solberg E; Skramstad M; Kvande T; Lodin J; Smith N; Syvertsen M; Kvithyld A. *Small scale oxidation experiments on AlMg alloys in various gas fired furnace atmospheres.* Light Metals 2020. The Minerals, Metals & Materials Society, Pittsburgh: Springer, New York.
7. Lodin J, Syvertsen M, Kvithyld A, Johansson A, Solberg E, Kvande T: *Study of the oxidation of an Al-5Mg alloy in various industrial melting furnace atmospheres.* Light Metals 2020. The Minerals, Metals & Materials Society, Pittsburgh: Springer, New York
8. OriginPro 9, Literature reference given at the bottom page of: https://www.originlab.com/doc/en/Origin-Help/Create-Contour-Graph#Adding.2FRemoving_Contour_Labels_and_Extracting_Contour_Lines

Table 1 Material properties used in the numerical model of the multi-chamber furnace

Parameter	Value	Unit
UBC emissivity	0.8	
UBC density	529	Kg/m^3
UBC heat capacity	1057	J/(Kg K)
UBC conductivity	5.3	W/ (m K)

equations with a k-ω turbulence model are solved in conjunction with the energy balance equation. The total mesh size is $6 \cdot 10^5$ hexahedral cells. The material properties used can be found in Table 1.

In the above, the thermal conductivity of the UBC bales is provided experimentally by using a single thermocouple and employing the analytical solution of the transient one-dimensional heat transfer equation [5]. In particular, a UBC bale was partly insulated by a ceramic fiber blanket in order to imitate one-dimensional heat transfer [5, 7]. Then, the thermal conductivity is estimated by using the method of maximum inclination under the condition of one-dimensional heat transfer [5].

The boundary condition for the furnace outside walls is set as convection and radiation heat transfer. The following account for "normal conditions" in our model: the recirculation outlet (No 6 in Fig. 1) is set to 0.8 kg/s and the recirculation inlet (No 7 in Fig. 1) is set to 0.4 kg/s. The windows between the two chambers (No 4 in Fig. 1) are considered open boundaries with an inflow temperature of 960 °C (average temperature of the main chamber of the furnace). Finally, the aluminum melt temperature is set to 730 °C. For simplicity, we assume that no chemical reactions take place in the furnace, thus, no organic gas is released from UBC decomposition (No 5 in Fig. 1). Thermolysis reactions will be incorporated in our ongoing research.

B. Lab scale experiments to derive temperature—delacquering degree correlation (TGA, electric resistance furnace) and incorporation of the above relationship into the model

A series of laboratory-scale experiments were also performed to study the delacquering process and feed the model with the relationship between the bales' temperature distribution and the achieved degree of delacquering. Polyester and epoxy-based lacquers were assumed as representative can lacquers and their thermal decomposition was studied by thermogravimetric analysis (TGA) using a "TA Instruments Q50" analyzer. Nine isothermal experiments were performed using 15–20 mg of each lacquer in the temperature range between 200 and 550 °C for 60 minholding time under air and low oxygen conditions (8% v/v O_2), considering the

later closer to industrial conditions. The delacquering degree was then determined as the weight loss of each lacquer at the particular temperature after 60 min of thermal treatment in the TGA furnace.

Lab-scale decoating experiments were also performed in UBC samples. In particular, 300 g of chopped UBCs were heated in a laboratory electric resistance furnace under atmospheric conditions for 60 min at four different temperatures. The achieved delacquering degrees were then compared to ones obtained by the can lacquers TGA tests.

C. Industrial validation of the calculated temperature distribution (thermocouples) and delacquering degree

Aiming to validate our model, an industrial-scale trial was performed by charging two UBC bundles in the scrap chamber of a multi-chamber furnace indented for experimental purposes. Each bundle consisted of stacked UBC bales (bundles) and remained for 30 min inside the furnace. The dimension of the two bundles were: (A) 0.8 × 2 × 1.1 m and (B) 1.2 × 2.1 × 1.1 m, respectively.

Results and Discussion

Lab-Scale Experiments

The curves in Fig. 2 were obtained from the average weight loss of both polymers in a TGA furnace. It is revealed that a pre-heating between 450 and 550 °C for half an hour under both air and low oxygen (8% v/v) atmosphere is sufficient for an almost complete decoating. The delacquering degree was also determined after heating chopped UBCs at four different temperatures (T = 270 °C, 350 °C, 400 °C, and 500 °C) in a laboratory electric resistance furnace. Images of "as-received" (ASR) UBCs and UBCs after decoating at each temperature are depicted in Fig. 3. It is revealed that UBCs after delacquering at 270 °C had a similar appearance to the ASR sample, while those at 350 °C were black/brown indicating carbon-containing tar residue formation and unreacted coating on their surface [8]. Moving to higher temperatures (400 °C), the metallic surface of UBCs started to be visible, and finally at 500 °C the UBCs were completely delacquered. By comparing TGA and electric

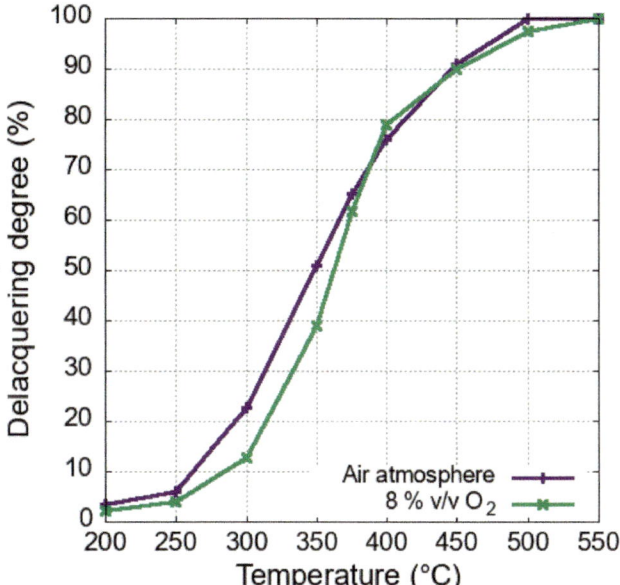

Fig. 2 Average estimated delacquering degree of polyester and epoxy-based lacquers versus temperature for t = 30 min (which is the holding time of the UBCs in the furnace), as determined by isothermal TGA experiments in air as well as in low oxygen (8% v/v) atmosphere

resistance furnace results, in Fig. 4, it is clear that the results obtained by TGA on lacquers agree with the ones obtained by UBC lab-scale decoating tests, thus they can be safely used to feed the model describing UBC delacquering process. The low oxygen (8% v/v) delacquering curve was incorporated into our numerical model for associating the temperature distribution with the delacquering degree.

Industrial Validation

Our numerical model is validated by performing an industrial-scale trial with two UBC bundles in the scrap chamber. In particular, in Fig. 5 we present the top view of the furnace scrap chamber as simulated by our numerical model. We observe that hot gas coming from the main chamber, through the windows, collides with the UBC bundles and heats them up. Moreover, the bundles are also heated by radiation from the aluminium melt as well as from the furnace refractory.

In the industrial-scale trial, the bundles' temperature was measured at specified positions by suitably applying thermocouples and was then compared with the computational results. In particular, as observed in Fig. 6, the numerical model is in satisfactory agreement with the experimental measurements. In both studied cases, it is revealed that the developed model slightly underestimates the UBCs temperature. Such a discrepancy can be attributed to the absence of chemical reactions (thermolysis) that would release volatile gases from the thermal decomposition of UBCs thus increasing the measured temperature. Another significant finding is the high-temperature difference between the surface and the center of UBC bales. In both bundles, the maximum recorded temperature from the thermocouples positioned on the UBC surface was 400–500 °C, implying a considerable coating removal. On the other hand, the temperature profile in the bundle center reached only ~ 100 °C, indicating that the developed heat was capable of removing solely moisture without contributing to polymer decomposition. The observed high-temperature difference between surface and center could be ascribed to the low UBC thermal conductivity as compared to bulk aluminium: 5.3 versus 93 W/(m K) (3104 alloy, at T $\approx$ 500 °C) [9], respectively.

Considering the correlation between the UBC temperature profile with the degree of organic removal (see Fig. 2), the overall delacquering degree, after 30 min in the scrap chamber, was calculated at $\sim 21\%$ w/w. The above result is in very good agreement with the experimentally obtained value since the weight difference of the bundles before and after the thermal treatment was $\sim 19\%$ w/w. It is useful to underline that the observed low overall delacquering degree is probably attributed to the dense arrangement of the UBC bales which limits the heat transfer efficiency. In the current

Fig. 3 Thermal treatment of chopped UBCs at four different temperatures (T = 270, 350, 400, and 500 °C) at a laboratory electric resistance furnace (holding time: 60 min). The "as-received" sample (ASR) is also depicted. The shiny metallic surface has been revealed in the case of the fully delacquered UBCs (T = 500 °C)

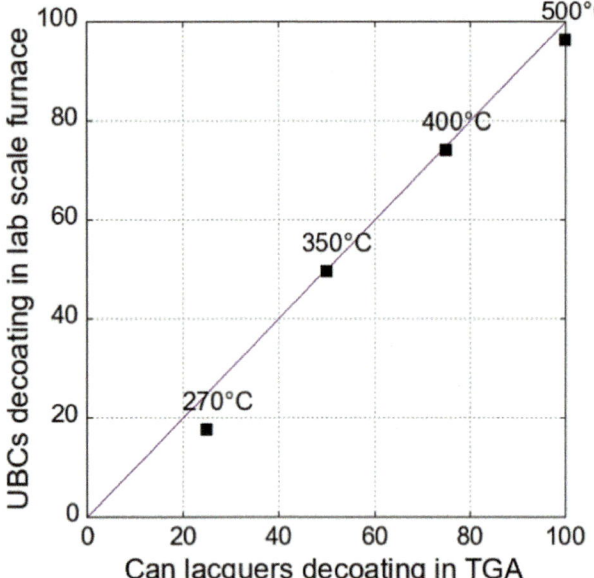

Fig. 4 UBC delacquering degrees (% wt.) derived by thermogravimetric analysis (TGA) in lacquers and UBCs decoating in laboratory electric resistance furnace under air atmosphere (holding time: 60 min in both cases)

case, it is revealed that a 30 min-preheating is not sufficient for complete delacquering. Moreover, the available oxygen is not capable of getting through internal can sheet layers and up to the bale center, thus the organic removal is hindered. In particular, only surface and near-surface sheets could exploit heat and actively contribute to the decoating process. Notice that the UBC stacking into large bundles is not a common production practice and is used here only for experimental purposes.

Process Parameters Scenarios

Our developed model can be safely applied to make predictions for the UBC temperature distribution when the process parameters are modified. In particular, the average UBC temperature profile for three scenarios with different process parameters is presented in Fig. 7.

- Scenario 1: The recirculation flow rate is increased from 0.8 to 1.6 kg/s (recirculation outlet, 6 in Fig. 1) and from 0.4 to 0.8 kg/s (recirculation inlet, 7 in Fig. 1),
- Scenario 2: The main chamber temperature is increased from 960 to 1160 °C,
- Scenario 3: Both bundles are moved closer to the molten aluminium by 1.2 m.

The reference case (normal conditions as described in the Methodology section) is also depicted.

In Fig. 7, it is revealed that the temperature increase in the main chamber (Scenario 2) has a significant impact on the UBC bundles preheating. The above conclusion is also confirmed by the calculation of the corresponding delacquering degrees, after 30 min stay in the scrap chamber. In particular, we calculated: 27, 36, and 24% w/w decoating for Scenarios 1, 2, and 3, respectively. Therefore, the delacquering degree has been increased by ~71%, in the case of Scenario 2, compared to the normal conditions (21% w/w). It is reasonable that a higher temperature in the main chamber would promote higher delacquering values, as a larger volume fraction of UBCs will be heated in the scrap chamber. Although such a higher temperature seems advantageous, other factors should be concurrently

Fig. 5 Visualization of gas flow inside the scrap chamber of the multi-chamber furnace as seen from the top. The flow pathlines are colored by the gas temperature shown in the color label (in K). Notice that the UBC inlet is on the left side of the figure (door side), whereas the windows between the two chambers are on the right side (burner side). The letters A and B indicate the two UBC bundles

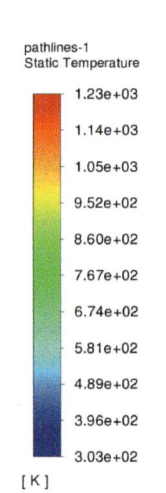

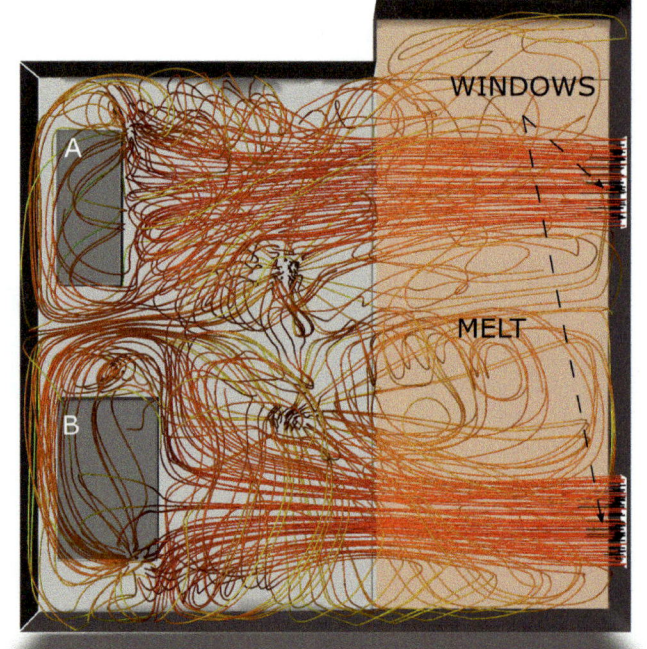

Fig. 6 Comparison between experimentally measured and calculated temperature variation of the UBC bundles at various positions: **a** Bundle A, door side (left side of Fig. 5), depth 20 mm, **b** Bundle A, door side, bundle surface, **c** Bundle B, door side, depth 15 mm, **d** Bundle B, burner side (right side of Fig. 5), bundle surface

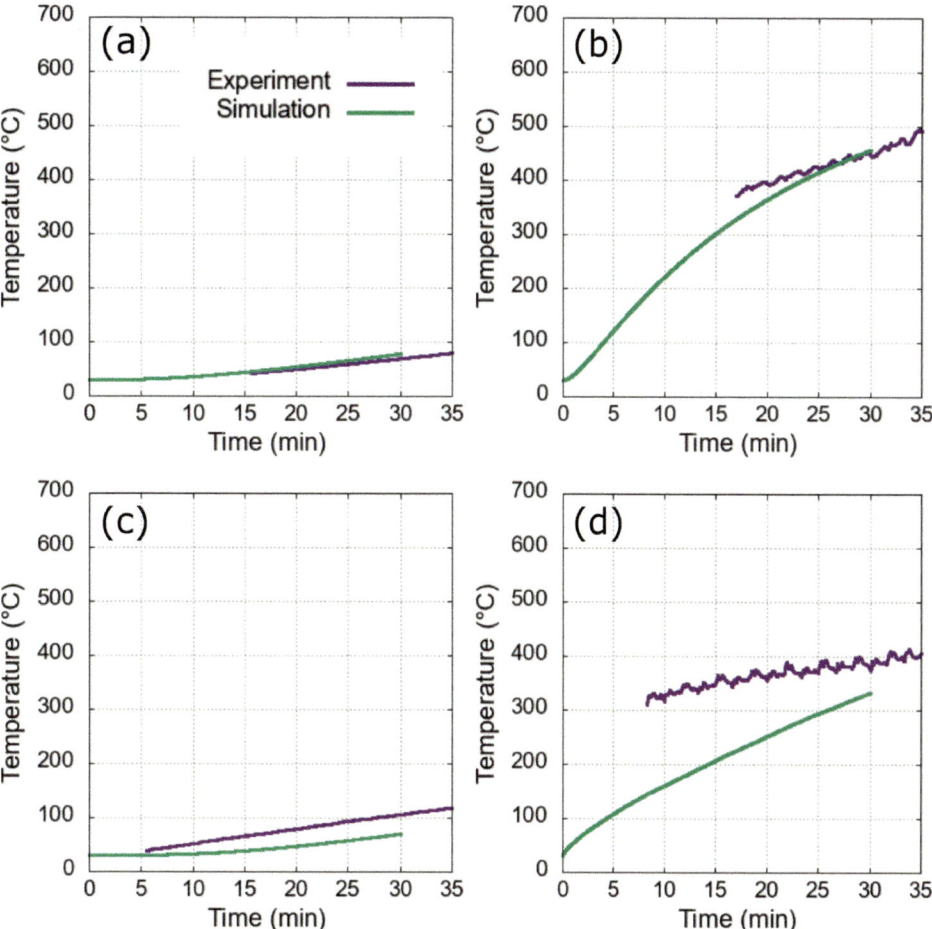

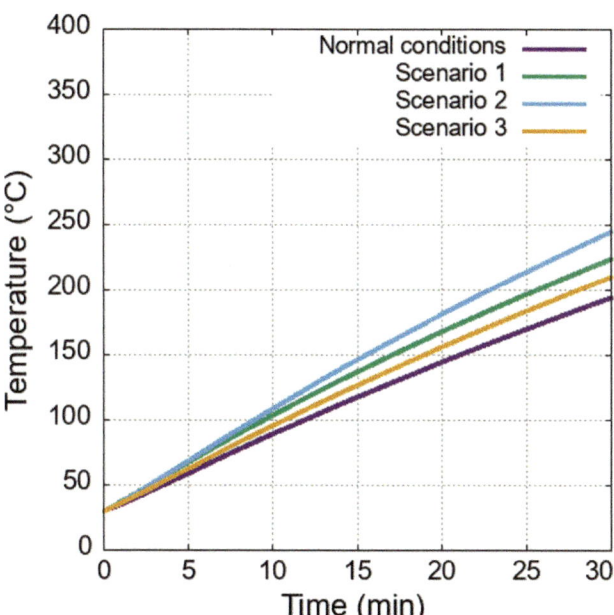

Fig. 7 Temporal evolution of the average UBC temperature for various process parameters scenarios: Scenario 1: Increase of recirculation flow rate, Scenario 2: Increase of main chamber temperature, Scenario 3: Displacement of the bundles closer to the melt

considered such as aluminium oxidation and dross formation, which directly reflect on the overall metal recovery (not examined in this study) on a basis of the industrial scale.

Conclusions and Future Work

The methodology presented in this work makes it possible to assess the degree of UBC delacquering when preheated in an industrial multi-chamber furnace. In coordination with TGA decoating experimental results, the accuracy of the numerical model is satisfactory and it can be used as a powerful tool for trials design and process optimization. The results obtained so far show that for achieving a higher degree of decoating under thermolysis conditions, a considerably larger volume fraction of UBCs should be above 450 °C. This may be realized by regulating various process parameters. Any process improvement achieved with this work will have a positive industrial, economic and environmental impact. This work is still in progress and will be used as a basis to promote the sustainable industrial implementation of circular economy practices.

References

1. Das S (2006) Designing Aluminum Alloys for a Recycle-Friendly World. Light Met. Age 64(3):26
2. Kvithyld A (2011) The recycling of contaminated Al scrap. Aluminium International Today 23(4):26
3. Schmitz C (2014) Handbook of aluminium recycling. Vulkan-Verlag GmbH, Germany
4. Evans P, Guest G (2002) The aluminium decoating handbook. Stein Atkinson Stordy Ltd, United Kingdom
5. Steglich, J et al. (2015) Conductive heat transfer in used beverage can scrap. Paper presented at the Advances in Materials and Processing Technologies Conference, Madrid, Spain, 14–17 December 2015
6. Steglich J, Dittrich R, Rosefort M, Friedrich B (2016) Pre-treatment of beverage can scrap to increase recycling efficiency. JMSE-A 3(3–4):57–65
7. Furu J (2013) An experimental and numerical study of heat transfer in aluminium melting and remelting furnaces. Doctoral thesis, Norwegian University of Science and Technology
8. Jaroni B, Flerus B, Friedrich B, Rombach G (2012) The Influence of Pyrolythic Reactions on the Aluminum Dross Formation during the Twin Chamber Remelting Process. In: Weiland H, Rollett AD, Cassada WA (ed) ICAA13. The Minerals, Metals & Materials Society, Pittsburgh; Springer, Cham p 1327–1336
9. Zhen F, Sun J, Li J (2016) Constitutive Equation for 3104 Alloy at High Temperatures in Consideration of Strain. High Temp. Mater. Process. 35(6):599–605

A Novel Green Melt Technology for Aluminum Alloys

Kaborson Ke, Xiyu Wen, and Dongjie Ke

Abstract

It is a novel green method of melt aluminum alloys that no chemical resolvent and fluxes were used. Currently, the harmful gases including Chlorine (Cl), Florine (F), and other harmful substances were exhausted during aluminum melt. The discharge of the solid dross including Cl and F has also been limited strictly due to environment protection. In order to reduce the discharges and resolve the problems of environmental pollution in aluminum industries, it is invented that the high pure argon gas by use of an artificial intelligence (AI) control system goes through a distinctive refractory structure with internal micropores, which distributed on certain regular patterns and included nanoparticles with rare earth elements in the bottom of melt furnace. Al liquid under the atmosphere is kept at a dynamic balance state with inert gas bubbles in microscale from starting melting to end of casting. The ingots from the melt processing can meet current requirements in aluminum industries. The hydrogen content was less than 0.3 ml/100 gAl, and the total dross reduced 50% or higher than 50%. Burning loss of metal decreased 0.2–0.8% and energy cost (natural gas) reduced 10% per ton aluminum alloy. If this system would be applied with a common degasser and filter system, H content in Al liquid could be controlled at less 0.05–0.10 ml/100 gAl level. The purity of the Al liquid arrives at a higher grade than some conventional melt processing and can realize the green melting and AI digital control characteristic under the industrial conditions.

Keywords

Aluminum alloy • Melt • No Chlorine and Florine • Gas bubbling degassing bed • Hydrogen content • Purity

Current Melt Processing of Aluminum Alloys

Currently, it is the first thing that aluminum and aluminum alloy liquid with high purity and quality can be obtained by different melt ways in aluminum industries. In general, the purity of aluminum liquid was improved and increased by chemical resolvents added with Cl and F and other chemical substance when melting aluminum alloys. The conventional processing of melting aluminum alloys has still a main problem, that is, the second contaminative probability and inclusions formed from the chemical resolvents. This will induce very more problems during production operation of high performance of aluminum alloys and lead to be exhausted of gas substance with Cl and F as well as harmful dross. In addition, the dross after melt aluminum alloys is also harmful for environment and will result in more difficulties for treating Al dross in aluminum industries.

To reduce environment contamination and exhaustion of harmful dross during melt aluminum alloys, the melt processing improved for aluminum alloys faces the challenge of a common issue, that is, source of the contamination. A lot of engineers and researchers developed and explored new technologies for decreasing the harmful dross formed from melting processing. Although, the granular flux, rotation degasser with the wall of furnace and rotation degasser on a vehicle as well as different on-line rotation degassers were applied to melt aluminum alloys to cut down dross, the chemical resolvents with Cl and F cannot still be got rid of. The true green melting of aluminum alloys has not been completely realized yet. It is very still difficult that the aluminum alloy melts with high purity and quality were obtained.

K. Ke · D. Ke (✉)
Fuzhou Metal-New High Temperature Technology Incorporation Limited, Fuzhou, Fujian 350119, P. R. China
e-mail: kdj@163.com

X. Wen
Center for Aluminum Technology, University of Kentucky, Lexington, KY 40514, USA

© The Minerals, Metals & Materials Society 2023
S. Broek (ed.), *Light Metals 2023*, The Minerals, Metals & Materials Series,
https://doi.org/10.1007/978-3-031-22532-1_126

To protect environment and improve operation and fluxing treatments in melt furnace as well as a high requirement for purity of aluminum alloy melts, the novel green technology has completed in industrial scale which no chemical resolvents added and the aluminum liquids were obtained with higher purity than conventional processing of melting aluminum alloys. It was invented that the high pure argon gas by use of an artificial intelligence (AI) control system goes though the distinctive refractory structures with micro poses on the surfaces including nanoparticles with rare earth elements, which distribute on a certain regular pattern in the bottom of the furnace. A porous bed on the bottom of furnace without chemical resolvents added and harmful substances to be exhausted has been used. It was realized that aluminum and aluminum alloys with high performance were produced under green environment by use of the technology.

Development of Green Melt Technology in Aluminum Industries

A lot of research has been carried out for green melt technologies of aluminum alloys due to needing improvement of the existing environment in the world, but some conventional aluminum melt processing used still mixed gas with Cl_2 and N_2 for removing out inclusions and lead to dross formed from melt aluminum alloys. To improve the work condition of aluminum plants, it was an effective way that powders of chemical resolvents were injected into melt aluminum alloy liquids by machines in aluminum industries. The cake-shaped chemical resolvents after removing out water prior to melt were put into aluminum liquid by steel cages and then stirred aluminum liquid to obtain pure aluminum alloys.

To improve melting processing, Alcan company applied initially granular flux of chemical resolvents with NaCl and KCl as well as KCl and $MgCl_2$ to purifying aluminum alloy liquids in the late 1980s [1]. So, a huge number of companies followed the method. The directly rolled/pressed into granules from the powder of some chemical resolvents with high-efficiency to reduce cost and atmosphere pollution can observably increase on the effect of aluminum liquids purifying.

In 2008, Metal-new High Temp. Tech. Co. Ltd (FMT) first obtained an inventing patent in P. R. China, a rotary flux injection degasser on the vehicle and a degasser on wall of furnace [2–5]. Subsequently, a company from Canada also product and sold the degassers in Chian market. Finally, the company sold widely the degassers to China after improvement and development underwent.

After porous bricks on bottom of furnaces used in the steel and iron industries, a company from Austria introduced the technology of porous bricks of a narrow slot type on

bottom of melt furnaces in aluminum industries in 1982. In addition, a porous plug technology also was developed in that time and put first porous plugs into a small die cast furnace pouring pistons for Ford Motor Company [6]. Currently, many furnaces with the porous bricks used in aluminum industries, but the porous bricks accompanied with N2 and Ar gases were only assistant to melt processing of aluminum alloys and some chemical resolvents must be added and used still. The green melting processing has no realization yet until now.

In addition, it was found that the technology was mainly applied to more vacuum furnaces. The method of no chemical resolvents added did not be reported to obtain aluminum alloy melt with high quality in industrial conditions. In general, plugging and leaking of aluminum liquid from the porous plugs on bottom of melt furnace often occurred. The technology in the application of aluminum industries has still a lot of difficulty.

Starting from 2005, FMT has developed and completed the technology about porous plugs on bottom of furnaces. The patents of "the porous bed technology" using porous bricks (or plugs) were approved in P. R. China and Germany [7–9]. By control of argon (Ar) gas bubbles going through the porous plugs, a dynamic balance of bubbling on the porous bed was formed on the bottom of furnace. The fine and disperse bubbles can singly move upward and reduce combining probability because of the porous construction with gas pulse control technique of special frequency. These bubbles can capture hydrogen and adsorb inclusions in aluminum liquids, which will purify aluminum alloy liquid and don't add any chemical resolvent. This technology has a very special point—not easily plugging. The content of hydrogen can be controlled range of 0.2–0.28 mL/(100gAl).

Keys of the Technology of Porous Plug Bed

Simulation of Al Melt Processing with the Porous Plug Bed

The flowing of gas of each nozzle of gas entered was simulatively analyzed in the Al liquid phase in a furnace by computation and water simulation technologies, and then thermokinetics of gas flow processing in the furnace was also simulated. Based on the result from the simulations, distribution of porous plugs and flow & temperature fields were optimized synergistically. A basic model with ideal conditions in melt furnace was built.

In the melt processing, the boiling and wave of aluminum liquids in furnace can appear due to advanced real-time control of volume and pressure of gas for each porous plug on the bottom of the furnace. The 3D mass transfer by convection and stirring by turbulent flow in aluminum liquids can also be completed by use of the control system. Finally, A dynamic balance of bubbling in the porous plugs on the porous bed came into being in the furnace, and then the gas bubbles upward from the bottom of furnace grabbed hydrogen and/or adsorbed the inclusion phases in aluminum liquid to move to surface of aluminum liquid in every location (dross in center and bottom reducing) in the furnace. The technology can keep high purity of aluminum liquid in the whole furnace.

Development of Materials and Design of the Porous Plug Bed

To obtain a porous bed with stable and long lifetime, the porous plugs on bottom of a furnace must have a similar coefficient of expansion with refractory bricks (materials) due to heating effect. So, no leaking and plugging can be realized in bottom of a furnace.

A new refractory material for the porous plugs (bricks) to use on bottom of melt furnace was developed by FMT. The performance of the refractory material has high thermostability of >1700 °C, softening temperature >1450 °C under loading, compression strength of >90 MPa and heating

shock of 25 times from 1100 °C to room temperature. In addition, the bricks with disperse 3D micropores has a special feature, that is, degree of porosity of >30%, and gas can go through with disperse micropores out of the bricks and the small bubbles were formed in Al liquid. The porous plug is not wetting with aluminum liquid and has a large wetting angle; The porous plug can prohibit aluminum liquid through (not easily plugging).

During the porous plug material manufactured, the nanoparticles with rare earth elements were added into size to improve bonding strength of interfaces and granules, so that the loose/porous bricks first were obtained. The bricks were put into high-temperature furnace with N2 atmosphere, and then SiC will convert into Si_xN_y in the sintering processing. The bricks with dispersal porous pores and no leaking and plugging for aluminum liquids were made.

Based on the result from the flow simulation, the construction of a furnace for melt aluminum alloys were designed as shown in Fig. 1a–c. The dispersal porous bricks (see Fig. 1) were put on bottom of the furnace, which formed a porous plug bed and the inert gas (Ar) can be blown through the porous plugs made by the bricks into the aluminum liquids. Programmable Logic Controller (PLC) was used to control volume and pressure of the gas as

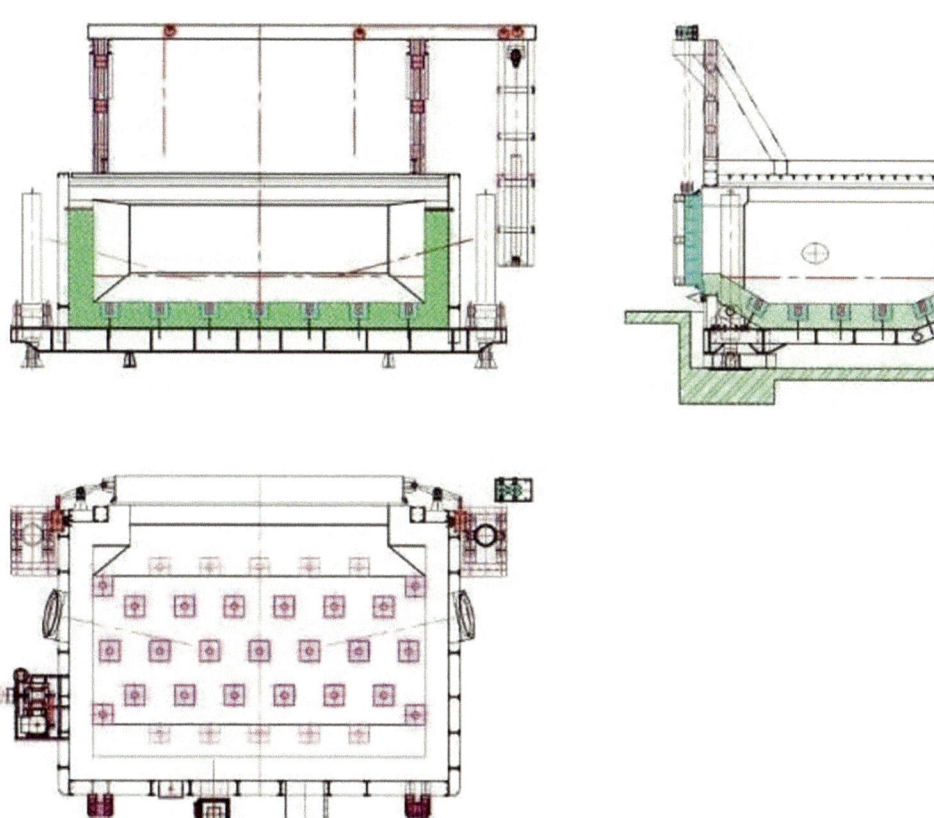

Fig. 1 Constructions of a furnace and porous plugs for melt aluminum alloys

well as the change of frequency and amplitude of pulses. The technology without chemical resolvent not only can decrease difference of temperatures in different locations in the furnace but also can make distribution of uniform alloying elements in melt furnace by heat and mass transfer in 3D convection.

Evaluation of Al Melt Quality

Measurement of Hydrogen Content

The HYCAL Hydrogen Determinator from EMC for on-line evaluation of hydrogen content during ingot casting trials was used. The trials were carried out by the FMT team at FMT's cast house on a melt furnace before and after degassing as well as after the FCC filter while running alloy AA5052. Content of hydrogen can finally be controlled range of 0.2–0.28 ml/(100 gAl) in the ingots.

Analysis and Determination of Inclusions in Aluminum Liquid

Because of the customer requested, melt trials were carried out by using the Prefil Footprinter of development from FMT for inclusion analysis. The metal flow distributions for casting drops from the processing without chemical resolvent were tested. All PODFA samples were analyzed at the lab of FTM. The Inclusions in Al liquid were at the level of less 0.282 mm^2/kg Al. Aluminum oxide films are thin and at a level of $\sim$No. of 60.

Homogeneity of Temperature Field in Furnace

Temperatures at 11 locations on the 100 mm depth to the surface of Al liquid and the 100 mm height to bottom of the furnace with 50-ton capacity were monitored. The locations in the furnace were set as shown in Fig. 2a. Measuring time kept at 3 min for each measurement. The measuring result was shown in Fig. 2b. The difference in temperatures at the locations is less than 2.0 °C.

Homogeneity of Composition of Aluminum Alloy Melts

Optical Emission Spectrometry (SPECTRO MMAXx Optical Emission Spectrometer) was used to determine approximate chemistry of samples taken at each location in the melt furnace according to ASTM 1251-11 Standard Test Method for Analysis of Aluminum and Aluminum Alloys by Spark Atomic Emission Spectrometry.

Composition at 11 locations on the 100 mm depth on the surface of Al liquid and the 100 mm height to bottom of the furnace with 50-tons capacity were also determined. The measuring result was listed below Table 1. The liquid of aluminum alloys has perfect homogeneity of the composition.

In addition, the measuring result of homogeneity of composition in casting start, medium, and ending were listed in Table 2. The good stability of chemical composition can be observed during casting.

Main Characteristic of the Green Project

The melting process of no chemical resolvent can avoid some residual harmfully chemical substance in dross of aluminum alloys. The door of a furnace often opened & closed did not need in melting processing because of no spraying and adding the resolvent as well as skimming operation. Heat lost remarkably reduced in duration of melt operation.

Each porous plug has an independent sub-system with pipes which connect to Ar gas station. The system was constituted by PLC in control cabinets and can provide independently intelligential parameters for each porous plug. The bubbling plug bed formed with all porous plugs and intensive bubbles (see Fig. 3) would catch hydrogen & inclusions and float upward based on the third law

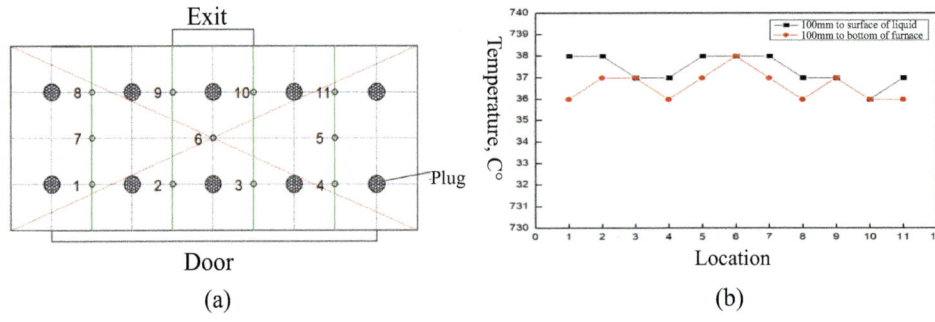

Fig. 2 Locations measured in the furnace in (**a**); measuring result in (**b**)

Table 1 compositions of different locations in the furnace (wt.%)

Location, top	Si	Fe	Cu	Mn	Mg	Cr	Ti	Location, bottom	Si	Fe	Cu	Mn	Mg	Cr	Ti
1	0.080	0.234	0.018	0.065	2.51	0.188	0.015	1	0.079	0.238	0.018	0.065	2.50	0.188	0.0102
2	0.082	0.241	0.018	0.066	2.53	0.187	0.015	2	0.081	0.244	0.0182	0.065	2.48	0.188	0.0101
3	0.082	0.237	0.018	0.065	2.52	0.189	0.015	3	0.081	0.239	0.0181	0.065	2.51	0.189	0.099
4	0.080	0.241	0.018	0.065	2.54	0.188	0.015	4	0.080	0.24	0.0183	0.065	2.53	0.186	0.0103
5	0.081	0.239	0.018	0.065	2.52	0.188	0.015	5	0.081	0.235	0.0182	0.065	2.51	0.187	0.0099
6	0.079	0.236	0.018	0.065	2.49	0.189	0.015	6	0.080	0.239	0.0184	0.065	2.48	0.189	0.0103
7	0.080	0.234	0.018	0.065	2.51	0.188	0.015	7	0.079	0.235	0.0182	0.065	2.50	0.188	0.0102
8	0.080	0.243	0.018	0.065	2.54	0.189	0.015	8	0.081	0.241	0.0183	0.065	2.53	0.189	0.0101
9	0.081	0.241	0.018	0.065	2.51	0.188	0.015	9	0.080	0.242	0.0181	0.065	2.50	0.187	0.01
10	0.081	0.237	0.018	0.065	2.49	0.187	0.015	10	0.081	0.235	0.0182	0.065	2.51	0.188	0.00990
11	0.081	0.242	0.018	0.065	2.48	0.189	0.015	11	0.080	0.241	0.0181	0.065	2.47	0.187	0102
12	0.079	0.239	0.018	0.065	2.51	0.187	0.015	12	0.079	0.243	0.0182	0.065	2.52	0.188	0.01
13	0.079	0.234	0.018	0.065	2.51	0.189	0.015	13	0.080	0.238	0.0182	0.065	2.50	0.189	0.0101

Table 2 Compositions and temperature starting, medium and ending casting

Location	Wt.% of main alloying elements							Temperature of Al liquid/°C
	Si	Fe	Cu	Mn	Mg	Cr	Ti	
Casting start	0.0788–0.0812	0.237–0.241	0.0180–0.0181	0.0651–0.0654	2.48–2.51	0.188–0.189	0.0146–0.0148	716–718
Medium of casting	0.0788–0.0812	0.240–0.242	0.0180–0.0181	0.0651–0.0654	2.47–2.49	0.188–0.189	0.0185–0.0187	712–713
Casting ending	0.0795–0.0805	0.241–0.243	0.0180–0.0181	0.0651–0.0654	2.45–2.47	0.186–0.188	0.0186–0.0188	701–703

Fig. 3 Porous plugs with intensive bubbles

(PV = constant) of thermodynamics. The process was realized and has advantages in industrial scale as follows:

(a) There are no evil-smelling odors due to use Ar gas instead of Cl, F, and ammonia. Those operators don't need to be afraid for health and environment. The cost of environmental protection will be saved.

(b) Total dross after the melting processing can reduce 50%. The dross easily meets requirements of industrial exhaust in any country. The dross can be treated without toxic and harmful substance.

The processing from melting start to casting end is in dynamic equilibrium state in all the time. The casting will start immediately if chemistry of aluminum alloys meets requirement of Al alloys because of no spraying powder, skimming and settling treatment, and so on. Total melt time will decrease by 1 h more. The processing can save natural gas per ton aluminum alloy than more $5m^3$/per cast drop in current conditions.

The Al liquid included hydrogen of about 0.5 ml/(100 gAl) before Ar gas going through the porous plugs. After treatment using the porous plug bed, the melting process can control H content in the range of 0.2–0.28 ml/(100 gAl), from current melting process not used chemical resolvents. It was seen that the melting process without chemical resolvents can obtain a lower level of H content in aluminum liquid.

The melting process shows that the dynamic equilibrium of aluminum alloy liquid under temperature and gravitational fields can obtain uniform alloying, that is, no segregation of composition. This is a foundation to produce high-quality aluminum alloy ingots.

No gases including Cl and F were produced and exhausted into the atmosphere, that is, the green melting process. The bottom of a furnace was kept flat and the porous plugs have no plugging or slog due to heavy compounds and inclusions with bubbles to float towards to surface of aluminum liquid.

Acknowledgements The authors thank the work of engineers of the Fuzhou Metal-new High Temp. Tech. Co. Ltd and support of Fuzhou University, Fujian Institute of Research on the Structure, Chalco Ruimin Co., Ltd, and Guangdong Hoshion Aluminum Co., Ltd. In addition, the authors thank support of EMC Hycal Limited in United Kingdom in hydrogen determination and MS & T Consulting in USA in PODFA measurement.

References

1. Beland, Guy, Dupuis, Claude and Riverin, Gaston. "Rotary Flux Injection: Chlorine-Free Technique for Furnace Preparation." Light Metals (1998) pp 843–847.
2. KE Dongjie, CHEN Qun, LI Yuhang, "Technical research on online degassing purification mechanism of molten aluminum", Light Alloy Fabrication Technology, 2012 Vol. 40, Issue 8, p 13-22.
3. WANG Shichong, KE Dongjie, "Study on bubble shape during water simulation experiment of rotary impeller spray process", Light Alloy Fabrication Technology, 2010,38(08) p 8-12.
4. WANG Shichong, KE Dongjie, CHEN Xiao, CHEN Qun. "Study on dynamic water simulation experiment for purification of molten aluminum alloys by rotary impeller", Light Alloy Fabrication Technology, 2010,38(03) p 18–24.
5. KE Dongjie, LU Guimin, Richard J., "Discussion on the development of green melting technology for aluminum alloys", Processings of the 13th National Light Alloy Processing Academic Exchange Conference,Hangzhou, Jiangsu province, P. R. China, 2005: p 7–13.
6. The Puur Gas Injection System – http://www.sparref.com/uploadedFiles/File/Puur_System_Brochure.pdf.
7. ZL201810880452.1, inventing patent of P. R. China, "A porous plug used bottom of melt furnace with gas stirring as well as a metal melt furnace".
8. ZL201810557056.5, inventing patent of P. R. China, "A metal melt method and furnace".
9. NR.112018002481, inventing patent of Germany, "A metal melt method and furnace".

MagPump

Oscar A. Perez and Eishin Takahashi

Abstract

TST is a California-based company leading the way in manufacturing aluminum ingot, billet, and slab worldwide for over 76 years. TST is committed to providing the highest quality products. TST is committed to continuous improvements in its casthouse operation and has installed the world's first and only permanent magnet-based pump, MagPump™, from Zmag (Japan). MagPump is designed to be a direct replacement for traditional mechanical pumps in side well (multi-chamber) furnaces without requiring furnace modification. MagPump is powered by zPMC™ (Zmag Permanent Magnetic Circuit) which generates virtual impellers to pump molten aluminum. Unlike mechanical pumps, MagPump does not have physical impellers and therefore is a nearly hands-free system, with no consumable parts, and reduced downtime. MagPump is also capable of other applications such as gas injection (e.g., Chlorine, Nitrogen, Argon) and scrap submersion. Operator safety has been increased as the necessity to approach the pump well has been minimized.

Keywords

Molten metal pumping • Magnetic stirring • Metal pump innovation

Introduction

TST Inc. has been leading the way since 1946 in manufacturing aluminum ingot, billet, and slab worldwide with five melting/holding furnaces. TST is committed to providing high-quality products that meet or exceed customer specifications. TST has set a goal to revolutionize how the company melts aluminum. This goal is achieved by adhering to an effective quality management system based on continual improvements and always searching for the technology to improve TST's quality production and manufacturing cost.

The MagPump technology addresses many of the industry challenges that companies are presently faced with. The design and operation of the MagPump maximizes operation and productivity and does not compromise quality.

O. A. Perez (✉)
TST Inc, 11601 S Etiwanda Ave, Fontana, CA 92337, USA
e-mail: operez@tst-inc.com

E. Takahashi
Zmag America, Ltd., 1011 NW Glisan Street, Unit 202, Portland, OR 97209, USA

© The Minerals, Metals & Materials Society 2023
S. Broek (ed.), *Light Metals 2023*, The Minerals, Metals & Materials Series,
https://doi.org/10.1007/978-3-031-22532-1_127

MagPump—The Design

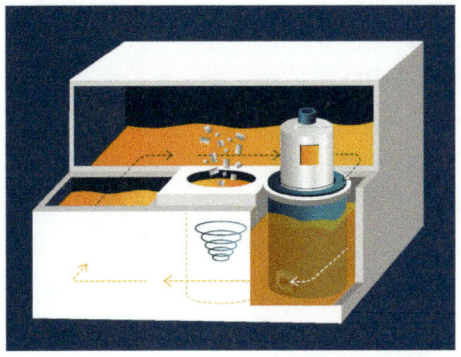

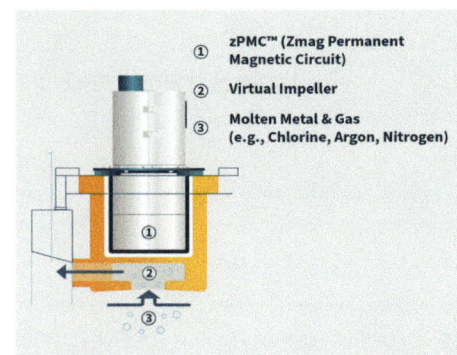

1. MagPump = MagPump Engine (zPMC) + MagPump Housing (refractory case)

MagPump Engine is powered by zPMC. zPMC is optimized to bring the best pump performance to a side well furnace. zPMC's magnetic fields are inherently existing (permanent) with the MagPump Engine. Since the field is permanent, there is no need to use electricity to generate magnetic fields as in the case of EMS (electromagnetic stirrer) and EMP (electromagnetic pump)—common technologies in the industry.

2. Virtual impeller using magnetic fields

There is no physical impeller with MagPump. Magnetic fields are used to pump molten aluminum, and this means that MagPump is a hands-free system.

3. MagPump Housing's pump room

The pump room is big and empty for increased performance. Molten aluminum is directed into the pump room from the bottom.

4. Gas injection

Gas can be injected from the bottom or in a path close to the discharge port to promote shearing and dissolution.

5. Air blower

A small industry-standard air blower is used to protect zPMC from molten aluminum heat. MagPump does not generate heat in its magnetic circuit (zPMC) as in the case of EMS and EMP.

6. Controls and electricity consumption

One VFD runs MagPump. The total electricity consumption of MagPump is about 10 kW.

7. No furnace modification

In most cases, MagPump does not require any furnace modification. "Plug and Play."

8. Vortex/chip submersion

MagPump is designed to work with a vortex system.

MagPump Installation

Zmag technology was truly plug-and-play; it was installed in the existing mechanical pump layout without the need for any modifications. TST also used the same VFD used by mechanical pumps with minor setting adjustments only for the MagPump™ motor to operate efficiently. Once MagPump was running to meet TST's operating standards, it was hands-free, with zero maintenance assistance after installation.

MagPump Operation

During the 6 weeks of trial, TST had six alloy changes. Even when alloy changes were not conducted, the bath was always nearly emptied on most casts (dry hearth or batch operation), TST cast an average of twice a day; however,

during the six weeks, the bath was emptied once per day due to TST's production schedule. 1 time per day * 5 days * 6 weeks = 30 times, 2.1 Million pounds of good metal was produced. TST has a second side well furnace next to the MagPump trial furnace; this furnace is a mirror image of the trial furnace. While TST had no issues at all with MagPump, TST did have challenges with the mechanical pump in the second furnace with wear and tear along with other unplanned issues. TST lost an average of 3.5–4 h of heat and production during the replacement and inspection of mechanical pumps which then required the addition of extra resources to assist in our second furnace. This lost production had the following impact:

Average 4 h downtime per event/failure × 4 events = 16 lost production hours.

If the average melt rate is 11,000 pounds per hour,

11,000 pounds per hour melt rate × 4 h downtime per event × 4 events = 176,000 pounds of lost production.

This is not including lost opportunity during alloy changes which required the raising of the mechanical pumps due to the metal level being too low for the mechanical pump to operate at an optimal RPM.

Extrapolating from the 6 week trial:

One year has 52 weeks. During the 6 weeks, mechanical pumps failed 4 times.

52/6 weeks × 4 events/failures = 35 events/failures expected per year.

11,000 pounds per hour melt rate × 4 h downtime × 35 events/failure per year = 1,540,000 pounds lost production.

In keeping with TST's goal of being environmentally responsible, MagPump also leads to a dramatic reduction in greenhouse gas:

At an average of 2,300 BTU per pound, TST would have used/wasted 2,300 BTU per pound × 1,540,000 pounds = 3,542,000,000 BTU.

LB—CO_2/mmBTU is 117, so 117 × 3,540,000,000/ 1,000,000 = 414,414 pounds or 188 tons per year of CO_2 reduced with MagPump.

Another benefit of MagPump is that the only inspection required was a simple check of the temperature display for MagPump's zPMC™ (Zmag Permanent Magnetic Circuit). A two second walk through and check of the displays was enough to let TST know MagPump was performing at peak performance. Additionally, Zmag's MagPump effectively minimized dross build-up in the pump well. Dross is typically a side effect of mechanical pumps. MagPump continuously ran at the set RPM of about 800. The metal level in the furnace did not affect MagPump's performance; MagPump did not need to be raised when the bath had a low heel since it pulled metal from underneath, not from the top. MagPump eliminated the risk of splashing when the metal level was low, which contributed to TST's commitment to providing a safe work environment for TST's associates.

MagPump Productivity

TST's present experience and most notable observation with operators of MagPump was observing the metal stirring movement that appeared to happen under the surface of the metal while maintaining a consistent temperature and constant charge rate. This resulted in a reduction of dross by not disturbing or breaking the metal surface. The effectiveness, reliability, and low maintenance of MagPump dramatically reduced the time needed by furnace operators to care for it, allowing operators to focus on other tasks.

MagPump Safety Contributions

MagPump has no moving impellers, and its low maintenance reduces the operator's exposure to molten metal. By having no moving parts submerged in the metal, it eliminates the risk of metal splashing. At TST, ensuring employee safety is always at the forefront of the decision-making process. MagPump provides peace of mind and priceless value, reducing the variable of uncertainty and provides stability that enhances productivity.

Zmag

Zmag was founded in Japan in 1990 and since its inception, it has committed itself to introducing cutting-edge magnetic engineering solutions. Zmag's original market was the sorting industry for scrap recycling, and its magnetic separators (e.g., eddy current separation) have been widely adopted by major companies in Asia. In the molten metal industry, Zmag introduced the world's first permanent magnet-based stirring system, MagStir™, which is powered by the Zmag Permanent Magnetic Circuit (zPMC™) technology. MagStir comes with varieties of designs, customised for different types of furnaces, and is widely adopted by casthouses around the world. Zmag also introduced Typhoon™, another zPMC-based chip submerging system widely adopted by automotive and die-cast companies in Japan, Asia, and other markets.

North American Market

In the North American market, the majority of casthouses operate side well furnaces equipped with (graphite) mechanical pumps. A mechanical pump comes with a rotating impeller shaft, which is submerged in molten aluminum and pumps aluminum in the side well and out to the main hearth. Flows from a mechanical pump help to quickly melt scrap that is charged in the charge well and/or chips that are charged in a vortex system next to a pump well. A mechanical pump is also used to inject gasses such as Argon, Nitrogen, and Chlorine to treat molten aluminum. Mechanical pumps have been the de facto standard technology in the industry for the past many decades.

Casthouse and Mechanical Pump Challenges

A mechanical pump is submerged in molten aluminum. Physical submersion in molten metal creates a great deal of unavoidable challenges for casthouses.

For example, some mechanical pump direct costs are:

1. Unpredictable running costs (consumable parts)

A rotating impeller shaft may be cracked or damaged due to dross balls, rocks, broken refractory, etc. A mechanical pump's support posts are also submerged in molten aluminum and oxidization thins and weakens the posts despite being protected by a special coating. The gas injection pipe and base are also required to be replaced when damaged or worn out. The timing of failure cannot be known and following Murphy's Law, failures usually occur at the worst possible time.

2. Maintenance shop and spare pumps

Since a mechanical pump needs to be repaired frequently, a multiple number of mechanical pumps are usually kept in the maintenance shop. Spare mechanical pumps need to be heated in a preheat chamber before being submerged in molten aluminum to avoid thermal shock.

3. Maintenance personnel

Typically, one or two dedicated maintenance personnel are required to maintain the plants' mechanical pumps.

4. Operator safety

When a mechanical pump fails, operators need to lift it out of molten aluminum and submerge a replacement pump. Safety is always a big concern for casthouses since any time operators are around molten aluminum, there is the possibility of an accident.

The industry's average annual mechanical pump budget per furnace is estimated to be about US \$200,000. This does not include the cost of spare mechanical pumps, which is about US \$27,000 per pump (on average, depending on size and other factors).

Hidden/indirect mechanical pump costs:

1. Lost production, revenue, and profit

A casthouse typically spends an average of 4 h to replace a mechanical pump. This long process is required because thermal shock needs to be avoided. Slow submergence is a must. The impact of lost production, revenue, and profit varies among casthouses; however, it can be calculated using a simple formula.

Loss of profit (US \$) = melt rate (T/h) × 4 h * # of downtime per year × aluminum profit (US \$/T).

2. Dross generation

A mechanical pump's rotating impeller shaft breaks the surface of molten aluminum and causes dross generation in the pump well. Cleaning dross around the rotating shaft and posts is not easy, safe, or operator friendly.

3. Metal splash and operator safety

A typical mechanical pump draws aluminum from the top of the base, and if a furnace sees a low heel and/or batch operation, a mechanical pump's RPM must be lowered so that it does not splash molten aluminum. Splashing is a serious safety concern. Furthermore, in a low heel condition, furnace production will suffer (which may be considered as downtime at some casthouses) because the mechanical pump runs at a lower RPM in order to reduce splashing, rather than at the pump's optimal RPM.

4. Wasted gas consumption and CO_2 emissions

When a mechanical pump is replaced, the furnace's burners keep running in order to maintain bath temperature over the full replacement operation (4 h on average). This wastes natural gas and emits CO_2.

5. CO_2 emissions related to graphite production

CO_2 is also emitted when graphite is produced. Reduction of use of graphite due to reduction in consumable parts is good for the environment and industry.

In order to solve the above-mentioned industry challenges, Zmag committed itself to invent MagPump™ that works as a direct replacement of a mechanical pump in a side well furnace. MagPump is designed to work as a pump rather than a stirrer.

Recycling of Aluminum from Aluminum Food Tubes

Sarina Bao, Anne Kvithyld, Gry Aletta Bjørlykke, and Kurt Sandaunet

Abstract

Aluminium is applied in food packaging due to its preservative capability. However, food residue and the fact that most packaging is thin gauge material, 50–250 μm, makes recycling challenging. In Northern Europe, processed cheese, caviar, and mayonnaise are popular items stored in toothpaste-shaped tubes. This paper focuses on the evaluation of the recyclability of these Al tubes from the aspect of tube thickness, user habits, food residue, benefits of pre-treatment, all in regard to yield. Food residue reduced the yield from around 88% (non-used empty) to 57% (with 3% food residue), and down to 34% (with 16% food residue). For comparison, the influence of beverage residue on yield was also studied. The influence of beverage residue is minimal, even neglectable after drying the can. The results also show that the influence of food residue on the yield is larger than that from decoating. The producer change in tube wall thickness did not influence the recycling yield considerably. However, a thinner tube makes it easier to be emptied. This together with that thinner tube uses less Al to protect the same amount of food implies that the thinner tube can be regarded as more environmentally friendly.

Keywords

Aluminium recycling • Aluminium food tubes • Metal yield • Decoating • Food residue

S. Bao (✉) · A. Kvithyld · K. Sandaunet
SINTEF Materials and Chemistry, N-7465 Trondheim, Norway
e-mail: sarina.bao@sintef.no

G. A. Bjørlykke
Kavli AS, Sandbrekkeveien 91, 5225 Nesttun, Norway

Introduction

Al is often applied in packaging due to its formability, strength, and protective quality. The Al packaging flows in Norway with an overview and challenges was reported in [1]. Also, a laboratory method for evaluating the quality and yield of the Al fraction from a Norwegian material recovery facility has been reported previously [2]. This work focus on the Al food tubes recycling.

In Northern Europe, processed cheese, caviar, and mayonnaise are stored in toothpaste like Al tube to extend food conservation. This tube is made from an Al slug, containing for example 1xxx alloy with Al > 99.5%, in a size of a coin by impact extrusion. The slug is pressed at a high velocity with extreme force into a desired form by a punch. Impact extrusion is often applied to manufacture tubes, cans, and technical parts in the automotive, food, cosmetics, pharmaceutical, and electrical industries.

After extrusion a protective barrier is sprayed inside the tube. Ink is dried on the outside of the tube on a white paint, which is resistant to most solvents and sun damage. Afterwards food can be inserted, and the gasket is sealed [3].

Analyses from disposed household glass and metal shows that there is 16.2% food residue left in tubes after use [4].

The coating is essential for food protection, but at the same time, it challenges the recycling process. If the coating is not removed before remelting, the content can hamper the recycling during remelting [5]. Thermal treatment is a standard industrial practice known as decoating [6]. However, burning of coating can increase oxidation and lead to environmentally unfriendly off-gas [7]. Oxidation can lead to metal loss and dross formation. Compaction [8] has been suggested as a method for preventing oxidation by reducing the surface area [9]. The fundamental processes of decoating is described in [10]. 10 min heat treatment at 550 °C has been chosen in the current work.

Al scrap can be remelted in reverberatory furnace, rotary furnace, crucible, or electrical furnaces in the industry [11].

Salt flux can be favorable when melting scraps of packaging. The salt layer can protect the metal underneath from further oxidation. It can extract impurities, inclusion, and oxides from the melt and extract metal entrapped in the dross. Beside reduced toxic properties and low vapor pressure, fluorides are often added into the salt to enhance coalescence. Fluorides help strip and break up the oxide layers so that imprisoned metal droplets can be accessed, and coalescence improved [12, 13].

After the sheet is milled, beverage cans are produced from forming and ironing operation [14]. Then washed, dried, painted externally, and coated internally before filling in the beverage [15]. Can lids (for example AA5182 with up to 5% Mg) are manufactured from a different alloy than can body (AA3004 or AA3104 with up to 1.4% Mg). In this work, the influence of beverage residue on recycling of Used Beverage Can (UBC) will be briefly discussed to compare with that of food residue in Al tube.

Classical studies performed by Rossel, [16] have demonstrated the role of the thickness on metal loss. The thinner the scrap the higher the metal loss in particular for thickness under 20 mm as shown in Fig. 1. As expected, metal loss increases with temperature (Fig. 2). Particularly alloy with higher Mg content (AlMg4.5) notes intensified metal loss.

Experimental

In this study Al tubes with two different gauge thicknesses were supplied by a food processor. The tubes were both non-used empty tubes and tubes filled with processed cheese as shown in Fig. 3. Filled beverage cans were provided by an Al company.

First, food was emptied from the used tubes by various methods, i.e., tube presser, hand roll, hand squeeze, squeezing key, and even more creative ways as stepping on, and scrolling on the table edge as shown in Fig. 4.

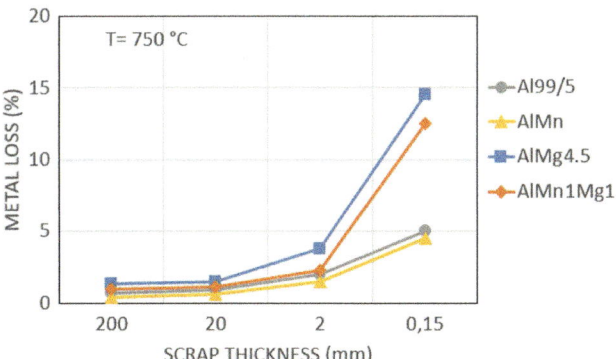

Fig. 1 The melt loss for scrap thickness at 750 °C. Original figure from [16]

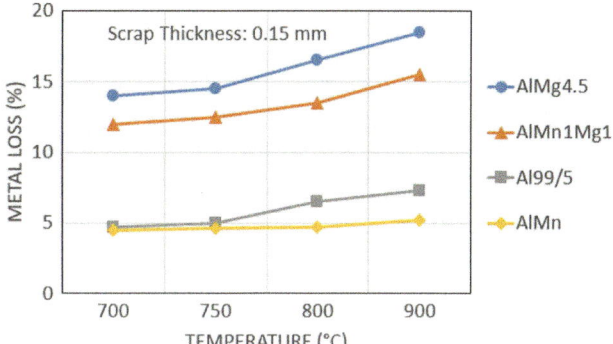

Fig. 2 The melt loss for various Al alloys as a function of temperature. Original figure from [16]

Fig. 3 Filled food tubes in the left picture (thicker gauged tube with white label), and empty non-used tubes in the right picture (thicker gauged tube to the left with blue color)

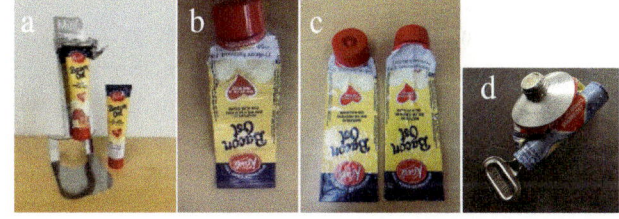

Fig. 4 Several ways to empty the food tubes. **a** tube presser, **b** tube emptied with tube presser, **c** tubes stepped on and rolled on the table edge, and **d** tubes rolled with squeezing key

Secondly, the tube's mass was registered, and gauge thickness was characterized by light microscopy.

Third, the Al tubes and cans were placed in a muffle furnace and heated up to 550 °C over approximately 30 min, and held at 550 °C for 10 min. After this decoating step was completed, white paint remained for tubes, while cans have changed the colour as shown in Fig. 5.

Fourth, to investigate the thermal decoating of lacquered Al scrap in detail, a Linkam TS 1500 hot stage with a Leica DMLM microscope was used. A 10 mm scrap piece

Fig. 5 Before and after decoating of Al tubes **a**, and before **b** and after **c** decoating of UBC

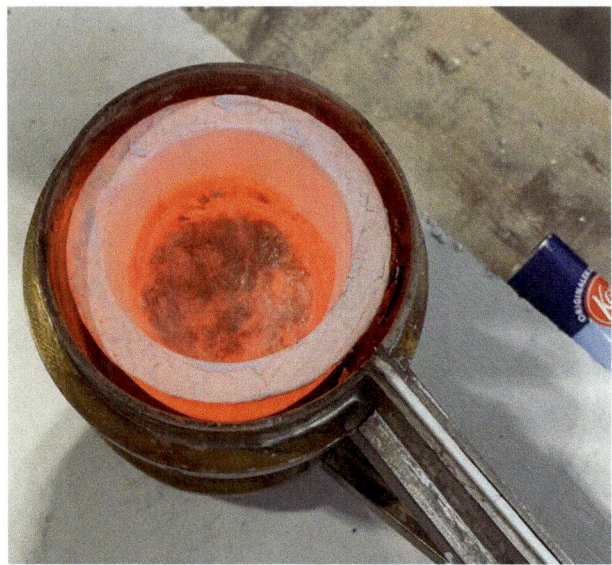

Fig. 6 Al tubes remelted in salt

from Al tube with the letter Ø was heated with 30 °C/min up to 700 °C and cooled to room temperature in 20 min in air. The same temperature profile was used in a thermogravimetric (TG) furnace in synthetic air, where the weight loss during decoating were measured as a function of time and temperature.

Fifth, as shown in Fig. 6, the Al samples were immersed into approximately 150 g melted salt, which contains 45% NaCl, 45% KCl, and 10% Na_3AlF_6, in a teacup-sized crucible in copper induction coil furnace. Both salt and the Al droplets were cast into a mould, after stirring and complete remelting. The salt was then crushed at cold conditions, and Al droplets were collected, soaked in water for overnight, dried, and weighted to determine the yield.

Fig. 7 Measurements of coatings and gauge thickness of the thin tube body. From left to the right, outer protective layer ink, (white) paint, Al, and inner protective barrier is 3.1 μm, 8.6 μm, 144.1 μm, and 8.3 μm, respectively

Table 1 Weight of Al tubes

	Used tube		Non-used tube	
	Thick	Thin	Thick	Thin
Weight (g)	14.0 ± 0.1	12.5 ± 0.1	13.5 ± 0.1	12.7 ± 0.1

Results and Discussion

Characterization of Al Food Tubes

From optical microscope analysis, as an example shown in Fig. 7, the shoulder and gauge thickness are 649 ± 7 μm and 145 ± 4 μm for thin tubes, and 841 ± 25 μm and 167 ± 7 μm for thick tubes. Thus, the thick tube has an extra 192 μm Al on the shoulder and 22 μm more Al on the body part of the tube.

After emptying the food in tubes by various methods, the tubes were washed, dried, and mass registered. As listed in Table 1, these two designs of thin tubes (Fig. 3) need 0.8–1.5 g less Al than the thick tubes.

Squeezing Out Food from Al Tube

Considering the variety of customers, a child of 5–7 years and an adult were engaged to squeeze out the cheese with a tube presser (Fig. 4a). As shown in Table 2, at least 5 g food residue would be left inside the tube, which is mostly located around the shoulder area. It is reasonable to see slightly more food residue for a child. Food residue with hand roll, hand squeeze, and squeezing key was bigger than when using a tube presser, stepping on, or rolling on the table edge. The food residue for thick and thin tubes is compared in Fig. 8. There is always slightly less, 0.4–4.3 g, food residue in thin

Table 2 Food residue inside the tube emptied with tube presser

	Man-force	Full mass (g)	After press (g)	Food residue (g)	Food residue[a] (%)
Thick tube	Child	188.2	20.8	6.7	3.8
	Adult	190.1	20.3	6.2	3.5
	Adult	188.2	19.5	5.6	3.2
Thin tube	Child	187.6	18.2	5.7	3.3
	Adult	186.8	17.9	5.3	3.0
	Adult	186.9	17.6	5.0	2.9

[a] Considering the net weight of cheese 175 g

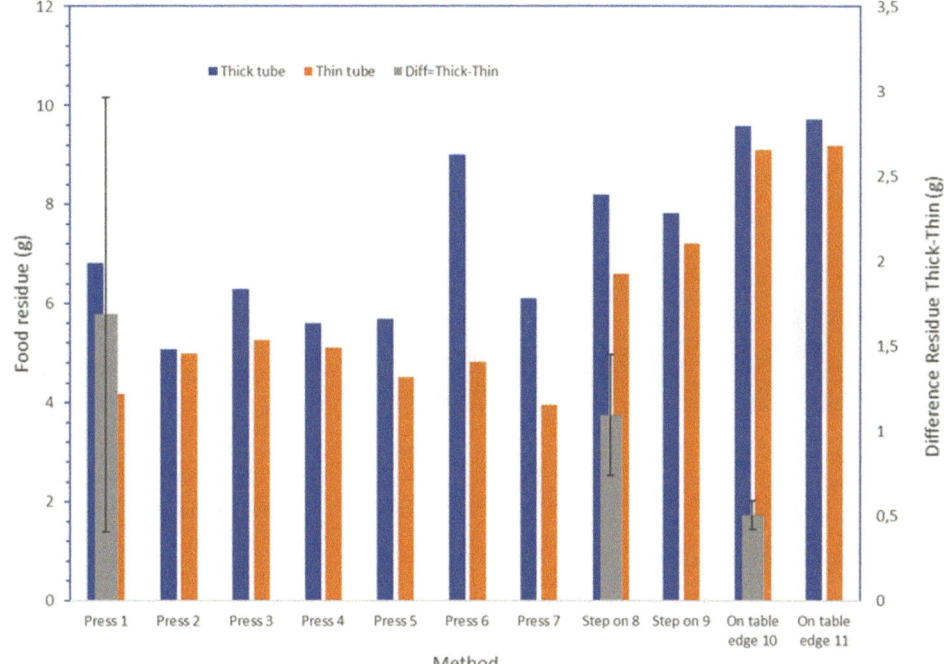

Fig. 8 Food residue in thick and thin used tubes with tube presser, stepping on, or rolling on the table edge

tubes with various methods. This means it is easier to squeeze out food from thin tubes. The highest food residue amount in this study is 6% and only a third of what was found in the analysis from household waste [4]. This means this study achieved likely the lowest limit of customer habit. More food residue should be expected in everyday life.

Thermal Decoating

The weight loss due to decoating is shown in Fig. 9 with 4.4, 3.8, and 1.5% for non-used thick tubes, non-used thin tubes, and UBC, respectively after preheating at 550 °C. The weight loss for the tubes is the same (within the uncertainty) and larger than that of the beverage can. The variation is due to coating components and thickness.

In Fig. 10, the fundamental processes described in the literature are confirmed for decoating. The first volatilization and pyrolysis of the coating (scission phase) has a peak

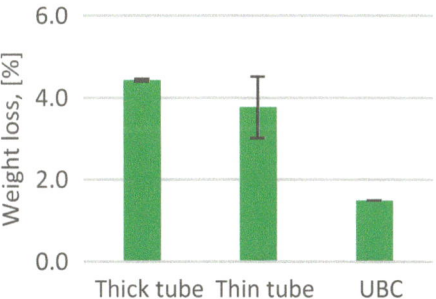

Fig. 9 Weight loss due to decoating

around 430 °C and the pigments and carbon residue are shown in the bottom left picture. The following combustion peak occurs around 550 °C (combustion phase) and only inorganic material remains on the surface as shown in the bottom right picture. The white residue is considered to be pigmented TiO_2. Pictures of starting materials and end are shown above the graphs in Fig. 10.

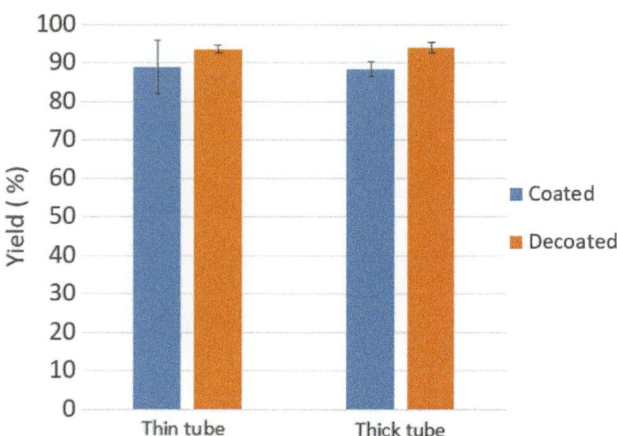

Fig. 12 Recycling yield of empty non-used tubes

Fig. 10 The weight loss with temperature during decoating and corresponding observation under a hot stage microscope

Remelting and Recycling

Remelting the tubes and beverage cans in salt with high content of Na_3AlF_6 had, as expected, enhanced the coalescence of the metal. Only 5 out of 23 experiments had more than one merged drop. Figures 11 and 12 present the recycling yield, which is calculated by metal out divided by

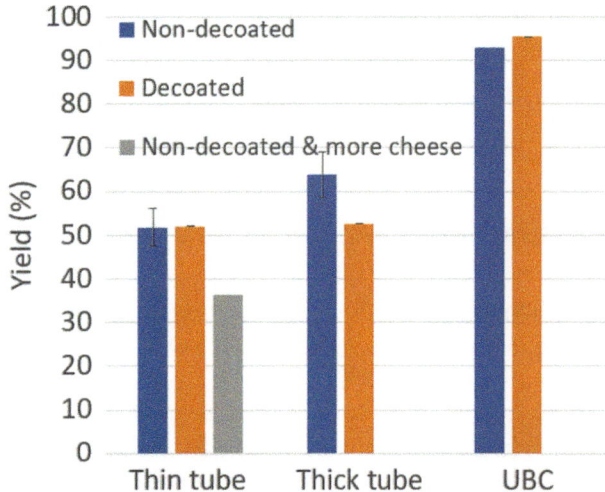

Fig. 11 Recycling yield of tubes and UBC with residue

material into the salt. Food residue in tube reduces the recycling yield from 88% (empty non-used tube) to around 57% (with 3% food residue), even down to 34% (with 16% food residue).

The result in Fig. 12 indicates that the recycling yield of the decoated tube is slightly higher than that of the non-decoated tube for non-used tube (94–88%). However, such an effect is not obvious for the used tube (Fig. 11).

Taking the coating into the denominator of yield calculation, the yield was slightly reduced for empty non-used tubes from 94 to 90%, and for used tubes with 3% residue from 52 to 49%. This indicates that decoating reduces the contamination and thus increases yield, 3–4% in the current case. However, food residue has decreased the yield from the 90 to the 50%, that is by as much as 40%.

With a thickness of several hundreds micrometer, when the shoulder and wall increase for 192 and 22 μm, it does not seem to change the recycling yield for non-used Al tubes (Fig. 12). But thick used tube has slightly larger recycling rate than thinner (Fig. 11).

Meanwhile UBC has much higher recycling yield than tubes (Fig. 11). This highlights that the influence of beverage residue is much less than that of food residue, even neglectable after drying.

Parameters Influence Al Tube Recycling

Thickness. From Fig. 1 we see that metal loss for alloy of 99.5% Al at 750 °C is around 5% for the gauge of 150 μm, and 4% for shoulder around 650 μm. The weight loss in this work is slightly larger, 6%, for non-used decoated tube.

Food residue One reason why recycling food tube is more difficult than that of UBC is food residue, which reduces the yield largely. Meanwhile, food residue produces large amount of gas, and has to be handled with care when recycled. Shredding, or slowly heating before entering molten metal is necessary.

The *lacquer thickness* is around 5% for tube body and 1% for shoulder for both thin and thick tubes, but slightly higher for thin tubes. Decoating reduces the contamination from lacquer and thus increases yield, but, as we observed, food residue reduces the yield more significantly.

Alloy. As known, alloying element Mg can be readily selectively oxidized during remelting due to its affinity for oxygen than Al. Thus, *UBC* alloys, AlMg4.5 (can top) and AlMn1Mg1(can body) give almost up to 3 times more weight loss than a 99.5% pure Al which is popular for *Al tube* extrusion (Fig. 2). However, the recycling yield is 30–40% lower for Al tubes in this work (Fig. 11), that is dominated by the food residue in Al tubes.

Recycling furnace. Different types of furnaces for melting Al scrap can be used, depending on the initial metal content in the scrap, type and content of impurities, geometry of the scrap, frequency of change in the alloy composition, operating condition, energy cost, and desired product quality [11]. For example, rotary furnace is more common in Europe due to high energy cost. While gas reverberatory furnace, which operate with a lower energy efficiency and requires lower capital cost, covers the 95% of Al scrap in the United States. Considering these features, the laboratory crucible furnace used in this work has less yield than reverberatory furnace.

The yield basically depends on the recycling method as described in the introduction. These experiments simulate part of a rotary furnace practice with salt. The industry uses less salt and has generally higher yields than what is usually obtained in small-scale laboratory experiments. It is experienced that the recycling yield in the industry of contaminated food packaging also can be low.

Conclusions

Al food tubes and UBC were remelted and recycled in a laboratory scale teacup-sized Cu coil induction furnace with salt.

Two Al tubes, thick and thin, were investigated. When the shoulder and wall thickness increase respectively 192 and 22 μm, thinner tubes require 0.8–1.5 g less Al alloy than the thick tubes. *This thickness change* does not seem to influence the recycling yield considerably. All the methods used to empty the Al tubes resulted in that the thinner tube had less food residue. The food left in the tubes lowers the recycling yield significantly. *Thus, a thinner Al tube* is more environmentally friendly.

Decoating reduces the contamination and thus increases yield, but food residue has a more dominating effect on yield.

UBC has a much higher recycling yield than Al tubes. The influence of beverage residue is less than that of food residue, even neglectable after drying.

Acknowledgements This research was carried out as part of the Norwegian Research Council (NRC)—funded IPN Project (296276) Alpakka-Circular Aluminium Packaging in Norway. It includes the following partners: Norsk Hydro, Metallco Aluminium, Norsk Metallgjenvinning, Infinitum, O. Kavli, NTNU, and SINTEF. Funding by the industrial partners and NRC is gratefully acknowledged. Ingrid Hansen and Kjersti Øverbø Schulte, SINTEF are acknowledged for hot microscopy study, and discussion and methods for emptying the tubes and packaging, respectively. Thanks are also given to Kavli and Nork Hydro for supplying Al tubes and UBC.

References

1. M. S. A. K. Magnus Skramstad, "Aluminium Packaging Flow in Norway," Trondheim, Apr. 2021.
2. S. Eggen, K. Sandaunet, L. Kolbeinsen, and A. Kvithyld, "Recycling of Aluminium from Mixed Household Waste," in *Minerals, Metals and Materials Series*, 2020, pp. 1091–1100. doi: https://doi.org/10.1007/978-3-030-36408-3_148.
3. *"How It's Made" featuring Montebello Packaging.* Accessed: Aug. 10, 2022. [Online Video]. Available: https://www.youtube.com/watch?v=8yOqFbygVAQ.
4. Sveinung Bjørnerud, "Plukkanalyser glass/metall 2020," 2020.
5. S. Capuzzi, A. Kvithyld, G. Timelli, A. Nordmark, E. Gumbmann, and T. A. Engh, "Coalescence of Clean, Coated, and Decoated Aluminum for Various Salts, and Salt–Scrap Ratios," *Journal of Sustainable Metallurgy*, vol. 4, no. 3, pp. 343–358, 2018.
6. T. A. Engh, G. K. Sigworth, and A. Kvithyld, *Principles of Metal Refining and Recycling.* Oxford University Press, 2021.
7. Solveig Høgåsen, "The effect of compaction and thermal treatment in the recovery of coated aluminium scrap through FTIR off-gas analysis and remelting in molten heel," Norwegian University of Science and technology, Trondheim, 2022.
8. A. Vallejo-Olivares, S. Høgåsen, A. Kvithyld, and G. Tranell, "Effect of Compaction and Thermal De-coating Pre-treatments on the Recyclability of Coated and Uncoated Aluminium," in *Light Metals 2022*, Springer, 2022, pp. 1029–1037.
9. C. Hamers and A. Jessberger, "Aluminium Cycle: Machining, Briquetting, Melting," *Global Recycling,* Mar. 2018. https://global-recycling.info/archives/2354 (accessed Aug. 10, 2022).
10. A. Kvithyld, C. E. M. Meskers, S. Gaal, M. Reuter, and T. A. Engh, "Recycling light metals: Optimal thermal de-coating," *JOM*, vol. 60, no. 8, pp. 47–51, 2008, doi: https://doi.org/10.1007/s11837-008-0107-y.
11. S. Capuzzi and G. Timelli, "Preparation and melting of scrap in aluminum recycling: A review," *Metals*, vol. 8, no. 4. MDPI AG, Apr. 01, 2018. doi: https://doi.org/10.3390/met8040249.
12. M. S. S. Besson, A. Pichat, E. Xolin, P. Chartrand, and B. Friedrich, "Improving Coalescence in Al-Recycling by Salt Optimization," in *Proceedings of EMC*, 2011, p. 1.

13. R. D. Peterson, "Effect of salt flux additives on aluminum droplet coalescence," in *Proceedings of the 2th International Symposium on Recycling of Metals and Engineered Materials*, 1990, pp. 69–84.

14. D. A. Doutre, "LiMCA and its Contribution to the Development of the Aluminum Beverage Container and UBC Recycling," 2011.

15. PE Americas, "Life Cycle Impact Assessment of Aluminum Beverage Cans," 2010. [Online]. Available: www.pe-americas.com.

16. H. Rossel, "Fundamental investigations about metal loss during remelting of extrusion and rolling fabrication scrap," *Light Metals*, pp. 721–729, 1990.

Recent Studies Using HR-TEM on the Fundamental Mechanism of Nucleation of *a*-Aluminum on TiB$_2$ in TiB D High-Efficiency Grain Refiners

John Courtenay and Yun Wang

Abstract

Recent studies using HR-TEM, (High-Resolution Transmission Electron Microscopy) at BCAST, Brunel University, London, on the mechanism of nucleation of *a*aluminium on TiB$_2$ in commercial high-efficiency grain refiners have shown that efficiency as measured by the Opticast Nucleation test can be directly related to the extent to which the TiB$_2$ particles have successfully formed a monoatomic layer of TiAl$_3$ on their basal plane. This factor was found to be predominant over other factors such as particle size distribution and average particle size. As a result of this research, it is now possible to produce a range of ultra-high efficiency grain refiners which enable addition rates to be further reduced to levels as low as 0.15 kg/t resulting in higher cost savings and the particle count being significantly reduced in the liquid metal leading to as cast metal quality being improved.

Keywords

Aluminum • Grain refinement • Efficiency

Introduction

Grain refinement during the casting of engineering alloys is usually desirable, since it results in not only a fine and equiaxed grain structure but also a significant reduction of casting defects, which in turn leads to an improved engineering performance [1]. As a common practice in the metal casting industry, grain refinement is usually achieved by chemical inoculation through the addition of grain refiners. Among a series of Al-Ti-B-based grain refiners developed for Al-alloys, Al-5Ti-1B (all compositions are in wt.% unless otherwise specified) master alloy has been the most widely used grain refiner, which contains potent TiB$_2$ particles for enhancing heterogeneous nucleation of a-Al grains [2]. Since the introduction of Al-Ti-B-based grain refiners at the beginning of the 1950s [3], significant effort has been made to optimize their performance largely through a trial-and-error approach and to understand the grain refining mechanisms, with both the theory and practice being extensively reviewed [4, 5]. Various researchers recognized that AI$_3$Ti was a much stronger nucleant than TiB$_2$; however, its presence could not be substantiated because it was not thermodynamically stable for the hypo peritectic case. For example, Johnsson [6] indicated that the aluminides probably take less than 1 min, but definitely less than 5 min to dissolve at a holding temperature of 775 °C. Jones and Pearson suggested that all aluminides dissolved in molten Al in less than 30 s. Although the dissolution time depends on both the holding temperature and the size of the AI$_3$Ti particles, the resulting consensus is that AI$_3$Ti dissolves rapidly above the Al liquidus and hence the peritectic reaction is not thermodynamically feasible.

For AI$_3$Ti to be responsible for grain refinement during the solidification of hypoperitectic alloys, other factors have to be operational. Vader et al. [7] and Backerud et al. [8] proposed the peritectic hulk theory in the early 1990s. This theory recognized that AI$_3$Ti is a more potent nucleant than TiB$_2$ and attempted to explain how the borides increase the stability of aluminides. It was suggested that the borides form a shell around the aluminides and slow down the dissolution of the aluminides. The aluminides eventually dissolve and leave a cell of liquid with approximately the peritectic composition. The peritectic reaction can then take place to form the α-Al.

J. Courtenay (✉)
MQP International Ltd, Solihull, UK
e-mail: john.courtenay@mqpltd.com

Y. Wang
BCAST Brunel University, Uxbridge, England

© The Minerals, Metals & Materials Society 2023
S. Broek (ed.), *Light Metals 2023*, The Minerals, Metals & Materials Series,
https://doi.org/10.1007/978-3-031-22532-1_129

However, a number of researchers (e.g., Refs. [1–13] investigated the Al-rich comer of the Al-Ti-B phase diagram and showed that boron had virtually no effect on the Al-Ti phase diagram. Therefore, the nucleation process cannot be explained purely based on theories that attempt to modify the conditions for the peritectic reaction to occur. Later in pioneering TEM carried out by Schumacher and Greer [10, 11] using a metallic-glass technique, it was claimed that the presence of an Al_3Ti coating phase on the TiB_2 was essential for them to be potent.

In the first phase of this work, experimental evidence for the existence of an atomic monolayer of (112) Al_3Ti twodimensional compound (2DC) on the (0001) TiB_2 surface of commercial Al-Ti-B-based grain refiners was reported [12, 13].

In the second phase of the work, HR-TEM studies of grain refiners with varying performance as measured by the Opticast Nucleation Test have been able to demonstrate that efficiency is related to the extent to which the TiB_2 particles have been successfully coated.

Study Using HR-TEM of the Extent of Layer Formation in Samples of Grain Refiner with 50 and 123% Efficiencies

Having established that nucleation can only occur when an Al_3Ti monolayer is present on the surface of the TiB_2 particles, the next phase of the work centered on examining the possibility that the wide variations in grain refiner performance observed using MQP's Opticast nucleation test measurements of efficiency might be due to differences in the extent to which the TiB_2 particles in the grain refiner had successfully formed an Al_3Ti layer.

Samples of commercial Optifine Al-3Ti-B production provided by MQP were examined by HR-TEM as follows:

Sample 1

This batch had been rejected for acceptance as Optifine because the specific efficiency measured by the Opticast Nucleation test was only 50% compared to the target efficiency of 100%.

Sample 2

This batch was from experimental production aimed at achieving substantially higher efficiency and had an efficiency measured at 123%.

To prepare thin foils for conventional HR-TEM, and HRSTEM examinations, slices from the rods of the grain refiner master alloys were mechanically ground and cut into

3 mm diameter discs. The discs were then hand ground to a thickness of less than 70 μm, followed by ion beam thinning using a Gatan PIPS facility.

Results

Sample 1 50% Efficiency

A total number of 37 TiB_2 particles were examined for the 50% efficiency grain refiner. Six particles were electronically thin enough and thus the status of their (0001) surfaces is clarified decisively by HR-TEM.

Five of the six particles had no Al_3Ti 2DC layer, and one exhibited a partial Al_3Ti 2DC layer on its basal plane (Fig. 1).

An orientation relationship (OR) between TiB_2 and Al grain is not observed in 50% sample, although limited particles were examined (Fig. 2).

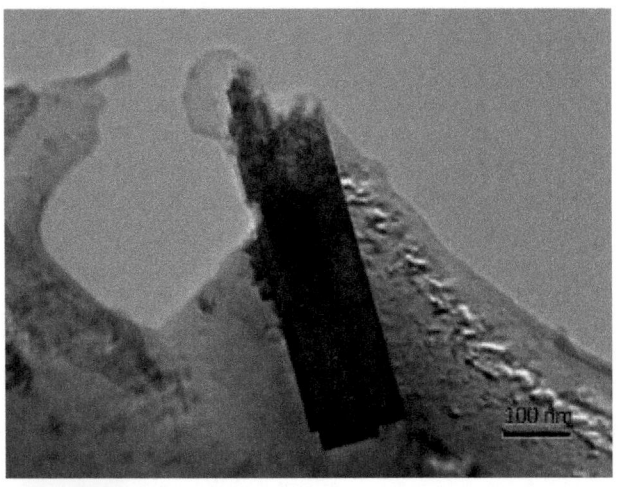

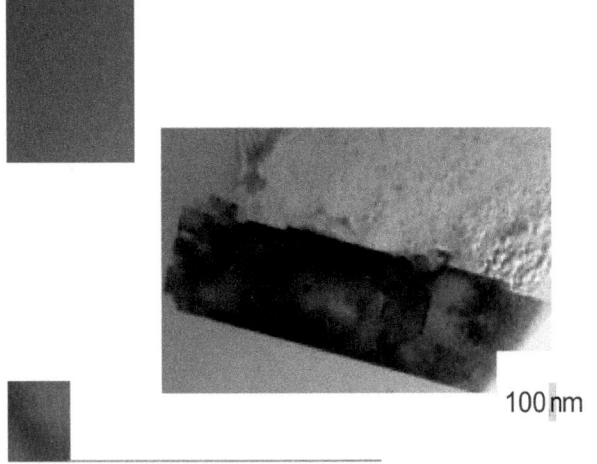

Fig. 1 TEM bright field images showing the morphology of the TiB_2 particles in the low-efficiency (50%) grain refiner master alloy. Both the particles are being viewed along their [1 1 −2 0] direction

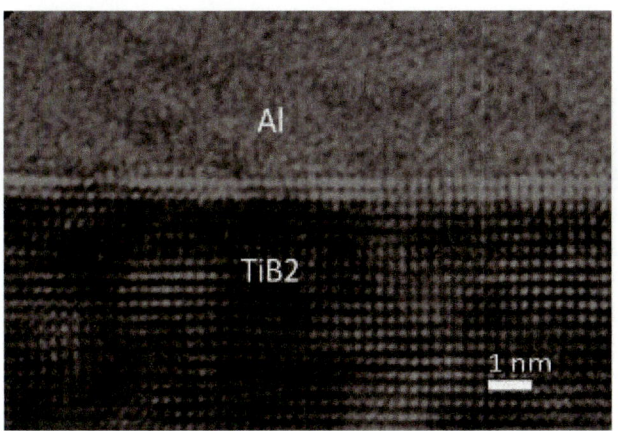

Fig. 2 No Al_3Ti 2DC layer is seen on the surface of TiB_2

Sample 2 123% Efficiency

A total of 33 samples were examined and among them, eight particles were electronically thin enough and thus the status of their (0001) surfaces was clarified by HR-TEM (Fig. 3).

Seven of the eight particles were confirmed to have the Al_3Ti 2DC layer, and one had no Al_3Ti 2DC layer.

In summary, in this phase, we have shown the following:

HR-TEM analysis shows the probability of TiB_2 particles which nucleated Al grains is significantly lower in 50% than in 123% sample.

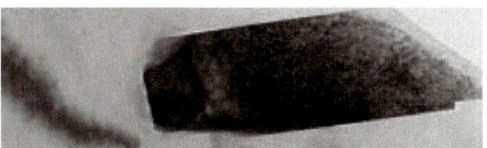

100 nrn

Fig. 3 TEM bright field images showing the morphology of the TiB_2 particles in the high-efficiency (123%) grain refiner master alloy. Both the particles are being viewed along their 1 **1** −2 0 direction

The extent of Al_3Ti 2DC on layer formation on TiB_2 particle is a priority factor in determining the refining efficiency.

The results can be shown in a histogram as follows:

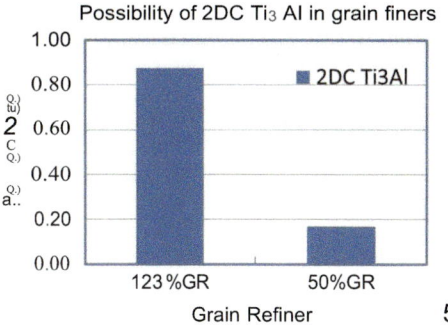

Development of New Super Efficiency Grain Refiner-Optifme 5:1125

Following promising results with experimental production of an Optifine 3:1 with higher efficiency, MQP decided to concentrate further development on a new version based on the Al-5Ti-B system.

(a)

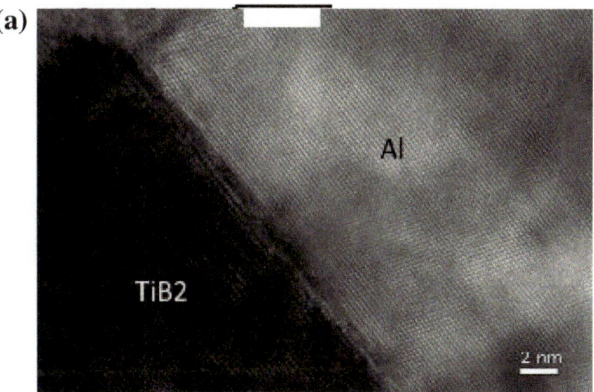

(b)

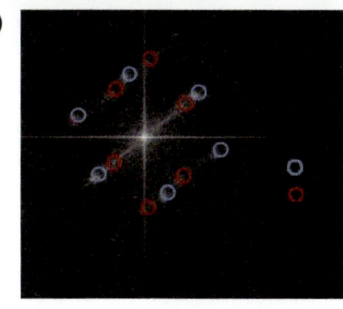

Fig. 4 **a** High-resolution TEM image showing a well-defined orientation relationship (OR) between TiB_2 and Al grain observed in high-efficiency (123%) sample Orientation Relationship (OR): (0001)[11 −20]TiB_2//(111)[0 −1 1]Al. **b** Fast Fourier transformation (FFT) pattern obtained from Fig. 4a showing the well-defined orientation relationship is (0 0 0 1)[1 1 −2 0]TiB_2//(1 1 1)[0 −**1** 1]Al

Samples as follows from trial production were provided for HR-TEM examination to quantify the degree of particle coating, SEM analysis to study morphology and particle size distribution, and macro analysis to review the rod structure.

- Batch 0417 W 5:1 (Not Optifine) 63% efficiency.
- Batch 6878 T Optifme 5:1 125 125% efficiency.

In sample 6878T, the much-expected orientation relationship (ORl) is observed in the Optifine 5:1 125, indicating its high grain refining efficiency (Fig. 5b).

Results of HR-TEM and SEM Analyses

HR-TEM Analysis—Batch 6878T Optifine 5:1125 120%

Nine TiB_2 particles from batch 6878T were successfully prepared for analysis and examination showed that all were fully covered with a Al_3Ti_2DC layer indicating 100% particle coating of the 125% efficiency Optifine 5:1 125.

Thus, it was concluded, albeit on a limited number of particles observed, that the degree of TiBD particle coating

with the Al_3Ti atomic monolayer was significantly higher in the new Optifine 5:1 125 product compared to the earlier experimental Optifine 3:1 which had an efficiency of 50%

Examination of Particle Morphology and Particle Size Distribution of Optifine 5:1125

Samples of the rod from trial production were examined at low magnification to compare the structure in the longitudinal section between the successful Optifine 5:1 125 (125% efficiency) batch used in the previous **HR** TEM examination with a rejected batch that had exhibited only 63% efficiency when measured on the Opticast Nucleation test.

A more uniform microstructure at low magnifications for the high-efficiency grain refiner batch 6878T was observed compared to that for the low-efficiency batch 0417TW.

Under examination by SEM at 20,00 k x, a clear difference in morphology can be observed. In the low-efficiency sample, evidence of significant amounts of salt reaction products can be seen around the periphery of the TiB_2 particles whilst in the high-efficiency sample, the TiB_2 particles are more clearly developed being clean and smooth with only minor amounts of salt reaction product present.

Measurement of Particle Size Distribution

From Figs. 9 and 10, it can be concluded that TiBD particles in the two samples are of similar average size dD and size distribution $a-$.

Alloy	Batch	Efficiency (%)	Size distribution (dD μm)	(J)	Particle coating $AlTi_3$
5:1	6878T	125	0.43	0.44	100%
5:1	0417	63	0.42	0.54	

(a)

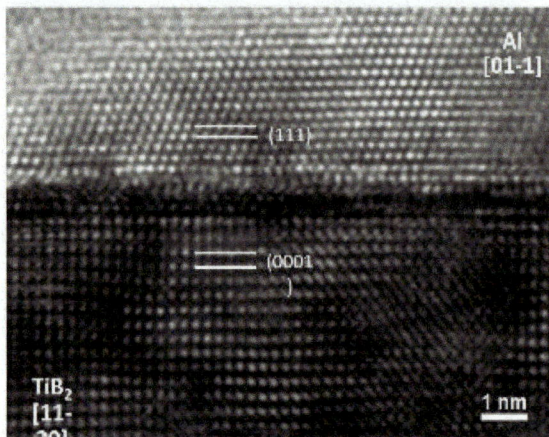

(b)

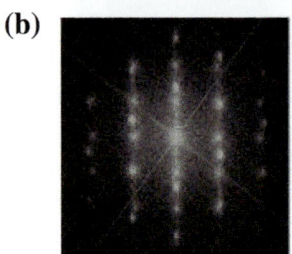
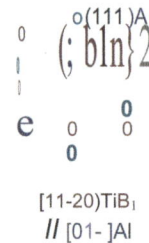

Fig. 5 **a** The much expected orientation relationship (OR1) is observed in the Al-5Ti-B, indicating its high grain refining efficiency. **b** The well-defined orientation relationship (OR): (0 0 0 1) [1 1 −2 O] TiB_2//(1 1 1)[0 −1 l] Al is again confirmed by the Fast Fourier transformation (FFT) pattern obtained from Fig. 5a

Fig. 6 HR-TEM lattice image indicates that the Al_3Ti 2DC layer is readily observed on the basal surface of TiB_2 particles in the Al-5Ti-1B master alloy (Optifine 5:1 125)

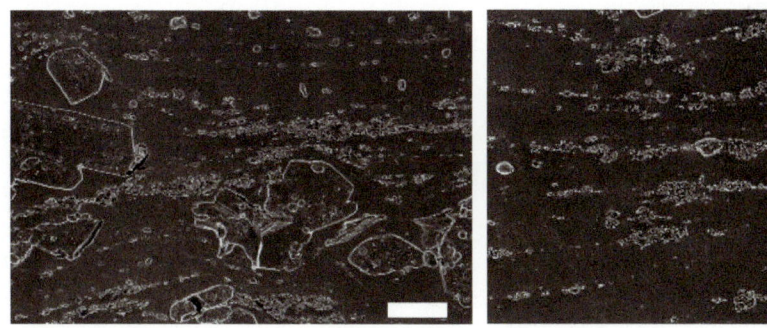

Fig. 7 Low-magnification examination of longitudinal sections from rod

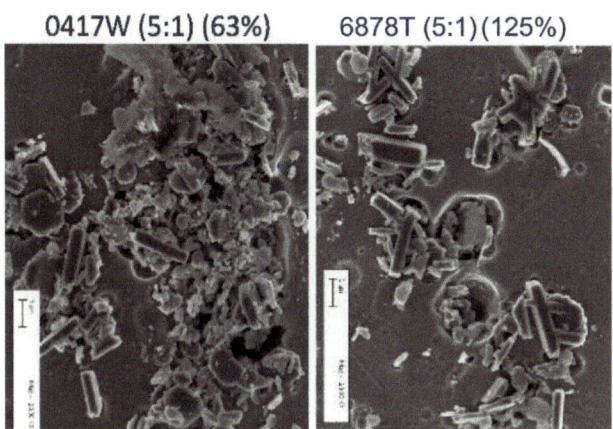

Fig. 8 High-magnification examination of longitudinal sections

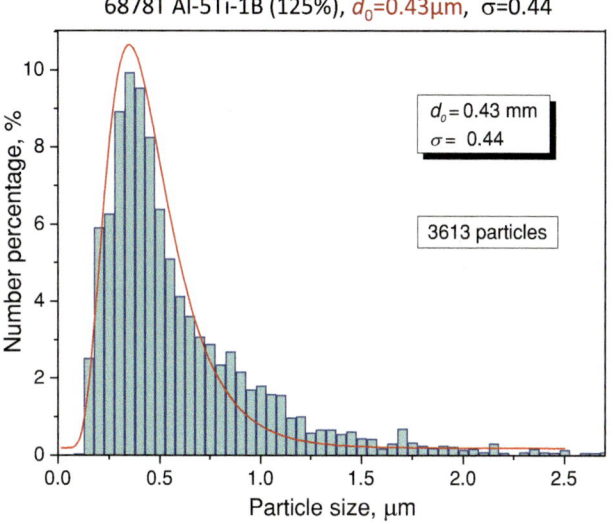

Fig. 10 Particle size distribution for 6878 T Al-5Ti-1B (125%), $d_0 = 0.43$ μm, cr = 0.44

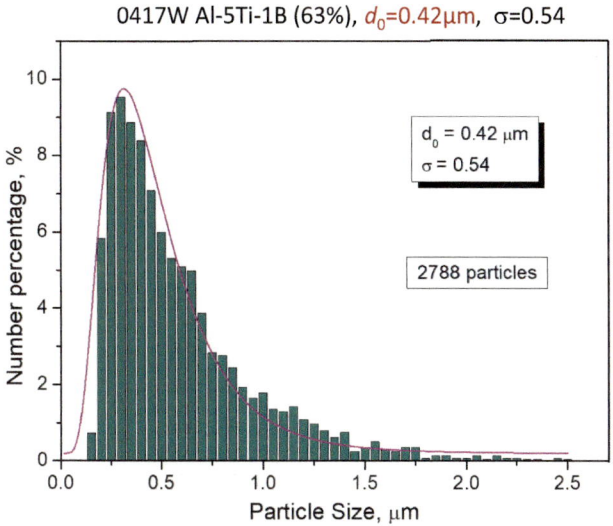

Fig. 9 Particle size distribution for 0417 W Al-5Ti-1B (63%), $d_0 = 0.42$ μm, cr = 0.54

Summary

- The microstructure of the high-efficiency Al-5Ti-1B alloy (6878 T, 125%) is more uniform than that of the low-efficiency one (0417 W, 63%).
- The low-efficiency alloy shows more severely agglomerated TiB_2 clusters than the high-efficiency one.
- Salt reaction products are readily found to remain in the low-efficiency alloy.
- In spite of their difference in refining efficiency, the TiB_2 particles in the three Al-5Ti-1B rods show similar average size and size distribution.

Table 1 Histogram showing % of particles coated with Al3Ti 2DC layer versus grain refiner efficiency

Grain refiner	Addition rate (kID'ton)	Alloy
Standard 3/1	1, 0.5, 0.2	AAA 1050
Optifine 3/1		
Optifine 5/1		

- TEM reveals that, so far, all the nine TiB$_2$ particles examined in the high-efficiency refiner are covered with Al$_3$Ti 2DC, indicating its high refining efficiency (125%).

This results in a reduction of the addition rate of 85% yielding even higher cost savings.

Discussion

In their 2015 paper on the "Grain refining mechanism in the Al-Ti-B system" Fan et al. [13] remarked in the summary of the discussion that "suitable size, size distribution, and adequate number density of TiB$_2$ particles in the Al-5Ti-1B grain refiner can also be contributing factors to effective grain refinement. The Al-Ti-B-based grain refiners were developed over 60 years, mainly by trial and error, and their degree of optimization is surprisingly high. It is anticipated that further improvement of grain refining efficiency will be very difficult, if at all possible".

Results for Super High-Efficiency Optifine 5:1125 Compared to Optifine 3:1

Laboratory trials were conducted using AA1050. In the trial, the performance of three grain refiners with significantly different efficiencies—including standard 3/1, Optifine 3.1, and Optifine 5.1:125—were tested at addition rates from 0.2 kg/t up to 1kg/t; see Table 1.

The alloy was melted in a resistance furnace and the temperature was held at 720 ± 5 °C throughout the testing period. The performance of each grain refiner was tested using the Opticast equipment.

Figure 11 shows the effect of addition rate versus grain size. The red hatched line indicates that Optifine 3:1 provides the same grain size at an addition rate of 0.28 kg/t compared to the standard grain refiner at an addition rate of 1 kg/t. This results in a reduction of the addition rate of more than 70%, which, in practice, will reduce the cost of grain refinement by 50%.

Respectively, the yellow hatched line shows that Optifine 5:1 125 provides the same grain size at an addition rate of 0.15 kg/t compared to the standard grain refiner at **1** kg/t.

This was the state of the art at that time. Although by using advanced techniques such as HR-TEM, HRSTEM, and super STEM, it had been possible to establish that in terms of the mechanism it was the formation of an ABTi 2DC monolayer that enables normally low nucleation ability TiB$_2$ particles to nucleate successfully; the quantitative effect of this and the other contributory factors such as particle size distribution on grain refiner efficiency had not been studied.

However, with the development of a rapid, reliable, accurate, and reproducible testing method capable of mass application in the production process, it is now possible to generate accurate data on how efficiency varies with a variety of factors. The TP-1 test has been the industry standard for testing for more than 50 years but without modification and the use of multiple samples, it has severe statistical limitations. It is also time-consuming to operate and requires a large melt volume. These practical limitations have hindered the measurement and control of efficiency.

In this study, the use of the Opticast Nucleation test has made it possible to correlate the enhanced understanding of the nucleation mechanism, in terms of the extent to which the Al$_3$Ti$_2$DC layer is present on TiB$_2$ particles, directly to the efficiency of the grain refiner expressed in grains/mm^3/ppm of boron, or for practical purposes kg/t of grain refiner addition to produce a given grain size for different alloys.

At the time of the 2015 study although it was considered that commercial grain refiners were quite well optimized, it has been demonstrated subsequently that this is not the case

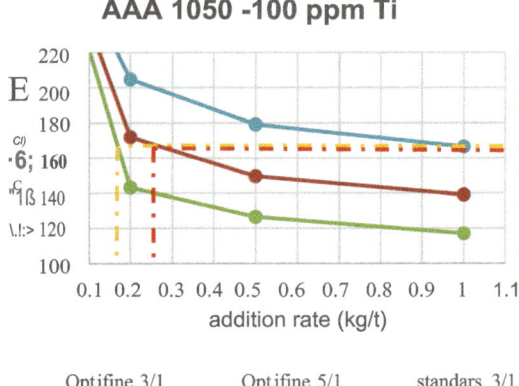

AAA 1050 -100 ppm Ti

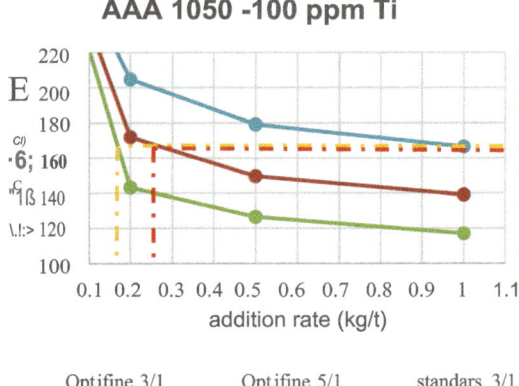

Fig. 11 Performance of Optifine 5:1125 on AAA 1050 compared to Optifine 3:**1** and standard grain refiner

and wide variations in efficiency exist between different batches of standard grain refiners supplied. If we take Optifine 3:**1** as the current standard for high-efficiency grain refiner with a relative efficiency of 100%, then although many batches of standard grain refiner may have relative efficiencies between 50 and 80%, there are always batches with efficiencies as low as 30% present and this means that application rates in practice have to be based on this value if cracking of cast product is to be avoided. Now with the application of the latest technology, it is possible to produce a commercial supply of grain refiner specified as either 100 or 120% efficiency representing in the best case an improvement of 85% in efficiency or in practical terms a reduction in addition rate from 1 kg/t down to 0.15 kg/t with strong benefits in terms of cost reduction and improvement of quality.

Conclusions

1. Whilst TiB_2 particles can nucleate a-Al, they have only low potency, and to nucleate effectively require the formation of a (112) Al_3Ti 2DC monolayer on the (0001) TiB_2 surface.

2. Al_3Ti 2DC is stable in concentrated Al-Ti solution but is unstable and dissolves in dilute Al-Ti solution.

3. Excess Ti a affects the effectiveness of Al-Ti-B grain refiner by the formation of Al_3Ti 2DC to increase the potency of TiB_2 particles and the provision of free Ti in the inoculated melt to cause columnar- to-equiaxed transition.

4. Studies of grain refiner batches with varying efficiencies have shown that efficiency is proportional to the extent to which the TiB_2 particles exhibit an Al_3Ti 2DC layer.

5. Specifically, in a 50% efficiency batch, only 20% of TiB_2 exhibit a Al_3Ti 2DC layer, whilst in a 123% batch, 80% of TiB_2 particles exhibited an Al_3Ti 2DC layer.

6. A new grain refiner has been developed, Optifine 5:1 125 based on the Al-5Ti-B system with a target of 120% efficiency. TEM examination of nine particles from a sample of batch 6878 T has revealed that all nine TiB_2 particles were covered with an Al_3Ti 2DC layer, indicating its high refining efficiency (125%).

7. The microstructure of the high-efficiency Al-5Ti-1B alloy (6878 T, 125%) is more uniform than that of the low-efficiency one (0417 W, 63%).

8. The low-efficiency alloy shows more severely agglomerated TiB_2 clusters than high-efficiency one.

9. Salt reaction products are readily found to remain in the low-efficiency alloy.

10. In spite of their difference in refining efficiency, the TiB_2 particles in both Al-5Ti-1B batches show similar average size and size distribution.

References

1. G.K. Sigworth, in: ASM Handbook, Casting. Metals Park, vol. 15, ASM, OH, 2008, p. 255.
2. B.S. Murty, S.A. Kori, **M.** Chakraborty, Int. Mater. Rev. 47(2002) 3.
3. A. Cibula, **J.** Inst. Met. 80 (1951) 1.
4. D.G. McCartney, Int. Mater. Rev. 34 (1989) 247.
5. G.P. Jones, **J.** Pearson, Metal!. Mater. Trans. B 7 (1976) 223.
6. **M.** Johnsson. A critical survey of the grain refining mecha- nisms of Al (PhD thesis), Stockholm University, 1993.
7. M. Vader, J. Noordegraaf, P.C. Van Wiggen, in: E.L. Rooy (Ed.), Light Metals, TMS, Warrendale, PA, 1991, pp. 1123–1130.
8. L. Backerud, P. Gustafson, M. Johnsson, Aluminum 67 (1991) 910.
9. G.K. Sigworth, Metal!. Mater. Trans. A 15 (1984) 277.
10. P. Schumacher, A.L. Greer, J. Worth, P.V. Evans, M.A. Kearns, P. Fisher, A.H. Green, Mater. Sci. Tech. 14 (1998) 394.
11. P. Schumacher, A.L. Greer, Mater. Sci. Eng. A 181 (1994) 1335.
12. Z. Fan, Y. Wang, M. Xia, S. Arumuganathar, Acta Mater. 57 (2009) 4891.
13. Z. Fan, Y. Wang, Y. Zhang, T. Oin, X.R.Zhou, G.E. Thompson, T. Pennycook and T. Hashimoto, Acta Mater. 84 (2015) 303.

A Cellular Automaton Model for Qualifying Current Grain Refiners and Prescribing Next-Generation Grain Refiners for Aluminium Alloys

G. Salloum-Abou-Jaoude, S. Sami, A. Jacot, and L. Rougier

Abstract

In cast aluminium products, small equiaxed grains reduce the risk of hot tears and shrinkage porosities by facilitating the liquid feeding in the interdendritic liquid. Although grain refinement in aluminium alloys is well known and widely used, grain size control is still not always guaranteed industrially. This depends largely on the nature and fabrication quality of the grain refiner rod. In this work, we developed a cellular automaton model to establish a clear link between the grain refiner type/nature and grain refining efficiency while accounting for the principal physical phenomena affecting grain refiner performance: grain refiner nature, nucleant size distribution, recalescence, solute suppressed nucleation zone. At TMS2020 [1] we experimentally highlighted the inconsistencies in grain refiner performance between different producers and in batches of the same producer. This model helps in qualifying grain refiners and would serve as a prescriber for designing next-generation grain refiners with superior efficiency.

Keywords

Grain refiner • Solidification • Aluminium alloys • Cellular automaton model

G. Salloum-Abou-Jaoude (✉) · S. Sami
Constellium Technology Center C-TEC, Parc Economique Centr'alp, 725 Rue Aristide Bergès, CS10027 Voreppe, France
e-mail: Georges.salloum-abou-jaoude@constellium.com

S. Sami · L. Rougier
ESI Group, 70 Rue Robert, 69006 Lyon, France

A. Jacot
Calcom ESI SA, Route Cantonale 105, 1025 St-Sulpice, Switzerland

Introduction

Grain size control during casting of aluminium alloys have always been essential to ensure sound cast quality and guarantee the desired properties of the final product. A non-reliable solidification microstructure control, reduces production yield, resulting in significant value loss. This also increases the plants runaround scrap in an industry where recycling is key. A plant is more than ever seeking to reduce its runaround scrap to maximise external scrap consumption resulting in reduced carbon footprint of the alloy. A well-controlled equiaxed grains size [1, 2], helps in increasing production yields by reducing the risk of hot tearing [3] and shrinkage porosities in cast products. This is a results from enhanced liquid feeding between the solidifying α-Al grains [4].

Although grain refinement in aluminium alloys is well known from the fundamental and experimental point of view, grain size control is still not always guaranteed at the industrial scale. This depends largely on the nature and fabrication quality of the grain refiner rod. There are still debates in the industrial community about what makes a grain refiner less or more efficient.

During the last decades there has been great interest to quantitatively predict solidification microstructures. Several modelling techniques have emerged such as multiphase-models [5–7] and cellular automata [8] which can provide fields of various microstructural quantities on the scale of an entire cast component. On a smaller scale, the phase-field method can now address the growth of columnar or equiaxed grains with a very high level of details about the solid morphology and the distribution of the solute elements in the microstructure including the effect of fluid flow [9, 10]. In spite of this progress, the prediction of grain size remains a modelling challenge. The reason is that the average grain size in a microstructure is the result of a complex competition between nucleation and growth and is governed by phenomena taking place at very different length scales.

© The Minerals, Metals & Materials Society 2023
S. Broek (ed.), *Light Metals 2023*, The Minerals, Metals & Materials Series,
https://doi.org/10.1007/978-3-031-22532-1_130

A major progress in the understanding of grain refinement and quantification of the grain size has been the free growth theory of Greer [11]. One of the outcomes of the theory is that the particle size distribution of the inoculant can be translated into a distribution of nucleation undercoolings. Combining this with a proper growth model, the grain size can be predicted [12]. Models predicting the grain size normally also include a calculation of a thermal recalescence, which will stop nucleation, leaving the less potent particles unused. Another phenomena to also take into consideration, is the solute suppressed nucleation zone (SSNZ), it is well described by Easton and StJohn [13].

Predicting the grain size using a direct representation of the individual grains requires a relatively large computation domain to have enough grains to carry out meaningful statistics. For grain refined alloys, computation domains must typically be in the order of 1 mm to correctly predict the grain size in a given region of the casting. While state-of-the art implementations of the phase-field method [9] would probably allow to address such volumes, the computation cost of the technique remains very high, and prevents any implementation in numerical schemes where such local microstructure calculations would be coupled with heat transfer on the process scale. Another choice would be mean-field approaches, which have been applied with success [8]. A direct description of the grains is however desired when non-random spatial distributions of nucleant particles are analysed, including the complex multiple interactions through their diffusion fields. For these reasons, an envelope model was considered. In this approach, only the grain envelope is directly represented whilst the internal solid morphology is simplified.

In 2020 an experimental method was published [1] allowing to establish a direct link between the nucleant particle size distribution and grain refining efficiency of AlTiB and AlTiC grain refiners during casting of 7xxx alloys. In this work we aim at developing a cellular automaton model that can use as input the experimentally measured grain refiner nucleant particle size distribution and predict the grain refining efficiency during casting of aluminium alloys.

Model Description

The model used in this work is based on the latest CAFE model renovation CAFE2G of ProCAST [15], originally developed at EPFL [8]. The model supports parallel computations, it uses a grid of cells to track the nucleation and growth of grains nucleating either at the mould surface or in the bulk. Nucleation is assumed to be heterogeneous and athermal. The model incorporates the main mechanisms that determine the final grain size: the size and spatial distribution of nucleant particles, the solute diffusion field around growing grains, thermal recalesence and site capture by SSNZ. The model has been designed to be compatible with a resolution of heat flow on the process scale.

For the purpose of this work, the model was revised and a new version was created allowing to take into consideration the principal physical phenomena affecting grain refiner performance:

1. Experimentally measured nucleant particle size distribution of TiB_2 and TiC based grain refiners
2. Growth kinetics
   a. Globular growth
   b. Dendritic growth
   c. Globular to dendritic transition

More details about the model can be found in [15]. The details for the physical laws used in the latest version of the model are described elsewhere [16].

In this paper, we focused on one 7xxx alloy refined at an addition rate of 1 kg/t with two grain refiners on AlTiB and another AlTiC.

The 7xxx alloy properties considered for the presented calculations are deduced from a pseudo binary approximation of the alloy taking into consideration all the alloying elements Zn, Mg, Cu, Zr, Fe, Si, Ti, Mn. The parameters introduced to the model are:

- Liquidus temperature
- Composition
- Diffusion coefficient of the elements in the liquid state
- Liquidus slope
- Partition coefficient
- Eutectic temperature
- Latent Heat
- Heat capacity at constant pressure
- Gibbs Thomson Coefficient

Experimental Data Analysis

To validate the model, we used the experimental procedure explained in TMS2020 [14]. Solidification experiments were carried out in the Cold Finger grain refiner test apparatus on both Al-5Ti-1B and Al-3Ti-0.15C grain refiners. In order to simulate precisely the solidification conditions, the characterisation of temperature distribution in the cold finger test during solidification of 7xxx alloys was also carried out (Fig. 1).

The TiB_2 and TiC particle size distributions were also quantified using a scanning electron microscope (SEM). Commercial grain refiner (cylindrical) rods were cut

Fig. 1 Image to the left shows the location of three thermocouples for thermal analysis of the Cold Finger solidification test. The image on the right shows the temperature curves during solidification of a 7xxx alloy with a liquidus at around 630 °C

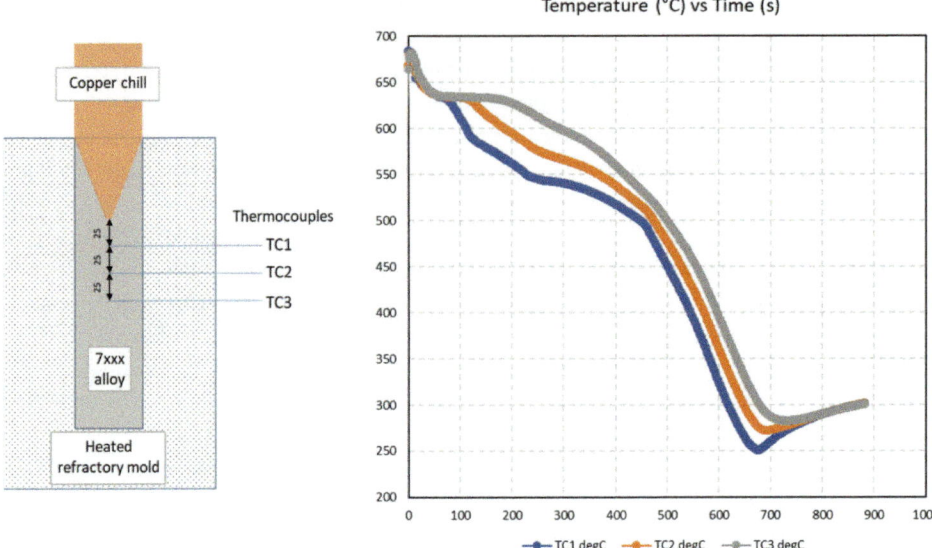

Fig. 2 Graph to the left shows size distribution of TiB_2 and TiC particles in commercial AlTiB and AlTiC grain refiners. The green rectangle shows the range identified using SEM on non-etched samples. The images to the right show examples of the microstructures and the segmentation of the quantified particles on the macro-etched samples

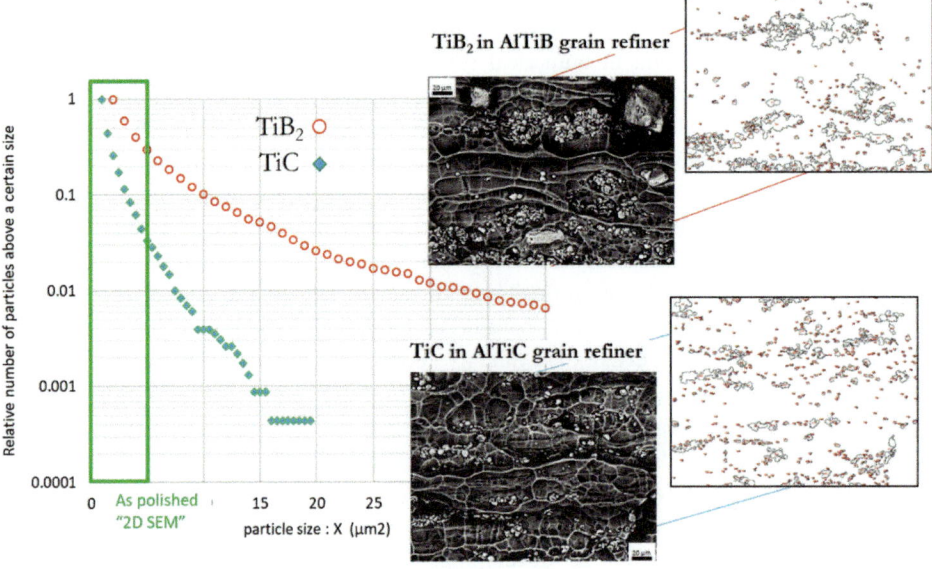

longitudinally along the central axis and polished, then examined under SEM. In order to have precise particle size description, the samples were quantified once under SEM in the as-polished state, and then a second under time after macro-etching in order to reveal the particle agglomerates. The procedure is detailed elsewhere [14], and the results are presented in Fig. 2.

The grain refiner particle size distributions were later fit with lognormal laws. Both the lognormal fit parameters and the temperature curves deduced experimentally were used as inputs in the grain refiner numerical model to predict the grain size after solidification.

Results

In this paper, we focused on calculations done under 0.8 K/s cooling rate simulating the Cold Finger test at a distance of 25 mm from the top. A 7xxx alloy was refined at an addition rate of 1 kg/t with two grain refiners, one AlTiB and another AlTiC.

First, the effect of grain refiner nature will be presented in a globular kinetic growth mode. Then a globular to dendritic criterion will be introduced into the model and its effect on grain size will be investigated and compared with measured experimental results. In the end, the effect of particle agglomerates will be discussed and its introduction into the model will be described and future perspectives are presented.

Globular Growth Kenetics

In this section, a round of numerical calculations is presented. The calculations are done considering globular growth kinetic for the grains. The simulated box is 0.3 cm³. Figure 3 presents the cooling curves (full lines) and the grain number (dashed lines) during solidification. One can see that a lower undercooling is reached with TiC grain refiner that that of TiB₂ grain refiner. This observation is awaited since TiC grain refiner has smaller average particle size and narrower distribution than that of TiB₂ Fig. 2.

The final grain size is deduced and compared to the experimentally obtained grain size in Fig. 4. It seems that the calculations predict well the tendency of grain refinement for TiB₂ and TiC. Furthermore, the TiB₂ calculation result seems to be an excellent prediction of the grain size since the calculated value is in the range of the experimental error bar. For TiC refinement, the calculated grain size is outside the experimental error bar.

Globular to Dendritic Transition

In order to be more precise, the authors investigated a criterion for switching the growth of the grains from globular to dendritic kinetics [16]. A spherical grain destabilizes when its radius exceeds a value multiple of the critical radius of free growth. Thus, based on this hypothesis, the model was

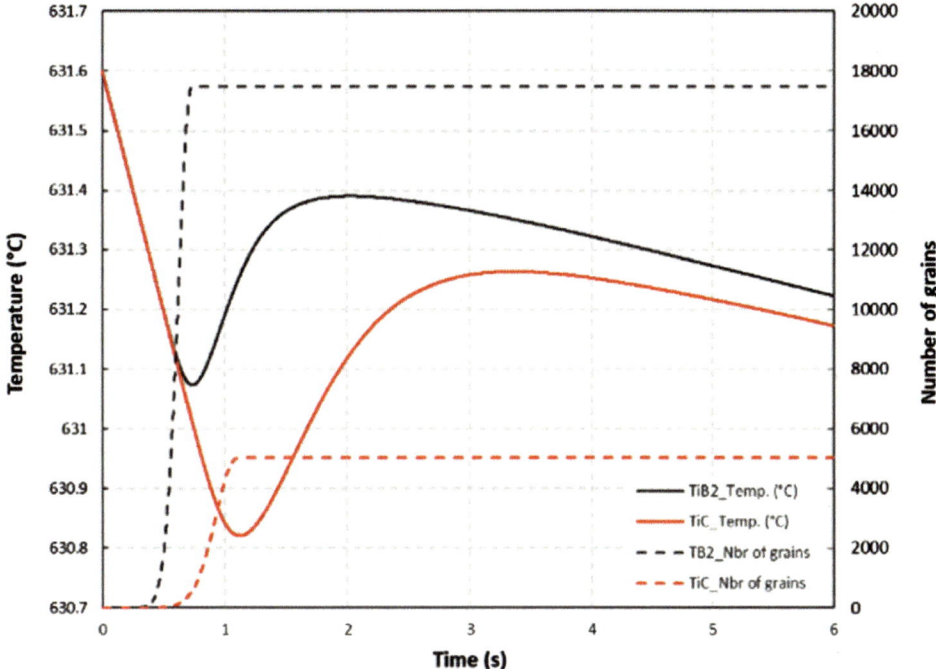

Fig. 3 Cooling curves (full lines) of the calculated 7xxx solidification refined with TiB₂ and TiC based grain refiners. The dashed lines indicate the number of grains as a function of time

We can see from Fig. 3 that the number of grains starts increasing just before reaching the lowest value of undercooling for both TiB₂ and TiC cases. The number of grains reaches its maximum at the maximum of undercooling at which point recalescence happens and no further nucleation events happen. This number of grains depicts the final grain size after solidification, we can already see that for this alloy at the calculated cooling rate using the globular growth kinetic, TiB₂ grain refinement seemed to be more efficient that TiC.

developed to take into consideration the transition from a globular to a dendritic form.

Figure 5 illustrates the adapted globular to dendritic transition criterion. The dotted and full lines represent respectively the critical radius of globular to dendritic transition (R_{gd}) for both TiB₂ and TiC based grain refiners. The dashed lines represent the maximum calculated grain size (R_{max}) during globular solidification. We can see that for the TiB₂ case, the R_{dg} and R_{max} curves never cross, that means that the Globular to Dendritic criterion is never met and the

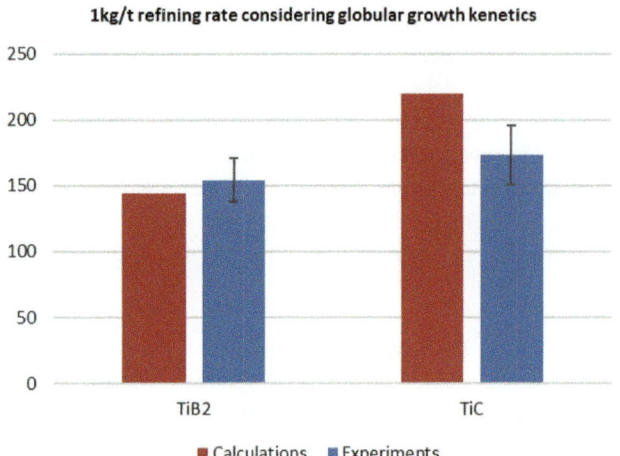

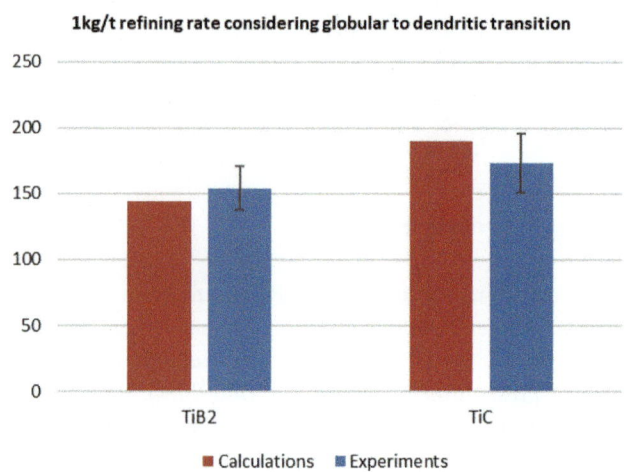

Fig. 4 Comparison between the calculated and the experimental final average grain size of both TiB_2 and TiC refined cases taking into consideration only globular growth kinetics

Fig. 6 Comparison between the calculated and the experimental final average grain size of both TiB_2 and TiC refined cases taking into consideration the Globular to Dendritic transition

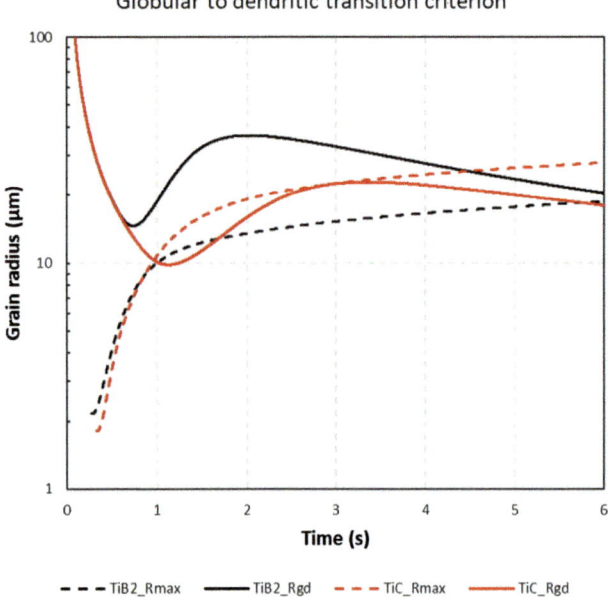

Fig. 5 Globular to dendritic transition criterion illustrated by the intersection of the R_{gd} curve with the Rmax for both grain refiners

growth will continue in globular fashion for the TiB_2 case. One the other hand, for the TiC case, the R_{gd} and the R_{max} curves meet just before the solidification time of 1 s. This means that the globular grains start to destabilise and grow following dendritic growth kinetics, this affects the nucleation phenomena still active since recalescence for the case of TiC did not yet start at this time (Fig. 3).

The model was adapted to include this transition and grain by grain the growth kinetic was switched from

globular to dendritic growth kinetics when the local conditions are met.

Figure 6 shows that the calculated grain size for the case of TiB_2 does not change when comparison with Fig. 4. This is expected since Fig. 5 shows that Globular to Dendritic transition is never attained for this calculated case with TiB_2. On the other hand, in the case of TiC, the calculated grain size drops down to 190 μm from the initially calculated 220 μm (Fig. 4) when only globular growth is considered. This shows that taking into consideration the globular to dendritic transition and switching the growth kinetic during growth is key for precise grain size prediction.

Conclusions

In this work, a new version of cellular automaton CAFE model was developed. This allowed to take into consideration:

- Experimentally measured grain refiner particle size distributions.
- The solidification conditions of the Cold Finger grain refiner test.
- The globular to dendritic transition criterion.

The numerical model showed that the predicted grain size of 7xxx alloy solidification experiments refined with TiB_2 and TiC based grain refiners have good match with experimental data. Including in the model a globular to dendritic transition improved the quantitative grain size prediction especially for the TiC based grain refiner.

The foreseen steps are:

- Adapt the model to take into consideration the physics behind nucleation and growth of grains on particle agglomerates
- Exploit this model to:

Discriminate between low and high efficiency grain refiners.

Design next-generation grain refiners with superior efficiency.

References

1. Jackson KA, Hunt JD, Uhlmann DR, Seward III TP (1966) Transactions of The Metallurgical Society of the American Institute of Mining, Metallurgical and Pet. Trans TMS-AIME 236:149–58
2. Nguyen-Thi H, Reinhart G, Mangelinck-Noël N, et al (2007) In-Situ and Real-Time Investigation of Columnar-to-Equiaxed Transition in Metallic Alloy. Metall Mater Trans A 38:1458–1464. https://doi.org/10.1007/s11661-007-9170-1
3. Rappaz M, Drezet J-M, Gremaud M (1999) A new hot-tearing criterion. Metall Mater Trans A 30:449–455. https://doi.org/10.1007/s11661-999-0334-z
4. Moldovan P (2007) Microporosity Formation in DC cast 5083 Alloy. TMS Light Met 733–737
5. Wang C, Beckermann C. Mat Sci Eng A 171:199
6. Appolaire B, Combeau H, Lesoult G. Mat Sci Eng A 487:33
7. Wu M, Ludwig A (2009). Acta Mater 57:5621
8. Gandin C-A, Desbiolles J-L, Rappaz M, Thevoz P (1999) A three-dimensional cellular automation-finite element model for the prediction of solidification grain structures. Metall Mater Trans A 30:3153–3165. https://doi.org/10.1007/s11661-999-0226-2
9. Boettinger WJ, Warren JA, Beckermann C, Karma A (2002) Phase-Field Simulation of Solidification. Annu Rev Mater Res 32:163–194. https://doi.org/10.1146/annurev.matsci.32.101901.155803
10. Takaki T, Sakane S, Ohno M, et al (2018) Competitive grain growth during directional solidification of a polycrystalline binary alloy: Three-dimensional large-scale phase-field study. Materialia 1:104–113. https://doi.org/10.1016/j.mtla.2018.05.002
11. Greer A, Bunn A, Tronche A, et al (2000) Modelling of inoculation of metallic melts: application to grain refinement of aluminium by Al–Ti–B. Acta Mater 48:2823–2835. https://doi.org/10.1016/S1359-6454(00)00094-X
12. Quested T, Greer A (2004) The effect of the size distribution of inoculant particles on as-cast grain size in aluminium alloys. Acta Mater 52:3859–3868. https://doi.org/10.1016/j.actamat.2004.04.035
13. Easton M, StJohn D (1999) Grain refinement of aluminum alloys: Part I. the nucleant and solute paradigms—a review of the literature. Metall Mater Trans A 30:1613–1623. https://doi.org/10.1007/s11661-999-0098-5
14. Salloum-Abou-Jaoude G, Jarry P, Celle P, Sarrazin E Effect of Nucleant Particle Size Distribution on the Grain Refining Efficiency of 7xxx Alloys | SpringerLink. https://link.springer.com/chapter/https://doi.org/10.1007/978-3-030-36408-3_134. Accessed 9 Sep 2022
15. Jacot A (2020) A cellular automaton approach for the prediction of grain size in grain refined alloys. IOP Conf Ser Mater Sci Eng 861:012061. https://doi.org/10.1088/1757-899X/861/1/012061
16. Rappaz M, Dantzig JA (2016) Chapter 11: Macro-and microstructures. In: Solidification, 2nd ed. pp 489–491

Modelling Contactless Ultrasound Treatment in a DC Casting Launder

Christopher Beckwith, Georgi Djambazov, and Koulis Pericleous

Abstract

Ultrasonic processing can be performed without a vibrating probe by electromagnetic induction with a suitable frequency where resonance conditions can be established. This contactless method is suitable for high-temperature or reactive metal alloys providing purity of the melt and durability of the equipment. Hydrogen bubbles coming out of solution grow by rectified diffusion, and larger bubbles escape from the top surface leading to degassing. Violent collapses of the remaining smaller bubbles help grain refinement. In this study, the application of a contactless 'top-coil' device to continuous casting via a launder is considered. Resonance is achieved by the positioning of baffles on either side of the coil. Electromagnetic forces also cause strong stirring, increasing residence time. The process is modelled using time domain and frequency domain methods, and results for the proposed setup are compared with data obtained for the immersed sonotrode. Accuracy and sensitivity to process and model parameters are discussed.

Keywords

Ultrasonic melt processing • Numerical modelling • Acoustic cavitation • Acoustic resonance

Introduction

Techniques of improving the quality of light alloy metal billets are of high importance, as reducing trapped hydrogen through degassing, grain refinement, and dispersion of metal clusters have been linked with improvements in mechanical properties including tensile strength and ductility [1]. One such method that has been the subject of a significant amount of recent research is the application of ultrasound (UST) while the melt is still in its liquid phase before casting [2]. Traditionally this is performed with the use of an immersed mechanical vibrating sonotrode [2, 3], and the high acoustic pressures directly next to the sonotrode surface result in the rapid growth and then collapse of bubbles through inertial cavitation. Unfortunately, inertially cavitating bubbles attenuate the sound field through thermal and viscous losses, in addition to acoustic radiation, which prevents cavitation from happening further away from the sonotrode surface. An alternative approach which has been the subject of recent research has been to replace the mechanical sonotrode with an AC induction coil [1, 4–6]. This approach relies on resonance to build suitable pressures, and as a result can trigger cavitation deep into the melt, far away from the liquid surface. This could have a number of advantages, including the repositioning of the active zone for maximum processing efficiency (for example, just above the liquid–solid interface during casting, or through the creation of multiple active zones at the antinodes of a standing wave). This is in addition to other, already well established benefits of contactless processing which include the reduction of contamination due to sonotrode erosion, which eventually also results in reduced cost as traditional processing requires frequent sonotrode tip replacements [6]. In addition, contactless treatment also allows for the processing of high temperature (Ni, Fe, Cu) or reactive (Ti, Zr) melts which cannot be processed using a mechanical sonotrode. However, existing work mostly implements contactless UST in small scale experiments in a crucible, and work implementing the treatment in a practical casting process is limited. One study [4] performed initial numerical simulations demonstrating how contactless UST might work with the coil placed directly in the hot top of a DC caster, and showed that it was possible to establish a fixed resonant frequency during casting, due to the impedance mismatch at

C. Beckwith (✉) · G. Djambazov · K. Pericleous
Computational Science and Engineering Group, University of Greenwich, 30 Park Row, London, 10 9LS, UK
e-mail: c.beckwith@greenwich.ac.uk

© The Minerals, Metals & Materials Society 2023
S. Broek (ed.), *Light Metals 2023*, The Minerals, Metals & Materials Series,
https://doi.org/10.1007/978-3-031-22532-1_131

the liquid–solid interface. Electromagnetic stirring then provides suitable mixing, transporting dendrite fragments which might lead to an evenly refined microstructure.

An alternative to processing directly on the hot top is to instead process further up in the launder. Previously, this has been investigated [3, 7, 8] for a mechanical sonotrode, and microstructure grain analysis has shown that this approach can result in more evenly refined grains. Processing in the launder is also less prone to macrosegregation caused by the acoustic streaming jet, and could potentially also be induced by the similar electromagnetic jet induced with the contactless method.

This work attempts to develop this idea further, with full 3D simulations of the fluid flow and solidification, and for the first time attempting to apply contactless UST further up in the casting process, directly in the launder instead of in the hot top.

Problem Description

DC casting of aluminium alloy billets has been carried out at the Advanced Metal Casting Centre (AMCC) of the Brunel Centre of Advanced Solidification Technology (BCAST). These experiments have so far used mechanical sonotrodes, as described in Fig. 1a. These casts used AA6XXX series aluminium with an addition of 0.25 wt% Zr but without the addition of an AlTiB grain refiner. The diameter of the cast billets measured 152 mm. Results from these experiments have previously been presented in [8] with a 5-kW magnetostrictive transducer (Reltec) driven at 17.3 kHz used to power the sonotrode, which has a diameter of 40 mm. The sonotrode tip was immersed 12 mm below the melt surface, oscillating with a peak-to-peak amplitude of 30 μm. Grain analysis showed that by applying UST in the launder, grain size decreased twofold and the presence of feathery grains was suppressed. The presence of partitions in the launders was also linked with increased acoustic pressures and more efficient UST, with the maximum acoustic pressure being twice as high as that without partitions, and RMS pressures increasing by 50%. This was linked with an additional grain refinement of approximately 10% compared to the same experiments without partitions.

For simulations investigating the potential effect of replacing the sonotrode with an AC Induction coil, the partitions also increase the potential for resonance by adding additional geometry for sound reflection, as the location of these partitions can be modified to target particular acoustic frequencies. Numerical simulations are carried out assuming the same experimental conditions and partition configuration as the experiments with a sonotrode, but with the induction coil immersed into the liquid as shown in Fig. 1b.

Modelling Approach

A number of authors have presented models for the frequency domain calculation of the acoustic field including the effect of inertially cavitating bubbles [9–11]. In our previous paper [5], the model of [9] was modified and a nonlinear Helmholtz equation for cavitation problems including the effect of a background source term $F^{\star}$ representing the Lorentz force was obtained. A description of the method is given in this section, but for a full description including derivation, please see the cited work. The equations governing the propagation of the sound field are given in Eqs. (1) and (2), where P represents the complex acoustic pressure field, and k_m^2 is a modified wave number due to the attenuation and change of speed of sound that exists in the presence of inertially cavitating bubbles. The nonlinear Helmholtz equation includes terms which allow for the variation in density at material boundaries (e.g. containing walls), in addition to the density change that occurs during solidification.

$$\nabla\left(\frac{1}{\rho}\nabla P - \frac{F^{\star}}{\rho}\right) + \frac{k_m^2}{\rho}P = 0 \tag{1}$$

$$k_m^2 = \left(\frac{\omega}{c}\right)^2 - \frac{\mathcal{A}(P)}{|P|} - i\frac{\mathcal{B}(P)}{|P|} \tag{2}$$

The dispersion coefficients $\mathcal{A}$ and $\mathcal{B}$ can then be calculated using Eq. (3) [9], which considers only the change in void faction over the last acoustic period.

$$\mathcal{A} = -\frac{\rho_l \omega^2}{\pi}\int_0^{2\pi}\frac{\partial\beta}{\partial\tau}\sin\tau d\tau, \mathcal{B} = -\frac{\rho_l \omega^2}{\pi}\int_0^{2\pi}\frac{\partial\beta}{\partial\tau}\cos\tau d\tau \tag{3}$$

where $\beta = 4/3\pi r^3 N$ is the void fraction and N is the number of bubbles. The void fraction must be computed with a bubble dynamics simulation, and many choices could be suitable for solving the time evolution of bubbles, for example [12, 13]. Here, the Keller-Miksis Equation (KME) as given in [14] is used due to first order compressibility and acoustic radiation terms and is shown here in Eq. (4).

$$\left(1 - \frac{\dot{R}}{C}\right)R\ddot{R} + \frac{3}{2}\dot{R}^2\left(1 - \frac{\dot{R}}{3c}\right)$$
$$= \frac{1}{\rho_l}\left(1 + \frac{\dot{R}}{c} + \frac{R}{c}\frac{d}{dt}\right)[p_l - p(t)] \tag{4}$$

where p_l represents the liquid pressure at the liquid gas interface and is defined by Eq. (5), where σ_e is the surface tension, μ is the liquid viscosity, and p_g is the pressure in the

Fig. 1 **a** A typical setup for ultrasonic melt treatment in the launder using a mechanical sonotrode. **b** The alternative setup uses an AC induction coil. Partitions are placed in the launder for both cases

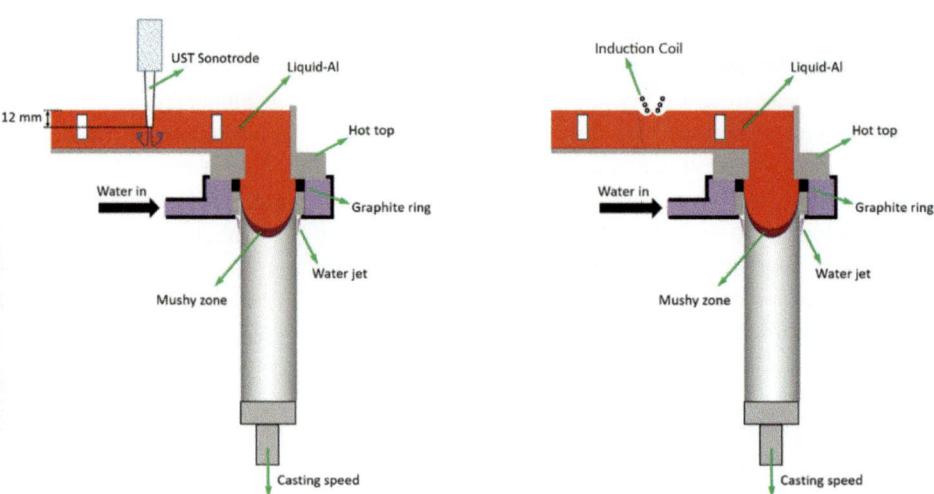

gas at the interface, which can be assumed to follow the adiabatic equation of state [15] given by Eq. (6).

$$p_l = p_g - \frac{2\sigma_e}{R} - \frac{4\mu\dot{R}}{R} \tag{5}$$

$$p_g = p_{g0}\left(\frac{R_0}{R}\right)^{3\gamma} \tag{6}$$

Here, $\gamma = 1.4$ is the polytropic exponent, and p_{g0} the initial gas pressure in the bubble. A background pressure $p(t) = p_0(1 - A\sin(\omega t))$ accounts for both the sinusoidal acoustic pressure with dimensionless amplitude A, and atmospheric pressure. It is important to run the single bubble model for more than one acoustic period so that simulation can converge to a harmonic solution, and the number of cycles needed increases with frequency and driving amplitude [9]. In the simulations in this paper, 500 cycles are chosen as the cut-off point at which if a harmonic solution has not been obtained, interpolation with cubic splines is used instead.

N is assumed to be a smoothed stepwise function $W(|P|)$ centred on the Blake pressure, with a smoothing distance equal to $P_{\text{blake}} - P_{\text{rd}}$, the difference between the Blake threshold and the rectified diffusion threshold, the acoustic pressure required for bubbles to begin growing. Under this pressure, hydrogen bubbles begin dissolving back into solution and do not significantly influence the acoustic field. Assuming that the driving frequency is far from the bubble resonant frequency, which is generally the case in metal processing. The resulting threshold is given by Eq. (7) [5], where C_i is the concentration of hydrogen in the bulk fluid, C_0 is the saturation concentration, η is the polytropic coefficient.

$$P_{\text{rd}}^2 = \frac{(\rho R_0^2 \omega_0^2)^2 \left[(1 - \omega^2/\omega_0^2)^2 + b^2(\omega^2/\omega_0^2)\right](1 + 2\sigma/R_0 P_\infty - C_i/C_0)}{(3 + 4K)(C_i/C_0) - \left\{\left[\frac{3}{4}(\eta - 1)(3\eta - 4)\right] + (4 - 3\eta)K\right\}(1 + 2\sigma/R_0 P_\infty)} \tag{7}$$

For the fluid flow simulation, Eqs. (8) and (9) describe the continuity and momentum equations. An additional term F_s is included to model the additional stirring due to the influence of a background field. For the top mounted induction coil, this term represents the induced Lorentz forces, the mean component of which drives the main fluid flow. For an immersed sonotrode, this term can be used to represent acoustic streaming, which has been shown in previous work [3, 7, 8]. The Lorentz forces induced by the coil are calculated from a separate simulation using Comsol's Magnetic Fields solver. Solidification at the mould is then included through the use of a continuum approach described in [16] and previously used in [7]. The continuum approach adds a Carman–Kozeny momentum sink term S_d that forces the fluid velocity to the background velocity v_{ref}. This was implemented using a custom OpenFOAM solver based on the included "buoyantPimpleFoam" solver (a combination of the PISO and SIMPLE algorithms). The electromagnetic source terms were exported from Comsol in CSV format and then interpolated onto the OpenFOAM mesh.

$$\nabla \cdot v = 0 \tag{8}$$

$$\rho_0 \frac{\partial v}{\partial t} + \rho_0 \nabla \cdot (vv) = -\nabla p + \mu_0 \nabla^2 v + \rho_0 g \beta_T (T - T_{\text{ref}}) \\ + S_d + F_s \tag{9}$$

where ρ_0 and μ_0 are the fluid density and the dynamic viscosity. Turbulence is included in the model, using the k-Omega-SST turbulence model. The system is closed by the energy balance, as given in Eq. (10).

$$\rho c_p \frac{\partial T}{\partial t} + \rho c_p \nabla \cdot (vT) = \nabla \cdot (k\nabla T) \\ - \rho_0 L_f \left[\frac{\partial g_l}{\partial t} + \nabla \cdot (vg_l)\right] \tag{10}$$

where c_p is the specific heat, T is the temperature, and k is the thermal conductivity, L_f the latent heat of fusion, and g_l the volume fraction of liquid. In the slurry, the effective dynamic viscosity μ_{eff} can be calculated from the Stefanescu formula [18, 19] given in Eqs. (11, 12), where μ_l is the liquid viscosity, f_s is the solid fraction, and f_c is the dendrite coherency point, chosen to be 0.3.

$$\mu_{eff} = \mu_l \left(\frac{1}{1 - F_\mu f_s / f_c} \right)^2 \qquad (11)$$

$$F_\mu = 0.5 - (1/\pi)\tan^{-1}(100(f_s - f_c)) \qquad (12)$$

At the water spray, the heat transfer can be described by a Fourier boundary condition with a heat flux function depending on the average temperature $\overline{T}$ between the surface of the billet and the bulk fluid [21]. Including the effect of nucleate boiling above a critical point $q_c = 3910\Delta T^{2.16}$ the heat transfer coefficient takes the form of Eq. (13).

$$h_c = \begin{cases} [-1.67 \times 10^5 + 704\overline{T}] \cdot Q'^{1/3}, & \text{if } q_c \geq q_i \\ [-1.67 \times 10^5 + 704\overline{T}] \cdot Q'^{1/3} + \frac{20.8(\Delta T_x)^3}{\Delta T}, & \text{if } q_c < q_i \end{cases} \qquad (13)$$

At the free surface, a surface radiation boundary condition is used and is given by Eq. (14), where $\epsilon = 0.3$ is the surface emissivity and $\sigma = 5.6708 \times 10^{-8}$ the Stefan–Boltzmann constant.

$$\nabla T = \frac{\epsilon \sigma (T_{amb}^4 - T^4)}{k} \qquad (14)$$

Results

Following the solution procedure described in [5], target acoustic Eigenfrequencies are first calculated from a linear model (achieved by setting N = 0), and are then used as an initial guess for the resonant frequency including the effect of cavitation. The frequency is then adjusted until resonance is achieved in the non-linear case. Target frequencies were taken only if they were below 30 kHz, to target specific modes that could be obtainable by the AC coil. A full list of material properties for the liquid aluminium is given in Table 1, and the properties of the copper induction coil are given in Table 2.

In the acoustic simulation, a sound soft ($P = 0$) boundary condition is used on the top surface, and sound hard boundaries are used elsewhere. At the inlet and outlet of the launder, a perfectly matched layer was used to prevent reflections, in an attempt to focus on resonant modes present due to partitions and to prevent unnatural resonant

Table 1 Model properties of liquid aluminium. Properties obtained from [15]

Electrical conductivity (S/m)	4e6
Relative permeability	1
Relative permittivity	1
Casting velocity (m s^{-1})	0.0023
Inlet temperature (K)	1003.15
Liquidus temperature (K)	929.2
Solidus temperature (K)	757.4
Latent heat (J kg^{-1})	375,696.0
Density (kg m^{-3})	2375
Speed of sound (m s^{-1})	4600
Thermal expansion coefficient (K^{-1})	2.3×10^{-5}
Kinematic viscosity (m^2 s^{-1})	5.5×10^{-7}

Table 2 Properties of the copper induction coil

Electrical conductivity (S/m)	5.998e7
Relative permeability	1
Relative permittivity	1
Density (kg m^{-3})	8700

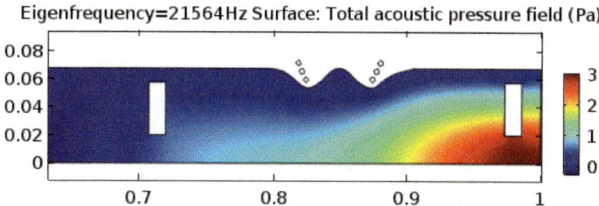

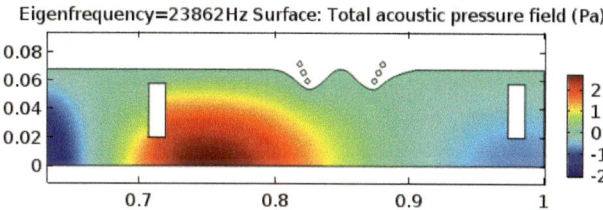

Fig. 2 Obtained Eigenfrequencies from linear theory potentially obtainable by the AC coil. **a, b** The alternative setup uses an AC induction coil. Partitions are placed in the launder for both cases

frequencies that might occur due to not simulating the entire launder. Figure 2 shows two chosen eigenfrequencies for the launder with partitions. The two Eigenfrequencies occurred at 21564 and 23862 Hz, corresponding to electrical frequencies of 10782 and 11931 Hz and high pressure regions were located close to the partitions. For the rest of the results in this work, the 21564 Hz mode will be the target mode that we will use.

The nonlinear Helmholtz solver is then used to compute the acoustic field in the presence of cavitating bubbles. The

Table 3 Hydrogen bubble properties [15]

Bubble density N	5e7
Surface tension (N m^{-1})	0.860
Vapour pressure (Pa)	0
Specific heat capacity (J kg^{-1} K^{-1})	717
Bulk temperature (K)	1013.15
Ambient bubble radius R_0 (m)	10×10^{-6}

properties of these hydrogen bubbles are given in Table 3. Due to the reduction in speed of sound that occurs with an increase in bubble volume fraction, the resonant frequency is slightly lower than that of the eigenfrequency study, and the closest frequency is calculated to be 21320 Hz, a shift of 244 Hz. Results for this case are shown in Fig. 3b. Peak pressures induced by the coil were approximately 180 kPa, well above the threshold for cavitation for 10 micron hydrogen bubbles in liquid aluminium (calculated to be approximately 154 kPa [5]). The traditional mechanical sonotrode is capable of reaching much higher pressures, up to 600 kPa directly below the sonotrode surface, but the active region is isolated to the area directly under the sonotrode, with acoustic shielding preventing higher pressures elsewhere in the domain. This can be seen in Fig. 3a. The top coil by comparison resulted in an active zone much deeper below the surface, located in the gap between the downstream partition and the base of the launder.

Figure 4 shows the induced magnetic field from the AC induction coil at an electrical frequency of 10660 Hz. The coil interacts with induction currents in the melt to repel the free surface, ensuring that no contact is made with the metal. At the electromagnetic skin layer, the magnetic flux density reaches a peak of 0.11 T, comparable to that in previous work in a crucible [5]. The interacting magnetic and electric fields result in a force on the aluminium $F = J \times B$. This force has a mean component given in Fig. 4b which drives

the bulk fluid flow, and is used as the source term F_s in Eq. (9), and a harmonic oscillating component given in Fig. 4c, which is responsible for the acoustic field source term $F^{\star}$. The magnitude of the mean part was found to be lower than the amplitude of the harmonic part, with peak amplitudes of the order 2e6 and 6e6 respectively.

The fluid flow induced by the sonotrode reached a peak velocity of approximately 1 m/s directly below the sonotrode, with the acoustic streaming jet having a velocity closer to 0.6 m/s. Flow patterns for this case can be seen in Fig. 5a. The top coil on the other hand, which does not rely on acoustic streaming but instead the electromagnetic Lorentz forces, reached a peak velocity of approximately 2.5 m/s, 2.5× what was obtained by the sonotrode. This results in very strong mixing and is likely to result in a very even distribution of processed particles, which has previously been linked with improved uniformity in grain size across the final billet [7]. This could then further be improved by designing a geometry such that the resonant mode, and therefore the active processing region, be located to increase residence time before mixing, as in these simulations the active zone primarily acts downstream, which could limit the effectiveness of treatment. The resulting sump profile showing the solidification of the billet in the mould and temperature contours are given in Fig. 5c.

Integrated Cylindrical Vessel

An alternative geometry using a cylindrical vessel integrated into the launder is presented in this section. This concept hopes to improve the efficiency of processing by encouraging a central resonant mode, with partitions placed to restrict flow in and out of the isolated cylindrical vessel. A diagram of this concept is given in Fig. 6.

Initial simulations were carried out on the cylindrical vessel, carried out in 3 stages. Initially, an electromagnetics solver computes the magnetic and electric fields, then the

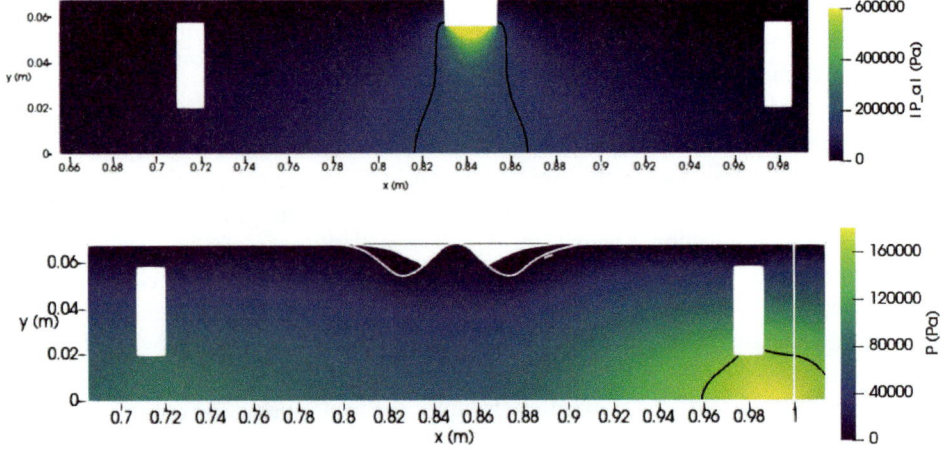

Fig. 3 The induced acoustic field including the effect of inertially cavitating bubbles, using an immersed sonotrode operating at 17300 Hz (**a**) and a top mounted induction coil operating at 10660 Hz (an acoustic frequency of 21320 Hz) (**b**). The black line in both figures indicates the active processing zone above the Blake threshold

Fig. 4 **a** Induced magnetic field from the AC induction coil at 10660 Hz, in addition to the magnitude of the resulting mean Lorentz force (**b**) and harmonic component (**c**)

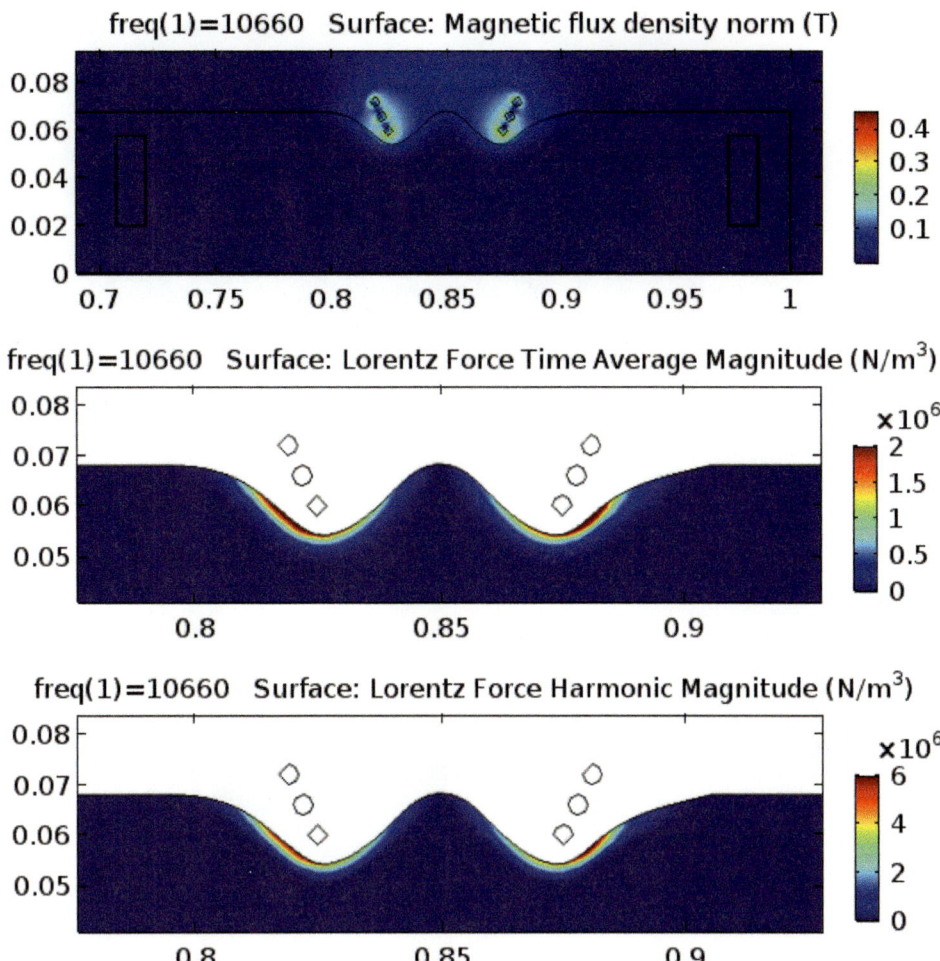

fluid flow, followed by the acoustics. For the electromagnetics and free surface model, Maxwell's equations in electromagnetic induction approximation are solved numerically via a formulation based on the electric field vectors (real and imaginary parts) and Biot-Savart integral [17]. The 3-turn top coil was modelled as 3 axisymmetric rings each carrying a current of 1700A at 14,580.4 Hz. This frequency was chosen as the lowest for which the periodic component of the Lorentz force induces acoustic resonance in the metal volume with a radius of 96.4 mm and a height of 100 mm. The coils are lowered from their initial positions where the lowest turn is 5 mm above the metal surface in steps of 1.25 mm and the computational mesh is deformed according to the mean pressure exerted by the Lorentz force until a total displacement of 17.5 mm (Fig. 7) is reached—just before the hydrostatic pressure of the liquid metal under the deformed free surface becomes too high for the generated Lorentz force to balance. Both the mean and the periodic components of the Lorentz force vectors are saved for the next stages.

The flow of the liquid metal is then caused by the mean component of the Lorentz force. Figure 8 shows the result of

a CFD simulation with a $k-\varepsilon$ turbulence model carried out in the PHYSICA code. The calculated maximum velocity in the downward jet is 0.38 m/s, comparable to typical flow patterns induced by an immersed sonotrode. This is less than the 2.5 m/s jet obtained by the frequency domain simulation in the previous section, but this can largely be explained by the coil operating at 1700A as opposed to 2000A, which also reduces the induced Lorentz forces.

Acoustic cavitation of any hydrogen bubbles formed in the melt is driven by the periodic part of the electromagnetic Lorentz force. It is simulated with a bespoke piece of software [5] capturing the time-dependent behaviour of both the acoustic field and the bubbles dispersed in the metal volume. 32,366 acoustic cycles were simulated with time step 1.75×10^{-7} s, taking about 5 h on a 32-processor workstation. The acoustic algorithm [18] needs a regular cartesian grid, so the depression of the free surface is not modelled within the acoustic part, but the total volume of liquid metal is retained. The equilibrium bubble radius, i.e. the radius without acoustic excitation, is assumed to be 7 μm and the bubble concentration is 1×10^7 m^{-1}. Acoustically soft (zero pressure amplitude) boundary conditions were

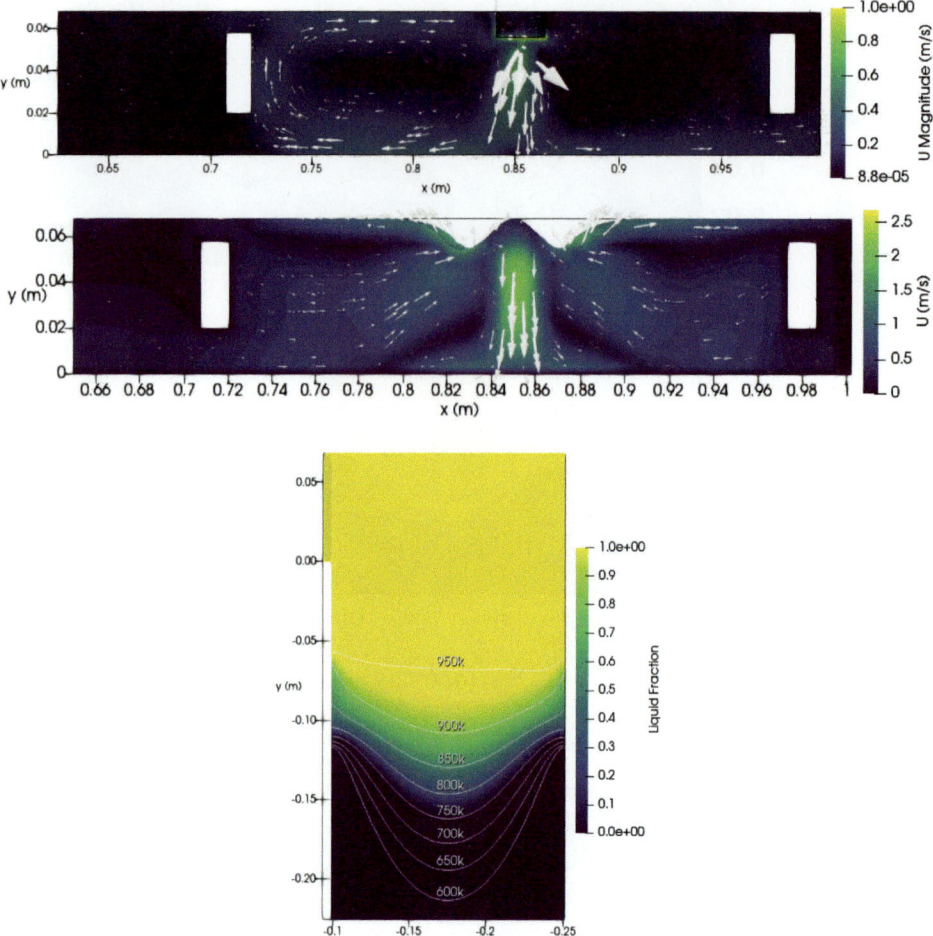

Fig. 5 The fluid flow induced by the acoustic streaming generated traditional mechanical sonotrode (**a**), Lorentz forces generated by the induction coil (**b**), and the sump profile in the mould with temperature contours (**c**)

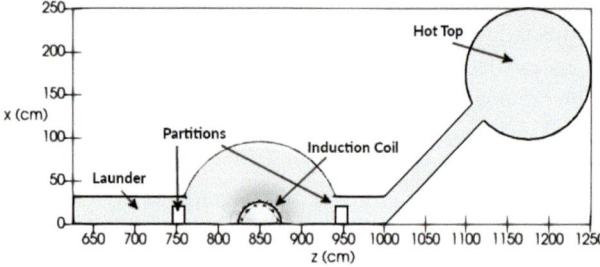

Fig. 6 Launder concept which aims to improve contactless UST processing efficiency by integrating an additional cylindrical vessel into the launder

assumed on all sides of the liquid metal volume. The simulated acoustic field as shown in Fig. 9 (for RMS pressure *p*) shows a spheroidal zone with an approximate radius of 2 cm around the centre of the domain which is above the Blake threshold for inertial cavitation (130 kPa RMS in the

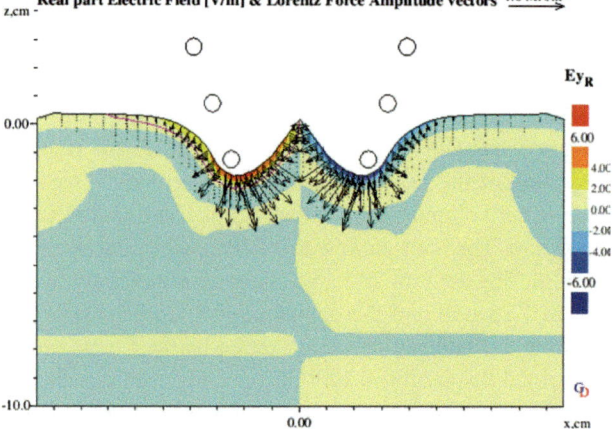

Fig. 7 Vertical central cross-section of the liquid metal volume and top coil showing the deformed free surface, electric field contours and amplitude of the periodic component of the Lorentz force vectors; 'real part' means in phase with the driving coil electrical current

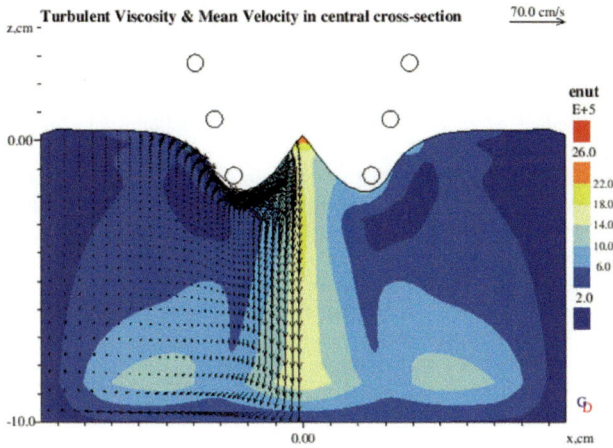

Fig. 8 CFD result for the mean flow driven by the steady component of the Lorentz force; contours—turbulent viscosity, vectors—flow velocity

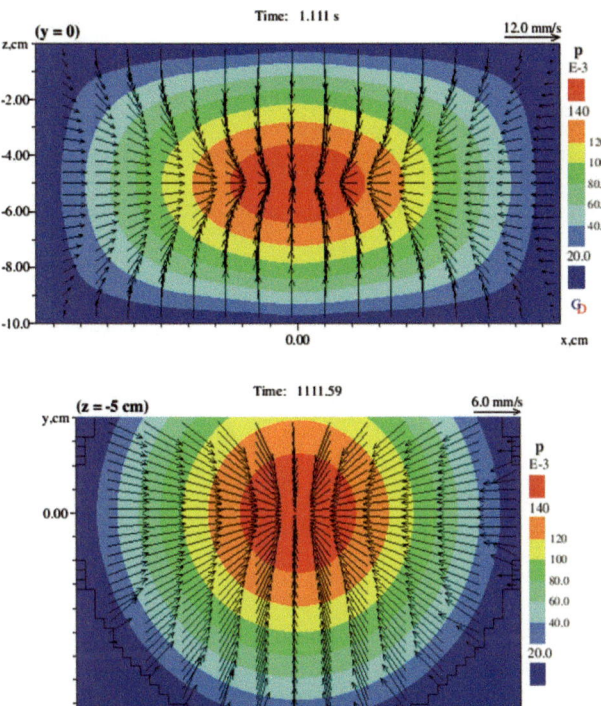

Fig. 9 Vertical and horizontal sections of the acoustic cavitation field showing the RMS acoustic pressure averaged over 0.1 s just before the end of the simulated 1.11 s

simulated case) and is the necessary condition for cavitation treatment of the melt. A Lorentz force frequency in the acoustic simulation was set at 29,132 Hz, but its amplitude was taken from the electromagnetic results. One can see the prescribed frequency is not exactly twice the electrical (as expected theoretically) and the slight shift is needed to

sustain resonance under the bubble action which alters the effective speed of sound in the liquid metal. This is important for practical applications as automatic frequency tuning will be needed.

Conclusions

A coupled model for simulating the acoustic field in the presence of bubbles, electromagnetic induction from an AC coil, and fluid flow has been presented and has been applied to the potential application of contactless UST processing in the launder. It has been shown numerically that it is possible to obtain pressures required for processing as the Blake threshold was exceeded deep into the launder close to the partitions. This mode was obtained at 10660 Hz electrical frequency, corresponding to an acoustic frequency of 21320 Hz, which is obtainable by the coil. In addition, induced flow velocities were 2.5× higher in the launder when using the Lorentz force, as opposed to traditional UST processing with a mechanical sonotrode. This could have benefits for enhanced mixing and could be used to evenly distribute processed particles that would result in an even distribution of refined grains in the final billet. However, as this was not tested further theoretical and experimental work needs to be carried out to determine the exact effect that this will have on grain refinement.

In addition, an alternative launder concept has been suggested that might be better suited for contactless UST. The new concept integrates a cylindrical vessel into the launder, and a resonant mode can then be calculated which results in a large active zone directly in the middle of the vessel. Induced flow velocities for this system are comparable to that obtained by traditional UST with a sonotrode but needed a higher acoustic frequency of 29132 Hz to obtain.

References

1. Dybalska, A.; Caden, A.; Griffiths, W.D.; Nashwan, Z.; Bojarevics, V.; Djambazov, G.; Tonry, C.E.H.; Pericleous, K.A. Enhancement of Mechanical Properties of Pure Aluminium through Contactless Melt Sonicating Treatment. *Materials* **2021**, *14*, 4479, doi:https://doi.org/10.3390/ma14164479.
2. Eskin, G.I.; Eskin, D.G. *Ultrasonic Treatment of Light Alloy Melts, Second Edition*; CRC Press, 2014; ISBN 978-1-4665-7798-5.
3. Subroto, T.; Eskin, D.G.; Beckwith, C.; Skalicky, I.; Roberts, D.; Tzanakis, I.; Pericleous, K. Structure Refinement Upon Ultrasonic Melt Treatment in a DC Casting Launder. *JOM* **2020**, *72*, 4071–4081, doi:https://doi.org/10.1007/s11837-020-04269-3.
4. Tonry, C.E.H.; Bojarevics, V.; Djambazov, G.; Pericleous, K. Contactless Ultrasonic Treatment in Direct Chill Casting. *JOM* **2020**, *72*, 4082–4091, doi:https://doi.org/10.1007/s11837-020-04370-7.

5. Beckwith, C.; Djambazov, G.; Pericleous, K.; Tonry, C. Comparison of Frequency Domain and Time Domain Methods for the Numerical Simulation of Contactless Ultrasonic Cavitation. *Ultrasonics Sonochemistry* **2022**, *89*, 106138, doi:https://doi.org/10.1016/j.ultsonch.2022.106138.

6. Pericleous, K.; Bojarevics, V.; Djambazov, G.; Dybalska, A.; Griffiths, W.D.; Tonry, C. Contactless Ultrasonic Cavitation in Alloy Melts. *Materials* **2019**, *12*, 3610, doi:https://doi.org/10.3390/ma12213610.

7. Beckwith, C.; Djambazov, G.; Pericleous, K.; Subroto, T.; Eskin, D.G.; Roberts, D.; Skalicky, I.; Tzanakis, I. Multiphysics Modelling of Ultrasonic Melt Treatment in the Hot-Top and Launder during Direct-Chill Casting: Path to Indirect Microstructure Simulation. *Metals* **2021**, *11*, 674, doi:https://doi.org/10.3390/met11050674.

8. Subroto, T.; Eskin, D.G.; Beckwith, C.; Roberts, D.; Tzanakis, I.; Pericleous, K. Effect of Flow Management on Ultrasonic Melt Processing in a Launder upon DC Casting. In Proceedings of the Light Metals 2022; Eskin, D., Ed.; Springer International Publishing: Cham, 2022; pp. 649–654.

9. Trujillo, F.J. A Strict Formulation of a Nonlinear Helmholtz Equation for the Propagation of Sound in Bubbly Liquids. Part I: Theory and Validation at Low Acoustic Pressure Amplitudes. *Ultrasonics Sonochemistry* **2018**, *47*, 75–98, doi:https://doi.org/10.1016/j.ultsonch.2018.04.014.

10. Louisnard, O. A Simple Model of Ultrasound Propagation in a Cavitating Liquid. Part I: Theory, Nonlinear Attenuation and Traveling Wave Generation. *Ultrasonics Sonochemistry* **2012**, *19*, 56–65, doi:https://doi.org/10.1016/j.ultsonch.2011.06.007.

11. Jamshidi, R.; Brenner, G. Dissipation of Ultrasonic Wave Propagation in Bubbly Liquids Considering the Effect of Compressibility to the First Order of Acoustical Mach Number. *Ultrasonics* **2013**, *53*, 842–848, doi:https://doi.org/10.1016/j.ultras.2012.12.004.

12. Plesset, M.S. The Dynamics of Cavitation Bubbles. *Journal of Applied Mechanics* **1949**, *16*, 277–282, doi:https://doi.org/10.1115/1.4009975.

13. Löfstedt, R.; Barber, B.P.; Putterman, S.J. Toward a Hydrodynamic Theory of Sonoluminescence. *Physics of Fluids A: Fluid Dynamics* **1993**, *5*, 2911–2928, doi:https://doi.org/10.1063/1.858700.

14. Keller, J.B.; Miksis, M. Bubble Oscillations of Large Amplitude. *The Journal of the Acoustical Society of America* **1980**, *68*, 628–633, doi:https://doi.org/10.1121/1.384720.

15. Harkin, A.; Nadim, A.; Kaper, T.J. On Acoustic Cavitation of Slightly Subcritical Bubbles. *Physics of Fluids* **1999**, *11*, 274–287, doi:https://doi.org/10.1063/1.869878.

16. Voller, V.R.; Prakash, C. A Fixed Grid Numerical Modelling Methodology for Convection-Diffusion Mushy Region Phase-Change Problems. *International Journal of Heat and Mass Transfer* **1987**, *30*, 1709–1719, doi:https://doi.org/10.1016/0017-9310(87)90317-6.

17. Djambazov, G.; Bojarevics, V.; Pericleous, K.; Croft, N. Finite Volume Solutions for Electromagnetic Induction Processing. *Applied Mathematical Modelling* **2015**, *39*, 4733–4745, doi:https://doi.org/10.1016/j.apm.2015.03.059.

18. Lebon, G.S.B.; Tzanakis, I.; Djambazov, G.; Pericleous, K.; Eskin, D.G. Numerical Modelling of Ultrasonic Waves in a Bubbly Newtonian Liquid Using a High-Order Acoustic Cavitation Model. *Ultrasonics Sonochemistry* **2017**, *37*, 660–668, doi:https://doi.org/10.1016/j.ultsonch.2017.02.031.

Numerical Analysis of Channel-Type Segregations in DC Casting Aluminum Slab

Keisuke Kamiya and Takuya Yamamoto

Abstract

In the direct chill casting of aluminum alloys, the stripe-shaped segregation called channel-type segregations is formed in the slab, but the mechanism of their formation is not clear, and the casting conditions under which the segregation is minimized have not been found. In this study, it is reported that a numerical simulation model for the segregation have been developed, and the segregation distribution in the Al–Mg alloy slab was numerically analyzed. As a result, the segregation similar to that observed in actual slabs was reproduced on numerical analysis. This simulation results showed that the channel-type segregations could be suppressed by colliding the strong down flow with the solidification front.

Keywords

Aluminum • DC casting • Segregation • Numerical analysis

Introduction

In the semi-continuous casting process called direct chill (DC) casting of aluminum alloys, the segregation of solute concentration occurs in the slab. Negative segregation is observed in the center region of the slab, and stripe-shaped segregation, called channel-type segregation is observed around the center region of the slab (Fig. 1). Since these segregations cause the changes in the mechanical and chemical properties of the final product, they have to be controlled. However, their formation mechanism remains unclear up to now.

In response to such problems, we have attempted to clear the formation mechanism of channel-type segregation by numerical simulation [1]. Figure 2 shows the results of numerical analysis of solute concentration in DC casting billet. This is the first example in which the channel-type segregation was reproduced numerically in the DC casting model for aluminum alloys. This simulation results allow us to explain in detail how the channel-type segregation is formed during the casting process.

Figure 3 shows the simulation results of Mg concentration around the solidification front in DC casting billet [1]. First, as the molten metal solidifies, a zone with high concentration of Mg (a in Fig. 3) is created at the solidification front, and the concentration of Mg causes occurrence of the upward solutal buoyancy flow along solidification front (b) resulting in the solidification delay (c). As a result, Mg is transferred from the low concentration zone toward the high concentration zone of delayed solidification due to Mg partition (d) between the solid and liquid phases. In this way, the low concentration zones (e) are formed below the high concentration zones. This segregation cycle is repeated, resulting in the formation of channel-type segregation.

As described, the mechanism of channel-type segregation has been cleared, however, casting conditions to suppress its occurrence have not yet been found. Therefore, the aim in this paper is to find a casting conditions to suppress the channel-type segregation in DC casting slab of aluminum alloys by the numerical simulation.

K. Kamiya (✉)
UACJ Corporation, 3-1-12 Minato-Ku, Chitose Nagoya, 455-8670, Aichi, Japan
e-mail: kamiya-keisuke@uacj.co.jp

T. Yamamoto
Tohoku University, 6-6-02 Aza-Aoba, Aramaki, Aoba-ku Sendai, 980-8579, Miyagi, Japan
e-mail: t-yamamoto@tohoku.ac.jp

© The Minerals, Metals & Materials Society 2023
S. Broek (ed.), *Light Metals 2023*, The Minerals, Metals & Materials Series,
https://doi.org/10.1007/978-3-031-22532-1_132

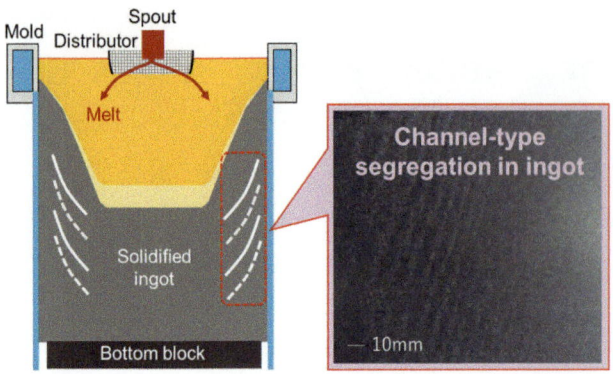

Fig. 1 Schematic view of DC casting and observed segregation on etched slab sample

Numerical Simulation Method

Governing Equation

The numerical simulation model in this study is based on the model proposed by Vreeman et al. [2], and the modified model by Fezi et al. [3], which basically consists of four basic governing equations for the balance of momentum, mass, enthalpy and chemical species concentration with solidification.

Fig. 2 Numerical analysis result of Al-2.5mass%Mg billet [1]

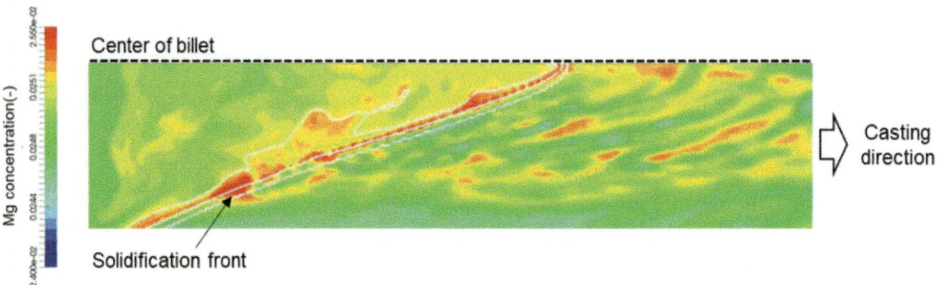

Fig. 3 Formation process of channel-type segregation shown by numerical analysis [1]

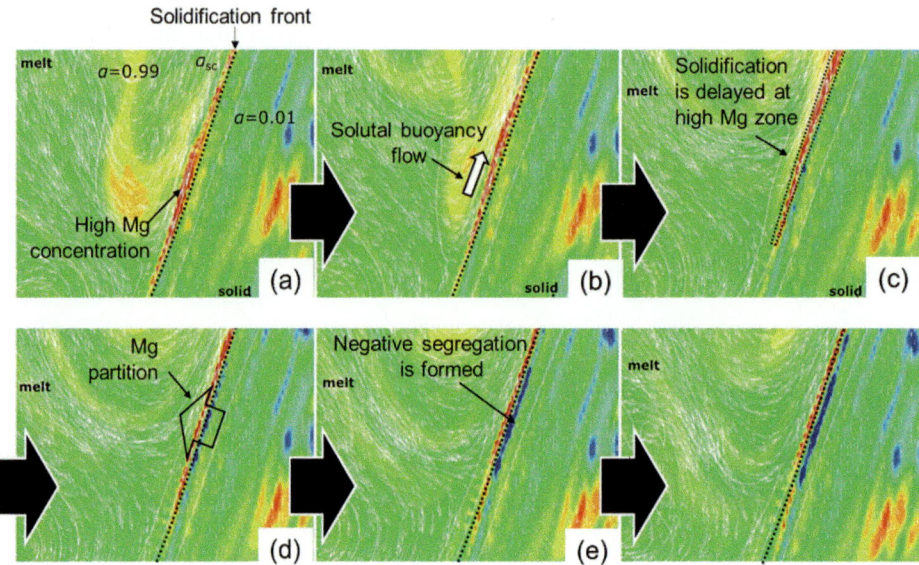

Each equation is given as

$$\frac{\partial(\rho \boldsymbol{u})}{\partial t} + \nabla(\rho \boldsymbol{u}\boldsymbol{u}) = -\nabla p + \nabla \cdot \left(\frac{\mu_l}{\rho_l}\nabla \boldsymbol{u}\right)$$
$$- \rho g(\beta_T(T - T_0) + \beta_C(C - C_0)) \quad (1)$$
$$+ (1 - P)S_{Slurry} + PS_{rigid}$$

$$\frac{\partial \rho}{\partial t} + \nabla \cdot (\rho \boldsymbol{u}) = 0 \quad (2)$$

$$\frac{\partial(\rho h)}{\partial t} + \nabla \cdot (\rho \boldsymbol{u}h) = \nabla \cdot \left(\frac{k}{c_{ps}}\nabla h\right)$$
$$+ \nabla \cdot \left(\frac{k}{c_{ps}}\nabla(h_s - h)\right) \quad (3)$$
$$- \nabla \cdot (\rho(\boldsymbol{u} - \boldsymbol{u}_s)(h_l - h))$$

$$\frac{\partial(\rho C)}{\partial t} + \nabla \cdot (\rho \boldsymbol{u}C) = \nabla \cdot (\alpha_l \rho_l D_l \nabla C)$$
$$+ \nabla \cdot (\alpha_l \rho_l D_l \nabla(C_l - C)) - \nabla \cdot (\rho(\boldsymbol{u} - \boldsymbol{u}_s)(C_l - C)) \quad (4)$$

where ρ is the density, $\boldsymbol{u}$ is the flow velocity, t is the time, p is the pressure, μ is the viscosity, $\boldsymbol{g}$ is the gravitational acceleration, β_T is the volume thermal expansion coefficient, β_C is the volume solutal expansion coefficient, T_0 is the reference temperature, C_0 is the reference composition, P is the packing variable, S_{Slurry} is the source term for mushy zones, S_{rigid} is the source term for solid zones, h is the enthalpy, k is the thermal conductivity, c_p is the heat capacity of solid, C is the composition, α is the volume fraction, D is the diffusion coefficient, the subscripts s and l indicate the solid and liquid phases, respectively.

The model used of the slurry and solid parts was that proposed by Plotkowski and Krane [4], Coleman and Krane [5]. The packing variable, P, can be written as

$$P = \min\left(\max\left(\left(1 - \frac{\alpha_s^c - \alpha_s}{\Delta\alpha_s}\right), 0\right), 1\right) \quad (5)$$

α_s^c is the critical volume fraction of solid phase, and $\Delta\alpha_s$ is the steepness of phase transition. If α_s is greater than 0 and less than α_s^c, then P is 0. If α_s is greater than α_s^c and less than 1, then P is 1.

The source term for mushy and solid zones can be written by

$$S_{Slurry} = -\nabla \cdot \left(\mu_l \frac{\rho f_s}{\rho_l}\nabla \boldsymbol{u}_s\right) + \nabla \cdot (\bar{\mu}_s(1 - \alpha)\nabla \boldsymbol{u}_s) - \nabla$$
$$\cdot \left[\left(\frac{\rho f_s}{\rho_l}\right)(\boldsymbol{u} - \boldsymbol{u}_s)(\boldsymbol{u} - \boldsymbol{u}_s)\right] \quad (6)$$

$$S_{rigid} = -\frac{\mu_l}{K}\frac{\rho}{\rho_l}(\boldsymbol{u} - \boldsymbol{u}_s) \quad (7)$$

where $\bar{\mu}_s$ is the effective solid viscosity, f is the mass fraction, and K is the permeability, which is described by the following Blake–Kozeny model [6].

The solid velocity $\boldsymbol{u}_s$ is modeled as

$$\boldsymbol{u}_s = (1 - P)\left(\boldsymbol{u} + f_l\left(\frac{\alpha(\rho_s - \rho_l)d_g^2}{18\mu_l}\boldsymbol{g}\right)\right) + P\boldsymbol{u}_{cast} \quad (8)$$

where d_g is the diameter of floating solid particle, and $\boldsymbol{u}_{cast}$ is the casting velocity.

Simulation Models

The physical properties and operating parameters are shown in Table 1. In this numerical simulation, Al-5.0mass%Mg alloy was used to simplify the numerical simulation model. Table 2 shows the calculation conditions. In this study, we focused on distributor geometry and casting speed and investigated the effect of casting speed in calculation #1–#3 and distributor in calculation #A–#C. Figure 4 shows the schematic view of the distributors, calculation domain, and boundary conditions. In this study, we devised three different distributor models. Distributor model (A) is the default distributor model that supply molten metal to the width direction of slab. The geometry of distributor in model (B) and (C) is square and supply molten metal to width, thickness, and depth directions. Distributor model (C) has a smaller diameter inlet into which molten metal inflows than the other models. The sizes of each distributor model are shown in Table 3.

As for the boundary conditions, in all distributor models, the distributor surface through which the molten metal passes was set to occur a pressure drop based on the Darcy law. Primary cooling and heat extraction from the bottom block were calculated based on Newton's cooling law, and both using a constant heat transfer coefficient. Secondary cooling with coolant was also calculated based on Newton's cooling law, and the heat transfer coefficient was set so that the value depends on the surface temperature of slab. Figure 5a shows the numerical grid used in this study. The slab size was 400 × 800 mm, to reduce the calculation time the calculation domain was limited to a quarter-symmetrical region of the slab. The size of grid cell in each direction was set to 5 mm. Figure 5b shows schematic drawing of dynamic grid motion. In DC casting, typical casting length of produced slabs is several meters, so in the present study, the dynamic grid motion was implemented by adding new grid cells.

Table 1 Physical properties and operating parameters for numerical simulation

Parameters	Marks	Values	Unit
Density of melt	ρ_l	2350	kg/m³
Density of solid	ρ_s	2650	kg/m³
Heat capacity of melt	c_{pl}	1180	J/kg K
Heat capacity of solid	c_{ps}	1000	J/kg K
Thermal conductivity of solid	k_s	140	W/m K
Kinematic viscosity	v	5.47×10^{-7}	m²/s
Thermal expansion coefficient	β_T	6.90×10^5	1/K
Solutal expansion coefficient	β_C	3.20×10^{-1}	–
Eutectic temperature	T_e	723	K
Eutectic concentration	C_e	0.38	–
Partition coefficient	k_p	0.47	–
Latent heat	L	389,000	J/kg
Gravitational acceleration	g	9.81	m/s²
Critical solid volume fraction	α_s^c	0.30	–
Smooth parameter of packing fraction	$\Delta\alpha_s$	0.05	–
Secondary dendrite arm spacing	λ	5.00×10^{-5}	m
Diameter of floating particle	d_g	7.50×10^{-5}	m
Averaged solid viscosity	$\overline{\mu_s}$	6.45×10^{-3}	Pa s
Casting speed	$\boldsymbol{u}_{cast}$	50, 70, 90	mm/min

Table 2 Calculation conditions

	Casting speed (mm/min)	Distributor model
Calculation #1	50	(A)
Calculation #2	70	(A)
Calculation #3	90	(A)
Calculation #A	70	(A)
Calculation #B	70	(B)
Calculation #C	70	(C)

Fig. 4 Schematic view of the distributors, calculation domain, and boundary conditions

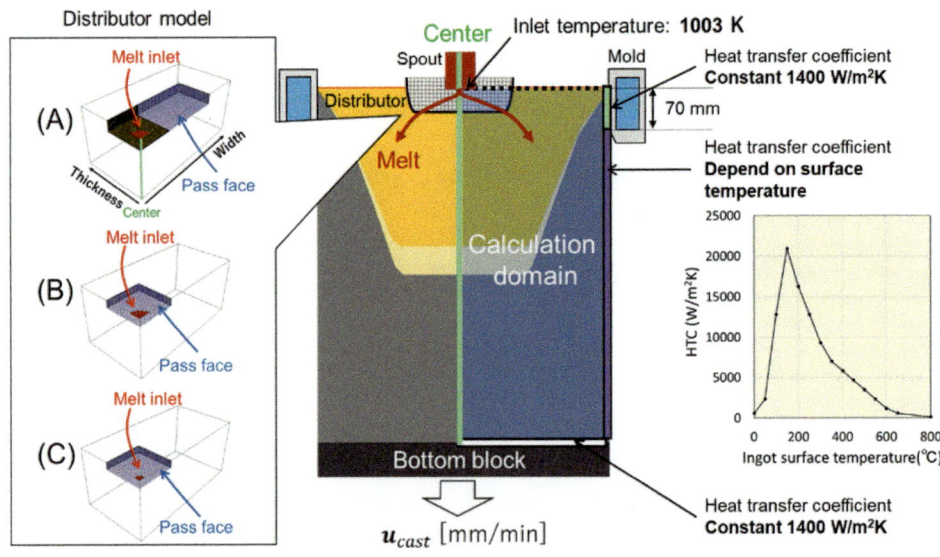

Table 3 Size of each distributor

Marks	Inlet (Φmm)	Width× thickness × depth (mm)
(A)	60	275 × 105 × 30
(B)	60	105 × 105 × 20
(C)	30	105 × 105 × 20

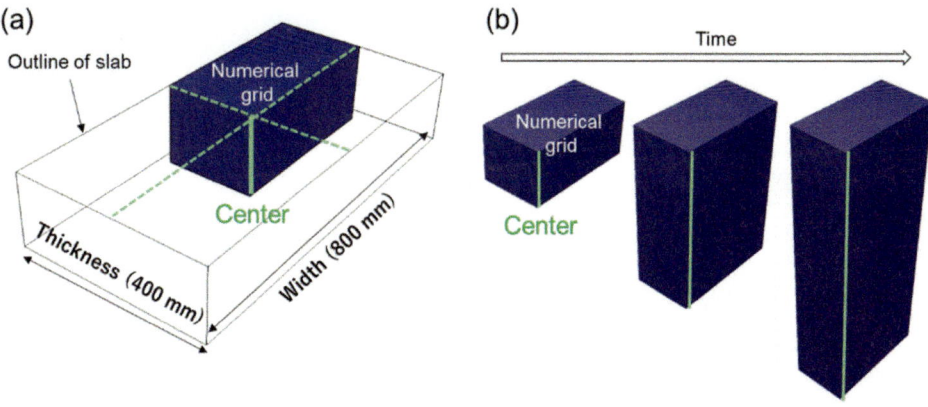

Fig. 5 Schematic view of **a** numerical grid and **b** dynamic grid motion

Results

Figure 6 shows the numerical simulation result of time variation in temperature and Mg concentration distribution within a slab at calculation #2. It can be seen that stripe-shaped negative segregation is formed successively along the solidification front in the thickness section as casting proceeds. Figures 7 and 8 show the results of Mg concentration distribution within a slab for the same casting length, and the graph plots the normalized Mg concentration by the average concentration in the thickness section of each slab. In Fig. 7, it is compared the difference of Mg concentration distribution each casting speed. As far as the standard deviation of Mg concentration is concerned, the casting speed and channel-type segregation strength is tiny correlations, and it is not expected to suppress the segregation by controlling casting speed.

In Fig. 8, the differences in Mg concentrations distribution for each distributor model are compared. The result of calculation #B has a smaller segregation concentration than calculation #A, but the segregation appears in a width cross section too. The result of calculation #C has an even smaller segregation concentration than calculation #B, and the segregation is suppressed in both the width and thickness cross sections.

Discussion

Figure 9 shows the concentration distribution of Mg and flow fields in the molten sump. At calculation #A, it can be seen that the molten metal forms circular flow that flows from distributor to width direction of slab, next flows downward along the solidification front to center of slab, finally flows upward along the solidification front to thickness direction of slab. This circular flow allows molten metal to flow in the same direction as the upward solutal buoyancy flow at the solidification front in thickness direction. Therefore, the high Mg concentration zone is widely distributed at the solidification front, and the partitioning of solute is promoted, resulting in pronouncing the channel-type segregation in the thickness cross section.

At calculation #B, molten metal flows in both the thickness and width directions, so circular flow as # A does not occur, but it can be seen that the downward flow separate at bottom of molten sump and exerts an upward flow in both the thickness and width direction. As a result, it is considered the high Mg concentration zone occur and the channel-type segregation was formed in both thickness and width cross section.

At calculation #C, inlet size is smaller than the other models, resulting in a higher inflow velocity.

Temperature distribution

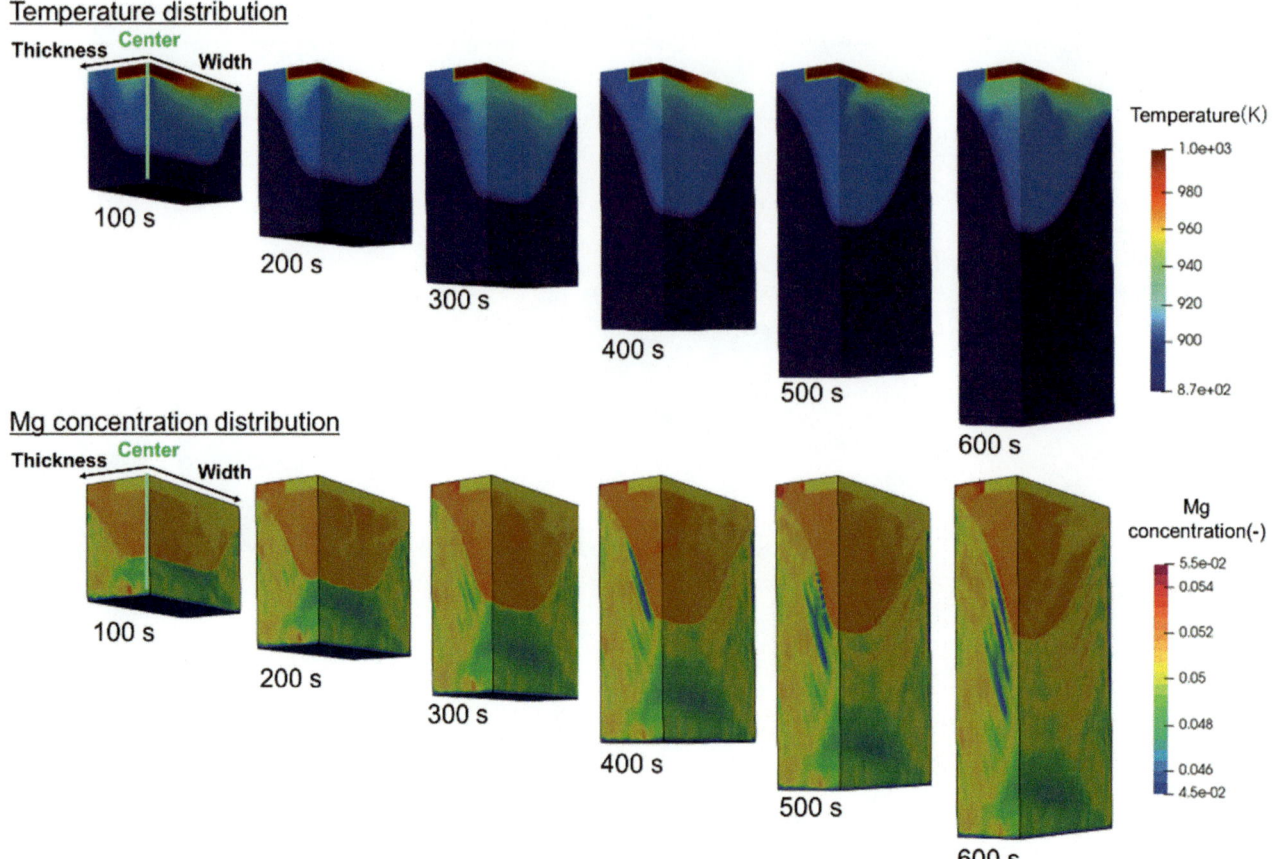

Mg concentration distribution

Fig. 6 Time variation in temperature and Mg concentration at calculation#2

Fig. 7 **a** Mg concentration distribution within a slab at each casting speed, and **b** the graphs of the normalized Mg concentration by the average concentration in the thickness section of each slab

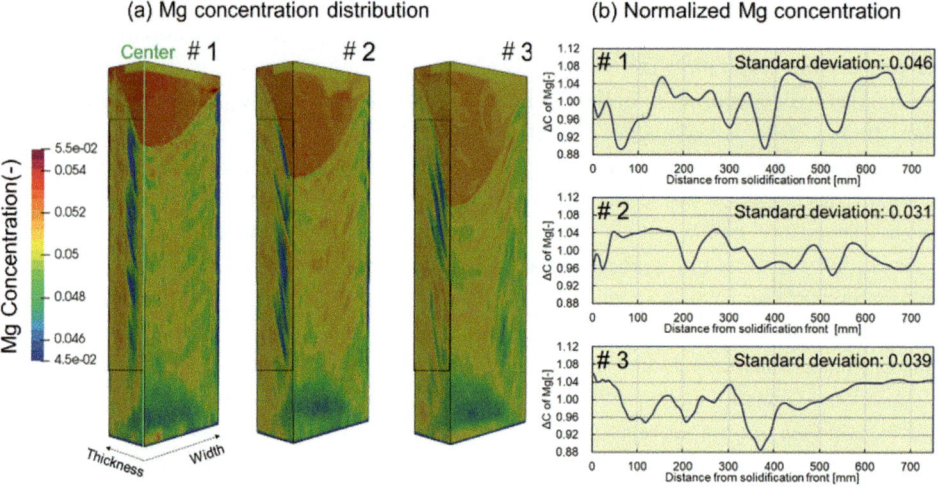

As a result, circular flow does not occur and downward flow against upward solutal buoyancy flow occurs throughout molten sump. Therefore, it is considered that the solute was diffuse in molten sump without collecting at the solidification front and the channel-type segregation did not occur in both thickness and width cross section.

Conclusion

In this study, channel-type segregation in DC casting of aluminum slabs was reproduced by numerical analysis, and the following results were obtained.

Fig. 8 **a** Mg concentration distribution within a slab at each distributor model, and **b** the graphs of the normalized Mg concentration by the average concentration in the thickness section of each slab

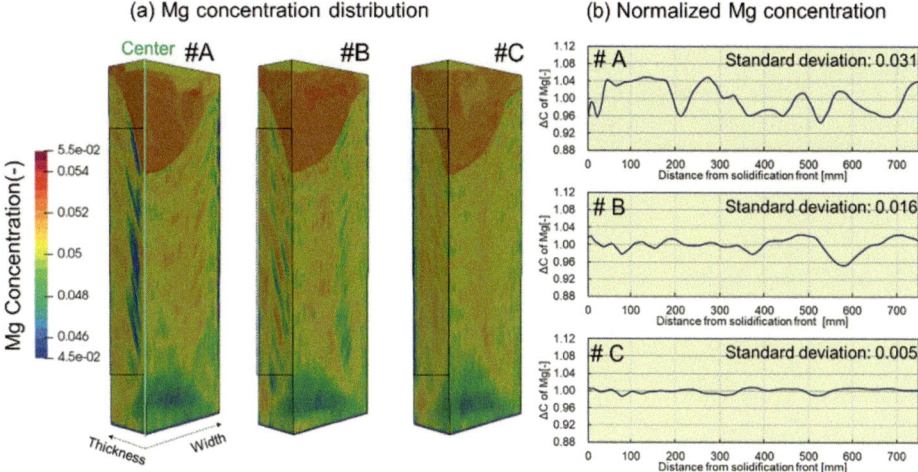

Fig. 9 **a** Flow field in molten sump and **b** Mg concentration distribution

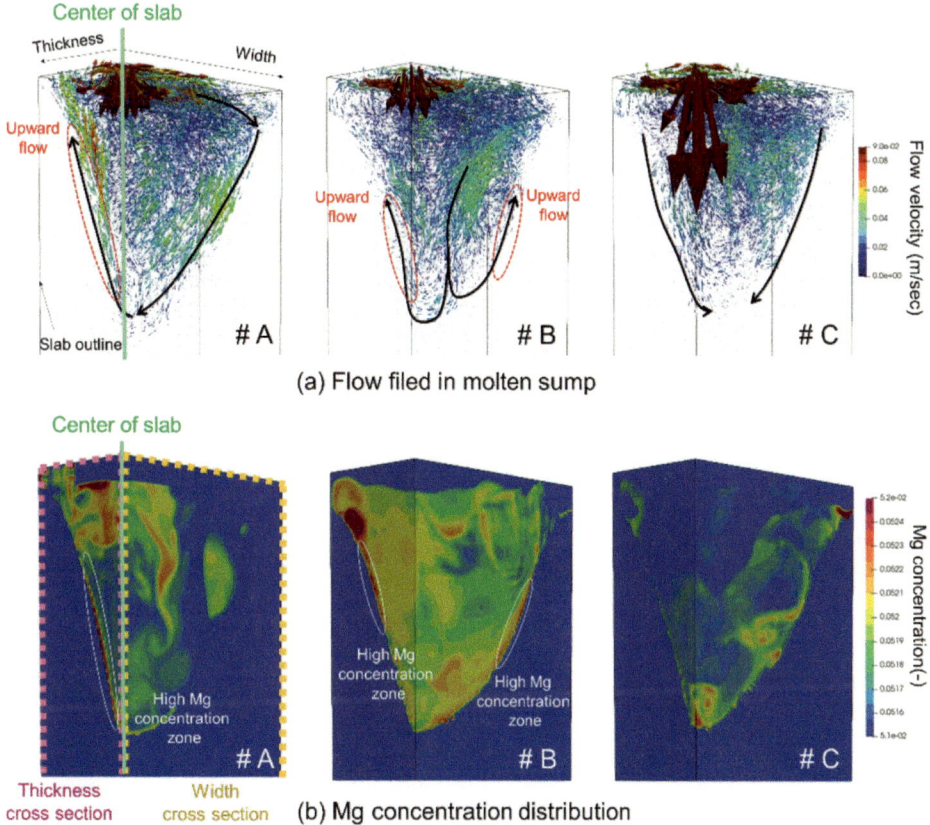

1. Channel-type segregation occurs at different locations in slab depending on the geometry of distributor.
2. Channel-type segregation is more pronounced in the high Mg zone formed by the molten metal upward flow at solidification front.
3. By reducing the area of molten metal inlet, the molten metal inflow velocity increase and strong downward flow occur throughout molten sump. Therefore, Mg concentration at the solidification front is suppressed and channel-type segregation can be reduced.

References

1. Yamamoto T, Kamiya K, Fukawa K, Yomogida S, Kubo T, Tsunekawa M, Komarov S (2021) Numerical Prediction of Channel-Type Segregation Formation in DC Casting of Al–Mg Billet. Metallurgical and Materials Transactions B. 52:4046–4060.
2. Christopher JV, Matthew JMK, Frank PI (2000) The effect of free-floating dendrites and convection on macrosegregation in direct chill cast aluminum alloys: Part I: model development. Int. J. Heat Mass Transf.43:677–686.

3. Fezi K, Plotkowski A, Krane MJM (2016) Macrosegregation modeling during direct-chill casting of aluminum alloy 7050. Numer. Heat Transf. A Appl.70:939–963.

4. Plotkowski A, Krane MJM (2016) A continuum grain attachment model for simulations of equiaxed solidification. Comput. Mater. Sci. 124:238–248.

5. Coleman J, Krane MJM (2020) Influence of liquid metal feeding on the flow and macrosegregation in DC casting. Mater. Sci. Technol 36:393–402.

6. Stefanescu DM (2008) Science and engineering of casting solidification. Springer, New York.

Effect of Casting Variables on Mechanical Properties of Direct Chill Cast Aluminium Alloy Billets

S. P. Mohapatra

Abstract

Microstructure of DC cast 6000 series Aluminium alloy billets developed during casting and homogenization largely determine the performance of end product. Apart from chemical composition, casting parameters like casting speed, casting temperature, and cooling water flow rate influence microstructure development during solidification. There is a need to understand the effect of each of these variables on the mechanical properties of a DC cast Aluminium alloy billet. In the present investigation, the effects of casting speed, cooling water flow rate, and casting temperature on Yield strength and work hardening behaviour of an AA6063 alloy billet have been discussed. An empirical relationship has been formulated to understand the impact of these operating parameters on mechanical properties of the casting.

Keywords

Aluminium • Microstructure • Mechanical properties

Introduction

Direct chill cast 6xxx series Aluminium alloys are widely used as extruded parts in construction and automotive sectors. Formation of the structure during casting and homogenization play a major role in determining the final product quality. Understanding various aspects of the manufacturing process and using them in developing qualitative and quantitative models has become a necessity. Such models must be able to predict the structure & properties of the cast product.

Microstructure of a DC cast billet is largely determined by three factors: chemical composition, casting parameters, and homogenization treatment. The microstructure variables of interest are: grain size, dendrite arm spacing (DAS), amount, and morphology of intermetallic particles. Each of these variables determines hot deformation behaviour and mechanical properties of the part. The mechanical properties of most cast alloys are seen to be strongly dependent on SDAS (Secondary Dendrite Arm Spacing). SDAS is controlled by solidification time whereas grain size is dependent on many different factors like composition, nucleation rate, grain refiner addition, etc. Billet manufacturers give more importance to SDAS at mid radius as a quality parameter. The effect of casting parameters on the variation of SDAS and consequently on mechanical properties of AA 6063 DC cast billet are analyzed in the present study.

The most vital factor for the formation of structure and quality of any casting is cooling rate. The spacing between secondary dendrite arms is a function of cooling rate alone. Cooling rate reflects the heat extraction rate during casting.

$$CR = \frac{Tl - Ts}{Tf} \qquad (1)$$

where Tl and Ts are liquidus and solidus temp, Tf is the time taken by the solidification front to move from the liquidus to the solidus line. The time taken by a single particle to travel in the mushy zone from liquidus and solidus isotherms can be traced from a thermal model.

Generally the structure is refined by increasing the heat extraction rate, i.e., by increasing cooling rate. A lot of studies have been made on the microstructure of binary alloys in order to determine the inter dependence of DAS and solidification parameters [1–4].

S. P. Mohapatra (✉)
DGM (R&D), Nalco Research & Technology Center, Bhubaneswar, Odisha, India
e-mail: satya.mohapatra@nalcoindia.co.in

© The Minerals, Metals & Materials Society 2023
S. Broek (ed.), *Light Metals 2023*, The Minerals, Metals & Materials Series,
https://doi.org/10.1007/978-3-031-22532-1_133

Secondary dendrite arm spacing (λ) which has an impact on the billet properties, is known to vary inversely with cooling rate during solidification as per formula [5, 6],

$$\lambda = AT_c^{-n} \qquad (2)$$

where 'A' is a constant for a particular casting technology, alloy, and size of casting. 'Tc' is the cooling rate. The exponent 'n' lies in between 0.2 and 0.4 for most of the Aluminium alloys [5, 6]. Many researchers have derived expressions in terms of local solidification time T_f as

$$\lambda = CT_f^m \qquad (3)$$

The exponent 'n' used in Eq. 2 has been summarized for a few Aluminium alloys in the literature [7].

A number of studies have pointed out the effect of microstructure and particularly of dendrite arm spacing upon mechanical properties [8–13]. P. R. Goulart et al. [14] developed experimental-based expressions which correlate the ultimate tensile strength with SDAS for Al-Si hypoeutectic alloys. Their experiment was conducted with a 100 mm × 190 mm × 140 mm steel mold in a laboratory by taking Al—9 wt % Si alloys. The study of Devdas et al. [15] on the prediction of SDAS has been significant. They have shown the variation of SDAS with respect to casting speed, super heat of metal and size of billet through mathematical modelling. There are hardly any studies carried out to illustrate the effect of casting variables on mechanical properties of AA6063 DC casting. Experimental investigation is adopted in this work to find out relationship between casting variables with SDAS. Subsequently experiments were carried out to model the effect of casting variables on mechanical properties of the AA 6063 DC cast billet.

Experimental

The experimental work for the study was carried out at the Billet Casting Facility of National Aluminium Company Ltd (Nalco) Smelter plant, India. The three important casting process parameters which significantly affect the structure of billet are casting speed, metal pouring temp & water flow rate. In order to understand the quantitative relation between casting parameters and SDAS, experimental studies were conducted. The parameters were varied one by one & influence of each variable was quantified.

Studying the effect of one process variable on structure could be possible only if other variables are kept constant during casting. The trials were conducted in different campaigns to generate 9 different cases for the complete study.

The pouring temp was varied from 695 °C to 725 °C, water flow was varied from 68 to 100 lpm, and casting speed from 102 to 126 mm/min during different casting campaigns, which were taken as experimental cases for this study. The following cases were experimented in order to identify the influence of process parameterTable 1.

A typical dendritic microstructure and SDAS of a 6063 DC cast billet are shown Fig. 1.

The sample for analysis of SDAS was taken from the mid radius point of a sound billet. The samples were prepared by following the standard procedures of metallography. Optical microscopic analysis was carried out to measure SDAS by line intercept method and individual counting method.

Average value of 10 numbers of readings was taken as the reference value. This is a standard procedure followed by Wislei Riuper Osorio and others and Eduardo Netto De Souza and others [16, 17] to calculate SDAS.

In order to understand the effect of casting parameters on mechanical properties of the billet five samples were collected from billets produced from five different sets of conditions mentioned in Table 2.

The samples were tested under compression to determine the properties of the materials. The specimens are of cylindrical in the shape of 30 mm diameter and 40 mm height. The specimens were cut to size by turning, using kerosene as cutting fluid. All specimens before experimentation were annealed in the boiling water for a period of two hours. The dimension was so chosen that the cylindrical specimen should avoid buckling during testing. The specimen had oil grooves turned on both ends to entrap lubricant during the compression process. After annealing the specimen in boiling water for two hours, its ends were adequately lubricated with lithium-based grease lubricant (commercially available SKF LGMT 3IN1 general purpose grease).

Experimental Determination of Stress–Strain Characteristics

In order to plot the stress–strain diagram, specimens were tested under uniaxial compression on the INSTRON®-600KN hydraulic pressing machine. To avoid the rate effect, the movement of the punch was adjusted to approximately 1 mm per minute. The compressive load is recorded at every 0.5 mm of punch travel. After compressing the specimen to about 10 mm it was taken out from the sub-press, re-machined to a cylindrical shape of diameter 30 mm with the oil grooves again turned on both ends and tested in compression. This process continued till the specimen was reduced to about 15 mm in height.

Table 1 Experimental cases for microstructural study

Cases	Water flow, (lpm)	Casting speed, (mm/min)	Pour temp, (° C)	
1	78	102	705	For different casting speed
2	78	110	706	
3	78	126	705	
4	78	126	705	For different pouring temp
5	78	125	712	
6	78	125	725	
7	68	126	706	For different water flow
8	78	126	705	
9	100	125	705	

Fig. 1 Micrograph of AA 6063 billet, **a** bright field SEM image identifying a grain, **b** dark field SEM image identifying SDAS

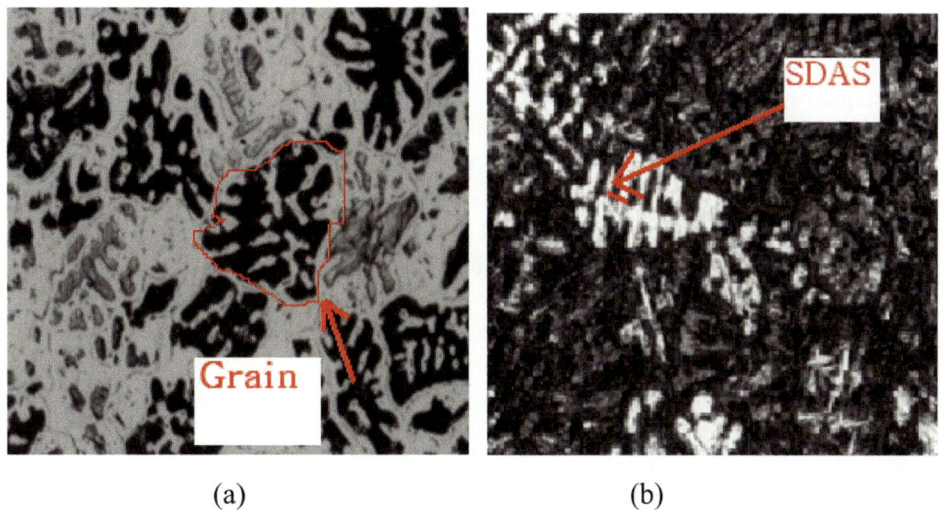

(a)   (b)

Table 2 Experimental cases for deformation study

Sample no	SDAS, (μm)	Speed V, (mm/min)	Water flow F (lpm)	Casting Temperature T, (° C)
1	13	138	135	690
2	15	137	120	702
3	16	127	130	702
4	18	126	100	698
5	20	126	78	705

Results and Discussion

The variation of measured SDAS values with individual casting parameters are depicted Fig. 2.

It is evident that as speed increases, cooling rate also increases proportionately. The rise in cooling rate raises heat extraction rate for a particular grain in the entire solidification zone. This prevents grain coarsening and ensures finer grain size. The dendrite arm spacing decreases as the grain size is finer.

Higher melt temperature will increase the size of SDAS as noticed from the above figure. If the casting temperature is very high, the liquid in the center of the billet will remain above the liquidus temperature for a long time. As a result, most of the crystals will melt again soon after they depart from the cast wall. Only the remaining crystals can extend near the cast wall to form the chill zone and a thin shell composed of randomly oriented small equiaxed grains will be formed on the surface of the billet. Consequently grains will remain longer time in liquid and coarsen.

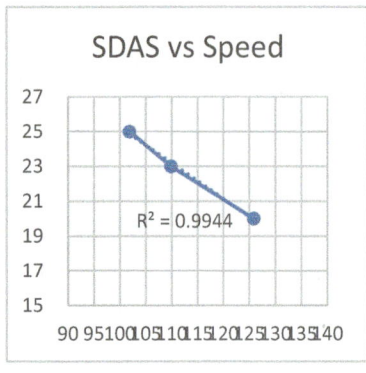
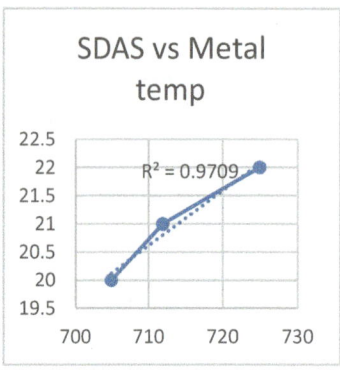
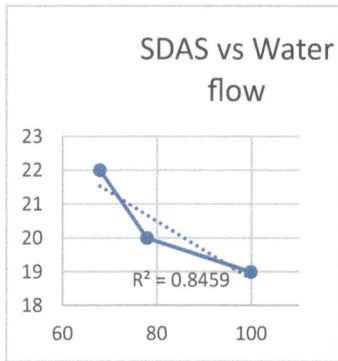

(a) SDAS vs Speed   (b) SDAS vs Temp   (c) SDAS vs Water flow

Fig. 2 Variation of SDAS with casting parameters

On the other hand, if the casting temperature is lower, all of the liquid will cool fast below the liquidus temperature and the crystals will move fast into the melt. Then a fine equiaxed billet structure will be produced. Again a caster will never want to run at higher melt temperature as it will increase the energy use and the risk of bleed out.

The water flow has less effect on the maximum heat that can be extracted, since after a certain flow rate, the heat transport mainly depends on the conduction through solid portion of the billet rather than to the outside water. It is also observed that increasing water flow has less impact on cooling rate beyond a certain flow rate. SDAS is found to be decreasing when water flow rate is increased as the time spent by a particle in the mushy zone is reduced.

Regression Model

It is seen from the relation between SDAS and casting parameters that SDAS varies linearly with all three parameters. The correlation coefficient in each case is around 0.9, which indicates a strong linear relationship. A multivariate linear regression analysis with a confidence level 95% has been taken up to correlate all the three casting variables with SDAS.

Considering the fact that the rate of change of each parameter is independent of the other variable, the composite regression model developed is as per the following equation:

$$\textbf{SDAS} = \textbf{17.47} - \textbf{0.169} * \textbf{V} - \textbf{0.0993} * \textbf{L} + \textbf{0.0464} * \textbf{T} \tag{4}$$

where V, L and T are casting speed in mm/min, water flow in lpm and T is in °C respectively.

The summary output of the analysis is presented in Table 3. The correlation coefficient of the analysis is 0.988. This indicates a strong relationship.

Table 3 Regression analysis summary output

Regression statistics	
Multiple R	0.9940463
R^2	0.9881281
Adjusted R^2	0.9810049
Standard error	0.2430962
Observations	9

The measured values of SDAS and values calculated from the model are compared in Table 4.

It can be seen from the table that the model is able to predict the SDAS value within a reasonable range of variation. This model is valid for a specific range of casting speed from 102 to 126 mm/min, pouring temp from 705 to 725 °C and water flow rate from 68 to 100 lpm. This study can be extended for large ranges of casting parameters. The casting speed is found to have the maximum impact on SDAS followed by the water flow and then metal casting temperature.

The model was also tested for five different cases undertaken to study the deformation behaviour of the billets with respect to casting variables (Table 5).

The standard errors are also seen to remain within acceptable range, so the model is validated.

In metal forming, the plastic region plays an important role because metal is plastically and permanently deformed in this process. The stress–strain relationship for a metal exhibits elasticity below the yield point and strain hardening after that. In the plastic region the metal's behavior is expressed by the equation;

$$\sigma = A \in^n \tag{5}$$

The mechanical properties of the specimens tested in compression can be characterized by their yield stress σ, their strength coefficient, (A) and work hardening exponent (n).

Table 4 Comparison between measurement and model prediction

Experimental cases	SDAS as per regression model output (equation-4), (µm)	SDAS as per Measurement values, (µm)	Standard error with respect to regression model, (%)
1	25.13	25	0.79
2	23.82	23	3.73
3	21.06	20	5.40
4	21.06	20	5.40
5	21.56	21	2.9
6	22.16	22	1.08
7	22.10	22	0.82
8	21.06	20	5.40
9	19.05	19	0.66

Table 5 Model validation

Experimental cases	SDAS as per regression model output (Eq. 4), (µm)	SDAS as per Measurement values, (µm)	Standard error with respect to regression model, (%)
1	12.7585	13	1.857692
2	14.9738	15	0.174667
3	15.6708	16	2.0575
4	18.6332	18	−3.51778
5	21.1426	20	−5.713

The result is shown in Table 6.

The characteristic equations for the different billets obtained from the compression test are summarized below.

$$\text{With SDAS 13 µm}: \sigma = 241.0\varepsilon^{0.19} \tag{6}$$

$$\text{With SDAS 15 µm}: \sigma = 229.0\varepsilon^{0.16} \tag{7}$$

$$\text{With SDAS 16 µm}: \sigma = 195.0\varepsilon^{0.14} \tag{8}$$

$$\text{With SDAS 18 µm}: \sigma = 193.0\varepsilon^{0.13} \tag{9}$$

$$\text{With SDAS 20 µm}: \sigma = 163.0\varepsilon^{0.10} \tag{10}$$

To determine the uniaxial yield stress of the billet material for any given reduction R, the corresponding strain imparted to the billet material during the compression process must be known beforehand. For the same reduction, the strain imparted to the billet during the practical forming processes is always higher than that imparted to the billet during compression because of the redundant work done during the forming processes. Following the principles laid by Johnson and Mellor [18], the above strain in the present case is calculated from the empirical relation,

$$\bar{\varepsilon} = 0.8 + 1.5 \ln\left(\frac{1}{1-R}\right) \tag{11}$$

Table 6 Compression test result

Sample no	SDAS, (µm)	Speed V, (mm/min)	Water flow F, (lpm)	Casting Temperature T, (°C)	Strength coefficient (A), (MPa)	Work hardening exponent, (n)	Yield strength, (MPa)
1	13	138	135	690	241	0.19	224.0
2	15	137	120	702	229	0.16	214.9
3	16	127	130	702	195	0.14	184.2
4	18	126	100	698	193	0.13	183.0
5	20	126	78	705	163	0.10	156.3

where R (for the present calculation, a severe situation, $R = 0.45$ was taken) is the fractional reduction. In Aluminum forming process, general reduction ratio 45% is considered as the limiting value. The average yield stress of the compressed billet is then calculated by dividing the area under the stress–strain curve with the corresponding value of the abscissa. The magnitude of the strain is calculated from Eq. (11) and the magnitude of the average yield stress σ_y for different specimens is determined in the above manner.

Figure 3 shows the variation of the mechanical properties as a function of SDAS. It can be seen from the figure that microstructural refining enhances the properties of the material. The yield strength and the strength coefficient are highest in the sample with SDAS of 13 μm. The sample with largest SDAS has shown a work hardening coefficient as 0.1 and yield strength of 156.3 MPa. The findings are matching with the findings of Q. Wang for Aluminum alloy A 356 and A 357 [19].

It is clearly seen that the yield strength decreases as SDAS is increased. The refining of spacings and consequently the grains enhance the blocking resistance to slipping of dislocations which causes deformation. So a material having less SDAS will exhibit higher strength.

The work hardening coefficient 'n' is found to be decreasing when SDAS is increased. This is also in agreement with the result of Defeng et al. [20] for different types of Aluminum alloys, A 319/356/357. Since 'n' defines the working range of a material, the deformation range of a material is shortened for a coarse grain size material. The formability of a material is dependent on strain gradients in the deformed metal. Strain gradient arises when deformation is not uniform. Since the most highly strained region will have the highest hardness, the load is passed on to the neighbouring elements for a metal with a high 'n'. This forces them to strain more and by doing so the strain

gradient is reduced [21]. As a result, higher formability can be achieved by the metal with higher 'n'. The ductility of the metal is also influenced by SDAS. In a coarse grain with large SDAS, the eutectic particles become coarser and longer. A particular homegenisation treatment will have less impact on a coarse grain than fine grain structure material as the diffusion distance is increased in a coarse grain material. So large and elongated particles in the metal will exhibit rapid development of particle stress and thus low ductility [22, 23].

So increasing the cooling rate refines the grain and also the eutectic particles, thereby improving the strength as well as formability of the material. It is seen that the work hardening exponent, 'n' which influences the formability of material is influenced by SDAS and strength coefficient A, which are in turn affected by the casting parameters.

Correlation between the deformation behavior of the material represented by the strain hardening coefficient and yield strength with its casting conditions, represented by SDAS is depicted in Fig. 4.

The variation of work hardening exponent with SDAS and yield strength for the present investigation can be expressed by an equation as follows:

$$n = \frac{\sigma_y^{0.799}}{56.3 \times \text{SDAS}^{0.764}} \tag{12}$$

In order to derive the value of yield strength with different SDAS and 'n' the following figure will be helpful (Fig. 5).

The relationship is expressed in the following equation:

$$\sigma_y = 347.2 \times \text{SDAS}^{0.379} \times n^{0.845} \tag{13}$$

The ultimate aim of a caster is to select his casting conditions for the desired quality of the product. In order to correlate the casting parameters like casting speed, pouring temp, and water flow rate to the final objective of the product, i.e., strength and formability, regression analysis was done. An empirical model is proposed for the strength hardening coefficient and yield strength of the product.

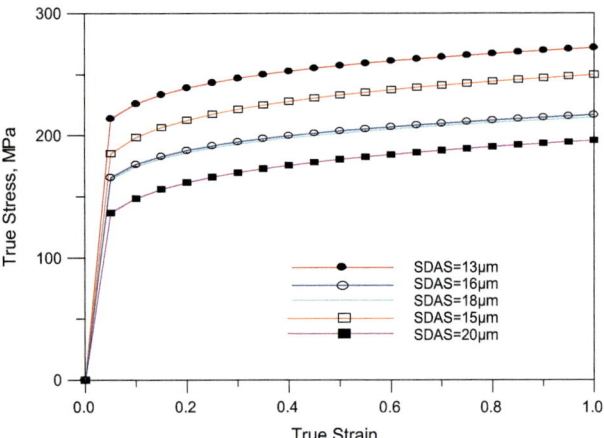

Fig. 3 Stress–strain curve for five different samples

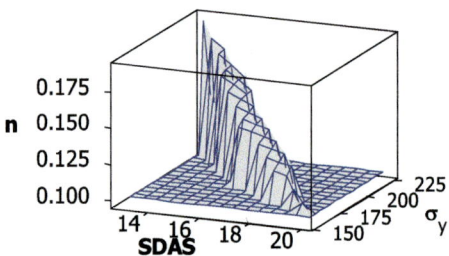

Fig. 4 Variation of work hardening coefficient with SDAS (μm) and Yield strength (MPa)

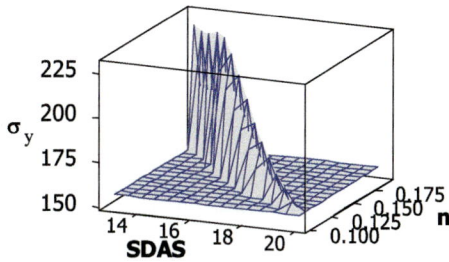

Fig. 5 Variation of σ_y (MPa) with SDAS (μm) and work hardening coefficient

Proposed Empirical Model

This model is applicable to only a specific technology (Wagstaff Air-slip) of billet casting and for a specific range of variables experimented. The applicability can be translated to any technology and any range of variables through experiments in line with the present work.

$$n = 1.3 \times 10^{15} \frac{V^{1.93} F^{0.555}}{T^{7.45}} \qquad (14)$$

$$\sigma_y = 2.96 \times 10^5 \frac{V^{1.74} F^{0.256}}{T^{2.6}} \qquad (15)$$

Conclusion

Increasing cooling rate reduces SDAS and thereby improves the yield strength of a material. The work hardening coefficient is decreased with an increase in SDAS, so that ductility of the material is diminished. The casting variables like casting speed and cooling water flow rate improves the formability of the metal whereas melt temperature has a negative impact. Similarly, the strength of the metal is improved by both casting speed and water flow and is lowered by metal casting temperature.

An attempt has been made to propose an empirical model for the prediction of work hardening coefficient and yield strength based on the three casting variables. It is envisaged that such a model could help the casting operators to fine tune the casting variables as per the requirement of product behavior (in line with casting technology permission values). The advantage of such a model is that the casting operator does not have to be knowledgeable on mathematical modeling. The control of dendritic as-cast microstructures by regulating operating parameters can be used as an alternative way to produce components with better mechanical properties.

Acknowledgements The author would like to sincerely thank Nalco management for allowing him to carry out the experiment in the plant and also giving permission to present the findings.

References

1. Measurement technique, Light Metals, TMS, USA: 883–889.
2. H. Jacobi and K. Schwerdtfeger (1976) Dendritic Morphology of Steady state Unidirectionally Solidified Steel, Metall. Trans. A, Vol. 7A: 811–820.
3. K. P. Young and D.H. Kirkwood (1975) The Dendritic Arm Spacing Arms Spacing of Aluminum-Copper Alloys Solidified Under Steady-State Conditions, Metall. Trans. A, Vol. 6A: 197–205.
4. J. A. Spittle and D. M. Lloy (1979) Dendritic Arms Spacing in Hypoeutectic Pb-Sb Alloys Directionally Solidified Under Steady and Non-Steady Conditions, Proc. Of International Conference on Solidification and Casting, The Metals Soceity, London: 15–20.
5. M. O. Shabani and Ali Mazahery (2011) Microstructural prediction of cast A256 alloy of a function of cooling rat, Journal of Materials, TMS, USA:132–136.
6. D. G. Eskin, V. I. Savran, and L. Katgerman (2008) Effects of Melt Temperature and Casting Speed on the Structure and Defect Formation during Direct-Chill Casting of an Al-Cu Alloy", Metallurgical And Materials Transactions A Volume 36A: 1965–1976.
7. J. Sengupta, D. Maijer, M. A. Wells, Cockroft and A. Laruch (2003) Mathematical Modelling Of Water Ejection and Water Incursion During The Start Up Phase Of The Dc Casting Process, Light Metals, TMS, USA: 841–847.
8. Wang Mengjun, Gan Chunlei, Liu Xinyu(2005) Deformation Behavior Of 6063 Aluminum Alloy During High-Speed Compression, Journal Of Wuhan University Of Technology, Vol. 20 No. 3: 40–43.
9. H. Cao And M. Wessén , "Effect of Microstructure on Mechanical Properties of As-Cast Mg-Al Alloys (2004) Metallurgical And Materials Transactions a Vol. 35 A, January: 309–319.
10. C. H. Caceres, C.J. Davidson, J. R. Griffiths,C. L. Newton (2002) Effects Of Solidification Rate And Ageing On The Microstructure And Mechanical Properties Of AZ91 Alloy" Materials Science and Engineering A325: 344–355.
11. Wislei R. OsoRio, Pedro R. Goulart, Givanildo A. Santos, Carlos Moura Neto, and Amauri Garcia (2006) Effect of Dendritic Arm Spacing on Mechanical Properties and Corrosion Resistance of Al 9 Wt Pct Si and Zn 27 Wt Pct Al Alloys, Metallurgical And Materials Transactions a Vol. 37A: 2525–2038.
12. Suyitno, D.G. Eskin, V.I. Savran, And L. Katgerman (2004) Effects of Alloy Composition and Casting Speed on Structure Formation and Hot Tearing during Direct-Chill Casting of Al-Cu Alloys", Metallurgical And Materials Transactions A Vol. 35 A: 3551–3561.
13. I. Ebrahimzadeh, G. H. Akbari (2007) Comparing Of Mechanical Behavior And Microstructure Of Continuous Cast And Hot Worked Cuzn40al1 Alloy", IJE Transactions B: Applications Vol. 20, No. 3, December: 249–256.
14. Pedro R. Goulart, Jos. E. Spinelli, Wislei R. Osrio, Amauri Garcia (2006) Mechanical Properties as a Function Of Microstructure And Solidification Thermal Variables Of Al–Si Castings", Material Science and Engineering A421: 245–253.
15. C. Devdas, Ivo Musulin, Olaf Celliers(1992) Prediction Of Microsrtucture Of Dc Cast 6063 Billets And Its Effect On Extrusion Processes, International Conference on Extrusion Technology: 121–128.

16. Doru M. Stefanescu, G. Upadhya, and D. Bandyopadhyay (1990) Heat Transfer-Solidification Kinetics Modelling of Solidification of Castings, Metallurgical Transactions A Volume 21: 997–1005.

17. M. J. Couper, B. Rinderer, M. Cooksey (2004) A Simplified Empirical Model of Precipitation Strengthening in 6000 Al-Mg-Si Extrusion Alloys, Materials Forum Volume 28: 159–164.

18. W. Johnson and P. B. Mellor (2003) Engineering Plasticity Van-Nostrand Reinhold, London.

19. Q. G. Warg, (2003) Microstructural effects on the tensile and fracture behavior of Aluminum casting alloys A356/357 inch, Metallurgical and Materials Transaction A, Vol-34A:2887–2889.

20. M. O. Defeng, H. Zhengfei, L. Xiaoshan (2010) Effect of Microstructure on Strain Hardening Exponent Prediction of Cast Aluminum Alloy, Acta Metallurgica Sinica, VOl-46, No-2:184–188.

21. G. E. Dieter (2001) Mechanical Metallurgy, McGraw-Hill Books, UK.

22. Q. G. Wang (2004) Plastic Deformation Behavior Of Aluminum Casting Alloys A356/357 Inch, Metallurgical and Materials Transaction A, Vol-35A: 2707–2718.

23. C. H. Caceres and J. R. Griffiths (1996) Damage by the Cracking of Silicon Particles in an 7Si-0.4Mg Casting Alloy, Acta Metallurgica, Vol. 44, No. 1: 25–33.

Stability of SiC and Al$_2$O$_3$ Reinforcement Particles in Thermomechanical Processed Direct Chill Cast Al MMnCs

Abdallah Abu Amara, Guangyu Liu, Dmitry Eskin, and Brian McKay

Abstract

In this study, aluminium alloys reinforced by alumina (Al$_2$O$_3$) and silicon carbide (SiC) were investigated. Lab-scale DC cast billets with a diameter of 80 mm were cast by ultrasound-assisted stir-casting technology. The billets were subsequently thermo-mechanically processed. The distribution of the particles in the matrix was analysed before and after thermomechanical processing. It was found that hot rolling improved the distribution of the reinforcement particles in the matrix. Reactions between the reinforcement particles and the matrix were also investigated, and their implications studied. It was found that alumina reinforcement reacted with magnesium (Mg) in the alloy to produce spinel and SiC reinforcement oxidised producing a layer of silicon oxide (SiO$_2$) around the particles. The consequences of these reactions are discussed. Whilst theoretical calculations of free energies for different carbides suggested that transition metals in the alloys should substitute the silicon in SiC, no reactions were observed in the physical experimentation. This was attributed to the weight percentages of the transition metals in the alloy being too small. These results are discussed and analysed further in this study, with plans for future experimentation.

Keywords

Al MMnC • DC casting • SiC particles • Al$_2$O$_3$ particles • Thermomechanical processing

A. Abu Amara · G. Liu (✉) · D. Eskin · B. McKay
Brunel Centre for Advanced Solidification Technology, Brunel University London, Uxbridge, UB8 3PH, Middlesex, UK
e-mail: guangyu.liu2@brunel.ac.uk

A. Abu Amara
e-mail: abdallah.abuamara@brunel.ac.uk

Introduction

The synthesis of lightweight materials with elevated mechanical properties has long been a challenge for engineers and material scientists. Materials with such characteristics are known to be highly desirable in transport applications, especially the aerospace and automotive industries. As legislative restrictions continue to grow around CO$_2$ emissions from vehicles and aircraft, Original Equipment Manufacturers (OEMs) are seeking to adopt the use of new materials in their designs. This comes with the knowledge that a 10% reduction in the vehicle weight can improve the fuel economy by 6–8% [1]. Furthermore, the electrical revolution is pushing new specifications and requirements for Battery Electric Vehicle (BEV) technology, whilst the aerospace industry has already begun implementing designs for hydrogen fuel cell aircraft [2]. The introduction of new technologies in the form of electric vehicles and hydrogenpowered aircraft will undoubtedly cause the OEMs to continue searching for new material technologies.

The addition of ceramic nano-particles as a reinforcement to aluminium alloys has proven to be a legitimate method for enhancing the material properties [3]. In recent years, there has been a growth in the interest of these materials due to their desirable mechanical properties. Many studies have shown that aluminium metal matrix composites (AlMMC) have superior mechanical properties when compared to their base alloys [3–9].

Nevertheless, there are still many challenges to overcome in the field of Metal Matrix Composites (MMCs). Reactions between the reinforcement and the matrix can occur, such as alumina reinforcement producing spinel phase (MgAl$_2$O$_4$) [10]. The growth of spinel causes the viscosity in the melt to increase which has negative implications on the castability of the material [10]. SiC reinforcement, on the other hand, has the potential to oxidise when in the melt and produces an SiO$_2$ layer around the particles. Unlike spinel phase, the SiO$_2$ has positive implications as this layer enhances the wettability, allowing for better incorporation of particles [11]. Moreover, the addition of carbide reinforcement particles to the aluminium alloys containing transition metals can lead to chemical reactions which substitute the metal in the reinforcement with the transition metal. For example, an addition of SiC reinforcement to an alloy containing Zr can lead to a chemical reaction between the reinforcement and the Zr in the alloy, resulting in the formation of ZrC particles rather than SiC [12].

Transition metals in alloys have been known to make substitutions in ceramic particle additions used for grain refinement. An example of this is the Zr poisoning effect. This effect which is reported in titanium carbide (TiC) particles reduces the grain refinement of TiC considerably [13]. The mechanism of the Zr poisoning effect is not completely agreed upon in the literature; however, it is generally accepted that the poisoning occurs as a result of Zr interacting with the nucleating particles such as Al_3Ti and TiC from the grain refiner [14].

The thermodynamic stability of the compounds can be used to determine whether or not a reaction will take place between the transition metals in the alloy and the carbide reinforcement. A carbide Ellingham diagram is presented in Fig. 1 which compares the Gibbs free energies of carbides at different temperatures [15]. It can be seen from the diagram that there are a number of transition metal carbides which exhibit lower free energies than silicon carbide, which is one of the reinforcement types used in this study. This means that these carbides are likely to appear in the material instead of the SiC particles under suitable conditions.

In this study, work is carried out to determine which reactions (if any) will occur between the matrix of an aluminium alloy and two types of reinforcement particles: Al_2O_3 and SiC. The base aluminium alloy and the composite materials are produced via DC casting then thermomechanically processed. Theoretical calculations of the free energies of the compounds are calculated to determine which carbides are likely to form in place of SiC. The implications of the reactions observed are discussed.

Methods

Experimental

Material Synthesis

6 kg of aluminium alloy was charged in an electric furnace and heated to 730 °C. At the same time, 150 g of Al-20 wt% SiC master-alloy powder was preheated to 200 °C in an electric oven. This amount corresponded to an addition level of 0.5 wt% SiC of the 6 kg base alloy. This master-alloy powder was produced by mechanical alloying and supplied by MBN Nanomaterialia (Italy). The master-alloy powder was introduced to the melt through mechanical stirring with a rotational speed of 400 RPM for a period of 10 min. Next, ultrasound was applied to the melt to further disperse the reinforcement particles. The same process was repeated for the base aluminium alloy, without any reinforcement addition. Also, it was repeated for the aluminium alloy with an addition of 150 g of Al-20 wt% Al_2O_3 master-alloy powder which corresponded to 0.5 wt% Al_2O_3.

Once the ultrasound was applied to the melt for 10 min at a frequency of 17 kHz and a power of 3.5 kW, the melt was direct chill (DC) cast through an 80 mm diameter mould, at a pouring temperature of 700 °C. The casting speed was set to 230 mm/min. The 80 mm diameter billet can be seen in Fig. 2b. Aluminium alloys, which are widely used in the automotive industry [16], are typically DC cast into round billets for the purpose of heat treatment and subsequent mechanical processing [17]. Lab-scale DC casting was performed in this study to replicate the process used in industry.

Thermo-Mechanical Processing

The homogenisation process was carried out at temperatures between 500 and 550 °C for a period of 3–4 h. The purpose of this step is to dissolve the Mg_2Si phase and to transform the β-AlFeSi (β-Al_5FeSi) to α-AlFeSi (α-$Al_{12}(FeMn)_3Si$) [18, 19]. Sarafoglou et al. [18] observed the dissolution of the Mg_2Si gradually over the course of 3 h at 540 °C in a 6xxx alloy.

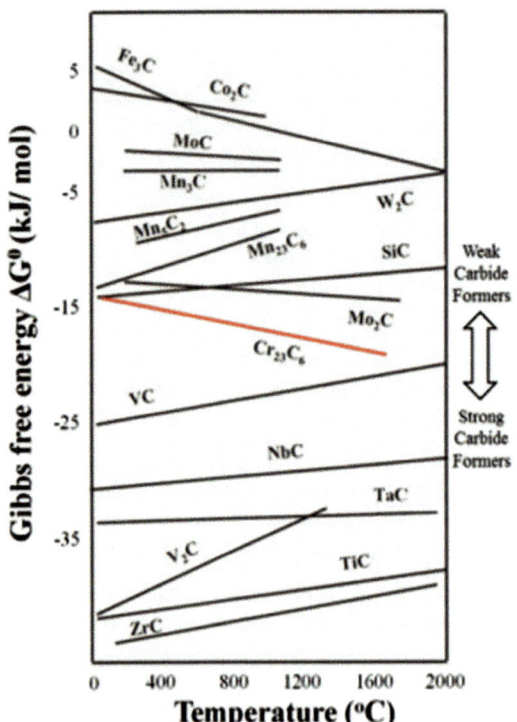

Fig. 1 Carbide Ellingham diagram

Fig. 2 **a** Casting setup, **b** 80 mm diameter billet, **c** rectangular samples obtained from billet, and **d** hot-rolled samples

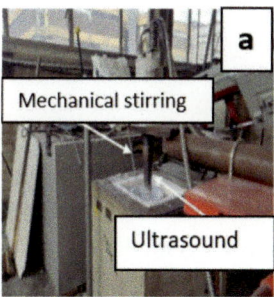

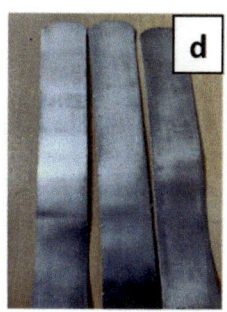

Rectangular samples were then obtained from the centres of the 80 mm diameter billets, for the purpose of hot rolling. These can be seen in Fig. 2c. The rectangular samples were rolled at a temperature of 350 °C, from a thickness of 13 mm down to a thickness of 2.5 mm ($\sim$80% reduction) as can be seen in Fig. 2d. This was achieved via eight passes through the rolling mill.

A T6 heat treatment was then performed. For the solution treatment step, the samples were heated to temperatures between 500 and 25 °C and then quenched in water. The samples were then left to naturally age (at room temperature) for one day. Finally, the samples were aged at temperatures in the range of 150–160 °C for 15 h. The homogenised-rolled samples were solution treated and aged according to temperatures and times taken from Smithells Metals Reference Book [20].

Characterisation

Samples were obtained from the as-cast and as-rolled homogenised castings. These samples were ground and polished in preparation for microstructural analysis. The microstructures of the castings were then studied via Optical Microscopy (OM). The use of Optical Microscopy allowed for the observation of nano-particle clusters in the matrix as well as secondary intermetallic phases. The measurement tool in the OM was also used for comparing the sizes of different clusters of reinforcement particles.

A Zeiss Supra 35 field-emission scanning electron microscope (FESEM), equipped with energy-dispersive X-ray spectroscopy (EDS), was also used to analyse the size of particles and microstructure of the cast materials. EDS was used to obtain the compositions of the reinforcement particles. This was important for confirming that these were added reinforcement. Compositions of the intermetallic phase were also analysed using EDS. The interfaces between the reinforcement and the matrix were studied to determine if there had any chemical reactions between the particles and the matrix.

Gibbs Free Energy Calculation

The Gibbs free energy was calculated using the following equation:

$$\Delta G = \Delta H - (T * \Delta S) \tag{1}$$

where ΔG is the change in the Gibbs free energy (kJ * mol^{-1}), ΔH is the change in the enthalpy (kJ * mol^{-1}), ΔS is the change in entropy (kJ * K^{-1}), and T is the temperature in Kelvin [21]. The change in enthalpy and entropy values was taken from the National Institute of Standards and Technology Database [22]. These were used to calculate the change in Gibbs free energy for SiC and the other carbides that were likely to form as a result of alloying elements in the alloy. These carbides are zirconium carbide (ZrC), titanium carbide (TiC), and chromium carbide ($Cr_{23}C_6$). An example calculation is shown below for the silicon carbide:

$$\Delta G = 29.02 - (1000 * 0.06404) = -35.02 \, \text{kJ/mol}$$

In this example, the ΔH is equal to 29.02 (kJ/mol) and the ΔS is 0.06404 (kJ/mol * K) [22]. The temperature used in the calculation is 1000 K which is approximately equivalent to 730 °C. This is the temperature at which the particles were added and processed in the aluminium melt (using mechanical stirring and ultrasonication). The remaining free energy values are presented in Results.

Results

Free Energies of Transition Metal Carbides

As previously mentioned in Sect. 2.2, the free energies of formation were calculated for ZrC, TiC, and $Cr_{23}C_6$ as presented in Table 1.

Table 1 shows that the three transition metals present in the alloy have lower free energies than SiC. This means that

Table 1 Calculated theoretical free energies of formation

Carbide	SiC	ZrC	TiC	$Cr_{23}C_6$
Free energy of formation (kJ/mol)	−35	−56	−46	−983

the reactions of all of the shown carbides are more spontaneous than the reaction of SiC. This suggests that, in theory, these carbides can form in place of the SiC particles.

Base Alloy's and Composite Materials' Microstructure

As-Cast Microstructure

The microstructure of the casting containing alumina reinforcement showed that the particles were spread across all of the matrix. In some areas, the particles can be seen clustering as in Fig. 3a, but generally the particles were not adhering to each other or agglomerating. Individual particles can be seen, as indicated, in Fig. 3b. The SiC castings however revealed large agglomerates greater than 100 µm in diameter as shown in Fig. 3c. It can be seen from Fig. 3d that the particles are not only clustering, but also coalescing, as if still bonded by the van der Waals forces of attraction.

The SEM micrographs and their corresponding EDS spectra shown in Fig. 4 were used to confirm that these particles are alumina and SiC through the use of EDX point analysis. For the alumina particles, two main aluminium and oxygen peaks can be seen with smaller magnesium and silicon peaks also visible in the spectra shown in Fig. 4d. This confirms that the particles present are Al_2O_3 with potential reactions with magnesium to produce spinel phase. This will be discussed in further detail below.

For the SiC particles, two main silicon and carbon peaks can be seen with a smaller aluminium peak also visible in the spectra in Fig. 4e. This confirms that the particles are silicon carbide. The existence of the aluminium peak may be a result of some of the backscattered electrons penetrating below the particle into the Al matrix.

There were no chemical reactions observed between the SiC reinforcement particles and the transition metals present in the alloy, even though the theoretical calculations suggested that some reactions should take place between the transition metals in the alloy and the reinforcement. Reasons for this will be outlined in the discussion section.

The alumina particles in Fig. 4a can be seen ranging in sizes between 0.2 and 2 µm, whilst the majority of the SiC particles in Fig. 4c are smaller than 100 nm in diameter. It is difficult to tell the exact sizes of these particles due to the limitation of the machine; however, TEM will be used in future studies to further characterise the SiC particles.

Rolled and Heat-Treated Microstructure

Figure 5a shows optical micrographs of the as-cast base alloy with 0.5 wt% alumina. It is clear that the particle dispersion in the rolled samples shown in Fig. 5b, c is better than that of the as-cast state (in Fig. 5a). It can be said that the rolling has caused the clustered alumina particles to disperse further in the matrix. A similar effect can be seen in Fig. 5d which represent the further dispersion of SiC particles due to rolling.

Fig. 3 Optical micrographs of **a** macro-distribution of alumina particles, **b** micro-distribution of alumina particles, **c** macro-distribution of SiC particles and **d** micro-distribution of SiC particles (magnified from **c**)

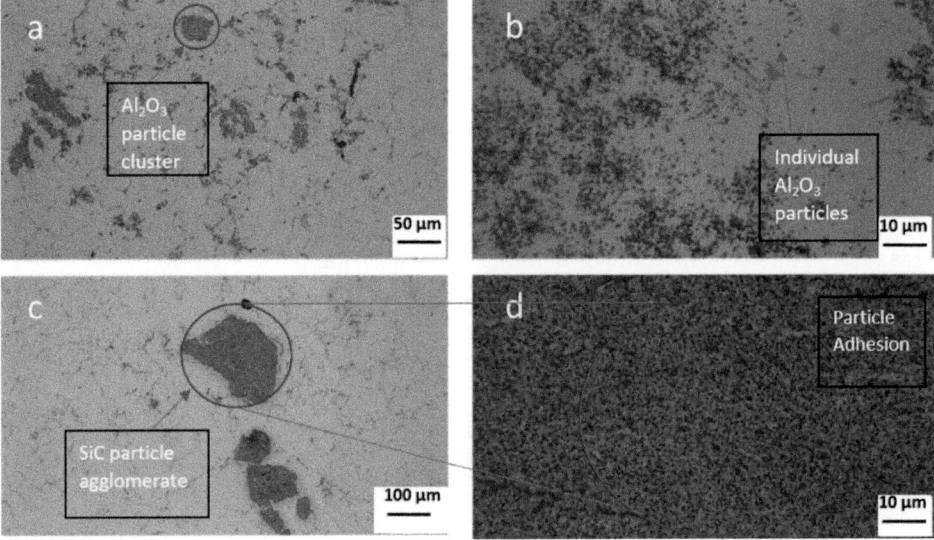

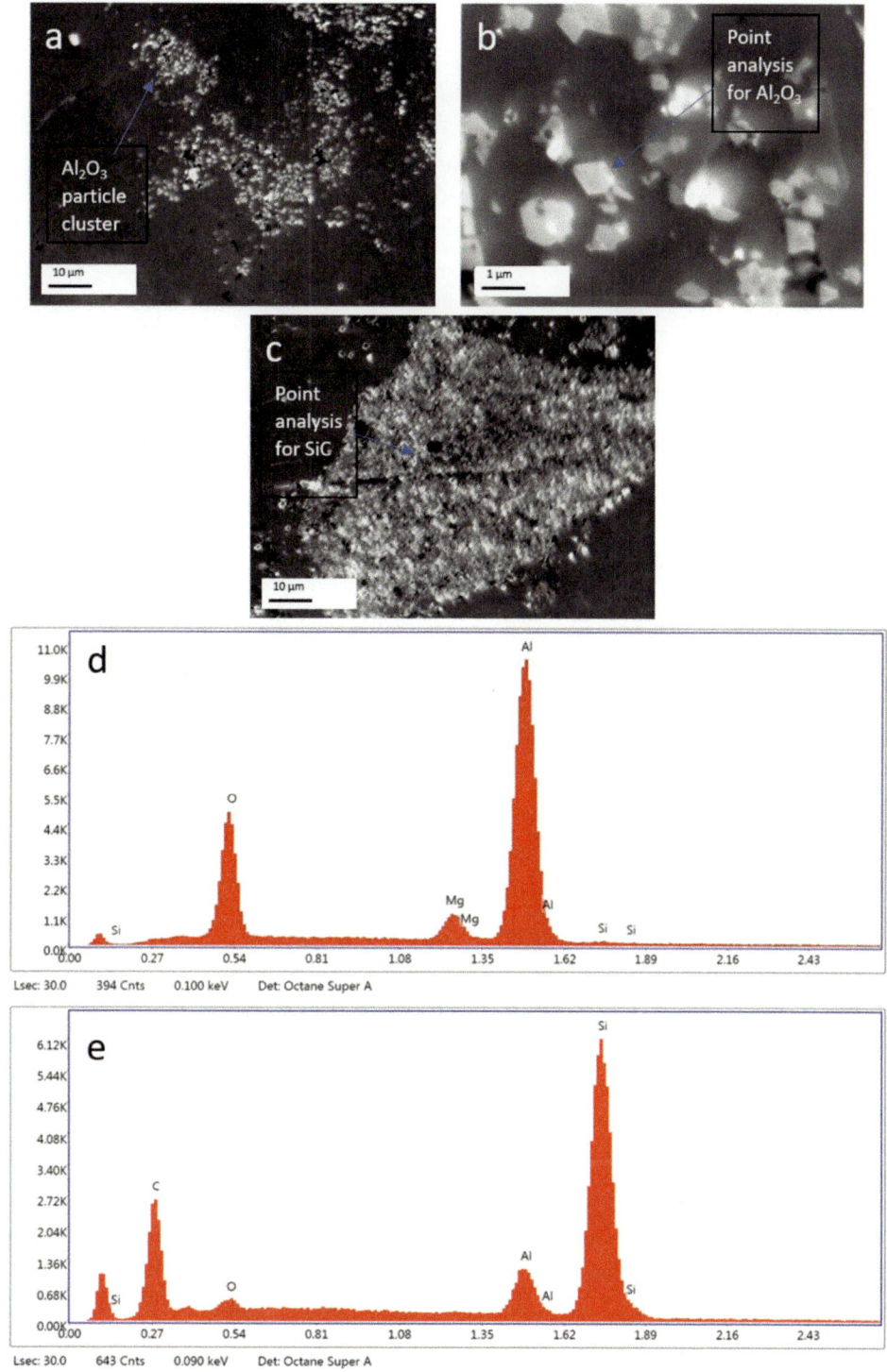

Fig. 4 **a** SEM micrograph of alumina clusters, **b** SEM micrograph of individual alumina particles, **c** SEM micrograph of a SiC agglomerate, **d** EDX spectra of an alumina particle, and **e** EDX spectra of a SiC particle

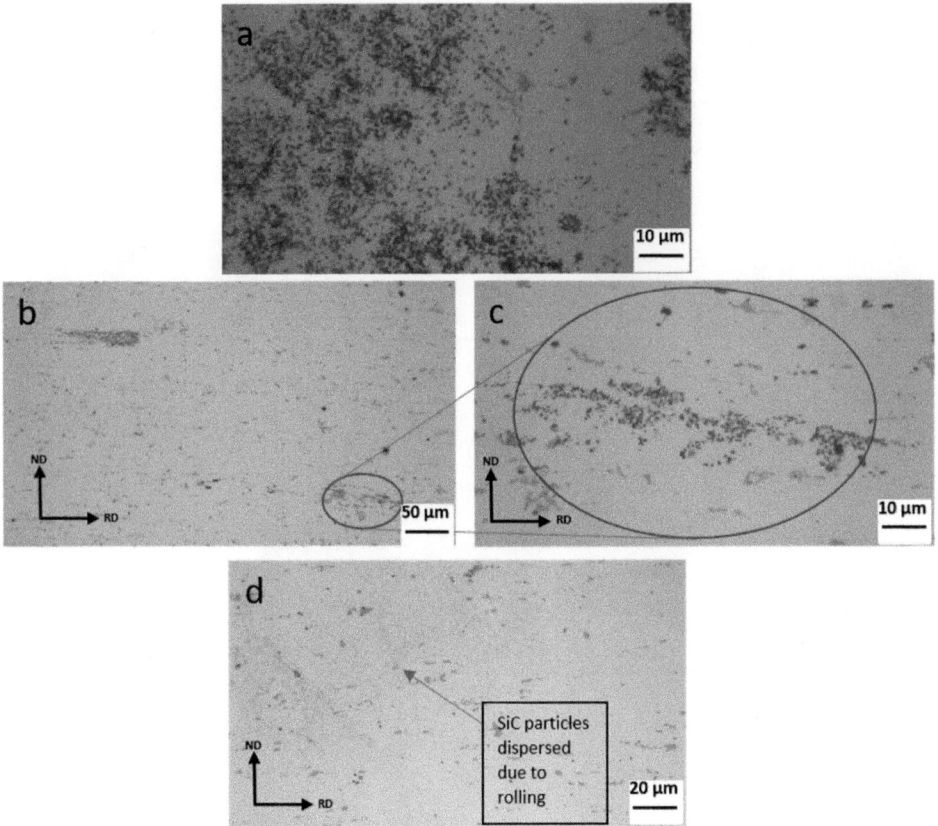

Fig. 5 Optical micrographs of composite materials, **a** as-cast base alloy with 0.5 wt% alumina before rolling, **b** rolled base alloy with 0.5 wt% alumina, **c** magnified from **b**, **e** base alloy with 0.5 wt% SiC

Homogenously distributed, fine-sized dispersoid particles were observed in the heat-treated rolled samples. The particles can be seen in Fig. 6. Particles can be seen ranging in size from approximately 100 nm up to almost 1 μm. The dispersoids were observed in the base alloy casting and both of the composite material castings (0.5 wt% alumina and 0.5 wt% SiC). The EDX spectra shown in Fig. 6c was taken at one of the dispersoid particles shown in Fig. 6a. There is a main peak and further smaller peaks at Mn, Fe, and Si. This suggests that the dispersoid particles observed are α-Al15 (MnFe)3Si2. This is a common dispersoid phase in the wrought aluminium alloys [23]. Another common dispersoid phase in the wrought aluminium alloys is the Al₃Zr; however, it is likely that these are the smaller particles observed in Fig. 6a, b [23]. Due to their relatively fine size, these nano-sized particles were difficult to characterise using the EDX on the SEM employed. It is planned to study these particles in further detail using TEM, in future work.

Discussion

Particle Distribution

The alumina reinforcement particles that showed a better distribution than the SiC particles can be attributed to the difference in size between the particles. The alumina particles exhibited diameters in the range of 0.2–1 μm. Even larger alumina particles were observed, in some cases reaching up to 2 μm in diameter. On the other hand, the SiC particles often had diameters of less than 100 nm. This means that these particles had larger van der Waals forces than the alumina particles, which made them more difficult to disperse in the matrix. It has been reported in previous studies that the stir-casting technique is not effective for the purpose of dispersing particles in the nano-meter range [24]. This is why the use of ultrasound is generally implemented

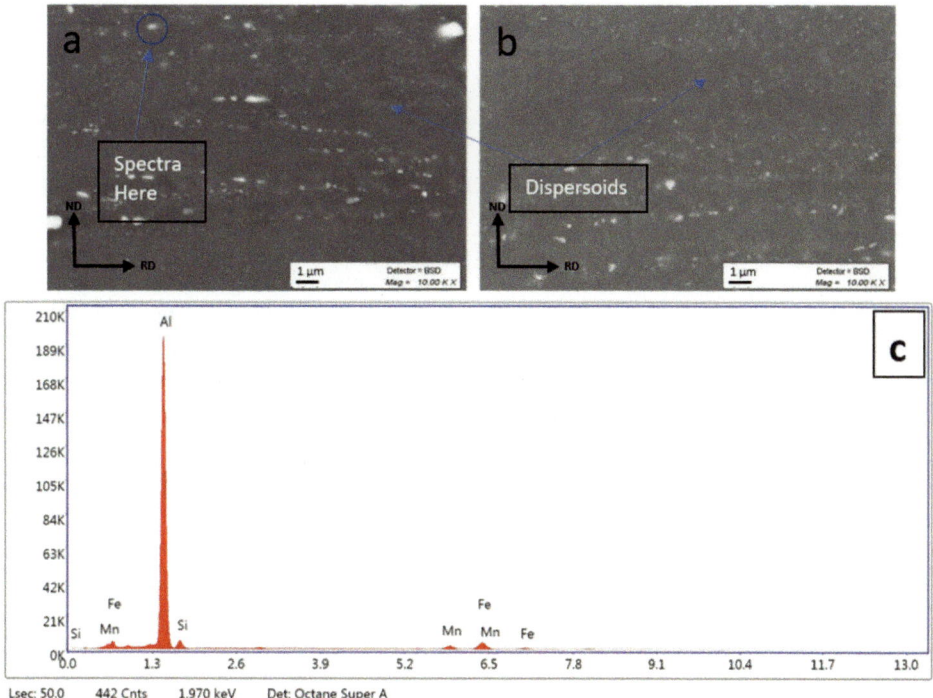

Fig. 6 SEM micrographs of rolled samples **a** base alloy dispersoids **b** composite material dispersoids and **c** spectra from dispersoid particle

in addition to mechanical stirring. However, the acoustic streaming and cavitation effect generally tend to occur directly below the ultrasonic sonotrode [7]. This means that in large volumes of melt, not all of the melt will benefit from the use of ultrasound.

Choi et al. [25] discussed the use of ultrasonication for the purpose of distributing nano-particles on a large scale. In the study, it was suggested that in order for the ultrasonication to be effective in large volumes of melt, all of the melt should be passed through a chamber with an active ultrasonic sonotrode. This method guarantees that almost all of the melt will at some point travel below the sonotrode and experience acoustic streaming and cavitation. This method may be experimented with in the future.

Previous studies have shown that alumina has poor wettability with aluminium [26] and this also contributed to the fact that there was no significant increase in mechanical properties. One of the key strengthening mechanisms mentioned is the load-bearing effect and this requires good adhesion between the particle and the matrix. Therefore, this strengthening mechanism has a very small effect when using alumina particles as a reinforcement due to the poor wettability. This means that using different types of particles as a reinforcement can result in greater increase in strength due to better wettability. An example of this is silicon carbide. Bao et al. [27] showed that the wetting angle of an Al–SiC system is approximately 79°, whilst the wetting angle of an Al–Al$_2$O$_3$ system is approximately 97°. However, the relatively small size of the silicon carbide particles used in this study caused them to agglomerate excessively which is likely to have had an impact on the properties in the as-cast state.

The improved distribution of particles in the matrix due to rolling is reported in the literature in several studies [28, 29]. Song et al. [28] found that 85% rolling reduction at 300 °C increased the distribution homogeneity of AlN reinforcement particles in the matrix. This in turn contributed to an increase in the strength and ductility of the material. However, Bisht et al. [30] reported observing cracks in the rolled materials due to agglomerated reinforcement particles. The material was rolled at 200 °C with 60% reduction. It is likely that the lower temperature contributed towards the cracking of the material, hence higher temperatures should be used in the cases of rolling composite materials with agglomerated particles.

Chemical Reactions Between Reinforcement and Matrix

The EDX spectra in Fig. 4d shows a small oxygen peak, which is likely due to the oxidation of the SiC particles during the synthesis process. The oxidation of SiC is used as a technique to treat SiC particles so that the processability of aluminium metal matrix composites containing SiC reinforcement can be improved [11]. The generation of a continuous SiO$_2$ layer allows for the incorporation of the particles in the molten aluminium [11]. This is because the degrading reaction of Al$_4$C$_3$ formation is replaced by alternative interfacial reaction which can enhance the wettability [11].

The EDX spectra in Fig. 4e shows the smaller Mg and Si peaks in addition to the main Al and O peaks, suggesting the appearance of spinel phase. The appearance of spinel has been reported when alumina reinforcement is used. For example, Forn et al. [31] reported observing a layer 1.5 μm in diameter with an irregular shape around the reinforcement particles. XRD was used to confirm the presence of MgAl$_2$O$_4$. The author pointed out that the appearance of this phase is detrimental to the mechanical properties as the decohesion between the matrix and the reinforcement phase, according to the fractography study, might be attributed to the spinel creation in the interface [31]. It was suggested that the cohesion between the reinforcement and the matrix can be improved through a T6 heat treatment. This was one of the reasons why a T6 heat treatment was carried out in this study.

No chemical reactions were observed between the SiC reinforcement and the transition metals present in the alloy. This can be attributed to two main contributing factors.

Firstly, the holding time of the material in the furnace may have been insufficient for the reactions to take place. It was mentioned previously that as the reinforcement was added to the melt, the melt was stirred continuously for 10 min. The melt was then left to sit for approximately 5 min before the ultrasonic treatment was applied for 10 min. This means that since the incorporation of the reinforcement, the total time in which the melt and the reinforcement was in the furnace was approximately 25 min. Considering that the reinforcement was added in the form of a master powder, some time would have been needed for the aluminium in the master powder to melt, releasing the SiC particles. So, it is likely that 25 min was not enough time for the reactions to take place.

Secondly, the compositions of transition metals in the alloy may have been too low for any reaction to take place. None of the transition metals present in the alloy had any wt % greater than 0.15%. In the case of Zr poisoning, the effect had been observed in 7xxx series alloys [32]. 7xxx series alloys typically contain more Zr than the aluminium alloy used in this study as will be discussed later. Furthermore, Bunn et al. [33] reported observing the Zr poisoning effect in an alloy containing 5 at% Zr. This suggests that the 0.15 wt % Zr in the alloy was also insufficient for any reaction to take place with the reinforcement.

The presence of transition metals such as Zr, V, and Cr in the wrought aluminium alloys is important for the purpose of solid-state reactions with the soluble elements during the homogenization process [23]. Typically, these reactions produce the nano-sized hard particles known as dispersoids [23]. The thermally introduced stable dispersoids (during homogenisation) are an effective way to retard the dislocation motion and inhibit recrystallisation [34]. Zr can be found in greater weight percentages in 7xxx alloys compared to other wrought alloys. Yet, the poisoning effect has been observed with 7xxx series alloys containing as little as 0.10% Zr [32]. Some other wrought alloys contain up to 0.2 wt% Zr, meaning that this certainly has the potential for producing an effect similar to Zr poisoning.

These substitution chemical reactions in the melt will be exothermic reactions due to negative free energies. This will cause the cooling rate to be slower in certain localised areas or pockets in the alloy. These areas will have coarser grain sizes and this risks the introduction of weak spots in the material, hence reducing mechanical properties.

Further experimentation will be carried out to determine whether the above-mentioned substitution reactions will occur with higher wt% of transition metals and/or longer processing times. This phenomenon is not mentioned extensively in the literature due to the limited number of studies on Al MMC in the wrought alloys (7xxx, in particular). These two series are the most popular in the automotive and aerospace fields. Hence, if Al MMC are to be used in these industries, it will be imperative to understand which reactions will occur when carbide reinforcements are incorporated in the mentioned alloys and how these will ultimately affect the final properties.

Conclusions

The present work studied the effect of addition of silicon carbide and aluminium oxide reinforcement particles to an aluminium alloy. The base aluminium alloy and composite materials were DC cast, homogenised, hot rolled, and heat treated. Characterisation was subsequently carried out. The study investigated the distribution of the particles in the matrix before and after hot rolling, as well as chemical reactions between the reinforcement and the alloy. The following key points can be concluded from the study:

- Whilst the mechanical stirring and ultrasonication did not fully disperse the particles in the matrix, the Al$_2$O$_3$ particles were more dispersed than the SiC. This is most likely as a result of the difference in size between the two types of reinforcement (the SiC being smaller). The rolling contributed to a more homogenous distribution of the particles in the matrix for both types of reinforcement.
- SiC reinforcement was found to produce a desirable oxidation reaction, in which an SiO$_2$ layer is formed around the particles. This will increase the wettability of the particles with the matrix. Al$_2$O$_3$ particles produced a

non-desirable spinel phase which can reduce the mechanical properties; however, according to existing literature, the negative effect of spinel can be reduced through a T6 heat treatment [31].

- Theoretical calculations of the free energies of formations of different carbides showed that some transition metals in the alloy should substitute the silicon in the SiC reinforcement. However, no such reactions or substitutions were observed in the physical experiments. This is likely as a result of the compositions of the transition metals in the alloy being relatively low, and the holding times of the melt being too short.

Acknowledgements The authors gratefully acknowledge the financial support of the European Union (FLAMINGo Grant Agreement No. 101007011) on this project. Constellium and MBN Nanomaterialia are also gratefully acknowledged for providing the aluminium alloys and the reinforcement nano-powder, respectively. Finally, the Experimental Techniques Centre at Brunel University is thanked for providing access to the equipment to conduct the characterisation work.

References

1. Energy.gov. 2020. *Lightweight Materials for Cars and Trucks.* [online] Available at: <https://www.energy.gov/eere/vehicles/lightweight-materials-cars-and-trucks> [Accessed 5 January 2020].
2. Professional Engineering, 2022. Three final FlyZero concepts show vision of the future for aerospace. [online] Available at: <https://www.imeche.org/news/news-article/three-final-flyzeroconcepts-show-vision-of-the-future-foraerospace#msdynttrid=K_LPLL2KQUnbzxeUVobarY3IpJJiawF1gw4HnuTNfQ0> [Accessed 2 February 2022].
3. Lazarova, R., Bojanova, N., Dimitrova, R. *et al.* Influence of Nanoparticles Introducing in the Melt of Aluminum Alloys on Castings Microstructure and Properties. *Inter Metalcast* **10,** 466–476 (2016).
4. Ramanathan, A., Krishnan, P.K., Muraliraja, R., A review on the production of metal matrix composites through stir casting – Furnace design, properties, challenges, and research opportunities, Journal of Manufacturing Processes, 2019, ISSN 1526-6125.
5. Ajith, C., M et al matrix composites: History, status, factors and future. Dissertation (2011).
6. Malaki, M., Xu, W., Kasar, A.K., Menezes, P.L., Dieringa, H., Varma, R.S., Gupta, M. Advanced Metal Matrix Nanocomposites. *Metals.* 2019.
7. Shinde, D.M., Sahoo, P. Fabrication of Aluminium Metal Matrix Nanocomposites: An Overview. In: Sahoo, S. (eds) Recent Advances in Layered Materials and Structures. Materials Horizons: From Nature to Nanomaterials. Springer, Singapore (2021).
8. Rasidhar, Lanka & Krishna, A. & Rao, Ch.Srinivasa. Fabrication and investigation on properties of Ilmenite (FeTiO3) based al-nanocomposite by stir casting process. International Journal of Bio-Science and Bio-Technology. 5. 193–199 (2013).
9. Ma Z.Y., Li Y.L., Liang, Y., Zheng, F., Bi, J., Tjong, S.C., Nanometric Si3N4 particulate-reinforced aluminum composite, Materials Science and Engineering: A, Volume 219, Issues 1–2, 1996, Pages 229–231, ISSN 0921-5093.
10. McLeod, A.D., Gabryel, C.M. Kinetics of the growth of spinel, MgAl$_2$O$_4$, on alumina particulate in aluminum alloys containing magnesium. *Metall Mater Trans A* **23**, 1279–1283 (1992). https://doi.org/10.1007/BF02665059.
11. Ureña, A., Martínez, E.E., Rodrigo, P., Gil, L., Oxidation treatments for SiC particles used as reinforcement in aluminium matrix composites, Composites Science and Technology, Volume 64, Issue 12, 2004, Pages 1843–1854, ISSN 0266-3538.
12. Bhanumurthy, K., Schmid-Fetzer, R., Interface reactions between SiC/Zr and development of zirconium base composites by in-situ solid state reactions, Scripta Materialia, Volume 45, Issue 5, 2001, Pages 547–553, ISSN 1359-462.
13. Yang, H., Qian, Z., Zhang, G., Nie, J., Liu, X., The grain refinement performance of B-doped TiC on Zr-containing Al alloys, Journal of Alloys and Compounds, Volume 731, 2018, Pages 774–783, ISSN 0925-8388.
14. Wang, Y., Fang, C.M., Zhou, L., Hashimoto, T., Zhou, X., Ramasse, Q.M., Fan, Z., Mechanism for Zr poisoning of Al-Ti-B based grain refiners, Acta Materialia, Volume 164, 2019, Pages 428–439, ISSN 1359-6454.
15. Park, J.H., Seo, H.S., Kim, K.Y., Alloy Design to Prevent Intergranular Corrosion of Low-Cr Ferritic Stainless Steel with Weak Carbide Formers. Journal of The Electrochemical Society, 162 (8) C412–C418 (2015).
16. Djukanovic, G., Aluminium Alloys in the Automotive Industry: A Handy Guide. In: *Aluminium Insider (2019).*
17. Tempelman, E., Shercliff, H., Ninaber van Eyben, B., Chapter 5 - Extrusion of Metals, Manufacturing and Design, Butterworth-Heinemann, 2014, Pages 69–83, ISBN 9780080999227.
18. Sarafoglou, PI., Serafeim, A., Fanikos, IA., Aristeidakis, J.S., Haidemenopoulos, G.N., Modeling of Microsegregation and Homogenization of 6xxx Al-Alloys Including Precipitation and Strengthening During Homogenization Cooling. Materials (Basel). 2019 May 1;12(9):1421. https://doi.org/10.3390/ma12091421.
19. Rinderer, B., The Metallurgy of Homogenisation. Materials Science Forum (Volume 693). https://doi.org/10.4028/www.scientific.net/MSF.693.264.
20. Smithells, C., Brandes, E., Brook, G. and Smithells, C., 1992. *[Metals reference book].* Oxford: Butterworth-Heinemann, pp. 29–19.
21. Shatynski, S.R. The thermochemistry of transition metal carbides. *Oxid Met* **13**, 105–118 (1979). https://doi.org/10.1007/BF00611975.
22. Webbook.nist.gov. 2022. *NIST Chemistry WebBook.* [online] Available at: <https://webbook.nist.gov/chemistry/> [Accessed 20 September 2022].
23. Lumley, R., 2010. *Fundamentals of aluminium metallurgy.* p. 370–374.
24. Mohanty, P., Mahapatra, R., Padhi, P., Ramana, CH.V.V., Mishra, D.K., Ultrasonic cavitation: An approach to synthesize uniformly dispersed metal matrix nanocomposites—A review, Nano-Structures & Nano-Objects, Volume 23, 2020, 100475, ISSN 2352-507X.
25. Choi, Hongseok & Cho, Woo-hyun & Hoefert, Daniel & Weiss, David & Li, Xiaochun. (2013). Scale-up Ultrasonic Processing System for Batch Production of Metallic Nanocomposites, American Foundry Society.
26. Bao, S., Tang, K., Kvithyld, A. *et al.* Wettability of Aluminum on Alumina. *Metall Mater Trans B* **42**, 1358–1366 (2011). https://doi.org/10.1007/s11663-011-9544-z.
27. Bao, S., Tang, K., Kvithyld, A., ENGH, T., Tangstad, M., *et al.* Wetting of pure aluminium on graphite, SiC and Al2O3 in aluminium filtration, Transactions of Nonferrous Metals Society of China, Volume 22, Issue 8, 2012, Pages 1930–1938, ISSN 1003-6326.

28. Song, L.; Lu, F.; Jin, F.; Nie, J.; Liu, G.; Zhao, Y. Effect of Cold and Warm Rolling on the Particle Distribution and Tensile Properties of Heterogeneous Structured AlN/Al Nanocomposites. *Materials* **2020**, *13*, 4001. https://doi.org/10.3390/ma13184001.

29. Nie J.F., Wang F., Li Y.S., Liu Y.F., Liu X.F., Zhao Y.H. Microstructure and mechanical properties of Al-TiB$_2$/TiC in situ composites improved via hot rolling. *Trans. Nonferr. Met. Soc.* 2017;27:2548–2554. https://doi.org/10.1016/S1003-6326(17)60283-8.

30. Bisht, A., Kumar, V., Li, L.H., Chen, Y., Agarwal, A., Lahiri, D., Effect of warm rolling and annealing on the mechanical properties of aluminum composite reinforced with boron nitride nanotubes, Materials Science and Engineering: A, 2018, Pages 366–373, ISSN 0921-5093.

31. Antonio Forn, M. Teresa Baile, Elisa Rupérez, Spinel effect on the mechanical properties of metal matrix composite AA6061/(Al2O3)

p, Journal of Materials Processing Technology, Volumes 143–144, 2003, Pages 58–61, ISSN 0924-0136.

32. Kearns, M.A., Cooper, P., Effects of solutes on grain refinement of selected wrought aluminium alloys Mater. Sci. Technol., 13 (1997), pp. 650–654.

33. Bunn, A.M. & Schumacher, P. & Kearns, Martin & Boothroyd, C. B. & Greer, A. (1999). Grain refinement by Al-Ti-B alloys in aluminium melts: A study of the mechanisms of poisoning by zirconium. Materials Science and Technology. 15. 1115–1123. https://doi.org/10.1179/026708399101505158.

34. Elasheri, A., Elgallad, E.M., Parson, N., & X.- G. Chen (2022) Improving the dispersoid distribution and recrystallization resistance of a Zr-containing 6xxx alloy using two-step homogenization, Philosophical Magazine, https://doi.org/10.1080/14786435.2022.2103597.

TRC Combi Box: A Compact Inline Melt Treatment Unit for Continuous Casting

M. Gorsunova-Balkenhol, M. Badowski, M. Betzing, J. Stotz, and Ø. Pedersen

Abstract

An efficient inline melt treatment process is a key prerequisite for aluminum producers to ensure compliance with today's product requirements. Modern DC casting lines use an equipment chain including Argon degassing, grain refiner addition, and mechanical filtration to reduce hydrogen and non-metallic inclusions to low levels. Individual units of this process chain have been adapted to twin roll casting (TRC) conditions with low melt flow rates, long casting campaigns, and high temperature sensitivity, while a holistic concept is pending. Speira and Drache initiated a project to develop a TRC degassing and filtration unit based on joint experience combining Argon degassing, CFF filtration and immersion heater temperature management into a user-friendly single box concept with small footprint and high metallurgical performance. This paper describes the technological approach for degassing, filtration and electrical heating, the integration into a single-unit concept and initial results of the pilot installation at Speira Karmøy.

Keywords

Combi box • Melt treatment • Continuous casting

M. Gorsunova-Balkenhol (✉) · M. Badowski
Speira GmbH, Research & Development, Georg-Von-Boeselager
Str. 21, 53117 Bonn, Germany
e-mail: margarita.gorshunova@speira.com

M. Betzing · J. Stotz
Drache Umwelttechnik GmbH, Werner-Von-Siemens-Straße
24-26, 65582 Diez, Germany

Ø. Pedersen
Speira Karmoy, Hydrovegen 160, 4265 Håvik, Norway

Background and Scope

Production of aluminum rolled products is a multistage process, where the quality of a final product is strongly affected by the cleanliness of the cast aluminum melt. Even minor impurities in the melt can cause formation of harmful defects in the solid product, which lead to loss of material in scrap. Expenses of such a failure may rise significantly, depending on which production stage it was detected. Today's quality requirements for aluminum products are very high, therefore an efficient melt treatment process has become a vital production step, as well as an important competing factor for aluminum manufactures. The main goal of the melt treatment is the elimination of harmful impurities, such as hydrogen, alkali metals, and non-metallic impurities. Whereas treatment of alkali metals is conducted in crucibles and furnaces, hydrogen and non-metallic inclusions can be most effectively removed during the inline treatment process chain.

Negative impact of hydrogen is well known in the aluminum industry. Excess of hydrogen in liquid aluminum leads to porosity in solidified ingots and formation of defects in plates or strips such as blister [1]. Various degassing aggregates for the removal of hydrogen have been developed in recent decades. Today the most common practice in Direct Chill (DC) casting is to remove hydrogen by injecting inert gas (mainly Argon) with a spinning rotor. In this process, dissolved monoatomic hydrogen from the melt, driven through the difference in partial pressure in the melt and in the gas bubbles, diffuses into the rising Argon gas bubbles. In the bubbles H_2 molecules are formed, which are then removed together with the inert gas bubbles on the melt surface. One of the most widely used apparatus of this kind is a box-degasser, such as Spinning Nozzle Inert Flotation (SNIF) or Aluminium Purification (Alpur) [2, 3]. It consists of a deep reactor, which is divided into several chambers—1 to 4 depending on the required capacity—and a lid, where rotors for each chamber are installed. The box-degasser

© The Minerals, Metals & Materials Society 2023
S. Broek (ed.), *Light Metals 2023*, The Minerals, Metals & Materials Series,
https://doi.org/10.1007/978-3-031-22532-1_135

provides a high hydrogen reduction efficiency and is quite simple in handling and maintenance. The main disadvantages of this reactor type are a large footprint and a relatively high melt volume remaining in the box between the drops, which does not allow fast alloy changes and requires additional heating.

Another concept developed to compensate disadvantages of the box-degasser is a launder degasser, such as Alcan Compact Degasser (ACD) [4]. It does not require heating and demands less space. A launder degasser consists of a wide launder segment and a lid, where rotors are installed. Due to a low reactive volume and smaller rotors, a higher number of rotors, typically 6–12, and higher inert gas flow rates are applied. Initially, the ACD is operated with addition of chlorine. When chlorine is eliminated a higher dross generation is reported. Hydrogen reduction rates are reported to be comparable with the box-degasser technology. The latest models of ACD are supplied with sealed lids, which is supposed to reduce dross generation and allow chlorine-free degassing.

The siphon inert reactor (SIR) represents another drain-free type of degasser. It was developed to combine the large reaction volume of the box-degasser with the flexibility of the launder degasser. Unique features of the SIR unit are the application of underpressure to rise the metal level in the box and a use of bottom-installed rotors for gas purging, which makes maintenance and handling more demanding. The hydrogen reduction efficiency of the SIR technology is quite good and is comparable with one of the convenient box-degassers [5]. Depending on the boundary conditions all mentioned equipment can reach in average a 50–70% reduction rate of hydrogen.

Non-metallic inclusions are another important aspect determining ingot quality. They vary strongly in form and origin, but all cause unwanted defects during processing of the cast ingot, such as tears and pin holes [1]. Although degassing is able to reduce the number of particles, mechanical filtration is more efficient and is essential for the assurance of the melt cleanliness. In DC casting the most commonly used technologies are ceramic foam and deep bed filtration. The principle of both systems is similar. They consist of a box divided into two chambers. In a bigger chamber a filter medium is positioned. Metal enters the box from the top and passes through the filter medium. Through the underflow in the lower part of the chamber, metal passes to the rising chamber and exits the box. In ceramic foam filtration (CFF) technology, the filtration medium consists of a thin ceramic foam plate made of alumina and some additives. The filtration efficiency of CFF depends on the porosity grade of the filter, a higher grade number corresponds to a finer pore size. The CFF technology is the most widely used due to a number of benefits such as easy maintenance and handling of the filter plates and low cost.

CFF is also very flexible regarding the change of an alloy because it is drained after each cast. Deep bed filtration (DBF) technology is known for higher filtration efficiency as CFF especially for the finer inclusion range. Filtration medium fills the whole chamber and consists of many layers of alumina balls and grits of different grain size, which can be stacked in different order. This construction provides a better efficiency but is also more demanding in maintenance. Due to a large metal volume remaining in the box between the drops, it also requires additional heating and is not flexible. Another filtration technology mostly used for high purity products is bonded particle filtration (BPF). BPF is also produced in the shape of a plate-like CFF but is made of silicon carbide grits. It is much denser and heavier than CFF, which makes handling more difficult. However, the filtration efficiency is very high. The cost of BPF is also much higher than CFF.

All the above-mentioned aggregates and technologies were mainly designed for the DC casting lines, where high flow rates and a batchwise casting mode are the main boundary conditions determining the process. In continuous casting, such as twin roll casting (TRC), opposite conditions with much lower flow rates and long casting campaigns are present. Besides, a much precise temperature management due to high sensitivity of the casting process has to be considered. So far, the main approach was to adapt existing DC casting equipment to continuous casting requirements, by reduction of the reaction volume. Several solutions are available on the market:

- Smaller SIR unit consists of one chamber with one rotor head, instead of double version of a standard SIR unit [6].
- There is also a version of ACD for the flow rates of 1,7 t/h available [4].
- Novelis PAE distributes smaller versions of the Alpur degasser and the Pechiney Deep Bed Filtration (PDBF) unit specially for continuous casting lines. Alpur model G3 uses a single or double rotor concept and a sealed lid [7]. The smallest version of PDBF is designed for flow-rates of 5 t/h by reduction of bed dimensions [8].

The main disadvantages of all existing concepts are still a quite oversized capacity and throughput, together with a large footprint and high operational and capital costs.

The Pyrotek Multicast filtration system represents an integrated solution, where mechanical filtration by means of a bonded particle filter and degassing with porous plugs are combined [9]. It was initially used in a continuous casting line in 80 s; however, for today's product requirements further advanced solutions are required. The main disadvantages identified by Speira Karmøy, where several units are installed, are the large footprint, inefficient temperature control, low hydrogen reduction efficiency of porous plugs, and high consumable cost for bonded particle filters. Moreover, a spare parts supply is very limited.

The Speira production site at Karmøy in Norway operates several TRC units which require a better technical solution to keep up with the increasing product demands and provide better operating conditions in compliance with modern safety requirements. In absence of a fitted solution, a project in cooperation with the company Drache Umwelttechnik GmbH has been initiated to develop a filtration unit specially tailored for the TRC line conditions and requirements. High particle and hydrogen removal efficiency as well as low temperature loss are the most critical requirements to be fulfilled under the following boundary conditions:

- Low metal flow rate of 1.5–2.5 t/h.
- Long casting campaign of 6–12 weeks and >2500 mt.
- Footprint of less than 3 m^2.

This paper summarizes basic considerations about the equipment concept and presents initial results of a pilot installation at the Speira Karmøy plant.

Concept Development

A concept of combining several melt treatment steps in one unit was inspired by previous work done at R&D Department of Speira (earlier Hydro Aluminium Rolled Products). In early 2000s, a longterm study on the filtration efficiency was conducted at Speira's R&D Department in Bonn. The results lead to a development of the XC-Filter concept, where ceramic foam filtration, deep bed filtration, grain refinement and heating elements were combined in one unit [10, 11]. The performance of the XC-Filter was investigated under different casting conditions on the dedicated production-scale casting center at Speira Rheinwerk plant. The key learnings from this development were that a ceramic foam filter can provide very high filtration efficiency and its capacity can substantially exceed the manufacturer specification. Furthermore, it was demonstrated that a combined filter box reduces the footprint of inline treatment, which also minimizes the heat loss. Additionally, it was learned that the application of immersion heaters can provide efficient temperature control for small metal volumes and throughputs.

Based on this experience, an integrated design concept combining degassing, mechanical filtration and temperature control was considered for the TRC lines in Karmøy. The design concept and functionality of each step are discussed in the following chapters.

Degassing

As mentioned before, in DC casting argon injection by several spinning rotors and a large reactor volume are used to ensure an efficient removal of Hydrogen for high flow rates of 20–50 t/h. The melt throughput in TRC is significantly lower, but an efficient reduction of hydrogen up to 75% is still required. Besides hydrogen reduction efficiency, low gas consumption, low surface turbulence, and impact of argon bubbles transport on heating and mechanical filtration had to be pursued. A water model study was performed at R&D Facility in Bonn to understand the degassing step for TRC and to determine the optimal process parameters. A cross section of 600 × 500 mm was selected as degasser chamber based on the footprint requirements and single-chamber hydrogen reduction rates at Speira sites. It was assumed that this cross section is still close to a square shape and supports a good argon bubble distribution. These dimensions were transferred into the model in 1:1 ratio. Different rotor head types, rotor speed, position of rotor head, and argon flow rates were tested in the model. The rotor head model Carré 140 from Gerken SA has performed the best regarding the bubble distribution and surface turbulence and was chosen for the pilot unit.

Based on the reactor volume and the highest flow rate, the minimum residence time of the melt in the degassing chamber was calculated to be about 8 min, from which a half-time period of approx. 4 min required to reduce the hydrogen concertation to 50% could be derived. A diagram on Fig. 1 demonstrates the monitored half-time period for oxygen reduction in the water model, depending on the rotor speed for different argon flow rates. To reach a desired hydrogen reduction of 75%, a double half-time period is required, which could be achieved by most of the parameter settings. Based on these results and the aim to minimize the energy and argon consumption, the lowest rotation speed of 200–250 rpm and a 10 l/min argon flow rate were set as a starting point for the pilot unit. Finally, even lower parameters were then defined in the commissioning phase.

The position of the rotor head was chosen to be 100 mm above the bottom with 75 mm underflow to ensure very low transfer of the gas bubbles into the heater chamber (which will be discussed in more detail below).

Filtration

Ceramic foam filters offer a significant advantage regarding cost and handling in comparison with other technologies such as bonded particle filters or bed filtration. As learned from XC-filter development, a high filtration performance

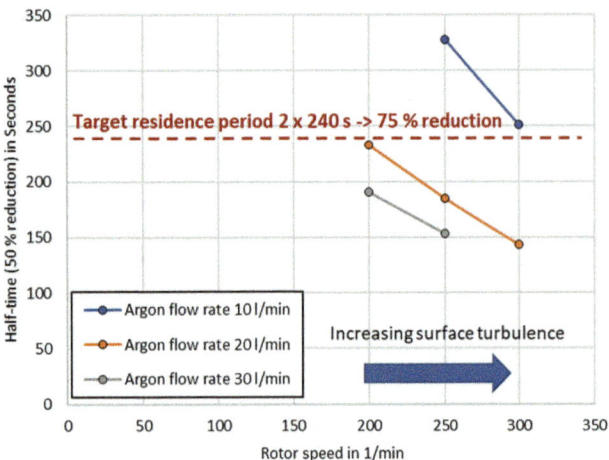

Fig. 1 Half-time period for oxygen reduction in the water model in dependence of rotor speed and argon flow rate for rotor head model Carré 140

can be achieved by CFF, and therefore it was decided to use this technology for the mechanical filtration step. The main parameters defining the filter layout are the flow rate and the filter capacity. A filter porosity of 30–40 ppi was chosen based on the quality of the melt and on the quality requirements for the final product. For the required filter porosity and TRC throughput of 1.5–2.5 t/h, a 7–9 inch CFF plate would be required. Based on the nominal filtration capacity more than forty 26-inch filter plates would be required for a standard casting campaign of 2500 mt. Even though XC-filter results showed that nominal capacity was significantly exceeded, the use of one filter plate for the whole casting campaign was not feasible.

A vertical gate CFF was chosen for the pilot unit. This solution provides more flexibility, since it can be exchanged during the cast if needed. The highest possible cross section of the plate close to 26 inch was chosen, which contributes to the extension of the filter plate life. Moreover, the vertical installation ensures a suitable priming height without increased launder depth by the height of the filter plate. Priming will likely start in the lower area of the filter plate and will move up during the use of the filter when particle deposition increases the local flow resistance. The sequential use of the filter increases the local melt velocity significantly above the average throughput and therefore closer to the recommended flow rates ensuring improved filtration performance. Nevertheless, the approach to exchange a CFF plate during running production is a new process and bears several risks, which need to be carefully assessed. A gripping tool was designed to ensure safe and ergonomic replacement.

Temperature Management

The TRC process demands a very precise temperature control, and therefore the temperature of the melt coming from the filter box has to be stable and easily adjustable. For this purpose, an immersion heating technology offers the best solution, which was already proven by the application in the XC-filter. An immersion heater consists of two main components. In the center, an electrical resistance heater embedded in a high-performance insulating material, which ensures electrical insulation and high heat transfer. This heating element is protected with a silicon nitride sheath. This material is insensitive to thermal shock, poses high mechanical resistance, excellent thermal conductivity and optimum resistance to corrosion.

Immersion heaters offer a whole range of benefits. The heating efficiency of immersion heaters is above 99%, which provides significant reduction of energy consumption in comparison with conventional heating systems. Temperature control accuracy lies in a range of ±2 °C. Heating up is fast and smooth. Immersion heaters are also easy to install and maintain. The main condition to assure all the benefits of the immersion heater technology is an unhindered heat transfer from the heater to the melt. Therefore, a perfect contact of the melt with the surface of the heater is essential for proper functionality and long life of the heater. In case gas bubbles or dross stick to the surface of the heater a local overheating occurs, which can cause damage of the immersion heater. To avoid the bubble transfer from the degasser chamber, the immersion heaters are positioned in a separate chamber. Both chambers are connected by an underflow the height of which was determined in the water model study for the degasser.

Considering the requirements of the TRC process, the heating system is supposed to provide a heating rate of 50 K/h in idle mode and an online temperature adjustment of 15 K at highest flow rate. A heat loss model was calculated to determine the amount of heat needed to maintain the required melt temperature. According to the model thermal inertia, the degassing process, radiation, and convection contribute to the total heat loss of approx. 15 kW (Fig. 2). An additional energy input of 12 kW would be required to fulfill the requirements of the plant. So, in total 27 kW should be generated by the heaters to ensure the proper heating conditions.

The heating power of the immersion heater can be calculated based on the heating zone length for a given diameter of the heater. Based on the lowest metal level in the box and required bottom clearance height a heating length of 310 mm was derived. Considering a specific heat flux of 20 W/cm² an

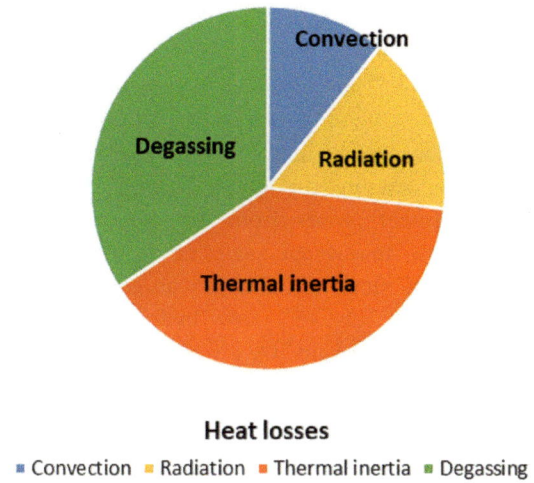

Fig. 2 Heat loss distribution in combi box calculated in a heat loss model

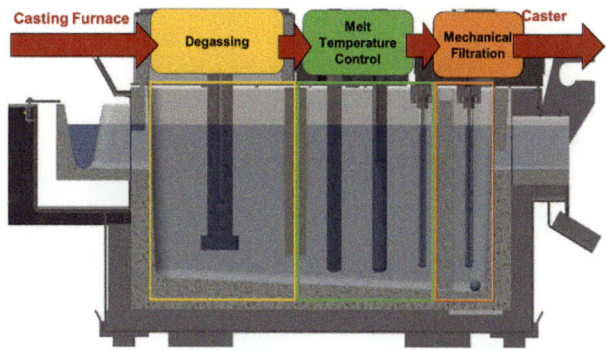

Fig. 3 A side view of the filter box showing the sequence of the chambers along the metal flow

immersion heater of 55 mm diameter can provide approx. 11 kW. Based on this evaluation two immersion heaters with the nominal capacity of 11 kW would be sufficient to compensate the heat loss of the box and a third heater would be necessary for heating up. Considering theoretical calculations three vertical immersion heaters with 12 kW heating power were chosen for the pilot unit. The heaters are arranged in a triangle to fit to the set dimensions and provide homogeneous heat distribution in the heating chamber. Two thermocouples are used for controlling of the heaters. One is for temperature control and another is a safety thermocouple for the melt level control, which prevents the heaters from overheating when the melt level is too low. Additionally, a thermocouple is positioned behind the CFF plate to track the temperature development in this area due to unknown heat conductivity of the CFF plate.

System Integration

After definition of a basic design of each process step, the chambers were set in the following sequence into a single unit:

- a first chamber for degassing,
- a second chamber for heating of the melt
- a third chamber for mechanical filtration

A cross section of a filter box along the metal flow is shown in Fig. 3. First metal enters the degassing chamber, through the underflow it comes to the next chamber, where it is heated up. After that metal passes through the vertical CFF plate and exits the box. The heater chamber and the CFF chamber are not physically separated. Afterwards the grain refiner is added in the trough.

This layout was chosen based on the following considerations:

- The addition of grain refiner after the box allows high filtration performance of the CFF independent of the position within the sequence.
- The CFF lifetime will be extended as the incoming inclusion load is already reduced by the degassing step.
- The immersion heaters are centrally collated with good heat transfer to degasser and CFF area. Lowest temperature expected downstream CFF.
- Low underflow between degasser and heater chamber inhibits a bubble transfer to the heater surface.
- The immersion heaters are located at the highest metal depth, which increases the heating length.

The final dimensions of the box are as follows 2000 × 1000 × 900 mm which fits to the space requirements on the casting line and provides enough space for operators to work. The metal volume in the box during the cast is approx. 1000 kg, and 700 kg has to be drained when the casting campaign is done.

Operational Concept

Besides the metallurgical performance, an improvement of working conditions and safety aspects was a very important aim for the new box. During the design phase various working situations were systematically analyzed to determine required conditions. The final design is presented on Fig. 4. The layout is organized in a way that all operations can be done from one side of the box ("operator side"). On the other side only a drainage bin is positioned.

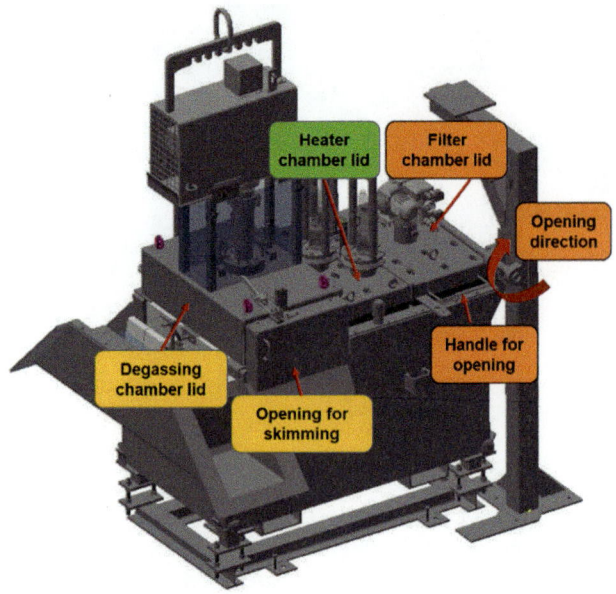

Fig. 4 An overview of the box design with the main components necessary for operation

The lid of the box is divided into three parts, one for each chamber. The lids were designed individually, and each lid can be lifted by crane. Dedrossing is one of the most frequent operations to be done on the box during the cast. In order to minimize dross formation, removable dams can be installed in the inlet and the outlet of the box. The dams are cut at the bottom which allows the melt flow in and out of the box but prevents entrance of air. An easy access for skimming is provided through the opening on the side of the degasser lid on the "operator side". A bottom part of the opening is lined with an inclined fused silica inlay which allows comfortable skimming and prevents sticking of the dross to the steel surface. Immersion heaters are mounted in the lid of the heater chamber and can be lifted together with the lid. Skimming of the heater chamber is supposed to be done through the opening of the filter chamber. The filter chamber has to be opened for skimming of the filter and heater chambers as well as for eventual changing of the filter plate. The lid of the filter chamber can be moved manually, for safe closing a counterweight is built in. Preheating of the box is done with hot air blowers from two positions. One is for the degasser chamber and another one is for the heater and filter chamber. The filter plate has to be pre-heated separately.

Initial Operational Results

The combi box was successfully installed and commissioned at Karmøy in January 2022. After the installation, several minor improvements had to be done, but the total concept is

functioning as expected. In April 2022, a first measurement campaign was conducted to evaluate a performance of the new filter concept. The results of the evaluation are presented below.

The hydrogen content in the melt before and after the combi box was measured by means of AlSCAN, with different measurement probes from Bomem and Accurity. Figure 5 demonstrates average values of measured hydrogen concentration and reduction efficiency in comparison with historical data. A relatively low hydrogen content of incoming melt under 0.3 ml/100 g was measured with both probes. The hydrogen concentration after the box was in average 0.087 ml/100 g measured with AlSCAN probe and 0.102 ml/100 g according to Accurity probe. The reduction efficiency was above 60%. These values mean a significant improvement of the degassing efficiency, since in previous years an average reduction efficiency of 37% was reached.

The filtration efficiency of the new box was evaluated by means of PoDFA. Sampling was done before and after the combi box and after addition of grain refiner. Figure 6 shows an overview of particle loading for measurements done in 2022 and in previous years from 2018 to 2020, when an old filter box was utilized. The particle concentration at the outlet of the combi box was very low in both samples (0.023 and 0.008 mm^2/kg) and a very high reduction efficiency up to 99% could be reached. However, the concentration of inclusions in the melt entering the box was higher than in years before, which could have promoted a better filtration efficiency through the cake formation. Based on the initial results it can be concluded that the reduction efficiency of the new filter concept is at least comparable with previously used bonded particle filters, which supports the statement about the high filtration efficiency of CFF technology.

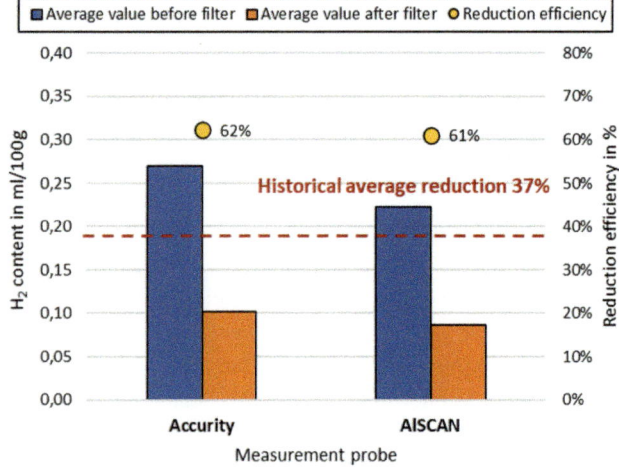

Fig. 5 Average hydrogen content in the melt before and after the combi box measured by AlSCAN and Accurity and corresponding reduction efficiency in comparison with historical data

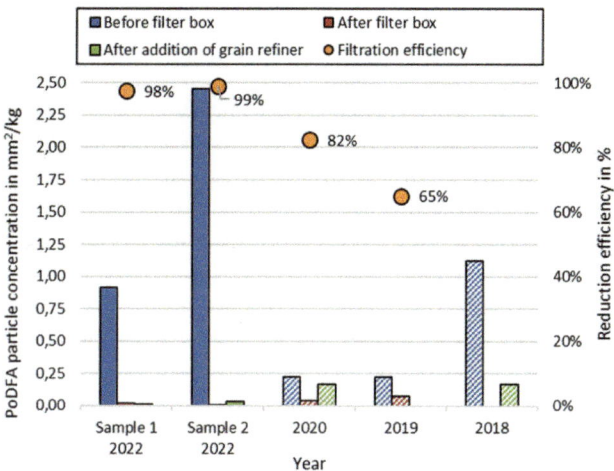

Fig. 6 PoDFA evaluation of particle concentration and reduction efficiency of the Combi Box measured in 2022 in comparison with previously used filtration unit from 2018 to 2020

Due to the ongoing adjusting of the process parameters, the CFF plate has been exchanged quite often and a real capacity has not been determined yet. At the time of PoDFA sampling, the CFF plate was 3 weeks in use and approx. 170 t of melt had gone through it, which is three times higher than the nominal capacity claimed by the manufacturer.

The evaluation of the temperature management is ongoing. The first results have demonstrated that the chosen configuration of the immersion heaters generates enough energy to maintain the melt temperature on the required level. In the steady state during the cast, the heaters use approx. 50% of their power capacity, which correlates with the data from the heat loss model quite well. The start of the new cast is the most demanding phase where the heaters are switched on full power. However, it is still in evaluation if the lower power input is also enough.

Very positive feedback is received from the casthouse personnel. The new box is reported to be more ergonomic and provides better working conditions. Much less lifting and manual work is required and the posture during the work is straight. The time required for the exchange of the filter plate has been significantly reduced from several hours to several minutes.

Summary and Conclusion

The new filter unit based on the specific requirement of the TRC process has been developed in cooperation with Speira GmbH and Drache Umwelttechnik GmbH. The base concept is a combination of three dedicated chambers for degassing, heating, and mechanical filtration in one unit. A pilot unit was successfully installed at Speira plant in Karmøy and has been running in a full-scale production mode since February 2022. No major changes in concept are considered. First results on the metallurgical performance of the box are very promising, although further long-term evaluations have to be done. The Karmøy casthouse plans to revamp another casting line with the Combi Box in 2023.

References

1. Schneider, W., Stranggießen von Al-Werkstoffen, VAW Aluminium AG 1991, 17–18
2. Frank, R.A., Mattocks, M.F., Recent developments in SNIF technology to provide improved refining capacity. Proceedings of the second international workshop on aluminium melt processing technologies, 1998, 119–128
3. Bildstein, J., Ventre, I., Alpur Technology – Present and Future, Light Metals 1990, TMS 1990, 755–763
4. Taylor, M.B., Maltais, B., Experience with the sealed Alcan compact degasser (ACD), Aluminium Cast House Technology 2005
5. Maland, G., Myrbostad, E., Venas, K., Hycast I-60 SIR – a new generation inline melt refining system., Light Metals 2002, TMS, 855–859
6. Hakonsen, A., Haugen, T., Fagerlie, J.O., Inline melt treatment for low to medium metal flow rates., Light Metals 2016, TMS 2016, 817–820
7. ALPUR® G3 cc Line degasser for continuous casting, http://www.novelispae.com/alpur-degasser/ Accessed 29 August 2022
8. Le Roy, G., Chaleau, J.-M., Charlier, P.h., PDBF: Proven filtration for high-end applications., Light Metals 2007, TMS 2007, 651–655
9. Bopp, J.T., Neff, D.V., Stankewicz, E.P., Degassing Multicast™ filtration system (DMC) – new technology for producing high quality molten metal., Light Metals 1987, TMS 1987, 729–736
10. Instone, S., Badowski, M., Schneider, W., Development of molten Metal filtration technology for aluminium., Light Metals 2005, TMS 2005
11. Counrtenay, J.H., Reusch, F., Instone, S., A review of the development of new filter technologies based on the principle of multistage filtration with grain refiner added in the intermediate stage., Light Metals 2011, TMS 2011, 769–774

CFD Modeling of Thin Sheet Product Using the Horizontal Single Belt Casting Method

Daniel R. Gonzalez-Morales, Mihaiela M. Isac, and Roderick I. L. Guthrie

Abstract

The Horizontal Single Belt Casting (HSBC) process involves casting a thin "river" of molten metal directly onto a fast-moving, continuously cooled, endless belt. By casting at higher cooling rates than is common, uniform solidification rates across unconstrained "thin sheets" of melt are maintained, yielding excellent cast properties and microstructures. A non-isothermal analysis of heat flows in the critical region where the melt first pours onto the water-cooled belt is made and compared with results from an experimental caster. For this, the CFD ANSYS-Fluent 19.1 code was run, using various combinations of melt processing parameters. It provided accurate predictions of the "windows" available for stable strips of AA6111, AA5182, and AA2024 alloys, to be cast. The HSBC's significant reductions in capital and operating costs, due to significantly reduced processing steps, lower energy consumption, and lower greenhouse gas emissions, make it a valuable alternative to current-day practices.

Keywords

CFD modelling • Horizontal Single Belt Casting (HSBC) • Meniscus • Thin strips

Introduction

Horizontal Single Belt Casting (HSBC) is one of the surviving Near Net Shape Casting Processes (NSSC), which were being developed since the 1980s, as an alternative to traditional casting technologies like Continuous Casting (CC) and Direct-Chill (DC) casting [5, 6]. The advantages of the HSBC process include very significant energy and cost savings with associated environmental advantages, since thin metal sheets are directly obtained from molten metal poured from a tundish or launder. The final sheets have shown excellent cast microstructures for final thermomechanical processing for a wide range of copper, aluminium, and steel alloys [2, 3, 13]. The HSBC was independently conceived and patented by R. Guthrie and J. Herbertson in 1988, in Canada [9].

The operational principle of the HSBC is that molten metal, contained within a tundish or launder, is poured with the help of gravity through a slot nozzle (~ 3 mm), directly onto a moving, water-cooled, horizontally moving belt. This belt provides for the progressive solidification of the liquid metal along it. A schematic of the pilot-scale caster used, along with its component's parts, is presented in Fig. 1. Extensive research of the HSBC process has been carried out, using both experimental (HSBC simulator and the HSBC pilot-scale system) and mathematical (computational fluid dynamics) routes, to determine the process parameters needed to obtain the desired properties of the final metal sheets. Some of the most important process parameters are the slot nozzle's liquid metal inlet velocity, the speed of the moving belt, the vertical clearance of the refractory backwall from the belt, and its inclination. The McGill Metals Processing Center (MMPC) has done extensive experimental work using the pilot-scale caster, processing a large variety of thin strips of steel (high-strength high-ductility steels, electrical transformer steels, TRIP and TWIP steels), various grades of aluminum (AA6111, AA2024, AA5182, etc.), and Cu–Sn–Ni alloys. The HSBC process has also been successfully implemented industrially in North America, where it is in commercial use for producing thin strips of copper alloys, by Materion Brush, USA. In Europe, a similar HSBC process has been in operation by Salzgitter Gmbh, at the steel plant in Peine, Germany, casting various grades of steel including high Mn-steel alloys [15].

At McGill/McGill Metals Processing Centre, the HSBC technology has been successfully used for processing thin

D. R. Gonzalez-Morales · M. M. Isac (✉) · R. I. L. Guthrie
Montreal, Canada
e-mail: Mihaiela.Isac@McGill.ca

© The Minerals, Metals & Materials Society 2023
S. Broek (ed.), *Light Metals 2023*, The Minerals, Metals & Materials Series,
https://doi.org/10.1007/978-3-031-22532-1_136

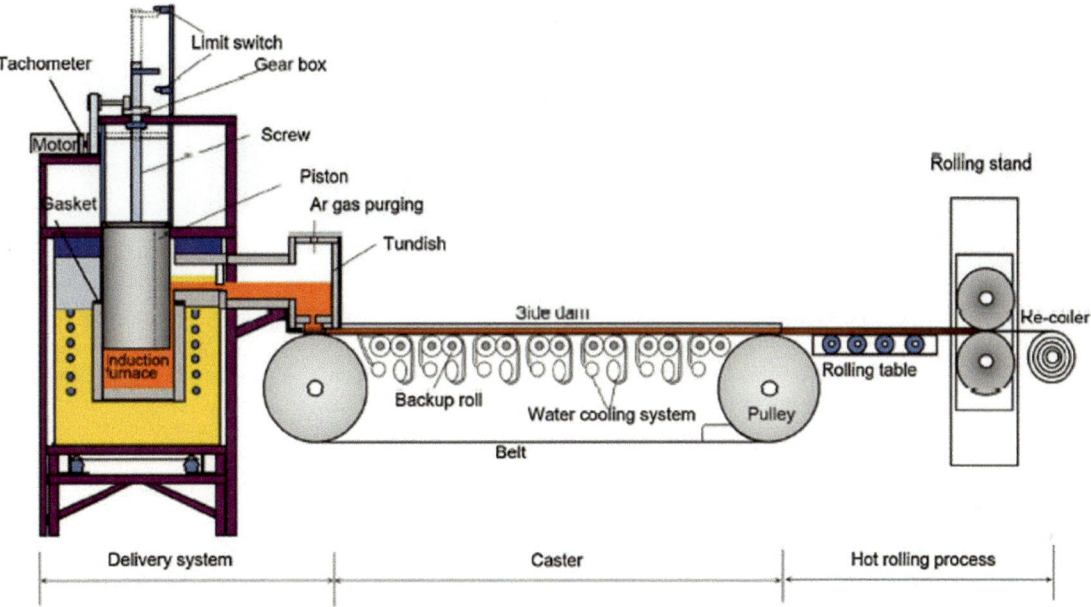

Fig. 1 Schematic diagram of a HSBC pilot-scale machine

strips of the aluminum AA2024, AA5182, and AA6111 alloys. The chemical composition and produced sheets dimensions are presented in Table 1. Figure 2 presents the pilot-scale HSBC system processing thin strips of AA6111 aluminum alloy, providing a view towards the feeding system (a) and a view towards the in-line pinch rolls-mini mill (b) in operation.

Optical microscopy of the three different aluminum alloys was then performed in order to evaluate the strips' microstructures. The microstructures thereby obtained are shown in Fig. 3 for the three cases, at 50 × magnification. For the AA2024 aluminum alloy [12], the final strip presented similar average roughness values on both top and bottom strip's surfaces. A fine, uniform equiaxed microstructure, Fig. 3a, was obtained across the thickness of as-cast thin strip product. For the AA5182 aluminum alloy [11], the resulting 5.0-mm-thick strip microstructure, Fig. 3 b, presented acceptable levels of porosity and an average grain size of 63.1 μm, practically half the grain size of the 123 μm found for a DC (Direct Chill) cast sample. Additionally, desired secondary phases were found within the grains, rather than at grain boundaries, due to the uniformly

high cooling rates across the strip, for the HSBC process. This translated into an improved strength and hardness compared to a DC cast product.

Regarding the use of the inclined feeding system, previous research has commented on the recommended practice for an isokinetic delivery of liquid metal onto the belt (inlet velocity = belt velocity) for reducing upper surface disturbances during casting. For the AA6111 alloy [13], HSBC pilot-scale experimental and CFD work was performed to compare the thin strip product quality when casting the AA6111 alloy (used in Ford F150 trucks), employing either single or double impingement feeding systems, with no side dams. The resulting strips again presented similar roughness values on both surfaces, with no surface indentations on the bottom surface. The strip microstructure, Fig. 3c, presented an average grain size of 85 μm with intermetallic phases distributed within the grains. Respective micrographs showing fine equiaxed grains, and some porosities (dark spots) were also detected, similarly to a DC cast product [13]. The resultant microstructures were suitable for rolling down directly into final sheet, without the need for performing additional long heat treatment annealing times

Table 1 Chemical composition and dimensions of aluminium alloys casted via HSBC pilot scale

	Al	Cu	Mg	Mn	Si	Zn	Thickness (mm)	Width (mm)
AA2024	Balance	3.8–4.9	1.2–1.8	0.3–0.9	<0.5	–	8.3	183
AA5182	Balance	–	4.0–5.0	0.2–0.5	<0.2	–	5.0	200, 250
AA6111	Balance	0.790	0.879	0.287	0.903	0.0007	6.0	250

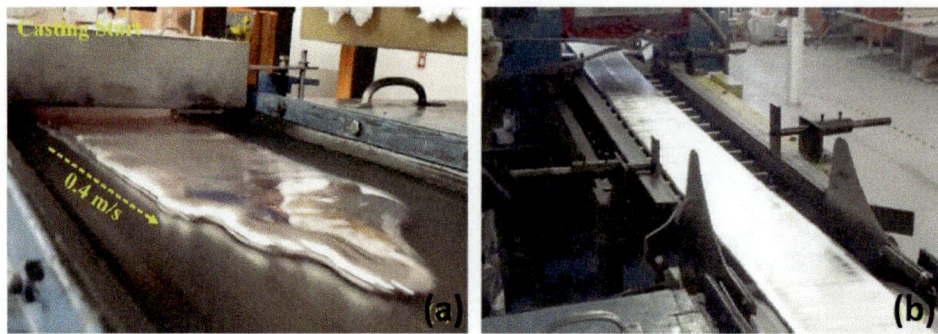

Fig. 2 Pilot-scale HSBC system casting thin strip of AA6111 aluminum alloy, **a** view towards the feeding system at the beginning of casting and **b** view towards the pinch roll/mini mill, during casting

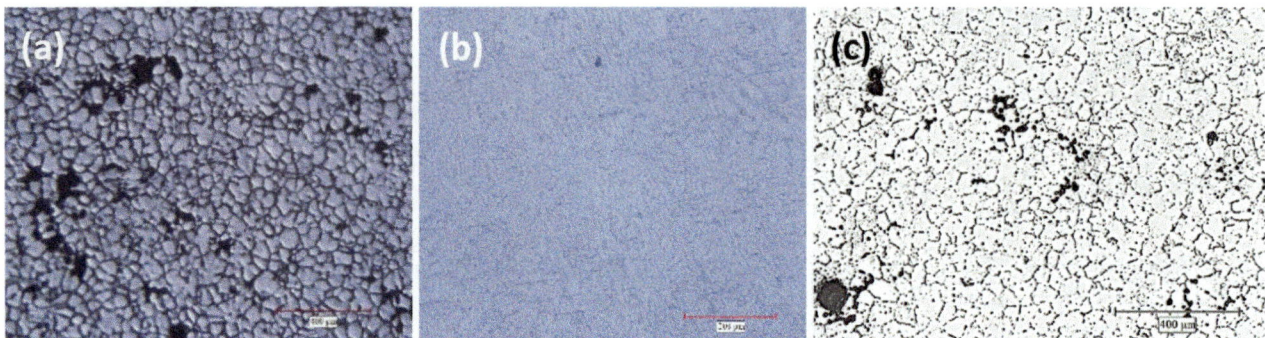

Fig. 3 HSBC obtained microstructures at 50 × for **a** AA2024 [12], **b** AA5182 [11], and **c** AA6111 [13] alloys

needed for DC cast material. This advantage for the HSBC process can be attributed to a more rapid and uniform rate of cooling across the thickness of the thin strip, plus measured higher initial heat fluxes ($\sim$10–20 MW/m^2) in the HSBC process, higher than typical values measured for conventional casting technologies ($\sim$1–3 MW/m^2) [1, 4].

Owing to the wide range of alloys cast via the HSBC process, several types of feeding systems have been developed to render the process more stable. Two main types of feeding systems have been researched: the single and double impingement metal feeding systems. Here, impingement refers to the number of times the flowing molten metal collides with an inclined wall or/and horizontal cooled belt. The double impingement system is normally achieved using a 45° inclination of a refractory backwall. In Fig. 4, a schematic of the single-impingement system is presented. Several analyses of the effects of the feeding system on the process have recently been reported [7, 8, 14, 16, 17].

More recent work on the HSBC process has focused on the isothermal modeling of the back meniscus stability during HCBC casting, using single [8] and double impingement [7] feeding systems, for an AA2024 alloy. The CFD work focused on analyzing the effect of the belt speed, back gap size, and contact angle, on the backflow and meniscus behavior. Back meniscus stability was identified as

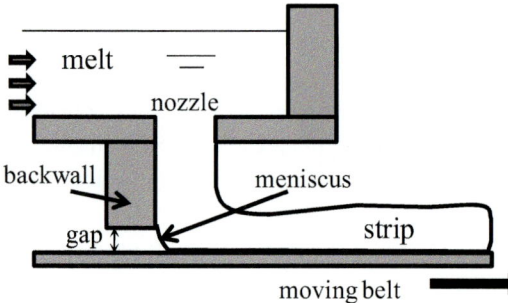

Fig. 4 Single impingement feeding systems for the HSBC process

being of paramount importance to obtain good bottom surface strip quality, since big flow instabilities, including backflows through the back gap between the moving belt and the stationary refractory backwall, will unavoidably lead to a poor-quality bottom surface. Optimal parameter combinations were predicted to promote a stable meniscus and finally a stable process, for avoiding product defects. The importance of the contact angle and its impact on the process stability were additionally studied. It was concluded that a non-wetting contact angle (>90°) is absolutely needed in avoiding backflows and providing a stable meniscus. Therefore, surface modification of the moving belt, or of the interfacial gas composition, is necessary for contact angle

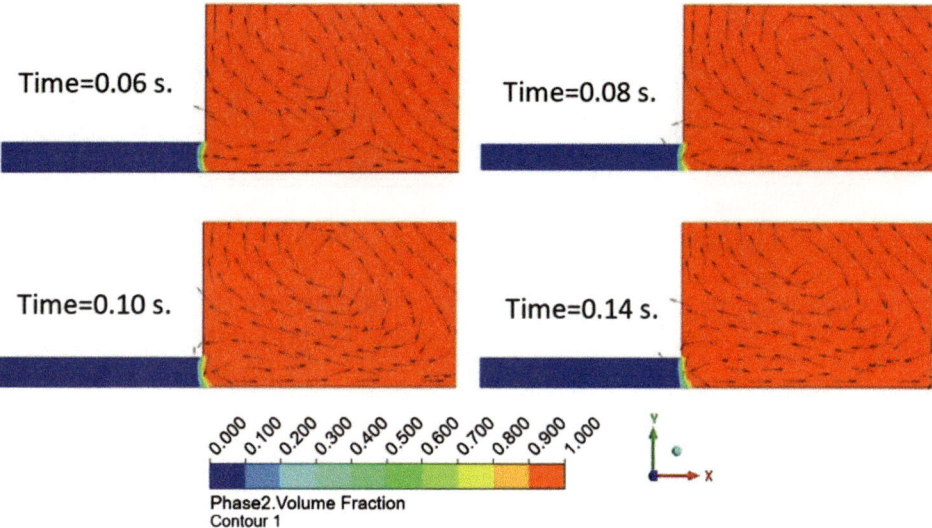

Fig. 5 Liquid aluminum volume fraction contours with velocity vectors map for the 0.5 mm back gap and 1.0 m/s belt speed at the meniscus zone case for a single-impingement feeding system, at different times [8]

modifications in the processing of some metals, or alloys, using HSBC. For a single-impingement system, it was concluded that a 0.5 mm back gap and a 1 m/s belt speed were needed to prevent significant backflows and provide for a stable back meniscus. These conditions were effective in promoting a stable HSBC process for the AA2024 alloy. Predicted isothermal liquid aluminum volume fraction contours with velocity vectors at different times are presented in Fig. 5 as an example of a predicted stable back meniscus [8].

Mathematical Modeling of the HSBC Process

For the present study, a 2D transient state, non-isothermal, multiphase model of the HSBC system for a single impingement was developed. For this, we included the addition of heat transfer and solidification. The mathematical study was performed for the three different commercial aluminum alloys, AA2024, AA5182, and AA6111, presented in Table 1, to review, and, if necessary, revise our

previous recommendations (back gap size of 0.5 mm and belt speed of 1.0 m/s) for promoting a stable back meniscus for the HSBC process. In the present study, the first 100 mm along the "x"-axis was considered to evaluate the potential for solidification of the liquid metal. The geometry of the systems was generated with the ANSYS SpaceClaim Software.

The geometry generated, together with important dimensions, are presented on Fig. 6. The mesh for the system was generated using ANSYS Meshing, with cells refinement in the impingements zones for better interphase accuracy. The generated meshes were mostly quadratic, with an average cell size of 1.5×10^{-4} m^2. The average number of cells was 120,000 cells, with an average orthogonality of 0.99, that being the characteristics needed for mesh-independent results. The boundary conditions are indicated, as well, in Fig. 6.

The melt inlet (3 mm) was defined with a Liquid Aluminum/AA2024 Alloy Volume Fraction of 1, an inlet velocity of 1.0 m/s and with a temperature of liquidus

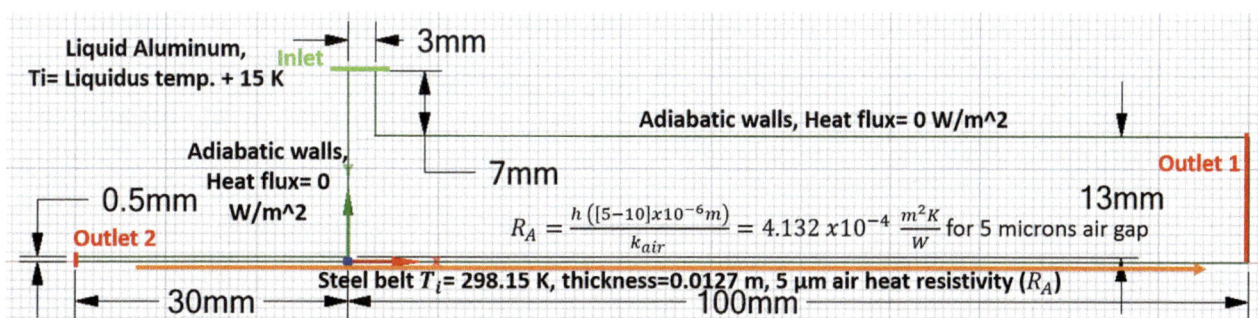

Fig. 6 Geometry generated in ANSYS SpaceClaim ©

Table 2 Materials properties, aluminum alloys

	AA2024	AA5182	AA6111
Density, kg/m³	2780	2600	2300
Viscosity, kg/m s	0.00122	0.00129	0.001338
Specific heat, J/kg K	875	904	1177
Thermal conductivity, W/m K	121	126	104
Standard state enthalpy, J/kg mol	1.1×10^7	1.1×10^7	1.1×10^7
Latent heat J/Kg	387,000	387,000	387,000
Solidus temperature, K	775.15	850.15	861.15
Liquidus temperature, K	911.15	911.15	923.15

Table 3 Material properties, air, and steel (belt material)

	Air	Steel (belt material)
Density, kg/m³	1.225	8030
Viscosity, kg/m s	1.7894×10^{-5}	–
Specific heat, J/kg K	1006.43	502.5
Thermal conductivity, W/m K	0.0242	202.4
Contact angle, °	120 with moving belt and 105 with adiabatic walls	
Surface tensions, N/m	0.914	

temperature +15 K. The moving belt conditions were defined as a moving wall with only an X-positive velocity component of 1.0 m/s for the three cases and an initial temperature of 298.15 K. A heat resistivity value was included, assuming a 5 μm air gap due to the texture and air entrainment, according to a previous detailed study [10]. All other walls were defined as being adiabatic, with non-slip condition. At time = 0, the system was assumed to be filled with air at 298.15 K. The properties of the liquid aluminum alloys AA2024, AA5182, and AA6111 are presented in Table 2 and the properties of the air and steel, belt material, are presented in Table 3.

The model was considered as transient state, with a variable time step starting at 1×10^{-6} s, maintaining a global Courant number of 0.25 for solution stability. The Volume of Fluid (VOF) multiphase model was applied to predict the two-phase interactions The k-ω Shear-Stress Transport (SST) Turbulence model was used to describe the effect of turbulence on the fluid flow. The equations and details of both models are described in a previous publication [8] and are omitted from this work, for brevity. However, the heat transfer and solidification modelling are now included in the present model, to confirm the functionality of

the previously defined optimal process parameters in the solidification phenomena. To model the heat transfer, the following equations were solved:

$$\frac{\partial}{\partial t}(\rho H) + \rho(\vec{v} \cdot \nabla)H = \nabla \cdot (k\nabla T) + S \qquad (1)$$

$$H = h_{\text{ref}} + \int_{T_{\text{ref}}}^{T} C_p dT + \beta LH \qquad (2)$$

$$\beta = \frac{T - T_{\text{solidus}}}{T_{\text{liquidus}} - T_{\text{solidus}}} \qquad (3)$$

Here H is the enthalpy, defined by Eq. (2), T is the temperature, k is the heat conductivity and S is a source term, defined more in detail in previous work [2]. The enthalpy is calculated via the h_{ref}, C_p, and LH, corresponding to the standard reference enthalpy, the specific heat, and the latent heat, respectively. β corresponds to the liquid fraction parameter and is defined by Eq. (3). The properties introduced into the momentum and heat transport equations were inputted according to the information provided in Table 1, for each simulation for each aluminum alloy.

Mathematical Modelling Results

The predicted liquid aluminum volume fraction contours for the three different alloys are presented in Fig. 7 at 0.66–0.70 s, when the process has reached a quasi-steady state. In general terms, no important differences are observed between the three cases, in the three cases no backflow was predicted, and a relatively stable back meniscus was also predicted. Under quasi-steady-state operation, no air entrainment is observed at the bottom surface and no discernible oscillations are observed on the top surface, meaning a stable process with a final good surface quality can be expected for the casting of the three different studied alloys (AA2024, AA5182, and AA6111). The volume fraction predictions agree well with the previous isothermal results [8] and confirm the effect of a 0.5 mm back gap and 1.0 m/s belt speed in the back meniscus stability, being an optimal set of parameters to promote a stable process after just around 0.70 s.

With the inclusion of the solidification process on the performed simulations, liquid aluminum mass fraction contours were calculated and are presented in Fig. 8 at 0.22 s, for the three aluminum alloys. Essentially, along the complete domain, no solidification is predicted, the molten metal is predicted to remain totally liquid during the first 100 mm of the HSBC process with a single-impingement feeding system for the three different cases. In the same predictions, some surface instabilities are observed at 0.22 s, especially for the AA6111 alloy. Said instabilities are expected to be marginal after 0.70 s, based on the previously discussed volume fraction contours. As well, these results indicate that the set of process parameters will not promote any solidification at the back meniscus part, which could curtail the process, due to the possible flow stagnation and solidification at the back meniscus. This was a concern of the previous study [8], owing to the anticipated low velocity at the first impingement zone, where a recirculation zone is formed. Additionally, this confirms that a 15 K superheat is adequate to avoid any solidification along the first 100 mm of the process using a single-impingement system.

Figure 9 depicts the predicted temperature contours for the three cases after 0.22 s of "start pour". The temperature scale is adjusted so as the low limit corresponds to the liquidus temperature of the AA2024 and AA5182 alloys, namely, 911.15 K. The respective temperature contours help us to detect where the melt will start to solidify. Based on the presented temperature contours, it is observed that solidification will start to take place around 80 mm from the back

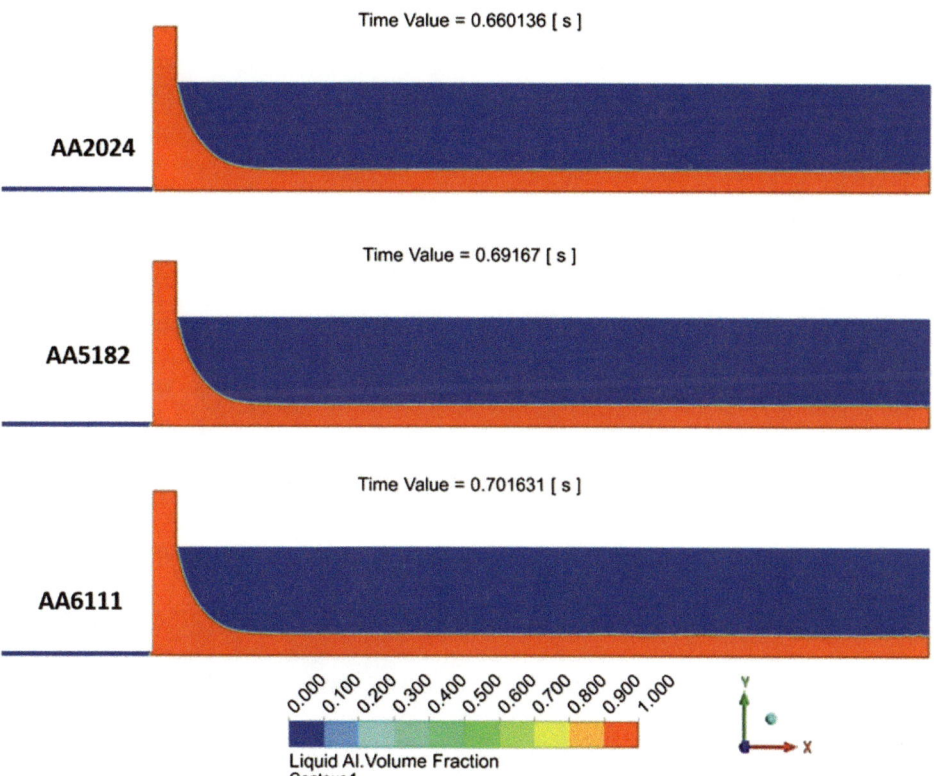

Fig. 7 Predicted liquid aluminum volume fraction contours for the three different alloys in quasi-steady-state operation

Fig. 8 Predicted liquid aluminum mass fraction contours for the three different alloys at quasi-steady-state operation

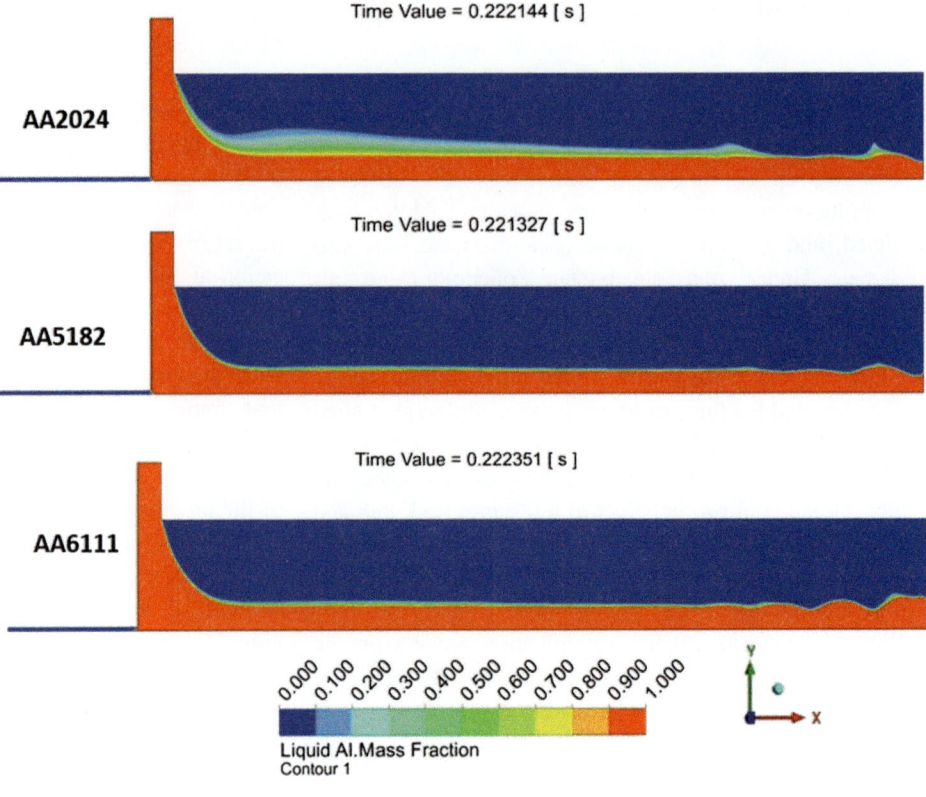

Fig. 9 Predicted temperature contours for the three different alloys at quasi-steady-state operation indicating distance along "x"-axis where solidification will start

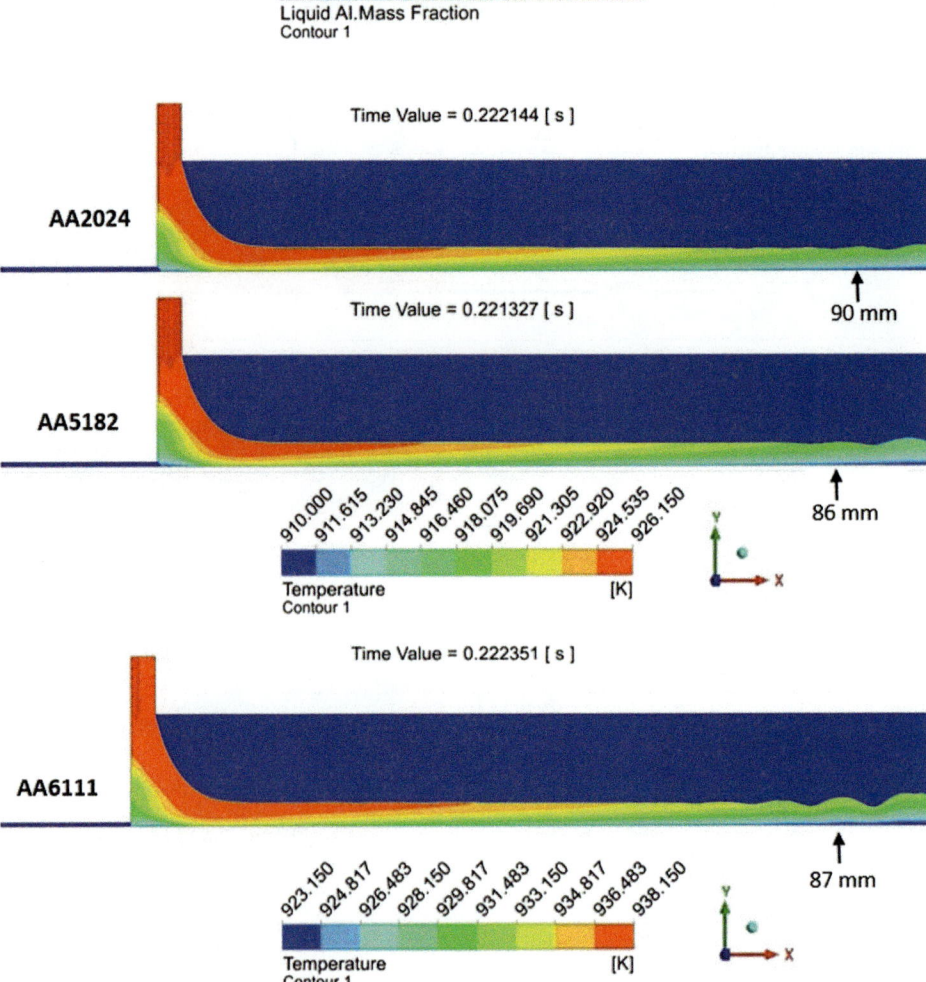

gap along the "x"-axis direction for the AA2024 and AA5182 alloys. The solidification will first start to take place from the bottom surface towards the top. However, since the temperature at the top surface is only about 10 degrees hotter than the bottom surface of the solidifying strip, and well within the solidus–liquidus temperature range, partial solidification occurs throughout the thickness of the strip, resulting in an equiaxed, as opposed to dendritic, microstructure. This is an inherent characteristic and advantage of the HSBC over conventional ingot casting methods. It can be observed that the already mentioned recirculation zone is responsible for an evident initial drop in temperature for the three cases of around 13 K for the AA2024 and AA5182 alloys and of around 15 K for the AA6111 alloy. Therefore, the alloys have lost practically all their superheat, but the hotter liquid above conducts heat downwards, such that solidification across the whole of the alloy, only initiates at about 90, 86, and 87 mm along the "x"-axis, for the AA2024, AA5182, and AA6111 alloys, respectively. It is

also worth remarking that a 15 K superheat is enough to promote a stable HSBC process window for a single-impingement feeding system. A detailed view of the first 20 mm on the "x"-axis of Fig. 9 results is presented in Fig. 10 and discussed below.

Figure 10 shows a magnified view of the temperature contours for the three different alloys, at quasi-steady-state operation, adjacent to the first point of contact with the belt. It is clearly seen that liquid temperatures drop near the back meniscus, due to the recirculation of the melt there, caused by a sudden change in the direction of the flow. The results for the AA2024 and AA5182 alloys are very similar, since they present the same liquidus temperature and other properties. For these two cases, it is observed that a tiny portion of the melt on the bottom surface reaches a temperature around 911.15 K, its liquidus temperature, but then the melt temperature again increases as it enters into contact with more of the top portion liquid at the inlet temperature of 926.15 K. The same process occurs for the AA6111 alloy,

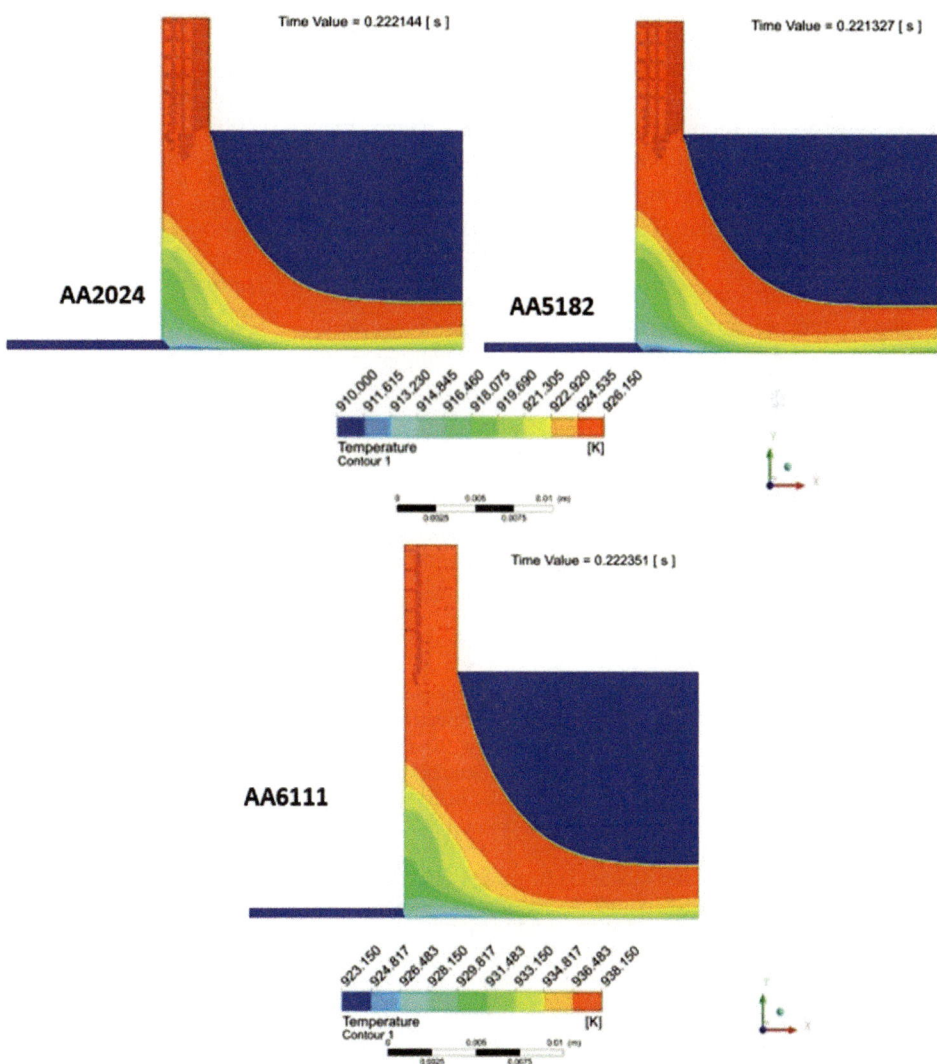

Fig. 10 Predicted temperature contours for the three different alloys, detail, at quasi-steady-state operation

but with its specific liquidus temperature of 923.15 K and inlet temperature of 938.15 K. For the three cases, merely a 5 K temperature difference can be observed between the top and bottom surfaces, 6 mm from the back wall, and this temperature difference will decrease along the "x"-axis. This fact explains why the HSBC process promotes minimal differences in the microstructure thickness-wise, together with homogeneous mechanical properties across the thickness of the strips, as previously observed experimentally [11–13].

Conclusions

- 2D multiphase simulations for three different aluminum alloys were performed, including the solidification process to assess the effect of a set of process parameters on the stability of the HSBC process.
- For the HSBC processing of thin strips of AA2024, AA5182, and AA6111 alloys, using a single-impingement feeding system, it was confirmed that a 0.5 mm back gap and a 1.0 m/s belt speed will avoid any backflows and important back meniscus instabilities, and will promote a stable process after 0.70 s of operation.
- No significant solidification was predicted along the first ~ 100 mm domain of the melt in contact with the belt as modelled, meaning that melt solidification does not interfere with the HSBC's process stability for the three cases studied. The solidification is predicted to start, around 87 mm from the back gap for the three different alloys, after all surface waves have subsided. This takes place after approximately 0.70 s for all three alloys.
- A 5 K temperature gradient was predicted across the thickness of the strip, during subsequent solidification along the belt. This will promote a uniform microstructure and properties across the thickness of the strip, thereby avoiding all casting defects associated with casting thicker strips, which require re-annealing and hot rolling, prior to hot rolling.

Acknowledgements The authors acknowledge the financial support received from the Natural Sciences and Engineering Research Council of Canada (NSERC) and the International Advisory Board of supporting companies of the McGill Metals Processing Centre (MMPC) in carrying out this research. The authors also acknowledge the support of the Quebec Centre of Research and Development of Aluminum (CQRDA), and the software licensing received from ANSYS Inc. to facilitate this research. Gonzalez-Morales D.R. would like to thank the Mexican National Consul of Science and Technology (CONACYT) for financial support for these graduate studies (CVU: 1036046).

References

1. Ge S, Celikin M, Isac M et al. (2015) Analysis and Evaluation of Novel Al-Mg-Sc-Zr Aerospace Alloy Strip Produced Using the Horizontal Single Belt Casting (HSBC) Process. Metallurgical and Materials Transactions B 46
2. Ge S, Celikin M, Isac M et al. (2014) Mathematical Modeling and Microstructure Analysis of Al–Mg–Sc–Zr Alloy Strips Produced by Horizontal Single Belt Casting (HSBC). ISIJ International 54:294–303
3. Ge S, Chang S, Wang T et al. (2016) Mathematical Modeling and Microstructure Analysis of Low Carbon Steel Strips Produced by Horizontal Single Belt Casting (HSBC). Metallurgical and Materials Transactions B 47:1893–1904
4. Ge S, Isac M, Guthrie R (2015) The Computational Fluid Dynamic (CFD) Modeling of the Horizontal Single Belt Casting (HSBC) Processing of Al-Mg-Sc-Zr Alloy Strips. Metallurgical and Materials Transactions B 46:2264
5. Ge S, Isac M, Guthrie RIL (2013) Progress in Strip Casting Technologies for Steel; Technical Developments. ISIJ International 53:729–742
6. Ge S, Isac M, Guthrie RIL (2012) Progress of Strip Casting Technology for Steel; Historical Developments. ISIJ International 52:2109–2122
7. Gonzalez-Morales DR, Isac MM, Guthrie RIL (2022) HSBC back-meniscus stability for a double impingement liquid metal feeding system In: Conference of Metallurgists Montreal, Canada
8. Gonzalez-Morales DR, Isac MM, Guthrie RIL (2022) Predicted Back-Meniscus Stability for Horizontal Single Belt Casting (HSBC) Using a Single Impingement Feeding System. In: Eskin D (ed) Light Metals 2022. Springer International Publishing, Cham, p 735–742
9. Guthrie RIL, Herbertson JG (1990) Continuous casting of thin metal strip. In, USA
10. Guthrie RIL, Isac M, Li D (2010) *Ab-initio* Predictions of Interfacial Heat Fluxes in Horizontal Single Belt Casting (HSBC), Incorporating Surface Texture and Air Gap Evolution. ISIJ International 50:1805–1813
11. Hsin C (2019) Production of automotive aluminum alloys AA5182 thin sheet material using the Horizontal Single Belt Casting (HSBC) process. In: McGill University, Montreal, Canada
12. Lee J, Isac M, Guthrie R (2018) Computational study and microstructural analysis of AA2024 strips processed via the horizontal single belt casting technique. In: COM and MS&T
13. Niaz U, Isac MM, Guthrie RIL (2020) Numerical Modeling of Transport Phenomena in the Horizontal Single Belt Casting (HSBC) Process for the Production of AA6111 Aluminum Alloy Strip. Processes 8:529
14. Niaz U, MineaIsac M, Guthrie RIL (2021) Horizontal Single Belt Casting of Thin Strips of an Advanced High Strength Steel (Fe–21%Mn–2.5%Al–2.8%Si–0.08%C wt%). steel research international 92:2000203
15. Wans J, Geerkens C, Cremers H et al. (2015) Belt casting technology experiences based on the worldwide first BCT caster. METEC & 2nd ESTAD
16. Xu M-G, Guthrie R, Isac M (2020) Transport phenomena in meniscus region of horizontal single belt casting. Transactions of Nonferrous Metals Society of China 30:3124–3132
17. Xu M, Isac M, Guthrie RIL (2018) A Numerical Simulation of Transport Phenomena During the Horizontal Single Belt Casting Process Using an Inclined Feeding System. Metallurgical and Materials Transactions B 49:1003–1013

Numerical and Experimental Investigation of Twin-Roll Casting of Aluminum–Lithium Strips

Olexandr Grydin, Kai-Uwe Garthe, Xueyang Yuan, Jette Broer, Olaf Keßler, Rostislav Králík, Miroslav Cieslar, and Mirko Schaper

Abstract

The main disadvantage of Al–Li alloys is their anisotropic behavior and limited formability. As conventional processes reach their limits, the formability of Al–Li alloys is to be improved using twin-roll casting. Twin-roll casting is an efficient process for the production of thin strips. The characteristic of this process is the combination of solidification and plastic deformation in a single unit. In this study, the twin-roll casting of Al–Li alloy using copper shells was represented by a numerical simulation through a thermofluid model implemented in ANSYS-Fluent. This simulation-based study investigates the influence of casting speed, casting temperature, strip thickness, and length of strip-forming zone on strip outlet temperature and specific size of deformation zone, thus optimizing the practical manufacturing process. The simulation determined parameters were used to manufacture strips, and the microstructure was analyzed.

Keywords

Twin-roll casting • Aluminum–lithium alloy • Numerical simulation • Process parameters • Thin strip • Microstructure

O. Grydin (✉) · K.-U. Garthe · X. Yuan · M. Schaper
Chair of Materials Science, Paderborn University, Paderborn, Germany
e-mail: grydin@lwk.upb.de

J. Broer · O. Keßler
Chair of Materials Science, Rostock University, Rostock, Germany

R. Králík · M. Cieslar
Faculty of Mathematics and Physics, Charles University, Prague, Czech Republic

Introduction

When scarce raw material resources, as well as the reduction of CO_2 emissions, are the focus of interest, the demands for future-oriented, individual, and economical manufacturing processes are constantly increasing [1]. A weight reduction of 100 kg already leads to fuel savings of 0.35 l/100 km and 8.4 g CO_2/km for gasoline engines [2], which is why the use of aluminum alloys and the associated weight reduction is of great importance. Since modern aluminum alloys have excellent specific stiffness and strength, steel parts can be replaced, reducing the weight by up to 40% [3, 4]. Research and development work continues to produce new or improved aluminum alloys, convinced by their outstanding performance potential. In particular, aluminum alloys with high lithium (Li) additions have unique properties superior to conventional aluminum alloys [5]. Recently, Al–Li alloys have thus attracted attention for their use in light construction and aerospace structures due to their impressive material properties, such as low density and high specific stiffness [6–8]. One wt.% Li reduces the density of the resulting Al alloy by about 3% and increases Young's modulus by about 6% [8]. Moreover, adding Li to Al results in fine precipitates that improve the strength of Al alloys [9]. However, the main disadvantage of Al–Li alloys is their anisotropic behavior with limited formability. It is due to different solidification phases and grain structure characteristics that occur during permanent mold casting and subsequent thermomechanical treatment [8]. As conventional methods reach their limits, processes for adapting the alloy chemistry and innovative casting and thermomechanical processing methods must be used. For example, the formability of Al–Li alloys is to be improved by combining twin-roll casting and thermomechanical treatment. Twin-roll casting is an energy-saving and efficient innovative process for producing thin strips [10, 11]. The characteristic of this process is the combination of solidification and plastic deformation of the metal in a single unit [12, 13]. This process produces near-net-shape strips

© The Minerals, Metals & Materials Society 2023
S. Broek (ed.), *Light Metals 2023*, The Minerals, Metals & Materials Series,
https://doi.org/10.1007/978-3-031-22532-1_137

directly from the molten metal [10]. In this study, the twin-roll casting of Al–Li alloys using copper shells was represented by numerical simulation using a thermofluid model implemented in ANSYS-Fluent. This simulation-based study investigates the influence of casting speed, casting temperature, strip thickness, and length of strip-forming zone on strip outlet temperature and specific size of deformation zone, thus optimizing the practical manufacturing process. The parameters determined through the simulation were used for the aluminum–lithium strip production, and the microstructure was analyzed on the basis of these parameters.

Numerical Finite Volume Model of the Twin-Roll Casting Process

The numerical finite volume model was developed. The theoretical analysis of the effects of different influencing variables of the twin-roll casting (casting speed, casting temperature, strip thickness, and length of the strip-forming zone) on the strip outlet temperature and the specific size of the deformation zone was performed. This model was implemented by using ANSYS-Fluent software. The processing of fluidic problems with the aid of fluidic simulation is divided into four work steps: preparation (model creation), pre-processing (meshing), solving (defining boundary conditions and solving model equations), and post-processing (graphical processing and presentation of results). The assumption that the melt was evenly distributed over the strip width and supplied to the strip-forming zone was made to accelerate calculations. Therefore, minor temperature and flow velocity changes along the strip width were not considered, and a two-dimensional model was developed.

Preparation

In the preparation step, the model of the twin-roll casting was created. The strip-forming zone and the copper shells were represented in a simplified form as a two-dimensional model. Figure 1 shows the dimensions of the twin-roll casting tool. The inner and outer copper shell radii are fixed, whereas the strip-forming zone length and the strip thickness are varied. The values in Fig. 1 correspond to basic parameters taken for the simulation, which amount to 50 mm for the strip-forming zone length and 3 mm for the strip thickness. Based on the drawing, a CAD model representing melt and copper shells was created.

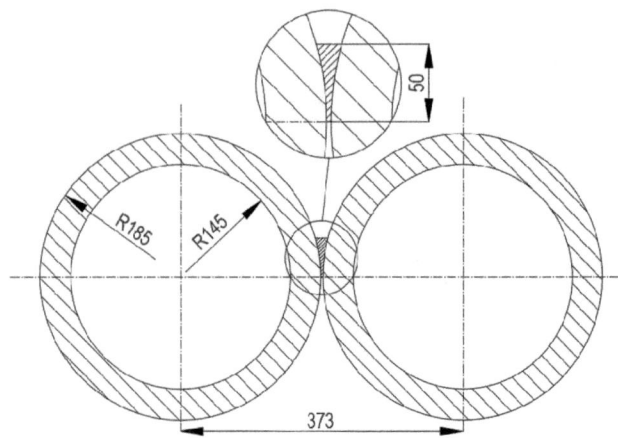

Fig. 1 Drawing of the twin-roll casting tool

Pre-processing

In this step, the previously created geometry was discretized (see Fig. 2). An advantage of Fluent is that it applies to unstructured meshes that can be easily created in ANSYS. However, the consequence of this is the loss of calculation accuracy. This study used structured meshes created with the finite element preprocessor HyperMesh to avoid this problem.

The discretization shown in Fig. 2 consists of 70,208 elements and 72,231 nodes. The discretized model was exported to ANSYS-Fluent to perform the following work step (solving).

Solving

In this step, the simulation was done in Fluent after the boundary conditions and material properties were adjusted. The melt temperature at the top of the strip-forming zone and the vertical component of the entry speed were set as boundary conditions. In addition, the speed of the copper shells was assumed as the rotation speed. Furthermore, it was considered that the copper shells are rigid; thus, the influence of their elastic deformation on the roll gap size can be neglected. The convective heat transfer from the copper shells to the coolant was described as a boundary condition by entering the heat transfer coefficient and the coolant temperature. Regarding the high turbulence caused by the lateral coolant supply and the rotation of the rolls, a heat transfer coefficient of 10,000 W/(m^2 K) results [14]. For the coolant temperature, 20 °C (293.15 K) was selected. The thermophysical properties of the aluminum–lithium alloy

Fig. 2 Discretization of the model

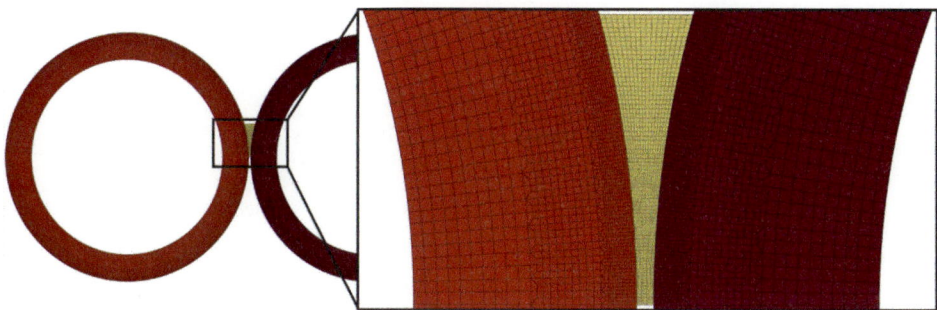

2099 and the copper shells are shown in Table 1, and the boundary conditions are in Table 2. The properties of the material in a semi-solid state were calculated using the mixture rule.

Subsequently, the boundary conditions and the material properties were set in the Fluent software, and the calculation was performed. The results of a complete analysis are shown in Fig. 3.

After the solving of the simulation was completed, the work step post-processing was carried out.

Post-processing

Post-processing is required to analyze the results. A well-known post-processor for analyzing Fluent's results is CFD. The influence of the process parameters on the casting process can be investigated by comparing the specific length of the deformation zone. The specific length is calculated using the following equation:

$$l_d' = \frac{l_d}{l_{SFZ}}$$

where l_d' is the specific length of the deformation zone, l_d is the length of the deformation zone, and l_{SFZ} is the length of the strip-forming zone. The length of the deformation zone can be calculated by the coordinates of the kissing point,

where the solidus temperature is reached. The melt is completely solidified in the deformation zone so that the liquid mass fraction in the deformation zone is zero. The influence of the process parameters can now be simulated and displayed by colorful pictures and curves using these points.

Experimental Procedure

After the numerical study, a trial on the twin-roll casting of an aluminum–lithium alloy 2099 was performed. The chemical composition of the cast strip measured with an optical emission spectrometer Q4 TASMAN from Bruker AXS GmbH is given in Table 3. The experiment was carried out using a lab-scaled vertical twin-roll caster with shells made of a copper alloy CuCr1Zr. The shell geometry corresponds with the specifications given in Fig. 1. Other characteristics of the utilized twin-roll casting unit are described in [15]. The numerical simulation results adjusted the twin-roll casting parameters to reach a moderate plastic reduction and avoid the overloading of the twin-roll caster. Casting speed of 7.2 m/min, casting temperature of 730 °C, strip thickness of 3 mm, and length of the strip-formation zone of 40 mm were used in the experiment. The melt was shielded by argon in the furnace and in the tundish during the pouring. The mean specific roll-separating force during the trial amounted to 1 kN/mm.

Table 1 Thermophysical material properties for the numerical simulation

Material	Melt	Solid	Copper shells
	Aluminum–lithium alloy		CuCr1Zr
Density in g/cm^3	2,3	2,59	8,91
Specific heat capacity in J/(kg °C)	1230	930	370
Thermal conductivity in W/(m K)	88	96	330
Viscosity in Pa s	0,012	100	
Phase transition temperature in °C	650	560	
Enthalpy of solidification J/kg	398,000		

Table 2 Boundary conditions

	Values
Heat transfer coefficient from shells to coolant in W/(m² K)	10,000
Coolant temperature in °C	20
Heat transfer coefficient from shells to air in W/(m² K)	20
Air temperature in °C	27
Heat transfer coefficient from melt to shell in W/(m² K)	12,000
Rotation speed in rad/s	0,72
Temperature of the melt in °C	730
Melt inlet speed in m/s	0,024
Pressure outlet in Pa	0

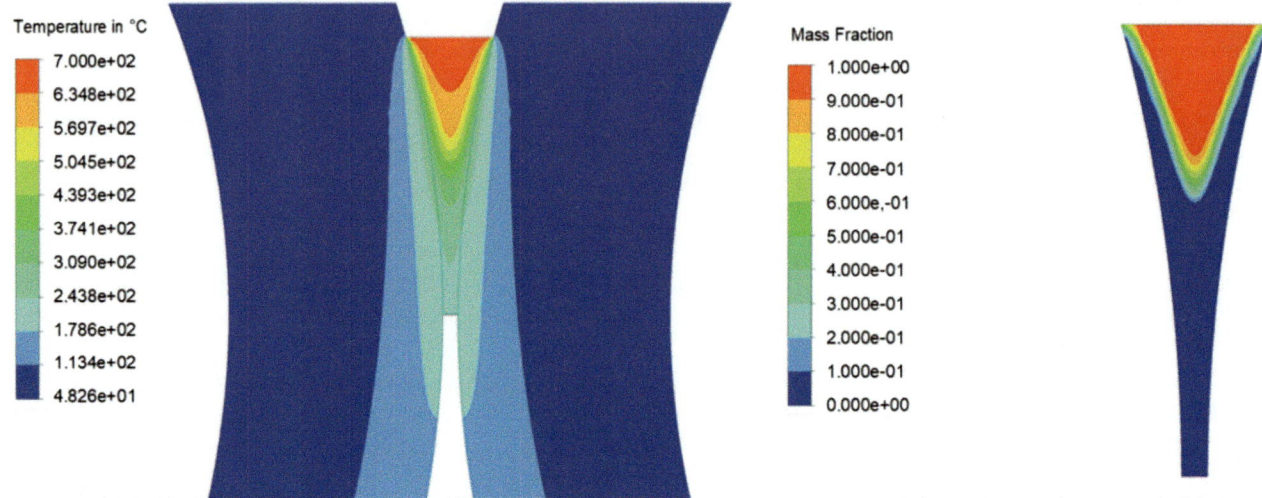

Fig. 3 Temperature (left) and liquid phase (right) distributions

Table 3 Chemical composition of the aluminum alloy 2099 used in the study

Element	Cu	Li	Zn	Mg	Mn	Zr	Al	Other
Content in %	2,82	1,783	0,63	0,388	0,279	0,126	93,37	balance

The microstructure of the as-cast material was studied by light optical microscope (LOM) observations using the Zeiss Axio Observer 7 microscope (Jena, Germany) in bright field and circular polarized light differential interference contrasts. These contrasts were used to observe and analyze precipitate dispersion and grain structures. Samples for these observations were polished with SiC papers from P800 up to P4000 FEPA grade. Struers OP-S suspension of colloidal silica was used to finalize polishing. Samples for precipitate dispersion observations were etched with an 0.5% solution of HF in ethanol for 10 s. Samples for grain structure observations were electrochemically polished by Barker's solution at 10 °C for 300 s. Observations were done from the rolling (RD) and transversal (TD) directions of the material. Grain size and secondary dendrite arm spacing (SDAS) were analyzed using the ImageJ software (National Institute of Health, USA).

Results and Discussion

The thermofluid model was used to determine the influence of casting speed, casting temperature, strip thickness, and length of strip-forming zone on strip outlet temperature and a specific length of the deformation zone. The results at a casting speed of 2–8 m/min (Fig. 4) have shown that the strip outlet temperature increases and the specific length of the deformation zone decreases with increasing casting speed. The contact between the shells and the melt is reduced with increasing casting speed; thus, the melt

Fig. 4 Influence of the casting speed on the twin-roll casting conditions at casting temperature of 730 °C, strip thickness of 3 mm, and length of strip-forming zone of 50 mm

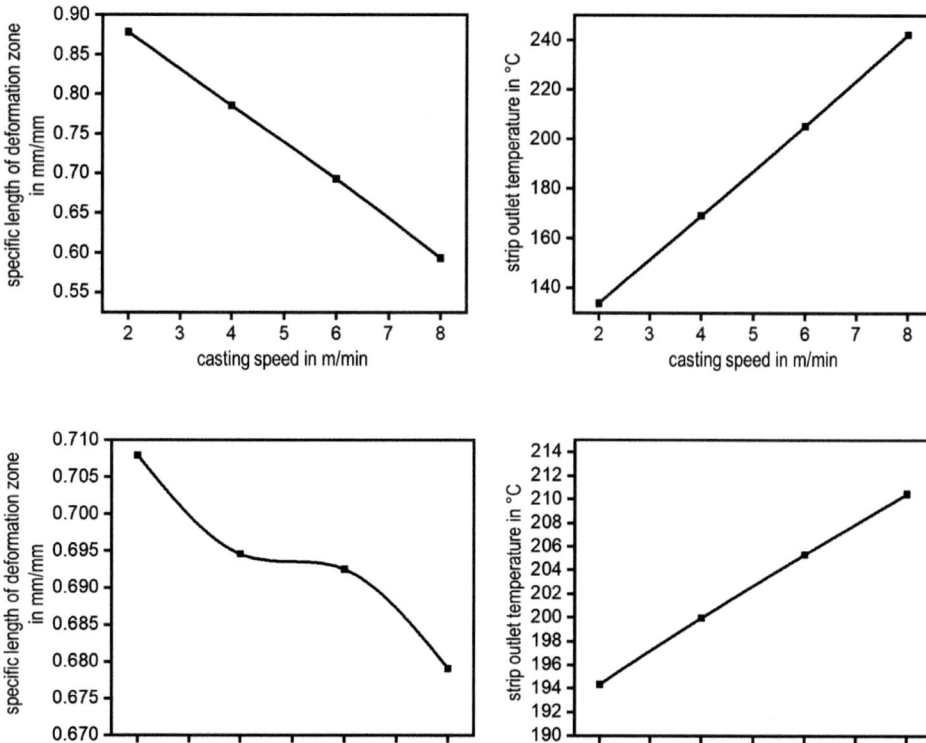

Fig. 5 Influence of the casting temperature on the twin-roll casting conditions at casting speed of 6 m/min, strip thickness of 3 mm, and length of strip-forming zone of 50 mm

solidifies more slowly. The cooling of aluminum–lithium alloy in copper shells is rapid, resulting in a relatively high deformation zone length and a low strip outlet temperature.

Furthermore, the results have shown that with increasing casting temperature (690–750 °C in steps of ten), the strip outlet temperature increases, and the specific length of the deformation zone becomes smaller (Fig. 5).

A variation of the strip thickness from 2 to 5 mm (Fig. 6) with otherwise constant parameters leads to an increase in strip outlet temperature and a decrease in the specific length of the deformation zone. The variation of the strip thickness has the most substantial effect on the twin-roll casting conditions changing.

A variation of the strip-forming zone length provides an inverted course. The results have shown that with the increasing length of the strip-forming zone (40–70 mm in steps of five) the strip outlet temperature decreases, and the length of the deformation zone increases (Fig. 7).

With the knowledge gained from the simulation, thin strips were produced and their microstructure investigated. As-cast material is relatively coarse-grained. The grain size of the material is two orders of magnitude higher than the solidified eutectic cells. Grains have a bimodal distribution, with equiaxed grains forming at the surface or in the center of the strip and columnar grains forming the bulk of the strip (Fig. 8). Equiaxed grains have a diameter of (260 ± 60) µm.

The columnar grains have an equiaxed cross section. They are (220 ± 20) µm wide and are (530 ± 160) µm high. The high scatter of grain height is mainly influenced by the presence or lack of center/surface grains. Such grain microstructure is typical for the strips of Al–Li alloys, which do not contain scandium and are twin-roll cast under the condition of a moderate roll-separating force [16].

Constituent particles formed during casting are concentrated into dendrites formed around eutectic cells. The size of the cells is characterized by SDAS. SDAS varies only slightly from surface to center of the strip, and the difference falls within the experimental scatter of both measurements (Fig. 9). The average SDAS is (5.8 ± 0.8) µm as evaluated from both observation directions. SDAS can be higher in some globular regions of the solidified strip. There is an apparent lack of major surface bleeding and centerline segregations in the observed strips. Some segregation occurs along grain boundaries close to the surface of the strip. The dendritic cells are elongated, and the primary axis of elongation differs from grain to grain but is consistent across individual grains.

On the one hand, a homogeneous distribution of finer grains over the cross section would be advantageous for the mechanical properties, especially for the ductility. On the other hand, a slight deformation during the process reduces the anisotropy of the mechanical properties. Analysis of the

Fig. 6 Influence of the strip thickness on the twin-roll casting conditions at casting temperature of 730 °C, casting speed of 6 m/min, and length of strip-forming zone of 50 mm

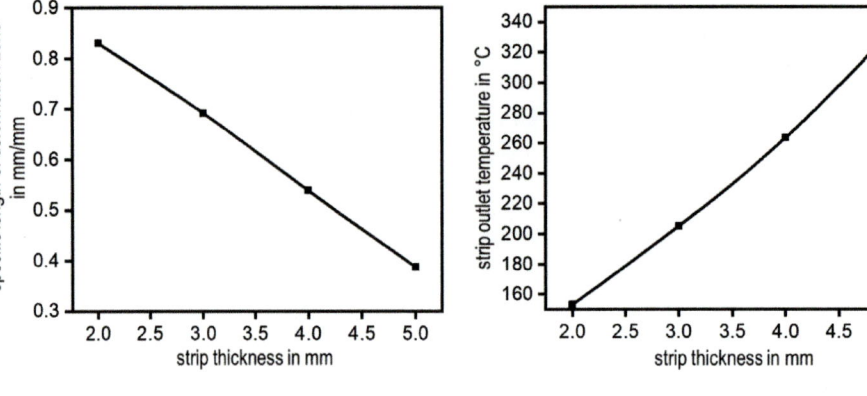

Fig. 7 Influence of the length of strip-forming zone on the twin-roll casting conditions at casting temperature of 730 °C, casting speed of 6 m/min, and strip thickness of 3 mm

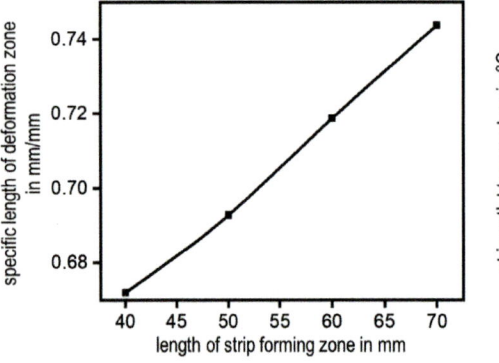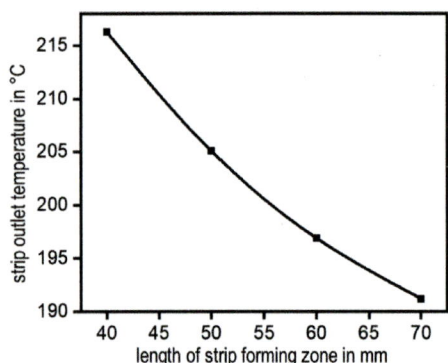

Fig. 8 Grain structure of the as-cast material. Observation from the RD (left) and TD (right)

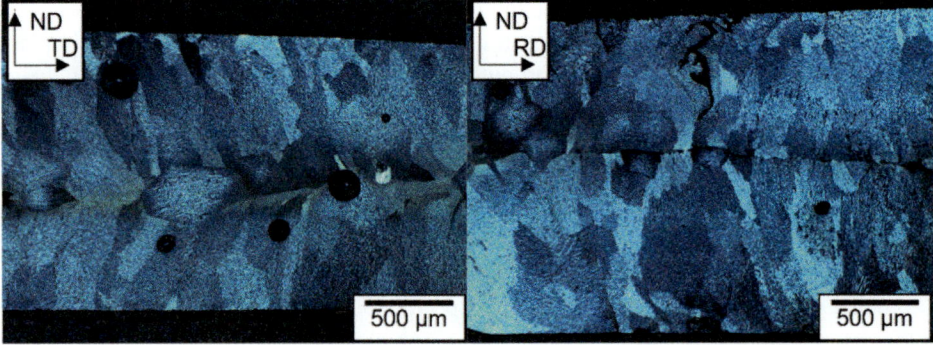

Fig. 9 Particle dispersion at the surface (left) and in the center (right) of the strip. Observation from the TD

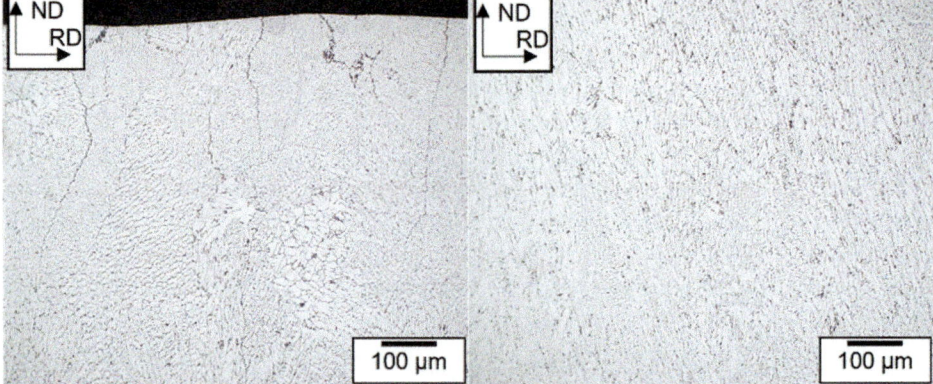

mechanical properties of the strips and their anisotropy as well as further process optimization for their improvement will be the subjects of further investigations.

Conclusions

Within the scope of the numerical study using the developed thermofluid model implemented in ANSYS-Fluent, dependencies of thermal and deformational states from the main technological process parameters such as casting speed, casting temperature, strip thickness, and length of strip-forming zone for the twin-roll casting of the aluminum–lithium alloy in copper shells were examined. Based on the established correlations, a laboratory twin-roll casting trial with moderate plastic reduction of the material was designed.

The aluminum–lithium strip produced using twin-roll casting has a coarse-grained structure with grain sizes ranging from 200 to 500 μm. Grain distribution is bimodal, with equiaxed grains in the center and at the surface and columnar grains forming the rest. Grain size is two orders of magnitude higher than the secondary dendrite arm spacing of the solidified eutectic cells. The observed dendrite arm spacing is the same across the strips. No significant segregates or dense particle clusters were detected.

Acknowledgements The authors thank the German Research Foundation (DFG) for the financial support of the project SCHA 1484/46-1 and the Czech Science Foundation for funding grant number 20-19170S.

References

1. Goede M, Stehlin M, Rafflenbeul L, Kopp G, Beeh E (2009) Super Light Car-lightweight construction thanks to a multi-material design and function integration. Eur. Transp. Res. Rev. 1: 5–10. https://doi.org/10.1007/s12544-008-0001-2
2. González Palencia J, Furubayashi T, Nakata T (2012) Energy use and CO2 emissions reduction potential in passenger car fleet using zero emission vehicles and lightweight materials. Energy 48: 548–565. https://doi.org/10.1016/j.energy.2012.09.041
3. Kleiner M, Geiger M, Klaus A (2003) Manufacturing of Lightweight Components by Metal Forming. CIRP Annals – Manu. Tech. 52: 521–542. https://doi.org/10.1016/S0007-8506(07)60202-9
4. Stolbchenko M, Grydin O, Schaper M (2015) Twin-roll casting and finishing treatment of thin strips of the hardening aluminum alloy EN AW-6082. Mater. Today. Proc. 2: 32–38. https://doi.org/10.1016/j.matpr.2015.05.014
5. Schlingmann T (2016) Einfluss verschiedener Legierungselemente auf Mikrostruktur und Werkstoffeigen-schaften von Al–Li–Legierungen. Ph.D. thesis, Berlin University
6. Abd El-Aty A (2017) Experimental investigation of tensile properties and anisotropy of 1420, 8090 and 2060 Al–Li alloys sheet undergoing different strain rates and fibre orientation: a comparative study. Procedia Eng. 207: 13–18. https://doi.org/10.1016/j.proeng.2017.10.730
7. Rioja R, Liu J (2012) The Evolution of Al–Li Base Products for Aerospace and Space Applications. Metall. Mater. Trans. A. 43(9): 3325-3337. https://doi.org/10.1007/s11661-012-1155-z
8. Lavernia E, Srivatsan T, Mohamed F (1990) Strength, deformation, fracture behaviour and ductility of aluminium-lithium alloys. J. Mater. Sci. 25(2): 1137–1158. https://doi.org/10.1007/bf00585420
9. Alexopoulos N, Migklis E (2013) Fatigue behavior of the aeronautical Al–Li 2198 aluminum alloy un-der constant amplitude loading. Int. J. Fat. 56: 95–105. https://doi.org/10.1016/j.ijfatigue.2013.07.009
10. Kammer C (1999) Continuous casting of Aluminium. Ph.D. thesis, Clausthal University
11. Watari H, Haga T (2005) Continuous Strip Casting of Magnesium Alloy by a Horizontal Twin Roll Caster. Con. Cast.: 70–76. https://doi.org/10.1002/9783527607969
12. Li B (1995) Producing Thin Strips by Twin-Roll Casting – Part I: Process Aspects and Quality Issues. JOM 47: 29–33. https://doi.org/10.1007/BF03221172
13. Sahoo S (2016) Review on vertical twin-roll strip casting: a key technology for quality strips. J. Metall. https://doi.org/10.1155/2016/1038950
14. Bondarenko S et al. (2018)Numerical analysis of twin-roll casting of strips with profiled cross-section. Mater. Res. https://doi.org/10.1590/1980-5373-MR-2017-1098
15. Grydin O et al. (2010) Experimental twin-roll casting equipment for production of thin strips. Metall. Min. Ind., 2(5): 348
16. Grydin O et al. (2020) New twin-roll cast Al–Li based alloys for high-strength applications. Metals. 10(8): 987. https://doi.org/10.3390/met10080987

Segregation Mechanisms and Their Effects on the Aluminium Flat Rolled Products (Sheet/Foil) Produced by Twin Roll Casting Technology

Onur Birbasar, Feyza Denizli, Eda Özkaya, Samet Sevinç, Ali Ulus, and Canan İnel

Abstract

Twin Roll Casting (TRC) technology has significantly higher solidification rates compared to conventional casting methods. The surface microstructure of the as-cast sheet produced with TRC typically shows a matrix supersaturated by alloying elements, while it also contains segregation along the centerline of the as-cast sheet. Surface and edge segregations also can be visible in as-cast materials. Segregation affects the final material properties in terms of the mechanical properties, corrosion properties, number of pinholes, and rolling performance. It is extremely important to eliminate or reduce the intensity of the segregation. In this study, the formation mechanisms and parameters for minimizing centerline segregation in 8xxx aluminum foil alloy cast with steel–steel and copper–copper shell pairs are investigated. Both surface and edge segregations are also studied. As-cast samples are prepared metallographically and investigated by using light microscope, scanning electron microscope equipped with EDS. Electrical conductivity measurements are also performed.

Keywords

Twin roll casting • Segregation • 8xxx wrought aluminum alloy

Introduction

Twin roll casting (TRC) has been popular for producing thin aluminum strips. The TRC process produces strips directly from liquid aluminum by feeding molten metal into the gap of two rotating water-cooled rolls. As a result, it solidifies and forms a solid shell on each roll surfaces with considerable amount of hot rolling before reaching the roll nip [1, 2]. TRC technique can provide lower operating costs, less energy consumption, and lower scrap rate compared to traditional DC casting method. During casting of aluminum alloy, process parameters such as liquid metal temperature, strip speed, water cooling, separating force, and amount of graphite affect the microstructural properties of the sheet such as grain morphology, distribution of alloying elements, the size and distribution of intermetallic particles, and the macro/micro-segregation [3]. Also, the alloy composition and the caster shell material (i.e. copper, steel) have a great influence on the surface properties of the cast material and its microstructure.

Due to the non-equilibrium solidification conditions, a supersaturated region of alloying elements near the surface appears and the centerline segregation reveals at the mid-plane of the as-cast sheet thickness. Such formations have different effects on the final sheet or foil material properties in terms of the corrosion properties, mechanical properties, surface defects, pinhole, and rolling performance. Therefore, it is important to eliminate or reduce the intensity of the macro- and micro-segregations for the downstream process efficiencies [4]. Steel caster rolls, which are widely used in the twin roll casting method, affect productivity and shell life, although they have different alloy grades, mechanical properties and thermal properties. Recently, due to having high thermal conductivity properties, copperalloy-based casting shell materials are used in twin roll casting. By using copper shells having high heat transfer capability from the liquid metal, different microstructural features are obtained [9]. In addition, high-temperature heat

O. Birbasar (✉) · F. Denizli · E. Özkaya · S. Sevinç · A. Ulus · C. İnel
Yazılıgürgen Mahallesi Fabrikalar Cad. No. 50, 54400 Karapürçek, Sakarya, Türkiye
e-mail: Onur.birbasar@asastr.com

© The Minerals, Metals & Materials Society 2023
S. Broek (ed.), *Light Metals 2023*, The Minerals, Metals & Materials Series,
https://doi.org/10.1007/978-3-031-22532-1_138

treatment of the cast material at different temperatures reduces the intensity of the centerline segregation, which usually remains until the final product. Another issue that is most problematic in the TRC process is edge cracking, which causes difficulties in downstream processing of strips. Edge cracking is caused by the rolling and solidification effect when the liquid metal comes in contact with the caster shell. During the twin roll casting process, edge cracks with different types can be seen depending on the alloy and casting parameters. Parameters such as liquid metal flow rate, liquid metal level, temperature, and graphite concentration applied to the shell surface must be kept under control to prevent edge cracking [1, 5–8].

In this study, the segregations encountered in the production of 8xxx series aluminum foil alloy produced by the twin roll casting method are investigated. Microstructural components and electrical conductivity measurements of cast sheets produced with copper and steel shells are compared. Edge cracks and surface segregations are investigated by scanning electron microscope (SEM) equipped with an energy dispersive spectrometer (EDS).

Experimental

In this study, 8xxx (Twin roll cast AlFeMnSi alloy) aluminum foil alloy is used and its chemical composition is given in Table 1. Pechiney Jumbo 3CM-type twin roll caster consisting of steel/steel pair and copper/copper pair rolls are used to produce strips with the thickness of 7,5 mm. Samples taken from the as-cast sheets produced by steel–steel and copper–copper pair shells are investigated for the microstructural characterization. In order to understand the effect of low- and high-temperature heat treatment on the centerline segregation, samples are heat treated at 480 °C for 8 h and 580 °C for 8 h, respectively. The surface quality of strips, namely, edge cracks and surface segregation defects are checked and samples are taken for the investigation. All samples of microstructure are metallographically prepared and investigated using light microscope (LM, Zeiss Scope A1-Vario) and SEM (Zeiss Evo MA15).

Table 1 Chemical composition of the samples

Weight-%				
Si	Fe	Cu	Mn	Ti
0,1–0,3	1,2–2,0	0,01–0,1	0,4–1,0	0,01–0,03

Results and Discussion

Figure 1 shows the microstructures of the as-cast sheets produced by steel–steel and copper–copper pair caster rolls. At the near-surface location of the solidification structure, both microstructures show both relatively higher density of secondary phases and smaller grains compared to the center of the cast material cross section. Although the center of solidified structure exhibits a equiaxed grain structure, grains on the surface appear in an elongated morphology in the direction of casting. Compared to the use of steel shell (Fig. 1f), relatively larger equiaxed grains and finer intermetallics are observed on the surface after the use of copper shell (Fig. 1e). This phenomenon is attributed to the rapid cooling effect due to the high thermal conductivity of copper.

As shown in Fig. 1, less centerline segregation encountered in production with copper shell is observed, although it is strongly affected by casting parameters (strip speed, tundish temperature, separating force, tip position, and cooling temperatures). This is related to the over-saturation of the matrix with alloying elements due to the rapid cooling effect. As-cast strip thickness of the material is 7,5 mm. With the data obtained from the electrical conductivity measurements, an inference can be made on the solidification characteristic. Electrical conductivity values of the material decrease with increasing solute elements dissolved in the matrix. While the conductivity values on the sheet produced with the copper shell are 21.3 MS/m (480 kHz) and 32,2 MS/m (240 kHz), this conductivity values on the sheet surface produced with the steel shell are determined as 21.0 MS/m (480 kHz) and 36,4 MS/m (240 kHz). For aluminum materials and a probe frequency of 480 kHz the penetration depth is 120 µm. A change in the probe frequency to 240 kHz leads to a penetration depth of 180 µm. Low electrical conductivity values obtained from the as-cast sheet surface produced with copper shell indicate that the matrix is more saturated with the alloying elements. It should always be taken into account that higher casting rates increase productivity, but can cause a macro-segregation in the middle of the casting material despite the use of a steel or copper shell.

The eutectic centerline segregation caused by the high casting speed in sheet material produced with steel shell is given in Fig. 2a, b. It is a fact that formed segregation within the solidified structure affects the final material properties in terms of the mechanical properties, corrosion properties, number of pinholes, and rolling performance. As it is well

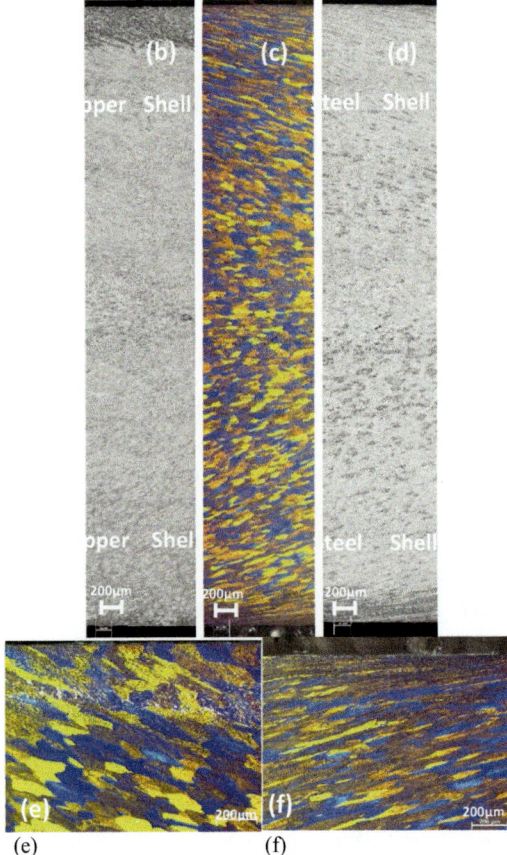

Fig. 1 LM micrographs showing the grain structures of a twin roll cast 8xxx series alloy produced with (**a, b, e**) copper shell and (**c, d, f**) steel shell

known, homogenization heat treatment is needed to eliminate in homogeneities, micro-segregations, residual stresses, and other defects that occur in the material during non-equilibrium solidification conditions that occur during the production process. Such a process is also desirable in industrial products that require increased deep drawability. However, although the density of mentioned discontinuities decreases with heat treatments, they do not disappear completely. After casting, the samples are subjected to homogenization heat treatment both at 480 and 580 °C for 8 h to observe the evolution of centerline segregation. SEM images showing the heat-treated matrices are given in Fig. 2c–f. EDS analysis indicates that the eutectic and intermetallic particles (Fig. 2b) are Fe–Mn–Si–Al based. As it can be seen in Fig. 2c–f, morphologies of the centerline segregation phases transformed into the spherical- or needle-like intermetallics. In addition, EDS analysis shows that the amount of Fe, Mn, and Si elements dissolved in the matrix decreases after heat treatment due to the formation of coarser intermetallic particles.

Casting parameters are extremely important in copper shell castings, as higher speeds are achieved compared to standard steel shell castings. As seen in Fig. 3a, as-cast sheet surface may have surface segregations due to the higher strip speed and different solidification conditions. In Fig. 3a, two different regions (marked with number 1 and 2) can be seen on the as-cast sheet surface. According to the SEM, SE, and BSE images, these two regions have different topographical features and different intermetallic phase morphologies. Eutectic intermetallic formation can be seen due to the high

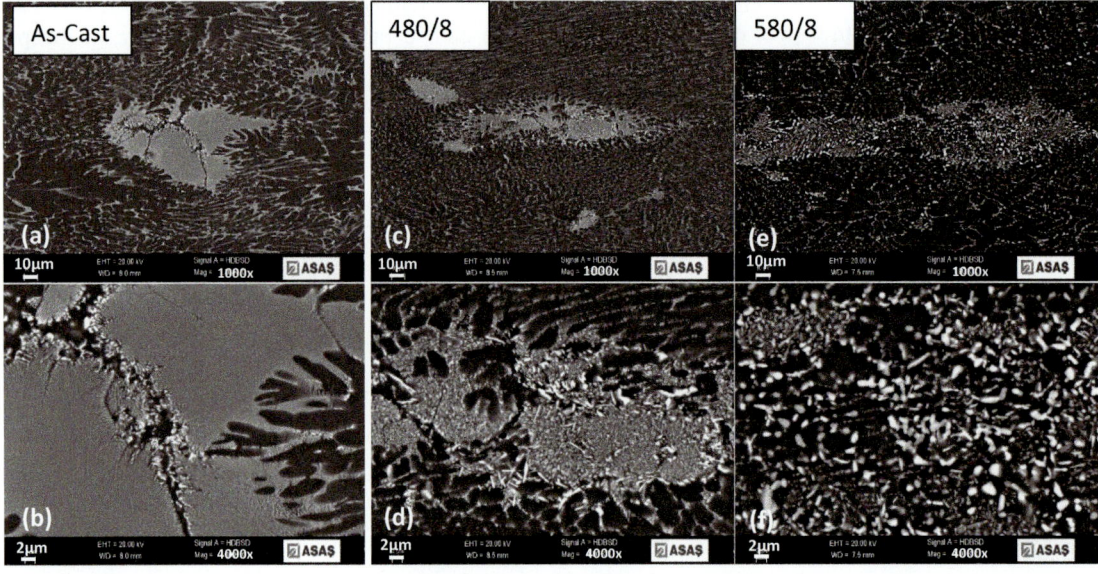

Fig. 2 SEM micrographs showing the structure of cast alloy (**a, b**), heat-treated alloy at 480 °C for 8 h (**c, d**) and 580 °C for 8 h (**e, f**)

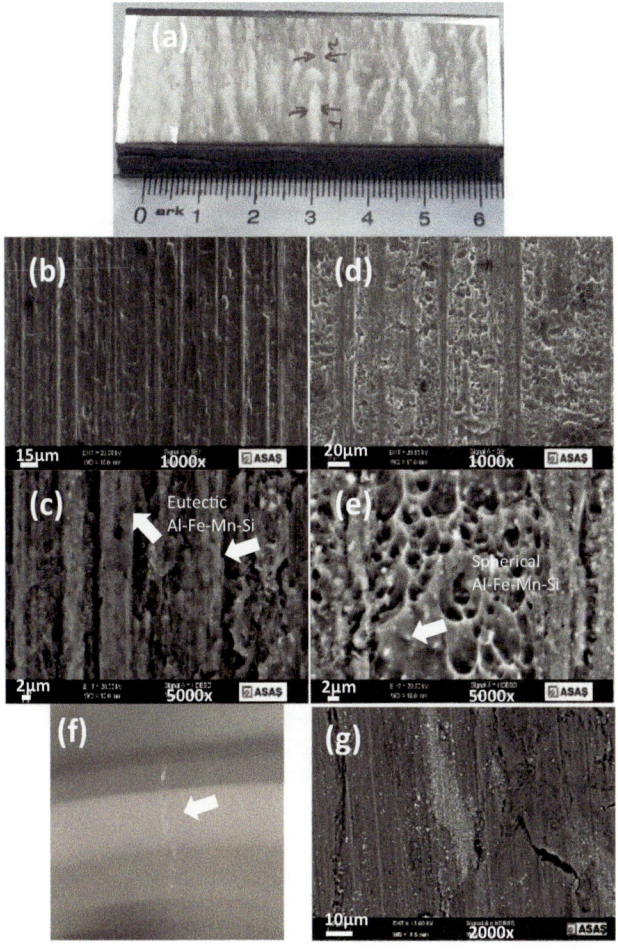

Fig. 3 **a** Macro-image of as-cast sheet material surface produced with copper shell; **b**, **c** SEM images of segregated area marked with number 1; **d**, **e** segregated area marked with number 2; **f** macro-image of surface defect on aluminum foil material; **g** BSE image of surface defect

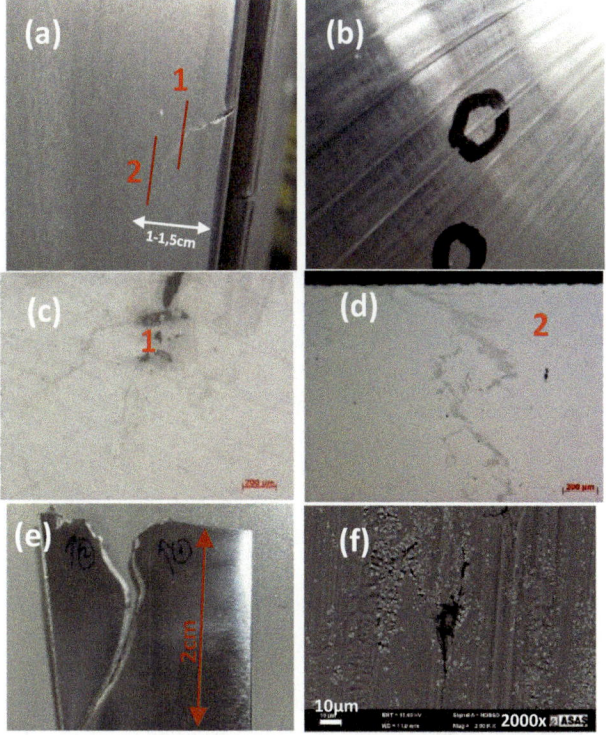

Fig. 4 Macro-images of; **a** as-cast sheet material edge, without any edge trimming. **b** Edge trimmed as-cast coil side surface. Optical microscope images of samples; **c** close to the crack tip marked with redline 1; **d** away from the crack tip marked with redline 2; **e** macro-image of crack formation on the edge of strip; **f** BSE images of crack area

solidification rate (Fig. 3c) and the particulated intermetallic phase formation appears due to the low solidification rate conditions (Fig. 3e). These kinds of segregations can cause surface defects (Fig. 3f, g) during cold rolling process to lower thicknesses. The inhomogeneity and different localized amounts of the graphite coating on the casting roll surface with the combined effect of the insufficient cooling conditions related to the casting parameters can cause surface segregations.

In twin roll strip casting process, the quality of the strips is an important issue. Regarding the surface quality of the strips another surface quality defect is edge cracking. The different microstructure at the edge is formed due to the casting speed, rolling effect, alloy composition, liquid metal flow, solidification behaviour, etc. Typical macro-image of

the crack formation on the edges is given in Fig. 4a It can be seen clearly a macro-crack formed in the edge of strip and extending along the TRC direction. Metallographic samples prepared from the edge of the crack area were investigated from the regions marked with 1 and 2 in Fig. 4a. Optical micrographs of the samples close to the crack tip marked with redline 1 and away from the crack tip marked with redline 2 are shown in Fig. 4c, d, respectively. It is clearly seen that a high volume of intermetallic phases (segregation) have formed in the vicinity of the crack area. Therefore, it can be concluded that the edge crack was initiated from the segregation line. During the casting process milling equipment enables accurate in-line milling of strip edges to eliminate solidification heterogeneities on the edges. However, crack roots may remain when edge milling operation is insufficient. In these cases, rolling defects may be encountered in downstream processes as seen in Fig. 4e. According to the BSE images of the problematic areas, the crack formation occurred due to the segregations related with the as-cast strip edge cracks.

Conclusions

Due to the high thermal conductivity of copper, relatively larger equiaxed grains and finer intermetallics are observed on copper shell which produced surface strip than that of steel shell produced surface strip. Although it is strongly affected by casting parameters (casting speed, separating force, temperatures, etc.), less centerline segregation was observed in strip produced with copper shell rollers.

Depending on the different solidification conditions related to the casting parameters and the high productivities achieved by using copper shell, segregation formations may occur on the as-cast sheet surfaces. If no precautions are taken, segregations can lead to quality defects in the downstream processes.

Most effective process parameters for the solution of segregation regardless of the alloy content are

- Decreasing the cooling water temperature to ensure uniform cooling condition on as-cast sheet surface.
- Reduce the casting speed.
- Reduce the casting temperature appropriately.
- Spraying homogenous graphite on to the caster shell surface.
- Promote fine equiaxed grain structure in both casting process and subsequent homogenization annealing

References

1. J. Huang, J. Li , C. Li, C. Huang, B. Friedrich, "Elimination of edge cracks and centerline segregation of twin-roll cast aluminum strip by ultrasonic melt treatment", Journal of Materials Research and Technology, vol. 9, 2020, 5034–5044.
2. M. Šlapáková, M. Zimina, S. Zaunschirm, J. Kastner, J. Baje, M. Cieslar, "3D analysis of macrosegregation in twin-roll cast AA3003 alloy", Materials Characterization, vol. 118, 2016, 44–49.
3. Z. Lv, F. Du, Z. An, H. Huang, Z. Xu, J. Sun, "Centerline segregation mechanism of twin-roll cast A3003 strip", Journal of Alloys and Compounds, vol. 643, 2015, 270–274.
4. Y. Li, C. He, J. Li, Z. Wang, D. Wu, G. Xu, "A Novel Approach to Improve the Microstructure and Mechanical Properties of Al–Mg–Si Aluminum Alloys during Twin-Roll Casting", Materials, vol. 13, 2020.
5. M.S. Kim, S.H. Kim, H.W. Kim, "Deformation-induced center segregation in twin-roll cast high-Mg Al–Mg strips", Scripta Materialia, vol. 152, 2018, 69–73.
6. J. T. Li, G. M. Xu, H. L. Yu, G. Chen, H. J. Li, C. Lu, J. Y. Guo, "Improvement of AA5052 sheet properties by electromagnetic twin-roll casting", The International Journal of Advanced Manufacturing Technology, vol. 85, 2016, 1007–1017.
7. C. M. G. Rodrigues, A. Ludwig, M. Wu, A. Kharicha, A. Vakhrushev, "A Comprehensive Analysis of Macrosegregation Formation During Twin-Roll Casting", Metallurgical and Materials Transactions B, vol. 50, 2019, 1334–1350.
8. B. Forbord, B. Andersson, F. Ingvaldsen, O. Austevik, J.A. Horst, I. Skauvik, "The formation of surface segregates during twin roll casting of aluminium alloys", Materials Science and Engineering A, vol. 4152, 2006, 12–20.
9. H. M. Altuner, C. Isıksacan, O Birbasar, M. Gunyuz, O. Meydanoglu, "Crystallographic texture development of as-cast 3105 alloy produced by St/Cu shell pair", Light Metals, 2016.

Novel Methods for Roll Texturing: EDT and Sandblast Applications for Aluminium Twin-Roll Cast and Cold Rolling

Yusuf Özçetin, Onur Birbaşar, Ali Ulus, Koray Dündar, Feyza Denizli, and Canan İnel

Abstract

In twin-roll continuous casting (TRC) and aluminum cold rolling, the rolls are conventionally roughened by grinding. However, this method may cause defects on the surface of the final sheet product. Structures called "shingles" are formed due to the nature of the TRC method. Because of the shingles, aluminum debris (SMUT) is carried on to the sheet surface during the manufacturing processes. With sandblasting, the surfaces of the rollers are roughened in a different topography, unlike the mill finish texture. Thanks to this topography, the shingle formation mechanism is prevented. With the EDT process applied to the surface of the cold rolling rolls, the resulting crater-like topography helps to eliminate the defective shingle structure from the casting. In this study, sandblasting of casting rolls; on the other hand, EDT process was applied to cold rolling rolls. The obtained results were evaluated by using surface characterization methods.

Keywords

EDT • Sandblasting • Texturing • Twin-roll casting

Y. Özçetin (✉) · O. Birbaşar · A. Ulus · K. Dündar · F. Denizli · C. İnel
Asaş Alüminyum San. Ve Tic. A.Ş, İstanbul, Turkey
e-mail: yusuf.ozcetin@asastr.com

O. Birbaşar
e-mail: onur.birbasar@asastr.com

A. Ulus
e-mail: ali.ulus@asastr.com

K. Dündar
e-mail: koray.dundar@asastr.com

F. Denizli
e-mail: feyza.denizli@asastr.com

C. İnel
e-mail: canan.inel@asastr.com

Introduction

The twin-roll continuous casting method is one of the most frequently used casting methods in aluminum sheet production. In this method, the liquid metal, which has passed through certain filtration processes, solidifies as soon as it passes between two counter-rotating steel rollers [1]. In the following process, depending on the final plate thickness, cold rolling processes are carried out. Finally, the production is terminated with the annealing process according to the desired final mechanical condition [2].

In both twin-roll casting and cold rolling the surface characteristic of the work rolls used is one of the important factors that directly affect the final product. Conventionally, casting and rolling rolls are roughened by a method called grinding. This method is one of the most established and sustainable in the industry. However, rolls roughened by grinding have some bad effects on the plate. Due to the rolling effect that occurs with the solidification during casting surface defects called shingles occur. These defects can still exist in the final product, even if a certain level is destroyed by rolling. This undesirable structure contains elevation differences at the micro-level. During all production processes, pollution is carried inside the shingles. This situation creates great negativities for some usage areas. These impurities are called "SMUT" [3].

In order to control the amount of SMUT in the final product the surface must be clear of shingles. Roughing the rolls on the casting side with a method called sandblasting and casting with these rolls has an effect on the shingle structure on the surface. Unlike traditional grinding, a different texture is formed on the plates in the casting thickness. The elevation difference of this texture is much less than the texture of the plates cast with the ground surface rolls. For this reason, it was proposed to apply the sandblasting process to the casting rolls.

The EDT (electron discharge texturing) method to roughen the cold rolling rolls gives a crater-like surface

© The Minerals, Metals & Materials Society 2023
S. Broek (ed.), *Light Metals 2023*, The Minerals, Metals & Materials Series,
https://doi.org/10.1007/978-3-031-22532-1_139

topography on the rolls which has a different surface texture than on the ground rolls. It is thought that it can break the surface structure that contains the shingle coming from the casting.

In this study, as separate experiments, sandblasting was applied to the casting rolls and EDT treatment was applied to the cold rolling rolls. Then, the surface characterization and inorganic pollution (SMUT) amounts were measured and the results were interpreted.

Materials and Methods

Sandblasting and EDT processes were carried out in two separate experimental processes. In the sandblasting process, only the casting rollers were sandblasted. In the ongoing cold rolling processes, the rolls were roughened with the conventional grinding process.

In the EDT process, while casting rolls are roughened by classical grinding, EDT process was applied to the cold rolling rolls for just first cold rolling pass. In ongoing cold rolling passes, rolling was continued with ground rolls. It was proposed that the undesired surface structure of the casting could be broken in the first pass. The applied process routes are shown in Fig. 1.

A 1050 aluminum alloy was used in the experiments. Rolling processes were carried out up to a final thickness of 1 mm. Finally, plate washing-stretching operations were carried out (Fig. 2).

Sandblasting is not a complex process. It is based on the principle of spraying sand on steel rollers while they are being rotated on rotating rotors in a closed atmosphere. This process can be provided in PLC-controlled manner with factors such as rotation speed and sandblasting flow. Below is the image of the device used during the process.

Sandblasted roll surface texture has significant differences compared to ground roll surface texture. Due to reasons such as high grinder disc speeds during the grinding process and the inability of the used disc to wear evenly in all parts of the roller corrugated structures may form between the grinding lines. However, surface roughness control can be done much more effectively than sandblasting. It is difficult to control the surface roughness in general in sandblasting. Before sandblasting, grinding is done as a pre-treatment. In this pre-treatment, the base ra is determined in grinding and surface texture is gained in sandblasting.

In Fig. 3, mobile-optical microscope images of standard ground and sandblasted casting rolls are shown. The sandblasted surface texture has significant differences compared to the ground texture. In the sandblasted surface texture, the topography shows almost no straight grind lines and has a sandy structure due to sand bombardment. While the surface roughness is around 1.10–1.20 μm for the ground roller surface, it was measured as 1.70–1.80 μm on the surface of the sandblasted roller. This difference can be explained by the fact that the sand texture has a more radical topography.

It is known that the sandblasted surface texture has a significant effect on reducing the effect of the "shingle"

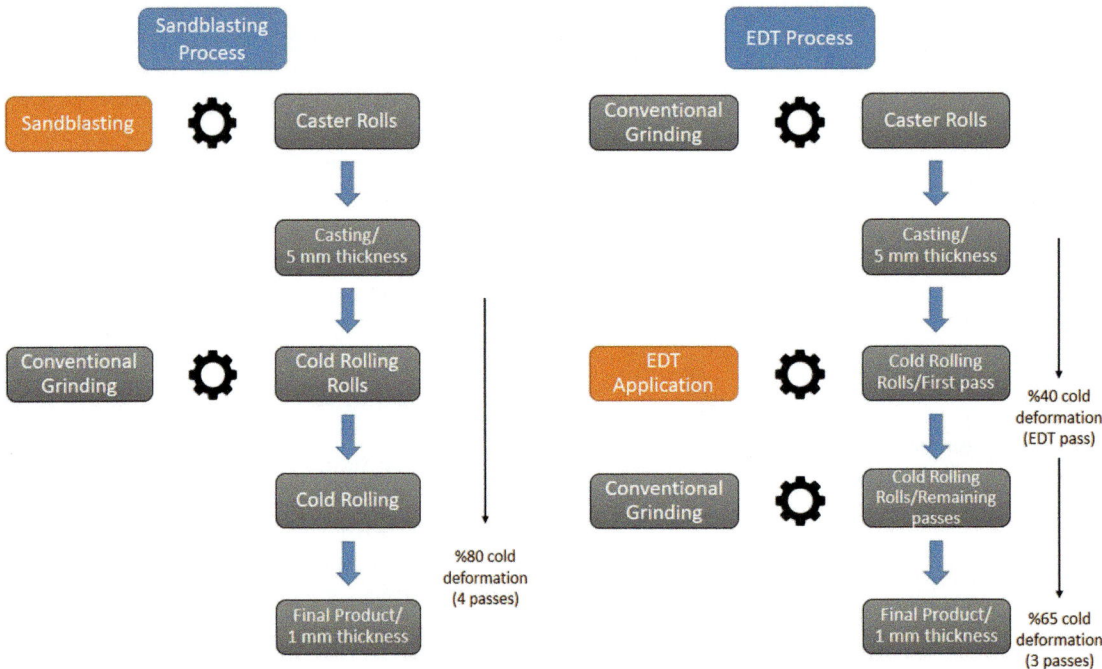

Fig. 1 Applied process routes

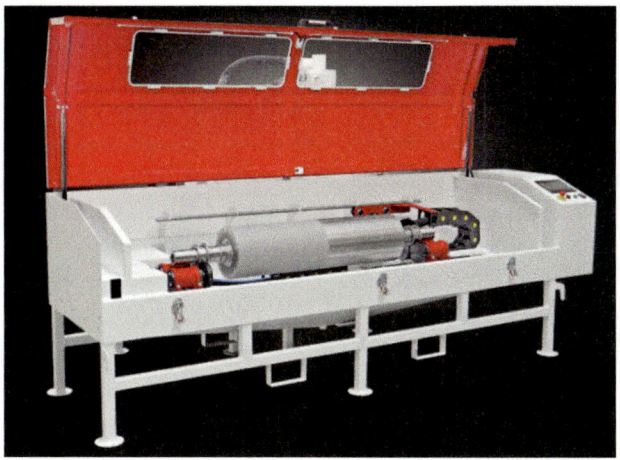

Fig. 2 Sandblasting machine [4]

structure formed by the continuous radial movement on the twin-roll continuous casting side. Due to the lines on the ground roller and the grooves with elevation differences between these lines, shingles are formed by the effect of radial force during casting. An SEM image of shingles is shown in Fig. 4.

The EDT process is a roll roughening method that has been the trend in recent years and is used by flat steel product manufacturers especially when producing for the automotive sector. EDT rollers are used in skin-pass lines to give the sheet surface superior paintability and deep drawability. Unlike grinding and sandblasting, no mechanical roughening is performed.

In this method, electrodes are placed close to the surface of the rotating roller (Fig. 5). In the gap between the electrodes, a dipole bridge is formed on account of the electric-discharge pulse passing through the dielectric. This results in an electrical breakdown. A small section of the roller surface melts, and a gas bubble forms in the dielectric. When the pulse is switched off, the bubble breaks down

under the action of the external pressure, with splashing of the molten metal [5, 6].

As the advantages of this method, roughness determination in a wide scale, homogeneous surface texture and high accuracy and similarity of the process can be said. As a disadvantage, high investment and operating costs can be specified.

Figure 6 shows mobile-optical microscope images of the surfaces of EDT treated and standard ground cold rolling work rolls. Based on the images, it can be said that the roller with EDT has a crater-like topography resulting from the discharge effect. Surface roughness was measured at the level of 2.3–2.5 μm. The standard ground surface work roll has the same grinding lines and grooves as the casting rolls surface. Ra was measured as 0.80–0.90 μm.

Certain surface characterization processes were applied to the plates produced from both processes. For plates cast with standard ground rollers and cast with sandblasted rollers, the surface morphology was investigated by scanning electron microscope (SEM) at casting thickness and at final thickness. In order to understand how the cold rolling work rolls prepared with EDT changes the surface, the first pass of cold rolling and final surface images after casting were examined. In addition, stylus profilometer images are shown in the casting thickness of the plates made by both processes. Finally, the inorganic contamination amounts of the final products of both processes at 1 mm were given and interpreted.

Experimental Studies

Sandblasting Process

SEM images of the surfaces from the casting thickness of the plates cast from standard ground and sandblasted rolls are shown in Fig. 7. Thin and long shingle structure dominates

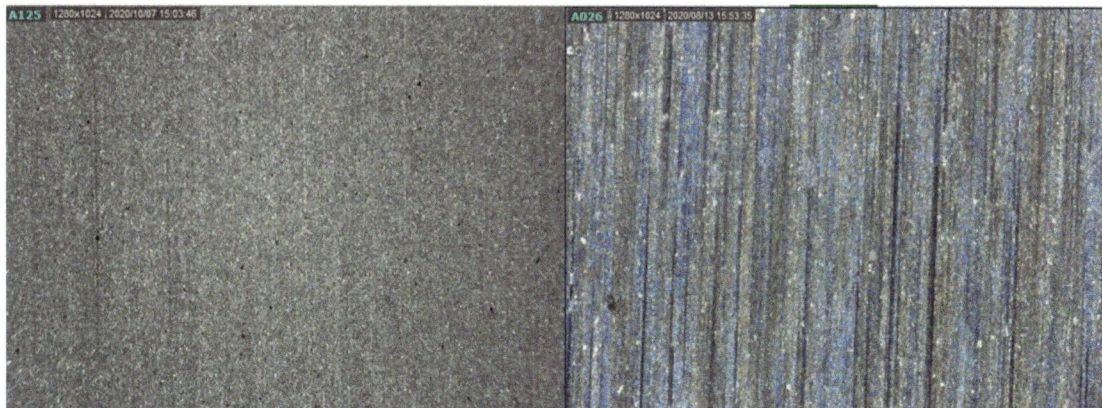

Fig. 3 Mobile-optical microscope images of sandblasted surface (left) and ground surface (right) rolls

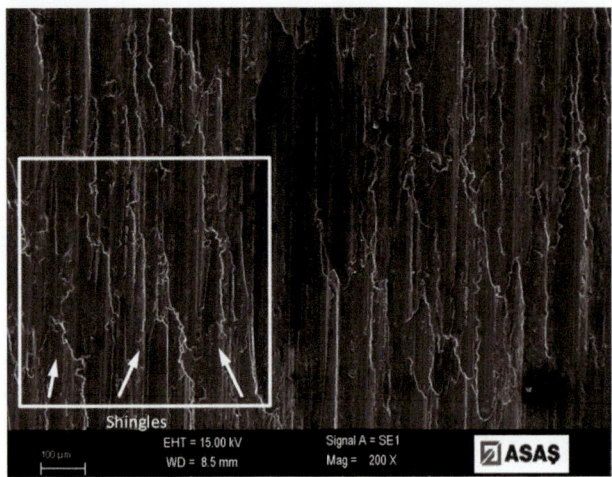

Fig. 4 Location of shingles on SEM image of a casted AA1050 plate (5 mm thickness)

the plate cast from a standard ground roller. In the plate cast from the sandblasted roller, the number of shingles is lower and the transverse dimension is larger. It is desirable that the number of shingles be low. Since the number of shingles will be low in the final thickness plate, it is expected that the amount of pollution carried through the processes will be less.

Figure 8 shows the stylus profilometer images in the casting thickness of the plates produced from ground and sandblasted rolls.

When the stylus profilometer data is examined it is apparent that the number of stigmas at radical height on the surface of the plate produced from the standard ground roller is high. This image supports the SEM image. On the plate produced by the sandblasted rollers, the elevation was more homogeneous and had low-throughout area. Likewise, the

consistency with the image in SEM was observed. The ra value was measured as 1.1 μm in the sandblasted process, while it was measured as 1.6 μm in the standard process.

When the final thickness plate surface images are examined in Fig. 9 the obvious difference between the surface lines can be seen. In the standard ground casting roll process, there are lines with a large number of elevation differences. In the sandblast casting roll process, it is apparently determined that there is a much smoother surface morphology compared to the other image. The surface of the plate produced with the sandblast cast rollers is close to the ideal surface.

In order to compare the final SMUT (inorganic pollution) amounts, analysis of residual oil and SMUT quantities from both processes was carried out. Table 1 shows that the quantity of SMUT in the sandblasting process was significantly lower.

EDT Process

Two different processes were applied to 1050 alloy plate cast from standard ground rollers. In the processes, the work rolls used in the cold rolling first passes were roughened by different methods. In one process, rolls were roughened by conventional grinding, while in the other process, the roughening process was carried out with the EDT method.

When the surface SEM images are examined after the cold rolling first pass at 2.8 mm (Fig. 10), it was observed that the faulty shingle surface in the casting continues in the standard ground surface cold rolling process. On the other hand, it was observed that almost all shingle structures were removed after the first pass in the cold rolling process with EDT prepared rolls. The difference is a significant one.

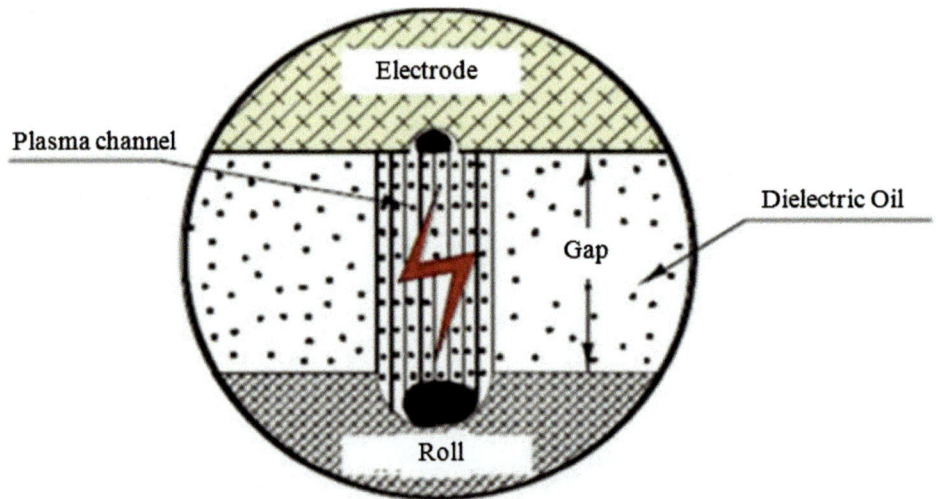

Fig. 5 Working principle of EDT method

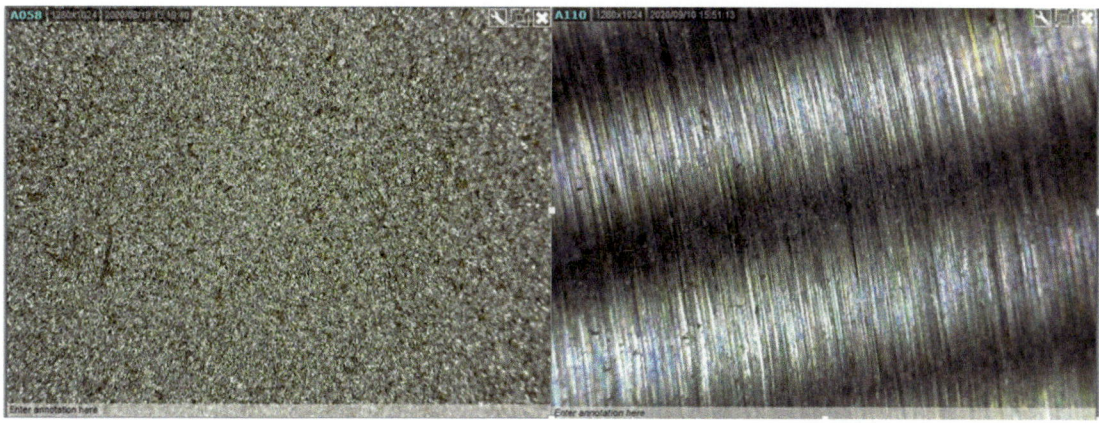

Fig. 6 Mobile-optical microscope images of EDT treated (left) and standard ground (right) cold rolling work rolls

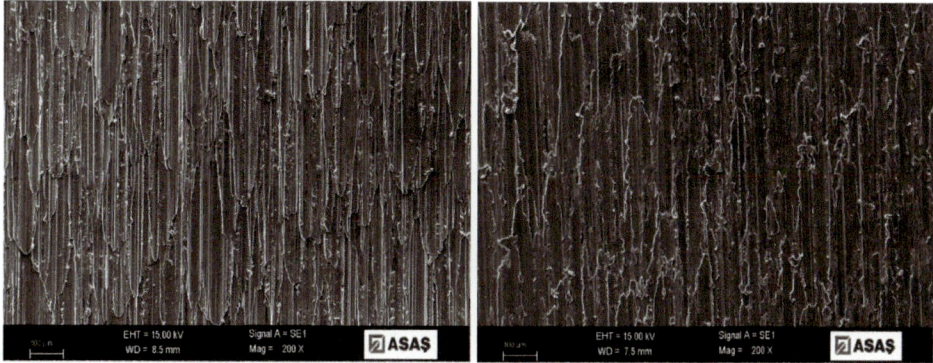

Fig. 7 SEM images of the casting thickness of the plates cast from standard ground (left) and sandblasted (right) rolls

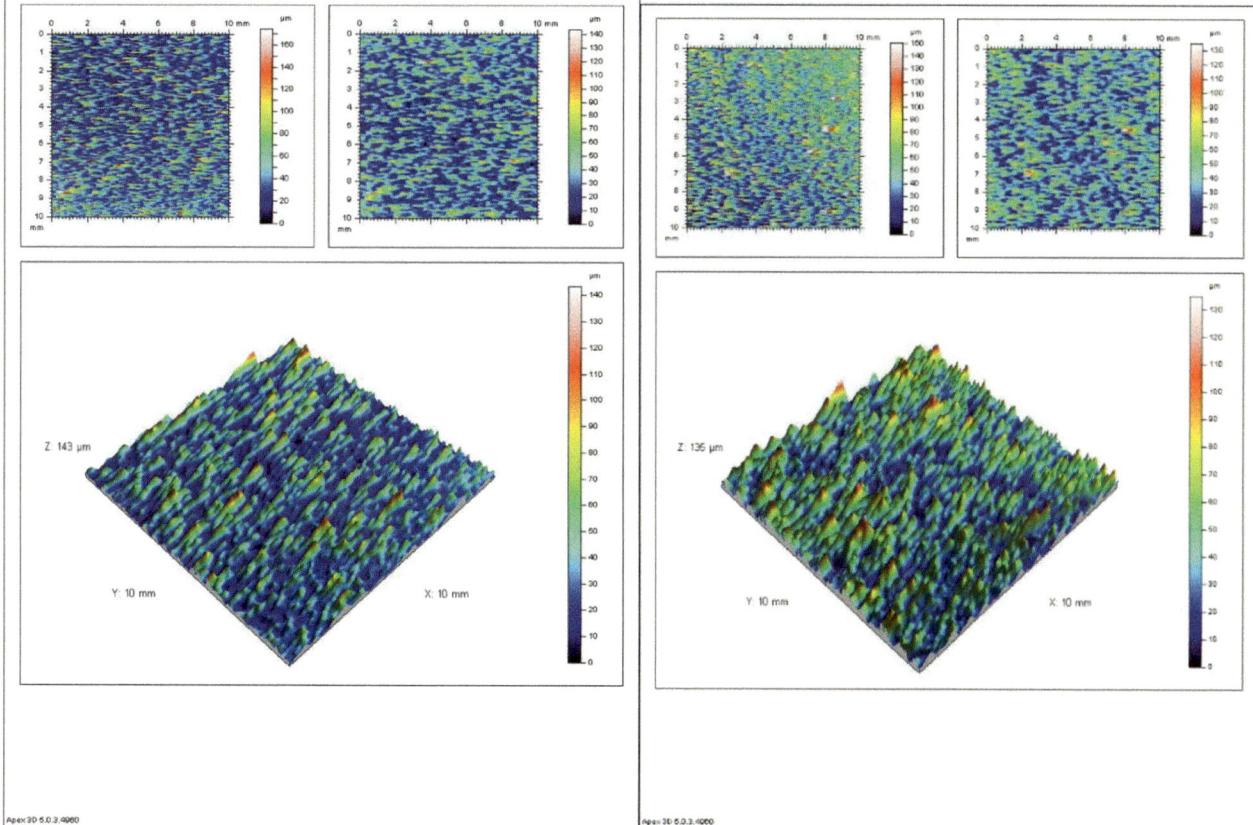

Fig. 8 Stylus profilometer images in the casting thickness of the plates produced from sandblasted (left) and ground (right) rolls

Fig. 9 SEM images of the final-thick(1 mm) plates. At left, standard process; at right, sandblasted process

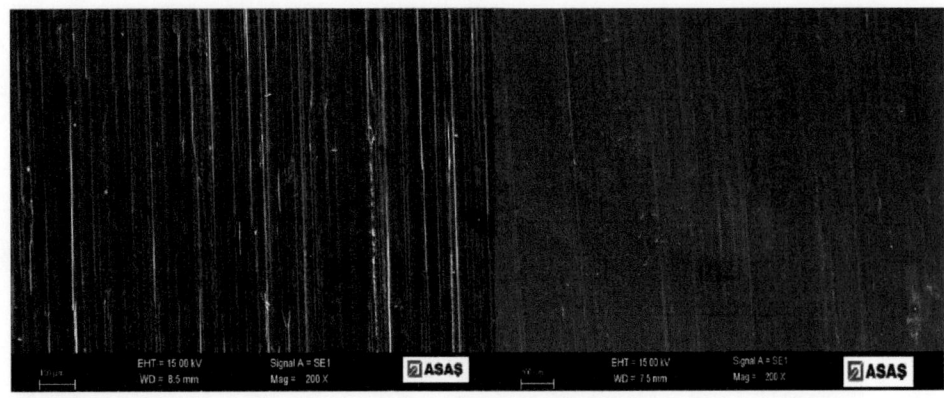

Table 1 Dirtiness analyses of sandblast and grinding processes for casting rollers

Process	Organic dirtiness max residual oil (mg/m^2)	Inorganic dirtiness SMUT (mg/m^2)	Total dirtiness (mg/m^2)
Sandblasting	3,33	2,67	6
Grinding	17,37	10,94	28,31

Fig. 10 SEM images after the first cold rolling pass (2.8 mm). At left, standard process; at right, EDT process

Fig. 11 SEM images of the final-thick plates. At left, standard process; at right, EDT process

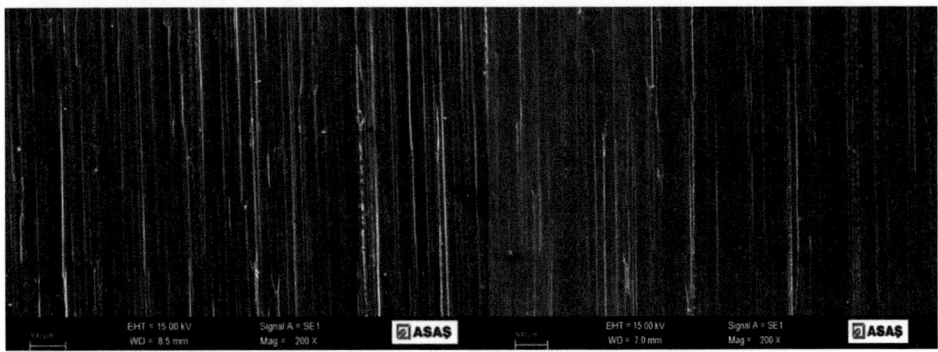

Table 2 Dirtiness analyses of EDT and grinding processes for cold working rollers

Process	Organic dirtiness max residual oil (mg/m^2)	Inorganic dirtiness SMUT (mg/m^2)	Total dirtiness (mg/m^2)
EDT	6,67	6,5	13,17
Grinding	17,37	10,94	28,31

SEM images of the final 1 mm thick plate samples are shown in Fig. 11. When the images are examined, it is seen that the topography in the EDT process is much cleaner compared to the standard process. It has been observed that groove-like structures in the standard plate continue to exist throughout the cold rolling passes.

Finally, the organic and inorganic pollution (SMUT) amounts in the final plate are shown in Table 2. In the process with EDT prepared rolls, 4–5 mg/m^2 SMUT was measured less than for the standard prepared rolls.

Conclusion

The following conclusions can be summed up based on this present work:

- It has been seen that both sandblasting of casting rolls and EDT treatment of cold rolling rolls are quite effective in eliminating casting welded shingle structure.
- It was observed that the quantity of inorganic (SMUT) pollution on the final thickness plate in the sandblasted casting roll process is lower at the level of 9 mg/m^2 compared to the standard ground casting roll process.
- It was observed that the quantity of final inorganic (SMUT) pollution on the final thickness plate in the EDT surface prepared cold rolling process was 5 mg/m2 less than the standard ground surface cold rolling process.
- At the same time, thanks to both sandblasting and EDT applications, there was a big decrease in the amount of residual oil on the surface.

- Sandblasting applied to cast rollers showed a greater reduction in the quantity of SMUT on the surface of the final thickness plate compared to EDT prepared surface on cold rolling rollers.
- More detailed studies should be carried out on the sandblasting process in areas of use with high expectation of surface cleaning.

Acknowledgements Authors thank Samet Sevinc regarding his efforts on metallographical operations and analysis.

References

1. Alper G, Alüminyum Sürekli Döküm Yöntemi ile Üretilmiş 5052–5182 Alüminyum Alaşımlarının Şekillendirilebililk Kabiliyetinin Belirlenmesi
2. Ünal. (2004). Continuous Casting of Aluminum, Patent No: US 6,672,368 B2
3. Ronald A. Reich, June M. Epp & David E. Gantzer (1996) A Mechanism for Generating Aluminum Debris in the Roll Bite and Its Partitioning between the Surface of the Work Roll and the Surface of the Sheet as Smudge, Tribology Transactions, 39:1, 23–32. https://doi.org/10.1080/10402009608983498
4. https://www.yenar.com.tr/urunler/machines2.php?dil=en
5. Sun Dale, Yao Lisong, Fan Qun, and Zhang Jian, Research on Service Performances between Different Textured Cold Rolls, AIS Tech., 2006, vol. 2, pp. 285–289
6. Salganik, V.M., Pivovarov, A.V., and Pivovarov, F.V., Improvement in Sheet Quality by Electrode-Discharge Texturing of Rollers, Stal, 2003, no. 4, pp. 46–47

Characterization of 8006 Aluminium Alloy Casted by TRC Technology with Steel–Steel and Copper–Copper Roll Pairs

Feyza Denizli, Onur Birbaşar, Koray Dündar, Yusuf Özçetin, Ali Ulus, and Canan İnel

Abstract

Aluminium alloys are widely used in many application areas due to the properties of high thermal conductivity, good formability, high corrosion resistance, and recyclability. Growing demand for aluminium materials leads the aluminium manufacturer to increase the casting productivity. One of the effective way to increase the casting speed in twin-roll casting (TRC) is changing the cooling conditions required for solidification. The shell component of a cast roll contacts with the molten metal ensures solidification by removing the heat from molten metal. Depending on the thermal conductivity of shell material, casting speed and productivity can be increased. However, casting microstructure and other properties of the material can change because of the solidification conditions. In this study, 8006 alloy is casted by TRC method using copper–copper roll pairs as an alternative to conventional steel–steel pairs. Microstructural analysis, electrical conductivity measurements and mechanical properties were investigated comparing with steel–steel casted material.

Keywords

8XXX • Solidification • Copper shell • Process technology • Twin-roll casting

Introduction

Aluminium alloys are widely used in many application areas such as automotive, construction, packaging, and food contact foil production due to the properties of high thermal conductivity, good formability, high corrosion resistance, and recyclability [1]. Twin-roll casting (TRC) method is one of the most preferred production methods of aluminium flat rolled products because of the low cost and short production cycle comparing to the Direct Chill (DC) casting method [2–4].

Growing demand for aluminium materials leads the aluminium manufacturer to increase the casting productivity. In the TRC process, the molten metal is transferred from the melting furnace through the tundish by the launder, and then it is distributed by the tip into the cast roll gap. When the molten metal contacts the water-cooled rotating rolls, it is solidified and aluminium sheet is produced. Aluminium alloys which have the low solidification range are produced with this method [2, 3, 5–7]. The cast roll comprises two components which are core and shell. The role of the shell is to aid the solidification by transferring the heat from the molten metal. Consequently, the casting speed and the productivity increase. Changing the cooling conditions required for solidification is one of the effective ways to increase the casting speed in TRC. For this reason, copper shell has a high thermal conductivity coefficient which is now being used as an alternative shell material [2–5, 7, 8]. However, since the cooling condition is changed, the microstructure and the distribution of intermetallics become different from the typical TRC casting microstructure. It's important to control the casting parameters and investigate the

F. Denizli (✉) · O. Birbaşar · K. Dündar · Y. Özçetin · A. Ulus · C. İnel
Asaş Alüminyum San. ve Tic. A.Ş, İstanbul, Turkey
e-mail: feyza.denizli@asastr.com

O. Birbaşar
e-mail: onur.birbasar@asastr.com

K. Dündar
e-mail: koray.dundar@asastr.com

Y. Özçetin
e-mail: yusuf.ozcetin@asastr.com

A. Ulus
e-mail: ali.ulus@asastr.com

C. İnel
e-mail: canan.inel@asastr.com

© The Minerals, Metals & Materials Society 2023
S. Broek (ed.), *Light Metals 2023*, The Minerals, Metals & Materials Series,
https://doi.org/10.1007/978-3-031-22532-1_140

microstructure of as-cast material and design the thermo-mechanical process accordingly [1, 3, 5].

In this study, 8006 alloy is casted by TRC method using copper–copper roll pairs as an alternative to conventional steel–steel pairs. Microstructural analysis, electrical conductivity measurements and mechanical properties are investigated comparing with steel–steel casted material.

Materials and Methods

In this study, 8006 aluminum alloy was casted to 7–7,5 mm thickness by a twin-roll casting machine on an industrial scale using copper–copper shell pairs. All the characterizations were repeated and compared to the standard steel–steel shell production. The chemical composition range of the samples is shown in Table 1. After the casting operation, samples were annealed at different temperatures between 520 and 580 °C/8 h in the laboratory-type furnace. Microstructural analysis of the copper–copper and steel–steel produced samples, hardness test, and electrical conductivity measurement were carried out by using optical microscope (ZEISS Axio Scope.A1), hardness tester (ZWICK ZV10), and electrical conductivity measurement device (FISCHER SIGMASCOPE SMP10), respectively. To investigate thermo-mechanical properties of the samples, they were rolled to a final thickness of 200 micron by using cold rolling testing machine on a laboratory scale. Tensile test, Erichsen cupping test, and macro- and microstructural analysis were carried out and reported.

Experimental Studies

Longitudinal cross-sectional microstructural analysis of as-cast samples is shown in Fig. 1. Contrary to the steel–steel grain structure, the grain structure at both surfaces of the copper–copper sample is equiaxial. It can be explained by the rapid heat transfer which ensures the homogeneous and equiaxed grain structure through the cross section due to the dynamic recrystallization [4, 5, 9].

Longitudinal cross-sectional microstructural analysis of annealed samples at 580 °C/8 h is shown in Fig. 2. After annealing, the grains have coarsened at the surfaces of the steel–steel sample by the mechanism of recrystallization and grain growth. However, it is observed that the grains have not still started to coarsen at the surfaces of the copper–copper sample because microstructure on the surfaces has

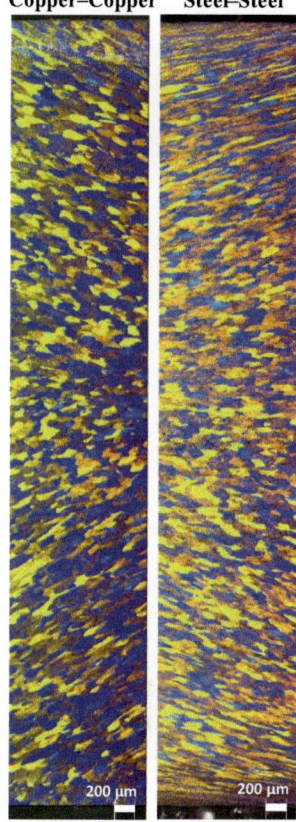

Fig. 1 As-cast longitudinal cross-sectional microstructure of the samples

already recrystallized during casting process (as shown in Fig. 1) [4, 5].

Electrical conductivity measurements are carried out on both as-cast and annealed samples (Table 2). The result shows that the electrical conductivity of the copper–copper sample is slightly lower than the steel–steel sample. Because of the rapid heat transfer, the intermetallics are not able to precipitate from the aluminium matrix and as a result of this, electrical conductivity decreases. By the help of annealing, intermetallics start to precipitate and the electrical conductivity of both samples increases [4, 5].

Hardness measurements were carried out on both as-cast and annealed samples (Table 3). The results are the average values measured from the top, bottom, and middle in the cross section. The result shows that the hardness value of the copper–copper sample is higher than the steel–steel sample. In addition, it's observed that while the annealing temperature increases, the hardness value of the samples decreases. The electrical conductivity measurements and hardness test result show that the values are very close to each after annealing [4] (Fig. 3).

Mechanical properties and Erichsen cupping test result of samples at final thickness are shown in Table 3. Copper–copper and steel–steel samples were rolled to the thickness of 200 μm by using cold rolling testing machine and then annealed at the same temperature and duration. It is observed

Table 1 The range of chemical composition of the samples

Si %	Fe %	Cu %	Mn %	Ti %
0,1–0,3	1,2–2,0	0,01–0,1	0,4–1,0	0,01–0,03

Fig. 2 Longitudinal cross-sectional microstructure of the samples after annealing at 580 °C/8 h

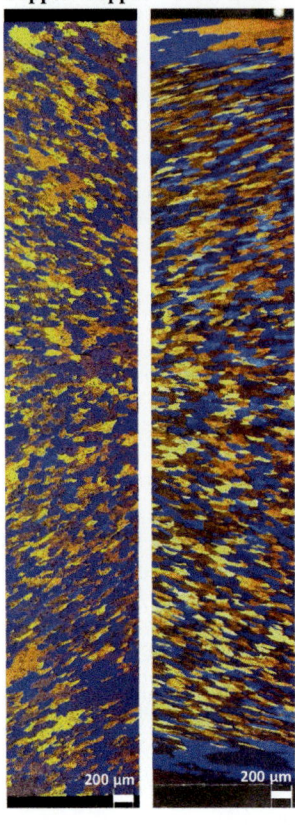

Copper–Copper   Steel–Steel

that the yield strength of copper–copper sample is higher than the steel–steel sample and similar in all directions. This result is in accordance with hardness test result which can be explained by the super-saturation during casting operation by the rapid heat transfer of copper–copper shell. In addition, it is observed that the deep-drawability of samples can be similar by the results of Erichsen values.

Figure 4 shows the longitudinal cross-sectional microstructure of the samples at final thickness. The grain growth is observed on the top and bottom surfaces of the steel–steel sample contrary to the copper–copper sample at a given temperature. Similarly, the grains at the surface of steel–steel samples are coarser than the copper–copper sample and recrystallized (Fig. 5). The difference between grain structures of final samples is the consequence of different solidification conditions during casting. While the steel–steel final sample shows the typical TRC material, copper–copper sample behaves differently because of the dynamic recrystallization which has occurred during the casting process and after annealing at as-cast thickness, the grain growth is not observed at surfaces.

Conclusion

- In this study, 8006 alloys are produced by TRC method using copper–copper roll pairs and its microstructural and mechanical properties are investigated comparing to the conventional steel–steel produced material.
- It is observed that the microstructure of copper–copper samples has more equiaxed and homogeneous grain structure along the longitudinal cross section compared to the steel–steel sample.
- Electrical conductivity of copper–copper as-cast material is lower than the steel–steel as-cast material. In addition, copper–copper as-cast material is harder than the steel–steel as-cast material because of the different cooling conditions. Intermetallics precipitate in the aluminium matrix at slow cooling rate and gives higher electrical conductivity.
- After annealing, the grains at the surfaces have not coarsened and recrystallized in the copper–copper sample which improves the grain structure of the final products.
- It is observed that since the copper–copper final product has not recrystallized yet, the temperature required for the final annealing can be higher than the steel–steel final sample in order to obtain similar mechanical properties.

Table 2 Electrical conductivity of the samples (mS/m)

Sample	As-cast	Annealed
Copper–copper	31,7	49,3
Steel–steel	36,2	49,1

Table 3 Mechanical properties of the samples at final thickness

Sample	Copper–copper			Steel–steel		
test direction (°)	0	45	90	0	45	90
Yield St (MPa)	83	80	81	73	69	72
Tensile St (MPa)	124	110	116	123	108	115
Elongation (%)	32,8	29,7	19,7	29,4	31,5	25
Erichsen (mm)	9,4			9,5		

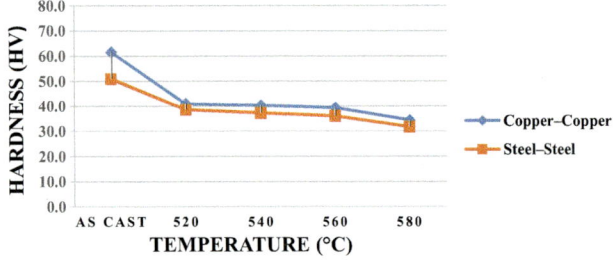

Fig. 3 Hardness results versus heat treatment temperature of the samples

Acknowledgements The authors are sincerely thankful to R&D laboratory specialist Samet Sevinç and technician Ahmet Bicat for their help in terms of preparation of the samples for metallographic examination and mechanical tests.

Copper–Copper

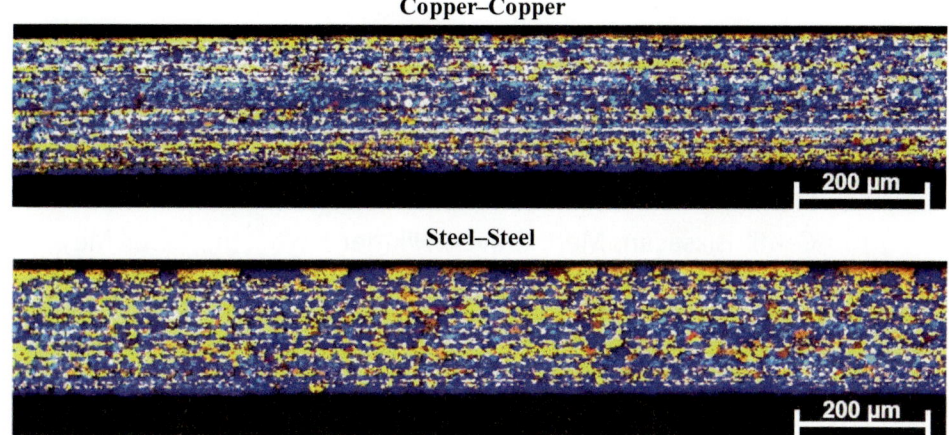

Fig. 4 Longitudinal cross-sectional microstructure of the samples at final thickness

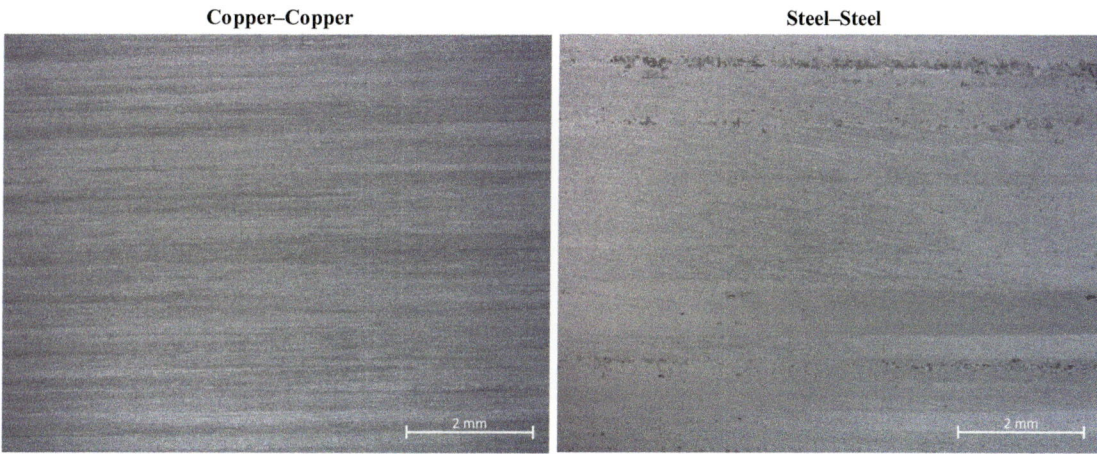

Fig. 5 Surface macrostructure of samples at final thickness

References

1. W. Chen, P. Zhao, Y. Zhou, Y. Pan, Effects of Homogenization Conditions on the Microstrucures of Twin-Roll Cast Foil Sock of AA8021 Aluminum Alloy, Materials Science Forum, ISSN: 1662-9752, Vol. 877, pp 296–302

2. R. E. Sanders, (2012) Continuous Casting for Aluminum Sheet: a Product Perspective, JOM, Vol. 64, No. 2, 2012, ©2012 TMS

3. C. W. Schmidt, A. Buchholz, K. Karhausen, The Importance of Heat Removal for Productivity in Industrial Twin Roll Casting of Aluminium, Light Metals, 2015, The Minerals, Metals & Materials Series, pp 1209–1213

4. D. Spathis, J. Tsiros, A. Arvanitis, H. Wobker, A. Clemente, The Use of Copper Shells by Twin Roll Strip Casters, Light Metals, 2010, The Minerals, Metals & Materials Series, pp 747–751

5. H. M. Altuner, C. Işıksaçan, O. Birbaşar, M. Günyüz, O. Meydanoğlu, Christallographic Texture Development of As-Cast

3105 Alloy Produced by St/Cu Shell Pair, Light Metals, 2016, The Minerals, Metals & Materials Series, pp 1025–1030

6. H. Westengen, K. Nes, Twin Roll Casting of Aluminium: The Occurance of Structure Inhomogenities and Defects in As Cast Strip, Light Metals 1984, The Minerals, Metals & Materials Society 2016, Essential Readings in Light Metals, pp 972–980

7. T. Haga, High Speed Roll Caster for Aluminium Alloy, Metals 2012, 11, 520. https://doi.org/10.3390/met11030520

8. R. E. Akdogan, H. M. Akdoğan, O. Birbaşar, M. Günyüz, Influence of Strip Thickness on As-Cast Material Properties of Twin-Roll Cast Aluminium Alloys, The Minerals, Metals & Materials Society 2019, C. Chesonis (ed.). Light Metals 2019, The Minerals, Metals & Materials Series, pp 1137–1141

9. X. Qian, L. Meng, X. Sheng, Y. Li, Z. Wang, The formation mechanism of the line segregation in twin-roll casting (TRC) 3003 aluminium alloy, Journal of Physics: Conference Series, AMCE-2021, 2194 (2022) 012007

Tailoring the As-Cast Microstructure of Twin-Roll Cast AA3105 Alloy Produced by St/Cu Shell Pair

Cemil Işıksaçan, Mert Gülver, Hikmet Kayaçetin, Onur Meydanoglu, and Erdem Atar

Abstract

Extracting the heat from the liquid metal with the help of the water-cooled caster rolls (designed with core and shrink-fitted shell) and then reaching as-cast strip in various thicknesses with different casting speeds generate the phenomena of the twin-roll casting (TRC) process. In order to increase productivity (combination of the casting speed and as-cast strip thickness) which is one of the most important outputs of the TRC, different shell materials having different heat transfer coefficient rate could be used. In case of using St/Cu (steel–copper) shell pair, higher productivity rate could be obtained. However, as-cast microstructure of the materials could have some differences such as behavior of the centerline segregation and grain structure compared to the sample produced by St/St (steel–steel) shell pair. In this study, casting parameters (setback, roll speed, strip speed, and thickness) are changed in order to observe the differences such as centerline segregation and grain structure close to the surfaces.

Keywords

Copper shell • Grain structure • Twin-roll casting

C. Işıksaçan · M. Gülver (✉) · H. Kayaçetin · O. Meydanoglu
Assan Alüminyum, R&D Center, DilovasıKocaeli, Turkey
e-mail: mert.gulver@assanaluminyum.com

C. Işıksaçan
e-mail: cemil.isiksacan@assanaluminyum.com

H. Kayaçetin
e-mail: hikmet.kayacetin@assanaluminyum.com

O. Meydanoglu
e-mail: onur.meydanoglu@assanaluminyum.com

C. Işıksaçan · E. Atar
Department of Materials Science and Engineering, Gebze
Technical University, GebzeKocaeli, Turkey
e-mail: atar@gtu.edu.tr

Introduction

In twin-roll casting (TRC), solidification of liquid metal is combined with hot deformation and it is very crucial to control various process parameters such as melt temperature, casting speed, setback, and strip thickness. The combination of the effects of these parameters plays an important role on solidification behavior of the liquid metal and dictates the microstructural features of the as-cast materials [1]. High cooling rates encountered in this technique lead to a highly super-saturated matrix and the change in the temperature gradient encountered through thickness of the solidifying metal results in heterogeneities in intermetallic particle sizes and distributions and leads to formation of macro-segregations, especially the centerline segregations (CLS). The CLS in TRC is due to the segregation of solute elements, which are swept to the center of the sheet by the two opposite solid/liquid interfaces [2–4]. TRC is well known as a relatively low productive process compared to traditional DC casting. Many attempts have been done to improve the TRC productivity; however, the most effective solution is the use of high thermal conductive materials as a sleeve in caster rolls [5]. Its high thermal conductivity makes copper alloys a preferred candidate for sleeves. However, copper possesses low mechanical properties under mechanical fatigue conditions. This drawback can be overcome with alloying which deteriorates thermal conductivity of the sleeves. Nevertheless, thermal conductivity of alloyed copper sleeves is still much higher than that of steel sleeves. Higher heat extraction of copper sleeve results in different microstructural constituents and material features [6].

In the light of this short review, the aim of this study is to investigate the microstructural behavior of AA3105 aluminum alloy produced by twin-roll casting with St/Cu shell pair at different as-cast strip thicknesses.

Experimental Studies

In this study, commercial-grade AA 3105 aluminum alloy was cast utilizing an industrial-scale twin-roll caster with steel/copper shell pair at two different strip gauges of 5.90 and 8.65 mm. Table 1 lists the chemical composition of alloy used in this study. As-cast strips were heat treated at 500 °C for 8 h in a laboratory-scale furnace to observe microstructural changes in the grain structures of as-cast strips.

Characterization of samples was made by microstructural surveys, mechanical tests, and electrical conductivity measurements. Microstructural characterization was carried out by optical microscope examinations on the cross sections of the samples under polarized light after preparing the samples according to standard metallographic methods and etching with Barker's solution. Mechanical properties of the samples were determined through thickness hardness measurements conducted on the polished cross sections of samples utilizing a micro-hardness tester with a Vickers indenter under an indentation load of 10 g. Electrical conductivity measurements were performed on as-cast strips as well as heat-treated samples at 60 kHz.

Table 1 Chemical composition of the AA 3105 alloy used in this study

Si	Fe	Cu	Mn	Mg	Ti
0.16	0.44	0.06	0.34	0.27	0.01

Results and Discussion

Typical etched cross-sectional optical microscope images of 5.90- and 8.65 mm-thick as-cast strips are shown in Fig. 1.

Due to the decreased cooling rates throughout the cross section in twin-roll casting, a symmetrical increase in grain size from outermost surfaces to mid-plane is expected with a super-saturated region at the outermost surfaces. The use of St/Cu shell pairs results in slightly different grain structures than those produced by the shell pair of the same material and alters the symmetrical change in grain structure. The centerline (the line where grains growing from rolls meet) shifts towards the steel roll. This observation can be attributed to the higher heat extraction of copper roll. On the other hand, increased as-cast strip thickness leads to quasi-equiaxed grains throughout the cross section.

Figure 2 shows through thickness hardness profiles of as-cast strips. Hardness profiles reveal that hardness of surface touching copper sleeve is higher than that of touching steel sleeve. Electrical conductivities of as-cast strips are given in Table 2. Electrical conductivity is an indication of saturation of aluminum matrix and thus amount of solute elements in the aluminum lattice. It has been reported that Mn and Fe when in solid solution have high negative impact on electrical conductivity of Al [7, 8]. When hardness profiles and electrical conductivities of as-cast strips are taken into consideration together, higher hardness of surfaces touching the copper sleeve can be attributed to the

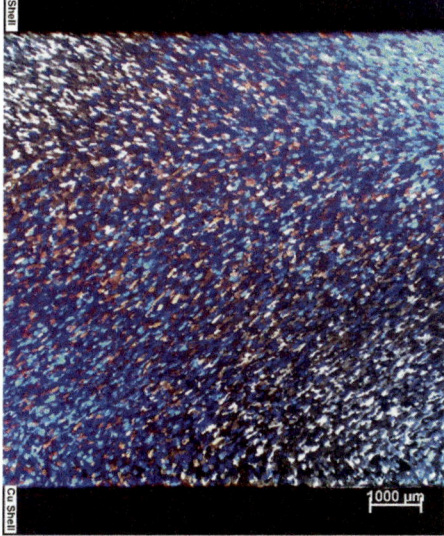

Fig. 1 Typical etched cross-sectional optical microscope images of as-cast strips, **a** 5.90 mm and **b** 8.65 mm thick

(a)   (b)

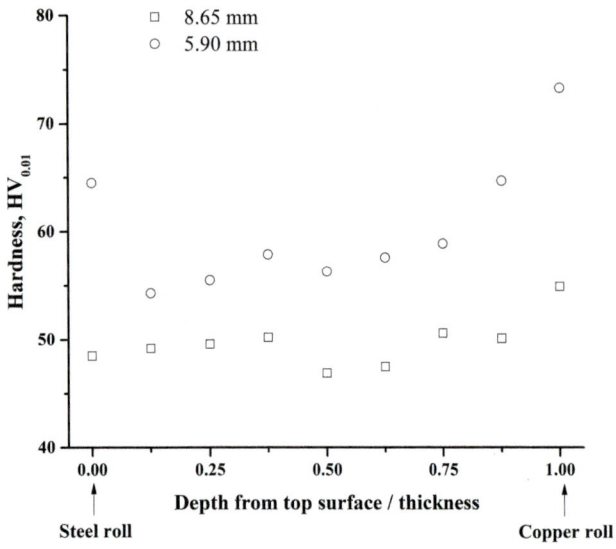

Fig. 2 Through thickness hardness profiles of as-cast strips

Table 2 Electrical conductivities of as-cast strips

As-cast strip thickness, mm	Electrical conductivity, MS/m	
	Steel sleeve	Copper sleeve
5.90	24.29	23.66
8.65	23.83	23.65

enrichment of aluminum matrix due to higher solidification rates provided by the copper sleeve. On the other hand, overall hardness of 8.65-mm-thick as-cast strip is lower and relatively homogeneous through the thickness as compared with 5.90-mm-thick as-cast strip.

Typical etched cross-sectional optical microscope images of 5.90-mm- and 8.65-mm-thick as-cast strips annealed at 500 °C for 8 h are shown in Fig. 3.

As can be seen in Fig. 3, heat treatment at 500 °C for 8 h leads to a grain growth starting at the outermost surfaces and penetrating through the thickness of 5.90 mm-thick as-cast strip where no morphological changes in grain structure of 8.65-mm-thick as-cast strip occur. Grain growth is more pronounced at the surface touching copper sleeve.

Figure 4 shows through thickness hardness profiles of as-cast strips annealed at 500 °C for 8 h. Heat treatment results in overall softening of as-cast samples. This is more pronounced for 5.90 mm as-cast strip. The hardness differences between copper side and steel side of as-cast strips can still be observed for 5.90 mm-thick as-cast strip where hardness differences between the surfaces disappear in 8.65-mm-thick as-cast strip.

Electrical conductivities of as-cast strips annealed at 500 °C for 8 h are given in Table 3. It was observed that there was an increase in electrical conductivity with heat treatment. Increase in electrical conductivity and decrease in hardness indicate precipitation of Mn- and Fe-bearing particles leading to decreased amount of Mn and Fe in solid solution.

Conclusions

The results of this study can be summarized as follows:

The use of steel/copper sleeve pair leads to shifted centerline towards steel sleeve due to higher heat extraction capability of copper as compared to that of steel.

Fig. 3 Typical etched cross-sectional optical microscope images of as-cast strips annealed at 500 °C for 8 h, **a** 5.90 mm, and **b** 8.65 mm thick

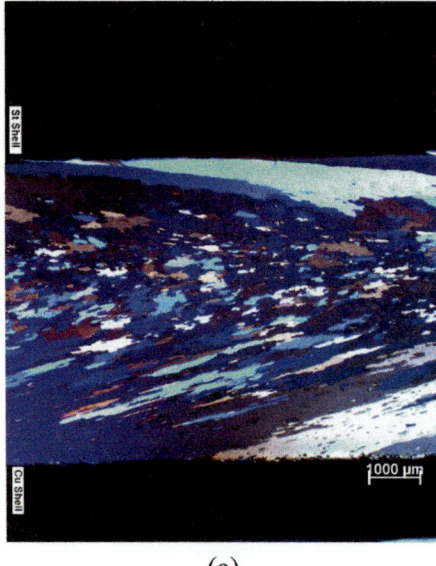

(a)

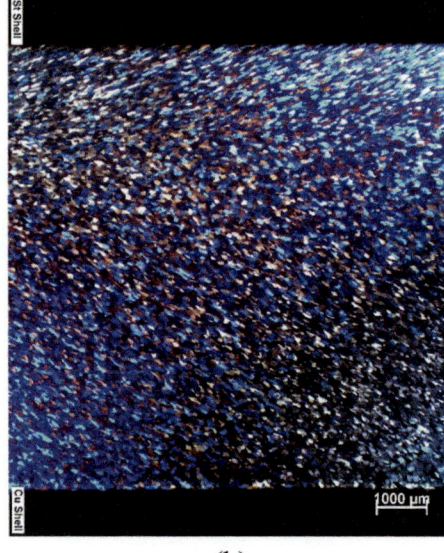

(b)

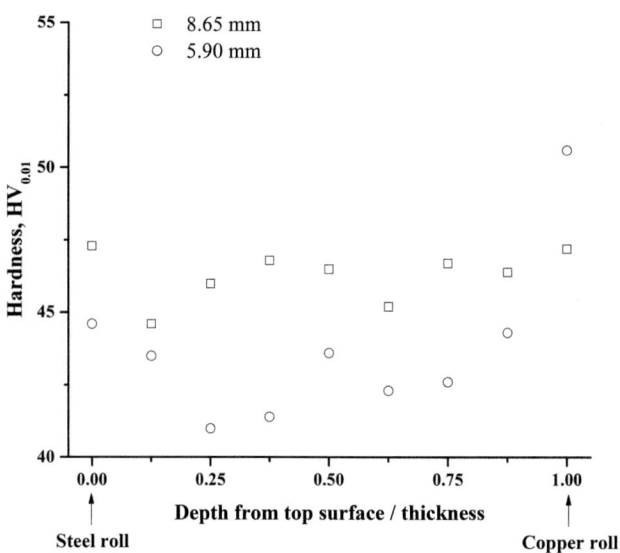

Fig. 4 Through thickness hardness profiles of as-cast strips annealed at 500 °C for 8 h

Table 3 Electrical conductivities of as-cast strips annealed at 500 °C for 8 h

As-cast strip thickness, mm	Electrical conductivity, MS/m	
	Steel sleeve	Copper sleeve
5.90	30.00	30.41
8.65	30.07	29.53

Increased as-cast strip thickness results in quasi-equiaxed grains.

Electrical conductivities of as-cast strips are lower on copper side as compared with steel side due to increased amount of solute element in the aluminum lattice.

Surface hardness of copper side is higher than that of steel side. This observation is more pronounced for thin gauge strip and increased as-cast strip thickness leads to relatively homogeneous hardness through the cross section.

Heat treatment at 500 °C for 8 h leads to a grain growth in thin gauge as-cast strip. However, increased as-cast strip thickness hinders morphological changes in grain structure.

Heat treatment at 500 °C for 8 h lowers the overall hardness and enhances the electrical conductivities of as-cast strips. Increased as-cast strip thickness provides relatively lower drop in overall hardness.

Acknowledgements The authors thank Mr. Nihat Yılmaz and Serdar Azaklıoğlu, technicians of Assan Alüminyum Laboratory, for their valuable help with metallographic studies.

References

1. Gras C., Meredith M., Hunt J. (2005) Microstructure and texture evolution after twin roll casting and subsequent cold rolling of Al-Mg-Mn aluminum alloys. Journal of Materials Processing Technology 169:156–163.
2. Birol Y. (2008) Recrystallization of a super-saturated Al-Mn alloy. Scripta Materialia 59:611–614.
3. Engler O., Laptyeva G., Wang N. (2013) Impact of homogenization on microchemistry and recrystallization of the Al-Fe-Mn alloy AA 8006. Material Characterization 79:60–75.
4. Lee Y., Kim H., Cho J. (2015) Process parameters and roll separation force in horizontal twin roll casting of aluminum alloys. Journal of Materials Processing Technology 218:48–56.
5. Clemente A., Tsiros J., Arvanitis A., Spathis D., Wobker H. (2010) The use of copper shells by twin roll strip casters. Paper presented at the 139th TMS Annual Meeting, Seattle, Washington, 14–18 February 2010.
6. Mollaoğlu Altuner H., Işıksaçan C., Birbaşar O., Günyüz M., Meydanoğlu O. (2016) Crystallographic texture development of as-cast 3105 Alloy produced by St/Cu shell pair. Paper presented at the 145th TMS Annual Meeting, Nashville, Tennessee, 14–18 February 2016.
7. Muggerud A. M. F., Mortsell E. A., Li Y., Holmestad R. (2013) Dispersoid strengthening in AA3xxx alloys with varying Mn and Si content during annealing at low temperatures. Materials Science and Engineering A 567:21–28.
8. Martins J. P., Carvalho A. L. M., Padilha A. F. (2009) Microstructure and texture assessment of Al-Mn-Fe-Si (3003) aluminum alloy produced by continuous and semicontinuous casting processes. J. Mater. Sci. 44:2966–2976.

Designing a Safe Casthouse

Alex W. Lowery

Abstract

The vast majority of casthouses built in our industry over the past decade were designed with production, not safety, in mind. This issue contributes to many incidents resulting in workers being injured and killed. This paper will discuss the common design mistakes that occur in casthouse design and construction. My paper will cover common hazards overlooked during casthouse design. The current design of casthouses focuses on a myriad of other goals excluding safety. I argue that only by focusing first on safety during the design of a casthouse can our industry stop these incidents from injuring and killing workers.

Keywords

Safety • Casthouse • Molten metal • Explosions

Disclaimer

Throughout this paper safety incidents involving injuries and/or fatalities are cited. It is not the intention of this paper to place blame upon companies or worker(s) who have suffered a workplace incident. Company names and locations regarding the incidents were purposely omitted so as not to identify the companies.

Introduction

Construction of a casthouse falls into one of two categories: brownfield or greenfield development. A brownfield project is one that carries constraints related to the current state of the site. It's a site that may have existing structures that architects need to remove or modify in some way. Many times a brownfield project is simply an expansion of an existing workplace. Greenfield developments occur on a completely vacant site. Architects can start completely from scratch.

After many tours of both types of casthouse projects, many common safety issues that are overlooked during the design and construction have been observed. The underlying issues are that architects or designers of casthouses have little or no experience in the aluminium industry. Provided with a budget and asked to design a building in an industry that they have little to no experience in. It is the hope of this author that this paper will be used to identify safety issues in a casthouse that are commonly overlooked. Each item described in this paper has resulted in serious injuries or fatalities.

Explosion Venting Panels

The aftermath of a casthouse after a molten metal explosion can be frightening. Twisted metal sheeting from the roof and walls are scattered a distance away after being blown off by the explosion. "The devastating impact an explosion can cause to an industrial facility, not just in terms of business interruption and lost productivity, but also the potential for loss of life or injury, (makes) the provision of a safe working environment is paramount to owners and operators of all businesses handling and storing potentially combustible materials," says Andy Moul, Technical Support Manager, Construction Specialties (UK) Ltd.

Explosion venting panels are an effective way to limit damage in the event of an explosion. These safety devices are most commonly seen in dust collectors or exhaust ducts. They have been installed in casthouse walls and roofs for decades. Unfortunately, for unknown reasons, they have been omitted over the past 20 years in most new casthouses

A. W. Lowery (✉)
Wise Chem LLC, P.O. Box 633 New Albany, OH 43054, USA
e-mail: alex.lowery@wisechem.net

© The Minerals, Metals & Materials Society 2023
S. Broek (ed.), *Light Metals 2023*, The Minerals, Metals & Materials Series,
https://doi.org/10.1007/978-3-031-22532-1_142

Photo 1 Casthouse explosion in 2011, three fatalities

being constructed. In their absence, architects have designed casthouse buildings that are enclosed except for man or truck doors. These doors are ineffective at dissipating the over pressure from an explosion because the force of the incident is directed up towards the roof. With no way for the pressure from the explosion to escape the over pressure results in sheet metal roofs and walls being blown out. If the walls and roof do blow out, a shockwave will rebound backwards toward the epicenter of the explosion, resulting in significant damage to machinery and workers. The force of an explosion and shockwave can be directed outside if explosion venting panels are installed on the upper walls and roof (Photo 1).

There are two types of explosion venting panels available to architects and designers; hinged or blow-out panels. "Blow-out panels are ejected from the building when safe pressures are exceeded; hinged panels are open by a top or bottom hinge and then return to a near closed position," explains Andy Moul. Casthouses used to be built with blow-out panels installed in the upper portions of the walls and the eves. This design was adequate but companies discovered that blow-out panels would have to be replaced. Since their shear bolts and fasteners detach when the over pressure of an explosion occurs. In response hinged explosion venting panels were developed and utilized in many other industries. They are reusable and can be calibrated and verified by non-destructive testing. Whereas blow-out panels calibration and verification are by destructive testing because of their sheered off bolts. When the internal pressure rises rapidly, the hinged panels release (open) quickly and once the over pressure has been reduced, return to a near closed position.

Casthouses should be designed with explosion venting hinged panels in the upper portions of the walls and the roof. Their placement and quantity can be determined by an engineer experienced in this field. In the simplest terms,

these safety devices should be above and around where molten metal is cast. It is that localized area (e.g., casting pit) where explosions could occur.

Pedestrian Walkways

A contributing root cause of fatalities every year in casthouses is the lack of defined and physically protected pedestrian walkways inside and outside of casthouses. A defined walkway can simply be markings on the floor where workers are trained to walk. Physically protected pedestrian walkways (PPPW) are defined walkways that are delineated with handrails or similar fixed installations that prevent pedestrians from leaving that protected area. More importantly, PPPW also prevent moveable equipment from entering that area.

At the beginning of a new casthouse project, the architect is provided with a list of machinery (e.g., furnaces, casting station, etc.) and a schematic of how they should be laid out in the workplace. Many architects design the smallest building possible to fit the production equipment due to budget limitations. This practice ignores the need for allocating areas for physically protected pedestrian walkways. Architects will instead include pedestrian walkways marked on the floor that commonly share pathways with moveable equipment. This decision to save money instead of expanding the building to accommodate pedestrian walkways unknowingly contributes to fatal incidents.

Not only have there been fatal incidents involving pedestrians inside but outside the casthouse too. Commonly the area surrounding the casthouse is filled with countless opportunities for pedestrians and moveable equipment to interact with deadly consequences. Vehicles operating outside a casthouse include incoming trucks with raw materials (e.g., scrap, sows, t-bar, etc.) and outgoing trucks hauling finished products away. Casthouses should install physically protected pedestrian walkways on the outside that force pedestrians to follow a path. If pedestrians are not physically constrained to a certain path, they will take shortcuts that increase their exposure to moveable equipment.

Storage of Consumables

Designing the smallest building also results in a lack of storage space for consumables. Consumables cover a wide variety of items that are needed for production in a casthouse. These items typically get delivered in cardboard boxes on wood pallets. It is not just one or two cardboard boxes or wooden pallets. It can be dozens of cardboard boxes on wooden pallets. These items ignite easily if contacted by molten metal.

Current casthouse building designs fail to allocate enough area for the storage of consumables. One casthouse placed their consumables in front of the furnace between two furnace charging stations. If a molten metal explosion occurred, at either furnace, the molten metal would be propelled onto the consumables resulting in a fire. Other casthouses store their material around the casting station. There have been incidents where a small molten metal explosion or a spill occurs igniting cardboard boxes or wood pallets. Architects need to design an area where consumables maybe properly stored and away from molten metal exposure.

Air Movement/Temperature

Casthouse building design ignores air movement in the workplace. This results in elevated temperatures in a workplace already known for high temperatures. An unintended consequence of poor air movement is that workers complain to management about their personnel protection equipment (e.g., safety clothing) being too hot to wear. Over time the casthouse management relaxes or disregards the industry's best practice toward safety regarding personal protective equipment when working around the molten metal.

Computer modeling is an ideal option for architects and designers of casthouses when looking at air movement and air temperature in a casthouse. Brian Bakowski, Principal with SLR International Corporation adds "The model can predict air movement by starting with solid inputs. Information such as ventilation volume, open doors, expected wind speeds, and even thermal currents contribute to how air moves within a building. The model can predict how the air will move and identify any areas where ventilation is an issue."

Casthouses with poor air movement and higher temperatures commonly allow their workers to wear secondary clothing instead of primary (aluminized) clothing when handling molten metal. These workplaces fail to realize that decision to allow secondary clothing results in their workers being exposed to more heat than if they wore secondary clothing. Because aluminized clothing reflects over 90% of the heat workers are exposed to.

The short and longterm health effects of wearing secondary clothing when exposed to high heat are unknown by many in the aluminum industry. The Canadian Centre for Occupational Health and Safety lists long-term heat exposure as a contribution to workers' heart, kidney, and liver damage. Temporary infertility in both women and men who are exposed to high heat work environments is also common.

Control Room Ventilation

Casthouses have a variety of control rooms manned by personnel overseeing a variety of tasks. The casting station control room is the most common. It´s where supervisors observe personnel during the casting process. The control rooms are normally air-conditioned. The concern here is whether or not the air being pumped into the control room is from inside or outside the casthouse building. The air inside the casthouse has micro particles of aluminum that are generated from the furnace, filter box, casting, etc. Peters et al. [9] state "exposure to aluminium dust may possibly increase the risk of cardiovascular disease and dementia of the Alzheimer's type." Any control room, office, breakroom, bathroom, etc. in a casthouse should have ventilation with fresh air extracted from outside the casthouse to minimize the exposure of workers breathing air that may contain aluminum fines.

Another concern is architects are unfamiliar with control room placement, their distance to molten metal, and the potential hazard of an explosion. This hazard became a catastrophe at a casthouse a few years ago. The furnace control room was located on the second floor in front of multiple furnaces. Six workers were in the control room when a rotary furnace suffered an explosion. The molten metal was projected like a shell from a cannon into the second floor control room. All six workers eventually succumbed to the injuries.

Thanks to reduced prices of video cameras their use has expanded inside casthouses. Control rooms no longer have to be close to potential hazards. Moving control rooms away from hazards protects workers.

Overhead Cranes

Architects who design casthouses rarely have any exposure to an existing casthouse prior to designing a new one. That became apparent on a tour of a casthouse when it was observed that the direct chill casting pit was built in the wrong orientation. The casthouse personnel had to attach a rope to the top of the billets and physically pull the load to rotate 90°. That was required because the overhead crane's hook block or trolley was fixed. This mistake in the orientation of the pit could have been easily identified if the architect understood the product (e.g., billet) being produced.

Another common mistake made in casthouse design is installing pendant operated cranes versus remote control or cab operated cranes. The hazard associated with pendant operated cranes is the close proximity the operator is to the load. The long length of billets or rolling ingots being lifted means the operator has a greater chance to be stuck if any of the products fall.

Lunch/Break Rooms

Many of the new casthouses have lunch rooms that are too small to accommodate the quantity of employees working per shift forcing workers either to eat outside or inside the casthouse. The practice of allowing workers to consume food or liquid on the casthouse production floor or any aluminum plant should be prohibited. One concern is that workers will unknowingly ingest aluminum fines that land on their meals prior to consumption.

Lunch and breakrooms should be attached to the casthouse building but not constructed inside. The health hazard is the airborne aluminum fines that are generated during normal casthouse operation. If lunch/break rooms are built outside of the main casthouse building, the air ventilation for these rooms can easily be connected to outside fresh air instead of interior plant air.

Fire Extinguishers

During the design of a casthouse, the architect will require a quantity and specific type(s) of extinguishers and place them throughout the workplace in accordance with any local fire regulations. Local fire regulations typically deal with the floor area of the building and the number of employees. The regulations rarely deal with specific hazards such as molten metal and aluminum fines that are found in casthouses. Therefore, a casthouse when it begins operation has an inadequate quantity of fire extinguishers.

Aluminium companies assume incorrectly that the local fire department would have sufficient quantity of fire extinguishers to handle any type of fire that may occur in the casthouse. This occurred when a fire occurred above an operating furnace in a casthouse. The aluminum company and local fire department did not have enough fire extinguishers on hand to put out the fire. In response, the responding fire department called for assistance from surrounding communities. Fifteen fire departments responded to the call for assistance. Over 65 class D extinguishers were brought to the fire by the responding fire departments and local citizens who opened closed factories to assist.

Class D fire extinguishers are required to put out any fire from a combustible metal. Casthouses need to have a sufficient quantity of Class D fire extinguishers and Class D extinguishing agents on hand for the local fire department to use if a fire breaks out.

Moveable Equipment

Casthouse building designs often overlook the importance of allocating space for indoor storage of furnace tending tools that are attached to moveable equipment. Dross skimmers, stirrers, and more need to be stored indoors. However, due to their long lengths varying 15–25 feet in length, the new casthouse design rarely affords the inside space for such long tools. Instead, casthouses are forced to store them near a truck door or outside. Either location exposes furnace tending tools to weather (e.g., rain, snow, etc.). There have been numerous incidents where unknowingly a wet furnace tending implement was placed into a molten bath resulting in a molten metal explosion. Architects should increase the interior space around the furnaces to allow for safe storage away from external doors for furnace tending tools.

Outside Storage

Outside storage of scrap, sows, etc. is a hidden hazard that many casthouses are oblivious to until an incident occurs. There are usually physical constraints on the areas designated for outside storage. Regardless, architects are unaware of the safety issues in this area and make numerous mistakes. Ground surrounding outside storage area should be level and completely covered in concrete. This will minimize the dirt and water that can be accidentally scooped up and charged into a furnace. A level concrete pad will also prevent stacks of t-bar, sows, etc. from leaning over on the soft ground (Photo 2).

The construction of bins to hold scrap is the issue that aluminum companies try to save money on. Some companies make the dangerous decision to stack large concrete blocks to form walls and make bins. This task is dangerous. Because there have been numerous incidents where these large concrete block walls topple over killing workers. A simple analogy is when a block building is being built. The masons use mortar or concrete to bond the block together because without it the walls could be easily pushed over.

Photo 2 Large concrete blocks stacked to form walls and bin

Location of Offices

There has been an ongoing trend to place offices inside the casthouse building. If possible, all offices should be attached to the exterior of the building. Air quality as previously mentioned in the offices located in a casthouse is a concern. Another hazard architects commonly overlook is the hazard of heat exposure as well as molten metal explosions. This lack of knowledge was apparent in a recent brownfield project where the architect placed offices on the first floor in front of furnaces. The office's plexiglass window that was originally installed started to warp because of the high temperatures when the furnace door was open.

Conduit Trenches

Conduit trenches are hidden hazards that many fail to acknowledge in a casthouse. In simplest terms, they are pathways under the floor where utilities run (e.g., electric, water natural gas, etc.,). Steel plates cover conduit trenches for easy access. A safety issue arises if a molten leak occurs nearby. If there are any gaps in the steel cover, the trench below will be breached by the molten metal. In addition to damaging any utilities in the trench, an explosion could occur if moisture is present. Some architects instruct that the conduit trenches be filled with sand. Sand retains moisture and explosions could still occur. In addition, maintenance workers need to remove sand to access the localized area where the work is. This time-consuming process is slowly going away because more and more countries have restricted the use of sand due to the health concerns of inhalation.

Molten metal explosions generated from the conduit trench trenches can be minimized by the application of an approved safety pit coating. The Aluminum Association's "Guidelines For Handling Molten Aluminum, Fourth Edition, May 2016" list Wise Chem E-212-F and E-115 as "identified as having the ability to prevent explosions from bleedouts .... during casting". The cover lids and the interior surfaces of the trenching should be coated with a Wise Chem coating to prevent an explosion from occurring.

Molten Metal Spills

Molten metal has the same viscosity as water and will travel to the lowest point possible if not stopped. The lowest points in a casthouse are normally the casting pit, adjacent maintenance pit, and pit under furnace and trenches. All of these locations need to be coated with an approved safety pit coating (e.g., Wise Chem E-212-F or E-115) to prevent an explosion from occurring.

Lighting

A frequent contributing root cause to incidents that occur during the night in casthouses is the lack of lighting. Workers have been injured or killed during the night time in or outside of casthouses because the moveable equipment operators could not see the pedestrians. No doubt there is a myriad of factors in every incident. However, lighting is an issue that can easily be corrected by architects during the design phase. Some architects have focused on the issue of poor lighting and found that specifying a white color for the inside sheeting of the building will better help reflect the lights.

Another lighting topic all aluminum plants deal with on an irregular basis is the loss of power. The loss of power can occur due to weather events or, in some regions, because of the rationing of electricity. Emergency backup lights should activate during a loss of electricity. Many new plants unfortunately have installed the minimum of emergency lights. They are installed just around exits and not around production equipment. One casthouse experienced a loss of power at dawn one morning at the end of a cast. Thankfully the sunrise provided natural light to safely abort the cast. Shockingly the casthouse personnel observed that the emergency lighting malfunctioned and did not work. One can imagine the dangers in a casthouse with no power and no visibility.

Conclusions

The vast majority of new casthouses built in the past decade were unintentionally designed and constructed with hidden hazards. These hidden hazards if not found and corrected may injure and kill workers.

Many of the hidden hazards can be eliminated from future projects if architects and designers focus first on safety and then production when planning a new casthouse. The majority of the hidden hazards discussed in this paper could be eliminated if the casthouse building was larger. A rough estimate on the additional square feet needed to eliminate a majority of the discussed hidden hazards would be less than 5%. For many projects that additional area could be realized by simply moving offices, break, or lunch rooms to the exterior of the casthouse building. Those freed-up areas would provide ample space to install physically protected pedestrian walkways inside and outside of the casthouse.

By educating the designers and architects of new casthouses, aluminium companies can not only make their workplaces safer but our industry too.

References

1. Bakowski, B., SLR International Corporation, Principal, personal communication September 1, 2022. Email: bbakowski@slrconsulting.com

2. Canadian Centre for Occupational Health & Safety (2022) OSH Answers Fact Sheets; Hot Environments - Health Effects and First Aid https://www.ccohs.ca/oshanswers/phys_agents/heat_health.html. Accessed: 2 September 2022

3. S. G Epstein, (1991) 'Causes and Prevention of Molten Aluminium Water Explosion," Light Metals 1991

4. "Guidelines for Handling Molten Aluminium, Fourth Edition," Aluminum Association, 2016

5. D.D. Leon, R.T. Richter, and T. L. Levendusky, (2001) "Investigation of Coatings Which Prevent Molten Aluminum/Water Explosions," Light Metals 2001

6. G. Long (1957) 'Explosions of Molten Aluminium and Water – Cause and Prevention", 107–112 Metal Progress

7. A.W. Lowery (2016) "Has Recent Advances in Direct Chill Casting Made Us Less Safe?" Light Metals 2016

8. Moul A (2021) Under pressure: the benefits of explosion venting panels. HazardEx: https://www.hazardexonthenet.net/article/187209/Under-pressure–the-benefits-of-explosion-venting-panels.aspx. Accessed: 1 September 2022

9. Peters et al (2013) "Long-term effects of aluminium dust inhalation" Occupational and Environmental Medicine Journal 2013

10. R.P. Taleyarkhan, V. Georgevich, and L. Nelson, (1997) "Fundamental Experimentation and Theoretical Modelling for Prevention of Molten Metal Explosion in Casting Pits, Light Metal Age 1997

11. Woloshyn1 et al (2017) "Overpressure due to a molten aluminum and water explosion in a casthouse." TMS 2017 – Light Metals – Casthouse Technology

Operations Assisting and Predictive Maintenance Tools in Cast Houses: Examples from AMAG Casting

Alexander Poscher, Martin Mönius, Eduard Faschang, and Bernd Prillhofer

Abstract

The interference-free functioning of cast house equipment such as moulds, starter blocks, metal level sensors, or simply the vehicle fleet is critical to a smooth and safe operation. Malfunctions can cause unforeseeable cast stops leading from downtime to safety issues and equipment damage due to bleed-outs, hang-ups, or metal freezing in the nozzles and they certainly reduce productivity. However, the wide variety of equipment used challenges production and maintenance employees responsible to keep an overview of the usage and condition of each single device at any point in time. Therefore, digital assistant systems seem to be a suitable way to support operations by tracking such devices on the shopfloor and predict at best equipment conditions depending on the past and future production program. This paper deals with a selection of digital support tools developed and used at AMAG casting to further improve condition-based maintenance and increased equipment availability.

Keywords

Predictive maintenance • Cast house • Mould maintenance • Operating hours • Equipment tracking • AMAG casting • Continuous casting • Starter blocks • RFID • DMC • MLS • Production vehicles

A. Poscher (✉) · M. Mönius · E. Faschang · B. Prillhofer
AMAG Casting GmbH, Ranshofen, Austria
e-mail: alexander.poscher@amag.at

M. Mönius
e-mail: martin.moenius@amag.at

E. Faschang
e-mail: eduard.faschang@amag.at

B. Prillhofer
e-mail: bernd.franz.prillhofer@amag.at

Introduction

Besides run to failure, time-based (preventive), proactive (includes potential re-design of machine parts to prevent such failure), and predictive maintenance are the maintenance philosophies used in the industry [1]. Modern data base systems, interlinking machines (IOT), enhanced sensor technologies, and analysis tools are used to obtain the actual operating conditions of critical plant systems and schedule maintenance activities before failure occurs [2, 3]. Misra et al. [4] state that vibration measurement, acoustic emission, oil analysis, infrared thermography, and motor current analysis are probably the most widely used examples for condition-monitoring systems as a basis for predictive maintenance.

According to the authors [1], the major advantage of predictive maintenance is given by the fact that machine downtimes can be scheduled in advance also taking lead-times of spare parts into account thus reducing the latter as well as productivity losses. Other advantages stated in [4, 5] point out that any failure has safety consequences, thus predefined actions can significantly reduce such risks.

Predictive maintenance is also seen to positively impact costs of maintenance because machine parts are replaced just before failure—thus on time. In contrast, preventive maintenance, due to predetermined time intervals, is limited to taking accelerated or retarded degradation of machine parts into account, thus the likelihood of machine part replacement prior to failure is significantly [5] higher compared to condition-based (predictive) maintenance. Moreover, costs of unplanned maintenance were estimated to be three times higher than in the case of a planned activity [6].

In the case of AMAG casting, condition-monitoring is currently widely used in, e.g., vibration monitoring of exhaust gas fans and thermography of furnaces. However, casting equipment has traditionally been maintained on predefined schedules or in case a defect has been detected (reactive). Nonetheless, complexity in the cast house grows as the quantity of moulds, starter blocks of various dimensions, metal level sensors, and the quantity of forklifts and wheel loaders increases. This fact makes it more and more difficult to keep the overview on the equipment condition, whereas first and foremost work safety can suffer in case of insufficiently maintained equipment [7, 8]. Every time a casting process is stopped due to an unplanned event, workers need to interact with the molten aluminium (e.g., draining the launders), to prevent further damage to the casting equipment or launder system. Not being prepared at this very moment, the situation itself increases the likelihood of a safety incident.

Vice versa a well-maintained equipment has a direct positive effect on reduced rates of cast abortions, correspondingly limited time losses caused by additional setup work, increased pit utilization and output, and thus higher productivity [8].

Since growing complexity does not necessarily go along with an increased workforce, available personal resources of production managers, maintenance employees, mould repair shop workers, or vehicle fleet managers just to name a few need to be allocated and prioritized more precisely to meet the strict requirements regarding work safety and productivity. This implies that operations assisting and condition-monitoring/prediction tools are required to be developed and implemented supporting the prioritization of maintenance work of critical cast house equipment. Among others such digital tools need to be able to process and analyse data from condition-monitoring sensors and provide aggregated and clear information to the previously mentioned departments and employees. There is an enhanced requirement that such tools be designed to support the exchange of data and information between departments in case equipment fails despite maintenance being carried out on time. Real-life examples have shown that the lack of information exchange of what the actual equipment problem was in production or the insufficient feedback what was found in the support facility (e.g., maintenance department) has caused the same defect to re-occur thus resulting in avoidable inefficiencies. Within this paper, the focus is set on AMAG casting's approach to implement condition-based maintenance tools for moulds, metal level sensors, and starter blocks and describes the implementation of a system automatically keeping track of a vehicle's operating hours.

One of the greatest challenges regarding moulds and metal level sensors is that condition data cannot be measured directly as it is the case for a vibration sensor mounted to an exhaust gas fan. The condition of moulds for example is given by the fact that water bores and screens have to be free of lime deposits or that contacting surfaces of EMC moulds do not exhibit corrosion layers influencing electrical resistance [10]. For metal level sensors, the situation is even more difficult, since gradual changes in accuracy are hard to detect

on the shopfloor, thus the development of a predictive maintenance strategy first needs to aim at the tracking of such equipment in production and second to determine mathematical relationships which describe the degradation of functionality by considering utilization and other influencing factors for each piece of casting equipment. Another challenge is that many available sensors for equipment tracking simply do not meet the requirements to withstand the harsh operating and environmental conditions in a cast house. Other problems to be solved are for instance low and/or varying lighting conditions in the cast house or casting pits, dust and the degree of contamination of sensors, corrosion, water vapor, visibility of machine parts to be detected by camera systems, and many more. Therefore, as one can imagine, there is no single answer or solution suitable to track critical components on a casting pit or the entire cast house. Instead, individual solutions needed to be developed or standard methods were adapted in a way to exhibit a high level of availability.

Currently, three different technologies are used at AMAG casting GmbH with the primary goal to track pieces of equipment in production and to gain distinct data on their level of utilization.

- Camera-based detection of moulds and metal level sensors (MLS).
- Ultra-high-frequency radio frequency identification (UHF-RFID) for starter blocks.
- Voltage-measuring sensors on production vehicles.

Identification of Moulds and Metal Level Sensors on the Casting Pit

As mentioned, moulds and MLS are two of the most critical components, which have a direct influence on the success or failure of the casting process.

The most sensitive moulds are those used for electromagnetic casting (EMC) [9], which is characterised by the fact that an electromagnetic field keeps the liquid aluminium in shape until a solidified shell is formed whereas the slab never has direct mould contact. Especially water screens, inverter cooling, and electric connectors need to be properly maintained to ensure stable casting conditions. For Low Head Composite Casting (LHC™) moulds, the most critical components are split jet valves, graphite liners, and water bores, whereas in the case of direct chill casting (DC) moulds defects on the water screens are most often found [11].

The MLS' task is to measure the level of liquid metal in the mould with the highest accuracy. The signal is used to control the liquid metal flow through the nozzles into the mould. Malfunction of either component (mould or the MLS) can directly result in an immediate and automated stop of the casting process. The fact that especially moulds and MLS are often exchanged between different casting pits needs to be addressed when developing a tracking tool as a prerequisite for a condition-prediction system.

For these specific applications, a solution was found, using an already installed camera system on the casting pits which has primarily been used to support the hands-free cast start [7]. Therefore, cameras mounted in the utility beam are aligned downwards, to allow for a direct view of the moulds and MLS.

To distinguish between the moulds and MLS, each component was given an individual designation, related to its characteristics (dimension, casting technology). All moulds and MLS were then equipped with data matrix code labels (DMC) containing an individual identity (ID). Each ID is linked to the component's designation in a data base where the relevant information for each cast is stored (date of cast, pit number, slab number, cast length, and casting technology).

During the start of the casting process, the installed cameras take several pictures of the casting equipment. Those pictures are then processed by a software, capable of identifying and reading the DMCs' IDs. The IDs are then compared to the entries in the data base and the dedicated components are selected. After the casting process, the entries for the identified equipment are updated in the data base.

Figure 1 shows the camera view of the moulds and MLS. Both are labelled with DMC codes.

Based on historic and current data stored for each single piece of casting equipment in conjunction with real-life degradation curves and the future production program, it is possible to predict equipment conditions and schedule maintenance activities. A more detailed description of the prediction tools used can be found later in this paper.

Starter Block Tracking

Starter blocks are used in vertical continuous casting to allow the first liquid metal poured into the moulds to solidify, thus forming the butt of the slab. Starter blocks are usually made of an aluminium alloy or steel and are connected via a starting base to the hydraulic casting cylinder. Similar to moulds, these pieces of equipment can be interchangeably used on various casting pits at AMAG casting GmbH, thus the development of a tracking tool to determine the utilization of each starter block represents the first step in the implementation of a condition-monitoring system.

During their operating lifetime, starter blocks undergo a variety of damages and deformation such as bending, shrinkage, and the formation of burrs. Figure 2 symbolises the directions of deformations.

Especially during the cast start such deformations can lead to deviating points of impact of cooling water leading to asymmetrical butt curl and its consequences. An increased likelihood of hang-ups or flashing is also found if starter blocks do not meet tight tolerance levels. Apart from this, notches close to the edges can lead to water spraying into the starter block's functional surface which can cause severe water vapor blow-ups in contact with liquid metal.

However, in contrast to moulds or MLS, the view on starter blocks is very often obstructed by other casting equipment, so that they cannot be identified using a downward aligned camera system. Cameras inside the casting pit, looking at the small end faces of the starter block, are not a

Fig. 1 Camera view onto the casting pit with DMC labels on MLS and moulds

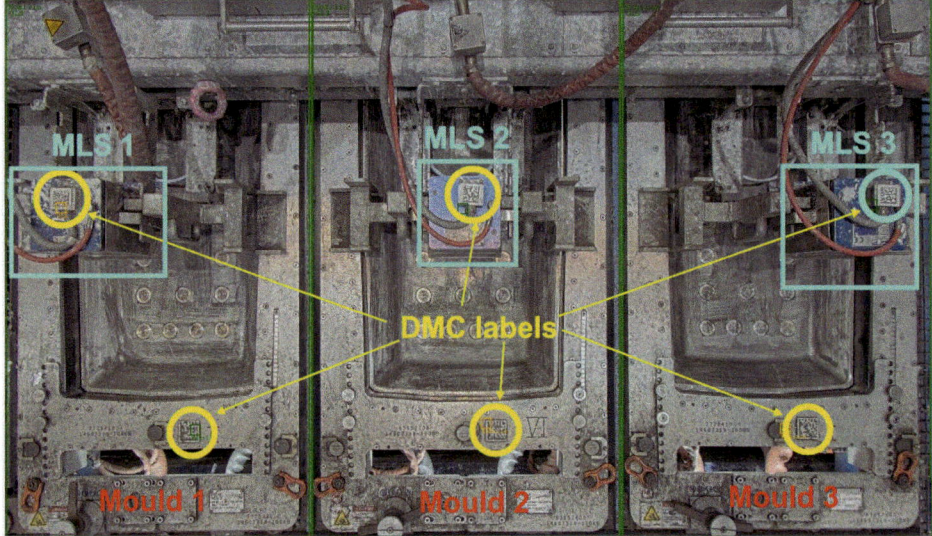

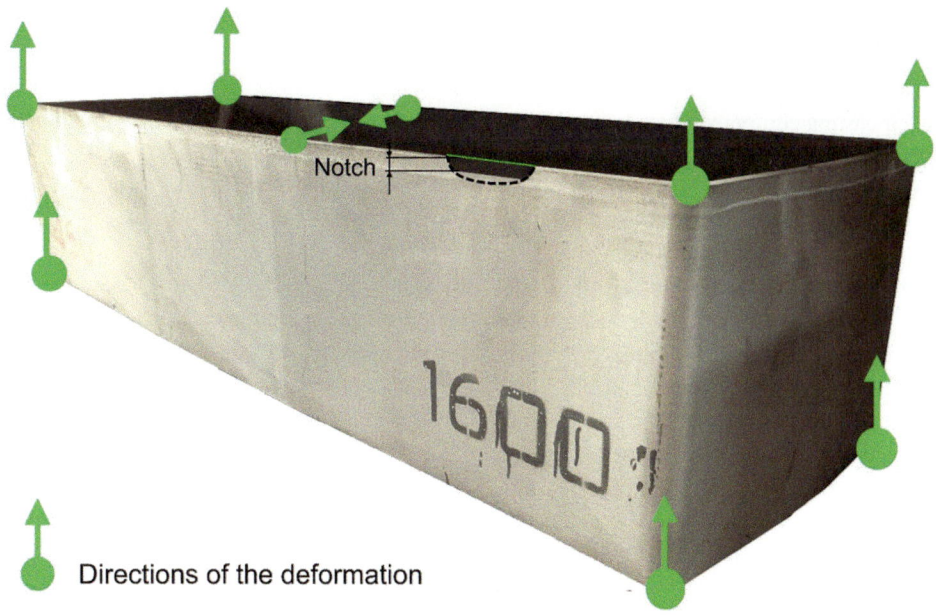

Notch

Directions of the deformation

Fig. 2 Main directions of deformation of a starter block during the use phase

viable option either, since water vapour and spray mist in conjunction with lighting conditions prevent any reliable optical identification method from proper functionality. Also, liquid metal from bleed-outs could damage camera systems in the casting pit or at least DMC codes mounted on starter blocks.

Therefore, an alternative system had to be found, capable of operating under the harsh conditions of a casting pit. UHF-RFID (Ultra-high-frequency RFID) tags were found to withstand the thermal impact of up to 100 °C (max. temperature of the starter block during the casting process), without losing their functionality or getting damaged, also

being water and steam resistant. After the selection of a suitable tag, an antenna and reader system had to be established at each position of the casting pit to read the information from the tag. By using this setup, it is possible to read the tag's information during the start of the casting process as the starter block passes the antenna.

The tag on each starter block contains an ID which is directly assigned to the starter block's designation stored in a data base. The linkage between the information on the tag and the data used for condition forecasting and scheduling of maintenance activities is designed in the same way as for moulds or MLS. Figure 3 depicts the described process.

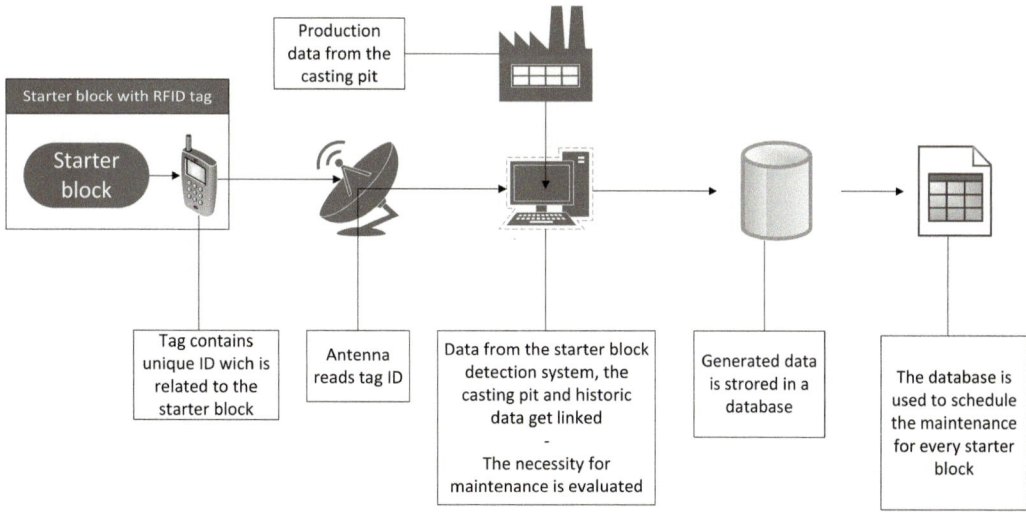

Fig. 3 Starter block tracking and maintenance scheduling

Automated Operating Hour Detection for Vehicles

Similar to the pieces of casting equipment already mentioned in this paper, the vehicle fleet represents another part a cast house needs to thoroughly take care of to secure its operations. Like passenger cars, production vehicles, such as forklifts, wheel loaders, or skimming manipulators, have defined intervals between service and maintenance works given by the manufacturer. The only difference is that these intervals are not based on the vehicle's mileage, but on actual operating hours. If these intervals are neglected, major damages to the vehicles can be the consequence. Regarding the quantity of transport equipment in a modern cast house (a total of over 100 in the AMAG casting GmbH), keeping track of the operating hours of every single vehicle is a difficult and time-consuming task, if done manually. Therefore, gathering these data without any human interaction seemed to be suitable in order not to bind valuable human resources to the search for vehicles, reading and documenting operating hours each week of the year.

So, a system was implemented which uses a sensor, that is directly linked to a component of the engine that is energized as soon as the vehicle's motor is running. In case a voltage is detected by the sensor, a signal is sent to a device that starts to measure the timespan in which the signal is active.

After the motor stops running, an automated event is triggered, which causes the time-measuring device to send the data via a wireless network connection (Wi-Fi) to a central server that updates the operating hour data in the data base.

The Right Maintenance Strategy for Every Type of Equipment

In the previous sections of this paper, mainly the tracking of the casting equipment and storage of relevant data in data bases were described. However, the derivation of suitable maintenance intervals based on actual utilization is the overall goal to achieve.

For casting equipment such as moulds, there may be service guidelines delivered by the manufacturer. However, it was found that simply focusing on the quantity of casts to determine the demand for maintenance for a specific mould is an insufficient approach, because more parameters seem to influence mould quality. Tests have shown that already after a small share of the suggested casts issues appeared, which were directly linked to a lack of maintenance. For this reason, additional parameters need to be introduced to be able to mathematically describe the condition of the mould.

The set of main parameters to describe the mould condition is listed as follows:

– Total cast length since the last maintenance.
– Time since the last service.
– Casting technology used.
– Frequency of usage.

The mentioned cast length describes the accumulated length of the produced slabs with a specific piece of equipment since its last maintenance, which is an indication of wear caused by thermal loads during the casting processes. This gives a measurement of potential corrosion or lime deposition due to residual humidity. Another very important factor in the evaluation of the necessity for maintenance is the casting technology. As mentioned earlier, the EMC technology is the most sensitive casting technology when it comes to issues due to a lack of service. In comparison, moulds of other casting technologies used at AMAG casting GmbH (LHC™ and DC casting) tolerate a higher cast length and time between service works compared to EMC moulds.

The frequency of usage describes how often a mould of a certain combination of format and technology is used. This information is utilized to prioritize maintenance tasks. Moulds with a high frequency of usage are getting prioritised in the mould repair shop, so that they have the highest possible availability for the production process.

The Mould Maintenance Factor (MMF) and the Starter Block Maintenance Factor (SBMF)

These four parameters (data for each piece of equipment is derived from tracking tools and the described data bases) are combined in an algorithm that calculates the so-called mould maintenance factor (MMF). One may compare the MMF with a wear margin which if depleted indicates that the respective part must be repaired or replaced. In combination with the forecast of the planned production program, service actions can be performed prior to the mould reaching its critical value of the MMF. The information provided in the AMAG casting's shopfloor management system is displayed in the order of highest to lowest priority (MMF). Automated reports are also distributed every evening to the different stakeholders giving a quick overview of the mould maintenance activity of the day and a forecast for upcoming service works.

Similar to the MMF, a factor to evaluate the necessity for maintenance for starter blocks is currently under development. This starter block maintenance factor (SBMF) takes an additional input parameter into account—the deformation (shrinkage and bending) of the equipment. Just as for moulds, the cast length since the last service is an indication of the wear caused by thermal loads. The casting technology also has a direct influence on the thermal load a starter block must withstand, because the distribution of cooling water varies with the technology used. Again, EMC starter blocks have the tendency to need a service more often compared to starter blocks for LHC™ or DC casting. However, in contrast to moulds, maintenance work of starter blocks is rather limited to the deburring of the edges which have direct contact with the liquid metal and to build-up welding of those edges. As one can imagine, deviations in dimensions cannot be corrected which is why internal specifications were developed defining critical values for necessary replacement. Therefore, a prediction model on the condition of a starter block needs to provide feedback on the likeliest point in time one of the quality criteria reaches a critical level. Once the development of the model is finished, it will then be used to monitor the stock of more than 1000 starter blocks at AMAG casting GmbH and enable an automated replacement strategy in case dimensions undercut internal threshold values.

For MLS it was found that three main factors have the biggest impact on the need for service. Those are the number of casts, the cast length, and the time span since the last maintenance activities. Based on this data, an improved model, similar to the one for the moulds, is under development.

Achieved Improvements and Outlook

For moulds, starter blocks and MLS tools were introduced to track each piece of equipment on the shopfloor when used on a casting pit. By interlinking these data with (semi-) quantitative data of measurements from the shopfloor (e.g., dimensional data of starter blocks over time or cast aborts which were caused by mould or MLS defects), it is possible to develop mathematical correlations which describe the degradation of the equipment, based on the parameters mentioned earlier in this paper. Since the implementation of the tracking and prediction tools for moulds, the rate of cast abortions was significantly reduced as depicted in Fig. 4. It is expected that when the MLS and starter block condition-prediction tool reaches the level of maturity, a further decrease in the cast abortion rate can be achieved.

In the authors' perspective, the positive impact of such tools is not only given by the fact that automatized scheduling of service activities is performed. Additionally, employees within the entire organization are encouraged to work together and to provide relevant information on

potential root causes or at least of phenomenological explanation of failures to maintenance departments.

Generally, it was found that the higher availability of information concerning the equipment's condition led to a positive synergetic effect between the personnel in the maintenance departments and the cast house production team. Since every maintenance activity is now documented digitally, the service history of every component is easily accessible throughout the organisation. These tools are also helpful if troubleshooting is needed and narrow down the possibilities in case a cast abortion occurs. The fully digitized maintenance protocol for MLS is an effective tool since it allows production managers to provide information of failures and vice versa to receive information on the detailed analysis. As a result, other root causes for cast abortions can be further identified and analysed.

In the case of production vehicles, the first step to an enhanced maintenance strategy has been taken. Each vehicle is currently getting equipped with the sensor and transmitting equipment mentioned above. We expect not only a significant reduction in human resources to keep track of vehicle operating hours but together with a set of our other measures, this system will result in a more precise view of the actual condition of the vehicle fleet and reduced costs for repairs. The newly installed system shall also be used to prove whether the manufacturers' suggested maintenance intervals are suitable for the harsh conditions in a cast house. With this information, an adopted maintenance plan can be created that shall reduce the number of repairs due to unsuitable standard service intervals.

Summary

As shown, predictive maintenance strategies can be applied to casting equipment and offer a variety of benefits in everyday production and work safety. A mould detection and tracking system in conjunction with a utilization-based prediction model for the upcoming necessary maintenance activity has been successfully implemented. The system uses cameras mounted on the utility beams of each casting pit. Moulds are identified using a special software tool and DMC codes mounted on each of them. Further actions need to be taken regarding the development of the starter block maintenance factor (SBMF). Starter blocks are equipped with Ultra-High-Frequency RFID tags, and casting pits are equipped with antennas to detect those. Currently, the SBMF is optimized to further improve the predictability of the condition of each starter block. Therefore, a model is generated, which describes the degradation of starter blocks throughout their lifetime as a function of the previously described influencing parameters. This model will then be used to monitor the stock of more than 1000 starter blocks at

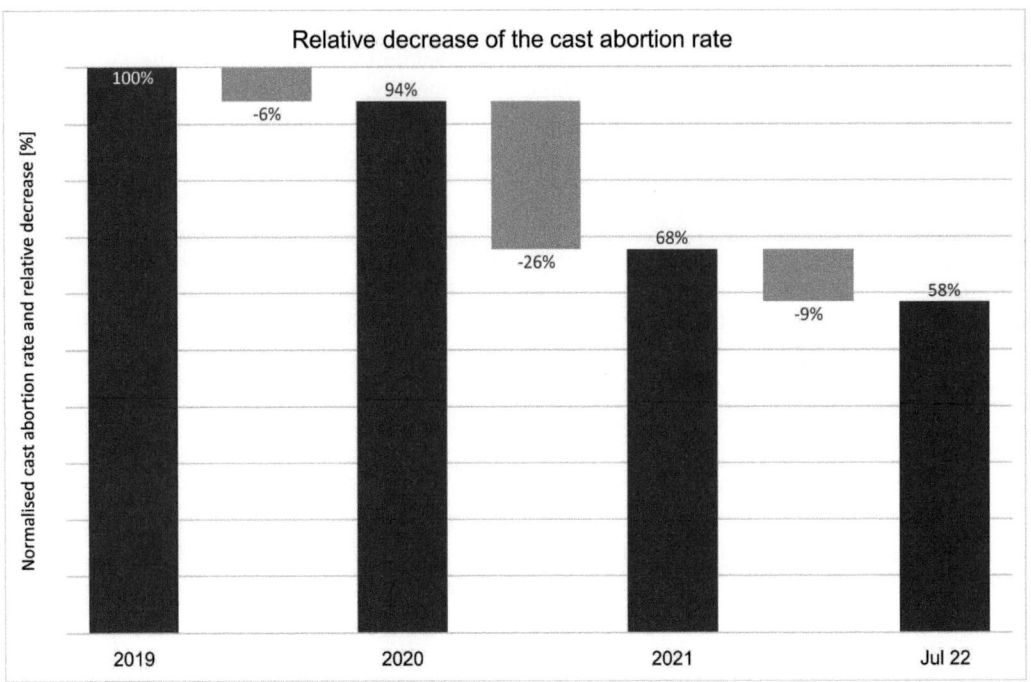

Fig. 4 Development of the cast abortion rate (2019–July 2022)

AMAG casting GmbH and enable an automated replacement strategy in case dimensions fall below internal threshold values.

For metal level sensors (MLS), it was found that the same identification system as for moulds can be used. Similar to the starter blocks, an improved condition-based model for MLS is currently in the phase of optimization.

For AMAG casting GmbH's vehicle fleet, the first steps towards an optimised maintenance strategy have been taken by integrating sensor and transmitting units for a fully automated vehicle operating hours tracking. The information will then be used for the development of an enhanced maintenance strategy considering the harsh conditions in a cast house.

References

1. Scheffer, C, Girdhar P (2004) Practical Machinery Vibration Analysis and Predictive Maintenance. Newnes, Oxford
2. Schlatt, A (2014) Predictive Maintenance Maschinenfehler und -ausfälle im Vorfeld erkennen. In: Genios Wirtschaftswissen Nr. 10 2014. GBI-Genios Deutsche Wirtschaftsdatenbank GmbH
3. Mobley, Keith R (2000) An Introduction to Predictive Maintenance. Second Edition. Elsevier, Woburn
4. Misra, K.B (2008) Maintenance Engineering and Maintainability: An Introduction. In: Misra K.B (ed) Handbook of Performability Engineering 2008. Springer, London p 755–772
5. CCPS (2005) Guidelines for Safe Handling of Powders and Bulk Solids. Center for Chemical Process Safety of the American Institute of Chemical Engineers, New York
6. Möller, N. et al. (2018) Handbook of Safety Principles. Wiley Essentials in Operations Research and Management Science. John Wiley & Sons
7. Prillhofer, B., Dobler, R., Mrnik, T. (2020). Hands-Free Casting at AMAG Casting GmbH—It Is Possible!. In: Tomsett, A. (eds) Light Metals 2020. The Minerals, Metals & Materials Series. Springer, Cham. https://doi.org/10.1007/978-3-030-36408-3_111
8. Steen, I.K., Håkonsen, A. (2011). Hycast™ Gas Cushion (GC) Billet Casting System. In: Lindsay, S.J. (eds) Light Metals 2011. Springer, Cham. https://doi.org/10.1007/978-3-319-48160-9_136
9. Xianshu Z., Zhaoxia L. & Yicheng W. (2001) Experimental study of concavity of large EMC slab, Science and Technology of Advanced Materials, 2:1, 113–116. https://doi.org/10.1016/S1468-6996(01)00036-5
10. Kharytonau D.,Zimowska M., Gurgul J., Mordarski G., Powalisz R., Rutowski A., Putynkowski G., Zięba A., Mokrzycki Ł., Socha R. (2022) Corrosion failure analysis of a cooling system of an injection mold, In Engineering Failure Analysis, Volume 135. https://doi.org/10.1016/j.engfailanal.2022.106118
11. Krone K., (2000) Aluminium Recycling. Vereinigung Deutscher Schmelzhütten e.V. (Hrsg.) Düsseldorf, p 382–390

Counter-Gravity Casting of Al Alloys: Microstructure and Properties

K. Georgarakis, J. Vian, D. Sgardelis, B. Souchon, Y. Chao, K. Konakoglou, M. Stiehler, and M. Jolly

Abstract

Counter-gravity casting can improve the structural integrity of castings by eliminating defects resulting from the turbulent flow of the molten metal during filling. The Constrained Rapid Induction Melting Single Shot Up-casting (CRIMSON) is an alternative counter-gravity casting approach designed for improving energy efficiency in castings in line with the concept of sustainable foundry. In this work, the microstructure and properties of the hypoeutectic 354 Al alloy produced by CRIMSON counter-gravity casting is investigated and compared with specimens produced by conventional gravity sand casting in the as-cast condition. The results indicate significant reduction in porosity, oxide inclusions and bifilms for the CRIMSON counter-gravity cast samples compared to that in conventional gravity cast samples in line with the controlled liquid flow of the counter-gravity filling. This can lead to significant improvements in terms of the casting yield, the expected fatigue resistance of the alloys as well as efficiency of the casting processes in terms of energy, materials use and greenhouse emissions. Furthermore, similar dendritic microstructure consisting of α-Al dendrites, Al–Si eutectic and intermetallic compounds in small volume fraction was observed for both CRIMSON and gravity cast samples. The mechanical behaviour was also evaluated using tensile tests and hardness tests.

Keywords

CRIMSON casting • Al alloys • Mechanical properties

Introduction

Metal casting is by nature an energy-intensive process as it typically involves the heating up and melting of metals and alloys at elevated temperatures. Most often the energy consumption, however, is much higher than what is thermodynamically necessary, which can be largely attributed to the low efficiency of conventional casting processes and foundry procedures. Furthermore, castings often suffer from various types of defects as a result of phenomena including shrinkage during solidification, entrainement, misruns, hot tears, and residual stresses [1, 2]. It is estimated that about 80% of the casting implications could be attributed to the entrained defects, mainly caused during filling. Non-controlled filling velocity, such as usually employed in gravity casting, can cause turbulence in the liquid flow resulting in defects such as oxide bifilms and air entrainment, that significantly deteriorate the structural integrity and engineering performance of castings [3–5]. To limit these defects, a maximum filling velocity criterion has been suggested by Campbell [6]; for Al alloys this equals to 0.5 ms^{-1}. Counter-gravity casting processes can actually provide such control in metal flow during filling improving the quality and structural integrity of castings [6, 7]. To this effect, the constrained rapid induction melting single shot up-casting (CRIMSON) has been designed to offer energy savings, increased casting yield, and reduce the propensity of casting defects such as double oxide films (DOF) and porosity [8]. CRIMSON concept suggests the use of high-quality raw materials and uses an induction furnace to rapidly melt the appropriate quantity of metal for filling a single mould thus making the traditional holding step redundant. Then the liquid is transferred to a computer-controlled casting unit that delivers the melt into the mould using counter-gravity filling. The mould including the running and feeding systems, as well as the liquid metal flows can be optimized by using casting computational fluid dynamic calculations [9].

K. Georgarakis (✉) · J. Vian · D. Sgardelis · B. Souchon · Y. Chao · K. Konakoglou · M. Stiehler · M. Jolly
School of Aerospace, Transport and Manufacturing, Cranfield University, Cranfield, UK
e-mail: k.georgarakis@cranfield.ac.uk

© The Minerals, Metals & Materials Society 2023
S. Broek (ed.), *Light Metals 2023*, The Minerals, Metals & Materials Series,
https://doi.org/10.1007/978-3-031-22532-1_144

Aluminum alloys are widely used as structural materials in numerous applications including in automotive and aerospace industries. Using appropriate alloy design such as addition of rare earth elements and appropriate manufacturing processes, aluminium alloys can reach mechanical (yield) strength as high as 1 GPa or higher [10–12]. More common industrial aluminium cast alloys however, can reach up to 320 MPa yield strength with appropriate processing. In particular, Al–Si–cast alloys show excellent castability, and are commonly used for the production of complex-shape parts with high specific strength due to their response to aging treatment [13, 14].

In this work, the hypoeutectic 354 Al alloy produced by CRIMSON counter-gravity casting and conventional gravity sand casting is investigated, in the as-cast condition, aiming to identify differences in the structural integrity and properties resulting from the difference in the casting process. In both processes, similar types of silica sand moulds were used in order to focus on the effect of the counter-gravity filling of CRIMSON casting.

Materials and Methods

The A354 Al alloy was used in this work. The raw material was in the form of direct chill (DC) billets, in a clean degassed condition from the metal supplier. Table 1 presents the chemical composition, typical for 354 Al alloy. Sand moulds were used to cast the alloy using (a) CRIMSON counter-gravity process and (b) conventional gravity casting for comparison. An assembly of six tensile bar specimens was produced with each casting as shown in Fig. 1. The

tensile bar specimens had a gauge length of 75 mm and diameter of 12.5 mm, respectively, in line with the ASTM E8/E8M standards. A SIEMENS D5005 diffractometer using $Cu - K_\alpha$ radiation was employed to study the crystal structure of the cast specimens. Measurements were taken at various areas of the castings as shown in Fig. 2c. The samples (cross sections) were observed with a Leica (DM 2700 M RL/TL) optical microscope (OM) equipped with a CCD camera. A TESCAN LYRA3 scanning electron microscope (SEM) equipped with energy-dispersive spectroscopy (EDS) was employed for the investigation of the microstructure. The metallographic preparation included chemical etching with Keller's agent. Tensile tests were performed using an INSTRON 5500R instrument and a strain rate of $2.22 \times 10^{-4}\ s^{-1}$. At least 3 measurements were conducted per casting method sample. Hardness measurements were performed on the as-cast vertically cut disks from the tensile bars, using a 10 kg applied load for a holding time of 12 s. Hardness values presented in this work are the average value of at least 10 measurements.

Results and Discussion

Figure 2a, b show XRD patterns from five different areas of CRIMSON and gravity cast specimens, respectively. Figure 2c indicates the position of these five areas on a tensile bar specimen; positions 1 and 2 are on the top part of the sample, 3 in the middle (gauge section) and position 4 in the lower part—close to the runner. At positions 1–4, XRD spectra were taken at the cross section of the specimens. Position 5 is found in the middle part (gauge section) and

Table 1 Chemical composition of an A354 Al alloy

	Si (wt.%)	Mg (wt.%)	Cu (wt.%)	Fe (wt.%)	Mn (wt.%)	Ti (wt.%)	Al
A354	9	0.5	1.8	0.2	0.1	0.2	Balance

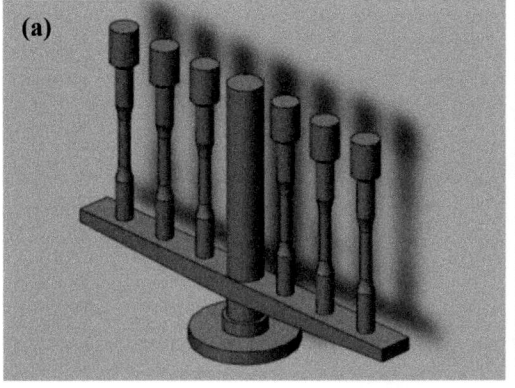
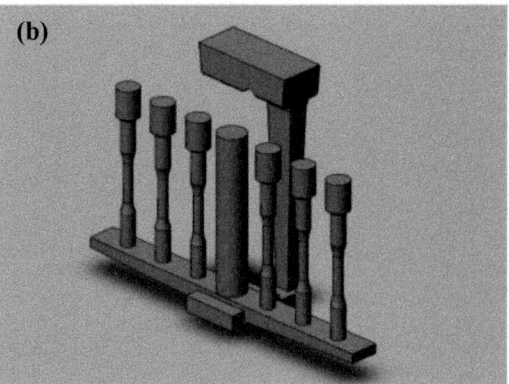

Fig. 1 Schematic illustrations of the mould designs; **a** CRIMSON casting and, **b** gravity casting [8]

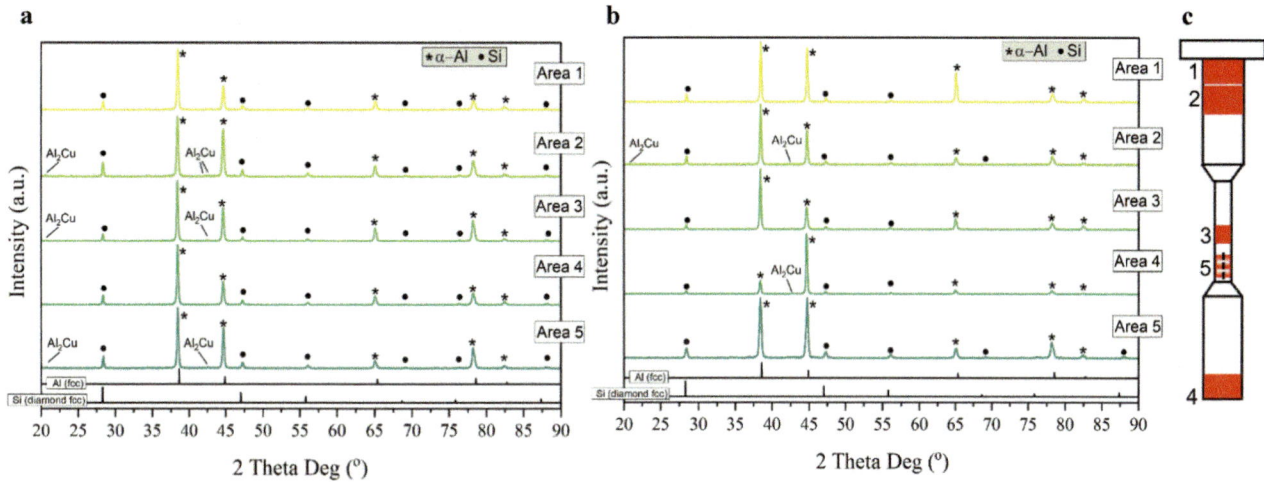

Fig. 2 XRD patterns of **a** the CRIMSON cast A354 alloy and, **b** the gravity-cast A354 alloy. **c** schematic illustration showing the five different areas/positions where XRD measurements were conducted. Areas 1–4 were cut across the solidification direction and area 5 vertical to the solidification direction

represents a longitudinal section, vertical to the solidification direction. As expected, the main reflections in the XRD spectra correspond to Al fcc and Si phases, whereas smaller reflections indicate the presence of θ-Al$_2$Cu phases in both CRIMSON and gravity-cast samples. These findings are typical for hypoeutectic aluminium–silicon–copper alloys. Comparing the XRD spectra for the two samples, no significant differences can be identified. The presence of other anticipated precipitated intermetallic phases, such as Q-Al$_5$Cu$_2$Mg$_8$Si$_6$, π-Al$_8$FeMg$_3$Si$_6$ and β-Al$_5$FeSi cannot be reliably identified with the XRD in this case due to their small volume fractions as well the overlapping of their reflections with those for Al and Si. However, evidence of such intermetallics were found during microscopical SEM analysis and will be discussed later on. A close observation of the pattern corresponding to position 4 of the gravity cast sample in Fig. 2b, one can notice a different intensity ratio between the (111) reflection at $37°$ $2-\theta$ angle and (200) reflection at $45°$ $2-\theta$ angle peak indicating a small increase of the (200) orientation compared to the other positions of the specimen. This must be related with a difference in the heat dissipation field during solidification of the gravity cast sample that is probably affected by the proximity of position 4 to the runner and the downsprue, Fig. 1b.

Optical micrographs taken from the centre and the mould-contacting edge of the cross-section at position 3 (Fig. 2c) of CRIMSON and gravity-cast samples are shown in Fig. 3. A clear difference in porosity can be observed. For the CRIMSON samples, Fig. 3a, b, some small pores mainly related to solidification shrinkage can be observed. On the other hand, for the gravity cast samples, Fig. 3c, d, in addition to the small pores related to shrinkage, pores significantly larger in size (up to 250 µm) with spherical and non-spherical shapes can be observed that may be resulting from air entrainment as well as oxide-induced defects [15, 16]. The microstructural features for both samples, Fig. 3, are typical for hypoeutectic Al–Si alloys with α-Al primary dendrites (light-coloured areas) surrounded by Al/Si eutectic phases (dark-coroured areas). The main microstructural features are comparable for both casting methods. Secondary dendrites arms spacing (SDAS) was found to be 31.8 ± 6.2 µm and 28.4 ± 4.7 µm for CRIMSON and gravity casting respectively.

Figure 4a shows an SEM micrograph from a gravity-cast A354 sample focusing on a pore wrapped up in an oxide that would possibly resemble a double oxide film (DOF). Figure 4b reveals the elemental content at various points of these defects numbered from 1 to 4. The high content of oxygen can be indicative of the presence of oxides such as Al$_2$O$_3$ supporting the hypothesis for the formation of double oxide films. DOF is a significant type of defect that is caused by surface turbulent flow during filling, which leads to the folding and breaking of the surface oxide film [15]. It is noted that these types of DOF defects have been mostly found in the gravity-cast samples suggesting that under the controlled liquid flow achieved by the counter-gravity filling of the CRIMSON casting can effectively reduce (or eliminate) the formation of double oxide films.

Figure 5 shows an SEM micrograph (accompanied by element maps) from a CRIMSON cast sample, depicting an area with Al dendrites, eutectic Al/Si colonies and intermetallic phases. The element mapping clearly indicates the distribution of Cu and Mg in different intermetallic phases. To further explore the formation of intermetallics in the cast alloys, Fig. 6 illustrates some intermetallic phases found in

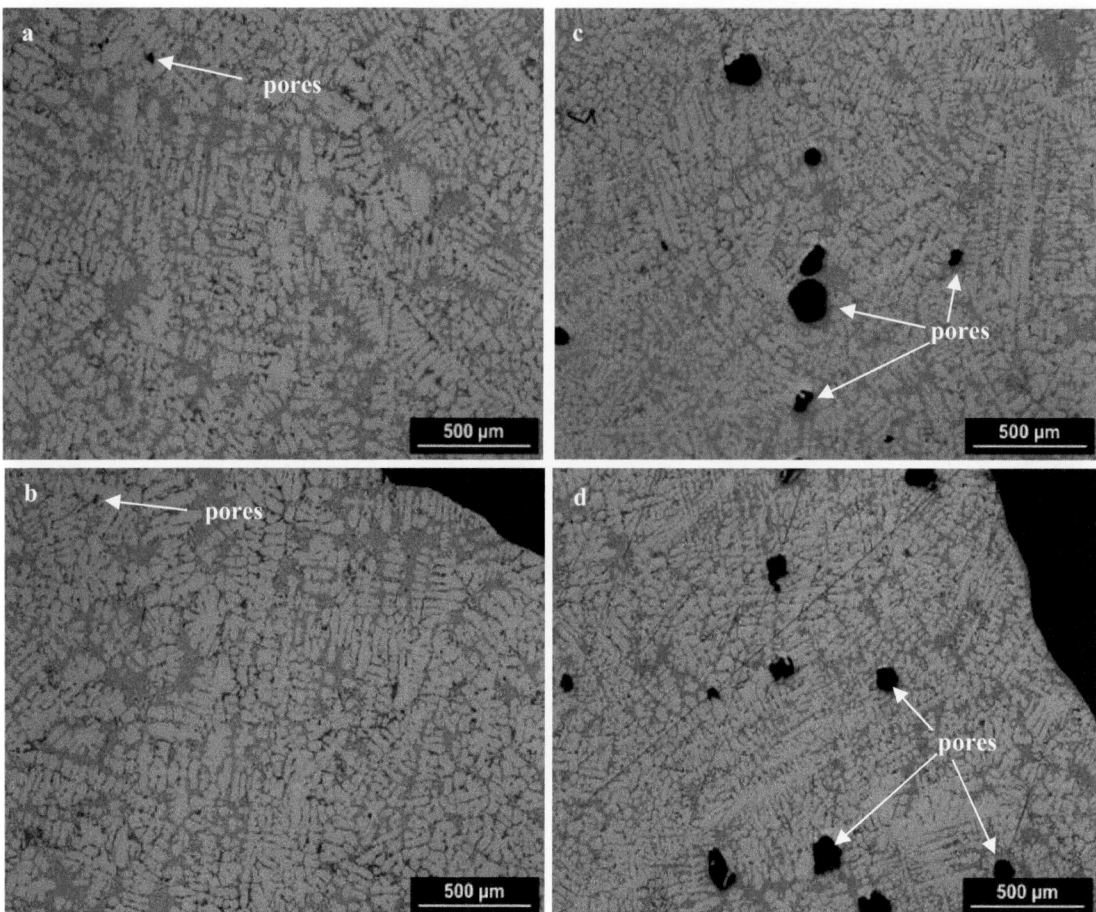

Fig. 3 Optical micrographs of the cross-section at the gage area (position 3): **a** CRIMSON casting at centre of the cross section, **b** CRIMSON casting at edge of the cross section, **c** gravity casting at the centre of the cross section and **d** gravity casting at the edge of the cross section

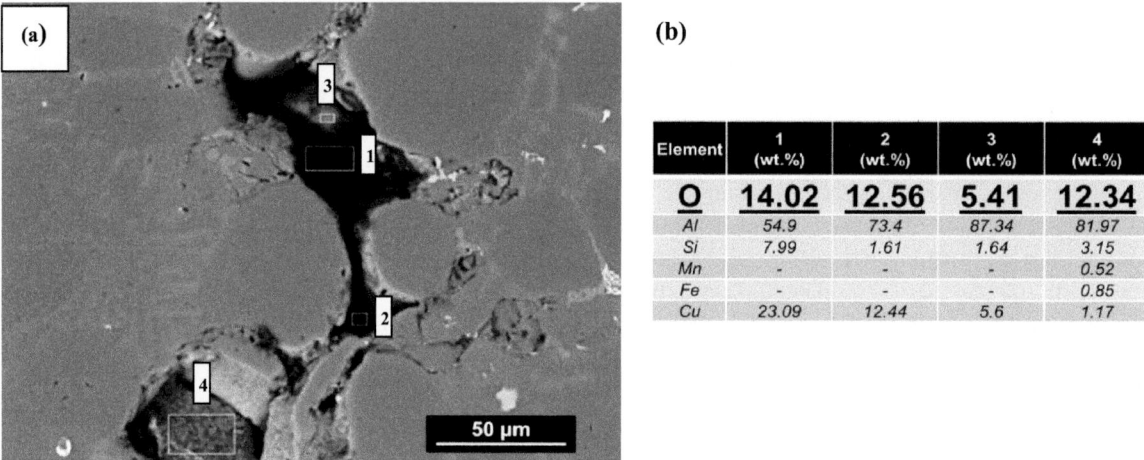

Element	1 (wt.%)	2 (wt.%)	3 (wt.%)	4 (wt.%)
O	**14.02**	**12.56**	**5.41**	**12.34**
Al	54.9	73.4	87.34	81.97
Si	7.99	1.61	1.64	3.15
Mn	-	-	-	0.52
Fe	-	-	-	0.85
Cu	23.09	12.44	5.6	1.17

Fig. 4 **a** SEM image of a DOF in the tensile gauge area gravity-cast microstructure. **b** EDS microanalysis of areas 1–4, as indicated in (**a**)

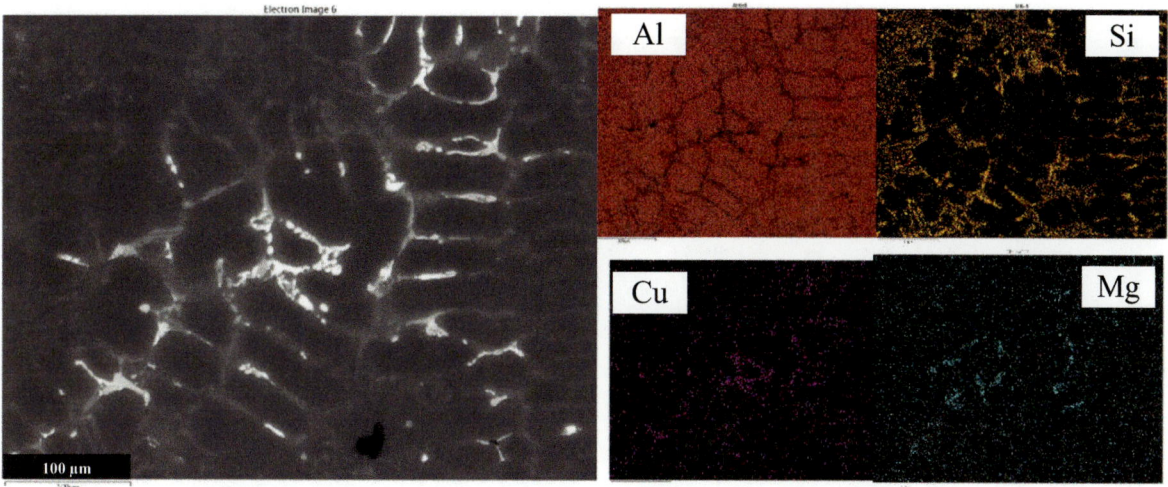

Fig. 5 Element mapping using SEM and EDS indicating the elemental distribution in secondary phases

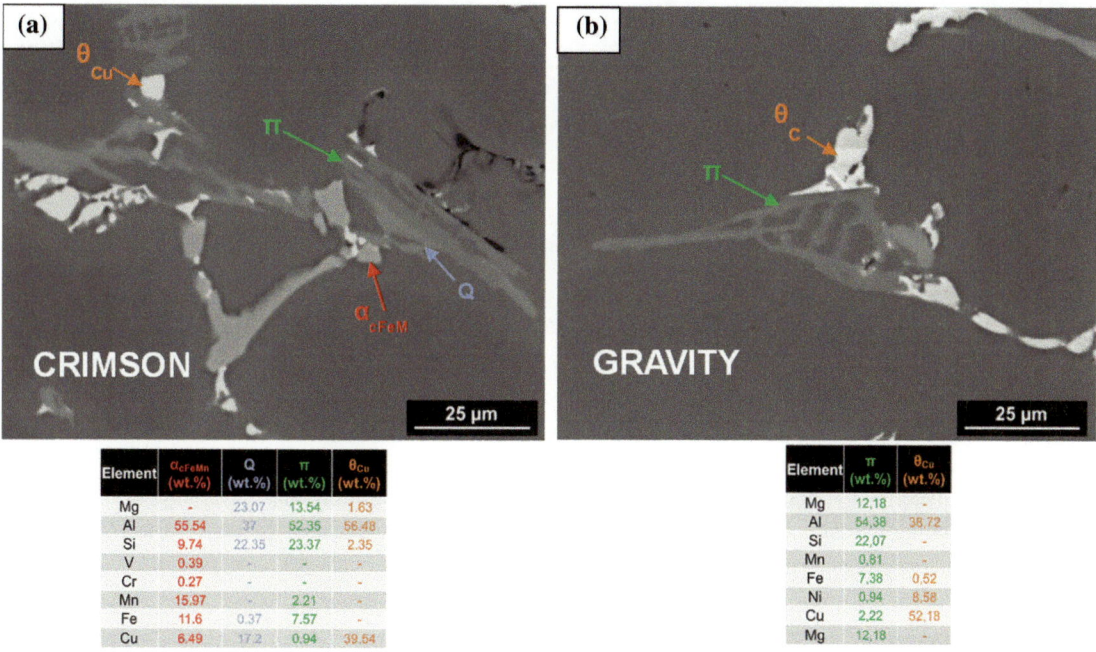

Element	α_{cFeMn} (wt.%)	Q (wt.%)	π (wt.%)	θ_{Cu} (wt.%)
Mg	-	23.07	13.54	1.63
Al	55.54	37	52.35	55.48
Si	9.74	22.35	23.37	2.35
V	0.39	-	-	-
Cr	0.27	-	-	-
Mn	15.97	-	2.21	-
Fe	11.6	0.37	7.57	-
Cu	6.49	17.2	0.94	39.54

Element	π (wt.%)	θ_{Cu} (wt.%)
Mg	12,18	-
Al	54,38	38,72
Si	22,07	-
Mn	0,81	-
Fe	7,38	0.52
Ni	0,94	8.58
Cu	2,22	52.18
Mg	12,18	-

Fig. 6 a SEM image of multiple intermetallic compounds clustered together as identified in CRIMSON castings. **b** SEM image of intermetallic compounds cluster found in gravity cast

both CRIMSON and gravity-cast samples revealing the presence of: (i) θ-Al_2Cu, (ii) Q–$Al_5Cu_2Mg_8Si_6$, (iii) π-$Al_8FeMg_3Si_6$, and (iv) α_cFeMn intermetallic compounds. The identification of the compounds was based on the chemical composition of the phases as well as their respective shapes [16]. The presence of the intermetallics, their morphology and size appear similar for both CRIMSON and gravity cast castings. As expected for aluminium–silicon–copper alloys the θ-Al_2Cu particles appear to form in spherical shapes, Q–$Al_5Cu_2Mg_8Si_6$ in irregularly rounded particles, π-$Al_8FeMg_3Si_6$ as Chinese script or branched polygons and α_cFeMn as branched polygons.

Hardness tests have shown an average value of 96.3 ± 11.2 HV and 88.0 ± 10.6 HV for the CRIMSON and gravity casting respectively. The standard deviation in both samples is at the anticipated range and it mainly derives from the hardness difference between the corresponding dendritic and eutectic areas. The yield strength (YS) and ultimate tensile strength (UTS) were found to be 151 MPa, 192 ± 22 MPa for CRIMSON sample, and 153 MPa,

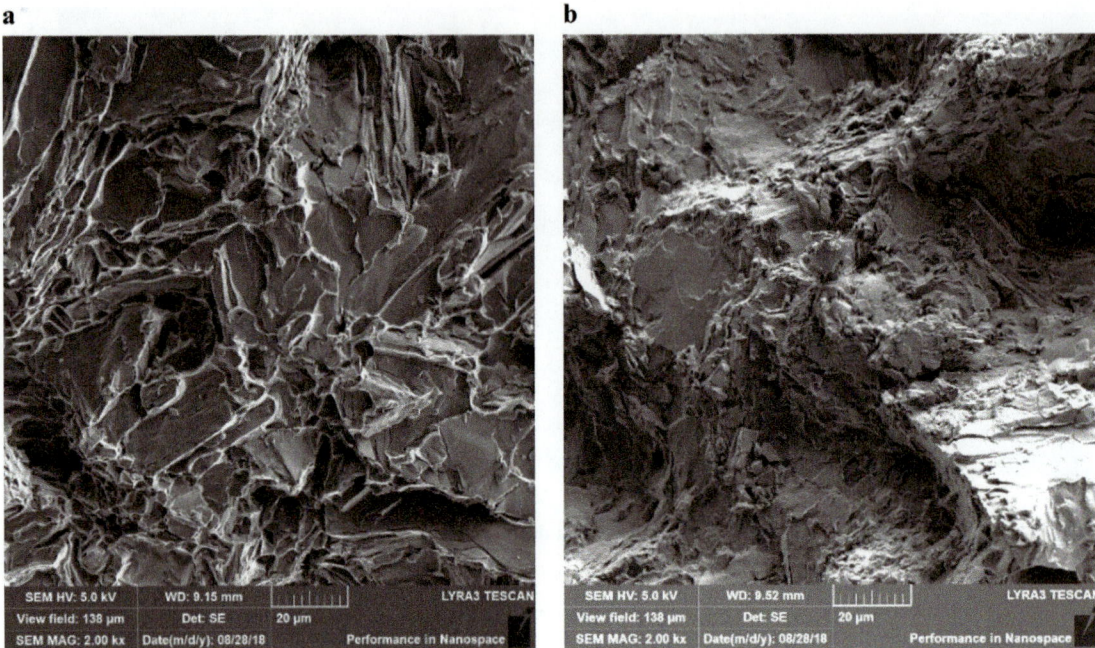

Fig. 7 Typical SEM images of the fractured surfaces after tensile tests: **a**, **b**, CRIMSON casting at low magnification, **c**, **d** gravity casting

195 ± 30 MPa, for gravity-cast samples, respectively, values comparable or better than expected for similar castings [17–19]. Plastic strain to failure was around 1% for CRIMSON and gravity castings. It is to be noted, that the mechanical strength and ductility usually improve with thermal treatment and ageing; this is the focus of an ongoing investigation. Figure 7 presents SEM images taken from typical fractured surfaces of the cast samples. Fracture analysis reveals a mixed fracture mode for the CRIMSON cast sample (Fig. 8a) with cleavage facets characteristics of (prevailing) brittle transgranular fracture coexisting with micro-dimples to a smaller extent. For the gravity-cast samples (Fig. 8b), the characteristics of brittle transgranular

fracture are even more prominent and the evidence of ductile fracture much more scarce.

Amongst the important observations in this study, is the premature failure of a gravity-cast sample during tensile tests; the sample fractured without any plastic deformation at only a fraction of the expected UTS (~ 138 MPa). This premature failure was directly related to a sizable entrapped oxide defect, as shown in Fig. 7, covering about a third of the cross-sectional area. Whereas, such a sizeable defect was not detected in other gravity samples, this observation highlights the unreliability of conventional gravity casting processes and underlines the benefits from processes such as CRIMSON that can reliably produce castings with high structural integrity.

Fig. 8 The fractured surfaces from a gravity cast specimen fractured prematurely at a fracture of the UTS (~ 138 MPa) without any plastic deformation. The arrows indicate the entrapped oxide defect

Conclusions

In this work, the structure and mechanical properties of Al castings produced by CRIMSON counter-gravity casting method was investigated and compared with gravity-cast specimens. The 354 Al alloy was used in this case in the as-cast condition. The microstructure was investigated using XRD, OM, and SEM revealing similar findings for the two castings processes. The microstructure was investigated using XRD, OM, and SEM revealing similar findings for the two casting processes. Dendritic microstructure was revealed mainly consisting of α-Al dendrites and Al–Si eutectic, as expected in Al–Si hypoeutectic alloys. Furthermore, the presence of the following intermetallic compounds has been confirmed: (i) θ-Al_2Cu, (ii) Q–$Al_5Cu_2Mg_8Si_6$, (iii) π-$Al_8FeMg_3Si_6$, and (iv) $\alpha_c FeMn$. The mechanical properties were found to be similar for the CRIMSON and gravity cast A354 alloy in terms of tensile strength and hardness. This is believed to be relevant to the similarity in the microstructure of the alloys. Interestingly, significant differences were found between the CRIMSON counter-gravity and the conventional gravity castings in terms of porosity and oxide bifilms. Microscopical examination revealed a significant reduction in the presence of pores relevant to air entrainment as well as oxide inclusions and bifilms for the CRIMSON counter-gravity cast samples, mainly achieved due to the controlled liquid flow during filing; in contrast to the turbulent flow during the conventional filing in gravity casting. The elimination of large pores and oxide defects in CRIMSON castings has a major impact on the structural integrity and the expected fatigue resistance of the castings, improving in parallel the casting yield of the process.

Acknowledgements Financial support from the EPSRC-DTP project "Low Energy Casting: Novel Induction Technologies for Energy Efficient Rapid Melting" (ref 2,541,199) in the framework of (EP/R513027/1) and the UK EPSRC project "Energy Resilient Manufacturing 2: Small is Beautiful Phase 2 (SIB2)" is gratefully acknowledged.

References

1. Campbell J (2003) Castings. Elsevier, Oxford
2. Jolly MR (2003) Castings, in: Milne I, Ritchie RO, Karihaloo B (Eds.), Comprehensive Structural Integrity, 1st ed., Elsevier, Oxford, 2003, p 377–466.
3. Jolly M, Katgerman L (2022) Modelling of defects in aluminium cast products. Progress in Materials Science 123: 100824
4. Divandari M, Campbell J (2004) Oxide film characteristics of Al-7Si-Mg alloy in dynamic conditions in casting. International Journal of Cast Metals Research 17(3):182-187. https://doi.org/10.1179/136404604225017546
5. Ktari A, El Mansori M (2021) Intelligent approach based on FEM simulations and soft computing techniques for filling system design optimisation in sand casting processes. The International Journal of Advanced Manufacturing Technology 114(8):981-995. https://doi.org/10.1007/s00170-021-06876-z
6. Campbell J (2012) Stop Pouring, Start Casting. Intern. Journal of Metal Casting 6: 7–18. https://doi.org/10.1007/BF03355529
7. Archer L, Hardin RA, Beckerman C (2018) Counter gravity sand casting of steel with pressurization during solidification, Intern J. of Metal Casting, 12 (3): 596-606
8. Dai X, Jolly M, Zeng B (2012) The improvement of aluminium casting process control by application of the new CRIMSON process. IOP Conference Series: Materials Science and Engineering 33:012009. https://doi.org/10.1088/1757-899X/33/1/012009
9. Papanikolaou M, Pagone E, Georgarakis K, Rogers K, Jolly Mark, Salonitis K. (2018) Design Optimisation of the Feeding System of a Novel Counter-Gravity Casting Process Metals, 8(10): 817; https://doi.org/10.3390/met8100817
10. Li Y, Georgarakis K, Pang S, Antonowicz J, Charlot F, LeMoulec A, Zhang T, Yavari AR (2009) AlNiY chill-zone alloys with good mechanical properties, J. Alloy. Compd., 477: p 346-349
11. Li Y, Georgarakis K, Pang S, Antonowicz J, Charlot F, LeMoulec A, Zhang T, Brice-Profeta S., Zhang T, Yavari AR (2009) Chill-zone aluminum alloys with GPa strength and good plasticity, J. Mater. Res., 24: p 1513-1521
12. Wang Z, Qu RT, Scudino S, Sun BA, Prashanth KG, Louzguine-Luzgin DV, Chen MW, Zhang ZF, Eckert J (2015) Hybrid nanostructured aluminum alloy with super-high strength. NPG Asia Mater. 7: p. e229
13. Dong X, Yang H, Zhu X, Ji S (2019) High strength and ductility aluminium alloy processed by high pressure die casting, J. Alloys Compds, 773: p 86-96
14. Toschi S, (2018) Optimization of A354 Al-Si-Cu-Mg Alloy Heat Treatment: Effect on Microstructure, Hardness, and Tensile Properties of Peak Aged and Overaged Alloy. Metals 8: p 961 https://doi.org/10.3390/met8110961
15. Campbell FC (2013) Metals Fabrication; Understanding the Basics. ASM International, Materials Park, Ohio
16. Brown, JR (2000) Foseco Non-Ferrous Foundryman's Handbook. Butterworth-Heinemann, Oxford.
17. Shabestari SG, Moemeni H (2004) Effect of copper and solidification conditions on the microstructure and mechanical properties of Al-Si-Mg alloys. Journal of Materials Processing Technology 153–154:193–198. https://doi.org/10.1016/j.jmatprotec.2004.04.302
18. Tiryakioglu, M. (2020) The effect of hydrogen on pore formation in aluminum alloy castings: Myth versus reality. Metals 10(3) p 368; https://doi.org/10.3390/met10030368.
19. Campbell, J (2018) Mini Casting Handbook. Aspect Design, Malvern.

Defect Minimisation in Vacuum-Assisted Plaster Mould Investment Casting Through Simulation of High-Value Aluminium Alloy Components

Emanuele Pagone, Christopher Jones, John Forde, William Shaw, Mark Jolly, and Konstantinos Salonitis

Abstract

Vacuum-assisted plaster mould investment casting is one of the best available processes to manufacture ultra-high complexity castings for the aerospace and defence sectors. In light of the emerging cross-sectoral manufacturing industry digitalisation, process simulation appears as a very important tool to improve casting yield, reduce metallurgical scrap, and reduce lead time to new product introduction. Considering the unique aspects and the level of customisation of the process system, this work will present a Computational Fluid Dynamics-based simulation tool with bespoke settings (that include thermophysical properties). Optimal fill and solidification parameters are identified for a representative geometry able to describe a variety of very complex, high-value aluminium alloy components through an iterative process.

Keywords

Aluminium · Shaping and forming · Modelling and simulation · Solidification

Introduction

Investment casting is a manufacturing process particularly useful to produce components with high precision and complex shape providing the designer with relatively large freedom. Although the process comprises a higher number of steps in comparison with other expendable mould alternatives, its products are near net shape, making it a very good choice for metals that are difficult to machine [1]. The typical main steps of investment casting can be summarised with the following sequence. The starting point is the production of a master pattern (i.e. a replica of the final geometry of the product) that is used to generate a master die filled with wax to produce wax patterns. Then, several of such patterns are arranged around a common sprue to form a so-called "tree". Furthermore, the "tree" is coated by investment material that is dried and hardened, where particular care is dedicated to the composition of the first layer that needs to be as smooth as possible. Subsequently, the wax patterns are removed by melting, whereas the obtained empty mould is heated at an optimal temperature before pouring liquid metal. Finally, the mould is broken and the castings obtained are separated from the sprue, while the small gate stubs are removed by machining [1].

The investment casting process studied in this work uses plaster as investment material and vacuum is generated within the mould before pouring liquid aluminium alloys. The objective of the study is to identify optimal fill and solidification parameters to minimise defects for a representative geometry using Computational Fluid Dynamics (CFD) simulations.

Although previous studies in this field can be identified, it can be observed that there is no previous literature covering the entire research aim considered. In particular, Kan et al. simulated numerically vacuum casting of Al-6061 quantifying the faster casting times under vacuum versus atmospheric conditions (0.95 s versus 1.2 s for the case considered). Furthermore, higher mould temperatures and an 8% increase in the heat transfer rate were reported during the vacuum casting process [2]. Although useful, this work does not consider plaster moulds. Ahmad et al. [3] used numerical simulations to improve the quality of product modelling the flow of resin into a five-axis impeller mould considering a vacuum casting process. Key parameters such as fluid temperature, flow rate, and filling time are considered reporting

E. Pagone (✉) · C. Jones · M. Jolly · K. Salonitis
Sustainable Manufacturing Systems Centre, School of Aerospace, Transport and Manufacturing, Cranfield University, Cranfield, MK43 0AL, UK
e-mail: e.pagone@cranfield.ac.uk

J. Forde · W. Shaw
Sylatech Limited, Kirkdale Road, Kirkbymoorside, YO62 6PX, North Yorkshire, UK

© The Minerals, Metals & Materials Society 2023
S. Broek (ed.), *Light Metals 2023*, The Minerals, Metals & Materials Series,
https://doi.org/10.1007/978-3-031-22532-1_145

the relevant optimal values obtained. However, also in this case, the material of the mould is not plaster and there is no comparison with atmospheric pressure processes. Tang et al. [4] developed a method to optimize the process parameters that have an impact on the quality of silicone rubber micro-mould cavities produced for vacuum casting. The authors show that the optimised parameters result in higher quality mould cavities quantified using dimensional analysis and white light interferometry. In particular, miniature gears with dimensional deviations of less than 3% in comparison with the benchmark product are reported. Zhang and Hu [5] combined a response surface method with an artificial fish swarm algorithm to optimise the filling process of a vacuum casting process improving the dimensional accuracy of castings. Suprapto et al. [6] developed a method to identify the location and the moment during a vacuum casting process when porosities are formed in aluminium alloys. Output parameters include morphological characteristics, number, and size of pores. The authors combine optical emission spectroscopy tests and fluid mechanic parameters and principles (e.g., Reynolds and Niyama numbers, Archimedes' principle, etc) pointing out that further experiments are recommended for validation purposes. Analysis of results recommends a limit of 0.15 cm^3 of hydrogen dissolved per 100 g of liquid metal as well as to minimise the presence of impurities with particular attention to copper compounds. Jin et al. [7] propose an improved design (with a novel overflow system) of a die for vacuum casting, considering two geometries. Results show that the proposed design prevents premature freezing of the metal (i.e., "cold runs") and reduce turbulence of liquid metal. However, also in this case, there is no comparison with a comparable non-vacuum process. On the other hand, Chaus et al. [8] compared the microstructure and properties of steel using gravity and vacuum casting reporting the surface roughness and impurity content alongside hardness and bending strength. Furthermore, it is reported that vacuum casting reduces the concentration of oxygen and hydrogen in the liquid metal by a factor of two and leads to better finishing and mechanical properties. Kuo et al. [9] used experimental design methods combined with the Vacuum Differential Pressure Casting (VDPC) technique to improve the tensile strength of castings with a focus on the production of the moulds. The mould cavity temperature is identified as the primary parameter (with an optimal cavity temperature of 35 °C) followed by the mixing time (with 40 s as optimal), differential pressure time (best at 8 s), and mixing chamber inlet valve angle (optimal at 60°). Ignaszak and Wojciechowski [10] validated plaster casting simulations of jewellery against experiments showing accurate prediction of the shrinkage porosity location. Lun Sin et al. [11] studied the reaction of molten magnesium alloys with plaster moulds under vacuum casting conditions using thermodynamics principles. The casting

temperature does not appear to affect the mould–metal reactivity within typical temperature ranges, while particles from the rough surface of the casting form a reaction layer between the plaster mould and the casting. Władysiak and Pawlak [12] studied the crystallization and cooling process of a silicon–aluminium alloy in a plaster mould using Topological Data Analysis parametrising the study with the casting temperature. The work shows that a casting temperature in the range of 725–800 °C prolongs the crystallization time and that 725 °C is the optimum for filling the mould and dimensional accuracy. Bogdanoff et al. [13] present a numerical procedure (based on CFD) to approximate the yield strength of A380 with plaster mould casting of A356.0 and A360.2. Yield strength of 140 and 210 MPa (respectively) is predicted after heat treatment. Brevick et al. [14] investigated the effect of different levels of silica sand and carbon fibre additions on the properties of plaster moulds and related aluminium alloy castings. Analysis of statistical data showed the relationship between wet and dry strength of plaster moulds with sand and carbon fibre content. An addition of 30% silica sand did not appear to reduce the strength of the plaster and it has the potential to reduce cost. Instead, since the addition of 0.5% carbon fibre significantly increased the strength of the plaster, the authors suggest adding both silica sand and carbon fibre to plaster moulds to significantly enhance their hardness.

Studies in the literature on the optimisation of casting parameters to minimise defects are not much focused on plaster mould vacuum investment casting. For example, Ahirrao and Marlapalle investigated the effect of several parameters (including pressure and die temperature) to minimise surface roughness and wear resistance of aluminium alloys with simulation but their study was conducted considering a high-pressure die casting process [15].

As it can be observed from this literature survey, although parts of the themes involved in this work have been studied extensively in the literature, not much has been investigated to minimise defects in plaster mould vacuum investment casting of aluminium alloys through CFD simulations.

Methods

A geometry representative of a family of products of interest (Fig. 1) has been investigated with CFD simulations using the commercial tool Flow-3D [16] where aluminium alloy A356 is the casting material.

The accuracy of insights gleamed from CFD simulations is governed by the accuracy and attention to detail in simulations setup. Broadly, the stages of this setup will include establishing the relevant physical models (with associated parameters), defining the properties of the materials present in the system, configuring a suitable meshing and applying

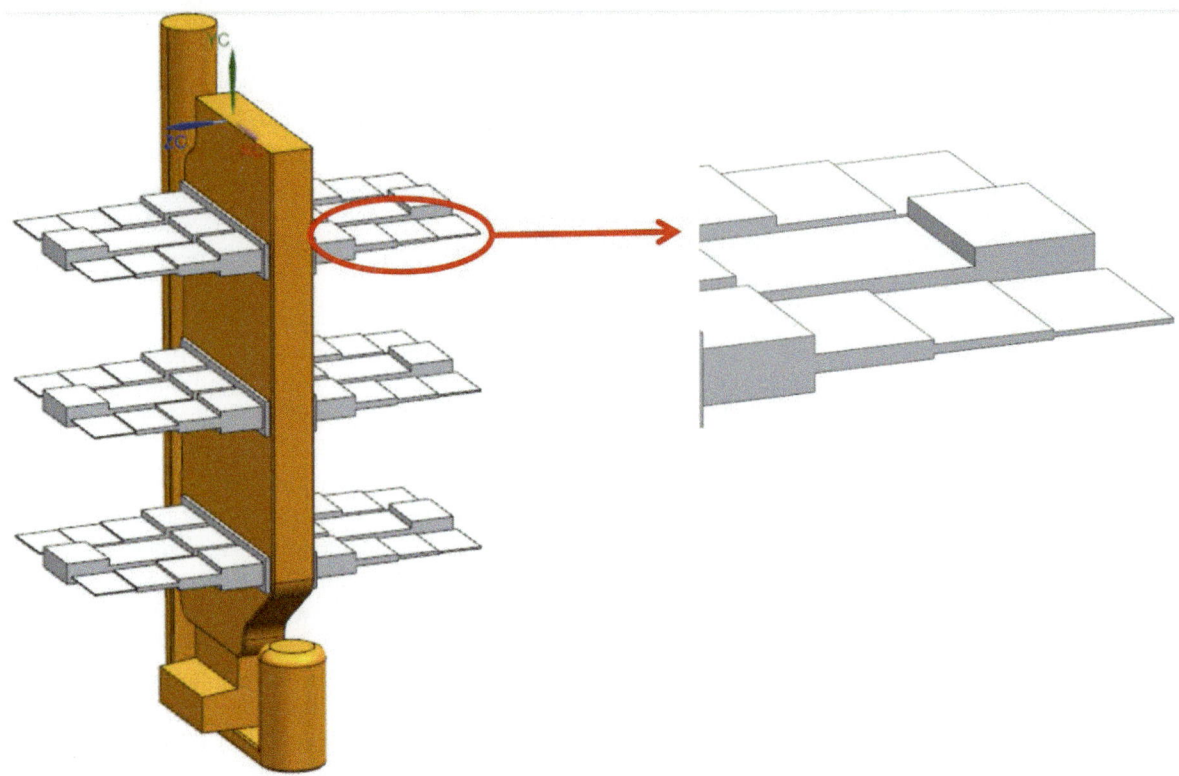

Fig. 1 Representation of the computational domain of the CFD simulations

physical boundary conditions with extremity limits. The decision of which outputs to monitor and at which frequency is a final consideration.

Second only to meshing, over-prescribing physical models has a significant impact on simulation run time. It is important that the aim of the simulations is clearly defined and only models in association with such aim are initiated. In this case, reducing the number of active models was not possible as all were relevant, however, alternative time-saving options are available in such situations. Therefore, the simulation was split into two halves—filling and solidification—in which flow effects were excluded in the latter part. With no flow effects, there was no need to solve the full momentum equations, presuming stationary fluid on which solidification calculations are made. In the initial (filling) simulation, the air entrainment, gravity, heat transfer and viscosity and turbulence models were activated whereas only the heat transfer and gravity models were active in the second (solidification) simulation in addition to activating the solidification model calculation. It is worth noting that during the second simulation, the solver for the heat transfer model was switched from explicit to implicit. Implicit solvers bypass the stability restrictions that are placed on explicit solvers allowing them to operate with a larger time step, ideal when working with computational demanding studies. Such solvers are not always recommended, requiring

a case-by-case sensitivity assessment ahead of the full study to confirm that it provides accurate results.

The properties of materials, or at least sufficiently comparable materials, are generally present in literature sources and provide adequate good accuracy in situations where exact material properties cannot be measured. Wu et al. [17] and Rahmanian and Wang [18] present data on the thermal conductivity and specific heat capacity of plaster as a function of temperature which have been employed in this study; the region of interest is between room temperature and 400 °C.

Definition of a suitable mesh is not a simple task, requiring a balance between resolution of complex geometries and simulation stability as well as run time. Flow-3D allows for multiple overlapping meshes to be generated to allow for local mesh refinement around areas of interest that helps in alleviating this dilemma. Figure 2 shows the grid lines of the mesh constructed around the casting geometry for this study. The mesh features a "global" mesh covering the whole domain with three additional mesh planes, visible in the left and centre panels of the figure, which grant local enhancement to resolution to the casting component. Mesh sizes are 0.0025 and 0.0003 m in the case of global and mesh planes, respectively.

Boundary conditions are applied to the outer faces of the global mesh block at the edge of the domain, meaning six

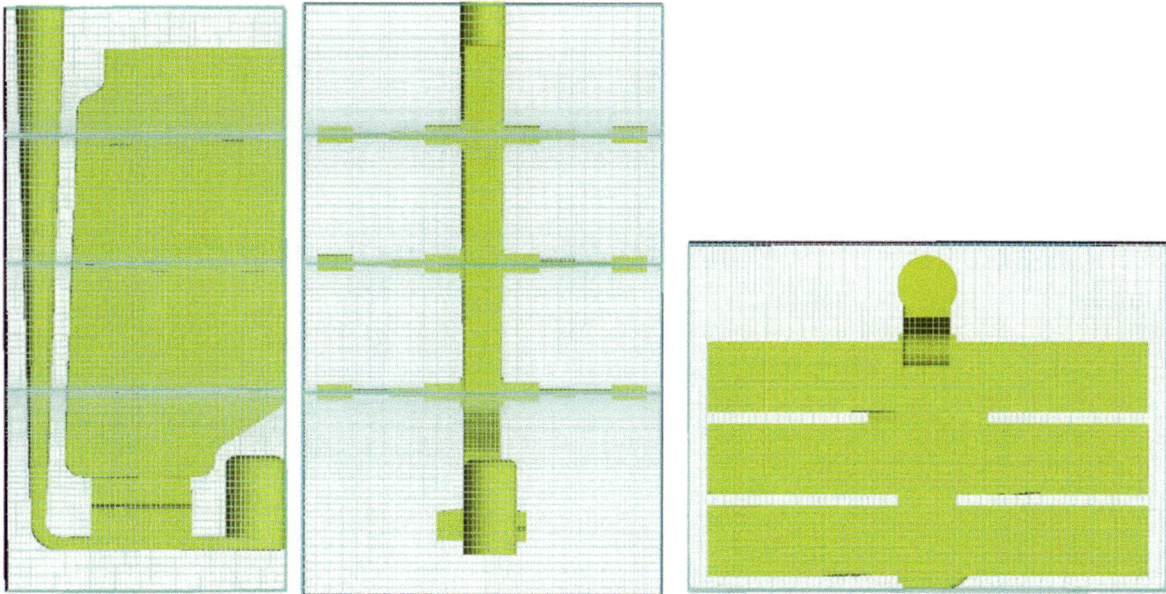

Fig. 2 Three-axis view of component mesh construction

boundaries in total given the cubic nature of the mesh space. To replicate conditions inside the vacuum chamber, in which air enters and leaves through a single valve, all boundaries except one were set to wall conditions with the final being set to a pressure boundary condition. Wall conditions apply a no-slip and zero velocity criteria at the boundary, preventing fluid from entering or leaving the domain although temperature and pressure information as initial or functions of time can be defined. The pressure condition at the final boundary facilities user-specified temperature and pressure as a function of time as well as allowing fluid to flow out of the domain.

Results and Discussion

During the filling of metal castings, the factors considered in this study to separate a sound casting from a sub-standard one are fluid velocity, fluid temperature, and evolution of air entrainment due to turbulent flow. Jin et al. [7] show that fluid velocities in excess of 0.5 m/s have a detrimental effect on filling and final casting quality. The predicted velocity profile in the computational domain at the final time step of the simulation (Fig. 3) shows that fluid flow throughout the running system has a velocity consistently in excess of 0.5 m/s which diminishes at the in-gate as fluid enters the main casting with the exception of a narrow vertical patch of excess velocity in the main body above the in-gate. It is believed this was a result of fluid inertia during filling.

The final distribution of temperature in the casting at the final time step of the filling simulation (Fig. 4) shows that

portions of the fluid furthest from the down sprue at 435 °C have begun solidifying (considering a melting temperature of the metal of 620 °C). Unlike the opposite side of the casting which was in proximity to the still hot alloy in the down sprue maintaining fluid temperature, this portion of the casting faces only the outside air leading to this accelerated cooling. Similarly, the horizontal aspects of the casting at the centre (right-hand panel of the figure) retain heat due to close contact with surrounding elevated temperature material, whereas alloy on the periphery sees relatively more rapid cooling. This pattern of solidification towards the centre of the casting has the potential to give rise to porosity in the casting. However, further investigations will be needed to determine the approximate position and extent of this porosity (as reported below). Importantly, as solidification has only just been initiated, the choice to divide the simulation into two parts with solidification modelling only in the second half is justifiable.

As would be expected, the volume fraction of air entrained (Fig. 5) has been forced to the extremities of the casting by the inbound fluid flow. In fact, there is a bubble trap at the end of the horizontal runner bar and at the tips of the horizontal elements of the computational domain. An additional explanation for this that could be presented is as a result of the thinning cross section of the casting towards the tips of these horizontal elements and, therefore, the alloy has difficulty reaching the extremity. This is considered the more likely scenario as a maximum porosity of 17.9% is higher than would be expected for a vacuum casting.

With the mould filled, discussion can turn to the behaviour of the casting during solidification. The solidification

Fig. 3 Fluid velocity (in m/s) contour plot of casting at final time step from three orthogonal orientations

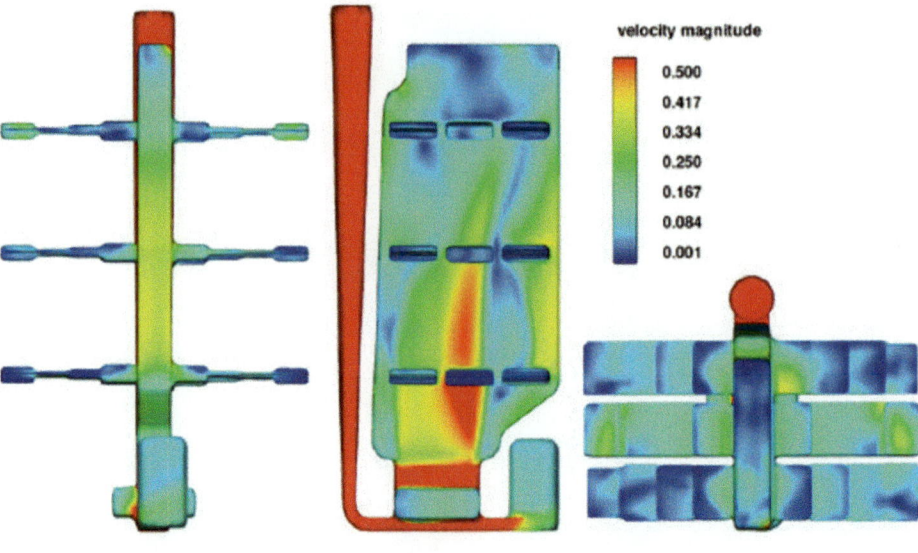

Fig. 4 Fluid temperature (in °C) contour plot of casting at final time step from three orthogonal orientations

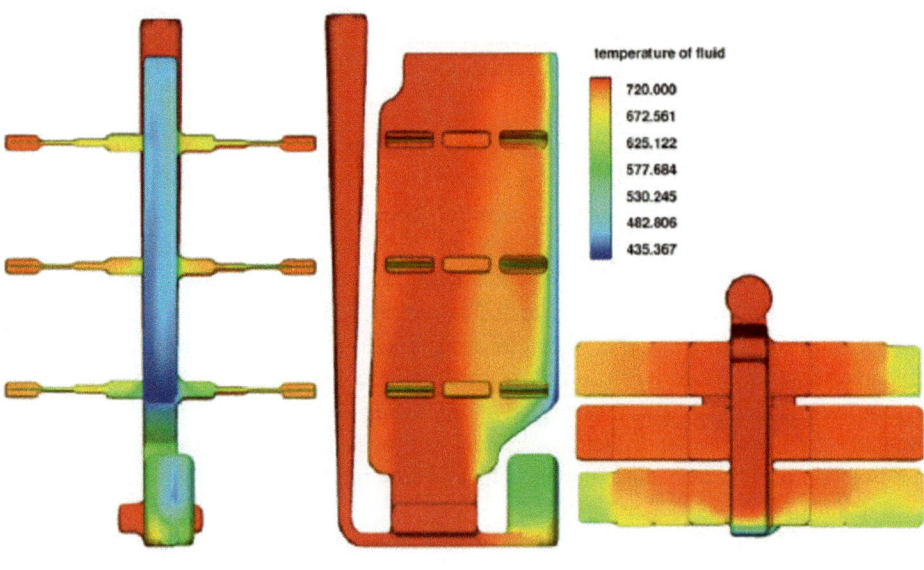

Fig. 5 Fluid air entrainment contour plot of casting at final time step from three orthogonal orientations

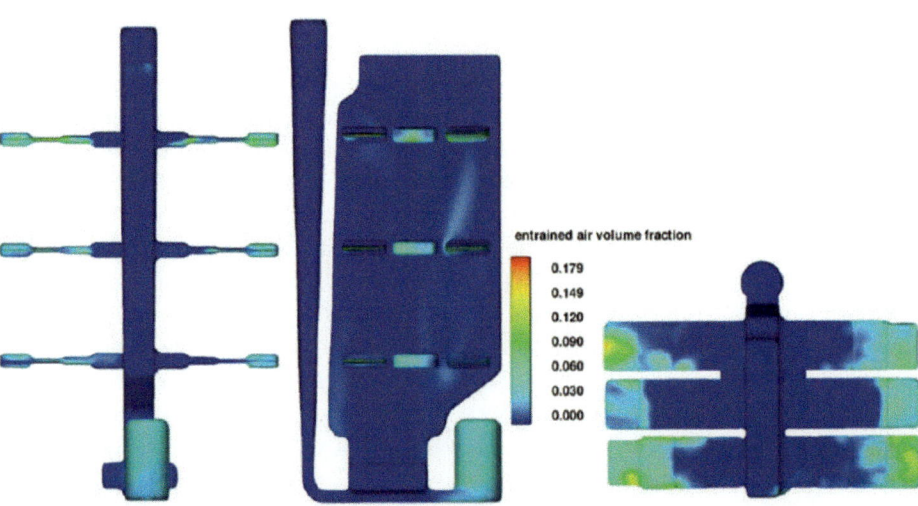

Fig. 6 Final temperature (in °C) distribution at the conclusion of solidification from three orthogonal orientations

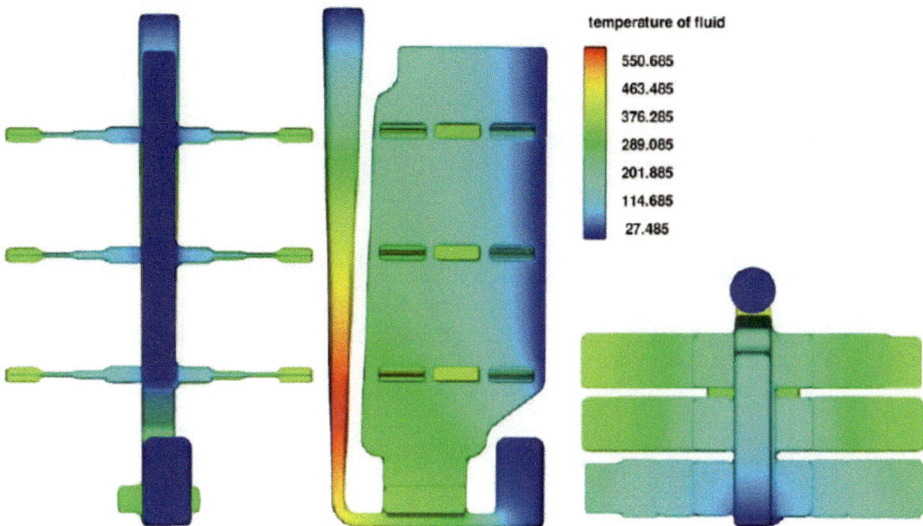

process as modelled was completed in 233 s, during which consideration of the fluid temperature and evolving porosity were selected. Considering the final temperature distribution in the computational domain at the end of solidification (Fig. 6), the outermost portion of the casting has almost reached room temperature with a rather steep gradient across the component to approximately 290 °C on the opposite face. A key feature of merit is that there are no isolated hot spots present in the casting that would otherwise lead to solidification defects.

Micro-porosity in the casting is the final element for discussion and, arguably, the most relevant as it relates directly to quality and operational performance in the context of this case. For components of this type, manufacturers typically do not allow more than 5% porosity, beyond which a casting would be rejected. As can be seen in Fig. 7, a maximum of 3.52% micro-porosity is predicted—a value comfortably below the maximum tolerable. It is interesting

that there is a difference between the predicted 3.52% micro-porosity and 17.9% entrained air suggesting that not all entrained air is contributing to micro-porosity. However, comparing the presented distributions of air entrainment and porosity there is significant overlap, implying air entrainment is a contributor to the porosity observed.

Given the importance placed on porosity in this casting due to the direct impact on component serviceability, this has been selected as the target parameter for an optimization assessment. Three parameters with the greatest influence on porosity have been selected: mould temperature, in-gate fluid temperature, and in-gate fluid velocity. An L9 orthogonal experimental design was constructed by selecting three levels for each of these parameters, as shown in Table 1. It should be noted that at least three levels are needed to appropriately identify non-linear patterns. An orthogonal test method has similarities with Taguchi methods [18] although there is a conceptual difference. Whilst the Taguchi method

Fig. 7 Micro-porosity map of casting at the conclusion of solidification from three orthogonal orientations

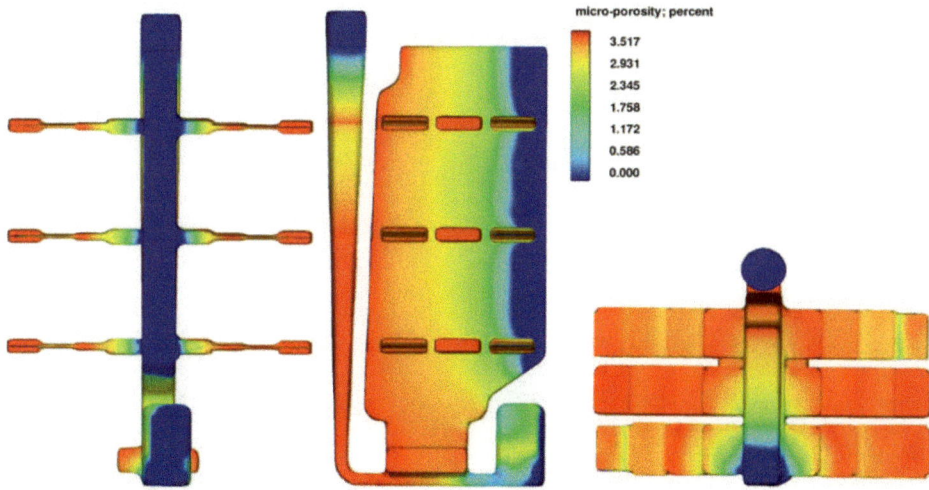

Table 1 Details of three levels used in optimization study

Parameter	Level 1	Level 2	Level 3
Mould temperature (°C)	350	400	450
In-gate fluid temperature (°C)	670	720	770
In-gate fluid velocity (m/s)	0.5	1	1.5

Table 2 Details of the parameter configuration used in each run

Simulation run	Mould temperature (°C)	In-gate fluid velocity (m/s)	In-gate fluid temperature (°C)
1	Level 1: 350	Level 1: 0.5	Level 1: 670
2	Level 1: 350	Level 2: 1.0	Level 2: 720
3	Level 1: 350	Level 3: 1.5	Level 3: 770
4	Level 2: 400	Level 1: 0.5	Level 2: 720
5	Level 2: 400	Level 2: 1.0	Level 3: 770
6	Level 2: 400	Level 3: 1.5	Level 1: 670
7	Level 3: 450	Level 1: 0.5	Level 3: 770
8	Level 3: 450	Level 2: 1.0	Level 1: 670
9	Level 3: 450	Level 3: 1.5	Level 2: 720

studies the stability of the entire system under a selection of different environments, the orthogonal method studies the responsiveness of target output parameters to the input control factors in the system.

Running these nine simulations according to a Design of Experiment (DoE) approach yields nine values of porosity from which conclusions can be drawn. The parameters used in each simulation run are detailed in Table 2, with the final porosity predicted by each simulation at the final time step (i.e., the end of solidification) displayed in a comparative histogram plot in Fig. 8.

It should be noted that this configuration of input parameters represents the second iteration of the investigation. The original selection of parameters was identified for extreme difference analysis upon the conclusion of which the best options were selected for further scrutiny. The micro-porosity predictions presented here were lower than those predicted in the original investigation, illustrating the success of this approach.

From this collection of porosity results, further assessments can be made to identify not only the most optimal setting for each parameter but also a measure of their relative importance. Considering each parameter in term, taking the average of the porosity results corresponding to each level in turn (all three L1s, L2s and L3s), 9 mean values corresponding to each level of each parameter can be obtained and have been denoted as M values. Plotting the three M values for each parameter (Fig. 9) the corresponding optimal value can be readily identified. As the goal is minimisation of porosity, the minimum M value for each parameter is sought. Hence, predictions suggest optimal values of 400 °C, 0.5 m/s and 770 °C for the mould temperature, fluid velocity and fluid temperature, respectively. These results are aligned with expectations based on the observation that a higher fluid temperature extends the onset of solidification and, hence, the formation of dendritic structure whilst filling is still occurring in thin sections. It is well known that a lower fluid velocity is beneficial in terms

Fig. 8 Histogram of percentage porosity predicted in the nine numerical trials

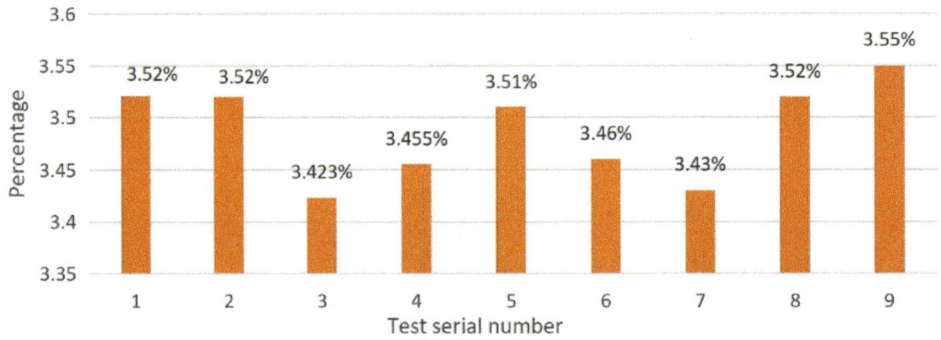

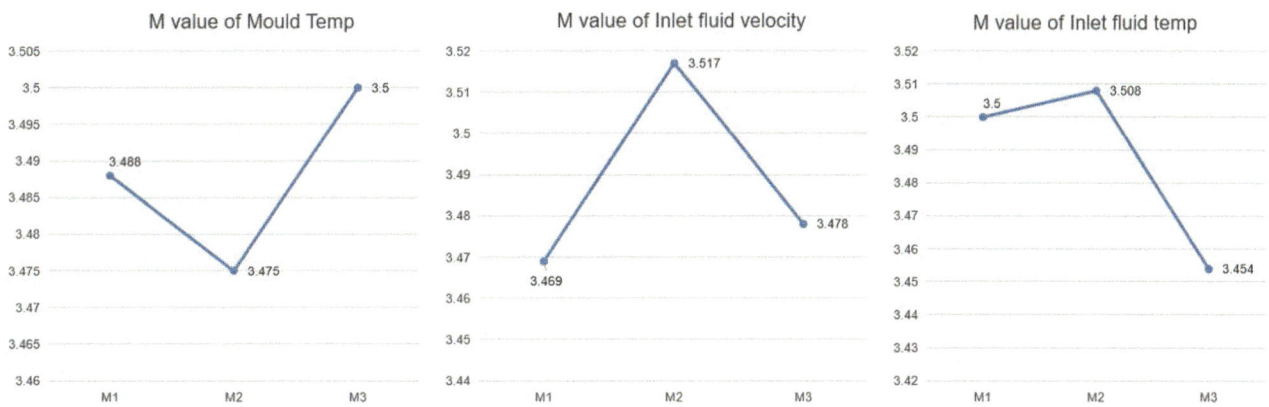

Fig. 9 M-value plots for each input parameter of the design of experiment

of lowering casting defects; however, an interesting observation is that it appears that there is a predicted reduction in micro-porosity when velocity is increased to 1.5 m/s. This would need further investigation to explain in full depth. Finally, the mould temperature must be high enough to ensure there is no premature solidification but not so high the excess heat is trapped within the system to fuel long-term porosity growth. The most optimal configuration, defined by the lowest M value, was a mould temperature of 400 °C, a fluid in-gate velocity of 0.5 m/s and an in-gate fluid temperature of 770 °C. It should be noted that the values of the input parameters were selected based on industry input. A broader range in these parameters could be investigated at a higher order orthogonal test to deepen understanding of the process beyond industry expectations; this however is a subject for further work.

As a final observation from this data, by considering the difference between the greatest and lowest value of the M values in each case, the relative importance of each parameter can be assessed. This relative importance will be denoted as the R value with R_1, R_2, and R_3 representing the relative importance of the mould temperature, fluid velocity, and fluid temperature, respectively. Making the calculation from this data set yields $R_1 = 0.025$, $R_2 = 0.048$, and $R_3 = 0.054$. The higher the R value, the greater the impact of that parameter, so this investigation reveals fluid temperature as the driving factor of the three investigated with fluid velocity and mould temperature taking second and third place.

There are of course limitations to such simulations. Optimization has suggested that a reduction in the filling rate would be beneficial, however, such a reduction would increase the filling time and hence the possibility of premature solidification and surface oxide formation. Equally, simulation models are not necessarily comprehensive. Depending on the number of models active and the extent of underpinning assumptions, aspects of the physical phenomena may not be fully captured.

Results obtained in this optimization study would readily lend themselves to a further optimization investigation on macro-shrinkage, air entrainment or volume fraction.

Conclusion and Further Work

This work presented a numerical study to minimise porosity formation in vacuum assisted plaster mould investment casting of aluminium alloy A356. A scientific literature survey revealed several studies on the broad topic but not many relevant, specific studies. A geometry designed to represent a family of products of interest has been modelled in the calculation domain of Computational Fluid Dynamics (CFD) simulations. The case considered showed a local maximum of 3.52% micro-porosity, a value well below the typical 5% threshold used in the industry to scrap high-value components. Results suggested also that not all entrained air is contributing to micro-porosity. A further optimisation based on a 9-level orthogonal Design of Experiment (DoE) based on mould temperature, in-gate fluid temperature, and in-gate fluid velocity has been carried out identifying optimal process parameter values. Furthermore, in-gate fluid temperature resulted the most important parameter to affect porosity, whereas the mould temperature the least important.

Future development aspects of this work can be summarised in the following areas.

- Further investigate the predicted reduction in micro-porosity when in-gate fluid velocity is increased to 1.5 m/s.

- A broader range of the input parameters considered in the DoE could investigate the design space beyond stringent industry expectations.
- Investigate a reduction of the filling rate (that has a beneficial effect towards micro-porosity) to identify cases of cold-runs or surface oxide formation.
- Further optimisation of macro-shrinkage, air entrainment, or volume fraction.

Acknowledgements The foundational work carried out by Mr Di Zhu as part of his MSc thesis project is gratefully acknowledged.

References

1. DeGarmo EP, Black JT, Kohser RA (2003) Material and Processes in Manufactuing, 9th ed., New York, Wiley
2. Kan M, Ipek O, Koru M Accessed 6 July 20172022) An Investigation into the Effect of Vacuum Conditions on the Filling Analysis of the Pressure Casting Process. International Journal of Metalcasting
3. Ahmad MN, Wahid MK, Maidin NA, Ab Rahman MH, Osman MH, Jumaidin R, Abu Hassan MA (2020) Flow Analysis of Five-Axis Impeller in Vacuum Casting by Computer Simulation. Journal of Advanced Research in Fluid Mechanics and Thermal Sciences, 61(2):181–189
4. Tang Y, Tan WK, Fuh JYH, Loh HT, Wong YS, Thian SCH, Lu L (2007) Micro-mould fabrication for a micro-gear via vacuum casting. Journal of Materials Processing Technology, 193:334–339.
5. Zhang HG, Hu QX (2016) Study of the filling mechanism and parameter optimization method for vacuum casting. Journal of Advanced Manufacturing Technology, 83:711–720
6. Suprapto W, Suharno B, Soedarsono JW, Priadi D (2011) Analytical and Experimental Models of Porosity Formation of Duralumin Cast in Vacuum Casting System. Advanced Materials Research, 271:76–83.
7. Jin CK, Jang CH, Kang CG (2013) Die design optimization of die casting for fabrication of fuel cell aluminum bipolar plate with micro-channel through casting simulation and experimental investigation. Advanced Materials Research, 27:2997-3003.
8. Chaus AS, Bračík M, Sahul M, Dománková M (2019) Microstructure and properties of M2 high-speed steel cast by the gravity and vacuum investment casting. Vacuum, 162:183–198.
9. Kuo CC, Liu HA, Chang CM (2020) Optimization of vacuum casting process parameters to enhance tensile strength of components using design of experiments approach. The International Journal of Advanced Manufacturing Technology, 106:3775–3785.
10. Ignaszak Z, Wojciechowski J (2020) Analysis and Validation of Database in Computer Aided Design of Jewellery Casting. Archives of Foundry Engineering, 20:9–16.
11. Lun Sin S, Dube D, Tremblay R (2006) Interfacial reactions between AZ91D magnesium alloy and plaster mould material during investment casting. Journal of Materials Science and Technology, 22:1456–1463
12. Władysiak R, Pawlak M (2009) Plaster mould casting process of AlSi11 alloy. Archives of Foundry Engineering, 9:225–232.
13. Bogdanoff T, Ghassemali E, Riestra M, Seifeddine S (2019) Prototyping of a High Pressure Die Cast Al-Si Alloy Using Plaster Mold Casting to Replicate Corresponding Mechanical Properties. In: Chesonis C. (eds) Light Metals 2019. The Minerals, Metals & Materials Series. Springer, Cham.
14. Brevick JR, Davis JW, Dincher C (1991) Towards improving the properties of plaster moulds and castings. Journal of Mechanical Engineering, 205(4).
15. Ahirrao RD, Marlapalle BG (2019) A review paper on analysis and optimization of aluminium casting parameters. International Research Journal of Engineering and Technology, 6(2):1129-1132.
16. FlowScience Flow-3D 2022R1, https://www.flow3d.com. Accessed 29 August 2022
17. Wu S, Li C, Mao C, Zhang J, Yang E, Mi G (2018) Process Design and Optimization of Thin-walled Shell Components by Gypsum Mould Investment Casting. Special Casting and Nonferrous Alloys, 38(3):332–335.
18. Rahmanian I, Wang YC (2012) A combined experimental and numerical method for extractingtemperature-dependent thermal conductivity of gypsum boards. Construction and Building Materials, 26(1):707-722.
19. Anastasiou KS (2002) Optimization of the aluminium die casting process based on the Taguchi method. Journal of Engineering Manufacture, 216(7).

Part VIII
Electrode Technology for Aluminum Production

Partial Replacement of Coke with Biocoke: Effect of Biocoke Production Temperature on Anode Quality

Belkacem Amara, Duygu Kocaefe, Yasar Kocaefe, Jules Côté, and André Gilbert

Abstract

Carbon anodes used in aluminum electrolysis are made of petroleum coke, coal tar pitch, recycled anodes, and butts. In order to reduce emissions of greenhouse gases (GHG), a group of researchers succeeded to replace a part of petroleum coke with biocoke modified using additives. Biocoke is obtained by the pyrolysis of wood chips at a temperature of about 1100 ºC, which is similar to the maximum temperature reached during anode baking. The focus of this study is the effect of biocoke production temperature on the anode quality. In this study, biocoke was produced at four different temperatures (600, 750, 950, and 1100 ºC). Anodes were produced without biocoke (standard anode) as well as with modified and non-modified biocokes. Then the anodes were characterized by measuring their densities, electrical resistivities, air and CO_2 reactivities, and bending strengths. The effect of biocoke production temperature on anode quality was determined by comparing the anode properties.

Keywords

Biocoke production • Biocoke calcination temperature • Biocoke modification • Anode properties • Petroleum coke

Introduction

The aluminum industry is one of Canada's main industries and 90% of aluminum is produced in Quebec. Canada is the fifth largest producer of aluminum in the world. But, it ranks first in the production of the greenest aluminum because of the use of renewable green energy (hydroelectric power) instead of fossil fuels. Nevertheless, the aluminum industry continues its efforts to achieve sustainable production and reduce greenhouse gas (GHG) emissions.

Aluminum is produced using an electrolytic process in which carbon anodes are regularly consumed [1]. Carbon anodes contain approximately 65% calcined petroleum coke. The production of this coke involves further treatment of the heaviest distillation product of petroleum. The calcination of green petroleum coke is also necessary to remove volatiles. Anode consumption during electrolysis results in the release of carbon as carbon dioxide, a GHG. If the calcined petroleum coke can be replaced (even partially) by another carbon-containing material coming from a renewable source, then the CO_2 produced is considered as non-contributing to the atmospheric carbon.

Improving anode quality is one of the priorities of the aluminum industry since this reduces production cost, energy and carbon consumption, and greenhouse gas (GHG) emissions, and also increases the production. Anode quality is strongly linked to those of raw materials [1, 2]. But, the quality of anode raw materials (petroleum coke and coal tar pitch) is decreasing. This requires the identification of new raw materials or the improvement of existing ones. One possibility is replacing a part of petroleum coke with biocoke, which is a renewable material [3, 4, 5, 6, 7, 8, 9]. The use of biocoke will reduce the GHG emissions of aluminum industry while adding value to the residue of wood industry, helping diversify its operations. Biocoke is known to be a reliable, low-cost, and low-sulphur material. The use of biocoke in the carbon anodes will make them more environment-friendly due to the reduction in GHG emissions.

B. Amara · D. Kocaefe (✉) · Y. Kocaefe
UQAC Research Chair On Industrial Materials (CHIMI), University of Quebec at Chicoutimi, 555, Boulevard de l'Université, Chicoutimi, QC G7H 2B1, Canada
e-mail: Duygu_Kocaefe@uqac.ca

J. Côté
Aluminerie Alouette, 400, Chemin de La Pointe-Noire, C.P. 1650, Sept-Îles, QC G4R 5M9, Canada

A. Gilbert
Boisaco Inc, 648, Chemin du Moulin, C.P. 250, Sacré-Cœur, QC G0T 1Y0, Canada

S. Broek (ed.), *Light Metals 2023*, The Minerals, Metals & Materials Series, https://doi.org/10.1007/978-3-031-22532-1_146

Since the trees consume CO_2 during photosynthesis producing O_2, CO_2 produced during the electrolysis by the biocoke does not add to GHG emissions.

Several researchers have studied the impact of using biocoke on the anode quality [3, 4, 5, 6, 7, 8, 9]. They observed a deterioration in all anode properties, and this was explained by the low density (high porosity) and the poor mechanical properties of biocoke. In one study, the carbon group at the University of Quebec at Chicoutimi (UQAC) showed that it is possible to replace 3% of the finest coke fraction with biocoke without degrading the anode quality [6]. But, in a second study, the same group found that this was not the case when another type of coke was used. In general, the reason for the poor quality of the anodes obtained when part of the coke is replaced with biocoke is the weak interactions (wettability) between pitch and biocoke due to the low concentration of functional groups.

A biocoke surface rich in heteroatoms promotes better wettability. Good wettability permits pitch to enter into the pores of the coke and the voids between the coke particles resulting in denser anodes. During wood calcination, many chemical functional groups are removed as volatile matter. This removal depletes the chemical groups present on the biocoke surface. The surface functional groups promote bonding and thus better interactions between biocoke and pitch. In order to improve the interaction of biocoke with pitch, biocoke surface was modified by the authors of the current study [8]. This enriched the functional groups on the biocoke surface, consequently improved biocoke–pitch interactions. Thus, the anodes made with modified biocoke had properties similar to those of the standard anode (anode without biocoke). The same authors also studied the effect of the coke type and the additive type on the anode properties [10, 11].

In all of the above studies, the final temperature of the calcination of biocoke was close to the anode baking temperature. In this article, the focus is on the final temperature of biocoke production. If biocoke produced at lower temperatures can be used to replace coke, the energy requirement for biocoke production will be reduced.

Materials and Methods

The same raw materials were used to produce all anodes. All raw materials (petroleum coke, coal tar pitch, green and baked anode rejects, and butts) for anode production were obtained from the industrial partner Aluminerie Alouette.

The wood chips used to produce biocoke were provided by the industrial partner Boisaco Inc. Biocokes were produced by heating them to different final temperatures (600, 760, 950, and 1100 °C) in the carbon laboratory of UQAC. These anodes are identified as T600, T760, T980, and T1100

in the figures. The percentage of biocoke used was 3%. The biocoke modification was performed using a chemical additive which belongs to the alkyl–phenyl–aldehyde class (Alfa Aesar). This additive is chosen because it is non-toxic, low-cost, and does not contaminate either anode or metal. The percentage of additive used was also 3% which was chosen based on the previous studies.

Laboratory Anode Production

In this study, laboratory-scale pilot anodes were produced. Coke/biocoke, anode butts, and rejected anodes were pre-heated separately and mixed with pitch in an intensive mixer to form an anode paste. Then the anode paste was vibro-compacted to obtain green anodes. Four cores with dimensions of 130 mm long and 50 mm diameter were taken from each anode. After, the cores were baked under the conditions which are similar to those of the industrial anodes up to 1100 °C.

Nine anodes were produced. The anode produced without biocoke was used as the reference. Eight of the nine anodes contain 3% unmodified biocoke or biocoke modified with 3% additive. Modified biocoke is designated as MOD and unmodified biocoke is designated as UNMOD in tables and figures. Table 1 shows the composition of each anode.

Density

The density measurement involves a non-destructive test for anode samples. This test is performed according to the standard ASTM D5502-00 [12]. The principle of this method is to determine the volume "V" and mass "m" of cylindrical samples (50 mm diameter and 130 mm height). Then, the density is calculated from these measurements.

The mean diameter is determined from eight diameters measured at eight different positions on the sample. The average height is calculated from four heights measured at different positions. The sample volume is calculated using Eq. 1. The anode density is calculated by dividing the sample mass by its volume. The anode density is the average of the densities of the four samples coming from the same anode.

$$V = \pi \cdot h \cdot \frac{d^2}{4} \qquad (1)$$

V   Sample volume (cm^3).
d   Average sample diameter (cm).
h   Average sample height (cm).
ρ Sample density (g/cm^3).

Table 1 Anode composition

Anode	Biocoke (%)	Additive (%)
Std*	0	0
T600-UNMOD	3	0
T600-MOD	3	3
T760-UNMOD	3	0
T760-MOD	3	3
T950-UNMOD	3	0
T950- MOD	3	3
T1100-UNMOD	3	0
T1100-MOD	3	3

where

$$\rho = \frac{m}{V} \tag{2}$$

m   Mass of the sample (g)

Electrical Resistivity

To measure the electrical resistivity of green and baked anodes, the ASTM D6120-97 standard [12] was used. The voltage was measured at different positions of the sample in order to determine the average voltage generated by the passage of an electrical current of 1 A. Equation 3 is used to calculate electrical resistivity. The electrical resistivity of the anode is the average resistivity of the four cores of the same anode.

$$RE = \frac{V \times A}{I \times L}. \tag{3}$$

RE   Electrical resistivity ($\mu\Omega$.m).
V   Voltage drop (mV).
A   Surface area perpendicular to the direction of electric current (I) passing through the anode sample (cm^2).
I   Electric current (A).
L   Distance between the two voltage drop measurement points (cm).

The surface "A" of each anode core is calculated using Eq. 4.

$$A = \frac{\pi \times d^2}{4} \tag{4}$$

d   Diameter (cm).

The diameter "d" is calculated from the eight measured diameters using Eq. 5.

$$d = \sqrt{\frac{8}{\frac{1}{d_1^2} + \frac{1}{d_2^2} + \cdots + \frac{1}{d_8^2}}} \tag{5}$$

Bending Strength

The principle of this test is based on the application of a load at the center on a cylindrical anode sample supported at two positions at the bottom. The standard used for this test is ISO 12986-1: 2014 [13]. The results of this test give an idea about the presence of cracks in the anode. If the anode has a low bending strength, this indicates that there are problems with the stability of the coke grains, mixing, and/or vibro-compaction conditions [7]. Equation 6 is used to calculate the maximum stress.

$$\sigma_{\max} = \frac{8 \cdot F_{\max} \cdot L}{\pi \cdot D^3} \tag{6}$$

$\sigma_{\max}$ Maximum bending stress [MPa].
$F_{\max}$ Maximum applied load [N].
L   Length between supports [mm].
d   Sample diameter [mm].

Air and CO$_2$ Reactivities

Air or CO$_2$ reactivity tests are destructive and are used to evaluate the anode consumption in the presence of air or CO$_2$ for a given period of time. The dimensions of the anode samples used for air and CO$_2$ reactivities are 50 mm diameter and 50 mm height (50 mm $\times$ 50 mm). According to the standard methods D6559 [14] and D6558 [15], these tests are conducted at the temperature of 525 °C for air (3 h) and 960 °C for CO$_2$ (7 h). The air and CO$_2$ reactivities are calculated using Eq. 7.

$$TR = \frac{1000(W_0 - W_t)}{t \cdot A} \tag{7}$$

TR   Total reactivity rate (mg/cm^2 h).
W_0 Initial mass of sample (g).
W_t Final mass of the sample (g).
t   Total test time (3 h for air and 7 h for CO_2).
A   Exposed surface of the sample (cm^2).

The sample area "A" is determined from Eq. 8.

$$A = \frac{\left(\pi dh + \frac{2\pi d^2}{4}\right)}{100} \qquad (8)$$

h   Sample height (mm).
d   Sample diameter (mm).

Results and Discussion

Density

The densities of the baked anodes produced with modified and unmodified biocoke as well as that of the standard anode, which does not contain any biocoke, are presented in Fig. 1. The results show that the densities of the anodes containing modified or unmodified biocoke are lower than the density of the standard anode, [6, 7]. Only the anode containing biocoke produced at 1100 °C and modified with 3% additive (T1100-MOD) has a density close to that of the standard anode. This can be explained by the final temperature of biocoke production. This is consistent with the findings of Huang et al. [6]. They found that biocoke prepared at a low temperature (425 °C) had a very low real density compared to that of biocoke produced at a coke calcination temperature used in industry. The biocoke modification improved the anode density compared to that of the anode containing unmodified biocoke. This shows that

the additive improved the interactions between biocoke and pitch by adding functional groups to biocoke surface. The combined effect of high-temperature biocoke calcination and biocoke chemical modification resulted in a higher density anode with a value quite close to that of the standard anode.

Electrical Resistivity

Figure 2 shows the electrical resistivities of the baked anodes for which the densities are presented in Fig. 1. It can be seen from this figure that the electrical resistivities of the anodes made with modified or unmodified biocoke are higher than that of the standard anode at the intermediate temperatures. This figure also shows that the anode made using biocoke produced at 1100 °C and modified with 3% additive (T1100-MOD) has an electrical resistivity similar to that of the standard anode. This is in an agreement with the density results of the anodes in Fig. 1. It is known that the higher the density of the anode is, the lower the electrical resistivity is [1, 2]. This is because low-density anodes have high porosity and cracks which increase the average resistivity of the anode. Again, the anode produced with biocoke calcined at high temperature followed by modification using an additive had an electrical resistivity similar to that of standard anode, which is consistent with the expectations.

Bending Strength

Bending strengths of the same anodes are presented in Fig. 3. As it can be seen in this figure, bending strengths of the anodes containing both modified and unmodified biocoke produced at temperatures lower than 1100 °C are much

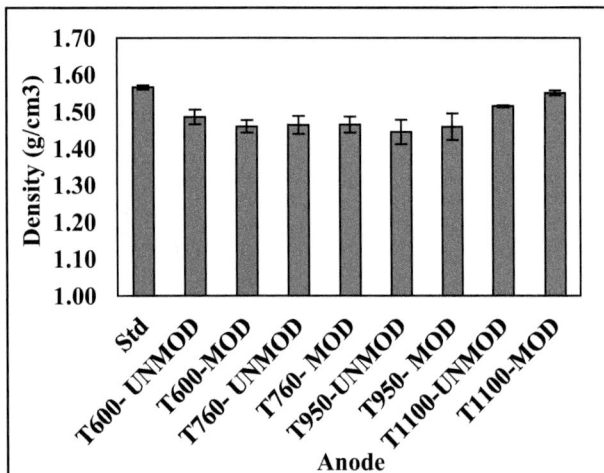

Fig. 1 Baked anode densities

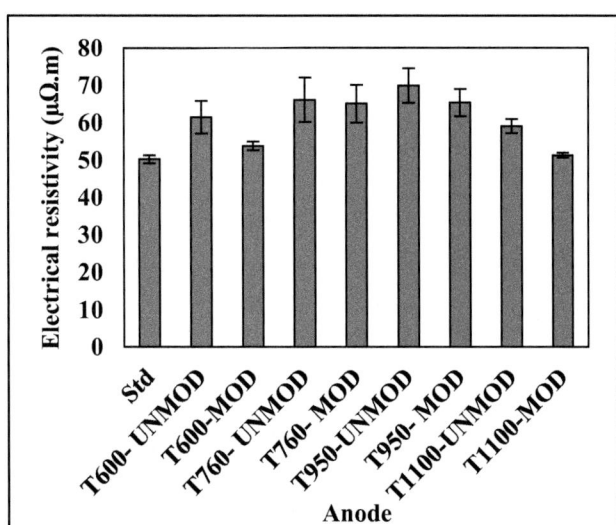

Fig. 2 Electrical resistivity of baked anodes

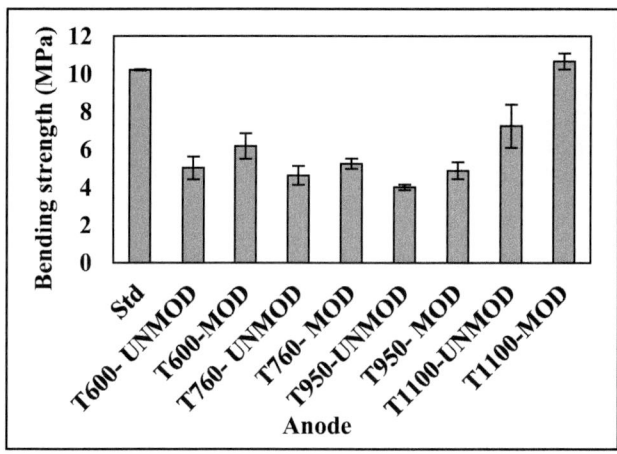

Fig. 3 Bending strength of the anodes

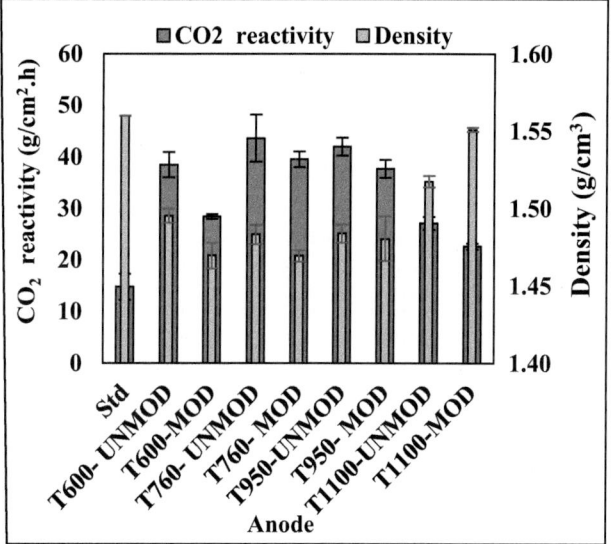

Fig. 5 CO_2 reactivity of the anodes

lower compared to the bending strength of standard anode. The anode produced with biocoke calcined at 1100 °C and modified with 3% additive (T1100-MOD) has a significantly higher bending strength than the others containing biocoke (modified or non-modified) and a slightly higher bending strength than that of the standard anode.

Air and CO₂ Reactivities

Air Reactivity

The air diffuses from the top of the electrolysis cell and reacts with the carbon anode $(C + O_2 \rightarrow CO_2)$. This causes the overconsumption of carbon, increases GHG emissions, and reduces the production [2]. The CO_2 reactivities of all anodes are presented in Fig. 4. As it can be seen from this figure, in general, the higher the density is, the lower the

reactivity is. The anode containing modified biocoke produced at 1100 °C has an air reactivity slightly lower than that of the standard anode.

CO₂ Reactivity

CO_2 produced during the electrolysis react with carbon anode $(CO_2 + C \rightarrow CO)$, decreasing the carbon available for aluminum production [2]. This also increases the GHG emissions. As shown in Fig. 5, the CO_2 reactivities of all anodes containing biocoke (modified or unmodified) are higher than that of the standard anode. Similar to the tendencies observed for electrical resistivity, CO_2 reactivity increases with decreasing density (increasing porosity) and bending strength (cracking). As reported in the literature, CO_2 reactivity is a diffusion-controlled reaction. It increases with increasing porosity since CO_2 can diffuse more easily into the anode and react with carbon. [2]. These results also show that only the anode produced with biocoke at 1100 °C and modified with the 3% additive (T1100-MOD) has a CO_2 reactivity closest to that of the standard anode.

Conclusions

Deterioration of the anode raw material quality and GHG emissions are among the major preoccupations of aluminium industry. Partially replacing coke with biocoke is one possible solution to reduce GHG emitted during aluminum production. Biocoke can also be considered as an alternative raw material. However, its use above a certain level deteriorates the anode properties. In this study, the focus was on

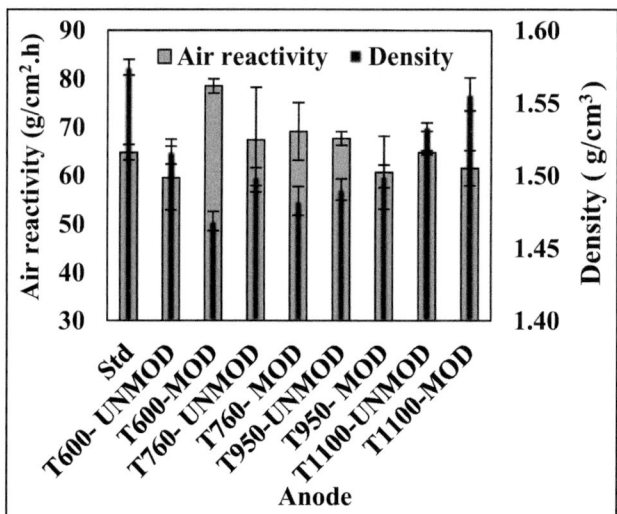

Fig. 4 Air reactivity of the anodes

the final calcination temperature during the biocoke preparation. Biocoke was produced by calcining wood at four different temperatures. Only one type of wood and one type of additive were used. The aim was to determine how biocoke produced at temperatures lower than the final anode baking temperature would affect the anode properties. If the effect is not significant, producing biocoke at lower temperatures would result in energy savings.

The replacement of 3% petroleum coke by unmodified and modified biocoke affects adversely the anode properties except at the highest calcination temperature. The anode properties were closest to those of the standard anode at this temperature when modified biocoke was used. The aim is to replace a part of the petroleum coke with biocoke without deteriorating the anode properties. The utilization of modified biocoke in anode production affects different anode properties differently. To decide if this replacement is feasible or not, other factors such as the GHG emission reduction due to the use of biocoke and the calcination of less petroleum coke, the utilization of a cheaper raw material, potential savings due to carbon tax, etc., should be considered.

Acknowledgements The financial and technical support of the Fonds de recherche du Québec-Nature et technologie (FRQNT), Aluminerie Alouette (AAI), Boisaco Inc., the University of Quebec at Chicoutimi (UQAC), and the Aluminium Research Center (REGAL) is greatly appreciated.

References

1. Charette A., Kocaefe Y. S., and Kocaefe D., Le carbone dans l'industrie de l'aluminium. Chicoutimi (Québec): Les presses de l'aluminium, 2012.

2. Hume SM. Anode reactivity: Influence of raw material properties, [2nd ed.]. ed. 1999, Sierre [Suisse]: R & D Carbon Ltd. xi, 433 p.

3. Huang X., Kocaefe D., Kocaefe Y., and Bhattacharyay D. (2016) *Wettability of bio-coke by coal tar pitch for its use in carbon anodes.* Colloids and Surfaces A: Physicochemical and Engineering Aspects vol. 490, p. 133-144.

4. Hussein A. (2014) *A bio-coke for anode production and the manufacturing method thereof.* PhD. Thesis, Université Laval, Québec, Canada.

5. Huang X., Kocaefe D., Kocaefe Y., and Bhattacharyay D. (2016) *Interaction of bio-coke with different coal tar pitches.* Fuel vol. 179, p. 179-192.

6. Huang X., Kocaefe D. and Kocaefe Y. (2018) *Utilization of biocoke as raw material for carbon anode production.* Energy & Fuels p. 1–20.

7. Hussein A., Fafard M., Ziegler D. and Alamdari H. (2017) *Effects of charcoal addition on the properties of carbon anodes.* Metals 7 (3), p.1-9.

8. Amara, B., Kocaefe, D., Faouzi, F-E., Kocaefe, Y., Bhattacharyay, D., Côté, J., Gilbert, A. (2021) *Modification of biocoke destined for the fabrication of anodes used in primary aluminum production.* Fuel 304, article 121352.

9. Elkasabi, Y., Yetunde Omolayo, Y., Spatari, S. (2021) *Continuous calcination of biocoke/petcoke blends in a rotary tube furnace.* ACS Sustainable Chem. *Eng.*, 9, p. 695–703

10. Amara, B, Kocaefe, D., Kocaefe, Y., Bhattacharyay, D., Côté, J., Gilbert, A. (2022) *Partial replacement of petroleum coke with modified biocoke during production of anodes used in aluminum industry: Effect of additive type.* Applied Sciences, published online, 12(7), https://doi.org/10.3390/app12073426.

11. Amara. B., Kocaefe, D., Kocaefe, Y., Bhattacharyay, D., Côté, J., Gilbert, A. (2022) *Effect of Coke Type on the Partial Replacement of Coke with Modified Biocoke in Anodes Used in Primary Aluminum Production.* Light Metals p. 818–825.

12. An American National Standard, ASTM D6120–97 (2007) *Standard test method for electrical resistivity of anode and cathode carbon material at room temperature.* 2010.

13. The International organization for standardization, ISO 12986–1:2014: *Carbonaceous materials used in the production of aluminium-Prebaked anodes and cathode blocks-Part 1: Determination of bending/shear strength by the three-point method.* 2014.

14. An American National Standard, ASTM D6559–00 (Reapproved 2005), *Standard test method for determination of TGA air reactivity of baked carbon anodes and cathode blocks.* 2010.

15. An American National Standard, ASTM D6558–00 (Reapproved 2005), *Standard test method for determination of TGA CO_2 reactivity of baked carbon anodes and cathode blocks.* 2010.

16. An American National Standard, ASTM D5502–00 (2005) *Standard test methodes for apparent density by physical measurements of manufactured anode and cathode carbon used by the aluminium industry.* 2010.

Method for Calcined Petroleum Coke Evaluation to Improve the Anode Quality

Sheetal Gupta, Suwarna Mahajan, Amit Gupta, and Vilas Tathavadkar

Abstract

Worldwide, the quality of petroleum coke, used for making carbon anode for aluminium smelters, is getting degraded with an increase in the impurities and higher presence of shot coke particles. Earlier studies reported that some of the anode problems like slabbing and vertical cracking are linked to the thermal shock resistance of coke material, which is majorly influenced by the structure of calcined petroleum coke (CPC). Therefore, a detailed method of optical microscopic analysis was established and successfully applied to evaluate CPC materials. The analysis of most of the samples was found to have the presence of sponge and isotropic structure. The findings of the optical microscopy have been correlated with the visual identification of shot, sponge and needle coke. The visual identification of different coke particles enables its integration with digital image processing techniques for ease of quantification of specific coke material.

Keywords

Calcined petroleum coke • Carbon anode • Microstructure • Aluminium smelter

Introduction

The calcined petroleum coke (CPC) is the main ingredient for manufacturing carbon anode. The quality and its grade play an important role in anode performance during smelter operation. CPC is produced by calcining the green petroleum coke (GPC) at temperatures greater than 1200 °C. GPC is a by-product of the petroleum refineries, and its properties such as chemical composition and structures are primarily influenced by the quality of crude oil feed to the delayed coker. A highly aromatic feedstock yields needle coke and a highly asphaltene feedstock produces shot coke. In between are other heavy hydrocarbons which, in the certain proportion with aromatics and asphaltenes, produce the sponge cokes [1].

Needle coke is a premium quality material with very low impurity levels. It has a low coefficient of thermal expansion (CTE) value less than 2.0×10^{-6}/K, making it the material of choice to produce the graphite electrodes used in steel manufacturing through electric arc furnaces. Sponge coke is the preferred material for carbon anode used in aluminium production and its CTE is typically in the range of 3.5–4.8×10^{-6}/K. Shot coke has a high level of sulfur and metal impurities with CTE value greater than 5.5×10^{-6}/K and is used as a fuel in different manufacturing operations [2].

The aluminum producers try to prevent the use of highly isotropic cokes, specifically shot cokes, because they have high CTE, low open porosities, and high level of impurities. Anodes made with these materials are more prone to thermal shock cracking such as slabbing (horizontal cracks) and vertical transversal cracks, during the rapid heat-up in aluminum electrolysis cells. Anodes can also suffer from lower mechanical strength and dusting problems during cell operation due to low levels of open macro-porosity for pitch penetration. This reduces the ability of pitch to interlock and bond the structure together during carbonization [3].

Hence, CPC structural information is crucial for assessing the anode quality. A routine characterization such as chemical composition and physical properties (densities, crystallinity, hardgrove grindability index, etc.) of CPC are being conducted at plant. However, these are not sufficient because they don't provide information on the thermal properties. To gain that insight, either we need to prepare lab scale anodes and characterize for CTE measurement or perform optical microscopic analysis.

S. Gupta (✉) · S. Mahajan · A. Gupta · V. Tathavadkar
Aditya Birla Science & Technology Company Pvt. Ltd., Navi Mumbai, Maharashtra, India
e-mail: sheetal.gupta@adityabirla.com

© The Minerals, Metals & Materials Society 2023
S. Broek (ed.), *Light Metals 2023*, The Minerals, Metals & Materials Series,
https://doi.org/10.1007/978-3-031-22532-1_147

Past studies [4–10] have reported optical microscopic analysis of CPC materials, however, little information is available about the techniques of sample preparation and the quantification of the structures. It has been also observed that obtaining a representative sample from a bulk is challenging for microscopic analysis as it requires a very small sample size. This limitation can be overcome using multiple samples but even then, microscopic analysis is very subjective in nature.

Therefore, this study focused on developing an easy way of quantifying specific CPC structure. To build on this, several microscopic analyses of CPC materials were performed and correlated them with their physical appearance. This has helped in identifying the structural difference of CPC materials through naked eye, thereby exploring the possibility of digital imaging techniques utilization with the developed knowledge of visual identification. After establishing this, a machine learning-based model would be developed and integrated with plant process.

Microscopic Analysis of CPC

A detailed sample preparation method starting from collecting a representative sample of CPC to molding and polishing has been described in the following section. The representative sample of CPC is encapsulated in resin by using vacuum molding technique. The molded sample is grounded and polished to flat surface for examination under polarized reflected light by optical microscope.

Sample Preparation

CPC materials come in a lot of 3200–3600 MT at plant through railway wagons, packed in bags of approx. 1.5 MT each. To open each bag and take out samples is challenging. Therefore, wagon was divided into four zones consisting of 800–900 MT material. From each zone, 30–40 bags were selected randomly and a sample weighing 1–2 kg was collected. This way 40–50 kg sample was collected from each zone of the wagon. Subsequently, Coning and Quartering (C&Q) method was used to bring the sample size to 100–200 gm, as a representative sample for microstructural analysis.

This CPC sample was selected (Fig. 1a) and placed in a mounting cup in such a way that it covers most of the inside area of a cup as shown in Fig. 1b. Epoxy resin-hardener mixture was poured in the cup to cover the sample completely and allowed some time to cure. In the case of porous material where resin must penetrate the sample, vacuum chamber was used to encapsulate as shown in Fig. 1c. Otherwise, resin would foam out of the cup during curing. The purpose of the present step is to degas the system and CPC sample to achieve complete penetration of any porosity in the sample. In case of CPC sample, vacuum chamber was operated at 0.3 bar pressure for 30 min to remove the entrapped air. Degassed encapsulated specimen from vacuum chamber was taken out and allowed to cure overnight at ambient temperature and pressure followed by sample polishing.

A polishing sequence involves grinding at around 200 rpm on a 320-grit silicon carbide disk exposing the CPC particles. This is followed by grinding on 500, 1000, and 2000 grit silicon carbide disk at same rpm, using water both as a lubricant and a coolant as shown in Fig. 1d. The polishing cloth was used for fine polishing by applying 3 μm high-performance diamond-based solution. The surface obtained after grinding and polishing, ensured to meet the following requirements:

- Enough material was removed to expose CPC surface.
- After each grinding or polishing step, sample was cleaned under running water while wiping gently with wet cotton wool.
- The surface should be free of grinding and polishing compounds.
- The surface should be free of scratches when examined at a magnification of 400X or 500X.

Final polished sample as shown in Fig. 2 is ready to be observed under optical microscope.

Fig. 1 **a** Calcined petroleum coke, **b** Cup mold with 15–25 mm size CPC sample, **c** Vacuum impregnated equipment, **d** Sample polishing machine

Fig. 2 Polished CPC sample

For microstructural analysis, any polarizing microscope with the capability of observing by reflected light can be used. The polarizer can be of the Nicol prism type. All optical components shall be of a quality to permit examination of the dry specimen at magnifications up to 400X to 500X under crossed polarizers. The analyzer should be oriented at 90° angle with respect to the polarizer for cross-polarizer examination. The optical microscope used for this study was from 'Carl Zeiss imager.A2m' as shown in Fig. 3.

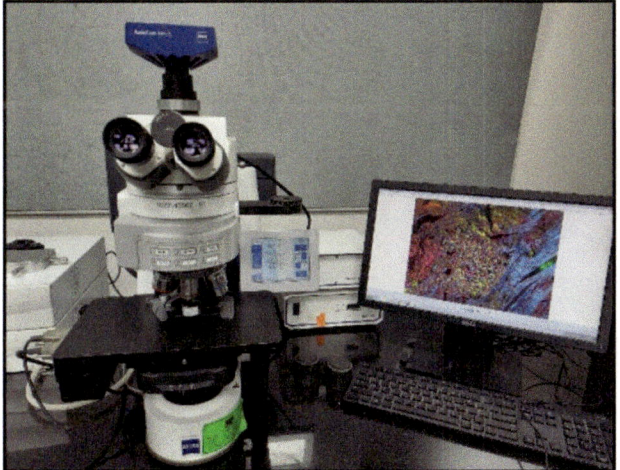

Fig. 3 Optical microscope coupled with a computer

Microstructural Analysis

The selection of coke material is a critical factor while controlling the anode quality because of the anisotropy and isotropy of CPC, which play a significant role in determining the strength, reactivity, and thermal properties of the anode. Hence, microscopic analysis of CPC is essential.

In this study, the optical microstructure analysis was performed while looking at polished cross-sections of CPC under polarized light. The working distance with 200X magnification was chosen for all microstructural analysis so that the entire sample can be evaluated with little to no refocusing. During the microstructure analysis, three kinds of structures were observed. Figure 4 shows the optical microstructures of needle, sponge, and shot coke particles whereas Fig. 5 shows a mix of all three structures present in one particle.

Figure 4a shows the anisotropic microstructure of needle coke particles which has large laminar porosities that are wider in the middle and tapering to a point toward the ends. Needle coke particles appear as a cluster of aligned needles while seeing through naked eyes. These particles are of coarser texture with layered structure and their properties change with orientations.

Figure 4b shows mixed microstructure of sponge coke particles which has irregular porosities and heterogeneous structure with a mixture of coarse and fine textures. Sponge coke particles have relatively high macro porosities and the pores are evident from visual examination of the coke. Sometimes these macro porosities are not visible due to the agglomeration of fine particles on the surface of coarser particles but can be seen after little hammering or crushing. The macro porosities in sponge coke allow good pitch penetration during mixing and lead to a mechanically strong interlocking structure that develops after anode baking.

Figure 4c shows isotropic microstructure of shot coke particles which is highly isotropic texture sometimes referred as a granular texture [2]. Isotropic structures tend to have irregular fine pores and exhibit similar properties in all directions. Shot coke particles look like small round spherical balls loosely bound together. Mostly, they agglomerate into ostrich egg-sized pieces. Shot coke particles up to 1 mm are easy to identify due to their unique shape and surface characteristics. These particles do not show any macro porosities and make a dense particle when agglomerated. Shot coke particles are also harder as compared to other kinds of coke particles.

Numerous numbers of CPC samples were prepared and analyzed under an optical microscope. It was observed that most of the samples were dominated by either sponge or

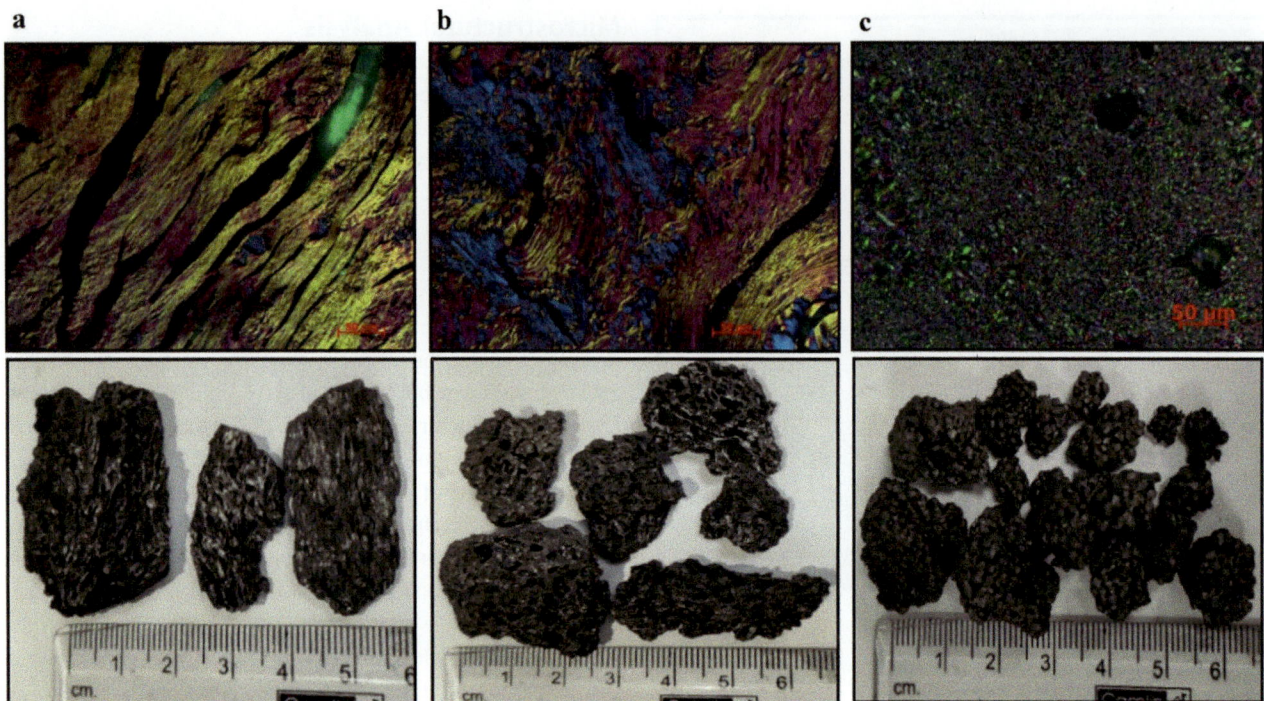

Fig. 4 **a** Needle coke (Anisotropic), **b** Sponge coke (Mixed), **c** Shot coke (isotropic) microstructures and their physical appearance

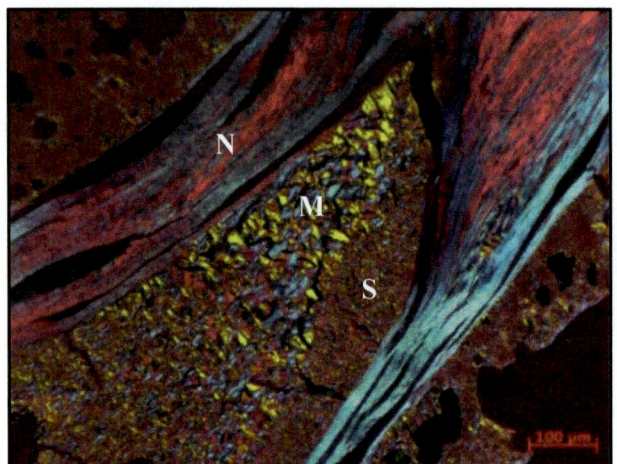

Fig. 5 Microstructure consists of needle (N), shot/isotropic (S), and mesophase (M)

During this analysis, it was observed that particles of size more than 5 mm are easy to be distinguished by naked eyes for their structure such as needle, sponge, and shot coke. Previous studies [5, 6] have described some methods to quantify CPC microstructure by counting specific sites for microstructure, measuring the size and shape of coke pores, and quantifying differences in reflected light intensity. But all these methods are either high time consuming or practically not possible at plant. In this study, microstructural analyses of the CPC material have been successfully correlated with the visual identification method which is less sophisticated as compared to cited work but enables a digital way of speedy quantification.

Digital Image Processing

In majority of the samples, the surface characteristics of CPC particle are identifiable through naked eyes, which facilitates the integration of digital image processing (DIP) techniques for faster quantification of specific coke material. The first step of DIP is to convert multicolor image into binary color image for ease of computational analysis. During this conversion, the higher light-reflecting surface would be shown as light blue colour, and a lesser light-reflecting surface as dark blue colour. The images were processed in MATLAB software. From Fig. 6, it can be observed that coke porosities are in dark blue color and solid/spherical/plateau

isotropic structure. It is also very likely that CPC will consist of a mixture of the anisotropic, isotropic, and mesophase (neither isotropic nor anisotropic) structure. Figure 5 shows the microstructural image captured at a lower magnification of 100X, so that all three phases are clearly visible in one frame.

Typical sponge cokes contain more mesophase microstructure than needle or isotropic cokes. However, a mix of each of the three microstructures can be found in most sponge cokes materials.

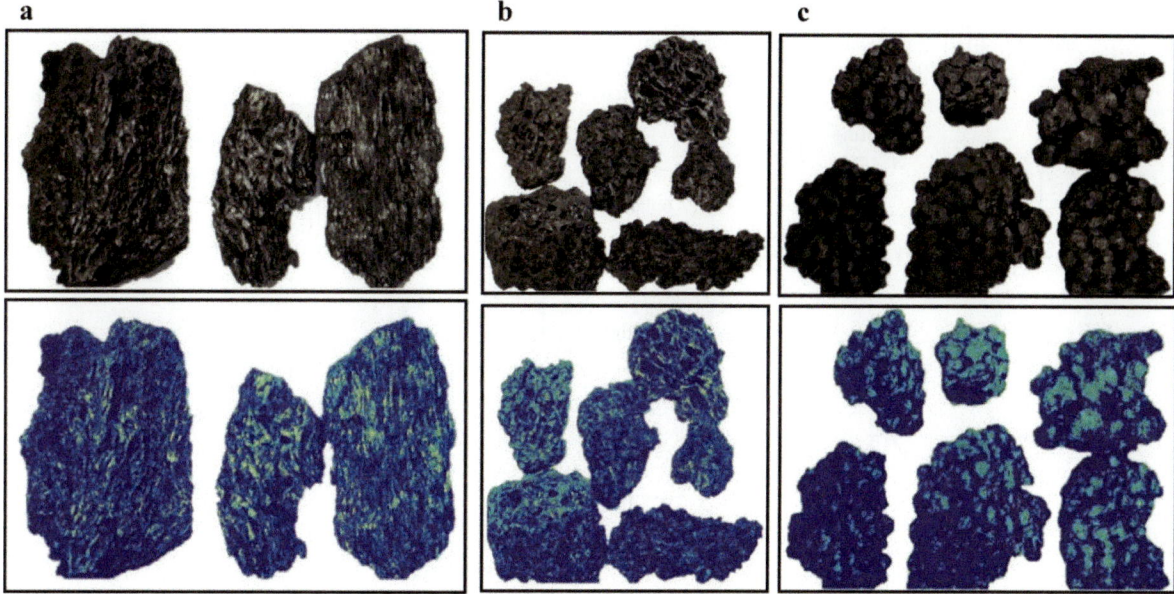

Fig. 6 Original and processed images of **a** Needle, **b** Sponge, and **c** Shot coke particles

surfaces are in light blue color since they reflect more light as compared to the porous/cavity surfaces. The following observations were made from the processed image of coke particles, which could be useful in developing a machine learning model for ease of identification.

- In case of needle coke particles shown in Fig. 6a, the presence of light blue color lines can be found surrounded by dark blue region. The aspect ratio (length to width ratio) of the light blue color region was found to be significantly higher than one.
- In case of sponge coke particles shown in Fig. 6b, the presence of dark blue region is surrounded by light blue color lines/curves representing the porous structure. In this case the aspect ratio of the dark blue color region was found close to one.
- Whereas, in case of shot coke particles shown in Fig. 6c, the presence of light blue region found to have more rounded shapes, representing the cluster of small spherical balls. In this case, the aspect ratio of the light blue color region was found close to one.

Applying the above method, the image analysis would help in identifying the type of particles in the total imaging area, which could potentially support the quantification of specific coke structure.

In the DIP analysis, there are few limitations which could adversely impact the analysis.

- The light source used while taking images can affect the identification of these structures, hence multiple diffused light sources would be advantageous for effective identification.
- The second limitation of this technique could be, in case of a sharp differentiation is missing through surface characteristic of particles. For example, in Fig. 7a, the fine dust particles agglomerating on the surface of porous structure of sponge coke particles, looks similar to Fig. 7b, where fine dust particles are covering the cluster of small balls of shot coke particles.

It can be overcome by taking images at a stage where CPC particles have been crushed/broken into a smaller

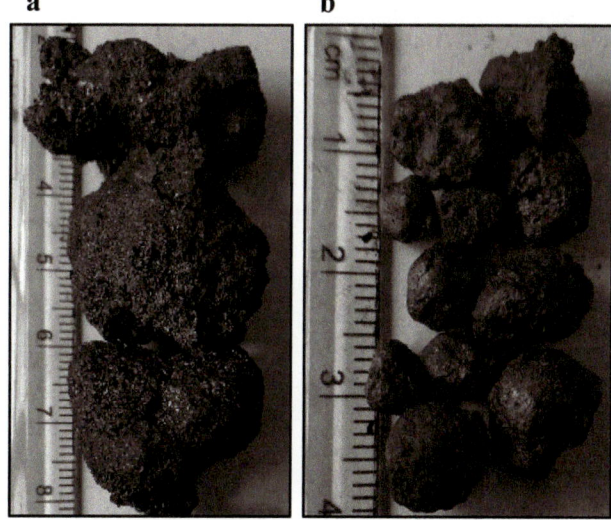

Fig. 7 a Sponge coke **b** Shot coke particles covered with fine dust particles

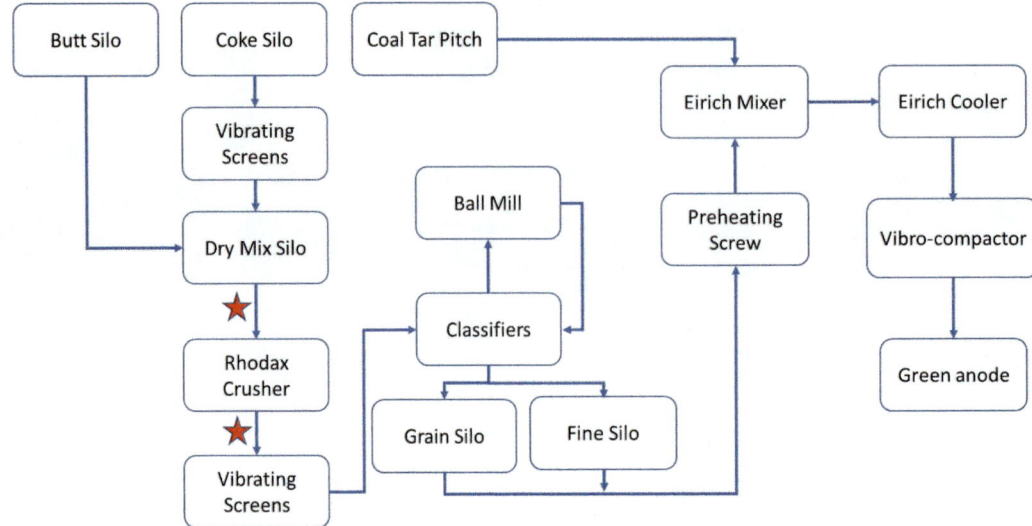

Fig. 8 Green anode plant process flow diagram

fraction. This would help in exposing the actual surface characteristic of sponge and shot coke particles.

For potential integration of DIP in the green anode manufacturing process, the availability of clear and noise-free images is crucial. Figure 8 shows the green anode plant process flow sheet along with the potential locations for placing digital imaging device (highlighted with stars).

To automate this analysis, the images from digital imaging device could be used to train the machine learning model for ease of identification and quantification of specific coke structure. This would be a future focus area of the development.

Conclusion

Sponge coke is the preferred material for making anodes for aluminium smelter, in comparison to needle/shot coke. The change in CPC quality over time, lead the focus on microscopic evaluation, which provides vital information about the coke structure. The microstructural analysis is crucial for identifying the presence of undesirable isotropic structure. This is very useful information to improve the anode quality and thereby performance of an aluminium smelter. The surface characteristics of CPC particles are correlated with microstructural analysis, which helped in faster image analysis. The digital image processing coupled with a machine learning model could have high potential for fast quantification of specific coke structure during the anode manufacturing process.

Acknowledgements The authors thank their colleagues in HINDALCO and Aditya Birla Science & Technology Company Pvt. Ltd. (ABSTCPL) for their support in conducting this study.

References

1. Les Edwards, Franz Vogt, Mike Robinette, (2009) Use of shot coke as an anode raw material. Paper presented at the TMS annual Meeting 2009
2. Les Edwards (2015) The history and future challenges of calcined petroleum coke production and use in aluminum smelting, Rain CII Carbon, JOM 67(2): 308-321
3. Les Edwards, Franz Vogt, Richard Love (2007) Electrode useful for molten salt electrolysis of aluminium oxide to aluminium. US patent 2007/0068800 A1, 29 March 2007.
4. J. A. Ross (2000) The use of petrographic techniques for evaluation of raw material and process changes in an aluminum smelter. Paper presented at TMS annual meeting 2000
5. Keith Neyrey, Les Edwards, J. A. Ross, Franz Vogt (2005) A tool for predicting anode performance of non-traditional calcined cokes. Paper presented at TMS annual meeting 2005
6. Andris Innus, Alain Jomphe, Hans Darmstadt (2013) A method for the rapid characterization of petroleum coke microstructure using polarized light microscopy. Paper presented at TMS annual meeting 2013
7. F. Vogt, R. Tonti, M. Hunt, and L. Edwards (2004) A preview of anode coke quality in 2007. Paper presented at TMS annual meeting 2004
8. M. Meier, W.K. Fischer, and R. Perruchoud (1994) Thermal shock of anodes – A solved problem?. Paper presented at TMS annual meeting 1994
9. S. Rorvik, M. Aanvik, M. Sorlie, H. Oye (2000) Characterization of optical texture in cokes by image analysis. Paper presented at TMS annual meeting 2000
10. Paul J. Rhedey (1977) Refinery feedstocks, coke structure and aluminium cell anodes. Paper published at TMS 1977

Influence of Crusher Type and Particle Shape on the Bulk Density of Blended Shaft and Hearth Calcined Anode Grade Petroleum Coke

Howard Childs, Mike Davidson, and Barry Sadler

Abstract

The bulk density of anode grade petroleum coke is an important property for aluminum smelting and has been extensively studied. This includes previous work by BP showing that particle shape was modified by different crushing technologies, and that changes in shape (i.e. sphericity and convexity) influenced coke bulk density. These laboratory studies indicated that the impact of particle shape on the Vibrated Bulk Density (VBD) of a range of cokes and butts was of similar magnitude to that of porosity. In ongoing customer support efforts, this work has been extended to study whether modifying particle shape can improve the VBD of blends of cokes calcined using different technologies—shaft and rotary hearth kilns. The results of this study are presented and the potential to enhance the VBD of shaft/non-shaft calcined coke blends by modifying particle shape with different crusher types determined at a laboratory scale.

Keywords

Carbon anodes • Aluminum smelting • Petroleum coke • Particle shape • Shaft coke • Blending • Crushing technology • Bulk density

H. Childs
BP Coke, Huntington Beach, CA, USA
e-mail: howard.childs@bp.com

M. Davidson
BP, San Diego, CA, USA
e-mail: mike.davidson@bp.com

B. Sadler (✉)
Net Carbon Consulting Pty Ltd, Melbourne, Australia
e-mail: barry.sadler@bigpond.com.au

Introduction

A key property of the carbon anodes used in aluminium smelting is baked anode geometric (or Apparent) Density (BAD), with plants generally aiming to maximise the BAD. As higher Green anode Apparent Densities (GAD's) normally lead to better BAD's, increasing GAD is a common objective of smelter green anode plants. One of the accepted ways of achieving this objective is to use higher Vibrated Bulk Density (VBD) cokes, such as some cokes calcined well in shaft calciners. Blending these cokes with others calcined in non-shaft calciners such as rotary kilns or hearth furnaces, is now quite widely practised in carbon plants.

BP Coke has a long established technical support program in which collaborative projects with Customers are aimed at maximising plant anode quality by better using the potential of available cokes. The objective of a key project over several years has been to gain a better understanding of how fundamental coke properties affect coke bulk density. An improved understanding of the influence of coke parameters such as the following may then lead to opportunities to maximise plant anode quality and performance:

- Porosity (Open and closed),
- Real Density (RD),
- Particle size and size distribution, and
- Particle shape

To date in this project, results have been published that showed how the shape of a range of anode filler materials (Including several cokes from different calcining technologies and plant butts) could be modified by using different laboratory crushing technologies, that changing particle shape parameters (i.e. sphericity and convexity, see definitions in [1]) has a significant influence on coke bulk density, and that this impact (within the range of materials tested) is of a similar magnitude as the influence of open porosity [1–3]. In this paper, the project work has been extended further to look at whether the ability to modify particle shape and hence packing has the potential to improve the bulk density of shaft and non-shaft calcined coke blends.

Background

Previous Work

A literature review of prior studies on the impact of coke parameters on bulk density and the previous work in this project on modifying particle shape and packing of a range of filler materials through different crushing technologies have been summarised in previous papers [1–3]. Laboratory "impact" type crushers (Plant equivalents would be impact crushers or hammer mills) were found to produce more regular shaped, less rough surfaced particles that packed better and gave a higher bulk density than crushers with a more "shear" like crushing action (e.g. disc mills or roll crushers).

The approach used in this project to clarify the impact of laboratory crushing/shape on VBD has been to screen Run of Kiln (RoK) coke to remove the fraction tested for shape and Bulk Density (Usually, -0.6 + 0.3 mm), and then crush the sample using the chosen crusher technology. The freshly generated fraction of interest is then screened out and tested. This means that all of the particles tested were crushed by the chosen crusher type. This is what happens in carbon plants with butts, however, for coke aggregate, the oversize material (e.g. +4.75 mm) is commonly screened out and only this is crushed.

To confirm that the methodology used in this project gave results aligned with the plant approach, a small laboratory study [3] showed that using the plant approach in the laboratory did reduce the gradient of the relationship between average particle shape and bulk density. Importantly, however, the relationship between crusher/shape and bulk density was still clearly evident.

Shaft Calcined Cokes

An approach used to increase anode BAD by a number of smelters is the on-site blending of high VBD shaft calcined cokes with non-shaft calcined cokes that generally have lower VBDs. The details of shaft calcining and the typical properties of these cokes have been well covered by others (e.g. [4, 5]).

Limited previous testwork with shaft cokes in this project [3] showed that a well-calcined shaft coke gave a higher Bulk Density for a given particle shape value than the non-shaft cokes tested (Fig. 1). This higher Bulk Density was associated with the lower porosity of the shaft coke. The shaft coke also appeared to respond better than some of the non-shaft cokes to "impact" type crushing to give more favourable particle shape properties.

The gradient of the trend between particle shape (convexity) and bulk density was found to be around the same for the shaft and non-shaft cokes/butts with the crushing technologies giving more regular shaped, higher bulk density shaft coke particles in the following order: "Puck" (impact) > Jaw > Roll crushing (Fig. 1).

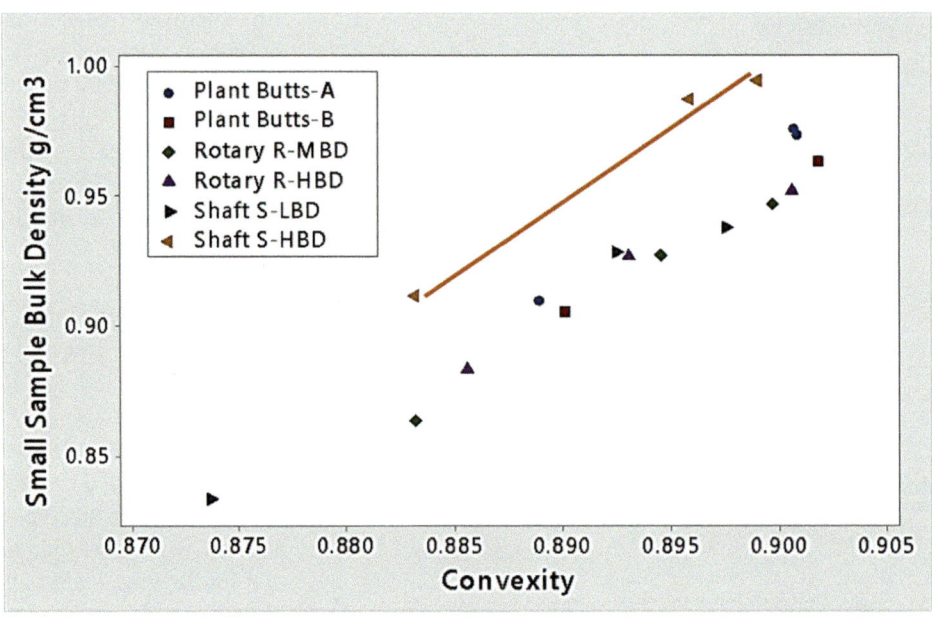

Fig. 1 Bulk Density versus Convexity for a range of anode filler materials. Although appearing to have a similar gradient, the typical shaft coke (S-HBD) gave a higher Bulk Density for a given convexity. This difference is associated with the lower porosity of this coke (Diagram from [3]. LBD—Low Bulk Density, MBD—Medium Bulk Density, and HBD—High Bulk Density. Full sample details in [3]. Small sample bulk density defined in [1])

Study Objectives

The objective of the study outlined in this paper was to determine if the understanding gained in previous work on the impact of crushers/particle shape on coke bulk density could be leveraged to improve the quality of shaft/non-shaft coke blends. More specifically, whether crushing of the shaft coke using specific crusher types to change the particle shape could improve blend packing and bulk density, and to identify any "sweet spot" synergistic blend ratios/crushing conditions that give better than expected results.

Experimental

Procedures Used

The experimental procedures used in the study were similar to those utilised previously [1, 3] except that the samples tested were blends of shaft and non-shaft cokes rather than unblended coke or butts. The coke blend contained a typical high-quality shaft calcined and a single source hearth calcined coke, with blend ratios of 0/100% Shaft/non-shaft, 25/75, 40/60, 50/50, 60/40, 75/25, and 100/0. These blends were selected to focus on the ratios typically used by plant operations.

Each blend ratio sample was crushed by either a jaw crusher (previously shown to give favourable particle shapes for packing) or a smooth roll crusher (i.e. there were no ridges nor teeth on the rolls), which tended to give less favourable particle shapes. Two particle size fractions were screened from the crushed samples and tested: −30 + 50 # (−0.6 + 0.3 mm) and a coarser fraction of −4 + 10 # (−4.75 + 2.0 mm). The samples were screened to remove the target fraction prior to crushing as per previous work so that all the coke tested was crushed by the selected crusher.

Particle shape was tested in a QICPIC instrument manufactured by Sympatec, following the method used previously [1, 6]. Vibrated Bulk Density was measured using the ASTM 8097 procedure for the Micromeritics GeoPyc instrument ("GeoPyc VBD") except that sample preparation was as defined by the experimental sample treatment (particle size and crusher) used.

Results

The GeoPyc VBD test results for the Shaft coke blend ratios crushed with the two crusher types (Jaw and Smooth roll) at the two target particle size fractions (−30 + 50 # [−0.6 + 0.3 mm], −4 + 10 # [−4.75 + 2.0 mm]) are shown in Fig. 2 below.

It can be seen from Fig. 2 that:

- At both the particle size ranges tested, the 100% **Jaw** crusher shaft coke (A Points) had a GeoPyc VBD **greater** than the RoK non-shaft coke (B Points), so increasing the

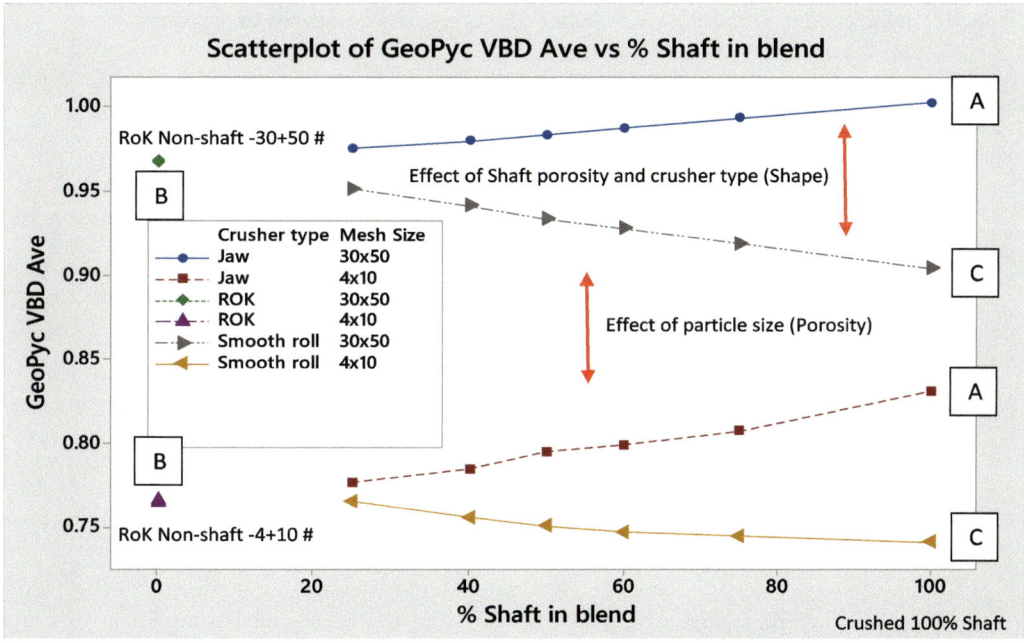

Fig. 2 Coke blend GeoPyc VBD (g/cm³) as a function of blend ratio (% Shaft coke in blend), particle size (−30 + 50 #, Top curves; −4 + 10 #, Bottom curves), and crusher used (Jaw or Smooth roll). "A" Points are 100% Shaft jaw crushed, "B" points are 100% Run of Kiln non-shaft coke, and "C" points are 100% Shaft smooth roll crushed

ratio of shaft coke in blend **increased** blend GeoPyc VBD. This was as expected, demonstrating the ability of shaft cokes to **improve** coke VBD and potentially anode BAD.

- However, at both the particle size ranges tested, 100% **Smooth roll** crushed shaft coke (C Points) had a GeoPyc VBD **less** than the RoK non-shaft coke, so increasing shaft coke in the blend **reduced** blend VBD. This was not expected.
- Whether the results are for blends with jaw-crushed shaft coke blends (i.e. Bulk Density increases at higher blend ratios) or smooth roll crushed shaft coke blends (i.e. Bulk Density decreases with higher blend ratios), the results were quite linear with no "sweet spot" blend ratio.
- The greater porosity of the larger particle size fraction was evident in the lower GeoPyc VBD of these fractions at all blend ratios.

The particle shape (Sphericity) test results for the blend ratios crushed with the two crusher types (Jaw and Smooth roll) at the two target particle size fractions (−30 + 50 # [−0.6 + 0.3 mm], −4 + 10 # [−4.75 + 2.0 mm]) are shown in Fig. 3 below.

It can be seen from Fig. 3 that:

- In general, sphericity had a significant influence on the GeoPyc VBD of the samples with more regular shapes (i.e. higher sphericity) giving the expected higher GeoPyc VBDs.
- The gradient of the relationship between bulk density and shape is similar for both particle sizes; however, better

shape and less porosity in the finer particle size material gave a higher VBD. The offset between the data sets was likely due to porosity differences.

- Blends of Jaw crusher shaft −30 + 50 # showed a different behaviour, forming a cluster rather than a line. The 100% jaw crusher shaft coke at −30 + 50 # had almost the same sphericity as the RoK non-shaft coke at the same particle size, so blending these materials did not significantly change sphericity. Since it was not shape, the increase in GeoPyc VBD with increasing jaw crusher shaft coke ratio of −30 + 50 # fraction would be due to the lower porosity of the shaft coke.

The convexity particle shape test results for the blend ratios crushed with the two crusher types (Jaw and Smooth roll) at the two target particle size fractions (−30 + 50 # [−0.6 + 0.3 mm], −4 + 10 # [−4.75 + 2.0 mm]) are shown in Fig. 4 below.

It can be seen from Fig. 4 that:

- The convexity versus GeoPyc VBD relationships across both the size fractions of Shaft coke blends were different to that of sphericity (Fig. 3). This had not been seen before in this project as previously sphericity and convexity were highly correlated and relationships with VBD similar [3].
- The reason why the smaller −30 + 50 # blend particles have a higher convexity but lower sphericity than the larger −4 + 10 # particles is not clear, but may be due to differing relationships between coke texture/porosity and convexity compared to sphericity.

Fig. 3 Coke blend GeoPyc VBD (g/cm³) as a function of particle shape (Sphericity), particle size (−30 + 50 #; −4 + 10 #), and crusher used (Jaw or Smooth roll)

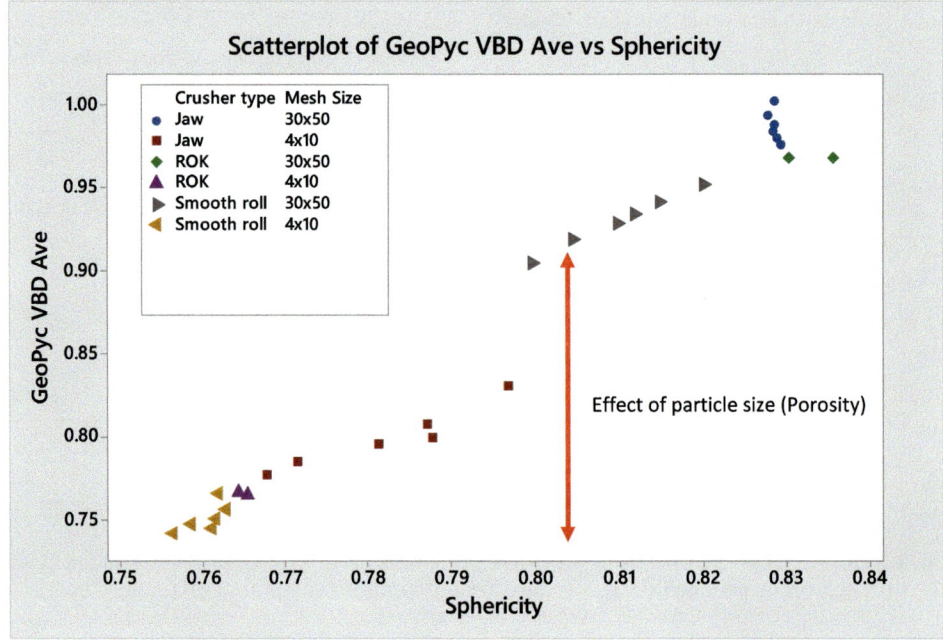

Fig. 4 Coke blend GeoPyc VBD (g/cm³) as a function of particle shape (Convexity, CV), particle size (−30 + 50 #; −4 + 10 #), and crusher used (Jaw or Smooth roll)

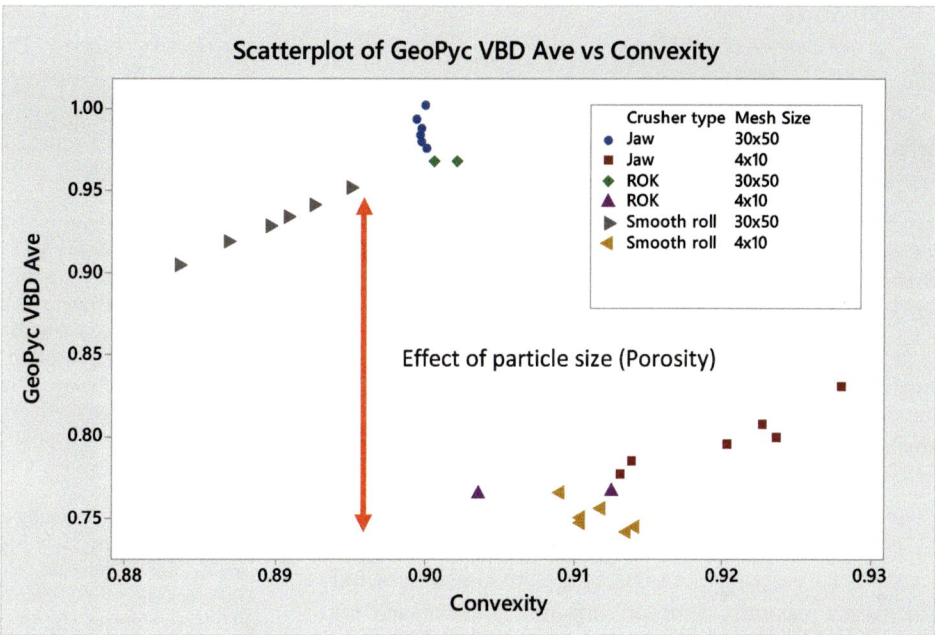

- The higher convexity of larger −4 + 10 # particles is better for packing than that of the smaller −30 + 50 # particles, however, the larger particles have a lower measured GeoPyc VBD. The greater porosity of the larger particles dominated the GeoPyc VBD, outweighing the positive effect of the higher convexity.

Discussion

The laboratory results outlined in the previous section indicate that:

- At both the particle size ranges tested (−30 + 50 #; −4 + 10 #), shaft coke particle shape and GeoPyc VBD were sensitive to crusher type (Jaw vs Smooth roll).
- It is evident that smooth roll crushers are not a good option for shaft cokes (At least on a laboratory scale) having a negative effect on coke Bulk Density.
- When all shaft coke particles are crushed, the Bulk Density of shaft and non-shaft coke blends can be significantly improved by crusher type selection (i.e. using jaw crushers instead of smooth roll crushers). If translated to plant operations, there may be potential to improve anode properties and performance in plants that do not currently use jaw or impact type crushers for anode aggregate and/or butts.
- The relationships between coke blend Bulk Density and Shaft coke blend ratio were generally quite linear, with no

indication of a "sweet spot" in blend ratio. This suggests that the impact of the (changing) shape of the shaft coke in the blend is to change the average shape of the blend; there was little or no indication of an "interaction" between the shapes of the shaft and non-shaft coke particles that affects the way the blend packs during the bulk density test.
- The impact of particle shape (Sphericity) and porosity on coke Bulk Density identified previous in this project was confirmed in this work, with evidence to show the impact of both on blend bulk density was significant.
- Although the mechanism is not yet clear, the impact of porosity was lower relative to sphericity than convexity, i.e. porosity appears to have a greater impact on bulk density than convexity.

Previous results from this project have suggested opportunities to improve plant aggregate and anode properties by modifying plant coke crushing principles. The current results again support this and show that, at least on a laboratory scale, using impact-type crushers, including jaw crushers can increase the bulk density of blends of shaft and non-shaft cokes. An optimal blend ratio of shaft to non-shaft, i.e. "sweet spot" was not identified when shaft coke shape was modified by crushing.

The current work has been extended further by looking at the effect of using different crushing technologies for both the shaft and non-shaft cokes in the blend on blend bulk density. The results of this will be published in a subsequent paper.

Conclusions

The findings outlined in this paper were an extension of previous studies that showed that when testing a single coke or butts, crushing method, particle shape, and bulk density are highly correlated, that using higher impact crushing methods gives particle shapes that pack better and increase particle Bulk Density, and that (within the range studied) changes in shape and open porosity have similar impacts on particle bulk density. These findings were generally confirmed in this work and extended to show that the bulk density of a blend of shaft and non-shaft cokes could be improved by using jaw crushing instead of smooth roll crushing. No evidence was found that there is a "sweet spot" blend ratio that optimised a shaft/non-shaft blend as the shape of the shaft coke is modified by crushing.

The previous results and the outcomes from this work suggest that designing and operating carbon plant crushing circuits to maximise "impact" crushing of coke and butt particles (including shaft and non-shaft coke blends) to achieve the target aggregate particle size could lead to better aggregate packing, higher bulk density values, and hence higher density anodes. This needs to be confirmed on a larger scale with an aggregate material.

References

1. Cannova FR, Davidson MD, Forte L, Sadler BA (2018) Influence of Crushing Technology and Particle Shape on the Bulk Density of Anode Grade Petroleum Coke. Light Metals 2018, TMS, 1169
2. Cannova FR, Davidson MD, Sadler BA, (2018) Opportunities to use particle technology to improve anode quality, Proc 12th Australasian Aluminium smelting technology conference, Queenstown, New Zealand, 7th - 11th December 2018, Paper 2a3
3. Cannova FR, Davidson MD, Sadler BA, (2020) Influence of particle shape and porosity on the bulk density of anode grade petroleum coke. Light Metals 2020, TMS, Page 1319
4. Edwards LE (2011) Quality and process performance of rotary kilns and shaft calciners. Light Metals 2011, TMS, 895
5. Ries K (2009) Enhancing Coke Bulk Density Through the Use of Alternate Calcining Technologies. Light Metals 2009, TMS, 945
6. QICPIC Instrument help file, Sympatec. http://www.sympatec.com/EN/ImageAnalysis/Fundamentals.html

Managing Green Petroleum Coke Properties Variations on Prebaked Anodes Quality in Aluminium Bahrain "Alba"

Hesham Buhazza, Vasantha Kumar Rangasamy, Nabeel Ebrahim Mohd Al Jallabi, Taleb Al Ansari, Abdulmohsin Hasan Radhi, Francois Morales, and Abdulla Habib

Abstract

Alba (Aluminium Bahrain) is one of the unique primary Aluminium smelters who has its own calciner plant to produce CPC, by converting different types of GPC to CPC. The quality of GPC has strong influence on the properties of the anodes. Due to market variations and price fluctuation, the properties of GPC purchased vary depending on the source. In addition, with latest expansion of Alba being the largest smelters outside China, hence, CPC requirements increased accordingly. Such changes bring up great challenges to both Calciner and Carbon operation team to manage these variations through precise blending and tuning process parameters to achieve optimum anodes quality. This paper discusses Alba strategy on managing the variations of GPC properties along with purchased CPC on quality of anodes in terms of desulphurization, anode density, baking level, stack emissions and others anode properties.

Keywords

Green Petroleum Coke (GPC) • Calcined Petroleum Coke (CPC) • Desulphurization • Anode quality • Baking level

Introduction

Aluminum industry uses blends of high- and low-sulfur green petroleum coke to balance between low production cost and suitable sulfur level to produce good anode quality.

Alba has its own rotary kilns to produce calcined petroleum coke by feeding a blend of several types of green petroleum coke. Nevertheless, with recent Alba expansions, CPC requirements increased accordingly which couldn't be met by Calciner plant and around 10% of CPC requirements has to be purchased annually.

Anode quality widely depends on the raw material quality. Calcined petroleum coke is the main raw material for preparing the prebaked anodes, which generally accounts for about 65% of the carbon anode weight. The quality of petroleum coke has a significant effect on the behavior of anodes in the electrolytic cell and, consequently, operational performance of the cell.

Sulfur content in green petroleum coke is being used for characterization in anode manufacturing and becomes the reference for anode production. Desulfurization has become a prominent issue for calciners and anode producers as the industry moves to blending higher sulfur cokes to achieve average coke sulfur levels and to comply with the SO_2 emission limit.

For smelters using a combination of shaft calcined coke and rotary kiln calcined coke, it is important that the smelter has an ability to blend these cokes. The porosities and bulk densities of these cokes are quite different and will result in large pitch level and anode density variations if not blended properly.

Differences in fines fraction and hardness of cokes from rotary kilns and shaft calciners can also drive additional variation in the paste plant if the cokes are not blended using relatively consistent ratios.

This paper presents and discusses the below:

- The GPC specifications and its blends by Alba calciner.
- Challenges encountered using high sulfur GPC.
- Challenges encountered using imported CPC.

H. Buhazza (✉) · V. K. Rangasamy · N. E. M. A. Jallabi ·
T. Al Ansari · A. H. Radhi · F. Morales · A. Habib
Aluminum Bahrain, Manama, Kingdom of Bahrain
e-mail: Hisham.Buhazza@alba.com.bh

© The Minerals, Metals & Materials Society 2023
S. Broek (ed.), *Light Metals 2023*, The Minerals, Metals & Materials Series,
https://doi.org/10.1007/978-3-031-22532-1_149

GPC Parameters and Blending Challenges

Alba has built Rotary Calciner plant as backward integration to its smelter functions. Calciner is receiving green petroleum coke from different suppliers. It has a storage capacity of 150,000 mt divided into different storage bays based on the green coke quality specifications.

Calcination of green petroleum coke is an exothermic reaction, which produces a significant amount of heat due to combustion of volatile matter hydrocarbon compounds present in GPC. This thermal up gradation process increases carbon to hydrogen ratio to improve electrical conductivity, real density, and oxidation characteristics. The main control proof variables are heating rate, VM/air ratio, calcinations temperature, tertiary air, and supplementary fuel (natural gas). The dynamics of calcination process are controlled with the draft to maintain automatic calcination temperature. Ranges have been set for various operating quality parameters which are continuously monitored to ensure meet the specified customer requirements.

Green Petroleum Coke Specifications

Alba Calciner procures green petroleum coke based on the following specifications (Table 1). However, the cokes are further divided into four categories based on sulphur content which is discussed more in the following sections.

Alba calciner was dealing with 17 different types of cokes from different sources in previous years but recently reduced to five GPC sources.

In recent years considering financial, economic challenges and scarcity in green coke qualities, Alba calciner team had to optimize the process to cope with these challenges.

Green Petroleum Coke Blending

Six blending silos (with a capacity of 1200 MT each) are available to fill different types of cokes and blend them at pre-set ratios, with the help of weigh feeders provided at bottom of each silo, to achieve targeted quality of calcined coke and consideration of economic feasibility as per the customer's specification.

Calciner is equipped with automatics samplers collecting samples at green coke feeding and outcome from coke cooler as per the set target timings. Samples are analyzed at dedicated laboratory at smelter for all the quality parameters as required for anode manufacturing. Intermediate samples are also collected to verify the process capability and adjust the process parameters as required.

Below Figs. 1 and 2 show the different percentages of green coke blend and its effect on calcined coke qualities. Figure 1 shows trend of calcined coke Lc and Fig. 2 shows trend of % sulphur in the calcined coke with respect to GPC blending. The variations occurred due to the coke used based on its inventory and availability. Carbon departments monitor and evaluate the changes in all quality parameters of calcined coke and adjust accordingly in order to achieve anode quality consistently.

Table 1 Green petroleum coke specifications

Green coke analysis	Units	Specification	
		Minimum	Maximum
Sulphur	%	0.6	4.5
Ash	%	0.1	0.3
Iron	ppm	150	200
Silicon	ppm	150	250
Vanadium	ppm	150	300
Nickel	ppm	130	200
Calcium	ppm	50	150
Sodium	ppm	50	100
Moisture (FOB Vessel)	%	–	8
Volatile Combustible Matter (VCM)	%	8	12
Hardgrove Grindability Index	No	70	110
+4 mesh	%	40	50
−20 mesh	%	15	20
Lump Size	mm	75	250

Fig. 1 Green petroleum coke blend and its effect on CPC Lc

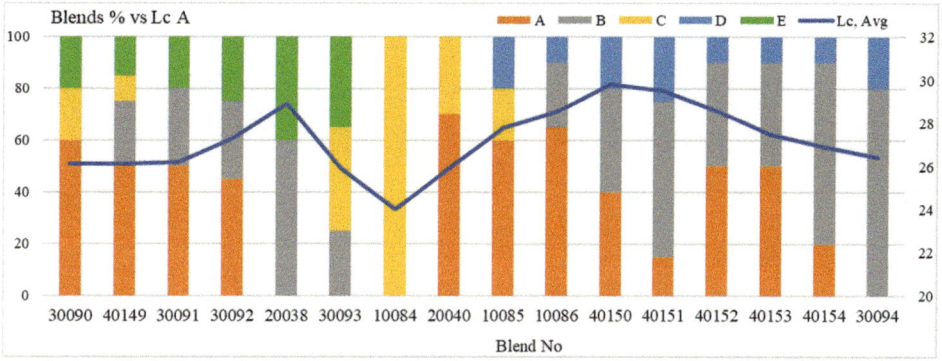

Fig. 2 Green petroleum coke blend and its effect on CPC sulphur

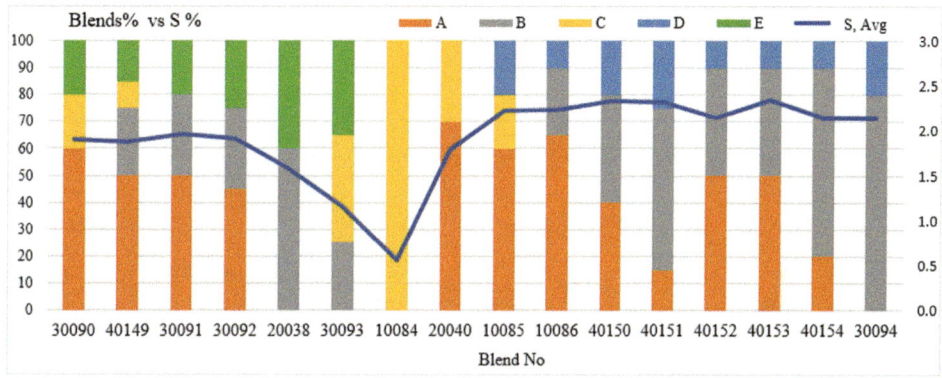

Challenges Encountered Using High-Sulfur Coke

Sulfur content is one of the important quality indexes of petroleum coke. Optimum sulfur content in baked anodes ∼2% must be achieved to avoid the negative impacts on the reduction cell performance.

The negative effects of high-sulfur petroleum coke on aluminum smelter are mainly in the following aspects:

- Decrease the baked anode density caused by desulfurization.
- Decrease the current efficiency in Potline due to the increase of porosity in the anode.
- Increase SO_x emission in Calciner and Anode Baking Furnace.

Alba has its own rotary kiln calciner to produce calcined petroleum coke (CPC) from different green petroleum coke sources (GPC). Based on sulfur content, these cokes being blended to achieve ∼2.5% Sulphur in CPC. Blending strategy shown in below Table 2. This strategy was established based on historical availability and economic feasibility of GPC in the market.

In the month of Apr'21, Alba procured GPC with high sulfur content (∼4.9%) and blended at the rate of 25% with other sources of GPC to produce CPC due to non-availability of GPC from other resources at that time.

With this blend, it has been observed that green anode density dropped and as consequence the baked anode density dropped as well by ∼0.02 g/cc. This drop can be explained due to the high sulfur loss under calcination temperature of >1350 °C, the sulfur from the coke releases out at faster rate which resulted in drop in vibrated bulk density (VBD), real density (RD), coke particle size (increase in fine particles), and increase in the coke micro-porosity and coke desulfurization (∼0.35%).

The SO_x emission also increased sharply\exceeding the legal limits of 500 mg/nm³ (Fig. 3). This can be explained due to the increase in the anode desulfurization by ∼59% at the baking furnace. (Fig. 4).

Alba decided to improve baked anode density, reduce the Sox emission below to the legal limit and reduce the anode desulphurization without reducing the peak gas temperature

Table 2 Green petroleum coke blending strategy

Sulfur content	0 to 1 S%	1 to 3 S%	3 to 4 S%	4 to 5 S%
Blending ratio	15	30	50	5

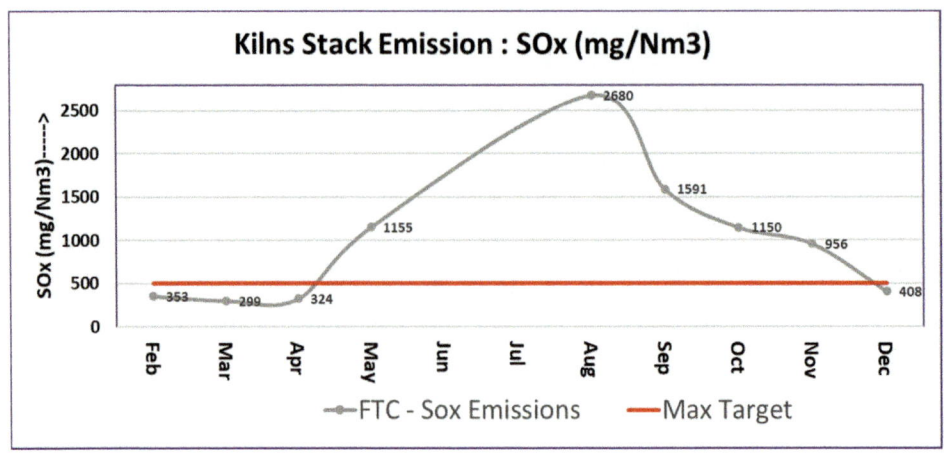

Fig. 3 SO$_x$ emission trend

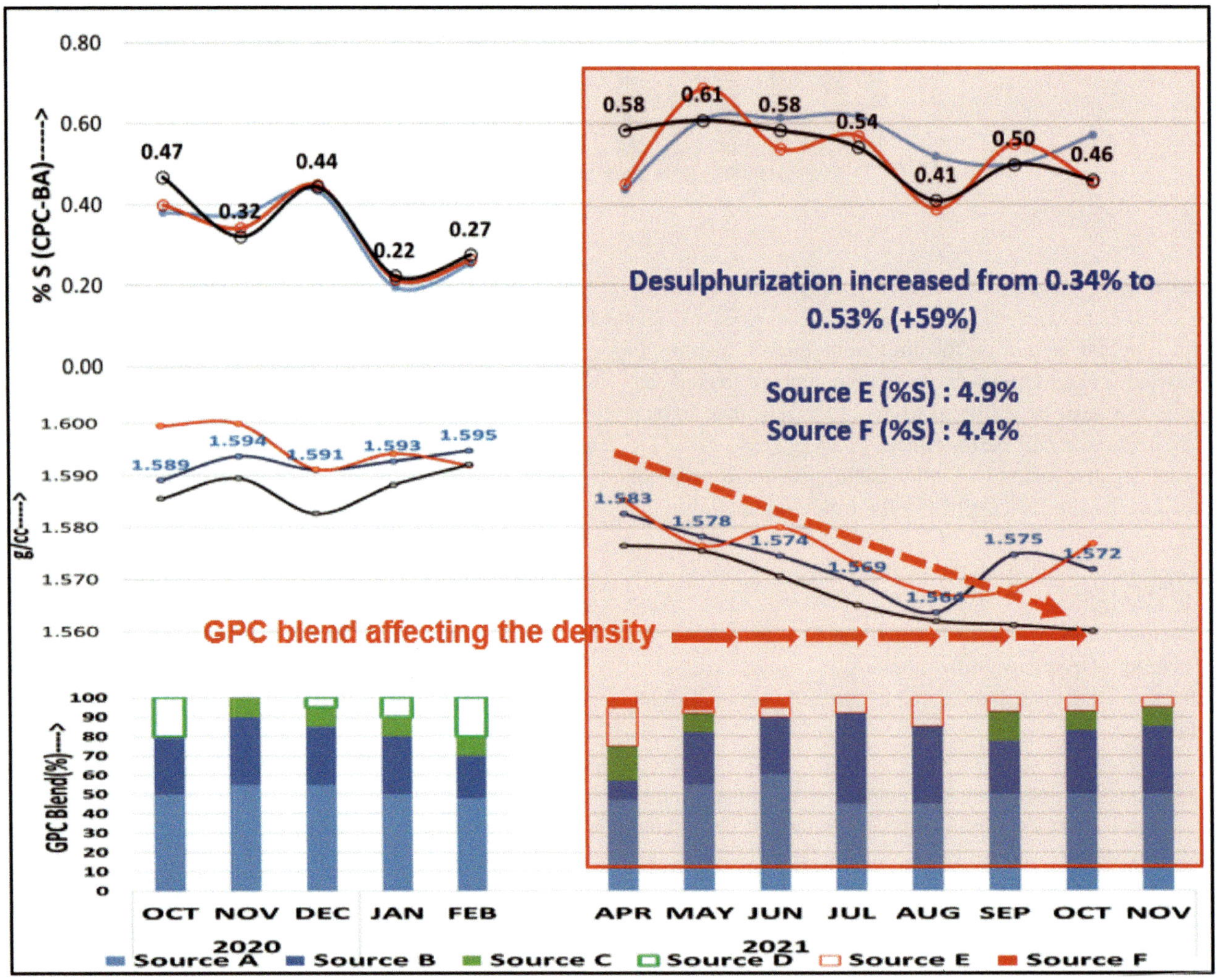

Fig. 4 Anode desulphurization, baked anode density, and GPC blend

Table 3 Sulfur % in GPC blend and its impact on BAD, SO$_x$, and desulphurization

High sulphur GPC blend (%)	Baked anode density (gm/cc)	Sox (mg/Nm^3)	Desulphurization (%)
25	1.576	1155	0.58
15	1.56	2680	0.5
10	1.571	1591	0.58
7	1.561	1150	0.4
5	1.564	408	0.3

in anode baking furnace in order to keep baking level Lc on target.

To improve the BAD, reduce the SO$_x$ and desulphurization, the high Sulphur GPC in the blend decreased from 25 to 5% gradually and it was observed that there was an improvement in So$_x$ emission and desulphurization as shown in Table 3.

Challenges of Imported CPC

The section below describes the challenges encountered when importing CPC from shaft and rotary calciners and their impact on the process.

Challenges Encountered Using Imported CPC (Source A)

Alba calciner can produces calcined petroleum coke at 90% from the total requirement for all the carbons plants in Alba,

the remaining requirement of 10% calcined petroleum coke is procured from the market.

In the year 2020 and 2021 Alba procured CPC from calciner with rotary Kilns to meet the requirement of carbon plant. The corresponding CPC parameters are listed in Table 4.

It has been observed that the fine fraction is low -100 mesh (<0.15 mm) and the sulfur is high in this coke.

Fractions of different-size particles are commonly used in a recipe designed to give a dense mix where the voids between the coarse particles are filled with medium-size particles and voids between coarse and medium particles are filled by small-size particles. Optimum quantity of coarser, medium, and finer fractions is required for anode formulation to obtain good anode quality.

Although the fine particles fill the gaps between the coarse or medium particles, there is a lot of inter-particle voids between the fine particles and the distance between the fine particles is far because fine particles to fill are not enough.

Due to the large proportion of fine particles, the number of small voids between fine particles increases significantly.

Table 4 Imported CPC parameters

Sl. No	Parameters	Unit	Target specs	Source A
1	Real density	g/cm^3	2.06–2.085	2.055–2.075
2	Vibrated bulk density	g/cm^3	0.87 Min	0.86–0.88
3	De dusting oil	%	0.4 Max	0.15–0.30
4	Moisture	%	0.05–0.25	0.05–0.25
5	Ash content	%	0.15 Max	0.05–0.25
6	CO$_2$ reactivity	%	12 Max	2.0–4.0
7	Air reactivity	%/	0.25 Max	0.1
8	Sulphur	%	3.0 Max	3.0–3.2
9	Vanadium	ppm	300 Max	190–320
10	Nickel	ppm	260 Max	70–120
11	Iron	ppm	300 Max	40–100
12	Silicon	ppm	200 Max	70–100
13	Calcium	ppm	150 Max	30–90
14	Sodium	ppm	100 Max	30–90
15	+4.75 mm	%	24 Min	28
16	−100 Mesh	%	3 Max	4

Fig. 5 Baked anode density with imported CPC

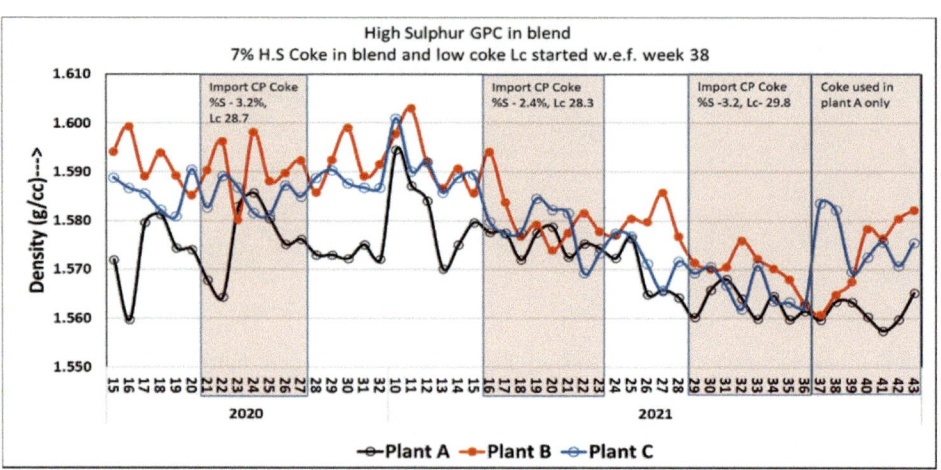

As voids between fine particles increase in CP Coke and high Sulfur green petroleum coke ($\sim$4.9%) used in the blend with the imported coke having high sulfur content (3.2%), which has resulted into drop to the green anode density as well as baked anode density (Fig. 5). CO_2 reactivity increased due to more porosity in baked anodes and as a result, carbon thickness under stub reduced (Fig. 6) and bath attack to stubs increased lead to iron in metal increase by $\sim$300 ppm (Fig. 7).

To minimize the effect of low baked anode density and increase of iron in metal, Alba decide to use imported CP coke to produce green anode in one carbon plant only.

It has been observed that baked anode density remained on the same low level for carbon plant where the imported coke used for anode production, and the baked anode density improved for other carbon plants where imported coke not used. (Fig. 5).

Challenges Encountered Using Imported CPC (Source B)

In Dec'21 and Feb'22, Alba procured CPC coke from a calciner using shaft kilns having VBD of >0.95gm/cc, vanadium at 350 ppm and Electrical Resistivity >500μΩ·cm. The imported shaft kiln CPC parameters are shown in Table 5 (Fig. 8).

In Apr'22, Alba procured shaft Kilns CP Coke having VBD of >0.95gm/cc, vanadium at 350 ppm and Electrical Resistivity > 500 μΩ·cm and blended with Alba CP Coke at a ratio from 30:70%, The blending was not consistent this time due to shortage of Alba CP Coke, This has resulted into pitch variation in recipe and increase in baked anode Electrical Resistivity (Fig. 9) and vanadium increased as well by 20 ppm.

During the same period, variations in paste plant process parameters observed which led to further increase of Electrical Resistivity as mentioned below.

Fig. 6 Carbon thickness under stub pin

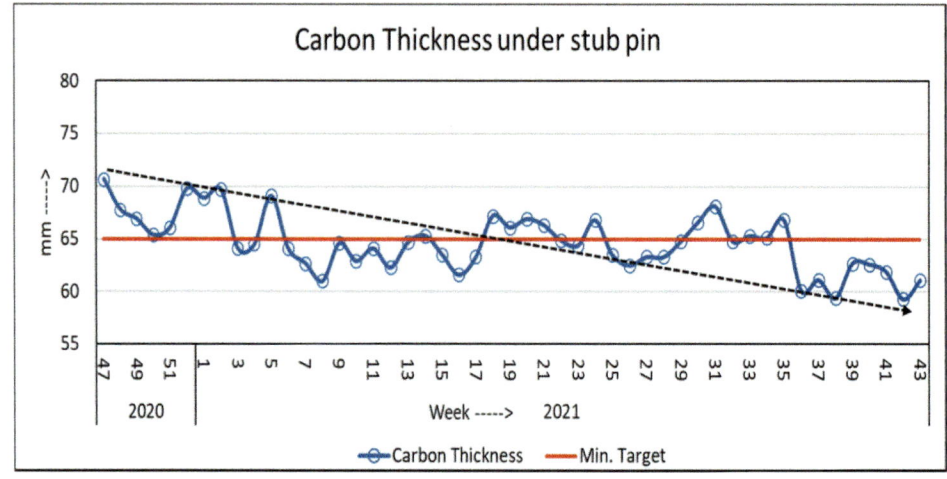

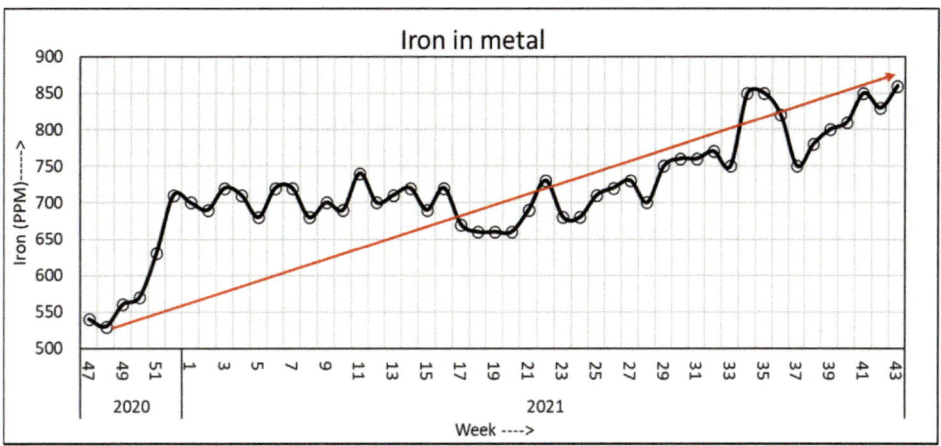

Fig. 7 Iron in metal

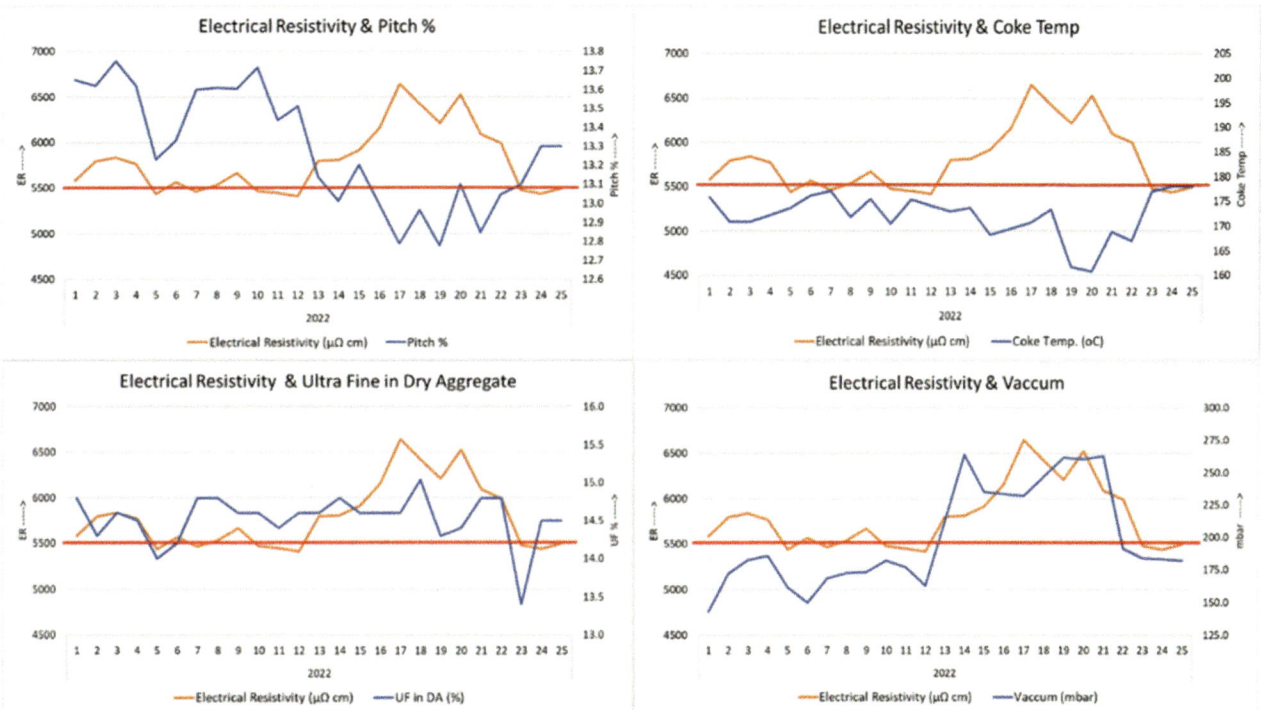

Fig. 8 Trend of paste plant parameters

- Variation in pitch addition from 12.9 to 13.5% because of variation in the blend of Shaft kilns and Alba rotary kilns CPC.
- Coke preheating temperature dropped from 175 to 160 °C because of HTM oil's low temperature due to malfunction in gas pressure regulator in boiler unit.
- Ultrafine in Dry Aggregate increased from 14 to 15% because of high percentage usage of shaft kiln CPC in the blend.
- Grain to Sand Ratio increased from 6 to 7 because of high percentage usage of shaft kiln CPC in blend.

- Vacuum decreased from 160 to 260 mbar to reduce baked anodes coke sticky tendency.

The increase in Electrical Resistivity of the baked anodes occurred at the worst possible period with the full impact of its recycled butts in the summer. Because of increase in Electrical Resistivity potline performance was negatively impacted resulting in increase in anodes spikes. From January 2022 to April 2022 the average anode failures (spikes) ahead of schedule changes was 1–4 nos/day, whereas during May 2022 to Jun 2022 the anodes increased up to 15 nos/day.

Table 5 CPC parameters

Sl. No	Parameters	Unit	Target specs	Source B
1	Real density	g/cm³	2.06–2.085	2.076
2	Vibrated bulk density	g/cm³	0.87 Min	0.952
3	Dedusting oil	%	0.4 Max	0.27
4	Moisture	%	0.05–0.25	0.07
5	Ash content	%	0.15 Max	0.17
6	CO_2 reactivity	%	12 Max	5.06
7	Sulphur	%	3.0 Max	2.55
8	Vanadium	ppm	300 Max	359
9	Nickel	ppm	260 Max	134
10	Ni + V	ppm	560 Max	493
11	Iron	ppm	300 Max	159
12	Silicon	ppm	200 Max	177
13	Calcium	ppm	150 Max	83
14	Sodium	ppm	100 Max	51
15	+4.75 mm	%	24 Min	38.1
16	−100 Mesh	%	3 Max	2.5
17	Electrical resistivity	μΩ cm	500 Max	539

Fig. 9 Monthly baked anode electrical resistivity

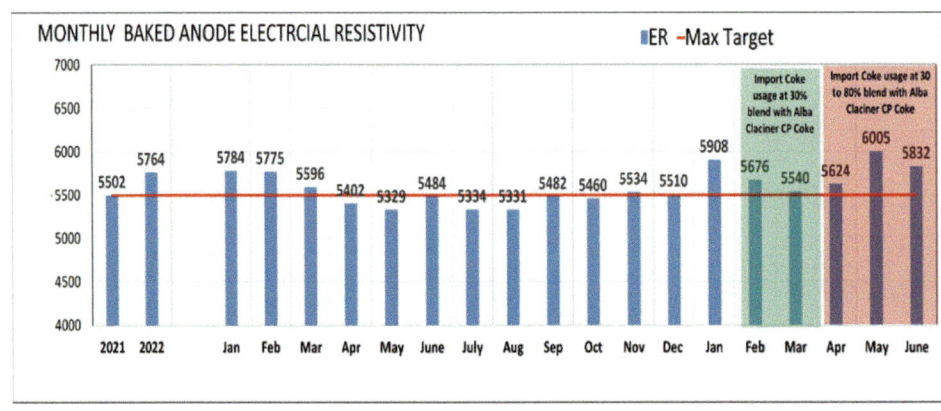

The increase of Electrical Resistivity in anodes decreased the anode to cathode distance (ACD), carbon dust increased from 45 to 56% (Fig. 11) and increase of spikes in potline (Fig. 10).

Due to the usage of 100% Alba CP coke for anode production and optimized paste plant process parameters resulted in declining trend of anodes electrical resistivity, which gives positive results in decrease of spikes in potline by 50% (Fig. 12).

Conclusion

Alba attempt to manage the high Sulfur GPC by blending with low-sulphur GPC in Alba calciner to reduce the impact on baked anode density, SO_x emission and anode desulphurization without reducing the peak gas temperature in baking furnace as it may result in an increase in carbon dusting and anode spike in potlines due to reduction in anode baking level (Lc).

To mitigate the impact of high sulphur GPC in the blend, high-sulfur GPC decreased from 25 to 5% gradually with low-sulfur GPC and observe a measurable improvement in SO_x emission and desulphurization.

The raw material change experienced by Alba was not planned, and in combination with other challenges, initially caused a significant negative impact on the smelter operation. The negative impact was swiftly reversed, through implementation of proper blending system. This serves as a useful case study of how a smelter can respond to such changes and demonstrates strong interaction between anode/cell performance and potline operations.

Fig. 10 Monthly spikes

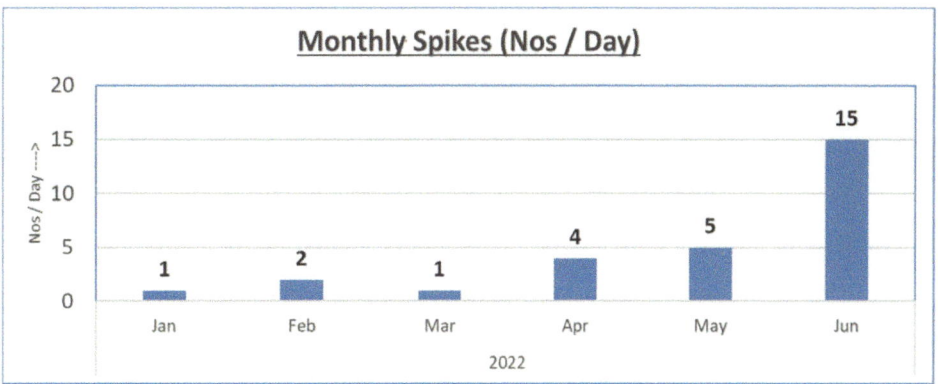

Fig. 11 Monthly carbon dust

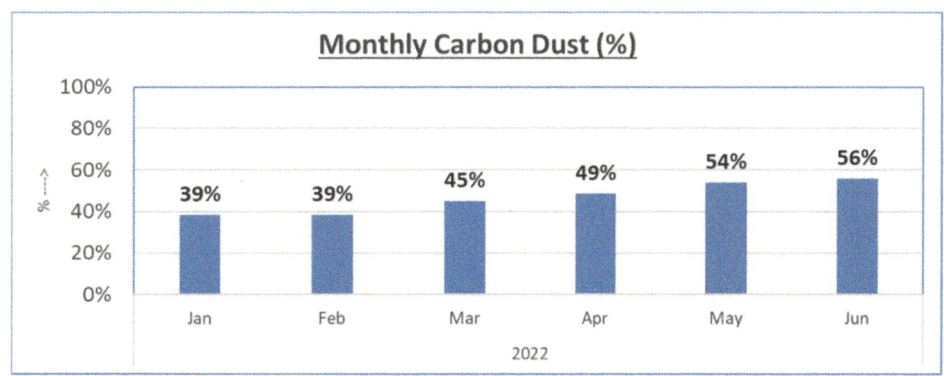

Fig. 12 Electrical resistivity versus no. of spikes per day in potline

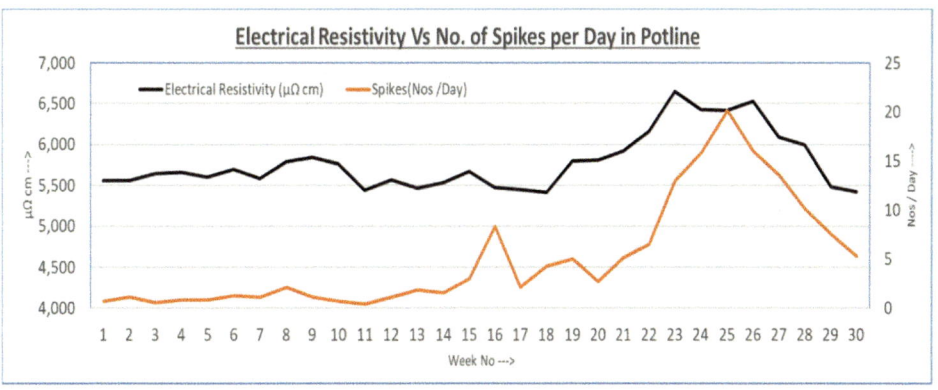

Some of these potential threats from CP Coke with high fines and high sulfur will no doubt evolve gradually and will be dealt with by the changes in practices and operations.

Alba faced the situation of low green anode density with the usage of CP Coke having high fines and high sulfur that led to low baked anode density. CO_2 reactivity increased due to more porosity in baked anodes and, as a result, carbon thickness under stub reduced and bath attack to stubs increased leading to iron in metal increase by ~ 300 ppm.

The effect of low baked anode density and increase of iron in metal was minimized by limiting the usage to only one carbon plant out of the four carbon plants in Alba.

Alba also faced the situation of using shaft Kilns CP Coke having VBD > 0.95gm/cc, vanadium at 350 ppm, and electrical resistivity > 500 μΩ·cm. Due to shortage of Alba CPC the blend was not consistent which resulted in pitch variation in recipe and, as a consequence, electrical resistivity increased in baked anodes and vanadium increased as well by 20 ppm which, led to ACD squeeze, an increase in carbon dust and anode spike cases. During the same period, variations in paste plant process parameters observed which led to further increase of Electrical Resistivity.

Due to usage of 100% Alba CP coke for anode production and optimized paste plant process parameters resulted in declining trend of anodes electrical resistivity, which gives positive results in a decrease in spikes in potline by 50%.

Pressure on raw material quality is likely to continue in future with regards to physical and chemical properties. This challenge will require further improvements in smelter operation which will come with a greater understanding of process dynamics, closer control of critical process variables.

References

1. Hai-Tao Jiang, Chang-Ting Tang, Zheng-Qing Ma, Ping Zhou, Yuan Li, and Pan-Pan Gao. "Effects of High-Sulfur Cokes on Physicochemical Properties of Prebaked Anodes in Aluminium Electrolysis". Light Metals 2018. The Minerals, Metals & Materials Series
2. S. Pietrzyk and J. Thonstad (2012) "Influence of The Sulphur Content in The Carbon Anodes in Aluminium Electrolysis – A Laboratory Study". Light Metals 2012. The Minerals, Metals & Materials Society
3. Les Charles Edwards, Keith J Neyrey, and Lorentz Petter Lossius, "A Review of Coke and Anode Desulfurization". Light Metals 2007.
4. Lorentz Petter Lossius, Keith J. Neyrey and Les Charles Edwards, "Coke and Anode Desulfurization Studies". Light Metals 2008. The Minerals, Metals & Materials Society
5. Dolfy Antonio Sinaga, Rainaldy Harahap, Edi Mugiono, Edwin El Ammar, and Al Rajak Sodikin "Higher CPC High Sulfur in Coke Blending Ratio to Optimize Production Cost Without Lowering Standard Anode Quality" Light Metals 2022, The Minerals, Metals & Materials Series,
6. Les Edwards, "Carbon Anode Raw Materials—Where Is the Cutting Edge?" Light Metals 2020, The Minerals, Metals & Materials Series.
7. Alexandre Gagnon, Nigel Backhouse, Hans Darmstadt, Esmé Ryan, Laurence Dyer, David G. Dixon, "Impurity Removal from Petroleum Coke" Light Metals 2013. The Minerals, Metals & Materials Society.
8. Les Edwards, "Quality and Process Performance of Rotary Kilns and Shaft Calciners" Light Metals 2011. The Minerals, Metals & Materials Society.
9. Les Edwards, Kevin Harp, and Christopher Kuhnt, "Use of Thermally Desulfurized Shaft CPC for Anode Production" Light Metals 2017, The Minerals, Metals & Materials Series
10. Gøril Jahrsengene, Stein Rørvik, Arne Petter Ratvik, Lorentz Petter Lossius, Richard G. Haverkamp, Ann Mari Svensson, "Reactivity of Coke in Relation to Sulfur Level and Microstructure" Light Metals 2019, The Minerals, Metals & Materials Series

Development of an Iron Aluminide Coating for Anticorrosion Protection of Anodic Pins

Henrique Santos, Roberto Seno, Antonio Couto, Alex Fukunaga, and Adriano Francisco

Abstract

The lifetime of an anodic pin in aluminum production process with Söderberg technology is around 3 years. The main consumption factor of the anodic pin is the corrosion of the steel by sulphide. To find a protective coating that will withstand the cycles of use and cleaning of anode pins is the main problem. The objective of this study was to investigate the formation and application of Iron Aluminide, due to its high resistance to sulphidation, as a coating for anodic pins. The application method chosen was the deposition of the mixture of aluminum and iron powders using laser cladding. Results showed the formation of iron aluminide, either in the form of FeAl or Fe_3Al during application, which makes this material promising for this purpose.

Keywords

Aluminum • Anode • Sulphidation • Coating • Pin

Introduction

Iron aluminides have, among many outstanding characteristics, excellent resistance to high temperature oxidation due to the formation of a protective alumina layer which favors its use in harsh environments [1]. These properties make this intermetallic a promising material to be used as a corrosion protection coating on Söderberg technology anodic pins (Fig. 1).

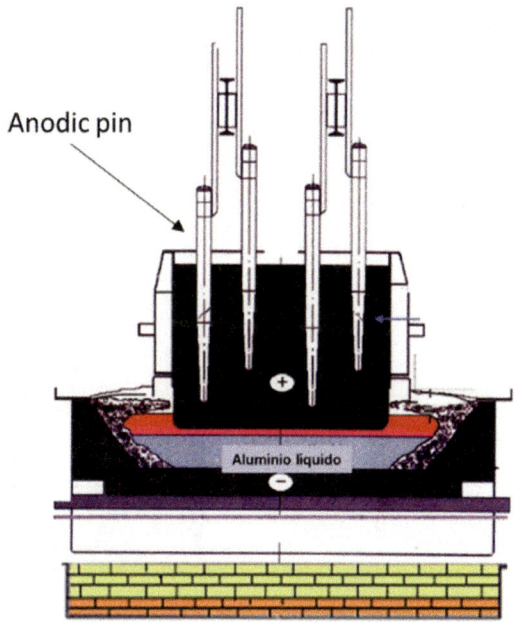

Fig. 1 Söderberg pot

H. Santos (✉) · R. Seno
CBA, Moraes Do Rego Str 347, Alumínio, 18125-000, Brazil
e-mail: henrique.correa@cba.com.br

A. Couto · A. Fukunaga
Universidade Presbiteriana Mackenzie, Rua da Consolação Str, São Paulo, 01302-907, Brazil

A. Francisco
HRC Laser Cladding, João Francisco Angeli Str 160, Piracicaba, 13413-087, Brazil

© The Minerals, Metals & Materials Society 2023
S. Broek (ed.), *Light Metals 2023*, The Minerals, Metals & Materials Series,
https://doi.org/10.1007/978-3-031-22532-1_150

The Fe-Al alloys can be described as a family of solid-solution substitutional alloys ordered at low temperatures and disordered at high temperatures. Table 1 lists the Fe–Al-based aluminides, their critical ordination temperature and melting temperature, among other parameters. Fe_3Al loses ordering at somewhat lower temperatures and passes through two ordered structures (DO3 and B2) before losing its crystalline ordering as showed in Figs. 2 and 3.

Among the characteristics of the material, corrosion resistance, especially in an environment with sulfides, is the main point to be observed. As shown in Fig. 4, the iron aluminide alloys exhibited corrosion rates lower than those of the best existing iron-base alloys (including coating material) by a couple of orders of magnitude when tested in a severe sulfidizing environment at 800 °C (1470 °F). In addition, the aluminides with more than 30% A1 are very resistant to corrosion in molten nitrate salt environments at 650 °C (1200 °F) [2].

Excellent oxidation and corrosion resistance, low density, and good fabricability make the aluminide alloys promising for structural use at temperatures up to 700 to 800 °C (1290–1470 °F). Potential applications include molten salt systems for chemical air separation, automotive exhaust systems, immersion heaters, heat exchangers, catalytic conversion vessels, chemical production systems, coal conversion systems, and so on [2].

When we look at the industrial reality, such as scale, cost and applicability, the challenge is even greater to make a coating based on iron aluminide feasible. Purchasing powdered intermetallic would be the simplest solution, but its cost and availability would make the entire process unfeasible. Then the study continued with the alternative of producing the intermetallic from the base metals during the coating application process [2].

Table 1 Intermetallic: their structures and related properties [2]

Alloy	Crystalline structure	Crystalline ordering temp. (°C)	Melting point (°C)	Density (g/cm³)	Yong module (GPa)
Fe_3Al	DO3 BCC	540	1540	6,72	141
Fe_3Al	B2 BCC	760	1540	–	–
FeAl	B2 BCC	1250	1250	5,56	161

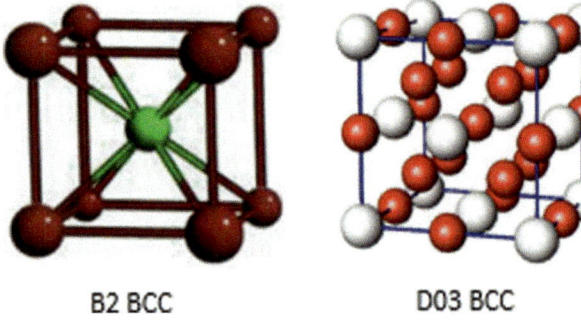

B2 BCC   DO3 BCC

Fig. 2 Crystalline structure DO3 BCC e B2 BCC

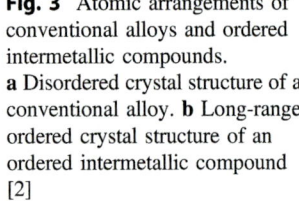

Fig. 3 Atomic arrangements of conventional alloys and ordered intermetallic compounds.
a Disordered crystal structure of a conventional alloy. **b** Long-range ordered crystal structure of an ordered intermetallic compound [2]

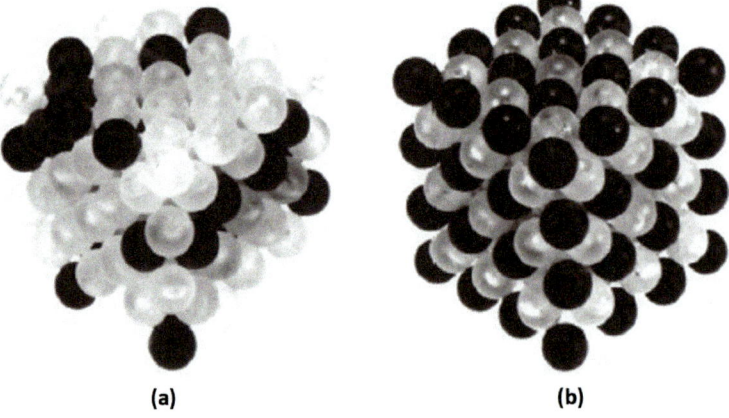

(a)   (b)

Fig. 4 Comparison of corrosion behavior of iron aluminides with that of conventional iron-base alloys Fig. 21 Fe–18Cr–6Al (coating material) and Fe–25Cr–20Ni. All materials were exposed to a severe sulfidizing environment at 800 °C (1470 °F) [2]

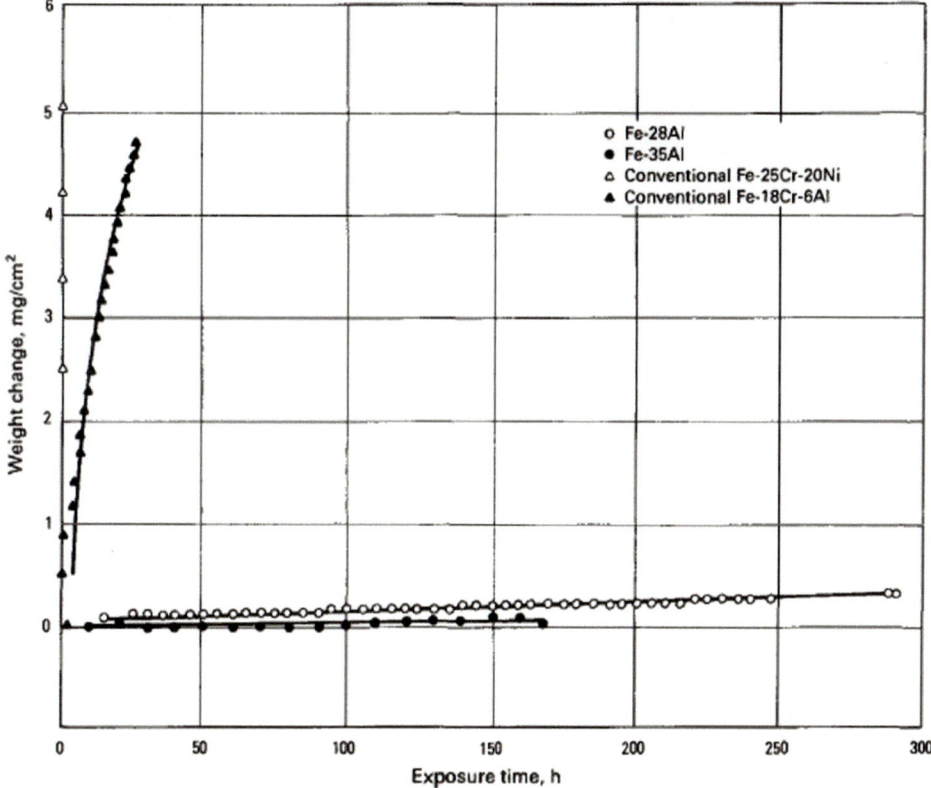

Technical Development

The first answer we need to go forward is whether it would be possible to produce iron aluminide from the powders of base metals.

Powder Preparation

The process chosen for this experiment was the deposition of the base metal powder mixture through Laser Cladding.

Some premises are necessary to be observed to obtain an adequate quality of the mixture for a good deposition:

a. Granulometric range;
b. Homogeneity;
c. Composition.

We use classification sieves to have an adequate distribution of particle size of aluminum and iron powders (Fig. 5).

Fig. 5 Vibrating sieves and MEV—EDS

 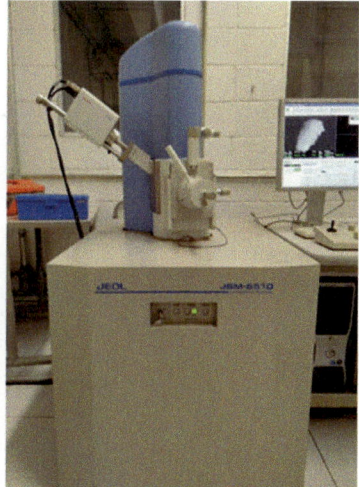

After obtaining the granulometry, a small portion of the Al and Fe powders were separated for SEM analysis (Scanning Electron Microscope) and EDS (Energy Dispersive Spectroscopy), in order to obtain the morphology and composition of the pure powders.

For the tests, we produced two samples of powders with 2 different compositions, 33%Al and 14%Al, calculated in weight, respectively, and homogenized in a cylindrical mixer (Fig. 6).

Fig. 6 Scale and mixer

Laser Cladding Application

Previous studies for this type of application were limited to low laser power, since there is a risk of projection and damage to the equipment lenses due to the high reactivity of aluminum powder. For an industrial approach, a high deposition rate is required to obtain productivity and this implies high laser power during coating application. Then we set off for a test on industrial equipment, in partnership with a local industry. The deposition will be made by the incidence of the laser in a fixed powder bed with predetermined size in a metal similar to the anodic pin. Several evaluation parameters were defined:

a. 2 powder mixture;
b. Bed height 2 mm;
c. 4 set points of laser power;
d. 3 different linear speed.

Results and Discussion

In all of the applications there weren´t projections or damage to the laser lens. Figure 7 shows the 6 coating tracks produced.

Fig. 7 Coating tracks produced

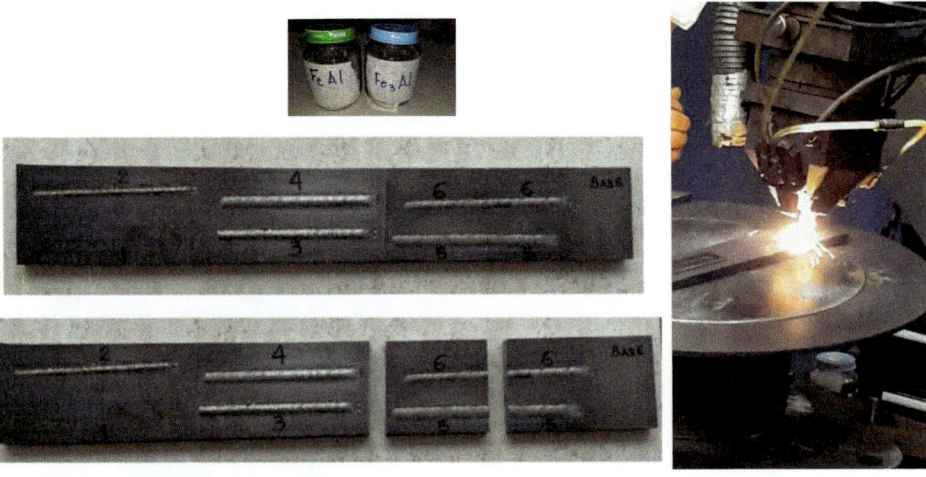

Track 1—33%Al

There was no material fused to the substrate, only the joining of the powders in small portions. There was no power required for melting.

Track 2—33%Al

There was no material fused to the substrate, but the substrate showed small roughness due to the increase of laser power.

Track 3 and 4—33%Al

There was the formation of the coating with dimensions similar to the dimensions of the mold (Fig. 8).

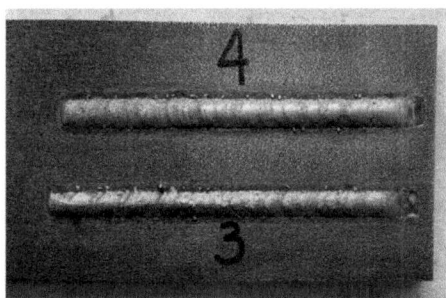

Fig. 8 Tracks 3 and 4

Track 5 and 6—14%Al

The coating was visually formed with a lower thickness than the other coatings and dimensions smaller with respect to the mold (Fig. 9).

Fig. 9 Tracks 5 and 6

XRD Analysis

To identify the iron aluminide it was necessary to compare the results of the XRD analysis with the existing standards. The main identifying elements are:

- The angular displacement of the Fe-α peak by the substitutional Al;
- The peak of Fe3Al—DO3;
- The peak of FeAl—B2;
- The peak of Al_2O_3.

The results show that in tracks 3 to 6 there was the formation of iron aluminide with some differences between them (Fig. 10).

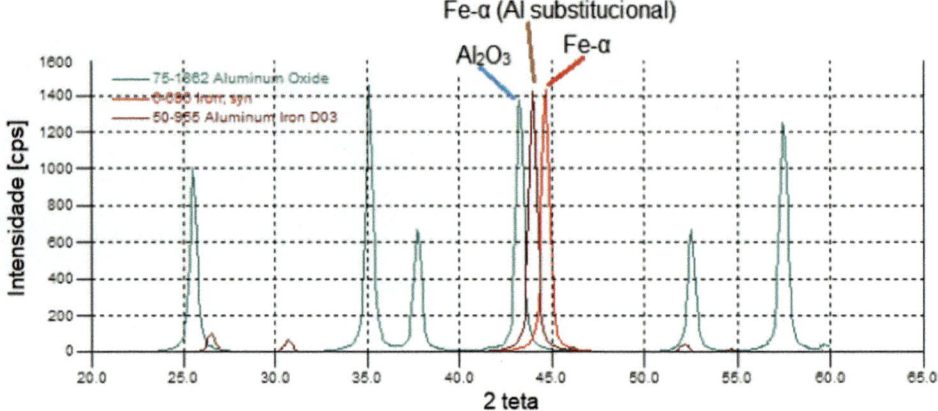

Fig. 10 JCPDS XRD Standard—Fe$_3$Al DO3, Fe-α e Al$_2$O$_3$

Track 3

As shown in Fig. 11, the displacement of the Fe-α angle with the insertion of substitutional Al. In addition, we can see small peaks of Fe$_3$Al DO3.

Track 4

As shown in Fig. 12, we can also see the displacement of the Fe-α angle with the insertion of substitutional Al, however in addition to the peak of Fe$_3$Al we have peaks of Al$_2$O$_3$.

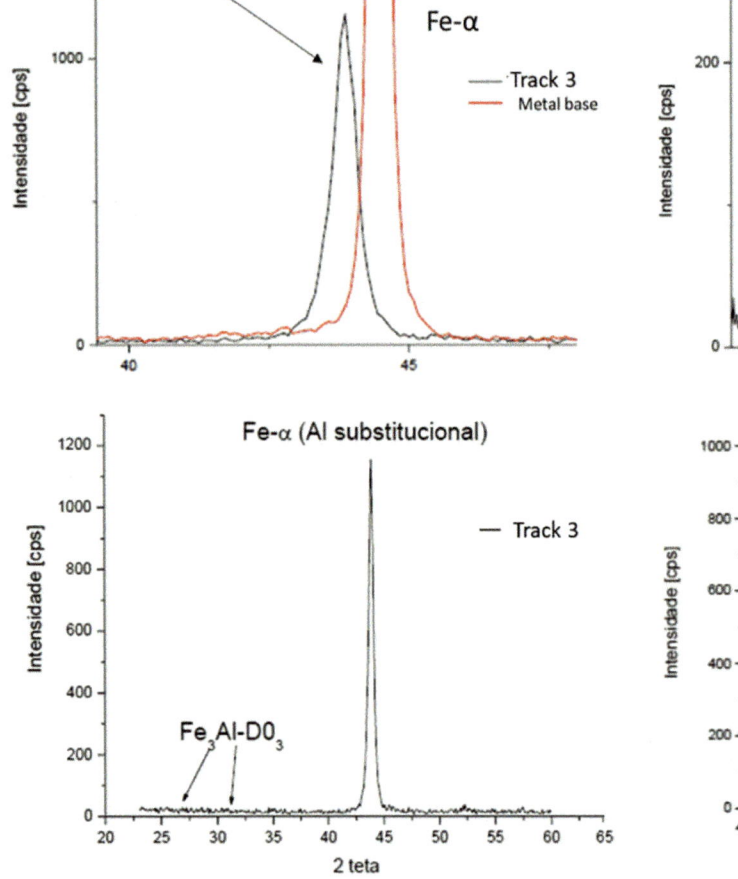

Fig. 11 Track 3 XRD analysis

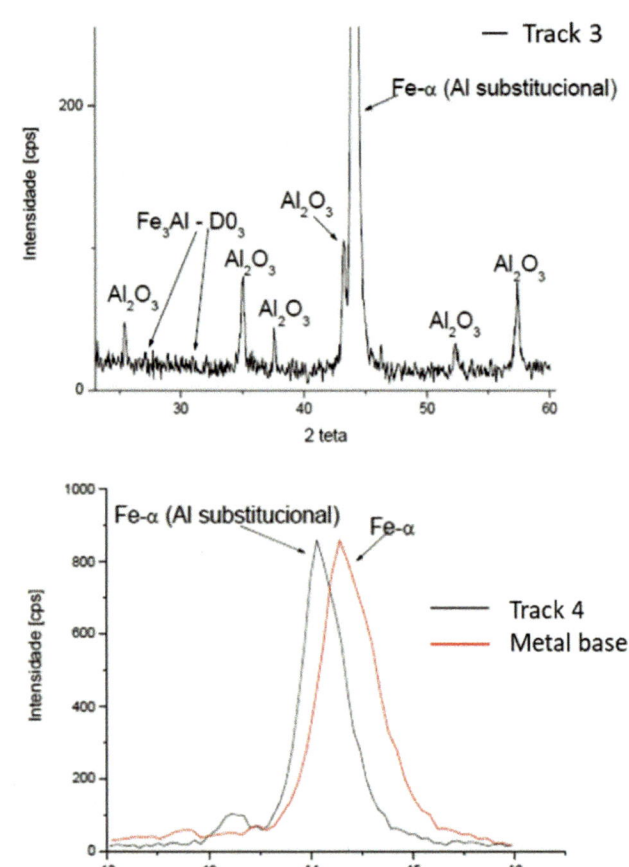

Fig. 12 Track 4 XRD analysis

Tracks 5 and 6

As shown in Fig. 13, as in the other tracks there is the displacement of the Fe-α angle with the insertion of substitutional Al, however, we can see the peak of FeAl B2 and peaks of Al_2O_3.

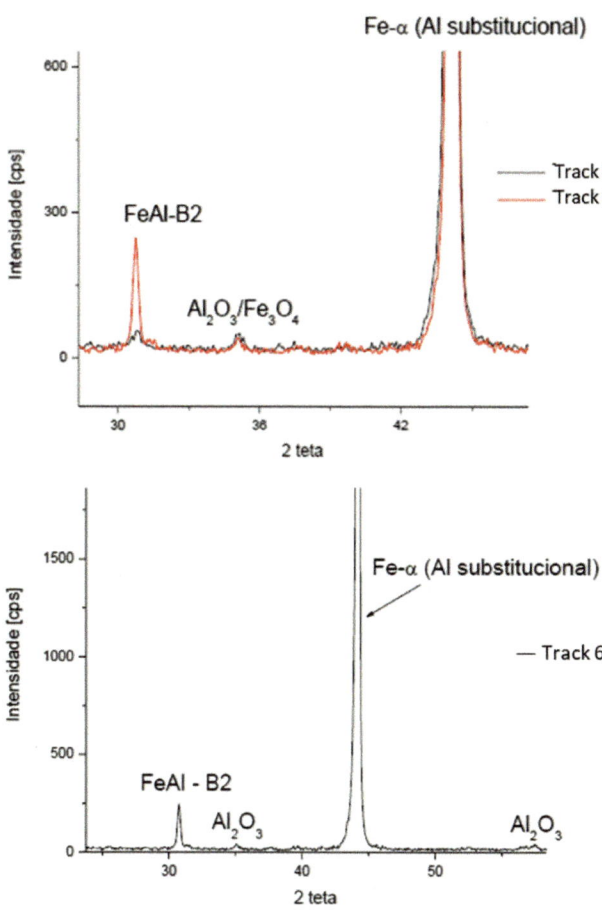

Fig. 13 Track 5 an 6 XRD analysis

Conclusion

The formation of the desired intermetallic was identified in 4 tested tracks, but we have significant differences between the tracks with different mixture contents. Morphology studies, mechanical and corrosion tests are being carried out. The main point of this paper concerns the formation of iron aluminide during laser deposition. This means that it is possible to develop an industrial scale process for the desired use of this coating from base metal powders. New challenges such as producing micro layers of coating, using a high-speed laser, adding the mixture or the powders separately will be studied.

References

1. Borges DF (2010) Processamento e caracterização de aluminetos de ferro obtidos a partir de matéria-prima reciclada. MsC theses, University of São Paulo
2. ASM Handbook, Volume 2: Properties and Selection: nonferrous Alloys and Special-Purpose Materials - ASM Handbook Committee, p 913–942
3. Bax B, Schäfer M, Pauly C, Mücklich F (2013) Coating and prototyping of single-phase iron aluminide by laser cladding. Surface & Coatings Technology 235 (2013) 773–777
4. Kovrov V, Zaikov Y, Tsvetov, Shtefanyuk Y, Pingin V, Golubev M (2017) Aluminide Coating Application for Protection of Anode Current-Supplying Pins in Soderberg Eletrolytic Cell for Aluminium Production. Materials Science Forum, Vol. 900, pp 141–145

New Methods to Determine PAH Emission Dynamics During Electrode Mass Processing

Ole Kjos, Thor Anders Aarhaug, Heiko Gaertner, Bente Håland, Jens Christian Fjelldal, Katarina Jakovljevic, Oscar Espeland, and Ida Kero

Abstract

New measurement techniques aiming to achieve a better understanding of the PAH emissions dynamics during production of electrode paste briquettes were tested. Traditional sampling method requires several hours of sampling to get enough PAH components on the adsorbent material to perform reliable analysis. As the process is changing from one batch to the next, emissions cannot be expected to be constant over time. Better time resolution is essential to increase the understanding of emission variations. Thermal desorption (TD) tubes allow for sampling times as short as 10 min. The SINTSENSE, a standalone photo ionization detector (PID) implementation allows for continuous emission monitoring providing online data second by second. However, the SINTSENSE will only report a total hydrocarbon estimate, with no differentiation between individual components.

O. Kjos (✉) · I. Kero
SINTEF Industry, 0314 Oslo, Norway
e-mail: ole.kjos@sintef.no

I. Kero
e-mail: ida.kero@sintef.no

T. A. Aarhaug · H. Gaertner
SINTEF Industry, 7034 Trondheim, Norway
e-mail: taarhaug@sintef.no

H. Gaertner
e-mail: heiko.gaertner@sintef.no

B. Håland · J. C. Fjelldal
Elkem Carbon, Oslo, Norway
e-mail: bente.sundby-haland@elkem.com

J. C. Fjelldal
e-mail: jens.christian.fjelldal@elkem.com

K. Jakovljevic
Norwegian University of Science and Technology, Trondheim, Norway
e-mail: Katarina.jakovljevic@ntnu.no

O. Espeland
Nemko Norlab, Porsgrunn, Norway
e-mail: oscar.espeland@nemkonorlab.no

Comparing results from these 3 methodologies gives interesting findings on the emission dynamics and variations in composition, in addition to benchmarking the individual measurement techniques.

Keywords

Aluminum • Environmental effects • Carbon materials

Introduction

Polycyclic Aromatic Hydrocarbons (PAH) are hydrocarbons consisting of two or more aromatic rings. The lightest PAH component is naphthalene, with only two connected rings. Due to their different sizes and molecular weights the PAH components have varying physical properties such as boiling point and vapor pressure. Emissions to air can be either as gas or condensed (liquid or solid) onto dust particles.

Almost all industrial processes involving the use of carbon materials at elevated temperatures have the potential to release PAH components, but PAH emissions are of particular concern when the carbon materials also contain pitch and tar. Classical metallurgical industries such as silicon, ferrosilicon and ferromanganese production [1, 2] as well as production and use of anodes in aluminum industry [3] are well-known sources of PAH emissions.

Conventional reporting of PAH emissions is usually based on a few hours of sampling once or twice every year, as full sampling in accordance with relevant standards such as EPA0010 or similar is complex and expensive. The yearly emissions are typically estimates of PAH content per Nm^3 air/off-gas, based on analysis of these samples, assuming that the emission profile during sampling is representative for the entire year. This assumption is known to be erroneous, and strange deviations from trends in reported emissions can be seen when studying yearly trends in open databases such as the Norwegian PRTR database [4]. It is very difficult to

© The Minerals, Metals & Materials Society 2023
S. Broek (ed.), *Light Metals 2023*, The Minerals, Metals & Materials Series,
https://doi.org/10.1007/978-3-031-22532-1_151

assess the degree to which a sample is actually representative with regards to variations in raw materials, production and process, or the impact of ambient conditions such as moisture, atmospheric pressure or wind.

Use of alternative sampling and measurement technologies, which can be conducted more frequently, or ideally continuously, can improve emission estimates and process understanding. Although these methodologies may be less accurate in themselves, the possibility to sample more frequently and hence better cover the process variations might improve the overall accuracy of the emission estimates [5]. In addition, the move towards industry 4.0, with more process control systems being digitized and requesting online data to be used as feedback, is providing insight in how and which sensors can be used in the industry. It is also desirable to combine the more frequent sampling by simple methodologies with the certified standard high precision sampling methods to establish correction factors for the simplified methodologies.

A production plant producing electrode paste briquettes and cylinders for Søderberg electrodes in metallurgical industries was used as a test site to evaluate different measurement strategies. The production of paste is a batch process and involves heating, mixing and casting operations of different pitch and raw materials like calcined anthracite and for some products also calcined coke materials. All of these processes may result in the release of PAH components. The amount and composition of the emissions will depend on the content of the coal-tar pitch, the softening point as well as the actual temperatures and time.

Two different emission points were selected to evaluate the performance of short-term sampling on thermal desorption (TD) tubes as well as online measurement with a photo ionization detector (PID) implementation. Conventional extractive sampling on filter and adsorbent (XAD2) was selected as reference measurements.

Measurement Principles

Three different measurement strategies were used, conventional extractive sampling on filter and XAD2, sampling on TD-tubes as well as online monitoring with a PID implementation. Each of these approaches has different sampling times, accuracies, and measurement challenges associated with them. Previous experience and historical, unpublished measurement data from various relevant industries have indicated that the gaseous PAH fraction tends to be significantly larger than the particle-bound fraction of the total PAH emissions. Hence, the measurement methods applied in this study were selected on the assumption that the gaseous fraction dominates the PAH picture.

Extractive Sampling

The extractive sampling was conducted using constant volumetric flow sampling. Two different setups with equipment from different manufacturers were used, but both had the same functionality and capability to sample gas and particulate matter in a representative manner in accordance with VDI 2066, EN 13,284–1, EPA5, and ISO 9096. The two setups were "ITES Automatic isokinetic sampler" (ITES) by Paul Gothe GmbH, Germany and "TPS4-ISOK4" by Environment AS, Germany (ISOK4). These setups were used to collect 2 and 24 h extractive samples, where the particulate matter was captured on glass microfiber thimble filters (MK160) with analytical purity, and gaseous PAH components by XAD2 adsorbent. The filter and sampling probe were kept at 120 °C to avoid condensation of PAH or water prior to the XAD2 adsorbent. Both the XAD2 and filter were then shipped to a commercial certified laboratory for analysis. A typical setup can be seen in Fig. 1, and it consists of several heavy components such as a high-capacity suction pump, control unit, filter, and sampling probe. Typical overall weight for a setup is 20–30 kg.

TD Tubes

Sampling by TD-tubes represents a significantly simplified sampling procedure compared to the standardized extractive sampling methods. A small flow, typically between 20 and 500 mL/min is sampled directly onto an adsorbent that is packed in a ¼" tube. Duration of a sample will vary depending on flow and expected concentrations but could be anything between a few minutes up to some hours.

TD-tubes have a widespread use for working environment monitoring but are not frequently used for emission monitoring. There is no separation between particles and gaseous components and the TD-tubes are not suitable for very dusty environments, although small amounts of particulate matter are usually not a problem. Analysis of TD-tubes is performed by heating the tube with the adsorbent and purging it by a clean gas to directly transfer the sample into a gas chromatograph (GC) for analysis. This is a simple 1-step procedure, and much simpler than the liquid extraction of PAH from XAD2 and filter as in the case of extractive sampling.

A typical kit with a battery-operated sampling pump, tubing, and a set of TD-tubes can easily fit in a backpack and will be lighter than 1 kg. For these experiments Carbograph TD1 from Markes International was used as an adsorbent due to availability. For better recovery of the heavier fraction of PAH components other adsorbents such as Tenax TA and XRO-440 [6], Carbopack C or glass beads. There are also

Fig. 1 Schematic view of the sampling system

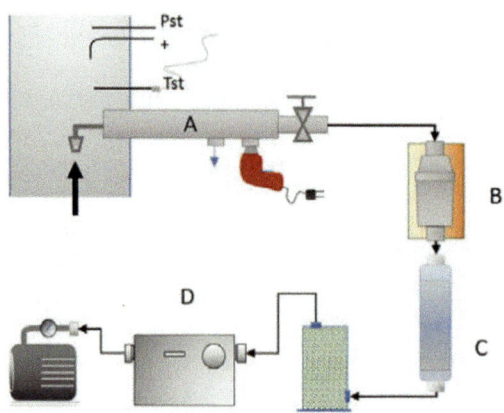

A. Heated sample lance
B. Heated filter-housing
C. XAD- 2
D. Flow control unit
   • Silica gel drying-tower
   • Control unit
   • Pump

purpose paced TD-tubes such as Markes PAH-tubes that are claimed to give good and linear response for the entire EPA-16 range of PAH components [7]. Sampling time was 10 min, and flow was 125 mL/min, controlled by a SKC PocketPump Touch.

Photo Ionization Detection, PID

SINTSENSE, an inhouse developed PID implementation is based on a Raspberry PI as a controller and logger and was created as a low-cost online monitoring system to follow trends in VOC and light PAH emissions. It does not have a selective response for each individual component but will show a trend for the total gaseous hydrocarbon emissions. It has been demonstrated to work satisfactory for up to 8 weeks in the off-gas of a Fe–Mn smelter [8].

A PID works by ionizing the targeted components in the gas mix by light and then measures the current generated when the ions hit a detection electrode. PID will not be selective as it will ionize all gaseous molecules that can be ionized by the output energy from the lamp. The PID used for these trials had a 10.6 eV lamp and will ionize all cyclic organic compounds. A PID has a fast response time, and by recording the value every 5 s we got a continuous trend. The PID outputs a voltage change as a response to changing hydrocarbon content, this voltage can be calibrated to a concentration, but only for a single hydrocarbon or a pre-defined mix of hydrocarbons. It is not possible to separate the contributions from each individual PAH component that is part of the signal. All results in this paper will be reported as naphthalene equivalents, which is the concentration of naphthalene that would give the recorded voltage of the sensing electrode. Since a PID electrode sensitivity generally increases with molecule size, a mixture of naphthalene with heavier PAH components would most likely result in a high estimate [9]

The results are mixed contributions from all PAH and other VOCs in gas phase calculated by the factory naphthalene calibration slope. For permanent use at a given PAH source this factor could be calibrated to increase accuracy.

The main aim for using the SINTSENSE in these trials is to get information about dynamics and how much hydrocarbon emission changes during the 2 and 24 h samples. By comparing PID results to the extractive sampling it is possible to get a lot of information about trends and time variations in the process emissions that cannot be captured by other means.

Results

In these studies, 2 different PAH sources on a carbon electrode paste plant were studied. The first source is the exhaust gas from a mixer, and the second is the draft from the casting line of this mixer. The emission points from the mixer have an adsorption system and dust filter where PAH is returned to the mixer by using the carbon (raw) material as a dry-scrubbing agent. The emission point from the casting does not have such a filter, but a wide stack with a condensation trap.

The ventilation from the casting had sufficient cross section to allow for all measurements to be conducted simultaneously. In Fig. 2 the complete setup at the stack over the castling line can be seen. The ISOK4 sampling line is on the left-hand side, while the ITES sampling line is seen just to the right of the center of the stack. TD-tubes and SINTSENSE sampling was conducted on a small hole at the back of the stack. There was no filtering of the gas at this stack, just a condensation trap at the bottom. The stack from the mixer had too small a diameter for simultaneous sampling, and sampling of 2 and 24 h samples had to be done in sequence.

Fig. 2 Picture of the casting emission point with all equipment installed

Reference Measurements

The fraction of PAH captured on the sampling filter (i.e. the particle-bound fraction) constitutes less than 1% of the overall emissions from the mixer, while from the casting around 10% of the overall PAH emissions were recovered from the filter. This confirms the assumption that the main part of PAH emissions is gaseous, and hence that our approaches to focus only on gaseous components are valid for this application.

Figure 3 shows the concentration estimates from the sampling on the casting stack. Measurements by all 3 methodologies are overlaid in the same figure. The width of the red, blue, and green bars represents the timeframe from which the samples are collected, and hence their value can be viewed as an average emission for that timeframe.

We can clearly see that there is a significant variation between each 2 h sample. The total PAH in the 3rd 2 h sample is only 26% compared to the 2nd sample. A typical routine sampling for reporting would consist of 2 such samples, and it is clear to see that two of these samples do not cover the full picture. Even for a 24 h timeframe, the representativity of two 2 h samples is poor.

Emissions from the mixer, shown in Fig. 4, had to be collected in sequence due to the stack diameter. The total PAH variation between the samples here is lower but

significant. The lowest 2 h sample is only 85% of the highest sample even though they were collected within a 40 h timeframe.

PID Measurements with SINTSENSE

In Figs. 3 and 4, the SINTSENSE signal is indicated by the black trendline, providing high time resolution data from which the dynamics in the emission profile can be studied. Events such as filling, mixing, and tapping can clearly be identified in the emissions from the mixer. It is also clear that the average PID signal represented by the magenta lines follows the same trend as the analysis of the total PAH from conventional sampling. Emissions from the casting line seem to have less variations from minute to minute and do not appear to be so affected by production events. However, the correlation between the averaged SINTSENSE signal and the conventional sampling is still strong.

Another finding is that the signal from the SINTSENSE indicates significant variation of PAH emissions between production batches. This is unexpected, as process estimations would have suggested that each batch of the same recipe should be identical when it comes to emissions. Possible explanations here are differences in mixing time before casting, and possibly also changes in the draft profile in the building due to operation of drive gates, natural

Fig. 3 Comparison of
naphthalene trends (upper) and
total PAH trends (lower) for all 3
methodologies during the 24 h
timeframe. Width of columns
represents the sampling time
covered

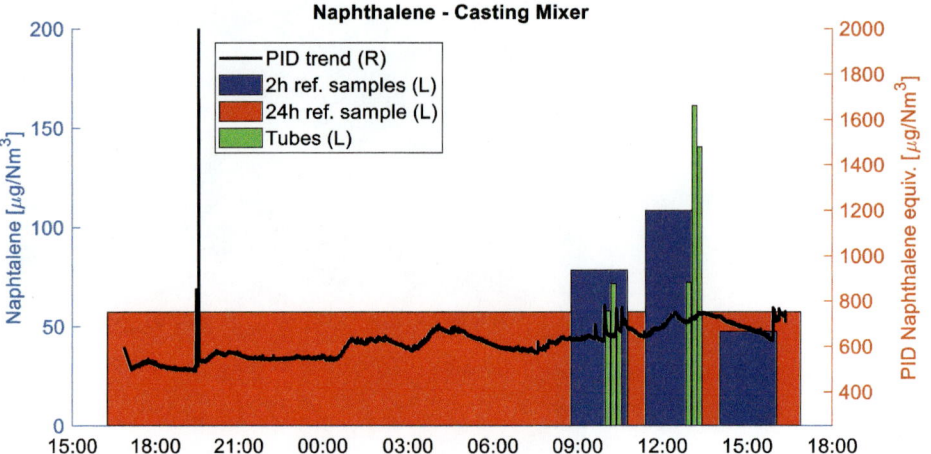

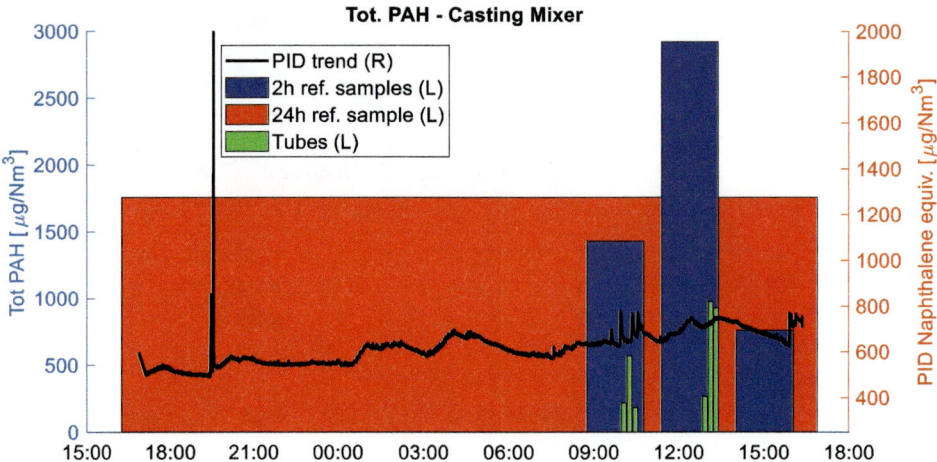

Fig. 4 Comparison of total PAH
trends for all 3 methodologies
during the collection of samples.
The planned 24 h sample had to
be slightly shortened for logistic
purposes and ended up around
20 h. The width of columns
represents the sampling time
covered

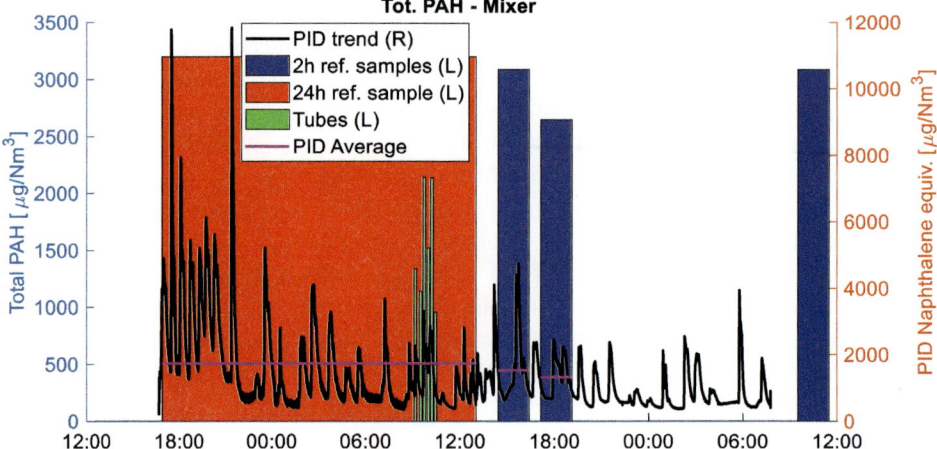

convection around high-temperature equipment (hall wind) or outside wind conditions and atmospheric pressure.

TD Tubes

Naphthalene estimates from TD-tubes and reference measurements using XAD2 and filter seen in upper part of Fig. 3 have a good agreement quantitatively. The TD-tube samples (green bars) vary within the expected range for the XAD2 sample (blue bars), at least when taking into consideration that the three 10 min samples only cover 25% of the sampling time of the 2 h sample.

For total PAH seen in the lower part of Fig. 3, the TD-tubes clearly underestimate the emissions compared to the reference samples. There are two explanations for this. Firstly, the TD-tubes have a limit for how much they can be heated. Hence it is challenging to completely desorb the heavy PAH components, especially in this case as the adsorbent is stronger than it ideally should be. This will impact recovery of heavy PAHs as to high heating will lead to decomposition of PAH. Secondly, the sampling is not isokinetic and the actual sampling itself will therefore under-sample particulate matter and the particle-born PAH is known to carry most of the heavy PAH components.

A comparison of the recovery of 6 PAH components from TD-tubes and reference measurements can be seen in Fig. 5. At least up to anthracene, the TD-tubes seems to produce results that are meaningful when comparing with the 2 and 24 h samples, especially when considering that there is not a complete overlap in sampling time between these samples. This is best seen in the samples from the casting (right) as these samples had a higher total content of heavier PAH due to the absence of any filtering, and hence naphthalene is not so dominant in the overall results.

An improved adsorbent such as Tenax in the TD-tubes would probably allow for better recovery of more heavy PAH components [6]. The gas chromatography mass spectroscopy (GCMS) methodology might also be limiting, depending on the setup of the laboratory.

To be able to estimate the total PAH emissions based on results from TD-tubes it would be necessary to establish a correlation factor between certain of the light PAHs and the heavier ones, especially in cases with more particulate matter (higher dust concentrations) since the particle-bound fraction is quite poorly recovered by this methodology. Such correlations could potentially be established by yearly reference measurements.

In Fig. 6 we can see an enlargement of the section in Fig. 4 where the TD-tubes are sampled. Note that SINT-SENSE response is plotted at the right axis, while the other values are on the left axis. The SINTSENSE response is given in naphthalene equivalents so a direct comparison of numbers is not possible, but the overall agreement is very good, nevertheless. The average PID signal from the time periods where the TD-tubes were sampled has a high degree of correlation. It is worth to keep in mind that both these methodologies have the same bias when it comes to only being sensitive towards gaseous PAH components. This bias is less of a concern for emission points equipped with a filter (such as the mixer case in this study) since the main part of PAH components would be expected to be in the gaseous state anyway. For emission points with unfiltered gas, and thus more dust particles present, the effect of this bias would be more significant.

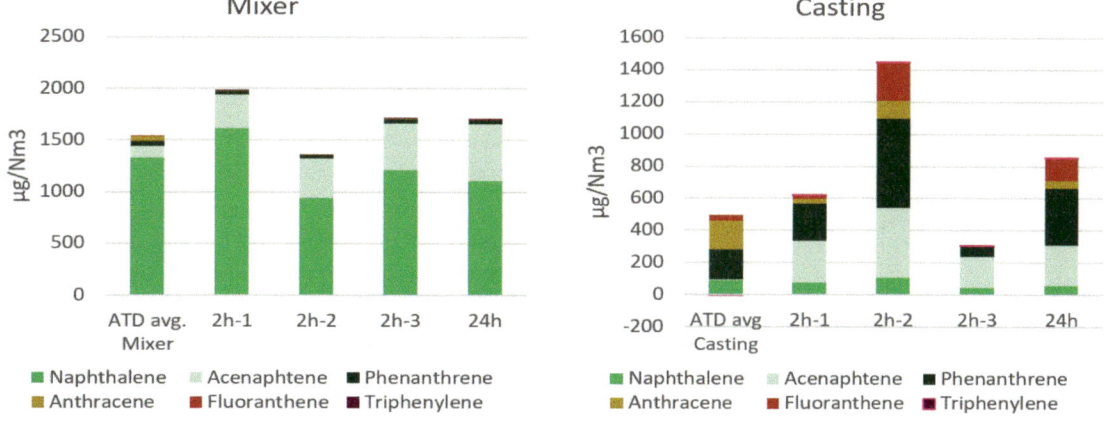

Fig. 5 Variations in the 6 most volatile PAH components in off-gas from the mixer (left) and casting from the mixer (right). For TD-tubes all 6 samples from each sampling point are averaged, giving a total sampling time of 60 min

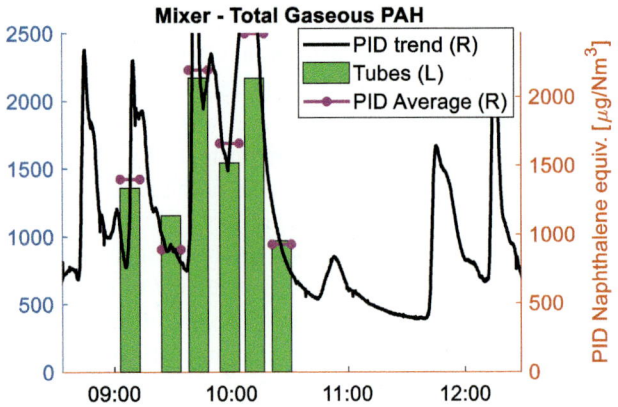

Fig. 6 Average PAH recovered from TD-tubes with SINTSENSE trend and averages overlayed

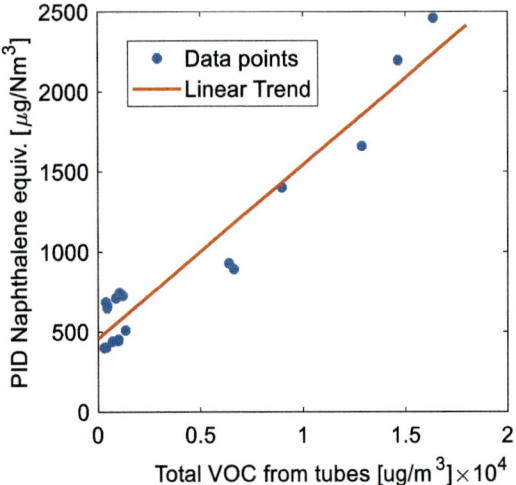

Fig. 7 Correlation between PID naphthalene equivalents estimated by the SINTSENSE and total VOC estimate from the TD-tubes gives $R^2 = 0.93$

In Fig. 7 we can see the overall agreement of the SINTSENSE signal and the TD-tubes for all simultaneous samplings conducted in this trial. The correlation is strong and linear over the entire concentration span and the actual slope seems to be very consistent. However, there appears to be some baseline shift in the data from the SINTSENSE resulting in a shift along the y-axis for one of the groups of data.

The rapid response from the SINTSENSE could be useful to improve or tune a process with regards to minimizing emissions, or to build an understanding of the process variations over time. Having a PID detector constantly monitoring the emissions in the weeks before and after a yearly certified extractive sampling could, for example, place that sample in an informative context, indicating if the

analysis were performed in a timeframe with average conditions or if the emission profile of the sampling day were deviating from normality.

Conclusions

There is no single way of measuring PAH emissions that are precise, easy, and cheap to conduct. A combination of certified reference measurements that can be supported by more frequent sampling by simplified low-cost measurements could improve the understanding of process variations and emission dynamics. It could also improve the representativity of the yearly emission estimates used in environmental reports.

The fraction of PAH captured on the sampling filter (i.e. the particle-bound fraction) constitutes less than 1% of the overall emissions from the mixer, and around 10% from the casting. Hence, the PAH emissions from the processes studied here are dominated by the gaseous PAH fraction. The simplified measuring methods focus primarily on gaseous components and are deemed valid for these applications. It is clear, however, that the dust concentration in the sampled off-gas is decisive for the suitability of these methods.

Both TD-tubes and the PID-based SINTSENSE show results that are in good agreement with expectations. Both correlate well with each other and with reference sampling using filter and XAD2. The results show that there are significant variations in the emissions both from second to second as well as between two 2 h samples taken back-to-back.

PID implementations such as SINTSENSE can capture trends and have short response time and low noise. Despite not being quantitative, it can give useful process information. This information has a significant potential to gain insight in process variations and planning of sampling strategies as well as for tuning and improvements of process control.

TD-tube sampling with Carbograph TD1 seems to give a representative and good analysis of anthracene and lighter PAH components, a weaker adsorbent would improve the method for heavier PAH-components. Sampling on TD-tubes can easily be conducted by plant personnel, and ease of sampling and comparably low cost of analysis could enable frequent sampling intervals with useful quantitative results as a way of capturing emission fluctuations between certified reference measurements.

Acknowledgements This work is part of the PAHssion project, funded by the Norwegian Research Council, grant number 295744.

References

1. I. Kero, P. Eidem, Y. Ma, H. Indersand, T. Aarhaug og S. Grådahl, «Airborne emissions from Mn ferroalloy production,» *JOM,* vol. 71, nr. 1, pp. 349–365, 2019.
2. I. Kero, S. Grådal og G. Tranell, «Airborne Emissions from Si / FeSi Production,» *JOM,* vol. 67, pp. 365–380, 2017.
3. J. Borgulat og T. Saszewski, «Fate of PAHs in the vincinity of aluminium smelter,» *Environ Sci Pollut Res Int.,* 2018.
4. «Norwegian PRTR,» [Internet]. Available: https://www.norskeutslipp.no/en. [Funnet 2022].
5. O. S. Kjos, T. Aarhaug, H. Gaertner og A. Brunsvik, «Dynamics of Anode Baking Furnace VOC Emissions Through a Firing Cycle,» i *Light Metals 2022,* 2022.
6. M. Wallace, J. Pleil, D. Whitaker og K. Oliver, «Recovery and reactivity of polycyclic aromatic hydrocarbons collected on selected sorbent tubes and analyzed by thermal desorption-gas chromatography / mass spectrometry,» *J. Chromatogra A.,* 2019.
7. M. International, *Application Note 115: Simple and reliable quantitation of ppt-level PAHs in air by TD-GC-MS,* Markes International, 2016.
8. T. A. Aarhaug, O. Kjos, I. T. Kero og E. Per Anders, «SINTSENSE: Low-cost Monitoring of VOC emissions from the Ferroally Industry,» *Proceedings of the 16th International Ferro-Alloys Congress (INFACON XVI),* 2021.
9. M. L. Langhorst, «Photoionization Detector Sensitivity of Organic Compounds,» *Journal of Chromatographic Science,* vol. 19, pp. 98–101, 1981.

Investigation of the Stacking Effects on the Electrical Resistivity of Industrial Baked Anodes

Thierno Saidou Barry, Donald Picard, Guillaume Gauvin, Julien Lauzon-Gauthier, and Houshang Alamdari

Abstract

Producing anodes with the highest uniformity and consistent properties is essential to improve the energy efficiency and productivity of the Hall-Héroult process. Achieving this objective is a great challenge, as it requires advanced investigations of all parameters related to process and materials which may influence the quality of the anode. Stacking anodes at the baking stage may be another parameter influencing their final properties. During the baking process, the anodes are stacked in the pits of the furnace and baked according to predefined temperature profiles. Stacking the anodes may generate mechanical stress on the lower anodes. This additional stress state could affect the rearrangement of the anode coke particles, leading to internal structure anisotropy. To investigate the effect of this parameter, we measured the electrical resistivity and density of the anodes baked at different positions in the furnace. The results show a statistically meaningful variation of the electrical resistivity as a function of anode position in the rows of the furnace.

Keywords

Anodes baking • Stacking effect • Mechanical stress • Electrical resistivity • Hall-Héroult • Aluminium

T. S. Barry (✉) · H. Alamdari
Department of Mining, Metallurgical and Materials Engineering, Laval University, Quebec City, QC G1V 0A6, Canada
e-mail: thierno-saidou.barry.2@ulaval.ca

D. Picard · G. Gauvin
NSERC-Alcoa Research Chair and RDC-UL Research Project, Quebec City, QC G1V 0A6, Canada

J. Lauzon-Gauthier
Continuous Improvement Smelting Technology, Alcoa, Deschambault, QC G0A 1S0, Canada

H. Alamdari
Aluminium Research Centre REGAL, Laval University, Quebec City, QC G1V 0A6, Canada

Introduction

The major challenges for the aluminium industry are to reduce its greenhouse gas emissions and its high energy consumption during the electrolysis of aluminum. Aluminum plants emit 5–20 tons of CO_2 [1] and consume 70.6 GJ of energy per ton of aluminum produced, of which 49.0 GJ are directly linked to the Hall-Héroult process [2]. Since the first application of the Hall-Héroult in the industry in 1914, several inventions have been adopted to reduce its energy consumption. These inventions have resulted in a considerable decrease in specific energy consumption with an increase in the amount of aluminum produced. From 1914 to 2000, the energy consumption in a modern cells has been reduced from 32.53 kWh to 13.10 kWh per kilogram of aluminium with an amount of 2308 kg of aluminium per day and with a current intensity of 300 kA [3].

Nevertheless, it remains until now a very energy-intensive process. To improve its energy efficiency, the use of anodes with uniform and consistent final properties is essential [4]. But, achieving this objective in aluminum smelters is a great challenge, as it requires advanced investigations into all parameters that can influence the manufacturing process and the quality of the anodes. Several investigations have more recently been carried out for this purpose and made it possible to understand some of these parameters. Most of these studies reported that the non-uniformity of the anodes final properties is due to the effect of the forming process (mixing and vibro-compaction/compaction) [5–8]; or it is due to the effect of their baking process (non-uniformity of baking temperatures) [9–12].

During baking, the anodes are stacked together in the pits and are baked according to predefined temperature profiles. The stacking may be another parameter influencing the final properties of anodes because it may generate additional mechanical stresses on the anode, especially on those lowest in the pit. This, combined with the thermochemical process of baking, could lead to internal structure anisotropy of these anodes. We hypothesize that the anode can deform under the stacking stress, resulting in the rearrangement of the coke particles and further reducing the distance between them in the direction parallel to the applied stress. In the literature and to the best of our knowledges, this effect on the final properties of anodes has never been investigated. In this study, we aim to analyze this effect on the uniformity of the final properties of baked anodes by mean of the electrical resistivity measurements and the apparent density determination. We have chosen the electrical resistivity method due to its simplicity and sensitivity [13]. It is also worth mentioning that the distributions of the temperatures inside the pits and inside the anodes were not known. Therefore, this effect on the subsequent analyses was not considered as the goal here was not to study a causal and effect phenomenon. But it was to check if there is a significant correlation between the electrical resistivities of the anodes and the stacking during the baking process.

Materials and Methods

Materials Description

The experimental measurements were performed on eighteen (18) carbon anode blocks manufactured and baked at an Alcoa smelter. Two different vibro-compactors were used for the forming process and all anodes were baked in the same furnace pit. We have denoted the two vibro-compactors as Vibro1 (V1) and Vibro2 (V2). Also, the position of each anode in the pit was known and they are labeled according to their row as top anodes (TA), middle anodes (MA), and bottom anodes (BA). Each row contained three anodes from V1 and three anodes from V2. An illustration of the anodes with the orientation of their stub holes and labeling in the pit is presented in Fig. 1.

Electrical Resistivity Measurements of Anodes Blocks: Four-Points Probe Method

The electrical resistivity of the anodes was measured using the Four-points probe method (4PP). It is a non-destructive method and allows to perform potential difference measurements at the surface of the anodes. It is also easier to perform measurements by this method at the same positions but in different measurement directions. According to the ASTM [14], a probe connected to a power supply DC HP/Agilent 6031A was used for the potential difference measurements. The probe was made up of four aligned probes including two outer ones through which the current is applied to the anode and two inner ones through which the potential difference relative to the supplied current is measured. The distance between the current conducting probes was 12.25 cm and the distance between the voltage probes was 10 cm (see Fig. 2).

Fig. 1 Illustration of the position of the anode blocks in the furnace

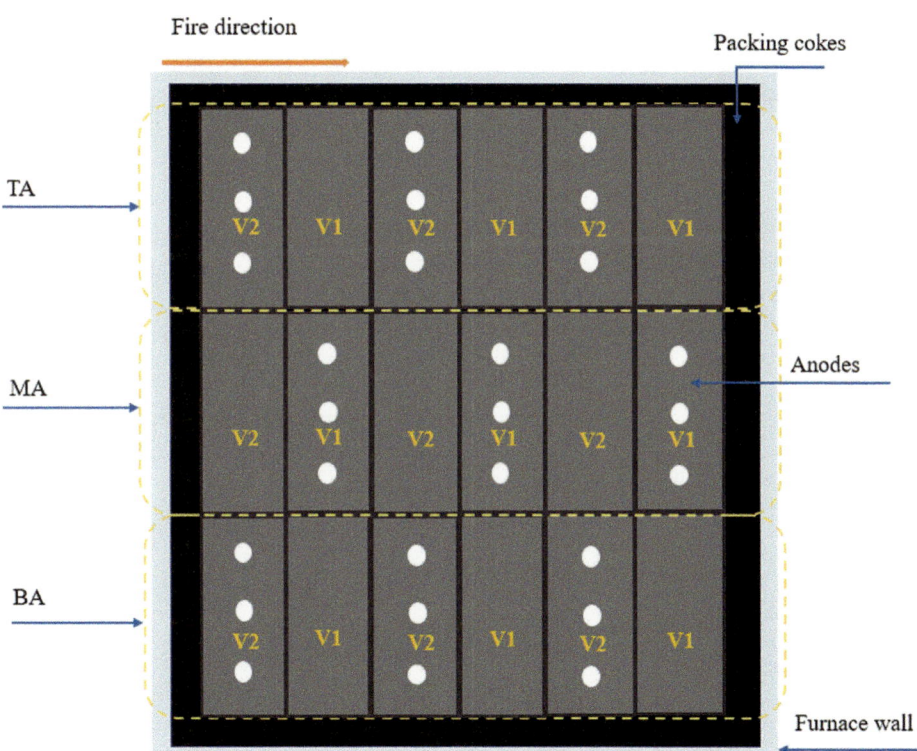

Fig. 2 Four-points probe apparatus for the electrical resistivity measurements

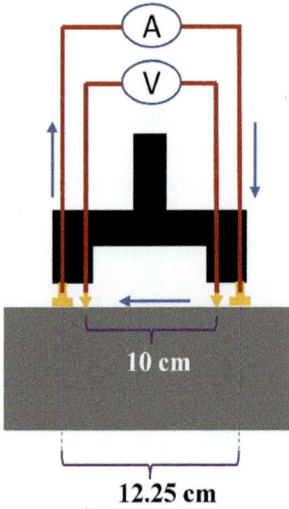

The potential difference measurements were carried out by placing the probe at specific locations (P1 to P6) on the anodes, as schematized in Fig. 3. These locations were on the two large lateral vertical surfaces of the anodes and the stub holes were chosen as positioning reference along the anode length. In each anode, 24 measurements were performed for a total of 12 measurements on each surface (see Fig. 3). Also, on each point, two measurements were taken in different directions. The directions of the measurements were vertical and horizontal. The horizontal and vertical direction references are in the compaction directions and the anode lying with its bottom towards the ground. As there were 18 anodes, a total of 432 measurements have been obtained. But 4 measurements from BA and 2 measurements from MA anodes are missing. These measurements could not be performed due to the presence of rigid packing coke on specific location of the surface of those anodes. In Table 1, the number of observations according to the number of anodes is presented. In Fig. 3, the anodes are illustrated according to their orientation during vibro-forming process. In this figure, the distance of 12.25 cm shows the distance between the two outer current probes of 4PP instrument, and the distance of 12 cm shows the distance where the

instrument is placed as a function of the upper and bottom layers of the anode.

To calculate the electrical resistivity from the resistance calculated with the 4PP measurements, a constant correction factor was determined by Finite Element Modeling (FEM) of the current flow path in the anode for each position and orientation. The formula to determine the electrical resistivity is presented in Eq. (1). Also, on one single anode, an external crack was detected at one measurement in the vertical direction. For this reason, an outlier detection test was performed only in this direction. In the horizontal direction, there was no physical reason obliging to perform outlier test and all measurements were recorded as it with the same method. The outliers were determined by calculating upper and lower limit bonds of all observations in vertical direction. Also, quantile–quantile diagrams have been plotted to verify the normal probability distribution law of observations with and without outliers. The quantile–quantile plots for both directions are presented in the results session.

$$\rho = k_\rho \frac{\Delta V}{I} \qquad (1)$$

In the formula ρ ($\mu\Omega m$) is the electrical resistivity, ΔV the potential difference (V), I the current intensity (A), k_ρ the correction factor of the resistivity ($\mu\Omega$-A-m)/(V).

Samples Coring and Cutting Methods

Three cylindrical core samples were obtained from each anode block, for a total of 54 samples. The cores were taken in the anode's stub holes (see Fig. 4). C1, C2 and C3 represent the core samples. For all anodes, C1, C2 and C3 were cored at the same locations. Their dimensions were 16 cm high and 5 cm in diameter. After coring, the samples were classified according to the position of each anode blocks in the furnace. They were then cut into smaller slices of 2 cm thickness and 5 cm in diameter with a crosscut saw. The slices selected for measurements were only taken from the top of the cores, due to the presence of microcracks on the

Fig. 3 Illustration of the locations of measurements on the surface of anodes according to horizontal direction and vertical direction. P1 to P6 indicate the locations of the probes

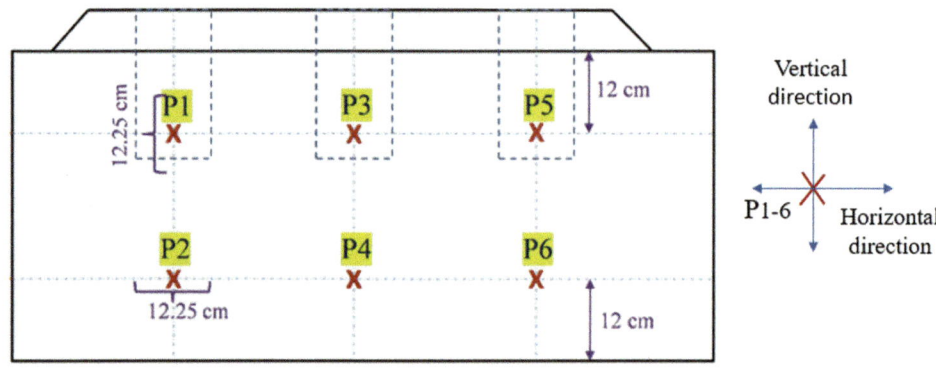

Table 1 Number of observations according to the number of anodes and rows in the furnace

Row of anodes in the furnace	TA	MA	BA
Number of anodes per row	6	6	6
Number of measurements per anode	24	24	24
Number of missing observations	0	2	4
Total of measurements per row	144	143	140

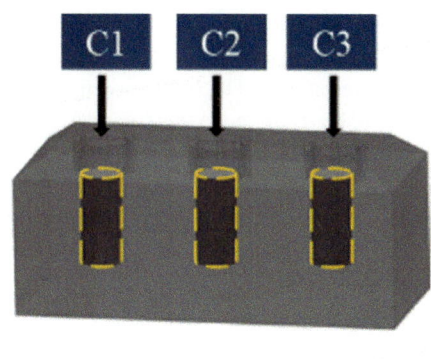

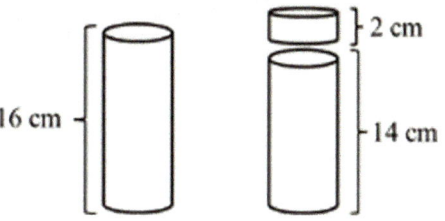

Fig. 4 Illustration of the location of cores on an anode block and the cut samples for electrical resistivity measurements

remaining part of the samples. The electrical resistivity and the density of 54 core samples (18 cores for each row of the anodes in the furnace) were analyzed (see Table 2). Also, quartile–quartile diagrams have been plotted to verify the normal probability distribution law of observations for electrical resistivity analysis and for density analysis.

Electrical Resistivity of Cores: Van Der Pauw Method

The Van der Pauw method works for any form of small samples which are much smaller in thickness than the diameter. The method was previously proposed to determine the electrical resistivity of carbon anodes samples by Rouget al. [15]. This method was used to obtain more information on the electrical resistivity of core samples of the

baked anode blocks. The setup used for the measurements was the same as that of [15] and it is presented in Fig. 5. It is constituted of four probes connected to four conducting copper bars. A small anode sample (2 cm thick and 5 cm in diameter) is placed between the bars. Two of the probes apply the current to the sample and the two others measure the potential difference. The setup was connected to a power supply DC HP/Agilent 6031A providing a current intensity of 1 A. The mathematical formula used to determine the electrical resistivity of the samples is presented in Eq. (2).

$$e^{-\pi.\frac{e}{\rho}.R_1} + e^{-\pi.\frac{e}{\rho}.R_2} = 1 \qquad (2)$$

In the formula, e is the thickness of the cores, R_1 and R_2 are the calculated electrical resistance, and ρ is the electrical resistivity.

Core Samples Apparent Density Determination Method

The baked apparent density of the slices (2 cm thick and 5 cm in diameter) from core samples was determined by the geometrical method, based on the ASTM [16] as presented in Fig. 6. This method consists of measuring the length (L) and the diameter (d) of the samples; and then calculating the ratio of the mass (m) of the samples to their volume (v). The mass was determined by an AND EJ-6100 of 6100 g capacity balance with an accuracy of 0.1 g. The measurements of the length and the diameter were carried out by a caliper Mitutoyo CD-12 "PSX with a measuring range of (0–300) mm and with a minimum indication of 0.01 mm.

Analysis Method

All results were analyzed by using descriptive statistics and plotting the mean and confidence interval of each group. The difference between the means was analyzed by using t-tests at a

Table 2 Number of measurements for different core samples and anodes rows in the furnace

Row of anodes in the furnace	TA	MA	BA
Number of anodes blocks per row	6	6	6
Number of cores per anode	3	3	3
Number of cores per row	18	18	18

Fig. 5 Set-up of Van der Pauw to determine the electrical resistivity of cores samples [15]

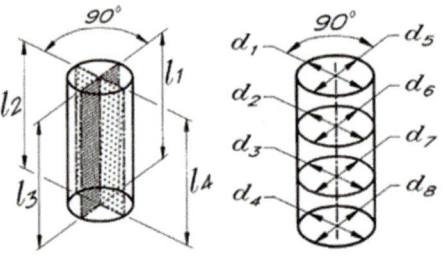

Fig. 6 Geometrical method to determine the apparent density of the core samples [16]

confidence level of 95%. Also, we have calculated separately the coefficient of variance (CV) of each group to check the dispersion of the observations by row. For 4PP method, all analysis were performed after removing outliers. The outlier detection method proposed in [17] chapter 7 was used to calculate upper limit (UL) and lower limit (LL) bonds using the first quartile (Q1), third quartile (Q3) and quartile range (QR) of observations using Eq. (3) and Eq. (4) respectively were determined. This was to check if the single data collected at the location of the crack in the vertical direction can be considered as an outlier. For cored samples, all observations were considered as there was no physical reason to remove outliers.

$$UL = Q3 + 1.5 \times QR \qquad (3)$$

$$LL = Q1 - 1.5 \times QR \qquad (4)$$

Interpretation of Data According to the Stacking

It is important to mention that the objective of this study was to analyze the electrical resistivity of a series of industrial anodes as a function of their stacking in the furnace during baking process. The results obtained are supposed to help in the identification of correlations that may influence the resistivity of anodes and to open new door for further investigations. These results do not allow to generalize the effect of stacking on all anodes baked in the industrial furnace. The principal reason is that the measurements were not gathered through a design of experiments, and we do not know exactly the temperature distribution in the furnace for each anode. The baking temperature and heat-up rates are known to have an influence on the electrical resistivity. The only known parameter is that the distribution of load applied to these anodes in the furnace is not uniform and the load applied on lower anodes is supposed to be higher than upper anodes. To help understanding the shortcoming of this study and how the results can be interpreted, the load applied on anodes according to their rows were roughly calculated by considering only the weight of 1000 kg per anode with a cross-sectional area of 0.38 m². The weights of packing cokes and refractories walls of the furnace have not been considered. For the cross-sectional area, we have considered only the small rectangular section of the anode. The formula to calculate the load is presented in Eq. (5) and (6) and illustrated in [18]. In this equation, the weight can change according to the row of the anode. Also, we have supposed that there was not applied load on TA. The calculated load on MA was equal to 25 kPa and the load on BA was equal to 50 kPa. Therefore, in the results section, the stacking effect of anodes refers only to the non-uniform distribution of the loads.

To this end analysis was performed as a function of the pit rows to:

1. determine the change of anodes electrical resistivity. In this part, all observations were considered.
2. evaluate the homogeneity of the electrical resistivity. In this part, the observations were grouped into upper and lower groups. The upper and lower groups refer to the top and bottom sections of the anodes respectively according to the vibro-compaction.
3. evaluate the presence of local anisotropy. The local anisotropy was determined by calculating the difference between horizontal and vertical observations using Eq. (7).

$$Load\,on\,MA = \frac{m_{anode} \times g}{A} = 25\,\text{kPa} \qquad (5)$$

$$Load\,on\,BA = \frac{2x m_{anode} \times g}{A} = 50\,\text{kPa} \qquad (6)$$

$$\Delta_{H,V} = ER_H - ER_V \qquad (7)$$

In these equations, m_{anode} is the weight of a single anode, g the gravitational acceleration, A the cross-sectional area of the anode, $\Delta_{H,V}$ the difference between horizontal and vertical measurements, ER_H and ER_V the electrical resistivities in horizontal and vertical directions respectively.

Results and Discussion

Quantile–Quantile Plots and Outlier Detection for Electrical Resistivity Measurements with 4PP

The quartile–quartile diagrams of each group (horizontal and vertical) before and after removing outlier are presented (Fig. 7) and the numerical values of Q1, Q3, QR, UL and LL are presented in Table 3. In Figs. 7a, b, it is possible to see that all measurements performed in the horizontal directions follow accurately a normal probability distribution law, but measurements performed in the vertical direction did not follow a normal distribution law. This can be explained by the effect of the data recorded in the location of an external crack. The resistivity value obtained at that location was visually higher than other measurements. After removing this outlier from the data, all remaining follow accurately a normal distribution law (Fig. 7c).

Electrical Resistivity of Anode Blocks as a Function of Vibro-Compactors

At first, the anode's electrical resistivity for each vibro-compactors has been compared to verify the potential effect of forming process on the results before comparing the anode as a function of their row in the furnace. In total, the electrical resistivity of 9 anodes per vibro-compactor were taken. The results were analyzed by calculating a t-test between the groups and the coefficient of variance (CV) for each group. The p-value was calculated to evaluate statistically the significant level between the means at 95% of the confidence interval. The coefficient of variation was used to evaluate the dispersion of the groups as all observations follow the law of normal distribution probability. The means and the confidence intervals of these values are plotted in Fig. 8 regardless of the position of the anodes in the furnace. The calculated p-value and CV are presented in Table 4. The analysis shows that there are no differences between the electrical resistivity of vibro 1 and vibro 2 anodes (Fig. 8).

Fig. 7 Quartile–Quartile plots of observations with outliers (**a** and **b**) and without outliers only in vertical direction (**c**)

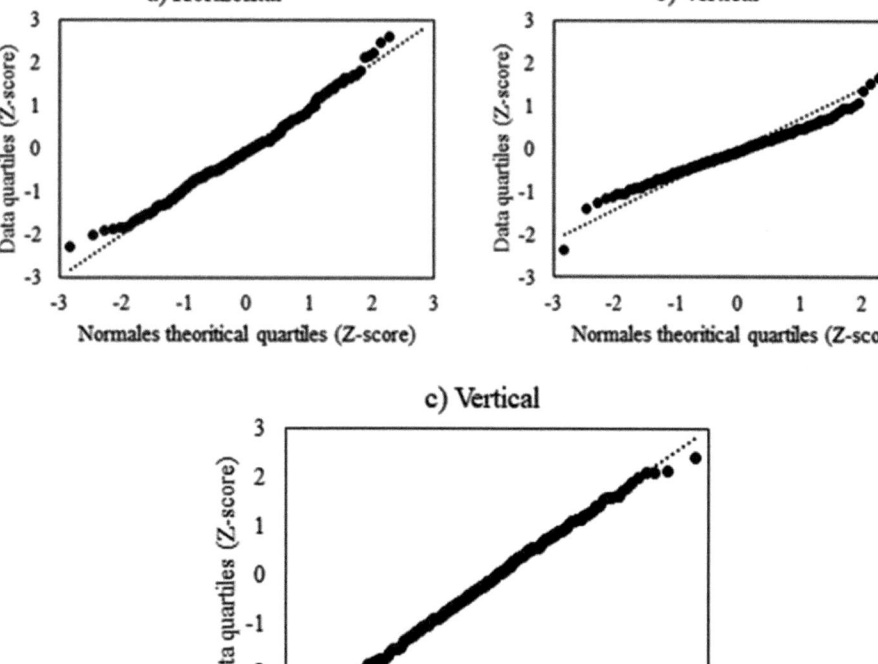

Table 3 Outliers of observations for both measurement directions

	Q1	Q3	QR	UL	LL
Vertical	49.613	54.972	5.359	63.010	41.574

Fig. 8 Electrical resistivity of anodes between vibro-compactors

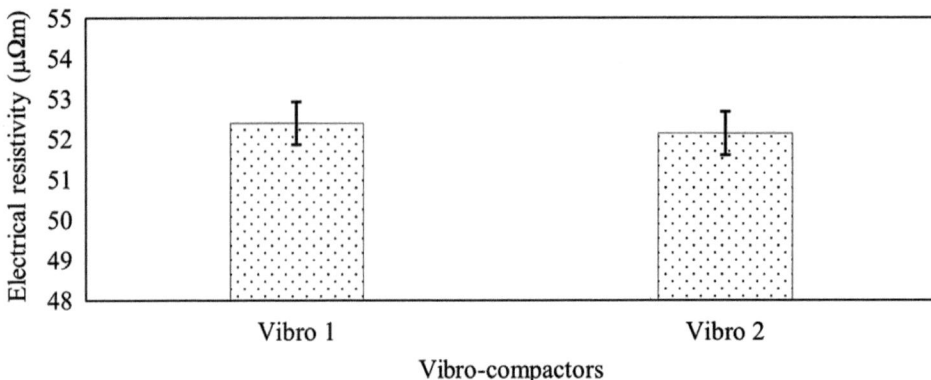

Table 4 CV and p-value between observations of vibro 1 and vibro 2

Vibros	Vibro 1	Vibro 2
CV	0.052	0.053
p-value	0.589	

Table 5 CV and p-value between observations according to the row of anodes in the furnace

p-value						CV		
BA	MA	BA	TA	MA	TA	BA	MA	TA
0.082		0.014		0.307		0.045	0.046	0.062

Stacking Effect on the Electrical Resistivity of Anode Blocks

In this part, the electrical resistivity of the anodes determined by 4PP method as described in the methodology section were analyzed according to their position in the furnace. The number of measurements in the anodes according to their rows was presented in Table 2. A t-tests on the differences between the means of each group pairs (TA-MA, TA-BA, MA-BA) with a confidence interval of 95% were performed to evaluate statistically the difference between means. Also, the CV of these groups was calculated to evaluate if they have similar variability. The results are presented in Table 5 and in Fig. 9. These results show a decrease of the electrical resistivity of anodes from top to the bottom of the furnace and those in the bottom (BA), were significantly less resistive than those in the top (TA). There was no significant difference between BA and MA at 95% of confidence interval. But at 90% confidence interval, the difference between BA and MA was significant as the p-value was only 0.08. These results were the first step in the confirmation of the basic hypothesis of this study, stipulating a correlation between the stacking effect (presence of additional mechanical stress during the

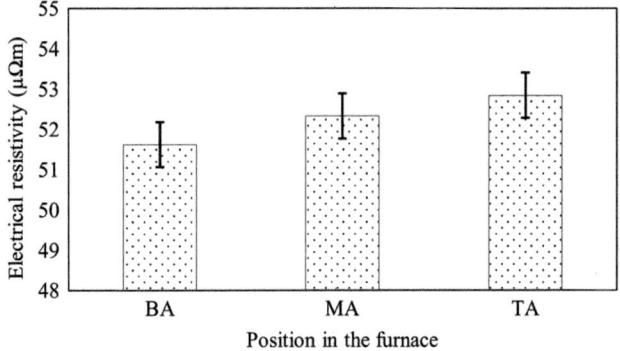

Fig. 9 Electrical resistivity of anodes as a function of their row in the furnace:

baking process) and the resistivity of anodes. However, due to the lack of information on the temperature distribution of these anodes more refined study should be done in controlled environment to reveal this effect on the change of anode electrical resistivity as a function of their baking position. To support this hypothesis, a work has been done by applying a load of 50 kPa under laboratory samples during baking to evaluate the effect of this load on their properties. The work was performed under controlled environment and the temperatures were kept uniform. The results are presented in [18] and show that the load of 50 kPa was enough to influence the properties of the samples by preventing the shrinkage of their diameter related to the baking, significantly decreasing their height due to the deformation of their internal structure by the rearrangement of the structure of capillary forces between the cokes particles and the binder matrix. This phenomenon has also influenced the volume and the baked apparent density of the samples.

Electrical Resistivity of Anode Blocks According to the Direction of Vibro-Forming Process

The stacking effect on the uniformity of the electrical resistivity of the anodes was investigated. The objective was to check if the heterogeneity in the anode could have an impact on the interpretation of differences between the rows. This is done by determining the electrical resistivity of the anodes as a function of the position of the probe on the height of the anodes versus the row of the anodes in the furnace. For the analysis, t-tests between each possible pairs of means and CV has been calculated for statistical evaluation (see Table 6). In these results presented below, upper layer denotes measurements performed in top section and bottom layer denote measurements performed in bottom section according to the orientation of anodes during vibro-forming process (see Fig. 10). Each layer in Fig. 10 contains measurements done in vertical and horizontal directions. The analysis was performed by comparing the electrical resistivity of P1, P3 and P5 positions to the electrical resistivity of P2, P4 and P6 positions. Each position has measurements performed both in the vertical and horizontal directions.

The analyzed results show a different behavior between the layers by row within the furnace. We can see in Fig. 11, that the electrical resistivity in the bottom layers of BA was significantly lower than the bottom layers of TA. For upper layers, there was no difference according to the row in the furnace. These results highlight that the electrical resistivity heterogeneity within the anode vary according to their row in the furnace. In one hand, the explanation on the difference between bottom layers of anodes may be attributed to the presence of an additional mechanical stress due to the loads and the effect of temperature profiles. The combination of these effects could lead to an internal structure deformation of bottom anodes. In the other hand, this difference may be linked to the initial density homogeneity of green anodes. In green anodes, there is known heterogeneity top to bottom of the anodes with the top layers having higher density and less resistive [19; 20] compared to the bottom layers. It is mentioned in [20] that this heterogeneity observed on green anodes may be attributable to the forming process due to the friction with the mold, unequal compact, etc. These new results presented in Fig. 11 are very interesting and show the possible correlation between the initial green density and the impact of load.

Electrical Resistivity of Anode Blocks Between Horizontal and Vertical Measurements

In this section, the electrical resistivity of the anodes was analyzed as a function of measurement directions to check local anisotropy with the 4PP resistivity. The local anisotropy can be defined as a property studying the change in the material internal structure due to sedimentary layers sequencing or existence of subvertical fault systems and

Table 6 Calculated p-value and CV between positions as a function of upper and bottom layer of anodes	Position in furnace	p-values						CV		
		BA	MA	BA	TA	MA	TA	BA	MA	TA
	Upper layer	0.341		0.333		0.894		0.050	0.048	0.061
	Bottom layer	0.124		0.010		0.180		0.040	0.045	0.062

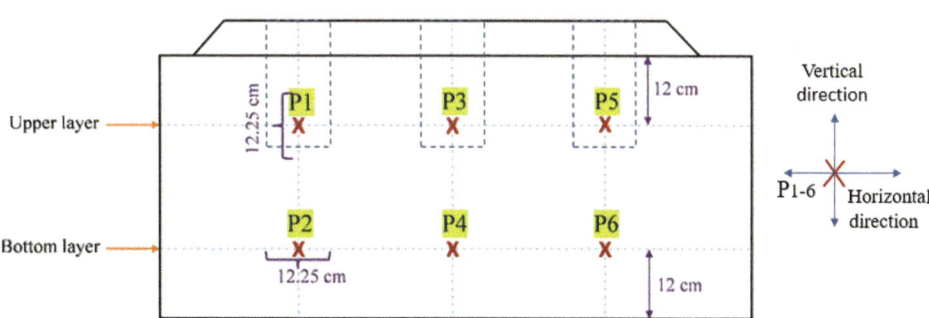

Fig. 10 Identification of upper layer and bottom layer on anode as a function of vibro-forming orientation

Fig. 11 Electrical resistivity of anodes between upper layer and bottom layer versus their row in the furnace

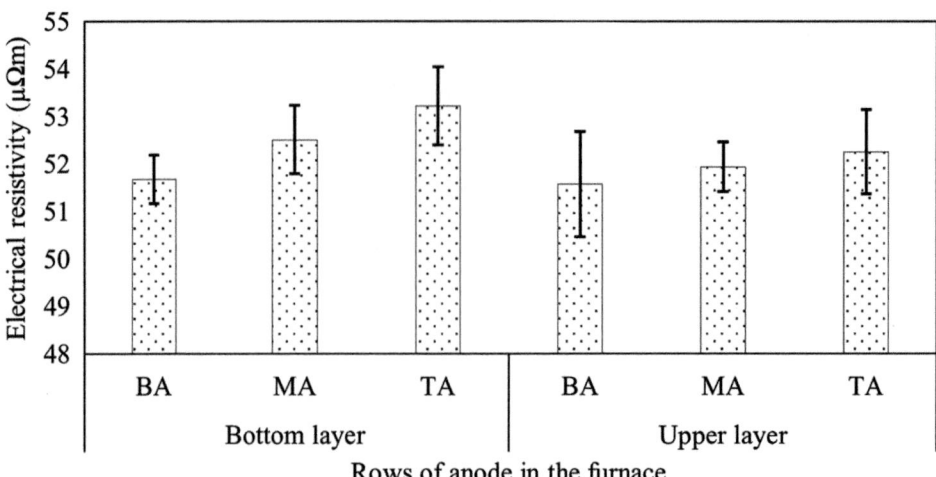

Fig. 12 Difference between the electrical resistivity (ER) of horizontal and vertical measurements directions as a function of anodes rows in the furnace

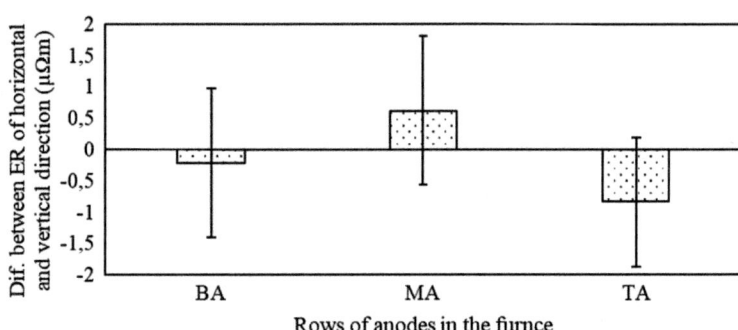

Table 7 CV and p-value of the difference between the electrical resistivity (ER) of horizontal and vertical direction

p-values						CV		
BA	MA	BA	TA	MA	TA	BA	MA	TA
0.324		0.431		0.068		-23.56	7.86	-5.14

depends to the direction in which the measurements is made, as well as the scale at which the measurements is made [21]. In the case of green carbon anodes, it has been reported that the causes of this local anisotropy may be related to the distribution of anode paste in the former before vibro-compaction causing pressure gradients, strain through the green anode body and the lack of intensity to compress the middle of the anode that is exposed to more vibration pressure than the top and the bottom part [20]. In this study, the local anisotropy was evaluated by determining the difference between horizontal and vertical directions at each measurement position to check if the stacking effect has led to the generation of structural deformation of the anodes at each location. The formula used to calculate this is presented in Eq. (7) and the obtained results are presented in Fig. 12 according to the row in the furnace. Since the confidence intervals for all the average differences by row include the zero value, there is no indication of local anisotropy

correlated to the row positions. Also, the analysis was supported by performing t-tests between each pair of row means. The calculated p-values obtained with t-tests and CV are presented in Table 7. The obtained results have not shown anisotropy in these anodes as function of their rows with the 4PP method. A more detailed investigation should be done by tracking and measuring the local anisotropy of the anodes before and after their baking process to validate if there is a change in anisotropy during baking.

Electrical Resistivity of Core Samples as a Function of Anodes Position in the Furnace

The Van der Pauw method was used to determine the specific electrical resistivity of the core samples. A total of 54 observations were analyzed using t-tests to evaluate statistically the difference between rows in the furnace. Each

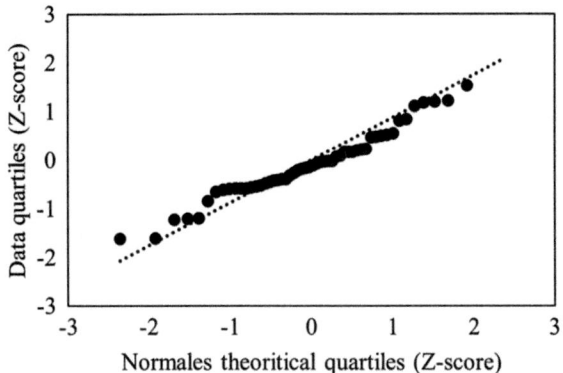

Fig. 13 Quartile–Quartile plots of all observations with outliers

Table 8 CV and p-value between observations according to the position of anodes in the furnace for the Van der Pauw electrical resistivity of core samples

p-values						CV		
BA	MA	BA	TA	MA	TA	BA	MA	TA
0.009		0.015		0.386		0.030	0.038	0.020

Table 9 CV and p-value between observations according to the position of anodes in the furnace for density of core samples

p-values						CV		
BA	MA	BA	TA	MA	TA	BA	MA	TA
0.193		0.156		0.618		0.032	0.052	0.011

row contains 18 observations of cored samples. To verify the normality of observations, quartile–quartile diagram has been plotted in Fig. 13. As the plotted quartile–quartile diagram showed good fit of normal distribution probability, all data were considered as it because there was not physical reason showing the possible presence of outliers during data collection.

In Table 8 and Fig. 14, the calculated p-value between rows, CV of each row, means and confidence interval of

observations at 95% are presented, respectively. From these results, it is possible to see that, the electrical resistivity of core samples from BA was significantly lower than the electrical resistivity of core samples from MA and TA. We have also observed a significant difference between BA and MA with Van der Pauw resistivity. Also, the CV values show that the dispersion of observations between groups was qualitatively similar. Then, it is possible to confirm that the electrical resistivity obtained from core samples was significantly different between the rows. These results were similar to those obtained with the 4PP method. The difference obtained between BA and MA with the Van der Pauw method may be explained by the fact that the measurement location was different (i.e. under the stub holes for the Van der Pauw method and on the large sides for the 4PP measurements).

Baked Density of Core Samples as a Function of Anodes Position in the Furnace

The baked apparent density of the core samples was also analyzed according to the position of the anode blocks in the furnace to evaluate the stacking effect on this property. A total of 54 observations were analyzed using t-tests to evaluate statistically the difference between each pair of rows in the furnace with 18 observations per row. Before analyzing the results, quartile–quartile plots have been plotted in Fig. 15 to check the normality.

The calculated p-values, CV and plots of observations according to the rows of anodes in the furnace are presented below. From these results, it was not possible to drawn conclusion on the densification of BA related to stacking effect. To confirm this effect on the densification as a function of anode stacking in the furnace, more anodes need to be investigated and other phenomena related to the baking process such as the effect of temperature during the baking and the effect of packing cokes surrounding the anodes. Also coring more samples on different locations of the anodes and

Fig. 14 Electrical resistivity (Van der Pauw method) of core samples as a function of the anode's row in the furnace

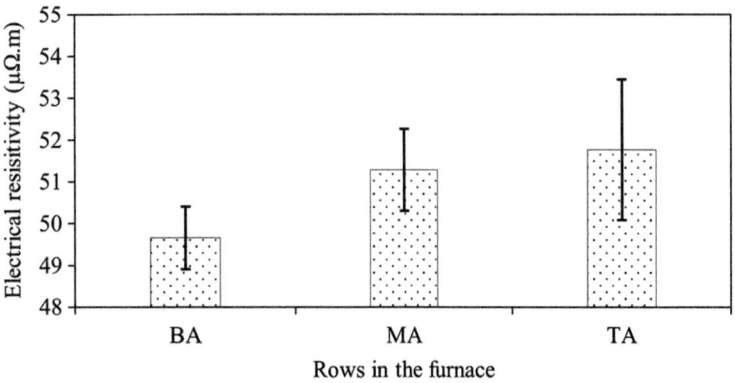

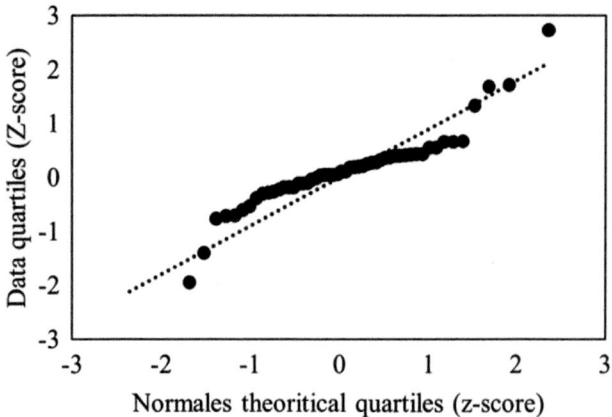

Fig. 15 Quartile–Quartile plots of all observations

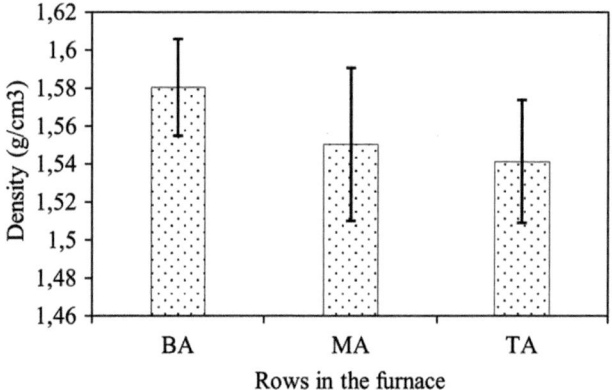

Fig. 16 Density of core samples as a function of the anode's row in the furnace

measuring the density on these locations or measuring the density of the full anodes could help to have a large comprehension on the possible densification of anodes according to the stacking effect. However, the work [18] realized in laboratory scale with additional load of 50 kPa during the baking of green anodes samples at 1100 °C has observed densification of these samples. The samples under the load have shrink on height and expand on diameter. Interested readers are invited to read [18].

Conclusion

In this study we have investigated the correlation of the stacking or loading for different rows in the furnace with the electrical resistivity and the baked density of anodes. The investigation was done on 18 anodes manufactured and baked at an Alcoa smelter. The results were analyzed as a function of the vibro-compactors, the position of the anodes in the furnace by row, the direction of measurements with respect to their rows and according to upper layer and

bottom layer of within the anode. All analyses were supported by statistical tests (t-test) to compare the significance level of the difference of means between the groups and the coefficient of variation (CV) has been calculated to evaluate the dispersion of measurements. For all analyses, obtained results revealed that the electrical resistivity of the anodes was significantly different according to their rows in the furnace. We have observed that lower anodes (BA) were less resistive compared to MA and TA. Also, the results shown that the bottom layer of BA was significantly lower compared to the bottom layer of MA and TA. This difference has an added value on the confirmation of the hypothesis of this study stipulating that there is a correlation between the loads due to stacking and the resistivity of anodes. It was also revealed that the impact of this load may be correlated to the initial green density of the anodes. For the baked density, no significant differences between rows of anodes in the furnace were observed. However, this study considers that, the only known parameter of these anodes is the distribution of load in the rows and the distribution of temperatures and heat-up rate between the rows was unknown. For this reason, it is impossible to generalize the stacking effect on all industrial anodes. Therefore, the need to have more refined investigation under temperatures-controlled environment and with varying green density. Other parameters inside the furnace like the effect of surrounding packing cokes which can generate the radial force on the anodes need also to be investigated.

Acknowledgements We thank Alcoa, the Natural Sciences and Engineering Research Council of Canada (NSERC), the Aluminium Research Centre (REGAL) and Mitacs-Acceleration for their financial support. We also thank Hugues Ferland, Expert Technician at REGAL, Dr. Hicham Chaouki, and Daniel De Araujo Costa Rodrigues for their exceptional help in this project.

References

1. Gielen D and van Dril T (1999), CO2 reduction strategies in the basic metals industry: A systems approach. Paper presented at the annual conference of the minerals, metals and materials society TMS, San Diego, 28 February - 4 March 1999
2. Worrell E et al. (2007), World best practice energy intensity values for selected industrial sectors. Environmental Energy Technologies Division Lawrence Berkeley National Laboratory, United States
3. Tarcy GP, Kvande H, and Tabereaux A (2011), Advancing the industrial aluminum process: 20th century breakthrough inventions and developments. JOM **63**(8):101
4. Mannweiler U and Keller F (1994), The design of a new anode technology for the aluminum industry. JOM **46**(2):15-21
5. Rebaine F. (2015), Étude de l'influence des paramètres de vibro-compaction sur les propriétés mécaniques des anodes crues en carbone. Ph.D. Thesis, Université du Québec a Chicoutimi
6. Chollier-Brym MJ et al. (2012), New method for representative measurement of anode electrical resistance. In: Suarez, CE (ed) Light Metals 2012. Springer, Cham, p 1299-1302

7. Léonard G et al. (2014), Anode electrical resistance measurements: learning and industrial on-line measurement equipment development. In: Grandfield, J (ed) Light Metals 2014. Springer, Cham., p 1269-1274

8. Rørvik S and Lossius LP (2019), Measurement of Anode Anisotropy by Micro X-Ray Computed Tomography. In: Chesonis, C (ed) Light Metals 2019. Springer, Cham., p 1293-1299

9. Severo DS, Gusberti V, and Pinto EC (2005), Advanced 3D modelling for anode baking furnaces. In: Kvande, H (ed) Light metals 2005. The Minerals, Metals & Material Society, p 697-702

10. Severo DS et al. (2011), Recent developments in anode baking furnace design. In: Lindsay, SJ (ed) Light Metals 2011. Springer. p. 853–858

11. Grégoire F, Gosselin L, and Alamdari H (23013), Sensitivity of carbon anode baking model outputs to kinetic parameters describing pitch pyrolysis. Industrial & Engineering Chemistry Research. 52(12): p 4465–4474

12. Holden I et al. (2016), Safe Operation of Anode Baking Furnaces, in Essential Readings. In: Tomsett, A and Johnson, J (ed) Light Metals 2016. Springer, Cham., p 396-402

13. Benzaoui A et al. (2017), A non-destructive technique for the on-line quality control of green and baked anodes. Metals 7(4): p 128

14. Standard Test Method for Electrical Resistivity of Anode and Cathode Carbon Material at Room Temperature. 2017

15. Rouget G et al., Electrical Resistivity Measurement of Carbon Anodes Using the Van der Pauw Method. Metals, 2017. 7(9): p. 369

16. Standard Test Method for Apparent Density by Physical Measurements of Manufactured Anode and Cathode Carbon Used by the Aluminum Industry, 2015

17. Hines WW, Montgomery DC, Goldsman DM, Borror CM (2017), Probabilités et statistiques pour ingénieurs, 3e édition, Chenelière Éducation, Montréal

18. Barry TS (2020), Effet de l'empilement des anodes de carbone pendant la cuisson sur leur densification et sur leur résistivité électrique. M. Sc. Thesis, Université Laval

19. Manolescu P et al (2022) Net carbon consumption in aluminum electrolysis: impact of anode properties and reduction cell-operation variables. J Sustain Metall https://doi.org/10.1007/s40831-022-00556-2

20. Ziegler DP and Secasan J. (2016), Methods for determining green electrode electrical resistivity and methods for making electrodes. U.S. Patent 9,416,458. 16 August 2016

21. Liu E and Martinez A. (2012), Fundamentals of seismic anisotropy. In: Liu, E and Matinez, A (ed), Seismic Fracture Characterization, EAGE (2012), p 29–57

New Generation Anode Baking Furnace: Use of Prefabrication for Additional Conversions at Bell-Bay Plant

Sandra Besson, David Deneef, Anthony Reeve, Youcef Nadjem, Meaghan Noonan, and Roy Cahill

Abstract

The anodes used for aluminium production are baked in order to reach the resistivity, mechanical resistance and reactivity adequate for the electrolysis process. The anodes are baked in pits that are usually separated from each other by fluewalls and headwalls made of dense refractory material. In 2017, an industrial prototype of the new generation anode baking technology was installed at the Bell-Bay smelter with 6 sections converted to this patented technology. The headwalls were partially removed to allow a productivity increase by 17% and gas consumption reduction by 30%. After the success of this first trial, the conversion of five additional 6-sections zones by the end of 2025 is planned for Bell-Bay's furnace. The use of prefabrication has been developed and implemented for the installation of the second generation of walls to:

1. shorten the construction duration and limit the production loss,
2. improve the ergonomics and safety for the construction workers and,
3. improve the overall brick assembly quality.

This paper reviews the process that led to an optimized second NG zone construction through prefabrication.

Keywords

Anodes in aluminium electrolysis • New Generation anode baking furnace • Rio Tinto Bell-Bay • Prefabrication

Introduction

In an open-type Anode Baking Furnace (ABF), anodes are placed in pits separated by hollow fluewalls, through which hot gases flow during the baking phase and air flows during the cooling phase. Sections are separated by headwalls through which flue walls in the same line are linked between one section and the next, thereby forming individual flue wall lines extending along the entire furnace.

The headwalls constitute 20 to 25% of the dense refractory mass installed in the furnace. They are heated and cooled during each fire cycle, which requires a large amount of gas and limits the ability to rapidly cool the anodes. The sealings between the fluewalls and the headwalls need regular maintenance, during which refractory maintenance operators enter the confined space of the pits and renew the fibers under difficult conditions: high temperature, working at height, presence of dust and fibers. Based on those considerations, a concept of New Generation (NG) anode baking furnace was proposed [1, 2].

After several developments and the installation of a pilot zone in one of the furnaces at Grande-Baie plant, the implementation of this technology on six sections of the Bell-Bay Aluminium (BBA) furnace was realized in 2017 [3, 4]. After the rebuild of this zone, a new one was built with prefabrication while the previous constructions took place *in situ*.

The purpose of this article is to describe the construction of the prefabrication of this new NG zone.

S. Besson (✉) · D. Deneef
Rio Tinto Aluminium Technology Solutions, Voreppe, France
e-mail: Sandra.besson@riotinto.com

A. Reeve · Y. Nadjem
Bell Bay Aluminium, TAS, Australia

M. Noonan · R. Cahill
Rio Tinto Aluminium Transformation and Technical Support - Pacific Operations, London, UK

© The Minerals, Metals & Materials Society 2023
S. Broek (ed.), *Light Metals 2023*, The Minerals, Metals & Materials Series,
https://doi.org/10.1007/978-3-031-22532-1_153

New Generation Anode Baking Furnace Concept

Concept of the New Generation (NG) Technology

The concept of the NG patented technology is to remove totally or partially the headwalls as presented in Fig. 1. It is applicable to both furnace revamping and new projects. For existing furnaces, it allows to increase the volume available for anode baking inside the pits. The number of headwalls to be removed should be selected depending on the anode dimensions to accommodate either an extra set of anodes or an increase in anode dimensions.

Headwall Functionalities and Challenges with Their Removal

The key role of the headwalls in the standard ABFs is the following:

- Allow for thermal expansion of the fluewalls
- Mechanically support fluewalls when pits are empty
- Separate sections (delimitation for operation sequence)

As the headwalls are removed in the NG design, it has been necessary to develop technical solutions and to adapt operational practices to assure the above functions. In particular, the NG design should allow the accommodation of the fluewall for thermal expansion which is usually allowed via the expansion gaps located at the junction between the headwall and the fluewalls in the standard design (see upper part of Fig. 2). As headwalls are no longer present, the internal design of the fluewall was modified so that the expansion and contraction can be handled by the fluewall itself through dedicated zones, called "breathing zones" (see lower part of Fig. 2). The design of bricks used in these areas has been modified so that relative movements between bricks is less inhibited in these locations compared with the rest of the fluewall.

New Generation Anode Baking Furnace Implementation at Bell-Bay

Bell-Bay Furnace and NG Technology Implementation

The anode baking furnace at BBA was constructed in 1989: the furnace has 3 fires with 48 sections, and 7 pits per section. Each pit is stacked with 7 layers of 4 anodes (horizontal loading).

With the necessity to rebuild some sections in the furnace in 2017 and the need for BBA to increase the anode production, an opportunity arose to use the NG design on one zone of this furnace. The NG design allowed to stack 7 anode sets instead of 6 for the same footprint (Fig. 3).

The first zone was installed in December 2017 with the support of Smelter Technical Support - Pacific Operations (Brisbane, Australia) and Aluminium Technology Solutions (Voreppe, France) teams.

Fig. 1 Concept of the New Generation Technology

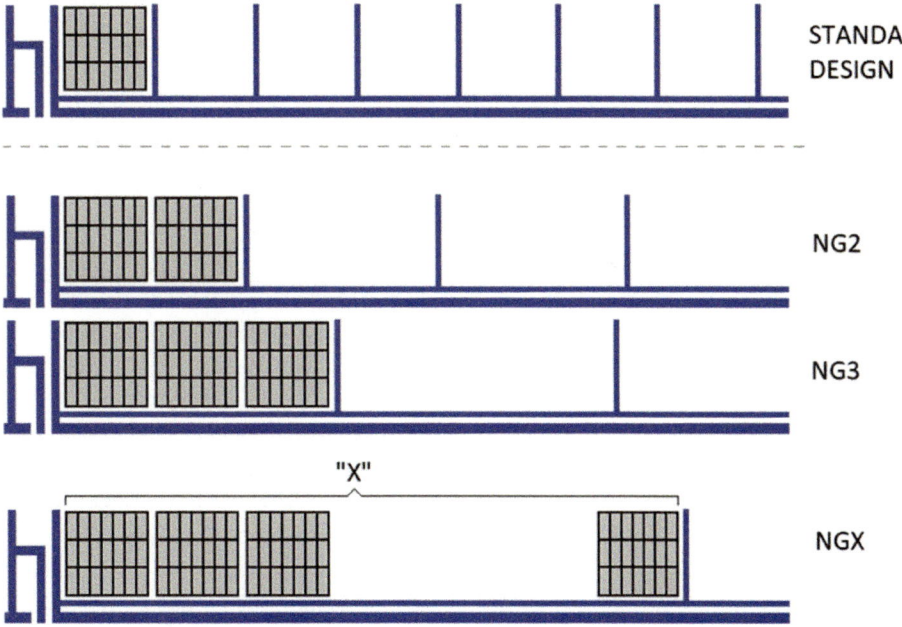

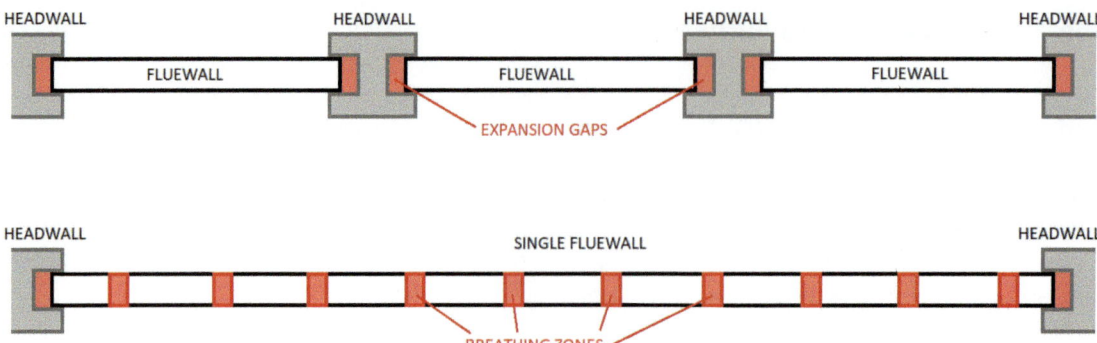

Fig. 2 Classical design (top) versus NG technology (bottom)

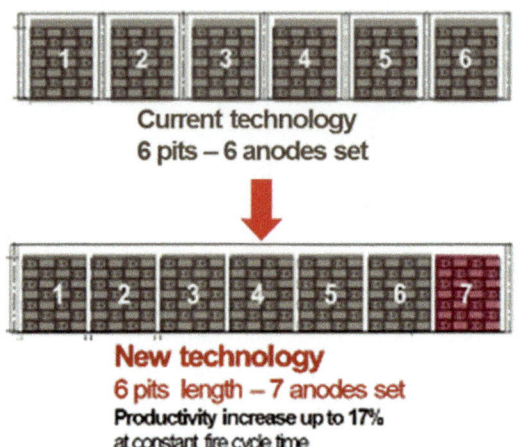

Fig. 3 Anode loading pattern in BBA furnace with current and NG Technology

Performance Achieved for the First Generation at BBA

The overall performance of this first 6-sections wide NG zone was beyond expectations [5].

Over the lifetime of 82 fire cycles, the NG zone allowed:

- 17% of productivity increase (+8759 anodes over the lifetime of the zone)
- 14% reduction of CAPEX
- A gas consumption reduction by 30% (equivalent of 613 kAUD gas savings)
- 2654 t CO_2 savings for 1/8th of the furnace converted
- 80% less sealing maintenance

Some pictures of the first NG zone at different ages are provided in Fig. 4.

Deployment of NG Technology on Other Sections of BBA's Furnace

Based on those results and BBA's business case, a program to deploy NG technology on 6 areas of the furnace by the end of 2025 has been validated. This means that ¾ of the furnace should be converted by this date.

After the rebuild of the first NG zone in 2021 with some design improvements, a second NG zone has been installed on the furnace in 2022. The choice was made to use prefabrication and the reasons for this choice are explained in the following paragraph.

Experience of Prefabrication for the NG Technology

Prefabrication in the NG Technology Context

Furnaces with standard technology and continuous fluewall replacement usually pre-build the fluewalls outside of the furnace and use the Furnace Tending Assembly (FTA) to move the fluewalls in one or two different parts. This allows to reduce the working time in the furnace and limits the impact on production, to reduce the co-activity in the furnace and to improve the ergonomics with the use of platforms to reduce the work at poor ergonomic heights.

Due to the length of the NG fluewalls at BBA (33 m long), the two first NG constructions in BBA were built *in situ*. Up to 33 people were mobilized to work simultaneously on the full length of the NG area. The first courses were difficult to build in terms of ergonomics because they were located below the operators' knee height. The speed of execution required to quickly install the fluewalls (despite the co-activity and despite the need to evacuate the work

Fig. 4 View of the NG zone after 21 rounds of fire (upper left), 41 rounds of fire (upper right) and at 75 rounds of fire shortly before the demolition (lower)

area during FTA runs). It could result in quality issues such as inhomogeneously sized degassing joints, lack of straightness or waves in the brick level.

When the decision was taken to generalize the NG technology for future rebuilds at BBA, a study for the prefabrication of the fluewalls was launched. The objective was to keep the original NG design, including the bricks design and the concept of breathing zones, while transporting the pre-built parts with the same methodology as for standard rebuilds (standard fluewalls are transported in two lifts at Bell-Bay). The cohesion between the different prefabricated parts was the subject of a specific design that could accommodate the NG design.

Fig. 5 Construction zone during the installation of the top lifts

Execution of Prefabrication for the 3rd NG Rebuild at BBA and Results

Twelve different parts had to be prefabricated for each fluewall prior to the construction. The prefabrication took place on the standard fluewalls building platforms and lasted for 10 weeks. The 96 prefabricated parts were stored on specific supports to facilitate their transportation from the storage area to the furnace end with a lifting crane vehicle. They were then handled by the FTA with the help of the same strongbacks that are usually used for standard fluewalls. Some minor modifications were required to the strongbacks, and removable weights were added to balance the loads and keep the non-symmetrical fluewall pieces levelled during installation.

The prefabricated method resulted in straighter fluewalls with better construction quality (Fig. 5). The number of people involved during the construction was reduced with an average of 17 people during 24 days instead of 24 people during 28 days. It resulted in labour savings of approximately 125 kAUD in comparison with the previous *in situ* construction. The ergonomics were also improved with a reduction of 95% in the tonnage of dense refractory installed at poor ergonomic height: 104 t of refractory were installed at chest height instead of below the knee for the *in situ* construction. The FTA utilization for the construction was also reduced by 96 h in comparison with the previous construction, which resulted in less co-activity between the workers inside the furnace and the FTA. In the future, the construction time could be reduced by 10 days with this methodology to limit the impact on production.

Conclusion and Perspectives

The prefabrication of the fluewalls for the NG zone at BBA has enabled improvements to ergonomics and construction quality, while reducing the number of workers required and usage of the FTA during the construction. The evolution of the refractory state of this zone will be followed-up closely in the coming months in comparison with the other NG zone that was built *in situ* 9 months before. Based on this successful experience, it is planned to use prefabrication for the future NG conversion on the BBA furnace.

References

1. Christian Jonville et al., Ring furnace including baking pits with a large horizontal aspect ratio and method of baking carbonaceous articles therein. *US Patent Number 8,684,727-B2,* 14 May 2007.
2. Pierre-Jean Roy et al., Ring furnace including baking pits with a large horizontal aspect ratio and method of baking carbonaceous articles therein. *European Patent EP 1 992 895 B1,* 14 May 2007.
3. Allan Graham et al., Increased Anode Production at Bell Bay Aluminium Using a New baking furnace design concept without headwall, *Proceedings 12th Australasian Aluminium Smelting Technology Conference, Queenstown, NZ 2018*
4. Arnaud Bourgier et al., Development of a new baking furnace design without headwall to increase anode production capacity, *Light Metals 2019*
5. Sandra Besson et al., New Generation Anode Baking Technology: a breakthrough technology to increase productivity and sustainability, *ICSOBA 2022*

AHEX Full Scale Experiences at TRIMET Aluminium SE

Anders Sørhuus, Vrauke Zeibig, Eivind Holmefjord, Ömer Mercan, and Elmar Sturm

Abstract

The new AHEX technology for treating the off gas fumes from the anode baking furnaces has been successfully demonstrated since 2018 for a complete new FTC at TRIMET Hamburg. With this new technology, the fumes are cooled in an indirect heat exchanger with water in a closed loop. Thereby, many issues with corrosion and deposits inside the traditional FTC based on the conditioning tower designs are avoided. Alumina is injected into the AHEX to keep the surfaces clean, and at the same time the alumina absorbs and removes part of the undesired components from the off gas stream. The less sticky and less humid gas due to the AHEX improves the operation conditions for the filter bags. The paper will present performance results achieved including operational and maintenance experiences over time after the start up of the first full scale AHEX FTC installed at TRIMET's Hamburg anode bake furnace.

Keywords

AHEX • FTC • Anode Bake

Introduction

The AHEX fume treatment process eliminates the use of water sprayed into the gas stream to cool and condense the volatile PAH components in the fumes. Cooling is instead achieved with the novel combined AHEX-heat exchanger/ alumina reactor to condense out the volatile hydrocarbons including the PAH components and tar. The alumina injected into the heat exchanger (AHEX) inlet not only removes the volatiles from the gas stream, but also keeps the surfaces clean to maintain good heat transfer over time [1].

As discussed in [1] the AHEX gives several benefits compared to the conventional cooling tower solution:

- Lower gas temperature can be achieved since the cooling is not limited by the water evaporation temperature of typically 105 °C to 110 °C in the conventional cooling tower.
- Elimination of the water spray reduces the corrosion issues, increases lifetime of the filterbags, and reduces problematic deposits in the inlet gas duct. In the traditional FTC due to the high hydrolysis caused by the high water content in the gas, the lifetime of filterbags can be roughly one year. At TRIMET four times longer life span for the filterbags has been observed.
- Full redundancy on the cooling is achieved since one AHEX is installed for each compartment.
- Heat can be recovered and used for useful purposes.

In addition there are other benefits as discussed in [1] including a simplified overall flow pattern of the gases with a downwards directed flow in the AHEX reactors, and thus there are no minimum lift velocity requirement. The minimum lift velocity in the conventional FTCs requires the installation of complex logics including recirculation of fumes that consequently are not needed for the operation of the AHEX based FTC.

In spring 2018 the first full scale AHEX FTC was commissioned to treat the fumes from the anode bake furnace at TRIMET Hamburg. This paper will discuss the results after approximately four years of operation.

A. Sørhuus (✉) · E. Holmefjord
REEL ECS, Drammensveien 165, 0277 Oslo, Norway
e-mail: anders.sorhuus@REEL-Norway.com

V. Zeibig · Ö. Mercan
TRIMET Aluminium SE, Aluminium Str. 1, 21129 Hamburg, Germany

E. Sturm
ESC-Consulting, Rispenweg 12, 21614 Buxtehude, Germany

© The Minerals, Metals & Materials Society 2023
S. Broek (ed.), *Light Metals 2023*, The Minerals, Metals & Materials Series,
https://doi.org/10.1007/978-3-031-22532-1_154

The Anode Bake Furnace at TRIMET

In the anode baking furnace, carbon anodes are baked to electrolysis readiness. For this purpose, the anodes are stacked in several layers in firing chambers with the aid of a crane and covered with packing coke. The individual combustion chambers are separated by hollow walls, the so-called flue walls.

Carbon anodes consist of calcined petroleum coke, coal tar pitch and anode residues contaminated by the electrolysis process. For the firing of the carbon anodes, the flue walls are heated up to 1200 °C by means of gas burners. In this way, the combustion chambers are indirectly heated in the firing process and the carbon anodes are baked. Natural gas is used as fuel. When the heating phase of the carbon anodes begins, the pitch in the anodes starts to soften. The volatile components of the pitch that are outgassed during the firing process are drawn into the flue wall and participate in the combustion.

The exhaust gas produced by the firing process in the flue walls is sucked through exhaust bridges into a ring gas line surrounding the furnace. Two radial blowers generate a vacuum of up to 1200 Pa in the ring gas line and draw the exhaust gas into the exhaust gas treatment system.

The exhaust gas temperatures immediately upstream of the exhaust gas treatment system range from 130 °C to 220 °C, depending on the progress of the fire. The exhaust gas contains, among others, hydrogen fluoride HF, sulfur dioxide SO_2, benzene, PAH, tar/ pitch residues and carbon monoxide CO.

Previous to 2018, the handling of the fumes included heat treatment in an RTO (Regenerative Thermal Oxidizer) upstream a dry scrubber. Since that time TRIMET has successfully improved the burner and furnace operations to reduce the most volatile hydrocarbons at the source. The improved furnace operation in combination with the improved treatment of the fumes in the AHEX FTC secures compliant operation without the RTO, and the high maintenance and energy costs of the RTO operation can thereby be avoided. This includes reduction of 1643 ton annually of CO_2 emitted from the RTO gas consumption of 8300 MWh pr year. In addition, 5000 MWh pr year electrical energy is saved in reduced fan power requirements (Fig. 1).

The Full Scale AHEX FTC

As shown in Fig. 2, the FTC comprise 3 compartments each with its integrated AHEX for removal of HF and tar components, and for cooling the anode baking fumes. The fumes from the furnace enter the top of the AHEX where it is mixed with alumina recycled from the hopper.

Primary alumina is injected into the filterbag nest according to the ABART principle, and thereby the pollutants are adsorbed on the alumina and removed through the overflow to the secondary silo. In Fig. 3, the outlet temperature of the AHEX can be seen to be below 90 °C, lower than the 105–110 °C limitation of the conventional conditioning spray towers. Therefore, due to the reduced temperatures, the AHEX-based FTC achieves lower emissions.

During the measurements in [1], the performance of the conditioning tower-based technology could be compared directly with the AHEX. The gas temperatures in the AHEX were intentionally adjusted to be comparable to those in the conditioning tower, and still lower emissions were realized. It is assumed that this is due to better mixing achieved in the AHEX reactor/HEX tubes. Especially as the fumes reach the low temperature boundary layer close to the inside surface of the HEX tubes, the tar starts to condense and will attach to the alumina particles.

Humidity from the water spray into the conditioning tower is prevented with the AHEX. With 100–150 °C cooling requirement of the gas, it can be estimated that AHEX has approx. 50–75 g H_2O / kg air less humid air compared to a conventional FTC. However, humidity from the air, fuel and anodes is also added to the fumes from the anode bake furnace and could already be at 30–50 g H_2O / kg air. With these assumptions, it can roughly be estimated that the less humidity in the AHEX contributes to a 10–20 °C lower acid dew point compared to a situation where SO_3 would form sulfuric acid at say 70–80 °C. It should be noted that this is a conservative approach for estimating the impact of the water spray on sulfuric acid formation. Locally, around the droplets of the water spray, very low temperatures are unavoidable, and once sulfuric acid is generated on these spots, it tends to remain as sulfuric acid, and not reverse its components back to SO_3 and H_2O once the temperature reaches the bulk at the conditioning tower outlet. In any case, the lower SO_3 dew point in the AHEX is a contributor to the lack of corrosion that has been observed. However, it is generally recommended that the gas ducting is insulated to prevent cold spots in the steel ducting.

Less humidity also reduces the hydrolysis of the filterbag materials, reduces the stickiness of the filtercake, and is expected to contribute to the long observed lifetime of the filterbags at TRIMET. Lower cost needlefelt materials based on polyester can also be considered. It should also be noted that the less humidity reduces the actual gas flow to be treated by up to 10%, which reduces the load and/or size requirements of the FTC proportionally.

As seen, the temperatures vary over the 24 h cycle of the furnace. In this figure, only 2 out of 3 fires were in operation. This reduce the overall temperatures and load on the FTC.

Fig. 1 Anode bake furnace (copyright TRIMET web page)

Fig. 2 The REEL AHEX FTC at TRIMET Hamburg

ABART 2- stage alumina injection dry scrubber with primary alumina injection onto the filterbags

Alumina hopper with alumina recyling to AHEX inlet

Water cooling and ORC/ clean gas duct

Mainanance lifting beams

Inlet duct

1 out of 3 AHEX

Fig. 3 Temperatures from the furnace, and out of AHEX over a 24 h anode bake cycle with 2 fires. The fume gas flow is also shown

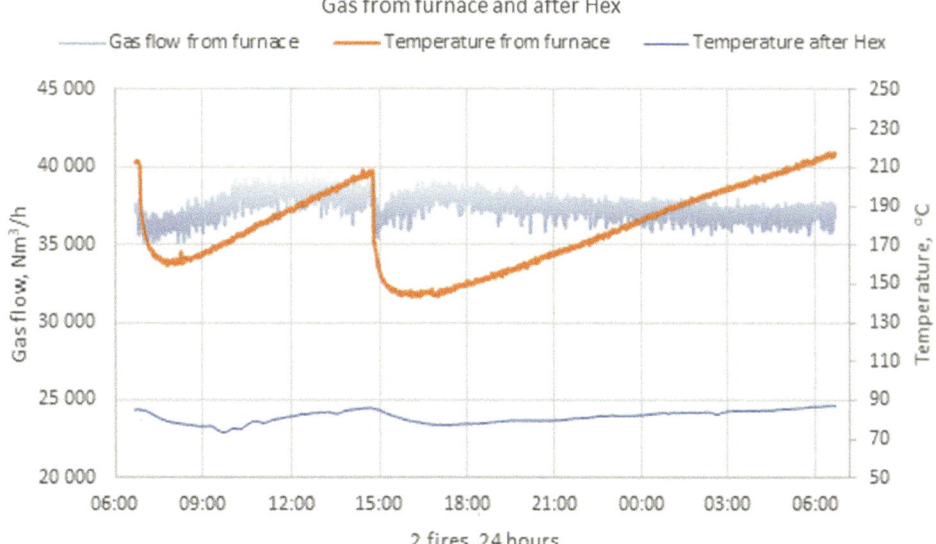

Table 1 Typical ORC operation data with 3 fires

Parameter	
Averaging time (h)	24
Average power generated kW	83
Accumulated power production (kWh over 24 h)	2000

The energy recovered from the AHEX can be used to produce electricity in the Organic Rankine Cycle (ORC) machine integrated with the cooling system for the AHEX cooling water. The ORC machine has some limitations in the range of temperatures and load when it can be used. Principally, the ORC machine needs to produce enough "steam" of organic fluid to keep the generator running at 50 Hz. The ORC will automatically start once the energy input is sufficient. Below are some data for the year of 2021 on the ORC operation (Table 1).

Other highlights after 4 years of operation are:

- More than four years lifetime of filterbags!
- No cleaning of AHEX surfaces
- Low pressure drop and energy consumption
- Minimum maintenance requirements after initial startup tuning and commissioning period

Erosion due to alumina injection is also known to be a challenge in traditional FTC designs. During week 34, 2022, one AHEX compartment was inspected inside, and no signs of erosion could be detected inside the filter compartment or inside the AHEX. Also, the AHEX inlet is protected with replaceable inserts to provide additional protection into each cooling tube, and so far, not one has required replacement. At the inlet alumina spreader above the AHEX, and at some venting pipes, erosion had been detected earlier during the pre-operation stage. With the great support of TRIMET operation, improvements were implemented, and now the plant has operated with almost four years of proven non-disturbed operation with considerably low maintenance requirements.

Measurement Results of HF, Benso(a)pyrene (BaP), and PAH 16

Dust, BaP and HF outlet measurements are performed regularly to check the stack emission compliance by TRIMET according to VDI 2470, and Table 2 show typical results.

As shown, inlet and outlet HF measurements for the AHEX reactor stage are provided. HF can be a convenient tracer to check the performance of the reactor stage versus the second polishing stage in the ABART 2-stage dry scrubber. Although the adsorption mechanisms of HF to the

alumina is different from the PAH components, some of the overall requirements for removal such as good mixing of the alumina and fumes remains. High removal efficiency of HF is therefore at least a good indication that PAH components will also have the possibility to adsorb to the alumina providing the temperature in the reactor stage is low enough to condense out the various PAH components.

With the measured numbers, the removal efficiencies for the first and overall dry scrubber stages can be estimated to 96,4% and 99,7% respectively.

To investigate further on the emissions of various PAH components, dedicated measurement campaigns were done in November 2019, and during week 34 of 2022 at the TRIMET AHEX FTC. PAH were extracted into XAD-2 (Fig. 4) and analyzed in the laboratory.

From measurements in November 2019, it was shown that more than 93% of the heavier PAH components were removed (see Fig. 5). This included Benzo(a)anthracene (160,7 °C), Chrysene(254 °C), Benzo(b)fluoranthene (168,3 °C), Benzo(k)fluoranthene (215,7 °C), Benzo(a)pyrene (179 °C), Indeno(1,2,3-cd)pyrene (163,3 °C), Dibenzo (a,h)anthracene (266,6 °C) and Benzo(g,h,i)perylene (278,3 °C). This is within the same range as calculated for the HF reactor above. However for the lighter components, lower removal efficiency is expected. Removal efficiencies of Naphthalene (78,2 °C), Acenaphtylene, Acenaphthene (95 °C), Fluorene (115 °C) and Phenanthrene (100,5 °C) where not determined accurately during the measurements in November 2019, but from measurements at the AHEX in Mosjøen [1], similar high efficiencies were measured for the light components.

Some melting temperatures [4] of the various PAH components are shown in the brackets above. It is interesting to note that many of the melting temperatures are above the temperatures in the AHEX which indicates that the main part of the PAH is present in solid state and that the main contributor to a low PAH emissions is good conditions and particle collection in the filter bags.

A second measurement campaign was done in week 34-2022 where some more alumina adsorbent was used (see Fig. 6). It should be noted that several operation conditions were different between the two campaigns, however by comparing the two sets of results, significantly better removal efficiencies can be calculated (average 99,1% versus 93,2% for the range Anthracene to Benzo(g,h,i)perylene).

Table 2 Dust, BaP, HF measurements at TRIMET Hamburg AHEX

Parameter	(mg/Nm³)
Dust	0,4
BaP	0,0001 (below detection)
HF inlet	50
HF outlet AHEX	1,8
HF outlet dry scrubber	0,14

 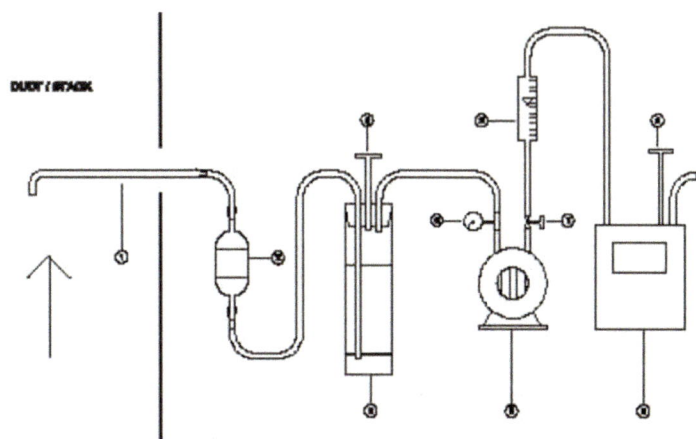

Fig. 4 REEL sampling train for PAH measurements based on NS 9815. Sampling probe (1), XAD-2 tube (2) Silica Gel bottle (3), Thermometer (4), Air tight pump (5), Vacuum Gauge (6), flow control valve (7), Rotameter (8), gas meter (9)

For the 16 PAH components measured in Fig. 6 an average removal efficiency of 93,1% is calculated including the lighter components- Naphthalene to Phenanthrene.

Discussion

The modern open furnace technology for baking anodes give much less emissions than in the past. Still, the raw gas fumes from the anode bake furnaces contain small fractions of partly un-combusted remains for the pitch and calcinated petrol coke including particulate and Polycyclic Aromatic Hydrocarbons (PAH) that must be treated before emitting the gas to the atmosphere. In addition recycled butts contain fluorides that generates significant amounts of HF into the raw gas. In some cases, the SO_2 present in the raw gas must also be treated.

In the US, EPA regulations emission factors are given for various aluminium production technologies based on 7 or 16 selected PAH components. Generally, the range of measured concentrations in the raw gas may vary considerably over the cycle time of the baking process with typically lighter volatiles escaping early and heavier components later as the baking temperature of the anode core increases. Typically,

the concentrations of PAH components tend to be higher at the start of the firing cycle.

Although the boiling temperature of most of the PAH components are higher than for the temperatures in the treatment process, the vapor pressure is still sufficient especially for the lighter components to reach the order of 100 mg/Nm3 gaseous saturated concentrations in the raw gas. With modern burner technology and open furnaces, the typical PAH 16 raw gas concentrations may be in the order of 1 mg/Nm³.

The EU BREF BAT 2016 describes several technologies for treating the anode bake fumes including for use of the alumina injection dry cleaning system. The recommendation is to treat Benzo(a)pyrene to less than 0,001 to 0,01 mg/Nm³ assuming it is a tracer for all the PAH components, particulate to less than 2–5 mg/Nm³ and HF to less than 0,3 to 0,5 mg/Nm³.

As seen from the values in Table 2 the measured emissions from the AHEX FTC are compliant with good margins (Table 3).

Since the alumina injection based FTCs generally rely on condensing the PAH components on the alumina surfaces, the lower temperature of the AHEX FTC is beneficial. However, the lighter components will tend to be adsorbed to

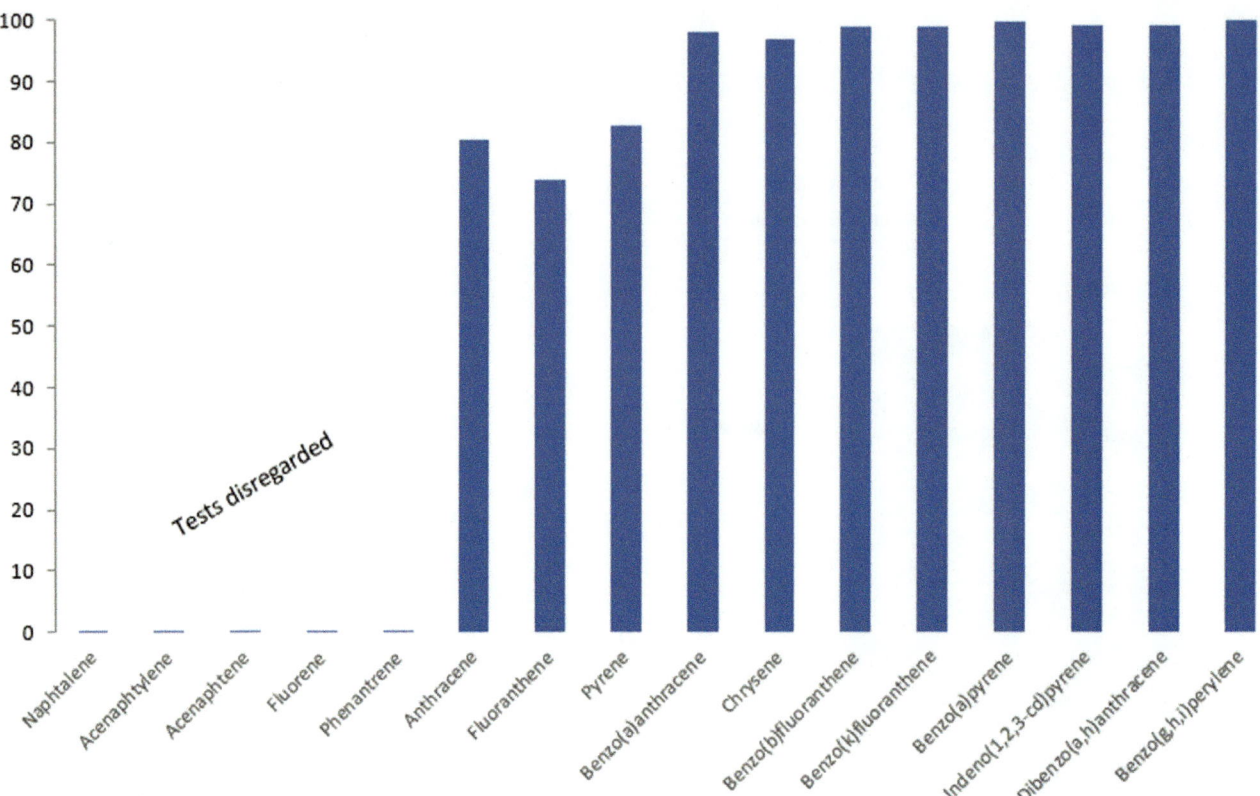

Fig. 5 Removal efficiency close to 100% for the heavier components with higher melting temperatures. Naphthalene to Phenanthrene analysis disregarded in the laboratory tests due too large inaccuracy

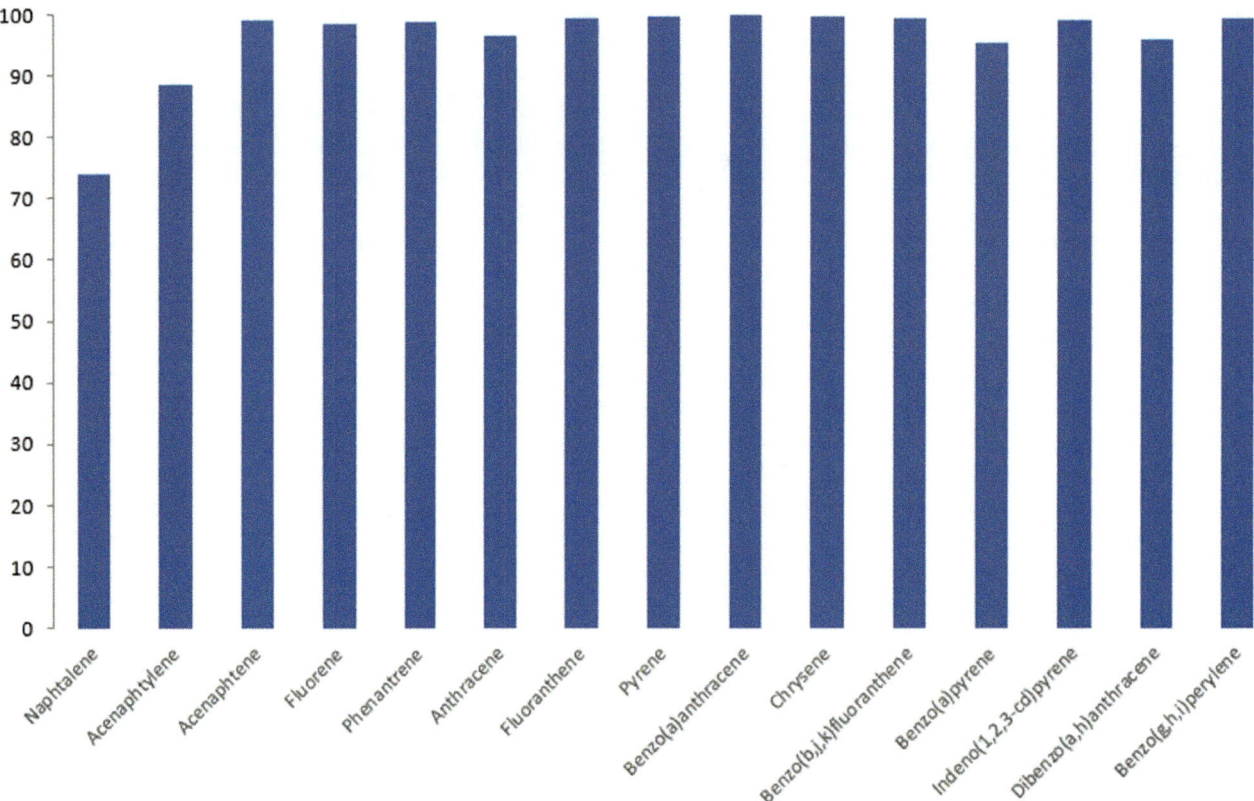

Fig. 6 Results from the second measurement campaign in week 34-2022

Table 3 Extract from EU BREF BAT 2016

Parameter	BAT-AEL (mg/Nm^3)
Dust	2–5
BaP	0,001–0,01
HF	0,3–0,5
Total fluorides	$\leq 0,8$

a lesser extent than the heavier components that will be more in solid particulate state and will be efficiently removed from a dry scrubber. Modern firing systems in the furnaces will tend to burn the light components almost 100%, and thus this combined with the AHEX FTC will make sure there is a good reduction of all the PAH components evolving from the anode bake anodes.

Conclusion

The first full scale AHEX FTC installed at TRIMET Hamburg has proven reliable and compliant operation for more than four years since the startup in 2018. Highlights are more than four years lifetime of filter bags with very low emissions of particulate, HF and BaP.

The overall efficiency for removal of PAHs show high efficiency, average 93,1% for the entire range of components analyzed (see Fig. 6). Differences were observed between the measurements in 2019 compared to 2022 for the heavier components (average 93,3% in 2019 and 99,1% in 2022). One of the differences was higher alumina consumption in 2022 compared to in 1999 which at least partly explains the differences observed.

As expected the more volatile components will typically be less efficiently adsorbed to the alumina. Good furnace operation with modern burning technology will, however, facilitate excellent combustion of the lighter volatiles, providing a perfect combination for treating the overall range of PAH components.

Typically 2000 kWh power is produced each day from the AHEX cooling water and improves the overall CO_2 footprint of the anode baking plant significantly.

Acknowledgements The measurements have been partly supported through the SFI metal production consortium. Measurements, support and feedback after 4 years of operation from TRIMET are highly appreciated.

References

1. Sørhuus, A. Ose,S. et al. "AHEX- a new Combined Waste Heat Recovery and Emission Control System for Anode Bake Furnaces. TMS. 2010.
2. Bøckman, O. "Method and Device for Removing Condensable Components from Warm Industrial Waste Gas". EP 0 870 529 A1, 1998.
3. Holmefjord, E. Trimet AHEX – PAH mesurements. GE/REEL internal report. 2019.
4. IARC Monographs Volume 92 Appendix "CHEMICAL AND PHYSICAL DATA FOR SOME NON-HETEROCYCLIC POLYCYCLIC AROMATIC HYDROCARBONS". 2007.

Inline Modal Detection System of Anodes and Cathodes Measuring Cracks and Physical Properties

Dag Herman Andersen and Ole Kristian Brandtzaeg

Abstract

An anode and cathode quality system based on modal analysis has been developed. The modal detection system finds the resonances, or "fingerprint", of the item tested by measuring the movements of the item caused by an input force from a hammer or shaker. The modal method is verified by modeling and measurements of anodes and cathodes to extract crack information and the elastic property, Young's Modulus (*YM*). Measurement times around $60\,s/item$ can be achieved. For a modal inline quality station to measure *YM*, the dimensions and weight of the item must be measured in advance. The resonances need also to be identified by mode shapes. Mode shapes are also important when it comes to extracting the crack plane orientation. The method has a significant potential of detecting anode and cathode cracks that can cause metal infiltration, unstable cell operation, poor cell performance, extra handling and early cathode failures. In addition, the modal test can provide feedback to production plants of anodes and cathodes to produce those items that are closest to the reference spectrum of resonances of a crack free homogenous product.

Keywords

Carbon anode • Anode fracture • Cathode crack • Young's modulus • Modal analysis • NDT

D. H. Andersen (✉)
Hydro Aluminium, Technology and Operational Support, 6885 Årdalstangen, Norway
e-mail: dag.herman.andersen@hydro.com

O. K. Brandtzaeg
Hydro Aluminium, Technology and Operational Support, 0283 Oslo, Norway
e-mail: ole.brandtzag@hydro.com

Introduction

In the anode manufacturing process for the aluminium industry, there is one final control of the green anodes in the paste plant (before baking) and one final control for the baked anode after the baking furnace. The final control in the paste plant measures bulk parameters including anode dimensions, density and pitch content, but for the final control of the baked anodes there is only limited physical measurements per anode item. However, the lack of baked anode bulk parameters is generally compensated by visual inspections and the core sample analysis on a percentage of the total anodes produced. Since anode tracking systems are on the edge of being realized [1] that will link baked anodes to green anode data, several bulk parameters at the final control of baked quality can be found if baked anode weight and dimensions are measured. Parameters, like baked density, dimensional- and volumetric shrinkages and baking loss are candidates in addition to the *YM* from the modal test.

The modal method sets the anode into resonances with larger wave lengths than with ultrasound methods and are therefore less sensitive to minor cracks in the anode structure. While ultrasound manages to detect both location and orientation of defects, the modal method can only detect the severity and a plane wise orientation of the defect. The orientation will be vertical along, vertical across or horizontal in the structure.

Both ultrasound and modal methods [2, 3] have earlier been tried out on baked anodes. The ultrasound method was found to require more advanced instrumentation and higher measurement times than the modal method [4], which makes the modal test favorable as an inline station in an anode production line.

Other crack detection systems based on electrical resistivity has been commercialized like the MIREA device, where the 2nd generation can achieve more than one anode measurement per minute in the future [5]. MIREA is the result of a technological development partnership between

S. Broek (ed.), *Light Metals 2023*, The Minerals, Metals & Materials Series,
https://doi.org/10.1007/978-3-031-22532-1_155

Rio Tinto, Fives ECL and Alouette Smelter and has been under commissioning at Alouette [6]. The system detects several resistivity values at different height levels in the anode. In the described "Proof of concept phase" [6], the measurement system was evaluated, including the degree of repeatability and how the repeatability changed over time.

The ratio a/F is measured in the modal test, where a is the acceleration from the accelerometers (receivers) and F is the force from the modal hammer- or shaker (exciter). The ratio is the transfer function, $H(f) = \frac{a}{F}(f)$, dependent of the frequency, f, and describes the process (the anode or cathode) rather than the input signals (F) and output signals (a) of the process. A variable F does not change the ratio since a will follow F and keep the ratio close to unchanged. This makes the measurement accurate, with the noise kept to a minimum. The measurement of a/F is resistant against anode variable surface hardness due to different heat treatment exposure from the baking furnace and the pitch level settings in the paste plant. Burn-off areas on the anode have a soft surface and the impact force from the exciter will be reduced. The soft impact will produce an energy distribution over a narrower frequency range which changes the force distribution over the frequencies of interest.

The left plot in Fig. 1 shows a measured magnitude of $H(f)$ at a position on the anode top surface along the measurement line shown orange in the right plot. The cursor (red crosshair) shows both real- and imaginary values of the chosen peak as $H(f)$ is a complex function, containing a real part, Re, and an imaginary part, Im.

The detected frequency of the resonance is close to identical in a repeatability test, regardless of the input force-magnitudes and positions of exciter and receivers. The uncertainty in the resonance is mainly due to the resolution, Δf, of the spectrum, $\frac{a}{F}(f)$, which is set to 0.25 Hz so that also closely spaced resonances can be detected. Since the height and width of anodes and cathodes are comparable ($h \approx w$),

some of the resonances (modes) in the spectrum will be closely spaced and can even reorder.

The modal method was modeled and tested on 5 baked anodes and 16 cathodes to find:

(1) The optimal foundation of a modal test station
(2) The optimal type-, direction- and positions of exciter and receivers
(3) The optimal signal treatment
(4) The Young's Modulus-equations from a modal test and verifications of the equations
(5) Crack identifiers from the modal parameters
(6) Developed software-procedures to automate the reading and detection of the modal parameters from spectrums.

Based on the above findings, a prototype is under development to verify measurement times (including on- and off loadings of exciter and receivers) and noise in the spectrums in the transition from fixed accelerometers to "clamp on"- or contactless principles.

In the next section, the measurement results from the 16 cathodes and 5 baked anodes are presented before the optimal instrumentation in a later section is described. The cathodes were not inspected by coring due to the cost and time constraints. The anodes were inspected by multiple core analyses and visual observations.

The Measurement Results

Figure 2 shows the resonances of 16 cathodes from the same producer, where each resonance (mode) is identified by a mode shape. The mode shapes are illustrated in Fig. 3.

The surprisingly low noise in the measured spectrums made it possible to detect the resonances and link with the associated mode shape.

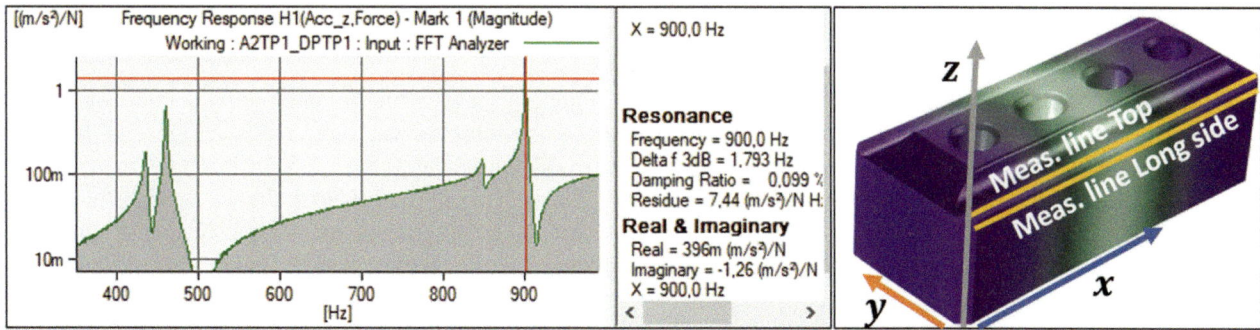

Fig. 1 Left plot shows the measured magnitude, $|H(f)| = \sqrt{Re(a/F)^2 + Im(a/F)^2}$, at a position in the z-direction at the orange measurement line at the anode top (right image). The peaks

in the spectrum are the resonances. The red crosshair-cursor is set on a resonance of 900 Hz, shown with a damping ratio of 0.099%, where $Re = 0.396$ and $Im = -1.26 \ \mathrm{m \cdot s^{-2}/N}$

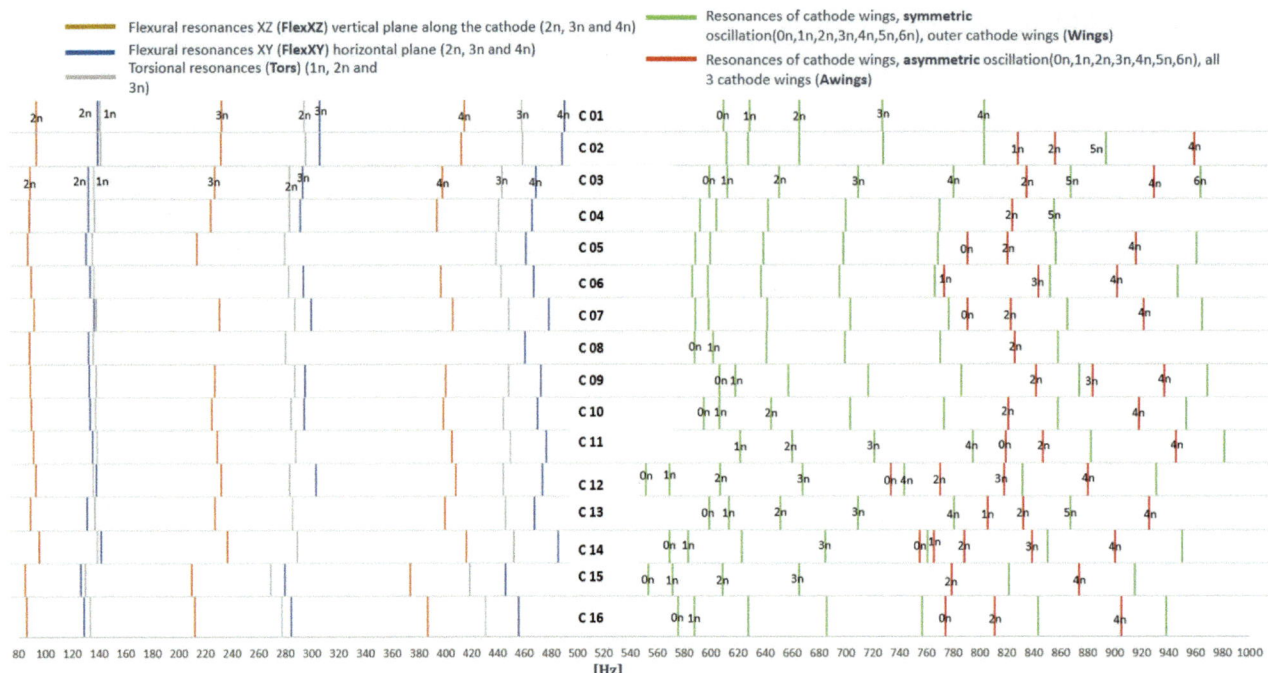

Fig. 2 The measured resonances of 16 cathode bars, not cast (denoted C01-C16) against the frequency. All lines of Cathode C01 and C03 are denoted with node numbers in the figure, while only part of the lines is denoted for the other cathodes to better see the resonance patterns. Node numbers are illustrated in Fig. 3

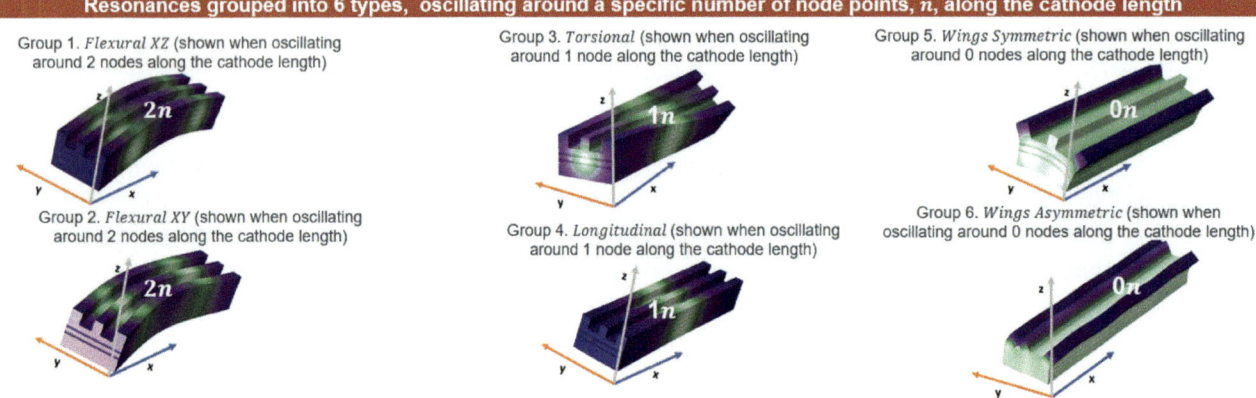

Fig. 3 The mode shapes of the cathode illustrated by a finite element model where each group of mode shape can oscillate at different numbers, n, of node points along the cathode length. Group 1–4 oscillates the whole body, while group 5–6 oscillates mainly the wings of the cathode

The crack identifiers from the modal test of anodes and cathodes

Modeled and measured spectrums have nearly the same ratio between resonances (mode ratios). An overall change of the *YM* in the anode or cathode transposes the same spectrum up or down in the frequency range, while it does not change the ratio between the resonances. Only inhomogeneous defects affect the ratios (within the same product type). For the measurements of, e.g. 4 resonances, f_1, f_2, f_3, f_4, there exist 6 mode ratios; $f_1/f_2, f_1/f_3, f_1/f_4, f_2/f_3, f_2/f_4, f_3/f_4$. Mode ratios can thereby be used as crack identifiers.

Figure 4 shows a "deviation of ratio"-diagram of an OK-cathode meaning the measured ratios are close to the modeled OK-cathode reference.

The mode, *WingsOn* (group 5, Figs. 2 and 3) is one of the modes which have decreased for cathode, C12 and C14, compared to the resonances belonging to group 1–4. A defect near the axis of oscillations near the outer wings of the cathodes is a possible and probable cause. Figure 5 summarizes the deviations of ratios for all the 16 cathodes, where C12 and C14 significantly deviates from the modeled OK reference.

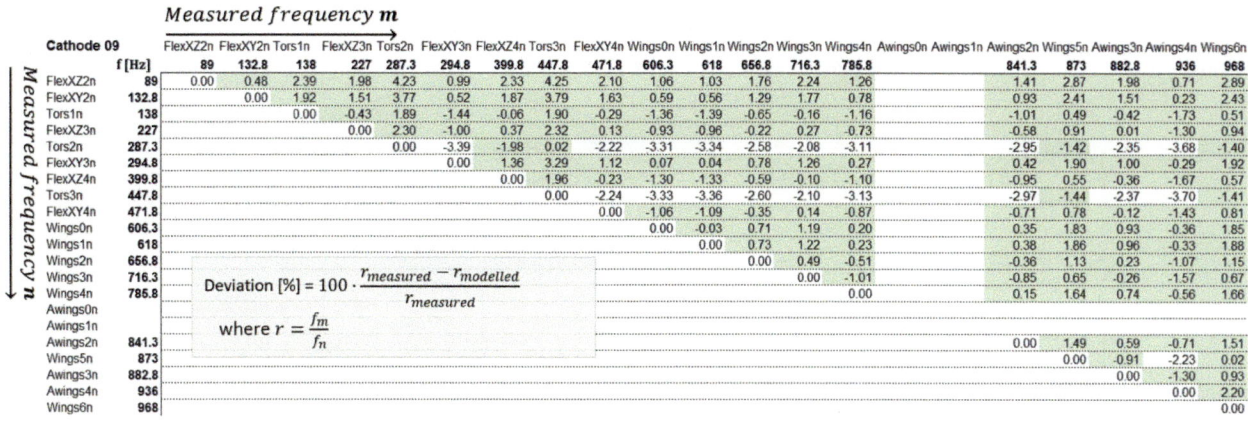

Measured frequency **m** →

Cathode 09	f [Hz]	FlexXZ2n 89	FlexXY2n 132.8	Tors1n 138	FlexXZ3n 227	Tors2n 287.3	FlexXY3n 294.8	FlexXZ4n 399.8	Tors3n 447.8	FlexXY4n 471.8	Wings0n 606.3	Wings1n 618	Wings2n 656.8	Wings3n 716.3	Wings4n 785.8	Awings0n	Awings1n	Awings2n 841.3	Wings5n 873	Awings3n 882.8	Awings4n 936	Wings6n 968
FlexXZ2n	89	0.00	0.48	2.39	1.98	4.23	0.99	2.33	4.25	2.10	1.06	1.03	1.76	2.24	1.26			1.41	2.87	1.98	0.71	2.89
FlexXY2n	132.8		0.00	1.92	1.51	3.77	0.52	1.87	3.79	1.63	0.59	0.56	1.29	1.77	0.78			0.93	2.41	1.51	0.23	2.43
Tors1n	138			0.00	-0.43	1.89	-1.44	-0.06	1.90	-0.29	-1.36	-1.39	-0.65	-0.16	-1.16			-1.01	0.49	-0.42	-1.73	0.51
FlexXZ3n	227				0.00	2.30	-1.00	0.37	2.32	0.13	-0.93	-0.96	-0.22	0.27	-0.73			-0.58	0.91	0.01	-1.30	0.94
Tors2n	287.3					0.00	-3.39	-1.98	0.02	-2.22	-3.31	-3.34	-2.58	-2.08	-3.11			-2.95	-1.42	-2.35	-3.68	-1.40
FlexXY3n	294.8						0.00	1.36	3.29	1.12	0.07	0.04	0.78	1.26	0.27			0.42	1.90	1.00	-0.29	1.92
FlexXZ4n	399.8							0.00	1.96	-0.23	-1.30	-1.33	-0.59	-0.10	-1.10			-0.95	0.55	-0.36	-1.67	0.57
Tors3n	447.8								0.00	-2.24	-3.33	-3.36	-2.60	-2.10	-3.13			-2.97	-1.44	-2.37	-3.70	-1.41
FlexXY4n	471.8									0.00	-1.06	-1.09	-0.35	0.14	-0.87			-0.71	0.78	-0.12	-1.43	0.81
Wings0n	606.3										0.00	-0.03	0.71	1.19	0.20			0.35	1.83	0.93	-0.36	1.85
Wings1n	618											0.00	0.73	1.22	0.23			0.38	1.86	0.96	-0.33	1.88
Wings2n	656.8												0.00	0.49	-0.51			-0.36	1.13	0.23	-1.07	1.15
Wings3n	716.3													0.00	-1.01			-0.85	0.65	-0.26	-1.57	0.67
Wings4n	785.8														0.00			0.15	1.64	0.74	-0.56	1.66
Awings0n																						
Awings1n																						
Awings2n	841.3																	0.00	1.49	0.59	-0.71	1.51
Wings5n	873																		0.00	-0.91	-2.23	0.02
Awings3n	882.8																			0.00	-1.30	0.93
Awings4n	936																				0.00	2.20
Wings6n	968																					0.00

Measured frequency n ↓

$$\text{Deviation [\%]} = 100 \cdot \frac{r_{measured} - r_{modelled}}{r_{measured}}$$

$$\text{where } r = \frac{f_m}{f_n}$$

Worst deviation:	-3.70 %	The ratio *Awings4n/Tors3n* is decreased by 3,70% compared to the modeled reference
Average Deviation of all ratios:	1.33 %	(from absolute values)
Average deviation Flex & Tors:	1.77 %	(from absolute values)

Fig. 4 Deviations of mode ratios between measured- and modeled resonances of the cathode, C09, from Fig. 3. Green ratios are those ratios where the measured deviation has not decreased more than −2.00% compared to the modeled OK cathode reference

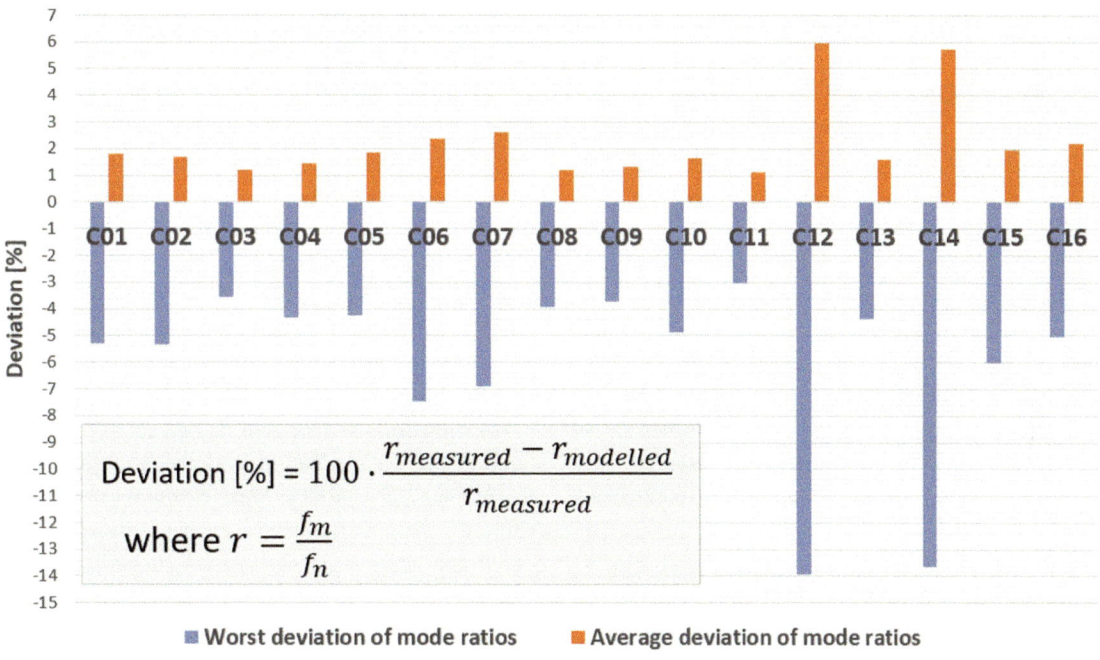

$$\text{Deviation [\%]} = 100 \cdot \frac{r_{measured} - r_{modelled}}{r_{measured}}$$

$$\text{where } r = \frac{f_m}{f_n}$$

■ Worst deviation of mode ratios   ■ Average deviation of mode ratios

Fig. 5 Average- and worst deviation of mode ratios of the 16 cathodes, C01–C16. The deviations of mode ratios are described in Fig. 4

Measured resonances of baked anodes have also been found with mode shapes. Figure 6 shows the measured resonances of 5 baked anodes, A1–A5, from one producer. The measured anodes have no slots and have therefore no measured oscillations dedicated for the "wings" or "legs" as the cathode mode shapes of group 5–6 in Fig. 3.

Interestingly, anode A2 and A3, Fig. 6, have been produced successively at the same vibrocompactor in the paste plant and are observed, as expected, to have almost the same spectrum. For anode A1 the spectrum has been "transposed"

downward due to a lower *YM* of the anode and not because of local defects (the mode ratios are preserved).

For anode, A5, the resonances have been reordered, where the 3rd mode has become the 2nd, and the 2nd has become the 3rd. In addition, the flexural *XY*-mode of 3 nodes (757 Hz, orange) has decreased significantly compared to the other modes. A5 is therefore a suspected anode with mode ratios not preserved. A4 is close to the same production time as A5. However, the flexural *XY*-mode (802 Hz) of A4 has not decreased to the same degree as A5.

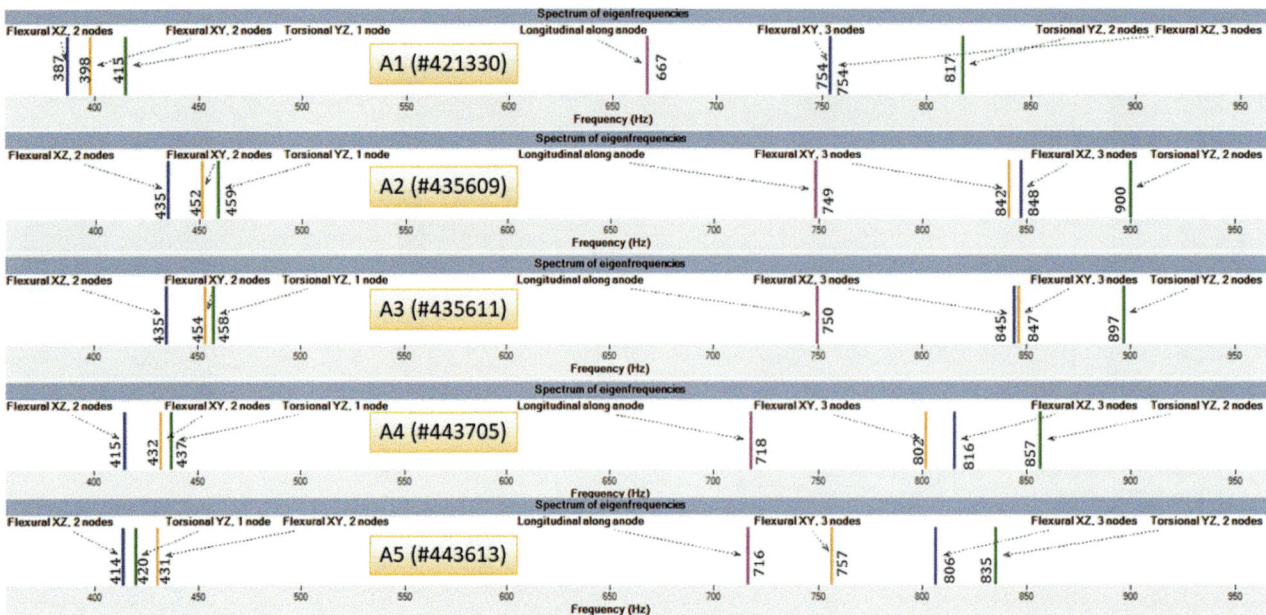

Fig. 6 The measured modes of 5 baked anodes, A1–A5 with ID-numbers, where each mode is found by a mode shape

A vertical defect along anode, A5, is expected as flexural modes oscillating in the horizontal XY-plane are sensitive to defects in the vertical plane along the anode (XZ). The acceptable "deviation of mode ratios" must be adjusted according to experiences from the plants, after a larger set of anodes have been measured.

Figure 7 shows the measurements of A5 for two of the seven identified modes, where the damping is found for each mode from the transfer function shown in Fig. 1. The damping signifies an increased energy loss in the material and will therefore be regarded as a 2nd crack identifier in the future measurements (in addition to the mode ratios).

The mode shape curves in Fig. 7 have been found from interpolated solid curves fitted to measured circle-points, $Im \frac{a}{F}(f)$, at positions along the measurement lines in Fig. 1, regardless of the value of $Re \frac{a}{F}(f)$. All three directions, x, y, z, are measured (blue, orange, grey lines respectively). The real value should, according to the theory, be zero at resonance, but rather are small values observed when measured. The dotted lines show the mode shape found from "estimated circles" in the Nyquist-plot, a method [7] which estimates $Im \frac{a}{F}(f)$ if $Re \frac{a}{F}(f) = 0$. From Mode 1, in Fig. 7, the solid curves cross the 0-line two times along the anode length meaning the mode shape has two node points along the anode length. In addition, the oscillation has no measured y-component (orange), and the z-component (grey) is the most sensitive. The automation software easily interprets mode 1 as the flexural mode oscillating in the XZ-plane. The automation software has, so far, criteria for detecting the first seven modes with mode shapes.

A third crack identifier can be found from the mode shapes, defined as *ModeOffset = ModeCenter − Anodecenter*. Ideally the mode shapes should be symmetric around the anode center, close to $l/2$ (grey cursor in the plots of Fig. 7). However, if the anode is inhomogeneous the symmetric point (mode center) of the oscillation (red cursor, Fig. 7) moves away from the anode center. Figure 8 summaries the measurement of mode offsets of the 5 anodes, except the mode offset of the longitudinal mode which were found to be inaccurate in this campaign.

The mode offsets (like the mode ratios) clearly point out anode A5 to have offsets significantly higher than for the other anodes. The modes, *FlexXZ2n* and *Tors1n*, see a mass center of the anode further away from $l/2$. The reason behind can be cracks, segregation of particles or inhomogeneous elasticity or density. Anode, A5, was from the core analyses the one with most of defects. The cracks were going vertically along the XZ-plane, mainly in the left part of the anode.

The Young's Modulus from the Modal Test

For the inline detection system to measure *YM*, baked dimensions and weight needs to me measured in advance. This is shown by a simple uniaxial mass (m)-spring-damper-system oscillating by a displacement u_x (in a direction of x). According to a second order system, the relation between stiffness, k, (elasticity) and resonance, f, is given by the well-known Eq. 1.

Fig. 7 Information panels for each identified mode, showing the two first modes of anode, A5

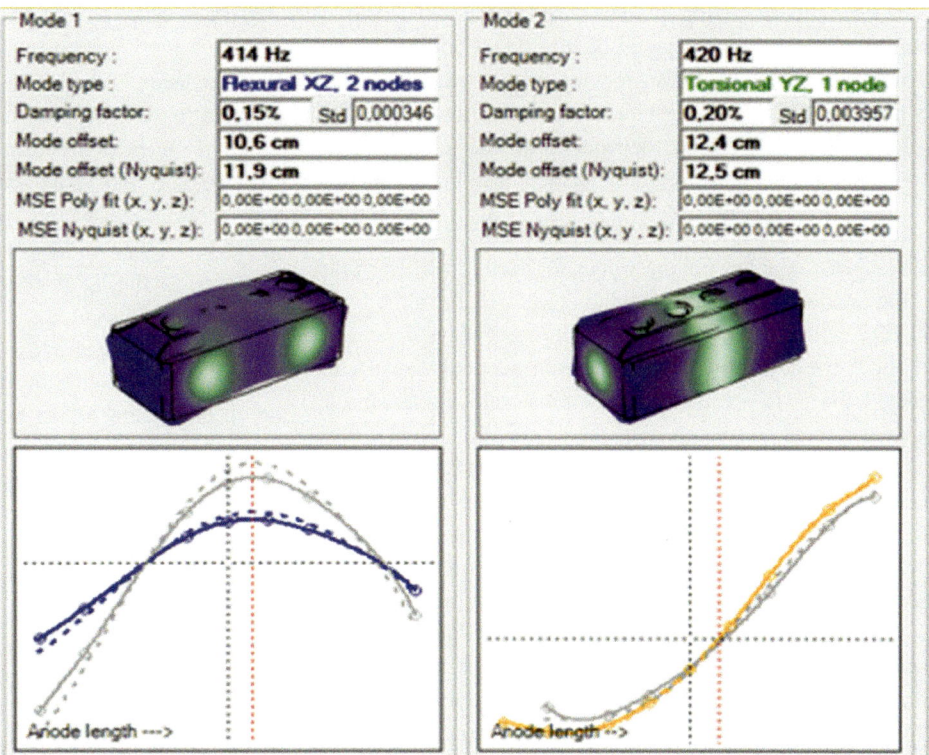

Fig. 8 Measured mode offsets for the anodes, A1–A5. The end caps of the error bars show the values of the mode offsets when calculated by the Nyquist-method, shown by dotted lines in Fig. 7

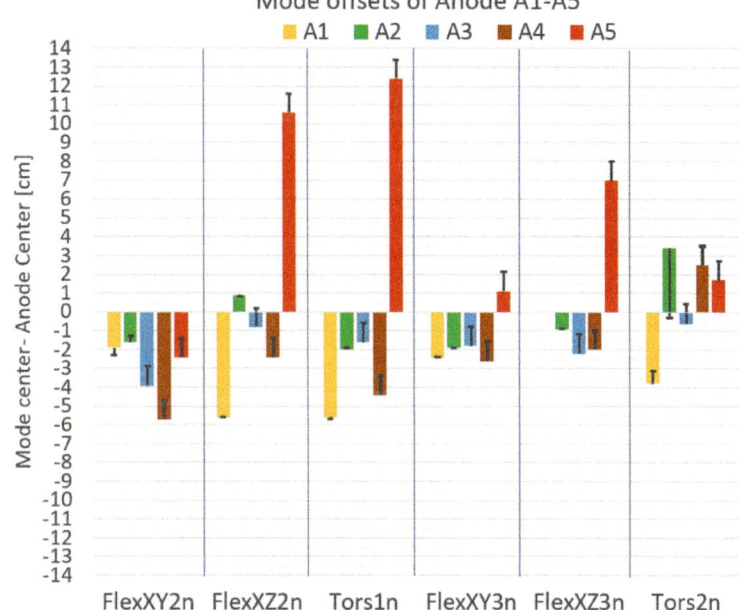

$$f = \frac{1}{2\pi} \sqrt{\frac{k}{m}} \qquad (1)$$

The stiffness, k, is defined by the spring force, $F_x = k u_x$ so that $k = F_x / u_x$.

By exchanging the spring with a material, having the same properties as the spring, Eq. 1 can be expressed by YM ($= E$) and ρ instead of k and m. Assuming a bar of length, l,

along x, with a cross section, $A = h \times w$, the modulus, E, can be written $E = \frac{l}{A} k$ and $\rho = m/V$. Equation 1 can thereby be expressed by E and ρ as in Eq. 2.

$$f = \frac{1}{l} \sqrt{\frac{E}{\rho}} \qquad (2)$$

The simple oscillating uniaxial case (Eqs. 1 and 2) shows that the measured resonance to be dependent both on anode density and elasticity. When moving away from an 1D-bar, to an anode of 3D, the relation between f and E will be different for each mode, dependent of the mode shape. A coefficient, C, must be implemented into Eq. 2, so that $f = C\frac{1}{l}\sqrt{\frac{E}{\rho}}$, where C is an expression of the anode dimensions, w, h, l, and has unique values for each identified mode of the anode or cathode. Both Eqs. (1 and 2) have the units of Hz, which lead to a coefficient term, $C(w, h, l)$, that should be non-dimensional. The finite element model calculated the 7 resonances for several combinations of the input parameters w, h, l, ρ, E of the model. This gave a data matrix for regression analysis to find $C(w, h, l)$ for each mode. The C-terms are therefore valid within the ranges of the input parameters of the model. With 7 identified modes, there are also 7 YM-equations which all will get the same value if the measured anode is crack free homogenous. However, since defects normally decrease some of the modes, the YM can be found by choosing the mode which carry the maximum value of YM (least affected by defects). Figure 9 shows the YM of the modal test when the longitudinal mode was chosen for all the 5 anodes.

The YM from the modal test highly correlates to the YM from loading tests. The selection of YM-equations among the modes can also be selected by taking the average YM from those modes which are involved in mode ratios closed to the modeled OK-reference. Equations are also derived for transversal isotropic materials as the cathode is specified with a vertical and horizontal YM of the product. Elasticity equations of Shear Modulus can also be derived since modes are found with mode shapes.

Optimal Instrumentation of the Modal Test

The foundation of the anode and cathode in the modal test was designed so that rigid body modes (RB-modes) were far below the frequencies of interest [8]. RB-modes are those which just oscillate and move the anode without deforming the body. Five anodes were tested by a modal hammer, and in an earlier test other anodes were tested with a shaker. In the hammer test the hammer moves from point to point at the measurement lines in Fig. 1 while the receivers are set at one position near the anode corner. With a shaker it will be opposite: the shaker is at the anode or cathode-corner while the receivers are at the actual measurement points at the measurement lines. In the future prototype the shaker solution will be chosen since this is more suited for automation and low measurement times.

For cathodes with wings or anodes with slots, the two measurement lines go along x close to the edge of an outer leg or wing, where one line is at the horizontal surface of the outer leg, and one is at the long side. For the anode with slots, it will be upside down in a test station.

With the hammer-solution, the recording time of the Fast Fourier Transform (FFT) needs to end before the next impact occurs. With a 0.25 Hz resolution in the spectrums, it will take 4 s for each impact. The measurement time (except on- and off-loading of exciters and receivers) will take 4 s multiplied with the number of measurement points along the measurement lines. With the anodes for testing, 9 points for each measurement line were defined, meaning $18\,points \times 4\,s = 72$ s. Choosing a coarser resolution of the FFT, like 0.5 Hz, will decrease the measurement time to the half (36 s).

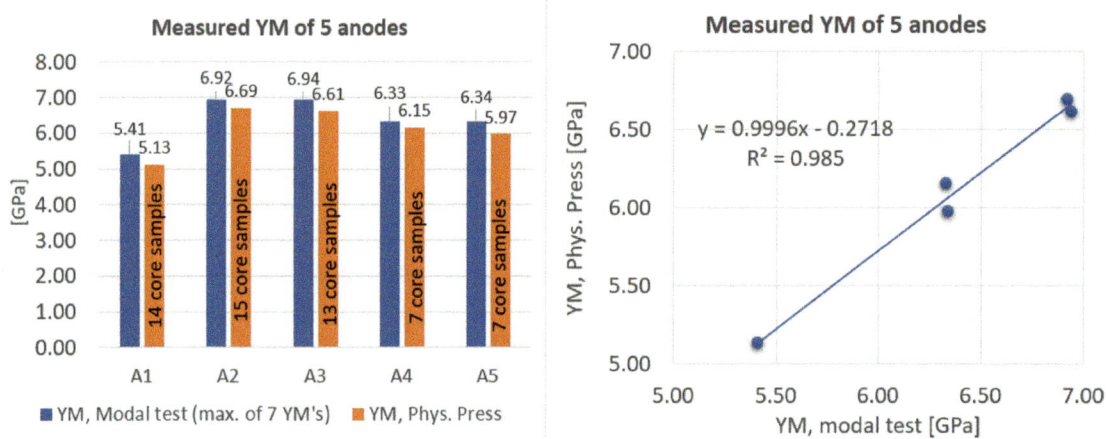

Fig. 9 Measured YM of the 5 baked anodes by the modal test (using the longitudinal mode), and by the physical loading test of sample cores of the same anodes

The mode shapes can be identified with fewer points, since it is not necessary to measure more than 2/3 of the length of the measurement line and still be able to identify the modes by mode shapes (since the mode shapes have symmetric behavior). This will reduce the above measurement times by around 30%.

The shaker solution is more time efficient since the force input are at only one position (the anode or cathode corner). The shaker will be set to deliver random noise in the relevant frequency range while the receivers are at positions along the two measurement lines. Spectrum averaging will be enabled, since each *FFT* recording time takes only 4 s (with 0.25 Hz resolution). The measurement time will be as low as 24 s if 6 spectrum-averages are carried out (spectrum averaging can also be carried out with overlapping, meaning that the number of averages can be higher). More than 35 s will then be available for on- and off-loading of the exciter and receivers.

The number of spectrum averages is decided by the noise in the measurements. The system measures online the coherent degree of linearity between a and F [8] which normally improves with the number of spectrum averages. The accuracy of the measurement can therefore be checked by the system itself; either it will be noise at a certain frequency range, or noise at a certain receiver along the measurement lines.

The coherence function is not available with hammer impacts unless the hammer hits at least two times at the same measurement point. This will double the number of impacts in the test and make the hammer-solution even more time consuming than the shaker-solution.

When it comes to the hardness of the cap of the hammer or shaker in contact with the anode or cathode, it should be chosen so that the power spectrum of the force does not decrease more than 20 dB over the frequency of interest [8]. The decrease has been measured to be around 4 dB in the frequency range of $100 - 1000$ Hz.

Time windowing of signals in the time domain are especially crucial for the hammer-solution, before the signals are transformed by *FFT* to the frequency domain. Transient time windows were chosen both for the impact force and the receivers for the 16 cathodes and the 5 anodes in the tests.

To search after a resonance, and to be sure that a certain peak in the spectrum is a resonance, certain threshold criteria were followed using information from the *magnitude* of $H(f)$ and the *phase* between a and F as $tan^{-1}\left(\frac{Im(a/F)}{Re(a/F)}\right)$. The peak under inspection had to be regarded as a resonance when: (1) The peak was higher than a certain *magnitude*, (2) The peak should have a defined 3 db-bandwidth, (3) The *phase* at the peak should decrease, (4) The *phase* at the peak should be within a range around $+90°$ or around $-90°$, (5) The same resonance among the measured spectrums of the tested item is found within a percentage of the frequency, set somewhat higher than the frequency resolution of the *FFT*.

The settings of the threshold criteria are set-up-parameters in the automation software. They can also be used as a guidance to manually detect peaks as resonances in the spectrum.

Summary

The modal test of anodes and cathodes was verified by modeling and measurements of 16 cathodes and 5 baked anodes. The instrumentation was set up in a robust and accurate way to suit a measurement time of $60\,s/item$ in an industrial line. The modal parameters were found to be sensitive to defects and gave the Young's Modulus of the anode and cathode. The noise in the measurement can be monitored online by the system itself. The modal system has a high degree of repeatability, especially the resonance, and will not be affected by varying impact forces and accelerations. The modal test is therefore fast and robust and is a candidate as an in-line station in the final control of an anode manufacturing plant.

Acknowledgements The authors are thankful for the financial support from the Norwegian Research Council (grant no. 309272) and for all the help from the Laboratory at Hydro Årdalstangen, Norway.

References

1. Bouché C, Genin X, Mahieu P, Georgel S (2021) AMELIOS Suite or the Fives Digital Package for Carbon 4.0. Light Metals, 940–950
2. Boubakar MB, Picard D, Duchesne C, Tessier J, Alamdari H, Farad M (2017) Non-Destructive Testing of Baked Anodes Based on Modal Analysis and Principal Component Analysis. Light Metals, 1289–1298
3. Boubakar MB (2017) Non-destructive quality control of carbon anodes using modal analysis, acoustoultrasonic and latent variable methods, Ph.D. Thesis, Université Laval
4. Rodrigues, D et al. (2020) A New Prototype for Acousto-Ultrasound Analysis of Carbon Anodes. Paper presented at the 38th conference and exhibition ICSOBA, online, 16–18 November 2020
5. Léonard G, Guérard S, Laroche D, Arnaud JC, Gourmaud S, Gagnon M, Marie-Chollier MJ, Perron Y (2014) Anode Electrical Resistance Measurements: Learning and Industrial On-line Measurement Equipment Development. Light Metals, 1269–1274
6. Gagnon M, Léonard G, Bernard A, Ghaoui YEl, Gourmaud S (2016) MIREA: An On-Line Quality Control Equipment Integration in an Opertional Context. Light Metals, 979–984
7. Miller CW (1978) Determination of Modal Parameters from Experimental Frequency Response Data, Master Thesis, University of Texas at Austin
8. Hewlett-Packard Co. (1997) The Fundamentals of Modal Testing, Application note 243-3, https://www.hpmemoryproject.org/an/pdf/an_243-3.pdf. Accessed 6 September 2022

Part IX
Scandium Extraction and Use in Aluminum Alloys

Investigations into Optimized Industrial Pilot Scale BR Leaching for Sc Extraction

Efthymios Balomenos, Panagiotis Davris, Alexandra Apostolopoulou,
Danai Marinos, Elena Mikeli, Aikaterini Toli, Dimitrios Kotsanis,
Grigoris Paschalis, and Dimitrios Panias

Abstract

Scandium is a critical metal with increasing demand in modern technologies, like solid oxide fuel cells and light-weight Al-Sc alloys. Sc is present at considerably high concentrations in various metallurgical by-products, including the bauxite residue (BR). Scandium extraction from the Greek bauxite residue (BR) that contains 70–100 mg/kg Sc has been demonstrated at an industrial pilot plant at MYTILINEOS. BR has been treated with sulfuric acid at conditions that allow for low acid consumption. In such conditions, high Sc leaching selectivity is achieved over Fe and Ti, which are the main impurities for the further purification of the leachate through the ion-exchange technique. This paper reports on the recent investigations on the effect of pH control during the leaching process, which can lead to optimized results. Moreover, the treatment of the produced leachate solution is also studied, in order to further refine it, prior to the ion-exchange.

Keywords

Red mud • Bauxite residue • Scandium • Recovery of scandium

Introduction

The primary production of aluminium requires approximately 2 tons of metallurgical grade alumina (Al_2O_3) for each ton of metal produced through molten salt electrolysis. To produce 2 tons of such alumina, alumina refineries need to process approximately 4 tons of bauxite ore through the Bayer process. The latter relies on the use of a caustic soda to selectively dissolve the aluminium content of the bauxite ore, which contains 40–60% aluminium oxides. The remaining ore, forms the solid by-product of the Bayer process which is termed *bauxite residue (BR)*. BR contains various undissolved metal oxides and hydroxides of Fe, Al, Ti and Si from the original bauxite ore as well as complex Al-Si-Na-Ca phases formed during the Bayer processing. Furthermore, BR contains in trace amounts metal oxides like V, Ga, REE/Sc and others, the concentration of which has practically doubled from the respective concentrations in the bauxite ore.

The acidic leaching of Sc from the BR faces two major challenges: (i) selectivity of leaching Sc against Fe and Ti which are both found in much greater concentrations in the BR and both tend to be co-extracted with Sc during ion exchange and (ii) acid consumption, as BR is naturally an alkaline material with a varying amount of Na, depending on the initial bauxite ore, processing conditions and process upsets in the alumina plant.

The selective leaching of Sc from Greek BR utilizing sulfuric acid to produce a suitable Sc containing solution for ion-exchange Sc extraction has been established in the previous work of the authors [1].

In the present paper, a method for automated control of the process based on the desired pulp pH is presented. Such a process should be able to optimize acid consumption and allow for the regulation of changes in the BR feed sodium chemical composition.

In addition, a lab scale study at the National Technical University of Athens (NTUA) of the produced PLS

E. Balomenos (✉) · P. Davris · A. Apostolopoulou · G. Paschalis · D. Panias
MYTILINEOS SA-Metallurgy BU, Ag. Nikolaos Plant, 320 03 Viotia, Greece
e-mail: efthymios.balomenos-external@alhellas.gr

D. Marinos · E. Mikeli · A. Toli · D. Kotsanis
NTUA, Laboratory of Metallurgy - TesMeT Group, Heroon Polytechniou 9, 157 80, Athens, Greece

© The Minerals, Metals & Materials Society 2023
S. Broek (ed.), *Light Metals 2023*, The Minerals, Metals & Materials Series,
https://doi.org/10.1007/978-3-031-22532-1_156

(pregnant liquid solution) is presented aiming at elucidating PLS composition and pre-treatment steps needed prior to ion-exchange.

Experimental

MYTILINEOS: The MYTILINEOS acid-leaching pilot plant consists of three polypropylene reactor tanks of up to 800 L capacity, with mechanical steering and heating/cooling through immersed coils for circulating steam and cooling water, respectively. Filter-pressed BR produced at MYTILINEOS alumina refinery is mixed with industrial water in the first reactor (100-TK-10) to produce pulp of specific density measured through an inline Coriolis Mass Flow Meter. The pulp is pumped to the second reactor (100-TK-30), where is heated and contacted with concentrated sulfuric acid. The pulp exiting the 100-TK-30 is driven to the cooling tank (200-TK-40), where it is cooled to 60 C and is subsequently passed to the filter press circuit. The filter press separates solids from liquid, generating the final PLS (Pregnant Liquid Solution) and the neutralized bauxite residue. The filter press used consists of 25 frames of 470 × 470 mm with 11 chamber plates and 12 membrane plates and a total filter area of 6.6 m^2. Inlet slurry is drawn via a diaphragm pump with a maximum working pressure of 15 bar. The filter cake washing was conducted directly on the filter press with fresh water inserted through the inlet slurry and directed to the washate-receiving tank. Cake squeezing and cake air blowing were also applied in the filter press.

To achieve better control of the leaching process, automation control was applied, utilizing a pH meter (Ceraliquid CPS41D, Endress and Hauser) in 100-TK-30 linked to the pilot's SCADA controlling the sulfuric acid peristaltic pump (Qdos, Watson-Marlow). The system was set to maintain the pH of the pulp at a value of 2.3 by controlling the flow of concentrated sulfuric acid in the 100-TK-30 reactor. Likewise, the temperature was controlled by automatically regulating the flow of pressurized steam in the heat exchange coils of the 100-TK-30 reactor, to achieve and maintain a temperature of 85 °C. The retention time of the BR pulp in the 100-TK-30 reactor was 30 min, and approximately 230 kg of dry BR was processed in this way per day. Two PLS samples were taken daily directly from the filtrate of the filterpress and analysed by ICP-OES (Avio 220, PerkinElmer) for their Sc, Fe, and Ti concentrations at MYTILINEOS. The PLS produced daily was homogenized in 5m3 tanks and samples were provided to NTUA for further analysis by ICP-OES and study.

NTUA: Chemical analysis—To analyze the chemical composition of the BR, alkaline fusion and wet chemical analysis using the inductively coupled plasma–optical emission spectrometry (ICP-OES) Optima 8000 by Perlin Elmer (Waltham, MA, USA) and Atomic Absorption Spectroscopy (AAS), (PinAAcle 900 T, Perkin Elmer, Waltham, MA, USA). Calibration standard solutions were prepared from commercially available ICP, standards obtained from Merck (Darmstadt, Germany). The standard solutions were prepared in a suitable concentration and diluted further with 1% v/v analytical grade nitric acid (65% wt.) as required for working standards. High-purity deionized water (18.2 MW/cm) and argon of special purity (99.999%) were used.

Measurements—The pH was measured by a Metrohm mobile pH meter 826 with an Ecotrode Plus electrode. The Oxidation/Reduction potential was measured by a Hach HQ40D Portable Multi Meter with a digital oxidation-reduction potential (ORP/Redox) probe IntelliCAL® MTC101. The results were transferred to the Eh using the equation Eh = $E_{m\ (redox)}$ + E_{ref}, where the E_{ref} was 207 mV, according to the probe used.

Physicochemical characterization—The X-ray diffraction analysis XRD pattern was recorded on a Rigaku Miniflex 600 benchtop diffractometer (Rigaku, Tokyo, Japan), with CuKa radiation (λ = 1.5405 Å) operating at 40 kV and 15 mA (the diffraction pattern was recorded between 3° and 75° (2θ), in 0.02° (2θ) steps with a scanning speed of 5° (2θ/min)).

Modeling of the PLS—The HSC software by Outotech (version 10.0.6.7) was utilized to model the PLS using both the Aqua and the Eh–pH modules. On the Eh–pH analysis, only aqueous and neutral species were considered. Also, the only S species that were considered were the SO_4^{-2} ones. Some iron species ($HFeO^-$, $HFeO_2^-$) were also not considered.

Results

Pilot Plant Leaching Campaign

The chemical composition of the BR used can be seen in Table 1. In total 9.8 t of BR were processed in the pilot over the course of 43 tests, producing 13.4 t of PLS. The pH control of the process proved effective as overall the pH values varied around 2.3 with ± 0.1 in each trial. Indicative variations of the pH during leaching are presented in Fig. 1, for two different trials, at conditions where a steady flow in reactor 100-TK-30 has been achieved. The BR pulp in the reactor tends to act as a buffer solution, neutralizing the incoming acidity; as a result the system constantly fluctuates the flow rate of the concentrated sulfuric acid stream. Depending on the variations in the initial pH levels in the BR pulp and the overall system response, differences in the total acid consumption are noted in each trial.

Table 1 Chemical composition of the BR from MYTILINEOS measured at NTUA

wt.%						ppm
Fe₂O₃	CaO	Al₂O₃	SiO₂	TiO₂	Na₂O	Sc
38.15	8.37	23.28	7.82	4.67	3.15	79.20

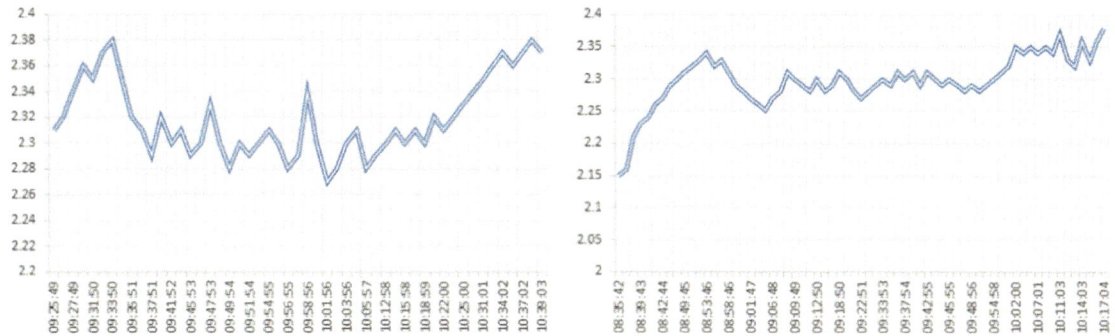

Fig. 1 Indicative pH profiles during leaching in two separate leaching tests

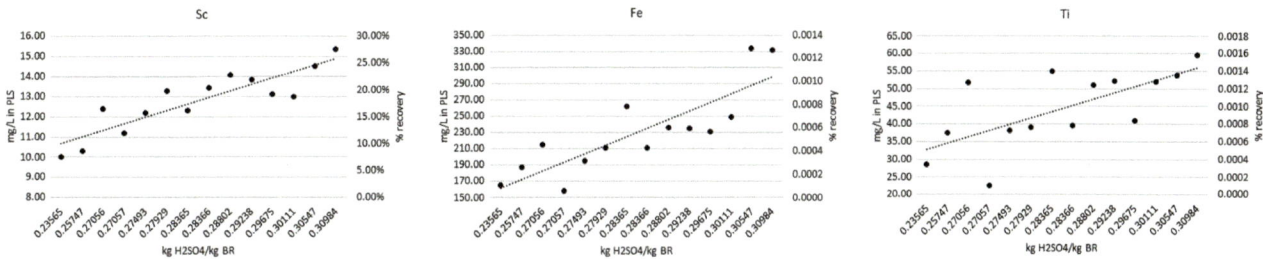

Fig. 2 Indicative Sc, Fe and Ti concentrations in the PLS at various kg H₂SO₄ /kg BR

Indicative results in terms of Sc, Fe and Ti extraction are presented in Fig. 2 as a function of the mass of concentrated sulfuric acid consumed per mass of BR treated. Dissolved Sc in the PLS ranged between 10–15 mg/L while Fe and Ti impurities were between 150–300 and 20–60 mg/L respectively.

From the results, a clear trend is confirmed between increasing the specific acid consumption and the recovery of Sc, Ti and Fe. This becomes more apparent when the specific acid consumption (kg acid/g Sc leached) is compared to the specific Sc leaching (gr Sc leached/ t BR), in Fig. 3. The potential for further optimization of the process can be seen if one moves towards the bottom right of the specific graph to achieve leaching of more than 20 g of Sc per t of BR with less than 15 kg of sulfuric acid per g of Sc.

The comparison of the homogenized PLS produced from this pH-driven campaign is presented in Table 2 and compared to the results achieved with the flowmeter campaign, where the pilot automation controlled the flow rate of the sulfuric acid pump at a set ratio with the BR pulp flow rate form 100-TK-10 [1]. The comparison of the two campaigns indicates a significant decrease in impurities levels as expressed by the dissolved Ti/Sc ratio in the PLS, for the pH-driven campaign.

PLS Treatment

A PLS sample produced from a single test in the pilot plant of MYTILINEOS was provided to NTUA for testing. The PLS sample was supersaturated, and thus, before the chemical analysis, solid to liquid separation had to be performed. The precipitate that was obtained from the solid to liquid separation was dissolved/measured by different chemical analysis methods as can be seen in Table 3.

The mineralogical composition of the solid precipitate was investigated through X-ray diffraction (XRD) analysis. The XRD profile of the solid precipitate (Fig. 4) exhibits four broad bands centered at 9.1 Å (9.7° 2θ), 3.4 Å (26.2° 2θ), 2.2 Å (40.7° 2θ) and 1.4 Å (66.7° 2θ), while it resembles that of allophane [2]. Besides the dominant occurrence of this poorly crystalline aluminosilicate, the sample also contains trace amounts of hematite, as it was possible to identify its two major peaks centered at 2.7 Å (33.1° 2θ) and 2.5 Å (35.7° 2θ), respectively.

The chemical analysis of the filtered PLS is seen in Table 4. The pH of the solution was measured at 25 °C and was found to be 3.36. In order to recover the Sc from this solution, an ion-exchange process will be used for

Fig. 3 Specific acid consumption versus specific Sc leaching

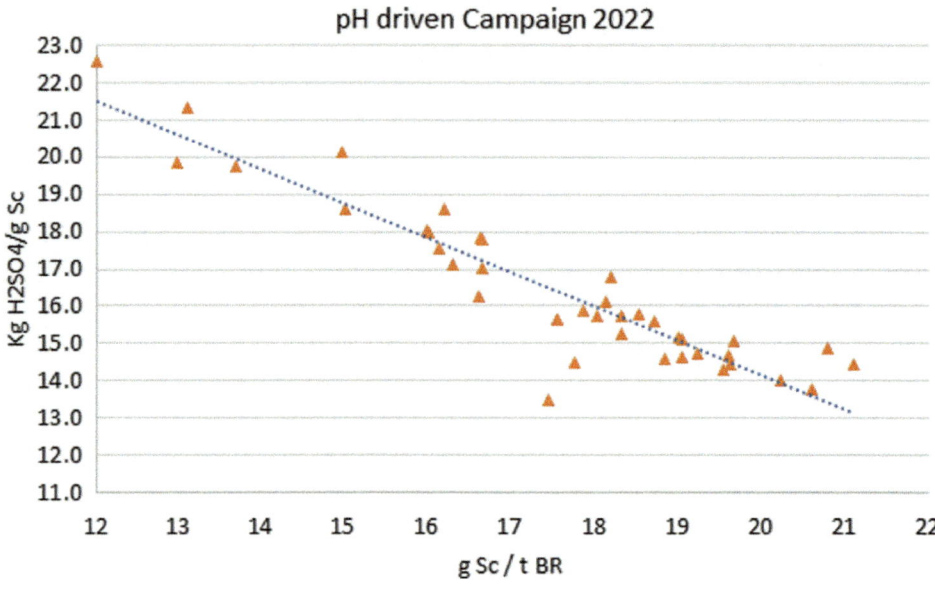

Table 2 Homogenized PLS results measured at MYTILINEOS

	t H_2SO_4/ t BR	Acid (kg)/ Sc (g)	g Sc/t BR	Sc	Ti	Fe	Ti/Sc ratio < 5
2012 pH-driven campaign	0.28	16.80	17.48	13.04	35.2	210.6	2.69
2011 Flowmeter-driven campaign	0.27	16.19	17.15	13.25	55.3	289.7	4.1

Table 3 Chemical analysis of the obtained precipitate

Chemical analysis	Fe_2O_3	Al_2O_3	TiO_2	SiO_2	CaO	Na_2O	Sc
XRF wt%	1.52	31.73	1.49	17.78	0.2	6.12	–
Fusion wt%	1.93	46.66	1.6	21.89	0.39	4.46	<0.05

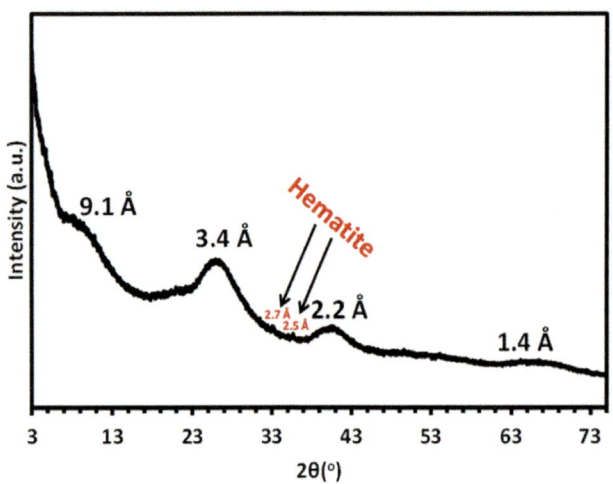

Fig. 4 XRD of amorphous aluminosilicate precipitate

purification. It is known, that the capacity of a weakly acidic resin, can significantly depend on the pH of the solution [3]. Most Ion-exchange resins are known for having the ability to first adsorb the higher valence ions [5]. The elements that are usually co-extracted together with Sc are Ti, Zr and Fe due to their high chemical affinity [4, 6]. If though, Fe^{+3} is reduced to Fe^{+2} by adding metallic iron according to reaction $2Fe^{+3} + Fe^0 = 3Fe^{+2}$, then Fe is usually not co-extracted increasing the ion-exchange resin's performance [6].

Before testing, the PLS solution was modeled, using the HSC software (ver: 10.0.6.7). At first, the aqua module was used to calculate the ionic strength of the solution based on its concentration. The ionic strength was found to be 3.57 mol/kg. The solution was also modeled using the Eh–pH module, using the ionic strength that was calculated from the aqua module, for the $H_2O/Fe/S$ system at 25 °C. The

Table 4 Chemical analysis of PLS sample (measured by ICP-OES or AAS at NTUA)

Elements	SO_4	Na	Al	Ca	Total Fe	Fe + 2	K	Si	Ti	Sc	Zr	V
Concentration (mg/L)	78,100*	12,800	9,800	577.5	280.9	172**	140	123	39.1	11.3	2.1	1

*Measured with a gravimetric method with the ignition of residue
**Measured by UV–VIS spectrophotometer, Fe(II) 1,10 Phenathrolin method

Eh–pH diagram produced can be seen in Fig. 5. From the graph, it can be seen that at a pH less than 4, the predominance area of Fe^{+3} species is always above 596 mV and the Fe^{+2} species predominance area is always below 538 mV depending on the pH.

To stabilize the PLS and to remove Fe interference in the ion-exchange, the PLS was studied for lowering the pH by adding sulfuric acid and metallic iron to reduce all the Fe^{+3} to Fe^{+2}.

The pH and the Oxidation/Reduction potential of the PLS were measured at room temperature. Since silica gel formation can also become a problem in such solutions, also the Si concentration was monitored in the experiments where sulfuric acid was added. The results showing the addition of sulfuric acid can be seen in Table 5 and the addition of metallic iron in Table 6. From the results, we can see that the addition of sulfuric acid did not have a great influence on the Eh or on the concentration of Fe^{+2} and Si. On the other hand, the addition of metallic iron changed greatly the Eh and the Fe^{+3} was totally reduced to Fe^{+2} from the first 0.315 g of Fe addition. Iron is reduced between 541.7 and 453.5 mV which corresponds with the theoretical calculations conducted with the HSC software.

Conclusions

The pH-driven BR leaching pilot campaign at MYTILINEOS has proven the capability to reproduce and even slightly improve the results of the previous flowmeter driven BR leaching campaign. This method of controlling the process is considered preferable as it allows for automatic corrections of the acid feed in light of changes in the BR's sodium content-pH level. Overall, the leached impurities of Fe and Ti were found to be fewer in comparison to the flowmeter campaign of 2021 with practically the same amounts of Sc recovery and acid consumption.

Lab scale study of the produced PLS revealed that as produced it is oversaturated in Al and Si, leading to precipitate formation over time. Sulfuric acid addition is required to stabilize the solution pH below 1.5. Approximately 60% of the dissolved iron is found to be Fe(II) cations. Addition of metallic iron as little as 3 kg/m³ of PLS can increase this amount to 95% allowing for Sc extraction through ion-exchange without significant Fe co-extraction.

The results presented here offer insight into further potential optimizations of the process that will allow the

Fig. 5 Eh–pH diagram of the PLS produced in the MYTILINEOS pilot plant calculated with HSC software

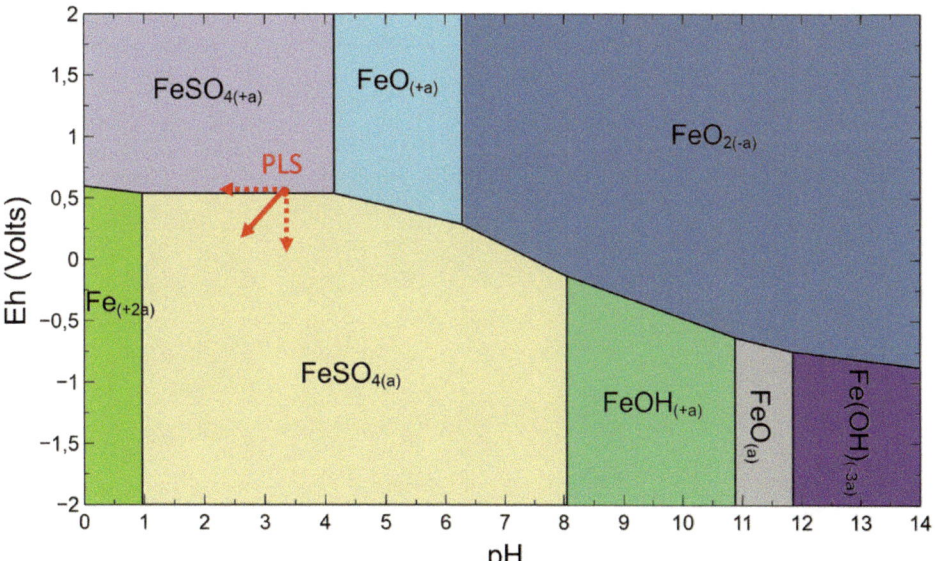

Table 5 Addition of concentrated sulfuric acid and its influence on the PLS

Added	2 h agitation		24 h contact		UV	ICP-OES	
ml of concentrated H_2SO_4 / L of PLS	pH	eh (mV)	pH	Eh (mV)	Fe(II) (mg/L)	Fe total (mg/L)	Si (mg/L)
0	3.36	541.7	3.35	559.2	172.3	272.2	123
5	1.54	617	1.6	612.6	176.28	270.8	121.9
10	1.05	612.1	1.09	612.5	176.53	266.1	118.4
15	0.8	607.8	0.84	606.7	174.54	269.3	121.5
15	0.83	607	0.88	608.7	171.31	265.2	117.3
20	0.6	605.7	0.64	603.2	168.32	249.3	114.2

Table 6 Addition of metallic iron and its influence on the PLS

Added			UV	ICP-OES	
Fe^0 (g/L)	pH	eh (mV)	Fe(II) mg/L	Fe total	Fe II%
0	3.36	541.7	172.3	272.200	63.3
0.315	0.76	453.5	559.9	587.300	95.34
0.617	0.76	438.3	846.8	899.200	94.18
1.014	0.78	400.2	1268.0	1400.000	90.57
1.51	0.81	382.9	1683.2	1800.000	93.51

leaching of more than 20 g/t of Sc from BR with less than 15 kg of acid per gr of Sc, by a small adjustment of the leaching pH. The specific acid consumption is expected to be further reduced, by recycling the raffinate from the subsequent ion-exchange process to extract Sc from the PLS.

Acknowledgements The research leading to these results has received funding by EIT-KIC, SCALE-Up Project, project number 21013. This publication reflects only the authors' views, exempting the SCALE-Up Consortium from any liability for the information presented herein.

References

1. Balomenos, E., Nazari, G., Davris, P., Abrenica, G., Pilihou, A., Mikeli, E., Panias, D., Patkar, S., & Xu, W.-Q. (2021). Scandium Extraction from Bauxite Residue Using Sulfuric Acid and a Composite Extractant-Enhanced Ion-Exchange Polymer Resin. Rare Metal Technology 2021, Cham.

2. Harsh, J., Chorover, J., & Nizeyimana, E. (2002). Allophane and Imogolite. In *Soil Mineralogy with Environmental Applications* (pp. 291–322). https://doi.org/10.2136/sssabookser7.c9

3. Helfferich, F. G. (1995). *Ion exchange*. Courier Corporation.

4. Mikeli, E., Marinos, D., Toli, A., Pilichou, A., Balomenos, E., & Panias, D. (2022). Use of Ion-Exchange Resins to Adsorb Scandium from Titanium Industry's Chloride Acidic Solution at Ambient Temperature. *Metals*, *12*(5). https://doi.org/10.3390/met12050864

5. Saha, B., & Streat, M. (2005). Adsorption of Trace Heavy Metals: Application of Surface Complexation Theory to a Macroporous Polymer and a Weakly Acidic Ion-Exchange Resin. *Industrial & Engineering Chemistry Research*, *44*(23), 8671–8681. https://doi.org/10.1021/ie048848

6. Wang, W., & Cheng, C. Y. (2011). Separation and purification of scandium by solvent extraction and related technologies: a review. *Journal of Chemical Technology & Biotechnology*, *86*(10), 1237-1246. https://doi.org/10.1002/jctb.2655

Solvent Extraction of Scandium from Titanium Process Solutions

Dimitrios Filippou, Michel Paquin, Yves Pépin, Mike Johnson, and Niels Verbaan

Abstract

Scandium is present in titanium dioxide feedstock (ilmenite, titania slags, etc.) and usually reports to process solutions and effluents either at the feedstock producer site or at the end user site (TiO$_2$ pigment or Ti metal plant). In this paper, we report on the refining of scandium from such solutions by solvent extraction. A mixture of D2EHPA and TBP was used to extract scandium from a scandium-rich intermediate aqueous solution, which had been previously recovered from a titania slag upgrading process. Scandium loaded organic was scrubbed with H$_2$SO$_4$ to remove titanium and thorium, and the scrubbed organic was stripped with NaOH to produce an impure scandium hydroxide precipitate. The precipitate was dissolved in H$_2$SO$_4$, and the resulting solution was treated further with Alamine® 336 and Aliquat® 336 to remove persistent impurities. The formation of solids in the D2EHPA stripping step was a significant process issue, but ultimately high purity scandium oxide was produced. For reasons not related to the work reported here, Rio Tinto eventually selected other processing methods for its scandium extraction plant in Sorel-Tracy, QC, Canada.

Keywords

Scandium • Solvent extraction • Chloride solutions • D2EHPA • Tributyl phosphate • TBP • Trioctylamine • Alamine 336 • Methyltrioctylammonium chloride • Aliquat 336

Introduction

Titanium ores and concentrates (ilmenite and rutile) may contain appreciable amounts of scandium, ranging between 10 and 60 g/t [1, 2] or even higher [3], as a substitute of titanium in the ilmenite or rutile crystal structure [3]. Upon smelting ilmenite, almost all scandium reports to the titania slag, and not to the iron metal [4]. The titania slag is used to produce TiO$_2$ pigment and titanium metal, and scandium ends up in process solutions and effluents of the latter sites.

For more than 60 years, Rio Tinto Fer et Titane (formerly QIT) has been smelting hard-rock ilmenite obtained from its Lac Tio Mine in Havre-Saint-Pierre, Quebec. In 1996, the company introduced a new pyro-hydrometallurgical process, coined as UGS ("UpGraded Slag"), for the upgrading of titania slag by leaching with hydrochloric acid [5, 6]. By sampling the main process streams in the UGS plant, it was observed that about 90% of the scandium is leached out of the slag and goes into the so-called "UGS spent acid." It was therefore decided to try to extract scandium from the UGS spent acid before this solution is sent to HCl regeneration (which is done by pyrohydrolysis).

The method selected for the extraction of scandium from UGS spent acid is using ion exchange resins, as described in [7]. The specific method has the advantage of removing scandium and few other minor impurities from the spent acid, without affecting the concentration of chloride salts such as FeCl$_2$, MgCl$_2$, and AlCl$_3$, which are necessary for the regeneration of hydrochloric acid. However, some impurities (zircon and thorium) proved to be persistent and follow scandium to the final scandium oxide product. For this reason, a two-stage process was chosen for further study, including (a) extraction by ion exchange resins and production of an intermediate scandium product, and (b) refining of the intermediate product for the removal of persistent impurities and production of high purity scandium oxide (Fig. 1).

D. Filippou (✉) · M. Paquin · Y. Pépin
Rio Tinto Iron and Titanium, Sorel-Tracy, QC J3R 1M6, Canada
e-mail: dimitrios.filippou@riotinto.com

M. Johnson · N. Verbaan
SGS Canada Inc, Lakefield, ON K0L 2H0, Canada

© The Minerals, Metals & Materials Society 2023
S. Broek (ed.), *Light Metals 2023*, The Minerals, Metals & Materials Series,
https://doi.org/10.1007/978-3-031-22532-1_157

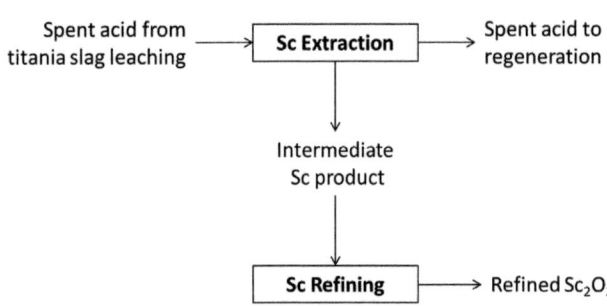

Fig. 1 Simplified block flow diagram of Rio Tinto's scandium recovery process

In this paper, we report on the second part of this process, i.e., the refining of an intermediate scandium product using solvent extraction. Solvent extraction is widely used in China, where most scandium is presently produced [8]. SGS Lakefield has also developed a solvent extraction process for scandium recovery from chloride solutions obtained by leaching niobium concentrate [9].

This study was part of the development work for Rio Tinto's first scandium recovery facility (demonstration plant) in Sorel-Tracy, Quebec. The objective of the work reported here was to validate whether solvent extraction could be an option for scandium refining.

Experimental

The feed material for this work was a chloride solution prepared from intermediate scandium precipitate supplied by Rio Tinto Fer et Titane, Sorel-Tracy, Quebec. The solution was acidified at pH 3.5 with a composition as shown in Table 1.

Based on SGS's previous experience [9], di-(2-ethylhexyl) phosphoric acid (D2EHPA, procured from Rhodia/Solvay), tributyl phosphate (TBP, from Millipore-Sigma and/or Fisher Scientific), trioctylamine Alamine® 336 (procured from BASF), and methyltrioctylammonium chloride Aliquat® 336 (from Millipore-Sigma and/or Fisher Scientific) were chosen for the solvent extraction work. These extractants were diluted

in Orfom® SX-80. All inorganic reagents used were of analytical/technical grade.

Small-scale extraction, scrubbing, and stripping contacts were conducted in an appropriately sized jacketed vessel equipped with a bottom drain, and an overhead mixer. Small-scale tests were carried out at ambient temperature, unless otherwise specified.

Larger scale tests ('bulk tests') were conducted within a 20 L vessel with a bottom drain and an overhead mixer, mixing the two phases for ten minutes before turning off the mixer and allowing the phases to separate overnight. All bulk contacts were at ambient temperature except for stripping, where the two phases were preheated separately to 50 °C before mixing.

Results and Discussion

Earlier Work

In earlier unpublished work, which is not presented in detail in this paper, the authors had determined that:

- Scandium extraction from acidic chloride solutions was very efficient using an organic mixture of 30% D2EHPA (as extractant) and 15% TBP (as modifier) diluted in Orfom® SX-80, at aqueous-to-organic (A/O) volume ratios less than one, and ambient temperature.
- The loaded organic was then treated with 3 mol/L H_2SO_4 (A/O = 1, ambient temperature) to scrub virtually all Th, 35 to 50% Ti, but only 25% Zr at best.
- The scrubbed organic was diluted with Orfom® SX-80 to decrease its loaded scandium concentration to approximately 1.0 g/L Sc, and then scandium was stripped by contacting with 100 g/L NaOH at 50 °C. Upon stripping, solid scandium precipitate was formed—most likely $Sc(OH)_3$.
- The precipitate, which settled at the bottom of the aqueous phase, was collected, filtered, and washed thoroughly to remove entrained impurities such as sodium, followed

Table 1 Assay of solution prepared from intermediate scandium precipitate

Element	Concentration (mg/L)	Element	Concentration (mg/L)
Sc	9630	Na	4720
Al	7	P	40
Ca	20	Th	359
Cr	6	Ti	80
Fe	7	U	1
Mg	5	V	1
Mn	0.1	Zr	63

by leaching with HCl for one hour at 80 °C to a terminal pH 1 and 14 g/L Sc.

- The leachate was treated with 10% w/w oxalic acid at 75 °C for about 30 min to precipitate scandium oxalate.
- The oxalate precipitate was calcined at 1000 °C for three hours to obtain scandium oxide (Sc_2O_3) containing only 60 g/t (ppm) Na, 26 g/t Th and less than 8 g/t Ti.

The above process was tested successfully at bench scale. However, the mass of the scandium oxide product was small and insufficient for full characterization. For example, it was not possible to analyse for Zr contamination.

Scoping Tests

In line with the results of the earlier work, the scandium solution of Table 1 was used in extraction testing with an organic mixture of 30% D2EHPA and 15% TBP in Orfom® SX-80, at three different A/O ratios: 5/7, 5/9, and 1/4. In the first two contacts at A/O 5/7 and 5/9, the phases did not separate at all, resulting in a persistent milky emulsion. The third extraction contact, while not ideal, eventually separated enough that the two phases could be analyzed, which showed that the scandium was loaded, as expected, alongside thorium, titanium, and zirconium. The organic from the third extraction contact was advanced to two scrubbing contacts, which proved that the original scrubbing regime with 3 mol/L H_2SO_4 could still handle the loaded thorium.

The poor physical separation in the three preliminary extraction contacts with 30% D2EHPA and 15% TBP suggested that the organic phase was being overloaded. To investigate further, the organic phase was modified to 15%

D2EHPA/7.5% TBP and 5% D2EHPA/2.5% TBP in the next set of tests. Also, the aqueous feed solution was diluted to 5 g/L Sc and adjusted to 20 g/L HCl. For each organic strength, two A/O ratios were tested at either ambient temperature or heated to 50 °C. The A/O ratios were selected based on the loading capacity of the organic mixture. The conditions and results of these two tests are summarized in Table 2.

By adjusting the phase ratios to stay away from potentially overloading the organic, phase separation was seen to improve (to under three minutes in most cases). Heating the contacts did not appear to offer any benefit, and some cases resulted in more stable emulsion forming. From a chemical standpoint, the contacts worked well in extracting scandium. These tests suggested that by keeping the scandium loading to less than 0.23 g/L/% D2EHPA in the organic, much of the emulsion issues could be prevented, although the organic phase remained hazy.

First Bulk Test

Based on the scoping test results, bulk processing was initiated using the 15% D2EHPA/7.5% TBP organic. The conditions and results of the work are summarized in Table 3.

The bulk extraction contact (E3) used the same conditions as E1-1, contacting the organic and diluted feed liquor at a ratio of 3/5 A/O. The contact was allowed to separate overnight, even though the interface between the organic and the aqueous was visible after 150 s. Some emulsion was formed, and although it was not as significant, the organic remained hazy and did not clear up after settling overnight.

Table 2 Conditions and results of scoping extraction tests with 15% D2EHPA/7.5% TBP and 5% D2EHPA/2.5% TBP. The aqueous feed solution was that of Table 1 diluted to 5 g/L Sc and adjusted to 20 g/L HCl

Test[a]	E1-1	E1-2	E1-3	E1-4	E2-1	E2-2	E2-3	E2-4
Organic feed	15% D2EHPA 7.5% TBP	15% D2EHPA 7.5% TBP	15% D2EHPA 7.5% TBP	15% D2EHPA 7.5% TBP	5% D2EHPA 2.5% TBP	5% D2EHPA 7.5% TBP	15% D2EHPA 2.5% TBP	5% D2EHPA 2.5% TBP
A/O ratio	3/5	3/5	2/7	2/7	1/5	1/5	1/10	1/10
Temperature	Ambient	50 °C	Ambient	50 °C	Ambient	50 °C	Ambient	50 °C
Final phase separation time	90 s	180 s	60 s	20 s	1 h	–	80 s	90 s
Emulsion	None	Minor	None	Minor	Moderate	Moderate	Minor	Minor
Loading to organic phase (%)								
Scandium	97	99	89	98	94	95	99	100
Thorium	75	92	87	99	89	91	98	99
Titanium	–	9	12	69	–	33	28	81
Zirconium	41	77	77	98	68	89	98	99

[a]All tests presented here are numbered as per the original lab books

Table 3 Conditions and results of the first bulk test. The aqueous feed solution for Test E3-1 was that of Table 1 diluted to 5 g/L Sc and adjusted to 20 g/L HCl

	Extraction	Scrubbing		Stripping
Test	E3-1	B1-1	B1-2	S2-1
Aqueous feed	As per E1-1	3 mol/L H$_2$SO$_4$	3 mol/L H$_2$SO$_4$	100 g/L NaOH
Organic feed	As per E1-1	E3-1, organic	B1-1, organic	B1-2, organic
A/O ratio	3/5	1/1	1/1	1/1
Temperature	Ambient	Ambient	Ambient	50 °C
Final phase separation time	150 s	600 s	420 s	600 s
Emulsion	Moderate	Haziness	Haziness	Minor
	Loading (%)	*Scrubbing (%)*		*Stripping (%)*
Scandium	100	–	–	99
Thorium	97	97	57	–
Titanium	31	78	66	–
Zirconium	94	4	5	–

Chemically, E3 worked very well, extracting essentially all scandium into the organic (3050 mg/L Sc). The loaded organic was then scrubbed in two successive contacts with 3 mol/L sulphuric acid at A/O ratio 1/1. Phase separation was adequate, although the haziness in the organic persisted. A small sample of the hazy organic was filtered for assay, and suspended crud recovered on the filter appeared to be responsible for the haziness. After successful small stripping tests confirmed past performance (not detailed in this paper), the whole volume of organic was advanced to stripping test S2, where it was contacted 1/1 with 100 g/L NaOH (both phases preheated to 50 °C) and then allowed to separate overnight. The solids produced in the stripping contact were split into two fractions after separation; the solids that settled within the aqueous phase (bottom solids) and solids that were trapped at the organic/aqueous interface and did not settle (top solids).

The solids generated in S2-1 were collected separately and washed using a laboratory scale centrifuge. The bottom solids separated relatively well in the initial centrifuge pass, but clarity suffered as washes were applied and the supernatant slowly became cloudy. The supernatant was eventually filtered via Millipore™ membrane filters, and the solids were recombined with the rest of the material. The top solids posed a greater challenge as a significant fraction of the precipitate resisted separation and remained floating in the residual organic. Repeatedly washing the top solids samples eventually improved the overall solids recovery, but as with the bottom solids, the washes became cloudy. The top solids washes were also filtered, and the recovered solids were combined with the rest of the top solids.

The top and bottom solids from S2-1 were leached separately in tests RL1 and RL2. In these tests, the solids were pulped in deionized water and heated to 85 °C. Hydrochloric acid was added to the slurry to reach a target pH of 1. The amount of water used was tailored to target a solution of 10 g/L Sc after dissolution. Once the target pH was reached, the slurry was held for approximately one hour before ending the test and filtering the pulp to remove any fine solids still suspended. The liquors from RL1 and RL2 were then treated with oxalic acid at 75 °C for half an hour, in tests ScP1 and ScP2, to precipitate scandium oxalate, targeting a 125% stoichiometric dose of oxalic acid relative to the scandium in solution. The precipitated scandium oxalate was then dried overnight, before calcining at 1000 °C to convert it to scandium oxide.

The assays of the two scandium oxide solids showed higher than desired thorium content (Table 4). Zirconium, another key impurity, was found to be elevated as well. The thorium content of ScP2 was higher than that of ScP1, and attributed to entrained crud/organic within the top solids (as organic crud was found after filtering RL2). Suspecting that something in the organic was to blame for the elevated thorium content, the small amount of crud that was recovered from the filtered sample of B1 scrubbed organic was submitted for assay, and revealed that the entrained solids causing the haziness in the organic after extraction contained 342 g/t Th and 3420 g/t Sc. As the organic remained hazy throughout extraction, scrubbing, and stripping, it was concluded that the crud was either partially stripped or physically carried through the centrifuge/washing step and dissolved in the releach, accounting for the high thorium in both final oxide products.

Table 4 Assays of scandium oxide samples produced in the first bulk test. The oxide samples were analyzed for many other elements but were found below detection limits

Component	Unit	Test	
		ScP1	ScP2
Al	g/t	11	11
Ca	g/t	2010	1770
Fe	g/t	462	261
Mg	g/t	2	3
Na	g/t	86	114
Sr	g/t	12	10
Ti	g/t	<0.4	<0.4
Zr	g/t	2620	1180
U	g/t	<0.5	<0.5
Th	g/t	407	1050
Sc	% wt	62.5	61.8
Sc_2O_3	% wt	95.8	94.8

Second Bulk Test

Knowing that the haziness in the extraction stage was in fact thorium-containing suspended crud, test E8-1 was a repeat of E3-1, but after allowing the phases to separate overnight, the organic was passed through a Whatman™ 1PS filter paper before advancing to scrubbing. The collected crud was discarded, and the clear organic was advanced to scrubbing B2. The organic remained clear throughout the scrubbing contacts in B2, which was increased from two to four to aggressively target the thorium on the organic.

Without the crud in the organic, the scrubbing contacts also separated very well, with clear phases and full separation within two minutes. The conditions and results of E8-1 and B2 are detailed Table 5. As expected, E8-1 worked very well chemically, extracting 99% of scandium and 74% of thorium. During scrubbing, the thorium content in the organic was reduced from 45 to less than 0.5 mg/L after the first contact alone (98.9% overall scrubbing efficiency). The remaining contacts served to polish the organic further, with little change in the scandium on the organic.

The B2-4 organic was advanced to stripping S7-1, which was run at 50 °C and is summarized in Table 5. Overall phase separation in S7-1 was good, with the bulk of the phases separated within two minutes. As seen in Fig. 2, the solids in the aqueous phase continued to settle out with more time, forming a visible band of solids at the bottom of the vessel after 50 min. This band of solids continued to

Table 5 Conditions and results of the second bulk test. The aqueous feed solution for Test E8-1 was that of Table 1 diluted to 5 g/L Sc and adjusted to 20 g/L HCl

	Extraction	Scrubbing				Stripping
Test	E8-1	B2-1	B2-2	B2-3	B2-4	S7-1
Aqueous feed	As per E1-1	3 mol/L H_2SO_4	3 mol/L H_2SO_4	3 mol/L H_2SO_4	3 mol/L H_2SO_4	100 g/L NaOH
Organic feed	As per E1-1	E8-1, organic	B2-1, organic	B2-2, organic	B2-3, organic	B2-4, organic
A/O ratio	3/5	1/1	1/1	1/1	1/1	1/1
Temperature	Ambient	Ambient	Ambient	Ambient	Ambient	50 °C
Final phase separation time	240 s	96 s	96 s	96 s	96 s	120 s
Emulsion	Moderate	None	None	None	None	Minor
	Loading (%)	*Scrubbing (%)*				*Stripping (%)*
Scandium	99	1	1	1	1	100
Thorium	74	98	85	41	7	–
Titanium	2	98	70	37	37	–
Zirconium	54	7	7	7	7	–

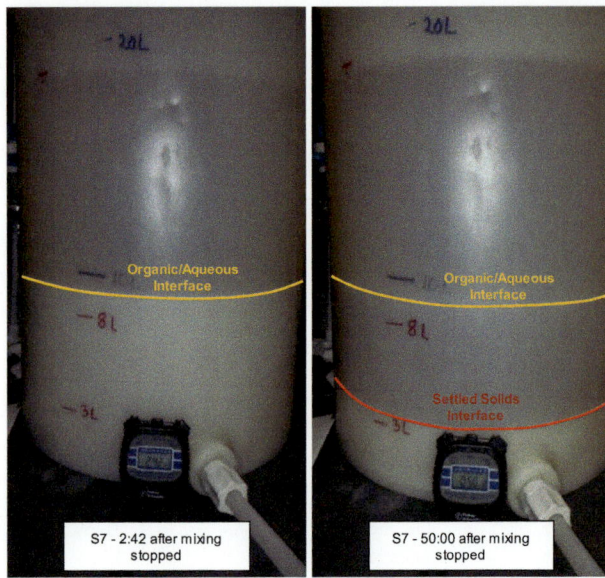

Fig. 2 Progression of solids settling in stripping S7-1

compress overnight, at which point the solids at the bottom of the vessel were collected. Like S2, a fraction of the precipitate was held up at the phase interface, but unlike S2, these solids were combined to form a single precipitate sample moving forward.

The solids were collected in a similar fashion as S2-1, draining the solids and using a laboratory centrifuge to dewater and wash the solids. A small amount of the solids was collected and leached unwashed for one hour at about 85 °C using HCl to a final pH of 1.03 (RL3). The remaining solids were washed/filtered through repeated cycles within the centrifuge and leached for one hour at about 85 °C using H_2SO_4 to a final pH of 0.67 (RL4). The unwashed sample

was leached to determine if the washing cycles were necessary for product purity, or if the material could be advanced as-produced. Conversely, the washed solids were leached in H_2SO_4 to accommodate subsequent downstream operations to target zirconium and niobium in the releach liquor. The filtrate from RL3 contained 7.04 g/L Sc and 0.5 mg/L Th. The filtrate of RL4 contained 20.8 g/L Sc and 1.3 mg/L Th. The thorium concentration in the filtrates of RL3 and RL4 were significantly lower than that of the filtrates of RL1 and RL2 (8.4 and 21.1 mg/L Th, respectively), and so crud control after scandium extraction was proven to be a key factor within the process flowsheet.

The liquor from RL4 was advanced to successive contacts with Alamine® 336 to remove trace zirconium, and Aliquat® 336 to remove trace niobium, at room temperature and an A/O of 1/1. Conditions and results from these two contacts are summarized in Table 6. There were no problems with phase separation in either contact and both reduced the target impurity metals to below the analytical detection limits while leaving the scandium untouched.

The liquors from RL3 and NbE1-1 were advanced to scandium precipitation via oxalic acid addition at 75 °C for half an hour. Both precipitates were dried overnight before calcining at 1000 °C.

Table 7 summarizes the calcine assays from ScP3 and ScP4. As expected, the final product from ScP3 contained higher levels of zirconium, but the thorium content was much lower than in ScP1 and ScP2 (only 46 g/t Th). On the other hand, ScP4 was low in thorium and in zirconium, meeting higher quality standards. In terms of purity, ScP3 was calculated to be 99.7% Sc_2O_3 based on 100% minus all impurities with grades above detection limits, and ScP4 was calculated to be 99.9% Sc_2O_3 on the same basis. ScP3 and ScP4 were also calculated to be 99.6% and 99.8% Sc_2O_3

Table 6 Conditions and results of zirconium and niobium removal in the second bulk test		Zr extraction	Nb extraction
Test		ZrE1-1	NbE1-1
Aqueous feed		RL4 filtrate	ZrE1-1, aqueous
Organic feed		5% Alamine 336 95% Orfom SX-80	5% Aliquat 336 2.5% Tridecanol 92.5% Orfom SX-80
A/O ratio		1/1	1/1
Temperature		Ambient	Ambient
Final phase separation time		573 s	240 s
Emulsion		None	None
Loading (%)			
Scandium		–	1
Thorium		26	27
Niobium		98	98
Zirconium		97	84

Table 7 Assays of scandium oxide samples produced in the second bulk test. The oxide samples were analyzed for many other elements but were found below detection limits

Component	Unit	Test	
		ScP3	ScP4
Al	g/t	18	14
Ca	g/t	615	<200
Fe	g/t	132	<10
Mg	g/t	<2	<2
Na	g/t	205	106
Nb	g/t	<40	<70
Sr	g/t	2	3
Ti	g/t	<0.4	<0.4
Zr	g/t	1080	150
U	g/t	<0.5	<0.5
Th	g/t	46	49
Sc	% wt	65.3	64.2
Sc_2O_3[a]	% wt	99.7	99.9

[a]Estimate by difference: 100%—all impurity oxides, taking anything below detection limit as equal to zero

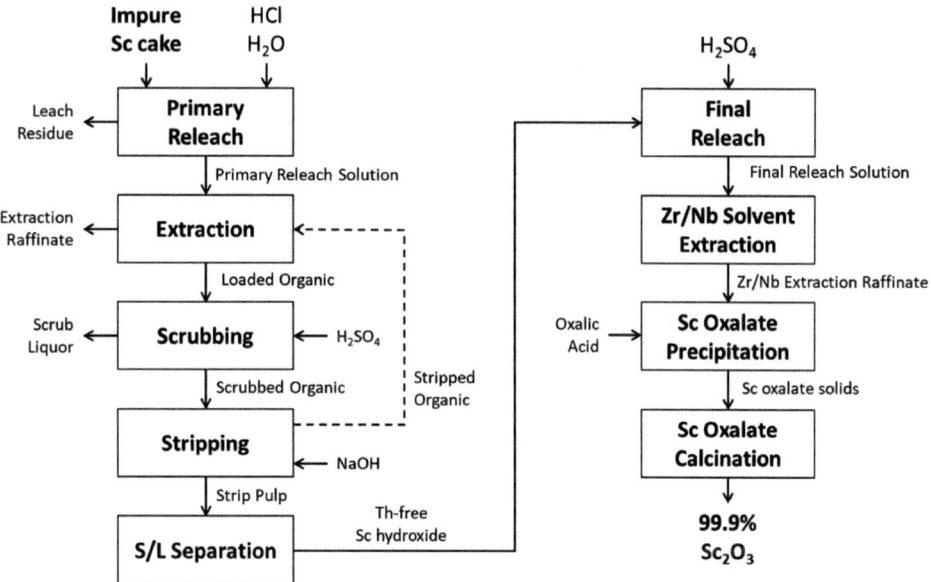

Fig. 3 Conceptual block flow diagram for refining impure scandium cake to Th and Zr-free high-quality scandium oxide via solvent extraction

based on 100% minus all impurities, taking anything below the detection limit as equal to the detection limit for a worst-case scenario.

Conclusions

Based on findings of the study detailed in this paper, the following conclusions and recommendations are made:

- A 99.9% pure scandium oxide product can be obtained using a flowsheet that incorporates D2EHPA solvent extraction to load scandium and a separate solvent

extraction process after releaching an intermediate scandium precipitation for zirconium (and niobium) control.

- The flowsheet shown in Fig. 3 requires further study and optimization to determine if the physical aspects of the solvent extraction process can be improved while still retaining the excellent scandium oxide product quality. Determining conditions to prevent or adequately control/remove crud during scandium loading is key, since the crud poisons the final product with thorium if carried through to stripping.

- Recovery of the precipitated solids after stripping scandium from D2EHPA poses a major solid/liquid separation challenge, as the material is slow to filter through traditional means and resists centrifugation in the presence of any entrained organic. Such problems have also been noted by other researchers [8]. Therefore, more work is needed to identify conditions that will prevent the precipitate from collecting at the A/O phase interface.

- The removal of zirconium and niobium from Th-free scandium hydroxide was successful, but requires further optimization.

Acknowledgements The authors want to thank the management of SGS Canada Inc. and Rio Tinto Iron & Titanium for their permission to publish this paper.It should be noted that, for reasons related to limited tolerance in organics in the overall UGS process, Rio Tinto eventually decided not to pursue the solvent extraction refining process, and to use instead ion exchange resins for the purification of the intermediate scandium product.

References

1. Borisenko, LF (1983) Raw-material resources of scandium. International Geology Review, 25(8):942–946. https://doi.org/10.1080/00206818309466787
2. Wang, W, Pranolo, Y, Cheng CY (2011) Metallurgical processes for scandium recovery from various resources: A review. Hydrometallurgy, 108(1–2): 100–108. https://doi.org/10.1016/j.hydromet.2011.03.001
3. Wang, Z, Li MYH, Liu, ZRR, Zhou, MF (2021) Scandium: Ore deposits, the pivotal role of magmatic enrichment and future exploration. Ore Geology Reviews, 128, 103906. https://doi.org/10.1016/j.oregeorev.2020.103906
4. Denisova, LV, Rogatkin, AA, Cherkashin, VI, Ilyushina, LM, Khlopkov, LP (1977) Vanadium-, niobium-, tantalum- and scandium distribution in the process of titanium slags melting. Tsvetnye Metally, (2):44–46.
5. Borowiec, K, Grau AE, Guéguin, M, Turgeon, JF (1998) Method to upgrade titania slag and resulting product. US Patent 5830420, 3 November 1998.
6. Borowiec, K, Grau AE, Guéguin, M, Turgeon, JF (2003) TiO_2 containing product including rutile, pseudo-brookite and ilmenite. US Patent 6531110, 11 March 2003.
7. Paquin, M, Filippou, D (2019) Process for polishing metal contaminants from an acidic solution comprising scandium. PCT World Patent Application WO2019213753, 14 November 2019.
8. Wang, W, Cheng, CY (2011) Separation and purification of scandium by solvent extraction and related technologies: a review. Journal of Chemical Technology and Biotechnology, 86: 1237–1246. https://doi.org/10.1002/jctb.2655.
9. Verbaan, N, Bourricaudy, E, Grammatikopoulos, T, Johnson, M, Larochelle, E, Honan, S, Smith, K, Sixberry, R (2017) A process flowsheet for the extraction of scandium from NioCorp's niobium/scandium Elk Creek deposit. In: COM2017–Hydrometallurgy of Nickel-Cobalt. Westmount, QC, Canada: Canadian Institute of Mining and Metallurgy (CIM). https://www.onemine.org/document/abstract.cfm?docid=236812.

State of the Art Technologies for Scandium Recovery, Purification, and Aluminum-Scandium Alloy Production

Anne Marie Reyes, Gomer Abrenica, and Ghazaleh Nazari

Abstract

Aluminum-based alloys containing minimal amount of scandium have been extensively investigated for their exceptional mechanical properties which make them particularly suitable for advanced engineering applications such as aerospace and defense by allowing design and construction of high-strength, lightweight structural parts. Understanding the challenges in scandium sourcing, recovery and purification processes, and alloying methods, a vertically integrated solution has been developed which includes a portfolio of technologies: (i) patented recovery of Sc from Sc-containing ores and industrial waste streams from ppm levels to 20–25% in the concentrate—validated in the pilot scale; (ii) purification of Sc from concentrates containing impurities such as titanium, iron and other rare earths to produce >99.5% pure scandium oxide—validated in the commercial scale; and (iii) patent-pending technology for production of the final alloy through a one-step alloying process using a vacuum induction furnace, refined casting technique, and suitable heat treatment processes—validated at demonstration scale.

Keywords

Additive manufacturing · Extraction and processing · Hydrometallurgy · Critical metal · Scandium · High-performance alloy

A. M. Reyes · G. Abrenica
Coherent, Cavite, Philippines

G. Nazari (✉)
Coherent, Saxonburg, PA, USA
e-mail: Ghazaleh.Nazari@coherent.com

Introduction

Scandium (Sc) is one of the highest-valued elements in the periodic table and a critical raw material essential for several emerging applications. Historically, supply of Sc has been limited due to its scarcity and high cost of production. Sc is sparsely distributed in the earth's crust and occurs in trace amount in many minerals. These minerals are concentrated in Australia, Philippines, China, Kazakhstan, Norway, Russia, Ukraine, and Madagascar. In Australia and the Philippines, Sc is found in nickel and cobalt ores, in Russia and Kazakhstan in uranium ores, in Madagascar and Norway in pegmatite rocks. In Ukraine, it is contained in iron ores and in China, it is contained in tin, tungsten, zirconium, and iron ores [1].

Sc is usually found as a by-product of other metal processing such as bauxite residue from alumina production, acid waste from titanium dioxide (TiO_2) pigment production, tailings from nickel/cobalt recovery processes, and tailings from tantalum, niobium, and tungsten metal recovery plants. Global generation of bauxite residues, highly alkaline waste residue from extraction of alumina from bauxite ore through the Bayer process, is estimated at 150 million tons per year [2] with an accumulated inventory of 3 billion tons [3]. This is equivalent to a potential Sc supply of about 270 thousand tons. Another substantial source of Sc is acid waste from TiO_2 pigment production plants, via chloride and sulfate routes, which is estimated to reach a total capacity of 8.4 million tons worldwide. Based on the average Sc concentration of 15–25 mg/kg, this waste stream can potentially provide about 200 tons of Sc per year [4].

The current world's supply and demand of Sc in the form of scandium oxide (Sc_2O_3) is only about 20–25 tons per year, and the demand is projected to increase to 1,000 tons per year in 2025. This sharp increase in Sc demand in near future is driven by major factors such as accelerating usage in solid oxide fuel cells (SOFCs) and the rising demand for aluminum-scandium alloys (Al-Sc).

© The Minerals, Metals & Materials Society 2023
S. Broek (ed.), *Light Metals 2023*, The Minerals, Metals & Materials Series,
https://doi.org/10.1007/978-3-031-22532-1_158

Sc is the most effective microalloying strengthener for Al and its alloys. Sc imparts substantial improvement in strength, even with small additions of between 0.05 and 1%, or 50 MPa per 0.1%. Aside from improved strength, other beneficial impacts of Sc on the properties of a wide range of Al alloys are improved resistance to hot cracking, resistance to crystallization, and reduced grain size. These exceptional properties make them particularly suitable for advanced engineering applications by allowing design and construction of strong and lightweight structural parts.

Technologies

Understanding the challenges in Sc sourcing, recovery and purification processes, and alloying methods, a vertically integrated solution (Fig. 1) has been developed which includes a portfolio of technologies:

Selective-Ion Recovery (SIR) Technology

Most Sc is produced today as a by-product of other mineral refining processes, such as bauxite residues from alumina production and acid wastes from TiO_2 pigment production. These sources have Sc concentrations that are at the levels of mg/kg and are considered exploitable but technologically and economically complicated to recover. In the pursuit of recovering Sc from such sources, SIR Technology was developed [5, 6] that includes the use of a composite extractant-enhanced ion-exchange resin to extract Sc from acidic solution or slurries, and its subsequent recovery as Sc concentrate. The factors that contribute to its performance have been studied to a great extent using pregnant leach solution (PLS) from various Sc-containing sources.

Bauxite residue contains various major metal oxides of iron (Fe), Al, Ti, calcium, silicon (Si), and sodium, as well as minor metal oxides like vanadium, gallium, Sc, and other rare-earth elements. The concentration of Sc in the bauxite residue varies depending on its concentration in the ore and the operation conditions of the Bayer process [7]. A sample of bauxite residue from Greece containing about 70 mg/kg Sc was subjected to a series of lab-scale experiments. A combination of parameters such as acid concentration, leaching temperature, residence time, and pulp density were evaluated. The initial acid concentrations were determined by acid–base titration. Leaching temperature and time were monitored during the entire duration of the experiments. Concentrations of Sc and critical elements such as Ti, Fe, and Si in the solution were measured using inductively-coupled plasma–optical emission spectrophotometry (ICP-OES). The loading tests were conducted in lab-fabricated columns.

High acid leaching experiments were carried out to increase Sc recovery. However, that resulted in generation of a PLS with high impurity levels that interfered with Sc loading. Low acid leaching tests were also carried out with the aim of reducing levels of impurities; however, it was found that Si concentration was very high, which led to silica precipitation in the columns and created pressure drop and fouling. Relatively low Sc loadings of 60 and 500 mg/L resin at low and high acid leaching were observed, respectively. To improve Sc loading, leaching conditions were optimized such that PLS contained maximum concentration of Sc while maintaining impurities to a minimum. The silica precipitation and fouling issue was also addressed in this process. The PLS produced under optimum conditions proceeded to the SIR for Sc recovery. As shown in Fig. 2, the SIR loading capacity increased dramatically to 5,000 mg/L, and the breakthrough point for Sc up to 200 bed-volumes (BVs) for the PLS generated from leaching BR at optimum conditions. This increase in the loading capacity corresponds to improvements in the economic viability of the process.

The patented SIR Technology was successfully demonstrated and scaled up in 2021 for recovery of Sc from bauxite residues [8–11]. This project received funding from the European Community's Horizon 2020 SCALE Program [12] and has won wide recognition and acclaim for its environmental benefits and as a new resource for rare-earth elements.

The applicability of the SIR process in recovery of Sc from the Ti tailings solution was evaluated [11, 13]. The acidic waste solution was obtained from a TiO_2 production plant employing carbothermal chlorination processing of Ti ores. As a preliminary evaluation on the effect of the other metal cations particularly Ti to the loading capacity of the resin for Sc, the solution containing about 80 g/L hydrochloric acid (HCl), 80 mg/L Sc, 5,000 mg/L Ti, and other transition metal ions was used. The adsorption capacity of the resin for Sc and Ti were 1,000 and 51,000 mg/L, respectively. Considering the high loading of Ti, a

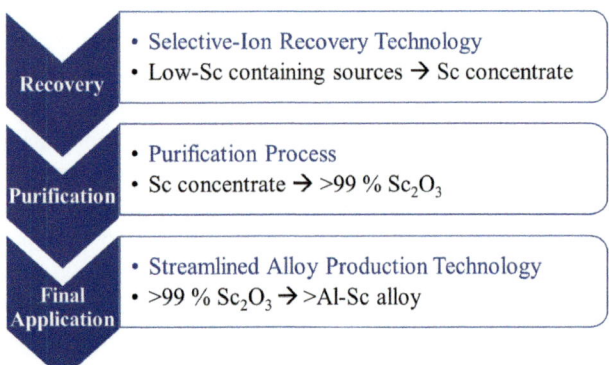

Fig. 1 Vertically integrated technologies

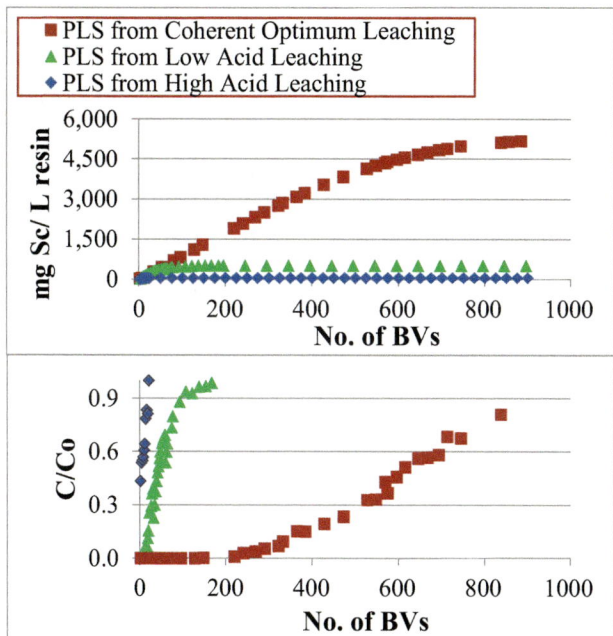

Fig. 2 Loading of Sc (top) and adsorption isotherm (bottom) on SIR from PLS generated by leaching bauxite residue

pre-treatment step for Ti removal was implemented to maximize Sc loading efficiency.

To separate Ti and Sc, precipitation experiments were conducted. It was found that the optimum pH for the preferential precipitation of Ti was at 0.70 ± 0.10 with more than 98% of the Ti precipitated while more than 80% of Sc was maintained in the solution. This solution was subjected to the succeeding adsorption experiments to evaluate the performance of the resin. As shown in Fig. 3, the resin was loaded to exhaustion using Ti tailings solution and the loading capacity of 20,000 mg Sc/L was obtained. The resin was eluted and regenerated according to the established parameters.

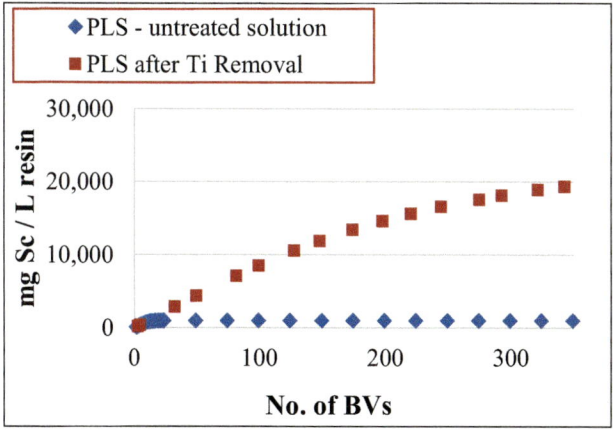

Fig. 3 Loading of Sc onto SIR from Ti Tailings solution with and without Ti removal

The eluate proceeded to the downstream treatment steps to precipitate Sc. The crude Sc concentrate contained 20–25 wt% Sc. Recovery of Sc in the elution and Sc hydroxide precipitation steps were 99% and 97%, respectively.

Purification of Crude Sc Concentrate

The crude Sc concentrate generated from the SIR process can be processed via the developed purification process. The purification process was developed and commercialized for recovery and purification of Sc from fluorite mine tailings. The facility, located in the Philippines, includes a rotary-tilting furnace, mixer-settlers, agitated tanks, and calciners and has a production capacity of 7,200 kg Sc_2O_3 per year [11].

The process involves leaching of the crude Sc concentrate to solubilize Sc. This resulting solution is then contacted with an organic extractant with which Sc forms an acid-insoluble complex. The Sc-containing organic is contacted with a series of scrubbing stages using various acid solutions that completely remove co-extracted rare-earth elements, Fe, thorium, uranium, and Ti, then with an alkaline solution to strip Sc in the form of Sc hydroxide precipitate. Depending on the type and concentrations of impurities, modifications in some process steps might be necessary. The Sc is dissolved from the said precipitates, further purified by oxalation, and calcined to produce >99% Sc_2O_3.

Streamlined Al-Sc Alloy Production Technology

Reduction of Sc and Alloying with Al

The conventional method of producing Al-Sc alloy is through the aluminothermic reduction of Sc from its oxide or halide compounds. Direct reduction of Sc from Sc_2O_3 is difficult because of its high thermodynamic stability, so it is often halogenated first to convert it to scandium fluoride (ScF_3) or scandium chloride ($ScCl_3$) prior to aluminothermic reduction [14]. These halides in combination with fluoride and/or chloride salts are preferred because Sc can be reduced at a relatively lower reduction temperature and shorter reaction time. The three reaction schemes shown in Table 1 were considered and tested.

Typically, ScF_3 is prepared by directly reacting Sc_2O_3 with hydrofluoric acid or contacting it with hot hydrogen fluoride gas at high temperature—both of which present challenges in safety and operations due to the chemicals' high toxicity. To significantly minimize the associated risks, the process employs preparing ScF_3 by dissolving Sc_2O_3 in HCl, then stoichiometrically precipitated using NaF.

Preliminary lab-scale tests were conducted using crucibles and a muffle furnace to develop the baseline

Table 1 Reaction schemes for Al₃Sc formation

Scheme	Reaction	Standard reduction potential (T = 298.15 K, P = 101.325 kPa) [15]	
1	$3Al + ScF_3 \rightarrow Al_3Sc + AlF_3$	$Al^{3+} + 3e^- \leftrightarrow Al$	$E^{\circ}(V) = -1.66$
		$Sc^{3+} + 3e^- \leftrightarrow Sc$	$E^{\circ}(V) = -2.07$
2	$2ScF_3 + 3Mg \rightarrow 2Sc + 3MgF_2$	$Sc^{3+} + 3e^- \leftrightarrow Sc$	$E^{\circ}(V) = -2.07$
	$3Al + Sc \rightarrow Al_3Sc$	$Mg^+ + e^- \leftrightarrow Mg$	$E^{\circ}(V) = -2.70$
3	$3NaF + ScF_3 \rightarrow Na_3ScF_6$		
	$Na_3ScF_6 + Al \rightarrow Na_3AlF_6 + Sc$		
	$3Al + Sc \rightarrow Al_3Sc$		

conditions for the process. Then, the succeeding experiments were conducted in vacuum induction furnace (VIF) at 25 kg alloy capacity to produce Al-Sc alloy containing about 2% Sc. The extent of reduction of Sc was determined by measuring the Sc content in the resulting alloy using ICP-OES.

The extent of reduction via reaction Scheme 1 at temperature of 800–850 °C under vacuum was only about 10%. The direct reduction of Sc from ScF₃ using Al is difficult considering the standard reduction potential of Al which is higher than that of Sc ion. For this reason, more electronegative metals such as Ca and Mg (Scheme 2) [14, 16, 17], or a mixture of molten salts (Scheme 3) are required to facilitate reduction. Once Sc is reduced, it is captured by Al forming Al-Sc alloy [18].

To evaluate Scheme 2, stoichiometric amount of the reactants was reacted with molten Al under vacuum at 800–850 °C. Under this condition, 84% of Sc was reduced. However, significant amount of the alloy was vaporized due to presence of Mg. Under atmospheric conditions, the ignition temperature of pure Mg is 623 °C [19] while in vacuum, Mg starts to volatilize at 300 °C.

The highest Sc reduction efficiency and alloy yield were obtained via reaction Scheme 3 at > 93% and 90%, respectively. This is due to the formation of an intermediate compound of sodium fluorine-scandate [18, 20] that facilitates the reaction. The reduced Sc then diffuses into the Al melt and forms the master alloy. However, this master alloy has no commercial application other than as a raw material for producing final alloys with lower %Sc.

Conventionally, the Al-Sc master alloy is remelted with other master alloys and diluted with pure Al to achieve the desired concentration in the final product. Alloying is done between 660 and 1000 °C depending on the required melting temperature of the master alloys being added. This two-stage practice leads to long production time, higher production cost, and higher material losses.

To circumvent the said issues, a series of experiments in the VIF was conducted to study the effects of various parameters such as temperature, time, and order of master alloy additions to reduction efficiency, product yield, and extent of alloying with other metals. The target composition of the final alloy containing Sc, magnesium (Mg), and zirconium (Zr) is shown in Table 2.

A summary of effects when the process is done beyond the optimum operating ranges is shown in Table 3.

From these experiments, a patent-pending technology has been developed to streamline Al-Sc alloy production [21]. This single-stage process combines the aluminothermic Sc reduction and the alloying steps with other metals. It utilizes the similarity in temperature required for reduction reaction and the temperature required to melt the other master alloys. The aluminothermic reduction of Sc is performed by melting ScF₃ together with NaF as flux and pure Al. This is done simultaneously with remelting and alloying of other metals or master alloys with high volatilization temperature at 800–850 °C under vacuum or at 900–1000 °C under atmospheric pressure, for 1–2 h. The melt is cooled to the temperature suitable for addition of the other metals or master alloys with low volatilization temperature. Any slags that formed are skimmed off. Vacuum can be induced to remove gases such as oxygen, hydrogen, and nitrogen from the melt. The melt is then casted in molds.

Casting

A casting procedure was also developed while evaluating the alloy production process. Experiments were carried out to determine the conditions at which casting defects are minimized, if not eliminated.

Segregation of alloying elements is usually observed during solidification due to differences in diffusivity and solubility of solutes in solid phase and in liquid phase. For castings where center portion is usually the last to solidify,

Table 2 Target chemical composition of final alloy

Element	Weight %
Al	Balance
Sc	0.20–0.40
Mg	4.00–6.00
Zr	0.10–0.30

Table 3 Effect of deviation from optimum process conditions

Parameters	Lower than optimum	Higher than optimum
Flux	↓Sc reduction efficiency	↑Sc reduction efficiency → ↑cost
Reduction Temperature	↓Sc reduction efficiency	↑Material loss to slag and Al volatilization
Reaction Time	↓Sc reduction	↑Material loss to slag and Al volatilization
Alloying Temperature	↓Alloying of Zr & Ti	↑Material loss due to Al and Mg volatilization

Table 4 Relative standard deviation of metal composition

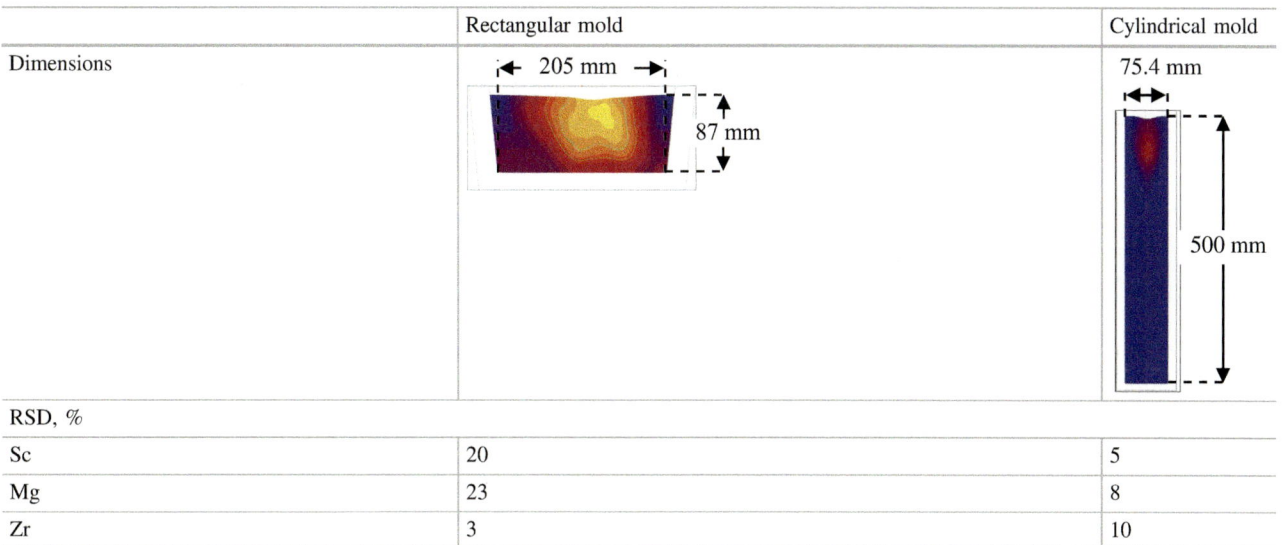

	Rectangular mold	Cylindrical mold
Dimensions	205 mm / 87 mm	75.4 mm / 500 mm
RSD, %		
Sc	20	5
Mg	23	8
Zr	3	10

concentrations of alloying elements Mn, Sc, Ti, and Zr are higher in the center due to their lower solubilities, while Mg is lower as it has a higher solubility [22]. To minimize segregation in the billet, solidification rate in all regions of the cast product must be as uniform as possible. Table 4 shows a simulation of the solidification gradient of the alloy for two different mold designs used in the experiments. In both tests, molten alloy was poured at T < 700 °C and was allowed to naturally cool down. Heat transfer was slowest at yellow regions which took the longest time to solidify while blue regions solidified immediately after pouring due to fast heat transfer. Casting in cylindrical mold has less solidification gradient for most parts of the billet, hence less segregation, which was confirmed by the RSD of the compositional analysis of the product.

To further improve homogeneity, one-directional cooling was employed to ensure that the top part is the last to solidify. The following casting procedures were implemented:

i. Placing a copper plate at the bottom of the mold to allow fast and uniform heat dissipation from the billet body due to its high heat conductivity

ii. Cooling copper plate with water to further increase the heat dissipation

Gas porosities and gas entrapment issues were resolved by reducing the pouring height, controlling pouring using a ladle, and by operating under vacuum. Additionally, micro-porosities found at the upper portion of the billet, as shown in Fig. 4, was eliminated by continuous heating of the top part of the mold which allows sufficient feeding of molten material to all parts of the solidifying billet. Pouring temperature was also controlled at 720–730 °C to minimize shrinkage defects.

The optimum parameters for Sc reduction, alloying, and casting were confirmed by conducting 6 alloying batches. The products were subsequently machined and polished to the required size and finish. Billets did not have any casting defects. Compositional variation between batches was low as shown in Table 5.

The developed streamlined alloying process is diagrammatically shown in Fig. 5. This has several advantages over the existing conventional process: (1) high reduction efficiency of Sc of up to 90%, (2) lower metal losses from volatilization due to lower reaction time and bypassing the double-step alloy production, (3) lower operating cost, (4) shorter cycle time, and (5) flexibility of the process to produce Al alloys with a wide range of Sc concentration from 0.05–2%.

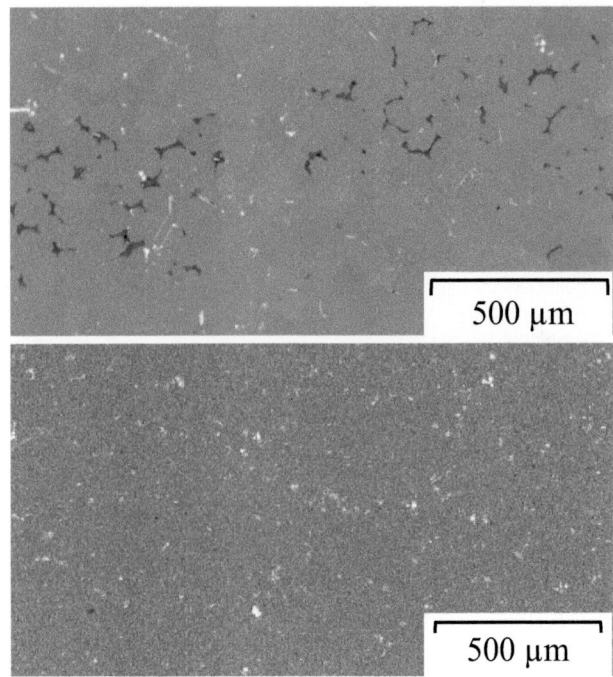

Fig. 4 SEM image of as-cast samples w/o continuous heating (top) w/continuous heating (bottom)

Heat Treatment, Primary and Secondary Al₃Sc Precipitates

Strengthening property of Sc is due to its ability to form nano-sized Al₃Sc precipitates that are coherently bound to the matrix. Depending on % Sc and cooling rate, two different types of precipitates can be formed. Primary precipitates are formed from liquid melt during alloy solidification which provide nucleation sites for Al grains causing grain refining. In an as-cast alloy, size of these precipitates can range from 20 to 400 μm and have a highly faceted morphology. For Sc concentrations greater than the eutectic composition, primary precipitates inevitably form from the liquid melt. Secondary precipitates, on the other hand, are formed from solid solution during specific heat treatment conditions. They are well-dispersed and spherical in shape, with average diameters of 20 nm. These precipitates strengthen the alloy by immobilizing dislocations and sub-grain boundaries and increases the shear stress required for dislocation slippage [23–27].

The goal was to dissolve and maintain the maximum amount of Sc in solid solution, then precipitate them out as secondary precipitates. In a binary Al-Sc system [25, 27], solubility of Sc in Al at the eutectic temperature is 0.3 wt%. High cooling rates reaching to 100 K/s can retain up to 0.5 wt%, while extremely high cooling rates of 10^7 K/s can retain as much as 5.22 wt% in solid solution [29–31]. Fast cooling followed by heat treatment ensures uniform distribution of the secondary precipitates that will be formed [32]. However, high cooling rates are only achievable by quenching small volumes of the alloy. Hence, with the casting setup and the quantity of final alloy, the presence of primary precipitates is expected.

Precipitation hardening of Al₃Sc particles are achieved during heat treatment at 300–350 °C, while temperature range of 350–450 °C is used for Al-Zr alloys [28]. For ternary Al-Sc-Zr systems, the combined effect of Al₃Sc and Al₃Zr precipitates were shown to retard decomposition of supersaturated solid solution and slow down coarsening of precipitates. Depending on the heat treatment regime, different forms of Al-Sc-Zr precipitate can be produced. If alloy is subjected to one-step heat treatment (isothermal) at temperatures 400–450 °C, Zr co-precipitates with Sc forming Al₃(Sc,Zr) phase. On the other hand, isochronal heat

Table 5 Compositional variation of the 6 billets

Element	RSD, %	Physical appearance
Sc	6	
Mg	2	
Zr	6	

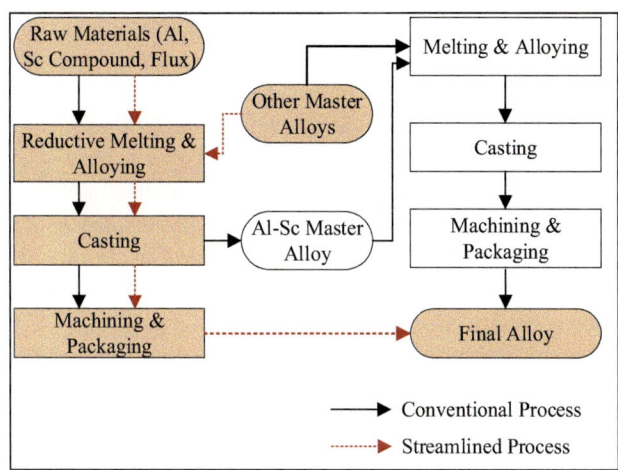

Fig. 5 Conventional and streamlined Al-Sc alloy production process

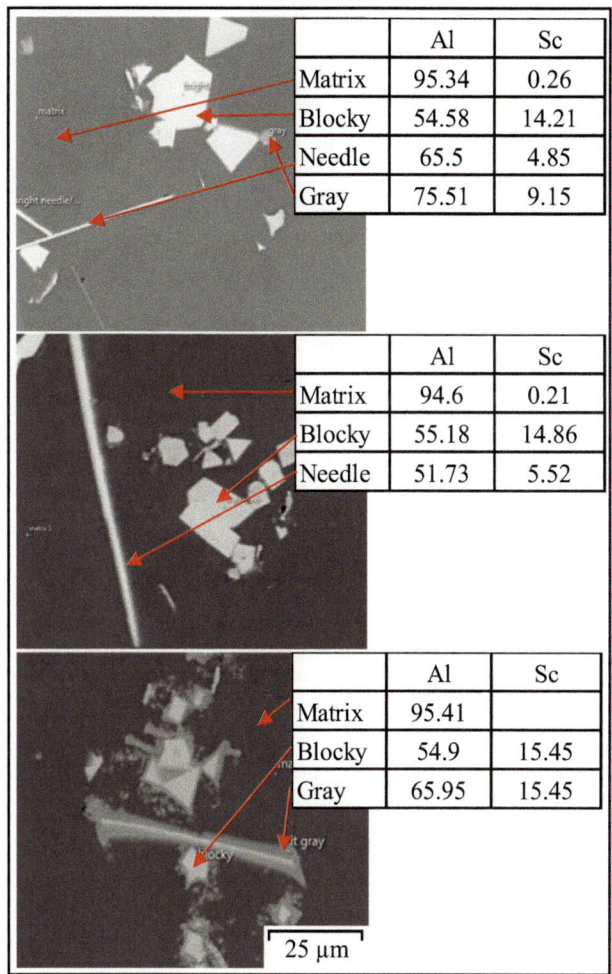

Fig. 6 SEM micrograph of as-cast structure (top), after 2-stage heat treatment at 300 °C-12 h then at 400C-12 h (middle), and after over-aging at 480C-8 h (bottom)

treatment forms precipitates with Al_3Sc-rich core and Al_3Zr-rich shell. This happens because Sc diffuses faster in Al forming Al_3Sc dispersoids which serve as nucleating sites for Zr when the alloy is subjected to a higher temperature. When the alloy reaches peak hardness, further exposure to higher temperature causes over-aging or the coarsening of precipitates, which decreases their strengthening effect [28].

To explore the effects of different heat treatment conditions, samples from one of the six billets were taken and subjected to a three-stage isochronal heat treatment as described below:

i. 300 °C-12 h—to form secondary Al_3Sc precipitates with a homogeneous distribution.
ii. 400 °C-12 h—to precipitate Al_3Zr around the Al_3Sc precipitates forming the core–shell structure. At this point, alloy is expected to reach peak hardness.
iii. 480 °C-8 h—to coarsen the precipitates and reduce the hardness.

Each sample was aged at the set time and temperature and then quenched in water. Conductivity measurement was done using Fischer Sigmascope SMP10 conductivity meter, while hardness was measured using Leco RT-240 hardness tester. Samples were mounted in epoxy and polished with 0.05 μm colloidal silica, in accordance with standard metallographic preparations for SEM imaging. Figure 6 shows that primary precipitate was present even in the as-cast sample. This is expected since the required cooling rate was not high enough to keep all Sc in solid solution Increase in conductivity after the first heat treatment step was observed, which corresponds to an increase in hardness, as shown in Fig. 7. This indicates formation of additional precipitates,

predictably Al_3Sc. After the second heat treatment stage, conductivity and hardness further increased, predictably due to the precipitation of Al_3Zr. After over-aging conductivity increased even further signifying that the remaining Sc precipitate out from solution, as also observed in the decrease in Sc in matrix. Hardness, however, decreased, which agrees with previous assessment that at higher temperatures, particles coalesce and lose coherency, causing a decrease in strengthening effect.

Summary

A vertically integrated solution has been developed which includes a portfolio of technologies that overcomes supply, economic, and technical challenges the current technologies are facing:

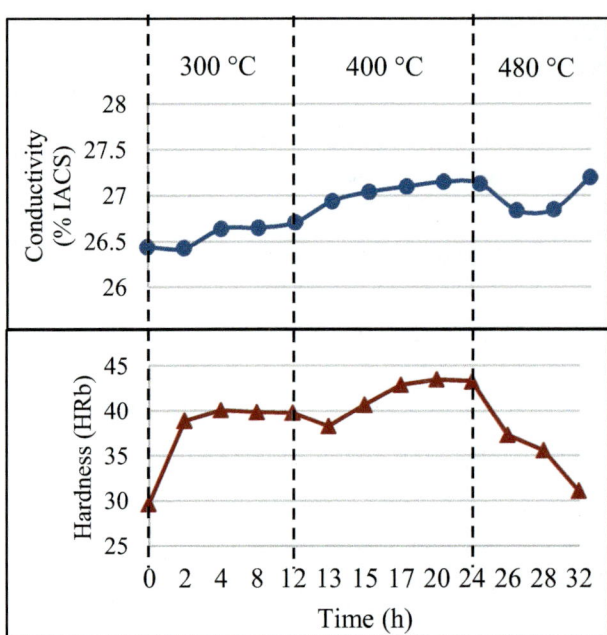

Fig. 7 Conductivity (top) and hardness (bottom) measurements on heat-treated samples

(i) A patented technology for recovery of Sc from Sc-containing ores and industrial waste streams. The applicability of the process in recovery of Sc from bauxite residue and Ti tailings solution was validated at pilot-scale. Leaching conditions and ion-exchange parameters were optimized to minimize impurities in solution, address silica precipitation and fouling issue, and increase Sc recovery, producing Sc concentrates containing 20–25%Sc. This significant increase in Sc content substantially reduces the sizing requirements for the downstream purification step and presents a pragmatic approach to the recovery of Sc from the said complex sources.

(ii) Purification of Sc from concentrates containing impurities such as Ti, Fe, and other rare earths. A purification process to produce >99% Sc_2O_3 has been validated at commercial scale. The Sc concentrates produced from SIR can be purified via this process.

(iii) A patent-pending technology for production of the final alloy through a one-step alloying process using a vacuum induction furnace, refined casting technique, and suitable heat treatment processes. This process which has been validated in the pilot scale has several advantages over the existing process: (1) high reduction efficiency of Sc of up to 90%, (2) lower metal losses from volatilization due to lower reaction time

and bypassing the double-step alloy production, (3) lower operating cost, (4) shorter cycle time, and (5) flexibility of the process to produce Al alloys with a wide range of Sc concentration from 0.05 to 2% Sc.

Our vision is to expand industrial applications for Sc and its alloys as key materials for high energy efficiency and mechanical strength. Our R&D is engaged in developing innovative ideas to address the current and potential opportunities to satisfy the demands of the growing world and is continually exploring different avenues for applying its refining and purification experience and expertise.

References

1. U.S. Geological Survey, Mineral Commodity Summaries, January 2021.
2. Evans K. (2016) The history, challenges, and new developments in the management and use of bauxite residue. J Sustainable Metal 2:316–331.
3. Binnemans K, Jones PT, Blanpain B, Van Gerven T and Pontikes Y. (2015). Towards zero-waste valorisation of rare-earth-containing industrial process residues: a critical review. J Clean Prod 99:17–38.
4. Yagmurlu B, Orberger B, Dittrich C, Croisé G, Scharfenberg R, Balomenos E, Panias D, Mikeli E, Maier C, Schneider R, Friedrich B, Dräger P, Baumgärtner F, Schmitz M, Letmathe P, Sakkas K, Georgopoulos C, and Van Den Laan H. (2021). Sustainable Supply of Scandium for the EU Industries from Liquid Iron Chloride Based TiO2 Plants; Materials Proceedings.
5. Xu, W-Q; Mattera VD, Abella MYR, Abrenica GM, Patkar S. (2019). Selective Recovery of Rare Earth Metals from an Acidic Slurry or Acidic Solution. US20190078175A1 and WO2019099859A1.
6. Xu, W-Q; Mattera VD, Abella MYR, Abrenica GM, Patkar S. (2017). Composite extractant-enhanced polymer resin, method of making the same, and its usage for extraction of valuable metal(s). WO 2017/074921A1.
7. Vind J, Malfliet A, Blanpain B, Tsakiridis P, Tkaczyk A, Vassiliadou V, and Panias D. (2018). Rare Earth Element Phases in Bauxite Residue.
8. Balomenos E, Nazari G, Davris P, Abrenica G, Pilihou A, Mikeli E, Panias D, Patkar S, & Xu WQ. (2021). Scandium Extraction from Bauxite Residue Using Sulfuric Acid and a Composite Extractant-Enhanced Ion-Exchange Polymer Resin. Rare Metal Technology.
9. Davris P, Balomenos P, Nazari G, Abrenica G, Patkar P, Xu WQ, and Karnachoritis Y. Viable Scandium Extraction from Bauxite Residue at Pilot Scale. Multidisciplinary Digital Publishing Institute.
10. Balomenos E, Davris P, Nazari G, Abrenica G, Martinez AM, Patkar S, Xu WQ, and Karnachoritis I. (2021). Developing a Parallel-to-Aluminium Value Chain for Scandium and Al-Sc Alloy Production. Pilot Scale Results under the SCALE Project; TRAVAUX 50, Proceedings of the 39 the International ICSOBA Conference.
11. Nazari G, Abrenica G, and Patkar S. (2022). II-VI's Capabilities in Scandium Recovery and Purification. COM 2022.

12. European Community's Horizon 2020 SCALE Program (H2020/2014–2020/No. 730105).

13. Nazari G, Abrenica G, Xu WQ, Patkar WQ, Benzing M. (2020). Recovery of Sc from Ti Tailing Solution using II-VI SIR Technology; ERES 2020 Conference.

14. Harata M, Nakamura T, Yakushiji H, Okabe TH. (2008). Production of scandium and Al-Sc alloy by metallothermic reduction. Mineral Processing and Extractive Metallurgy Vol 117-2, 95-99.

15. Lide DR. (2005). CRC Handbook of Chemistry and Physics, 85th ed. Internet version 2005

16. Brinkmann F, Mazurek C, Friedrich B. (2019). Metallothermic Al-Sc co-reduction by vacuum induction melting using Ca. Metals 2019, 9, 1223.

17. Xiao J, Ding W, Peng Y, Chen T, Zou K (2020). Preparing Sc-bearing master alloy using aluminum-magnesium thermoreduction method. Metals 2020 10 960.

18. Suzdaltsev AV, Nikolaev, AY, Zaikov YP. (2019). A modern ways for Al-Sc master alloys production. M. Zinigrad, & L. Leontiev (Eds.). Optimization of the composition, structure and properties of metals, oxides, composites, nano and amorphous materials (pp. 161–167). Ariel University Press.

19. Fassel WM Jr, Gulbransen, LB, Lewis JR, Hamilton JH. (1951). Ignition temperatures of magnesium and magnesium alloys. Journal of Metals. 522–528.

20. Kulikov BP, Baranov VN, Bezrukikh AI. (2018). Preparation of aluminum-scandium master alloys by aluminothermal reduction of scandium fluoride extracted from Sc2O3. Metallurgist Vol.61 Nos. 11-12, 1115-1121.

21. Nazari, G, Abrenica, G, Reyes, Anne Marie. Streamlined Process for Aluminum-Scandium Alloy Production. (2022). US Patent Application No 63/375,717.

22. Ghosh, A. (2001). Segregation in cast products. Sadhana Vol. 26 Parts 1&2, 5–24.

23. Lohar AK, Mondal BN, Panigrahi, SC (2010). Influence of cooling rate on the microstructure and ageing behavior of as-cast Al-Sc-Zr alloy. Journal of Materials Processing Technology 210, 2135-2141.

24. Ahmad, Z. (2003). The properties and application of scandium-reinforced aluminum. JOM.

25. Milman, Y. (2006). Scandium effect on increasing mechanical properties of aluminum alloys. Vol 25 Nos 1–2.

26. Ren L, Huimin G, Wang W, Wang S, Li C, Wang Z, Zhai Y, Ma P. (2020). The microstructure and properties of an Al-Mg-0.3Sc alloy deposited by wire arc additive manufacturing. Metals 2020, 10, 320.

27. Royset J, Ryum N, (2005). Scandium in aluminium alloys. International Materials Reviews 2005 Vol 50 No1.

28. Belov, N., Alabin, A., Eskin, D., & Istomin-Kastrovskii, V. (2006). Optimization of hardening of Al-Zr-Sc cast alloys. J Mater Sci (2006) 41, 5890-5899.

29. Johnson, ND. (2011). Processing and mechanical properties of cast aluminum containing scandium, zirconium, and ytterbium. Master's Thesis, Michigan Technological University.

30. Toropova LS, Eskin DG, Kharakterova ML, Dobatkina TV. (1998). Advanced Aluminum Alloys Containing Scandium. 1st. Milton Park, Abingdon, Oxon: Taylor & Francis.

31. Babic E, Zlatic V. (1976) Changes of the lattice parameter in Al 3d alloys due to the virtual bound state. Solid State Communications. 18(6):705-708.

32. Kavalco PM, Canale CF, Totten GE (2009). Quenching of aluminum alloys: Cooling rate, strength, and intergranular corrosion. Heat Treating Progress 9(7): 25-30.

FEA Materials—Aluminum Scandium Master Alloy Production Technology

Rick Salvucci, Brian Hunt, and Eugene Prahin

Abstract

The production of Aluminum Scandium (AlSc) master alloy ranging between 2 and 8wt % is of interest because this alloy family has outstanding material properties, such as strength-to-weight ratio, weldability and corrosion resistance. Despite its attractive properties, commercialization of Scandium-containing aluminum alloys has been hindered by the lack of availability and high cost of Scandium (Sc). This has resulted in very little development in this space. However, access to this material is projected to increase in the coming years. The increase in supply will lower the cost of Sc, enabling commercialization of AlSc, as long as low-cost production techniques are also available. FEA Materials LLC (FEAM) has developed a hybrid metallothermic/electrolysis process (FEAM Process) that enables production of Al-2%Sc below \$40/kg at scandium oxide market price of \$1000/kg or lower. The FEAM Process can theoretically achieve Sc concentrations as high as 8wt % in the alloy, with consistent concentrations of up to 4wt % observed in experiments to-date. The system is designed for continuous operation and maintains a material yield above 90%. The FEAM Process uses a proprietary molten salt composition that is both robust and self-correcting, overcoming the challenges with traditional electrolysis processes. The cell platform is modular, allowing for tuned production rates and flexible infrastructure requirements. The FEAM Process runs under a controlled atmosphere to keep tight control of output streams and to ensure a safe work environment (no perfluorocarbon emissions). FEAM is currently producing upwards of 16 kg Al 2%Sc per shift with the capability to expand to 50 kg/shift within its facility in Westborough, Massachusetts.

Keywords

Electrolysis • Metallothermic • AlSc • Master alloy • Scandium • Low-cost production • Modular platform • Self-correcting

Introduction

Scandium (Sc) is the best strengthener for aluminum (Al) alloys per mole of alloy addition [1]. Al_3Sc coherent precipitates are very fine and stable at high temperature, making these alloys particularly suited to welding or sintering, such as in 3-D printing. Indeed, with yield stress of 525 MPa [2]—twice that of leading powder alloy AlSi10Mg [3]—the strength-to-density ratio $\sigma y/\rho$ of sintered Scalmalloy (3.6% Mg 1% Sc) powder at 1.94×10^5 m^2/s^2 is 20% higher than that of Ti-6-4 [4]. Tensile and bending stiffness/density E/ρ and $E^{1/3}/\rho$ are 3% and 40% higher, respectively, than titanium alloys. At approximately \$3500/kg, Sc is expensive but many alloys benefit from just 0.2 wt % Sc. The typical commercial master alloy consists of 2 wt % Sc-balance Al, with a market price of about \$65–70/kg, which corresponds to \$3064–\$3314 per kg Sc. This market price assumes scandium oxide (Sc_2O_3) price of approximately \$1000/kg.

Though more abundant than lead in the Earth's crust, Sc is very dispersed and hard to isolate, such that worldwide production of Sc_2O_3 is only 20–25 tonnes per year. Sc_2O_3 is the lowest-cost form of Sc available in the market today. With a price of approximately \$1000/kg for 98–99% pure material [5], the metals basis cost is \$1533/kg Sc. Scandium fluoride (ScF_3) is a precursor to making pure Sc metal via calciothermic reduction. ScF_3 is trading at approximately \$1300/kg [6], with metals basis cost of \$2947/kg.

There are three common production techniques for AlSc:

R. Salvucci · B. Hunt · E. Prahin (✉)
400 Madison Avenue, Suite 11A, New York, NY 10017, USA
e-mail: eprahin@feamaterials.com

© The Minerals, Metals & Materials Society 2023
S. Broek (ed.), *Light Metals 2023*, The Minerals, Metals & Materials Series,
https://doi.org/10.1007/978-3-031-22532-1_159

(1) Sc dissolution relies on superheated aluminum metal to mix with high-cost Sc metal, followed by melt crystallization. This method is time intensive and very costly.

(2) Aluminothermic reduction takes Sc_2O_3 or high cost ScF_3 and mixes it into molten Al, yielding an AlSc master alloy [7, 8]. Oxide based reduction has typically been limited to around 2wt % Sc. While this process is simple, there is a high likelihood for oxygen impurities to enter the finished alloy due to the processing method. ScF_3-based reductions can enable higher concentration AlSc alloys. However, the cost of Sc on a metals basis from fluoride results in an expensive alloy.

(3) Lastly, electrolysis has been demonstrated with cryolite or cryolite derivative molten salts similar to the Hall-Heroult process. However, controlling the molten salt composition and maintaining high material yield on the input Sc_2O_3 is one challenge with this production method. This, combined with the hazardous nature of the molten salts, shortened cell lifetime and potential hazardous emissions (perfluorinated compounds) of the system, makes commercialization difficult.

The FEAM Process was developed and optimized to overcome these challenges seen in the Hall-Heroult process. The FEAM Process leverages fast aluminothermic reduction and low-cost electrolysis to produce AlSc alloys at concentrations as high as 8wt % from Sc_2O_3. The molten salt was engineered to be robust and stable during cell operation, enabling high material yields and consistent production. The operating window for these systems has been tailored to prevent any hazardous emissions from being generated during the process. Figure 1 describes the metallothermic and electrolytic reactions—the former occurring continuously and the latter with applied potential.

During electrolysis, the concentration of Sc in the bath far exceeds that reduced into the cathode. This makes sense due to the lower electronegativity of Sc than Al. However, this makes it difficult to precisely characterize the ratio of metallothermic to electrolytic reaction, nor to establish or measure a long-term steady-state bath composition. Looking only at the input Al and Sc_2O_3 showed the ability to produce AlSc using low-cost input materials. This demonstrates an electrolysis intermediate between co-reduction of both oxides to make all of the alloy and reduction of Sc_2O_3 to pure Sc for later mixing with Al.

Experimental

For all experimentation, FEAM used its versatile low volume/high value cell infrastructure which is capable of cell current upwards of 580A and crucible diameters of 12″. Heat to the cell is supplied via external resistance heaters. Located in the center of the cell is a corrosion resistant retort allowing for controlled atmosphere and continuous operation. The external retort is equipped with fixed feedthroughs for all electrodes and sampling equipment. A crucible was loaded into the retort of 12″ OD, capable of holding upwards of 20 kg of electrolyte. The cathode, in this case, was a liquid Al charge located at the bottom of the crucible. Electrical connection to the cathode was made through a current collector. Anodes were comprised of graphite and submersed into the electrolyte. DC power was supplied to the electrodes via an HP power supply capable of 580A. During operation of the cell, oxide and other powders can be added to the electrolyte in a batch process or continuously through an atmosphere controlled automated powder feeding system. The cell is continuously purged with an inert gas and process

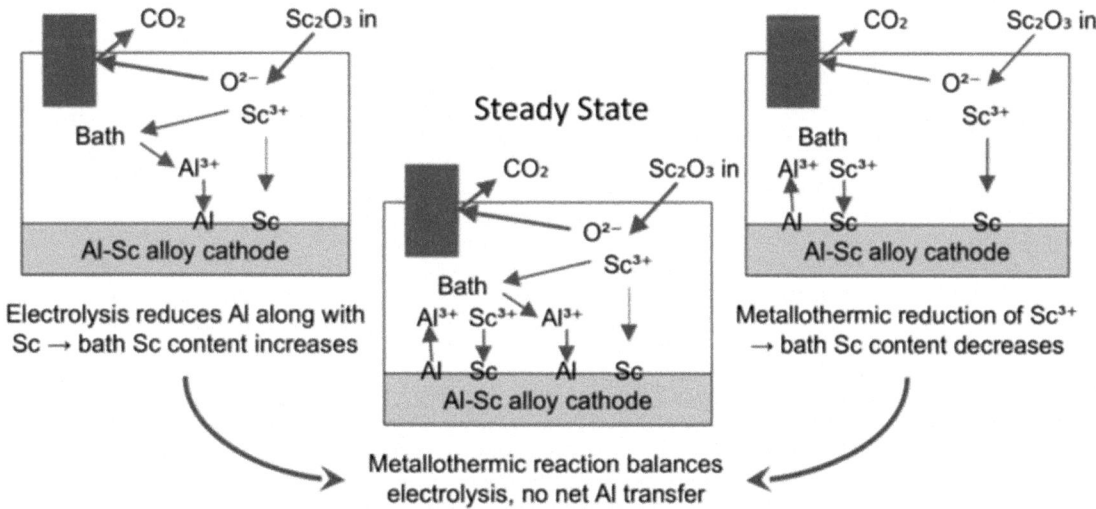

Fig. 1 Hybrid Electrolysis/Metallothermic reaction

byproducts are removed via an external exhaust system. The inert gas prevents moisture from entering the system, which could lead to harmful hydrogen fluoride (HF) emissions. At scale, the inert gas could be substituted with cell off gases to prevent moisture ingress and HF formation.

For data acquisition, the cell is equipped with a variety of sensors to study the electrolytic properties and heat management, including:

- Crucible and heater thermocouples,
- DC current sensors,
- Voltage sense leads to record both operating voltage of the cell and reference voltages to study electrochemical potentials,
- Pressure transducers,
- Current and voltage monitor for external heaters, and
- Sampling port for RGA analysis.

The system was charged with a ScF_3 containing cryolite derivative. Analytical grade chemicals were used for all fluorides and oxides except for AlF_3 which was industrial smelters grade. All chemicals were dried in a separate furnace for several days at approximately 350 °C to decompose any hydroxides and drive off all water from hydrates. Following the drying process, all chemicals were either added to the cell or stored in an oven to prevent any water from attaching.

Experimental Procedure

Experiments at FEAM follow a standard experimental procedure described below:

1. Heat cell to desired operating temperature and allow system to equilibrate
2. Measure electrolyte height in the cell and extract salt sample
3. Measure salt composition and adjust salt chemistry/bath height as needed
4. Load Al charge, electrodes, and current collectors
5. Allow system to equilibrate
6. Measure salt height
7. Extract alloy sample
8. Begin electrolysis at desired Sc_2O_3 feed
9. Run electrolysis for desired Amp-hrs
10. Extract alloy via manual ladling process.

Results

The FEAM cell was operated in a semi-continuous state for several days to observe process stability, yields and total Sc concentration in the achieved master alloy. Alloy samples were measured via energy-dispersive X-ray spectroscopy (EDS) to check for Sc concentration and the presence of segregated Al_3Sc based phases. During each run, AlSc samples were extracted at a fixed time interval to explore how Sc concentration increased over the course of the run. Following equilibration, electrolysis was conducted for a total of 450 Amp-hours at an operating current of ∼200A.

Figure 2 shows the progression of Sc in the alloy over the course of a run. The results show a rapid increase in Sc concentration prior to the start of electrolysis. During subsequent electrolysis, a gradual increase in Sc can be observed. It is hypothesized that during the temperature

Fig. 2 Scandium concentration in the produced Alloy over 450Ahr Run

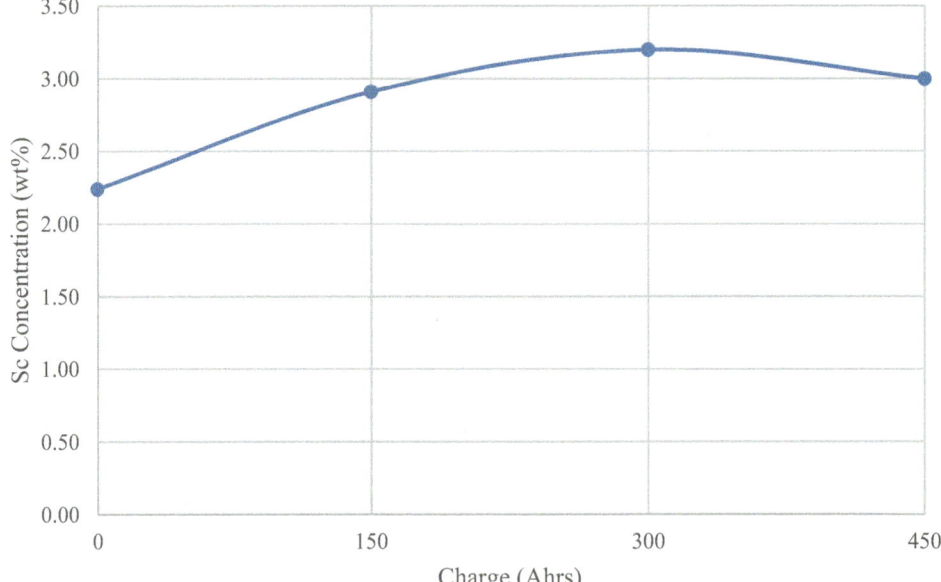

equilibration step the Sc in the bath is aluminothermically reducing from Sc_2O_3 and/or ScF_3. The reactions are hypothesized as follows:

$$Sc_2O_3 + Al = AlSc + Al_2O_3 \qquad (1)$$

$$ScF_3 + Al = AlSc + AlF_3 \qquad (2)$$

In either case, the byproducts are readily dissolved into the electrolyte and the Sc product is dissolved into the molten Al. Given that only 0.12 kg of Sc is produced during this initial step and the bath is ~ 20 kg, we can assume the byproducts have a negligible impact on electrolyte composition. Once DC power is applied to the cell, Sc_2O_3 feeds begin. Figure 2 shows that over the course of electrolysis, there is a large amount of current passed with a relatively small increase in Sc in the product alloy. Therefore, it can be assumed that a large portion of the electrolytic process is reducing Al_2O_3 in the electrolyte.

Al_2O_3 can be present in the electrolyte from two separate reactions. The first is described by Eq. 1 (the aluminothermic reduction of Sc_2O_3). While this can account for some of the alumina, it is likely not the only source. AlSc alloys >3% have been demonstrated via the initial aluminothermic reaction in the FEAM Process which exceeds the typical limit of oxide based metallothermic reductions ($\sim 2\%$). Therefore, one can infer the high Sc concentration is enabled by the presence of ScF_3 and thus creating a surplus of AlF_3. In the presence of excess AlF_3, the Sc_2O_3 feed can react according to the following reaction:

$$Sc_2O_3 + AlF_3 = ScF_3 + Al_2O_3 \qquad (3)$$

This reaction replenishes the spent ScF_3 and electrolysis consumes the generated Al_2O_3. This creates an environment where the relative concentration of Al and Sc in the electrolyte can be easily controlled as long as the process is operating at high material yield and Al_2O_3 concentration is not increasing in the electrolyte. The net result is that the concentrations of reactive cations in the bath are in equilibrium despite external perturbations due to reactions 1,2,3. The combination of these reactions creates a self-balancing environment within the electrolyte.

Figure 3 shows the material yield of the process as calculated by measuring the final alloy mass and concentration of Sc in the product. The total charge (Ahrs), Sc_2O_3 addition and initial aluminum mass were kept constant across all the runs. The results show an exceptionally high material yield over several days with some runs exceeding 92% on a Sc basis. Material yields do show some cyclic-like pattern, oscillating between periods of lower and higher material yields. This is likely related to minor changes caused by periodic in situ salt adjustments to restore proper salt heights and chemistries during operation.

Samples are extracted at operating temperature and rapidly cooled to limit phase segregation of the Sc. Samples are then polished and subjected to both EDS and inductively coupled plasma–optical emission spectroscopy (ICP-OES) measurements to quantify Sc concentration. Using the concentration of Sc in the sample and the initial Al charge, the approximate mass of Sc produced can be calculated. This method does slightly underreport the material yield and mass of Sc produced because Al produced during electrolysis is unaccounted for.

As per Fig. 4, Sc percentages in the final alloy are routinely exceeding 3.0%, with a few runs approaching 4.0%. Summing the reactions from above, the FEAM Process produces high concentration AlSc from an initial aluminothermic reduction from both Sc_2O_3 and ScF_3. The

Fig. 3 Material yield over 9 runs

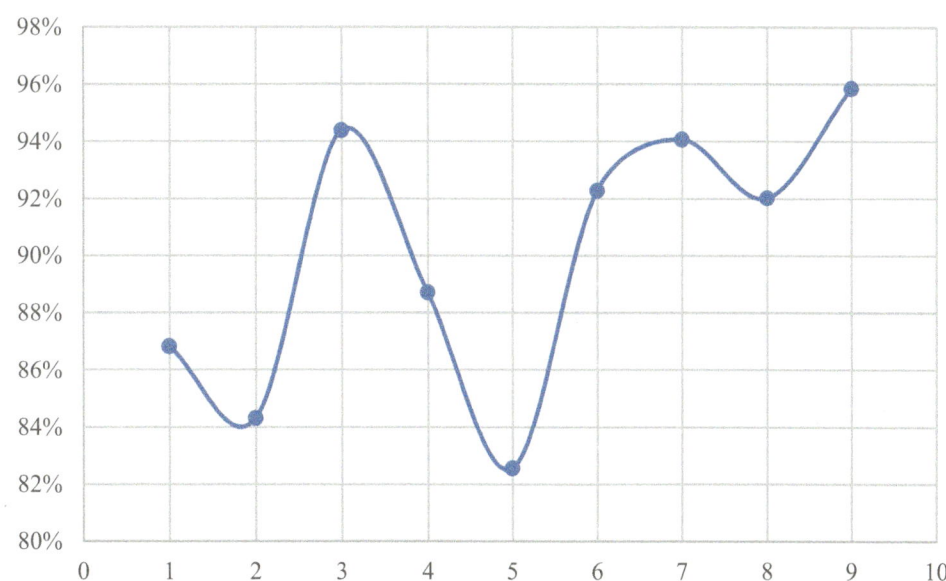

Fig. 4 Scandium content (%) in final alloy over 9 runs

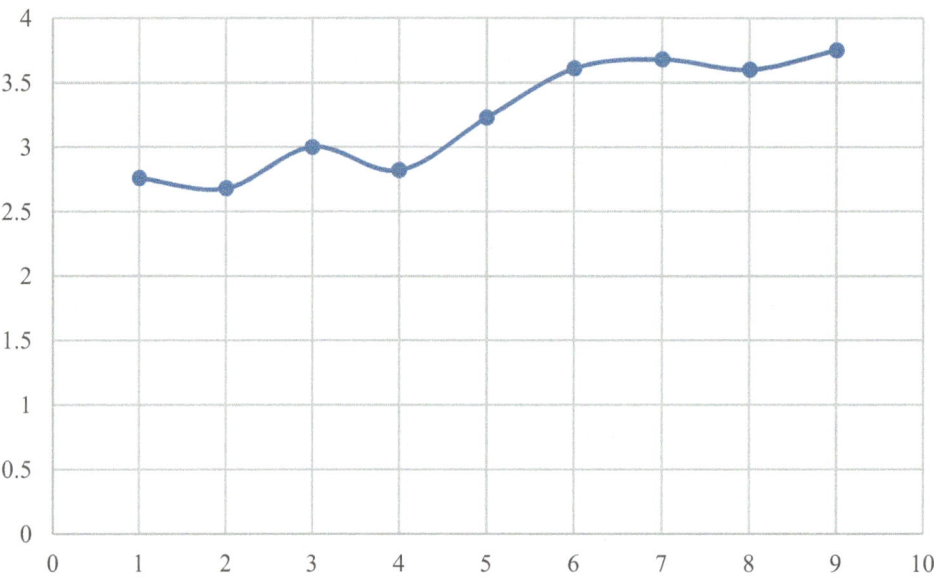

subsequent biproducts are then dissolved into the electrolyte. Electrolyte based reactions with Sc_2O_3 result in a net reaction where AlSc alloys are produced from Sc_2O_3 leveraging a fluoride reduction as an intermediary step.

Conclusions

Typical commercial processes to produce AlSc master alloys are often limited to Al-2%Sc. The FEAM Process leverages a combination of oxide and fluoride aluminothermic reduction followed by subsequent electrolysis to continue reducing Sc and replenish spent Sc from the electrolyte. The result is a process which can operate at both high material yields (upward of 95% demonstrated) and produce high Sc loadings in the final product. The product produced by the FEAM Process enables a reduction in required cell size, shipping cost and operating cost by reducing the volume of Al required to produce high quantities of Sc. Combined with desirable product characteristics, the FEAM Process leverages a self-balancing salt chemistry which enables stable production capabilities and industry adaptation.

Through the combination of the aforementioned advances, the FEAM Process-combined metallothermic and electrolytic approach can enable significant price reduction of scandium containing master alloys. On a comparable basis, the current implied market price of Sc within Al-2%Sc ($65–70/kg) is in the $3064–3314/kg range [5] while the FEAM Process provides Sc within the alloy at $1813/kg, a cost reduction of 40.8–45.3% [9].

References

1. Z. Ahmad, "The Properties and application of Scandium-Reinforced Aluminum," JOM February 2003, p. 35.
2. Source: RSP Technology RSA-501 AE on MatWeb, URL: https://www.matweb.com/search/datasheet.aspx?matguid=436f7a56c98446778a8fb769dff7e99d. Accessed 1 October 2022.
3. Source: EOS data sheet, URL: https://www.eos.info/03_system-related-assets/material-related-contents/metal-materials-and-examples/metal-material-datasheet/aluminium/material_datasheet_eos_aluminium-alsi10mg_en_web.pdf. Accessed 1 October 2022.
4. Source: MatWeb, URL: https://www.matweb.com/search/datasheet.aspx?matguid=b978fc9e4f194184b038fdb3c9bae53a&n=1. Accessed 1 October 2022.
5. Source: Metals.com, URL: https://www.metal.com/Rare-Earth-Oxides/202104090003. Accessed 1 October 2022.
6. Source: https://www.chemicalbook.com/Price/Scandium-trifluoride.htm. Accessed 1 October 2022.
7. F.H. Spedding et al., "Preparation and Properties of High Purity Scandium Metal," Trans. Met. Soc. AIME, 218:608, 1960.
8. A.B. Shubin and K.Yu. Shunyaev, "Thermodynamic Calculations of the Interaction of Scandium Halides with Aluminum," Russ. J. Phys. Chem. A 84(12):2011-2016, 2010.
9. Source: "Al-Sc Alloy Production - TEA Summary Report" by Sidi Deng, Purdue Laboratory for Sustainable Manufacturing

Scandium Master Alloy Production Via Sulfidation and Vacuum Aluminothermic Reduction

Caspar Stinn, Ethan Benderly-Kremen, and Antoine Allanore

Abstract

Scandium is a critical component for high strength aluminum products, yet manufacture is burdened by challenges in metal reduction and alloying. Current best practice begins with generation of an aluminum-scandium master alloy from oxide or halide precursors. However, this approach is characterized by high costs and large environmental impacts. Recent results have shown that employing a metal sulfide feedstock for aluminum master alloy production increases metal yield and improves process economics. Herein, we conduct the sulfidation of scandium oxide using elemental sulfur to generate a scandium sulfide intermediate, which we experimentally confirm to be amenable for reduction to metal. We demonstrate production of aluminum-scandium master alloy at the hundred-gram scale from scandium sulfide using aluminothermic reduction via reactive vacuum distillation. Chemical analysis is conducted to determine product purity and yield. Operating conditions to manufacture master alloys with scandium contents of 2 wt% and higher are tested.

Keywords

Scandium • Master alloy • Sulfidation • Aluminothermic reduction • Vacuum treatment • Induction melting

Introduction

Scandium alloying additions to aluminum provide numerous benefits, including inhibition of recrystallization during heat treatment, high specific strengthening, grain refinement, reduced hot cracking during welding, and improved fatigue resistance [1]. These attributes have placed scandium containing aluminum alloys among desired materials for applications requiring good manufacturability and ultra-high strength to weight ratios [2]. Examples are currently limited to niche uses however, such as sporting equipment and some aerospace applications. Utilization in more general applications is currently hindered by high costs stemming from challenges in extraction, reduction, and alloying [3].

Scandium is a rare earth element that is produced as a by product of titanium dioxide pigment production [4], nickel laterite processing [5] and rare earth mineral processing [3]. Other sources such as red mud from the Bayer process have been considered but are currently not utilized industrially [6]. Scandium is generally recovered in hydrometallurgical solvent extraction circuits and calcined to an oxide [7]. Several processing pathways have been proposed for reduction of scandium oxide or its halogenated compounds to produce aluminum master alloys, including molten salt electrolysis [8, 9] and metallothermic reduction [10–12]. Industrial-scale operation using these methods is challenging however due to burdens associated with halogenation, high costs, low yields, and undesirable byproduct formation.

An attractive avenue for sustainable and economical metal production is sulfide chemistry [13, 14]. Sulfidation of oxides with elemental sulfur is more chemically benign than fluorination chemistry [15], and has been practiced industrially in nickel laterite processing for half a century [16]. Recently, aluminothermic reduction of manganese sulfide via reactive vacuum distillation was shown to produce high purity aluminum-manganese alloys with yields over 95% [17].

Scandium sulfide may be produced from the oxide via roasting the oxide with sulfur gas via the following reaction in a graphite crucible [14]:

$$Sc_2O_3(s) + \frac{9}{4}S_2(g) = Sc_2S_3(s) + \frac{3}{2}SO_2(g) \qquad (1)$$

C. Stinn · E. Benderly-Kremen · A. Allanore (✉)
Department of Materials Science and Engineering, Massachusetts Institute of Technology, Cambridge, MA, USA
e-mail: allanore@mit.edu

© The Minerals, Metals & Materials Society 2023
S. Broek (ed.), *Light Metals 2023*, The Minerals, Metals & Materials Series,
https://doi.org/10.1007/978-3-031-22532-1_160

In the presence of graphite, product sulfur dioxide reacts to form carbon monoxide and regenerate sulfur [14]:

$$SO_2(g) + 2C(s) = 2CO(g) + \frac{1}{2}S_2(g) \qquad (2)$$

By le Chatelier's principle, the *carbothermically driven sulfur reflux* of Reaction 2 within the reactor shifts the equilibrium of Reaction 1 to the right to favor scandium sulfide production.

Aluminothermic reduction of scandium sulfide may be described by the following reaction:

$$Sc_2S_3 + 2Al(l) = 2Sc(s,l) + Al_2S_3(l,g) \qquad (3)$$

Assuming unit activities or partial pressures for all components and a reference state temperature of 25 °C and pressure of 1 atm, Reaction 3 is thermodynamically non-spontaneous at all temperatures below 4500 °C. When excess aluminum is utilized, the activity of the scandium is lowered in the alloy product phase. In a system open to aluminum sulfide, vacuum may be applied to distill aluminum sulfide and lower its activity or partial pressure. By le Chatelier's principle, these forces may shift the equilibrium of Reaction 3 toward the right in order to aluminothermically reduce scandium sulfide to aluminum-scandium alloy via reactive vacuum distillation. This approach is similar thermodynamically to the Pidgeon process [18] for silicothermic reduction of magnesium oxide with ferrosilicon to magnesium via the following reaction:

$$FeSi(s) + 2MgO(s) = Fe(s) + SiO_2(s) + 2Mg(g) \qquad (4)$$

Assuming unit activities or partial pressures for all components and a reference state temperature of 25 °C and pressure of 1 atm, Reaction 4 is also thermodynamically non-spontaneous. However, by continuously distilling product magnesium from the system and reacting product silica with calcium to form calcium silicate, the equilibrium of Reaction 4 is driven to the right. Aluminothermic reduction of scandium sulfide via reactive vacuum distillation differs operationally from the Pidgeon process in that aluminum-scandium alloy is the condensed product whereas magnesium is the distilled product.

Herein, we synthesize scandium sulfide via sulfidation of scandium oxide with elemental sulfur. We then produce an aluminum-scandium alloy at the 100 g scale through aluminothermic reduction of the scandium sulfide via reactive vacuum distillation. 6061 aluminum alloy is utilized as a reductant, enabling analysis of impurity management and propagation in the process. These results present a path forward toward employing environmentally sustainable and economically competitive sulfide chemistry for aluminum-scandium master alloy production.

Methods

Scandium Sulfide Synthesis

For synthesis of scandium sulfide, scandium oxide (Sc_2O_3, 99.99%, Sumitomo Metal Mining) was used as a precursor for sulfidation. The scandium oxide was reacted with sulfur gas (99.5%, sublimed, Acros Organics) at a partial pressure of 0.1 atm to 0.5 atm in a graphite packed bed reactor at a temperature of 1350 °C for 2 h using published apparatuses and procedures [14]. The packed bed reactor was fabricated in-house from 63.5 mm diameter graphite rod stock (EC-12 grade, Graphite Store) and had an outer diameter of 56 mm, inner diameter of 50 mm, depth of 26 mm, bottom thickness of 10 mm, with 3 mm holes spaced in a grid 3 mm apart throughout the bottom. Holes 10 mm in diameter lined the wall 8 mm from the top with 15 mm spacing radially between them. A piece of tissue paper (Kimwipes Delicate Task Wipers, Kimtech) was placed in the bottom of the reactor to limit scandium oxide loss through the bottom holes. 30 g of scandium oxide powder was loaded into the tray without any packing or compression. The crystalline phase composition of the scandium sulfide product was analyzed via x-ray diffraction following sulfidation. The scandium sulfide product was then utilized as a feedstock for aluminum-scandium alloy production.

Aluminum-Scandium Alloy Synthesis

Aluminum-scandium alloy was produced from scandium sulfide through aluminothermic reduction via reactive vacuum distillation in a vacuum induction furnace using apparatuses and procedures described previously [17]. 8 g of scandium sulfide and 200 g of 6061 aluminum alloy (McMaster-Carr) were mixed in a graphite crucible with an outer diameter of 65 mm, inner diameter of 45 mm, depth of 120 mm, and bottom thickness of 30 mm, machined in house from 75 mm diameter graphite rod stock (EC-12 grade, Graphite Store). The starting composition of the 6061 alloy is reported in Table 1. The scandium sulfide and aluminum in the graphite crucible were induction heated from 25 °C to 1550 °C at a vacuum of 10^{-3} atm in 25 min then held at 1550 °C and 10^{-3} atm for 25 min. The furnace power was then shut off and the crucible containing the metal product was allowed to furnace quench under vacuum, dropping from 1550 °C to 600 °C in 10 min.

Following aluminothermic reduction, the crucible containing the metal product was vacuum cast in epoxy and cross sectioned in half using a reciprocating saw. Metal product was extracted from half of the sample for external chemical composition analysis by spark optical emission

Table 1 Composition on a mass fraction basis of 6061 reductant and aluminum-scandium alloy product determined via spark-OES, LECO, ICP-MS, and SEM-EDS analysis

	Al	Sc	Mg	Si	Fe	Cu	Cr	Zn	Ti	Mn	S	O	C
6061 (OES[a], LECO[b], ICP[c])	rem	nil[c]	0.8[a]	0.7[a]	0.41[a]	0.33[a]	0.05[a]	0.04[a]	0.02[a]	0.05[a]	nil[b]	0.001[b]	0.007[b]
Al-Sc, product (OES[a], LECO[b], ICP[c])	rem	0.38[c]	0.02[a]	0.57[a]	0.39[a]	0.37[a]	0.05[a]	0.01[a]	0.06[a]	0.04[a]	0.12[b]	3.18[b]	0.39[b]
Al-Sc, product (SEM-EDS)	94.4 (4.0)	3.3 (4.0)	na	0.4 (0.1)	0.4 (0.1)	0.5 (0.2)	0.1 (0.1)	0.1 (0.1)	0.1 (0.1)	0.1 (0.1)	0.5 (0.1)	na	na
Al-Sc, ScAl₃ (SEM-EDS)	76.4 (0.3)	21.4 (0.3)	na	0.2 (0.1)	0.3 (0.1)	0.2 (0.1)	0.4 (0.1)	0.3 (0.1)	0.4 (0.1)	0.1 (0.1)	0.3 (0.2)	na	na
Al-Sc, Al (SEM-EDS)	97.8 (0.1)	0.3 (0.2)	na	0.2 (0.1)	0.3 (0.1)	0.6 (0.1)	0.1 (0.1)	0.1 (0.1)	0.1 (0.1)	0.1 (0.1)	0.3 (0.2)	na	na
Al-Sc, metallic distillate (SEM-EDS)	96.1 (0.9)	0.3 (0.1)	na	0.5 (0.2)	0.5 (0.1)	0.5 (0.2)	0.1 (0.1)	0.2 (0.1)	0.1 (0.1)	0.1 (0.1)	1.6 (0.7)	na	na

Values in parenthesis correspond to errors of ± one standard deviation, "nil" denotes values below the detection limit, "rem" denotes a component that constitutes the remainder of the composition, and "na" denotes species not analyzed herein via SEM-EDS

spectroscopy (OES, ATS, Marietta, GA, USA) for magnesium, silicon, iron, copper, chromium, zinc, titanium, and manganese, light element combustion analysis (LECO, ATS, Marietta, GA, USA) for carbon, oxygen, and sulfur, and inductively coupled plasma mass spectroscopy (ICP-MS, ATS, Marietta, GA, USA) for scandium. The other half of the cross sectioned sample was polished for element mapping and quantification of composition via a scanning electron microscope equipped with an energy-dispersive X-ray spectrometer (SEM-EDS, JEOL JSM-6610LV, JEOL). For quantification of bulk alloy and distillate compositions, representative samplings of 2 mm × 2 mm EDS area spectrums were conducted across the surfaces of the product. Crystalline phase identification was conducted via XRD of the product surface.

Results and Discussion

Scandium Sulfide Synthesis

Sulfidation of scandium sesquioxide to scandium sesquisulfide at 1350 °C for 2 h proceeded to approximately 70% conversion. The powder bed exhibited a yellow shell on the outside with white unreacted powder on the inside, illustrated in Fig. 1. The yellow shell had sintered, allowing it to be easily removed from the loose unreacted powder on the inside. The yellow shell was then crushed for phase identification via XRD. The XRD spectrum, illustrated in Fig. 2, revealed that scandium sesquisulfide was the only crystalline phase present at detectable levels within the outer

shell. Further analysis, such as LECO, would be necessary to determine the level of dissolved oxygen in the sulfide product. The homogenous nature and crystalline composition of the scandium sulfide product in the outer shell suggests that complete conversion of oxide to sulfide was achieved there. Incomplete conversion in the center of the bed was likely due to mass transport limitations.

The absence of oxide or oxysulfide phases in the scandium sulfide product indicates that the reaction was not limited by *intragrain* diffusion effects within individual scandium oxide particles. Instead, sintering of the scandium sulfide likely hindered *intergrain* diffusion through the product shell to the core of the powder bed [14]. These limitations to oxide sulfidation conversion may be resolved through optimization of reaction time and temperature, reactor mass transport, and the powder/pellet density. For sulfidation in a packed bed reactor, the thickness of the scandium sulfide shell provides insight into the maximum bed thickness able to support complete conversion of oxide to sulfide at the temperature, pressures, and powder bed morphology employed herein. Likewise, for sulfidation in a fluidized bed reactor, the observed scandium sulfide product shell provides insight into the maximum pellet diameter able to support complete conversion of oxide to sulfide under the utilized feedstock and reaction conditions. Established mass transport and kinetic relations are available to design a sulfidation reactor for complete conversion of scandium oxide to sulfide [14]. Reactors with stacked beds or multiple hearths may be employed to increase the productivity of sulfidation for a given reactor cross sectional area [19].

Fig. 1 Scandium sulfide synthesized via sulfidation of scandium oxide with elemental sulfur gas in the presence of carbon. The yellow outer shell, shown via XRD in Fig. 2 to be homogeneous scandium sesquisulfide, sintered during the reaction and limited mass transport of sulfur gas into the center of the powder bed. Color figures available online

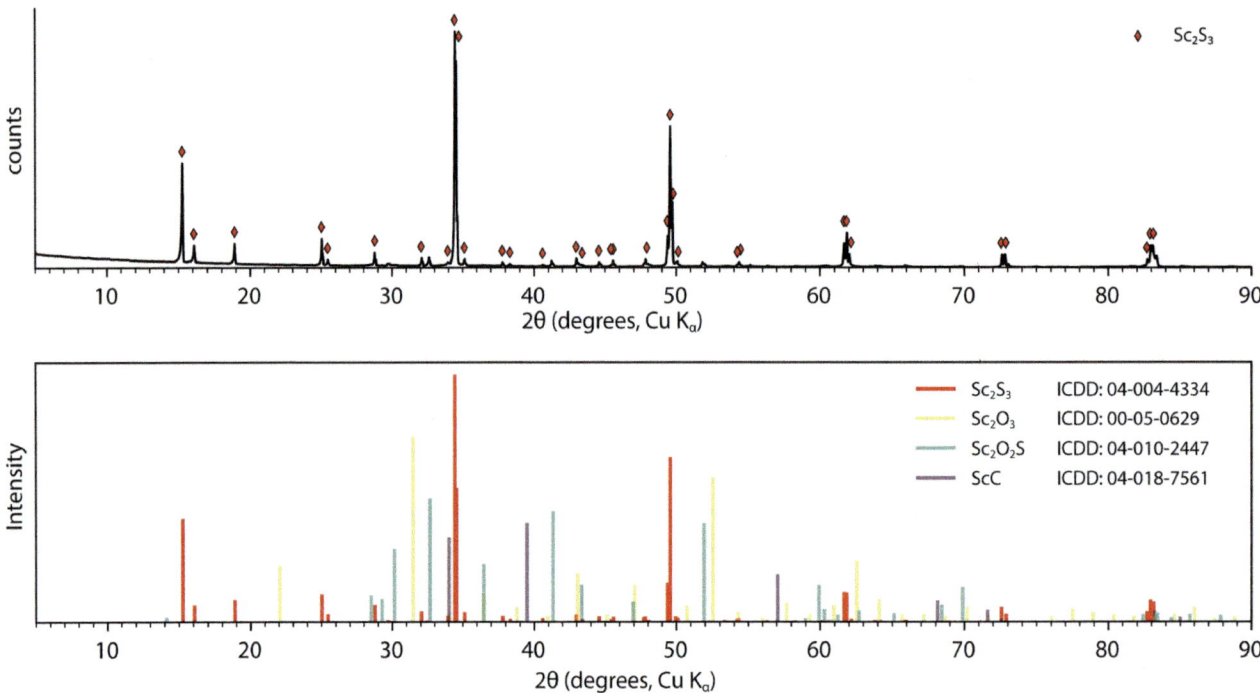

Fig. 2 Qualitative XRD results of sulfidized scandium oxide compared with published XRD spectrums, revealing a homogenous scandium sesquisulfide product in the yellow sulfidized shell (Fig. 1) free of crystalline oxide, oxysulfide, and carbide impurities

Aluminum-Scandium Alloy Synthesis

The scandium sesquisulfide product obtained from the outside of the bed was utilized for subsequent aluminum-scandium alloy production through aluminothermic reduction via reactive vacuum distillation. As a result of the aluminothermic reduction of scandium sesquisulfide, a metal product with dispersed needle structures, a metallic distillate, and non-metal distillate were formed. These regions are depicted in the cross section illustrated in Fig. 3. Needle structures were observed to protrude inward from the wall of the susceptor, forming a web within the center of the solidified melt that was interrupted by micron-scale and millimeter-scale porosity. Potential implications of inhomogeneous morphology on sampling for chemical analysis are discussed later.

The compositions of the metallic products from SEM-EDS, spark-OES, LECO, and ICP-MS are reported in Table 1. From SEM-EDS analysis, a total scandium content of approximately 3.3 wt% was achieved in the metal product, with some surveyed 2 mm × 2 mm sections such as in Fig. 3b exhibiting total scandium contents above 6 wt%. The

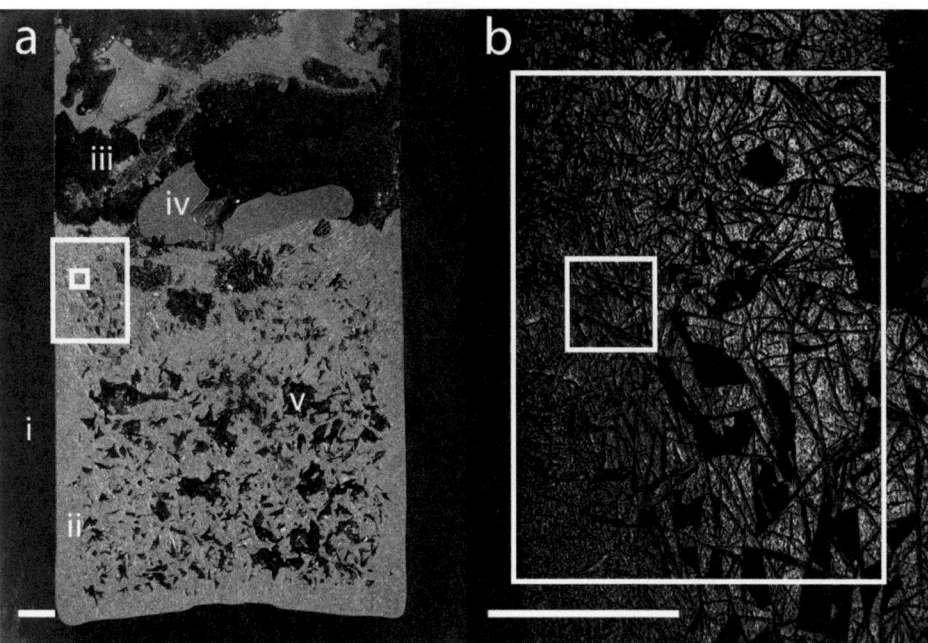

Fig. 3 Aluminum-scandium alloy produced via reactive vacuum distillation. Panel a corresponds to a cross section of the graphite crucible (i) containing the aluminum-scandium alloy product (ii), sulfide distillate (iii), and aluminum metal distillate (iv). Significant porosity (v) is observed in the aluminum-scandium alloy following cooling. The region marked by the outer box is detailed in panel b, revealing the existence of needle-shaped intermetallic phases throughout the aluminum-scandium alloy product. The chemical composition of the region marked by the inner box is quantified using SEM/EDS element maps in Fig. 4. Scale bars: 5 mm

scandium content of the distillate metal product was significantly lower at 0.3 wt% and less variable. Composition of the nonmetallic distillate requires further analysis. These results indicate that sulfidation followed by aluminothermic reduction via reactive vacuum distillation constitutes a promising avenue for aluminum-scandium master alloy production with high scandium yields. For commercial scale production of aluminum-scandium master alloy using sulfide chemistry, challenges with product homogeneity and macrosegregation of scandium aluminide intermetallics must be resolved. To recover residual scandium from distillate products, they may be recycled into sulfidation or reduction steps.

The observed scandium content herein of 3.3 wt% exceeds that of commercial master alloy, marketed at a grade of 2 wt% scandium. Lower amounts of scandium sulfide may be introduced during reduction to produce 2 wt% scandium aluminum alloy if desired. To test the viability of more scandium-rich metal products for subsequent alloying, such as those manufactured herein via sulfide chemistry, their melting and mixing behaviors require exploration. The morphology and surface area of the aluminum-scandium master alloy will require optimization to maximize dissolution rate and yield while minimizing oxidative loss of the scandium-containing phases.

Within the metallic product, SEM/EDS element mapping illustrated in Fig. 4 revealed the needle phases to be a scandium aluminide intermetallic compound dispersed in aluminum. SEM-EDS compositions of needle phases and the bulk aluminum in the metallic product are reported in Table 1. The scandium content of the needle-shaped intermetallic phases was 21.4 wt%, whereas the scandium content of aluminum bulk was 0.3 wt%. Aluminum sulfide and aluminum-silicon-iron precipitants were also present in the bulk aluminum. The aluminum sulfide precipitants were observed to be present at the interface of the scandium aluminide and bulk aluminum, whereas aluminum-silicon-iron precipitants were dispersed throughout. XRD analysis confirmed the scandium aluminide to be $ScAl_3$, illustrated in Fig. 5. The aluminum sulfide and iron-silicon aluminide precipitants were found to be Al_2S_3 and $Fe_3(Al_{0.3}Si_{0.7})$ respectively. Trace amounts of aluminum carbide were also observed. Scandium-containing carbide, oxide, or oxysulfide byproducts were not observed via SEM/EDS or XRD, indicating that complete conversion of scandium sulfide to metal is possible.

Analysis of the aluminum scandium alloy product via ICP-MS yielded a scandium content in the metallic product of 0.38 wt%. This result was inconsistent with those obtained via SEM/EDS and XRD. This discrepancy is likely due to sampling error, and may have explanations rooted in thermodynamics or mass transport. As shown in Fig. 3, the distillate metal phase is in direct contact with the aluminum-scandium product, yet they remain unmixed. During reactive

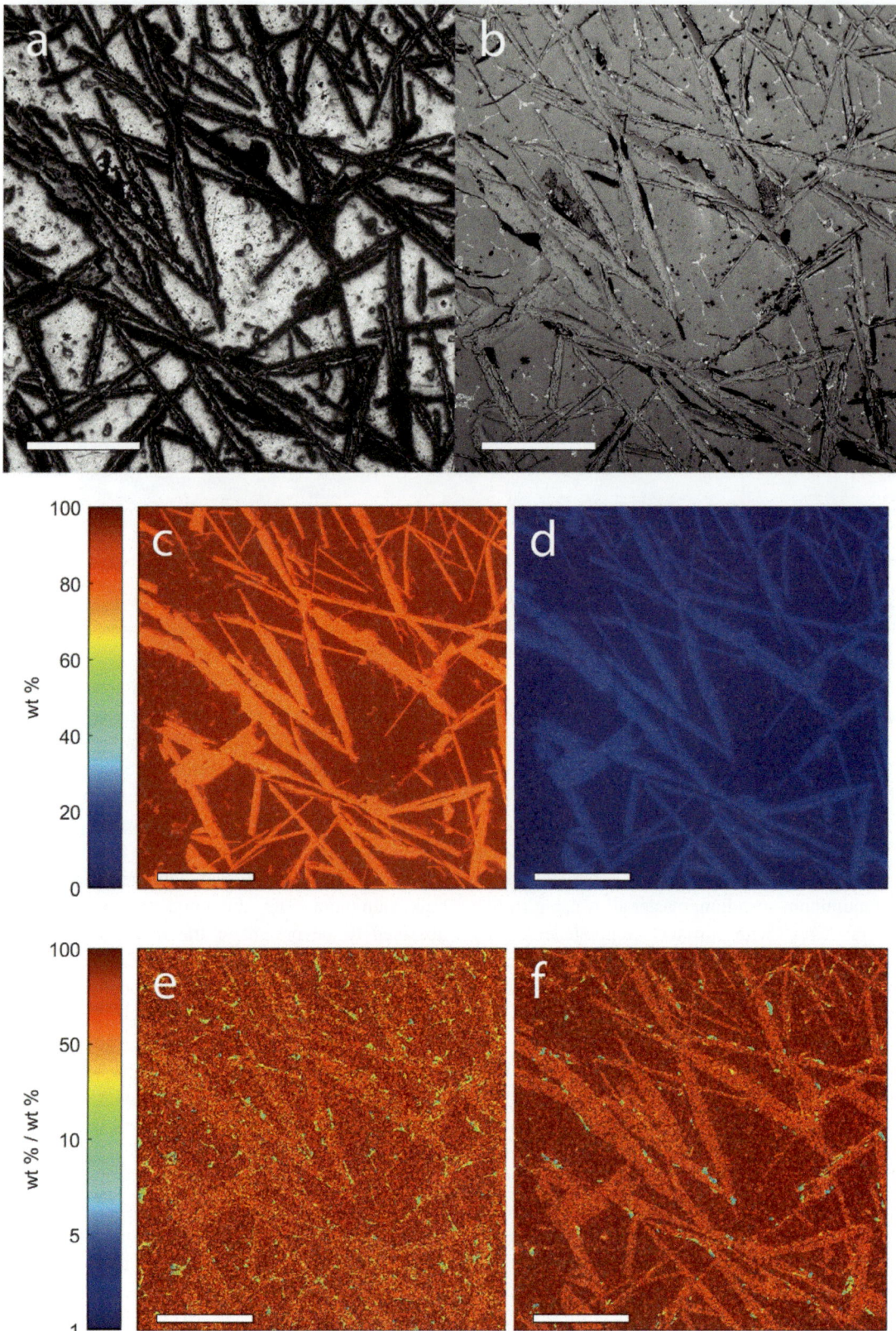

Fig. 4 Spatial element mapping of aluminum-scandium alloy components via SEM/EDS. Panels a and b correspond to optical and SEM/BEC images respectively of the region analyzed. Panels c and d report aluminum and scandium spatial compositions respectively in the alloy. Needle phases are observed via XRD to correspond to ScAl₃ intermetallic compound. Panels e and f report aluminum to iron and aluminum to sulfur mass ratios respectively. Iron present from the 6061 aluminum alloy feedstock is shown to form smaller iron-silicon aluminide intermetallic phases dispersed in the aluminum bulk within the aluminum-scandium alloy product. Residual sulfur is found to be concentrated in aluminum sulfide phases at the interface between ScAl₃ phases and the aluminum bulk. Scale bars: 500 μm

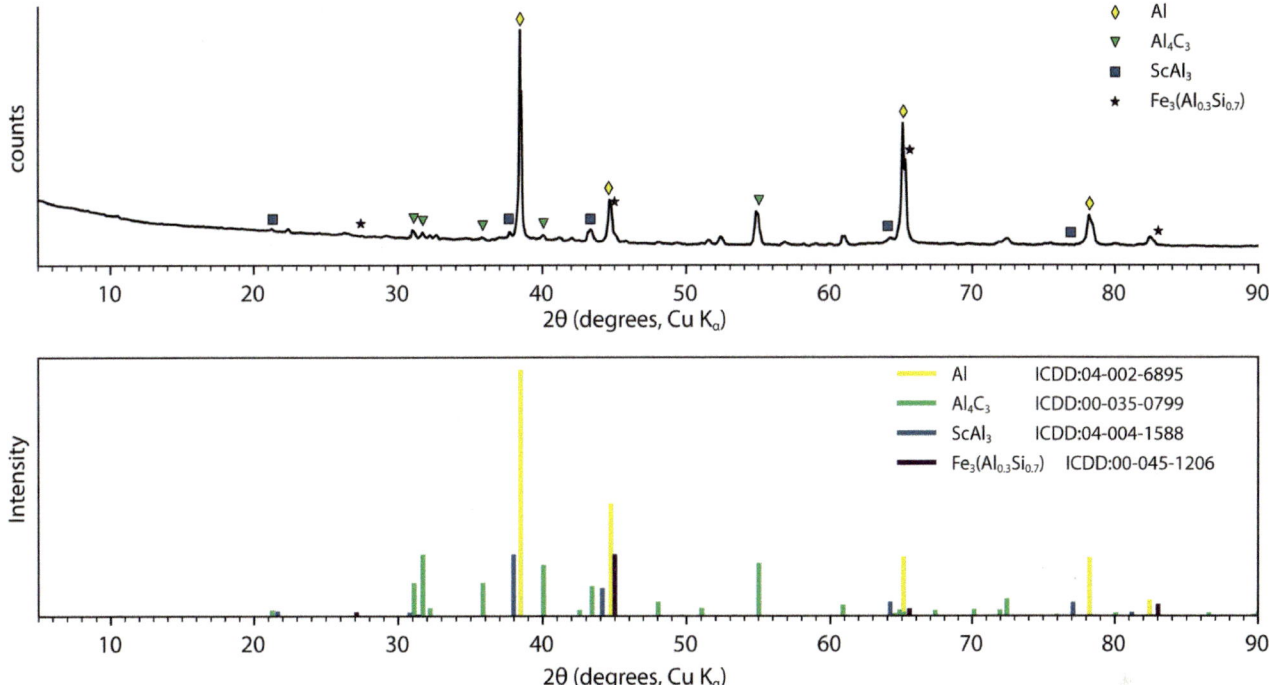

Fig. 5 Qualitative XRD results for the aluminum scandium alloy product compared with published XRD spectrums, revealing the presence of aluminum, scandium aluminide (ScAl₃), aluminum carbide (Al_4C_3), and iron silicon aluminide ($Fe_3(Al_{0.3}Si_{0.7})$) phases. Analysis was performed on the bulk as-crystalized sample, resulting in texturing of the spectrum data

vacuum distillation, aluminum vapors recondensed within the graphite susceptor and flowed back down toward the aluminum-scandium alloy melt, separating from more volatile magnesium and zinc species that recondensed elsewhere as discussed later. Due to highly enthalpically-favorable mixing and compound formation between aluminum and scandium, aluminum-scandium complexes and short range ordering likely persist hundreds of degrees above the alloy's melting point [20]. The presence of short range ordering may slow diffusion and remixing of aluminum-scandium alloy with aluminum distillate [21], possibly leading to local macroscopic variations in scandium content that were preserved when the product was quenched. During sampling for ICP-MS, one of these scandium-deficient regions may have been analyzed.

Alternatively, oxidation of scandium aluminide phases during handling may have hindered analysis. From XRD, oxide phases were not observed in the metal product. However, in subsequent LECO analysis, significant oxidation was observed. Scandium is known to exhibit a higher affinity for oxygen than aluminum in both pure [14] and alloyed [22] forms, indicating that preferential oxidation of aluminum-scandium intermetallic phases may have occurred during the time between in-house SEM-EDS/XRD and external spark-OES/LECO/ICP-MS. Scandium-aluminum oxide coatings on the surface of the intermetallic may have subsequently hindered their aqueous dissolution and analysis

via ICP-MS. The ICP-MS results for overall scandium content were comparable to the scandium content observed via SEM-EDS analysis of the bulk aluminum phases found between intermetallics.

Results from SEM-EDS and spark-OES for the partitioning of alloying agents from the 6061 reductant between metal product and distillate phases are in good agreement, suggesting that analytical and sampling challenges were limited to scandium quantification and ICP-MS. Magnesium, zinc, and some silicon impurities introduced as alloying agents in the 6061 feedstock were removed from the system during reactive vacuum distillation. Magnesium and zinc were also depleted in the condensed aluminum distillate, suggesting they were condensed in the nonmetallic distillate or outside the graphite susceptor. Iron, copper, manganese, and chromium impurities showed a fairly uniform partitioning between metallic product and metallic distillate phases. Meanwhile, reactive vacuum distillation was shown to enrich the metal product in titanium. Within the metal product, residual copper preferentially partitioned to the aluminum bulk, whereas residual chromium, zinc, and titanium preferentially partitioned to scandium aluminide intermetallic phases.

Overall these results indicate that impure or scrap aluminum reductants may potentially be used for the aluminothermic reduction of sulfides. Magnesium, silicon, and zinc impurities in the reduction feedstock are readily volatilized and have a limited impact on product purity. Iron impurities reacted with silicon in the metal product, lowering the extent of silicon distillation. Meanwhile, other 6061 alloying agents showed limited selectivity in distillation and may accumulate in the metal product. Depending on the specifications required for the aluminum-scandium master alloy product, aluminum with lower levels of deleterious impurities may be employed. The use of unsorted scrap in reduction remains another path of exploration.

Conclusions

Scandium sulfide was synthesized from scandium oxide via sulfidation with elemental sulfur using a packed bed reactor. Intergrain diffusion limitations through sintered, fully converted scandium sulfide products were observed to likely hinder conversion in the center of the particle bed, informing the design of industrial processes for scandium sulfide production. Pathways exist to optimize mass transport effects to support full conversion of oxide to sulfide on commercial scales.

Product scandium sulfide was successfully aluminothermically reduced to aluminum-scandium alloy via reactive vacuum distillation. A scandium content of approximately 3.3 wt% was achieved in the bulk aluminum, existing in the form of $ScAl_3$ intermetallic needles dispersed throughout. Achieving homogenous distribution of $ScAl_3$ intermetallics throughout the alloy product will require process optimization. 6061 aluminum alloy was employed as a reductant and behaviors of scrap impurities during aluminothermic reduction of sulfides were explored. Magnesium, zinc and some silicon impurities were expelled during the reactive vacuum distillation, while iron, copper, chromium, titanium, and manganese impurities remained or accumulated in the melt. Iron reacted with residual silicon to form iron-silicon aluminide precipitants in the alloy product. Residual sulfur was found in small Al_2S_3 phases on the edges of $ScAl_3$ phases. The alloying behavior of aluminum-scandium products following reduction and the optimal choice for aluminum reductants both require further analysis.

Acknowledgements The authors thank Dr. Hiro Higuchi and Sumitomo Metal Mining for graciously providing the sample of scandium oxide.

References

1. Ahmad Z (2003) The Properties and Application of Scandium-Reinforced Aluminum. J. Miner. Met. Mater. Soc. 55:35–39
2. Riva S, Yusenko K V., Lavery N P, Jarvis D J, & Brown S G R (2016) The scandium effect in multicomponent alloys. Int. Mater. Rev. 61(3):203–228. https://doi.org/10.1080/09506608.2015.1137692
3. Botelho Junior A B, Espinosa D C R, Vaughan J, & Tenório J A S (2021) Recovery of scandium from various sources: A critical review of the state of the art and future prospects. Miner. Eng. 172:107148. https://doi.org/10.1016/j.mineng.2021.107148
4. Zhang L, Zhang T A, Lv G, Zhang W, Li T, & Cao X (2021) Separation and Extraction of Scandium from Titanium Dioxide Waste Acid. JOM 73(5):1301–1309. https://doi.org/10.1007/s11837-021-04629-7
5. Kaya Ş, Dittrich C, Stopic S, & Friedrich B (2017) Concentration and separation of scandium from Ni laterite ore processing streams. Metals (Basel). 7(12):0–6. https://doi.org/10.3390/met7120557
6. Wang W, Pranolo Y, & Cheng C Y (2011) Metallurgical processes for scandium recovery from various resources: A review. Hydrometallurgy 108(1–2):100–108. https://doi.org/10.1016/j.hydromet.2011.03.001
7. Zhao B, Zhang J, & Schreiner B (2016) Separation Hydrometallurgy of Rare Earth Elements. Springer International Publishing AG Switzerland,
8. Nikolaev A Y, Suzdaltsev A V., & Zaikov Y P (2019) Electrowinning of Aluminum and Scandium from KF-AlF3-Sc2O3 Melts for the Synthesis of Al-Sc Master Alloys. J. Electrochem. Soc. 166(8):D252–D257. https://doi.org/10.1149/2.0231908jes
9. Harata M, Yasuda K, Yakushiji H, & Okabe T H (2009) Electrochemical production of Al-Sc alloy in CaCl2-Sc2O3 molten salt. J. Alloys Compd. 474(1–2):124–130. https://doi.org/10.1016/j.jallcom.2008.06.110
10. Brinkmann F, Mazurek C, & Friedrich B (2019) Metallothermic Al-Sc co-reduction by vacuum induction melting using Ca. Metals (Basel). 9(11). https://doi.org/10.3390/met9111223
11. Kulikov B P, Baranov V N, Bezrukikh A I, Deev V B, & Motkov M M (2018) Preparation of Aluminum-Scandium Master Alloys by Aluminothermal Reduction of Scandium Fluoride Extracted from Sc2O3. Metallurgist 61(11–12):1115–1121. https://doi.org/10.1007/s11015-018-0614-1
12. Xiao J, Ding W, Peng Y, Chen T, & Zou K (2020) Preparing Sc-bearing master alloy using aluminum–magnesium thermoreduction method. Metals (Basel). 10(7):1–14. https://doi.org/10.3390/met10070960
13. Daehn K E, Stinn C, Rush L, Benderly-Kremen E, Wagner M E, Boury C, Chmielowiec B, Gutierrez C, & Allanore A (2022) Liquid Copper and Iron Production from Chalcopyrite, in the Absence of Oxygen. Metals (Basel). 12(9):1440. https://doi.org/10.3390/met12091440
14. Stinn C & Allanore A (2022) Selective sulfidation of metal compounds. Nature 602:78–83. https://doi.org/10.1038/s41586-021-04321-5
15. Ahmadi E & Suzuki R O (2021) Tantalum Metal Production Through High-Efficiency Electrochemical Reduction of TaS2 in Molten CaCl2. J. Sustain. Metall. 7(2):437–447. https://doi.org/10.1007/s40831-021-00347-1
16. Diaz C M, Landolt C A, Vahed A, Warner A E M, & Taylor J C (1988) A Review of Nickel Pyrometallurgical Operations. JOM 40(9):28–33. https://doi.org/10.1007/BF03258548
17. Stinn C, Toll S, & Allanore A (2022) Aluminothermic Reduction of Sulfides via Reactive Vacuum Distillation. Light Metals 2022. p 681–688
18. Halmann M, Frei A, & Steinfeld A (2008) Magnesium production by the pidgeon process involving dolomite calcination and MgO silicothermic reduction: Thermodynamic and environmental analyses. Ind. Eng. Chem. Res. 47(7):2146–2154. https://doi.org/10.1021/ie071234v

19. Allanore A & Stinn C (2021) Selective Sulfidation and Desulfidation. US2021/0277531A1. 2021

20. Shevchenko M O, Kudin V G, & Berezutskii V V (2014) Thermodynamic Poerperties of Al-Sc Alloys. Powder Metall. Met. Ceram. 53(3):151–157

21. Stinn C & Allanore A (2018) Thermodynamic and Structural Study of the Copper-Aluminum System by the Electrochemical Method Using a Copper-Selective Beta″ Alumina Membrane. Metall. Mater. Trans. B 48(6):2922–2929. https://doi.org/10.1007/s11663-018-1400-y

22. Obidov Z R, Amonova A V., & Ganiev I N (2013) Effect of scandium doping on the oxidation resistance of Zn5Al and Zn55Al alloys. Russ. J. Phys. Chem. A 87(4):702–703. https://doi.org/10.1134/S0036024413040201

European Scandium for a Lighter and Greener Future

Henk van der Laan, Beate Orberger, Carsten Dittrich,
Robin Scharfenberg, Edward Peters, Georges Croisé, Pierre Feydi,
Carolin Maier, Richard Schneider, Bernd Friedrich, Yashvi Baria,
Konstantinos Sakkas, and Christos Georgopolous

Abstract

Scandium is a soft silvery metal, with an atomic number of 21 it is the lightest of the transition metals. The melting point is 1.541 °C, the boiling point is 2.836 °C and with a density of 2.985 g/cm^3 is slightly heavier than Aluminium. Scandium is actually not rare—it is more abundant than precious metals and commercial metals like cobalt, lead and mercury. Scandium is primarily produced as a byproduct from the mining of other metals or minerals like Bauxite, Coal, Rare Earth Elements (REE), Iron, Tungsten, Uranium, Zirconia or Titaniumdioxide. Scandium and scandium compounds have unique properties for many advanced technological applications. Scandium is considered as a Strategic metal by the EU and by the US government due to the current limited Western supply situation. Scandium is increasingly used in energy storage systems such as solid oxygen fuel cells (SOFC) and for green hydrogen production in solid oxide electrolyser cells (SOEC). AlScN piezoelectric films for energy generation are important compounds for 5G applications. Today, the EU imports 100%, mainly from China. Therefore, a continuous supply of scandium at reasonable prices must be ensured in and for Europe, and the dependency on China must be reduced. Europe is leading in the development of green technologies and has sufficient scandium resources. This paper will briefly explain the status and potential of Scandium compound production in Europe, in particular providing an update on the ScaVanger project.

Scandium Properties

Scandium (Sc) is a soft silvery metal, with an atomic number of 21, it is the lightest of the transition metals. The melting point is 1.541 °C, the boiling point is 2.836 °C and with a density of 2.985 g/cm^3 is slightly heavier than Aluminium. Thermal conductivity of Scandium is 15.8 W/(m·K). The linear thermal expansion coefficient of Scandium is 10.2 μm/(m·K).

The high melting point makes Sc interesting for many applications, especially for high strength light alloys.

Scandium Resources

Scandium is actually not rare—it is more abundant [1] than precious metals and commercial metals like cobalt, lead and mercury. Scandium rarely concentrates in nature [2, 4], therefore a primary scandium mine is not available. Scandium is primarily produced as a byproduct from the mining of other metals or minerals like Bauxite [5], Coal, Nickel laterites, Rare Earth Elements (REE), Iron, Tungsten, Uranium, Zirconia or Titaniumdioxide. Scandium occurs in aqueous acidic residual solutions from metallurgical processing.

Today, Europe imports 100% scandium oxide (Sc_2O_3). Major suppliers are located in China ($\sim 67\%$), Russia and Kazakhstan.

Scandium Applications and Markets

Scandium and its compounds have unique properties [3] for many advanced technological applications. Scandium is considered as a strategic metal by the EU and by the US government due to the current limited Western supply situation. Scandium is increasingly used in energy storage systems such as solid oxygen fuel cells (SOFC) and for green

H. van der Laan (✉) · B. Orberger · C. Dittrich · R. Scharfenberg ·
E. Peters · G. Croisé · P. Feydi · C. Maier · R. Schneider ·
B. Friedrich · Y. Baria · K. Sakkas · C. Georgopolous
Rotterdam, The Netherlands
e-mail: henkvanderlaan@yahoo.com

© The Minerals, Metals & Materials Society 2023
S. Broek (ed.), *Light Metals 2023*, The Minerals, Metals & Materials Series,
https://doi.org/10.1007/978-3-031-22532-1_161

hydrogen production in solid oxide electrolyzer cells (SOEC). Mg-Sc alloys are mentioned as a promising light weight application for hydrogen storage.

For aluminium and its alloys, AlSc2 is a strong grain refiner, it is also an alloying ingredient resulting in improved mechanical properties in Al–Mg-Zr alloys. It has a positive effect on the weldability of Al–Mg alloys [6], these alloys are extremely suitable for marine and aerospace applications. The "Green" aerospace alloy 5024 (AlMg4.5Sc0,2) has ideal properties to be used in existing commercial planes for fuselage panels as a replacement of the traditional 2024 aircraft alloy resulting in a 4–5% weight saving. Other promising applications of Al–Mg-Sc alloys are LPG tankers and windmill rotor blades. Al-Sc-N piezoelectric thin film depositions in the semicon industry are important compounds for 5-G and 6-G applications and for power electronic applications.

Summarising, Aluminium-scandium alloys, scandium oxides and fluorides can play a major role in our energy transition and innovative technologies to reduce CO_2 footprints.

Future European Production

Today, the EU imports 100%, mainly from China ($\sim 67\%$) as scandium oxide (Sc_2O_3). This lack of a reliable and sustainable supply chain for scandium prevents a broader use of Sc-applications. Therefore, a continuous supply of scandium at reasonable prices must be ensured in and for Europe, furthermore, the dependency on China must be reduced.

Europe can become a leader in the growing market applications covering the entire value chain when a continuous scandium supply is ensured. Europe's tradition in extractive metallurgy provides the know-how and experience of scandium production.

The lack of a primary scandium mine in Europe opens the door for industrial symbiosis processes for the winning of Scandium precursor from industrial residues like scandium containing waste acid from other industries (Titanium, Nickel and Aluminium). Europe is leading in the development of green technologies and has sufficient scandium resources. These resources indeed are presently locked in metallurgical residues. They are hosted in residual iron-chloride or sulphate solutions from titanium oxide (TiO_2) pigment production. Each year, about 1.5 million tons of TiO_2 are manufactured in Belgium, Germany, the Netherlands, Slovenia and the United Kingdom. The waste acids are neutralised and either landfilled at high costs and causing

environmental risks, or, in some plants, sold as a product to the chemical industry.

ScaVanger Production

The two EIT-Rawmaterials projects; ScaVanger (2021–2024) and Scaleup (2022–2024)* will bring Scandium compounds and a Scandium Aluminium alloy to the market. Scandium, Niobium and Vanadium will be extracted through hydrometallurgical processing "in-line" from acid residual solutions during metallurgical processing of the Titanium dioxide. ScaVanger is an industrial symbiosis (IS) project for the winning of Sc, Nb and V from industrial residues out of the Titaniumdioxide pigment industry. Scaleup is an IS project for the winning of Sc, Ga and V from red mud, which is industrial waste from Alumina production.

The ScaVanger's ecological and economical processing includes acid and water recycling, as well as metal recoveries during the cleaning process. This industrial symbiosis process converts residue to valuable products—economically and ecologically-friendly. ScaVanger's business model is focussed on Scandium hydroxide first. ScaVanger intends to produce a Scandium precursor in the form of a Scandium hydroxide filtercake. This feedstock will be converted into several Scandium compounds to serve different Sc markets.

1. Scandium hydroxide.
2. Scandium oxide and scandium fluoride corresponding to about 45 tons per year equivalent to scandium oxide for the European market.
3. AlSc alloys in mass production to supply to specialized Al-Sc-Mg producers.

Upscaling of metallurgical processing to preindustrial production will be performed for 3 years from both resources: iron-rich chloride solutions (TiO_2 production in the EU) and residue (alumina production at MYTILINEOS S.A, Greece). This scandium production will start up-ramping in 2025. The concepts can be adapted and exported to other TiO_2 and Al processing plants in and outside the EU (Fig. 1).

Conclusion

Sc-compounds (Aluminium-scandium alloys, scandium-hydroxide, -oxide and -fluoride) production in Europe can play a major role in the energy transition and development of innovative technologies to reduce CO_2 footprints.

European Scandium
For a Lighter and Greener Future

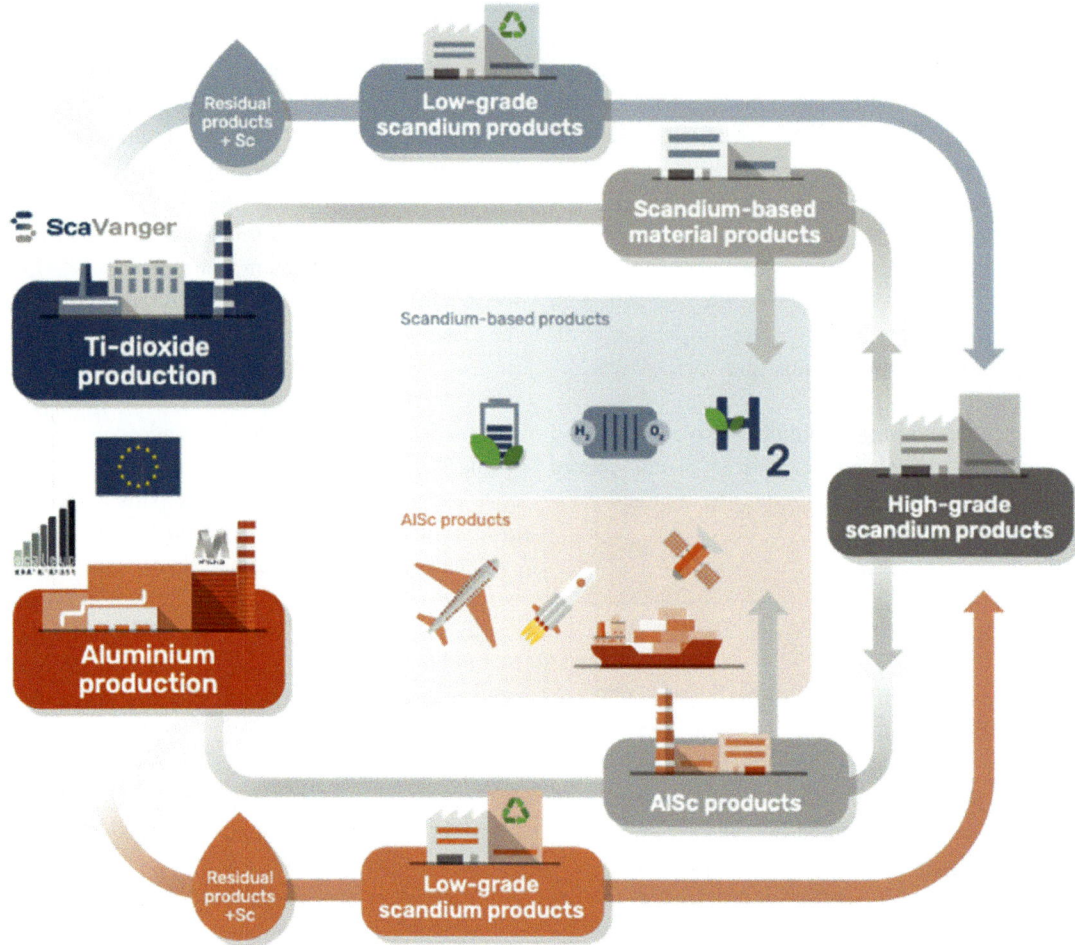

Fig. 1 Schematic of The ScaVanger and Scaleup projects, which are co-funded by EIT Raw Materials (https://eitrawmaterials.eu/). Contact: Beate Orberger, Catura Geoprojects: beate.orberger@catura.eu

Acknowledgements The Scavanger and Scaleup projects are co-funded by EIT Raw Materials GmbH.

References

1. `Scandium.` Los Alamos National Laboratory. https://periodic.lanl.gov/21.shtml
2. The Scandium content of some Norwegian Minerals. https://foreninger.uio.no/ngf/ngt/pdfs/NGT_41_2-4_197-210.pdf
3. The mineralogy of Scandium-Mindat.org. https://www.mindat.org/element/Scandium
4. Bibliography of the Geology and Mineralogy of the Rare Earths and Scandium to 1971, Geological Survey Bulletin 1366 https://pubs.usgs.gov/bul/1366/report.pdf
5. Occurence of Scandium and Rare Earth elemenyts in Jamaican bauxite waste. Willard Pinnock. https://www.researchgate.net/publication/247862176
6. "White Paper – SCANDIUM; A review of the element, its characteristics, and current and emerging commercial applications W. P. C. Duyvesteyn and G. F. Putnam, EMC Metals Corporation (TSX: EMC.TO)," May 2014. https://scandiummining.com/site/assets/files/5740/scandium-white-paperemc-website-june-2014-.pdf
7. Scandium: Ore deposits, the pivotal of magmatic enrichment and future exploration https://www.sciencedirect.com/science/article/abs/pii/S016913682031091X
8. Solvent extraction behaviour of scandium from lateritic nickel-cobalt ores using different organic reagents. Serif Kaya, Ece Ferizoglu, Yavuz A. Topkaya. ppmp1855.pdf (pwr.wroc.pl)

Formation of Al$_3$Sc Dispersoids and Associated Strengthening

Thomas Dorin, Lu Jiang, and Timothy Langan

Abstract

The exceptional strengthening potential of scandium additions to aluminum was first discovered in the 1960s. The impact that scandium additions have on the alloys' properties varies with parameters of the thermo-mechanical process and alloy composition. The benefits from scandium come from the various influences it has on the alloys' microstructure which can be summarized as four main effects, grain refinement, increased recrystallisation temperature, dispersoid strengthening and nucleation of strengthening phase. The strength improvement from Sc additions is mainly provided by the formation of fine dispersions of Al$_3$Sc L1$_2$ dispersoids. The level of strengthening depends on the size and number density of these dispersoids. Temperatures in the range of 250 to 350 °C are usually preferred to nucleate a high number of small Al$_3$Sc. The use of higher temperatures will most commonly generate coarser particles with less strengthening potential. This paper reviews the heat treatment strategies to generate fine dispersions of dispersoids, the origins of dispersoid strengthening, the impact of strengthening in different alloy series and finally the effect of adding Sc in combination with Zr.

Keywords

Al$_3$Sc • Strengthening • Scandium

T. Dorin (✉) · L. Jiang
Institute for Frontier Materials, Deakin University, Geelong, VIC 3216, Australia
e-mail: thomas.dorin@deakin.edu.au

T. Langan
Sunrise Energy Metals, Melbourne, VIC 3000, Australia

Introduction

Aluminium in its pure form has an inherently low strength. Additions of alloying elements combined with specific thermo-mechanical treatments are efficient ways to enhance the strength of these alloys. The highest strength aluminum alloys are from the "age-hardenable" category and require a high temperature solution treatment (i.e. 500–600 °C) and water quench followed by a low temperature ageing treatment (<200 °C) [1]. The addition of slow diffusing transition elements, such as Sc, usually stays supersaturated upon solidification when added below their maximum solubility. Dispersoids can then be formed at relatively high temperatures (compared to traditional ageing treatment) usually in the range of 250–450 °C. These dispersoids have an L1$_2$ structure and an Al$_3$M (where M is a transition or rare earth element) stoichiometry.

The formation of trialuminide compounds of the type Al$_3$M are particularly attractive as they lead to significant strength additions with minimal impact on other properties such as corrosion performance and electrical conductivity. Other attractive characteristics of Al$_3$M particles are: low density (their composition is 75at.% Al), high specific strength, excellent thermal stability and oxidation resistance. Elements that can form an Al$_3$M structure are Sc, Y, Ti, Zr, Hf, V, Nb, Ta, Er, Tm, Yb, and Lu [2]. Among all these elements, Sc has attracted particular attention as it can achieve exceptional strengthening for relatively low additions. Zr is commonly used in combination with Sc as it leads to the formation of highly thermally stable dispersoids with a core–shell morphology. The core–shell structure is a consequence of the different diffusivities of Sc and Zr which lead to a core enriched in Sc and a shell enriched in Zr. In reality, the core–shell particles have an Al$_3$(Sc$_x$,Zr$_{1-x}$) composition with x going from 1 to 0 from the center of the particle to the interface with the Al matrix [3].

S. Broek (ed.), *Light Metals 2023*, The Minerals, Metals & Materials Series,
https://doi.org/10.1007/978-3-031-22532-1_162

This paper will give a brief overview of the origins of strengthening from Al$_3$Sc dispersoids and discuss heat treatment strategies that yield optimal strengthening.

Heat Treatment Strategies

The Al$_3$Sc dispersoids form at an intermediate temperature which is not accommodated in traditional heat treatment cycles employed for wrought aluminum products. Indeed, conventional homogenization and solution treatment temperature are usually higher, in the range of 450–600 °C and artificial ageing treatments are usually below 200 °C. This suggests that the thermal cycles of traditional alloy series have to be revamped when Sc is added. As mentioned earlier the Al$_3$Sc have to be formed during a heat treatment at a temperature in the range of 250–350 °C [4].

The two common strategies to form Al$_3$Sc are:

1. Add an additional homogenization step at the very start of the thermo-mechanical schedule.
2. Increase the final ageing temperature.

Solution 2 is often not practical for age-hardenable alloys as it will result in over-ageing of the microstructure. Indeed the final ageing temperature would have to be raised significantly above 250 °C to activate the formation of Al$_3$Sc. This would result in the formation of coarse precipitates. This strategy has been used in Al-Cu-Sc alloys with a final ageing at 250 °C for 8 h [5]. In that case, Sc was shown to segregate at the interfaces between the θ' precipitates and the Al matrix restricting their coarsening. The authors of [5] reported a strength increase of ∼60 MPa from 0.3wt.% Sc addition. The interfacial segregation has also the advantage of stabilizing the θ' precipitates when using these alloys at elevated temperatures [6].

If solution 1 is selected, there then needs to be a strategy in place to ensure that the dispersoids will remain stable throughout the rest of thermo-mechanical treatment. The most common solution is the use of Zr which can form an Al$_3$Zr shell at the Al$_3$Sc/Al interface. The thermal stability of Al$_3$Zr prevents the dispersoids to coarsen or dissolve so that they can survive the higher temperature heat treatment such as the solution treatment.

In the case of Sc-containing age-hardenable wrought alloys, the core–shell dispersoids will have to withstand the solution treatment which is commonly used after the forming process and before the final artificial ageing stage. An example of a tailored thermo-mechanical treatment of an Sc and Zr containing alloy is displayed in Fig. 1. The homogenization often consists of three steps (the two first step for the core–shell dispersoids and the third higher temperature step to homogenize the microstructure. Hot forming is then conducted and results in the formation of a strong grain texture. The presence of core–shell dispersoids is very efficient at grain boundary pinning which can prevent recrystallisation during the solution heat treatment. Finally, a stress relief test can be used as dislocations sometimes act as nucleation sites for the main strengthening phase and artificial ageing is conducted. Figure 1 also presents a schematic of the evolution of the microstructure during the full treatment.

The use of a combination of Sc and Zr requires further tailoring of the homogenization treatment to form an optimal distribution of fine and numerous core–shell Al$_3$Sc/Al$_3$Zr dispersoids. This can be done by either using a 2-step treatment where the core will mostly form in the first step and the shell can then form in the second step. The use of a temperature ramp can also be used to reach the same effect. The core–shell morphology is a direct result of the significantly different diffusivities of these species. Ideally, a fine dispersion of Al$_3$Sc would nucleate in the first step and the second step activates the diffusion of Zr to form a shell around the Al$_3$Sc before they start to dissolve and coarsen. Figure 2 shows an example of fine core–shell dispersoids formed during a tailored two-step homogenization such as

Fig. 1 Example of thermos-mechanical treatment tailored for the case of a wrought aluminum alloy containing Sc. A schematic of the evolution of the microstructure is also provided

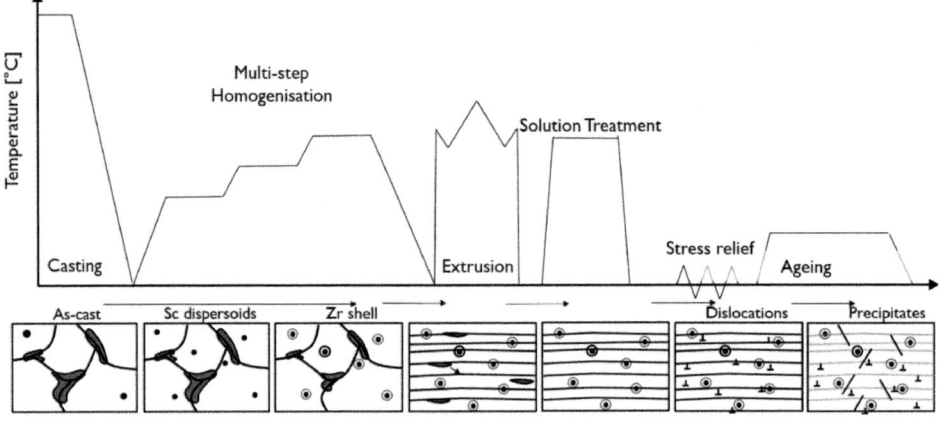

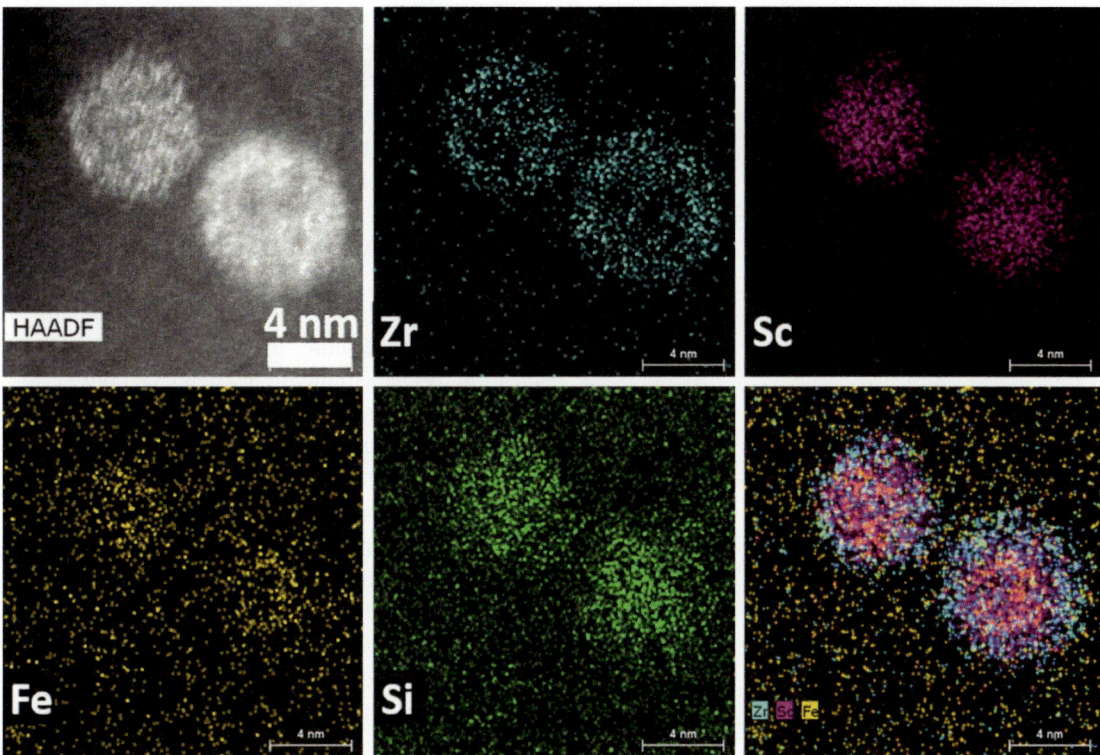

Fig. 2 HAADF-STEM and EDX of core–shell dispersoids particles. The presence of a predominantly Al₃Sc core and Al₃Zr shell is clearly depicted. The core–shell is clearly visible is highlighted in the composite image at the bottom right. The presence of Si and Fe in the core can also clearly be identified

the one shown in Fig. 1. The Al₃Sc-core and Al₃Zr-shell morphology is clearly depicted. The presence of Si and Fe inside the core can also be seen. Si impurities have been shown to have a significant effect on the formation kinetics of the dispersoids [7]. The effect of Fe impurities on the formation kinetics has not been quantified but is very important as Fe is present as an impurity or alloying element in all aluminum alloys. Some recent work revealed that some anisotropic Fe-rich plate like precipitate can form in the presence of Sc [8].

Origins of Al₃Sc Strengthening

Overall, Scandium is often regarded as the element from the periodic table with the highest strengthening potential when added to aluminum. One of the most studied precipitation system in physical metallurgy is Al-Cu and the precipitation of the θ' precipitates. Cu additions would typically result in around 5 to 10 MPa strength improvement per 0.1 wt.% Cu addition, an order of magnitude lower than for Sc additions. The main difference is that the maximum solubility of Cu is much larger than that of Sc, allowing for larger amounts of

Cu (up to 5–6 wt.%) to be added. This in turn results in alloys with superior strength. Comparatively the maximum solubility of Sc is 0.33 wt.% which gives less freedom to add large amounts of Sc and limits the maximum strength achievable. However, one of the key benefits of Sc is the potential of developing very lean Al alloys with exceptional combination of properties, such as strength, conductivity, corrosion resistance and formability. The increment in yield strength is plotted as a function of Sc content in Fig. 3. The average increment experienced when adding Sc below its maximum solubility is ~50 MPa increase per 0.1 wt.%Sc added. This strengthening value varies depending on the alloy system with the maximum strengthening obtained in binary Al-Sc systems. This strengthening has been shown to be affected in the presence of other alloying elements such as Si or Cu. This is due to the formation of alternative Sc-rich phases such AlSiSc or AlCuSc that do not provide strengthening and deplete the Sc solute. In 6xxx-series alloys, some literature even reports negative effects of Sc additions. Recent studies have shown, that the use of specifically tailored heat treatments (such as the one in Fig. 1) allows for a positive effect of Sc in both Al-Cu [6, 9–11] and Al-Mg-Si [12–18] alloys.

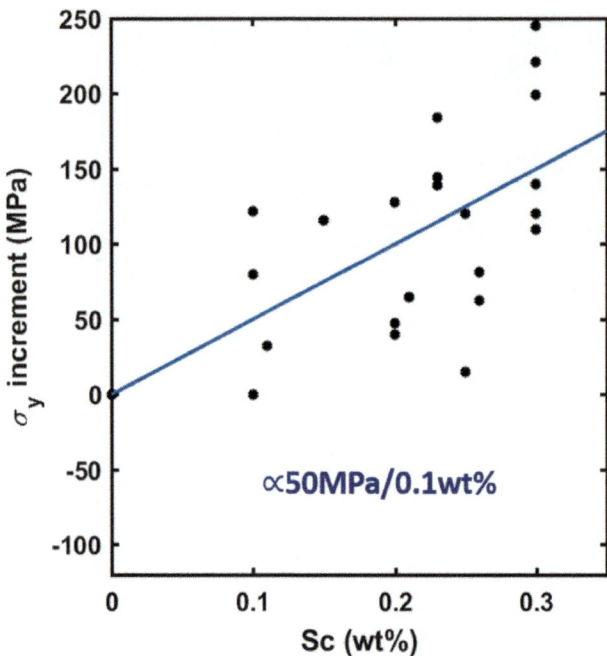

Fig. 3 Yield strength increment plotted as a function of Sc content. The data used to plot the graph was extracted from [4, 19–21]. The average increment is 50 MPa per 0.1 wt.%Sc addition. This increment varies greatly depending on the alloy type and heat treatment applied as can be seen from the scattered data in the graph showing up to ~150 MPa increase for 0.1 wt.%Sc addition down to 0 MPa increment in some cases

and number density of these dispersoids. Temperatures in the range of 250–350 °C are usually preferred to nucleate a high number of small Al_3Sc. The use of higher temperatures will most commonly generate coarser particles with less strengthening potential.

The increment of yield strength and corresponding dislocation mechanisms related to Al_3Sc precipitates was systematically investigated by Seidman et al. [22]. When deforming at room temperature, the dislocation/precipitate interaction mechanisms (precipitate shearing or precipitate by-passing) determine the level of strengthening. The shearing mechanism involves chemical, coherency, modulus mismatch, and anti-phase boundary (APB) strengthening. For larger precipitates, a moving dislocation will bow all the way around the precipitates and leave a dislocation loop [23]. Seidman et al. systematically generated microstructures with different precipitate sizes at approximately constant volume fractions in an Al-0.3wt.%Sc alloy. Maximum strength increment is found to occur for a precipitate radius of ~1.5–2 nm. The transition between shearing and by-passing was confirmed through analytical models and a transition radius of 2.1 nm was predicted, confirming the experimental results. The maximum strengthening achieved in Seidman's study was about 500 MPa (HV) which translates to a yield strength increment of ~167 MPa for 0.3wt. % Sc addition.

Nucleation of Other Strengthening Phases

The presence of second phase particles in an aluminum matrix is known to influence the breakdown of solid solutions. Various mechanisms have been proposed to explain the role of the interface in modifying nucleation, including vacancies sinks, solute segregation at precipitate interfaces and reduced surface energy for nucleation. Sc and Al_3Sc dispersoids have been reported to influence precipitation mechanisms in age-hardenable alloy systems with most of the reports found for additions of Sc in the 2xxx and 6xxx-series of aluminum alloys.

In the case of 2xxx-series Al alloys, the addition of Sc has been reported to have a negative [24] or a positive [6, 9–11] impact. These different results come from the different thermo-mechanical treatments used in the respective studies. The use of a traditional heat treatment for 2xxx-series alloys results in the formation of the W-AlCuSc phase depleting the solid solution from both Cu and Sc and hence resulting in a negative impact on the mechanical properties. Contrastingly, the use of a tailored heat treat (such as the one described in Fig. 1) results in a positive impact from Sc additions. In that case, the use of a low temperature first step heat treatment during homogenisation results in the formation of Al_3Sc and W-AlCuSc could be kept to a minimum [25]. The formation

The sources of strengthening from scandium additions are the following:

- Solid solution strengthening
- Precipitation strengthening
- Grain refinement
- Work hardening from piling up of dislocations
- Nucleation of other strengthening phases.

The two most important contributors to the yield strength of an Sc containing Al alloy are the strengthening from precipitation and from nucleation of other strengthening phases. As a result, these two main contributors will be reviewed in greater detail.

Precipitation Strengthening

The exceptional strengthening potential of scandium additions to aluminum was first discovered in the 1960s. A majority of the early work came out from Russia and is summarized in a comprehensive book by Toropova et al. [4]. The strength improvement from Sc additions is mainly provided by the formation of fine dispersions of Al_3Sc $L1_2$ dispersoids. The level of strengthening depends on the size

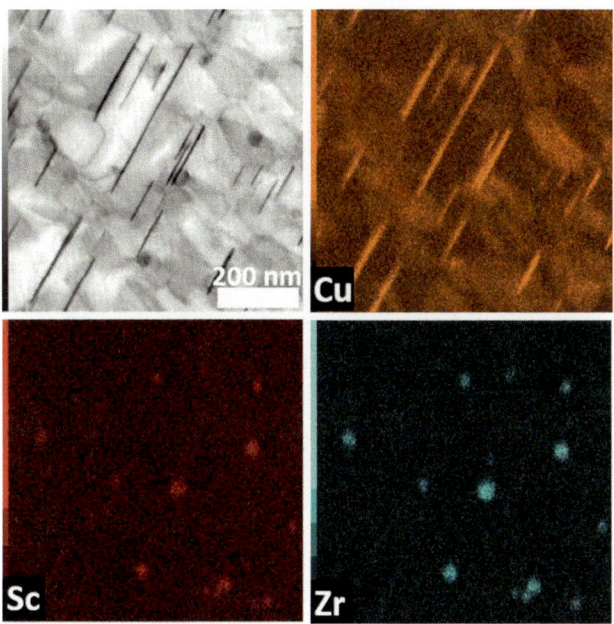

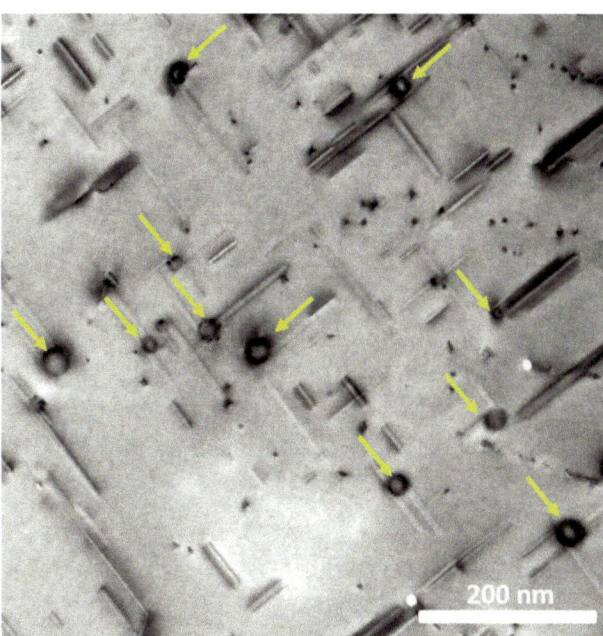

Fig. 4 Bright field TEM of a peak aged Al-4wt.%Cu-0.1wt.% Sc-0.1wt.%Zr which was prepared following a thermo-mechanical treatment akin to the one in Fig. 1. The EDX reveals the presence of the core–shell dispersoids which are systematically associated with the Cu-rich θ' precipitates

Fig. 5 Bright field TEM of a peak aged Al-0.5wt.%Mg-0.5wt.% Si-0.05wt.%Sc-0.1wt.%Zr which was prepared following a thermo-mechanical treatment akin to the one on Fig. 1. The yellow arrows point to associations between the spherical Al₃(Sc,Zr) dispersoids and β" precipitates

of an Al₃Zr shell was then efficient to entrap the Sc to prevent the formation of the W-AlCuSc during further heat treatments. The dispersoids were then shown to act as preferential nucleation sites for the main strengthening θ' phase resulting in a significant refinement of these precipitates. This refinement was shown to contribute significant strength to the resulting alloys. This association is evident in Fig. 4 which shows the microstructure of an Al-4wt.% Cu-0.1wt.%Sc-0.1wt.%Zr.

Similarly, in the 6xxx-series Al-Mg-Si alloys, there has been reports of negative [26–29] or positive [12–18] impact from Sc additions. The negative impact was again due to the formation of a ternary AlSiSc phase (referred as τ or V-phase [30]) that could be avoided by using a tailored multi-step homogenization treatment. In the right heat treatment conditions, Sc and Zr have been shown to result in significant strength increment in Al-Mg-Si alloys. The rod-like β"-precipitates have been shown to be systematically associated with Al₃Sc precipitates in atom probe tomography results [31]. Systematic TEM characterisations have confirmed these associations between dispersoids and β"-precipitates [32]. This nucleation effect was found to result in a refinement of the β" precipitates and hence a higher strengthening from the rod precipitates. The β" precipitates were also reported to have greater thermal stability in the presence of Sc [29] (Fig. 5).

Conclusions

In this short paper, we discuss the importance of tailoring the thermo-mechanical heat treatment when using Sc in wrought aluminium alloys. The Al₃Sc dispersoids provide significant strengthening from dispersion strengthening. It is often important to use Zr in combination with Sc to form Al₃(Sc, Zr) dispersoids with high thermal stability. This is particularly important in alloys that require the use of high temperature solution heat treatment so that the dispersoids can remain stable.

The presence of fine Al₃(Scₓ,Zr₁₋ₓ) particles in the microstructure has recently been reported to act as nucleation sites for other precipitates in 2xxx-series and 6xxx-series alloys. This nucleation effect was shown to result in significant strength contributions. Understanding the origins of strengthening from Al₃(Scₓ,Zr₁₋ₓ) particles is key to design suitable heat treatments to maximise the benefits from Sc additions when used in different alloy classes.

Acknowledgements Dr Thomas Dorin is the recipient of a Discovery Early Career Award (project number DE190100614) funded by the Australian Research Council. University of Manchester's Electron Microscopy centre is acknowledged for use of FEI Titan G2 80-200 scanning transmission electron microscope. Deakin University's Advanced Characterisation Facility is acknowledged for use of the TEM-FEG JEOL 2100F.

References

1. I. Polmear, D. StJohn, J.-F. Nie, and M. Qian, *Light alloys: metallurgy of the light metals.* Butterworth-Heinemann, 2017.

2. T. Dorin, M. Ramajayam, A. Vahid, and T. Langan, "Aluminium Scandium Alloys," in *Fundamentals of Aluminium Metallurgy*, Elsevier, 2018, pp. 439–494. https://doi.org/10.1016/B978-0-08-102063-0.00012-6.

3. T. Dorin *et al.*, "Stability and stoichiometry of L12 Al3(Sc,Zr) dispersoids in Al-(Si)-Sc-Zr alloys," *Acta Materialia*, vol. 216, p. 117117, Sep. 2021, https://doi.org/10.1016/j.actamat.2021.117117.

4. L. S. Toropova, D. G. Eskin, and M. L. Kharakterova, Eds., *Advanced aluminum alloys containing scandium: structure and properties.* Amsterdam: Gordon & Breach, 1998.

5. C. Yang *et al.*, "The influence of Sc solute partitioning on the microalloying effect and mechanical properties of Al-Cu alloys with minor Sc addition," *Acta Materialia*, vol. 119, pp. 68–79, 2016, https://doi.org/10.1016/j.actamat.2016.08.013.

6. L. Jiang, B. Rouxel, T. Langan, and T. Dorin, "Coupled segregation mechanisms of Sc, Zr and Mn at θ′ interfaces enhances the strength and thermal stability of Al-Cu alloys," *Acta Materialia*, vol. 206, p. 116634, Mar. 2021, https://doi.org/10.1016/j.actamat.2021.116634.

7. C. Booth-Morrison, Z. Mao, M. Diaz, D. C. Dunand, C. Wolverton, and D. N. Seidman, "Role of silicon in accelerating the nucleation of Al3(Sc,Zr) precipitates in dilute Al–Sc–Zr alloys," *Acta Mat.*, vol. 60, no. 12, pp. 4740–4752, 2012.

8. T. Dorin, S. Babaniaris, L. Jiang, A. Cassel, A. Eggeman, and J. D. Robson, "Precipitation Sequence in Al-Sc-Zr alloys revisited," *Materialia*, vol. 26, p. 101608, 2022, https://doi.org/10.1016/j.mtla.2022.101608.

9. T. Dorin, M. Ramajayam, and T. J. Langan, "Impact of Scandium and Zirconium on extrudability, microstructure and hardness of a binary Al-Cu alloy," *Materials Today: Proceedings*, In Press 2017.

10. B. Rouxel, M. Ramajayam, T. J. Langan, J. Lamb, P. G. Sanders, and T. Dorin, "Effect of dislocations, Al3(Sc,Zr) distribution and ageing temperature on θ′ precipitation in Al-Cu-(Sc)-(Zr) alloys," *Materialia*, vol. 9, p. 100610, Mar. 2020, https://doi.org/10.1016/j.mtla.2020.100610.

11. B. A. Chen *et al.*, "Effect of solution treatment on precipitation behaviors and age hardening response of Al–Cu alloys with Sc addition," *Materials Science and Engineering: A*, vol. 530, pp. 607–617, 2011.

12. L. Lityńska-Dobrzyńska, "Effect of heat treatment on the sequence of phases formation in Al-Mg-Si alloy with Sc and Zr additions," *Archives of Metallurgy and Materials*, vol. 51, no. 4, pp. 555–560, 2006.

13. L. Lityńska-Dobrzyńska, "Precipitation of phases in Al-Mg-Si-Cu alloy with Sc and Zr additions during heat treatment," in *Solid State Phenomena*, 2007, vol. 130, pp. 163–166.

14. L. L. Rokhlin, N. R. Bochvar, and I. E. Tarytina, "Joint effect of scandium and zirconium on the structure and the strength properties of Al-Mg₂Si–Based alloys," *Russian Metallurgy (Metally)*, vol. 2015, no. 9, pp. 726–731, 2015.

15. T. Dorin, M. Ramajayam, S. Babaniaris, L. Jiang, and T. J. Langan, "Precipitation sequence in Al–Mg–Si–Sc–Zr alloys during isochronal aging," *Materialia*, vol. 8, p. 100437, 2019.

16. S. Babaniaris, M. Ramajayam, L. Jiang, T. Langan, and T. Dorin, "Tailored precipitation route for the effective utilisation of Sc and Zr in an Al-Mg-Si alloy," *Materialia*, p. 100656, 2020.

17. S. Babaniaris, M. Ramajayam, L. Jiang, T. Langan, and T. Dorin, "Developing an Optimized Homogenization Process for Sc and Zr Containing Al-Mg-Si Alloys," in *Light Metals 2019*, 2019, pp. 1445–1453.

18. T. Dorin, M. Ramajayam, and T. J. Langan, "Effects of Mg, Si, and Cu on the formation of the Al3Sc/Al3Zr dispersoids," in *International Conference on Aluminium Alloys*, 2018, vol. 16.

19. J. Røyset and N. Ryum, "Scandium in aluminium alloys," *International Materials Reviews*, vol. 50, no. 1, pp. 19–44, Feb. 2005, https://doi.org/10.1179/174328005X14311.

20. V. V. Zakharov, "Effect of scandium on the structure and properties of aluminum alloys," *Metal Science and Heat Treatment*, vol. 45, no. 7–8, pp. 246–253, 2003.

21. Z. Ahmad, "The properties and application of scandium-reinforced aluminum," *Jom*, vol. 55, no. 2, pp. 35–39, 2003.

22. D. N. Seidman, E. A. Marquis, and D. C. Dunand, "Precipitation strengthening at ambient and elevated temperatures of heat-treatable Al(Sc) alloys," *Acta Materialia*, vol. 50, no. 16, pp. 4021–4035, Sep. 2002, https://doi.org/10.1016/S1359-6454(02)00201-X.

23. E. Nembach, *Particle strengthening of metals and alloys.* New York, NY: John Wiley and Sons, 1997.

24. V. V. Zakharov and T. D. Rostova, "On the possibility of scandium alloying of copper-containing aluminum alloys," *Metal Science and Heat Treatment*, vol. 37, no. 2, pp. 65–69, Feb. 1995, https://doi.org/10.1007/BF01157047.

25. S. K. Kairy *et al.*, "Simultaneous improvement in corrosion resistance and hardness of a model 2xxx series Al-Cu alloy with the microstructural variation caused by Sc and Zr additions," *Corrosion Science*, vol. 158, p. 108095, Sep. 2019, https://doi.org/10.1016/j.corsci.2019.108095.

26. J. Royset, U. Tundal, S. R. Skjervold, G. Waterloo, C. Braathen, and O. Reiso, "An Investigation of Sc Addition on the Extrudability, Recrystallisation Resistance and Mechanical Properties of a 6082- and a 7108-Alloy," in *Materials Forum*, 2004, vol. 28, pp. 246–251.

27. M. Vlach, B. Smola, I. Stulíková, and V. Očenášek, "Microstructure and mechanical properties of the AA6082 aluminium alloy with small additions of Sc and Zr," *International Journal of Materials Research*, vol. 100, no. 3, pp. 420–423, 2009.

28. T. Nakamura, T. Matsuo, M. Ikeda, and S. Komatsu, "Effect of Scandium Addition on Aging Behavior of 6061 Aluminum Alloy," in *Advanced Materials Research*, 2007, vol. 15, pp. 7–12.

29. E. P. Kwon, K. D. Woo, S. H. Kim, D. S. Kang, K. J. Lee, and J. Y. Jeon, "The effect of an addition of Sc and Zr on the precipitation behavior of AA6061 alloy," *Met. Mater. Int.*, vol. 16, no. 5, pp. 701–707, Oct. 2010, https://doi.org/10.1007/s12540-010-1002-y.

30. J. Dumbre *et al.*, "Understanding the formation of (Al, Si) 3Sc and V-phase (AlSc2Si2) in Al-Si-Sc alloys via ex situ heat treatments and in situ transmission electron microscopy studies," *Journal of Alloys and Compounds*, p. 158511, 2020.

31. T. Dorin, M. Ramajayam, S. Babaniaris, and T. J. Langan, "Micro-segregation and precipitates in as-solidified Al-Sc-Zr-(Mg)-(Si)-(Cu) alloys," *Materials Characterization*, vol. 154, pp. 353–362, Aug. 2019, https://doi.org/10.1016/j.matchar.2019.06.021.

32. S. Babaniaris, M. Ramajayam, L. Jiang, T. Langan, and T. Dorin, "Tailored precipitation route for the effective utilisation of Sc and Zr in an Al-Mg-Si alloy," *Materialia*, vol. 10, p. 100656, May 2020, https://doi.org/10.1016/j.mtla.2020.100656.

Use of Sc to Improve the Properties of AA5083 Cast and Rolled Products

Paul Rometsch, Jerome Fourmann, Emad Elgallad, and X.-Grant Chen

Abstract

The properties of 5xxx aluminum alloys can be improved with small additions of Sc. When Sc and Zr are added to 5xxx alloys, the alloys become heat-treatable as $Al_3(Sc, Zr)$ nanoprecipitates form at 300–400 °C. However, the heat treatment and thermo-mechanical processing (TMP) need to be adapted to maximize the value of the Sc addition. In this work, AA5083 slabs were DC cast with and without minor additions of Sc and/or Zr. The room/elevated temperature mechanical properties of the cast materials were evaluated after various annealing and homogenization treatments. Samples were then hot and cold rolled with different TMP treatments. The microstructure, room temperature mechanical properties and corrosion performance were evaluated for selected tempers. It was evident that the properties could be improved significantly with small additions of Sc, but that this depends very much on the amount of Sc and on the processing parameters.

Keywords

Aluminum alloy AA5083 • Scandium • Rolling • Heat treatment • Mechanical properties

P. Rometsch (✉)
Arvida Research and Development Center, Rio Tinto Aluminium, Saguenay, QC G7S 4K8, Canada
e-mail: paul.rometsch@riotinto.com

J. Fourmann
Technical Marketing, Rio Tinto Aluminium, Chicago, IL 60601-7329, USA
e-mail: jerome.fourmann@riotinto.com

E. Elgallad · X.-G. Chen
Department of Applied Science, University of Quebec at Chicoutimi, Saguenay, QC G7H 2B1, Canada
e-mail: Emad_Elgallad@uqac.ca

X.-G. Chen
e-mail: XGrant_Chen@uqac.ca

Introduction

Among the wrought aluminum alloys, scandium (Sc) is known to work particularly well in the Al–Mg system (usually together with Zr), and there is already a wealth of information about how Sc behaves in the 5xxx series alloys compared to other alloy systems [1–7]. It has been reported that Sc can increase the yield strength of aluminum alloys by an average of about 50 MPa per 0.1 wt% Sc addition [1]. However, this increment in yield strength can vary significantly depending on the alloy, the Sc content and how the material has been processed.

Since the work hardening 5xxx alloys additionally become precipitation hardening when Sc is added, it is evident that the properties can vary widely depending on how the material is rolled and heat treated. For example, recent work has shown that the $Al_3(Sc,Zr)$ hardening precipitates in AA5083 coarsen during hot rolling at 500 °C and has suggested that a lower hot rolling temperature of 400–425 °C could result in a higher strength [8, 9]. Furthermore, $Al_3(Sc,Zr)$ precipitates can inhibit recrystallization and cause increased precipitation hardening at typical annealing temperatures, thereby changing the properties and perhaps even the meaning of a standard O temper. There is also evidence to indicate that for AA5083, a greater increment in yield strength can be achieved with the first 0.1 wt% Sc addition than with each further addition of 0.1 wt% Sc [8–10]. Considering the cost of Sc, it is therefore important to understand how the properties of 5xxx alloys can be maximised with a minimal Sc addition.

This paper reports on a study evaluating the effects of 0.05 and 0.10 wt% Sc additions on the microstructure and property evolution of AA5083 in different heat treatment and thermo-mechanical processing (TMP) conditions. The first part of the work explored the effects of Sc and heat treatment on the room temperature and elevated temperature mechanical properties of cast slabs. The second part of the work

S. Broek (ed.), *Light Metals 2023*, The Minerals, Metals & Materials Series,
https://doi.org/10.1007/978-3-031-22532-1_163

explored the effects of Sc and TMP on the microstructure and property evolution during hot and cold rolling in different tempers.

Materials and Methods

Four slabs were DC cast on the casting simulator at the Rio Tinto Arvida Research and Development Centre. Small amounts of Zr and Sc were added to an AA5083 base alloy as shown in Table 1. The concentrations of all the other alloying elements were kept at a constant level for all four alloys so that the relative effects of the Zr and Sc additions could be isolated. The slabs were approximately 590 mm wide $\times$ 185 mm thick $\times$ 950 mm long. Test samples were cut from between the slab center and surface after scrapping some material at surface, head, and butt locations.

For *the first part of the work*, as-cast test samples were exposed to the following heat treatments:

- 2 h at 300 °C
- 16 h at 300 °C
- 16 h at 350 °C
- 12 h at 400 °C
- 16 h at 300 °C +6 h at 425 °C.

In each case, the heating rate was 50 °C/h and the cooling rate was 100 °C/h. Some of the samples soaked for 12 h at 400 °C were subsequently subjected to a thermal exposure of 100 h at 200 or 300 °C to evaluate their thermal stability.

Room temperature tensile tests were conducted on an Intron machine according to ASTM E8 using full size rectangular samples with a gauge length of 50 mm. Elevated temperature tensile testing was conducted on a Gleeble 3800 machine at 200 and 300 °C both without and with prior thermal exposure at these temperatures for 100 h. Samples with a gauge length of 25 mm were heated to the test temperature at a heating rate of 2 °C/s and held for 3 min before testing with a strain rate fixed at 1×10^{-3} s^{-1}. The electrical conductivity was measured at room temperature using an eddy current device for the purpose of assessing the relative

effects on the thermal conductivity, which is linearly proportional to the electrical conductivity at a given temperature according to the Wiedemann–Franz law.

For *the second part of the work*, the same four alloys were subjected to a series of hot and cold rolling trials. Test pieces with dimensions of 190 mm $\times$ 48 mm $\times$ 16 mm were machined from the slabs after a fixed identical homogenization treatment had been applied to all. The pieces were then hot rolled from 16 to 4 mm thickness in multiple passes. Two separate trials were conducted to assess the effect of hot rolling temperature:

- Test 1: Pieces were repeatedly heated to and hot rolled at 475–500 °C,
- Test 2: Pieces were repeatedly heated to and hot rolled at 400–425 °C.

Cold rolling and additional processing were then applied as shown in Table 2. After this, the materials were evaluated using the following methods:

- Room temperature tensile testing using full size samples with a gauge length of 50 mm (ASTM E8)
- VDA 3-point bend testing to a 30 N load drop (VDA 238–100 Plate Bending Test for Metallic Materials: Verband der Automobilindustrie, www.vda.de, 2017-04)
- Electrical conductivity (eddy current device)
- Rockwell hardness (HRF)
- 10 days acidified salt spray corrosion test (ASTM G85-A3 SWAAT) for Test 1 samples
- 24 h HNO$_3$ intergranular corrosion test (ASTM G67 NAMLT) for Test 2 samples
- Metallography for corrosion evaluation, and general grain structure with Barker's etch.

Results and Discussion

This section is divided into two parts of the work that were described in the previous section.

Table 1 Alloy compositions based on AA5083, with actual OES analysis results shown for Zr and Sc (wt %)

Alloy	Si	Fe	Cu	Mn	Mg	Cr	Ti	Zr	Sc
#1 (Base AA5083)	≤ 0.40	≤ 0.40	≤ 0.10	0.40–1.0	4.0–4.9	0.05–0.25	≤ 0.15	–	–
#2 (Base + 0.08Zr)	≤ 0.40	≤ 0.40	≤ 0.10	0.40–1.0	4.0–4.9	0.05–0.25	≤ 0.15	**0.08**	–
#3 (Base + 0.08Zr + 0.05Sc)	≤ 0.40	≤ 0.40	≤ 0.10	0.40–1.0	4.0–4.9	0.05–0.25	≤ 0.15	**0.08**	**0.05**
#4 (Base + 0.08Zr + 0.10Sc)	≤ 0.40	≤ 0.40	≤ 0.10	0.40–1.0	4.0–4.9	0.05–0.25	≤ 0.15	**0.08**	**0.10**

Table 2 TMP treatments applied to all four alloys

Test No	Hot rolling	Cold rolling	Annealing	Cold rolling	Thermal stabilization	Final thickness	Final temper
1	16 to 4 mm at 475–500 °C	4 to 1 mm	–	–	–	1 mm	~H18
1	16 to 4 mm at 475–500 °C	4 to 1 mm	0.5 h at 400 °C	–	–	1 mm	~O
2	16 to 4 mm at 400–425 °C	4 to 1 mm	–	–	–	1 mm	~H18
2	16 to 4 mm at 400–425 °C	4 to 1 mm	0.5 h at 400 °C	–	–	1 mm	~O
2	16 to 4 mm at 400–425 °C	4 to 1 mm	0.5 h at 400 °C	1 to 0.85 mm	–	0.85 mm	~H116[*]
2	16 to 4 mm at 400–425 °C	4 to 1 mm	0.5 h at 400 °C	1 to 0.80 mm	1 h at 185 °C	0.80 mm	~H321

[*]This TMP treatment resulted in a higher strength than is usual for AA5083-H116

Effect of Sc on AA5083 Cast Slab

The room temperature tensile properties of the four cast alloys are shown for different heat treatment conditions in Fig. 1. It is evident that after heat treatment at 300–400 °C, a 0.10 wt% Sc addition (Alloy #4) increased the yield strength (YS) and ultimate tensile strength (UTS) over that of the base alloy (Alloy #1) by about 60 MPa and 10–37 MPa, respectively, while the elongation decreased as the strength increased.

On the other hand, a 0.08 wt% Zr addition on its own (Alloy #2) had a rather small effect on the strength, while a 0.05 wt% Sc addition (Alloy #3) increased the yield strength by up to 13–15 MPa (e.g. after 16 h at 350 °C and 12 h at 400 °C) but with some losses in UTS and elongation. In the as-cast condition, the alloys with Sc had similar properties as the base alloy.

It can be deduced from these results that most, if not all, of the Sc remained in solution during the DC casting and then precipitated out during heat treatment at 300–425 °C, thereby increasing the yield strength by about 60 MPa per 0.1 wt% Sc addition. The highest strengths (YS of 195 MPa and UTS of 287 MPa) were achieved for the alloy with 0.10 wt% Sc (Alloy #4) at the optimal heat treatment condition after 12 h at 400 °C. However, since heat treatments at 300, 350 and 425 °C also resulted in high yield strengths of 182–186 MPa, it is evident that the yield strength of this alloy is not very sensitive to heat treatment conditions over this range of temperatures.

The elevated temperature tensile testing results in Fig. 2 indicate that after a thermal exposure of 100 h at 200 °C, the addition of 0.10 wt% Sc increased the yield strength at 200 °C over the base alloy by 32% from 137 to 181 MPa. After a thermal exposure of 100 h at 300 °C, the addition of 0.10 wt% Sc increased the yield strength at 300 °C by 18% from 120 to 142 MPa. The addition of 0.05 wt% Sc also increased the yield strength, but to a lesser extent. Again, the effects of Sc on UTS were much smaller and the elongation mostly increased as the yield strength decreased.

It is important to note here that Sc improves not only the elevated temperature yield strength, but also the thermal stability at 200–300 °C. The tensile strengths measured at those elevated temperatures remained almost unchanged before and after the thermal exposure. If a precipitation hardening aluminum alloy such as a 7xxx series alloy had been exposed to 100 h at 200–300 °C, it would have over-aged and softened dramatically and irreversibly. By contrast, a Sc and Zr containing 5xxx alloy (e.g. Alloy #4) that was already precipitation hardened at a higher temperature of 300–425 °C would be able to endure very long thermal exposures at 200–300 °C without loss in room temperature strength, while also maintaining its elevated temperature strength. As a result, such alloys could be attractive for applications where machined aluminum tooling such as dies for the plastic/rubber molding industries need to withstand elevated temperature operating conditions for extended periods of time.

Furthermore, the thermal conductivity is expected to be negligibly impacted by Sc as the electrical conductivity after 12 h at 400 °C was measured to be 25.9 and 26.2% IACS for Alloy #4 and Alloy #1, respectively, or 26.0 and 26.4% IACS respectively after an additional 100 h at 300 °C. This also confirms that these alloys are relatively stable during the thermal exposure of 100 h at 300 °C.

Effect of Sc on Hot and Cold Rolled AA5083

Figure 3 shows a yield strength comparison for all four alloy variants processed to the H18 and O tempers with both high and low hot rolling temperatures as outlined in Table 2. It is evident that for the Sc-containing alloys, the low hot rolling temperature of 400–425 °C in Test 2 gave clearly higher yield strengths than the high hot rolling temperature of 475–500 °C in Test 1. By contrast, the alloys without Sc were much less sensitive to the hot rolling temperature, especially in the O temper. Furthermore, the results in Fig. 3 show that a 0.08 wt% Zr addition on its own (Alloy #2) has a rather

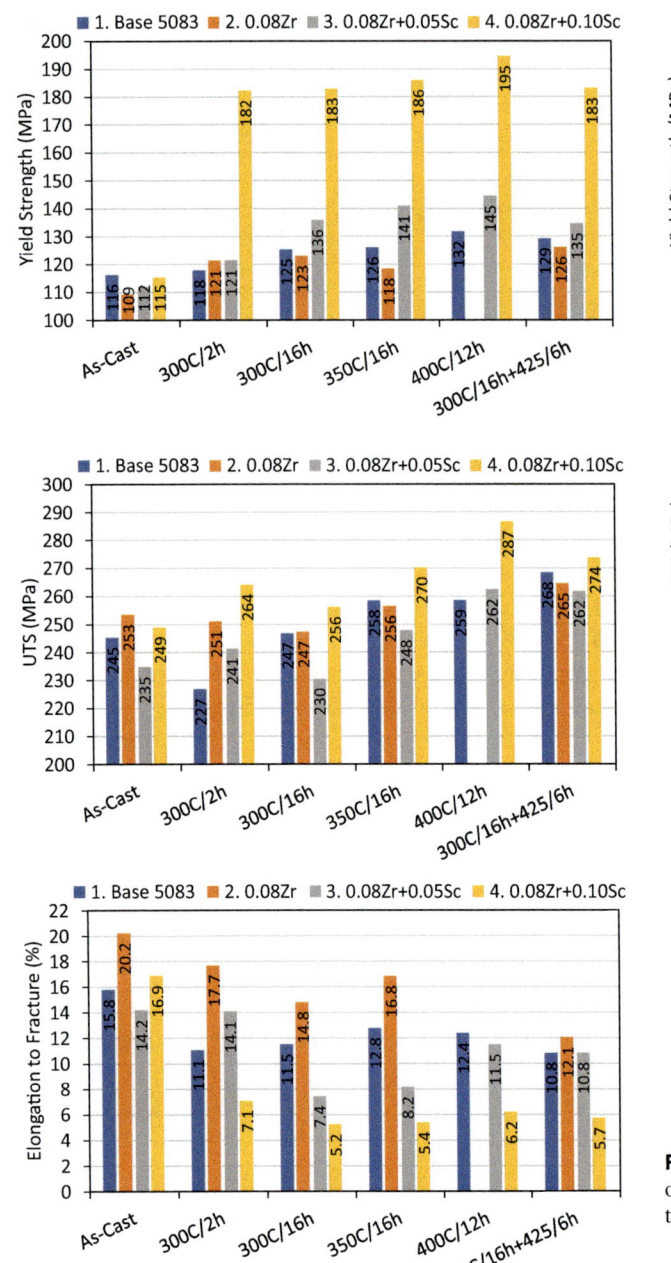

Fig. 1 Effects of Zr, Sc and heat treatment on room temperature tensile properties of cast AA5083

negligible effect, resulting in the same yield strength as the base AA5083 and an insensitivity to the hot rolling temperature. By contrast, the 0.05 wt% Sc addition (Alloy #3) started to increase the yield strength with the low hot rolling temperature in Test 2.

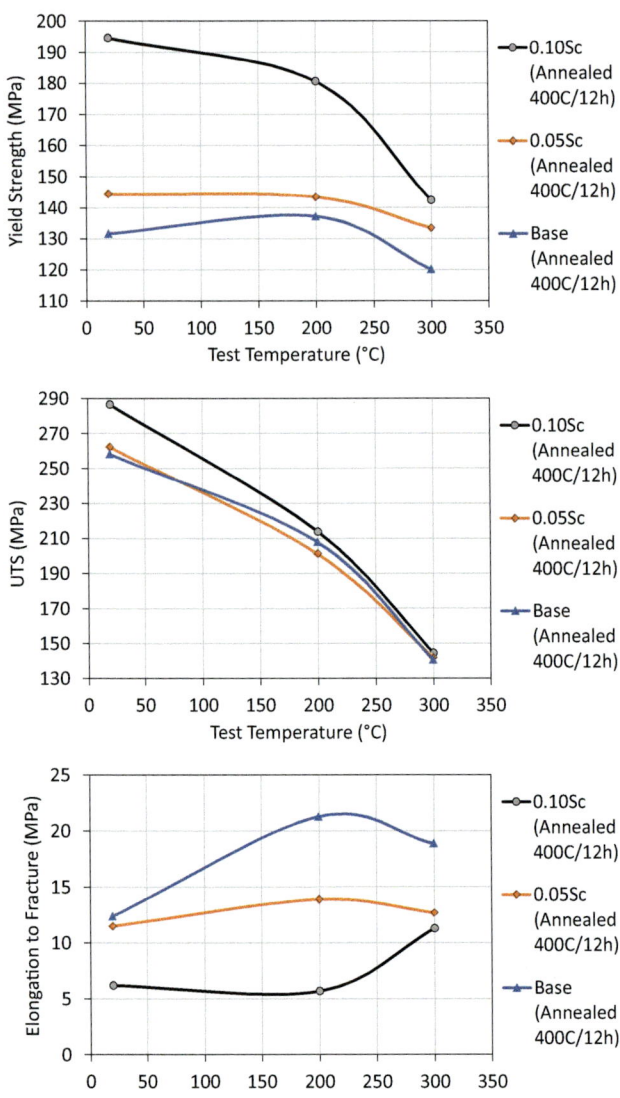

Fig. 2 Effect of Sc on room and elevated temperature tensile properties of cast and annealed AA5083 after 100 h of thermal exposure at the test temperature (for the samples tested at 200 and 300 °C)

When using the low hot rolling temperature, Alloy #4 with 0.10 wt% Sc clearly reached the highest yield strength out of all the alloys i.e. 450 MPa in the H18 temper and 291 MPa in the O temper. With the higher hot rolling temperature, the yield strengths were approximately 80 MPa lower than this in each temper. This confirms the hypothesis of Algendy et al. [9] that a lower hot rolling temperature of 400–425 °C is more likely to resist the $Al_3(Sc,Zr)$ precipitate coarsening (i.e. over-aging) effect that was observed upon rolling at 500 °C.

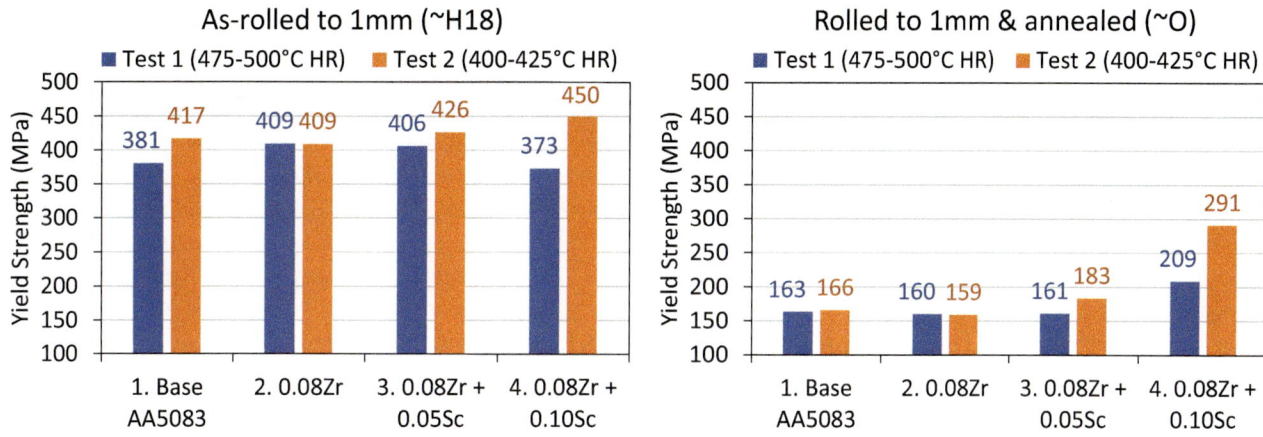

Fig. 3 Effects of Zr, Sc and hot rolling temperature on yield strength in H18 and O tempers

Considering that better results were obtained with the low hot rolling temperature, most of the remaining work focused on Test 2 with TMP to approximate H18, O, H116 and H321 tempers as shown in Table 2. Based on the tensile property data in Fig. 4, it is evident that Alloy #4 with 0.10 wt% Sc clearly produced the highest yield strength (YS) and ultimate tensile strength (UTS) for each temper, while the elongation decreased with increasing strength. Compared to the base AA5083 (Alloy #1), the addition of 0.10 wt% Sc (Alloy #4):

- Increased the YS by 8%, 75%, 25% and 33% for the H18, O, H116 and H321 tempers, respectively,
- Increased the UTS by 5%, 26%, 19% and 16% for the H18, O, H116 and H321 tempers, respectively.

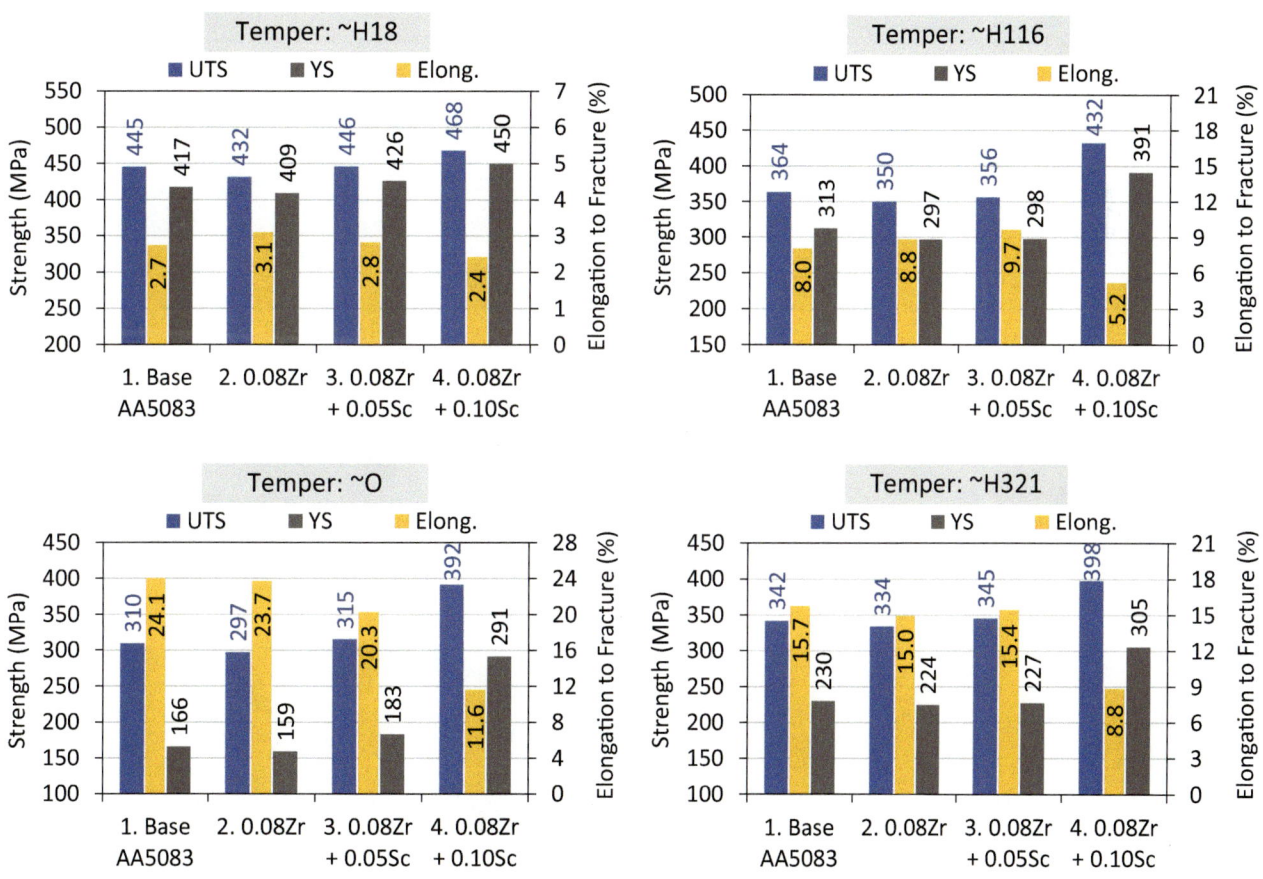

Fig. 4 Effects of Zr and Sc on tensile properties in different tempers for Test 2 with TMP as shown in Table 2

With the addition of 0.05 wt% Sc (Alloy #3), the strength was much closer to that of the base AA5083 (Alloy #1) for all tempers except for the O temper, where the 0.05 wt% Sc addition increased the yield strength by 10%. On the other hand, the alloy with only a 0.08 wt% Zr addition (Alloy #2) exhibited a small drop in strength for each temper compared to the base AA5083 (Alloy #1).

The effects of Zr and Sc on the grain structure evolution in the different tempers are shown in Fig. 5. As expected, all alloys exhibited an unrecrystallized deformed grain structure in the cold rolled condition ($\sim$H18). Upon annealing for 0.5 h at 400 °C ($\sim$O), the base AA5083 (Alloy #1) and the alloy with only a 0.08 wt% Zr addition (Alloy #2) recrystallized completely whereas Alloy #3 with 0.05 wt% Sc was partially recrystallized and Alloy #4 with 0.10 wt% Sc remained completely unrecrystallized. The impressive 125 MPa (or 75%) increase in yield strength with the 0.10 wt% Sc addition (i.e. 291 MPa for Alloy #4 versus 166 MPa for Alloy #1 in the O temper) may therefore be attributed to a combination of precipitation hardening and substructural hardening effects.

Whilst the annealing treatment decreased the yield strength of Alloy #4 from 450 to 291 MPa due to some recovery effects (vs. 417 to 166 MPa for Alloy #1 due to complete recrystallization), the temperature was not high enough to coarsen or dissolve the Al$_3$(Sc,Zr) precipitates to the extent that they would cease to inhibit recrystallization and lose their precipitation hardening effect. As a result, the elongation increased only from 2.4 to 11.6% upon annealing (vs. 2.7 to 24.1% for Alloy #1). Nevertheless, the desired balance of O temper strength and ductility could easily be achieved for Alloy #4 by modifying the annealing temperature as needed [7]. For Alloy #3 it might also be possible to increase the O temper strength by decreasing the annealing temperature to further inhibit recrystallization.

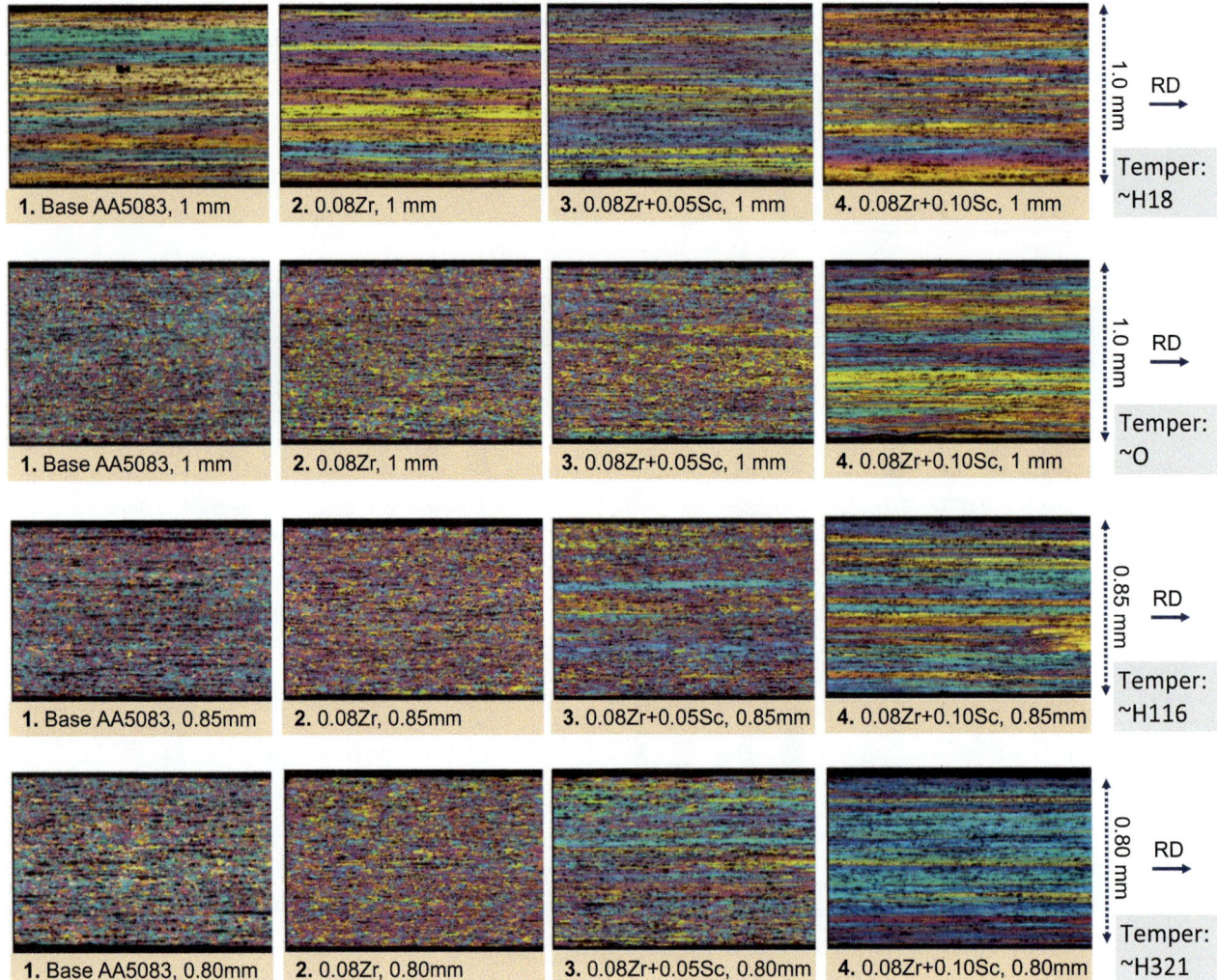

Fig. 5 Effects of Zr and Sc on grain structure in different tempers for Test 2 with TMP as shown in Table 2

The grain structures for the H116 and H321 tempers in Fig. 5 show some additional signs of deformation as might be expected from the extra cold rolling that was applied. The cold rolling reduction from 1.0 to 0.85 mm thickness turned out to be excessive as the strength increased by more than might be expected for AA5083-H116. Upon further cold rolling to 0.80 mm thickness and then applying a thermal stabilization treatment of 1 h at 185 °C, the strength of each alloy decreased while the elongation increased as might be expected for AA5083-H321. Whilst Alloy #4 exhibited a clearly higher strength and lower elongation compared to the other alloys in these two tempers, any desired combination of strength and ductility could potentially be achieved by tuning the annealing conditions.

Figure 6 shows ductility results in the form of VDA bend angle measured in the longitudinal (L) and transverse (T) orientations. Since Alloy #4 had the highest yield strength in each temper, it also exhibited the lowest ductility in bending. However, it is interesting to note that the 0.10 wt % Sc addition in Alloy #4 eliminated the L versus T anisotropy in these 3-point bend testing results. Again, it should be possible to achieve any desired combination of strength and ductility by tuning the TMP/annealing conditions and the Sc content.

Figure 7 shows hardness and electrical conductivity comparisons for Alloy #1 and Alloy #4 in all four tempers. As expected, the hardness trends are similar to the strength trends that have already been described, with Alloy #4 exhibiting the highest hardness for each temper. The electrical conductivity results show that whilst the addition of 0.10 wt% Sc along with 0.08 wt% Zr decreases the H18 conductivity slightly from 29 to 28.5%IACS, it has a

negligible effect for all the other tempers. It is therefore possible that the high dislocation content after cold rolling enabled some more Sc and Zr to precipitate out during the annealing treatment.

Since 5xxx series alloys are often used for marine and transport applications, it is important to understand how Sc additions might affect the corrosion performance. Figure 8 shows salt water acetic acid test (SWAAT) corrosion results for Alloy #1 and Alloy #4 in as-cast, H18 and O tempers from Test 1 after 10 days in the salt spray chamber. The results show similar average pit depths for both alloys, with deeper pits being observed in the as-cast condition. However, the area ratio of pits was 15–33% lower for Alloy #4. Further analysis showed that the 0.10 wt% Sc addition resulted in fewer pits, narrower pits and a lower pit volume for all the evaluated conditions. Interestingly, both alloys also exhibited some intergranular corrosion (IGC) attack, but only after annealing to the O temper.

Figure 9 shows nitric acid mass loss test (NAMLT) results for Alloy #1 and Alloy #4 in the O, H116 and H321 tempers for Test 2 with TMP conditions as shown in Table 2. It is immediately evident that Alloy #1 exhibited severe IGC in the H321 temper whereas all other results may be considered to be IGC-resistant (i.e. well within a mass loss of 1–15 mg/cm^2 as described in ASTM G67). Interestingly, the 0.10 wt% Sc addition resulted in the lowest mass loss for each temper, thereby confirming again that Sc acts to improve the corrosion resistance of AA5083. It appears that the unrecrystallized grain structure of Alloy #4 may have helped to improve its resistance to IGC. As shown in Fig. 9, the base AA5083-H321 exhibited an almost continuous network of grain boundary precipitation that would

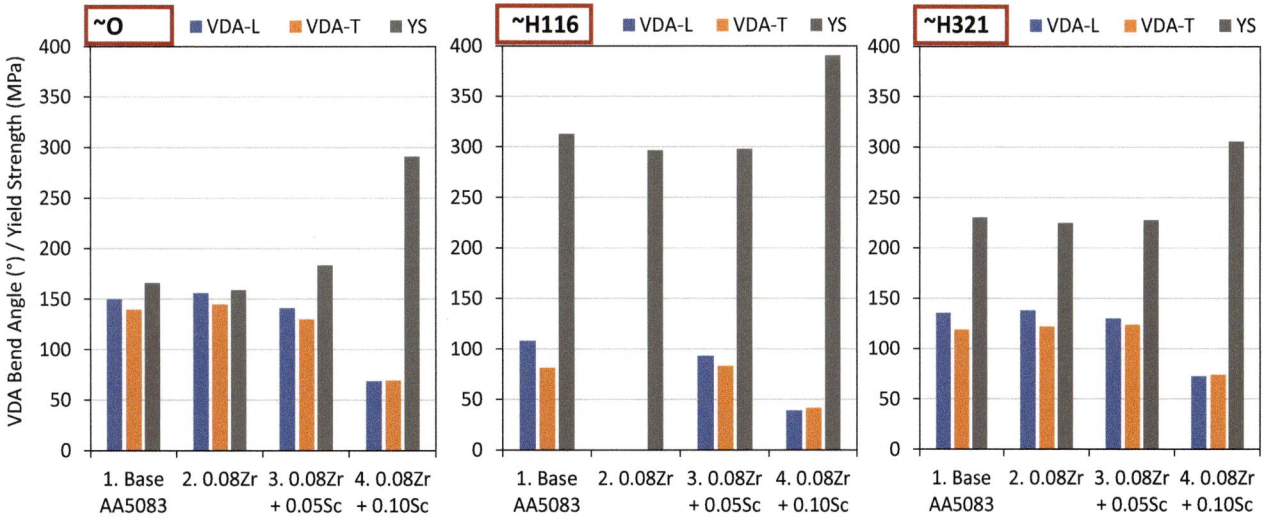

Fig. 6 Effects of Zr and Sc on VDA bendability in different tempers for Test 2 with TMP as shown in Table 2

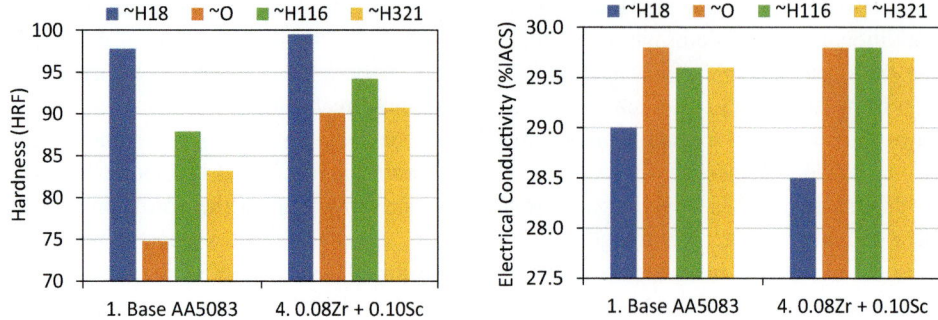

Fig. 7 Hardness and electrical conductivity results for Alloy #1 and Alloy #4 in different tempers for Test 2 with TMP as shown in Table 2

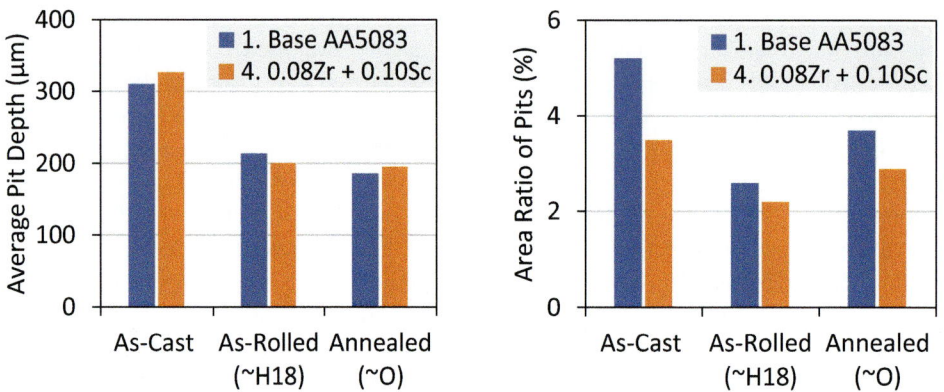

Fig. 8 10-day SWAAT corrosion results for Alloy #1 and Alloy #4 in different conditions (Test 1)

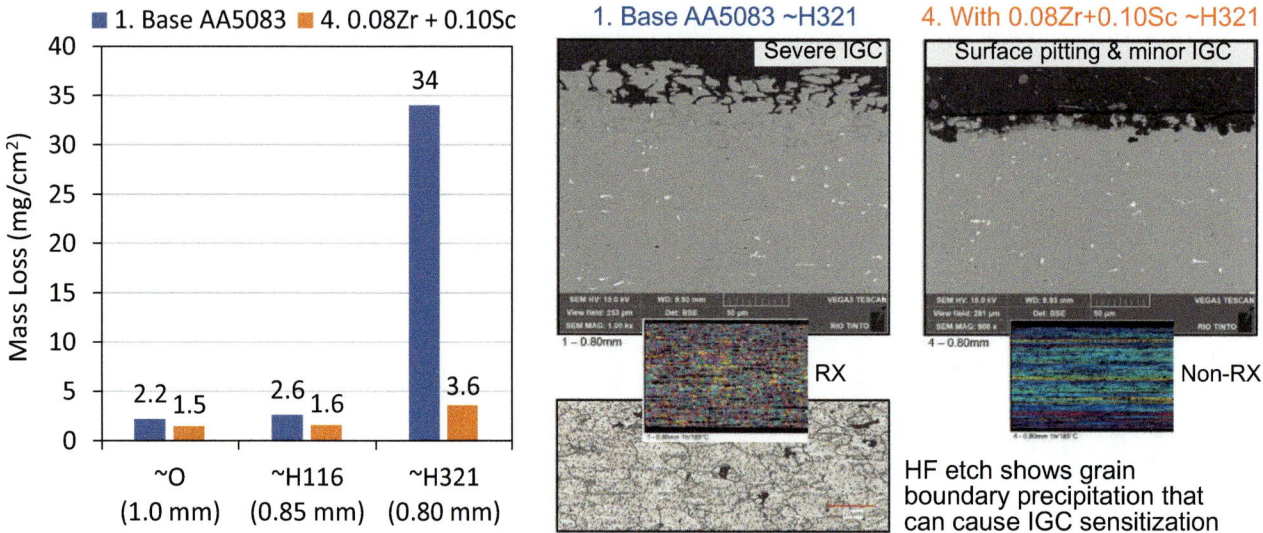

Fig. 9 NAMLT intergranular corrosion results for Alloy #1 and Alloy #4 in different tempers (Test 2)

have formed and sensitized the fully recrystallized material during the thermal stabilization treatment of 1 h at 185 °C. By contrast, the recrystallization inhibition effect of Sc could make it harder for the IGC to penetrate far below the surface due to the very long pancake-shaped grains providing fewer grain boundary pathways. This is not the first paper showing that Sc and Zr can improve the corrosion resistance of 5xxx alloys [11, 12].

Overall, the findings in this work fit very well with recent work at Rusal and elsewhere, which has demonstrated that the remarkable improvements in properties achieved by microalloying 5xxx alloys such as AA5083 with ~0.1 wt% Sc can provide compelling solutions for some marine/transport applications and welded structures [3, 10, 13–15].

Conclusions

The work on DC-cast AA5083 slabs revealed that after heat treatment at 300–425 °C:

- 0.10 wt% Sc with 0.08 wt% Zr increased the yield strength by ~60 MPa.
- 0.05 wt% Sc with 0.08 wt% Zr increased the yield strength by up to 15 MPa.
- Sc increased the ultimate tensile strength to a smaller extent, while the elongation decreased with increasing strength.
- The highest strengths (YS of 195 MPa and UTS of 287 MPa) were achieved for the alloy with 0.10 wt% Sc and 0.08 wt% Zr.
- 0.10 wt% Sc with 0.08 wt% Zr increased the elevated temperature yield strength by 32% at 200 °C and by 18% at 300 °C (relative to the base alloy), while also maintaining the thermal stability at 200–300 °C.

After hot and cold rolling, it was found that:

- AA5083 with 0.10 wt% Sc and 0.08 wt% Zr is sensitive to the hot rolling temperature such that hot rolling at 400–425 °C resulted in a ~80 MPa higher yield strength than hot rolling at 475–500 °C.
- The addition of 0.08 wt% Zr on its own had a negligible effect on the properties, while 0.05 wt% Sc together with 0.08 wt% Zr started to inhibit recrystallization and increase the yield strength for some tempers.
- The addition of 0.10 wt% Sc together with 0.08 wt% Zr had dramatic effects on the microstructure and properties of AA5083: it increased the yield strength by 8, 75, 25 and 33% for the H18, O, H116 and H321 tempers,

respectively; it increased the ultimate tensile strength by up to 26%; it decreased the ductility as it increased the strength, but the anisotropy in bending was eliminated; it inhibited recrystallization; it improved the corrosion performance and gave intergranular corrosion resistance for the O, H116 and H321 tempers.

- It should be possible to achieve almost any desired combination of microstructure and properties by tuning the processing/annealing conditions and the Sc content.

Acknowledgements The authors thank Dr Julie Lévesque at the Centre de Métallurgie du Québec (CMQ) for help with the rolling trials. Mr Pierre-Luc Privé and other technicians at Rio Tinto's Arvida Research and Development Centre are thanked for their laboratory work contributions.

References

1. Dorin, T, Ramajayam, M, Vahid, A, Langan, T (2018) Aluminium scandium alloys, chapter 12 in: Lumley, RN (ed) Fundamentals of aluminium metallurgy – recent advances. Woodhead Publishing, Elsevier, p 439–494.
2. Eskin, DG (2018) Sc applications in aluminum alloys: overview of Russian research in the 20th century, in: Martin, O (ed) Light metals 2018. The Minerals, Metals & Materials Society, p 1565–1572. https://doi.org/10.1007/978-3-319-72284-9_204.
3. Eskin DG (2020) The scandium story – part II, impact on aluminum alloys and their applications. Light Metal Age, August 2020, p 40–44.
4. Røyset J, Ryum N (2005) Scandium in aluminium alloys. International Materials Reviews 50(1):19–44.
5. Riva S, Yusenko KV, Lavery NP, Jarvis DJ, Brown SGR (2016) The scandium effect in multicomponent alloys. International Materials Reviews 61(3):203–228.
6. Ahmad Z (2003) The properties and application of scandium-reinforced aluminum. JOM, February 2003, p 35–39.
7. Li J, Yang X, Xiang S, Zhang Y, Shi J, Qiu Y, Sanders RE (2021) Effects of Sc and Zr addition on microstructure and mechanical properties of AA5182. Materials 14:4753, p 1–13. https://doi.org/10.3390/ma14164753.
8. Algendy AY, Liu K, Rometsch P, Parson N, Chen XG (2022) Evolution of discontinuous/continuous Al₃(Sc,Zr) precipitation in Al-Mg-Mn 5083 alloy during thermomechanical process and its impact on tensile properties. Materials Characterization 192:112241. https://doi.org/10.1016/j.matchar.2022.112241
9. Algendy AY, Liu K, Rometsch P, Parson N, Chen XG (2022) Effects of AlMn dispersoids and Al₃(Sc,Zr) precipitates on the microstructure and ambient/elevated-temperature mechanical properties of hot-rolled AA5083 alloys. Materials Science and Engineering A 855:143950. https://doi.org/10.1016/j.msea.2022.143950.
10. Mann V, Krokhin A, Alabin A, Fokin D, Valchuk S (2019) New Al-Mg-Sc alloys for shipbuilding and marine applications. Light Metal Age, June 2019, p 29–30 and 49.
11. Peng Y, Li S, Deng Y, Zhou H, Xu G, Yin Z (2016) Synergetic effects of Sc and Zr microalloying and heat treatment on

mechanical properties and exfoliation corrosion behavior of Al-Mg-Mn alloys. Materials Science and Engineering A 666:61–71.

12. Tang Z, Jiang F, Long M, Jiang J, Liu H, Tong M (2020) Effect of annealing temperature on microstructure, mechanical properties and corrosion behavior of Al-Mg-Mn-Sc-Zr alloy. Applied Surface Science 514:146081.

13. Krokhin AY, Mann VK, Ryabov DK, Babitskiy NA (2018) Effect of treatment parameters on structure, mechanical and corrosion properties of Al-Mg-Sc alloy forgings with reduced concentration of scandium, in: Martin, O (ed) Light metals 2018. The Minerals, Metals & Materials Society, p 1573–1580. https://doi.org/10.1007/978-3-319-72284-9_205.

14. Zakharov VV (2018) Prospects of creation of aluminum alloys sparingly alloyed with scandium. Metal Science and Heat Treatment 60(3-4):172–176. https://doi.org/10.1007/s11041-018-0256-8.

15. Yuryev PO, Baranov VN, Orelkina TA, Bezrukikh AI, Voroshilov DS, Murashkin MY, Partyko EG, Konstantinov IL, Yanov VV, Stepanenko NA (2021) Investigation the structure in cast and deformed states of aluminum alloy, economically alloyed with scandium and zirconium. The International Journal of Advanced Manufacturing Technology 115:263–274. https://doi.org/10.1007/s00170-021-07206-z.

Efficiency of Sc for Strengthening and Formability Improvement of 5XXX BIW Sheets

Margarita Nikitina, Aleksandr Gradoboev, Dmitry Ryabov, Roman Vakhromov, Viktor Mann, and Aleksandr Krokhin

Abstract

The use of aluminum makes it possible to reduce the weight of the car, thereby improving the environmental impact of transport by reducing CO_2 emissions. Rolling sheets made of 5182 alloy are widely used for the manufacture of BIW parts using deep drawing. However, to increase the weight efficiency of the car it is necessary to develop an alloy with higher mechanical properties without losing processability. The results of previous studies of 5XXX series alloys showed that additives of Scandium, Zirconium and other REE (rare earth elements) and TM (transition metals) lead to a significant increase in strength characteristics. The work presents the results of studies of the influence of small additives of Scandium and other REE on the mechanical and technological properties of alloys based on 5182, as well as the dependencies of changes in the level of mechanical and technological properties on rolling and heat treatment parameters are established.

Keywords

Aluminum-scandium alloys • Influence of small additives of Scandium and other REE • Rolling • Influence heat treatment • BIW • Established

Introduction

Reducing the weight of the car is one of the priority directions of the development of vehicles, thereby improving the environmental friendliness of transport by reducing CO_2 emissions. One of the means of reducing the weight of the car is the use of aluminum alloys. Rolling sheets made of 5182 alloy are widely used for the manufacture of BIW parts manufactured by deep drawing. However, to increase the weight efficiency of the car it is necessary to develop an alloy with higher mechanical properties without losing processability. Scandium, Zirconium and other rare earth elements (REE) is the most effective additive for improving the strength characteristics of aluminum alloys. The greatest efficiency from the introduction of Scandium is achieved in 5XXX series aluminum alloys [1].

Despite the high efficiency in hardening aluminum alloys of the 5XXX series, scandium still remains an expensive additive. The market value of the $AlSc_2$ master alloy is approximately $42/kg. Therefore, in order to ensure the introduction of scandium-containing alloys instead of those traditionally used, it is necessary that the increase in the cost of the alloy is compensated by reducing the required mass of semi-finished products due to their higher strength characteristics. That is, for example, a car manufacturer purchasing more expensive rolled Al-Mg-Sc alloy will have the effect of reducing the weight of the structure, but, at the same time, the cost of the purchased semi-finished products will remain the same due to the difference in weight due to the use of smaller thicknesses. In this regard, in our work, in order to replace alloy 5182 in BIW, we have limited ourselves to the Scandium content of 0.02wt%.

The work presents the results of studies of the effect of small additives of zirconium scandium and other REE as well as changes in the content of magnesium and other additives on the mechanical and technological properties of alloys based on 5182. The dependencies of changes in the

M. Nikitina (✉) · A. Gradoboev · D. Ryabov · R. Vakhromov
LLC Light Materials and Technologies Institute UC RUSAL,
Moscow, Russia
e-mail: Margarita.Nikitina@rusal.com

V. Mann · A. Krokhin
JSC RUSAL Management, Moscow, Russia

© The Minerals, Metals & Materials Society 2023
S. Broek (ed.), *Light Metals 2023*, The Minerals, Metals & Materials Series,
https://doi.org/10.1007/978-3-031-22532-1_164

level of mechanical and technological properties on the rolling and heat treatment modes had been also investigated.

Methods

Experiments on the effect of changes in chemical composition, degree of deformation and heat treatment were carried out on sheets with a thickness of 1 mm thick sheets with of alloy 5182 sheets were also produced under the same laboratory conditions as all other compositions. Chemical compositions described in the present work are presented in Table 1.

The casting of the compositions was carried out on a small casting machine using casting equipment manufactured by Wagstaff. Slabs were cast with a cross section of 120 × 340 mm. The slabs were scalped in order to obtain blanks with dimensions of 100 × 320 × 3 40 mm.

The slabs were rolled on a reversible DUO 400 mill with a maximum rolling force of 2500 KN. Hot rolling was carried out down to 7 mm followed by cold rolling down to 1 mm. All compositions were filled with a deformation degree of 50%, and the compositions that demonstrated good mechanical properties were rolled with 3 types of deformation degrees: 30%, 50%, 65%. The width of the resulting strips was 300 mm.

The mechanical properties of the sheets were determined after stretching. The mechanical tests were carried out on the MTS Criterion 40 universal testing machine. The test temperature was 25 °C. The tests were carried out on samples after rolling, as well as after various annealing practices. The main task of annealing was to obtain complete recrystallization.

Annealing of the samples were carried out in a laboratory furnace Nabertherm GmbH 30/65. After annealing, the mechanical properties, structure, as well as plastic characteristics were studied by the Eriksen test.

The microstructure of the alloys was studied using a Zeiss Axio Observer microscope. The granular structure of the samples was studied after anodic oxidation in an electrolyte consisting of an aqueous solution of hydrofluoric and boric

acids. The anode film on the samples were obtained using the electrolytic polishing and etching unit Struers LectroPol-5. The samples were observed in polarized light after oxidation.

Results and Discussion

The effect of scandium and zirconium additives on the mechanical properties of alloy 5182 is shown in Fig. 1. Mechanical properties are evaluated on sheets after rolling with a final deformation degree of 50%.

The additives of scandium and zirconium in the composition of 5182 lead to an increase in UTS and YS, which indicates the expediency of using these additives of zirconium and scandium up to 0.02wt.% to increase mechanical properties.

It is well known that alloys with a high magnesium content are not technologically advanced in manufacturing. The option of compensating for magnesium reduction with scandium and zirconium additives is considered. The composition contains 3 percent magnesium, unlike alloy 5182, the magnesium content in which is at the level of 4.5wt.%, and the addition of scandium and zirconium, a comparison of properties in the recrystallized state with a final deformation of 50% is shown in Fig. 2 and in the recrystallized state in Fig. 3.

As can be seen from Figs. 2 and 3, scandium and zirconium additives successfully compensate the decrease in Magnesium content.

In addition to scandium and zirconium additives, cerium and yttrium additives are also effective. Two compositions were considered in which Scandium content was reduced by 0.01wt.%, and 0.1wt.% Ce was added to one of them, and

Table 1 Chemical compositions

Alloy number	Chemical composition
1 alloy	5182
2 alloy	Zr + Sc
3 alloy	3 Mg + Zr + Sc
4 alloy	Zr + 0,01Sc + Ce
5 alloy	Zr + 0,01Sc + Y
6 alloy	5 Mg + Zr + Sc

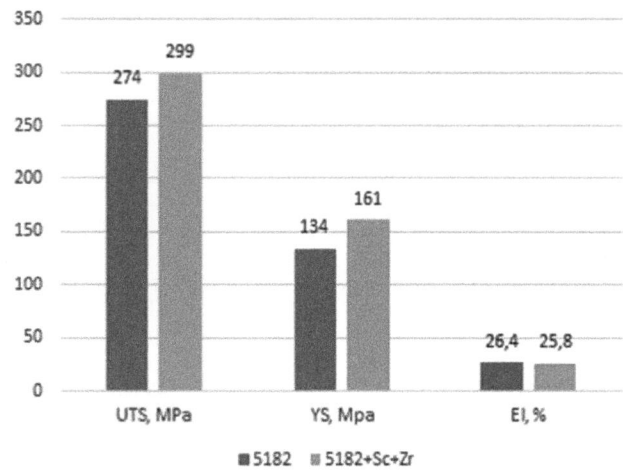

Fig. 1 Comparison of mechanical properties of 5182 and 5182 with additives Sc and Zr in the recrystallized state

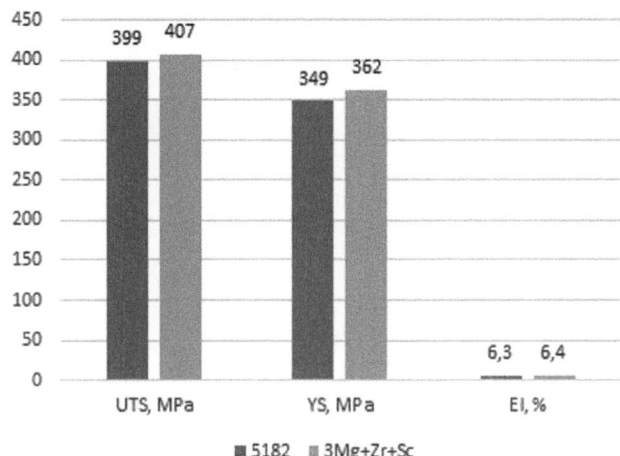

Fig. 2 Comparison of mechanical properties of 5182 and 5182 with 3% Mg and Sc and Zr additives after rolling

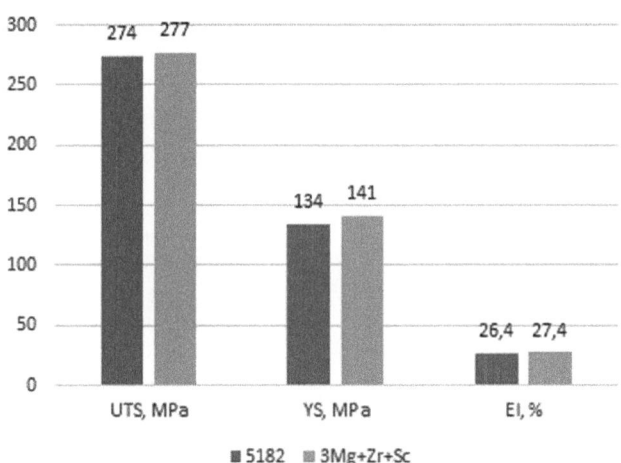

Fig. 3 Comparison of mechanical properties of 5182 and 5182 with 3% Mg and Sc and Zr additives in the recrystallized state

0.06wt.%Y—in the other one. Comparison of the mechanical properties are shown in Fig. 4.

The introduction of Ce and Y additives instead of reducing Sc did not lead to an increase in properties relative to the 5182 composition after rolling with 50% deformation degree. However, in the fully crystallized state, the addition of Ce showed some increase in YS. A comparison of the properties of the compositions in the recrystallized state are shown in Fig. 5.

In order to obtain higher strength characteristics, the option of adding Sc, Zr and an increase of Mg from 4.5wt.% to 5wt.% is considered. A comparison of the mechanical properties in the state after rolling with a deformation degree of 50% are shown in Fig. 6.

An increase in the magnesium content by 0.5 wt.% gives an increase in mechanical properties after rolling, however,

after the recrystallization annealing (Fig. 7), it practically does not give any advantages, which may indicate that the addition of magnesium stimulates the ability to obtain high characteristics by deformation.

The effect of additives Zr and Sc, as well as Mg on formability was determined by the Eriksen test. The test results are presented in Table 2.

According to the data obtained, these additives Sc and Zr, as well as Mg increases by 0.5%, have practically no effect on the formability of the material.

To investigate the influence of the degree of deformation on the mechanical properties the following compositions were chosen: 5182, 5182 + Zr + Sc and 5 Mg + Zr + Sc. Figures 8 and 9 show comparisons of UTS and YS dependences at different strains with subsequent recrystallization annealing.

The above dependences show that with increasing degree of strain, after recrystallization annealing the alloys with Scandium content are observed to decrease UTS and YS, and the alloy without Scandium tends to increase UTS and YS. The typical structures of the alloys after rolling with different degrees of deformation are shown in Figs. 10 and 11.

As can be seen in Figs. 10 and 11, as the degree of deformation increases, the structure elaboration improves. While at 30% deformation both elongated and rounded grains are observed, with increasing deformation of 50% the structure becomes banded, at 65% deformation a denser structure, elongated and finer grains are observed.

Conclusions

1. The impact of scandium and other REE additives on mechanical properties has been studied
2. Zirconium and Scandium additives, up to 0.02 percent, have been found to effectively increase UTS and YS
3. Scandium and zirconium additives were found to successfully compensate for the decrease in Magnesium
4. The introduction of additives Cerium and Yttrium, with a reduction of scandium content to 0.01wt.% has almost no effect on the mechanical properties compared to the base alloy
5. A 0.5wt.% increase in magnesium gives a negligible increase in UTS and YS after recrystallization, but improves the hardening ability of the alloy due to deformation
6. It was found that with increasing degree of deformation, after recrystallization annealing the alloys with Scandium content are observed to decrease UTS and YS, and the alloy without Scandium tends to increase UTS and YS.

Fig. 4 The effect of Sc
substitution with REE additives
on mechanical properties of 5182
alloy after rolling

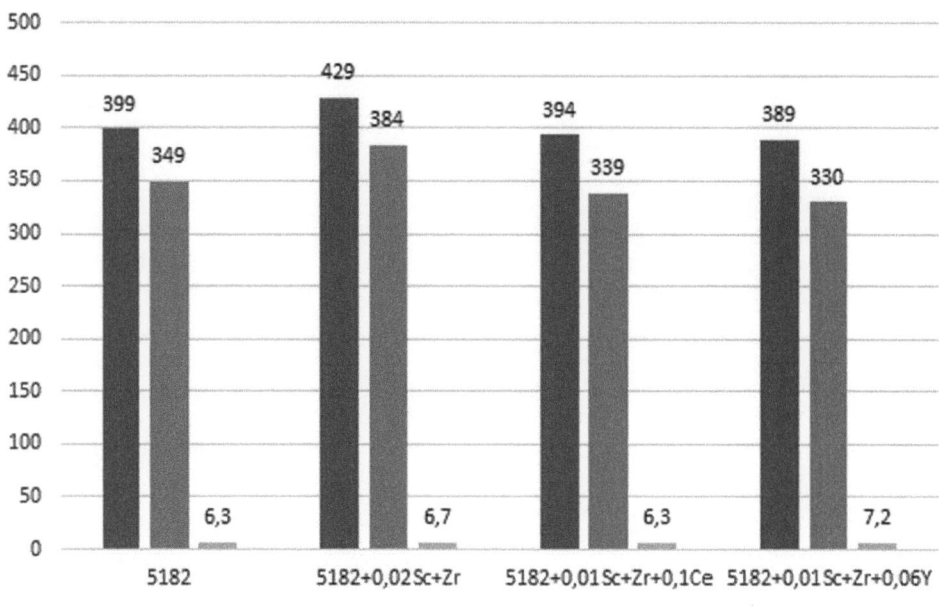

Fig. 5 The effect of Sc
substitution with REE additives
on the mechanical properties of
5182 alloy in the recrystallized
state

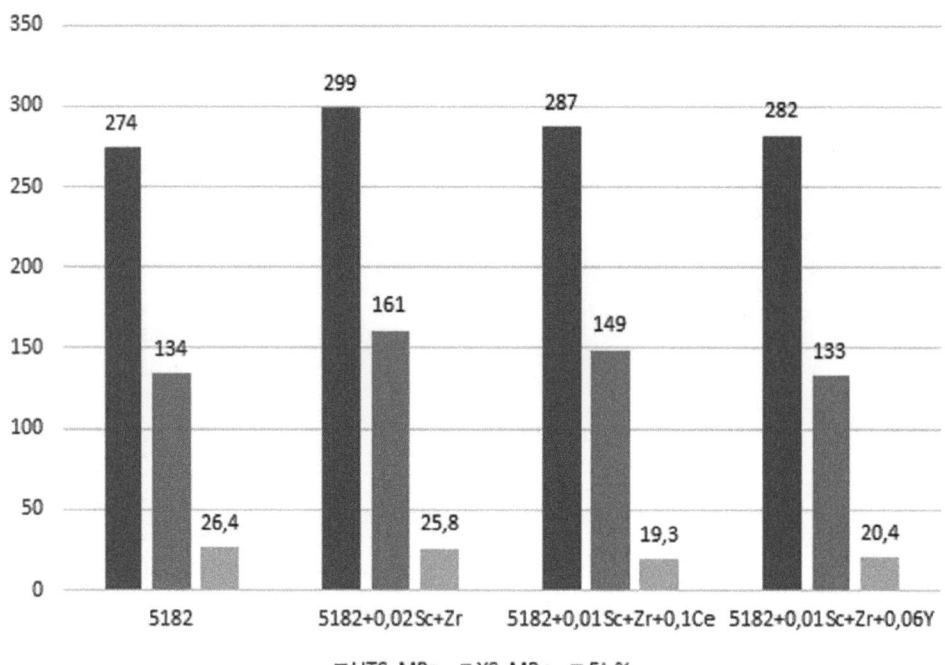

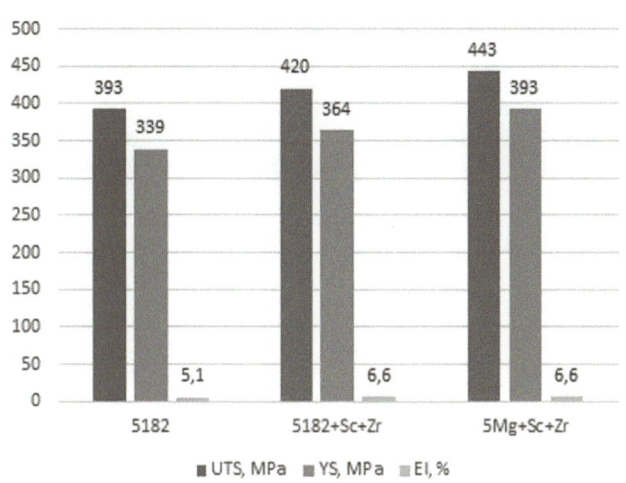

Fig. 6 The effect of increasing Mg by 0.5% on the properties of rolled semi-finished products after rolling

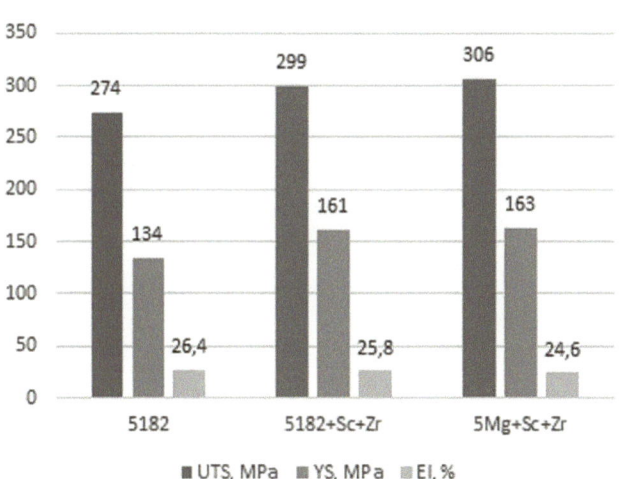

Fig. 7 The effect of increasing Mg by 0.5% on the properties of rolled semi-finished products in the recrystallized state

Table 2 Eriksen test

Chemical composition	Hole depth, mm
5182	8.0
5182 + Sc + Zr	7.9
5 Mg + Sc + Zr	8.0

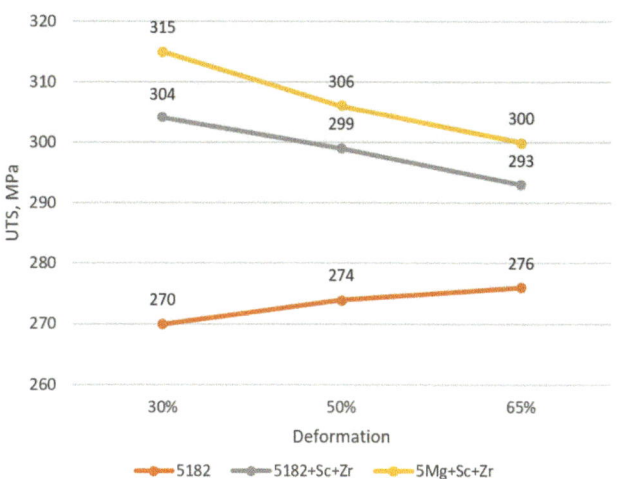

Fig. 8 Dependence of UTS on the degree of deformation

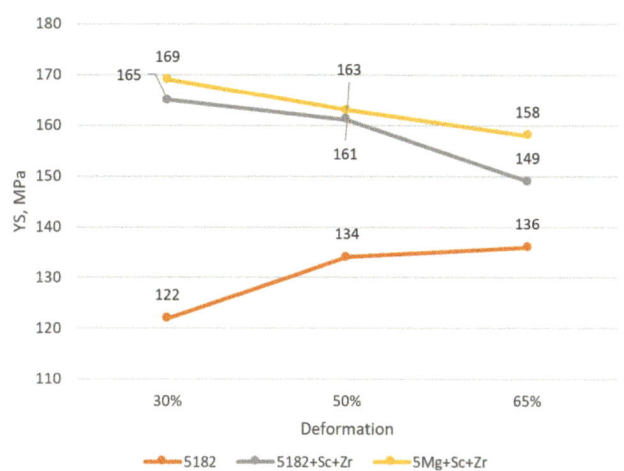

Fig. 9 Dependence of YS on the degree of deformation

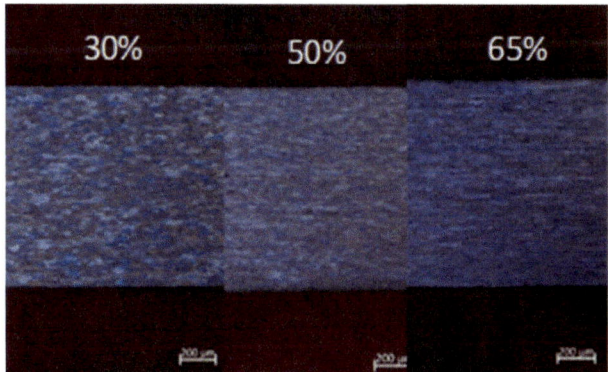

Fig. 10 Dependence of the sheet structure on the degree of deformation of the studied compositions 5182 + Sc + Zr

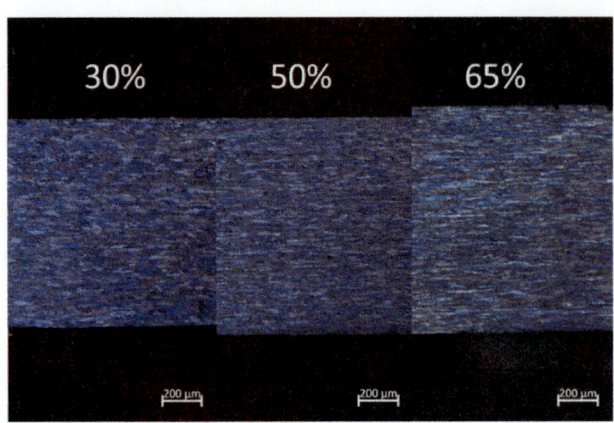

Fig. 11 Dependence of the sheet structure on the degree of deformation of the studied compositions 5 Mg + Sc + Zr

References

1. Eskin D, (2020) The Scandium Story–Part II: Impact on Aluminum Alloys and Their Applications. Light Metal Age 78(5):40-44;

2. Fuller, C. Temporal evolution of the nanostructure of Al(Sc,Zr) alloys: Part I – Chemical compositions of Al(ScZr) precipitates / C. Fuller, J. Murray, D. Seidman // Acta Mater. – 2005. – V. 53. – P. 5401–5413;

3. Tolley, A. Segregation in Al3(Sc,Zr) precipitates in Al–Sc–Zr alloys / A. Tolley, V. Radmilovic, U. Dahmen // Scripta Mater. – 2005. – V. 52. – P. 621–625;

4. Knipling, K. E.Criteria for developing castable, creep-resistant aluminum-based alloys – A review / K.E. Knipling, D.C. Dunand, D.N. Seidman // Zeitschrift für Metallkunde. – 2006. – V. 97. – P. 246–265;

5. Mann V, Krokhin A, Alabin A, Fokin D, Valchuk S (2019) New Al-Mg-Sc Alloys for Shipbuilding and Marine Applications. Light Metal Age 77(3):29–30,49

6. Kaufman, JG (2000) Introduction to Aluminum Alloys and Tempers. ASM International

Effect of Sc and Zr Additions on Dispersoid Microstructure and Mechanical Properties of Hot-Rolled AA5083

Ahmed Y. Algendy, Kun Liu, Paul Rometsch, Nick Parson, and X.-Grant Chen

Abstract

5xxx aluminum alloys are traditionally considered non-heat-treatable. With the addition of Sc/Zr and multistep heat treatment, two kinds of dispersoids (AlMn and $Al_3(Sc,Zr)$) were formed. The effect of Sc additions (0.08–0.43 wt.%) on dispersoid formation and mechanical properties of hot-rolled sheets was investigated. The results showed that tensile properties initially increased with increasing Sc addition. The yield strength (YS) and ultimate tensile strength (UTS) of the alloy with 0.16 wt. % Sc reached 295 and 411 MPa, respectively, showing improvements of 28% in YS and 8% in UTS compared to the base alloy. However, with a further increase of Sc, the tensile properties declined owing to the formation of a line/fan-shaped microstructure associated with discontinuous $Al_3(Sc,Zr)$ precipitation during solidification. The evolution of $Al_3(Sc,Zr)$ and AlMn dispersoids during heat treatment and hot rolling was characterized using scanning and transmission electron microscopies. Their influence on the mechanical properties of hot-rolled AA5083 alloys was discussed.

Keywords

Aluminum 5083 alloy • Sc and Zr addition • Dispersoids • Mechanical properties

A. Y. Algendy · K. Liu · X.-G. Chen (✉)
Department of Applied Science, University of Quebec at Chicoutimi, Saguenay, QC G7H 2B1, Canada
e-mail: xgrant_chen@uqac.ca

P. Rometsch · N. Parson
Arvida Research and Development Center, Rio Tinto Aluminium, Saguenay, QC G7S 4K8, Canada

Introduction

Owing to their excellent combination of high strength-to-weight ratio, good formability, high toughness, excellent weldability and corrosion resistance, Al-Mg-Mn 5xxx alloys are considered excellent candidates for transportation and construction industries [1, 2]. Traditionally, 5xxx aluminum alloys are classified as non-heat-treatable alloys. Therefore, the achievable strength of this alloy series is more limited than that of heat-treatable high-strength aluminum alloys. Microalloying with Sc can significantly improve the mechanical properties, preserve the work hardening and enhance the recrystallization resistance of Al-Mg-Mn 5xxx alloys [3, 4]. Owing to its cost-effectiveness, Zr is often added together with Sc. The combination of Sc and Zr can form core–shell $Al_3(Sc_{1-x},Zr_x)$ precipitates with the same $L1_2$-crystal structure of Al_3Sc while improving coarsening resistance [5].

In addition to work hardening and solid solution strengthening, dispersoid strengthening can provide a significant potential for improving the mechanical strength of non-heat-treatable aluminum alloys, such as 3xxx [6, 7] and 5xxx [8] alloys. Recent studies in AA3004 and AA5083 alloys demonstrated that a multistep heat treatment can be used to promote the precipitation of submicron-sized Mn-bearing dispersoids and nano-sized $Al_3(Sc,Zr)$ dispersoids, resulting in enhanced mechanical properties [8–10]. In Al-Mg-Mn 5xxx alloys, the precipitation temperatures of AlMn dispersoids and $Al_3(Sc,Zr)$ precipitates are similar, and both phases are coarsening-resistant, which provides a common basis for improving the mechanical properties during heat treatment. However, there is little information on the synergetic effects of these strengthening phases in 5xxx alloys.

With additions of Sc and Sc + Zr, $Al_3Sc/Al_3(Sc,Zr)$ precipitates can be formed either continuously or discontinuously [11, 12]. Continuous precipitation usually occurs during aging (300–400 °C), in which nano-sized $Al_3Sc/$

© The Minerals, Metals & Materials Society 2023
S. Broek (ed.), *Light Metals 2023*, The Minerals, Metals & Materials Series,
https://doi.org/10.1007/978-3-031-22532-1_165

Al$_3$(Sc,Zr) precipitates effectively retard the dislocation and grain boundary movements and significantly improve the alloy properties [5, 13]. However, discontinuous precipitation can also occur during solidification with rod- and lamellae-like morphologies [14, 15]. The tendency of Al-Sc/Zr alloys to form Al$_3$Sc/Al$_3$(Sc,Zr) discontinuous precipitates is strongly related to their Sc level (often in high-Sc hypereutectic alloys) [12, 15] and solidification rate [14, 16]. Most studies considered Al$_3$Sc/Al$_3$(Sc,Zr) discontinuous precipitates as an undesirable microstructure feature, which negatively affects the mechanical properties of the alloy.

This study investigated the effect of Sc additions (0.08–0.43 wt.%) on the precipitation behavior of dispersoids and mechanical properties in a typical Al-Mg-Mn AA5083 alloy during thermomechanical processing (heat treatment and hot rolling). To clarify the roles of micro-alloyed Sc and Zr, the microstructural evolution after heat treatment and hot rolling was characterized by scanning electron microscopy (SEM) and transmission electron microscopy (TEM).

Methodology

Four Al-Mg-Mn alloys were prepared according to the typical AA5083 chemical composition with various Sc and Zr levels, denoted as the Base, Sc08, Sc15, and Sc43 following their Sc levels. The chemical compositions of the experimental alloys, as analyzed by optical emission spectroscopy, are shown in Table 1. The alloys were prepared in an electrical resistance furnace, and the melt was cast in a permanent steel mold preheated at 250 °C with a cooling rate of 2 °C/s to produce cast ingots with a dimension of 30 × 40 × 80 mm.

After casting, all alloys were subjected to a three-step heat treatment (275 °C/12 h + 375 °C/48 h + 425 °C/12 h) to promote the precipitation of AlMn-dispersoids and Al$_3$(Sc, Zr) precipitates and to improve the processability of rolling [8, 10]. After heat treatment, the samples were hot-rolled at 500 ± 20 °C to a final thickness of 3.2 mm (87% reduction) using a laboratory-scale rolling mill. After rolling and before mechanical testing, all rolled sheets were annealed at 300 °C for 5 h to relieve the rolling-induced thermal stress.

The Vickers microhardness was measured at room temperature with a load of 10 g and a 20 s dwell time. The average value was calculated from 20 measurements for each sample. Uniaxial tensile tests were conducted using an Instron 8801 servo-hydraulic testing unit at a strain rate of 0.5 mm/min. The tensile samples were machined according to ASTM E8/ E8M-16a in the rolling direction with a gauge length of 32 mm and gauge area of 3 × 6 mm. Average results were obtained from three repeated tests.

The microstructure evolution was characterized using scanning electron microscopy (SEM, JSM-6480LV) and transmission electron microscopy (TEM, JEM-2100). In addition, the cast samples were etched with Keller's reagent for 30 s to highlight the discontinuous precipitates. Image analysis with ImageJ software was used for measuring the area fraction of intermetallic particles, the size and number density of the AlMn dispersoids and Al$_3$(Sc,Zr) precipitates. The number density was measured according to Eq. 1 [8, 10]

$$ND = \frac{N}{A * (D + t)} \tag{1}$$

where; N is the number of particles in the TEM image, A is the total area, D is the equivalent diameter, and t is the thickness of the TEM foil. Finally, the grain structure was mapped and analyzed after rolling and annealing using the Electron Back-Scatter Diffraction (EBSD) technique with a step size of 1 μm.

Results and Discussion

Evolution of the Microstructure in the As-Cast and Heat Treatment

Figure 1 displays the as-cast microstructure of the experimental alloys after Keller etching. The microstructure is mainly composed of the α-Al matrix surrounded by intermetallic phases distributed along the interdendritic regions. The three major intermetallic compounds (IMC) in the experimental alloys were Fe/Mn-rich phases (α-Al$_{15}$(Fe, Mn)$_3$Si$_2$ and Al$_6$(Fe,Mn)), and primary Mg$_2$Si as identified by SEM-EDS. A small amount of the low melting point eutectic phases, τ-Al$_6$CuMg$_4$ and β-Al$_5$Mg$_3$, was also

Table 1 Chemical composition of experimental alloys

Alloys	Elements, wt.%								
	Mg	Mn	Si	Fe	Cu	Cr	Ti	Sc	Zr
Base	4.78	0.79	0.26	0.31	0.12	0.14	0.09	–	–
Sc08	4.76	0.79	0.26	0.35	0.10	0.15	0.10	**0.08**	**0.08**
Sc15	4.75	0.81	0.31	0.31	0.11	0.15	0.09	**0.16**	**0.17**
Sc43	4.76	0.75	0.30	0.33	0.10	0.15	0.10	**0.43**	**0.15**

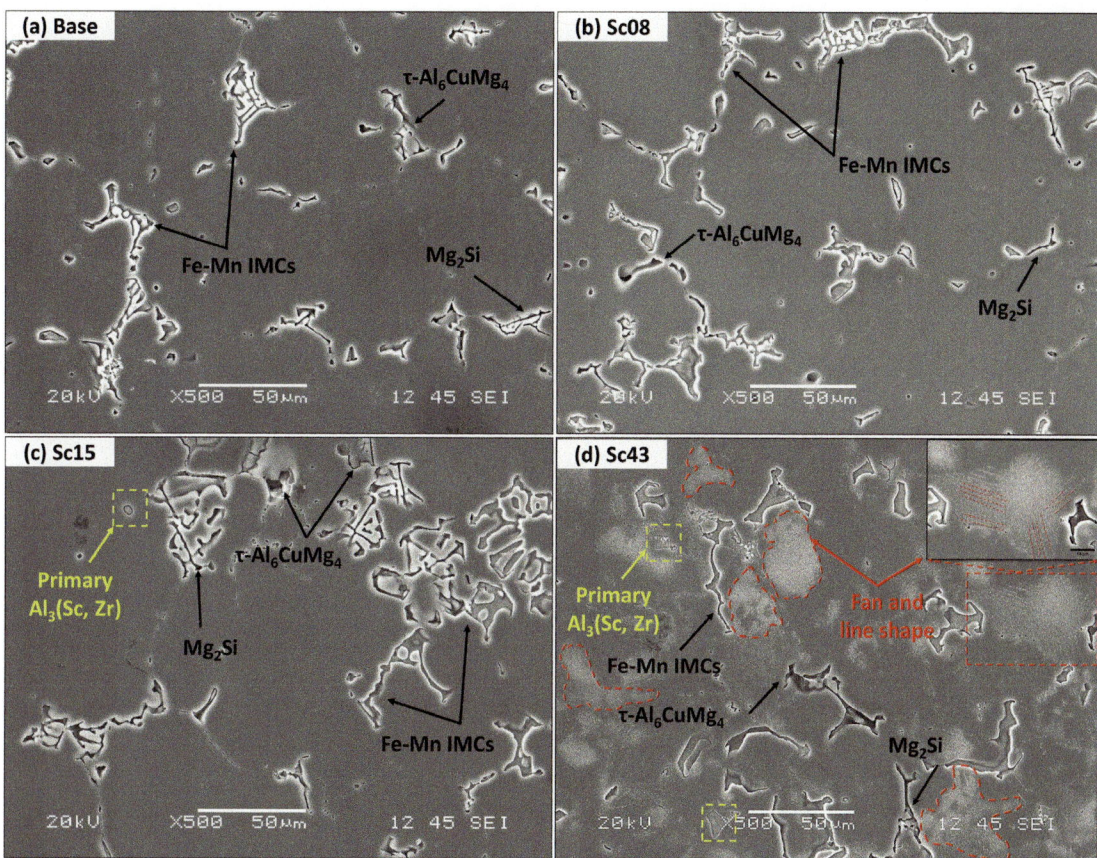

Fig. 1 SEM images showing the as-cast microstructure evolution with Sc and Zr content after etching with Keller's reagent

detected along the interdendritic regions. Furthermore, the area fraction of Fe/Mn-rich intermetallic and Mg_2Si remarkably increased with the Sc and Zr addition. For instance, the area fractions of Fe/Mn-rich and Mg_2Si IMCs increased from 1.98%, 0.7% in the base alloy to 2.31%, 0.77% in alloy Sc08, and further to 3.14%, 0.9% in alloy Sc15. This could be attributed to the reduction of Mn and Mg solubility associated with the addition of Sc and Zr [17, 18].

In addition, in the Sc-containing alloys (Sc15 and Sc43), a few primary $Al_3(Sc,Zr)$ particles were detected inside the aluminum dendrite cells and their fraction increased with increasing Sc content (Fig. 1 c and d). Furthermore, in the alloy Sc43 with the highest Sc content (0.43%), several areas exhibited the formation of fan- and line-shaped aggregates (marked by red dashed lines in Fig. 1d), which was rarely observed in the Sc08/Sc15 alloy. This type of precipitate was related to the decomposition of the supersaturated solid solution to form discontinuous $Al_3(Sc,Zr)$ precipitates (DCP) at a moving grain boundary. More details on the discontinuous precipitation mechanism were reported in our previous study [19].

After heat treatment, the low-melting-point eutectic phases (τ-Al_6CuMg_4/β-Al_5Mg_3) were entirely dissolved in the aluminum matrix with a partial dissolution of Mg_2Si in all cases. Meanwhile, a large number of duplex precipitation populations (AlMn dispersoids and $Al_3(Sc,Zr)$ precipitates) were formed, as shown in Fig. 2. Bright-field TEM images in Fig. 2 a-d show the precipitation of submicron-sized Mn-dispersoids; their size and number density are quantified in Table 2. The types of AlMn dispersoids were identified as Al_4Mn and Al_6Mn based on SADP and TEM-EDS analysis in our previous work [8]. With increasing Sc and Zr contents, the size of the dispersoids increased, and the number density decreased. For instance, the equivalent diameter of AlMn dispersoids in the base alloy was ∼25 nm and the number density was ∼56×10^{20} m^{-3}, while the diameter of dispersoids in the Sc15 alloy increased to ∼33.5 nm and the number density decreased to 30×10^{20} m^{-3} (Table 2). There are two possible reasons for increasing the size and decreasing the number density of AlMn dispersoids in Sc-containing alloys. First, adding Sc and Zr reduces the solubility of Mn, Mg, and Si in the aluminum matrix [17, 18]. Secondly, the addition decreases the nucleation efficiency of AlMn dispersoids during heat treatment by decreasing the nucleation sites in the first stage (275 °C/12 h) [20, 21].

In addition to the AlMn-dispersoids, a large number of nano-sized, spherical $Al_3(Sc,Zr)$ precipitates were also

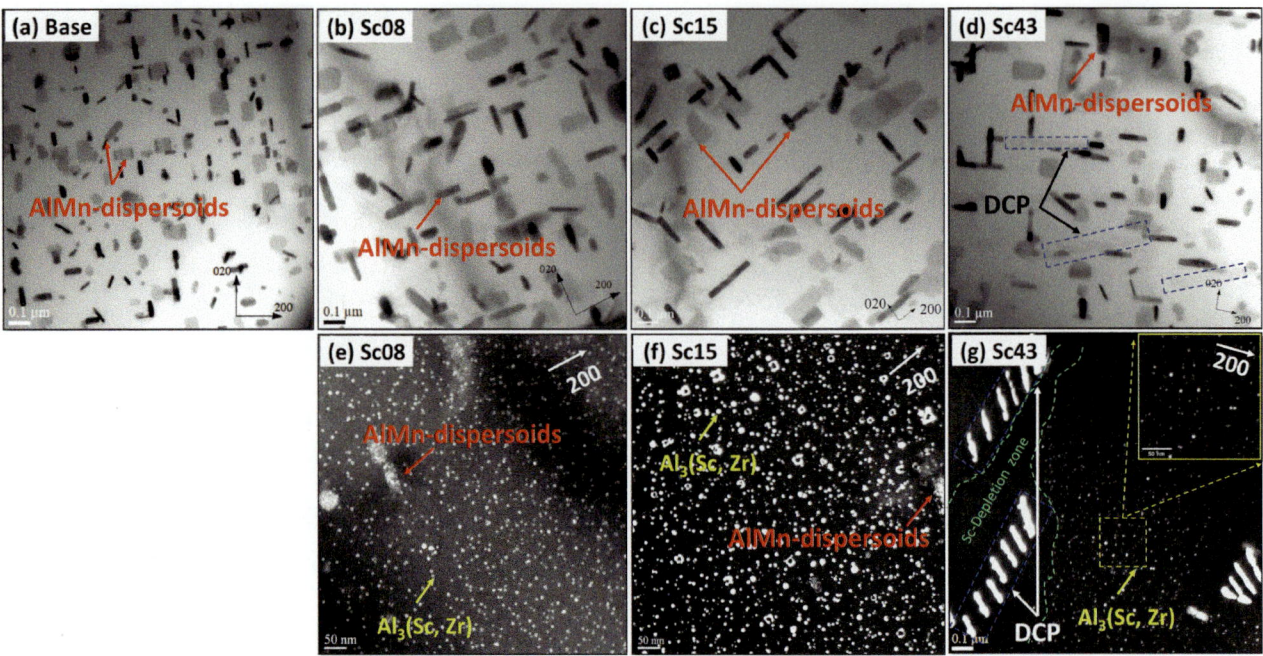

Fig. 2 Bright-field TEM images (**a, b, c, d**) showing the distribution of AlMn dispersoids for Base, Sc08, Sc15 and Sc43; dark-field TEM images (**e, f, g**) showing the distribution of $Al_3(Sc,Zr)$ precipitates for Sc08, Sc15 and Sc43

Table 2 Quantitative measurements of AlMn dispersoids and $Al_3(Sc,Zr)$ after heat treatment and rolling

Alloy	After heat treatment				After rolling and annealing			
	AlMn dispersoids		$Al_3(Sc,Zr)$		AlMn dispersoids		$Al_3(Sc,Zr)$	
	d (nm)	N_d (10^{20}/m³)	d (nm)	N_d (10^{22}/m³)	d (nm)	N_d (10^{20}/m³)	d (nm)	N_d (10^{22}/m³)
Base	24.8 ± 1.7	55.9 ± 22.2	–	–	32.4 ± 0.2	23.1 ± 46	–	–
Sc08	29.6 ± 5	30.9 ± 64.0	5.3 ± 0.5	11.3 ± 683	57.2 ± 0.6	6.01 ± 9.6	11.9 ± 0.5	2.7 ± 385
Sc15	33.5 ± 2.6	30.1 ± 56.7	6.1 ± 1.2	19.7 ± 741	56.0 ± 0.6	6.48 ± 13.3	12.6 ± 0.4	5.7 ± 297
Sc43	33.6 ± 1.6	28.7 ± 21.2	8.3 ± 1.4	5.2 ± 741	64.6 ± 1.7	6.80 ± 11.3	37.8 ± 2.6	0.7 ± 177

formed in the Sc-containing alloys, as shown in the dark-field TEM images in Fig. 2e–g. The $Al_3(Sc,Zr)$ precipitates were much finer with a higher number density than the AlMn-dispersoids, and were uniformly distributed in the Al matrix. These precipitates filled the space between the AlMn dispersoids, lowering the inter-particle distance, and resulting in more obstacles for dislocation movements. In addition, Fig. 2d, g revealed a number of rod-like precipitates, non-uniformly distributed in the Al matrix, which originally came from the as-cast fan- and line-shaped microstructure (Fig. 1d) but were partially dissolved during the heat treatment. Those rod-like precipitates were discontinuous precipitates of $Al_3(Sc,Zr)$ (DCP) formed during solidification [15, 19], and their size was much larger than the spherical ones. Furthermore, as shown in Fig. 2g (marked by green dashed lines), there were almost no spherical $Al_3(Sc,Zr)$ precipitates in the vicinity of the rod-like discontinuous precipitates.

The image analysis results in Table 2 show that the number density of spherical $Al_3(Sc,Zr)$ precipitates initially increased with increasing Sc content to 0.16 wt.% (Sc15) and then decreased with increasing Sc to 0.43 wt.% (Sc43). The reason for this effect is that the formation of discontinuous precipitates consumed large amounts of Sc and Zr solutes, decreasing the Sc and Zr content in the solid solution. Furthermore, it created a depleted zone surrounding the rod-like precipitates, preventing further precipitation of the spherical $Al_3(Sc,Zr)$ precipitates.

Microstructure Evolution During Hot Rolling

Prior to hot rolling, all samples were preheated at 500 °C/1.5 h and then hot-rolled at 500 °C. Typical microstructures after hot rolling are shown in Fig. 3. Because of the high reduction ratio (87%), the intermetallic particles were

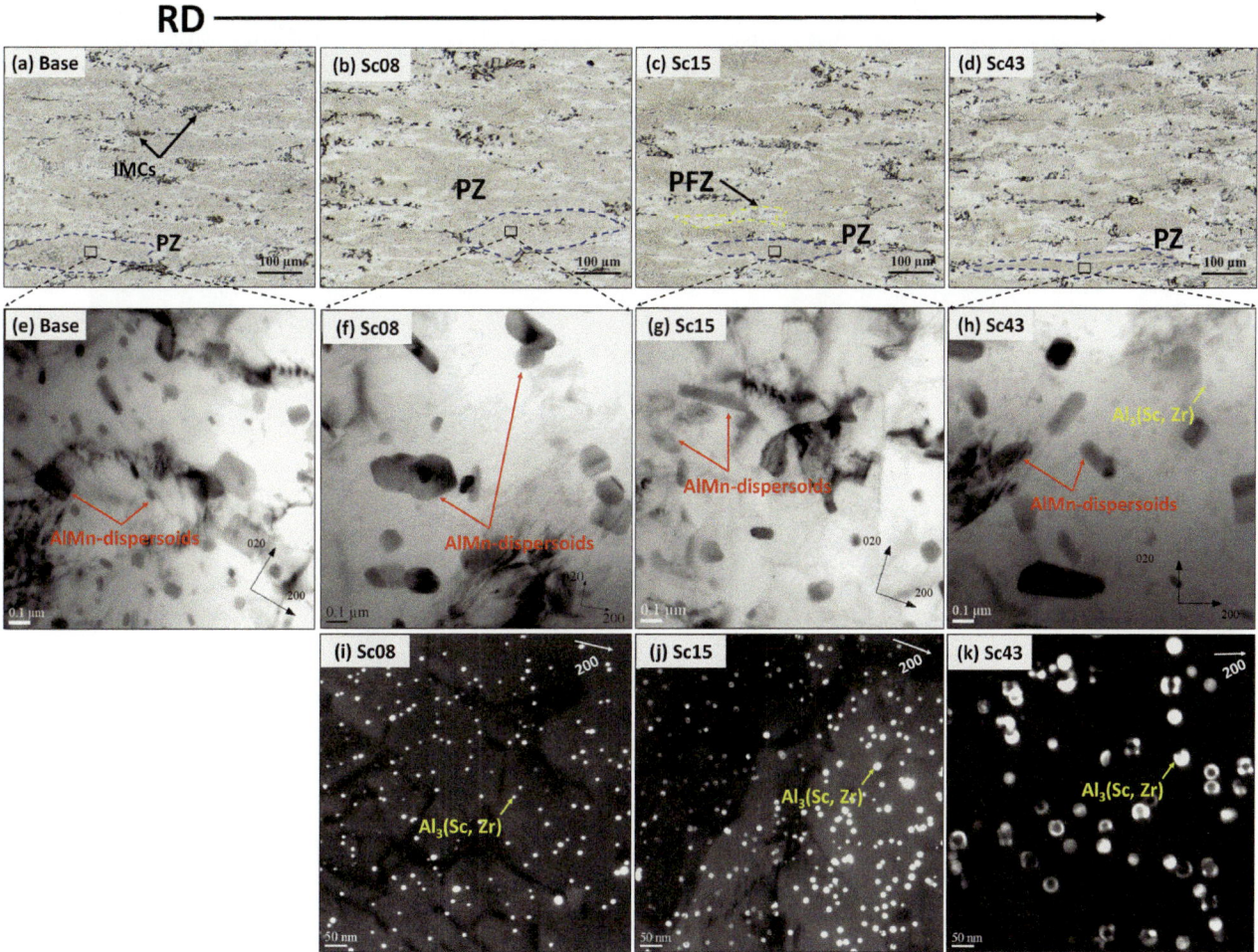

Fig. 3 Typical hot-rolled microstructures: (**a**, **b**, **c**, **d**) OM images and (**e**, **f**, **g**, **h**) bright-field TEM images showing the distribution of AlMn-dispersoids for Base, Sc08, Sc15, and Sc43; (**i**, **j**, **k**) dark-field TEM images showing the distribution of $Al_3(Sc,Zr)$ precipitates for Sc08, Sc15 and Sc43, respectively

fragmented, and the grains were elongated in the rolling direction, as shown in Fig. 3a–d. In addition, the line- and fan-shaped structures related to DCP in Sc43 completely disappeared. This could be attributed to the accelerated dissolution of the DCP rod-like precipitates due to the high rolling temperature (500 °C) and the accelerated diffusion rates during hot deformation [7, 22, 23].

Bright- and dark-field TEM images (Fig. 3e–h) and (Fig. 2i–k) revealed that both AlMn dispersoids and spherical $Al_3(Sc,Zr)$ precipitate coarsened during preheating and hot rolling as compared to the heat-treated alloys. This resulted in a size increase, and a significant decrease in number density (see Fig. 3 and Table 2). The quantitative results in Table 2 show that the number density of AlMn dispersoids decreased by 59%, 79%, and 76% during hot rolling for the Base, Sc15, and Sc43 alloys, respectively. Although the AlMn dispersoids coarsened, the number density in the base alloy was still higher than in Sc-containing alloys. On the other hand, the fine and

spherical $Al_3(Sc,Zr)$ particles coarsened and exhibited a reduction in number density, particularly in the Sc43 alloy. For instance, the $Al_3(Sc,Zr)$ precipitate size increased during rolling by 104% and 107% in the Sc08 and Sc15 alloys, compared to 355% in the Sc43 alloy.

The reduced particle density and increased size of both AlMn dispersoids and $Al_3(Sc,Zr)$ precipitates after hot rolling are predominantly attributed to two factors. Firstly, the rolling temperature (500 °C) is higher than their formation temperature [7, 22]. Secondly, the high number density of dislocations generated during rolling accelerates the diffusion of alloying elements in the matrix, resulting in the growth and coarsening of both types of particles [23]. The most significant coarsening of spherical $Al_3(Sc,Zr)$ precipitates occurred in the Sc43 alloy (Fig. 3h), and the dissolved DCP $Al_3(Sc,Zr)$ further accelerates the coarsening of the spherical $Al_3(Sc,Zr)$ through the diffusion/growth mechanism, indicating their negative effect on the characteristics of the spherical $Al_3(Sc,Zr)$ precipitates [16].

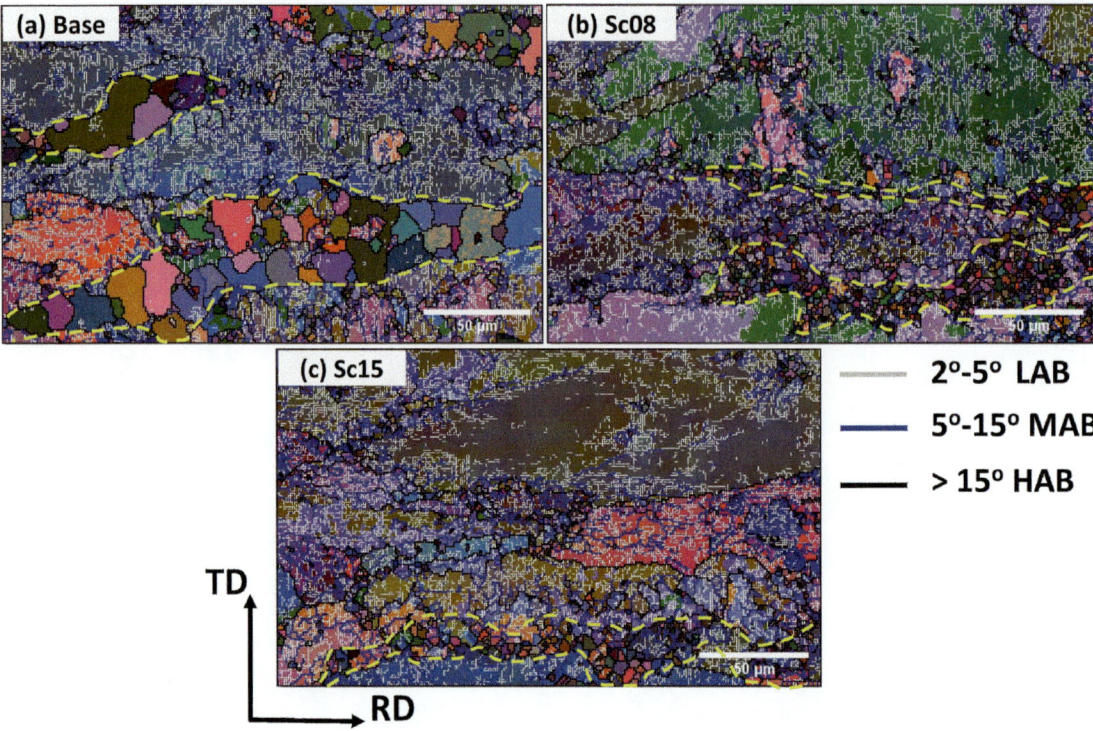

Fig. 4 All Euler orientation maps showing the grain structure after hot rolling and annealing

Figure 4 illustrates the grain structure of the base Sc08 and Sc15 alloys via Euler orientation maps after hot rolling and annealing. It can be seen that all samples exhibited a mixture of deformed grains in the rolling direction and recrystallized grains along the grain boundaries. In the base alloy, the recrystallized grains were coarser with an average size of 25 μm (marked by yellow dashed lines in Fig. 4a) and were randomly grown in the deformed grains. However, in the Sc-containing alloys, a chain of fine equiaxed and recrystallized grains (marked by yellow lines in Fig. 4b, c) was distributed along the grain boundaries with an average size of 3–8 μm. The lower fraction and the smaller size of recrystallized grains in the Sc-containing alloys can be explained by the effective pinning effect of $Al_3(Sc,Zr)$ precipitates [3, 24, 25]. The predominant fibrous deformed grains along with the fine recrystallized grains in the Sc-containing alloys could provide additional strengthening at room temperature by inhibiting dislocation movement and grain rotation [26].

Mechanical Properties of Hot-Rolled Sheets

Figure 5 shows the microhardness of the experimental alloys after heat treatment and rolling. In general, the microhardness of the Sc-containing alloys was substantially higher than the base alloy, indicating the significant strengthening

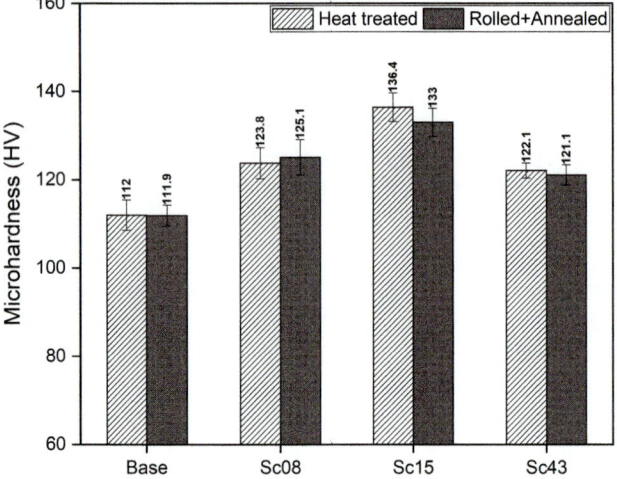

Fig. 5 Microhardness after heat treatment and hot rolling/annealing

effect of the nano-sized, spherical $Al_3(Sc,Zr)$ precipitates. It can be seen that the microhardness value first increased with increasing Sc content, from the base alloy via Sc08 up to the Sc15 alloy, and then decreased significantly to the Sc43 alloy. The microhardness evolution after rolling and annealing showed a similar trend to the as heat-treated condition, corresponding to the substantial strengthening effect of the fine and spherical $Al_3(Sc,Zr)$ precipitates (Fig. 3).

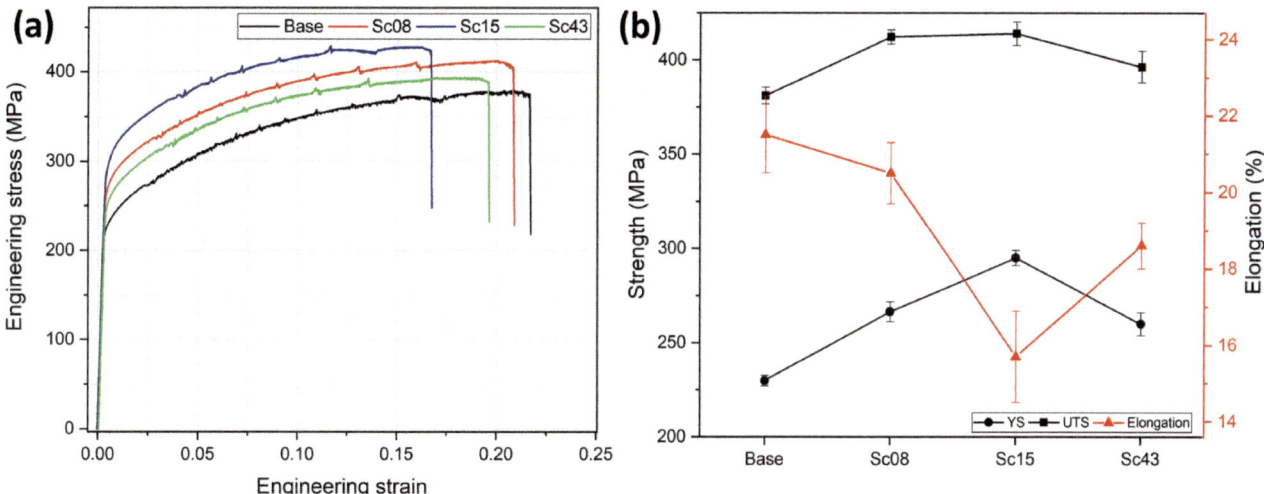

Fig. 6 **a** Typical engineering stress–strain curves of the rolled/annealed samples, and **b** the mean tensile properties versus Sc content

Figure 6 shows the ambient-temperature tensile properties of all four hot-rolled samples. Figure 6a demonstrates the typical engineering stress–strain curves. As shown in Fig. 6a, the stress level significantly increased with the addition of Sc and Zr, confirming the contribution of Sc towards improving the tensile properties. Figure 6b presents the tensile property data (YS, UTS and elongation). Similar to the microhardness trend, the YS and UTS initially increased with the Sc addition. The Sc15 alloy achieved the highest YS and UTS at 295 and 411 MPa, representing an improvement of 28% in YS and 8% in UTS relative to the base alloy. However, further increasing the Sc content to 0.43 wt.% significantly decreased the strength. For example, the Sc43 alloy gave a YS of 260 MPa, corresponding to an improvement of 13% compared to the base alloy, which was slightly lower than that of the Sc08 alloy.

As shown in Figs. 5 and 6, both the hardness and strength initially increased with increasing Sc addition up to 0.16% Sc, and then gradually degraded with further addition of Sc to 0.43%. These changes in mechanical properties are mainly attributed to the evolution of microstructures, especially the evolution of both AlMn dispersoids and spherical Al$_3$(Sc,Zr) precipitates. As illustrated in Fig. 2 and Table 2, with the addition of Sc and Zr, although the size of AlMn dispersoids increased and the number density decreased, the precipitation of a large number of fine and spherical Al$_3$(Sc, Zr) more than compensated for the loss of AlMn dispersoids, leading to the higher hardness in Sc-containing alloys (Fig. 5). During hot-rolling, although both AlMn dispersoids and spherical Al$_3$(Sc,Zr) coarsened (Fig. 3), the spherical Al$_3$(Sc,Zr) still effectively improved the mechanical properties due to their fine size and high number density. Meanwhile, the number density of fine and spherical Al$_3$(Sc, Zr) increased with the increasing Sc up to ∼0.16%, which

gave the highest mechanical properties. However, a decrease of both hardness and mechanical properties was observed with further addition of Sc (0.43%). This was principally associated with the formation of coarse discontinuous Al$_3$(Sc,Zr) in the as-cast condition, consuming more Sc and Zr from the solid solution. This resulted in a low number density of coarser spherical Al$_3$(Sc,Zr) precipitates after heat treatment (Fig. 3 and Table 2), resulting in the decreased hardness of Sc043 after heat treatment in Fig. 4. During hot-rolling, the released Sc and Zr solutes from the dissolution of discontinuous Al$_3$(Sc,Zr) further accelerated the coarsening of spherical Al$_3$(Sc,Zr) precipitates (Fig. 3k), leading to lower mechanical properties compared to the Sc15 alloy. However, the mechanical properties of the Sc43 alloy were still higher than the Sc-free base alloy, further confirming the positive contribution of nano-sized, spherical Al$_3$(Sc,Zr) precipitates on mechanical properties. However, due to the high cost of Sc, the addition of Sc should be well controlled to achieve the optimal mechanical properties for industrial applications, which is ∼0.16 wt.% Sc based on the strength evolution in the context of this work.

Conclusions

(1) During multistep heat treatment, two strengthening phases (AlMn dispersoids and Al$_3$(Sc,Zr) precipitates) precipitated in the Sc/Zr-containing Al-Mg-Mn based alloys. However, the addition of Sc and Zr caused a reduction in number density of AlMn dispersoids. At high Sc levels (0.43 wt.%), the presence of coarse rod-like discontinuous Al$_3$(Sc,Zr) reduced the number density of fine and spherical Al$_3$(Sc,Zr) precipitates.

(2) During hot-rolling, both AlMn dispersoids and spherical $Al_3(Sc,Zr)$ exhibited coarsening. More significant coarsening of spherical $Al_3(Sc,Zr)$ particles occurred with the 0.43% Sc addition due to the dissolution of discontinuous $Al_3(Sc,Zr)$.

(3) Hardness and strength increased with Sc additions up to 0.16 wt.%, followed by a decrease with further addition of Sc to 0.43 wt.%.

(4) In the hot-rolled and annealed condition, the highest strength was achieved with 0.16 wt.% Sc. The YS and UTS values of 295 and 411 MPa, respectively, represented improvements of 28% in YS and 8% in UTS compared to the base alloy free of Sc.

Acknowledgements The authors acknowledge the financial support from the Natural Sciences and Engineering Research Council of Canada (NSERC) and Rio Tinto Aluminum under the Grant No. CRDPJ 514651-17, through the Research Chair in Metallurgy of Aluminum Transformation at the University of Quebec at Chicoutimi.

References

1. J.R. Davis, (2001) Aluminum and aluminum alloys, light metals and alloys p 351–416. https://doi.org/10.1361/autb2001p351

2. J.A.V.D. Hoeven, L. Zhuang, (2002) A New 5xxx Series Alloy Developed for Automotive Applications, SAE TECHNICAL PAPER SERIES (724): 1-8. https://doi.org/10.4271/2002-01-2128

3. J. Jiang, F. Jiang, M. Zhang, Z. Tang, M. Tong, (2020) Recrystallization behavior of Al-Mg-Mn-Sc-Zr alloy based on two different deformation ways, Materials Letters 265: 127455. https://doi.org/10.1016/j.matlet.2020.127455

4. Y. Peng, S. Li, Y. Deng, H. Zhou, G. Xu, Z. Yin, (2016) Synergetic effects of Sc and Zr microalloying and heat treatment on mechanical properties and exfoliation corrosion behavior of Al-Mg-Mn alloys, Materials Science and Engineering A 666: 61-71. https://doi.org/10.1016/j.msea.2016.04.029

5. C.B. Fuller, J.L. Murray, D.N. Seidman, (2005) Temporal evolution of the nanostructure of Al(Sc ,Zr) alloys:Part I - Chemical compositions of Al3(Sc1-xZrx) precipitates, Acta Materialia 53(20): 5401-5413. https://doi.org/10.1016/j.actamat.2005.08.016

6. Y. Li, A. Muggerud, A. Olsen, T. Furu, (2012) Precipitation of partially coherent α-Al (Mn, Fe) Si dispersoids and their strengthening effect in AA 3003 alloy, Acta Materialia 60(3): 1004-1014. https://doi.org/10.1016/j.actamat.2011.11.003

7. K. Liu, X.G. Chen, (2015) Development of Al-Mn-Mg 3004 alloy for applications at elevated temperature via dispersoid strengthening, Materials and Design 84: 340-350. https://doi.org/10.1016/j.matdes.2015.06.140

8. A.Y. Algendy, K. Liu, X.G. Chen, (2021) Evolution of dispersoids during multistep heat treatments and their effect on rolling performance in an Al-5 % Mg-0 . 8 % Mn alloy, Materials Characterization 181:111487. https://doi.org/10.1016/j.matchar.2021.111487

9. Z. Li, Z. Zhang, X.G. Chen, (2018) Improvement in the mechanical properties and creep resistance of Al-Mn-Mg 3004 alloy with Sc and Zr addition, Materials Science and Engineering A 729: 196-207. https://doi.org/10.1016/j.msea.2018.05.055

10. A.Y. Algendy, K. Liu, P. Rometsch, N. Parson, X.G. Chen, (2022) Effects of AlMn dispersoids and Al3(Sc,Zr) precipitates on the microstructure and ambient/elevated-temperature mechanical properties of hot-rolled AA5083 alloys, Materials Science and Engineering: A 855: 143950. https://doi.org/10.1016/j.msea.2022.143950

11. A.K. Lohar, B. Mondal, D. Rafaja, V. Klemm, S.C. Panigrahi, (2009) Microstructural investigations on as-cast and annealed Al-Sc and Al-Sc-Zr alloys, Materials Characterization 60(11): 1387-1394. https://doi.org/10.1016/j.matchar.2009.06.012

12. J. Røyset, N. Ryum, (2005) Scandium in aluminium alloys, International Materials Reviews 50(1): 19-44. https://doi.org/10.1179/174328005X14311

13. D.N. Seidman, E.A. Marquis, D.C. Dunand, (2002) Precipitation strengthening at ambient and elevated temperatures of heat-treatable Al (Sc) alloys, Acta Materialia 50(16): 4021-4035. https://doi.org/10.1016/S1359-6454(02)00201-X

14. Q. Dong, A. Howells, D.J. Lloyd, M. Gallerneault, V. Fallah, (2020) Effect of solidification cooling rate on kinetics of continuous/discontinuous Al3(Sc,Zr) precipitation and the subsequent age-hardening response in cold-rolled AlMgSc(Zr) sheets, Materials Science and Engineering A 772: 138693. https://doi.org/10.1016/j.msea.2019.138693

15. Y. Sun, Q. Pan, Y. Luo, S. Liu, W. Wang, J. Ye, Y. Shi, Z. Huang, S. Xiang, Y. Liu, (2021) The effects of scandium heterogeneous distribution on the precipitation behavior of Al3(Sc, Zr) in aluminum alloys, Materials Characterization 174: 110971. https://doi.org/10.1016/j.matchar.2021.110971

16. A.K. Lohar, B.N. Mondal, S.C. Panigrahi, (2010) Influence of cooling rate on the microstructure and ageing behavior of as-cast Al–Sc–Zr alloy, Journal of Materials Processing Tech. 210(15): 2135-2141. https://doi.org/10.1016/j.jmatprotec.2010.07.035

17. K. Liu, E. Elgallad, C. Li, X.G. Chen, (2021) Effects of Zr and Sc additions on precipitation of α-Al(FeMn)Si dispersoids under various heat treatments in Al-Mg-Si AA6082 alloys, International Journal of Materials Research 112(9): 706-716. https://doi.org/10.1515/ijmr-2021-8283

18. L.L. Rokhlin, N.R. Bochvar, I.E. Tarytina, N.P. Leonova, (2010) Phase composition and recrystallization of Al-based Al-Sc-Mn-Zr alloys, Russian Metallurgy (Metally) 2010(3): 241-247. https://doi.org/10.1134/S0036029510030158

19. A.Y. Algendy, K. Liu, P. Rometsch, N. Parson, X.G. Chen, (2022) Evolution of discontinuous/continuous Al3(Sc, Zr) precipitation in Al-Mg-Mn 5083 alloy during thermomechanical process and its impact on tensile properties, Materials Characterization 192: 112241. https://doi.org/10.1016/j.matchar.2022.112241

20. Z. Li, Z. Zhang, X.G. Chen, (2018) Effect of Metastable Mg2Si and Dislocations on α-Al(MnFe)Si Dispersoid Formation in Al-Mn-Mg 3xxx Alloys, Metallurgical and Materials Transactions A: Physical Metallurgy and Materials Science 49(11): 5799-5814. https://doi.org/10.1007/s11661-018-4852-4

21. M.J. Starink, N. Gao, N. Kamp, S.C. Wang, P.D. Pitcher, I. Sinclair, (2006) Relations between microstructure, precipitation, age-formability and damage tolerance of Al-Cu-Mg-Li (Mn, Zr, Sc) alloys for age forming, Materials Science and Engineering A 418(1-2): 241-249. https://doi.org/10.1016/j.msea.2005.11.023

22. P. Xu, F. Jiang, M. Tong, Z. Tang, J. Jiang, N. Yan, Y. Peng, (2019) Precipitation characteristics and morphological transitions of Al3Sc precipitates, Journal of Alloys and Compounds 790: 509-516. https://doi.org/10.1016/j.jallcom.2019.03.256

23. M. Cabibbo, E. Evangelista, M. Vedani, (2005) Influence of severe plastic deformations on secondary phase precipitation in a 6082

Al-Mg-Si alloy, Metallurgical and Materials Transactions A 36(5): 1353-1364. https://doi.org/10.1007/s11661-005-0226-9

24. M. Li, Q. Pan, Y. Shi, X. Sun, H. Xiang, (2017) High strain rate superplasticity in an Al–Mg–Sc–Zr alloy processed via simple rolling, Materials Science and Engineering A 687 298-305. https://doi.org/10.1016/j.msea.2017.01.091

25. J. Jiang, F. Jiang, M. Zhang, Z. Tang, M. Tong, (2020) Al3(Sc, Zr) precipitation in deformed Al-Mg-Mn-Sc-Zr alloy: effect of annealing temperature and dislocation density, Journal of Alloys and Compounds 831: 154856. https://doi.org/10.1016/j.jallcom.2020.154856

26. A.H. Chokshi, (2020) Grain boundary processes in strengthening, weakening, and superplasticity, Advanced Engineering Materials 22(1): 1900748. https://doi.org/10.1002/adem.201900748

Effect of Cooling Rate on W-Phase Formation in Al-Cu-Sc Alloys

Austin DePottey, Lu Jiang, Thomas Dorin, Thomas Wood, Timothy Langan, and Paul Sanders

Abstract

Aluminum-copper-scandium alloys show significant potential for high-strength applications; however, the formation of the detrimental W-phase (nominally Al_8Cu_4Sc) has prevented commercial adoption. There is not a strong consensus as to what conditions lead to the formation of W-phase, but two key factors are the cooling rate during solidification and the homogenization heat treatment. In this work, the effect of cooling rate on the formation of W-phase in Al-Cu-Sc alloys is investigated utilizing wedge molds that produce solidification rates from ~ 0.25 to 100 K/s. Samples are examined in both the as-cast state and following homogenization and aging treatments.

Keywords

W-phase • Al-Cu-Sc alloy • Al8Cu4Sc

Background

With increasing regulatory and customer demands, design engineers are increasingly turning to materials with high strength-to-weight ratios. This is especially true in the aerospace and automotive industries. Aluminum alloys are commonly used for lightweighting, due to good strength-to-weight ratios and ease of manufacturing. Most aluminum alloys however, have poor high-temperature strength, which limits the applications where they may be used [1]. Thus, there is demand for lightweight, high-strength materials that have improved high-temperature properties over commercially available materials.

One potential candidate for higher temperatures is aluminum-copper-scandium alloys. Recent work shows that scandium additions to aluminum-copper alloys result in a more refined θ' precipitates, which in turn increases the strength [2]. The alloy utilizes a heat treatment (Fig. 1) which first precipitates out scandium as Al_3Sc dispersoid cores, and then forms a shell by activating zirconium diffusion at a higher temperature. A final, still higher temperature step is used to solutionize any remaining θ phase. The interface of the $Al_3(Sc, Zr)$ dispersoids then acts as nucleation sites for θ' precipitates during natural or artificial aging. Due to the homogeneous distribution of the $Al_3(Sc, Zr)$ dispersoids, the θ' precipitates nucleate in larger numbers and therefore grow to smaller average size [2, 3]. Several recent studies show that scandium preferentially segregates to coherent θ' boundaries, which impedes copper diffusion [4, 5] and restricts coarsening of the θ' precipitates, slowing high-temperature deformation.

One of the primary reasons that aluminum-copper-scandium alloys have not been thoroughly investigated until recently is due to formation of the ternary W-phase. This phase, with a composition $Al_{8-x}Cu_{4+x}Sc$ ($0 \leq x \leq \sim 2.15$) [6], is stable throughout the temperature range for used homogenization/solutionizing treatments [7]. The W-phase consumes copper and scandium, leading to lower matrix concentrations and fewer strengthening precipitates, such that scandium-containing aluminum-copper alloys may have lower mechanical properties than scandium-free alloys [8–10].

Literature on the formation of W-phase during non-equilibrium thermal processing is inconsistent. Table 1 summarizes the formation of W-phase in Al-Cu-Sc alloys. As predicted by thermodynamics [11], W-phase tends not to form in alloys that are low in copper (approximately 2.5 wt %) [7]. Additionally, samples that experience longer times at high temperatures, either during homogenization or due to

A. DePottey (✉) · T. Wood · P. Sanders
Michigan Technological University, Houghton, USA
e-mail: amdepott@mtu.edu

L. Jiang · T. Dorin
Deakin University, Melbourne, Australia

T. Langan
Sunrise Energy Metals, Melbourne, Australia

© The Minerals, Metals & Materials Society 2023
S. Broek (ed.), *Light Metals 2023*, The Minerals, Metals & Materials Series,
https://doi.org/10.1007/978-3-031-22532-1_166

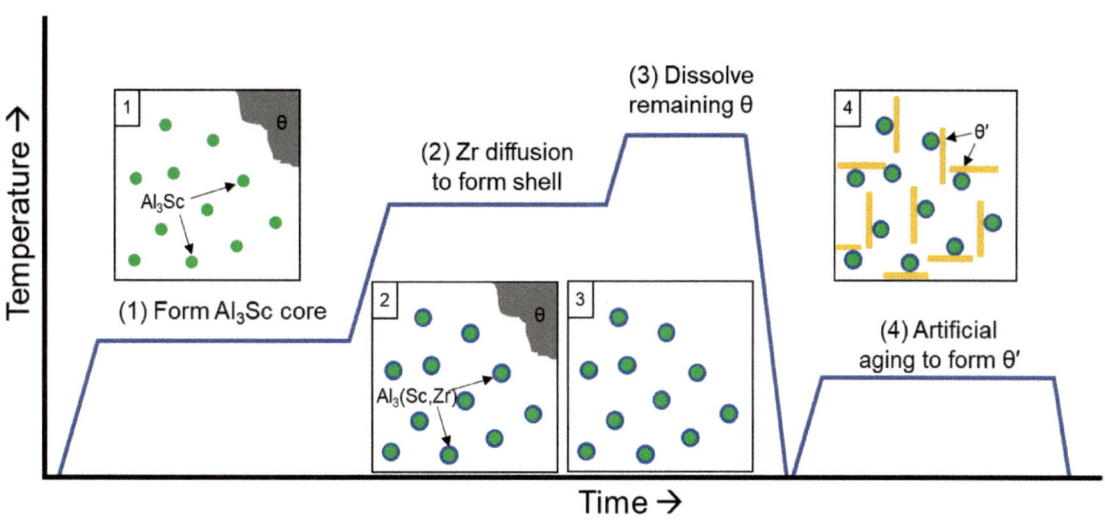

Fig. 1 Schematic of heat treatment utilized by Dorin et al. that produces refinement of θ′ after aging [2]

Table 1 Summary of literature results on W-phase formation

Chemistry	Heat treatment	W-phase present?	References
Al-5.9Cu-0.10Sc-0.15Zr-0.3 Mg-0.3Mn-0.09Ti	As-cast, metal mold preheated to 200 °C	No	[15]
Al-5.9Cu-0.30Sc-0.13Zr-0.3 Mg-0.3Mn-0.09Ti	As-cast, metal mold preheated to 200 °C	Yes	[15]
Al-5.6Cu-0.17Sc-0.12Zr-0.72 Mg-0.5Ag-0.32Mn-0.07Ti-0.1Ge-0.02Ni-0.03Fe-0.01Si-0.02 V	As-cast, chilled copper mold, ~ 70 K/s solidification rate	No	[8]
Al-5.6Cu-0.17Sc-0.12Zr-0.72 Mg-0.5Ag-0.32Mn-0.07Ti-0.1Ge-0.02Ni-0.03Fe-0.01Si-0.02 V	360 °C / 6 h + 510 ° C / 24 h	Yes	[8]
Al-4.18Cu-0.063Sc-0.1Zr-0.92Cu-0.3 Mg-0.38Ag	As-cast, chilled copper mold	No	[16]
Al-4.18Cu-0.063Sc-0.1Zr-0.92Cu-0.3 Mg-0.38Ag	470 °C / 8 h + 520 ° C / 24 h	Yes	[16]
Al-4.5Cu-0.4Sc	Solidified in DSC, 0.1–0.8 K/s solidification rate	Yes	[14]
Al-4.5Cu-0.4Sc	Atomized powder, ~ 800–10,000 K/s solidification rate	No	[13]
Al-4.5Cu-0.4Sc	Atomized powder, 535 °C / 18.5 h + water quench + 240 ° C/2 h	Yes	[13]
Al-4.5Cu-0.6Sc-1.5 Mg	As-cast, 10–100 K/s solidification rate	No	[12]
Al-4.5Cu-0.6Sc-1.5 Mg	Solidified in DSC, 0.16 K/s solidification rate	Yes	[12]
Al-2.57Cu-0.24Sc	As-cast, steel mold preheated to 200 °C	No	[17]

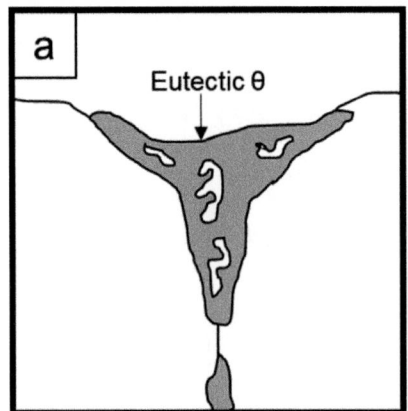 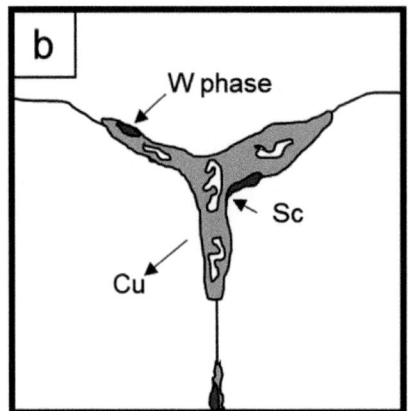 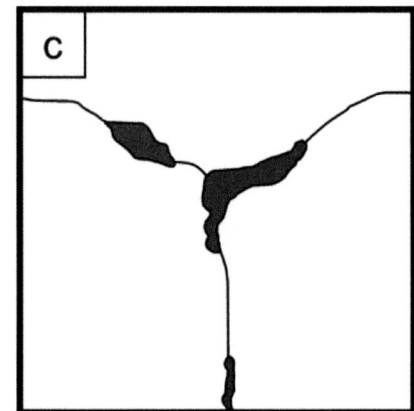

Fig. 2 A mechanism for formation of W-phase: **a** an eutectic θ phase that contains scandium forms during solidification, **b** at high temperatures, copper diffuses out of θ phase and scandium diffuses in, with W-phase nucleating at the incoherent θ phase interface, and **c** at long times, the matrix is depleted of scandium and W-phase remains in the eutectic areas

slow solidification rates, are likely to form the W-phase [12–14].

A proposed mechanism for the formation of W-phase begins with scandium enrichment in the eutectic θ phase during solidification. When held at high temperatures, copper diffuses out of the θ phase, while additional scandium diffuses into the eutectic region, nucleating W-phase at the incoherent interface. Scandium diffusion into the eutectic region continues, resulting in coarse particles of W-phase and a matrix depleted of scandium [8, 16] (Fig. 2).

Given the limited understanding of W-phase formation, and the significant potential of aluminum-copper-scandium alloys, this study was undertaken to better control the formation of W-phase. Key factors include solidification rates in the range of commercial processes such as direct chill casting (~1–20 K/s) [18, 19], and the comparison of traditional heat treatments with those used in more recent studies [2, 5, 20, 21].

Experimental

Two alloys (Table 2) were prepared using commercially pure aluminum ingot (99.9%), with master alloys used for copper (50 wt%), scandium (2 wt%), and zirconium (10 wt %). Alloys were heated in a graphite crucible in an electric resistance furnace to 760 °C, then degassed for 15 min using argon gas. The chemistry was verified using a Bruker Q4 optical emission spectrometer.

The alloys were poured into wedge-shaped copper and chemically bonded sand molds. The wedge geometry for both mold types was 267 mm in length × 100 mm in width × 32 mm thick at the top tapering down to a point (Fig. 3). Type-K thermocouples were inserted evenly along the centerline of the mold in five locations to collect cooling rate data. The castings were cooled in the mold until all thermocouples read below 200 °C.

Wedges were sectioned and heat treated (Fig. 4). To avoid any phase formation during heating and more closely simulate an isothermal treatment, homogenization was performed in a salt bath, and artificial aging was performed in a fluidized bed. The time between quenching and artificial aging was less than 10 min. Following heat treatment, the wedges were sectioned to the centerline of the mold and Vickers hardness (10 kg load, 15 s dwell) measurements were made at regular intervals along the length of the wedge (at least three replicates per location).

Optical metallography was performed on selected areas of each wedge under each treatment condition. Dilute Keller's reagent (0.5% HF, 0.75% HCl, 1.25% mL HNO3, balance distilled water) was applied for 20 s. Scanning electron microscopy (SEM) was performed using a FEI Philips XL 40 SEM equipped with an Oxford Ultimax EDS detector. Transmission electron microscopy (TEM) samples were prepared by mechanically polishing to 120 μm followed by electropolishing in an electrolyte of 10% nitric acid in methanol at −25 °C with a voltage of 30 V. TEM was performed on a FEI Titan Themis operated at 200 kV.

Table 2 Compositions produced in this work (wt%)

Alloy	Cu	Sc	Zr	Fe	Si
Cu6	6.25	< 0.01	0.003	0.05	0.04
Cu6ScZr	5.63	0.114	0.088	0.09	0.04

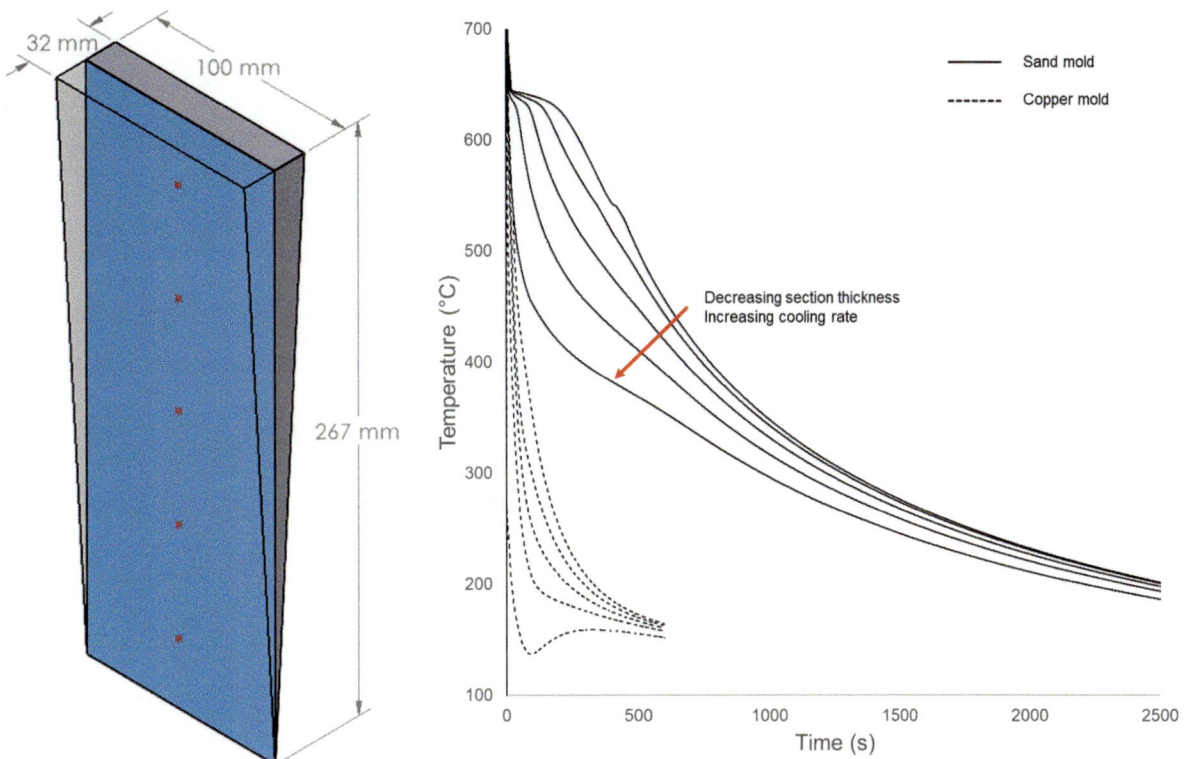

Fig. 3 (Left) Wedge mold geometry. Red points are thermocouple locations. Blue plane is the plane that all samples were analyzed on. (Right) Cooling curves collected from the wedges. Solidification rates are taken as an average between the liquidus, and solidus temperatures observed from these cooling curves.

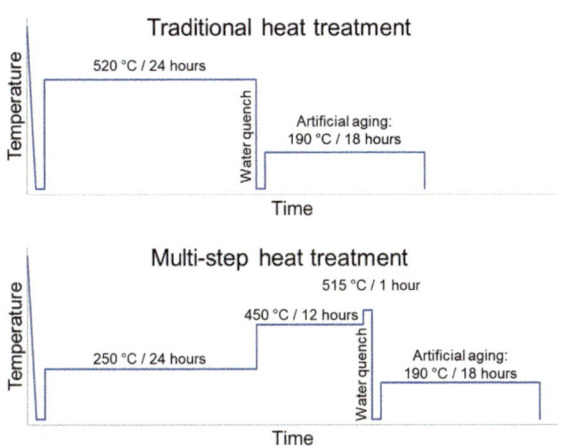

Fig. 4 Heat treatments examined in this work

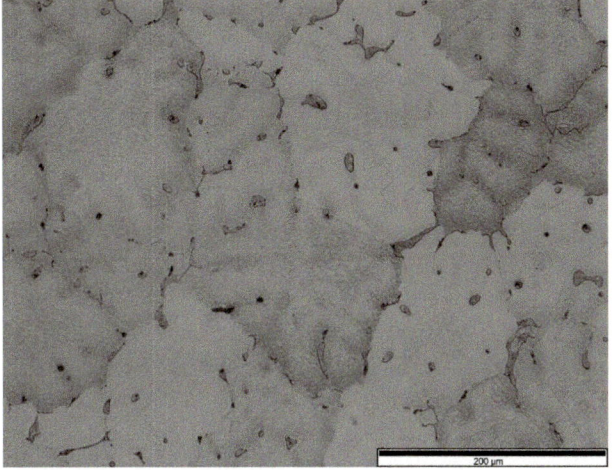

Fig. 5 Typical microstructure in the as-cast state (Cu6 at 5 K/s solidification rate) exhibiting visible coring

Results

As-Cast Condition

For both alloys, the observed microstructure exhibited dendritic solidification with coring of copper observed for all solidification rates (Fig. 5). In the SEM, the scandium-containing samples exhibited scandium segregation to the θ phase (Fig. 6), as has previously been reported [8, 22].

The grain size of both alloys studied was approximately the same for a given solidification rate, and there is very little effect of solidification rate on hardness. The scandium-

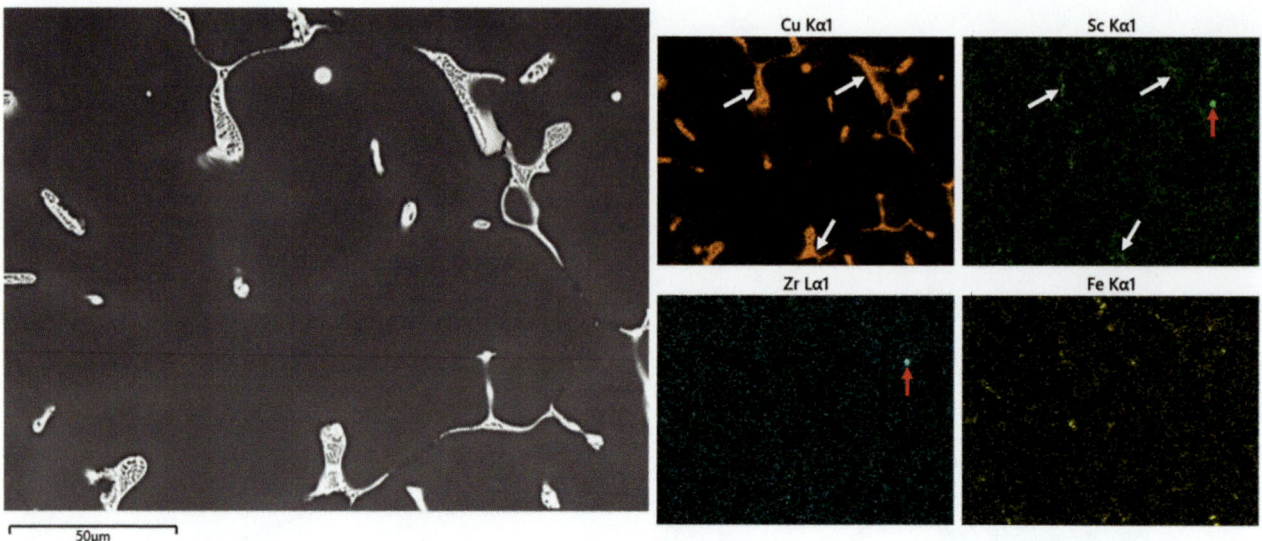

Fig. 6 SEM image of Cu6ScZr at 5 K/s solidification rate in the as-cast state, with copper and scandium segregation observed in the eutectic region (white arrows). Red arrows indicate what appears to be Al$_3$(Sc,Zr) formed upon solidification

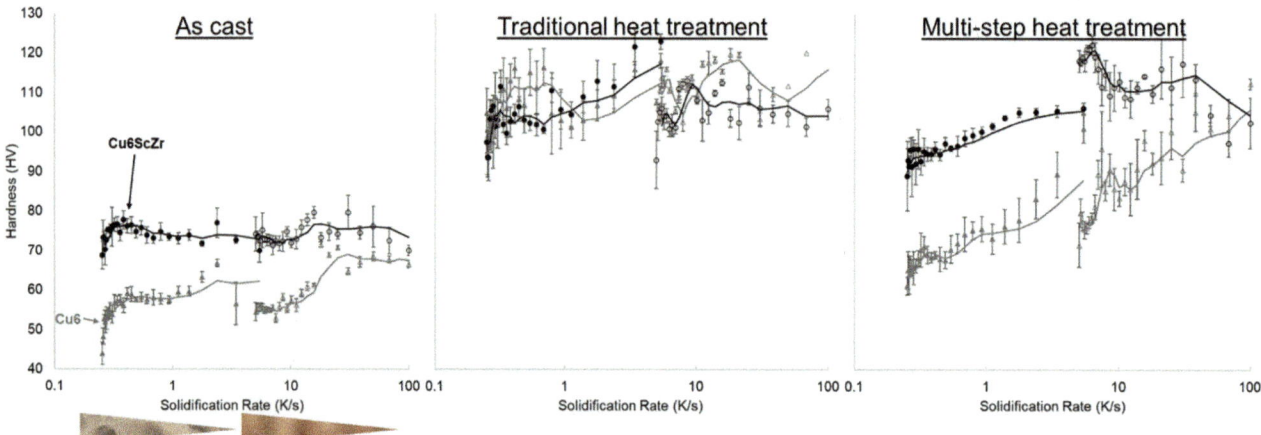

Fig. 7 Hardness of samples in each of the three states analyzed. Lower hardness at slower solidification rates is due to coarse porosity within top of the castings

containing sample exhibits higher hardness than the scandium-free samples (Fig. 7). Optical microscopy reveals coarse shrinkage porosity in slowly solidifying cast sections, leading to the lower local hardness.

Closer examination of the scandium-containing samples in slowly cooled eutectic regions reveals the presence of two phases (Fig. 8). The scandium-rich regions are likely W-phase that are formed during solidification. These phases are not observed in sections with solidification rates above ∼ 7 K/s, due to less time for W-phase formation and/or finer microstructure.

Traditional Heat Treatment

There was little hardness difference between scandium-containing and scandium-free alloys following the traditional heat treatment, and there was little change in response with cooling rate. The microstructure of scandium-free alloys was as expected, consisting mostly of copper-rich aluminum grains, with some undissolved θ present at prior eutectic regions (Fig. 9). Plate-like iron-rich intermetallics were observed in sections with slow solidification rates.

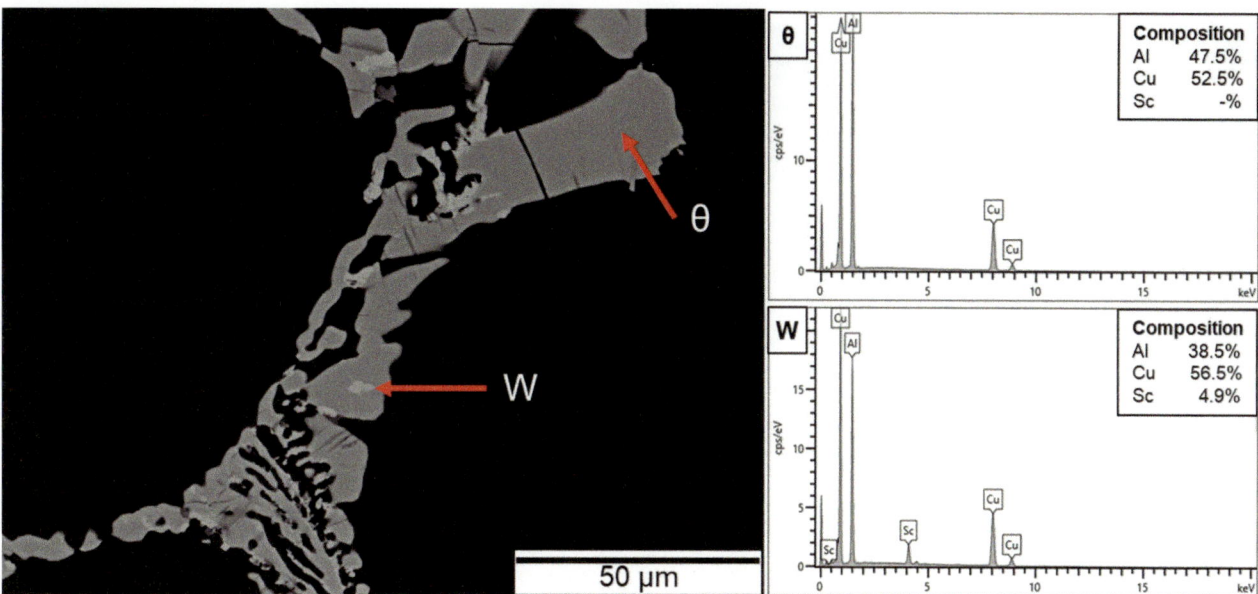

Fig. 8 BSE-SEM image of as-cast structure of Cu6ScZr at 0.25 K/s solidification rate, in the eutectic region. EDS shows that the majority of the region is θ phase, while the brighter phases are scandium-enriched and likely W-phase

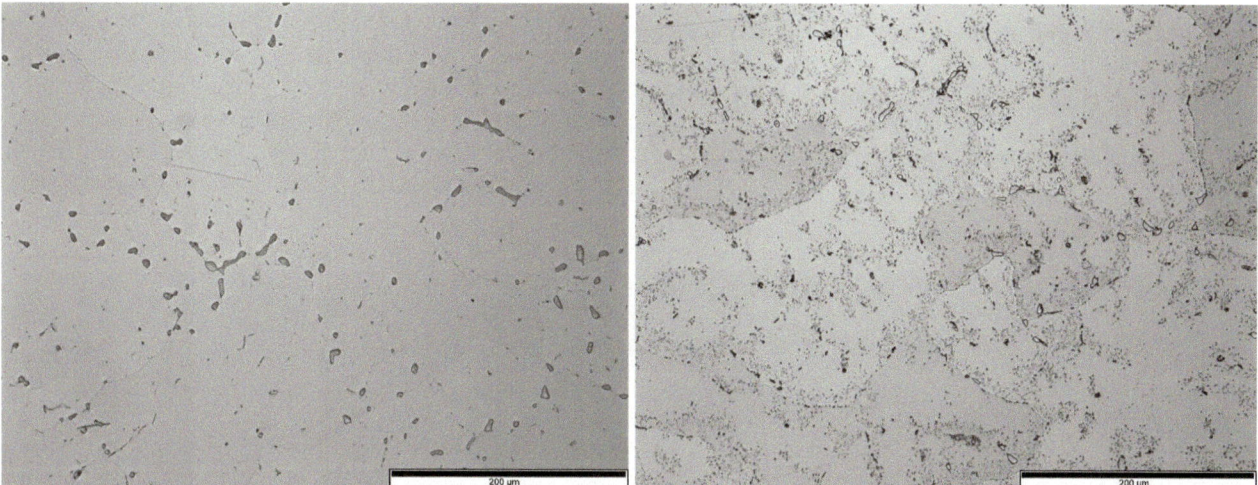

Fig. 9 (Left) Cu6 at 5 K/s solidification rate, showing primarily an α-Al matrix with small amounts of undissolved θ phase. (Right) Cu6ScZr at 5 K/s solidification rate, showing the prior dendritic structure with small phases present near the dendrite boundaries.

The scandium alloys contain a large number of small (∼0.5 μm) phases at the prior dendrite boundaries, enabled by enrichment of Cu due to coring in the as-cast structure. This phenomenon was observed in the entire range of solidification rates but was difficult to resolve at the fastest solidification rates. These phases are observed to be copper- and scandium-rich, indicating that they may be W-phase.

Multi-Step Heat Treatment

Following the multi-step heat treatment, the scandium-free alloys had an increase in hardness with increasing solidification rate, likely due to incomplete dissolution of the coarser θ phase at slower solidification rates given the final short solutionizing step. The scandium-containing alloys did

Fig. 10 Microstructure of Cu6ScZr at 5 K/s solidification rate, after multi-step heat treatment

exhibit a small increase in hardness, although it was lower in magnitude than the scandium-free alloys.

Similar to the traditional heat treatment, the scandium-containing alloys exhibited a large number of small ($\sim$0.5 µm) phases present at the prior dendrite boundaries or copperenriched regions (Fig. 10). This was less pronounced than in the traditional heat treatment, likely in part due to the multi-step heat treatment spending less time at high temperatures. The formation of coarsening resistant $Al_3(Sc,Zr)$ dispersoids, which tie up scandium, may reduce

the presence of the scandium-rich phases following a multi-step heat treatment.

TEM

To better understand the microstructure observed with optical microscopy and SEM, TEM was performed on the small phases observed in the scandium-containing alloys (Fig. 11). While diffraction in the small phases was unable to confirm the crystallographic structure, EDS measurements indicate that the phases are enriched in copper and scandium. Precipitate-free zones were also observed surrounding the small phases. In the areas away from the small phases, the θ' precipitate count and size closely matched what was previously reported in literature [2].

Discussion

As-Cast Condition

The W-phase formation in the as-cast microstructure was in good agreement with the literature. Previously, solidification rates that led to the formation of W-phase were on the order of 0.1–1 K/s [12, 14]. At faster solidification rates, >10 K/s, W-phase was generally not observed prior to heat treatment [8, 12, 13]. The transition at 7 K/s observed experimentally

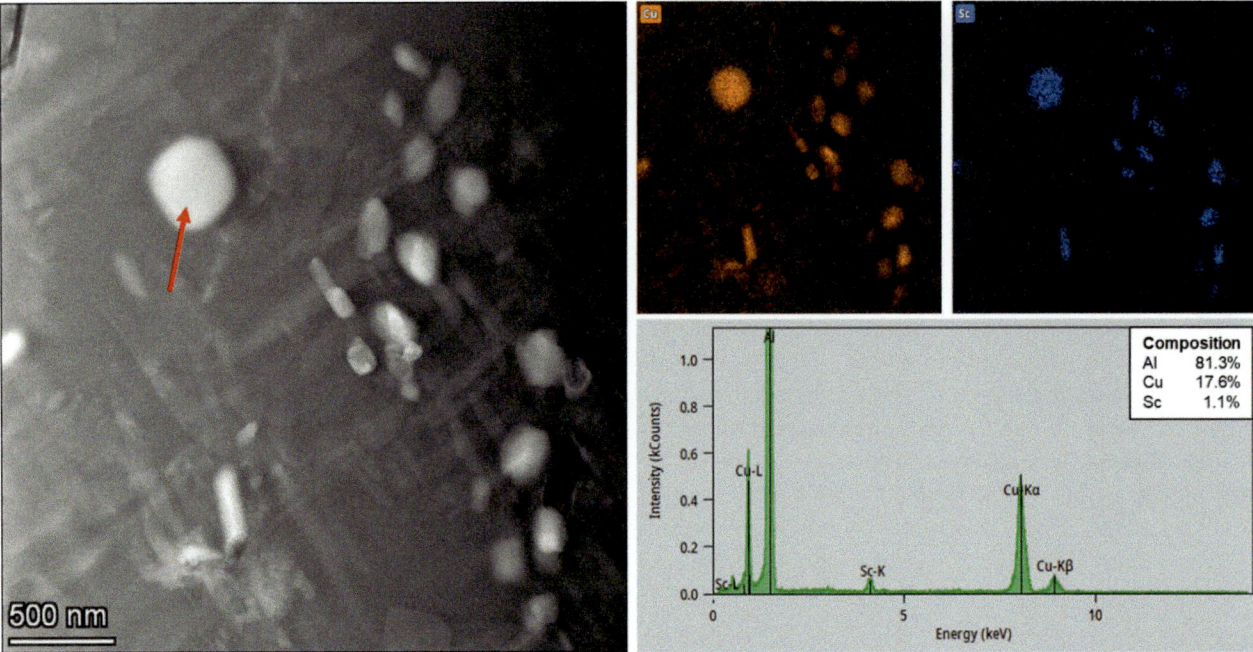

Fig. 11 HAADF-TEM micrograph of Cu6ScZr at 0.25 K/s solidification rate following the multi-step heat treatment. The bright phases are the small, dark particles observed in optical microscopy. EDS spectrum for the bright phases indicates they are enriched in copper and scandium

in this work is in good agreement with the previous findings. Additionally, the observation of scandium segregation into the eutectic θ phase agrees with previous literature findings [8, 22].

Heat-Treated Conditions

The presence of W-phase in heat-treated aluminum-copper-scandium alloys was previously observed as only occurring at the prior eutectic region [8, 16], which leads to the theory that W-phase only nucleates on existing θ interfaces. However, in this work, it was observed that small (∼0.5 μm) phases precipitated out within the matrix, which has not previously been described in literature. While the crystal structure has not been determined, they are rich in copper and scandium, and are likely W-phase. These phases were observed in all cooling rates, which explains the relatively flat hardening response in respect to solidification rate. These phases exhibited large (approximately 2–3 μm) precipitate-free zones surrounding them, which confirms the detrimental nature of W-phase.

The formation mechanism of these phases is not fully understood, but they tend to form in regions that were previously high in copper due to solidification coring. One possible explanation is that an intermediate θ phase is nucleating in early stages of the heat treatments, due to the increased copper concentration, and W-phase is nucleating on the resulting interfaces, eventually transforming the θ into W-phase. The other possible explanation is that W-phase is nucleating directly in the matrix, due to the higher copper and scandium concentration in those areas. Additional work is necessary to fully understand the formation sequence of these phases and how to avoid them.

Conclusions

Aluminum-copper-(scandium) alloys were studied at solidification rates from ∼ 0.25–100 K/s utilizing wedge molds. As-cast and two heat-treated conditions were examined for hardening response and formation of the ternary W-phase.

1: Minimal W-phase is formed during solidification at rates above ∼7 K/s, so it may be possible to manufacture billets (especially at smaller diameters) utilizing direct chill casting to avoid W-phase formation in the as-cast state.

2: In both heat treatment conditions, copper- and scandium-rich phases were observed to form at the regions where copper segregated during solidification between secondary dendrite arms and near the dendrite boundaries. The mechanism for the formation of these phases is not yet understood.

References

1. Davis JR (1993) ASM Specialty Handbook: Aluminum and Aluminum Alloys. ASM International.
2. Dorin T, Ramajayam M, Lamb J, Langan T (2017) Effect of Sc and Zr additions on the microstructure/strength of Al-Cu binary alloys. Mater Sci Eng A 707:58–64. https://doi.org/https://doi.org/10.1016/j.msea.2017.09.032
3. Deane K, Sanders P (2018) Effect of Zr Additions on Thermal Stability of Al-Cu Precipitates in As-Cast and Cold Worked Samples. Metals 8:331. https://doi.org/https://doi.org/10.3390/met8050331
4. Shin D, Shyam A, Lee S, et al (2017) Solute segregation at the Al/θ′-Al2Cu interface in Al-Cu alloys. Acta Mater 141:327–340. https://doi.org/https://doi.org/10.1016/j.actamat.2017.09.020
5. Jiang L, Rouxel B, Langan T, Dorin T (2021) Coupled segregation mechanisms of Sc, Zr and Mn at θ′ interfaces enhances the strength and thermal stability of Al-Cu alloys. Acta Mater 206:116634. https://doi.org/https://doi.org/10.1016/j.actamat.2021.116634
6. Suski W, Cichorek T, Wochowski K, et al (1997) Low-temperature electrical resistance of the U(Cu,Ni)4Al8 system and magnetic and electrical properties of ScCu4+xAl8−x. Phys B Condens Matter 230–232:324–326. https://doi.org/10.1016/S0921-4526(96)00704-1
7. Bo H, Liu LB, Jin ZP (2010) Thermodynamic analysis of Al-Sc, Cu-Sc and Al-Cu-Sc system. J Alloys Compd 490:318–325. https://doi.org/https://doi.org/10.1016/j.jallcom.2009.10.003
8. Gazizov M, Teleshov V, Zakharov V, Kaibyshev R (2011) Solidification behaviour and the effects of homogenisation on the structure of an Al–Cu–Mg–Ag–Sc alloy. J Alloys Compd 509:9497–9507. https://doi.org/https://doi.org/10.1016/j.jallcom.2011.07.050
9. Lee S-L, Wu C-T, Chen Y-D (2015) Effects of Minor Sc and Zr on the Microstructure and Mechanical Properties of Al-4.6Cu-0.3Mg-0.6Ag Alloys. J Mater Eng Perform 24:1165–1172. https://doi.org/https://doi.org/10.1007/s11665-014-1364-2
10. Zakharov VV, Rostova TD (1995) On the possibility of scandium alloying of copper-containing aluminum alloys. Met Sci Heat Treat 37:65–69. https://doi.org/https://doi.org/10.1007/BF01157047
11. Thermo-Calc Software TCAL8 Al-alloys database. (accessed 1 Feb 2022).
12. Norman AF, Hyde K, Costello F, et al (2003) Examination of the effect of Sc on 2000 and 7000 series aluminium alloy castings: for improvements in fusion welding. Mater Sci Eng A 354:188–198. https://doi.org/https://doi.org/10.1016/S0921-5093(02)00942-5
13. Bogno A-A, Henein H, Ivey DG, et al (2020) Effects of scandium on rapid solidified hypo-eutectic aluminium copper. Can Metall Q 59:101–115. https://doi.org/https://doi.org/10.1080/00084433.2019.1696606
14. Bogno A-A, Valloton J, Henein H, et al (2018) Effects of scandium on hypoeutectic aluminium copper microstructures under low solidification rate conditions. Can Metall Q 57:148–159. https://doi.org/https://doi.org/10.1080/00084433.2017.1403106

15. Qin J, Ma M, Tan P, et al (2022) Effects of Sc alloying on the evolution of solidification microstructure and formation of W phase in as-cast 2519 aluminum alloys. J Alloys Compd 898:162764. https://doi.org/https://doi.org/10.1016/j.jallcom.2021.162764

16. Jia M, Zheng Z, Gong Z (2014) Microstructure evolution of the 1469 Al–Cu–Li–Sc alloy during homogenization. J Alloys Compd 614:131–139. https://doi.org/https://doi.org/10.1016/j.jallcom.2014.06.033

17. Gao YH, Kuang J, Liu G, Sun J (2019) Effect of minor Sc and Fe co-addition on the microstructure and mechanical properties of Al-Cu alloys during homogenization treatment. Mater Sci Eng A 746:11–26. https://doi.org/https://doi.org/10.1016/j.msea.2018.12.099

18. Eskin DG, Katgerman L (2009) Solidification phenomena related to direct chill casting of aluminium alloys: fundamental studies and future challenges. Mater Technol 24:152–156. https://doi.org/https://doi.org/10.1179/106678509X12489478523537

19. Das SK (2006) Modeling and Optimization of Direct Chill Casting to Reduce Ingot Cracking. 74.

20. Rouxel B, Ramajayam M, Langan TJ, et al (2020) Effect of dislocations, Al3(Sc,Zr) distribution and ageing temperature on θ′ precipitation in Al-Cu-(Sc)-(Zr) alloys. Materialia 9:100610. https://doi.org/https://doi.org/10.1016/j.mtla.2020.100610

21. Lamb J, Rouxel B, Langan T, Dorin T (2020) Novel Al-Cu-Mn-Zr-Sc Compositions Exhibiting Increased Mechanical Performance after a High-Temperature Thermal Exposure. J Mater Eng Perform 29:5672–5684. https://doi.org/https://doi.org/10.1007/s11665-020-05040-0

22. Røyset J, Leinum JR, Øverlie HG, Reiso O (2006) An Investigation of the Solubility of Scandium in Iron-Bearing Constituent Particles in Aluminium Alloys. Mater Sci Forum 519–521:531–536. https://doi.org/https://doi.org/10.4028/www.scientific.net/MSF.519-521.531

Solute Clustering During Natural Ageing in Al-Cu-(Sc)-(Zr) Alloys

Lu Jiang, Kathleen Wood, Anna Sokolova, Robert Knott, Timothy Langan, and Thomas Dorin

Abstract

Solute clustering during natural ageing in Al-4wt.%Cu alloys with and without Sc and Zr has been studied. The alloys were heat treated to form Al_3(Sc, Zr) dispersoids before being solutionised at 500 °C followed by a water quench. Scanning transmission electron microscopy (STEM) and atom probe tomography (APT) were used to characterize Al_3(Sc, Zr) dispersoids in the Sc- and Zr-containing alloy. The APT results show that Cu segregates at the interfaces between Al_3(Sc, Zr) dispersoids and the Al matrix. Solute clustering during natural ageing was quantified in situ with small-angle X-ray scattering (SAXS). The in situ SAXS results show that the presence of Al_3(Sc, Zr) dispersoids significantly retards the solute clustering formation during natural ageing in the Al-Cu alloys. This could be related to the Cu segregation at the Al_3(Sc, Zr) dispersoids/Al matrix interfaces, which could deplete the solid solution from Cu solute which is no more available for solute clustering. Additionally, Sc solutes could bind with vacancies which reduces the free vacancies for the diffusion of Cu.

Keywords

Solute clustering · Natural ageing · Al-Cu alloys · Al_3(Sc · Zr) dispersoids

L. Jiang (✉) · T. Dorin
Institute for Frontier Materials, Deakin University, Geelong, Australia
e-mail: l.jiang@deakin.edu.au

K. Wood · A. Sokolova · R. Knott
Australian Nuclear Science and Technology Organisation, Sydney, Australia

T. Langan
Sunrise Energy Metals, Melbourne, Australia

Introduction

Reducing the weight of transportation vehicles is among the top priorities to achieve the goal of net-zero emissions. This potential can be realised through various means. The use of advanced, lightweight materials is one of the key methods to achieve optimal vehicle performance and the lightest weight. Al-Cu alloys have been used widely due to their outstanding mechanical properties [1, 2]. However, there is always a compromise in properties of aluminium alloys, such as strength, weldability, corrosion resistance, and thermal stability in harsh environments. For example, some alloys obtain their strength from precipitation hardening; however, this leads to relatively poor weldability and increased risk of stress corrosion cracking. Therefore, to reach and enhance lightweighting goals, there is a pressing need to develop new aluminium alloys that can enable even higher strength than current alloys and maintain other main properties, such as corrosion performance.

Sc is the most potent strength enhancer for Al alloys, providing up to ~ 100 MPa increase per 0.1wt.% addition without compromising other bulk properties [3]. Sc-modified Al alloys have an equivalent yield and tensile strength to some steel and titanium alloys, but are 1/3 the weight of steel and 40% lighter than titanium. Recently, a minor amount of Sc and Zr has been added to an Al-4wt.% Cu alloy to increase the yield strength by 150 MPa due to the formation of nano-sized Al_3(Sc, Zr) dispersoids which enhance the precipitation of strengthening phase, θ', at artificial ageing at elevated temperatures [4].

There remains, however, technical challenges related to understand microstructural evolution during room temperature storage, which limits the further development and manufacturing of new Sc-containing Al-Cu alloys. For Al-Cu alloys, room temperature storage after quenching from high-temperature solutionisation treatment and before elevated temperature artificial ageing is essential. During room temperature storage, the hardness/strength of the alloy

© The Minerals, Metals & Materials Society 2023
S. Broek (ed.), *Light Metals 2023*, The Minerals, Metals & Materials Series,
https://doi.org/10.1007/978-3-031-22532-1_167

increases due to the formation of nano-sized unordered accumulations of solute atoms, so-called solute clusters. This phenomenon is known as natural ageing (NA). However, to date, we are still lacking a basic understanding of how $Al_3(Sc, Zr)$ dispersoids affect the solute clustering during natural ageing in Al-Cu alloys. Therefore, the aim of this work is to investigate the effect of the presence of $Al_3(Sc, Zr)$ dispersoids on the solute clustering during natural ageing in Al-Cu alloys using small-angle X-ray scattering (SAXS). Hardness was also monitored during natural ageing. Scanning transmission electron microscopy (STEM) and atom probe tomography (APT) were used to characterise the formation of $Al_3(Sc, Zr)$ dispersoids.

Materials and Experimental Methods

The alloys used for this work were Al-4wt.%Cu with and without 0.1wt.%Sc and 0.1wt.%Zr, which are named as Al-Cu-Sc-Zr and Al-Cu alloys, respectively. Both alloys were cast at 720 °C in a small induction furnace with a ~ 4 kg capacity for aluminium and poured into a cylindrical steel mold. The alloys were cast from master alloys to ensure a maximum homogeneity of the billets. The master alloys were Al-33wt.%Cu, Al-2wt.%Sc, and Al-5wt.%Zr. After casting, both alloys were heat treated at 250 °C for 24 h, 450 °C for 12 h, and 500 °C for 8 h, in order to form $Al_3(Sc, Zr)$ dispersoids and dissolve the coarse intermetallics left over from the casting. After that, a traditional solution treatment was conducted at 500 °C for 1 h, followed by quenching at cold water. Then, natural ageing was conducted at room temperature.

Hardness of both alloys during natural ageing was monitored using HWDV-75 Vickers Hardness Tester under 1 kg load with a 30 s loading time at room temperature. At least seven indentations were measured. The maximum and minimum values were removed when calculating the average hardness.

Scanning transmission electron microscopy (STEM) foils were electropolished in a twin-jet Tenupol instrument in a solution of 33% nitric acid in methanol at -25 °C at a voltage of 13 V. STEM was performed on a JEOL 2100F TEM at 200 kV voltage.

Atom probe tomography (APT) samples were prepared by a standard two-step electropolishing. APT was conducted on a CAMECA Local Electrode Atom Probe (LEAP) 5000XR instrument. Data collected was performed at 25 K with a pulse fraction of 20%, a pulse repetition rate of 200 kHz, and a detection rate of 0.5%. APT data reconstruction and analyses were performed on the Integrated Visualization and Analysis Software (IVAS) version 3.8.10.

Small-angle neutron scattering (SAXS) was carried out at room temperature on a Bruker NANOSTAR SAXS instrument at the Australian Nuclear Science and Technology Organisation (ANSTO). This instrument is equipped with a rotating Cu K_α anode that emits photons with a wavelength of 0.1541 nm. The beam was focused through a three-point collimation setup. The presented experiments were conducted in a vacuum chamber fitted with a multi-sample stage. Samples for SAXS were homogenised foils with a thickness of ~ 100 micron. The foils were solutionised at 500 °C for 1 h followed by water quenching, before being transferred to SAXS chamber within 15 min. Samples were exposed to SAXS beam for 600 s. The data analyses were performed using SasView software.

Results

Hardness

Figure 1 shows the hardness evolution of the Al-Cu and Al-Cu-Sc-Zr alloys during natural ageing after quenching from the solutionisation treatment at 500 °C for 1 h. It can be seen that the Al-Cu-Sc-Zr alloy starts at 62.6 ± 1.2 HV, ~ 10 HV higher than the Al-Cu alloy (53.2 ± 1.6 HV), evidencing the existence of $Al_3(Sc, Zr)$ dispersoids. As the natural ageing time increases, the hardness of both alloys increases during natural ageing. However, the alloy without Sc and Zr shows a significant increase in hardness compared to the alloy containing Sc and Zr, which suggests that the presence of $Al_3(Sc, Zr)$ dispersoids significantly delays the kinetics of natural ageing in the Al-Cu alloy. For example, after natural ageing for 24 h, the hardness of the Al-Cu alloy is 87.6 ± 5.0 HV with an increment of hardness of ~ 34

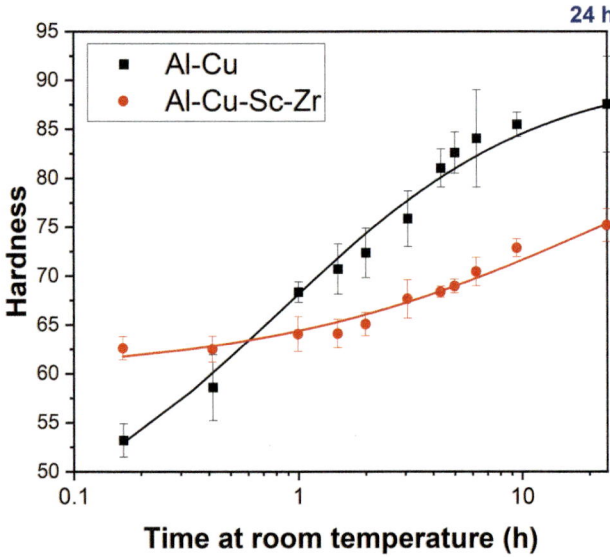

Fig. 1 Hardness evolution of the two alloys during natural ageing up to 24 h

HV. However, the hardness of the Al-Cu-Sc-Zr alloy after natural ageing for 24 h is 75.2 ± 1.7 HV with an increment of hardness of only ~12.6 HV.

Charaterisation of Al₃(Sc, Zr) Dispersoids

Figure 2 shows both STEM and APT characterisation of the Al-Cu-Sc-Zr alloy after the solutionisation treatment at 500 °C for 1 h. The STEM image in Fig. 2a confirms that formation of $Al_3(Sc, Zr)$ dispersoids. The average diameter of the dispersoids is ~16.8 ± 3.2 nm. Figure 2b shows a 1.0 at.% Zr iso-concentration surface and the corresponding proximity histograms, obtained by APT. It is revealed that the $Al_3(Sc, Zr)$ dispersoids have a core–shell structure with a Sc-rich core and a Zr-rich shell, which is consistent to the previous work [4–7].

Charaterisation of Solute Clusters

Figure 3 shows the SAXS one-dimensional scattering profiles for both alloys after natural ageing for 15 min and 24 h. It is shown that the Al-Cu alloy shows a significant difference in SAXS scattering intensity between natural ageing for 15 min and natural ageing for 24 h, suggesting a large amount of clusters formed during natural ageing. However, the Al-Cu alloy containing Sc and Zr shows a relatively smaller difference in SAXS scattering intensity between natural ageing for 15 min and natural ageing for 24 h, evidencing the less clustering occurred during natural ageing compared to the Al-Cu alloy.

Table 1 gives the average radius and volume fraction of the solute clusters in the two alloys after natural ageing for 24 h, extracted from the SAXS scattering profiles in Fig. 3. After natural ageing for 24 h, the average size of the solute clusters in the Al-Cu alloy is ~1.15 nm, which is about two times larger than that in the Al-Cu-Sc-Zr alloy (0.67 nm). In addition, the Al-Cu alloy shows a volume fraction of the solute clusters (~1.8%), which is clearly higher than that in the Al-Cu-Sc-Zr alloy (~0.7%).

Discussion

The present work shows that a minor addition of Sc and Zr can remarkably slow down the kinetics of natural ageing in an Al-Cu alloy. The SAXS results have demonstrated that the presence of nano-sized $Al_3(Sc, Zr)$ dispersoids can retard the formation of solute clusters in Al-Cu alloys during natural ageing.

It is known that solute clustering is a diffusion-driven process which strongly depends on the availability of solute atoms, the presence of vacancies, and their interaction with the solute atoms [8–10]. Therefore, there are two possible mechanisms to explain the delay of solute clustering in the

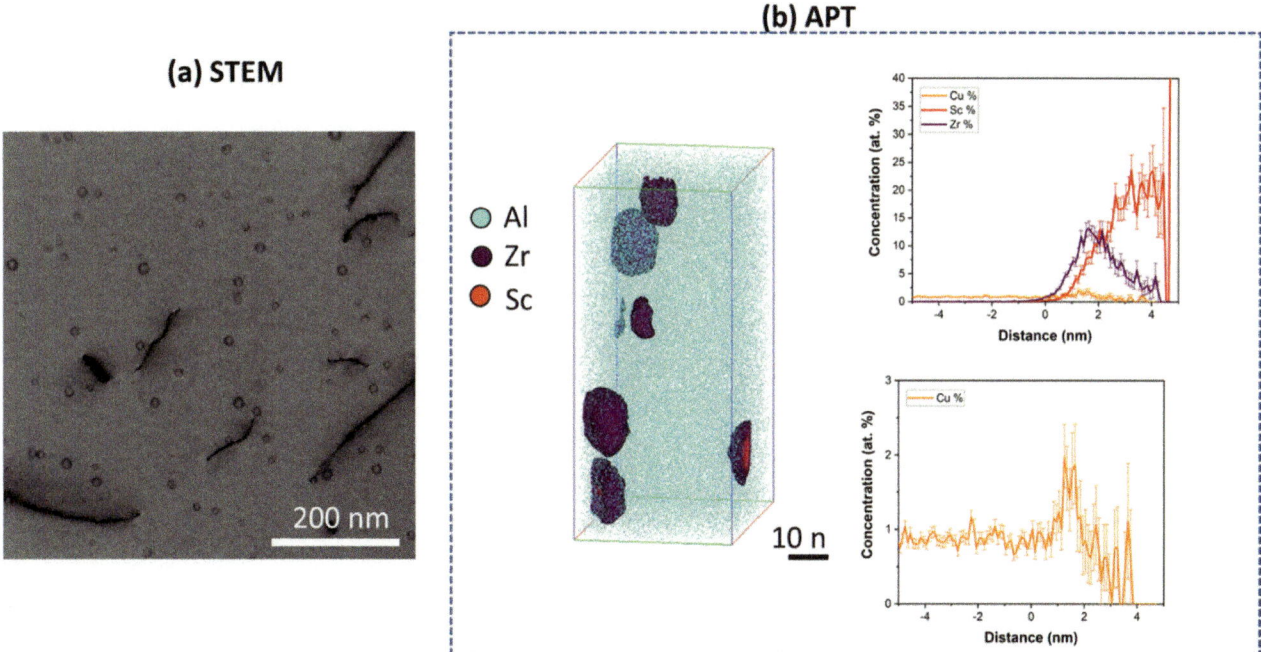

Fig. 2 a STEM image of the Al-Cu-Sc-Zr alloy after the solutionisation treatment at 500 °C for 1 h, taken from [110]_Al zone axis; **b** 3D atom map of Al, Zr, and Sc in the Al-Cu-Sc-Zr alloy after > 3-day natural ageing containing a 1.0 at.% Zr iso-concentration surfaces highlighting the presence of Al₃(Sc, Zr) dispersoids and the corresponding proximity histogram

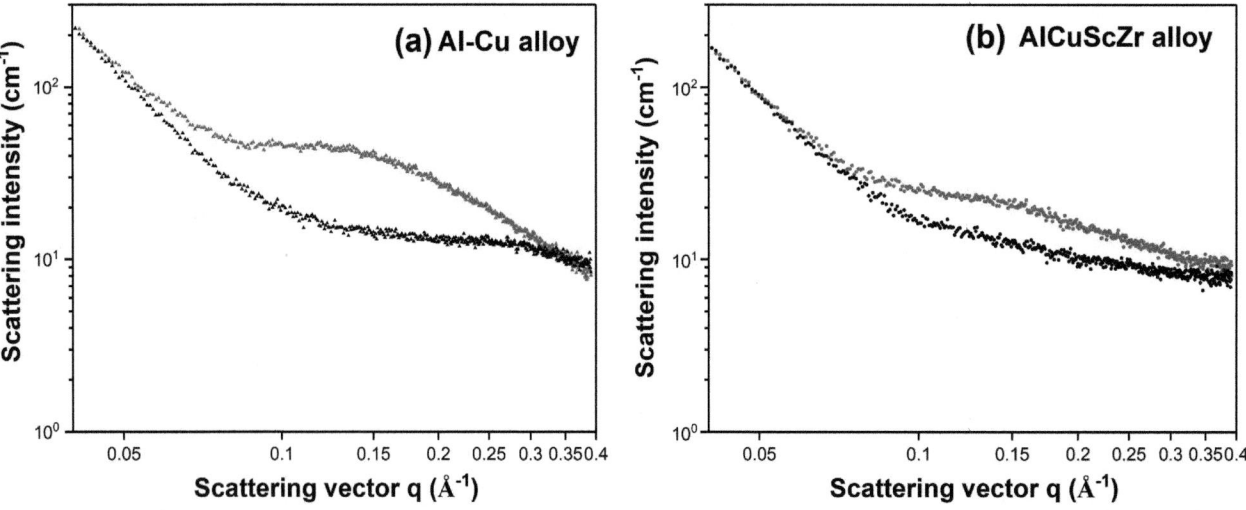

Fig. 3 SAXS one-dimensional scattering profiles for **a** Al-Cu alloy and **b** Al-Cu-Sc-Zr alloy after natural ageing for 15 min and 24 h

Table 1 The size and volume fraction of the solute clusters in the Al-Cu and Al-Cu-Sc-Zr alloys after natural ageing for 24 h		Average radius (nm)	Volume fraction (%)
	Al-Cu alloy	1.15	1.8
	Al-Cu-Sc-Zr alloy	0.67	0.7

Al-Cu-Sc-Zr alloy during natural ageing. Firstly, the APT results in Fig. 2b clearly reveal that Cu atoms segregate at the interfaces between $Al_3(Sc, Zr)$ dispersoids and the Al matrix. This can deplete the solid solution from Cu solutes which is no more available from solute clusters during natural ageing, which could lead to a lower concentration of free Cu solute atoms for clustering during natural ageing. Secondly, the kinetics of solute clustering is directly related to the solute-vacancy-binding energy. Table 2 lists the solute-vacancy-binding energies E_b for Cu, Sc, and Zr. It can be seen that the vacancy-binding energies E_b for Sc and Zr are significantly higher than that for Cu, which suggests that vacancies in the matrix can be trapped by Sc and Zr solutes. This results in less vacancies that are available for Cu diffusion during natural ageing. In order to validate these explanations, future work will focus on the effect of solute treatment and Cu concentrations on the solute clustering during natural ageing in Al-Cu alloys containing Sc and Zr.

In terms of the cluster strengthening during natural ageing, Friedel [12] proposed a strengthening model to estimate the strengthening contribution from clusters:

$$\sigma_{clusters} = k^{3/2} M G \sqrt{3/2\pi b} (f_v R)^{1/2} \qquad (1)$$

where M is the Taylor factor (~ 3), f_v and R are the volume fraction and radius (nm) of the clusters obtained using SAXS (Table 1), k is a fitting parameter and estimated to be 0.09 for a similar alloying system, and G and b are the shear modulus (25 GPa) and the burgers vector (0.286), respectively [13]. Based on the equation above, the calculated cluster strengthening in the Al-Cu and Al-Cu-Sc-Zr alloys after natural ageing for 24 h is given in Table 3. It can be seen that the calculated cluster strengthening in both alloys after natural ageing for 24 h is comparable to the strength increment that is converted from hardness increment

Table 2 Solute-vacancy-binding energies E_b for Cu, Sc, and Zr [11]

Element	E_b [eV]
Cu	0.0 ± 0.12
Sc	0.25–0.5
Zr	0.24 ± 0.02

Table 3 Calculated cluster strengthening ($\sigma_{clusters}$) in the two alloys after natural ageing for 24 h, together with the strength increment ($\Delta\sigma_y$) converted from hardness increment

	$\sigma_{clusters}$, calculated cluster strengthening	$\Delta\sigma_y$, converted from hardness increment (MPa)
Al-Cu alloy	125.5	120.7
Al-Cu-Sc-Zr alloy	59.7	44.73

($\Delta\sigma_y = 3.55 \times \Delta HV$ [14]). This indicates that the formation of clusters is the main factor that leads to the increase of hardness during natural ageing. The lower cluster strengthening in the Al-Cu-Sc-Zr alloy after natural ageing for 24 h is due to the smaller cluster size and the lower volume fraction of cluster.

Conclusions

The current research has investigated the effect of $Al_3(Sc, Zr)$ dispersoids on the solute clustering during natural ageing in an Al-4wt.%Cu alloy. The main results are summarised as follows:

(1) The kinetics of the natural ageing is delayed in the Sc- and Zr-containing Al-Cu alloy.
(2) The presence of $Al_3(Sc, Zr)$ dispersoids retards the formation of Cu-rich clusters during natural ageing.
(3) Two main possible mechanisms: (1) Cu atoms segregate at the $Al_3(Sc, Zr)$/matrix interfaces, which depletes the solid solution from Cu solute which is no more available for solute clustering; (2) Sc and Zr solutes bind with vacancies which reduces the free vacancies for the diffusion of Cu.

Acknowledgements Dr. Thomas Dorin is the recipient of an Australian Research Council Australian Discovery Early Career Award (DE190100614) funded by the Australian Government. The authors acknowledge Sunrise Energy Metals for providing constant support for this study. Deakin University's Advanced Characterisation Facility is acknowledged for use of the TEM-FEG JEOL 2100F and LEAP 5000XR APT instruments. The authors also acknowledge Australian Nuclear Science and Technology Organisation for allocating SANS and SAXS beamtime session at Australian Centre for Neutron Scattering (Proposal No: P9585).

References

1. J.R. Davis, Aluminum and aluminum alloys, ASM international1993.
2. I. Polmear, D. StJohn, J.-F. Nie, M. Qian, Light alloys: metallurgy of the light metals, Butterworth-Heinemann2017.
3. [3] S.K. Kairy, B. Rouxel, J. Dumbre, J. Lamb, T.J. Langan, T. Dorin, N. Birbilis, Simultaneous improvement in corrosion resistance and hardness of a model 2xxx series Al-Cu alloy with the microstructural variation caused by Sc and Zr additions, Corrosion Science 158 (2019) 108095.
4. [4] T. Dorin, M. Ramajayam, J. Lamb, T. Langan, Effect of Sc and Zr additions on the microstructure/strength of Al-Cu binary alloys, Mater. Sci. Eng. A 707 (2017) 58-64.
5. [5] T. Dorin, M. Ramajayam, S. Babaniaris, L. Jiang, T.J. Langan, Precipitation sequence in Al–Mg–Si–Sc–Zr alloys during isochronal aging, Materialia 8 (2019) 100437.
6. [6] L. Jiang, T. Langan, T. Wood, P. Sanders, T. Dorin, Isotropy of precipitate distribution in pre-stretched Al-Cu-(Sc)-(Zr) alloys, Scripta Materialia 210 (2022) 114452.
7. [7] L. Jiang, B. Rouxel, T. Langan, T. Dorin, Coupled segregation mechanisms of Sc, Zr and Mn at θ' interfaces enhances the strength and thermal stability of Al-Cu alloys, Acta Materialia 206 (2021) 116634.
8. Y. Zhang, Z. Zhang, N.V. Medhekar, L. Bourgeois, Vacancy-tuned precipitation pathways in Al-1.7 Cu-0.025In-0.025Sb (at.%) alloy, Acta Materialia 141 (2017) 341–351.
9. [9] Y. Nagai, M. Murayama, Z. Tang, T. Nonaka, K. Hono, M. Hasegawa, Role of vacancy–solute complex in the initial rapid age hardening in an Al–Cu–Mg alloy, Acta Materialia 49(5) (2001) 913-920.
10. [10] J. Banhart, M.D.H. Lay, C.S.T. Chang, A.J. Hill, Kinetics of natural aging in Al-Mg-Si alloys studied by positron annihilation lifetime spectroscopy, Physical Review B 83(1) (2011) 014101.
11. [11] H. Kimura, A. Kimura, R.R. Hasiguti, A resistometric study on the role of quenched-in vacancies in ageing of Al-Cu alloys, Acta Metallurgica 10 (1962) 607-619.
12. J. Friedel, Dislocations Pergamon Press, New York (1964).
13. H. FROST, M. ASHBY, Deformation-mechanism maps: The plasticity and creep of metals and ceramics(Book), Oxford, Pergamon Press, 1982, 175 p (1982).
14. [14] J.T. Busby, M.C. Hash, G.S. Was, The relationship between hardness and yield stress in irradiated austenitic and ferritic steels, Journal of Nuclear Materials 336(2) (2005) 267-278.

Effect of Zr and Sc on Intermetallic Morphology and Hardening of an Al–Fe Alloy

Suwaree Chankitmunkong, Dmitry G. Eskin, Chaowalit Limmaneevichitr, Phromphong Pandee, and Onnjira Diewwanit

Abstract

We studied the effect of zirconium and scandium on an Al-7 wt% Fe cast alloy with potential heat- and wear-resistant applications. An addition of 0.2% Zr resulted in thinning of primary Al_3Fe particles, while an addition of 0.15% Zr and 0.15% Sc changed the morphology of primary intermetallics from needles to flower-like shape. While the addition of Zr did not affect the properties, the Zr + Sc joint additions increased the hardness of the as-cast Al–Fe alloy. The hardness of the base alloy increased upon annealing from 40 to 80 HV (450 °C, 2 h) and to 110 HV (350 °C, 20 h). The wear resistance of the Al–Fe alloy was also improved by Zr and Sc addition, especially after annealing. The observed effects are likely to be linked to the supersaturation of Zr and Sc in the aluminum solid solution during solidification and precipitation of dispersoids during annealing.

Keywords

Aging • Aluminium–iron alloy • Scandium • Zirconium • Wear

S. Chankitmunkong
Department of Industrial Engineering, School of Engineering, King Mongkut's Institute of Technology Ladkrabang, Chalongkrung Road, Ladkrabang, Bangkok, 10520, Thailand
e-mail: suwaree.ch@kmitl.ac.th

D. G. Eskin
Brunel University London, BCAST, Uxbridge, UB8 3PH, Middlesex, UK
e-mail: dmitry.eskin@brunel.ac.uk

D. G. Eskin
Tomsk State University, Tomsk, 634050, Russian Federation

C. Limmaneevichitr · P. Pandee
Department of Production Engineering, Faculty of Engineering, King Mongkut's University of Technology Thonburi, Pracha-Utid Road, Bangmod, Tungkhru, Bangkok, 10140, Thailand
e-mail: chaowalit.lim@mail.kmutt.ac.th

P. Pandee
e-mail: phrompong.pan@mail.kmutt.ac.th

O. Diewwanit (✉)
Department of Tool and Materials Engineering, Faculty of Engineering, King Mongkut's University of Technology Thonburi, Pracha-Utid Road, Bangmod, Tungkhru, Bangkok, 10140, Thailand
e-mail: onnjira.tha@kmutt.ac.th

Introduction

A new class of Al–Fe-based alloys is a group of materials which may be widely used for structural and automobile applications [1]. Their main advantages are low-density, good corrosion resistance, and good mechanical properties [2]. Hypereutectic Al–Fe alloys have received a great deal of attention over the last decade [3, 4], as they contain hard and stable Al_3Fe particles, both of primary and eutectic origins, in Al matrix [5].

Previous work showed that the addition of Ce and La rare earth metals into an Al-5Fe alloy resulted in the refinement of the primary phase [6], which led to the improved tensile strength of the alloy. It is well known that the simultaneous addition of Zr and Sc in Al provides better mechanical properties than the separate additions due to the formation of thermally stable core–shell $L1_2$ $Al_3(Sc_x,Zr_{1-x})$ nanoprecipitates created upon annealing [7, 8], in which the core is enriched with Sc and the shell enriched with Zr [8, 9]. These nanoprecipitates have benefits of rapid precipitation and coherency from Al_3Sc and slow coarsening due to the formation of a highly stable Zr-rich shell [10, 11]. It is also beneficial in the economical aspect since the costly Sc can be replaced by much less costly Zr, while it can maintain good mechanical properties at both room and elevated temperature [12].

Reinforcement of ceramic whiskers, fibers, and particles in aluminum/aluminum alloys has been extensively explored as tribo-materials [13, 14]. The formation of hard primary and eutectic phases in aluminum alloys could be

© The Minerals, Metals & Materials Society 2023
S. Broek (ed.), *Light Metals 2023*, The Minerals, Metals & Materials Series,
https://doi.org/10.1007/978-3-031-22532-1_168

successfully used in typical wear applications such as pistons and cylinder liners of automobile engines [15, 16].

This work is aimed at studying the effects of Sc and Zr on the microstructure and properties of hypereutectic Al–7 wt% Fe alloys. The formation of intermetallics and their morphology have been examined. The alloys have been subjected to different annealing temperatures to characterize dispersion hardening by Vickers microhardness. Wear properties have been analyzed and related to the microstructure.

Experimental

Three experimental Al–Fe-based alloys were prepared in an induction furnace at a temperature of 750 °C and then poured into a permanent mold (width 100 mm, height 200 mm, and 17 mm in thickness). 99.85 pure Al, Al–20% Fe, Al–8% Zr, and Al–2% Sc master alloys were used as starting materials. The cast alloy was cut into small pieces of 25 mm × 25 mm × 17 mm in dimensions before annealing at 350 °C and 450 °C for 20 h. The compositions of the alloys as obtained from a spark optical emission spectrometer (ARL 3460 model) and are given in Table 1.

Specimens were cut 20 mm in the middle of the casting to examine the as-cast microstructure. These specimens were prepared by a standard metallographic method, using Keller's reagent as an etchant to reveal the intermetallics by using a light microscope (Carl Zeiss, model Zeiss Axioscope 5). The specimens were polished down to 1 μm for Vickers microhardness measurement. At least ten times measurements were made on each specimen with a 30 N load and 5-s dwell time, using a microhardness FM-700e, FUTURE-TECH, Tokyo, Japan.

Specimens of 20 mm × 20 mm were mechanically polished and ultrasonically cleaned with acetone prior to the wear testing using a pin on disc high-temperature tribometer (THT tribometer CSM instruments, Switzerland) with sliding motion against alumina balls of 6 mm diameter for the cycles test from 5000 to 30,000. The weight loss was measured through 3D image analysis of worn surface by using digital microscope (Olympus DSX1000 microscope). Worn surface of the alloys were analyzed by a scanning electron microscope (SEM, JSM-6610 LV, Joel, Tokyo Japan).

Results and Discussion

Figure 1 shows that the initial microstructure of as-cast Al–Fe alloys consisted of primary and eutectic Al_3Fe particles in the eutectic matrix. Typical primary intermetallic phases are present in all three alloys as a dispersion of elongated crystals (random cross sections of coarse plate dendrites) as can be seen in Fig. 1a. However, the addition of 0.2% Zr resulted in thinning of primary Al_3Fe particles (Fig. 1b), while an addition of 0.15% Zr and 0.15% Sc changed the morphology of primary intermetallics from needle-like to flower-like shape as shown in Fig. 1c. It has been previously reported that the primary Al_3Fe can be modified into fine needle-like and fibrous shape by the addition of Zr and Sc [1] and some other rare earth metals [17, 18].

Figure 2 shows the average hardness of the Al–Fe cast alloy, which is about 38 HV, and there is no significant difference in the Al-7Fe-0.15Zr cast alloy, while the hardness increased in the Al-7Fe-0.15Zr-0.15Sc alloy up to 55 HV. This may be due to the supersaturation of Zr and Sc in the aluminum solid solution during solidification [19]. A similar trend is demonstrated in the annealed alloys as measured after annealed at 350 °C and 450 °C for 20 h, Fig. 3. Interestingly, the hardness of the binary Al–Fe alloy increased from 38 to 50 HV after annealing at 350 °C for 20 h, probably due to the precipitates of Al_6Fe as had been reported elsewhere [20, 21].

It is interesting to see that the hardness of Al-7Fe with Zr and Sc addition alloy is the highest after aging at 350 °C for 20 h, reaching 110 HV. This significant increase in hardening effect is apparently due to the precipitation of nanometer-scale core–shell $Al_3(Sc,Zr)$ particles similar to other aluminum alloys with Zr and Sc [1, 7–11]. Annealing at 450 °C results in the decrease in hardness, most probably due to the coarsening of the precipitates [10, 11].

Figure 4 shows the wear volume losses of Al–Fe alloy as a function of cycles at a load of 2 N. The wear losses increase with increasing the number of cycles. The results demonstrated that the volume losses of the alloys were similar after initial number of cycles, while on further increasing the cycles the changes in volume loss are clearly dependent on the composition and heat treatment of the alloys. The as-cast binary Al-7Fe alloy had the

Table 1 Chemical compositions of the experimental alloys

Alloys	Chemical composition (wt%)			
	Al	Fe	Zr	Sc
Al-7Fe	Bal	7.155	–	–
Al-7Fe-0.2Zr	Bal	7.156	0.198	–
Al-7Fe-0.15Zr-0.15Sc	Bal	7.142	0.149	0.159

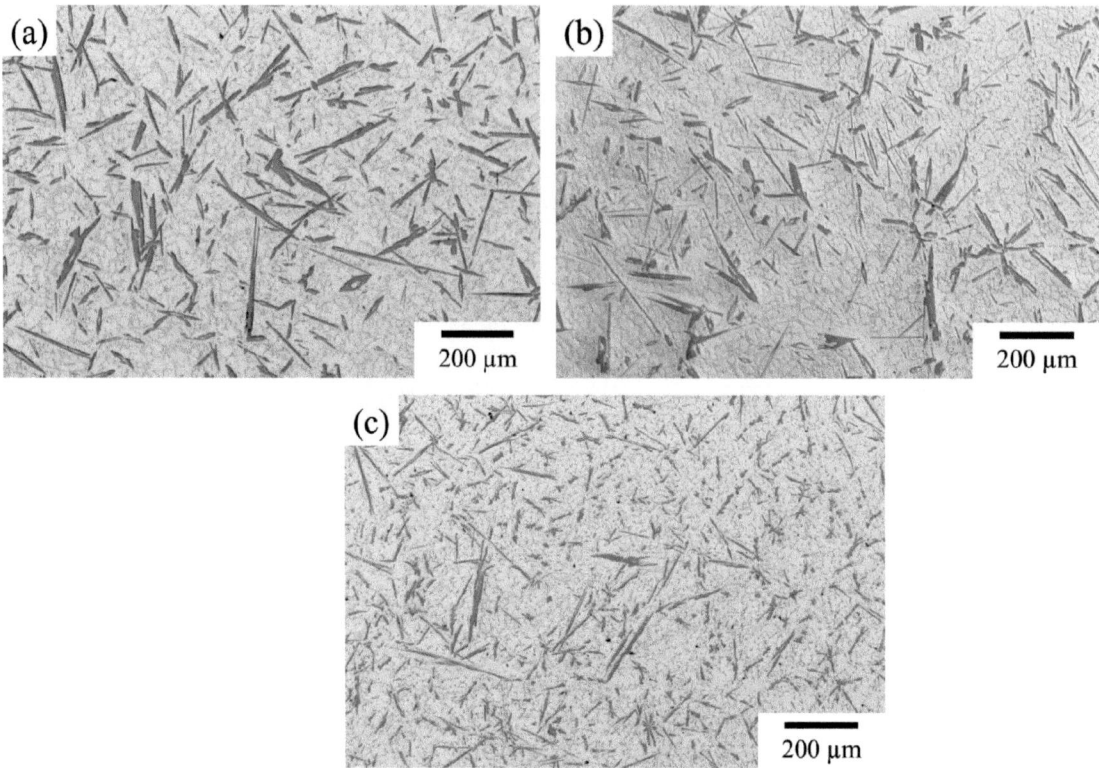

Fig. 1 Microstructure of the alloys with and without Zr and Sc additions: **a** Al-7Fe, **b** Al-7Fe-0.15Zr, and **c** Al-7Fe-0.15Zr-0.15Sc

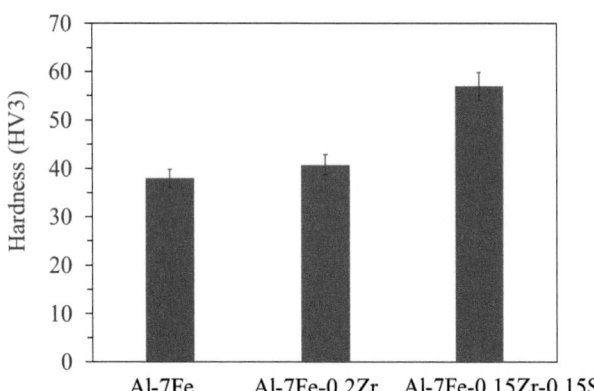

Fig. 2 Hardness of Al–Fe alloys in as-cast condition

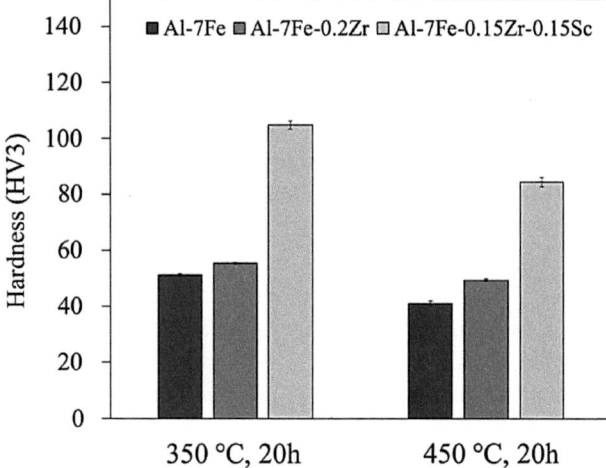

Fig. 3 Hardness of Al–Fe alloys after annealing at 350 °C and 450 °C for 20 h

highest volume loss when the cycles increased up to 10,000. This was similar to the previous work, which investigated the wear volume loss as a function of the number of fretting cycles [16]. The results also showed that the annealing consistently improved wear resistance, even in the binary alloy. The addition of Zr and, especially, Zr + Sc improved the wear resistance, especially after annealing. This is likely to be due to the more uniformly distributed refined hard intermetallics of Al$_3$Fe (Fig. 1) and also precipitates of hard nanoparticles of Al$_3$(Sc,Zr) and Al$_6$Fe in the Al matrix [22, 23].

An extensive study was conducted to understand the mechanism of wear by observing the worn surface in an SEM as shown in Fig. 5. The wear track surface of the as-cast Al–Fe-Zr-Sc alloy was highly deformed with deep grooves and cracks and gross delamination occurred, leading to larger volume loss as can be seen in Fig. 5a and c. Such kind of delamination wear was also reported in an Al–Fe-V alloy [24]. On the other hand, the alloy annealed at 350 °C

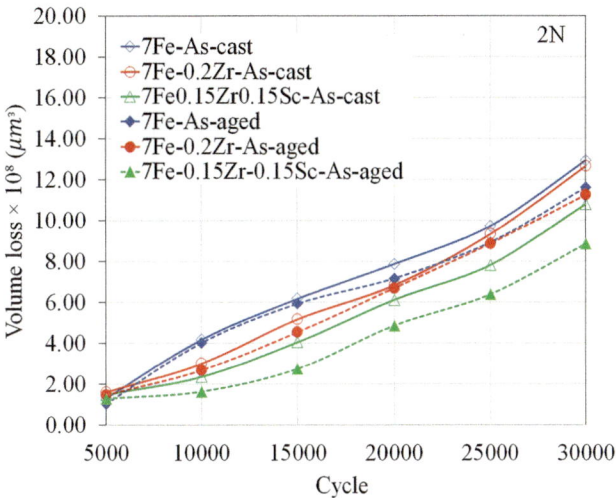

Fig. 4 Volume loss of the tested Al–Fe alloys in as-cast and as-annealed conditions at 350 °C for 20 h after wear testing with applied load 2 N from 5000 to 30,000 cycles

for 20 h was largely covered with oxide layer and had smooth worn surface as shown in Fig. 5b and d, which agrees with the previous results obtained on Al–Fe and Al–Si alloys [25, 26]. This is a result of much harder Al matrix that provided better resistance to wear.

Conclusions

1. The addition of 0.2%Zr to an Al–7Fe alloy resulted in thinning of primary Al_3Fe particles, while the morphology of intermetallic is changed from needles to flower-like shape with the addition of 0.15% Zr and 0.15% Sc.

2. Joint Zr and Sc additions provided substantial hardening ability of the Al–7Fe alloys with the hardness increasing from 40 to 80 HV (450 °C, 2 h) and to 110 HV (350 °C, 20 h). There was also some hardening associated with the possible precipitation of Al_6Fe in the binary alloy at 350 °C.

3. The wear of Al–Fe alloy decreased by Zr and Sc additions, especially after annealing. The observed effects are likely to be linked to the refinement of primary Al_3Fe intermetallics and supersaturation of Zr, Sc, and Fe in the aluminum solid solution during solidification and precipitation of dispersoids during annealing.

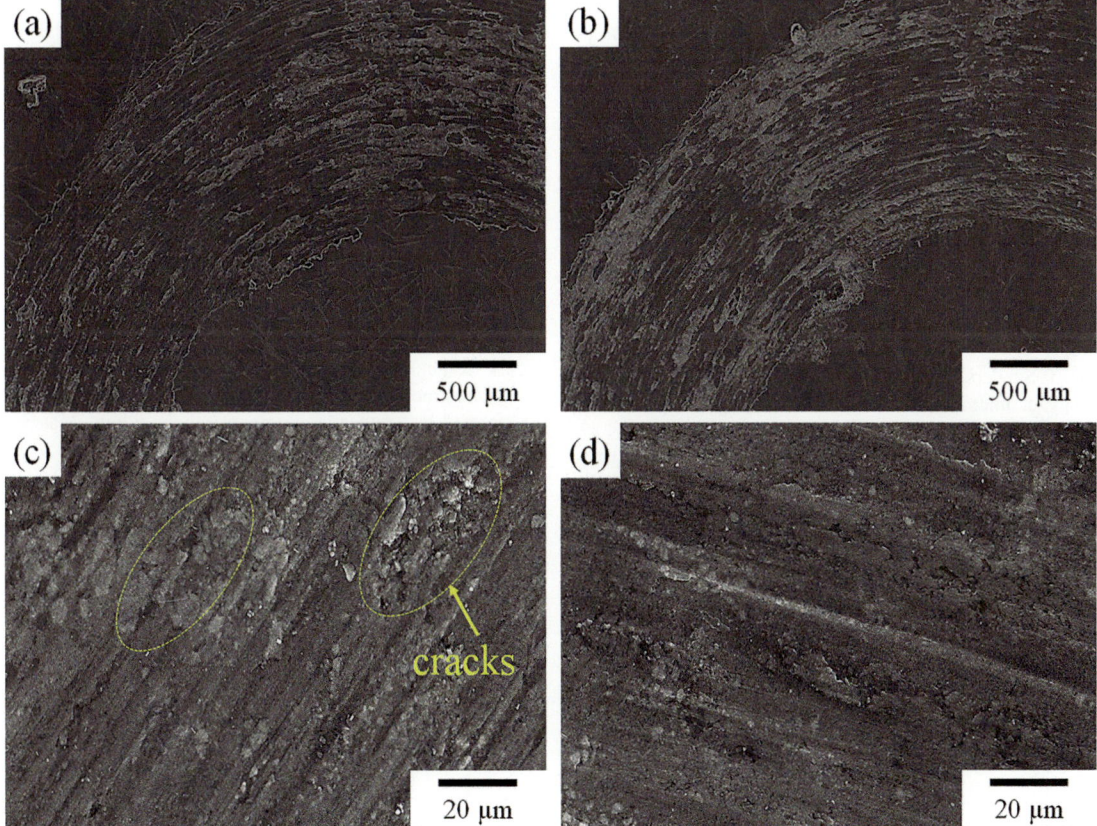

Fig. 5 Worn surface image of Al-7Fe-0.15Zr-0.15Sc addition under load test of 2 N with 30,000 cycle: **a, c** as-cast specimen and **b, d** as-aged at specimen at 350 °C for 20 h

Acknowledgements The authors are grateful for grant number KREF046409 provided by King Mongkut's Institute of Technology Ladkrabang. The authors also would like to thank the Department of Tool specimen and Materials Engineering and Production Engineering Department, King Mongkut's University of Technology Thonburi laboratory facilities. FE-SEM center at school of engineering, King Mongkut's Institute of Technology Ladkrabang.

References

1. K. Dai, J. Ye, Z. Wang, M. Gao, J. Chen, R. Guan, Effects of Sc and Zr addition on the solidification and mechanical properties of Al–Fe alloys, Journal of Materials Research and Technology 18 (2022) 112-121.

2. X. Wang, R. Guan, R. Misra, Y. Wang, H. Li, Y. Shang, The mechanistic contribution of nanosized Al_3Fe phase on the mechanical properties of Al-Fe alloy, Materials Science and Engineering: A 724 (2018) 452-460.

3. D. Liang, H. Jones, Morphologies of primary Al3Fe in Bridgman solidification and TIG weld traversing of hypereutectic Al-Fe alloys, Materials Science and Engineering: A 173(1-2) (1993) 109-114.

4. O. Senkov, F. Froes, V. Stolyarov, R. Valiev, J. Liu, Microstructure and microhardness of an Al-Fe alloy subjected to severe plastic deformation and aging, Nanostructured Materials 10(5) (1998) 691-698.

5. M.I.R. Basariya, N.K. Mukhopadhyay, Structural and mechanical behaviour of Al-Fe intermetallics, Intermetallic compounds–formation and applications. Rijeka: IntechOpen (2018) 97–122.

6. Y. Liang, Z. Shi, G. Li, R. Zhang, M. Li, Effects of rare earth modification on microstructure refinement and mechanical properties of Al-2 wt% Fe alloys, Materials Research Express 6(10) (2019) 106504.

7. C.B. Fuller, D.N. Seidman, D.C. Dunand, Mechanical properties of Al(Sc, Zr) alloys at ambient and elevated temperatures, Acta Materialia 51(16) (2003) 4803-4814.

8. A. Tolley, V. Radmilovic, U. Dahmen, Segregation in Al_3(Sc, Zr) precipitates in Al–Sc–Zr alloys, Scripta Materialia 52(7) (2005) 621-625.

9. C.B. Fuller, J.L. Murray, D.N. Seidman, Temporal evolution of the nanostructure of Al(Sc, Zr) alloys: Part I–Chemical compositions of $Al_3(Sc_{1-x}Zr_x)$ precipitates, Acta Materialia 53(20) (2005) 5401-5413.

10. Y. Riddle, T. Sanders, A study of coarsening, recrystallization, and morphology of microstructure in Al-Sc-(Zr)-(Mg) alloys, Metallurgical and Materials Transactions A 35(1) (2004) 341-350.

11. C.B. Fuller, D.N. Seidman, Temporal evolution of the nanostructure of Al(Sc,Zr) alloys: Part II-coarsening of $Al_3(Sc_{1-x}Zr_x)$ precipitates, Acta Materialia 53(20) (2005) 5415-5428.

12. A. De Luca, D.C. Dunand, D.N. Seidman, Scandium-Enriched Nanoprecipitates in Aluminum Providing Enhanced Coarsening and Creep Resistance, Light Metals 2018, Springer, 2018, pp. 1589-1594.

13. Z. Zhang, J. Zhang, Y. Mai, Wear behaviour of SiCp/Al-Si composites, Wear 176 (1994) 231-7.

14. P. Chokemorh, P. Pandee, S. Chankitmunkong, U. Patakham, C. Limmaneevichitr, Primary Si refinement and eutectic Si modification in Al-20Si via P-Ce addition, Materials Research Express (2022).

15. Z. Nouri, R. Taghiabadi, Tribological properties improvement of conventionally-cast Al-8.5 Fe-1.3 V-1.7 Si alloy by multi-pass friction stir processing, Transactions of Nonferrous Metals Society of China 31(5) (2021) 1262–1275.

16. X. Niu, L. Froyen, L. Delaey, C. Peytour, Fretting wear of mechanically alloyed Al-Fe and Al-Fe-Mn alloys, Wear 193(1) (1996) 78-90.

17. W. Wang, N. Takata, A. Suzuki, M. Kobashi, M. Kato, Formation of multiple intermetallic phases in a hypereutectic Al–Fe binary alloy additively manufactured by laser powder bed fusion, Intermetallics 125 (2020) 106892.

18. S. Luo, Z. Shi, N. Li, Y. Lin, Y. Liang, Y. Zeng, Crystallization inhibition and microstructure refinement of Al-5Fe alloys by addition of rare earth elements, Journal of Alloys and Compounds 789 (2019) 90-99.

19. V. Zakharov, Stability of the solid solution of scandium in aluminum, Metal Science and Heat Treatment 39(2) (1997) 61-66.

20. H. Qu, W.D. Liu, Study on the stability and phase transition of precipitated phases Al_6Fe and Al_3Fe in Al-Fe Alloy, In Advanced Materials Research, 2011, vol. 306, pp. 438-442.

21. H. Tsubakino, A. Yamamoto, T. Kato, A. Suehiro, Precipitation in deformed Al-Fe and Al-Fe-Si dilute alloys, In Materials science forum, vol. 331, pp. 951–956. Trans Tech Publications Ltd, 2000.

22. X. Niu, A. Mulaba-Bafubiandi, L. Froyen, L. Delaey, C. Peytour, Mössbauer study of phase formation in mechanically alloyed Al-Fe and Al-Fe-Mn powders, Scripta Metallurgica et Materialia 31(9) (1994) 1157-1162.

23. Y. Wang, R. Li, T. Yuan, L. Zou, M. Wang, H. Yang, Microstructure and mechanical properties of Al-Fe-Sc-Zr alloy additively manufactured by selective laser melting, Materials Characterization 180 (2021) 111397.

24. K. Sahoo, C. Krishnan, A. Chakrabarti, Studies on wear characteristics of Al–Fe–V–Si alloys, Wear 239(2) (2000) 211-218.

25. S. Mohan, S. Srivastava, Surface behaviour of as-cast Al–Fe intermetallic composites, Tribology Letters 22(1) (2006) 45-51.

26. N. Raghukiran, R. Kumar, Effect of scandium addition on the microstructure, mechanical and wear properties of the spray formed hypereutectic aluminum–silicon alloys, Materials Science and Engineering: A 641 (2015) 138-147.

Effect of Sc, Zr, and Other REE on the 1XXX Conductive Aluminum Alloy Properties

Ruslan Aliev, Alexander Gradoboev, Dmitry Ryabov, Roman Vakhromov, Aleksandr Krokhin, and Viktor Mann

Abstract

The modern trend of transport is electric vehicles. This is due to the high environmental friendliness of electric vehicles and international regulations that regulate carbon footprint. Aluminum alloys are good alternative to copper for use in bus bars and wires. In order to replace copper for the manufacture of bus bars for electric vehicles, aluminum alloys should have high electrical conductivity at the level of technical grades (AA 1350) and at the same time have increased strength and thermal stability. The most promising method of improving strength properties is doping with small additives Sc, Zr, and REE, which form fine dispersed precipitates, providing an increase in strength properties, significantly increasing recrystallization temperature and thermal stability. In the present paper, a study of the effect of additives of a small amount of Sc, Zr, and other REE on mechanical properties and electrical conductivity alloys of aluminum of 1XXX series was carried out.

Keywords

Aluminum alloy • Sc • Rear-earth elements • Busbar • Wiring

Introduction

The modern trend in transport engineering is electric vehicles. Annual growth of their production and consumption is about 30%. This is due to the high environmental

R. Aliev (✉) · A. Gradoboev · D. Ryabov · R. Vakhromov
LLC Light Materials and Technologies Institute UC RUSAL, Moscow, Russia
e-mail: Ruslan.Aliev2@rusal.com

A. Krokhin · V. Mann
JSC RUSAL Management, Moscow, Russia

friendliness of this type of transport, lower noise levels, lower operating and maintenance costs, etc.

Electric vehicles use electric motor drives, high-capacity batteries, power inverters, and power distribution systems from the charging source to the battery and then throughout the vehicle. Electrically conductive busbars and cables—a system of electrical conductors for distributing current—connect all of this.

Copper is currently widely used as the material for the conductive buses that power the electric car. Despite its main advantage—high electrical conductivity, copper has two main disadvantages—high weight and cost. Aluminum and its alloys have long proved to be a positive substitute for copper in wires, cables, and busbars.

The higher specific resistance of aluminum leads to an increase in the cross-sectional area of the conductor. However, even with the increased cross section, the weight of aluminum wire is lower than copper wire, and the cost of aluminum wire is 30–50% that of copper wire for equal conductivity. In addition, the stock of copper is decreasing and with the high demand for electric vehicles, its stock may be insufficient. In this regard, the use of aluminum alloys for electric vehicle wiring is a promising direction.

Certain requirements are imposed on the materials of conductive elements in terms of electrical conductivity, mechanical properties, and fatigue curve. Currently, aluminum alloys 1350 (ASTM B236) and 6101 (ASTM B317) are used for making busbars. Alloy 1350, while having high electrical conductivity of 61–62%IACS, has insufficient strength and fatigue resistance. Alloy 6101 is more than twice as strong, but its main disadvantage is the lower electrical conductivity of 55–58%IACS [1].

Taking into account the limitation of alloying for aluminum materials used in electrical engineering to increase the strength of the 1xxx series such as 1350 and 1370, we can consider the introduction of small addition of scandium and other rare-earth elements that provide the effect of

© The Minerals, Metals & Materials Society 2023
S. Broek (ed.), *Light Metals 2023*, The Minerals, Metals & Materials Series,
https://doi.org/10.1007/978-3-031-22532-1_169

solid-solution hardening and getting nonrecrystallized structure.

Sc is one of the most effective alloying elements that increase the mechanical properties of aluminum alloys. The maximum solubility of Sc in aluminum is 0.38% wt. at 660 °C, [2]. The greatest effect of hardening Sc is observed in the alloying of aluminum alloys series 3xxx and 5xxx [3, 4].

Zirconium additives are also used to solve the problem of heat resistance of aluminum wires. Maximum concentration of Zr in solid solution is 0.28 wt. % at 660.8 °C [2]. The study of the effect of zirconium content on the electrical conductivity and mechanical properties of electrical wire rods is devoted to dissertation work [5]. Wire rod is produced by combined casting, rolling and extrusion process (CREP), at a temperature of the melt on the wheel crystallizer at least 720 °C at 0.2% Zr and not less than 760 °C at 0.3% Zr. Under the recommended modes of annealing, the structure in the rod is formed by the presence of fibrous uncrystallized grains and nanoparticles of Al_3Zr phase ($L1_2$) with size up to 20 nm. The author achieved the highest electrical conductivity of the rods about 28.5 µOh-mm (60.5% IACS) due to its multi-stage annealing at the temperature of the last stage in the range of 420–450 °C.

A group of researchers [6] reported the results of studying the effect of Zr concentration in the range from 0.18 to 0.476% wt. when it is added to the technical aluminum of type 1370. The authors explain the greatest increase in hardness with increasing zirconium concentration as well as the lowest values of electrical resistance which are almost independent of zirconium concentration after 450 °C annealing by the fact that zirconium from solid solution becomes bound to nanoparticles of $L1_2$ phase with an average size of not more than 10 nm. Heating above 450 °C results in coarsening of the particles and their subsequent transformation into the stable $D0_{23}$ phase, which grows to 2 µm after annealing at 600 °C.

High electric conductivity and strength are achieved by joint alloying with Sc and Zr. Two-stage heat treatment for complete precipitation of nanodispersed $Al_3(Sc, Zr)$ particles, where Sc is segregated from a solid solution at temperatures above 250 °C, and Zr at temperatures above 400 °C [7, 8].

The authors [9] show a study of Al-0.06% Sc-0.23% Zr alloy based on high-purity aluminum of 99.999%. They selected modes of one-stage aging at 300 °C and two-stage aging at temperatures of 300 and 400 °C. As a result, they achieved the highest strength properties up to UTS 190 MPa with an electrical conductivity of 57.6% IACS.

The team of authors [10] shows a comparative study of mechanical properties and electrical conductivity of aluminum wires from pure aluminum (99.996%) and its alloying with zirconium and scandium on alloys Al-0.16% Zr, Al-0.16% Sc, and Al-0.12%Sc-0.04% Zr. The authors concluded that alloying of pure aluminum with 0.16% Sc and 0.16% Zr increases the ultimate tensile strength but reduces the electrical conductivity. The joint alloying with 0.12% Sc and 0.04% Zr with optimal heat treatment resulted in the highest tensile strength up to 160 MPa and high electrical conductivity at the level of technical aluminum grades.

A large lineup of authors [11] shows a study of economically alloyed material based on pure aluminum (99.99%) with less scandium and the addition of iron Al-0.05% Sc-0.1% Zr-0.2 Fe. The authors concluded that the alloying with Sc and Zr improves the shape of iron-containing particles of cast samples. Aging at 300 °C with an exposure time of up to 24 h yielded spherical Al_3Fe particles less than 1 µm and spherical nanoparticles of $Al_3(Sc,Zr)$. This also achieved a high strength of 135 MPa and raised the electrical conductivity to 61% IACS. A subsequent cold drawing process of the aged samples resulted in an increase in tensile strength to 165.7 MPa with a slight decrease in conductivity to 60% IACS.

The high strengthening efficiency of aluminum alloys with scandium is limited by the very high cost of scandium itself. One of the elements used in the manufacture of electrically conductive aluminum alloys is cerium. With the same content of scandium and cerium in the Earth's crust, their cost and consumption differ by three to four orders of magnitude [12]. For cerium, the low cost is accompanied by rather limited hardening effects: the insignificant solubility of cerium in the solid state in aluminum makes the hardening during aging not so effective. At the same time, the alloying of aluminum with cerium still allows obtaining alloys with improved high-temperature characteristics [13, 14].

In addition to cerium, yttrium is used as a hardening element. Like scandium, yttrium forms an intermetallic compound with aluminum [2]. In work [15] based on technical aluminum (99.7% wt.), the alloys Al-0.2Ce, Al-0.2Ce-0.2Sc, Al-0.2Ce-0.1Y, Al-0.2Ce-0.12Zr, and Al-0.2Ce-0.2Sc-0.1Y were obtained. The alloy Al-0,2Ce-0,2Sc-0,1Y has the best complex characteristics. After cold deformation (drawing) and annealing, the samples have a tensile strength of 190–200 MPa, elongation of 8%, and electrical conductivity of 61%IACS.

Other rare-earth elements (REE) capable of increasing strength properties without a significant decrease in electrical conductivity are Er and Yb. Both elements like Sc form with aluminum a metastable phase $Al_3(Er, Yb)$ with a cubic lattice of type $L1_2$ [2].

The authors [16] consider the dissolution kinetics of small additions of Yb and Er in aluminum (99.99% wt.) at different temperatures, evaluated by measuring the electrical conductivity, among other things. Electrical conductivity of 60.5%IACS is achieved with Yb 0.064% wt and Er 0.074% wt doping.

Article [17] provides data on the partial replacement of scandium by Er and Yb in low-alloy alloys. The authors studied the alloy Al-0,141Yb-0,062Er-0,047Sc-0,121Zr % wt. based on pure aluminum (99.99%wt.). The addition of Er and Yb allowed after the selected aging modes to obtain a larger number of nanodispersed precipitations $Al_3(Sc_{0,56}Yb_{0,14}Er_{0,10}Zr_{0,20})$ smaller in radius to 3.5 nm, i.e., to achieve a more dispersed structure. In this case, as shown by atomic probe tomography measurements, Yb and Er are concentrated in the center of the dispersoids, while the shell is enriched in Sc and Zr. The addition of Er and Yb markedly increases the fatigue characteristics of low-alloyed alloys with Sc and Zr.

Most of the given examples of the effect of Sc and other rare-earth elements alloying were carried out on small laboratory samples of ingots, which were also subjected to high-temperature homogenization of 620–640 °C followed by quenching. In large-scale production and melting of industrial batches of ingots or slabs, hardening operations on ingots are not possible, and the properties can be greatly separated.

In this work, the influence of alloying with Sc, Zr and rare-earth elements such as Ce, Y, Er, and Yb in the production of semi-industrial slabs of 1xxx series electrical-grade aluminum and rolled products from them was investigated.

Experimental Procedure and Material

Three model alloy compositions based on electrical aluminum 99.7% (Table 1) with different contents of alloying elements (% wt.) were made Al-Sc (up to 0.03%)—Ce + Y (up to 1.5%) and Al-Sc + Zr (up to 0.09%) and Al-Sc-Zr + Er, Yb (up to 0.1%). Alloy 1350 was also fabricated as a reference to compare mechanical properties and electrical conductivity.

The alloys were prepared in an induction melting furnace. The melt was modified with Al-5Ti-1B bar modifier. Casting of slabs was carried out on the existing Wagstaff Pilot Casting Complex production facilities (in Krasnoyarsk). As a result, slabs of 200 × 400 mm size and 1000–1100 mm length were obtained (Fig. 1a).

After trimming the defective ends and scalping from slabs, the sheets were rolled (Fig. 1b). Hot rolling was applied down to 10 mm with preheating temperatures from 380 to 420 °C. Cold rolling was applied from 10 to 3 mm.

The sheets at the final thickness of 3 mm were annealed at temperatures of 150–400 °C for 3 h in a convection furnace. Mechanical properties were measured according to ASTM E8 on samples with a 6 mm subsize specimen. The conductivity of the sheets after heat treatment was measured by eddy-current method with recalculation in % IACS (are based on the International Annealed Copper Standard adopted by IEC in 1913, which is 58· × 10^3 S/m at 20 °C for 100% conductivity).

Results and Discussion

Casting Slabs

Analysis of the macrostructure of the ingots of each of the three model compositions showed the absence of porosity. The structure is uniform with silicon and iron phases separated along the borders of dendrites (Fig. 2). The presence of Sc, Zr, and other REE did not change the morphology of the structure and did not lead to fragmentation of dendrites.

On the one hand, there are no mandatory homogenization requirements for electrical aluminum grades of the 1xxx series and it is possible to process slabs (rolling) without using this operation. On the other hand, when rolling large slabs, the hot-rolling stage is indispensable, where to achieve the required deformations, it is necessary to heat the billets up to 500 °C to reduce the forces and increase the plasticity of the material. In this case, Sc precipitates from the solid solution at 220–300 °C, the diffusion rate of Sc in Al is relatively high and Al_3Sc dispersoids can coarsen relatively quickly with increasing temperature of heat treatment. When the heat treatment temperature increases to 500 °C, Zr is also released as coarse and large particles with a DO_{23} structure. This can all lead to a significant degradation of the mechanical properties.

Given the experience of heat treatment of large-scale slabs of scandium alloys like 5081 and 5181 to obtain a uniform separation of particles containing Sc, Zr, and other REE before the hot-rolling stage, studies of changes in hardness and electrical conductivity on ingot samples were conducted.

As the first stage of heat treatment the temperature interval 270–360 °C with holding times up to 6 h was chosen. The highest hardness is achieved at 300–330 °C with a maximum exposure time of 6 h. At the same time at 360 °C the hardness decreases (Fig. 3). The electrical

Table 1 Electrical-grade aluminum (% wt.)

Si	Fe	Cu	Mg	Zn	Ga	Cr	Mn + Ti + V	Al
0,07 ± 0,01	0,18 ± 0,02	≤ 0,01	≤ 0,01	≤ 0,04	≤ 0,03	≤ 0,005	≤ 0,01	99.7

Fig. 1 Casting slabs and 3 mm rolling sheet from new alloys

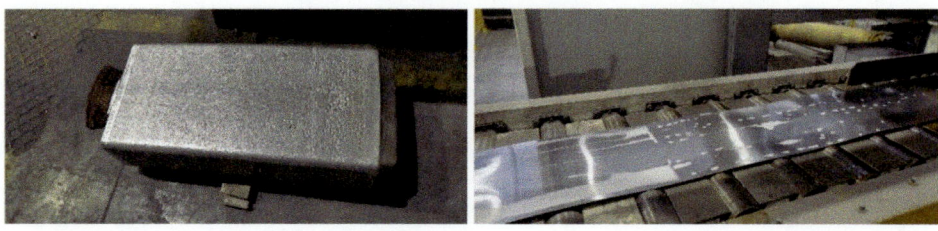

Fig. 2 Microstructure of slabs: **a** 1350, **b** Al-Sc-Ce, Y; **c** Al-Sc, Zr; and **d** Al-Sc, Zr-Er, Yb

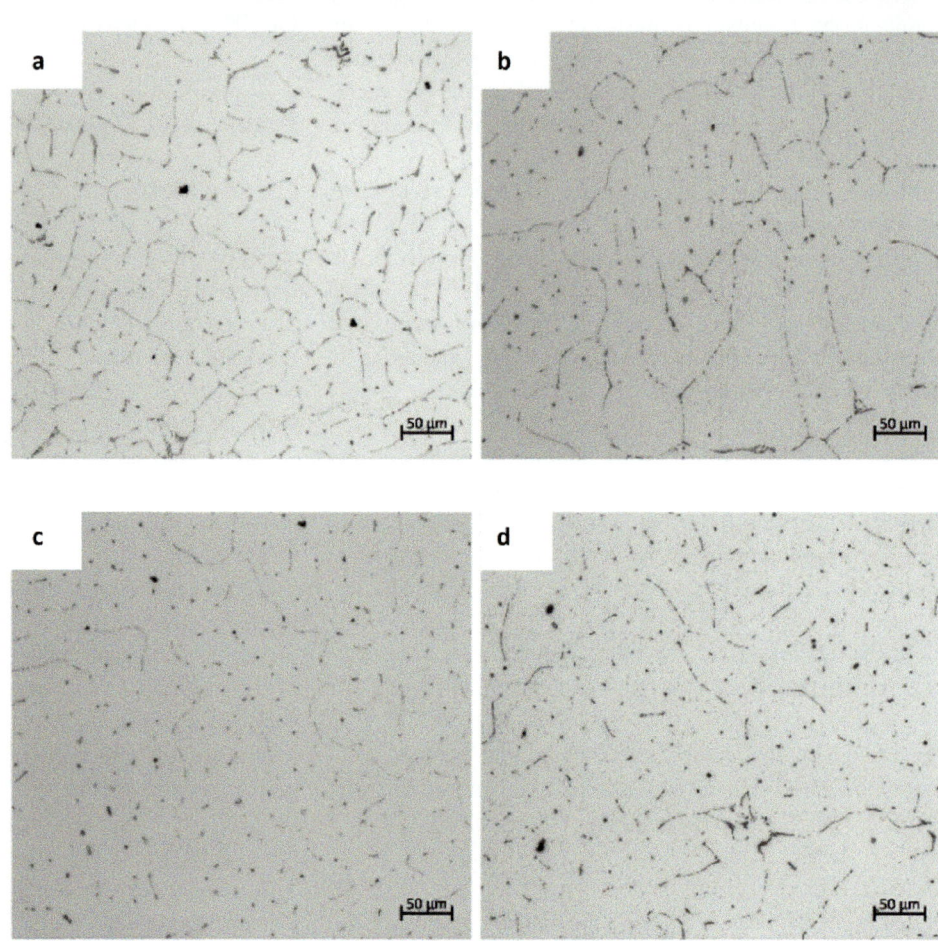

conductivity (Fig. 4) reaches a maximum already at 3 h, which should indicate scandium release from the solid solution.

The temperature of 420–425 °C was chosen as the second stage for the extraction of scandium, REE and separation of most of the zirconium-containing particles. The hardness increases for the compositions Al-Sc, Zr-Er-Yb, practically does not change for the compositions Al-Sc, Zr, and for the compositions Al-Sc-Ce, Y (without Zr) hardness drops sharply (Fig. 5) at the beginning of annealing. The electrical conductivity at this temperature increases almost to a maximum at 3 h of annealing.

In connection with the data obtained by measuring the electrical resistance and hardness on the annealed samples,

the ingots were heat treated according to optimal regimes to achieve maximum strength.

Rolling Sheets

The change in mechanical properties of the rolled sheets with a thickness of 3 mm of model compositions in comparison with sheets from 1350 is shown in Fig. 6. The strength of the 1350 sheets begins to decrease after annealing at 150 °C and already at 300 °C the sheets are in a fully annealed state with a minimum UTS of 70 MPa. For the compositions Al-Sc, Zr and Al-Sc-Ce, Y a similar dependence is observed, but the strength reduction begins at 250 °C.

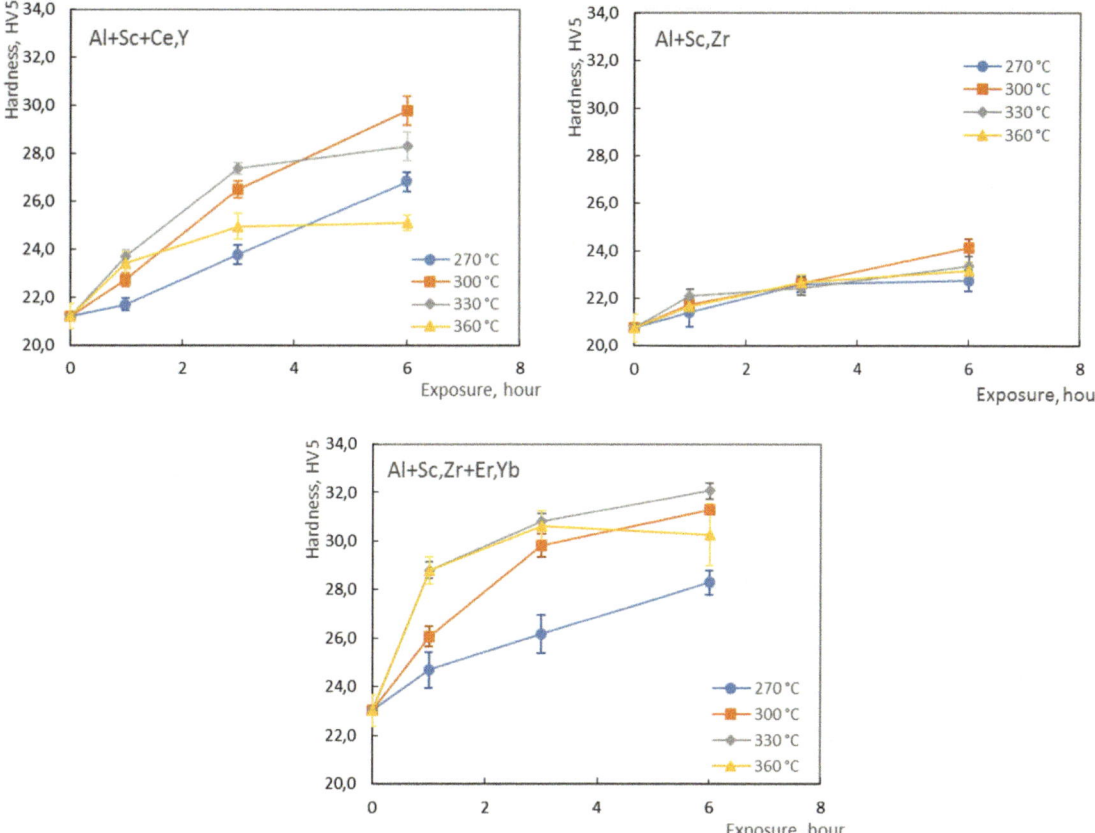

Fig. 3 Changes in the hardness of slabs samples after first stage of heat treatment with different temperatures

At the same time, the annealing at 400 °C for composition Al-Sc, Zr does not lead to complete softening of the material (UTS 110, YS 95 MPa). On the contrary, the Al-Sc-Ce, Y sheets are in a fully annealed state after heat treatment with 400 °C and the mechanical properties of them are comparable with those of 1350. This seems to indicate that the Al₃Sc precipitates are enlarging and are no longer effectively hardening the material. The highest mechanical properties have Al-Sc, Zr-Er, Yb sheets with UTS 186, YS 178 MPa and elongation of 11%. During heat treatment of sheets of this composition, the strength is practically unchanged up to 250 °C. A smooth decrease in strength occurs only starting at 300 °C, and after annealing at 400 °C the sheets still have high UTS 135, YS 120 MPa.

The metallographic analysis of the structure is shown in Fig. 7. After rolling, the sheets of all compositions have a deformation fibrous structure with elongated grains. At the same time, the sheets of compositions with Zr have a more strongly hardened structure, which apparently indicates the suppression of the processes of dynamic recrystallization at the stage of hot rolling. Annealing at 300 °C leads to recrystallization of 1350 sheets, which is typical for all alloys of the 1xxx series. The alloying of Sc in combination with Ce and Y suppresses recrystallization, which only begins at

300 °C, and is complete at 400 °C. For both compositions Al-Sc, Zr and Al-Sc, Zr-Er, Yb, the absence of recrystallization of hardened sheets is observed even after 400 °C (with annealing time of 3 h), which also explains the high mechanical properties after this annealing.

In order to stabilize the mechanical properties of the highly cold worked sheets, an annealing at 150 °C for 3 h was applied. This condition according to DIN EN 515 corresponds most closely to the H38 temper. The mechanical properties and electrical conductivity for the rolling sheets of the model alloys in comparison with those of 1350 are shown in Fig. 8. The greatest strength is possessed by Al-Sc, Zr-Er, Yb (UTS 172 MPa, YS 165 MPa), but the electrical conductivity decreases to 60%IACS. In contrast, the alloy alloyed with Sc, Ce, and Y has slightly lower strength (UTS 160 MPa, YS 150 MPa), but the electrical conductivity is at the level of electrical-grade aluminum at 62% IACS.

The results of comparative fatigue tests with a tensile-compression loading scheme (with asymmetry coefficient −1) of 1350 and Al-Sc, Zr-Er, Yb sheets in the H38 temper are shown in Fig. 9. The rare-earth elements Sc, Er, and Yb as well as Zr form fine precipitates of Al₃M, and, on the one hand, increase the strength, and on the other hand, the small size does not lead to material embrittlement, which

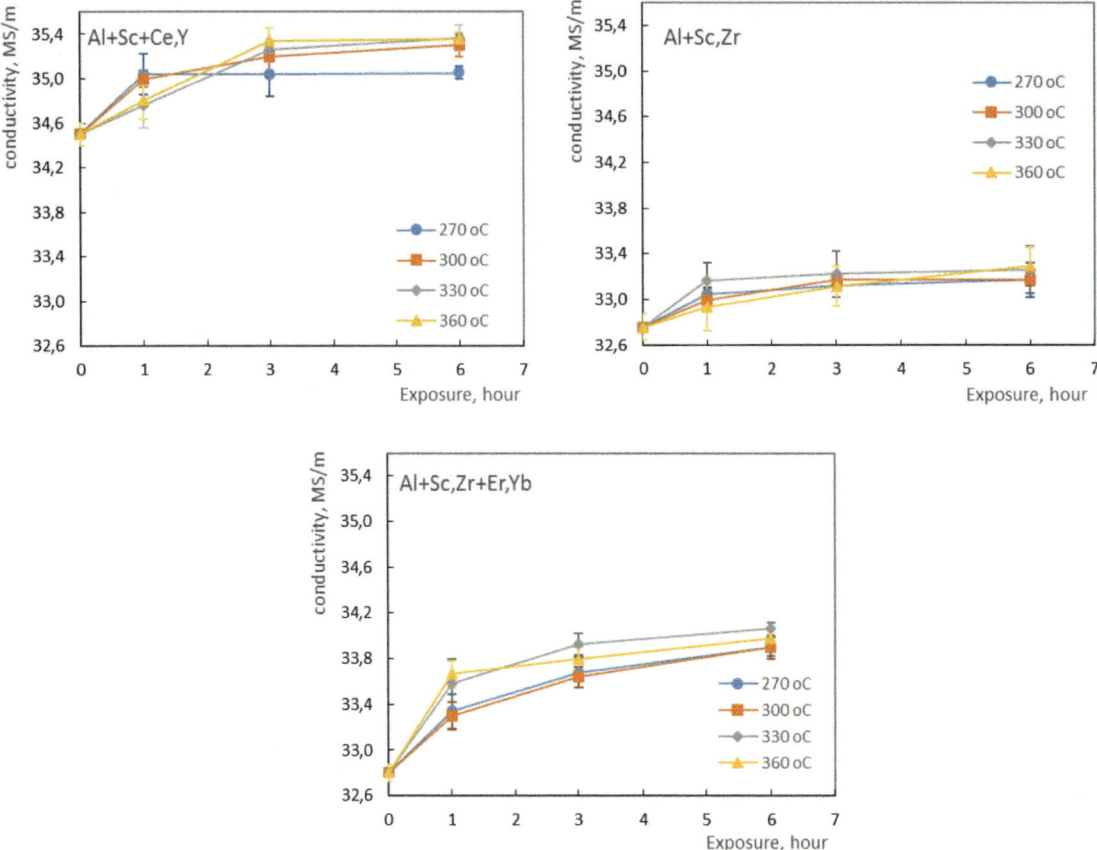

Fig. 4 Changes in the conductivity of slabs samples after first stage of heat treatment with different temperatures

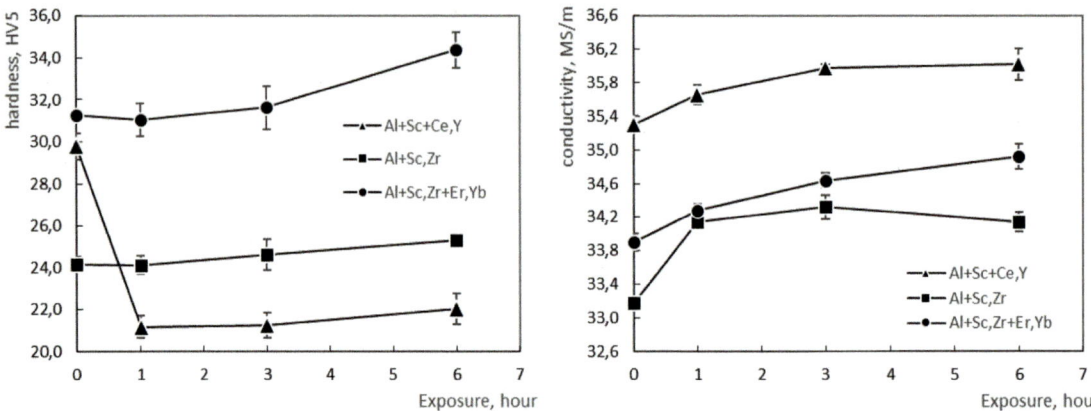

Fig. 5 Changes of hardness and conductivity of slabs samples after second stage of heat treatment with 420 °C (first stage was 300 °C/6 h)

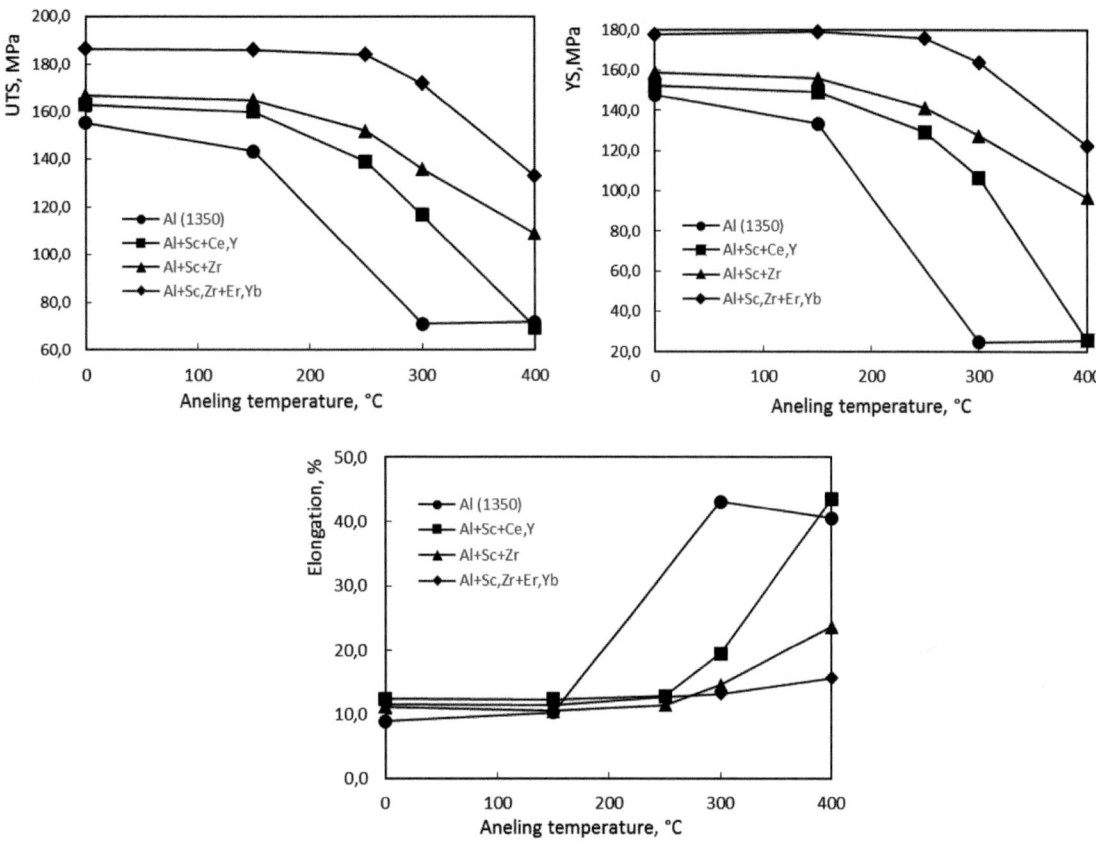

Fig. 6 Mechanical properties of 3 mm rolled sheets after annealing with different temperatures (3 h exposure time)

was confirmed by cyclic tests. The limit of fatigue of samples at 2×10^7 cycles is about 70 MPa, which is about 20% higher than that for the alloy 1350.

Conclusion

In summary, the effects of adding Sc, Zr, and other REE on the 1XXX conductive aluminum alloys:

- Alloying of Sc and other REE of 1xxx series alloys requires selection of regimes of heat treatment of industrial slabs and hot-rolling temperature. Using heats above 420 °C on compositions with Sc without additional alloying with Zr leads to degradation of strength properties.
- The alloying of Sc together with Ce and Y of electrical-grade aluminum type such as 1350 leads to an increase in the strength in hardened and stabilized state up to UTS 160 MPa, YS 150 MPa, elongation 12% and practically does not reduce the electrical conductivity, which remains at 61–62% IACS.

- The Al-Sc-Ce, Y composition in the fully annealed state after heat treatment at 400 °C has mechanical properties comparable with those of 1350. Apparently, at this temperature, the Al$_3$Sc precipitates coarsen and cease to harden the material.
- The addition of Zr completely suppresses recrystallization even of heavily hardened sheets of compositions Al-Sc, Zr and Al-Sc, Zr-Er, Yb up to annealing temperature 400 °C, which, among other things, contributes to the preservation of high mechanical properties when thermally influenced.
- The addition of Zr as well as Er and Yb leads to a slight decrease in conductivity, but in general allows to obtain a product with electrical conductivity of about 60%IACS.
- The composition Al-Sc, Zr-Er, Yb in H38 temper (after stabilizing annealing at 150 °C) with UTS 172 MPa, YS 165 MPa, elongation 11% has the highest mechanical properties.
- The alloying of Sc, Zr, Er, and Yb in addition to mechanical properties also leads to an increased fatigue tests by more than 20%.

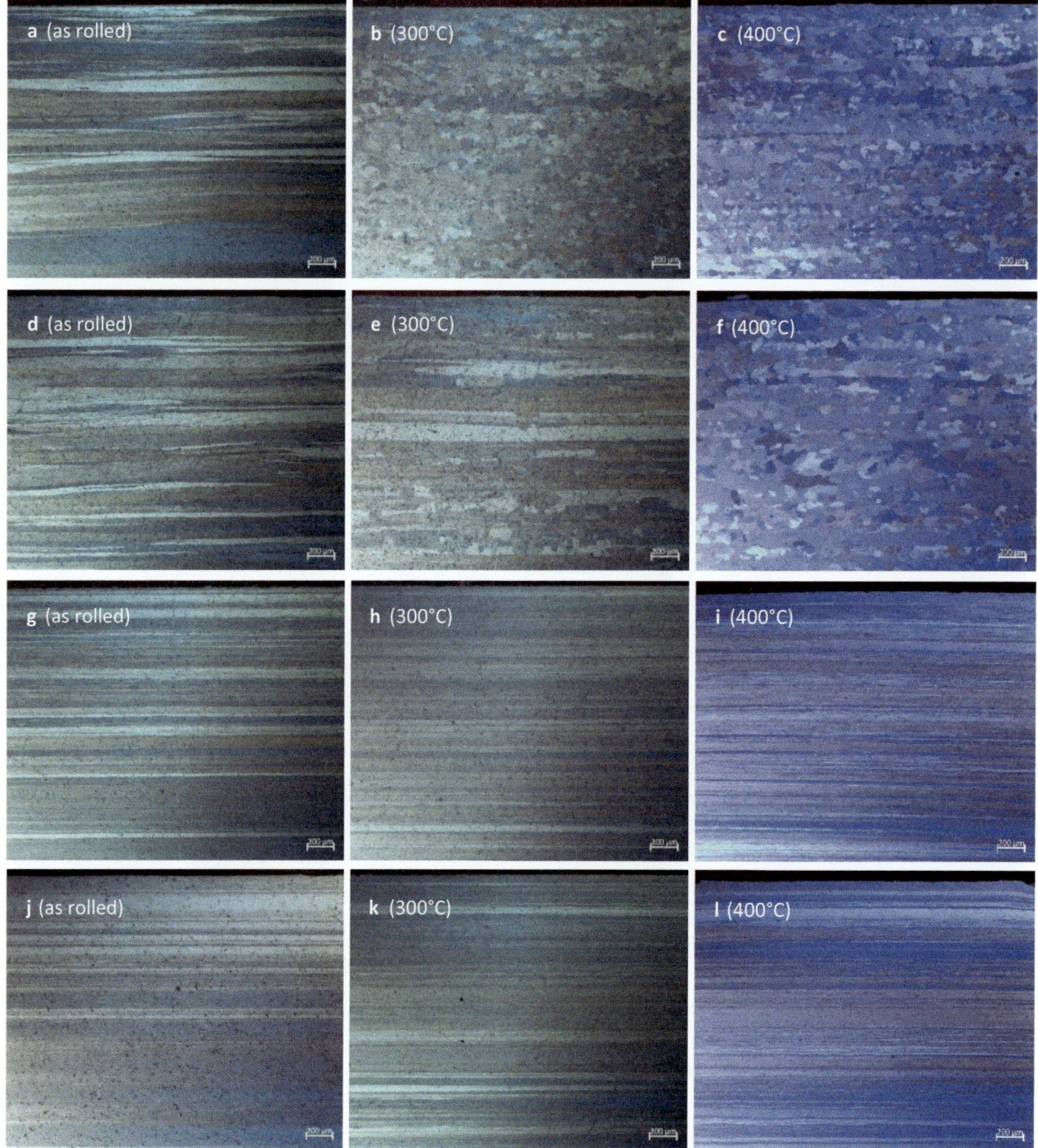

Fig. 7 Microstructure of sheets in longitudinal rolling direction: **a–c** 1350; **d–f** Sc-Ce, Y; **g–i** Sc-Zr; **j–l** Sc-Zr-Er-Yb

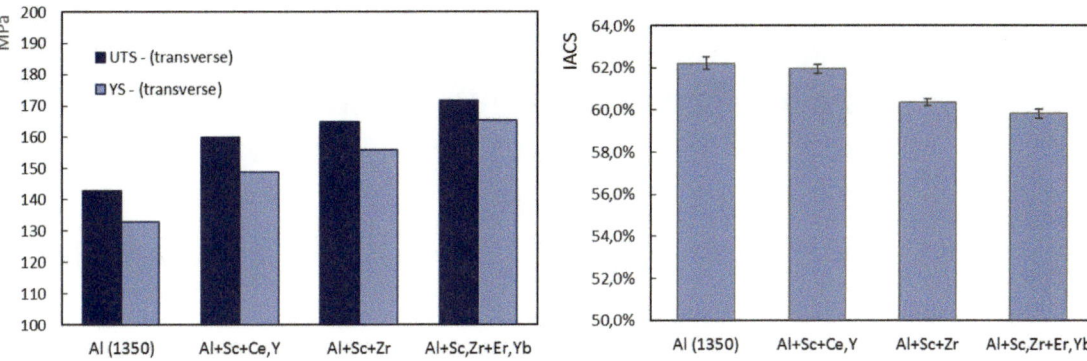

Fig. 8 Mechanical properties and conductivity for 1350 and new alloys after stabilized treatment

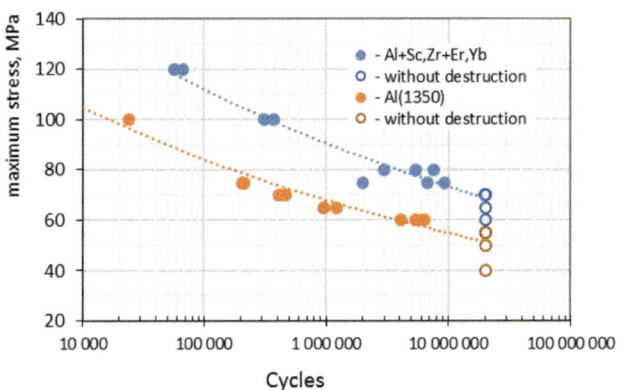

Fig. 9 Fatigue curve for 1350 and Al-Sc, Zr-Er, Yb after stabilized treatment

References

1. Aluminum Electrical Conductor Handbook. Section IV Bus Conductors Chapter 13. 1989.
2. K. E. Knipling et al.: Criteria for developing castable, creep-resistant aluminum-based alloys – A review. Zeitschrift für Metallkunde. 97 (2006).
3. Dorin, Thomas & Ramajayam, Mahendra & Vahid, Alireza & Langan, Timothy. (2018). Aluminium Scandium Alloys. https://doi.org/10.1016/B978-0-08-102063-0.00012-6.
4. [4] Mann V, Krokhin A, Alabin A, Fokin D, Valchuk S (2019) New Al-Mg-Sc Alloys for Shipbuilding and Marine Applications. Light Metal Age 77(3):29-30.
5. Matveeva I.A. Research and development of production technology of aluminum wire rod with zirconium addition by continuous casting and rolling in order to produce heat-resistant power transmission lines. Dissertation of PhD in Technical Sciences. M. 2014.
6. Belov N.A., Dostaeva A.M., Shurkin P.K., Korotkova N.O., Yakovlev A.A. Impact of annealing on electrical resistivity and hardness of hot-rolled alloyed aluminum sheets with Zr content up to 0,5 wt.%. Izvestiya Vuzov. Tsvetnaya Metallurgiya 2016.
7. K.E. Knipling, R.A. Karnesky, C.P. Lee, D.C. Dunand, D.N. Seidman, Acta Mater.58 (2010) 5184–5195.».
8. [8] K.E. Knipling, D.N. Seidman, D.C. Dunand, Acta Mater. 59 (2011) 943–954.
9. Liu, Li & Jiang, Jian-Tang & Zhang, Bo & Shao, Wen-Zhu & Zhen, Liang. Enhancement of strength and electrical conductivity for a dilute Al-Sc-Zr alloy via heat treatments and cold drawing. Journal of Materials Science & Technology. 35. (2018).
10. Chao, Run-Ze & Guan, Xi-Hua & Guan, Ren-guo & Tie, Di & Lian, Chao & Wang, Xiang & Zhang, Jian. (2014). Effect of Zr and Sc on mechanical properties and electrical conductivities of Al wires. Transactions of Nonferrous Metals Society of China.
11. Di Tie, Yu Wang, Xiang Wang et al. Microstructure Evolution and Properties Tailoring of Rheo-Extruded Al-Sc-Zr-Fe Conductor via Thermo-Mechanical Treatment. Materials. 13. 845. (2020).
12. [12] Frank Czerwinski. Critical Assessment 36: Assessing differences between the use of cerium and scandium in aluminium alloying. Materials science and technology. 2020, Vol. 36, No. 3, 255–263.
13. Zachary Sims, Orlando Rios, David Weiss et all. High Performance Aluminum-Cerium Alloys for High-Temperature Applications. Materials Horizons. 4. (2017).
14. Dobatkin V.I., Elagin V.I., Fedorov V.M. Rapidly crystallized aluminum alloys. VILS, 1995.
15. Microstructure and properties of novel Al-Ce-Sc, Al-Ce-Y, Al-Ce-Zr and Al-Ce-Sc-Y alloy conductors processed by die casting, hot extrusion and cold drawing / W. Wang, Q. Pan, Y. Sun [et al.] // Journal of Materials Science and Technology. – 2020. – Vol. 5.
16. Zhang Yi, Gao Kunyuan, Wen Shengping et all. Determination of Er and Yb solvuses and trialuminide nucleation in Al-Er and Al-Yb alloys. Journal of Alloys and Compounds. (2014).
17. Nhon Q. Vo, Davaadorj Bayansan, Amirreza Sanaty-Zadeh et all. Effect of Yb microadditions on creep resistance of a dilute Al-Er-Sc-Zr alloy. Materialia, Volume 4, 2018, Pages 65–69.

Developing Al-Zr-Sc Alloys as High-Temperature-Resistant Conductors for Electric Overhead Line Applications

Quan Shao, Emad Elgallad, Alexandre Maltais, and X.-Grant Chen

Abstract

The effect of a minor Sc addition (within 0.1 wt.%) to improve the mechanical properties and electrical conductivity of Al-Zr-Sc conductor wires was investigated. The thermal-resistance properties of Sc-containing wires after thermal exposure at 310 °C and 400 °C were evaluated according to the international standard IEC 62,004. The results show that a simultaneous improvement of mechanical properties and electrical conductivity while maintaining outstanding thermal resistance was achieved by microalloying with Sc in comparison with the base Al-Zr alloy. This was attributed to the precipitation of a high number density of Al₃(Sc, Zr) nanoparticles. Excellent combination of ultimate tensile strength and electrical conductivity (188–197 MPa and 58.0–59.5% IACS) was obtained in Sc-containing alloys using conventional thermomechanical process (e.g., casting, rolling and wire drawing) to fulfill the particular requirements of different standard conductor grades. The newly developed conductor alloys provide a much-needed outstanding performance for overhead line applications in the electric industry.

Keywords

Al-Zr-Sc alloys • Mechanical properties • Electrical conductivity • Thermal resistance properties

Introduction

With the massive societal shift toward electrification, the industrial demand for high strength and high thermal resistance of aluminum conductors is continuously growing due to their lightweight and the low cost [1, 2]. Al-Mg-Si 6xxx alloys are commonly used for the production of All-Aluminium Alloy Conductors (AAACs), offering two categories: 1) high strength of 315 MPa with a relatively low electrical conductivity of 52.5% IACS and 2) moderate strength of 240 MPa with a high electrical conductivity of 59.0% IACS [3]. However, the poor thermal-resistance property of Al-Mg-Si 6xxx alloys limits their operating temperature below 100 °C [4, 5]. This, in turn, restricts the usage and current-carrying capability of these alloys, taking into consideration that increasing this capability generally raises the wire temperature.

Thermal-resistant aluminium alloy wires for overhead line conductors have been introduced to adopt at much higher operating temperatures (180–310 °C) than the maximum allowable operating temperature of 6xxx alloy wires. They are commonly Al-Zr-based alloy wires and are designated as AT1, AT2, AT3, and AT4 according to different combinations of electrical conductivity, strength, and thermal-resistant property based on the International Standard IEC 62,004 [6]. Table 1 lists the electrical, mechanical, and thermal-resistant properties of Al alloy wires in this standard. The improved thermal stability of Al-Zr-based alloy wires arises from the precipitation of the Al₃Zr nanoparticles, which are thermally stable upon heating at high temperatures due to the low diffusivity of Zr atoms in α-Al matrix [7–10]. However, the low strength of Al-Zr-based alloys limits their use in the production of electrical overhead line conductors [11, 12].

Q. Shao · E. Elgallad · X.-G. Chen (✉)
Department of Applied Science, University of Quebec at Chicoutimi, Saguenay, QC G7H 2B1, Canada
e-mail: xgrant_chen@uqac.ca

A. Maltais
Arvida Research and Development Center, Rio Tinto Aluminum, Saguenay, QC G7S 4K8, Canada

© The Minerals, Metals & Materials Society 2023
S. Broek (ed.), *Light Metals 2023*, The Minerals, Metals & Materials Series,
https://doi.org/10.1007/978-3-031-22532-1_170

Table 1 Electrical, mechanical, and thermal-resistant properties of Al alloy wires [6]

Category	Conductivity (%IACS)	*Tensile strength (MPa)	*Elongation (%)	**Thermal-resistant temperature (°C)	
				1 h heating	400 h heating
AT1	60.0	159	2.0	230	180
AT2	55.0	225	2.0	230	180
AT3	60.0	159	2.0	280	240
AT4	58.0	159	2.0	400	310

* Tensile properties of 4.0–4.5 mm diameter wires
** A minimum strength of 90% of the initial tensile strength after heating should be maintained to fulfil the thermal-resistant property of the wires

The addition of Sc to Al-Zr alloys can result in higher mechanical properties and better thermal resistance [13–15] due to the precipitation of the coherent and thermally stable $L1_2$-Al_3(Sc, Zr) nanoparticles, which are composed of a Sc-enriched core surrounded by a Zr-enriched shell. Therefore, they possess slow coarsening kinetics and can effectively impede dislocation movement and grain boundary migration [15, 16]. Several studies on Al-Sc-Zr as the targeted electrical conductors have been reported in literature [11, 12, 16]. It is reported that a high Sc-content Al-0.35Sc-0.2Zr alloy provided a strength of 210 MPa and an electrical conductivity of 60.2% IACS with a sluggish strength degradation after prolonged exposure at 400 °C for 100 h [12]. It is also reported a dilute Al-0.06Sc-0.23Zr alloy with optimized process exhibited a strength of 194 MPa with an electrical conductivity of 61% IACS [11].

Although Al-Sc alloys have recently been introduced as promising candidates for high thermal-resistant aluminium alloys, the impact of Sc under the thermomechanical process on comprehensive properties of Al-Zr-Sc conductors (mechanical, electrical, and thermal-resistance) has not been systematically assessed. The present study aims at developing a series of Al-Zr-Sc thermal-resistant conductors that could yield different combinations of mechanical properties, thermal-resistant properties, and electrical conductivities, satisfying the specifications of AT1, AT2, AT3, and AT4 grades. The approach adopted here is based on microalloying of Sc at a low range (~0.10 wt.%) for cost-effectiveness and using a conventional thermomechanical processing route.

Experimental Procedures

Table 2 lists the actual chemical compositions of the four experimental alloys, analyzed by optical emission spectroscopy. The alloy A represents the benchmark Al-Zr-based material. Alloys B and C focus on a low Sc range (~0.10 wt. %) with different combinations of Si and Fe. To examine the feasibility of using higher Sc levels (~0.2 wt.%), alloy D is included for comparison with the low Sc alloys B and C. The melts were batched using an electrical resistance furnace and cast into a Y-shaped permanent mold to obtain rectangular cast ingots with dimensions of 30 × 40 × 80 mm. The melting temperature was kept at 750 °C, i.e., above the liquidus temperature of the Al-0.2%Zr alloys (720 °C).

The thermomechanical processing of the cast ingots consists of hot rolling, solution heat treatment, cold wire drawing, and aging treatment. Hot rolling was carried out on a laboratory-scale rolling mill with multiple passes at a temperature of 375 °C for alloy A and 550 °C for alloys B to D. The rolling began from the 26.5-mm-thick ingot and ended with 8-mm-thick strip (~70% reduction ratio). The hot-rolled strips were cut and machined into square bars with dimensions of 6.7 × 6.7 mm for the subsequent wire drawing. Cold wire drawing was performed on a laboratory-scale wire drawing mill to draw the square bars into wires with a diameter of 4.3 mm (~68% reduction ratio). Both solution and aging treatments were conducted in a programmable air circulating electric furnace, and their parameters are listed in Table 3.

Table 2 The chemical compositions of the alloys prepared (wt.%)

Alloy	Si	Fe	Zr	Sc	Al
A	0.26	0.27	0.14	0.00	Bal.
B	0.03	0.27	0.08	0.07	Bal.
C	0.24	0.28	0.11	0.07	Bal.
D	0.27	0.30	0.11	0.17	Bal.

Table 3 Heat treatment parameters

Alloy	Solution treatment	Aging treatment
A	N/A	400 °C for 72 h
B, C and D	600 °C for 8 h	350 °C for 48 h

Vickers microhardness measurements were performed using an NG-1000 CCD microhardness tester with a load of 25 g and a dwell time of 15 s. The electrical conductivity (EC, % IACS) was measured using a Sigmascope SMP10 electrical conductivity measurement instrument based on the eddy current method. The microhardness and EC measurements were conducted at room temperature on the polished cross sections of 4.3 mm diameter wire samples.

The tensile properties of 4.3 mm diameter wires were measured at room temperature using an Instron 8801 servo-hydraulic unit with a strain rate of 25 mm/min. Three wires were tested per each condition, and the average values of ultimate tensile strength (UTS) and percentage elongation (El) were calculated. To assess the thermal-resistant property, 4.3 diameter wires were subjected to thermal exposure at 400 °C for 1 h and 310 °C for 400 h. The UTSs of the thermally exposed wires were measured and then compared with the initial values before thermal exposure.

Samples for transmission electron microscope (TEM) observations were prepared from the cross sections of 4.3 mm diameter wires. The nanosize Al_3Zr and $Al_3(Zr, Sc)$ precipitates were observed in the dark-field mode in a Jeol JEM-2100 TEM operating at 200 kV. The average number density of these precipitates was measured based on the image analysis of TEM images.

Results

As-Cast Microstructure

The typical as-cast microstructures of alloys B and C (as examples) are shown in Fig. 1. In general, the microstructures of the alloys are composed of primary α-Al grains along with AlFe and AlFeSi intermetallic phases (gray particles) distributed at the aluminum dendrite boundaries. The addition of Si to alloy C slightly increased the fraction of AlFeSi phase compared with alloy B. No primary intermetallic phases containing Zr and/or Sc were detected in both alloys. This implies that Zr and Sc solutes were highly dissolved in the α-Al matrix.

Effect of Alloy Composition on the Microhardness and Electrical Conductivity

The effect of alloy composition on the EC and microhardness of 4.3 mm diameter wires is shown in Fig. 2. The base A alloy containing Zr only exhibited the lowest hardness of 42 HV. The other three alloys containing Sc exhibited much higher hardness ranging from 65.3 to 68.5 HV, resulting in a hardness increase of 56–63% compared with the base alloy. The hardness of alloy C was higher than that of alloy B (68.5 HV vs. 65.3 HV), even though both alloys contained the same level of Sc (0.07 wt.%), which is attributed to the higher Si level in the former relative to the latter (0.24 wt.% vs. 0.03 wt.%). It was reported that Si could promote the decomposition of the Zr-containing α-Al solid solution [17, 18]. Solute atoms of Si remained in the α-Al solid solution could also increase the strength via solid solution strengthening. Although alloy D contained higher Sc level compared to alloys B and C (0.17 wt.% vs. 0.07 wt.%), its hardness (67.8 HV) is still lower than that of alloy C (68.5 HV). This indicates that the use of high Sc levels would not be helpful with the adopted thermomechanical process to further improve the strength. The EC of the experimental alloys ranged from 57.8 to 59.5% IACS, and the highest EC was achieved by alloy B followed by alloys A, C, and D. This is

Fig. 1 SEM micrographs showing the as-cast microstructure of alloys **a** B and **b** C

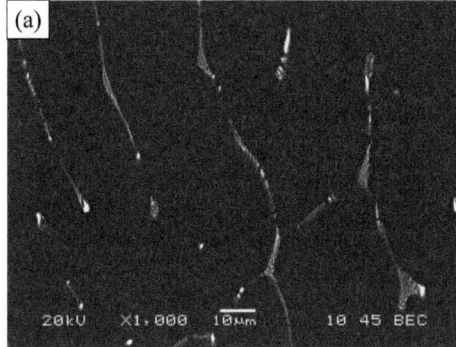
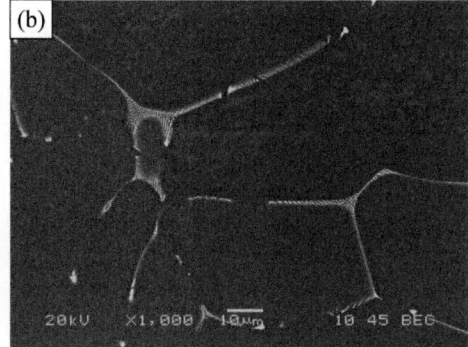

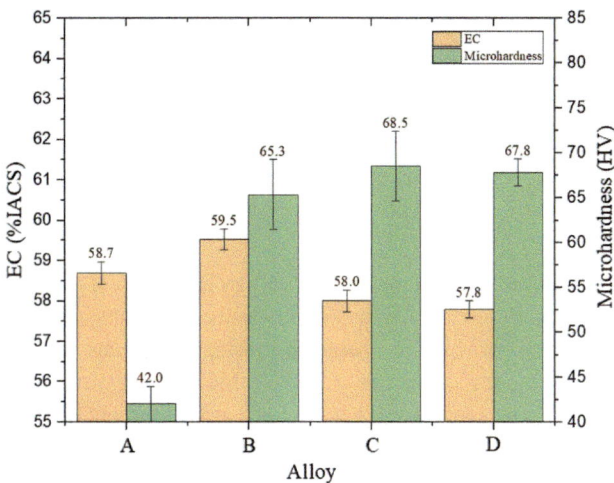

Fig. 2 Microhardness and EC of the experimental alloys

consistent with the levels of alloying elements contained in these alloys; the lower the level of alloying elements, the higher is the EC of the alloy, and vice versa.

Effect of Alloy Composition on Tensile Properties

The UTS and El of 4.3 mm diameter wires are shown in Fig. 3. The variation of the UTS with the alloy composition agrees well with the results of microhardness (Fig. 2). The UTS of the alloy A containing Zr only was 105 MPa, which is too low to match the minimum required UTS of the thermal-resistant Al alloy conductors (159 MPa, Table 1). The UTS of the other Sc-containing alloys ranged from 186 to 197 MPa, which well exceeded the minimum specified

UTS of the thermal-resistant AT1, AT3, and AT4 grades (159 MPa, Table 1). It is evident that microalloying with Sc resulted in an outstanding increase in the strength (77–88%) compared to the base alloy A, revealing the great potential of the minor Sc addition to improve the strength of thermal-resistant conductors. The elongations of the Sc-containing alloys ranged from 5.7 to 9.3%, greatly exceeding the minimum required elongation of the thermal-resistant Al alloy conductors (2.0%, Table 1).

Thermal-Resistant Properties

From the results of Figs. 2 and 3, it can be seen that only alloys B and C can meet wire specifications defined in the standard IEC 62,004. The wire samples of alloys B and C were therefore selected to evaluate thermal-resistant properties after thermal exposures at 400 °C for 1 h and 310 °C for 400 h, as specified for AT4 grade (Table 1), which is the highest standard of thermal resistance among the four grades. The UTS of both alloys before and after both thermal exposures is compared in Fig. 4. It is apparent that both thermal exposures did not cause a remarkable reduction in the UTS. In both exposure conditions, the reductions were 0.4 and 2.7% for alloy B, and 3.5 and 4.0% for alloy C, respectively. These slight reductions are much less than 10%, implying that alloys B and C can fulfill the highest thermal-resistant property of the AT4 grade, and consequently the lower thermal-resistant properties of the AT1, AT2, and AT3 grades (Table 1). The outstanding thermal resistance of alloys B and C is undoubtedly attributed to the addition of Sc that resulted in the precipitation of the thermally stable $Al_3(Sc, Zr)$ nanoparticles.

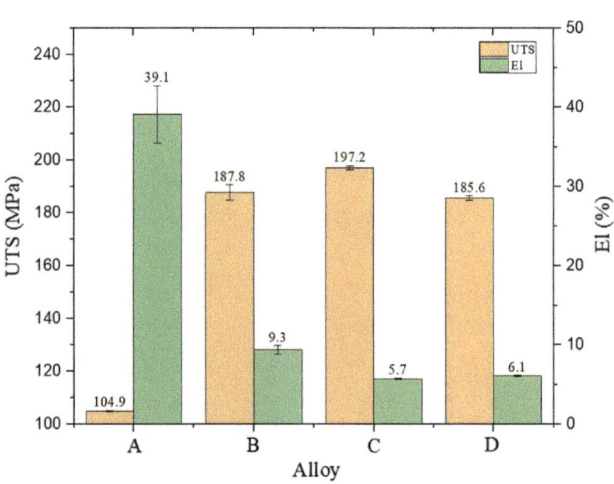

Fig. 3 Tensile strength (UTS) and elongation (El) of the experimental alloys

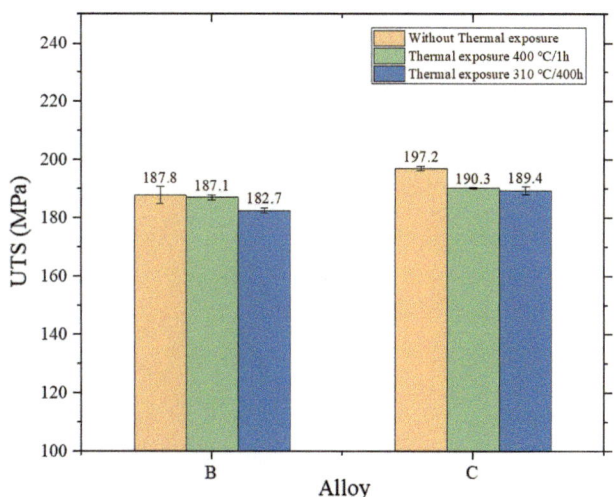

Fig. 4 UTS of alloys B and C before and after thermal exposures at 400 °C/1 h and 310 °C/400 h

Precipitate Microstructure

Figure 5 shows the nanosized precipitates formed in alloy A containing Zr only and alloy C (as an example of Sc-containing alloys). Very few Al_3Zr precipitates with a relatively large size were formed in alloy A (Fig. 5a). On the other hand, fine uniformly distributed $Al_3(Sc, Zr)$ precipitates were found in alloy C (Fig. 5b). The characteristics of the nanosized precipitates in alloys A, B, and C are listed in Table 4. As can be seen, the $Al_3(Sc, Zr)$ precipitates formed in alloys B and C had higher number densities and smaller sizes in comparison with the Al_3Zr precipitates formed in alloy A, explaining the large difference in the strength without and with Sc addition even with a low Sc level (Fig. 3). The higher number density of the $Al_3(Sc, Zr)$ precipitates in alloy C relative to alloy B explains the relatively higher hardness and strength of alloy C (Figs. 2 and 3) and can be attributed to the role of Si in stimulating the precipitation of $Al_3(Sc, Zr)$ precipitates [17, 18].

Discussion

The results obtained above reveal that the Al-Zr base alloy could hardly achieve the required mechanical properties of the thermal-resistant aluminium conductors. The maximum achievable strength of the Al-Zr alloy (alloy A, 0.14 wt.% Zr) is 105 MPa, while the minimum specified strength of thermal-resistant aluminium conductors is 159 MPa. Due to low solubility of Zr in Al at the liquidus temperature (~ 0.12 wt.%), adding a high level of Zr could not increase the quantity of Zr solutes in the solid Al matrix and hence would not be much helpful for improving the strength. On the other hand, microalloy with Sc in Al-Zr alloys resulted in a substantial increase in the mechanical properties due to the precipitation of a large number $Al_3(Sc, Zr)$ nanoparticles. The strength and elongation of the Sc-containing alloys ranged from 186 to 197 MPa and from 5.7 to 9.3%, respectively, highly exceeding specifications of thermal-resistant aluminium conductors (159 MPa and 1.5%). Simultaneously, the Sc-containing alloys provided high electrical conductivities of 57.8–59.5% IACS and excellent thermal-resistant properties; the maximum strength reduction after thermal exposures at 400 °C for 1 h and 310° for 400 h is limited to 4%, while the maximum specified reduction is 10% [6]. These improvements were achieved by a low Sc level (less than 0.1 wt.%), which could be costly affordable for the production of high-temperature-resistant conductors with such promising properties. On the other hand, the use of a higher level of Sc (0.17 wt.%, alloy D) did not further improve the strength of the wires and it, therefore, seems not feasible with the approach adopted in the present study.

Based on these results, Sc-containing B and C alloys show a significant potential for producing thermal-resistant aluminium conductors. Because the thermal-resistant-properties of these alloys fulfill the particular requirements of all AT1, AT2, AT3, and AT4 grades, the possible assignment of the alloys to these grades is made according to the UTS and EC values in comparison with the specified standard values of the four grades (Table 1), as shown in Table 5. As can be seen, both AT1 and AT3 grades with extra high electrical conductivity are almost achievable by alloy B. A very slight increase in the electrical conductivity of this alloy is required to reach the specified value (59.5% IACS vs. 60.0% IACS), which is expected to be easily attained by increasing the aging time beyond 48 h or by the Boron treatment of the melt. The AT4 grade with extra high thermal resistance is fully achievable by alloy C with a much higher strength (197 MPa vs. 159 MPa) and a typical EC (58.0% IACS). Because grade AT2 requires a relatively low EC (55% IACS) but relatively high strength (225 MPa), an increase of other alloying elements in alloy C, such as Fe and Si, with optimized aging treatment could satisfy the strength requirement of this grade.

Fig. 5 Dark-field TEM images showing Al_3Zr and $Al_3(Sc, Zr)$ precipitates formed in **a** A and **b** C alloys, respectively

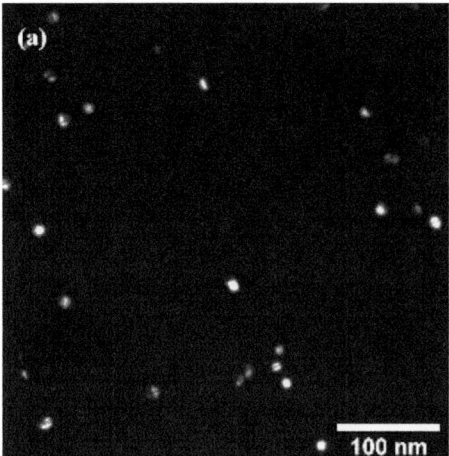
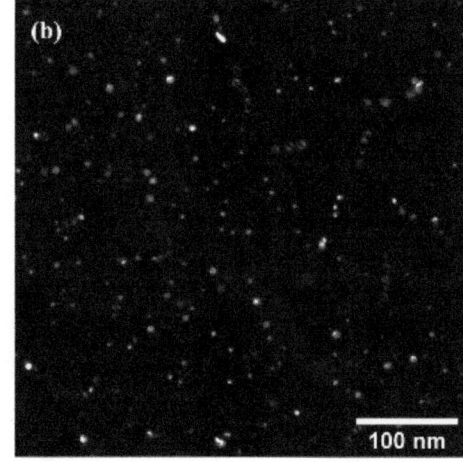

Table 4 Characteristics of the precipitates formed in alloys A, B, and C

Alloys	Number density, $\times 10^{22}$ m^{-3}	Average diameter, nm
A	0.03 ± 0.04	8.84 ± 2.48
B	1.17 ± 0.27	4.17 ± 1.25
C	1.51 ± 0.26	4.19 ± 1.57

Table 5 Assignment of alloys B and C to standard grades of thermal-resistant aluminium conductors

Alloy	EC (%IACS)	UTS (MPa)	Standard grade	Note
B	59.5	187.8	AT1 and AT3	Almost achieved
C	58.0	197.2	AT4	Completely achieved

Conclusions

The effect of microalloying with Sc to improve the mechanical properties and electrical conductivity of Al-Zr-Sc conductor wires was investigated. From the results obtained, the following conclusions can be drawn:

The Al-Zr base alloy could hardly achieve the minimum required strength of the thermal-resistant aluminium conductors. A minor addition of 0.07 wt.% of Sc resulted in a significant increase of 88% in the strength compared to the base Al-Zr alloy through the precipitation of a large number of Al_3(Sc, Zr) nanoparticles.

The Sc-containing alloys provided superior strength highly exceeding the specified strength of standard grades (188–197 MPa vs. 159 MPa) while maintaining high electrical conductivities ranging between 58.0 and 59.5% IACS.

The Sc-containing alloys exhibited excellent thermal-resistant properties at 310 °C and 400 °C, satisfying the highest standard among the four grades of thermal-resistant aluminium conductors. The maximum strength reduction after thermal exposure at these temperatures was limited to 4%, which was lower than the allowable reduction of 10%.

With an excellent combination of mechanical strength, electrical conductivity, and thermal-resistant properties, the Sc-containing alloys developed in the present study could be deemed as promising candidates for producing thermal-resistant aluminium overhead lines for the electric industry with an affordable cost and a conventional thermomechanical manufacturing process.

Acknowledgements The authors acknowledge the financial support of the Natural Sciences and Engineering Research Council of Canada (NSERC) under the Grant No. CRDPJ 514651-17 and Rio Tinto Aluminum through the Research Chair in the Metallurgy of Aluminum Transformation at University of Quebec at Chicoutimi.

References

1. Murashkin MY, Sabirov I, Sauvage X, Valiev RZ (2016) Nanostructured Al and Cu Alloys with Superior Strength and Electrical Conductivity. J. Mater Sci. 51:33–49. https://doi.org/10.1007/s10853-015-9354-9
2. Blackburn, JL (2017) Symmetrical Components for Power Systems Engineering. CRC Press, Boca Raton.
3. Al-Mg-Si alloy wire for overhead line conductor JB/T 8134-1997.
4. Yuan W, Liang Z, Zhang C, Wei L (2012) Effects of La addition on the mechanical properties and thermal-resistant properties of Al–Mg–Si–Zr alloys based on AA 6201. Mater. Des. 34:788–792. https://doi.org/10.1016/j.matdes.2011.07.003
5. Yuan W, Liang Z (2011) Effect of Zr addition on properties of Al–Mg–Si aluminum alloy used for all aluminum alloy conductor. Mater. Des. 32(8-9):4195–4200.
6. Thermal-resistant aluminium alloy wire for overhead line conductor International Standard IEC 62004, 2007-2,
7. Li H, Bin J, Liu J, Gao Z, Lu X (2012) Precipitation evolution and coarsening resistance at 400 °C of Al microalloyed with Zr and Er. Scr. Mater. 67(1):73–76. https://doi.org/10.1016/j.scriptamat.2012.03.026
8. Muthaiah VS, Mula S (2016) Effect of zirconium on thermal stability of nanocrystalline aluminium alloy prepared by mechanical alloying. J. Alloys Compd. 688:571–580. https://doi.org/10.1016/j.jallcom.2016.07.038
9. Robson J, Prangnell P (2001) Dispersoid precipitation and process modelling in zirconium containing commercial aluminium alloys. Acta Mater. 49(4):599–613. https://doi.org/10.1016/S1359-6454(00)00351-7
10. Knipling KE, Dunand DC, Seidman DN (2008) Precipitation evolution in Al–Zr and Al–Zr–Ti alloys during aging at 450–600 °C. Acta Mater. 56(6):1182–1195. https://doi.org/10.1016/j.actamat.2007.11.011
11. Liua L, Jianga J.-T, Zhanga B, Shaoa W-Z, Zhen L (2019) Enhancement of strength and electrical conductivity for a dilute Al-Sc-Zr alloy via heat treatments and cold drawing. J. Mater. Sci. Tech. 35(6):962–971. https://doi.org/10.1016/j.jmst.2018.12.023
12. Guan R, Shen Y, Zhao Z, Wang X (2017) A high-strength, ductile Al-0.35Sc-0.2Zr alloy with good electrical conductivity strengthened by coherent nanosized-precipitates. J. Mater. Sci. Tech. 33(3):215–223. https://doi.org/10.1016/j.jmst.2017.01.017

13. Fuller CB, Seidman DN, Dunand DC (2003) Mechanical properties of Al(Sc,Zr) alloys at ambient and elevated temperatures. Acta Mater. 51(16):4803–4814. https://doi.org/10.1016/S1359-6454 (03)00320-3

14. Knipling KE, Karnesky RA, Lee CP, Dunand DC, Seidman DN (2010) Precipitation evolution in Al–0.1Sc, Al–0.1Zr and Al–0.1Sc–0.1Zr (at.%) alloys during isochronal aging. Acta Mater. 58 (15):5184–5195. https://doi.org/10.1016/j.actamat.2010.05.054.

15. Knipling KE, Seidman DN, Dunand DC (2011) Ambient- and high-temperature mechanical properties of isochronally aged Al–0.06Sc, Al–0.06Zr and Al–0.06Sc–0.06Zr (at.%) alloys. Acta Mater. 59(3):943–954. https://doi.org/10.1016/j.actamat.2010.10.017.

16. Zhou W, Cai B, Li W, Liu Z, Yang S, Heat-resistant Al–0.2Sc–0.04Zr electrical conductor (2012) Mater. Sci. Eng. A 552:353–358. https://doi.org/10.1016/j.msea.2012.05.051.

17. Alabin AN, Belov NA, Korotkova NO, Samoshinal ME (2017) Effect of annealing on the electrical resistivity and strengthening of low-alloy Alloys of the Al – Zr – Si system. Met. Sci. Heat Treat. 58(9):527–531. https://doi.org/10.1007/s11041-017-0048-6.

18. Morozova A, Mogucheva A, Bukin D, Lukianova O, Korotkova N, Belov N, Kaibyshev R (2017) Effect of Si and Zr on the microstructure and properties of Al-Fe-Si-Zr alloys. Metals 7(11):495. https://doi.org/10.3390/met7110495

The Development of New Aluminum Alloys for the Laser-Powder Bed Fusion Process

Nathan Andrew Smith, Mostafa Yakout, Mohamed Elbestawi, Phil Chataigneau, and Peter Cashin

Abstract

The laser-powder bed fusion (L-PBF) processing environment poses significant challenges for the expansion of materials available for usage as a result of the drastic cooling rates native to this manufacturing method. First introduced in the casting manufacturing ecosystem, the incorporation of small amounts of grain refining rare-earth metals, particularly scandium, into aluminum alloys (Al alloys) provided an avenue to resist the detrimental effects imposed by extreme thermal gradients, while also imparting a not insignificant increase in material strength. Through the analysis of solidification behaviour, complementary element pairings, and owing to the supersaturation potential in these manufacturing conditions, a material design strategy is outlined based upon observations made through experimentation and from the literature. Results show that an apparent extensive processing window exists for the material studied, with many opportunities for further improvement of print results available. Further work objectives are outlined for work involving the synthesis of new materials using scandium as a significant additive, owing to its grain refining qualities.

Keywords

Additive manufacturing • Alloy design • Aluminum alloys • Laser beam melting • Laser-powder bed fusion • Rare-earth element • Scandium • Selective laser melting

Introduction

A prominent school of thought when employing additive manufacturing (AM), particularly for components that have been historically produced using conventional manufacturing methods, is that the component shall preferably be redesigned to accommodate the new manufacturing process. Even with intricate part designs, the process of designing for AM stems from the idea that, while a design may be ideal for the application, it must pass through the filter of whether it is cost effective to manufacture through available methods.

AM allows a high degree of design freedom for components, entire mechanical assemblies, or even monolithic compliant mechanisms that might otherwise be impossible to fabricate via traditional means of manufacture. As a result, the possibilities for shape and weight optimized component designs need not be constrained through requiring the use of complex moulds [1] or by the inability to remove excess material along complex internal pathways.

While designing specifically for AM bears an appropriate intent to design parts in a fashion that best suites the application and fabrication process, it does neglect a key factor that is frequently overlooked: the material selected for a component based upon the application requirements now needs to accommodate the AM process itself. Aluminum, steel, and titanium alloys have been historically processed using subtractive and formative manufacturing processes, and while they have experienced significant uptake in AM [2], they are not necessarily well-suited (in their current alloy compositions) to the AM process environment, which can vary between several discrete types of processes.

N. A. Smith (✉) · M. Elbestawi
Department of Mechanical Engineering, McMaster University, Hamilton, ON, Canada
e-mail: smithna2@mcmaster.ca

M. Yakout (✉)
Department of Mechanical Engineering, University of Alberta, Edmonton, AB, Canada
e-mail: yakout@ualberta.ca

P. Chataigneau
Independent Consultant, Toronto, ON, Canada

P. Cashin
Imperial Mining Group Ltd, Montreal, QC, Canada

© The Minerals, Metals & Materials Society 2023
S. Broek (ed.), *Light Metals 2023*, The Minerals, Metals & Materials Series,
https://doi.org/10.1007/978-3-031-22532-1_171

Since there exists only a limited number of commercially available materials that are specifically designed or optimized for AM processes, there has been a significant rise in the volume of publications regarding AM-specific materials in recent years [2, 3]. Particularly, there is a need for the development of aluminum alloys (Al alloys) that are specific to AM [4, 5]. This is largely a consequence of the susceptibility to hot cracking in many Al alloys [5, 6], tied to the characteristic of weldability in several alloy series (namely 2xxx, 6xxx, and 7xxx series) being poor [7].

The main objective of the present study is identifying the key factors that contribute to the effective development of new Al alloys and processing parameters that are specifically optimized for the laser-powder bed fusion (L-PBF) manufacturing process, with a primary focus upon Sc-modified alloy formulations. The problematic aspects of the L-PBF process are summarized, in addition to the availability of printable materials and their interactions with this particular AM process. Following this introductory content is a brief review of the literature with respect to developing these new Al alloys, followed by details and observations from preliminary experiments on Scalmalloy® serving as a baseline material for future investigations.

L-PBF Process

The L-PBF process itself can vary slightly depending upon the machine (powder hoppers versus reservoirs, powder rollers versus recoater blades, etc.), but is generally governed by a common suite of parameters. In addition, many of the issues that exist within the L-PBF process remain fairly universal to the available materials for printing, details of which are covered in the following sections.

Parameters

Although most evaluations of material performance in the L-PBF manufacturing ecosystem incorporate several core parameters, there are a plethora of parameters that are at the disposal of the machine operator [8], controlling everything from pre- and post-exposure values, skin and core printing aspects, and the amount of powder used to prepare each new layer. There is also a degree to which some parameters (or properties) are difficult to tune, narrowing the scope of what can be considered controllable [9]. The primary parameters include the laser power (P, in W), scanning speed (v, in mm/s), hatch spacing (h, in mm), scan strategy (raster, chessboard, etc.), scan rotation (in degrees), and layer thickness (t, in mm), some of which are combined in Eq. (1) to define the volumetric laser energy density, (E_v, in J/mm^3).

$$E_v = \frac{P}{v \times h \times t} \tag{1}$$

This equation is expressed frequently in the literature [9–16] for the assessment of the amount of energy imparted to the material, despite there being significant factors, such as material properties, that are neglected. Thomas et al. [17] refined this relation to include properties such as the material reflectivity and specific heat capacity, although there exists criticism on the accuracy of values used, particularly with respect to the reflectivity of Al not representing the actual characteristics of the powder substrate [18].

The impacts of altering these parameters, or any others not mentioned, are perceived to predominantly contribute to the success or failure of a print sequence. Depending upon the desirable traits for the output, such as minimal surface roughness, each parameter can have an impact that is assigned a weight, as shown by Mahmoodkhani et al. [8] in their survey of 23 process parameters.

Process Issues

Specific to Al alloys, the issues that arise in the L-PBF process are comprised of the ability of the material to absorb the incoming laser energy, the rapid solidification rates resulting from high thermal gradients, and defects that are native to powder-based AM processes.

Laser Absorption

The generation of a melt pool is the initial step towards fusing the powder substrate, either to the build platform or prior layers, and requires an adequate injection of energy to allow the substrate to transition to the liquid state. In the case of Al, the amount of energy is disproportionately high with respect to its melting point as a result of a reduction in energy absorption from an intrinsically elevated reflectivity [1, 9]. In this respect, the frequency (and laser spot size) of the laser beam determines the absorption, which can vary between L-PBF machines and accompanying laser types [19].

Cooling Rates

Cooling rates in L-PBF are significant (10^3–10^6 K/s) by nature [1–3, 20, 21], with equally extreme thermal gradients, being simulated within the melt pool by Lopez-Botello et al. [22] (attaining a maximum of 3.5×10^6 K/m in their study). By comparison, the cooling rates native to casting are typically orders of magnitude less than L-PBF [23], marking a significant departure from the gradients present in conventional manufacturing methods. These extreme thermal gradients are largely a consequence of the process of rapidly heating a minute region of the substrate, with a near-ambient temperature region (unfused powder or previously cooled layers) in close proximity.

Defects

Some of the prominent defects in AM processes (particularly in L-PBF) are a consequence of the extreme cooling rates, in addition to the nature of melting a powder substrate. Delamination of layers [24], crack propagation along columnar inter-layer grains [25], and vaporization of the substrate material are examples of such issues. In the absence of finetuned process parameters, over-melting (Fig. 1a) or lack of fusion voids (Fig. 1b) can result, in negatively impacting part relative density. When the melting mode shifts from conductive to the keyhole, trapped gas pores (Fig. 1c) can form well below the present layer being fused [24]. And at lower energy levels, the powder adjacent to the part exterior can adhere in an incomplete melted state, known as balling (Fig. 1d).

Constituent Element Losses

The high thermal energy imparted by the laser in order to generate a melt pool of sufficient size to fuse the newly deposited powder to previous layers can exceed the boiling point of some alloy constituents. In the context of Al alloys, Cd, Zn, Mg, and Li represent the low end of thresholds for boiling points, most of which are exceeded during L-PBF fabrication. Depending upon the amount of energy imparted and the concentration of one of these elements (or others, in cases where extremely high melt pool temperatures are experienced, such as with high laser power and low scanning speed parameter combinations), the losses can be substantial.

In the literature, Mg has been removed entirely [4, 26], reducing the possibility for element loss (but not necessarily for this express purpose). While this reduces the presence of this variability, it may have promoted the retention of a coarse grain microstructure. This would be a result of the solubility inhibition qualities of Mg when paired with potent grain refiners such as Zr [27] and Sc [28].

Another action taken includes reducing the amount of Mg to a level that exerts a lower impact upon the final chemistry [21]. While this strategy was used in combination with increasing the amount of Mn for improving strength, the concentrations of the grain refiners (Sc and Zr) were higher as well, leading to a typical microstructure, but at the expense of the weight reducing the quality of Mg (which was substituted with Mn in the same proportion).

Most recently, an attempt has been made to retain the high concentration of Mg in an Sc-containing Al alloy (Scalmalloy®) through the incorporation of a vaporization inhibitor. The minor addition of Ca (roughly one tenth the amount of Mg present) reduced the amount of Mg vaporized by at least a factor of three [29]. This method of managing selective element losses may be the most interesting, since alloy additions are carefully selected to perform specific functions (increase strength, promote improved solidification, decrease specific density, etc.) that may not be achievable with any alternatives.

Developing New Aluminum Alloys

The qualities that constituent elements impart to an alloy are generally framed according to contributions to the physical properties (tensile strength, corrosion resistance, conductivity, etc.) of the final component, as opposed to the mechanisms that allow for a tolerance of the processing environment. The availability of these elements and their impact upon the microstructure to resist process-related faults are significant aspects and are discussed in the following sections.

Material Availability Versus Application Needs

A limited number of Al alloys are commercially available for the L-PBF process. This is in contrast to the many

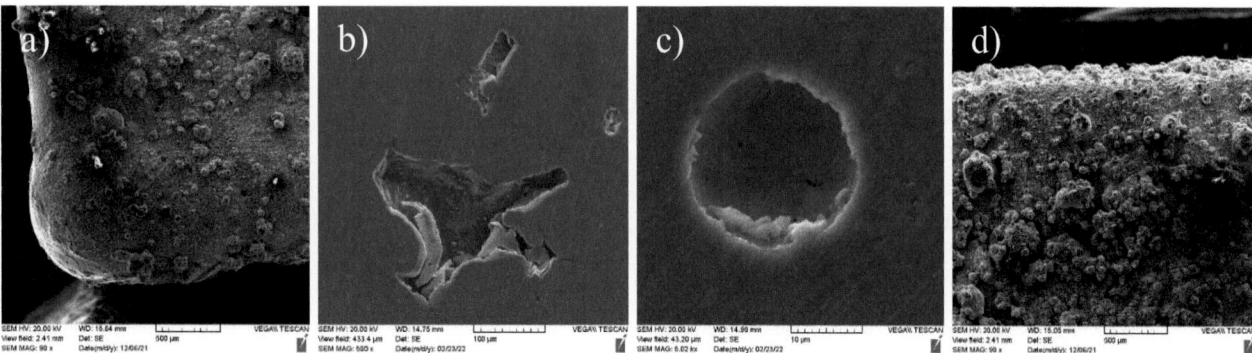

Fig. 1 SEM images taken by the author exemplifying common defects: **a** over-melting, causing rounded corners; **b** lack of fusion voids; **c** a special gas pore; and **d** balling on the sample exterior

options that exist for conventional manufacturing methods, which range across many series of alloy compositions. Despite this wealth of options, many of these traditional alloys are simply not suitable for the L-PBF process, either because of their thermal properties or their interaction with the high cooling rates. As a result, there is a substantial opportunity for growth in this respect, as posited by Palm et al. [4] as part of their characterization of ScanCromAl®.

For example, while not a particularly new metric, flowability (as a category under powder properties) assumes an important role in the ability spread powder layers in an L-PBF machine [9] but is otherwise not of significant importance in traditional manufacturing practices. While a complete inventory of implications upon printed part performance tied to powder properties (such as flowability and spreadability measurements) continue to be assessed in the literature [30], Al alloys typically exhibit poor performance with respect to these metrics [18]. In contrast, the light-weight aspect of Al alloys is a significant benefit to their usage, particularly for structural components that prioritize mass minimization.

In addition to the need to resolve serious defect-related issues, tailored materials for the L-PBF process open the opportunity for enhanced material properties that are not possible through other means of part production. Namely, the ability to customize the microstructure (assuming the presence of a generous processing window) is made available as a feature [1–3]. This beneficial aspect of a material engineered to perform well in the L-PBF environment is not likely to be realized through material selection or modification with simply the finished component application in mind, but yields the ability to design a range of components with intentional variable microstructure, which may present the chance to expand the applicability of topology optimization-like methods for part design.

Grain Refinement

When investigating the ability to modify AA7075, Martin et al. [31] observed when printing the unmodified AA7075 an immediate failure under tensile loading. This is indicative of the extremes of the L-PBF process rendering typical high-strength alloys (albeit with poor weldability) unusable as a result of the processing conditions not being compatible with the material. In the same study, the solution to this issue was grain refinement through the incorporation of Zr. This addition promoted heterogeneous nucleation early in the solidification process (see Fig. 2a, showing the result of Sc addition to Al, a similar grain refiner), generating barriers to the propagation of cracks. While this is not a complete solution [6], as it does not completely eliminate cracking, it

is effective at drastically reducing the impacts of solidification cracking on the physical performance of printed components.

Scandium Usage

Sc was first introduced in Al castings after it was discovered that it imparted a significant positive impact upon the alloy. Namely, Sc additions have the highest contribution to alloy strengthening through precipitation hardening [4, 5, 32–34]. This is in part a result of the exceptional grain refining ability of this metal [2, 6, 35]. Once the solubility limit in aluminum is exceeded, Al_3Sc dispersoids form readily, serving as potent nucleation sites during solidification. The typical solubility of Sc in Al (with much lower thermal gradients than typical for L-PBF) is roughly 0.4 wt. % [32, 33, 36]. This threshold increases to around 0.6 wt. % in the L-PBF process, owing to the ability to maintain an amount of Sc in a metastable dissolved form within the Al base [36, 37].

In the initial investigation of Scalmalloy® by its developers [38], the mechanical performance was characterized in comparison to a wrought Al alloy and another common L-PBF material (namely AA7050 and AlSi10Mg, respectively). Following heat treatment, Scalmalloy exhibited mechanical properties that exceeded those of the comparative wrought material. In contrast, while AlSi10Mg produces quality components, the level of mechanical performance is substantially lower. As such, Scalmalloy® is able to reach or exceed benchmark performance values of common aerospace Al alloys, while retaining advantages in material efficiency (reduction of losses compared to subtractive manufacturing methods) and enhanced part complexity. This ability to greatly enhance the alloy strength [21] while also serving to remedy hot cracking susceptibility [6] through enhanced nucleation when exceeding a relatively low solubility threshold outlines the need to investigate Al alloy systems specific to L-PBF that incorporate Sc as a pivotal constituent alloying element.

A key component in the application of Sc, in addition to pairings with Zr (which reduces the amount of Sc needed to achieve the desired grain refining effect), is the low lattice misfit of their intermetallic products in the aluminum matrix [6, 39].

Although its qualities when combined with Al are formidable, there is a hesitancy to use Sc in Al on a large scale due to prior difficulties in extracting an adequate supply [35], although the process of mining and separating the metal has improved in recent years [4]. In addition, newly extracted Sc is in limited supply [34, 40], contributing to its continued extraordinary cost. For these reasons, maintaining an optimal minimum of high-value additives, while still obtaining the desired microstructural outcome, is paramount. Future research efforts are directed towards minimizing the amount of Sc needed to achieve this ideal microstructure, aided by complimentary element pairings to control the substrate costs and increase the practicality of such process-optimized modified alloys.

Near-Eutectic Solidification

As the material cools rapidly, and in the case where some liquid portion still persists, the liquid material is insufficient to fill the intervening space between adjacent dendrites, leading to solidification cracking [6, 31]. This can be the result in alloys that have a solidification curve terminus analogous to Fig. 2b, where the solidification temperature drops substantially due to the undissolved Mg content. As opposed to the grain refinement solution, the constituent elements selected can instead contribute to a plateau in the final stage of solidification. This end stage flat response defines these additive elements as near-eutectic solidifiers. Successful Al alloy formulations for L-PBF, such as AlSi10Mg, owe this status to the near-eutectic solidification behaviour, avoiding the severe cracking susceptibility that is native to other Al alloys [2, 6].

Experimental Methods

Material

The material used to carry out the following experiments is based upon AA5083, modified to include Sc and Zr. The commercial name for this alloy is Scalmalloy®, which was first developed and examined by Schmidtke et al. [38] as an ideal candidate for use in AM processes. For the current work, the supplied and measured composition information is

contained in Table 1. The format of the material was a powder substrate with a prescribed particle size range of 20–63 μm, produced via gas atomization.

Samples

A series of sample sets (see Table 2) were produced that correspond to a full factorial design of experiment (DOE) with two (2) factors (laser power and scanning speed) having three (3) levels, one (1) factor (scan rotation) having two (2) levels, with all permutations having three (3) replicates (producing a total of 54 samples across a single build platform). The hatch spacing (100 μm, corresponding roughly to the laser spot size) and scan strategy (bidirectional raster) remained constant, along with the layer thickness (at 30 μm). The laser power (W) and scanning speed (mm/s) ranged 220–350 and 800–1600, respectively. As for the scan rotation between layers, 67° and 90° were the analyzed conditions. Within the print chamber, the atmosphere consisted of argon gas, with less than 0.1% oxygen content throughout the build process.

The samples were cubes with a side length of 10 mm, dimensions selected to optimize for a minimum amount of material used while also allowing for enough volume to elicit a representative crack density [18] and sufficient mass to carry out relative density measurements, as per the ATSM standard B962-17 [42].

The labelling convention (shown with all artifacts engraved in Fig. 3a, an example label in schematic form in Fig. 3b, and the corresponding printed cube in Fig. 3c) was based upon binary coding with several rows across adjacent sides to identify the batch and sample location, accompanied by an arrow to give the orientation and serve as a standard artifact across all samples.

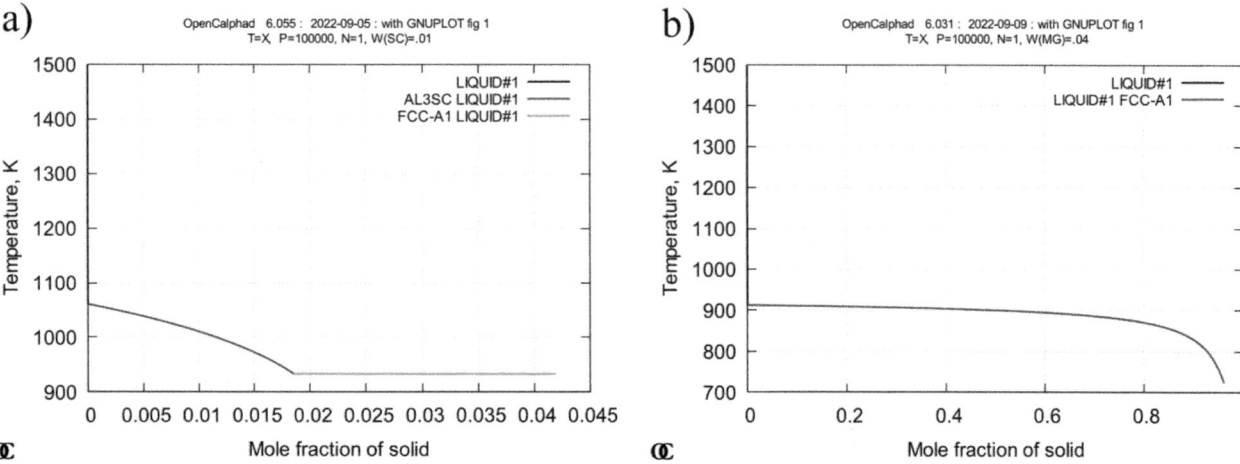

Fig. 2 Solidification curves for: **a** Sc, showing the initial occurring within the first 2 % of solidification; and **b** Mg, showing the detrimental decline in solidification temperature that leads to cracking.

Table 1 Composition information for Scalmalloy® from Carpenter Additive specifications and subsequently following ICP-OES measurements, averaged across the virgin and first-round recycled powder. All values are in wt. % [41]

	Al	Mg	Mn	Sc	Zr	Fe	Si	Ti	Cu	Zn	V	O
Nominal	Bal	4.20–5.10	0.30–0.80	0.60–0.88	0.20–0.50	0.40	0.40	0.15	0.10	0.25	0.10	0.05
Actual	Bal	4.62	0.56	0.70	0.29	0.11	0.05	0.02	–	–	–	–

Table 2 Sample processing parameter sets, with the levels existing within the range of 220–350 W for laser power and 800–1600 mm/s for scanning speed. Sets are duplicated for the 90° and 67° scan rotations

Parameters Sample Sets	Laser Power (−, 0, +)	Scanning Speed (−, 0, +)
S1	−	+
S3	−	0
S6	−	−
S2	0	+
S5	0	0
S8	0	−
S4	+	+
S7	+	0
S9	+	−

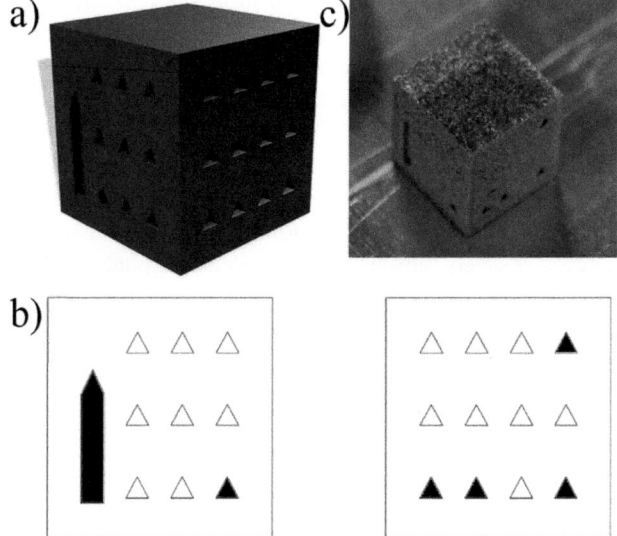

Fig. 3 Samples, with **a** showing the initial model, **b** showing an example identifying code, and **c** showing the printed sample

Equipment

All samples were produced using an EOS EOSINT M280 with a 400 W laser module. Scanning electron microscope (SEM) images were captured at multiple stages following the printing of the samples using a TESCAN VEGA-II LSU, located at the Canadian Centre for Electron Microscopy (CCEM). Relative density measurements were obtained through the Archimedes method. Composition measurements, as reported in this work, were obtained through induction-coupled plasma optical emission spectrometry (ICP-OES) using a Varian Vista-PRO ICP-OES system.

Results

For the initial stage of investigation in Sc-modified Al alloys, samples were analyzed to determine the relative density in response to changes in laser power and scanning speed and the resulting processing window, potential confounding factors such as selective vaporization of constituent elements in Scalmalloy®, and the implications upon material design strategies. The work documented below is meant to serve as a baseline for future work, characterizing what is available (material-wise) and tailored for the L-PBF process, with consideration for what can be improved upon to further the range of useable alloys.

Density

The relative density of the samples produced was assessed using the Archimedes method, with individual sample masses averaging in the range of 2.4 g and a constant specific density of 2.67 g cm^{-3} assumed for the material [41]. The density results corresponding to the sample sets (Fig. 4, with sets organized in ascending order according to E_v) exhibit a high number of samples that average above 99%, for both scan rotation permutations. As Scalmalloy® is designed specifically for L-PBF, this result was not unexpected. In the case of S1, defects dominate the cause for lower relative density (corresponding to the lack of fusion voids shown in Fig. 2b), as opposed to predominantly fine porosities being present in higher relative density samples (Fig. 1c, from sample within the parameter set S4).

The processing window is not shown in its entirety, since there is no notable decline at the end due to excess vaporization to increase porosity, as can be observed in the work performed by Koutny et al. [43] with Scalmalloy®. That being said, these results are similar to results obtained by Spierings et al. [10, 11], where the decline is not present, with an upturn as energy density increases (although the values shown in those experiments involve lower laser

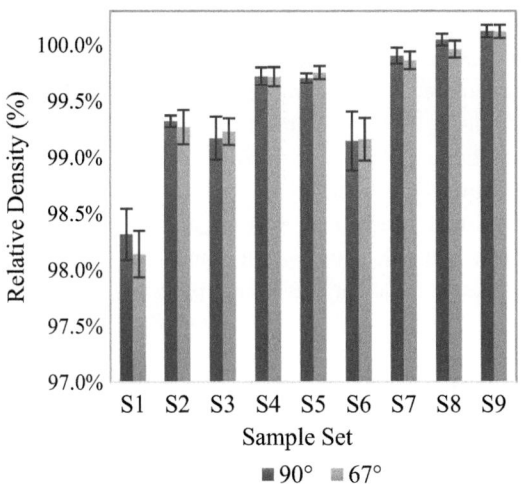

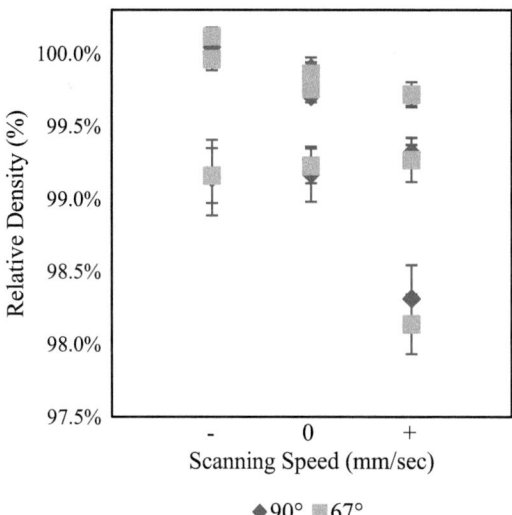

Fig. 4 Relative density results, plotted with respect to the sample set, arranged in ascending order of volumetric energy density. Error bars correspond to the standard deviation of multiple measurements of individual samples, with replicates grouped together according to the parameters and scan rotation

Fig. 6 Relative density results, sorted with respect to scanning speed, with the lowest speed level from DOE on left (-)

power, balanced with a lower scanning speed to reach higher energy values).

Although there appears to be a complex curve following the density results, there are a couple of factors that are likely to be influencing the observed trend. Primarily, the influence of the laser power on relative density, namely the lowest level (220 W), corresponds to the lowest results (for S1, S3, and S6 in Fig. 5). Isolating these results, there is a simple curve that remains, with a steady increase in density with higher energy input.

Interestingly, there is another trend that appears in both the isolated laser power (Fig. 5) and scanning speed (Fig. 6)

graphs, where there is a convergence with the change in parameters. In the case of laser power, a higher laser power corresponds to a tighter distribution at the upper limit of the power range in these experiments. With respect to the scanning speed, there appears to be a region of the minimal variation in the mid-range, which in principle would contribute to more consistent print results.

Alloying Element Vaporization

As alluded to in discussion of relative density results, another factor that could be influencing the complex nature of the density curve with respect to energy input is the selective vaporization of constituent elements. With boiling points that exist below the melt pool temperature, some elements are likely to be lost in varying amounts following the initial fusion stage (less likely with subsequent passes and lower thermal load). Material loss has been observed in the literature for Mg [4, 7, 21], Mg along with Zn [5, 44–46], and even Mg along with Mn [26] (where exceedingly high melt pool temperatures were present).

Specific to this analysis, select concentrations of Mg in pre- and post-print samples are summarized in Table 3. These results show a marked decrease in the amount of Mg in the printed parts with respect to the initial state, to the degree that the specific density of the material is likely to be affected. The impact would be an increase in specific density, leading to a potential inflation of the relative density results, scaled with increasing energy input. Although not explicitly quantified in this work, the result would be values that are lower, but also with less deviation between energy levels (leading to a more uniform print performance across a

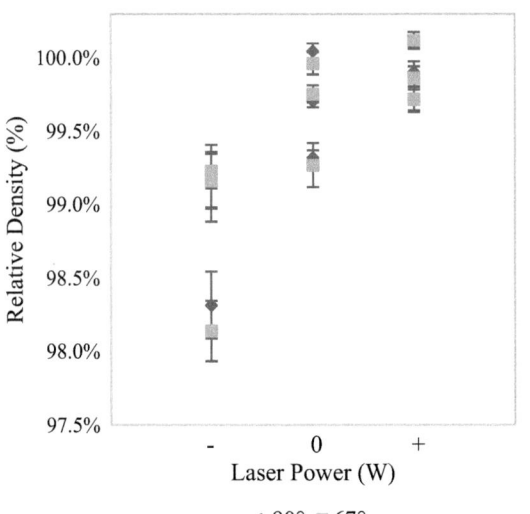

Fig. 5 Relative density results, sorted with respect to laser power, with lowest power level from DOE on left (-)

Table 3 Amount of Mg present before and after processing, including the upper/lower limits of the parameter sets (S9 and S1, respectively)

Processing stage of substrate	Mg amount (wt. %)
Mg concentration in powder	**4.62**
Mg concentration, **LOW** energy	**4.20**
Mg concentration, **HIGH** energy	**3.35**

wide processing range). This may translate to wider than observed applicable processing window for this material, with less variation than can be observed through measuring relative density alone. Despite this probable influence, the specific density of the material was assumed as constant as part of the relative density calculations, which corresponds to previous research efforts in the literature relating to the use of similar modified Al alloys [10, 11, 46]. While the amount of Mg lost will impact the value of the specific gravity, the amount of Zn lost and the uptake (or decline) of oxygen resulting from the printing process were not captured in the composition analysis, both of which remain important components that could also impact the specific density to a comparable degree. The authors acknowledge that the selective elemental losses are of consequence in the determination of relative density due to the direct impact upon the alloy specific density, and future work will involve investigation into the characterization and scope of this variation.

Material Design Implications

The grain refining quality of this Sc-modified Al alloy, with documented high strength relative to some wrought Al alloys, establishes Scalmalloy® as a success in the L-PBF domain. Despite the advantages, there are aspects that can be improved with future material design. Namely, either eliminating (through element exclusion) or controlling the relative material loss is an important factor. Coupled with the Mg mass loss is the increase in solidification temperature range from this element, which in itself increases the risk of crack formation.

If selective vaporizing elements are necessary to include in the composition of new Al alloys, then characterizing the material performance will likely need to incorporate methods of measuring changes to specific density, and the impact upon porosity within printed components. As mentioned, the use of these elements can either be tailored to correspond to acceptable losses at certain processing parameters, limited to exist within stable solubility limits, or retained in the near-desired final concentration in the presence of inhibitors to selective vaporization.

Regarding the amount of Sc required, complimentary element pairings are critical to manage the amount of grain refiners required, functioning as solubility inhibitors ideally [27, 47], as is the case with Sc and Zr, where Zr additions can potentially reduce the amount of Sc required by nearly a factor of two [37]. In addition, complimentary near-eutectic solidifiers will need to be investigated to operate in conjunction with the Sc and Zr grain refiners to approach a more appropriate material for the L-PBF process.

Discussion

The results reported show that there is a significant opportunity for new alloy compositions that employ constituent elements and element groups that cater to the L-PBF process. Despite this, there are equally substantial hurdles to the continued investigation and implementation of these new formulations, largely tied to the physical process of making the powder material.

Notwithstanding the harsh conditions native to L-PBF or the aspects of element selection and portioning, for testing a range of new alloy designs, the scale of powder required from the perspective of a small research initiative is minimal, amounting to only tens of kilograms. While these low thresholds for conducting research appear manageable, the minimum threshold of production for a powder producer making these new alloys may be situated at a much higher level, starting in the range of hundreds of kilograms.

In the instance that small batches for experimentation are feasible, there may exist significant differences in batch-to-batch quality and composition, owing to the low level of homogenization over a limited sample size, as opposed to a typical production line which would process an order of magnitude greater amount of powder. The small batch size, not having a robust market, may also carry a proportionately large cost per kilogram, making the cost equivalent to a larger batch regardless of the reduced amount of powder requested. This may also be coupled with a more intensive process required to generate the desired alloy, whether for the bulk master alloy or in the conversion to the finished powder.

Manually testing alloy formulations on an incremental basis is therefore likely to be time and cost prohibitive, lending an advantage to automated methods of simulating the behaviour of these materials. While these automated tools and specific criteria are available for narrowing the list of appropriate alloy compositions, the timeline for testing and qualifying these alloy designs can still be extensive and are hindered in part by the lack of data available (in comparison to the scale of alloys developed for conventional manufacturing methods). As such, developing materials using automated means to optimize for ideal performance (emphasizing wide processing windows, as opposed to single characteristics such as peak strength) in the L-PBF ecosystem, in conjunction with methods to reduce the barriers to obtaining the new alloy powders (potentially through the investigating low-waste, low-cost powder production methods, which are often portrayed as poor or unusable due to non-spherical powder morphology), will allow for a greater inventory of materials to be available as a baseline in AM for future material finetuning for specific applications.

Conclusion

Despite the limitations imposed by cost, rarity of some constituent materials, and some less-than-ideal properties (high coefficient of thermal expansion, reduced mechanical performance at elevated temperatures, and selective vaporization of a major alloying element), Scalmalloy® and related Sc-modified Al alloys hold significant promise for AM and aerospace applications. Sc-modified Al alloys show improvement over available AM materials and parity with many conventional wrought materials. Since experimentation and development appear to still be within the early stages, the potential for growth in its applicability to more circumstances is evident.

In summary, the following points were investigated and discussed:

- Density measurements were taken for an L-PBF optimized alloy, but need to be refined to account for the change in material composition, with future material investigations considering this aspect, both in the design of alloys and observation of relative densities;
- A combined approach of incorporating grain refiners and near-eutectic solidifiers into alloy compositions is of considerable importance for ensuring viable alloy solutions, emphasizing the need for simulation of the solidification behaviour as a component of the alloy modification and development processes; and
- Barriers to the testing of new alloys must be addressed, with alternative powder production methods being pursued to reduce the initial cost of these investigative efforts.

Future efforts will be directed towards experimentation with modified high-strength Al alloys and the development of new alloys that are well-suited to the L-PBF process, ideally with a wide processing window that features defects that are resolvable through parametric adjustments.

Acknowledgements This project is partially supported by the Natural Sciences and Engineering Research Council (NSERC) of Canada [Alliance Program].

References

1. Li XP et al. (2017) Selective laser melting of nano-TiB2 decorated AlSi10Mg alloy with high fracture strength and ductility. Acta Materialia. 129:183-193. doi: https://doi.org/10.1016/j.actamat.2017.02.062
2. Aversa A et al. (2019) New Aluminum Alloys Specifically Designed for Laser Powder Bed Fusion: A Review. Materials. 12(7):1007. doi: https://doi.org/10.3390/ma12071007
3. Wang P et al. (2018) Microstructure and mechanical properties of a heat-treatable Al-3.5 Cu-1.5 Mg-1Si alloy produced by selective laser melting. Materials Science and Engineering: A. 711:562-570. doi: https://doi.org/10.1016/j.msea.2017.11.063
4. Palm F, Bärtl M, Schimbäck D, Maier A (2020) New tailored high strength & ductile Al-alloys for laser powderbed fusion (LPB-F). Paper presented at the 11th CIRP Conference on Photonic Technologies, LANE 2020.
5. Kotadia H, Gibbons G, Das A, Howes P (2021) A review of Laser Powder Bed Fusion Additive Manufacturing of aluminium alloys: Microstructure and properties. Additive Manufacturing.102155. doi: https://doi.org/10.1016/j.addma.2021.102155.
6. Mishra RS, Thapliyal S (2021) Design approaches for printability-performance synergy in Al alloys for laser-powder bed additive manufacturing. Materials & Design. 204:109640. doi: https://doi.org/10.1016/j.matdes.2021.109640
7. Croteau JR et al. (2018) Microstructure and mechanical properties of Al-Mg-Zr alloys processed by selective laser melting. Acta Materialia. 153:35-44. doi: https://doi.org/10.1016/j.actamat.2018.04.053
8. Mahmoodkhani Y et al. (2018) Determination of the most contributing laser powder bed fusion process parameters on the surface roughness quality of Hastelloy X components. Glob. Power Propuls. Soc., Montreal.1–8.
9. Galy C, Le Guen E, Lacoste E, Arvieu C (2018) Main defects observed in aluminum alloy parts produced by SLM: from causes to consequences. Additive manufacturing. 22:165-175. doi: https://doi.org/10.1016/j.addma.2018.05.005
10. Spierings AB, Dawson K, Voegtlin M, Palm F, Uggowitzer PJ (2016) Microstructure and mechanical properties of as-processed scandium-modified aluminium using selective laser melting. CIRP Annals. 65(1):213-216. doi: https://doi.org/10.1016/j.cirp.2016.04.057
11. Spierings AB, Dawson K, Uggowitzer PJ, Wegener K (2018) Influence of SLM scan-speed on microstructure, precipitation of Al3Sc particles and mechanical properties in Sc- and Zr-modified Al-Mg alloys. Materials & Design. 140:134-143. doi: https://doi.org/10.1016/j.matdes.2017.11.053
12. Bi J et al. (2019) Microstructure and mechanical properties of a novel Sc and Zr modified 7075 aluminum alloy prepared by selective laser melting. Materials Science and Engineering: A. 768:138478. doi: https://doi.org/10.1016/j.msea.2019.138478
13. Spierings AB, Dawson K, Kern K, Palm F, Wegener K (2017) SLM-processed Sc-and Zr-modified Al-Mg alloy: Mechanical properties and microstructural effects of heat treatment. Materials Science and Engineering: A. 701:264-273. doi: https://doi.org/10.1016/j.msea.2017.06.089

14. Zhang H et al. (2018) Selective laser melting of rare earth element Sc modified aluminum alloy: Thermodynamics of precipitation behavior and its influence on mechanical properties. Additive Manufacturing. 23:1-12. doi: https://doi.org/10.1016/j.addma.2018.07.002

15. Shi Y, Yang K, Kairy SK, Palm F, Wu X, Rometsch PA (2018) Effect of platform temperature on the porosity, microstructure and mechanical properties of an Al–Mg–Sc–Zr alloy fabricated by selective laser melting. Materials Science and Engineering: A. 732:41-52. doi: https://doi.org/10.1016/j.msea.2018.06.049

16. Herzog D, Seyda V, Wycisk E, Emmelmann C (2016) Additive manufacturing of metals. Acta Materialia. 117:371-392. doi: https://doi.org/10.1016/j.actamat.2016.07.019

17. Thomas M, Baxter GJ, Todd I (2016) Normalised model-based processing diagrams for additive layer manufacture of engineering alloys. Acta Materialia. 108:26-35. doi: https://doi.org/10.1016/j.actamat.2016.02.025

18. Aboulkhair NT, Simonelli M, Parry L, Ashcroft I, Tuck C, Hague R (2019) 3D printing of Aluminium alloys: Additive Manufacturing of Aluminium alloys using selective laser melting. Progress in Materials Science. 106:100578. doi: https://doi.org/10.1016/j.pmatsci.2019.100578

19. Spierings AB (2018) Powder Spreadability and Characterization of Sc- and Zr-modified Aluminium Alloys processed by Selective Laser Melting. Quality Management System for additive manufacturing. Ph.D. thesis, ETH Zurich. Available: http://hdl.handle.net/20.500.11850/253924.

20. Tradowsky U, White J, Ward R, Read N, Reimers W, Attallah M (2016) Selective laser melting of AlSi10Mg: Influence of post-processing on the microstructural and tensile properties development. Materials & Design. 105:212-222. doi: https://doi.org/10.1016/j.matdes.2016.05.066

21. Bayoumy D, Schliephake D, Dietrich S, Wu X, Zhu Y, Huang A (2021) Intensive processing optimization for achieving strong and ductile Al-Mn-Mg-Sc-Zr alloy produced by selective laser melting. Materials & Design. 198:109317. doi: https://doi.org/10.1016/j.matdes.2020.109317

22. Lopez-Botello O, Martinez-Hernandez U, Ramírez J, Pinna C, Mumtaz K (2017) Two-dimensional simulation of grain structure growth within selective laser melted AA-2024. Materials & Design. 113:369-376. doi: https://doi.org/10.1016/j.matdes.2016.10.031

23. Qin H, Fallah V, Dong Q, Brochu M, Daymond MR, Gallerneault M (2018) Solidification pattern, microstructure and texture development in Laser Powder Bed Fusion (LPBF) of Al10SiMg alloy. Materials Characterization. 145:29-38. doi: https://doi.org/10.1016/j.matchar.2018.08.025

24. DebRoy T et al. (2018) Additive manufacturing of metallic components – Process, structure and properties. Progress in Materials Science. 92:112-224. doi: https://doi.org/10.1016/j.pmatsci.2017.10.001

25. Zhou L et al. (2019) Microstructure and mechanical properties of Zr-modified aluminum alloy 5083 manufactured by laser powder bed fusion. Additive Manufacturing. 28:485-496. doi: https://doi.org/10.1016/j.addma.2019.05.027

26. Bärtl M, Xiao X, Brillo J, Palm F (2022) Influence of Surface Tension and Evaporation on Melt Dynamics of Aluminum Alloys for Laser Powder Bed Fusion. Journal of Materials Engineering and Performance.1-13. doi: https://doi.org/10.1007/s11665-022-06592-z.

27. Sigli C (2004) Zirconium solubility in aluminum alloys. Paper presented at the Materials Forum. 28:1353-1358.

28. Spierings AB et al. (2017) Microstructural features of Sc-and Zr-modified Al-Mg alloys processed by selective laser melting. Materials & Design. 115:52-63.

29. Deillon L, Jensch F, Palm F, Bambach M (2022) A new high strength Al–Mg–Sc alloy for laser powder bed fusion with calcium addition to effectively prevent magnesium evaporation. Journal of Materials Processing Technology. 300:117416. doi: https://doi.org/10.1016/j.jmatprotec.2021.117416

30. Vock S, Klöden B, Kirchner A, Weißgärber T, Kieback B (2019) Powders for powder bed fusion: a review. Progress in Additive Manufacturing. 4(4):383-397. doi: https://doi.org/10.1007/s40964-019-00078-6

31. Martin JH, Yahata BD, Hundley JM, Mayer JA, Schaedler TA, Pollock TM (2017) 3D printing of high-strength aluminium alloys. Nature. 549(7672):365-369. doi: https://doi.org/10.1038/nature23894

32. Ahmad Z (2003) The properties and application of scandium-reinforced aluminum. Jom. 55(2):35-39. doi: https://doi.org/10.1007/s11837-003-0224-6

33. Zakharov V (2003) Effect of scandium on the structure and properties of aluminum alloys. Metal science and heat treatment. 45(7):246-253. doi: https://doi.org/10.1023/A:1027368032062

34. Czerwinski F (2020) Critical Assessment 36: Assessing differences between the use of cerium and scandium in aluminium alloying. Materials Science and Technology. 36(3):255-263. doi: https://doi.org/10.1080/02670836.2019.1702775

35. Thapliyal S et al. (2021) Design of heterogeneous structured Al alloys with wide processing window for laser-powder bed fusion additive manufacturing. Additive Manufacturing. 42:102002. doi: https://doi.org/10.1016/j.addma.2021.102002

36. Bi J et al. (2020) Densification, microstructure and mechanical properties of an Al-14.1 Mg-0.47 Si-0.31 Sc-0.17 Zr alloy printed by selective laser melting. Materials Science and Engineering: A. 774:138931. doi: https://doi.org/10.1016/j.msea.2020.138931.

37. Zakharov V, Filatov YA, Fisenko I (2020) Scandium alloying of aluminum alloys. Metal Science and Heat Treatment. 62 (7):518-523. doi: https://doi.org/10.1007/s11041-020-00595-0

38. Schmidtke K, Palm F, Hawkins A, Emmelmann C (2011) Process and Mechanical Properties: Applicability of a Scandium modified Al-alloy for Laser Additive Manufacturing. Physics Procedia. 12:369-374. doi: https://doi.org/10.1016/j.phpro.2011.03.047

39. Janghorban A, Antoni-Zdziobek A, Lomello-Tafin M, Antion C, Mazingue T, Pisch A (2013) Phase equilibria in the aluminium-rich side of the Al–Zr system. Journal of thermal analysis and calorimetry. 114(3):1015-1020.

40. Popov VV et al. (2021) Powder Bed Fusion Additive Manufacturing Using Critical Raw Materials: A Review. Materials. 14 (4):909. doi: https://doi.org/10.3390/ma14040909

41. (2021) Scalmalloy - Technical Datasheet. Carpenter Additive. Available: https://f.hubspotusercontent10.net/hubfs/6205315/Resources/Data%20Sheets/20210624–Scalmalloy_Datasheet_Digital_F.pdf. Accessed: 15 September 2021.

42. Committee B (2017) ASTM B962–17 Standard Test Methods for Density of Compacted or Sintered Powder Metallurgy (PM) Products Using Archimedes Principle. ASTM International, nd. https://doi.org/10.1520/B0962-17.

43. Koutny D et al. (2018) Processing of Al-Sc aluminum alloy using SLM technology. Procedia CIRP. 74:44-48. doi: https://doi.org/10.1016/j.procir.2018.08.027

44. Mauduit A, Pillot S, Gransac H (2017) Study of the suitability of aluminum alloys for additive manufacturing by laser powder bed fusion. Sci. Bull. 79(4):219-238.

45. Babu A, Kairy S, Huang A, Birbilis N (2020) Laser powder bed fusion of high solute Al-Zn-Mg alloys: Processing, characterisation and properties. Materials & Design. 196:109183. doi: https://doi.org/10.1016/j.matdes.2020.109183

46. Montero-Sistiaga ML et al. (2016) Changing the alloy composition of Al7075 for better processability by selective laser melting. Journal of Materials Processing Technology. 238:437-445. doi: https://doi.org/10.1016/j.jmatprotec.2016.08.003.

47. Zakharov V (2017) About alloying of aluminum alloys with transition metals. Metal Science and Heat Treatment. 59(1):67-71. doi: https://doi.org/10.1007/s11041-017-0104-2

Sustainable Scandium Recovery Method from Metallic 3D Printing Powders

Bengi Yagmurlu and Carsten Dittrich

Abstract

With the upcoming limitations on gasoline-based vehicles, especially in Europe under the Paris Accord, research on electrical/hydrogen powered vehicles and lightweight alloys for lower CO_2 emission has peaked. Scandium (Sc) became one of the key elements for advanced Al-based lightweight alloys due to major improvements in physical properties. Due to extreme prices and rare primary sources, recovery of Sc from available end-of-life (EoL) products has clearly became the focus of research. Al-Sc-containing 3D printing powders are one of the promising products to be utilized in the industry. To recover the critical metals from these EoL products or from the production wastes, hydrometallurgical processes are great alternatives due to possibility of elemental separation with low energy consumption. However, conventional leaching of electronegative metallic particles and alloys with a strong mineral acid can result in generation of H_2 gas and an aggressive exothermic reaction. While such a leaching process can be made safe at laboratory scale there are significant challenges at commercial scale from a control and zero harm perspective. Hence, there is a need for greener and sustainable process with easier control, lower chemical use, energy, and CO_2 emission without any H_2 gas evolution. Thus, an innovative leaching approach was applied with a metal salt solution with higher reduction potential to dissolve the metallic powders with cementation of the metal from the leaching solution. Then, the target was separated and purified from the impurities via refining operations. The flexibility of end product is another advantage since metal carbonates, metal sulfates, metal oxides, metal fluorides and metal oxides could be produced via this processing route. The process is also scaled up and tested in a continuous mini-pilot scale.

Keywords

Recycling • Scandium • End-of-life • Solvent extraction • Leaching • 3D printing

Introduction

Research on the lightweight alloys and the greener energy sources are spiked due to upcoming limitations on gasoline-based vehicles and strict restrictions in CO_2 emission with the Paris Accord [1] in Europe. Since aluminium (Al) is the most abundant metal in the earth crust with no accessibility problems and low price, Al alloys take the lead on design of lightweight alloys. Even though it has several disadvantages such as low strength, incompatibility for welding and low thermal stability, these physical and mechanical properties could be enhanced through alloying.

Scandium (Sc), which is an extremely expensive element for large-scale industrial use at the moment, has attracted attention in the last decade due to property improvements in the alloy and in the energy sector and was hence classified as a critical metal for the future [2−6]. A new generation of advanced materials is represented by Sc reinforced Al alloys as a result of various advantages over high-strength Al alloys. When used as a tuning metal, Sc can significantly improve the properties of Al alloys, majorly as an outcome of the formation of Al_3Sc phase. Additionally, the mechanical properties of the material could be improved as a result of recrystallization resistance by Sc addition to Al alloys [7]. Many welding related problems of Al alloys, such as soft spots and loss of strength in heat-affected zones, could be eliminated via grain refinement and anti-recrystallization effect during welding. Hot cracking in Al alloy

B. Yagmurlu (✉)
Institute of Mineral and Waste Processing, Recycling and Circular Economy Systems, TU Clausthal, Clausthal-Zellerfeld, Germany
e-mail: bengi.yagmurlu@tu-clausthal.de

C. Dittrich
MEAB Chemie Technik GmbH, Aachen, Germany

© The Minerals, Metals & Materials Society 2023
S. Broek (ed.), *Light Metals 2023*, The Minerals, Metals & Materials Series,
https://doi.org/10.1007/978-3-031-22532-1_172

welding could also be eliminated by the use of Sc-modified filler material [8].

Sc is not the rarest element on earth crust and its rarity comparable with that of Co and Pb, even though the direct minerals containing concentrated Sc are extremely infrequent. Hence, Sc is produced majorly as a by-product of Ti, U, REEs, Ni, and bauxite residue and the estimated production of Sc_2O_3 reaches 15 tonnes/year [9−17]. End-of-life products containing Sc could be another important source in the future. However, there is currently no reported recycling of these materials industrially [18, 19] and limited amount of studies presenting feasible recycling routes [19, 20]. Therefore, a complete and sustainable route for Sc recovery from Al-Sc alloys, Scalmalloy in this case, was proposed through hydrometallurgical methods in this study.

When electronegative metallic particles and alloys are leached with mineral acids, H_2 gas generation and an aggressive exothermic reaction occur. Such problems could be safe at smaller scale, but they would be challenging to deal in larger scales. Therefore, we propose an alternative method for handling such metallic powders in an innovative way, which is based on conventional cementation process [21].

Experimental Procedure

Scalmalloy® powder, which was designed to be used in additive layer manufacturing, was used as the source material, and the composition was given in Table 1.

It contains aluminium, magnesium, manganese, scandium, and zirconium. The particle size of the powder was between 10 and 120 μm. Series of leaching tests were applied to the same source without any pre-treatment.

Leaching experiments were carried out in a glass reactor and heating stirring was controlled by use of a hot plate with magnetic stirrer. $CuSO_4$ was selected as the lixiviant in this study. Stoichiometric amount of $CuSO_4$ solution was prepared and used in the leaching tests. For the preparation of the solution, reagent-grade $CuSO_4$ was used. Leaching experiments were performed under atmospheric pressure for

Table 1 Elemental composition of Scalmalloy fine powder

Scalmalloy® fine powder concentration in g/kg	
Al	930
Mg	48
Sc	7
Mn	5
Zr	4

5, 15 30, 60, and 120 min. Leaching experiments were repeated three times independently to ensure the accuracy of the calculations and the analysis. Samples from leach liquor were vacuum filtered and leach residues were dried overnight at 90 °C. In order to evaluate leaching efficiency, leachates were analysed by ICP-OES technique.

The produced pregnant leach solution (PLS) was further treated with solvent extraction for the separation of Sc in a mini-pilot scale. For that purpose, di-(2-ethylhexyl) phosphoric acid (D2EHPA) was used as the organic reagent. The concentration of the D2EHPA was 10 vol.% and the diluent was D80. H_2SO_4 was used as the scrubbing solution to minimize the co-extracted impurities together with Sc in the organic phase. The organic is then washed with dilute salt solution to eliminate any entrained aqueous phase before the stripping step. Sc in the organic phase is then stripped via NH_4F to produce $(NH_4)_3ScF_6$ in the strip liquor. MEAB MSU 0.5-type mixer settlers were used for the mini-pilot-scale purification tests. In total, 9 mixer settlers having 0.5 L of settling volume were used in the tests.

Results and Discussion

As was discussed in the previous section, the alloy powder has a fine particle size, which makes the processing via pyrometallurgical operations hard, although it is more common to process metallic wastes via re-smelting or re-fusing. Hydrometallurgical operations on the other hand, fit this alloy powder better since Sc could be separated with high purity while consuming less energy compared to the pyrometallurgical options. Moreover, relatively pure source with significant Sc amount makes this waste a promising source for Sc.

The use of conventional mineral acid on this alloy powder to leach an uncontrollable reaction occurs. Significant H_2 evolution was observed when metallic particles reacted with strong mineral acids and these acids have a higher reduction potential and propensity to yield hydrogen gas with its capacity to form explosive mixtures with oxygen. The equation of the leaching reaction with sulfuric acid is given below:

$$Al\ (s)\ +\ H_2SO_4\ =\ Al_2(SO_4)_3 +\ H_2(g)$$

Additionally, the temperature control was extremely difficult due to the occurrence of this intense exothermic reaction. Although almost complete dissolution of the constituent metals can be achieved employing conventional mineral acids (HCl or H_2SO_4), the reaction was impossible to control. Thus, an alternative method is used to leach this reactive alloy powder and to obtain a leach solution containing targeted metals. For this purpose, $CuSO_4$ solution was used as the lixiviant which yielded the following reactions upon addition:

$$2\,Me + 3\,Cu^{2+} = 2\,Me^{3+} + 3\,Cu(s)\,(Me = Al,\ Sc)$$
$$Me + Cu^{2+} = Me^{2+} + Cu(s)\,(Me = Mg,\ Mn)$$
$$Me + 2\,Cu^{2+} = Me^{4+} + 2\,Cu(s)\,(Me = Zr)$$

In this case, Cu^{2+} reacted with the metal particles in the alloy powder instead of H^+ and Cu^{2+} is reduced into Cu(s) while the metallic constituents of the alloy formed the respective ions in the aqueous solution. The reaction became a non-violent and controllable reaction where Cu(s) is obtained as the by-product while dissolving the alloy powder. The cemented Cu metal then could be recycled in the process and can be used in cycles upon addition of sulfuric acid and oxidation agent. By doing so, the process becomes a sustainable alternative for the recycling of this alloy powder.

When the Scalmalloy powder was reacted with stoichiometric amount of $CuSO_4$ solution, immediate formation of metallic Cu was observed as was hypothesized. 50, 70, and 90 °C were tested to observe the effect of the temperature on the process. Optimum leaching temperature was determined as 90 °C. The effect of temperature and the duration was shown in Table 2 by comparing the leaching efficiencies of the constituent metals at 50 and 90 °C for 15 and 60 min.

As can be seen from the Table 2, successful dissolution of the target elements was achieved in both cases, although the efficiencies were lower when 50 °C was applied. While 97% of the Al was leached from the alloy powder, 82 and 83% of Sc and Zr were dissolved, respectively. Upon increasing the temperature to 90 °C, almost complete Al, Sc, and Zr were observed. In addition, the reaction was much faster since an additional energy was sustained to the leaching system. Mg and Mn leaching stayed at lower levels and similar leaching efficiencies were found in both temperatures. The duration also had a negligible impact on the leaching of both Mn and Mg.

After successful leaching of the Scalmalloy alloy powder by $CuSO_4$, the PLS obtained was further processed with solvent extraction to isolate Sc from the impurities in the solution. Due to low amount of solution for the mini-pilot-scale test, a synthetic solution with the complete characteristics of the real solution was prepared. As the

organic phase of the SX process, 10 vol. % D2EHPA in D80 kerosene was selected as the extractant. 50 L of this synthetic PLS was processed in 12 h of operation time. The initial solution was containing approximately 32 g/L Al, 0.5 g/L Sc, 0.1 g/L Zr, and 0.05 g/L Mg and Mn in its composition. Since all Cu was consumed during the dissolution in the real trial, no Cu was added to the synthetic solution.

Sc was almost completely extracted from the PLS in the extraction step while half of the Zr and minor amount of Al is co-extracted together with the target metal. Negligible amount of Mg and Mn was also found to be co-extracted to the organic phase. After the extraction step, the loaded organic was scrubbed by the use of 4 M H_2SO_4 in order to eliminate the majority of the co-extracted impurities. In this step, no Sc was lost to the scrubbing while majority of the Al was removed from the organic. Moreover, Zr remained as the major impurity and only 36% of the co-extracted Zr was scrubbed out from the loaded organic. The negligible amount that was co-extracted was completely removed from the organic phase during the scrubbing. Then, the scrubbed organic was washed with dilute salt solution to eliminate any entrained acid solution. Since, 3 M NH_4F was utilized in the stripping step as the stripping solution, any entrained acid from the scrubbing phase could result in the formation of HF. Therefore, the washing step was applied to eliminate the risk of HF formation. The washed organic was mixed with NH_4F in the stripping step to take out all the loaded Sc from the organic. One of the reasons to use NH_4F in the stripping step is to directly form $(NH_4)_3ScF_6$ in the strip liquor. After crystallization, this can be further calcined into ScF_3 which can directly be used in the Al-Sc alloy production by the molten salt electrolysis, or can be further calcined into Sc_2O_3 to be used in the solid-oxide fuel cell production.

The average extraction, scrubbing, and the stripping efficiencies of the mini-pilot-scale test were given in Table 3. In addition, the mini-pilot test setup can be found in Fig. 1.

As a result of the complete operation, Sc was almost completely isolated from Al, Mg, and Mn in the solution while some inclusion of the Zr occurs in the strip liquor. Process was found to be very promising to separate and purify electronegative metals from the alloys via hydrometallurgical operations. The continuous process flow diagram is presented in Fig. 2.

Table 2 Leaching efficiencies of the metals in Scalmalloy powder with $CuSO_4$ solution at 50 and 90 °C for 10 and 60 min

Temperature (°C)	Duration (min)	Al (%)	Sc (%)	Zr (%)	Mg (%)	Mn (%)
50	15	86	77	74	4	16
	60	97	82	83	5	19
90	15	98	89	89	5	18
	60	100	98	95	5	20

Table 3 Extraction, scrubbing, and stripping efficiencies of the solvent extraction process

Stage	Al (%)	Sc (%)	Zr (%)	Mg (%)	Mn (%)
Extraction	<1	99	56	<1	<1
Scrubbing	67	<1	36	99	99
Stripping	70	99	91	<1	<1

Fig. 1 Mini-pilot testwork which was conducted by MEAB MSU 0.5 mixer settlers

Fig. 2 Basic flow diagram of the dissolution and solvent extraction process

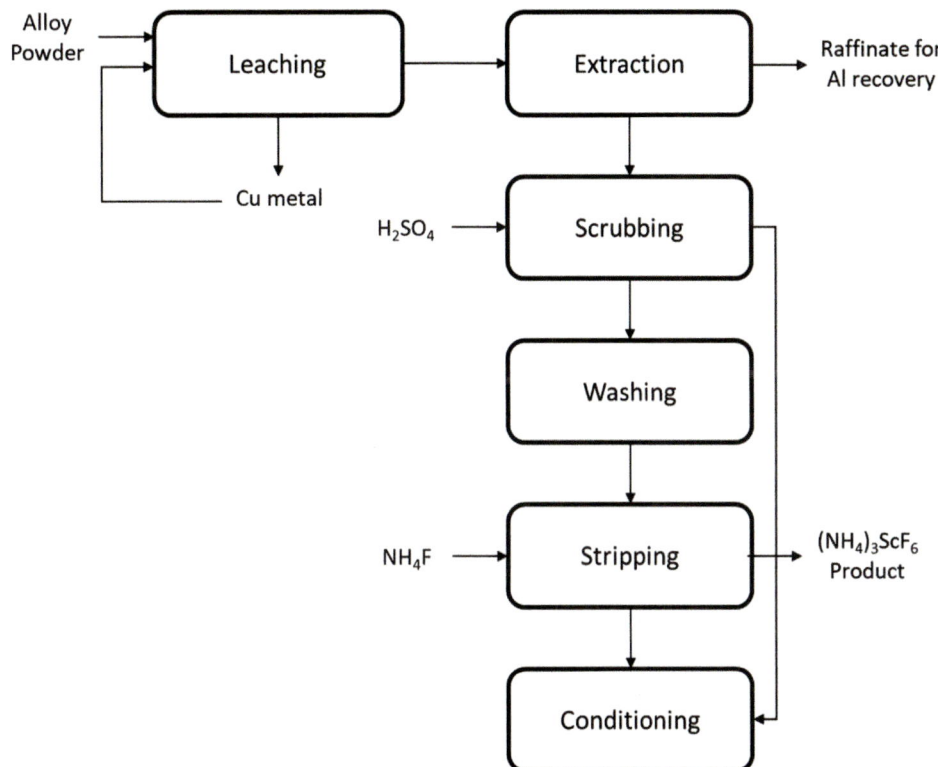

Conclusions

In this study, a hydrometallurgical concept for Sc-containing alloy powders that are inappropriate for pyrometallurgical recycling options is presented. It was shown that it is safe to treat extremely electronegative metals by this sustainable hydrometallurgical process. The generation of hydrogen during leaching with mineral acids and the extreme exothermic reaction was bypassed via the use of a more electropositive metal (Cu^{2+}) to hydrogen was employed. Additionally, the precipitated metallic Cu can also be recycled during the processing in cycles without leaving the circuit and thus enhances the sustainability of this leach process.

In the second part of the process, Sc was almost completely recovered and isolated from most of the impurities. As a result, a strip liquor which has more than 95% purity was achieved. The solution can be further treated by antisolvent crystallization which may enable both the antisolvent and the strip solution recycling [22, 23]. In this case, only sulfuric acid would be consumed during the processing both for Cu dissolution and the scrubbing step.

Acknowledgements The research leading to these results has received funding from the European Community's Horizon 2020 Programme SCALE (H2020/2014-2020/No. 730105). This publication reflects only the author's view, exempting the Community from any liability.

References

1. UNFCCC. Adoption of the Paris Agreement, Report No. FCCC/CP/2015/L.9/Rev.1. (http://unfccc.int/resource/docs/2015/cop21/eng/l09r01.pdf, 2015).
2. European Comission. Study on the EU's list of Critical Raw Materials: final report. 1–153.
3. Zhang, H. et al. Selective laser melting of rare earth element Sc modified aluminum alloy: Thermodynamics of precipitation behavior and its influence on mechanical properties. Additive Manufacturing 23, 1-12 (2018).
4. Røyset, J. & Ryum, N. Scandium in aluminium alloys. International Materials Reviews 50, 19-44 (2005).
5. Avdibegović, D. et al. Combined multi-step precipitation and supported ionic liquid phase chromatography for the recovery of rare earths from leach solutions of bauxite residues. Hydrometallurgy 180, 229-235 (2018).
6. Yagmurlu, B. et al. Synthesis of Scandium Phosphate after Peroxide Assisted Leaching of Iron Depleted Bauxite Residue (Red Mud) Slags. Scientific reports 9, 1-10 (2019).
7. Ma, K. et al. Mechanical behavior and strengthening mechanisms in ultrafine grain precipitation-strengthened aluminum alloy. Acta Materialia 62, 141-155 (2014).
8. Irving, B. Scandium places Aluminium Welding on a new Plateau. Welding journal 76 (1997).
9. Kaya, Ş. & Topkaya, Y. in Rare Earths Industry 171–182 (Elsevier, 2016).
10. Wang, W., Pranolo, Y. & Cheng, C. Y. Metallurgical processes for scandium recovery from various resources: A review. Hydrometallurgy 108, 100-108 (2011).
11. Li, Y. et al. Separation and recovery of scandium and titanium from spent sulfuric acid solution from the titanium dioxide production process. Hydrometallurgy 178, 1-6 (2018).
12. Alkan, G. et al. Novel Approach for Enhanced Scandium and Titanium Leaching Efficiency from Bauxite Residue with Suppressed Silica Gel Formation. Scientific Reports 8, 5676, doi: https://doi.org/10.1038/s41598-018-24077-9 (2018).
13. Yagmurlu, B., Dittrich, C. & Friedrich, B. Precipitation Trends of Scandium in Synthetic Red Mud Solutions with Different Precipitation Agents. Journal of Sustainable Metallurgy 3, 90-98 (2017).
14. Yagmurlu, B., Dittrich, C. & Friedrich, B. Effect of Aqueous Media on the Recovery of Scandium by Selective Precipitation. Metals 8, 314 (2018).
15. Yagmurlu, B., Zhang, W., Heikkilä, M. J., Koivula, R. T. & Friedrich, B. Solid-State Conversion of Scandium Phosphate into Scandium Oxide with Sodium Compounds. Industrial & Engineering Chemistry Research 58, 14609-14620 (2019).
16. Hedwig, S. et al. Nanofiltration-Enhanced Solvent Extraction of Scandium from TiO2 Acid Waste. ACS Sustainable Chemistry and Engineering 10 (18), 6063-6071 (2022).
17. Yagmurlu, B. et al. Sustainable Supply of Scandium for the EU Industries from Liquid Iron Chloride Based TiO2 Plants, Materials Proceedings 5 (1), 86 (2021)
18. Cordier, D.J. USGS Minerals Information: Scandium. (U.S. Geological Survey, 2022).
19. Binnemans, K. et al. Recycling of rare earths: a critical review. Journal of cleaner production 51, 1-22 (2013).
20. Ditze, A. & Kongolo, K. Recovery of scandium from magnesium, aluminium and iron scrap. Hydrometallurgy 44, 179-184 (1997).
21. Yagmurlu, B., Dittrich, C., Dunn, G. Innovative Scandium Recovery Method from Metallic End Life Products, Proceedings of 3rd European Rare Earth Resources Conference, Online (October 2020).
22. Peters, E. M., Dittrich, C., Yagmurlu, B., & Forsberg, K. Co-precipitation of Impurity (Ti, Fe, Al, Zr, U, Th) Phases during the Recovery of (NH4)3ScF6 from Strip Liquors by Anti-solvent Crystallization. In Rare Metal Technology 2020 (pp. 177–189). Springer, Cham. (2020).
23. Peters, E. M., Svärd, M., & Forsberg, K. Phase equilibria of ammonium scandium fluoride phases in aqueous alcohol mixtures for metal recovery by anti-solvent crystallization. Separation and Purification Technology, 252, 117449 (2020).

New Aluminium–Scandium Welding Wires for Additive Manufacturing

Thomas Dorin, Lu Jiang, and Andrew Sales

Abstract

This work explores new welding wire compositions tailored for the emerging technology of Wire Additive Manufacturing (WAM). The 5183 alloy is a commonly used welding wire as it provides strength through the presence of a high content of Mg through solid solution strengthening mechanism. These alloys experience a number of shortcomings such as low strength compared to age-hardenable alloys and are prone to sensitisation when used in marine environment. This work uses co-additions of Sc, Zr, and other transition elements to a 5xxx-series alloy to provide additional strengthening through the formation of L1$_2$ Al$_3$X particles. The formation of the Al$_3$X during a tailored heat treatment will be discussed. Tensile tests are conducted to evaluate the impact on the mechanical properties. Atom probe tomography and transmission electron microscopy are used to better explain the role of the different elements on the formation kinetics of Al$_3$X. The newly developed alloys will enable the creation of high strength, corrosion resistant WAM structures.

Keywords

WAAM • Al3Sc • 5xxx

T. Dorin (✉) · L. Jiang
Institute for Frontier Materials, Deakin University, Geelong, VIC 3216, Australia
e-mail: Thomas.dorin@deakin.edu.au

A. Sales
AML3D Ltd, 35 Woomera Ave, Edinburgh, SA 5111, Australia

Introduction

Novel additive manufacturing (AM) technologies will play a key role in achieving the light weighting of modern vehicle structures which is key to reducing greenhouse gas emissions. AM technologies remove limits on design and are hence key to the design of lighter structures. The wire arc additive manufacturing (WAAM) technology offers a number of benefits over other common AM such as scalability, low waste, low running cost, and short production cycle. WAAM consists in the deposition of successive layers of metal by welding in which the metal wire is fed at a controlled rate into the welding arc resulting in a fully dense component [1].

The beginning of WAAM can be traced back to the 1920s making it the oldest AM technology. It is only recently over the past two decades that WAAM has been optimised for industrial AM applications. WAAM can be used with any type of weldable material. A great amount of research has been conducted on WAAM of Titanium, Aluminium and Ferrous alloys. Aluminium is particularly suited to WAAM for its great weldability and ability to be welded in air. A number of recent review papers describe the WAAM technique in greater details [2–4].

Aluminium alloys for most AM technologies will be exposed to extreme heating and cooling cycles (i.e., powder and wire-based AM) and extreme strain (friction stir AM) which are not suited for common high strength age-hardenable alloys used in the automotive and aerospace industries. Age-hardenable alloys require long heat treatments (up to 24 h) of the final part, which is not economically feasible for the case of large-scale AM parts. There is also great scope of use of WAAM for the repair of structures where post-heat treatments would not be possible. We hence need to re-think common alloy design concepts to design a new generation of alloys for the growing AM sector. 5xxx-series alloys gain their strength through solid solution hardening from the presence of high levels of Mg solute.

These alloys are hence ideal for welding and are commonly used in traditional welding applications such as in the marine sector due to their exceptional corrosion resistance. However 5xxx-series alloys have the disadvantage of having very low strength in the as-welded condition with a yield strength of around 150 MPa.

Scandium is the most potent strengthening element when added to aluminium. This means that only a very low amount of scandium is needed to provide significant strengthening effects, with an average yield strength increase of 50 MPa per 0.1 wt.%Sc addition [5]. The fact that only small amounts are required means that the negative effect on corrosion is minor compared to other elements. This gives the potential of substantial strength improvements, with a reduced impact on corrosion resistance. The Sc strengthening effect stems from the presence of homogeneously dispersed Al_3Sc spherical precipitates with a $L1_2$ structure. Sc usually supersaturates during solidification and the precipitates can be formed during a post-heat treatment at ~ 300 °C. Other transition elements are known to substitute to Sc in Al_3Sc affecting their formation kinetics. The ideal balance between Sc and these other transition elements to obtain maximum strength and minimum post-ageing requirements of welded 5xxx-series alloys is currently unknown. In this work, a range of minor alloying additions to a 5183 type alloy are explored with the aim of maximising the strength of the produced WAAM parts and reduce the need for post-heat treatments.

Alloy Selection

From literature review and using machine learning, a number of elements were identified to be used in combination with Sc in a 5xxx-series aluminium alloys. A range of transition elements were trialled. A number of welding wire compositions were prepared in small batches using casting of a 60 mm diameter billet, extrusion to a 9.5 mm rod and then wire drawn to the final diameter of 1.2 mm. The exact compositions are proprietary but the composition ranges are given in Table 1. The other minor additions of transition elements are grouped into the category "Other $L1_2$ formers" as they will substitute to Sc in Al_3Sc. The identity of these elements cannot be revealed at this time.

This work did not optimise the process parameters and hence the same process parameters were used for all compositions. In these trials, the deposits were aged after

deposition. We are also exploring the variation of process parameters to improve the as-deposited properties and remove the need of post-heat treatments. Most of the aging results will be presented on 5 compositions using Vickers hardness compositions. The discussion will incorporate additional results coming from our full study which was conducted on a total of 15 compositions.

Temperature Selection Via Isochronal Ageing

Isochronal aging consists in heat treating a sample for a range of temperatures at constant duration in an iterative manner and recording the evolution of hardness as a function of temperature to identify the best temperatures for an isothermal heat treatment. The principle of isochronal aging is depicted in Fig. 1.

In the present work, each sample was heat treated from 100 °C to 450 °C in 25 °C steps with a constant holding time of 3 h at each temperature. The hardness was measured after each temperature step and the same sample was put back in the furnace for the next temperature step.

The evolution of hardness as a function of temperature is reported in Fig. 2. The increase in hardness is linked with the formation of Al_3X particles. The peak hardness is observed at temperatures anywhere between 300 and 375 °C. As a result three isothermal treatments were trialled at 300, 335, and 375 °C.

Isothermal Heat Treatments

Ageing was conducted at three temperatures, 300, 335 and 375 °C and hardness evolution was recorded for all 5 alloys as displayed in Fig. 3.

At 300 °C, alloys 1, 2, and 4 reached peak hardness of over 100 Hv after 2 h of ageing. Alloys 3 and 5 reached peak hardness around 90 Hv. At 335 °C, Alloys 1 and 2 reached a peak hardness of 95–100 Hv after only 1 h of ageing. All other alloys reached peak hardness below 90 Hv. At 375 °C, Alloy 2 reached a peak around 90 Hv after only 30 min of ageing. All other alloys reached peak hardness below 90 Hv. The best hardness was reached at 300 °C but a faster ageing can be conducted at 335 °C.

Tensile tests were conducted on the three most promising compositions (alloys 1, 2 and 4) and tensile test was also conducted on a reference alloy 6 (5183 alloy). The

Table 1 Composition ranges used in this study in wt.%

Wt.%	Mg	Si	Sc	Zr	Other L12 formers	Zn	Mn	Cr	Ti	Fe
min	3	0.05	0	0	0	0	0	0	0	0
max	6	0.15	0.3	0.2	0.3	0.1	0.5	0.05	0.05	0.1

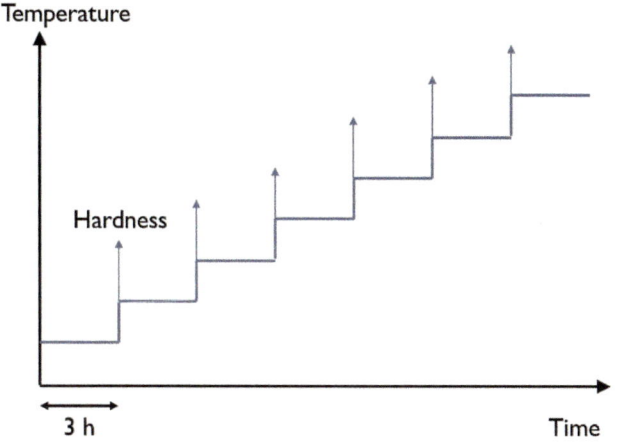

Fig. 1 Schematic showing the principle of an isochronal aging treatment

Microstructure Characterisation—Effect of the Minor Elements

Transmission electron microscopy was conducted on Alloy 2 after 2 h of heat treatment at 300 °C. The TEM results are shown in Fig. 5 and reveal a fine dispersion of Al_3X particles (Fig. 5b). The presence of fine sub-grains can clearly be seen in Fig. 5a. Higher magnification micrograph and corresponding diffraction pattern confirm that the particles imaged display an $L1_2$ structure, Fig. 5c and 5d. Finally the EDX confirms that these precipitates are rich in Sc, Fig. 5e and 5f. The average size of precipitates in this sample was 6.9 nm.

stress/strain curves are reported in Fig. 4. Three samples were tested for each alloy to ensure reproducibility. One can notice that Alloys 1, 2, and 4 have a higher strength than the reference 5183 alloy with no impact on ductility. The highest YS and UTS of, respectively, 275 and 400 MPa were reached on Alloy 1.

Discussion

This work explored a number of welding wire compositions for the WAAM technology. These compositions were based on the 5183 alloy with minor additions of transition elements including Sc. Overall, the properties were found to be significantly improved over a reference 5183 alloy but the improvement was found to depend on the composition. All

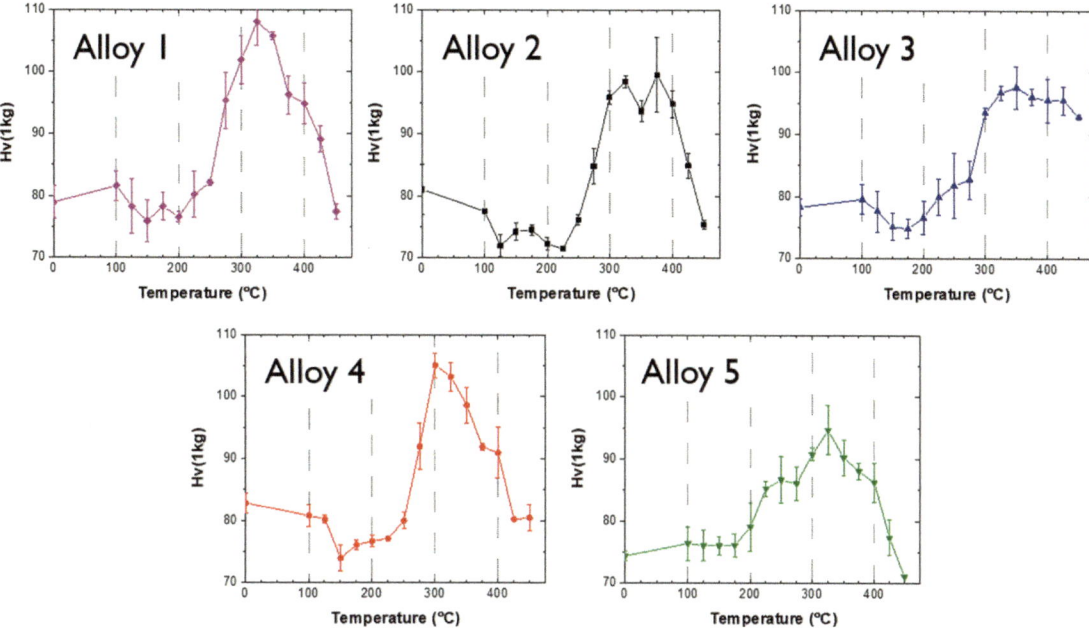

Fig. 2 Hardness evolution as a function of temperatures during isochronal ageing from 100 to 450 °C in 25 °C steps and a holding time of 3 h at each temperature

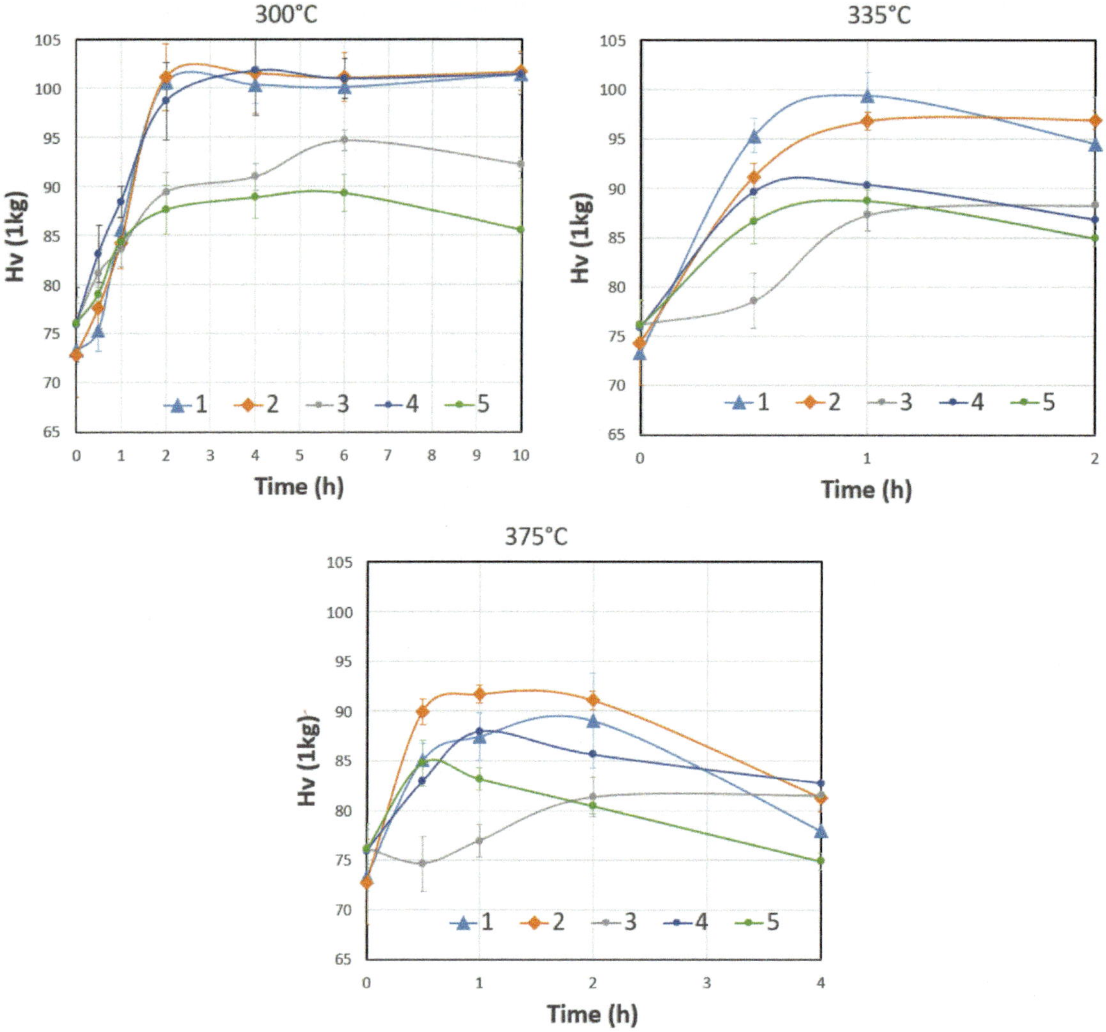

300°C

335°C

375°C

Fig. 3 Hardness evolution during ageing at 300 °C, 335 °C and 375 °C for all 5 alloys

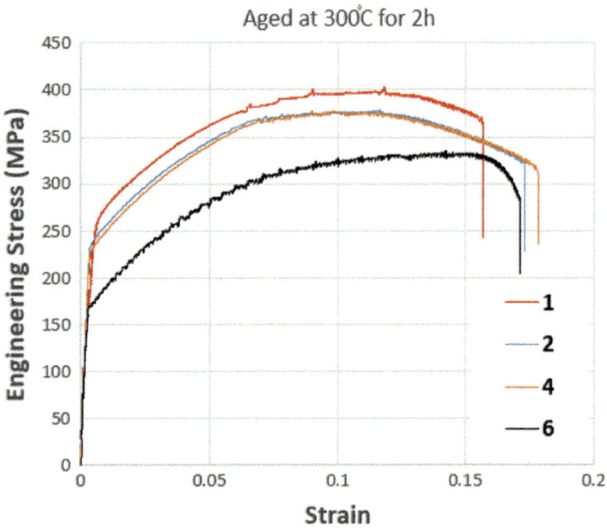

Fig. 4 Stress/Strain curves for alloys 1, 2, 4, and 6 after 2 h at 300 °C

the minor elements used in this study result in the formation of $L1_2$ Al_3X type precipitates that result in substantial strengthening after applying a heat treatment at 300 °C.

In order to better understand the variations observed in our study, the yield strength was plotted as a function of the total content of $L1_2$ forming elements in Fig. 6a. Overall the YS was found to be linearly dependent on the total amount of $L1_2$ formers apart from three outliers (light blue points). The YS was then plotted as a function of Si content and it was found that higher Si levels had a negative effect on the YS even when disregarding any other changes to the composition. As a result, we found that the YS was directly proportional to the total amount of $L1_2$ formers divided by the amount of Si in at.%, Fig. 6c. As a result, this study reveals the polluting effect of Si especially when present in amounts larger than 0.1 at.% in 5xxx-series alloys containing Sc and other transition elements.

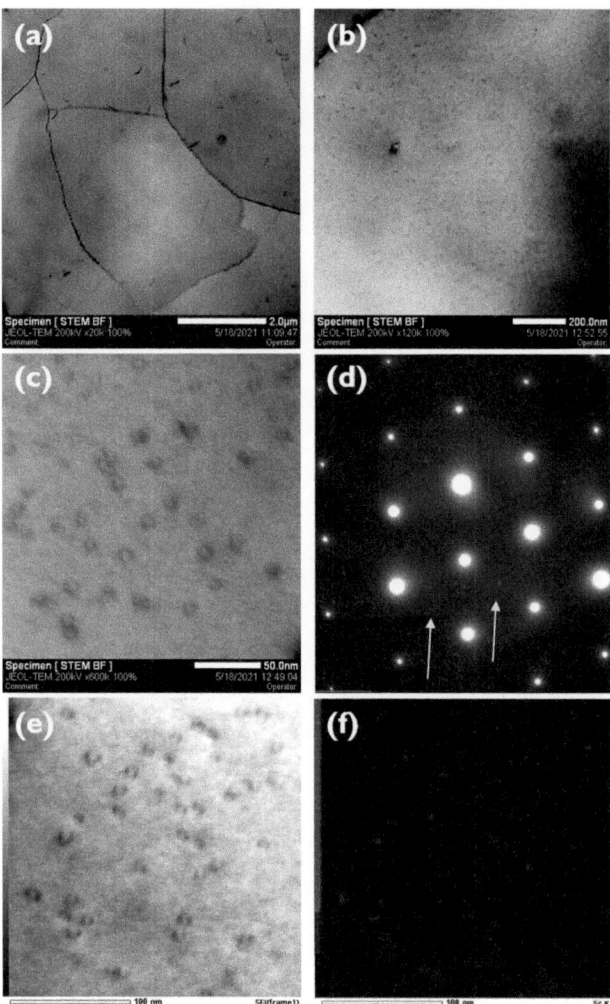

Fig. 5 TEM of alloy 2 aged for 2 h at 300 °C at different magnifications. **a** low magnification reveals fine sub-grains, **b** higher magnification displays a high number density of L12 particles, **c** higher magnification of dispersoids and **d** corresponding selected area diffraction pattern, the sub-lattice diffraction spots are highlighted with yellow arrows and confirms the presence of the L12 structured dispersoids. **e** and **f** micrograph and corresponding EDX of Sc

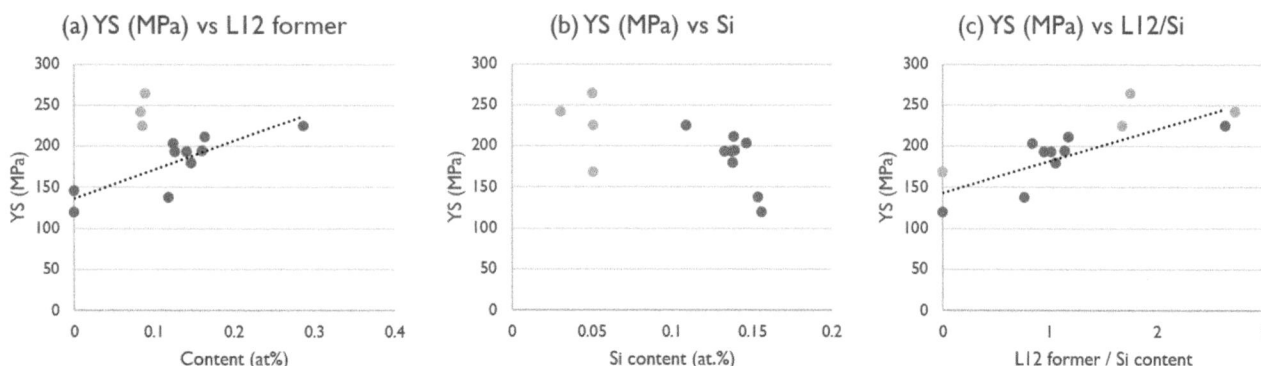

Fig. 6 YS in MPa plotted as a function of **a** the total content of $L1_2$ forming elements in at.%, **b** the Si content in at.% and **c** the total content of $L1_2$ forming elements in at.% divided by the Si content in at.%

Conclusions

This study explored the use of minor alloying additions in welding wire for WAAM. The minor alloying elements used in this study were selected by machine learning based on prior available data. A significant improvement in Vickers Hardness, YS and UTS was observed in all the studied alloys. Microstructure characterisation revealed the presence of a fine dispersion of $L1_2$ Al_3X particles which are responsible for the strength increase. Further analysis of the results revealed that the yield strength was generally higher when larger amounts of $L1_2$ forming elements were added. However, it was clearly revealed that low amounts of Si should be targeted as Si was found to have a polluting effect resulting in decreased strength.

Acknowledgements Dr. Thomas Dorin is the recipient of a Discovery Early Career Award (project number DE190100614) funded by the Australian Research Council. Deakin University's Advanced Characterisation Facility is acknowledged for use of the TEM-FEG JEOL 2100F and LEAP 5000XR APT instruments. AML3D materials and production engineer Magdalen Tan is warmly thanked for her kind help during various visits to the AML3D facilities in Edinburgh, SA, Australia.

References

1. S. W. Williams, F. Martina, A. C. Addison, J. Ding, G. Pardal, and P. Colegrove, "Wire + Arc Additive Manufacturing," *null*, vol. 32, no. 7, pp. 641–647, May 2016, doi: https://doi.org/10.1179/1743284715Y.0000000073.
2. C. R. Cunningham, J. M. Flynn, A. Shokrani, V. Dhokia, and S. T. Newman, "Invited review article: Strategies and processes for high quality wire arc additive manufacturing," *Additive Manufacturing*, vol. 22, pp. 672–686, Aug. 2018, doi: https://doi.org/10.1016/j.addma.2018.06.020.
3. K. S. Derekar, "A review of wire arc additive manufacturing and advances in wire arc additive manufacturing of aluminium," *null*, vol. 34, no. 8, pp. 895–916, May 2018, doi: https://doi.org/10.1080/02670836.2018.1455012.
4. C. Xia *et al.*, "A review on wire arc additive manufacturing: Monitoring, control and a framework of automated system," *Journal of Manufacturing Systems*, vol. 57, pp. 31–45, Oct. 2020, doi: https://doi.org/10.1016/j.jmsy.2020.08.008.
5. T. Dorin, M. Ramajayam, A. Vahid, and T. Langan, "Aluminium Scandium Alloys," in *Fundamentals of Aluminium Metallurgy*, Elsevier, 2018, pp. 439–494.

Comparative Study of Al-Mg-Ti(-Sc-Zr) Alloys Fabricated by Cold Metal Transfer and Electron Beam Additive Manufacturing

Jiangqi Zhu, Xingchen Yan, Timothy Langan, and Jian-Feng Nie

Abstract

In this work, Al-5 Mg-0.1Ti and Al-5 Mg-0.1Ti-0.2Sc-0.1Zr (wt.%) alloys fabricated by cold metal transfer (CMT) and electron beam additive manufacturing (EBAM) are investigated. The Sc and Zr addition improves the processability of Al–Mg-Ti alloys by eliminating cracks in their as-deposited microstructure. Additionally, the Sc and Zr addition promotes columnar to equiaxed transition in the as-deposited Al–Mg-Ti alloys, exhibiting a fully equiaxed microstructure with grain sizes decreased by at least 70% compared to the unmodified alloys in both cases. Grain refinement in the as-deposited Al–Mg-Ti-Sc-Zr alloy samples is attributable to the inoculation effect of primary Al_3(Sc, Zr, Ti) phase and Zener pinning effect of the solid-state precipitates of Al_3Sc. Moreover, the Sc and Zr addition improves the hardness of Al–Mg-Ti alloys in the as-deposited state, achieved by grain size strengthening and precipitate strengthening.

Keywords

Cold metal transfer • Electron beam additive manufacturing • Al-Mg-Sc-Zr-Ti alloy

J. Zhu · J.-F. Nie (✉)
Department of Materials Science and Engineering, Monash University, Melbourne, VIC 3800, Australia
e-mail: jianfeng.nie@monash.edu

X. Yan
National Engineering Laboratory for Modern Materials Surface Engineering Technology, The Key Lab of Guangdong for Modern Surface Engineering Technology, Institute of New Materials, Guangdong Academy of Sciences, Guangzhou, 510651, P.R. China

T. Langan
Sunrise Energy Metals, 12/21 Howleys Rd, Notting Hill, VIC 3168, Australia

Introduction

Additive manufacturing (AM) of aluminum alloys has received increasing attention and emerged as a hot topic because Al alloys are commonly used for lightweight applications. However, there is a printability-properties trade-off for the AM of Al alloys, as most Al alloys, especially high-strength Al alloys, are not printable with the AM technologies. To date, the most studied and used Al alloys for AM are based on Al-Si casting alloys due to their excellent castability and printability [1–3]. However, mechanical properties of the Al-Si alloys are often inadequate for some specific applications [4].

In contrast, the printability of most high-strength Al alloys is insufficiently good to be fabricated reliably by existing AM technologies due to solidification cracking [5–7]. The microstructure of additively manufactured high-strength Al alloys usually consists of a large fraction of columnar grains, and solidification cracks tend to form along columnar grain boundaries. One exception is the Sc-containing Al alloys, especially the Al-Mg-Sc(-Zr) alloys [8, 9]. The Al-Mg-Sc(-Zr) alloys have excellent printability and mechanical properties. During the AM process of the Al–Mg-Sc(-Zr) alloys, the primary Al_3Sc and/or Al_3(Sc, Zr) particles form before the nucleation of Al grains during solidification. These primary Al_3Sc and/or Al_3(Sc, Zr) particles provide heterogeneous nucleating sites for Al grains and thus cause remarkable grain refinement [10, 11]. Additionally, the Al-Mg-Sc(-Zr) alloys exhibit a strong age-hardening response that is accomplished by the solid-state precipitation of Al_3Sc or Al_3(Sc, Zr) during the post-AM ageing treatment [10]. As a result, mechanical properties of Sc-containing Al alloys fabricated by the AM technologies are approaching to those of the strongest Al alloys fabricated by conventional processing methods [7, 12].

Up to date, there are few studies on Sc-containing Al alloys fabricated by wire-based AM. The influence of the Sc and Zr addition on the microstructure of the Al-Mg alloy

S. Broek (ed.), *Light Metals 2023*, The Minerals, Metals & Materials Series,
https://doi.org/10.1007/978-3-031-22532-1_174

components fabricated by wire-based AM is still not well established. In this study, the Al-5 Mg-0.1Ti and Al-5 Mg-0.1Ti-0.2Sc-0.1Zr (wt.%) alloys are selected for bulk product fabrication using CMT and EBAM. The aim is to study the influence of the Sc and Zr addition on the microstructure and mechanical properties of the Al-Mg-Ti alloys fabricated by the CMT and EBAM.

Experimental Methodology

The compositions of the Al-Mg-Ti and Al-Mg-Ti-Sc-Zr alloy filler wires used in this study are listed in Table 1. Four types of bulk samples were fabricated from (1) Al-Mg-Ti alloy by CMT, (2) Al-Mg-Ti-Sc-Zr alloy by CMT, (3) Al-Mg-Ti alloy by EBAM, and (4) Al-Mg-Ti-Sc-Zr alloy by EBAM. A Fronius CMT system and a self-developed EBAM system were used in this study.

The single-track multi-layer bulk components were produced. The geometric sizes of the CMT-deposited bulk samples of the Al-Mg-Ti and Al-Mg-Ti-Sc-Zr alloys are 154 mm (length) × 213 mm (height) × 8 mm (width) and EBAM-deposited bulk samples of the Al-Mg-Ti and Al-Mg-Ti-Sc-Zr alloys are 140 mm (length) × 90 mm (height) × 15 mm (width). All bulk samples were deposited on an AA5083 Al alloy sheet. The processing parameters are shown in Table 2. A dwell time of 30 s was set between the deposition of each layer. The bulk samples were sectioned, mounted, and polished for optical microscopy and scanning electron microscopy (SEM) characterization. Transmission electron microscopy (TEM) was used to characterize the nano-sized particles. The hardness of the as-deposited bulk samples was measured using a Duramin A-300 Hardness tester with a load of 0.05 kg.

Results and Discussion

The CMT- and EBAM-deposited bulk sample of the Al-Mg-Ti alloy showed a layered structure comprising columnar grains at the bottom and equiaxed grains at the top

of a layer (Fig. 1a and b). The grains were much smaller in the CMT- and EBAM-deposited bulk samples of the Al-Mg-Ti-Sc-Zr alloy, indicating a grain refinement effect caused by the Sc and Zr addition (Fig. 1c and d). In addition, the CMT- and EBAM-deposited bulk samples of the Al-Mg-Ti-Sc-Zr alloy showed a fully equiaxed grain structure. It is noteworthy that the grains of CMT-deposited bulk samples of the Al-Mg-Ti and Al-Mg-Ti-Sc-Zr alloys were smaller than that in the EBAM-deposited bulk samples of Al-Mg-Ti and Al-Mg-Ti-Sc-Zr alloy, respectively.

Several types of particles with different contrast were visible in the back-scattered electron (BSE) images of the CMT- and EBAM-deposited bulk samples of the Al-Mg-Ti-Sc-Zr alloy. The bright particles were primary $Al_3(Sc, Zr, Ti)$ particles and $Al_6(Fe, Mn)$. The dark particles were Mg_2Si (Fig. 2a and b). The primary $Al_3(Sc, Zr, Ti)$ particles form before nucleation of Al grains during solidification and hence act as nucleation sites for Al grains, leading to grain refinement in the as-deposited bulk samples of the Al-Mg-Ti-Sc-Zr alloy that were fabricated by both CMT and EBAM.

In addition to the coarse particles, the bright field (BF) TEM images showed many nano-sized Al_3Sc precipitates in the CMT- and EBAM-deposited bulk samples of the Al-Mg-Ti-Sc-Zr alloy (Fig. 2c and d). The Al_3Sc precipitates were finer in the CMT-deposited bulk samples of the Al-Mg-Ti-Sc-Zr alloys. The hardness values of the CMT- and EBAM-deposited bulk samples of the Al-Mg-Ti-Sc-Zr alloy were 120 $HV_{0.05}$ and 85 $HV_{0.05}$, respectively, which were higher than those of counterpart CMT- and EBAM-deposited Al-Mg-Ti alloy, respectively (Table 3). The higher hardness of the as-deposited bulk samples of Al-Mg-Ti-Sc-Zr alloy is attributable to precipitate strengthening, via solid-state precipitation of the nano-sized Al_3Sc, and grain size strengthening. Additionally, the hardness was lower in the EBAM-deposited bulk samples, which is attributable to the larger grain size and the severe Mg evaporation during the EBAM fabrication process. The Mg concentration decreased by nearly 70% after the EBAM printing process. The presence of the nano-sized Al_3Sc in the microstructure of the as-deposited CMT and EBAM bulk

Table 1 Nominal compositions of Al-Mg-Ti and Al-Mg-Ti-Sc-Zr wires (wt.%)

	Al	Mg	Mn	Fe	Si	Ti	Zn	Sc	Zr
Al-Mg-Ti	Bal	5.00	0.15	0.10	0.04	0.07	–	–	–
Al-Mg-Ti-Sc-Zr	Bal	5.50	0.08	0.09	0.20	0.10	–	0.21	0.10

Table 2 Processing parameters of Al-Mg-Ti and Al-Mg-Ti-Sc-Zr alloys fabricated by CMT and EBAM

	Current (A)	Voltage (V)	Transverse speed (m/min)	Wire feeding rate (m/min)
CMT-deposited Al-Mg-Ti and Al-Mg-Ti-Sc-Zr	157	30	1.0	9
EBAM-deposited Al-Mg-Ti and Al-Mg-Ti-Sc-Zr	0.03	60,000	0.5	5

Fig. 1 Polarised light optical micrographs of **a** CMT-deposited Al-Mg-Ti alloy, **b** EBAM-deposited Al-Mg-Ti alloy, **c** CMT-deposited Al-Mg-Ti-Sc-Zr alloy, and **d** EBAM-deposited Al-Mg-Ti-Sc-Zr alloy

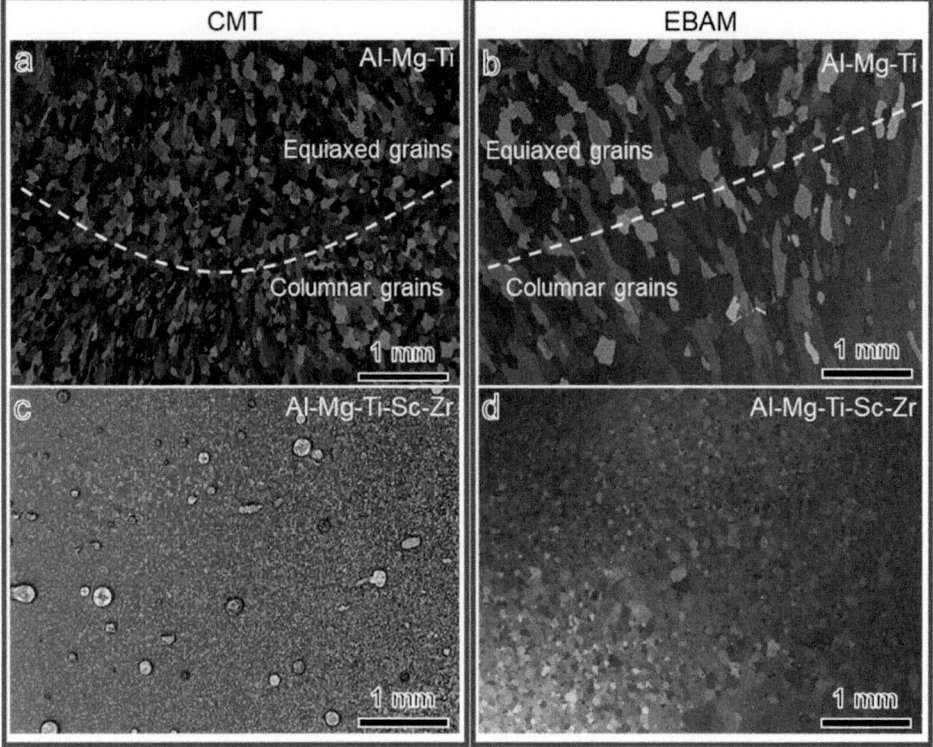

Fig. 2 **a**, **b** BSE SEM images and **c**, **d** BF TEM images of bottom parts of CMT- and EBAM-deposited bulk samples of the Al-Mg-Ti-Sc-Zr alloy

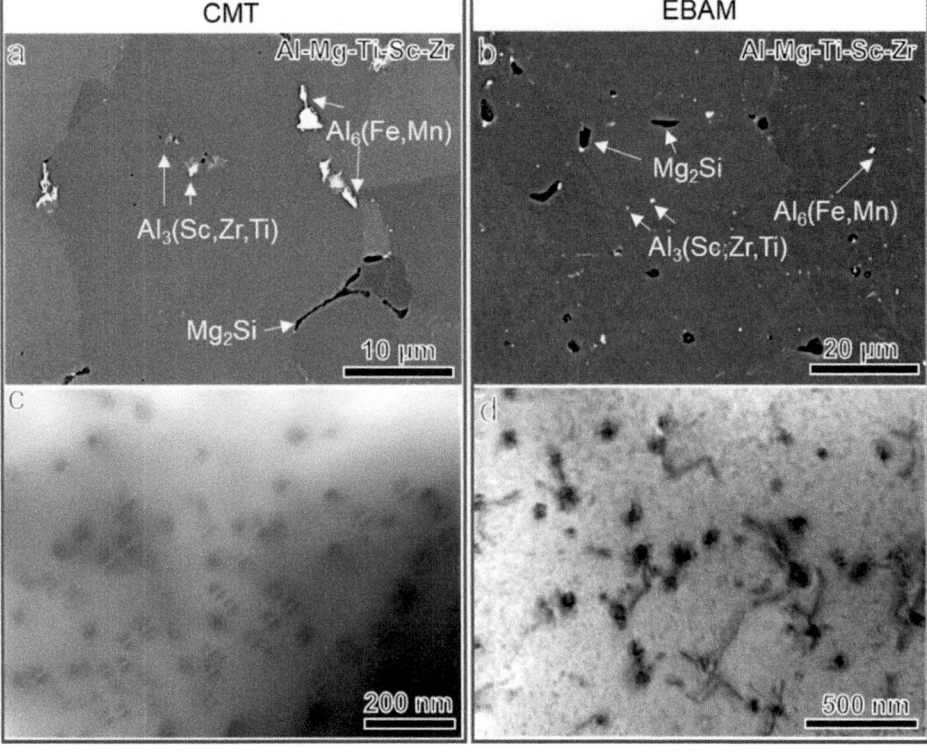

Table 3 Hardness values of as-deposited bulk samples of Al-Mg-Ti and Al-Mg-Ti-Sc-Zr alloys

		Average hardness ($HV_{0.05}$)
CMT	Al-Mg-Ti	100
	Al-Mg-Ti-Sc-Zr	112
EBAM	Al-Mg-Ti	71
	Al-Mg-Ti-Sc-Zr	85

samples of the Al-Mg-Ti-Sc-Zr alloy indicated the occurrence of heat treatment of a previously deposited layer from the heat introduced by the deposition of a subsequently deposited layer. The solid-state precipitation of the nano-sized Al_3Sc during the deposition process might reduce the age-hardening response in the post-deposition heat treatment. The investigation of the aged hardening response of the as-deposited bulk samples of the Al-Mg-Ti-Sc-Zr alloy fabricated in this study during post-deposition heat treatment in the temperature range 300–350 °C is ongoing. Such results will be reported in a separate paper.

Conclusions

In this study, bulk samples were produced from the Al-5 Mg-0.1Ti and Al-5.5 Mg-0.1Ti-0.2Sc-0.1Zr (wt.%) alloys by the CMT and EBAM. The main conclusions are summarised as follows:

1. The CMT- and EBAM-deposited bulk samples of the Al-Mg-Ti alloy have a layered structure comprising columnar grains at the bottom and equiaxed grains at the top of the layer.
2. The Sc and Zr addition causes an effective grain refinement as the grain size is much smaller in the as-deposited bulk samples of the Al-Mg-Ti-Sc-Zr alloy fabricated by EBAM and CMT. Additionally, the as-deposited bulk samples of the Al-Mg-Ti-Sc-Zr alloy produced by CMT and EBAM show a fully equiaxed grain structure. The grain refinement effect is attributable to the inoculation effect of the primary Al_3(Sc, Zr, Ti) phase.
3. In addition to grain refinement, the Sc and Zr addition also increases the hardness of the as-deposited bulk samples of the Al-Mg-Ti alloy fabricated by CMT and EBAM. The increase in the hardness is attributable to the smaller grain size and the solid-state precipitation of Al_3Sc.

References

1. Haselhuhn AS, Buhr MW, Wijnen B, Sanders PG, Pearce JM (2016) Structure-property relationships of common aluminum weld alloys utilized as feedstock for GMAW-based 3-D metal printing. Mater. Sci. Eng. A. 673:511–523. https://doi.org/10.1016/j.msea.2016.07.099.
2. Wang D, Lu J, Tang S, Yu L, Fan H, Ji L, Liu C (2018) Reducing porosity and refining grains for arc additive manufacturing aluminum alloy by adjusting arc pulse frequency and current, Materials (Basel) 11:1344. https://doi.org/10.3390/ma11081344.
3. Langelandsvik G, Horgar A, Furu T, Roven HJ, Akselsen OM (2020) Comparative study of eutectic Al-Si alloys manufactured by WAAM and casting, Int. J. Adv. Manuf. Technol. 110:935–947. https://doi.org/10.1007/s00170-020-05735-7.
4. Rometsch PA, Zhu Y, Wu X, Huang A (2022) Review of high-strength aluminium alloys for additive manufacturing by laser powder bed fusion, Mater. Des. 219:110779. https://doi.org/10.1016/j.matdes.2022.110779.
5. Martin JH, Yahata BD, Hundley JM, Mayer JA, Schaedler TA, Pollock TM (2017) 3D printing of high-strength aluminium alloys, Nature 549:365–369. https://doi.org/10.1038/nature23894.
6. Croteau JR, Griffiths S, Rossell MD, Leinenbach C, Kenel C, Jansen V, Seidman DN, Dunand DC, Vo NQ (2018) Microstructure and mechanical properties of Al-Mg-Zr alloys processed by selective laser melting, Acta Mater. 153:35–44. https://doi.org/10.1016/j.actamat.2018.04.053.
7. Jia Q, Rometsch P, Kürnsteiner P, Chao Q, Huang A, Weyland M, Bourgeois L, Wu X (2019) Selective laser melting of a high strength Al-Mn-Sc alloy: Alloy design and strengthening mechanisms, Acta Mater. 171:108–118. https://doi.org/10.1016/j.actamat.2019.04.014.
8. Spierings AB, Dawson K, Voegtlin M, Palm F, Uggowitzer PJ (2016) Microstructure and mechanical properties of as-processed scandium-modified aluminium using selective laser melting, CIRP Ann. Manuf. Technol. 65:213–216. https://doi.org/10.1016/j.cirp.2016.04.057.
9. Schmidtke K, Palm F, Hawkins A, Emmelmann C (2011) Process and mechanical properties: Applicability of a scandium modified Al-alloy for laser additive manufacturing, Phys. Procedia 12:369–374. https://doi.org/10.1016/j.phpro.2011.03.047.
10. Røyset J, Ryum N (2005) Scandium in aluminium alloys, Int. Mater. Rev. 50:19–44. https://doi.org/10.1179/174328005X14311.

11. Spierings AB, Dawson K, Heeling T, Uggowitzer PJ, Schäublin R, Palm F, Wegener K (2017) Microstructural features of Sc- and Zr-modified Al-Mg alloys processed by selective laser melting, Mater. Des. 115:52–63. https://doi.org/10.1016/j.matdes.2016.11.040.

12. Rometsch P, Jia Q, Yang KV, Wu X (2019) Aluminum alloys for selective laser melting - towards improved performance. In: Froes, F and Boyer, R (ed) Additive manufacturing for the aerospace industry, Elsevier, pp. 301–325. https://doi.org/10.1016/B978-0-12-814062-8.00016-9.

Dissolution and Development of Al₃(Sc, Zr) Dispersoids in Structures Produced Via Wire Arc Additive Manufacturing

Sonja Blickley, Tori Nizzi, Anna Palmcook, Austin Schaub, Erico Freitas, Tim Langan, Carson Williams, and Paul Sanders

Abstract

Aluminum alloys are of growing interest in fields requiring lightweight structural materials and are common in additive manufacturing applications. Aluminum wire-arc additive manufacturing (WAAM) as a method of production is also increasing in development and use, predominantly using available welding filler alloys (i.e. 5356, 5183, etc.) as feedstock. As a manufacturing method, there is an interest in pursuing new aluminum alloy development to produce superior strengths to currently available materials. Alloys of aluminum and scandium offer increased strength as well as post-processing heat treatability that makes them well suited as a feedstock for WAAM applications, however, the complex thermo-mechanical history of wire production, as well as the thermal input during WAAM processing, has a significant impact on the precipitation, dissolution, structure, and morphology of Al₃Sc dispersoids. In this work, aluminum alloy 5025 is used to produce WAAM structures, and the resulting material properties and characteristics are analyzed. Transmission electron microscopy is used to study Sc dissolution, Al₃X dispersoid morphology, and dispersoid chemistry in WAAM samples in both the feedstock and as-printed conditions. Suggestions are made for optimizing Al-Sc alloys in WAAM applications.

Keywords

Aluminum • Scandium • Additive manufacturing

Introduction

Aluminum and its alloys have been used in welding and brazing applications for decades, with 4xxx and 5xxx series alloys filler wire produced for these applications [1]. In the growing field of additive manufacturing (AM), aluminum is an attractive choice for lightweight, moderate strength applications. High strength alloys and metal matrix composites have been produced in powder-based AM processes, but there is growing use of wire-based AM applications, specifically in wire-arc additive manufacturing (WAAM), resulting in demand for aluminum alloys specifically designed for WAAM that may utilize alloying elements not well suited for traditional welding environments. To this end, aluminum-scandium alloys are being explored.

Aluminum filler alloys used for gas metal-arc welding (GMAW) need to be welded without solidification cracking. This excludes certain high strength systems such as 2xxx and 7xxx series alloys [1]. Furthermore, most aluminum GMAW applications involve welding base metal parts with specific tempers, which means that post-weld heat treatment is difficult given the presence of both wrought and solidified microstructures. WAAM applications are more suited for post-print heat treatment, as the entire structure shares the same chemistry and does not have prior wrought mechanical processing that would be negatively impacted by thermal treatments.

Aluminum alloys with Sc additions form Al₃Sc dispersoids, which resist hot cracking, refine the grains, and provide post-print precipitate strengthening via heat treatment. Scandium is known to be a potent grain refiner [2] in aluminum, which adds strength and toughness to welded or printed structures. Proper heat treatment of Al-Sc alloys

S. Blickley · T. Nizzi · A. Palmcook · A. Schaub · E. Freitas · P. Sanders (✉)
Michigan Technological University, Houghton, MI, USA
e-mail: sanders@mtu.edu

T. Langan
Sunrise Metals, Notting Hill, Victoria, Australia

C. Williams
Hobart Brothers LLC, Traverse City, MI, USA

© The Minerals, Metals & Materials Society 2023
S. Broek (ed.), *Light Metals 2023*, The Minerals, Metals & Materials Series,
https://doi.org/10.1007/978-3-031-22532-1_175

produces a fine distribution of Al_3Sc dispersoids which strengthens the material. After welding, a fraction of the scandium is supersaturated in the matrix, and when heat treated forms dispersoids on the scale of 2–10 nm that confer significant strengthening. However, the thermal history during wire production can lead to dispersoids significantly larger than 2–10 nm, and these overaged dispersoids are ineffective strengtheners. Dissolution of these particles must occur during weld processing to optimize strengthening through a post-weld heat treatment. Dispersoid dissolution during the WAAM processes is not fully understood, but is necessary for efficient utilization of expensive Sc additions.

Previous work welding with aluminum filler alloys containing scandium has confirmed reduced crack sensitivity, improved grain refinement, and significant age hardening response in post-weld heat treatments [3, 4]. Similar results have been observed in wire-arc additive manufacturing [5], [6]. The presence of a post-weld hardening response implies that a Sc supersaturated solid solution is achieved during the welding process, suggesting that large Al_3Sc dispersoids initially present in the filler alloy are dissolved during welding.

This study traces the scandium evolution during WAAM processing, and examines dispersoid morphology before and after the printing process. Al5025, which is essentially Al5356 with Sc and Zr additions, was chosen as the aluminum filler wire alloy. Scanning transmission electron microscopy (STEM) is used to characterize Al_3Sc dispersoid size and composition in the wire feedstock and in WAAM structures. The as-printed and heat treated WAAM materials hardness and electrical conductivity are quantified.

Methods

Rod with the nominal composition of Al5025 was drawn into 1.2 mm wire. The composition of the rod and subsequent WAAM material was measured by optical emission spectroscopy (OES/Bruker Q4 Tasman/3 replicates). The Zr OES calibration was incorrect, but Zr was known to be to specification. All tabular data in this paper are rounded to the least significant figure.

Gas metal-arc welding (GMAW) WAAM was performed on a Fanuc Robot R-1000iA/80F using a Lincoln Power Wave R450 with STT Module in short circuit mode at 14.5 V. Walls 100 mm long by 4–5 mm wide were printed on a 3 mm thick 5000 series substrate with 1.2 mm diameter Al5025 wire. The travel speed was 5 mm/s, wire feed was 53 mm/s, and layer height was 2 mm with a 15 s pause between layers. A Type K thermocouple sampled at 1 Hz was used to assess the weld pool temperature by positioning it upon the third WAAM layer and printing multiple layers

above it. After printing, a portion of the WAAM material was aged at 340 °C for 4 h in air.

The as-printed and aged WAAM materials were sectioned, ground, and polished to 0.5 μm colloidal silica. Conductivity was measured at the centerline of a polished longitudinal section with a Fischer SigmaScope SMP10 (13 mm probe diameter/3 replicates by 4 operators). Vickers microhardness was performed at the center of a mounted transverse section with a LECO M-400-G1 using a 100 g load with a 15 s dwell time (4 replicates by 4 operators). Conductivity and microhardness measurements each passed a gage repeatability and reproducibility study, with percent contribution measurement errors of 0.1% for conductivity and 6% for microhardness.

Transmission electron microscopy samples were cut from the center of transverse sections of the rod prior to drawing and the WAAM material after printing. Samples were ground using silicon carbide paper to less than 100 μm thick and punched into 3 mm diameter disks. These disks were electropolished in nitric acid and methanol at -30 °C. Prior to STEM imaging, the disks were plasma cleaned. A probe aberration-corrected FEI Titan Themis STEM, at 200 kV, was used to image in the high angle annular dark field (HAADF) mode with EDS to determine phase compositions.

Results

Alloys in the 5000 system are dominated by the nominal 5 wt% Mg in solution. The 5025 alloy system includes 0.3 wt% and 0.15 wt% Zr, and the solubilities of these elements, which are already low, are reduced by the presence of Mg. For example, the maximum Sc solubility at the eutectic temperature is reduced from 0.32 wt% in the Al-Sc binary to 0.23 wt% in the Al–Mg-Sc ternary to 0.15 wt% the Al–Mg-Sc-Zr quaternary (Fig. 1). The addition of Zr raises the temperature required to completely melt the alloy to above 750 °C and also makes it possible to first form primary Al_3Zr upon cooling. Al_3Sc dispersoids may start to form around 650 °C. In practice, faster cooling rates (typical in WAAM) limit the formation of primary dispersoids, leading to a matrix supersaturated in Sc and Zr [7] which during heat treatment typically nucleates and grows as $L1_2$ dispersoids.

Single-bead width WAAM prints of various heights are shown in Fig. 2a. A cross section shows good density for most of the height with some porosity at the top (Fig. 2b). Subsequent analysis was performed at the middle of the height and the middle of the wall thickness.

A Type K thermocouple was positioned at the center of the 3rd layer (Fig. 2a) to measure the temperature during the WAAM process. The temperature trace (Fig. 3) for the 4th layer shows a rapid rise to 1350 °C followed by logarithmic

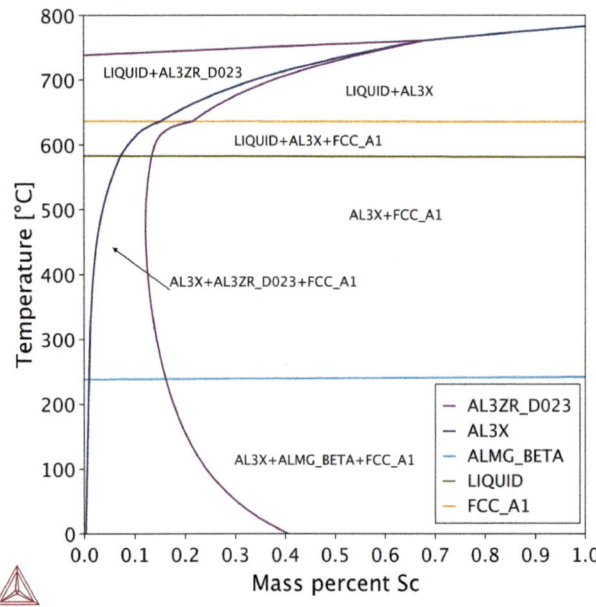

Fig. 1 Pseudo binary plotted versus Sc with 4.6 wt% Mg and 0.15 wt % Zr (Table 1). The DO_{23} Al_3Zr phase is stable to 750 °C, which is about 100 °C higher than Al_3Sc (Thermo-Calc TCAL8.1)

cooling to 100 °C. Print time per layer was 20 s, and combined with a 15 s pause, this leads to 35 s between print traverses over the thermocouple. Subsequent weld passes lead to small (<10 °C) temperature rises in the 3rd print layer. According to OMEGA, the maximum useful (measurable) range of a Type K thermocouple is 1260 °C, while the maximum material range is 1372 °C [8], so the thermocouple was at the limits of its capability.

Analysis of the rod feedstock and the as-printed WAAM material in the STEM was used to determine the dispersoid size and composition (Fig. 4). The rod had a wide distribution of large dispersoids (70 nm) (Fig. 5a). In addition there appears to be a correlation between the Sc and Zr, possibly suggesting a core–shell Al_3X morphology (Fig. 4 top). After WAAM, most of the large dispersoids are no

longer present and there is a fine dispersion of Sc (Fig. 4 center bottom) with a still finer dispersion of Zr (Fig. 4 right bottom). Upon closer inspection, Al_3Sc dispersoids are present with a size of 2.2 nm (Fig. 5b). Several large dispersoids are present, with the one in the top center appearing to be more Sc on the left and Zr on the right.

The electrical conductivity is inversely proportional to the amount of solute in the matrix. Given the high Mg levels in Al5025 (nominally 5 wt%), the conductivity of the as-printed WAAM material is about 14 MS/m (Table 2), which is significantly lower than pure Al (35 MS/m). If precipitation hardening occurs during aging, the conductivity should rise as solute precipitates from the matrix. After heat treatment, the WAAM conductivity rises from 13.8 MS/m to 14.7 MS/m, and given the precision of this measurement (0.1 MS/m) this is a statistically significant increase (Table 2). The wrought-processed rod had an initial hardness of 900 MPa, which was reduced to 780 MPa by the WAAM process (Table 2). There was a significant rise in hardness after a 4 h/340 °C heat treatment targeting Al_3Sc precipitation, indicating that some Sc remained in solution after WAAM processing and was able to precipitate to provide additional strength.

Discussion

Due to commercial concerns, the details of the rod manufacturing and wire drawing processes cannot be disclosed. However, it can be reasonably assumed that dispersoids present in the rod feedstock (Fig. 4 top) did not get smaller during wire drawing. It appears that melt temperatures reaching 1350 °C for a duration on the order of a second (Fig. 3) are sufficient to dissolve 70 nm $Al_3(Sc, Zr)$ core–shell dispersoids (Fig. 4 bottom).

The WAAM process produces a fine dispersion Al_3Sc dispersoids (Fig. 5) but not measurable Al_3Zr dispersoids. To assess the ability to form Al_3Sc dispersoids during

 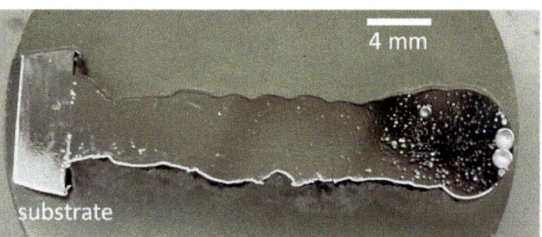

Fig. 2 Representative single-bead wall sections (4 by 100 mm) with various heights (a). Note thermocouple (TC) location. A cross section of a wall showing substrate, high-density wall section, and some porosity at the top (b)

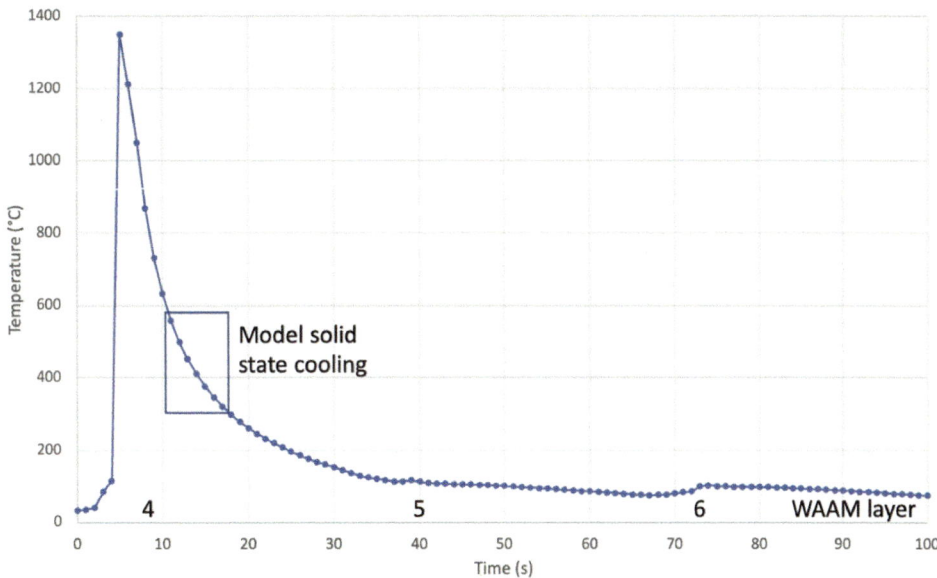

Fig. 3 Type K thermocouple placed upon the 3rd print layer (Fig. 2a) indicates a rapid temperature rise when printed on directly but little effect from subsequent printed layers. Box indicates the region for microstructure modeling during the solid state cooling portion of the process

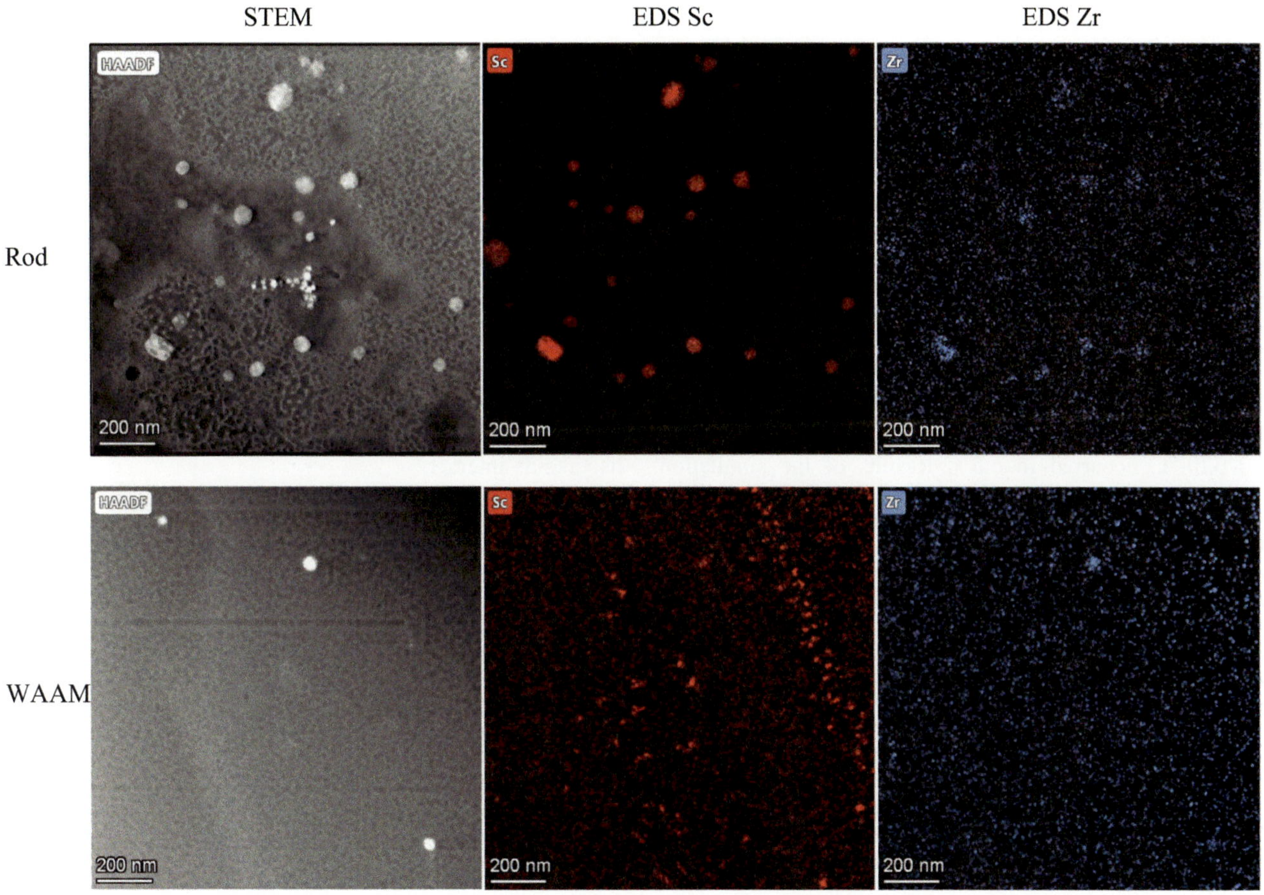

	STEM	EDS Sc	EDS Zr
Rod			
WAAM			

Fig. 4 Representative HAADF STEM for rod before wire drawing (top) and as-printed WAAM (bottom)

Fig. 5 Dispersoid diameter size distributions from the rod before wire drawing (70 ± 3 nm; n = 79) (**a**) and the as-printed WAAM (2.2 ± 0.1 nm; n = 64) (**b**). Note the factor of 30 difference in x-axis scale

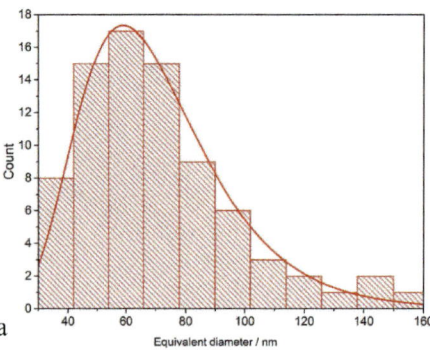
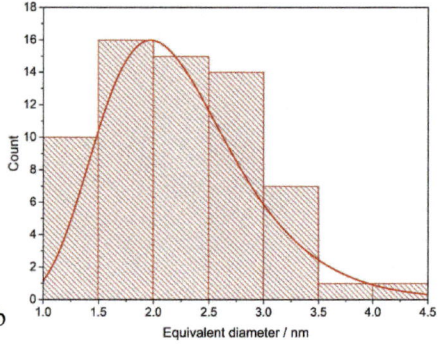

Table 1 Specification of Al5025 compared to OES measurement of rod and WAAM (wt%)

	Si	Fe	Cu	Mn	Mg	Cr	Zn	Ti	Sc	Zr
Al5025	0.25	0.40	0.10	0.05–0.20	4.5–5.5	0.05–0.20	0.10	0.06–0.20	0.3	0.15
Rod	0.03	0.07	0.04	0.19	4.58	0.002	0.002	0.05	0.32	n/a
WAAM	0.04	0.08	0.04	0.19	4.67	0.002	0.004	0.05	0.32	n/a

Table 2 Electrical conductivity and Vickers microhardness

	Conductivity MS/m	Vickers hardness MPa
Rod		900
WAAM	13.8	780
WAAM + 340 °C 4 h	14.7	1080

WAAM, a Thermo-Calc Precipitation (TCAL8.1/MOBAL7) simulation was performed with 4.6 wt% Mg, 0.3 wt% Sc and 0.15 wt% Zr using the experimental cooling data in the box inside Fig. 2. The upper temperature was near the solidus for this alloy of 590 °C. The lower temperature was selected as 320 °C, which is approximately the optimal range for Al$_3$Sc precipitation in isothermal heat treatments, and too low a temperature for significant changes on the timescale of seconds. Logarithmic cooling observed over the range of 590–320 °C occurred in 6.5 s. Results of the simulation suggest that the particles would grow in size and volume fraction up to about 3 s (450 °C) (Fig. 6). After this time and temperature, no significant changes in size or volume fractions are observed. Heating from subsequent WAAM layers was insignificant (Fig. 3), so no further changes in dispersoid morphology are expected. The disperoid diameter was 1.5 nm, which is slightly smaller than the measured diameter of 2 nm, while the volume fraction was about 0.5%. In the Thermo-Calc results, it is unclear if and how the Sc and Zr dispersoids interact.

Fig. 6 Thermo-Calc Precipitation calculation results for 4.6 wt% Mg, 0.3 wt% Sc, and 0.15 wt% Zr during WAAM cooling from 590 °C to 320 °C (box in Fig. 3)

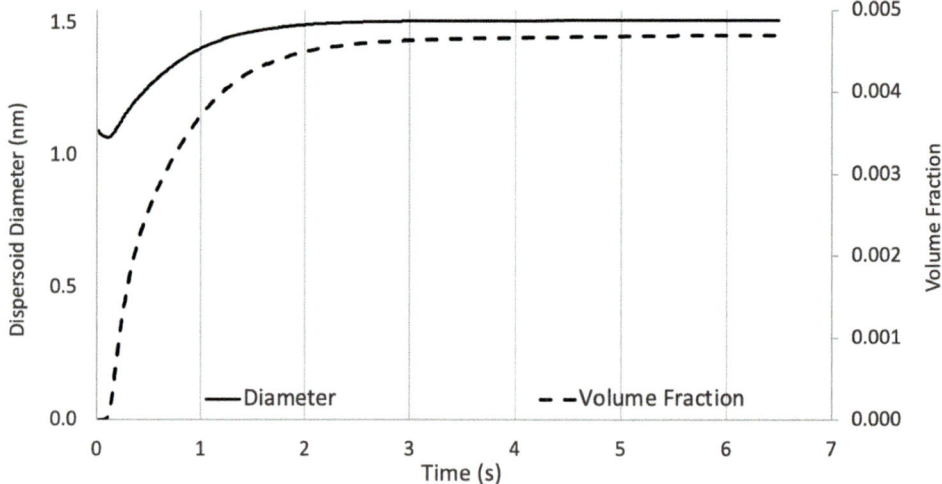

A heat treatment after WAAM targeted to precipitate Al$_3$Sc led to decreases in electrical conductivity and increases in hardness, implying that further precipitation was possible from a matrix still supersaturated with Sc, consistent with other post-WAAM strengthening responses observed in the literature [5]. As homogeneous nucleation of Sc is commonly observed, it is anticipated that the current 2.2 nm dispersoids (Fig. 5b) would grow further, while additional Al$_3$Sc would nucleate and grow.

Development of aluminum alloys for WAAM should consider the whole process from direct chill casting through extrusion, rod rolling, drawing, printing. For moderate temperature applications (under 250 °C), including elements that can only be solutionized by melting (e.g., Ti, Zr) complicates thermo-mechanical processing while not providing tangible benefits.

References

1. *Welding Aluminum: Theory and Practice*, 4th ed. The Aluminum Association, Inc., 2002.
2. V. Fallah *et al.*, "Atomic-scale pathway of early-stage precipitation in Al–Mg–Si alloys," *Acta Mater.*, vol. 82, pp. 457–467, 2015.
3. A. F. Norman, S. S. Birley, and P. B. Prangnell, "Development of new high strength Al – Sc filler wires for fusion welding 7000 series aluminium aerospace alloys," *Sci. Technol. Weld. Join.*, vol. 8, no. 4, pp. 235–245, Aug. 2003, doi: https://doi.org/10.1179/136217103225010989.
4. Z. Ahmad, "The properties and application of scandium-reinforced aluminum," *JOM*, vol. 55, no. 2, pp. 35–39, Feb. 2003, doi: https://doi.org/10.1007/s11837-003-0224-6.
5. T. Ponomareva, M. Ponomarev, K. Arseniy, and M. Ivanov, "Wire Arc Additive Manufacturing of Al-Mg Alloy with the Addition of Scandium and Zirconium," *Materials*, vol. 14, no. 13, pp. 36–65, 2021.

Author Index

Subject Index